PHYSICS

PHYSICS

FIFTH EDITION

JAMES S. WALKER

Western Washington University

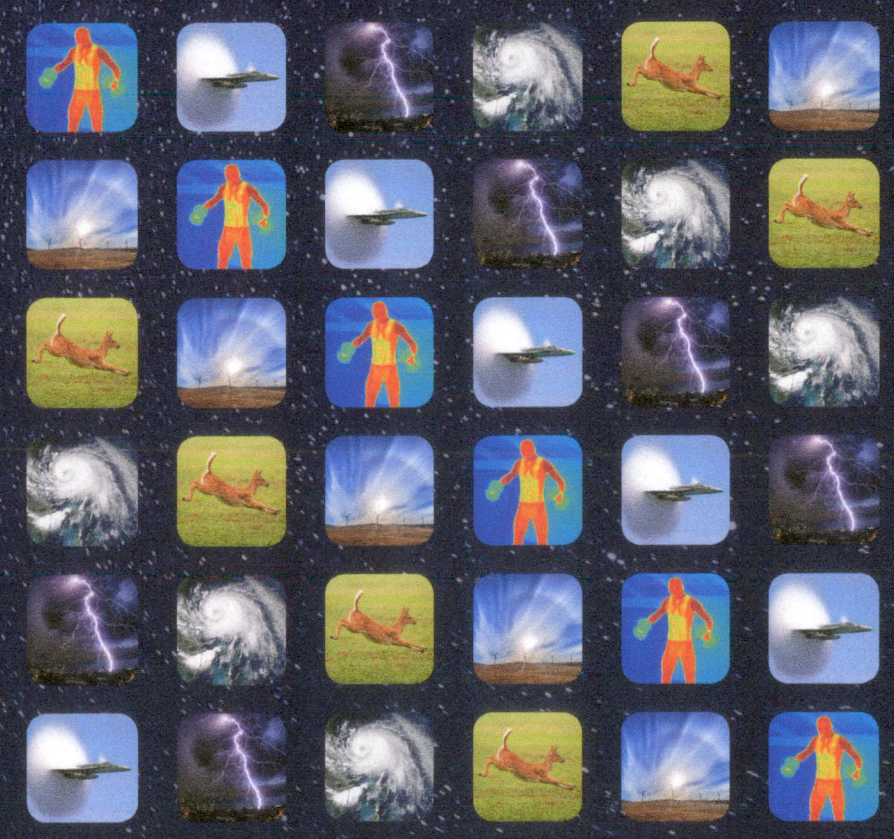

PEARSON

Editor-in-Chief: Jeanne Zalesky
Acquisitions Editor: Darien Estes
Director of Marketing: Christy Lesko
Marketing Manager: Elizabeth Ellsworth
Project Manager: Chandrika Madhavan
Program Manager: Katie Conley
Development Manager: Cathy Murphy
Program and Project Management Team Lead: Kristen Flathman
Copyeditor: Carol Reitz
Compositor: Cenveo Publisher Services
Vendor Production Manager: Rose Kernan
Illustrations: Rolin Graphics
Design Manager: Derek Bacchus
Interior and Cover Designer: Gary Hespenheide
Rights & Permissions Project Manager: Tim Nicholls
Photo Researcher: Eric Schrader
Media Producer: Sarah Kaubisch
Manufacturing Buyer: Maura Zaldivar-Garcia
Printer and Binder: LSC Communications
Cover Printer: Phoenix Color

THIS BOOK IS DEDICATED TO MY
PARENTS, IVAN AND JANET WALKER,
AND TO MY WIFE, BETSY.

Cover Photo Credits: The Corona on grassland: David Hu/Getty Images;
spotted deer: Nishant Shah/Getty Images; thermal image of young
male athlete training: Joseph Giacomin/Getty Images; Sydney summer
lightning strike: Australian Land, City, People Scape Photographer/Getty
Images; F/A-18 Hornet: US Navy; Hurricane Arthur's cloud-covered eye:
NASA Goddard MODIS Rapid Response Team

Library of Congress Cataloging-in-Publication Data
Names: Walker, James S., 1950- author.
Title: Physics/James S. Walker, Western Washington University.
Description: Fifth edition. | San Francisco : Pearson Addison-Wesley,
[2016]
 | Includes index.
Identifiers: LCCN 2015044999| ISBN 9780321976444 | ISBN 0321976444
Subjects: LCSH: Physics--Textbooks.
Classification: LCC QC23.2 .W35 2016 | DDC 530--dc23
LC record available at http://lccn.loc.gov/2015044999

ISBN 10: 0-321-97644-4; ISBN 13: 978-0-321-97644-4 (Student edition)
ISBN 10: 0-134-03124-5; ISBN 13: 978-0-134-03124-8 (Volume One)
ISBN 10: 0-134-03125-3; ISBN 13: 978-0-134-03125-5 (Volume Two)

JAMES S. WALKER

James Walker obtained his Ph.D. in theoretical physics from the University of Washington in 1978. He subsequently served as a post-doc at the University of Pennsylvania, the Massachusetts Institute of Technology, and the University of California at San Diego before joining the physics faculty at Washington State University in 1983. Professor Walker's research interests include statistical mechanics, critical phenomena, and chaos. His many publications on the application of renormalization-group theory to systems ranging from absorbed monolayers to binary-fluid mixtures have appeared in *Physical Review, Physical Review Letters, Physica,* and a host of other publications. He has also participated in observations on the summit of Mauna Kea, looking for evidence of extra-solar planets.

Jim Walker likes to work with students at all levels, from judging elementary school science fairs to writing research papers with graduate students, and has taught introductory physics for many years. Through his enjoyment of this course and his empathy for students, Jim has earned a reputation as an innovative, enthusiastic, and effective teacher. Jim's educational publications include "Reappearing Phases" (*Scientific American,* May 1987) as well as articles in the *American Journal of Physics* and *The Physics Teacher.* In recognition of his contributions to the teaching of physics at Washington State University, Jim was named the Boeing Distinguished Professor of Science and Mathematics Education for 2001–2003.

When he is not writing, conducting research, teaching, or developing new classroom demonstrations and pedagogical materials, Jim enjoys amateur astronomy, eclipse chasing, bird and dragonfly watching, photography, juggling, unicycling, boogie boarding, and kayaking. Jim is also an avid jazz pianist and organist. He has served as ballpark organist for a number of Class A minor league baseball teams, including the Bellingham Mariners, an affiliate of the Seattle Mariners, and the Salem-Keizer Volcanoes, an affiliate of the San Francisco Giants. He can play "Take Me Out to the Ball Game" in his sleep.

BRIEF CONTENTS

1 **Introduction to Physics** 1

PART I MECHANICS

2 **One-Dimensional Kinematics** 19

3 **Vectors in Physics** 60

4 **Two-Dimensional Kinematics** 88

5 **Newton's Laws of Motion** 117

6 **Applications of Newton's Laws** 155

7 **Work and Kinetic Energy** 195

8 **Potential Energy and Conservation of Energy** 223

9 **Linear Momentum and Collisions** 257

10 **Rotational Kinematics and Energy** 299

11 **Rotational Dynamics and Static Equilibrium** 334

12 **Gravity** 381

13 **Oscillations About Equilibrium** 418

14 **Waves and Sound** 456

15 **Fluids** 502

PART II THERMAL PHYSICS

16 **Temperature and Heat** 544

17 **Phases and Phase Changes** 579

18 **The Laws of Thermodynamics** 617

PART III ELECTROMAGNETISM

19 **Electric Charges, Forces, and Fields** 657

20 **Electric Potential and Electric Potential Energy** 695

21 **Electric Current and Direct-Current Circuits** 730

22 **Magnetism** 769

23 **Magnetic Flux and Faraday's Law of Induction** 807

24 **Alternating-Current Circuits** 845

PART IV LIGHT AND OPTICS

25 **Electromagnetic Waves** 880

26 **Geometrical Optics** 913

27 **Optical Instruments** 954

28 **Physical Optics: Interference and Diffraction** 983

PART V MODERN PHYSICS

29 **Relativity** 1017

30 **Quantum Physics** 1051

31 **Atomic Physics** 1082

32 **Nuclear Physics and Nuclear Radiation** 1119

An accessible, problem-solving approach to physics, grounded in real-world applications

PHYSICS

FIFTH EDITION

JAMES S. WALKER

What's the Big Idea?

James Walker's *Physics*, Fifth Edition engages students by connecting conceptual and quantitative physics to the world around them, making complex concepts understandable, and helping students build problem-solving skills. New "just in time" learning aids, such as the "Big Ideas," quickly orient students to the overarching principles of each chapter, while new Real-World Physics and Biological Applications expose students to physics they can observe in their own lives. A revised problem-solving pedagogy allows students to build a deep understanding of the relationship between the conceptual and the quantitative.

Ample "just in time" learning aids help students where they need it, when they need it

NEW! BIG IDEAS appear on each chapter-opening page to quickly orient students to the overarching principles of each chapter. Details highlighting key takeaways for students appear in the chapter margins next to where these ideas are introduced in the narrative.

Big Ideas

1 Work is force times distance.

2 Kinetic energy is one-half mass times velocity squared.

3 Power is the rate at which wo

Big Idea 1 Work is force times distance when force and displacement are in the same direction. More generally, work is the component of force in the direction of displacement times the distance.

NEW! PHYSICS IN CONTEXT calls attention to a related or supporting concept covered in a previous chapter ("Physics in Context: Looking Back"); or alerts students to a concept to be covered in a future chapter that relates to what they're reading ("Physics in Context: Looking Ahead").

PHYSICS IN CONTEXT
Looking Back

Conservation of energy, first introduced in Chapter 8, is just as important in rotational motion as it is in linear motion.

PHYSICS IN CONTEXT
Looking Ahead

In Chapter 5 we will introduce one of the most important concepts in all of physics—force. It is a vector quantity. Other important vector quantities to be introduced in later chapters include linear momentum (Chapter 9), angular momentum (Chapter 11), electric field (Chapter 19), and magnetic field (Chapter 22).

NEW! REAL-WORLD PHYSICS AND BIOLOGICAL APPLICATIONS woven into the text have been updated with fresh topics relevant to today's student. These applications bring abstract physics principles to life with real-life examples from the world around us.

Enhance Your Understanding
(Answers given at the end of the chapter)

5. An object moves along the brown path in **FIGURE 3-30** in the direction indicated. Which physical quantity (position, acceleration, velocity) is represented by the following vectors: (a) $\vec{A}$, (b) $\vec{B}$, (c) $\vec{C}$?

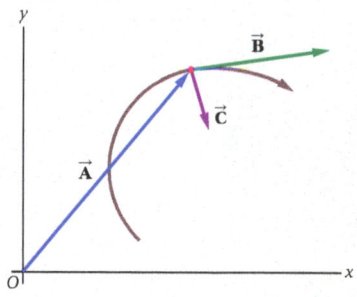

▲ **FIGURE 3-30** Enhance Your Understanding 5

Section Review

- Many physical quantities are represented by vectors. Among these are position, displacement, velocity, and acceleration.

NEW! ENHANCE YOUR UNDERSTANDING qualitative multiple-choice and ranking questions appear before each Section Review to give students an opportunity to practice what they've just learned. Answers are listed at the end of the chapter.

NEW! SECTION REVIEWS briefly synthesize the key ideas covered in the preceding section for a quick at-a-glance summary.

Thorough problem-solving instruction and multiple opportunities for practice

The Fifth Edition continues the Walker tradition of providing ample opportunity for students to develop problem-solving skills with a greater variety of example types and more thoroughly stepped out explanations and guidance.

NEW! KNOWNS AND UNKNOWNS have been added to worked examples to model how scientists think about setting up a problem before they solve it.

NEW! 50% UPDATED AND REVISED FULLY WORKED EXAMPLES provide a systematic process for solving problems:

- **Picture the Problem** encourages students to visualize the situation, identify and label important quantities, and set up a coordinate system. This step is accompanied by a figure and free-body diagram when appropriate.
- **UPDATED! Reasoning and Strategy** helps students learn to analyze the problem, identify the key physical concepts, and map a plan for the solution, including the Known and Unknown quantities.
- **Solution** is presented in a *two-column format* to help students translate the words of the problem on the left to the equations they will use to solve it on the right.
- **Insight** points out interesting or significant features of the problem, solution process, or the result.
- **Practice Problems** give students the opportunity to test their understanding and skills on similar problems to the one just worked.

NEW! END-OF-CHAPTER HOMEWORK set includes 10 new problems per chapter and more than 15% revised problems. Hundreds of additional problems are available in Mastering as an alternate problem set as well as the Instructor's Solutions Manual.

NEW! BIO PASSAGE PROBLEMS have been thoroughly rewritten to better reflect the new MCAT exam, released in 2015. With a focus on skills and core competencies, rather than rote knowledge, each Bio Passage Problem offers opportunities to practice the types of questions pre-meds will encounter on the exam.

EXAMPLE 7-2 · HEADING FOR THE ER

An intern pushes an 87-kg patient on an 18-kg gurney, producing an acceleration of 0.55 m/s². **(a)** How much work does the intern do in pushing the patient and gurney through a distance of 1.9 m? Assume the gurney moves without friction. **(b)** How far must the intern push the gurney to do 140 J of work?

PICTURE THE PROBLEM
Our sketch shows the physical situation for this problem. Notice that the force exerted by the intern is in the same direction as the displacement of the gurney; therefore, we know that the work is $W = Fd$.

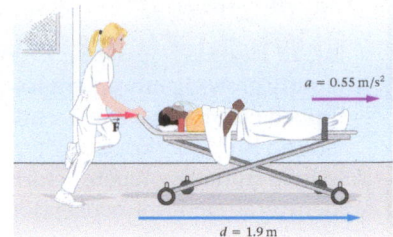

$a = 0.55 \text{ m/s}^2$

$d = 1.9 \text{ m}$

REASONING AND STRATEGY
We are not given the magnitude of the force, F, so we cannot apply Equation 7-1 ($W = Fd$) directly. However, we are given the mass and acceleration of the patient and gurney, and from them we can calculate the force with $F = ma$. The work done by the intern is then $W = Fd$, where $d = 1.9$ m.

Known Mass of patient, 87 kg; mass of gurney, 18 kg; acceleration, $a = 0.55$ m/s²; **(a)** pushing distance, $d = 1.9$ m; **(b)** work, $W = 140$ J.

Unknown **(a)** Work done, $W = ?$ **(b)** Pushing distance, $d = ?$

SOLUTION

Part (a)

1. First, find the force F exerted by the intern: $\qquad F = ma = (87 \text{ kg} + 18 \text{ kg})(0.55 \text{ m/s}^2) = 58 \text{ N}$
2. The work done by the intern, W, is the force times the distance: $\qquad W = Fd = (58 \text{ N})(1.9 \text{ m}) = 110 \text{ J}$

Part (b)

3. Use $W = Fd$ to solve for the distance d: $\qquad d = \dfrac{W}{F} = \dfrac{140 \text{ J}}{58 \text{ N}} = 2.4 \text{ m}$

INSIGHT
You might wonder whether the work done by the intern depends on the speed of the gurney. The answer is no. The work done on an object, $W = Fd$, doesn't depend on whether the object moves through the distance d quickly or slowly. What does depend on the speed of the gurney is the *rate* at which work is done, which we discuss in detail in Section 7-4.

PRACTICE PROBLEM — PREDICT/CALCULATE
(a) If the total mass of the gurney plus patient is halved and the acceleration is doubled, does the work done by the intern increase, decrease, or remain the same? Explain. **(b)** Determine the work in this case. [**Answer: (a)** The work remains the same because the two changes offset one another; that is, $F = ma = (m/2)(2a)$. **(b)** The work is 110 J, as before.]

Some related homework problems: Problem 3, Problem 4

PASSAGE PROBLEMS

Bam!—*Apollo 15* Lands on the Moon

The first word spoken on the surface of the Moon after *Apollo 15* landed on July 30, 1971, was "Bam!" This was James Irwin's involuntary reaction to their rather bone-jarring touchdown. "We did hit harder than any of the other flights!" says Irwin. "And I was startled, obviously, when I said, 'Bam!'"

The reason for the "firm touchdown" of *Apollo 15*, as pilot David Scott later characterized it, was that the rocket engine was shut off a bit earlier than planned, when the lander was still 4.30 ft above the lunar surface and moving downward with a speed of 0.500 ft/s. From that point on the lander descended in lunar free fall, with an acceleration of 1.62 m/s². As a result, the landing speed of *Apollo 15* was by far the largest of any of the *Apollo* missions. In comparison, Neil Armstrong's landing speed on *Apollo 11* was the lowest at 1.7 ft/s—he didn't shut off the engine until the footpads were actually on the surface. *Apollos 12, 14,* and *17* all landed with speeds between 3.0 and 3.5 ft/s.

To better understand the descent of *Apollo 15*, we show its trajectory during the final stages of landing in FIGURE 2-47 (a). In FIGURE 2-47 (b) we show a variety of speed-versus-time plots.

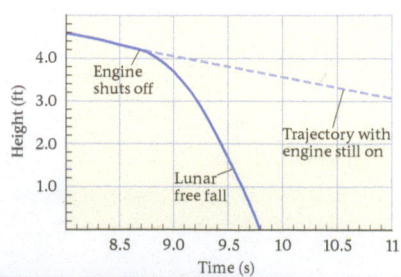

Personalize learning with MasteringPhysics®

BEFORE CLASS

NEW! INTERACTIVE PRELECTURE VIDEOS address the rapidly growing movement toward prelecture teaching and flipped classrooms. These whiteboard-style animations provide an introduction to key topics with embedded assessment to help students prepare before lecture and to help professors identify student misconceptions.

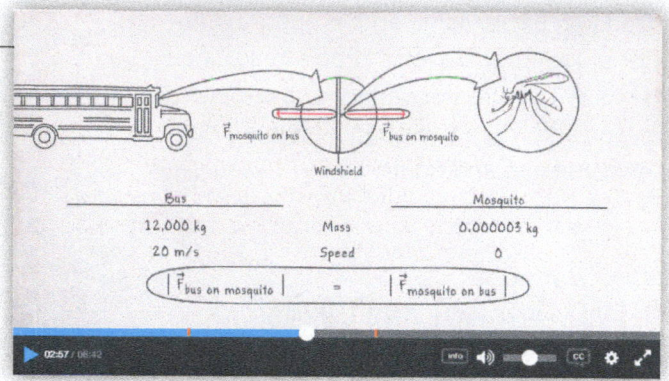

NEW! DYNAMIC STUDY MODULES (DSMs) leverage research from the fields of cognitive psychology, neurobiology, and game studies to help students study on their own by continuously assessing their activity and performance, then using data and analytics to provide personalized content in real time to reinforce concepts that target each student's particular strengths and weaknesses. Assignable for Pre-Class Prep or Self-Study, physics DSMs include the mathematics of physics, conceptual definitions, relationships for topics across all of mechanics and electricity and magnetism, and more.

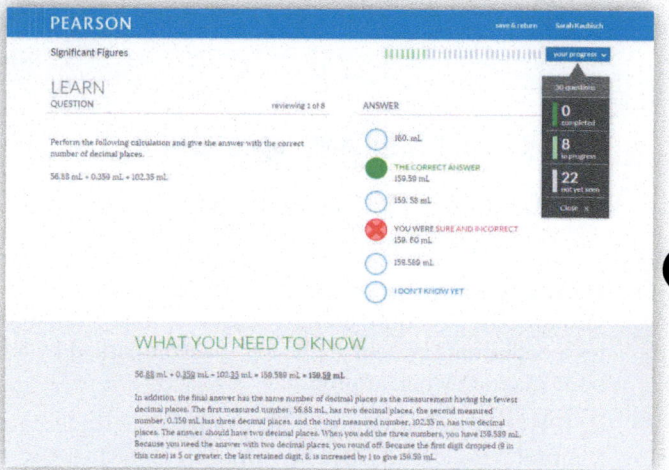

DURING CLASS

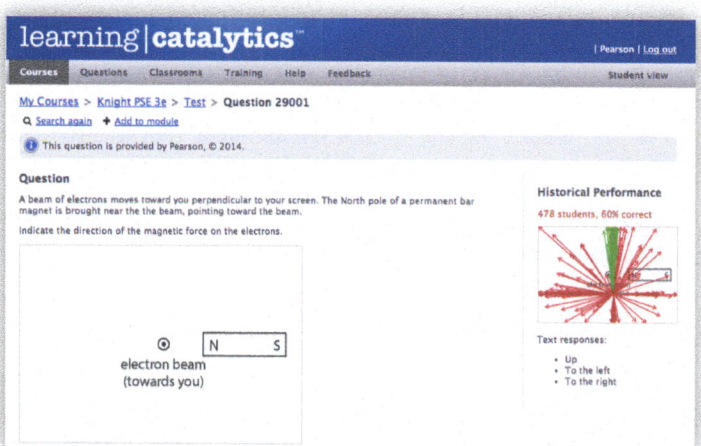

NEW! LEARNING CATALYTICS™ is an interactive classroom tool that uses students' devices to engage them in class and provide "just in time" analytics to inform your lecture/active classroom. A new math palette for the mathematical expression question type allows for Greek symbols, PI, Euler's number, Logarithm, Exponent, Trigonometric functions, Absolute value, Square root, Nth square root, and Fractions with the exception of vector or unit vector.

AFTER CLASS

NEW! **ENHANCED END-OF-CHAPTER QUESTIONS** offer students instructional support when and where they need it including links to the eText, math remediation, and wrong-answer specific feedback for an improved student experience and greater learning gains.

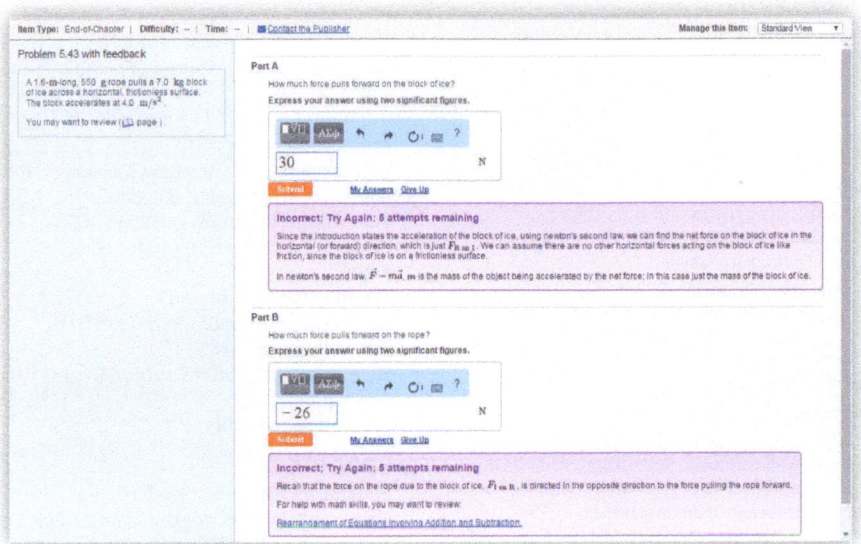

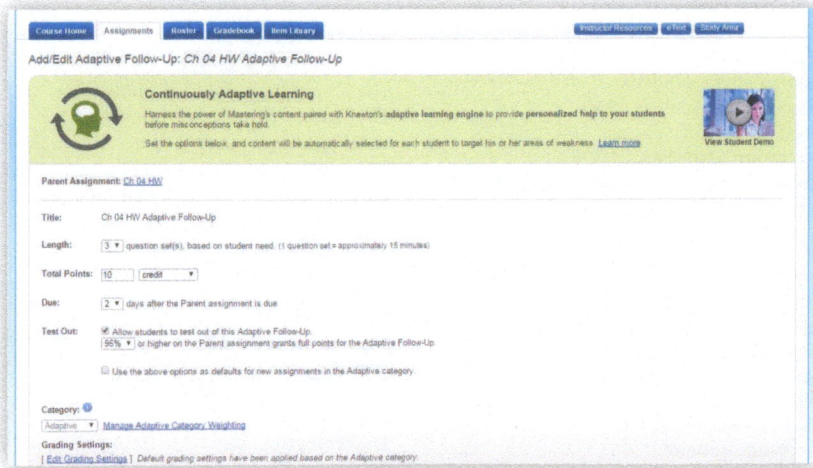

ADAPTIVE FOLLOW-UPS are personalized assignments that pair Mastering's powerful content with Knewton's adaptive learning engine to provide individualized help to students before misconceptions take hold. These adaptive follow-ups address topics students struggled with on assigned homework, including core prerequisite topics.

APPLICATIONS IN THE TEXT

Note: This list includes applied topics that receive significant discussion in the chapter text or a worked Example, as well as topics that are touched on in - end-of-chapter Conceptual Questions, Conceptual Exercises, and Problems. Topics of particular relevance to the life sciences or medicine are marked **BIO**. Topics related to Passage Problems are marked **PP**.

CHAPTER 1

Estimation: How many raindrops in a storm   11
The strike of a Mantis shrimp   17 **BIO**
Mosquito courtship   18 **BIO**
Using a cricket as a thermometer   18 **PP**, **BIO**

CHAPTER 2

Takeoff distance for an airliner   38
The stopping distance of a car   40
Calculating the speed of a lava bomb   46
Apollo 15 lands on the Moon   59 **PP**

CHAPTER 3

Crossing a river   78
Food in digestive system   85 **BIO**
The watch that won the Longitude Prize   86
Motion camouflage in dragonflies   87 **PP**, **BIO**

CHAPTER 4

Traffic collision avoidance system   90
The parabolic trajectory of projectiles   96
Golf on the Moon   104
How an archerfish hunts   106 **BIO**
Punkin Chunkin   112
Spitting llamas   113 **BIO**
Volcanoes on Io   113
Landing rovers on Mars   115 **PP**

CHAPTER 5

How walking affects your height   123 **BIO**
Astronaut jet packs   127
Stopping an airplane with Foamcrete   127
Simulating weightlessness   140 **BIO**
The force of running   152 **BIO**
A leafhopper's leap   153 **BIO**
Increasing safety in a collision   154 **PP**, **BIO**

CHAPTER 6

Antilock braking systems   163
Setting a broken leg with traction   165 **BIO**
Tethered astronauts   173
Skids and banked roadways   180
Centrifuges and ultracentrifuges   182 **BIO**
Spiderweb forces   187 **BIO**
A human centrifuge   190 **BIO**
Bubble net fishing   190 **BIO**
Nasal strips   193 **PP**, **BIO**

CHAPTER 7

Human power output and flight   213 **BIO**
Peristaltic work   217 **BIO**
The reentry of Skylab   218
Human-powered flight   219 **BIO**
Power output of the human brain   221 **BIO**
The biplane dinosaur   221 **PP**, **BIO**

CHAPTER 8

Converting food energy to mechanical energy   230 **BIO**
Jump of a flea   231 **BIO**
Strike of the mantis shrimp   233 **BIO**
The wing of the hawkmoth   250 **BIO**
Mantis shrimp smasher   251 **BIO**
Bird tendons   254 **BIO**
Nasal strips   254
The jump of a flea   255 **BIO**
The flight of dragonflies   256 **PP**, **BIO**

CHAPTER 9

The force between a ball and a bat   263
The ballistocardiograph   269 **BIO**
Concussions   269 **BIO**
Heartbeat detectors   269 **BIO**
The ballistic pendulum   275
Analyzing a traffic accident   276
The center of mass of the arm   284 **BIO**
An exploding rocket   288
The Saturn V rocket   289
Delta IV Rocket   289
Concussion impulse   293 **BIO**
Concussion Recoil   293 **BIO**
Escaping octopus   297 **BIO**

CHAPTER 10

The operation of a CD   308
Types of centrifuges   309
The microhematocrit centrifuge   309 **BIO**
Moment of inertia of the Earth   316
Flywheel energy recovery system   317
Dental drills, the world's fastest turbines   327
Spin-dry dragonflies   329 **BIO**
Dolphin tricks   331 **BIO**
Human-powered centrifuge   332 **PP**, **BIO**

CHAPTER 11

Starfish attack   336 **BIO**
Applying the brakes   347
Forces required for structural stability   347
An arm in a sling   348 **BIO**
Hurricanes and tornadoes   360
The angular speed of a pulsar   360
The precession of the Earth   368
Gyroscopes in navigation and space   368
Correcting torsiversion   379 **PP**, **BIO**

CHAPTER 12

The dependence of gravity on altitude   386
Weighing the Earth   388
The internal structure of the Earth and Moon   389
The Sun and Mercury   394
Geosynchronous satellites   396
The Global Positioning System (GPS)   395
Maneuvering spacecraft   397
Planetary atmospheres   405
Black holes   407
Gravitational lensing   407
Tides   407
Tidal locking   408
The Roche limit and Saturn's rings   409

CHAPTER 13

Woodpecker impact   420 **BIO**
Measuring the mass of a "weightless" astronaut   431

The pendulum clock and pulsilogium   438
Adjusting a grandfather clock   441
Walking speed   442 **BIO**
Resonance: radio tuners   445
Resonance: spider webs   446
Resonance: bridges   446
Slow-motion dragonfly   449 **BIO**
A cricket thermometer   454 **PP**, **BIO**

CHAPTER 14

Calculating distance to a lightning strike   465
Ultrasonic sounds in nature   466
Ultrasonic scans   467 **BIO**
Shock wave lithotripsy   467 **BIO**
Infrasonic communication in elephants and whales   467
Infrasound produced by meteors   467
Echolocation   467 **BIO**
Human perception of sound intensity   468 **BIO**
Measuring the speed of blood flow   477 **BIO**
The red shift of distant galaxies   477
Connecting speakers in phase   480
Active noise reduction   481
The shape of a piano   485
Human perception of pitch   485 **BIO**
Frets on a guitar   485
Human hearing and the ear canal   487 **BIO**
Organ pipes   488
Chelyabinsk meteor   496

CHAPTER 15

Walking on lily pads   505 **BIO**
Pressure at the wreck of the *Titanic*   508
The barometer   509
The hydraulic lift   511
Measuring the body's density   515 **BIO**
Measuring body fat   515 **BIO**
The swim bladder in fish   518 **BIO**
Diving sea mammals   518 **BIO**
Maximum load indicator on a ship   519
The tip of the iceberg   519
Hoses and nozzles   521
The lift produced by an airplane wing   527
The force on a roof in a windstorm   528
Ventilation in a prairie dog burrow   528 **BIO**
Blood speed in the pulmonary artery   531 **BIO**
Breathing, alveoli, and premature birth   533 **BIO**
Vasodilation   539 **BIO**
Occlusion in an artery   540 **BIO**
Cooking doughnuts   542 **PP**

CHAPTER 16

Bimetallic strips   552
Antiscalding devices   553
Thermal expansion joints   553
Floating icebergs   555
The ecology of lakes   556 **BIO**
Bursting water pipes   556
Water and the climate   560
Insulated windows   566
Countercurrent exchange   566 **BIO**
Cold hands and feet   566 **BIO**
Convection on the Earth and Sun   567
Using color to measure temperature   567
Temperatures of the stars   567
Thermos bottles and the Dewar   569

CHAPTER 17
Take a deep breath   581 **BIO**
Stretching a bone   593 **BIO**
The autoclave   598 **BIO**
The pressure cooker   598
Adding salt to boiling water   598
Ice melting under pressure   599
Frost wedging and frost heaving   599
Biological antifreeze   600 **BIO**
Cooling the body with evaporation   600 **BIO**
Stability of planetary atmospheres   601
Homemade ice cream   605
Diving in the bathysphere   616 **PP**

CHAPTER 18
Diesel engines   627
Using adiabatic heating to start a fire   627
Rain shadows   631
The steam engine   632
Refrigerators   637
Air conditioners   638
Heat pumps   639
Heat death of the universe   646
Entropy and life   646 **BIO**
Energy from the ocean   656 **PP**
Seismographs   805
Electric guitar pickups   805
Magnetic disk drives and credit card readers   806
T coils and induction loops   806
Magnetic braking and speedometers   809
Induction stove   810
Magnetic antitheft devices   813
Tracking the movement of insects   813 **BIO**
Tracking the motion of the human eye   813 **BIO**
Electric generators   813

CHAPTER 19
Bacterial infection from endoscopic
   surgery   662 **BIO**
Electrodialysis for water purification   672
Electric fish   675 **BIO**
Electrical shark repellent   675 **BIO**
Solar Panels   678
Electrical shielding   679
Lightning rods and Saint Elmo's fire   680
Electrostatic precipitation   681
Bumblebees and static cling   693 **PP**, **BIO**

CHAPTER 20
The electrocardiograph   710 **BIO**
The electroencephalograph   711 **BIO**
Computer keyboards   717
The theremin—a musical instrument you play
   without touching   717
The electronic flash   719
The defibrillator   719 **BIO**
Automatic external defibrillator   726 **BIO**
The electric eel   729 **PP**, **BIO**

CHAPTER 21
The bolometer   737
Thermistors and fever thermometers   737
Superconductors and high-temperature
   superconductivity   737
"Battery check" meters   740
Three-way lightbulbs   743
"Touch-sensitive" lamps   752
Delay circuits in windshield wipers and turn
   signals   757
Pacemakers   757 **BIO**
Footwear safety   767 **PP**, **BIO**

CHAPTER 22
Refrigerator magnets   771
The Earth's magnetic field   771
The electromagnetic flowmeter   778 **BIO**
The mass spectrometer   780
Synchrotron Undulator   781
The aurora borealis and aurora australis   782
The galvanometer   787
MRI instruments   794
Magnetite in living organisms   796 **BIO**
Magnetism and the brain   796 **BIO**
Magnetic Recycling   796
Magnetic levitation   796
Magnetoencephalography   806 **PP**, **BIO**

CHAPTER 23
Dynamic microphones and seismographs   813
Electric guitar pickups   813
Magnetic disk drives and credit card readers   814
Magnetic braking and speedometers   817
Induction stove   817
Magnetic antitheft devices   821
Tracking the movement of insects   821 **BIO**
Tracking the motion of the human eye   821 **BIO**
Electric generators   822
Motion Electric Generators   823
Electric motors   824
Energy recovery technology in cars   824
High-voltage electric power transmission   834
Loop detectors on roadways   844 **PP**

CHAPTER 24
Electric shock hazard   850 **BIO**
Polarized plugs and grounded plugs   850
Ground fault circuit interrupter   851
Light dimmers   862
Tuning a radio or television   870
Metal detectors   870
Persistence of vision   877 **BIO**
Human Impedance   879 **PP**, **BIO**

CHAPTER 25
Radio and television communications   883
Doppler radar   887
Nexrad   888
Infrared receptors in pit vipers   890 **BIO**
Biological effects of ultraviolet light   891 **BIO**
Irradiated food   891 **BIO**
Photoelastic stress analysis   900
Liquid crystal diplays (LCDs)   901
Navigating using polarized light from the
   sky   902 **BIO**
Why the sky is blue   902
How Polaroid sunglasses cut reflected
   glare   902
Visible-light curing in dentistry   912 **PP**

CHAPTER 26
Micromirror devices and digital movie
   projection   915
Corner reflectors and the Earth-Moon
   distance   919
Parabolic mirrors   921
Apparent depth   931
Mirages   932
Porro prisms in binoculars   934
Optical fibers and endoscopes   934 **BIO**
Underwater vision   938 **BIO**
The rainbow   942
The focal length of a lens   953 **PP**

CHAPTER 27
Optical properties of the human eye   955 **BIO**
Speed and aperture settings on a camera   958
Extended vision: Correcting
   nearsightedness   961 **BIO**
Intracorneal rings   963 **BIO**
Radial keratotomy   963 **BIO**
Correcting farsightedness   964 **BIO**
Keratometers   965 **BIO**
Achromatic lenses   974
Cataracts and intraocular lenses   982 **PP**, **BIO**

CHAPTER 28
Newton's rings   993
Soap bubbles and oil slicks   993
Nonreflective coating   996
Reading the information on a CD   996
Pointillism and painting   1005
Color television images   1005
Acousto-optic modulation   1007
X-ray diffraction and the structure of DNA   1007
Grating spectroscopes   1007
Iridescence in nature   1008 **BIO**
Resolving lines on an HDTV   1016 **PP**

CHAPTER 29
Nuclear power—converting mass to energy   1034
The energy of the Sun   1034
Positron-emission tomography   1036 **BIO**
Gravitational lensing   1041
Black Holes   1041
The search for gravity waves   1042
Relativity in a TV set   1049 **PP**

CHAPTER 30
Measuring the temperature of a star   1053
Dark-adapted vision   1057 **BIO**
Photocells   1060
Solar energy panels   1060
Sailing on a beam of light   1062
Optical tweezers   1062
Electron microscopes   1067
Scanning tunneling microscopy   1073
Owl vision/Human vision   1078 **BIO**
Hydrothermal vents   1081 **PP**

CHAPTER 31
Medical X-ray tubes   1104
Computerized axial tomography   1107 **BIO**
Helium-neon laser   1107
Laser eye surgery   1108 **BIO**
Photodynamic therapy   1109 **BIO**
Holography   1109
Fluorescent light bulbs   1110
Applications of fluorescence in forensics   1111
Detecting scorpions at night   1111 **BIO**
The GFP Bunny   1111 **BIO**
Welding a detached retina   1118 **PP**, **BIO**

CHAPTER 32
Smoke detector   1128
Dating the Iceman   1136
Nuclear reactors   1142
Powering the Sun: the proton-proton cycle   1143
Manmade fusion   1143
Radiation and cells   1144
Radioactive tracers   1146
Positron-emission tomography   1146
Magnetic resonance imaging (MRI)   1147
Higgs Boson   1150
Treating a hyperactive thyroid   1157 **PP**, **BIO**

Teaching introductory algebra-based physics can be a most challenging—and rewarding—experience. Students enter the course with a wide range of backgrounds, interests, and skills and we, the instructors, strive not only to convey the basic concepts and fundamental laws of physics, but also to give students an appreciation for its relevance and appeal.

I wrote this book to help with that task. It incorporates a number of unique and innovative pedagogical features that evolved from years of teaching experience. The materials have been tested extensively in the classroom and in focus groups, and refined based on comments from students and teachers who used the earlier editions of the text. The enthusiastic response I've received from users of the first four editions is both flattering and motivating. The Fifth Edition has been improved in direct response to this feedback with new and revised examples; modern biological and real-world physics applications woven into the text; a refreshed homework problem set, including new biological passage problems that align with the new MCAT exam; and "just in time" learning aids throughout the text.

Learning Tools in the Text

A key goal of this text is to help students make the connection between a conceptual understanding of physics and the various skills necessary to solve quantitative problems. One of the chief means to that end is an integrated system of learning tools—including fully worked Examples with solutions in two-column format, two-column Quick Examples, Conceptual Examples, and Exercises. Each of these tools is specialized to meet the needs of students at a particular point in the development of a chapter.

These needs are not always the same. Sometimes students benefit from a detailed explanation of how to tackle a particular problem; at other times it is helpful for them to explore a key idea in a conceptual context. And sometimes, all that is required is practice using a new equation or definition.

This text emulates the teaching style of successful instructors by providing the right tool at the right time and place. This "just in time" approach helps students master the new ideas and concepts as they are encountered.

In a similar spirit, two new features appear at the end of every section in each chapter to give students immediate, timely feedback on the material just covered. These features are the **Enhance Your Understanding** and the **Section Review**. The Enhance Your Understanding feature is a conceptual question designed to solidify the concepts presented in the section. Answers to the Enhance Your Understanding questions are given at the end of each chapter. The Section Review gives a brief review of the key concepts covered in that section.

WORKED EXAMPLES WITH SOLUTIONS IN TWO-COLUMN FORMAT

Examples provide students with a complete and detailed method of solving a particular type of problem. The Examples in this text are presented in a format that focuses on the basic strategies and thought processes involved in problem solving. This focus on the intimate connection between conceptual insights and problem-solving techniques encourages students to view the ability to solve problems as a logical outgrowth of conceptual understanding. In addition, the Examples encourage students to think of solving physics problems as an opportunity to exercise their innate creativity.

Each **Example** has the same basic structure:

- **Picture the Problem** This first step discusses how the physical situation can be represented visually and what such a representation can tell us about how to analyze and solve the problem. At this step, we set up a coordinate system where appropriate, and label important quantities.

- **Reasoning and Strategy** The Reasoning and Strategy section addresses the commonly asked question, "How do I get started?" It does this by providing a clear overview of the problem and helping students to identify the relevant physical principles. It then guides the student in using known relationships to map out a step-by-step path to the solution. In the Fifth Edition I've fleshed out this step to give students more thorough guidance.   **UPDATED**

- **Knowns and Unknowns** Before the process of solving the problem begins, we list the quantities that are known (given), and those that are to be found. This new feature in the Reasoning and Strategy step helps students organize their thoughts, and sets a clear goal for their calculations.   **NEW**

- **Solution in a Two-Column Format** In the step-by-step Solution of the problem, each step is presented with a prose statement in the left-hand column and the corresponding mathematical implementation in the right-hand column. Each step clearly translates the idea described in words into the appropriate equations.

- **Insight** Each Example wraps up with an Insight—a comment regarding the solution just obtained. Some Insights deal with possible alternative solution techniques, others with new ideas suggested by the results.

- **Practice Problem** Following the Insight is a Practice Problem, which gives the student a chance to practice the type of calculation just presented. The Practice Problems are always accompanied by answers, and provide students with a valuable check on their understanding of the material. **Some of the Practice Problems are of the new Predict/Calculate type** in the end-of-chapter homework section. These problems ask for a prediction based on physical concepts, and then present a numerical problem to verify the prediction. Finally, each Example ends with a reference to some related end-of-chapter Problems to allow students to test their skills further.   **UPDATED**

QUICK EXAMPLES

Quick Examples are streamlined versions of the full Examples. By streamlining the process, this new feature allows for more sample problems to be covered in the text without taking up too much space or becoming redundant in the details.   **NEW**

CONCEPTUAL EXAMPLES

Conceptual Examples help students sharpen their insight into key physical principles. A typical Conceptual Example presents a thought-provoking question that can be answered by logical reasoning based on physical concepts rather than by numerical calculations. The statement of the question is followed by a detailed discussion and analysis in the section titled Reasoning and Discussion, and the Answer is given at the end of the checkpoint for quick and easy reference.

NEW to this edition are **Conceptual Examples that prepare students to solve Predict/Explain** problems in the end-of-chapter homework section. These problems ask for a prediction, and then ask the student to pick the best explanation from those provided.

EXERCISES

Exercises present brief one-step calculations designed to illustrate the application of important new relationships, without the expenditure of time and space required by a fully worked Example or Quick Example. Exercises generally give students an opportunity to practice the use of a new equation, become familiar with the units of a new physical quantity, and get a feeling for typical magnitudes.

PROBLEM-SOLVING NOTES

Each chapter includes a number of Problem-Solving Notes presented in a "just in time" fashion in the margin. These practical hints are designed to highlight useful problem-solving methods while helping students avoid common pitfalls and misconceptions.

End-of-Chapter Learning Tools

The end-of-chapter material in this text also includes a number of innovations, along with refinements of more familiar elements.

- Each chapter concludes with a **Chapter Summary** presented in an easy-to-use outline style. Key concepts, equations, and important figures are organized by topic for convenient reference.

- The homework for each chapter begins with a section of **Conceptual Questions**. Answers to the odd-numbered Questions can be found in the back of the book. Answers to even-numbered Conceptual Questions are available in the Instructor's Solutions Manual.

- Following the Conceptual Questions is a complete set of **Problems and Conceptual Exercises. Conceptual Exercises (CE)** consist of multiple-choice and ranking questions. Answers to the odd-numbered Exercises can be found in the back of the book. Answers to even-numbered Conceptual Exercises are available in the online Instructor's Solutions Manual.

UPDATED
- The Fifth Edition boasts a refreshed set of homework problems: at least 10 new problems per chapter and more than 15% of the previous edition problems have been revised and improved. **Problems** are divided into sections, with increasing difficulty levels indicated with one, two, or three colored bullets. Problems of particular real-world interest are indicated with titles. In addition, a section titled **General Problems** presents a variety of problems that use material from two or more sections within the chapter, or refer to material covered in earlier chapters.

- **Bio/Med Problems** are homework problems of special biological or medical relevance. These problems are indicated with the symbol **BIO**.

NEW
- **Predict/Calculate** is a new type of problem that combines a conceptual question (prediction) with a numerical problem (calculation). Problems of this type, which stress the importance of reasoning from basic principles, show how conceptual insight and numerical calculation go hand in hand in physics.

- **Predict/Explain** problems ask students to predict what will happen in a given physical situation, and then to choose the best explanation for their prediction.

UPDATED
- **Passage Problems** are based on an extended multi-paragraph description of a physical situation. These problems are similar to those found on MCAT exams, and have associated multiple-choice questions. All of the Passage Problems in the book are either new or revised and support the new MCAT released in 2015.

Perspective Across Chapters

It's easy for students to miss the forest for the trees—to overlook the unifying concepts that are central to physics and make the details easier to learn, understand, and retain. To address this difficulty, the Fifth Edition adds a **NEW** feature called **Physics In Context**. This feature, which appears in the margin at appropriate locations in the chapters, comes in two varieties—**Looking Back** and **Looking Ahead**. The Looking Back variety connects material just developed to related material from earlier chapters. This helps students apply their understanding of earlier material to a new situation, and provides a greater perspective on physics as a whole. The Looking Ahead variety gives students a "heads up" that the material presented in this chapter is important at a specific point later in the text. With these two varieties working together, Physics In Context helps students develop connections between different topics in physics that share a common central theme.

Scope and Organization

The presentation of physics in this text follows the standard practice for introductory courses, with only a few well-motivated refinements.

PHYSICS IN CONTEXT
Looking Back

Conservation of energy, first introduced in Chapter 8, is just as important in rotational motion as it is in linear motion.

PHYSICS IN CONTEXT
Looking Ahead

In Chapter 5 we will introduce one of the most important concepts in all of physics—force. It is a vector quantity. Other important vector quantities to be introduced in later chapters include linear momentum (Chapter 9), angular momentum (Chapter 11), electric field (Chapter 19), and magnetic field (Chapter 22).

First, note that Chapter 3 is devoted to **vectors and their application to physics**. My experience has been that students benefit greatly from a full discussion of vectors early in the course. Most students have seen vectors and trigonometric functions before, but rarely from the point of view of physics. Thus, including a chapter on vectors sends a message that this is important material, and gives students an opportunity to develop and improve their understanding of vectors in a physics context.

Note also that **additional time is given to some of the more fundamental aspects of physics**, such as Newton's laws and energy. Presenting such material in two chapters gives the student a better opportunity to assimilate and master these crucial topics.

REAL-WORLD AND BIOLOGICAL PHYSICS

Since physics applies to everything in nature, it's only reasonable to point out applications of physics that students may encounter in the real world. Each chapter presents a number of discussions focusing on "Real-World Physics." Those of general interest are designated by **RWP** at the start of the paragraph. Applications that pertain more specifically to biology or medicine are indicated by **BIO**.

UPDATED

These applications have been thoughtfully updated with topics that are current and relevant to today's student. For example, new **RWP** features include timely discussions of accelerometers in cell phones and game controllers (Chapter 2), traffic collision avoidance systems (Chapter 4), the tension in a tether connecting astronauts in the movie *Gravity* (Chapter 6), and the use of magnetic forces to improve recycling (Chapter 22). Updated **BIO** topics include the deadly blow delivered by a mantis shrimp (Chapter 8), the 1000-rpm mid-air spinning behavior of dragonflies (Chapter 10), how starfish use torque to open clamshells (Chapter 11), and how radioactivity is used to treat hyperactive thyroids (Chapter 32).

These new applications are in addition to the many classic, and often surprising, examples of physics that have always been an important part of Walker *Physics*. Examples like these engage a student's interest, and help motivate them to think more deeply about physics concepts. Students are often intrigued, for example, to discover that they are shorter at the end of the day than when they get up in the morning (Chapter 5), that humming next to a spider web can cause a resonance effect that sends the spider into a tizzy (Chapter 13), and that scorpions in the nighttime desert are brightly fluorescent when illuminated by an ultraviolet flashlight (Chapter 31). With real-world applications like these, it's easy to show students that physics is relevant to their everyday lives.

VISUALS

One of the most fundamental ways in which we learn is by comparing and contrasting. A new feature called **Visualizing Concepts** helps with this process by presenting a selection of photos that illustrate a physical concept in a variety of different contexts. Grouping carefully chosen photographs in this way helps students to see the universality of physics. We have also included a number of **demonstration photos** that use high-speed time-lapse photography to dramatically illustrate topics, such as standing waves, static versus kinetic friction, and the motion of center of mass, in a way that reveals physical principles in the world around us.

Resources

The Fifth Edition is supplemented by an ancillary package developed to address the needs of both students and instructors.

FOR THE INSTRUCTOR

Instructor's Solutions Manual by Kenneth L. Menningen (University of Wisconsin–Stevens Point) is available online at the Instructor's Resource Center: www.pearsonhighered.com/educator, on the Instructor's Resource DVD, and in the Instructor's Resource Area on Mastering. You will find detailed, worked solutions to every Problem and Conceptual Exercise in the text, all solved using the step-by-step problem-solving strategy of the in-chapter Examples (Picture the Problem, Reasoning and Strategy, two-column Solutions, and Insight). The solutions also contain answers to the even-numbered Conceptual Questions as well as problem statements and solutions for the hundreds of problems in the alternate problem set in Mastering.

The cross-platform **Instructor's Resource DVD** (ISBN 978-0-321-76570-3) provides a comprehensive library of applets from ActivPhysics OnLine as well as PhET simulations. All line figures, photos, and examples from the textbook are provided in JPEG format and the key equations are available in editable Word files. A revised set of Lecture Outlines and Clicker questions, both in PowerPoint, are included for use in lecture. Assets available in Mastering are provided here, too: Pause and Predict Video Tutor Demonstrations and Author Demonstration Videos. And it includes the Test Bank in Word and TestGen formats and the Instructor's Solutions Manual in Word and pdf.

MasteringPhysics® (www.masteringphysics.com) is the most advanced, educationally effective, and widely used physics homework and tutorial system in the world. Ten years in development, it provides instructors with a library of extensively pre-tested end-of-chapter problems and rich multipart, multistep tutorials that incorporate a wide variety of answer types, wrong answer feedback, individualized help (comprising hints or simpler sub-problems upon request), all driven by the largest metadatabase of student problem-solving in the world.

New to MasteringPhysics:

NEW   **Interactive Prelecture Videos** address the rapidly growing movement toward prelecture teaching and flipped classrooms. These whiteboard-style animations provide an introduction to key topics with embedded assessment to help students prepare before lecture and to help professors identify student misconceptions.

NEW   **Dynamic Study Modules (DSMs)** leverage research from the fields of cognitive psychology, neurobiology, and game studies to help students study on their own by continuously assessing their activity and performance, then using data and analytics to provide personalized content in real-time to reinforce concepts that target each student's particular strengths and weaknesses. Assignable for Pre-Class Prep or Self-Study, physics DSMs include the mathematics of physics, conceptual definitions, relationships for topics across all of mechanics and electricity and magnetism, and more.

Learning Catalytics™ is an inter-   **NEW** active classroom tool that uses students' devices to engage them in class and provide "just in time" analytics to inform your lecture/active classroom. A new math palette for the mathematical expression question type allows for Greek symbols, PI, Euler's number, Logarithm, Exponent, Trigonometric functions, Absolute value, Square root, Nth square root, and Fractions with the exception of vector or unit vector.

Enhanced End-of-Chapter Ques-   **NEW** **tions** offer students instructional support when and where they need it including links to the eText, math remediation, and wrong-answer specific feedback for an improved student experience and greater learning gains.

Adaptive Follow-Ups are personalized assignments that pair Mastering's powerful content with Knewton's adaptive learning engine to provide individualized help to students before misconceptions take hold. These adaptive follow-ups address topics students struggled with on assigned homework, including core prerequisite topics.

An **alternate set of hundreds of**   **NEW** **homework problems** not included in the textbook gives instructors more assignable homework options than before. Solutions and problem statements are available in the Instructor's Solutions Manual.

MCAT questions are a set of multi-   **NEW** part passage problems and standalone problems in a biological context that cover the key topics of the new 2015 MCAT format.

Video Tutor Demomonstration   **NEW** **coaching activities** give brief demonstrations that include "Pause and Predict" questions with Mastering assessment containing wrong answer feedback and hints.

The **Test Bank** contains more than 2000 high-quality problems assignable as auto-graded items in MasteringPhysics. Test files are provided both in TestGen (an easy-to-use, fully networkable program for creating and editing quizzes and exams) and Word formats. Available in the

MasteringPhysics Instructor's Area, on the Instructor's Resource Center (www.pearsonhighered.com/irc), and on the Instructor's Resource DVD.

FOR THE STUDENT

MasteringPhysics® (www.masteringphysics.com) is the most advanced physics homework and tutorial system available. This online homework and tutoring system guides students through the most important topics in physics with self-paced tutorials that provide individualized coaching. These assignable, in-depth tutorials are designed to coach students with hints and feedback specific to their individual errors. Instructors can also assign end-of-chapter problems from every chapter including multiple-choice questions, section-specific exercises, and general problems. Quantitative problems can be assigned with numerical answers and randomized values (with sig fig feedback) or solutions. Mastering is available either bundled with the textbook, as an online download—with or without the eText—and as a standalone item available for purchase at the campus bookstore.

Pearson eText is available through MasteringPhysics®, either automatically when MasteringPhysics® is packaged with new books or as a purchased upgrade online. Allowing students access to the text wherever they have access to the Internet, Pearson eText comprises the full text with additional interactive features. Users can search for words or phrases, create notes, highlight text, bookmark sections, and click on definitions to key terms as they read.

Acknowledgments

I would like to express my sincere gratitude to colleagues at Washington State University and Western Washington University, as well as to many others in the physics community, for their contributions to this project.

In particular, I would like to thank Professor Ken Menningen of the University of Wisconsin–Stevens Point for his painstaking attention to detail in producing the Instructor's Solutions Manual. Ken has also contributed a number of new excellent Real-World Applications, Biological Applications, and end-of-chapter homework problems to the Fifth Edition.

My thanks are also due to the many wonderful and talented people at Pearson who have been such a pleasure to work with during the development of the Fifth Edition, and especially to Katie Conley, Jeanne Zalesky, Darien Estes, Cathy Murphy and Chandrika Madhavan.

In addition, I am grateful for the dedicated efforts of Rose Kernan, who choreographed a delightfully smooth production process.

Finally, I owe a great debt to all my students over the years. My interactions with them provided the motivation and inspiration that led to this book.

We are grateful to the following instructors for their thoughtful comments on the revision of this text.

REVIEWERS OF THE FIFTH EDITION

Ugur Akgun
University of Iowa

Kapila Castoldi
Oakland University

Joshua Colwell
University of Central Florida

Diana Driscoll
Case Western Reserve University

Daniel Giese
University of Wisconsin, Milwaukee

Klaus Honscheid
Ohio State University

Darrin Johnson
University of Minnesota, Duluth

Weining Man
San Francisco State University

Kin Mon
University of Georgia

Kriton Papavasiliou
Virginia Polytechnic Institute

Arnd Pralle
SUNY, University at Buffalo

Andrew Richter
Valparaiso University

Andreas Shalchi
University of Manitoba

Douglas Sherman
San Jose State University

Maxim Sukharev
Arizona State University

Michael Tammaro
University of Rhode Island

Orhan Yenen
University of Nebraska, Lincoln

Ataollah Zamani
Houston Community College

REVIEWERS OF THE FOURTH EDITION

Raymond Benge
Tarrant County College–NE Campus

Matthew Bigelow
Saint Cloud University

Edward J. Brash
Christopher Newport University

Michaela Burkardt
New Mexico State University

Jennifer Chen
University of Illinois at Chicago

Eugenia Ciocan
Clemson University

Shahida Dar
University of Delaware

Joseph Dodoo
University of Maryland, Eastern Shore

Thomas Dooling
University of Northern Carolina at Pembroke

Hui Fang
Sam Houston State University

Carlos E. Figueroa
Cabrillo College

Lyle Ford
University of Wisconsin, Eau Claire

Darrin Johnson
University of Minnesota, Duluth

Paul Lee
California State University, Northridge

Sheng-Chiang (John) Lee
Mercer University

Nilanga Liyanage
University of Virginia

Michael Ottinger
Missouri Western State University

Melodi Rodrigue
University of Nevada

Claudiu Rusu
Richland College of DCCCD

Mark Sprague
East Carolina University

Richard Szwerc
Montgomery College

Lisa Will
San Diego City College

Guanghua Xu
University of Houston

Bill Yen
University of Georgia

REVIEWERS OF PREVIOUS EDITIONS

Daniel Akerib, *Case Western Reserve University*
Richard Akerib, *Queens College*
Alice M. Hawthorne Allen, *Virginia Tech*
Barbara S. Andereck, *Ohio Wesleyan University*
Eva Andrei, *Rutgers University*
Bradley C. Antanaitis, *Lafayette College*
Michael Arnett, *Kirkwood Community College*
Robert W. Arts, *Pikeville College*

David Balogh, *Fresno City College*
David T. Bannon, *Oregon State University*
Rama Bansil, *Boston University*
Anand Batra, *Howard University*
Paul Beale, *University of Colorado–Boulder*
Mike Berger, *Indiana University*
David Berman, *University of Iowa*
S. M. Bhagat, *University of Maryland*

James D. Borgardt, *Juniata College*
James P. Boyle, *Western Connecticut State University*
David Branning, *Trinity College*
Jeff Braun, *University of Evansville*
Matthew E. Briggs, *University of Wisconsin-Madison*
Jack Brockway, *State University of New York-Oswego*
Neal Cason, *University of Notre Dame*
Thomas B. Cobb, *Bowling Green State University*
Lattie Collins, *Eastern Tennessee State University*
James Cook, *Middle Tennessee State University*
Stephen Cotanch, *North Carolina State University*
David Craig, *LeMoyne College*
David Curott, *University of North Alabama*
William Dabby, *Edison Community College*
Robert Davie, *St. Petersburg Junior College*
Steven Davis, *University of Arkansas-Little Rock*
N. E. Davison, *University of Manitoba*
Duane Deardorff, *University of North Carolina at Chapel Hill*
Edward Derringh, *Wentworth Institute of Technology*
Martha Dickinson, *Maine Maritime Academy*
Anthony DiStefano, *University of Scranton*
David C. Doughty, Jr., *Christopher Newport University*
F. Eugene Dunnam, *University of Florida*
John J. Dykla, *Loyola University-Chicago*
Eldon Eckard, *Bainbridge College*
Donald Elliott, *Carroll College*
David Elmore, *Purdue University*
Robert Endorf, *University of Cincinnati*
Raymond Enzweiler, *Northern Kentucky University*
John Erdei, *University of Dayton*
David Faust, *Mt. Hood Community College*
Frank Ferrone, *Drexel University*
John Flaherty, *Yuba College*
Curt W. Foltz, *Clarion University*
Lewis Ford, *Texas A&M University*
Armin Fuchs, *Florida Atlantic University*
Joseph Gallant, *Kent State University, Trumbull*
Asim Gangopadhyaya, *Loyola University-Chicago*
Thor Garber, *Pensacola Junior College*
David Gerdes, *University of Michigan*
John D. Gieringer, *Alvernia College*
Karen Gipson, *Grand Valley State University*
Barry Gilbert, *Rhode Island College*
Fred Gittes, *Washington State University*
Michael Graf, *Boston College*
William Gregg, *Louisiana State University*
Rainer Grobe, *Illinois State University*
Steven Hagen, *University of Florida*
Mitchell Haeri, *Saddleback College*
Parameswar Hari, *California State University-Fresno*
Xiaochun He, *Georgia State University*
Timothy G. Heil, *University of Georgia*
J. Erik Hendrickson, *University of Wisconsin-Eau Claire*

Scott Holmstrom, *University of Tulsa*
John Hopkins, *The Pennsylvania State University*
Manuel A. Huerta, *University of Miami*
Zafar Ismail, *Daemen College*
Adam Johnston, *Weber State University*
Gordon O. Johnson, *Washington State University*
Nadejda Kaltcheva, *University of Wisconsin-Oshkosh*
William Karstens, *Saint Michael's College*
Sanford Kern, *Colorado State University*
Dana Klinck, *Hillsborough Community College*
Ilkka Koskelo, *San Francisco State University*
Laird Kramer, *Florida International University*
R. Gary Layton, *Northern Arizona University*
Kevin M. Lee, *University of Nebraska-Lincoln*
Michael Lieber, *University of Arkansas*
Ian M. Lindevald, *Truman State University*
Mark Lindsay, *University of Louisville*
Jeff Loats, *Fort Lewis College*
Daniel Ludwigsen, *Kettering University*
Lorin Matthews, *Baylor University*
Hilliard Macomber, *University of Northern Iowa*
Trecia Markes, *University of Nebraska-Kearny*
William McNairy, *Duke University*
Kenneth L. Menningen, *University of Wisconsin-Stevens Point*
Joseph Mills, *University of Houston*
Anatoly Miroshnichenko, *University of Toledo*
Wouter Montfrooij, *University of Missouri*
Gary Morris, *Valparaiso University*
Paul Morris, *Abilene Christian University*
David Moyle, *Clemson University*
Ashok Muthukrishnan, *Texas A&M University*
K. W. Nicholson, *Central Alabama Community College*
Robert Oman, *University of South Florida*
Michael Ottinger, *Missouri Western State College*
Larry Owens, *College of the Sequoias*
A. Ray Penner, *Malaspina University*
Francis Pichanick, *University of Massachusetts, Amherst*
Robert Piserchio, *San Diego State University*
Anthony Pitucco, *Pima Community College*
William Pollard, *Valdosta State University*
Jerry Polson, *Southeastern Oklahoma State University*
Robert Pompi, *Binghamton University*
David Procopio, *Mohawk Valley Community College*
Earl Prohofsky, *Purdue University*
Jia Quan, *Pasadena City College*
David Raffaelle, *Glendale Community College*
Michele Rallis, *Ohio State University*
Michael Ram, *State University of New York-Buffalo*
Prabha Ramakrishnan, *North Carolina State University*
Rex Ramsier, *University of Akron*
John F. Reading, *Texas A&M University*
Lawrence B. Rees, *Brigham Young University*
M. Anthony Reynolds, *Embry Riddle University*
Dennis Rioux, *University of Wisconsin-Oshkosh*
John A. Rochowicz, Jr., *Alvernia College*

Bob Rogers, *San Francisco State University*
Gaylon Ross, *University of Central Arkansas*
Lawrence G. Rowan, *University of North Carolina at Chapel Hill*
Gerald Royce, *Mary Washington College*
Wolfgang Rueckner, *Harvard University*
Misa T. Saros, *Viterbo University*
C. Gregory Seab, *University of New Orleans*
Mats Selen, *University of Illinois*
Bartlett Sheinberg, *Houston Community College*
Peter Shull, *Oklahoma State University*
Christopher Sirola, *Tri-County Technical College*
Daniel Smith, *South Carolina State University*
Leigh M. Smith, *University of Cincinnati*
Soren Sorensen, *University of Tennessee-Knoxville*
Mark W. Sprague, *East Carolina University*
George Strobel, *University of Georgia*
Carey E. Stronach, *Virginia State University*
Irina Struganova, *Barry University*

Daniel Stump, *Michigan State University*
Leo Takahashi, *Penn State University–Beaver*
Harold Taylor, *Richard Stockton College*
Frederick Thomas, *Sinclair Community College*
Jack Tuszynski, *University of Alberta*
Lorin Vant Hull, *University of Houston*
John A. Underwood, *Austin Community College, Rio Grande*
Karl Vogler, *Northern Kentucky University*
Desmond Walsh, *Memorial University of Newfoundland*
Toby Ward, *College of Lake County*
Richard Webb, *Pacific Union College*
Lawrence Weinstein, *Old Dominion University*
Jeremiah Williams, *Illinois Wesleyan University*
Linda Winkler, *Moorhead State University*
Lowell Wood, *University of Houston*
Robert Wood, *University of Georgia*
Jeffrey L. Wragg, *College of Charleston*

STUDENT REVIEWERS

We wish to thank the following students at New Mexico State University and Chemetka Community College for providing helpful feedback during the development of the fourth edition of this text. Their comments offered us valuable insight into the student experience.

Teresa M. Abbott
Rachel Acuna
Sonia Arroyos
Joanna Beeson
Carl Bryce
Jennifer Currier
Juan Farias
Mark Ferra
Bonnie Galloway

Cameron Haider
Gina Hedberg
Kyle Kazsinas
Ty Keeney
Justin Kvenzi
Tannia Lau
Ann MaKarewicz
Jasmine Pando
Jenna Painter

Jonathan Romero
Aaron Ryther
Sarah Salaido
Ashley Slape
Christina Timmons
Christopher Torrez
Charmaine Vega
Elisa Wingerd

We would also like to thank the following students at Boston University, California State University–Chico, the University of Houston, Washington State University, and North Carolina State University for providing helpful feedback via review or focus group for the first three editions of this text:

Ali Ahmed
Joel Amato
Max Aquino
Margaret Baker
Tynisa Bennett
Joshua Carmichael
Sabrina Carrie
Suprna Chandra
Kara Coffee
Tyler Cumby
Rebecca Currell
Philip Dagostino
Andrew D. Fisher
Shadi Miri Ghomizadea

Colleen Hanlon
Jonas Hauptmann
Parker Havron
Jamie Helms
Robert Hubbard
Tamara Jones
Bryce Lewis
Michelle Lim
Candida Mejia
Roderick Minogue
Ryan Morrison
Hang Nguyen
Mary Nguyen
Julie Palakovich

Suraj Parekh
Scott Parsons
Peter Ploewski
Darren B. Robertson
Chris Simons
Tiffany Switzer
Steven Taylor
Monique Thomas
Khang Tran
Michael Vasquez
Jerod Williams
Nathan Witwer
Alexander Wood
Melissa Wright

ATTENDEE: Lynda Klein, *California State University–Chico*

As a student preparing to take an algebra-based physics course, you are probably aware that physics applies to absolutely everything in the natural world, from raindrops and people, to galaxies and atoms. Because physics is so wide-ranging and comprehensive, it can sometimes seem a bit overwhelming. This text, which reflects nearly two decades of classroom experience, is designed to help you deal with a large body of information and develop a working understanding of the basic concepts in physics. Now in its fifth edition, it incorporates many refinements that have come directly from interacting with students using the first four editions. As a result of these interactions, I am confident that as you develop a deeper understanding of physics, you will also enrich your experience of the world in which you live.

Now, I must admit that I like physics, and so I may be a bit biased in this respect. Still, the reason I teach and continue to study physics is that I enjoy the insight it gives into the physical world. I can't help but notice—and enjoy—aspects of physics all around me each and every day. I would like to share some of this enjoyment and delight in the natural world with you. It is for this reason that I undertook the task of writing this book.

To assist you in the process of studying physics, this text incorporates a number of learning aids, including two-column Examples, Quick Examples, Conceptual Examples, and Exercises. These and other elements work together in a unified way to enhance your understanding of physics on both a conceptual and a quantitative level—they have been developed to give you the benefit of what we know about how students learn physics, and to incorporate strategies that have proven successful to students over the years. The pages that follow will introduce these elements to you, describe the purpose of each, and explain how they can help you.

As you progress through the text, you will encounter many interesting and intriguing applications of physics drawn from the world around you. Some of these, such as magnetically levitated trains or the Global Positioning System (GPS), are primarily technological in nature. Others focus on explaining familiar or not-so-familiar phenomena, such as why the Moon has no atmosphere, how sweating cools the body, or why flying saucer shaped clouds often hover over mountain peaks even on a clear day. Still others, such as countercurrent heat exchange in animals and humans, or the use of sound waves to destroy kidney stones, are of particular relevance to students of biology and those interested in pursuing a career in medicine.

In many cases, you may find the applications to be a bit surprising. Did you know, for example, that you are shorter at the end of the day than when you first get up in the morning? (This is discussed in Chapter 5.) That an instrument called the ballistocardiograph can detect the presence of a person hiding in a truck, just by registering the minute recoil from the beating of the stowaway's heart? (This is discussed in Chapter 9.) That if you hum next to a spider's web at just the right pitch you can cause a resonance effect that sends the spider into a tizzy? (This is discussed in Chapter 13.) That powerful magnets can exploit the phenomenon of diamagnetism to levitate living creatures? (This is discussed in Chapter 22.) That scorpions in the nighttime desert are brightly fluorescent when illuminated by ultraviolet light? (This is discussed in Chapter 31.) The natural world truly is filled with marvelous things, and physics applies to all of them.

Writing this textbook was a rewarding and enjoyable experience for me. I hope using it will prove equally rewarding to you, and that it will inspire an appreciation for physics that will give you a lifetime of enjoyment.

James S. Walker

DETAILED CONTENTS

1 INTRODUCTION TO PHYSICS 1

1-1 Physics and the Laws of Nature 2

1-2 Units of Length, Mass, and Time 2

1-3 Dimensional Analysis 5

1-4 Significant Figures, Scientific Notation, and Round-Off Error 6

1-5 Converting Units 9

1-6 Order-of-Magnitude Calculations 11

1-7 Scalars and Vectors 12

1-8 Problem Solving in Physics 13

Chapter Summary 15

Conceptual Questions 16

Problems and Conceptual Exercises 16

PART I MECHANICS

2 ONE-DIMENSIONAL KINEMATICS 19

2-1 Position, Distance, and Displacement 20

2-2 Average Speed and Velocity 22

2-3 Instantaneous Velocity 26

2-4 Acceleration 28

2-5 Motion with Constant Acceleration 33

2-6 Applications of the Equations of Motion 39

2-7 Freely Falling Objects 43

Chapter Summary 49

Conceptual Questions 50

Problems and Conceptual Exercises 51

3 VECTORS IN PHYSICS 60

3-1 Scalars Versus Vectors 61

3-2 The Components of a Vector 61

3-3 Adding and Subtracting Vectors 65

3-4 Unit Vectors 70

3-5 Position, Displacement, Velocity, and Acceleration Vectors 71

3-6 Relative Motion 76

Chapter Summary 79

Conceptual Questions 81

Problems and Conceptual Exercises 81

4 TWO-DIMENSIONAL KINEMATICS 88

4-1 Motion in Two Dimensions 89

4-2 Projectile Motion: Basic Equations 92

4-3 Zero Launch Angle 94

4-4 General Launch Angle 99

4-5 Projectile Motion: Key Characteristics 103

Chapter Summary 108

Conceptual Questions 109

Problems and Conceptual Exercises 109

5 NEWTON'S LAWS OF MOTION 117

5-1 Force and Mass 118

5-2 Newton's First Law of Motion 118

5-3 Newton's Second Law of Motion 120

5-4 Newton's Third Law of Motion 129

5-5 The Vector Nature of Forces: Forces in Two Dimensions 134

5-6 Weight 137

5-7 Normal Forces 141

Chapter Summary 145

Conceptual Questions 146

Problems and Conceptual Exercises 147

6 APPLICATIONS OF NEWTON'S LAWS 155

6-1 Frictional Forces 156

6-2 Strings and Springs 164

6-3 Translational Equilibrium 169

6-4 Connected Objects 172

6-5 Circular Motion 177

Chapter Summary 182

Conceptual Questions 184

Problems and Conceptual Exercises 184

7 WORK AND KINETIC ENERGY 195

7-1 Work Done by a Constant Force 196

7-2 Kinetic Energy and the Work–Energy Theorem 203

7-3 Work Done by a Variable Force 207

7-4 Power 212

Chapter Summary 215

Conceptual Questions 216

Problems and Conceptual Exercises 217

8 POTENTIAL ENERGY AND CONSERVATION OF ENERGY 223

8-1 Conservative and Nonconservative Forces 224

8-2 Potential Energy and the Work Done by Conservative Forces 228

8-3 Conservation of Mechanical Energy 233

8-4 Work Done by Nonconservative Forces 241

8-5 Potential Energy Curves and Equipotentials 246

Chapter Summary 248

Conceptual Questions 249

Problems and Conceptual Exercises 249

9 LINEAR MOMENTUM AND COLLISIONS 257

9-1 Linear Momentum 258

9-2 Momentum and Newton's Second Law 261

9-3 Impulse 262

9-4 Conservation of Linear Momentum 266

9-5 Inelastic Collisions 271

9-6 Elastic Collisions 277

9-7 Center of Mass 283

9-8 Systems with Changing Mass: Rocket Propulsion 288

Chapter Summary 290

Conceptual Questions 291

Problems and Conceptual Exercises 292

10 ROTATIONAL KINEMATICS AND ENERGY 299

10-1 Angular Position, Velocity, and Acceleration 300

10-2 Rotational Kinematics 304

10-3 Connections Between Linear and Rotational Quantities 307

10-4 Rolling Motion 311

10-5 Rotational Kinetic Energy and the Moment of Inertia 313

10-6 Conservation of Energy 318

Chapter Summary 323

Conceptual Questions 325

Problems and Conceptual Exercises 326

11 ROTATIONAL DYNAMICS AND STATIC EQUILIBRIUM 334

11-1 Torque 335

11-2 Torque and Angular Acceleration 338

11-3 Zero Torque and Static Equilibrium 343

11-4 Center of Mass and Balance 350

11-5 Dynamic Applications of Torque 354

11-6 Angular Momentum 356

11-7 Conservation of Angular Momentum 360

11-8 Rotational Work and Power 365

11-9 The Vector Nature of Rotational Motion 366

Chapter Summary 368

Conceptual Questions 370

Problems and Conceptual Exercises 371

12 GRAVITY 381

12-1 Newton's Law of Universal Gravitation 382

12-2 Gravitational Attraction of Spherical Bodies 385

12-3 Kepler's Laws of Orbital Motion 391

12-4 Gravitational Potential Energy 398

12-5 Energy Conservation 401

12-6 Tides 407

Chapter Summary 409

Conceptual Questions 411

Problems and Conceptual Exercises 411

13 OSCILLATIONS ABOUT EQUILIBRIUM 418

13-1 Periodic Motion 419

13-2 Simple Harmonic Motion 420

13-3 Connections Between Uniform Circular Motion and Simple Harmonic Motion 424

13-4 The Period of a Mass on a Spring 430

13-5 Energy Conservation in Oscillatory Motion 435

13-6 The Pendulum 438

13-7 Damped Oscillations 444

13-8 Driven Oscillations and Resonance 445

Chapter Summary 446

Conceptual Questions 448

Problems and Conceptual Exercises 448

14 WAVES AND SOUND 456

14-1 Types of Waves 457

14-2 Waves on a String 460

14-3 Harmonic Wave Functions 463

14-4 Sound Waves 464

14-5 Sound Intensity 468

14-6 The Doppler Effect 472

14-7 Superposition and Interference 478

14-8 Standing Waves 482

14-9 Beats 489

Chapter Summary 492

Conceptual Questions 494

Problems and Conceptual Exercises 494

15 FLUIDS 502

15-1 Density 503

15-2 Pressure 504

15-3 Static Equilibrium in Fluids: Pressure and Depth 507

15-4 Archimedes' Principle and Buoyancy 513

15-5 Applications of Archimedes' Principle 514

15-6 Fluid Flow and Continuity 521

15-7 Bernoulli's Equation 523

15-8 Applications of Bernoulli's Equation 527

15-9 Viscosity and Surface Tension 530

Chapter Summary 534

Conceptual Questions 535

Problems and Conceptual Exercises 536

PART II THERMAL PHYSICS

16 TEMPERATURE AND HEAT 544

16-1 Temperature and the Zeroth Law of Thermodynamics 545

16-2 Temperature Scales 546

16-3 Thermal Expansion 550

16-4 Heat and Mechanical Work 556

16-5 Specific Heats 558

16-6 Conduction, Convection, and Radiation 562

Chapter Summary 570

Conceptual Questions 572

Problems and Conceptual Exercises 572

17 PHASES AND PHASE CHANGES 579

17-1   Ideal Gases 580

17-2   The Kinetic Theory of Gases 586

17-3   Solids and Elastic Deformation 592

17-4   Phase Equilibrium and Evaporation 597

17-5   Latent Heats 602

17-6   Phase Changes and Energy Conservation 606

   Chapter Summary 608

   Conceptual Questions 610

   Problems and Conceptual Exercises 610

18 THE LAWS OF THERMODYNAMICS 617

18-1   The Zeroth Law of Thermodynamics 618

18-2   The First Law of Thermodynamics 618

18-3   Thermal Processes 620

18-4   Specific Heats for an Ideal Gas: Constant Pressure, Constant Volume 628

18-5   The Second Law of Thermodynamics 631

18-6   Heat Engines and the Carnot Cycle 632

18-7   Refrigerators, Air Conditioners, and Heat Pumps 637

18-8   Entropy 640

18-9   Order, Disorder, and Entropy 645

18-10 The Third Law of Thermodynamics 647

   Chapter Summary 648

   Conceptual Questions 650

   Problems and Conceptual Exercises 650

PART III ELECTROMAGNETISM

19 ELECTRIC CHARGES, FORCES, AND FIELDS 657

19-1   Electric Charge 658

19-2   Insulators and Conductors 662

19-3   Coulomb's Law 663

19-4   The Electric Field 670

19-5   Electric Field Lines 675

19-6   Shielding and Charging by Induction 679

19-7   Electric Flux and Gauss's Law 682

   Chapter Summary 686

   Conceptual Questions 687

   Problems and Conceptual Exercises 688

20 ELECTRIC POTENTIAL AND ELECTRIC POTENTIAL ENERGY 695

20-1   Electric Potential Energy and the Electric Potential 696

20-2   Energy Conservation 700

20-3   The Electric Potential of Point Charges 703

20-4   Equipotential Surfaces and the Electric Field 707

20-5   Capacitors and Dielectrics 712

20-6   Electrical Energy Storage 718

   Chapter Summary 720

   Conceptual Questions 722

   Problems and Conceptual Exercises 722

21   ELECTRIC CURRENT AND DIRECT-CURRENT CIRCUITS   730

21-1   Electric Current   731

21-2   Resistance and Ohm's Law   734

21-3   Energy and Power in Electric Circuits   738

21-4   Resistors in Series and Parallel   741

21-5   Kirchhoff's Rules   747

21-6   Circuits Containing Capacitors   751

21-7   *RC* Circuits   754

21-8   Ammeters and Voltmeters   758

Chapter Summary   759

Conceptual Questions   761

Problems and Conceptual Exercises   761

22   MAGNETISM   769

22-1   The Magnetic Field   770

22-2   The Magnetic Force on Moving Charges   772

22-3   The Motion of Charged Particles in a Magnetic Field   777

22-4   The Magnetic Force Exerted on a Current-Carrying Wire   783

22-5   Loops of Current and Magnetic Torque   785

22-6   Electric Currents, Magnetic Fields, and Ampère's Law   788

22-7   Current Loops and Solenoids   792

22-8   Magnetism in Matter   795

Chapter Summary   797

Conceptual Questions   799

Problems and Conceptual Exercises   800

23   MAGNETIC FLUX AND FARADAY'S LAW OF INDUCTION   807

23-1   Induced Electromotive Force   808

23-2   Magnetic Flux   809

23-3   Faraday's Law of Induction   812

23-4   Lenz's Law   814

23-5   Mechanical Work and Electrical Energy   818

23-6   Generators and Motors   821

23-7   Inductance   825

23-8   *RL* Circuits   828

23-9   Energy Stored in a Magnetic Field   830

23-10   Transformers   832

Chapter Summary   834

Conceptual Questions   837

Problems and Conceptual Exercises   837

24   ALTERNATING-CURRENT CIRCUITS   845

24-1   Alternating Voltages and Currents   846

24-2   Capacitors in ac Circuits   851

24-3   *RC* Circuits   855

24-4   Inductors in ac Circuits   859

24-5   *RLC* Circuits   863

24-6   Resonance in Electric Circuits   867

Chapter Summary   872

Conceptual Questions   874

Problems and Conceptual Exercises   874

PART IV   LIGHT AND OPTICS

25   ELECTROMAGNETIC WAVES   880

25-1   The Production of Electromagnetic Waves   881

25-2   The Propagation of Electromagnetic Waves   884

25-3   The Electromagnetic Spectrum   889

25-4   Energy and Momentum in Electromagnetic Waves   892

25-5   Polarization   896

Chapter Summary   904

Conceptual Questions   905

Problems and Conceptual Exercises   906

26   GEOMETRICAL OPTICS   913

26-1   The Reflection of Light   914

26-2   Forming Images with a Plane Mirror   916

26-3   Spherical Mirrors   919

26-4   Ray Tracing and the Mirror Equation   922

26-5   The Refraction of Light   928

26-6   Ray Tracing for Lenses   936

26-7   The Thin-Lens Equation   939

26-8   Dispersion and the Rainbow   941

Chapter Summary   943

Conceptual Questions   945

Problems and Conceptual Exercises   946

27   OPTICAL INSTRUMENTS   954

27-1   The Human Eye and the Camera   955

27-2   Lenses in Combination and Corrective Optics   958

27-3   The Magnifying Glass   966

27-4   The Compound Microscope   969

27-5   Telescopes   971

27-6   Lens Aberrations   974

Chapter Summary   975

Conceptual Questions   976

Problems and Conceptual Exercises   977

28   PHYSICAL OPTICS: INTERFERENCE AND DIFFRACTION   983

28-1   Superposition and Interference   984

28-2   Young's Two-Slit Experiment   986

28-3   Interference in Reflected Waves   990

28-4   Diffraction   998

28-5   Resolution   1002

28-6   Diffraction Gratings   1005

Chapter Summary   1009

Conceptual Questions   1010

Problems and Conceptual Exercises   1011

PART V   MODERN PHYSICS

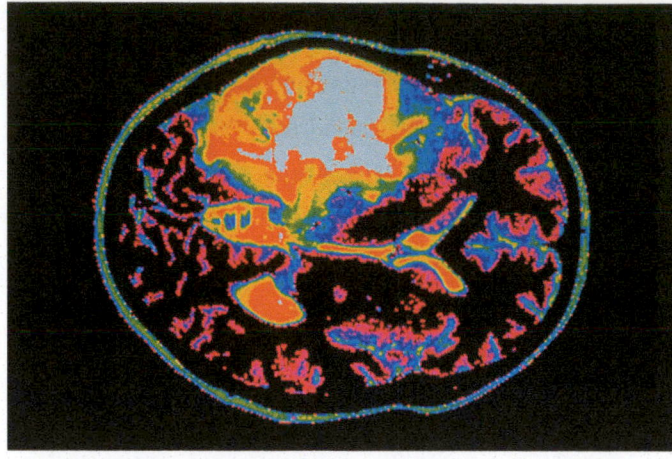

29   RELATIVITY   1017

29-1   The Postulates of Special Relativity   1018

29-2   The Relativity of Time and Time Dilation   1020

29-3   The Relativity of Length and Length Contraction   1025

29-4   The Relativistic Addition of Velocities   1028

29-5   Relativistic Momentum   1031

29-6   Relativistic Energy and $E = mc^2$ 1033

29-7   The Relativistic Universe   1038

29-8   General Relativity   1039

Chapter Summary   1043

Conceptual Questions   1045

Problems and Conceptual Exercises   1045

30   QUANTUM PHYSICS   1051

30-1   Blackbody Radiation and Planck's Hypothesis of Quantized Energy   1052

30-2   Photons and the Photoelectric Effect   1056

30-3   The Mass and Momentum of a Photon   1061

30-4   Photon Scattering and the Compton Effect   1062

30-5   The de Broglie Hypothesis and Wave–Particle Duality   1065

30-6   The Heisenberg Uncertainty Principle   1069

30-7   Quantum Tunneling   1073

Chapter Summary   1074

Conceptual Questions   1076

Problems and Conceptual Exercises   1076

31   ATOMIC PHYSICS   1082

31-1   Early Models of the Atom   1083

31-2   The Spectrum of Atomic Hydrogen   1084

31-3   Bohr's Model of the Hydrogen Atom   1088

31-4   de Broglie Waves and the Bohr Model   1094

31-5   The Quantum Mechanical Hydrogen Atom   1096

31-6   Multielectron Atoms and the Periodic Table   1100

31-7   Atomic Radiation   1104

Chapter Summary   1112

Conceptual Questions   1114

Problems and Conceptual Exercises   1114

32   NUCLEAR PHYSICS AND NUCLEAR RADIATION   1119

32-1   The Constituents and Structure of Nuclei   1120

32-2   Radioactivity   1125

32-3   Half-Life and Radioactive Dating   1132

32-4   Nuclear Binding Energy   1137

32-5   Nuclear Fission   1139

32-6   Nuclear Fusion   1142

32-7   Practical Applications of Nuclear Physics   1144

32-8   Elementary Particles   1147

32-9   Unified Forces and Cosmology   1150

Chapter Summary   1152

Conceptual Questions   1153

Problems and Conceptual Exercises   1154

APPENDICES

Appendix A   Basic Mathematical Tools   A-0

Appendix B   Typical Values   A-9

Appendix C   Planetary Data   A-10

Appendix D   Elements of Electrical Circuits   A-11

Appendix E   Periodic Table of the Elements   A-12

Appendix F   Properties of Selected Isotopes   A-13

Answers to Odd-Numbered End-of-Chapter Questions   A-16

Answers to Odd-Numbered Conceptual Questions   A-28

Credits   C-1

Index   I-1

PHYSICS

Big Ideas

1 Physics is the study of the laws of nature.

2 The basic units in physics are length, mass, and time.

3 Valid physics equations must be dimensionally consistent.

4 Some physical quantities are scalars; others are vectors.

Introduction to Physics

▲ Physics is a quantitative science, based on careful measurements of quantities such as mass, length, and time. In the measurement shown here, a naturalist determines the mass of a cooperative penguin. The length of the penguin, and its age, are also important data in monitoring the penguin's health.

The goal of physics is to gain a deeper understanding of the world in which we live. For example, the laws of physics allow us to predict the behavior of everything from rockets sent to the Moon, to integrated chips in computers, to lasers used to perform eye surgery. In short, everything in nature—from atoms and subatomic particles to solar systems and galaxies—obeys the laws of physics. In this chapter, we develop a common "language" of physics that will be used throughout this book.

Big Idea 1 Physics is the study of the laws of nature. These laws determine the behavior of all physical objects and processes in the universe.

1-1 Physics and the Laws of Nature

Physics is the study of the fundamental laws of nature, which, simply put, are the laws that underlie all physical phenomena in the universe. Remarkably, we have found that these laws can be expressed in terms of mathematical equations. As a result, it is possible to make precise, quantitative comparisons between the predictions of theory—derived from the mathematical form of the laws—and the observations of experiments. Physics, then, is a science rooted equally firmly in both theory and experiment. As physicists make new observations, they constantly test and—if necessary—refine the present theories.

What makes physics particularly fascinating is the fact that it relates to everything in the universe. There is a great beauty in the vision that physics brings to our view of the universe—namely, that all the complexity and variety that we see in the world around us, and in the universe as a whole, are manifestations of a few fundamental laws and principles. That we can discover and apply these basic laws of nature is both astounding and exhilarating.

For those not familiar with the subject, physics may seem to be little more than a confusing mass of formulas. Sometimes, in fact, these formulas can be the trees that block the view of the forest. For a physicist, however, the many formulas of physics are simply different ways of expressing a few fundamental ideas. It is the forest—the basic laws and principles of physical phenomena in nature—that is the focus of this text.

Enhance Your Understanding (Answers given at the end of the chapter)

1. The laws of physics apply to which of the following: **(a)** gravity, **(b)** electricity, **(c)** magnetism, **(d)** light, **(e)** atoms, or **(f)** all of these?

Section Review

- Physics combines both theory and experiment in the study of the laws of nature.

1-2 Units of Length, Mass, and Time

Big Idea 2 The basic units of physical quantities in the first part of this book are length, mass, and time. Later in the book, other units, like temperature and electric current, are introduced.

To make quantitative comparisons between the laws of physics and our experience of the natural world, certain basic physical quantities must be measured. The most common of these quantities are **length** (L), **mass** (M), and **time** (T). In fact, in the next several chapters these are the only quantities that arise. Later in the text, additional quantities, such as temperature and electric current, will be introduced as needed.

We begin by defining the units in which each of these quantities is measured. Once the units are defined, the values obtained in specific measurements can be expressed as multiples of them. For example, our unit of length is the **meter** (m). It follows, then, that a person who is 1.5 m tall has a height 1.5 times this unit of length. Similar comments apply to the unit of mass, the **kilogram**, and the unit of time, the **second**.

The detailed system of units used in this book was established in 1960 at the Eleventh General Conference on Weights and Measures in Paris, France, and goes by the name Système International d'Unités, or SI for short. Thus, when we refer to **SI units**, we mean units of meters (m), kilograms (kg), and seconds (s). Taking the first letter from each of these units leads to an alternate name that is often used—the **mks system**.

In the remainder of this section we define each of the SI units.

Length

Early units of length were often associated with the human body. For example, the Egyptians defined the cubit to be the distance from the elbow to the tip of the middle finger. Similarly, the foot was at one time defined to be the length of the royal

TABLE 1-1 Typical Distances

Distance from Earth to the nearest large galaxy (the Andromeda galaxy, M31)	2×10^{22} m
Diameter of our galaxy (the Milky Way)	1×10^{21} m
Distance from Earth to the nearest star (other than the Sun)	4×10^{16} m
One light-year	9.46×10^{15} m
Average radius of Pluto's orbit	6×10^{12} m
Distance from Earth to the Sun	1.5×10^{11} m
Radius of Earth	6.37×10^6 m
Length of a football field	10^2 m
Height of a person	2 m
Diameter of a CD	0.12 m
Diameter of the aorta	0.018 m
Diameter of a period in a sentence	5×10^{-4} m
Diameter of a red blood cell	8×10^{-6} m
Diameter of the hydrogen atom	10^{-10} m
Diameter of a proton	2×10^{-15} m

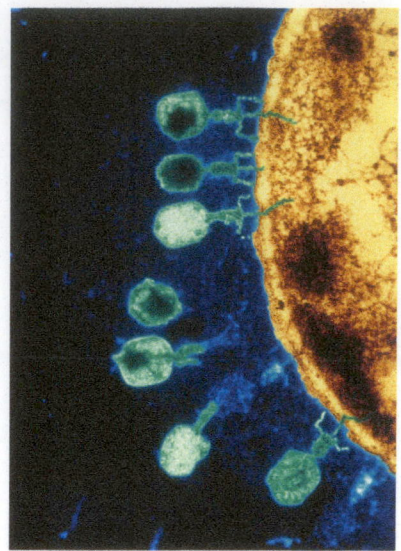

(a)

(b)

▲ **FIGURE 1-1 (a)** The size of these viruses, seen here attacking a bacterial cell, is about 10^{-7} m. **(b)** The diameter of this typical galaxy is about 10^{21} m. (How many viruses would it take to span the galaxy?)

foot of King Louis XIV. As colorful as these units may be, they are not particularly reproducible—at least not to great precision.

In 1793 the French Academy of Sciences, seeking a more objective and reproducible standard, decided to define a unit of length equal to one ten-millionth the distance from the North Pole to the equator. This new unit was named the metre (from the Greek *metron* for "measure"). The preferred spelling in the United States is *meter*. This definition was widely accepted, and in 1799 a "standard" meter was produced. It consisted of a platinum-iridium alloy rod with two marks on it one meter apart.

Since 1983 we have used an even more precise definition of the meter, based on the speed of light in a vacuum:

> One meter is defined to be the distance traveled by light in a vacuum in 1/299,792,458 of a second.

No matter how its definition is refined, however, a meter is still about 3.28 feet, which is roughly 10 percent longer than a yard. A list of typical lengths is given in Table 1-1, and an illustration of the range of lengths is given in **FIGURE 1-1**.

Mass

In SI units, mass is measured in kilograms. Unlike the meter, the kilogram is not based on any natural physical quantity. By convention, the kilogram is defined as follows:

> The kilogram is the mass of a particular platinum-iridium alloy cylinder at the International Bureau of Weights and Measures in Sèvres, France. The cylinder, referred to as the standard kilogram, is shown in **FIGURE 1-2**.

To put the kilogram in everyday terms, a quart of milk has a mass slightly less than 1 kilogram. Additional masses, in kilograms, are given in Table 1-2.

It's important to note that we do not define the kilogram to be the *weight* of the platinum-iridium cylinder. In fact, weight and mass are quite different quantities, even though they are often confused in everyday language. Mass is an intrinsic, unchanging property of an object. Weight, in contrast, is a measure of the gravitational force acting on an object, which can vary depending on the object's location. For example, if you are fortunate enough to travel to Mars someday, you will find that your weight is less than on Earth, though your mass is unchanged. The force of gravity will be discussed in detail in Chapter 12.

TABLE 1-2 Typical Masses

Galaxy (Milky Way)	4×10^{41} kg
Sun	2×10^{30} kg
Earth	5.97×10^{24} kg
Space shuttle	2×10^6 kg
Elephant	5400 kg
Automobile	1200 kg
Human	70 kg
Baseball	0.15 kg
Honeybee	1.5×10^{-4} kg
Red blood cell	10^{-13} kg
Bacterium	10^{-15} kg
Hydrogen atom	1.67×10^{-27} kg
Electron	9.11×10^{-31} kg

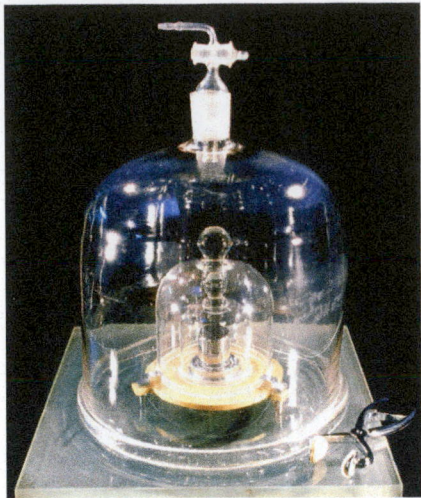

▲ **FIGURE 1-2** The standard kilogram, a cylinder of platinum and iridium 0.039 m in height and diameter, is kept under carefully controlled conditions in Sèvres, France. Exact replicas are maintained in other laboratories around the world.

**PHYSICS
IN CONTEXT
Looking Ahead**

The three physical dimensions introduced in this chapter—mass, length, and time—are the only ones we'll use until Chapter 16, when temperature is introduced. Other quantities found in the next several chapters, like force, momentum, and energy, are combinations of these three basic dimensions.

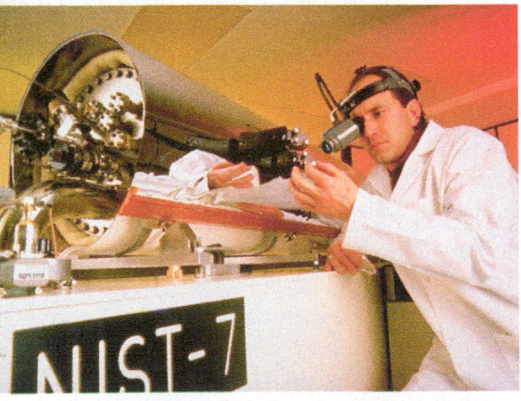

▲ **FIGURE 1-3** This atomic clock, which keeps time on the basis of radiation from cesium atoms, is accurate to about two ten-millionths of a second per year. (How much time would it take for it to gain or lose an hour?)

Time

Nature has provided us with a fairly accurate timepiece in the revolving Earth. In fact, prior to 1956 the mean solar day was defined to consist of 24 hours, with 60 minutes per hour, and 60 seconds per minute, for a total of $(24)(60)(60) = 86,400$ seconds. Even the rotation of the Earth is not completely regular, however.

Today, the most accurate timekeepers are "atomic clocks," like the one in **FIGURE 1-3**, which are based on characteristic frequencies of radiation emitted by certain atoms. These clocks have typical accuracies of about 1 second in 6 million years. The atomic clock used for defining the second operates with cesium-133 atoms. In particular, the second is defined as follows:

> One second is defined to be the time it takes for radiation from a cesium-133 atom to complete 9,192,631,770 cycles of oscillation.

A range of characteristic time intervals is given in Table 1-3.

Atomic Clocks and Official Time The nation's time and frequency standard is determined by a *cesium fountain atomic clock* developed at the National Institute of Standards and Technology (NIST) in Boulder, Colorado. The fountain atomic clock, designated NIST-F2, produces a "fountain" of cesium atoms that are projected upward in a vacuum to a height of about a meter. It takes roughly a second for the atoms to rise and fall through this height (as we shall see in the next chapter), and during this relatively long period of time, the frequency of their oscillation can be measured with great precision. In fact, the NIST-F2 will gain or lose no more than one second in every 300 million years of operation.

Atomic clocks are almost commonplace these days. For example, the satellites that participate in the Global Positioning System (GPS) actually carry atomic clocks with them in orbit. This allows them to make the precise time measurements that are needed for an equally precise determination of a GPS user's position and speed. Similarly, the "atomic clocks" that are advertised for use in the home, while not atomic in their operation, nonetheless get their time from radio signals sent out from the atomic clocks at NIST in Boulder. You can access the official U.S. time on your computer by going to time.gov on the Web.

Other Systems of Units and Standard Prefixes

Although SI units are used throughout most of this book, and are used almost exclusively in scientific research and in industry, we will occasionally refer to other systems that you may encounter from time to time.

For example, a system of units similar to the mks system, though comprised of smaller units, is the **cgs system**, which stands for centimeter (cm), gram (g), and second (s). In addition, the British engineering system is often encountered in everyday usage in the United States. Its basic units are the slug for mass, the foot (ft) for length, and the second (s) for time.

Finally, multiples of the basic units are common no matter which system is used. Standard prefixes are used to designate common multiples in powers of ten. For example, the prefix *kilo* means one thousand, or, equivalently, 10^3. Thus, 1 kilogram is 10^3 grams, and 1 kilometer is 10^3 meters. Similarly, *milli* is the prefix for one thousandth, or 10^{-3}. Thus, a millimeter is 10^{-3} meter, and so on. The most common prefixes are listed in Table 1-4.

EXERCISE 1-1 CONVERTING UNITS

   a. A baseball has a diameter of 75 millimeters. Give the diameter in meters and kilometers.

   b. A house sells for 450,000 dollars. Express the price of the house in kilodollars and megadollars.

REASONING AND SOLUTION
Refer to Table 1-4 for a list of prefixes and their corresponding powers of ten.

   a. 0.075 m, 0.000075 km

   b. 450 kilodollars, 0.450 megadollars

Enhance Your Understanding　　　(Answers given at the end of the chapter)

2. Rank the following lengths (A, B, C, and D) in order of increasing value. Indicate ties where appropriate. A = 2 millimeters; B = 10 micrometers; C = 0.1 kilometers; D = 10 centimeters.

Section Review

- The SI units of length, mass, and time are the meter, kilogram, and second, respectively.

TABLE 1-3　Typical Times

Age of the universe	4×10^{17} s
Age of the Earth	1.4×10^{17} s
Existence of human species	6×10^{12} s
Human lifetime	2×10^{9} s
One year	3×10^{7} s
One day	8.6×10^{4} s
Time between heart-beats	0.8 s
Human reaction time	0.1 s
One cycle of a high-pitched sound wave	5×10^{-5} s
One cycle of an AM radio wave	10^{-6} s
One cycle of a visible light wave	2×10^{-15} s

1-3 Dimensional Analysis

In physics, when we speak of the **dimension** of a physical quantity, we refer to the *type* of quantity in question, regardless of the units used in the measurement. For example, a distance measured in cubits and another distance measured in light-years both have the same dimension—length. The same is true of compound units such as velocity, which has the dimensions of length per unit time (length/time). A velocity measured in miles per hour has the same dimensions—length/time—as a velocity measured in inches per century.

Now, any valid formula in physics must be **dimensionally consistent**; that is, each term in the equation must have the same dimensions. It simply doesn't make sense to add a distance to a time, for example, any more than it makes sense to add apples and oranges. They are different things.

Checking Dimensional Consistency and Dimensional Analysis To check the dimensional consistency of an equation, it's convenient to introduce a special notation for the dimension of a quantity. We will use square brackets, [ ], for this purpose. Thus, if x represents a distance, which has dimensions of length [L], we write this as $x = [L]$. Similarly, a velocity, v, has dimensions of length [L] per time [T]; thus we write $v = [L]/[T]$ to indicate its dimensions. Acceleration, a, which is the change in velocity per time, has the dimensions $a = ([L]/[T])/[T] = [L]/[T^2]$. The dimensions of some common physical quantities are listed in Table 1-5.

Let's use this notation to check the dimensional consistency of a simple equation. Consider the following formula:

$$x = x_0 + vt$$

In this equation, x and x_0 represent distances, v is a velocity, and t is time. Writing out the dimensions of each term, we have

$$[L] = [L] + \frac{[L]}{[T]}[T]$$

It might seem that the last term has different dimensions than the other two. However, dimensions obey the same rules of algebra as other quantities. Thus the dimensions of time cancel in the last term:

$$[L] = [L] + \frac{[L]}{[T]}[T] = [L] + [L]$$

As a result, we see that each term in this formula has the same dimension, [L]. This type of calculation with dimensions is referred to as **dimensional analysis.**

TABLE 1-4　Common Prefixes

Power	Prefix	Abbreviation
10^{15}	peta	P
10^{12}	tera	T
10^{9}	giga	G
10^{6}	mega	M
10^{3}	kilo	k
10^{2}	hecto	h
10^{1}	deka	da
10^{-1}	deci	d
10^{-2}	centi	c
10^{-3}	milli	m
10^{-6}	micro	μ
10^{-9}	nano	n
10^{-12}	pico	p
10^{-15}	femto	f

Big Idea 3 Valid equations in physics must be dimensionally consistent. This means that each term in the equation must have the same dimensions.

EXERCISE 1-2　ANALYZING DIMENSIONS

Show that $x = x_0 + v_0 t + \frac{1}{2}at^2$ is dimensionally consistent. The quantities x and x_0 are distances, v_0 is a velocity, and a is an acceleration.

CONTINUED

TABLE 1-5 Dimensions of Some Common Physical Quantities

Quantity	Dimension
Distance	$[L]$
Area	$[L^2]$
Volume	$[L^3]$
Velocity	$[L]/[T]$
Acceleration	$[L]/[T^2]$
Energy	$[M][L^2]/[T^2]$

PHYSICS IN CONTEXT
Looking Ahead

Dimensional analysis is used frequently in the coming chapters to verify that each term in an equation has the correct dimensions. We will also use dimensional analysis to help derive some results, such as the speed of waves on a string in Chapter 14.

REASONING AND SOLUTION

Use the dimensions given in Table 1-5 for each term in the equation. Cancel dimensions where appropriate.

Original equation: $\quad x = x_0 + v_0 t + \frac{1}{2}at^2$

Dimensions: $\quad [L] = [L] + \frac{[L]}{[T]}[T] + \frac{[L]}{[T^2]}[T^2] = [L] + [L] + [L]$

Notice that $\frac{1}{2}$ is ignored in this analysis because it is simply a numerical factor, and has no dimensions.

Later in this text you will derive your own formulas from time to time. As you do so, it is helpful to check dimensional consistency at each step of the derivation. If at any time the dimensions don't agree, you will know that a mistake has been made, and you can go back and look for it. If the dimensions check, however, it's not a guarantee the formula is correct—after all, dimensionless numerical factors, like $\frac{1}{2}$ or 2, don't show up in a dimensional check.

Enhance Your Understanding
(Answers given at the end of the chapter)

3. Give the dimensions of each of the following quantities (A, B, C, D), given that x is a distance, v is a velocity, a is an acceleration, and t is a time. A = at; B = v/t; C = $0.5x/t^2$; D = $2ax$.

Section Review

- The dimension of a quantity is the type of quantity it is—such as length, mass, or time.
- Each term in a valid physics equation must have the same dimensions.

▲ **FIGURE 1-4** Every measurement has some degree of uncertainty associated with it. How precise would you expect this measurement to be?

1-4 Significant Figures, Scientific Notation, and Round-Off Error

In this section we discuss a few issues regarding the numerical values that arise in scientific measurements and calculations.

Significant Figures

When a length, a mass, or a time is measured in a scientific experiment, the result is known only to within a certain accuracy. The inaccuracy or uncertainty can be caused by a number of factors, as illustrated in **FIGURE 1-4**, ranging from limitations of the measuring device itself to limitations associated with the senses and the skill of the person performing the experiment. In any case, the fact that observed values of experimental quantities have inherent uncertainties should always be kept in mind when performing calculations with those values.

The level of uncertainty in a numerical value is indicated by the number of **significant figures** it contains. We define significant figures as follows:

Number of Significant Figures
The number of significant figures ("sig figs" for short) in a physical quantity is equal to the number of digits in it that are known with certainty.

As an example, suppose you want to determine the walking speed of your pet tortoise. To do so, you measure the time, t, it takes for the tortoise to walk a distance, d, and then you calculate the quotient, d/t. When you measure the distance with a ruler, which has one tick mark per millimeter, you find that $d = 21.4$ cm. Each of these three digits is known with certainty, though the digit that follows the 4 is uncertain. Hence, we say that d has *three* significant figures. Similarly, you measure the time with an old pocket watch, and determine that $t = 8.5$ s. It follows that t is known to only *two* significant figures, because you are not certain of the digit that follows the 5.

We would now like to calculate the speed of the tortoise. Using the above values for d and t, and a calculator with eight digits in its display, we find $(21.4 \text{ cm})/(8.5 \text{ s}) = 2.5176470$ cm/s. Clearly, such an accurate value for the speed is unjustified, considering the limitations of our measurements. In general, the number of significant figures that result when we multiply or divide physical quantities is given by the following rule of thumb:

Significant Figures When Multiplying or Dividing
The number of significant figures after multiplication or division is equal to the number of significant figures in the *least* accurately known quantity.

In our speed calculation, for example, we know the distance to three significant figures, but we know the time to only two significant figures. As a result, the speed should be given with just two significant figures: $d/t = (21.4 \text{ cm})/(8.5 \text{ s}) = 2.5$ cm/s. Thus, 2.5 cm/s is our best estimate for the tortoise's speed, given the uncertainty in our measurements.

EXAMPLE 1-3   IT'S THE TORTOISE BY A HARE

A tortoise races a rabbit by walking with a constant speed of 2.51 cm/s for 12.27 s. How much distance does the tortoise cover?

PICTURE THE PROBLEM
The race between the rabbit and the tortoise is shown in our sketch. The rabbit pauses to eat a carrot while the tortoise walks with a constant speed.

REASONING AND STRATEGY
The distance covered by the tortoise is the speed of the tortoise multiplied by the time during which it walks.

Known   Constant speed = 2.51 cm/s; walking time = 12.27 s.
Unknown Distance, d = ?

SOLUTION

1. Multiply the speed by the time to find the distance d:

$$d = (\text{speed})(\text{time})$$
$$= (2.51 \text{ cm/s})(12.27 \text{ s}) = 30.8 \text{ cm}$$

INSIGHT
If we simply multiply 2.51 cm/s by 12.27 s, we obtain 30.7977 cm. We don't give all of these digits in our answer, however. In particular, because the quantity that is known with the least accuracy (the speed) has only three significant figures, we give a result with three significant figures. In addition, notice that the third digit in our answer has been rounded up from 7 to 8 because the number that follows 7 is greater than or equal to 5.

PRACTICE PROBLEM
How much time does it take for the tortoise to walk 17 cm?   [**Answer:** $t = (17 \text{ cm})/(2.51 \text{ cm/s}) = 6.8$ s]

Some related homework problems: Problem 14, Problem 18

The distance of 17 cm in the above Practice Problem has only two significant figures, because we don't know the digits to the right of the decimal place. If the distance were given as 17.0 cm, on the other hand, it would have three significant figures. The role of significant figures in a real-world situation is illustrated in **FIGURE 1-5**.

When physical quantities are added or subtracted, we use a slightly different rule of thumb. In this case, the rule involves the number of decimal places in each of the terms:

Significant Figures When Adding or Subtracting
The number of decimal places after addition or subtraction is equal to the smallest number of decimal places in any of the individual terms.

▲ **FIGURE 1-5** The time measurements for a race like this must be accurate to several significant figures. This level of accuracy is needed to determine both the winning time, and the order in which the other runners finished.

Thus, if you make a time measurement of 16.74 s and then a subsequent time measurement of 5.1 s, the total time of the two measurements should be given as 16.74 s + 5.1 s = 21.8 s, rather than 21.84 s.

EXERCISE 1-4   FINDING THE TOTAL WEIGHT

You and a friend go fishing. You catch a fish that weighs 10.7 lb, and your friend catches a fish that weighs 8.35 lb. What is the combined weight of the two fish?

REASONING AND SOLUTION
Simply adding the two numbers gives 19.05 lb. According to our rule of thumb, however, the final result must have only a single decimal place (corresponding to the term with the smallest number of decimal places). Rounding off to one decimal place, then, gives 19.1 lb as the accepted result.

Scientific Notation

The number of significant figures in a given quantity may be ambiguous due to the presence of zeros at the beginning or end of the number. For example, if a distance is stated to be 2500 m, the two zeros could be significant figures, or they could be zeros that simply show where the decimal point is located. If the two zeros are significant figures, the uncertainty in the distance is roughly a meter; if they are not significant figures, however, the uncertainty is about 100 m.

To remove this type of ambiguity, we can write the distance in **scientific notation**—that is, as a number of order unity times an appropriate power of ten. Thus, in this example, we would express the distance as 2.5×10^3 m if there are only two significant figures, or as 2.500×10^3 m to indicate four significant figures. Likewise, a time given as 0.000036 s has only two significant figures—the preceding zeros only serve to fix the decimal point. If this quantity were known to three significant figures, we would write it as 0.0000360 s, or equivalently as 3.60×10^{-5} s, to remove any ambiguity. See Appendix A for a more detailed discussion of scientific notation.

EXERCISE 1-5   SIGNIFICANT FIGURES

How many significant figures are there in (a) 0.0210, (b) 5.060, (c) 7.10×10^{-4}, (d) 1.0×10^5?

REASONING AND SOLUTION
Preceding zeros don't count when determining the number of significant figures, but trailing zeros and zeros within a number do count. As a result, the number of significant figures is as follows: (a) 3, (b) 4, (c) 3, (d) 2.

Round-Off Error

Finally, even if you perform all your calculations to the same number of significant figures as in the text, you may occasionally obtain an answer that differs in its last digit from that given in the book. In most cases this is not an issue as far as understanding the physics is concerned—usually it is due to **round-off error**.

Round-off error occurs when numerical results are rounded off at different times during a calculation. To see how this works, let's consider a simple example. Suppose you are shopping for knickknacks, and you buy one item for $2.21, plus 8 percent sales tax. The total price is $2.3868 or, rounded off to the nearest penny, $2.39. Later, you buy another item for $1.35. With tax this becomes $1.458 or, again to the nearest penny, $1.46. The total expenditure for these two items is $2.39 + $1.46 = $3.85.

Now, let's do the rounding off in a different way. Suppose you buy both items at the same time for a total before-tax price of $2.21 + $1.35 = $3.56. Adding in the 8 percent tax gives $3.8448, which rounds off to $3.84, one penny different from the previous amount. This type of discrepancy can also occur in physics problems. In general, it's a good idea to keep one extra digit throughout your calculations whenever possible, rounding off only the final result. But, while this practice can help to reduce the likelihood of round-off error, there is no way to avoid it in every situation.

4. Rank the following numbers in order of increasing number of significant figures. Indicate ties where appropriate. A = 1; B = 1001; C = 10.01; D = 123.

Section Review

- The number of significant figures in a physical quantity is the number of digits known with certainty. The greater the number of significant figures, the more accurately the quantity is known.

- Round-off error occurs when numbers are rounded off at different points in a calculation. The small discrepancy that results is not significant.

1-5 Converting Units

It's often convenient to convert from one set of units to another. For example, suppose you want to convert 316 ft to its equivalent in meters. Looking at the conversion factors on the inside front cover of the text, we see that

$$1\ \text{m} = 3.281\ \text{ft}$$

Equivalently,

$$\frac{1\ \text{m}}{3.281\ \text{ft}} = 1$$

Now, to make the conversion, we simply multiply 316 ft by this expression, which is equivalent to multiplying by 1:

$$(316\ \cancel{\text{ft}})\left(\frac{1\ \text{m}}{3.281\ \cancel{\text{ft}}}\right) = 96.3\ \text{m}$$

We write the conversion factor in this particular way, as 1 m divided by 3.281 ft, so that the units of feet cancel out, leaving the final result in the desired units of meters.

  Of course, we can just as easily convert from meters to feet if we use the reciprocal of this conversion factor—which is also equal to 1:

$$1 = \frac{3.281\ \text{ft}}{1\ \text{m}}$$

For example, a distance of 26.4 m is converted to feet by canceling out the units of meters, as follows:

$$(26.4\ \cancel{\text{m}})\left(\frac{3.281\ \text{ft}}{1\ \cancel{\text{m}}}\right) = 86.6\ \text{ft}$$

Thus, we see that converting units is as easy as multiplying by 1—because that's really what you're doing. A real-world example of unit conversion is shown in **FIGURE 1-6**.

▲ **FIGURE 1-6** From this sign, you can calculate factors for converting miles to kilometers and vice versa. (Why do you think the conversion factors seem to vary for different destinations?)

EXAMPLE 1-6   A HIGH-VOLUME ITEM

The interior of a popular microwave oven has a width W = 15.5 in., a depth D = 14.5 in., and a height H = 9.25 in. What is the interior volume of the oven in SI units?

PICTURE THE PROBLEM
In our sketch we picture the oven, and indicate the dimensions given in the problem statement.

REASONING AND STRATEGY
We begin by converting the width, depth, and height of the oven to meters. Once this is done, the volume in SI units is simply the product of the three dimensions.

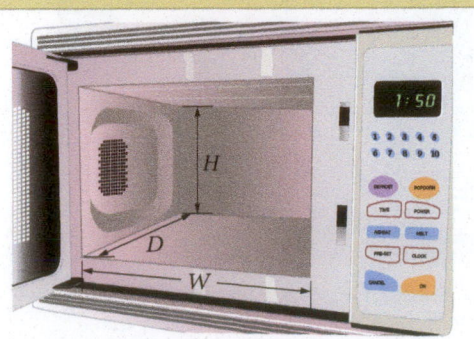

CONTINUED

| **Known** | Width, $W = 15.5$ in.; depth, $D = 14.5$ in.; height, $H = 9.25$ in. |
| **Unknown** | Volume, $V = ?$ |

SOLUTION

1. Convert the width of the oven to meters:

$$W = (15.5 \text{ in.})\left(\frac{1 \text{ ft}}{12 \text{ in.}}\right)\left(\frac{1 \text{ m}}{3.281 \text{ ft}}\right) = 0.394 \text{ m}$$

2. Convert the depth to meters:

$$D = (14.5 \text{ in.})\left(\frac{1 \text{ ft}}{12 \text{ in.}}\right)\left(\frac{1 \text{ m}}{3.281 \text{ ft}}\right) = 0.368 \text{ m}$$

3. Convert the height to meters:

$$H = (9.25 \text{ in.})\left(\frac{1 \text{ m}}{39.4 \text{ in.}}\right) = 0.235 \text{ m}$$

4. Calculate the volume of the oven in cubic meters:

$$V = W \times D \times H = (0.394 \text{ m})(0.368 \text{ m})(0.235 \text{ m}) = 0.0341 \text{ m}^3$$

INSIGHT

This volume is just 3.41 percent of a cubic meter. In terms of cubic feet, which is the unit used by manufacturers, the volume is 1.20 cu ft. (Note: We used a different conversion factor in Step 3, just to illustrate that conversions can be carried out in different ways.)

PRACTICE PROBLEM

What is the volume of the oven if its width is one-half of a yard, and the other dimensions are unchanged?
[**Answer:** $V = 0.0395 \text{ m}^3$]

Some related homework problems: Problem 21, Problem 22

Converting More Than One Unit at a Time   The procedure outlined above can be applied to conversions involving any number of units. For instance, if you walk at 3.00 mi/h, how fast is that in m/s? In this case we need the following additional conversion factors:

$$1 \text{ mi} = 5280 \text{ ft} \qquad 1 \text{ h} = 3600 \text{ s}$$

With these factors at hand, we carry out the conversion as follows:

$$(3.00 \text{ mi/h})\left(\frac{5280 \text{ ft}}{1 \text{ mi}}\right)\left(\frac{1 \text{ m}}{3.281 \text{ ft}}\right)\left(\frac{1 \text{ h}}{3600 \text{ s}}\right) = 1.34 \text{ m/s}$$

In each conversion factor, the numerator is equal to the denominator. In addition, each conversion factor is written in such a way that the unwanted units cancel, leaving just meters per second in our final result.

QUICK EXAMPLE 1-7   FIND THE SPEED OF BLOOD

Blood in the human aorta (**FIGURE 1-7**) can have speeds of 39.7 cm/s. How fast is this in **(a)** ft/min and **(b)** mi/h?

REASONING AND SOLUTION

To convert a speed to different units it may be necessary to convert both the units of length and those of time.

Part (a) Convert centimeters to feet, and seconds to minutes:

$$39.7 \text{ cm/s} = (39.7 \text{ cm/s})\left(\frac{1 \text{ m}}{100 \text{ cm}}\right)\left(\frac{3.281 \text{ ft}}{1 \text{ m}}\right)\left(\frac{60 \text{ s}}{1 \text{ min}}\right)$$

$$= 78.2 \text{ ft/min}$$

Part (b) Convert centimeters to miles, and seconds to hours:

$$39.7 \text{ cm/s} = (39.7 \text{ cm/s})\left(\frac{1 \text{ m}}{100 \text{ cm}}\right)\left(\frac{3.281 \text{ ft}}{1 \text{ m}}\right)\left(\frac{1 \text{ mi}}{5280 \text{ ft}}\right)\left(\frac{3600 \text{ s}}{1 \text{ h}}\right)$$

$$= 0.888 \text{ mi/h}$$

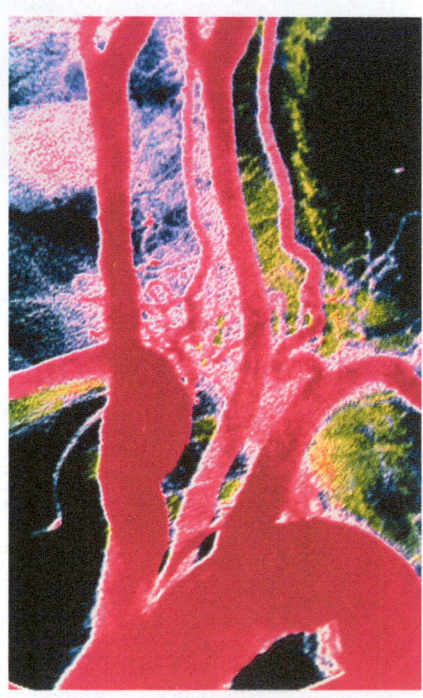

▲ **FIGURE 1-7** Major blood vessels branch from the aorta (bottom), the artery that receives blood directly from the heart.

5. To convert a length of 6 ft to its equivalent in meters, which of the following conversion factors (A, B, C, D) should we use? $A = \left(\dfrac{1 \text{ mi}}{5280 \text{ ft}}\right)$; $B = \left(\dfrac{3.281 \text{ ft}}{1 \text{ m}}\right)$; $C = \left(\dfrac{3.281 \text{ m}}{1 \text{ ft}}\right)$; $D = \left(\dfrac{1 \text{ m}}{3.281 \text{ ft}}\right)$.

Section Review

• To convert units, multiply by a conversion factor that cancels one unit and replaces it with another.

1-6 Order-of-Magnitude Calculations

An **order-of-magnitude** calculation is a rough "ballpark" estimate designed to be accurate to within a factor of about 10. One purpose of such a calculation is to give a quick idea of what order of magnitude should be expected from a more complete calculation. If an order-of-magnitude calculation indicates that a distance should be on the order of 10^4 m, for example, and your calculator gives an answer on the order of 10^7 m, then there is an error somewhere that needs to be resolved.

For example, suppose you would like to estimate the speed of a cliff diver on entering the water. First, the cliff may be 20 or 30 feet high; thus in SI units we would say that the order of magnitude of the cliff's height is 10 m—certainly not 1 m or 10^2 m. Next, the diver hits the water something like a second later—certainly not 0.1 s later or 10 s later. Thus, a reasonable order-of-magnitude estimate of the diver's speed is 10 m/1 s = 10 m/s, or roughly 20 mi/h. If you do a detailed calculation and your answer is on the order of 10^4 m/s, you probably entered one of your numbers incorrectly. Order-of-magnitude calculations like this are often referred to as "Fermi problems" in honor of Italian physicist Enrico Fermi (**FIGURE 1-8**), who was a master of such calculations.

Another reason for doing an order-of-magnitude calculation is to get a feeling for what size numbers we are talking about in situations where a precise count is not possible. This is illustrated in the following Example.

▲ **FIGURE 1-8** Enrico Fermi (1901–1954) was renowned for his ability to pose and solve interesting order-of-magnitude problems. A winner of the 1938 Nobel Prize in physics, Fermi would ask his classes to obtain order-of-magnitude estimates for questions such as "How many piano tuners are there in Chicago?"

EXAMPLE 1-8 ESTIMATION: HOW MANY RAINDROPS IN A STORM?

A thunderstorm drops half an inch (~ 0.01 m) of rain on Washington, DC, which covers an area of about 70 square miles ($\sim 10^8$ m^2). Estimate the number of raindrops that fell during the storm.

PICTURE THE PROBLEM
Our sketch shows an area $A = 10^8$ m^2 covered to a depth $d = 0.01$ m by rainwater from the storm. Each drop of rain is approximated by a small sphere with a radius of $r = (4 \text{ mm})/2 = 2$ mm.

REASONING AND STRATEGY
To find the number of raindrops, we first calculate the volume of water required to cover 10^8 m^2 to a depth of 0.01 m. Next, we calculate the volume of an individual drop of rain, recalling that the volume of a sphere of radius r is $4\pi r^3/3$. We estimate the radius of a raindrop to be about 2 mm. Finally, dividing the volume of a drop into the volume of water that fell during the storm gives the number of drops.

Known   Area of rainfall, $A = 10^8$ m^2; depth of rain, $d = 0.01$ m; radius of raindrop, $r = 2$ mm.
Unknown Number of raindrops = ?

CONTINUED

SOLUTION

1. Calculate the order of magnitude of the volume of water, V_{water}, that fell during the storm:

$$V_{water} = Ad = (10^8 \text{ m}^2)(0.01 \text{ m}) \approx 10^6 \text{ m}^3$$

2. Calculate the order of magnitude of the volume of a drop of rain, V_{drop}. Recall that if the diameter of a drop is 4 mm, its radius is $r = 2 \text{ mm} = 0.002 \text{ m}$:

$$V_{drop} = \frac{4}{3}\pi r^3 \approx \frac{4}{3}\pi(0.002 \text{ m})^3 \approx 10^{-8} \text{ m}^3$$

3. Divide V_{drop} into V_{water} to find the order of magnitude of the number of drops that fell during the storm:

$$number\ of\ raindrops \approx \frac{V_{water}}{V_{drop}} \approx \frac{10^6 \text{ m}^3}{10^{-8} \text{ m}^3} = 10^{14}$$

INSIGHT

Thus, the number of raindrops in this one small storm is roughly 10,000 times greater than the current population of the Earth.

PRACTICE PROBLEM

If a storm pelts Washington, DC, with 10^{15} raindrops, how many inches of rain fall on the city?   [**Answer:** About 5 inches]

Some related homework problems: Problem 35, Problem 38

Appendix B provides a number of interesting "typical values" for length, mass, speed, acceleration, and many other quantities. You may find these to be of use in making your own order-of-magnitude estimates.

Enhance Your Understanding   (Answers given at the end of the chapter)

6. Give an order-of-magnitude estimate for the speed at which fingernails grow. Express your answer in meters per second.

Section Review

*   Order-of-magnitude estimates are rough calculations that give answers accurate to within a factor of 10.

1-7 Scalars and Vectors

Physical quantities are sometimes defined solely in terms of a number and the corresponding unit, like the volume of a room or the temperature of the air it contains. Other quantities require both a numerical value *and* a direction. For example, suppose a car is traveling at a rate of 25 m/s in a direction that is due north. Both pieces of information—the rate of travel (25 m/s) and the direction (north)—are required to fully specify the motion of the car. The rate of travel is given the name **speed**; the rate of travel combined with the direction is referred to as the **velocity**.

In general, quantities that are specified by a numerical value only are referred to as **scalars**; quantities that require both a numerical value and a direction are called **vectors**:

*   A scalar is a numerical value, expressed in terms of appropriate units. An example is the temperature of a room or the speed of a car.

*   A vector is a physical quantity with both a numerical value and a direction. An example is the velocity of a car.

All the physical quantities discussed in this text are either vectors or scalars. The properties of numbers (scalars) are well known, but the properties of vectors are sometimes less well known—though no less important. For this reason, you will find that Chapter 3 is devoted entirely to a discussion of vectors in two and three dimensions and, more specifically, to how they are used in physics.

Big Idea 4 Some physical quantities are described by scalars, which are simply numbers. Other physical quantities are described by vectors, which combine both a numerical value and a direction.

Vectors in One Dimension The rather straightforward special case of vectors in one dimension is discussed in Chapter 2. There, we shall see that the direction of a velocity vector, for example, can be only to the left or to the right, up or down, and so on. That is, only two choices are available for the direction of a vector in one dimension.

This is illustrated in **FIGURE 1-9**, where we see two cars, each traveling with a speed of 25 m/s. Notice that the cars are traveling in opposite directions, with car 1 moving to the right and car 2 moving to the left. We indicate the direction of travel with a plus sign for motion to the right, and a negative sign for motion to the left. Thus, the velocity of car 1 is written $v_1 = +25$ m/s, and the velocity of car 2 is $v_2 = -25$ m/s. The speed of each car is the absolute value, or **magnitude,** of the velocity; that is, speed $= |v_1| = |v_2| = 25$ m/s.

Whenever we deal with one-dimensional vectors, we shall indicate their direction with the appropriate sign. Many examples are found in Chapter 2 and, again, in later chapters where the simplicity of one dimension can again be applied.

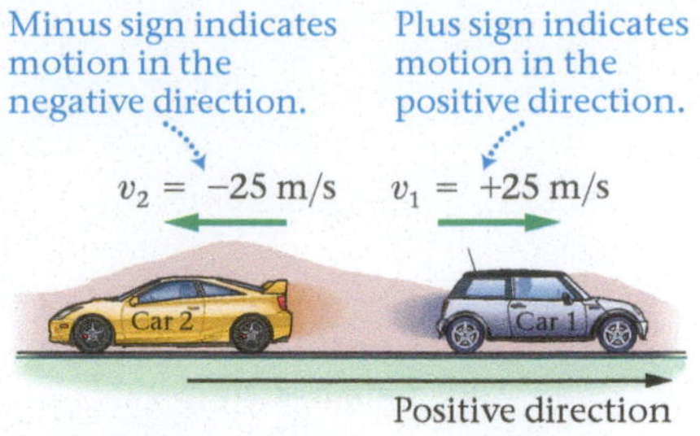

Enhance Your Understanding (Answers given at the end of the chapter)

7. **(a)** Can two cars have the same speed but different velocities? Explain. **(b)** Can two cars have the same velocity but different speeds? Explain.

Section Review

- Scalars are numbers. Vectors have both a numerical value and a direction.
- In one dimension, the direction of a vector is designated by its sign.

Minus sign indicates motion in the negative direction.

Plus sign indicates motion in the positive direction.

$v_2 = -25$ m/s $v_1 = +25$ m/s

Positive direction

▲ **FIGURE 1-9 Velocity vectors in one dimension** The two cars shown in this figure have equal speeds of 25 m/s but are traveling in opposite directions. To indicate the direction of travel, we first choose a positive direction (to the right in this case), and then give appropriate signs to the velocity of each car. For example, car 1 moves to the right, and hence its velocity is positive, $v_1 = +25$ m/s; the velocity of car 2 is negative, $v_2 = -25$ m/s, because it moves to the left.

1-8 Problem Solving in Physics

Physics is a lot like swimming—you have to learn by doing. You could read a book on swimming and memorize every word in it, but when you jump into a pool the first time you are going to have problems. Similarly, you could read this book carefully, memorizing every formula in it, but when you finish, you still haven't learned physics. To learn physics, you have to go beyond passive reading; you have to interact with physics and experience it by doing problems.

Problem-Solving Tips In what follows, we present a general overview of problem solving in physics. The suggestions given below, which apply to problems in all areas of physics, should help to develop a systematic approach.

We should emphasize at the outset that there is no recipe for solving problems in physics—it is a creative activity. In fact, the opportunity to be creative is one of the attractions of physics. The following suggestions, then, are not intended as a rigid set of steps that must be followed like the steps in a computer program. Rather, they provide a general guideline that experienced problem solvers find to be effective.

Problem-Solving Suggestions

- **Read the problem carefully** Before you can solve a problem, you need to know exactly what information it gives and what it asks you to determine. Some information is given explicitly, as when a problem states that a person has a mass of 70 kg. Other information is implicit; for example, saying that a ball is dropped from rest means that its initial speed is zero. Clearly, a *careful* reading is the essential first step in problem solving.

- **Sketch the system** This may seem like a step you can skip—but don't. A sketch helps you to acquire a physical feeling for the system. It also provides an opportunity to label those quantities that are known and those that are to be determined.

- **Visualize the physical process** Try to visualize what is happening in the system as if you were watching it in a movie. Your sketch should help. This step ties in closely with the next step.

- **Strategize** This may be the most difficult, but at the same time the most creative, part of the problem-solving process. From your sketch and visualization, try to identify the physical processes at work in the system. Ask yourself what concepts or principles are involved in this situation. Then, develop a strategy—a game plan—for solving the problem.

PHYSICS IN CONTEXT
Looking Ahead

In this chapter we introduced the idea of a vector in one dimension and showed how the direction of the vector can be indicated by its sign. These concepts are developed in more detail in Chapter 2.

PHYSICS IN CONTEXT
Looking Ahead

Vectors are extended to two and three dimensions in Chapter 3. Vectors are a standard tool throughout the remaining chapters on mechanics, and they appear again in electricity and magnetism.

- **Identify appropriate equations** Once a strategy has been developed, find the specific equations that are needed to carry it out.

- **Solve the equations** Use basic algebra to solve the equations identified in the previous step. Work with symbols such as x or y for the most part, substituting numerical values near the end of the calculations. Working with symbols will make it easier to go back over a problem to locate and identify mistakes, if there are any, and to explore limits and special cases.

- **Check your answer** Once you have an answer, check to see if it makes sense: (i) Does it have the correct dimensions? (ii) Is the numerical value reasonable?

- **Explore limits/special cases** Getting the correct answer is nice, but it's not all there is to physics. You can learn a great deal about physics and about the connection between physics and mathematics by checking various limits of your answer. For example, if you have two masses in your system, m_1 and m_2, what happens in the special case where $m_1 = 0$, or in the case where $m_1 = m_2$? Check to see whether your answer and your physical intuition agree.

Following these suggestions will help to strengthen your ability to solve problems—in physics and in other disciplines as well.

Pedagogical Features The pedagogical features in this text are designed to deepen your understanding of physics concepts, and at the same time to develop your problem-solving skills. Perhaps the most important of these are the **Examples,** which present a complete solution to a substantial problem and include these helpful sections:

- A sketch of the system, along with a **Picture the Problem** discussion to help you visualize the problem.

- A **Reasoning and Strategy** section to develop a "game plan" before carrying out the calculations.

- A **Known** and **Unknown** section to specify what is given and what is to be calculated.

- A two-column, step-by-step **Solution** with the reasoning given in words in the left-hand column and the corresponding math given in the right-hand column.

- An **Insight** section to point out interesting aspects of the calculation and its results.

- A **Practice Problem** to give you a chance to try out the calculations just given on a similar problem.

These sections work together to give a thorough understanding of a problem and its solution. You will also find **Quick Examples,** which are streamlined versions of the full Examples, and **Exercises,** which present simple one-line calculations.

In addition, **Conceptual Examples** are designed to strengthen your understanding of the basic concepts of physics, and to develop your skill in using these concepts to "make sense" of the numerical aspects of physics. Some of the homework problems help with making these connections as well. **Predict/Calculate** problems ask for a prediction based on physical concepts, followed by a calculation to verify the prediction. **Predict/Explain** problems ask for a prediction and the corresponding explanation. You will also find problems of these types in some of the Examples and Conceptual Examples in each chapter.

Finally, it's often tempting to look for shortcuts when doing a problem—to look for a formula that seems to fit and some numbers to plug into it. It may seem harder to think ahead, to be systematic as you solve the problem, and then to think back over what you've done at the end of the problem. The extra effort is worth it, however. It will help you develop powerful problem-solving skills that can be applied to unexpected problems you may encounter on exams—and in life in general.

Section Review

- Problem solving is an important part of learning physics. Practice is required, as in learning to ride a bicycle.

CHAPTER 1 REVIEW

CHAPTER SUMMARY

1–1 PHYSICS AND THE LAWS OF NATURE
Physics is based on a small number of fundamental laws and principles.

1–2 UNITS OF LENGTH, MASS, AND TIME

Length
One meter is defined as the distance traveled by light in a vacuum in 1/299,792,458 second.

Mass
One kilogram is the mass of a metal cylinder kept at the International Bureau of Weights and Measures.

Time
One second is the time required for a particular type of radiation from cesium-133 to undergo 9,192,631,770 oscillations.

1–3 DIMENSIONAL ANALYSIS

Dimension
The dimension of a quantity is the type of quantity it is—for example, length [L], mass [M], or time [T].

Dimensional Consistency
An equation is dimensionally consistent if each term in it has the same dimensions. All valid physical equations are dimensionally consistent.

Dimensional Analysis
A calculation based on the dimensional consistency of an equation.

1–4 SIGNIFICANT FIGURES, SCIENTIFIC NOTATION, AND ROUND-OFF ERROR

Significant Figures
The number of digits reliably known, excluding digits that simply indicate the decimal place.

Round-off Error
Discrepancies caused by rounding off numbers in intermediate results.

1–5 CONVERTING UNITS
Multiply by the ratio of two units to convert from one to another.

1–6 ORDER-OF-MAGNITUDE CALCULATIONS
A ballpark estimate designed to be accurate to within the nearest power of ten.

1–7 SCALARS AND VECTORS
A physical quantity that can be represented by a numerical value only is called a scalar. Quantities that require a direction in addition to the numerical value are called vectors.

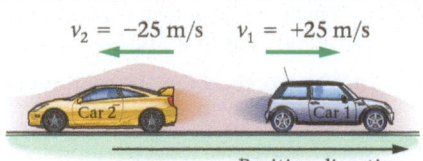

$v_2 = -25 \text{ m/s}$ $v_1 = +25 \text{ m/s}$

Positive direction

1–8 PROBLEM SOLVING IN PHYSICS
A good general approach to problem solving is as follows: read, sketch, visualize, strategize, identify equations, solve, check, explore limits.

ANSWERS TO ENHANCE YOUR UNDERSTANDING QUESTIONS

1. (**f**) all of these.

2. $B < A < D < C$.

3. $A = [L]/[T]; B = [L]/[T^2]$;
   $C = [L]/[T^2]; D = [L^2]/[T^2]$.

4. $A < D < B = C$.

5. D.

6. 10^{-9} m/s.

7. (**a**) Yes. They can travel equally fast in different directions.
   (**b**) No. Both the speed and direction must be the same.

CONCEPTUAL QUESTIONS

For instructor-assigned homework, go to www.masteringphysics.com. (MP)

Conceptual Questions are meant to stimulate discussion of significant physics concepts. Answers to odd-numbered Conceptual Questions are given in the back of the book.

1. Can dimensional analysis determine whether the area of a circle is πr^2 or $2\pi r^2$? Explain.

2. If a distance d has units of meters, and a time T has units of seconds, does the quantity $T + d$ make sense physically? What about the quantity d/T? Explain in both cases.

3. Is it possible for two quantities to (**a**) have the same units but different dimensions or (**b**) have the same dimensions but different units? Explain.

4. The frequency of a pendulum is the number of oscillations it makes per second. Is frequency a scalar or a vector quantity? Explain.

5. Albert runs 34 m to the right in 10 s. Isaac runs 17 m to the left in 5 s. (**a**) Is Albert's speed greater than, less than, or equal to Isaac's speed? (**b**) Is Albert's velocity equal to Isaac's velocity? Explain.

6. Give an order-of-magnitude estimate for the time in seconds of the following: (**a**) a year; (**b**) a baseball game; (**c**) a heartbeat; (**d**) the age of the Earth; (**e**) the age of a person.

7. Give an order-of-magnitude estimate for the length in meters of the following: (**a**) a person; (**b**) a fly; (**c**) a car; (**d**) a 747 airplane; (**e**) an interstate freeway stretching coast-to-coast.

PROBLEMS AND CONCEPTUAL EXERCISES

Answers to odd-numbered Problems and Conceptual Exercises can be found in the back of the book. **BIO** *identifies problems of biological or medical interest;* **CE** *indicates a conceptual exercise.* Predict/Explain *problems ask for two responses: (a) your prediction of a physical outcome, and (b) the best explanation among three provided; and* Predict/Calculate *problems ask for a prediction of a physical outcome, based on fundamental physics concepts, and follow that with a numerical calculation to verify the prediction. On all problems, bullets (•, ••, •••) indicate the level of difficulty.*

SECTION 1-2 UNITS OF LENGTH, MASS, AND TIME

1. • **The Hunger Games** The movie *The Hunger Games* brought in about $152,000,000 in its opening weekend. Express this amount in (**a**) gigadollars and (**b**) teradollars.

2. • **BIO The Thickness of Hair** A human hair has a thickness of about 85 μm. What is this in (**a**) meters and (**b**) millimeters?

3. • The speed of light in a vacuum is approximately 0.3 Gm/s. Express the speed of light in meters per second.

4. • **A Fast Computer** IBM has a computer it calls the Blue Gene/L that can do 136.8 teracalculations per second. How many calculations can it do in a nanosecond?

SECTION 1-3 DIMENSIONAL ANALYSIS

5. • **CE** Which of the following equations are dimensionally consistent? (**a**) $x = \frac{1}{2}at^2$, (**b**) $t = v/x$, (**c**) $t = (2x/a)^{1/2}$.

6. • **CE** Which of the following quantities have the dimensions of a time? (**a**) x/v, (**b**) a/v, (**c**) $(2x/a)^{1/2}$, (**d**) v^2/a.

7. • **CE** Which of the following quantities have the dimensions of a distance? (**a**) vt, (**b**) $\frac{1}{2}at^2$, (**c**) $2at$, (**d**) v^2/a.

8. • **CE** Which of the following quantities have the dimensions of a speed? (**a**) $\frac{1}{2}at^2$, (**b**) at, (**c**) $(2x/a)^{1/2}$, (**d**) $(2ax)^{1/2}$.

9. •• Velocity is related to acceleration and distance by the following expression: $v^2 = 2ax^p$. Find the power p that makes this equation dimensionally consistent.

10. •• Acceleration is related to distance and time by the following expression: $a = 2xt^p$. Find the power p that makes this equation dimensionally consistent.

11. •• The time t required for an object to fall from rest through a height h is given by

$$t = h^p \sqrt{\frac{2}{g}}$$

In this expression, g is the acceleration due to gravity. Find the power p that makes this equation dimensionally consistent.

12. •• Newton's second law (to be discussed in Chapter 5) states that the acceleration a of an object is proportional to the force F acting on it, and inversely proportional to its mass m. That is,

$$a = \frac{F}{m}$$

What are the dimensions of force?

13. •• The time T required for one complete oscillation of a mass m on a spring of force constant k is

$$T = 2\pi \sqrt{\frac{m}{k}}$$

Find the dimensions k must have for this equation to be dimensionally correct.

SECTION 1-4 SIGNIFICANT FIGURES, SCIENTIFIC NOTATION, AND ROUND-OFF ERROR

14. • The speed of light to five significant figures is 2.9979×10^8 m/s. What is the speed of light to three significant figures?

15. • A parking lot is 124.3 m long and 41.06 m wide. What is the perimeter of the lot?

16. • On a fishing trip you catch a 2.77-lb bass, a 14.3-lb rock cod, and a 13.43-lb salmon. What is the total weight of your catch?

17. •• How many significant figures are there in (a) 0.0000303 and (b) 6.201×10^5?

18. •• What is the area of a circle of radius (a) 11.37 m and (b) 6.8 m?

19. •• The first several digits of π, the ratio of the circumference to the diameter of a circle, are $\pi = 3.14159265358979\ldots$. What is π to (a) three significant figures, (b) five significant figures, and (c) seven significant figures?

SECTION 1-5 CONVERTING UNITS

20. • Rank the following speeds in order of increasing magnitude. Indicate ties where appropriate. (a) 0.25 m/s, (b) 0.75 km/h, (c) 12 ft/s, (d) 16 cm/s.

21. • The Ark of the Covenant is described as a chest of acacia wood 2.5 cubits in length and 1.5 cubits in width and height. Given that a cubit is equivalent to 17.7 in., find the volume of the ark in cubic feet.

22. • A car drives on a highway with a speed of 68 mi/h. What is this speed in km/h?

23. • **Angel Falls** Water going over Angel Falls, in Venezuela, the world's highest waterfall, drops through a distance of 3212 ft. What is this distance in km?

24. • An electronic advertising sign repeats a message every 9 seconds, day and night, for a week. How many times did the message appear on the sign?

25. • **BIO Blue Whales** The blue whale (*Balaenoptera musculus*) is thought to be the largest animal ever to inhabit the Earth. The longest recorded blue whale had a length of 108 ft. What is this length in meters?

Angel Falls (Problem 23)

26. • **The Star of Africa** The Star of Africa, a diamond in the royal scepter of the British crown jewels, has a mass of 530.2 carats, where 1 carat = 0.20 g. Given that 1 kg has an approximate weight of 2.21 lb, what is the weight of this diamond in pounds?

27. • **BIO Woodpecker Impact** When red-headed woodpeckers (*Melanerpes erythrocephalus*) strike the trunk of a tree, they can experience an acceleration ten times greater than the acceleration of gravity, or about 98.1 m/s². What is this acceleration in ft/s²?

28. •• **Predict/Calculate** Many highways have a speed limit of 55 mi/h. (a) Is this speed greater than, less than, or equal to 55 km/h? Explain. (b) Find the speed limit in km/h that corresponds to 55 mi/h.

29. •• **BIO Mantis Shrimp** Peacock mantis shrimps (*Odontodactylus scyllarus*) feed largely on snails. They shatter the shells of their prey by delivering a sharp blow with their front legs, which have been observed to reach peak speeds of 23 m/s. What is this speed in (a) feet per second and (b) miles per hour?

30. •• **A Jiffy** The American physical chemist Gilbert Newton Lewis (1875–1946) proposed a unit of time called the "jiffy." According to Lewis, 1 jiffy = the time it takes light to travel one centimeter. (a) If you perform a task in a jiffy, how much time has it taken in seconds? (b) How many jiffys are in one minute? (Use the fact that the speed of light is approximately 2.9979×10^8 m/s.)

31. •• **The Mutchkin and the Noggin** (a) A mutchkin is a Scottish unit of liquid measure equal to 0.42 L. How many mutchkins are required to fill a container that measures one foot on a side? (b) A noggin is a volume equal to 0.28 mutchkin. What is the conversion factor between noggins and gallons?

32. •• Suppose 1.0 cubic meter of oil is spilled into the ocean. Find the area of the resulting slick, assuming that it is one molecule thick, and that each molecule occupies a cube 0.50 μm on a side.

33. •• The acceleration of gravity is approximately 9.81 m/s² (depending on your location). What is the acceleration of gravity in feet per second squared?

34. •• **BIO Squid Nerve Impulses** Nerve impulses in giant axons of the squid can travel with a speed of 25.0 m/s. How fast is this in (a) ft/s and (b) mi/h?

SECTION 1-6 ORDER-OF-MAGNITUDE CALCULATIONS

35. • Give a ballpark estimate of the number of seats in a typical major-league ballpark.

36. • Estimate the speed at which your hair grows. Give your answer in meters per second.

37. •• Milk is often sold by the gallon in plastic containers. (a) Estimate the number of gallons of milk that are purchased in the United States each year. (b) What approximate weight of plastic does this represent?

38. •• New York is roughly 3000 miles from Seattle. When it is 10:00 A.M. in Seattle, it is 1:00 P.M. in New York. Using this information, estimate (a) the rotational speed of the surface of Earth, (b) the circumference of Earth, and (c) the radius of Earth.

GENERAL PROBLEMS

39. • **CE** Which of the following equations are dimensionally consistent? (a) $v = at$, (b) $v = \frac{1}{2}at^2$, (c) $t = a/v$, (d) $v^2 = 2ax$.

40. • **CE** Which of the following quantities have the dimensions of an acceleration? (a) xt^2, (b) v^2/x, (c) x/t^2, (d) v/t.

41. • **BIO Photosynthesis** The light that plants absorb to perform photosynthesis has a wavelength that peaks near 675 nm. Express this distance in (a) millimeters and (b) inches.

42. • **Glacial Speed** On June 9, 1983, the lower part of the Variegated Glacier in Alaska was observed to be moving at a rate of 210 feet per day. What is this speed in meters per second?

Alaska's Variegated Glacier (Problem 42)

43. • One liter of pure gold weighs 42.5 lb. If you discovered 1.0 ft³ of pure gold, how much would it weigh in pounds?

44. • What is the speed in miles per second of a beam of light traveling at 3.00×10^8 m/s?

45. • **BIO** **Rattlesnake Frequency** A timber rattlesnake (*Crotalus horridus*) shakes its rattle at a characteristic frequency of about 3300 shakes per minute. What is this frequency in shakes per second?

46. •• **BIO** A single human red blood cell has a mass of about 27 pg. What is this mass in **(a)** kilograms and **(b)** nanograms?

47. •• **BIO** **Corn Growth** Sweet corn plants can grow 4.1 cm in height per day. What is this speed in **(a)** millimeters per second and **(b)** feet per week?

48. •• **BIO** **Mosquito Courtship** Male mosquitoes in the mood for mating find female mosquitoes of their own species by listening for the characteristic "buzzing" frequency of the female's wing beats. This frequency is about 605 wing beats per second. **(a)** How many wing beats occur in one minute? **(b)** How many cycles of oscillation does the radiation from a cesium-133 atom complete during one mosquito wing beat?

49. •• Predict/Calculate A Porsche sports car can accelerate at 14 m/s². **(a)** Is this acceleration greater than, less than, or equal to 14 ft/s²? Explain. **(b)** Determine the acceleration of a Porsche in ft/s².

50. •• **A Speeding Bullet** The fastest commercially available rifle bullet claims a muzzle velocity of 4225 ft/s. **(a)** What is this speed in miles per hour? **(b)** How far (in meters) can this bullet travel in 5.0 ms?

51. •• **BIO** **Human Nerve Fibers** Type A nerve fibers in humans can conduct nerve impulses at speeds up to 140 m/s. **(a)** How fast are the nerve impulses in miles per hour? **(b)** How far (in meters) can the impulses travel in 5.0 ms?

52. •• **BIO** **Brain Growth** The mass of a newborn baby's brain has been found to increase by about 1.6 mg per minute. **(a)** How much does the brain's mass increase in one day? **(b)** How much time does it take for the brain's mass to increase by 0.0075 kg?

53. •• **The Huygens Probe** NASA's *Cassini* mission to Saturn released a probe on December 25, 2004, that landed on the Saturnian moon Titan on January 14, 2005. The probe, which was named *Huygens*, was released with a gentle relative speed of 31 cm/s. As *Huygens* moved away from the main spacecraft, it rotated at a rate of seven revolutions per minute. **(a)** How many revolutions had *Huygens* completed when it was 150 yards from the mother ship? **(b)** How far did *Huygens* move away from the mother ship during each revolution? Give your answer in feet.

54. •• **BIO** **Spin-Dry Dragonflies** Some dragonflies are observed to splash into the water to bathe, and then spin rapidly to shed the water. Slow-motion video shot at 240 frames per second shows that dragonflies complete one spin revolution every 14 frames. What is the spin rate of the dragonfly in revolutions per minute (rpm)?

55. ••• Acceleration is related to velocity and time by the following expression: $a = v^p t^q$. Find the powers p and q that make this equation dimensionally consistent.

56. ••• The period T of a simple pendulum is the amount of time required for it to undergo one complete oscillation. If the length of the pendulum is L and the acceleration of gravity is g, then T is given by $T = 2\pi L^p g^q$. Find the powers p and q required for dimensional consistency.

57. ••• Driving along a crowded freeway, you notice that it takes a time t to go from one mile marker to the next. When you increase your speed by 7.9 mi/h, the time to go one mile decreases by 13 s. What was your original speed?

PASSAGE PROBLEMS

BIO **Using a Cricket as a Thermometer**

All chemical reactions, whether organic or inorganic, proceed at a rate that depends on temperature—the higher the temperature, the higher the rate of reaction. This can be understood in terms of molecules moving with increased energy as the temperature is increased, and colliding with other molecules more frequently. In the case of organic reactions, the result is that metabolic processes speed up with increasing temperature.

An increased or decreased metabolic rate can manifest itself in a number of ways. For example, a cricket trying to attract a mate chirps at a rate that depends on its overall rate of metabolism. As a result, the chirping rate of crickets depends directly on temperature. In fact, some people even use a pet cricket as a thermometer.

The cricket that is most accurate as a thermometer is the snowy tree cricket (*Oecanthus fultoni* Walker). Its rate of chirping is described by the following formula:

$$N = \text{number of chirps per 13.0 seconds} = T - 40.0$$

In this expression, T is the temperature in degrees Fahrenheit.

58. • Which plot in **FIGURE 1-10**—A, B, C, D, or E—represents the chirping rate of the snowy tree cricket?

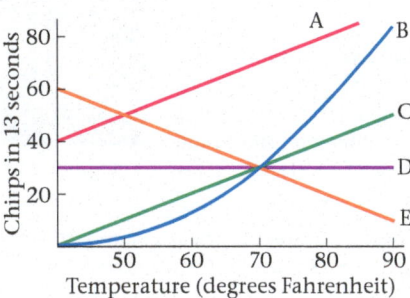

Chirping rate versus temperature

FIGURE1-10 Problem 58

59. •• If the temperature is 43 °F, how much time does it take for the cricket to produce 12 chirps?

A. 12 s   B. 24 s   C. 43 s   D. 52 s

60. •• Your pet cricket chirps 112 times in one minute (60.0 s). What is the temperature in degrees Fahrenheit?

A. 41.9   B. 47.0   C. 64.3   D. 74.7

61. •• Suppose a snowy tree cricket is chirping when the temperature is 65.0 °F. How many oscillations does the radiation from a cesium-133 atom complete between successive chirps?

A. 7.98×10^7 B. 3.68×10^8
C. 4.78×10^9 D. 9.58×10^9

Big
Ideas

1 Position is measured relative to a coordinate system.

2 Velocity is the rate of change of position with time.

3 Acceleration is the rate of change of velocity with time.

4 Speed increases (decreases) when velocity and acceleration are in the same (opposite) direction.

5 Equations of motion relate position, velocity, acceleration, and time.

6 Free fall is motion with a constant downward acceleration of magnitude *g*.

One-Dimensional
Kinematics

▲ These sprinters provide a good illustration of one-dimensional, straight-line motion. They accelerated as they left the starting blocks, and now they maintain a constant velocity as they approach the finish line.

We begin our study of physics with **mechanics**, the area of physics most apparent to us in our everyday lives. Every time you raise an arm, stand up or sit down, throw a ball, or open a door, your actions are governed by the laws of mechanics. In this chapter we focus on kinematics—the basic properties of motion—in one dimension.

2-1 Position, Distance, and Displacement

In physics, the terms *position*, *distance*, and *displacement* have specific meanings. This section gives the physics definitions of these terms and shows how they are used to describe the motion of a particle.

Position The first step in describing the motion of a particle is to set up a **coordinate system** that defines its location—that is, its **position**. An example of a coordinate system in one dimension is shown in **FIGURE 2-1**. This is simply an x axis, with an origin (where $x = 0$) and an arrow pointing in the positive direction—the direction in which x increases. In setting up a coordinate system, we are free to choose the origin and the positive direction as we like, but once we make a choice, we must be consistent with it throughout any calculations that follow.

▶ **FIGURE 2-1 A one-dimensional coordinate system** You are free to choose the origin and positive direction as you like, but once your choice is made, stick with it.

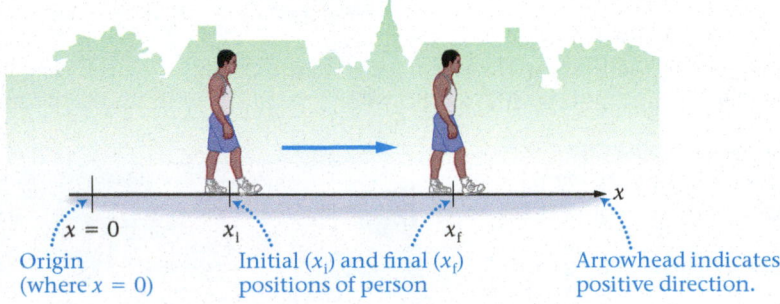

Origin (where $x = 0$)   Initial (x_i) and final (x_f) positions of person   Arrowhead indicates positive direction.

The "particle" in Figure 2-1 is a person who has moved to the right from an initial position, x_i, to a final position, x_f. Because the positive direction is to the right, it follows that x_f is greater than x_i; that is, $x_f > x_i$.

Distance Now that we've seen how to set up a coordinate system, let's use one to investigate the situation shown in **FIGURE 2-2**. Suppose you leave your house, drive to the grocery store, and then return home. The **distance** you've covered in your trip is 4.3 mi + 4.3 mi = 8.6 mi. In general, distance is defined as the length of a trip:

Definition of Distance

distance = total length of travel

SI unit: meter, m

Using SI units, we find that the distance in this case is

$$8.6 \text{ mi} = (8.6 \text{ mi})\left(\frac{1609 \text{ m}}{1 \text{ mi}}\right) = 1.4 \times 10^4 \text{ m}$$

▶ **FIGURE 2-2 One-dimensional coordinates** The locations of your house, your friend's house, and the grocery store in terms of a one-dimensional coordinate system.

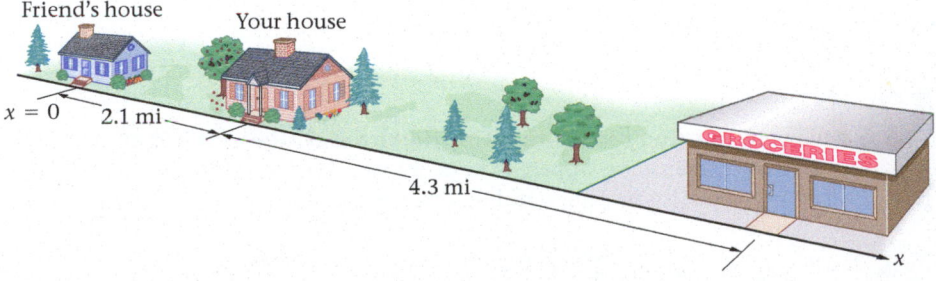

Friend's house   Your house

$x = 0$ 2.1 mi

4.3 mi

In a car, the distance traveled is indicated by the odometer. Distance is always positive, and is a scalar quantity (see Chapter 1) with no associated direction.

Displacement Another useful way to characterize a particle's motion is in terms of its **displacement**, Δx, which is simply the change in position:

Definition: Displacement, Δx

displacement = change in position = final position − initial position

displacement = $\Delta x = x_f - x_i$ 2-1

SI unit: meter, m

The SI units of displacement are meters—the same as for distance—but displacement and distance are really quite different. For example, in the round trip from your house to the grocery store and back, the distance traveled is 8.6 mi, whereas the displacement is zero because $x_f = x_i = 2.1$ mi, and hence $\Delta x = x_f - x_i = 0$.

Notice that we use the delta notation, Δx, as a convenient shorthand for the quantity $x_f - x_i$. (See Appendix A for a complete discussion of delta notation.) Also, notice that Δx can be positive (if the final position is to the right of the initial position, $x_f > x_i$), negative (if the final position is to the left of the initial position, $x_f < x_i$), or zero (if the final and initial positions are the same, $x_f = x_i$). In fact, the displacement is a one-dimensional vector, as defined in Chapter 1, and its direction (right or left) is given by its sign (positive or negative, respectively).

For example, suppose you go from your house to the grocery store and then to your friend's house in Figure 2-2. On this trip the distance is 10.7 mi, but the displacement is

$$\Delta x = x_f - x_i = (0) - (2.1\,\text{mi}) = -2.1\,\text{mi}$$

The minus sign means your displacement is in the negative direction; that is, to the left.

QUICK EXAMPLE 2-1 DISPLACEMENT AND DISTANCE

Predict/Calculate

Referring to Figure 2-2, suppose you take a trip from your friend's house to the grocery store and then to your house. **(a)** Is the displacement for this trip positive, negative, or zero? Explain. **(b)** Find the displacement for this trip. **(c)** What is the distance covered in this trip?

REASONING AND SOLUTION

Displacement is the final position minus the initial position; it can be positive or negative. Distance is the length of travel, which is always positive.

1. **Part (a)** The displacement is positive because the final position is to the right (positive direction) of the initial position.

2. **Part (b)** Determine the initial $x_i = 0$
   position for the trip, using Figure 2-2:

3. Determine the final position for the $x_f = 2.1$ mi
   trip, using Figure 2-2:

4. Subtract x_i from x_f to find the $\Delta x = x_f - x_i = 2.1\,\text{mi} - 0 = 2.1\,\text{mi}$
   displacement. Notice that the result
   is positive, as expected:

5. **Part (c)** Add the distances for the $2.1\,\text{mi} + 4.3\,\text{mi} + 4.3\,\text{mi} = 10.7\,\text{mi}$
   various parts of the trip:

Enhance Your Understanding   (Answers given at the end of the chapter)

1. For each of the following questions, give an example if your answer is yes. Explain why not if your answer is no. **(a)** Is it possible to take a trip in which the distance covered is less than the magnitude of the displacement? **(b)** Is it possible to take a trip in which the distance covered is greater than the magnitude of the displacement?

Section Review

- Distance is the total length of a trip.
- Displacement is the change in position; displacement = $\Delta x = x_f - x_i$.

PHYSICS IN CONTEXT
Looking Back

In this chapter we make extensive use of the sign conventions for one-dimensional vectors introduced in Chapter 1—positive for one direction, negative for the opposite direction.

Big Idea 1 Position and displacement are measured relative to a coordinate system. The coordinate system must include an origin and a positive direction. Displacement has a direction and a magnitude, and hence it is a vector. Distance has only a numerical value, and hence it is a scalar.

2-2 Average Speed and Velocity

The next step in describing motion is to consider how rapidly an object moves. For example, how much time does it take for a major-league fastball to reach home plate? How far does an orbiting satellite travel in one hour? These are examples of some of the most basic questions regarding motion, and in this section we learn how to answer them.

Average Speed The simplest way to characterize the rate of motion is with the **average speed**:

$$\text{average speed} = \frac{\text{distance}}{\text{elapsed time}} \qquad 2\text{-}2$$

The dimensions of average speed are distance per time or, in SI units, meters per second, m/s. Both distance and elapsed time are positive; thus average speed is always positive.

EXAMPLE 2-2   DOG RUN

A dog trots back to its owner with an average speed of 1.40 m/s from a distance of 2.3 m. How much time does it take for the dog to reach its owner?

PICTURE THE PROBLEM
The dog moves in a straight line through a distance of $d = 2.3$ m. The average speed of the dog is $v = 1.40$ m/s.

REASONING AND STRATEGY
Equation 2-2 $\left(\text{average speed} = \dfrac{\text{distance}}{\text{elapsed time}}\right)$ relates average speed, distance, and elapsed time. We can solve for the elapsed time by rearranging this equation.

Known   Average speed, $v = 1.40$ m/s; distance, $d = 2.3$ m.
Unknown   Elapsed time = ?

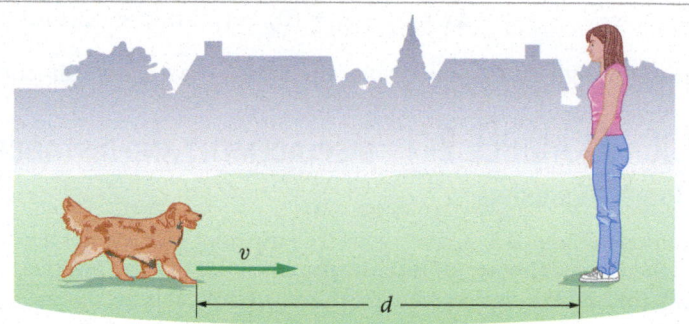

SOLUTION

1. Rearrange Equation 2-2 to solve for the elapsed time:
$$\text{elapsed time} = \frac{\text{distance}}{\text{average speed}} = \frac{d}{v}$$

2. Substitute numerical values to find the time:
$$\text{elapsed time} = \frac{2.3 \text{ m}}{1.40 \text{ m/s}} = \frac{2.3}{1.40}\text{ s} = 1.6 \text{ s}$$

INSIGHT
As this example shows, Equation 2-2 is not just a formula for calculating the average speed. It relates speed, time, and distance. Any one of these quantities can be determined if the other two are known.

PRACTICE PROBLEM
A dog trots with an average speed of 1.44 m/s for 1.9 s. What is the distance it covers?
[**Answer:** distance = (average speed) (elapsed time) = (1.44 m/s) (1.9 s) = 2.7 m]

Some related homework problems: Problem 11, Problem 13

Next, we calculate the average speed for a trip consisting of two parts of equal length, each traveled with a different speed.

CONCEPTUAL EXAMPLE 2-3   AVERAGE SPEED   PREDICT/EXPLAIN

You drive 4.00 mi at 30.0 mi/h and then another 4.00 mi at 50.0 mi/h. **(a)** Is your average speed for the 8.00-mi trip greater than, less than, or equal to 40.0 mi/h? **(b)** Which of the following is the *best explanation* for your prediction?

  I. The average of 30.0 mi/h and 50.0 mi/h is 40.0 mi/h, and hence this is the average speed for the trip.

 II. You go farther during the 50.0 mi/h part of the trip, and hence your average speed is greater than 40.0 mi/h.

III. You spend more time during your trip traveling at 30.0 mi/h, and hence your average speed is less than 40.0 mi/h.

CONTINUED

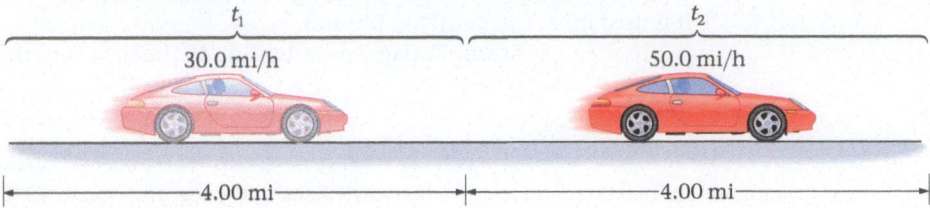

REASONING AND DISCUSSION

At first glance it might seem that the average speed is definitely 40.0 mi/h. On further reflection, however, it is clear that it takes more time to travel 4.00 mi at 30.0 mi/h than it does to travel 4.00 mi at 50.0 mi/h. Therefore, you will be traveling at the lower speed for a greater period of time, and hence your average speed will be *less* than 40.0 mi/h—that is, closer to 30.0 mi/h than to 50.0 mi/h.

ANSWER

(a) The average speed is less than 40.0 mi/h. **(b)** The best explanation is III.

To confirm the conclusion of Conceptual Example 2-3, we simply apply the definition of average speed to find its value for the entire trip. We already know that the distance traveled is 8.00 mi; what we need now is the total elapsed time. The elapsed time on the first 4.00 mi is

$$t_1 = \frac{4.00 \text{ mi}}{30.0 \text{ mi/h}} = \left(\frac{4.00}{30.0}\right) \text{h} = 0.133 \text{ h}$$

The time required to cover the second 4.00 mi is

$$t_2 = \frac{4.00 \text{ mi}}{50.0 \text{ mi/h}} = \left(\frac{4.00}{50.0}\right) \text{h} = 0.0800 \text{ h}$$

Therefore, the elapsed time for the entire trip is

$$t_1 + t_2 = 0.133 \text{ h} + 0.0800 \text{ h} = 0.213 \text{ h}$$

This gives the following average speed:

$$\text{average speed} = \frac{8.00 \text{ mi}}{0.213 \text{ h}} = 37.6 \text{ mi/h} < 40.0 \text{ mi/h}$$

It's important to note that a "guess" will never give a detailed result like 37.6 mi/h; a systematic, step-by-step calculation is required.

Average Velocity There is another physical quantity that is often more useful than the average speed. It is the **average velocity**, v_{av}, and it is defined as displacement per time:

Definition: Average Velocity, v_{av}
$$\text{average velocity} = \frac{\text{displacement}}{\text{elapsed time}}$$
$$v_{av} = \frac{\Delta x}{\Delta t} = \frac{x_f - x_i}{t_f - t_i} \qquad \text{2-3}$$
SI unit: meter per second, m/s

Not only does the average velocity tell us, on average, how fast something is moving, it also tells us the *direction* the object is moving. For example, if an object moves in the positive direction, then $x_f > x_i$ and the average velocity is positive, $v_{av} > 0$. On the other hand, if an object moves in the negative direction, then $x_f < x_i$ and $v_{av} < 0$. As with displacement, the average velocity is a one-dimensional vector, and its direction

Balls Take High and Low Tracks

PROBLEM-SOLVING NOTE

"Coordinate" the Problem

The first step in solving a physics problem is to produce a simple sketch of the system. Your sketch should include a coordinate system, along with an origin and a positive direction. Next, you should identify quantities that are given in the problem, such as initial position, initial velocity, acceleration, and so on. These preliminaries will help you produce a mathematical representation of the problem.

is given by its sign. Average velocity gives more information than average speed; hence it is used more frequently in physics.

In the next Example, pay close attention to the sign of each quantity.

EXAMPLE 2-4   SPRINT TRAINING

An athlete sprints in a straight line for 50.0 m in 8.00 s, and then walks slowly back to the starting line in 40.0 s. If the "sprint direction" is taken to be positive, what are (a) the average sprint velocity, (b) the average walking velocity, and (c) the average velocity for the complete round trip?

PICTURE THE PROBLEM

In our sketch we set up a coordinate system with the sprint going in the positive x direction, as described in the problem. For convenience, we choose the origin to be at the starting line. The finish line, then, is at $x = 50.0$ m.

REASONING AND STRATEGY

In each part of the problem we are asked for the average velocity, and we are given information for times and distances. All that is needed, then, is to determine $\Delta x = x_f - x_i$ and $\Delta t = t_f - t_i$ in each case, and to apply Equation 2-3 $\left(v_{av} = \frac{\Delta x}{\Delta t}\right)$.

Known   Sprint distance = 50.0 m; sprint time = 8.00 s; walking distance = 50.0 m; walking time = 40.0 s.

Unknown   (a) Average sprint velocity, $v_{av} = ?$ (b) Average walking velocity, $v_{av} = ?$ (c) Average velocity for round trip, $v_{av} = ?$

SOLUTION

Part (a)

1. Apply $v_{av} = \dfrac{\Delta x}{\Delta t}$ to the sprint, with $x_f = 50.0$ m, $x_i = 0$, $t_f = 8.00$ s, and $t_i = 0$:

$$v_{av} = \frac{\Delta x}{\Delta t} = \frac{x_f - x_i}{t_f - t_i} = \frac{50.0 \text{ m} - 0}{8.00 \text{ s} - 0} = \frac{50.0}{8.00}\text{ m/s} = 6.25 \text{ m/s}$$

Part (b)

2. Apply $v_{av} = \dfrac{\Delta x}{\Delta t}$ to the walk. In this case, $x_f = 0$, $x_i = 50.0$ m, $t_f = 48.0$ s, and $t_i = 8.00$ s:

$$v_{av} = \frac{x_f - x_i}{t_f - t_i} = \frac{0 - 50.0 \text{ m}}{48.0 \text{ s} - 8.00 \text{ s}} = -\frac{50.0}{40.0}\text{ m/s} = -1.25 \text{ m/s}$$

Part (c)

3. For the round trip, $x_f = x_i = 0$; thus $\Delta x = 0$:

$$v_{av} = \frac{\Delta x}{\Delta t} = \frac{0}{48.0 \text{ s}} = 0$$

INSIGHT

The sign of the velocities in parts (a) and (b) indicates the direction of motion: positive for motion to the right, negative for motion to the left. In addition, notice that the average *speed* for the entire 100.0-m trip (100.0 m/48.0 s = 2.08 m/s) is nonzero, even though the average velocity vanishes.

PRACTICE PROBLEM

If the average velocity during the walk is −1.50 m/s, how much time does it take the athlete to walk back to the starting line?
[**Answer:** $\Delta t = \Delta x/v_{av} = (-50.0 \text{ m})/(-1.50 \text{ m/s}) = 33.3$ s]

Some related homework problems: Problem 9, Problem 14, Problem 16

Graphical Interpretation of Average Velocity

TABLE 2-1 Time and Position Values for Figure 2-3

t (s)	x (m)
0	1
1	3
2	4
3	2
4	−1

It's often useful to "visualize" a particle's motion by sketching its position as a function of time. For example, suppose a particle moves back and forth along the x axis, with the positions and times listed in Table 2-1. This data is plotted in **FIGURE 2-3 (a)**, which is certainly a better way to "see" the motion than a table of numbers.

Even so, this way of showing a particle's position and time is a bit messy, so let's replot the same information with a different type of graph. In **FIGURE 2-3 (b)** we again plot the motion shown in Figure 2-3 (a), but this time with the vertical axis representing the position, x, and the horizontal axis representing time, t. This is referred to as an **x-versus-t graph**, and it makes it much easier to visualize a particle's motion, as we shall see.

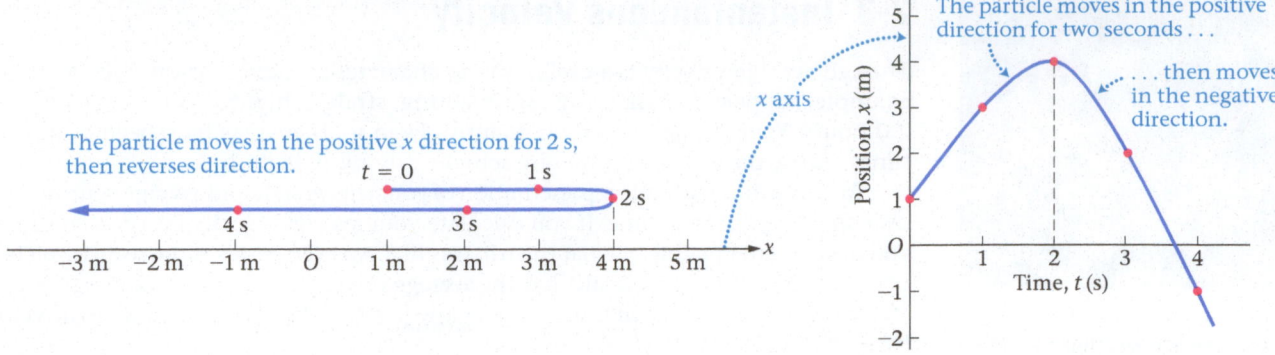

(a) The particle's path shown on a coordinate axis

(b) The same path as a graph of position x versus time t

▲ **FIGURE 2-3 Two ways to visualize one-dimensional motion (a)** Position plotted for different times. Although the path is shown with a "U" shape for clarity, the particle actually moves straight back and forth along the x axis. **(b)** Plot of position (vertical) versus time (horizontal).

Slope Is Equal to Average Velocity An x-versus-t plot leads to a particularly useful interpretation of average velocity. To see how, suppose you would like to know the average velocity of the particle in Figure 2-3 from $t = 0$ to $t = 3$ s. From our definition of average velocity in Equation 2-3, we know that $v_{av} = \Delta x/\Delta t$, or $v_{av} = (2\ m - 1\ m)/(3\ s - 0) = +0.3$ m/s for this particle. To relate this to the x-versus-t plot, draw a straight line connecting the position at $t = 0$ (call this point A) and the position at $t = 3$ s (point B). The result is shown in **FIGURE 2-4 (a)**.

The slope of the straight line from A to B is equal to the rise over the run, which in this case is $\Delta x/\Delta t$. But $\Delta x/\Delta t$ is the average velocity. Thus, we conclude the following:

> The *slope* of a line connecting two points on an x-versus-t plot is equal to the *average velocity* during that time interval.

As an additional example, let's calculate the average velocity of this particle between times $t = 2$ s and $t = 3$ s. A line connecting the corresponding points is shown in **FIGURE 2-4 (b)**. The first thing we notice about this line is that it has a negative slope; thus $v_{av} < 0$ and the particle is moving to the left. We can also see that this line is inclined more steeply than the line in Figure 2-4 (a); hence the magnitude of its slope (its speed) is greater. In fact, if we calculate the slope of this line we find that $v_{av} = -2$ m/s for this time interval.

Thus, connecting points on an x-versus-t plot gives an immediate "feeling" for the average velocity over a given time interval. This type of graphical analysis will be particularly useful in the next section.

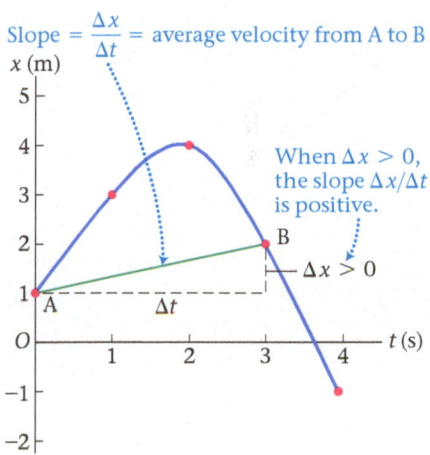

(a) Average velocity between $t = 0$ and $t = 3$ s

Enhance Your Understanding

(Answers given at the end of the chapter)

2. The position of an object as a function of time is shown in **FIGURE 2-5**. For the time intervals between **(a)** points 1 and 2, **(b)** points 2 and 3, and **(c)** points 3 and 4, state whether the average velocity is positive, negative, or zero.

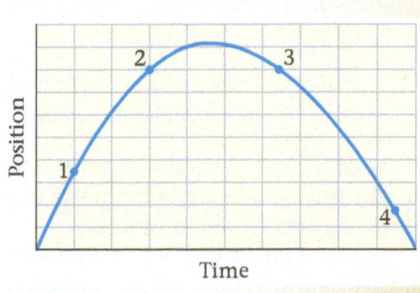

▲ **FIGURE 2-5**

Section Review

- Average speed is distance divided by time.
- Average velocity is displacement divided by time; $v_{av} = \dfrac{\Delta x}{\Delta t}$.
- The slope of a line connecting two points on an x-versus-t plot is the average velocity between those two points.

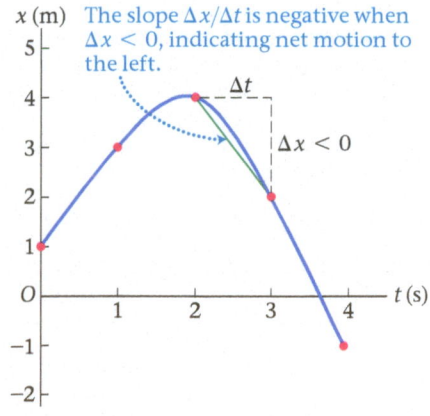

(b) Average velocity between $t = 2$ s and $t = 3$ s

▲ **FIGURE 2-4 Average velocity on an x-versus-t graph** The slope of a straight line between any two points on an x-versus-t graph equals the average velocity between those points. Positive slopes indicate net motion to the right; negative slopes indicate net motion to the left.

▲ **FIGURE 2-6** A speedometer indicates a car's instantaneous speed but gives no information about its *direction*. Thus, the speedometer is truly a "speed meter," not a velocity meter.

Big Idea **2** Velocity is the rate of change of position with time, and speed is the magnitude of velocity. Velocity has a magnitude and a direction, and hence it is a vector; speed has only a magnitude, and hence it is a scalar.

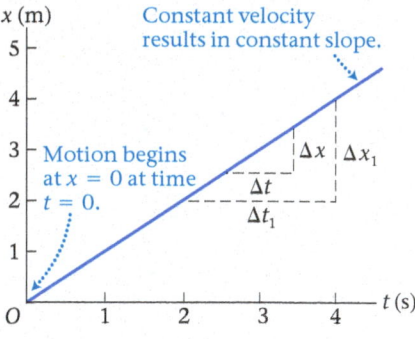

▲ **FIGURE 2-7** **Constant velocity corresponds to constant slope on an x-versus-t graph** The slope $\Delta x_1/\Delta t_1$ is equal to $(4\,\text{m} - 2\,\text{m})/(4\,\text{s} - 2\,\text{s}) = (2\,\text{m})/(2\,\text{s}) = 1\,\text{m/s}$. Because the x-versus-t plot is a straight line, the slope $\Delta x/\Delta t$ is also equal to 1 m/s for any value of Δt.

TABLE 2-2 x-versus-t Values for Figure 2-8

$t(s)$	$x(m)$
0	0
0.25	9.85
0.50	17.2
0.75	22.3
1.00	25.6
1.25	27.4
1.50	28.1
1.75	28.0
2.00	27.4

2-3 Instantaneous Velocity

Though average velocity is a useful way to characterize motion, it can miss a lot. For example, suppose you travel by car on a long, straight highway, covering 92 mi in 2.0 hours. Your average velocity is 46 mi/h. Even so, there may have been only a few times during the trip when you were actually driving at 46 mi/h.

To have a more accurate representation of your trip, you should average your velocity over shorter periods of time. If you calculate your average velocity every 15 minutes, you have a better picture of what the trip was like. An even better, more realistic picture of the trip is obtained if you calculate the average velocity every minute or every second. Ideally, when you deal with the motion of any particle, it's desirable to know the velocity of the particle at each *instant* of time.

This idea of a velocity corresponding to an instant of time is what is meant by the **instantaneous velocity**. Mathematically, we define the instantaneous velocity as follows:

Definition: Instantaneous Velocity, v

$$v = \lim_{\Delta t \to 0} \frac{\Delta x}{\Delta t} \qquad\qquad 2\text{-}4$$

SI unit: meter per second, m/s

In this expression the notation $\lim_{\Delta t \to 0}$ means "evaluate the average velocity, $\Delta x/\Delta t$, over shorter and shorter time intervals, approaching zero in the limit." Notice that the instantaneous velocity can be positive, negative, or zero, just like the average velocity—and just like the average velocity, the instantaneous velocity is a one-dimensional vector. The magnitude of the instantaneous velocity is called the **instantaneous speed**. In a car, the speedometer gives a reading of the vehicle's instantaneous speed. **FIGURE 2-6** shows a speedometer reading 60 mi/h.

Calculating the Instantaneous Velocity As Δt becomes smaller, Δx becomes smaller as well, but the ratio $\Delta x/\Delta t$ approaches a constant value. To see how this works, consider first the simple case of a particle moving with a constant velocity of $+1$ m/s. If the particle starts at $x = 0$ at $t = 0$, then its position at $t = 1$ s is $x = 1$ m, its position at $t = 2$ s is $x = 2$ m, and so on. Plotting this motion in an x-versus-t plot gives a straight line, as shown in **FIGURE 2-7**.

Now, suppose we want to find the instantaneous velocity at $t = 3$ s. To do so, we calculate the average velocity over small intervals of time centered at 3 s, and let the time intervals become arbitrarily small, as shown in the Figure. Because the x-versus-t plot is a straight line, it's clear that $\Delta x/\Delta t = \Delta x_1/\Delta t_1$, no matter how small the time interval Δt. As Δt becomes smaller, so does Δx, but the ratio $\Delta x/\Delta t$ is simply the slope of the line, 1 m/s. Thus, the instantaneous velocity at $t = 3$ s is 1 m/s.

Of course, in this example the instantaneous velocity is 1 m/s for any instant of time, not just $t = 3$ s. Therefore:

> When the velocity is constant, the average velocity over any time interval is equal to the instantaneous velocity at any time.

In general, a particle's velocity varies with time, and the x-versus-t plot is not a straight line. An example is shown in **FIGURE 2-8**, with the corresponding numerical values of x and t given in Table 2-2.

In this case, what is the instantaneous velocity at, say, $t = 1.00$ s? As a first approximation, let's calculate the average velocity for the time interval from $t_i = 0$ to $t_f = 2.00$ s. Notice that this time interval is centered at $t = 1.00$ s. From Table 2-2 we see that $x_i = 0$ and $x_f = 27.4$ m; thus $v_{av} = 13.7$ m/s. The corresponding straight line connecting these two points is the lowest straight line in Figure 2-8.

The next three lines, in upward progression, refer to time intervals from 0.250 s to 1.75 s, 0.500 s to 1.50 s, and 0.750 s to 1.25 s, respectively. The corresponding average velocities, given in Table 2-3, are 12.1 m/s, 10.9 m/s, and 10.2 m/s. Table 2-3 also gives results for even smaller time intervals. In particular, for the interval from 0.900 s to 1.10 s

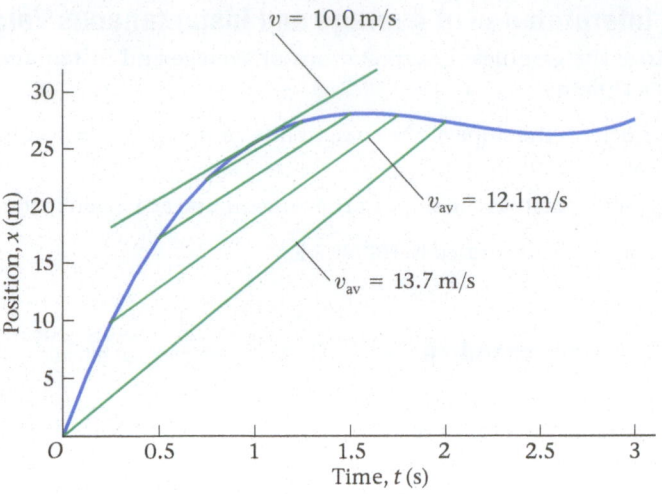

◄ **FIGURE 2-8 Instantaneous velocity** An x-versus-t plot for motion with variable velocity. The instantaneous velocity at $t = 1$ s is equal to the slope of the tangent line at that time. The average velocity for a small time interval centered on $t = 1$ s approaches the instantaneous velocity at $t = 1$ s as the time interval goes to zero.

TABLE 2-3 Calculating the Instantaneous Velocity at $t = 1$ s for Figure 2-8

t_i (s)	t_f (s)	Δt (s)	x_i (m)	x_f (m)	Δx (m)	$v_{av} = \Delta x/\Delta t$ (m/s)
0	2.00	2.00	0	27.4	27.4	13.7
0.250	1.75	1.50	9.85	28.0	18.2	12.1
0.500	1.50	1.00	17.2	28.1	10.9	10.9
0.750	1.25	0.50	22.3	27.4	5.10	10.2
0.900	1.10	0.20	24.5	26.5	2.00	10.0
0.950	1.05	0.10	25.1	26.1	1.00	10.0

the average velocity is 10.0 m/s. Smaller intervals also give 10.0 m/s. Thus, we can conclude that the instantaneous velocity at $t = 1.00$ s is $v = 10.0$ m/s.

Tangent Lines and the Instantaneous Velocity The uppermost straight line in Figure 2-8 is the tangent line to the x-versus-t curve at the time $t = 1.00$ s; that is, it is the line that touches the curve at just a single point. Its slope is 10.0 m/s. Clearly, the average-velocity lines have slopes that approach the slope of the tangent line as the time intervals become smaller. This is an example of the following general result:

> The instantaneous velocity at a given time is equal to the slope of the tangent line at that point on an x-versus-t graph.

Thus, a visual inspection of an x-versus-t graph gives information not only about the location of a particle, but also about its velocity.

CONCEPTUAL EXAMPLE 2-5 INSTANTANEOUS VELOCITY

Referring to Figure 2-8, is the instantaneous velocity at $t = 0.500$ s **(a)** greater than, **(b)** less than, or **(c)** the same as the instantaneous velocity at $t = 1.00$ s?

REASONING AND DISCUSSION
From the x-versus-t graph in Figure 2-8 it is clear that the slope of a tangent line drawn at $t = 0.500$ s is greater than the slope of the tangent line at $t = 1.00$ s. It follows that the particle's velocity at 0.500 s is greater than its velocity at 1.00 s.

ANSWER
(a) The instantaneous velocity is greater at $t = 0.500$ s.

In the remainder of the book, when we say *velocity* it is to be understood that we mean *instantaneous* velocity. If we want to refer to the average velocity, we will specifically say average velocity.

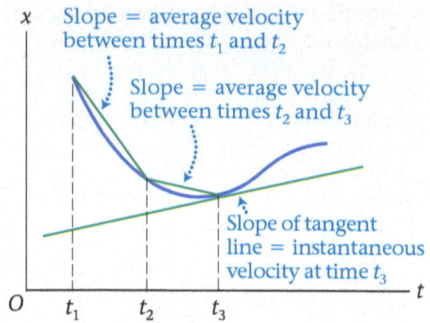

▲ FIGURE 2-9 Graphical interpretation of average and instantaneous velocity
Average velocities correspond to the slope of straight-line segments connecting different points on an x-versus-t graph. Instantaneous velocities are given by the slope of the tangent line at a given time.

Graphical Interpretation of Average and Instantaneous Velocity

Let's summarize the graphical interpretations of average and instantaneous velocity on an x-versus-t graph:

- Average velocity is the slope of the straight line connecting two points corresponding to a given time interval.

- Instantaneous velocity is the slope of the tangent line at a given instant of time.

These relationships are illustrated in **FIGURE 2-9**.

Enhance Your Understanding   (Answers given at the end of the chapter)

3. **FIGURE 2-10** shows the position-versus-time graph for an object. Rank the instantaneous velocity of the object at points A, B, C, and D in order of increasing velocity, from most negative to most positive. Indicate ties where appropriate.

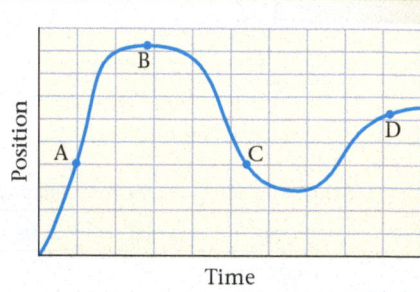

▲ FIGURE 2-10

Section Review

- Instantaneous velocity is the limit of average velocity over shorter and shorter time intervals; $v = \lim_{\Delta t \to 0} \frac{\Delta x}{\Delta t}$. The magnitude of the instantaneous velocity is the instantaneous speed.

- The instantaneous velocity on an x-versus-t plot is the slope of a tangent line at a given time.

2-4 Acceleration

Just as velocity is the rate of change of *displacement* with time, **acceleration** is the rate of change of *velocity* with time. Thus, an object accelerates whenever its velocity *changes,* no matter what the change—it accelerates when its velocity increases, it accelerates when its velocity decreases. Of all the concepts discussed in this chapter, perhaps none is more central to physics than acceleration. Galileo, for example, showed that falling bodies move with constant acceleration. Newton showed that acceleration and force are directly related, as we shall see in Chapter 5. Thus, it is particularly important to have a clear, complete understanding of acceleration before leaving this chapter.

Average Acceleration We begin with the definition of **average acceleration**, which is the change in velocity divided by the change in time:

Definition: Average Acceleration, a_{av}

$$a_{av} = \frac{\Delta v}{\Delta t} = \frac{v_f - v_i}{t_f - t_i}$$

2-5

SI unit: meter per second per second, m/s^2

Notice that the dimensions of average acceleration are the dimensions of velocity per time, or (meters per second) per second:

$$\frac{\text{meters per second}}{\text{second}} = \frac{m/s}{s} = \frac{m}{s^2}$$

This is generally spoken as "meters per second squared." For example, the acceleration of gravity on the Earth's surface is approximately 9.81 m/s^2, which means that the velocity of a falling object changes by 9.81 meters per second (m/s) every second (s). In addition, we see that the average acceleration can be positive, negative, or zero. In fact, it is a one-dimensional vector, just like displacement, average velocity, and instantaneous velocity. Typical magnitudes of acceleration are given in Table 2-4. A real-world example of linear acceleration is shown in **FIGURE 2-11**.

EXERCISE 2-6   AVERAGE ACCELERATION

a. A certain car can go from 0 to 60.0 mi/h in 7.40 s. What is the average acceleration of this car in meters per second squared?

b. An airplane has an average acceleration of 2.19 m/s^2 during takeoff. If the airplane starts at rest, how much time does it take for it to reach a speed of 174 mi/h?

REASONING AND SOLUTION

(a) Average acceleration is the change in velocity divided by the elapsed time, $a_{\text{av}} = \Delta v/\Delta t$.

(b) The equation for average acceleration can be rearranged to solve for the elapsed time, $\Delta t = \Delta v/a_{\text{av}}$.

a. average acceleration $= a_{\text{av}} = (60.0 \text{ mi/h})/(7.40 \text{ s})$

$$= (60.0 \text{ mi/h})\left(\frac{1609 \text{ m}}{1 \text{ mi}}\right)\left(\frac{1 \text{ h}}{3600 \text{ s}}\right)/(7.40 \text{ s})$$

$$= (26.8 \text{ m/s})/(7.40 \text{ s}) = 3.62 \text{ m/s}^2$$

b. $\Delta t = \Delta v/a_{\text{av}} = (174 \text{ mi/h})\left(\dfrac{1609 \text{ m}}{1 \text{ mi}}\right)\left(\dfrac{1 \text{ h}}{3600 \text{ s}}\right)/(2.19 \text{ m/s}^2)$

$$= (77.8 \text{ m/s})/(2.19 \text{ m/s}^2) = 35.5 \text{ s}$$

Instantaneous Acceleration   We considered the limit of smaller and smaller time intervals to find an instantaneous velocity, and we can do the same to define the **instantaneous acceleration:**

> **Definition: Instantaneous Acceleration, a**
>
> $$a = \lim_{\Delta t \to 0} \frac{\Delta v}{\Delta t} \qquad \text{2-6}$$
>
> SI unit: meter per second per second, m/s^2

As you might expect, the instantaneous acceleration is a vector, just like the average acceleration, and its direction in one dimension is given by its sign. For simplicity, when we say acceleration in this text, we are referring to the instantaneous acceleration.

One final note before we go on to some examples. If the acceleration is constant, it has the same value at all times. Therefore:

> When acceleration is constant, the instantaneous and average accelerations are the same.

We shall make use of this fact when we return to the special case of constant acceleration in the next section.

Graphical Interpretation of Acceleration

To see how acceleration can be interpreted graphically, suppose a particle has a constant acceleration of -0.50 m/s^2. This means that the velocity of the particle *decreases* by 0.50 m/s each second. Thus, if its velocity is 1.0 m/s at $t = 0$, then at $t = 1$ s its velocity is 0.50 m/s, at $t = 2$ s its velocity is 0, at $t = 3$ s its velocity is -0.50 m/s, and so on. This is illustrated by curve I in **FIGURE 2-12**, where we see that a plot of v versus

▲ **FIGURE 2-11** The space shuttle *Columbia* accelerates upward on the initial phase of its journey into orbit. The astronauts on board experienced an approximately linear acceleration that may have been as great as 20 m/s^2.

Big Idea 3 Acceleration is the rate of change of velocity with time. Acceleration has both a magnitude and a direction, and hence it is a vector.

TABLE 2-4 Typical Accelerations (m/s^2)

Ultracentrifuge	3×10^6
Bullet fired from a rifle	4.4×10^5
Batted baseball	3×10^4
Click beetle righting itself	400
Acceleration required to deploy air bags	60
Bungee jump	30
High jump	15
Acceleration of gravity on Earth	9.81
Emergency stop in a car	8
Airplane during takeoff	5
An elevator	3
Acceleration of gravity on the Moon	1.62

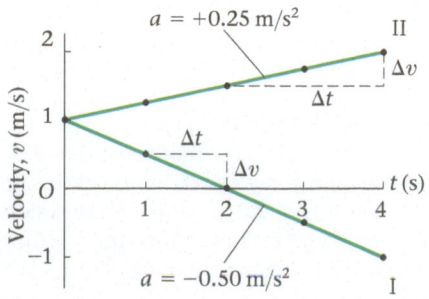

▶ **FIGURE 2-12** *v*-versus-*t* plots for motion with constant acceleration Curve I represents the movement of a particle with constant acceleration $a = -0.50 \text{ m/s}^2$. Curve II represents the motion of a particle with constant acceleration $a = +0.25 \text{ m/s}^2$.

t results in a straight line with a negative slope. Curve II in Figure 2-12 has a positive slope, corresponding to a constant acceleration of $+0.25 \text{ m/s}^2$. Thus, in terms of a *v*-versus-*t* plot, a constant acceleration results in a straight line with a slope equal to the acceleration.

CONCEPTUAL EXAMPLE 2-7 SPEED AS A FUNCTION OF TIME

The speed of a particle with the *v*-versus-*t* graph shown by curve II in Figure 2-12 increases steadily with time. Consider, instead, a particle whose *v*-versus-*t* graph is given by curve I in Figure 2-12. As a function of time, does the speed of this particle **(a)** increase, **(b)** decrease, or **(c)** decrease and then increase?

REASONING AND DISCUSSION
Recall that speed is the *magnitude* of velocity. In curve I of Figure 2-12 the speed starts out at 1.0 m/s, then *decreases* to 0 at *t* = 2 s. After *t* = 2 s the speed *increases* again. For example, at *t* = 3 s the speed is 0.50 m/s, and at *t* = 4 s the speed is 1 m/s.

Did you realize that the particle represented by curve I in Figure 2-12 changes direction at *t* = 2 s? It certainly does. Before *t* = 2 s the particle moves in the positive direction; after *t* = 2 s it moves in the negative direction. At precisely *t* = 2 s the particle is momentarily at rest. However, regardless of whether the particle is moving in the positive direction, moving in the negative direction, or instantaneously at rest, it still has the same constant acceleration. Acceleration has to do only with the way the velocity is *changing* at a given moment.

ANSWER
(c) The speed decreases and then increases.

The graphical interpretations for velocity presented in Figure 2-9 apply equally well to acceleration, with just one small change: Instead of an *x*-versus-*t* graph, we use a *v*-versus-*t* graph, as in **FIGURE 2-13**. Thus, the average acceleration in a *v*-versus-*t* plot is the slope of a straight line connecting points corresponding to two different times. Similarly, the instantaneous acceleration is the slope of the tangent line at a particular time.

▶ **FIGURE 2-13 Graphical interpretation of average and instantaneous acceleration** Average accelerations correspond to the slopes of straight-line segments connecting different points on a *v*-versus-*t* graph. Instantaneous accelerations are given by the slope of the tangent line at a given time.

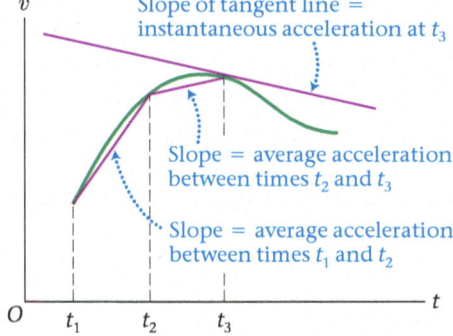

EXAMPLE 2-8 ACCELERATION OF A BICYCLE

A cyclist riding in a straight line has an initial velocity of 3.5 m/s, and accelerates at -1.0 m/s^2 for 2.0 s. The cyclist then coasts with zero acceleration for 3.0 s, and finally accelerates at 1.5 m/s^2 for an additional 2.0 s. **(a)** What is the final velocity of the cyclist? **(b)** What is the average acceleration of the cyclist for these seven seconds?

PICTURE THE PROBLEM
We begin by sketching a *v*-versus-*t* plot for the cyclist. The basic idea is that each interval of constant acceleration is represented by a straight line of the appropriate slope. Therefore, we draw a straight line with the slope -1.0 m/s^2 from *t* = 0 to *t* = 2.0 s, a line with zero slope from *t* = 2.0 s to *t* = 5.0 s, and a line with the slope 1.5 m/s^2 from *t* = 5.0 s to *t* = 7.0 s. The line connecting the initial and final points determines the average acceleration.

CONTINUED

REASONING AND STRATEGY

During each period of constant acceleration the change in velocity is $\Delta v = a_{av}\Delta t = a\Delta t$.

a. Adding the individual changes in velocity gives the total change, $\Delta v = v_f - v_i$. Because v_i is known, this expression can be solved for the final velocity, $v_f = \Delta v + v_i$.

b. The average acceleration can be calculated using Equation 2-5, $a_{av} = \Delta v/\Delta t$. Notice that Δv has been obtained in part (a), and the total time interval is $\Delta t = 7.0$ s.

Known   Initial velocity, $v_i = 3.5$ m/s; accelerations, $a_1 = -1.0$ m/s^2, $a_2 = 0$, $a_3 = 1.5$ m/s^2.

Unknown (a) Final velocity, $v_f = ?$ (b) Average acceleration, $a_{av} = ?$

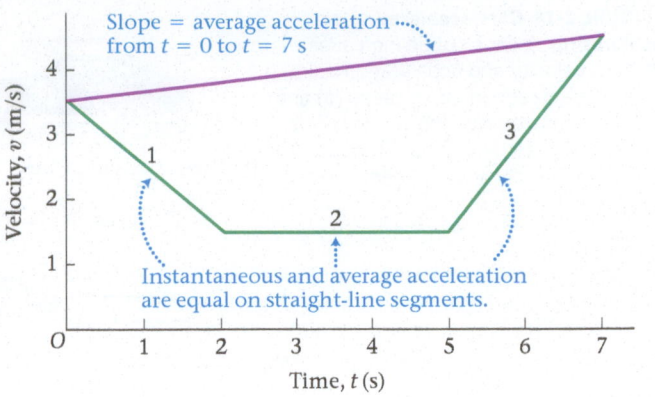

SOLUTION

Part (a)

1. Find the change in velocity during each of the three periods of constant acceleration:

$$\Delta v_1 = a_1\Delta t_1 = (-1.0 \text{ m/s}^2)(2.0 \text{ s}) = -2.0 \text{ m/s}$$
$$\Delta v_2 = a_2\Delta t_2 = (0)(3.0 \text{ s}) = 0$$
$$\Delta v_3 = a_3\Delta t_3 = (1.5 \text{ m/s}^2)(2.0 \text{ s}) = 3.0 \text{ m/s}$$

2. Sum the change in velocity for each period to obtain the total Δv:

$$\Delta v = \Delta v_1 + \Delta v_2 + \Delta v_3$$
$$= -2.0 \text{ m/s} + 0 + 3.0 \text{ m/s} = 1.0 \text{ m/s}$$

3. Use Δv to find v_f, recalling that $v_i = 3.5$ m/s:

$$\Delta v = v_f - v_i$$
$$v_f = \Delta v + v_i = 1.0 \text{ m/s} + 3.5 \text{ m/s} = 4.5 \text{ m/s}$$

Part (b)

4. The average acceleration is $\Delta v/\Delta t$:

$$a_{av} = \frac{\Delta v}{\Delta t} = \frac{1.0 \text{ m/s}}{7.0 \text{ s}} = 0.14 \text{ m/s}^2$$

INSIGHT

The average acceleration for these seven seconds is *not* the average of the individual accelerations, -1.0 m/s^2, 0, and 1.5 m/s^2. The reason is that different amounts of time are spent with each acceleration.

PRACTICE PROBLEM

What is the average acceleration of the cyclist between $t = 3.0$ s and $t = 6.0$ s?
[**Answer:** $a_{av} = \Delta v/\Delta t = (3.0 \text{ m/s} - 1.5 \text{ m/s})/(6.0 \text{ s} - 3.0 \text{ s}) = 0.50 \text{ m/s}^2$]

Some related homework problems: Problem 32, Problem 34

Relating the Signs of Velocity and Acceleration to the Change in Speed   In one dimension, nonzero velocities and accelerations are either positive or negative, depending on whether they point in the positive or negative direction of the coordinate system chosen. Thus, the velocity and acceleration of an object may have the same or opposite signs. (Of course, in two or three dimensions the relationship between velocity and acceleration can be much more varied, as we shall see in the next several chapters.) This leads to the following two possibilities in one dimension:

- When the velocity and acceleration of an object have the same sign (point in the same direction), the speed of the object increases.

- When the velocity and acceleration of an object have opposite signs (point in opposite directions), the speed of the object decreases.

These two possibilities are illustrated in **FIGURE 2-14**. Notice that when a particle's speed increases, it means either that its velocity becomes more positive, as in **FIGURE 2-14 (a)**, or more negative, as in **FIGURE 2-14 (d)**. In either case, it is the magnitude of the velocity—the speed—that increases. **FIGURE 2-15** shows an example where the velocity and the acceleration are definitely in the same direction.

When a particle's speed decreases, it is often said to be *decelerating*. A common misconception is that deceleration implies a negative acceleration. This is not true. Deceleration can be caused by a positive or a negative acceleration, depending on the direction of the initial velocity. For example, the car in **FIGURE 2-14 (b)** has a positive velocity and

PHYSICS IN CONTEXT
Looking Ahead

The distinctions developed in this chapter between velocity and acceleration play a key role in our understanding of Newton's laws of motion in Chapters 5 and 6.

Big Idea 4 Speed increases when velocity and acceleration are in the same direction; speed decreases when velocity and acceleration are in opposite directions.

▶ **FIGURE 2-14 Cars accelerating or decelerating** A car's speed increases when its velocity and acceleration point in the same direction, as in cases **(a)** and **(d)**. When the velocity and acceleration point in opposite directions, as in cases **(b)** and **(c)**, the car's speed decreases.

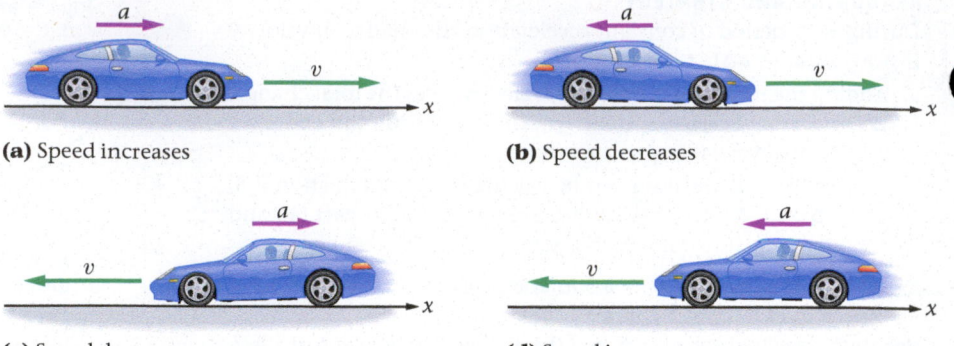

(a) Speed increases **(b)** Speed decreases

(c) Speed decreases **(d)** Speed increases

▶ **FIGURE 2-15** The winner of this race was traveling at a speed of 313.91 mi/h at the end of the quarter-mile course. The winning time was just 4.607 s, and hence the *average* acceleration during this race was approximately three times the acceleration of gravity (which is covered in Section 2-7).

a negative acceleration, while the car in **FIGURE 2-14 (c)** has a negative velocity and a positive acceleration. In both cases, the speed of the car decreases. Again, all that is required for deceleration in one dimension is that the velocity and acceleration have *opposite signs*; that is, they must point in *opposite directions*, as in parts (b) and (c) of Figure 2-14.

Velocity-versus-time plots for the four situations shown in Figure 2-14 are presented in **FIGURE 2-16**. In each of the four plots in Figure 2-16 we assume constant acceleration. Be sure to understand clearly the connection between the *v*-versus-*t* plots in Figure 2-16 and the corresponding physical motions indicated in Figure 2-14.

EXAMPLE 2-9   THE FERRY DOCKS

A ferry makes a short run between two docks: one in Anacortes, Washington, the other on Guemes Island. As the ferry approaches the Guemes dock (traveling in the positive *x* direction), its speed is 6.8 m/s. **(a)** If the ferry slows to a stop in 13.3 s, what is its average acceleration? **(b)** As the ferry returns to the Anacortes dock, its speed is 6.1 m/s. If it comes to rest in 12.9 s, what is its average acceleration?

PICTURE THE PROBLEM

Our sketch shows the locations of the two docks and the positive direction indicated in the problem. The distance between docks is not given, nor is it needed.

REASONING AND STRATEGY

We are given the initial and final velocities (the ferry comes to a stop in each case, so its final speed is zero) and the relevant times. Therefore, we can find the average acceleration using $a_{av} = \Delta v/\Delta t$, being careful to get the signs right.

Known   **(a)** Initial velocity, $v_i = 6.8$ m/s; time to stop, $\Delta t = 13.3$ s. **(b)** Initial velocity, $v_i = -6.1$ m/s; time to stop, $\Delta t = 12.9$ s.

Unknown   **(a)** Average acceleration, $a_{av} = ?$ **(b)** Average acceleration, $a_{av} = ?$

SOLUTION

Part (a)

1. Calculate the average acceleration, noting that $v_i = 6.8$ m/s and $v_f = 0$:

$$a_{av} = \frac{\Delta v}{\Delta t} = \frac{v_f - v_i}{\Delta t} = \frac{0 - 6.8 \text{ m/s}}{13.3 \text{ s}} = -0.51 \text{ m/s}^2$$

CONTINUED

Part (b)

2. In this case, $v_1 = -6.1$ m/s and $v_f = 0$:

$$a_{av} = \frac{\Delta v}{\Delta t} = \frac{v_f - v_1}{\Delta t} = \frac{0 - (-6.1 \text{ m/s})}{12.9 \text{ s}} = 0.47 \text{ m/s}^2$$

INSIGHT

In each case, the acceleration of the ferry is opposite in sign to its velocity; therefore the ferry decelerates.

PRACTICE PROBLEM

When the ferry leaves Guemes Island, its speed increases from 0 to 4.8 m/s in 9.05 s. What is its average acceleration?
[**Answer:** $a_{av} = -0.53$ m/s^2]

Some related homework problems: Problem 30, Problem 36

RWP* The ability to detect and measure acceleration has become an important capability of many new technologies. A device that measures acceleration is called an *accelerometer*. In recent years a number of accelerometers have been developed that are contained in tiny integrated circuit chips, such as the one shown in **FIGURE 2-17**. These accelerometers allow devices such as smartphones, video game consoles, automotive air bag sensors, aircraft flight stabilization systems, and numerous others to detect acceleration and respond accordingly.

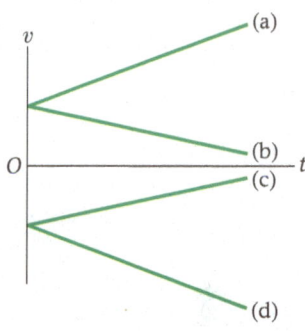

▲ **FIGURE 2-16** *v-versus-t* **plots with constant acceleration** Four plots of *v* versus *t* corresponding to the four situations shown in Figure 2-14. Notice that the speed increases in cases **(a)** and **(d)**, but decreases in cases **(b)** and **(c)**.

Enhance Your Understanding   (Answers given at the end of the chapter)

4. At a certain time, object 1 has an initial velocity of -2 m/s and an acceleration of 3 m/s^2. At the same time, object 2 has an initial velocity of 5 m/s and an acceleration of -1 m/s^2. **(a)** Is the speed of object 1 increasing or decreasing? Explain. **(b)** Is the velocity of object 1 increasing or decreasing? Explain. **(c)** Is the speed of object 2 increasing or decreasing? Explain. **(d)** Is the velocity of object 2 increasing or decreasing? Explain.

Section Review

- The average acceleration is the change in velocity divided by the time; $a_{av} = \dfrac{\Delta v}{\Delta t}$.

- The instantaneous acceleration is the average acceleration over shorter and shorter time intervals; $a = \lim\limits_{\Delta t \to 0} \dfrac{\Delta v}{\Delta t}$.

- Average acceleration is the slope between two points on a *v-versus-t* plot; instantaneous acceleration is the slope of a tangent line on a *v-versus-t* plot.

2-5 Motion with Constant Acceleration

In this section, we derive equations that describe the motion of particles moving with **constant acceleration**. These "equations of motion" can be used to describe a wide range of everyday phenomena. For example, in an idealized world with no air resistance, falling bodies have constant acceleration.

Velocity as a Function of Time   As mentioned in the previous section, if a particle has constant acceleration—that is, the same acceleration at every instant of time—then its instantaneous acceleration, *a*, is equal to its average acceleration, a_{av}. Recalling the definition of average acceleration (Equation 2-5), we have

$$a_{av} = \frac{v_f - v_1}{t_f - t_1} = a$$

▲ **FIGURE 2-17 Integrated circuit accelerometer** Microscopic image of the ADXL330, the accelerometer incorporated into the Nintendo *Wii* remote controller. When the chip is accelerated in any direction, it generates an electrical signal that is proportional to the magnitude of the acceleration. It also indicates the direction of acceleration.

---

*Real World Physics applications are denoted by the acronym RWP.

The initial and final times may be chosen arbitrarily in this equation. For example, let $t_i = 0$ for the initial time, and let $v_i = v_0$ denote the velocity at time zero.* For the final time and velocity we drop the subscripts to simplify the notation; thus we let $t_f = t$ and $v_f = v$. With these identifications we have

$$a_{av} = \frac{v - v_0}{t - 0} = a$$

With a slight rearrangement we find

$$v - v_0 = a(t - 0) = at$$

This yields our first equation of motion:

Constant-Acceleration Equation of Motion: Velocity as a Function of Time

$$v = v_0 + at \qquad \qquad 2\text{-}7$$

Equation 2-7 describes a straight line on a v-versus-t plot. The line crosses the velocity axis at the value v_0 and has a slope a, in agreement with the graphical interpretations discussed in the previous section. For example, in curve I of Figure 2-12, the equation of motion is $v = v_0 + at = (1 \text{ m/s}) + (-0.5 \text{ m/s}^2)t$. Also, notice that $(-0.5 \text{ m/s}^2)t$ has the units $(\text{m/s})^2$ $(\text{s}) = \text{m/s}$; thus each term in Equation 2-7 has the same dimensions (as it must to be a valid physical equation).

PHYSICS IN CONTEXT
Looking Back

We are careful to check the dimensional consistency of our equations in this chapter. This concept was introduced in Chapter 1.

EXERCISE 2-10 VELOCITY WITH CONSTANT ACCELERATION

A ball is thrown straight upward with an initial velocity of $+7.3 \text{ m/s}$. If the acceleration of the ball is downward, with the value -9.81 m/s^2, find the velocity of the ball after **(a)** 0.45 s and **(b)** 0.90 s.

REASONING AND SOLUTION
For constant acceleration, velocity is related to acceleration and time by Equation 2-7 $(v = v_0 + at)$. Given the initial velocity $(v_0 = 7.3 \text{ m/s})$ and the acceleration $(a = -9.81 \text{ m/s}^2)$, the final velocity can be found by substituting the desired time.

a. Substituting $t = 0.45$ s in $v = v_0 + at$ yields

$$v = 7.3 \text{ m/s} + (-9.81 \text{ m/s}^2)(0.45 \text{ s}) = 2.9 \text{ m/s}$$

The ball is still moving upward at this time.

b. Similarly, using $t = 0.90$ s in $v = v_0 + at$ gives

$$v = 7.3 \text{ m/s} + (-9.81 \text{ m/s}^2)(0.90 \text{ s}) = -1.5 \text{ m/s}$$

The ball is moving downward (negative direction) at this time.

Position as a Function of Time and Velocity How far does a particle move in a given time if its acceleration is constant? To answer this question, recall the definition of average velocity:

$$v_{av} = \frac{\Delta x}{\Delta t} = \frac{x_f - x_i}{t_f - t_i}$$

Using the same identifications given previously for initial and final times, and letting $x_i = x_0$ and $x_f = x$, we have

$$v_{av} = \frac{x - x_0}{t - 0}$$

Multiplying through by $(t - 0)$ gives

$$x - x_0 = v_{av}(t - 0) = v_{av}t$$

Finally, a simple rearrangement results in

$$x = x_0 + v_{av}t \qquad \qquad 2\text{-}8$$

---

*We often use the subscript i to denote the initial value of a quantity, as in v_i. In the special case where the initial value corresponds to zero time, $t = 0$, we use the subscript 0 to be more specific, as in v_0.

◀ FIGURE 2-18 **Average velocity** **(a)** When acceleration is constant, the velocity varies linearly with time. As a result, the average velocity, v_{av}, is simply the average of the initial velocity, v_0, and the final velocity, v. **(b)** The velocity curve for nonconstant acceleration is nonlinear. In this case, the average velocity is no longer midway between the initial and final velocities.

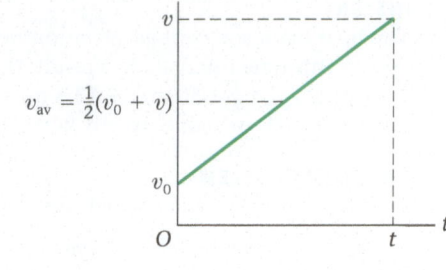

(a)

Now, Equation 2-8 is fine as it is. In fact, it applies whether the acceleration is constant or not. A more useful expression for the case of constant acceleration is obtained by writing v_{av} in terms of the initial and final velocities. This can be done by referring to **FIGURE 2-18 (a)**. Here the velocity changes linearly (since a is constant) from v_0 at $t = 0$ to v at some later time t. The average velocity during this period of time is simply the average of the initial and final velocities—that is, the sum of the two velocities divided by two:

Constant-Acceleration Equation of Motion: Average Velocity

$$v_{av} = \tfrac{1}{2}(v_0 + v) \qquad\qquad 2\text{-}9$$

The average velocity is indicated in the figure. Notice that if the acceleration is not constant, as in **FIGURE 2-18 (b)**, this simple averaging of initial and final velocities is no longer valid.

Substituting the expression for v_{av} from Equation 2-9 into Equation 2-8 yields

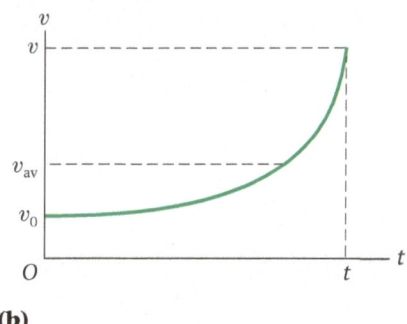

(b)

Constant-Acceleration Equation of Motion: Position as a Function of Time

$$x = x_0 + \tfrac{1}{2}(v_0 + v)t \qquad\qquad 2\text{-}10$$

This equation, like Equation 2-7, is valid *only* for constant acceleration. The utility of Equations 2-7 and 2-10 is illustrated in the next Example.

EXAMPLE 2-11   FULL SPEED AHEAD

A boat moves slowly inside a marina (so as not to leave a wake) with a constant speed of 1.50 m/s. As soon as it passes the breakwater, leaving the marina, it throttles up and accelerates at 2.40 m/s². **(a)** How fast is the boat moving after accelerating for 5.00 s? **(b)** How far has the boat traveled in these 5.00 s?

PICTURE THE PROBLEM
In our sketch we choose the origin to be at the breakwater, and the positive x direction to be the direction of motion. With this choice the initial position is $x_0 = 0$, and the initial velocity is $v_0 = 1.50$ m/s.

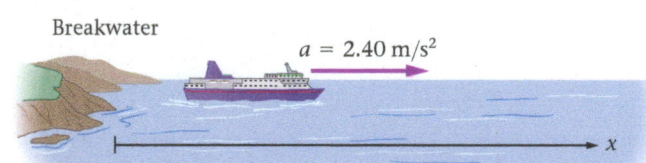

REASONING AND STRATEGY
The acceleration is constant, so we can use Equations 2-7 to 2-10. In part (a) we want to relate velocity to time, so we use Equation 2-7, $v = v_0 + at$. In part (b) our knowledge of the initial and final velocities allows us to relate position to time using Equation 2-10, $x = x_0 + \tfrac{1}{2}(v_0 + v)t$.

Known   Velocity at time $t = 0$, $v_0 = 1.50$ m/s; acceleration, $a = 2.40$ m/s².
Unknown   **(a)** Velocity at $t = 5.00$ s, $v = ?$ **(b)** Distance traveled at $t = 5.00$ s, $x = ?$

SOLUTION

Part (a)

1. Use Equation 2-7 ($v = v_0 + at$) to find the final velocity, with $v_0 = 1.50$ m/s and $a = 2.40$ m/s²:

$\quad v = v_0 + at = 1.50 \text{ m/s} + (2.40 \text{ m/s}^2)(5.00 \text{ s})$
$\quad\quad = 1.50 \text{ m/s} + 12.0 \text{ m/s} = 13.5 \text{ m/s}$

Part (b)

2. Apply Equation 2-10 $\left(x = x_0 + \tfrac{1}{2}(v_0 + v)t\right)$ to find the distance covered, using the result for v obtained in part (a):

$\quad x = x_0 + \tfrac{1}{2}(v_0 + v)t$
$\quad\quad = 0 + \tfrac{1}{2}(1.50 \text{ m/s} + 13.5 \text{ m/s})(5.00 \text{ s})$
$\quad\quad = (7.50 \text{ m/s})(5.00 \text{ s}) = 37.5 \text{ m}$

CONTINUED

INSIGHT
The boat has a constant acceleration between $t = 0$ and $t = 5.00$ s, and hence its velocity-versus-time curve is linear during this time interval. As a result, the average velocity for these 5.00 seconds is the average of the initial and final velocities, $v_{av} = \frac{1}{2}(1.50$ m/s $+ 13.5$ m/s$) = 7.50$ m/s. Multiplying the average velocity by the time, 5.00 s, gives the distance traveled—which is exactly what Equation 2-10 does in Step 2.

PRACTICE PROBLEM
At what time is the boat's speed equal to 10.0 m/s?   [**Answer:** $t = 3.54$ s]

Some related homework problems: Problem 40, Problem 43

Position as a Function of Time and Acceleration   The velocity of the boat in Example 2-11 is plotted as a function of time in **FIGURE 2-19**, with the acceleration starting at time $t = 0$ and ending at $t = 5.00$ s. We will now show that the *distance* traveled by the boat from $t = 0$ to $t = 5.00$ s *is equal to the corresponding area under the velocity-versus-time curve*. This is a general result, valid for any velocity curve and any time interval:

The distance traveled by an object from a time t_1 to a time t_2 is equal to the area under the velocity curve between those two times.

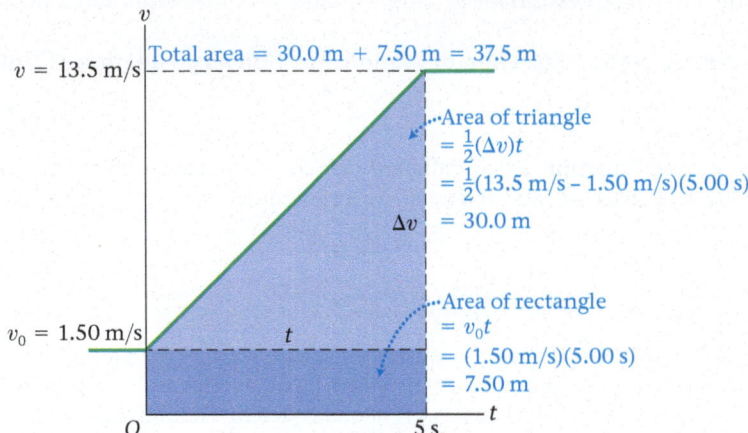

▲ **FIGURE 2-19 Velocity versus time for the boat in Example 2-11** The distance traveled by the boat between $t = 0$ and $t = 5.00$ s is equal to the corresponding area under the velocity curve.

In this case, the area is the sum of the areas of a rectangle and a triangle. The rectangle has a base of 5.00 s and a height of 1.50 m/s, which gives an area of $(5.00$ s$)(1.50$ m/s$) = 7.50$ m. Similarly, the triangle has a base of 5.00 s and a height of $(13.5$ m/s $- 1.50$ m/s$) = 12.0$ m/s, for an area of $\frac{1}{2}(5.00$ s$)(12.0$ m/s$) = 30.0$ m. Clearly, the total area is 37.5 m, which is the same distance found in Example 2-11.

Staying with Example 2-11 for a moment, let's repeat the calculation of part (b), only this time for the general case. First, we use the final velocity from part (a), calculated with $v = v_0 + at$, in the expression for the average velocity, $v_{av} = \frac{1}{2}(v_0 + v)$. Symbolically, this gives the following result:

$$\tfrac{1}{2}(v_0 + v) = \tfrac{1}{2}\left[v_0 + (v_0 + at)\right] = v_0 + \tfrac{1}{2}at \qquad \text{(constant acceleration)}$$

Next, we substitute this result into Equation 2-10 $\left(x = x_0 + \tfrac{1}{2}(v_0 + v)t\right)$, which yields

$$x = x_0 + \tfrac{1}{2}(v_0 + v)t = x_0 + \left(v_0 + \tfrac{1}{2}at\right)t$$

Simplifying this expression yields the following result:

Constant-Acceleration Equation of Motion: Position as a Function of Time

$x = x_0 + v_0t + \frac{1}{2}at^2$ 2-11

Here we have an expression for position versus time that is explicitly in terms of the acceleration, a.

Notice that each term in Equation 2-11 has the same dimensions, as they must. For example, the velocity term, $v_0 t$, has the units $(\text{m/s})(\text{s}) = \text{m}$. Similarly, the acceleration term, $\frac{1}{2}at^2$, has the units $(\text{m/s}^2)(\text{s}^2) = \text{m}$.

EXERCISE 2-12   POSITION WITH CONSTANT ACCELERATION

Repeat part (b) of Example 2-11 using Equation 2-11, $x = x_0 + v_0 t + \frac{1}{2}at^2$.

REASONING AND SOLUTION

Given the initial position, initial velocity, and acceleration, the position at any given time can be found with $x = x_0 + v_0 t + \frac{1}{2}at^2$. In this case, we have

$$x = x_0 + v_0 t + \frac{1}{2}at^2 = 0 + (1.50\,\text{m/s})(5.00\,\text{s}) + \frac{1}{2}(2.40\,\text{m/s}^2)(5.00\,\text{s})^2 = 37.5\,\text{m}$$

The next Example gives further insight into the physical meaning of Equation 2-11.

EXAMPLE 2-13   PUT THE PEDAL TO THE METAL

A drag racer starts from rest and accelerates at $7.40\,\text{m/s}^2$. How far has it traveled in (a) 1.00 s, (b) 2.00 s, (c) 3.00 s?

PICTURE THE PROBLEM

We set up a coordinate system in which the drag racer starts at the origin and accelerates in the positive x direction. With this choice, it follows that $x_0 = 0$ and $a = +7.40\,\text{m/s}^2$. Also, the racer starts from rest, and hence its initial velocity is zero, $v_0 = 0$. Incidentally, the positions of the racer in the sketch have been drawn to scale.

$t = 0.00$ $t = 1.00\,\text{s}$ $t = 2.00\,\text{s}$ $t = 3.00\,\text{s}$

REASONING AND STRATEGY

This problem gives the acceleration, which is constant, and asks for a relationship between position and time. Therefore, we use Equation 2-11, $x = x_0 + v_0 t + \frac{1}{2}at^2$.

Known   Velocity at time $t = 0$, $v_0 = 0$; acceleration, $a = 7.40\,\text{m/s}^2$.
Unknown   Distance traveled, $x = ?$, at (a) $t = 1.00\,\text{s}$, (b) $t = 2.00\,\text{s}$, (c) $t = 3.00\,\text{s}$

SOLUTION

Part (a)

1. Evaluate $x = x_0 + v_0 t + \frac{1}{2}at^2$ with $a = 7.40\,\text{m/s}^2$ and $t = 1.00\,\text{s}$:

$$x = x_0 + v_0 t + \frac{1}{2}at^2 = 0 + 0 + \frac{1}{2}at^2 = \frac{1}{2}at^2$$
$$x = \frac{1}{2}(7.40\,\text{m/s}^2)(1.00\,\text{s})^2 = 3.70\,\text{m}$$

Part (b)

2. From the calculation in part (a), we see that $x = x_0 + v_0 t + \frac{1}{2}at^2$ reduces to $x = \frac{1}{2}at^2$ in this situation. Evaluate $x = \frac{1}{2}at^2$ at $t = 2.00\,\text{s}$:

$$x = \frac{1}{2}at^2 = \frac{1}{2}(7.40\,\text{m/s}^2)(2.00\,\text{s})^2 = 14.8\,\text{m} = 4(3.70\,\text{m})$$

Part (c)

3. Evaluate $x = \frac{1}{2}at^2$ at $t = 3.00\,\text{s}$:

$$x = \frac{1}{2}at^2$$
$$= \frac{1}{2}(7.40\,\text{m/s}^2)(3.00\,\text{s})^2 = 33.3\,\text{m} = 9(3.70\,\text{m})$$

INSIGHT

This Example illustrates one of the key features of accelerated motion—position does not change uniformly with time when an object accelerates. In this case, the distance traveled in the first two seconds is 4 times the distance traveled in the first second, and the distance traveled in the first three seconds is 9 times the distance traveled in the first second. This kind of behavior is a direct result of the fact that x depends on t^2 when the acceleration is nonzero.

PRACTICE PROBLEM

We've seen that in one second the racer travels 3.70 m. How much time does it take for the racer to travel twice this distance, $2(3.70\,\text{m}) = 7.40\,\text{m}$?   [**Answer:** $t = \sqrt{2}\,\text{s} = 1.41\,\text{s}$]

Some related homework problems: Problem 43, Problem 46

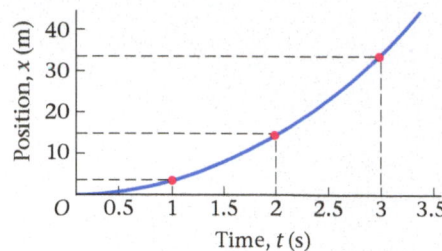

▲ **FIGURE 2-20 Position versus time for Example 2-13** The upward-curving, parabolic shape of this *x*-versus-*t* plot indicates a positive, constant acceleration. The dots on the curve show the position of the drag racer in Example 2-13 at the times 1.00 s, 2.00 s, and 3.00 s.

PHYSICS IN CONTEXT
Looking Ahead

The equations developed for motion with constant acceleration in this chapter are used again with slightly different symbols when we study motion in two dimensions in Chapter 4 and rotational motion in Chapter 10.

FIGURE 2-20 shows a graph of *x* versus *t* for Example 2-13. Notice the parabolic shape of the *x*-versus-*t* curve, which is due to the $\frac{1}{2}at^2$ term and is characteristic of constant acceleration. In particular, if acceleration is positive ($a > 0$), then a plot of *x* versus *t* curves upward; if acceleration is negative ($a < 0$), a plot of *x* versus *t* curves downward. The greater the magnitude of *a*, the greater the curvature. In contrast, if a particle moves with constant velocity ($a = 0$), the t^2 dependence vanishes, and the *x*-versus-*t* plot is a straight line.

Velocity as a Function of Position   Our final equation of motion with constant acceleration relates velocity to position. We start by solving for the time, *t*, in Equation 2-7 ($v = v_0 + at$):

$$v = v_0 + at \quad \text{or} \quad t = \frac{v - v_0}{a}$$

Next, we substitute this result into Equation 2-10, $x = x_0 + \frac{1}{2}(v_0 + v)t$, thus eliminating *t*:

$$x = x_0 + \frac{1}{2}(v_0 + v)t = x_0 + \frac{1}{2}(v_0 + v)\left(\frac{v - v_0}{a}\right)$$

Noting that $(v_0 + v)(v - v_0) = v_0 v - v_0^2 + v^2 - vv_0 = v^2 - v_0^2$, we have

$$x = x_0 + \frac{v^2 - v_0^2}{2a}$$

Finally, a straightforward rearrangement of terms yields the following:

Constant-Acceleration Equation of Motion: Velocity in Terms of Displacement

$$v^2 = v_0^2 + 2a(x - x_0) = v_0^2 + 2a\Delta x \qquad \qquad 2\text{-}12$$

This equation allows us to relate the velocity at one position to the velocity at another position, without knowing how much time is involved. The next Example shows how Equation 2-12 is used.

EXAMPLE 2-14   TAKEOFF DISTANCE FOR AN AIRLINER

RWP Jets at John F. Kennedy International Airport accelerate from rest at one end of a runway, and must attain takeoff speed before reaching the other end of the runway. **(a)** Plane A has acceleration *a* and takeoff speed v_{to}. What is the minimum length of runway, Δx_A, required for this plane? Give a symbolic answer. **(b)** Plane B has the same acceleration as plane A, but requires twice the takeoff speed. Find Δx_B and compare with Δx_A. **(c)** Find the minimum runway length for plane A if $a = 2.20 \text{ m/s}^2$ and $v_{to} = 95.0 \text{ m/s}$. (These values are typical for a 747 jetliner. For purposes of comparison, the shortest runway at JFK International Airport is 04R/22L, which has a length of 2560 m.)

PICTURE THE PROBLEM
In our sketch, we choose the positive *x* direction to be the direction of motion. With this choice, it follows that the acceleration of the plane is positive, $a = +2.20 \text{ m/s}^2$. Similarly, the takeoff velocity is positive as well, $v_{to} = +95.0 \text{ m/s}$.

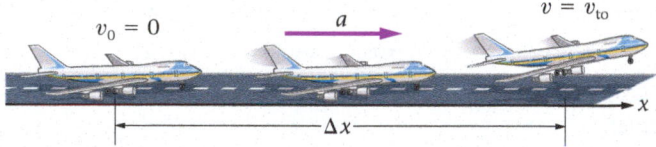

REASONING AND STRATEGY
From the sketch it's clear that we want to express Δx, the distance the plane travels in attaining takeoff speed, in terms of the acceleration, *a*, and the takeoff speed, v_{to}. Equation 2-12 ($v^2 = v_0^2 + 2a\Delta x$), which relates distance to velocity, allows us to do this.

Known   Acceleration, $a = 2.20 \text{ m/s}^2$; takeoff velocity, $v_{to} = 95.0 \text{ m/s}$.
Unknown   Takeoff distance, $\Delta x = ?$

SOLUTION

Part (a)

1. Solve $v^2 = v_0^2 + 2a(x - x_0) = v_0^2 + 2a\Delta x$ for Δx.
   To find Δx_A, set $v_0 = 0$ and $v = v_{to}$:

$$\Delta x = \frac{v^2 - v_0^2}{2a}$$

$$\Delta x_A = \frac{v_{to}^2}{2a}$$

CONTINUED

Part (b)

2. To find Δx_B, simply change v_{to} to $2v_{to}$ in part (a):

$$\Delta x_B = \frac{(2v_{to})^2}{2a} = \frac{4v_{to}^2}{2a} = 4\Delta x_A$$

Part (c)

3. Substitute numerical values into the result found in part (a):

$$\Delta x_A = \frac{v_{to}^2}{2a} = \frac{(95.0 \text{ m/s})^2}{2(2.20 \text{ m/s}^2)} = 2050 \text{ m}$$

INSIGHT

This Example illustrates the fact that there are many advantages to obtaining symbolic results before substituting numerical values. In this case, we found that the takeoff distance is proportional to v^2; hence, we conclude immediately that doubling v results in a fourfold increase of Δx.

PRACTICE PROBLEM

Find the minimum acceleration needed for a takeoff speed of $v_{to} = (95.0 \text{ m/s})/2 = 47.5 \text{ m/s}$ on a runway of length $\Delta x = (2050 \text{ m})/4 = 513 \text{ m}$.   [**Answer:** $a = v_{to}^2/2\Delta x = 2.20 \text{ m/s}^2$]

Some related homework problems: Problem 40, Problem 90

Enhance Your Understanding   (Answers given at the end of the chapter)

5. The equation of motion for an object moving with constant acceleration is $x = 6 \text{ m} - (5 \text{ m/s})t + (4 \text{ m/s}^2)t^2$. **(a)** What is the position of this object at $t = 0$? **(b)** What is the velocity of this object at $t = 0$? **(c)** What is the acceleration of this object?

Section Review

- Motion with constant acceleration can be described by equations of motion relating quantities like position, velocity, time, and acceleration.

2-6 Applications of the Equations of Motion

We devote this section to a variety of examples that further illustrate the use of the constant-acceleration equations of motion. For convenience, all of our constant-acceleration equations of motion are collected in Table 2-5.

Big Idea 5 Equations of motion for constant acceleration relate position, velocity, acceleration, and time.

TABLE 2-5 Constant-Acceleration Equations of Motion

Variables Related	Equation	Number
velocity, time, acceleration	$v = v_0 + at$	2-7
initial, final, and average velocity	$v_{av} = \frac{1}{2}(v_0 + v)$	2-9
position, time, velocity	$x = x_0 + \frac{1}{2}(v_0 + v)t$	2-10
position, time, acceleration	$x = x_0 + v_0 t + \frac{1}{2} at^2$	2-11
velocity, position, acceleration	$v^2 = v_0^2 + 2a(x - x_0) = v_0^2 + 2a\Delta x$	2-12

In our first Example, we consider the distance and time needed to brake a vehicle to a complete stop.

EXAMPLE 2-15   HIT THE BRAKES!

A park ranger driving on a back country road suddenly sees a deer "frozen" in the headlights. The ranger, who is driving at 11.4 m/s, immediately applies the brakes and slows with an acceleration of 3.80 m/s^2. **(a)** If the deer is 20.0 m from the ranger's vehicle when the brakes are applied, how close does the ranger come to hitting the deer? **(b)** How much time is needed for the ranger's vehicle to stop?

CONTINUED

PICTURE THE PROBLEM

We choose the positive x direction to be the direction of motion. With this choice it follows that $v_0 = +11.4$ m/s. In addition, the fact that the ranger's vehicle is slowing down means its acceleration points in the *opposite* direction to that of the velocity [see Figure 2-14 (b) and (c)]. Therefore, the vehicle's acceleration is $a = -3.80$ m/s^2. Finally, when the vehicle comes to rest its velocity is zero, $v = 0$.

REASONING AND STRATEGY

The acceleration is constant, so we can use the equations listed in Table 2-5. In part (a) we want to find a distance when we know the velocity and acceleration, so we use a rearranged version of Equation 2-12 ($v^2 = v_0^2 + 2a\Delta x$). In part (b) we want to find a time when we know the velocity and acceleration, so we use a rearranged version of Equation 2-7 ($v = v_0 + at$).

Known Velocity at time $t = 0$, $v_0 = 11.4$ m/s; acceleration, $a = -3.80$ m/s^2.
Unknown (a) Distance required to stop, $\Delta x = ?$ (b) Time required to stop, $t = ?$

SOLUTION

Part (a)

1. Solve $v^2 = v_0^2 + 2a\Delta x$ for Δx:

$$\Delta x = \frac{v^2 - v_0^2}{2a}$$

2. Set $v = 0$, and substitute numerical values:

$$\Delta x = -\frac{v_0^2}{2a} = -\frac{(11.4 \text{ m/s})^2}{2(-3.80 \text{ m/s}^2)} = 17.1 \text{ m}$$

3. Subtract Δx from 20.0 m to find the distance between the stopped vehicle and the deer:

$$20.0 \text{ m} - 17.1 \text{ m} = 2.9 \text{ m}$$

Part (b)

4. Set $v = 0$ in $v = v_0 + at$ and solve for t:

$$v = v_0 + at = 0$$

$$t = -\frac{v_0}{a} = -\frac{11.4 \text{ m/s}}{(-3.80 \text{ m/s}^2)} = 3.00 \text{ s}$$

INSIGHT

Notice the different ways that t and Δx depend on the initial speed. If the initial speed is doubled, for example, the time needed to stop also doubles. On the other hand, the distance needed to stop increases by a factor of four. This is one reason speed on the highway has such a great influence on safety.

PRACTICE PROBLEM

Show that using $t = 3.00$ s in Equation 2-11 $\left(x = x_0 + v_0 t + \frac{1}{2}at^2\right)$ results in the same distance needed to stop.
[**Answer:** $x = x_0 + v_0 t + \frac{1}{2}at^2 = 0 + (11.4 \text{ m/s})(3.00 \text{ s}) + \frac{1}{2}(-3.80 \text{ m/s}^2)(3.00 \text{ s})^2 = 17.1$ m, as expected]

Some related homework problems: Problem 55, Problem 56

In Example 2-15, we calculated the distance necessary for a vehicle to come to a complete stop. But how does the speed v vary with distance as the vehicle slows down? The next Conceptual Example deals with this topic.

CONCEPTUAL EXAMPLE 2-16   STOPPING DISTANCE

The ranger in Example 2-15 brakes for 17.1 m. After braking for only half that distance, $\frac{1}{2}(17.1 \text{ m}) = 8.55$ m, is the ranger's speed **(a)** equal to $\frac{1}{2}v_0$, **(b)** greater than $\frac{1}{2}v_0$, or **(c)** less than $\frac{1}{2}v_0$?

REASONING AND DISCUSSION

As pointed out in the Insight for Example 2-15, the fact that the stopping distance, Δx, depends on v_0^2 means that this distance increases by a factor of four when the speed is doubled. For example, the stopping distance with an initial speed of v_0 is four times the stopping distance when the initial speed is $v_0/2$.

CONTINUED

To apply this observation to the ranger, suppose that the stopping distance with an initial speed of v_0 is Δx. It follows that the stopping distance for an initial speed of $v_0/2$ is $\Delta x/4$. This means that as the ranger slows from v_0 to 0, it takes a distance $\Delta x/4$ to slow from $v_0/2$ to 0, and the remaining distance, $3\Delta x/4$, to slow from v_0 to $v_0/2$. Thus, at the halfway point the ranger has not yet slowed to half of the initial velocity—the speed at this point is greater than $v_0/2$.

ANSWER

(b) The ranger's speed is greater than $\frac{1}{2}v_0$.

Clearly, v does not decrease uniformly with distance. A plot showing v as a function of x for Example 2-15 is shown in **FIGURE 2-21**. As we can see from the graph, v changes more in the second half of the braking distance than in the first half.

We close this section with a familiar everyday example: a police car accelerating to overtake a speeder. This is the first time we use two equations of motion for two different objects to solve a problem—but it won't be the last. Problems of this type are often more interesting than problems involving only a single object, and they relate to many types of situations in everyday life.

PROBLEM-SOLVING NOTE

Strategize

Before you attempt to solve a problem, it is a good idea to have some sort of plan, or "strategy," for how to proceed. It may be as simple as saying, "The problem asks me to relate velocity and time; therefore I will use Equation 2-7." In other cases the strategy is a bit more involved. Producing effective strategies is one of the most challenging—and creative—aspects of problem solving.

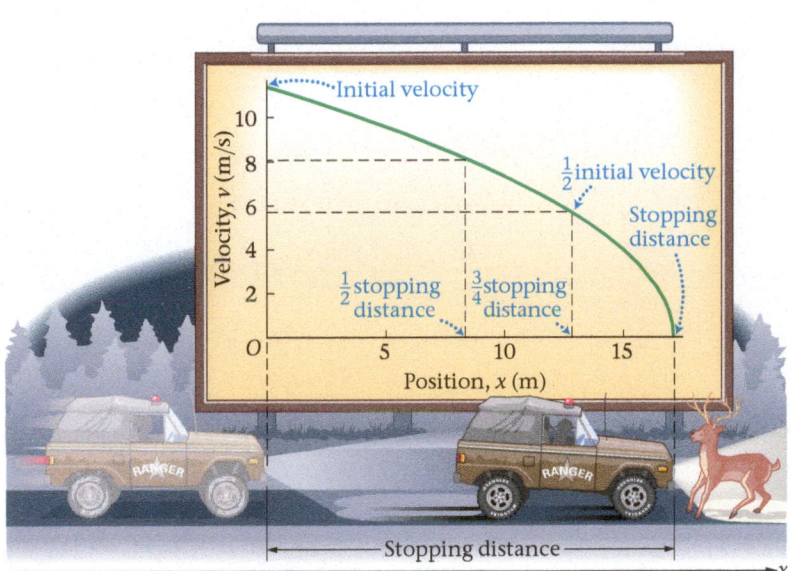

◄ **FIGURE 2-21 Velocity as a function of position for the ranger in Example 2-15** The ranger's vehicle comes to rest with constant acceleration, which means that its velocity decreases uniformly with time. The velocity *does not* decrease uniformly with distance, however. In particular, note how rapidly the velocity decreases in the final one-quarter of the stopping distance.

EXAMPLE 2-17 CATCHING A SPEEDER

A speeder doing 40.0 mi/h (about 17.9 m/s) in a 25 mi/h zone approaches a parked police car. The instant the speeder passes the police car, the police begin their pursuit. If the speeder maintains a constant velocity, and the police car accelerates with a constant acceleration of 4.51 m/s², **(a)** how much time does it take for the police car to catch the speeder, **(b)** how far have the two cars traveled in this time, and **(c)** what is the velocity of the police car when it catches the speeder?

PICTURE THE PROBLEM

Our sketch shows the two cars at the moment the speeder passes the resting police car. At this instant, which we take to be $t = 0$, both the speeder and the police car are at the origin, $x = 0$. In addition, we choose the positive x direction to be the direction of motion; therefore, the speeder's initial velocity is given by $v_s = +17.9$ m/s, and the police car's initial velocity is zero. The speeder's acceleration is zero, but the police car has an acceleration given by $a = +4.51$ m/s². Finally, our figure shows the linear x-versus-t plot for the speeder, and the parabolic x-versus-t plot for the police car.

REASONING AND STRATEGY

To solve this problem, we first write a position-versus-time equation $\left(x = x_0 + v_0 t + \frac{1}{2}at^2\right)$ for the police car, x_p, and a separate equation for the speeder, x_s. Next, we find the time it takes the police car to catch the speeder by setting $x_p = x_s$ and solving the resulting equation for t. Once the catch time is determined, it's straightforward to calculate the distance traveled and the velocity of the police car.

CONTINUED

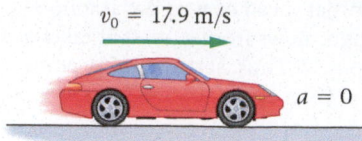

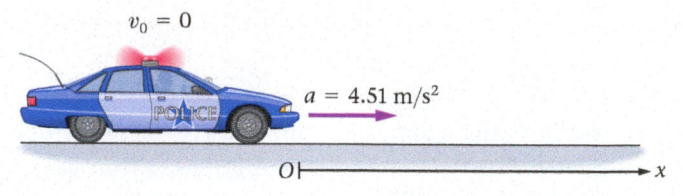

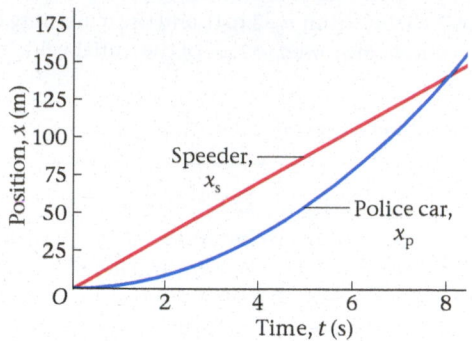

Known Speeder velocity, $v_s = +17.9$ m/s; initial police velocity $= 0$; police acceleration, $a = +4.51$ m/s^2.
Unknown (a) Time to catch speeder, $t = $? (b) Distance to catch speeder? (c) Velocity of police when speeder is caught, $v_p = $?

SOLUTION

Part (a)

1. Use $x = x_0 + v_0 t + \frac{1}{2}at^2$ to write equations of motion for the two vehicles. For the police car, $v_0 = 0$ and $a = 4.51$ m/s^2. For the speeder, $v_0 = 17.9$ m/s $= v_s$ and $a = 0$:

$$x_p = \frac{1}{2}at^2$$
$$x_s = v_s t$$

2. Set $x_p = x_s$ and solve for the time:

$$\frac{1}{2}at^2 = v_s t \quad \text{or} \quad \left(\frac{1}{2}at - v_s\right)t = 0$$

$$\text{two solutions: } t = 0 \quad \text{and} \quad t = \frac{2v_s}{a}$$

3. Clearly, $t = 0$ corresponds to the initial conditions, because both vehicles started at $x = 0$ at that time. The time of interest is obtained by substituting numerical values into the other solution:

$$t = \frac{2v_s}{a} = \frac{2(17.9 \text{ m/s})}{4.51 \text{ m/s}^2} = 7.94 \text{ s}$$

Part (b)

4. Substitute $t = 7.94$ s into the equations of motion for x_p and x_s. Notice that $x_p = x_s$, as expected:

$$x_p = \frac{1}{2}at^2 = \frac{1}{2}(4.51 \text{ m/s}^2)(7.94 \text{ s})^2 = 142$$
$$x_s = v_s t = (17.9 \text{ m/s})(7.94 \text{ s}) = 142 \text{ m}$$

Part (c)

5. To find the velocity of the police car use Equation 2-7 ($v = v_0 + at$), which relates velocity to time:

$$v_p = v_0 + at = 0 + (4.51 \text{ m/s}^2)(7.94 \text{ s}) = 35.8 \text{ m/s}$$

INSIGHT

When the police car catches up with the speeder, its velocity is 35.8 m/s, which is exactly twice the velocity of the speeder. A coincidence? Not at all. When the police car catches the speeder, both have traveled the same distance (142 m) in the same time (7.94 s); therefore, they have the same average velocity. Of course, the average velocity of the speeder is simply v_s. The average velocity of the police car is $\frac{1}{2}(v_0 + v)$, since its acceleration is constant, and thus $\frac{1}{2}(v_0 + v) = v_s$. Noting that $v_0 = 0$ for the police car, we see that $v = 2v_s$. This result is independent of the acceleration of the police car, as we show in the following Practice Problem.

PRACTICE PROBLEM

Repeat this Example for the case where the acceleration of the police car is $a = 3.25$ m/s^2. [**Answer: (a)** $t = 11.0$ s, **(b)** $x_p = x_s = 197$ m, **(c)** $v_p = 35.8$ m/s]

Some related homework problems: Problem 62, Problem 64

Enhance Your Understanding (Answers given at the end of the chapter)

6. A submerged alligator swims directly toward two unsuspecting ducks. The equation of motion of the alligator is $x_a = (0.25 \text{ m/s}^2)t^2$, the equation of motion of duck 1 is $x_{d1} = 3 \text{ m} - (1.5 \text{ m/s})t$, and the equation of motion of duck 2 is $x_{d2} = 3 \text{ m} + (1 \text{ m/s})t$. Which duck does the alligator encounter first?

CONTINUED

Section Review

- Equations of motion for constant acceleration are listed in Table 2-5. As an example, the equation relating velocity and time is $v = v_0 + at$.

2-7 Freely Falling Objects

The most famous example of motion with constant acceleration is **free fall**—the motion of an object falling freely under the influence of gravity. It was Galileo (1564–1642) who first showed, with his own experiments, that falling bodies move with constant acceleration. His conclusions were based on experiments done by rolling balls down inclines of various steepness. By using an incline, Galileo was able to reduce the acceleration of the balls, thus producing motion slow enough to be timed with the instruments available at the time.

Galileo also pointed out that objects of different weight fall with the *same* constant acceleration—provided air resistance is small enough to be ignored. Whether he dropped objects from the Leaning Tower of Pisa to demonstrate this fact, as legend has it, will probably never be known for certain, but we do know that he performed extensive experiments to support his claim.

Today it is easy to verify Galileo's assertion by dropping objects in a vacuum chamber, where the effects of air resistance are essentially removed. In a standard classroom demonstration, a feather and a coin are dropped in a vacuum, and both fall at the same rate. In 1971, a novel version of this experiment was carried out on the Moon by astronaut David Scott. In the near-perfect vacuum on the Moon's surface he dropped a feather and a hammer and showed a worldwide television audience that they fell to the ground in the same time. Examples of free fall in different contexts are shown in **FIGURE 2-22**.

To illustrate the effect of air resistance in everyday terms, drop a sheet of paper and a rubber ball from the same height (**FIGURE 2-23**). The paper drifts slowly to the ground, taking much more time to fall than the ball. Now, wad the sheet of paper into a tight ball and repeat the experiment. This time the ball of paper and the rubber ball reach the ground in nearly the same time. What was different in the two experiments? Clearly, when the sheet of paper was wadded into a ball, the effect of air resistance on it was greatly reduced, so that both objects fell almost as they would in a vacuum.

Characteristics of Free Fall   Before considering a few examples, let's discuss exactly what is meant by "free fall." To begin, the word *free* in free fall means free from any effects other than gravity. For example, in free fall we assume that an object's motion is not influenced by any form of friction or air resistance.

Free fall is the motion of an object subject *only* to the influence of gravity.

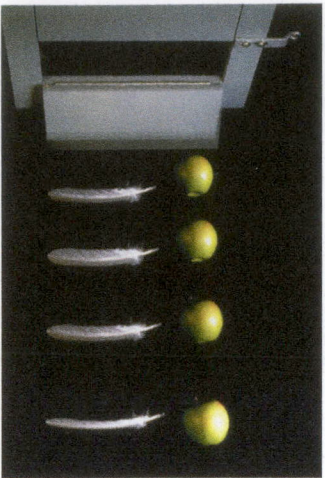

(a)

(b)

(c)

▲ **FIGURE 2-22 Visualizing Concepts Free fall (a)** In the absence of air resistance, all objects fall with the same acceleration, regardless of their mass. **(b)** Whether she is on the way up, at the peak of her flight, or on the way down, this girl is in free fall, accelerating downward with the acceleration of gravity. Only when she is in contact with the blanket does her acceleration change. **(c)** In the absence of air resistance, these lava bombs from the Kilauea volcano on the Big Island of Hawaii would strike the water with the same speed they had when they were blasted into the air. It's an example of the symmetry of free fall.

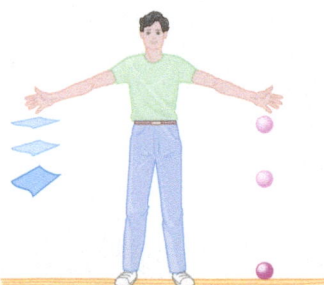

(a) Dropping a sheet of paper and a rubber ball

(b) Dropping a wadded-up sheet of paper and a rubber ball

▲ **FIGURE 2-23 Free fall and air resistance**

Big Idea 6 Free fall is motion with a constant downward acceleration of magnitude g, where g is the acceleration due to gravity.

Though free fall is an idealization—which does not apply to many real-world situations—it is still a useful approximation in many other cases. In the following examples we assume that the motion may be considered as free fall.

Next, it should be realized that the word *fall* in free fall does not mean the object is necessarily moving downward. By free fall, we mean *any* motion under the influence of gravity alone. If you drop a ball, it is in free fall. If you throw a ball upward or downward, it is in free fall as soon as it leaves your hand.

> An object is in free fall as soon as it is released, whether it is dropped from rest, thrown downward, or thrown upward.

Finally, the acceleration produced by gravity on the Earth's surface (sometimes called the gravitational strength) is denoted with the symbol g. As a shorthand name, we will frequently refer to g simply as "the acceleration due to gravity." In fact, as we shall see in Chapter 12, the value of g varies according to one's location on the surface of the Earth, as well as one's altitude above it. Table 2-6 shows how g varies with latitude.

In all the calculations that follow in this book, we shall use $g = 9.81 \text{ m/s}^2$ for the acceleration due to gravity. Note, in particular, that g always stands for $+9.81 \text{ m/s}^2$, never -9.81 m/s^2. For example, if we choose a coordinate system with the positive direction upward, the acceleration in free fall is $a = -g$. If the positive direction is downward, then free-fall acceleration is $a = g$.

With these comments in mind, we're ready to explore a variety of free-fall examples.

TABLE 2-6 Values of g at Different Locations on Earth (m/s^2)

Location	Latitude	g
North Pole	90° N	9.832
Oslo, Norway	60° N	9.819
Hong Kong	30° N	9.793
Quito, Ecuador	0°	9.780

EXAMPLE 2-18 | LEMON DROP

A lemon drops from a tree and falls to the ground 3.15 m below. **(a)** How much time does it take for the lemon to reach the ground? **(b)** What is the lemon's speed just before it hits the ground?

PICTURE THE PROBLEM
In our sketch we choose the origin to be at the drop height of the lemon, and we let the positive x direction be downward. With these choices, $x_0 = 0$, $a = g$, and the ground is at $x = h = 3.15 \text{ m}$. Of course, the initial velocity is zero, $v_0 = 0$, because the lemon drops from rest out of the tree.

REASONING AND STRATEGY
We can ignore air resistance in this case and model the motion as free fall. This means we can assume a constant acceleration equal to g and use the equations of motion in Table 2-5. For part (a) we want to find the time of fall when we know the distance and acceleration, so we use Equation 2-11 $\left(x = x_0 + v_0 t + \frac{1}{2}at^2\right)$. For part (b) we can relate velocity to time by using Equation 2-7 ($v = v_0 + at$), or we can relate velocity to position by using Equation 2-12 ($v^2 = v_0^2 + 2a\Delta x$). We will implement both approaches and show that the results are the same.

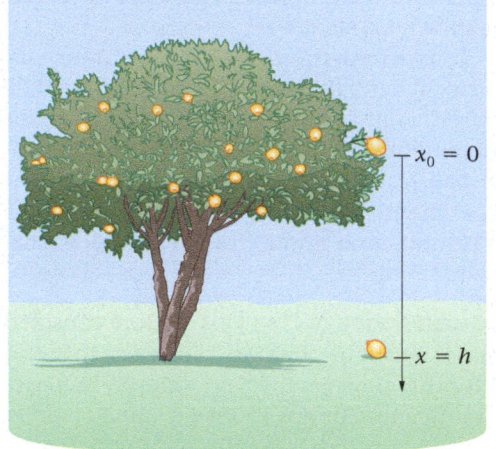

$x_0 = 0$

$x = h$

Known   Initial position of lemon, $x_0 = 0$; final position of lemon, $x = h = 3.15 \text{ m}$; initial velocity of lemon, $v_0 = 0$; acceleration of lemon, $a = g = 9.81 \text{ m/s}^2$.

Unknown   **(a)** Drop time, $t = ?$ **(b)** Landing speed, $v = ?$

SOLUTION

Part (a)

1. Write $x = x_0 + v_0 t + \frac{1}{2}at^2$, with $x_0 = 0$, $v_0 = 0$, and $a = g$:

$$x = x_0 + v_0 t + \frac{1}{2}at^2 = 0 + 0 + \frac{1}{2}gt^2 = \frac{1}{2}gt^2$$

2. Solve for the time, t, and set $x = h = 3.15 \text{ m}$:

$$t = \sqrt{\frac{2x}{g}} = \sqrt{\frac{2h}{g}} = \sqrt{\frac{2(3.15 \text{ m})}{9.81 \text{ m/s}^2}} = 0.801 \text{ s}$$

Part (b)

3. To find the velocity, use the time found in part (a), $t = 0.801 \text{ s}$, in $v = v_0 + at$:

$$v = v_0 + gt = 0 + (9.81 \text{ m/s}^2)(0.801 \text{ s}) = 7.86 \text{ m/s}$$

4. We can also find the velocity without knowing the time by using $v^2 = v_0^2 + 2a\Delta x$ with $\Delta x = 3.15 \text{ m}$:

$$v^2 = v_0^2 + 2a\Delta x = 0 + 2g\Delta x$$
$$v = \sqrt{2g\Delta x} = \sqrt{2(9.81 \text{ m/s}^2)(3.15 \text{ m})} = 7.86 \text{ m/s}$$

CONTINUED

INSIGHT
We could just as well solve this problem with the origin at ground level, the drop height at $x = h$, and the positive x direction upward, which means that $a = -g$. The results are the same, of course.

PRACTICE PROBLEM — PREDICT/CALCULATE
Consider the distance the lemon drops in half the time required to reach the ground—that is, in the time $t = (0.801 \text{ s})/2$. **(a)** Is this distance greater than, less than, or equal to half the distance to the ground? Explain. **(b)** Find the distance the lemon drops in this time.   [**Answer: (a)** The distance is less than half the distance to the ground, because the average speed of the lemon in the first half of its drop is less than its average speed in the second half of its drop. **(b)** The distance dropped is

$$x = \tfrac{1}{2}gt^2 = \tfrac{1}{2}(9.81 \text{ m/s}^2)\left(\frac{0.801 \text{ s}}{2}\right)^2 = 0.787 \text{ m, which is only one-quarter the distance to the ground.}]$$

Some related homework problems: Problem 70, Problem 73

Free Fall from Rest The special case of free fall from rest occurs so frequently, and in so many different contexts, that it deserves special attention. If we take x_0 to be zero, and positive to be downward, then position as a function of time is $x = x_0 + v_0 t + \tfrac{1}{2}gt^2 = 0 + 0 + \tfrac{1}{2}gt^2$, or

$$x = \tfrac{1}{2}gt^2 \qquad\qquad 2\text{-}13$$

Similarly, velocity as a function of time is

$$v = gt \qquad\qquad 2\text{-}14$$

In addition, velocity as a function of position is

$$v = \sqrt{2gx} \qquad\qquad 2\text{-}15$$

The behavior of these functions is illustrated in **FIGURE 2-24**. Notice that position increases with time squared, whereas velocity increases linearly with time.

Next we consider two objects that drop from rest, one after the other, and discuss how their separation varies with time.

$t = 0$	$v = 0$	$x = 0$
$t = 1\text{ s}$	$v = 9.81 \text{ m/s}$	$x = 4.91 \text{ m}$
$t = 2\text{ s}$	$v = 19.6 \text{ m/s}$	$x = 19.6 \text{ m}$
$t = 3\text{ s}$	$v = 29.4 \text{ m/s}$	$x = 44.1 \text{ m}$
$t = 4\text{ s}$	$v = 39.2 \text{ m/s}$	$x = 78.5 \text{ m}$

▶ **FIGURE 2-24 Free fall from rest** Position and velocity are shown as functions of time. It is apparent that velocity depends linearly on t, whereas position depends on t^2.

CONCEPTUAL EXAMPLE 2-19   FREE-FALL SEPARATION

Drops of water detach from the tip of an icicle and fall from rest. When one drop separates, the drop ahead of it has already fallen through a distance d, as shown below. As these two drops continue to fall, does their separation **(a)** increase, **(b)** decrease, or **(c)** stay the same?

REASONING AND DISCUSSION
It might seem that the separation between the drops will remain the same, because both are in free fall. This is not so. The drop that has a head start always has a greater velocity than the one that comes next. Therefore, the first drop covers a greater distance in any interval of time, and as a result, the separation between the drops increases.

ANSWER
(a) The separation between the drops increases.

An erupting volcano shooting out fountains of lava is an impressive sight. In the next Example we show how a simple timing experiment can determine the initial velocity of the erupting lava.

EXAMPLE 2-20 BOMBS AWAY: CALCULATING THE SPEED OF A LAVA BOMB

RWP A volcano shoots out blobs of molten lava, called lava bombs, from its summit. A geologist observing the eruption uses a stopwatch to time the flight of a particular lava bomb that is projected straight upward. If the time for the bomb to rise and fall back to its launch height is 4.75 s, and its acceleration is 9.81 m/s² downward, what is its initial speed?

PICTURE THE PROBLEM

Our sketch shows a coordinate system with upward as the positive x direction. For clarity, we offset the upward and downward trajectories slightly in our sketch. In addition, we choose $t = 0$ to be the time at which the lava bomb is launched. With these choices it follows that $x_0 = 0$ and the acceleration is $a = -g = -9.81$ m/s². The initial speed to be determined is v_0.

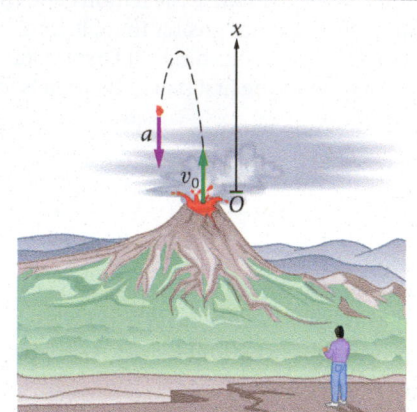

REASONING AND STRATEGY

Once again, we can ignore air resistance and model the motion of the lava bomb as free fall—this time with an initial upward velocity. We know that the lava bomb starts at $x = 0$ at the time $t = 0$ and returns to $x = 0$ at the time $t = 4.75$ s. This means that we know the bomb's position, time, and acceleration ($a = -g$), from which we would like to determine the initial velocity. A reasonable approach is to use Equation 2-11 $\left(x = x_0 + v_0 t + \frac{1}{2} at^2 \right)$ and solve it for the one unknown it contains, v_0.

Known Flight time, $t = 4.75$ s; acceleration, $a = -g = -9.81$ m/s².
Unknown Initial velocity of the lava bomb, $v_0 = ?$

SOLUTION

1. Write out $x = x_0 + v_0 t + \frac{1}{2} at^2$ with $x_0 = 0$ and $a = -g$. Factor out a time, t, from the two remaining terms:

$$x = x_0 + v_0 t + \frac{1}{2} at^2 = v_0 t - \frac{1}{2} gt^2 = \left(v_0 - \frac{1}{2} gt \right)t$$

2. Set x equal to zero, because this is the position of the lava bomb at $t = 0$ and $t = 4.75$ s:

$$x = \left(v_0 - \frac{1}{2} gt \right)t = 0$$

Two solutions: (i) $t = 0$ (ii) $v_0 - \frac{1}{2} gt = 0$

3. The first solution is simply the initial condition; that is, $x = 0$ at $t = 0$. Solve the second solution for the initial speed:

$$v_0 - \frac{1}{2} gt = 0 \quad \text{or} \quad v_0 = \frac{1}{2} gt$$

4. Substitute numerical values for g and the time the lava bomb lands:

$$v_0 = \frac{1}{2} gt = \frac{1}{2}(9.81 \text{ m/s}^2)(4.75 \text{ s}) = 23.3 \text{ m/s}$$

INSIGHT

A geologist can determine a lava bomb's initial speed by simply observing its flight time. Knowing the lava bomb's initial speed can help geologists determine how severe a volcanic eruption will be, and how dangerous it might be to people in the surrounding area.

PRACTICE PROBLEM

A second lava bomb is projected straight upward with an initial speed of 25 m/s. What is its flight time?
[**Answer:** $t = 5.1$ s]

Some related homework problems: Problem 71, Problem 78

PROBLEM-SOLVING NOTE

Check Your Solution

Once you have a solution to a problem, check to see whether it makes sense. First, make sure the units are correct; m/s for speed, m/s² for acceleration, and so on. Second, check the numerical value of your answer. If you are solving for the speed of a diver dropping from a 3.0-m diving board and you get an unreasonable value like 200 m/s (≈ 450 mi/h), chances are good that you've made a mistake.

What is the speed of a lava bomb when it returns to Earth; that is, when it returns to the same level from which it was launched? Physical intuition might suggest that, in the absence of air resistance, it should be the same as its initial speed. To show that this hypothesis is indeed correct, write out Equation 2-7 ($v = v_0 + at$) for this case:

$$v = v_0 - gt$$

Substituting numerical values, we find

$$v = v_0 - gt = 23.3 \text{ m/s} - (9.81 \text{ m/s}^2)(4.75 \text{ s}) = -23.3 \text{ m/s}$$

Thus, the velocity of the lava bomb when it lands is just the negative of the velocity it had when launched upward. Or, put another way, when the lava bomb lands, it has the same speed as when it was launched; it's just traveling in the opposite direction.

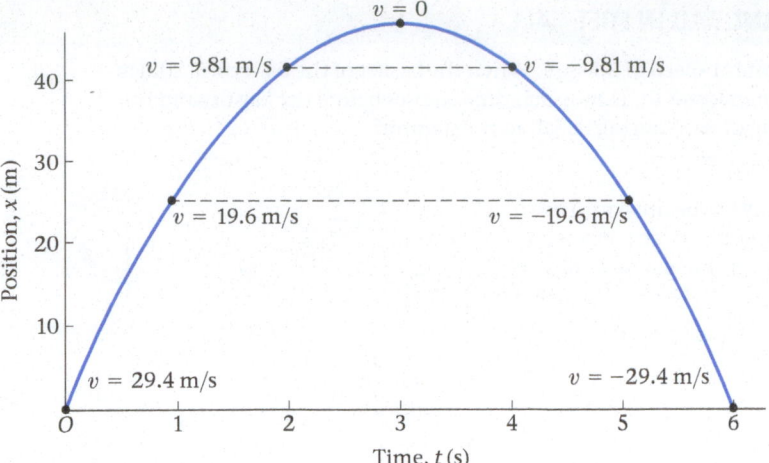

It's instructive to verify this result symbolically. Recall from Example 2-20 that $v_0 = \frac{1}{2}gt$, where t is the time the bomb lands. Substituting this result into Equation 2-7, we find

$$v = \frac{1}{2}gt - gt = -\frac{1}{2}gt = -v_0$$

The advantage of the symbolic solution lies in showing that the result is not a fluke—no matter what the initial velocity, no matter what the acceleration, the bomb lands with the velocity $-v_0$.

The Symmetry of Free Fall The preceding results suggest a symmetry relating the motion on the way up to the motion on the way down. To make this symmetry more apparent, we first solve for the time when the lava bomb lands. Using the result $v_0 = \frac{1}{2}gt$ from Example 2-20, we find

$$t = \frac{2v_0}{g} \qquad \textit{(time of landing)}$$

Next, we find the time when the velocity of the lava bomb is zero, which is at its highest point. Setting $v = 0$ in Equation 2-7 ($v = v_0 + at$), we have $v = v_0 - gt = 0$, or

$$t = \frac{v_0}{g} \qquad \textit{(time when $v = 0$)}$$

This is exactly half the time required for the lava to make the round trip. Thus, the velocity of the lava bomb is zero and the height of the bomb is greatest exactly halfway between launch and landing.

This symmetry is illustrated in **FIGURE 2-25**. In this case we consider a lava bomb that is in the air for 6.00 s, moving without air resistance. Note that at $t = 3.00$ s the lava bomb is at its highest point and its velocity is zero. At times equally spaced before and after $t = 3.00$ s, the lava bomb is at the same height and has the same speed, but is moving in opposite directions. As a result of this symmetry, a movie of the lava bomb's flight would look the same whether run forward or in reverse.

FIGURE 2-26 shows the time dependence of position, velocity, and acceleration for an object in free fall without air resistance after being thrown upward. As soon as the object is released, it begins to accelerate downward—as indicated by the negative slope of the velocity-versus-time plot—though it isn't necessarily moving downward. For example, if you throw a ball *upward*, it begins to accelerate *downward* the moment it leaves your hand. It continues moving upward, however, until its speed diminishes to zero. Because gravity is causing the downward acceleration, and gravity doesn't turn off just because the ball's velocity goes through zero, the ball continues to accelerate downward even when it is momentarily at rest.

Similarly, in the next Example we consider a sandbag that falls from an ascending hot-air balloon. This means that before the bag is in free fall it was moving upward—just like a ball thrown upward. And just like the ball, the sandbag continues moving upward for a brief time before momentarily stopping and then moving downward.

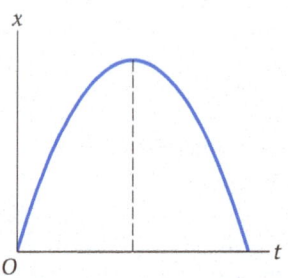

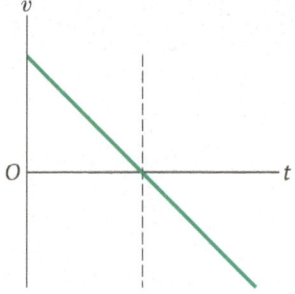

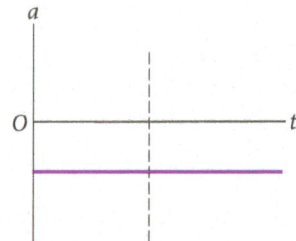

▲ **FIGURE 2-26 Position, velocity, and acceleration of a lava bomb as functions of time** The fact that the x-versus-t plot is curved indicates an acceleration; the downward curvature shows that the acceleration is negative. This is also clear from the v-versus-t plot, which has a negative slope. The constant slope of the straight line in the v-versus-t plot indicates a constant acceleration, as shown in the a-versus-t plot.

EXAMPLE 2-21   LOOK OUT BELOW! A SANDBAG IN FREE FALL

A hot-air balloon is rising straight upward with a constant speed of 6.5 m/s. When the basket of the balloon is 20.0 m above the ground, a bag of sand tied to the basket comes loose. **(a)** How much time elapses before the bag of sand hits the ground? **(b)** What is the greatest height of the bag of sand during its fall to the ground?

PICTURE THE PROBLEM
We choose the origin to be at ground level and positive to be upward. This means that, for the bag, we have $x_0 = 20.0$ m, $v_0 = 6.5$ m/s, and $a = -g$. Our sketch also shows snapshots of the balloon and bag of sand at three different times, starting at $t = 0$ when the bag comes loose. Notice that the bag is moving upward with the balloon at the time it comes loose. It therefore continues to move upward for a short time after it separates from the basket, exactly as if it had been thrown upward.

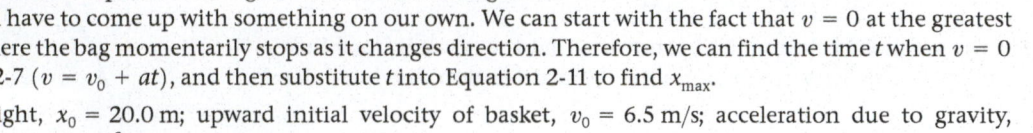

REASONING AND STRATEGY
The effects of air resistance on the sandbag can be ignored. As a result, we can use the equations in Table 2-5 with a constant acceleration $a = -g$.

In part (a) we want to relate position and time—knowing the initial position and initial velocity—so we use Equation 2-11 $\left(x = x_0 + v_0 t + \frac{1}{2}at^2\right)$. To find the time the bag hits the ground, we set $x = 0$ and solve for t.

For part (b) we have no expression that gives the maximum height of a particle—so we will have to come up with something on our own. We can start with the fact that $v = 0$ at the greatest height, since it is there the bag momentarily stops as it changes direction. Therefore, we can find the time t when $v = 0$ by using Equation 2-7 $(v = v_0 + at)$, and then substitute t into Equation 2-11 to find x_{max}.

Known   Drop height, $x_0 = 20.0$ m; upward initial velocity of basket, $v_0 = 6.5$ m/s; acceleration due to gravity, $a = -g = -9.81$ m/s².

Unknown   **(a)** Time to reach the ground, $t = ?$ **(b)** Maximum height of sandbag, $x_{max} = ?$

SOLUTION

Part (a)

1. Apply $x = x_0 + v_0 t + \frac{1}{2}at^2$ to the bag of sand. Do this by letting $a = -g$ and setting $x = 0$, which corresponds to ground level:

$$x = x_0 + v_0 t - \frac{1}{2}gt^2 = 0$$

2. Notice that we have a quadratic equation for t, with $A = -\frac{1}{2}g = -\frac{1}{2}(9.81 \text{ m/s}^2)$, $B = v_0 = 6.5$ m/s, and $C = x_0 = 20.0$ m. Solve this equation for t. The positive solution, 2.78 s, applies to this problem: (Quadratic equations and their solutions are discussed in Appendix A. In general, one can expect two solutions to a quadratic equation.)

$$t = \frac{-v_0 \pm \sqrt{v_0^2 - 4\left(-\frac{1}{2}g\right)(x_0)}}{2\left(-\frac{1}{2}g\right)}$$

$$= \frac{-(6.5 \text{ m/s}) \pm \sqrt{(6.5 \text{ m/s})^2 + 2(9.81 \text{ m/s}^2)(20.0 \text{ m})}}{(-9.81 \text{ m/s}^2)}$$

$$= \frac{-(6.5 \text{ m/s}) \pm 20.8 \text{ m/s}}{(-9.81 \text{ m/s}^2)} = 2.78 \text{ s}, -1.46 \text{ s}$$

Part (b)

3. Apply $v = v_0 + at$ to the bag of sand, and solve for the time when the velocity equals zero:

$$v = v_0 + at = v_0 - gt$$

$$v_0 - gt = 0 \quad \text{or} \quad t = \frac{v_0}{g} = \frac{6.5 \text{ m/s}}{9.81 \text{ m/s}^2} = 0.66 \text{ s}$$

4. Use $t = 0.66$ s in $x = x_0 + v_0 t + \frac{1}{2}at^2$ to find the maximum height:

$$x_{max} = 20.0 \text{ m} + (6.5 \text{ m/s})(0.66 \text{ s}) - \frac{1}{2}(9.81 \text{ m/s}^2)(0.66 \text{ s})^2$$
$$= 22 \text{ m}$$

INSIGHT
Thus, the bag of sand continues to move upward for 0.66 s after it separates from the basket, reaching a maximum height of 22 m above the ground. It then begins to move downward, and hits the ground 2.78 s after detaching from the basket.

PRACTICE PROBLEM
What is the velocity of the sandbag just before it hits the ground? [**Answer:** $v = v_0 - gt = (6.5 \text{ m/s}) - (9.81 \text{ m/s}^2) \times (2.78 \text{ s}) = -20.8$ m/s; the minus sign indicates the bag is moving downward, as expected.]

Some related homework problems: Problem 83, Problem 84

7. On a distant, airless planet, an astronaut drops a rock to test the planet's gravitational pull. The astronaut finds that in the first second of falling (from $t = 0$ to $t = 1$ s) the rock drops a distance of 1 m. How far does the rock drop from $t = 1$ s to $t = 2$ s? **(a)** 1 m, **(b)** 2 m, **(c)** 3 m, **(d)** 4 m.

Section Review

* Free fall is motion with a constant downward acceleration of magnitude $g = 9.81$ m/s^2.

CHAPTER 2 REVIEW

CHAPTER SUMMARY

2-1 POSITION, DISTANCE, AND DISPLACEMENT

Position
Position is the location of an object as measured on a coordinate system.

Distance
Distance is the total length of travel, from beginning to end. Distance is always positive.

Displacement
Displacement, Δx, is the change in position; that is, $\Delta x = x_f - x_i$.

Positive and Negative Displacement
The *sign* of the displacement indicates the *direction* of motion.

2-2 AVERAGE SPEED AND VELOCITY

Average Speed
Average speed is *distance* divided by elapsed time.

Average Velocity
Average velocity is *displacement* divided by elapsed time. Average velocity is positive for motion in the positive direction, negative for motion in the negative direction.

Graphical Interpretation
In an *x*-versus-*t* plot, the average velocity is the slope of a line connecting two points.

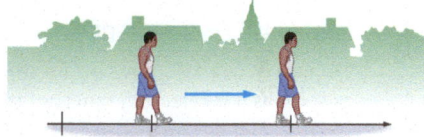
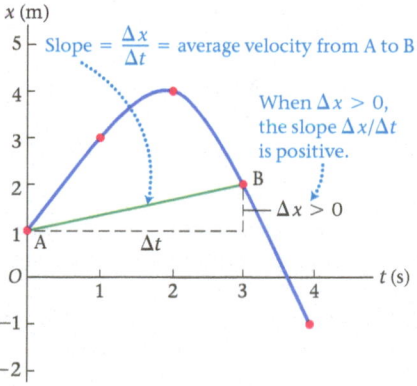

2-3 INSTANTANEOUS VELOCITY

The velocity at an instant of time, v, is the average velocity over shorter and shorter time intervals: $v = \lim\limits_{\Delta t \to 0} \frac{\Delta x}{\Delta t}$.

Graphical Interpretation
In an *x*-versus-*t* plot, the instantaneous velocity is the slope of a tangent line at a given instant of time.

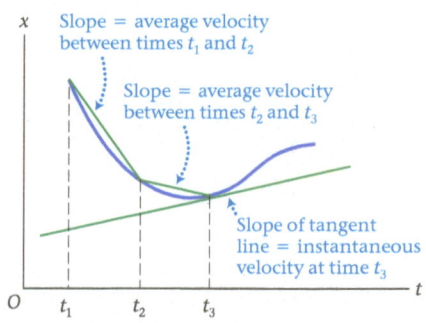

2-4 ACCELERATION

Average Acceleration

Average acceleration is the change in velocity divided by the elapsed time.

Instantaneous Acceleration

The acceleration at an instant of time, a, is the limit of the average acceleration over shorter and shorter time intervals: $a = \lim\limits_{\Delta t \to 0} \dfrac{\Delta v}{\Delta t}$.

Graphical Interpretation

In a v-versus-t plot, the instantaneous acceleration is the slope of a tangent line at a given instant of time.

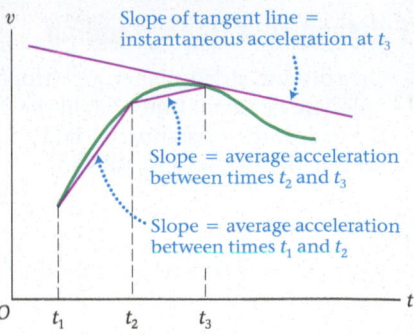

2-5 MOTION WITH CONSTANT ACCELERATION

Several different equations are used to describe the motion of objects moving with constant acceleration. These "equations of motion" are listed in Table 2-5. The ones that are used most often are (1) velocity as a function of time ($v = v_0 + at$), (2) position as a function of time $\left(x = x_0 + v_0 t + \frac{1}{2}at^2\right)$, and (3) velocity as a function of position ($v^2 = v_0^2 + 2a\Delta x$).

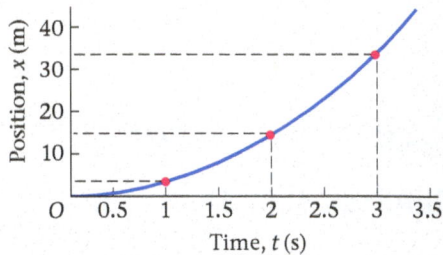

2-7 FREELY FALLING OBJECTS

Objects in free fall move under the influence of gravity alone. An object is in free fall as soon as it is released, whether it is thrown upward, thrown downward, or released from rest. As a result, objects in free fall move with a constant downward acceleration of magnitude $g = 9.81 \text{ m/s}^2$.

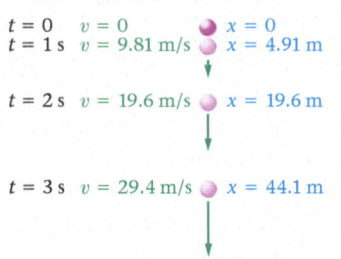

ANSWERS TO ENHANCE YOUR UNDERSTANDING QUESTIONS

1. **(a)** No. Distance is always increasing on a trip, whereas the magnitude of displacement can increase or decrease. **(b)** Yes. For example, in a 5-mi round trip the distance is 5 mi but the magnitude of the displacement is zero.

2. **(a)** Positive. **(b)** Zero. **(c)** Negative.

3. C < B < D < A.

4. **(a)** The speed is decreasing because velocity and acceleration have opposite signs. **(b)** The velocity is increasing (becoming less negative) because the acceleration is positive. **(c)** The speed is decreasing because velocity and acceleration have opposite signs. **(d)** The velocity is decreasing (becoming less positive) because the acceleration is negative.

5. **(a)** $x = 6$ m. **(b)** $v = -5$ m/s. **(c)** $a = 8 \text{ m/s}^2$.

6. The alligator encounters duck 1 first.

7. **(c)** 3 m.

CONCEPTUAL QUESTIONS

For instructor-assigned homework, go to www.masteringphysics.com. (MP)

(Answers to odd-numbered Conceptual Questions can be found in the back of the book.)
(The effects of air resistance are to be ignored in this chapter.)

1. You take your dog on a walk to a nearby park. On the way, your dog takes many short side trips to chase squirrels, examine fire hydrants, and so on. When you arrive at the park, do you and your dog have the same displacement from home? Have you and your dog traveled the same distance? Explain.

2. Does an odometer in a car measure distance or displacement? Explain.

3. An astronaut orbits Earth in the space shuttle. In one complete orbit, is the magnitude of the displacement the same as the distance traveled? Explain.

4. After a tennis match the players dash to the net to congratulate one another. If they both run with a speed of 3 m/s, are their velocities equal? Explain.

5. Does a speedometer measure speed or velocity? Explain.

6. Is it possible for a car to circle a racetrack with constant velocity? Can it do so with constant speed? Explain.

7. For what kinds of motion are the instantaneous and average velocities equal?

8. Assume that the brakes in your car create a constant deceleration, regardless of how fast you are going. If you double your driving speed, how does this affect **(a)** the time required to come to a stop, and **(b)** the distance needed to stop?

9. The velocity of an object is zero at a given instant of time. **(a)** Is it possible for the object's acceleration to be zero at this time? Explain. **(b)** Is it possible for the object's acceleration to be nonzero at this time? Explain.

10. If the velocity of an object is nonzero, can its acceleration be zero? Give an example if your answer is yes; explain why not if your answer is no.

11. Is it possible for an object to have zero average velocity over a given interval of time, yet still be accelerating during the interval? Give an example if your answer is yes; explain why not if your answer is no.

12. A batter hits a pop fly straight up. (a) Is the acceleration of the ball on the way up different from its acceleration on the way down? (b) Is the acceleration of the ball at the top of its flight different from its acceleration just before it lands?

13. A person on a trampoline bounces straight upward with an initial speed of 4.5 m/s. What is the person's speed when she returns to her initial height?

14. A volcano shoots a lava bomb straight upward. Does the displacement of the lava bomb depend on (a) your choice of origin for your coordinate system, or (b) your choice of a positive direction? Explain in each case.

PROBLEMS AND CONCEPTUAL EXERCISES

Answers to odd-numbered Problems and Conceptual Exercises can be found in the back of the book. **BIO** *identifies problems of biological or medical interest;* **CE** *indicates a conceptual exercise.* **Predict/Explain** *problems ask for two responses: (a) your prediction of a physical outcome, and (b) the best explanation among three provided; and* **Predict/Calculate** *problems ask for a prediction of a physical outcome, based on fundamental physics concepts, and follow that with a numerical calculation to verify the prediction. On all problems, bullets (•, ••, •••) indicate the level of difficulty.*

SECTION 2-1 POSITION, DISTANCE, AND DISPLACEMENT

1. • Referring to **FIGURE 2-27**, you walk from your home to the library, then to the park. (a) What is the distance traveled? (b) What is your displacement?

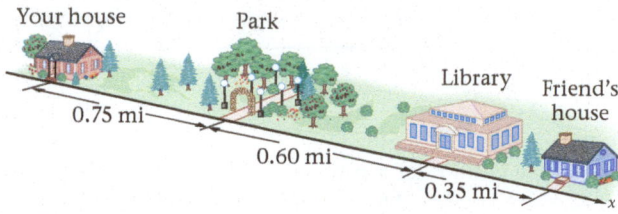

FIGURE 2-27 Problems 1 and 2

2. • In Figure 2-27, you walk from the park to your friend's house, then back to your house. What are your (a) distance traveled, and (b) displacement?

3. • The two tennis players shown in **FIGURE 2-28** walk to the net to congratulate one another. (a) Find the distance traveled and the displacement of player A. (b) Repeat for player B.

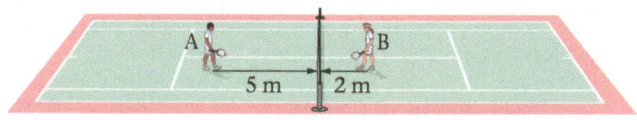

FIGURE 2-28 Problem 3

4. • The golfer in **FIGURE 2-29** sinks the ball in two putts, as shown. What are (a) the distance traveled by the ball, and (b) the displacement of the ball?

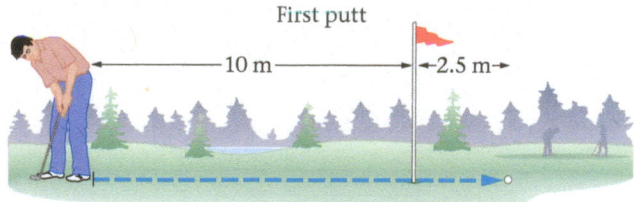

First putt

Second putt

FIGURE 2-29 Problem 4

5. • A jogger runs on the track shown in **FIGURE 2-30**. Neglecting the curvature of the corners, (a) what are the distance traveled and the displacement in running from point A to point B? (b) Find the distance and displacement for a complete circuit of the track.

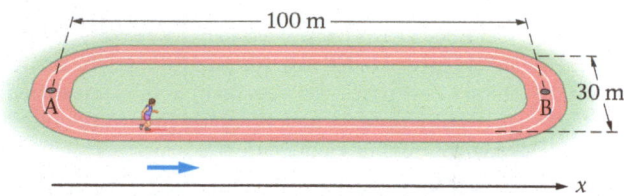

FIGURE 2-30 Problem 5

6. •• **Predict/Calculate** A child rides a pony on a circular track whose radius is 5.25 m. (a) Find the distance traveled and the displacement after the child has gone halfway around the track. (b) Does the distance traveled increase, decrease, or stay the same when the child completes one circuit of the track? Explain. (c) Does the displacement increase, decrease, or stay the same when the child completes one circuit of the track? Explain. (d) Find the distance and displacement after a complete circuit of the track.

SECTION 2-2 AVERAGE SPEED AND VELOCITY

7. • **CE** **Predict/Explain** You drive your car in a straight line at 15 m/s for 10 kilometers, then at 25 m/s for another 10 kilometers. (a) Is your average speed for the entire trip greater than, less than, or equal to 20 m/s? (b) Choose the *best* explanation from among the following:
   - I. More time is spent at 15 m/s than at 25 m/s.
   - II. The average of 15 m/s and 25 m/s is 20 m/s.
   - III. Less time is spent at 15 m/s than at 25 m/s.

8. • **CE** **Predict/Explain** You drive your car in a straight line at 15 m/s for 10 minutes, then at 25 m/s for another 10 minutes. (a) Is your average speed for the entire trip greater than, less than, or equal to 20 m/s? (b) Choose the *best* explanation from among the following:
   - I. More time is required to drive at 15 m/s than at 25 m/s.
   - II. Less distance is covered at 25 m/s than at 15 m/s.
   - III. Equal time is spent at 15 m/s and 25 m/s.

9. • Usain Bolt of Jamaica set a world record in 2009 for the 200-meter dash with a time of 19.19 seconds. What was his average speed? Give your answer in meters per second and miles per hour.

10. • **BIO** Kangaroos have been clocked at speeds of 65 km/h. (a) How far can a kangaroo hop in 3.2 minutes at this speed? (b) How much time will it take a kangaroo to hop 0.25 km at this speed?

11. • **Rubber Ducks** A severe storm on January 10, 1992, caused a cargo ship near the Aleutian Islands to spill 29,000 rubber ducks

and other bath toys into the ocean. Ten months later hundreds of rubber ducks began to appear along the shoreline near Sitka, Alaska, roughly 1600 miles away. What was the approximate average speed of the ocean current that carried the ducks to shore in **(a)** m/s and **(b)** mi/h? (Rubber ducks from the same spill began to appear on the coast of Maine in July 2003.)

12. • Radio waves travel at the speed of light, approximately 186,000 miles per second. How much time does it take for a radio message to travel from Earth to the Moon and back? (See the inside back cover for the necessary data.)

13. • It was a dark and stormy night, when suddenly you saw a flash of lightning. Six and a half seconds later you heard the thunder. Given that the speed of sound in air is about 340 m/s, how far away was the lightning bolt? (Ignore the travel time for the flash of light.)

14. • **BIO Nerve Impulses** The human nervous system can propagate nerve impulses at about 10^2 m/s. Estimate the time it takes for a nerve impulse to travel 2 m from your toes to your brain.

15. •• A finch rides on the back of a Galapagos tortoise, which walks at the stately pace of 0.060 m/s. After 1.5 minutes the finch tires of the tortoise's slow pace, and takes flight in the same direction for another 1.5 minutes at 11 m/s. What was the average speed of the finch for this 3.0-minute interval?

16. •• You jog at 9.1 km/h for 5.0 km, then you jump into a car and drive an additional 13 km. With what average speed must you drive your car if your average speed for the entire 18 km is to be 25 km/h?

17. •• A dog runs back and forth between its two owners, who are walking toward one another (**FIGURE 2-31**). The dog starts running when the owners are 8.2 m apart. If the dog runs with a speed of 2.7 m/s, and the owners each walk with a speed of 1.3 m/s, how far has the dog traveled when the owners meet?

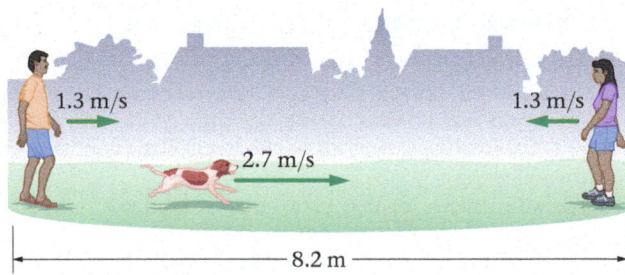

FIGURE 2-31 Problem 17

18. •• **BIO Predict/Calculate** Blood flows through a major artery at 1.0 m/s for 0.50 s, then at 0.60 m/s for another 0.50 s through a smaller artery. **(a)** Is the average speed of the blood 0.80 m/s, greater than 0.80 m/s, or less than 0.80 m/s? Explain. **(b)** Verify your answer to part (a) by calculating the average speed.

19. •• **BIO Predict/Calculate** Blood flows through a major artery at 1.0 m/s over a distance of 0.50 m, then at 0.60 m/s for another 0.50 m through a smaller artery. **(a)** Is the average speed of the blood 0.80 m/s, greater than 0.80 m/s, or less than 0.80 m/s? Explain. **(b)** Verify your answer to part (a) by calculating the average speed.

20. •• In heavy rush-hour traffic you drive in a straight line at 12 m/s for 1.5 minutes, then you have to stop for 3.5 minutes, and finally you drive at 15 m/s for another 2.5 minutes. **(a)** Plot a position-versus-time graph for this motion. Your plot should extend from $t = 0$ to $t = 7.5$ minutes. **(b)** Use your plot from part (a) to calculate the average velocity between $t = 0$ and $t = 7.5$ minutes.

21. •• **Predict/Calculate** An expectant father paces back and forth, producing the position-versus-time graph shown in **FIGURE 2-32**. Without performing a calculation, indicate whether the father's velocity is positive, negative, or zero on each of the following segments of the graph: **(a)** A, **(b)** B, **(c)** C, and **(d)** D. Calculate the numerical value of the father's velocity for the segments **(e)** A, **(f)** B, **(g)** C, and **(h)** D, and show that your results verify your answers to parts (a)–(d).

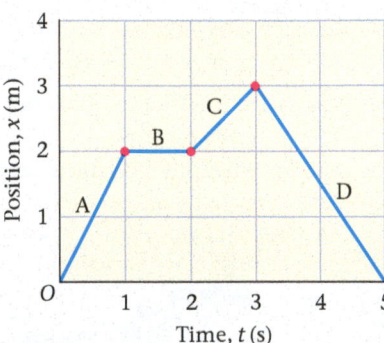

FIGURE 2-32 Problem 21

22. •• The position of a particle as a function of time is given by $x = (-5 \text{ m/s})t + (3 \text{ m/s}^2)t^2$. **(a)** Plot x versus t for $t = 0$ to $t = 2$ s. **(b)** Find the average velocity of the particle from $t = 0$ to $t = 1$ s. **(c)** Find the average speed from $t = 0$ to $t = 1$ s.

23. •• The position of a particle as a function of time is given by $x = (6 \text{ m/s})t + (-2 \text{ m/s}^2)t^2$. **(a)** Plot x versus t for $t = 0$ to $t = 2$ s. **(b)** Find the average velocity of the particle from $t = 0$ to $t = 1$ s. **(c)** Find the average speed from $t = 0$ to $t = 1$ s.

24. •• **Predict/Calculate** A tennis player moves back and forth along the baseline while waiting for her opponent to serve, producing the position-versus-time graph shown in **FIGURE 2-33**. **(a)** Without performing a calculation, indicate on which of the segments of the graph, A, B, or C, the player has the greatest speed. Calculate the player's speed for **(b)** segment A, **(c)** segment B, and **(d)** segment C, and show that your results verify your answers to part (a).

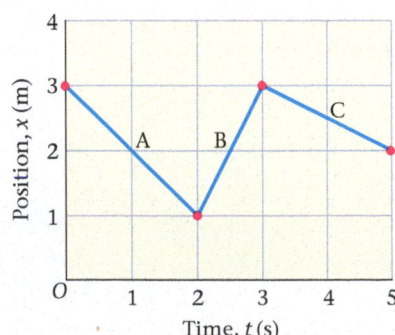

FIGURE 2-33 Problem 24

25. ••• On your wedding day you leave for the church 30.0 minutes before the ceremony is to begin, which should be plenty of time since the church is only 10.0 miles away. On the way, however, you have to make an unanticipated stop for construction work on the road. As a result, your average speed for the first 15 minutes is only 5.0 mi/h. What average speed do you need for the rest of the trip to get you to the church on time?

SECTION 2-3 INSTANTANEOUS VELOCITY

26. • **CE** The position-versus-time plot of a boat positioning itself next to a dock is shown in **FIGURE 2-34**. Rank the six points indicated in the plot in order of increasing value of the velocity v, starting with the most negative. Indicate a tie with an equal sign.

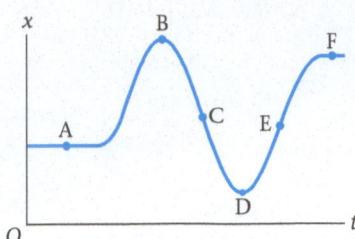

FIGURE 2-34 Problem 26

27. •• The position of a particle as a function of time is given by $x = (2.0 \text{ m/s})t + (-3.0 \text{ m/s}^3)t^3$. (a) Plot x versus t for time from $t = 0$ to $t = 1.0$ s. (b) Find the average velocity of the particle from $t = 0.35$ s to $t = 0.45$ s. (c) Find the average velocity from $t = 0.39$ s to $t = 0.41$ s. (d) Do you expect the instantaneous velocity at $t = 0.40$ s to be closer to 0.54 m/s, 0.56 m/s, or 0.58 m/s? Explain.

28. •• The position of a particle as a function of time is given by $x = (-2.00 \text{ m/s})t + (3.00 \text{ m/s}^3)t^3$. (a) Plot x versus t for time from $t = 0$ to $t = 1.00$ s. (b) Find the average velocity of the particle from $t = 0.150$ s to $t = 0.250$ s. (c) Find the average velocity from $t = 0.190$ s to $t = 0.210$ s. (d) Do you expect the instantaneous velocity at $t = 0.200$ s to be closer to -1.62 m/s, -1.64 m/s, or -1.66 m/s? Explain.

SECTION 2-4 ACCELERATION

29. • **CE Predict/Explain** On two occasions you accelerate uniformly from rest along an on-ramp in order to merge into interstate traffic at 65 mi/h. On-ramp A is 200 m long and on-ramp B is 400 m long. (a) Is your acceleration along on-ramp A greater than, less than, or equal to your acceleration along ramp B? (b) Choose the *best explanation* from among the following:

  I. The shorter acceleration distance on ramp A requires a greater acceleration.

  II. In each case, you start from rest and have the same final speed.

  III. The greater distance traveled on ramp B requires a greater acceleration.

30. • A 747 airliner reaches its takeoff speed of 156 mi/h in 35.2 s. What is the magnitude of its average acceleration?

31. • At the starting gun, a runner accelerates at 1.9 m/s² for 5.2 s. The runner's acceleration is zero for the rest of the race. What is the speed of the runner (a) at $t = 2.0$ s, and (b) at the end of the race?

32. • A jet makes a landing traveling due east with a speed of 70.6 m/s. If the jet comes to rest in 13.0 s, what are the magnitude and direction of its average acceleration?

33. • A car is traveling due north at 23.6 m/s. Find the velocity of the car after 7.10 s if its acceleration is (a) 1.30 m/s² due north, or (b) 1.15 m/s² due south.

34. •• A motorcycle moves according to the velocity-versus-time graph shown in **FIGURE 2-35**. Find the average acceleration of the motorcycle during each of the following segments of the motion: (a) A, (b) B, and (c) C.

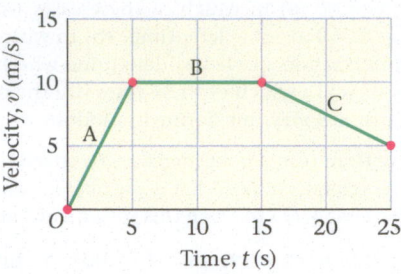

FIGURE 2-35 Problem 34

35. •• A person on horseback moves according to the velocity-versus-time graph shown in **FIGURE 2-36**. Find the displacement of the person for each of the following segments of the motion: (a) A, (b) B, and (c) C.

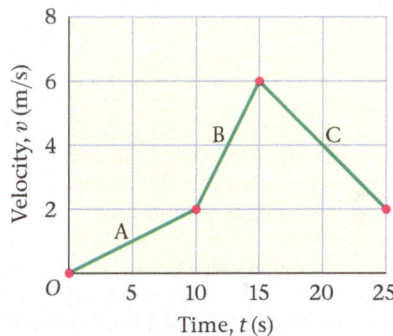

FIGURE 2-36 Problem 35

36. •• Running with an initial velocity of $+9.2$ m/s, a horse has an average acceleration of -1.81 m/s². How much time does it take for the horse to decrease its velocity to $+5.5$ m/s?

37. •• **Predict/Calculate** Assume that the brakes in your car create a constant deceleration of 4.2 m/s² regardless of how fast you are driving. If you double your driving speed from 18 m/s to 36 m/s, (a) does the time required to come to a stop increase by a factor of two or a factor of four? Explain. Verify your answer to part (a) by calculating the stopping times for initial speeds of (b) 18 m/s and (c) 36 m/s.

38. •• As a train accelerates away from a station, it reaches a speed of 4.7 m/s in 5.0 s. If the train's acceleration remains constant, what is its speed after an additional 5.0 s has elapsed?

39. •• A particle has an acceleration of $+6.24$ m/s² for 0.450 s. At the end of this time the particle's velocity is $+9.31$ m/s. What was the particle's initial velocity?

SECTION 2-5 MOTION WITH CONSTANT ACCELERATION

40. • Landing with a speed of 71.4 m/s, and traveling due south, a jet comes to rest in 949 m. Assuming the jet slows with constant acceleration, find the magnitude and direction of its acceleration.

41. • When you see a traffic light turn red, you apply the brakes until you come to a stop. If your initial speed was 18 m/s, and you were heading due west, what was your average velocity during braking? Assume constant deceleration.

42. **CE** •• A ball is released at the point $x = 2$ m on an inclined plane with a nonzero initial velocity. After being released, the ball moves with constant acceleration. The acceleration and initial velocity of the ball are described by one of the following four cases: case 1, $a > 0$, $v_0 > 0$; case 2, $a > 0$, $v_0 < 0$; case 3, $a < 0$, $v_0 > 0$; case

4, $a < 0$, $v_0 < 0$. **(a)** In which of these cases will the ball definitely pass $x = 0$ at some later time? **(b)** In which of these cases is more information needed to determine whether the ball will cross $x = 0$? **(c)** In which of these cases will the ball come to rest momentarily at some time during its motion?

43. •• Starting from rest, a boat increases its speed to 4.82 m/s with constant acceleration. **(a)** What is the boat's average speed? **(b)** If it takes the boat 4.77 s to reach this speed, how far has it traveled?

44. •• The position of a car as a function of time is given by $x = (50 \text{ m}) + (-5.0 \text{ m/s})t + (-10 \text{ m/s}^2)t^2$. **(a)** What are the initial position, initial velocity, and acceleration of the car? **(b)** Plot x versus t for $t = 0$ to $t = 2.0$ s. **(c)** What distance does the car travel during the first 1.0 s? **(d)** What is the average velocity of the car between $t = 1.0$ s and $t = 2.0$ s?

45. •• The position of a ball as a function of time is given by $x = (5.0 \text{ m/s})t + (-10 \text{ m/s}^2)t^2$. **(a)** What are the initial position, initial velocity, and acceleration of the ball? **(b)** Plot x versus t for $t = 0$ to $t = 2.0$ s. **(c)** Find the average velocity of the ball from $t = 0$ to $t = 1.0$ s. **(d)** Find the average speed of the ball between $t = 1.0$ s and $t = 2.0$ s.

46. •• **BIO** A cheetah can accelerate from rest to 25.0 m/s in 6.22 s. Assuming constant acceleration, **(a)** how far has the cheetah run in this time? **(b)** After sprinting for just 3.11 s, is the cheetah's speed 12.5 m/s, greater than 12.5 m/s, or less than 12.5 m/s? Explain. **(c)** What is the cheetah's average speed for the first 3.11 s of its sprint? For the second 3.11 s of its sprint? **(d)** Calculate the distance covered by the cheetah in the first 3.11 s and the second 3.11 s.

47. •• A sled slides from rest down an icy slope. Measurements taken from a video show that the distance from the starting point to the sled at various times is as follows:

Time (s)	Distance from starting point (m)
0	0
0.50	0.36
1.0	1.3
1.5	3.2
2.0	5.7
2.5	8.8

(a) Plot the position of the sled versus time. **(b)** From your plot, determine the approximate average speed of the sled from $t = 0.25$ s to $t = 1.3$ s. **(c)** Determine the approximate acceleration of the sled, assuming it to be constant.

SECTION 2-6 APPLICATIONS OF THE EQUATIONS OF MOTION

48. • A child slides down a hill on a toboggan with an acceleration of 1.6 m/s². If she starts at rest, how far has she traveled in **(a)** 1.0 s, **(b)** 2.0 s, and **(c)** 3.0 s?

49. • **The Detonator** On a ride called the Detonator at Worlds of Fun in Kansas City, passengers accelerate straight downward from rest to 45 mi/h in 2.2 seconds. What is the average acceleration of the passengers on this ride?

The Detonator (Problem 49)

50. • **Jules Verne** In his novel *From the Earth to the Moon* (1865), Jules Verne describes a spaceship that is blasted out of a cannon, called the *Columbiad*, with a speed of 12,000 yards/s. The *Columbiad* is 900 ft long, but part of it is packed with explosives, so the spaceship accelerates over a distance of only 700 ft. Estimate the acceleration experienced by the occupants of the spaceship during launch. Give your answer in m/s². (Verne realized that the "travelers would . . . encounter a violent recoil," but he probably didn't know that people generally lose consciousness if they experience accelerations greater than about $7g \sim 70 \text{ m/s}^2$.)

51. •• **BIO Bacterial Motion** Approximately 0.1% of the bacteria in an adult human's intestines are *Escherichia coli*. These bacteria have been observed to move with speeds up to 15 μm/s and maximum accelerations of 166 μm/s². Suppose an *E. coli* bacterium in your intestines starts at rest and accelerates at 156 μm/s². How much **(a)** time and **(b)** distance are required for the bacterium to reach a speed of 12 μm/s?

52. •• Two cars drive on a straight highway. At time $t = 0$, car 1 passes mile marker 0 traveling due east with a speed of 20.0 m/s. At the same time, car 2 is 1.0 km east of mile marker 0 traveling at 30.0 m/s due west. Car 1 is speeding up with an acceleration of magnitude 2.5 m/s², and car 2 is slowing down with an acceleration of magnitude 3.2 m/s². **(a)** Write x-versus-t equations of motion for both cars, taking east as the positive direction. **(b)** At what time do the cars pass next to one another?

53. •• **A Meteorite Strikes** On October 9, 1992, a 27-pound meteorite struck a car in Peekskill, NY, leaving a dent 22 cm deep in the trunk. If the meteorite struck the car with a speed of 130 m/s, what was the magnitude of its deceleration, assuming it to be constant?

54. •• A rocket blasts off and moves straight upward from the launch pad with constant acceleration. After 2.8 s the rocket is at a height of 91 m. **(a)** What are the magnitude and direction of the rocket's acceleration? **(b)** What is its speed at this time?

55. •• **Predict/Calculate** You are driving through town at 12.0 m/s when suddenly a ball rolls out in front of you. You apply the brakes and begin decelerating at 3.5 m/s². **(a)** How far do you travel before stopping? **(b)** When you have traveled only half the distance in part (a), is your speed 6.0 m/s, greater than 6.0 m/s, or less than 6.0 m/s? **(c)** Find the speed of your car when you have traveled half the distance in part (a).

56. •• **Predict/Calculate** You are driving through town at 16 m/s when suddenly a car backs out of a driveway in front of you. You apply the brakes and begin decelerating at 3.2 m/s². **(a)** How much time does it take to stop? **(b)** After braking half the time found in part (a), is your speed 8.0 m/s, greater than 8.0 m/s, or less than 8.0 m/s? **(c)** Find your speed after braking half the time found in part (a).

57. •• **BIO** **Predict/Calculate** **A Tongue's Acceleration** When a chameleon captures an insect, its tongue can extend 16 cm in 0.10 s. **(a)** Find the magnitude of the tongue's acceleration, assuming it to be constant. **(b)** In the first 0.050 s, does the tongue extend 8.0 cm, more than 8.0 cm, or less than 8.0 cm? **(c)** Find the extension of the tongue in the first 0.050 s.

It's not polite to reach! (Problem 57)

58. •• **BIO** **Surviving a Large Deceleration** On July 13, 1977, while on a test drive at Britain's Silverstone racetrack, the throttle on David Purley's car stuck wide open. The resulting crash subjected Purley to the greatest "g-force" ever survived by a human—he decelerated from 173 km/h to zero in a distance of only about 0.66 m. Calculate the magnitude of the acceleration experienced by Purley (assuming it to be constant), and express your answer in units of the acceleration of gravity, $g = 9.81$ m/s².

59. •• A boat is cruising in a straight line at a constant speed of 2.6 m/s when it is shifted into neutral. After coasting 12 m the engine is engaged again, and the boat resumes cruising at the reduced constant speed of 1.6 m/s. Assuming constant acceleration while coasting, **(a)** how much time did it take for the boat to coast the 12 m? **(b)** What was the boat's acceleration while it was coasting? **(c)** What was the speed of the boat when it had coasted for 6.0 m? Explain.

60. •• A model rocket rises with constant acceleration to a height of 4.2 m, at which point its speed is 26.0 m/s. **(a)** How much time does it take for the rocket to reach this height? **(b)** What was the magnitude of the rocket's acceleration? **(c)** Find the height and speed of the rocket 0.10 s after launch.

61. •• The infamous chicken is dashing toward home plate with a speed of 5.7 m/s when he decides to hit the dirt. The chicken slides for 1.2 s, just reaching the plate as he stops (safe, of course). **(a)** What are the magnitude and direction of the chicken's acceleration? **(b)** How far did the chicken slide?

62. •• A bicyclist is finishing his repair of a flat tire when a friend rides by with a constant speed of 3.5 m/s. Two seconds later the bicyclist hops on his bike and accelerates at 2.4 m/s² until he catches his friend. **(a)** How much time does it take until he catches his friend? **(b)** How far has he traveled in this time? **(c)** What is his speed when he catches up?

63. •• A car in stop-and-go traffic starts at rest, moves forward 22 m in 8.0 s, then comes to rest again. The velocity-versus-time plot for this car is given in **FIGURE 2-37**. What distance does the car cover in **(a)** the first 4.0 seconds of its motion and **(b)** the last 2.0 seconds of its motion? **(c)** What is the constant speed V that characterizes the middle portion of its motion?

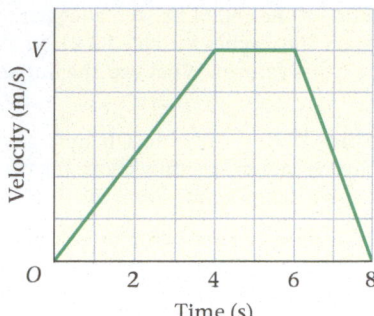

FIGURE 2-37 Problem 63

64. ••• A car and a truck are heading directly toward one another on a straight and narrow street, but they avoid a head-on collision by simultaneously applying their brakes at $t = 0$. The resulting velocity-versus-time graphs are shown in **FIGURE 2-38**. What is the separation between the car and the truck when they have come to rest, given that at $t = 0$ the car is at $x = 15$ m and the truck is at $x = -35$ m? (Note that this information determines which line in the graph corresponds to which vehicle.)

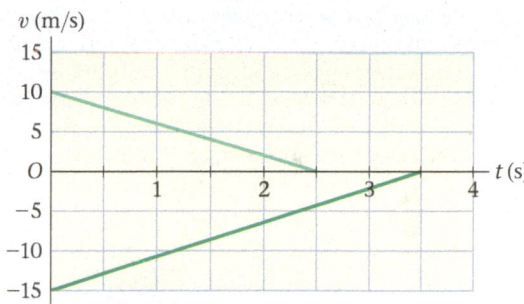

FIGURE 2-38 Problem 64

65. ••• Suppose you use videos to analyze the motion of penguins as they slide from rest down three different frictionless ramps (A, B, C) and into the water. The distances they travel and the times elapsed are recorded in the following table:

Ramp	A	B	C
distance, m	4.09	1.96	1.08
time, s	2.19	1.08	0.663

(a) Find the magnitude of the penguin's acceleration for each ramp. Assume the acceleration is uniform. **(b)** Determine the inclination angle θ above horizontal for each ramp, assuming that the accelerations for part (a) are in close agreement with the formula, $a = g \sin \theta$. (We will derive this formula in Chapter 5.)

SECTION 2-7 FREELY FALLING OBJECTS

66. • **CE** At the edge of a roof you throw ball 1 upward with an initial speed v_0; a moment later you throw ball 2 downward with the same initial speed. The balls land at the same time. Which of the following statements is true for the instant just before the balls hit the ground? A. The speed of ball 1 is greater than the speed of ball

2; **B.** The speed of ball 1 is equal to the speed of ball 2; **C.** The speed of ball 1 is less than the speed of ball 2.

67. • A cliff diver drops from rest to the water below. How many seconds does it take for the diver to go from 0 to 60; that is, to go from 0 mph to 60 mph? (For comparison, it takes about 3.5 s to 4.0 s for a powerful car to go from 0 to 60.)

68. • For a flourish at the end of her act, a juggler tosses a single ball high in the air. She catches the ball 3.2 s later at the same height from which it was thrown. What was the initial upward speed of the ball?

69. • **Soaring Shaun** During the 2014 Olympic games, snowboarder Shaun White rose 6.4 m vertically above the rim of the half-pipe. With what speed did he launch from the rim?

70. • **BIO** Gulls are often observed dropping clams and other shellfish from a height to the rocks below, as a means of opening the shells. If a seagull drops a shell from rest at a height of 17 m, how fast is the shell moving when it hits the rocks?

71. • A volcano launches a lava bomb straight upward with an initial speed of 28 m/s. Taking upward to be the positive direction, find the speed and direction of motion of the lava bomb **(a)** 2.0 seconds and **(b)** 3.0 seconds after it is launched.

72. • **An Extraterrestrial Volcano** The first active volcano observed outside the Earth was discovered in 1979 on Io, one of the moons of Jupiter. The volcano was observed to be ejecting material to a height of about 3.00×10^5 m. Given that the acceleration of gravity on Io is 1.80 m/s², find the initial velocity of the ejected material.

73. • **BIO** **Measure Your Reaction Time** Here's something you *can* try at home—an experiment to measure your reaction time. Have a friend hold a ruler by one end, letting the other end hang down vertically. At the lower end, hold your thumb and index finger on either side of the ruler, ready to grip it. Have your friend release the ruler without warning. Catch it as quickly as you can. If you catch the ruler 5.2 cm from the lower end, what is your reaction time?

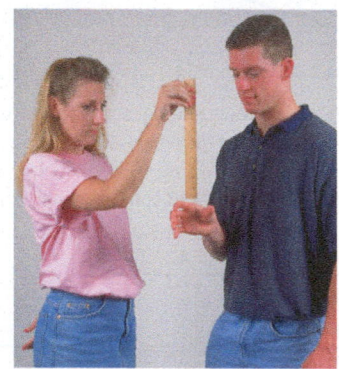

How fast are your reactions? (Problem 73)

74. •• **CE** **Predict/Explain** A carpenter on the roof of a building accidentally drops her hammer. As the hammer falls it passes two windows of equal height, as shown in **FIGURE 2-39**. **(a)** Is the *increase* in speed of the hammer as it drops past window 1 greater than, less than, or equal to the *increase* in speed as it drops past window 2? **(b)** Choose the *best explanation* from among the following:
    I. The greater speed at window 2 results in a greater increase in speed.
    II. Constant acceleration means the hammer speeds up the same amount for each window.
    III. The hammer spends more time dropping past window 1.

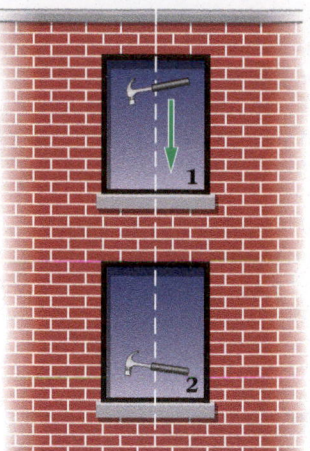

FIGURE 2-39 Problem 74

75. •• **CE** **Predict/Explain** **FIGURE 2-40** shows a *v*-versus-*t* plot for the hammer dropped by the carpenter in Problem 74. Notice that the times when the hammer passes the two windows are indicated by shaded areas. **(a)** Is the area of the shaded region corresponding to window 1 greater than, less than, or equal to the area of the shaded region corresponding to window 2? **(b)** Choose the *best explanation* from among the following:

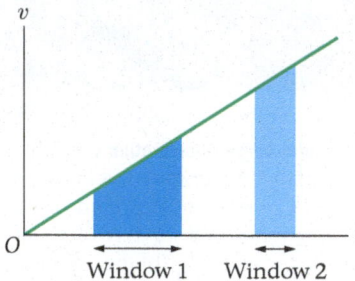

FIGURE 2-40 Problem 75

    I. The shaded area for window 2 is higher than the shaded area for window 1.
    II. The windows are equally tall.
    III. The shaded area for window 1 is wider than the shaded area for window 2.

76. •• **CE** A ball is thrown straight upward with an initial speed v_0. When it reaches the top of its flight at height h, a second ball is thrown straight upward with the same initial speed. Do the balls cross paths at height $\frac{1}{2}h$, above $\frac{1}{2}h$, or below $\frac{1}{2}h$?

77. •• On a hot summer day in the state of Washington while kayaking, I saw several swimmers jump from a railroad bridge into the Snohomish River below. The swimmers stepped off the bridge, and I estimated that they hit the water 1.5 s later. **(a)** How high was the bridge? **(b)** How fast were the swimmers moving when they hit the water? **(c)** What would the swimmers' drop time be if the bridge were twice as high?

78. •• **Highest Water Fountain** The USA's highest fountain of water is located, appropriately enough, in Fountain Hills, Arizona. The fountain rises to a height of 560 ft (5 feet higher than the Washington Monument). **(a)** What is the initial speed of the water? **(b)** How much time does it take for water to reach the top of the fountain?

79. •• Wrongly called for a foul, an angry basketball player throws the ball straight down to the floor. If the ball bounces straight up and returns to the floor 3.2 s after first striking it, what was the ball's greatest height above the floor?

80. •• To celebrate a victory, a pitcher throws her glove straight upward with an initial speed of 6.5 m/s. **(a)** How much time does it take for the glove to return to the pitcher? **(b)** How much time does it take for the glove to reach its maximum height?

81. •• **Predict/Calculate** Standing at the edge of a cliff 30.5 m high, you drop a ball. Later, you throw a second ball downward with an initial speed of 11.2 m/s. **(a)** Which ball has the greater *increase* in speed when it reaches the base of the cliff, or do both balls speed up by the same amount? **(b)** Verify your answer to part (a) with a calculation.

82. •• You shoot an arrow into the air. Two seconds later (2.00 s) the arrow has gone straight upward to a height of 30.0 m above its launch point. **(a)** What was the arrow's initial speed? **(b)** How much time did it take for the arrow to first reach a height of 15.0 m above its launch point?

83. •• While riding on an elevator descending with a constant speed of 3.0 m/s, you accidentally drop a book from under your arm. **(a)** How much time does it take for the book to reach the elevator floor, 1.2 m below your arm? **(b)** What is the book's speed relative to you when it hits the elevator floor?

84. •• A hot-air balloon is descending at a rate of 2.3 m/s when a passenger drops a camera. If the camera is 41 m above the ground when it is dropped, **(a)** how much time does it take for the camera to reach the ground, and **(b)** what is its velocity just before it lands? Let upward be the positive direction for this problem.

85. •• A model rocket blasts off and moves upward with an acceleration of 12 m/s² until it reaches a height of 29 m, at which point its engine shuts off and it continues its flight in free fall. **(a)** What is the maximum height attained by the rocket? **(b)** What is the speed of the rocket just before it hits the ground? **(c)** What is the total duration of the rocket's flight?

86. •• **BIO** The southern flying squirrel (*Glaucomys volans*) can reduce its free-fall acceleration by extending a furry membrane called a *patagium* and gliding through the air. **FIGURE 2-41** shows the vertical position-versus-time plot of one squirrel's flight. Use the information given in the plot to determine the acceleration of the squirrel. Assume that its initial height is $x_0 = 4.0$ m, its initial vertical speed is $v_0 = 0$, and the elapsed time is $t = 1.10$ s.

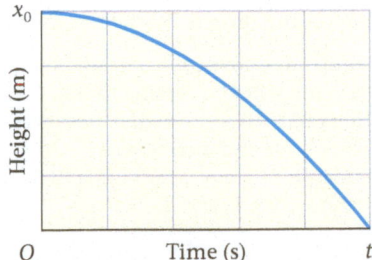

FIGURE 2-41 Problem 86

87. ••• **Hitting the "High Striker"** A young woman at a carnival steps up to the "high striker," a popular test of strength where the contestant hits one end of a lever with a mallet, propelling a small metal plug upward toward a bell. She gives the mallet a mighty swing and sends the plug to the top of the striker, where it rings the bell. **FIGURE 2-42** shows the corresponding position-versus-time plot for the plug. Using the information given in the plot, answer the following questions: **(a)** What is the average speed of the plug during its upward journey? **(b)** By how much does the speed of the plug decrease during its upward journey? **(c)** What is the initial speed of the plug? (Assume the plug is in free fall during its upward motion, with no effects of air resistance or friction.)

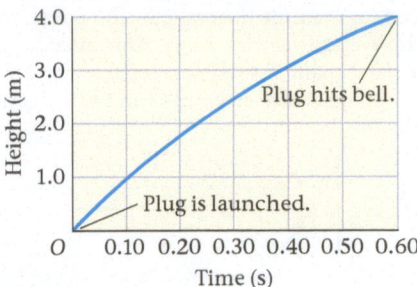

FIGURE 2-42 Problem 87

88. ••• While sitting on a tree branch 10.0 m above the ground, you drop a chestnut. When the chestnut has fallen 2.5 m, you throw a second chestnut straight down. What initial speed must you give the second chestnut if they are both to reach the ground at the same time?

GENERAL PROBLEMS

89. • An astronaut on the Moon drops a rock straight downward from a height of 1.25 m. If the acceleration of gravity on the Moon is 1.62 m/s², what is the speed of the rock just before it lands?

90. • **Taipei 101** An elevator in the Taipei 101 skyscraper can start from rest and accelerate to a maximum speed of 16 m/s with a constant acceleration of 1.1 m/s². Through what distance does the elevator move as it accelerates to its maximum speed?

91. • **A Supersonic Waterfall** Geologists have learned of periods in the past when the Strait of Gibraltar closed off, and the Mediterranean Sea dried out and become a desert. Later, when the strait reopened, a massive saltwater waterfall was created. According to geologists, the water in this waterfall was supersonic; that is, it fell with speeds in excess of the speed of sound. Ignoring air resistance, what is the minimum height necessary to create a supersonic waterfall? (The speed of sound may be taken to be 340 m/s.)

92. • A juggler throws a ball straight up into the air. If the ball returns to its original position in the juggler's hand in 0.98 s, what was its maximum height above the juggler's hand?

93. •• **CE** At the edge of a roof you drop ball A from rest, and at the same instant your friend throws ball B upward from the ground with an initial velocity of v_0 sufficient to reach your position on the roof. What are the signs (positive or negative) of the velocity and the acceleration of each ball at a point halfway between the ground and the roof? Assume upward is the positive direction.

94. •• **CE** Two balls start their motion at the same time, with ball A dropped from rest and ball B thrown upward with an initial speed v_0. Which of the five plots shown in **FIGURE 2-43** corresponds to **(a)** ball A and **(b)** ball B?

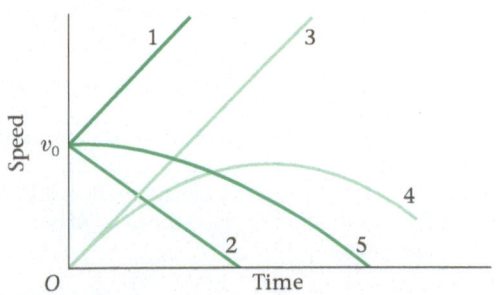

FIGURE 2-43 Problem 94

95. •• **CE** Refer to the position-versus-time plot in **FIGURE 2-44** for the following questions: **(a)** Is the average speed for the time interval between points 1 and 2 greater than, less than, or equal to the average speed for the time interval between points 1 and 3? Explain. **(b)** Is the average velocity for the time interval between points 2 and 4 greater than, less than, or equal to the average velocity for the time interval between points 3 and 4? Explain.

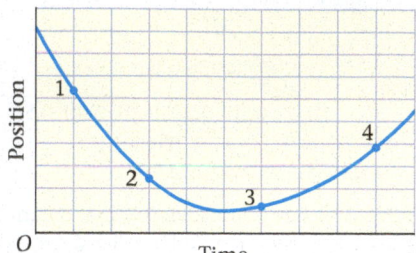

FIGURE 2-44 Problem 95

96. •• **Drop Tower** NASA operates a 2.2-second drop tower at the Glenn Research Center in Cleveland, Ohio. At this facility, experimental packages are dropped from the top of the tower, on the 8th floor of the building. During their 2.2 seconds of free fall, the packages experience a microgravity environment similar to that of a spacecraft in orbit. **(a)** What is the drop distance of a 2.2-s tower? **(b)** How fast are the packages traveling when they hit the air bags at the bottom of the tower? **(c)** If an experimental package comes to rest over a distance of 0.75 m upon hitting the air bags, what is the average stopping acceleration?

97. •• The velocity-versus-time graph for an object moving in a straight line is given in **FIGURE 2-45**. **(a)** What is the acceleration of this object? **(b)** If this object is at $x = 12.0$ m at time $t = 0$, what is its position at $t = 1.00$ s? **(c)** Assuming the acceleration of the object remains constant, what is its position at $t = 5.00$ s?

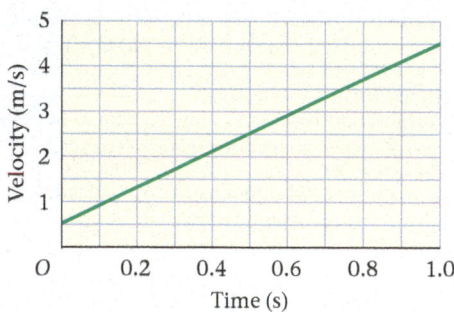

FIGURE 2-45 Problem 97

98. •• At the 18th green of the U.S. Open you need to make a 23.5-ft putt to win the tournament. When you hit the ball, giving it an initial speed of 1.54 m/s, it stops 6.00 ft short of the hole. **(a)** Assuming the deceleration caused by the grass is constant, what should the initial speed have been to just make the putt? **(b)** What initial speed do you need to make the remaining 6.00-ft putt?

99. •• A glaucous-winged gull, ascending straight upward at 5.20 m/s, drops a shell when it is 12.5 m above the ground. **(a)** What are the magnitude and direction of the shell's acceleration just after it is released? **(b)** Find the maximum height above the ground reached by the shell. **(c)** How much time does it take for the shell to reach the ground? **(d)** What is the speed of the shell just before it hits the ground?

100. •• A doctor, preparing to give a patient an injection, squirts a small amount of liquid straight upward from a syringe. If the liquid emerges with a speed of 1.5 m/s, **(a)** how much time does it take for it to return to the level of the syringe? **(b)** What is the maximum height of the liquid above the syringe?

101. •• A hot-air balloon has just lifted off and is rising at the constant rate of 2.0 m/s. Suddenly one of the passengers realizes she has left her camera on the ground. A friend picks it up and tosses it straight upward with an initial speed of 13 m/s. If the passenger is 2.5 m above her friend when the camera is tossed, how high is she when the camera reaches her?

102. •• Astronauts on a distant planet throw a rock straight upward and record its motion with a video camera. After digitizing their video, they are able to produce the graph of height, y, versus time, t, shown in **FIGURE 2-46**. **(a)** What is the acceleration of gravity on this planet? **(b)** What was the initial speed of the rock?

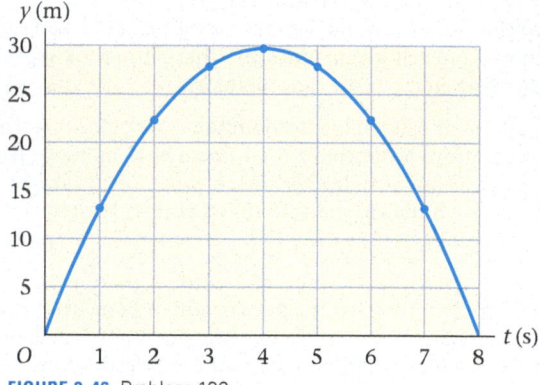

FIGURE 2-46 Problem 102

103. •• **BIO A Jet-Propelled Squid** Squids can move through the water using a form of jet propulsion. Suppose a squid jets forward from rest with constant acceleration for 0.170 s, moving through a distance of 0.179 m. The squid then turns off its jets and coasts to rest with constant acceleration. The total time for this motion (from rest to rest) is 0.400 s, and the total distance covered is 0.421 m. What is the magnitude of the squid's acceleration during **(a)** the time it is jetting and **(b)** the time it coasts to a stop?

104. ••• A ball, dropped from rest, covers three-quarters of the distance to the ground in the last second of its fall. **(a)** From what height was the ball dropped? **(b)** What was the total time of fall?

105. ••• You drop a ski glove from a height h onto fresh snow, and it sinks to a depth d before coming to rest. **(a)** In terms of g and h, what is the speed of the glove when it reaches the snow? **(b)** What are the magnitude and direction of the glove's acceleration as it moves through the snow, assuming it to be constant? Give your answer in terms of g, h, and d.

106. ••• To find the height of an overhead power line, you throw a ball straight upward. The ball passes the line on the way up after 0.75 s, and passes it again on the way down 1.5 s after it was tossed. What are the height of the power line and the initial speed of the ball?

107. ••• Sitting in a second-story apartment, a physicist notices a ball moving straight upward just outside her window. The ball is visible for 0.25 s as it moves a distance of 1.05 m from the bottom to the top of the window. **(a)** How much time does it take before the ball reappears? **(b)** What is the greatest height of the ball above the top of the window?

PASSAGE PROBLEMS

Bam!—*Apollo 15* Lands on the Moon

The first word spoken on the surface of the Moon after *Apollo 15* landed on July 30, 1971, was "Bam!" This was James Irwin's involuntary reaction to their rather bone-jarring touchdown. "We did hit harder than any of the other flights!" says Irwin. "And I was startled, obviously, when I said, 'Bam!'"

The reason for the "firm touchdown" of *Apollo 15*, as pilot David Scott later characterized it, was that the rocket engine was shut off a bit earlier than planned, when the lander was still 4.30 ft above the lunar surface and moving downward with a speed of 0.500 ft/s. From that point on the lander descended in lunar free fall, with an acceleration of 1.62 m/s². As a result, the landing speed of *Apollo 15* was by far the largest of any of the *Apollo* missions. In comparison, Neil Armstrong's landing speed on *Apollo 11* was the lowest at 1.7 ft/s—he didn't shut off the engine until the footpads were actually on the surface. *Apollos 12, 14,* and *17* all landed with speeds between 3.0 and 3.5 ft/s.

To better understand the descent of *Apollo 15,* we show its trajectory during the final stages of landing in **FIGURE 2-47 (a).** In **FIGURE 2-47 (b)** we show a variety of speed-versus-time plots.

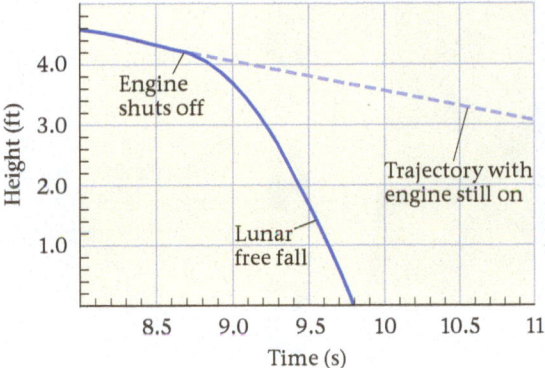

(a)

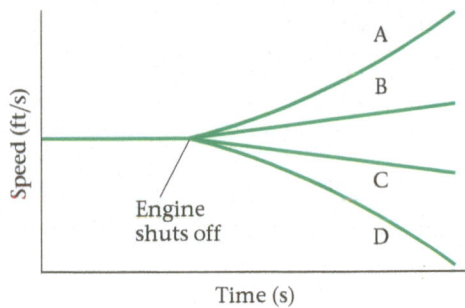

(b)
FIGURE 2-47 Problems 108, 109, 110, and 111

108. • How much time did it take for the lander to drop the final 4.30 ft to the Moon's surface?

A. 1.18 s	**B.** 1.37 s
C. 1.78 s	**D.** 2.36 s

109. •• What was the impact speed of the lander when it touched down? Give your answer in feet per second (ft/s), the same units used by the astronauts.

A. 2.41 ft/s	**B.** 6.78 ft/s
C. 9.95 ft/s	**D.** 10.6 ft/s

110. • Which of the speed-versus-time plots in Figure 2-48 (b) correctly represents the speed of the *Apollo 15* lander?

A   B   C   D

111. • Suppose, instead of shutting off the engine, the astronauts had increased its thrust, giving the lander a small, but constant, upward acceleration. Which speed-versus-time plot in Figure 2-48 (b) would describe this situation?

A   B   C   D

112. •• **Referring to Example 2-17** Suppose the speeder (red car) is traveling with a constant speed of 25 m/s, and the maximum acceleration of the police car (blue car) is 3.8 m/s². If the police car is to start from rest and catch the speeder in 15 s or less, what is the maximum head-start distance the speeder can have? Measure time from the moment the police car starts.

113. •• **Referring to Example 2-17** The speeder passes the position of the police car with a constant speed of 15 m/s. The police car immediately starts from rest and pursues the speeder with constant acceleration. What acceleration must the police car have if it is to catch the speeder in 7.0 s? Measure time from the moment the police car starts.

114. •• **Predict/Calculate** **Referring to Example 2-21** **(a)** In Example 2-21, the bag of sand is released at 20.0 m and reaches a maximum height of 22 m. If the bag had been released at 30.0 m instead, with everything else remaining the same, would its maximum height be 32 m, greater than 32 m, or less than 32 m? **(b)** Find the speed of the bag just before it lands when it is released from 30.0 m.

115. •• **Referring to Example 2-21** Suppose the balloon is descending with a constant speed of 4.2 m/s when the bag of sand comes loose at a height of 35 m. **(a)** What is the amount of time the bag is in the air? **(b)** What is the speed of the bag when it is 15 m above the ground?

3

Vectors in Physics

Big Ideas

1 A vector has both a magnitude and a direction.

2 The component of a vector is a measure of how much it extends in a certain direction.

3 The most precise way to add or subtract vectors is to add or subtract their components.

4 Unit vectors are dimensionless, have a magnitude of 1, and point in a specific direction.

5 Position, displacement, velocity, and acceleration are all vectors.

6 Relative-motion problems are an application of vector addition and subtraction.

▲ The points of the compass have long been used as a framework for indicating directions. In physics, we more frequently indicate directions in terms of *x* and *y* rather than N, S, E, and W. Either way, specifying a direction as well as a magnitude is essential to defining one of the physicist's most useful tools, the vector.

O f all the mathematical tools used in this book, perhaps none is more important than the vector. In this chapter we discuss what a vector is, how it differs from a scalar, and how it can describe a physical quantity.

3-1 Scalars Versus Vectors

Numbers can represent many quantities in physics. For example, a numerical value, together with the appropriate units, can specify the volume of a container, the temperature of the air, or the time of an event. In physics, a number with its units is referred to as a **scalar**:

> A scalar is a number with units. It can be positive, negative, or zero.

Sometimes, however, a scalar isn't enough to adequately describe a physical quantity—in many cases, a direction is needed as well. For example, suppose you're walking in an unfamiliar city and you want directions to the library. You ask a passerby, "Do you know where the library is?" If the person replies "Yes," and walks on, he hasn't been too helpful. If he says, "Yes, it is half a mile from here," that is more helpful, but you still don't know where it is. The library could be anywhere on a circle of radius one-half mile, as shown in **FIGURE 3-1**. To pin down the location, you need a reply such as, "Yes, the library is half a mile northwest of here." With both a distance *and* a direction, you know the location of the library.

Thus, if you walk northwest for half a mile you arrive at the library, as indicated by the upper left arrow in Figure 3-1. The arrow points in the direction traveled, and its **magnitude,** 0.5 mi in this case, represents the distance covered. In general, a quantity that is specified by both a *magnitude* and a *direction* is represented by a **vector**:

> A vector is a mathematical quantity with both a magnitude and a direction.

In the example of walking to the library, the vector corresponding to the trip is the displacement vector. Additional examples of displacement vectors are given on the directional signpost shown in **FIGURE 3-2**.

Symbols Used to Represent Vectors   When we indicate a vector on a diagram or a sketch, we draw an arrow, as in Figure 3-1. To indicate a vector with a written symbol, we use **boldface** for the vector itself, with a small arrow above it to remind us of its vector nature, and we use *italics* to indicate its magnitude. For example, the upper-left vector in Figure 3-1 is designated by the symbol $\vec{\mathbf{r}}$, and its magnitude is $r = 0.5$ mi. (When we represent a vector in a graph, we sometimes label it with the corresponding boldface symbol, and sometimes with the appropriate magnitude.) It is common in handwritten material to draw a small arrow over the vector's symbol, which is very similar to the way vectors are represented in this text.

Enhance Your Understanding   (Answers given at the end of the chapter)

1. Is it possible for two vectors to be different from one another, but still have the same magnitude? State why if your answer is no; give an example if your answer is yes.

Section Review

- A scalar is a number with a magnitude and units, but no direction.
- A vector has both a magnitude and a direction.

3-2 The Components of a Vector

When we discussed the directions for finding a library in the previous section, we pointed out that knowing the magnitude and direction angle—0.5 mi northwest—gives its precise location. We left out one key ingredient in actually *getting* to the library, however. In most cities it would not be possible to simply walk in a straight line for 0.5 mi directly to the library, because to do so would take you through buildings where there are no doors, through people's backyards, and through all kinds of other obstructions. In fact, if the city streets are laid out along north-south and east-west directions, you might instead walk west for a certain distance, then turn and proceed north an equal distance until you reach the library, as illustrated in **FIGURE 3-3**. What you have

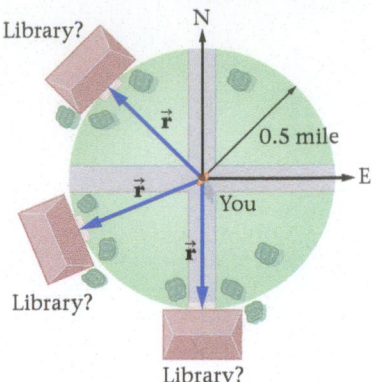

▲ **FIGURE 3-1 Distance and direction** If you know only that the library is 0.5 mi from you, it could lie anywhere on a circle of radius 0.5 mi. If, instead, you are told that the library is 0.5 mi northwest, you know its precise location.

Big Idea 1 A vector has both a magnitude and a direction; a scalar is simply a number with appropriate units.

▲ **FIGURE 3-2** This directional signpost gives both a distance and a direction to each city. In effect, the sign defines a displacement vector for each of these destinations.

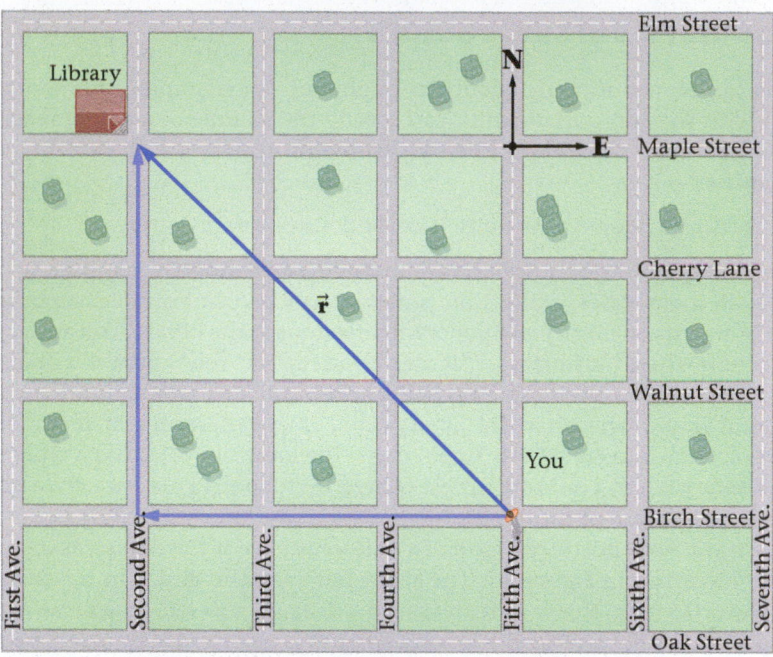

▲ **FIGURE 3-3 A walk along city streets to the library** By taking the indicated path, we have "resolved" the vector $\vec{\mathbf{r}}$ into east-west and north-south components.

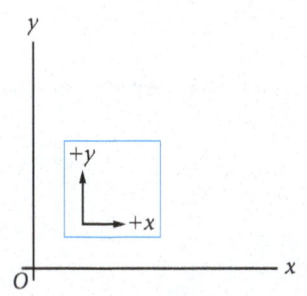

▲ **FIGURE 3-4 A two-dimensional coordinate system** The positive x and y directions are indicated in this shorthand form.

Big Idea 2 The component of a vector is a measure of how much it extends in a certain direction.

just done is "resolved" the displacement vector $\vec{\mathbf{r}}$, between you and the library into east-west and north-south "components."

Defining the Components of a Vector In general, to find the components of a vector we need to set up a coordinate system. In two dimensions we choose an origin, O, and a positive direction for both the x and the y axes, as in **FIGURE 3-4**. If the system were three-dimensional, we would also indicate a z axis.

Now, a vector is defined by its magnitude (indicated by the length of the arrow representing the vector) and its direction. For example, suppose an ant leaves its nest at the origin and, after foraging for some time, is at the location given by the vector $\vec{\mathbf{r}}$, in **FIGURE 3-5 (a)**. This vector has a magnitude $r = 1.10$ m and points in a direction $\theta = 25.0°$ above the x axis. Equivalently, $\vec{\mathbf{r}}$, can be defined by saying that it extends a distance r_x in the x direction and a distance r_y in the y direction, as shown in **FIGURE 3-5 (b)**. The quantities r_x and r_y are referred to as the x and y **scalar components** of the vector $\vec{\mathbf{r}}$.

Converting from Magnitude and Angle to Components We can find r_x and r_y by using standard trigonometric relationships, as summarized in the Problem-Solving Note on this page. Referring to Figure 3-5 (b), we see that r_x and r_y have the following values:

$$r_x = r \cos 25.0° = (1.10 \text{ m})(0.906) = 0.997 \text{ m}$$

$$r_y = r \sin 25.0° = (1.10 \text{ m})(0.423) = 0.465 \text{ m}$$

► **FIGURE 3-5 A vector and its scalar components (a)** The vector $\vec{\mathbf{r}}$ is defined by its length ($r = 1.10$ m) and its direction angle ($\theta = 25.0°$) measured counterclockwise from the positive x axis. **(b)** Alternatively, the vector $\vec{\mathbf{r}}$ can be defined by its x component, $r_x = 0.997$ m, and its y component, $r_y = 0.465$ m.

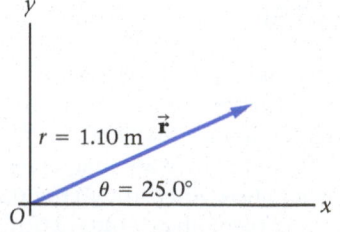

(a) A vector defined in terms of its length and direction angle

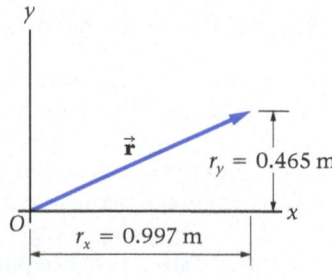

(b) The same vector defined in terms of its x and y components

Thus, we can say that the ant's final displacement is equivalent to what it would be if the ant had simply walked 0.997 m in the x direction and then 0.465 m in the y direction.

Converting from Components to Magnitude and Direction   Sometimes you are given the components of a vector, and from them you would like to find the corresponding magnitude and direction. As an example, let's start with the components $r_x = 0.997$ m and $r_y = 0.465$ m, just determined above, and verify that $r = 1.10$ m and $\theta = 25.0°$. First, notice that r_x, r_y, and r form a right triangle with r as the hypotenuse. Thus, we can use the Pythagorean theorem (see Appendix A) to find r in terms of r_x and r_y. This gives the expected result:

$$r = \sqrt{r_x^2 + r_y^2} = \sqrt{(0.997\,\text{m})^2 + (0.465\,\text{m})^2} = \sqrt{1.21\,\text{m}^2} = 1.10\,\text{m}$$

Second, we can use any two sides of the triangle to obtain the angle θ, as shown in the next three calculations:

$$\theta = \sin^{-1}\left(\frac{0.465\,\text{m}}{1.10\,\text{m}}\right) = \sin^{-1}(0.423) = 25.0°$$

$$\theta = \cos^{-1}\left(\frac{0.997\,\text{m}}{1.10\,\text{m}}\right) = \cos^{-1}(0.906) = 25.0°$$

$$\theta = \tan^{-1}\left(\frac{0.465\,\text{m}}{0.997\,\text{m}}\right) = \tan^{-1}(0.466) = 25.0°$$

The final expression, the one involving the inverse tangent ($\tan^{-1}$), is the one that is most commonly used.

In some situations we know a vector's magnitude and direction; in other cases we are given the vector's components. You will find it useful to be able to convert quickly and easily from one description of a vector to the other using trigonometric functions and the Pythagorean theorem.

PROBLEM-SOLVING NOTE

A Vector and Its Components

Given the magnitude and direction of a vector, find its components:

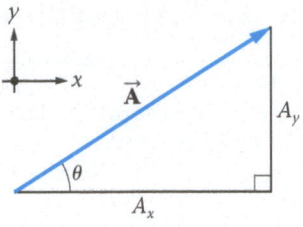

$$A_x = A\cos\theta$$
$$A_y = A\sin\theta$$

Given the components of a vector, find its magnitude and direction:

$$A = \sqrt{A_x^2 + A_y^2}$$
$$\theta = \tan^{-1}\frac{A_y}{A_x}$$

PHYSICS IN CONTEXT
Looking Back

In Chapter 2 we indicated the direction of a vector with a plus sign ($+$) or a minus sign ($-$), because only two directions were possible. With the results from this chapter we can now deal with vectors that point in any direction.

EXAMPLE 3-1   ZIP-A-DEE ZIPLINE

RWP* A neighborhood zipline is set up from a large tree to the ground. The zipline makes an angle of $\theta = 18°$ with the horizontal, and is anchored to the ground 12 m from the base of the tree. **(a)** What is the height of the top of the zipline above the ground? **(b)** What length of wire is needed to make the zipline?

PICTURE THE PROBLEM
Our sketch shows an adventurer heading toward the ground on the zipline. The angle of the zipline relative to the horizontal is $\theta = 18°$, and the horizontal distance from the base of the tree to the anchor point is $r_x = 12$ m. The opposite side of the zipline triangle is the height to the top of the zipline, r_y, and the hypotenuse is the length of the zipline, r.

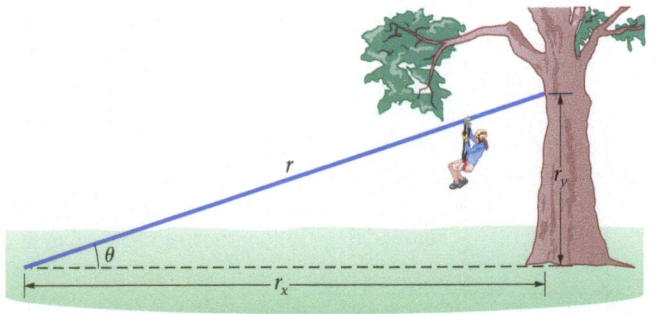

REASONING AND STRATEGY
The tangent of θ is the height of the zipline triangle divided by the base: $\tan\theta = r_y/r_x$. We know both θ and the base of the triangle, r_x, and hence we can rearrange this relationship to find the height r_y. Similarly, the length of the zipline can be obtained by solving $\cos\theta = r_x/r$ for r.

Known   Angle of zipline, $\theta = 18°$; horizontal distance to the tree, $r_x = 12$ m.
Unknown (a) Height of zipline, $r_y = ?$ (b) Length of zipline, $r = ?$

CONTINUED

---

*Real World Physics applications are denoted by the acronym RWP.

SOLUTION

Part (a)

1. Use $\tan\theta = r_y/r_x$ to solve for the height of the zipline, r_y:

$$r_y = r_x \tan\theta = (12\text{ m})\tan 18° = 3.9\text{ m}$$

Part (b)

2. Use $\cos\theta = r_x/r$ to solve for the length of the zipline:

$$r = \frac{r_x}{\cos\theta} = \frac{12\text{ m}}{\cos 18°} = 13\text{ m}$$

INSIGHT

Another way to solve part (b) is to use the Pythagorean theorem: $r = \sqrt{r_x^2 + r_y^2} = \sqrt{(12\text{ m})^2 + (3.9\text{ m})^2} = 13$ m. Thus, if we let $\vec{\mathbf{r}}$ denote the vector from the bottom to the top of the zipline, its magnitude is 13 m and its direction above the horizontal is 18°. Equivalently, the x component of $\vec{\mathbf{r}}$ is 12 m and its y component is 3.9 m.

PRACTICE PROBLEM — PREDICT/CALCULATE

(a) If the zipline is anchored to the ground 24 m from the base of the tree, but the top is at the same location, is the angle to the top of the zipline greater than, less than, or equal to 18°? Explain. (b) Find the angle to the top of the zipline in this case. [**Answer: (a)** The angle is less than 18° because the base of the triangle is greater, while the height remains the same. **(b)** The angle is $\theta = \tan^{-1}\left(\dfrac{3.9\text{ m}}{24\text{ m}}\right) = 9.2°$, which is less than 18°, as expected.]

Some related homework problems: Problem 5, Problem 19

EXERCISE 3-2 COMPONENTS, MAGNITUDE, AND DIRECTION

a. Find the components A_x and A_y for the vector $\vec{\mathbf{A}}$, whose magnitude and direction are given by $A = 3.5$ m and $\theta = 66°$, respectively.

b. Find B and θ for the vector $\vec{\mathbf{B}}$ with components $B_x = 75.5$ m and $B_y = 6.20$ m.

REASONING AND SOLUTION

(a) Multiply the magnitude A by $\cos\theta$ to find the x component; multiply the magnitude by $\sin\theta$ to find the y component. (b) Use the Pythagorean theorem to find the magnitude B, and the inverse tangent to find the angle θ.

a. $A_x = (3.5\text{ m})\cos 66° = 1.4\text{ m}$, $A_y = (3.5\text{ m})\sin 66° = 3.2\text{ m}$

b. $B = \sqrt{(75.5\text{ m})^2 + (6.20\text{ m})^2} = 75.8\text{ m}$, $\theta = \tan^{-1}\left(\dfrac{6.20\text{ m}}{75.5\text{ m}}\right) = 4.69°$

Determining the Sign of Vector Components How do we determine the correct signs for the x and y components of a vector? This is done by considering the right triangle formed by A_x, A_y, and $\vec{\mathbf{A}}$, as shown in **FIGURE 3-6**. To determine the sign of A_x, start at the tail of the vector and move along the x axis toward the right angle. If you are moving in the positive x direction, then A_x is positive ($A_x > 0$); if you are moving in the negative x direction, then A_x is negative ($A_x < 0$). For the y component, start at the right angle and move toward the tip of the arrow. A_y is positive or negative depending on whether you are moving in the positive or negative y direction.

For example, consider the vector shown in **FIGURE 3-7 (a)**. In this case, $A_x > 0$ and $A_y < 0$, as indicated in the figure. Similarly, the signs of A_x and A_y are given in **FIGURE 3-7 (b, c, d)** for the vectors shown there. Be sure to verify each of these cases by applying the rules just given. As we continue our study of physics, it's important to be able to find the components of a vector *and* to assign to them the correct signs.

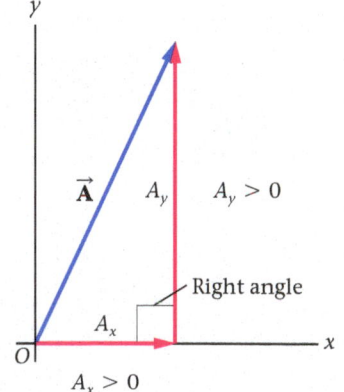

▲ **FIGURE 3-6** A vector whose x and y components are positive

EXERCISE 3-3 FIND THE COMPONENTS

The vector $\vec{\mathbf{A}}$ has a magnitude of 9.75 m. Find its components for direction angles of

a. $\theta = 7.25°$ **c.** $\theta = 235°$

b. $\theta = 115°$ **d.** $\theta = 355°$

CONTINUED

REASONING AND SOLUTION

Multiplying the magnitude A by $\cos\theta$ gives the x component; multiplying the magnitude by $\sin\theta$ gives the y component.

a. $A_x = (9.75\text{ m})\cos 7.25° = 9.67\text{ m}, A_y = (9.75\text{ m})\sin 7.25° = 1.23\text{ m}$

b. $A_x = (9.75\text{ m})\cos 115° = -4.12\text{ m}, A_y = (9.75\text{ m})\sin 115° = 8.84\text{ m}$

c. $A_x = (9.75\text{ m})\cos 235° = -5.59\text{ m}, A_y = (9.75\text{ m})\sin 235° = -7.99\text{ m}$

d. $A_x = (9.75\text{ m})\cos 355° = 9.71\text{ m}, A_y = (9.75\text{ m})\sin 355° = -0.850\text{ m}$

Using Care with a Calculator Be careful when using your calculator to determine the direction angle, θ, because you may need to add 180° to get the correct angle. For example, if $A_x = -0.50$ m and $A_y = 1.0$ m, your calculator will give the following result:

$$\theta = \tan^{-1}\left(\frac{1.0\text{ m}}{-0.50\text{ m}}\right) = \tan^{-1}(-2.0) = -63°$$

Does this angle correspond to the specified vector? The way to check is to sketch $\vec{\mathbf{A}}$. When you do, your drawing is similar to Figure 3-7 (c), and thus the direction angle of $\vec{\mathbf{A}}$ should be between 90° and 180°. To obtain the correct angle, add 180° to the calculator's result:

$$\theta = -63° + 180° = 117°$$

This, in fact, is the direction angle for the vector $\vec{\mathbf{A}}$. Why the need to add 180°? Well, the tangents of θ and $(\theta + 180°)$ are the same for any θ. A calculator doesn't know which of these answers is the one that's appropriate for your situation—you have to make that decision.

EXERCISE 3-4 DIRECTION ANGLE

The vector $\vec{\mathbf{B}}$ has components $B_x = -2.10$ m and $B_y = -1.70$ m. Find the direction angle, θ, for this vector.

REASONING AND SOLUTION

The direction angle is found using the inverse tangent. If the resulting angle is not appropriate for the vector, it may be necessary to add 180°. This is illustrated in the following calculations:

$$\tan^{-1}[(-1.70\text{ m})/(-2.10\text{ m})] = \tan^{-1}(1.70/2.10) = 39.0°, \theta = 39.0° + 180° = 219°$$

Enhance Your Understanding   (Answers given at the end of the chapter)

2. A vector has the components $A_x = 5$ m and $A_y = 3$ m. **(a)** Is the magnitude of $\vec{\mathbf{A}}$ greater than, less than, or equal to 8 m? Explain. **(b)** Is the direction angle of $\vec{\mathbf{A}}$ greater than, less than, or equal to 45°? Explain.

Section Review

- The x and y components of a vector tell how far the vector extends in those directions. The components are found using the sine and cosine functions.

- The magnitude of a vector is found from its components by using the Pythagorean theorem. The angle is found from the components by using the inverse tangent.

3-3 Adding and Subtracting Vectors

One important reason for determining the components of a vector is that they are useful in adding and subtracting vectors. In this section we begin by defining vector addition graphically, and then show how the same addition can be performed more concisely and accurately with components.

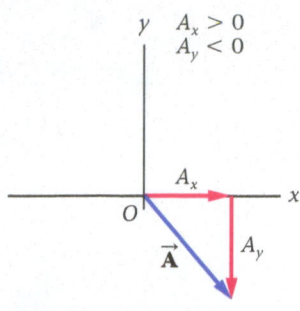

(a)

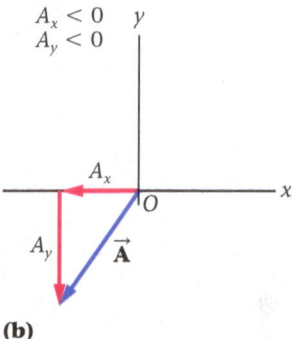

(b)

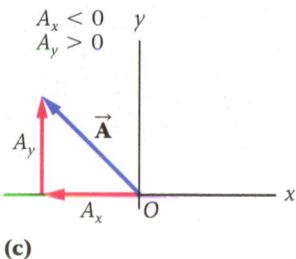

(c)

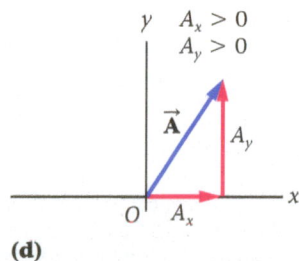

(d)

▲ **FIGURE 3-7 Examples of vectors with components of different signs** To determine the signs of a vector's components, it is only necessary to observe the direction in which they point. For example, in part (a) the x component points in the positive direction; hence $A_x > 0$. Similarly, the y component in part (a) points in the negative y direction; therefore $A_y < 0$.

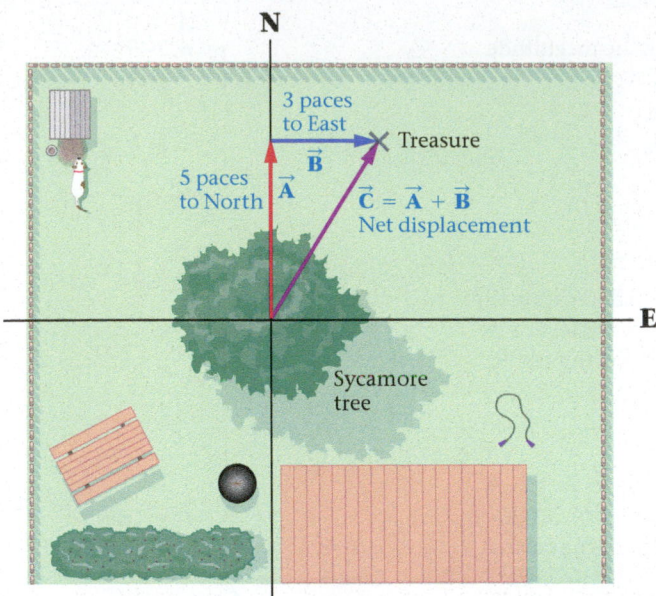

▲ **FIGURE 3-8 The sum of two vectors** To go from the sycamore tree to the treasure, one must first go 5 paces north ($\vec{A}$) and then 3 paces east ($\vec{B}$). The net displacement from the tree to the treasure is $\vec{C} = \vec{A} + \vec{B}$.

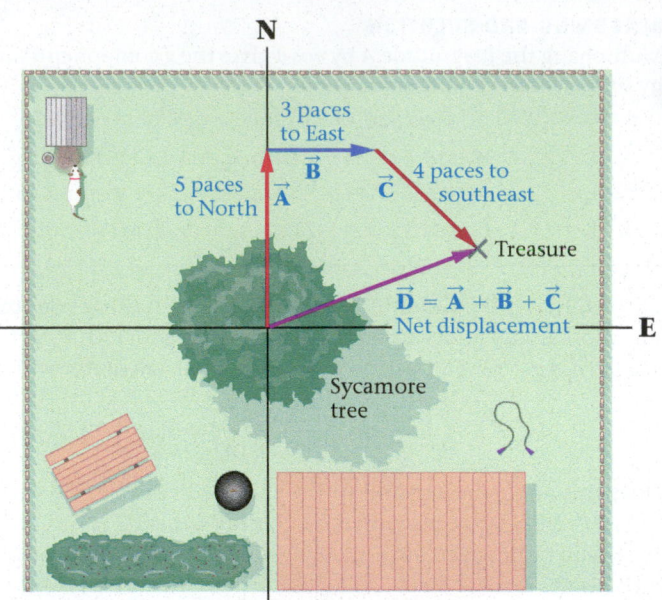

▲ **FIGURE 3-9 Adding several vectors** Searching for a treasure that is 5 paces north ($\vec{A}$), 3 paces east ($\vec{B}$), and 4 paces southeast ($\vec{C}$) of the sycamore tree. The net displacement from the tree to the treasure is $\vec{D} = \vec{A} + \vec{B} + \vec{C}$.

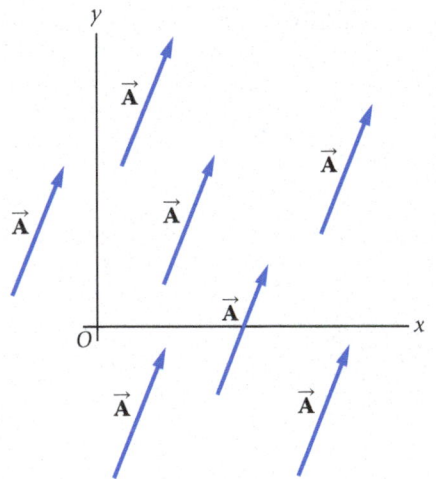

▲ **FIGURE 3-10 Identical vectors $\vec{A}$ at different locations** A vector is defined by its direction and length; its location is immaterial.

◀ **FIGURE 3-11** To a good approximation, these snow geese are all moving in the same direction with the same speed. As a result, their velocity vectors are equal, even though their positions are different.

Adding Vectors Graphically

One day you open an old chest in the attic and find a treasure map inside. To locate the buried treasure, the map says that you must "Go to the sycamore tree in the backyard, march 5 paces north, then 3 paces east, and dig there." If these two displacements are represented by the vectors $\vec{A}$ and $\vec{B}$ in **FIGURE 3-8**, the total displacement from the tree to the treasure is given by the vector $\vec{C}$. We say that $\vec{C}$ is the *vector sum* of $\vec{A}$ and $\vec{B}$; that is, $\vec{C} = \vec{A} + \vec{B}$. In general, vectors are added graphically according to the following rule:

> To add the vectors $\vec{A}$ and $\vec{B}$, place the tail of $\vec{B}$ at the head of $\vec{A}$. The sum, $\vec{C} = \vec{A} + \vec{B}$, is the vector extending from the tail of $\vec{A}$ to the head of $\vec{B}$.

If the instructions to find the treasure were a bit more complicated—5 paces north, 3 paces east, then 4 paces southeast, for example—the path from the sycamore tree to the treasure would be like that shown in **FIGURE 3-9**. In this case, the total displacement, $\vec{D}$, is the sum of the three vectors $\vec{A}$, $\vec{B}$, and $\vec{C}$; that is, $\vec{D} = \vec{A} + \vec{B} + \vec{C}$. It follows that to add more than two vectors, we just keep placing the vectors head-to-tail, head-to-tail, and then draw a vector from the tail of the first vector to the head of the last vector, as in Figure 3-9.

In order to place a given pair of vectors head-to-tail, it may be necessary to move the corresponding arrows. This is fine, as long as you don't change their length or their direction. After all, a vector is defined by its length and direction; if these are unchanged, so is the vector.

> A vector is defined by its magnitude and direction, regardless of its location.

For example, in **FIGURE 3-10** all of the vectors are the same, even though they are at different locations on the graph. Similarly, the geese in **FIGURE 3-11** all have the same velocity vectors, even though they are in different locations.

As an example of moving vectors to different locations, consider two vectors, $\vec{A}$ and $\vec{B}$, and their vector sum, $\vec{C}$, as illustrated in **FIGURE 3-12 (a)**:

$$\vec{C} = \vec{A} + \vec{B}$$

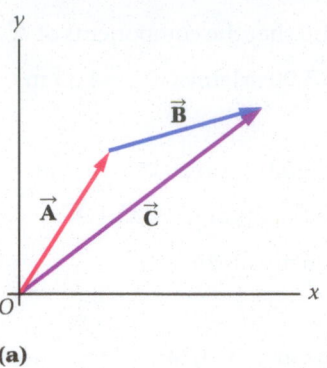

(a)

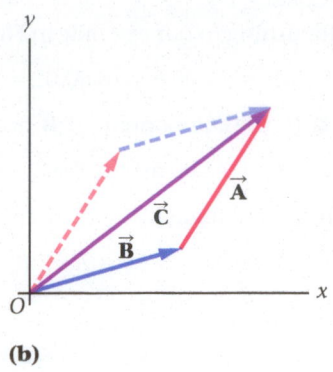

(b)

◄ **FIGURE 3-12** $\vec{A} + \vec{B} = \vec{B} + \vec{A}$ The vector $\vec{C}$ is equal to **(a)** $\vec{A} + \vec{B}$ and **(b)** $\vec{B} + \vec{A}$. Note also that $\vec{C}$ is the diagonal of the parallelogram formed by the vectors $\vec{A}$ and $\vec{B}$. For this reason, this method of vector addition is referred to as the "parallelogram method."

By moving the arrow representing $\vec{B}$ so that its tail is at the origin, and moving the arrow for $\vec{A}$ so that its tail is at the head of $\vec{B}$, we obtain the construction shown in **FIGURE 3-12 (b)**. From this graph we see that $\vec{C}$, which is $\vec{A} + \vec{B}$, is also equal to $\vec{B} + \vec{A}$:

$$\vec{C} = \vec{A} + \vec{B} = \vec{B} + \vec{A}$$

That is, the sum of vectors is independent of the order in which the vectors are added.

Now, suppose that $\vec{A}$ has a magnitude of 5.00 m and a direction angle of 60.0° above the x axis, and that $\vec{B}$ has a magnitude of 4.00 m and a direction angle of 20.0° above the x axis. These two vectors and their sum, $\vec{C}$, are shown in **FIGURE 3-13**. The question is: What are the length and direction angle of $\vec{C}$?

A graphical way to answer this question is to simply measure the length and direction of $\vec{C}$ in Figure 3-13. With a ruler, we find the length of $\vec{C}$ to be approximately 1.75 times the length of $\vec{A}$, which means that $\vec{C}$ is roughly 1.75 (5.00 m) = 8.75 m. Similarly, with a protractor we measure the angle θ to be about 45.0° above the x axis.

Adding Vectors Using Components

The graphical method of adding vectors yields approximate results, limited by the accuracy with which the vectors can be drawn and measured. In contrast, exact results can be obtained by adding $\vec{A}$ and $\vec{B}$ in terms of their components. To see how this is done, consider **FIGURE 3-14 (a)**, which shows the components of $\vec{A}$ and $\vec{B}$, and **FIGURE 3-14 (b)**, which shows the components of $\vec{C}$. Clearly,

$$C_x = A_x + B_x$$

and

$$C_y = A_y + B_y$$

Thus, to add vectors, you simply add the components.

▲ **FIGURE 3-13 Graphical addition of vectors** The vector $\vec{A}$ has a magnitude of 5.00 m and a direction angle of 60.0°; the vector $\vec{B}$ has a magnitude of 4.00 m and a direction angle of 20.0°. The magnitude and direction of $\vec{C} = \vec{A} + \vec{B}$ can be measured on the graph with a ruler and a protractor.

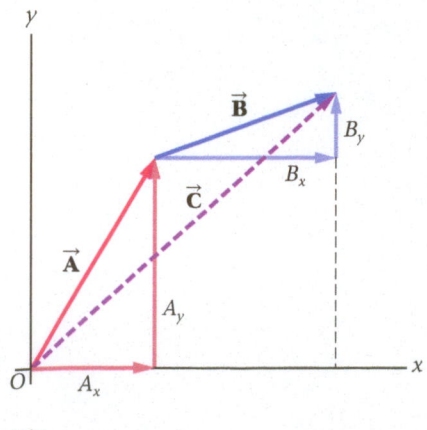

(a)

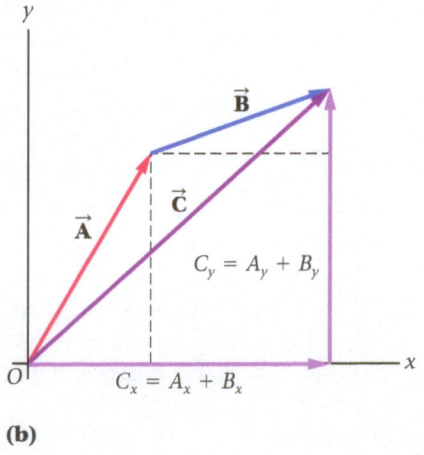

(b)

◄ **FIGURE 3-14 Component addition of vectors (a)** The x and y components of $\vec{A}$ and $\vec{B}$. **(b)** The x and y components of $\vec{C}$. Notice that $C_x = A_x + B_x$ and $C_y = A_y + B_y$.

Returning to our example in Figure 3-13, we find that the components of $\vec{\mathbf{A}}$ are

$$A_x = (5.00 \text{ m}) \cos 60.0° = 2.50 \text{ m}, \quad A_y = (5.00 \text{ m}) \sin 60.0° = 4.33 \text{ m}$$

Similarly, the components of $\vec{\mathbf{B}}$ are

$$B_x = (4.00 \text{ m}) \cos 20.0° = 3.76 \text{ m}, \quad B_y = (4.00 \text{ m}) \sin 20.0° = 1.37 \text{ m}$$

Adding component by component yields the components of $\vec{\mathbf{C}} = \vec{\mathbf{A}} + \vec{\mathbf{B}}$:

$$C_x = A_x + B_x = 2.50 \text{ m} + 3.76 \text{ m} = 6.26 \text{ m}$$

and

$$C_y = A_y + B_y = 4.33 \text{ m} + 1.37 \text{ m} = 5.70 \text{ m}$$

With these results, we can now find *precise* values for C, the magnitude of vector $\vec{\mathbf{C}}$, and its direction angle θ. In particular,

$$C = \sqrt{C_x{}^2 + C_y{}^2} = \sqrt{(6.26 \text{ m})^2 + (5.70 \text{ m})^2} = \sqrt{71.7 \text{ m}^2} = 8.47 \text{ m}$$

and

$$\theta = \tan^{-1}\left(\frac{C_y}{C_x}\right) = \tan^{-1}\left(\frac{5.70 \text{ m}}{6.26 \text{ m}}\right) = \tan^{-1}(0.911) = 42.3°$$

Big Idea **3** The most precise way to add or subtract vectors is to add or subtract their components.

Note that the approximate results found by graphical addition are in rough agreement with these exact results.

In the future, we will always add vectors using components—graphical addition is useful primarily as a rough check on the results obtained with components.

EXAMPLE 3-5   TREASURE HUNT: FIND THE DIRECTION AND MAGNITUDE

What are the magnitude D and direction angle θ of the net displacement for the treasure hunt illustrated in the figure?

PICTURE THE PROBLEM
Our drawing shows the direction and magnitude for each of the three vectors, $\vec{\mathbf{A}}$, $\vec{\mathbf{B}}$, and $\vec{\mathbf{C}}$, that lead to the location of the buried treasure. The net displacement is the vector sum, $\vec{\mathbf{D}} = \vec{\mathbf{A}} + \vec{\mathbf{B}} + \vec{\mathbf{C}}$.

REASONING AND STRATEGY
We'll start by writing out the components for $\vec{\mathbf{A}}$, $\vec{\mathbf{B}}$, and $\vec{\mathbf{C}}$. Notice that $\vec{\mathbf{A}}$ has only a positive y component (in the N direction), $\vec{\mathbf{B}}$ has only a positive x component (in the E direction), and $\vec{\mathbf{C}}$ has both a positive x component and a negative y component. Adding the vectors component by component yields the components of $\vec{\mathbf{D}}$. We then convert the components of $\vec{\mathbf{D}}$ to the magnitude D and angle θ.

Known   The lengths of the segments of the path are A = 3.75 m, B = 2.25 m, and C = 3.00 m. The directions are given in the figure.

Unknown   Length of final displacement, D = ? Direction angle of final displacement, θ = ?

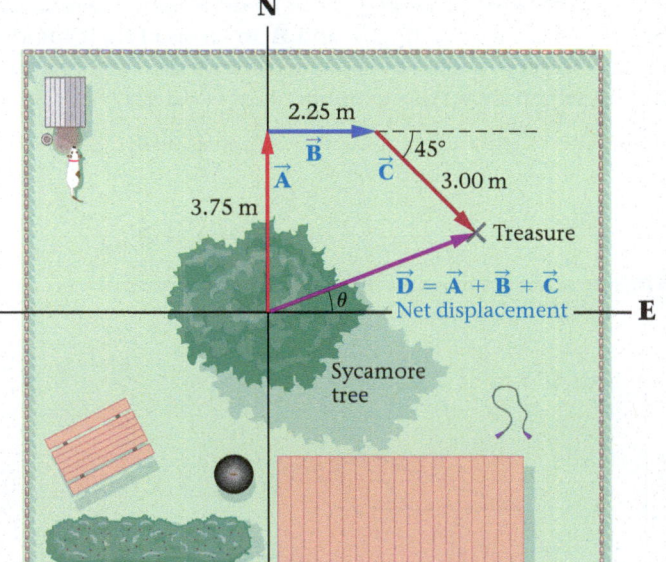

SOLUTION

1. Find the components of $\vec{\mathbf{A}}$: $A_x = 0, A_y = 3.75 \text{ m}$

2. Find the components of $\vec{\mathbf{B}}$: $B_x = 2.25 \text{ m}, B_y = 0$

3. Find the components of $\vec{\mathbf{C}}$: $C_x = (3.00 \text{ m}) \cos(-45°) = 2.12 \text{ m}$
                                          $C_y = (3.00 \text{ m}) \sin(-45°) = -2.12 \text{ m}$

4. Sum the components of $\vec{\mathbf{A}}$, $\vec{\mathbf{B}}$, and $\vec{\mathbf{C}}$ to find the components of $\vec{\mathbf{D}}$: $D_x = 0 + 2.25 \text{ m} + 2.12 \text{ m} = 4.37 \text{ m}$
                                          $D_y = 3.75 \text{ m} + 0 - 2.12 \text{ m} = 1.63 \text{ m}$

CONTINUED

5. Use the Pythagorean theorem to find D and the inverse tangent to find θ:

$$D = \sqrt{(4.37 \text{ m})^2 + (1.63 \text{ m})^2} = 4.66 \text{ m}$$

$$\theta = \tan^{-1}\left(\frac{1.63 \text{ m}}{4.37 \text{ m}}\right) = 20.5°$$

INSIGHT

A direct route to the treasure would be to walk 4.66 m from the tree in a direction 20.5° above the *x* axis.

PRACTICE PROBLEM — PREDICT/CALCULATE

(a) If the magnitudes of $\vec{A}$, $\vec{B}$, and $\vec{C}$ are all decreased by a factor of two, by what factors do you expect D and θ to change? **(b)** Verify your predictions with a numerical calculation.   [**Answer: (a)** The distance D will be halved; the angle θ will stay the same. **(b)** $D = (4.66 \text{ m})/2 = 2.33 \text{ m}, \theta = 20.5°$.]

Some related homework problems: Problem 22, Problem 28, Problem 74

Subtracting Vectors

Next, how do we subtract vectors? Suppose, for example, that we would like to determine the vector $\vec{D}$, where

$$\vec{D} = \vec{A} - \vec{B}$$

As an example, we'll consider the vectors $\vec{A}$ and $\vec{B}$ that are shown in Figure 3-13. To find $\vec{D}$, we start by rewriting the equation for it as follows:

$$\vec{D} = \vec{A} + (-\vec{B})$$

That is, $\vec{D}$ is the sum of $\vec{A}$ and $-\vec{B}$. Now the negative of a vector has a very simple graphical interpretation:

> The *negative* of a vector is represented by an arrow of the same length as the original vector, but pointing in the *opposite* direction. That is, multiplying a vector by minus one *reverses its direction*.

For example, the vectors $\vec{B}$ and $-\vec{B}$ are indicated in **FIGURE 3-15 (a)**. Thus, to subtract $\vec{B}$ from $\vec{A}$, simply reverse the direction of $\vec{B}$ and add it to $\vec{A}$, as indicated in **FIGURE 3-15 (b)**.

In terms of components, you subtract vectors by simply subtracting the components. For example, consider the following subtraction:

$$\vec{D} = \vec{A} - \vec{B}$$

The components of $\vec{D}$ are given by subtracting the components of $\vec{B}$ from the components of $\vec{A}$:

$$D_x = A_x - B_x$$

$$D_y = A_y - B_y$$

Once the components of $\vec{D}$ are found, its magnitude and direction angle can be calculated as usual.

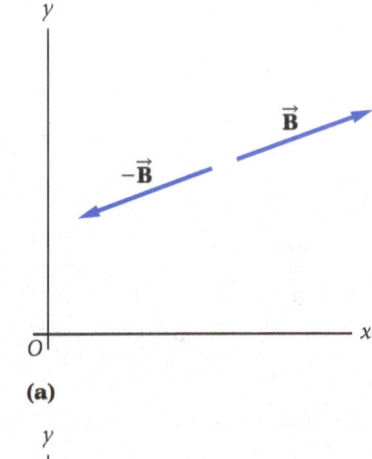

(a)

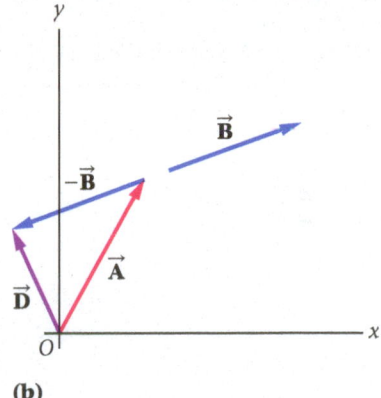

(b)

▲ **FIGURE 3-15 Vector subtraction (a)** The vector $\vec{B}$ and its negative $-\vec{B}$. **(b)** A vector construction for $\vec{D} = \vec{A} - \vec{B}$.

EXERCISE 3-6   SUBTRACTING VECTORS

a. For the vectors $\vec{A}$ and $\vec{B}$ shown in Figure 3-13, find the components of $\vec{D} = \vec{A} - \vec{B}$.

b. Find D and θ for the vector in part (a), and compare your answer with the vector $\vec{D}$ shown in Figure 3-15 (b).

REASONING AND SOLUTION

(a) Subtract the components of $\vec{B}$ from the components of $\vec{A}$ to find the components of $\vec{D}$:

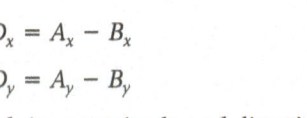

$$D_x = A_x - B_x = (2.50 \text{ m}) - (3.76 \text{ m}) = -1.26 \text{ m},$$

$$D_y = A_y - B_y = (4.33 \text{ m}) - (1.37 \text{ m}) = 2.96 \text{ m}$$

(b) Converting components to magnitude and direction, using the Pythagorean theorem and the inverse tangent, we find $D = 3.22$ m and $\theta = -66.9° + 180° = 113°$. In Figure 3-15 (b) we see that $\vec{D}$ is shorter than $\vec{B}$, which has a magnitude of 4.00 m, and its direction angle is somewhat greater than 90°, in agreement with our numerical results.

Enhance Your Understanding   (Answers given at the end of the chapter)

3. Several vectors are shown in **FIGURE 3-16**. The red vectors have a magnitude equal to a diagonal across the square formed by the blue vectors. Which of the given vectors represents the following quantities: (a) $\vec{A} + \vec{B}$, (b) $\vec{D} - \vec{A}$, (c) $\vec{G} - \vec{C}$, and (d) $\vec{B} + \vec{H} - \vec{D}$?

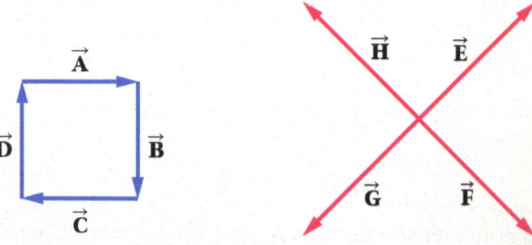

▲ **FIGURE 3-16** Enhance Your Understanding 3

Section Review

• Vectors are added graphically by placing them head-to-tail.

• Vectors are added with components by adding the components.

3-4 Unit Vectors

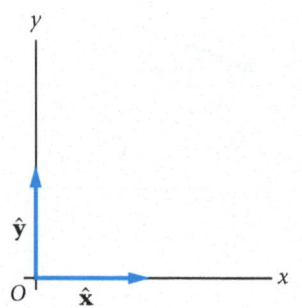

▲ **FIGURE 3-17 Unit vectors** The unit vectors $\hat{\mathbf{x}}$ and $\hat{\mathbf{y}}$ point in the positive x and y directions, respectively.

Big Idea 4 Unit vectors are dimensionless, have a magnitude of 1, and point in a specific direction. The most commonly used unit vectors point in the positive x or positive y direction.

Unit vectors provide a convenient way of expressing an arbitrary vector in terms of its components, as we shall see. But first, let's define what we mean by a unit vector. In particular, the unit vectors $\hat{\mathbf{x}}$ and $\hat{\mathbf{y}}$ are defined to be dimensionless vectors of unit magnitude pointing in the positive x and y directions, respectively:

• The x unit vector, $\hat{\mathbf{x}}$, is a dimensionless vector of unit length pointing in the positive x direction.

• The y unit vector, $\hat{\mathbf{y}}$, is a dimensionless vector of unit length pointing in the positive y direction.

FIGURE 3-17 shows $\hat{\mathbf{x}}$ and $\hat{\mathbf{y}}$ on a two-dimensional coordinate system. Unit vectors have no physical dimensions—like mass, length, or time—and hence they are used to specify direction only.

Multiplying Unit Vectors by Scalars

To see the utility of unit vectors, consider the effect of multiplying a vector by a scalar. For example, multiplying a vector by 3 increases its magnitude by a factor of 3 but does not change its direction, as shown in **FIGURE 3-18**. Multiplying by -3 increases the magnitude by a factor of 3 *and* reverses the direction of the vector. This is also shown in Figure 3-18. In the case of unit vectors—which have a magnitude of 1 and are dimensionless—multiplication by a scalar results in a vector with the same magnitude and dimensions as the scalar.

For example, if a vector $\vec{A}$ has the scalar components $A_x = 5$ m and $A_y = 3$ m, we can write it as

$$\vec{A} = (5\text{ m})\hat{\mathbf{x}} + (3\text{ m})\hat{\mathbf{y}}$$

We refer to the quantities $(5\text{ m})\hat{\mathbf{x}}$ and $(3\text{ m})\hat{\mathbf{y}}$ as the x and y **vector components** of the vector $\vec{A}$. In general, an arbitrary two-dimensional vector $\vec{A}$ can always be written as the sum of a vector component in the x direction and a vector component in the y direction:

$$\vec{A} = A_x\hat{\mathbf{x}} + A_y\hat{\mathbf{y}}$$

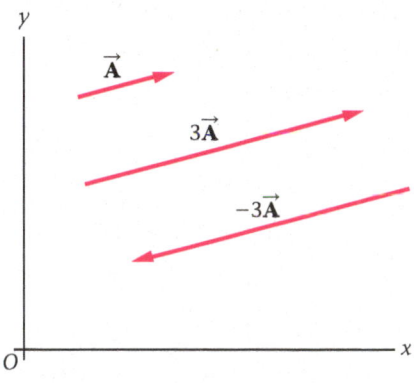

▲ **FIGURE 3-18 Multiplying a vector by a scalar** Multiplying a vector by a positive scalar different from 1 will change the length of the vector but leave its direction the same. If the vector is multiplied by a negative scalar its direction is reversed.

This is illustrated in **FIGURE 3-19 (a)**. An equivalent way of representing the vector components of a vector is illustrated in **FIGURE 3-19 (b)**. In this case we see that the vector components are the *projection* of a vector onto the x and y axes. The sign of the vector components is positive if they point in the positive x or y direction, and negative if they point in the negative x or y direction. This is how vector components will generally be shown in later chapters.

Finally, notice that vector addition and subtraction are straightforward with unit vector notation. For example,

$$\vec{\mathbf{C}} = \vec{\mathbf{A}} + \vec{\mathbf{B}} = (A_x + B_x)\hat{\mathbf{x}} + (A_y + B_y)\hat{\mathbf{y}}$$

and

$$\vec{\mathbf{D}} = \vec{\mathbf{A}} - \vec{\mathbf{B}} = (A_x - B_x)\hat{\mathbf{x}} + (A_y - B_y)\hat{\mathbf{y}}$$

Clearly, unit vectors provide a useful way to keep track of the x and y components of a vector.

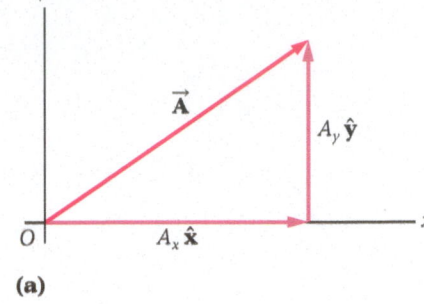

(a)

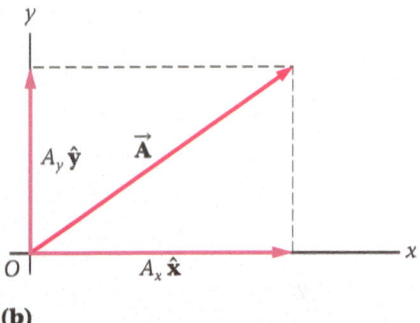

(b)

▲ **FIGURE 3-19 Vector components (a)** A vector $\vec{\mathbf{A}}$ can be written in terms of unit vectors as $\vec{\mathbf{A}} = A_x\hat{\mathbf{x}} + A_y\hat{\mathbf{y}}$. **(b)** Vector components can be thought of as the projection of the vector onto the x and y axes. This method of representing vector components will be used frequently in subsequent chapters.

Enhance Your Understanding   (Answers given at the end of the chapter)

4. Which of the vectors shown in
    FIGURE 3-20 has the same direction as
    (a) $(1\text{ m})\,\hat{\mathbf{x}} + (1\text{ m})\,\hat{\mathbf{y}}$,
    (b) $(1\text{ m})\,\hat{\mathbf{x}} - (1\text{ m})\,\hat{\mathbf{y}}$,
    (c) $-(1\text{ m})\,\hat{\mathbf{x}} + (1\text{ m})\,\hat{\mathbf{y}}$,
    (d) $-(1\text{ m})\,\hat{\mathbf{x}} - (1\text{ m})\,\hat{\mathbf{y}}$?

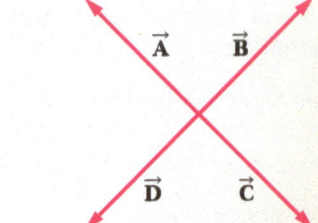

▲ **FIGURE 3-20** Enhance Your Understanding 4

Section Review

- Unit vectors provide a convenient way to represent the vector components of a vector.

3-5 Position, Displacement, Velocity, and Acceleration Vectors

In Chapter 2 we discussed four different one-dimensional vectors: position, displacement, velocity, and acceleration. Each of these quantities had a direction associated with it, indicated by its sign; positive meant in the positive direction, negative meant in the negative direction. Now we consider these vectors again, this time in two dimensions, where the possibilities for direction are not so limited.

Position Vectors

To begin, imagine a two-dimensional coordinate system, as in **FIGURE 3-21**. Position is indicated by a vector from the origin to the location in question. We refer to the position vector as $\vec{\mathbf{r}}$; its units are meters, m.

Definition: Position Vector, $\vec{\mathbf{r}}$

position vector $= \vec{\mathbf{r}}$ 3-1

SI unit: meter, m

In terms of unit vectors, the position vector is simply $\vec{\mathbf{r}} = x\,\hat{\mathbf{x}} + y\,\hat{\mathbf{y}}$.

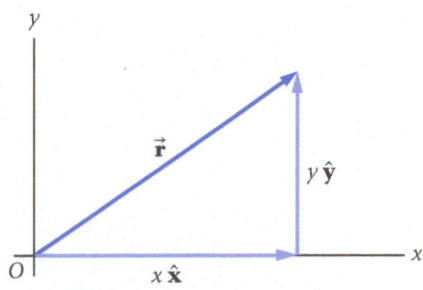

▲ **FIGURE 3-21 Position vector** The position vector $\vec{\mathbf{r}}$ points from the origin to the current location of an object. The x and y vector components of $\vec{\mathbf{r}}$ are $x\hat{\mathbf{x}}$ and $y\hat{\mathbf{y}}$, respectively.

Displacement Vectors

Now, suppose that initially you are at the location indicated by the position vector $\vec{\mathbf{r}}_i$ (subscript i stands for initial), and that later you are at the final position represented by

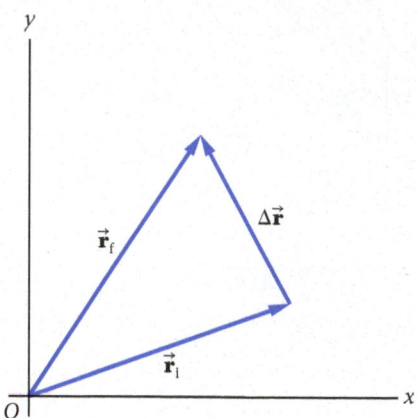

▲ **FIGURE 3-22 Displacement vector** The displacement vector $\Delta\vec{\mathbf{r}}$ is the change in position. It points from the head of the initial position vector $\vec{\mathbf{r}}_i$ to the head of the final position vector $\vec{\mathbf{r}}_f$. Thus $\vec{\mathbf{r}}_f = \vec{\mathbf{r}}_i + \Delta\vec{\mathbf{r}}$ or $\Delta\vec{\mathbf{r}} = \vec{\mathbf{r}}_f - \vec{\mathbf{r}}_i$.

▲ **FIGURE 3-23** A map can be used to determine the direction and magnitude of the displacement vector from your initial position to your desired final position.

the position vector $\vec{\mathbf{r}}_f$ (subscript f stands for final). Your displacement vector, $\Delta\vec{\mathbf{r}}$, is the change in position:

> **Definition: Displacement Vector, $\Delta\vec{\mathbf{r}}$**
>
> $$\Delta\vec{\mathbf{r}} = \vec{\mathbf{r}}_f - \vec{\mathbf{r}}_i \qquad \text{3-2}$$
>
> SI unit: meter, m

Rearranging this definition slightly, we see that

$$\vec{\mathbf{r}}_f = \vec{\mathbf{r}}_i + \Delta\vec{\mathbf{r}}$$

That is, the final position is equal to the initial position plus the change in position. This is illustrated in **FIGURE 3-22**, where we see that $\Delta\vec{\mathbf{r}}$ extends from the head of $\vec{\mathbf{r}}_i$ to the head of $\vec{\mathbf{r}}_f$. This is similar to how a map and compass can be used to determine the displacement vector from your present position to your destination, as in **FIGURE 3-23**.

Velocity Vectors

Next, the average velocity vector is defined as the displacement vector $\Delta\vec{\mathbf{r}}$ divided by the elapsed time Δt:

> **Definition: Average Velocity Vector, $\vec{\mathbf{v}}_{av}$**
>
> $$\vec{\mathbf{v}}_{av} = \frac{\Delta\vec{\mathbf{r}}}{\Delta t} \qquad \text{3-3}$$
>
> SI unit: meter per second, m/s

Since $\Delta\vec{\mathbf{r}}$ is a vector, it follows that $\vec{\mathbf{v}}_{av}$ is also a vector; it is the vector $\Delta\vec{\mathbf{r}}$ multiplied by the scalar $(1/\Delta t)$. Thus $\vec{\mathbf{v}}_{av}$ is parallel to $\Delta\vec{\mathbf{r}}$ and has the units m/s.

QUICK EXAMPLE 3-7 A DRAGONFLY'S AVERAGE VELOCITY

A dragonfly is initially at the position $\vec{\mathbf{r}}_i = (4.60\,\text{m})\hat{\mathbf{x}} + (0.960\,\text{m})\hat{\mathbf{y}}$, as shown in the sketch. After 3.00 s, the dragonfly is at the position $\vec{\mathbf{r}}_f = (0.520\,\text{m})\hat{\mathbf{x}} + (2.40\,\text{m})\hat{\mathbf{y}}$. What was the dragonfly's average velocity during this time?

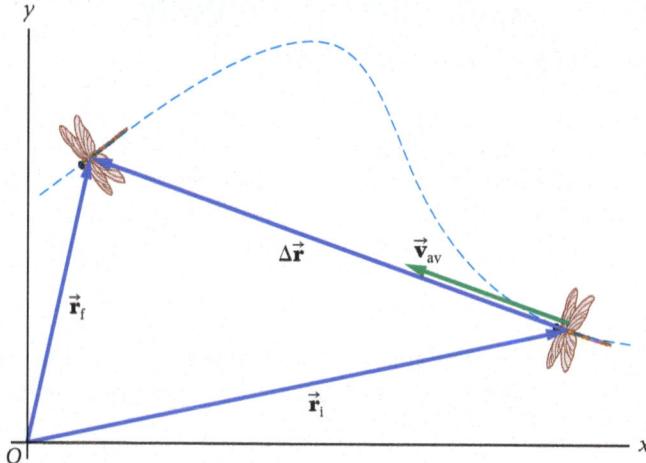

REASONING AND SOLUTION
We show the initial and final positions in our sketch, labeled $\vec{\mathbf{r}}_i$ and $\vec{\mathbf{r}}_f$, respectively. The corresponding change in position is $\Delta\vec{\mathbf{r}} = \vec{\mathbf{r}}_f - \vec{\mathbf{r}}_i$. Dividing $\Delta\vec{\mathbf{r}}$ by the time $\Delta t = 3.00\,\text{s}$ yields the average velocity.

1. Calculate the change in position:
$$\Delta\vec{\mathbf{r}} = \vec{\mathbf{r}}_f - \vec{\mathbf{r}}_i$$
$$= (0.520\,\text{m} - 4.60\,\text{m})\hat{\mathbf{x}} + (2.40\,\text{m} - 0.960\,\text{m})\hat{\mathbf{y}}$$
$$= (-4.08\,\text{m})\hat{\mathbf{x}} + (1.44\,\text{m})\hat{\mathbf{y}}$$

CONTINUED

2. Divide by the elapsed time to find the average velocity:

$$\vec{v}_{av} = \frac{\Delta \vec{r}}{\Delta t} = \frac{(-4.08 \text{ m})\hat{x} + (1.44 \text{ m})\hat{y}}{3.00 \text{ s}}$$
$$= (-1.36 \text{ m/s})\hat{x} + (0.480 \text{ m/s})\hat{y}$$

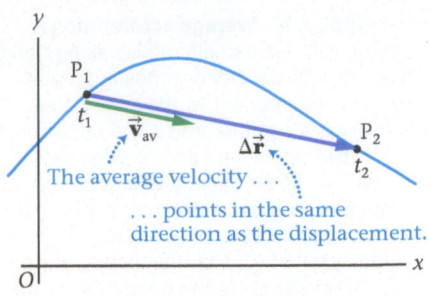

▲ **FIGURE 3-24 Average velocity vector** The average velocity, $\vec{v}_{av}$, points in the same direction as the displacement, $\Delta \vec{r}$, for any given interval of time.

To help visualize $\vec{v}_{av}$, imagine a particle moving in two dimensions along the blue path shown in **FIGURE 3-24**. If the particle is at point P_1 at time t_1, and at P_2 at time t_2, its displacement is indicated by the vector $\Delta \vec{r}$. The average velocity is parallel to $\Delta \vec{r}$, as indicated in Figure 3-24. It makes sense physically that $\vec{v}_{av}$ is parallel to $\Delta \vec{r}$; after all, on *average* you have moved in the direction of $\Delta \vec{r}$ during the time from t_1 to t_2. To put it another way, a particle that starts at P_1 at the time t_1 and moves with the velocity $\vec{v}_{av}$ until the time t_2 will arrive in precisely the same location as the particle that follows the blue path.

If we consider smaller and smaller time intervals, as in **FIGURE 3-25**, it is possible to calculate the instantaneous velocity vector:

Definition: Instantaneous Velocity Vector, $\vec{v}$

$$\vec{v} = \lim_{\Delta t \to 0} \frac{\Delta \vec{r}}{\Delta t} \qquad \text{3-4}$$

SI unit: meter per second, m/s

As can be seen in Figure 3-25, the instantaneous velocity at a given time is tangential to the path of the particle at that time. In addition, the magnitude of the velocity vector is the instantaneous speed of the particle. Thus, the instantaneous velocity vector tells you both how fast a particle is moving and in what direction.

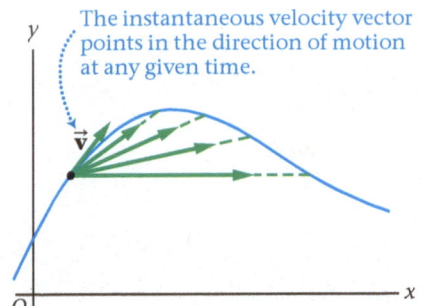

▲ **FIGURE 3-25 Instantaneous velocity vector** The instantaneous velocity vector $\vec{v}$ is obtained by calculating the average velocity vector over smaller and smaller time intervals. In the limit of vanishingly small time intervals, the average velocity approaches the instantaneous velocity, which points in the direction of motion.

EXERCISE 3-8   A TRAVELING TROUT

Find the speed and direction of motion for a rainbow trout whose velocity is $\vec{v} = (2.6 \text{ m/s})\hat{x} + (-1.9 \text{ m/s})\hat{y}$.

REASONING AND SOLUTION

The velocity components are given, and hence we can find the magnitude of the velocity (the speed) using the Pythagorean theorem. Therefore, the speed of the trout is

$$v = \sqrt{(2.6 \text{ m/s})^2 + (-1.9 \text{ m/s})^2} = 3.2 \text{ m/s}$$

We can find the direction of motion using the inverse tangent, as follows:

$$\theta = \tan^{-1}\left(\frac{-1.9 \text{ m/s}}{2.6 \text{ m/s}}\right) = -36°$$

We conclude that the trout moves at the rate of 3.2 m/s in a direction 36° *below* the x axis.

Acceleration Vectors

Finally, the average acceleration vector over an interval of time, Δt, is defined as the change in the velocity vector, $\Delta \vec{v}$, divided by the elapsed time Δt:

Definition: Average Acceleration Vector, $\vec{a}_{av}$

$$\vec{a}_{av} = \frac{\Delta \vec{v}}{\Delta t} \qquad \text{3-5}$$

SI unit: meter per second per second, m/s²

An example is given in **FIGURE 3-26 (a)**, where we show the initial and final velocity vectors corresponding to two different times. The change in velocity is defined as

$$\Delta \vec{v} = \vec{v}_f - \vec{v}_i$$

It follows that

$$\vec{v}_f = \vec{v}_i + \Delta \vec{v}$$

▶ **FIGURE 3-26 Average acceleration vector (a)** As a particle moves along the blue path its velocity changes in magnitude and direction. At the time t_i the velocity is $\vec{\mathbf{v}}_i$; at the time t_f the velocity is $\vec{\mathbf{v}}_f$. **(b)** The average acceleration vector $\vec{\mathbf{a}}_{av} = \Delta\vec{\mathbf{v}}/\Delta t$ points in the direction of the change in velocity vector $\Delta\vec{\mathbf{v}}$. We obtain $\Delta\vec{\mathbf{v}}$ by moving $\vec{\mathbf{v}}_f$ so that its tail coincides with the tail of $\vec{\mathbf{v}}_i$, and then drawing the arrow that connects the head of $\vec{\mathbf{v}}_i$ to the head of $\vec{\mathbf{v}}_f$. Note that $\vec{\mathbf{a}}_{av}$ need not point in the direction of motion, and in general it doesn't.

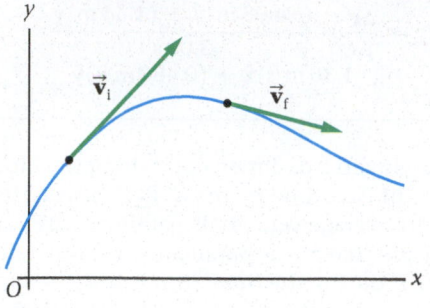

(a) The instantaneous velocity at two different times

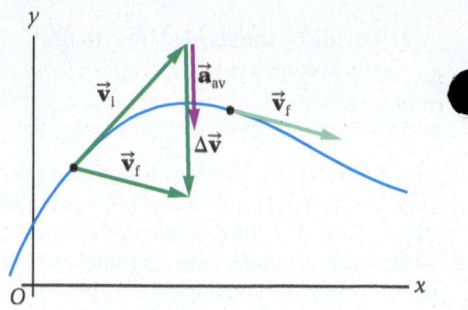

(b) The average acceleration vector points in the same direction as the change in velocity.

This is indicated in **FIGURE 3-26 (b)**. Notice that $\Delta\vec{\mathbf{v}}$ is the vector extending from the head of $\vec{\mathbf{v}}_i$ to the head of $\vec{\mathbf{v}}_f$, just as $\Delta\vec{\mathbf{r}}$ extends from the head of $\vec{\mathbf{r}}_i$ to the head of $\vec{\mathbf{r}}_f$ in Figure 3-22. The direction of $\vec{\mathbf{a}}_{av}$ is the direction of $\Delta\vec{\mathbf{v}}$, as shown in Figure 3-26 (b).

Can an object accelerate if its speed is constant? Absolutely—if its direction changes. Consider a car traveling with a constant speed on a circular track, as shown in **FIGURE 3-27**. Suppose that the initial velocity of the car is $\vec{\mathbf{v}}_i = (12\ \text{m/s})\hat{\mathbf{x}}$, and that 10.0 s later its final velocity is $\vec{\mathbf{v}}_f = (-12\ \text{m/s})\hat{\mathbf{y}}$. The speed is 12 m/s in each case, but the velocity is different because the *direction* has changed. Calculating the average acceleration, we find a nonzero acceleration:

$$\vec{\mathbf{a}}_{av} = \frac{\Delta\vec{\mathbf{v}}}{\Delta t} = \frac{\vec{\mathbf{v}}_f - \vec{\mathbf{v}}_i}{10.0\ \text{s}}$$

$$= \frac{(-12\ \text{m/s})\hat{\mathbf{y}} - (12\ \text{m/s})\hat{\mathbf{x}}}{10.0\ \text{s}} = (-1.2\ \text{m/s}^2)\hat{\mathbf{x}} + (-1.2\ \text{m/s}^2)\hat{\mathbf{y}}$$

PHYSICS IN CONTEXT
Looking Ahead

In Chapter 5 we will introduce one of the most important concepts in all of physics—force. It is a vector quantity. Other important vector quantities to be introduced in later chapters include linear momentum (Chapter 9), angular momentum (Chapter 11), electric field (Chapter 19), and magnetic field (Chapter 22).

Thus, a change in direction is just as important as a change in speed in producing an acceleration. We shall study circular motion in detail in Chapter 6.

Finally, by going to infinitesimally small time intervals, $\Delta t \rightarrow 0$, we can define the instantaneous acceleration vector:

Definition: Instantaneous Acceleration Vector, $\vec{\mathbf{a}}$

$$\vec{\mathbf{a}} = \lim_{\Delta t \rightarrow 0} \frac{\Delta\vec{\mathbf{v}}}{\Delta t} \qquad \qquad 3\text{-}6$$

SI unit: meter per second per second, m/s²

▶ **FIGURE 3-27 Average acceleration for a car traveling in a circle with constant speed** Although the speed of this car never changes, it is still accelerating—due to the change in its direction of motion. For the time interval depicted, the car's average acceleration is in the direction of $\Delta\vec{\mathbf{v}}$, which is toward the center of the circle. (As we shall see in Chapter 6, the car's acceleration is always toward the center of the circle.)

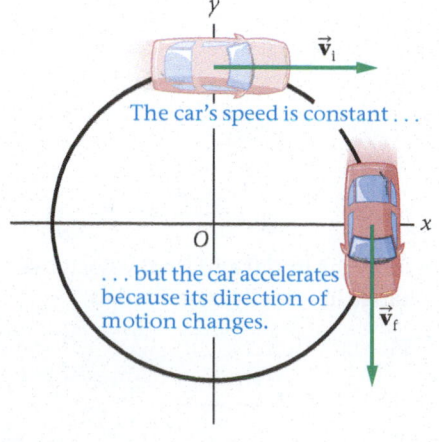

The car's speed is constant . . .

. . . but the car accelerates because its direction of motion changes.

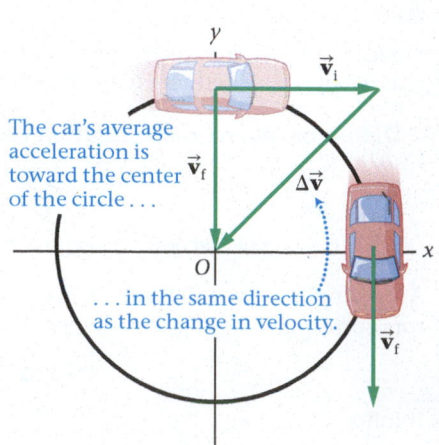

The car's average acceleration is toward the center of the circle . . .

. . . in the same direction as the change in velocity.

QUICK EXAMPLE 3-9   FIND THE AVERAGE ACCELERATION

A car is traveling with an initial velocity of $\vec{\mathbf{v}}_i = (-6.36 \text{ m/s})\hat{\mathbf{x}} + (6.36 \text{ m/s})\hat{\mathbf{y}}$. Eight seconds later, after following a bend in the road, its final velocity is $\vec{\mathbf{v}}_f = (15.00 \text{ m/s})\hat{\mathbf{y}}$. What are the magnitude and direction of its average acceleration during those 8.00 seconds?

REASONING AND SOLUTION

Given the initial and final velocities, we can find the corresponding change in velocity, $\Delta\vec{\mathbf{v}} = \vec{\mathbf{v}}_f - \vec{\mathbf{v}}_i$. Dividing $\Delta\vec{\mathbf{v}}$ by the time $\Delta t = 8.00$ s yields the average acceleration vector. We can find its magnitude and direction using the Pythagorean theorem and the inverse tangent, respectively.

1. Calculate the change in velocity, $\Delta\vec{\mathbf{v}}$:

$$\begin{aligned}\Delta\vec{\mathbf{v}} &= \vec{\mathbf{v}}_f - \vec{\mathbf{v}}_i \\ &= (15.00 \text{ m/s})\hat{\mathbf{y}} - [(-6.36 \text{ m/s})\hat{\mathbf{x}} + (6.36 \text{ m/s})\hat{\mathbf{y}}] \\ &= (6.36 \text{ m/s})\hat{\mathbf{x}} + (8.64 \text{ m/s})\hat{\mathbf{y}}\end{aligned}$$

2. Divide the change in velocity by the time to determine the average acceleration, $\vec{\mathbf{a}}_{av}$:

$$\begin{aligned}\vec{\mathbf{a}}_{av} &= \frac{\Delta\vec{\mathbf{v}}}{\Delta t} = \frac{(6.36 \text{ m/s})\hat{\mathbf{x}} + (8.64 \text{ m/s})\hat{\mathbf{y}}}{(8.00 \text{ s})} \\ &= (0.795 \text{ m/s}^2)\hat{\mathbf{x}} + (1.08 \text{ m/s}^2)\hat{\mathbf{y}}\end{aligned}$$

3. Find the magnitude, a_{av}, and direction angle, θ, of the average acceleration:

$$\begin{aligned}a_{av} &= \sqrt{(0.795 \text{ m/s}^2)^2 + (1.08 \text{ m/s}^2)^2} \\ &= 1.34 \text{ m/s}^2\end{aligned}$$

$$\theta = \tan^{-1}\left(\frac{1.08 \text{ m/s}^2}{0.795 \text{ m/s}^2}\right)$$

$$= 53.6° \quad \text{(above the positive } x \text{ axis)}$$

It's important to recognize the following distinctions between the velocity vector and the acceleration vector:

- The velocity vector, $\vec{\mathbf{v}}$, always points in the direction of a particle's motion.

- The acceleration vector, $\vec{\mathbf{a}}$, can point in directions other than the direction of motion, and in general it does.

In one dimension, the acceleration is either in the same direction as the velocity or in the opposite direction. In two dimensions, the acceleration can be in any direction relative to the velocity.

An example of a particle's motion, showing the velocity and acceleration vectors at various times, is presented in **FIGURE 3-28**. Notice that in all cases the velocity is tangential to the motion, though the acceleration points in various directions. When the acceleration is perpendicular to the velocity of an object, as at points (2) and (3) in Figure 3-28, the speed of the object remains constant while its direction of motion changes. At points (1) and (4) in Figure 3-28 the acceleration is antiparallel (opposite) or parallel to the velocity of the object, respectively. In such cases, the direction

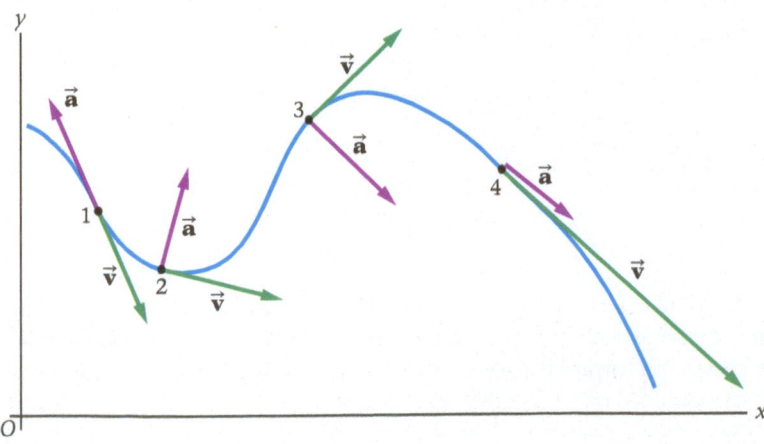

◄ **FIGURE 3-28 Velocity and acceleration vectors for a particle moving along a winding path** The acceleration of a particle need not point in the direction of motion. At point (1) the particle is slowing down, at (2) it is turning to the left, at (3) it is turning to the right, and, finally, at point (4) it is speeding up.

▶ **FIGURE 3-29** The velocities of these cyclists change in both magnitude and direction as they slow to negotiate a series of sharp curves and then speed up again. Both kinds of velocity change involve an acceleration.

Big Idea 5 Position, displacement, velocity, and acceleration are all vectors; each has a specific direction and magnitude.

PHYSICS IN CONTEXT
Looking Ahead

The vectors developed in this section play a key role in Chapter 4, where we study two-dimensional kinematics. In particular, vectors and their components will allow us to analyze two-dimensional motion as a combination of two completely independent one-dimensional motions.

of motion remains the same while the speed changes. A similar real-world example is provided in **FIGURE 3-29**. Throughout the next chapter we shall see further examples of motion in which the velocity and acceleration are in different directions.

Enhance Your Understanding   (Answers given at the end of the chapter)

5. An object moves along the brown path in **FIGURE 3-30** in the direction indicated. Which physical quantity (position, acceleration, velocity) is represented by the following vectors: (a) $\vec{A}$, (b) $\vec{B}$, (c) $\vec{C}$?

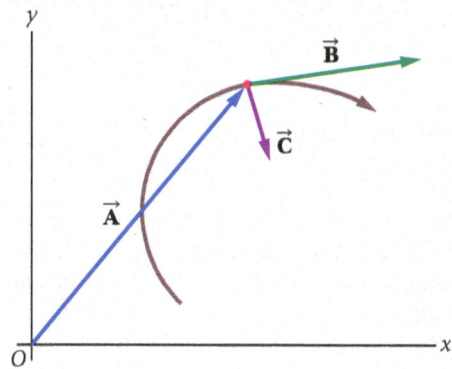

▲ **FIGURE 3-30** Enhance Your Understanding 5

Section Review

- Many physical quantities are represented by vectors. Among these are position, displacement, velocity, and acceleration.

3-6 Relative Motion

A good example of the use of vectors is in the description of relative motion. Suppose, for example, that you are standing on the ground as a train goes by at 15.0 m/s, as shown in **FIGURE 3-31**. Inside the train, a person is walking in the forward direction at 1.2 m/s relative to the train. How fast is the person moving relative to you? Clearly, the answer is 1.2 m/s + 15.0 m/s = 16.2 m/s. What if the person had been walking with the same speed, but toward the back of the train? In this case, you would see the person going by with a speed of −1.2 m/s + 15.0 m/s = 13.8 m/s.

Let's generalize these results. Call the velocity of the *train* relative to the *ground* $\vec{v}_{tg}$, the velocity of the *passenger* relative to the *train* $\vec{v}_{pt}$, and the velocity of the *passenger*

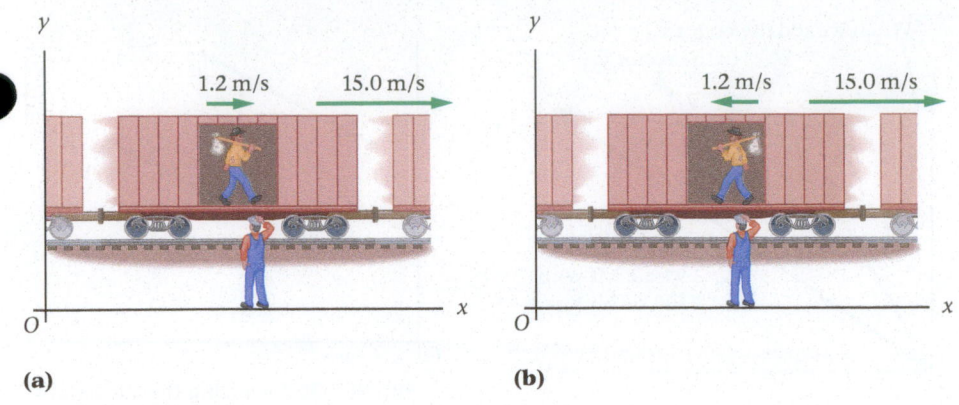

(a)

(b)

◄ **FIGURE 3-31 Relative velocity of a passenger on a train with respect to a person on the ground (a)** The passenger walks toward the front of the train. **(b)** The passenger walks toward the rear of the train.

relative to the ground $\vec{\mathbf{v}}_{pg}$. As we saw in the previous paragraph, the velocity of the passenger relative to the ground is

$$\vec{\mathbf{v}}_{pg} = \vec{\mathbf{v}}_{pt} + \vec{\mathbf{v}}_{tg} \qquad 3\text{-}7$$

This vector addition is illustrated in **FIGURE 3-32** for the two cases we discussed.

Though this example dealt with one-dimensional motion, Equation 3-7 is valid for velocity vectors pointing in any directions. For example, instead of walking on the car's floor, the person might be climbing a ladder to the roof of the car, as in **FIGURE 3-33**. In this case $\vec{\mathbf{v}}_{pt}$ is vertical, $\vec{\mathbf{v}}_{tg}$ is horizontal, and $\vec{\mathbf{v}}_{pg}$ is simply the vector sum $\vec{\mathbf{v}}_{pt} + \vec{\mathbf{v}}_{tg}$.

Big Idea 6 Relative motion is a simple application of vector addition and subtraction.

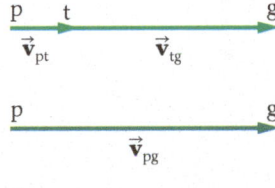

(a)

QUICK EXAMPLE 3-10 RELATIVE MOTION

Suppose the person in Figure 3-33 is climbing a vertical ladder with a speed of 0.24 m/s, and the train is slowly coasting forward at 0.68 m/s. Find the speed and direction of motion of the person relative to the ground.

REASONING AND SOLUTION

The velocity of the person relative to the ground (pg) is equal to the velocity of the person relative to the train (pt) plus the velocity of the train relative to the ground (tg). The velocity of the person relative to the train is in the positive y direction, $\vec{\mathbf{v}}_{pt} = (0.24\,\text{m/s})\hat{\mathbf{y}}$, and the velocity of the train relative to the ground is in the positive x direction, $\vec{\mathbf{v}}_{tg} = (0.68\,\text{m/s})\hat{\mathbf{x}}$. Thus, the desired velocity is

$$\vec{\mathbf{v}}_{pg} = \vec{\mathbf{v}}_{pt} + \vec{\mathbf{v}}_{tg}$$
$$= (0.24\,\text{m/s})\hat{\mathbf{y}} + (0.68\,\text{m/s})\hat{\mathbf{x}}$$

In the more usual order of writing things, with the x component first, we have

$$\vec{\mathbf{v}}_{pg} = (0.68\,\text{m/s})\hat{\mathbf{x}} + (0.24\,\text{m/s})\hat{\mathbf{y}}$$

It follows that $v_{pg} = \sqrt{(0.68\,\text{m/s})^2 + (0.24\,\text{m/s})^2} = 0.72\,\text{m/s}$ and $\theta = \tan^{-1}(0.24/0.68) = 19°$.

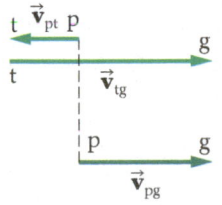

(b)

▲ **FIGURE 3-32 Adding velocity vectors** Vector addition to find the velocity of the passenger with respect to the ground for **(a)** Figure 3-31 (a) and **(b)** Figure 3-31 (b).

◄ **FIGURE 3-33 Relative velocity in two dimensions** A person climbs up a ladder on a moving train with velocity $\vec{\mathbf{v}}_{pt}$ relative to the train. If the train moves relative to the ground with a velocity $\vec{\mathbf{v}}_{tg}$, the velocity of the person on the train relative to the ground is $\vec{\mathbf{v}}_{pg} = \vec{\mathbf{v}}_{pt} + \vec{\mathbf{v}}_{tg}$.

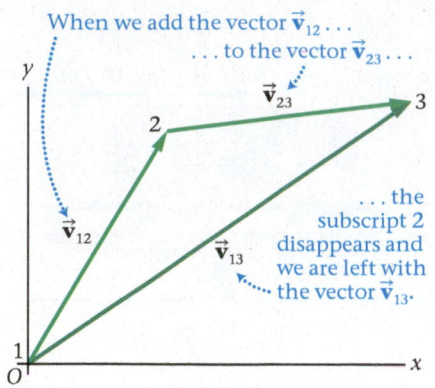

▲ FIGURE 3-34 **Vector addition used to determine relative velocity**

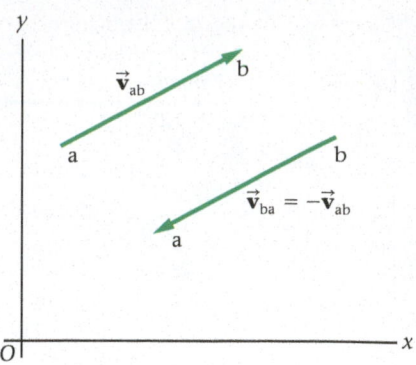

▲ FIGURE 3-35 **Reversing the subscripts of a velocity reverses the corresponding velocity vector**

Subscripts Are the Key to Relative Motion Notice that the subscripts in Equation 3-7 and Exercise 3-10 follow a definite pattern. On the left-hand side of the equation we have the subscripts pg. On the right-hand side we have two sets of subscripts, pt and tg; notice that a pair of t's has been inserted between the p and the g. This pattern always holds for any relative-motion problem, though the subscripts will be different when referring to different objects. Thus, we can say quite generally that

$$\vec{\mathbf{v}}_{13} = \vec{\mathbf{v}}_{12} + \vec{\mathbf{v}}_{23} \qquad \text{3-8}$$

In the train example, we can identify 1 as the *passenger*, 2 as the *train*, and 3 as the *ground*.

The vector addition in Equation 3-8 is shown in **FIGURE 3-34**. For convenience in seeing how the subscripts are ordered in the equation, we have labeled the tail of each vector with its first subscript and the head of each vector with its second subscript.

One final note about velocities and their subscripts: Reversing the subscripts reverses the velocity. This is indicated in **FIGURE 3-35**, where we see that

$$\vec{\mathbf{v}}_{ba} = -\vec{\mathbf{v}}_{ab}$$

Physically, what we are saying is that if you are riding in a car due *north* at 20 m/s relative to the ground, then the ground, relative to you, is moving due *south* at 20 m/s.

Let's apply these results to a real-world, two-dimensional example.

EXAMPLE 3-11 CROSSING A RIVER

RWP You ride in a boat whose speed relative to the water is 6.1 m/s. The boat points at an angle of 25° upstream on a river flowing at 1.4 m/s. **(a)** What is the velocity of the boat relative to the ground? **(b)** Suppose the speed of the boat relative to the water remains the same, but the direction in which it points is changed. What angle is required for the boat to go straight across the river—that is, for the boat to move perpendicular to the shores of the river?

PICTURE THE PROBLEM
We choose the x axis to be perpendicular to the river, and the y axis to point upstream. With these choices the velocity of the water relative to the ground has a magnitude of 1.4 m/s and points in the negative y direction. In part (a), the velocity of the boat relative to the water has a direction angle of $\theta = 25°$ above the x axis. In part (b), the angle θ is to be determined.

REASONING AND STRATEGY
If the water were still, the boat would move in the direction in which it is pointed. With the water flowing downstream, as shown, the boat will move in a direction closer to the x axis. **(a)** To find the velocity of the boat we use $\vec{\mathbf{v}}_{13} = \vec{\mathbf{v}}_{12} + \vec{\mathbf{v}}_{23}$ with 1 referring to the boat (b), 2 referring to the water (w), and 3 referring to the ground (g). **(b)** To go "straight across the river" means that the velocity of the boat relative to the ground should be in the x direction. Thus, we choose the angle θ that cancels the y component of velocity.

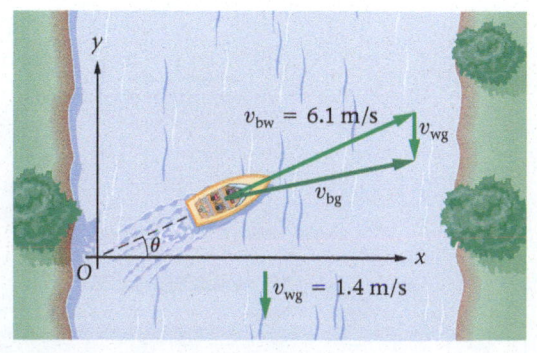

CONTINUED

Known   The speed of the boat relative to the water, $v_{bw} = 6.1$ m/s; the angle of the boat above the x axis, $\theta = 25°$; the speed of the water relative to the ground, $v_{wg} = 1.4$ m/s.

Unknown   (a) Velocity of the boat relative to the ground, $\vec{\mathbf{v}}_{bg} = ?$
(b) Angle of motion to go straight across the river, $\theta = ?$

SOLUTION

Part (a)

1. Rewrite $\vec{\mathbf{v}}_{13} = \vec{\mathbf{v}}_{12} + \vec{\mathbf{v}}_{23}$ with $1 \rightarrow b$, $2 \rightarrow w$, and $3 \rightarrow g$:

$$\vec{\mathbf{v}}_{bg} = \vec{\mathbf{v}}_{bw} + \vec{\mathbf{v}}_{wg}$$

2. From our sketch we see that the water flows at 1.4 m/s in the negative y direction relative to the ground:

$$\vec{\mathbf{v}}_{wg} = (-1.4 \text{ m/s})\hat{\mathbf{y}}$$

3. The velocity of the boat relative to the water is given in the problem statement:

$$\vec{\mathbf{v}}_{bw} = (6.1 \text{ m/s}) \cos 25° \, \hat{\mathbf{x}} + (6.1 \text{ m/s}) \sin 25° \, \hat{\mathbf{y}}$$
$$= (5.5 \text{ m/s})\hat{\mathbf{x}} + (2.6 \text{ m/s})\hat{\mathbf{y}}$$

4. Carry out the vector sum in Step 1 to find $\vec{\mathbf{v}}_{bg}$:

$$\vec{\mathbf{v}}_{bg} = (5.5 \text{ m/s})\hat{\mathbf{x}} + (2.6 \text{ m/s} - 1.4 \text{ m/s})\hat{\mathbf{y}}$$
$$= (5.5 \text{ m/s})\hat{\mathbf{x}} + (1.2 \text{ m/s})\hat{\mathbf{y}}$$

Part (b)

5. To cancel the y component of $\vec{\mathbf{v}}_{bg}$, we choose the angle that gives 1.4 m/s for the y component of $\vec{\mathbf{v}}_{bw}$:

$$(6.1 \text{ m/s}) \sin \theta = 1.4 \text{ m/s}$$

6. Solve for θ. With this angle, we see that the y component of $\vec{\mathbf{v}}_{bg}$ in Step 4 will be zero:

$$\theta = \sin^{-1}(1.4/6.1) = 13°$$

INSIGHT

(a) The speed of the boat relative to the ground is $\sqrt{(5.5 \text{ m/s})^2 + (1.2 \text{ m/s})^2} = 5.6$ m/s, and the direction angle of its motion is $\theta = \tan^{-1}(1.2/5.5) = 12°$ upstream. (b) The speed of the boat relative to the ground in this case is equal to the x component of its velocity, because the y component is zero. Therefore, its speed is $(6.1 \text{ m/s}) \cos 13° = 5.9$ m/s.

PRACTICE PROBLEM

Find the speed and direction of the boat relative to the ground in part (a) of this Example if the river flows at 4.5 m/s. Everything else remains the same.   [**Answer:** $v_{bg} = 5.8$ m/s, $\theta = -19°$. In this case, a person on the ground sees the boat going slowly downstream, even though the boat itself points upstream.]

Some related homework problems: Problem 50, Problem 55, Problem 57

Enhance Your Understanding   (Answers given at the end of the chapter)

6. Suppose the speed of the boat is increased in part (a) of Example 3-11, while everything else remains the same. In this case, is the angle of the boat's path across the river greater than, less than, or equal to 12°? Explain.

Section Review

- Relative motion is analyzed with vector addition. For example, the velocity of object 1 relative to object 3 is $\vec{\mathbf{v}}_{13} = \vec{\mathbf{v}}_{12} + \vec{\mathbf{v}}_{23}$, where object 2 can be any other object.

CHAPTER 3 REVIEW

CHAPTER SUMMARY

3-1   SCALARS VERSUS VECTORS

Scalar
A number with appropriate units. Examples of scalar quantities include time and length.

Vector
A quantity with both a magnitude and a direction. Examples include displacement, velocity, and acceleration.

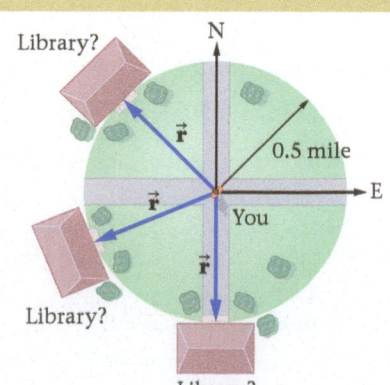

3-2 THE COMPONENTS OF A VECTOR

x Component of Vector $\vec{A}$
$A_x = A\cos\theta$, where θ is measured relative to the x axis.

y Component of Vector $\vec{A}$
$A_y = A\sin\theta$, where θ is measured relative to the x axis.

Magnitude of Vector $\vec{A}$
The magnitude of $\vec{A}$ is $A = \sqrt{A_x{}^2 + A_y{}^2}$.

Direction Angle of Vector $\vec{A}$
The direction angle of $\vec{A}$ is $\theta = \tan^{-1}(A_y/A_x)$, where θ is measured relative to the x axis.

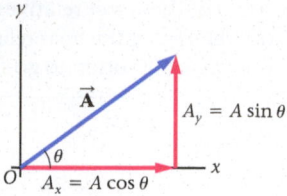

3-3 ADDING AND SUBTRACTING VECTORS

Graphical Method
To add $\vec{A}$ and $\vec{B}$, place them so that the tail of $\vec{B}$ is at the head of $\vec{A}$. The sum $\vec{C} = \vec{A} + \vec{B}$ is the arrow from the tail of $\vec{A}$ to the head of $\vec{B}$.

Component Method
If $\vec{C} = \vec{A} + \vec{B}$, then $C_x = A_x + B_x$ and $C_y = A_y + B_y$. If $\vec{C} = \vec{A} - \vec{B}$, then $C_x = A_x - B_x$ and $C_y = A_y - B_y$.

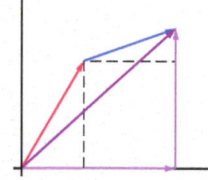

3-4 UNIT VECTORS

x Unit Vector
Written $\hat{\mathbf{x}}$, the x unit vector is a dimensionless vector of unit length in the positive x direction.

y Unit Vector
Written $\hat{\mathbf{y}}$, the y unit vector is a dimensionless vector of unit length in the positive y direction.

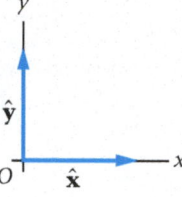

3-5 POSITION, DISPLACEMENT, VELOCITY, AND ACCELERATION VECTORS

Position Vector
The position vector $\vec{r}$ points from the origin to a particle's location.

Displacement Vector
The displacement vector $\Delta\vec{r}$ is the change in position; $\Delta\vec{r} = \vec{r}_f - \vec{r}_i$.

Velocity Vector
The velocity vector $\vec{v}$ points in the direction of motion and has a magnitude equal to the speed.

Acceleration Vector
The acceleration vector $\vec{a}$ indicates how quickly and in what direction the velocity is changing. It need not point in the direction of motion.

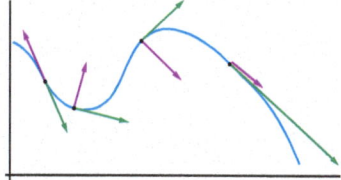

3-6 RELATIVE MOTION

Velocity of Object 1 Relative to Object 3
$\vec{v}_{13} = \vec{v}_{12} + \vec{v}_{23}$, where object 2 can be anything.

Reversing the Subscripts on a Velocity
$\vec{v}_{12} = -\vec{v}_{21}$.

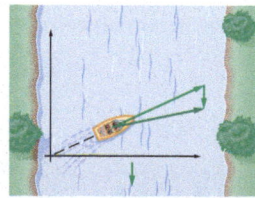

ANSWERS TO ENHANCE YOUR UNDERSTANDING QUESTIONS

1. Yes, if the vectors point in different directions.

2. **(a)** Less than 8 m because the x and y components are at right angles to one another. **(b)** Less than 45° because the vertical side of the corresponding triangle is shorter than the horizontal side.

3. **(a)** $\vec{F}$, **(b)** $\vec{H}$, **(c)** $\vec{B}$, **(d)** $\vec{G}$.

4. **(a)** $\vec{B}$, **(b)** $\vec{C}$, **(c)** $\vec{A}$, **(d)** $\vec{D}$.

5. **(a)** Position. **(b)** Velocity. **(c)** Acceleration.

6. Greater than 12°. The angle will be closer to the boat's heading of 25° because as the boat speeds up the river has less effect.

CONCEPTUAL QUESTIONS

For instructor-assigned homework, go to www.masteringphysics.com. (MP)

(Answers to odd-numbered Conceptual Questions can be found in the back of the book.)

1. For the following quantities, indicate which is a scalar and which is a vector: **(a)** the time it takes for you to run the 100-yard dash; **(b)** your displacement after running the 100-yard dash; **(c)** your average velocity while running; **(d)** your average speed while running.

2. Which, if any, of the vectors shown in **FIGURE 3-36** are equal?

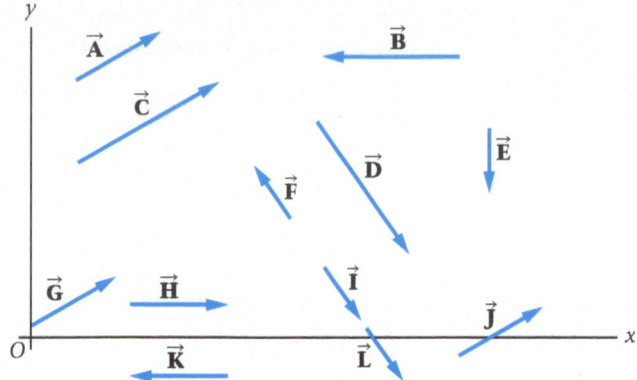

FIGURE 3-36 Conceptual Question 2

3. Given that $\vec{A} + \vec{B} = 0$, **(a)** how does the magnitude of $\vec{B}$ compare with the magnitude of $\vec{A}$? **(b)** How does the direction of $\vec{B}$ compare with the direction of $\vec{A}$?

4. Can a component of a vector be greater than the vector's magnitude?

5. Suppose that $\vec{A}$ and $\vec{B}$ have nonzero magnitude. Is it possible for $\vec{A} + \vec{B}$ to be zero?

6. Can a vector with zero magnitude have one or more components that are nonzero? Explain.

7. Given that $\vec{A} + \vec{B} = \vec{C}$, and that $A^2 + B^2 = C^2$, how are $\vec{A}$ and $\vec{B}$ oriented relative to one another?

8. Given that $\vec{A} + \vec{B} = \vec{C}$, and that $A + B = C$, how are $\vec{A}$ and $\vec{B}$ oriented relative to one another?

9. Given that $\vec{A} + \vec{B} = \vec{C}$, and that $A - B = C$, how are $\vec{A}$ and $\vec{B}$ oriented relative to one another?

10. Vector $\vec{A}$ has x and y components of equal magnitude. What can you say about the possible directions of $\vec{A}$?

11. The components of a vector $\vec{A}$ satisfy the relation $A_x = -A_y \neq 0$. What are the possible directions of $\vec{A}$?

12. Use a sketch to show that two vectors of unequal magnitude cannot add to zero, but that three vectors of unequal magnitude can.

13. Rain is falling vertically downward and you are running for shelter. To keep driest, should you hold your umbrella vertically, tilted forward, or tilted backward? Explain.

14. When sailing, the wind feels stronger when you sail upwind ("beating") than when you are sailing downwind ("running"). Explain.

PROBLEMS AND CONCEPTUAL EXERCISES

Answers to odd-numbered Problems and Conceptual Exercises can be found in the back of the book. **BIO** identifies problems of biological or medical interest; **CE** indicates a conceptual exercise. **Predict/Explain** problems ask for two responses: (a) your prediction of a physical outcome, and (b) the best explanation among three provided; and **Predict/Calculate** problems ask for a prediction of a physical outcome, based on fundamental physics concepts, and follow that with a numerical calculation to verify the prediction. On all problems, bullets (•, ••, •••) indicate the level of difficulty.

SECTION 3-2 THE COMPONENTS OF A VECTOR

1. • **CE** Suppose that each component of a certain vector is doubled. **(a)** By what multiplicative factor does the magnitude of the vector change? **(b)** By what multiplicative factor does the direction angle of the vector change?

2. • **CE** Rank the vectors in **FIGURE 3-37** in order of increasing magnitude.

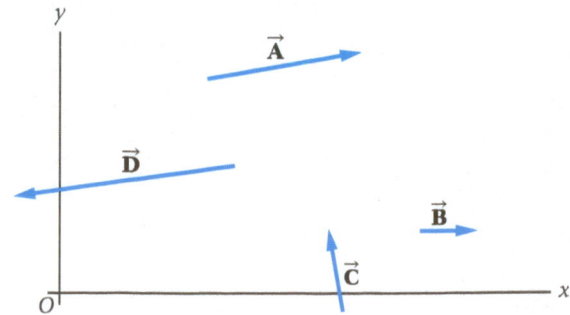

FIGURE 3-37 Problems 2, 3, and 4

3. • **CE** Rank the vectors in Figure 3-37 in order of increasing value of their x component.

4. • **CE** Rank the vectors in Figure 3-37 in order of increasing value of their y component.

5. • The press box at a baseball park is 44.5 ft above the ground. A reporter in the press box looks at an angle of 13.4° below the horizontal to see second base. What is the horizontal distance from the press box to second base?

6. • You are driving up a long, inclined road. After 1.2 miles you notice that signs along the roadside indicate that your elevation has increased by 530 ft. **(a)** What is the angle of the road above the horizontal? **(b)** How far do you have to drive to gain an additional 150 ft of elevation?

7. • **A One-Percent Grade** A road that rises 1 ft for every 100 ft traveled horizontally is said to have a 1% grade. Portions of the Lewiston grade, near Lewiston, Idaho, have a 6% grade. At what angle is this road inclined above the horizontal?

8. • You walk in a straight line for 95 m at an angle of 162° above the positive x axis. What are the x and y components of your displacement?

9. •• Find the x and y components of a position vector $\vec{r}$ of magnitude $r = 88$ m, if its angle relative to the x axis is **(a)** 32.0° and **(b)** 64.0°.

10. •• A vector has the components $A_x = 22$ m and $A_y = 13$ m. **(a)** What is the magnitude of this vector? **(b)** What angle does this vector make with the positive x axis?

11. •• A vector has the components $A_x = -36$ m and $A_y = 43$ m. **(a)** What is the magnitude of this vector? **(b)** What angle does this vector make with the positive x axis?

12. •• A baseball "diamond" (**FIGURE 3-38**) is a square with sides 90 ft in length. If the positive x axis points from home plate to first base, and the positive y axis points from home plate to third base, find the displacement vector of a base runner who has just hit **(a)** a double, **(b)** a triple, or **(c)** a home run.

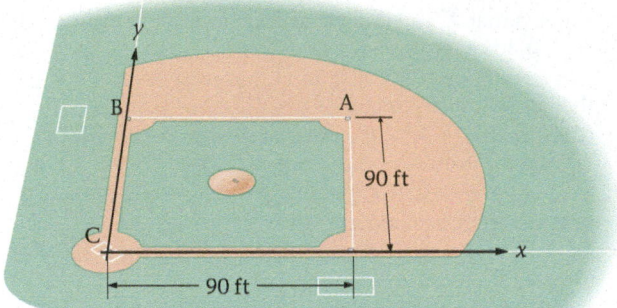

FIGURE 3-38 Problem 12

13. •• A lighthouse that rises 49 ft above the surface of the water sits on a rocky cliff that extends 19 ft from its base, as shown in **FIGURE 3-39**. A sailor on the deck of a ship sights the top of the lighthouse at an angle of 30.0° above the horizontal. If the sailor's eye level is 14 ft above the water, how far is the ship from the rocks?

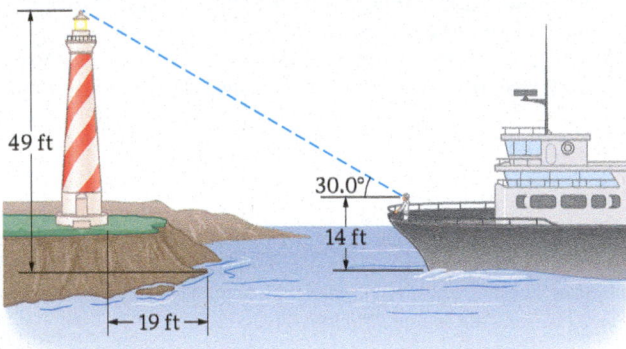

FIGURE 3-39 Problem 13

14. •• **H₂O** A water molecule is shown schematically in **FIGURE 3-40**. The distance from the center of the oxygen atom to the center of a hydrogen atom is 0.96 Å, and the angle between the hydrogen atoms is 104.5°. Find the center-to-center distance between the hydrogen atoms. (1 Å = 10^{-10} m.)

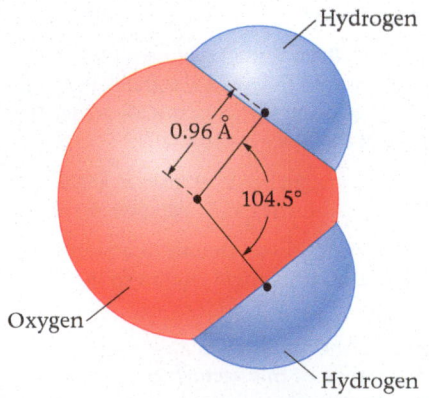

FIGURE 3-40 Problem 14

15. •• **Predict/Calculate** The x and y components of a vector $\vec{r}$ are $r_x = 14$ m and $r_y = -9.5$ m, respectively. Find **(a)** the direction and **(b)** the magnitude of the vector $\vec{r}$. **(c)** If both r_x and r_y are doubled,

how do you predict your answers to parts (a) and (b) will change? **(d)** Verify your prediction in part (c) by calculating the magnitude and direction for this new case.

16. •• You drive a car 680 ft to the east, then 340 ft to the north. **(a)** What is the magnitude of your displacement? **(b)** Using a sketch, estimate the direction of your displacement. **(c)** Verify your estimate in part (b) with a numerical calculation of the direction.

17. •• Vector $\vec{A}$ has a magnitude of 50 units and points in the positive x direction. A second vector, $\vec{B}$, has a magnitude of 120 units and points at an angle of 70° below the x axis. Which vector has **(a)** the greater x component, and **(b)** the greater y component?

18. •• A treasure map directs you to start at a palm tree and walk due north for 15.0 m. You are then to turn 90° and walk 22.0 m; then turn 90° again and walk 5.00 m. Give the distance from the palm tree, and the direction relative to north, for each of the four possible locations of the treasure.

19. •• A whale comes to the surface to breathe and then dives at an angle of 20.0° below the horizontal (**FIGURE 3-41**). If the whale continues in a straight line for 215 m, **(a)** how deep is it, and **(b)** how far has it traveled horizontally?

FIGURE 3-41 Problem 19

SECTION 3-3   ADDING AND SUBTRACTING VECTORS

20. • **CE** Consider the vectors $\vec{A}$ and $\vec{B}$ shown in **FIGURE 3-42**. Which of the other four vectors in the figure ($\vec{C}$, $\vec{D}$, $\vec{E}$, and $\vec{F}$) best represents the *direction* of **(a)** $\vec{A} + \vec{B}$, **(b)** $\vec{A} - \vec{B}$, and **(c)** $\vec{B} - \vec{A}$?

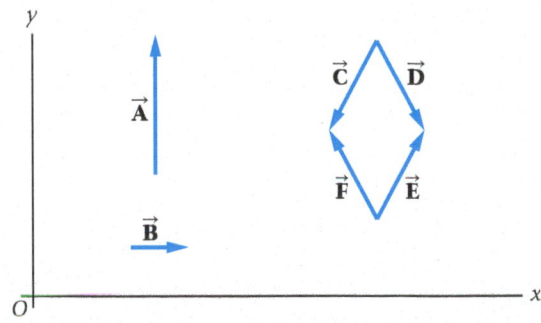

FIGURE 3-42 Problems 20 and 21

21. • **CE** Refer to Figure 3-42 for the following questions: **(a)** Is the magnitude of $\vec{A} + \vec{D}$ greater than, less than, or equal to the magnitude of $\vec{A} + \vec{E}$? **(b)** Is the magnitude of $\vec{A} + \vec{E}$ greater than, less than, or equal to the magnitude of $\vec{A} + \vec{F}$?

22. •• A vector $\vec{A}$ has a magnitude of 40.0 m and points in a direction 20.0° below the positive x axis. A second vector, $\vec{B}$, has a magnitude of 75.0 m and points in a direction 50.0° above the positive x axis. **(a)** Sketch the vectors $\vec{A}$, $\vec{B}$, and $\vec{C} = \vec{A} + \vec{B}$. **(b)** Using the component method of vector addition, find the magnitude and direction of the vector $\vec{C}$.

23. •• An air traffic controller observes two airplanes approaching the airport. The displacement from the control tower to plane 1 is given by the vector $\vec{A}$, which has a magnitude of 220 km and points in a direction 32° north of west. The displacement from the control tower to plane 2 is given by the vector $\vec{B}$, which has a magnitude of 140 km and points 65° east of north. (a) Sketch the vectors $\vec{A}$, $-\vec{B}$, and $\vec{D} = \vec{A} - \vec{B}$. Notice that $\vec{D}$ is the displacement from plane 2 to plane 1. (b) Find the magnitude and direction of the vector $\vec{D}$.

24. •• The initial velocity of a car, $\vec{v}_i$, is 45 km/h in the positive x direction. The final velocity of the car, $\vec{v}_f$, is 66 km/h in a direction that points 75° above the positive x axis. (a) Sketch the vectors $-\vec{v}_i$, $\vec{v}_f$, and $\Delta\vec{v} = \vec{v}_f - \vec{v}_i$. (b) Find the magnitude and direction of the change in velocity, $\Delta\vec{v}$.

25. •• Vector $\vec{A}$ points in the positive x direction and has a magnitude of 75 m. The vector $\vec{C} = \vec{A} + \vec{B}$ points in the positive y direction and has a magnitude of 95 m. (a) Sketch $\vec{A}$, $\vec{B}$, and $\vec{C}$. (b) Estimate the magnitude and direction of the vector $\vec{B}$. (c) Verify your estimate in part (b) with a numerical calculation.

26. •• Vector $\vec{A}$ points in the negative x direction and has a magnitude of 22 units. The vector $\vec{B}$ points in the positive y direction. (a) Find the magnitude of $\vec{B}$ if $\vec{A} + \vec{B}$ has a magnitude of 37 units. (b) Sketch $\vec{A}$ and $\vec{B}$.

27. •• Vector $\vec{A}$ points in the negative y direction and has a magnitude of 5 units. Vector $\vec{B}$ has twice the magnitude and points in the positive x direction. Find the direction and magnitude of (a) $\vec{A} + \vec{B}$, (b) $\vec{A} - \vec{B}$, and (c) $\vec{B} - \vec{A}$.

28. •• A basketball player runs down the court, following the path indicated by the vectors $\vec{A}$, $\vec{B}$, and $\vec{C}$ in **FIGURE 3-43**. The magnitudes of these three vectors are $A = 10.0$ m, $B = 20.0$ m, and $C = 7.0$ m. Find the magnitude and direction of the net displacement of the player using (a) the graphical method and (b) the component method of vector addition. Compare your results.

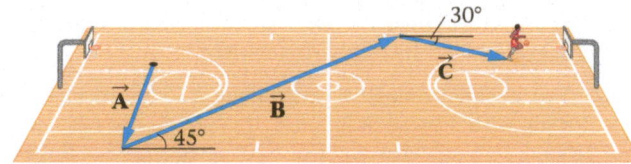

FIGURE 3-43 Problem 28

SECTION 3-4   UNIT VECTORS

29. • A particle undergoes a displacement $\Delta\vec{r}$ of magnitude 54 m in a direction 42° below the x axis. Express $\Delta\vec{r}$ in terms of the unit vectors $\hat{x}$ and $\hat{y}$.

30. • A vector has a magnitude of 3.50 m and points in a direction that is 145° counterclockwise from the x axis. Find the x and y components of this vector.

31. • A vector $\vec{A}$ has a length of 6.1 m and points in the negative x direction. Find (a) the x component and (b) the magnitude of the vector $-3.7\,\vec{A}$.

32. • The vector $-5.2\,\vec{A}$ has a magnitude of 34 m and points in the positive x direction. Find (a) the x component and (b) the magnitude of the vector $\vec{A}$.

33. • Find the direction and magnitude of the vectors.
 (a) $\vec{A} = (5.0\,\text{m})\hat{x} + (-2.0\,\text{m})\hat{y}$,
 (b) $\vec{B} = (-2.0\,\text{m})\hat{x} + (5.0\,\text{m})\hat{y}$, and (c) $\vec{A} + \vec{B}$.

34. • Find the direction and magnitude of the vectors.
 (a) $\vec{A} = (22\,\text{m})\hat{x} + (-14\,\text{m})\hat{y}$,
 (b) $\vec{B} = (2.5\,\text{m})\hat{x} + (13\,\text{m})\hat{y}$, and (c) $\vec{A} + \vec{B}$.

35. • For the vectors given in Problem 34, express (a) $\vec{A} - \vec{B}$ and (b) $\vec{B} - \vec{A}$ in unit vector notation.

36. • Express each of the vectors in **FIGURE 3-44** in unit vector notation.

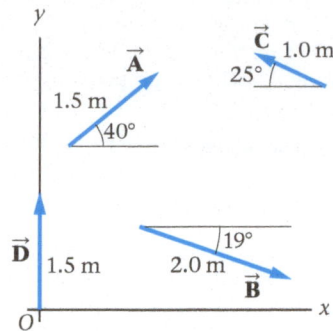

FIGURE 3-44 Problems 36 and 37

37. •• Referring to the vectors in Figure 3-44, express the sum $\vec{A} + \vec{B} + \vec{C}$ in unit vector notation.

SECTION 3-5   POSITION, DISPLACEMENT, VELOCITY, AND ACCELERATION VECTORS

38. • **CE** The blue curves shown in **FIGURE 3-45** display the constant-speed motion of two different particles in the x-y plane. For each of the eight vectors in Figure 3-45, state whether it is (a) a position vector, (b) a velocity vector, or (c) an acceleration vector for the particles.

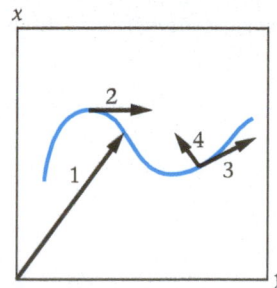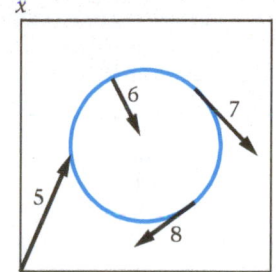

FIGURE 3-45 Problem 38

39. • What are the direction and magnitude of your total displacement if you have traveled due west with a speed of 23 m/s for 175 s, then due south at 12 m/s for 285 s?

40. •• **Predict/Calculate** **Moving the Knight** Two of the allowed chess moves for a knight are shown in **FIGURE 3-46**. (a) Is the magnitude of displacement 1 greater than, less than, or equal to the magnitude of displacement 2? Explain. (b) Find the magnitude and direction of the knight's displacement for each of the two moves. Assume that the checkerboard squares are 3.5 cm on a side.

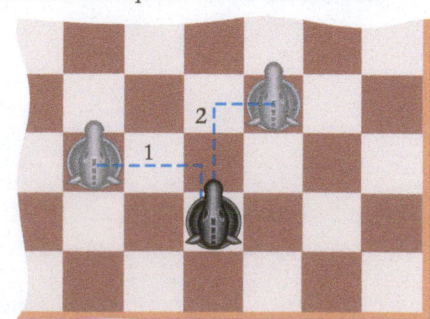

FIGURE 3-46 Problem 40

41. •• To visit your favorite ice cream shop, you must travel 490 m west on Main Street and then 950 m south on Division Street. **(a)** Find the total distance you traveled. **(b)** Write your displacement vector in unit vector notation, taking the $\hat{\mathbf{x}}$ direction to be east and the $\hat{\mathbf{y}}$ direction to be north. **(c)** Write the displacement vector required to return to your starting point in unit vector notation.

42. •• Referring to Problem 41, suppose you take 44 s to complete the 490-m displacement and 73 s to complete the 950-m displacement. **(a)** What are the magnitude and direction of your average velocity during this 117-second period of time? **(b)** What is your average speed during the trip?

43. •• You drive a car 1500 ft to the east, then 2500 ft to the north. If the trip took 3.0 minutes, what were the direction and magnitude of your average velocity?

44. •• Predict/Calculate A jogger runs with a speed of 3.25 m/s in a direction 30.0° above the x axis. **(a)** Find the x and y components of the jogger's velocity. **(b)** How will the velocity components found in part (a) change if the jogger's speed is halved? **(c)** Calculate the components of the velocity for the case where the jogger's speed is halved.

45. •• You throw a ball upward with an initial speed of 4.5 m/s. When it returns to your hand 0.92 s later, it has the same speed in the downward direction (assuming air resistance can be ignored). What was the average acceleration vector of the ball?

46. •• Consider a skateboarder who starts from rest at the top of a ramp that is inclined at an angle of 17.5° to the horizontal. Assuming that the skateboarder's acceleration is $g \sin 17.5°$, find his speed when he reaches the bottom of the ramp in 3.25 s.

47. ••• In a soccer game a midfielder kicks the ball from one touchline directly toward the other with a velocity of $(14.0 \text{ m/s}) \hat{\mathbf{x}}$. After the ball travels for 2.50 s, a striker makes an acrobatic kick that gives the ball a new velocity in the $\hat{\mathbf{y}}$ direction, and after 0.900 s the ball passes the diving goalkeeper and goes into the net. If the total displacement of the ball is 40.2 m at 29.5° above the x axis, with what speed did the striker kick the ball?

SECTION 3-6 RELATIVE MOTION

48. • CE The accompanying photo shows a KC-10A Extender using a boom to refuel an aircraft in flight. If the velocity of the KC-10A is 125 m/s due east relative to the ground, what is the velocity of the aircraft being refueled relative to **(a)** the ground, and **(b)** the KC-10A?

Air-to-air refueling. (Problem 48)

49. • As an airplane taxies on the runway with a speed of 16.5 m/s, a flight attendant walks toward the tail of the plane with a speed

of 1.22 m/s. What is the flight attendant's speed relative to the ground?

50. • Referring to part (a) of Example 3-11, find the time it takes for the boat to reach the opposite shore if the river is 35 m wide.

51. •• A police car travels at 38.0 m/s due east while in pursuit of a vehicle that is traveling at 33.5 m/s due east. **(a)** What is the velocity of the vehicle relative to the police car? **(b)** What is the velocity of the police car relative to the vehicle?

52. •• Consider the river crossing problem in Example 3-11. Suppose we would like the boat to move directly across the river (in the positive x direction) with a speed of 5.0 m/s. What is the corresponding speed and direction of the boat's velocity relative to the water?

53. •• As you hurry to catch your flight at the local airport, you encounter a moving walkway that is 61 m long and has a speed of 2.2 m/s relative to the ground. If it takes you 49 s to cover 61 m when walking on the ground, how much time will it take you to cover the same distance on the walkway? Assume that you walk with the same speed on the walkway as you do on the ground.

54. •• In Problem 53, how much time would it take you to cover the 61-m length of the walkway if, once you get on the walkway, you immediately turn around and start walking in the opposite direction with a speed of 1.3 m/s relative to the walkway?

55. •• Predict/Calculate The pilot of an airplane wishes to fly due north, but there is a 65-km/h wind blowing toward the east. **(a)** In what direction relative to north should the pilot head her plane if its speed relative to the air is 340 km/h? **(b)** Draw a vector diagram that illustrates your result in part (a). **(c)** If the pilot decreases the air speed of the plane, but still wants to head due north, should the angle found in part (a) be increased or decreased?

56. •• A passenger walks from one side of a ferry to the other as it approaches a dock. If the passenger's velocity is 1.50 m/s due north relative to the ferry, and 4.50 m/s at an angle of 30.0° west of north relative to the water, what are the direction and magnitude of the ferry's velocity relative to the water?

57. •• You are riding on a Jet Ski at an angle of 35° upstream on a river flowing with a speed of 2.8 m/s. If your velocity relative to the ground is 9.5 m/s at an angle of 20.0° upstream, what is the speed of the Jet Ski relative to the water? (*Note*: Angles are measured relative to the x axis shown in Example 3-11.)

58. ••• Predict/Calculate In Problem 57, suppose the Jet Ski is moving at a speed of 11 m/s relative to the water. **(a)** At what angle must you point the Jet Ski if your velocity relative to the ground is to be perpendicular to the shore of the river? **(b)** If you increase the speed of the Jet Ski relative to the water, does the angle in part (a) increase, decrease, or stay the same? Explain. **(c)** Calculate the new angle if you increase the Jet Ski speed to 15 m/s. (*Note*: Angles are measured relative to the x axis shown in Example 3-11.)

59. ••• Predict/Calculate Two people take identical Jet Skis across a river, traveling at the same speed relative to the water. Jet Ski A heads directly across the river and is carried downstream by the current before reaching the opposite shore. Jet Ski B travels in a direction that is 35° upstream and arrives at the opposite shore directly across from the starting point. **(a)** Which Jet Ski reaches the opposite shore in the least amount of time? **(b)** Confirm your answer to part (a) by finding the ratio of the time it takes for the two Jet Skis to cross the river. (*Note:* Angles are measured relative to the x axis shown in Example 3-11.)

GENERAL PROBLEMS

60. • **CE** Predict/Explain Consider the vectors $\vec{A} = (1.2\,\text{m})\hat{x}$ and $\vec{B} = (-3.4\,\text{m})\hat{x}$. **(a)** Is the magnitude of vector $\vec{A}$ greater than, less than, or equal to the magnitude of vector $\vec{B}$? **(b)** Choose the *best explanation* from among the following:

  I. The number 3.4 is greater than the number 1.2.
  II. The component of $\vec{B}$ is negative.
  III. The vector $\vec{A}$ points in the positive x direction.

61. • **CE** Predict/Explain Two vectors are defined as follows: $\vec{A} = (-3.8\,\text{m})\hat{x}$ and $\vec{B} = (2.1\,\text{m})\hat{y}$. **(a)** Is the magnitude of $\vec{A} + \vec{B}$ greater than, less than, or equal to the magnitude of $\vec{A} - \vec{B}$? **(b)** Choose the *best explanation* from among the following:

  I. A vector sum is always greater than a vector difference.
  II. $\vec{A} + \vec{B}$ and $\vec{A} - \vec{B}$ produce similar triangles because $\vec{A}$ is perpendicular to $\vec{B}$.
  III. The vector $-\vec{B}$ is in the same direction as $\vec{A}$.

62. • To be compliant with regulations the inclination angle of a wheelchair ramp must not exceed 4.76°. If the wheelchair ramp must have a vertical height of 1.22 m, what is its minimum horizontal length?

63. • Find the direction and magnitude of the vector $2\vec{A} + \vec{B}$, where $\vec{A} = (12.1\,\text{m})\hat{x}$ and $\vec{B} = (-32.2\,\text{m})\hat{y}$.

64. •• **CE** The components of a vector $\vec{A}$ satisfy $A_x < 0$ and $A_y < 0$. Is the direction angle of $\vec{A}$ between 0° and 90°, between 90° and 180°, between 180° and 270°, or between 270° and 360°?

65. •• **CE** The components of a vector $\vec{B}$ satisfy $B_x > 0$ and $B_y < 0$. Is the direction angle of $\vec{B}$ between 0° and 90°, between 90° and 180°, between 180° and 270°, or between 270° and 360°?

66. •• It is given that $\vec{A} - \vec{B} = (-51.4\,\text{m})\hat{x}$, $\vec{C} = (62.2\,\text{m})\hat{x}$, and $\vec{A} + \vec{B} + \vec{C} = (13.8\,\text{m})\hat{x}$. Find the vectors $\vec{A}$ and $\vec{B}$.

67. •• You pilot an airplane with the intent to fly 392 km in a direction 55.0° south of east from the starting airport. Air traffic control requires you to first fly due south for 85.0 km. **(a)** After completing the 85.0-km flight due south, how far are you from your destination? **(b)** In what direction should you point your airplane to complete the trip? Give your answer as an angle relative to due east. **(c)** If your airplane has an average ground speed of 485 km/h, what estimated time of travel should you give to your passengers?

68. •• Find the x, y, and z components of the vector $\vec{A}$ shown in **FIGURE 3-47**, given that $A = 65$ m.

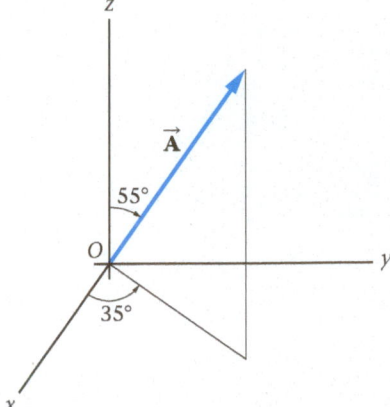

FIGURE 3-47 Problem 68

69. •• Observer 1 rides in a car and drops a ball from rest straight downward, relative to the interior of the car. The car moves horizontally with a constant speed of 3.50 m/s relative to observer 2 standing

on the sidewalk. **(a)** What is the speed of the ball 1.00 s after it is released, as measured by observer 2? **(b)** What is the direction of travel of the ball 1.00 s after it is released, as measured relative to the horizontal by observer 2?

70. •• A person riding in a subway train drops a ball from rest straight downward, relative to the interior of the train. The train is moving horizontally with a constant speed of 6.7 m/s. A second person standing at rest on the subway platform observes the ball drop. From the point of view of the person on the platform, the ball is released at the position $x = 0$ and $y = 1.2$ m. Make a plot of the position of the ball for the times $t = 0$, 0.1 s, 0.2 s, 0.3 s, and 0.4 s. (Your plot is parabolic in shape, as we shall see in Chapter 4.)

71. •• A football is thrown horizontally with an initial velocity of $(16.6\,\text{m/s})\hat{x}$. Ignoring air resistance, the average acceleration of the football over any period of time is $(-9.81\,\text{m/s}^2)\hat{y}$. **(a)** Find the velocity vector of the ball 1.75 s after it is thrown. **(b)** Find the magnitude and direction of the velocity at this time.

72. •• As a function of time, the velocity of the football described in Problem 71 can be written as $\vec{v} = (16.6\,\text{m/s})\hat{x} - [(9.81\,\text{m/s}^2)t]\hat{y}$. Calculate the average acceleration vector of the football for the time periods **(a)** $t = 0$ to $t = 1.00$ s, **(b)** $t = 0$ to $t = 2.50$ s, and **(c)** $t = 0$ to $t = 5.00$ s.

73. •• Two airplanes taxi as they approach the terminal. Plane 1 taxies with a speed of 12 m/s due north. Plane 2 taxies with a speed of 7.5 m/s in a direction 20° north of west. **(a)** What are the direction and magnitude of the velocity of plane 1 relative to plane 2? **(b)** What are the direction and magnitude of the velocity of plane 2 relative to plane 1?

74. •• A shopper at the supermarket follows the path indicated by vectors $\vec{A}$, $\vec{B}$, $\vec{C}$, and $\vec{D}$ in **FIGURE 3-48**. Given that the vectors have the magnitudes $A = 51$ ft, $B = 45$ ft, $C = 35$ ft, and $D = 13$ ft, find the total displacement of the shopper using **(a)** the graphical method and **(b)** the component method of vector addition. Give the direction of the displacement relative to the direction of vector $\vec{A}$.

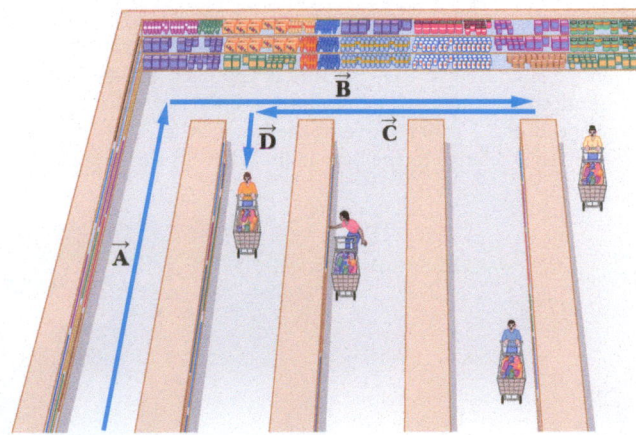

FIGURE 3-48 Problem 74

75. •• **BIO** A food particle from your breakfast takes a circuitous path through your digestive system. Suppose its motion over a period of time can be represented by the four displacement vectors depicted in **FIGURE 3-49**. Let the vectors have magnitudes $A = 8.0$ cm, $B = 16$ cm, $C = 23$ cm, and $D = 5.6$ cm. **(a)** What is the total displacement of the food particle? Give its direction relative to the direction of vector $\vec{A}$. **(b)** If the average speed of the particle is 0.010 mm/s, what is its average velocity over the time interval of the four displacements?

FIGURE 3-49 Problem 75

76. •• Initially, a particle is moving at 4.10 m/s at an angle of 33.5° above the horizontal. Two seconds later, its velocity is 6.05 m/s at an angle of 59.0° below the horizontal. What was the particle's average acceleration during these 2.00 seconds?

77. •• A passenger on a stopped bus notices that rain is falling vertically just outside the window. When the bus moves with constant velocity, the passenger observes that the falling raindrops are now making an angle of 15° with respect to the vertical. (a) What is the ratio of the speed of the raindrops to the speed of the bus? (b) Find the speed of the raindrops, given that the bus is moving with a speed of 18 m/s.

78. •• **Predict/Calculate** Suppose we orient the x axis of a two-dimensional coordinate system along the beach at Waikiki, as shown in **FIGURE 3-50**. Waves approaching the beach have a velocity relative to the shore given by $\vec{\mathbf{v}}_{ws} = (-1.3 \text{ m/s})\hat{\mathbf{y}}$. Surfers move more rapidly than the waves, but at an angle θ to the beach. The angle is chosen so that the surfers approach the shore with the same speed as the waves along the y direction. (a) If a surfer has a speed of 7.2 m/s relative to the water, what is his direction of motion θ relative to the beach? (b) What his the surfer's velocity

relative to the wave? (c) If the surfer's speed is increased, will the angle in part (a) increase or decrease? Explain.

79. •• **Predict/Calculate The Longitude Problem** In 1759, John Harrison (1693–1776) completed his fourth precision chronometer, the H4, which eventually won the celebrated Longitude Prize. (For the human drama behind the Longitude Prize, see *Longitude*, by Dava Sobel.) When the minute hand of the H4 indicates 10 minutes past the hour, it extends 3.0 cm in the horizontal direction. (a) How long is the H4's minute hand? (b) At 10 minutes past the hour, is the extension of the minute hand in the vertical direction more than, less than, or equal to 3.0 cm? Explain. (c) Calculate the vertical extension of the minute hand at 10 minutes past the hour.

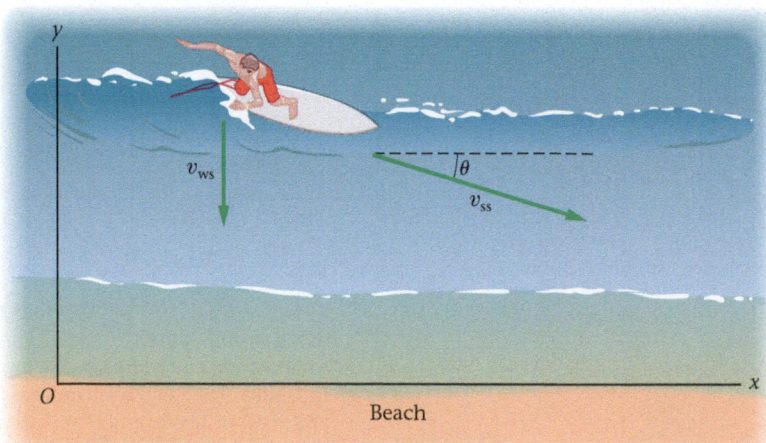

Not just a watch! The Harrison H4.
(Problem 79)

80. ••• Referring to Example 3-11, (a) what heading must the boat have if it is to land directly across the river from its starting point? (b) How much time is required for this trip if the river is 25.0 m wide? (c) Suppose the speed of the boat is increased, but it is still desired to land directly across from the starting point. Should the boat's heading be more upstream, more downstream, or the same as in part (a)? Explain.

81. ••• Vector $\vec{\mathbf{A}}$ points in the negative x direction. Vector $\vec{\mathbf{B}}$ points at an angle of 30.0° above the positive x axis. Vector $\vec{\mathbf{C}}$ has a magnitude of 15 m and points in a direction 40.0° below the positive x axis. Given that $\vec{\mathbf{A}} + \vec{\mathbf{B}} + \vec{\mathbf{C}} = 0$, find the magnitudes of $\vec{\mathbf{A}}$ and $\vec{\mathbf{B}}$.

FIGURE 3-50 Problem 78

82. ••• As two boats approach the marina, the velocity of boat 1 relative to boat 2 is 2.15 m/s in a direction 47.0° east of north. If boat 1 has a velocity that is 0.775 m/s due north, what is the velocity (magnitude and direction) of boat 2?

PASSAGE PROBLEMS

BIO **Motion Camouflage in Dragonflies**

Dragonflies, whose ancestors were once the size of hawks, have prowled the skies in search of small flying insects for over 250 million years. Faster and more maneuverable than any other insect, they even fold their front two legs in flight and tuck them behind their head to be as streamlined as possible. They also employ an intriguing stalking strategy known as "motion camouflage" to approach their prey almost undetected.

   The basic idea of motion camouflage is for the dragonfly to move in such a way that the line of sight from the prey to the dragonfly is always in the same direction. Moving in this way, the dragonfly appears almost motionless to its prey, as if it were an object at infinity. Eventually the prey notices the dragonfly has grown in size and is therefore closer, but by that time it's too late for the prey to evade capture.

   A typical capture scenario is shown in **FIGURE 3-51**, where the prey moves in the positive y direction with the constant speed $v_p = 0.750$ m/s, and the dragonfly moves at an angle $\theta = 48.5°$ to the x axis with the constant speed v_d. If the dragonfly chooses its speed correctly, the line of sight from the prey to the dragonfly will always be in the same direction—parallel to the x axis in this case.

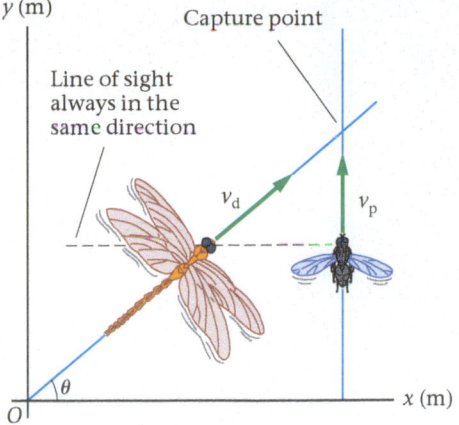

FIGURE 3-51 Problems 83, 84, 85, and 86

83. • What speed must the dragonfly have if the line of sight, which is parallel to the x axis initially, is to remain parallel to the x axis?

   A. 0.562 m/s   B. 0.664 m/s
   C. 1.00 m/s   D. 1.13 m/s

84. • Suppose the dragonfly now approaches its prey along a path with $\theta > 48.5°$, but it still keeps the line of sight parallel to the x axis. Is the speed of the dragonfly in this new case greater than, less than, or equal to its speed in Problem 83?

85. •• If the dragonfly approaches its prey with a speed of 0.950 m/s, what angle θ is required to maintain a constant line of sight parallel to the x axis?

   A. 37.9°   B. 38.3°
   C. 51.7°   D. 52.1°

86. •• In a similar situation, the dragonfly is observed to fly at a constant speed, but its angle to the x axis gradually increases from 20° to 40°. Which of the following is the best explanation for this behavior?

   A. The prey is moving in the positive y direction and maintaining a constant speed.
   B. The prey is moving in the positive y direction and increasing its speed.
   C. The prey is moving in the positive y direction and decreasing its speed.
   D. The prey is moving in the positive x direction and maintaining a constant speed.

87. •• **Referring to Example 3-11** Suppose the speed of the boat relative to the water is 7.0 m/s. **(a)** At what angle to the x axis must the boat be headed if it is to land directly across the river from its starting position? **(b)** If the speed of the boat relative to the water is increased, will the angle needed to go directly across the river increase, decrease, or stay the same? Explain.

88. ••• **Referring to Example 3-11** Suppose that the boat has a speed of 6.7 m/s relative to the water, and that the dock on the opposite shore of the river is at the location $x = 55$ m and $y = 28$ m relative to the starting point of the boat. **(a)** At what angle relative to the x axis must the boat be pointed in order to reach the other dock? **(b)** With the angle found in part (a), what is the speed of the boat relative to the ground?

4

Two-Dimensional Kinematics

Big Ideas

1 Two-dimensional motion consists of independent horizontal and vertical motions.

2 Objects in free fall move under the influence of gravity alone.

3 Objects in free fall follow parabolic paths.

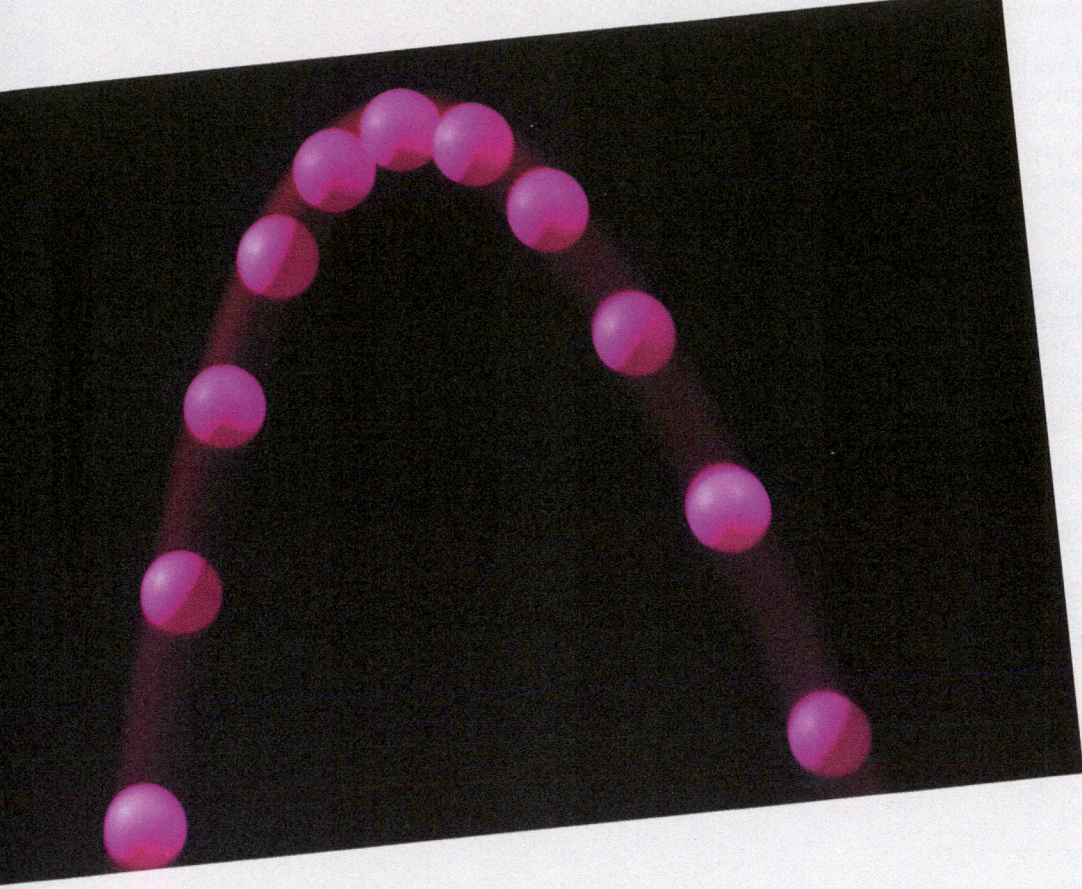

▲ When you hear the word projectile, you probably think of an artillery shell or a home run into the upper deck. But the term projectile applies to any object moving under the influence of gravity alone. For example, a juggling ball undergoes projectile motion—and follows a parabolic path—as it moves from one hand to the other. In this chapter we explore the physical laws that govern projectile motion.

The main idea in this chapter is quite simple: Horizontal and vertical motions are independent. That's it. This chapter develops and applies the idea of independence of motion to many common physical systems.

4-1 Motion in Two Dimensions

In this section we develop equations of motion to describe objects moving in two dimensions. First, we consider motion with constant velocity, determining x and y as functions of time. Later, we investigate motion with constant acceleration. In both cases, we show that the one-dimensional kinematic equations of Chapter 2 can be extended in a straightforward way to two dimensions.

Constant Velocity

To begin, consider the simple situation shown in **FIGURE 4-1**. A turtle starts at the origin at $t = 0$ and moves with a constant speed $v_0 = 0.26$ m/s in a direction 25° above the x axis. How far has the turtle moved in the x and y directions after 5.0 seconds?

First, notice that the turtle moves in a straight line a distance given by speed multiplied by time:

$$d = v_0 t = (0.26 \text{ m/s})(5.0 \text{ s}) = 1.3 \text{ m}$$

This is indicated in Figure 4-1 (a). From the definitions of sine and cosine given in the previous chapter, we see that the horizontal (x) and vertical (y) distances are given by

$$x = d \cos 25° = 1.2 \text{ m}$$
$$y = d \sin 25° = 0.55 \text{ m}$$

An alternative way to approach this problem is to treat the x and y motions separately. First, we determine the speed of the turtle in each direction. Referring to Figure 4-1 (b), we see that the x component of velocity is

$$v_{0x} = v_0 \cos 25° = 0.24 \text{ m/s}$$

Similarly, the y component of velocity is

$$v_{0y} = v_0 \sin 25° = 0.11 \text{ m/s}$$

Next, we find the distance traveled by the turtle in the x and y directions by multiplying the speed in each direction by the time:

$$x = v_{0x}t = (0.24 \text{ m/s})(5.0 \text{ s}) = 1.2 \text{ m}$$
$$y = v_{0y}t = (0.11 \text{ m/s})(5.0 \text{ s}) = 0.55 \text{ m}$$

This is in agreement with our previous results. To summarize, we can think of the turtle's actual motion as a combination of separate x and y motions.

Basic Equations for Constant Velocity   In general, the turtle (or any object) might start at a position $x = x_0$ and $y = y_0$ at time $t = 0$. In this case, the equations of motion are

$$x = x_0 + v_{0x}t \qquad\qquad\text{4-1}$$

and

$$y = y_0 + v_{0y}t \qquad\qquad\text{4-2}$$

These equations give x and y as functions of time.

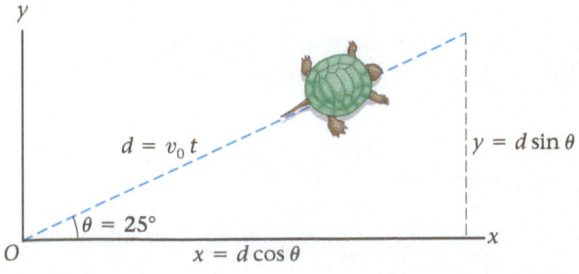

(a)

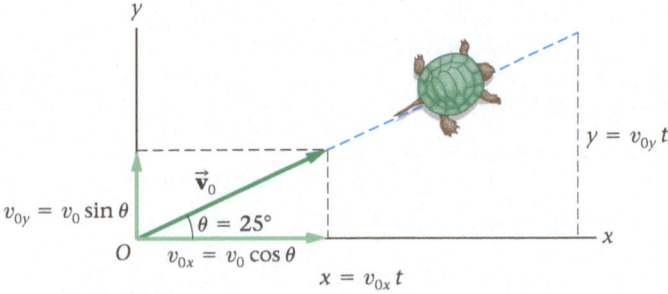

(b)

▲ **FIGURE 4-1 Constant velocity** A turtle walks from the origin with a speed of $v_0 = 0.26$ m/s. **(a)** In a time t the turtle moves through a straight-line distance of $d = v_0 t$; thus the x and y displacements are $x = d \cos \theta$ and $y = d \sin \theta$. **(b)** Equivalently, the turtle's x and y components of velocity are $v_{0x} = v_0 \cos \theta$ and $v_{0y} = v_0 \sin \theta$; hence $x = v_{0x}t$ and $y = v_{0y}t$.

Compare these equations with Equation 2-11, $x = x_0 + v_0 t + \frac{1}{2}at^2$, which gives position as a function of time in one dimension. When the acceleration is zero, as it is for the turtle, Equation 2-11 reduces to $x = x_0 + v_0 t$. Replacing v_0 with the x component of the velocity, v_{0x}, yields Equation 4-1. Similarly, replacing each x in Equation 4-1 with y converts it to Equation 4-2, the y equation of motion.

A situation illustrating the use of $x = x_0 + v_{0x}t$ and $y = y_0 + v_{0y}t$ is given in the following Example.

EXAMPLE 4-1   THE EAGLE DESCENDS

An eagle perched on a tree limb 19.5 m above the water spots a fish swimming near the surface. The eagle pushes off from the branch and descends toward the water. By adjusting its body in flight, the eagle maintains a constant speed of 3.10 m/s at an angle of 20.0° below the horizontal. **(a)** How much time does it take for the eagle to reach the water? **(b)** How far has the eagle traveled in the horizontal direction when it reaches the water?

PICTURE THE PROBLEM

We set up our coordinate system so that the eagle starts at $x_0 = 0$ and $y_0 = h = 19.5$ m. The water level is $y = 0$. As indicated in our sketch, $v_{0x} = v_0 \cos\theta$ and $v_{0y} = -v_0 \sin\theta$, where $v_0 = 3.10$ m/s and $\theta = 20.0°$. The eagle's velocity is constant, and therefore its equations of motion are given by $x = x_0 + v_{0x}t$ and $y = y_0 + v_{0y}t$.

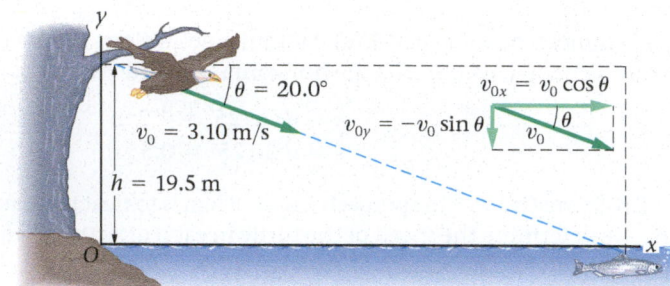

REASONING AND STRATEGY

As usual in such problems, it's best to treat the eagle's flight as a combination of separate x and y motions. Since we are given the constant speed of the eagle, and the angle at which it descends, we can find the x and y components of its velocity. We then use the y equation of motion, $y = y_0 + v_{0y}t$, to find the time t when the eagle reaches the water. Next, we use this value of t in the x equation of motion, $x = x_0 + v_{0x}t$, to find the horizontal distance the bird travels.

Known   Height, $h = 19.5$ m; speed, $v_0 = 3.10$ m/s; angle, $\theta = 20.0°$.
Unknown   **(a)** Time, $t = $? **(b)** Horizontal distance, $x = $?

SOLUTION

Part (a)

1. Begin by determining the x and y components of the eagle's velocity, v_{0x} and v_{0y}:

$$v_{0x} = v_0 \cos\theta = (3.10\,\text{m/s}) \cos 20.0° = 2.91\,\text{m/s}$$
$$v_{0y} = -v_0 \sin\theta = -(3.10\,\text{m/s}) \sin 20.0° = -1.06\,\text{m/s}$$

2. Now, set $y = 0$ in $y = y_0 + v_{0y}t$ and solve for t:

$$y = y_0 + v_{0y}t = h + v_{0y}t = 0$$
$$t = -\frac{h}{v_{0y}} = -\frac{19.5\,\text{m}}{(-1.06\,\text{m/s})} = 18.4\,\text{s}$$

Part (b)

3. Substitute $t = 18.4$ s into $x = x_0 + v_{0x}t$ to find x:

$$x = x_0 + v_{0x}t = 0 + (2.91\,\text{m/s})(18.4\,\text{s}) = 53.5\,\text{m}$$

INSIGHT

Notice how the two minus signs in Step 2 combine to give a positive time. One minus sign comes from setting $y = 0$, the other from the fact that v_{0y} is negative. No matter where we choose the origin, or what direction we choose to be positive, the time will always have the same value.

As mentioned in the problem statement, the eagle cannot travel in a straight line by simply dropping from the tree limb—it has to adjust its wings and tail to produce enough lift and drag to balance the force of gravity. Airplanes do the same thing when they adjust their flight surfaces to make a smooth landing.

PRACTICE PROBLEM

What is the location of the eagle 2.00 s after it takes flight?   [**Answer:** $x = 5.82$ m, $y = 17.4$ m]

Some related homework problems: Problem 2, Problem 3

RWP* One interesting application of the mathematical description of motion in *three* dimensions is the Traffic Collision Avoidance System (TCAS), which uses information from radio signals exchanged between equipped aircraft to warn pilots of a

---

*Real World Physics applications are denoted by the acronym RWP.

possible mid-air collision. When a possible collision is detected, the system automatically negotiates a mutual avoidance maneuver and gives the pilots instructions on a cockpit display as well as by a synthesized voice. The TCAS has prevented numerous mid-air collisions and has saved many lives.

Constant Acceleration

To study motion with constant acceleration in two dimensions we repeat what was done in one dimension in Chapter 2, but with separate equations for both x and y. For example, to obtain x as a function of time we start with $x = x_0 + v_0 t + \frac{1}{2} a t^2$ (Equation 2-11) and replace both v_0 and a with the corresponding x components, v_{0x} and a_x. This gives

$$x = x_0 + v_{0x} t + \frac{1}{2} a_x t^2 \qquad \text{4-3(a)}$$

To obtain y as a function of time, we write y in place of x in Equation 4-3(a):

$$y = y_0 + v_{0y} t + \frac{1}{2} a_y t^2 \qquad \text{4-3(b)}$$

These are the position-versus-time equations of motion for two dimensions. (In three dimensions we introduce a third coordinate direction and label it z. We would then simply replace x with z in Equation 4-3(a) to obtain z as a function of time.)

The same approach gives velocity as a function of time. Start with Equation 2-7, $v = v_0 + at$, and write it in terms of x and y components. This yields

$$v_x = v_{0x} + a_x t \qquad \text{4-4(a)}$$
$$v_y = v_{0y} + a_y t \qquad \text{4-4(b)}$$

Notice that we simply repeat everything we did for one dimension, only now with separate equations for the x and y components.

Finally, we can write $v^2 = v_0^2 + 2a\Delta x$ in terms of components as well:

$$v_x^2 = v_{0x}^2 + 2a_x \Delta x \qquad \text{4-5(a)}$$
$$v_y^2 = v_{0y}^2 + 2a_y \Delta y \qquad \text{4-5(b)}$$

Our results are listed in Table 4-1.

PHYSICS IN CONTEXT
Looking Back

The equations of one-dimensional kinematics derived in Chapter 2 are used again in this chapter, even though we are now studying kinematics in two dimensions. For example, the equations in Table 4-1 are the same as those used in Chapter 2, only now they are applied individually to the x and y directions.

TABLE 4-1 Constant-Acceleration Equations of Motion

Position as a function of time	Velocity as a function of time	Velocity as a function of position
$x = x_0 + v_{0x} t + \frac{1}{2} a_x t^2$	$v_x = v_{0x} + a_x t$	$v_x^2 = v_{0x}^2 + 2a_x \Delta x$
$y = y_0 + v_{0y} t + \frac{1}{2} a_y t^2$	$v_y = v_{0y} + a_y t$	$v_y^2 = v_{0y}^2 + 2a_y \Delta y$

Big Idea 1 Two-dimensional motion is a combination of horizontal and vertical motions. The key concept behind two-dimensional motion is that the horizontal and vertical motions are completely independent of one another; each can be considered separately as one-dimensional motion.

The Same Equations Are Used Throughout This Chapter The fundamental equations in Table 4-1 are used to obtain *all* of the results presented throughout the rest of this chapter. Though it may *appear* sometimes that we are writing new sets of equations for different problems, the equations aren't new at all—what we're actually doing is simply writing these same equations again, but with different specific values substituted for the constants that appear in them.

EXAMPLE 4-2   A CHOPPER ACCELERATES VERTICALLY

A helicopter is in level flight with a constant horizontal speed of 11 m/s. At $t = 0$ it begins to accelerate vertically at 0.96 m/s^2. Assuming the helicopter's acceleration remains constant for the time interval of interest, find **(a)** the horizontal and vertical distances through which it moves in 5.3 s and **(b)** its x and y velocity components at $t = 5.3$ s.

PICTURE THE PROBLEM

In our sketch we have placed the origin of a two-dimensional coordinate system at the location of the helicopter at the initial time, $t = 0$. In addition, we have chosen the initial direction of motion to be in the positive x direction, and the direction of acceleration to be in the positive y direction. As a result, it follows that $x_0 = y_0 = 0$, $v_{0x} = 11$ m/s, $v_{0y} = 0$, $a_x = 0$, and $a_y = 0.96$ m/s^2. As the helicopter moves upward, its y component of velocity increases, resulting in a curved path, as shown.

CONTINUED

REASONING AND STRATEGY

a. We want to relate position and time, and hence we find the horizontal position of the helicopter with $x = x_0 + v_{0x}t + \frac{1}{2}a_xt^2$, and the vertical position with $y = y_0 + v_{0y}t + \frac{1}{2}a_yt^2$. In each case we substitute $t = 5.3$ s.

b. The velocity components at $t = 5.3$ s can be found using $v_x = v_{0x} + a_xt$ and $v_y = v_{0y} + a_yt$.

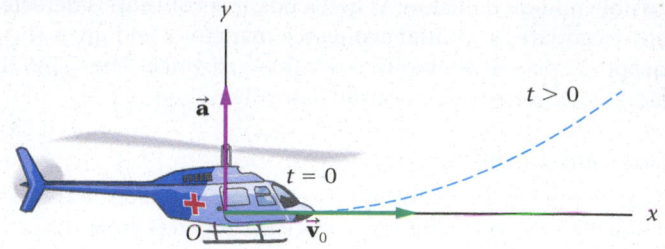

Known   Horizontal speed, $v_{0x} = 11$ m/s; vertical acceleration, $a_y = 0.96$ m/s².

Unknown   (a) Horizontal distance, $x = ?$; vertical distance, $y = ?$ (b) Horizontal component of velocity, $v_x = ?$; vertical component of velocity, $v_y = ?$

SOLUTION

Part (a)

1. Use $x = x_0 + v_{0x}t + \frac{1}{2}a_xt^2$ to find x at $t = 5.3$ s: $x = x_0 + v_{0x}t + \frac{1}{2}a_xt^2 = 0 + (11 \text{ m/s})(5.3 \text{ s}) + 0 = 58 \text{ m}$

2. Use $y = y_0 + v_{0y}t + \frac{1}{2}a_yt^2$ to find y at $t = 5.3$ s: $y = y_0 + v_{0y}t + \frac{1}{2}a_yt^2 = 0 + 0 + \frac{1}{2}(0.96 \text{ m/s}^2)(5.3 \text{ s})^2 = 13 \text{ m}$

Part (b)

3. Use $v_x = v_{0x} + a_xt$ to find v_x at $t = 5.3$ s: $v_x = v_{0x} + a_xt = 11 \text{ m/s} + 0 = 11 \text{ m/s}$

4. Use $v_y = v_{0y} + a_yt$ to find v_y at $t = 5.3$ s: $v_y = v_{0y} + a_yt = 0 + (0.96 \text{ m/s}^2)(5.3 \text{ s}) = 5.1 \text{ m/s}$

INSIGHT

If we assume constant acceleration, the vertical position of the helicopter will eventually increase more rapidly with time than the horizontal position, due to the t^2 dependence of y as compared with the t dependence of x. This results in a curved, parabolic path for the helicopter, as shown in our sketch. The helicopter's velocity at 5.3 s is $v = \sqrt{v_x^2 + v_y^2} = \sqrt{(11 \text{ m/s})^2 + (5.1 \text{ m/s})^2} = 12 \text{ m/s}$ at an angle of $\theta = \tan^{-1}(v_y/v_x) = \tan^{-1}[(5.1 \text{ m/s})/(11 \text{ m/s})] = 25°$ above the x axis. It's clear that the angle of flight must be less than 45° at this time because the x component of velocity is greater than the y component.

PRACTICE PROBLEM

How much time is required for the helicopter to move 18 m vertically from its initial position?   [**Answer:** $t = 6.1$ s]

Some related homework problems: Problem 4, Problem 5, Problem 6.

Enhance Your Understanding   (Answers given at the end of the chapter)

1. The equations of motion of an object are $x = (1 \text{ m/s}^2)t^2$ and $y = (2 \text{ m}) - (3 \text{ m/s})t$. What are the x and y components of the object's (a) initial position, (b) initial velocity, and (c) acceleration?

Section Review

- Motion in two dimensions is a combination of one-dimensional motion in the x and y directions. The basic equations used in this chapter are a straightforward extension of the equations of motion in Chapter 2. They are listed in Table 4-1.

4-2 Projectile Motion: Basic Equations

We now apply the independence of horizontal and vertical motions to projectiles. Just what do we mean by a projectile? Well, a **projectile** is an object that is thrown, kicked, batted, or otherwise launched into motion and then allowed to follow a path determined solely by the influence of gravity. As you might expect, this covers a wide variety of physical systems.

In studying projectile motion we make the following assumptions:

- Air resistance is ignored.

- The acceleration due to gravity is constant, downward, and has a magnitude equal to $g = 9.81$ m/s².

- The Earth's rotation is ignored.

Air resistance can be significant if a projectile moves with relatively high speed or if it encounters a strong wind. In many everyday situations, however, like tossing a ball to a friend or dropping a book, air resistance is relatively insignificant. As for the acceleration due to gravity, $g = 9.81$ m/s², this value varies slightly from place to place on the Earth's surface and decreases with increasing altitude. In addition, the rotation of the Earth can be significant when we consider projectiles that cover great distances. Little error is made in ignoring the variation of g or the rotation of the Earth, however, in the examples of projectile motion considered in this chapter.

Equations of Motion for Projectiles Let's incorporate the preceding assumptions into the equations of motion given in the previous section. Suppose, as in **FIGURE 4-2**, that the x axis is horizontal and the y axis is vertical, with the positive direction upward. Noting that downward is the negative direction, it follows that

$$a_y = -9.81 \text{ m/s}^2 = -g$$

Gravity causes no acceleration in the x direction. Thus, the x component of acceleration is zero:

$$a_x = 0$$

With these acceleration components substituted into the fundamental constant-acceleration equations of motion (Table 4-1) we find:

Projectile Motion ($a_x = 0$, $a_y = -g$)

$$x = x_0 + v_{0x}t \qquad v_x = v_{0x} \qquad v_x^2 = v_{0x}^2$$
$$y = y_0 + v_{0y}t - \tfrac{1}{2}gt^2 \qquad v_y = v_{0y} - gt \qquad v_y^2 = v_{0y}^2 - 2g\Delta y$$

4-6

In these expressions, the positive y direction is upward and the quantity g is positive. *All* of our studies of projectile motion use Equations 4-6 as our fundamental equations—again, special cases simply correspond to substituting different specific values for the constants.

Demonstrating Independence of Motion A simple demonstration illustrates the independence of horizontal and vertical motions in projectile motion. First, while standing still, drop a rubber ball to the floor and catch it on the rebound. Notice that the ball goes straight down, lands near your feet, and returns almost to the level of your hand in about a second.

Next, walk—or roller skate—with constant speed before dropping the ball, then observe its motion carefully. To you, its motion looks the same as before: It goes straight down, lands near your feet, bounces straight back up, and returns in about one second. This is illustrated in **FIGURE 4-3**. The fact that you were moving in the horizontal direction the whole time had no effect on the ball's vertical motion—the motions are independent.

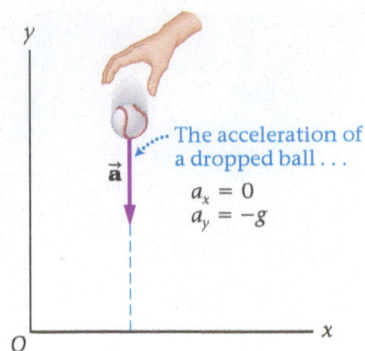

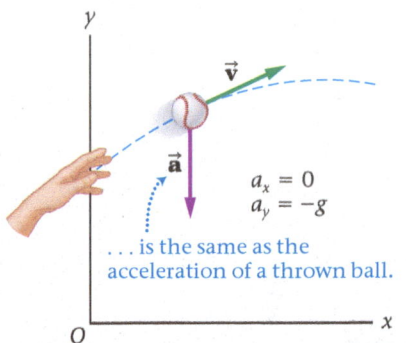

▲ **FIGURE 4-2 Acceleration in free fall** All objects in free fall have acceleration components $a_x = 0$ and $a_y = -g$ when the coordinate system is chosen as shown here. This is true regardless of whether the object is dropped, thrown, kicked, or otherwise set into motion.

Big Idea 2 Projectiles are objects that move under the influence of gravity alone. Projectiles can be dropped from rest or thrown at some angle to the horizontal. Once they are launched, they have all the characteristics of projectile motion, irrespective of how their motion started.

PHYSICS IN CONTEXT
Looking Ahead

The basic idea behind projectile motion is used again in Chapter 12, when we consider orbital motion.

PROBLEM-SOLVING NOTE

Acceleration of a Projectile

When the x axis is chosen to be horizontal and the y axis points vertically upward, it follows that the acceleration of an ideal projectile is $a_x = 0$ and $a_y = -g$.

▲ **FIGURE 4-3 Independence of vertical and horizontal motions** When you drop a ball while walking, running, or skating with constant velocity, it appears to you to drop straight down from the point where you released it. To a person at rest, the ball follows a curved path that combines horizontal and vertical motions.

(a)

(b)

(c)

▲ **FIGURE 4-4 Visualizing Concepts Independence of Motion (a)** An athlete jumps upward from a moving skateboard. The athlete retains his initial horizontal velocity, and hence remains directly above the skateboard at all times. **(b)** The pilot ejection seat of a jet fighter is being ground-tested. The horizontal and vertical motions are independent, and hence the test dummy is still almost directly above the cockpit from which it was ejected. (Notice that air resistance is beginning to reduce the dummy's horizontal velocity.) **(c)** This rollerblader may not be thinking about independence of motion, but the ball she released illustrates the concept perfectly as it falls directly below her hand.

To an observer who sees you walking by, the ball follows a curved path, as shown. The precise shape of this curved path—a parabola—is verified in the next section. Additional examples of this principle are shown in **FIGURE 4-4**.

Ball Fired Upward from Moving Cart

Ball Fired Upward from Accelerating Cart

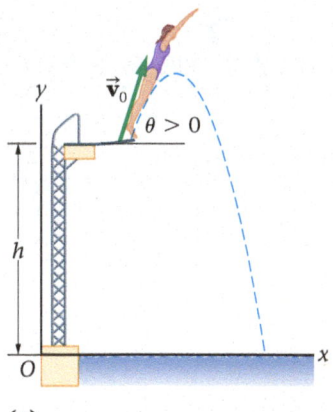

(a)

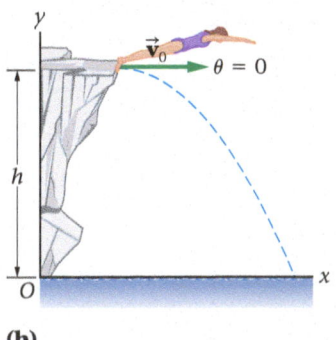

(b)

◀ **FIGURE 4-5 Launch angle of a projectile (a)** A projectile launched at an angle above the horizontal, $\theta > 0$. A launch below the horizontal would correspond to $\theta < 0$. **(b)** A projectile launched horizontally, $\theta = 0$. In this section we consider $\theta = 0$. The next section deals with $\theta \neq 0$.

Enhance Your Understanding

(Answers given at the end of the chapter)

2. A sailor drops a pair of binoculars from the "crow's nest" at the top of a sailing ship's mast. When the ship is at rest, the binoculars land at the base of the mast. If the ship is moving forward with constant velocity, and air resistance can be ignored, do the binoculars land in front of the mast (toward the front of the ship), at the base of the mast, or behind the mast (toward the rear of the ship)? Explain.

Section Review

• Projectiles move under the influence of gravity alone, with a constant downward acceleration. The basic equations describing projectile motion are summarized in Equations 4-6.

4-3 Zero Launch Angle

A special case of some interest is a projectile launched horizontally, so that the angle between the initial velocity and the horizontal is $\theta = 0$. We devote this section to a brief look at this type of motion.

Equations of Motion

Suppose you are walking with a speed v_0 when you release a ball from a height h. If we choose ground level to be $y = 0$, and the release point to be directly above the origin, the initial position of the ball is given by

$$x_0 = 0$$

▶ **FIGURE 4-6 Trajectory of a projectile launched horizontally** In this plot, the projectile was launched from a height of 9.5 m with an initial speed of 5.0 m/s. The positions shown in the plot correspond to the times $t = 0.20\,\text{s}, 0.40\,\text{s}, 0.60\,\text{s}, \ldots$. Notice the uniform motion in the x direction, and the accelerated motion in the y direction.

and

$$y_0 = h$$

This is illustrated in Figure 4-3.

The initial velocity is horizontal in this case, corresponding to $\theta = 0$ in **FIGURE 4-5**. As a result, the x component of the initial velocity is simply the initial speed:

$$v_{0x} = v_0 \cos 0° = v_0$$

The y component of the initial velocity is zero:

$$v_{0y} = v_0 \sin 0° = 0$$

Substituting these specific values into our fundamental equations for projectile motion (Equations 4-6) gives the following simplified results for zero launch angle ($\theta = 0$):

$$x = v_0 t \qquad v_x = v_0 = \text{constant} \qquad v_x^2 = v_0^2 = \text{constant}$$
$$y = h - \tfrac{1}{2}gt^2 \qquad v_y = -gt \qquad v_y^2 = -2g\Delta y \qquad \text{4-7}$$

Snapshots of this motion at equal time intervals are shown in **FIGURE 4-6**.

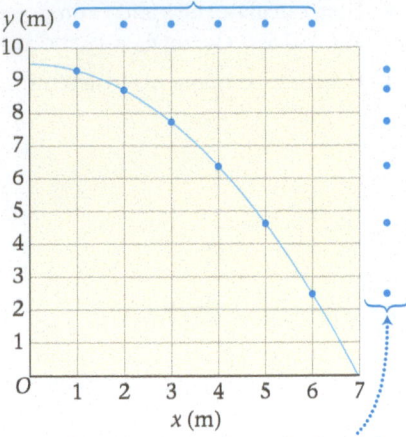

Horizontal motion is uniform—equal distance in equal time.

Vertical motion is accelerated—the object goes farther in each successive interval.

   Dropped and Thrown Balls

PROBLEM-SOLVING NOTE

Identify Initial Conditions

The launch point of a projectile determines x_0 and y_0. The initial velocity of a projectile determines v_{0x} and v_{0y}.

EXAMPLE 4-3   DROPPING A BALL

A person skateboarding with a constant speed of 1.30 m/s releases a ball from a height of 1.25 m above the ground. Given that $x_0 = 0$ and $y_0 = h = 1.25\,\text{m}$, find x and y for **(a)** $t = 0.250\,\text{s}$ and **(b)** $t = 0.500\,\text{s}$. **(c)** Find the velocity, speed, and direction of motion of the ball at $t = 0.500\,\text{s}$.

PICTURE THE PROBLEM

The ball starts at $x_0 = 0$ and $y_0 = h = 1.25\,\text{m}$. Its initial velocity is horizontal; therefore $v_{0x} = v_0 = 1.30\,\text{m/s}$ and $v_{0y} = 0$. In addition, it accelerates with the acceleration due to gravity in the negative y direction, $a_y = -g$, and moves with constant speed in the x direction, $a_x = 0$.

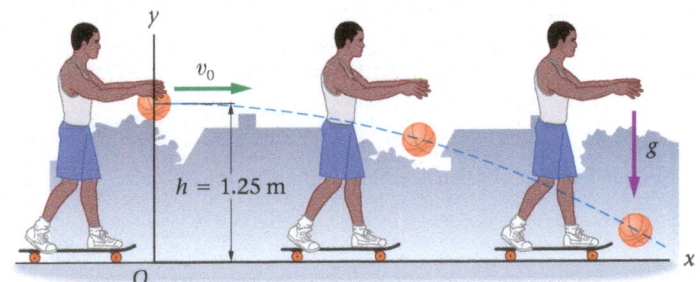

REASONING AND STRATEGY

The x and y positions are given by $x = v_0 t$ and $y = h - \tfrac{1}{2}gt^2$, respectively. We simply substitute time into these expressions. Similarly, the velocity components are $v_x = v_0$ and $v_y = -gt$.

Known   Speed, $v_0 = 1.30\,\text{m/s}$; height, $h = 1.25\,\text{m}$.
Unknown **(a)** $x = ?$ and $y = ?$ at $t = 0.250\,\text{s}$; **(b)** $x = ?$ and $y = ?$ at $t = 0.500\,\text{s}$;
             (c) velocity, $\vec{\mathbf{v}} = ?$, speed, $v = ?$, direction angle, $\theta = ?$, at $t = 0.500\,\text{s}$

SOLUTION

Part (a)

1. Substitute $t = 0.250\,\text{s}$ into the x and y equations of motion:

$$x = v_0 t = (1.30\,\text{m/s})(0.250\,\text{s}) = 0.325\,\text{m}$$
$$y = h - \tfrac{1}{2}gt^2$$
$$= 1.25\,\text{m} - \tfrac{1}{2}(9.81\,\text{m/s}^2)(0.250\,\text{s})^2 = 0.943\,\text{m}$$

Part (b)

2. Substitute $t = 0.500\,\text{s}$ into the x and y equations of motion. Notice that the ball is only about an inch above the ground at this time:

$$x = v_0 t = (1.30\,\text{m/s})(0.500\,\text{s}) = 0.650\,\text{m}$$
$$y = h - \tfrac{1}{2}gt^2$$
$$= 1.25\,\text{m} - \tfrac{1}{2}(9.81\,\text{m/s}^2)(0.500\,\text{s})^2 = 0.0238\,\text{m}$$

CONTINUED

Part (c)

3. First, calculate the x and y components of the velocity at $t = 0.500$ s using $v_x = v_0$ and $v_y = -gt$:

$$v_x = v_0 = 1.30 \text{ m/s}$$
$$v_y = -gt = -(9.81 \text{ m/s}^2)(0.500 \text{ s}) = -4.91 \text{ m/s}$$

4. Use these components to determine $\vec{\mathbf{v}}$, v, and θ:

$$\vec{\mathbf{v}} = (1.30 \text{ m/s})\hat{\mathbf{x}} + (-4.91 \text{ m/s})\hat{\mathbf{y}}$$
$$v = \sqrt{v_x^2 + v_y^2}$$
$$= \sqrt{(1.30 \text{ m/s})^2 + (-4.91 \text{ m/s})^2} = 5.08 \text{ m/s}$$
$$\theta = \tan^{-1}\frac{v_y}{v_x} = \tan^{-1}\frac{(-4.91 \text{ m/s})}{1.30 \text{ m/s}} = -75.2°$$

INSIGHT
It's interesting to note that the x position of the ball doesn't depend on the acceleration of gravity, g, and its y position doesn't depend on the initial horizontal speed of the ball, v_0. For example, if the person is running when he drops the ball, the ball is moving faster in the horizontal direction, and it keeps up with the person as it falls. Its vertical motion doesn't change at all, however.

PRACTICE PROBLEM
How much time does it take for the ball to land?   [**Answer:** From the results of part (b), it's clear that the time of landing is slightly greater than 0.500 s. Setting $y = 0$ gives a precise answer: $t = \sqrt{2h/g} = 0.505$ s.]

Some related homework problems: Problem 17, Problem 18, Problem 21

CONCEPTUAL EXAMPLE 4-4   COMPARE SPLASHDOWN SPEEDS

Two youngsters dive off an overhang into a lake. Diver 1 drops straight down, and diver 2 runs off the cliff with an initial horizontal speed v_0. Is the splashdown speed of diver 2 **(a)** greater than, **(b)** less than, or **(c)** equal to the splashdown speed of diver 1?

REASONING AND DISCUSSION
Neither diver has an initial y component of velocity, and both fall with the same vertical acceleration—the acceleration due to gravity. Therefore, the two divers fall for the same amount of time, and their y components of velocity are the same at splashdown. However, because diver 2 also has a nonzero x component of velocity, unlike diver 1, the speed of diver 2 is greater.

ANSWER
(a) The speed of diver 2 is greater than that of diver 1.

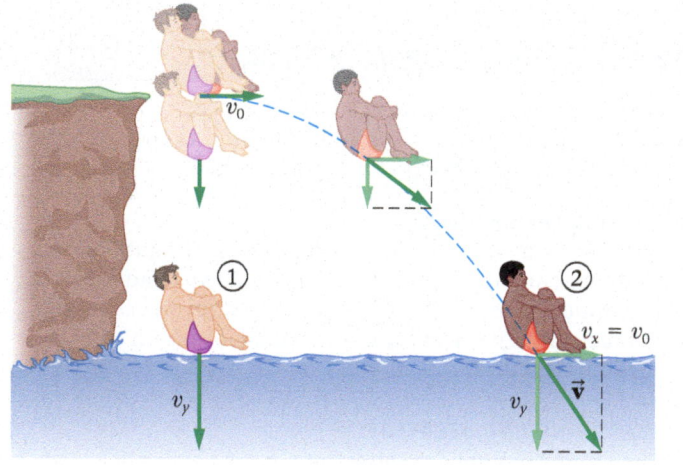

Parabolic Path

RWP Just what is the shape of the curved path followed by a projectile launched horizontally? This can be found by combining $x = v_0 t$ and $y = h - \frac{1}{2}gt^2$, which allows us to express y in terms of x. First, solve for time using the x equation. This gives

$$t = \frac{x}{v_0}$$

Next, substitute this result into the y equation to eliminate t:

$$y = h - \frac{1}{2}g\left(\frac{x}{v_0}\right)^2 = h - \left(\frac{g}{2v_0^2}\right)x^2 \qquad 4\text{-}8$$

It follows that y has the form

$$y = a + bx^2$$

◄ **FIGURE 4-7** Visualizing Concepts
Parabolic Trajectories Lava bombs (left) and fountain jets (right) trace out parabolic paths, as is typical in projectile motion. The trajectories are only slightly altered by air resistance.

In this expression, $a = h$ = constant and $b = -g/2v_0^2$ = constant. This is the equation of a *parabola* that curves downward, a characteristic shape in projectile motion. Examples of parabolas produced by real-world projectiles are shown in **FIGURE 4-7**.

Landing Site

Where does a projectile land if it is launched horizontally with a speed v_0 from a height h?

The most direct way to answer this question is to set $y = 0$ in Equation 4-8, since $y = 0$ corresponds to ground level. This gives

$$0 = h - \left(\frac{g}{2v_0^2}\right)x^2$$

Solving for x yields the landing site:

$$x = v_0\sqrt{\frac{2h}{g}} \qquad \text{4-9}$$

We've chosen the positive sign for the square root because the projectile was launched in the positive x direction, and hence it lands at a positive value of x.

PROBLEM-SOLVING NOTE

Use Independence of Motion

Projectile problems can be solved by breaking the problem into its x and y components, and then solving for the motion of each component separately.

EXAMPLE 4-5   JUMPING A CREVASSE

A mountain climber encounters a crevasse in an ice field. The opposite side of the crevasse is 2.74 m lower, and is separated horizontally by a distance of 4.10 m. To cross the crevasse, the climber gets a running start and jumps in the horizontal direction. **(a)** What is the minimum speed needed by the climber to safely cross the crevasse? **(b)** If, instead, the climber's speed is 6.00 m/s, where does the climber land?

PICTURE THE PROBLEM
The mountain climber jumps from $x_0 = 0$ and $y_0 = h = 2.74$ m. The landing site for part (a) is $x = w = 4.10$ m and $y = 0$. As for the initial velocity, we are given that $v_{0x} = v_0$ and $v_{0y} = 0$. Finally, with our choice of coordinates it follows that $a_x = 0$ and $a_y = -g$.

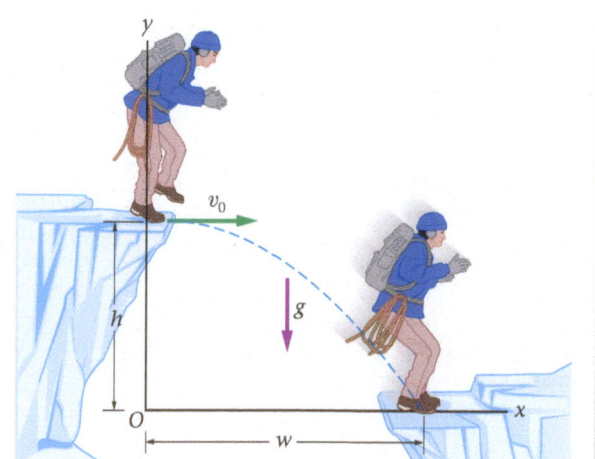

REASONING AND STRATEGY
We can model the climber as a projectile, and apply our equations for projectile motion with a horizontal launch.

a. From Equations 4-7 we have that $x = v_0 t$ and $y = h - \frac{1}{2}gt^2$. Setting $y = 0$ determines the time of landing. Using this time in the x equation gives the horizontal landing position in terms of the initial speed.

b. We can now use the relationship from part (a) to find x in terms of $v_0 = 6.00$ m/s.

Known Height, $h = 2.74$ m; width, $w = 4.10$ m.
Unknown (a) Minimum initial speed, $v_0 = ?$ (b) Landing distance, $x = ?$, for $v_0 = 6.00$ m/s

SOLUTION

Part (a)

1. Set $y = h - \frac{1}{2}gt^2$ equal to zero (landing condition) and solve for the corresponding time t:

$$y = h - \frac{1}{2}gt^2 = 0$$

$$t = \sqrt{\frac{2h}{g}}$$

CONTINUED

2. Substitute this expression for t into the x equation of motion, $x = v_0 t$, and solve for the speed, v_0:

$$x = v_0 t = v_0 \sqrt{\frac{2h}{g}} \quad \text{or} \quad v_0 = x \sqrt{\frac{g}{2h}}$$

3. Substitute numerical values in this expression:

$$v_0 = x \sqrt{\frac{g}{2h}} = (4.10 \text{ m}) \sqrt{\frac{9.81 \text{ m/s}^2}{2(2.74 \text{ m})}} = 5.49 \text{ m/s}$$

Part (b)

4. Substitute $v_0 = 6.00$ m/s into the expression for x obtained in Step 2, $x = v_0 \sqrt{2h/g}$:

$$x = v_0 \sqrt{\frac{2h}{g}} = (6.00 \text{ m/s}) \sqrt{\frac{2(2.74 \text{ m})}{9.81 \text{ m/s}^2}} = 4.48 \text{ m}$$

INSIGHT

The minimum speed to safely cross the crevasse is 5.49 m/s. If the initial horizontal speed is 6.00 m/s, the climber will land 4.48 m − 4.10 m = 0.38 m beyond the edge of the crevasse.

PRACTICE PROBLEM

(a) When the climber's speed is the minimum needed to cross the crevasse, $v_0 = 5.49$ m/s, for what amount of time is the climber in the air? (b) For what amount of time is the climber in the air when $v_0 = 6.00$ m/s?   [**Answer:** (a) $t = x/v_0 = (4.10 \text{ m})/(5.49 \text{ m/s}) = 0.747$ s.   (b) $t = x/v_0 = (4.48 \text{ m})/(6.00 \text{ m/s}) = 0.747$ s. The times are exactly the same! The answer to both parts is simply the time needed to fall through a height h: $t = \sqrt{2h/g} = 0.747$ s.]

Some related homework problems: Problem 13, Problem 14, Problem 19

CONCEPTUAL EXAMPLE 4-6   MINIMUM SPEED   PREDICT/EXPLAIN

(a) If the height h is increased in the Example 4-5, but the width w remains the same, does the minimum speed needed to cross the crevasse increase, decrease, or stay the same? (b) Which of the following is the *best explanation* for your prediction?

  I. The time of fall is greater, so the minimum speed decreases.

 II. The width remains the same, so the minimum speed stays the same as well.

III. Increasing the height increases the minimum speed.

REASONING AND DISCUSSION

If the height is greater, the time of fall is also greater. This means that the climber is in the air for a greater time, and hence the horizontal distance covered for a given initial speed is also greater. It follows that a lower initial speed allows for a safe crossing.

ANSWER

(a) The minimum speed decreases. (b) The best explanation is I.

**PHYSICS
IN CONTEXT**
Looking Ahead

Two-dimensional kinematics comes up again when we study the motion of charged particles (like electrons) in electric fields. For example, compare Figure 19-41 and Figure 22-10 with the person jumping a crevasse in Example 4-5. The same basic principles apply in each case.

Enhance Your Understanding   (Answers given at the end of the chapter)

3. Two objects, A and B, are launched horizontally, as indicated in **FIGURE 4-8**. The horizontal distance covered before landing is smaller for object A than for object B. (a) Is the initial speed of object A greater than, less than, or equal to the initial speed of object B? Explain. (b) Is the amount of time object A is in the air greater than, less than, or equal to the time object B is in the air? Explain.

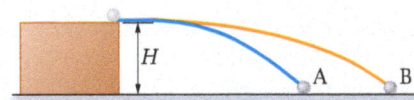

▲ **FIGURE 4-8** Enhance Your Understanding 3

Section Review

• Projectiles follow a parabolic path.

• A projectile launched horizontally falls to the ground in the same time as one that is dropped from rest from the same height.

4-4 General Launch Angle

We now consider the case of a projectile launched at an arbitrary angle with respect to the horizontal. As always, we return to our basic equations for projectile motion (Equations 4-6), and this time we simply let θ be nonzero.

FIGURE 4-9 (a) shows a projectile launched with an initial speed v_0 at an angle θ above the horizontal. Since the projectile starts at the origin, the initial x and y positions are zero:

$$x_0 = y_0 = 0$$

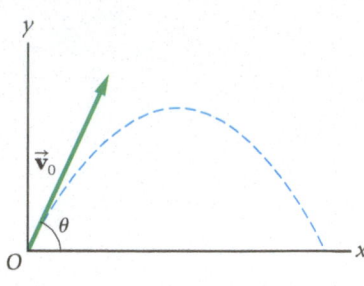
(a)

The components of the initial velocity are determined as indicated in **FIGURE 4-9 (b)**:

$$v_{0x} = v_0 \cos \theta$$
$$v_{0y} = v_0 \sin \theta$$

As a quick check, notice that if $\theta = 0$, then $v_{0x} = v_0$ and $v_{0y} = 0$, as expected. Similarly, if $\theta = 90°$, we find $v_{0x} = 0$ and $v_{0y} = v_0$. These checks are depicted in **FIGURE 4-9 (c)**.

General Equations of Motion Substituting the preceding results into the basic equations for projectile motion yields the following results for a general launch angle:

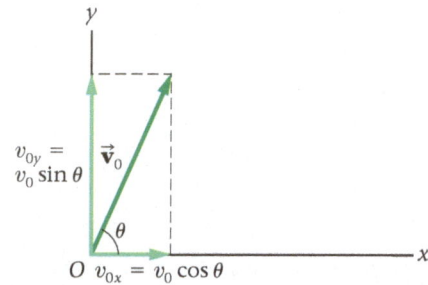
(b)

$$x = (v_0 \cos \theta)t \qquad v_x = v_0 \cos \theta \qquad v_x{}^2 = v_0{}^2 \cos^2 \theta$$
$$y = (v_0 \sin \theta)t - \tfrac{1}{2}gt^2 \quad v_y = v_0 \sin \theta - gt \quad v_y{}^2 = v_0{}^2 \sin^2 \theta - 2g\Delta y$$

4-10

These equations, which are valid for any launch angle, reduce to the simpler Equations 4-7 when we set $\theta = 0$ and $y_0 = h$. In the next two Examples, we use Equations 4-10 to calculate a projectile's position and velocity for three equally spaced times.

QUICK EXAMPLE 4-7   POSITION OF A PROJECTILE

A projectile is launched from the origin with an initial speed of 20.0 m/s at an angle of 35.0° above the horizontal. Find the x and y positions of the projectile at the times **(a)** $t = 0.500$ s, **(b)** $t = 1.00$ s, and **(c)** $t = 1.50$ s.

REASONING AND SOLUTION
Use $x = (v_0 \cos \theta)t$ with $v_0 = 20.0$ m/s and $\theta = 35.0°$ to find the x position of the projectile. Similarly, use $y = (v_0 \sin \theta)t - \tfrac{1}{2}gt^2$ with $g = 9.81$ m/s^2 to find the y position.

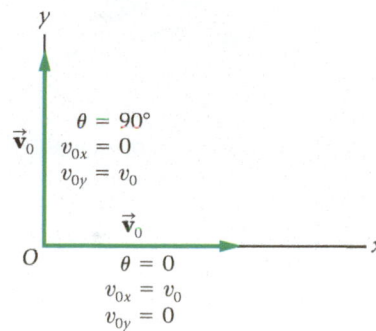
(c)

▲ **FIGURE 4-9 Projectile with an arbitrary launch angle (a)** A projectile launched from the origin at an angle θ above the horizontal. **(b)** The x and y components of the initial velocity. **(c)** Velocity components in the limits $\theta = 0$ and $\theta = 90°$.

a. $x = (20.0 \text{ m/s})(\cos 35.0°)(0.500 \text{ s}) = 8.19$ m
$y = (20.0 \text{ m/s})(\sin 35.0°)(0.500 \text{ s}) - \tfrac{1}{2}(9.81 \text{ m/s}^2)(0.500 \text{ s})^2 = 4.51$ m

b. Repeating the calculation in part (a), but with 1.00 s for the time, we find $x = 16.4$ m and $y = 6.57$ m.

c. Using 1.50 s for the time in part (a), we find $x = 24.6$ m and $y = 6.17$ m.

Notice that x increases steadily with time. In contrast, y first increases and then decreases.

PHYSICS IN CONTEXT
Looking Back

The vector methods developed in Chapter 3 are used in this section to resolve velocity vectors into their x and y components.

QUICK EXAMPLE 4-8   VELOCITY OF A PROJECTILE

Find the velocity of the projectile in Quick Example 4-7 at the times **(a)** $t = 0.500$ s, **(b)** $t = 1.00$ s, and **(c)** $t = 1.50$ s.

REASONING AND SOLUTION
The x component of the velocity is constant, with the value $v_x = v_0 \cos \theta = (20.0 \text{ m/s}) \cos 35.0° = 16.4$ m/s. The y component of the velocity changes with time according to the relationship $v_y = v_0 \sin \theta - gt$.

a. $v_x = v_0 \cos \theta = (20.0 \text{ m/s}) \cos 35.0° = 16.4$ m/s
$v_y = (20.0 \text{ m/s}) \sin 35.0° - (9.81 \text{ m/s}^2)(0.500 \text{ s}) = 6.57$ m/s
$\vec{v} = (16.4 \text{ m/s})\hat{x} + (6.57 \text{ m/s})\hat{y}$

CONTINUED

b. Repeating part (a) with 1.00 s for the time, we find $\vec{\mathbf{v}} = (16.4\ \text{m/s})\hat{\mathbf{x}} + (1.66\ \text{m/s})\hat{\mathbf{y}}$.

c. Repeating part (a) with 1.50 s for the time, we find $\vec{\mathbf{v}} = (16.4\ \text{m/s})\hat{\mathbf{x}} + (-3.24\ \text{m/s})\hat{\mathbf{y}}$.

Notice that the projectile is moving upward in parts (a) and (b), but is moving downward (negative v_y) in part (c).

FIGURE 4-10 shows the projectile referred to in the previous Quick Examples for a series of times spaced by 0.10 s. Notice that the points in Figure 4-10 are not evenly spaced in terms of position, even though they are evenly spaced in time. In fact, the points bunch closer together at the top of the trajectory, showing that a comparatively large fraction of the flight time is spent near the highest point. This is why it seems that a basketball player soaring toward a slam dunk, or a ballerina performing a grand jeté, is "hanging in air." This concept is explored further in FIGURE 4-11.

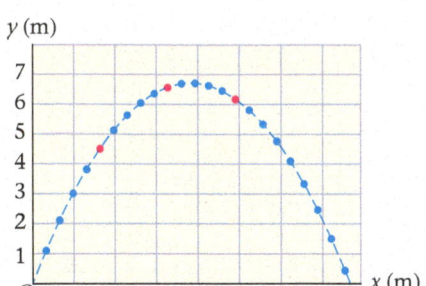

◄ **FIGURE 4-10 Snapshots of a trajectory** This plot shows a projectile launched from the origin with an initial speed of 20.0 m/s at an angle of 35.0° above the horizontal. The positions shown in the plot correspond to the times $t = 0.1\ \text{s}, 0.2\ \text{s}, 0.3\ \text{s}, \ldots$. Red dots mark the positions considered in Quick Examples 4-7 and 4-8.

▲ **FIGURE 4-11 Visualizing Concepts Hang Time** "Hanging in air" near the peak of a jump requires no special knack—in fact, it's an unavoidable consequence of the laws of physics. This phenomenon, which makes big leapers (such as deer and dancers) look particularly graceful, can also make life dangerous for salmon fighting their way upstream to spawn.

EXAMPLE 4-9   A ROUGH SHOT

Chipping from the rough, a golfer sends the ball over a 3.00-m-high tree that is 14.0 m away. The ball lands at the same level from which it was struck after traveling a horizontal distance of 17.8 m—on the green, of course. The ball leaves the club at 54.0° above the horizontal and lands on the green 2.24 s later. **(a)** What was the initial speed of the ball? **(b)** How high was the ball when it passed directly over the tree?

PICTURE THE PROBLEM

Our sketch shows the ball taking flight from the origin, $x_0 = y_0 = 0$, with a launch angle of 54.0°, and arcing over the tree. The individual points along the parabolic trajectory correspond to equal time intervals.

REASONING AND STRATEGY

a. The projectile moves with constant speed in the x direction, and hence the x component of velocity is simply the horizontal distance divided by time. Knowing v_x and θ, we can find v_0 from $v_x = v_0 \cos\theta$.

b. We can use $x = (v_0 \cos\theta)t$ to find the time when the ball is at $x = 14.0\ \text{m}$. Substituting this time into $y = (v_0 \sin\theta)t - \frac{1}{2}gt^2$ gives the height.

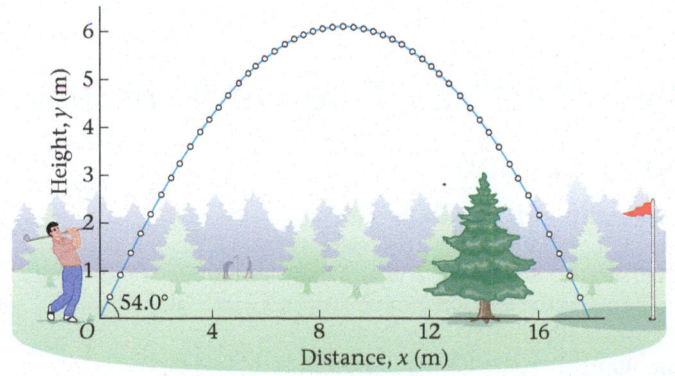

Known Horizontal distance, $d = 17.8\ \text{m}$; launch angle, $\theta = 54.0°$; time of flight, $t = 2.24\ \text{s}$.
Unknown (a) Initial speed, $v_0 = ?$ (b) Height of ball at tree, $y = ?$

CONTINUED

SOLUTION

Part (a)

1. Divide the horizontal distance, d, by the time of flight, t, to obtain v_x:

$$v_x = \frac{d}{t} = \frac{17.8\,\text{m}}{2.24\,\text{s}} = 7.95\,\text{m/s}$$

2. Use $v_x = v_0 \cos\theta$ to find v_0, the launch speed:

$$v_x = v_0 \cos\theta \quad\text{or}\quad v_0 = \frac{v_x}{\cos\theta} = \frac{7.95\,\text{m/s}}{\cos 54.0°} = 13.5\,\text{m/s}$$

Part (b)

3. Use $x = (v_0 \cos\theta)t$ to find the time when $x = 14.0\,\text{m}$. Recall that $x_0 = 0$:

$$x = (v_0 \cos\theta)t \quad\text{or}\quad t = \frac{x}{v_0 \cos\theta} = \frac{14.0\,\text{m}}{7.95\,\text{m/s}} = 1.76\,\text{s}$$

4. Evaluate $y = (v_0 \sin\theta)t - \frac{1}{2}gt^2$ at the time found in Step 3. Recall that $y_0 = 0$:

$$y = (v_0 \sin\theta)t - \frac{1}{2}gt^2$$
$$= [(13.5\,\text{m/s})\sin 54.0°](1.76\,\text{s}) - \frac{1}{2}(9.81\,\text{m/s}^2)(1.76\,\text{s})^2$$
$$= 4.03\,\text{m}$$

INSIGHT

The ball clears the top of the tree by $4.03\,\text{m} - 3.00\,\text{m} = 1.03\,\text{m}$, and lands on the green $2.24\,\text{s} - 1.76\,\text{s} = 0.48\,\text{s}$ later. Just before landing, its speed (in the absence of air resistance) is again 13.5 m/s—the same as when it was launched. This result will be verified in the next section.

PRACTICE PROBLEM

What are the speed and direction of the ball when it passes over the tree? [**Answer:** To find the ball's speed and direction, notice that $v_x = 7.95\,\text{m/s}$ and $v_y = v_0 \sin\theta - gt = -6.34\,\text{m/s}$. It follows that $v = \sqrt{v_x^2 + v_y^2} = 10.2\,\text{m/s}$ and $\theta = \tan^{-1}(v_y/v_x) = -38.6°$.]

Some related homework problems: Problem 32, Problem 38

QUICK EXAMPLE 4-10   AN ELEVATED GREEN

Ball Fired from Cart on Incline

A golfer hits a ball from the origin with an initial speed of 15.0 m/s at an angle of 62.0° above the horizontal. The ball is in the air 2.58 s, and lands on a green that is above the level where the ball was struck.

   a. What is the level of the green?

   b. How far has the ball traveled in the horizontal direction when it lands?

   c. What is the ball's y component of velocity just before it lands?

REASONING AND SOLUTION

The equations of projectile motion (Equations 4-10) can be applied to the golf ball, with $\theta = 62.0°$, $v_0 = 15.0\,\text{m/s}$, and the landing time equal to $t = 2.58\,\text{s}$. **(a)** The level of the green is given by y as a function of time, $y = (v_0 \sin\theta)t - \frac{1}{2}gt^2$. **(b)** The horizontal distance is given by x as a function of time, $x = (v_0 \cos\theta)t$. **(c)** The y component of velocity is given by $v_y = v_0 \sin\theta - gt$.

   a. Evaluate $y = (v_0 \sin\theta)t - \frac{1}{2}gt^2$ at $t = 2.58\,\text{s}$:

$$y = (v_0 \sin\theta)t - \frac{1}{2}gt^2$$
$$= (15.0\,\text{m/s})(\sin 62.0°)(2.58\,\text{s}) - \frac{1}{2}(9.81\,\text{m/s}^2)(2.58\,\text{s})^2$$
$$= 1.52\,\text{m (above the level where the ball was struck)}$$

   b. Substitute $t = 2.58\,\text{s}$ into $x = (v_0 \cos\theta)t$:

$$x = (v_0 \cos\theta)t$$
$$= (15.0\,\text{m/s})(\cos 62.0°)(2.58\,\text{s}) = 18.2\,\text{m}$$

   c. Substitute $t = 2.58\,\text{s}$ into $v_y = v_0 \sin\theta - gt$ to find v_y:

$$v_y = v_0 \sin\theta - gt$$
$$= (15.0\,\text{m/s})\sin 62.0° - (9.81\,\text{m/s}^2)(2.58\,\text{s}) = -12.1\,\text{m/s}$$

The negative value of v_y indicates that the ball is moving downward at $t = 2.58\,\text{s}$, as expected.

Colliding Projectiles The next Example presents a classic situation in which two projectiles collide. One projectile is launched from the origin, and thus its equations of motion are given by Equations 4-10. The second projectile is simply dropped from a height, which is a special case of the equations of motion in Equations 4-7 with $v_0 = 0$.

EXAMPLE 4-11 A LEAP OF FAITH

A trained dolphin leaps from the water with an initial speed of 12.0 m/s. It jumps directly toward a ball held by the trainer a horizontal distance of 5.50 m away and a vertical distance of 4.10 m above the water. In the absence of gravity, the dolphin would move in a straight line to the ball and catch it, but because of gravity the dolphin follows a parabolic path well below the ball's initial position, as shown in the sketch. If the trainer releases the ball the instant the dolphin leaves the water, show that the dolphin and the falling ball meet.

PICTURE THE PROBLEM

In our sketch we have the dolphin leaping from the water at the origin $x_0 = y_0 = 0$ with an angle above the horizontal given by $\theta = \tan^{-1}(h/d)$. The initial position of the ball is $x_0 = d = 5.50$ m and $y_0 = h = 4.10$ m, and its initial velocity is zero. The ball drops straight down with the acceleration due to gravity, $a_y = -g$.

REASONING AND STRATEGY

We want to show that when the dolphin is at $x = d$, its height above the water is the same as the height of the ball above the water. To do this we first find the time when the dolphin is at $x = d$, then calculate y for the dolphin at this time. Next, we calculate y of the ball at the same time and then check to see if they are equal. Because the ball drops from rest from a height h, its y equation of motion is $y = h - \frac{1}{2}gt^2$, as in Equations 4-7 in Section 4-3.

Known   Initial speed, $v_0 = 12.0$ m/s; initial horizontal distance to ball, $d = 5.50$ m; initial vertical distance to ball, $h = 4.10$ m; acceleration of ball and dolphin, $a_y = -g$.

Unknown   Height at which dolphin and ball meet, $y_d = y_b = ?$

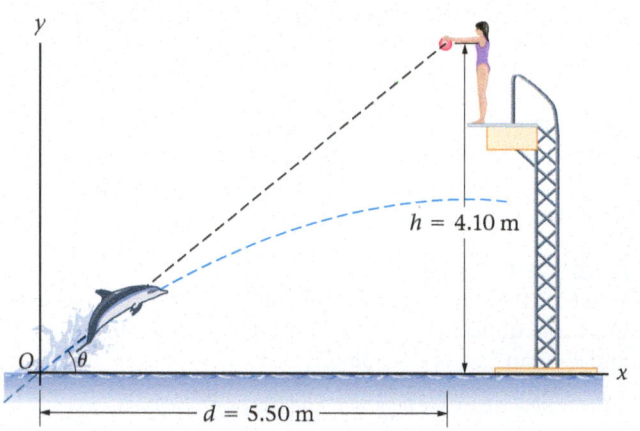

$h = 4.10$ m

$d = 5.50$ m

SOLUTION

1. Calculate the angle at which the dolphin leaves the water:

$$\theta = \tan^{-1}\left(\frac{h}{d}\right) = \tan^{-1}\left(\frac{4.10\,\text{m}}{5.50\,\text{m}}\right) = 36.7°$$

2. Use this angle and the initial speed to find the time t when the x position of the dolphin, x_d, is equal to 5.50 m. The x equation of motion is $x_d = (v_0 \cos\theta)t$:

$$x_d = (v_0 \cos\theta)t = [(12.0\,\text{m/s})\cos 36.7°]t = (9.62\,\text{m/s})t$$
$$= 5.50\,\text{m}$$
$$t = \frac{5.50\,\text{m}}{9.62\,\text{m/s}} = 0.572\,\text{s}$$

3. Evaluate the y position of the dolphin, y_d, at $t = 0.572$ s. The y equation of motion is $y_d = (v_0 \sin\theta)t - \frac{1}{2}gt^2$:

$$y_d = (v_0 \sin\theta)t - \frac{1}{2}gt^2$$
$$= [(12.0\,\text{m/s})\sin 36.7°](0.572\,\text{s}) - \frac{1}{2}(9.81\,\text{m/s}^2)(0.572\,\text{s})^2$$
$$= 4.10\,\text{m} - 1.60\,\text{m} = 2.50\,\text{m}$$

4. Evaluate the y position of the ball, y_b, at $t = 0.572$ s. The ball's equation of motion is $y_b = h - \frac{1}{2}gt^2$:

$$y_b = h - \frac{1}{2}gt^2 = 4.10\,\text{m} - \frac{1}{2}(9.81\,\text{m/s}^2)(0.572\,\text{s})^2$$
$$= 4.10\,\text{m} - 1.60\,\text{m} = 2.50\,\text{m}$$

INSIGHT

In the absence of gravity, both the dolphin and the ball are at $x = 5.50$ m and $y = 4.10$ m at $t = 0.572$ s. Because of gravity, however, the dolphin and the ball fall below their zero-gravity positions—and by the *same amount*, 1.60 m. In fact, from the point of view of the dolphin, the ball is always at the same angle of 36.7° above the horizontal until it is caught.

This is shown in the accompanying plot, where the red dots show the position of the ball at ten equally spaced times, and the blue dots show the position of the dolphin at the corresponding times. In addition, the dashed lines from the dolphin to the ball all make the same angle with the horizontal, 36.7°.

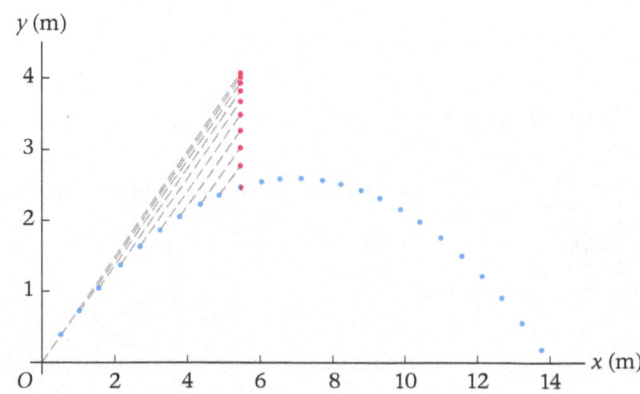

PRACTICE PROBLEM

At what height does the dolphin catch the ball if it leaves the water with an initial speed of 8.22 m/s? [**Answer:** $y_d = y_b = 0.680$ m. If the dolphin's initial speed is less than 7.50 m/s, it reenters the water before it catches the ball.]

Some related homework problems: Problem 69, Problem 79

4. A projectile is launched and lands at the same level, as shown in **FIGURE 4-12**. Three points during the flight of the projectile are indicated in the figure. **(a)** Which of the four velocity vectors (shown in green) corresponds to points 1, 2, and 3? **(b)** Which of the four acceleration vectors (shown in purple) corresponds to points 1, 2, and 3?

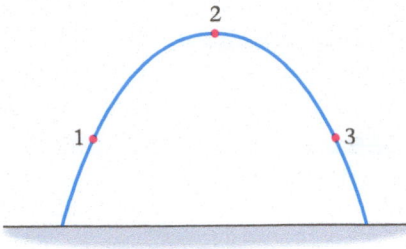

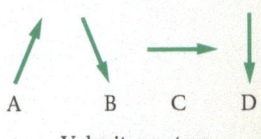

A   B   C   D

Velocity vectors

E   F   G   H

Acceleration vectors

▲ **FIGURE 4-12** Enhance Your Understanding 4

Section Review

- The basic equations for projectile motion are given in Equations 4-10. They combine constant-acceleration motion in the vertical direction with constant-speed motion in the horizontal direction.

4-5 Projectile Motion: Key Characteristics

We conclude this chapter with a brief look at some additional characteristics of projectile motion that are both interesting and useful. In all cases, our results follow as a direct consequence of the fundamental kinematic equations (Equations 4-10) describing projectile motion.

Range

The **range**, R, of a projectile is the horizontal distance it travels before landing. We consider the case shown in **FIGURE 4-13**, where the initial and final elevations are the same ($y = 0$). The initial elevation occurs at $t = 0$, when the projectile is launched, and the projectile returns to the same elevation at a later time when it lands. One way to obtain the range, then, is as follows:

(i) Find the later time ($t > 0$) when the projectile lands by setting $y = 0$ in the expression

$$y = (v_0 \sin \theta)t - \tfrac{1}{2}gt^2.$$

(ii) Substitute the time found in (i) into the x equation of motion. Carrying out the first part of the calculation yields the following:

$$(v_0 \sin \theta)t - \tfrac{1}{2}gt^2 = 0 \quad \text{or} \quad (v_0 \sin \theta)t = \tfrac{1}{2}gt^2$$

Clearly, $t = 0$ is a solution to this equation—it corresponds to the initial condition—but the solution we seek is a time that is greater than zero. We can find the desired time by dividing both sides of the equation by t. This gives

$$(v_0 \sin \theta) = \tfrac{1}{2}gt \quad \text{or} \quad t = \left(\frac{2v_0}{g}\right)\sin \theta \qquad \text{4-11}$$

This is the time when the projectile lands—also known as the *time of flight*.

  Now, substitute this time into $x = (v_0 \cos \theta)t$ to find the value of x when the projectile lands:

$$x = (v_0 \cos \theta)t = (v_0 \cos \theta)\left(\frac{2v_0}{g}\right)\sin \theta = \left(\frac{2v_0{}^2}{g}\right)\sin \theta \cos \theta$$

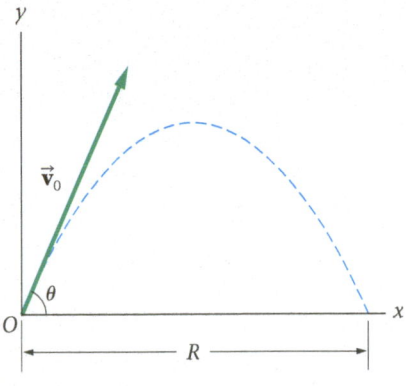

▲ **FIGURE 4-13 Range of a projectile** The range R of a projectile is the horizontal distance it travels between its takeoff and landing positions.

Use the Same Math Regardless of the Initial Conditions

Once an object is launched, its trajectory follows the kinematic equations of motion, regardless of specific differences in the initial conditions. Thus, our equations of motion can be used to derive any desired characteristic of projectile motion, including range, symmetry, and maximum height.

This value of x is the range, R; thus

$$R = \left(\frac{2v_0^2}{g}\right)\sin\theta\cos\theta$$

Using the trigonometric identity $\sin 2\theta = 2\sin\theta\cos\theta$, as given in Appendix A, we can write this more compactly as follows:

$$R = \left(\frac{v_0^2}{g}\right)\sin 2\theta \quad \text{(same initial and final elevation)} \qquad 4\text{-}12$$

EXERCISE 4-12 FIND THE INITIAL SPEED

A football game begins with a kickoff in which the ball travels a horizontal distance of 37 m (approximately 40 yd) and lands at the same level from which it was kicked. If the ball was kicked at an angle of 44° above the horizontal, what was its initial speed?

REASONING AND SOLUTION

We know the range ($R = 37$ m) and the initial angle ($\theta = 44°$.) With this information we can rearrange the equation $R = (v_0^2/g)\sin 2\theta$ to solve for the initial speed, v_0:

$$v_0 = \sqrt{gR/\sin 2\theta}$$
$$= \sqrt{(9.81 \text{ m/s}^2)(37 \text{ m})/\sin 88°} = 19 \text{ m/s}$$

We choose the positive square root because we are interested in the *speed* of the ball, which is always positive.

RWP Notice that R depends inversely on the acceleration due to gravity, g—thus the smaller g, the larger the range. For example, a projectile launched on the Moon, where the acceleration due to gravity is only about one-sixth that on Earth, travels about six times as far as it would on Earth. It was for this reason that astronaut Alan Shepard simply couldn't resist the temptation of bringing a golf club and ball with him on the third lunar landing mission in 1971. He ambled out onto the Fra Mauro Highlands and became the first person to hit a golf shot on the Moon. His distance was undoubtedly respectable—unfortunately, his ball landed in a sand trap.

Maximum Range Now, what launch angle gives the greatest range? From Equation 4-12 we see that R varies with angle as $\sin 2\theta$; thus R is maximum when $\sin 2\theta$ is greatest—that is, when $\sin 2\theta = 1$. Since $\sin 90° = 1$, it follows that $\theta = 45°$ gives the maximum range. Thus

$$R_{\text{max}} = \frac{v_0^2}{g} \qquad 4\text{-}13$$

As expected, the range (Equation 4-12) and maximum range (Equation 4-13) depend strongly on the initial speed of the projectile—they are both proportional to v_0^2.

These results are specifically for the case where a projectile lands at the same level from which it was launched. If a projectile lands at a higher level, for example, the launch angle that gives the maximum range is greater than 45°, and if it lands at a lower level, the angle for the maximum range is less than 45°.

Finally, the range given here applies only to the ideal case of no air resistance. In cases where air resistance is significant, as in the flight of a rapidly moving golf ball, for example, the overall range of the ball is reduced. In addition, the maximum range occurs for a launch angle less than 45° (**FIGURE 4-14**). The reason is that with a smaller launch angle the golf ball is in the air for less time, giving air resistance less time to affect its flight.

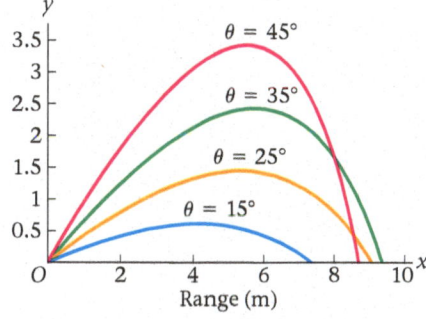

▲ **FIGURE 4-14 Projectiles with air resistance** Projectiles with the same initial speed but different launch angles showing the effects of air resistance. Notice that the maximum range occurs for a launch angle less than 45°, and that the projectiles return to the ground at a steeper angle than the launch angle.

Symmetry in Projectile Motion

There are many striking symmetries in projectile motion, beginning with the graceful symmetry of the parabola itself. As an example, recall that the time when a projectile lands at the same level from which it was launched is

$$t = \left(\frac{2v_0}{g}\right)\sin\theta$$

This result was obtained in Equation 4-11. Now, recall that when a projectile is launched, its y component of velocity is $v_y = v_0 \sin \theta$. When the projectile lands, at time $t = (2v_0/g) \sin \theta$, its y component of velocity is

$$v_y = v_0 \sin \theta - gt = v_0 \sin \theta - g\left(\frac{2v_0}{g}\right) \sin \theta = -v_0 \sin \theta$$

This is exactly the opposite of the y component of the velocity when it was launched. Since the x component of velocity is always the same, it follows that the projectile lands with the same *speed* it had when it was launched. The *velocities* are different, however, because the direction of motion is different at launch and landing. In fact, there's a nice symmetry with the direction of motion: If the initial velocity is *above* the horizontal by the angle θ, the landing velocity is *below* the horizontal by the same angle θ.

So far, these results have referred to launching and landing, which both occur at $y = 0$. The same symmetry occurs at any level, though. That is, at a given height the speed of a projectile is the same on the way up as on the way down. In addition, the angle of the velocity above the horizontal on the way up is the same as the angle below the horizontal on the way down. This is illustrated in **FIGURE 4-15** and in the next Conceptual Example.

> **Big Idea** **3** Projectiles follow parabolic paths that curve downward toward the ground. The motion of a projectile on the way up is mirrored by its motion on the way down.

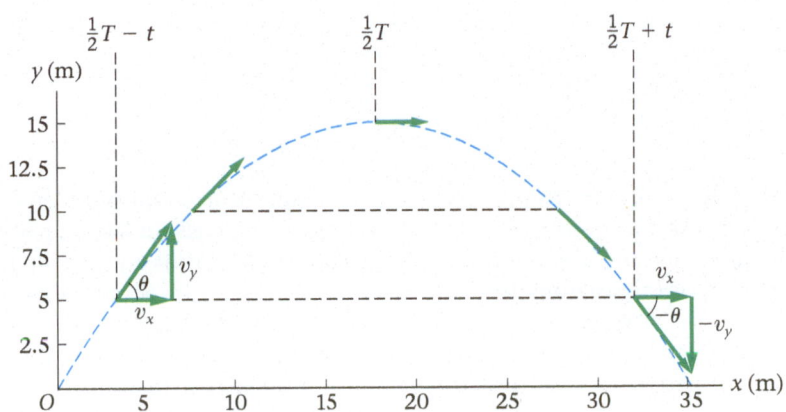

◄ **FIGURE 4-15 Velocity vectors for a projectile launched at the origin** At a given height the speed (length of velocity vector) is the same on the way up as on the way down. The direction of motion on the way up is above the horizontal by the same amount that it is below the horizontal on the way down. In this case, the total time of flight is T, and the greatest height is reached at the time $T/2$. Notice that the speed is the same at the time $(T/2) - t$ as it is at the time $(T/2) + t$.

CONCEPTUAL EXAMPLE 4-13   COMPARE LANDING SPEEDS

Two cats are sitting side-by-side on a fence when they're startled by a dog. The cats jump off the fence with equal speeds, but different initial angles relative to the horizontal. Cat 1 jumps at an angle of 45° above the horizontal; cat 2 jumps at an angle of 45° below the horizontal. Is the landing speed of cat 1 greater than, less than, or equal to the landing speed of cat 2?

REASONING AND DISCUSSION

One consequence of symmetry in projectile motion is that when cat 1 returns to the level of the jump, its speed is the same as its initial speed. In addition, it is now moving downward at 45° below the horizontal. From that point on its motion is the same as that of cat 2. Therefore, cat 1 lands with the same speed as cat 2.

What if cat 1 jumps horizontally? Or suppose it jumps at an angle of 80° above the horizontal? In either case, the final speed is unchanged! In fact, for a given initial speed, the speed on landing simply doesn't depend on the direction at which the cat jumps. This is shown in Homework Problems 34 and 77. We return to this point in Chapter 8 when we discuss energy conservation.

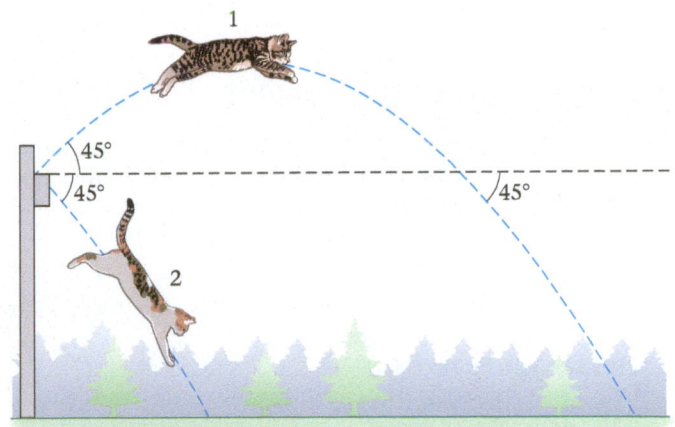

ANSWER
The cats have equal speeds on landing.

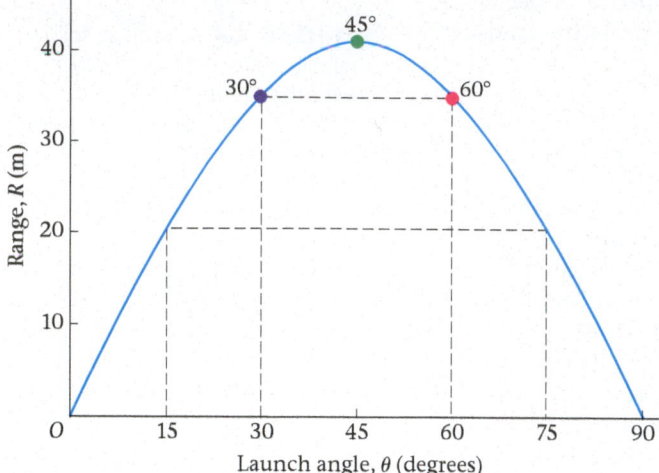

(a) Launch angles that are greater or less than 45° by the same amount give the same range.

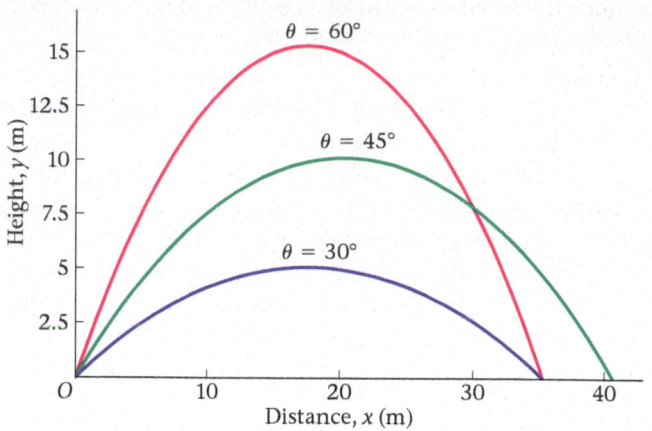

(b) Projectiles with $\theta = 30°$ and $\theta = 60°$ follow different paths but have the same range.

Range of a Gun at Two Firing Angles

▲ **FIGURE 4-16 Range and launch angle in the absence of air resistance (a)** A plot of range versus launch angle for a projectile launched with an initial speed of 20 m/s. Note that the maximum range occurs at $\theta = 45°$. Launch angles equally greater than or less than 45°, such as 30° and 60°, give the same range. **(b)** Trajectories of projectiles with initial speeds of 20 m/s and launch angles of 60°, 45°, and 30°. The projectiles with launch angles of 30° and 60° land at the same location.

Range Versus Launch Angle   As our final example of projectile symmetry, consider the range R. A plot of R versus launch angle θ is shown in **FIGURE 4-16 (a)** for $v_0 = 20$ m/s. Note that in the absence of air resistance, R is greatest at $\theta = 45°$, as pointed out previously. In addition, we can see from the figure that the range for angles equally above or below 45° is the same. For example, if air resistance is negligible, the range for $\theta = 30°$ is the same as the range for $\theta = 60°$, as we can see in both parts (a) and (b) of Figure 4-16.

Symmetries such as these are just some of the many reasons why physicists find physics to be "beautiful" and "aesthetically pleasing." Discovering such patterns and symmetries in nature is really what physics is all about. A physicist does not consider the beauty of projectile motion to be diminished by analyzing it in detail. Just the opposite—detailed analysis reveals deeper, more subtle, and sometimes unexpected levels of beauty.

EXAMPLE 4-14   WHAT A SHOT!

The archerfish hunts by dislodging an unsuspecting insect from its resting place with a stream of water expelled from the fish's mouth. Suppose the archerfish squirts water with an initial speed of 2.30 m/s at an angle of 19.5° above the horizontal. When the stream of water reaches a beetle on a leaf at height h above the water's surface, it is moving horizontally.

a. How much time does the beetle have to react?

b. What is the height h of the beetle?

c. What is the horizontal distance d between the fish and the beetle when the water is launched?

PICTURE THE PROBLEM

Our sketch shows the fish squirting water from the origin, $x_0 = y_0 = 0$, and the beetle at $x = d$, $y = h$. The stream of water starts off with a speed $v_0 = 2.30$ m/s at an angle $\theta = 19.5°$ above the horizontal. The water is moving horizontally when it reaches the beetle.

REASONING AND STRATEGY

a. Because the stream of water is moving horizontally when it reaches the beetle, it is at the top of its parabolic trajectory, as can be seen in Figure 4-15. This means that its y component of velocity is zero. Therefore, we can set $v_y = 0$ in $v_y = v_0 \sin \theta - gt$ and solve for the time t.

CONTINUED

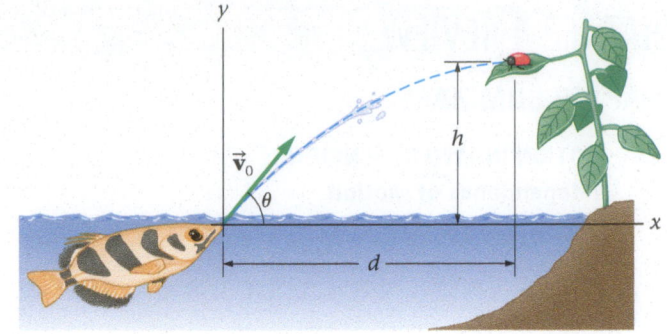

b. To find the maximum height of the stream of water, and of the beetle, we substitute the time found in part (a) into $y = (v_0 \sin\theta)t - \frac{1}{2}gt^2$.

c. Similarly, we can find the horizontal distance d by substituting the time from part (a) into $x = (v_0 \cos\theta)t$.

Known Initial speed, $v_0 = 2.30$ m/s; angle above the horizontal, $\theta = 19.5°$.

Unknown **(a)** Reaction time, $t = ?$ **(b)** Height of beetle, $h = ?$ **(c)** Horizontal distance to beetle, $d = ?$

SOLUTION

Part (a)

1. Set $v_y = v_0 \sin\theta - gt$ equal to zero and solve for the corresponding time t:

$$v_y = v_{0y} - gt = v_0 \sin\theta - gt = 0$$
$$t = \frac{v_0 \sin\theta}{g}$$

2. Substitute numerical values to determine the reaction time:

$$t = \frac{v_0 \sin\theta}{g} = \frac{(2.30 \text{ m/s}) \sin 19.5°}{9.81 \text{ m/s}^2} = 0.0783 \text{ s}$$

Part (b)

3. To calculate the height, we can substitute $t = (v_0 \sin\theta)/g$ or $t = 0.0783$ s into $y = (v_0 \sin\theta)t - \frac{1}{2}gt^2$. For practice, let's work symbolically:

$$y = (v_0 \sin\theta)\left(\frac{v_0 \sin\theta}{g}\right) - \frac{1}{2}g\left(\frac{v_0 \sin\theta}{g}\right)^2 = \frac{(v_0 \sin\theta)^2}{2g}$$

4. Now, substitute numerical values into our symbolic result to find the height, $y = h$:

$$h = \frac{(v_0 \sin\theta)^2}{2g} = \frac{[(2.30 \text{ m/s}) \sin 19.5°]^2}{2(9.81 \text{ m/s}^2)} = 0.0300 \text{ m}$$

Part (c)

5. We can find the horizontal distance d using x as a function of time, $x = (v_0 \cos\theta)t$:

$$x = (v_0 \cos\theta)t$$
$$d = [(2.30 \text{ m/s}) \cos 19.5°](0.0783 \text{ s}) = 0.170 \text{ m}$$

INSIGHT

To hit the beetle, the fish aims 19.5° above the horizontal. For comparison, the straight-line angle to the beetle is $\tan^{-1}(0.0300/0.170) = 10.0°$. Therefore, the fish cannot aim directly at its prey if it wants a meal. An example of an archerfish scoring a direct hit is shown in **FIGURE 4-17**.

Finally, notice that by working symbolically in Step 3 we derived a general result for the maximum height of a projectile. In particular, we found that $y_{max} = (v_0 \sin\theta)^2/2g$, a result that is valid for any launch speed and angle.

PRACTICE PROBLEM

How far does the stream of water go if it happens to miss the beetle? [**Answer:** By symmetry, the distance d is half the range. Thus the stream of water travels a distance $R = 2d = 0.340$ m.]

Some related homework problems: Problem 43, Problem 82

▶ **FIGURE 4-17** An archerfish would have trouble procuring its lunch without an instinctive grasp of projectile motion.

Enhance Your Understanding   (Answers given at the end of the chapter)

5. A baseball player throws a ball to another player. The initial speed of the ball is v_0, and it is thrown at an angle of 45° above the horizontal. Is there ever a time during its flight to the other player when its speed is $\frac{1}{2}v_0$? Explain.

Section Review

• Many symmetries characterize projectile motion. For example, the speed of a projectile at a given height is the same whether the projectile is on its way up or on its way down.

CHAPTER 4 REVIEW

CHAPTER SUMMARY

4-1   MOTION IN TWO DIMENSIONS

Independence of Motion
Motion in the x and y directions are independent of one another.

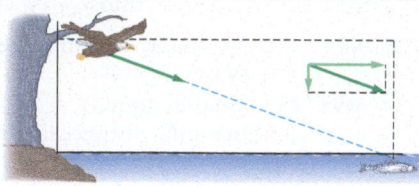

4-2   PROJECTILE MOTION: BASIC EQUATIONS

Projectile motion refers to the path of an object after it is thrown, kicked, batted, or otherwise launched into the air.

x and y as Functions of Time
The x and y equations of motion are

$$x = x_0 + v_{0x}t$$
$$y = y_0 + v_{0y}t - \tfrac{1}{2}gt^2$$

4-6

v_x and v_y as Functions of Time
The velocity components vary with time as follows:

$$v_x = v_{0x}$$
$$v_y = v_{0y} - gt$$

4-6

v_x and v_y as Functions of Displacement
v_x and v_y vary with displacement as

$$v_x{}^2 = v_{0x}{}^2$$
$$v_y{}^2 = v_{0y}{}^2 - 2g\Delta y$$

4-6

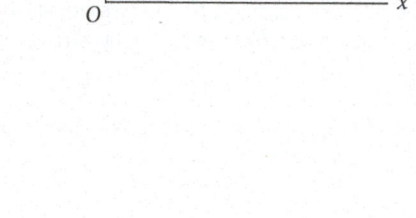

$$a_x = 0$$
$$a_y = -g$$

4-3   ZERO LAUNCH ANGLE

Equations of Motion
A projectile launched horizontally from $x_0 = 0$, $y_0 = h$ with an initial speed v_0 has the following equations of motion:

$$x = v_0 t \qquad v_x = v_0 \qquad v_x{}^2 = v_0{}^2$$
$$y = h - \tfrac{1}{2}gt^2 \qquad v_y = -gt \qquad v_y{}^2 = -2g\Delta y$$

4-7

Parabolic Path
The parabolic path followed by a projectile launched horizontally with an initial speed v_0 is described by

$$y = h - \left(\frac{g}{2v_0{}^2}\right)x^2$$

4-8

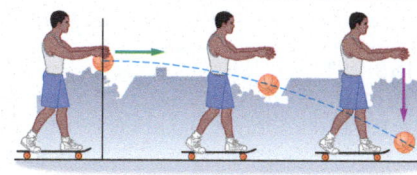

4-4   GENERAL LAUNCH ANGLE

Launch from the Origin
The equations of motion for a launch from the origin with an initial speed v_0 at an angle of θ with respect to the horizontal are

$$x = (v_0 \cos\theta)t \qquad v_x = v_0 \cos\theta \qquad v_x{}^2 = v_0{}^2 \cos^2\theta$$
$$y = (v_0 \sin\theta)t - \tfrac{1}{2}gt^2 \qquad v_y = v_0 \sin\theta - gt \qquad v_y{}^2 = v_0{}^2 \sin^2\theta - 2g\Delta y$$

4-10

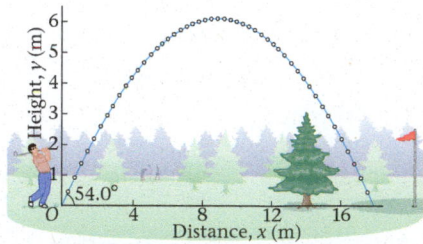

4-5   PROJECTILE MOTION: KEY CHARACTERISTICS

Range
The range of a projectile launched with an initial speed v_0 and a launch angle θ is

$$R = \left(\frac{v_0{}^2}{g}\right)\sin 2\theta \quad \text{(same initial and final elevation)}$$

4-12

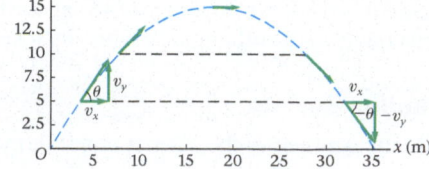

Symmetry
The motion of a projectile on the way down mirrors its motion on the way up.

ANSWERS TO ENHANCE YOUR UNDERSTANDING QUESTIONS

1. **(a)** $x_0 = 0$, $y_0 = 2$ m. **(b)** $v_{0x} = 0$, $v_{0y} = 3$ m/s. **(c)** $a_x = 2$ m/s^2, $a_y = 0$.

2. At the base of the mast. The binoculars continue to move forward with constant velocity as they fall.

3. **(a)** Less, because object A covers a smaller horizontal distance. **(b)** The times are equal because the objects fall through the same vertical distance.

4. **(a)** Point 1, vector A; point 2, vector C; point 3, vector B. **(b)** Points 1 through 3, vector G.

5. No. The minimum speed is $v_x = v_0 \cos 45° = 0.7v_0$, which is greater than $0.5v_0$.

CONCEPTUAL QUESTIONS

For instructor-assigned homework, go to www.masteringphysics.com.

(Answers to odd-numbered Conceptual Questions can be found in the back of the book.)
(The effects of air resistance are to be ignored in this chapter.)

1. What is the acceleration of a projectile when it reaches its highest point? What is its acceleration just before and just after reaching this point?

2. A projectile is launched with an initial speed of v_0 at an angle θ above the horizontal. It lands at the same level from which it was launched. What was its average velocity between launch and landing? Explain.

3. A projectile is launched from level ground. When it lands, its direction of motion has rotated clockwise through 60°. What was the launch angle? Explain.

4. In a game of baseball, a player hits a high fly ball to the outfield. **(a)** Is there a point during the flight of the ball where its velocity is parallel to its acceleration? **(b)** Is there a point where the ball's velocity is perpendicular to its acceleration? Explain in each case.

5. A projectile is launched with an initial velocity of $\vec{\mathbf{v}} = (4\text{ m/s})\hat{\mathbf{x}} + (3\text{ m/s})\hat{\mathbf{y}}$. What is the velocity of the projectile when it reaches its highest point? Explain.

6. A projectile is launched from a level surface with an initial velocity of $\vec{\mathbf{v}} = (2\text{ m/s})\hat{\mathbf{x}} + (4\text{ m/s})\hat{\mathbf{y}}$. What is the velocity of the projectile just before it lands? Explain.

7. Do projectiles for which air resistance is nonnegligible, such as a bullet fired from a rifle, have maximum range when the launch angle is greater than, less than, or equal to 45°? Explain.

8. Two projectiles are launched from the same point at the same angle above the horizontal. Projectile 1 reaches a maximum height twice that of projectile 2. What is the ratio of the initial speed of projectile 1 to the initial speed of projectile 2? Explain.

9. A child rides on a pony walking with constant velocity. The boy leans over to one side and a scoop of ice cream falls from his ice cream cone. Describe the path of the scoop of ice cream as seen by **(a)** the child and **(b)** his parents standing on the ground nearby.

10. Driving down the highway, you find yourself behind a heavily loaded tomato truck. You follow close behind the truck, keeping the same speed. Suddenly a tomato falls from the back of the truck. Will the tomato hit your car or land on the road, assuming you continue moving with the same speed and direction? Explain.

11. A projectile is launched from the origin of a coordinate system where the positive x axis points horizontally to the right and the positive y axis points vertically upward. What was the projectile's launch angle with respect to the x axis if, at its highest point, its direction of motion has rotated **(a)** clockwise through 50° or **(b)** counterclockwise through 30°? Explain.

PROBLEMS AND CONCEPTUAL EXERCISES

Answers to odd-numbered Problems and Conceptual Exercises can be found in the back of the book. **BIO** *identifies problems of biological or medical interest;* **CE** *indicates a conceptual exercise.* **Predict/Explain** *problems ask for two responses: (a) your prediction of a physical outcome, and (b) the best explanation among three provided; and* **Predict/Calculate** *problems ask for a prediction of a physical outcome, based on fundamental physics concepts, and follow that with a numerical calculation to verify the prediction. On all problems, bullets (•, ••, •••) indicate the level of difficulty.* (The effects of air resistance can be ignored in this chapter)

SECTION 4-1   MOTION IN TWO DIMENSIONS

1. • **CE Predict/Explain** As you walk briskly down the street, you toss a small ball into the air. **(a)** If you want the ball to land in your hand when it comes back down, should you toss the ball straight upward, in a forward direction, or in a backward direction, relative to your body? **(b)** Choose the *best explanation* from among the following:
   I. If the ball is thrown straight up you will leave it behind.
   II. You have to throw the ball in the direction you are walking.
   III. The ball moves in the forward direction with your walking speed at all times.

2. • A sailboat runs before the wind with a constant speed of 4.2 m/s in a direction 32° north of west. How far **(a)** west and **(b)** north has the sailboat traveled in 25 min?

3. • As you walk to class with a constant speed of 1.75 m/s, you are moving in a direction that is 18.0° north of east. How much time does it take to change your position by **(a)** 20.0 m east or **(b)** 30.0 m north?

4. • Starting from rest, a car accelerates at 2.0 m/s^2 up a hill that is inclined 5.5° above the horizontal. How far **(a)** horizontally and **(b)** vertically has the car traveled in 12 s?

5. •• **Predict/Calculate** A particle passes through the origin with a velocity of $(6.2\text{ m/s})\hat{\mathbf{y}}$. If the particle's acceleration is $(-4.4\text{ m/s}^2)\hat{\mathbf{x}}$, **(a)** what are its x and y positions after 5.0 s? **(b)** What are v_x and v_y at this time? **(c)** Does the speed of this particle increase with time, decrease with time, or increase and then decrease? Explain.

6. •• A skateboarder travels on a horizontal surface with an initial velocity of 3.8 m/s toward the south and a constant acceleration of 2.2 m/s^2 toward the east. Let the x direction be eastward and the y direction be northward, and let the skateboarder be at the origin at $t = 0$. **(a)** What are her x and y positions at $t = 0.80$ s? **(b)** What are her x and y velocity components at $t = 0.80$ s?

7. •• A hot-air balloon is drifting in level flight due east at 2.5 m/s due to a light wind. The pilot suddenly notices that the balloon must gain 24 m of altitude in order to clear the top of a hill 120 m

to the east. **(a)** How much time does the pilot have to make the altitude change without crashing into the hill? **(b)** What minimum, constant, upward acceleration is needed in order to clear the hill? **(c)** What are the horizontal and vertical components of the balloon's velocity at the instant that it clears the top of the hill?

8. •• An electron in a cathode-ray tube is traveling horizontally at 2.10×10^9 cm/s when deflection plates give it an upward acceleration of 5.30×10^{17} cm/s². **(a)** How much time does it take for the electron to cover a horizontal distance of 6.20 cm? **(b)** What is its vertical displacement during this time?

9. •• Two canoeists start paddling at the same time and head toward a small island in a lake, as shown in **FIGURE 4-18**. Canoeist 1 paddles with a speed of 1.10 m/s at an angle of 45° north of east. Canoeist 2 starts on the opposite shore of the lake, a distance of 1.5 km due east of canoeist 1. **(a)** In what direction relative to north must canoeist 2 paddle to reach the island? **(b)** What speed must canoeist 2 have if the two canoes are to arrive at the island at the same time?

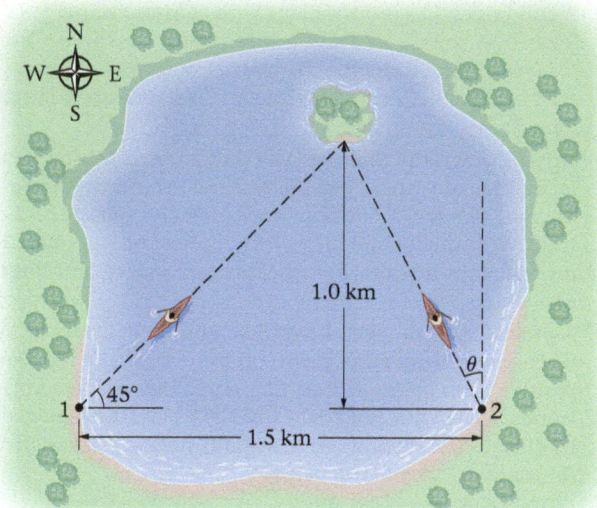

FIGURE 4-18 Problem 9

SECTION 4-3   ZERO LAUNCH ANGLE

10. • **CE** Predict/Explain Two divers run horizontally off the edge of a low cliff. Diver 2 runs with twice the speed of diver 1. **(a)** When the divers hit the water, is the horizontal distance covered by diver 2 twice as much as, four times as much as, or equal to the horizontal distance covered by diver 1? **(b)** Choose the *best explanation* from among the following:
   I. The drop time is the same for both divers.
   II. Drop distance depends on t^2.
   III. All divers in free fall cover the same distance.

11. • **CE** Predict/Explain Two youngsters dive off an overhang into a lake. Diver 1 drops straight down, and diver 2 runs off the cliff with an initial horizontal speed v_0. **(a)** Is the splashdown speed of diver 2 greater than, less than, or equal to the splashdown speed of diver 1? **(b)** Choose the *best explanation* from among the following:
   I. Both divers are in free fall, and hence they will have the same splashdown speed.
   II. The divers have the same vertical speed at splashdown, but diver 2 has the greater horizontal speed.
   III. The diver who drops straight down gains more speed than the one who moves horizontally.

12. • An archer shoots an arrow horizontally at a target 15 m away. The arrow is aimed directly at the center of the target, but it hits 52 cm lower. What was the initial speed of the arrow?

13. • **Victoria Falls** The great, gray-green, greasy Zambezi River flows over Victoria Falls in south central Africa. The falls are approximately 108 m high. If the river is flowing horizontally at 3.60 m/s just before going over the falls, what is the speed of the water when it hits the bottom? Assume the water is in free fall as it drops.

14. • A diver runs horizontally off the end of a diving board with an initial speed of 1.85 m/s. If the diving board is 3.00 m above the water, what is the diver's speed just before she enters the water?

15. • An astronaut on the planet Zircon tosses a rock horizontally with a speed of 6.95 m/s. The rock falls through a vertical distance of 1.40 m and lands a horizontal distance of 8.75 m from the astronaut. What is the acceleration due to gravity on Zircon?

16. •• Predict/Calculate **Pitcher's Mounds** Pitcher's mounds are raised to compensate for the vertical drop of the ball as it travels a horizontal distance of 18 m to the catcher. **(a)** If a pitch is thrown horizontally with an initial speed of 32 m/s, how far does it drop by the time it reaches the catcher? **(b)** If the speed of the pitch is increased, does the drop distance increase, decrease, or stay the same? Explain. **(c)** If this baseball game were to be played on the Moon, would the drop distance increase, decrease, or stay the same? Explain.

17. •• Playing shortstop, you pick up a ground ball and throw it to second base. The ball is thrown horizontally, with a speed of 18 m/s, directly toward point A (**FIGURE 4-19**). When the ball reaches the second baseman 0.54 s later, it is caught at point B. **(a)** How far were you from the second baseman? **(b)** What is the distance of vertical drop, AB?

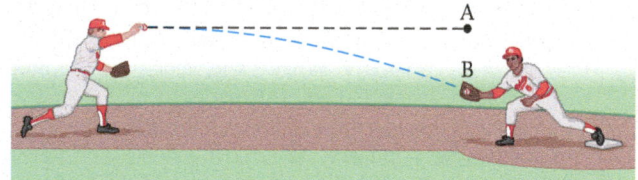

FIGURE 4-19 Problem 17

18. •• Predict/Calculate A crow is flying horizontally with a constant speed of 2.70 m/s when it releases a clam from its beak (**FIGURE 4-20**). The clam lands on the rocky beach 2.10 s later. Just before the clam lands, what are **(a)** its horizontal component of velocity, and **(b)** its vertical component of velocity? **(c)** How would your answers to parts (a) and (b) change if the speed of the crow were increased? Explain.

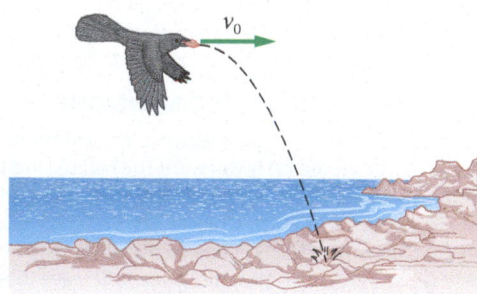

FIGURE 4-20 Problem 18

19. •• A mountain climber jumps a 2.8-m-wide crevasse by leaping horizontally with a speed of 7.8 m/s. **(a)** If the climber's direction of motion on landing is −45°, what is the height difference between the two sides of the crevasse? **(b)** Where does the climber land?

20. •• Predict/Calculate A white-crowned sparrow flying horizontally with a speed of 1.80 m/s folds its wings and begins to drop in free fall. **(a)** How far does the sparrow fall after traveling a horizontal

distance of 0.500 m? **(b)** If the sparrow's initial speed is increased, does the distance of fall increase, decrease, or stay the same?

21. •• **Pumpkin Toss** In Denver, children bring their old jack-o-lanterns to the top of a tower and compete for accuracy in hitting a target on the ground (**FIGURE 4-21**). Suppose that the tower is 9.0 m high and that the bull's-eye is a horizontal distance of 3.5 m from the launch point. If the pumpkin is thrown horizontally, what is the launch speed needed to hit the bull's-eye?

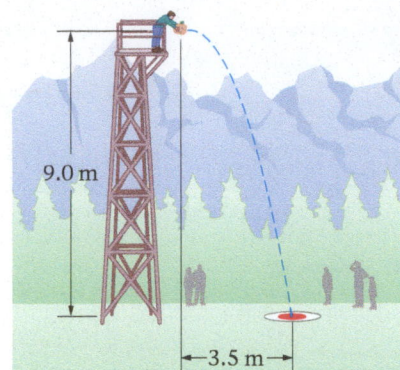

FIGURE 4–21 Problem 21

22. •• Fairgoers ride a Ferris wheel with a radius of 5.00 m (**FIGURE 4-22**). The wheel completes one revolution every 32.0 s. **(a)** What is the average speed of a rider on this Ferris wheel? **(b)** If a rider accidentally drops a stuffed animal at the top of the wheel, where does it land relative to the base of the ride? (*Note:* The bottom of the wheel is 1.75 m above the ground.)

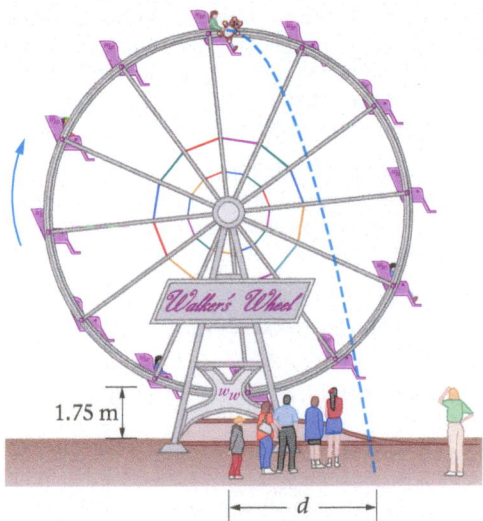

FIGURE 4-22 Problem 22

23. •• **Predict/Calculate** A swimmer runs horizontally off a diving board with a speed of 3.32 m/s and hits the water a horizontal distance of 1.78 m from the end of the board. **(a)** How high above the water was the diving board? **(b)** If the swimmer runs off the board with a reduced speed, does it take more, less, or the same time to reach the water?

24. •• **Baseball and the Washington Monument** On August 25, 1894, Chicago catcher William Schriver caught a baseball thrown from the top of the Washington Monument (555 ft, 898 steps). **(a)** If the ball was thrown horizontally with a speed of 5.00 m/s, where did it land? **(b)** What were the ball's speed and direction of motion when caught?

25. ••• A basketball is thrown horizontally with an initial speed of 4.20 m/s (**FIGURE 4-23**). A straight line drawn from the release point to the landing point makes an angle of 30.0° with the horizontal. What was the release height?

FIGURE 4-23 Problem 25

26. ••• **Predict/Calculate** A ball rolls off a table and falls 0.75 m to the floor, landing with a speed of 4.0 m/s. **(a)** What is the acceleration of the ball just before it strikes the ground? **(b)** What was the initial speed of the ball? **(c)** What initial speed must the ball have if it is to land with a speed of 5.0 m/s?

SECTION 4-4 GENERAL LAUNCH ANGLE

27. • **CE** A certain projectile is launched with an initial speed v_0. At its highest point its speed is $\frac{1}{2}v_0$. What was the launch angle of the projectile?

   A. 30°   **B.** 45°   **C.** 60°   **D.** 75°

28. • **CE** Three projectiles (A, B, and C) are launched with the same initial speed but with different launch angles, as shown in **FIGURE 4-24**. Rank the projectiles in order of increasing **(a)** horizontal component of initial velocity and **(b)** time of flight. Indicate ties where appropriate.

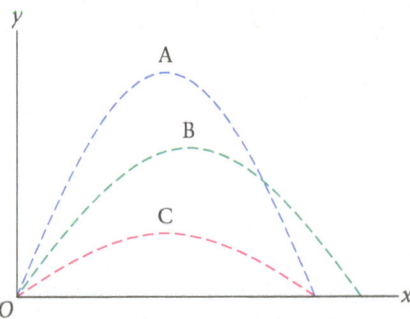

FIGURE 4-24 Problewm 28

29. • **CE** Three projectiles (A, B, and C) are launched with different initial speeds so that they reach the same maximum height, as shown in **FIGURE 4-25**. Rank the projectiles in order of increasing **(a)** initial speed and **(b)** time of flight. Indicate ties where appropriate.

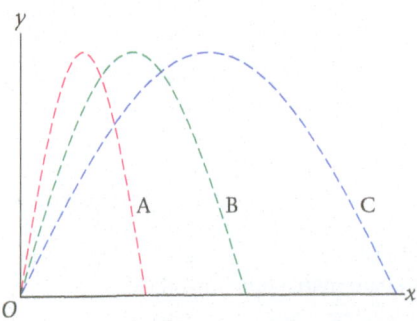

FIGURE 4-25 Problem 29

30. • A cannonball is launched at an angle above level ground, giving it an initial vertical and horizontal velocity that are each positive. The cannonball lands at a time T after it is launched. **(a)** Which plot (A, B, C, D, or E) in **FIGURE 4-26** best represents the horizontal component of the cannonball's velocity? **(b)** Which plot best represents the vertical component of the cannonball's velocity?

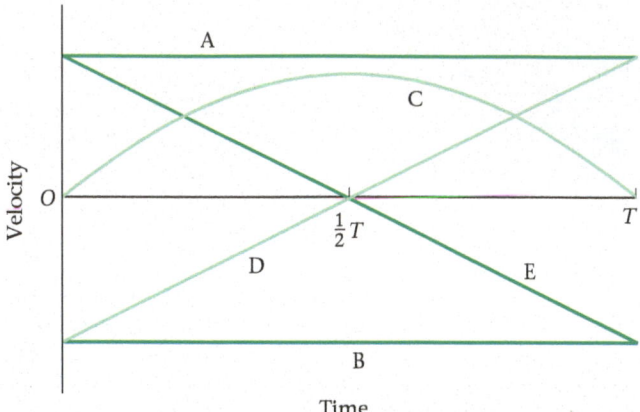

FIGURE 4-26 Problem 30

31. • A second baseman tosses the ball to the first baseman, who catches it at the same level from which it was thrown. The throw is made with an initial speed of 18.0 m/s at an angle of 37.5° above the horizontal. **(a)** What is the horizontal component of the ball's velocity just before it is caught? **(b)** For what amount of time is the ball in the air?

32. • A soccer ball is kicked with a speed of 15.6 m/s at an angle of 32.5° above the horizontal. If the ball lands at the same level from which it was kicked, for what amount of time was it in the air?

33. •• In a game of basketball, a forward makes a bounce pass to the center. The ball is thrown with an initial speed of 4.3 m/s at an angle of 15° below the horizontal. It is released 0.80 m above the floor. What horizontal distance does the ball cover before bouncing?

34. •• **Predict/Calculate** Snowballs are thrown with a speed of 13 m/s from a roof 7.0 m above the ground. Snowball A is thrown straight downward; snowball B is thrown in a direction 25° above the horizontal. **(a)** Is the landing speed of snowball A greater than, less than, or the same as the landing speed of snowball B? Explain. **(b)** Verify your answer to part (a) by calculating the landing speed of both snowballs.

35. •• In Problem 34, find the direction of motion of the two snowballs just before they land.

36. •• A golfer gives a ball a maximum initial speed of 51.4 m/s. **(a)** What is the longest possible hole-in-one for this golfer? Neglect any distance the ball might roll on the green and assume that the tee and the green are at the same level. **(b)** What is the minimum speed of the ball during this hole-in-one shot?

37. •• What is the highest tree the ball in the previous problem could clear on its way to the longest possible hole-in-one?

38. •• The "hang time" of a punt is measured to be 4.50 s. If the ball was kicked at an angle of 63.0° above the horizontal and was caught at the same level from which it was kicked, what was its initial speed?

39. •• In a friendly game of handball, you hit the ball essentially at ground level and send it toward the wall with a speed of 16 m/s at an angle of 37° above the horizontal. **(a)** What amount of time is required for the ball to reach the wall if it is 3.3 m away? **(b)** How high is the ball when it hits the wall?

40. •• On a hot summer day, a young girl swings on a rope above the local swimming hole (**FIGURE 4-27**). When she lets go of the rope her initial velocity is 2.25 m/s at an angle of 35.0° above the horizontal. If she is in flight for 0.616 s, how high above the water was she when she let go of the rope?

FIGURE 4-27 Problem 40

41. •• A certain projectile is launched with an initial speed v_0. At its highest point its speed is $v_0/4$. What was the launch angle?

SECTION 4-5 PROJECTILE MOTION: KEY CHARACTERISTICS

42. • **Punkin Chunkin** In Dover, Delaware, a post-Halloween tradition is "Punkin Chunkin," in which contestants build cannons, catapults, trebuchets, and other devices to launch pumpkins and compete for the greatest distance. Though hard to believe, pumpkins have been projected a distance of 4694 feet in this contest. What is the minimum initial speed needed for such a shot?

43. • A dolphin jumps with an initial velocity of 12.0 m/s at an angle of 40.0° above the horizontal. The dolphin passes through the center of a hoop before returning to the water. If the dolphin is moving horizontally when it goes through the hoop, how high above the water is the center of the hoop?

44. • A player passes a basketball to another player who catches it at the same level from which it was thrown. The initial speed of the ball is 7.1 m/s, and it travels a distance of 4.6 m. What were **(a)** the initial direction of the ball and **(b)** its time of flight?

45. • A golf ball is struck with a five iron on level ground. It lands 92.2 m away 4.30 s later. What were **(a)** the direction and **(b)** the magnitude of the initial velocity?

46. •• **CE** **Predict/Explain** You throw a ball into the air with an initial speed of 10 m/s at an angle of 60° above the horizontal. The ball returns to the level from which it was thrown in the time T. **(a)** Referring to **FIGURE 4-28**, which of the plots (A, B, or C) best represents the speed of the ball as a function of time? **(b)** Choose the *best explanation* from among the following:
I. Gravity causes the ball's speed to increase during its flight.
II. The ball has zero speed at its highest point.
III. The ball's speed decreases during its flight, but it doesn't go to zero.

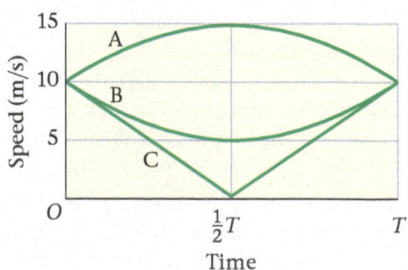

FIGURE 4-28 Problem 46

47. •• A football quarterback shows off his skill by throwing a pass 45.70 m downfield and into a bucket. The quarterback consistently launches the ball at 38.00° above horizontal, and the bucket is placed at the same level from which the ball is thrown. **(a)** What initial speed is needed so that the ball lands in the bucket? **(b)** By how much would the launch speed have to be increased if the bucket is moved to 46.50 m downfield?

48. •• A clever inventor has created a device that can launch water balloons with an initial speed of 85.0 m/s. Her goal is to pass a balloon through a small hoop mounted on the observation platform at the top of the Eiffel Tower, 276 m above the ground. If the balloon is to pass through the hoop at the peak of its flight, at what angle above horizontal should she launch the balloon?

49. •• **Predict/Calculate** **Volcanoes on Io** Astronomers have discovered several volcanoes on Io, a moon of Jupiter. One of them, named Loki, ejects lava to a maximum height of 2.00×10^5 m. **(a)** What is the initial speed of the lava? (The acceleration of gravity on Io is $1.80 \, \text{m/s}^2$.) **(b)** If this volcano were on Earth, would the maximum height of the ejected lava be greater than, less than, or the same as on Io? Explain.

50. •• **Predict/Calculate** A soccer ball is kicked with an initial speed of 10.2 m/s in a direction 25.0° above the horizontal. Find the magnitude and direction of its velocity **(a)** 0.250 s and **(b)** 0.500 s after being kicked. **(c)** Is the ball at its greatest height before or after 0.500 s? Explain.

51. •• A soccer ball is kicked with an initial speed of 8.25 m/s. After 0.750 s it is at its highest point. What was its initial direction of motion?

52. •• An archer shoots an arrow over a castle wall by launching it with an initial speed of 28 m/s at 65° above horizontal. Assume the arrow lands on the other side of the castle wall at the same elevation as the launch point. **(a)** What maximum height does the arrow attain? **(b)** What is the range of the arrow's flight? **(c)** What is the vertical component of the arrow's velocity just before it lands on the other side of the castle wall?

GENERAL PROBLEMS

53. • **CE** Child 1 throws a snowball horizontally from the top of a roof; child 2 throws a snowball straight down. Once in flight, is the acceleration of snowball 2 greater than, less than, or equal to the acceleration of snowball 1?

54. • **CE** The penguin to the left in the accompanying photo is about to land on an ice floe. Just before it lands, is its speed greater than, less than, or equal to its speed when it left the water?

This penguin behaves much like a projectile from the time it leaves the water until it touches down on the ice. (Problem 54)

55. • **CE** Dolphins may leap from the water just for the fun of it. At the instant a leaping dolphin lands, is its speed greater than, less than, or equal to its speed when it left the water?

56. • **CE** **Predict/Explain** A person flips a coin into the air and it lands on the ground a few feet away. **(a)** If the person were to perform an identical coin flip on an elevator rising with constant speed, would the coin's time of flight be greater than, less than, or equal to its time of flight when the person was at rest? **(b)** Choose the *best explanation* from among the following:

I. The floor of the elevator is moving upward, and hence it catches up with the coin in mid flight.

II. The coin has the same upward speed as the elevator when it is tossed, and the elevator's speed doesn't change during the coin's flight.

III. The coin starts off with a greater upward speed because of the elevator, and hence it reaches a greater height.

57. • **CE** **Predict/Explain** Suppose the elevator in the previous problem is rising with a constant upward acceleration, rather than constant velocity. **(a)** In this case, would the coin's time of flight be greater than, less than, or equal to its time of flight when the person was at rest? **(b)** Choose the *best explanation* from among the following:

I. The coin has the same acceleration once it is tossed, whether the elevator accelerates or not.

II. The elevator's upward speed increases during the coin's flight, and hence it catches up with the coin at a greater height than before.

III. The coin's downward acceleration is less than before because the elevator's upward acceleration partially cancels it.

58. • A train moving with constant velocity travels 170 m north in 12 s and an undetermined distance to the west. The speed of the train is 32 m/s. **(a)** Find the direction of the train's motion relative to north. **(b)** How far west has the train traveled in this time?

59. • A tennis ball is struck in such a way that it leaves the racket with a speed of 4.87 m/s in the horizontal direction. When the ball hits the court, it is a horizontal distance of 1.95 m from the racket. Find the height of the tennis ball when it left the racket.

60. • A person tosses a ball for her puppy to retrieve. The ball leaves her hand horizontally with a speed of 4.6 m/s. If the initial height of the ball is 0.95 m above the ground, how far does it travel in the horizontal direction before landing?

61. •• An osprey flies horizontally with a constant speed of 6.8 m/s when it drops the fish it was carrying. How much time elapses after the fish is dropped before the speed of the fish doubles?

62. •• **Predict/Calculate** A hot-air balloon rises from the ground with a velocity of $(2.00 \, \text{m/s})\hat{\mathbf{y}}$. A champagne bottle is opened to celebrate takeoff, expelling the cork horizontally with a velocity of $(5.00 \, \text{m/s})\hat{\mathbf{x}}$ relative to the balloon. When opened, the bottle is 6.00 m above the ground. **(a)** What is the initial velocity of the cork, as seen by an observer on the ground? Give your answer in terms of the x and y unit vectors. **(b)** What are the speed of the cork and its initial direction of motion as seen by the same observer? **(c)** Determine the maximum height above the ground attained by the cork. **(d)** For what amount of time does the cork remain in the air?

63. •• In a friendly neighborhood squirt gun contest a participant runs at 7.8 m/s horizontally off the back deck and fires her squirt gun in the plane of her motion but 45° above horizontal. The gun can shoot water at 11 m/s relative to the barrel, and she fires the gun 0.42 s after leaving the deck. **(a)** What is the initial velocity of the water particles as seen by an observer on the ground? Give your answer in terms of the horizontal and vertical components. **(b)** At the instant she fires, the gun is 1.9 m above the level ground. How far will the water travel horizontally before landing?

64. •• **BIO** **Spitting Llamas** An agitated llama may spit to assert dominance, or to ward off threats. Llamas can spit a considerable distance, and people handling them need to keep this in mind. If the spittle from a llama is launched from an initial height of 1.8 m with a speed of 6.1 m/s, and at an angle of 12° above horizontal, how far will it travel horizontally?

65. •• A particle leaves the origin with an initial velocity $\vec{v} = (2.40 \, \text{m/s})\hat{x}$, and moves with constant acceleration $\vec{a} = (-1.90 \, \text{m/s}^2)\hat{x} + (3.20 \, \text{m/s}^2)\hat{y}$. (a) How far does the particle move in the x direction before turning around? (b) What is the particle's velocity at this time? (c) Plot the particle's position at $t = 0.500$ s, 1.00 s, 1.50 s, and 2.00 s. Use these results to sketch position versus time for the particle.

66. •• **BIO** When the dried-up seed pod of a scotch broom plant bursts open, it shoots out a seed with an initial velocity of 2.62 m/s at an angle of 60.5° above the horizontal. If the seed pod is 0.455 m above the ground, (a) what amount of time does it take for the seed to land? (b) What horizontal distance does it cover during its flight?

67. •• **Trick Shot** In an Internet video an athlete launches a basketball from a stadium platform that is 16.2 m higher than the hoop. He makes the basket by launching the ball at an angle of 14.5° above horizontal with a speed of 11.6 m/s. What horizontal distance does the ball travel before passing through the hoop?

68. •• A shot-putter throws the shot with an initial speed of 12.2 m/s from a height of 5.15 ft above the ground. What is the range of the shot if the launch angle is (a) 20.0°, (b) 30.0°, or (c) 40.0°?

69. •• Two marbles are launched at $t = 0$ in the experiment illustrated in **FIGURE 4-29**. Marble 1 is launched horizontally with a speed of 4.20 m/s from a height $h = 0.950$ m. Marble 2 is launched from ground level with a speed of 5.94 m/s at an angle $\theta = 45.0°$ above the horizontal. (a) Where would the marbles collide in the absence of gravity? Give the x and y coordinates of the collision point. (b) Where do the marbles collide given that gravity produces a downward acceleration of $g = 9.81 \, \text{m/s}^2$? Give the x and y coordinates.

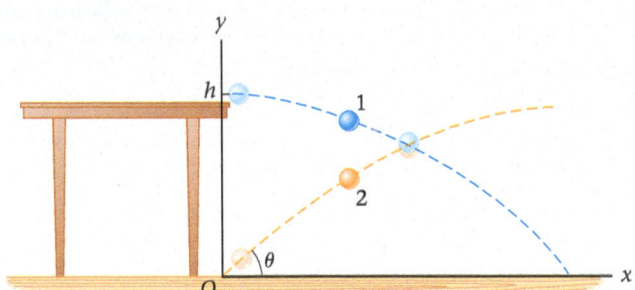

FIGURE 4-29 Problem 69

70. •• **Rescue Swimmers** Coast Guard rescue swimmers are trained to leap from helicopters into the sea to save boaters in distress. The rescuers like to step off their helicopter when it is "ten and ten," which means that it is *ten* feet above the water and moving forward horizontally at *ten* knots. What are (a) the speed and (b) the direction of motion as a rescue swimmer enters the water following a ten and ten jump?

71. •• A football player kicks a field goal, launching the ball at an angle of 48° above the horizontal. During the kick, the ball is in contact with the player's foot for 0.045 s, and the ball's acceleration is 290 m/s². What is the range of the football?

72. •• A ball thrown straight upward returns to its original level in 2.75 s. A second ball is thrown at an angle of 40.0° above the

horizontal. What is the initial speed of the second ball if it also returns to its original level in 2.75 s?

73. •• **Predict/Calculate** To decide who pays for lunch, a passenger on a moving train tosses a coin straight upward with an initial speed of 5.25 m/s and catches it again when it returns to its initial level. From the point of view of the passenger, then, the coin's initial velocity is $(5.25 \, \text{m/s})\hat{y}$. The train's velocity relative to the ground is $(12.1 \, \text{m/s})\hat{x}$. (a) What is the minimum speed of the coin relative to the ground during its flight? At what point in the coin's flight does this minimum speed occur? Explain. (b) Find the initial speed and direction of the coin as seen by an observer on the ground. (c) Use the expression for y_{max} derived in Example 4-14 to calculate the maximum height of the coin, as seen by an observer on the ground. (d) What is the maximum height of the coin from the point of view of the passenger, who sees only one-dimensional motion?

74. •• **Predict/Calculate** A cannon is placed at the bottom of a cliff 61.5 m high. If the cannon is fired straight upward, the cannonball just reaches the top of the cliff. (a) What is the initial speed of the cannonball? (b) Suppose a second cannon is placed at the top of the cliff. This cannon is fired horizontally, giving its cannonballs the same initial speed found in part (a). Show that the range of this cannon is the same as the maximum range of the cannon at the base of the cliff. (Assume the ground at the base of the cliff is level, though the result is valid even if the ground is not level.)

75. •• A golfer hits a shot to an elevated green. The ball leaves the club with an initial speed of 16 m/s at an angle of 58° above the horizontal. If the speed of the ball just before it lands is 12 m/s, what is the elevation of the green above the point where the ball is struck?

76. •• **Shot Put Record** A men's world record for the shot put, 23.12 m, was set by Randy Barnes of the United States on May 20, 1990. If the shot was launched from 6.00 ft above the ground at an initial angle of 42.0°, what was its initial speed?

77. •• Referring to Conceptual Example 4-13, suppose the two cats jump from an elevation of 2.5 m with an initial speed of 3.0 m/s. What is the speed of each cat when it is 1.0 m above the ground?

78. •• **A "Lob" Pass Versus a "Bullet"** A quarterback can throw a receiver a high, lazy "lob" pass or a low, quick "bullet" pass. These passes are indicated by curves 1 and 2, respectively, in **FIGURE 4-30**. (a) The lob pass is thrown with an initial speed of 21.5 m/s and its time of flight is 3.97 s. What is its launch angle? (b) The bullet pass is thrown with a launch angle of 25.0°. What is the initial speed of this pass? (c) What is the time of flight of the bullet pass?

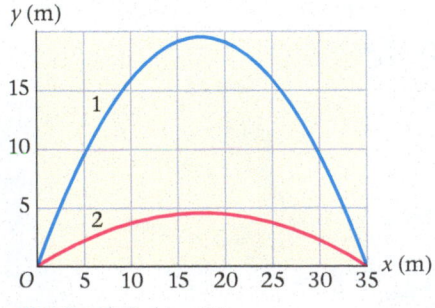

FIGURE 4-30 Problem 78

79. •• For summertime fun, you decide to combine diving from a board with shooting a basketball through a hoop. (a) During several practice runs you stand at the end of a diving board and launch the basketball horizontally from a position 4.50 m above

the water. If the average landing spot is 6.25 m horizontally from your initial position, what is the average launch speed? **(b)** Now you step off the diving board and launch the ball in order to make a basket in a hoop that is 3.75 m horizontally from the ball and 1.00 m above the water as shown in **FIGURE 4-31**. At what time after stepping off the board should you launch the ball? *Hint:* Consider the time required for the ball to travel the required horizontal distance. **(c)** What are the horizontal and vertical components of the ball's velocity at the instant of launch?

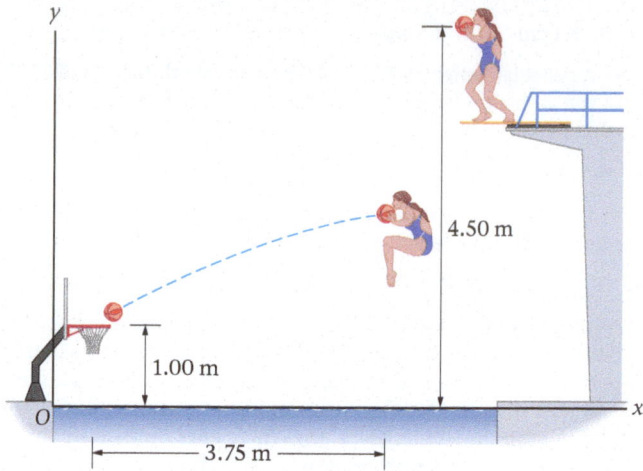

FIGURE 4-31 Problem 79

80. •• **Landing on Mars** When the twin exploration rovers, *Spirit* and *Opportunity*, landed on Mars, their method of landing was unique and elaborate. After initial braking with rockets and parachutes to a virtual standstill several meters above the ground, the rovers inflated four air bags with six lobes each. The rovers were then detached from the parachutes and allowed to drop in free fall (3.72 m/s^2) to the surface, where they bounced about 12 times before coming to rest. They then deflated their air bags, righted themselves, and began to explore the surface. **FIGURE 4-32** shows a rover with its surrounding cushion of air bags making its first contact with the Martian surface. Assume that the first bounce of the rover is with an initial speed of 9.92 m/s at an angle of 75.0° above the horizontal. **(a)** What is the maximum height of a rover between its first and second bounces? **(b)** How much time elapses between the first and second bounces? **(c)** How far does a rover travel in the horizontal direction between its first and second bounces?

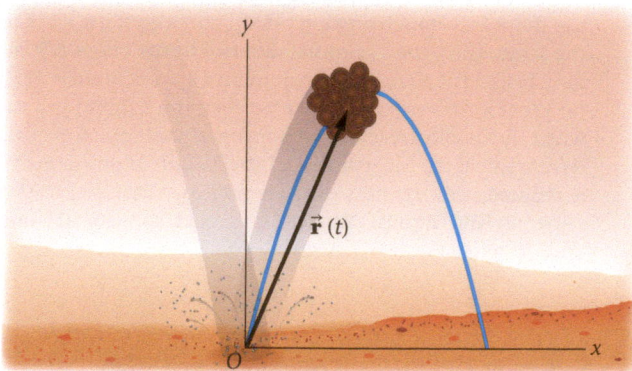

FIGURE 4-32 Problem 80

81. ••• **Collision Course** A useful rule of thumb in piloting is that if the heading from your airplane to a second airplane remains constant, the two airplanes are on a collision course. Consider the two airplanes shown in **FIGURE 4-33**. At time $t = 0$, airplane 1 is at the location $(X, 0)$ and moving in the positive y direction; airplane 2 is at $(0, Y)$ and moving in the positive x direction. The speed of airplane 1 is v_1. **(a)** What speed must airplane 2 have if the airplanes are to collide at the point (X, Y)? **(b)** Assuming airplane 2 has the speed found in part (a), calculate the displacement from airplane 1 to airplane 2, $\Delta\vec{r} = \vec{r}_2 - \vec{r}_1$. **(c)** Use your results from part (b) to show that $(\Delta r)_y/(\Delta r)_x = -Y/X$, independent of time. This shows that $\Delta\vec{r} = \vec{r}_2 - \vec{r}_1$ maintains a constant direction until the collision, as specified in the rule of thumb.

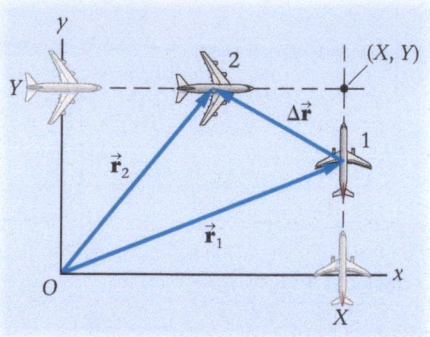

FIGURE 4-33 Problem 81

82. ••• As discussed in Example 4-14, the archerfish hunts by dislodging an unsuspecting insect from its resting place with a stream of water expelled from the fish's mouth. Suppose the archerfish squirts water with a speed of 2.15 m/s at an angle of 52.0° above the horizontal, and aims for a beetle on a leaf 3.00 cm above the water's surface. **(a)** At what horizontal distance from the beetle should the archerfish fire if it is to hit its target in the least time? **(b)** How much time will the beetle have to react?

83. ••• Find the launch angle for which the range and maximum height of a projectile are the same.

84. ••• A mountain climber jumps a crevasse of width W by leaping horizontally with speed v_0. **(a)** If the height difference between the two sides of the crevasse is h, what is the minimum value of v_0 for the climber to land safely on the other side? **(b)** In this case, what is the climber's direction of motion on landing?

85. ••• **Landing on a Different Level** A projectile fired from $y = 0$ with initial speed v_0 and initial angle θ lands on a different level, $y = h$. Show that the time of flight of the projectile is

$$T = \tfrac{1}{2}T_0\left(1 + \sqrt{1 - \frac{h}{H}}\right)$$

In this expression, T_0 is the time of flight for $h = 0$, and H is the maximum height of the projectile.

86. ••• A mountain climber jumps a crevasse by leaping horizontally with speed v_0. If the climber's direction of motion on landing is θ below the horizontal, what is the height difference h between the two sides of the crevasse?

87. ••• **Projectiles: Coming or Going?** Most projectiles continually move farther from the origin during their flight, but this is not the case if the launch angle is greater than $\cos^{-1}(\tfrac{1}{3}) = 70.5°$. For example, the projectile shown in **FIGURE 4-34** has a launch angle of 75.0° and an initial speed of 10.1 m/s. During the portion of its motion shown in red, it is moving closer to the origin—it is moving away on the blue portions. Calculate the distance from the origin to the projectile **(a)** at the start of the red portion, **(b)** at the end of the

red portion, and **(c)** just before the projectile lands. Notice that the distance for part **(b)** is the smallest of the three.

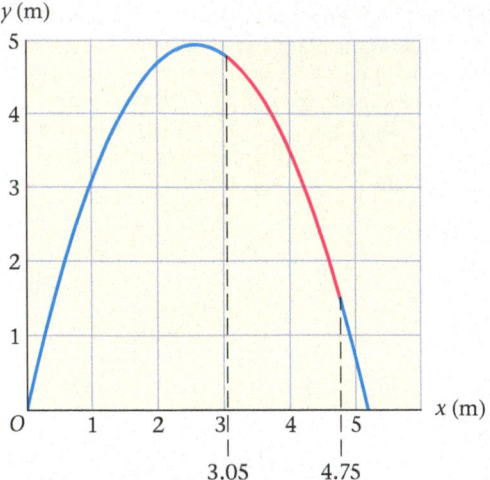

FIGURE 4-34 Problem 87

PASSAGE PROBLEMS

Caterpillar Pellets

The larvae (caterpillars) of certain species of butterflies and moths construct shelters on a host plant out of folded or rolled leaves secured with silk threads. Many of these types of caterpillars eliminate waste by ejecting fecal pellets (frass) at high speeds so that the pellets are projected far away from the caterpillar and its shelter. Various explanations for this behavior have been proposed, but some studies have shown that it may help keep predatory insects such as wasps from locating the caterpillars by homing in on the odor of caterpillar frass.

Video microscopy has been used to study the pellet ejection process. In one study, the video images reveal that a group of Brazilian skipper caterpillars that are about 50 mm long eject pellets at angles between 10° and 40° above the horizontal. There is no correlation between the size of the pellet and the angle at which it is ejected, but larger pellets are shot with lower speeds, as shown in **FIGURE 4-35**. The pellets are small and dense enough that at the speeds at which they travel, air resistance is negligible.

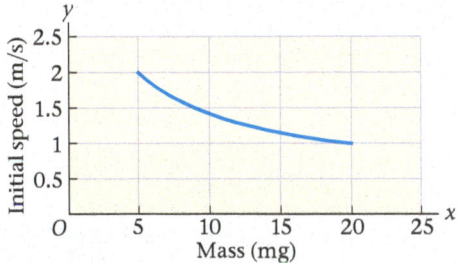

FIGURE 4-35 Problems 88, 89, 90, and 91

88. • If pellets shot at the same angle are compared, which of the following is true?

    A. A lighter pellet travels farther because of its higher ejection speed.
    B. A heavier pellet travels farther because it is launched more horizontally.
    C. Pellets of all masses travel equally far because the launch angle is the same.

89. •• What is the maximum height above its launch site of a 20-mg pellet that's launched at an angle of 20° above the horizontal?

    A. 5.1 cm   **B.** 4.5 cm   **C.** 1.2 cm   **D.** 6.0 mm

90. •• A pellet launched at 30° above the horizontal that spends 0.15 s in the air before returning to its original level has approximately what mass?

    A. 2 mg   **B.** 9 mg   **C.** 14 mg   **D.** 20 mg

91. •• What is the maximum range of a 5-mg pellet that lands at the same height from which it is ejected?

    A. 14 cm   **B.** 20 cm   **C.** 40 cm   **D.** 78 cm

92. •• **REFERRING TO EXAMPLE 4-9 (a)** At what launch angle *greater* than 54.0° does the golf ball just barely miss the top of the tree in front of the green? Assume that the ball has an initial speed of 13.5 m/s, and that the tree is 3.00 m high and is a horizontal distance of 14.0 m from the launch point. **(b)** Where does the ball land in the case described in part **(a)**? **(c)** At what launch angle *less* than 54.0° does the golf ball just barely miss the top of the tree in front of the green? **(d)** Where does the ball land in the case described in part **(c)**?

93. •• **REFERRING TO EXAMPLE 4-9** Suppose that the golf ball is launched with a speed of 15.0 m/s at an angle of 57.5° above the horizontal, and that it lands on a green 3.50 m above the level where it was struck. **(a)** What horizontal distance does the ball cover during its flight? **(b)** What increase in initial speed would be needed to increase the horizontal distance in part **(a)** by 7.50 m? Assume everything else remains the same.

94. •• **REFERRING TO EXAMPLE 4-11** Suppose the ball is dropped at the horizontal distance of 5.50 m, but from a new height of 5.00 m. The dolphin jumps with the same speed of 12.0 m/s. **(a)** What launch angle must the dolphin have if it is to catch the ball? **(b)** At what height does the dolphin catch the ball in this case? **(c)** What is the minimum initial speed the dolphin must have to catch the ball before it hits the water?

95. •• **Predict/Calculate REFERRING TO EXAMPLE 4-11** Suppose we change the dolphin's launch angle to 45.0°, but everything else remains the same. Thus, the horizontal distance to the ball is 5.50 m, the drop height is 4.10 m, and the dolphin's launch speed is 12.0 m/s. **(a)** What is the vertical distance between the dolphin and the ball when the dolphin reaches the horizontal position of the ball? We refer to this as the "miss distance." **(b)** If the dolphin's launch speed is reduced, will the miss distance increase, decrease, or stay the same? **(c)** Find the miss distance for a launch speed of 10.0 m/s.

Big Ideas

1 A force is a push or a pull.

2 An object at rest stays at rest if the net force acting on it is zero.

3 A net force causes an object to accelerate.

4 All forces come in equal but opposite pairs.

Newton's Laws of Motion

▲ This motorist is learning firsthand that it takes a force to accelerate an object—the greater the force, the greater the acceleration. What she may not realize, however, is that the car is also pushing back on her with a force of equal magnitude. These observations follow directly from Newton's laws of motion, the subject of this chapter.

We are all subject to Newton's laws of motion, whether we know it or not. In this chapter we present Newton's three laws and show how they can be applied to everyday situations.

5-1 Force and Mass

Two important elements in Newton's laws, and all of physics, are force and mass. In this section we discuss the physical meaning of these concepts.

Force Simply put, a **force** is a push or a pull. When you push on a box to slide it across the floor, for example, or pull on the handle of a wagon to give a child a ride, you are exerting a force. Similarly, when you hold this book in your hand, you exert an upward force to oppose the downward pull of gravity. If you set the book on a table, the table exerts the same upward force you exerted a moment before. Forces are truly all around us.

Now, when you push or pull on something, there are two quantities that characterize the force you are exerting. The first is the strength or **magnitude** of your force; the second is the **direction** in which you are pushing or pulling. Because a force is determined by both a magnitude and a direction, it is a vector. We consider the vector properties of forces in more detail in Section 5-5.

In general, an object has several forces acting on it at any given time. In the previous example, a book at rest on a table experiences a downward force due to gravity and an upward force due to the table. If you push the book across the table, it also experiences a horizontal force due to your push. The total, or net, force exerted on the book is the vector sum of the individual forces acting on it.

Mass A second key ingredient in Newton's laws is the **mass** of an object, which is a measure of how difficult it is to change its velocity. For example, mass determines how hard it is to start an object moving if it is at rest, to bring it to rest if it is moving, or to change its direction of motion. For example, if you throw a baseball or catch one thrown to you, the force required is not too great. But if you want to start a car moving or stop a car that is coming at you, the force involved is much greater. It follows that the mass of a car is greater than the mass of a baseball.

In agreement with everyday usage, mass can also be thought of as a measure of the quantity of matter in an object. Thus, it is clear that the mass of an elephant, for example, is much greater than the mass of a mouse, but much less than the mass of Earth. We measure mass in units of kilograms (kg), where one kilogram is defined as the mass of a standard cylinder of platinum-iridium, as discussed in Chapter 1. A list of typical masses is given in Table 5-1.

The properties of force and mass presented here are developed in detail in the next three sections.

Big Idea **1** A force is a push or a pull in a given direction. Most objects are acted on by a number of forces at any given time.

TABLE 5-1 Typical Masses in Kilograms (kg)

Earth	5.97×10^{24}
Space shuttle	2,000,000
Blue whale (largest animal on Earth)	105,000
Whale shark (largest fish)	18,000
Elephant (largest land animal)	5400
Automobile	1200
Human (adult)	70
Gallon of milk	3.9
Quart of milk	1.0
Baseball	0.145
Honeybee	0.00015
Bacterium	10^{-15}

Enhance Your Understanding   (Answers given at the end of the chapter)

1. Two forces have magnitudes F_1 and F_2. If these forces are added together, which of the following statements about the net force is correct? **A:** The net force must have a magnitude equal to $F_1 + F_2$. **B:** The net force can have a magnitude less than $F_1 + F_2$. **C:** The net force can have a magnitude greater than $F_1 + F_2$. **D:** The net force has a magnitude greater than F_1.

Section Review

• A force is a push or a pull. Forces are vectors, with specific directions and magnitudes.

• Mass is the quantity of matter contained in an object.

5-2 Newton's First Law of Motion

If you've ever stood in line at an airport, pushing your bags forward a few feet at a time, you know that as soon as you stop pushing the bags, they stop moving. Observations such as this often lead to the erroneous conclusion that a force is required for an object to move. In fact, according to Newton's first law of motion, a force is required only to *change* an object's motion.

Side view   **End view**

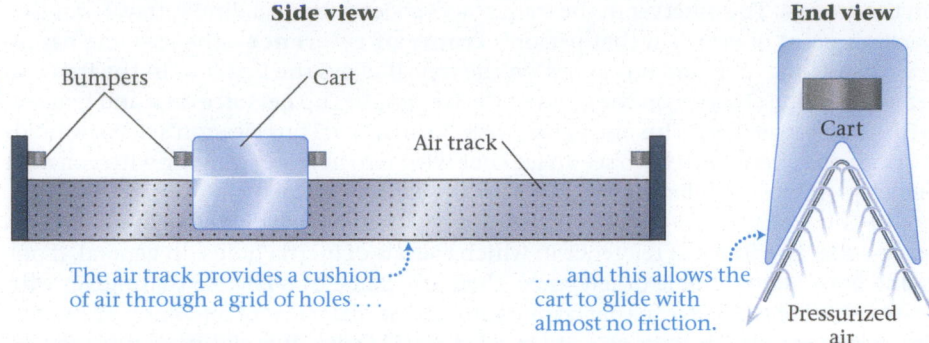

▲ **FIGURE 5-1 The air track** An air track provides a cushion of air on which a cart can ride with virtually no friction.

What's missing in this analysis is the force of friction between the bags and the floor. When you stop pushing the bags, it is not true that they stop moving because they no longer have a force acting on them. On the contrary, there is a rather large *frictional force* between the bags and the floor. It is this force that causes the bags to come to rest.

Reducing Friction   To see how motion is affected by reducing friction, imagine that you slide on dirt into second base during a baseball game. You won't slide very far before stopping. On the other hand, if you slide with the same initial speed on a sheet of ice—where the friction is much less than on a ball field—you slide considerably farther. If you could reduce the friction more, you would slide even farther.

In the classroom, air tracks allow us to observe motion with practically no friction. An example of such a device is shown schematically in **FIGURE 5-1** and in real life in **FIGURE 5-2**. Notice that air is blown through small holes in the track, creating a cushion of air for a small "cart" to ride on. A cart placed at rest on a level track remains at rest—unless you push on it to get it started.

Once set in motion, the cart glides along with constant velocity—constant speed in a straight line—until it hits a bumper at the end of the track. The bumper exerts a force on the cart, causing it to change its direction of motion. After bouncing off the bumper, the cart again moves with constant velocity. If the track could be extended to infinite length, and could be made perfectly frictionless, the cart would simply keep moving with constant velocity forever.

Newton's first law of motion summarizes these observations in the following statements:

▲ **FIGURE 5-2** An air track provides a nearly frictionless environment for experiments involving linear motion.

> **Newton's First Law**
>
> An object at rest remains at rest as long as no net force acts on it.
>
> An object moving with constant velocity continues to move with the same speed and in the same direction as long as no net force acts on it.

Notice the recurring phrase, "no net force," in these statements. It's important to realize that this can mean one of two things: (i) no force acts on the object; or (ii) forces act on the object, but they sum to zero. We shall see examples of the second possibility later in this chapter and again in the next chapter.

Newton's first law, which was first enunciated by Galileo, is also known as the **law of inertia**. This is an appropriate name because the literal meaning of the word *inertia* is "laziness." Speaking loosely, we can say that matter is "lazy," in that it won't change its motion unless forced to do so. For example, if an object is at rest, it won't start moving on its own. If an object is already moving with constant velocity, it won't alter its speed *or* direction, unless a force causes the change. We call this property of matter its inertia.

Frames of Reference   According to Newton's first law, being at rest and moving with constant velocity are actually equivalent. To see this, imagine two observers: one is in a train moving with constant velocity; the second is standing next to the tracks, at rest

Big Idea **2** An object at rest stays at rest if the net force acting on it is zero. An object moving with constant velocity continues to move with constant velocity if the net force acting on it is zero.

Suspended Balls: Which String Breaks?

on the ground. The observer in the train places an ice cube on a dinner tray. From that person's point of view—in that person's **frame of reference**—the ice cube has no net force acting on it and it is at rest on the tray. It obeys the first law. In the frame of reference of the observer on the ground, the ice cube has no net force on it and it moves with constant velocity. This also agrees with the first law. Thus Newton's first law holds for both observers: They both see an ice cube with zero net force moving with constant velocity—it's just that for the first observer the constant velocity happens to be zero.

In this example, we say that each observer is in an **inertial frame of reference**—that is, a frame of reference in which the law of inertia holds. In general, if one frame is an inertial frame of reference, then any frame of reference that moves with constant velocity relative to that frame is also an inertial frame of reference. Thus, if an object moves with constant velocity in one inertial frame, it is always possible to find another inertial frame in which the object is at rest. It is in this sense that there really isn't any difference between being at rest and moving with constant velocity. It's all relative—relative to the frame of reference the object is viewed from.

This gives us a more compact statement of Newton's first law:

> If the net force on an object is zero, its velocity is constant.

As an example of a frame of reference that is not inertial, imagine that the train carrying the first observer suddenly comes to a halt. From the point of view of the observer in the train, there is still no net force on the ice cube. However, because of the rapid braking, the ice cube flies off the tray. In fact, the ice cube simply continues to move forward with the same constant velocity it had before, while the *train* comes to rest. To the observer on the train, it appears that the ice cube has accelerated forward, even though no force acts on it, which is in violation of Newton's first law. Thus the *accelerating* train is *not* an inertial frame of reference; we refer to it as a **noninertial frame of reference**.

In general, any frame of reference that accelerates relative to an inertial frame is a noninertial frame. The surface of the Earth accelerates slightly, due to its rotational and orbital motions, but because the acceleration is so small, it may be considered an excellent approximation to an inertial frame of reference. Unless specifically stated otherwise, we will always consider the surface of the Earth to be an inertial frame.

Enhance Your Understanding   (Answers given at the end of the chapter)

2. Which of the following statements is correct? **A:** A net force is required to keep an object moving with constant velocity. **B:** A net force is required to make an object accelerate. **C:** When you hit the brakes in a car, a force pushes you toward the front of the car. **D:** When you accelerate in a car, a force pushes you back into your seat. **E:** An object that is moving will continue to move with the same velocity unless a net force acts on it.

Section Review

- **Newton's First Law of Motion** An object at rest remains at rest, as long as no net force acts on it. An object moving with constant velocity maintains its constant velocity, as long as no net force acts on it.

- An inertial frame of reference is one in which Newton's first law is valid. The surface of the Earth is a good approximation to an inertial frame of reference.

5-3 Newton's Second Law of Motion

To hold an object in your hand, you have to exert an upward force to oppose, or "balance," the force of gravity. If you suddenly remove your hand so that the only force acting on the object is gravity, the object accelerates downward, as discussed in Chapter 2. This is one example of Newton's second law, which states, basically, that unbalanced forces cause accelerations.

To explore this in more detail, consider a spring scale of the type used to weigh fish. The scale gives a reading of the force, F, exerted by the spring contained within it. If we hang one weight from the scale, it gives a reading that we will call F_1. If two identical weights are attached, the scale reads $F_2 = 2F_1$, as indicated in **FIGURE 5-3**. With these two forces marked on the scale, we are ready to perform some force experiments.

First, attach the scale to an air-track cart, as in **FIGURE 5-4**. If we pull with a force F_1, we observe that the cart accelerates at the rate a_1. If we now pull with a force $F_2 = 2F_1$, the acceleration we observe is $a_2 = 2a_1$. Thus, the acceleration is proportional to the force—the greater the force, the greater the acceleration.

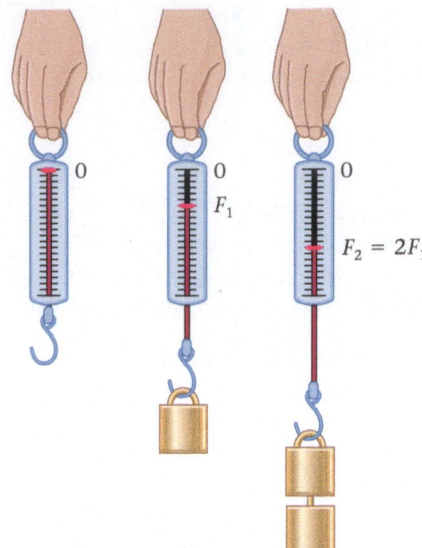

▲ **FIGURE 5-3 Calibrating a "force meter"** With two weights, the force exerted by the scale is twice the force exerted when only a single weight is attached.

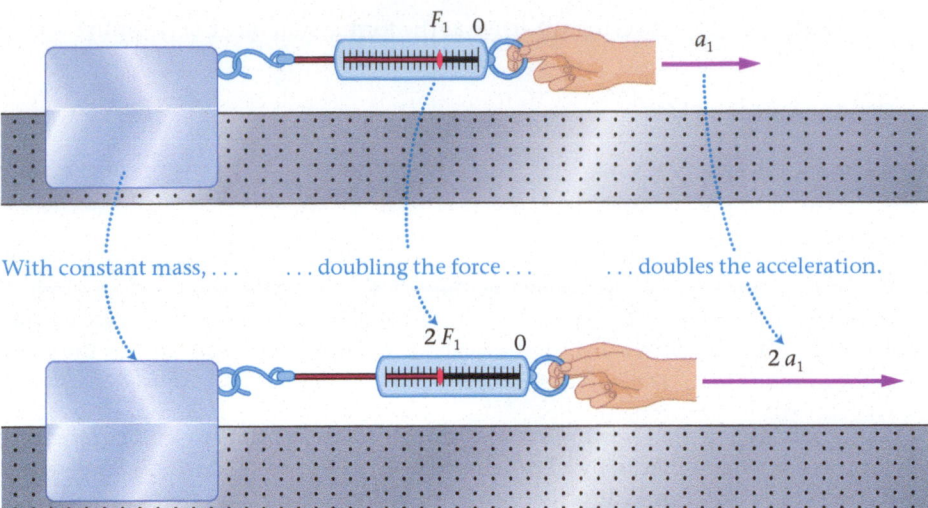

With constant mass, ... doubling the force . . .   . . . doubles the acceleration.

▲ **FIGURE 5-4 Acceleration is proportional to force** The spring calibrated in Figure 5-3 is used to accelerate a mass on a "frictionless" air track. If the force is doubled, the acceleration is also doubled.

Second, instead of doubling the force, let's double the mass m of the cart by connecting two together, as in **FIGURE 5-5**. In this case, if we pull a mass $2m$ with a force F_1 we find an acceleration equal to $\frac{1}{2}a_1$. Thus, the acceleration is inversely proportional to mass—the greater the mass, the less the acceleration.

◄ **FIGURE 5-5 Acceleration is inversely proportional to mass** If the mass of an object is doubled but the force remains the same, the acceleration is halved.

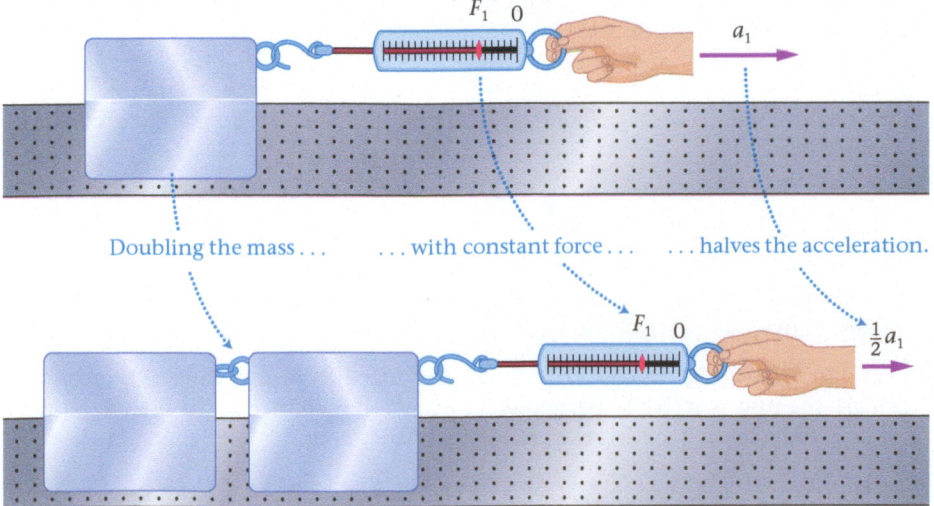

Doubling the mass . . .   . . . with constant force . . .   . . . halves the acceleration.

Combining these results, we find that in this simple case—with just one force in just one direction—the acceleration is given by

$$a = \frac{F}{m}$$

Big Idea 3 A net force acting on an object causes it to accelerate.

PHYSICS
IN CONTEXT
Looking Back

The fact that a constant force produces a constant acceleration gives special significance to the discussion of constant acceleration in Chapters 2 and 4.

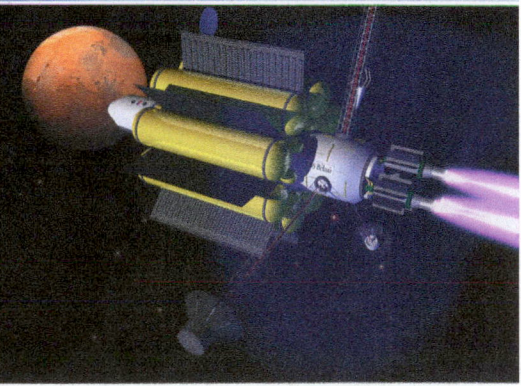

Rearranging the equation yields the form of Newton's second law that is perhaps best known, $F = ma$. The dependence of acceleration on force and mass is illustrated in **FIGURE 5-6**.

Net Force and Newton's Second Law In general, there may be several forces acting on a given mass, and these forces may be in different directions. Thus, we replace F with the sum of the force vectors acting on a mass. We call this the net force:

$$\text{sum of force vectors} = \vec{F}_{net} = \sum \vec{F}$$

The notation $\sum \vec{F}$, which uses the Greek letter sigma (Σ), is read "sum of $\vec{F}$." Recalling that acceleration is also a vector, we arrive at the formal statement of Newton's second law of motion:

Newton's Second Law

$$\vec{a} = \frac{\sum \vec{F}}{m} \quad \text{or} \quad \sum \vec{F} = m\vec{a} \qquad \text{5-1}$$

In words:

> If an object of mass m is acted on by a net force $\sum \vec{F}$, it will experience an acceleration $\vec{a}$ that is equal to the net force divided by the mass. Because the net force is a vector, the acceleration is also a vector. In fact, the direction of an object's acceleration is the *same* as the direction of the net force acting on it.

One should note that Newton's laws cannot be derived from anything more basic. In fact, this is what we mean by a law of nature. The validity of Newton's laws, and all other laws of nature, comes directly from comparisons with experiment.

Newton's Second Law in Component Form In terms of vector components, an equivalent statement of the second law is

$$\sum F_x = ma_x \qquad \sum F_y = ma_y \qquad \sum F_z = ma_z \qquad \text{5-2}$$

Notice that Newton's second law holds independently for each coordinate direction. This component form of the second law is particularly useful when we solve problems involving more than one direction.

Let's pause for a moment to consider an important special case of the second law. Suppose an object has zero net force acting on it. This may be because no forces act on it at all, or because it is acted on by forces whose vector sum is zero. In either case, we can state this mathematically as

$$\sum \vec{F} = 0$$

Now, according to Newton's second law, we conclude that the acceleration of this object must be zero:

$$\vec{a} = \frac{\sum \vec{F}}{m} = \frac{0}{m} = 0$$

But if an object's acceleration is zero, its velocity must be constant. In other words, if the net force on an object is zero, the object moves with constant velocity. This is Newton's first law. Thus we see that Newton's first and second laws are consistent with one another.

Units for Measuring Forces Forces are measured in a unit called, appropriately enough, the **newton (N)**. In particular, one newton is defined as the force required to give one kilogram of mass an acceleration of 1 m/s^2. Thus,

$$1 \text{ N} = (1 \text{ kg})(1 \text{ m/s}^2) = 1 \text{ kg} \cdot \text{m/s}^2 \qquad \text{5-3}$$

◄ **FIGURE 5-6 Visualizing Concepts Force, Mass, and Acceleration** Even though the tugboats exert a large force on this ship, the ship's acceleration is small. This is because the acceleration of an object is inversely proportional to its mass, and the mass of the ship is enormous. The force exerted on the unfortunate hockey player is much smaller than the force exerted on the ship, but his resulting acceleration is much larger, due to his relatively small mass. The acceleration caused by an ion-thrust rocket is even smaller than that of the hockey player, but when it acts for a long period of time the final velocity of the rocket can be quite large.

In everyday terms, a newton is roughly a quarter of a pound. A force in newtons divided by a mass in kilograms has the units of acceleration:

$$\frac{1\,\text{N}}{1\,\text{kg}} = \frac{1\,\text{kg} \cdot \text{m/s}^2}{1\,\text{kg}} = 1\,\text{m/s}^2 \qquad\qquad 5\text{-}4$$

Other common units for force are presented in Table 5-2. Typical forces and their magnitudes in newtons are listed in Table 5-3.

TABLE 5-2 Units of Mass, Acceleration, and Force

System of units	Mass	Acceleration	Force
SI	kilogram (kg)	m/s²	newton (N)
cgs	gram (g)	cm/s²	dyne (dyn)
British	slug	ft/s²	pound (lb)

(*Note:* 1 N = 10^5 dyne = 0.225 lb)

TABLE 5-3 Typical Forces in Newtons (N)

Main engines of space shuttle	31,000,000
Pulling force of locomotive	250,000
Thrust of jet engine	75,000
Force to accelerate a car	7000
Weight of adult human	700
Weight of an apple	1
Weight of a rose	0.1
Weight of an ant	0.001

EXERCISE 5-1 FIND THE ACCELERATION

The net force acting on a Cessna 172 airplane has a magnitude of 1900 N and points in the positive x direction. If the plane has a mass of 860 kg, what is its acceleration?

REASONING AND SOLUTION

Only x components are involved in this system, so we use the x component of Newton's second law, $\Sigma F_x = ma_x$. Solving this equation for the acceleration yields

$$a_x = \frac{\Sigma F_x}{m} = \frac{1900\,\text{N}}{860\,\text{kg}} = 2.2\,\text{m/s}^2$$

The plane accelerates in the positive x direction, as expected. The magnitude of the acceleration is about one-fourth the acceleration due to gravity.

PHYSICS IN CONTEXT
Looking Back

All forces are vectors, and therefore the ability to use and manipulate vectors is essential to a full and complete understanding of forces. Again, we see the importance of the vector material presented in Chapter 3.

The following Conceptual Example presents a situation in which both Newton's first and second laws play an important role.

CONCEPTUAL EXAMPLE 5-2   TIGHTENING A HAMMER

The metal head of a hammer is loose. To tighten it, you drop the hammer down onto a table. Should you **(a)** drop the hammer with the handle end down, **(b)** drop the hammer with the head end down, or **(c)** do you get the same result either way?

REASONING AND DISCUSSION

It might seem that since the same hammer hits against the same table in either case, there shouldn't be a difference. Actually, there is.

In case (a) the handle of the hammer comes to rest when it hits the table, but the head—with its large inertia—continues downward until a force acts on it to bring it to rest. The force that acts on it is supplied by the handle, which results in the head being wedged more tightly onto the handle. Because the metal head is heavy, the force wedging it onto the handle is great. In case (b) the head of the hammer comes to rest, but the handle continues to move until a force brings it to rest. The handle is lighter than the head, however; thus the force acting on it is less, resulting in less tightening.

ANSWER

(a) Drop the hammer with the handle end down.

RWP* A similar effect occurs when you walk—with each step you tamp your head down onto your spine, as when you drop a hammer handle end down. This causes you to grow shorter during the day! Try it. Measure your height first thing in the morning, then again before going to bed. If you're like many people, you'll find that you have shrunk by an inch or so during the day.

---

*Real World Physics applications are denoted by the acronym RWP.

PROBLEM-SOLVING NOTE

External Forces

External forces acting on an object fall into two main classes: (i) forces at the point of contact with another object, and (ii) forces exerted by an external agent, such as gravity.

Free-Body Diagrams

When we solve problems involving forces and Newton's laws, it's essential to begin by making a sketch that indicates *each and every external force* acting on a given object. This type of sketch is referred to as a **free-body diagram**. If we are concerned only with nonrotational motion, as is the case in this and the next chapter, we treat the object of interest as a point particle and apply each of the forces acting on the object to that point, as **FIGURE 5-7** shows. Once the forces are drawn, we choose a coordinate system and resolve each force into components. At this point, Newton's second law can be applied to each coordinate direction separately.

For example, in Figure 5-7 there are three external forces acting on the chair. One is the force $\vec{F}$ exerted by the person. In addition, gravity exerts a downward force, $\vec{W}$, which is simply the weight of the chair. Finally, the floor exerts an upward force on the chair that prevents it from falling toward the center of the Earth. This force is referred to as the *normal force*, $\vec{N}$, because it is perpendicular (that is, normal) to the surface of the floor. We will consider the weight and the normal force in greater detail in Sections 5-6 and 5-7, respectively.

(a) Sketch the forces

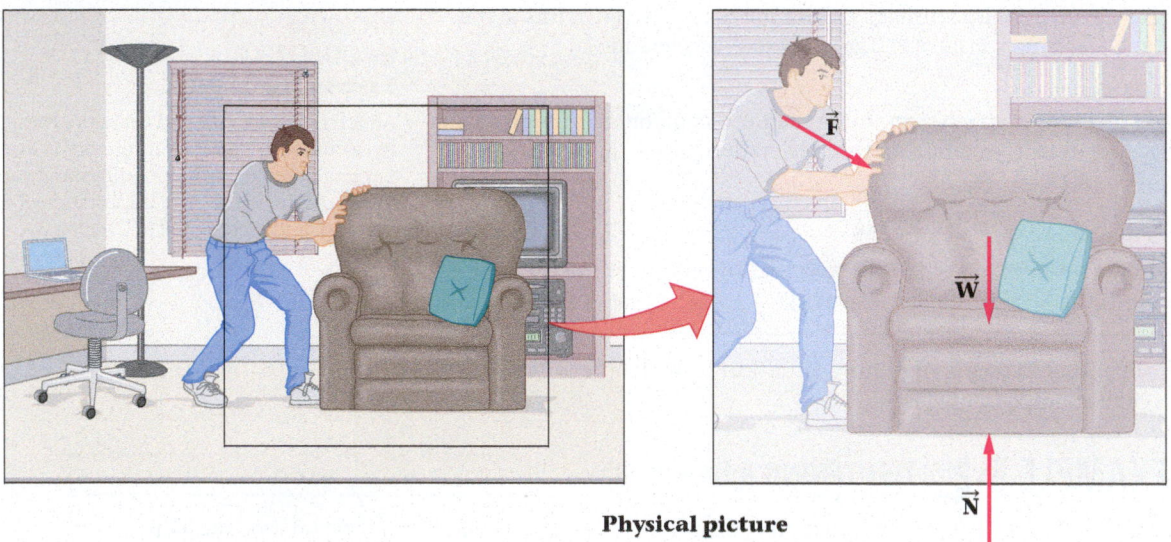

Physical picture

(b) Isolate the object of interest

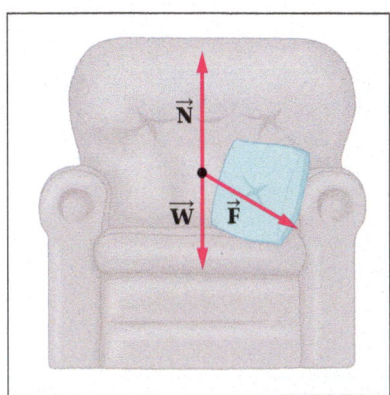

(c) Choose a convenient coordinate system

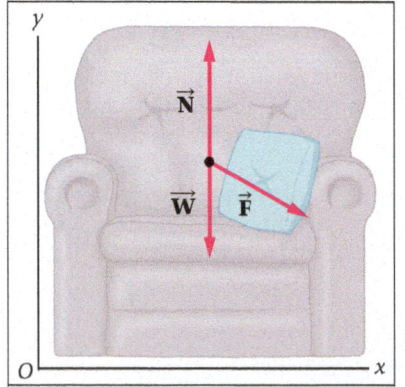

(d) Resolve forces into their components

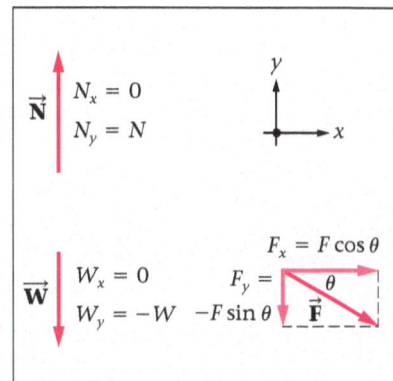

Free-body diagram

▲ **FIGURE 5-7 Constructing and using a free-body diagram** The four basic steps in constructing and using a free-body diagram. **(a)** Sketch all of the external forces acting on an object of interest. Note that only forces acting *on* the object are shown; none of the forces exerted *by* the object are included. **(b)** Isolate the object and treat it as a point particle. **(c)** Choose a convenient coordinate system. This will often mean aligning a coordinate axis to coincide with the direction of one or more forces in the system. **(d)** Resolve each of the forces into components using the coordinate system of part (c).

We can summarize the steps involved in constructing a free-body diagram as follows:

Sketch the Forces

Identify and sketch all of the external forces acting on an object. Sketching the forces roughly to scale will help in estimating the direction and magnitude of the net force.

Isolate the Object of Interest

Replace the object with a point particle of the same mass. Apply each of the forces acting on the object to that point.

Choose a Convenient Coordinate System

Any coordinate system will work; however, if the object moves in a known direction, it is often convenient to pick that direction for one of the coordinate axes. Otherwise, it is reasonable to choose a coordinate system that aligns with one or more of the forces acting on the object.

Resolve the Forces into Components

Determine the components of each force in the free-body diagram.

Apply Newton's Second Law to Each Coordinate Direction

Analyze motion in each coordinate direction using the component form of Newton's second law, as given in Equation 5-2.

These basic steps are illustrated in Figure 5-7. Notice that the figures in this chapter use the labels "Physical picture" to indicate a sketch of the physical situation and "Free-body diagram" to indicate a free-body sketch.

One-Dimensional Systems   In this section, we apply free-body diagrams to one-dimensional examples; we'll save two-dimensional systems for Section 5-5. Suppose, for instance, that you hold a book at rest in your hand. What is the magnitude of the upward force that your hand must exert to keep the book at rest? From everyday experience, we expect that the upward force must be equal in magnitude to the weight of the book, but let's see how this result can be obtained directly from Newton's second law.

We begin with a sketch of the physical situation, as shown in **FIGURE 5-8 (a)**. The corresponding free-body diagram, in **FIGURE 5-8 (b)**, shows just the book, represented by a point, and the forces acting on it. Notice that two forces act on the book: (i) the downward force of gravity, $\vec{W}$, and (ii) the upward force, $\vec{F}$, exerted by your hand. Only the forces acting *on* the book are included in the free-body diagram.

Now that the free-body diagram is drawn, we indicate a coordinate system so that the forces can be resolved into components. In this case all the forces are vertical, thus we draw a y axis in the vertical direction. Note that we have chosen upward to be the positive direction. With this choice, the y components of the forces are $F_y = F$ and $W_y = -W$. It follows that

$$\sum F_y = F - W$$

Using the y component of the second law ($\sum F_y = ma_y$), we find

$$F - W = ma_y$$

The book remains at rest, and hence its acceleration is zero. Thus, $a_y = 0$, which gives

$$F - W = ma_y = 0 \quad \text{or} \quad F = W$$

This is the expected result, but we had to be careful with the signs of components to obtain it.

Next, we consider a situation where the net force acting on an object is nonzero, meaning that its acceleration is also nonzero.

PROBLEM-SOLVING NOTE

Picture the Problem

In problems involving Newton's laws, it is important to begin with a free-body diagram and to identify all the external forces that act on an object. Once these forces are identified and resolved into their components, Newton's laws can be applied in a straightforward way. It is crucial, however, that only external forces acting on the object be included, and that none of the external forces be omitted.

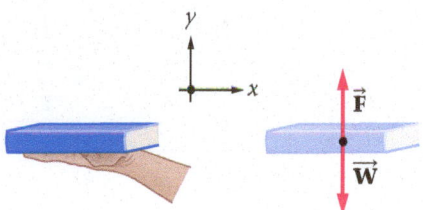

(a) Physical picture   **(b) Free-body diagram**

▲ **FIGURE 5-8 A book supported in a person's hand (a)** The physical situation. **(b)** The free-body diagram for the book, showing the two external forces acting on it. We also indicate our choice for a coordinate system.

EXAMPLE 5-3   THREE FORCES

Moe, Larry, and Curly push on a 752-kg boat that floats next to a dock. They each exert an 80.5-N force parallel to the dock. **(a)** What is the acceleration of the boat if they all push in the same direction? Give both direction and magnitude. **(b)** What are the magnitude and direction of the boat's acceleration if Larry and Curly push in the opposite direction to Moe's push?

PICTURE THE PROBLEM
In our sketch we indicate the three relevant forces acting on the boat: $\vec{F}_M$, $\vec{F}_L$, and $\vec{F}_C$. Notice that we have chosen the positive x direction to the right, in the direction that all three push for part (a). Therefore, all three forces have a positive x component in part (a). In part (b), however, the forces exerted by Larry and Curly have negative x components.

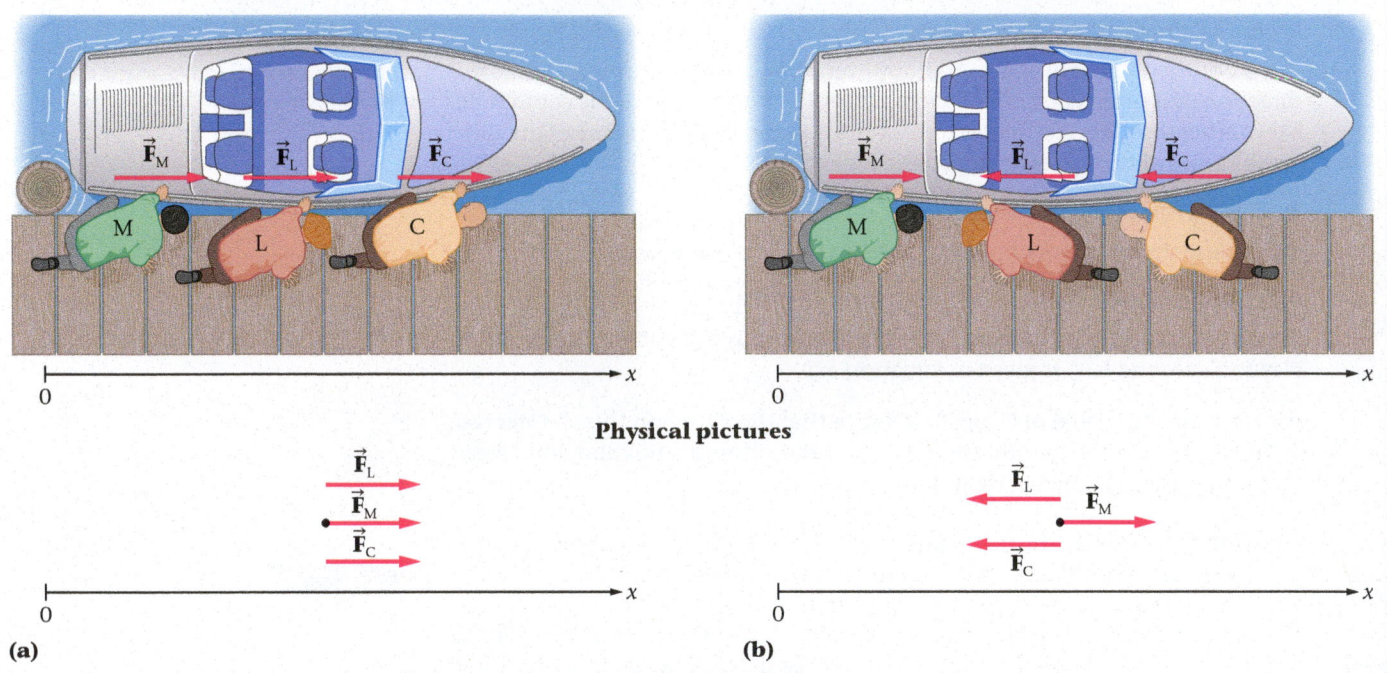

Physical pictures

(a)   **(b)**

Free-body diagrams

REASONING AND STRATEGY
We know the mass of the boat and the forces acting on it, and hence we can find the acceleration using $\sum F_x = ma_x$. Even though this problem is one-dimensional, it's important to think of it in terms of vector components. For example, when we sum the x components of the forces, we are careful to use the appropriate signs—just as we always do when dealing with vectors.

Known   Mass of the boat, $m = 752$ kg; force of each person $= 80.5$ N.
Unknown **(a)** Acceleration of the boat, $a_x = ?$, when all forces are in the same direction
             (b) Acceleration of the boat, $a_x = ?$, when forces are in different directions

SOLUTION

Part (a)

1. Write out the x component for each of the three forces:

$$F_{M,x} = F_{L,x} = F_{C,x} = 80.5 \text{ N}$$

2. Sum the x components of force and set equal to ma_x:

$$\sum F_x = F_{M,x} + F_{L,x} + F_{C,x} = 241.5 \text{ N} = ma_x$$

3. Divide by the mass to find a_x. Because a_x is positive, the acceleration is to the right, as expected:

$$a_x = \frac{\sum F_x}{m} = \frac{241.5 \text{ N}}{752 \text{ kg}} = 0.321 \text{ m/s}^2$$

Part (b)

4. Again, start by writing the x component for each force:

$$F_{M,x} = 80.5 \text{ N}$$
$$F_{L,x} = F_{C,x} = -80.5 \text{ N}$$

5. Sum the x components of force and set equal to ma_x:

$$\sum F_x = F_{M,x} + F_{L,x} + F_{C,x}$$
$$= 80.5 \text{ N} - 80.5 \text{ N} - 80.5 \text{ N} = -80.5 \text{ N} = ma_x$$

6. Solve for a_x. In this case a_x is negative, indicating an acceleration to the left:

$$a_x = \frac{\sum F_x}{m} = \frac{-80.5 \text{ N}}{752 \text{ kg}} = -0.107 \text{ m/s}^2$$

CONTINUED

Combining Newton's Laws with Kinematics   In some problems, we are given information that allows us to calculate an object's acceleration using the kinematic equations of Chapters 2 and 4 . Once the acceleration is known, the second law can be used to find the net force that caused the acceleration.

RWP For example, suppose an astronaut uses a jet pack to push a satellite toward the space shuttle. These jet packs, which are known to NASA as Manned Maneuvering Units, or MMUs, are basically small "one-person rockets" strapped to the back of an astronaut's spacesuit. An MMU contains pressurized nitrogen gas that can be released through varying combinations of 24 nozzles spaced around the unit, producing a force of about 10 pounds. The MMUs contain enough propellant for a six-hour EVA (extra-vehicular activity).

In **FIGURE 5-9 (a)** we show a physical situation where an astronaut pushes on a 655-kg satellite. The corresponding free-body diagram for the satellite is shown in **FIGURE 5-9 (b)**. Notice that we've chosen the x axis to point in the direction of the push. Now, if the satellite starts at rest and moves 0.675 m after 5.00 seconds of pushing, what is the force, F, exerted on it by the astronaut?

Clearly, we would like to use Newton's second law (basically, $\vec{\mathbf{F}} = m\vec{\mathbf{a}}$) to find the force, but we know only the mass of the satellite, not its acceleration. We can find the acceleration, however, by assuming constant acceleration (after all, the force is constant) and using the kinematic equation relating position to time: $x = x_0 + v_{0x}t + \frac{1}{2}a_x t^2$. We can choose the initial position of the satellite to be $x_0 = 0$, and we are given that it starts at rest; thus $v_{0x} = 0$. Hence,

$$x = \tfrac{1}{2}a_x t^2$$

Since we know the distance covered in a given time, we can solve for the acceleration:

$$a_x = \frac{2x}{t^2} = \frac{2(0.675 \text{ m})}{(5.00 \text{ s})^2} = 0.0540 \text{ m/s}^2$$

Now that kinematics has provided the acceleration, we use the x component of the second law to find the force. Only one force acts on the satellite, and its x component is F; thus,

$$\sum F_x = F = ma_x$$
$$F = ma_x = (655 \text{ kg})(0.0540 \text{ m/s}^2) = 35.4 \text{ N}$$

This force corresponds to a push of about 8 pounds.

Another system in which we use kinematics to find the acceleration is presented in the next Example.

(a) Physical picture

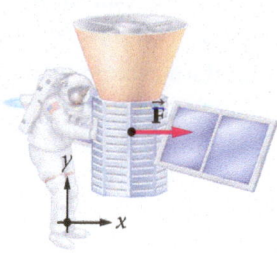

(b) Free-body diagram

▲ **FIGURE 5-9 An astronaut using a jet pack to push a satellite (a)** The physical situation. **(b)** The free-body diagram for the satellite. Only one force acts on the satellite, and it is in the positive x direction.

PHYSICS IN CONTEXT
Looking Ahead

Forces are a central theme throughout physics. In particular, we shall see in Chapters 7 and 8 that a force acting on an object over a distance changes its energy.

EXAMPLE 5-4   THE FORCE EXERTED BY FOAMCRETE

RWP Foamcrete is a substance designed to stop an airplane that has run off the end of a runway, without causing injury to passengers. It is solid enough to support a car, but it crumbles under the weight of a large airplane, as shown in **FIGURE 5-10**. By crumbling, Foamcrete slows the plane to a safe stop. For example, suppose a 747 jetliner with a mass of 1.75×10^5 kg and an initial speed of 26.8 m/s is slowed to a stop in 122 m. What is the magnitude of the retarding force exerted by the Foamcrete on the plane?

CONTINUED

PICTURE THE PROBLEM

The airplane and its stopping distance of $\Delta x = 122$ m are shown in our sketch. The initial velocity is $v_0 = 26.8$ m/s in the positive x direction, and the final velocity of the plane is $v = 0$. The retarding force is in the negative x direction, and hence we can write it as $\vec{F} = (-f)\hat{x}$, where $-f$ is the x component of the force, and f is the magnitude.

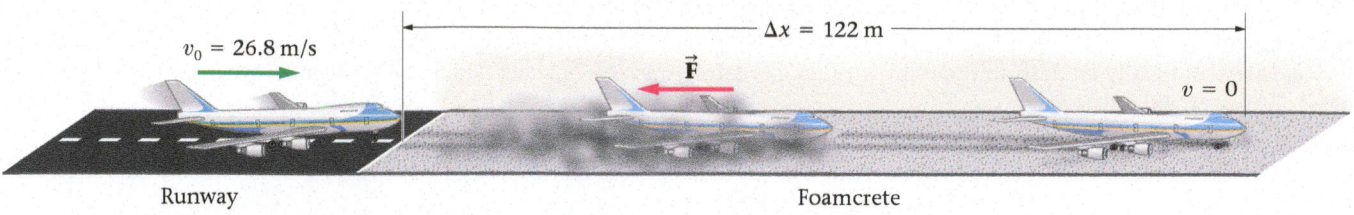

REASONING AND STRATEGY

First, use the kinematics equation that relates displacement to velocity to solve for the acceleration of the plane. Knowing the acceleration and mass, we can determine the retarding force acting on the plane.

Known Mass of the airplane, $m = 1.75 \times 10^5$ kg; initial speed of airplane, $v_0 = 26.8$ m/s; stopping distance, $\Delta x = 122$ m.
Unknown Magnitude of the retarding force, $f = ?$

SOLUTION

1. Solve $v^2 = v_0^2 + 2a_x \Delta x$ to find the plane's acceleration. Recall that $v = 0, v_0 = 26.8$ m/s, and $\Delta x = 122$ m:

$$a_x = \frac{v^2 - v_0^2}{2\Delta x} = \frac{-(26.8 \text{ m/s})^2}{2(122 \text{ m})}$$
$$= -2.94 \text{ m/s}^2$$

2. Sum the forces in the x direction (there's only one force in the x direction in this system):

$$\sum F_x = -f$$

3. Set the sum of the forces equal to mass times acceleration:

$$-f = ma_x \quad \text{or, equivalently,} \quad f = -ma_x$$

4. Solve for the magnitude of the retarding force, f:

$$f = -ma_x = -(1.75 \times 10^5 \text{ kg})(-2.94 \text{ m/s}^2)$$
$$= 5.15 \times 10^5 \text{ N}$$

INSIGHT

The plane moves in the positive direction, but the retarding force exerted on it, and its acceleration, are in the negative direction. As a result, the plane's speed decreases with time.

PRACTICE PROBLEM

What is the stopping distance if the magnitude of the retarding force is $f = 2.25 \times 10^5$ N? **[Answer: 278 m]**

Some related homework problems: Problem 9, Problem 12

▶ **FIGURE 5-10** A technician inspects the landing gear of an airliner in a test of Foamcrete.

EXAMPLE 5-5   PITCH MAN: ESTIMATE THE FORCE ON THE BALL

A pitcher throws a 0.15-kg baseball, accelerating it from rest to a speed of about 90 mi/h. Estimate the force exerted by the pitcher on the ball.

PICTURE THE PROBLEM

We choose the x axis to point in the direction of the pitch. Also indicated in the sketch is the distance over which the pitcher accelerates the ball, Δx. Because we are interested only in the pitch, and not in the subsequent motion of the ball, we ignore the effects of gravity.

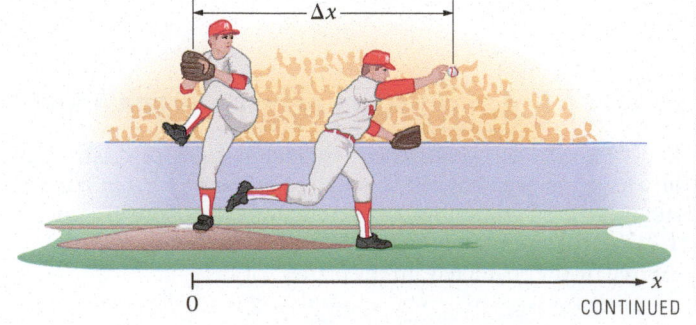

CONTINUED

REASONING AND STRATEGY

We know the mass, so we can find the force with $F_x = ma_x$ if we can estimate the acceleration. To find the acceleration, we start with the fact that $v_0 = 0$ and $v \approx 90$ mi/h. In addition, we can see from the sketch that a reasonable estimate for Δx is about 2.0 m. Combining these results with the kinematic equation $v^2 = v_0^2 + 2a_x \Delta x$ yields the acceleration, which we then use to find the force.

Known   Mass of baseball, $m = 0.15$ kg; initial speed, $v_0 = 0$; final speed, $v = 90$ mi/h.
Unknown   Force exerted by pitcher, $F_x = ?$

SOLUTION

1. Starting with the fact that 60 mi/h = 1 mi/min, perform a rough back-of-the-envelope conversion of 90 mi/h to meters per second:

$$v \approx 90 \text{ mi/h} = \frac{1.5 \text{ mi}}{\text{min}} \approx \frac{2400 \text{ m}}{60 \text{ s}} = 40 \text{ m/s}$$

2. Solve $v^2 = v_0^2 + 2a_x \Delta x$ for the acceleration, a_x. Use the estimates $\Delta x \approx 2.0$ m and $v \approx 40$ m/s:

$$a_x = \frac{v^2 - v_0^2}{2 \Delta x} \approx \frac{(40 \text{ m/s})^2 - 0}{2(2.0 \text{ m})} = 400 \text{ m/s}^2$$

3. Find the corresponding force with $F_x = ma_x$:

$$F_x = ma_x \approx (0.15 \text{ kg})(400 \text{ m/s}^2) = 60 \text{ N} \approx 10 \text{ lb}$$

INSIGHT

On the one hand, this is a sizable force, especially when you consider that the ball itself weighs only about 1/3 lb. Thus, the pitcher exerts a force on the ball that is about 30 times greater than the force exerted by Earth's gravity. It follows that ignoring gravity during the pitch is a reasonable approximation.

On the other hand, you might say that 10 lb isn't that much force for a person to exert. That's true, but this force is being exerted with an average speed of about 20 m/s, which means that the pitcher is actually generating about 1.5 horsepower—a sizable power output for a person. We will cover power in detail in Chapter 7, and relate it to human capabilities.

PRACTICE PROBLEM

What is the approximate speed of the ball if the force exerted by the pitcher is $\frac{1}{2}(60 \text{ N}) = 30$ N?   **[Answer: 30 m/s or 60 mi/h]**

Some related homework problems: Problem 5, Problem 9

Enhance Your Understanding   (Answers given at the end of the chapter)

3. The acceleration of an object has a magnitude a. What is the magnitude of the acceleration in the following cases? **(a)** All the forces acting on the object are doubled. **(b)** The mass and the net force acting on the object are doubled. **(c)** The net force acting on the object is doubled, and its mass is halved. **(d)** The mass of the object is doubled, and the net force acting on it is halved.

Section Review

- **Newton's Second Law of Motion** The acceleration of an object is proportional to the force acting on it, and inversely proportional to its mass.

- A free-body diagram shows all the external forces acting on a given object.

5-4 Newton's Third Law of Motion

Nature never produces just one force at a time—*forces always come in pairs*. In addition, the forces in a pair always act on *different objects*, are *equal in magnitude*, and point in *opposite directions*. This is Newton's third law of motion.

Newton's Third Law

For every force that acts on an object, there is a reaction force acting on a different object that is equal in magnitude and opposite in direction.

In a somewhat more specific form:

> If object 1 exerts a force $\vec{F}$ on object 2, then object 2 exerts a force $-\vec{F}$ on object 1.

This law, more commonly known by its abbreviated form, "For every action there is an equal and opposite reaction," completes Newton's laws of motion.

Big Idea 4 If object 1 exerts a force on object 2, then object 2 exerts an equal but opposite force on object 1. In short, all forces come in equal but opposite pairs.

Weighing a Hovering Magnet

▶ **FIGURE 5-11 Examples of action-reaction force pairs**

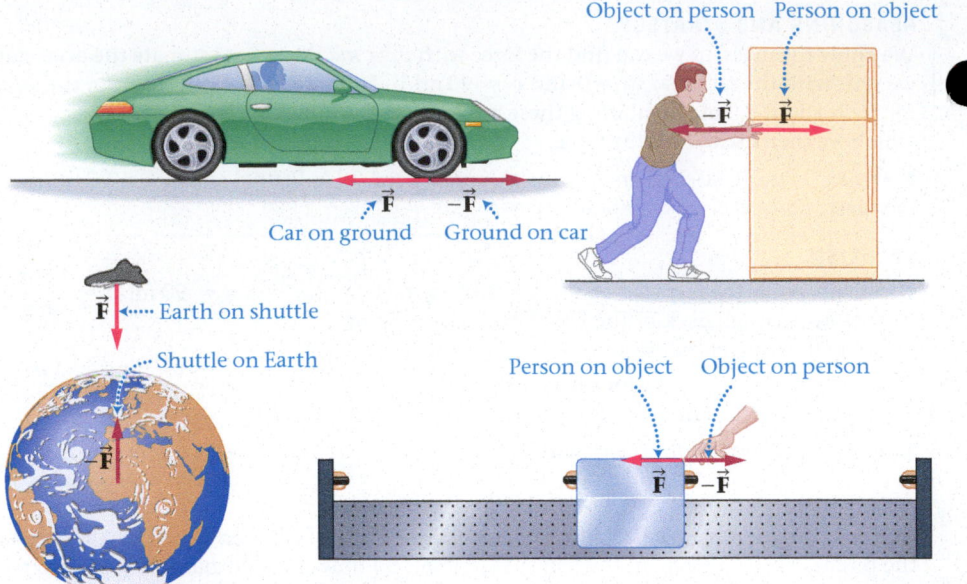

FIGURE 5-11 illustrates some action-reaction pairs. Notice that there is always a reaction force, whether the action force pushes on something hard to move, like a refrigerator, or on something that moves with no friction, like an air-track cart. In some cases, the reaction force tends to be overlooked, as when the Earth exerts a *downward* gravitational force on the space shuttle, and the shuttle exerts an equal and opposite *upward* gravitational force on the Earth. Still, the reaction force always exists.

Third-Law Forces Act on Different Objects Another important aspect of the third law is that the action-reaction forces always act on *different* objects. This, again, is illustrated in Figure 5-11. Thus, in drawing a free-body diagram, we draw only one of the action-reaction pair of forces for a given object. The other force in the pair would appear in the free-body diagram of a different object. As a result:

> The two forces in an action-reaction pair do not cancel in the free-body diagram for an object.

For example, consider a car accelerating from rest, as in Figure 5-11. As the car's engine turns the wheels, the tires exert a force on the road. By the third law, the road exerts an equal and opposite force on the car's tires. It is this second force—which acts on the car through its tires—that propels the car forward. The force exerted by the tires on the road does not accelerate the car.

Because the action-reaction forces act on different objects, they generally produce different accelerations. This is the case in the next Example.

EXAMPLE 5-6   TIPPY CANOE

Two groups of canoeists meet in the middle of a lake. After a brief visit, a person in canoe 1 pushes on canoe 2 with a force of 46 N to separate the canoes. The mass of canoe 1 and its occupants is $m_1 = 150$ kg, and the mass of canoe 2 and its occupants is $m_2 = 250$ kg. **(a)** Find the acceleration the push gives to each canoe. **(b)** What is the separation of the canoes after 1.2 s of pushing?

PICTURE THE PROBLEM

We have chosen the positive x direction to point from canoe 1 to canoe 2. With this choice, the force exerted on canoe 2 is $\vec{\mathbf{F}}_2 = (+46 \text{ N})\hat{\mathbf{x}}$. By Newton's third law, the force exerted on the person in canoe 1, and thus on canoe 1 itself if the person is firmly seated, is $\vec{\mathbf{F}}_1 = (-46 \text{ N})\hat{\mathbf{x}}$. For convenience, we have placed the origin at the point where the canoes touch.

REASONING AND STRATEGY

From Newton's third law, the force on canoe 1 is equal in magnitude to the force on canoe 2. The masses of the canoes are different, however, and therefore their accelerations are different as well. **(a)** We can find the acceleration of each

CONTINUED

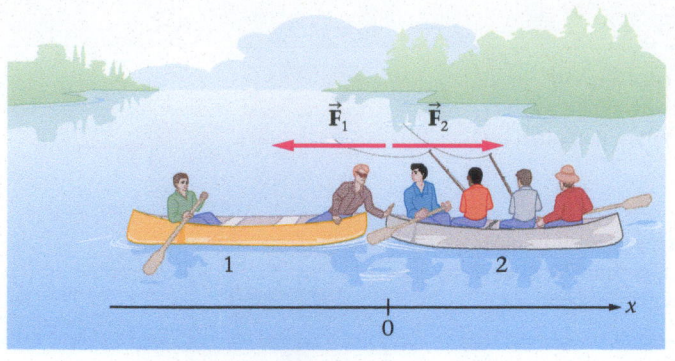
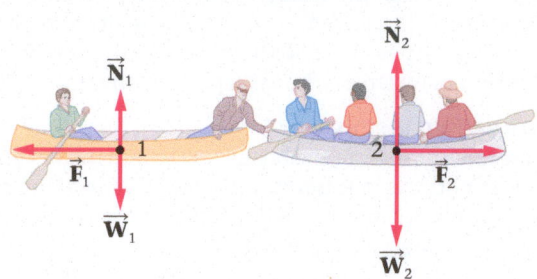

Physical picture   **Free-body diagrams**

canoe by solving $\Sigma F_x = ma_x$ for a_x. **(b)** The kinematic equation relating position to time, $x = x_0 + v_{0x}t + \frac{1}{2}a_xt^2$, can then be used to find the displacement of each canoe.

Known Force of separation, 46 N; mass of canoe 1, $m_1 = 150$ kg; mass of canoe 2, $m_2 = 250$ kg; time of pushing, $t = 1.2$ s.

Unknown **(a)** Acceleration of canoe 1, $a_{1,x} = ?$ Acceleration of canoe 2, $a_{2,x} = ?$ **(b)** Separation of canoes at $t = 1.2$ s, $x_2 - x_1 = ?$

SOLUTION

Part (a)

1. Use Newton's second law to find the acceleration of canoe 2:

$$a_{2,x} = \frac{\Sigma F_{2,x}}{m_2} = \frac{46\,\text{N}}{250\,\text{kg}} = 0.18\,\text{m/s}^2$$

2. Do the same calculation for canoe 1. Notice that the acceleration of canoe 1 is in the negative direction:

$$a_{1,x} = \frac{\Sigma F_{1,x}}{m_1} = \frac{-46\,\text{N}}{150\,\text{kg}} = -0.31\,\text{m/s}^2$$

Part (b)

3. Use $x = x_0 + v_{0x}t + \frac{1}{2}a_xt^2$ to find the position of canoe 2 at $t = 1.2$ s. From the problem statement, we know the canoes start at the origin ($x_0 = 0$) and at rest ($v_{0x} = 0$):

$$x_2 = \frac{1}{2}a_{2,x}t^2 = \frac{1}{2}(0.18\,\text{m/s}^2)(1.2\,\text{s})^2 = 0.13\,\text{m}$$

4. Repeat the calculation for canoe 1:

$$x_1 = \frac{1}{2}a_{1,x}t^2 = \frac{1}{2}(-0.31\,\text{m/s}^2)(1.2\,\text{s})^2 = -0.22\,\text{m}$$

5. Subtract the two positions to find the separation of the canoes:

$$x_2 - x_1 = 0.13\,\text{m} - (-0.22\,\text{m}) = 0.35\,\text{m}$$

INSIGHT

The same magnitude of force acts on each canoe; hence the lighter one has the greater acceleration and the greater displacement. If the heavier canoe were replaced by a large ship of great mass, both vessels would still accelerate as a result of the push. However, the acceleration of the large ship would be so small as to be practically imperceptible. In this case, it would appear as if only the canoe moved, whereas, in fact, both vessels move.

PRACTICE PROBLEM — PREDICT/CALCULATE

(a) If the mass of canoe 2 is increased, does its acceleration increase, decrease, or stay the same? Explain. **(b)** Check your prediction by calculating the acceleration for the case where canoe 2 is replaced by a 25,000-kg ship. [**Answer:** **(a)** Acceleration is inversely proportional to mass, so the acceleration of canoe 2 decreases. **(b)** In this case, $a = 0.0018$ m/s^2. This is one hundred times smaller than the acceleration found in this Example, because the mass is one hundred times larger.]

Some related homework problems: Problem 20, Problem 21

Contact Forces When objects are touching one another, the action-reaction forces are often referred to as **contact forces**. The behavior of contact forces is explored in the following Conceptual Example.

CONCEPTUAL EXAMPLE 5-7   CONTACT FORCES

Two boxes—one large and heavy, the other small and light—rest on a smooth, level floor. You push with a force $\vec{\mathbf{F}}$ on either the small box or the large box. Is the contact force between the two boxes **(a)** the same in either case, **(b)** larger when you push on the large box, or **(c)** larger when you push on the small box?

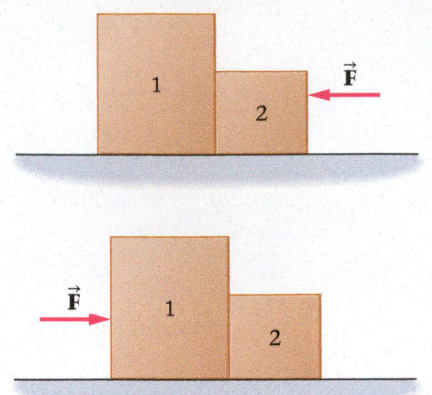

REASONING AND DISCUSSION
The same force pushes on the boxes, and so you might think the force of contact is the same in both cases. It is not. What we can conclude, however, is that the boxes have the same acceleration in either case—the same net force acts on the same total mass, so the same acceleration, a, results.

    To find the contact force between the boxes, we focus our attention on each box individually, and note that *Newton's second law must be satisfied for each of the boxes, just as it is for the entire two-box system.* For example, when the external force is applied to the small box, the only force acting on the *large* box (mass m_1) is the contact force; hence, the contact force must have a magnitude equal to m_1a. In the second case, the only force acting on the *small* box (mass m_2) is the contact force, and so the magnitude of the contact force is m_2a. Noting that m_1 is greater than m_2, we conclude that the force of contact is larger when you push on the small box, m_1a, than when you push on the large box, m_2a.

    To put it another way, the contact force is larger when *it* must push the larger box.

ANSWER
(c) The contact force is larger when you push on the small box.

In the next Example, we calculate a numerical value for the contact force in a system similar to that described in Conceptual Example 5-7. We also show explicitly that Newton's third law is required for a full analysis of this system.

EXAMPLE 5-8   WHEN PUSH COMES TO SHOVE

A box of mass $m_1 = 10.0\,\text{kg}$ rests on a smooth, horizontal floor in contact with a box of mass $m_2 = 5.00\,\text{kg}$. You now push on box 1 with a horizontal force of magnitude $F = 30.0\,\text{N}$. **(a)** What is the acceleration of the boxes? **(b)** What is the force of contact between the boxes?

PICTURE THE PROBLEM
We choose the x axis to be horizontal and pointing to the right. Thus, the force pushing the boxes is $\vec{\mathbf{F}} = (30.0\,\text{N})\hat{\mathbf{x}}$. The contact forces are labeled as follows: $\vec{\mathbf{F}}_1$ is the contact force exerted on box 1; $\vec{\mathbf{F}}_2$ is the contact force exerted on box 2. By Newton's third law, the contact forces have the same magnitude, f, but point in opposite directions. With our coordinate system, we have $\vec{\mathbf{F}}_1 = -f\hat{\mathbf{x}}$ and $\vec{\mathbf{F}}_2 = f\hat{\mathbf{x}}$.

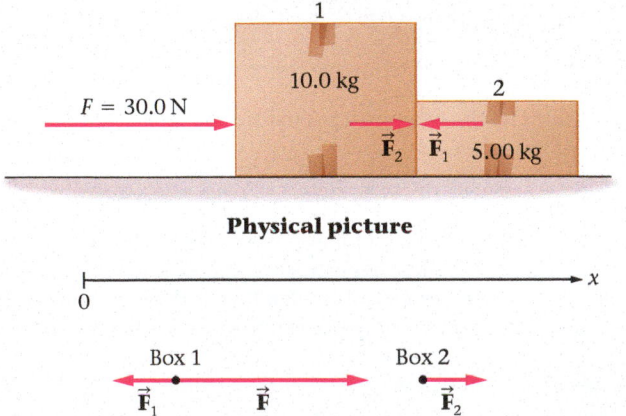

REASONING AND STRATEGY

a. The two boxes are in contact, and hence they have the same acceleration. We find this acceleration with Newton's second law; that is, we divide the net horizontal force by the total mass of the two boxes.

b. In this case, we consider the system to consist solely of box 2. The mass in this case is 5.00 kg, and the only horizontal force acting on the system is $\vec{\mathbf{F}}_2$. Thus, we can find f, the magnitude of $\vec{\mathbf{F}}_2$, by requiring that box 2 has the acceleration found in part (a).

Known   Mass of box 1, $m_1 = 10.0\,\text{kg}$; mass of box 2, $m_2 = 5.00\,\text{kg}$; pushing force, $F = 30.0\,\text{N}$.
Unknown   **(a)** Acceleration of the boxes, $a_x = ?$ **(b)** Force of contact, $f = ?$

SOLUTION

Part (a)

1. Find the net horizontal force acting on the two boxes. Notice that $\vec{\mathbf{F}}_1$ and $\vec{\mathbf{F}}_2$ are equal in magnitude but opposite in direction. Hence, they sum to zero; $\vec{\mathbf{F}}_1 + \vec{\mathbf{F}}_2 = 0$:

$$\sum_{\substack{\text{both}\\\text{boxes}}} F_x = F = 30.0\,\text{N}$$

CONTINUED

2. Divide the net force by the total mass, $m_1 + m_2$, to find the acceleration of the boxes:

$$a_x = \frac{\sum F_x}{m_1 + m_2}$$

$$= \frac{30.0\,\text{N}}{(10.0\,\text{kg} + 5.00\,\text{kg})} = \frac{30.0\,\text{N}}{15.0\,\text{kg}} = 2.00\,\text{m/s}^2$$

Part (b)

3. Find the net horizontal force acting on box 2, and set it equal to the mass of box 2 times its acceleration:

$$\sum_{\text{box2}} F_x = F_{2,x} = f = m_2 a_x$$

4. Determine the magnitude of the contact force, f, by substituting numerical values for m_2 and a_x:

$$f = m_2 a_x = (5.00\,\text{kg})(2.00\,\text{m/s}^2) = 10.0\,\text{N}$$

INSIGHT

The net horizontal force acting on box 1 is $F - f = 30.0\,\text{N} - 10.0\,\text{N} = 20.0\,\text{N}$, and hence its acceleration is $(20.0\,\text{N})/(10.0\,\text{kg}) = 2.00\,\text{m/s}^2$. Thus, as expected, box 1 and box 2 have precisely the same acceleration.

If box 2 were not present, the 30.0-N force acting on box 1 would give it an acceleration of $3.00\,\text{m/s}^2$. As it is, the contact force between the boxes slows box 1, so that its acceleration is less than $3.00\,\text{m/s}^2$. At the same time, the contact force gives box 2 an acceleration that is greater than zero. The precise value of the contact force is simply the value that gives *both boxes* the same acceleration.

PRACTICE PROBLEM

Suppose the relative positions of the boxes are reversed, so that F pushes on the small box, as shown here. Calculate the contact force for this case, and show that the force is greater than 10.0 N, as expected from Conceptual Example 5-7.   [**Answer:** The contact force in this case is 20.0 N, double its previous value. This follows because the box being pushed by the contact force has twice the mass of the box that was pushed by the contact force originally.]

Some related homework problems: Problem 22, Problem 23

BIO Living animals utilize Newton's third law of motion to propel themselves. Perhaps the champion of propulsion is the flea (*Ctenocephalides felis*), which can jump 33 cm horizontally, a distance that is more than 150 times its 2-mm body length. To propel itself, the flea's legs exert a force on the surface on which it sits. The surface, in turn, exerts an equal and opposite force on the flea, in accordance with Newton's third law. This force then accelerates the flea according to Newton's second law of motion. All biological propulsion can be explained in similar terms. In Chapter 8 we will use energy concepts to reexamine the flea's amazing jumping ability.

Enhance Your Understanding   (Answers given at the end of the chapter)

4. A force F pushes on three boxes that slide without friction on a horizontal surface, as indicated in **FIGURE 5-12**. Box 1 has a mass of 1 kg, box 2 has a mass of 2 kg, and box 3 has a mass of 3 kg. The boxes are in contact, and slide as a group with an acceleration a. Is the magnitude of the force of contact between boxes 1 and 2 greater than, less than, or equal to the magnitude of the force of contact between boxes 2 and 3? Explain.

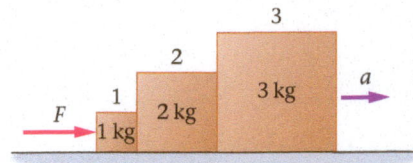

▲ **FIGURE 5-12** Enhance Your Understanding 4

Section Review

- **Newton's Third Law of Motion** Forces always come in pairs that are equal in magnitude but opposite in direction. The two forces in a pair always act on different objects.

5-5 The Vector Nature of Forces: Forces in Two Dimensions

When we presented Newton's second law in Section 5-3, we said that an object's acceleration is equal to the net force acting on it divided by its mass. If more than one force acts on an object, experiments show that its acceleration is in the direction of the vector sum of the forces.

As an example of the vector nature of forces, suppose two astronauts are using jet packs to push a 940-kg satellite toward a space station, as shown in **FIGURE 5-13**. With the coordinate system indicated in the figure, astronaut 1 pushes in the positive x direction and astronaut 2 pushes in a direction 52° above the x axis. If astronaut 1 pushes with a force of magnitude $F_1 = 26$ N and astronaut 2 pushes with a force of magnitude $F_2 = 41$ N, what are the magnitude and direction of the satellite's acceleration?

▶ **FIGURE 5-13 Two astronauts pushing a satellite with forces that differ in magnitude and direction** The acceleration of the satellite can be found by calculating a_x and a_y separately, then combining these components to find a and θ.

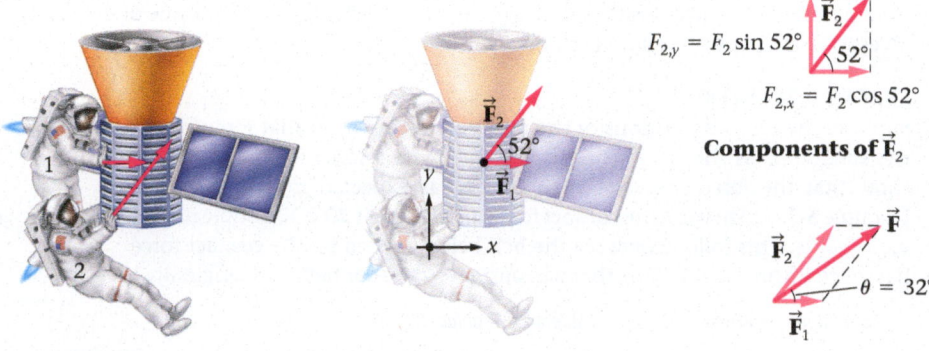

(a) Physical picture   (b) Free-body diagram   Total force

Different Coordinate Directions Can Be Treated Independently   The easiest way to solve a problem like this is to treat each coordinate direction independently of the other, just as we did many times when studying two-dimensional kinematics in Chapter 4. Thus, we first resolve each force into its x and y components. Referring to Figure 5-13, we see that for the x direction

$$F_{1,x} = F_1$$
$$F_{2,x} = F_2 \cos 52°$$

For the y direction

$$F_{1,y} = 0$$
$$F_{2,y} = F_2 \sin 52°$$

Next, we find the acceleration in the x direction by using the x component of Newton's second law:

$$\sum F_x = ma_x$$

Applied to this system, we have

$$\sum F_x = F_{1,x} + F_{2,x} = F_1 + F_2 \cos 52° = 26\,\text{N} + (41\,\text{N}) \cos 52° = 51\,\text{N}$$
$$= ma_x$$

Solving for the acceleration yields

$$a_x = \frac{\sum F_x}{m} = \frac{51\,\text{N}}{940\,\text{kg}} = 0.054\,\text{m/s}^2$$

Similarly, in the y direction we start with

$$\sum F_y = ma_y$$

This gives

$$\sum F_y = F_{1,y} + F_{2,y} = 0 + F_2 \sin 52° = (41\,\text{N}) \sin 52° = 32\,\text{N}$$
$$= ma_y$$

As a result, the y component of acceleration is

$$a_y = \frac{\sum F_y}{m} = \frac{32 \text{ N}}{940 \text{ kg}} = 0.034 \text{ m/s}^2$$

Thus, the satellite accelerates in both the x and the y directions. Its total acceleration has a magnitude of

$$a = \sqrt{a_x{}^2 + a_y{}^2} = \sqrt{(0.054 \text{ m/s}^2)^2 + (0.034 \text{ m/s}^2)^2} = 0.064 \text{ m/s}^2$$

From Figure 5-13 we expect the total acceleration to be in a direction above the x axis but at an angle less than 52°. Straightforward calculation yields

$$\theta = \tan^{-1}\left(\frac{a_y}{a_x}\right) = \tan^{-1}\left(\frac{0.034 \text{ m/s}^2}{0.054 \text{ m/s}^2}\right) = \tan^{-1}(0.63) = 32°$$

This is the same direction as the total force in Figure 5-13, as expected.

The next two Examples give further practice with resolving force vectors and using Newton's second law in component form.

PHYSICS IN CONTEXT
Looking Back

As with two-dimensional kinematics in Chapter 4, where motions in the x and y directions were seen to be independent, the x and y components of force are independent. In particular, acceleration in the x direction depends on only the x component of force, and acceleration in the y direction depends on only the y component of force.

EXAMPLE 5-9 BARELY OUT OF REACH

Campers lift a food cooler with a weight of $W = 192.0$ N out of reach of bears using two ropes hung over tree branches. Rope 1 exerts a force $\vec{F}_1$ with a magnitude of 166 N at an angle of 47.4° above the horizontal, as shown in the sketch. Rope 2 exerts a force $\vec{F}_2$ at an angle of 36.1° above the horizontal. **(a)** Find the magnitude of the force $\vec{F}_2$ that produces a net force on the cooler that is straight upward. **(b)** Determine the net force acting on the cooler, given that $\vec{F}_2$ has the magnitude determined in part (a).

PICTURE THE PROBLEM

Our physical picture and free-body diagram show the cooler and the three forces acting on it, as well as the angles of $\vec{F}_1$ and $\vec{F}_2$ relative to the horizontal. We also show the x and y components of $\vec{F}_1$ and $\vec{F}_2$. The weight has only a negative y component; that is, $\vec{W} = -(192.0 \text{ N})\hat{y}$.

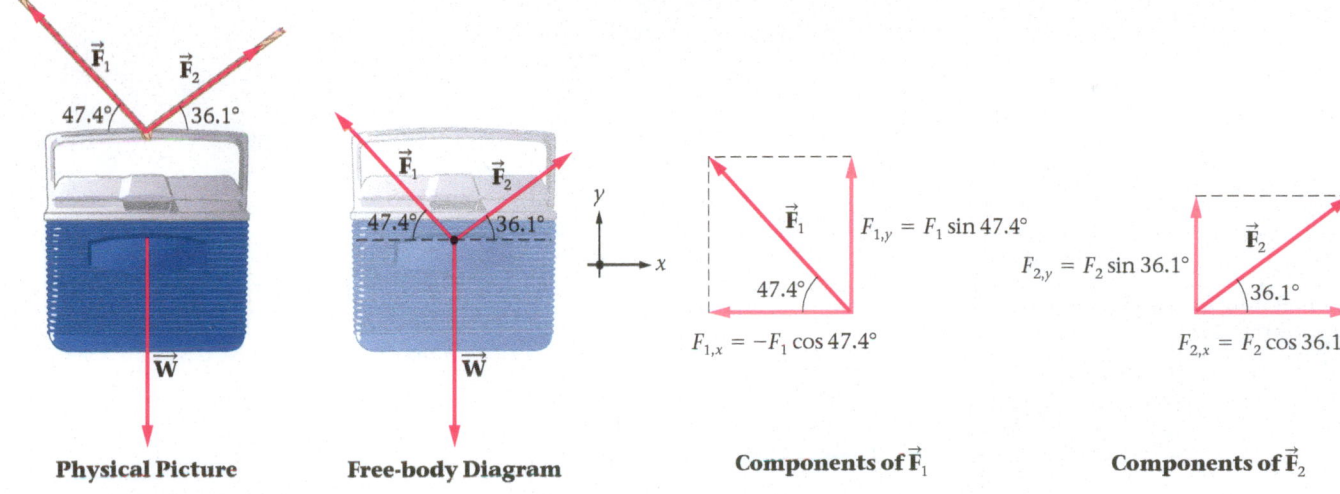

| Physical Picture | Free-body Diagram | Components of $\vec{F}_1$ | Components of $\vec{F}_2$ |

REASONING AND STRATEGY

a. We want the net force to be purely vertical. This means that the x component of the net force must be zero; that is, $\sum F_x = 0$. This condition can be solved to yield F_2, the magnitude of $\vec{F}_2$.

b. The x component of the net force is zero if we use F_2 from part (a). All that remains to find the net force on the cooler is to sum the y components of all three forces.

Known   Weight of cooler, $W = 192.0$ N; force exerted by rope 1, 166 N at 47.4° above horizontal; force exerted by rope 2, 36.1° above horizontal.

Unknown (a) Magnitude of force exerted by rope 2 to produce a straight upward net force, $F_2 = ?$
(b) Net force acting on cooler, $\sum F_y = ?$

SOLUTION

Part (a)

1. Begin by writing the x component of each force. Notice that $\vec{W}$ has no x component and that the x component of $\vec{F}_1$ points in the negative x direction:

$F_{1,x} = -F_1 \cos 47.4°$
$F_{2,x} = F_2 \cos 36.1°$
$W_x = 0$

CONTINUED

2. Sum the x components of force and set equal to zero. The magnitude F_2 is the only unknown in this equation:

$$\sum F_x = -F_1 \cos 47.4° + F_2 \cos 36.1° + 0 = 0 \quad \text{or}$$
$$F_1 \cos 47.4° = F_2 \cos 36.1°$$

3. Solve for F_2:

$$F_2 = \frac{F_1 \cos 47.4°}{\cos 36.1°} = \frac{(166\,\text{N}) \cos 47.4°}{\cos 36.1°} = 139\,\text{N}$$

Part (b)

4. Determine the y component of each force:

$$F_{1,y} = F_1 \sin 47.4° = (166\,\text{N}) \sin 47.4° = 122\,\text{N}$$
$$F_{2,y} = F_2 \sin 36.1° = (139\,\text{N}) \sin 36.1° = 81.9\,\text{N}$$
$$W_y = -W = -192.0\,\text{N}$$

5. Sum the y components of force:

$$\sum F_y = F_1 \sin 47.4° + F_2 \sin 36.1° - W$$
$$= 122\,\text{N} + 81.9\,\text{N} - 192.0\,\text{N} = 11.9\,\text{N}$$

INSIGHT

Thus, the net force acting on the cooler is 11.9 N in the positive y direction; that is, $\vec{F}_{net} = (11.9\,\text{N})\hat{y}$. As a double check that the x component of force is zero, we have $\sum F_x = -F_1 \cos 47.4° + F_2 \cos 36.1° + 0 = -(166\,\text{N})\cos 47.4° + (139\,\text{N})\cos 36.1° = -112\,\text{N} + 112\,\text{N} = 0$.

PRACTICE PROBLEM — PREDICT/CALCULATE

Suppose we change the angle of $\vec{F}_1$ to be 37.4° above the horizontal. **(a)** Is the magnitude F_2 that gives a zero x component of the net force greater than, less than, or equal to 139 N? Explain. **(b)** Find the new value of F_2.   [**Answer:** **(a)** Changing the angle to 37.4° increases the magnitude of the x component of $\vec{F}_1$. This means that F_2 is greater than 139 N. **(b)** $F_2 = 163\,\text{N}$]

Some related homework problems: Problem 29, Problem 31, Problem 33

QUICK EXAMPLE 5-10   FIND THE SPEED OF THE SLED

A 4.60-kg sled is pulled across a smooth ice surface. The force acting on the sled has a magnitude of $F = 6.20$ N and points in a direction $\theta = 35.0°$ above the horizontal. If the sled starts at rest, how fast is it moving after being pulled for 1.15 s?

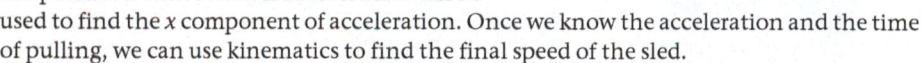

REASONING AND SOLUTION

The physical system is shown in our sketch, along with a coordinate system. In terms of this coordinate system, the force is 35.0° above the positive x axis. Newton's second law can be used to find the x component of acceleration. Once we know the acceleration and the time of pulling, we can use kinematics to find the final speed of the sled.

1. Find the x component of $\vec{F}$:

$$F_x = F \cos \theta = (6.20\,\text{N}) \cos (35.0°) = 5.08\,\text{N}$$

2. Apply Newton's second law to the x direction:

$$\sum F_x = F_x = ma_x$$

3. Solve for the x component of acceleration:

$$a_x = \frac{F_x}{m} = \frac{5.08\,\text{N}}{4.60\,\text{kg}} = 1.10\,\text{m/s}^2$$

4. Use $v_x = v_{0x} + a_x t$ to find the speed of the sled:

$$v_x = v_{0x} + a_x t$$
$$= 0 + (1.10\,\text{m/s}^2)(1.15\,\text{s}) = 1.27\,\text{m/s}$$

The y component of $\vec{F}$ has no effect on the acceleration of the sled.

FIGURE 5-14 shows a dramatic example of a case where the net force acting on a system is zero. As a result, neither climber is accelerating. Further examples of this type will be discussed in Chapter 6.

▲ **FIGURE 5-14** At the moment this picture was taken, the acceleration of both climbers was zero because the net force acting on them was zero. In particular, the upward forces exerted on the lower climber by the other climber and the ropes exactly cancel the downward force that gravity exerts on her.

Enhance Your Understanding   (Answers given at the end of the chapter)

5. An object is acted on by a single force that is at an angle of 35° above the positive x axis. The object's x component of acceleration is 1 m/s². Is the y component of the object's acceleration greater than, less than, or equal to 1 m/s²? Explain.

CONTINUED

Section Review

• Newton's second law can be applied to each coordinate direction independently.

5-6 Weight

When you step onto a scale to weigh yourself, the scale gives a measurement of the pull of Earth's gravity. This is your weight, W. Similarly, the weight of any object on the Earth's surface is simply the gravitational force exerted on it by the Earth.

> The weight, W, of an object on the Earth's surface is the gravitational force exerted on it by the Earth.

As we know from everyday experience, the greater the mass of an object, the greater its weight. For example, if you put a brick on a scale and weigh it, you might get a reading of 9.0 N. If you put a second, identical brick on the scale—which doubles the mass—you will find a weight of $2(9.0 \text{ N}) = 18$ N. Clearly, there must be a simple connection between weight, W, and mass, m.

Relating Weight and Mass   To see exactly what this connection is, consider taking one of the bricks just mentioned and letting it drop in free fall. As indicated in **FIGURE 5-15,** the only force acting on the brick is its weight, W, which is downward. If we choose upward to be the positive direction, we have

$$\sum F_y = -W$$

In addition, we know from Chapter 2 that the brick moves downward with an acceleration of $g = 9.81 \text{ m/s}^2$ regardless of its mass. Thus,

$$a_y = -g$$

Using these results in Newton's second law, we find

$$\sum F_y = ma_y$$
$$-W = -mg$$

Therefore, the weight of an object of mass m is $W = mg$:

> **Definition: Weight, W**
>
> $W = mg$ 5-5
>
> SI unit: newton, N

Distinguishing Weight and Mass   There is a clear distinction between weight and mass. Weight is a gravitational force, measured in newtons; mass is a measure of the inertia of an object, and it is given in kilograms. For example, if you were to travel to the Moon, your mass would not change—you would have the same amount of matter in you, regardless of your location. On the other hand, the gravitational force on the Moon's surface is less than the gravitational force on the Earth's surface. As a result, you would weigh less on the Moon than on the Earth, even though your mass is the same.

To be specific, on Earth an 81.0-kg person has a weight given by

$$W_{\text{Earth}} = mg_{\text{Earth}} = (81.0 \text{ kg})(9.81 \text{ m/s}^2) = 795 \text{ N}$$

In contrast, the same person on the Moon, where the acceleration of gravity is 1.62 m/s², weighs only

$$W_{\text{Moon}} = mg_{\text{Moon}} = (81.0 \text{ kg})(1.62 \text{ m/s}^2) = 131 \text{ N}$$

This is roughly one-sixth the weight on Earth. If, sometime in the future, there is a Lunar Olympics, the Moon's low gravity would be a boon for pole-vaulters, gymnasts, and others.

Finally, because weight is a force—which is a vector quantity—it has both a magnitude and a direction. Its magnitude, of course, is mg, and its direction is simply the

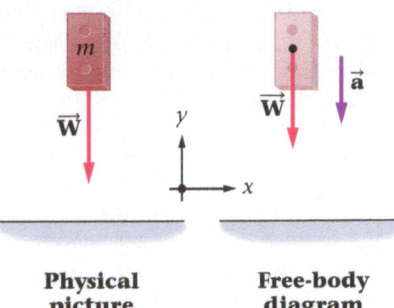

▲ **FIGURE 5-15 Weight and mass** A brick of mass m has only one force acting on it in free fall—its weight, $\vec{W}$. The resulting acceleration has a magnitude $a = g$; hence $W = mg$.

PHYSICS IN CONTEXT
Looking Ahead

Another important application of forces is in the study of collisions. Central to this topic is the concept of momentum, a physical quantity that is changed when a force acts on an object over a period of time. We shall study momentum in Chapter 9.

PHYSICS IN CONTEXT
Looking Ahead

In this chapter we introduced the force law for gravity near the Earth's surface, $F = mg$. The more general law of gravity, valid at any location, is introduced in Chapter 12.

direction of gravitational acceleration. Thus, if $\vec{g}$ denotes a vector of magnitude g, pointing in the direction of free-fall acceleration, the weight of an object can be written in vector form as follows:

$$\vec{W} = m\vec{g}$$

We use the weight vector and its magnitude, mg, in the next Example.

EXAMPLE 5-11   WHERE'S THE FIRE?

The fire alarm goes off, and a 97-kg fireman slides 3.0 m down a pole to the ground floor. Suppose the fireman starts from rest, slides with constant acceleration, and reaches the ground floor in 1.2 s. What was the upward force $\vec{F}$ exerted by the pole on the fireman?

PICTURE THE PROBLEM

Our sketch shows the fireman sliding down the pole and the two forces acting on him: the upward force exerted by the pole, $\vec{F}$, and the downward force of gravity, $\vec{W}$. We choose the positive y direction to be upward; therefore $\vec{F} = F\hat{y}$ and $\vec{W} = (-mg)\hat{y}$. In addition, we choose $y = 0$ to be at ground level.

REASONING AND STRATEGY

The basic idea in approaching this problem is to apply Newton's second law to the y direction: $\Sigma F_y = ma_y$. The acceleration is not given directly, but we can find it using the kinematic equation $y = y_0 + v_{0y}t + \frac{1}{2}a_yt^2$. Substituting the result for a_y into Newton's second law, along with $W_y = -W = -mg$, allows us to solve for the unknown force, $\vec{F}$.

Known   Mass of fireman, $m = 97$ kg; sliding distance, $y_0 = 3.0$ m; sliding time starting from rest, $t = 1.2$ s.

Unknown   Upward force exerted on the fireman, $\vec{F}$ = ?

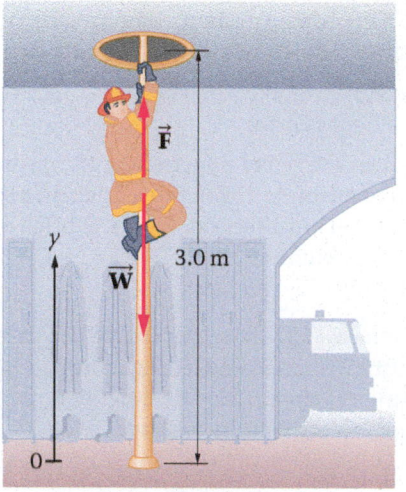

Physical picture

Free-body diagram

SOLUTION

1. Solve $y = y_0 + v_{0y}t + \frac{1}{2}a_yt^2$ for a_y, using the fact that $v_{0y} = 0$:

$$y = y_0 + v_{0y}t + \tfrac{1}{2}a_yt^2 = y_0 + \tfrac{1}{2}a_yt^2$$
$$a_y = \frac{2(y - y_0)}{t^2}$$

2. Substitute $y = 0, y_0 = 3.0$ m, and $t = 1.2$ s to find the acceleration:

$$a_y = \frac{2(0 - 3.0\,\text{m})}{(1.2\,\text{s})^2} = -4.2\,\text{m/s}^2$$

3. Sum the forces in the y direction:

$$\Sigma F_y = F - mg$$

4. Set the sum of the forces equal to mass times acceleration:

$$F - mg = ma_y$$

5. Solve for F, the y component of the force exerted by the pole. Use the result for F to write the force vector $\vec{F}$:

$$F = mg + ma_y = m(g + a_y)$$
$$= (97\,\text{kg})(9.81\,\text{m/s}^2 - 4.2\,\text{m/s}^2) = 540\,\text{N}$$
$$\vec{F} = (540\,\text{N})\hat{y}$$

INSIGHT

How is it that the pole exerts a force on the fireman? Well, by wrapping his arms and legs around the pole as he slides, the fireman exerts a downward force on the pole. By Newton's third law, the pole exerts an upward force of equal magnitude on the fireman. These forces are due to friction, which we shall study in detail in Chapter 6.

PRACTICE PROBLEM

What is the fireman's acceleration if the force exerted on him by the pole is 650 N?   [**Answer:** $a_y = -3.1$ m/s^2]

Some related homework problems: Problem 36, Problem 68

Apparent Weight

We've all had the experience of riding in an elevator and feeling either heavy or light, depending on its motion. For example, when an elevator accelerates upward from rest, we feel heavier. On the other hand, we feel lighter when an elevator that is moving upward comes to rest by accelerating downward. In short, the motion of an elevator can give rise to an **apparent weight** that differs from our true weight. Why?

▶ **FIGURE 5-16 Apparent weight** A person rides in an elevator that is accelerating upward. Because the acceleration is upward, the net force must also be upward. As a result, the force exerted on the person by the floor of the elevator, $\vec{W}_a$, must be greater than the person's weight, $\vec{W}$. This means that the person feels heavier than normal.

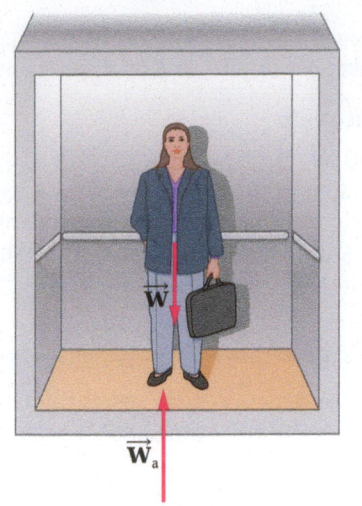

Physical picture

The reason is that our sensation of weight in this case is due to the force exerted on our feet by the floor of the elevator. If this force is greater than our weight, mg, we feel heavy; if it is less than mg, we feel light.

Riding in an Accelerating Elevator As an example, imagine you are in an elevator that is moving with an upward acceleration a, as indicated in **FIGURE 5-16**. Two forces act on you: (i) your weight, W, acting downward; and (ii) the upward normal force exerted on your feet by the floor of the elevator. Let's call the second force W_a, since it represents your apparent weight—that is, W_a is the force that pushes upward on your feet and gives you the sensation of your "weight" pushing down on the floor. We can find W_a by applying Newton's second law to the vertical direction.

To be specific, the sum of the forces acting on you is

$$\sum F_y = W_a - W$$

By Newton's second law, this sum must equal ma_y. Setting $a_y = a$, we find

$$W_a - W = ma$$

Solving for the apparent weight, W_a, yields

$$W_a = W + ma$$
$$= mg + ma = m(g + a) \qquad \text{5-6}$$

Notice that W_a is greater than your weight, mg, and hence you feel heavier. In fact, your apparent weight is precisely what it would be if you were suddenly "transported" to a planet where the acceleration due to gravity is $g + a$ instead of g.

On the other hand, if the elevator accelerates downward, so that $a_y = -a$, your apparent weight is found by simply replacing a with $-a$ in Equation 5-6:

$$W_a = W - ma$$
$$= mg - ma = m(g - a) \qquad \text{5-7}$$

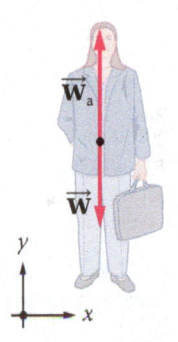

Free-body diagram

In this case you feel lighter than usual.

We explore these results in the next Example, in which we consider weighing a fish on a scale. The reading on the scale is equal to the upward force it exerts on an object. Thus, the upward force exerted by the scale is the apparent weight, W_a.

EXAMPLE 5-12 SOMETHING FISHY

As part of an attempt to combine physics and biology in the same class, an instructor asks students to weigh a 5.2-kg salmon by hanging it from a fish scale attached to the ceiling of an elevator. What is the apparent weight of the salmon, $\vec{W}_a$, if the elevator **(a)** is at rest, **(b)** moves with an upward acceleration of 3.5 m/s², or **(c)** moves with a downward acceleration of 2.3 m/s²?

PICTURE THE PROBLEM
The free-body diagram for the salmon shows the weight of the salmon, $\vec{W}$, and the force exerted by the scale, $\vec{W}_a$, which is the apparent weight. Notice that upward is the positive direction. Therefore, the y component of the weight is $-W = -mg$ and the y component of the apparent weight is W_a.

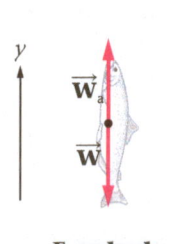

Physical picture   **Free-body diagram**

REASONING AND STRATEGY
We know the magnitude of the weight, $W = mg$, and the acceleration, a_y. To find the apparent weight, W_a, we use Newton's second law applied to the y component:
$\sum F_y = W_a - W = ma_y$. **(a)** Set $a_y = 0$. **(b)** Set $a_y = 3.5 \text{ m/s}^2$. **(c)** Set $a_y = -2.3 \text{ m/s}^2$.

CONTINUED

Known Mass of salmon, $m = 5.2$ kg; acceleration of elevator, (a) $a_y = 0$, (b) $a_y = 3.5$ m/s^2, (c) $a_y = -2.3$ m/s^2.
Unknown (a), (b), and (c) Apparent weight, $\vec{W}_a = ?$

SOLUTION

Part (a)

1. Sum the y components of the forces and set equal to mass, m, times the y component of acceleration, a_y:

$$\sum F_y = W_a - W = ma_y$$

2. Solve for W_a and set $a_y = 0$. Use the numerical result to write the vector $\vec{W}$:

$$W_a = W + ma_y$$
$$= mg + 0 = (5.2\text{ kg})(9.81\text{ m/s}^2) = 51\text{ N}$$
$$\vec{W}_a = (51\text{ N})\hat{y}$$

Part (b)

3. Use $W_a = W + ma_y$ again, this time with $a_y = 3.5$ m/s^2. Use the numerical result to write the vector $\vec{W}_a$:

$$W_a = W + ma_y$$
$$= mg + ma_y = 51\text{ N} + (5.2\text{ kg})(3.5\text{ m/s}^2) = 69\text{ N}$$
$$\vec{W}_a = (69\text{ N})\hat{y}$$

Part (c)

4. Use $W_a = W + ma_y$ again, this time with $a_y = -2.3$ m/s^2. Use the numerical result to write the vector $\vec{W}_a$:

$$W_a = W + ma_y$$
$$= mg + ma_y = 51\text{ N} + (5.2\text{ kg})(-2.3\text{ m/s}^2) = 39\text{ N}$$
$$\vec{W}_a = (39\text{ N})\hat{y}$$

INSIGHT

When the salmon is at rest, or moving with constant velocity, its acceleration is zero and the apparent weight is equal to the actual weight, mg. In part (b) the apparent weight is greater than the actual weight because the scale must exert an upward force capable not only of supporting the salmon, but of accelerating it upward as well. In part (c) the apparent weight is less than the actual weight. In this case the net force acting on the salmon is downward, and hence its acceleration is downward.

PRACTICE PROBLEM

What is the elevator's acceleration if the scale gives a reading of (a) 52 N, (b) 42 N? [**Answer:** (a) $a_y = 0.19$ m/s^2, (b) $a_y = -1.7$ m/s^2]

Some related homework problems: Problem 38, Problem 40, Problem 41

▲ **FIGURE 5-17** Astronaut candidates pose for a floating class picture during weightlessness training aboard the "vomit comet."

Apparent Weight and Weightlessness Let's return for a moment to Equation 5-7:

$$W_a = m(g - a)$$

This result indicates that a person feels lighter than normal when riding in an elevator with a downward acceleration a. In particular, if the elevator's downward acceleration is g—that is, if the elevator is in free fall—it follows that $W_a = m(g - g) = 0$. Thus, a person feels "weightless" (zero apparent weight) in a freely falling elevator!

RWP NASA uses this effect when training astronauts. Trainees are sent aloft in a KC-135 airplane, as shown in **FIGURE 5-17**. This plane is affectionately known as the "vomit comet" because many trainees experience nausea along with the weightlessness. To generate an experience of weightlessness, the plane flies on a parabolic path—the same path followed by a projectile in free fall. Each round of weightlessness lasts about half a minute, after which the plane pulls up to regain altitude and start the cycle again. On a typical flight, trainees experience about 40 cycles of weightlessness. Many scenes in the movie *Apollo 13* were shot in 30-second takes aboard the vomit comet.

RWP Many amusement parks have popular rides, like drop towers and roller coasters, that give visitors a thrill by providing brief periods of weightlessness. For example, the Zumanjaro: Drop of Doom ride at Six Flags Great Adventure in Jackson, New Jersey, is a drop tower that is 126 m (415 ft) high, providing passengers with a few seconds of downward acceleration at 9.81 m/s^2, so that they are weightless relative to the gondola in which they are riding. Similarly, the Outlaw Run roller coaster at Silver Dollar City in Branson, Missouri, features a 49.4-m (162 ft) drop at a 81° inclination during its 87-second ride in which passengers briefly feel weightless relative to their seats.

This idea of free-fall weightlessness applies to more than just the vomit comet. In fact, astronauts in orbit experience weightlessness for the same reason—they and their craft are actually in free fall. As we shall see in detail in Chapter 12, orbital motion is just a special case of free fall.

CONCEPTUAL EXAMPLE 5-13   **ELEVATOR RIDE** PREDICT/EXPLAIN

(a) If you ride in an elevator moving upward with constant speed, is your apparent weight greater than, less than, or equal to your actual weight, *mg*? **(b)** Which of the following is the *best explanation* for your prediction?

  I. The elevator is moving upward. This will cause a person to press down harder on the floor, giving an apparent weight greater than *mg*.

  II. The elevator moves with zero acceleration, and therefore the apparent weight is equal to *mg*.

  III. Because the elevator is moving upward it is partially lifting the person, resulting in an apparent weight less than *mg*.

REASONING AND DISCUSSION

If the elevator is moving in a straight line with constant speed, its acceleration is zero. Now, if the acceleration is zero, the net force must also be zero. Hence, the upward force exerted by the floor of the elevator, W_a, must equal the downward force of gravity, *mg*. As a result, your apparent weight is equal to *mg*.

Notice that this conclusion agrees with Equations 5-6 and 5-7, with $a = 0$.

ANSWER

(a) Your apparent weight is equal to *mg*. **(b)** The best explanation is II.

Enhance Your Understanding   (Answers given at the end of the chapter)

6. When a certain person steps onto a scale on solid ground, the reading is 750 N. When the same person rides in an elevator that accelerates, the reading on the scale is different. Rank the three cases (A, B, and C) shown in **FIGURE 5-18** in order of increasing acceleration (from most negative to most positive) in the *y* direction. Indicate ties where appropriate.

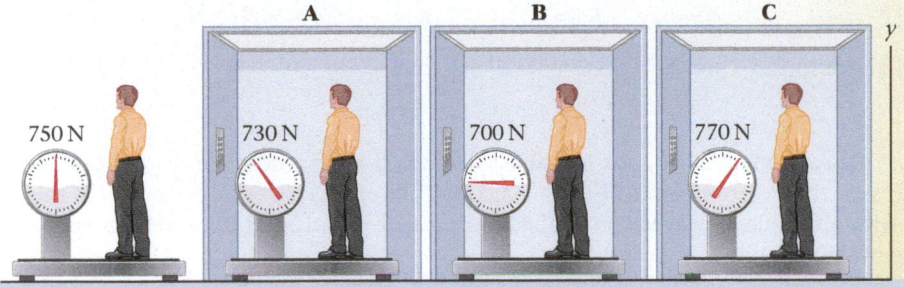

▲ **FIGURE 5-18** Enhance Your Understanding 6

Section Review

• Weight is the force of gravity exerted on an object.

• The apparent weight of an object depends on the acceleration of its frame of reference.

5-7 Normal Forces

As you get ready for lunch, you take a can of soup from the cupboard and place it on the kitchen counter. The can is now at rest, which means that its acceleration is zero, so the net force acting on it is also zero. Thus, you know that the downward force of gravity is being opposed by an upward force exerted by the counter, as shown in **FIGURE 5-19**. As we've mentioned before, this force is referred to as the **normal force**, $\vec{N}$. The reason the force is called normal is that it is *perpendicular to the surface,* and in mathematical terms, *normal* simply means perpendicular.

    The origin of the normal force is the interaction between atoms in a solid that acts to maintain its shape. When the can of soup is placed on the countertop, for example, it causes an imperceptibly small compression of the surface of the counter. This is similar to compressing a spring, and just like a spring, the countertop exerts a force to oppose the compression. Therefore, the greater the weight placed on the countertop, the greater the normal force it exerts to oppose being compressed.

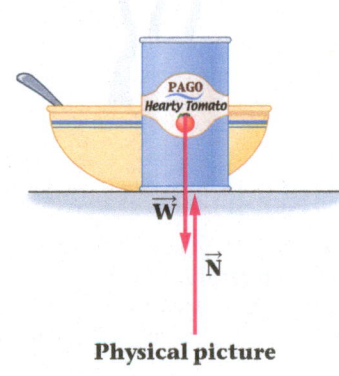

Physical picture

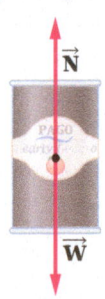

Free-body diagram

▲ **FIGURE 5-19 The normal force is perpendicular to a surface** A can of soup rests on a kitchen counter, which exerts a normal (perpendicular) force, $\vec{N}$, to support it. In the special case shown here, the normal force is equal in magnitude to the weight, $W = mg$, and opposite in direction.

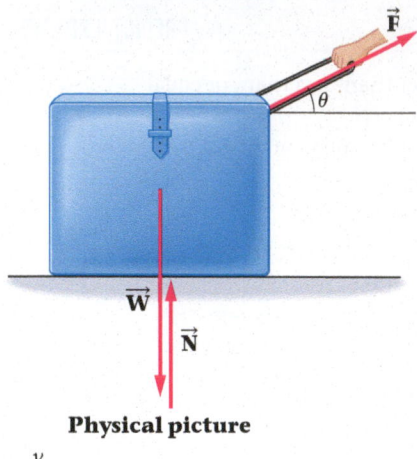

Physical picture

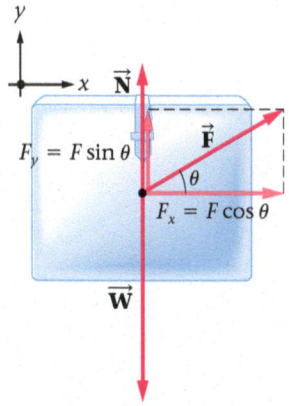

Free-body diagram

▲ **FIGURE 5-20** **The normal force may differ from the weight** A suitcase is pulled across the floor by an applied force of magnitude F, directed at an angle θ above the horizontal. As a result of the upward component of $\vec{F}$, the normal force $\vec{N}$ will have a magnitude less than the weight of the suitcase.

The Normal Force Can Be Greater Than, Less Than, or Equal to the Weight   In the example of the soup can and the countertop, the magnitude of the normal force is equal to the weight of the can. This is a special case, however. In general, the normal force can also be greater than or less than the weight of an object.

To see how this can come about, consider pulling a 12.0-kg suitcase across a smooth floor by exerting a force, $\vec{F}$, at an angle θ above the horizontal. The weight of the suitcase is $mg = (12.0\,\text{kg})(9.81\,\text{m/s}^2) = 118\,\text{N}$. The normal force will have a magnitude less than this, however, because the force $\vec{F}$ has an upward component that supports part of the suitcase's weight. To be specific, suppose that $\vec{F}$ has a magnitude of 45.0 N and that $\theta = 20.0°$. What is the normal force exerted by the floor on the suitcase?

The situation is illustrated in **FIGURE 5-20**, where we show the three forces acting on the suitcase: (i) the weight of the suitcase, $\vec{W}$, (ii) the force $\vec{F}$, and (iii) the normal force, $\vec{N}$. We also indicate a typical coordinate system in the figure, with the x axis horizontal and the y axis vertical. Now, the key to solving a problem like this is to realize that since the suitcase does not move in the y direction, its y component of acceleration is zero; that is, $a_y = 0$. It follows, from Newton's second law, that the sum of the y components of force must also equal zero; that is, $\Sigma F_y = ma_y = 0$. Using this condition, we can solve for the one force that is unknown, $\vec{N}$.

To find $\vec{N}$, then, we start by writing out the y component of each force. For the weight we have $W_y = -mg = -118\,\text{N}$; for the applied force, $\vec{F}$, the y component is $F_y = F \sin 20.0° = (45.0\,\text{N}) \sin 20.0° = 15.4\,\text{N}$; finally, the y component of the normal force is $N_y = N$. Setting the sum of the y components of force equal to zero yields

$$\Sigma F_y = W_y + F_y + N_y = -mg + F\sin 20.0° + N = 0$$

Solving for N gives

$$N = mg - F \sin 20.0° = 118\,\text{N} - 15.4\,\text{N} = 103\,\text{N}$$

In vector form,

$$\vec{N} = N_y \hat{y} = (103\,\text{N})\hat{y}$$

Thus, as mentioned, the normal force has a magnitude less than $mg = 118\,\text{N}$ because the y component of $\vec{F}$, $F_y = F \sin 20.0°$, supports part of the weight. In the following Example, however, the applied forces cause the normal force to be greater than the weight.

EXAMPLE 5-14   A MOVEABLE FEAST

A 4.53-kg holiday turkey is moved toward the center of a smooth table by two people at the feast. Person 1 exerts a force $\vec{F}_1$ with a magnitude of 35.5 N at an angle of 34.6° below the horizontal, as indicated in the sketch. Person 2 exerts a force $\vec{F}_2$ with a magnitude of 39.7 N at an angle of 21.2° above the horizontal. What is the magnitude N of the normal force exerted by the table on the turkey?

PICTURE THE PROBLEM
The sketch shows our choice of coordinate system, as well as all the forces acting on the turkey. Notice that $\vec{F}_1$ has a positive x component and a negative y component; $\vec{F}_2$ has positive x and y components. The weight $\vec{W}$ and the normal force $\vec{N}$ have only y components. For example, noting that the mass of the turkey is $m = 4.53$ kg, we have for the components of the weight, $W_x = 0$ and $W_y = -mg = -(4.53\,\text{kg})(9.81\,\text{m/s}^2) = -44.4\,\text{N}$. The components of the normal force are $N_x = 0$ and $N_y = N$. For the moment we ignore friction with the table, which we will cover in the next chapter.

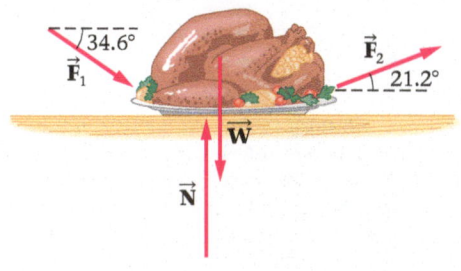

Components of $\vec{F}_1$

$$F_{1,x} = F_1 \cos 34.6°$$

$$F_{1,y} = -F_1 \sin 34.6°$$

Components of $\vec{F}_2$

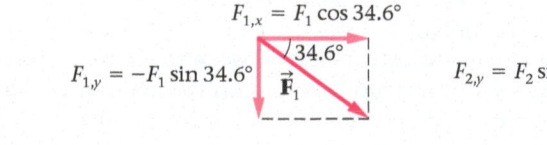

$$F_{2,y} = F_2 \sin 21.2°$$

$$F_{2,x} = F_2 \cos 21.2°$$

CONTINUED

REASONING AND STRATEGY

The basic idea in this problem is to apply Newton's second law to the y direction. Specifically, the fact that there is no motion in the y direction means that the acceleration in the y direction is zero. Hence, we can find the magnitude N of the normal force by setting $\sum F_y = ma_y = 0$.

Known   Mass of turkey, $m = 4.53$ kg; force 1, 35.5 N at 34.6° below the horizontal; force 2, 39.7 N at 21.2° above the horizontal.

Unknown   Magnitude of the normal force exerted by the table, $N = ?$

SOLUTION

1. Write out the y component of each force. The only force we don't know is the normal. We represent its magnitude by N:

$$F_{1,y} = -F_1 \sin 34.6° = -(35.5 \text{ N}) \sin 34.6° = -20.2 \text{ N}$$
$$F_{2,y} = F_2 \sin 21.2° = (39.7 \text{ N}) \sin 21.2° = 14.4 \text{ N}$$
$$N_y = N \quad W_y = -mg = -44.4 \text{ N}$$

2. Sum the y components of force:

$$\sum F_y = F_{1,y} + F_{2,y} + N_y + W_y$$
$$= -20.2 \text{ N} + 14.4 \text{ N} + N - 44.4 \text{ N}$$

3. Set this sum equal to 0, since the acceleration in the y direction is zero, and solve for N:

$$-20.2 \text{ N} + 14.4 \text{ N} + N - 44.4 \text{ N} = 0$$
$$N = 20.2 \text{ N} - 14.4 \text{ N} + 44.4 \text{ N} = 50.2 \text{ N}$$

INSIGHT

The normal force is greater in magnitude than the weight, $mg = 44.4$ N. This is because the force $\vec{F}_1$ pushes down with a y component that is greater than the upward y component of $\vec{F}_2$, even though $\vec{F}_2$ has the greater magnitude.

In general, the normal force exerted by a surface is only as large as is necessary to prevent the motion of an object into the surface. If the required force is larger than the material can provide, the surface will break.

PRACTICE PROBLEM

What is the magnitude of the normal force if $\vec{F}_2$ is at an angle of 42.5° above the horizontal?   [**Answer:** In this case, $N = 37.8$ N, which is less than the weight of the turkey.]

Some related homework problems: Problem 45, Problem 52

Normal Forces on Tilted Surfaces   To this point, we have considered surfaces that are horizontal, in which case the normal force is vertical. When a surface is inclined, the normal force is still at right angles to the surface, even though it is no longer vertical. This is illustrated in **FIGURE 5-21**. (If friction is present, a surface may also exert a force that is parallel to its surface. This will be considered in detail in Chapter 6.)

When we choose a coordinate system for an inclined surface, it is generally best to have the x and y axes of the system parallel and perpendicular to the surface, respectively, as in **FIGURE 5-22**. One can imagine the coordinate system to be "bolted down" to the surface, so that when the surface is tilted, the coordinate system tilts along with it.

With this choice of coordinate system, there is no motion in the y direction, even on the inclined surface, and the normal force points in the positive y direction. Thus, the condition that determines the normal force is still $\sum F_y = ma_y = 0$ as before. In addition, if the object slides on the surface, its motion is purely in the x direction.

Finally, if the surface is inclined by an angle θ, notice that the weight—which is still vertically downward—is at the same angle θ with respect to the negative y axis, as shown in Figure 5-22. As a result, the x and y components of the weight are

$$W_x = W \sin \theta = mg \sin \theta \qquad \text{5-8}$$

and

$$W_y = -W \cos \theta = -mg \cos \theta \qquad \text{5-9}$$

Let's quickly check some special cases of these results. First, if $\theta = 0$ the surface is horizontal, and we find $W_x = 0$, $W_y = -mg$, as expected. Second, if $\theta = 90°$ the surface is vertical; therefore, the weight is parallel to the surface, pointing in the positive x direction. In this case, $W_x = mg$ and $W_y = 0$.

The next Example shows how to use the weight components to find the acceleration of an object on an inclined surface.

The normal force is perpendicular to the surface that produces it... $\vec{N}$ ...and hence it may or may not be vertical.

▲ **FIGURE 5-21 An object on an inclined surface** The normal force $\vec{N}$ is always at right angles to the surface; hence, it is not always in the vertical direction.

▶ **FIGURE 5-22 Components of the weight on an inclined surface** Whenever a surface is tilted by an angle θ, the weight $\vec{W}$ makes the same angle θ with respect to the negative y axis. This is proven in part (b), where we show that $\theta + \phi = 90°$ and that $\theta' + \phi = 90°$. From these results it follows that $\theta' = \theta$. The component of the weight perpendicular to the surface is $W_y = -W\cos\theta$; the component parallel to the surface is $W_x = W\sin\theta$.

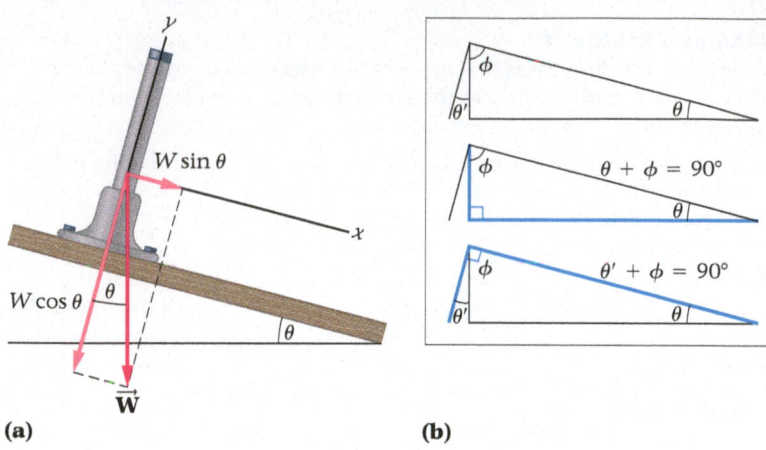

(a)

(b)

EXAMPLE 5-15   TOBOGGAN TO THE BOTTOM

A child of mass m rides on a toboggan down a slick, ice-covered hill inclined at an angle θ with respect to the horizontal. **(a)** What is the acceleration of the child? **(b)** What is the normal force exerted on the child by the toboggan?

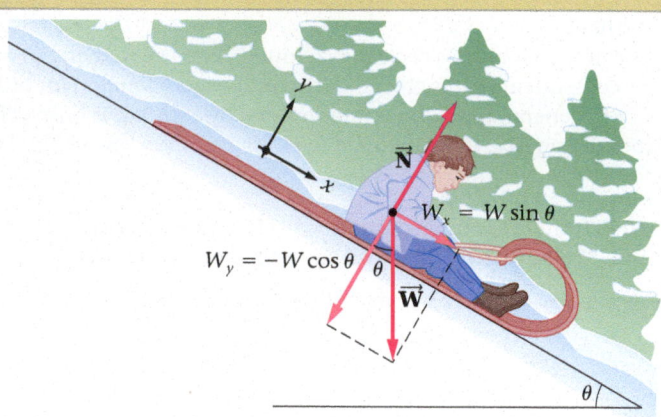

PICTURE THE PROBLEM

We choose the x axis to be parallel to the slope, with the positive direction pointing downhill. Similarly, we choose the y axis to be perpendicular to the slope, pointing up and to the right. With these choices, the x component of $\vec{W}$ is positive, $W_x = W\sin\theta$, and its y component is negative, $W_y = -W\cos\theta$. Finally, the x component of the normal force is zero, $N_x = 0$, and its y component is positive, $N_y = N$.

REASONING AND STRATEGY

Only two forces act on the child: (i) the weight, $\vec{W}$, and (ii) the normal force, $\vec{N}$. **(a)** We find the child's acceleration by solving $\Sigma F_x = ma_x$ for a_x. **(b)** Because there is no motion in the y direction, the y component of acceleration is zero. Therefore, we can find the normal force by setting $\Sigma F_y = ma_y = 0$.

Known   Mass of child, m; angle of incline, θ.
Unknown   **(a)** Acceleration of the child, $a_x = ?$ **(b)** Normal force exerted on the child, $\vec{N} = ?$

SOLUTION

Part (a)

1. Write out the x components of the forces acting on the child:

$$N_x = 0$$
$$W_x = W\sin\theta = mg\sin\theta$$

2. Sum the x components of the forces and set equal to ma_x:

$$\Sigma F_x = N_x + W_x = mg\sin\theta = ma_x$$

3. Divide by the mass m to find the acceleration in the x direction:

$$a_x = \frac{\Sigma F_x}{m} = \frac{mg\sin\theta}{m} = g\sin\theta$$

Part (b)

4. Write out the y components of the forces acting on the child:

$$N_y = N$$
$$W_y = -W\cos\theta = -mg\cos\theta$$

5. Sum the y components of the forces and set the sum equal to zero, since $a_y = 0$:

$$\Sigma F_y = N_y + W_y = N - mg\cos\theta$$
$$= ma_y = 0$$

6. Solve for the magnitude of the normal force, N:

$$N - mg\cos\theta = 0 \quad \text{or} \quad N = mg\cos\theta$$

7. Write the normal force in vector form:

$$\vec{N} = (mg\cos\theta)\hat{y}$$

INSIGHT

Notice that for θ between 0 and 90° the acceleration of the child is *less* than the acceleration of gravity. This is because only a *component* of the weight is causing the acceleration.

CONTINUED

Let's check some special cases of our general result, $a_x = g \sin \theta$. First, let $\theta = 0$. In this case, we find zero acceleration: $a_x = g \sin 0 = 0$. This makes sense because with $\theta = 0$ the hill is actually level, and we don't expect an acceleration. Second, let $\theta = 90°$. In this case, the hill is vertical, and the toboggan should drop straight down in free fall. This also agrees with our general result: $a_x = g \sin 90° = g$.

PRACTICE PROBLEM

What is the child's acceleration if his mass is doubled to $2m$? [**Answer:** The acceleration is still $a_x = g \sin \theta$. As in free fall, the acceleration produced by gravity is independent of mass.]

Some related homework problems: Problem 46, Problem 51

Enhance Your Understanding   (Answers given at the end of the chapter)

7. **FIGURE 5-23** shows four identical bricks that are at rest on surfaces tilted at different angles. Rank the bricks in order of increasing normal force exerted on them by the surface. Indicate ties where appropriate.

A   B   C   D

▲ **FIGURE 5-23** Enhance Your Understanding 7

Section Review

• A normal force is perpendicular (normal) to a surface.

CHAPTER 5 REVIEW

CHAPTER SUMMARY

5-1   FORCE AND MASS

Force
A push or a pull.

Mass
A measure of the difficulty in accelerating an object. Equivalently, a measure of the quantity of matter in an object.

5-2   NEWTON'S FIRST LAW OF MOTION

First Law (Law of Inertia)
If the net force on an object is zero, its velocity is constant.

Inertial Frame of Reference
A frame of reference in which the first law holds. All inertial frames of reference move with constant velocity relative to one another.

5-3   NEWTON'S SECOND LAW OF MOTION

Second Law
An object of mass m has an acceleration $\vec{a}$ given by the net force $\Sigma \vec{F}$ divided by m. That is,

$$\vec{a} = \Sigma \vec{F}/m \qquad 5\text{-}1$$

Component Form

$$\Sigma F_x = ma_x \quad \Sigma F_y = ma_y \quad \Sigma F_z = ma_z \qquad 5\text{-}2$$

SI Unit: Newton (N)

$$1\,\text{N} = 1\,\text{kg} \cdot \text{m/s}^2 \qquad 5\text{-}3$$

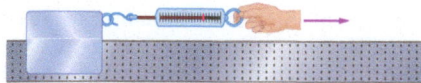

Free-Body Diagram
A sketch showing all external forces acting on an object.

5-4 NEWTON'S THIRD LAW OF MOTION

Third Law
For every force that acts on an object, there is a reaction force acting on a different object that is equal in magnitude and opposite in direction.

Contact Forces
Action-reaction pair of forces produced by the physical contact of two objects.

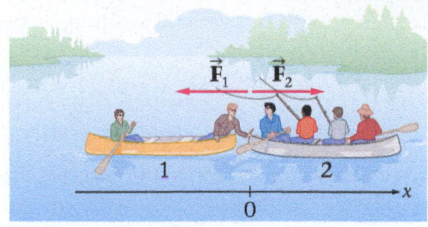

5-5 THE VECTOR NATURE OF FORCES: FORCES IN TWO DIMENSIONS

Forces are vectors.

Newton's second law can be applied to each component of force separately and independently.

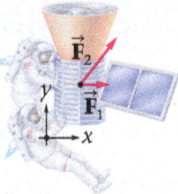

5-6 WEIGHT

Gravitational force exerted *by the Earth on an object*.

On the surface of the Earth the weight, W, of an object of mass m has the magnitude

$$W = mg \qquad\qquad 5\text{-}5$$

Apparent Weight
The force felt from contact with the floor or a scale in an accelerating system—for example, the sensation of feeling heavier or lighter in an accelerating elevator.

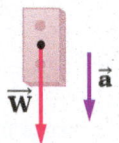

5-7 NORMAL FORCES

A force exerted by a surface that is *perpendicular* to the surface.

ANSWERS TO ENHANCE YOUR UNDERSTANDING QUESTIONS

1. Statement B is correct. The net force can have a magnitude with any value from $|F_1 - F_2|$ to $(F_1 + F_2)$ depending on the direction of the two forces.

2. Statements B and E are correct—a net force is required to make an object accelerate (change its velocity).

3. **(a)** $2a$. **(b)** a. **(c)** $4a$. **(d)** $a/4$.

4. The force of contact between boxes 1 and 2 is greater than the force of contact between boxes 2 and 3. The reason is that the force between 1 and 2 must give an acceleration a to a mass of 2 kg + 3 kg = 5 kg, but the force between 2 and 3 must give an acceleration a to a mass of only 3 kg.

5. Less than 1 m/s² because the y component of a vector is less than the x component when the vector is only 35° above the x axis.

6. B < A < C.

7. B = C < D < A.

CONCEPTUAL QUESTIONS

For instructor-assigned homework, go to www.masteringphysics.com.

(Answers to odd-numbered Conceptual Questions can be found in the back of the book.)

1. Driving down the road, you hit the brakes suddenly. As a result, your body moves toward the front of the car. Explain, using Newton's laws.

2. You've probably seen pictures of someone pulling a tablecloth out from under glasses, plates, and silverware set out for a formal dinner. Perhaps you've even tried it yourself. Using Newton's laws of motion, explain how this stunt works.

3. As you read this, you are most likely sitting quietly in a chair. Can you conclude, therefore, that you are at rest? Explain.

4. When a dog gets wet, it shakes its body from head to tail to shed the water. Explain, in terms of Newton's first law, why this works.

5. A young girl slides down a rope. As she slides faster and faster she tightens her grip, increasing the force exerted on her by the rope. What happens when this force is equal in magnitude to her weight? Explain.

6. A block of mass m hangs from a string attached to a ceiling, as shown in **FIGURE 5-24**. An identical string hangs down from the bottom of the block. Which string breaks if **(a)** the lower string is pulled with a slowly increasing force or **(b)** the lower string is jerked rapidly downward? Explain.

7. An astronaut on a space walk discovers that his jet pack no longer works, leaving him stranded 50 m from the spacecraft. If the jet

pack is removable, explain how the astronaut can still use it to return to the ship.

8. Two untethered astronauts on a space walk decide to take a break and play catch with a baseball. Describe what happens as the game of catch progresses.

9. In **FIGURE 5-25** Wilbur asks Mr. Ed, the talking horse, to pull a cart. Mr. Ed replies that he would like to, but the laws of nature just won't allow it. According to Newton's third law, he says, if he pulls on the wagon it pulls back on him with an equal force. Clearly, then, the net force is zero and the wagon will stay put. How should Wilbur answer the clever horse?

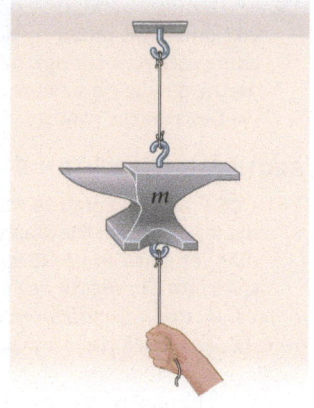

FIGURE 5-24 Conceptual Question 6

FIGURE 5-25 Conceptual Question 9

10. A whole brick has more mass than half a brick, thus the whole brick is harder to accelerate. Why doesn't a whole brick fall more slowly than half a brick? Explain.

11. The force exerted by gravity on a whole brick is greater than the force exerted by gravity on half a brick. Why, then, doesn't a whole brick fall faster than half a brick? Explain.

12. Is it possible for an object at rest to have only a single force acting on it? If your answer is yes, provide an example. If your answer is no, explain why not.

13. Is it possible for an object to be in motion and yet have zero net force acting on it? Explain.

14. A bird cage, with a parrot inside, hangs from a scale. The parrot decides to hop to a higher perch. What can you say about the reading on the scale (a) when the parrot jumps, (b) when the parrot is in the air, and (c) when the parrot lands on the second perch? Assume that the scale responds rapidly so that it gives an accurate reading at all times.

15. Suppose you jump from the cliffs of Acapulco and perform a perfect swan dive. As you fall, you exert an upward force on the Earth equal in magnitude to the downward force the Earth exerts on you. Why, then, does it seem that you are the one doing all the accelerating? Since the forces are the same, why aren't the accelerations?

16. A friend tells you that since his car is at rest, there are no forces acting on it. How would you reply?

17. Since all objects are "weightless" in orbit, how is it possible for an orbiting astronaut to tell if one object has more mass than another object? Explain.

18. To clean a rug, you can hang it from a clothesline and beat it with a tennis racket. Use Newton's laws to explain why beating the rug should have a cleansing effect.

19. If you step off a high board and drop to the water below, you plunge into the water without injury. On the other hand, if you were to drop the same distance onto solid ground, you might break a leg. Use Newton's laws to explain the difference.

20. Is it possible for an object to be moving in one direction while the net force acting on it is in another direction? If your answer is yes, provide an example. If your answer is no, explain why not.

21. Since a bucket of water is "weightless" in space, would it hurt to kick the bucket? Explain.

22. In the movie *The Rocketeer*, a teenager discovers a jet-powered backpack in an old barn. The backpack allows him to fly at incredible speeds. In one scene, however, he uses the backpack to rapidly accelerate an old pickup truck that is being chased by "bad guys." He does this by bracing his arms against the cab of the pickup and firing the backpack, giving the truck the acceleration of a drag racer. Is the physics of this scene "Good," "Bad," or "Ugly"? Explain.

23. List three common objects that have a weight of approximately 1 N.

PROBLEMS AND CONCEPTUAL EXERCISES

Answers to odd-numbered Problems and Conceptual Exercises can be found in the back of the book. **BIO** *identifies problems of biological or medical interest;* **CE** *indicates a conceptual exercise.* Predict/Explain *problems ask for two responses: (a) your prediction of a physical outcome, and (b) the best explanation among three provided; and* Predict/Calculate *problems ask for a prediction of a physical outcome, based on fundamental physics concepts, and follow that with a numerical calculation to verify the prediction. On all problems, bullets (•, ••, •••) indicate the level of difficulty.*

SECTION 5-3 NEWTON'S SECOND LAW OF MOTION

1. • **CE** An object of mass m is initially at rest. After a force of magnitude F acts on it for a time T, the object has a speed v. Suppose the mass of the object is doubled, and the magnitude of the force acting on it is quadrupled. In terms of T, how much time does it take for the object to accelerate from rest to a speed v now?

2. • On a planet far, far away, an astronaut picks up a rock. The rock has a mass of 5.00 kg, and on this particular planet its weight is 40.0 N. If the astronaut exerts an upward force of 46.2 N on the rock, what is its acceleration?

3. • In a grocery store, you push a 15.4-kg shopping cart horizontally with a force of 13.4 N. If the cart starts at rest, how far does it move in 2.50 s?

4. • You are pulling your little sister on her sled across an icy (frictionless) surface. When you exert a constant horizontal force of 120 N, the sled has an acceleration of 2.5 m/s². If the sled has a mass of 7.4 kg, what is the mass of your little sister?

5. • A 0.53-kg billiard ball initially at rest is given a speed of 12 m/s during a time interval of 4.0 ms. What average force acted on the ball during this time?

6. • A 92-kg water skier floating in a lake is pulled from rest to a speed of 12 m/s in a distance of 25 m. What is the net force exerted on the skier, assuming his acceleration is constant?

7. • A 0.5-kg object is acted on by a force whose x component varies with time as shown in **FIGURE 5-26**. Draw the corresponding a_x-versus-time graph for this object.

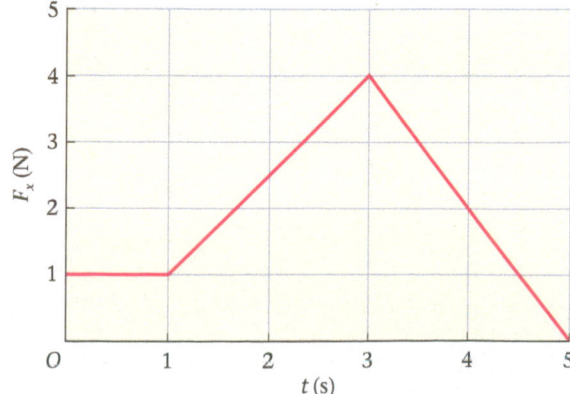

FIGURE 5-26 Problem 7

8. •• **CE Predict/Explain** You drop two balls of equal diameter from the same height at the same time. Ball 1 is made of metal and has a greater mass than ball 2, which is made of wood. The upward force due to air resistance is the same for both balls. **(a)** Is the drop time of ball 1 greater than, less than, or equal to the drop time of ball 2? **(b)** Choose the *best explanation* from among the following:

   I. The acceleration of gravity is the same for all objects, regardless of mass.

   II. The more massive ball is harder to accelerate.

   III. Air resistance has less effect on the more massive ball.

9. •• **Predict/Calculate** A 42.0-kg parachutist is moving straight downward with a speed of 3.85 m/s. **(a)** If the parachutist comes to rest with constant acceleration over a distance of 0.750 m, what force does the ground exert on her? **(b)** If the parachutist comes to rest over a shorter distance, is the force exerted by the ground greater than, less than, or the same as in part (a)? Explain.

10. •• **Predict/Calculate** In baseball, a pitcher can accelerate a 0.15-kg ball from rest to 98 mi/h in a distance of 1.7 m. **(a)** What is the average force exerted on the ball during the pitch? **(b)** If the mass of the ball is increased, is the force required of the pitcher increased, decreased, or unchanged? Explain.

11. •• A major-league catcher gloves a 92 mi/h pitch and brings it to rest in 0.15 m. If the force exerted by the catcher is 803 N, what is the mass of the ball?

12. •• Driving home from school one day, you spot a ball rolling out into the street (**FIGURE 5-27**). You brake for 1.20 s, slowing your 950-kg car from 16.0 m/s to 9.50 m/s. **(a)** What was the average force exerted on your car during braking? **(b)** How far did you travel while braking?

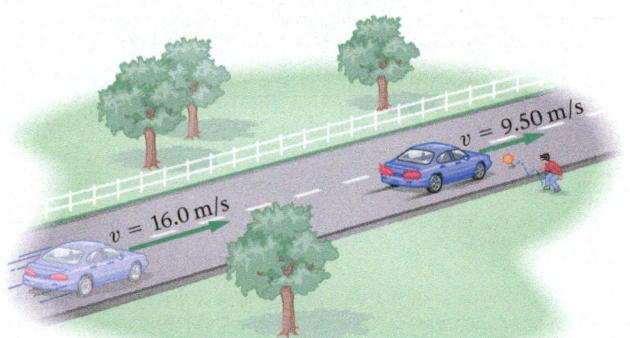

FIGURE 5-27 Problem 12

13. •• **Stopping a 747** A 747 jetliner lands and begins to slow to a stop as it moves along the runway. If its mass is 3.50×10^5 kg, its speed is 75.0 m/s, and the net braking force is 7.25×10^5 N, **(a)** what is its speed 10.0 s later? **(b)** How far has it traveled in this time?

14. •• The v_x-versus-time graph for a 1.8-kg object is shown in **FIGURE 5-28**. A single force acts on the object, and the force has only an x component. What is F_x at **(a)** $t = 0.50$ s, **(b)** $t = 2.0$ s, and **(c)** $t = 4.0$ s?

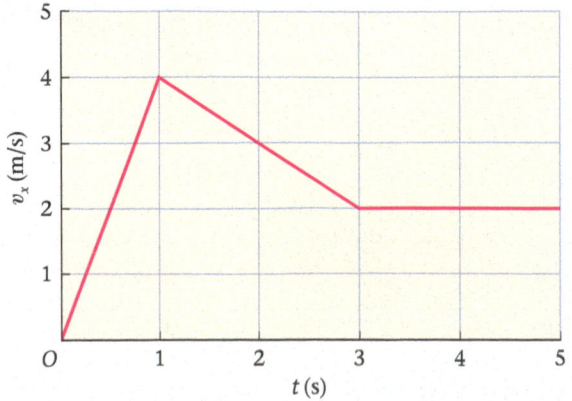

FIGURE 5-28 Problem 14

15. •• **Predict/Calculate** A drag racer crosses the finish line doing 202 mi/h and promptly deploys her drag chute (the small parachute used for braking). **(a)** What force must the drag chute exert on the 891-kg car to slow it to 45.0 mi/h in a distance of 185 m? **(b)** Describe the strategy you used to solve part (a).

SECTION 5-4 NEWTON'S THIRD LAW OF MOTION

16. • **CE Predict/Explain** A small car collides with a large truck. **(a)** Is the magnitude of the force experienced by the car greater than, less than, or equal to the magnitude of the force experienced by the truck? **(b)** Choose the *best explanation* from among the following:

   I. Action-reaction forces always have equal magnitude.

   II. The truck has more mass, and hence the force exerted on it is greater.

   III. The massive truck exerts a greater force on the lightweight car.

17. • **CE Predict/Explain** A small car collides with a large truck. **(a)** Is the acceleration experienced by the car greater than, less than, or equal to the acceleration experienced by the truck? **(b)** Choose the *best explanation* from among the following:

   I. The truck exerts a larger force on the car, giving it the greater acceleration.

   II. Both vehicles experience the same magnitude of force; therefore the lightweight car experiences the greater acceleration.

   III. The greater force exerted on the truck gives it the greater acceleration.

18. • As you catch a 0.14-kg ball it accelerates at -320 m/s^2 and comes to rest in your hand. What force does the ball exert on your hand?

19. • **BIO Woodpecker Concussion Prevention** A woodpecker exerts a 230-N force on a tree trunk during its effort to find hidden insects to eat. If its head has an effective mass of 0.025 kg, what is the acceleration of its head away from the tree trunk? The woodpecker's skull has a special spongy bone plate to protect itself against brain concussions.

20. •• On vacation, your 1400-kg car pulls a 560-kg trailer away from a stoplight with an acceleration of 1.85 m/s^2. **(a)** What is the net force exerted on the trailer? **(b)** What force does the trailer exert on the car? **(c)** What is the net force acting on the car?

21. •• **Predict/Calculate** An 85-kg parent and a 24-kg child meet at the center of an ice rink. They place their hands together and push. **(a)** Is the force experienced by the child greater than, less than, or the same as the force experienced by the parent? **(b)** Is the acceleration of the child greater than, less than, or the same as the acceleration of the parent? Explain. **(c)** If the acceleration of the child is 3.3 m/s^2 in magnitude, what is the magnitude of the parent's acceleration?

22. •• A force of magnitude 7.50 N pushes three boxes with masses $m_1 = 1.30$ kg, $m_2 = 3.20$ kg, and $m_3 = 4.90$ kg, as shown in **FIGURE 5-29**. Find the magnitude of the contact force **(a)** between boxes 1 and 2, and **(b)** between boxes 2 and 3.

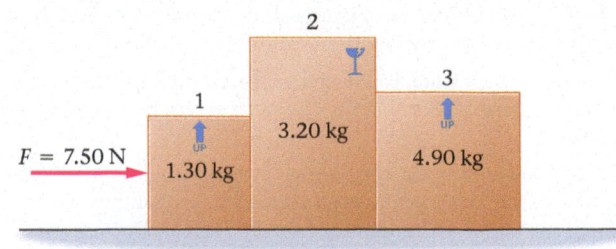

FIGURE 5-29 Problem 22

23. •• A force of magnitude 7.50 N pushes three boxes with masses $m_1 = 1.30$ kg, $m_2 = 3.20$ kg, and $m_3 = 4.90$ kg, as shown in **FIGURE 5-30**. Find the magnitude of the contact force **(a)** between boxes 1 and 2, and **(b)** between boxes 2 and 3.

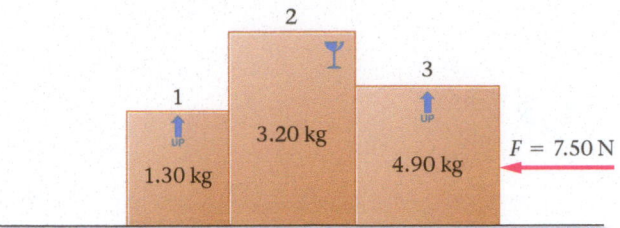

FIGURE 5-30 Problem 23

24. •• **Predict/Calculate** Two boxes sit side-by-side on a smooth horizontal surface. The lighter box has a mass of 5.2 kg; the heavier box has a mass of 7.4 kg. **(a)** Find the contact force between these boxes when a horizontal force of 5.0 N is applied to the light box. **(b)** If the 5.0-N force is applied to the heavy box instead, is the contact force between the boxes the same as, greater than, or less than the contact force in part (a)? Explain. **(c)** Verify your answer to part (b) by calculating the contact force in this case.

SECTION 5-5 THE VECTOR NATURE OF FORCES

25. • **CE** A skateboarder on a ramp is accelerated by a nonzero net force. For each of the following statements, state whether it is always true, never true, or sometimes true. **(a)** The skateboarder is moving in the direction of the net force. **(b)** The acceleration of the skateboarder is at right angles to the net force. **(c)** The acceleration of the skateboarder is in the same direction as the net force. **(d)** The skateboarder is instantaneously at rest.

26. • **CE** Three objects, A, B, and C, have x and y components of velocity that vary with time as shown in **FIGURE 5-31**. What is the direction of the net force acting on **(a)** object A, **(b)** object B, and **(c)** object C, as measured from the positive x axis? (All of the nonzero slopes have the same magnitude.)

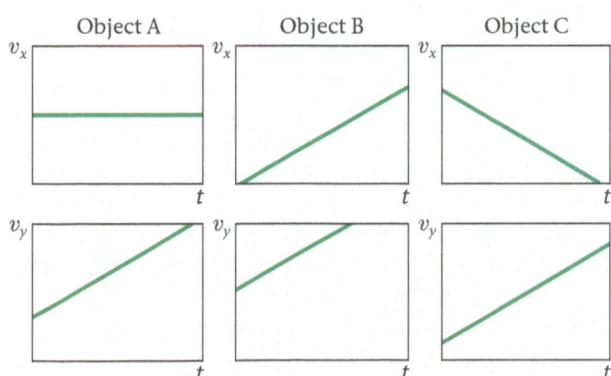

FIGURE 5-31 Problem 26

27. • A farm tractor tows a 3300-kg trailer up a 14° incline with a steady speed of 2.8 m/s. What force does the tractor exert on the trailer? (Ignore friction.)

28. • A shopper pushes a 7.5-kg shopping cart up a 13° incline, as shown in **FIGURE 5-32**. Find the magnitude of the horizontal force, $\vec{F}$, needed to give the cart an acceleration of 1.41 m/s².

FIGURE 5-32 Problem 28

29. • Two crewmen pull a raft through a lock, as shown in **FIGURE 5-33**. One crewman pulls with a force of 130 N at an angle of 34° relative to the forward direction of the raft. The second crewman, on the opposite side of the lock, pulls at an angle of 45°. With what force should the second crewman pull so that the net force of the two crewmen is in the forward direction?

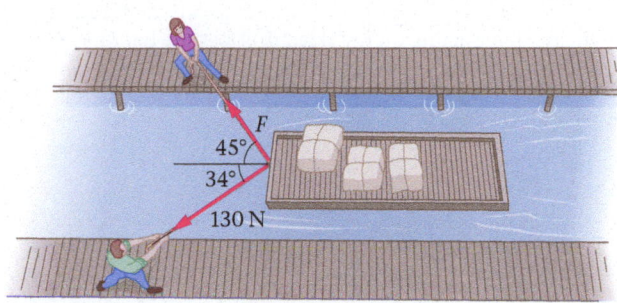

FIGURE 5-33 Problem 29

30. •• **CE** A hockey puck is acted on by one or more forces, as shown in **FIGURE 5-34**. Rank the four cases, A, B, C, and D, in order of the magnitude of the puck's acceleration, starting with the smallest. Indicate ties where appropriate.

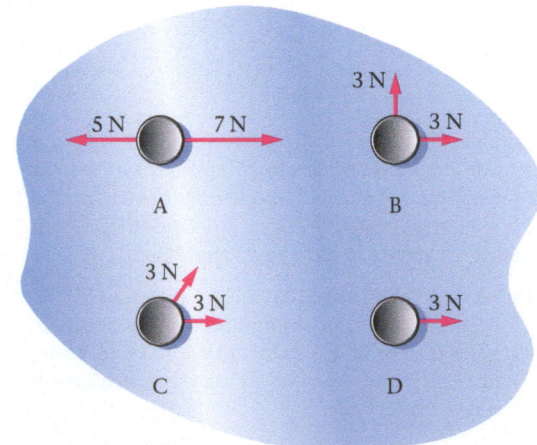

FIGURE 5-34 Problem 30

31. •• To give a 19-kg child a ride, two teenagers pull on a 3.7-kg sled with ropes, as indicated in **FIGURE 5-35**. Both teenagers pull with a force of 55 N at an angle of 35° relative to the forward direction, which is the direction of motion. In addition, the snow exerts a retarding force on the sled that points opposite to the direction of motion, and has a magnitude of 57 N. Find the acceleration of the sled and child.

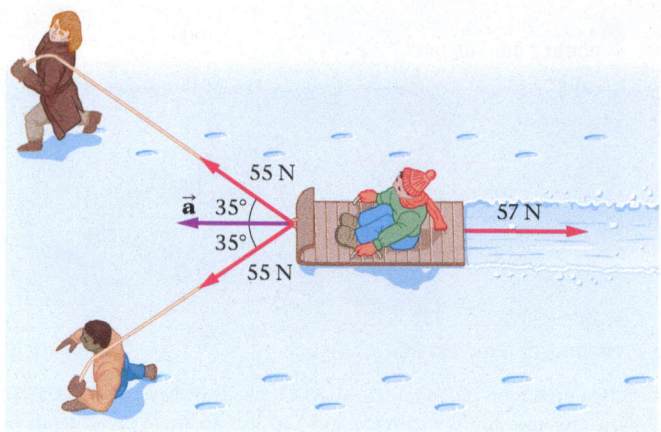

FIGURE 5-35 Problem 31

32. •• **Predict/Calculate** A 65-kg skier speeds down a trail, as shown in **FIGURE 5-36**. The surface is smooth and inclined at an angle of 22° with the horizontal. **(a)** Find the direction and magnitude of the net force acting on the skier. **(b)** Does the net force exerted on the skier increase, decrease, or stay the same as the slope becomes steeper? Explain.

FIGURE 5-36 Problems 32 and 46

33. •• An object acted on by three forces moves with constant velocity. One force acting on the object is in the positive x direction and has a magnitude of 6.5 N; a second force has a magnitude of 4.4 N and points in the negative y direction. Find the direction and magnitude of the third force acting on the object.

34. •• A train is traveling up a 2.88° incline at a speed of 4.31 m/s when the last car breaks free and begins to coast without friction. **(a)** How much time does it take for the last car to come to rest momentarily? **(b)** How far did the last car travel before momentarily coming to rest?

35. •• **The Force Exerted on the Moon** In **FIGURE 5-37** we show the Earth, Moon, and Sun (not to scale) in their relative positions at the time when the Moon is in its third-quarter phase. Though few people realize it, the force exerted on the Moon by the Sun is actually greater than the force exerted on the Moon by the Earth. In fact, the force exerted on the Moon by the Sun has a magnitude of $F_{SM} = 4.34 \times 10^{20}$ N, whereas the force exerted by the Earth has a magnitude of only $F_{EM} = 1.98 \times 10^{20}$ N. These forces are indicated to scale in Figure 5-37. Find **(a)** the direction

and **(b)** the magnitude of the net force acting on the Moon. **(c)** Given that the mass of the Moon is $M_M = 7.35 \times 10^{22}$ kg, find the magnitude of its acceleration at the time of the third-quarter phase.

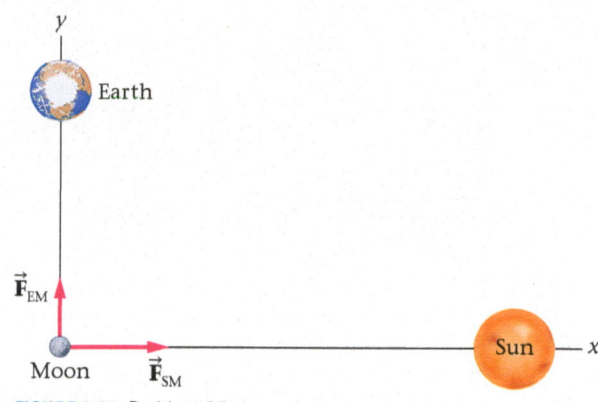

FIGURE 5-37 Problem 35

SECTION 5-6 WEIGHT

36. • You pull upward on a stuffed suitcase with a force of 105 N, and it accelerates upward at 0.725 m/s². What are **(a)** the mass and **(b)** the weight of the suitcase?

37. • **BIO Brain Growth** A newborn baby's brain grows rapidly. In fact, it has been found to increase in mass by about 1.6 mg per minute. **(a)** How much does the brain's weight increase in one day? **(b)** How much time does it take for the brain's weight to increase by 0.15 N?

38. • Suppose a rocket launches with an acceleration of 30.5 m/s². What is the apparent weight of a 92-kg astronaut aboard this rocket?

39. • During an episode of turbulence in an airplane you feel 190 N heavier than usual. If your mass is 78 kg, what are the magnitude and direction of the airplane's acceleration?

40. • At the bow of a ship on a stormy sea, a crewman conducts an experiment by standing on a bathroom scale. In calm waters, the scale reads 182 lb. During the storm, the crewman finds a maximum reading of 225 lb and a minimum reading of 138 lb. Find **(a)** the maximum upward acceleration and **(b)** the maximum downward acceleration experienced by the crewman.

41. •• **Predict/Calculate** As part of a physics experiment, you stand on a bathroom scale in an elevator. Though your normal weight is 610 N, the scale at the moment reads 730 N. **(a)** Is the acceleration of the elevator upward, downward, or zero? Explain. **(b)** Calculate the magnitude of the elevator's acceleration. **(c)** What, if anything, can you say about the velocity of the elevator? Explain.

42. •• When you weigh yourself on good old *terra firma* (solid ground), your weight is 142 lb. In an elevator your apparent weight is 121 lb. What are the direction and magnitude of the elevator's acceleration?

43. •• **Predict/Calculate BIO Flight of the Samara** A 1.21-g samara— the winged fruit of a maple tree—falls toward the ground with a constant speed of 1.1 m/s (**FIGURE 5-38**). **(a)** What is the force of air resistance exerted on the samara? **(b)** If the constant speed of descent is greater than 1.1 m/s, is the force of air resistance greater than, less than, or the same as in part (a)? Explain.

FIGURE 5-38 Problem 43

44. •••When you lift a bowling ball with a force of 82 N, the ball accelerates upward with an acceleration a. If you lift with a force of 92 N, the ball's acceleration is $2a$. Find **(a)** the weight of the bowling ball, and **(b)** the acceleration a.

SECTION 5-7 NORMAL FORCES

45. • A 23-kg suitcase is pulled with constant speed by a handle at an angle of 25° above the horizontal. If the normal force exerted on the suitcase is 180 N, what is the force F applied to the handle?

46. • **(a)** Draw a free-body diagram for the skier in Problem 32. **(b)** Determine the normal force acting on the skier.

47. • A 9.3-kg child sits in a 3.7-kg high chair. **(a)** Draw a free-body diagram for the child, and find the normal force exerted by the chair on the child. **(b)** Draw a free-body diagram for the chair, and find the normal force exerted by the floor on the chair.

48. •• **FIGURE 5-39** shows the normal force N experienced by a rider of weight $W = mg$ on a Ferris wheel as a function of time. **(a)** At what time, A, B, C, or D, is the rider accelerating upward? **(b)** What is the magnitude of the maximum acceleration experienced by the rider?

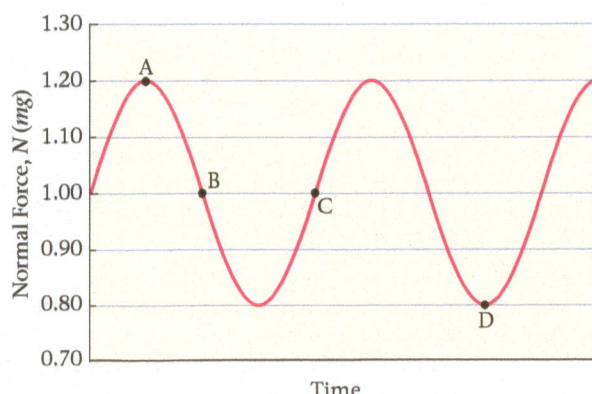

FIGURE 5-39 Problem 48

49. •• **FIGURE 5-40** shows the normal force N as a function of the angle θ for the suitcase shown in Figure 5-20. Determine the magnitude of the force $\vec{F}$ for each of the three curves shown in Figure 5-40. Give your answer in terms of the weight of the suitcase, mg.

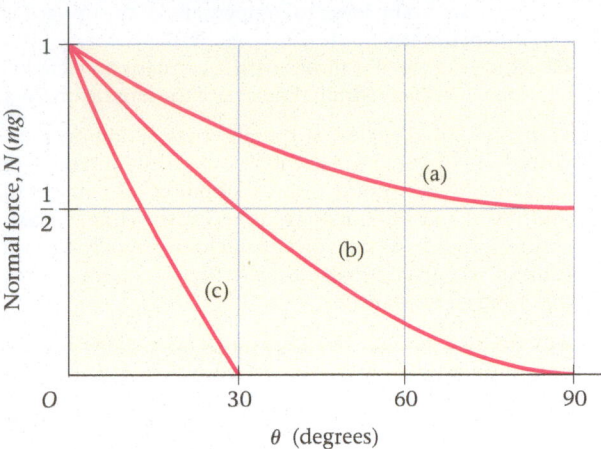

FIGURE 5-40 Problem 49

50. •• A 5.0-kg bag of potatoes sits on the bottom of a stationary shopping cart. **(a)** Sketch a free-body diagram for the bag of potatoes. **(b)** Now suppose the cart moves with a constant velocity. How does this affect your free-body diagram? Explain.

51. •• Predict/Calculate **(a)** Find the normal force exerted on a 2.9-kg book resting on a surface inclined at 36° above the horizontal. **(b)** If the angle of the incline is reduced, do you expect the normal force to increase, decrease, or stay the same? Explain.

52. •• Predict/Calculate A gardener mows a lawn with an old-fashioned push mower. The handle of the mower makes an angle of 32° with the surface of the lawn. **(a)** If a 68-N force is applied along the handle of the 17-kg mower, what is the normal force exerted by the lawn on the mower? **(b)** If the angle between the surface of the lawn and the handle of the mower is increased, does the normal force exerted by the lawn increase, decrease, or stay the same? Explain.

53. •••An ant walks slowly away from the top of a bowling ball, as shown in **FIGURE 5-41**. If the ant starts to slip when the normal force on its feet drops below one-half its weight, at what angle θ does slipping begin?

FIGURE 5-41 Problem 53

GENERAL PROBLEMS

54. • **CE** Predict/Explain Riding in an elevator moving upward with constant speed, you begin a game of darts. **(a)** Do you have to aim your darts higher than, lower than, or the same as when you play darts on solid ground? **(b)** Choose the *best explanation* from among the following:
I. The elevator rises during the time it takes for the dart to travel to the dartboard.
II. The elevator moves with constant velocity. Therefore, Newton's laws apply within the elevator in the same way as on the ground.
III. You have to aim lower to compensate for the upward speed of the elevator.

55. • **CE** Predict/Explain Riding in an elevator moving with a constant upward acceleration, you begin a game of darts. **(a)** Do you have to aim your darts higher than, lower than, or the same as when you play darts on solid ground? **(b)** Choose the *best explanation* from among the following:
I. The elevator accelerates upward, giving its passengers a greater "effective" acceleration due to gravity.

II. You have to aim lower to compensate for the upward acceleration of the elevator.

III. Since the elevator moves with a constant acceleration, Newton's laws apply within the elevator the same as on the ground.

56. • **CE** Give the direction of the net force acting on each of the following objects. If the net force is zero, state "zero." **(a)** A car accelerating northward from a stoplight. **(b)** A car traveling southward and slowing down. **(c)** A car traveling westward with constant speed. **(d)** A skydiver parachuting downward with constant speed. **(e)** A baseball during its flight from pitcher to catcher (ignoring air resistance).

57. • **CE** Predict/Explain You jump out of an airplane and open your parachute after an extended period of free fall. **(a)** To decelerate your fall, must the force exerted on you by the parachute be greater than, less than, or equal to your weight? **(b)** Choose the *best explanation* from among the following:
    I. Parachutes can only exert forces that are less than the weight of the skydiver.
    II. The parachute exerts a force exactly equal to the skydiver's weight.
    III. To decelerate after free fall, the net force acting on a skydiver must be upward.

58. • In a tennis serve, a 0.070-kg ball can be accelerated from rest to 36 m/s over a distance of 0.75 m. Find the magnitude of the average force exerted by the racket on the ball during the serve.

59. • **BIO Human Heart Force** The left ventricle of the human heart expels about 0.070 kg of oxygenated blood straight upward into the ascending aorta with a force of 5.5 N. What is the acceleration of the blood as it enters the aorta?

60. • A 51.5-kg swimmer with an initial speed of 1.25 m/s decides to coast until she comes to rest. If she slows with constant acceleration and stops after coasting 2.20 m, what was the force exerted on her by the water?

61. • The a_x-versus-time graph for a 2.0-kg object is shown in **FIGURE 5-42**. Draw the corresponding F_x-versus-time graph, assuming only a single force acts on the object.

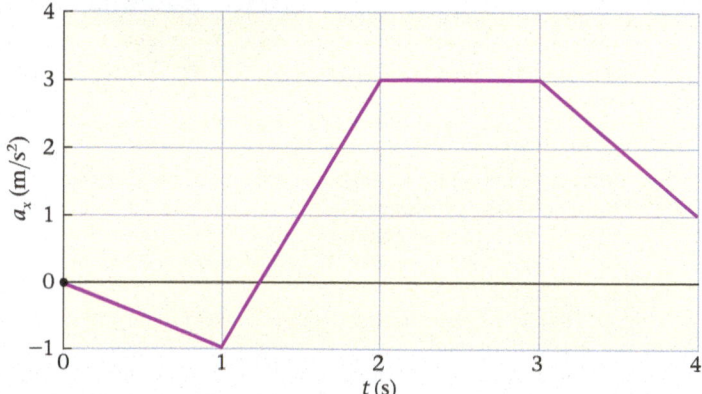

FIGURE 5-42 Problem 61

62. • A skateboarder starts from rest and rolls down a ramp that is inclined at an angle of 15° below the horizontal. Friction can be ignored. **(a)** What is the acceleration of the skateboader? **(b)** What is the skateboarder's speed 1.5 s after starting?

63. •• The rotors of a 15,200-kg heavy-lift helicopter exert a downward force of 322,000 N in order to accelerate itself and its external cargo upward at 0.50 m/s². By itself, what upward force does the external cargo exert on the Earth?

64. •• As it pulls itself up to a branch, a chimpanzee accelerates upward at 2.2 m/s² at the instant it exerts a 260-N force downward on the branch. The situation is shown in **FIGURE 5-43**. **(a)** In what direction is (i) the force the branch exerts on the chimpanzee; (ii) the force the tree exerts on the branch; and (iii) the force the chimpanzee exerts on the Earth? **(b)** Find the magnitude of the force the chimpanzee exerts on the Earth.

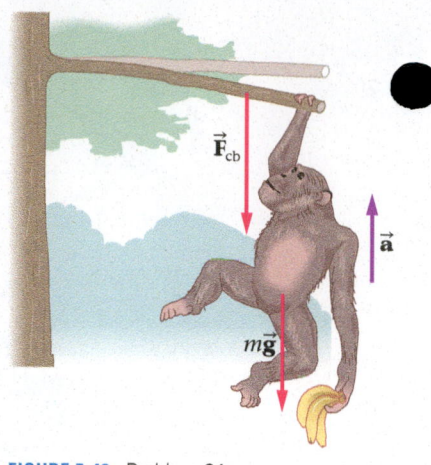

FIGURE 5-43 Problem 64

65. •• **CE** Each of the three identical hockey pucks shown in **FIGURE 5-44** is acted on by a 3-N force. Puck A moves with a speed of 7 m/s in a direction opposite to the force; puck B is instantaneously at rest; puck C moves with a speed of 7 m/s at right angles to the force. Rank the three pucks in order of the magnitude of their acceleration, starting with the smallest. Indicate ties with an equal sign.

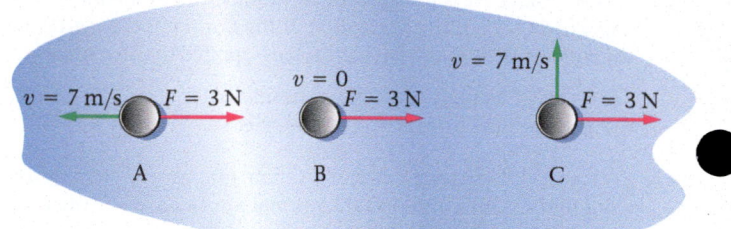

FIGURE 5-44 Problem 65

66. •• Predict/Calculate **The VASIMR Rocket** NASA plans to use a new type of rocket, a Variable Specific Impulse Magnetoplasma Rocket (VASIMR), on future missions. If a VASIMR can produce 6.0 N of thrust (force), and has a mass of 620 kg, **(a)** what acceleration will it experience? Assume that the only force acting on the rocket is its own thrust, and that the mass of the rocket is constant. **(b)** Over what distance must the rocket accelerate from rest to achieve a speed of 9500 m/s? **(c)** When the rocket has covered one-quarter the acceleration distance found in part (b), is its average speed 1/2, 1/3, or 1/4 its average speed during the final three-quarters of the acceleration distance? Explain.

67. •• An object of mass $m = 5.95$ kg has an acceleration $\vec{a} = (1.17 \text{ m/s}^2)\hat{x} + (-0.664 \text{ m/s}^2)\hat{y}$. Three forces act on this object: $\vec{F}_1$, $\vec{F}_2$, and $\vec{F}_3$. Given that $\vec{F}_1 = (3.22 \text{ N})\hat{x}$ and $\vec{F}_2 = (-1.55 \text{ N})\hat{x} + (2.05 \text{ N})\hat{y}$, find $\vec{F}_3$.

68. •• At the local grocery store, you push a 14.5-kg shopping cart. You stop for a moment to add a bag of dog food to your cart. With a force of 12.0 N, you now accelerate the cart from rest through a distance of 2.29 m in 3.00 s. What was the mass of the dog food?

69. •• **BIO** Predict/Calculate **The Force of Running** Biomechanical research has shown that when a 67-kg person is running, the force exerted on each foot as it strikes the ground can be as great as 2300 N. **(a)** What is the ratio of the force exerted on the foot by the ground to the person's body weight? **(b)** If the only forces acting on the person are (i) the force exerted by the ground and (ii) the person's weight, what are the magnitude and direction of the person's acceleration?

(c) If the acceleration found in part (b) acts for 10.0 ms, what is the resulting change in the vertical component of the person's velocity?

70. •• **BIO Predict/Calculate Grasshopper Liftoff** To become airborne, a 2.0-g grasshopper requires a takeoff speed of 2.7 m/s. It acquires this speed by extending its hind legs through a distance of 3.7 cm. **(a)** What is the average acceleration of the grasshopper during takeoff? **(b)** Find the magnitude of the average net force exerted on the grasshopper by its hind legs during takeoff. **(c)** If the mass of the grasshopper increases, does the takeoff acceleration increase, decrease, or stay the same? **(d)** If the mass of the grasshopper increases, does the required takeoff force increase, decrease, or stay the same? Explain.

71. •• **Takeoff from an Aircraft Carrier** On an aircraft carrier, a jet can be catapulted from 0 to 155 mi/h in 2.00 s. If the average force exerted by the catapult is 9.35×10^5 N, what is the mass of the jet?

72. •• A 79-kg person stands on a bathroom scale and rides in an elevator whose v_y-versus-time graph is shown in **FIGURE 5-45**. What is the reading on the scale at **(a)** $t = 1.0$ s, **(b)** $t = 4.0$ s, and **(c)** $t = 8.0$ s?

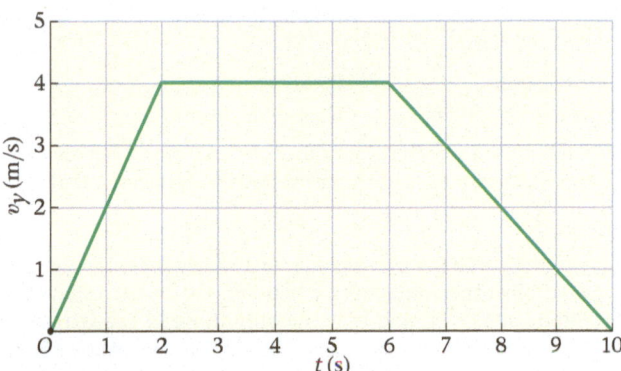

FIGURE 5-45 Problem 72

73. •• **Predict/Calculate** An archer shoots a 0.024-kg arrow at a target with a speed of 54 m/s. When it hits the target, it penetrates to a depth of 0.083 m. **(a)** What was the average force exerted by the target on the arrow? **(b)** If the mass of the arrow is doubled, and the force exerted by the target on the arrow remains the same, by what multiplicative factor does the penetration depth change? Explain.

74. •• An apple of mass $m = 0.13$ kg falls out of a tree from a height $h = 3.2$ m. **(a)** What is the magnitude of the force of gravity, mg, acting on the apple? **(b)** What is the apple's speed, v, just before it lands? **(c)** Show that the force of gravity times the height, mgh, is equal to $\frac{1}{2}mv^2$. (We shall investigate the significance of this result in Chapter 8.) Be sure to show that the dimensions are in agreement as well as the numerical values.

75. •• **BIO The Fall of T. rex** Paleontologists estimate that if a *Tyrannosaurus rex* had tripped and fallen, the ground would have exerted a normal force of approximately 260,000 N on its torso as it landed. Assuming the torso has a mass of 3800 kg, **(a)** find the magnitude of the torso's upward acceleration as it comes to rest. (For comparison, humans lose consciousness with an acceleration of about 7g.) **(b)** Assuming the torso of a T. rex is in free fall for a distance of 1.46 m as it falls to the ground, how much time is required for the torso to come to rest once it contacts the ground?

76. •• **Deep Space 1** The NASA spacecraft *Deep Space 1* was shut down on December 18, 2001, following a three-year journey to the asteroid Braille and the comet Borrelly. This spacecraft used a solar-powered ion engine to produce 0.021 pounds of thrust (force) by stripping electrons from xenon atoms and accelerating the resulting ions to 70,000 mi/h. The thrust was only as much as the weight of a couple sheets of paper, but the engine operated continuously for 16,000 hours. As a result, the speed of the spacecraft increased by

7900 mi/h. What was the mass of *Deep Space 1*? (Assume that the mass of the xenon gas is negligible.)

77. •• Your groceries are in a bag with paper handles. The two handles together will tear off if a force greater than 51.5 N is applied to them. What is the greatest mass of groceries that can be lifted safely with this bag, given that the bag is raised **(a)** with constant speed, or **(b)** with an acceleration of 1.25 m/s²?

78. •• **BIO A Leafhopper's Leap** The motion of jumping insects is of interest not only to biologists, but also to those who design robots. **FIGURE 5-46** shows the speed-versus-time graph for two different takeoff jumps of a 19-mg green leafhopper, as determined from high-speed videos. **(a)** What is the magnitude of the average net force acting on the leafhopper's body during jump 1? **(b)** What is the ratio of the force found in part (a) to the leafhopper's weight? **(c)** Is the average force during jump 2 greater than, less than, or equal to the average force during jump 1? Explain.

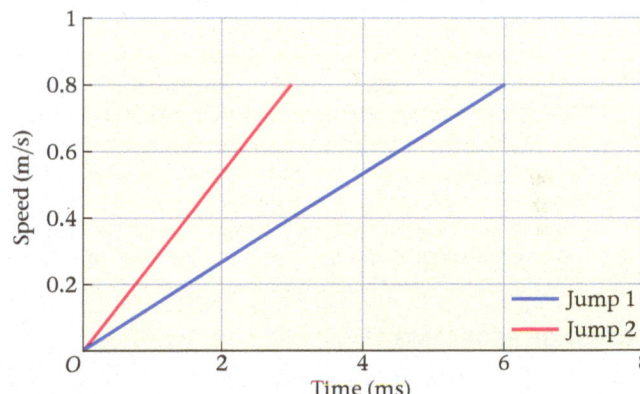

FIGURE 5-46 Problem 78

79. •• **Predict/Calculate** At the airport, you observe some of the jets as they take off. With your watch you find that it takes about 35 seconds for a plane to go from rest to takeoff speed. In addition, you estimate that the distance required is about 1.5 km. **(a)** If the mass of a jet is 1.70×10^5 kg, what force is needed for takeoff? **(b)** Describe the strategy you used to solve part (a).

80. •• **BIO Gecko Feet** Researchers have found that a gecko's foot is covered with hundreds of thousands of small hairs (*setae*) that allow it to walk up walls and even across ceilings. A single foot pad, which has an area of 1.0 cm², can attach to a wall or ceiling with a force of 11 N. **(a)** How many 250-g geckos could be suspended from the ceiling by a single foot pad? **(b)** Estimate the force per square centimeter that your body exerts on the soles of your shoes, and compare with the 11 N/cm² of the sticky gecko foot.

81. •• Two boxes are at rest on a smooth, horizontal surface. The boxes are in contact with one another. If box 1 is pushed with a force of magnitude $F = 12.00$ N, the contact force between the boxes is 8.50 N; if, instead, box 2 is pushed with the force F, the contact force is 12.00 N − 8.50 N = 3.50 N. In either case, the boxes move together with an acceleration of 1.70 m/s². What is the mass of **(a)** box 1 and **(b)** box 2?

82. ••• You have been hired to help improve the material movement system at a manufacturing plant. Boxes containing 22 kg of tomato sauce in glass jars must slide from rest down a frictionless roller ramp to the loading dock, but they must not accelerate at a rate that exceeds 2.5 m/s² because of safety concerns. **(a)** What is the maximum angle of inclination of the ramp? **(b)** If the vertical distance the ramp must span is 1.2 m, with what speed will the boxes exit the bottom of the ramp? **(c)** What is the normal force on a box as it moves down the ramp?

83. ••• You and some friends take a hot-air balloon ride. One friend is late, so the balloon floats a couple of feet off the ground as you wait. Before this person arrives, the combined weight of the basket and people is 1220 kg, and the balloon, which contains 5980 kg of hot air, is neutrally buoyant. When the late arrival climbs up into the basket, the balloon begins to accelerate downward at 0.10 m/s^2. What was the mass of the last person to climb aboard?

84. ••• A baseball of mass m and initial speed v strikes a catcher's mitt. If the mitt moves a distance Δx as it brings the ball to rest, what is the average force it exerts on the ball?

85. ••• When two people push in the same direction on an object of mass m they cause an acceleration of magnitude a_1. When the same people push in opposite directions, the acceleration of the object has a magnitude a_2. Determine the magnitude of the force exerted by each of the two people in terms of m, a_1, and a_2.

86. ••• An air-track cart of mass $m_1 = 0.14 \text{ kg}$ is moving with a speed $v_0 = 1.3 \text{ m/s}$ to the right when it collides with a cart of mass $m_2 = 0.25 \text{ kg}$ that is at rest. Each cart has a wad of putty on its bumper, and hence they stick together as a result of their collision. Suppose the average contact force between the carts is $F = 1.5 \text{ N}$ during the collision. (a) What is the acceleration of cart 1? Give direction and magnitude. (b) What is the acceleration of cart 2? Give direction and magnitude. (c) How much time does it take for both carts to have the same speed? (Once the carts have the same speed the collision is over and the contact force vanishes.) (d) What is the final speed of the carts, v_f? (e) Show that $m_1 v_0$ is equal to $(m_1 + m_2)v_f$.(We shall investigate the significance of this result in Chapter 9.)

PASSAGE PROBLEMS

BIO Increasing Safety in a Collision

Safety experts say that an automobile accident is really a succession of three separate collisions: (1) the automobile collides with an obstacle and comes to rest; (2) people within the car continue to move forward until they collide with the interior of the car, or are brought to rest by a restraint system like a seatbelt or an air bag; and (3) organs within the occupants' bodies continue to move forward until they collide with the body wall and are brought to rest. Not much can be done about the third collision, but the effects of the first two can be mitigated by increasing the distance over which the car and its occupants are brought to rest.

For example, the severity of the first collision is reduced by building collapsible "crumple zones" into the body of a car, and by placing compressible collision barriers near dangerous obstacles like bridge supports. The second collision is addressed primarily through the use of seatbelts and air bags. These devices reduce the force that acts on an occupant to survivable levels by increasing the distance over which he or she comes to rest. This is illustrated in **FIGURE 5-47**, where we see the force exerted on a 65.0-kg driver who slows from an initial speed of 18.0 m/s (lower curve) or 36.0 m/s (upper curve) to rest in a distance ranging from 5.00 cm to 1.00 m.

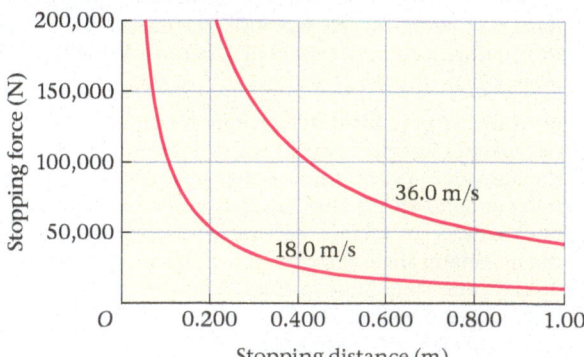

FIGURE 5-47 Problems 87, 88, 89, and 90

87. •• The combination of "crumple zones" and air bags/seatbelts might increase the distance over which a person stops in a collision to as great as 1.00 m. What is the magnitude of the force exerted on a 65.0-kg driver who decelerates from 18.0 m/s to 0.00 m/s over a distance of 1.00 m?

    A. 162 N   B. 585 N
    C. 1.05×10^4 N   D. 2.11×10^4 N

88. •• A driver who does not wear a seatbelt continues to move forward with a speed of 18.0 m/s (due to inertia) until something solid like the steering wheel is encountered. The driver now comes to rest in a much shorter distance—perhaps only a few centimeters. Find the magnitude of the net force acting on a 65.0-kg driver who is decelerated from 18.0 m/s to rest in 5.00 cm.

    A. 3240 N   B. 1.17×10^4 N
    C. 2.11×10^5 N   D. 4.21×10^5 N

89. • Suppose the initial speed of the driver is doubled to 36.0 m/s. If the driver still has a mass of 65.0 kg, and comes to rest in 1.00 m, what is the magnitude of the force exerted on the driver during this collision?

    A. 648 N   B. 1170 N
    C. 2.11×10^4 N   D. 4.21×10^4 N

90. • If both the speed and stopping distance of a driver are doubled, by what factor does the force exerted on the driver change?

    A. 0.5   B. 1   C. 2   D. 4

91. •• Predict/Calculate **Referring to Example 5-8** Suppose that we would like the contact force between the boxes to have a magnitude of 5.00 N, and that the only thing in the system we are allowed to change is the mass of box 2; the mass of box 1 is 10.0 kg and the applied force is 30.0 N. (a) Should the mass of box 2 be increased or decreased? Explain. (b) Find the mass of box 2 that results in a contact force of magnitude 5.00 N. (c) What is the acceleration of the boxes in this case?

92. •• **Referring to Example 5-8** Suppose the force of 30.0 N pushes on two boxes of unknown mass. We know, however, that the acceleration of the boxes is 1.20 m/s^2 and the contact force has a magnitude of 4.45 N. Find the mass of (a) box 1 and (b) box 2.

93. •• Predict/Calculate **Referring to Figure 5-13** Suppose the magnitude of $\vec{F}_2$ is increased from 41 N to 55 N, and that everything else in the system remains the same. (a) Do you expect the direction of the satellite's acceleration to be greater than, less than, or equal to 32°? Explain. Find (b) the direction and (c) the magnitude of the satellite's acceleration in this case.

94. •• Predict/Calculate **Referring to Figure 5-13** Suppose we would like the acceleration of the satellite to be at an angle of 25°, and the only quantity we can change in the system is the magnitude of $\vec{F}_1$. (a) Should the magnitude of $\vec{F}_1$ be increased or decreased? Explain. (b) What is the magnitude of the satellite's acceleration in this case?

Applications of Newton's Laws

Big Ideas

1 Friction is a force exerted between surfaces in contact.

2 Springs exert a force proportional to the amount of stretch or compression.

3 A system is in translational equilibrium when the total force acting on it is zero.

4 An object in circular motion accelerates toward the center of the circle.

▲ The climber in this photograph may not be thinking about Newton's laws, but they are involved in every aspect of his endeavor. He relies on the forces transmitted through the ropes to support his weight, and on the pulleys through which the ropes are threaded to give those forces the desired directions. His forward progress is made possible by the substantial frictional force between his hands and the rope. These are but a few of the many real-world applications of Newton's laws that are explored in this chapter.

Newton's laws of motion can be applied to an immense variety of systems. This chapter shows how Newton's laws are applied to friction, springs, connected objects, and circular motion.

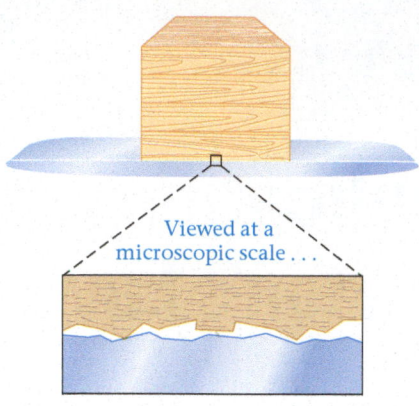

Viewed at a
microscopic scale . . .

. . . even a "smooth" surface is rough.

▲ **FIGURE 6-1** **The origin of friction** Even "smooth" surfaces have irregularities at the microscopic level that lead to friction.

Big Idea 1 Friction is a force exerted between surfaces in contact. The force of friction is different depending on whether the surfaces are in static contact, or slide relative to one another.

6-1 Frictional Forces

In Chapter 5 we assumed that surfaces are smooth and that objects slide without resistance to their motion. No surface is perfectly smooth, however. When viewed on the atomic level, even the "smoothest" surface is actually rough and jagged, as indicated in **FIGURE 6-1**. To slide one such surface across another requires a force large enough to overcome the resistance of microscopic hills and valleys bumping together. This is the origin of the force we call **friction.**

We often think of friction as something that should be reduced, or even eliminated if possible. For example, roughly 20% of the gasoline you buy does nothing but overcome friction within your car's engine. Clearly, reducing that friction would be desirable.

On the other hand, friction can be helpful—even indispensable—in other situations. Suppose, for example, that you are standing still and then decide to begin walking forward. The force that accelerates you is the force of friction between your shoes and the ground. We simply couldn't walk or run without friction—it's hard enough when friction is merely reduced, as on an icy sidewalk. Similarly, starting and stopping a car, and even turning a corner, all require friction. Examples of systems with small and large amounts of friction are shown in **FIGURE 6-2**.

Since friction is caused by the random, microscopic irregularities of a surface, and since it is greatly affected by other factors such as the presence of lubricants, there is no simple "law of nature" for friction. There are, however, some very useful rules of thumb that give accurate results for calculating frictional forces. In what follows, we describe these rules of thumb for the two most common types of friction—kinetic friction and static friction.

Kinetic Friction

As its name implies, kinetic friction is the friction produced when surfaces slide against one another. The force generated by this friction, which we designate with the symbol f_k, acts to oppose the sliding.

A series of simple experiments illustrates the main characteristics of kinetic friction. First, imagine attaching a spring scale to a rough object, like a brick, and pulling it across a table, as shown in **FIGURE 6-3**. If the brick moves with constant velocity, Newton's second law tells us that the net force on the brick must be zero. Hence, the force indicated on the scale, F, has the same magnitude as the force of kinetic friction, f_k. Now, if we repeat the experiment, but this time put a second brick on top of the first, we find that the force needed to pull the brick with constant velocity is doubled, to $2F$.

From this experiment we see that when we double the normal force—by stacking up two bricks, for example—the force of kinetic friction is also doubled. In general,

▶ **FIGURE 6-2 Visualizing Concepts**
Minimizing and Maximizing Friction
Friction plays an important role in almost everything we do. Sometimes it is desirable to reduce friction; in other cases we want as much friction as possible. For example, it is more fun to ride on a water slide (left) if the friction is low. When running, however, we need friction to help us speed up, slow down, and make turns. The sole of this running shoe (right), like a car tire, is designed to maximize friction.

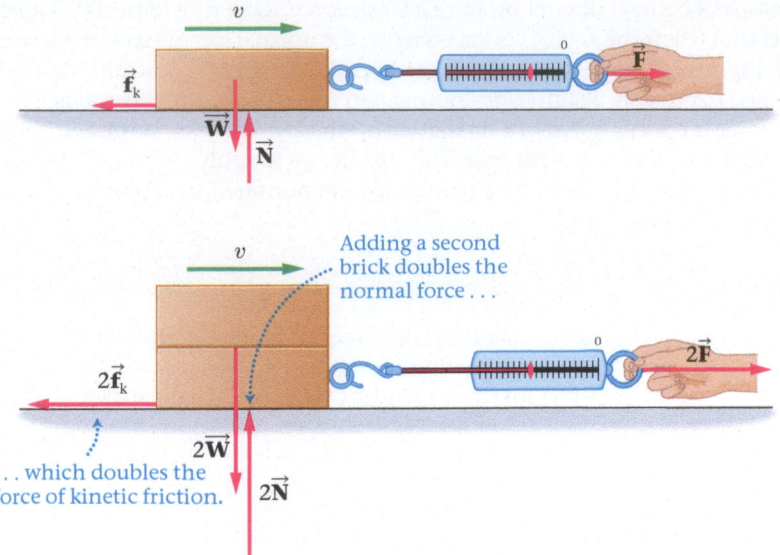

◀ **FIGURE 6-3 The force of kinetic friction depends on the normal force** In the top part of the figure, a force F is required to pull the brick with constant speed v. Thus the force of kinetic friction is $f_k = F$. In the bottom part, the normal force has been doubled, and so has the force of kinetic friction, to $f_k = 2F$.

the magnitude of the force of kinetic friction is proportional to the magnitude of the normal force, N. This is stated mathematically as follows:

$$f_k = \mu_k N \qquad\qquad 6\text{-}1$$

The constant of proportionality, μ_k (pronounced "mew sub k"), is referred to as the **coefficient of kinetic friction.** In Figure 6-3 the normal force is equal to the weight of the bricks, but this is a special case. The normal force is greater than the weight if someone pushes down on the bricks, and this would cause more friction, or less than the weight of the bricks if they are placed on an incline or someone pulls upward on them.

Because f_k and N are both forces, and hence have the same units, it follows that μ_k is a dimensionless number. The coefficient of kinetic friction is always positive, and typical values range between 0 and 1, as indicated in Table 6-1. The interpretation of μ_k is simple: If $\mu_k = 0.1$, for example, the force of kinetic friction is one-tenth of the normal force. Simply put, the greater μ_k, the greater the friction.

As we know from everyday experience, the force of kinetic friction tends to oppose relative motion, as shown in Figure 6-3. Thus, $f_k = \mu_k N$ is not a vector equation, because N is perpendicular to the direction of motion. When doing calculations with the force of kinetic friction, we use $f_k = \mu_k N$ to find its magnitude, and we draw its direction so that it is opposite to the direction of motion.

There are two more friction experiments of particular interest. First, suppose that when we pull a brick, we initially pull it at the speed v, then later at the speed $2v$. What forces do we measure? It turns out that the force of kinetic friction is approximately the same in each case. Second, let's try standing the brick on end, so that it has a smaller area in contact with the table. If this smaller area is half the previous area, is the force halved? No! It turns out that the force of kinetic friction remains essentially the same, regardless of the area of contact.

We summarize our observations with the following three rules of thumb for kinetic friction:

Rules of Thumb for Kinetic Friction
The force of kinetic friction between two surfaces is:

1. Proportional to the magnitude of the normal force, N, between the surfaces:

$$f_k = \mu_k N$$

2. Independent of the relative speed of the surfaces.
3. Independent of the area of contact between the surfaces.

Before we show how to use f_k in calculations, we should make a comment regarding rule 3. This rule often seems rather surprising and counterintuitive. How is it that

TABLE 6-1 Typical Coefficients of Friction

Materials	Kinetic, μ_k	Static, μ_s
Rubber on concrete (dry)	0.80	1–4
Steel on steel	0.57	0.74
Glass on glass	0.40	0.94
Wood on leather	0.40	0.50
Copper on steel	0.36	0.53
Rubber on concrete (wet)	0.25	0.30
Steel on ice	0.06	0.10
Waxed ski on snow	0.05	0.10
Teflon on Teflon	0.04	0.04
Synovial joints in humans	0.003	0.01

PHYSICS IN CONTEXT
Looking Back

The equations of kinematics from Chapters 2 and 4 are useful in this chapter, especially in Examples 6-1 and 6-2.

Choice of Coordinate System: Incline

On an incline, align one axis (x) parallel to the surface, and the other axis (y) perpendicular to the surface. That way the motion is in the x direction. Since no motion occurs in the y direction, we know that $a_y = 0$.

a larger area of contact doesn't produce a larger force? One way to think about this is to consider that when the area of contact is large, the normal force is spread out over a large area, giving a small force per area, F/A. As a result, the microscopic hills and valleys are not pressed too deeply against one another. On the other hand, if the area is small, the normal force is concentrated in a small region, which presses the surfaces together more firmly, due to the large force per area. The net effect is roughly the same in either case.

The next Example considers a commonly encountered situation in which kinetic friction plays a decisive role.

EXAMPLE 6-1 PASS THE SALT—PLEASE

Someone at the other end of the table asks you to pass the salt. Feeling quite dashing, you slide the 50.0-g salt shaker in that direction, giving it an initial speed of 1.15 m/s. **(a)** If the shaker comes to rest with constant acceleration in 0.840 m, what is the coefficient of kinetic friction between the shaker and the table? **(b)** How much time is required for the shaker to come to rest if you slide it with an initial speed of 1.32 m/s?

PICTURE THE PROBLEM

We choose the positive x direction to be the direction of motion, and the positive y direction to be upward. Two forces act in the y direction: the shaker's weight, $\vec{W} = -W\hat{y} = -mg\hat{y}$, and the normal force, $\vec{N} = N\hat{y}$. Only one force acts in the x direction: the force of kinetic friction, $\vec{f}_k = -\mu_k N\hat{x}$. Notice that the shaker moves through a distance of 0.840 m with an initial speed $v_{0x} = 1.15$ m/s.

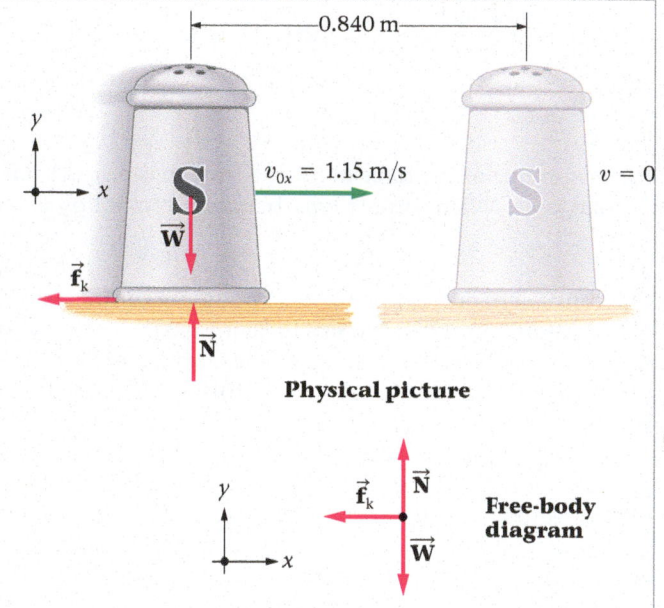

Physical picture

Free-body diagram

REASONING AND STRATEGY

a. The frictional force has a magnitude of $f_k = \mu_k N$, and hence it follows that $\mu_k = f_k/N$. Therefore, we need to find the magnitudes of the frictional force, f_k, and the normal force, N, to determine μ_k. To find f_k we set $\Sigma F_x = ma_x$, and find a_x with the kinematic equation $v_x{}^2 = v_{0x}{}^2 + 2a_x\Delta x$. To find N we set $a_y = 0$ (since there is no motion in the y direction) and solve for N using $\Sigma F_y = ma_y = 0$.

b. The coefficient of kinetic friction is independent of the sliding speed, and hence the acceleration of the shaker is also independent of the speed. As a result, we can use the acceleration from part (a) in the equation $v_x = v_{0x} + a_x t$ to find the sliding time.

Known Mass of salt shaker, $m = 50.0$ g; initial speed, $v_0 = 1.15$ m/s or 1.32 m/s; sliding distance, $\Delta x = 0.840$ m.

Unknown **(a)** Coefficient of kinetic friction, $\mu_k = ?$ **(b)** Time to come to rest, $t = ?$

SOLUTION

Part (a)

1. Set $\Sigma F_x = ma_x$ to find f_k in terms of a_x:
$$\Sigma F_x = -f_k = ma_x \quad \text{or} \quad f_k = -ma_x$$

2. Determine a_x by using the kinematic equation relating velocity to position, $v_x{}^2 = v_{0x}{}^2 + 2a_x\Delta x$:
$$v_x{}^2 = v_{0x}{}^2 + 2a_x\Delta x$$
$$a_x = \frac{v_x{}^2 - v_{0x}{}^2}{2\Delta x} = \frac{0 - (1.15 \text{ m/s})^2}{2(0.840 \text{ m})} = -0.787 \text{ m/s}^2$$

3. Set $\Sigma F_y = ma_y = 0$ to find the normal force, N:
$$\Sigma F_y = N + (-W) = ma_y = 0 \quad \text{or} \quad N = W = mg$$

4. Substitute $N = mg$ and $f_k = -ma_x$ (with $a_x = -0.787$ m/s^2) into $\mu_k = f_k/N$ to find μ_k:
$$\mu_k = \frac{f_k}{N} = \frac{-ma_x}{mg} = \frac{-a_x}{g} = \frac{-(-0.787 \text{ m/s}^2)}{9.81 \text{ m/s}^2} = 0.0802$$

Part (b)

5. Use $a_x = -0.787$ m/s^2, $v_{0x} = 1.32$ m/s, and $v_x = 0$ in $v_x = v_{0x} + a_x t$ to solve for the time, t:
$$v_x = v_{0x} + a_x t \quad \text{or}$$
$$t = \frac{v_x - v_{0x}}{a_x} = \frac{0 - (1.32 \text{ m/s})}{-0.787 \text{ m/s}^2} = 1.68 \text{ s}$$

INSIGHT

Notice that m canceled in Step 4, so our result for the coefficient of friction is independent of the shaker's mass. For example, if we were to slide a shaker with twice the mass, but with the same initial speed, it would slide the same distance. It's unlikely this independence would have been apparent if we had worked the problem numerically rather than symbolically. Part (b) shows that the same comments apply to the sliding time—it too is independent of the shaker's mass.

CONTINUED

PRACTICE PROBLEM

Given an initial speed of 1.15 m/s and a coefficient of kinetic friction equal to 0.120, what are (a) the acceleration of the shaker and (b) the distance it slides? [**Answer: (a)** $a_x = -1.18$ m/s², **(b)** 0.560 m]

Some related homework problems: Problem 3, Problem 13

In the next Example, the system is inclined at an angle θ relative to the horizontal. As a result, the normal force responsible for the kinetic friction is *less* than the weight of the object.

EXAMPLE 6-2 MAKING A BIG SPLASH

A trained sea lion slides from rest with constant acceleration down a 3.0-m-long ramp into a pool of water. If the ramp is inclined at an angle of $\theta = 23°$ above the horizontal and the coefficient of kinetic friction between the sea lion and the ramp is 0.26, how much time does it take for the sea lion to make a splash in the pool?

PICTURE THE PROBLEM

As is usual with inclined surfaces, we choose one axis to be parallel to the surface and the other to be perpendicular to it. In our sketch, the sea lion accelerates in the positive x direction ($a_x > 0$), having started from rest, $v_{0x} = 0$. We are free to choose the initial position of the sea lion to be $x_0 = 0$. There is no motion in the y direction, and therefore $a_y = 0$. Finally, we note from the free-body diagram that $\vec{N} = N\hat{y}$, $\vec{f}_k = -\mu_k N\hat{x}$, and $\vec{W} = (mg \sin \theta)\hat{x} + (-mg \cos \theta)\hat{y}$.

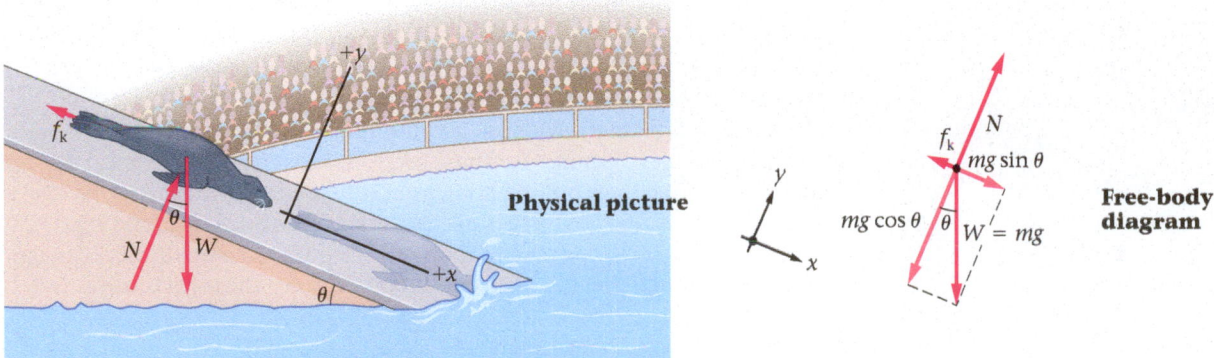

Physical picture

Free-body diagram

REASONING AND STRATEGY

We can use the kinematic equation relating position to time, $x = x_0 + v_{0x}t + \frac{1}{2}a_x t^2$, to find the time of the sea lion's slide. It will be necessary, however, to first determine the acceleration of the sea lion in the x direction, a_x.

To find a_x we apply Newton's second law to the sea lion. First, we can find N by setting $\Sigma F_y = ma_y$ equal to zero (since $a_y = 0$). It is important to start by finding N because we need it to find the force of kinetic friction, $f_k = \mu_k N$. Using f_k in the sum of forces in the x direction, $\Sigma F_x = ma_x$, allows us to solve for a_x and, finally, for the time.

Known Length of ramp, $x = 3.0$ m; angle of incline, $\theta = 23°$; coefficient of kinetic friction, $\mu_k = 0.26$.
Unknown Sliding time, $t = ?$

SOLUTION

1. We begin by resolving each of the three force vectors into x and y components:

$$N_x = 0 \qquad\qquad N_y = N$$
$$f_{k,x} = -f_k = -\mu_k N \qquad f_{k,y} = 0$$
$$W_x = mg \sin \theta \qquad W_y = -mg \cos \theta$$

2. Set $\Sigma F_y = ma_y = 0$ to find N. We see that N is less than the weight, mg:

$$\Sigma F_y = N - mg \cos \theta = ma_y = 0$$
$$N = mg \cos \theta$$

3. Next, set $\Sigma F_x = ma_x$. Notice that the mass cancels in this equation:

$$\Sigma F_x = mg \sin \theta - f_k$$
$$= mg \sin \theta - \mu_k mg \cos \theta = ma_x$$

4. Solve for the acceleration in the x direction, a_x:

$$a_x = g(\sin \theta - \mu_k \cos \theta)$$
$$= (9.81 \text{ m/s}^2)[\sin 23° - (0.26) \cos 23°]$$
$$= 1.5 \text{ m/s}^2$$

CONTINUED

5. Use $x = x_0 + v_{0x}t + \frac{1}{2}a_xt^2$ to find the time when the sea lion reaches the bottom. We choose $x_0 = 0$, and we are given that $v_{0x} = 0$; hence we set $x = \frac{1}{2}a_xt^2 = 3.0$ m and solve for t:

$$x = \frac{1}{2}a_xt^2 = 3.0 \text{ m}$$

$$t = \sqrt{\frac{2x}{a_x}} = \sqrt{\frac{2(3.0 \text{ m})}{1.5 \text{ m/s}^2}} = 2.0 \text{ s}$$

INSIGHT

We don't need the sea lion's mass to find the time. On the other hand, if we wanted the magnitude of the force of kinetic friction, $f_k = \mu_k N = \mu_k mg \cos\theta$, the mass would be needed.

It is useful to compare the sliding salt shaker in Example 6-1 with the sliding sea lion in this Example. In the case of the salt shaker, friction is the only force acting along the direction of motion (opposite to the direction of motion, in fact), and it brings the object to rest. Because of the slope on which the sea lion slides, however, it experiences both a component of its weight in the forward direction and the frictional force opposite to the motion. Because the component of the weight is the larger of the two forces, the sea lion accelerates down the slope—friction only acts to slow its progress.

PRACTICE PROBLEM

How much time would it take the sea lion to reach the water if there were no friction in this system?   [**Answer:** 1.3 s]

Some related homework problems: Problem 4, Problem 58

Static Friction

Static friction tends to keep two surfaces from moving relative to one another. Like kinetic friction, it is due to the microscopic irregularities of surfaces that are in contact. In fact, static friction is typically stronger than kinetic friction because when surfaces are in static contact, their microscopic hills and valleys nestle down deeply into one another, forming a strong connection between the surfaces. In kinetic friction, the surfaces bounce along relative to one another and don't become as firmly enmeshed. Static friction also depends on lubrication, as illustrated in **FIGURE 6-4**.

As we did with kinetic friction, let's use the results of some simple experiments to determine the rules of thumb for static friction. We start with a brick at rest on a table, with no horizontal force pulling on it, as in **FIGURE 6-5**. Of course, in this case the force of static friction is zero; no force is needed to keep the brick from sliding.

Next, we attach a spring scale to the brick and pull with a small force of magnitude F_1, a force small enough that the brick doesn't move. Since the brick is still at rest, it follows

▲ **FIGURE 6-4** The coefficient of static friction between two surfaces depends on many factors, including whether the surfaces are dry or wet. On the desert floor of Death Valley, California, occasional rains can reduce the friction between rocks and the sandy ground to such an extent that strong winds can move the rocks over considerable distances. This results in linear "rock trails," which record the direction of the winds at different times.

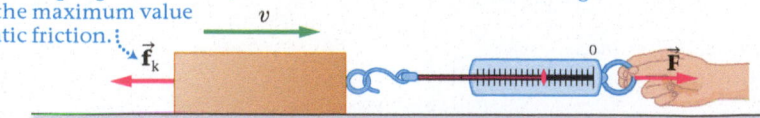

▲ **FIGURE 6-5 The maximum limit of static friction** As the force applied to an object increases, so does the force of static friction—up to a certain point. Beyond this maximum value, static friction can no longer hold the object, and it begins to slide. Now kinetic friction takes over.

that the force of static friction, f_s, is equal in magnitude to the applied force; that is, $f_s = F_1$. Now, we increase the applied force to a new value, F_2, which is still small enough that the brick stays at rest. In this case, the force of static friction has also increased so that $f_s = F_2$. If we continue increasing the applied force, we eventually reach a value beyond which the brick starts to move and kinetic friction takes over, as shown in the figure. It follows that there is a maximum force, $f_{s,max}$, that can be exerted by static friction. Thus, the force of static friction, f_s, can have any value between 0 and $f_{s,max}$:

$$0 \le f_s \le f_{s,max} \qquad\qquad 6\text{-}2$$

Imagine repeating the experiment, only now with a second brick on top of the first. This doubles the normal force and it also doubles the maximum force of static friction. Thus, the maximum force is proportional to the magnitude of the normal force; that is,

$$f_{s,max} = \mu_s N \qquad\qquad 6\text{-}3$$

The constant of proportionality is called μ_s (pronounced "mew sub s"), the **coefficient of static friction.** Notice that μ_s is dimensionless, just like μ_k. Typical values are given in Table 6-1. In most cases, μ_s is greater than μ_k, indicating that the force of static friction is greater than the force of kinetic friction. In some cases, like rubber in contact with dry concrete, μ_s is greater than 1.

Two additional experimental results regarding static friction are important: (i) Static friction, like kinetic friction, is independent of the area of contact. (ii) The force of static friction is parallel to the surface of contact, and opposite to the direction the object would move if there were no friction. All of these observations are summarized in the following rules of thumb:

Rules of Thumb for Static Friction
The force of static friction between two surfaces has the following properties:

1. It takes on any value between zero and the maximum possible force of static friction, $f_{s,max} = \mu_s N$:

$$0 \le f_s \le \mu_s N$$

2. It is independent of the area of contact between the surfaces.
3. It is parallel to the surface of contact, and in the direction that opposes relative motion.

The next Example presents a practical method of determining the coefficient of static friction in a real-world system.

EXAMPLE 6-3   SLIGHTLY TILTED

A flatbed truck slowly tilts its bed upward to dispose of a 95.0-kg crate. For small angles of tilt the crate stays put, but when the tilt angle exceeds $\theta = 23.3°$, the crate begins to slide. What is the coefficient of static friction between the bed of the truck and the crate?

PICTURE THE PROBLEM
We align our coordinate system with the incline, and choose the positive x direction to point down the slope. Three forces act on the crate: the normal force, $\vec{\mathbf{N}} = N\hat{\mathbf{y}}$, the force of static friction, $\vec{\mathbf{f}}_s = -\mu_s N\hat{\mathbf{x}}$, and the weight, $\vec{\mathbf{W}} = (mg \sin \theta)\hat{\mathbf{x}} + (-mg \cos \theta)\hat{\mathbf{y}}$.

Physical picture

Free-body diagram

CONTINUED

REASONING AND STRATEGY

When the crate is on the verge of slipping, but has not yet slipped, its acceleration is zero in both the x and y directions. In addition, "on the verge of slipping" means that the magnitude of the static friction is at its maximum value, $f_s = f_{s,max} = \mu_s N$. Thus, we set $\Sigma F_y = ma_y = 0$ to find N, and then use $\Sigma F_x = ma_x = 0$ to find μ_s.

Known   Mass of crate, $m = 95.0$ kg; critical tilt angle, $\theta = 23.2°$.
Unknown   Coefficient of static friction, $\mu_s = ?$

SOLUTION

1. Resolve the three force vectors acting on the crate into x and y components:

$$N_x = 0 \qquad\qquad N_y = N$$
$$f_{s,x} = -f_{s,max} = -\mu_s N \qquad f_{s,y} = 0$$
$$W_x = mg \sin \theta \qquad W_y = -mg \cos \theta$$

2. Set $\Sigma F_y = ma_y = 0$, since $a_y = 0$. Solve for the normal force, N:

$$\Sigma F_y = N_y + f_{s,y} + W_y = N + 0 - mg \cos \theta = ma_y = 0$$
$$N = mg \cos \theta$$

3. Set $\Sigma F_x = ma_x = 0$, since the crate is at rest, and use the result for N obtained in Step 2:

$$\Sigma F_x = N_x + f_{s,x} + W_x = ma_x = 0$$
$$= 0 - \mu_s N + mg \sin \theta$$
$$= 0 - \mu_s mg \cos \theta + mg \sin \theta$$

4. Solve the expression for the coefficient of static friction, μ_s:

$$\mu_s mg \cos \theta = mg \sin \theta$$
$$\mu_s = \frac{mg \sin \theta}{mg \cos \theta} = \tan \theta = \tan 23.2° = 0.429$$

INSIGHT

In general, if an object is on the verge of slipping when a surface is tilted at a critical angle θ_c, the coefficient of static friction between the object and the surface is $\mu_s = \tan \theta_c$. This result is independent of the mass of the object. For example, the critical angle for this crate is precisely the same whether it is filled with feathers or lead bricks. Real-world examples of the critical angle are found in hourglasses and talus slopes, as shown in **FIGURE 6-6**.

PRACTICE PROBLEM

Find the magnitude of the force of static friction acting on the crate just before it starts to slide.
[**Answer:** $f_{s,max} = \mu_s N = 367$ N]

Some related homework problems: Problem 9, Problem 11

▶ **FIGURE 6-6 Visualizing Concepts Talus Slope** The angle that the sloping sides of a sand pile (left) make with the horizontal is determined by the coefficient of static friction between the grains of sand, in much the same way that static friction determines the angle at which the crate in Example 6-3 begins to slide. The same basic mechanism determines the angle of the cone-shaped mass of rock debris at the base of a cliff—known as a talus slope (right).

Recall that static friction can have a magnitude that is less than its maximum possible value. This point is emphasized in the next Example.

QUICK EXAMPLE 6-4   THE FORCE OF STATIC FRICTION

In Example 6-3, what is the magnitude of the force of static friction acting on the crate when the truck bed is tilted at an angle of 20.0°?

REASONING AND SOLUTION

The crate is at rest, and hence its acceleration is zero in the x direction: $a_x = 0$. This condition determines the magnitude of the static friction force necessary to hold the crate in place.

1. Sum the x components of force acting on the crate: $\qquad \Sigma F_x = 0 - f_s + mg \sin \theta$

CONTINUED

2. Set this sum equal to zero (since $a_x = 0$) $f_s = mg \sin \theta$
and solve for the magnitude of the
static friction force, f_s:

3. Substitute numerical values, $f_s = (95.0 \, \text{kg})(9.81 \, \text{m/s}^2) \sin (20.0°) = 319 \, \text{N}$
including $\theta = 20.0°$:

The force of static friction in this case has a magnitude of 319 N, which is less than the
value of 367 N found in the Practice Problem in Example 6-3, even though the coefficient
of static friction is precisely the same.

Finally, friction often enters into problems that involve vehicles with rolling wheels.
In Conceptual Example 6-5, we consider which type of friction is appropriate in such cases.

CONCEPTUAL EXAMPLE 6-5   FRICTION FOR ROLLING TIRES

A car drives with its tires rolling freely. Is the friction between the tires and the road **(a)** kinetic
or **(b)** static?

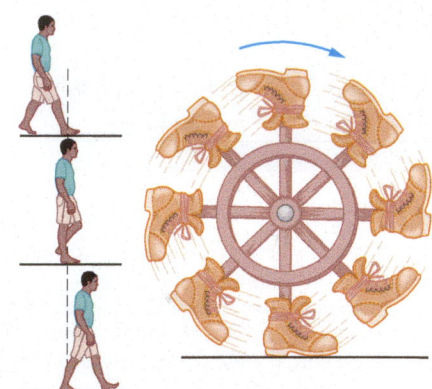

REASONING AND DISCUSSION

A reasonable-sounding answer is that because the car is moving, the friction between its
tires and the road must be kinetic friction—but this is not the case.

Actually, the friction is static because the bottom of the tire is in static contact with the
road. To understand this, watch your feet as you walk. Even though you are moving, each
foot is in static contact with the ground once you step down on it. Your foot doesn't move
again until you lift it up and move it forward for the next step. A tire can be thought of as a
succession of feet arranged in a circle, each momentarily in static contact with the ground.

ANSWER

(b) The friction between rolling tires and the road is static friction.

RWP* If a car skids, the friction acting on it is kinetic; if its wheels are rolling, the
friction is static. Static friction is generally greater than kinetic friction, however, and
hence it follows that a car can be stopped in a shorter distance if its wheels are rolling (static friction) than if its wheels are locked up (kinetic friction). This is the idea
behind the antilock braking systems (ABS) that are available on many cars. When the
brakes are applied in a car with ABS, an electronic rotation sensor at each wheel detects whether the wheel is about to start skidding. To prevent skidding, a small computer automatically begins to pulse the hydraulic pressure in the brake lines in short
bursts, causing the brakes to release and then reapply in rapid succession. This allows
the wheels to continue rotating, even in an emergency stop, and for static friction to
determine the stopping distance. **FIGURE 6-7** shows a comparison of braking distances
for cars with and without ABS. An added benefit of ABS is that a driver is better able to
steer and control a braking car if its wheels are rotating. An illustration of the danger
posed by locked wheels is presented in **FIGURE 6-8**.

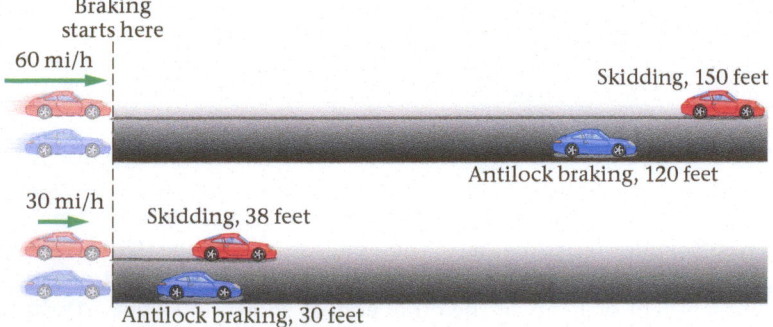

◄ **FIGURE 6-7 Stopping distance with
and without ABS** Antilock braking systems (ABS) allow a car to stop with static
friction rather than kinetic friction—even
in a case where a person slams on the
brakes. As a result, the braking distance is
reduced. Professional drivers can beat the
performance of ABS by carefully adjusting
the force they apply to the brake pedal
during a stop, but ABS provides essentially the same performance—within a few
percent—for a person who simply pushes
the brake pedal to the floor and holds it
there.

---

*Real World Physics applications are denoted by the acronym RWP.

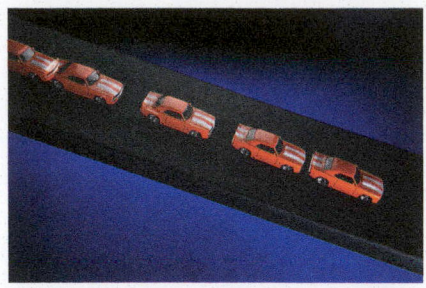

(a)

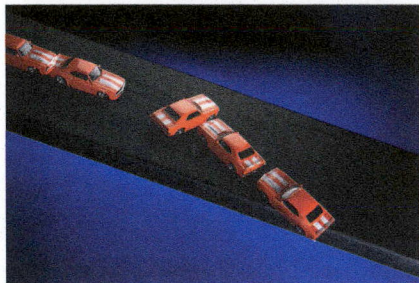

(b)

▲ **FIGURE 6-8 Visualizing Concepts Static Versus Kinetic Friction** The two time-lapse images show a toy car as it slides down an inclined surface. In **(a)** the front wheels are locked and skid on the surface, but the rear wheels roll without slipping. The front wheels experience kinetic friction and the rear wheels experience static friction. Because the force of kinetic friction is usually less than the force of static friction, the front wheels go down the incline first, pulling the rear wheels behind. The situation is reversed in **(b)**, where the rear wheels are the ones that skid and experience a smaller frictional force. As a result, the rear wheels slide down the incline more quickly than the front wheels, causing the car to spin around. This change in behavior could be dangerous in a real-life situation.

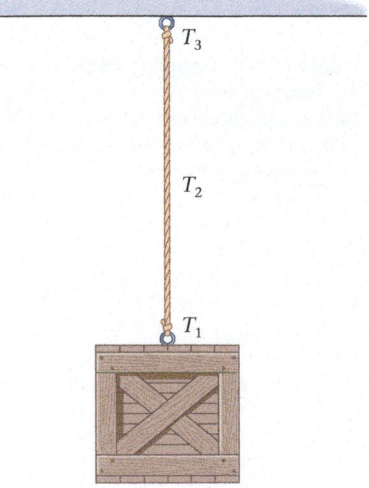

1. A block rests on a rough, horizontal surface, as shown in **FIGURE 6-9**. A force will be applied to the block along one of the four indicated directions (A, B, C, D). Rank the directions in order of increasing force required to make the block move. Indicate ties where appropriate.

▲ **FIGURE 6-9** Enhance Your Understanding 1

Section Review

- Frictional forces are due to the microscopic roughness of surfaces in contact.

- The force of kinetic friction is proportional to the magnitude of the normal force. That is, $f_k = \mu_k N$, where μ_k is the coefficient of kinetic friction.

- The maximum force of static friction is proportional to the magnitude of the normal force. That is, $f_{s,max} = \mu_s N$, where μ_s is the coefficient of static friction. The force of static friction can always be less than the maximum amount.

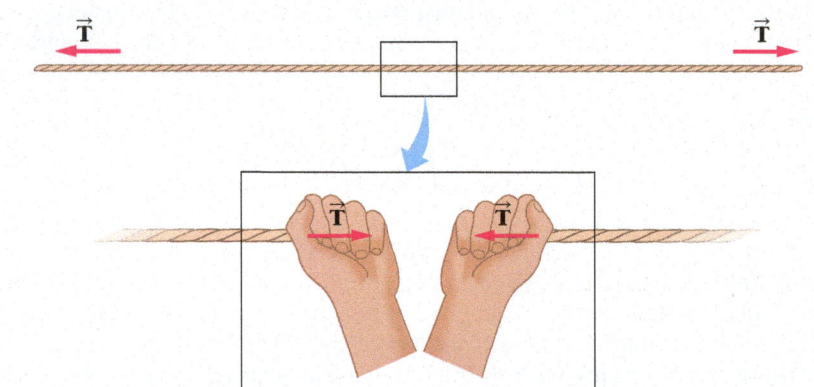

▲ **FIGURE 6-10 Tension in a string** A string, pulled from either end, has a tension, T. If the string were to be cut at any point, the force required to hold the ends together is T.

6-2 Strings and Springs

A common way to exert a force on an object is to pull on it with a string, a rope, or a wire. Similarly, you can push or pull on an object if you attach it to a spring. In this section we discuss the basic features of strings and springs and how they transmit forces.

Strings and Tension

Imagine picking up a light string and holding it with one end in each hand. If you pull to the right with your right hand with a force T and to the left with your left hand with a force T, the string becomes taut. In such a case, we say that there is a **tension** T in the string. To be more specific, if your friend were to cut the string at some point, the tension T is the force pulling the ends apart, as illustrated in **FIGURE 6-10**—that is, T is the force your friend would have to exert with each hand to hold the cut ends together. At any given point in the string, the tension pulls equally to the right and to the left.

As an example, consider a rope that is attached to the ceiling at one end, and to a box with a weight of 105 N at the other end, as shown in **FIGURE 6-11**. In addition,

◄ **FIGURE 6-11 Tension in a heavy rope** Because of the weight of the rope, the tension is noticeably different at points 1, 2, and 3. In the limit of a rope of zero mass, the tension is the same throughout the rope.

suppose that the rope is uniform and has a total weight of 2.00 N. What is the tension in the rope (i) where it attaches to the box, (ii) at its midpoint, and (iii) where it attaches to the ceiling?

First, the rope holds the box at rest; thus, the tension where the rope attaches to the box is simply the weight of the box, $T_1 = 105$ N. At the midpoint of the rope, the tension supports the weight of the box, plus half the weight of the rope. Thus, $T_2 = 105$ N $+ \frac{1}{2}(2.00$ N$) = 106$ N. Similarly, at the ceiling the tension supports the box plus all of the rope, giving a tension of $T_3 = 107$ N.

From this discussion, we can see that the tension in the rope changes slightly from top to bottom because of the mass of the rope. If the rope had less mass, the difference in tension between its two ends would also be less. In particular, if the rope's mass were to be vanishingly small, the difference in tension would vanish as well. In this text we will assume that all ropes, strings, wires, and so on are essentially massless—unless specifically stated otherwise—and hence the tension is the same throughout their length.

Pulleys Change the Direction of the Tension Pulleys are often used to redirect a force exerted by a string, as indicated in **FIGURE 6-12**. In the ideal case, a pulley has no mass and no friction in its bearings. Thus, the following statement applies:

> An ideal pulley changes the direction of the tension in a string, without changing its magnitude.

If a system contains more than one pulley, however, it's possible to arrange them in such a way as to "magnify a force," even if each pulley itself merely redirects the tension in a string. The traction device considered in the next Example shows one way this can be accomplished in a system that uses three ideal pulleys.

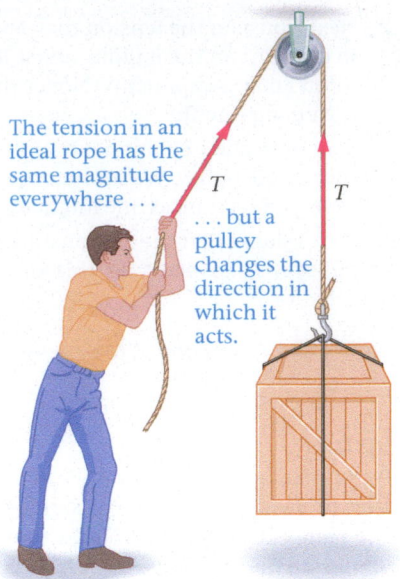

The tension in an ideal rope has the same magnitude everywhere... ...but a pulley changes the direction in which it acts.

▲ **FIGURE 6-12 A pulley changes the direction of a tension**

EXAMPLE 6-6 SETTING A BROKEN LEG WITH TRACTION

A traction device employing three pulleys is applied to a broken leg, as shown in the sketch. The middle pulley is attached to the sole of the foot, and a mass m supplies the tension in the ropes. Find the value of the mass m if the force exerted on the sole of the foot by the middle pulley is to be 165 N.

PICTURE THE PROBLEM
Our sketch shows the physical picture as well as the tension forces acting on the middle pulley. Notice that on the upper portion of the rope the tension is $\vec{T}_1 = (T\cos 40.0°)\hat{x} + (T\sin 40.0°)\hat{y}$; on the lower portion it is $\vec{T}_2 = (T\cos 40.0°)\hat{x} + (-T\sin 40.0°)\hat{y}$.

REASONING AND STRATEGY
We begin by noting that the rope supports the hanging mass m. As a result, the tension in the rope, T, must be equal in magnitude to the weight of the mass: $T = mg$.

Next, the pulleys simply change the direction of the tension without changing its magnitude. Therefore, the net force exerted on the sole of the foot is the sum of the tension T at 40.0° above the horizontal plus the tension T at 40.0° below the horizontal. We will calculate the net force component by component.

Once we calculate the net force acting on the foot, we set it equal to 165 N and solve for the tension T. Finally, we find the mass using the relationship $T = mg$.

Known Force exerted on foot, 165 N.
Unknown Suspended mass, $m = ?$

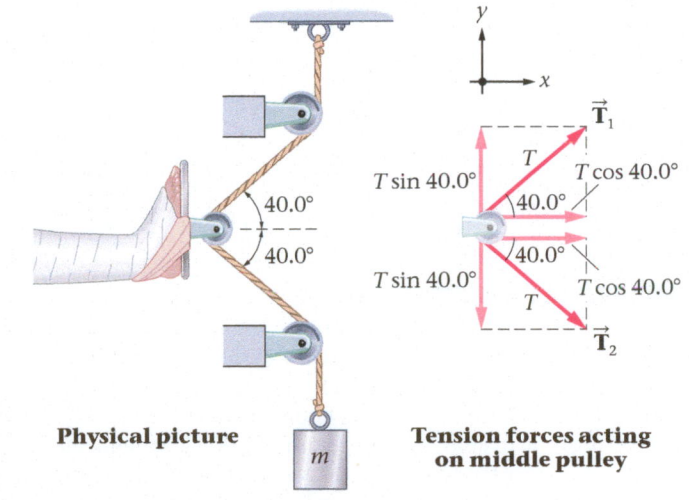

Physical picture

Tension forces acting on middle pulley

SOLUTION

1. First, consider the tension that acts upward and to the right on the middle pulley. Resolve this tension into x and y components:

$$T_{1,x} = T\cos 40.0° \qquad T_{1,y} = T\sin 40.0°$$

CONTINUED

2. Next, consider the tension that acts downward and to the right on the middle pulley. Resolve this tension into x and y components. Notice the minus sign in the y component:

$$T_{2,x} = T\cos 40.0° \qquad T_{2,y} = -T\sin 40.0°$$

3. Sum the x and y components of the force acting on the middle pulley. We see that the net force acts only in the x direction, as one might expect from symmetry:

$$\sum F_x = T\cos 40.0° + T\cos 40.0° = 2T\cos 40.0°$$
$$\sum F_y = T\sin 40.0° - T\sin 40.0° = 0$$

4. Step 3 shows that the net force acting on the middle pulley is $2T\cos 40.0°$. Set this force equal to 165 N and solve for T:

$$2T\cos 40.0° = 165 \text{ N}$$
$$T = \frac{165 \text{ N}}{2\cos 40.0°} = 108 \text{ N}$$

5. Solve for the mass, m, using $T = mg$:

$$T = mg$$
$$m = \frac{T}{g} = \frac{108 \text{ N}}{9.81 \text{ m/s}^2} = 11.0 \text{ kg}$$

INSIGHT

As pointed out earlier, this pulley arrangement "magnifies the force" in the sense that a 108-N weight attached to the rope produces a 165-N force exerted on the foot by the middle pulley. Notice that the tension in the rope always has the same value—$T = 108$ N—as expected with ideal pulleys, but because of the arrangement of the pulleys the force applied to the foot by the rope is $2T\cos 40.0° > T$.

The force exerted on the foot by the middle pulley produces an opposing force in the leg that acts in the direction of the head (a cephalad force), as desired to set a broken leg and keep it straight as it heals.

PRACTICE PROBLEM — PREDICT/CALCULATE

(a) Would the required mass m increase or decrease if the angles in this device were changed from 40.0° to 30.0°?
(b) Find the mass m for an angle of 30.0°.   [**Answer:** (a) m would decrease. (b) 9.71 kg]

Some related homework problems: Problem 26, Problem 34

CONCEPTUAL EXAMPLE 6-7   COMPARE THE READINGS ON THE SCALES

The vertical scale shown in the sketch reads 9.81 N. Is the reading of the horizontal scale (a) greater than, (b) less than, or (c) equal to 9.81 N?

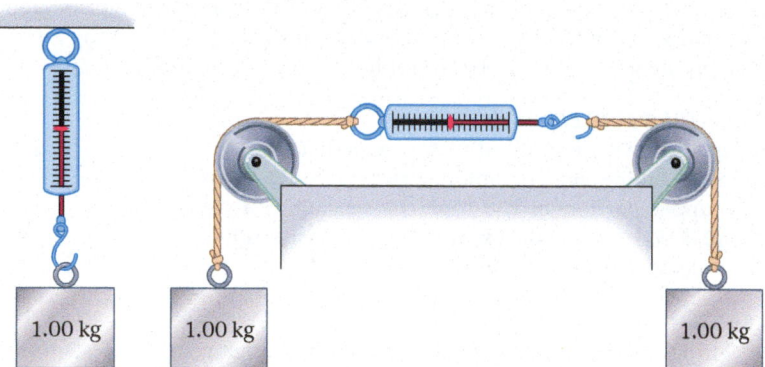

1.00 kg   1.00 kg   1.00 kg

REASONING AND DISCUSSION

Because a pulley simply changes the direction of the tension in a string without changing its magnitude, it's clear that the vertical scale attached to the ceiling reads the same as the horizontal scale shown in the sketch to the right.

There is no difference, however, between attaching the left end of the horizontal scale to something rigid or attaching it to another 1.00-kg hanging mass, as in the problem statement. In either case the scale is at rest, and hence a force of 9.81 N must be exerted to the left to balance the 9.81-N force exerted by the mass on the right end of the scale. As a result, the two horizontal scales read the same.

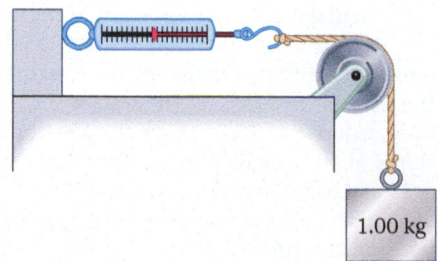

1.00 kg

ANSWER

(c) The reading of the horizontal scale is equal to 9.81 N.

Tension in String between Hanging Weights

Springs and Hooke's Law

Suppose you take a spring of length L, as shown in **FIGURE 6-13 (a)**, and attach it to a block. If you pull on the spring, causing it to stretch to a length $L + x$, the spring *pulls*

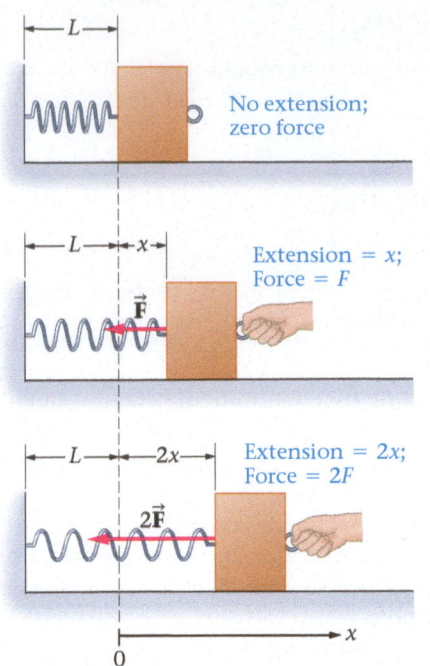

(a) Doubling the extension doubles the force.

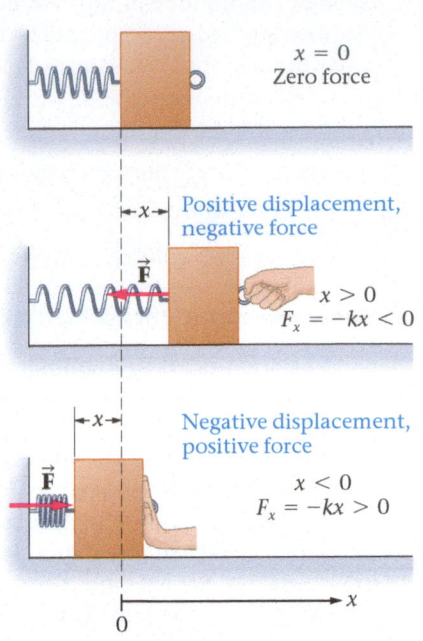

(b) The spring force is opposite to the displacement from equilibrium.

◀ **FIGURE 6-13 The force exerted by a spring** When we are dealing with a spring, it is convenient to choose the origin at the equilibrium (zero force) position.

on the block with a force of magnitude F. If you increase the length of the spring to $L + 2x$, the force exerted by the spring increases to $2F$. Similarly, if you compress the spring to a length $L - x$, the spring *pushes* on the block with a force of magnitude F, where F is the same force given previously. As you might expect, compression to a length $L - 2x$ results in a push of magnitude $2F$.

As a result of these experiments, we can say that a spring exerts a force that is proportional to the amount, x, by which it is stretched or compressed. Thus, if F is the magnitude of the spring force, it follows that

$$F = kx$$

In this expression, k is a constant of proportionality, referred to as the **force constant.** Because F has units of newtons and x has units of meters, it follows that k has units of newtons per meter, or N/m. The larger the value of k, the stiffer the spring.

To be more precise, consider the spring shown in **FIGURE 6-13 (b)**. Notice that we have placed the origin of the x axis at the equilibrium length of the spring—that is, at the position of the spring when no force acts on it. Now, if we stretch the spring so that the end of the spring is at a positive value of x ($x > 0$), we find that the spring exerts a force of magnitude kx in the negative x direction. Thus, the spring force (which has only an x component) can be written as

$$F_x = -kx$$

Similarly, consider compressing the spring so that its end is at a negative value of x ($x < 0$). In this case, the force exerted by the spring is of magnitude kx, and points in the positive x direction, as is shown in Figure 6-13 (b). Again, we can write the spring force as

$$F_x = -kx$$

To see that this is correct—that is, that F_x is positive in this case—recall that x is negative, which means that $(-x)$ is positive.

This result for the force of a spring is known as Hooke's law, after Robert Hooke (1635–1703). It's really just a good rule of thumb rather than a law of nature. Clearly, it can't work for any amount of stretching. For example, we know that if we stretch a spring far enough, it will be permanently deformed and will never return to its original length. Still, for small stretches or compressions, Hooke's law is quite accurate.

Big Idea **2** Springs exert a force proportional to the amount of stretch or compression.

PHYSICS IN CONTEXT
Looking Ahead

Our discussion of springs, and Hooke's law in particular, is important when we consider oscillations in Chapter 13.

▲ **FIGURE 6-14 Visualizing Concepts**
Springs Springs come in a variety of sizes and shapes. The large springs on a railroad car (top) are so stiff and heavy that you can't compress or stretch them by hand. Still, three of them are needed to smooth the ride of this car. In contrast, the delicate spiral spring inside a watch (bottom) flexes with even the slightest touch. It exerts enough force, however, to power the equally delicate mechanism of the watch.

Rules of Thumb for Springs (Hooke's Law)

A spring stretched or compressed by the amount x from its equilibrium length exerts a force whose x component is given by

$$F_x = -kx \; (\textit{gives magnitude and direction}) \qquad \text{6-4}$$

If we are interested only in the magnitude of the force associated with a given stretch or compression, we use the somewhat simpler form of Hooke's law:

$$F = kx \; (\textit{gives magnitude only}) \qquad \text{6-5}$$

In this text, we consider only **ideal springs**—that is, springs that are massless, and that are assumed to obey Hooke's law exactly. A variety of real-world springs is shown in Visualizing Physics in **FIGURE 6-14**.

Hooke's Law Finds Many Applications Recalling that the stretch of a spring and the force it exerts are proportional, we can see how a spring scale operates. In particular, pulling on the two ends of a scale stretches the spring inside it by an amount proportional to the applied force. Once the scale is calibrated—by stretching the spring with a known, or reference, force—we can use it to measure other unknown forces.

It's also useful to note that Hooke's law, which we've introduced in the context of ideal springs, is particularly important in physics because it applies to so much more than just springs. For example, the forces that hold atoms together are often modeled by Hooke's law—that is, as "interatomic springs"—and these are the forces that are ultimately responsible for the normal force (Chapter 5), vibrations and oscillations (Chapter 13), wave motion (Chapter 14), and even the thermal expansion of solids (Chapter 16). And this just scratches the surface—Hooke's law comes up in one form or another in virtually every field of physics, as well as chemistry, geology, and biology.

BIO Elastic Lungs For instance, your lungs obey a version of Hooke's law. The spring-like property of your lungs arises from surface tension and tissue elasticity, which assists with exhalation and opposes inhalation. The greater the volume of air that you inhale (corresponding to the spring stretch distance x), the greater the force required to do so, and the greater the force your lungs exert when you exhale. Various diseases can change the effective force constant of lungs, and hamper their function. For example, *pulmonary fibrosis* causes scarring of tissues and an increase in the effective force constant of the lungs—this makes it more difficult for a person to inhale. *Emphysema* destroys elastic tissue and results in lungs with a lower force constant—this makes it difficult for a person to exhale fully. In the following Quick Example, we present another biomedical application of Hooke's law associated with breathing.

QUICK EXAMPLE 6-8 NASAL STRIPS

BIO An increasingly popular device for improving air flow through nasal passages is the nasal strip, which consists of two flat, polyester springs enclosed by an adhesive tape covering. Measurements show that a nasal strip can exert an outward force of 0.22 N on the nose, causing it to expand by 3.6 mm. **(a)** Treating the nose as an ideal spring, find its force constant in newtons per meter. **(b)** How much force would be required to expand the nose by 4.1 mm?

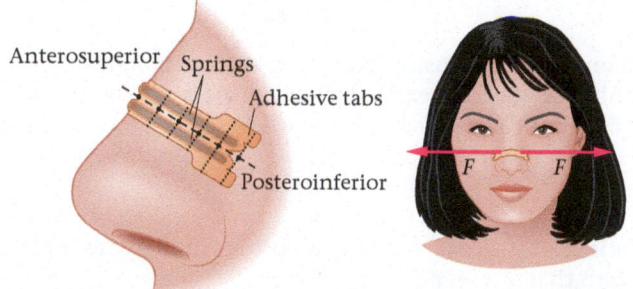

REASONING AND SOLUTION

Using Hooke's law, along with the applied force and resulting stretch, we can solve for the force constant. Once the force constant is determined, we can use it to solve for the force corresponding to any given stretch.

CONTINUED

Part (a)

1. Solve the magnitude form of Hooke's law, $F = kx$, for the force constant, k. Then substitute numerical values for F and x:

$$k = \frac{F}{x}$$
$$= \frac{0.22\text{ N}}{0.0036\text{ m}} = 61\text{ N/m}$$

Part (b)

2. Use $F = kx$ to find the required force:

$$F = kx = (61\text{ N/m})(0.0041\text{ m}) = 0.25\text{ N}$$

The human nose is certainly not an ideal spring, but Hooke's law is still a useful way to model its behavior when dealing with forces and the stretches they cause. In addition, notice the linear nature of Hooke's law—a 14% increase in the stretch distance (from 3.6 mm to 4.1 mm) requires a 14% increase in force (from 0.22 N to 0.25 N).

Enhance Your Understanding (Answers given at the end of the chapter)

2. When a mass is attached to a certain spring, the spring stretches by the amount x. Suppose we now connect two of these springs together in series, as shown in **FIGURE 6-15**. If we attach the same mass to this system, is the amount of stretch greater than, less than, or equal to x? Explain.

Section Review

- The force transmitted through a string is referred to as the tension.

- The force exerted by a spring stretched by the amount x has a magnitude $F = kx$. This is referred to as Hooke's law.

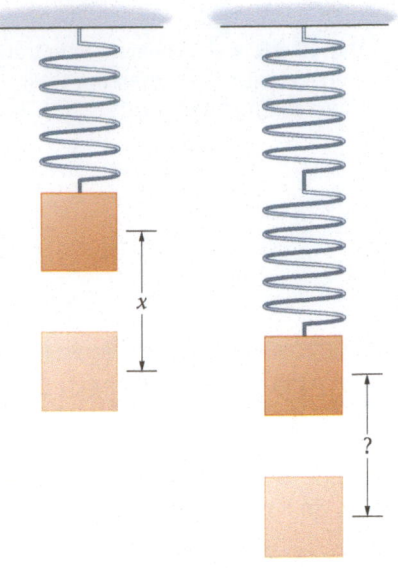

▲ **FIGURE 6-15** Enhance Your Understanding 2

6-3 Translational Equilibrium

When we say that an object is in **translational equilibrium,** we mean that the net force acting on it is zero. We state this mathematically as follows:

$$\sum \vec{\mathbf{F}} = 0 \qquad\qquad 6\text{-}6$$

From Newton's second law, this is equivalent to saying that the object's acceleration is zero. In two-dimensional systems, translational equilibrium implies two independent conditions, one for each coordinate direction: $\sum F_x = 0$ and $\sum F_y = 0$. In one dimension, only one of these conditions applies.

As a first example, consider the one-dimensional situation illustrated in **FIGURE 6-16.** Here we see a person lifting a bucket of water from a well by pulling down on a rope that passes over a pulley. If the bucket's mass is m, and it is rising with constant speed v, what is the tension T_1 in the rope attached to the bucket? In addition, what is the tension T_2 in the chain that supports the pulley?

To answer these questions, we first note that both the bucket and the pulley are in equilibrium; that is, they both have zero acceleration. As a result, the net force on each of them must be zero.

Let's start with the bucket. In Figure 6-16, we see that just two forces act on the bucket, both along the y direction: (i) its weight, $W = mg$ downward, and (ii) the tension in the rope, T_1 upward. If we take upward to be the positive direction, we can write $\sum F_y = 0$ for the bucket as follows:

$$T_1 - mg = 0$$

Therefore, the tension in the rope is $T_1 = mg$. Notice that this is also the force the person must exert downward on the rope, as expected.

Next, we consider the pulley. In Figure 6-16, we see that three forces act on it: (i) the tension in the chain, T_2 upward, (ii) the tension in the part of the rope

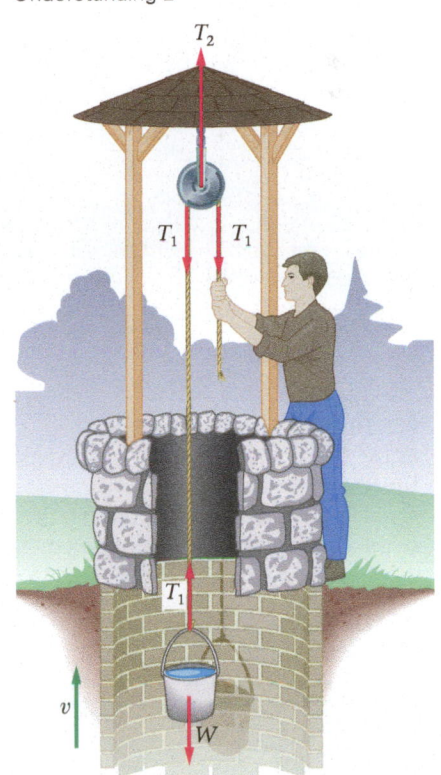

Physical picture

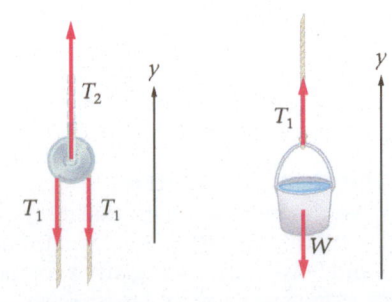

Forces acting on the pulley   **Forces acting on the bucket**

▶ **FIGURE 6-16 Raising a bucket** A person lifts a bucket of water from the bottom of a well with a constant speed, v. Because the speed is constant, the net force acting on the bucket is zero.

Big Idea 3 A system is in translational equilibrium when the total force acting on it is zero.

leading to the bucket, T_1 downward, and (iii) the tension in the part of the rope leading to the person, T_1 downward. We don't include the weight of the pulley, since we consider it to be ideal—that is, massless and frictionless. If we again take upward to be positive, the statement that the net force acting on the pulley is zero ($\Sigma F_y = 0$) can be written as

$$T_2 - T_1 - T_1 = 0$$

It follows that the tension in the chain is $T_2 = 2T_1 = 2mg$, twice the weight of the bucket of water!

In the next Conceptual Example we consider a slight variation of this situation.

CONCEPTUAL EXAMPLE 6-9 COMPARING TENSIONS

A person hoists a bucket of water from a well and holds the rope, keeping the bucket at rest, as in (1). Next, the person ties the rope to the bucket handle so the rope holds the bucket in place, as in (2). In this case, is the tension in the rope **(a)** greater than, **(b)** less than, or **(c)** equal to the tension in the first case?

(1)   (2)   (3)   (4)

REASONING AND DISCUSSION

In the first case (1), the only upward force exerted on the bucket is the tension in the rope. The bucket is at rest, and hence the tension must be equal in magnitude to the weight of the bucket (3). In the second case (2), the two ends of the rope exert equal upward forces on the bucket; therefore, the tension in the rope is only half the weight of the bucket. To see this more clearly, imagine cutting the bucket in half so that each end of the rope supports half the weight, as indicated in (4).

ANSWER

(b) The tension in the second case is less than in the first.

In the next two Examples, we consider two-dimensional systems in which forces act at various angles with respect to one another. Hence, our first step is to resolve the relevant vectors into their x and y components. Following that, we apply the conditions for translational equilibrium: $\Sigma F_x = 0$ and $\Sigma F_y = 0$.

EXAMPLE 6-10 SUSPENDED VEGETATION

To hang a 6.20-kg pot of flowers, a gardener uses two wires—one attached horizontally to a wall, the other sloping upward at an angle of $\theta = 40.0°$ and attached to the ceiling. Find the tension in each wire.

PICTURE THE PROBLEM

We choose a typical coordinate system, with the positive x direction to the right and the positive y direction upward. With this choice, tension 1 is in the positive x direction, $\vec{\mathbf{T}}_1 = T_1\hat{\mathbf{x}}$; the weight is in the negative y direction, $\vec{\mathbf{W}} = -mg\,\hat{\mathbf{y}}$; and tension 2 has a negative x component and a positive y component, $\vec{\mathbf{T}}_2 = (-T_2\cos\theta)\hat{\mathbf{x}} + (T_2\sin\theta)\hat{\mathbf{y}}$.

CONTINUED

REASONING AND STRATEGY

The pot is at rest, and therefore the net force acting on it is zero. As a result, we can say that (i) $\Sigma F_x = 0$ and (ii) $\Sigma F_y = 0$. These two conditions allow us to determine the magnitude of the two tensions, T_1 and T_2.

Known   Mass of flower pot, $m = 6.20$ kg; angle of upper wire, $\theta = 40.0°$.

Unknown   Tension in each wire: $T_1 = ?$, $T_2 = ?$

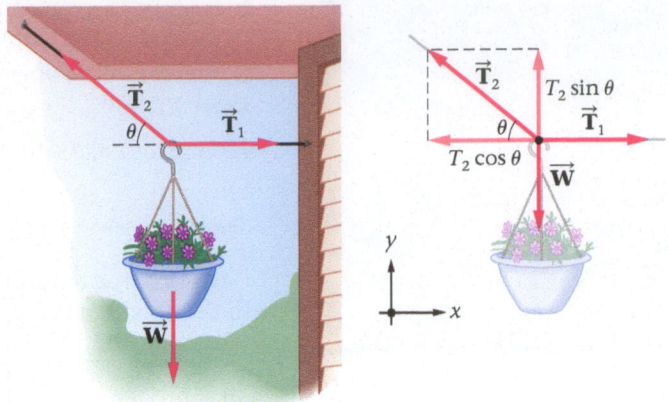

Physical picture   **Free-body diagram**

SOLUTION

1. First, resolve each of the forces acting on the pot into x and y components:

$$T_{1,x} = T_1 \qquad\qquad T_{1,y} = 0$$
$$T_{2,x} = -T_2 \cos\theta \qquad T_{2,y} = T_2 \sin\theta$$
$$W_x = 0 \qquad\qquad W_y = -mg$$

2. Now, set $\Sigma F_x = 0$. Notice that this condition gives a relationship between T_1 and T_2:

$$\Sigma F_x = T_{1,x} + T_{2,x} + W_x = T_1 + (-T_2 \cos\theta) + 0 = 0$$
$$T_1 = T_2 \cos\theta$$

3. Next, set $\Sigma F_y = 0$. This time, the resulting condition determines T_2 in terms of the weight, mg:

$$\Sigma F_y = T_{1,y} + T_{2,y} + W_y = 0 + T_2 \sin\theta + (-mg) = 0$$
$$T_2 \sin\theta = mg$$

4. Use the relationship obtained in Step 3 to find T_2:

$$T_2 = \frac{mg}{\sin\theta} = \frac{(6.20\,\text{kg})(9.81\,\text{m/s}^2)}{\sin 40.0°} = 94.6\,\text{N}$$

5. Finally, use the connection between the two tensions (obtained from $\Sigma F_x = 0$ in Step 2) to find T_1:

$$T_1 = T_2 \cos\theta = (94.6\,\text{N})\cos 40.0° = 72.5\,\text{N}$$

INSIGHT

Notice that even though two wires suspend the pot, they both have tensions *greater* than the pot's weight, $mg = 60.8$ N. This is an important point for architects and engineers to consider when designing structures.

PRACTICE PROBLEM

Find T_1 and T_2 if the second wire slopes upward at the angle **(a)** $\theta = 20°$, **(b)** $\theta = 60.0°$, or **(c)** $\theta = 90.0°$. [**Answer: (a)** $T_1 = 167$ N, $T_2 = 178$ N; **(b)** $T_1 = 35.1$ N, $T_2 = 70.2$ N; **(c)** $T_1 = 0$, $T_2 = mg = 60.8$ N]

Some related homework problems: Problem 30, Problem 32 , Problem 72

QUICK EXAMPLE 6-11   THE FORCES IN A LOW-TECH LAUNDRY

A 1.84-kg bag of clothespins hangs in the middle of a clothesline, causing it to sag at an angle $\theta = 3.50°$. Find the tension, T, in the clothesline.

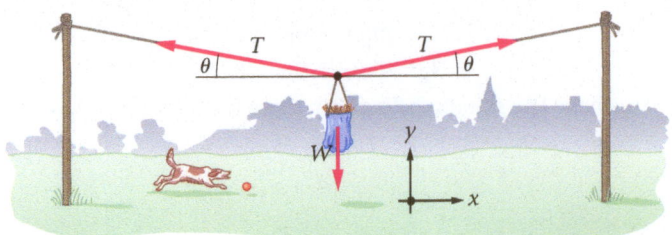

REASONING AND SOLUTION

The tension in the clothesline must result in a net vertical component of force equal in magnitude to the weight of the clothespins. The horizontal components of the tension cancel because there is no horizontal acceleration.

CONTINUED

▲ **FIGURE 6-17** Like the bag of clothespins in Quick Example 6-11, this mountain climber is in static equilibrium. Since the ropes suspending the climber are nearly horizontal, the tension in them is significantly greater than the climber's weight.

1. Find the y component for the tension on each side of the bag:

   $T_y = T \sin \theta$

2. Find the y component of the bag's weight:

   $W_y = -mg$

3. Set $\Sigma F_y = 0$:

   $T \sin \theta + T \sin \theta - mg = 0$

4. Solve for T and substitute numerical values:

   $T = \dfrac{mg}{2 \sin \theta} = \dfrac{(1.84 \text{ kg})(9.81 \text{ m/s}^2)}{2 \sin 3.50°}$

   $= 148 \text{ N}$

At 148 N the tension in the clothesline is quite large, especially when compared with the weight of the clothespin bag itself, which is only 18.1 N. The reason for such a large value is that the vertical component of the two tensions is $2T \sin \theta$, which, for $\theta = 3.50°$, is only $(0.122)T$. If $(0.122)T$ is to equal the weight of the bag, it is clear that T must be roughly eight times the bag's weight. On the other hand, if $\theta = 90°$ the tension is just $mg/2$, as expected. A similar situation involving large tensions and a rock climber is shown in **FIGURE 6-17**.

If you and a friend were to pull on the two ends of the clothesline, in an attempt to straighten it out, you would find that no matter how hard you pulled, the line would still sag. You may be able to reduce θ to quite a small value, but as you do so the corresponding tension increases rapidly. In principle, it would take an infinite force to completely straighten the line and reduce θ to zero.

Enhance Your Understanding   (Answers given at the end of the chapter)

3. Suppose the tension in the clothesline in Quick Example 6-11 is 120 N. Is the sag angle greater than, less than, or equal to 3.50°? Explain.

Section Review

- An object in translational equilibrium has zero net force acting on it. This means it won't have a translational (left/right, up/down) acceleration.

6-4 Connected Objects

Interesting applications of Newton's laws arise when we consider accelerating objects that are tied together. Suppose, for example, that a force of magnitude F pulls two boxes—connected by a string—along a frictionless surface, as in **FIGURE 6-18**. In such a case, the string has a certain tension, T, and the two boxes have the same acceleration, a. Given the masses of the boxes and the applied force F, we would like to determine both the tension in the string and the acceleration of the boxes.

First, we sketch the free-body diagram for each box. Box 1 has two horizontal forces acting on it: (i) the tension T to the left, and (ii) the force F to the right. Box 2 has only a single horizontal force, the tension T to the right. If we take the positive direction to be to the right, Newton's second law for the two boxes can be written as follows:

$$F - T = m_1 a_1 = m_1 a \quad \text{box 1}$$
$$T = m_2 a_2 = m_2 a \quad \text{box 2}$$

6-7

Because the boxes have the same acceleration, a, we have set $a_1 = a_2 = a$.

▼ **FIGURE 6-18 Two boxes connected by a string** The string ensures that the two boxes have the same acceleration. This physical connection results in a mathematical connection, as shown in Equation 6-7. In this case we treat each box as a separate system.

Box 2   Box 1

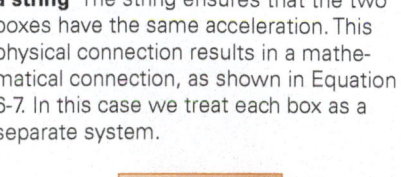

Physical picture

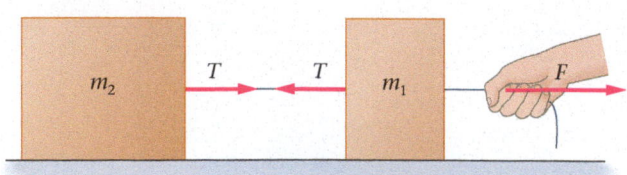

Free-body diagrams
(horizontal components only)

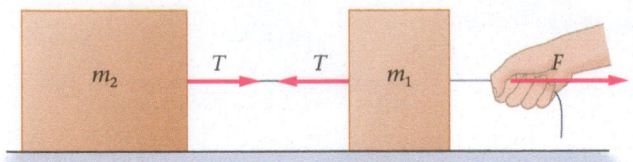

Physical picture

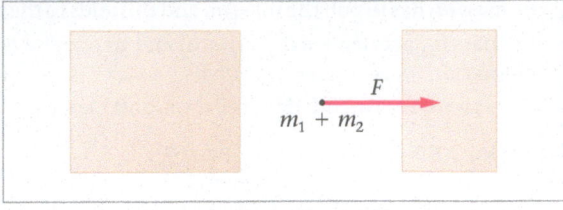

Free-body diagram
(horizontal components only)

▲ **FIGURE 6-19 Two boxes, one system** In this case we consider the two boxes together as a single system of mass $m_1 + m_2$. The only external horizontal force acting on this system is $\vec{F}$; hence the horizontal acceleration of the system is $a = F/(m_1 + m_2)$, in agreement with Equation 6-8.

Next, we can eliminate the tension T by adding the two equations:

$$F - T = m_1 a$$
$$\underline{T = m_2 a}$$
$$F = (m_1 + m_2)a$$

With this result, it is straightforward to solve for the acceleration in terms of the applied force F:

$$a = \frac{F}{m_1 + m_2} \qquad\qquad \text{6-8}$$

Finally, we substitute this expression for a into either of the original equations to find the tension. The algebra is simpler if we use the equation for box 2. We find

$$T = m_2 a = \left(\frac{m_2}{m_1 + m_2}\right)F \qquad\qquad \text{6-9}$$

It is left as an exercise to show that the equation for box 1 gives the same expression for T.

A second way to approach this problem is to treat both boxes together as a single system with a mass $m_1 + m_2$, as shown in **FIGURE 6-19**. The only *external* horizontal force acting on this system is the applied force F—the two tension forces are now *internal* to the system, and internal forces are not included when applying Newton's second law. As a result, the horizontal acceleration is simply $F/(m_1 + m_2)$, as given in Equation 6-8. This is certainly a quick way to find the acceleration a, but to find the tension T we must still use one of the relationships given in Equations 6-7.

In general, we are always free to choose the "system" any way we like. We can choose any individual object, as when we considered box 1 and box 2 separately, or we can choose all the objects together. The important point is that Newton's second law is equally valid no matter what choice we make for the system, as long as we remember to include only forces *external* to *that system* in the corresponding free-body diagram.

RWP A dramatic example of tension in a rope is depicted in the 2013 movie *Gravity*, in which two astronauts orbit the Earth while tethered by a long rope (**FIGURE 6-20**). Whenever astronaut 1 fires a jetpack to go forward, the tether became taut. The tension T in the tether pulls backward on astronaut 1 and forward on astronaut 2. The fact that the astronauts are joined by a tether demands that the acceleration of astronaut 1, $a_1 = (F_{\text{jetpack}} - T)/m_1$, be exactly the same as the acceleration of astronaut 2, $a_2 = T/m_2$. When the jetpack is turned off, the tension in the tether becomes zero and both astronauts drift at constant velocity. The movie depicts the laws of physics realistically amid the drama of running out of oxygen while attempting to travel a great distance!

▲ **FIGURE 6-20 Tethered astronauts** A dramatic scene from the movie *Gravity* illustrates the physics of objects connected by a rope.

CONCEPTUAL EXAMPLE 6-12   TENSION IN THE STRING   PREDICT/EXPLAIN

Two masses, m_1 and m_2, are connected by a string that passes over a pulley. Mass m_1 slides without friction on a horizontal tabletop, and mass m_2 falls vertically downward. Both masses move with a constant acceleration of magnitude a. **(a)** Is the tension in the string greater than, less than, or equal to the weight of block 2, $m_2 g$? **(b)** Which of the following is the *best explanation* for your prediction?

CONTINUED

I. The mass m_2 has to pull the mass m_1 and this adds to the tension, making it greater than m_2g.

II. The mass m_2 accelerates downward, and hence the tension is less than m_2g so that the net force on m_2 is downward.

III. The mass m_2 is suspended by the rope, and hence the tension is equal to m_2g.

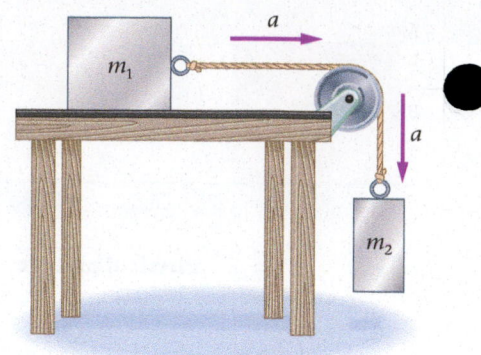

REASONING AND DISCUSSION

First, note that m_2 accelerates downward, which means that the net force acting on it is downward. Only two forces act on m_2, however: the tension in the string (upward) and its weight (downward). Because the net force is downward, the tension in the string must be less than the weight, m_2g. This is explanation II.

A common misconception is that since m_2 has to pull m_1 behind it, the tension in the string must be greater than m_2g. Certainly, attaching the string to m_1 has an effect on the tension. If the string were not attached, its tension would be zero. Hence, m_2 pulling on m_1 increases the tension to a value greater than zero, but still less than m_2g.

ANSWER

(a) The tension in the string is less than m_2g. (b) The best explanation is II.

Choose a Coordinate System That Follows the Motion In the next Example, we verify the qualitative conclusions given in Conceptual Example 6-12 with a detailed calculation. But first, a note about choosing a coordinate system for a problem such as this. Rather than apply the same coordinate system to both masses, it is useful to take into consideration the fact that a pulley simply changes the direction of the tension in a string. With this in mind, we choose a set of axes that "follow the motion" of the string, so that both masses accelerate in the positive x direction with accelerations of equal magnitude. Example 6-13 illustrates the use of this type of coordinate system.

PROBLEM-SOLVING NOTE

Choice of Coordinate System: Connected Objects

If two objects are connected by a string passing over a pulley, let the coordinate system follow the direction of the string. With this choice, both objects have accelerations of the same magnitude and in the same coordinate direction.

EXAMPLE 6-13   CONNECTED BLOCKS

A block of mass m_1 slides on a frictionless tabletop. It is connected to a string that passes over a pulley and suspends a mass m_2. Find **(a)** the acceleration of the masses and **(b)** the tension in the string.

PICTURE THE PROBLEM

Our coordinate system follows the motion of the string so that both masses move in the positive x direction. Since the masses are connected, their accelerations have the same magnitude. Thus, $a_{1,x} = a_{2,x} = a$. In addition, notice that the tension, $\vec{T}$, is in the positive x direction for mass 1, but in the negative x direction for mass 2. Its magnitude, T, is the same for each mass, however. Finally, the weight of mass 2, W_2, acts in the positive x direction, whereas the weight of mass 1 is offset by the normal force, N.

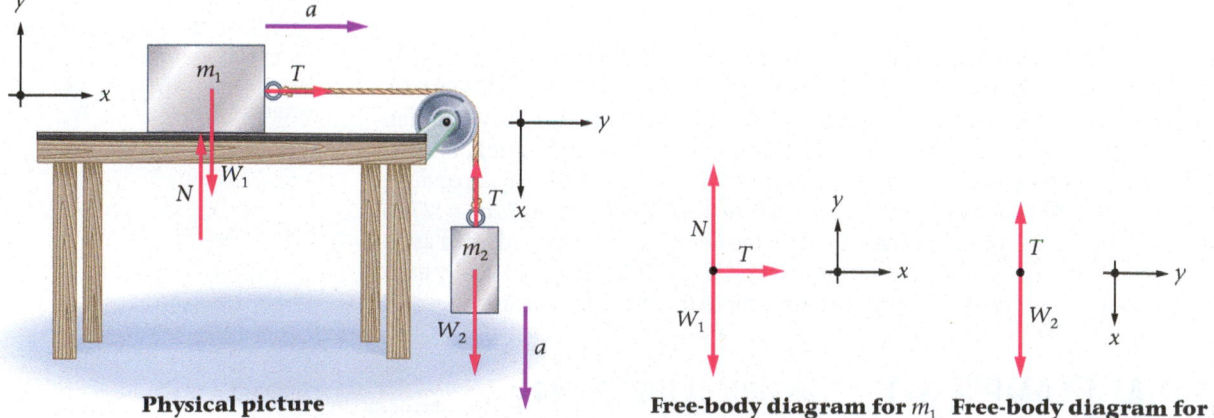

Physical picture   Free-body diagram for m_1 Free-body diagram for m_2

REASONING AND STRATEGY

Applying Newton's second law to the two masses yields the following relationships: for mass 1, $\Sigma F_{1,x} = T = m_1 a_{1,x} = m_1 a$, and for mass 2, $\Sigma F_{2,x} = m_2g - T = m_2 a_{2,x} = m_2 a$. These two equations can be solved for the two unknowns, a and T.

CONTINUED

Known   Symbolic masses of the blocks, m_1, m_2.
Unknown   (a) Acceleration of the blocks, $a = ?$ (b) Tension in the string, $T = ?$

SOLUTION

Part (a)

1. First, write $\sum F_{1,x} = m_1 a$. Note that the only force acting on m_1 in the x direction is T:

$$\sum F_{1,x} = T = m_1 a$$
$$T = m_1 a$$

2. Next, write $\sum F_{2,x} = m_2 a$. In this case, two forces act in the x direction: $W_2 = m_2 g$ (positive direction) and T (negative direction):

$$\sum F_{2,x} = m_2 g - T = m_2 a$$
$$m_2 g - T = m_2 a$$

3. Sum the two relationships obtained to eliminate T:

$$T = m_1 a$$
$$\underline{m_2 g - T = m_2 a}$$
$$m_2 g = (m_1 + m_2)a$$

4. Solve for a:

$$a = \left(\frac{m_2}{m_1 + m_2}\right)g$$

Part (b)

5. Substitute a into the first equation ($T = m_1 a$) to find T:

$$T = m_1 a = \left(\frac{m_1 m_2}{m_1 + m_2}\right)g$$

INSIGHT

We could just as well have determined T using $m_2 g - T = m_2 a$, though the algebra is a bit messier. Also, notice that $a = 0$ if $m_2 = 0$, and that $a = g$ if $m_1 = 0$, as expected. Similarly, $T = 0$ if either m_1 or m_2 is zero. This type of check, where you connect equations with physical situations, is one of the best ways to increase your understanding of physics.

PRACTICE PROBLEM

Find the acceleration and tension for the case $m_1 = 1.50$ kg and $m_2 = 0.750$ kg, and compare the tension to $m_2 g$. [**Answer:** $a = 3.27$ m/s², $T = 4.91$ N $< m_2 g = 7.36$ N]

Some related homework problems: Problem 38, Problem 41

Conceptual Example 6-12 shows that the tension in the string is less than $m_2 g$. Let's rewrite our solution for T from Example 6-13 to show that this is indeed the case. We have

$$T = \left(\frac{m_1 m_2}{m_1 + m_2}\right)g = \left(\frac{m_1}{m_1 + m_2}\right)m_2 g$$

Noting that the ratio $m_1/(m_1 + m_2)$ is always less than 1 (as long as m_2 is nonzero), it follows that $T < m_2 g$, as expected.

A Method for Measuring the Acceleration due to Gravity   We conclude this section with a classic system that can be used to measure the acceleration due to gravity. It is referred to as Atwood's machine, and it is basically two blocks of different mass connected by a string that passes over a pulley. The resulting acceleration of the blocks is related to the acceleration due to gravity by a relatively simple expression, which we derive in the following Example.

EXAMPLE 6-14   ATWOOD'S MACHINE

Atwood's machine consists of two masses connected by a string that passes over a pulley, as shown in the sketch. Find the acceleration of the masses for general m_1 and m_2, and evaluate for the specific case $m_1 = 3.1$ kg, $m_2 = 4.4$ kg.

PICTURE THE PROBLEM

Our sketch shows Atwood's machine, along with our choice of coordinate directions for the two blocks. Note that both blocks accelerate in the positive x direction with accelerations of equal magnitude, a. From the free-body diagrams we can see that for mass 1 the weight is in the negative x direction and the tension is in the positive x direction. For mass 2, the tension is in the negative x direction and the weight is in the positive x direction. The tension has the same magnitude T for both masses, but their weights are different.

CONTINUED

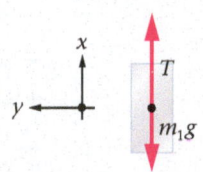
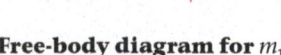
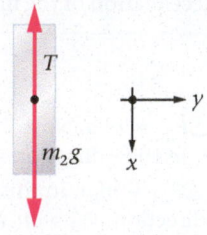

Physical picture

Free-body diagram for m_1 **Free-body diagram for** m_2

REASONING AND STRATEGY

To find the acceleration of the blocks, we follow the same strategy given in the previous Example. In particular, we start by applying Newton's second law to each block individually, using the fact that $a_{1,x} = a_{2,x} = a$. This gives two equations, both involving the tension T and the acceleration a. Eliminating T allows us to solve for the acceleration.

Known   Masses, $m_1 = 3.1$ kg, $m_2 = 4.4$ kg.
Unknown   Acceleration of the masses, $a = ?$

SOLUTION

1. Begin by writing out the expression $\sum F_{1,x} = m_1 a$.
   Note that two forces act in the x direction: T
   (positive direction) and $m_1 g$ (negative direction):

   $$\sum F_{1,x} = T - m_1 g = m_1 a$$

2. Next, write out $\sum F_{2,x} = m_2 a$. The two forces acting in
   the x direction in this case are $m_2 g$ (positive direction)
   and T (negative direction):

   $$\sum F_{2,x} = m_2 g - T = m_2 a$$

3. Sum the two relationships obtained above to eliminate T:

   $$
   \begin{aligned}
   T - m_1 g &= m_1 a \\
   m_2 g - T &= m_2 a \\
   \hline
   (m_2 - m_1)g &= (m_1 + m_2)a
   \end{aligned}
   $$

4. Solve for a:

   $$a = \left(\frac{m_2 - m_1}{m_1 + m_2}\right)g$$

5. To evaluate the acceleration, substitute numerical
   values for the masses and for g:

   $$a = \left(\frac{m_2 - m_1}{m_1 + m_2}\right)g$$
   $$= \left(\frac{4.4 \text{ kg} - 3.1 \text{ kg}}{3.1 \text{ kg} + 4.4 \text{ kg}}\right)(9.81 \text{ m/s}^2) = 1.7 \text{ m/s}^2$$

INSIGHT

Because m_2 is greater than m_1, we find that the acceleration is positive, meaning that the masses accelerate in the positive x direction. On the other hand, if m_1 were greater than m_2, we would find that a is negative, indicating that the masses accelerate in the negative x direction. Finally, if $m_1 = m_2$, we have $a = 0$, as expected.

PRACTICE PROBLEM — PREDICT/CALCULATE

(a) If m_1 is increased by a small amount, does the acceleration of the blocks increase, decrease, or stay the same? Explain. **(b)** Check your answer to part (a) by evaluating the acceleration for $m_1 = 3.3$ kg.   [**Answer: (a)** The acceleration decreases because the masses are more nearly balanced. **(b)** $a = 1.4$ m/s^2]

Some related homework problems: Problem 40, Problem 41, Problem 43

Enhance Your Understanding   (Answers given at the end of the chapter)

4. Three boxes are connected by ropes
   and pulled across a smooth, horizontal surface with an acceleration a, as
   shown in **FIGURE 6-21**. The force applied
   to the first box on the right is F, the
   tension in the rope connecting the first

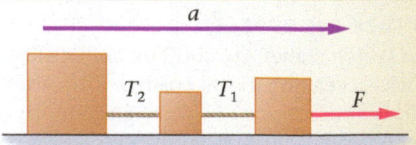

▲ **FIGURE 6-21** Enhance Your Understanding 4

CONTINUED

and second boxes is T_1, and the tension in the rope connecting the second and third boxes is T_2. Rank the three forces (F, T_1, T_2) in order of increasing magnitude. Indicate ties where appropriate.

Section Review

- Objects connected by strings have the same acceleration. This implies certain values for the tension in the strings.

Big Idea **4** When an object moves in a circular path, it accelerates toward the center of the path. As a result, circular motion requires a force directed toward the center.

6-5 Circular Motion

According to Newton's second law, if no force acts on an object, it will move with constant speed in a constant direction. A force is required to change the speed, the direction, or both. For example, if you drive a car with constant speed on a circular track, the direction of the car's motion changes continuously. A force must act on the car to cause this change in direction. We would like to know two things about a force that causes circular motion: (i) What is its direction, and (ii) What is its magnitude?

First, let's consider the direction of the force. Imagine swinging a ball tied to a string in a circle above your head, as shown in **FIGURE 6-22**. As you swing the ball, you feel a tension in the string pulling outward. Of course, on the other end of the string, where it attaches to the ball, the tension pulls inward, toward the center of the circle. Thus, the force the ball experiences is a force that is always directed toward the center of the circle. In summary:

> To make an object move in a circle with constant speed, a force must act on it that is directed toward the center of the circle.

Because the ball is acted on by a *force* toward the center of the circle, it follows that it must be *accelerating* toward the center of the circle. This might seem odd at first: How can a ball that moves with constant speed have an acceleration? The answer is that acceleration is produced whenever the speed or direction of the velocity changes—and in circular motion, the direction changes continuously. The resulting center-directed acceleration is called **centripetal acceleration** (centripetal is from the Latin for "center seeking").

Calculating the Centripetal Acceleration   Let's calculate the magnitude of the centripetal acceleration, a_{cp}, for an object moving with a constant speed v in a circle of radius r. **FIGURE 6-23** shows the circular path of an object, with the center of the circle at the origin. To calculate the acceleration at the top of the circle, at point P, we first calculate the average acceleration from point 1 to point 2:

$$\vec{\mathbf{a}}_{av} = \frac{\Delta \vec{\mathbf{v}}}{\Delta t} = \frac{\vec{\mathbf{v}}_2 - \vec{\mathbf{v}}_1}{\Delta t} \qquad 6\text{-}10$$

The instantaneous acceleration at P is the limit of $\vec{\mathbf{a}}_{av}$ as points 1 and 2 move closer to P.

Referring to Figure 6-23, we see that $\vec{\mathbf{v}}_1$ is at an angle θ above the horizontal, and $\vec{\mathbf{v}}_2$ is at an angle θ below the horizontal. Both $\vec{\mathbf{v}}_1$ and $\vec{\mathbf{v}}_2$ have a magnitude v. Therefore, we can write the two velocities in vector form as follows:

$$\vec{\mathbf{v}}_1 = (v \cos \theta)\hat{\mathbf{x}} + (v \sin \theta)\hat{\mathbf{y}}$$
$$\vec{\mathbf{v}}_2 = (v \cos \theta)\hat{\mathbf{x}} + (-v \sin \theta)\hat{\mathbf{y}}$$

Substituting these results into $\vec{\mathbf{a}}_{av}$ gives

$$\vec{\mathbf{a}}_{av} = \frac{\vec{\mathbf{v}}_2 - \vec{\mathbf{v}}_1}{\Delta t} = \frac{-2v \sin \theta}{\Delta t}\hat{\mathbf{y}} \qquad 6\text{-}11$$

Notice that $\vec{\mathbf{a}}_{av}$ points in the negative y direction—which, at point P, is toward the center of the circle.

To complete the calculation, we need to know the time, Δt, it takes the object to go from point 1 to point 2. Since the object's speed is v, and the distance from point

▲ **FIGURE 6-22 Swinging a ball in a circle** The tension in the string pulls outward on the person's hand and pulls inward on the ball.

Ball Leaves Circular Track

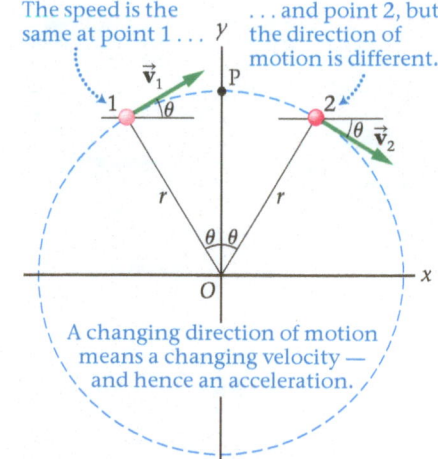

▲ **FIGURE 6-23 A particle moving with constant speed in a circular path centered on the origin** The speed of the particle is constant, but its velocity is constantly changing direction. Because the velocity changes, the particle is accelerating.

TABLE 6-2 $\dfrac{\sin\theta}{\theta}$ **for Values of θ Approaching Zero**

θ, radians	$\dfrac{\sin\theta}{\theta}$
1.00	0.841
0.500	0.959
0.250	0.990
0.125	0.997
0.0625	0.999

PHYSICS IN CONTEXT
Looking Back

The derivation of the direction and magnitude of centripetal acceleration makes extensive use of our knowledge of vectors from Chapter 3, and especially how to resolve vectors into components.

PHYSICS IN CONTEXT
Looking Ahead

Circular motion comes up again in a number of physical systems, especially when we consider gravitational orbital motion in Chapter 12 and the Bohr model of the hydrogen atom in Chapter 31.

PROBLEM-SOLVING NOTE

Choice of Coordinate System: Circular Motion

In circular motion, it is convenient to choose the coordinate system so that one axis points toward the center of the circle. Then, we know that the acceleration in that direction must be $a_{cp} = v^2/r$.

1 to point 2 is $d = r(2\theta)$, where θ is measured in radians (see Appendix, page A-2 for a discussion of radians and degrees), we find

$$\Delta t = \frac{d}{v} = \frac{2r\theta}{v} \qquad \text{6-12}$$

Combining this result for Δt with the previous result for $\vec{a}_{av}$ gives

$$\vec{a}_{av} = \frac{-2v\sin\theta}{(2r\theta/v)}\hat{y} = -\frac{v^2}{r}\left(\frac{\sin\theta}{\theta}\right)\hat{y} \qquad \text{6-13}$$

To find $\vec{a}$ at point P, we let points 1 and 2 approach P, which means letting θ go to zero. Table 6-2 shows that as θ goes to zero ($\theta \rightarrow 0$), the ratio $(\sin\theta)/\theta$ goes to 1:

$$\lim_{\theta\to0}\frac{\sin\theta}{\theta} = 1$$

It follows, then, that the instantaneous acceleration at point P is

$$\vec{a} = -\frac{v^2}{r}\hat{y} = -a_{cp}\hat{y} \qquad \text{6-14}$$

As mentioned, the direction of the acceleration is toward the center of the circle, and now we see that its magnitude is

$$a_{cp} = \frac{v^2}{r} \qquad \text{6-15}$$

We can summarize these results as follows:

- When an object moves in a circular path of radius r with constant speed v, its centripetal acceleration has a magnitude given by $a_{cp} = v^2/r$.

- A centripetal force must be applied to an object to give it circular motion. For an object of mass m, the net force acting on it must have a magnitude given by

$$f_{cp} = ma_{cp} = m\frac{v^2}{r} \qquad \text{6-16}$$

This force must be directed toward the center of the circle.

The **centripetal force,** f_{cp}, can be produced in any number of ways. For example, f_{cp} might be the tension in a string, as in the example with the ball, or it might be due to friction between tires and the road, as when a car turns a corner. In addition, f_{cp} could be the force of gravity that causes a satellite, or the Moon, to orbit the Earth. Thus, f_{cp} is a force that must be present to cause circular motion, but the specific cause of f_{cp} varies from system to system. The centripetal force in a carnival ride is illustrated in **FIGURE 6-24**.

We now show how these results for centripetal force and centripetal acceleration can be applied in practice.

◄ **FIGURE 6-24** The people enjoying this carnival ride are experiencing a centripetal acceleration of roughly 10 m/s² directed inward, toward the axis of rotation. The force needed to produce this acceleration, which keeps the riders moving in a circular path, is provided by the horizontal component of the tension in the chains.

EXAMPLE 6-15   ROUNDING A CORNER

A 1200-kg car rounds a corner of radius $r = 45$ m. If the coefficient of static friction between the tires and the road is $\mu_s = 0.82$, what is the greatest speed the car can have in the corner without skidding?

PICTURE THE PROBLEM

In the first sketch we show a bird's-eye view of the car as it moves along its circular path. The next sketch shows the car moving directly toward the observer. Notice that we have chosen the positive x direction to point toward the center of the circular path, and the positive y axis to point vertically upward. We also indicate the three forces acting on the car: gravity, $\vec{W} = -W\hat{y} = -mg\hat{y}$; the normal force, $\vec{N} = N\hat{y}$; and the force of static friction, $\vec{f}_s = \mu_s N\hat{x}$.

CONTINUED

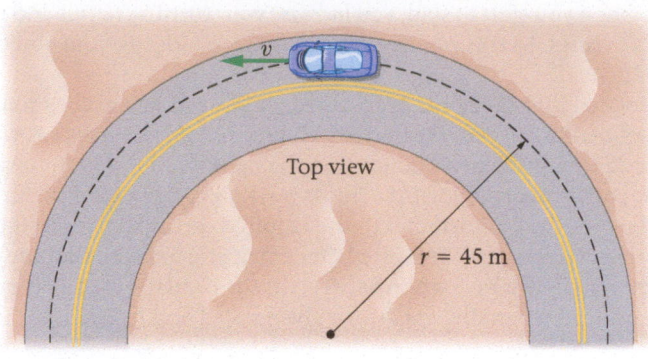

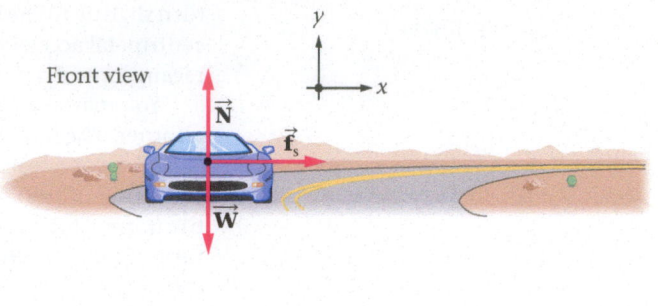

REASONING AND STRATEGY

In this system, the force of static friction provides the centripetal force required for the car to move in a circular path. That is why the force of friction is at right angles to the car's direction of motion; it is directed toward the center of the circle. In addition, the friction in this case is static because the car's tires are rolling without slipping—always making static contact with the ground. Finally, if the car moves faster, more centripetal force (i.e., more friction) is required. Thus, the greatest speed for the car corresponds to the maximum static friction, $f_s = \mu_s N$. Hence, by setting $\mu_s N$ equal to the centripetal force, $ma_{cp} = mv^2/r$, we can solve for v.

Known Mass of car, $m = 1200$ kg; corner radius, $r = 45$ m; coefficient of static friction, $\mu_s = 0.82$.
Unknown Maximum speed without skidding, $v = ?$

SOLUTION

1. Sum the x components of force to relate the force of static friction to the centripetal acceleration of the car:

$$\sum F_x = f_s = ma_x$$

2. The car moves in a circular path, with the center of the circle in the x direction, and hence it follows that $a_x = a_{cp} = v^2/r$. Make this substitution, along with $f_s = \mu_s N$ for the force of static friction:

$$\mu_s N = ma_{cp} = m\frac{v^2}{r}$$

3. Next, set the sum of the y components of force equal to zero, since $a_y = 0$:

$$\sum F_y = N - W = ma_y = 0$$

4. Solve for the normal force:

$$N = W = mg$$

5. Substitute the result $N = mg$ in Step 2 and solve for v. Notice that the mass of the car cancels:

$$\mu_s mg = m\frac{v^2}{r}$$
$$v = \sqrt{\mu_s rg}$$

6. Substitute numerical values to determine v:

$$v = \sqrt{(0.82)(45 \text{ m})(9.81 \text{ m/s}^2)} = 19 \text{ m/s}$$

INSIGHT

The maximum speed is less if the radius is smaller (tighter corner) or if μ_s is smaller (slick road). The mass of the vehicle, however, is irrelevant. For example, the maximum speed is precisely the same for a motorcycle rounding this corner as it is for a large, heavily loaded truck.

PRACTICE PROBLEM — PREDICT/CALCULATE

Suppose the situation described in this Example takes place on the Moon, where the acceleration due to gravity is less than it is on Earth. **(a)** If a lunar rover goes around this same corner, is its maximum speed greater than, less than, or the same as the result found in Step 4? Explain. **(b)** Find the maximum speed for a lunar rover when it rounds a corner with $r = 45$ m and $\mu_s = 0.82$. (On the Moon, $g = 1.62$ m/s^2.) [Answer: **(a)** The maximum speed is less, because gravity holds the rover to the road with a smaller force. **(b)** On the Moon we find $v = 7.7$ m/s, considerably less than the maximum speed of 19 m/s on Earth.]

Some related homework problems: Problem 47, Problem 74

RWP If you try to round a corner too rapidly, you may experience a skid; that is, your car may begin to slide sideways across the road. A common bit of road wisdom is that you should turn in the direction of the skid to regain control—which, to most people, seems counterintuitive. The advice is sound, however. Suppose you are turning to the left and begin to skid to the right. If you turn more sharply to the left to try to correct for the skid, you simply reduce the turning radius of your car, r. The result is that the centripetal acceleration, v^2/r, becomes larger, and an even larger force is required from the road to make the turn. The tendency to skid is therefore increased. On the other hand, if you turn slightly to the right when you start to skid, you *increase* your turning radius and the centripetal acceleration *decreases*. In this case your car may stop skidding, and you can regain control of your vehicle.

You may also have noticed that many roads are tilted, or banked, when they round a corner. The same type of banking is observed on many automobile racetracks as well. Next time you drive around a banked curve, notice that the banking tilts you in toward the center of the circular path you are following. This is by design. On a banked curve, the normal force exerted by the road contributes to the required centripetal force. If the tilt angle is just right, the normal force provides all of the centripetal force, so that the car can negotiate the curve even if there is no friction between its tires and the road. The next Example determines the optimum banking angle for a given speed and radius of turn.

EXAMPLE 6-16   BANK ON IT

If a roadway is banked at the proper angle, a car can round a corner without any assistance from friction between the tires and the road. Find the appropriate banking angle for a 900-kg car traveling at 20.5 m/s in a turn of radius 85.0 m.

PICTURE THE PROBLEM

We choose the positive y axis to point vertically upward and the positive x direction to point toward the center of the circular path. Since $\vec{N}$ is perpendicular to the banked roadway, it is at an angle θ to the y axis. Therefore, $\vec{N} = (N\sin\theta)\hat{x} + (N\cos\theta)\hat{y}$ and $\vec{W} = -W\hat{y} = -mg\hat{y}$.

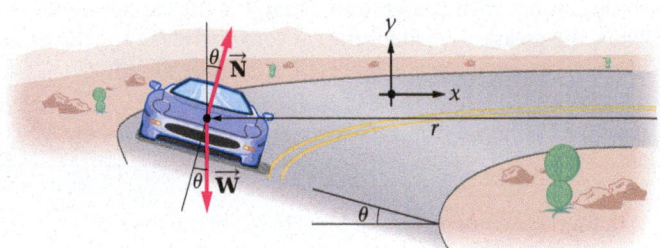

REASONING AND STRATEGY

In order for the car to move in a circular path, there must be a force acting on it in the positive x direction. Because the weight $\vec{W}$ has no x component, it follows that the normal force $\vec{N}$ must supply the needed centripetal force. Thus, we find N by setting $\Sigma F_y = ma_y = 0$, since there is no motion in the y direction. Then we use N in $\Sigma F_x = ma_x = mv^2/r$ to find the angle θ.

Known   Mass of car, m = 900 kg; speed, v = 20.5 m/s; turn radius, r = 85.0 m.
Unknown   Optimum banking angle, θ = ?

SOLUTION

1. Start by determining N from the condition $\Sigma F_y = 0$:

$$\Sigma F_y = N\cos\theta - W = 0$$
$$N = \frac{W}{\cos\theta} = \frac{mg}{\cos\theta}$$

2. Next, set $\Sigma F_x = mv^2/r$:

$$\Sigma F_x = N\sin\theta$$
$$= ma_x = ma_{cp} = m\frac{v^2}{r}$$

3. Substitute $N = mg/\cos\theta$ (from $\Sigma F_y = 0$, Step 1) and solve for θ, using the fact that $\sin\theta/\cos\theta = \tan\theta$. Notice that, once again, the mass of the car cancels:

$$N\sin\theta = \frac{mg}{\cos\theta}\sin\theta = m\frac{v^2}{r}$$
$$\tan\theta = \frac{v^2}{gr} \quad \text{or} \quad \theta = \tan^{-1}\left(\frac{v^2}{gr}\right)$$

4. Substitute numerical values to determine θ:

$$\theta = \tan^{-1}\left[\frac{(20.5 \text{ m/s})^2}{(9.81 \text{ m/s}^2)(85.0 \text{ m})}\right] = 26.7°$$

INSIGHT

The symbolic result in Step 3 shows that the banking angle increases with increasing speed and decreasing radius of turn, as one would expect.

From the point of view of a passenger, the experience of rounding a properly banked corner is basically the same as riding on a level road—there are no "sideways forces" to make the turn uncomfortable. There is one small difference, however: The passenger feels heavier due to the increased normal force. The Visualizing Concepts feature in **FIGURE 6-25** shows a variety of situations where banking in a turn is useful.

PRACTICE PROBLEM

A turn of radius 65 m is banked at 30.0°. What speed should a car have in order to make the turn with no assistance from friction?   [**Answer:** v = 19 m/s]

Some related homework problems: Problem 49, Problem 78

If you've ever driven through a dip in the road, you know that you feel momentarily heavier near the bottom of the dip. This change in apparent weight is due to the approximately circular motion of the car, as we show next.

QUICK EXAMPLE 6-17   NORMAL FORCE IN A DIP

Predict/Calculate

While driving along a country lane with a constant speed of 17.0 m/s you encounter a dip in the road. The dip can be approximated as a circular arc, with a radius of 65.1 m. **(a)** At the bottom of the dip, is the magnitude of the normal force exerted by a car seat on a passenger in the car greater than, less than, or equal to the passenger's weight? Explain. **(b)** What is the normal force exerted by a car seat on an 80.0-kg passenger when the car is at the bottom of the dip?

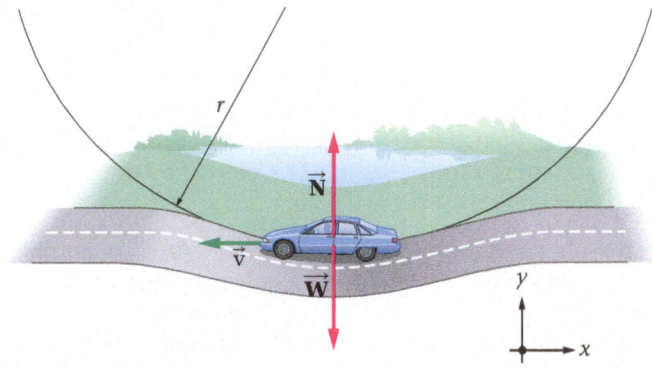

REASONING AND SOLUTION

Part (a) At the bottom of the dip, the normal force is greater than the weight of the passenger. This is because the normal force must not only counter the weight of the passenger, but also provide the centripetal force toward the center of the circular path needed for circular motion.

Part (b) At the bottom of the dip the passenger is moving along a circular path and accelerating vertically upward—with an acceleration of v^2/r toward the center of the circle. Set the sum of the vertical forces equal to the mass times the vertical acceleration of the passenger.

1. Write $\sum F_y = ma_y$ for the passenger: $N - mg = ma_y$

2. Replace a_y with the centripetal acceleration: $a_y = v^2/r$

3. Solve for N: $N = mg + mv^2/r$

4. Substitute numerical values: $N = (80.0\text{ kg})(9.81\text{ m/s}^2) + (80.0\text{ kg})(17.0\text{ m/s})^2/(65.1\text{ m})$
$$= 785\text{ N} + 355\text{ N}$$
$$= 1140\text{ N}$$

The same physics applies to a jet pilot who pulls a plane out of a high-speed dive. In that case, the magnitude of the effect can be much larger, resulting in a decrease of blood flow to the brain and eventually to loss of consciousness. Here's a case where basic physics really can be a matter of life and death!

A similar calculation can be applied to a car going over the top of a bump. In that case, circular motion results in a reduced apparent weight. With just the right speed you can even experience momentary "weightlessness."

RWP Next, we determine the acceleration produced in a **centrifuge,** a common device in biological and medical laboratories that uses large centripetal accelerations to perform such tasks as separating red and white blood cells from serum. A simplified top view of a centrifuge is shown in **FIGURE 6-26**, and a laboratory centrifuge is shown in **FIGURE 6-27**.

▲ **FIGURE 6-25 Visualizing Concepts**
Banked Turns The steeply banked track at the Talladega Superspeedway in Alabama (top) helps to keep the rapidly moving cars from skidding off along a tangential path. Even when there is no solid roadway, however, banking can still help. Airplanes bank when making turns (center) to keep from "skidding" sideways. Banking is beneficial in another way as well. Occupants of cars on a banked roadway or of a banking airplane feel no sideways force when the banking angle is just right, so turns become a safer and more comfortable experience. For this reason, some trains use hydraulic suspension systems to bank when rounding corners (bottom), even though the tracks themselves are level.

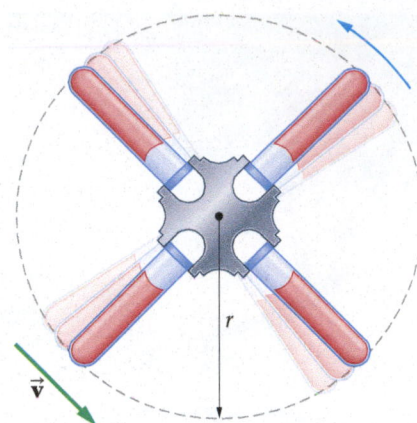

▲ **FIGURE 6-26 Simplified top view of a centrifuge in operation**

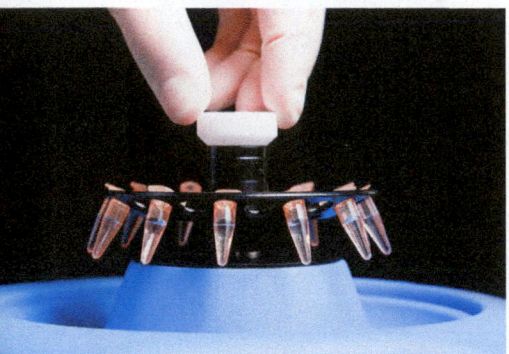

▲ **FIGURE 6-27** A laboratory centrifuge of the kind commonly used to separate blood components.

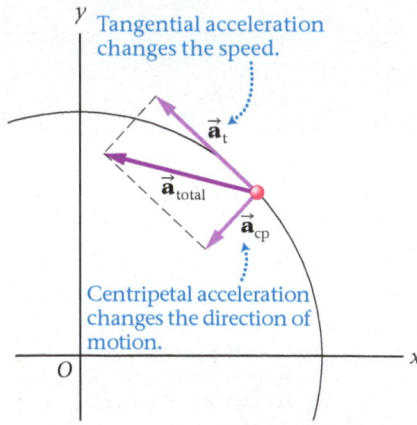

◀ **FIGURE 6-28 A particle moving in a circular path with tangential acceleration** In this case, the particle's speed is increasing at the rate given by a_t.

EXERCISE 6-18   CENTRIFUGE ACCELERATION

The centrifuge in Figure 6-26 rotates at a rate that gives the bottom of the test tube a linear speed of 90.6 m/s. If the bottom of the test tube is 9.25 cm from the axis of rotation, what is the centripetal acceleration experienced there?

REASONING AND SOLUTION

Applying the relationship $a_{cp} = v^2/r$ yields

$$a_{cp} = \frac{v^2}{r} = \frac{(90.6\ \text{m/s})^2}{0.0925\ \text{m}} = 88{,}700\ \text{m/s}^2 = 9040\,g$$

In this expression, g is the acceleration due to gravity, $9.81\ \text{m/s}^2$.

Thus, a centrifuge can produce centripetal accelerations that are many thousand times greater than the acceleration due to gravity. In fact, devices referred to as **ultracentrifuges** can produce accelerations as great as 1 million g. Even in the relatively modest case considered in Exercise 6-18, the forces involved in a centrifuge can be quite significant. For example, if the contents of the test tube have a relatively modest mass of 12.0 g, the centripetal force that must be exerted by the bottom of the tube is $(0.0120\ \text{kg})(9040\,g) = 1064\ \text{N}$, or about 239 lb!

Circular Motion with Variable Speed   Supppose the speed of an object moving in a circular path increases or decreases. In such a case, the object has both an acceleration tangential to its path that changes its speed, $\vec{a}_t$, and a centripetal acceleration perpendicular to its path that changes its direction of motion, $\vec{a}_{cp}$. Such a situation is illustrated in **FIGURE 6-28**. The total acceleration of the object is the vector sum of $\vec{a}_t$ and $\vec{a}_{cp}$. We will explore this case more fully in Chapter 10.

Enhance Your Understanding   (Answers given at the end of the chapter)

5. A system consists of an object with mass m and speed v that moves on a circular path with a radius of curvature r. Rank the following systems (A, B, C, D) in order of increasing centripetal force required to cause the motion. Indicate ties where appropriate. A: $m = 1$ kg, $v = 5$ m/s, $r = 5$ m; B: $m = 6$ kg, $v = 1$ m/s, $r = 2$ m; C: $m = 6$ kg, $v = 2$ m/s, $r = 2$ m; D: $m = 10$ kg, $v = 2$ m/s, $r = 5$ m.

Section Review

*   An object moving along a circular path of radius r with a speed v experiences a centripetal (cp) acceleration toward the center of the circle of magnitude $a_{cp} = v^2/r$.

*   The centripetal (cp) force that must be exerted on an object of mass m to make it move in a circular path of radius r with a speed v is $f_{cp} = ma_{cp} = mv^2/r$.

CHAPTER 6 REVIEW

CHAPTER SUMMARY

6-1   FRICTIONAL FORCES

Frictional rforces are due to the microscopic roughness of surfaces in contact. As a rule of thumb, friction is independent of the area of contact and the relative speed of the surfaces.

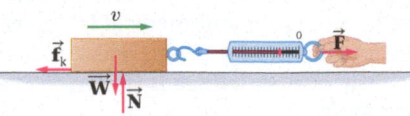

Kinetic Friction

Kinetic friction is experienced by surfaces that are in contact and moving relative to one another. The force of kinetic friction is given by

$$f_k = \mu_k N \qquad\qquad 6\text{-}1$$

In this expression, μ_k is the coefficient of kinetic friction and N is the magnitude of the normal force.

Static Friction

Static friction is experienced by surfaces that are in static contact. The maximum force of static friction is given by

$$f_{s,max} = \mu_s N \qquad\qquad 6\text{-}3$$

In this expression, μ_s is the coefficient of static friction and N is the magnitude of the normal force. The force of static friction can have any magnitude between zero and its maximum value.

6-2   STRINGS AND SPRINGS

Strings and springs provide a common way of exerting forces on objects. Ideal strings and springs are massless.

Tension

Tension is the force transmitted through a string. The tension is the same throughout the length of an ideal string.

Hooke's Law

Hooke's law states that the force exerted by an ideal spring stretched by the amount x is proportional to the stretch. Specifically,

$$F_x = -kx \qquad\qquad 6\text{-}4$$

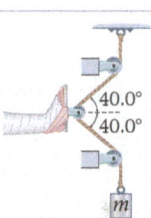

6-3   TRANSLATIONAL EQUILIBRIUM

An object is in translational equilibrium if the net force acting on it is zero.

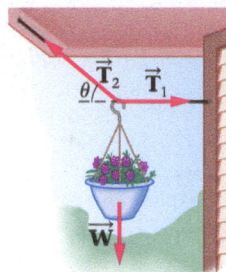

6-4   CONNECTED OBJECTS

Connected objects are linked physically and mathematically. For example, objects connected by strings have the same magnitude of acceleration.

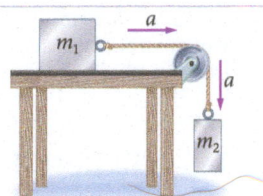

6-5   CIRCULAR MOTION

An object moving with speed v in a circle of radius r has a centripetal acceleration of magnitude $a_{cp} = v^2/r$ directed toward the center of the circle. If the object has a mass m, the force required for the circular motion is

$$f_{cp} = ma_{cp} = mv^2/r \qquad\qquad 6\text{-}16$$

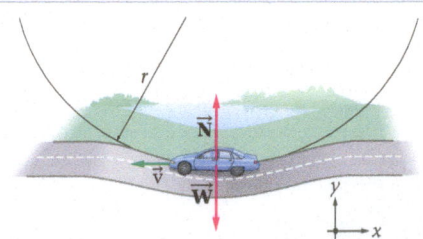

ANSWERS TO ENHANCE YOUR UNDERSTANDING QUESTIONS

1. $A < B = D < C$.

2. Greater than x. Each of the springs stretches by the amount x, and hence the total stretch is $2x$.

3. The sag angle is greater, because with a smaller tension ($120\ N$ versus $148\ N$) a larger vertical component is required to support the weight. This in turn requires a larger angle.

4. $T_2 < T_1 < F$.

5. $B < A < D < C$.

CONCEPTUAL QUESTIONS

For instructor-assigned homework, go to www.masteringphysics.com. MP

(Answers to odd-numbered Conceptual Questions can be found in the back of the book.)

1. A clothesline always sags a little, even if nothing hangs from it. Explain.

2. In the *Jurassic Park* sequel, *The Lost World*, a man tries to keep a large vehicle from going over a cliff by connecting a cable from his Jeep to the vehicle. The man then puts the Jeep in gear and spins the rear wheels. Do you expect that spinning the tires will increase the force exerted by the Jeep on the vehicle? Why or why not?

3. When a traffic accident is investigated, it is common for the length of the skid marks to be measured. How could this information be used to estimate the initial speed of the vehicle that left the skid marks?

4. In a car with rear-wheel drive, the maximum acceleration is often less than the maximum deceleration. Why?

5. A train typically requires a much greater distance to come to rest, for a given initial speed, than does a car. Why?

6. Give some everyday examples of situations in which friction is beneficial.

7. At the local farm, you buy a flat of strawberries and place them on the backseat of the car. On the way home, you begin to brake as you approach a stop sign. At first the strawberries stay put, but as you brake a bit harder, they begin to slide off the seat. Explain.

8. It is possible to spin a bucket of water in a vertical circle and have none of the water spill when the bucket is upside down. How would you explain this to members of your family?

9. Water sprays off a rapidly turning bicycle wheel. Why?

10. Can an object be in translational equilibrium if it is moving? Explain.

11. In a dramatic circus act, a motorcyclist drives his bike around the inside of a vertical circle. How is this possible, considering that the motorcycle is upside down at the top of the circle?

12. The gravitational attraction of the Earth is only slightly less at the altitude of an orbiting spacecraft than it is on the Earth's surface. Why is it, then, that astronauts feel weightless?

13. A popular carnival ride has passengers stand with their backs against the inside wall of a cylinder. As the cylinder begins to spin, the passengers feel as if they are being pushed against the wall. Explain.

14. Referring to Question 13, after the cylinder reaches operating speed, the floor is lowered away, leaving the passengers "stuck" to the wall. Explain.

15. Your car is stuck on an icy side street. Some students on their way to class see your predicament and help out by sitting on the trunk of your car to increase its traction. Why does this help?

16. The parking brake on a car causes the rear wheels to lock up. What would be the likely consequence of applying the parking brake in a car that is in rapid motion? (*Note*: Do *not* try this at home.)

17. **BIO** The foot of your average gecko is covered with billions of tiny hair tips—called spatulae—that are made of keratin, the protein found in human hair. A subtle shift of the electron distribution in both the spatulae and the wall to which a gecko clings produces an adhesive force by means of the van der Waals interaction between molecules. Suppose a gecko uses its spatulae to cling to a vertical windowpane. If you were to describe this situation in terms of a coefficient of static friction, μ_s, what value would you assign to μ_s? Is this a sensible way to model the gecko's feat? Explain.

18. Discuss the physics involved in the spin cycle of a washing machine. In particular, how is circular motion related to the removal of water from the clothes?

19. The gas pedal and the brake pedal are capable of causing a car to accelerate. Can the steering wheel also produce an acceleration? Explain.

20. In the movie *2001: A Space Odyssey*, a rotating space station provides "artificial gravity" for its inhabitants. How does this work?

21. When rounding a corner on a bicycle or a motorcycle, the driver leans inward, toward the center of the circle. Why?

PROBLEMS AND CONCEPTUAL EXERCISES

Answers to odd-numbered Problems and Conceptual Exercises can be found in the back of the book. **BIO** *identifies problems of biological or medical interest;* **CE** *indicates a conceptual exercise.* **Predict/Explain** *problems ask for two responses: (a) your prediction of a physical outcome, and (b) the best explanation among three provided; and* **Predict/Calculate** *problems ask for a prediction of a physical outcome, based on fundamental physics concepts, and follow that with a numerical calculation to verify the prediction. On all problems, bullets* (•, ••, •••) *indicate the level of difficulty.*

SECTION 6-1 FRICTIONAL FORCES

1. • **CE** **Predict/Explain** You push two identical bricks across a table-top with constant speed, v, as shown in **FIGURE 6-29**. In case 1, you place the bricks end to end; in case 2, you stack the bricks one on top of the other. **(a)** Is the force of kinetic friction in case 1 greater than, less than, or equal to the force of kinetic friction in case 2? **(b)** Choose the *best explanation* from among the following:

   I. The normal force in case 2 is larger, and hence the bricks press down more firmly against the tabletop.

   II. The normal force is the same in the two cases, and friction is independent of surface area.

   III. Case 1 has more surface area in contact with the tabletop, and this leads to more friction.

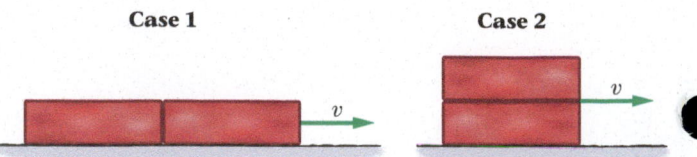

Case 1　　　　　　　**Case 2**

FIGURE 6-29 Problem 1

2. • **CE Predict/Explain** Two drivers traveling side-by-side at the same speed suddenly see a deer in the road ahead of them and begin braking. Driver 1 stops by locking up his brakes and screeching to a halt; driver 2 stops by applying her brakes just to the verge of locking, so that the wheels continue to turn until her car comes to a complete stop. **(a)** All other factors being equal, is the stopping distance of driver 1 greater than, less than, or equal to the stopping distance of driver 2? **(b)** Choose the *best explanation* from among the following:
  I. Locking up the brakes gives the greatest possible braking force.
  II. The same tires on the same road result in the same force of friction.
  III. Locked-up brakes lead to sliding (kinetic) friction, which is less than rolling (static) friction.

3. • A 1.8-kg block slides on a horizontal surface with a speed of $v = 0.80$ m/s and an acceleration of magnitude $a = 2.8$ m/s², as shown in **FIGURE 6-30**. **(a)** What is the coefficient of kinetic friction between the block and the surface? **(b)** When the speed of the block slows to 0.40 m/s, is the magnitude of the acceleration greater than, less than, or equal to 2.8 m/s²? Explain.

FIGURE 6-30 Problem 3

4. • A child goes down a playground slide with an acceleration of 1.26 m/s². Find the coefficient of kinetic friction between the child and the slide if the slide is inclined at an angle of 33.0° below the horizontal.

5. • What is the minimum horizontal force F needed to make the box start moving in **FIGURE 6-31**? The coefficients of kinetic and static friction between the box and the floor are 0.27 and 0.38, respectively.

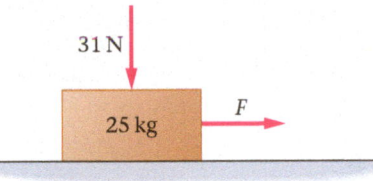

FIGURE 6-31 Problem 5

6. •• What is the minimum horizontal force F needed to make the box start moving in **FIGURE 6-32**? The coefficients of kinetic and static friction between the box and the floor are 0.24 and 0.41, respectively.

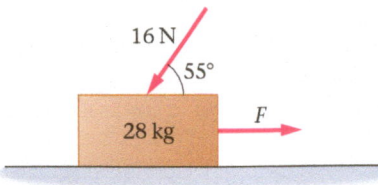

FIGURE 6-32 Problem 6

7. •• **CE** The three identical boxes shown in **FIGURE 6-33** remain at rest on a rough, horizontal surface, even though they are acted on by two different forces, $\vec{F}_1$ and $\vec{F}_2$. All of the forces labeled $\vec{F}_1$ have the same magnitude; all of the forces labeled $\vec{F}_2$ are identical to one another. Rank the boxes in order of increasing magnitude of the force of static friction between them and the surface. Indicate ties where appropriate.

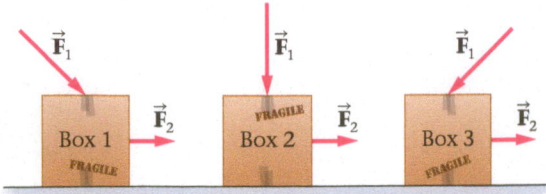

FIGURE 6-33 Problem 7

8. •• To move a large crate across a rough floor, you push on it with a force F at an angle of 21° below the horizontal, as shown in **FIGURE 6-34**. Find the force necessary to start the crate moving, given that the mass of the crate is 32 kg and the coefficient of static friction between the crate and the floor is 0.57.

FIGURE 6-34 Problem 8

9. •• **Predict/Calculate** A 37-kg crate is placed on an inclined ramp. When the angle the ramp makes with the horizontal is increased to 23°, the crate begins to slide downward. **(a)** What is the coefficient of static friction between the crate and the ramp? **(b)** At what angle does the crate begin to slide if its mass is doubled?

10. •• **Coffee To Go** A person places a cup of coffee on the roof of his car while he dashes back into the house for a forgotten item. When he returns to the car, he hops in and takes off with the coffee cup still on the roof. **(a)** If the coefficient of static friction between the coffee cup and the roof of the car is 0.24, what is the maximum acceleration the car can have without causing the cup to slide? Ignore the effects of air resistance. **(b)** What is the smallest amount of time in which the person can accelerate the car from rest to 15 m/s and still keep the coffee cup on the roof?

11. •• A mug rests on an inclined surface, as shown in **FIGURE 6-35**. **(a)** What is the magnitude of the frictional force exerted on the mug? **(b)** What is the minimum coefficient of static friction required to keep the mug from sliding?

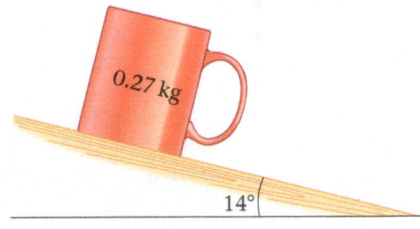

FIGURE 6-35 Problem 11

12. •• **Predict/Calculate Force Times Distance** At the local hockey rink, a puck with a mass of 0.12 kg is given an initial speed of $v = 5.3$ m/s. **(a)** If the coefficient of kinetic friction between the ice and the puck is 0.11, what distance d does the puck slide before coming to rest? **(b)** If the mass of the puck is doubled, does the frictional force F exerted on the puck increase, decrease, or stay the same? Explain. **(c)** Does the stopping distance of the puck increase, decrease, or stay the same when its mass is doubled? Explain. **(d)** For the situation considered in part (a), show that $Fd = \frac{1}{2}mv^2$. (In Chapter 7 we will see $\frac{1}{2}mv^2$ is the kinetic energy of an object.)

13. ••• **Predict/Calculate** The coefficient of kinetic friction between the tires of your car and the roadway is μ. **(a)** If your initial speed is v and you lock your tires during braking, how far do you skid? Give your answer in terms of v, μ, and m, the mass of your car. **(b)** If you double your speed, what happens to the stopping distance? **(c)** What is the stopping distance for a truck with twice the mass of your car, assuming the same initial speed and coefficient of kinetic friction?

SECTION 6-2 STRINGS AND SPRINGS

14. • **CE** A certain spring has a force constant k. **(a)** If this spring is cut in half, does the resulting half spring have a force constant that is greater than, less than, or equal to k? **(b)** If two of the original full-length springs are connected end to end, does the resulting double spring have a force constant that is greater than, less than, or equal to k?

15. • **CE** A certain spring has a force constant k. **(a)** If two such identical springs are connected in parallel to each other, as shown in part (a) of **FIGURE 6-36**, does the resulting double spring have an effective force constant that is greater than, less than, or equal to k? Explain. **(b)** If two of the original springs are connected as shown in part (b) of Figure 6-36, does the resulting double spring have an effective force constant that is greater than, less than, or equal to k? Explain.

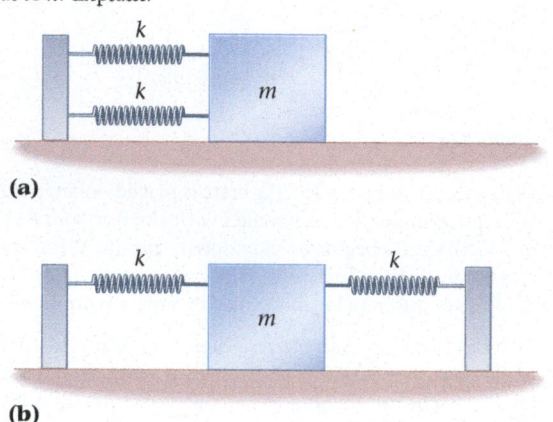

(a)

(b)

FIGURE 6-36 Problem 15

16. • Pulling up on a rope, you lift a 7.27-kg bucket of water from a well with an acceleration of 1.24 m/s^2. What is the tension in the rope?

17. • When a 9.09-kg mass is placed on top of a vertical spring, the spring compresses 4.18 cm. Find the force constant of the spring.

18. • **Predict/Calculate** A backpack full of books weighing 52.0 N rests on a table in a physics laboratory classroom. A spring with a force constant of 150 N/m is attached to the backpack and pulled horizontally, as indicated in **FIGURE 6-37**. **(a)** If the spring is pulled until it stretches 2.00 cm and the pack remains at rest, what is the force of friction exerted on the backpack by the table? **(b)** Does your answer to part (a) change if the mass of the backpack is doubled? Explain.

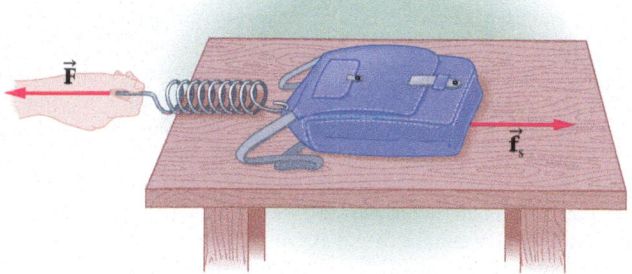

FIGURE 6-37 Problem 18

19. •• Two springs, with force constants $k_1 = 150 \text{ N/m}$ and $k_2 = 250 \text{ N/m}$, are connected in series, as shown in **FIGURE 6-38**. When a mass $m = 0.90$ kg is attached to the springs, what is the amount of stretch, x?

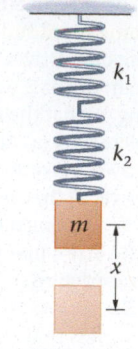

FIGURE 6-38 Problem 19

20. •• **Predict/Calculate** Illinois Jones is being pulled from a snake pit with a rope that breaks if the tension in it exceeds 755 N. **(a)** If Illinois Jones has a mass of 70.0 kg and the snake pit is 3.40 m deep, what is the minimum time that is required to pull our intrepid explorer from the pit? **(b)** Explain why the rope breaks if Jones is pulled from the pit in less time than that calculated in part (a).

21. •• **Predict/Calculate** A spring with a force constant of 120 N/m is used to push a 0.27-kg block of wood against a wall, as shown in **FIGURE 6-39**. **(a)** Find the minimum compression of the spring needed to keep the block from falling, given that the coefficient of static friction between the block and the wall is 0.46. **(b)** Does your answer to part (a) change if the mass of the block of wood is doubled? Explain.

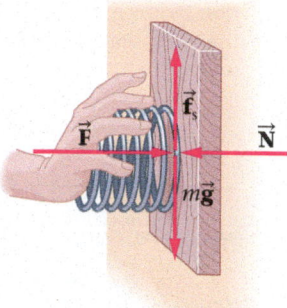

FIGURE 6-39 Problem 21

22. •• A spring is suspended vertically from the ceiling of a room. When a mass is attached to the bottom of the spring, the amount of stretch is recorded. A plot of stretch versus mass is given in **FIGURE 6-40**. Use the data given in the plot to determine the force constant of the spring.

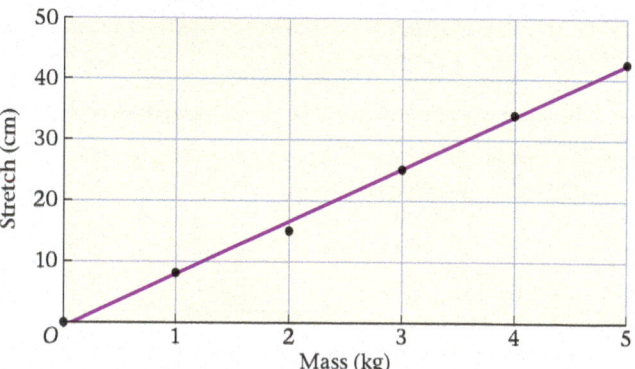

FIGURE 6-40 Problem 22

23. •• **Mechanical Advantage** The pulley system shown in **FIGURE 6-41** is used to lift a $m = 67$-kg crate. Note that one chain connects the upper pulley to the ceiling and a second chain connects the lower pulley to the crate. Assuming the masses of the chains, pulleys, and ropes are negligible, determine **(a)** the force $\vec{F}$ required to lift the crate with constant speed, **(b)** the tension in the upper chain, and **(c)** the tension in the lower chain.

FIGURE 6-41 Problem 23

SECTION 6-3 TRANSLATIONAL EQUILIBRIUM

24. • Pulling the string on a bow back with a force of 28.7 lb, an archer prepares to shoot an arrow. If the archer pulls in the center of the string, and the angle between the two halves is 138°, what is the tension in the string?

25. • In **FIGURE 6-42** we see two blocks connected by a string and tied to a wall. The mass of the lower block is 1.0 kg; the mass of the upper block is 2.0 kg. Given that the angle of the incline is 31°, find the tensions in **(a)** the string connecting the two blocks and **(b)** the string that is tied to the wall.

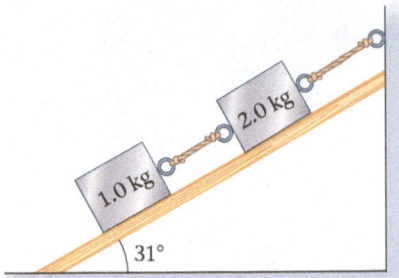

FIGURE 6-42 Problem 25

26. • **BIO Traction** After a skiing accident, your leg is in a cast and supported in a traction device, as shown in **FIGURE 6-43**. Find the magnitude of the force $\vec{F}$ exerted by the leg on the small pulley. (By Newton's third law, the small pulley exerts an equal and opposite force on the leg.) Let the mass m be 2.27 kg.

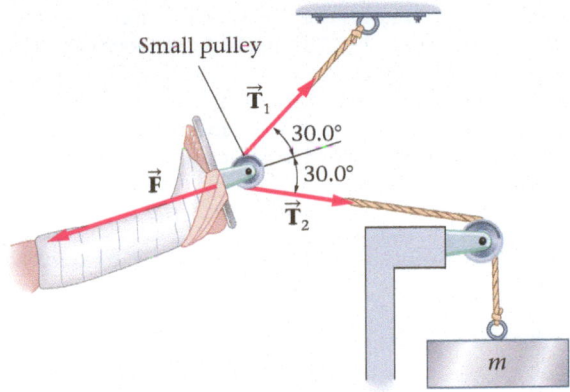

FIGURE 6-43 Problem 26

27. • Two blocks are connected by a string, as shown in **FIGURE 6-44**. The smooth inclined surface makes an angle of 42° with the horizontal, and the block on the incline has a mass of 6.7 kg. Find the mass of the hanging block that will cause the system to be in equilibrium. (The pulley is assumed to be ideal.)

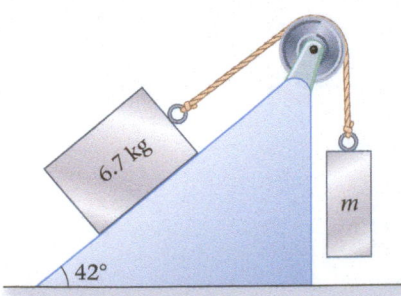

FIGURE 6-44 Problem 27

28. • **Predict/Calculate** The system shown in **FIGURE 6-45** consists of two pulleys and two blocks. The pulleys each have a mass of 2.0 kg, the block on the left has a mass of 16 kg, and the block on the right has a mass m. The system is in equilibrium, and the string is massless. **(a)** Is the mass m greater than, less than, or equal to 16 kg? Explain. **(b)** Determine the mass m.

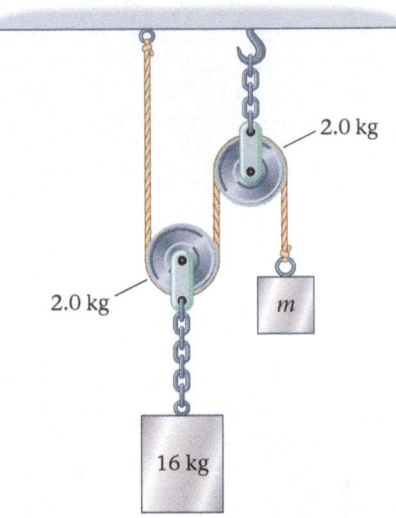

FIGURE 6-45 Problem 28

29. • **CE Predict/Explain** **(a)** Referring to the hanging planter in Example 6-10, which of the three graphs (A, B, or C) in **FIGURE 6-46** shows an accurate plot of the tensions T_1 and T_2 as a function of the angle θ? **(b)** Choose the *best explanation* from among the following:

  I. The two tensions must be equal at some angle between $\theta = 0$ and $\theta = 90°$.

  II. T_2 is greater than T_1 at all angles, and is equal to mg at $\theta = 90°$.

  III. T_2 is less than T_1 at all angles, and is equal to 0 at $\theta = 90°$.

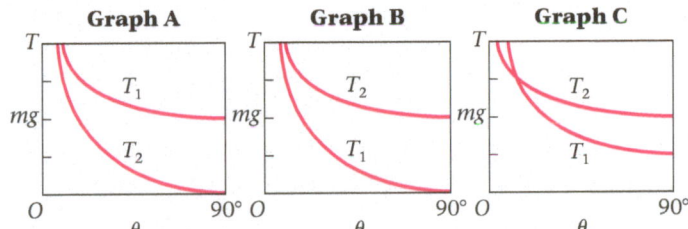

FIGURE 6-46 Problem 29

30. •• **BIO Spiderweb Forces** An orb-weaver spider sits in her web that is supported by three radial strands, as shown in **FIGURE 6-47**. Assume that only the radial strands contribute to supporting the weight of the spider. If the mass of the spider is 5.2×10^{-4} kg, and the radial strands are all under the same tension, find the magnitude of the tension, T.

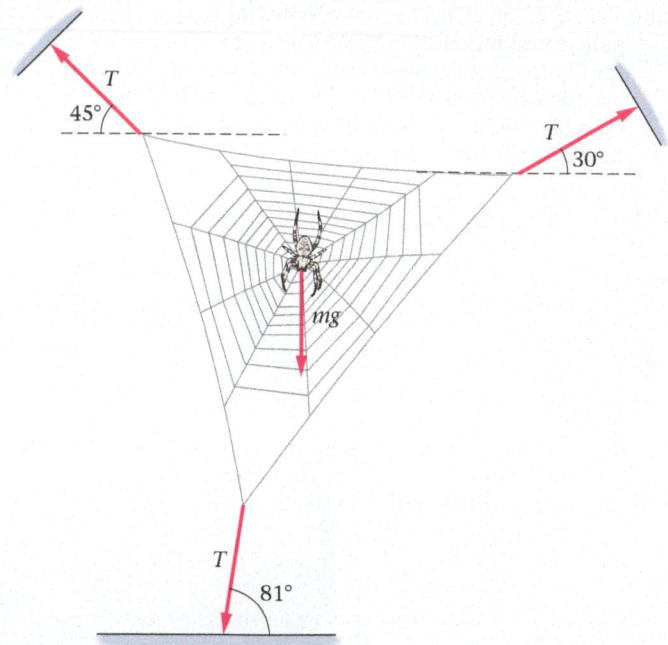

FIGURE 6-47 Problem 30

31. •• A 0.15-kg ball is placed in a shallow wedge with an opening angle of 120°, as shown in **FIGURE 6-48**. For each contact point between the wedge and the ball, determine the force exerted on the ball. Assume the system is frictionless.

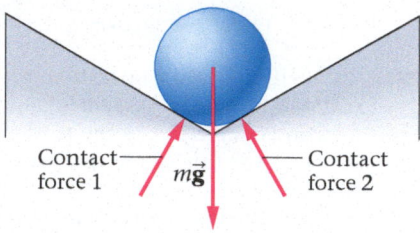

FIGURE 6-48 Problem 31

32. •• **Predict/Calculate** A picture hangs on the wall suspended by two strings, as shown in **FIGURE 6-49**. The tension in string 1 is 1.7 N. **(a)** Is the tension in string 2 greater than, less than, or equal to 1.7 N? Explain. **(b)** Verify your answer to part (a) by calculating the tension in string 2. **(c)** What is the weight of the picture?

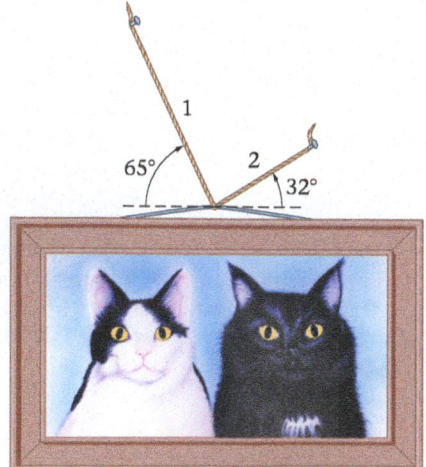

FIGURE 6-49 Problem 32

33. •• **Predict/Calculate** You want to nail a 1.6-kg board onto the wall of a barn. To position the board before nailing, you push it against

the wall with a horizontal force $\vec{\mathbf{F}}$ to keep it from sliding to the ground (**FIGURE 6-50**). **(a)** If the coefficient of static friction between the board and the wall is 0.79, what is the least force you can apply and still hold the board in place? **(b)** What happens to the force of static friction if you push against the wall with a force greater than that found in part (a)?

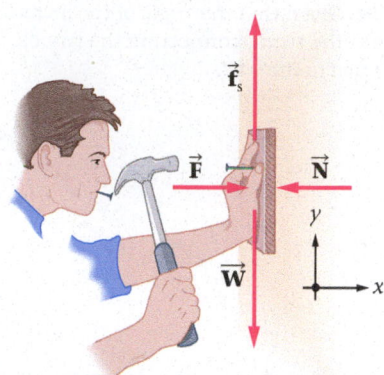

FIGURE 6-50 Problem 33

34. ••• **BIO The Russell Traction System** To immobilize a fractured femur (the thigh bone), doctors often utilize the Russell traction system illustrated in **FIGURE 6-51**. Notice that one force is applied directly to the knee, $\vec{\mathbf{F}}_1$, while two other forces, $\vec{\mathbf{F}}_2$ and $\vec{\mathbf{F}}_3$, are applied to the foot. The latter two forces combine to give a force $\vec{\mathbf{F}}_2 + \vec{\mathbf{F}}_3$ that is transmitted through the lower leg to the knee. The result is that the knee experiences the total force $\vec{\mathbf{F}}_{\text{total}} = \vec{\mathbf{F}}_1 + \vec{\mathbf{F}}_2 + \vec{\mathbf{F}}_3$. The goal of this traction system is to have $\vec{\mathbf{F}}_{\text{total}}$ directly in line with the fractured femur, at an angle of 20.0° above the horizontal. Find **(a)** the angle θ required to produce this alignment of $\vec{\mathbf{F}}_{\text{total}}$ and **(b)** the magnitude of the force, $\vec{\mathbf{F}}_{\text{total}}$, that is applied to the femur in this case. (Assume the pulleys are ideal.)

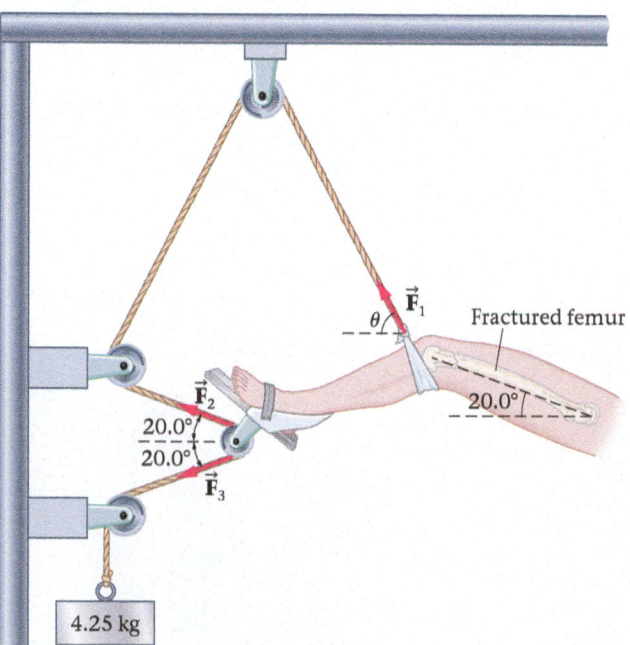

FIGURE 6-51 Problem 34

SECTION 6-4 CONNECTED OBJECTS

35. • **CE** In Example 6-13 (Connected Blocks), suppose m_1 and m_2 are both increased by a factor of 2. **(a)** Does the acceleration of the blocks increase, decrease, or stay the same? **(b)** Does the tension in the string increase, decrease, or stay the same?

36. • **CE** **Predict/Explain** Suppose m_1 and m_2 in Example 6-14 (Atwood's Machine) are both increased by 1 kg. Does the acceleration of the blocks increase, decrease, or stay the same? **(b)** Choose the *best explanation* from among the following:
    I. The net force acting on the blocks is the same, but the total mass that must be accelerated is greater.
    II. The difference in the masses is the same, and this is what determines the net force on the system.
    III. The force exerted on each block is greater, leading to an increased acceleration.

37. • **CE** Three boxes of masses m, $2m$, and $3m$ are connected as shown in **FIGURE 6-52**. A force is applied to accelerate the boxes toward the right. **(a)** Rank the tensions in the ropes, T_1, T_2, and T_3, from smallest to largest, when the boxes are in motion and there is no friction between the boxes and the horizontal surface. **(b)** Rank the tensions in the ropes, from smallest to largest, when the boxes are in motion and there is a coefficient of kinetic friction μ_k between the boxes and the horizontal surface.

FIGURE 6-52 Problem 37

38. •• Find the acceleration of the masses shown in **FIGURE 6-53**, given that $m_1 = 1.0$ kg, $m_2 = 2.0$ kg, and $m_3 = 3.0$ kg. Assume the table is frictionless and the masses move freely.

FIGURE 6-53 Problem 38

39. •• **Predict/Calculate** **(a)** If the hanging mass m_3 in Figure 6-53 is decreased, will the acceleration of the system increase, decrease, or stay the same? Explain. **(b)** Find the acceleration of the system given that $m_1 = 1.0$ kg, $m_2 = 2.0$ kg, and $m_3 = 0.50$ kg.

40. •• Two blocks are connected by a string, as shown in **FIGURE 6-54**. The smooth inclined surface makes an angle of 35° with the horizontal, and the block on the incline has a mass of 5.7 kg. The mass of the hanging block is $m = 3.2$ kg. Find **(a)** the direction and **(b)** the magnitude of the hanging block's acceleration.

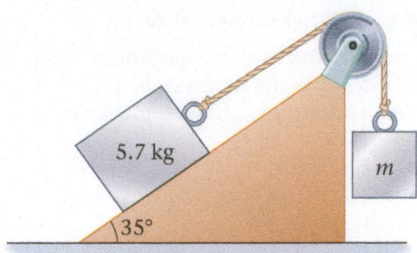

FIGURE 6-54 Problem 40

41. •• **Predict/Calculate** A 3.50-kg block on a smooth tabletop is attached by a string to a hanging block of mass 2.80 kg, as shown in **FIGURE 6-55**. The blocks are released from rest and allowed to move freely. **(a)** Is the tension in the string greater than, less than, or equal to the weight of the hanging mass? Find **(b)** the acceleration of the blocks and **(c)** the tension in the string.

FIGURE 6-55 Problem 41

42. •• **Predict/Calculate** A 7.7-N force pulls horizontally on a 1.6-kg block that slides on a smooth horizontal surface. This block is connected by a horizontal string to a second block of mass $m_2 = 0.83$ kg on the same surface. **(a)** What is the acceleration of the blocks? **(b)** What is the tension in the string? **(c)** If the mass of block 1 is increased, does the tension in the string increase, decrease, or stay the same?

43. •• **Predict/Calculate** **(a)** Find the magnitude of the acceleration of the two blocks in **FIGURE 6-56 (a)**. Assume the top block slides without friction on the horizontal surface. **(b)** If the 12-N block in Figure 6-56 (a) is replaced with a 12-N force pulling downward, as in **FIGURE 6-56 (b)**, is the acceleration of the sliding block greater than, less than, or equal to the acceleration calculated in part (a)? Explain. **(c)** Calculate the acceleration of the block in Figure 6-56 (b), assuming it slides without friction.

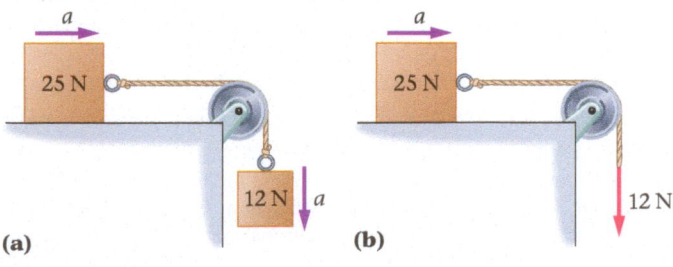

FIGURE 6-56 Problem 43

SECTION 6-5 CIRCULAR MOTION

44. • **CE** A car drives with constant speed on an elliptical track, as shown in **FIGURE 6-57**. Rank the points A, B, and C in order of increasing likelihood that the car might skid. Indicate ties where appropriate.

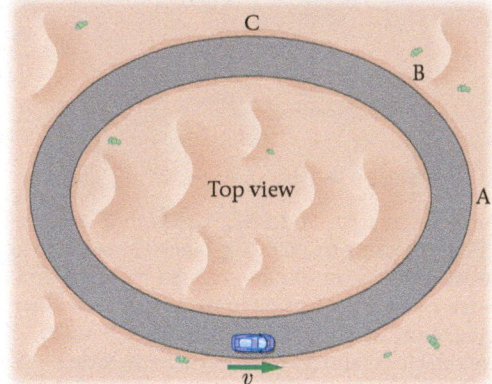

FIGURE 6-57 Problem 44

45. • **CE** A puck attached to a string undergoes circular motion on an air table. If the string breaks at the point indicated in **FIGURE 6-58**, is the subsequent motion of the puck best described by path A, B, C, or D?

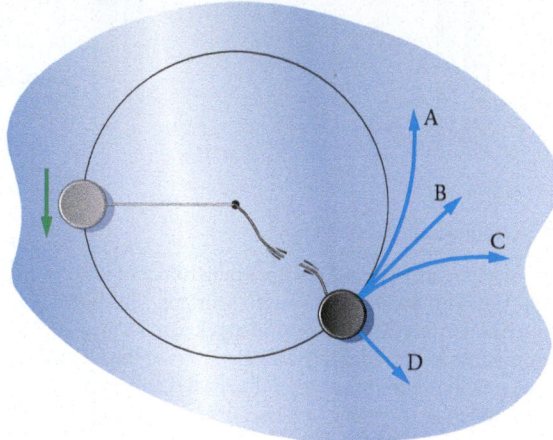

FIGURE 6-58 Problem 45

46. • **BIO Bubble Net Fishing** Humpback whales sometimes catch fish by swimming rapidly in a circle, blowing a curtain of bubbles that confuses a school of fish and traps it in a small area, where the whales can easily catch and eat them. Suppose a 28,000-kg humpback whale swims at 2.2 m/s in a circle of radius 8.5 m. What centripetal force must the whale generate?

47. • When you take your 1900-kg car out for a spin, you go around a corner of radius 53 m with a speed of 13 m/s. The coefficient of static friction between the car and the road is 0.88. Assuming your car doesn't skid, what is the force exerted on it by static friction?

48. • **BIO A Human Centrifuge** To test the effects of high acceleration on the human body, the National Aeronautics and Space Administration (NASA) constructed a large centrifuge at the Manned Spacecraft Center in Houston. In this device, astronauts were placed in a capsule that moved in a circular path with a radius of 15 m. If the astronauts in this centrifuge experienced a centripetal acceleration 9.0 times that of gravity, what was the linear speed of the capsule?

49. • A car goes around a curve on a road that is banked at an angle of 21.5°. Even though the road is slick, the car will stay on the road without any friction between its tires and the road when its speed is 19.8 m/s. What is the radius of the curve?

50. • **Clearview Screen** Large ships often have circular structures in their windshields, as shown in **FIGURE 6-59**. What is their purpose? Called clearview screens, they consist of a glass disk that is rotated at high speed by a motor (in the center of the circle) to disperse rain and spray. If the screen has a diameter of 0.39 m and rotates at 1700 rpm, the speed at the rim of the screen is 35 m/s. What is the centripetal acceleration at the rim of the screen? (A large centripetal acceleration keeps liquid from remaining on the screen. For comparison, recall that the acceleration due to gravity is 9.81 m/s².)

FIGURE 6-59 Problem 50

51. •• **Predict/Calculate** (a) As you ride on a Ferris wheel, your apparent weight is different at the top than at the bottom. Explain. (b) Calculate your apparent weight at the top and bottom of a Ferris wheel, given that the radius of the wheel is 7.2 m, it completes one revolution every 28 s, and your mass is 55 kg.

52. •• Driving in your car with a constant speed of $v = 22$ m/s, you encounter a bump in the road that has a circular cross section, as indicated in **FIGURE 6-60**. If the radius of curvature of the bump is $r = 52$ m, find the apparent weight of a 67-kg person in your car as you pass over the top of the bump.

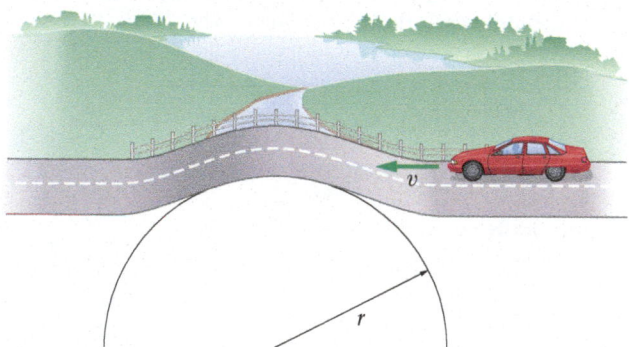

FIGURE 6-60 Problem 52

GENERAL PROBLEMS

53. • **CE** If you weigh yourself on a bathroom scale at the equator, is the reading you get greater than, less than, or equal to the reading you get if you weigh yourself at the North Pole?

54. • **CE BIO Maneuvering a Jet** Humans lose consciousness if exposed to prolonged accelerations of more than about 7g. This is of concern to jet fighter pilots, who may experience centripetal accelerations of this magnitude when making high-speed turns. Suppose we would like to decrease the centripetal acceleration of a jet. Rank the following changes in flight path in order of how effective they are in decreasing the centripetal acceleration, starting with the least effective: A, decrease the turning radius by a factor of two; B, decrease the speed by a factor of three; or C, increase the turning radius by a factor of four.

55. • **CE BIO Gravitropism** As plants grow, they tend to align their stems and roots along the direction of the gravitational field. This tendency, which is related to differential concentrations of plant hormones known as auxins, is referred to as *gravitropism*. As an illustration of gravitropism, experiments show that seedlings placed in pots on the rim of a rotating turntable do not grow in the vertical direction. Do you expect their stems to tilt inward—toward the axis of rotation—or outward—away from the axis of rotation?

56. • **BIO Human-Powered Centrifuge** One of the hazards of prolonged weightlessness on long space missions is bone and muscle loss. To combat this problem, NASA is studying a human-powered centrifuge, like the one shown in **FIGURE 6-61**. In this device, one astronaut pedals a suspended, bicycle-like contraption, while a second astronaut rides standing up in a similarly suspended platform. If the standing astronaut moves with a speed of 2.7 m/s at a distance of 2.9 m from the center of the device, what is the astronaut's centripetal acceleration? Give your answer as a multiple of g, the acceleration due to gravity on the surface of the Earth.

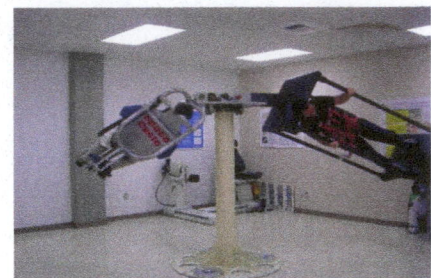

FIGURE 6-61 Problem 56 A human-powered centrifuge

57. •• **Predict/Calculate** A 9.3-kg box slides across the floor of an elevator with a coefficient of kinetic friction equal to 0.38. **(a)** What is the force of kinetic friction acting on the box when the elevator is at rest? **(b)** If the elevator moves upward with constant acceleration, is the force of kinetic friction acting on the box greater than, less than, or equal to the value found in part (a)? Explain. **(c)** Determine the force of kinetic friction acting on the sliding box when the elevator moves upward with a constant acceleration of 1.4 m/s².

58. •• A child goes down a playground slide that is inclined at an angle of 26.5° below the horizontal. Find the acceleration of the child given that the coefficient of kinetic friction between the child and the slide is 0.315.

59. •• **Spin-Dry Dragonflies** Some dragonflies splash down onto the surface of a lake to clean themselves. After this dunking, the dragonflies gain altitude, and then spin rapidly at about 1000 rpm to spray the water off their bodies. When the dragonflies do this "spin-dry," they tuck themselves into a "ball." If they spin with a linear speed of 2.3 m/s, and curl into a ball with a radius of 1.1 cm, what is the centripetal acceleration they produce?

FIGURE 6-62 Problem 59

60. •• **The da Vinci Code** Leonardo da Vinci (1452–1519) is credited with being the first to perform quantitative experiments on friction, though his results weren't known until centuries later, due in part to the secret code (mirror writing) he used in his notebooks. Leonardo would place a block of wood on an inclined plane and measure the angle at which the block begins to slide. He reports that the coefficient of static friction was 0.25 in his experiments. At what angle did Leonardo's blocks begin to slide?

61. •• A 4.5-kg sled is pulled with constant speed across a rough, horizontal surface with various amounts of weight added to the sled. Data showing the required force for several different amounts of added weight are given in the table.

Added weight (kg)	Force (N)
0	11
1	15
2	17
3	21
4	22

Plot the data and determine the coefficient of kinetic friction between the sled and the horizontal surface.

62. •• A 0.045-kg golf ball hangs by a string from the rearview mirror of a car, as shown in **FIGURE 6-63**. The car accelerates in the forward direction with a magnitude of 2.1 m/s². **(a)** What angle θ does the string make with the vertical? **(b)** What is the tension in the string?

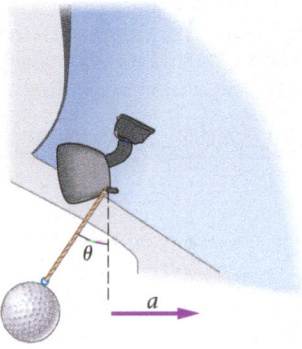

FIGURE 6-63 Problem 62

63. •• A physics textbook weighing 22 N rests on a desk. The coefficient of static friction between the book and the desk is $\mu_s = 0.60$; the coefficient of kinetic friction is $\mu_k = 0.40$. You push horizontally on the book with a force that gradually increases from 0 to 15 N, and then slowly decreases to 5.0 N, as indicated in the table below. For each value of the applied force given in the table, give the magnitude of the force of friction and state whether the book is accelerating, decelerating, at rest, or moving with constant speed.

Applied force	Friction force	Motion
0		
5.0 N		
11 N		
15 N		
11 N		
8.0 N		
5.0 N		

64. •• **Predict/Calculate** The blocks shown in **FIGURE 6-64** are at rest. **(a)** Find the frictional force exerted on block A given that the mass of block A is 8.82 kg, the mass of block B is 2.33 kg, and the coefficient of static friction between block A and the surface on which it rests is 0.320. **(b)** If the mass of block A is doubled, does the frictional force exerted on it increase, decrease, or stay the same? Explain.

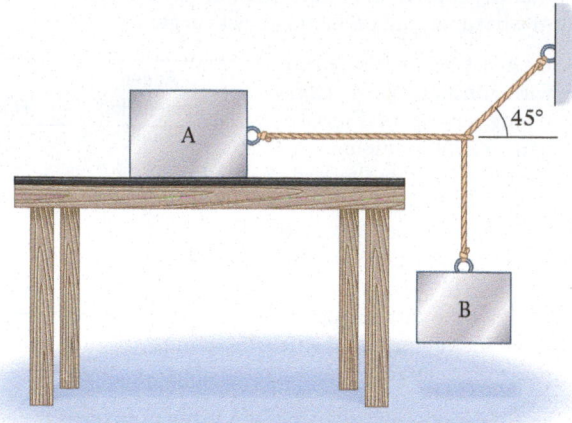

FIGURE 6-64 Problem 64

65. •• **A Conical Pendulum** A 0.075-kg toy airplane is tied to the ceiling with a string. When the airplane's motor is started, it moves with a constant speed of 4.47 m/s in a horizontal circle of radius 1.28 m, as illustrated in **FIGURE 6-65**. Find **(a)** the angle the string makes with the vertical and **(b)** the tension in the string.

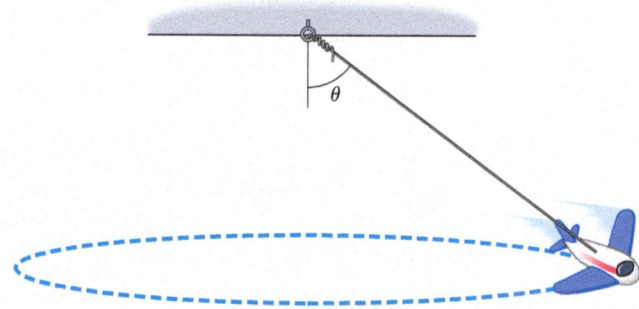

FIGURE 6-65 Problem 65

66. •• A tugboat tows a barge at constant speed with a 3500-kg cable, as shown in **FIGURE 6-66**. If the angle the cable makes with the horizontal where it attaches to the barge and the tugboat is 22°, find the force the cable exerts on the barge in the forward direction.

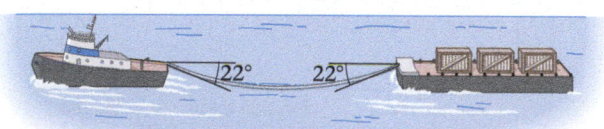

FIGURE 6-66 Problem 66

67. •• **Predict/Calculate** Two blocks, stacked one on top of the other, can move without friction on the horizontal surface shown in **FIGURE 6-67**. The surface between the two blocks is rough, however, with a coefficient of static friction equal to 0.47. **(a)** If a horizontal force F is applied to the 5.0-kg bottom block, what is the maximum value F can have before the

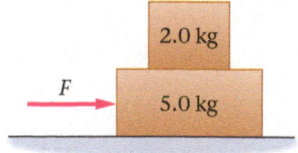

FIGURE 6-67 Problem 67

2.0-kg top block begins to slip? **(b)** If the mass of the top block is increased, does the maximum value of F increase, decrease, or stay the same? Explain.

68. •• **Predict/Calculate** In a daring rescue by helicopter, two men with a combined mass of 172 kg are lifted to safety. **(a)** If the helicopter lifts the men straight up with constant acceleration, is the tension in the rescue cable greater than, less than, or equal to the combined weight of the men? Explain. **(b)** Determine the tension in the cable if the men are lifted with a constant acceleration of 1.10 m/s².

69. •• **Predict/Calculate** A light spring with a force constant of 13 N/m is connected to a wall and to a 1.2-kg toy bulldozer, as shown in **FIGURE 6-68**. When the electric motor in the bulldozer is turned on, it stretches the spring for a distance of 0.45 m before its tread begins to slip on the floor. **(a)** Which coefficient of friction (static or kinetic) can be determined from this information? Explain. **(b)** What is the numerical value of this coefficient of friction?

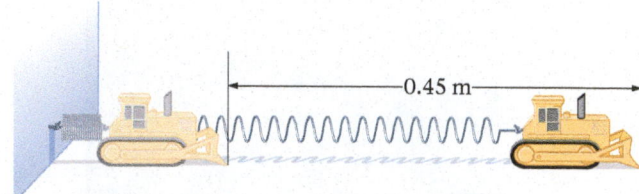

FIGURE 6-68 Problem 69

70. •• **Predict/Calculate** The blocks in **FIGURE 6-69** have an acceleration of magnitude $a = 1.1$ m/s². **(a)** Is the tension in the string greater than, less than, or equal to (1.5 kg) × (9.81 m/s²)? Explain. **(b)** What is the tension in the string connecting the blocks? **(c)** What is the magnitude of the force of kinetic friction acting on the upper block? **(d)** What is the coefficient of kinetic friction between the upper block and the horizontal surface?

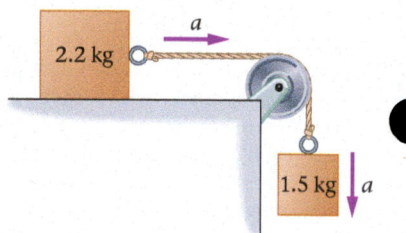

FIGURE 6-69 Problem 70

71. •• **Predict/Calculate Playing a Violin** The tension in a violin string is 2.7 N. When pushed down against the neck of the violin, the string makes an angle of 4.1° with the horizontal. **(a)** With what force must you push down on the string to bring it into contact with the neck? **(b)** If the angle were less than 4.1°, would the required force be greater than, less than, or the same as in part (a)? Explain.

72. •• **Predict/Calculate** A 9.8-kg monkey hangs from a rope, as shown in **FIGURE 6-70**. The tension on the left side of the rope is T_1 and the tension on the right side is T_2. **(a)** Which tension is largest in magnitude, or are they the same? Explain. **(b)** Determine the tensions T_1 and T_2.

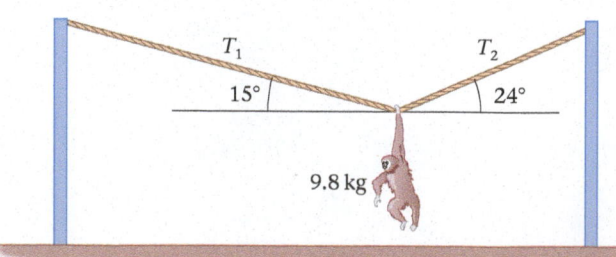

FIGURE 6-70 Problem 72

73. •• As your plane circles an airport, it moves in a horizontal circle of radius 2300 m with a speed of 390 km/h. If the lift of the airplane's wings is perpendicular to the wings, at what angle should the plane be banked so that it doesn't tend to slip sideways?

74. •• At a playground, a 22-kg child sits on a spinning merry-go-round, as shown from above in **FIGURE 6-71**. The merry-go-round completes one revolution every 6.2 s, and the child sits at a radius of $r = 1.4$ m. **(a)** What is the force of static friction acting on the child? **(b)** What is the minimum coefficient of static friction between the child and the merry-go-round to keep the child from slipping?

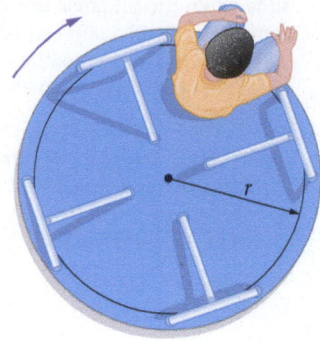

FIGURE 6-71 Problem 74

75. ••• A 2.0-kg box rests on a plank that is inclined at an angle of 65° above the horizontal. The upper end of the box is attached to a spring with a force constant of 360 N/m, as shown in **FIGURE 6-72**. If the coefficient of static friction between the box and the plank is 0.22, what is the maximum amount the spring can be stretched and the box remain at rest?

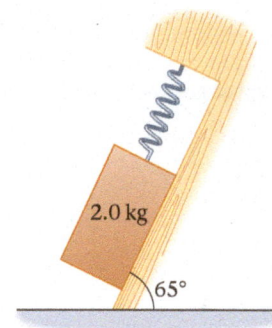

FIGURE 6-72 Problem 75

76. ••• A wood block of mass m rests on a larger wood block of mass M that rests on a wooden table. The coefficients of static and kinetic friction between all surfaces are μ_s and μ_k, respectively. What is the minimum horizontal force, F, applied to the lower block that will cause it to slide out from under the upper block?

77. ••• A hockey puck of mass m is attached to a string that passes through a hole in the center of a table, as shown in **FIGURE 6-73**. The hockey puck moves in a circle of radius r. Tied to the other end of the string, and hanging vertically beneath the table, is a mass M. Assuming the tabletop is perfectly smooth, what speed must the hockey puck have if the mass M is to remain at rest?

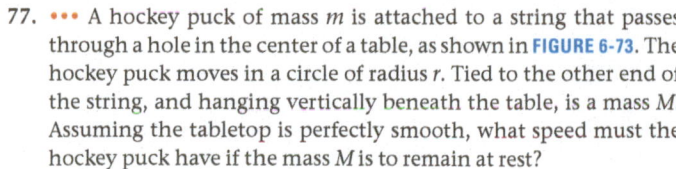

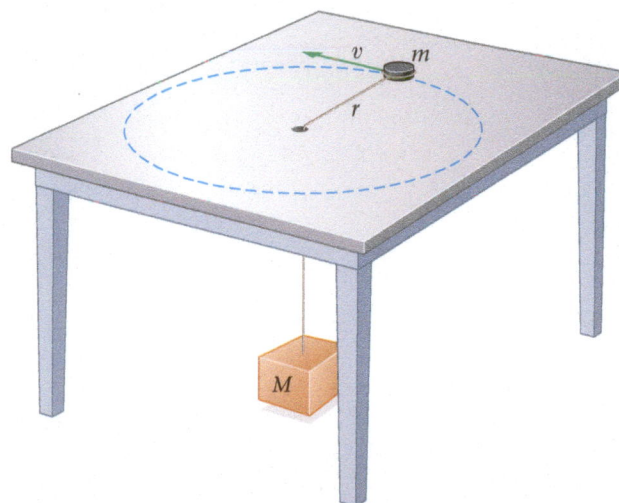

FIGURE 6-73 Problem 77

78. ••• **Predict/Calculate** A popular ride at amusement parks is illustrated in **FIGURE 6-74**. In this ride, people sit in a swing that is suspended from a rotating arm. Riders are at a distance of 12 m from the axis of rotation and move with a speed of 25 mi/h. **(a)** Find the centripetal acceleration of the riders. **(b)** Find the angle θ the supporting wires make with the vertical. **(c)** If you observe a ride like that in Figure 6-74, or as shown in the photo on page 178, you will notice that all the swings are at the same angle θ to the vertical, regardless of the weight of the rider. Explain.

FIGURE 6-74 Problem 78

79. ••• **A Conveyor Belt** A box is placed on a conveyor belt that moves with a constant speed of 1.25 m/s. The coefficient of kinetic friction between the box and the belt is 0.780. **(a)** How much time does it take for the box to stop sliding relative to the belt? **(b)** How far does the box move in this time?

80. ••• As part of a circus act, a person drives a motorcycle with constant speed v around the inside of a vertical track of radius r, as indicated in **FIGURE 6-75**. If the combined mass of the motorcycle and rider is m, find the normal force exerted on the motorcycle by the track at the points **(a)** A, **(b)** B, and **(c)** C.

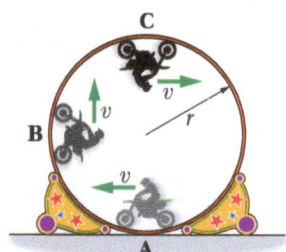

FIGURE 6-75 Problem 80

PASSAGE PROBLEMS

BIO Nasal Strips

People in all walks of life use nasal strips, or external nasal dilator strips (ENDS), to alleviate a number of respiratory problems. First introduced to eliminate snoring, they are now finding use in a number of other areas. For example, dentists have found that nasal strips help patients breathe better during dental procedures, making the experience considerably more pleasant for both doctor and patient.

One of the great advantages of ENDS is that no drugs are involved; the strips are a purely mechanical device, consisting of two flat, polyester springs enclosed by an adhesive tape covering. When applied to the nose, they exert an outward force that enlarges the nasal passages and reduces the resistance to air flow (see the illustration in Quick Example 6-8). The mechanism shown in **FIGURE 6-76 (a)** is used to measure the behavior of these strips. For example, if a 30-g weight is placed on the movable platform (of negligible mass), the strip is found to compress from an initial length of 50 mm to a reduced length of 19 mm, as can be seen in **FIGURE 6-76 (b)**.

81. • On the straight-line segment II in Figure 6-76 (b) we see that increasing the applied mass from 20 g to 25 g results in a reduction of the end-to-end distance from 42 mm to 24 mm. What is the force constant in N/m on segment II?

    A. 0.28 N/m   B. 0.37 N/m
    C. 2.7 N/m   D. 3.6 N/m

82. • Rank the straight segments I, II, and III in order of increasing "stiffness" of the nasal strips.

83. • In use on a typical human nose, the end-to-end distance is measured to be 30 mm. What force does the strip exert on the nasal passages in this situation?

    A. 0.23 N   B. 0.69 N
    C. 9.0 N   D. 2 N

84. •• Predict/Calculate REFERRING TO EXAMPLE 6-3 Suppose the coefficients of static and kinetic friction between the crate and the truck bed are 0.415 and 0.382, respectively. **(a)** Does the crate begin to slide at a tilt angle that is greater than, less than, or equal to 23.2°? **(b)** Verify your answer to part (a) by determining the angle at which the crate begins to slide. **(c)** Find the length of time it takes for the crate to slide a distance of 2.75 m when the tilt angle has the value found in part (b).

85. •• Predict/Calculate REFERRING TO EXAMPLE 6-3 The crate begins to slide when the tilt angle is 17.5°. When the crate reaches the bottom of the flatbed, after sliding a distance of 2.75 m, its speed is 3.11 m/s. Find **(a)** the coefficient of static friction and **(b)** the coefficient of kinetic friction between the crate and the flatbed.

86. •• REFERRING TO EXAMPLE 6-13 Suppose that the mass on the frictionless tabletop has the value $m_1 = 2.45$ kg. **(a)** Find the value of m_2 that gives an acceleration of 2.85 m/s². **(b)** What is the corresponding tension, T, in the string? **(c)** Calculate the ratio $T/m_2 g$ and show that it is less than 1, as expected.

87. •• REFERRING TO EXAMPLE 6-15 **(a)** At what speed will the force of static friction exerted on the car by the road be equal to half the weight of the car? The mass of the car is $m = 1200$ kg, the radius of the corner is $r = 45$ m, and the coefficient of static friction between the tires and the road is $\mu_s = 0.82$. **(b)** Suppose that the mass of the car is now doubled, and that it moves with a speed that again makes the force of static friction equal to half the car's weight. Is this new speed greater than, less than, or equal to the speed in part (a)?

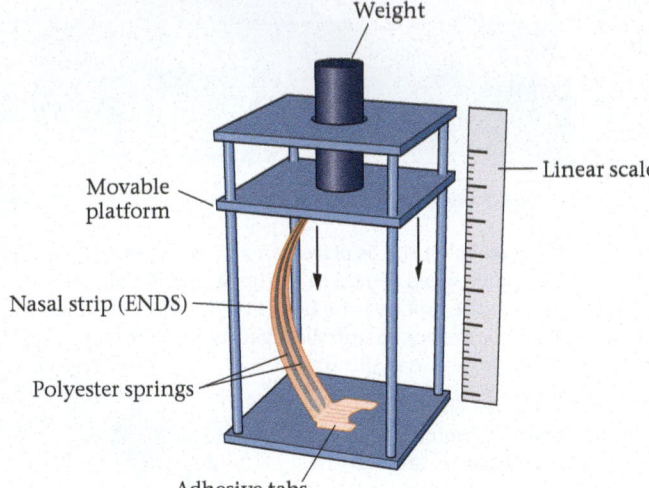

(a)

(b)

FIGURE 6-76 Problems 81, 82, and 83

Big Ideas

1 Work is force times distance.

2 Kinetic energy is one-half mass times velocity squared.

3 Power is the rate at which work is done.

Work and Kinetic Energy

▲ The work done by this cyclist can increase his energy of motion. The relationship between work (force times distance) and energy of motion (kinetic energy) is developed in detail in this chapter.

In this chapter we introduce the idea that force times distance is an important new physical quantity, which we refer to as *work*. Closely related to work is the energy of motion, or *kinetic* *energy*. Together, work and kinetic energy set the stage for a deeper understanding of the physical world.

7-1 Work Done by a Constant Force

In this section we define work—in the physics sense of the word—and apply our definition to a variety of physical situations. We start with the simplest case—namely, the work done when force and displacement are in the same direction.

Force in the Direction of Displacement

When we push a shopping cart in a store or pull a suitcase through an airport, we do work. The greater the force, the greater the work; the greater the distance, the greater the work. These simple ideas form the basis for our definition of work.

To be specific, suppose we push a box with a constant force $\vec{F}$, as shown in **FIGURE 7-1**. If we move the box *in the direction of* $\vec{F}$ through a displacement $\vec{d}$, the **work** W we have done is Fd:

> **Definition of Work, W, When a Constant Force Is in the Direction of Displacement**
>
> $$W = Fd \qquad\qquad 7\text{-}1$$
>
> SI unit: newton-meter (N · m) = joule, J

Work is the product of two magnitudes, and hence it is a scalar. In addition, notice that a small force acting over a large distance gives the same work as a large force acting over a small distance. For example, $W = (1\,\text{N})(400\,\text{m}) = (400\,\text{N})(1\,\text{m})$.

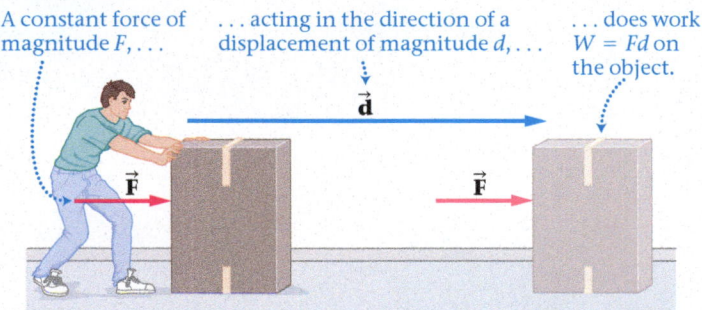

A constant force of magnitude F,... ... acting in the direction of a displacement of magnitude d,... ... does work $W = Fd$ on the object.

▲ **FIGURE 7-1 Work: constant force in the direction of motion** A constant force $\vec{F}$ pushes a box through a displacement $\vec{d}$. In this special case, where the force and displacement are in the *same* direction, the work done on the box by the force is $W = Fd$.

The dimensions of work are newtons (force) times meters (distance), or N·m. This combination of dimensions is called the **joule** (rhymes with "school," as commonly pronounced) in honor of James Prescott Joule (1818–1889), a dedicated physicist who is said to have conducted physics experiments even while on his honeymoon. We define a joule as follows:

> **Definition of the joule, J**
>
> $$1\ \text{joule} = 1\,\text{J} = 1\,\text{N} \cdot \text{m} = 1(\text{kg} \cdot \text{m/s}^2) \cdot \text{m} = 1\,\text{kg} \cdot \text{m}^2/\text{s}^2 \qquad\qquad 7\text{-}2$$

To get a better feeling for work and the associated units, suppose you exert a force of 82.0 N on the box in Figure 7-1 and move it in the direction of the force through a distance of 3.00 m. The work you have done is

$$W = Fd = (82.0\,\text{N})(3.00\,\text{m}) = 246\,\text{N} \cdot \text{m} = 246\,\text{J}$$

Similarly, if you do 5.00 J of work to lift a book through a vertical distance of 0.750 m, the force you have exerted on the book is

$$F = \frac{W}{d} = \frac{5.00\,\text{J}}{0.750\,\text{m}} = \frac{5.00\,\text{N} \cdot \text{m}}{0.750\,\text{m}} = 6.67\,\text{N}$$

EXERCISE 7-1   WORK DONE

One species of Darwin's finch, *Geospiza magnirostris*, can exert a force of 217 N with its beak as it cracks open a *Tribulus* seed case. If its beak moves through a distance of 0.42 cm during this operation, how much work does the finch do to get the seed?

REASONING AND SOLUTION

The work the finch does is force times distance. In this case, the force is 217 N and the distance is 0.42 cm = 0.0042 m. Thus,

$$W = Fd = (217\text{ N})(0.0042\text{ m}) = 0.91\text{ J}$$

This is a relatively small amount of work for a human, but significant for a small bird.

Just how much work is a joule, anyway? Well, you do one joule of work when you lift a gallon of milk through a height of about an inch, or lift an apple a meter. One joule of work lights a 100-watt lightbulb for 0.01 second or heats a glass of water 0.00125 degree Celsius. Clearly, a joule is a modest amount of work in everyday terms. Additional examples of work are listed in Table 7-1.

TABLE 7-1 Typical Values of Work

Activity	Equivalent work (J)
Annual U.S. energy use	8×10^{19}
Mt. St. Helens eruption	10^{18}
Burning 1 gallon of gas	10^{8}
Human food intake/day	10^{7}
Melting an ice cube	10^{4}
Lighting a 100-W bulb for 1 minute	6000
Heartbeat	0.5
Turning a page of a book	10^{-3}
Hop of a flea	10^{-7}
Breaking a bond in DNA	10^{-20}

EXAMPLE 7-2   HEADING FOR THE ER

An intern pushes an 87-kg patient on an 18-kg gurney, producing an acceleration of 0.55 m/s². **(a)** How much work does the intern do in pushing the patient and gurney through a distance of 1.9 m? Assume the gurney moves without friction. **(b)** How far must the intern push the gurney to do 140 J of work?

PICTURE THE PROBLEM

Our sketch shows the physical situation for this problem. Notice that the force exerted by the intern is in the same direction as the displacement of the gurney; therefore, we know that the work is $W = Fd$.

REASONING AND STRATEGY

We are not given the magnitude of the force, F, so we cannot apply Equation 7-1 ($W = Fd$) directly. However, we are given the mass and acceleration of the patient and gurney, and from them we can calculate the force with $F = ma$. The work done by the intern is then $W = Fd$, where $d = 1.9$ m.

Known Mass of patient, 87 kg; mass of gurney, 18 kg; acceleration, $a = 0.55$ m/s²; **(a)** pushing distance, $d = 1.9$ m; **(b)** work, $W = 140$ J.

Unknown **(a)** Work done, $W = ?$ **(b)** Pushing distance, $d = ?$

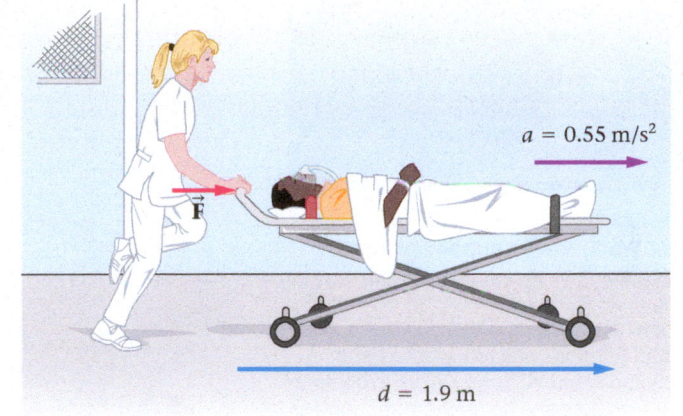

SOLUTION

Part (a)

1. First, find the force F exerted by the intern: $F = ma = (87\text{ kg} + 18\text{ kg})(0.55\text{ m/s}^2) = 58\text{ N}$
2. The work done by the intern, W, is the force times the distance: $W = Fd = (58\text{ N})(1.9\text{ m}) = 110\text{ J}$

Part (b)

3. Use $W = Fd$ to solve for the distance d: $d = \dfrac{W}{F} = \dfrac{140\text{ J}}{58\text{ N}} = 2.4\text{ m}$

INSIGHT

You might wonder whether the work done by the intern depends on the speed of the gurney. The answer is no. The work done on an object, $W = Fd$, doesn't depend on whether the object moves through the distance d quickly or slowly. What does depend on the speed of the gurney is the *rate* at which work is done, which we discuss in detail in Section 7-4.

PRACTICE PROBLEM — PREDICT/CALCULATE

(a) If the total mass of the gurney plus patient is halved and the acceleration is doubled, does the work done by the intern increase, decrease, or remain the same? Explain. **(b)** Determine the work in this case.   [**Answer: (a)** The work remains the same because the two changes offset one another; that is, $F = ma = (m/2)(2a)$. **(b)** The work is 110 J, as before.]

Some related homework problems: Problem 3, Problem 4

▲ **FIGURE 7-2 Visualizing Concepts Force, Work, and Displacement** The person doing push ups does positive work as he moves upward, but does no work when he holds himself in place. The weightlifter does positive work as she raises the weights, but no work is done on the weights as she holds them motionless above her head.

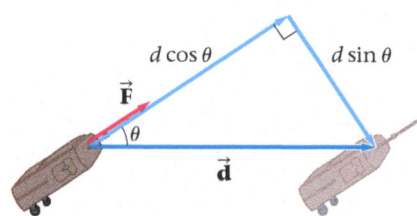

▲ **FIGURE 7-4 Force at an angle to direction of motion: another look** The displacement of the suitcase in Figure 7-3 is equivalent to a displacement of magnitude $d \cos \theta$ in the direction of the force $\vec{F}$, plus a displacement of magnitude $d \sin \theta$ perpendicular to the force. Only the displacement parallel to the force results in nonzero work; hence the total work done is $F(d \cos \theta)$ as expected.

Zero Distance Implies Zero Work   Before moving on, let's note an interesting point about our definition of work. It's clear from Equation 7-1 that *the work W is zero if the distance d is zero*—and this is true regardless of how great the force might be, as illustrated in **FIGURE 7-2**. For example, if you push against a solid wall, you do no work on it, even though you may become tired from your efforts. Similarly, if you stand in one place holding a 50-pound suitcase in your hand, you do no work on the suitcase, even though you soon feel worn out. The fact that we become tired when we push against a wall or hold a heavy object is due to the repeated contraction and expansion of individual cells within our muscles. Thus, even when we are "at rest," our muscles are doing mechanical work on the microscopic level.

Force at an Angle to the Displacement

In **FIGURE 7-3** we see a person pulling a suitcase on a level surface with a strap that makes an angle θ with the horizontal—in this case the force is at an angle to the direction of motion. How do we calculate the work now? Well, instead of force times distance, we say that work is the *component* of force in the *direction* of displacement times the magnitude of the displacement. In Figure 7-3, the component of force in the direction of the displacement is $F \cos \theta$ and the magnitude of the displacement is d. Therefore, the work is $F \cos \theta$ times d:

> **Definition of Work When the Angle Between a Constant Force and the Displacement Is θ**
>
> $$W = (F \cos \theta)d = Fd \cos \theta$$ 7-3
>
> SI unit: joule, J

Of course, in the case where the force is in the direction of motion, the angle θ is zero; then $W = Fd \cos 0 = Fd \cdot 1 = Fd$, in agreement with Equation 7-1.

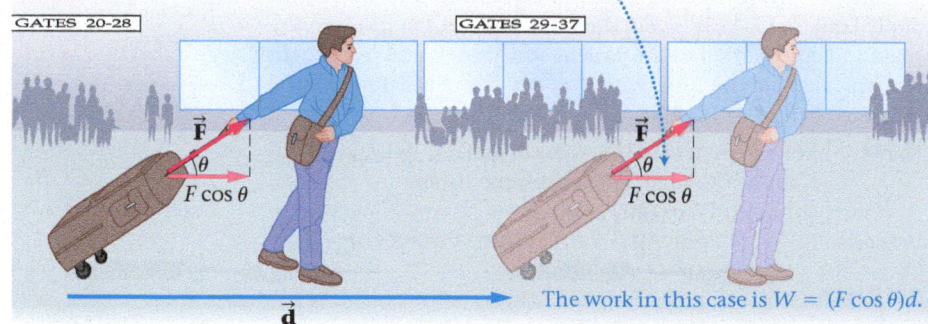

The component of force in the direction of displacement is $F \cos \theta$. This is the only component of the force that does work.

The work in this case is $W = (F \cos \theta)d$.

▲ **FIGURE 7-3 Work: force at an angle to the direction of motion** A person pulls a suitcase with a strap at an angle θ to the direction of motion. The component of force in the direction of motion is $F \cos \theta$, and the work done by the person is $W = (F \cos \theta)d$.

Equally interesting is a situation in which the force and the displacement are at right angles to one another. In this case $\theta = 90°$ and the work done by the force F is zero; $W = Fd \cos 90° = 0$.

This result leads naturally to an alternative way to think about the expression $W = Fd \cos \theta$. In **FIGURE 7-4** we show the displacement and the force for the suitcase in Figure 7-3. Notice that the displacement is equivalent to a displacement in the direction of the force of magnitude $(d \cos \theta)$ *plus* a displacement at right angles to the force of magnitude $(d \sin \theta)$. Since the displacement at right angles to the force corresponds to zero work and the displacement in the direction of the force corresponds to a work $W = F(d \cos \theta)$, it follows that the work done in this case is $Fd \cos \theta$, as given in Equation 7-3. Thus, the work done by a force can be thought of in the following two *equivalent* ways:

(i) *Work is the component of force in the direction of the displacement times the magnitude of the displacement.*

(ii) *Work is the component of displacement in the direction of the force times the magnitude of the force.*

In both interpretations, the mathematical expression for work is exactly the same, $W = Fd\cos\theta$, where θ is the angle between the force vector and the displacement vector when they are placed tail-to-tail. This definition of θ is illustrated in Figure 7-4.

Finally, we can also express work as the **dot product** of the vectors $\vec{F}$ and $\vec{d}$; that is, $W = \vec{F} \cdot \vec{d} = Fd\cos\theta$. Notice that the dot product, which is always a scalar, is simply the magnitude of one vector times the magnitude of the second vector times the cosine of the angle between them. We discuss the dot product in greater detail in Appendix A.

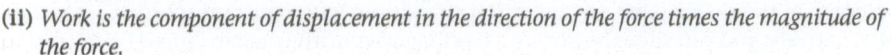

Big Idea 1 Work is force times distance when force and displacement are in the same direction. More generally, work is the component of force in the direction of displacement times the distance.

PHYSICS IN CONTEXT
Looking Back

Even though work is a scalar quantity, notice that vector components (Chapter 3) are used in the definition of work.

EXAMPLE 7-3   GRAVITY ESCAPE SYSTEM

RWP* In a gravity escape system (GES), an enclosed lifeboat on a large ship is deployed by letting it slide down a ramp and then continue in free fall to the water below. Suppose a 4970-kg lifeboat slides a distance of 5.00 m on a ramp, dropping through a vertical height of 2.50 m. How much work does gravity do on the boat?

PICTURE THE PROBLEM

From our sketch, we see that the force of gravity $m\vec{g}$ and the displacement $\vec{d}$ are at an angle θ relative to one another when placed tail-to-tail, and that θ is also the angle the ramp makes with the vertical. In addition, we note that the vertical height of the ramp is $h = 2.50$ m and the length of the ramp is $d = 5.00$ m.

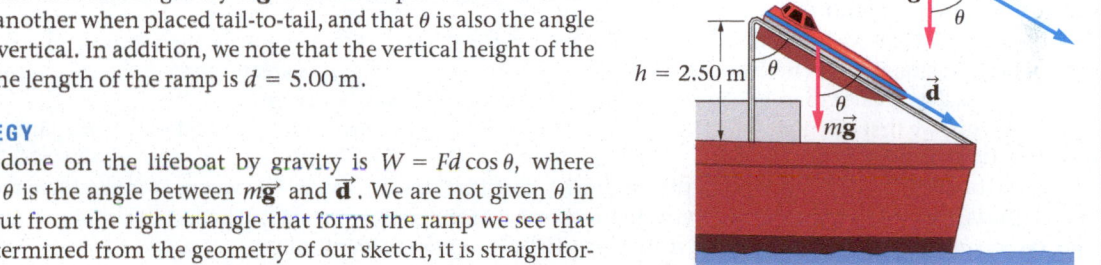

REASONING AND STRATEGY

By definition, the work done on the lifeboat by gravity is $W = Fd\cos\theta$, where $F = mg$, $d = 5.00$ m, and θ is the angle between $m\vec{g}$ and $\vec{d}$. We are not given θ in the problem statement, but from the right triangle that forms the ramp we see that $\cos\theta = h/d$. Once θ is determined from the geometry of our sketch, it is straightforward to calculate W.

Known   Mass of lifeboat, $m = 4970$ kg; sliding distance, $d = 5.00$ m; vertical height, $h = 2.50$ m.
Unknown   Work done by gravity, $W = ?$

SOLUTION

1. First, find the component of $\vec{F} = m\vec{g}$ in the direction of motion:

$$F\cos\theta = (mg)\left(\frac{h}{d}\right)$$
$$= (4970\text{ kg})(9.81\text{ m/s}^2)\left(\frac{2.50\text{ m}}{5.00\text{ m}}\right) = 24{,}400\text{ N}$$

2. Multiply by distance to find the work:

$$W = (F\cos\theta)d = (24{,}400\text{ N})(5.00\text{ m}) = 122{,}000\text{ J}$$

3. Alternatively, cancel d algebraically before substituting numerical values:

$$W = Fd\cos\theta = (mg)(d)\left(\frac{h}{d}\right)$$
$$= mgh = (4970\text{ kg})(9.81\text{ m/s}^2)(2.50\text{ m}) = 122{,}000\text{ J}$$

INSIGHT

The work is simply $W = mgh$, exactly the same as if the lifeboat had fallen straight down through the height h.

Working the problem symbolically, as in Step 3, results in two distinct advantages. First, it makes for a simpler expression for the work. Second, and more important, it shows that the distance d cancels; hence the work depends on the height h but not on the distance. Such a result is not apparent when we work solely with numbers, as in Steps 1 and 2.

PRACTICE PROBLEM

Suppose the lifeboat slides halfway to the water, gets stuck for a moment, and then starts up again and continues to the end of the ramp. What is the work done by gravity in this case?   [**Answer:** The work done by gravity is exactly the same, $W = mgh$, independent of how the boat moves down the ramp.]

Some related homework problems: Problem 10, Problem 11

Next, we present a Conceptual Example that compares the work required to move an object along two different paths.

---

*Real World Physics applications are denoted by the acronym RWP.

CONCEPTUAL EXAMPLE 7-4 PATH DEPENDENCE OF WORK

You want to load a box into the back of a truck, as shown in the sketch. One way is to lift it straight up with constant speed through a height h, as shown, doing a work W_1. Alternatively, you can slide the box up a loading ramp with constant speed a distance L, doing a work W_2. Assuming the box slides on the ramp without friction, which of the following statements is correct: **(a)** $W_1 < W_2$, **(b)** $W_1 = W_2$, **(c)** $W_1 > W_2$?

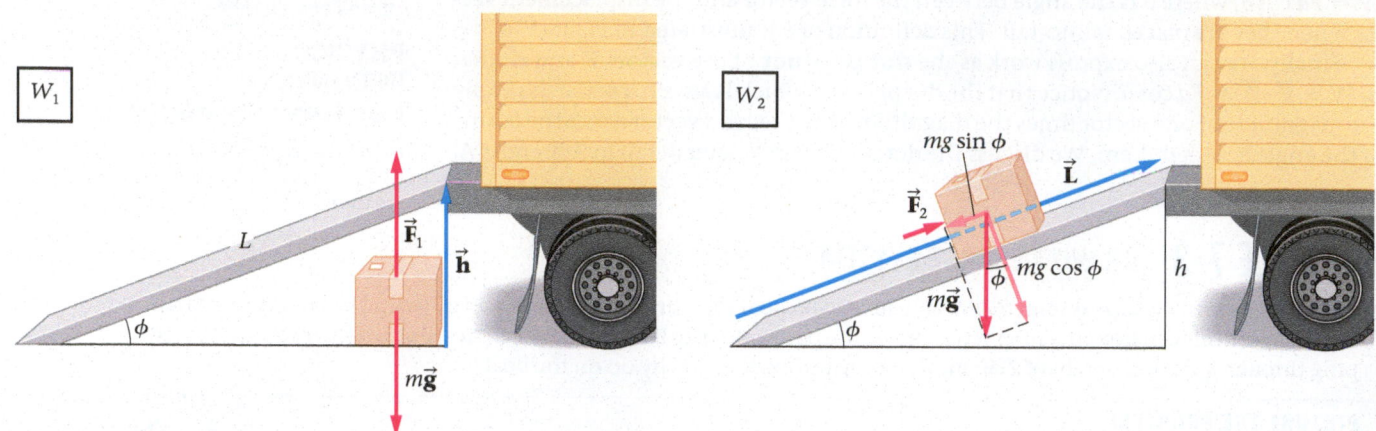

REASONING AND DISCUSSION

You might think that W_2 is less than W_1, because the force needed to slide the box up the ramp, F_2, is less than the force needed to lift it straight up. On the other hand, the distance up the ramp, L, is greater than the vertical distance, h, so perhaps W_2 should be greater than W_1. In fact, these two effects cancel exactly, giving $W_1 = W_2$.

To see this, we first calculate W_1. The force needed to lift the box with constant speed is $F_1 = mg$, and the height is h; therefore $W_1 = mgh$.

Next, the work to slide the box up the ramp with constant speed is $W_2 = F_2 L$, where F_2 is the force required to push against the tangential component of gravity. In the sketch we see that $F_2 = mg \sin \phi$. The sketch also shows that $\sin \phi = h/L$; thus $W_2 = (mg \sin \phi) L = (mg)(h/L)L = mgh = W_1$: The two works are identical!

Clearly, the ramp is a useful device—it reduces the *force* required to move the box upward from $F_1 = mg$ to $F_2 = mg(h/L)$. Even so, it doesn't decrease the amount of *work* we need to do. As we have seen, the reduced force on the ramp is offset by the increased distance.

ANSWER

(b) $W_1 = W_2$

Negative Work and Total Work

Work depends on the angle between the force, $\vec{F}$, and the displacement (or direction of motion), $\vec{d}$. This dependence gives rise to three distinct possibilities, as shown in **FIGURE 7-5**:

(i) *Work is positive if the force has a component in the direction of motion* $(-90° < \theta < 90°)$.

(ii) *Work is zero if the force has no component in the direction of motion* $(\theta = \pm 90°)$.

(iii) *Work is negative if the force has a component opposite to the direction of motion* $(90° < \theta < 270°)$.

Thus, whenever we calculate work, we must be careful about its sign and not just assume it is positive.

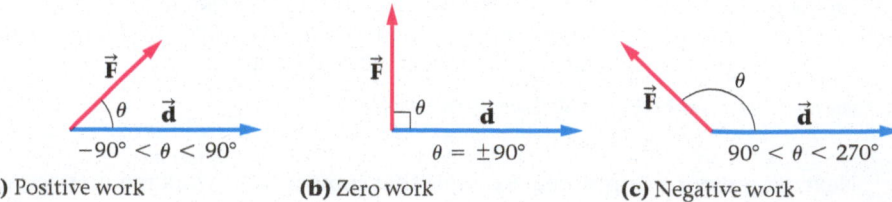

(a) Positive work **(b)** Zero work **(c)** Negative work

▲ **FIGURE 7-5 Positive, negative, and zero work** Work is positive when the force is in the same general direction as the displacement and is negative if the force is generally opposite to the displacement. Zero work is done if the force is at right angles to the displacement.

When more than one force acts on an object, the total work is the sum of the work done by each force separately. Thus, if force $\vec{\mathbf{F}}_1$ does work W_1, force $\vec{\mathbf{F}}_2$ does work W_2, and so on, the total work is

$$W_{\text{total}} = W_1 + W_2 + W_3 + \cdots = \sum W \qquad 7\text{-}4$$

Equivalently, the total work can be calculated by first performing a vector sum of all the forces acting on an object to obtain $\vec{\mathbf{F}}_{\text{total}}$ and then using our basic definition of work:

$$W_{\text{total}} = (F_{\text{total}} \cos \theta)d = F_{\text{total}}\, d \cos \theta \qquad 7\text{-}5$$

In this expression, θ is the angle between $\vec{\mathbf{F}}_{\text{total}}$ and the displacement $\vec{\mathbf{d}}$. In the next two Examples we calculate the total work in each of these ways.

EXAMPLE 7-5   A COASTING CAR I

A car of mass m coasts down a hill inclined at an angle ϕ below the horizontal. The car is acted on by three forces: (i) the normal force $\vec{\mathbf{N}}$ exerted by the road, (ii) a force due to air resistance, $\vec{\mathbf{F}}_{\text{air}}$, and (iii) the force of gravity, $m\vec{\mathbf{g}}$. Find the total work done on the car as it travels a distance d along the road.

PICTURE THE PROBLEM

Because ϕ is the angle the slope makes with the horizontal, it is also the angle between $m\vec{\mathbf{g}}$ and the downward normal direction, as was shown in Figure 5-22. It follows that the angle between $m\vec{\mathbf{g}}$ and the displacement $\vec{\mathbf{d}}$ is $\theta = 90° - \phi$. Our sketch also shows that the angle between $\vec{\mathbf{N}}$ and $\vec{\mathbf{d}}$ is $\theta = 90°$, and the angle between $\vec{\mathbf{F}}_{\text{air}}$ and $\vec{\mathbf{d}}$ is $\theta = 180°$.

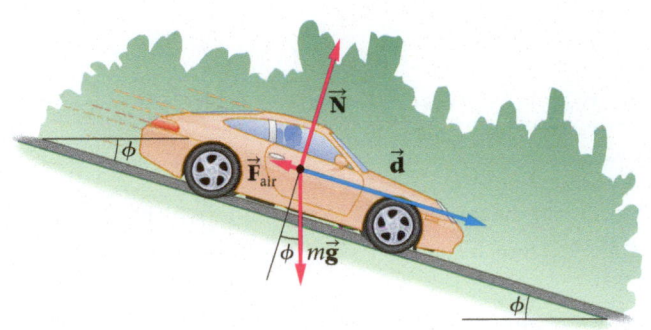
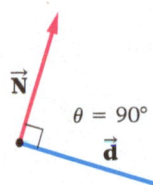
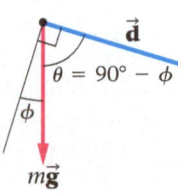

REASONING AND STRATEGY

For each force we calculate the work using $W = Fd \cos \theta$, where θ is the angle between that particular force and the displacement $\vec{\mathbf{d}}$. The total work is the sum of the work done by each of the three forces.

Known   Mass of car, m; angle of incline, ϕ; distance, d; forces acting on car, $\vec{\mathbf{N}}$, $\vec{\mathbf{F}}_{\text{air}}$, and $m\vec{\mathbf{g}}$.
Unknown Work done on car, $W = ?$

SOLUTION

1. We start with the work done by the normal force, $\vec{\mathbf{N}}$. From the figure we see that $\theta = 90°$ for this force:

$$W_N = Nd \cos \theta = Nd \cos 90° = Nd(0) = 0$$

2. For the force of air resistance, $\theta = 180°$:

$$W_{\text{air}} = F_{\text{air}}d \cos 180° = F_{\text{air}}d(-1) = -F_{\text{air}}d$$

3. For gravity the angle θ is $\theta = 90° - \phi$, as indicated in the sketch. Recall that $\cos(90° - \phi) = \sin \phi$ (see Appendix A):

$$W_{mg} = mgd \cos (90° - \phi) = mgd \sin \phi$$

4. The total work is the sum of the individual works:

$$W_{\text{total}} = W_N + W_{\text{air}} + W_{mg} = 0 - F_{\text{air}}d + mgd \sin \phi$$

INSIGHT

The normal force is perpendicular to the motion of the car, and thus does no work. Air resistance points in a direction that opposes the motion, so it does negative work. On the other hand, gravity has a component in the direction of motion; therefore, its work is positive. The physical significance of positive, negative, and zero work will be discussed in detail in the next section.

PRACTICE PROBLEM

Calculate the total work done on a 1550-kg car as it coasts 20.4 m down a hill with $\phi = 5.00°$. Let the force due to air resistance be 15.0 N.
[**Answer:** $W_{\text{total}} = W_N + W_{\text{air}} + W_{mg} = 0 - F_{\text{air}}d + mgd \sin \phi = 0 - 306\,\text{J} + 2.70 \times 10^4\,\text{J} = 2.67 \times 10^4\,\text{J}$]

Some related homework problems: Problem 15, Problem 63

In the next Example, we first sum the forces acting on the car to find F_{total}. Once the total force is determined, we calculate the total work using $W_{total} = F_{total}d \cos \theta$.

EXAMPLE 7-6 · A COASTING CAR II

Consider the car described in Example 7-5. Calculate the total work done on the car using $W_{total} = F_{total}d \cos \theta$.

PICTURE THE PROBLEM

First, we choose the x axis to point down the slope, and the y axis to be at right angles to the slope. With this choice, there is no acceleration in the y direction, which means that the total force in that direction must be zero. As a result, the total force acting on the car is in the x direction. The magnitude of the total force is $mg \sin \phi - F_{air}$, as can be seen in our sketch.

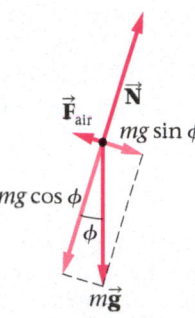
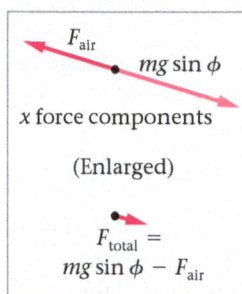

REASONING AND STRATEGY

We begin by finding the x component of each force vector, and then we sum them to find the total force acting on the car. As can be seen from the sketch, the total force points in the positive x direction—that is, in the same direction as the displacement. Therefore, the angle θ in $W = F_{total}d \cos \theta$ is zero.

Known Mass of car, m; angle of incline, ϕ; distance, d; forces acting on car, $\vec{N}$, $\vec{F}_{air}$, and $m\vec{g}$.
Unknown Work done on car, $W = ?$

SOLUTION

1. Referring to the sketch, we see that the *magnitude* of the total force is $mg \sin \phi$ minus F_{air}:

$$F_{total} = mg \sin \phi - F_{air}$$

2. The *direction* of $\vec{F}_{total}$ is the same as the direction of $\vec{d}$; thus $\theta = 0°$. We can now calculate W_{total}:

$$W_{total} = F_{total}d \cos \theta = (mg \sin \phi - F_{air})d \cos 0°$$
$$= mgd \sin \phi - F_{air}d$$

INSIGHT

Notice that we were careful to calculate both the magnitude and the direction of the total force. The magnitude (which is always positive) gives F_{total} and the direction gives $\theta = 0°$, allowing us to use $W_{total} = F_{total}d \cos \theta$.

PRACTICE PROBLEM

Suppose the total work done on a 1620-kg car as it coasts 25.0 m down a hill with $\phi = 6.00°$ is $W_{total} = 3.75 \times 10^4$ J. Find the magnitude of the force due to air resistance. [**Answer:** $F_{air}d = -W_{total} + mgd \sin \phi = 4030$ J; thus $F_{air} = (4030\,\text{J})/d = 161$ N]

Some related homework problems: Problem 15, Problem 63

PROBLEM-SOLVING NOTE

Work Can Be Positive, Negative, or Zero

When you calculate work, be sure to keep track of whether it is positive or negative. The distinction is important, since positive work increases speed, whereas negative work decreases speed. Zero work, of course, has no effect on speed.

Enhance Your Understanding (Answers given at the end of the chapter)

1. A block slides a distance d to the right on a horizontal surface, as indicated in **FIGURE 7-6**. Rank the four forces that act on the block (F, N, mg, f_k) in order of the amount of work they do, from most negative to most positive. Indicate ties where appropriate.

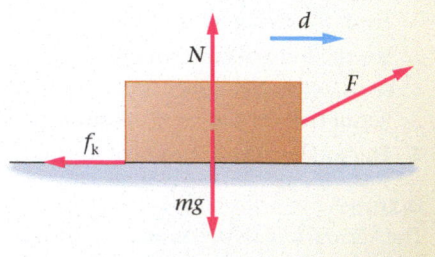

▲ **FIGURE 7-6** Enhance Your Understanding 1.

Section Review

- The work done by a force is $W = (F \cos \theta)d = Fd \cos \theta$, where F is the force, d is the distance, and θ is the angle between the force and displacement.

CONTINUED

- Work is positive when the force and displacement are in the same direction, zero if the force is perpendicular to the displacement, and negative if the force and displacement are in opposite directions.

7-2 Kinetic Energy and the Work–Energy Theorem

Suppose you drop an apple. As it falls, gravity does positive work on it, as indicated in **FIGURE 7-7**, and its speed increases. If you toss the apple upward, gravity does negative work on it, and the apple slows down. In general, whenever the total work done on an object is positive, its speed increases; whenever the total work is negative, its speed decreases. In this section we derive an important result, the **work–energy theorem**, which makes this connection between work and change in speed precise.

Work and Kinetic Energy   To begin, consider an apple of mass m falling through the air, and suppose that two forces act on the apple—gravity, $m\vec{g}$, and the average force of air resistance, $\vec{F}_{air}$. The total force acting on the apple, $\vec{F}_{total}$, gives the apple a constant downward acceleration of magnitude

$$a = \frac{F_{total}}{m}$$

Since the total force is downward and the motion is downward, the work done on the apple is positive.

Now, suppose that the initial speed of the apple is v_i, and that after falling a distance d its speed increases to v_f. The apple falls with constant acceleration a; hence constant-acceleration kinematics (Equation 2–12) gives

$$v_f{}^2 = v_i{}^2 + 2ad$$

With a slight rearrangement we find

$$2ad = v_f{}^2 - v_i{}^2$$

Next, we substitute $a = F_{total}/m$ into this equation:

$$2\left(\frac{F_{total}}{m}\right)d = v_f{}^2 - v_i{}^2$$

Multiplying both sides by m and dividing by 2 yields

$$F_{total}d = \tfrac{1}{2}mv_f{}^2 - \tfrac{1}{2}mv_i{}^2$$

In this expression, $F_{total}d$ is simply the total work done on the apple. Thus we find

$$W_{total} = \tfrac{1}{2}mv_f{}^2 - \tfrac{1}{2}mv_i{}^2$$

This shows that the total work is directly related to the change in speed. Notice that $W_{total} > 0$ means $v_f > v_i$, $W_{total} < 0$ means $v_f < v_i$, and $W_{total} = 0$ means $v_f = v_i$.

The quantity $\tfrac{1}{2}mv^2$ in the equation for W_{total} has a special significance in physics. We call it the **kinetic energy**, K:

Definition of Kinetic Energy, K

$K = \tfrac{1}{2}mv^2$ 7-6

SI unit: $kg \cdot m^2/s^2$ = joule, J

In general, the kinetic energy of an object is the energy due to its motion. We measure kinetic energy in joules, the same units as work, and both kinetic energy and work are scalars. Unlike work, however, kinetic energy is never negative. Instead, K is always greater than or equal to zero, independent of the direction of motion or the direction of any forces.

Force is in the direction of displacement . . .

. . . so positive work is done on the apple. $\vec{d}$ $m\vec{g}$

This causes the apple to speed up.

Apple falling:
$W > 0$, speed increases

(a)

Negative work done on the apple . . .

$\vec{d}$ $m\vec{g}$

. . . causes it to slow down.

Apple tossed upward:
$W < 0$, speed decreases

(b)

▲ **FIGURE 7-7 Gravitational work (a)** The work done by gravity on an apple that moves downward is positive. If the apple is in free fall, this positive work will result in an increase in speed. **(b)** In contrast, the work done by gravity on an apple that moves upward is negative. If the apple is in free fall, the negative work done by gravity will result in a decrease of speed.

**PHYSICS
IN CONTEXT
Looking Back**

The kinematic equations of motion for constant acceleration (Chapters 2 and 4) are used in the derivation of kinetic energy.

**PHYSICS
IN CONTEXT
Looking Ahead**

In Chapter 8 we introduce the concept of potential energy. The combination of kinetic and potential energy is referred to as mechanical energy, which will play a central role in our discussion of the conservation of energy.

TABLE 7-2 Typical Kinetic Energies

Source	Approximate kinetic energy (J)
Jet aircraft at 500 mi/h	10^9
Car at 60 mi/h	10^6
Home-run baseball	10^2
Person at walking speed	50
Housefly in flight	10^{-4}

Big Idea 2 Kinetic energy is one-half mass times velocity squared. The total work done on an object is equal to the change in its kinetic energy.

**PHYSICS
IN CONTEXT
Looking Ahead**

The kinetic energy before and after a collision is an important characterizing feature, as we shall see in Chapter 9. In addition, look for kinetic energy to reappear when we discuss rotational motion in Chapter 10, and again when we study ideal gases in Chapter 17.

To get a feeling for typical values of kinetic energy, consider your kinetic energy when jogging. If we assume a mass of about 62 kg and a speed of 2.5 m/s, your kinetic energy is $K = \frac{1}{2}(62 \text{ kg})(2.5 \text{ m/s})^2 = 190 \text{ J}$. Additional examples of kinetic energy are given in Table 7-2.

EXERCISE 7-7 KINETIC ENERGY TAKES OFF

A small airplane moving along the runway during takeoff has a mass of 690 kg and a kinetic energy of 25,000 J. **(a)** What is the speed of the plane? **(b)** By what multiplicative factor does the kinetic energy of the plane change if its speed is tripled?

REASONING AND SOLUTION
(a) The kinetic energy of an object is given by $K = \frac{1}{2}mv^2$. We know that $K = 25{,}000 \text{ J}$ and $m = 690 \text{ kg}$; therefore the speed is

$$v = \sqrt{\frac{2K}{m}} = \sqrt{\frac{2(25{,}000 \text{ J})}{690 \text{ kg}}} = 8.5 \text{ m/s}$$

(b) Kinetic energy depends on the speed squared, and hence tripling the speed increases the kinetic energy by a factor of nine.

The Work–Energy Theorem   In the preceding discussion we used a calculation of work to derive an expression for the kinetic energy of an object. The precise connection we derived between these quantities is known as the **work–energy theorem.** This theorem can be stated as follows:

Work–Energy Theorem
The *total work* done on an object is equal to the *change* in its kinetic energy:

$$W_{\text{total}} = \Delta K = \tfrac{1}{2}mv_f{}^2 - \tfrac{1}{2}mv_i{}^2 \qquad \text{7-7}$$

Thus, the work–energy theorem says that when a force acts on an object over a distance—doing work on it—the result is a change in the speed of the object, and hence a change in its energy of motion. Equation 7-7 is the quantitative expression of this connection.

We have derived the work–energy theorem for a force that is constant in direction and magnitude, but it is valid for any force. In fact, the work–energy theorem is completely general, making it one of the most important and fundamental results in physics. It is also a very handy tool for problem solving, as we shall see many times throughout this text.

EXERCISE 7-8 WORK TO ACCELERATE

A 220-kg motorcycle is cruising at 14 m/s. What is the total work that must be done on the motorcycle to increase its speed to 19 m/s?

REASONING AND SOLUTION
From the work–energy theorem, we know that the total work required to change an object's kinetic energy is $W_{\text{total}} = \Delta K = \frac{1}{2}mv_f{}^2 - \frac{1}{2}mv_i{}^2$. Given that $m = 220 \text{ kg}$, $v_i = 14 \text{ m/s}$, and $v_f = 19 \text{ m/s}$, we find

$$W_{\text{total}} = \tfrac{1}{2}mv_f{}^2 - \tfrac{1}{2}mv_i{}^2$$
$$= \tfrac{1}{2}(220 \text{ kg})(19 \text{ m/s})^2 - \tfrac{1}{2}(220 \text{ kg})(14 \text{ m/s})^2 = 18{,}000 \text{ J}$$

This much energy could lift a 90-kg person through a vertical distance of about 20 m.

We now present Examples showing how the work–energy theorem is used in practical situations.

EXAMPLE 7-9   HIT THE BOOKS

A 4.10-kg box of books is lifted vertically from rest a distance of 1.60 m with a constant, upward applied force of 52.7 N. Find **(a)** the work done by the applied force, **(b)** the work done by gravity, and **(c)** the final speed of the box.

PICTURE THE PROBLEM

Our sketch shows that the direction of motion of the box is upward. In addition, we see that the applied force, $\vec{F}_{app}$, is upward and the force of gravity, $m\vec{g}$, is downward. Finally, the box is lifted from rest ($v_i = 0$) through a distance $\Delta y = 1.60$ m.

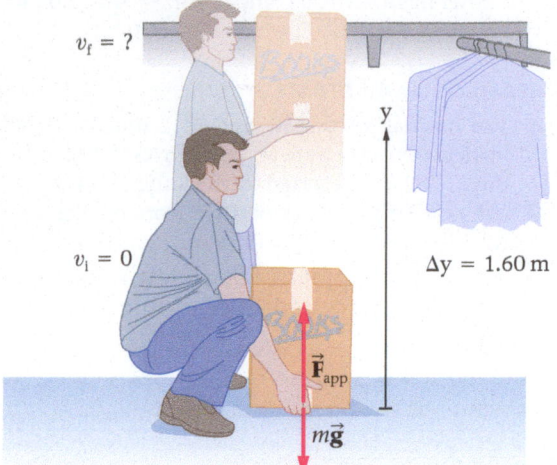

REASONING AND STRATEGY

The applied force is in the direction of motion, so the work it does, W_{app}, is positive. Gravity is opposite in direction to the motion; thus its work, W_{mg}, is negative. The total work is the sum of W_{app} and W_{mg} and the final speed of the box is found by applying the work–energy theorem, $W_{total} = \Delta K$.

Known   Mass of box, $m = 4.10$ kg; vertical distance, $\Delta y = 1.60$ m; applied force, $F_{app} = 52.7$ N.

Unknown   **(a)** Work done by applied force, $W_{app} = ?$ **(b)** work done by gravity, $W_{mg} = ?$ **(c)** Final speed of box, $v_f = ?$

SOLUTION

Part (a)

1. First we find the work done by the applied force. In this case, $\theta = 0°$ and the distance is $\Delta y = 1.60$ m:

$$W_{app} = F_{app} \cos 0° \, \Delta y = (52.7 \text{ N})(1)(1.60 \text{ m}) = 84.3 \text{ J}$$

Part (b)

2. Next, we calculate the work done by gravity. The distance is $\Delta y = 1.60$ m, as before, but now $\theta = 180°$:

$$W_{mg} = mg \cos 180° \, \Delta y$$
$$= (4.10 \text{ kg})(9.81 \text{ m/s}^2)(-1)(1.60 \text{ m}) = -64.4 \text{ J}$$

Part (c)

3. The total work done on the box, W_{total}, is the sum of W_{app} and W_{mg}:

$$W_{total} = W_{app} + W_{mg} = 84.3 \text{ J} - 64.4 \text{ J} = 19.9 \text{ J}$$

4. To find the final speed, v_f, we apply the work–energy theorem. Recall that the box started at rest; thus $v_i = 0$:

$$W_{total} = \tfrac{1}{2}mv_f^2 - \tfrac{1}{2}mv_i^2 = \tfrac{1}{2}mv_f^2$$
$$v_f = \sqrt{\frac{2W_{total}}{m}} = \sqrt{\frac{2(19.9 \text{ J})}{4.10 \text{ kg}}} = 3.12 \text{ m/s}$$

INSIGHT

As a check on our result, we can find v_f in a completely different way. First, calculate the acceleration of the box with the result $a = (F_{app} - mg)/m = 3.04 \text{ m/s}^2$. Next, use this result in the kinematic equation $v^2 = v_0^2 + 2a\Delta y$. With $v_0 = 0$ and $\Delta y = 1.60$ m, we find $v = 3.12$ m/s, in agreement with the results using the work–energy theorem.

PRACTICE PROBLEM — PREDICT/CALCULATE

(a) If the box is lifted only a quarter of the distance, $\Delta y = (1.60 \text{ m})/4 = 0.400$ m, is the final speed 1/8, 1/4, or 1/2 of the value found in this Example? Explain. **(b)** Determine v_f in this case.   [**Answer: (a)** Work depends linearly on Δy, and v_f depends on the square root of the work. As a result, it follows that the final speed is $\sqrt{1/4} = \tfrac{1}{2}$ the value found in Step 4. **(b)** We find $v_f = \tfrac{1}{2}(3.12 \text{ m/s}) = 1.56 \text{ m/s}$.]

Some related homework problems: Problem 28, Problem 29, Problem 65

In the previous Example the initial speed was zero. This is not always the case, of course. The next Example considers a case with a nonzero initial speed.

EXAMPLE 7-10   GOING FOR A STROLL

When a father takes his daughter out for a spin in a stroller, he exerts a force $\vec{F}$ with a magnitude of 44.0 N at an angle of 32.0° below the horizontal, as shown in the sketch. The stroller and child together have a mass $m = 22.7$ kg. **(a)** How much work does the father do on the stroller when he pushes it through a horizontal distance $d = 1.13$ m? **(b)** If the initial speed of the stroller is $v_i = 1.37$ m/s, what is its final speed? (Assume the stroller rolls with negligible friction.)

CONTINUED

PICTURE THE PROBLEM

Our sketch shows the direction of motion and the directions of each of the forces. The normal forces and the force due to gravity are vertical, whereas the displacement is horizontal. The force exerted by the father has a vertical component, $-F \sin 32.0°$, and a horizontal component, $F \cos 32.0°$, where $F = 44.0$ N.

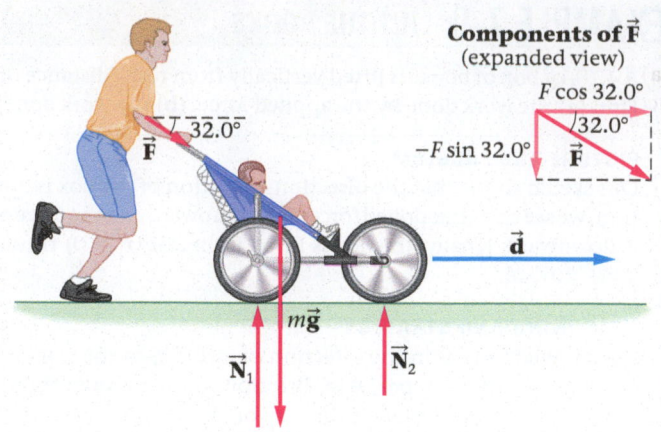

REASONING AND STRATEGY

a. The normal forces ($\vec{N}_1$ and $\vec{N}_2$) and the weight ($m\vec{g}$) do no work because they are at right angles to the horizontal displacement. The force exerted by the father, however, has a horizontal component that does positive work on the stroller. Therefore, the total work is simply the work done by the father.

b. After calculating the work in part (a), we can find the final speed v_f by applying the work–energy theorem with $v_i = 1.37$ m/s.

Known   Force, $F = 44.0$ N; angle below horizontal, $\theta = 32.0°$; mass, $m = 22.7$ kg; distance, $d = 1.13$ m; initial speed, $v_i = 1.37$ m/s.

Unknown   (a) Work done by the father, $W_{father} = ?$ (b) Final speed, $v_f = ?$

SOLUTION

Part (a)

1. The work done by the father is the horizontal component of his force times the distance, $(F \cos \theta)d$. This is also the total work done on the stroller:

$$W_{father} = (F \cos \theta)d$$
$$= (44.0 \, \text{N})(\cos 32.0°)(1.13 \, \text{m})$$
$$= 42.2 \, \text{J} = W_{total}$$

Part (b)

2. Use the work–energy theorem to solve for the final speed:

$$W_{total} = \Delta K = \tfrac{1}{2}mv_f^2 - \tfrac{1}{2}mv_i^2$$
$$\tfrac{1}{2}mv_f^2 = W_{total} + \tfrac{1}{2}mv_i^2$$
$$v_f = \sqrt{\frac{2W_{total}}{m} + v_i^2}$$

3. Substitute numerical values to get the desired result:

$$v_f = \sqrt{\frac{2(42.2 \, \text{J})}{22.7 \, \text{kg}} + (1.37 \, \text{m/s})^2}$$
$$= 2.37 \, \text{m/s}$$

INSIGHT

Notice that the speed of the stroller increased by 1.00 m/s, from 1.37 m/s to 2.37 m/s. If the stroller had started from rest, would its speed increase from zero to 1.00 m/s? No. The work–energy theorem depends on the *square* of the speeds, and hence speeds don't simply add or subtract in an intuitive "linear" way. The final speed in this case is greater than 1.00 m/s, as we show in the following Practice Problem.

PRACTICE PROBLEM

Suppose the stroller starts at rest. What is its final speed in this case?   [**Answer:** $v_f = \sqrt{2W_{total}/m} = 1.93$ m/s, almost twice what simple "linear reasoning" would suggest]

Some related homework problems: Problem 29, Problem 62

PROBLEM-SOLVING NOTE

Be Careful About Linear Reasoning

Though some relationships are linear—if you *double* the mass, you *double* the kinetic energy—others are not. For example, if you *double* the speed, you *quadruple* the kinetic energy. Be careful not to jump to conclusions based on linear reasoning.

The final speeds in the previous Examples could have been found using Newton's laws and the constant-acceleration kinematics of Chapter 2, as indicated in the Insight in Example 7-9. The work–energy theorem provides an alternative method of calculation that is often much easier to apply than Newton's laws. We return to this point in Chapter 8.

CONCEPTUAL EXAMPLE 7-11   COMPARE THE WORK

(a) To accelerate a certain car from rest to the speed v requires the work W_1, as shown in the sketch. The work needed to accelerate the car from v to $2v$ is W_2. Which of the following statements is correct: $W_2 = W_1$, $W_2 = 2W_1$, $W_2 = 3W_1$, $W_2 = 4W_1$? **(b)** Which of the following is the *best explanation* for your prediction?

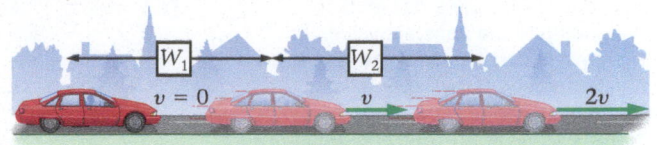

I.   The increase in speed is the same, so the work is also the same.

II. To double the speed requires double the work.

III. Kinetic energy depends on v^2, and hence it takes four times as much work to increase the speed to $2v$.

IV. Four times as much work is required to go from 0 to $2v$ as to go from 0 to v. Therefore, the work required to increase the speed from v to $2v$ is three times the original work.

REASONING AND DISCUSSION

A common misconception is to reason that because we increase the speed by the same amount in each case, the work required is the same. It is not, and the reason is that work depends on the speed squared rather than on the speed itself.

To see how this works, first calculate W_1, the work needed to go from rest to a speed v. From the work–energy theorem, with $v_i = 0$ and $v_f = v$, we find $W_1 = \frac{1}{2}mv_f^2 - \frac{1}{2}mv_i^2 = \frac{1}{2}mv^2$. Similarly, the work W_2 needed to go from $v_i = v$ to $v_f = 2v$ is $W_2 = \frac{1}{2}m(2v)^2 - \frac{1}{2}mv^2 = 3\left(\frac{1}{2}mv^2\right) = 3W_1$.

ANSWER

(a) The required work is $W_2 = 3W_1$. **(b)** The best explanation is IV.

Enhance Your Understanding   (Answers given at the end of the chapter)

2. An object has an initial kinetic energy of 100 J. A few minutes later, after a single external force has acted on the object, its kinetic energy is 200 J. Is the work done by the force positive, negative, or zero? Explain.

Section Review

• The kinetic energy of an object is one-half its mass times its velocity squared.

• The total work done on an object is equal to the change in its kinetic energy.

7-3 Work Done by a Variable Force

Thus far we have calculated work only for constant forces, yet most forces in nature vary with position. For example, the force exerted by a spring depends on how far the spring is stretched, and the force of gravity between planets depends on their separation. In this section we show how to calculate the work for a force that varies with position.

A Graphical Interpretation of Work   First, let's review briefly the case of a constant force, and develop a graphical interpretation of work. **FIGURE 7-8** shows a constant force plotted versus position, x. If the force acts in the positive x direction and moves an object a distance d, from x_1 to x_2, the work it does is $W = Fd = F(x_2 - x_1)$. Referring to the figure, we see that the work is equal to the shaded area[1] between the force line and the x axis.

Next, consider a force that has the value F_1 from $x = 0$ to $x = x_1$ and a different value F_2 from $x = x_1$ to $x = x_2$, as in **FIGURE 7-9 (a)**. The work in this case is the sum of the works done by F_1 and F_2. Therefore, $W = F_1 x_1 + F_2(x_2 - x_1)$. This, again, is the area between the force lines and the x axis. Clearly, this type of calculation can be extended to a force with any number of different values, as indicated in **FIGURE 7-9 (b)**.

If a force varies continuously with position, we can approximate it with a series of constant values that follow the shape of the curve, as shown in **FIGURE 7-10 (a)**. It follows that the work done by the continuous force is approximately equal to the area of the corresponding

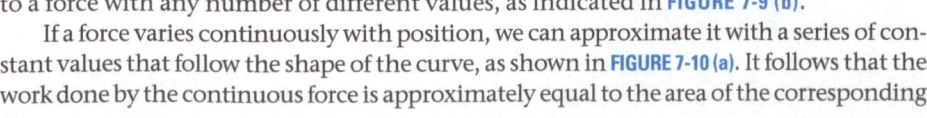

▲ **FIGURE 7-8 Graphical representation of the work done by a constant force** A constant force F acting through a distance d does the work $W = Fd$. Note that Fd is also equal to the shaded area between the force line and the x axis.

---

[1]Usually, area has the dimensions of (length) × (length), or length^2. In this case, however, the vertical axis is force and the horizontal axis is distance. As a result, the dimensions of area are (force) × (distance), which in SI units is N·m = J.

▼ **FIGURE 7-10 Work done by a continuously varying force** (a) A continuously varying force can be approximated by a series of constant values that follow the shape of the curve. **(b)** The work done by the continuous force is approximately equal to the area of the small rectangles corresponding to the constant values of force shown in part (a). **(c)** In the limit of an infinite number of vanishingly small rectangles, we see that the work done by the force is equal to the area between the force curve and the *x* axis.

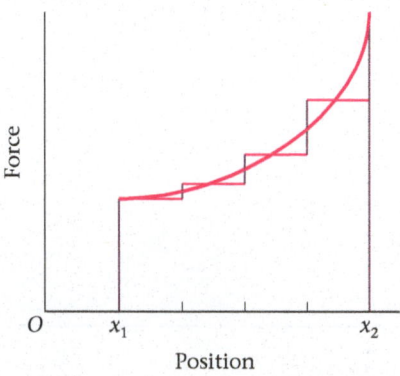

(a) Approximating a continuous force

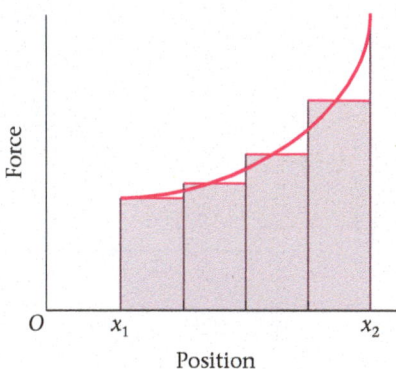

(b) Approximating the work done by a continuous force

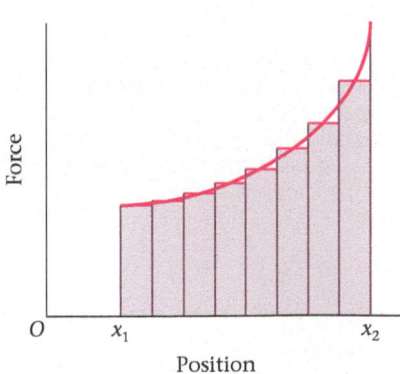

(c) A better approximation

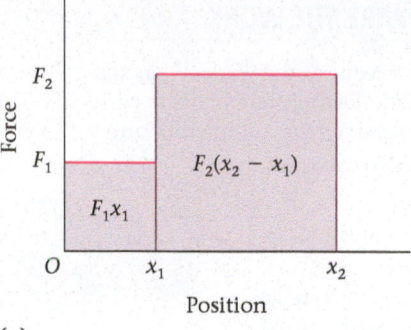

(a)

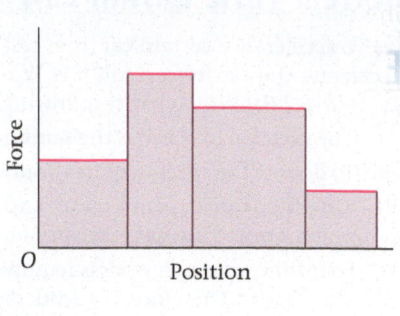

(b)

▲ **FIGURE 7-9 Work done by a nonconstant force** (a) A force with a value F_1 from 0 to x_1 and a value F_2 from x_1 to x_2 does the work $W = F_1 x_1 + F_2(x_2 - x_1)$. This is simply the combined area of the two shaded rectangles. **(b)** If a force takes on a number of different values, the work it does is still the total area between the force lines and the *x* axis, just as in part (a).

rectangles, as **FIGURE 7-10 (b)** shows. The approximation can be made better by using more rectangles, as illustrated in **FIGURE 7-10 (c)**. In the limit of an infinite number of vanishingly small rectangles, the area of the rectangles becomes identical to the area under the force curve. Hence this area is the work done by the continuous force. To summarize:

> The work done by a force in moving an object from one position to another is equal to the corresponding area between the force curve and the *x* axis.

Work for the Spring Force A case of particular interest is a spring. If we recall that the force exerted by a spring is given by $F_x = -kx$ (Section 6-2), it follows that the force we must exert to hold the spring at the position *x* is $+kx$. This is illustrated in **FIGURE 7-11**, where we also show that the corresponding force curve is a straight line extending from the origin. Therefore, the work we do in stretching a spring from $x = 0$ (equilibrium) to the general position *x* is the shaded triangular area shown in **FIGURE 7-12**. This area is equal to $\frac{1}{2}(\text{base})(\text{height})$, where in this case the base is *x* and the height is *kx*. As a result, the work is $\frac{1}{2}(x)(kx) = \frac{1}{2}kx^2$. Similar reasoning shows that the work needed to compress a spring a distance *x* is also $\frac{1}{2}kx^2$. Therefore:

Work to Stretch or Compress a Spring a Distance *x* from Equilibrium

$$W = \frac{1}{2}kx^2 \qquad\qquad\qquad 7\text{-}8$$

SI unit: joule, J

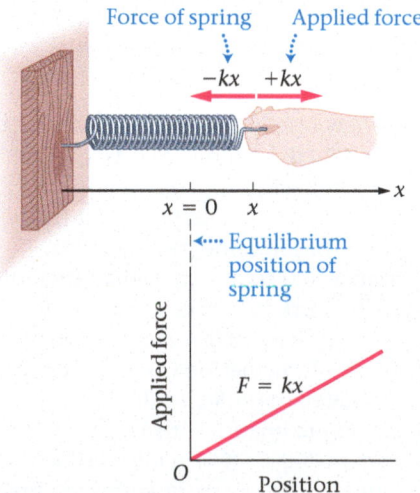

▲ **FIGURE 7-11 Stretching a spring** The force we must exert on a spring to stretch it a distance *x* is $+kx$. Thus, applied force versus position for a spring is a straight line of slope *k*.

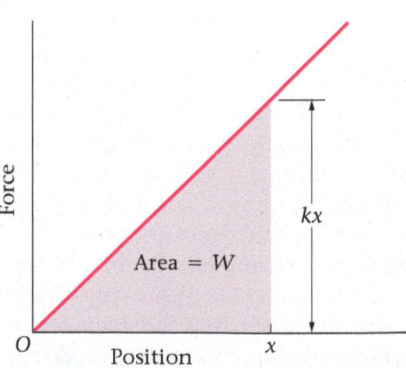

▲ **FIGURE 7-12 Work needed to stretch a spring a distance *x*** The work done is equal to the shaded area, which is a right triangle. The area of the triangle is $\frac{1}{2}(x)(kx) = \frac{1}{2}kx^2$.

We can get a feeling for the amount of work required to compress a typical spring in the following Exercise.

EXERCISE 7-12   SPRING WORK

(a) A toy snake has a spring with a force constant of 230 N/m. How much work is required to stretch this spring 2.5 cm? **(b)** The coiled suspension spring in a car has a force constant of 21,000 N/m. If 3.5 J of work is done to compress this spring, what is the compression distance?

REASONING AND SOLUTION

Work is related to the force constant and compression (or stretch) by $W = \frac{1}{2}kx^2$. We can apply this relationship to both questions. **(a)** Substitute $k = 230$ N/m and $x = 0.025$ m into $W = \frac{1}{2}kx^2$:

$$W = \tfrac{1}{2}kx^2 = \tfrac{1}{2}(230 \text{ N/m})(0.025 \text{ m})^2 = 0.072 \text{ J}$$

(b) Solve $W = \frac{1}{2}kx^2$ for x, then substitute $W = 3.5$ J and $k = 21,000$ N/m:

$$x = \sqrt{\frac{2W}{k}} = \sqrt{\frac{2(3.5 \text{ J})}{21,000 \text{ N/m}}} = 0.018 \text{ m}$$

The work done in compressing or expanding a spring varies with the second power of x, the displacement from equilibrium. The consequences of this dependence are explored throughout the rest of this section.

"Springs" in Other Contexts   Our results for a spring apply to more than just the classic case of a helical coil of wire. In fact, any flexible structure satisfies the relationships $F_x = -kx$ and $W = \frac{1}{2}kx^2$, given the appropriate value of the force constant, k, and small enough displacements, x. Several examples were mentioned in Section 6-2.

Here we consider an example from the field of nanotechnology—namely, the cantilevers used in **atomic-force microscopy** (AFM). As we show in Example 7-13, a typical atomic-force cantilever is basically a thin silicon bar about 250 μm in length, supported at one end like a diving board, with a sharp, hanging point at the other end. When the point is pulled across the surface of a material—like an old-fashioned phonograph needle in the groove of a record—individual atoms on the surface cause the point to move up and down, deflecting the cantilever. These deflections, which can be measured by reflecting a laser beam from the top of the cantilever, are then converted into an atomic-level picture of the surface, as shown in **FIGURE 7-13**.

A typical force constant for an AFM cantilever is on the order of 1 N/m, much smaller than the 100–500-N/m force constant of a common lab spring. The implications of this are discussed in the following Example.

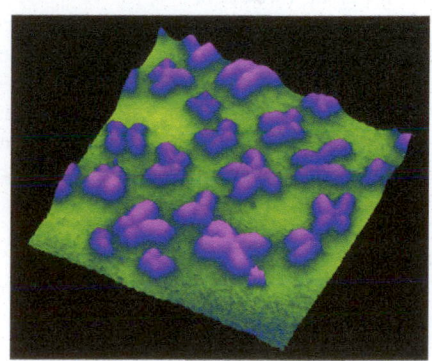

▲ **FIGURE 7-13** Human chromosomes, as imaged by an atomic-force microscope.

EXAMPLE 7-13   FLEXING AN AFM CANTILEVER

The work required to deflect a typical AFM cantilever by 0.10 nm is 1.2×10^{-20} J. **(a)** What is the force constant of the cantilever, treating it as an ideal spring? **(b)** How much work is required to increase the deflection of the cantilever from 0.10 nm to 0.20 nm?

PICTURE THE PROBLEM

The upper sketch shows the cantilever and its sharp point being dragged across the surface of a material. In the lower sketch, we show an exaggerated view of the cantilever's deflection, and indicate that it is equivalent to the stretch of an "effective" ideal spring with a force constant k.

REASONING AND STRATEGY

a. Given that $W = 1.2 \times 10^{-20}$ J for a deflection of $x = 0.10$ nm, we can find the effective force constant k using $W = \frac{1}{2}kx^2$.

b. To find the work required to deflect from $x = 0.10$ nm to $x = 0.20$ nm, $W_{1\rightarrow 2}$, we calculate the work to deflect from $x = 0$ to $x = 0.20$ nm, $W_{0\rightarrow 2}$, and then subtract the work needed to deflect from $x = 0$ to $x = 0.10$ nm, $W_{0\rightarrow 1}$. (Notice that we *cannot* simply assume that the work to go from $x = 0.10$ nm to $x = 0.20$ nm is the same as the work to go from $x = 0$ to $x = 0.10$ nm.)

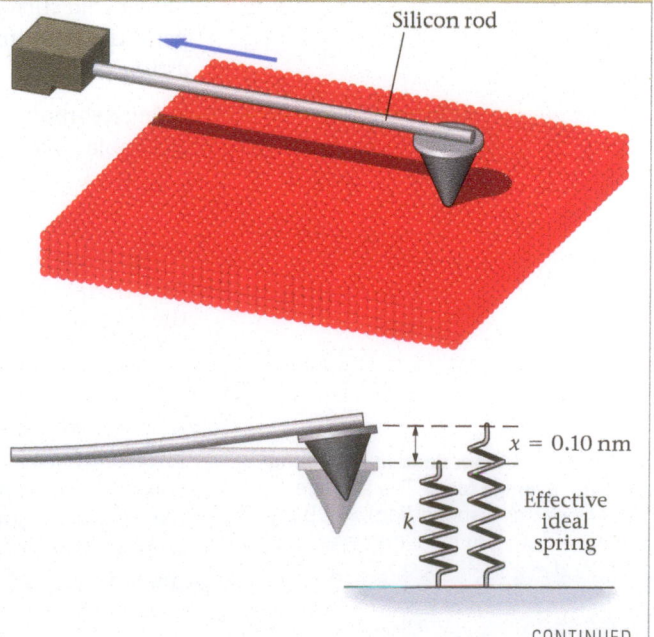

CONTINUED

Known Deflection distance, $x = 0.10 \text{ nm} = 0.10 \times 10^{-9} \text{ m}$; work to deflect, $W = 1.2 \times 10^{-20} \text{ J}$.
Unknown (a) Force constant, $k = ?$ (b) Work to increase deflection, $W_{1 \to 2} = ?$

SOLUTION

Part (a)

1. Solve $W = \frac{1}{2}kx^2$ for the force constant k:

$$k = \frac{2W}{x^2} = \frac{2(1.2 \times 10^{-20} \text{ J})}{(0.10 \times 10^{-9} \text{ m})^2} = 2.4 \text{ N/m}$$

Part (b)

2. First, calculate the work needed to deflect the cantilever from $x = 0$ to $x = 0.20 \text{ nm}$:

$$W_{0 \to 2} = \frac{1}{2}kx^2$$
$$= \frac{1}{2}(2.4 \text{ N/m})(0.2 \times 10^{-9} \text{ m})^2 = 4.8 \times 10^{-20} \text{ J}$$

3. Subtract from the above result the work to deflect from $x = 0$ to $x = 0.10 \text{ nm}$, which the problem statement gives as $1.2 \times 10^{-20} \text{ J}$:

$$W_{1 \to 2} = W_{0 \to 2} - W_{0 \to 1}$$
$$= 4.8 \times 10^{-20} \text{ J} - 1.2 \times 10^{-20} \text{ J} = 3.6 \times 10^{-20} \text{ J}$$

INSIGHT

Our results show that more energy is needed to deflect the cantilever the second 0.10 nm than to deflect it the first 0.10 nm. Why? The reason is that the force of the cantilever increases with distance; thus, the average force over the second 0.10 nm is greater than the average force over the first 0.10 nm. In fact, we can see from the graph that the average force between 0.10 nm and 0.20 nm (0.36 nN) is three times the average force between 0 and 0.10 nm (0.12 nN). It follows that the work required for the second 0.10 nm is three times the work required for the first 0.10 nm.

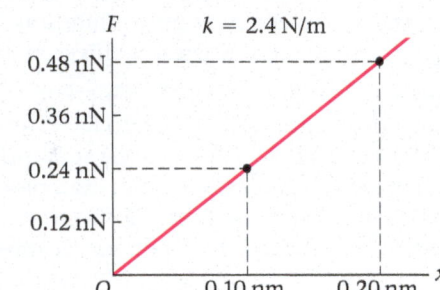

PRACTICE PROBLEM — PREDICT/CALCULATE

A second cantilever has half the force constant of the cantilever in this Example. (a) Is the work required to deflect the second cantilever by 0.20 nm greater than, less than, or equal to the work required to deflect the cantilever in this Example by 0.10 nm? Explain. (b) Determine the work required to deflect the second cantilever 0.20 nm. [Answer: (a) Halving the force constant halves the work, but doubling the deflection quadruples the work. The net effect is that the work increases by a factor of two. (b) The work required is $2.4 \times 10^{-20} \text{ J}$.]

Some related homework problems: Problem 31, Problem 36

BIO Another example of the work required to stretch a flexible structure is the work done inside your eye. The human eye can *accommodate*, or focus on objects at different distances, by using *ciliary muscles* to alter the shape of the flexible lens behind the pupil of the eye. (The physiology and optical properties of the eye are explored in more detail in Chapter 27.) The shorter the distance you attempt to focus your eye, the harder your ciliary muscles must work to change the shape of the lens. This is why long periods of viewing close-up objects, such as a book or a computer screen, can lead to *asthenopia*, or eye strain.

Using Average Force to Calculate Work An equivalent way to calculate the work for a variable force is to multiply the average force, F_{av}, by the distance, d:

$$W = F_{av}d \qquad \qquad 7\text{-}9$$

For a spring that is stretched a distance x from equilibrium, the force varies linearly from 0 to kx. Thus, the average force is $F_{av} = \frac{1}{2}kx$, as indicated in **FIGURE 7-14**. Therefore, the work is

$$W = \frac{1}{2}kx(x) = \frac{1}{2}kx^2$$

As expected, our result agrees with Equation 7-8.

Finally, when you stretch or compress a spring from its equilibrium position, the work you do is always positive. The work done *by* a spring, however, may be positive or negative, depending on the situation. For example, consider a block sliding to the right with an initial speed v_0 on a smooth, horizontal surface, as shown in **FIGURE 7-15 (a)**. When the block begins to compress the spring, as in **FIGURE 7-15 (b)**, the spring exerts a force

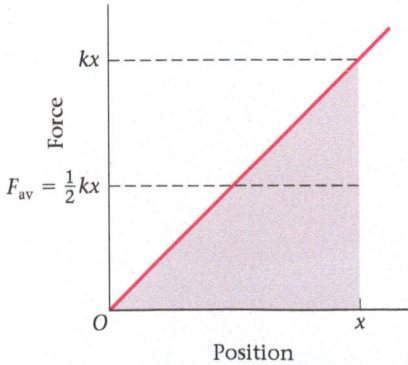

▲ **FIGURE 7-14 Work done in stretching a spring: average force** The average force to stretch a spring from $x = 0$ to x is $F_{av} = \frac{1}{2}kx$, and the work done is $W = F_{av}d = \frac{1}{2}kx(x) = \frac{1}{2}kx^2$.

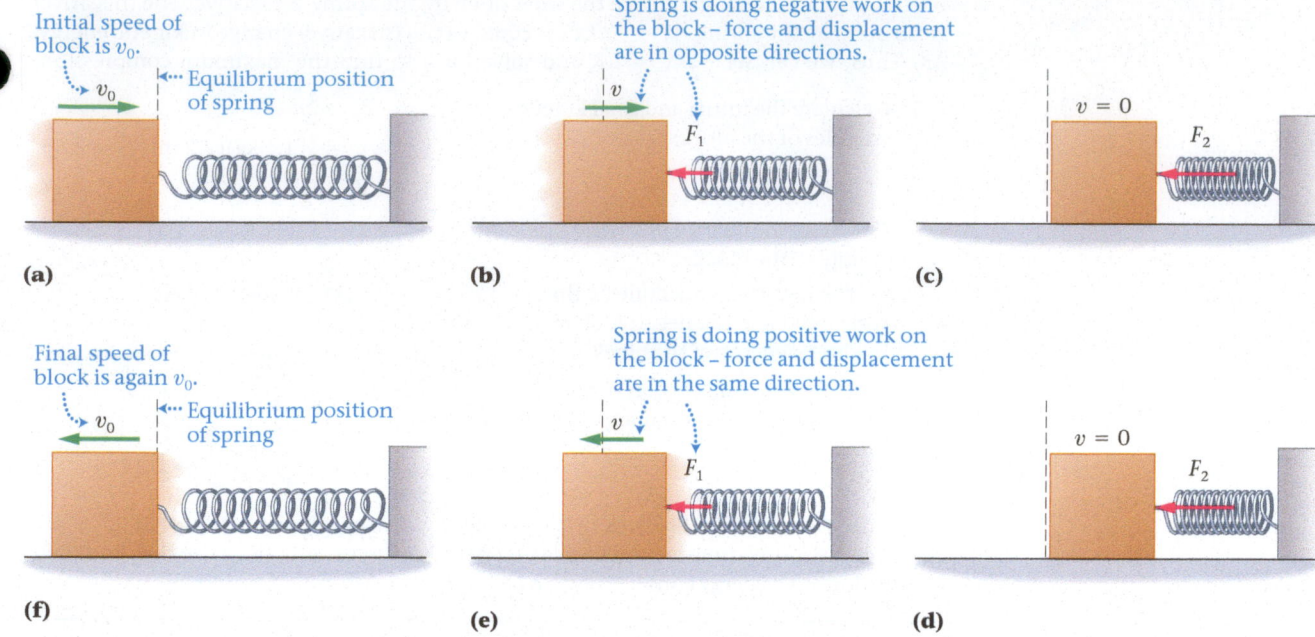

▲ **FIGURE 7-15 The work done by a spring can be positive or negative** **(a)** A block slides to the right on a frictionless surface with a speed v_0 until it encounters a spring. **(b)** The spring now exerts a force to the left—opposite to the block's motion—and hence it does negative work on the block. This causes the block's speed to decrease. **(c)** The negative work done by the spring eventually is equal in magnitude to the block's initial kinetic energy, at which point the block comes to rest momentarily. As the spring expands, **(d)** and **(e)**, it does positive work on the block and increases its speed. **(f)** When the block leaves the spring, its speed is again equal to v_0.

on the block to the left—that is, opposite to the block's direction of motion. As a result, the spring does *negative* work on the block, which causes the block's speed to decrease. Eventually the negative work done by the spring, $W = -\frac{1}{2}kx^2$, is equal in magnitude to the initial kinetic energy of the block. At this point, **FIGURE 7-15 (c)**, the block comes to rest momentarily, and $W = \Delta K = K_f - K_i = 0 - K_i = -K_i = -\frac{1}{2}mv_0^2 = -\frac{1}{2}kx^2$.

After the block comes to rest, the spring expands back to its equilibrium position, as we see in **FIGURE 7-15 (d)–(f)**. During this expansion the force exerted by the spring is in the same direction as the block's motion, and hence it does *positive* work in the amount $W = \frac{1}{2}kx^2$. As a result, the block leaves the spring with the same speed it had initially.

QUICK EXAMPLE 7-14 A BLOCK COMPRESSES A SPRING

A block with a mass of 1.5 kg and an initial speed of $v_i = 2.2$ m/s slides on a frictionless, horizontal surface. The block comes into contact with a spring that is in its equilibrium position, and compresses it until the block comes to rest momentarily. Find the maximum compression of the spring, assuming its force constant is 475 N/m.

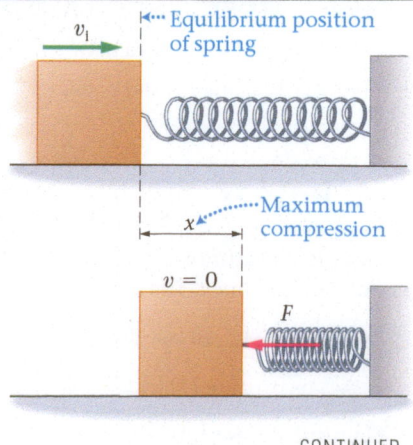

REASONING AND SOLUTION

Our sketch shows the block just as it comes into contact with the spring with an initial speed of $v_i = 2.2$ m/s. We also show the spring at maximum compression, x, when the block is momentarily at rest. As the block comes to rest, the force of the spring is opposite to the

CONTINUED

block's displacement, and hence the work done by the spring is negative. The (negative) work done by the spring, $W = -\frac{1}{2}kx^2$, is equal to the (negative) change in kinetic energy, ΔK. Thus, we can set $-\frac{1}{2}kx^2 = \Delta K$ and solve for x to find the maximum compression.

1. Calculate the initial and final kinetic energies of the block:

$$K_i = \frac{1}{2}mv_i^2$$
$$= \frac{1}{2}(1.5 \text{ kg})(2.2 \text{ m/s})^2 = 3.6 \text{ J}$$
$$K_f = 0$$

2. Calculate the change in kinetic energy of the block:

$$\Delta K = K_f - K_i = -3.6 \text{ J}$$

3. Set the negative work done by the spring equal to the negative change in kinetic energy of the block:

$$-\frac{1}{2}kx^2 = \Delta K = -3.6 \text{ J}$$

4. Solve for the compression, x, and substitute numerical values:

$$x = \sqrt{\frac{-2\Delta K}{k}}$$
$$= \sqrt{\frac{-2(-3.6 \text{ J})}{475 \text{ N/m}}} = 0.12 \text{ m}$$

When the spring brings the block to rest, the kinetic energy it had initially is not lost—it is stored in the spring itself and can be released later. We discuss situations like this, and their connection with energy conservation, in Chapter 8.

Enhance Your Understanding   (Answers given at the end of the chapter)

3. As an object moves along the positive x axis, the force shown in **FIGURE 7-16** acts on it. Is the work done by the force from $x = 0$ to $x = 4$ m greater than, less than, or equal to the work done by the force from $x = 5$ m to $x = 9$ m? Explain.

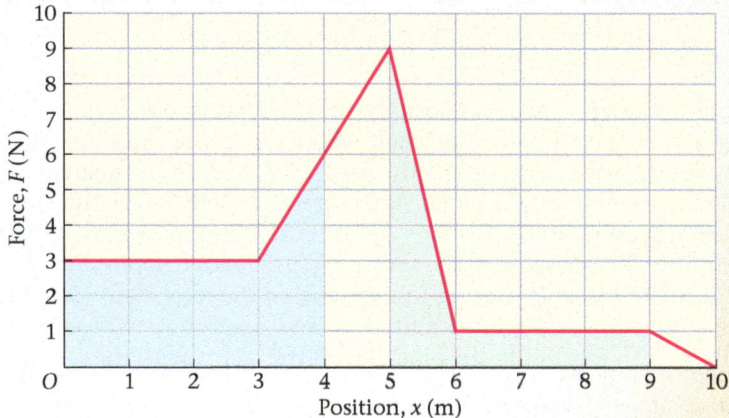

▲ **FIGURE 7-16** Enhance Your Understanding 3.

Section Review

- Work is equal to the area between a force curve and the x axis.
- Work is also equal to the average force times displacement.

7-4 Power

Power is a measure of how *quickly* work is done. To be precise, suppose the work W is performed in the time t. The average power delivered during this time is defined as follows:

Definition of Average Power, P

$$P = \frac{W}{t}$$

7-10

SI unit: J/s = watt, W

For simplicity of notation we drop the usual subscript av for an average quantity and simply understand that the power P refers to an average power unless stated otherwise.

The Dimensions of Power   The dimensions of power are joules (work) per second (time). We define one joule per second to be a watt (W), after James Watt (1736–1819), the Scottish engineer and inventor who played a key role in the development of practical steam engines:

$$1 \text{ watt} = 1 \text{ W} = 1 \text{ J/s} \qquad\qquad 7\text{-}11$$

Of course, the watt is the unit of power used to rate the output of lightbulbs. Another common unit of power is the horsepower (hp), which is used to rate the output of car engines. It is defined as follows:

$$1 \text{ horsepower} = 1 \text{ hp} = 746 \text{ W} \qquad\qquad 7\text{-}12$$

Though it sounds like a horse should be able to produce one horsepower, in fact, a horse can generate only about 2/3 hp for sustained periods. The reason for the discrepancy is that when James Watt defined the horsepower—as a way to characterize the output of his steam engines—he purposely chose a unit that was overly generous to the horse, so that potential investors couldn't complain he was overstating the capability of his engines.

Human Power   To get a feel for the magnitude of the watt and the horsepower, consider the power you might generate when walking up a flight of stairs. Suppose, for example, that an 80.0-kg person walks up a flight of stairs in 20.0 s, and that the altitude gain is 12.0 ft (3.66 m). Referring to Example 7-3 and Conceptual Example 7-4, we find that the work done by the person is $W = mgh = (80.0 \text{ kg})(9.81 \text{ m/s}^2)(3.66 \text{ m}) = 2870 \text{ J}$. To find the power, we simply divide the work by the time: $P = W/t = (2870 \text{ J})/(20.0 \text{ s}) = 144 \text{ W} = 0.193 \text{ hp}$. Thus, a leisurely stroll up the stairs requires about 1/5 hp or 150 W. Similarly, the power produced by a sprinter bolting out of the starting blocks is about 1 hp, and the greatest power most people can produce for sustained periods of time is roughly 1/3 to 1/2 hp. Further examples of power are given in Table 7-3.

Human-powered flight is a feat just barely within our capabilities, since the most efficient human-powered airplanes require a steady power output of about 1/3 hp. In 1979 the *Gossamer Albatross* became the first (and so far the only) human-powered aircraft to fly across the English Channel. This 22.25-mile flight—from Folkestone, England, to Cap Gris-Nez, France—took 2 hours 49 minutes and required a total energy output roughly equivalent to climbing to the top of the Empire State Building 10 times. The *Gossamer Albatross* is shown in midflight in **FIGURE 7-17**.

RWP Power output is also an important factor in the performance of a car. For example, suppose it takes a certain amount of work, W, to accelerate a car from 0 to 60 mi/h. If the average power provided by the engine is P, then according to Equation 7-10, the amount of time required to reach 60 mi/h is $t = W/P$. Clearly, the greater the power P, the less the time required to accelerate. Thus, in a loose way of speaking, we can say that of "how fast it can go fast."

Big Idea **3** Power is the rate at which work is done. The more work that is done in a shorter time, the greater the power.

TABLE 7-3 Typical Values of Power

Source	Approximate power (W)
Hoover Dam	1.34×10^9
Car moving at 40 mi/h	7×10^4
Home stove	1.2×10^4
Sunlight falling on one square meter	1380
Refrigerator	615
Television	200
Person walking up stairs	150
Human brain	20

▲ **FIGURE 7-17** The *Gossamer Albatross* Twice the aircraft touched the surface of the water, but the pilot was able to maintain control.

EXAMPLE 7-15   PASSING FANCY

To pass a slow-moving truck, you want your fancy 1.30×10^3-kg car to accelerate from 13.4 m/s (30.0 mi/h) to 17.9 m/s (40.0 mi/h) in 3.00 s. What is the minimum power required for this pass?

PICTURE THE PROBLEM
Our sketch shows the car accelerating from an initial speed of $v_i = 13.4$ m/s to a final speed of $v_f = 17.9$ m/s. We assume the road is level, so that no work is done against gravity, and that friction and air resistance may be ignored.

CONTINUED

REASONING AND STRATEGY

Power is work divided by time, and work is equal to the change in kinetic energy as the car accelerates. We can determine the change in kinetic energy from the given mass of the car and its initial and final speeds. With this information at hand, we can determine the power with the relationship $P = W/t = \Delta K/t$.

Known Mass of car, $m = 1.30 \times 10^3$ kg; initial speed, $v_i = 13.4$ m/s; final speed, $v_f = 17.9$ m/s; time, $t = 3.00$ s.

Unknown Minimum power, $P = ?$

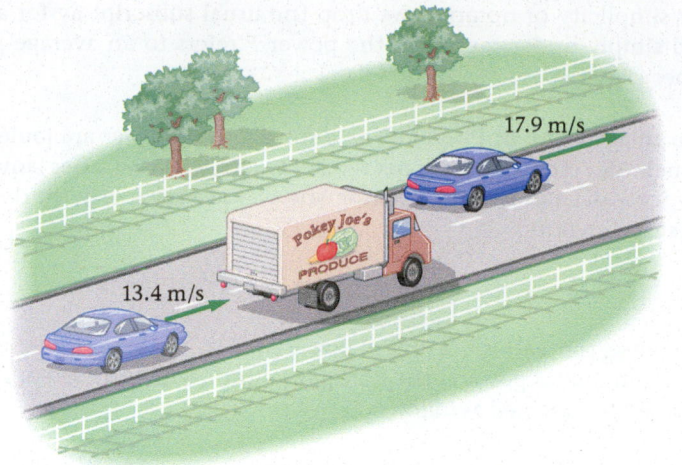

SOLUTION

1. First, calculate the change in kinetic energy:

$$\Delta K = \tfrac{1}{2}mv_f^2 - \tfrac{1}{2}mv_i^2 = \tfrac{1}{2}(1.30 \times 10^3 \text{ kg})(17.9 \text{ m/s})^2$$
$$-\tfrac{1}{2}(1.30 \times 10^3 \text{ kg})(13.4 \text{ m/s})^2$$
$$= 9.16 \times 10^4 \text{ J}$$

2. Divide by time to find the minimum power (the actual power would have to be greater to overcome frictional losses):

$$P = \frac{W}{t} = \frac{\Delta K}{t} = \frac{9.16 \times 10^4 \text{ J}}{3.00 \text{ s}} = 3.05 \times 10^4 \text{ W} = 40.9 \text{ hp}$$

INSIGHT

Suppose that your fancy car continues to produce the same 3.05×10^4 W of power as it accelerates from $v = 17.9$ m/s (40.0 mi/h) to $v = 22.4$ ms (50.0 mi/h). Is the time required more than, less than, or equal to 3.00 s? It will take more than 3.00 s. The reason is that ΔK is greater for a change in speed from 40.0 mi/h to 50.0 mi/h than for a change in speed from 30.0 mi/h to 40.0 mi/h, because K depends on speed squared. Since ΔK is greater, the time $t = \Delta K/P$ is also greater.

PRACTICE PROBLEM

Find the time required to accelerate from 40.0 mi/h to 50.0 mi/h with 3.05×10^4 W of power.
[**Answer:** First, $\Delta K = 1.18 \times 10^5$ J. Next, $P = \Delta K/t$ can be solved for time to give $t = \Delta K/P$. Thus, $t = 3.87$ s.]

Some related homework problems: Problem 43, Problem 56, Problem 72

Power and Speed Finally, consider a system in which a car, or some other object, is moving with a constant speed v. For example, a car might be traveling uphill on a road inclined at an angle θ above the horizontal. To maintain a constant speed, the engine must exert a constant force F equal to the combined effects of friction, gravity, and air resistance, as indicated in **FIGURE 7-18**. Now, as the car travels a distance d, the work done by the engine is $W = Fd$, and the power it delivers is

$$P = \frac{W}{t} = \frac{Fd}{t}$$

▼ **FIGURE 7-18 Driving up a hill** A car traveling uphill at constant speed requires a constant force, F, of magnitude $mg \sin \theta + F_{\text{air res}} + F_{\text{friction}}$ applied in the direction of motion.

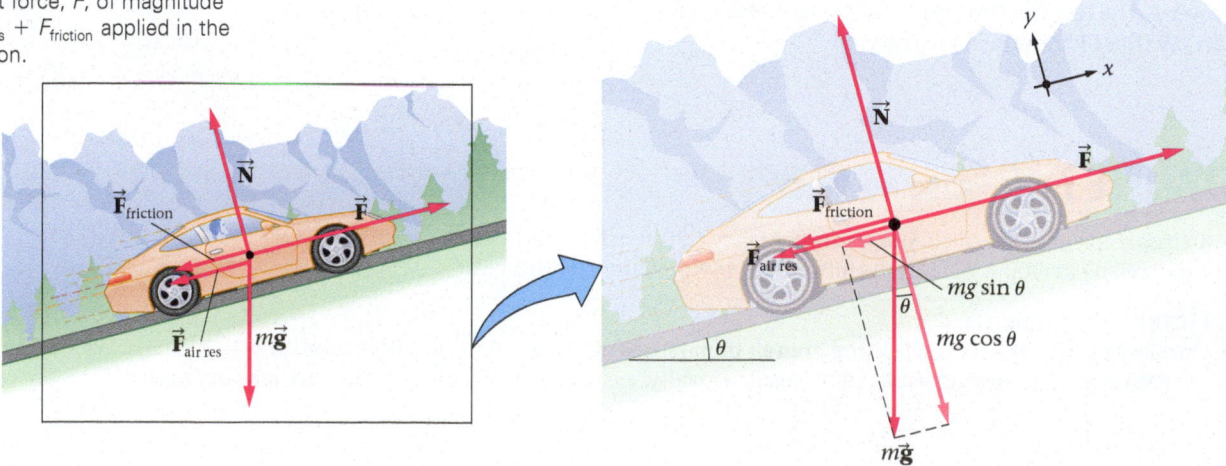

Since the car has a constant speed, $v = d/t$, it follows that

$$P = \frac{Fd}{t} = F\left(\frac{d}{t}\right) = Fv \qquad\qquad \text{7-13}$$

Notice that power is directly proportional to both the force and the speed. For example, suppose you push a heavy shopping cart with a force F. You produce twice as much power when you push at 2 m/s than when you push at 1 m/s, even though you are pushing no harder. It's just that the amount of work you do in a given time period is doubled.

EXERCISE 7-16 UPHILL SPEED

It takes a force of 1450 N to keep a 1300-kg car moving with constant speed up a slope of 5.25°. If the engine delivers 4.10×10^4 W (55 hp) to the drive wheels, what is the maximum speed of the car?

REASONING AND SOLUTION
Solve the power relationship, $P = Fv$, for the speed v, and substitute $P = 4.10 \times 10^4$ W and $F = 1450$ N:

$$v = \frac{P}{F} = \frac{4.10 \times 10^4 \text{ W}}{1450 \text{ N}} = 28.3 \text{ m/s}$$

This is approximately 63 mi/h. The power delivered by the engine overcomes friction and air resistance, and also provides the force needed to lift the car up the hill.

Enhance Your Understanding   (Answers given at the end of the chapter)

4. Engine A generates the power P and operates for 10 s; engine B generates the power $2P$ and operates for 5 s; engine C generates the power $2P$ and operates for 20 s; and engine D generates the power $P/2$ and operates for 40 s. Rank the engines in order of increasing work done. Indicate ties where appropriate.

Section Review

- Power is the rate at which work is done.
- A force F pushing or pulling an object at a speed v produces the power $P = Fv$.

CHAPTER 7 REVIEW

CHAPTER SUMMARY

7-1   WORK DONE BY A CONSTANT FORCE

A force exerted through a distance performs mechanical work.

Force in Direction of Motion
In this, the simplest case, work is force times distance:

$$W = Fd \qquad\qquad \text{7-1}$$

Force at an Angle θ to Motion
Work is the component of force in the direction of motion, $F \cos\theta$, times distance, d:

$$W = (F \cos\theta)d = Fd \cos\theta \qquad\qquad \text{7-3}$$

Negative and Total Work
Work is negative if the force opposes the motion. If more than one force does work, the total work is the sum of the works done by each force separately:

$$W_{\text{total}} = W_1 + W_2 + W_3 + \cdots \qquad\qquad \text{7-4}$$

Equivalently, sum the forces first to find F_{total}, then

$$W_{\text{total}} = (F_{\text{total}} \cos\theta)d = F_{\text{total}}\, d \cos\theta \qquad\qquad \text{7-5}$$

Units

The SI unit of work and energy is the joule, J:

$$1\,\text{J} = 1\,\text{N}\cdot\text{m} \qquad\qquad 7\text{-}2$$

7-2 KINETIC ENERGY AND THE WORK–ENERGY THEOREM

Kinetic energy is one-half mass times speed squared:

$$K = \tfrac{1}{2}mv^2 \qquad\qquad 7\text{-}7$$

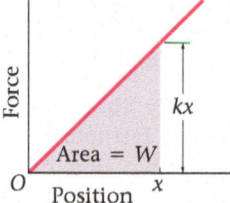

Total work is equal to the change in kinetic energy:

$$W_{\text{total}} = \Delta K = \tfrac{1}{2}mv_{\text{f}}^2 - \tfrac{1}{2}mv_{\text{i}}^2 \qquad\qquad 7\text{-}6$$

7-3 WORK DONE BY A VARIABLE FORCE

Work is equal to the area between the force curve and the displacement on the x axis. For the case of a spring force, the work to stretch or compress a distance x from equilibrium is

$$W = \tfrac{1}{2}kx^2 \qquad\qquad 7\text{-}8$$

7-4 POWER

Average power is the work divided by the time required to do the work:

$$P = \frac{W}{t} \qquad\qquad 7\text{-}10$$

Equivalently, power is force times speed:

$$P = Fv \qquad\qquad 7\text{-}13$$

Units

The SI unit of power is the watt, W:

$$1\,\text{W} = 1\,\text{J/s} \qquad\qquad 7\text{-}11$$

$$746\,\text{W} = 1\,\text{hp} \qquad\qquad 7\text{-}12$$

ANSWERS TO ENHANCE YOUR UNDERSTANDING QUESTIONS

1. Work done by $f_{\text{k}} <$ work done by $N =$ work done by $mg <$ work done by F.

2. The kinetic energy of the object has increased, and hence the total work done on it is positive.

3. The work done from $x = 0$ to $x = 4$ m is greater than the work done from $x = 5$ m to $x = 9$ m because the blue area is larger than the green area.

4. $A = B < D < C$.

CONCEPTUAL QUESTIONS

(Answers to odd-numbered Conceptual Questions can be found in the back of the book.)

For instructor-assigned homework, go to www.masteringphysics.com.

1. Is it possible to do work on an object that remains at rest?

2. A friend makes the statement, "Only the total force acting on an object can do work." Is this statement true or false? If it is true, state why; if it is false, give a counterexample.

3. A friend makes the statement, "A force that is always perpendicular to the velocity of a particle does no work on the particle." Is this statement true or false? If it is true, state why; if it is false, give a counterexample.

4. The net work done on a certain object is zero. What can you say about its speed?

5. Give an example of a frictional force doing positive work.

6. A ski boat moves with constant velocity. Is the net force acting on the boat doing work? Explain.

7. A package rests on the floor of an elevator that is rising with constant speed. The elevator exerts an upward normal force on the package, and hence does positive work on it. Why doesn't the kinetic energy of the package increase?

8. An object moves with constant velocity. Is it safe to conclude that no force acts on the object? Why, or why not?

9. Engine 1 does twice the work of engine 2. Is it correct to conclude that engine 1 produces twice as much power as engine 2? Explain.

10. Engine 1 produces twice the power of engine 2. Is it correct to conclude that engine 1 does twice as much work as engine 2? Explain.

PROBLEMS AND CONCEPTUAL EXERCISES

Answers to odd-numbered Problems and Conceptual Exercises can be found in the back of the book. **BIO** *identifies problems of biological or medical interest;* **CE** *indicates a conceptual exercise.* Predict/Explain *problems ask for two responses: (a) your prediction of a physical outcome, and (b) the best explanation among three provided; and* Predict/Calculate *problems ask for a prediction of a physical outcome, based on fundamental physics concepts, and follow that with a numerical calculation to verify the prediction. On all problems, bullets (•, ••, •••) indicate the level of difficulty.*

SECTION 7-1 WORK DONE BY A CONSTANT FORCE

1. • **CE** A pendulum bob swings from point I to point II along the circular arc indicated in **FIGURE 7-19**. (a) Is the work done on the bob by gravity positive, negative, or zero? Explain. (b) Is the work done on the bob by the string positive, negative, or zero? Explain.

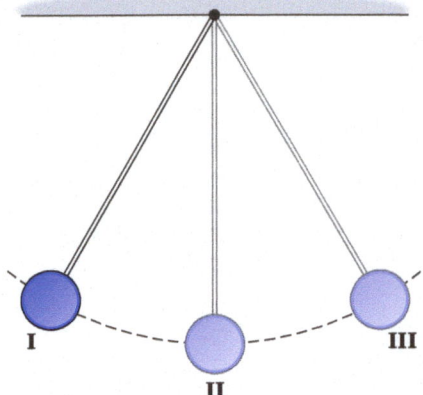

FIGURE 7-19 Problems 1 and 2

2. • **CE** A pendulum bob swings from point II to point III along the circular arc indicated in Figure 7-19. (a) Is the work done on the bob by gravity positive, negative, or zero? Explain. (b) Is the work done on the bob by the string positive, negative, or zero? Explain.

3. • A farmhand pushes a 26-kg bale of hay 3.9 m across the floor of a barn. If she exerts a horizontal force of 88 N on the hay, how much work has she done?

4. • Children in a tree house lift a small dog in a basket 3.85 m up to their house. If it takes 201 J of work to do this, what is the combined mass of the dog and basket?

5. • Early one October, you go to a pumpkin patch to select your Halloween pumpkin. You lift the 3.2-kg pumpkin to a height of 1.2 m, then carry it 50.0 m (on level ground) to the check-out stand. (a) Calculate the work you do on the pumpkin as you lift it from the ground. (b) How much work do you do on the pumpkin as you carry it from the field?

6. • The coefficient of kinetic friction between a suitcase and the floor is 0.272. If the suitcase has a mass of 71.5 kg, how far can it be pushed across the level floor with 642 J of work?

7. • **BIO Peristaltic Work** The human small intestine moves food along its 6.9-m length by means of *peristalsis*, a contraction and relaxation of muscles that propagate in a wave along the digestive tract. If the average force exerted by peristalsis is 0.18 N, how much work does the small intestine do to move food along its entire length?

8. •• Predict/Calculate A tow rope, parallel to the water, pulls a water skier directly behind the boat with constant velocity for a distance of 72 m before the skier falls. The tension in the rope is 140 N. (a) Is the work done on the boat by the rope positive, negative, or zero? Explain. (b) Calculate the work done by the rope on the boat.

9. •• A child pulls a friend in a little red wagon with constant speed. If the child pulls with a force of 16 N for 10.0 m, and the handle of the wagon is inclined at an angle of 25° above the horizontal, how much work does the child do on the wagon?

10. •• A 57-kg packing crate is pulled with constant speed across a rough floor with a rope that is at an angle of 37° above the horizontal. If the tension in the rope is 142 N, how much work is done on the crate to move it 6.1 m?

11. •• Predict/Calculate To clean a floor, a janitor pushes on a mop handle with a force of 50.0 N. (a) If the mop handle is at an angle of 55° above the horizontal, how much work is required to push the mop 0.50 m? (b) If the angle the mop handle makes with the horizontal is increased to 65°, does the work done by the janitor increase, decrease, or stay the same? Explain.

12. •• A small plane tows a glider at constant speed and altitude. If the plane does 2.00×10^5 J of work to tow the glider 145 m and the tension in the tow rope is 2560 N, what is the angle between the tow rope and the horizontal?

13. •• As a snowboarder descends a mountain slope, gravity, the normal force, and friction combine to exert a net force of 320 N in the downhill direction. The net force does 4700 J of work on her during one segment of her descent. If her path cuts across the face of the mountain at a 65° angle from the direction of the net force, over what distance does she travel?

14. •• A young woman on a skateboard is pulled by a rope attached to a bicycle. The velocity of the skateboarder is $\vec{\mathbf{v}} = (4.1 \text{ m/s})\hat{\mathbf{x}}$ and the force exerted on her by the rope is $\vec{\mathbf{F}} = (17 \text{ N})\hat{\mathbf{x}} + (12 \text{ N})\hat{\mathbf{y}}$. (a) Find the work done on the skateboarder by the rope in 25 seconds. (b) Assuming the velocity of the bike is the same as that of the skateboarder, find the work the rope does on the bicycle in 25 seconds.

15. •• To keep her dog from running away while she talks to a friend, Susan pulls gently on the dog's leash with a constant force given by $\vec{\mathbf{F}} = (2.2 \text{ N})\hat{\mathbf{x}} + (1.1 \text{ N})\hat{\mathbf{y}}$. How much work does she do on the dog if its displacement is (a) $\vec{\mathbf{d}} = (0.25 \text{ m})\hat{\mathbf{x}}$, (b) $\vec{\mathbf{d}} = (0.25 \text{ m})\hat{\mathbf{y}}$, or (c) $\vec{\mathbf{d}} = (-0.50 \text{ m})\hat{\mathbf{x}} + (-0.25 \text{ m})\hat{\mathbf{y}}$?

16. •• Water skiers often ride to one side of the center line of a boat, as shown in **FIGURE 7-20**. In this case, the ski boat is traveling at 12 m/s and the tension in the rope is 91 N. If the boat does 3800 J of work on the skier in 50.0 m, what is the angle θ between the tow rope and the center line of the boat?

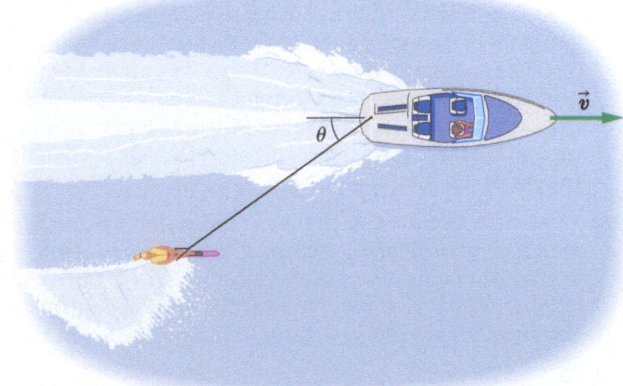

FIGURE 7-20 Problems 16 and 62

SECTION 7-2 KINETIC ENERGY AND THE WORK–ENERGY THEOREM

17. • **CE** A pitcher throws a ball at 90 mi/h and the catcher stops it in her glove. (a) Is the work done on the ball by the pitcher positive, negative, or zero? Explain. (b) Is the work done on the ball by the catcher positive, negative, or zero? Explain.

18. • How much work is needed for a 73-kg runner to accelerate from rest to 7.7 m/s?

19. • **Skylab's Reentry** When Skylab reentered the Earth's atmosphere on July 11, 1979, it broke into a myriad of pieces. One of the largest fragments was a 1770-kg lead-lined film vault, and it landed with an estimated speed of 120 m/s. What was the kinetic energy of the film vault when it landed?

20. • **Predict/Calculate** A 9.50-g bullet has a speed of 1.30 km/s. (a) What is its kinetic energy in joules? (b) What is the bullet's kinetic energy if its speed is halved? (c) If its speed is doubled?

21. •• The energy required to increase the speed of a certain car from 18 m/s to 24 m/s is 190 kJ. (a) What is the mass of the car? (b) At what speed is the car's kinetic energy equal to 190 kJ?

22. •• **CE Predict/Explain** The work W_0 accelerates a car from 0 to 50 km/h. (a) Is the work required to accelerate the car from 50 km/h to 150 km/h equal to $2W_0$, $3W_0$, $8W_0$, or $9W_0$? (b) Choose the *best explanation* from among the following:
    I. The work to accelerate the car depends on the speed squared.
    II. The final speed is three times the speed that was produced by the work W_0.
    III. The increase in speed from 50 km/h to 150 km/h is twice the increase in speed from 0 to 50 km/h.

23. •• **CE** Car A has a mass m and a speed v, car B has a mass $m/2$ and a speed $3v$, car C has a mass $3m$ and a speed $v/2$, and car D has a mass $4m$ and a speed $v/2$. Rank the cars in order of increasing kinetic energy. Indicate ties where appropriate.

24. •• **Predict/Calculate** A 0.14-kg pinecone falls 16 m to the ground, where it lands with a speed of 13 m/s. (a) With what speed would the pinecone have landed if there had been no air resistance? (b) Did air resistance do positive work, negative work, or zero work on the pinecone? Explain.

25. •• In the previous problem, (a) how much work was done on the pinecone by air resistance? (b) What was the average force of air resistance exerted on the pinecone?

26. •• At $t = 1.0$ s, a 0.55-kg object is falling with a speed of 6.5 m/s. At $t = 2.0$ s, it has a kinetic energy of 38 J. (a) What is the kinetic energy of the object at $t = 1.0$ s? (b) What is the speed of the object at $t = 2.0$ s? (c) How much work was done on the object between $t = 1.0$ s and $t = 2.0$ s?

27. •• After hitting a long fly ball that goes over the right fielder's head and lands in the outfield, the batter decides to keep going past second base and try for third base. The 62.0-kg player begins sliding 3.40 m from the base with a speed of 4.35 m/s. If the player comes to rest at third base, (a) how much work was done on the player by friction? (b) What was the coefficient of kinetic friction between the player and the ground?

28. •• **Predict/Calculate** A 1100-kg car coasts on a horizontal road with a speed of 19 m/s. After crossing an unpaved, sandy stretch of road 32 m long, its speed decreases to 12 m/s. (a) Was the net work done on the car positive, negative, or zero? Explain. (b) Find the magnitude of the average net force on the car in the sandy section.

29. •• A 65-kg bicyclist rides his 8.8-kg bicycle with a speed of 14 m/s. (a) How much work must be done by the brakes to bring the bike and rider to a stop? (b) How far does the bicycle travel if it takes 4.0 s to come to rest? (c) What is the magnitude of the braking force?

SECTION 7-3 WORK DONE BY A VARIABLE FORCE

30. • **CE** A block of mass m and speed v collides with a spring, compressing it a distance Δx. What is the compression of the spring if the force constant of the spring is increased by a factor of four?

31. • A spring with a force constant of 3.5×10^4 N/m is initially at its equilibrium length. (a) How much work must you do to stretch the spring 0.050 m? (b) How much work must you do to compress it 0.050 m?

32. • Initially sliding with a speed of 4.1 m/s, a 1.5-kg block collides with a spring and compresses it 0.27 m before coming to rest. What is the force constant of the spring?

33. • The force shown in **FIGURE 7-21** moves an object from $x = 0$ to $x = 0.75$ m. (a) How much work is done by the force? (b) How much work is done by the force if the object moves from $x = 0.15$ m to $x = 0.60$ m?

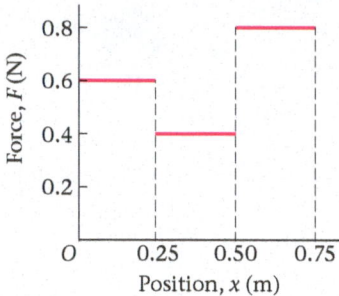

FIGURE 7-21 Problem 33

34. • An object is acted on by the force shown in **FIGURE 7-22**. What is the final position of the object if its initial position is $x = 0.40$ m and the work done on it is equal to (a) 0.21 J, or (b) −0.19 J?

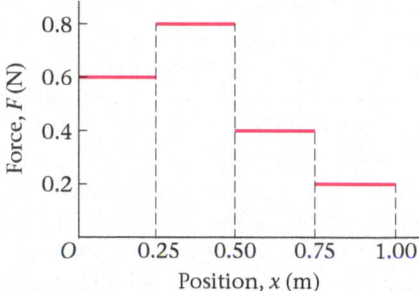

FIGURE 7-22 Problems 34 and 37

35. •• To compress spring 1 by 0.20 m takes 150 J of work. Stretching spring 2 by 0.30 m requires 210 J of work. Which spring is stiffer?

36. •• **Predict/Calculate** It takes 180 J of work to compress a certain spring 0.15 m. (a) What is the force constant of this spring? (b) To compress the spring an additional 0.15 m, does it take 180 J, more than 180 J, or less than 180 J? Verify your answer with a calculation.

37. •• The force shown in Figure 7-22 acts on a 1.3-kg object whose initial speed is 0.35 m/s and initial position is $x = 0.27$ m. (a) Find

the speed of the object when it is at the location $x = 0.99$ m. **(b)** At what location would the object's speed be 0.25 m/s?

38. ••• A block is acted on by a force that varies as $(2.0 \times 10^4 \, \text{N/m})x$ for $0 \le x \le 0.21$ m, and then remains constant at 4200 N for larger x. How much work does the force do on the block in moving it **(a)** from $x = 0$ to $x = 0.30$ m, or **(b)** from $x = 0.10$ m to $x = 0.40$ m?

SECTION 7-4   POWER

39. • **CE** Force F_1 does 5 J of work in 10 seconds, force F_2 does 3 J of work in 5 seconds, force F_3 does 6 J of work in 18 seconds, and force F_4 does 25 J of work in 125 seconds. Rank these forces in order of increasing power they produce. Indicate ties where appropriate.

40. • **BIO Climbing the Empire State Building** A new record for running the stairs of the Empire State Building was set on February 4, 2003. The 86 flights, with a total of 1576 steps, was run in 9 minutes and 33 seconds. If the height gain of each step was 0.20 m, and the mass of the runner was 70.0 kg, what was his average power output during the climb? Give your answer in both watts and horsepower.

41. • Calculate the power output of a 14-mg fly as it walks straight up a windowpane at 2.3 cm/s.

42. • An ice cube is placed in a microwave oven. Suppose the oven delivers 105 W of power to the ice cube and that it takes 32,200 J to melt it. How much time does it take for the ice cube to melt?

43. • Your car produces about 34 kW of power to maintain a constant speed of 31 m/s on the highway. What average force does the engine exert?

44. • You raise a bucket of water from the bottom of a deep well. If your power output is 108 W, and the mass of the bucket and the water in it is 5.00 kg, with what speed can you raise the bucket? Ignore the weight of the rope.

45. •• **BIO Salmon Migration** As Chinook salmon swim upstream in the Yukon River to spawn, they must travel about 3000 km in about 90 days, and in the process each fish must do about 1.7×10^6 J of work. **(a)** What is the average power output of a Chinook salmon as it swims upstream? **(b)** What average force does the salmon exert as it swims?

46. •• In order to keep a leaking ship from sinking, it is necessary to pump 12.0 lb of water each second from below deck up a height of 2.00 m and over the side. What is the minimum horsepower motor that can be used to save the ship?

47. •• **Predict/Calculate** A kayaker paddles with a power output of 35.0 W to maintain a steady speed of 2.24 m/s. **(a)** Calculate the resistive force exerted by the water on the kayak. **(b)** If the kayaker doubles her power output, and the resistive force due to the water remains the same, by what factor does the kayaker's speed change?

48. •• **BIO Human-Powered Flight** Human-powered aircraft require a pilot to pedal, as in a bicycle, and produce a sustained power output of about 0.30 hp. The *Gossamer Albatross* flew across the English Channel on June 12, 1979, in 2 h 49 min. **(a)** How much energy did the pilot expend during the flight? **(b)** What is the minimum number of Snickers bars (280 Cal per bar) the pilot would have to consume to be "fueled up" for the flight? [*Note:* The nutritional calorie, 1 Cal, is equivalent to 1000 calories (1000 cal) as defined in physics. In addition, the conversion factor between calories and joules is as follows: 1 Cal = 1000 cal = 1 kcal = 4184 J.]

49. •• **Predict/Calculate Beating to Windward** A sailboat can be propelled into the wind by a maneuver called *beating to windward*. Beating requires the sailboat to travel in a zigzag pattern at an angle to the wind that is greater than the *no-go zone*, which is shaded red in **FIGURE 7-23**. When a sailboat is just outside the no-go zone (boats B in the figure) the wind exerts a force $\vec{\mathbf{F}}$ on the sail that has a component in the direction of motion $\vec{\mathbf{v}}$. Similar comments apply to boats C. The work done by the wind on the sail is $W = Fd\cos\theta$, and because $v = d/t$, the propulsion power $P = W/t$ delivered to the sailboat is $Fv\cos\theta$, where θ is the angle between the sail force and the direction of motion. **(a)** Assuming that F and v have the same magnitudes for each sailboat, will the propulsion power delivered to sailboats B be greater than, less than, or the same as the propulsion power delivered to sailboats C? Explain. **(b)** If $F = 870$ N and $v = 11$ m/s, what propulsion power is delivered to sailboats B, for which $\theta = 79°$? **(c)** What propulsion power is delivered to sailboats C, for which $\theta = 56°$?

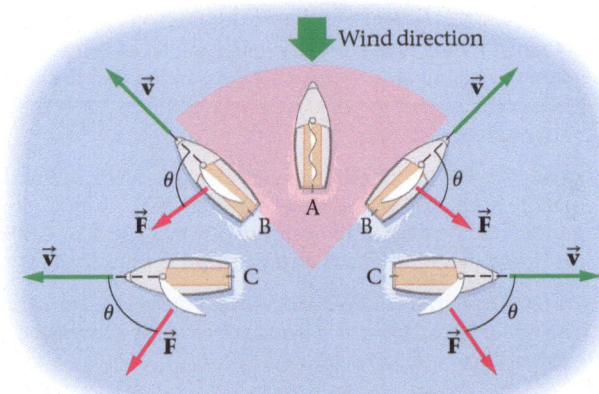

FIGURE 7-23 Problem 49

50. •• **Predict/Calculate** A grandfather clock is powered by the descent of a 4.35-kg weight. **(a)** If the weight descends through a distance of 0.760 m in 3.25 days, how much power does it deliver to the clock? **(b)** To increase the power delivered to the clock, should the time it takes for the mass to descend be increased or decreased? Explain.

51. ••• **Predict/Calculate** A certain car can accelerate from rest to the speed v in T seconds. If the power output of the car remains constant, **(a)** how much time does it take for the car to accelerate from v to $2v$? **(b)** How fast is the car moving at $2T$ seconds after starting?

GENERAL PROBLEMS

52. • **CE** As the three small sailboats shown in **FIGURE 7-24** drift next to a dock, because of wind and water currents, students pull on a line attached to the bow and exert forces of equal magnitude F. Each boat drifts through the same distance d. Rank the three boats (A, B, and C) in order of increasing work done on the boat by the force F. Indicate ties where appropriate.

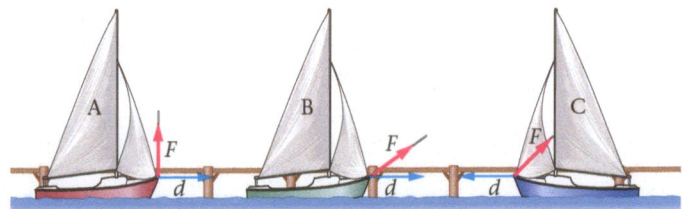

FIGURE 7-24 Problem 52

53. • **CE** Predict/Explain A car is accelerated by a constant force, F. The distance required to accelerate the car from rest to the speed v is Δx. **(a)** Is the distance required to accelerate the car from the speed v to the speed $2v$ equal to Δx, $2\Delta x$, $3\Delta x$, or $4\Delta x$? **(b)** Choose the *best explanation* from among the following:

    I. The final speed is twice the initial speed.

    II. The increase in speed is the same in each case.

    III. Work is force times distance, and work depends on the speed squared.

54. • **CE** Car 1 has four times the mass of car 2, but they both have the same kinetic energy. If the speed of car 2 is v, is the speed of car 1 equal to $v/4$, $v/2$, $2v$, or $4v$? Explain.

55. • **BIO** **Muscle Cells** Biological muscle cells can be thought of as nanomotors that use the chemical energy of ATP to produce mechanical work. Measurements show that the active proteins within a muscle cell (such as myosin and actin) can produce a force of about 1.5 pN and displacements of 8.0 nm. How much work is done by such proteins?

56. • A small motor runs a lift that raises a load of bricks weighing 836 N to a height of 10.7 m in 23.2 s. Assuming that the bricks are lifted with constant speed, what is the minimum power the motor must produce?

57. • You push a 67-kg box across a floor where the coefficient of kinetic friction is $\mu_k = 0.55$. The force you exert is horizontal. **(a)** How much power is needed to push the box at a speed of 0.50 m/s? **(b)** How much work do you do if you push the box for 35 s?

58. • A 1300-kg elevator is lifted at a constant speed of 1.3 m/s through a height of 22 m. How much work is done **(a)** by the tension in the elevator cable and **(b)** by gravity?

59. •• **CE** The work W_0 is required to accelerate a car from rest to the speed v_0. How much work is required to accelerate the car **(a)** from rest to the speed $v_0/2$ and **(b)** from $v_0/2$ to v_0?

60. •• After a tornado, a 0.55-g straw was found embedded 2.3 cm into the trunk of a tree. If the average force exerted on the straw by the tree was 65 N, what was the speed of the straw when it hit the tree?

61. •• You throw a glove straight upward to celebrate a victory. Its initial kinetic energy is K and it reaches a maximum height h. What is the kinetic energy of the glove when it is at the height $h/2$?

62. •• The water skier in Figure 7-20 is at an angle of 32.5° with respect to the center line of the boat, and is being pulled at a constant speed of 15.6 m/s. If the tension in the tow rope is 111 N, **(a)** how much work does the rope do on the skier in 10.0 s? **(b)** How much work does the resistive force of water do on the skier in the same time?

63. •• Predict/Calculate A sled with a mass of 5.80 kg is pulled along the ground through a displacement given by $\vec{d} = (4.55 \text{ m})\hat{x}$. (Let the x axis be horizontal and the y axis be vertical.) **(a)** How much work is done on the sled when the force acting on it is $\vec{F} = (2.89 \text{ N})\hat{x} + (0.131 \text{ N})\hat{y}$? **(b)** How much work is done on the sled when the force acting on it is $\vec{F} = (2.89 \text{ N})\hat{x} + (0.231 \text{ N})\hat{y}$? **(c)** If the mass of the sled is increased, does the work done by the forces in parts **(a)** and **(b)** increase, decrease, or stay the same? Explain.

64. •• Predict/Calculate A 0.19-kg apple falls from a branch 3.5 m above the ground. **(a)** Does the power delivered to the apple by gravity increase, decrease, or stay the same during the time the apple falls to the ground? Explain. Find the power delivered by gravity to the apple when the apple is **(b)** 2.5 m and **(c)** 1.5 m above the ground.

65. •• A boy pulls a bag of baseball bats across a ball field toward the parking lot. The bag of bats has a mass of 6.80 kg, and the boy exerts a horizontal force of 24.0 N on the bag. As a result, the bag accelerates from rest to a speed of 1.12 m/s in a distance of 5.25 m. What is the coefficient of kinetic friction between the bag and the ground?

66. •• At the instant it leaves the player's hand after a jump shot, a 0.62-kg basketball is traveling 8.4 m/s. When it passes through the hoop its speed is 6.8 m/s. **(a)** How much work has gravity done on the ball? **(b)** What is the vertical component of the ball's displacement? **(c)** If the magnitude of the total displacement of the ball is 4.6 m, what is the angle between the displacement vector and the gravitational force?

67. •• The force shown in **FIGURE 7-25** acts on an object that moves along the x axis. How much work is done by the force as the object moves from **(a)** $x = 0$ to $x = 2.0$ m, **(b)** $x = 1.0$ m to $x = 4.0$ m, and **(c)** $x = 3.5$ m to $x = 1.2$ m?

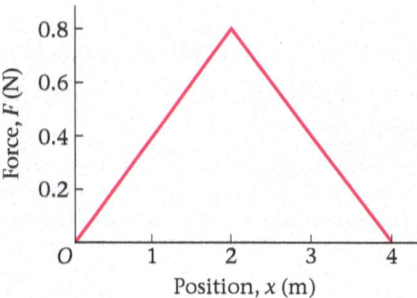

FIGURE 7-25 Problem 67

68. •• **A Compound Bow** A compound bow in archery allows the user to hold the bowstring at full draw with considerably less force than the maximum force exerted by the string. The draw force as a function of the string position x for a particular compound bow is shown in **FIGURE 7-26**. **(a)** How much work does the archer do on the bow in order to draw the string from $x = 0$ to $x = 0.60$ m? **(b)** If all of this work becomes the kinetic energy of a 0.065-kg arrow, what is the speed of the arrow?

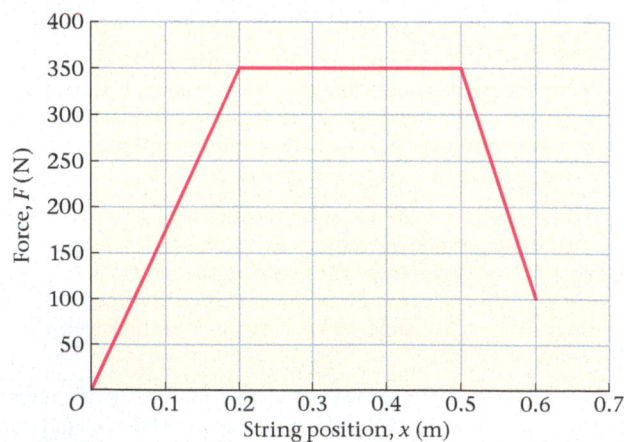

FIGURE 7-26 Problem 68

69. •• **A Compound Versus a Simple Bow** The compound bow in the previous problem requires 152 J of work to draw the string back to $x = 0.540$ m, at which point the archer need only exert 250 N of force to hold the string in place. **(a)** Suppose a simple bow is used instead of a compound bow. What should its force constant k (in N/m) be so that it requires the same amount of work to draw the

string a distance of $x = 0.540$ m? Assume the simple bow behaves the same as an ideal spring. **(b)** How much force must the archer exert to draw the simple bow to $x = 0.540$ m? Compare this to the 250-N force required by the compound bow.

70. •• Calculate the power output of a 0.42-g spider as it walks up a windowpane at 2.3 cm/s. The spider walks on a path that is at 25° to the vertical, as illustrated in **FIGURE 7-27**.

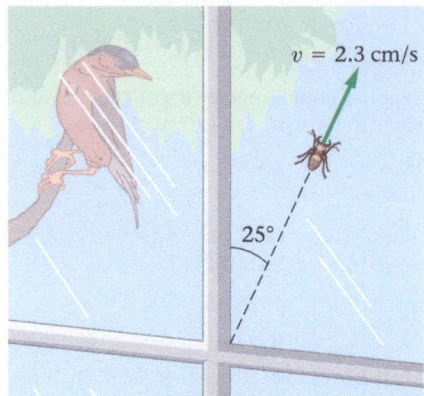

FIGURE 7-27 Problem 70

71. •• **Cookie Power** To make a batch of cookies, you mix half a bag of chocolate chips into a bowl of cookie dough, exerting a 28-N force on the stirring spoon. Assume that your force is always in the direction of motion of the spoon. **(a)** What power is needed to move the spoon at a speed of 0.44 m/s? **(b)** How much work do you do if you stir the mixture for 1.5 min?

72. •• **Predict/Calculate** A pitcher accelerates a 0.14-kg hardball from rest to 42.5 m/s in 0.060 s. **(a)** How much work does the pitcher do on the ball? **(b)** What is the pitcher's power output during the pitch? **(c)** Suppose the ball reaches 42.5 m/s in less than 0.060 s. Is the power produced by the pitcher in this case more than, less than, or the same as the power found in part **(b)**? Explain.

73. •• **BIO Brain Power** The human brain consumes about 22 W of power under normal conditions, though more power may be required during exams. **(a)** For what amount of time can one Snickers bar (see the note following Problem 48) power the normally functioning brain? **(b)** At what rate must you lift a 3.6-kg container of milk (one gallon) if the power output of your arm is to be 22 W? **(c)** How much time does it take to lift the milk container through a distance of 1.0 m at this rate?

74. •• **Meteorite** On October 9, 1992, a 27-pound meteorite struck a car in Peekskill, NY, creating a dent about 22 cm deep. If the initial speed of the meteorite was 550 m/s, what was the average force exerted on the meteorite by the car?

75. ••• **BIO Powering a Pigeon** A pigeon in flight experiences a force of air resistance given approximately by $F = bv^2$, where v is the flight speed and b is a constant. **(a)** What are the units of the constant b? **(b)** What is the greatest possible speed of the pigeon if its maximum power output is P? **(c)** By what factor does the greatest possible speed increase if the maximum power is doubled?

76. ••• **Springs in Series** Two springs, with force constants k_1 and k_2, are connected in series, as shown in **FIGURE 7-28**. How much work is required to stretch this system a distance x from the equilibrium position?

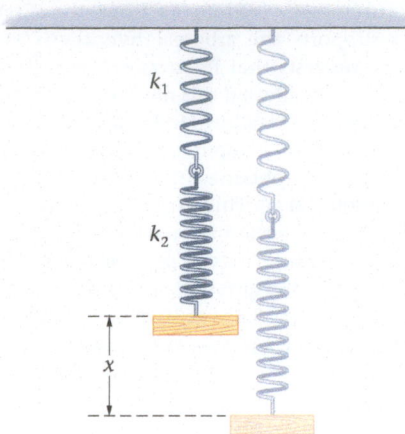

FIGURE 7-28 Problem 76

77. ••• **Springs in Parallel** Two springs, with force constants k_1 and k_2, are connected in parallel, as shown in **FIGURE 7-29**. How much work is required to stretch this system a distance x from the equilibrium position?

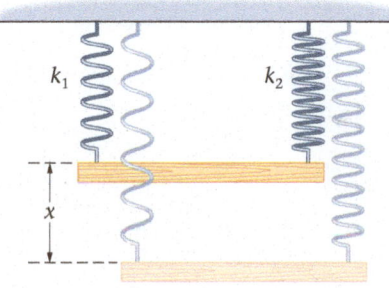

FIGURE 7-29 Problem 77

78. ••• A block rests on a horizontal frictionless surface. A string is attached to the block, and is pulled with a force of 45.0 N at an angle θ above the horizontal, as shown in **FIGURE 7-30**. After the block is pulled through a distance of 1.50 m, its speed is 2.60 m/s, and 50.0 J of work has been done on it. **(a)** What is the angle θ? **(b)** What is the mass of the block?

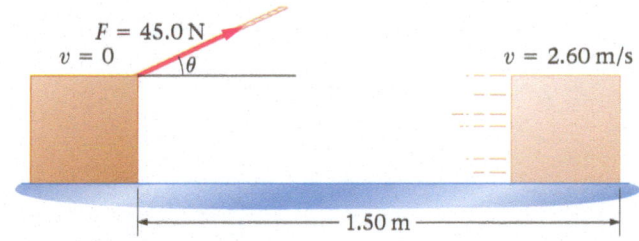

FIGURE 7-30 Problem 78

PASSAGE PROBLEMS

BIO *Microraptor gui*: **The Biplane Dinosaur**

The evolution of flight is a subject of intense interest in paleontology. Some subscribe to the "cursorial" (or ground-up) hypothesis, in which flight began with ground-dwelling animals running and jumping after prey. Others favor the "arboreal" (or trees-down) hypothesis, in which tree-dwelling animals, like modern-day flying squirrels, developed flight as an extension of gliding from tree to tree.

A recently discovered fossil from the Cretaceous period in China supports the arboreal hypothesis and adds a new element—it suggests that feathers on both the wings and the lower legs and feet allowed this dinosaur, *Microraptor gui*, to glide much like a biplane, as shown in **FIGURE 7-31 (a)**. Researchers have produced a detailed computer simulation of *Microraptor*, and with its help have obtained the power-versus-speed plot presented in **FIGURE 7-31 (b)**. This curve shows how much power is required for flight at speeds between 0 and 30 m/s. Notice that the power increases at high speeds, as expected, but it is also high for low speeds, where the dinosaur is almost hovering. A minimum of 8.1 W is needed for flight at 10 m/s. The lower horizontal line shows the estimated 9.8-W power output of *Microraptor*, indicating the small range of speeds for which flight would be possible. The upper horizontal line shows the wider range of flight speeds that would be available if *Microraptor* were able to produce 20 W of power.

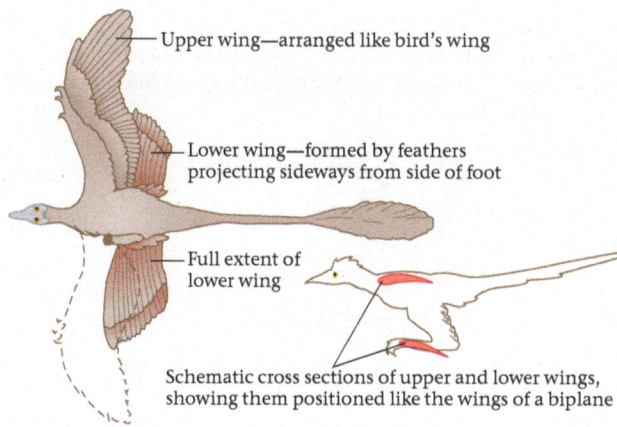

— Upper wing—arranged like bird's wing

— Lower wing—formed by feathers projecting sideways from side of foot

— Full extent of lower wing

Schematic cross sections of upper and lower wings, showing them positioned like the wings of a biplane

(a) Possible reconstruction of *Microraptor gui* **in flight**

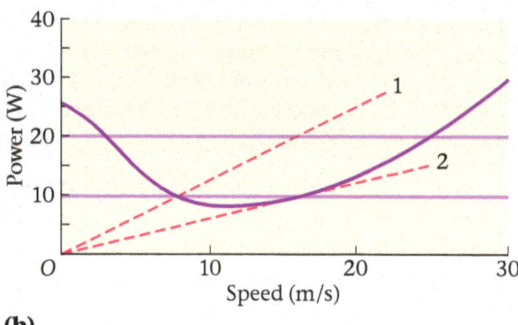

(b)

FIGURE 7-31 Problems 79, 80, 81, and 82

Also of interest are the two dashed, straight lines labeled 1 and 2. These lines represent constant ratios of power to speed—that is, a constant value for P/v. Referring to Equation 7-13, we see that $P/v = Fv/v = F$, so lines 1 and 2 correspond to lines of constant force. Line 2 is interesting in that it has the smallest slope that still touches the power-versus-speed curve.

79. • Estimate the range of flight speeds for *Microraptor gui* if its power output is 9.8 W.

     A. 0–7.7 m/s   B. 7.7–15 m/s
     C. 15–30 m/s   D. 0–15 m/s

80. • What approximate range of flight speeds would be possible if *Microraptor gui* could produce 20 W of power?

     A. 0–25 m/s   B. 25–30 m/s
     C. 2.5–25 m/s   D. 0–2.5 m/s

81. •• How much energy would *Microraptor* have to expend to fly with a speed of 10 m/s for 1.0 minute?

     A. 8.1 J   B. 81 J
     C. 490 J   D. 600 J

82. • Estimate the minimum force that *Microraptor* must exert to fly.

     A. 0.65 N   B. 1.3 N
     C. 1.0 N   D. 10 N

83. •• **REFERRING TO FIGURE 7-15** Suppose the block has a mass of 1.4 kg and an initial speed of 0.62 m/s. **(a)** What force constant must the spring have if the maximum compression is to be 2.4 cm? **(b)** If the spring has the force constant found in part (a), find the maximum compression if the mass of the block is doubled *and* its initial speed is halved.

84. •• **Predict/Calculate** **REFERRING TO FIGURE 7-15** In the situation shown in Figure 7-15 (d), a spring with a force constant of 750 N/m is compressed by 4.1 cm. **(a)** If the speed of the block in Figure 7-15 (f) is 0.88 m/s, what is its mass? **(b)** If the mass of the block is doubled, is the final speed greater than, less than, or equal to 0.44 m/s? **(c)** Find the final speed for the case described in part (b).

85. •• **Predict/Calculate** **REFERRING TO EXAMPLE 7-15** Suppose the car has a mass of 1400 kg and delivers 48 hp to the wheels. **(a)** How much time does it take for the car to increase its speed from 15 m/s to 25 m/s? **(b)** Would the time required to increase the speed from 5.0 m/s to 15 m/s be greater than, less than, or equal to the time found in part (a)? **(c)** Determine the time required to accelerate from 5.0 m/s to 15 m/s.

Potential Energy and Conservation of Energy

Big Ideas

1 Forces are either conservative (conserve energy) or nonconservative (dissipate it to other forms).

2 Potential energy is a storage mechanism for energy.

3 Mechanical energy is the sum of the potential and kinetic energies of a system.

▲ As this athlete hangs motionless in midair for a moment, he might wonder what happened to the kinetic energy he had just a few moments before? Where did that energy go? And how does it reappear as he begins to fall back toward the ground? The answers to these questions involve new types of energy that are the subject of this chapter.

One of the greatest accomplishments of physics is the concept of energy and its conservation. As we shall see, the universe has a certain amount of energy, and this energy can ebb and flow from one form to another, but the total amount remains fixed. Conservation of energy is of central importance in physics, and is also an important practical tool in problem solving.

8-1 Conservative and Nonconservative Forces

Work done by person = mgh

Work done by gravity = mgh

▲ **FIGURE 8-1 Work against gravity** Lifting a box against gravity with constant speed takes work mgh. When the box is released, gravity does the same work on the box as it falls. Gravity is a conservative force.

PHYSICS IN CONTEXT
Looking Back

The concept of work, first introduced in Chapter 7, is important in this chapter as well. For example, we use work (force times distance) in this section to illustrate the difference between conservative and non-conservative forces.

In physics, we classify forces according to whether they are *conservative* or *nonconservative*. The key distinction is that when a conservative force acts, the work it does is stored in a form that can be released as kinetic energy at a later time. In contrast, nonconservative forces dissipate energy to other forms, such as thermal energy, that cannot be recovered. In this section, we sharpen this distinction and explore some examples of conservative and nonconservative forces.

Gravity Is a Conservative Force   Perhaps the simplest case of a conservative force is gravity. Imagine lifting a box of mass m from the floor to a height h, as in **FIGURE 8-1**. To lift the box with constant speed, the force you must exert against gravity is mg. Since the upward distance is h, the work you do on the box is $W = mgh$. If you now release the box and allow it to drop back to the floor, gravity does the same work, $W = mgh$, and in the process gives the box an equivalent amount of kinetic energy. Thus, the work done by a **conservative force** can be recovered later as kinetic energy.

Friction Is a Nonconservative Force   Contrast gravity with the force of kinetic friction, which is nonconservative. To slide a box of mass m across the floor with constant speed, as shown in **FIGURE 8-2**, you must exert a force of magnitude $\mu_k N = \mu_k mg$. After the box has moved a distance d, the work you have done is $W = \mu_k mgd$. In this case, when you release the box it simply stays put—friction does no work on it after you let go. Thus, the work done by a **nonconservative force** cannot be recovered later as kinetic energy; instead, it is converted to other forms of energy, such as a slight warming of the floor and box in our example. A visual comparison of conservative and nonconservative forces is presented in **FIGURE 8-3**.

Work = $\mu_k mgd$

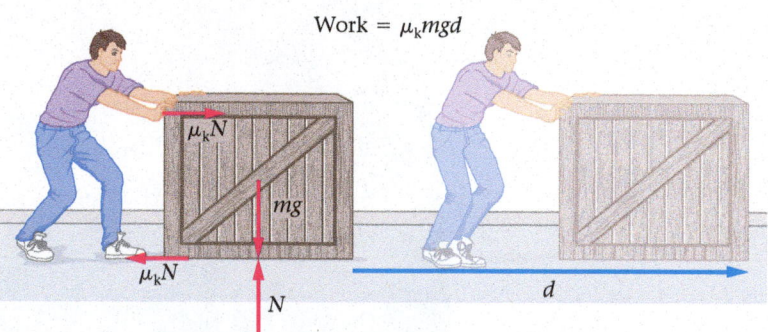

▲ **FIGURE 8-2 Work against friction** Pushing a box with constant speed against friction takes the work $\mu_k mgd$. When the box is released, it quickly comes to rest and friction does no further work. Friction is a nonconservative force.

Forces on a Closed Path   The differences between conservative and nonconservative forces are even more apparent for an object moving around a closed path. Consider, for example, the path shown in **FIGURE 8-4**. If we move a box of mass m along this path, the total work done by gravity is the sum of the work done on each segment of the path; that is, $W_{\text{total}} = W_{\text{AB}} + W_{\text{BC}} + W_{\text{CD}} + W_{\text{DA}}$. The work done by gravity from A to B and from C to D is zero, because the force is at right angles to the displacement on these segments. Thus $W_{\text{AB}} = W_{\text{CD}} = 0$. On the segment from B to C, gravity does negative work (displacement and force are in opposite directions), but it does positive work from D to A (displacement and force are in the same direction). Hence, $W_{\text{BC}} = -mgh$ and $W_{\text{DA}} = mgh$. As a result, the total work done by gravity is zero:

$$W_{\text{total}} = 0 + (-mgh) + 0 + mgh = 0$$

With friction, the results are quite different. If we push the box around the closed horizontal path shown in **FIGURE 8-5**, the total work done by friction does not vanish. In fact, friction does the negative work $W = -f_k d = -\mu_k mgd$ on each segment. Therefore, the total work done by kinetic friction is

$$W_{\text{total}} = (-\mu_k mgd) + (-\mu_k mgd) + (-\mu_k mgd) + (-\mu_k mgd) = -4\mu_k mgd$$

▲ **FIGURE 8-3** **Visualizing Concepts** **Conservative Versus Nonconservative Forces** Because gravity is a conservative force, the work done against gravity in lifting these logs (left) can, in principle, all be recovered. If the logs are released, for example, they will acquire an amount of kinetic energy exactly equal to the work done to lift them. Friction, by contrast, is a nonconservative force. Some of the work done by this spinning grind-stone (right) goes into removing material from the object being ground, while the rest is transformed into sound energy and (especially) heat. Most of this work can never be recovered as kinetic energy.

These results lead to the following definition of a conservative force:

Conservative Force: Definition 1

A conservative force does zero total work on a closed path.

A roller coaster provides a good illustration of this definition. A car on a roller coaster has speed v at point A in **FIGURE 8-6**, it speeds up as it drops to point B, it slows down as it approaches point C, and so on. When the car returns to its original height, at point D, it will again have the speed v, as long as friction and other nonconservative forces can be neglected. Similarly, if the car completes a circuit of the track and returns to point A, it will again have the speed v. Hence, a roller-coaster car's kinetic energy is unchanged ($\Delta K = 0$) after *any* complete circuit of the track. From the work–energy theorem, $W_{\text{total}} = \Delta K$, it follows that the work done by gravity is zero for the closed path of the car, as expected for a conservative force.

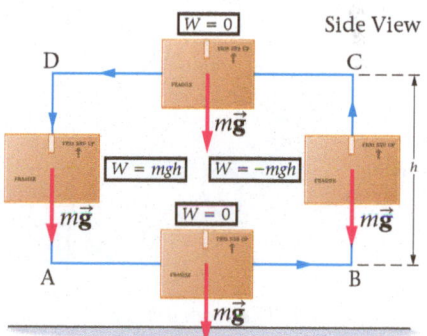

▲ **FIGURE 8-4** **Work done by gravity on a closed path is zero** Gravity does no work on the two horizontal segments of the path. On the two vertical segments, the amounts of work done are equal in magnitude but opposite in sign. Therefore, the total work done by gravity on this—or any—closed path is zero.

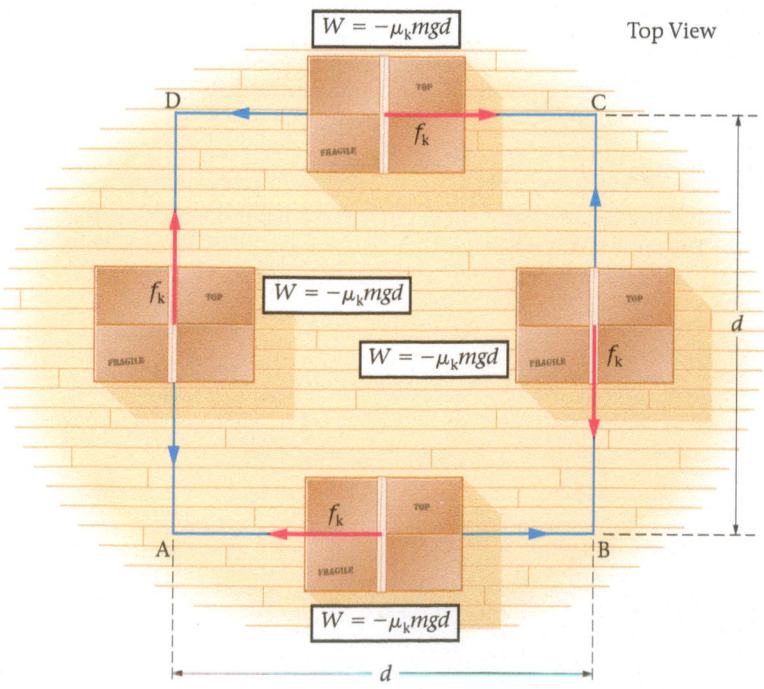

◀ **FIGURE 8-5** **Work done by friction on a closed path is nonzero** The work done by friction when an object moves through a distance d is $-\mu_k mgd$. Thus, the total work done by friction on a closed path is nonzero. In this case, it is equal to $-4\mu_k mgd$.

▶ **FIGURE 8-6 Gravity is a conservative force** If frictional forces can be ignored, a roller-coaster car will have the same speed at points A and D, since they are at the same height. Hence, after any complete circuit of the track, the speed of the car returns to its initial value. It follows that the change in kinetic energy is zero for a complete circuit, and therefore the work done by gravity is also zero.

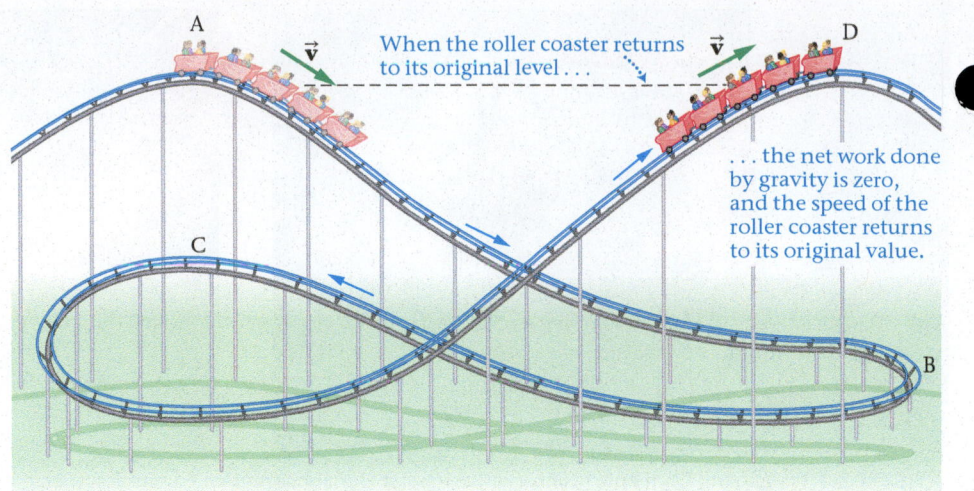

When the roller coaster returns to its original level . . .

. . . the net work done by gravity is zero, and the speed of the roller coaster returns to its original value.

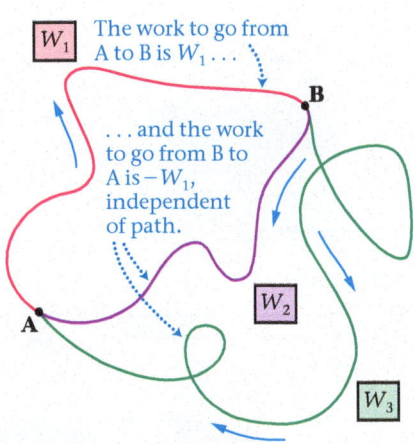

W_1 The work to go from A to B is W_1 . . .

. . . and the work to go from B to A is $-W_1$, independent of path.

W_2

W_3

▲ **FIGURE 8-7 The work done by a conservative force is independent of path** Considering paths 1 and 2, we see that $W_1 + W_2 = 0$, or $W_2 = -W_1$. From paths 1 and 3, however, we see that $W_1 + W_3 = 0$, or $W_3 = -W_1$. It follows, then, that $W_3 = W_2$, since they are both equal to $-W_1$; hence the work done in going from B to A is independent of the path.

The Work Done by a Conservative Force Is Path Independent The property of zero total work on a closed path has interesting consequences for conservative forces. For instance, consider the closed paths shown in **FIGURE 8-7**. On each of these paths, we know that the work done by a conservative force is zero. Thus, it follows from paths 1 and 2 that $W_{\text{total}} = W_1 + W_2 = 0$, or

$$W_2 = -W_1$$

Similarly, using paths 1 and 3 we have $W_{\text{total}} = W_1 + W_3 = 0$, or

$$W_3 = -W_1$$

As a result, we see that the work done on path 3 is the same as the work done on path 2:

$$W_3 = W_2$$

But paths 2 and 3 are arbitrary, as long as they start at point B and end at point A. This leads to an equivalent definition of a conservative force:

Conservative Force: Definition 2
The work done by a conservative force in going from point A to point B is *independent of the path* from A to B.

This definition is given an explicit check in Example 8-1.

EXAMPLE 8-1   DIFFERENT PATHS, DIFFERENT FORCES

(a) A 4.57-kg box is moved with constant speed from A to B along the two paths shown in part (a) of our sketch. Calculate the work done by gravity on each of these paths. (b) The same box is pushed across a floor from A to B along path 1 and path 2 in part (b) of our sketch. If the coefficient of kinetic friction between the box and the surface is $\mu_k = 0.63$, how much work is done by friction along each path?

Side View

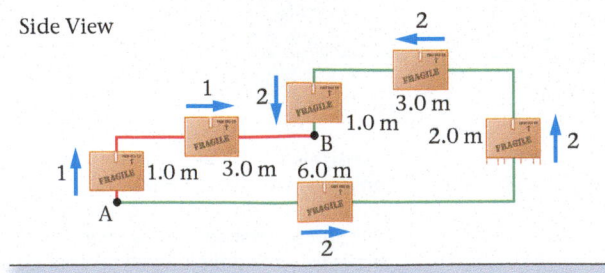

Top View

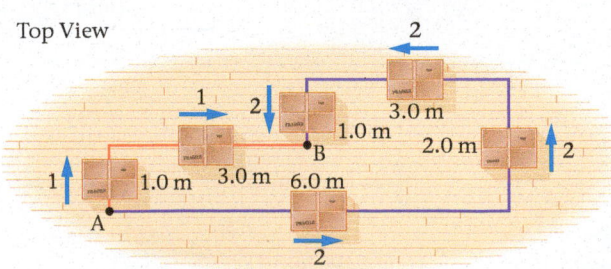

(a)   **(b)**   CONTINUED

PICTURE THE PROBLEM

Part (a) of our sketch shows two different paths a box might be taken through in going from point A to point B. Path 1 is indicated by two red lines, indicating a vertical displacement of 1.0 m and a horizontal displacement of 3.0 m. Path 2, indicated in green, consists of two horizontal and two vertical displacements. In this case, we are interested in the work done by gravity. Part (b) shows the same basic paths—path 1 in orange and path 2 in purple—only this time on a rough floor. Here it is the force of kinetic friction that is of interest.

REASONING AND STRATEGY

To calculate the work done for each path, we break it down into segments. Path 1 is made up of two segments; path 2 has four segments.

a. For gravity, the work is zero on horizontal segments. On vertical segments, the work done by gravity is positive when the motion is downward and negative when the motion is upward.

b. The work done by kinetic friction is negative on all segments of both paths.

Known Mass of box, $m = 4.57$ kg; coefficient of kinetic friction, $\mu_k = 0.63$; the paths are shown in the sketch.
Unknown (a) Work done by gravity on two paths: $W_1 = ?; W_2 = ?$
(b) Work done by friction on two paths: $W_1 = ?; W_2 = ?$

SOLUTION

Part (a)

1. Using $W = Fd = mgy$, calculate the work done by gravity along the two segments of path 1:

$$W_1 = -(4.57 \text{ kg})(9.81 \text{ m/s}^2)(1.0 \text{ m}) + 0 = -45 \text{ J}$$

2. In the same way, calculate the work done by gravity along the four segments of path 2:

$$W_2 = 0 - (4.57 \text{ kg})(9.81 \text{ m/s}^2)(2.0 \text{ m})$$
$$+ 0 + (4.57 \text{ kg})(9.81 \text{ m/s}^2)(1.0 \text{ m}) = -45 \text{ J}$$

Part (b)

3. Using $F = \mu_k N$, calculate the work done by kinetic friction along the two segments of path 1:

$$W_1 = -(0.63)(4.57 \text{ kg})(9.81 \text{ m/s}^2)(1.0 \text{ m})$$
$$- (0.63)(4.57 \text{ kg})(9.81 \text{ m/s}^2)(3.0 \text{ m}) = -110 \text{ J}$$

4. Similarly, calculate the work done by kinetic friction along the four segments of path 2:

$$W_2 = -(0.63)(4.57 \text{ kg})(9.81 \text{ m/s}^2)(6.0 \text{ m})$$
$$- (0.63)(4.57 \text{ kg})(9.81 \text{ m/s}^2)(2.0 \text{ m})$$
$$- (0.63)(4.57 \text{ kg})(9.81 \text{ m/s}^2)(3.0 \text{ m})$$
$$- (0.63)(4.57 \text{ kg})(9.81 \text{ m/s}^2)(1.0 \text{ m}) = -340 \text{ J}$$

INSIGHT

As expected, the conservative force of gravity gives the same work in going from A to B, regardless of the path. The work done by kinetic friction, however, has a larger magnitude on the path of greater length.

PRACTICE PROBLEM — PREDICT/CALCULATE

Suppose the work done by gravity when the box is moved from point B to a new point C is 140 J. (a) Is point C above, below, or at the same height as point B? Explain. (b) Find the vertical distance between points B and C.
[**Answer:** (a) The work done by gravity is positive. This means that the displacement of the box is downward—in the direction of the force of gravity. Thus point C is *below* point B. (b) Point C is 3.1 m below point B.]

Some related homework problems: Problem 2, Problem 3

Table 8-1 summarizes the different kinds of conservative and nonconservative forces we have encountered thus far in this text.

Enhance Your Understanding (Answers given at the end of the chapter)

1. In **FIGURE 8-8**, the work done by a conservative force along path AB is 15 J, and the work the force does along path BC is −5 J. What is the work done by this force along the paths (a) CDA and (b) AEC?

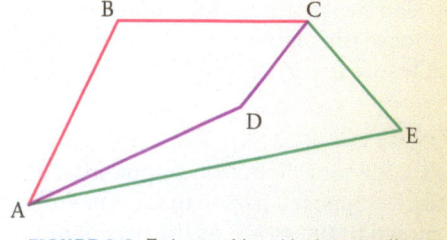

▲ **FIGURE 8-8** Enhance Your Understanding 1.

CONTINUED

TABLE 8-1 Conservative and Nonconservative Forces

Force	Section
Conservative forces	
Gravity	5-6
Spring force	6-2
Nonconservative forces	
Friction	6-1
Tension in a rope, cable, etc.	6-2
Forces exerted by a motor	7-4
Forces exerted by muscles	5-3

Section Review

- The work done by a conservative force is zero along a closed path, and is independent of the path between any two points.

- The work done by a conservative force can be recovered later. The work done by a nonconservative force cannot be recovered.

Big Idea 2 Potential energy is a storage mechanism for energy. Each type of conservative force has its own form of potential energy. There is no potential energy associated with nonconservative forces.

PHYSICS IN CONTEXT
Looking Back

Work (Chapter 7) is used in this section to introduce the concept of potential energy, U, and to define its change.

PROBLEM-SOLVING NOTE

Zero of Potential Energy

When working potential energy problems it is important to make a definite choice for the location where the potential energy is to be set equal to zero. Any location can be chosen, but once the choice is made, it must be used consistently.

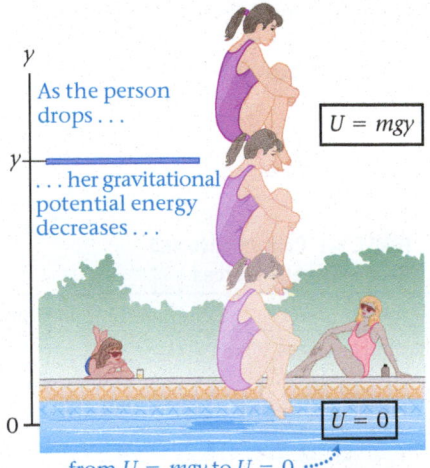

▲ **FIGURE 8-9 Gravitational Potential Energy** A person drops from a diving board into a swimming pool. The diving board is at the height y, and the surface of the water is at $y = 0$. We choose the gravitational potential energy to be zero at $y = 0$; hence, the potential energy is mgy at the diving board.

8-2 Potential Energy and the Work Done by Conservative Forces

Work must be done to lift a bowling ball from the floor to a shelf. Once on the shelf, the bowling ball has zero kinetic energy, just as it did on the floor. Even so, the work done in lifting the ball has not been lost. If the ball is allowed to fall from the shelf, gravity does the same amount of work on it as you did to lift it in the first place. As a result, the work you did is "recovered" in the form of kinetic energy. Thus, we say that when the ball is lifted to a new position, there is an increase in **potential energy,** U, and that this potential energy can be converted to kinetic energy when the ball falls.

In a sense, potential energy is a storage system for energy. When we increase the separation between the ball and the ground, the work we do is stored in the form of an increased potential energy. Not only that, but the storage system is perfect, in the sense that the energy is never lost as long as the separation remains the same. The ball can rest on the shelf for a million years, and still, when it falls, it gains the same amount of kinetic energy.

Work done against friction, however, is not "stored" as potential energy. Instead, it is dissipated into other forms of energy such as heat or sound. The same is true of other nonconservative forces. Only conservative forces have the potential-energy storage system.

Before proceeding, we should point out an interesting difference between kinetic and potential energy. Kinetic energy is given by the expression $K = \frac{1}{2}mv^2$, no matter what force might be involved. On the other hand, each different conservative force has a different expression for its potential energy. To see how this comes about, we turn now to a precise definition of potential energy.

Potential Energy, U

When a conservative force does an amount of work W_c (the subscript c stands for conservative), the corresponding potential energy U is changed according to the following definition:

Definition of the Change in Potential Energy, U

$$W_c = U_i - U_f = -(U_f - U_i) = -\Delta U \qquad \text{8-1}$$

SI unit: joule, J

In words, the work done by a conservative force is equal to the negative of the change in the corresponding potential energy. For example, when an object falls, gravity does *positive* work on it and the gravitational potential energy *decreases.* Similarly, when an object is lifted, gravity does *negative* work and the gravitational potential energy *increases.*

Work is a scalar with the units of joules, and hence the same is true of potential energy. In addition, our definition determines only the *change* in potential energy between two points, not the actual value of the potential energy. As a result, we are free to choose the place where the potential energy is zero ($U = 0$) in much the same way that we are free to choose the location of the origin of a coordinate system.

Gravity

Let's apply our definition of potential energy to the force of gravity near the Earth's surface. Suppose a person of mass m drops a distance y from a diving board into a pool, as shown in **FIGURE 8-9**. As the person drops, gravity does the work

$$W_c = mgy$$

When we apply the definition given in Equation 8-1, the corresponding change in potential energy is

$$-\Delta U = U_i - U_f = W_c = mgy$$

In this expression, U_i is the potential energy when the diver is on the board, and U_f is the potential energy when the diver enters the water. Rearranging slightly, we have

$$U_i = mgy + U_f \qquad\qquad 8\text{-}2$$

Notice that the initial potential energy, U_i, is greater than the final potential energy, U_f.

As mentioned above, we are free to choose $U = 0$ anywhere we like; *only the change in U is important*. For example, if you slip and fall to the ground, you hit with the same thud whether you fall in Denver (altitude 1 mile) or in Honolulu (at sea level). It's the difference in height that matters, not the height itself. This is illustrated in **FIGURE 8-10**. The only issue to be careful about when choosing a location for $U = 0$ is to be consistent with the choice once it is made.

In general, we choose $U = 0$ in a convenient location. In Figure 8-9, a reasonable place for $U = 0$ is the surface of the water, where $y = 0$; that is, $U_f = 0$. Then, Equation 8-2 becomes $U_i = mgy$. If we omit the subscript on U_i, letting U stand for the potential energy at the arbitrary height y, we have:

Gravitational Potential Energy (Near Earth's Surface)

$$U = mgy \qquad\qquad 8\text{-}3$$

The gravitational potential energy depends only on the height, y, and is independent of the horizontal position.

▲ **FIGURE 8-10** The $U = 0$ level for the gravitational potential energy of this system can be assigned to the point where the diver starts his dive, to the water level, or to any other level. Regardless of the choice, however, his kinetic energy when he strikes the water will be exactly equal to the difference in gravitational potential energy between his launch and splashdown points.

EXERCISE 8-2   GRAVITATIONAL POTENTIAL ENERGY

Find the gravitational potential energy associated with a 7.25-kg bowling ball that is held 1.45 m above the ground. Let $U = 0$ be at ground level.

REASONING AND SOLUTION
The gravitational potential energy is given by $U = mgh$. Substituting $m = 7.25$ kg and $h = 1.45$ m yields

$$U = mgy = (7.25\ \text{kg})(9.81\ \text{m/s}^2)(1.45\ \text{m}) = 103\ \text{J}$$

The next Example considers the change in gravitational potential energy associated with a mountain climber, given different choices for the location of $U = 0$.

EXAMPLE 8-3   PIKES PEAK OR BUST

An 82.0-kg mountain climber is in the final stage of the ascent of 4301-m-high Pikes Peak. What is the change in gravitational potential energy as the climber gains the last 100.0 m of altitude? Let $U = 0$ be **(a)** at sea level or **(b)** at the top of the peak.

PICTURE THE PROBLEM
Our sketch shows the mountain climber and the last 100.0 m of altitude to be climbed. We choose a typical coordinate system, with the positive y axis upward and the positive x axis to the right.

REASONING AND STRATEGY
The gravitational potential energy of the Earth–climber system depends only on the height y; the path followed in gaining the last 100.0 m of altitude is unimportant. The change in potential energy is $\Delta U = U_f - U_i = mgy_f - mgy_i$, where y_f is the altitude of the peak and y_i is 100.0 m less than y_f.

Known   Mass of climber, $m = 82.0$ kg; difference in altitude, $y_f - y_i = 100.0$ m; **(a)** $U = 0$ at sea level; **(b)** $U = 0$ at top of peak.

Unknown   **(a)** Change in potential energy, $\Delta U = ?$ **(b)** Change in potential energy, $\Delta U = ?$

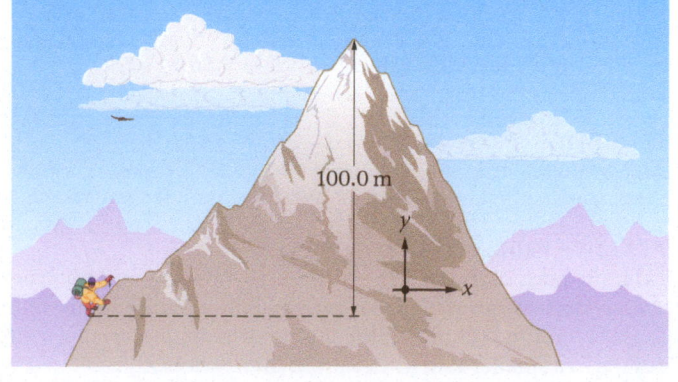

CONTINUED

SOLUTION

Part (a)

1. Calculate ΔU with $y_f = 4301$ m and $y_i = 4201$ m:

$$\Delta U = mgy_f - mgy_i$$
$$= (82.0 \text{ kg})(9.81 \text{ m/s}^2)(4301 \text{ m})$$
$$- (82.0 \text{ kg})(9.81 \text{ m/s}^2)(4201 \text{ m}) = 80{,}400 \text{ J}$$

Part (b)

2. Calculate ΔU with $y_f = 0$ and $y_i = -100.0$ m:

$$\Delta U = mgy_f - mgy_i$$
$$= (82.0 \text{ kg})(9.81 \text{ m/s}^2)(0)$$
$$- (82.0 \text{ kg})(9.81 \text{ m/s}^2)(-100.0 \text{ m}) = 80{,}400 \text{ J}$$

INSIGHT

As expected, the *change* in gravitational potential energy does not depend on where we choose $U = 0$. Nor does it depend on the path taken between the initial and final points.

PRACTICE PROBLEM

Find the altitude of the climber for which the gravitational potential energy of the Earth–climber system is 1.00×10^5 J less than it is when the climber is at the summit.   [**Answer: 4180 m**]

Some related homework problems: Problem 10, Problem 17

A single item of food can be converted into a surprisingly large amount of potential energy. This is shown for the case of a candy bar in the next Example.

EXAMPLE 8-4   CONVERTING FOOD ENERGY TO MECHANICAL ENERGY

BIO A bicyclist eats a 195 Cal = 195 kcal candy bar. The energy equivalent of the candy is 8.16×10^5 J, and the mass of the bicyclist and bicycle together is 70.3 kg. If the bicyclist could convert all of the candy bar's energy to gravitational potential energy, what gain in altitude would she achieve?

PICTURE THE PROBLEM

The bicyclist pedals up a slope, gaining altitude. She starts at the origin, $y = 0$, and rises in altitude to the final height $y = h$.

REASONING AND STRATEGY

The initial gravitational potential energy of the system is $U = 0$; the final potential energy is $U = mgh$. To find the altitude gain, we set $U = mgh$ equal to the energy provided by the candy bar, 8.16×10^5 J, and solve for the height h.

Known   Energy of candy bar, $U = 8.16 \times 10^5$ J; mass of bicyclist and bicycle, $m = 70.3$ kg.
Unknown   Gain in altitude, $h = ?$

SOLUTION

1. Solve $U = mgh$ for h:

$$U = mgh$$
$$h = \frac{U}{mg}$$

2. Substitute numerical values, with $U = 8.16 \times 10^5$ J:

$$h = \frac{U}{mg} = \frac{8.16 \times 10^5 \text{ J}}{(70.3 \text{ kg})(9.81 \text{ m/s}^2)} = 1180 \text{ m}$$

INSIGHT

This is almost 3/4 of a mile in elevation. Even if we take into account the fact that metabolic efficiency is only about 20%, the height is still 236 m, or nearly 0.15 mi. It's remarkable just how much our bodies can do with so little.

PRACTICE PROBLEM — PREDICT/CALCULATE

(a) If the mass of the bicyclist is increased—by adding a backpack, for example—does the possible elevation gain increase, decrease, or stay the same? Explain. (b) Find the elevation gain if the bicyclist and bicycle have a combined mass of 85.5 kg.   [**Answer: (a)** The elevation gain is inversely proportional to the mass. Hence, an increase in mass results in a smaller gain in elevation. **(b)** $h = 973$ m]

Some related homework problems: Problem 9, Problem 11

We have been careful *not* to say that the potential energy of a mountain climber—or any object—increases when its height increases. The reason is that the potential energy is a property of an entire system, not of its individual parts. The correct statement is that if an object is lifted, the potential energy of the Earth–object system is increased.

Springs

Consider a spring that is stretched from its equilibrium position a distance x. According to Equation 7-8, the work required to cause this stretch is given by $W = \frac{1}{2}kx^2$. Therefore, if the spring is released—and allowed to return to the equilibrium position—it will do the same work, $\frac{1}{2}kx^2$. From our definition of potential energy, then, we see that

$$W_c = \tfrac{1}{2}kx^2 = U_i - U_f \qquad \text{8-4}$$

In this case, U_f is the potential energy when the spring is at $x = 0$ (equilibrium position), and U_i is the potential energy when the spring is stretched by the amount x.

A convenient choice for $U = 0$ is the equilibrium position of the spring. With this choice we have $U_f = 0$, and Equation 8-4 becomes $U_i = \frac{1}{2}kx^2$. Omitting the subscript i, so that U represents the potential energy of the spring for an arbitrary amount of stretch x, we have:

Potential Energy of a Spring

$U = \frac{1}{2}kx^2$ $\qquad$ 8-5

Since U depends on x^2, which is positive even if x is negative, the potential energy of a spring is always greater than or equal to zero. Thus, a spring's potential energy increases whenever it is displaced from equilibrium. A dramatic example of a stretched spring is shown in **FIGURE 8-11**.

▲ **FIGURE 8-11** Because springs, and bungee cords, exert conservative forces, they can serve as energy storage devices. In this case, the stretched bungee cord is beginning to give up the energy it has stored, and to convert that potential energy into kinetic energy as the jumper is pulled rapidly skyward.

EXERCISE 8-5   SPRING POTENTIAL ENERGY

(a) Find the potential energy of a spring with force constant $k = 740$ N/m if it is stretched by 2.1 cm. **(b)** Find the force constant of a spring whose potential energy is 1.7 J when it is compressed by 6.4 cm.

REASONING AND SOLUTION

(a) The spring potential energy is $U = \frac{1}{2}kx^2$. We substitute $k = 740$ N/m and $x = 0.021$ m:

$$U = \tfrac{1}{2}(740\,\text{N/m})(0.021\,\text{m})^2 = 0.16\,\text{J}$$

(b) We can rearrange $U = \frac{1}{2}kx^2$ to find the force constant, $k = \dfrac{2U}{x^2}$. Then we substitute $U = 1.7$ J and $x = -0.064$ m:

$$k = \frac{2U}{x^2} = \frac{2(1.7\,\text{J})}{(-0.064\,\text{m})^2} = 830\,\text{N/m}$$

Notice that the potential energies for gravity $(U = mgy)$ and a spring $\left(U = \frac{1}{2}kx^2\right)$ are given by different expressions. As mentioned earlier, each conservative force has its own form of potential energy.

EXAMPLE 8-6   COMPRESSED ENERGY AND THE JUMP OF A FLEA

When a force of 120.0 N is applied to a certain spring, it causes a stretch of 2.25 cm. What is the potential energy of this spring when it is **(a)** compressed by 3.50 cm or **(b)** expanded by 7.00 cm?

PICTURE THE PROBLEM

The top sketch shows the spring stretched 2.25 cm by the force $F_1 = 120.0$ N. The lower sketch shows the same spring compressed by a second force, F_2, which causes a compression of 3.50 cm. An expansion of the spring by 7.00 cm would look similar to the top sketch.

CONTINUED

REASONING AND STRATEGY

From the first piece of information—a certain force causes a certain stretch—we can calculate the force constant using $F = kx$. Once we know k, we can find the potential energy for either a compression or an expansion with $U = \frac{1}{2}kx^2$.

Known   Applied force, $F = 120.0$ N; resulting stretch of spring, $x = 0.0225$ m; **(a)** compression, $x = -0.0350$ m; **(b)** expansion, $x = 0.0700$ m.

Unknown   **(a)** Potential energy of spring, $U = ?$ **(b)** Potential energy of spring, $U = ?$

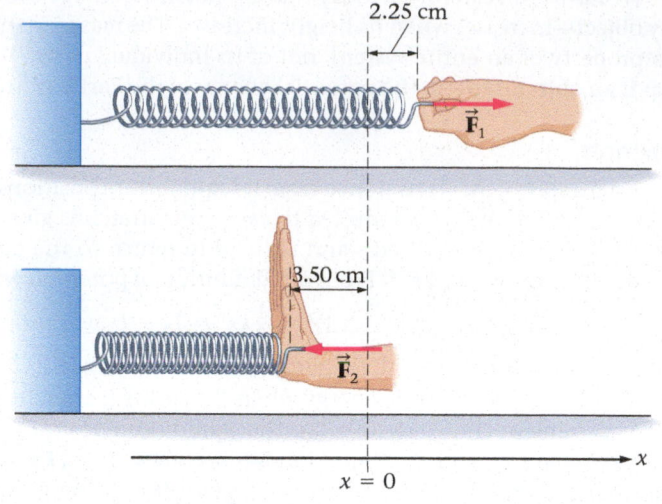

SOLUTION

1. Solve $F = kx$ for the spring constant, k:

$$F = kx$$
$$k = \frac{F}{x} = \frac{120.0\,\text{N}}{0.0225\,\text{m}} = 5330\,\text{N/m}$$

Part (a)

2. Substitute $k = 5330$ N/m and $x = -0.0350$ m into the potential energy expression, $U = \frac{1}{2}kx^2$:

$$U = \tfrac{1}{2}kx^2 = \tfrac{1}{2}(5330\,\text{N/m})(-0.0350\,\text{m})^2 = 3.26\,\text{J}$$

Part (b)

3. Substitute $k = 5330$ N/m and $x = 0.0700$ m into the potential energy expression, $U = \frac{1}{2}kx^2$:

$$U = \tfrac{1}{2}kx^2 = \tfrac{1}{2}(5330\,\text{N/m})(0.0700\,\text{m})^2 = 13.1\,\text{J}$$

INSIGHT

Though this Example deals with ideal springs, the same basic physics applies to many other real-world situations. A case in point is the jump of a flea, in which a flea can propel itself up to 100 times its body length. The physics behind this feat is the slow accumulation of energy in a "springy" strip of resilin in the thigh and coxa of the flea, as shown in the accompanying sketches, and the sudden release of this energy at a later time. Specifically, as the flea's muscles flex the leg, the resilin strip is compressed, storing the work done by the muscles in the form of potential energy, $U = \frac{1}{2}kx^2$. Later, when a trigger mechanism unlocks the flexed leg, the energy stored in the resilin is released explosively—rapidly extending the leg and propelling the flea upward. See Problem 69 for a calculation using the force constant of resilin.

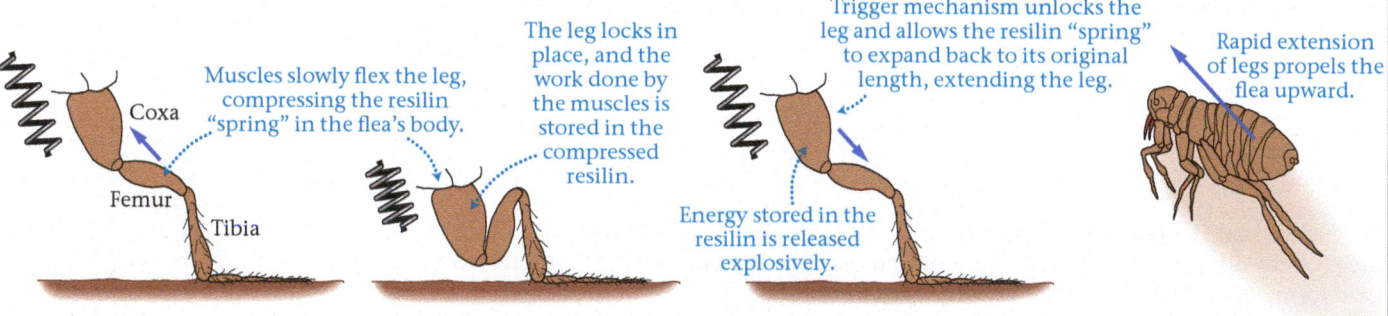

PRACTICE PROBLEM

What stretch is necessary for the spring in this Example to have a potential energy of 5.00 J?   [**Answer:** 4.33 cm]

Some related homework problems: Problem 12, Problem 14, and Problem 16

The jump of a flea is similar in many respects to the operation of a bow and arrow. In the latter case, the work done in slowly pulling the string back is stored in the flex of the bow. The string is held in place while aim is taken, and then released to allow the bow to return to its original shape. This propels the arrow forward with great speed—a speed many times faster than could be obtained by simply throwing the arrow with the same arm muscles that pulled back on the string. An analogous process occurs in the flash unit of a camera, as we shall see in Chapter 20.

BIO Another example of a creature that has the ability to store energy in a spring is the peacock mantis shrimp (*Odontodactylus scyllarus*), shown in **FIGURE 8-12**. This small crustacean can kill its prey by smashing it with a club-like appendage at amazing speeds. By storing energy in a spring-like portion of exoskeleton on its appendage, it can move the club from rest to 23 m/s with an acceleration equivalent to 10,600*g*, similar to the acceleration of a 0.22-caliber bullet when fired. The club strikes with a force as large as 1500 N—two thousand times the weight of the shrimp—and is capable of shattering snails, crabs, mollusks, and rock oysters, cracking their shells into pieces. Mantis shrimp have even been known to break the glass walls of the aquarium in which they are kept! The material from which its club is made is exceptionally tough, and has inspired the design of new carbon fiber–epoxy composite materials.

Enhance Your Understanding   (Answers given at the end of the chapter)

2. The work done by a conservative force on a system is 10 J, and the initial potential energy of the system is 30 J. **(a)** Is the system's final potential energy greater than, less than, or equal to 30 J? Explain. **(b)** What is the final potential energy of the system?

Section Review

* Potential energy is an energy storage mechanism.
* The work done by a conservative force is equal to the negative of the change in potential energy.

▲ **FIGURE 8-12** A peacock mantis shrimp is a formidable predator with its spring-loaded, club-like appendages.

8-3 Conservation of Mechanical Energy

In this section, we show how potential energy can be used as a powerful tool in solving a variety of problems and in gaining greater insight into the workings of physical systems. To do so, we begin by defining the **mechanical energy,** *E,* as the sum of the potential and kinetic energies of an object:

$$E = U + K \qquad \text{8-6}$$

The significance of mechanical energy is that it is **conserved** in systems that involve only conservative forces. By conserved, we mean that its value never changes; that is, E = constant. (In situations where nonconservative forces are involved, the mechanical energy can change, as when friction converts mechanical energy to thermal energy. When *all* possible forms of energy are considered, like mechanical and thermal energy, energy is always found to be conserved.)

Mechanical Energy and Conservative Forces   To show that the mechanical energy E is conserved for conservative forces, we start with the work–energy theorem from Chapter 7:

$$W_{\text{total}} = \Delta K = K_{\text{f}} - K_{\text{i}}$$

Suppose for a moment that the system has only a single force and that the force is conservative. If this is the case, then the total work, W_{total}, is the work done by the conservative force, W_{c}:

$$W_{\text{total}} = W_{\text{c}}$$

Big Idea 3 Mechanical energy is the sum of the potential and kinetic energies of a system. The total amount of mechanical energy is conserved—remains constant—for systems that have only conservative forces.

▲ **FIGURE 8-13 Visualizing Concepts**
Conservation of Energy A roller coaster (top) illustrates the conservation of energy. With every descent, gravitational potential energy is converted into kinetic energy; with every rise, kinetic energy is converted back into gravitational potential energy. If friction is neglected, the total mechanical energy of the car remains constant. The same principle is exploited at a pumped-storage facility, such as this one at the Mormon Flat Dam near Phoenix, Arizona (bottom). When surplus electrical power is available, it is used to pump water uphill into the reservoir. This process, in effect, stores electrical energy as gravitational potential energy. When power demand is high, the stored water is allowed to flow back downhill through the electrical generators in the dam, converting the gravitational energy to kinetic energy and the kinetic energy to electrical energy.

PHYSICS IN CONTEXT
Looking Ahead

Conservation of energy is one of the key elements in the study of elastic collisions in Section 9-6.

PROBLEM-SOLVING NOTE

Conservative Systems

A convenient approach to problems involving energy conservation is to first sketch the system, and then label the initial and final points with i and f, respectively. To apply energy conservation, write out the energy at these two points and set $E_i = E_f$.

From the definition of potential energy, we know that $W_c = -\Delta U = U_i - U_f$. Combining these results, we have

$$W_{total} = W_c$$
$$K_f - K_i = U_i - U_f$$

With a slight rearrangement we find

$$U_f + K_f = U_i + K_i$$

Recalling that $E = U + K$, we have

$$E_f = E_i$$

The initial and final points can be chosen arbitrarily, and hence it follows that E is conserved:

$$E = \text{constant}$$

If the system has more than one conservative force, the only change to these results is to replace U with the sum of the potential energies of all the forces.

To summarize:

Conservation of Mechanical Energy
In systems with conservative forces only, the mechanical energy E is conserved; that is,
$E = U + K = \text{constant}$.

In terms of physical systems, conservation of mechanical energy means that energy can be converted between potential and kinetic forms, but that the sum remains the same. As an example, in the roller coaster shown in Figure 8-6, the gravitational potential energy decreases as the car approaches point B; as it does, the car's kinetic energy increases by the same amount. Similar examples of energy conservation are presented in **FIGURE 8-13**. From a practical point of view, conservation of mechanical energy means that many physics problems can be solved by what amounts to simple bookkeeping.

Applying Energy Conservation Consider a key chain of mass m that is dropped to the floor from a height h, as illustrated in **FIGURE 8-14**. The question is, how fast are the keys moving just before they land? We know how to solve this problem using Newton's laws and kinematics, but let's see how energy conservation can be used instead.

First, we note that the only force acting on the keys is gravity—ignoring air resistance, of course—and that gravity is a conservative force. As a result, we can say that $E = U + K$ is constant during the entire time the keys are falling. To solve the problem, then, we pick two points on the motion of the keys, say i and f in Figure 8-14, and we set the mechanical energy equal at these points:

$$E_i = E_f \qquad 8\text{-}7$$

Writing this out in terms of potential and kinetic energies, we have

$$U_i + K_i = U_f + K_f \qquad 8\text{-}8$$

This one equation—which is nothing but bookkeeping—can be used to solve for the one unknown, the final speed.

To be specific, in Figure 8-14 (a) we choose $y = 0$ at ground level, which means that $U_i = mgh$. In addition, the fact that the keys are released from rest means that $K_i = 0$. Similarly, at point f—just before hitting the ground—the energy is all kinetic, and the potential energy is zero; that is, $U_f = 0$, $K_f = \frac{1}{2}mv^2$. Substituting these values into $U_i + K_i = U_f + K_f$, we find

$$mgh + 0 = 0 + \frac{1}{2}mv^2$$

Canceling m and solving for v yields the same result we get with kinematics:

$$v = \sqrt{2gh}$$

Suppose, instead, that we had chosen $y = 0$ to be at the release point of the keys, as in Figure 8-14 (b), so that the keys land at $y = -h$. Now, when the keys are released,

$U_i = mgh$
$K_i = 0$

$v = 0$

h

$U_f = 0$
$K_f = \frac{1}{2}mv^2$

$v = ?$

(a)

$U_i = 0$
$K_i = 0$

$v = 0$

h

$U_f = -mgh$
$K_f = \frac{1}{2}mv^2$

$v = ?$

(b)

◄ **FIGURE 8-14 Solving a kinematics problem using conservation of energy (a)** A set of keys falls to the floor. Ignoring frictional forces, we know that the mechanical energy at points i and f must be equal: $E_i = E_f$. Using this condition, we can find the speed of the keys just before they land. **(b)** The same physical situation as in part (a), except this time we have chosen $y = 0$ to be at the point where the keys are dropped. As before, we set $E_i = E_f$ to find the speed of the keys just before they land. The result is the same.

we have $U_i = 0$ and $K_i = 0$, and when they land, $U_f = -mgh$ and $K_f = \frac{1}{2}mv^2$. Substituting these results in $U_i + K_i = U_f + K_f$ yields

$$0 + 0 = -mgh + \frac{1}{2}mv^2$$

Solving for v gives the same result:

$$v = \sqrt{2gh}$$

Thus, as expected, changing the zero level has no effect on the physical results.

EXAMPLE 8-7   GRADUATION FLING

At the end of a graduation ceremony, graduates fling their caps into the air. Suppose a 0.120-kg cap is thrown straight upward with an initial speed of 7.85 m/s, and that frictional forces can be ignored. **(a)** Use kinematics to find the speed of the cap when it is 1.18 m above the release point. **(b)** Show that the mechanical energy at the release point is the same as the mechanical energy 1.18 m above the release point.

PICTURE THE PROBLEM
In our sketch, we choose $y = 0$ to be at the level where the cap is released with an initial speed of 7.85 m/s, and we choose upward to be the positive direction. In addition, we designate the release point as i (initial) and the point at which $y = 1.18$ m as f (final). It is the speed at point f that we wish to find.

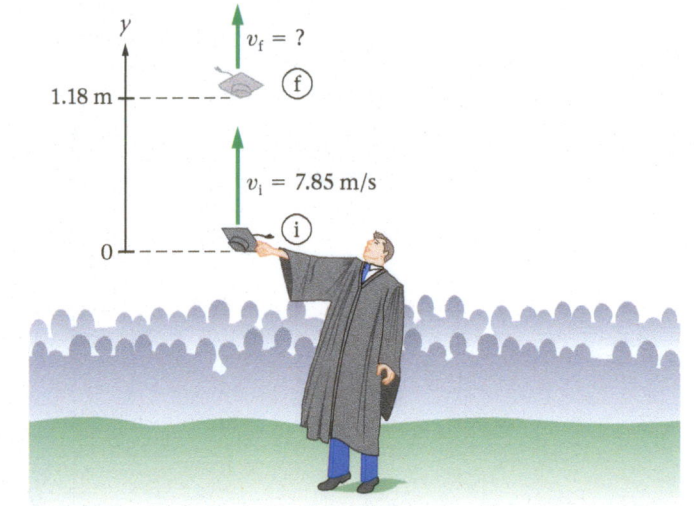

REASONING AND STRATEGY

a. The cap is in free fall, which justifies the use of constant-acceleration kinematics. In addition, the cap is thrown straight upward, and hence the motion is one dimensional along the y direction. We want to relate velocity to position; thus we use $v_f^2 = v_i^2 + 2a_y\Delta y$ (Section 2-5). In this case, $v_i = 7.85$ m/s, $\Delta y = 1.18$ m, and $a_y = -g$. Substituting these values gives v_f.

b. At the points i and f we calculate the mechanical energy, $E = U + K$, with $U = mgy$ for the gravitational potential energy. The kinetic energy, as always, is $K = \frac{1}{2}mv^2$.

Known   Mass of cap, $m = 0.120$ kg; initial speed, $v_i = 7.85$ m/s; height above release point, $\Delta y = 1.18$ m.
Unknown   **(a)** Speed of cap, $v_f = ?$ **(b)** Mechanical energy at initial and final points, $E_i = ?, E_f = ?$

CONTINUED

SOLUTION

Part (a)

1. Use kinematics to solve for v_f:

$$v_f^2 = v_i^2 + 2a_y \Delta y$$
$$v_f = \pm\sqrt{v_i^2 + 2a_y \Delta y}$$

2. Substitute $v_i = 7.85 \text{ m/s}$, $\Delta y = 1.18 \text{ m}$, and $a_y = -g$ to find v_f. Choose the plus sign, since we are interested only in the speed:

$$v_f = \sqrt{v_i^2 + 2a_y \Delta y}$$
$$= \sqrt{(7.85 \text{ m/s})^2 + 2(-9.81 \text{ m/s}^2)(1.18 \text{ m})} = 6.20 \text{ m/s}$$

Part (b)

3. Calculate E_i. At this point $y_i = 0$ and $v_i = 7.85 \text{ m/s}$:

$$E_i = U_i + K_i = mgy_i + \tfrac{1}{2}mv_i^2$$
$$= 0 + \tfrac{1}{2}(0.120 \text{ kg})(7.85 \text{ m/s})^2 = 3.70 \text{ J}$$

4. Calculate E_f. At this point $y_f = 1.18 \text{ m}$ and $v_f = 6.20 \text{ m/s}$:

$$E_f = U_f + K_f = mgy_f + \tfrac{1}{2}mv_f^2$$
$$= (0.120 \text{ kg})(9.81 \text{ m/s}^2)(1.18 \text{ m}) + \tfrac{1}{2}(0.120 \text{ kg})(6.20 \text{ m/s})^2$$
$$= 1.39 \text{ J} + 2.31 \text{ J} = 3.70 \text{ J}$$

INSIGHT

As expected, E_f is equal to E_i. In the remaining Examples in this section we turn this process around; we start with $E_f = E_i$, and use this relationship to find a final speed or a final height. As we shall see, this procedure of using energy conservation is a more powerful approach—it actually makes the calculations much simpler.

PRACTICE PROBLEM

Use energy conservation to find the height at which the speed of the cap is 5.00 m/s.   [**Answer:** 1.87 m]

Some related homework problems: Problem 27, Problem 29

Finding the Speed at a Given Height An interesting extension of Example 8-7 is shown in **FIGURE 8-15**. In this case, we are given that the speed of the cap is v_i at the height y_i, and we would like to know its speed v_f when it is at the height y_f.

To find v_f, we apply energy conservation to the points i and f:

$$U_i + K_i = U_f + K_f$$

Writing out U and K specifically for these two points yields the following:

$$mgy_i + \tfrac{1}{2}mv_i^2 = mgy_f + \tfrac{1}{2}mv_f^2$$

As before, we cancel m and solve for the unknown speed, v_f:

$$v_f = \sqrt{v_i^2 + 2g(y_i - y_f)}$$

This result is in agreement with the kinematic equation, $v_y^2 = v_{0y}^2 + 2a_y \Delta y$.

▶ **FIGURE 8-15 Speed is independent of path** If the speed of the cap is v_i at the height y_i, its speed is v_f at the height y_f, independent of the path between the two heights. This assumes, of course, that frictional forces can be neglected.

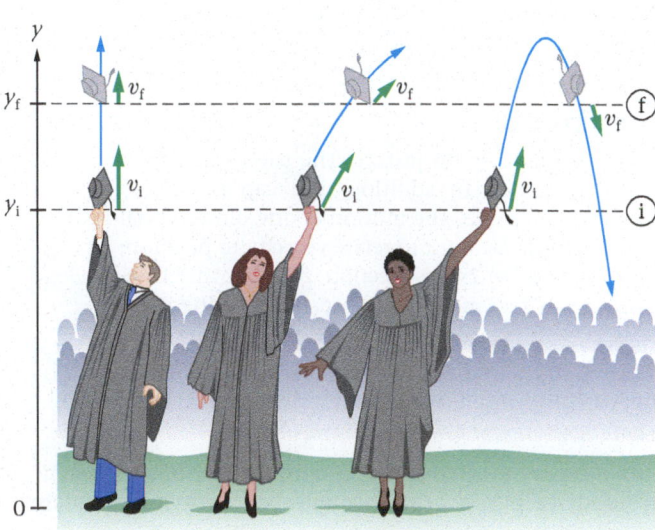

Notice that v_f depends only on y_i and y_f, not on the path connecting them. This is because conservative forces such as gravity do work that is path independent. What this means physically is that the cap has the same speed v_f at the height y_f, whether it goes straight upward or follows some other trajectory, as in Figure 8-15. All that matters is the height difference.

EXAMPLE 8-8   CATCHING A HOME RUN

In the bottom of the ninth inning, a player hits a 0.15-kg baseball over the outfield fence. The ball leaves the bat with a speed of 36 m/s, and a fan in the bleachers catches it 7.2 m above the point where it was hit. Assuming frictional forces can be ignored, find **(a)** the kinetic energy of the ball and **(b)** its speed just before it's caught.

PICTURE THE PROBLEM
Our sketch shows the ball's trajectory. We label the hit point i and the catch point f. At point i we choose $y_i = 0$; at point f, then, $y_f = h = 7.2$ m. In addition, we are given that $v_i = 36$ m/s. The final speed v_f is to be determined.

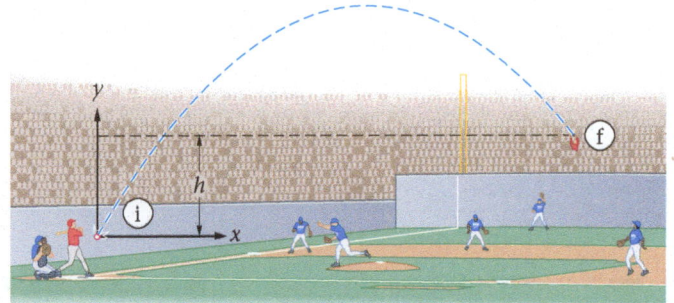

REASONING AND STRATEGY

a. Because frictional forces can be ignored, it follows that the initial mechanical energy is equal to the final mechanical energy; that is, $U_i + K_i = U_f + K_f$. Use this relationship to find K_f.

b. Once K_f is determined, use $K_f = \frac{1}{2}mv_f^2$ to find v_f.

Known   Mass of baseball, $m = 0.15$ kg; initial speed of ball, $v_i = 36$ m/s; catch height, $h = 7.2$ m.
Unknown   **(a)** Kinetic energy of ball just before being caught, $K_f = ?$ **(b)** Speed of ball just before being caught, $v_f = ?$

SOLUTION

Part (a)

1. Begin by writing U and K for point i:

$$U_i = 0$$
$$K_i = \tfrac{1}{2}mv_i^2 = \tfrac{1}{2}(0.15 \text{ kg})(36 \text{ m/s})^2 = 97 \text{ J}$$

2. Next, write U and K for point f:

$$U_f = mgh = (0.15 \text{ kg})(9.81 \text{ m/s}^2)(7.2 \text{ m}) = 11 \text{ J}$$
$$K_f = \tfrac{1}{2}mv_f^2$$

3. Set the total mechanical energy at point i, $E_i = U_i + K_i$, equal to the total mechanical energy at point f, $E_f = U_f + K_f$, and solve for K_f:

$$U_i + K_i = U_f + K_f$$
$$0 + 97 \text{ J} = 11 \text{ J} + K_f$$
$$K_f = 97 \text{ J} - 11 \text{ J} = 86 \text{ J}$$

Part (b)

4. Use $K_f = \frac{1}{2}mv_f^2$ to find v_f:

$$K_f = \tfrac{1}{2}mv_f^2$$
$$v_f = \sqrt{\frac{2K_f}{m}} = \sqrt{\frac{2(86 \text{ J})}{0.15 \text{ kg}}} = 34 \text{ m/s}$$

INSIGHT
To find the ball's speed just before being caught, we need to know the height of point f, but we don't need to know any details about the ball's trajectory. For example, it is not necessary to know the angle at which the ball leaves the bat or its maximum height.

The histograms show the values of U and K at the points i and f. Notice that the energy of the system is still mostly kinetic just before the ball is caught.

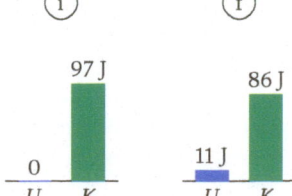

PRACTICE PROBLEM
If the mass of the ball were increased, would the catch speed be greater than, less than, or the same as the value we just found?   [**Answer:** The same. U and K depend on mass in the same way; hence the mass cancels.]

Some related homework problems: Problem 25, Problem 27

The connection between height difference and speeds is explored further in the following two Examples. The first Example deals with the speed of swimmers on water slides, like those pictured in **FIGURE 8-16**.

CONCEPTUAL EXAMPLE 8-9 | COMPARE THE FINAL SPEEDS

PREDICT/EXPLAIN

Swimmers at a water park can enter a pool using one of two frictionless slides of equal height. Slide 1 approaches the water with a uniform slope; slide 2 dips rapidly at first, then levels out. **(a)** Is the speed at the bottom of slide 1 greater than, less than, or the same as the speed at the bottom of slide 2? **(b)** Which of the following is the *best explanation* for your prediction?

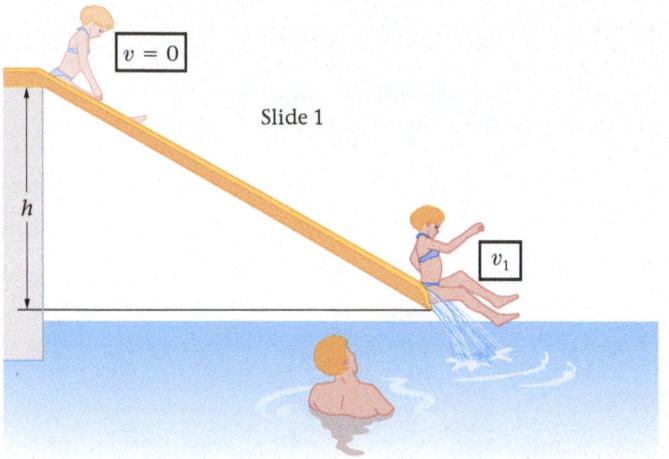
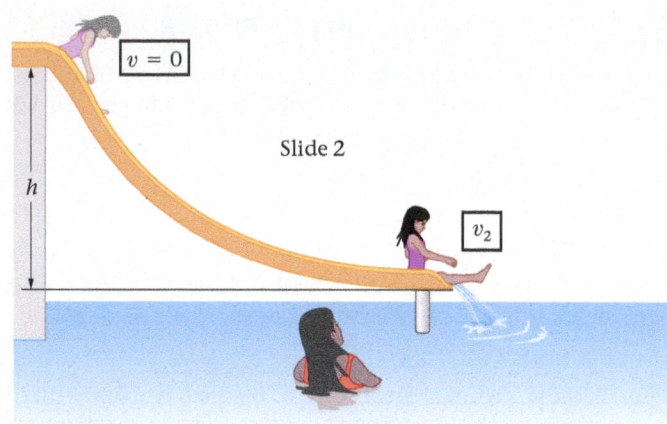

I. The same amount of potential energy is converted to kinetic energy in each case, so the speeds are the same.

II. Slide 2 dips rapidly, resulting in a greater speed for that slide.

III. The uniform drop of slide 1 results in a steady increase in speed, and a greater final speed for that slide.

REASONING AND DISCUSSION

In both cases, the same amount of potential energy, mgh, is converted to kinetic energy. Because the conversion of gravitational potential energy to kinetic energy is the *only* energy transaction taking place, it follows that the speed is the same for each slide.

Interestingly, although the final speeds are the same, the time required to reach the water is less for slide 2. The reason is that swimmer 2 reaches a high speed early and maintains it, whereas the speed of swimmer 1 increases slowly and steadily.

ANSWER

(a) The speeds are the same for the two slides. **(b)** The best explanation is I.

EXAMPLE 8-10 | SKATEBOARD EXIT RAMP

A 55-kg skateboarder enters a ramp moving horizontally with a speed of 6.5 m/s and leaves the ramp moving vertically with a speed of 4.1 m/s. Find the height of the ramp, assuming no energy loss to frictional forces.

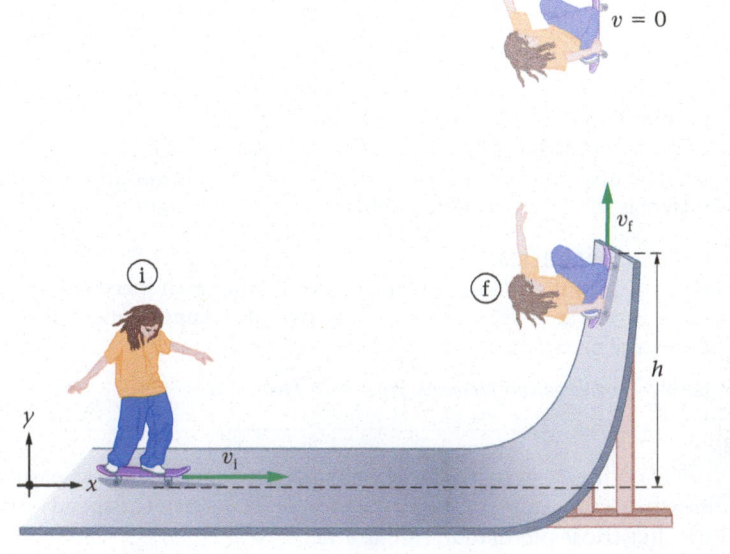

PICTURE THE PROBLEM

We choose $y = 0$ to be the level of the bottom of the ramp; thus the gravitational potential energy is zero there. Point i indicates the skateboarder entering the ramp with a speed of 6.5 m/s; point f is the top of the ramp, where the speed is 4.1 m/s.

REASONING AND STRATEGY

To find h, we simply set the initial energy, $E_i = U_i + K_i$, equal to the final energy, $E_f = U_f + K_f$.

Known Mass of skateboarder, $m = 55$ kg; initial speed, $v_i = 6.5$ m/s; final speed, $v_f = 4.1$ m/s.

Unknown Height of ramp, $h = ?$

CONTINUED

SOLUTION

1. Write expressions for U_i and K_i:

$$U_i = mg \cdot 0 = 0 \qquad K_i = \tfrac{1}{2}mv_i^2$$

2. Write expressions for U_f and K_f:

$$U_f = mgh \qquad K_f = \tfrac{1}{2}mv_f^2$$

3. Set the total mechanical energy at point i, $E_i = U_i + K_i$, equal to the total mechanical energy at point f, $E_f = U_f + K_f$:

$$U_i + K_i = U_f + K_f$$
$$0 + \tfrac{1}{2}mv_i^2 = mgh + \tfrac{1}{2}mv_f^2$$

4. Solve for h. Notice that m cancels:

$$mgh = \tfrac{1}{2}mv_i^2 - \tfrac{1}{2}mv_f^2$$

$$h = \frac{v_i^2 - v_f^2}{2g}$$

5. Substitute numerical values:

$$h = \frac{(6.5 \text{ m/s})^2 - (4.1 \text{ m/s})^2}{2(9.81 \text{ m/s}^2)} = 1.3 \text{ m}$$

INSIGHT

Our result for h is independent of the shape of the ramp—it is equally valid for one with the shape shown here, or one that simply inclines upward at a constant angle. In addition, the height does not depend on the person's mass, as we see in Step 4 where the mass cancels.

The histograms show U and K to scale at the points i and f, as well as at the maximum height, where $K = 0$.

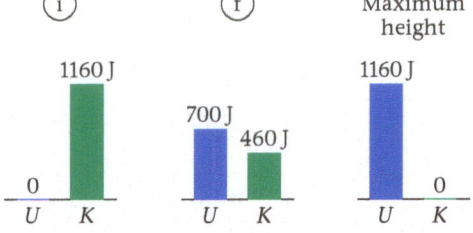

PRACTICE PROBLEM

What is the skateboarder's maximum height above the bottom of the ramp?   [**Answer: 2.2 m**]

Some related homework problems: Problem 23, Problem 29, Problem 60

It's useful to express the equation in Step 3 from Example 8-10 $\left(0 + \tfrac{1}{2}mv_i^2 = mgh + \tfrac{1}{2}mv_f^2\right)$ in words. First, the left side of the equation is the initial kinetic energy of the skateboarder, $\tfrac{1}{2}mv_i^2$. This is the initial energy content of the system. At point f the system still has the same amount of energy, only now part of it, mgh, is in the form of gravitational potential energy. The remainder is the final kinetic energy, $\tfrac{1}{2}mv_f^2$.

Conceptual Example 8-11 considers the effect of a slight change in the initial speed of an object.

▶ **FIGURE 8-16** Does the shape of the slide matter? (See Conceptual Example 8-9.)

CONCEPTUAL EXAMPLE 8-11   WHAT'S THE FINAL SPEED

A snow boarder coasts on a smooth track that rises from one level to another. If the snowboarder's initial speed is 4 m/s, the snowboarder just makes it to the upper level and comes to rest. With a slightly greater initial speed of 5 m/s, the snowboarder is still moving to the right on the upper level. Is the snowboarder's final speed in this case **(a)** 1 m/s, **(b)** 2 m/s, or **(c)** 3 m/s?

REASONING AND DISCUSSION

A plausible-sounding answer is that since the initial speed is greater by 1 m/s in the second case, the final speed should be greater by 1 m/s as well. Therefore, the answer should be 0 + 1 m/s = 1 m/s. This is incorrect, however.

As surprising as it may seem, an increase in the initial speed from 4 m/s to 5 m/s results in an increase in the final speed from 0 to 4 m/s. This is due to the fact that kinetic energy depends on v^2 rather than v; thus, it is the difference in v^2 that counts. In this case, the initial value of v^2 increases from 16 m²/s² to 25 m²/s², for a total increase of 25 m²/s² − 16 m²/s² = 9 m²/s². The final value of v^2 must increase by the same amount, 9 m²/s² = (3 m/s)². As a result, the final speed is 3 m/s.

ANSWER

(c) The final speed of the snowboarder in the second case is 3 m/s.

PHYSICS IN CONTEXT
Looking Ahead

The wide-ranging importance of energy conservation is illustrated by its use in the following disparate topics: rotational motion (Section 10-6), gravitation (Section 12-5), oscillatory motion (Section 13-5), fluid dynamics (Section 15-7), and phase changes (Section 17-6).

Let's check the results of the previous Conceptual Example with a specific numerical example. Suppose the snowboarder has a mass of 74.0 kg. It follows that in the first case the initial kinetic energy is $K_i = \frac{1}{2}(74.0\text{ kg})(4.00\text{ m/s})^2 = 592\text{ J}$. At the top of the hill all of this kinetic energy is converted to gravitational potential energy, mgh.

In the second case, the initial speed of the snowboarder is 5.00 m/s; thus, the initial kinetic energy is $K_i = \frac{1}{2}(74.0\text{ kg})(5.00\text{ m/s})^2 = 925\text{ J}$. When the snowboarder reaches the top of the hill, 592 J of this kinetic energy is converted to gravitational potential energy, leaving the snowboarder with a final kinetic energy of 925 J − 592 J = 333 J. The corresponding speed is given by

$$\frac{1}{2}mv^2 = 333\text{ J}$$

$$v = \sqrt{\frac{2(333\text{ J})}{m}} = \sqrt{\frac{2(333\text{ J})}{74.0\text{ kg}}} = \sqrt{9.00\text{ m}^2/\text{s}^2} = 3.00\text{ m/s}$$

Thus, as expected, the snowboarder in the second case has a final speed of 3.00 m/s. We conclude this section with two Examples involving springs.

EXAMPLE 8-12   SPRING TIME

A 1.70-kg block slides on a horizontal, frictionless surface until it encounters a spring with a force constant of 955 N/m. The block comes to rest after compressing the spring a distance of 4.60 cm. Find the initial speed of the block. (Ignore air resistance and any energy lost when the block initially contacts the spring.)

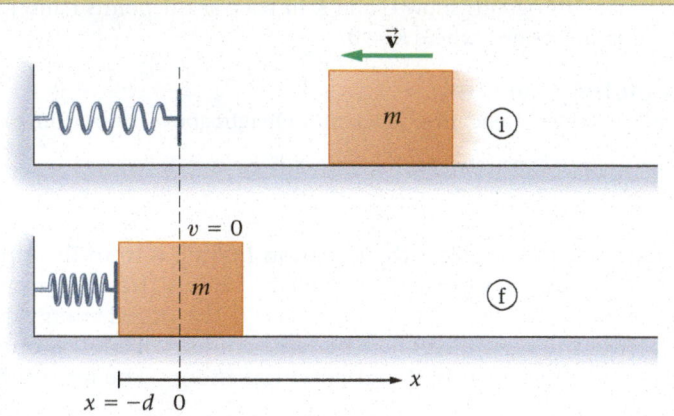

PICTURE THE PROBLEM
Point i refers to times before the block makes contact with the spring, which means the block has a speed v_i and the end of the spring is at $x = 0$. Point f refers to the time when the block has come to rest, which means $v_f = 0$, and the spring is compressed to $x = -d = -4.60$ cm.

    We can choose the center of the block to be the $y = 0$ level. With this choice, the gravitational potential energy of the sytem is zero at all times.

REASONING AND STRATEGY
Set E_i equal to E_f to find the one unknown, v_i. Notice that the initial energy, E_i, is the kinetic energy of the block before it reaches the spring. The final energy, E_f, is the potential energy of the compressed spring.

Known   Mass of block, $m = 1.70$ kg; force constant, $k = 955$ N/m; compression distance, $d = 0.0460$ m; final speed, $v_f = 0$.
Unknown   Initial speed of block, $v_i = ?$

SOLUTION

1. Write expressions for U_i and K_i. For U, we consider only the potential energy of the spring, $U = \frac{1}{2}kx^2$:
$$U_i = \frac{1}{2}k \cdot 0^2 = 0 \qquad K_i = \frac{1}{2}mv_i^2$$

2. Write expressions for U_f and K_f:
$$U_f = \frac{1}{2}k(-d)^2 = \frac{1}{2}kd^2 \qquad K_f = \frac{1}{2}m \cdot 0^2 = 0$$

3. Set the initial mechanical energy, $E_i = U_i + K_i$, equal to the final mechanical energy, $E_f = U_f + K_f$, and solve for v:
$$U_i + K_i = U_f + K_f$$
$$0 + \frac{1}{2}mv_i^2 = \frac{1}{2}kd^2 + 0$$
$$v_i = d\sqrt{\frac{k}{m}}$$

4. Substitute numerical values:
$$v_i = d\sqrt{\frac{k}{m}} = (0.0460\text{ m})\sqrt{\frac{955\text{ N/m}}{1.70\text{ kg}}} = 1.09\text{ m/s}$$

INSIGHT
After the block comes to rest, the spring expands again, converting its potential energy back into the kinetic energy of the block. When the block leaves the spring, moving to the right, its speed is once again 1.09 m/s.

PRACTICE PROBLEM
What is the compression distance, d, if the block's initial speed is 0.500 m/s? [**Answer:** 2.11 cm]

Some related homework problems: Problem 26, Problem 28

QUICK EXAMPLE 8-13   FIND THE SPEED OF THE BLOCK

Suppose the spring and block in Example 8-12 are oriented vertically. Initially, the spring is compressed 4.60 cm and the block is at rest. When the block is released, it accelerates upward. Find the speed of the block when the spring has returned to its equilibrium position.

REASONING AND SOLUTION

The physical system is shown in our sketch, with the positive y direction upward, and $y = 0$ at the equilibrium position of the spring. The initial condition corresponds to the block at rest at a negative value of y and the spring compressed. Thus, the initial mechanical energy is the positive spring potential energy $(U = \frac{1}{2}kd^2)$, plus the negative gravitational potential energy $(U = -mgd)$, plus zero kinetic energy $(K = 0)$. The final condition corresponds to the block leaving the uncompressed spring, where the both the spring potential energy and gravitational potential energy are zero. Therefore, the final mechanical energy is simply the kinetic energy of the block. We set E_i equal to E_f to find the one unknown, which in this case is v_f.

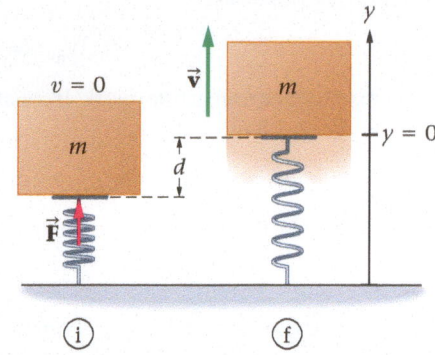

1. Write an expression for the initial mechanical energy E_i:

$$E_i = U_i + K_i = \tfrac{1}{2}kd^2 - mgd + 0$$

2. Write an expression for the final mechanical energy E_f:

$$E_f = U_f + K_f = 0 + 0 + \tfrac{1}{2}mv_f^2$$

3. Set E_i equal to E_f and solve for v_f:

$$\tfrac{1}{2}kd^2 - mgd = \tfrac{1}{2}mv_f^2$$

$$v_f = \sqrt{\frac{kd^2}{m} - 2gd}$$

4. Substitute numerical values:

$$v_f = \sqrt{\frac{(955 \text{ N/m})(0.0460 \text{ m})^2}{1.70 \text{ kg}} - 2(9.81 \text{ m/s}^2)(0.0460 \text{ m})}$$

$$= 0.535 \text{ m/s}$$

In this system, part of the initial potential energy of the spring $\left(\frac{1}{2}kd^2\right)$ goes into increasing the gravitational potential energy of the block (mgd). The remainder of the initial energy, $\frac{1}{2}kd^2 - mgd$, is converted into the block's kinetic energy.

Enhance Your Understanding   (Answers given at the end of the chapter)

3. A system with only conservative forces acting on it has a mechanical energy of 50 J and an initial kinetic energy of 25 J. If the final potential energy of the system is 15 J, what is its final kinetic energy?

Section Review

• If a system is acted on by conservative forces only, its mechanical energy, $E = U + K$, is conserved.

8-4 Work Done by Nonconservative Forces

Nonconservative forces change the amount of mechanical energy in a system. They might decrease the mechanical energy by converting it to thermal energy, or increase it by converting muscular work to kinetic or potential energy. In some systems, both types of processes occur at the same time.

Connecting Nonconservative Work and Energy Change   To see the connection between the work done by a nonconservative force, W_{nc}, and the mechanical energy, E, we return once more to the work–energy theorem, which says that the *total* work is equal to the change in kinetic energy:

$$W_{total} = \Delta K$$

Suppose, for instance, that a system has one conservative and one nonconservative force. In this case, the total work is the sum of the conservative work W_c and the nonconservative work W_{nc}:

$$W_{total} = W_c + W_{nc}$$

Recalling that conservative work is related to the change in potential energy by $W_c = -\Delta U$, we have

$$W_{total} = -\Delta U + W_{nc} = \Delta K$$

Solving this relationship for the nonconservative work yields

$$W_{nc} = \Delta U + \Delta K$$

Finally, since the total mechanical energy is $E = U + K$, it follows that the *change* in mechanical energy is $\Delta E = \Delta U + \Delta K$. As a result, *the nonconservative work is simply the change in mechanical energy:*

$$W_{nc} = \Delta E = E_f - E_i \qquad\qquad \text{8-9}$$

If more than one nonconservative force acts, we simply add the nonconservative work done by each such force to obtain W_{nc}.

At this point it may be useful to collect the three "working relationships" that have been introduced in the last two chapters:

$$\begin{aligned} W_{total} &= \Delta K \\ W_c &= -\Delta U \\ W_{nc} &= \Delta E \end{aligned} \qquad\qquad \text{8-10}$$

PROBLEM-SOLVING NOTE

Nonconservative Systems

Start by sketching the system and labeling the initial and final points with i and f, respectively. The initial and final mechanical energies are related to the nonconservative work by $W_{nc} = E_f - E_i$.

Notice that positive nonconservative work increases the total mechanical energy of a system, while negative nonconservative work decreases the mechanical energy—and converts it to other forms. In the next Example, for instance, part of the initial mechanical energy of a leaf is converted to heat and other forms of energy by air resistance as the leaf falls to the ground.

EXAMPLE 8-14   A LEAF FALLS IN THE FOREST: FIND THE NONCONSERVATIVE WORK

Deep in the forest, a 17.0-g leaf falls from a tree and drops straight to the ground. If its initial height was 5.30 m and its speed just before landing was 1.3 m/s, how much nonconservative work was done on the leaf?

PICTURE THE PROBLEM
In our sketch, the initial and final points are labeled i and f, respectively. At the initial point, the leaf drops from rest, $v_i = 0$, from a height $y_i = h = 5.30$ m. At the final point, just before landing at $y_f = 0$, the leaf's speed is $v_f = 1.3$ m/s.

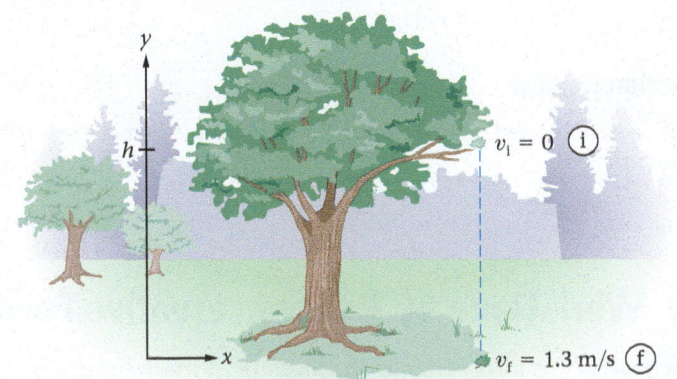

REASONING AND STRATEGY
To begin, calculate the initial mechanical energy, E_i, and the final mechanical energy, E_f. Once these energies have been determined, the nonconservative work is the change in energy, $W_{nc} = \Delta E = E_f - E_i$.

Known   Mass of leaf, $m = 0.0170$ kg; initial height, $y_i = h = 5.30$ m; final height, $y_f = 0$; initial speed, $v_i = 0$; final speed, $v_f = 1.3$ m/s.

Unknown   Nonconservative work done on leaf, $W_{nc} = ?$

CONTINUED

SOLUTION

1. Evaluate U_i, K_i, and E_i:

$U_i = mgh = (0.0170 \text{ kg})(9.81 \text{ m/s}^2)(5.30 \text{ m}) = 0.884 \text{ J}$
$K_i = \frac{1}{2}m \cdot 0^2 = 0$
$E_i = U_i + K_i = 0.884 \text{ J}$

2. Next, evaluate U_f, K_f, and E_f.

$U_f = mg \cdot 0 = 0$
$K_f = \frac{1}{2}mv_f^2 = \frac{1}{2}(0.0170 \text{ kg})(1.3 \text{ m/s})^2 = 0.014 \text{ J}$
$E_f = U_f + K_f = 0.014 \text{ J}$

3. Use $W_{nc} = \Delta E$ to find the nonconservative work:

$W_{nc} = \Delta E = E_f - E_i = 0.014 \text{ J} - 0.884 \text{ J} = -0.870 \text{ J}$

INSIGHT

Most of the leaf's initial mechanical energy is dissipated as it falls. This is indicated in the histograms. The small amount that remains (only about 1.6%) appears as the kinetic energy of the leaf just before it lands.

PRACTICE PROBLEM

What was the average nonconservative force exerted on the leaf as it fell?
[**Answer:** $W_{nc} = -Fh$, $F = -W_{nc}/h = 0.164 \text{ N}$, upward]

Some related homework problems: Problem 34, Problem 36, Problem 38

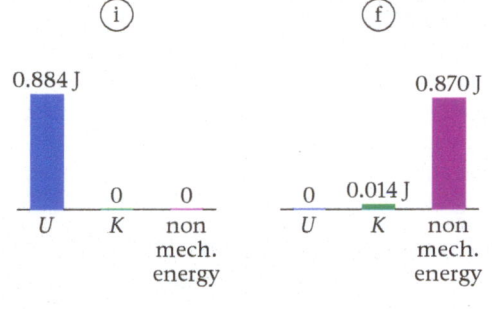

In the following Example, we use nonconservative work to find the drop height of a ski glove.

EXAMPLE 8-15   SNOW JOB

A skier drops her 0.312-kg glove from rest at a height h into fresh powder snow. The glove falls to a depth $d = 0.147 \text{ m}$ below the surface of the snow before coming to rest. If the nonconservative work done on the glove is $W_{nc} = -1.44 \text{ J}$, what is the drop height of the glove?

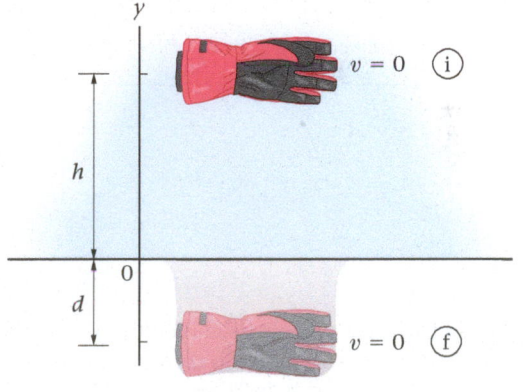

PICTURE THE PROBLEM

We choose the positive y direction to be upward. In addition, we set $y = 0$ at the level of the snow. The glove starts with zero speed at $y = h$, and ends with zero speed at $y = -d$.

REASONING AND STRATEGY

We start by calculating the initial and final mechanical energies, E_i and E_f. Given these energies, and the fact that $W_{nc} = -1.44 \text{ J}$, we can use $W_{nc} = \Delta E = E_f - E_i$ to solve for the one unknown, h.

Known   Mass of glove, $m = 0.312 \text{ kg}$; depth in snow, $d = 0.147 \text{ m}$; nonconservative work, $W_{nc} = -1.44 \text{ J}$.
Unknown   Drop height, $h = ?$

SOLUTION

1. Write the initial mechanical energy, E_i:

$E_i = mgh + 0 = mgh$

2. Write the final mechanical energy, E_f:

$E_f = mg(-d) + 0 = -mgd$

3. Set W_{nc} equal to ΔE:

$W_{nc} = \Delta E = E_f - E_i = -mgd - mgh$

4. Solve for h:

$h = -\dfrac{(W_{nc} + mgd)}{mg}$

5. Substitute numerical values:

$h = -\dfrac{[-1.44 \text{ J} + (0.312 \text{ kg})(9.81 \text{ m/s}^2)(0.147 \text{ m})]}{(0.312 \text{ kg})(9.81 \text{ m/s}^2)} = 0.323 \text{ m}$

INSIGHT

Another way to write Step 3 is $E_f = E_i + W_{nc}$. In words, this equation says that the final mechanical energy is the initial mechanical energy plus the nonconservative work done on the system. In this case, $W_{nc} < 0$; hence the final mechanical energy is less than the initial mechanical energy.

CONTINUED

We now present a Conceptual Example that further examines the relationship between nonconservative work and distance.

CONCEPTUAL EXAMPLE 8-16   JUDGING A PUTT

A golfer badly misjudges a putt, sending the ball only one-quarter of the distance to the hole. The original putt gave the ball an initial speed of v_i. If the force of resistance due to the grass is constant, would an initial speed of **(a)** $2v_i$, **(b)** $3v_i$, or **(c)** $4v_i$ be needed to get the ball to the hole from its original position?

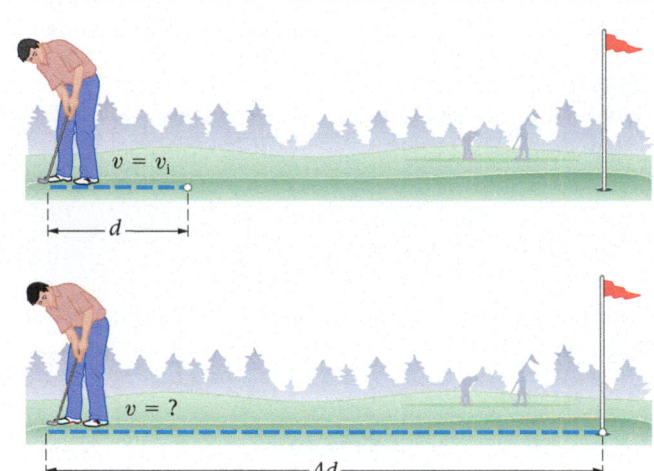

REASONING AND DISCUSSION
In the original putt, the ball started with a kinetic energy of $\frac{1}{2}mv_i^2$ and came to rest in the distance d. The kinetic energy was dissipated by the nonconservative force due to grass resistance, F, which does the work $W_{nc} = -Fd$. Since the change in mechanical energy is $\Delta E = 0 - \frac{1}{2}mv_i^2 = -\frac{1}{2}mv_i^2$, it follows from $W_{nc} = \Delta E$ that $Fd = \frac{1}{2}mv_i^2$. Therefore, to go four times the distance, $4d$, we need to give the ball four times as much kinetic energy. Noting that kinetic energy is proportional to v^2, we see that the initial speed need only be doubled.

ANSWER
(a) The initial speed should be doubled to $2v_i$.

A common example of a nonconservative force is kinetic friction. In the next Example, we show how to include the effects of friction in a system that also includes kinetic energy and gravitational potential energy.

EXAMPLE 8-17   LANDING WITH A THUD

A block of mass $m_1 = 2.40$ kg is connected to a second block of mass $m_2 = 1.80$ kg, as shown in the sketch. When the blocks are released from rest, they move through a distance $d = 0.500$ m, at which point m_2 hits the floor. Given that the coefficient of kinetic friction between m_1 and the horizontal surface is $\mu_k = 0.450$, find the speed of the blocks just before m_2 lands.

PICTURE THE PROBLEM
We choose $y = 0$ to be at floor level; therefore, the gravitational potential energy of m_2 is zero when it lands. The potential energy of m_1 doesn't change during this process; it is always m_1gh. Thus, it isn't necessary to know the value of h. We've labeled the initial and final points with i and f, respectively.

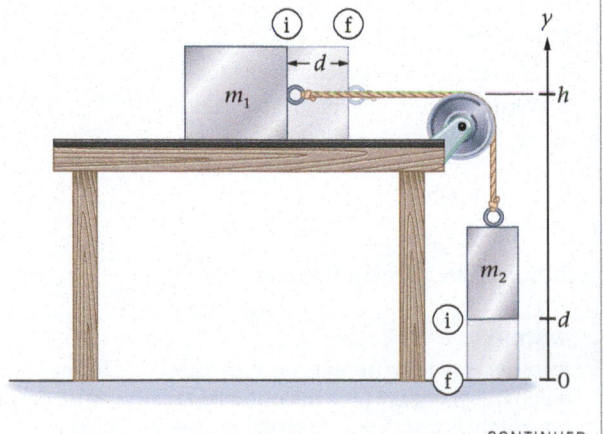

REASONING AND STRATEGY
A nonconservative force (friction) is doing work in this system, and hence we use $W_{nc} = \Delta E = E_f - E_i$. We can solve this relationship for the final mechanical energy, $E_f = E_i + W_{nc}$, and from this determine the final speed, v_f.

Known   Mass of block 1, $m_1 = 2.40$ kg; mass of block 2, $m_2 = 1.80$ kg; distance, $d = 0.500$ m; coefficient of kinetic friction, $\mu_k = 0.450$.
Unknown   Final speed (just before landing), $v_f = ?$

CONTINUED

SOLUTION

1. Evaluate U_i, K_i, and E_i. Be sure to include contributions from both masses:

$$U_i = m_1gh + m_2gd$$
$$K_i = \tfrac{1}{2}m_1 \cdot 0^2 + \tfrac{1}{2}m_2 \cdot 0^2 = 0$$
$$E_i = U_i + K_i = m_1gh + m_2gd$$

2. Next, evaluate U_f, K_f, and E_f. Note that E_f depends on the final speed, v_f:

$$U_f = m_1gh + 0$$
$$K_f = \tfrac{1}{2}m_1v_f^2 + \tfrac{1}{2}m_2v_f^2$$
$$E_f = U_f + K_f = m_1gh + \tfrac{1}{2}m_1v_f^2 + \tfrac{1}{2}m_2v_f^2$$

3. Calculate the nonconservative work, W_{nc}. Recall that the force of friction is $f_k = \mu_k N = \mu_k m_1 g$, and that it points opposite to the displacement of distance d:

$$W_{nc} = -f_k d = -\mu_k m_1 gd$$

4. Set W_{nc} equal to $\Delta E = E_f - E_i$ and solve for E_f. Notice that m_1gh cancels because it occurs in both E_i and E_f:

$$E_f = E_i + W_{nc}$$
$$\cancel{m_1gh} + \tfrac{1}{2}m_1v_f^2 + \tfrac{1}{2}m_2v_f^2 = \cancel{m_1gh} + m_2gd - \mu_k m_1 gd$$

5. Solve for v_f:

$$v_f = \sqrt{\frac{2(m_2 - \mu_k m_1)gd}{m_1 + m_2}}$$

6. Substitute numerical values:

$$v_f = \sqrt{\frac{2[1.80\,\text{kg} - (0.450)(2.40\,\text{kg})](9.81\,\text{m/s}^2)(0.500\,\text{m})}{2.40\,\text{kg} + 1.80\,\text{kg}}}$$
$$= 1.30\,\text{m/s}$$

INSIGHT

Step 4 can be simplified to the following: $\tfrac{1}{2}m_1v_f^2 + \tfrac{1}{2}m_2v_f^2 = m_2gd - \mu_k m_1 gd$. Translating this into words, we can say that the final kinetic energy of the blocks is equal to the initial gravitational potential energy of m_2, minus the energy dissipated by friction on m_1.

PRACTICE PROBLEM

Find the coefficient of kinetic friction if the final speed of the blocks is 0.950 m/s.   [**Answer:** $\mu_k = 0.589$]

Some related homework problems: Problem 40, Problem 42, Problem 64

Next, we present an Example for the common situation of a system in which two different nonconservative forces do work.

QUICK EXAMPLE 8-18   MARATHON MAN: FIND THE HEIGHT OF THE HILL

An 80.0-kg jogger starts from rest and runs uphill into a stiff breeze. At the top of the hill the jogger has done the work $W_{nc1} = +1.80 \times 10^4$ J, air resistance has done the work $W_{nc2} = -4420$ J, and the jogger's final speed is 3.50 m/s. Find the height of the hill.

REASONING AND SOLUTION

We choose $y = 0$ to be the level at the base of the hill. We also let the positive y direction be upward. The jogger starts with the speed $v_i = 0$ and ends with the speed $v_f = 3.50$ m/s. Nonconservative forces are present in this system, and hence we can use $W_{nc} = \Delta E = E_f - E_i$, where in this case $W_{nc} = W_{nc1} + W_{nc2}$. We can solve this relationship for the final energy, $E_f = E_i + W_{nc}$, and from this determine the height, h.

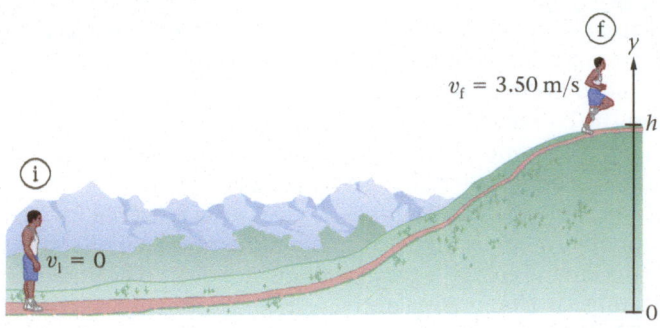

1. Calculate the initial mechanical energy, E_i:

$$E_i = U_i + K_i = 0 + 0 = 0$$

2. Calculate the final mechanical energy, E_f:

$$E_f = U_f + K_f = mgh + \tfrac{1}{2}mv_f^2$$

CONTINUED

3. Rearrange $W_{nc} = E_f - E_i$ and set E_f equal to $E_i + W_{nc}$:

$$E_f = E_i + W_{nc}$$
$$mgh + \tfrac{1}{2}mv_f^2 = 0 + W_{nc}$$

4. Rearrange to solve for h:

$$h = \frac{\left(W_{nc} - \tfrac{1}{2}mv_f^2\right)}{mg}$$

5. Calculate the total nonconservative work:

$$W_{nc} = W_{nc1} + W_{nc2} = 13{,}600\,\text{J}$$

6. Substitute numerical values to determine h:

$$h = \frac{\left[13{,}600\,\text{J} - \tfrac{1}{2}(80.0\,\text{kg})(3.50\,\text{m/s})^2\right]}{(80.0\,\text{kg})(9.81\,\text{m/s}^2)} = 16.7\,\text{m}$$

As usual when we deal with energy calculations, our final result is independent of the shape of the hill.

A dramatic example of the conservation of energy in the presence of nonconservative forces is shown in **FIGURE 8-17**.

Enhance Your Understanding   (Answers given at the end of the chapter)

4. A system is acted on by more than one force, resulting in the following amounts of work: $W_{total} = 100\,\text{J}$, $W_c = 35\,\text{J}$, and $W_{nc} = 65\,\text{J}$. Find the final values of the mechanical energy, kinetic energy, and potential energy of this system, given that their initial values are $E_i = 15\,\text{J}$, $K_i = 5\,\text{J}$, and $U_i = 10\,\text{J}$.

Section Review

* The work done by nonconservative forces on a system is equal to the change in the system's mechanical energy.

8-5   Potential Energy Curves and Equipotentials

FIGURE 8-18 shows a metal ball rolling on a roller coaster–like track. Initially the ball is at rest at point A. The height at A is $y = h$, and hence the ball's initial mechanical energy is $E_0 = mgh$. If friction and other nonconservative forces can be ignored, the ball's mechanical energy remains fixed at E_0 throughout its motion. Thus,

$$E = U + K = E_0$$

As the ball moves, its potential energy falls and rises in the same way as the track. After all, the gravitational potential energy, $U = mgy$, is directly proportional to the height of the track, y. In a sense, then, the track itself represents a graph of the corresponding potential energy.

This is shown explicitly in **FIGURE 8-19**, where we plot energy on the vertical axis and x on the horizontal axis. The potential energy U looks just like the track in Figure 8-18. In addition, we plot a horizontal line at the value E_0, indicating the constant energy of the ball. Since the potential energy plus the kinetic energy must always add up to E_0, it follows that K is the amount of energy from the potential energy curve up to the horizontal line at E_0. This is also shown in Figure 8-19.

Examining an energy plot like Figure 8-19 gives a great deal of information about the motion of an object. For example, at point B the potential energy has its lowest value, and thus the kinetic energy is greatest there. At point C the potential energy has increased, indicating a corresponding decrease in kinetic energy. As the ball continues to the right, the potential energy increases until, at point D, it is again equal to the total energy, E_0. At this point the kinetic energy is zero, and the ball comes to rest momentarily. It then "turns around" and begins to move to the left, eventually returning to point A where it again stops, changes direction, and begins a new cycle. Points A and D, then, are referred to as **turning points** of the motion.

▲ **FIGURE 8-17** Steep highways often have "runaway truck" ramps that enable truck drivers whose brakes fail to bring their rigs to a safe stop. These ramps provide a perfect illustration of the conservation of energy. From a physics point of view, the driver's problem is to get rid of an enormous amount of kinetic energy in the safest possible way. The ramps run uphill, so some of the kinetic energy is simply converted back into gravitational potential energy. In addition, the ramps are typically surfaced with sand or gravel, allowing much of the initial kinetic energy to be dissipated by friction into other forms of energy, such as sound and heat.

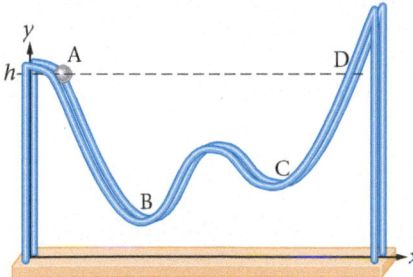

▲ **FIGURE 8-18** **A ball rolling on a frictionless track** The ball starts at A, where $y = h$, with zero speed. Its greatest speed occurs at B. At D, where $y = h$ again, its speed returns to zero.

Chin Basher?

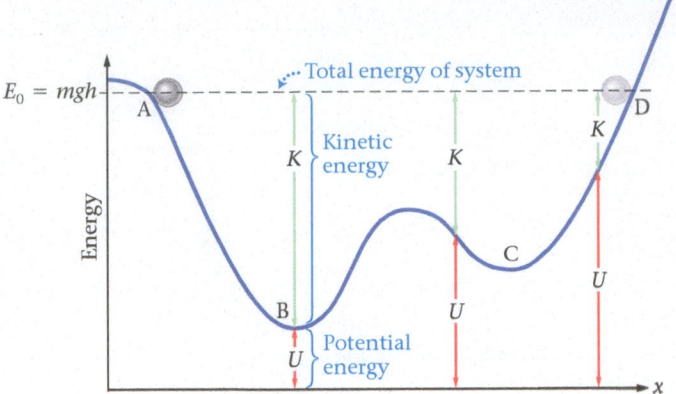

◄ **FIGURE 8-19 Gravitational potential energy versus position for the track shown in Figure 8-18** The shape of the potential energy curve is exactly the same as the shape of the track. In this case, the total mechanical energy is fixed at its initial value, $E_0 = U + K = mgh$. Because the height of the curve is U, by definition, it follows that K is the distance from the curve up to the dashed line at $E_0 = mgh$. Note that K is largest at B. In addition, K vanishes at A and D, which are turning points of the motion.

Turning points are also seen in the motion of a mass on a spring, as indicated in **FIGURE 8-20**. Figure 8-20 (a) shows a mass pulled to the position $x = A$ and then released from rest; Figure 8-20 (b) shows the potential energy of the system, $U = \frac{1}{2}kx^2$. Starting the system this way gives it an initial energy $E_0 = \frac{1}{2}kA^2$, shown by the horizontal line in Figure 8-20 (b). As the mass moves to the left, its speed increases, reaching a maximum where the potential energy is lowest, at $x = 0$. If no nonconservative forces act, the mass continues to $x = -A$, where it stops momentarily before returning to $x = A$. This type of **oscillatory motion** will be studied in detail in Chapter 13.

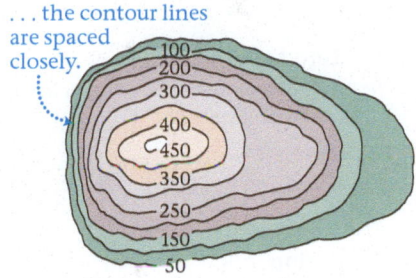

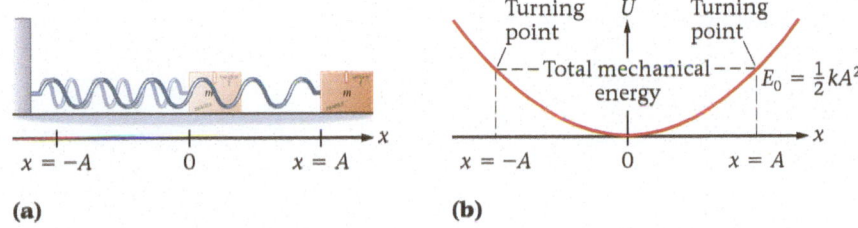

▲ **FIGURE 8-20 A mass on a spring (a)** A spring is stretched by an amount A, giving it a potential energy of $U = \frac{1}{2}kA^2$. **(b)** The potential energy curve, $U = \frac{1}{2}kx^2$, for the spring in (a). Because the mass starts at rest, its initial mechanical energy is $E_0 = \frac{1}{2}kA^2$. The mass oscillates between $x = A$ and $x = -A$.

In some cases, a two-dimensional plot of potential energy contours is useful. For instance, **FIGURE 8-21** shows a contour map of a hill. Each contour corresponds to a given altitude and, hence, to a given value of the gravitational potential energy. In general, lines corresponding to constant values of potential energy are called **equipotentials.** Since the altitude changes by equal amounts from one contour to the next, it follows that when gravitational equipotentials are packed close together, the corresponding terrain is steep. In contrast, when the equipotentials are widely spaced, the ground is nearly flat, since a large horizontal distance is required for a given change in altitude. We shall see similar plots with similar interpretations when we study electric potential energy in Chapter 20.

Contour map of mountain (from above)

▲ **FIGURE 8-21 A contour map** A small mountain (top, in side view) is very steep on the left, more gently sloping on the right. A contour map of this mountain (bottom) shows a series of equal-altitude contour lines from 50 ft to 450 ft. Notice that the contour lines are packed close together where the terrain is steep, but are widely spaced where it is more level.

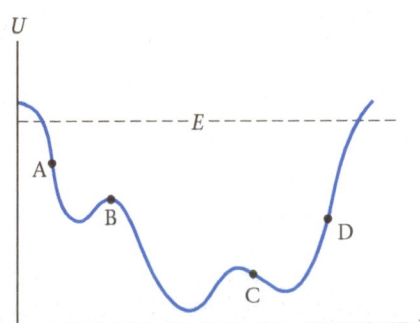

▲ **FIGURE 8-22** Enhance Your Understanding 5.

CHAPTER 8 REVIEW

CHAPTER SUMMARY

8-1   CONSERVATIVE AND NONCONSERVATIVE FORCES

Conservative forces conserve the mechanical energy of a system. Thus, in a conservative system the total mechanical energy remains constant.

Nonconservative forces convert mechanical energy into other forms of energy, or convert other forms of energy into mechanical energy.

Conservative Force, Definition

A conservative force does zero total work on any closed path. In addition, the work done by a conservative force in going from point A to point B is *independent of the path* from A to B. Examples of conservative forces are given in Table 8-1.

Nonconservative Force, Definition

The work done by a nonconservative force on a closed path is nonzero. The work is also path dependent. Examples of nonconservative forces are given in Table 8-1.

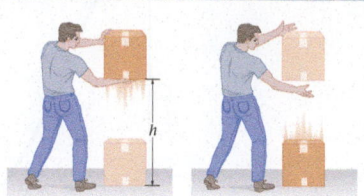

8-2   POTENTIAL ENERGY AND THE WORK DONE BY CONSERVATIVE FORCES

Potential energy, U, can "store" energy in a system. Energy in the form of potential energy can be converted to kinetic or other forms of energy.

Potential Energy, Definition

The work done by a conservative force is the negative of the change in potential energy:

$$W_c = -\Delta U = U_i - U_f \qquad \text{8-1}$$

Zero Level

Any location can be chosen for $U = 0$. Once the choice is made, however, it must be used consistently.

Gravity

Choosing $y = 0$ to be the zero level near Earth's surface,

$$U = mgy \qquad \text{8-3}$$

Spring

Choosing $x = 0$ (the equilibrium position) to be the zero level,

$$U = \tfrac{1}{2}kx^2 \qquad \text{8-5}$$

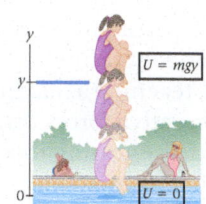

8-3   CONSERVATION OF MECHANICAL ENERGY

Mechanical energy, E, is conserved in systems with conservative forces only.

Mechanical Energy, Definition

Mechanical energy is the sum of the potential and kinetic energies of a system:

$$E = U + K \qquad \text{8-6}$$

8-4   WORK DONE BY NONCONSERVATIVE FORCES

Nonconservative forces can change the mechanical energy of a system.

Change in Mechanical Energy

The work done by a nonconservative force is equal to the change in the mechanical energy of a system:

$$W_{nc} = \Delta E = E_f - E_i \qquad \text{8-9}$$

8-5   POTENTIAL ENERGY CURVES AND EQUIPOTENTIALS

A potential energy curve plots U as a function of position.

An equipotential plot shows contours corresponding to constant values of U.

Turning Points

Turning points occur where an object stops momentarily before reversing direction. At turning points the kinetic energy is zero.

Oscillatory Motion

An object moving back and forth between two turning points is said to have oscillatory motion.

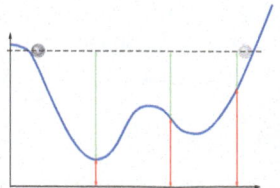

ANSWERS TO ENHANCE YOUR UNDERSTANDING QUESTIONS

1. (a) −10 J. (b) 10 J.
2. (a) Less than 30 J because positive work by a conservative force decreases the potential energy. (b) 20 J.
3. 35 J.
4. $E_f = 80$ J, $K_f = 105$ J, $U_f = -25$ J.
5. A < B < D < C.

CONCEPTUAL QUESTIONS

For instructor-assigned homework, go to www.masteringphysics.com.

(Answers to odd-numbered Conceptual Questions can be found in the back of the book.)

1. Is it possible for the kinetic energy of an object to be negative? Is it possible for the gravitational potential energy of an object to be negative? Explain.

2. If the stretch of a spring is doubled, the force it exerts is also doubled. By what factor does the spring's potential energy increase?

3. When a mass is placed on top of a vertical spring, the spring compresses and the mass moves downward. Analyze this system in terms of its mechanical energy.

4. If a spring is stretched so far that it is permanently deformed, its force is no longer conservative. Why?

5. An object is thrown upward to a person on a roof. At what point is the object's kinetic energy at maximum? At what point is the potential energy of the system at maximum? At what locations do these energies have their minimum values?

6. It is a law of nature that the total energy of the universe is conserved. What do politicians mean, then, when they urge "energy conservation"?

7. Discuss the various energy conversions that occur when a person performs a pole vault. Include as many conversions as you can, and consider times before, during, and after the actual vault itself.

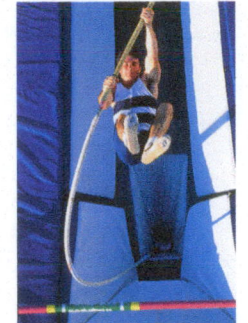

How many energy conversions can you identify? (Conceptual Question 7)

8. Discuss the nature of the work done by the equipment shown in this photo. What types of forces are involved?

Conservative or nonconservative? (Conceptual Question 8)

9. If the force on an object is zero, does that mean the potential energy of the system is zero? If the potential energy of a system is zero, is the force zero?

10. When a ball is thrown upward, its mechanical energy, $E = mgy + \frac{1}{2}mv^2$, is constant with time if air resistance can be ignored. How does E vary with time if air resistance cannot be ignored?

11. When a ball is thrown upward, it spends the same amount of time on the way up as on the way down—as long as air resistance can be ignored. If air resistance is taken into account, is the time on the way down the same as, greater than, or less than the time on the way up? Explain.

PROBLEMS AND CONCEPTUAL EXERCISES

Answers to odd-numbered Problems and Conceptual Exercises can be found in the back of the book. **BIO** *identifies problems of biological or medical interest;* **CE** *indicates a conceptual exercise.* **Predict/Explain** *problems ask for two responses: (a) your prediction of a physical outcome, and (b) the best explanation among three provided; and* **Predict/Calculate** *problems ask for a prediction of a physical outcome, based on fundamental physics concepts, and follow that with a numerical calculation to verify the prediction. On all problems, bullets (•, ••, •••) indicate the level of difficulty.*

SECTION 8-1 CONSERVATIVE AND NONCONSERVATIVE FORCES

1. • **CE** The work done by a conservative force is indicated in **FIGURE 8-23** for a variety of different paths connecting the points A and B. What is the work done by this force **(a)** on path 1 and **(b)** on path 2?

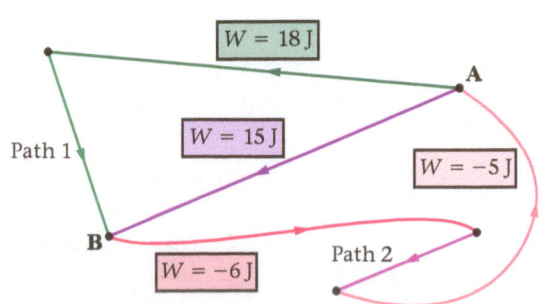

FIGURE 8-23 Problem 1

2. • Calculate the work done by gravity as a 3.2-kg object is moved from point A to point B in **FIGURE 8-24** along paths 1, 2, and 3.

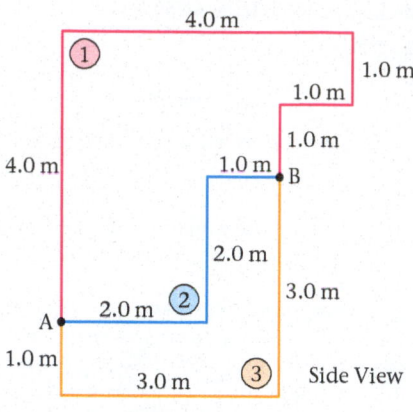

FIGURE 8-24 Problem 2

3. • Calculate the work done by friction as a 3.7-kg box is slid along a floor from point A to point B in **FIGURE 8-25** along paths 1, 2, and 3. Assume that the coefficient of kinetic friction between the box and the floor is 0.26.

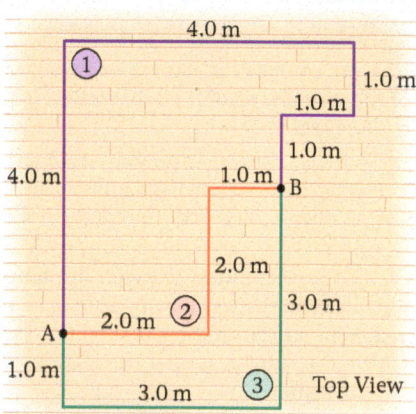

FIGURE 8-25 Problem 3

4. • Predict/Calculate A 2.8-kg block is attached to a spring with a force constant of 375 N/m, as shown in **FIGURE 8-26**. (a) Find the work done by the spring on the block as the block moves from A to B along paths 1 and 2. (b) How do your results depend on the mass of the block? Specifically, if you increase the mass, does the work done by the spring increase, decrease, or stay the same? (Assume the system is frictionless.)

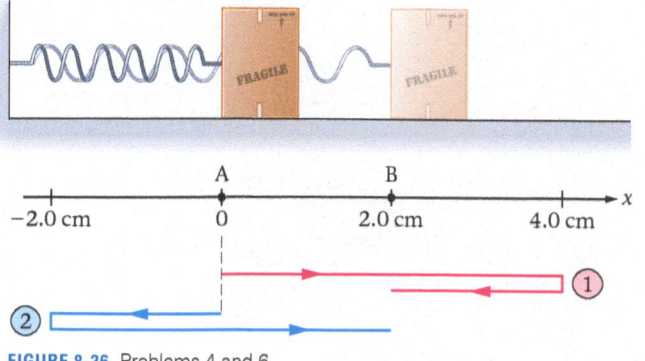

FIGURE 8-26 Problems 4 and 6

5. • Predict/Calculate (a) Calculate the work done by gravity as a 5.2-kg object is moved from A to B in **FIGURE 8-27** along paths 1 and 2. (b) How do your results depend on the mass of the block? Specifically, if you increase the mass, does the work done by gravity increase, decrease, or stay the same?

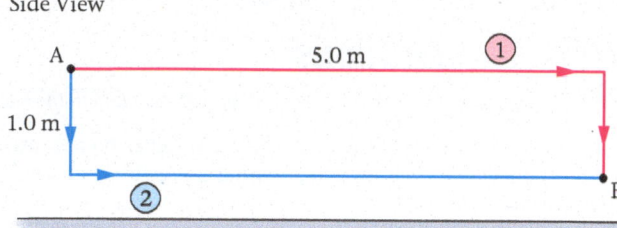

FIGURE 8-27 Problem 5

6. •• In the system shown in Figure 8-26, suppose the block has a mass of 2.7 kg, the spring has a force constant of 480 N/m, and the coefficient of kinetic friction between the block and the floor is 0.16. (a) Find the work done on the block by the spring and by friction as the block is moved from point A to point B along path 2. (b) Find the work done on the block by the spring and by friction if the block is moved directly from point A to point B.

SECTION 8-2 POTENTIAL ENERGY AND THE WORK DONE BY CONSERVATIVE FORCES

7. • **CE** Predict/Explain Ball 1 is thrown to the ground with an initial downward speed; ball 2 is dropped to the ground from rest. Assuming the balls have the same mass and are released from the same height, is the change in gravitational potential energy of ball 1 greater than, less than, or equal to the change in gravitational potential energy of ball 2? (b) Choose the *best explanation* from among the following:
   I. Ball 1 has the greater total energy, and therefore more energy can go into gravitational potential energy.
   II. The gravitational potential energy depends only on the mass of the ball and the drop height.
   III. All of the initial energy of ball 2 is gravitational potential energy.

8. • **CE** A mass is attached to the bottom of a vertical spring. This causes the spring to stretch and the mass to move downward. (a) Does the potential energy of the spring increase, decrease, or stay the same during this process? Explain. (b) Does the gravitational potential energy of the Earth–mass system increase, decrease, or stay the same during this process? Explain.

9. • Find the gravitational potential energy of an 88-kg person standing atop Mt. Everest at an altitude of 8848 m. Use sea level as the location for $y = 0$.

10. • A student lifts a 1.42-kg book from her desk to a bookshelf. If the gravitational potential energy of the book–Earth system increases by 9.08 J, how high is the bookshelf above the desk?

11. • At the local ski slope, an 82.0-kg skier rides a gondola to the top of the mountain. If the lift has a length of 2950 m and makes an angle of 13.1° with the horizontal, what is the change in the gravitational potential energy of the skier–Earth system?

12. •• **BIO** The Wing of the Hawkmoth Experiments performed on the wing of a hawkmoth (*Manduca sexta*) show that it deflects by a distance of $x = 4.8$ mm when a force of magnitude $F = 3.0$ mN is applied at the tip, as indicated in **FIGURE 8-28**. Treating the wing as an ideal spring, find (a) the force constant of the wing and (b) the energy stored in the wing when it is deflected. (c) What force must be applied to the tip of the wing to store twice the energy found in part (b)?

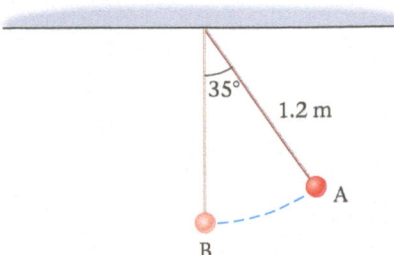

FIGURE 8-28 Problem 12

13. •• **Predict/Calculate** A vertical spring stores 0.962 J in spring potential energy when a 3.5-kg mass is suspended from it. **(a)** By what multiplicative factor does the spring potential energy change if the mass attached to the spring is doubled? **(b)** Verify your answer to part (a) by calculating the spring potential energy when a 7.0-kg mass is attached to the spring.

14. •• Pushing on the pump of a soap dispenser compresses a small spring. When the spring is compressed 0.50 cm, its potential energy is 0.0025 J. **(a)** What is the force constant of the spring? **(b)** What compression is required for the spring potential energy to equal 0.0084 J?

15. •• **BIO Mantis Shrimp Smasher** A peacock mantis shrimp smashes its prey with a hammer-like appendage by storing energy in a spring-like section of exoskeleton on its appendage. The spring has a force constant of 5.9×10^4 N/m. **(a)** If the spring stores 0.0103 J of energy, over what distance must the shrimp compress it? **(b)** What power does the shrimp generate if it unleashes the energy in 0.0018 s? (The power per mass of the shrimp appendage is more than 20 times greater than that generated by muscle alone, showcasing the shrimp's ability to harness the potential energy of its spring.)

16. •• **Predict/Calculate** The work required to stretch a certain spring from an elongation of 4.00 cm to an elongation of 5.00 cm is 30.5 J. **(a)** Is the work required to increase the elongation of the spring from 5.00 cm to 6.00 cm greater than, less than, or equal to 30.5 J? Explain. **(b)** Verify your answer to part (a) by calculating the required work.

17. •• A 0.33-kg pendulum bob is attached to a string 1.2 m long. What is the change in the gravitational potential energy of the system as the bob swings from point A to point B in **FIGURE 8-29**?

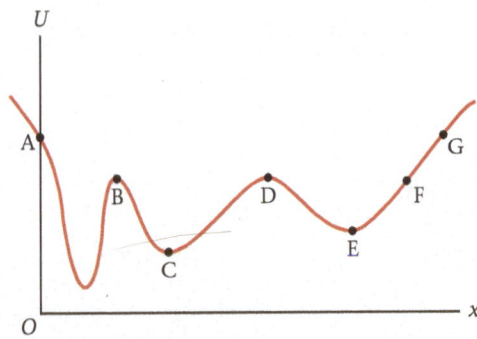

FIGURE 8-29 Problems 17 and 30

SECTION 8-3   CONSERVATION OF MECHANICAL ENERGY

18. • **CE Predict/Explain** You throw a ball upward and let it fall to the ground. Your friend drops an identical ball straight down to the ground from the same height. Is the change in kinetic energy of your ball greater than, less than, or equal to the change in kinetic energy of your friend's ball? **(b)** Choose the *best explanation* from among the following:
   I. Your friend's ball converts all its initial energy into kinetic energy.

II. Your ball is in the air longer, which results in a greater change in kinetic energy.
III. The change in gravitational potential energy is the same for each ball, which means the change in kinetic energy must be the same also.

19. • **CE Predict/Explain** When a ball of mass *m* is dropped from rest from a height *h*, its kinetic energy just before landing is *K*. Now, suppose a second ball of mass 4*m* is dropped from rest from a height *h*/4. **(a)** Just before ball 2 lands, is its kinetic energy 4*K*, 2*K*, *K*, *K*/2, or *K*/4? **(b)** Choose the *best explanation* from among the following:
   I. The two balls have the same initial energy.
   II. The more massive ball will have the greater kinetic energy.
   III. The reduced drop height results in a reduced kinetic energy.

20. • **CE** For an object moving along the *x* axis, the potential energy of the frictionless system is shown in **FIGURE 8-30**. Suppose the object is released from rest at the point A. Rank the other points in the figure in increasing order of the object's speed. Indicate ties where appropriate.

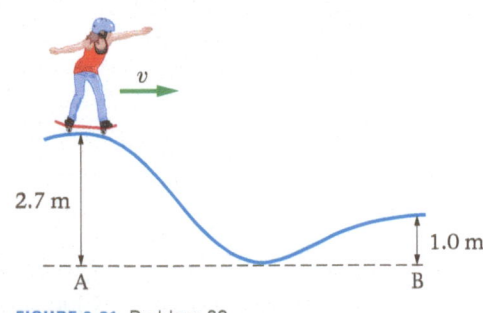

FIGURE 8-30 Problem 20

21. • At an amusement park, a swimmer uses a water slide to enter the main pool. If the swimmer starts at rest, slides without friction, and descends through a vertical height of 2.31 m, what is her speed at the bottom of the slide?

22. • **Predict/Calculate** A player passes a 0.600-kg basketball downcourt for a fast break. The ball leaves the player's hands with a speed of 8.30 m/s and slows down to 7.10 m/s at its highest point. **(a)** Ignoring air resistance, how high above the release point is the ball when it is at its maximum height? **(b)** How would doubling the ball's mass affect the result in part (a)? Explain.

23. • A skateboarder at a skate park rides along the path shown in **FIGURE 8-31**. If the speed of the skateboarder at point A is *v* = 1.4 m/s, what is her speed at point B? Assume that friction is negligible.

FIGURE 8-31 Problem 23

24. •• **CE** Three balls are thrown upward with the same initial speed v_0, but at different angles relative to the horizontal, as shown in **FIGURE 8-32**. Ignoring air resistance, indicate which of the following statements is correct: At the dashed level, (A) ball 3 has the lowest speed; (B) ball 1 has the lowest speed; (C) all three balls have the same speed; (D) the speed of the balls depends on their mass.

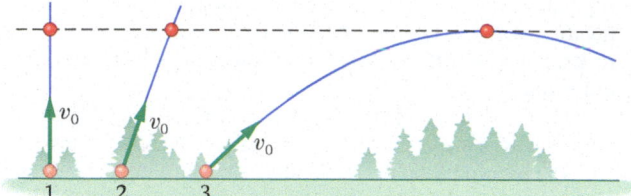

FIGURE 8-32 Problem 24

25. •• A 0.21-kg apple falls from a tree to the ground, 4.0 m below. Ignoring air resistance, determine the apple's kinetic energy, K, the gravitational potential energy of the system, U, and the total mechanical energy of the system, E, when the apple's height above the ground is (a) 4.0 m, (b) 3.0 m, (c) 2.0 m, (d) 1.0 m, and (e) 0 m. Take ground level to be $y = 0$.

26. •• **Predict/Calculate** A 2.9-kg block slides with a speed of 2.1 m/s on a frictionless horizontal surface until it encounters a spring. (a) If the block compresses the spring 5.6 cm before coming to rest, what is the force constant of the spring? (b) What initial speed should the block have to compress the spring by 1.4 cm?

27. •• A 0.26-kg rock is thrown vertically upward from the top of a cliff that is 32 m high. When it hits the ground at the base of the cliff, the rock has a speed of 29 m/s. Assuming that air resistance can be ignored, find (a) the initial speed of the rock and (b) the greatest height of the rock as measured from the base of the cliff.

28. •• A 1.40-kg block slides with a speed of 0.950 m/s on a frictionless horizontal surface until it encounters a spring with a force constant of 734 N/m. The block comes to rest after compressing the spring 4.15 cm. Find the spring potential energy, U, the kinetic energy of the block, K, and the total mechanical energy of the system, E, for compressions of (a) 0 cm, (b) 1.00 cm, (c) 2.00 cm, (d) 3.00 cm, and (e) 4.00 cm.

29. •• A 5.76-kg rock is dropped and allowed to fall freely. Find the initial kinetic energy, the final kinetic energy, and the change in kinetic energy for (a) the first 2.00 m of fall and (b) the second 2.00 m of fall.

30. •• **Predict/Calculate** Suppose the pendulum bob in Figure 8-29 has a mass of 0.33 kg and is moving to the right at point B with a speed of 2.4 m/s. Air resistance is negligible. (a) What is the change in the system's gravitational potential energy when the bob reaches point A? (b) What is the speed of the bob at point A? (c) If the mass of the bob is increased, does your answer to part (a) increase, decrease, or stay the same? Explain. (d) If the mass of the bob is increased, does your answer to part (b) increase, decrease, or stay the same? Explain.

31. ••• The two masses in the Atwood's machine shown in **FIGURE 8-33** are initially at rest at the same height. After they are released, the large mass, m_2, falls through a height h and hits the floor, and the small mass, m_1, rises through a height h. (a) Find the speed of the masses just before m_2 lands, giving your answer in terms of m_1, m_2, g, and h. Assume the ropes and pulley have negligible mass and that friction can be ignored. (b) Evaluate your answer to part (a) for the case $h = 1.2$ m, $m_1 = 3.7$ kg, and $m_2 = 4.1$ kg.

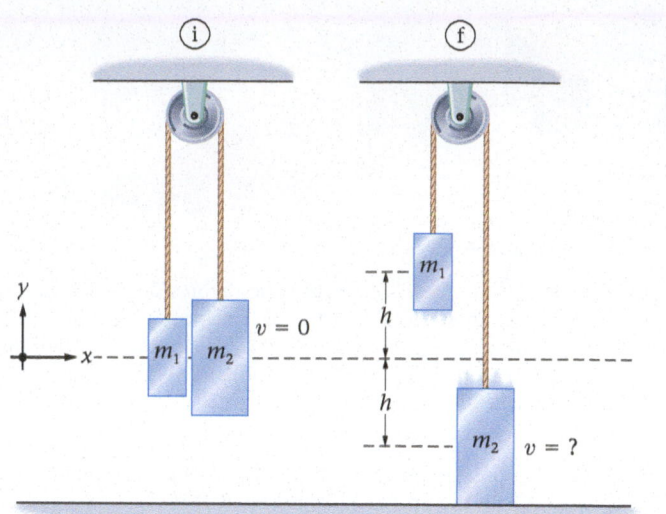

FIGURE 8-33 Problems 31, 32, and 63

32. ••• In the previous problem, suppose the masses have an initial speed of 0.20 m/s, and that m_2 is moving upward. How high does m_2 rise above its initial position before momentarily coming to rest, given that $m_1 = 3.7$ kg and $m_2 = 4.1$ kg?

SECTION 8-4 WORK DONE BY NONCONSERVATIVE FORCES

33. • **CE** **Predict/Explain** On reentry, the Space Shuttle's protective heat tiles became extremely hot. (a) Was the mechanical energy of the Shuttle–Earth system when the Shuttle landed greater than, less than, or the same as when it was in orbit? (b) Choose the *best explanation* from among the following:

   I. Dropping out of orbit increases the mechanical energy of the Shuttle.
   II. Gravity is a conservative force.
   III. A portion of the mechanical energy has been converted to heat energy.

34. • Catching a wave, a 77-kg surfer starts with a speed of 1.3 m/s, drops through a height of 1.65 m, and ends with a speed of 8.2 m/s. How much nonconservative work was done on the surfer?

35. • At a playground, a 19-kg child plays on a slide that drops through a height of 2.3 m. The child starts at rest at the top of the slide. On the way down, the slide does a nonconservative work of −361 J on the child. What is the child's speed at the bottom of the slide?

36. • Starting at rest at the edge of a swimming pool, a 72.0-kg athlete swims along the surface of the water and reaches a speed of 1.20 m/s by doing the work $W_{nc1} = +161$ J. Find the nonconservative work, W_{nc2}, done by the water on the athlete.

37. • A 22,000-kg airplane lands with a speed of 64 m/s on a stationary aircraft carrier deck that is 115 m long. Find the work done by nonconservative forces in stopping the plane.

38. •• A 78-kg skateboarder grinds down a hubba ledge that is 2.5 m long and inclined at 19° below the horizontal. Kinetic friction dissipates half of her initial potential energy to thermal and sound energies. What is the coefficient of kinetic friction between her skateboard and the ledge surface?

39. •• **CE** You ride your bicycle down a hill, maintaining a constant speed the entire time. (a) As you ride, does the gravitational potential energy of the you–bike–Earth system increase, decrease, or stay the same? Explain. (b) Does the kinetic energy of you and your bike increase, decrease, or stay the same? Explain. (c) Does the mechanical energy of the you–bike–Earth system increase, decrease, or stay the same? Explain.

40. •• A 111-kg seal at an amusement park slides from rest down a ramp into the pool below. The top of the ramp is 1.75 m higher than the surface of the water, and the ramp is inclined at an angle of 26.5° above the horizontal. If the seal reaches the water with a speed of 4.25 m/s, what are **(a)** the work done by kinetic friction and **(b)** the coefficient of kinetic friction between the seal and the ramp?

41. •• A 1.9-kg rock is released from rest at the surface of a pond 1.8 m deep. As the rock falls, a constant upward force of 4.6 N is exerted on it by water resistance. Calculate the nonconservative work, W_{nc}, done by water resistance on the rock, the gravitational potential energy of the system, U, the kinetic energy of the rock, K, and the total mechanical energy of the system, E, when the depth of the rock below the water's surface is **(a)** 0 m, **(b)** 0.50 m, and **(c)** 1.0 m. Let $y = 0$ be at the bottom of the pond.

42. •• A 1250-kg car drives up a hill that is 16.2 m high. During the drive, two nonconservative forces do work on the car: (i) the force of friction, and (ii) the force generated by the car's engine. The work done by friction is -3.11×10^5 J; the work done by the engine is $+6.44 \times 10^5$ J. Find the change in the car's kinetic energy from the bottom of the hill to the top of the hill.

43. ••• The Outlaw Run roller coaster in Branson, Missouri, features a track that is inclined at 81° below the horizontal and that spans a 49.4-m (162 ft) change in height. **(a)** If the coefficient of friction for the roller coaster is $\mu_k = 0.29$, what is the magnitude of the friction force on a 1700-kg coaster train? **(b)** How much work is done by friction during the time the coaster train travels along the incline? **(c)** What is the speed of the coaster train at the bottom of the incline, assuming it started from rest at the top?

44. ••• A 1.80-kg block slides on a rough horizontal surface. The block hits a spring with a speed of 2.00 m/s and compresses it a distance of 11.0 cm before coming to rest. If the coefficient of kinetic friction between the block and the surface is $\mu_k = 0.560$, what is the force constant of the spring?

SECTION 8-5   POTENTIAL ENERGY CURVES AND EQUIPOTENTIALS

45. • **FIGURE 8-34** shows a potential energy curve as a function of x. In qualitative terms, describe the subsequent motion of an object that starts at rest at point A.

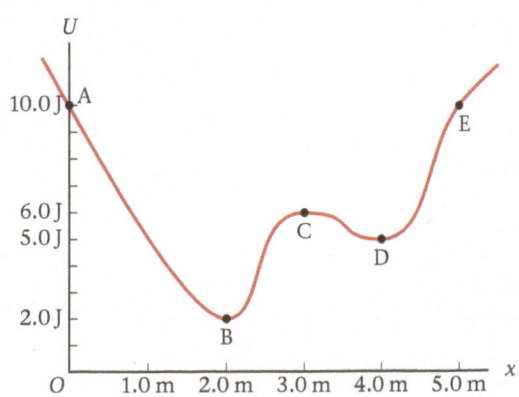

FIGURE 8-34 Problems 45, 46, 47, and 48

46. • An object moves along the x axis, subject to the potential energy shown in Figure 8-34. The object has a mass of 1.1 kg and starts at rest at point A. **(a)** What is the object's speed at point B? **(b)** At point C? **(c)** At point D? **(d)** What are the turning points for this object?

47. • A 1.34-kg object moves along the x axis, subject to the potential energy shown in Figure 8-34. If the object's speed at point C is 1.25 m/s, what are the approximate locations of its turning points?

48. •• The potential energy of a particle moving along the x axis is shown in Figure 8-34. When the particle is at $x = 1.0$ m it has 3.6 J of kinetic energy. Give approximate answers to the following questions. **(a)** What is the total mechanical energy of the system? **(b)** What is the smallest value of x the particle can reach? **(c)** What is the largest value of x the particle can reach?

49. •• A block of mass $m = 0.88$ kg is connected to a spring of force constant $k = 845$ N/m on a smooth, horizontal surface. **(a)** Plot the potential energy of the spring from $x = -5.00$ cm to $x = 5.00$ cm. **(b)** Determine the turning points of the block if its speed at $x = 0$ is 1.4 m/s.

50. •• A ball of mass $m = 0.75$ kg is thrown straight upward with an initial speed of 8.9 m/s. **(a)** Plot the gravitational potential energy of the ball from its launch height, $y = 0$, to the height $y = 5.0$ m. Let $U = 0$ correspond to $y = 0$. **(b)** Determine the turning point (maximum height) of the ball.

51. •• **FIGURE 8-35** depicts the potential energy of a 350-kg pallet that slides from rest down a frictionless roller conveyer ramp, then moves across a frictionless horizontal conveyer, and finally encounters a horizontal spring at $x = 4.0$ m. When the pallet encounters the spring, it momentarily comes to rest at $x = 6.0$ m. **(a)** What is the force constant of the spring? **(b)** At what height h above the horizontal conveyer did the pallet begin sliding from rest?

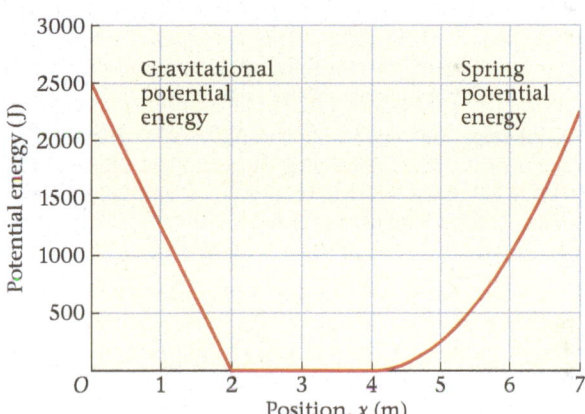

FIGURE 8-35 Problems 51 and 52

52. •• Figure 8-35 depicts the potential energy of a 655-kg pallet that slides from rest down a frictionless roller conveyer ramp, then moves across a frictionless horizontal conveyer, and finally encounters a horizontal spring at $x = 4.0$ m. **(a)** What is the angle of inclination of the conveyer ramp? **(b)** What will be the speed of the pallet at $x = 6.0$ m if it begins sliding from rest at $x = 0$?

GENERAL PROBLEMS

53. • **CE** You and a friend both solve a problem involving a skier going down a slope. When comparing solutions, you notice that your choice for the $y = 0$ level is different than the $y = 0$ level chosen by your friend. Will your answers agree or disagree on the following quantities: **(a)** the skier's potential energy; **(b)** the skier's change in potential energy; **(c)** the skier's kinetic energy?

54. • **CE** A particle moves under the influence of a conservative force. At point A the particle has a kinetic energy of 12 J; at point B the particle is momentarily at rest, and the potential energy of the system is 25 J; at point C the potential energy of the system is 5 J. **(a)** What is the potential energy of the system when the particle is at point A? **(b)** What is the kinetic energy of the particle at point C?

55. •• A sled slides without friction down a small, ice-covered hill. On its first run down the hill, the sled starts from rest at the top of the hill and its speed at the bottom is 7.50 m/s. On a second run, the sled starts with a speed of 1.50 m/s at the top. What is the speed of the sled at the bottom of the hill on the second run?

56. •• A 74-kg skier encounters a dip in the snow's surface that has a circular cross section with a radius of curvature of 12 m. If the skier's speed at point A in **FIGURE 8-36** is 7.1 m/s, what is the normal force exerted by the snow on the skier at point B? Ignore frictional forces.

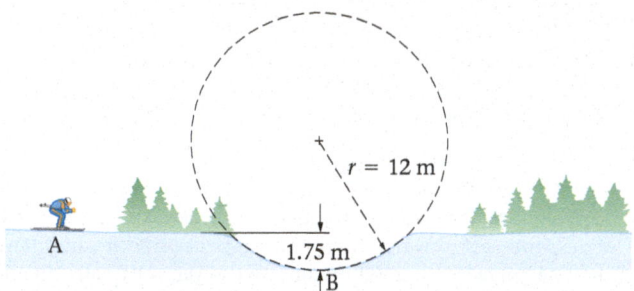

FIGURE 8-36 Problem 56

57. •• **Running Shoes** The soles of a popular make of running shoe have a force constant of 2.0×10^5 N/m. Treat the soles as ideal springs for the following questions. **(a)** If a 62-kg person stands in a pair of these shoes, with her weight distributed equally on both feet, how much does she compress the soles? **(b)** How much energy is stored in the soles of her shoes when she's standing?

58. •• **Nasal Strips** The force required to flex a nasal strip and apply it to the nose is 0.25 N; the energy stored in the strip when flexed is 0.0022 J. Assume the strip to be an ideal spring for the following calculations. Find **(a)** the distance through which the strip is flexed and **(b)** the force constant of the strip.

59. •• The water slide shown in **FIGURE 8-37** ends at a height of 1.50 m above the pool. If the person starts from rest at point A and lands in the water at point B, what is the height h of the water slide? (Assume the water slide is frictionless.)

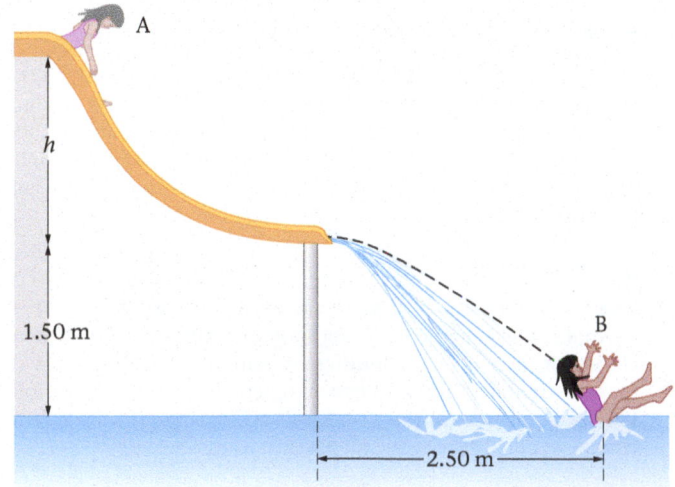

FIGURE 8-37 Problems 59

60. •• A skateboarder starts at point A in **FIGURE 8-38** and rises to a height of 2.64 m above the top of the ramp at point B. What was the skateboarder's initial speed at point A?

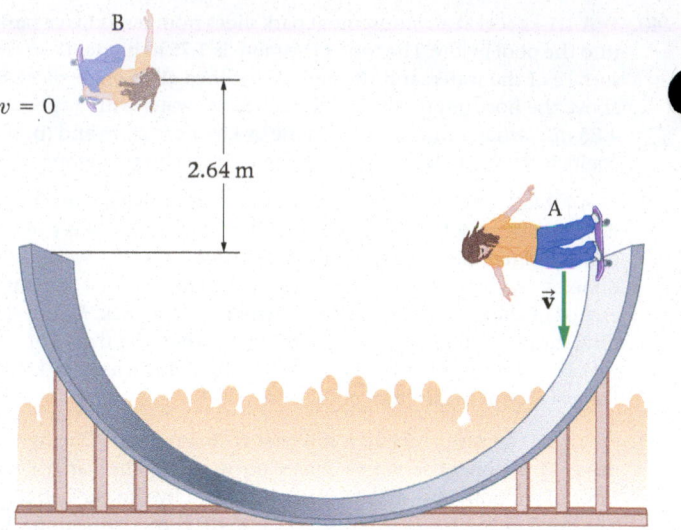

FIGURE 8-38 Problem 60

61. •• **The Crash of Skylab** NASA's *Skylab*, the largest spacecraft ever to fall back to the Earth, reentered the Earth's atmosphere on July 11, 1979, and broke into a myriad of pieces. One of the largest fragments was a 1770-kg, lead-lined film vault, which landed with an estimated speed of 120 m/s. If its speed at an altitude of 148 km was 7950 m/s, how much nonconservative work was done on the film vault during reentry?

62. •• **BIO Bird Tendons** Several studies indicate that the elastic properties of tendons can change in response to exercise. In one study, guinea fowl were divided into a group that ran for 30 minutes a day on a treadmill, and a control group that was kept in their cages. After 12 weeks, the researchers measured the mechanical properties of the gastrocnemius tendon in each bird. The results for one control bird and one treadmill bird are shown in **FIGURE 8-39**. **(a)** Treating the tendon as an ideal spring, what is the force constant for the control bird? **(b)** By how much does the control bird's tendon need to be stretched in order to have the same amount of stored energy as the tendon of the treadmill bird at a deformation of 1.0 mm?

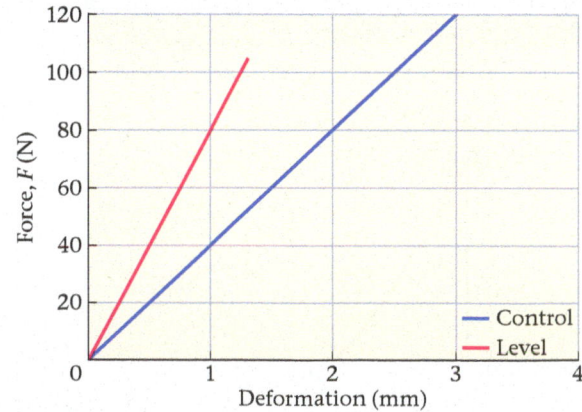

FIGURE 8-39 Problem 62

63. •• In the Atwood's machine of Problem 31, the mass m_2 remains at rest once it hits the floor, but the mass m_1 continues moving upward. How much higher does m_1 go after m_2 has landed? Give your answer for the case $h = 1.2$ m, $m_1 = 3.7$ kg, and $m_2 = 4.1$ kg.

64. •• A 6.60-kg block slides with an initial speed of 2.24 m/s up a ramp inclined at an angle of 18.4° with the horizontal. The coefficient of kinetic friction between the block and the ramp is 0.44. Use energy conservation to find the distance the block slides before coming to rest.

65. •• Jeff of the Jungle swings on a 7.6-m vine that initially makes an angle of 37° with the vertical. If Jeff starts at rest and has a mass of 78 kg, what is the tension in the vine at the lowest point of the swing?

66. •• A 1.9-kg block slides down a frictionless ramp, as shown in **FIGURE 8-40**. The top of the ramp is 1.5 m above the ground; the bottom of the ramp is 0.25 m above the ground. The block leaves the ramp moving horizontally, and lands a horizontal distance d away. Find the distance d.

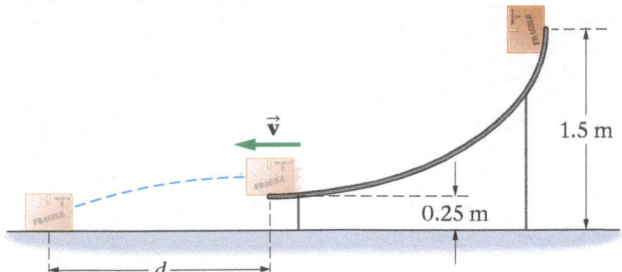

FIGURE 8-40 Problems 66 and 67

67. •• Suppose the ramp in Figure 8-40 is not frictionless. Find the distance d for the case in which friction on the ramp does -9.7 J of work on the block before it becomes airborne.

68. •• **BIO** **Compressing the Ground** A running track at Harvard University uses a surface with a force constant of 2.5×10^5 N/m. This surface is compressed slightly every time a runner's foot lands on it. The force exerted by the foot, according to the Saucony shoe company, has a magnitude of 2700 N for a typical runner. Treating the track's surface as an ideal spring, find **(a)** the amount of compression caused by a foot hitting the track and **(b)** the energy stored briefly in the track every time a foot lands.

69. •• **BIO** **A Flea's Jump** The resilin in the body of a flea has a force constant of about 24 N/m, and when the flea cocks its jumping legs, the resilin associated with each leg is compressed by approximately 0.12 mm. Given that the flea has a mass of 0.56 mg, and that two legs are used in a jump, estimate the maximum height a flea can attain by using the energy stored in the resilin. (Assume the resilin to be an ideal spring.)

70. ••• **Predict/Calculate** **Tension at the Bottom** A ball of mass m is attached to a string of length L and released from rest at the point A in **FIGURE 8-41**. **(a)** Show that the tension in the string when the ball reaches point B is $3mg$, independent of the length l. **(b)** Give a detailed physical explanation for the fact that the tension at point B is independent of the length l.

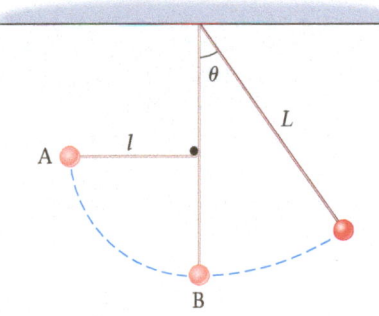

FIGURE 8-41 Problem 70

71. ••• An ice cube is placed on top of an overturned spherical bowl of radius r, as indicated in **FIGURE 8-42**. If the ice cube slides downward from rest at the top of the bowl, at what angle θ does it separate from the bowl? In other words, at what angle does the normal force between the ice cube and the bowl go to zero?

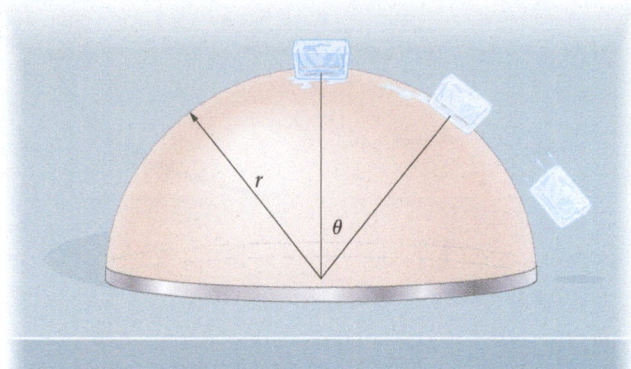

FIGURE 8-42 Problem 71

72. ••• **Predict/Calculate** The two blocks shown in **FIGURE 8-43** are moving with an initial speed v. **(a)** If the system is frictionless, find the distance d the blocks travel before coming to rest. (Let $U = 0$ correspond to the initial position of block 2.) **(b)** Is the work done on block 2 by the rope positive, negative, or zero? Explain. **(c)** Calculate the work done on block 2 by the rope.

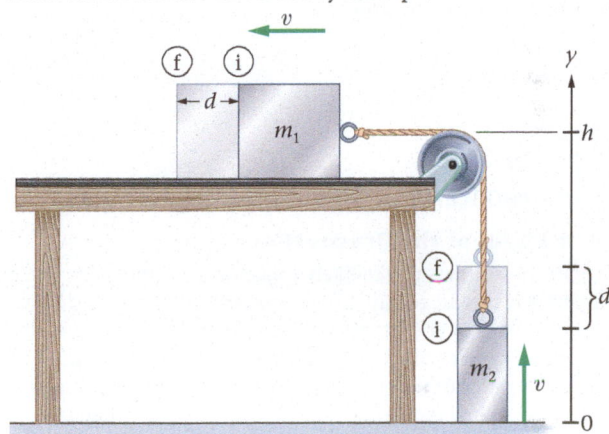

FIGURE 8-43 Problem 72

73. ••• **Predict/Calculate** **Loop-the-Loop** **(a)** A block of mass m slides from rest on a frictionless loop-the-loop track, as shown in **FIGURE 8-44**. What is the minimum release height, h, required for the block to maintain contact with the track at all times? Give your answer in terms of the radius of the loop, r. **(b)** Explain why the release height obtained in part (a) is independent of the block's mass.

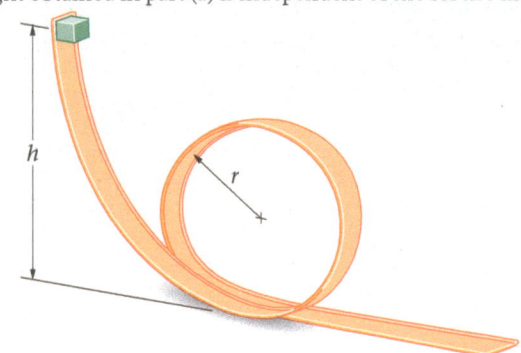

FIGURE 8-44 Problem 73

74. ••• **FIGURE 8-45** shows a 1.75-kg block at rest on a ramp of height h. When the block is released, it slides without friction to the bottom of the ramp, and then continues across a surface that is frictionless except for a rough patch of width 10.0 cm that has a coefficient of kinetic friction $\mu_k = 0.640$. Find h such that the block's speed after crossing the rough patch is 3.50 m/s.

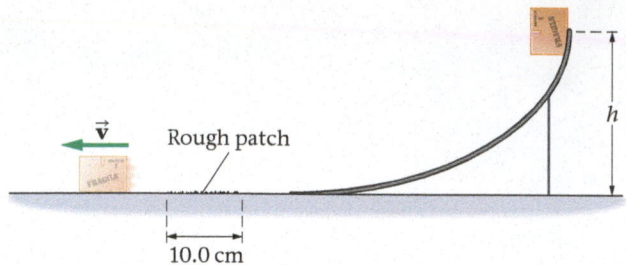

FIGURE 8-45 Problem 74

75. ••• In **FIGURE 8-46** a 1.2-kg block is held at rest against a spring with a force constant $k = 730 \, \text{N/m}$. Initially, the spring is compressed a distance d. When the block is released, it slides across a surface that is frictionless except for a rough patch of width 5.0 cm that has a coefficient of kinetic friction $\mu_k = 0.44$. Find d such that the block's speed after crossing the rough patch is 2.3 m/s.

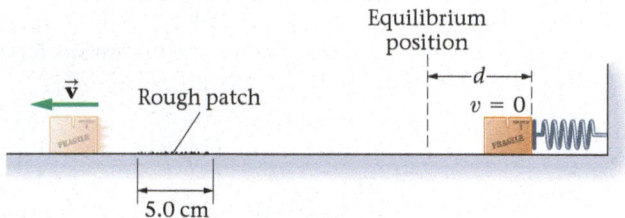

FIGURE 8-46 Problem 75

PASSAGE PROBLEMS

BIO The Flight of the Dragonflies

Of all the animals you're likely to see on a summer day, the most ancient is the dragonfly. In fact, the fossil record for dragonflies extends back over 250 million years, more than twice as long as for birds. Ancient dragonflies could be as large as a hawk, and were surely buzzing around the heads of both *T. Rex* and *Triceratops*.

Dragonflies belong to the order Odonata ("toothed jaws") and the suborder Anisoptera ("different wings"), a reference to the fact that their hindwings are wider front-to-back than their forewings. (Damselflies, in contrast, have forewings and hindwings that are the same width.) Although ancient in their lineage, dragonflies are among the fastest flying and most acrobatic of all insects; some of their maneuvers subject them to accelerations as great as 20g.

The properties of dragonfly wings, and how they account for such speed and mobility, have been of great interest to biologists. **FIGURE 8-47 (a)** shows an experimental setup designed to measure the force constant of Plexiglas models of wings, which are used in wind tunnel tests. A downward force is applied to the model wing at the tip (1 for hindwing, 2 for forewing) or at two-thirds the distance to the tip (3 for hindwing, 4 for forewing). As the force is varied in magnitude, the resulting deflection of the wing is measured. The results are shown in **FIGURE 8-47 (b)**. Notice that significant differences are seen between the hindwings and forewings, as one might expect from their different shapes.

76. • Treating the model wing as an ideal spring, what is the force constant of the hindwing when a force is applied to its tip?

    A. 94 N/m   **B.** 130 N/m
    C. 290 N/m   **D.** 330 N/m

77. • What is the force constant of the hindwing when a force is applied at two-thirds the distance from the base of the wing to the tip?

    A. 94 N/m   **B.** 130 N/m
    C. 290 N/m   **D.** 330 N/m

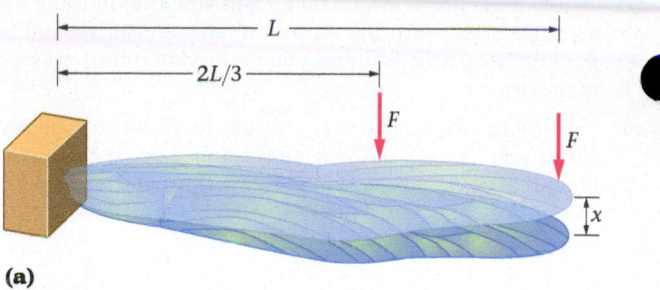

(a)

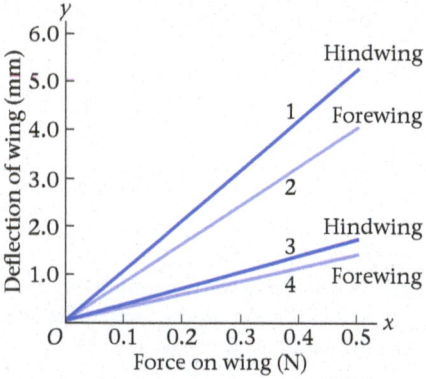

(b)

FIGURE 8-47 Problems 76, 77, 78, and 79

78. • Which of the wings is "stiffer"?

    A. The hindwing
    B. The forewing
    C. Depends on where the force is applied
    D. They are equally "stiff"

79. •• How much energy is stored in the forewing when a force at the tip deflects it by 3.5 mm?

    A. 0.766 mJ   **B.** 49.0 mJ
    C. 0.219 J   **D.** 1.70 kJ

80. •• **Predict/Calculate REFERRING TO EXAMPLE 8-12** Consider a spring with a force constant of 955 N/m. **(a)** Suppose the mass of the block is 1.70 kg, but its initial speed can be varied. What initial speed is required to give a maximum spring compression of 4.00 cm? **(b)** Suppose the initial speed of the block is 1.09 m/s, but its mass can be varied. What mass is required to give a maximum spring compression of 4.00 cm?

81. •• **REFERRING TO EXAMPLE 8-12** Suppose the block is released from rest with the spring compressed 5.00 cm. The mass of the block is 1.70 kg and the force constant of the spring is 955 N/m. **(a)** What is the speed of the block when the spring expands to a compression of only 2.50 cm? **(b)** What is the speed of the block after it leaves the spring?

82. •• **REFERRING TO EXAMPLE 8-17** Suppose we would like the landing speed of block 2 to be increased to 1.50 m/s. **(a)** Should the coefficient of kinetic friction between block 1 and the tabletop be increased or decreased? **(b)** Find the required coefficient of kinetic friction for a landing speed of 1.50 m/s. Note that $m_1 = 2.40 \, \text{kg}$, $m_2 = 1.80 \, \text{kg}$, and $d = 0.500 \, \text{m}$.

Big Ideas

1 Momentum is mass times velocity.

2 Impulse is force multiplied by time.

3 Momentum is conserved when the net force is zero.

4 Collisions conserve momentum, but may or may not conserve mechanical energy.

5 Systems behave as if all their mass were concentrated at the center of mass.

Linear Momentum and Collisions

▲ A bowling ball hitting a group of pins is just one example of the many collisions that occur all around us everyday. As we shall see in this chapter, collisions transfer a new physical quantity—momentum—between the colliding objects, while keeping the total amount of momentum constant. In this chapter, we show how momentum is related to Newton's second law, and how it can be used to analyze a wide range of collisions.

Conservation laws play a central role in physics. In this chapter we introduce the concept of *momentum* and show that it, like energy, is a conserved quantity. Nothing we can do—in fact, nothing that can occur in nature—can change the total energy or total momentum of the universe.

9-1 Linear Momentum

Imagine standing at rest on a skateboard that rolls without friction on a horizontal surface. If you catch a heavy, slow-moving ball tossed to you by a friend, you begin to move. If, on the other hand, your friend tosses you a lightweight, yet fast-moving ball, the net effect may be the same—that is, catching a lightweight ball moving fast enough causes you to move with the same speed as when you caught the heavy ball.

In physics, these observations are made precise with a quantity called the **linear momentum, $\vec{p}$**, which is defined as the product of the mass m and velocity $\vec{v}$ of an object:

Definition of Linear Momentum, $\vec{p}$

$$\vec{p} = m\vec{v} \qquad\qquad 9\text{-}1$$

SI unit: $\text{kg} \cdot \text{m/s}$

Similarly, the magnitude of the linear momentum is simply the object's mass times its speed:

Magnitude of Linear Momentum, p

$$p = mv \qquad\qquad 9\text{-}2$$

SI unit: $\text{kg} \cdot \text{m/s}$

Thus, if the heavy ball mentioned above has twice the mass of the light ball, but the light ball has twice the speed of the heavy ball, the momenta of the two balls are equal in magnitude; that is, $(2m)v = m(2v)$.

It follows from $\vec{p} = m\vec{v}$ and $p = mv$ that the units of linear momentum are those of mass times velocity—that is, $\text{kg} \cdot \text{m/s}$. No special shorthand name is given to this combination of units. In addition, notice that *linear momentum*—or just *momentum* for short—is the momentum of an object moving in a straight line. In Chapter 11 we introduce *angular momentum* for objects that are rotating.

The following Exercise gives a feeling for the magnitude of the momentum of everyday objects.

EXERCISE 9-1 MOMENTUM

(a) A 72.3-kg jogger runs with a speed of 3.13 m/s in the positive x direction. What is the jogger's momentum vector? **(b)** A 0.170-kg hockey puck has a momentum with a magnitude of $1.52 \text{ kg} \cdot \text{m/s}$. What is the speed of the puck? **(c)** The momentum of an airplane flying with a speed of 54.9 m/s has a magnitude of $3.93 \times 10^4 \text{ kg} \cdot \text{m/s}$. What is the mass of the airplane?

REASONING AND SOLUTION

We use the vector relationship, $\vec{p} = m\vec{v}$, when a direction is specified. In other cases, we work with magnitudes, and use the relationship $p = mv$.

  a. Using $\vec{p} = m\vec{v}$, and noting that the direction of the velocity is $\hat{x}$, we find

$$\vec{p} = m\vec{v} = (72.3 \text{ kg})(3.13 \text{ m/s})\hat{x} = (226 \text{ kg} \cdot \text{m/s})\hat{x}$$

  b. Solving $p = mv$ for v yields

$$v = \frac{p}{m} = \frac{1.52 \text{ kg} \cdot \text{m/s}}{0.170 \text{ kg}} = 8.94 \text{ m/s}$$

  c. Solving $p = mv$ for m yields

$$m = \frac{p}{v} = \frac{3.93 \times 10^4 \text{ kg} \cdot \text{m/s}}{54.9 \text{ m/s}} = 716 \text{ kg}$$

Calculating Changes in Momentum Because the velocity $\vec{v}$ is a vector with both a magnitude and a direction, so too is the momentum, $\vec{p} = m\vec{v}$. In problems that are strictly one-dimensional, the vectors simplify to a single component. This makes the calculations much easier—we just have to be careful to use the correct sign of the component.

As an illustration of the vector nature of momentum, and its importance in analyzing real-world situations, consider the situations shown in **FIGURE 9-1**. In Figure 9-1 (a), a 0.10-kg beanbag bear is dropped to the floor, where it hits with a speed of 4.0 m/s and sticks. In Figure 9-1 (b), a 0.10-kg rubber ball also hits the floor with a speed of 4.0 m/s, but in this case the ball bounces upward off the floor. If we assume an ideal rubber ball, its initial upward speed is 4.0 m/s. Now the question in each case is, "What is the change in momentum?"

To approach the problem systematically, we introduce a coordinate system as shown in Figure 9-1. With this choice, it follows that neither object has momentum in the x direction; thus, the system is one-dimensional. We need only consider the y component of momentum, p_y, and be careful about its sign.

We begin with the beanbag. Just before hitting the floor, it moves downward (that is, in the negative y direction) with a speed of $v = 4.0$ m/s. Letting m stand for the mass of the beanbag, we find that the initial momentum is

$$p_{y,i} = m(-v)$$

After landing on the floor, the beanbag is at rest; hence, its final momentum is zero:

$$p_{y,f} = m(0) = 0$$

Therefore the change in momentum for the beanbag is

$$\Delta p_y = p_{y,f} - p_{y,i} = 0 - m(-v) = mv$$
$$= (0.10 \text{ kg})(4.0 \text{ m/s}) = 0.40 \text{ kg} \cdot \text{m/s}$$

Notice that the change in momentum is positive—that is, in the upward direction. This makes sense because, before the bag landed, it had a negative (downward) momentum in the y direction. In order to increase the momentum from a negative value to zero, it is necessary to add a positive (upward) momentum.

Next, consider the rubber ball in Figure 9-1 (b). Before bouncing, its momentum is

$$p_{y,i} = m(-v)$$

This is the same initial momentum as for the beanbag. After bouncing, the ball is moving in the upward (positive) direction and its momentum is

$$p_{y,f} = mv$$

As a result, the change in momentum for the rubber ball is

$$\Delta p_y = p_{y,f} - p_{y,i} = mv - m(-v) = 2mv$$
$$= 2(0.10 \text{ kg})(4.0 \text{ m/s}) = 0.80 \text{ kg} \cdot \text{m/s}$$

This is *twice* the change in momentum of the beanbag! The reason is that in this case the momentum in the y direction must first be increased from $-mv$ to 0, and then increased again from 0 to mv. For the beanbag, the change was simply from $-mv$ to 0.

Notice how important it is to be careful about the vector nature of the momentum and to use the correct sign for p_y. Otherwise, we might have concluded—erroneously— that the rubber ball had zero change in momentum, since the *magnitude* of its momentum was unchanged by the bounce. In fact, its momentum does change due to the change in its *direction* of motion.

Calculating the Momentum for a System of Objects

Momentum is a vector, and hence the total momentum of a system of objects is the *vector* sum of the momenta of all the objects. That is,

$$\vec{\mathbf{P}}_{\text{total}} = \vec{\mathbf{P}}_1 + \vec{\mathbf{P}}_2 + \vec{\mathbf{P}}_3 + \cdots \qquad \text{9-3}$$

This is illustrated for the case of three objects in the following Example.

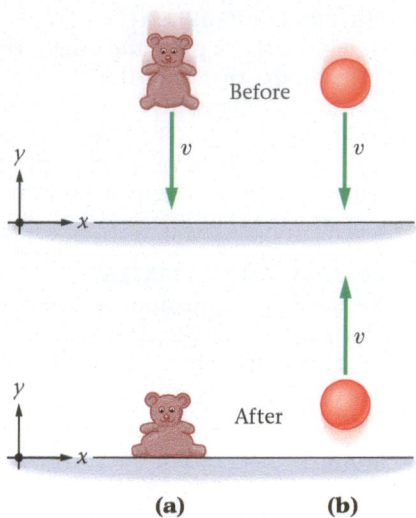

▲ **FIGURE 9-1 Change in momentum**
A beanbag bear and a rubber ball, with the same mass m and the same downward speed v, hit the floor. **(a)** The beanbag bear comes to rest on hitting the floor. Its change in momentum is mv upward. **(b)** The rubber ball bounces upward with a speed v. Its change in momentum is $2mv$ upward.

 Happy/Sad Pendulums

PROBLEM-SOLVING NOTE

Coordinate Systems

Be sure to draw a coordinate system for momentum problems, even if the problem is only one-dimensional. It is important to use the coordinate system to assign the correct sign to velocities and momenta in the system.

EXAMPLE 9-2   DUCK, DUCK, GOOSE

At a city park, a person throws some bread into a duck pond. Two 4.00-kg ducks and a 9.00-kg goose paddle rapidly toward the bread, as shown in our sketch. If the ducks swim at 1.10 m/s, and the goose swims with a speed of 1.30 m/s, find the magnitude and direction of the total momentum of the three birds.

CONTINUED

PICTURE THE PROBLEM

In our sketch we place the origin where the bread floats on the water. Notice that duck 1 swims in the positive x direction, duck 2 swims in the negative y direction, and the goose swims in the positive y direction. Therefore, $\vec{\mathbf{p}}_{d1} = m_d v_d \hat{\mathbf{x}}$, $\vec{\mathbf{p}}_{d2} = -m_d v_d \hat{\mathbf{y}}$, and $\vec{\mathbf{p}}_g = m_g v_g \hat{\mathbf{y}}$, where $v_d = 1.10$ m/s, $m_d = 4.00$ kg, $v_g = 1.30$ m/s, and $m_g = 9.00$ kg. The total momentum, $\vec{\mathbf{p}}_{total}$, points at an angle θ relative to the positive x axis, as indicated.

REASONING AND STRATEGY

We write the momentum of each bird as a vector, using unit vectors in the x and y directions. Next, we sum these vectors component by component to find the total momentum. Finally, we use the components of the total momentum to calculate its magnitude and direction.

Known   Mass of the ducks, $m_d = 4.00$ kg; mass of the goose, $m_g = 9.00$ kg; speed of the ducks, $v_d = 1.10$ m/s; speed of the goose, $v_g = 1.30$ m/s.

Unknown   Magnitude of total momentum, $p_{total} = ?$ Direction of total momentum, $\theta = ?$

SOLUTION

1. Use x and y unit vectors to express the momentum of each bird in vector form:

$$\vec{\mathbf{p}}_{d1} = m_d v_d \hat{\mathbf{x}} = (4.00 \text{ kg})(1.10 \text{ m/s})\hat{\mathbf{x}}$$
$$= (4.40 \text{ kg} \cdot \text{m/s})\hat{\mathbf{x}}$$

$$\vec{\mathbf{p}}_{d2} = -m_d v_d \hat{\mathbf{y}} = -(4.00 \text{ kg})(1.10 \text{ m/s})\hat{\mathbf{y}}$$
$$= -(4.40 \text{ kg} \cdot \text{m/s})\hat{\mathbf{y}}$$

$$\vec{\mathbf{p}}_g = m_g v_g \hat{\mathbf{y}} = (9.00 \text{ kg})(1.30 \text{ m/s})\hat{\mathbf{y}}$$
$$= (11.7 \text{ kg} \cdot \text{m/s})\hat{\mathbf{y}}$$

2. Sum the momentum vectors to obtain the total momentum:

$$\vec{\mathbf{p}}_{total} = \vec{\mathbf{p}}_{d1} + \vec{\mathbf{p}}_{d2} + \vec{\mathbf{p}}_g$$
$$= (4.40 \text{ kg} \cdot \text{m/s})\hat{\mathbf{x}} + (-4.40 \text{ kg} \cdot \text{m/s} + 11.7 \text{ kg} \cdot \text{m/s})\hat{\mathbf{y}}$$
$$= (4.40 \text{ kg} \cdot \text{m/s})\hat{\mathbf{x}} + (7.30 \text{ kg} \cdot \text{m/s})\hat{\mathbf{y}}$$

3. Calculate the magnitude of the total momentum:

$$p_{total} = \sqrt{p_{total,x}^2 + p_{total,y}^2}$$
$$= \sqrt{(4.40 \text{ kg} \cdot \text{m/s})^2 + (7.30 \text{ kg} \cdot \text{m/s})^2}$$
$$= 8.52 \text{ kg} \cdot \text{m/s}$$

4. Calculate the direction of the total momentum:

$$\theta = \tan^{-1}\left(\frac{p_{total,y}}{p_{total,x}}\right) = \tan^{-1}\left(\frac{7.30 \text{ kg} \cdot \text{m/s}}{4.40 \text{ kg} \cdot \text{m/s}}\right) = 58.9°$$

INSIGHT

Notice that the momentum of each bird depends only on its mass and velocity; it is independent of the bird's location. In addition, we observe that the magnitude of the total momentum is less than the sum of the magnitudes of each individual bird's momentum. This is generally the case when dealing with vector addition—the only exception is when all vectors point in the same direction.

PRACTICE PROBLEM — PREDICT/CALCULATE

(a) Suppose we can change the speed of the goose. If the total momentum of the three birds is to have a y component equal to zero, should the speed of the goose be increased or decreased? Explain. **(b)** Verify your prediction by calculating the required speed.   [**Answer: (a)** The goose's speed must be decreased, because its current y component of momentum is greater in magnitude than that of duck 2. **(b)** Setting the magnitude of the goose's momentum equal to 4.40 kg · m/s, we find $v_g = 0.489$ m/s.]

Some related homework problems: Problem 1, Problem 3, Problem 4

PHYSICS IN CONTEXT
Looking Ahead

Linear momentum plays an important role in quantum physics. For example, the momentum of a particle is related to its de Broglie wavelength in Section 30-5, and to the uncertainty principle in Section 30-6.

Enhance Your Understanding   (Answers given at the end of the chapter)

1. A block of wood is struck by a bullet. **(a)** Is the block more likely to be knocked over if the bullet is metal and embeds itself in the wood, or if the bullet is rubber and bounces off the wood? **(b)** Choose the *best explanation* for your prediction from among the following:

  I. The change in momentum when a bullet rebounds is larger than when it is brought to rest.

CONTINUED

    II. The metal bullet does more damage to the block.

    III. The rubber bullet bounces off, and hence it has little effect.

Section Review

- The linear momentum of an object is $\vec{\mathbf{p}} = m\vec{\mathbf{v}}$.
- The magnitude of the linear momentum is $p = mv$.
- The total linear momentum for a system of objects is $\vec{\mathbf{p}}_{total} = \vec{\mathbf{p}}_1 + \vec{\mathbf{p}}_2 + \vec{\mathbf{p}}_3 + \cdots$.

9-2 Momentum and Newton's Second Law

In Chapter 5 we introduced Newton's second law:

$$\sum \vec{\mathbf{F}} = m\vec{\mathbf{a}}$$

We mentioned at the time that this expression is valid only for objects that have constant mass. Newton's original statement of the second law, which is valid even if the mass changes, is as follows:

Newton's Second Law

$$\sum \vec{\mathbf{F}} = \frac{\Delta \vec{\mathbf{p}}}{\Delta t} \qquad\qquad 9\text{-}4$$

That is, the net force acting on an object is equal to the rate of change of its momentum.

    To show the connection between these two statements of the second law, consider the change in momentum, $\Delta\vec{\mathbf{p}}$. Recalling that $\vec{\mathbf{p}} = m\vec{\mathbf{v}}$, we have

$$\Delta\vec{\mathbf{p}} = \vec{\mathbf{p}}_f - \vec{\mathbf{p}}_i = m_f\vec{\mathbf{v}}_f - m_i\vec{\mathbf{v}}_i$$

If the mass is constant, so that $m_f = m_i = m$, it follows that the change in momentum is simply m times $\Delta\vec{\mathbf{v}}$:

$$\Delta\vec{\mathbf{p}} = m_f\vec{\mathbf{v}}_f - m_i\vec{\mathbf{v}}_i = m(\vec{\mathbf{v}}_f - \vec{\mathbf{v}}_i) = m\Delta\vec{\mathbf{v}}$$

As a result, Newton's second law for objects of constant mass is

$$\sum \vec{\mathbf{F}} = \frac{\Delta\vec{\mathbf{p}}}{\Delta t} = m\frac{\Delta\vec{\mathbf{v}}}{\Delta t}$$

If we recall that acceleration is the rate of change of velocity with time, $\vec{\mathbf{a}} = \Delta\vec{\mathbf{v}}/\Delta t$, the above equation becomes

$$\sum \vec{\mathbf{F}} = \frac{\Delta\vec{\mathbf{p}}}{\Delta t} = m\vec{\mathbf{a}} \qquad\qquad 9\text{-}5$$

Hence, the two statements of Newton's second law are equivalent if the mass is constant—as is often the case. The statement in terms of momentum is always valid, however. In the remainder of this chapter, we shall use the general form of Newton's second law to investigate the connections between forces and changes in momentum.

**PHYSICS
IN CONTEXT**
Looking Ahead

In Chapter 11 we will see that the rate of change of angular momentum is related to the "rotational force," referred to as torque.

Enhance Your Understanding   (Answers given at the end of the chapter)

2. Object 1 has a mass of 1 kg and is acted on by a force of 100 N. Object 2 has a mass of 100 kg and is also acted on by a force of 100 N. **(a)** Is the change in momentum in one second greater for object 1, greater for object 2, or the same for the two objects? Explain. **(b)** Is the change in velocity in one second greater for object 1, greater for object 2, or the same for the two objects? Explain.

Section Review

- The net force acting on an object is equal to the rate of change of its momentum.

► **FIGURE 9-2 The average force during a collision** The force between two objects that collide, as when a bat hits a baseball, rises rapidly to very large values, then drops again to zero in a matter of milliseconds. Rather than try to describe the complex behavior of the force, we focus on its average value, F_{av}. Notice that the area under the F_{av} rectangle is the same as the area under the actual force curve.

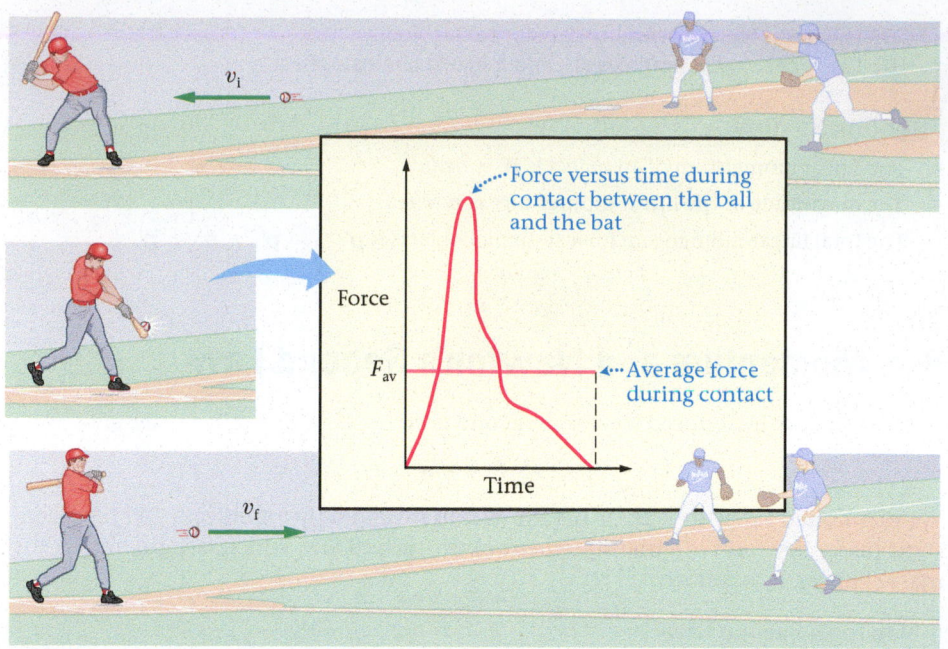

9-3 Impulse

The pitcher delivers a fastball, the batter takes a swing, and with a crack of the bat the ball that was approaching home plate at 95.0 mi/h is now heading toward the pitcher at 115 mi/h. In the language of physics, we say that the bat has delivered an **impulse**, $\vec{\mathbf{I}}$, to the ball.

During the brief time the ball and bat are in contact—perhaps as little as a thousandth of a second—the force between them rises rapidly to a large value, as shown in **FIGURE 9-2**, then falls back to zero as the ball takes flight. It would be almost impossible, of course, to describe every detail of the way the force varies with time. Instead, we focus on the average force exerted by the bat, $\vec{\mathbf{F}}_{av}$, which is also shown in Figure 9-2. The impulse delivered by the bat is defined to be $\vec{\mathbf{F}}_{av}$ times the amount of time, Δt, that the ball and bat are in contact:

Definition of Impulse, $\vec{\mathbf{I}}$

$$\vec{\mathbf{I}} = \vec{\mathbf{F}}_{av}\,\Delta t \qquad \text{9-6}$$

SI unit: $N \cdot s = kg \cdot m/s$

The magnitude of the impulse is simply the area under the force-versus-time curve.

Notice that impulse is a vector and that it points in the same direction as the average force. In addition, its units are $N \cdot s = (kg \cdot m/s^2) \cdot s = kg \cdot m/s$, the same as the units of momentum. It's no accident that impulse and momentum have the same units. In fact, if we rearrange Newton's second law, $\sum \vec{\mathbf{F}} = \vec{\mathbf{F}}_{total} = \dfrac{\Delta \vec{\mathbf{p}}}{\Delta t}$, we can see that the total force times Δt is simply the change in momentum:

$$\vec{\mathbf{F}}_{total} = \frac{\Delta \vec{\mathbf{p}}}{\Delta t}$$

$$\vec{\mathbf{F}}_{total}\,\Delta t = \Delta \vec{\mathbf{p}}$$

The same applies to impulse if we replace $\vec{\mathbf{F}}_{total}$ with $\vec{\mathbf{F}}_{av}$. Thus, we find:

Momentum–Impulse Theorem

$$\vec{\mathbf{I}} = \vec{\mathbf{F}}_{av}\,\Delta t = \Delta \vec{\mathbf{p}} \qquad \text{9-7}$$

For instance, if we know the impulse delivered to an object—that is, its change in momentum, $\Delta \vec{\mathbf{p}}$—and the time interval during which the change occurs, the average force that caused the impulse is simply $\Delta \vec{\mathbf{p}}/\Delta t$.

PHYSICS IN CONTEXT
Looking Back

In this chapter we see that force multiplied by the *time* over which it acts is related to a change in momentum. In Chapters 7 and 8 we saw that force multiplied by the *distance* over which it acts leads to a change in energy.

Impulse and Force As an example of the connection between impulse and force, let's calculate the impulse given to the baseball considered at the beginning of this section. We will also calculate the average force between the ball and the bat.

First, we set up a coordinate system with the positive x axis pointing from home plate toward the pitcher's mound, as indicated in **FIGURE 9-3**. If the ball's mass is 0.145 kg, its initial momentum—which is in the negative x direction—is

$$\vec{\mathbf{p}}_i = -mv_i\hat{\mathbf{x}} = -(0.145 \text{ kg})(95.0 \text{ mi/h})\left(\frac{0.447 \text{ m/s}}{1 \text{ mi/h}}\right)\hat{\mathbf{x}} = -(6.16 \text{ kg} \cdot \text{m/s})\hat{\mathbf{x}}$$

Immediately after the hit, the ball's final momentum is in the positive x direction:

$$\vec{\mathbf{p}}_f = mv_f\hat{\mathbf{x}} = (0.145 \text{ kg})(115 \text{ mi/h})\left(\frac{0.447 \text{ m/s}}{1 \text{ mi/h}}\right)\hat{\mathbf{x}} = (7.45 \text{ kg} \cdot \text{m/s})\hat{\mathbf{x}}$$

The impulse, then, is

$$\vec{\mathbf{I}} = \Delta\vec{\mathbf{p}} = \vec{\mathbf{p}}_f - \vec{\mathbf{p}}_i = [7.45 \text{ kg} \cdot \text{m/s} - (-6.16 \text{ kg} \cdot \text{m/s})]\hat{\mathbf{x}} = (13.6 \text{ kg} \cdot \text{m/s})\hat{\mathbf{x}}$$

If the ball and bat are in contact for $1.20 \text{ ms} = 1.20 \times 10^{-3}$ s, a typical time, then the average force is

$$\vec{\mathbf{F}}_{av} = \frac{\Delta\vec{\mathbf{p}}}{\Delta t} = \frac{\vec{\mathbf{I}}}{\Delta t} = \frac{(13.6 \text{ kg} \cdot \text{m/s})\hat{\mathbf{x}}}{1.20 \times 10^{-3} \text{ s}} = (1.13 \times 10^4 \text{ N})\hat{\mathbf{x}}.$$

RWP* The average force is in the positive x direction—that is, toward the pitcher, as expected. In addition, the magnitude of the average force is remarkably large, more than 2500 pounds! This explains why high-speed photographs show the ball flattening out during a hit, as in **FIGURE 9-4**. In addition, notice that the weight of the ball, which is only about 0.3 lb, is negligible compared to the forces involved during the hit. On the other hand, extending the time of impact can reduce the force, as in the cushion breaking the pole vaulter's fall in Figure 9-4.

Big Idea 2 Impulse is force multiplied by the time over which it acts. Impulse is equal to the change in momentum of a system.

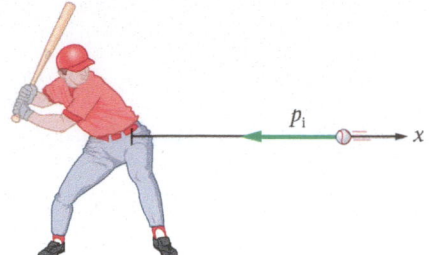

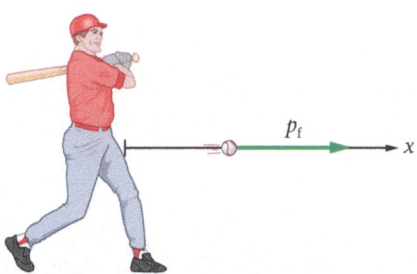

▲ **FIGURE 9-3 Hitting a baseball** The impulse delivered to the ball by the bat changes the ball's momentum from $-p_i\hat{\mathbf{x}}$ to $p_f\hat{\mathbf{x}}$.

◄ **FIGURE 9-4 Visualizing Concepts Impulse** When a baseball is hit by a bat (left), an enormous force (thousands of newtons) acts for a very short period of time—perhaps only a few ms. During this time, the ball is dramatically deformed by the impact. To keep the same thing from happening to a pole vaulter (right), a deeply padded landing area is provided. The change in the pole vaulter's momentum as he is brought to a stop, $mv = F\Delta t$, is the same whether he lands on a mat or on concrete. However, the padding yields, thereby greatly prolonging the time Δt during which he is in contact with the mat. The corresponding force on the vaulter is thus markedly decreased.

QUICK EXAMPLE 9-3   FIND THE FINAL SPEED OF THE BALL

A 0.144-kg baseball is moving toward home plate with a speed of 43.0 m/s when it is bunted (hit softly). The bat exerts an average force of 6.50×10^3 N on the ball for 1.30 ms. The average force is directed toward the pitcher, which we take to be the positive x direction. What is the final speed of the ball?

REASONING AND SOLUTION

We know the force and the time over which it acts. This gives the impulse, which is the change in momentum. We also know the initial momentum of the ball, from its mass

CONTINUED

*Real World Physics applications are denoted by the acronym RWP.

and initial speed. Therefore, we can find the final momentum, and from that the final speed.

1. Use $\vec{I} = \vec{F}_{av}\Delta t = \Delta\vec{p}$ to relate the change in momentum to the average force:

$$\vec{I} = \vec{F}_{av}\Delta t = \Delta\vec{p}$$
$$= \vec{p}_f - \vec{p}_i$$

2. Rearrange to solve for the final momentum:

$$\vec{p}_f = \vec{I} + \vec{p}_i$$

3. Calculate the initial momentum, $\vec{p}_i$. Recall that the ball's initial velocity is in the negative x direction:

$$\vec{p}_i = m\vec{v}_i$$
$$= (0.144\ \text{kg})[(-43.0\ \text{m/s})\hat{\mathbf{x}}]$$
$$= (-6.19\ \text{kg}\cdot\text{m/s})\hat{\mathbf{x}}$$

4. Calculate the impulse, $\vec{I}$. Recall that the average force exerted by the bat is in the positive x direction:

$$\vec{I} = \vec{F}_{av}\Delta t$$
$$= [(6.50 \times 10^3\ \text{N})\hat{\mathbf{x}}](1.30 \times 10^{-3}\ \text{s})$$
$$= (8.45\ \text{kg}\cdot\text{m/s})\hat{\mathbf{x}}$$

5. Substitute the results for impulse and initial momentum into Step 2 to find the final momentum, $\vec{p}_f$:

$$\vec{p}_f = \vec{I} + \vec{p}_i$$
$$= (8.45\ \text{kg}\cdot\text{m/s})\hat{\mathbf{x}} + (-6.19\ \text{kg}\cdot\text{m/s})\hat{\mathbf{x}}$$
$$= (2.26\ \text{kg}\cdot\text{m/s})\hat{\mathbf{x}}$$

6. Divide the final momentum by the mass to find the final velocity, $\vec{v}_f$. The final speed, v_f, is the magnitude of the final velocity:

$$\vec{v}_f = \frac{\vec{p}_f}{m} = \frac{(2.26\ \text{kg}\cdot\text{m/s})\hat{\mathbf{x}}}{0.144\ \text{kg}} = (15.7\ \text{m/s})\hat{\mathbf{x}}$$
$$v_f = 15.7\ \text{m/s}$$

With our choice of coordinate system, we see that the initial momentum of the ball was in the negative x direction. The impulse applied to the ball, however, resulted in a final momentum (and velocity) in the positive x direction.

Everyday Examples of Impulse and Force We saw in Section 9-1 that the change in momentum is different for an object that hits something and sticks compared with an object that hits and bounces off. This means that the impulse, and hence the force, is different in the two cases. We explore this in the following Conceptual Example.

CONCEPTUAL EXAMPLE 9-4   RAIN VERSUS HAIL   PREDICT/EXPLAIN

A person stands under an umbrella during a rain shower. A few minutes later the raindrops turn to hail—though the number of "drops" hitting the umbrella per time remains the same, as does their speed and mass. **(a)** Is the force required to hold the umbrella in the hail greater than, less than, or equal to the force required in the rain? **(b)** Which of the following is the *best explanation* for your prediction?

 I. The force required is greater because the hail bounces off the umbrella, which causes a greater change in momentum and a greater impulse.

 II. Less force is required because the hail bounces off the umbrella, and hence has little effect on it. In contrast, the raindrops hit the umbrella and come to rest.

III. The force is the same in either case because the same mass of water (liquid or ice) hits the umbrella with the same speed.

REASONING AND DISCUSSION

When raindrops strike the umbrella, they tend to splatter and run off; when hailstones hit the umbrella, they bounce back upward. As a result, the change in momentum is greater for the hail—just as the change in momentum is greater for a rubber ball bouncing off the floor than it is for a beanbag landing on the floor. Hence, the impulse and the force are greater with hail.

ANSWER

(a) The force is greater in the hail. **(b)** The best explanation is I.

We conclude this section with an additional calculation involving impulse.

EXAMPLE 9-5 JOGGING A THOUGHT

A force platform measures the horizontal force exerted on a person's foot by the ground. The sketch shows force-versus-time data taken by a force platform for a 71-kg person who is jogging with increasing speed. The first part of the data ($t = 0$ to 0.30 s) indicates a negative force as the person's foot comes to rest. The impulse (area under the curve) during this time period is $I_1 = -2.7$ kg · m/s. The second part of the data ($t = 0.30$ s $-$ 0.80 s) shows a positive force as the person pushes off against the ground to gain speed. The impulse during this time period is $I_2 = 26.2$ kg · m/s. For the time period from $t = 0$ to $t = 0.80$ s, what are (a) the average horizontal force exerted on the jogger and (b) the jogger's change in speed?

PICTURE THE PROBLEM

The plot of data from the force platform shows the force exerted on a person's foot as it first comes to rest (from $t = 0$ to $t = 0.30$ s) and then pushes off against the ground (from $t = 0.30$ s to $t = 0.80$ s). The impulse is negative at first ($I_1 < 0$), but then becomes positive ($I_2 > 0$). The net impulse delivered to the person is positive, $I_{net} = I_1 + I_2 > 0$, indicating an increase in speed.

REASONING AND STRATEGY

a. The net impulse delivered to the jogger from $t = 0$ to $t = 0.80$ s is $I_{net} = I_1 + I_2$. The average horizontal force exerted during this time interval is the net impulse divided by the elapsed time, $F_{av} = I_{net}/\Delta t$.

b. The net impulse I_{net} delivered to the person from $t = 0$ to $t = 0.80$ s is equal to the change in momentum $\Delta p = m\Delta v$ during that time period. The corresponding change in speed is given by the change in momentum divided by the mass m of the jogger; that is, $\Delta v = \Delta p/m = I_{net}/m$.

Known Mass of jogger, $m = 71$ kg; impulse from $t = 0$ to $t = 0.30$ s, $I_1 = -2.7$ kg · m/s; impulse from $t = 0.30$ s to $t = 0.80$ s, $I_2 = 26.2$ kg · m/s.

Unknown (a) Average force, $F_{av} = ?$ (b) Change in speed, $\Delta v = ?$

SOLUTION

Part (a)

1. Calculate the net impulse in the horizontal direction, I_{net}:

$$I_{net} = I_1 + I_2$$
$$= -2.7 \text{ kg} \cdot \text{m/s} + 26.2 \text{ kg} \cdot \text{m/s} = 23.5 \text{ kg} \cdot \text{m/s}$$

2. Divide the net impulse by the time period, $\Delta t = 0.80$ s, to find the average force, F_{av}:

$$F_{av} = \frac{I_{net}}{\Delta t} = \frac{23.5 \text{ kg} \cdot \text{m/s}}{0.80 \text{ s}} = 29 \text{ N}$$

Part (b)

3. Divide the net impulse by the jogger's mass to find the change in speed:

$$\Delta v = \frac{\Delta p}{m} = \frac{I_{net}}{m} = \frac{23.5 \text{ kg} \cdot \text{m/s}}{71 \text{ kg}} = 0.33 \text{ m/s}$$

INSIGHT

In this case, the floor delivers a *horizontal* impulse to the jogger, causing an increase in speed. In **FIGURE 9-5**, the ground delivers a *vertical* impulse to a vampire bat, propelling it into the air.

PRACTICE PROBLEM

For the time period from $t = 0$ to $t = 0.30$ s, when the jogger's foot is coming to rest on the floor, what is the jogger's change in speed? [**Answer:** The change in speed is small, only -0.038 m/s.]

Some related homework problems: Problem 14, Problem 15, Problem 17

Enhance Your Understanding   (Answers given at the end of the chapter)

3. A steel ball and a rubber ball collide. Is the impulse received by the steel ball greater than, less than, or equal to the impulse received by the rubber ball? Explain.

CONTINUED

▲ **FIGURE 9-5** Most bats take off by simply dropping from their perch, but vampire bats like this one must leap from the ground to become airborne. They do so by rocking forward onto their front limbs and then pushing off, using their extremely strong pectoral muscles. As the bat pushes downward, the ground exerts an upward reaction force on the bat, with a corresponding impulse sufficient to propel it upward a considerable distance. In fact, a vampire bat can launch itself 1 m or more into the air in less than a tenth of a second.

Big Idea **3** Momentum is conserved when the net force acting on a system is zero.

Section Review

- The impulse delivered to an object is $\vec{\mathbf{I}} = \vec{\mathbf{F}}_{av}\,\Delta t$.
- The impulse delivered to an object is equal to the change in its momentum: $\vec{\mathbf{I}} = \vec{\mathbf{F}}_{av}\,\Delta t = \Delta\vec{\mathbf{p}}$.
- The impulse is a vector that is proportional to the force vector.

9-4 Conservation of Linear Momentum

In this section we turn to perhaps the most important aspect of linear momentum—the fact that it is a conserved quantity. In this respect, it plays a fundamental role in physics similar to that of energy. We shall also see that momentum conservation leads to simplifications in our calculations, making it of great practical significance.

First, recall that the net force acting on an object is equal to the rate of change of its momentum:

$$\sum\vec{\mathbf{F}} = \frac{\Delta\vec{\mathbf{p}}}{\Delta t}$$

Rearranging this expression, we find that the change in momentum during a time interval Δt is

$$\Delta\vec{\mathbf{p}} = \left(\sum\vec{\mathbf{F}}\right)\Delta t \qquad 9\text{-}8$$

Clearly, then, if the net force acting on an object is zero, $\sum\vec{\mathbf{F}} = 0$, its change in momentum is also zero:

$$\Delta\vec{\mathbf{p}} = \left(\sum\vec{\mathbf{F}}\right)\Delta t = 0$$

Writing the change of momentum in terms of its initial and final values, we have

$$\Delta\vec{\mathbf{p}} = \vec{\mathbf{p}}_f - \vec{\mathbf{p}}_i = 0$$

This means that the initial and final momenta are equal:

$$\vec{\mathbf{p}}_f = \vec{\mathbf{p}}_i \qquad 9\text{-}9$$

The momentum does not change in this case, and hence we say that it is **conserved.** To summarize:

Conservation of Momentum

If the net force acting on an object is zero, its momentum is conserved; that is,

$$\vec{\mathbf{p}}_f = \vec{\mathbf{p}}_i$$

In some cases, the force is zero in one direction and nonzero in another. For example, an object in free fall has no force in the x direction, $F_x = 0$, but has a nonzero component of force in the y direction, $F_y \neq 0$. As a result, the object's x component of momentum remains constant, while its y component of momentum changes with time. Therefore, in applying momentum conservation, we must remember that both the force and the momentum are vector quantities and that the momentum conservation principle applies separately to each coordinate direction.

Internal Versus External Forces

The net force acting on a system of objects is the sum of forces applied from outside the system (external forces, $\vec{\mathbf{F}}_{ext}$) and forces acting between objects within the system (internal forces, $\vec{\mathbf{F}}_{int}$). Thus, we can write

$$\vec{\mathbf{F}}_{net} = \sum\vec{\mathbf{F}} = \sum\vec{\mathbf{F}}_{ext} + \sum\vec{\mathbf{F}}_{int}$$

As we shall see, internal and external forces play very *different* roles in terms of how they affect the momentum of a system.

◄ **FIGURE 9-6 Separating two canoes**
A system composed of two canoes and
their occupants. The forces $\vec{F}_1$ and $\vec{F}_2$ are
internal to the system, and hence they
sum to zero.

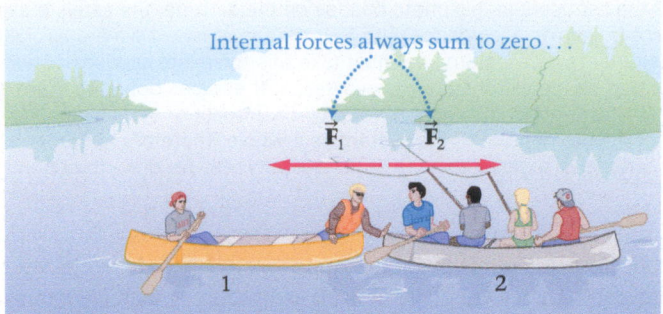

Internal forces always sum to zero...

$\vec{F}_1$ $\vec{F}_2$

1   2

...and hence they have no effect on the net momentum of the system.

To illustrate the distinction, consider the case of two canoes floating at rest next to one another on a lake, as described in Example 5-6 and shown in **FIGURE 9-6**. Let's consider the "system" to be the two canoes and the people inside them. When a person in canoe 1 pushes on canoe 2, a force $\vec{F}_2$ is exerted on canoe 2. By Newton's third law, an equal and opposite force, $\vec{F}_1 = -\vec{F}_2$, is exerted on the person in canoe 1. We say that $\vec{F}_1$ and $\vec{F}_2$ are internal forces, because they act between objects that are part of the same system. Notice, also, that these two forces sum to zero:

$$\vec{F}_1 + \vec{F}_2 = (-\vec{F}_2) + \vec{F}_2 = 0$$

This is a special case, of course, but it demonstrates the following general principles:

- Internal forces, like all forces, occur in action-reaction pairs. Both members of an action-reaction pair are within the system when the forces are internal.

- The forces in action-reaction pairs are equal and opposite—due to Newton's third law—and hence internal forces must *always* sum to zero:

$$\sum \vec{F}_{int} = 0$$

The fact that internal forces always cancel means that the net force acting on a system of objects is simply the sum of the *external* forces acting on it:

$$\vec{F}_{net} = \sum \vec{F}_{ext} + \sum \vec{F}_{int} = \sum \vec{F}_{ext} + 0 = \sum \vec{F}_{ext}$$

The external forces, on the other hand, may or may not sum to zero—it all depends on the particular situation. For example, if the system consists of the two canoes in Figure 9-6, the external forces are the weights of the people and the canoes acting downward, and the upward, normal force exerted by the water to keep the canoes afloat. These forces sum to zero, and there is no acceleration in the vertical direction. In other cases, like an object in free fall, the net external force is nonzero.

Finally, how do external and internal forces affect the momentum of a system? To see the connection, first note that Newton's second law gives the change in the net momentum for a given time interval Δt:

$$\Delta \vec{p}_{net} = \vec{F}_{net} \Delta t$$

Because the internal forces cancel, however, the change in the net momentum is directly related to the net *external* force:

$$\Delta \vec{p}_{net} = \left(\sum \vec{F}_{ext} \right) \Delta t \qquad 9\text{-}10$$

Therefore, the key distinction between internal and external forces is the following:

Conservation of Momentum for a System of Objects

- *Internal* forces have absolutely no effect on the net momentum of a system.

- If the *net external* force acting on a system is zero, its net momentum is conserved. That is,

$$\vec{p}_{1,f} + \vec{p}_{2,f} + \vec{p}_{3,f} + \cdots = \vec{p}_{1,i} + \vec{p}_{2,i} + \vec{p}_{3,i} + \cdots$$

◄ **FIGURE 9-7** If the astronaut in this photo pushes on the satellite, the satellite exerts an equal but opposite force on him, in accordance with Newton's third law. If we are calculating the change in the astronaut's momentum, we must take this force into account. However, if we define the system to be the astronaut *and* the satellite, the forces between them are internal to the system. Whatever effect they may have on the astronaut or the satellite individually, they do not affect the momentum of the system as a whole. Therefore, whether a particular force counts as internal or external depends entirely on where we draw the boundaries of the system.

It's important to point out that these statements apply only to the *net* momentum of a system, and not to the momentum of each individual object. For example, suppose that a system consists of two objects, 1 and 2, and that the net external force acting on the system is zero. As a result, the net momentum must remain constant:

$$\vec{\mathbf{P}}_{net} = \vec{\mathbf{p}}_1 + \vec{\mathbf{p}}_2 = \text{constant}$$

This does not mean, however, that $\vec{\mathbf{p}}_1$ is constant or that $\vec{\mathbf{p}}_2$ is constant. All we can say is that the *sum* of $\vec{\mathbf{p}}_1$ and $\vec{\mathbf{p}}_2$ does not change. An example of this is the astronaut pushing off from a satellite in **FIGURE 9-7**.

PROBLEM-SOLVING NOTE

Internal Versus External Forces

It is important to keep in mind that internal forces cannot change the momentum of a system—only a net external force can do that.

Applying Momentum Conservation As a specific example of momentum conservation, consider the two canoes floating on a lake, as described previously. Initially the momentum of the system is zero, because the canoes are at rest. After a person pushes the canoes apart, they are both moving, and hence both have nonzero momentum. Thus, the momentum of each canoe has changed. On the other hand, because the net external force acting on the system is zero, the sum of the canoes' momenta must still vanish. We show this in the next Example.

EXAMPLE 9-6 TIPPY CANOE

Two groups of canoeists meet in the middle of a lake. After a brief visit, a person in canoe 1 pushes on canoe 2 with a force of 46 N to separate the canoes. If the mass of canoe 1 and its occupants is 130 kg, and the mass of canoe 2 and its occupants is 250 kg, find the momentum of each canoe after 1.20 s of pushing.

PICTURE THE PROBLEM

We choose the positive *x* direction to point from canoe 1 to canoe 2. With this choice, the force exerted on canoe 2 is $\vec{\mathbf{F}}_2 = (46\text{ N})\hat{\mathbf{x}}$ and the force exerted on canoe 1 is $\vec{\mathbf{F}}_1 = (-46\text{ N})\hat{\mathbf{x}}$.

REASONING AND STRATEGY

First, we find the acceleration of each canoe using $a_x = F_x/m$. Next, we use kinematics, $v_x = v_{0x} + a_x t$, to find the velocity at time *t*. Notice that the canoes start at rest, and hence $v_{0x} = 0$. Finally, the momentum is calculated using $p_x = mv_x$.

Known Mass of canoe 1, $m_1 = 130$ kg; mass of canoe 2, $m_2 = 250$ kg; magnitude of pushing force, $F = 46$ N; pushing time, $t = 1.20$ s.

Unknown Final momentum of canoe 1, $\vec{\mathbf{p}}_{1,f} = ?$ Final momentum of canoe 2, $\vec{\mathbf{p}}_{2,f} = ?$

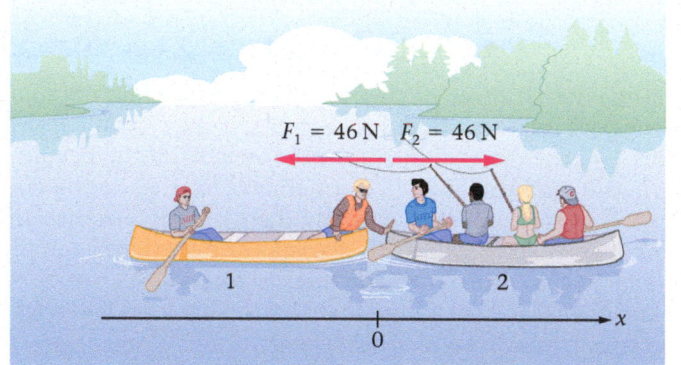

SOLUTION

1. Use Newton's second law to find the acceleration of canoe 1. Note that the acceleration of canoe 1 is in the negative *x* direction:

$$a_{1,x} = \frac{\sum F_{1,x}}{m_1} = \frac{-46\text{ N}}{130\text{ kg}} = -0.35\text{ m/s}^2$$

2. Do the same calculation for canoe 2:

$$a_{2,x} = \frac{\sum F_{2,x}}{m_2} = \frac{46\text{ N}}{250\text{ kg}} = 0.18\text{ m/s}^2$$

3. Calculate the velocity of each canoe at $t = 1.20$ s:

$$v_{1,x} = a_{1,x}t = (-0.35\text{ m/s}^2)(1.20\text{ s}) = -0.42\text{ m/s}$$
$$v_{2,x} = a_{2,x}t = (0.18\text{ m/s}^2)(1.20\text{ s}) = 0.22\text{ m/s}$$

4. Calculate the momentum of each canoe at $t = 1.20$ s:

$$p_{1,x} = m_1 v_{1,x} = (130\text{ kg})(-0.42\text{ m/s}) = -55\text{ kg}\cdot\text{m/s}$$
$$p_{2,x} = m_2 v_{2,x} = (250\text{ kg})(0.22\text{ m/s}) = 55\text{ kg}\cdot\text{m/s}$$

CONTINUED

The sum of the momenta of the two moving canoes is zero. This is just what one would expect: The canoes start at rest with zero momentum, there is zero net external force acting on the system, and hence the final momentum must also be zero. The final velocities *do not* add to zero; it is momentum ($m\vec{v}$) that is conserved, not velocity ($\vec{v}$).

Finally, we solved this problem using one-dimensional kinematics so that we could clearly see the distinction between velocity and momentum. An alternative way to calculate the final momentum of each canoe is to use $\Delta\vec{p} = \vec{p}_f - \vec{p}_i = \vec{F}\Delta t$. For canoe 1 we have $\vec{p}_{1,f} = \vec{F}_1\Delta t + \vec{p}_{1,i} = [(-46\,\text{N})\hat{x}](1.20\,\text{s}) + 0 = (-55\,\text{kg}\cdot\text{m/s})\hat{x}$, in agreement with our results above. A similar calculation yields $\vec{p}_{2,f} = (55\,\text{kg}\cdot\text{m/s})\hat{x}$ for canoe 2.

PRACTICE PROBLEM
What are the final momenta of the canoes if they are pushed apart with a force of 65 N?   [**Answer:** $p_{1,x} = -78\,\text{kg}\cdot\text{m/s}$; $p_{2,x} = 78\,\text{kg}\cdot\text{m/s}$]

Some related homework problems: Problem 23, Problem 24, Problem 26

BIO The Physics of Recoil

In a situation like the one described in Example 9-6, the person in canoe 1 pushes canoe 2 away. At the same time, canoe 1 begins to move in the opposite direction. This is referred to as **recoil**. It is essentially the same as the recoil one experiences when firing a gun or when turning on a strong stream of water.

Participants in a number of athletic sports, including American football, soccer, hockey, and of course boxing, can experience a traumatic brain injury called a *concussion*. A concussion is an example of an injury caused by recoil. The human brain is surrounded by protective cerebrospinal fluid, which allows it to move a little bit inside the skull. Initially the brain and the skull are at rest relative to each other and the momentum of the system is zero. If the skull is suddenly accelerated forward, due to an impact at the back of the head as indicated in **FIGURE 9-8**, the brain remains at rest for a brief moment until it collides with the back of the skull. The brain then recoils from the skull and moves forward—in fact, it may even collide again, this time with the front of the skull. Such collision and recoil events cause the concussive brain injury.

Another interesting example of recoil involves the human body. Perhaps you've noticed, when resting quietly in a rocking or reclining chair, that the chair wobbles back and forth slightly about once a second. The reason for this movement is that each time your heart pumps blood in one direction (from the atria to the ventricles, then to the aorta and pulmonary arteries, and so on) your body recoils in the opposite direction. Because the recoil depends on the force exerted by your heart on the blood and the volume of blood expelled from the heart with each beat, it is possible to gain valuable medical information regarding the health of your heart by analyzing the recoil it produces.

The medical instrument that employs the physical principle of recoil is called the *ballistocardiograph*. It is a completely noninvasive technology that simply requires the patient to sit comfortably in a chair fitted with sensitive force sensors under the seat and behind the back. Sophisticated bathroom scales also utilize this technology. A ballistocardiographic (BCG) scale detects the recoil vibrations of the body as a person stands on the scale. This allows the BCG scale to display not only the person's body weight but his or her heart rate as well.

A more dramatic application of heartbeat recoil is currently being used at the Riverbend Maximum Security Institution in Tennessee. The only successful breakout from this prison occurred when four inmates hid in a secret compartment in a delivery truck that was leaving the facility. The institution now uses a heartbeat recoil detector that would have foiled this escape. Each vehicle leaving the prison must stop at a checkpoint where a small motion detector is attached to it with a suction cup. Any persons hidden in the vehicle will reveal their presence by the very beating of their hearts. These heartbeat detectors have proved to be 100% effective, even though the recoil of the heart may displace a large truck by only a few millionths of an inch. Similar systems are being used at other high-security installations and border crossings.

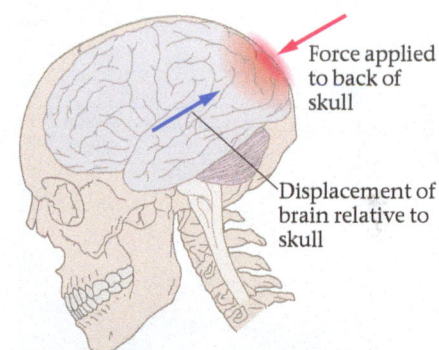

Force applied to back of skull

Displacement of brain relative to skull

▲ **FIGURE 9-8** A sharp force applied to the back of the head can accelerate the skull forward relative to the brain. The brain then collides with the back of the skull and rebounds off of it, which can cause a concussion.

CONCEPTUAL EXAMPLE 9-7 MOMENTUM VERSUS KINETIC ENERGY

The final momentum of the two canoes and their occupants in Example 9-6 is equal to their initial momentum. Is the final kinetic energy of the system (a) greater than, (b) less than, or (c) equal to its initial kinetic energy?

REASONING AND DISCUSSION

The final momentum of the two canoes is zero because one canoe has a positive momentum and the other has a negative momentum of the same magnitude. The two momenta, then, sum to zero. Kinetic energy, which is $\frac{1}{2}mv^2$, cannot be negative; hence no such cancellation is possible. Both canoes have positive kinetic energies, and therefore the final kinetic energy is greater than the initial kinetic energy, which is zero.

Where does the increase in kinetic energy come from? It comes from the muscular work done by the person who pushes the canoes apart.

ANSWER

(a) K_f is greater than K_i.

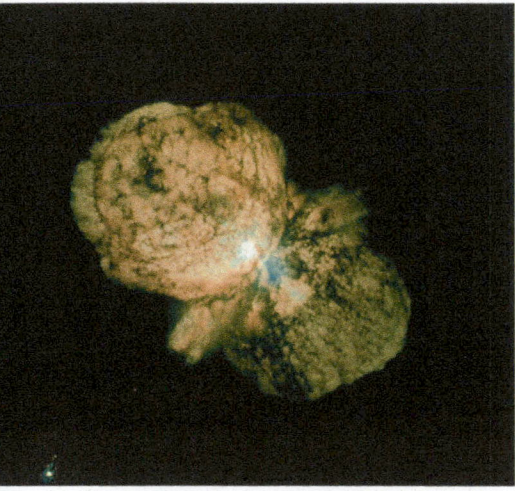

▲ **FIGURE 9-9** This Hubble Space Telescope photograph shows the aftermath of a violent explosion of the star Eta Carinae. The explosion produced two bright lobes of matter spewing outward in opposite directions. In this photograph, these lobes have expanded to about the size of our solar system. The momentum of the star before the explosion must be the same as the total momentum of the star and the bright lobes after the explosion. Since the lobes are roughly symmetrical and move in opposite directions, their net momentum is essentially zero. Thus, we conclude that the momentum of the star itself was virtually unchanged by the explosion.

RWP A special case of some interest is the universe. Since there is nothing external to the universe—by definition—it follows that the net external force acting on it is zero. Therefore, its net momentum is conserved. No matter what happens—a comet collides with the Earth, a star explodes and becomes a supernova, a black hole swallows part of a galaxy—the total momentum of the universe simply cannot change. A particularly vivid illustration of momentum conservation in our own galaxy is provided by the exploding star Eta Carinae. As can be seen in the Hubble Space Telescope photograph in **FIGURE 9-9**, jets of material are moving away from the star in opposite directions, just like the canoes moving apart from one another in Example 9-6.

Conservation of momentum also applies to the more down-to-Earth situation described in the next Quick Example.

QUICK EXAMPLE 9-8 THE SPEED OF A LOG

Predict/Calculate

A 61-kg lumberjack stands on a 320-kg log that floats on a pond. Both the log and the lumberjack are at rest initially. In addition, the log points directly toward the shore, which we take to be the positive x direction. (a) If the lumberjack trots along the log toward the shore with a speed v relative to the shore, the log moves away from the shore. Is the speed of the log relative to the shore greater than, less than, or equal to v? Explain. (b) If the lumberjack trots with a velocity of 1.8 m/s relative to the shore, what is the velocity of the log relative to the shore? Ignore friction between the log and the water.

REASONING AND SOLUTION

Part (a) The speed of the log relative to the shore is less than v. To see this, notice that before the lumberjack begins trotting toward the shore, the system has zero momentum. Only internal forces are involved as the lumberjack trots, and so the total momentum remains equal to zero. Therefore, the momentum of the lumberjack moving toward the shore is balanced by the momentum of the log moving away from the shore. The log has the greater mass, and hence the speed of the log is less than v.

Part (b)

1. Set the total momentum of the lumberjack (lj) plus the log equal to zero:

$$\vec{\mathbf{p}}'_{lj} + \vec{\mathbf{p}}'_{log} = 0$$

CONTINUED

2. Solve for the momentum of the log in terms of the momentum of the lumberjack:

$$\vec{\mathbf{p}}'_{\text{log}} = -\vec{\mathbf{p}}'_{\text{lj}}$$

3. Calculate the momentum of the trotting lumberjack. Recall that the direction toward the shore is the positive x direction:

$$\vec{\mathbf{p}}'_{\text{lj}} = m_{\text{lj}} v_{\text{lj}} \hat{\mathbf{x}} = (61 \text{ kg})(1.8 \text{ m/s})\hat{\mathbf{x}}$$
$$= (110 \text{ kg} \cdot \text{m/s})\hat{\mathbf{x}}$$

4. Calculate the momentum of the log:

$$\vec{\mathbf{p}}'_{\text{log}} = -\vec{\mathbf{p}}'_{\text{lj}} = -(110 \text{ kg} \cdot \text{m/s})\hat{\mathbf{x}}$$

5. Divide the log's momentum by its mass to find its velocity:

$$\vec{\mathbf{p}}'_{\text{log}} = m_{\text{log}} \vec{\mathbf{v}}'_{\text{log}}$$

$$\vec{\mathbf{v}}'_{\text{log}} = \frac{\vec{\mathbf{p}}'_{\text{log}}}{m_{\text{log}}} = \frac{-(110 \text{ kg} \cdot \text{m/s})\hat{\mathbf{x}}}{320 \text{ kg}} = -(0.34 \text{ m/s})\hat{\mathbf{x}}$$

As expected, the log moves away from the shore—in the minus x direction—and with a speed (0.34 m/s) that is less than the speed of the lumberjack (1.8 m/s). With these speeds, the magnitude of the log's momentum is equal to the magnitude of the lumberjack's momentum.

Enhance Your Understanding   (Answers given at the end of the chapter)

4. An object initially at rest breaks into two pieces as the result of an explosion. Piece 1 has twice the kinetic energy of piece 2. **(a)** If piece 1 has a momentum of magnitude p, what is the magnitude of the momentum of piece 2? Explain. **(b)** If the mass of piece 1 is m, what is the mass of piece 2? Explain.

Section Review

- Momentum is conserved when the net force acting on a system is zero.
- Internal forces have no effect on the net momentum of a system.

9-5 Inelastic Collisions

In this section we consider **collisions**. By a collision we mean a situation in which two objects strike one another, and in which the net external force is either zero or negligibly small. For example, if two train cars roll on a level track and hit one another, this is a collision. In this case, the net external force—the weight downward plus the normal force exerted by the tracks upward—is zero. As a result, the momentum of the two-car system is conserved.

Another example of a collision is a baseball that is struck by a bat. In this case, the external forces are not zero because the weight of the ball is not balanced by any other force. However, as we have seen in Section 9-3, the forces exerted during the hit are much larger than the weight of the ball or the bat. Hence, to a good approximation, we may neglect the external forces (the weight of the ball and bat) in this case, and say that the momentum of the ball–bat system is conserved.

Now it may seem surprising at first, but the fact that the momentum of a system is conserved during a collision does not necessarily mean that the system's kinetic energy is conserved. In fact most, or even all, of a system's kinetic energy may be converted to other forms during a collision. In general, collisions are categorized according to what happens to the kinetic energy of the system. There are two possibilities. After a collision, the final kinetic energy, K_f, is either equal to the initial kinetic energy, K_i, or it is not. If $K_f = K_i$, the collision is said to be **elastic**. We consider elastic collisions in the next section.

On the other hand, the kinetic energy may change during a collision. Usually it decreases due to losses associated with sound, heat, and deformation. Sometimes it increases—if the collision sets off an explosion, for instance. In any event, collisions in which the kinetic energy is not conserved are referred to as **inelastic**:

Big Idea 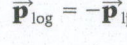 Collisions conserve momentum, but may or may not conserve mechanical energy. Mechanical energy is conserved in elastic collisions, but not in inelastic collisions.

PHYSICS IN CONTEXT
Looking Back

Kinetic energy (Chapter 7) plays a key role in analyzing collisions, leading to the distinction between elastic and inelastic collisions.

Inelastic Collisions

In an inelastic collision, the momentum of a system is conserved:

$$\vec{p}_f = \vec{p}_i$$

The kinetic energy of the system is not conserved in an inelastic collision:

$$K_f \neq K_i$$

In the special case where objects stick together after the collision, we say that the collision is **completely inelastic**:

Completely Inelastic Collisions

When objects stick together after colliding, the collision is completely inelastic.

In a completely inelastic collision, the maximum amount of kinetic energy is lost. If the total momentum of the system is zero, this means that all of the kinetic energy is lost. For systems with nonzero total momentum, however, some kinetic energy remains after the collision. Examples of elastic and inelastic collisions are shown in **FIGURE 9-10**.

Inelastic Collisions in One Dimension

Consider a system consisting of two train cars of mass m_1 and m_2 on a smooth, level track. The cars have initial velocities $v_{1,i}$ and $v_{2,i}$, as shown in **FIGURE 9-11**. When the cars collide, the coupling mechanism latches, causing the cars to stick together and move as a unit. What is the speed of the cars after the collision?

To answer this question, we begin by noting that the system has zero net external force acting on it, and hence its momentum is conserved. The initial momentum is

$$p_i = m_1 v_{1,i} + m_2 v_{2,i}$$

After the collision, the objects move together with a common velocity v_f. Therefore, the final momentum is

$$p_f = (m_1 + m_2)v_f$$

Equating the initial and final momenta yields $m_1 v_{1,i} + m_2 v_{2,i} = (m_1 + m_2)v_f$, or

$$v_f = \frac{m_1 v_{1,i} + m_2 v_{2,i}}{m_1 + m_2} \qquad \text{9-11}$$

Thus, the final velocity is equal to the initial momentum divided by the total mass. If the initial momentum is zero, as when two identical train cars approach one another with equal speed, the final velocity is zero and the system is at rest after the collision.

▲ **FIGURE 9-10 Visualizing Concepts**
Collisions In both elastic and inelastic collisions, momentum is conserved. The same is not true of kinetic energy, however. In the largely inelastic collision at top, much of the hockey players' initial kinetic energy is transformed into work—rearranging the players' anatomies and shattering the glass at the rink. In the highly elastic collision at bottom, the ball rebounds with very little diminution of its kinetic energy (though a little energy is lost as sound and heat).

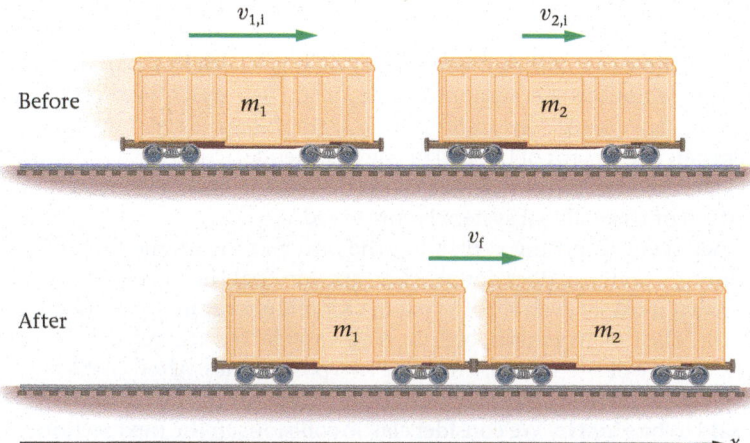

▲ **FIGURE 9-11 Railroad cars collide and stick together** After the collision, the cars stick together and move with the same speed.

EXERCISE 9-9 SPEED AFTER THE COLLISION

A 1400-kg car moving at 3.2 m/s is struck in the rear by a 2900-kg truck moving in the same direction at 5.3 m/s. If the vehicles stick together after the collision, what is their speed immediately after colliding? (Assume external forces may be ignored.)

REASONING AND SOLUTION

The vehicles stick together, and hence their collision is completely inelastic. As a result, we can find the final velocity with $v_f = (m_1v_{1,i} + m_2v_{2,i})/(m_1 + m_2)$ (Equation 9-11). In this case, $m_1 = 1400$ kg, $v_{1,i} = 3.2$ m/s, $m_2 = 2900$ kg, and $v_{2,i} = 5.3$ m/s; hence the final velocity is

$$v_f = \frac{m_1v_{1,i} + m_2v_{2,i}}{m_1 + m_2} = \frac{(1400\ \text{kg})(3.2\ \text{m/s}) + (2900\ \text{kg})(5.3\ \text{m/s})}{1400\ \text{kg} + 2900\ \text{kg}} = 4.6\ \text{m/s}$$

Let's apply $v_f = (m_1v_{1,i} + m_2v_{2,i})/(m_1 + m_2)$ to the simple case of two railroad cars of equal mass m, one of which is at rest before the collision. If the other car has an initial speed of v, what is the final speed of the two cars? Letting $m_1 = m_2 = m$, $v_{1,i} = v$, and $v_{2,i} = 0$, we find

$$v_f = \frac{mv + m \cdot 0}{m + m} = \frac{m}{2m}v = \frac{1}{2}v \qquad \text{9-12}$$

As you might have guessed, the final speed is one-half the initial speed.

No momentum is lost in this collision of railroad cars, but some of the initial kinetic energy is converted to other forms. Some propagates away as sound, some is converted to heat, and some creates permanent deformations in the metal of the latching mechanism. The precise amount of kinetic energy that is lost is addressed in the following Conceptual Example.

PROBLEM-SOLVING NOTE

Momentum Versus Energy Conservation

Be sure to distinguish between momentum conservation and energy conservation. A common error is to assume that kinetic energy is conserved just because the momentum is conserved.

CONCEPTUAL EXAMPLE 9-10 HOW MUCH KINETIC ENERGY IS LOST?

A railroad car of mass m and speed v collides with and sticks to an identical railroad car that is initially at rest. After the collision, is the kinetic energy of the system (a) 1/2, (b) 1/3, or (c) 1/4 of its initial kinetic energy?

REASONING AND DISCUSSION

Before the collision, the kinetic energy of the system is

$$K_i = \frac{1}{2}mv^2$$

After the collision, the mass is doubled and the speed is halved. Hence, the final kinetic energy is

$$K_f = \frac{1}{2}(2m)\left(\frac{v}{2}\right)^2 = \frac{1}{2}(2m)\left(\frac{v^2}{4}\right) = \frac{1}{2}\left(\frac{1}{2}mv^2\right) = \frac{1}{2}K_i$$

Therefore, one-half of the initial kinetic energy is converted to other forms of energy.

ANSWER

(a) The final kinetic energy is one-half the initial kinetic energy.

It's interesting that we know the precise amount of kinetic energy that was lost during the collision of the train cars, even though we don't know just how much went into sound, how much went into heat, and so on. It's not necessary to know all of the details to determine how much kinetic energy was lost.

We also know how much momentum was lost—none.

EXAMPLE 9-11   GOAL-LINE STAND

On a touchdown attempt, a 95.0-kg running back runs toward the end zone at 3.75 m/s. A 111-kg linebacker moving at 4.10 m/s meets the runner in a head-on collision. If the two players stick together, **(a)** what is their velocity immediately after the collision? **(b)** What are the initial and final kinetic energies of the system?

PICTURE THE PROBLEM

In our sketch, we let subscript 1 refer to the red-and-gray running back who carries the ball, and subscript 2 refers to the blue-and-gold linebacker who makes the tackle. The direction of the running back's initial motion is taken to be in the positive x direction. Therefore, the initial velocities of the players are $\vec{v}_1 = (3.75 \text{ m/s})\hat{x}$ and $\vec{v}_2 = (-4.10 \text{ m/s})\hat{x}$.

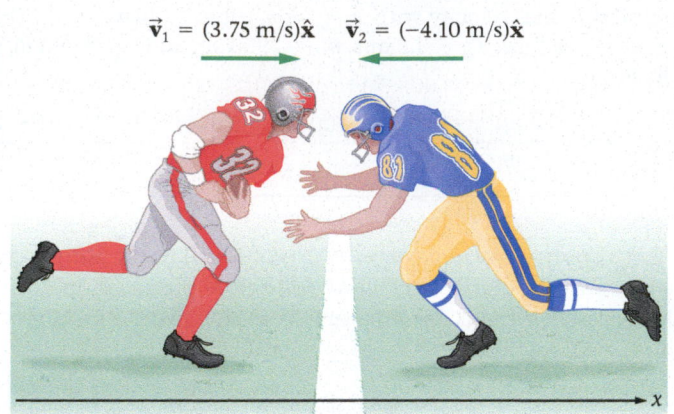

$\vec{v}_1 = (3.75 \text{ m/s})\hat{x} \quad \vec{v}_2 = (-4.10 \text{ m/s})\hat{x}$

REASONING AND STRATEGY

a. The final velocity can be found by applying momentum conservation to the system consisting of the two players. Initially, the players have momenta in opposite directions. After the collision, the players move together with a combined mass $m_1 + m_2$ and a velocity $\vec{v}_f$.

b. The kinetic energies can be found by applying $\frac{1}{2}mv^2$ to the players individually to obtain the initial kinetic energy, and then to their combined motion for the final kinetic energy.

Known   Player 1: $m_1 = 95.0 \text{ kg}$, $\vec{v}_1 = (3.75 \text{ m/s})\hat{x}$; Player 2: $m_2 = 111 \text{ kg}$, $\vec{v}_2 = (-4.10 \text{ m/s})\hat{x}$.
Unknown   **(a)** $\vec{v}_f = ?$ **(b)** $K_i = ? K_f = ?$

SOLUTION

Part (a)

1. Set the initial momenta of the two players equal to their final momentum:

$$m_1\vec{v}_1 + m_2\vec{v}_2 = (m_1 + m_2)\vec{v}_f$$

2. Solve for the final velocity and substitute numerical values, being careful to use the appropriate signs:

$$\vec{v}_f = \frac{m_1\vec{v}_1 + m_2\vec{v}_2}{m_1 + m_2}$$

$$= \frac{(95.0 \text{ kg})(3.75 \text{ m/s})\hat{x} + (111 \text{ kg})(-4.10 \text{ m/s})\hat{x}}{95.0 \text{ kg} + 111 \text{ kg}}$$

$$= (-0.480 \text{ m/s})\hat{x}$$

Part (b)

3. Calculate the initial kinetic energy of the two players:

$$K_i = \frac{1}{2}m_1v_1^2 + \frac{1}{2}m_2v_2^2$$

$$= \frac{1}{2}(95.0 \text{ kg})(3.75 \text{ m/s})^2 + \frac{1}{2}(111 \text{ kg})(4.10 \text{ m/s})^2$$

$$= 1600 \text{ J}$$

4. Calculate the final kinetic energy of the players, noting that they both move with the same velocity after the collision:

$$K_f = \frac{1}{2}(m_1 + m_2)v_f^2$$

$$= \frac{1}{2}(95.0 \text{ kg} + 111 \text{ kg})(0.480 \text{ m/s})^2$$

$$= 23.7 \text{ J}$$

INSIGHT

After the collision, the two players are moving in the negative x direction—that is, away from the end zone—with a speed of 0.480 m/s. This is because the linebacker had more negative momentum than the running back had positive momentum. As for the kinetic energy, of the original 1600 J, only 23.7 J is left after the collision. This means that more than 98% of the original kinetic energy is converted to other forms. Even so, *none* of the momentum is lost.

PRACTICE PROBLEM — PREDICT/CALCULATE

(a) If the final speed of the two players is to be zero, should the speed of the running back (red and gray) be increased or decreased? Explain. **(b)** Check your prediction by calculating the required speed for the running back.
[**Answer: (a)** The running back's speed should be increased, so that the magnitude of his momentum will match that of the linebacker. **(b)** The desired speed is 4.79 m/s.]

Some related homework problems: Problem 31, Problem 34

EXAMPLE 9-12   BALLISTIC PENDULUM

In a ballistic pendulum, an object of mass m is fired with an initial speed v_i at the bob of a pendulum. The bob has a mass M, and is suspended by a rod of negligible mass. After the collision, the object and the bob stick together and swing through an arc, eventually gaining a height h. Find the height h in terms of m, M, v_i, and g.

PICTURE THE PROBLEM

Our sketch shows the physical setup of a ballistic pendulum. Initially, only the object of mass m is moving, and it moves in the positive x direction with a speed v_i. Immediately after the collision, the bob and object move together with a new speed, v_f, which is determined by momentum conservation. Finally, the pendulum continues to swing to the right until its speed decreases to zero and it comes to rest at the height h.

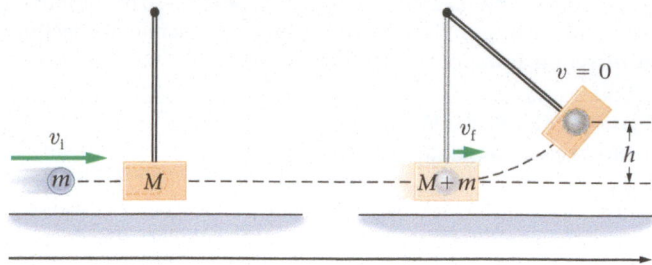

REASONING AND STRATEGY

There are two distinct physical processes at work in the ballistic pendulum. The first is a completely inelastic collision between the bob and the object. Momentum is conserved during this collision, but kinetic energy is not. After the collision, the remaining kinetic energy is converted to gravitational potential energy, which determines how high the bob and object will rise.

Known   Mass of object, m; mass of bob, M; initial speed of object, v_i.
Unknown   Height after collision, $h = ?$

SOLUTION

1. Set the momentum just before the bob–object collision equal to the momentum just after the collision. Let v_f be the speed just after the collision:

$$mv_i = (M + m)v_f$$

2. Solve for the speed just after the collision, v_f:

$$v_f = \left(\frac{m}{M + m}\right)v_i$$

3. Calculate the kinetic energy just after the collision:

$$K_f = \frac{1}{2}(M + m)v_f^2 = \frac{1}{2}(M + m)\left(\frac{m}{M + m}\right)^2 v_i^2$$

$$= \frac{1}{2}mv_i^2\left(\frac{m}{M + m}\right)$$

4. Set the kinetic energy after the collision equal to the gravitational potential energy of the bob–object combination at the height h:

$$\frac{1}{2}mv_i^2\left(\frac{m}{M + m}\right) = (M + m)gh$$

5. Solve for the height, h:

$$h = \left(\frac{m}{M + m}\right)^2\left(\frac{v_i^2}{2g}\right)$$

INSIGHT

A ballistic pendulum is often used to measure the speed of a rapidly moving object, such as a bullet. If a bullet were shot straight up, it would rise to the height $v_i^2/2g$, which can be thousands of feet. On the other hand, if a bullet of mass m is fired into a ballistic pendulum, in which M is much greater than m, the bullet will reach only a small fraction of this height. Thus, the ballistic pendulum makes for a more convenient and practical measurement.

PRACTICE PROBLEM

A 7.00-g bullet is fired into a ballistic pendulum whose bob has a mass of 0.950 kg. If the bob rises to a height of 0.220 m, what was the initial speed of the bullet?   [**Answer:** $v_i = 284$ m/s. If this bullet were fired straight up, it would rise 4.11 km $\approx$ 13,000 ft in the absence of air resistance.]

Some related homework problems: Problem 32, Problem 33

Inelastic Collisions in Two Dimensions

Next we consider collisions in two dimensions, where we must conserve the momentum component by component. To do this, we set up a coordinate system and resolve the initial momentum into x and y components. Next, we demand that the final momentum have precisely the same x and y components as the initial momentum. That is,

$$p_{x,i} = p_{x,f} \quad \text{and} \quad p_{y,i} = p_{y,f}$$

PROBLEM-SOLVING NOTE

Sketch the System Before and After the Collision

In problems involving collisions, it is useful to draw the system before and after the collision. Be sure to label the relevant masses, velocities, and angles.

The following Example shows how to carry out such a calculation in a practical situation.

EXAMPLE 9-13   ANALYZING A TRAFFIC ACCIDENT

RWP A car with a mass of 950 kg and a speed of 16 m/s approaches an intersection, as shown in the sketch. A 1300-kg minivan traveling at 21 m/s is heading for the same intersection. The car and minivan collide and stick together. Find the direction and speed of the wrecked vehicles just after the collision, assuming external forces can be ignored.

PICTURE THE PROBLEM
In our sketch, we align the x and y axes with the crossing streets. With this choice, $\vec{\mathbf{v}}_1$ (the car's initial velocity) is in the positive x direction, and $\vec{\mathbf{v}}_2$ (the minivan's initial velocity) is in the positive y direction. In addition, the problem statement indicates that $m_1 = 950$ kg and $m_2 = 1300$ kg. After the collision, the two vehicles move together (as a unit) with a speed v_f in a direction θ with respect to the positive x axis.

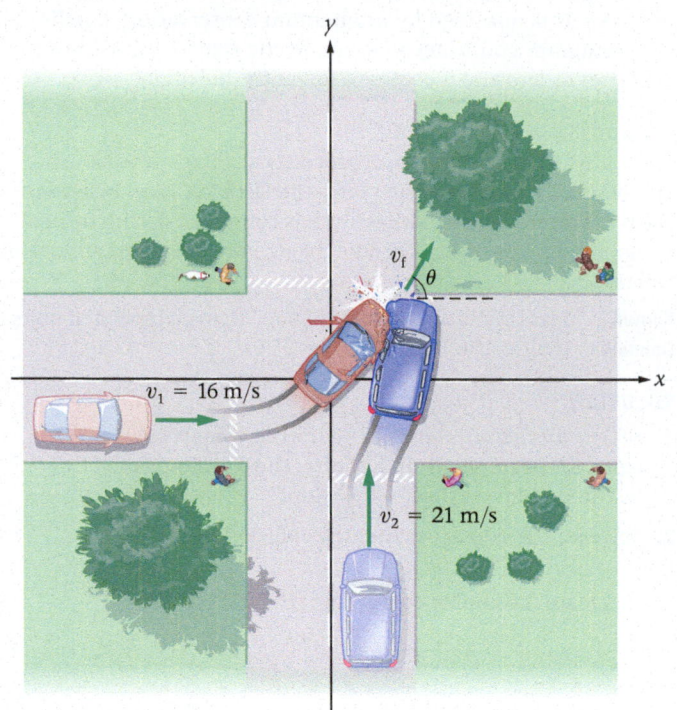

REASONING AND STRATEGY
Because external forces can be ignored, the total momentum of the system must be conserved during the collision. This is really two conditions: (i) The x component of momentum is conserved, and (ii) the y component of momentum is conserved. These two conditions determine the two unknowns: the final speed, v_f, and the final direction, θ.

Known   Mass of car, $m_1 = 950$ kg; initial velocity of car, $\vec{\mathbf{v}}_1 = (16 \text{ m/s})\hat{\mathbf{x}}$; mass of minivan, $m_2 = 1300$ kg; initial velocity of minivan, $\vec{\mathbf{v}}_2 = (21 \text{ m/s})\hat{\mathbf{y}}$.
Unknown   Final speed, $v_f = ?$ Final direction, $\theta = ?$

SOLUTION

1. Set the initial x component of momentum equal to the final x component of momentum. Note that the x component of $\vec{\mathbf{v}}_f$ is $v_f \cos\theta$:

$$m_1 v_1 = (m_1 + m_2) v_f \cos\theta$$

2. Set the initial y component of momentum equal to the final y component of momentum. Note that the y component of $\vec{\mathbf{v}}_f$ is $v_f \sin\theta$:

$$m_2 v_2 = (m_1 + m_2) v_f \sin\theta$$

3. Divide the y momentum equation by the x momentum equation. This eliminates v_f, giving an equation involving θ alone:

$$\frac{m_2 v_2}{m_1 v_1} = \frac{(m_1 + m_2) v_f \sin\theta}{(m_1 + m_2) v_f \cos\theta} = \frac{\sin\theta}{\cos\theta} = \tan\theta$$

4. Solve for θ:

$$\theta = \tan^{-1}\left(\frac{m_2 v_2}{m_1 v_1}\right) = \tan^{-1}\left[\frac{(1300 \text{ kg})(21 \text{ m/s})}{(950 \text{ kg})(16 \text{ m/s})}\right]$$
$$= \tan^{-1}(1.8) = 61°$$

5. The final speed can be found using either the x or the y momentum equation. Here we use the x equation:

$$v_f = \frac{m_1 v_1}{(m_1 + m_2) \cos\theta} = \frac{(950 \text{ kg})(16 \text{ m/s})}{(950 \text{ kg} + 1300 \text{ kg}) \cos 61°}$$
$$= 14 \text{ m/s}$$

INSIGHT
As a check, you should verify that the y momentum equation gives the same value for v_f.

When a collision occurs in the real world, a traffic-accident investigation team measures skid marks at the scene of the crash and uses this information—along with some basic physics—to determine the initial speeds and directions of the vehicles. This information is often presented in court, where it can lead to a clear identification of the driver at fault.

CONTINUED

Enhance Your Understanding   (Answers given at the end of the chapter)

5. Two objects travel in opposite directions along a straight line. They have momenta of equal magnitude, and they stick together when they collide. **(a)** What is the final speed of the two objects? Explain. **(b)** What percentage of the initial kinetic energy is converted to other forms as a result of the collision?

Section Review

- In an inelastic collision, the final kinetic energy of a system is not equal to the initial kinetic energy of the system.

- A completely inelastic collision is one in which objects hit and stick together.

- In a two-dimensional collision, momentum conservation must be applied to both the x and y directions.

9-6 Elastic Collisions

In this section we consider collisions in which both momentum and kinetic energy are conserved. As mentioned in the previous section, such collisions are referred to as elastic:

Elastic Collisions

In an elastic collision, momentum and kinetic energy are conserved. That is,

$$\vec{P}_f = \vec{P}_i$$

and

$$K_f = K_i$$

Most collisions in everyday life are rather poor approximations to being elastic—usually a significant amount of energy is converted to other forms. However, the collision of objects that bounce off one another with little deformation—billiard balls, for example—provides a reasonably good approximation to an elastic collision. In the subatomic world, on the other hand, elastic collisions are common.

Elastic Collisions in One Dimension

Consider a head-on collision of two carts on an air track, as pictured in **FIGURE 9-12**. The carts are provided with bumpers that give an elastic bounce when the carts collide.

**PHYSICS
IN CONTEXT**
Looking Ahead

In Chapter 15 we shall see that elastic collisions between molecules and the wall of a container lead to the pressure exerted by a gas.

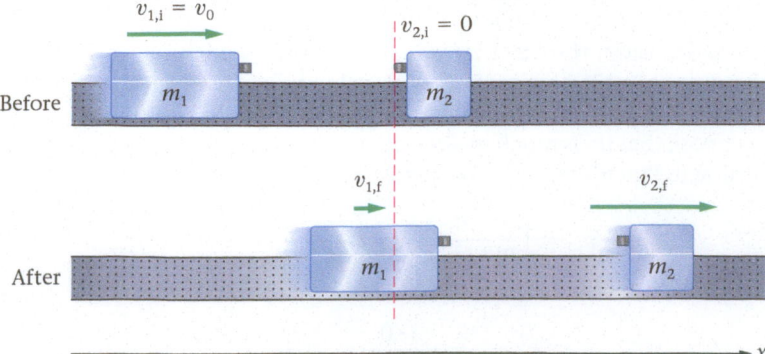

◄ **FIGURE 9-12 An elastic collision between two air carts** In the case pictured, $v_{1,f}$ is to the right (positive), which means that m_1 is greater than m_2. In fact, we have chosen $m_1 = 2m_2$ for this plot; therefore, $v_{1,f} = v_0/3$ and $v_{2,f} = 4v_0/3$ as given by Equations 9-13. If m_1 were less than m_2, cart 1 would bounce back toward the left, meaning that $v_{1,f}$ would be negative.

Let's suppose that initially cart 1 is moving to the right with a speed v_0, and cart 2 is at rest. If the masses of the carts are m_1 and m_2, respectively, then momentum conservation can be written as follows:

$$m_1 v_0 = m_1 v_{1,f} + m_2 v_{2,f}$$

In this expression, $v_{1,f}$ and $v_{2,f}$ are the final velocities of the two carts. Notice that we say velocities, not speeds, because it is possible for cart 1 to reverse direction, in which case $v_{1,f}$ would be negative.

Next, the fact that this is an elastic collision means the final velocities must also satisfy energy conservation:

$$\frac{1}{2}m_1 v_0^2 = \frac{1}{2}m_1 v_{1,f}^2 + \frac{1}{2}m_2 v_{2,f}^2$$

Thus, we now have two equations for the two unknowns, $v_{1,f}$ and $v_{2,f}$. Straightforward—though messy—algebra yields the following results:

$$v_{1,f} = \left(\frac{m_1 - m_2}{m_1 + m_2}\right)v_0$$

$$v_{2,f} = \left(\frac{2m_1}{m_1 + m_2}\right)v_0$$

9-13

Notice that the final velocity of cart 1 can be positive, negative, or zero, depending on whether m_1 is greater than, less than, or equal to m_2, respectively. The final velocity of cart 2, however, is always positive.

EXERCISE 9-14   BOUNCING BUMPER CARS

At an amusement park, a 98.5-kg bumper car moving with a speed of 1.18 m/s bounces elastically off a second 122-kg bumper car at rest. Find the final velocities of the two cars.

REASONING AND SOLUTION
The situation is just like the one described by Equations 9-13, where object 1 moves with an initial speed v_0 and collides elastically with object 2 at rest. Using m_1 = 98.5 kg, m_2 = 122 kg, and v_0 = 1.18 m/s, we can calculate the final velocities as follows:

$$v_{1,f} = \left(\frac{m_1 - m_2}{m_1 + m_2}\right)v_0 = \left(\frac{98.5 \text{ kg} - 122 \text{ kg}}{98.5 \text{ kg} + 122 \text{ kg}}\right)(1.18 \text{ m/s}) = -0.126 \text{ m/s}$$

$$v_{2,f} = \left(\frac{2m_1}{m_1 + m_2}\right)v_0 = \left(\frac{2(98.5 \text{ kg})}{98.5 \text{ kg} + 122 \text{ kg}}\right)(1.18 \text{ m/s}) = 1.05 \text{ m/s}$$

The direction of travel of car 1 has been reversed, as indicated by its negative velocity.

Let's check a few special cases of our results. First, consider the case where the two carts have equal masses: $m_1 = m_2 = m$. Substituting into Equations 9-13, we find

$$v_{1,f} = \left(\frac{m - m}{m + m}\right)v_0 = 0$$

and

$$v_{2,f} = \left(\frac{2m}{m + m}\right)v_0 = v_0$$

Thus, after the collision, the cart that was moving with velocity v_0 is now at rest, and the cart that was at rest is now moving with velocity v_0. In effect, the carts have "exchanged" velocities. This case is illustrated in **FIGURE 9-13 (a)**.

Next, suppose that m_2 is much greater than m_1 or, equivalently, that m_1 approaches zero. Returning to Equations 9-13, and setting $m_1 = 0$, we find

$$v_{1,f} = \left(\frac{0 - m_2}{0 + m_2}\right)v_0 = \left(\frac{-m_2}{m_2}\right)v_0 = -v_0$$

and

$$v_{2,f} = \frac{2 \cdot 0}{0 + m_2}v_0 = 0$$

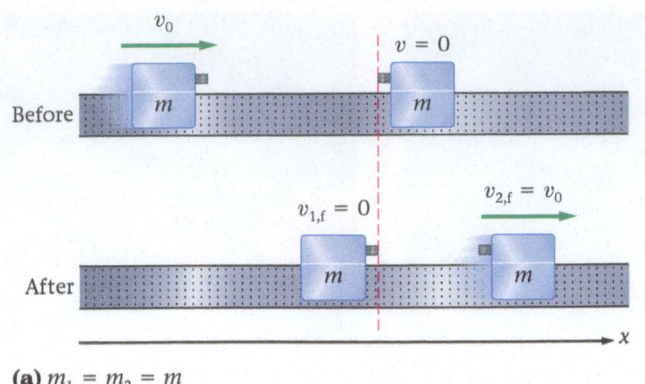

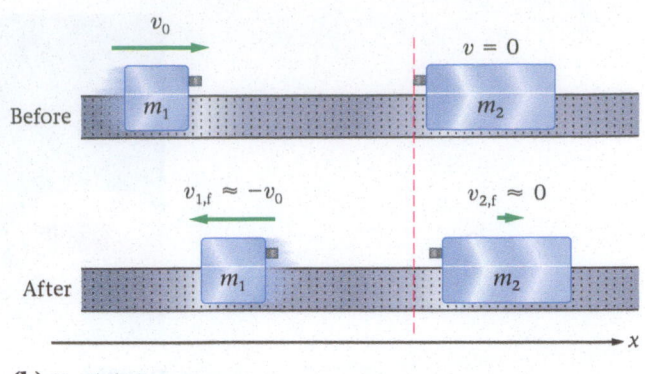

(a) $m_1 = m_2 = m$

(b) $m_1 << m_2$

▶ **FIGURE 9-13 Elastic collisions between air carts of various masses** (a) Carts of equal mass exchange velocities when they collide. (b) When a light cart collides with a stationary heavy cart, its direction of motion is reversed. Its speed is practically unchanged. (c) When a heavy cart collides with a stationary light cart, it continues to move in the same direction with essentially the same speed. The light cart moves off with a speed that is roughly twice the initial speed of the heavy cart.

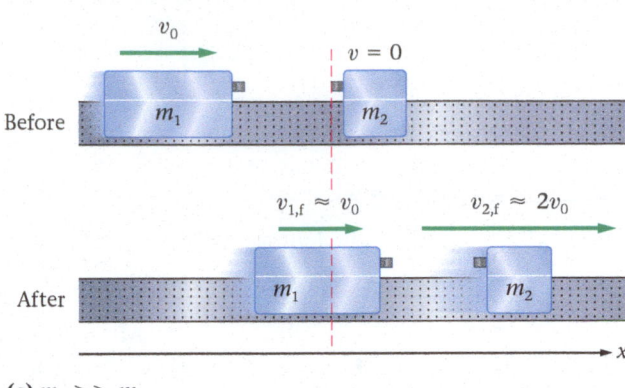

(c) $m_1 >> m_2$

Physically, we interpret these results as follows: A very light cart collides with a heavy cart that is at rest. The heavy cart hardly budges, but the light cart is reflected, heading *backward* (remember the minus sign in $-v_0$) with the same speed it had initially. For example, if you throw a ball against a wall, the wall is the very heavy object and the ball is the light object. The ball bounces back with the same speed it had initially (assuming an ideal elastic collision). We show a case in which m_1 is much less than m_2 in **FIGURE 9-13 (b)**.

Finally, what happens when m_1 is much greater than m_2? To check this limit we can set m_2 equal to zero. We consider the results in the following Conceptual Example.

CONCEPTUAL EXAMPLE 9-15   THE HOVERFLY AND THE ELEPHANT

A hoverfly is happily maintaining a fixed position about 3 m above the ground when an elephant charges out of the bush and collides with it. The fly bounces elastically off the forehead of the elephant. If the initial speed of the elephant is v_0, is the speed v of the fly after the collision equal to (a) v_0, (b) $2v_0$, or (c) $3v_0$?

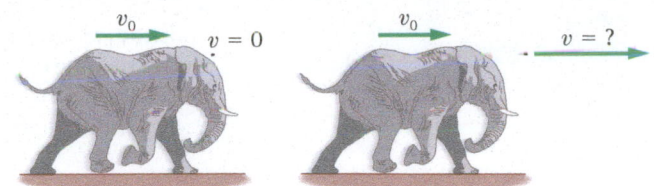

REASONING AND DISCUSSION

We can use Equations 9-13 to find the final speeds of the fly and the elephant. First, let m_1 be the mass of the elephant, and m_2 be the mass of the fly. Clearly, m_2 is vanishingly small compared with m_1, and hence we can evaluate Equations 9-13 in the limit $m_2 \rightarrow 0$. For the elephant, this yields

$$v_{1,f} = \left(\frac{m_1 - m_2}{m_1 + m_2}\right)v_0 \xrightarrow{m_2 \rightarrow 0} \left(\frac{m_1}{m_1}\right)v_0 = v_0$$

Similarly, for the fly we have

$$v_{2,f} = \left(\frac{2m_1}{m_1 + m_2}\right)v_0 \xrightarrow{m_2 \rightarrow 0} \left(\frac{2m_1}{m_1}\right)v_0 = 2v_0$$

As expected, the speed of the elephant is unaffected. The fly, however, rebounds with *twice* the speed of the elephant. **FIGURE 9-13 (c)** illustrates this case with air carts.

ANSWER

(b) The speed of the fly is $2v_0$.

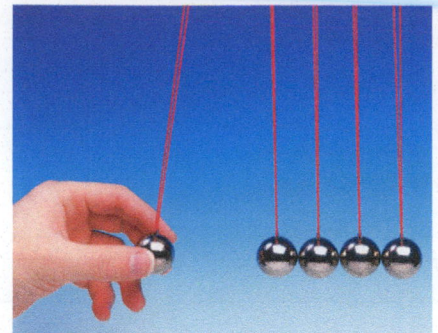 

▲ **FIGURE 9-14** This device, known as a Newton's cradle, illustrates some of the basic features of elastic collisions between objects of equal mass. When the end ball is pulled out to the side and then released, it creates a rapid succession of elastic collisions among the balls. In each collision, one ball comes to rest while the next one begins to move with the original speed, just as with the air carts in Figure 9-13 (a). When the collisions reach the other end of the apparatus, the last ball swings out to the same height from which the first ball was released.

If two balls are pulled out and released, two balls swing out at the other side, and so on. To see why this must be so, imagine that the two balls swing in with a speed v and a single ball swings out at the other side with a speed v'. What value must v' have (a) to conserve momentum and (b) to conserve kinetic energy? Since the required speed is $v' = 2v$ for (a) and $v' = \sqrt{2}v$ for (b), it follows that it is not possible to conserve both momentum and kinetic energy with two balls swinging in and one ball swinging out.

After the collision discussed above, the fly is separating from the elephant with the speed $2v_0 - v_0 = v_0$. Before the collision, the elephant was approaching the fly with the same speed, v_0. This is a special case of the following general result:

> The speed of separation after a head-on elastic collision is equal to the speed of approach before the collision.

An excellent example of elastic collisions is provided by the familiar apparatus shown in **FIGURE 9-14**. Another example with a rather surprising result is provided in **FIGURE 9-15**.

Elastic Collisions in Two Dimensions

In a two-dimensional elastic collision, if we are given the final speed and direction of one of the objects, we can find the speed and direction of the other object using energy conservation and momentum conservation. For example, consider the collision of two 7.00-kg curling stones, as depicted in **FIGURE 9-16**. One stone is at rest initially; the other approaches with a speed $v_{1,i} = 1.50$ m/s. The collision is not head-on, and after the collision, stone 1 moves with a speed of $v_{1,f} = 0.610$ m/s in a direction 66.0° away from the initial line of motion. What are the speed and direction of stone 2?

First, let's find the speed of stone 2. The easiest way to do this is to simply require that the final kinetic energy be equal to the initial kinetic energy. Initially, the kinetic energy is

$$K_i = \tfrac{1}{2}m_1 v_{1,i}^2 = \tfrac{1}{2}(7.00 \text{ kg})(1.50 \text{ m/s})^2 = 7.88 \text{ J}$$

After the collision stone 1 has a speed of 0.610 m/s and stone 2 has the speed $v_{2,f}$. Hence, the final kinetic energy is

$$K_f = \tfrac{1}{2}m_1 v_{1,f}^2 + \tfrac{1}{2}m_2 v_{2,f}^2 = \tfrac{1}{2}(7.00 \text{ kg})(0.610 \text{ m/s})^2 + \tfrac{1}{2}m_2 v_{2,f}^2$$
$$= 1.30 \text{ J} + \tfrac{1}{2}m_2 v_{2,f}^2 = K_i$$

Solving for the speed of stone 2, we find

$$v_{2,f} = 1.37 \text{ m/s}$$

Next, we can find the direction of motion of stone 2 by requiring that the momentum be conserved. For example, initially there is no momentum in the y direction. This must be true after the collision as well. Hence, we have the following condition:

$$0 = m_1 v_{1,f} \sin 66.0° - m_2 v_{2,f} \sin \theta$$

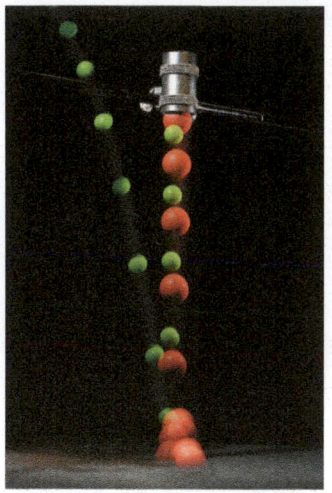

▲ **FIGURE 9-15 Visualizing Concepts**
Height Amplification In a collision between two objects of different mass, like the small and large balls in this photo, a significant amount of momentum can be transferred from the large object to the small object. Even though the total momentum is conserved, the small object can be given a speed that is significantly higher than any of the initial speeds. This is illustrated in the photo by the height to which the small ball bounces. A similar process occurs in the collapse of a star during a supernova explosion. The resulting collision can send jets of material racing away from the supernova at nearly the speed of light, just like the small ball that takes off with such a high speed in this collision.

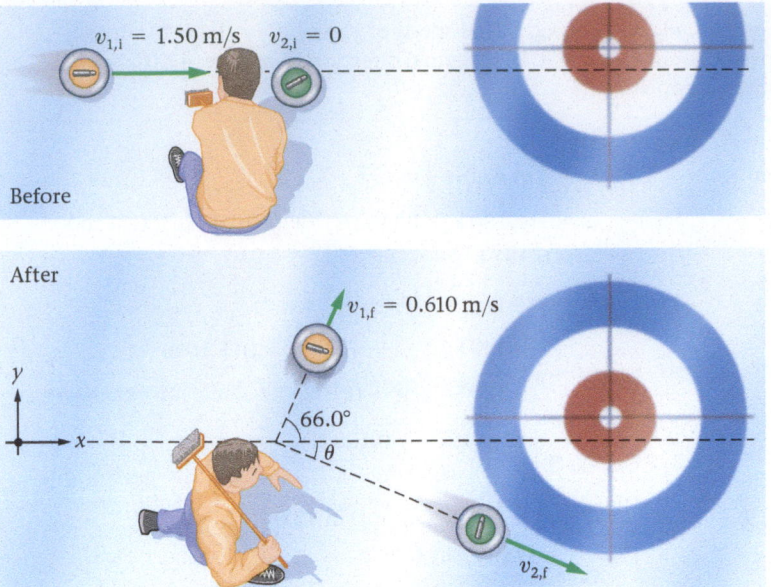

◄ **FIGURE 9-16 Two curling stones undergo an elastic collision** The speed of curling stone 2 after this collision can be determined using energy conservation; its direction of motion can be found using momentum conservation in either the x or the y direction.

Solving for the angle θ we find

$$\theta = 24.0°$$

Conserving momentum in the x direction gives the same result.

EXAMPLE 9-16   TWO FRUITS IN TWO DIMENSIONS: ANALYZING AN ELASTIC COLLISION

Two astronauts on opposite ends of a spaceship are comparing lunches. One has an apple, the other has an orange. They decide to trade. Astronaut 1 tosses the 0.130-kg apple toward astronaut 2 with a speed of 1.11 m/s. The 0.160-kg orange is tossed from astronaut 2 to astronaut 1 with a speed of 1.21 m/s. Unfortunately, the fruits collide, sending the orange off with a speed of 1.16 m/s at an angle of 42.0° with respect to its original direction of motion. Find the final speed and direction of the apple, assuming an elastic collision. Give the apple's direction relative to its original direction of motion.

PICTURE THE PROBLEM

In our sketch we label the apple as object 1 and the orange as object 2. We also choose the positive x direction to be in the initial direction of motion of the apple. We shall describe the "Before" and "After" sketches separately:

BEFORE

Initially, the apple moves in the positive x direction with a speed of 1.11 m/s, and the orange moves in the negative x direction with a speed of 1.21 m/s. There is no momentum in the y direction before the collision.

AFTER

After the collision, the orange moves with a speed of 1.16 m/s in a direction 42° below the negative x axis. As a result, the orange now has momentum in the negative y direction. To cancel this y momentum, the apple must move in a direction that is above the positive x axis, as indicated in the sketch.

REASONING AND STRATEGY

As described in the text, we first find the speed of the apple by demanding that the initial and final kinetic energies be the same. Next, we find the angle θ by conserving momentum in either the x or the y direction—the results are the same whichever direction is chosen.

CONTINUED

Known   Mass of apple, $m_1 = 0.130$ kg; mass of orange, $m_2 = 0.160$ kg; initial speed of apple, $v_{1,i} = 1.11$ m/s; initial speed of orange, $v_{2,i} = 1.21$ m/s; final speed of orange, $v_{2,f} = 1.16$ m/s; final direction of orange, 42.0°.

Unknown   Final speed of apple, $v_{1,f} = ?$ Final direction of apple, $\theta = ?$

SOLUTION

1. Calculate the initial kinetic energy of the system:

$$K_i = \tfrac{1}{2}m_1v_{1,i}^2 + \tfrac{1}{2}m_2v_{2,i}^2$$
$$= \tfrac{1}{2}(0.130\text{ kg})(1.11\text{ m/s})^2 + \tfrac{1}{2}(0.160\text{ kg})(1.21\text{ m/s})^2$$
$$= 0.197\text{ J}$$

2. Calculate the final kinetic energy of the system in terms of $v_{1,f}$:

$$K_f = \tfrac{1}{2}m_1v_{1,f}^2 + \tfrac{1}{2}m_2v_{2,f}^2$$
$$= \tfrac{1}{2}(0.130\text{ kg})v_{1,f}^2 + \tfrac{1}{2}(0.160\text{ kg})(1.16\text{ m/s})^2$$
$$= \tfrac{1}{2}(0.130\text{ kg})v_{1,f}^2 + 0.108\text{ J}$$

3. Set $K_f = K_i$ to find $v_{1,f}$:

$$v_{1,f} = \sqrt{\frac{2(0.197\text{ J} - 0.108\text{ J})}{0.130\text{ kg}}} = 1.17\text{ m/s}$$

4. Set the final y component of momentum equal to zero and solve for $\sin\theta$:

$$0 = m_1v_{1,f}\sin\theta - m_2v_{2,f}\sin 42.0°$$
$$\sin\theta = \frac{m_2v_{2,f}\sin 42.0°}{m_1v_{1,f}}$$

5. Substitute numerical values:

$$\sin\theta = \frac{(0.160\text{ kg})(1.16\text{ m/s})\sin 42.0°}{(0.130\text{ kg})(1.17\text{ m/s})} = 0.817$$
$$\theta = \sin^{-1}(0.817) = 54.8°$$

INSIGHT

The x momentum equation gives the same value for θ, as expected.

PRACTICE PROBLEM

Suppose that after the collision the apple moves in the positive y direction with a speed of 1.27 m/s. What are the final speed and direction of the orange in this case?   [**Answer:** The orange moves with a speed of 1.07 m/s in a direction 74.6° below the negative x axis.]

Some related homework problems: Problem 35, Problem 40

Enhance Your Understanding   (Answers given at the end of the chapter)

6. The 1-kg block in **FIGURE 9-17** slides to the right on a frictionless surface and undergoes an elastic collision with a 2-kg block that is initially at rest. **(a)** What is the direction of travel of the 1-kg block after the collision? **(b)** Which block has the greater speed after the collision?

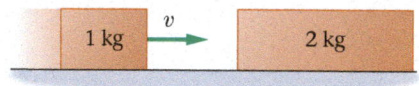

▲ **FIGURE 9-17** Enhance Your Understanding 6.

Section Review

- In elastic collisions, the final kinetic energy of a system is equal to the initial kinetic energy of the system.

- In a one-dimensional collision between a mass m_1 with an initial velocity of v_0, and a mass m_2 at rest, the final velocities are

$$v_{1,f} = \left(\frac{m_1 - m_2}{m_1 + m_2}\right)v_0 \quad \text{and} \quad v_{2,f} = \left(\frac{2m_1}{m_1 + m_2}\right)v_0.$$

- In a two-dimensional collision, kinetic energy is conserved, and so are the x and y components of momentum.

9-7 Center of Mass

In this section we introduce the concept of the center of mass. We begin by defining its location for a given system of masses. Next we consider the motion of the center of mass, and show how it is related to the net external force acting on the system.

Location of the Center of Mass

There is one point in any system of objects that has special significance—the **center of mass (CM)**. One of the reasons the center of mass is so special is that, in many ways, a system behaves as if all of its mass were concentrated there. As a result, a system can be balanced at its center of mass:

> The center of mass of a system of masses is the point where the system can be balanced in a uniform gravitational field.

For example, suppose you are making a mobile. At one stage in its construction, you want to balance a light rod with objects of mass m_1 and m_2 connected to either end, as indicated in **FIGURE 9-18**. To make the rod balance, you should attach a string to the center of mass of the system, just as if all its mass were concentrated at that point. In a sense, you can think of the center of mass as the "average" location of the system's mass. An elaborate example of a balanced mobile is shown in **FIGURE 9-19**.

To be more specific, suppose the two objects connected to the rod have the same mass. In this case the center of mass is at the midpoint of the rod, because this is where it balances. On the other hand, if one object has more mass than the other, the center of mass is closer to the heavier object, as indicated in **FIGURE 9-20**. In general, if a mass m_1 is on the x axis at the position x_1, and a mass m_2 is at the position x_2, as in Figure 9-18, then the location of the center of mass, X_{cm}, is defined as the "weighted" average of the two positions:

Center of Mass for Two Objects

$$X_{cm} = \frac{m_1 x_1 + m_2 x_2}{m_1 + m_2} = \frac{m_1 x_1 + m_2 x_2}{M} \qquad \text{9-14}$$

Notice that we have used $M = m_1 + m_2$ for the total mass of the two objects, and that the two positions, x_1 and x_2, are multiplied—or weighted—by their respective masses.

To see that this definition of X_{cm} agrees with our expectations, consider first the case where the masses are equal: $m_1 = m_2 = m$. In this case, $M = m_1 + m_2 = 2m$ and $X_{cm} = (mx_1 + mx_2)/2m = \frac{1}{2}(x_1 + x_2)$. Thus, as expected, if two masses are equal, their center of mass is halfway between them. On the other hand, if one mass is larger than the other, the center of mass is closer to the larger mass.

EXERCISE 9-17 CENTER OF MASS

Suppose that the masses in Figure 9-18 are separated by 0.800 m, and that $m_1 = 0.325$ kg and $m_2 = 0.166$ kg. What is the distance from m_1 to the center of mass of the system?

REASONING AND SOLUTION
Let $x_1 = 0$ and $x_2 = 0.800$ m in Figure 9-18. This yields

$$X_{cm} = \frac{m_1 x_1 + m_2 x_2}{m_1 + m_2} = \frac{(0.325 \text{ kg}) \cdot 0 + (0.166 \text{ kg})(0.800 \text{ m})}{0.325 \text{ kg} + 0.166 \text{ kg}} = 0.270 \text{ m}$$

Thus, the center of mass is 0.270 m from x_1, the larger of the two masses. Because the midpoint between the masses is 0.400 m from x_1, it follows that the center of mass is closer to the larger mass, as expected.

Center of Mass for Multiple or Continuous Objects in Two Dimensions
To extend the definition of X_{cm} to more general situations, first consider a system that contains many objects, not just two. In that case, X_{cm} is the sum of m times x for each object,

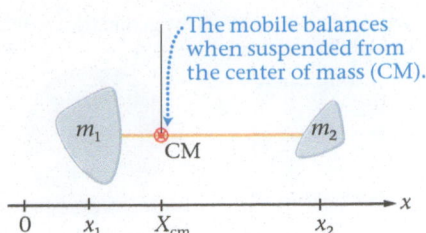

▲ **FIGURE 9-18 Balancing a mobile** Consider a portion of a mobile with masses m_1 and m_2 at the locations x_1 and x_2, respectively. The object balances when a string is attached at the center of mass. Since the center of mass is closer to m_1 than to m_2, it follows that m_1 is greater than m_2.

▲ **FIGURE 9-19** This mobile by Alexander Calder illustrates the concept of center of mass with artistic flair. Each arm of the mobile is in balance because it is suspended at its center of mass.

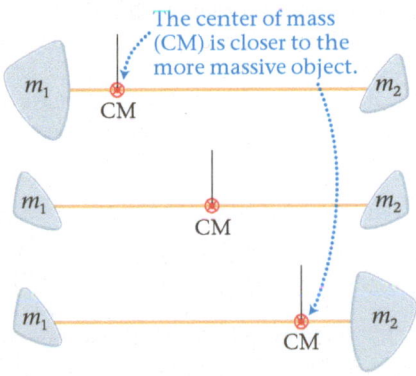

▲ **FIGURE 9-20 The center of mass of two objects** The center of mass is closer to the larger mass, or equidistant between the masses if they are equal.

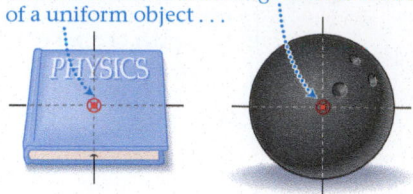

The center of mass is at the geometric center of a uniform object . . .

. . . even if there is no mass at that location.

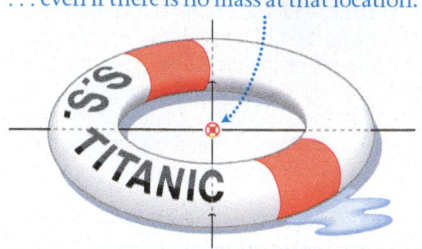

▲ **FIGURE 9-21 Locating the center of mass** In an object of continuous, uniform mass distribution, the center of mass is located at the geometric center of the object. In some cases, this means that the center of mass is not located within the object.

divided by the total mass of the system, M. If, in addition, the objects in the system are not in a line, but are distributed in two dimensions, the center of mass has both an x coordinate, X_{cm}, and a y coordinate, Y_{cm}. As one would expect, Y_{cm} is simply the sum of m times y for each object, divided by M. Thus, the x coordinate of the center of mass is

X Coordinate of the Center of Mass

$$X_{cm} = \frac{m_1 x_1 + m_2 x_2 + \cdots}{m_1 + m_2 + \cdots} = \frac{\sum mx}{M}$$

9-15

Similarly, the y coordinate of the center of mass is

Y Coordinate of the Center of Mass

$$Y_{cm} = \frac{m_1 y_1 + m_2 y_2 + \cdots}{m_1 + m_2 + \cdots} = \frac{\sum my}{M}$$

9-16

In systems with a continuous, uniform distribution of mass, the center of mass is at the geometric center of the object, as illustrated in **FIGURE 9-21**. Notice that it is common for the center of mass to be located in a position where no mass exists, as in a life preserver, where the center of mass is precisely in the center of the hole.

EXAMPLE 9-18 CENTER OF MASS OF THE ARM

RWP A person's arm is held with the upper arm vertical, the lower arm and hand horizontal. **(a)** Find the center of mass of the arm in this configuration, given the following data: The upper arm has a mass of 2.5 kg and a center of mass 0.18 m above the elbow; the lower arm has a mass of 1.6 kg and a center of mass 0.12 m to the right of the elbow; the hand has a mass of 0.64 kg and a center of mass 0.40 m to the right of the elbow. **(b)** A 0.14-kg baseball is placed on the palm of the hand. If the radius of the ball is 3.7 cm, find the center of mass of the arm–ball system.

PICTURE THE PROBLEM

We place the origin at the elbow, with the x and y axes pointing along the lower and upper arms, respectively. In the sketch, the center of mass of each of the three parts of the arm is indicated by an x; the center of mass of the entire arm is at the point labeled CM. The inset shows the baseball on the palm of the hand.

REASONING AND STRATEGY

a. Using the information given in the problem statement, we can treat the arm as a system of three point masses placed as follows: 2.5 kg at (0, 0.18 m); 1.6 kg at (0.12 m, 0); 0.64 kg at (0.40 m, 0). We substitute these masses and locations into Equations 9-15 and 9-16 to find the x and y coordinates of the center of mass, respectively.

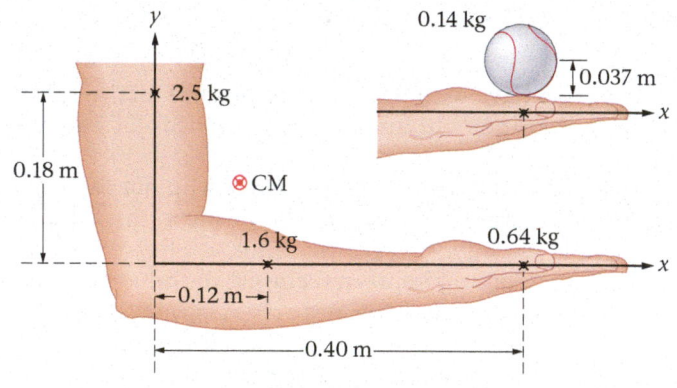

b. Treat the center of mass found in part (a) as a point particle with a mass 2.5 kg + 1.6 kg + 0.64 kg = 4.7 kg at the location (X_{cm}, Y_{cm}). The baseball can be treated as a point particle of mass 0.14 kg at the location (0.40 m, 0.037 m).

Known   Mass and location of parts of the arm and baseball given in drawing.
Unknown   **(a)** $X_{cm} = ?$ and $Y_{cm} = ?$ for arm **(b)** $X_{cm} = ?$ and $Y_{cm} = ?$ for arm–ball system

SOLUTION

Part (a)

1. Calculate the x coordinate of the center of mass:

$$X_{cm} = \frac{(2.5\ \text{kg})(0) + (1.6\ \text{kg})(0.12\ \text{m}) + (0.64\ \text{kg})(0.40\ \text{m})}{2.5\ \text{kg} + 1.6\ \text{kg} + 0.64\ \text{kg}}$$

$$= 0.095\ \text{m}$$

2. Do the same calculation for the y coordinate of the center of mass:

$$Y_{cm} = \frac{(2.5\ \text{kg})(0.18\ \text{m}) + (1.6\ \text{kg})(0) + (0.64\ \text{kg})(0)}{2.5\ \text{kg} + 1.6\ \text{kg} + 0.64\ \text{kg}}$$

$$= 0.095\ \text{m}$$

CONTINUED

Part (b)

3. Calculate the new x coordinate of the center of mass:

$$X_{cm} = \frac{(4.7 \text{ kg})(0.095 \text{ m}) + (0.14 \text{ kg})(0.40 \text{ m})}{4.7 \text{ kg} + 0.14 \text{ kg}}$$

$$= 0.10 \text{ m}$$

4. Calculate the new y coordinate of the center of mass:

$$Y_{cm} = \frac{(4.7 \text{ kg})(0.095 \text{ m}) + (0.14 \text{ kg})(0.037 \text{ m})}{4.7 \text{ kg} + 0.14 \text{ kg}}$$

$$= 0.093 \text{ m}$$

INSIGHT

As is often the case, the center of mass of an arm held in this position is in a location where no mass exists—you might say the center of mass is having an out-of-body experience. This effect can sometimes be put to good use, as when the center of mass of a high jumper passes beneath the horizontal bar while the body passes above it. (See Conceptual Question 16 for a photo of this technique in action, in the famous "Fosbury flop.")

PRACTICE PROBLEM — PREDICT/CALCULATE

Suppose the mass of the baseball is increased to 0.25 kg. **(a)** Does X_{cm} increase, decrease, or stay the same? **(b)** Does Y_{cm} increase, decrease, or stay the same? **(c)** Check your answers to parts (a) and (b) by finding the center of mass of the arm–ball system in this case.   **[Answer: (a)** The ball adds mass at a large value of x, and hence X_{cm} increases. **(b)** The ball adds mass at a small value of y, and hence Y_{cm} decreases. **(c)** $X_{cm} = 0.11$ m, $Y_{cm} = 0.092$ m]

Some related homework problems: Problem 47, Problem 48, Problem 50

Motion of the Center of Mass

Another reason the center of mass is important is that its motion often displays a remarkable simplicity when compared with the motion of other parts of a system. To analyze this motion, we consider both the velocity and acceleration of the center of mass. Each of these quantities is defined in analogy with the definition of the center of mass itself.

For example, to find the velocity of the center of mass, we first multiply the mass of each object in a system, m, by its velocity, $\vec{v}$, to give $m_1\vec{v}_1, m_2\vec{v}_2$, and so on. Next, we add all these products together, $m_1\vec{v}_1 + m_2\vec{v}_2 + \cdots$, and divide by the total mass, $M = m_1 + m_2 + \cdots$. The result, by definition, is the velocity of the center of mass, $\vec{V}_{cm}$:

Velocity of the Center of Mass

$$\vec{V}_{cm} = \frac{m_1\vec{v}_1 + m_2\vec{v}_2 + \cdots}{m_1 + m_2 + \cdots} = \frac{\sum m\vec{v}}{M} \qquad \text{9-17}$$

Comparing with Equation 9-15, we see that $\vec{V}_{cm}$ is the same as X_{cm} with each position x replaced with a velocity vector $\vec{v}$. In addition, note that the total mass of the system, M, times the velocity of the center of mass, $\vec{V}_{cm}$, is simply the total momentum of the system:

$$M\vec{V}_{cm} = m_1\vec{v}_1 + m_2\vec{v}_2 + \cdots = \vec{p}_1 + \vec{p}_2 + \cdots = \vec{p}_{total}$$

To gain more information on how the center of mass moves, we next consider its acceleration, $\vec{A}_{cm}$. As expected by analogy with $\vec{V}_{cm}$, the acceleration of the center of mass is defined as follows:

Acceleration of the Center of Mass

$$\vec{A}_{cm} = \frac{m_1\vec{a}_1 + m_2\vec{a}_2 + \cdots}{m_1 + m_2 + \cdots} = \frac{\sum m\vec{a}}{M} \qquad \text{9-18}$$

The vector $\vec{A}_{cm}$ contains the terms $m_1\vec{a}_1, m_2\vec{a}_2$, and so on, for each object in the system. From Newton's second law, however, we know that $m_1\vec{a}_1$ is simply $\vec{F}_1$, the net force acting on mass 1. The same conclusion applies to each of the masses. Therefore, we find that the total mass of the system, M, times the acceleration of the center of mass, $\vec{\mathbf{A}}_{cm}$, is simply the total force acting on the system:

$$M\vec{\mathbf{A}}_{cm} = m_1\vec{\mathbf{a}}_1 + m_2\vec{\mathbf{a}}_2 + \cdots = \vec{\mathbf{F}}_1 + \vec{\mathbf{F}}_2 + \cdots = \vec{\mathbf{F}}_{total}$$

Recall, however, that the total force acting on a system is the same as the net external force, $\vec{\mathbf{F}}_{net,ext}$, because the internal forces cancel. Therefore, $M\vec{\mathbf{A}}_{cm}$ is the net external force acting on the system:

Newton's Second Law for a System of Particles

$$M\vec{\mathbf{A}}_{cm} = \vec{\mathbf{F}}_{net,ext}$$

9-19

Zero Net External Force For systems in which $\vec{\mathbf{F}}_{net,ext}$ is zero, it follows that the acceleration of the center of mass is zero. Hence, if the center of mass is initially at rest, it remains at rest. Similarly, if the center of mass is moving initially, it continues to move with the same velocity. For example, in a collision between two air-track carts, the velocity of each cart changes as a result of the collision. The velocity of the center of mass of the two carts, however, is the same before and after the collision. We explore this situation in the following Example.

EXAMPLE 9-19   CRASH OF THE AIR CARTS

An air cart with a mass m and speed v_0 moves toward a second, identical air cart that is at rest. When the carts collide they stick together and move as one. Find the velocity of the center of mass of this system **(a)** before and **(b)** after the carts collide.

PICTURE THE PROBLEM
We choose the positive x direction to be the direction of motion of the incoming cart, whose initial speed is v_0. Notice that the carts have wads of putty on their bumpers; this ensures that they stick together when they collide and thereafter move as a unit. Their final speed is v_f.

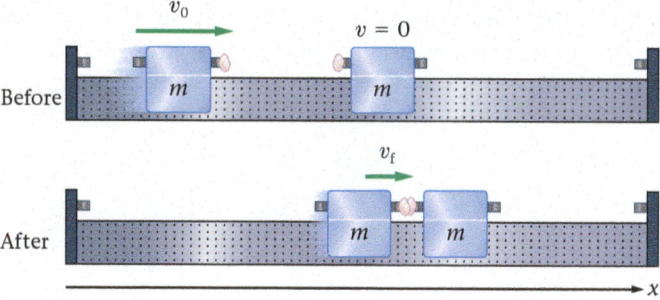

REASONING AND STRATEGY

a. We can find the velocity of the center of mass by applying Equation 9-17 to the case of just two masses; $\vec{\mathbf{V}}_{cm} = (m_1\vec{\mathbf{v}}_1 + m_2\vec{\mathbf{v}}_2)/M$. In this case, $\vec{\mathbf{v}}_1 = v_0\hat{\mathbf{x}}$, $\vec{\mathbf{v}}_2 = 0$, and $m_1 = m_2 = m$.

b. After the collision the two masses move with the same velocity, $\vec{\mathbf{v}}_f = v_f\hat{\mathbf{x}}$, which is given by momentum conservation (Equation 9-11, $v_f = (m_1v_{1,i} + m_2v_{2,i})/(m_1 + m_2)$). Hence, $\vec{\mathbf{V}}_{cm} = (m_1\vec{\mathbf{v}}_f + m_2\vec{\mathbf{v}}_f)/M$.

Known   Mass of air carts, m; initial speed of one air cart, v_0; initial speed of other air cart, $v = 0$.
Unknown Velocity of center of mass, $\vec{\mathbf{V}}_{cm} = ?$ **(a)** before and **(b)** after collision

---

SOLUTION

Part (a)

1. Use $\vec{\mathbf{V}}_{cm} = (m_1\vec{\mathbf{v}}_1 + m_2\vec{\mathbf{v}}_2)/M$ to find the velocity of the center of mass before the collision:

$$\vec{\mathbf{V}}_{cm} = \frac{(m_1\vec{\mathbf{v}}_1 + m_2\vec{\mathbf{v}}_2)}{m_1 + m_2} = \frac{(mv_0\hat{\mathbf{x}} + m\cdot 0)}{m + m} = \tfrac{1}{2}v_0\hat{\mathbf{x}}$$

Part (b)

2. Use momentum conservation in the x direction to find the speed of the carts after the collision:

$$mv_0 = mv_f + mv_f$$
$$v_f = \tfrac{1}{2}v_0$$

3. Calculate the velocity of the center of mass of the two carts after the collision:

$$\vec{\mathbf{V}}_{cm} = \frac{(m_1\vec{\mathbf{v}}_1 + m_2\vec{\mathbf{v}}_2)}{m_1 + m_2} = \frac{(mv_f\hat{\mathbf{x}} + mv_f\hat{\mathbf{x}})}{m + m} = v_f\hat{\mathbf{x}} = \tfrac{1}{2}v_0\hat{\mathbf{x}}$$

CONTINUED

INSIGHT

As expected, the velocity of each cart changes when they collide: The velocity of cart 1 changes from v_0 to $\frac{1}{2}v_0$, and the velocity of cart 2 changes from 0 to $\frac{1}{2}v_0$. On the other hand, the velocity of the center of mass is completely unaffected by the collision.

This finding is illustrated in the figure, where we show a sequence of equal-time snapshots of the system just before and just after the collision. First, we note that the incoming cart moves two distance units for every time interval until it collides with the second cart. From that point on, the two carts are locked together, and move one distance unit per time interval. In contrast, the center of mass (CM), which is centered between the two equal-mass carts, progresses uniformly throughout the sequence, advancing one unit of distance for each time interval.

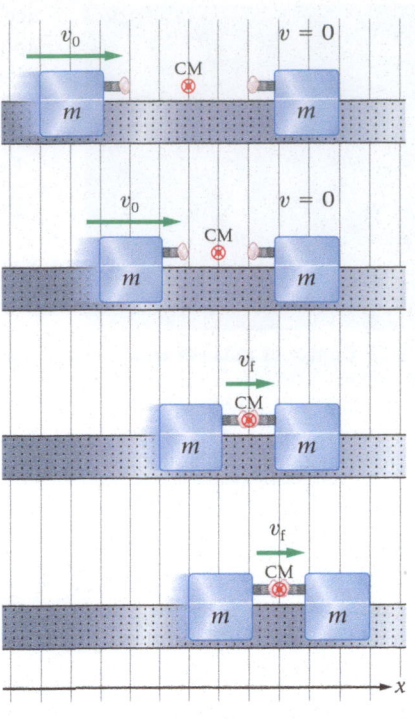

PRACTICE PROBLEM — PREDICT/CALCULATE

(a) If the mass of the cart that is moving initially is doubled to $2m$, does the velocity of the center of mass increase, decrease, or stay the same? Explain.
(b) Verify your answer to part (a) by calculating the velocity of the center of mass in this case.   [**Answer: (a)** More mass is moving, and hence V_{cm} increases. **(b)** The velocity of the center of mass is $\vec{V}_{cm} = (2v_0/3)\hat{x}$, both before and after the collision.]

Some related homework problems: Problem 53, Problem 64, Problem 73

Nonzero Net External Force   Recall that Newton's second law, as applied to the center of mass, states that the acceleration of the center of mass is related to the net external force as follows:

$$M\vec{A}_{cm} = \vec{F}_{net,ext}$$

This is completely analogous to the relationship between the acceleration of an object of mass m and the net force $\vec{F}_{net}$ applied to it:

$$m\vec{a} = \vec{F}_{net}$$

Therefore, when $\vec{F}_{net,ext}$ is nonzero, we can conclude the following:

> The center of mass of a system accelerates precisely as if it were a point particle of mass M acted on by the force $\vec{F}_{net,ext}$.

Big Idea **5** The center of mass of a system behaves as if all the mass were concentrated there.

For this reason, the motion of the center of mass can be quite simple compared to the motion of its constituent parts. For example, a hammer tossed into the air with a rotation is shown in **FIGURE 9-22**. The motion of one part of the hammer—the yellow tip of the handle, let's say—follows a complicated path in space. On the other hand, the path of the center of mass (red dot) is a simple parabola, precisely the same path that a point mass would follow.

RWP Similarly, consider a fireworks rocket launched into the sky, as illustrated in **FIGURE 9-23**. The center of mass of the rocket follows a parabolic path, if air resistance is ignored. At some point in its path it explodes into numerous individual pieces. The explosion is due to internal forces, however, which must therefore sum to zero. Hence, the net external force acting on the pieces of the rocket is the same before, during, and after the explosion. As a result, the center of mass has a constant downward

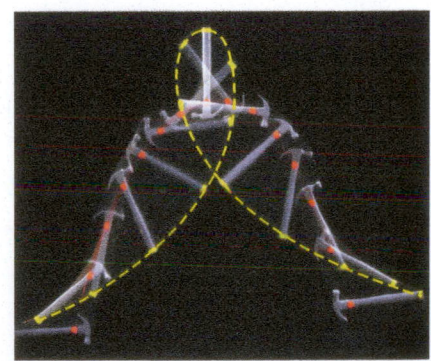

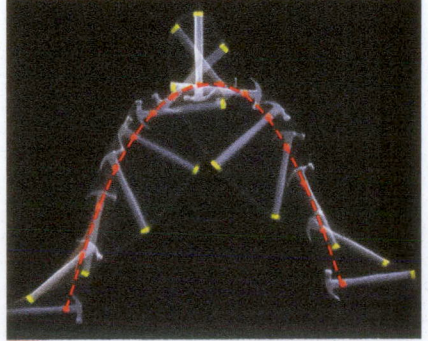

▶ **FIGURE 9-22 Simple Motion of the Center of Mass** As this hammer flies through the air, its motion is quite complex. Some parts of the hammer follow wild trajectories with strange loops and turns (yellow path on the top). There is one point on the hammer, however, that travels on a smooth, simple parabolic path—the center of mass. The center of mass (red path on the bottom) travels as if all the mass of the hammer were concentrated there.

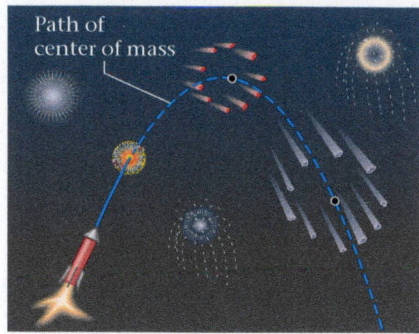

▲ **FIGURE 9-23 Center of mass of an exploding rocket** A fireworks rocket follows a parabolic path, ignoring air resistance, until it explodes. After exploding, its center of mass continues on the same parabolic path until some of the fragments start to land.

acceleration and continues to follow the original parabolic path. It is only when an additional external force acts on the system, as when one of the pieces of the rocket hits the ground, that the path of the center of mass changes.

7. A piece of sheet metal of mass M is cut into the shape of a right triangle, as shown in **FIGURE 9-24**.

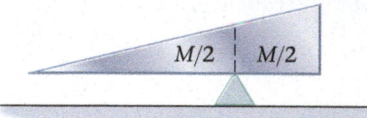

▲ **FIGURE 9-24** Enhance Your Understanding 7.

A vertical dashed line is drawn on the sheet at the point where the mass on either side of the line is $M/2$. The sheet is now placed on a fulcrum just under the dashed line and released from rest. **(a)** Does the metal sheet remain level, rotate counterclockwise (left end down), or rotate clockwise (right end down)? **(b)** Choose the *best explanation* for your prediction from among the following:

   I. Equal mass on either side will keep the metal sheet level.
  II. The metal sheet extends for a greater distance to the left, which shifts the center of mass to the left of the dashed line.
 III. The center of mass is to the right of the dashed line because the sheet of metal is thicker there.

Section Review

- The location of the center of mass is given by $X_{cm} = \dfrac{m_1 x_1 + m_2 x_2 + \cdots}{m_1 + m_2 + \cdots} = \dfrac{\sum mx}{M}$ and $Y_{cm} = \dfrac{m_1 y_1 + m_2 y_2 + \cdots}{m_1 + m_2 + \cdots} = \dfrac{\sum my}{M}$.

- The velocity of the center of mass is defined as $\vec{\mathbf{V}}_{cm} = \dfrac{m_1 \vec{\mathbf{v}}_1 + m_2 \vec{\mathbf{v}}_2 + \cdots}{m_1 + m_2 + \cdots} = \dfrac{\sum m\vec{\mathbf{v}}}{M}$.

- The acceleration of the center of mass is defined as $\vec{\mathbf{A}}_{cm} = \dfrac{m_1 \vec{\mathbf{a}}_1 + m_2 \vec{\mathbf{a}}_2 + \cdots}{m_1 + m_2 + \cdots} = \dfrac{\sum m\vec{\mathbf{a}}}{M}$. It is related to the net external force as follows: $M\vec{\mathbf{A}}_{cm} = \vec{\mathbf{F}}_{net,ext}$.

9-8 Systems with Changing Mass: Rocket Propulsion

We close this chapter with systems whose mass does change. The mass of a rocket, for example, changes as its engines operate because it ejects part of the fuel as it burns. The burning process is produced by internal forces; hence the total momentum of the rocket and its fuel remains constant.

Consider, then, a rocket at rest in outer space, far from any large, massive objects. When the rocket's engine is fired, it expels a certain mass of fuel out the back with a speed v. If the mass of the ejected fuel is Δm, then the momentum of the ejected fuel has a magnitude equal to $(\Delta m)v$. The total momentum of the system must still be zero, and therefore the rocket acquires an equivalent amount of momentum in the forward direction. Hence, the momentum of the rocket increases by the amount

$$\Delta p = (\Delta m)v$$

If this increase in momentum occurs in the time Δt, the force exerted on the rocket is the change in its momentum divided by the time interval (Equation 9-4); that is,

$$F = \frac{\Delta p}{\Delta t} = \left(\frac{\Delta m}{\Delta t}\right)v$$

Water Rocket

The force exerted on the rocket by the ejected fuel is referred to as **thrust**:

Thrust

$$\text{thrust} = \left(\frac{\Delta m}{\Delta t}\right)v \qquad \qquad \text{9-20}$$

SI unit: newton, N

RWP By $\Delta m/\Delta t$, we simply mean the amount of mass per time coming out of the rocket. For example, on the Delta IV Heavy rocket, among the most powerful rockets in operation today, the main engines eject fuel at the rate of 2510 kg/s with a speed of 3710 m/s. As a result, the thrust produced by these engines is

$$\text{thrust} = \left(\frac{\Delta m}{\Delta t}\right)v = (2510 \text{ kg/s})(3710 \text{ m/s}) = 9.31 \times 10^6 \text{ N}$$

This is equivalent to 2.09 million pounds, a small amount compared to the 7.60 million pounds generated by the Saturn V rocket that was used for the manned missions to the Moon. However, because the Delta IV Heavy rocket and its cargo weigh only 1.68 million pounds, there is sufficient thrust to give the rocket an upward acceleration. The acceleration increases as the rocket's mass decreases due to the expelled fuel. Real-world examples of rocket propulsion are shown in FIGURE 9-25.

EXERCISE 9-20   LEAVING THE MOON

The ascent stage of the lunar lander was designed to produce 15,600 N of thrust at liftoff. If the fuel was burned at the rate of 6.40 kg/s, what was the speed of the ejected fuel?

REASONING AND SOLUTION
The thrust, fuel consumption rate ($\Delta m/\Delta t$), and speed of ejected fuel v, are related by thrust $= (\Delta m/\Delta t)v$. We can rearrange this equation to solve for the speed:

$$v = \frac{\text{thrust}}{(\Delta m/\Delta t)} = \frac{15,600 \text{ N}}{6.40 \text{ kg/s}} = 2440 \text{ m/s}$$

This is about seven times greater than the speed of sound on Earth.

A common question regarding rockets is, "How can a rocket accelerate in outer space when it has nothing to push against?" The answer is that rockets, in effect, push against their own fuel. The situation is similar to firing a gun. When a bullet is ejected by the internal combustion of the gunpowder, the person firing the gun feels a recoil. If the person were in space, or standing on a frictionless surface, the recoil would give him or her a speed in the direction opposite to the bullet. The burning of a rocket engine provides a continuous recoil, almost as if the rocket were firing a steady stream of bullets out the back.

▲ **FIGURE 9-25 Visualizing Concepts**
Rocket Propulsion A Delta Heavy IV rocket (top) makes use of the principle of conservation of momentum: Mass (the products of explosive burning of fuel) is ejected at high speed in one direction, causing the rocket to move in the opposite direction. The same method of propulsion has evolved in octopi (bottom) and some other animals. When danger threatens and a quick escape is needed, powerful muscles contract to create a jet of water that propels the animal to safety. (A smokescreen of ink provides additional security.)

Enhance Your Understanding   (Answers given at the end of the chapter)

8. You place a bucket on a scale and pour water into it at a constant rate. When the weight of the bucket and the water in it is 20 N, and you're still pouring water into the bucket, is the reading on the scale greater than, less than, or equal to 20 N? Explain.

Section Review

• The thrust (force) produced by a rocket engine is thrust $= (\Delta m/\Delta t)v$.

CHAPTER 9 REVIEW

CHAPTER SUMMARY

9-1 LINEAR MOMENTUM

The linear momentum of an object of mass m moving with velocity $\vec{v}$ is

$$\vec{p} = m\vec{v} \qquad\qquad 9\text{-}1$$

Momentum of a System of Objects
In a system of several objects, the total linear momentum is the vector sum of the individual momenta:

$$\vec{P}_{\text{total}} = \vec{P}_1 + \vec{P}_2 + \vec{P}_3 + \cdots \qquad\qquad 9\text{-}3$$

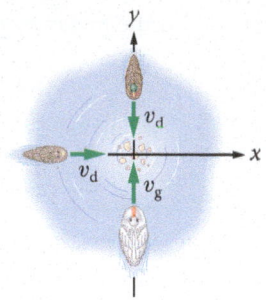

9-2 MOMENTUM AND NEWTON'S SECOND LAW

In terms of momentum, Newton's second law is

$$\sum \vec{F} = \frac{\Delta \vec{p}}{\Delta t} \qquad\qquad 9\text{-}4$$

9-3 IMPULSE

The impulse delivered to an object by an average force $\vec{F}_{\text{av}}$ acting for a time Δt is

$$\vec{I} = \vec{F}_{\text{av}}\, \Delta t \qquad\qquad 9\text{-}6$$

9-4 CONSERVATION OF LINEAR MOMENTUM

The momentum of an object is conserved (remains constant) if the net force acting on it is zero.

9-5 INELASTIC COLLISIONS

In an inelastic collision, the final kinetic energy is different from the initial kinetic energy. The kinetic energy is usually less after a collision, but it can also be more than the initial kinetic energy.

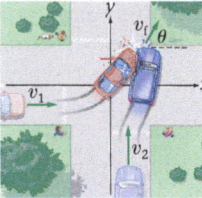

9-6 ELASTIC COLLISIONS

In an elastic collision, the final kinetic energy is equal to the initial kinetic energy.

9-7 CENTER OF MASS

The location of the center of mass of a two-dimensional system of objects is defined as follows:

$$X_{\text{cm}} = \frac{m_1 x_1 + m_2 x_2 + \cdots}{m_1 + m_2 + \cdots} = \frac{\sum mx}{M} \qquad\qquad 9\text{-}15$$

and

$$Y_{\text{cm}} = \frac{m_1 y_1 + m_2 y_2 + \cdots}{m_1 + m_2 + \cdots} = \frac{\sum my}{M} \qquad\qquad 9\text{-}16$$

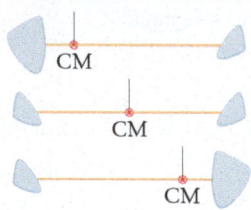

Motion of the Center of Mass
The center of mass accelerates as if the net external force acted on a single object of mass $M = m_1 + m_2 + \cdots$.

9-8 SYSTEMS WITH CHANGING MASS: ROCKET PROPULSION

The mass of a rocket changes because its engines expel fuel when they are fired. If fuel is expelled with the speed v and at the rate $\Delta m/\Delta t$, the thrust experienced by the rocket is

$$\text{thrust} = \left(\frac{\Delta m}{\Delta t}\right)v \qquad 9\text{-}20$$

ANSWERS TO ENHANCE YOUR UNDERSTANDING QUESTIONS

1. **(a)** The rubber bullet is more likely to topple the block. **(b)** The best explanation is I.

2. **(a)** The change in momentum is the same for the two objects because the force (momentum change per time) acting on them is the same. **(b)** The change in velocity is greater for the less massive object 1 because momentum is the product of mass and velocity.

3. The impulse is the same for the two balls because the force each experiences has the same magnitude (due to Newton's third law) and acts for the same amount of time.

4. **(a)** The total momentum must be zero after the explosion, and hence the momentum of piece 2 must have a magnitude p and point in a direction opposite to the momentum of piece 1. **(b)** The mass of piece 2 is $2m$, which, with $v_2 = \frac{1}{2}v_1$, satisfies both $p_1 = p_2$ and $K_1 = 2K_2$.

5. **(a)** The objects are at rest ($v = 0$) after the collision, because the total momentum of the system is zero. **(b)** 100% of the initial kinetic energy is converted to other forms, because the final kinetic energy is zero.

6. **(a)** The 1-kg mass rebounds from the 2-kg mass and travels to the left after the collision. **(b)** After the collision, the 2-kg mass has a greater speed ($2v/3$) than the 1-kg mass ($v/3$).

7. **(a)** The metal sheet will rotate counterclockwise (left end down). **(b)** The best explanation is II.

8. The reading is greater than 20 N because an additional upward force is needed to stop the water as it comes to rest in the bucket.

CONCEPTUAL QUESTIONS

For instructor-assigned homework, go to www.masteringphysics.com

(Answers to odd-numbered Conceptual Questions can be found in the back of the book.)

1. If you drop your keys, their momentum increases as they fall. Why is the momentum of the keys not conserved? Does this mean that the momentum of the universe increases as the keys fall? Explain.

2. By what factor does an object's kinetic energy change if its speed is doubled? By what factor does its momentum change?

3. A system of particles is known to have zero kinetic energy. What can you say about the momentum of the system?

4. A system of particles is known to have zero momentum. Does it follow that the kinetic energy of the system is also zero? Explain.

5. On a calm day you connect an electric fan to a battery on your sailboat and generate a breeze. Can the wind produced by the fan be used to power the sailboat? Explain.

6. Crash statistics show that it is safer to be riding in a heavy car in an accident than in a light car. Explain in terms of physical principles.

7. **(a)** As you approach a stoplight, you apply the brakes and bring your car to rest. What happened to your car's initial momentum? **(b)** When the light turns green, you accelerate until you reach cruising speed. What force was responsible for increasing your car's momentum?

8. An object at rest on a frictionless surface is struck by a second object. Is it possible for both objects to be at rest after the collision? Explain.

9. **(a)** Can two objects on a horizontal frictionless surface have a collision in which all the initial kinetic energy of the system is lost? Explain, and give a specific example if your answer is yes. **(b)** Can two such objects have a collision in which all the initial momentum of the system is lost? Explain, and give a specific example if your answer is yes.

10. Two cars collide at an intersection. If the cars do not stick together, can we conclude that their collision was elastic? Explain.

11. At the instant a bullet is fired from a gun, the bullet and the gun have equal and opposite momenta. Which object—the bullet or the gun—has the greater kinetic energy? Explain. How does your answer apply to the observation that it is safe to hold a gun while it is fired, whereas the bullet is deadly?

12. An hourglass is turned over, and the sand is allowed to pour from the upper half of the glass to the lower half. If the hourglass is resting on a scale, and the total mass of the hourglass and sand is M, describe the reading on the scale as the sand runs to the bottom.

13. In the classic movie *The Spirit of St. Louis*, Jimmy Stewart portrays Charles Lindbergh on his history-making transatlantic flight. Lindbergh is concerned about the weight of his fuel-laden airplane. As he flies over Newfoundland he notices a fly on the dashboard. Speaking to the fly, he wonders aloud, "Does the plane weigh less if you fly inside it as it's flying? Now that's an interesting question." What do you think?

14. A tall, slender drinking glass with a thin base is initially empty. **(a)** Where is the center of mass of the glass? **(b)** Suppose the glass is now filled slowly with water until it is completely full. Describe the position and motion of the center of mass during the filling process.

15. Lifting one foot into the air, you balance on the other foot. What can you say about the location of your center of mass?

16. In the "Fosbury flop" method of high jumping, named for the track and field star Dick Fosbury, an athlete's center of mass may pass under the bar while the athlete's body passes over the bar. Explain how this is possible.

The "Fosbury flop." (Conceptual Question 16)

PROBLEMS AND CONCEPTUAL EXERCISES

Answers to odd-numbered Problems and Conceptual Exercises can be found in the back of the book. **BIO** *identifies problems of biological or medical interest;* **CE** *indicates a conceptual exercise.* **Predict/Explain** *problems ask for two responses: (a) your prediction of a physical outcome, and (b) the best explanation among three provided; and* **Predict/Calculate** *problems ask for a prediction of a physical outcome, based on fundamental physics concepts, and follow that with a numerical calculation to verify the prediction. On all problems, bullets (•, ••, •••) indicate the level of difficulty.*

SECTION 9-1 LINEAR MOMENTUM

1. • What is the mass of a mallard duck whose speed is 8.9 m/s and whose momentum has a magnitude of 11 kg·m/s?

2. • **(a)** What is the magnitude of the momentum of a 0.0053-kg marble whose speed is 0.65 m/s? **(b)** What is the speed of a 0.132-kg baseball whose momentum has a magnitude of 3.28 kg·m/s?

3. • A 54-kg person walks due north with a speed of 1.2 m/s, and her 6.9-kg dog runs directly toward her, moving due south, with a speed of 1.7 m/s. What is the magnitude of the total momentum of this system?

4. •• A 26.2-kg dog is running northward at 2.70 m/s, while a 5.30-kg cat is running eastward at 3.04 m/s. Their 74.0-kg owner has the same momentum as the two pets taken together. Find the direction and magnitude of the owner's velocity.

5. •• **Predict/Calculate** Two air-track carts move toward one another on an air track. Cart 1 has a mass of 0.28 kg and a speed of 0.97 m/s. Cart 2 has a mass of 0.64 kg. **(a)** What speed must cart 2 have if the total momentum of the system is to be zero? **(b)** Since the momentum of the system is zero, does it follow that the kinetic energy of the system is also zero? **(c)** Verify your answer to part (b) by calculating the system's kinetic energy.

6. •• A 0.145-kg baseball is dropped from rest. If the magnitude of the baseball's momentum is 1.55 kg·m/s just before it lands on the ground, from what height was it dropped?

7. •• A 285-g ball falls vertically downward, hitting the floor with a speed of 2.5 m/s and rebounding upward with a speed of 2.0 m/s. **(a)** Find the magnitude of the change in the ball's momentum. **(b)** Find the change in the magnitude of the ball's momentum. **(c)** Which of the two quantities calculated in parts (a) and (b) is more directly related to the net force acting on the ball during its collision with the floor? Explain.

8. ••• Object 1 has a mass m_1 and a velocity $\vec{v}_1 = (2.80 \text{ m/s})\hat{x}$. Object 2 has a mass m_2 and a velocity $\vec{v}_2 = (3.10 \text{ m/s})\hat{y}$. The total momentum of these two objects has a magnitude of 17.6 kg·m/s and points in a direction 66.5° above the positive x axis. Find m_1 and m_2.

SECTION 9-3 IMPULSE

9. • **CE** Your car rolls slowly in a parking lot and bangs into the metal base of a light pole. In terms of safety, is it better for your collision with the light pole to be elastic, inelastic, or is the safety risk the same for either case? Explain.

10. • **CE** **Predict/Explain** A net force of 200 N acts on a 100-kg boulder, and a force of the same magnitude acts on a 100-g pebble. **(a)** Is the change of the boulder's momentum in one second greater than, less than, or equal to the change of the pebble's momentum in the same time period? **(b)** Choose the *best explanation* from among the following:
    I. The large mass of the boulder gives it the greater momentum.
    II. The force causes a much greater speed in the 100-g pebble, resulting in more momentum.
    III. Equal force means equal change in momentum for a given time.

11. • **CE** **Predict/Explain** Referring to the previous question, **(a)** is the change in the boulder's speed in one second greater than, less than, or equal to the change in speed of the pebble in the same

time period? **(b)** Choose the *best explanation* from among the following:
    I. The large mass of the boulder results in a small acceleration.
    II. The same force results in the same change in speed for a given time.
    III. Once the boulder gets moving it is harder to stop than the pebble.

12. • **CE** **Predict/Explain** Two identical cars, each traveling at 16 m/s, slam into a concrete wall and come to rest. In car A the air bag does not deploy and the driver hits the steering wheel; in car B the driver contacts the deployed air bag. **(a)** Is the impulse delivered by the steering wheel to driver A greater than, less than, or equal to the impulse delivered by the air bag to driver B? **(b)** Choose the *best explanation* from among the following:
    I. The larger force delivered by the steering wheel delivers a greater impulse.
    II. The longer interaction time with the air bag delivers a greater impulse.
    III. The change in momentum of each driver is the same.

13. • **CE** Force A has a magnitude F and acts for the time Δt, force B has a magnitude $2F$ and acts for the time $\Delta t/3$, force C has a magnitude $5F$ and acts for the time $\Delta t/10$, and force D has a magnitude $10F$ and acts for the time $\Delta t/100$. Rank these forces in order of increasing impulse. Indicate ties where appropriate.

14. • Find the magnitude of the impulse delivered to a soccer ball when a player kicks it with a force of 1250 N. Assume that the player's foot is in contact with the ball for 5.95×10^{-3} s.

15. • A 0.45-kg croquet ball is initially at rest on the grass. When the ball is struck by a mallet, the average force exerted on it is 270 N. If the ball's speed after being struck is 5.4 m/s, for what amount of time was the mallet in contact with the ball?

16. • When spiking a volleyball, a player changes the velocity of the ball from 4.2 m/s to −24 m/s along a certain direction. If the impulse delivered to the ball by the player is −9.3 kg·m/s, what is the mass of the volleyball?

17. •• **Force Platform** A force platform measures the horizontal force exerted on a person's foot by the ground. In **FIGURE 9-26** we see force-versus-time data taken by a force platform for a person who starts jogging at time $t = 0$. The impulse imparted to the jogger from $t = 0$ to $t = 0.6$ s is 198 N·s. **(a)** What is the average force exerted on the jogger from $t = 0$ to $t = 0.6$ s? **(b)** If the jogger's change in speed during this time period is 2.69 m/s, what is the jogger's mass?

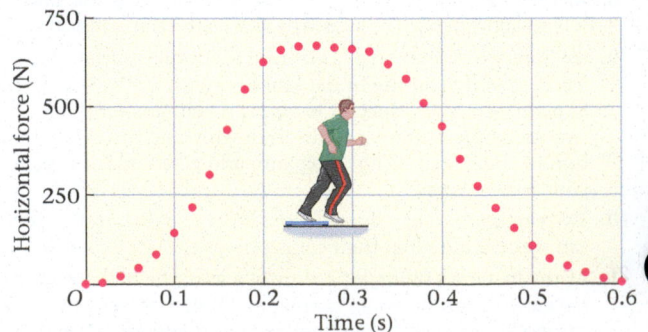

FIGURE 9-26 Problem 17

18. •• **Air Bag Safety** If a driver makes contact with a steering wheel during a 35-mph crash, she comes to rest in about $\Delta t = 15$ ms. If, during an identical crash, the driver makes contact with an air bag, she comes to rest in about $\Delta t = 110$ ms. What is the ratio F_{sw}/F_{ab} of the force exerted by the steering wheel to the force exerted by the air bag on the driver?

19. •• To make a bounce pass, a player throws a 0.60-kg basketball toward the floor. The ball hits the floor with a speed of 5.4 m/s at an angle of 65° to the vertical. If the ball rebounds with the same speed and angle, what was the impulse delivered to it by the floor?

20. •• **BIO Concussion Impulse** One study suggests that a brain concussion can occur if a person's head is subjected to an acceleration as small as $60g$, where g is the acceleration of gravity. **(a)** If the mass of the person's head is 4.8 kg, what is the magnitude of the force? **(b)** If the impulse delivered to the person's head is 18 kg · m/s, over what time interval is the force applied?

21. •• **Predict/Calculate** A 0.14-kg baseball moves toward home plate with a velocity $\vec{v}_i = (-36 \text{ m/s})\hat{x}$. After striking the bat, the ball moves vertically upward with a velocity $\vec{v}_f = (18 \text{ m/s})\hat{y}$. **(a)** Find the direction and magnitude of the impulse delivered to the ball by the bat. Assume that the ball and bat are in contact for 1.5 ms. **(b)** How would your answer to part (a) change if the mass of the ball were doubled? **(c)** How would your answer to part (a) change if the mass of the bat were doubled instead?

22. •• A player bounces a 0.43-kg soccer ball off her head, changing the velocity of the ball from $\vec{v}_i = (8.8 \text{ m/s})\hat{x} + (-2.3 \text{ m/s})\hat{y}$ to $\vec{v}_f = (5.2 \text{ m/s})\hat{x} + (3.7 \text{ m/s})\hat{y}$. If the ball is in contact with the player's head for 6.7 ms, what are **(a)** the direction and **(b)** the magnitude of the impulse delivered to the ball?

SECTION 9-4 CONSERVATION OF LINEAR MOMENTUM

23. • Two ice skaters stand at rest in the center of an ice rink. When they push off against one another the 45-kg skater acquires a speed of 0.62 m/s. If the speed of the other skater is 0.89 m/s, what is this skater's mass?

24. • A 0.042-kg pet lab mouse sits on a 0.35-kg air-track cart, as shown in **FIGURE 9-27**. The cart is at rest, as is a second cart with a mass of 0.25 kg. The lab mouse now jumps to the second cart. After the jump, the 0.35-kg cart has a speed of $v_1 = 0.88$ m/s. What is the speed v_2 of the mouse and the 0.25-kg cart?

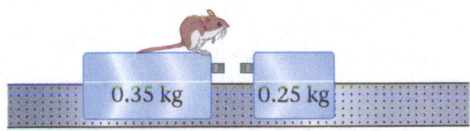

(a) Before the jump

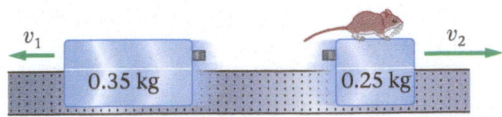

(b) After the jump

FIGURE 9-27 Problem 24

25. •• An object initially at rest breaks into two pieces as the result of an explosion. One piece has twice the kinetic energy of the other piece. What is the ratio of the masses of the two pieces? Which piece has the larger mass?

26. •• A 92-kg astronaut and a 1200-kg satellite are at rest relative to the space shuttle. The astronaut pushes on the satellite, giving it a speed of 0.14 m/s directly away from the shuttle. Seven and a half seconds later the astronaut comes into contact with the shuttle. What was the initial distance from the shuttle to the astronaut?

27. •• The recoil of a shotgun can be significant. Suppose a 3.3-kg shotgun is held tightly by an arm and shoulder with a combined mass of 15.0 kg. When the gun fires a projectile with a mass of 0.038 kg and a speed of 385 m/s, what is the recoil velocity of the shotgun and arm–shoulder combination?

28. ••• A plate drops onto a smooth floor and shatters into three pieces of equal mass. Two of the pieces go off with equal speeds v at right angles to one another. Find the speed and direction of the third piece.

SECTION 9-5 INELASTIC COLLISIONS

29. • Suppose the car in Example 9-13 has an initial speed of 20.0 m/s and that the direction of the wreckage after the collision is 40.0° above the x axis. Find the initial speed of the minivan and the final speed of the wreckage.

30. • Two 78.5-kg hockey players skating at 4.47 m/s collide and stick together. If the angle between their initial directions was 105°, what is their speed after the collision?

31. • An air-track cart with mass $m_1 = 0.32$ kg and initial speed $v_0 = 0.85$ m/s collides with and sticks to a second cart that is at rest initially. If the mass of the second cart is $m_2 = 0.44$ kg, how much kinetic energy is lost as a result of the collision?

32. •• **Predict/Calculate** A bullet with a mass of 4.0 g and a speed of 650 m/s is fired at a block of wood with a mass of 0.095 kg. The block rests on a frictionless surface, and is thin enough that the bullet passes completely through it. Immediately after the bullet exits the block, the speed of the block is 23 m/s. **(a)** What is the speed of the bullet when it exits the block? **(b)** Is the final kinetic energy of this system equal to, less than, or greater than the initial kinetic energy? Explain. **(c)** Verify your answer to part (b) by calculating the initial and final kinetic energies of the system.

33. •• **BIO Concussion Recoil** The human head can be considered as a 3.3-kg cranium protecting a 1.5-kg brain, with a small amount of cerebrospinal fluid that allows the brain to move a little bit inside the cranium. Suppose a cranium at rest is subjected to a force of 2800 N for 6.5 ms in the forward direction. **(a)** What is the final speed of the cranium? **(b)** The back of the cranium then collides with the back of the brain, which is still at rest, and the two move together. What is their final speed? **(c)** The cranium now hits an external object and suddenly comes to rest, but the brain continues to move forward. If the front of the brain interacts with the front of the cranium over a period of 15 ms before coming to rest, what average force is exerted on the brain by the cranium?

34. ••• Two objects moving with a speed v travel in opposite directions in a straight line. The objects stick together when they collide, and move with a speed of $v/4$ after the collision. **(a)** What is the ratio of the final kinetic energy of the system to the initial kinetic energy? **(b)** What is the ratio of the mass of the more massive object to the mass of the less massive object?

SECTION 9-6 ELASTIC COLLISIONS

35. • In the apple-orange collision in Example 9-16, suppose the final velocity of the orange is 1.03 m/s in the negative y direction. What are the final speed and direction of the apple in this case?

36. • A 732-kg car stopped at an intersection is rear-ended by a 1720-kg truck moving with a speed of 15.5 m/s. If the car was in neutral and its brakes were off, so that the collision is approximately elastic, find the final speed of both vehicles after the collision.

37. •• The collision between a hammer and a nail can be considered to be approximately elastic. Calculate the kinetic energy acquired by a 7.2-g nail when it is struck by a 450-g hammer moving with an initial speed of 8.8 m/s.

38. •• **Predict/Calculate** A charging bull elephant with a mass of 5240 kg comes directly toward you with a speed of 4.55 m/s. You toss a 0.150-kg rubber ball at the elephant with a speed of 7.81 m/s. **(a)** When

the ball bounces back toward you, what is its speed? **(b)** How do you account for the fact that the ball's kinetic energy has increased?

39. •• **Moderating a Neutron** In a nuclear reactor, neutrons released by nuclear fission must be slowed down before they can trigger additional reactions in other nuclei. To see what sort of material is most effective in slowing (or moderating) a neutron, calculate the ratio of a neutron's final kinetic energy to its initial kinetic energy, K_f/K_i, for a head-on elastic collision with each of the following stationary target particles. (*Note:* The mass of a neutron is $m = 1.009$ u, where the atomic mass unit, u, is defined as follows: $1 \text{ u} = 1.66 \times 10^{-27}$ kg.) **(a)** An electron ($M = 5.49 \times 10^{-4}$ u). **(b)** A proton ($M = 1.007$ u). **(c)** The nucleus of a lead atom ($M = 207.2$ u).

40. •• The three air carts shown in **FIGURE 9-28** have masses, reading from left to right, of $4m$, $2m$, and m, respectively.

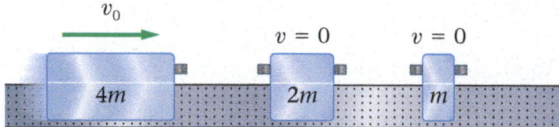

FIGURE 9-28 Problem 40

The most massive cart has an initial speed of v_0; the other two carts are at rest initially. All carts are equipped with spring bumpers that give elastic collisions. **(a)** Find the final speed of each cart. **(b)** Verify that the final kinetic energy of the system is equal to the initial kinetic energy. (Assume the air track is long enough to accommodate all collisions.)

41. •• An air-track cart with mass $m = 0.25$ kg and speed $v_0 = 1.2$ m/s approaches two other carts that are at rest and have masses $2m$ and $3m$, as indicated in **FIGURE 9-29**. The carts have bumpers that make all the collisions elastic. Find the final speed of each of the carts, assuming the air track extends indefinitely in either direction.

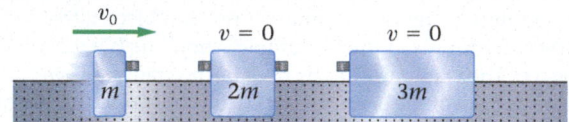

FIGURE 9-29 Problem 41

SECTION 9-7 CENTER OF MASS

42. • **CE Predict/Explain** A stalactite in a cave has drops of water falling from it to the cave floor below. The drops are equally spaced in time and come in rapid succession, so that at any given moment there are many drops in midair. **(a)** Is the center of mass of the midair drops higher than, lower than, or equal to the half-way distance between the tip of the stalactite and the cave floor? **(b)** Choose the *best explanation* from among the following:
   I. The drops bunch up as they near the floor of the cave.
   II. The drops are equally spaced as they fall, since they are released at equal times.
   III. Though equally spaced in time, the drops are closer together higher up.

43. • **CE** A baseball bat balances when placed on a fulcrum in the position shown in **FIGURE 9-30**. Is the mass of the bat to the left of the fulcrum greater than, less than, or equal to the mass of the bat to the right of the fulcrum? Explain.

FIGURE 9-30 Problem 43

44. • Find the x coordinate of the center of mass of the bricks shown in **FIGURE 9-31**.

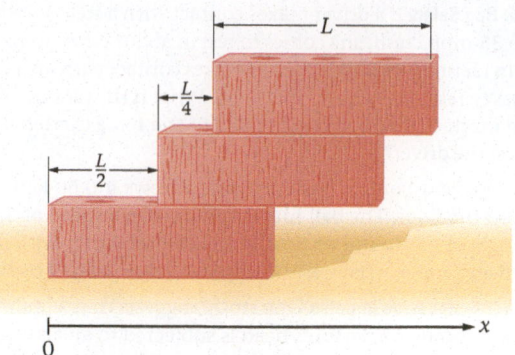

FIGURE 9-31 Problem 44

45. • You are holding a shopping basket at the grocery store with two 0.62-kg cartons of cereal at the left end of the basket. The basket is 0.61 m long. Where should you place a 1.9-kg half gallon of milk, relative to the left end of the basket, so that the center of mass of your groceries is at the center of the basket?

46. •• **CE** A pencil standing upright on its eraser end falls over and lands on a table. As the pencil falls, its eraser does not slip. The following questions refer to the contact force exerted on the pencil by the table. Let the positive x direction be in the direction the pencil falls, and the positive y direction be vertically upward. **(a)** During the pencil's fall, is the x component of the contact force positive, negative, or zero? Explain. **(b)** Is the y component of the contact force greater than, less than, or equal to the weight of the pencil? Explain.

47. •• **BIO Human Center of Mass** Suppose the human body can be reduced to individual point masses as shown in **FIGURE 9-32**. If $m_L = 14.0$ kg, $y_L = 0.65$ m, $m_A = 9.00$ kg, $y_A = 1.20$ m, $m_T = 24.5$ kg, $y_T = 1.25$ m, $m_H = 6.50$ kg, and $y_H = 1.65$ m, what is the vertical height of the center of mass?

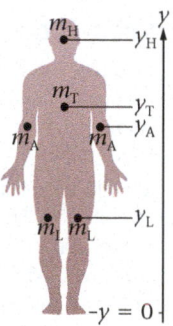

48. •• The location of the center of mass of the partially eaten, 12-inch-diameter pizza shown in **FIGURE 9-33** is $X_{cm} = -1.4$ in. and $Y_{cm} = -1.4$ in. Assuming each quadrant of the pizza to be the same, find the center of mass of the uneaten pizza above the x axis (that is, the portion of the pizza in the second quadrant).

FIGURE 9-32 Problem 47

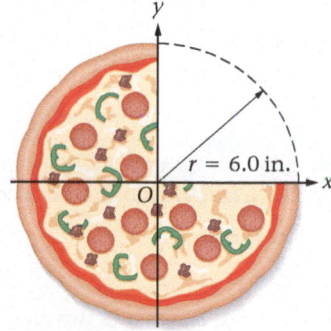

FIGURE 9-33 Problem 48

49. •• **The Center of Mass of Sulfur Dioxide** Sulfur dioxide (SO_2) consists of two oxygen atoms (each of mass 16 u, where u is defined in Problem 39) and a single sulfur atom (of mass 32 u). The center-to-center distance between the sulfur atom and either of the oxygen atoms is 0.143 nm, and the angle formed by the three atoms is 120°, as shown in **FIGURE 9-34**. Find the x and y coordinates of the center of mass of this molecule.

FIGURE 9-34 Problem 49

50. •• **Predict/Calculate** Three uniform metersticks, each of mass m, are placed on the floor as follows: stick 1 lies along the y axis from $y = 0$ to $y = 1.0$ m, stick 2 lies along the x axis from $x = 0$ to $x = 1.0$ m, stick 3 lies along the x axis from $x = 1.0$ m to $x = 2.0$ m. **(a)** Find the location of the center of mass of the metersticks. **(b)** How would the location of the center of mass be affected if the mass of the metersticks were doubled?

51. •• A 0.726-kg rope 2.00 meters long lies on a floor. You grasp one end of the rope and begin lifting it upward with a constant speed of 0.710 m/s. Find the position and velocity of the rope's center of mass from the time you begin lifting the rope to the time the last piece of rope lifts off the floor. Plot your results. (Assume the rope occupies negligible volume directly below the point where it is being lifted.)

52. •• Consider the system shown in **FIGURE 9-35**. Assume that after the string breaks the ball falls through the liquid with constant speed. If the mass of the bucket *and* the liquid is 1.20 kg, and the mass of the ball is 0.150 kg, what is the reading on the scale **(a)** before and **(b)** after the string breaks?

FIGURE 9-35 Problem 52

53. ••• A metal block of mass m is attached to the ceiling by a spring. Connected to the bottom of this block is a string that supports a second block of the same mass m, as shown in **FIGURE 9-36**. The string connecting the two blocks is now cut. **(a)** What is the net force acting on the two-block system immediately after the string is cut? **(b)** What is the acceleration of the center of mass of the two-block system immediately after the string is cut?

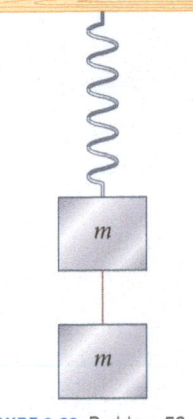

FIGURE 9-36 Problem 53

SECTION 9-8 SYSTEMS WITH CHANGING MASS: ROCKET PROPULSION

54. • **Helicopter Thrust** During a rescue operation, a 5300-kg helicopter hovers above a fixed point. The helicopter blades send air downward with a speed of 62 m/s. What mass of air must pass through the blades every second to produce enough thrust for the helicopter to hover?

55. • **Rocks for a Rocket Engine** A child sits in a wagon with a pile of 0.34-kg rocks. If she can throw each rock with a speed of 9.5 m/s relative to the ground, causing the wagon to move, how many rocks must she throw per minute to maintain a constant average speed against a 3.4-N force of friction?

56. • A 57.8-kg person holding two 0.880-kg bricks stands on a 2.10-kg skateboard. Initially, the skateboard and the person are at rest. The person now throws the two bricks at the same time so that their speed relative to the person is 17.0 m/s. What is the recoil speed of the person and the skateboard relative to the ground, assuming the skateboard moves without friction?

57. • A fire hose can expel water at a rate of 9.5 kg/s (150 gallons/minute) with a speed of 24 m/s. How much force must the firefighters exert on the hose in order to hold it steady?

58. •• A 0.540-kg bucket rests on a scale. Into this bucket you pour sand at the constant rate of 56.0 g/s. If the sand lands in the bucket with a speed of 3.20 m/s, **(a)** what is the reading of the scale when there is 0.750 kg of sand in the bucket? **(b)** What is the weight of the bucket and the 0.750 kg of sand?

59. •• **Predict/Calculate** Holding a long rope by its upper end, you lower it onto a scale. The rope has a mass of 0.13 kg per meter of length, and is lowered onto the scale at the constant rate of 1.4 m/s. **(a)** Calculate the thrust exerted by the rope as it lands on the scale. **(b)** At the instant when the amount of rope at rest on the scale has a weight of 2.5 N, does the scale read 2.5 N, more than 2.5 N, or less than 2.5 N? Explain. **(c)** Check your answer to part (b) by calculating the reading on the scale at this time.

GENERAL PROBLEMS

60. • **CE** Object A has a mass m, object B has a mass $2m$, and object C has a mass $m/2$. Rank these objects in order of increasing kinetic energy, given that they all have the same momentum. Indicate ties where appropriate.

61. • **CE** Object A has a mass m, object B has a mass $4m$, and object C has a mass $m/4$. Rank these objects in order of increasing momentum, given that they all have the same kinetic energy. Indicate ties where appropriate.

62. • **CE** A juggler performs a series of tricks with three bowling balls while standing on a bathroom scale. Is the average reading of the scale greater than, less than, or equal to the weight of the juggler plus the weight of the three balls? Explain.

63. • A golfer attempts a birdie putt, sending the 0.045-kg ball toward the hole with a speed of 1.2 m/s. Unfortunately, the ball does a "horseshoe" and goes halfway around the rim of the hole. The ball ends up heading back toward the golfer with a speed of 1.1 m/s. If we take the positive direction to be from the golfer to the hole, what is the impulse delivered to the ball by the hole?

64. • **Predict/Calculate** Two trucks drive directly away from one another on a straight line, as indicated in **FIGURE 9-37**. The trucks drive with their speeds adjusted precisely so that the center of mass (CM) of the two trucks remains stationary—that is, the velocity of the center of mass is zero. The mass and speed of truck 1 are $m_1 = 8400$ kg and $v_1 = 12$ m/s, respectively. The mass of truck 2 is $m_2 = 10,200$ kg. **(a)** Is the speed of truck 2 greater than, less than, or equal to the speed of truck 1? Explain. **(b)** Find the speed of truck 2.

FIGURE 9-37 Problem 64

65. •• **CE** Predict/Explain **FIGURE 9-38** shows a block of mass 2*m* at rest on a horizontal, frictionless table. Attached to this block by a string that passes over a pulley is a second block, with a mass *m*. The initial position of the center of mass of the blocks is indicated by the point i.

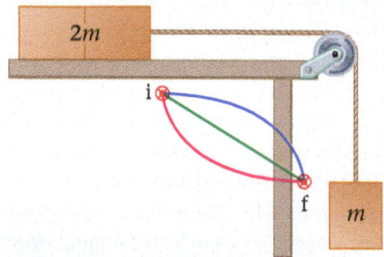

FIGURE 9-38 Problem 65

The blocks are now released and allowed to accelerate; a short time later their center of mass is at the point f. **(a)** Did the center of mass follow the red path, the green path, or the blue path? **(b)** Choose the *best explanation* from among the following:

   I. The center of mass must always be closer to the 2*m* block than to the *m* block.

   II. The center of mass starts at rest, and moves in a straight line in the direction of the net force.

   III. The masses are accelerating, which implies parabolic motion.

66. •• A 1.35-kg block of wood sits at the edge of a table, 0.782 m above the floor. A 0.0105-kg bullet moving horizontally with a speed of 715 m/s embeds itself within the block. What horizontal distance does the block cover before hitting the ground?

67. •• In a stunt, three people jump off a platform and fall 9.9 m onto a large air bag. A fourth person at the other end of the air bag, the "flier," is launched 17 m vertically into the air. Assume all four people have a mass of 75 kg. **(a)** What fraction of the mechanical energy of the three jumpers is transferred to the mechanical energy of the flier? **(b)** What is the total momentum of the three jumpers just before they land on the air bag? Let upward be the positive direction. **(c)** What is the momentum of the flier just after launch? **(d)** Notice that momentum is not conserved during this stunt, unless you include the Earth. What impulse is delivered to the jumpers–flier system by the Earth at the instant of the launch?

68. •• Predict/Calculate The carton of eggs shown in **FIGURE 9-39** is filled with a dozen eggs, each of mass *m*. Initially, the center of mass of the eggs is at the center of the carton. **(a)** Does the location of the center of mass of the eggs change more if egg 1 is removed or if egg 2 is removed? Explain. **(b)** Find the center of mass of the eggs when egg 1 is removed. **(c)** Find the center of mass of the eggs if egg 2 is removed instead.

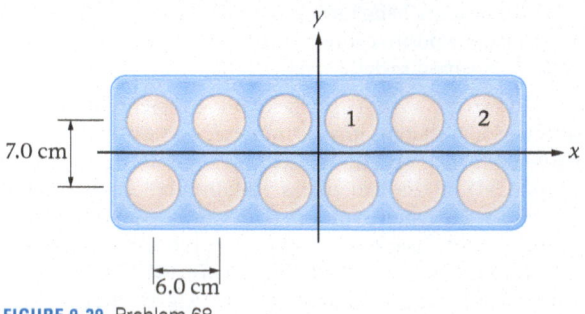

FIGURE 9-39 Problem 68

69. •• **The Force of a Storm** During a severe storm in Palm Beach, FL, on January 2, 1999, 31 inches of rain fell in a period of nine hours. Assuming that the raindrops hit the ground with a speed of 10 m/s, estimate the average upward force exerted by one square meter of ground to stop the falling raindrops during the storm. (*Note:* One cubic meter of water has a mass of 1000 kg.)

70. •• An experiment is performed in which two air carts collide on an air track. Position-versus-time data from a video of the experiment is shown in **FIGURE 9-40**. **(a)** Is the collision elastic or inelastic? Explain. **(b)** What is the speed of the center of mass of the two carts?

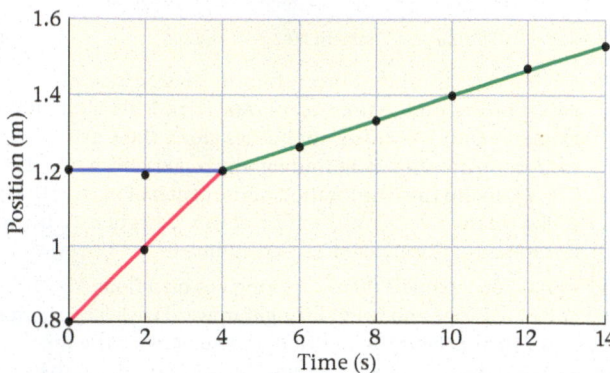

FIGURE 9-40 Problems 70 and 71

71. •• Figure 9-40 shows position-versus-time data from an experiment in which two air carts collide on an air track. If the mass of the air cart that is moving initially is 0.24 kg, what is the mass of the other cart?

72. •• To balance a 35.5-kg automobile tire and wheel, a mechanic must place a 50.2-g lead weight 25.0 cm from the center of the wheel. When the wheel is balanced, its center of mass is exactly at the center of the wheel. How far from the center of the wheel was its center of mass before the lead weight was added?

73. •• A hoop of mass *M* and radius *R* rests on a smooth, level surface. The inside of the hoop has ridges on either side, so that it forms a track on which a ball can roll, as indicated in **FIGURE 9-41**. If a ball of mass 2*M* and radius *r* = *R*/4 is released as shown, the system rocks back and forth until it comes to rest with the ball at the bottom of the hoop. When the ball comes to rest, what is the *x* coordinate of its center?

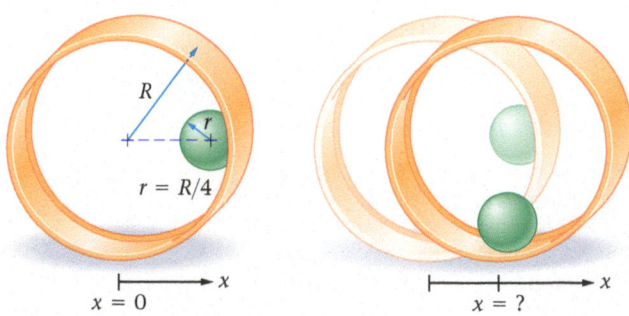

FIGURE 9-41 Problem 73

74. •• Predict/Calculate A 63-kg canoeist stands in the middle of her 22-kg canoe. The canoe is 3.0 m long, and the end that is closest to land is 2.5 m from the shore. The canoeist now walks toward the shore until she comes to the end of the canoe. **(a)** When the canoeist stops at the end of her canoe, is her distance from the shore equal to, greater than, or less than 2.5 m? Explain. **(b)** Verify your answer to part (a) by calculating the distance from the canoeist to shore.

75. •• A honeybee with a mass of 0.150 g lands on one end of a floating 4.75-g popsicle stick, as shown in **FIGURE 9-42**. After sitting at rest for a moment, it runs toward the other end with a velocity $\vec{v}_b$ relative to the still water. The stick moves in the opposite direction with a speed of 0.120 cm/s. What is the velocity of the bee? (Let the direction of the bee's motion be the positive x direction.)

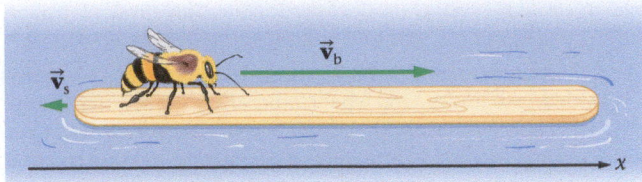

FIGURE 9-42 Problem 75

76. •• A young hockey player stands at rest on the ice holding a 1.3-kg helmet. The player tosses the helmet with a speed of 6.5 m/s in a direction 11° above the horizontal, and recoils with a speed of 0.25 m/s. Find the mass of the hockey player.

77. •• Predict/Calculate A 0.341-kg block of wood hangs from the ceiling by a string, and a 0.0567-kg wad of putty is thrown straight upward, striking the bottom of the block with a speed of 8.05 m/s. The wad of putty sticks to the block. (a) Is the mechanical energy of this system conserved? (b) How high does the putty–block system rise above the original position of the block?

78. •• A 0.454-kg block is attached to a horizontal spring that is at its equilibrium length, and whose force constant is 25.0 N/m. The block rests on a frictionless surface. A 0.0500-kg wad of putty is thrown horizontally at the block, hitting it with a speed of 8.94 m/s and sticking. How far does the putty–block system compress the spring?

79. •• BIO Escaping Octopus The giant Pacific octopus (*Enteroctopus dofleini*) has a typical mass of 15 kg and can propel itself by means of a water jet when frightened. (a) If it expels water at a mass rate of 2.4 kg/s and with a speed of 27 m/s, what thrust can it generate? (b) If the seawater exerts a resistive force of 18 N on the octopus, what acceleration can it achieve? (c) If the octopus maintains its jet propulsion for 3.0 s, what maximum velocity can it attain?

80. •• The Center of Mass of Water Find the center of mass of a water molecule, referring to **FIGURE 9-43** for the relevant angles and distances. The mass of a hydrogen atom is 1.0 u, and the mass of an oxygen atom is 16 u, where u is the atomic mass unit (see Problem 39). Use the center of the oxygen atom as the origin of your coordinate system.

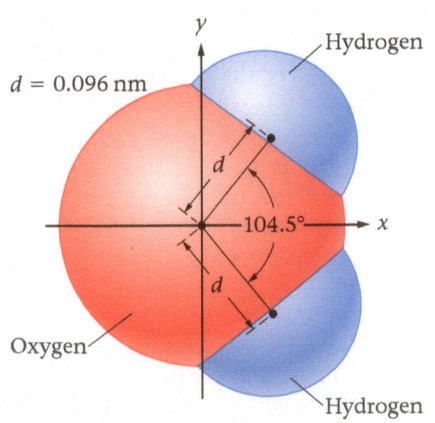

FIGURE 9-43 Problem 80

81. •• The three air carts shown in **FIGURE 9-44** have masses, reading from left to right, of m, $2m$, and $4m$, respectively. Initially, the cart on the right is at rest, whereas the other two carts are moving to the right with a speed v_0. All carts are equipped with putty bumpers that give completely inelastic collisions. (a) Find the final speed of the carts. (b) Calculate the ratio of the final kinetic energy of the system to the initial kinetic energy.

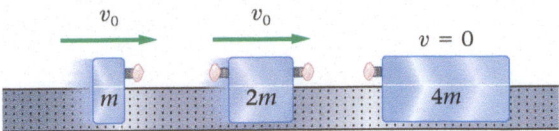

FIGURE 9-44 Problem 81

82. •• Unlimited Overhang Four identical textbooks, each of length L, are stacked near the edge of a table, as shown in **FIGURE 9-45**. The books are stacked in such a way that the distance they overhang the edge of the table, d, is maximized. Find the maximum overhang distance d in terms of L. In particular, show that $d > L$; that is, the top book is completely to the right of the table edge. (In principle, the overhang distance d can be made as large as desired simply by increasing the number of books in the stack.)

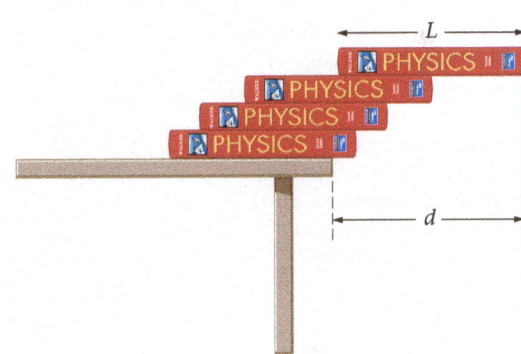

FIGURE 9-45 Problem 82

83. ••• Consider a one-dimensional, head-on elastic collision. One object has a mass m_1 and an initial velocity v_1; the other has a mass m_2 and an initial velocity v_2. Use momentum conservation and energy conservation to show that the final velocities of the two masses are

$$v_{1,f} = \left(\frac{m_1 - m_2}{m_1 + m_2}\right)v_1 + \left(\frac{2m_2}{m_1 + m_2}\right)v_2$$

$$v_{2,f} = \left(\frac{2m_1}{m_1 + m_2}\right)v_1 + \left(\frac{m_2 - m_1}{m_1 + m_2}\right)v_2$$

84. ••• Two air carts of mass $m_1 = 0.84$ kg and $m_2 = 0.42$ kg are placed on a frictionless track. Cart 1 is at rest initially, and has a spring bumper with a force constant of 690 N/m. Cart 2 has a flat metal surface for a bumper, and moves toward the bumper of the stationary cart with an initial speed $v = 0.68$ m/s. (a) What is the speed of the two carts at the moment when their speeds are equal? (b) How much energy is stored in the spring bumper when the carts have the same speed? (c) What is the final speed of the carts after the collision?

85. ••• Golden Earrings and the Golden Ratio A popular earring design features a circular piece of gold of diameter D with a circular cutout of diameter d, as shown in **FIGURE 9-46**. If this earring is to balance at the point P, show that the diameters must satisfy the condition $D = \phi d$, where $\phi = \left(1 + \sqrt{5}\right)/2 = 1.61803$ is the famous "golden ratio."

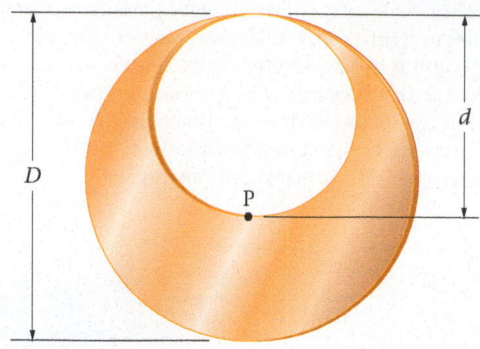

FIGURE 9-46 Problem 85

86. ••• **Amplified Rebound Height** Two small rubber balls are dropped from rest at a height h above a hard floor. When the balls are released, the lighter ball (with mass m) is directly above the heavier ball (with mass M). Assume the heavier ball reaches the floor first and bounces elastically; thus, when the balls collide, the ball of mass M is moving upward with a speed v and the ball of mass m is moving downward with essentially the same speed. In terms of h, find the height to which the ball of mass m rises after the collision. (Use the results given in Problem 83, and assume the balls collide at ground level.)

87. ••• **Predict/Calculate** **Weighing a Block on an Incline** A wedge of mass m_1 is firmly attached to the top of a scale, as shown in **FIGURE 9-47**. The inclined surface of the wedge makes an angle θ with the horizontal. Now, a block of mass m_2 is placed on the inclined surface of the wedge and allowed to accelerate without friction down the slope. **(a)** Show that the reading on the scale while the block slides is

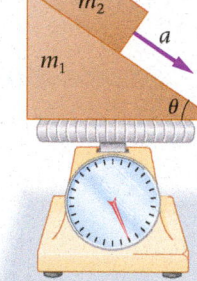

$$(m_1 + m_2\cos^2\theta)g$$

(b) Explain why the reading on the scale is less than $(m_1 + m_2)g$. **(c)** Show that the expression in part (a) gives the expected results for $\theta = 0$ and $\theta = 90°$.

FIGURE 9-47 Problem 87

88. ••• **Predict/Calculate** A uniform rope of length L and mass M rests on a table. **(a)** If you lift one end of the rope upward with a constant *speed*, v, show that the rope's center of mass moves upward with constant *acceleration*. **(b)** Next, suppose you hold the rope suspended in air, with its lower end just touching the table. If you now lower the rope with a constant speed, v, onto the table, is the acceleration of the rope's center of mass upward or downward? Explain your answer. **(c)** Find the magnitude and direction of the acceleration of the rope's center of mass for the case described in part (b). Compare with part (a).

PASSAGE PROBLEMS

BIO North Atlantic right whales (*Eubalaena glacialis*) are one of the largest and rarest marine mammals (adults can weigh up to 70 tons). Right whales move slowly near the surface to feed, and their thick blubber gives them enough buoyancy to float at the surface after they're killed, which explains why whalers called them the "right" whales to hunt. Given this, it's not surprising that centuries of whaling, which ended only in 1949, drove the species nearly to extinction. It's estimated that currently only about 400 individuals remain.

One of the reasons the population has been slow to recover is that right whales spend much of their time in shallow coastal waters, where they are susceptible to collisions with ships and entanglement in fishing gear. The United States has taken several steps to protect right whales from shipping-related injuries. For instance, ships weighing more than 300 tons are required to report when they enter specific regions off the U.S. coast that are important right-whale habitats. In addition, ships longer than 65 feet are required to slow down to 10 knots or less in certain critical areas, such as whale calving grounds. (Ship speeds are measured in knots, or nautical miles per hour; one nautical mile is 1.151 miles.)

In order to guide whale conservation policies, researchers are working to understand what factors affect the severity of injury to whales in a collision. Unfortunately, it's not always easy to gather information about the extent of whale injuries as a result of a particular collision. One approach is to mathematically model the effects of collisions. In one model of a collision between a ship and a whale, the duration of the impact decreases from 0.50 s for a ship moving at 5.0 knots to 0.38 s at 15 knots. Because the collision is this short, forces on the whale from water drag during the collision can be neglected.

89. •• The lightest ship that's subject to the mandatory reporting regulations, traveling at 5.0 knots, collides with a large stationary adult right whale of maximum size. As sometimes happens, the whale remains stuck on the prow of the ship after the collision. By how much does the speed of the ship change as a result of the collision? Give your answer in knots.

    A. 0.95   **B.** 1.1   **C.** 2.5   **D.** 4.1

90. •• By how much does the whale's speed change (in knots) as a result of this collision?

    A. 0.95   **B.** 1.1   **C.** 2.5   **D.** 4.1

91. •• What is the magnitude of the average force (in kN) the ship exerts on the whale during the collision? (*Note:* 1 ton = 907 kg)

    A. 62   **B.** 270   **C.** 520   **D.** 620

92. •• If the ship was traveling at 5.0 knots but was 10 times more massive, how would the force of the collision on the whale compare?

    A. It would be 100 times larger.
    B. It would be 10 times larger.
    C. It would be larger, but not as much as 10 times larger.
    D. It would be the same.

93. •• **REFERRING TO EXAMPLE 9-12** Suppose a bullet of mass $m = 6.75$ g is fired into a ballistic pendulum whose bob has a mass of $M = 0.675$ kg. **(a)** If the bob rises to a height of 0.128 m, what was the initial speed of the bullet? **(b)** What was the speed of the bullet–bob combination immediately after the collision took place?

94. •• **REFERRING TO EXAMPLE 9-12** A bullet with a mass $m = 8.10$ g and an initial speed $v_i = 320$ m/s is fired into a ballistic pendulum. What mass must the bob have if the bullet–bob combination is to rise to a maximum height of 0.125 m after the collision?

95. •• **REFERRING TO EXAMPLE 9-19** Suppose that cart 1 has a mass of 3.00 kg and an initial speed of 0.250 m/s. Cart 2 has a mass of 1.00 kg and is at rest initially. **(a)** What is the final speed of the carts? **(b)** How much kinetic energy is lost as a result of the collision?

96. •• **REFERRING TO EXAMPLE 9-19** Suppose the two carts have equal masses and are both moving to the right before the collision. The initial speed of cart 1 (on the left) is v_0, and the initial speed of cart 2 (on the right) is $v_0/2$. **(a)** What is the speed of the center of mass of this system? **(b)** What percentage of the initial kinetic energy is lost as a result of the collision? **(c)** Suppose the collision is elastic. What are the final speeds of the two carts in this case?

Big Ideas

1 Rotational quantities are analogous to the corresponding linear quantities.

2 Rolling motion is a combination of linear and rotational motions.

3 A rotating object has kinetic energy, just like an object in linear motion.

Rotational Kinematics and Energy

▲ Because of the earth's rotation about its axis, the stars in this long-exposure photograph appear to follow circular paths across the sky—with Polaris, the north star, very near the axis of rotation. Notice that each star moves through the same angle in the course of the exposure. However, the farther a star is from the axis of rotation, the longer the arc it traces out. This is just one example of the connections among angle, radius, and arc length that we explore in this chapter.

Rotational motion is a part of everyday life, from the spinning planet on which we live to the spinning wheels of an automobile. In this chapter we present the basic concepts of rotational motion, and show the close connection between rotational and linear motions.

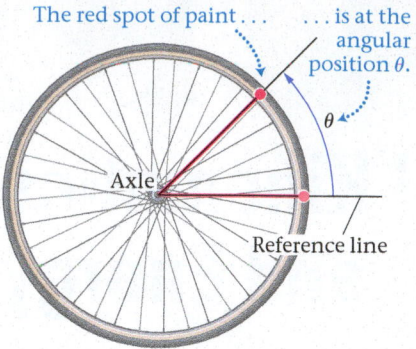

The red spot of paint . . .   . . . is at the angular position θ.

Axle

Reference line

▲ **FIGURE 10-1 Angular position** The angular position, θ, of a spot of paint on a bicycle wheel. The reference line, where θ = 0, is drawn horizontal here but any direction can be chosen.

10-1 Angular Position, Velocity, and Acceleration

To describe the motion of an object moving in a straight line, it's useful to establish a coordinate system with a definite origin and positive direction. In terms of this coordinate system we can measure the object's position, velocity, and acceleration.

Similarly, to describe rotational motion, we define "angular" quantities that are analogous to the linear position, velocity, and acceleration. These angular quantities, which are defined in terms of an angular "coordinate system," form the basis of our study of rotation. We begin by defining the most basic angular quantity—the angular position.

Angular Position, θ

Consider a bicycle wheel that is free to rotate about its axle, as shown in **FIGURE 10-1**. We say that the axle is the **axis of rotation** for the wheel. As the wheel rotates, each and every point on it moves in a circular path centered on the axis of rotation.

Now, suppose there is a small spot of red paint on the tire, and we want to describe the rotational motion of the spot. The **angular position** of the spot is defined to be the angle, θ, that a line from the axle to the spot makes with a reference line, as indicated in Figure 10-1.

Definition of Angular Position, θ

θ = angle measured from reference line   10-1

SI unit: radian (rad), which is dimensionless

The reference line simply defines θ = 0; it is analogous to the origin in a linear coordinate system. The reference line begins at the axis of rotation, and may be chosen to point in any direction—just as an origin may be placed anywhere along a coordinate axis. Once chosen, however, the reference line must be used consistently.

The spot of paint in Figure 10-1 is rotated counterclockwise from the reference line by the angle θ. By convention, we say that this angle is positive. Clockwise rotations correspond to negative angles.

Sign Convention for Angular Position

By convention:

$$\theta > 0 \text{ counterclockwise rotation from reference line}$$
$$\theta < 0 \text{ clockwise rotation from reference line}$$

Examples of rotational motion are shown in **FIGURE 10-2**.

Units of Angular Position   Now that we have established a reference line (for θ = 0) and a positive direction of rotation (counterclockwise), we must choose units in which to measure angles. Common units are degrees (°) and revolutions (rev), where one revolution—that is, going completely around a circle—corresponds to 360°:

$$1 \text{ rev} = 360°$$

▶ **FIGURE 10-2 Visualizing Concepts Rotational Motion** Rotational motion is everywhere in our universe, on every scale of length and time. A galaxy like the one at left may take millions of years to complete a single rotation about its center, while the skater in the middle photo spins several times in a second. The bacterium at right moves in a corkscrew path by rapidly twirling its flagella (the fine projections at either end of the cell) like whips.

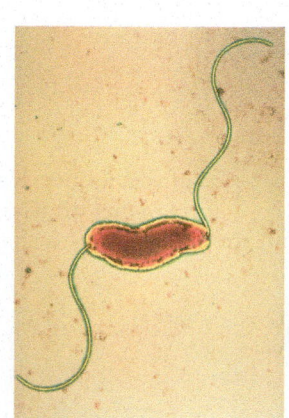

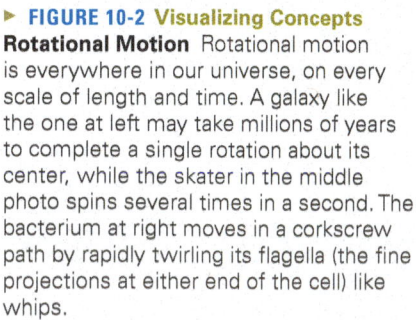

The most convenient units for scientific calculations, however, are radians. A **radian** (rad) is defined as follows:

> A radian is the angle for which the arc length on a circle of radius r is equal to the radius of the circle.

This definition is useful because it establishes a particularly simple relationship between an angle measured in radians and the corresponding arc length, as illustrated in **FIGURE 10-3**. For example, it follows from our definition that for an angle of one radian, the arc length s is equal to the radius: $s = r$. Similarly, an angle of two radians corresponds to an arc length of two radii, $s = 2r$, and so on. Thus, the arc length s for an arbitrary angle θ measured in radians is given by the following relationship:

$$s = r\theta \qquad \text{10-2}$$

This simple and straightforward relationship does not hold for degrees or revolutions—additional conversion factors are needed in those cases. An example of the connection among arc length, radius, and angle is shown in **FIGURE 10-4**.

In one complete revolution, the arc length is the circumference of a circle, $C = 2\pi r$. Comparing with $s = r\theta$, we see that a complete revolution corresponds to 2π radians:

$$1 \text{ rev} = 360° = 2\pi \text{ rad}$$

Equivalently,

$$1 \text{ rad} = \frac{360°}{2\pi} = 57.3°$$

One final note on the units for angles: Radians, as well as degrees and revolutions, are dimensionless. In the relationship $s = r\theta$, for example, the arc length and the radius both have SI units of meters. For the equation to be dimensionally consistent, it is necessary that θ have no dimensions. Still, if an angle θ is, let's say, three radians, we will write it as $\theta = 3$ rad to remind us of the angular units being used.

Angular Velocity, ω

As the bicycle wheel in Figure 10-1 rotates, the angular position of the spot of red paint changes. This is illustrated in **FIGURE 10-5**. The **angular displacement** of the spot, $\Delta\theta$, is

$$\Delta\theta = \theta_f - \theta_i$$

If we divide the angular displacement by the time, Δt, during which the displacement occurs, the result is the **average angular velocity**, ω_{av}.

Definition of Average Angular Velocity, ω_{av}

$$\omega_{av} = \frac{\Delta\theta}{\Delta t} \qquad \text{10-3}$$

SI unit: radian per second (rad/s) = s⁻¹

This is analogous to the definition of the average linear velocity: $v_{av} = \Delta x / \Delta t$. Notice that the units of linear velocity are m/s, whereas the units of angular velocity are rad/s.

In addition to the average angular velocity, we can define an **instantaneous angular velocity** as the limit of ω_{av} as the time interval, Δt, approaches zero. The instantaneous angular velocity, then, is:

Definition of Instantaneous Angular Velocity, ω

$$\omega = \lim_{\Delta t \to 0} \frac{\Delta\theta}{\Delta t} \qquad \text{10-4}$$

SI unit: rad/s = s⁻¹

Generally, we shall refer to the instantaneous angular velocity simply as the angular velocity.

Notice that we call ω the angular velocity, not the angular speed. The reason is that ω can be positive or negative, depending on the sense of rotation. For example, if the

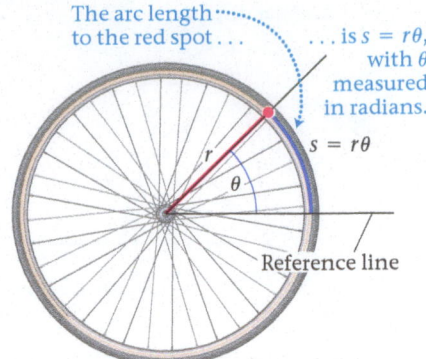

▲ **FIGURE 10-3 Arc length** The arc length, s, from the reference line to the spot of paint is given by $s = r\theta$ if the angular position θ is measured in radians.

PROBLEM-SOLVING NOTE

Radians

Remember to measure angles in radians when using the relationship $s = r\theta$.

▲ **FIGURE 10-4** The rotating carnival ride in this time exposure illustrates the connection between radius and arc length. As the wheel rotates through a given angle, the lights on the spokes trace out luminous "wedges" that are wider the greater the distance from the axis of rotation—as one would expect from $s = r\theta$.

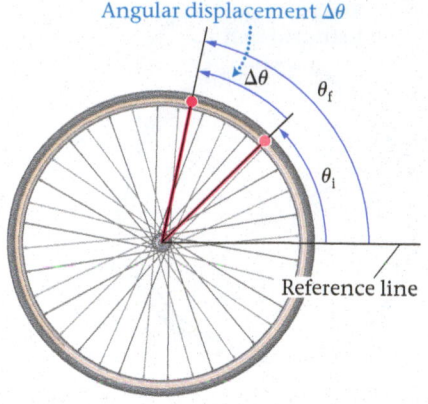

Angular displacement $\Delta\theta$

▲ **FIGURE 10-5 Angular displacement** As the wheel rotates, the spot of paint undergoes an angular displacement, $\Delta\theta = \theta_f - \theta_i$.

red paint spot rotates in the counterclockwise sense, the angular position, θ, increases. As a result, $\Delta\theta$ is positive and, therefore, so is ω. Similarly, clockwise rotation corresponds to a negative $\Delta\theta$ and hence a negative ω.

> **Sign Convention for Angular Velocity**
> By convention:
>
> $$\omega > 0 \text{ counterclockwise rotation}$$
> $$\omega < 0 \text{ clockwise rotation}$$

The sign convention for angular velocity is illustrated in **FIGURE 10-6**. In analogy with linear motion, the sign of ω indicates the *direction* of the angular velocity *vector*, as we shall see in detail in Chapter 11. Similarly, the magnitude of the angular velocity is the **angular speed,** just as in the one-dimensional case.

In the following Exercise we utilize the definitions and conversion factors presented so far in this section.

EXERCISE 10-1 ANGULAR VELOCITY AND SPEED

(a) A wind turbine for generating electricity rotates clockwise at the rate of 17.0 rpm (revolutions per minute). What is its angular velocity in rad/s? **(b)** A CD rotates through an angle of 106° in 0.0860 s. What is the angular speed of the CD in rad/s?

REASONING AND SOLUTION

A rotation rate in revolutions per minute can be converted to rad/s by noting that 1 revolution = 2π rad. In addition, knowing the angle $\Delta\theta$ through which an object rotates in a time Δt allows us to calculate the angular velocity with $\theta = \Delta\theta/\Delta t$. Recall that clockwise rotation corresponds to the negative direction, and that 2π rad = 360°.

a. Convert from rpm to rad/s. Use the fact that clockwise rotation corresponds to a negative angular velocity:

$$\omega = -17.0 \text{ rpm} = \left(-17.0\frac{\text{rev}}{\text{min}}\right)\left(\frac{2\pi \text{ rad}}{1 \text{ rev}}\right)\left(\frac{1 \text{ min}}{60 \text{ s}}\right) = -1.78 \text{ rad/s}$$

b. Convert $\Delta\theta$ from degrees (°) to radians (rad) in the relationship $\omega = \Delta\theta/\Delta t$. This gives the following result for the angular speed:

$$\omega = \frac{\Delta\theta}{\Delta t} = \frac{106°\left(\dfrac{2\pi \text{ rad}}{360°}\right)}{0.0860 \text{ s}} = 21.5 \text{ rad/s}$$

The same symbol, ω, is used for both angular velocity and angular speed in Exercise 10-1. Which quantity is meant in a given situation will be clear from the context in which it is used.

Angular Velocity and Period As a simple application of angular velocity, consider the following question: If an object rotates with a constant angular velocity, ω, how much time, T, is required for it to complete one full revolution?

To solve this problem, first note that because ω is constant, the instantaneous angular velocity is equal to the average angular velocity. That is,

$$\omega = \omega_{av} = \frac{\Delta\theta}{\Delta t}$$

In one revolution, we know that $\Delta\theta = 2\pi$ and $\Delta t = T$. Therefore,

$$\omega = \frac{\Delta\theta}{\Delta t} = \frac{2\pi}{T}$$

Finally, solving for T we find

$$T = \frac{2\pi}{\omega}$$

The time to complete one revolution, T, is referred to as the **period.**

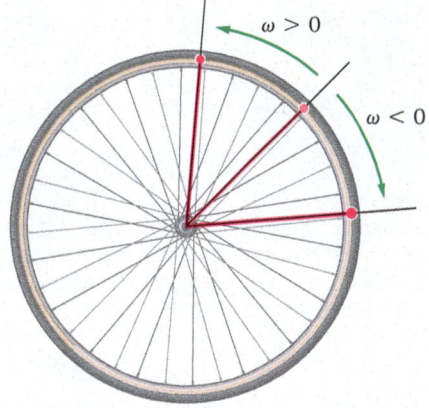

▲ **FIGURE 10-6 Angular speed and velocity** Counterclockwise rotation is defined to correspond to a positive angular velocity, ω. Clockwise rotation corresponds to a negative angular velocity. The magnitude of the angular velocity is referred to as the angular speed.

PHYSICS IN CONTEXT
Looking Ahead

Just as linear speed is related to linear momentum (Chapter 9), angular speed is related to angular momentum. This is discussed in detail in Chapter 11.

Definition of Period, *T*

$$T = \frac{2\pi}{\omega}$$ 10-5

SI unit: second, s

The angular speed in Equation 10-5 must be measured in rad/s for the relationship to be valid.

EXERCISE 10-2   PERIOD OF ROTATION

Find the period of a microwave tray that is rotating with an angular speed of 5.0 rpm.

REASONING AND SOLUTION
To apply $T = 2\pi/\omega$ we must first express ω in terms of rad/s:

$$\omega = 5.0 \text{ rpm} = \left(5.0\frac{\text{rev}}{\text{min}}\right)\left(\frac{2\pi \text{ rad}}{1 \text{ rev}}\right)\left(\frac{1 \text{ min}}{60 \text{ s}}\right) = 0.52 \text{ rad/s}$$

Now we can calculate the period:

$$T = \frac{2\pi}{\omega} = \frac{2\pi \text{ rad}}{0.52 \text{ rad/s}} = 12 \text{ s}$$

Angular Acceleration, α

If the angular velocity of the rotating bicycle wheel we've been considering increases or decreases with time, we say that the wheel experiences an **angular acceleration, α**. The average angular acceleration is the change in angular velocity, $\Delta\omega$, in a given interval of time, Δt:

Definition of Average Angular Acceleration, α_{av}

$$\alpha_{av} = \frac{\Delta\omega}{\Delta t}$$ 10-6

SI unit: radian per second per second $(\text{rad/s}^2) = \text{s}^{-2}$

The SI units of α are rad/s^2, which, since rad is dimensionless, is simply s^{-2}. If the angular acceleration α is constant, which is often the case, then the average angular acceleration is simply equal to the angular acceleration itself, $\alpha_{av} = \alpha$.

As expected, the instantaneous angular acceleration is the limit of α_{av} as the time interval, Δt, approaches zero:

Definition of Instantaneous Angular Acceleration, α

$$\alpha = \lim_{\Delta t \to 0} \frac{\Delta\omega}{\Delta t}$$ 10-7

SI unit: $\text{rad/s}^2 = \text{s}^{-2}$

When referring to the instantaneous angular acceleration, we will usually just say angular acceleration.

The sign of the angular acceleration is determined by whether the change in the angular velocity is positive or negative. For example, if ω is becoming more positive, so that ω_f is greater than ω_i, it follows that α is positive. Similarly, if ω is becoming more negative, so that ω_f is less than ω_i, it follows that α is negative. Therefore, if ω and α have the same sign, the speed of rotation is increasing. If ω and α have opposite signs, the speed of rotation is decreasing. This is illustrated in **FIGURE 10-7**.

EXERCISE 10-3   ANGULAR ACCELERATION

A ceiling fan that was rotating with an angular velocity of 12.6 rad/s begins to slow down when it is turned off. How much time does it take for the fan to come to rest if it has a constant angular acceleration of -0.168 rad/s^2?

CONTINUED

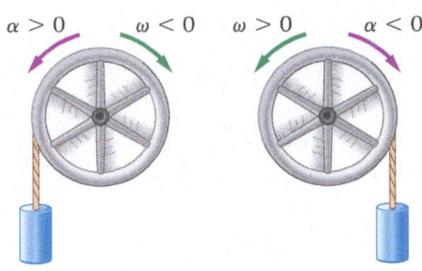

(a) Angular speed increases   **(b)** Angular speed increases

(c) Angular speed decreases   **(d)** Angular speed decreases

▲ **FIGURE 10-7 Angular acceleration and angular speed** When angular velocity and acceleration have the same sign, as in **(a)** and **(b)**, the angular speed increases. When angular velocity and angular acceleration have opposite signs, as in **(c)** and **(d)**, the angular speed decreases.

PHYSICS IN CONTEXT
Looking Back

Notice that our definitions of position, velocity, and acceleration from Chapter 2 are generalized here to apply to rotational motion.

PHYSICS IN CONTEXT
Looking Ahead

In Chapter 11, we relate force to angular acceleration in much the same way that force and acceleration are related in linear motion. This results in the concept of torque.

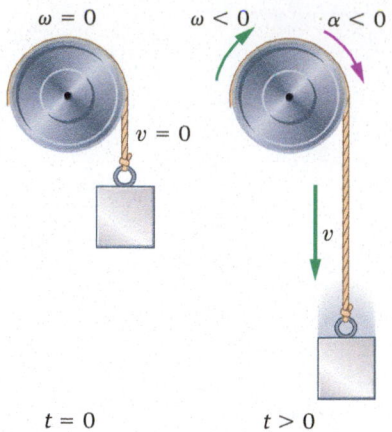

$\omega = 0$ $\omega < 0$ $\alpha < 0$

$v = 0$

v

$t = 0$ $t > 0$

▲ **FIGURE 10-8 A pulley with constant angular acceleration** A mass is attached to a string wrapped around a pulley. As the mass falls, it causes the pulley to increase its angular speed with a constant angular acceleration.

REASONING AND SOLUTION

We know the initial and final angular velocities, as well as the angular acceleration. The fact that the angular acceleration is constant means that $\alpha_{av} = \alpha = -0.168$ rad/s². We can then rearrange Equation 10-6 ($\alpha_{av} = \Delta\omega/\Delta t$) to find the elapsed time, Δt:

$$\Delta t = \frac{\Delta\omega}{\alpha_{av}} = \frac{\omega_f - \omega_i}{\alpha} = \frac{0 - 12.6 \text{ rad/s}}{-0.168 \text{ rad/s}^2} = 75.0 \text{ s}$$

Enhance Your Understanding   (Answers given at the end of the chapter)

1. Rank the following systems (A, B, C, D) in order of increasing **(a)** angular speed and **(b)** angular velocity. Indicate ties where appropriate. **A:** 2 rad/s clockwise; **B:** 3 rpm counterclockwise; **C:** 100 rpm counterclockwise; **D:** 10 rad/s clockwise.

Section Review

- Angular position is the angle relative to a reference line, angular velocity is the rate of change of angular position, and angular acceleration is the rate of change of angular velocity.

- Rotation in the counterclockwise direction is positive; rotation in the clockwise direction is negative.

10-2 Rotational Kinematics

Just as the kinematics of Chapter 2 described linear motion, rotational kinematics describes rotational motion. In this section, as in Chapter 2, we concentrate on the important special case of constant acceleration.

Angular Velocity with Constant Angular Acceleration As an example of a system with constant angular acceleration, consider the pulley shown in **FIGURE 10-8**. Wrapped around the circumference of the pulley is a string, with a mass attached to its free end. When the mass is released, the pulley begins to rotate—slowly at first, but then faster and faster. As we will see in Chapter 11, the pulley is accelerating with constant angular acceleration.

Because α is constant, it follows that the average and instantaneous angular accelerations are equal. Hence,

$$\alpha = \alpha_{av} = \frac{\Delta\omega}{\Delta t}$$

Suppose that the pulley starts with the initial angular velocity ω_0 at time $t = 0$, and that at the later time t its angular velocity is ω. Substituting these values into the preceding expression for α yields

$$\alpha = \frac{\Delta\omega}{\Delta t} = \frac{\omega - \omega_0}{t - 0} = \frac{\omega - \omega_0}{t}$$

Rearranging, we see that the angular velocity, ω, varies with time as follows:

$$\omega = \omega_0 + \alpha t \qquad \text{10-8}$$

EXERCISE 10-4 ANGULAR KINEMATICS

If the angular velocity of the pulley in Figure 10-8 is -6.43 rad/s at a given time, and its constant angular acceleration is -0.491 rad/s², what is the angular velocity of the pulley 1.25 s later?

REASONING AND SOLUTION

The angular velocity, ω, is found by applying $\omega = \omega_0 + \alpha t$:

$$\omega = \omega_0 + \alpha t = -6.43 \text{ rad/s} + (-0.491 \text{ rad/s}^2)(1.25 \text{ s}) = -7.04 \text{ rad/s}$$

The angular speed has *increased*, as expected, because ω and α have the *same sign*.

PROBLEM-SOLVING NOTE

Sign of the Angular Acceleration

The sign of the angular acceleration (by itself) doesn't determine whether a rotating system speeds up or slows down. What really matters is whether the signs of the angular acceleration and the angular velocity are the same (speeds up) or different (slows down).

Rotational and Linear Analogies Notice the close analogy between $\omega = \omega_0 + \alpha t$ for angular velocity and the corresponding relationship for linear velocity, Equation 2-7:

$$v = v_0 + at$$

Clearly, the equation for angular velocity can be obtained from our previous equation for linear velocity by replacing v with ω and replacing a with α. This type of analogy between linear and angular quantities can be most useful both in deriving angular equations—by starting with linear equations and using analogies—and in obtaining a better physical understanding of angular systems. Several linear-angular analogs are listed in the accompanying table.

Using these analogies, we can rewrite all the kinematic equations in Chapter 2 in angular form. The following table gives both the linear kinematic equations and their angular counterparts.

Linear Quantity	Angular Quantity
x	θ
v	ω
a	α

Big Idea **1** Rotational quantities, like angular position, angular velocity, and angular acceleration, are analogous to the corresponding linear quantities. In addition, the kinematics of rotational motion is essentially the same as the kinematics of linear motion.

Linear Equation (a = constant)		Angular Equation (α = constant)	
$v = v_0 + at$	2-7	$\omega = \omega_0 + \alpha t$	10-8
$x = x_0 + \frac{1}{2}(v_0 + v)t$	2-10	$\theta = \theta_0 + \frac{1}{2}(\omega_0 + \omega)t$	10-9
$x = x_0 + v_0 t + \frac{1}{2}at^2$	2-11	$\theta = \theta_0 + \omega_0 t + \frac{1}{2}\alpha t^2$	10-10
$v^2 = v_0^2 + 2a(x - x_0)$	2-12	$\omega^2 = \omega_0^2 + 2\alpha(\theta - \theta_0)$	10-11

In solving kinematic problems involving rotation, we apply these angular equations in the same way that the linear equations were applied in Chapter 2. In a sense, then, this material is a review—after all, the mathematics is essentially the same. The only difference comes in the physical interpretation of the results. We will emphasize the rotational interpretations throughout the chapter.

PROBLEM-SOLVING NOTE

Rotational Kinematics

Using analogies between linear and angular quantities often helps when solving problems involving rotational kinematics.

EXAMPLE 10-5 THROWN FOR A CURVE

To throw a curve ball, a pitcher gives the ball an initial angular speed of 36.0 rad/s. Just before the catcher gloves the ball 0.595 s later, its angular speed has decreased (due to air resistance) to 34.2 rad/s. **(a)** What is the ball's angular acceleration, assuming it to be constant? **(b)** How many revolutions does the ball make before being caught?

PICTURE THE PROBLEM
We choose the ball's initial direction of rotation to be positive. As a result, the angular acceleration (which decreases the angular speed) will be negative. We can also identify the initial angular velocity to be $\omega_0 = 36.0$ rad/s, and the final angular velocity to be $\omega = 34.2$ rad/s.

REASONING AND STRATEGY
The problem states that the angular acceleration of the ball is constant. It follows that the equations of angular kinematics (Equations 10-8 to 10-11) apply.

a. To relate angular velocity to time, we use $\omega = \omega_0 + \alpha t$. This can be solved for α.

b. To relate angle to time we use $\theta = \theta_0 + \omega_0 t + \frac{1}{2}\alpha t^2$. The angular displacement of the ball is $\theta - \theta_0$.

Known Initial angular speed, $\omega_0 = 36.0$ rad/s; final angular speed, $\omega = 34.2$ rad/s; elapsed time, $t = 0.595$ s.
Unknown **(a)** Angular acceleration, $\alpha = ?$ **(b)** Number of revolutions, $\theta - \theta_0 = ?$

SOLUTION

Part (a)

1. Solve $\omega = \omega_0 + \alpha t$ for the angular acceleration, α:

$$\omega = \omega_0 + \alpha t$$
$$\alpha = \frac{\omega - \omega_0}{t}$$

CONTINUED

2. Substitute numerical values to find α:

$$\alpha = \frac{\omega - \omega_0}{t}$$

$$= \frac{34.2 \text{ rad/s} - 36.0 \text{ rad/s}}{0.595 \text{ s}} = -3.03 \text{ rad/s}^2$$

Part (b)

3. Use $\theta = \theta_0 + \omega_0 t + \frac{1}{2}\alpha t^2$ to calculate the angular displacement of the ball:

$$\theta - \theta_0 = \omega_0 t + \frac{1}{2}\alpha t^2$$
$$= (36.0 \text{ rad/s})(0.595 \text{ s}) + \frac{1}{2}(-3.03 \text{ rad/s}^2)(0.595 \text{ s})^2$$
$$= 20.9 \text{ rad}$$

4. Convert the angular displacement to revolutions:

$$\theta - \theta_0 = 20.9 \text{ rad} = 20.9 \text{ rad}\left(\frac{1 \text{ rev}}{2\pi \text{ rad}}\right) = 3.33 \text{ rev}$$

INSIGHT

The ball rotates through three-and-one-third revolutions during its time in flight.

An alternative method of solution is to use the kinematic relationship given in Equation 10-9. This procedure yields $\theta - \theta_0 = \frac{1}{2}(\omega_0 + \omega)t = 20.9 \text{ rad}$, in agreement with our previous result.

PRACTICE PROBLEM

(a) What is the angular velocity of the ball 0.500 s after it is thrown? **(b)** What is the ball's angular velocity after it completes its first full revolution?   [**Answer: (a)** Use $\omega = \omega_0 + \alpha t$ to find $\omega = 34.5 \text{ rad/s}$. **(b)** Use $\omega^2 = \omega_0^2 + 2\alpha(\theta - \theta_0)$ to find $\omega = 35.5 \text{ rad/s}$.]

Some related homework problems: Problem 16, Problem 19

EXAMPLE 10-6   WHEEL OF MISFORTUNE

On a certain game show, contestants spin a wheel when it is their turn. One contestant gives the wheel an initial angular speed of 3.40 rad/s. It then rotates through one-and-one-quarter revolutions and comes to rest on the BANKRUPT space. **(a)** Find the angular acceleration of the wheel, assuming it to be constant. **(b)** How much time does it take for the wheel to come to rest?

PICTURE THE PROBLEM

We choose the initial angular velocity to be positive, $\omega_0 = +3.40 \text{ rad/s}$, and indicate it with a counterclockwise rotation in our sketch. Because the wheel is slowing down, its angular acceleration must have the opposite sign of its angular velocity, which means it is negative (clockwise) in this case. After a rotation of 1.25 rev the wheel will read BANKRUPT.

REASONING AND STRATEGY

As in Example 10-5, we can use the kinematic equations for constant angular acceleration, Equations 10-8 to 10-11.

a. To begin, we are given the initial angular velocity, $\omega_0 = +3.40 \text{ rad/s}$, the final angular velocity, $\omega = 0$ (the wheel comes to rest), and the angular displacement, $\theta - \theta_0 = 1.25 \text{ rev}$. We can find the angular acceleration using $\omega^2 = \omega_0^2 + 2\alpha(\theta - \theta_0)$.

b. Knowing the initial angular velocity and acceleration, we can find the time to stop rotating with $\omega = \omega_0 + \alpha t$.

Known   Initial angular speed, $\omega_0 = +3.40 \text{ rad/s}$; final angular speed, $\omega = 0$; angle of rotation, $\theta - \theta_0 = 1.25 \text{ rev}$.

Unknown   **(a)** Angular acceleration, $\alpha = ?$ **(b)** Time to stop, $t = ?$

SOLUTION

Part (a)

1. Solve $\omega^2 = \omega_0^2 + 2\alpha(\theta - \theta_0)$ for the angular acceleration, α:

$$\omega^2 = \omega_0^2 + 2\alpha(\theta - \theta_0)$$
$$\alpha = \frac{\omega^2 - \omega_0^2}{2(\theta - \theta_0)}$$

2. Convert $\theta - \theta_0 = 1.25 \text{ rev}$ to radians:

$$\theta - \theta_0 = 1.25 \text{ rev} = 1.25 \text{ rev}\left(\frac{2\pi \text{ rad}}{1 \text{ rev}}\right) = 7.85 \text{ rad}$$

CONTINUED

3. Substitute numerical values to find α:

$$\alpha = \frac{\omega^2 - \omega_0^2}{2(\theta - \theta_0)} = \frac{0 - (3.40 \text{ rad/s})^2}{2(7.85 \text{ rad})} = -0.736 \text{ rad/s}^2$$

Part (b)

4. Solve $\omega = \omega_0 + \alpha t$ for the time, t:

$$\omega = \omega_0 + \alpha t$$

$$t = \frac{\omega - \omega_0}{\alpha}$$

5. Substitute numerical values to find t:

$$t = \frac{\omega - \omega_0}{\alpha} = \frac{0 - 3.40 \text{ rad/s}}{(-0.736 \text{ rad/s}^2)} = 4.62 \text{ s}$$

INSIGHT

Notice that we converted the angular displacement in revolutions, 1.25 rev, to radians, 7.85 rad, in Step 2. This ensures that all angular quantities are measured with the same angular units. In addition, notice that it is not necessary to define a reference line—that is, a direction for $\theta = 0$. All we need to know is the angular displacement, $\theta - \theta_0$, not the individual angles θ and θ_0.

PRACTICE PROBLEM

What is the angular speed of the wheel after one complete revolution?   [**Answer:** $\omega = 1.52$ rad/s]

Some related homework problems: Problem 15, Problem 17

Enhance Your Understanding   (Answers given at the end of the chapter)

2. An object at rest begins to rotate at $t = 0$ with constant angular acceleration. After a time t the object has rotated through an angle θ and has an angular velocity ω. **(a)** Through what angle has the object rotated in the time $2t$? **(b)** What is the angular velocity of the object at time $2t$?

Section Review

- The basic analogies between linear and angular quantities are as follows: position, x analogous to θ; velocity, v analogous to ω; acceleration, a analogous to α.

- Using the linear-angular analogies, we can convert the kinematic equations of Chapter 2 to kinematic equations for angular motion.

10-3 Connections Between Linear and Rotational Quantities

At a local county fair a child rides on a merry-go-round. The ride completes one revolution, 2π radians, every $T = 7.50$ s. Therefore, the angular speed of the child, from Equation 10-5 ($\omega = 2\pi/T$), is

$$\omega = \frac{2\pi}{T} = \frac{2\pi}{7.50 \text{ s}} = 0.838 \text{ rad/s}$$

The path followed by the child is circular, with the center of the circle at the axis of rotation of the merry-go-round. In addition, at any instant of time the child is moving in a direction that is *tangential* to the circular path, as **FIGURE 10-9** shows. What is the tangential speed, v_t, of the child? In other words, what is the speed of the wind in the child's face?

We can find the child's tangential speed by dividing the distance traveled—the circumference of the circular path, $2\pi r$—by the time required to complete one revolution, T. That is,

$$v_t = \frac{2\pi r}{T} = r\left(\frac{2\pi}{T}\right)$$

Because $2\pi/T$ is simply the angular speed, ω, we can express the tangential speed as follows:

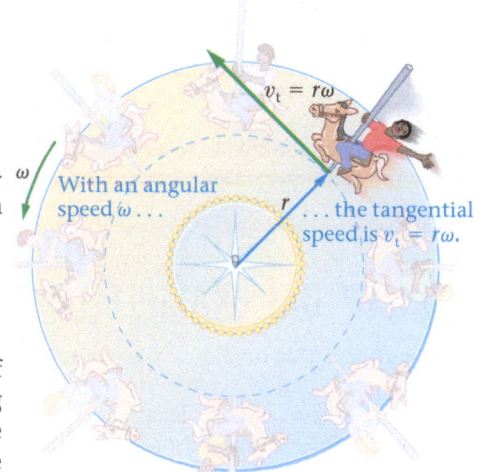

$v_t = r\omega$

With an angular speed ω...

r ...the tangential speed is $v_t = r\omega$.

▲ **FIGURE 10-9 Angular and linear speed** Overhead view of a child riding on a merry-go-round. The child's path is a circle centered on the merry-go-round's axis of rotation. At any given time the child is moving tangential to the circular path with a speed $v_t = r\omega$.

▲ **FIGURE 10-10 Visualizing Concepts Tangential Speed** In the top photo, the letters on the label are blurred by the rotational motion of the record. Notice that the blurring is greater the greater the distance from the axis of rotation—showing that the linear speed of the record increases with radius, as expected. Similarly, the boy in the bottom photo is moving faster than the girl, because his path is at a greater distance from the axis of rotation.

PHYSICS IN CONTEXT
Looking Ahead

Though it is a bit surprising at first, rotational motion is directly related to the motion of a pendulum swinging back and forth, and to the motion of a mass oscillating up and down on a spring. These connections are established in Section 13-3.

Tangential Speed of a Rotating Object

$$v_t = r\omega$$

SI unit: m/s

10-12

It is important to note that ω must be given in rad/s for this relationship to be valid.

In the case of the merry-go-round, if the radius of the child's circular path is $r = 4.25$ m, the tangential speed is $v_t = r\omega = (4.25\text{ m})(0.838\text{ rad/s}) = 3.56$ m/s. Examples of the dependence of linear speed on radius are shown in **FIGURE 10-10**. When it is clear that we are referring to the tangential speed, we will often drop the subscript t, and simply write $v = r\omega$.

An interesting application of the relationship between linear and angular speeds is provided in the operation of a compact disk (CD). As you may know, a CD is played by shining a laser beam onto the disk, and then converting the pattern of reflected light into a pattern of sound waves. For proper operation, however, the linear speed of the disk where the laser beam shines on it must be maintained at the constant value of 1.25 m/s. As the CD is played, the laser beam scans the disk in a spiral track from near the center outward to the rim. In order to maintain the required linear speed, the angular speed of the disk must decrease as the beam scans outward to larger distances, r. The required angular speeds are determined in the next Exercise.

EXERCISE 10-7 ANGULAR SPEED OF A CD

Find the angular speed a CD must have to give a linear speed of 1.25 m/s when the laser beam shines on the disk **(a)** 2.50 cm and **(b)** 6.00 cm from its center.

REASONING AND SOLUTION

a. Using $v = 1.25$ m/s and $r = 0.0250$ m in $v_t = r\omega$, we find

$$\omega = \frac{v}{r} = \frac{1.25\text{ m/s}}{0.0250\text{ m}} = 50.0\text{ rad/s} = 477\text{ rpm}$$

b. Similarly, with $r = 0.0600$ m we find

$$\omega = \frac{v}{r} = \frac{1.25\text{ m/s}}{0.0600\text{ m}} = 20.8\text{ rad/s} = 199\text{ rpm}$$

Thus, a CD slows from about 500 rpm to roughly 200 rpm as it plays.

RWP* How do the angular and tangential speeds of an object vary from one point to another? We explore this question in the following Conceptual Example.

*Real World Physics applications are denoted by the acronym RWP.

CONCEPTUAL EXAMPLE 10-8 COMPARE THE SPEEDS   PREDICT/EXPLAIN

Two children ride on a merry-go-round, with child 1 at a greater distance from the axis of rotation than child 2, as shown in our sketch. **(a)** Is the angular speed of child 1 greater than, less than, or the same as the angular speed of child 2? **(b)** Which of the following is the *best explanation* for your prediction?

I. Child 1 is moving faster than child 2, and hence has the greater angular speed.

II. Child 2 is closer to the axis of rotation, which results in a greater angular speed.

III. Both children complete a revolution in the same time, so their angular speeds are the same.

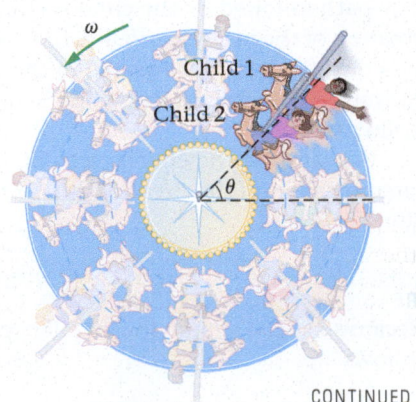

REASONING AND DISCUSSION

At any given time, the angle θ for child 1 is the same as the angle for child 2, as shown. Therefore, when the angle for child 1 has gone through 2π, for example, so has the angle for child 2. As a result, they have the same angular speed. In fact, *each and every point on the merry-go-round has exactly the same angular speed.*

CONTINUED

The tangential speeds are different, however. Child 1 has the greater tangential speed because he travels around a larger circle in the same time that child 2 travels around a smaller circle. This is in agreement with the relationship $v = r\omega$, because the radius to child 1 is greater than the radius to child 2. That is, $v_1 = r_1\omega > v_2 = r_2\omega$.

ANSWER
(a) The angular speeds are the same for the two children. **(b)** The best explanation is III.

Centripetal Acceleration   Because the children on the merry-go-round move in a circular path, they experience a centripetal acceleration, a_{cp} (Section 6-5). The centripetal acceleration is always directed toward the axis of rotation and has a magnitude given by

$$a_{cp} = \frac{v^2}{r}$$

The speed v in this expression is the tangential speed, $v = v_t = r\omega$, and therefore the centripetal acceleration in terms of ω is

$$a_{cp} = \frac{(r\omega)^2}{r}$$

Canceling one power of r, we have:

Centripetal Acceleration of a Rotating Object

$a_{cp} = r\omega^2$ 	10-13

SI unit: m/s^2

If the radius of a child's circular path on the merry-go-round is 4.25 m, and the angular speed of the ride is 0.838 rad/s, the centripetal acceleration of the child is $a_{cp} = r\omega^2 = (4.25\text{ m})(0.838\text{ rad/s})^2 = 2.98\text{ m/s}^2$.

   RWP Although the centripetal acceleration of a merry-go-round is typically only a fraction of the acceleration of gravity, rotating devices called **centrifuges** can produce extremely large centripetal accelerations. Two examples of centrifuges of different sizes are shown in **FIGURE 10-11**. The world's most powerful research centrifuge, operated by the U.S. Army Corps of Engineers, can subject 2.2-ton payloads to accelerations as high as $350g$ (350 times greater than the acceleration of gravity). This centrifuge is used to study earthquake engineering and dam erosion. The Air Force uses centrifuges to subject prospective jet pilots to the accelerations they will experience during rapid flight maneuvers, and in the future NASA may even use a human-powered centrifuge for gravity studies aboard the International Space Station.

   The centrifuges most commonly encountered in everyday life are those found in virtually every medical laboratory in the world. These devices, which can produce centripetal accelerations in excess of $13,000g$, are used to separate blood cells from blood plasma. They do this by speeding up the natural tendency of cells to settle out of plasma from days to minutes. The ratio of the packed cell volume to the total blood volume gives the *hematocrit value*, which is a useful clinical indicator of blood quality. In the next Example we consider the operation of a *microhematocrit centrifuge*, which measures the hematocrit value of a small (micro) sample of blood.

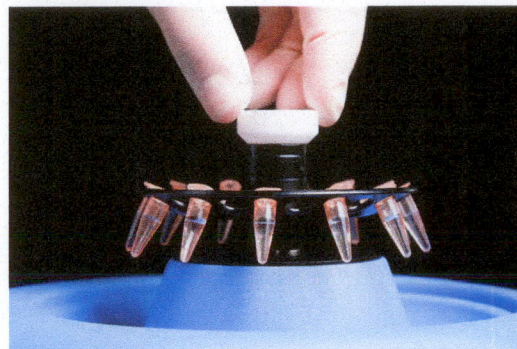

▲ **FIGURE 10-11 Visualizing Concepts Centripetal Acceleration** The large centrifuge (top), at the Gagarin Cosmonaut Training Center, is used to train Russian cosmonauts for space missions. This device, which rotates at 36 rpm, can produce a centripetal acceleration of over 290 m/s^2, 30 times the acceleration of gravity. The microcentrifuge with Eppendorf tubes (bottom) is used to separate blood cells from plasma. The volume of red blood cells in a given quantity of whole blood is a major factor in determining the oxygen-carrying capacity of the blood.

QUICK EXAMPLE 10-9   THE MICROHEMATOCRIT

BIO In a microhematocrit centrifuge, small samples of blood are placed in heparinized capillary tubes (heparin is an anticoagulant). The tubes are rotated at 11,500 rpm, with the bottoms of the tubes 9.07 cm from the axis of rotation. **(a)** Find the linear speed of the bottom of the tubes. **(b)** What is the centripetal acceleration at the bottom of the tubes?

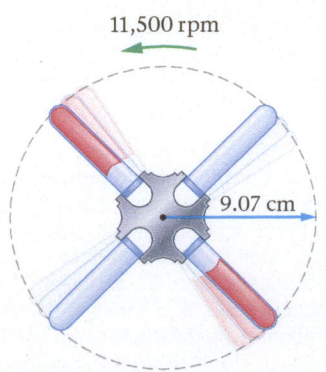

11,500 rpm

9.07 cm

REASONING AND SOLUTION
Our sketch shows a top view of the centrifuge, with the capillary tubes rotating at 11,500 rpm. Notice that the bottoms of the tubes move in a circular path of radius 9.07 cm.

CONTINUED

a. Linear and angular speeds are related by $v = r\omega$. Once we convert the angular speed to rad/s, we can use this relationship to determine the linear speed v.

b. The centripetal acceleration is $a_{cp} = r\omega^2$. Using ω from part (a) yields the desired result.

Part (a)

1. Convert the angular speed, ω, to radians per second:

$$\omega = (11{,}500 \text{ rev/min})\left(\frac{2\pi \text{ rad}}{1 \text{ rev}}\right)\left(\frac{1 \text{ min}}{60 \text{ s}}\right)$$
$$= 1.20 \times 10^3 \text{ rad/s}$$

2. Use $v = r\omega$ to calculate the linear speed:

$$v = r\omega = (0.0907 \text{ m})(1.20 \times 10^3 \text{ rad/s})$$
$$= 109 \text{ m/s}$$

Part (b)

3. Calculate the centripetal acceleration using $a_{cp} = r\omega^2$:

$$a_{cp} = r\omega^2 = (0.0907 \text{ m})(1.20 \times 10^3 \text{ rad/s})^2$$
$$= 131{,}000 \text{ m/s}^2$$

4. As a check, calculate the centripetal acceleration using $a_{cp} = v^2/r$:

$$a_{cp} = \frac{v^2}{r} = \frac{(109 \text{ m/s})^2}{0.0907 \text{ m}} = 131{,}000 \text{ m/s}^2$$

Every point on a centrifuge tube has the same angular speed. As a result, points near the top of a tube have smaller linear speeds and centripetal accelerations than do points near the bottom of a tube. In this case, the bottoms of the tubes experience a centripetal acceleration about 13,400 times greater than the acceleration of gravity on the surface of the Earth; that is, $a_{cp} = 131{,}000 \text{ m/s}^2 = 13{,}400g$. This means that the force exerted on the bottom of a tube by a 1-ounce sample would be about 800 lb.

Tangential Acceleration If the angular speed of the merry-go-round in Conceptual Example 10-8 changes, the tangential speed of the children changes as well. It follows, then, that the children will experience a tangential acceleration, a_t. We can determine a_t by considering the relationship $v_t = r\omega$. If ω changes by the amount $\Delta\omega$, with r remaining constant, the corresponding change in the tangential speed is

$$\Delta v_t = r\Delta\omega$$

If this change in ω occurs in the time Δt, the tangential acceleration is

$$a_t = \frac{\Delta v_t}{\Delta t} = r\frac{\Delta\omega}{\Delta t}$$

Recalling that $\Delta\omega/\Delta t$ is the angular acceleration, α, we find that

Tangential Acceleration of a Rotating Object

$$a_t = r\alpha \qquad\qquad 10\text{-}14$$

SI unit: m/s²

As with the tangential speed, we will often drop the subscript t in a_t when no confusion will arise.

In general, the children on the merry-go-round may experience both tangential and centripetal accelerations at the same time. Recall that a_t is due to a changing tangential speed, and that a_{cp} is caused by a changing direction of motion, even if the tangential speed remains constant. To summarize:

Tangential Versus Centripetal Acceleration

$a_t = r\alpha$ \qquad due to changing angular speed

$a_{cp} = r\omega^2$ \qquad due to changing direction of motion

As the names suggest, the tangential acceleration is always tangential to an object's path; the centripetal acceleration is always perpendicular to its path.

In cases in which both the centripetal and tangential accelerations are present, the total acceleration is the vector sum of the two, as indicated in **FIGURE 10-12**. Notice that

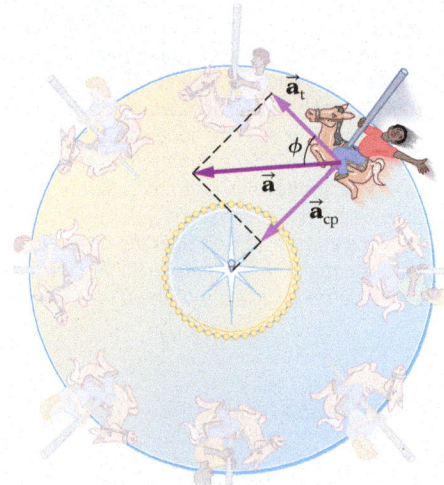

▲ **FIGURE 10-12 Centripetal and tangential accelerations** If the angular speed of the merry-go-round is increased, the child will experience two accelerations: (i) a tangential acceleration, $\vec{a}_t$, and (ii) a centripetal acceleration, $\vec{a}_{cp}$. The child's total acceleration, $\vec{a}$, is the vector sum of $\vec{a}_t$ and $\vec{a}_{cp}$.

$\vec{a}_t$ and $\vec{a}_{cp}$ are at right angles to one another, and hence the magnitude of the total acceleration is given by the Pythagorean theorem:

$$a = \sqrt{a_t{}^2 + a_{cp}{}^2}$$

The direction of the total acceleration, measured relative to the tangential direction, is

$$\phi = \tan^{-1}\left(\frac{a_{cp}}{a_t}\right)$$

This angle is shown in Figure 10-12.

Enhance Your Understanding   (Answers given at the end of the chapter)

3. Disk 1 has a radius r and rotates with an angular velocity ω. A point on the rim of disk 1 has a tangential speed of 1 m/s and a centripetal acceleration of 3 m/s². Disk 2 has a radius $2r$ and rotates with an angular velocity 2ω. What are **(a)** the tangential speed and **(b)** the centripetal acceleration of a point on the rim of disk 2?

Section Review

- The tangential speed of a point a distance r from the axis of a rotating object is $v_t = r\omega$.

- The magnitude of the centripetal acceleration in terms of angular speed is $a_{cp} = r\omega^2$, and the magnitude of the tangential acceleration in terms of angular acceleration is $a_t = r\alpha$.

10-4 Rolling Motion

We began this chapter with a bicycle wheel rotating about its axle. In that case, the axle was at rest and every point on the wheel, such as the spot of red paint, moved in a circular path about the axle. We would like to consider a different situation now. Suppose the bicycle wheel is rolling freely, as indicated in **FIGURE 10-13**, with no slipping between the tire and the ground. The wheel still rotates about the axle, but the axle itself is moving in a straight line. As a result, the motion of the wheel is a combination of both rotational motion and linear (or **translational**) motion.

Connecting Rotational and Tangential Motions in a Rolling Wheel   To see the connection between the wheel's rotational and translational motions, we show one full rotation of the wheel in Figure 10-13. During this rotation, the axle translates (moves) forward through a distance equal to the circumference of the wheel, $2\pi r$. Because the time required for one rotation is the period, T, the translational speed of the axle is

$$v = \frac{2\pi r}{T}$$

Recalling that $\omega = 2\pi/T$, we find

$$v = r\omega = v_t \qquad\qquad 10\text{-}15$$

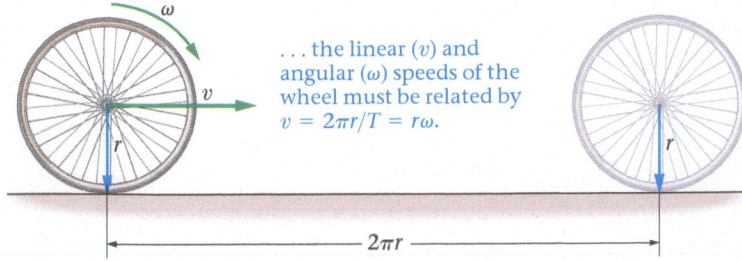

To roll without slipping . . .

ω

. . . the linear (v) and angular (ω) speeds of the wheel must be related by $v = 2\pi r/T = r\omega$.

v

r

r

$2\pi r$

◀ **FIGURE 10-13 Rolling without slipping** A wheel of radius r rolling without slipping. During one complete revolution, the center of the wheel moves forward through a distance $2\pi r$.

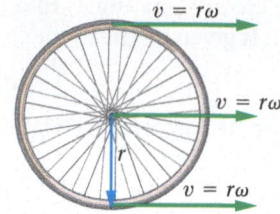

(a) Pure rotational motion   (b) Pure translational motion

▲ **FIGURE 10-14 Rotational and translational motions of a wheel (a)** In pure rotational motion, the velocities at the top and bottom of the wheel are in opposite directions. **(b)** In pure translational motion, each point on the wheel moves with the same speed in the same direction.

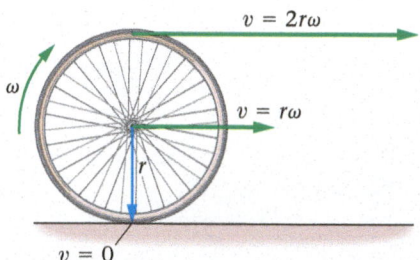

▲ **FIGURE 10-15 Velocities in rolling motion** In a wheel that rolls without slipping, the point in contact with the ground is instantaneously at rest. The center of the wheel moves forward with the speed $v = r\omega$, and the top of the wheel moves forward with twice that speed, $v = 2r\omega$.

Hence, the translational speed of the axle is equal to the tangential speed of a point on the rim of a wheel spinning with angular speed ω.

A rolling object, then, combines rotational motion with angular speed ω, and translational motion with linear speed $v = r\omega$, where r is the radius of the object. Let's consider these two motions one at a time. First, in **FIGURE 10-14 (a)** we show pure rotational motion with angular speed ω. In this case, the axle is at rest, and points at the top and bottom of the wheel have tangential velocities that are equal in magnitude, $v = r\omega$, but point in opposite directions.

Next, we consider translational motion with speed $v = r\omega$. This is illustrated in **FIGURE 10-14 (b)**, where we see that each point on the wheel moves in the same direction with the same velocity. If this were the only motion the wheel had, it would be skidding across the ground, instead of rolling without slipping.

Finally, we combine these two motions by simply adding the velocity vectors in Figures 10-14 (a) and (b). The result is shown in **FIGURE 10-15**. At the top of the wheel the two velocity vectors are in the same direction, so they sum to give a speed of $2r\omega$. At the axle, the velocity vectors sum to give a speed $r\omega$. Finally, at the bottom of the wheel, the velocity vectors from rotation and translation have equal magnitude, but are in opposite directions. As a result, these velocities cancel, giving a speed of zero where the wheel is in contact with the ground. An example of a bicycle wheel rolling without slipping is shown in **FIGURE 10-16**. Notice how the spokes are blurry at the top of the wheel but sharp at the bottom.

The fact that the bottom of the wheel is instantaneously at rest, so that it is in static contact with the ground, is precisely what is meant by "rolling without slipping." Thus, a wheel that rolls without slipping is just like the situation when you are walking—even though your body as a whole moves forward, the soles of your shoes are momentarily at rest every time you place them on the ground. This point was discussed in detail in Conceptual Example 6-5.

◄ **FIGURE 10-16** This photograph of a rolling wheel gives a visual indication of the speed of its various parts. The bottom of the wheel is at rest at any instant, so the image there is sharp. The top of the wheel has the greatest speed, and the image there shows the most blurring. (Compare Figure 10-15.)

EXERCISE 10-10   ROLLING TIRES

A car with tires of radius 31 cm drives on the highway at 65 mph. **(a)** What is the angular speed of the tires? **(b)** What is the linear speed of the tops of the tires?

REASONING AND SOLUTION

a. We know the linear speed and the radius of the tires. Thus, we can rearrange $v = r\omega$ to solve for the angular speed, ω. (A useful conversion factor here is 1 mph = 0.447 m/s.)

$$\omega = \frac{v}{r} = \frac{(65 \text{ mph})\left(\dfrac{0.447 \text{ m/s}}{1 \text{ mph}}\right)}{0.31 \text{ m}} = 94 \text{ rad/s}$$

This is about 15 revolutions per second.

b. The tops of the tires have a speed of $2v = 130$ mph.

Enhance Your Understanding

(Answers given at the end of the chapter)

4. In a popular pirate movie, two pirates find themselves on a runaway water wheel, as shown in **FIGURE 10-17**. The wheel rolls without slipping to the right with a linear speed v. Pirate 1 at the bottom of the wheel runs to the right with a speed v_1, and pirate 2 on top of the wheel runs to the left with a speed v_2. Each pirate runs with the speed necessary to maintain a fixed position relative to the axle of the wheel as it rolls. What are the running speeds of **(a)** pirate 1 and **(b)** pirate 2 in terms of v? **(c)** What are the speeds of the two pirates relative to the ground?

Running with speed v_2 to the left

Running with speed v_1 to the right

▲ **FIGURE 10-17** Enhance Your Understanding 4.

Section Review

- A wheel that rolls without slipping has zero speed at the bottom of the wheel, speed v at the axle, and speed $2v$ at the top of the wheel.

10-5 Rotational Kinetic Energy and the Moment of Inertia

An object in motion has kinetic energy, whether that motion is translational, rotational, or a combination of the two. In translational motion, for example, the kinetic energy of a mass m moving with a speed v is $K = \frac{1}{2}mv^2$. We cannot use this expression for a rotating object, however, because the speed v of each particle within a rotating object varies with its distance r from the axis of rotation. Thus, there is no unique value of v for an entire rotating object. On the other hand, there *is* a unique value of ω, the angular speed, and it applies to all particles in the object.

Kinetic Energy in Terms of Angular Speed To see how the kinetic energy of a rotating object depends on its angular speed, we start with a particularly simple system consisting of a rod of negligible mass and length r rotating about one end with an angular speed ω. Attached to the other end of the rod is a point mass m, as **FIGURE 10-18** shows. To find the kinetic energy of the mass, recall that its linear speed is $v = r\omega$ (Equation 10-12). Therefore, the translational kinetic energy of the mass m is

$$K = \tfrac{1}{2}mv^2 = \tfrac{1}{2}m(r\omega)^2 = \tfrac{1}{2}(mr^2)\omega^2 \qquad \text{10-16}$$

Notice that the kinetic energy of the mass depends not only on the angular speed squared (analogous to the way the translational kinetic energy depends on the linear speed squared), but also on the radius squared—that is, the kinetic energy depends on the *distribution* of mass in the rotating object. To be specific, mass near the axis of rotation contributes little to the kinetic energy because its speed ($v = r\omega$) is small. On the other hand, the farther a mass is from the axis of rotation, the greater its speed v for a given angular velocity, and thus the greater its kinetic energy.

You've probably noticed that the kinetic energy in Equation 10-16 is similar in form to the translational kinetic energy. Instead of $\frac{1}{2}(m)v^2$, we now have $\frac{1}{2}(mr^2)\omega^2$. Clearly, then, the quantity mr^2 plays the role of the mass for the rotating object. This "rotational mass" is given a special name in physics: the **moment of inertia**, I. Thus, in general, the kinetic energy of an object rotating with an angular speed ω can be written as:

Rotational Kinetic Energy

$$K = \tfrac{1}{2}I\omega^2 \qquad \text{10-17}$$

SI unit: J

Big Idea **2** Rolling motion is a combination of linear and rotational motions. When a wheel rolls without slipping with a linear speed v, the rim of the wheel has an instantaneous speed of zero where it touches the ground and an instantaneous speed of $2v$ at the top of the wheel.

Big Idea **3** A rotating object has kinetic energy, just like an object in linear motion. The kinetic energy of rotation can be used in energy conservation in the same way as linear kinetic energy.

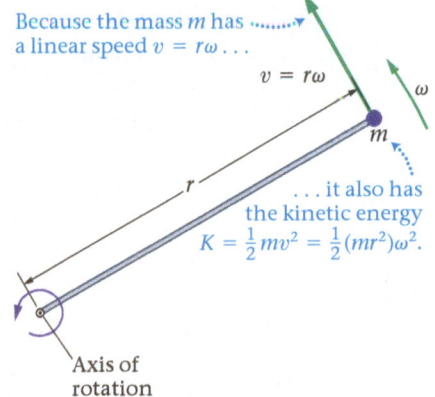

Because the mass m has a linear speed $v = r\omega$ ...

$v = r\omega$

m

... it also has the kinetic energy $K = \frac{1}{2}mv^2 = \frac{1}{2}(mr^2)\omega^2$.

Axis of rotation

▲ **FIGURE 10-18** **Kinetic energy of a rotating object** As this rod rotates about the axis of rotation with an angular speed ω, the mass has a speed of $v = r\omega$. It follows that the kinetic energy of the mass is $K = \frac{1}{2}mv^2 = \frac{1}{2}(mr^2)\omega^2$.

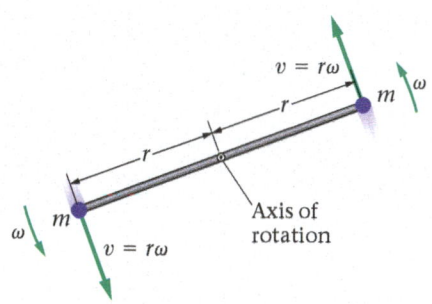

▲ **FIGURE 10-19 Kinetic energy of a rotating object of arbitrary shape** To calculate the kinetic energy of an object of arbitrary shape as it rotates about an axis with angular speed ω, imagine dividing it into small mass elements, m_i. The total kinetic energy of the object is the sum of the kinetic energies of all the mass elements.

**PHYSICS
IN CONTEXT
Looking Back**

The kinetic energy, first defined in Chapter 7, plays a key role in defining the moment of inertia.

The greater the moment of inertia, the greater an object's rotational kinetic energy. As we have just seen, in the special case of a point mass m a distance r from the axis of rotation, the moment of inertia is simply $I = mr^2$.

Determining the Moment of Inertia We now show how to find the moment of inertia for an object of arbitrary, fixed shape, as in **FIGURE 10-19**. Suppose, for example, that this object rotates about the axis indicated in the figure with an angular speed ω. To calculate the kinetic energy of the object, we first imagine dividing it into a collection of small mass elements, m_i. We then calculate the kinetic energy of each element and sum over all elements.

Following this plan, we find that the total kinetic energy of an arbitrary rotating object is

$$K = \sum\left(\tfrac{1}{2}m_iv_i^2\right)$$

In this expression, m_i is the mass of one of the small mass elements and v_i is its speed. If m_i is at the radius r_i from the axis of rotation, as indicated in Figure 10-19, its speed is $v_i = r_i\omega$. Notice that it is not necessary to write a separate angular speed, ω_i, for each element, because *all* mass elements of the object have exactly the same angular speed, ω. Therefore,

$$K = \sum\left(\tfrac{1}{2}m_ir_i^2\omega^2\right) = \tfrac{1}{2}\left(\sum m_ir_i^2\right)\omega^2$$

Now, in analogy with our results for the single mass, we can define the moment of inertia, I, as follows:

Definition of Moment of Inertia, I

$$I = \sum m_ir_i^2 \qquad\qquad 10\text{-}18$$

SI unit: kg · m^2

The precise value of I for a given object depends on its distribution of mass. A simple example of this dependence is given in the following Exercise.

EXERCISE 10-11   MOMENT OF INERTIA

Use the general definition of the moment of inertia, $I = \sum m_ir_i^2$, to find the moment of inertia for the dumbbell-shaped object shown in **FIGURE 10-20**. Notice that the axis of rotation goes through the center of the object and points out of the page. In addition, assume that the masses may be treated as point masses.

REASONING AND SOLUTION
Referring to Figure 10-20, we see that $m_1 = m_2 = m$ and $r_1 = r_2 = r$. Therefore, the moment of inertia is

$$I = \sum m_ir_i^2 = m_1r_1^2 + m_2r_2^2 = mr^2 + mr^2 = 2mr^2$$

▲ **FIGURE 10-20 A dumbbell-shaped object rotating about its center**

The connection between rotational kinetic energy and the moment of inertia is explored in more detail in the following Example.

EXAMPLE 10-12   NOSE TO THE GRINDSTONE

A grindstone with a radius of 0.610 m is being used to sharpen an ax. **(a)** If the linear speed of the stone relative to the ax is 1.50 m/s, and the stone's rotational kinetic energy is 13.0 J, what is its moment of inertia? **(b)** If the linear speed is doubled to 3.00 m/s, what is the corresponding kinetic energy of the grindstone?

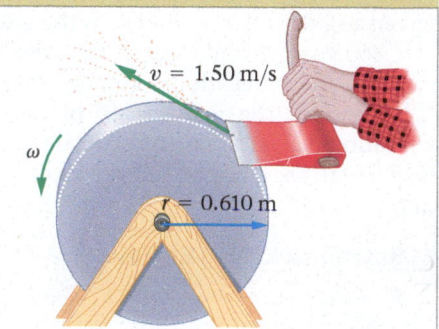

PICTURE THE PROBLEM
Our sketch shows the grindstone spinning with an angular speed ω, which is not given in the problem statement. We do know, however, that the linear speed of the grindstone at its rim is $v = 1.50$ m/s and that its radius is $r = 0.610$ m. At this rate of rotation, the stone has a kinetic energy of 13.0 J.

CONTINUED

REASONING AND STRATEGY

a. Recall that rotational kinetic energy and moment of inertia are related by $K = \frac{1}{2}I\omega^2$; thus $I = 2K/\omega^2$. We are not given ω, but we can find it from the connection between linear and angular speed, $v = r\omega$. Thus, we begin by finding ω. We then use ω, along with the kinetic energy K, to find I.

b. We can find the new angular speed with $\omega = v/r$. We can then use I from part (a), along with $K = \frac{1}{2}I\omega^2$, to find the new kinetic energy.

Known   Radius, $r = 0.610$ m; linear speed, $v = 1.50$ m/s; kinetic energy, $K = 13.0$ J.
Unknown   **(a)** Moment of inertia, $I = ?$ **(b)** $K = ?$ for $v = 3.00$ m/s

SOLUTION

Part (a)

1. Find the angular speed of the grindstone:

$$\omega = \frac{v}{r} = \frac{1.50 \text{ m/s}}{0.610 \text{ m}} = 2.46 \text{ rad/s}$$

2. Solve for the moment of inertia in terms of kinetic energy:

$$K = \frac{1}{2}I\omega^2 \text{ or } I = \frac{2K}{\omega^2}$$

3. Substitute numerical values for K and ω:

$$I = \frac{2K}{\omega^2} = \frac{2(13.0 \text{ J})}{(2.46 \text{ rad/s})^2} = 4.30 \text{ J} \cdot \text{s}^2 = 4.30 \text{ kg} \cdot \text{m}^2$$

Part (b)

4. Find the angular speed of the grindstone corresponding to $v = 3.00$ m/s:

$$\omega = \frac{v}{r} = \frac{3.00 \text{ m/s}}{0.610 \text{ m}} = 4.92 \text{ rad/s}$$

5. Determine the kinetic energy, K, using the moment of inertia, I, from part (a):

$$K = \frac{1}{2}I\omega^2 = \frac{1}{2}(4.30 \text{ kg} \cdot \text{m}^2)(4.92 \text{ rad/s})^2 = 52.0 \text{ J}$$

INSIGHT

(a) We found I by relating it to the rotational kinetic energy of the grindstone. Later in this section we show how to calculate the moment of inertia of a disk directly, given its radius and mass. **(b)** Doubling the linear speed, v, results in a doubling of the angular speed, ω. The kinetic energy K depends on ω^2; therefore doubling ω increases K by a factor of 4—from 13.0 J to 4(13.0 J) = 52.0 J.

PRACTICE PROBLEM — PREDICT/CALCULATE

When the ax is pressed firmly against the grindstone for sharpening, the angular speed of the grindstone decreases. **(a)** If the rotational kinetic energy of the grindstone is cut in half to 6.50 J, is the corresponding angular speed greater than, less than, or equal to one-half the original angular speed of 2.46 rad/s? Explain. **(b)** Find the new angular speed. [**Answer: (a)** The rotational kinetic energy depends on the square of the angular speed. Therefore, reducing the kinetic energy by a factor of 2 corresponds to reducing the angular speed by a factor of $\sqrt{2}$. Thus, the new angular speed is greater than ½ the initial angular speed. **(b)** The new angular speed (using $K = 6.50$ J) is $\omega = \sqrt{2K/I} = 1.74$ rad/s, which is greater than ½(2.46 rad/s) = 1.23 rad/s.]

Some related homework problems: Problem 47, Problem 48, Problem 51

The Moment of Inertia for Various Shapes   We return now to the dependence of the moment of inertia on the particular shape, or mass distribution, of an object. Suppose, for example, that a mass M is formed into the shape of a *hoop* of radius R. In addition, consider the case where the axis of rotation is perpendicular to the plane of the hoop and passes through its center, as shown in **FIGURE 10-21**. This is similar to a bicycle wheel rotating about its axle, if one ignores the spokes. In terms of small mass elements, we can write the moment of inertia as

$$I = \sum m_i r_i^2$$

Each mass element of the hoop, however, is at the same radius R from the axis of rotation; that is, $r_i = R$. Hence, the moment of inertia in this case is

$$I = \sum m_i r_i^2 = \sum m_i R^2 = \left(\sum m_i\right)R^2$$

Clearly, the sum of all the elementary masses is simply the total mass of the hoop, $\sum m_i = M$. Therefore, the moment of inertia of a hoop of mass M and radius R is

$$I = MR^2 \text{ (hoop)}$$

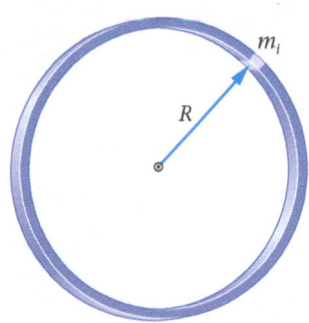

▲ **FIGURE 10-21 The moment of inertia of a hoop** Consider a hoop of mass M and radius R. Each small mass element is at the same distance, R, from the center of the hoop. The moment of inertia in this case is $I = MR^2$.

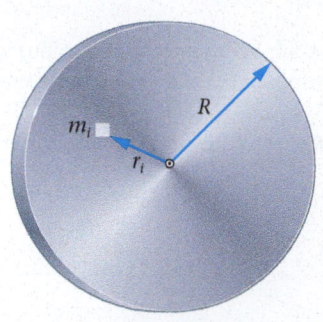

In contrast, if the same mass, M, is formed into a uniform *disk* of the same radius, R, the moment of inertia is different. To see this, notice that it is no longer true that $r_i = R$ for all mass elements. In fact, most of the mass elements are closer to the axis of rotation than was the case for the hoop, as indicated in **FIGURE 10-22**. Thus, most of the r_i are less than R, and hence the moment of inertia will be smaller for the disk than for the hoop. A detailed calculation, summing over all mass elements, yields the following result:

$$I = \tfrac{1}{2}MR^2 \text{ (disk)}$$

As expected, I is less for the disk than for the hoop.

▲ **FIGURE 10-22 The moment of inertia of a disk** Consider a disk of mass M and radius R. Mass elements for the disk are at distances from the center ranging from 0 to R. The moment of inertia in this case is $I = \tfrac{1}{2}MR^2$.

EXERCISE 10-13　UNIFORM DISK

If the grindstone in Example 10-12 is a uniform disk, what is its mass?

REASONING AND SOLUTION

Solving $I = \tfrac{1}{2}MR^2$ for the mass yields

$$M = \frac{2I}{R^2} = \frac{2(4.30 \text{ kg} \cdot \text{m}^2)}{(0.610 \text{ m})^2} = 23.1 \text{ kg}$$

Thus, the grindstone has a weight of roughly 51 lb.

Table 10-1 collects moments of inertia for a variety of objects. Notice that in all cases the moment of inertia is of the form $I = (\text{constant})MR^2$. It is only the numerical constant in front of MR^2 that changes from one object to another.

Note also that objects of the same general shape but with different mass distributions—such as solid and hollow spheres—have different moments of inertia. In particular, a hollow sphere has a larger I than a solid sphere of the same mass, for the same reason that a hoop's moment of inertia is greater than a disk's: More of its mass is at a greater distance from the axis of rotation. Thus, I is a measure of both the shape *and* the mass distribution of an object.

RWP Moment of Inertia of the Earth　For a good example of how the moment of inertia depends on the distribution of mass, consider the moment of inertia of the Earth. If the Earth were a uniform sphere of mass M_E and radius R_E, its moment of inertia

---

TABLE 10-1　Moments of Inertia for Uniform, Rigid Objects of Various Shapes and Total Mass M

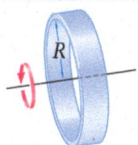 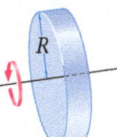 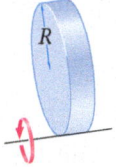 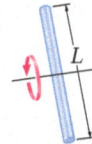 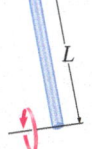

| Hoop or cylindrical shell $I = MR^2$ | Disk or solid cylinder $I = \tfrac{1}{2}MR^2$ | Disk or solid cylinder (axis at rim) $I = \tfrac{3}{2}MR^2$ | Long thin rod (axis through midpoint) $I = \tfrac{1}{12}ML^2$ | Long thin rod (axis at one end) $I = \tfrac{1}{3}ML^2$ |

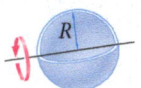 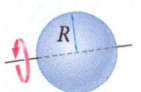 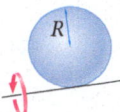 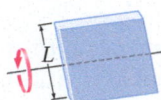 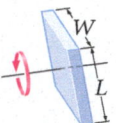

| Hollow sphere $I = \tfrac{2}{3}MR^2$ | Solid sphere $I = \tfrac{2}{5}MR^2$ | Solid sphere (axis at rim) $I = \tfrac{7}{5}MR^2$ | Solid plate (axis through center, in plane of plate) $I = \tfrac{1}{12}ML^2$ | Solid plate (axis perpendicular to plane of plate) $I = \tfrac{1}{12}M(L^2 + W^2)$ |

would be $\frac{2}{5}M_E R_E{}^2 = 0.4 M_E R_E{}^2$. In fact, the Earth's moment of inertia is only $0.331 M_E R_E{}^2$, considerably less than for a uniform sphere. This is due to the fact that the Earth is not homogeneous, but instead has a dense inner core surrounded by a less dense outer core and an even less dense mantle, as shown in **FIGURE 10-23**. The resulting concentration of mass near its axis of rotation gives the Earth a much smaller moment of inertia than it would have if its mass were uniformly distributed.

On the other hand, if the polar ice caps were to melt and release their water into the oceans, the Earth's moment of inertia would increase. This is because mass that had been near the axis of rotation (in the polar ice) would now be distributed more or less uniformly around the Earth (in the oceans). With more of the Earth's mass at greater distances from the axis of rotation, the moment of inertia would increase. If such an event were to occur, not only would the moment of inertia increase, but the length of the day would increase as well. We will discuss the reasons for this—which have to do with the conservation of angular momentum—in the next chapter.

The moment of inertia of an object also depends on the location and orientation of the axis of rotation. If the axis of rotation is moved, all of the r_i change, leading to a different result for I. This is investigated for a dumbbell system in the following Conceptual Example.

▲ **FIGURE 10-23** The distribution of mass in the Earth is not uniform. Dense materials, like iron and nickel, have concentrated near the center, while less dense materials, like silicon and aluminum, have risen to the surface. This concentration of mass near the axis of rotation lowers the Earth's moment of inertia.

CONCEPTUAL EXAMPLE 10-14   COMPARE THE MOMENTS OF INERTIA

If the dumbbell-shaped object in Figure 10-20 is rotated about one end, is its moment of inertia greater than, less than, or the same as the moment of inertia about its center? As before, assume that the masses can be treated as point masses.

REASONING AND DISCUSSION

As we saw in Exercise 10-11, the moment of inertia about the center of the dumbbell is $I = 2mR^2$. When the axis is at one end, that mass is at the radius $r = 0$, and the other mass is at $r = 2R$. Therefore, the moment of inertia is

$$I = \sum m_i r_i{}^2 = m \cdot 0 + m(2R)^2 = 4mR^2$$

Thus, the moment of inertia doubles when the axis of rotation is moved from the center to one end.

The reason I increases is that the moment of inertia depends on the radius squared. Hence, even small increases in r can cause significant increases in I. By moving the axis to one end, the radius to the other mass is increased to its greatest possible value. As a result, I increases.

ANSWER
The moment of inertia is greater about one end than about the center.

Extending Linear-Rotational Analogies to Mass and Kinetic Energy   In the accompanying table, we summarize the similarities between the translational kinetic energy, $K = \frac{1}{2}mv^2$, and the rotational kinetic energy, $K = \frac{1}{2}I\omega^2$. As expected, we see that the linear speed, v, has been replaced with the angular speed, ω. In addition, notice that the mass, m, has been replaced with the moment of inertia, I.

As suggested by these analogies, the moment of inertia, I, plays the same role in rotational motion that mass plays in translational motion. For example, the larger I, the more resistant an object is to any change in its angular velocity—an object with a large I is difficult to start rotating, and once it is rotating, it is difficult to stop. We shall see further applications of this analogy in the next chapter when we consider angular momentum.

RWP   Rotational kinetic energy can be used to improve the energy efficiency of an automobile. Whenever the car slows down, its kinetic energy can be temporarily stored by rotating a flywheel at a high angular speed. When the driver releases the brake, the energy stored in the flywheel's rotation is transferred to the rear wheels by a specially designed transmission. An example of this type of energy-recovery system is shown in **FIGURE 10-24**. Volvo Car Group has extensively tested the technology on public roads with an experimental version of their S60 car. The measurements indicate that the

Linear Quantity	Angular Quantity
v	ω
m	I
$\frac{1}{2}mv^2$	$\frac{1}{2}I\omega^2$

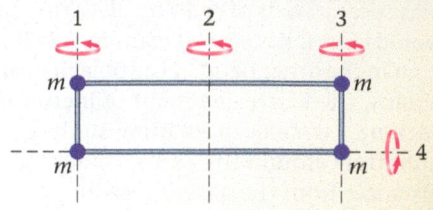

▲ **FIGURE 10-24 Flywheel technology for energy recovery** A rapidly spinning flywheel in the circular structure on the passenger side of the car stores energy from braking and uses it to help accelerate the car from a stop.

flywheel energy storage can reduce fuel consumption by up to 25%, especially in urban start-and-stop traffic.

Enhance Your Understanding   (Answers given at the end of the chapter)

5. Four masses m are connected by rods of negligible mass in a rectangular arrangement, as shown in **FIGURE 10-25**. The system can be rotated about one of four different axes of rotation (1, 2, 3, or 4). Rank the four axes of rotation in order of increasing moment of inertia of the rotating system. Indicate ties where appropriate.

▲ **FIGURE 10-25** Enhance Your Understanding 5.

Section Review

- Rotational kinetic energy is $\frac{1}{2}I\omega^2$, in close analogy to linear kinetic energy $\frac{1}{2}mv^2$.

- The moment of inertia I is the rotational analog to the mass in a linear system. The greater the mass, and the greater its distance from the axis of rotation, the greater the moment of inertia.

10-6 Conservation of Energy

In this section, we consider the mechanical energy of objects that roll without slipping, and show how to apply energy conservation to such systems. In addition, we consider objects that rotate as a string or rope unwinds: for example, a pulley with a string wrapped around its circumference, or a yo-yo with a string wrapped around its axle. As long as the unwinding process and the rolling motion occur without slipping, the two situations are basically the same—at least as far as energy considerations are concerned.

Energy Conservation for Rolling Objects To apply energy conservation to rolling objects, we first need to determine the kinetic energy of rolling motion. In Section 10-4 we saw that rolling motion is a combination of rotation and translation. It follows, then, that the kinetic energy of a rolling object is simply the sum of its translational kinetic energy, $\frac{1}{2}mv^2$, and its rotational kinetic energy, $\frac{1}{2}I\omega^2$:

Kinetic Energy of Rolling Motion

$$K = \tfrac{1}{2}mv^2 + \tfrac{1}{2}I\omega^2$$

10-19

The moment of inertia I in this expression is the moment of inertia about the center of the rolling object.

We can simplify the expression for the kinetic energy of a rolling object by using the fact that linear and angular speeds are related. In fact, recall that $v = r\omega$ (Equation 10-12), which can be rewritten as $\omega = v/r$. Substituting this into our expression for the rolling kinetic energy yields

Kinetic Energy of Rolling Motion: Alternative Form

$$K = \tfrac{1}{2}mv^2 + \tfrac{1}{2}I\left(\frac{v}{r}\right)^2 = \tfrac{1}{2}mv^2\left(1 + \frac{I}{mr^2}\right)$$

10-20

Noting that $I = (\text{constant})mr^2$, we see that the last term in Equation 10-20 is a constant that depends on the shape and mass distribution of the rolling object.

A special case of some interest is the point particle. In this case, by definition, all of the mass is at a single point. Therefore, $r = 0$, and hence $I = 0$. Substituting $I = 0$ in either Equation 10-19 or Equation 10-20 yields $K = \frac{1}{2}mv^2$, as expected.

Next, we apply Equations 10-19 and 10-20 to a hollow sphere (a basketball) that rolls without slipping.

PROBLEM-SOLVING NOTE

Energy Conservation with Rotational Motion

When applying energy conservation to a system with rotational motion, be sure to include the rotational kinetic energy, $\frac{1}{2}I\omega^2$.

**PHYSICS
IN CONTEXT
Looking Back**

Conservation of energy, first introduced in Chapter 8, is just as important in rotational motion as it is in linear motion.

EXAMPLE 10-15 · LIKE A ROLLING BASKETBALL

A basketball is basically a hollow sphere with a mass of 0.650 kg and a radius of 12.1 cm. If a basketball rolls without slipping with a linear speed of 1.33 m/s, find **(a)** the translational kinetic energy, **(b)** the rotational kinetic energy, and **(c)** the total kinetic energy of the ball.

PICTURE THE PROBLEM

The ball rolls without slipping, and hence its angular speed and linear speed are related by $v = r\omega$. The linear speed is $v = 1.33$ m/s and the radius is $r = 12.1$ cm. We are also given that the mass of the ball is 0.650 kg.

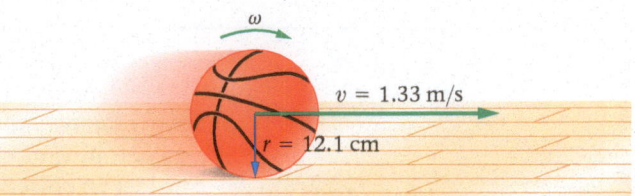

REASONING AND STRATEGY

We calculate each contribution to the kinetic energy separately. The translational kinetic energy, of course, is simply $\frac{1}{2}mv^2$. For the rotational kinetic energy, $\frac{1}{2}I\omega^2$, we use the fact that the moment of inertia for a hollow sphere is $I = \frac{2}{3}mr^2$. Finally, the ball rolls without slipping, and hence its angular speed is $\omega = v/r$. The total kinetic energy is the sum of the translational and rotational kinetic energies.

Known Mass of ball, $m = 0.650$ kg; radius, $r = 12.1$ cm; linear speed, $v = 1.33$ m/s; moment of inertia, $I = \frac{2}{3}mr^2$.
Unknown **(a)** Translational kinetic energy = ? **(b)** Rotational kinetic energy = ? **(c)** Total kinetic energy, K = ?

SOLUTION

Part (a)

1. Calculate the translational kinetic energy, $\frac{1}{2}mv^2$:

$$\tfrac{1}{2}mv^2 = \tfrac{1}{2}(0.650 \text{ kg})(1.33 \text{ m/s})^2 = 0.575 \text{ J}$$

Part (b)

2. Calculate the rotational kinetic energy symbolically, using $I = \frac{2}{3}mr^2$ and $\omega = v/r$. Notice that the radius cancels:

$$\tfrac{1}{2}I\omega^2 = \tfrac{1}{2}\left(\tfrac{2}{3}mr^2\right)\left(\tfrac{v}{r}\right)^2 = \tfrac{2}{3}\left(\tfrac{1}{2}mv^2\right) = \tfrac{1}{3}mv^2$$

3. Substitute the numerical values:

$$\tfrac{1}{2}I\omega^2 = \tfrac{1}{3}mv^2$$
$$= \tfrac{1}{3}(0.650 \text{ kg})(1.33 \text{ m/s})^2 = 0.383 \text{ J}$$

Part (c)

4. Sum the kinetic energies obtained in parts (a) and (b):

$$K = 0.575 \text{ J} + 0.383 \text{ J} = 0.958 \text{ J}$$

5. The same result is obtained using $K = \frac{1}{2}mv^2\left(1 + \frac{I}{mr^2}\right)$:

$$K = \tfrac{1}{2}mv^2\left(1 + \tfrac{I}{mr^2}\right) = \tfrac{1}{2}mv^2\left(1 + \tfrac{2}{3}\right)$$
$$= \tfrac{5}{3}\left(\tfrac{1}{2}mv^2\right) = \tfrac{5}{3}(0.575 \text{ J}) = 0.958 \text{ J}$$

INSIGHT

The symbolic result in Step 2 shows that the rotational kinetic energy of a hollow sphere rolling without slipping is precisely two-thirds the sphere's translational kinetic energy. This result is independent of the sphere's radius, as we can see by the cancellation of the radius r in Step 2.

PRACTICE PROBLEM — PREDICT/CALCULATE

Suppose the hollow sphere in this Example is replaced with a uniform disk of the same mass and radius. **(a)** If the disk rolls without slipping with the same speed as the hollow sphere, is its rotational kinetic energy greater than, less than, or equal to the rotational kinetic energy of the sphere? Explain. **(b)** Find the rotational kinetic energy of the disk. [**Answer: (a)** The disk has a smaller moment of inertia than the hollow sphere, and hence its rotational kinetic energy is smaller as well. **(b)** The disk's rotational kinetic energy is 0.287 J, which is less than the hollow sphere's rotational kinetic energy of 0.383 J.]

Some related homework problems: Problem 54, Problem 68, Problem 71

CONCEPTUAL EXAMPLE 10-16 · COMPARE KINETIC ENERGIES

A solid sphere and a hollow sphere of the same mass and radius roll without slipping at the same speed. Is the kinetic energy of the solid sphere greater than, less than, or the same as the kinetic energy of the hollow sphere?

CONTINUED

REASONING AND DISCUSSION
Both spheres have the same translational kinetic energy, since they have the same mass and speed. The rotational kinetic energy, however, is proportional to the moment of inertia. The hollow sphere has the greater moment of inertia, and hence it has the greater kinetic energy.

ANSWER
The solid sphere has less kinetic energy than the hollow sphere.

▲ **FIGURE 10-26 An object rolls down an incline** An object starts at rest at the top of an inclined plane and rolls without slipping to the bottom. The speed of the object at the bottom depends on its moment of inertia—a larger moment of inertia results in a lower speed.

Energy Conservation with Rolling Motion   Now that we can calculate the kinetic energy of rolling motion, we show how to apply it to energy conservation. For example, consider an object of mass m, radius r, and moment of inertia I at the top of a ramp, as shown in **FIGURE 10-26**. The object is released from rest and allowed to roll to the bottom, a vertical height h below the starting point. What is the object's speed on reaching the bottom?

The simplest way to solve this problem is to use energy conservation. To do so, we set the initial mechanical energy at the top (i) equal to the final mechanical energy at the bottom (f). That is,

$$K_i + U_i = K_f + U_f$$

Because we are dealing with rolling motion, the kinetic energy is

$$K = \tfrac{1}{2}mv^2\left(1 + \frac{I}{mr^2}\right)$$

The potential energy is simply that due to the uniform gravitational field. Therefore,

$$U = mgy$$

With $y = h$ at the top of the ramp and the object starting at rest, we have

$$K_i + U_i = 0 + mgh = mgh$$

Similarly, with $y = 0$ at the bottom of the ramp and the object rolling with a speed v, we find

$$K_f + U_f = \tfrac{1}{2}mv^2\left(1 + \frac{I}{mr^2}\right) + 0 = \tfrac{1}{2}mv^2\left(1 + \frac{I}{mr^2}\right)$$

Setting the initial and final energies equal yields

$$mgh = \tfrac{1}{2}mv^2\left(1 + \frac{I}{mr^2}\right)$$

Solving for v, we find

$$v = \sqrt{\frac{2gh}{1 + \dfrac{I}{mr^2}}}$$

Let's quickly check one special case—namely, $I = 0$. With this substitution, we find

$$v = \sqrt{2gh}$$

This is the speed an object would have after falling straight down with no rotation through a distance h. Thus, setting $I = 0$ means there is no rotational kinetic energy, and hence the result is the same as for a point particle. As I becomes larger, the speed at the bottom of the ramp is smaller.

Canned Food Race

CONCEPTUAL EXAMPLE 10-17   WHICH OBJECT WINS THE RACE?

A disk and a hoop of the same mass and radius are released at the same time at the top of an inclined plane. Does the disk reach the bottom of the plane before, after, or at the same time as the hoop?

CONTINUED

REASONING AND DISCUSSION

As we've just seen, the larger the moment of inertia, I, the smaller the speed, v. Hence the object with the larger moment of inertia (the hoop in this case) loses the race to the bottom, because its speed is less than the speed of the disk at any given height.

Another way to think about this is to recall that both objects have the same mechanical energy to begin with—namely, mgh. For the hoop, more of this initial potential energy goes into rotational kinetic energy, because the hoop has the larger moment of inertia; therefore, less energy is left for translational motion. As a result, the hoop moves more slowly and loses the race.

ANSWER

The disk wins the race by reaching the bottom before the hoop.

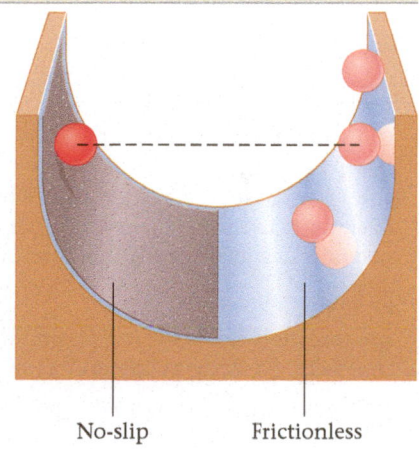

In the next Conceptual Example, we consider the effects of a surface that changes from nonslip to frictionless.

CONCEPTUAL EXAMPLE 10-18   COMPARE HEIGHTS

A ball is released from rest on a no-slip surface, as shown. After reaching its lowest point, the ball begins to rise again, this time on a frictionless surface. When the ball reaches its maximum height on the frictionless surface, is it at a greater height, at a lesser height, or at the same height as when it was released?

REASONING AND DISCUSSION

As the ball descends on the no-slip surface, it begins to rotate, increasing its angular speed until it reaches the lowest point of the surface. When it begins to rise again, there is no friction to slow the rotational motion; thus, the ball continues to rotate with the same angular speed it had at its lowest point. Therefore, some of the ball's initial gravitational potential energy remains in the form of rotational kinetic energy. As a result, less energy is available to be converted back into gravitational potential energy, and the height is less.

ANSWER

The height on the frictionless side is less.

Energy Conservation with Pulleys   We can also apply energy conservation to the case of a pulley, or similar object, with a string that winds or unwinds without slipping. In such cases, the relationship $v = r\omega$ is valid and we can follow the same methods applied to an object that rolls without slipping.

EXAMPLE 10-19   SPINNING WHEEL

A block of mass m is attached to a string that is wrapped around the circumference of a wheel of radius R and moment of inertia I. The wheel rotates freely about its axis and the string wraps around its circumference without slipping. Initially the wheel rotates with an angular speed ω, causing the block to rise with a linear speed v. To what height does the block rise before coming to rest? Give a symbolic answer.

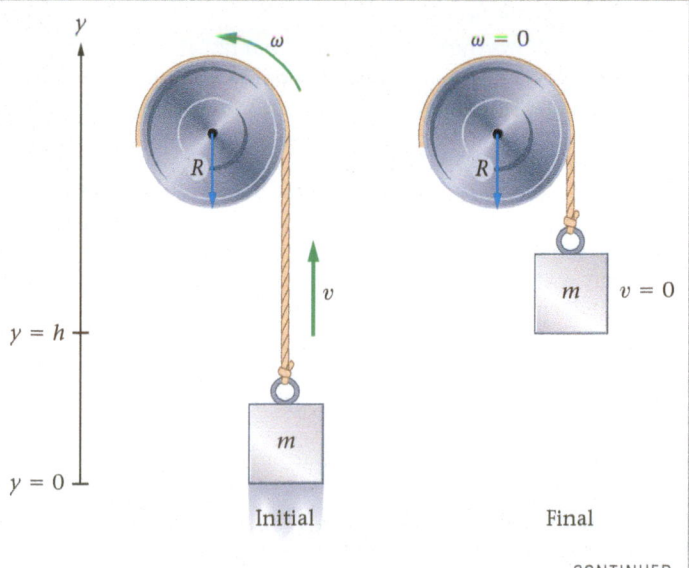

PICTURE THE PROBLEM

In our sketch, we choose the origin of the y axis to be at the initial height of the block. The positive y direction, as usual, is chosen to be upward. When the block comes to rest, then, it is at the height $y = h > 0$, where h is to be determined from the initial speed of the block and the properties of the wheel.

REASONING AND STRATEGY

The problem statement gives two key pieces of information. First, the string wraps onto the disk without slipping; therefore, $v = R\omega$. Second, the wheel rotates freely, which means that the

CONTINUED

mechanical energy of the system is conserved. Thus, at the height h the initial kinetic energy of the system has been converted to gravitational potential energy. This condition can be used to find h.

Before we continue, notice that the mechanical energy of the system includes the following contributions: (i) linear kinetic energy for the block, (ii) rotational kinetic energy for the wheel, and (iii) gravitational potential energy for the block. We do not include the gravitational potential energy of the wheel because its height does not change.

Known Mass of block, m; radius of wheel, R; moment of inertia of wheel, I; initial angular speed, ω; initial linear speed, v.

Unknown Height gained by the block, $h = ?$

SOLUTION

1. Write an expression for the initial mechanical energy of the system, E_i, including all three contributions mentioned in the Strategy:

$$E_i = \tfrac{1}{2}mv^2 + \tfrac{1}{2}I\omega^2 + mgy$$
$$= \tfrac{1}{2}mv^2 + \tfrac{1}{2}I\left(\frac{v}{R}\right)^2 + 0$$

2. Write an expression for the final mechanical energy of the system, E_f:

$$E_f = \tfrac{1}{2}mv^2 + \tfrac{1}{2}I\omega^2 + mgy$$
$$= 0 + 0 + mgh$$

3. Set the initial and final mechanical energies equal to one another, $E_i = E_f$:

$$E_i = \tfrac{1}{2}mv^2 + \tfrac{1}{2}I\left(\frac{v}{R}\right)^2 = \tfrac{1}{2}mv^2\left(1 + \frac{I}{mR^2}\right) = mgh = E_f$$

4. Solve for the height, h:

$$h = \left(\frac{v^2}{2g}\right)\left(1 + \frac{I}{mR^2}\right)$$

INSIGHT

If the block were moving upward with speed v on its own—not attached to anything—it would rise to the height $h = v^2/2g$. We recover this result if $I = 0$, because in that case it is as if the wheel were not there. If the wheel is present, and I is nonzero, the block rises to a height that is *greater* than $v^2/2g$. The reason is that the wheel has kinetic energy, in addition to the kinetic energy of the block, and the sum of these kinetic energies must be converted to gravitational potential energy before the block and the wheel stop moving.

PRACTICE PROBLEM

Suppose the wheel is a disk with a mass equal to the mass m of the block. Find an expression for the height h in this case. [**Answer:** The moment of inertia of the wheel is $I = \tfrac{1}{2}mR^2$. Therefore, $h = \left(\tfrac{3}{2}\right)(v^2/2g)$.]

Some related homework problems: Problem 57, Problem 59, Problem 62

EXAMPLE 10-20 FIND THE YO-YO'S SPEED

Yo-Yo man releases a yo-yo from rest and allows it to drop, as he keeps the top end of the string stationary. The mass of the yo-yo is 0.059 kg, its moment of inertia is 2.5×10^{-5} kg·m², and the radius r of the axle the string wraps around is 0.0068 m. What is the linear speed v of the yo-yo after it has dropped through a height of $h = 0.50$ m?

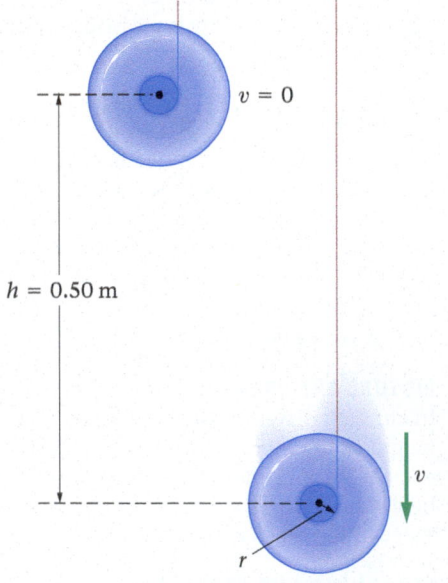

PICTURE THE PROBLEM

Our sketch shows the yo-yo released from rest and descending through a height of 0.50 m. As the yo-yo drops, it gains both linear and angular speed. The linear speed is related to the angular speed by $v = r\omega$, where r is the radius of the yo-yo's axle.

REASONING AND STRATEGY

The basic idea in this system is that the decrease in gravitational potential energy as the yo-yo drops is equal to the increase in the yo-yo's linear and rotational kinetic energies. That is, the final mechanical energy is equal to the initial mechanical energy. We start, then, by writing expressions for the initial and final mechanical energies. Setting these quantities equal yields a relationship that can be solved for the speed v.

Known Mass of yo-yo, $m = 0.059$ kg; moment of inertia, $I = 2.5 \times 10^{-5}$ kg·m²; radius of axle, $r = 0.0068$ m; drop height, $h = 0.50$ m.

Unknown Linear speed of yo-yo, $v = ?$

SOLUTION

1. Write the initial energy of the system, which is equal to its initial gravitational potential energy:

$$E_i = mgh$$

CONTINUED

2. Write the final energy of the system, which is equal to its final kinetic energy:

$$E_f = \frac{1}{2}mv^2\left(1 + \frac{I}{mr^2}\right)$$

3. Set the initial and final mechanical energies equal to one another, $E_i = E_f$, and solve for v:

$$v = \sqrt{\frac{2gh}{\left(1 + \dfrac{I}{mr^2}\right)}}$$

4. Substitute numerical values:

$$v = \sqrt{\frac{2(9.81 \text{ m/s}^2)(0.50 \text{ m})}{\left(1 + \dfrac{(2.5 \times 10^{-5} \text{ kg} \cdot \text{m}^2)}{(0.059 \text{ kg})(0.0068 \text{ m})^2}\right)}} = 0.98 \text{ m/s}$$

INSIGHT

The linear speed of the yo-yo is $v = r\omega$, where r is the radius of the axle from which the string unwraps without slipping. Therefore, the r in the term I/mr^2 is the radius of the axle. The outer radius of the yo-yo affects its moment of inertia, but I is given to us in the problem statement, and hence the outer radius is not needed.

PRACTICE PROBLEM — PREDICT/CALCULATE

(a) If the yo-yo's moment of inertia is increased, does the final linear speed increase, decrease, or stay the same? Explain. **(b)** Find the final linear speed for the case $I = 3.5 \times 10^{-5} \text{ kg} \cdot \text{m}^2$. [**Answer: (a)** Increasing the moment of inertia increases the rotational kinetic energy, and leaves less energy for the translational kinetic energy. As a result, the linear speed decreases. **(b)** $v = 0.84 \text{ m/s}$]

Some related homework problems: Problem 62, Problem 82, Problem 84

Enhance Your Understanding   (Answers given at the end of the chapter)

6. A hoop, a disk, a solid sphere, and a hollow sphere are raced down an incline, as in Conceptual Example 10-17. The objects all have the same mass and radius. Rank the objects in the order in which they finish the race, from first to last. Indicate ties where appropriate.

Section Review

*   To apply conservation of energy to a rotating system, it is necessary to include the kinetic energy of rotation.

CHAPTER 10 REVIEW

CHAPTER SUMMARY

10-1 ANGULAR POSITION, VELOCITY, AND ACCELERATION

To describe rotational motion, rotational analogs of position, velocity, and acceleration are defined.

Angular Position

Angular position, θ, is the angle measured from an arbitrary reference line:

$$\theta \text{ (in radians)} = \text{arc length/radius} = s/r \qquad 10\text{-}2$$

Angular Velocity

Angular velocity, ω, is the rate of change of angular position. The average angular velocity is

$$\omega_{av} = \frac{\Delta\theta}{\Delta t} \qquad 10\text{-}3$$

The instantaneous angular velocity is the limit of ω_{av} as Δt approaches zero:

$$\omega = \lim_{\Delta t \to 0}\frac{\Delta\theta}{\Delta t} \qquad 10\text{-}4$$

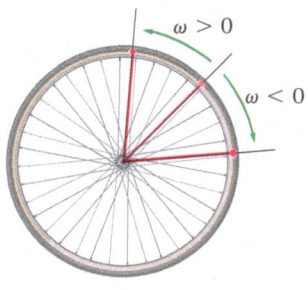

Angular Acceleration

Angular acceleration, α, is the rate of change of angular velocity. The average angular acceleration is

$$\alpha_{\text{av}} = \frac{\Delta\omega}{\Delta t} \qquad \text{10-6}$$

The instantaneous angular acceleration is the limit of α_{av} as Δt approaches zero:

$$\alpha = \lim_{\Delta t \to 0} \frac{\Delta\omega}{\Delta t} \qquad \text{10-7}$$

Period of Rotation

The period, T, is the time required to complete one full rotation. If the angular velocity is constant, T is related to ω as follows:

$$T = \frac{2\pi}{\omega} \qquad \text{10-5}$$

Sign Convention

Counterclockwise rotations are positive; clockwise rotations are negative.

10-2 ROTATIONAL KINEMATICS

Rotational kinematics is the description of angular motion, in the same way that linear kinematics describes linear motion. In both cases, we assume constant acceleration.

Linear–Angular Analogs

Rotational kinematics is related to linear kinematics by the following linear–angular analogs:

Linear Quantity	Angular Quantity
x	θ
v	ω
a	α

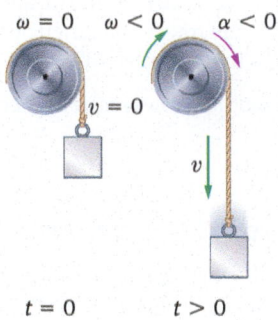

10-3 CONNECTIONS BETWEEN LINEAR AND ROTATIONAL QUANTITIES

A point on a rotating object follows a circular path. At any instant of time, the point is moving in a direction tangential to the circle, with a linear speed and acceleration.

Tangential Speed

The tangential speed, v_t, of a point on a rotating object is

$$v_t = r\omega \qquad \text{10-12}$$

Centripetal Acceleration

The centripetal acceleration, a_{cp}, of a point on a rotating object is

$$a_{\text{cp}} = r\omega^2 \qquad \text{10-13}$$

Centripetal acceleration is due to a change in the direction of motion.

Tangential Acceleration

The tangential acceleration, a_t, of a point on a rotating object is

$$a_t = r\alpha \qquad \text{10-14}$$

Tangential acceleration is due to a change in speed.

Total Acceleration

The total acceleration of a point on a rotating object is the vector sum of its tangential and centripetal accelerations.

10-4 ROLLING MOTION

Rolling motion is a combination of translational and rotational motions. An object of radius r, rolling without slipping, translates with linear speed v and rotates with angular speed

$$\omega = v/r \qquad \text{10-15}$$

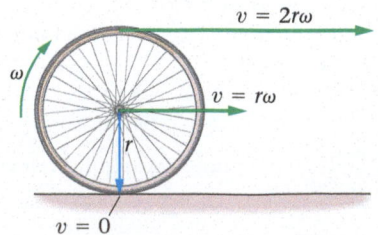

10-5 ROTATIONAL KINETIC ENERGY AND THE MOMENT OF INERTIA

Rotating objects have kinetic energy, just as objects in linear motion have kinetic energy.

Rotational Kinetic Energy

The kinetic energy of a rotating object is

$$K = \tfrac{1}{2}I\omega^2 \qquad\qquad \text{10-17}$$

The quantity I is the moment of inertia.

Moment of Inertia, Discrete Masses

The moment of inertia, I, of a collection of masses, m_i, at distances r_i from the axis of rotation is

$$I = \sum m_i r_i^2 \qquad\qquad \text{10-18}$$

10-6 CONSERVATION OF ENERGY

Energy conservation can be applied to a variety of rotational systems in the same way that it is applied to translational systems.

Kinetic Energy of Rolling Motion

The kinetic energy of an object that rolls without slipping is

$$K = \tfrac{1}{2}mv^2 + \tfrac{1}{2}I\omega^2 \qquad\qquad \text{10-19}$$

Since rolling without slipping implies that $\omega = v/r$, the kinetic energy can be written as follows:

$$K = \tfrac{1}{2}mv^2 + \tfrac{1}{2}I\left(\frac{v}{r}\right)^2 = \tfrac{1}{2}mv^2\left(1 + \frac{I}{mr^2}\right) \qquad\qquad \text{10-20}$$

ANSWERS TO ENHANCE YOUR UNDERSTANDING QUESTIONS

1. **(a)** B < A < D < C. **(b)** D < A < B < C.
2. **(a)** 4θ. **(b)** 2ω.
3. **(a)** 4 m/s. **(b)** 24 m/s^2.
4. **(a)** v. **(b)** v. **(c)** v.
5. 4 < 2 < 1 = 3.
6. First = solid sphere; second = disk; third = hollow sphere; last = hoop.

CONCEPTUAL QUESTIONS

For instructor-assigned homework, go to www.masteringphysics.com

(Answers to odd-numbered Conceptual Questions can be found in the back of the book.)

1. A rigid object rotates about a fixed axis. Do all points on the object have the same angular speed? Do all points on the object have the same linear speed? Explain.

2. Can you drive your car in such a way that your tangential acceleration is zero while at the same time your centripetal acceleration is nonzero? Give an example if your answer is yes; state why not if your answer is no.

3. Can you drive your car in such a way that your tangential acceleration is nonzero while at the same time your centripetal acceleration is zero? Give an example if your answer is yes; state why not if your answer is no.

4. The fact that the Earth rotates gives people in New York a linear speed of about 750 mi/h. Where should you stand on the Earth to have the smallest possible linear speed?

5. At the local carnival you and a friend decide to take a ride on the Ferris wheel. As the wheel rotates with a constant angular speed, your friend poses the following questions: **(a)** Is my linear velocity constant? **(b)** Is my linear speed constant? **(c)** Is the magnitude of my centripetal acceleration constant? **(d)** Is the direction of my centripetal acceleration constant? What is your answer to each of these questions?

6. Why should changing the axis of rotation of an object change its moment of inertia, given that its shape and mass remain the same?

7. Give a common, everyday example for each of the following: **(a)** An object that has zero rotational kinetic energy but nonzero translational kinetic energy. **(b)** An object that has zero translational kinetic energy but nonzero rotational kinetic energy. **(c)** An object that has nonzero rotational and translational kinetic energies.

8. Two spheres have identical radii and masses. How might you tell which of these spheres is hollow and which is solid?

9. At the grocery store you pick up a can of beef broth and a can of chunky beef stew. The cans are identical in diameter and weight. Rolling both of them down the aisle with the same initial speed, you notice that the can of chunky stew rolls much farther than the can of broth. Why?

10. Suppose we change the race shown in Conceptual Example 10-17 so that a hoop of radius R and mass M races a hoop of radius R and mass $2M$. **(a)** Does the hoop with mass M finish before, after, or at the same time as the hoop with mass $2M$? Explain. **(b)** How would your answer to part (a) change if the hoops had different radii? Explain.

PROBLEMS AND CONCEPTUAL EXERCISES

Answers to odd-numbered Problems and Conceptual Exercises can be found in the back of the book. **BIO** *identifies problems of biological or medical interest;* **CE** *indicates a conceptual exercise.* **Predict/Explain** *problems ask for two responses: (a) your prediction of a physical outcome, and (b) the best explanation among three provided; and* **Predict/Calculate** *problems ask for a prediction of a physical outcome, based on fundamental physics concepts, and follow that with a numerical calculation to verify the prediction. On all problems, bullets (•, ••, •••) indicate the level of difficulty.*

SECTION 10-1 ANGULAR POSITION, VELOCITY, AND ACCELERATION

1. • The following angles are given in degrees. Convert them to radians: 30°, 45°, 90°, 180°.

2. • The following angles are given in radians. Convert them to degrees: $\pi/6$, 0.70π, 1.5π, 5π.

3. • Express the angular velocity of the second hand on a clock in the following units: **(a)** rev/h, **(b)** deg/min, and **(c)** rad/s.

4. • Rank the following in order of increasing angular speed: an automobile tire rotating at 2.00×10^3 deg/s, an electric drill rotating at 400.0 rev/min, and an airplane propeller rotating at 40.0 rad/s.

5. • A spot of paint on a bicycle tire moves in a circular path of radius 0.29 m. When the spot has traveled a linear distance of 2.73 m, through what angle has the tire rotated? Give your answer in radians.

6. • **The Crab Nebula** One of the most studied objects in the night sky is the Crab nebula, the remains of a supernova explosion observed by the Chinese in 1054. In 1968 it was discovered that a pulsar—a rapidly rotating neutron star that emits a pulse of radio waves with each revolution—lies near the center of the Crab nebula. The period of this pulsar is 33 ms. What is the angular speed (in rad/s) of the Crab nebula pulsar?

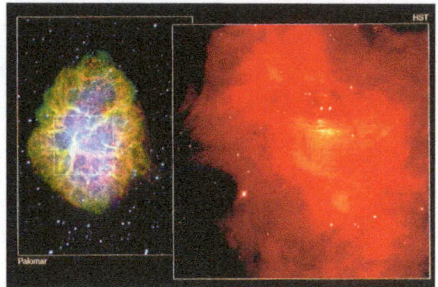

The photo is a true-color visible light image of the Crab nebula. In the false-color breakout, the pulsar can be seen as the left member of the pair of stars just above the center of the frame. (Problem 6)

7. • **BIO Hummingbird Flapping** The Ruby-throated Hummingbird (*Archilochus colubris*) can flap its wings with an angular speed of 346 rad/s. What is the period of its flapping motion?

8. • The following table gives the angular speed of a rotating fan blade at various times as it slows to a stop. Find the average angular acceleration for the times **(a)** $t = 0$ to $t = 2.0$ s, **(b)** $t = 0$ to $t = 4.0$ s, and **(c)** $t = 2.0$ s to $t = 4.0$ s.

Time (s)	Angular speed (rad/s)
0	5.0
2.0	4.1
4.0	3.0

9. • Four objects, a, b, c, and d, have the initial angular velocities and angular accelerations given in the following table. Determine the angular velocity of each object after an elapsed time of 3.5 s.

	Initial angular velocity (rad/s)	Angular acceleration (rad/s²)
(a)	15	4.6
(b)	16	−2.2
(c)	−17	3.6
(d)	−18	−7.2

10. •• **Predict/Calculate** A 3.5-inch floppy disk in a computer rotates with a period of 2.00×10^{-1} s. What are **(a)** the angular speed of the disk and **(b)** the linear speed of a point on the rim of the disk? **(c)** Does a point near the center of the disk have an angular speed that is greater than, less than, or the same as the angular speed found in part (a)? Explain. (*Note*: A 3.5-inch floppy disk is 3.5 inches in diameter.)

11. •• The angle an airplane propeller makes with the horizontal as a function of time is given by $\theta = (125 \text{ rad/s})t + (42.5 \text{ rad/s}^2)t^2$. **(a)** Estimate the instantaneous angular velocity at $t = 0.00$ by calculating the average angular velocity from $t = 0.00$ to $t = 0.010$ s. **(b)** Estimate the instantaneous angular velocity at $t = 1.000$ s by calculating the average angular velocity from $t = 1.000$ s to $t = 1.010$ s. **(c)** Estimate the instantaneous angular velocity at $t = 2.000$ s by calculating the average angular velocity from $t = 2.000$ s to $t = 2.010$ s. **(d)** Based on your results from parts (a), (b), and (c), is the angular acceleration of the propeller positive, negative, or zero? Explain. **(e)** Calculate the average angular acceleration from $t = 0.00$ to $t = 1.00$ s and from $t = 1.00$ s to $t = 2.00$ s.

SECTION 10-2 ROTATIONAL KINEMATICS

12. • **CE** An object at rest begins to rotate with a constant angular acceleration. If this object rotates through an angle θ in the time t, through what angle did it rotate in the time $t/2$?

13. • **CE** An object at rest begins to rotate with a constant angular acceleration. If the angular speed of the object is ω after the time t, what was its angular speed at the time $t/2$?

14. • The angular speed of a propeller on a boat increases with constant acceleration from 11 rad/s to 39 rad/s in 3.0 revolutions. What is the angular acceleration of the propeller?

15. •• After fixing a flat tire on a bicycle you give the wheel a spin. **(a)** If its initial angular speed was 6.35 rad/s and it rotated 14.2 revolutions before coming to rest, what was its average angular acceleration? **(b)** For what length of time did the wheel rotate?

16. •• **Predict/Calculate** A ceiling fan is rotating at 0.96 rev/s. When turned off, it slows uniformly to a stop in 2.4 min. **(a)** How many revolutions does the fan make in this time? **(b)** Using the result from part (a), find the number of revolutions the fan must make for its speed to decrease from 0.96 rev/s to 0.48 rev/s.

17. •• A discus thrower starts from rest and begins to rotate with a constant angular acceleration of 2.2 rad/s². **(a)** How many revolutions does it take for the discus thrower's angular speed to reach 6.3 rad/s? **(b)** How much time does this take?

18. •• **Half Time** At 3:00 the hour hand and the minute hand of a clock point in directions that are 90.0° apart. What is the first time after

3:00 that the angle between the two hands has decreased by half to 45.0°?

When the little hand is on the
3 and the big hand is on the
12 . . . . (Problem 18)

19. •• **BIO** A centrifuge is a common laboratory instrument that separates components of differing densities in solution. This is accomplished by spinning a sample around in a circle with a large angular speed. Suppose that after a centrifuge in a medical laboratory is turned off, it continues to rotate with a constant angular deceleration for 10.2 s before coming to rest. (a) If its initial angular speed was 3850 rpm, what is the magnitude of its angular acceleration? (b) How many revolutions did the centrifuge complete after being turned off?

20. •• **The Slowing Earth** The Earth's rate of rotation is constantly decreasing, causing the day to increase in duration. In the year 2006 the Earth took about 0.840 s longer to complete 365 revolutions than it did in the year 1906. What was the average angular acceleration, in rad/s², of the Earth during this time?

21. •• When a carpenter shuts off his circular saw, the 10.0-inch-diameter blade slows from 3984 rpm to 0 in 1.25 s. (a) What is the angular acceleration of the blade? (b) What is the distance traveled by a point on the rim of the blade during the deceleration? (c) What is the magnitude of the net displacement of a point on the rim of the blade during the deceleration?

22. •• **The World's Fastest Turbine** The drill used by most dentists today is powered by a small air turbine that can operate at angular speeds of 350,000 rpm. These drills, along with ultrasonic dental drills, are the fastest turbines in the world—far exceeding the angular speeds of jet engines. Suppose a drill starts at rest and comes up to operating speed in 2.1 s. (a) Find the angular acceleration produced by the drill, assuming it to be constant. (b) How many revolutions does the drill bit make as it comes up to speed?

SECTION 10-3 CONNECTIONS BETWEEN LINEAR AND ROTATIONAL QUANTITIES

23. • **CE** Predict/Explain Two children, Jason and Betsy, ride on the same merry-go-round. Jason is a distance R from the axis of rotation; Betsy is a distance $2R$ from the axis. (a) Is the rotational period of Jason greater than, less than, or equal to the rotational period of Betsy? (b) Choose the *best explanation* from among the following:
   I. The period is greater for Jason because he moves more slowly than Betsy.
   II. The period is greater for Betsy since she must go around a circle with a larger circumference.
   III. It takes the same amount of time for the merry-go-round to complete a revolution for all points on the merry-go-round.

24. • The hour hand on a certain clock is 8.2 cm long. Find the tangential speed of the tip of this hand.

25. • The outer edge of a rotating Frisbee with a diameter of 29 cm has a linear speed of 3.7 m/s. What is the angular speed of the Frisbee?

26. • A carousel at the local carnival rotates once every 45 seconds. (a) What is the linear speed of an outer horse on the carousel, which is 2.75 m from the axis of rotation? (b) What is the linear speed of an inner horse that is 1.75 m from the axis of rotation?

27. • A chainsaw is shown in **FIGURE 10-27**. When the saw is in operation, the chain moves with a linear speed of $v = 5.5$ m/s. At the end of the saw, the chain follows a semicircular path with a radius of $r = 0.044$ m. (a) What is the angular speed of the chain as it goes around the end of the saw? (b) What is the centripetal acceleration of the chain at the end of the saw?

FIGURE 10-27 Problem 27

28. •• Predict/Calculate Jeff of the Jungle swings on a vine that is 7.20 m long (**FIGURE 10-28**). At the bottom of the swing, just before hitting the tree, Jeff's linear speed is 8.50 m/s. (a) Find Jeff's angular speed at this time. (b) What centripetal acceleration does Jeff experience at the bottom of his swing? (c) What exerts the force that is responsible for Jeff's centripetal acceleration?

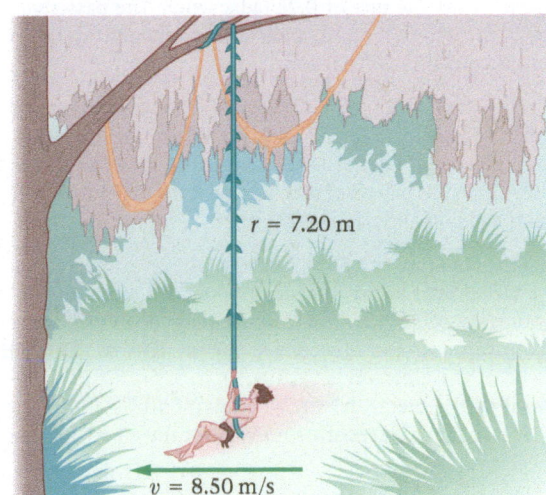

$r = 7.20$ m

$v = 8.50$ m/s

FIGURE 10-28 Problem 28

29. •• A compact disk, which has a diameter of 12.0 cm, speeds up uniformly from 0.00 to 4.00 rev/s in 3.00 s. What is the tangential acceleration of a point on the outer rim of the disk at the moment when its angular speed is (a) 2.00 rev/s and (b) 3.00 rev/s?

30. •• Predict/Calculate When a compact disk with a 12.0-cm diameter is rotating at 34.6 rad/s, what are (a) the linear speed and (b) the centripetal acceleration of a point on its outer rim? (c) Consider a point on the CD that is halfway between its center and its outer rim. Without repeating all of the calculations required for parts (a) and (b), determine the linear speed and the centripetal acceleration of this point.

31. •• **Predict/Calculate** As Tony the fisherman reels in a "big one," he turns the spool on his fishing reel at the rate of 2.0 complete revolutions every second (**FIGURE 10-29**). **(a)** If the radius of the reel is 2.3 cm, what is the linear speed of the fishing line as it is reeled in? **(b)** How would your answer to part (a) change if the radius of the reel were doubled?

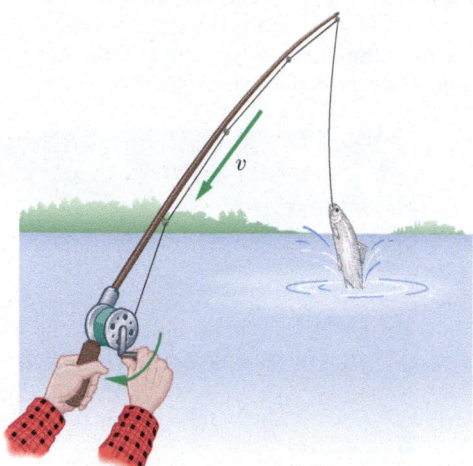

FIGURE 10-29 Problem 31

32. •• A Ferris wheel with a radius of 9.5 m rotates at a constant rate, completing one revolution every 36 s. Find the direction and magnitude of a passenger's acceleration when **(a)** at the top and **(b)** at the bottom of the wheel.

33. •• Suppose the Ferris wheel in the previous problem begins to decelerate at the rate of 0.22 rad/s² when the passenger is at the top of the wheel. Find the direction and magnitude of the passenger's acceleration at that time.

34. •• To polish a filling, a dentist attaches a sanding disk with a radius of 3.20 mm to the drill. **(a)** When the drill is operated at 2.15×10^4 rad/s, what is the tangential speed of the rim of the disk? **(b)** What period of rotation must the disk have if the tangential speed of its rim is to be 275 m/s?

35. •• **The Bohr Atom** The Bohr model of the hydrogen atom pictures the electron as a tiny particle moving in a circular orbit about a stationary proton. In the lowest-energy orbit the distance from the proton to the electron is 5.29×10^{-11} m, and the linear speed of the electron is 2.18×10^6 m/s. **(a)** What is the angular speed of the electron? **(b)** How many orbits about the proton does it make each second? **(c)** What is the electron's centripetal acceleration?

36. ••• A wheel of radius R starts from rest and accelerates with a constant angular acceleration α about a fixed axis. At what time t will the centripetal and tangential accelerations of a point on the rim have the same magnitude?

SECTION 10-4 ROLLING MOTION

37. • **CE Microwave Tray** Most microwave ovens have a glass tray that sits on top of a circular ring with three small wheels, as shown in **FIGURE 10-30**. A small motor in the base of the microwave rotates the glass tray to provide uniform heating. When the glass tray completes one revolution, does the circular ring below it rotate through 0.5 rev, 1 rev, 2 rev, or 3 rev? Explain.

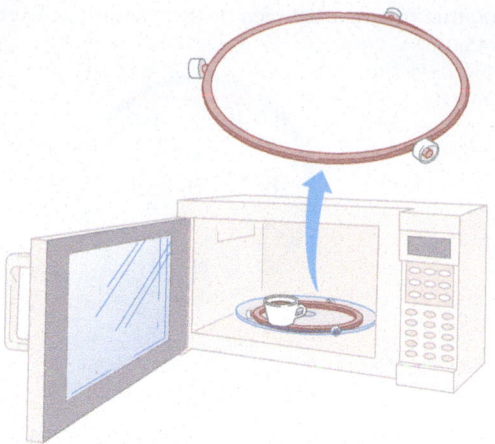

FIGURE 10-30 Problem 37

38. • The tires on a car have a radius of 31 cm. What is the angular speed of these tires when the car is driven at 15 m/s?

39. • A child pedals a tricycle, giving the driving wheel an angular speed of $\omega = 0.875$ rev/s (**FIGURE 10-31**). If the radius of the wheel is $r = 0.260$ m, what is the child's linear speed, v?

FIGURE 10-31 Problem 39

40. • A soccer ball, which has a circumference of 70.0 cm, rolls 14.0 yards in 3.35 s. What was the average angular speed of the ball during this time?

41. •• You drive down the road at 31 m/s (70 mi/h) in a car whose tires have a radius of 34 cm. **(a)** What is the period of rotation of the tires? **(b)** Through what angle does a tire rotate in one second?

42. •• The wheels of a skateboard roll without slipping as it accelerates at 0.35 m/s² down an 85-m-long hill. If the skateboarder travels at 1.8 m/s at the top of the hill, what is the average angular speed of the 2.6-cm-radius wheels during the entire trip down the hill?

43. •• The page feed roller of a computer printer grips each 11-inch-long sheet of paper and feeds it through the print mechanism. If the roller has a radius of 6.4 mm and the drive motor has a maximum angular speed of 480 rpm, what is the maximum number of pages that the printer can print each minute?

44. •• As you drive down the road at 17 m/s, you press on the gas pedal and speed up with a uniform acceleration of 1.12 m/s² for 0.65 s. If the tires on your car have a radius of 33 cm, what is their angular displacement during this period of acceleration?

SECTION 10-5 ROTATIONAL KINETIC ENERGY AND THE MOMENT OF INERTIA

45. • **CE Predict/Explain** The minute and hour hands of a clock have a common axis of rotation and equal mass. The minute hand is long, thin, and uniform; the hour hand is short, thick, and uniform. **(a)** Is the moment of inertia of the minute hand greater than, less than, or equal to the moment of inertia of the hour hand? **(b)** Choose the *best explanation* from among the following:

I. The hands have equal mass, and hence equal moments of inertia.

II. Having mass farther from the axis of rotation results in a greater moment of inertia.

III. The more compact hour hand concentrates its mass and has the greater moment of inertia.

46. **CE • Predict/Explain** Suppose a bicycle wheel is rotated about an axis through its rim and parallel to its axle. **(a)** Is its moment of inertia about this axis greater than, less than, or equal to its moment of inertia about its axle? **(b)** Choose the *best explanation* from among the following:

I. The moment of inertia is greatest when an object is rotated about its center.

II. The mass and shape of the wheel remain the same.

III. Mass is farther from the axis when the wheel is rotated about the rim.

47. **•** The moment of inertia of a 0.98-kg bicycle wheel rotating about its center is 0.13 kg·m². What is the radius of this wheel, assuming the weight of the spokes can be ignored?

48. **•** An electric fan spinning with an angular speed of 13 rad/s has a kinetic energy of 4.6 J. What is the moment of inertia of the fan?

49. **•• BIO Spin-Dry Dragonflies** Some dragonflies splash down onto the surface of a lake to clean themselves. After this dunking, the dragonflies gain altitude, and then spin rapidly at about 1000 rpm to spray the water off their bodies. When the dragonflies do this "spin-dry," they tuck themselves into a "ball" with a moment of inertia of 2.6×10^{-7} kg·m². How much energy must the dragonfly generate to spin itself at this rate?

50. **•• CE** The L-shaped object in **FIGURE 10-32** can be rotated in one of the following three ways: case 1, rotation about the x axis; case 2, rotation about the y axis; and case 3, rotation about the z axis (which passes through the origin perpendicular to the plane of the figure). Rank these three cases in order of increasing moment of inertia. Indicate ties where appropriate.

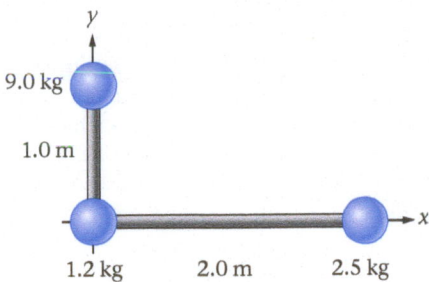

FIGURE 10-32 Problem 50

51. **•• Predict/Calculate** A 12-g CD with a radius of 6.0 cm rotates with an angular speed of 34 rad/s. **(a)** What is its kinetic energy? **(b)** If its kinetic energy is to be quadrupled, should its angular speed be doubled, quadrupled, or increased by a factor of 16? **(c)** Calculate the angular speed required to quadruple the kinetic energy the CD had in part (a).

52. **••** The engine of a model airplane must both spin a propeller and push air backward to propel the airplane forward. **(a)** Model the propeller as three 0.20-m-long thin rods of mass 0.040 kg each, with the rotation axis at one end. What is the moment of inertia of the propeller? **(b)** How much energy is required to rotate the propeller at 5500 rpm? Ignore the energy required to push the air.

53. **•• Flywheel Energy Storage** One way to store energy is in the rotational motion of a flywheel, and some have proposed using such technology to power automobiles. One unit is based on a 6.0-kg flywheel in the shape of a hoop of radius 0.10 m that spins as fast as 60,000 rpm. **(a)** How much kinetic energy is stored by the flywheel when it is rotating at its maximum rate? **(b)** If the vehicle requires an average power of 11 kW under normal driving conditions, for how much time can it operate with the energy stored in the flywheel?

54. **••** When a pitcher throws a curve ball, the ball is given a fairly rapid spin. If a 0.15-kg baseball with a radius of 3.7 cm is thrown with a linear speed of 48 m/s and an angular speed of 42 rad/s, how much of its kinetic energy is translational and how much is rotational? Assume the ball is a uniform, solid sphere.

55. **••** A lawn mower has a flat, rod-shaped steel blade that rotates about its center. The mass of the blade is 0.65 kg and its length is 0.55 m. **(a)** What is the rotational energy of the blade at its operating angular speed of 3500 rpm? **(b)** If all of the rotational kinetic energy of the blade could be converted to gravitational potential energy, to what height would the blade rise?

SECTION 10-6 CONSERVATION OF ENERGY

56. **• CE** Consider the physical situation shown in Conceptual Example 10-18. Suppose this time a ball is released from rest on the frictionless surface. When the ball comes to rest on the no-slip surface, is its height greater than, less than, or equal to the height from which it was released?

57. **•** Suppose the block in Example 10-19 has a mass of 2.1 kg and an initial upward speed of 0.33 m/s. Find the moment of inertia of the wheel if its radius is 8.0 cm and the block rises to a height of 7.4 cm before momentarily coming to rest.

58. **••** Calculate the speeds of **(a)** the disk and **(b)** the hoop at the bottom of the inclined plane in Conceptual Example 10-17 if the height of the incline is 0.55 m.

59. **•• Predict/Calculate Atwood's Machine** The two masses ($m_1 = 5.0$ kg and $m_2 = 3.0$ kg) in the Atwood's machine shown in **FIGURE 10-33** are released from rest, with m_1 at a height of 0.75 m above the floor. When m_1 hits the ground its speed is 1.8 m/s. Assuming that the pulley is a uniform disk with a radius of 12 cm, **(a)** outline a strategy that allows you to find the mass of the pulley. **(b)** Implement the strategy given in part (a) and determine the pulley's mass.

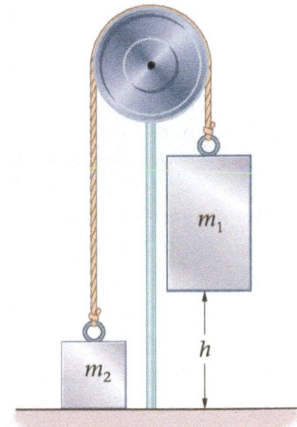

FIGURE 10-33 Problem 59

60. **••** In Conceptual Example 10-18, assume the ball is a solid sphere of radius 2.9 cm and mass 0.14 kg. If the ball is released from rest at a height of 0.78 m above the bottom of the track on the no-slip side, **(a)** what is its angular speed when it is on the frictionless side of the track? **(b)** How high does the ball rise on the frictionless side?

61. **•• Predict/Calculate** After you pick up a spare, your bowling ball rolls without slipping back toward the ball rack with a linear

speed of $v_i = 2.62$ m/s (**FIGURE 10-34**). To reach the rack, the ball rolls up a ramp that rises through a vertical height of $h = 0.47$ m. **(a)** What is the linear speed of the ball when it reaches the top of the ramp? **(b)** If the radius of the ball were increased, would the speed found in part (a) increase, decrease, or stay the same? Explain.

FIGURE 10-34 Problem 61

62. •• **Predict/Calculate** A 1.3-kg block is tied to a string that is wrapped around the rim of a pulley of radius 7.2 cm. The block is released from rest. **(a)** Assuming the pulley is a uniform disk with a mass of 0.31 kg, find the speed of the block after it has fallen through a height of 0.50 m. **(b)** If a small lead weight is attached near the rim of the pulley and this experiment is repeated, will the speed of the block increase, decrease, or stay the same? Explain.

63. •• After doing some exercises on the floor, you are lying on your back with one leg pointing straight up. If you allow your leg to fall freely until it hits the floor (**FIGURE 10-35**), what is the tangential speed of your foot just before it lands? Assume the leg can be treated as a uniform rod 0.95 m long that pivots freely about the hip.

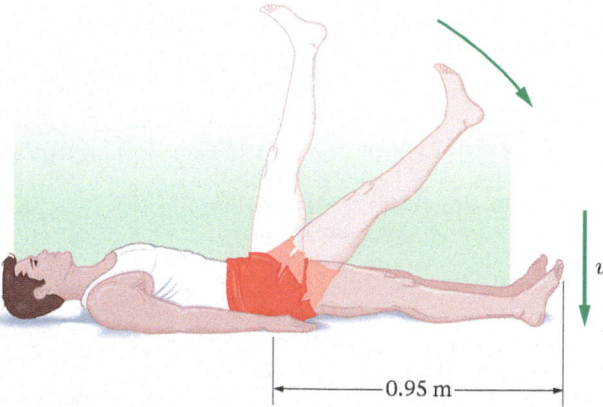

FIGURE 10-35 Problem 63

64. ••• A 2.0-kg solid cylinder (radius = 0.10 m, length = 0.50 m) is released from rest at the top of a ramp and allowed to roll without slipping. The ramp is 0.75 m high and 5.0 m long. When the cylinder reaches the bottom of the ramp, what are **(a)** its total kinetic energy, **(b)** its rotational kinetic energy, and **(c)** its translational kinetic energy?

65. ••• A 2.5-kg solid sphere (radius = 0.10 m) is released from rest at the top of a ramp and allowed to roll without slipping. The ramp is 0.75 m high and 5.6 m long. When the sphere reaches the bottom of the ramp, what are **(a)** its total kinetic energy, **(b)** its rotational kinetic energy, and **(c)** its translational kinetic energy?

GENERAL PROBLEMS

66. • **CE** As you switch a fan setting from its slowest speed to its highest, its angular speed ω triples. **(a)** Does the period of rotation T of the fan blade increase, decrease, or stay the same? Explain. **(b)** What is the ratio of the new rotation period to the old value, T_{new}/T_{old}?

67. • **CE** When you stand on the observation deck of the Empire State Building in New York, is your linear speed due to the Earth's rotation greater than, less than, or the same as when you were waiting for the elevators on the ground floor?

68. • What linear speed must a 0.065-kg hula hoop have if its total kinetic energy is to be 0.12 J? Assume the hoop rolls on the ground without slipping.

69. • **BIO Losing Consciousness** A pilot performing a horizontal turn will lose consciousness if she experiences a centripetal acceleration greater than 7.00 times the acceleration of gravity. What is the minimum radius turn she can make without losing consciousness if her plane is flying with a constant speed of 245 m/s?

70. • The angular velocity of a rotating wheel as a function of time is shown in the graph in **FIGURE 10-36**. What is the angular displacement of the wheel, $\theta - \theta_0$, between the times $t = 0$ and $t = 5$ s?

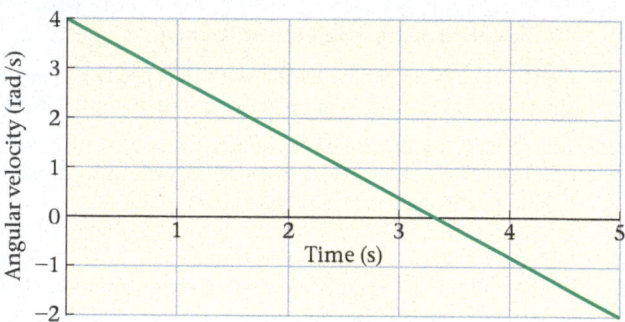

FIGURE 10-36 Problem 70

71. •• A 68-kg commuter rides on an 8.1-kg bicycle. Each bicycle wheel is a hoop of radius 33 cm and mass 1.4 kg. **(a)** What energy must the biker supply to move at 7.2 m/s? Include the kinetic energy of the biker, the kinetic energy of the bike frame, and the kinetic energy of the rotating wheels. **(b)** What percentage of the energy she expends goes into the rotation of the wheels? **(c)** What percentage goes into her own kinetic energy?

72. •• **CE** Place two quarters on a table with their rims touching, as shown in **FIGURE 10-37**. While holding one quarter fixed, roll the other one—without slipping—around the circumference of the fixed quarter until it has completed one round trip. How many revolutions has the rolling quarter made about its center?

FIGURE 10-37 Problem 72

73. •• **CE** The object shown in **FIGURE 10-38** can be rotated in three different ways: case 1, rotation about the x axis; case 2, rotation about the y axis; and case 3, rotation about the z axis. Rank these three cases in order of increasing moment of inertia. Indicate ties where appropriate.

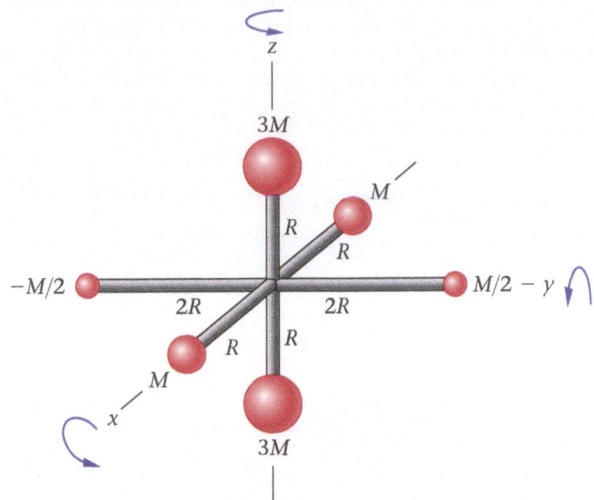

FIGURE 10-38 Problem 73

74. •• **Predict/Calculate** **When the Hands Align** A mathematically inclined friend e-mails you the following instructions: "Meet me in the cafeteria the first time after 2:00 P.M. today that the hands of a clock point in the same direction." **(a)** Is the desired meeting time before, after, or equal to 2:10 P.M.? Explain. **(b)** Is the desired meeting time before, after, or equal to 2:15 P.M.? Explain. **(c)** When should you meet your friend?

75. •• **BIO** **Spinning Dragonflies** Measurements show that when a dragonfly does a "spin-dry" (Problem 49), it can "spin up" from $\omega_0 = 0$ to $\omega = 104$ rad/s in 0.250 s. **(a)** Assuming constant angular acceleration, how many revolutions does the dragonfly make in this time? **(b)** How many revolutions has the dragonfly made when its angular speed is 51.2 rad/s?

76. •• **Predict/Calculate** A potter's wheel of radius 6.8 cm rotates with a period of 0.52 s. What are **(a)** the linear speed and **(b)** the centripetal acceleration of a small lump of clay on the rim of the wheel? **(c)** How do your answers to parts (a) and (b) change if the period of rotation is doubled?

77. •• **Predict/Calculate** **Playing a CD** The record in an old-fashioned record player always rotates at the same angular speed. With CDs, the situation is different. For a CD to play properly, the point on the CD where the laser beam shines must have a linear speed $v_t = 1.25$ m/s, as indicated in **FIGURE 10-39**. **(a)** As the CD plays from the center outward, does its angular speed increase, decrease, or stay the same? Explain. **(b)** Find the angular speed of a CD when the laser beam is 2.50 cm from its center. **(c)** Repeat part (b) for the laser beam 6.00 cm from the center. **(d)** If the CD plays for 66.5 min, and the laser beam moves from 2.50 cm to 6.00 cm during this time, what is the CD's average angular acceleration?

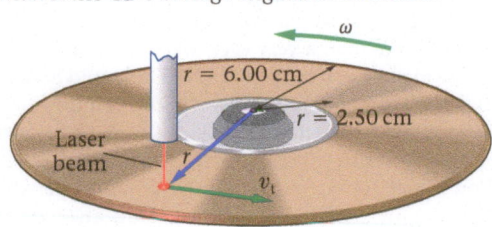

FIGURE 10-39 Problem 77

78. •• **BIO** **Roller Pigeons** Pigeons are bred to display a number of interesting characteristics. One breed of pigeon, the "roller," is remarkable for the fact that it does a number of backward somersaults as it drops straight down toward the ground. Suppose a roller pigeon drops from rest and free falls downward for a distance of 14 m. If the pigeon somersaults at the rate of 12 rad/s, how many revolutions has it completed by the end of its fall?

79. •• **BIO** **Dolphin Tricks** A bottlenose dolphin (*Tursiops truncatus*) is about 2.5 m long and has a mass of 250 kg. It can jump 3.5 m above the surface of the water while flipping nose-to-tail at 5.6 rad/s, fast enough to complete 1.5 rotations before landing in the water. **(a)** How much energy must the dolphin generate to jump 3.5 m above the surface of the water? **(b)** If the dolphin's moment of inertia about its rotation axis is 225 kg · m², how much energy must the dolphin generate to rotate its body in this way?

Jumping dolphin. (Problem 79)

80. •• As a marble with a diameter of 1.6 cm rolls down an incline, its center moves with a linear acceleration of 3.3 m/s². **(a)** What is the angular acceleration of the marble? **(b)** What is the angular speed of the marble after it rolls for 1.5 s from rest?

81. •• A rubber ball with a radius of 3.2 cm rolls along the horizontal surface of a table with a constant linear speed v. When the ball rolls off the edge of the table, it falls 0.66 m to the floor below. If the ball completes 0.37 revolution during its fall, what was its linear speed, v?

82. •• **Predict/Calculate** A yo-yo moves downward until it reaches the end of its string, where it "sleeps." As it sleeps—that is, spins in place—its angular speed decreases from 35 rad/s to 25 rad/s. During this time it completes 120 revolutions. **(a)** How much time did it take for the yo-yo to slow from 35 rad/s to 25 rad/s? **(b)** How much time does it take for the yo-yo to slow from 25 rad/s to 15 rad/s? Assume a constant angular acceleration as the yo-yo sleeps.

83. •• **Predict/Calculate** **(a)** An automobile with tires of radius 32 cm accelerates from 0 to 45 mph in 9.1 s. Find the angular acceleration of the tires. **(b)** How does your answer to part (a) change if the radius of the tires is halved?

84. •• **A Yo-Yo with a Brain** Yomega ("The yo-yo with a brain") is constructed with a clever clutch mechanism in its axle that allows it to rotate freely and "sleep" when its angular speed is greater than a certain critical value. When the yo-yo's angular speed falls below this value, the clutch engages, causing the yo-yo to climb the string to the user's hand. If the moment of inertia of the yo-yo is 7.4×10^{-5} kg $\cdot$ m^2, its mass is 0.11 kg, and the string is 1.0 m long, what is the smallest angular speed that will allow the yo-yo to return to the user's hand?

85. •• The rotor in a centrifuge has an initial angular speed of 430 rad/s. After 8.2 s of constant angular acceleration, its angular speed has increased to 550 rad/s. During this time, what were **(a)** the angular acceleration of the rotor and **(b)** the angle through which it turned?

86. •• The Sun, with Earth in tow, orbits about the center of the Milky Way galaxy at a speed of 137 miles per second, completing one revolution every 240 million years. **(a)** Find the angular speed of the Sun relative to the center of the Milky Way. **(b)** Find the distance from the Sun to the center of the Milky Way.

87. •• A person walks into a room and switches on the ceiling fan. The fan accelerates with constant angular acceleration for 15 s until it reaches its operating angular speed of 1.9 rotations/s—after that its speed remains constant as long as the switch is "on." The person stays in the room for a short time; then, 5.5 minutes after turning the fan on, she switches it off again and leaves the room. The fan now decelerates with constant angular acceleration, taking 2.4 minutes to come to rest. What is the total number of revolutions made by the fan, from the time it was turned on until the time it stopped?

88. •• **BIO Preventing Bone Loss in Space** When astronauts return from prolonged space flights, they often suffer from bone loss, resulting in brittle bones that may take weeks for their bodies to rebuild. One solution may be to expose astronauts to periods of substantial "g forces" in a centrifuge carried aboard their spaceship. To test this approach, NASA conducted a study in which four people spent 22 hours each in a compartment attached to the end of a 28-foot arm that rotated with an angular speed of 10.0 rpm. **(a)** What centripetal acceleration, in terms of g, did these volunteers experience? **(b)** What was their linear speed?

89. ••• A thin, uniform rod of length L and mass M is pivoted about one end, as shown in **FIGURE 10-40**. The rod is released from rest in a horizontal position, and allowed to swing downward without friction or air resistance. When the rod is vertical, what are **(a)** its angular speed ω and **(b)** the tangential speed v_t of its free end?

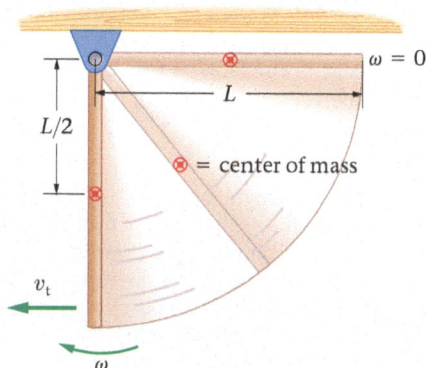

FIGURE 10-40 Problem 89

90. ••• **Center of Percussion** In the previous problem, suppose a small metal ball of mass $m = 2M$ is attached to the rod a distance d from the pivot. The rod and ball are released from rest in the horizontal position. **(a)** Show that when the rod reaches the vertical position, the speed of its tip is

$$v_t = \sqrt{3gL}\sqrt{\frac{1 + 4(d/L)}{1 + 6(d/L)^2}}$$

(b) At what finite value of d/L is the speed of the rod the same as it is for $d = 0$? (This value of d/L is the **center of percussion**, or "sweet spot," of the rod.)

91. ••• A wooden plank rests on two soup cans laid on their sides. Each can has a diameter of 6.5 cm, and the plank is 3.0 m long. Initially, one can is placed 1.0 m inward from either end of the plank, as **FIGURE 10-41** shows. The plank is now pulled 1.0 m to the right, and the cans roll without slipping. **(a)** How far does the center of each can move? **(b)** How many rotations does each can make?

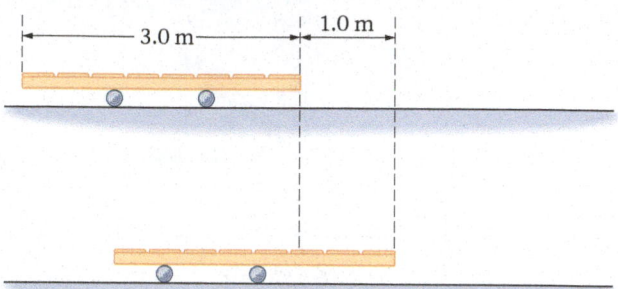

FIGURE 10-41 Problem 91

92. ••• A person rides on a 12-m-diameter Ferris wheel that rotates at the constant rate of 8.1 rpm. Calculate the magnitude and direction of the force that the seat exerts on a 65-kg person when he is **(a)** at the top of the wheel, **(b)** at the bottom of the wheel, and **(c)** halfway up the wheel.

93. ••• **Predict/Calculate** A solid sphere with a diameter of 0.17 m is released from rest; it then rolls without slipping down a ramp, dropping through a vertical height of 0.61 m. The ball leaves the bottom of the ramp, which is 1.22 m above the floor, moving horizontally (**FIGURE 10-42**). **(a)** Through what horizontal distance d does the ball move before landing? **(b)** How many revolutions does the ball make during its fall? **(c)** If the ramp were to be made frictionless, would the distance d increase, decrease, or stay the same? Explain.

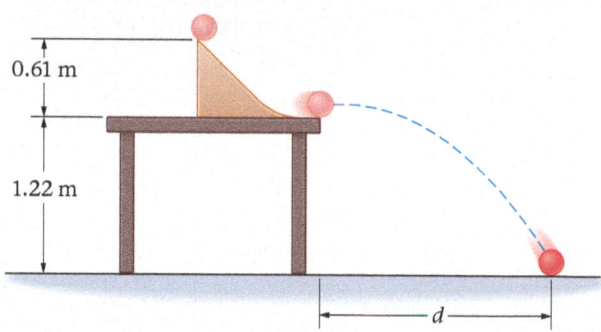

FIGURE 10-42 Problem 93

PASSAGE PROBLEMS

BIO Human-Powered Centrifuge

Space travel is fraught with hazards, not the least of which are the many side effects of prolonged weightlessness, including weakened muscles, bone loss, decreased coordination, and unsteady balance. If you are fortunate enough to go on a trip to Mars, which could take more than a year each way, you might be a bit "weak in the knees" by the time you arrive. This could lead to problems when you try to take your first "small step" on the surface.

To counteract these effects, NASA is looking into ways to provide astronauts with "portable gravity" on long space flights. One method under consideration is the human-powered centrifuge, which not only subjects the astronauts to artificial gravity, but also gives them aerobic exercise. The device is basically a rotating, circular platform on which two astronauts lie supine along a diameter, head-to-head at the center, with their feet at opposite rims, as shown in the accompanying photo. The radius of the platform in this test model is 6.25 ft. As one astronaut pedals to rotate the platform, the astronaut facing the other direction can exercise in the artificial gravity. Alternatively, a third astronaut on a stationary bicycle can provide the rotation for the other two. While the astronauts' feet are at the outer rim of the platform, their heads are near the center of the platform, and their hearts are 4.50 feet from the rim, which means that different parts of the astronauts' bodies will experience different "gravitational" accelerations.

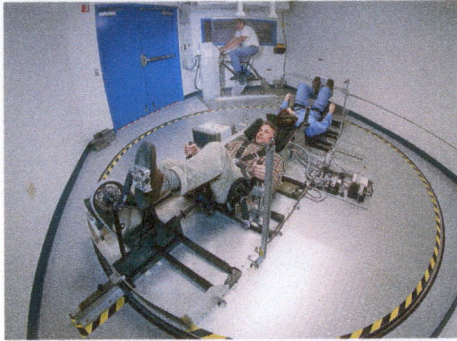

Human-powered centrifuge.

FIGURE 10-43 shows the centripetal acceleration (in g) produced by a rotating platform at four different radii. Notice that the acceleration increases as the square of the angular speed. Also indicated in Figure 10-43 are acceleration levels corresponding to 1, 3, and 5 gs. It is thought that enhanced gravitational effects may be desirable because the astronauts will experience the artificial gravity for only relatively brief periods of time during the flight.

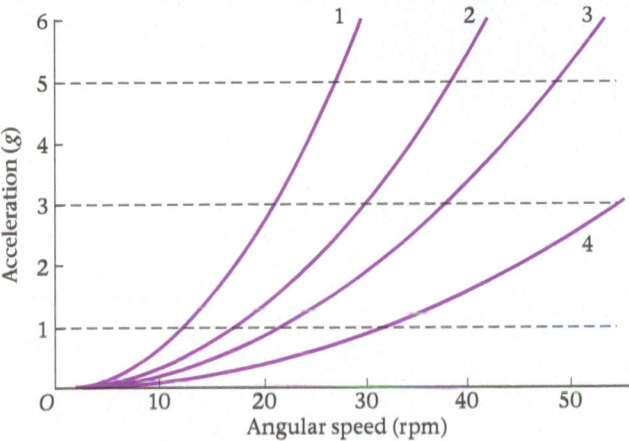

FIGURE 10-43 Problems 94, 95, 96, and 97

94. • Rank the four curves shown in Figure 10-43 in order of increasing radius. Indicate ties where appropriate.

95. • A particular experimental protocol calls for measuring the physiological parameters during exercise with 2.00g at the feet. What angular speed in rpm must the platform in this test model have in order to produce this centripetal acceleration at the feet?

A. 3.20 rpm   B. 26.5 rpm
C. 30.6 rpm   D. 57.8 rpm

96. • If the acceleration is 2.00g at the feet, what is the acceleration at the heart?

A. 0   B. 0.560g
C. 1.40g D. 2.00g

97. • Which of the curves shown in Figure 10-43 corresponds to the location of the feet in this test model?

A. 1   B. 2
C. 3   D. 4

98. •• **REFERRING TO CONCEPTUAL EXAMPLE 10-17** Suppose we race a disk and a hollow spherical shell, like a basketball. The spherical shell has a mass M and a radius R; the disk has a mass $2M$ and a radius $2R$. **(a)** Which object wins the race? If the two objects are released at rest, and the height of the ramp is $h = 0.75$ m, find the speed of **(b)** the disk and **(c)** the spherical shell when they reach the bottom of the ramp.

99. •• **REFERRING TO CONCEPTUAL EXAMPLE 10-17** Consider a race between the following three objects: object 1, a disk; object 2, a solid sphere; and object 3, a hollow spherical shell. All objects have the same mass and radius. **(a)** Rank the three objects in the order in which they finish the race. Indicate a tie where appropriate. **(b)** Rank the objects in order of increasing kinetic energy at the bottom of the ramp. Indicate a tie where appropriate.

11

Rotational Dynamics and Static Equilibrium

Big Ideas

1 Torque is the rotational analog of force in a linear system.

2 A constant torque causes a constant angular acceleration.

3 A system is in static equilibrium if the net force and net torque are equal to zero.

4 Angular momentum is the rotational analog of linear momentum.

5 Angular momentum is conserved in systems that have zero net torque.

▲ Equilibrium, and the sense of serenity that comes with it, requires more than just forces that add to zero. To keep from falling, for example, this acrobat's hands must exert forces that add up to his total weight. But the total weight must be shared between his two hands in just the right way, or else his body will rotate and the pose will be lost. To ensure equilibrium, a new physical quantity—the torque—must also be zero. In this chapter we introduce the torque and show that equilibrium occurs only when both the net force and the net torque are zero. We will also consider the consequences of nonzero torque.

In this chapter we develop the concept of torque, and show that it is the rotational equivalent of force. We also introduce the notion of *angular momentum*, and show that it is conserved when the net torque acting on a system is zero.

11-1 Torque

Suppose you want to loosen a nut by rotating it counterclockwise with a wrench. If you have ever used a wrench in this way, you probably know that the nut is more likely to turn if you apply your force as far from the nut as possible, as indicated in **FIGURE 11-1** and **FIGURE 11-2 (a)**. Applying a force near the nut would not be very effective—you could still get the nut to turn, but it would require considerably more effort! Similarly, it is much easier to open a revolving door if you push far from the axis of rotation, as indicated in **FIGURE 11-2 (b)**. Clearly, then, the tendency for a force to cause a rotation increases with the distance, r, from the axis of rotation to the force. As a result, it's useful to define a quantity called the **torque**, τ, that takes into account both the magnitude of the force, F, *and* the distance from the axis of rotation, r:

▲ **FIGURE 11-1** The long handle of this wrench enables the user to produce a large torque without having to exert a very great force.

Definition of Torque, τ, for a Tangential Force

$$\tau = rF \qquad\qquad 11\text{-}1$$

SI unit: N · m

With this definition, we see that torque increases with both the force and the distance.

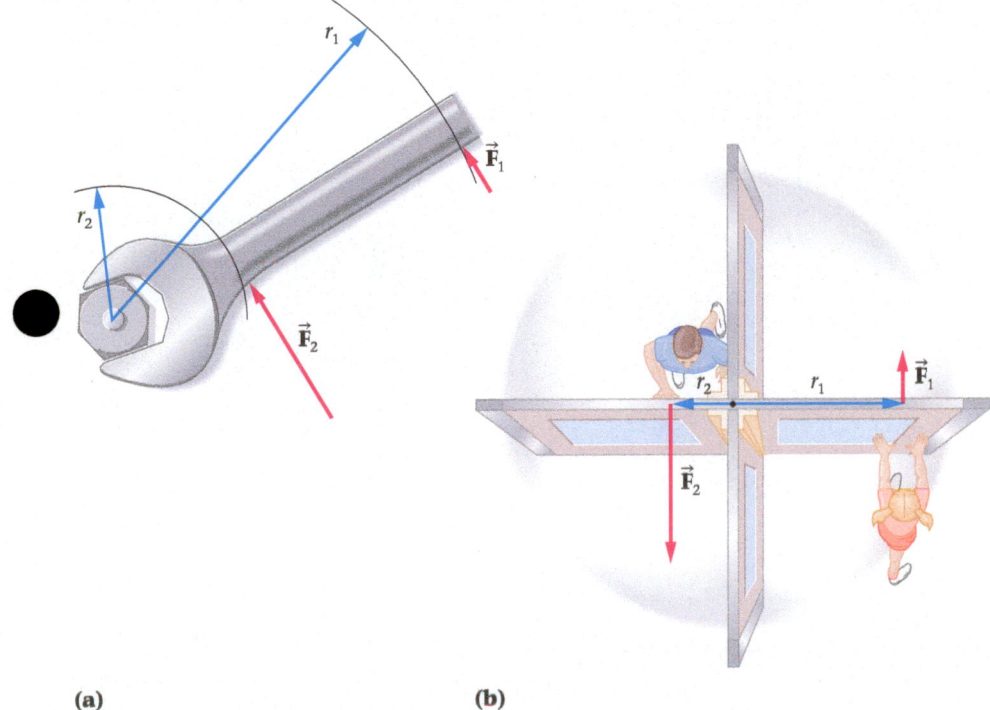

(a)   (b)

◄ **FIGURE 11-2 Applying a torque**
(a) When a wrench is used to loosen a nut, less force is required if it is applied far from the nut. **(b)** Similarly, less force is required to open a revolving door if it is applied far from the axis of rotation.

Equation 11-1 is valid only when the applied force is *tangential* to a circle of radius r centered on the axis of rotation, as in Figure 11-2. The more general case is considered later in this section. First, we use $\tau = rF$ to determine how much force is needed to open a swinging door, depending on where we apply the force.

EXERCISE 11-1 TORQUE AND FORCE

To open the door in Figure 11-2 (b) a tangential force F is applied at a distance r from the axis of rotation. Suppose the minimum torque required to open the door is 3.4 N · m. **(a)** What force must be applied to open the door if r is 0.91 m? **(b)** At what minimum distance must a force of 6.9 N be applied to open the door?

REASONING AND SOLUTION

a. Setting $\tau = rF = 3.4$ N · m, we can solve for the required force:

$$F = \frac{\tau}{r} = \frac{3.4\ \text{N} \cdot \text{m}}{0.91\ \text{m}} = 3.7\ \text{N}$$

CONTINUED

PROBLEM-SOLVING NOTE

The Units of Torque

Notice that the units of torque are N · m, the same as the units of work. Though their units are the same, torque, τ, and work, W, represent different physical quantities and should not be confused with one another.

▲ **FIGURE 11-3 Starfish attack** A starfish attempts to open a clam by exerting a torque on the two halves of its shell. The clam uses its adductor muscles to exert an opposite torque on the shell.

▶ **FIGURE 11-4 Only the tangential component of a force causes a torque** **(a)** A radial force causes no rotation. In this case, the force $\vec{F}$ is opposed by an equal and opposite force exerted by the axle of the merry-go-round. The merry-go-round does not rotate. **(b)** A force applied at an angle θ with respect to the radial direction. The radial component of this force, $F \cos \theta$, causes no rotation; the tangential component, $F \sin \theta$, can cause a rotation.

Big Idea 1 Torque is the rotational analog of force in a linear system. Torque depends on the direction and magnitude of the force applied to an object, and also on the distance from the point of application of the force to the axis of rotation.

PHYSICS IN CONTEXT
Looking Back

The concept of force (Chapters 5 and 6) has been extended to torque, its rotational equivalent. We apply Newton's laws to rotational motion in Sections 11-2 and 11-5, just as was done for linear motion in Chapters 5 and 6.

PHYSICS IN CONTEXT
Looking Ahead

Torque arises in the discussion of magnetic fields and the forces they exert in Chapter 22. The torques due to magnetic fields are also the key element in the operation of electric motors, as we show in Chapter 23.

b. This time, we solve $\tau = rF = 3.4 \, \text{N} \cdot \text{m}$ for the required distance:

$$r = \frac{\tau}{F} = \frac{3.4 \, \text{N} \cdot \text{m}}{6.9 \, \text{N}} = 0.49 \, \text{m}$$

As expected, the required distance is smaller when the applied force is greater.

BIO Many human inventions make use of torque and rotation, but there are also examples in nature—from the hinged jaws of the Venus flytrap to those of the alligator. One classic example of a struggle between torques is the clam and the starfish, as shown in **FIGURE 11-3**. The hard clam (*Mercenaria mercenaria*) keeps its two halves partly open for filter feeding, but it also has very strong adductor muscles to hold the shell closed when it senses danger. A starfish such as *Asterias forbesi* feeds by prying open the two halves of the clamshell and then inserting its stomach into the clam in order to digest it *in situ*. The starfish uses hydraulic suction in its tube feet to exert a large torque on the clamshell. If the starfish can exert sufficient torque, and for a sufficient length of time, it can wear out the clam's adductor muscles and enjoy a meal.

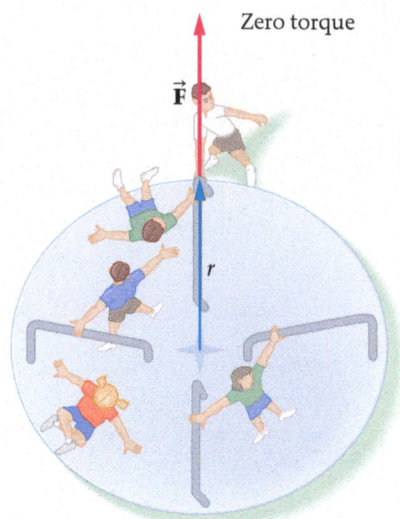

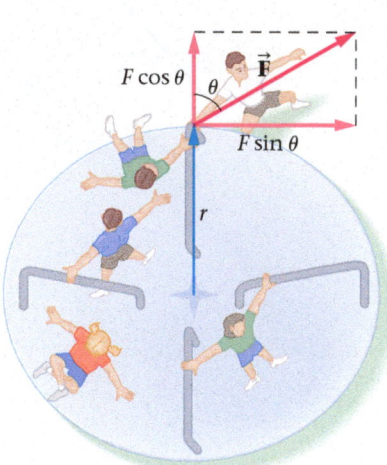

(a) A radial force produces zero torque

(b) Only the tangential component of force causes a torque

Torque with Nontangential Forces   So far we have considered tangential forces only. What happens if you exert a force in a direction that is not tangential? Suppose, for example, that you pull on a playground merry-go-round in a direction that is radial—that is, along a line that extends through the axis of rotation—as in **FIGURE 11-4 (a)**. In this case, your force has no tendency to cause a rotation. Instead, the axle of the merry-go-round simply exerts an equal and opposite force, and the merry-go-round remains at rest. Similarly, if you were to push or pull in a radial direction on a swinging door it would not rotate. We conclude that a *radial force produces zero torque*.

On the other hand, suppose your force is at an angle θ relative to a radial line, as shown in **FIGURE 11-4 (b)**. What is the torque now? To analyze this case, we first resolve the force vector $\vec{F}$ into radial and tangential components. Referring to the figure, we see that the radial component has a magnitude of $F \cos \theta$, and the tangential component has a magnitude of $F \sin \theta$. Because it is the tangential component alone that causes rotation, we define the torque to have a magnitude of $r(F \sin \theta)$. That is:

General Definition of the Magnitude of Torque, τ

$$\tau = r(F \sin \theta) \tag{11-2}$$

SI units: N · m

(More generally, the torque can be defined as the **cross product** of the vectors $\vec{r}$ and $\vec{F}$; that is, $\vec{\tau} = \vec{r} \times \vec{F}$. The cross product is discussed in detail in Appendix A.)

As a quick check, notice that a radial force corresponds to $\theta = 0$. In this case, $\tau = r(F \sin 0) = 0$, as expected. If the force is tangential, however, it follows that $\theta = \pi/2$. This gives $\tau = r(F \sin \pi/2) = rF$, in agreement with our original definition in Equation 11-1.

Torque and the Moment Arm An equivalent way to define torque is in terms of the **moment arm**, $r_\perp$. The idea here is to extend a line through the force vector, as in **FIGURE 11-5**, and then draw a second line from the axis of rotation perpendicular to the line of the force. The perpendicular distance from the axis of rotation to the line of the force is defined to be $r_\perp$. From the figure, we see that

$$r_\perp = r \sin \theta$$

In addition, we note that a simple rearrangement of the torque expression in Equation 11-2 yields

$$\tau = r(F \sin \theta) = (r \sin \theta)F$$

Thus, the torque can be written as the moment arm times the force:

$$\tau = r_\perp F \qquad \text{11-3}$$

Sign Convention for Torque Just as a force applied to an object gives rise to a linear acceleration, a torque applied to an object gives rise to an angular acceleration. Specifically, if a torque acts on an object at rest, the object will begin to rotate; if a torque acts on a rotating object, the object's angular velocity will change. In fact, the greater the torque applied to an object, the greater its angular acceleration, as we shall see in the next section. For this reason, the sign of the torque is determined by the same convention used previously in Section 10-1 for angular acceleration:

Sign Convention for Torque
By convention, if a torque τ acts alone, then

$\tau > 0$ if the torque causes a counterclockwise angular acceleration
$\tau < 0$ if the torque causes a clockwise angular acceleration

In a system with more than one torque, as in **FIGURE 11-6**, the sign of each torque is determined by the type of angular acceleration *it alone* would produce. The net torque acting on the system, then, is the sum of the individual torques, taking into account the proper signs. This is illustrated in the following Example.

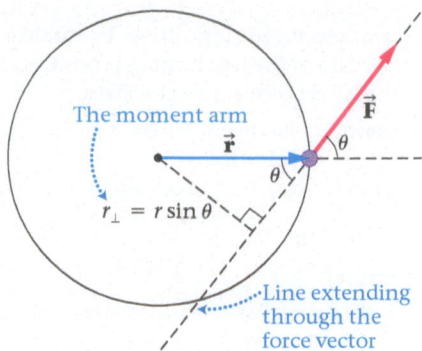

▲ **FIGURE 11-5 The moment arm** To find the moment arm, $r_\perp$, for a given force, first extend a line through the force vector. This line is sometimes referred to as the "line of action." Next, drop a perpendicular line from the axis of rotation to the line of the force. The perpendicular distance is $r_\perp = r \sin \theta$.

PROBLEM-SOLVING NOTE

The Sign of Torques

The sign of a torque is determined by the direction of rotation it would cause if it were the only torque acting in the system.

▲ **FIGURE 11-6** The net torque on the wheel of this ship is the sum of the torques exerted by the two helmsmen.

EXAMPLE 11-2   TORQUES TO THE LEFT AND TORQUES TO THE RIGHT

Two helmsmen, in disagreement about which way to turn a ship, exert the forces shown in the sketch of a ship's wheel. The wheel has a radius of 0.74 m, and the two forces have the magnitudes $F_1 = 72$ N and $F_2 = 58$ N. Find (a) the torque caused by $\vec{F}_1$ and (b) the torque caused by $\vec{F}_2$. (c) In which direction does the wheel turn as a result of these two forces?

PICTURE THE PROBLEM
Our sketch shows that both forces are applied at the distance $r = 0.74$ m from the axis of rotation. However, $F_1 = 72$ N is at an angle of 50.0° relative to the radial direction, whereas $F_2 = 58$ N is tangential, which means that its angle relative to the radial direction is 90.0°.

REASONING AND STRATEGY
For each force, we find the magnitude of the corresponding torque, using $\tau = rF \sin \theta$. As for the signs of the torques, we must consider the angular acceleration each force alone would cause. $\vec{F}_1$ acting alone would cause the wheel to accelerate counterclockwise, and

CONTINUED

hence its torque is positive. $\vec{F}_2$ would accelerate the wheel clockwise if it acted alone, and hence its torque is negative. If the sum of the two torques is positive, the wheel accelerates counterclockwise; if the sum of the two torques is negative, the wheel accelerates clockwise.

Known   Radius of wheel, $r = 0.74$ m; force 1, $F_1 = 72$ N; force 2, $F_2 = 58$ N; direction of force 1, 50.0°; direction of force 2, 90.0°.

Unknown   (a) Torque caused by $\vec{F}_1$, $\tau_1 = ?$ (b) Torque caused by $\vec{F}_2$, $\tau_2 = ?$ (c) Direction of rotation?

SOLUTION

Part (a)

1. Use $\tau = rF \sin \theta$ to calculate the torque due to $\vec{F}_1$. $\tau_1 = rF_1 \sin 50.0° = (0.74 \text{ m})(72 \text{ N}) \sin 50.0° = 41 \text{ N} \cdot \text{m}$
    Recall that this torque is positive:

Part (b)

2. Similarly, calculate the torque due to $\vec{F}_2$. $\tau_2 = -rF_2 \sin 90.0° = -(0.74 \text{ m})(58 \text{ N}) = -43 \text{ N} \cdot \text{m}$
    Recall that this torque is negative:

Part (c)

3. Sum the torques from parts (a) and (b) to find the net torque: $\tau_{net} = \tau_1 + \tau_2 = 41 \text{ N} \cdot \text{m} - 43 \text{ N} \cdot \text{m} = -2 \text{ N} \cdot \text{m}$

INSIGHT

Because the net torque is negative, the wheel accelerates clockwise. Thus, even though $\vec{F}_2$ is the smaller force, it has the greater effect in determining the wheel's direction of acceleration. This is because $\vec{F}_2$ is applied tangentially, whereas $\vec{F}_1$ is applied in a direction that is partially radial.

PRACTICE PROBLEM

What magnitude of $\vec{F}_2$ would yield zero net torque on the wheel, assuming everything else remains the same?   **[Answer: $F_2 = 55$ N]**

Some related homework problems: Problem 1, Problem 2

Enhance Your Understanding   (Answers given at the end of the chapter)

1. A bicycle wheel is mounted on an axle, as shown in **FIGURE 11-7**. Rank the four forces of equal magnitude (F_1, F_2, F_3, F_4) in order of the torque they produce, from most negative to most positive. Indicate ties where appropriate.

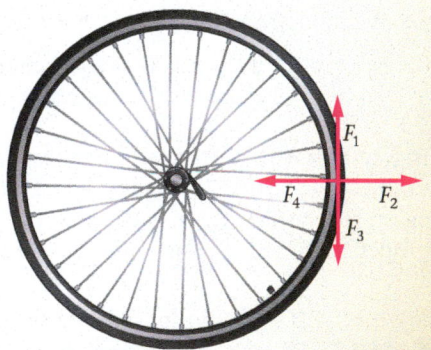

▲ **FIGURE 11-7** Enhance Your Understanding 1.

Section Review

- Torque has a magnitude given by $\tau = rF \sin \theta$, where F is the force, r is the distance to the axis of rotation, and θ is the angle between the force and the radial direction.

- A torque is positive if it alone would cause a counterclockwise rotation, and negative if it alone would cause a clockwise rotation.

11-2 Torque and Angular Acceleration

In the previous section, we indicated that a torque causes a change in the rotational motion of an object. To be more precise, a single torque, τ, acting on an object causes the object to have an angular acceleration, α. In this section we develop the specific relationship between τ and α.

Torque and Angular Acceleration Consider a small object of mass m connected to an axis of rotation by a light rod of length r, as in **FIGURE 11-8**. If a tangential force of magnitude F is applied to the mass, it will move with an acceleration given by Newton's second law:

$$a = \frac{F}{m}$$

From Equation 10-14, we know that the linear and angular accelerations are related by

$$\alpha = \frac{a}{r}$$

Combining these results yields the following expression for the angular acceleration:

$$\alpha = \frac{a}{r} = \frac{(F/m)}{r} = \frac{F}{mr}$$

Multiplying both numerator and denominator by r gives

$$\alpha = \left(\frac{r}{r}\right)\frac{F}{mr} = \frac{rF}{mr^2}$$

Now this last result is rather interesting, because the numerator and denominator have simple interpretations. First, the numerator is the torque, $\tau = rF$, for the case of a tangential force (Equation 11-1). Second, the denominator is the moment of inertia of a single mass m rotating at a radius r; that is, $I = mr^2$. Therefore, we find that

$$\alpha = \frac{rF}{mr^2} = \frac{\tau}{I}$$

Thus, the angular acceleration is directly proportional to the torque, and inversely proportional to the moment of inertia. Rearranging this slightly, we find

$$\tau = I\alpha$$

Notice the similarity between this result and Newton's second law for linear systems, $F = ma$.

Newton's Second Law for Rotational Motion The relationship $\tau = I\alpha$ was derived for the special case of a tangential force and a single mass rotating at a radius r. However, the result is completely general. For example, in a system with more than one torque, the relationship $\tau = I\alpha$ is replaced with $\tau_{net} = \Sigma\tau = I\alpha$, where τ_{net} is the net torque acting on the system. This gives the *rotational* version of Newton's second law:

Newton's Second Law for Rotational Motion

$$\sum \tau = I\alpha \qquad\qquad\qquad 11\text{-}4$$

If only a single torque acts on a system, we simply write $\tau = I\alpha$.

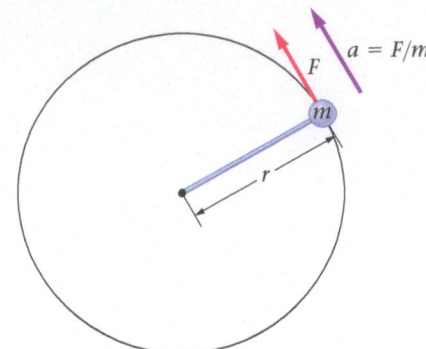

▲ **FIGURE 11-8 Torque and angular acceleration** A tangential force F applied to a mass m gives it a linear acceleration of magnitude $a = F/m$. The corresponding angular acceleration is $\alpha = \tau/I$, where $\tau = rF$ and $I = mr^2$.

QUICK EXAMPLE 11-3 ANGULAR ACCELERATION

A light rope wrapped around a disk-shaped pulley is pulled tangentially with a force of 0.51 N. Find the angular acceleration of the pulley, given that its mass is 2.4 kg and its radius is 0.13 m.

REASONING AND SOLUTION
To find the angular acceleration of the pulley, we first calculate the torque exerted on it by the rope ($\tau = rF$ for a tangential force). Next, we calculate the pulley's moment of inertia $\left(I = \frac{1}{2}mr^2 \text{ for a disk}\right)$. The angular acceleration of the pulley is the torque exerted on it divided by its moment of inertia.

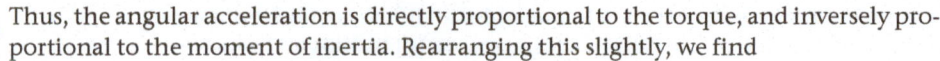

1. Calculate the torque applied to $\tau = rF = (0.13\text{ m})(0.51\text{ N}) = 0.066\text{ N}\cdot\text{m}$
   the disk:

CONTINUED

2. Determine the moment of inertia $I = \frac{1}{2}mr^2 = \frac{1}{2}(2.4 \text{ kg})(0.13 \text{ m})^2 = 0.020 \text{ kg} \cdot \text{m}^2$
of the disk-shaped pulley:

3. Divide the torque by the moment $\alpha = \dfrac{\tau}{I} = \dfrac{0.066 \text{ N} \cdot \text{m}}{0.020 \text{ kg} \cdot \text{m}^2} = 3.3 \text{ rad/s}^2$
of inertia to find the angular
acceleration of the pulley:

How would the angular acceleration change if the radius of the pulley were doubled, with everything else remaining the same? Well, doubling the radius doubles the torque (see Step 1), but quadruples the moment of inertia (see Step 2). Hence, the angular acceleration of the pulley is halved if the radius of the pulley is doubled.

Big Idea 2 A constant torque applied to an object causes a constant angular acceleration, just as a constant force causes a constant linear acceleration.

It's easy to remember the rotational version of Newton's second law, $\Sigma\tau = I\alpha$, by using analogies between rotational and linear quantities. We have already seen that I is the analog of m, and that α is the analog of a. Similarly, τ, which causes an angular acceleration, is the analog of F, which causes a linear acceleration. To summarize:

Linear Quantity	Angular Quantity
m	I
a	α
F	τ

PHYSICS IN CONTEXT
Looking Back

The connection between rotational and linear quantities (Chapter 10) is used here to relate torque to angular acceleration.

Thus, just as $\Sigma F = ma$ describes linear motion, $\Sigma\tau = I\alpha$ describes rotational motion.

EXAMPLE 11-4   GETTING WINDED

A wind turbine has a radius of 37 m, a moment of inertia of $8.4 \times 10^6 \text{ kg} \cdot \text{m}^2$, and rotates with an initial angular speed of 1.1 rad/s. When the wind speed increases, the turbine experiences a net torque of $1.6 \times 10^5 \text{ N} \cdot \text{m}$. **(a)** How many rotations does it take for the angular speed to increase to 1.8 rad/s? **(b)** What is the linear speed of the tips of the turbine blades 27 s after the acceleration begins?

PICTURE THE PROBLEM
Our sketch shows the wind turbine rotating with an initial angular speed of $\omega_0 = 1.1$ rad/s, and accelerating with an angular acceleration α in the same direction as the angular velocity. The result will be an increase in the angular speed.

REASONING AND STRATEGY
This is basically an angular kinematics problem, as in Chapter 10, but in this case we must first calculate the angular acceleration using $\alpha = \tau/I$. **(a)** Once α is known, we can find the angular displacement, $\Delta\theta$, using $\omega^2 = \omega_0^2 + 2\alpha\,\Delta\theta$. We divide $\Delta\theta$ by 2π to convert from radians to revolutions. **(b)** We can find the angular speed at $t = 27$ s by using $\omega = \omega_0 + \alpha t$. The linear speed at the tip of the blades is then $v = r\omega$.

$\omega_0 = 1.1$ rad/s

Known   Radius of turbine, $r = 37$ m; moment of inertia, $I = 8.4 \times 10^6 \text{ kg} \cdot \text{m}^2$; initial angular speed, $\omega_0 = 1.1$ rad/s; net torque, $\tau = 1.6 \times 10^5 \text{ N} \cdot \text{m}$.
Unknown   **(a)** Angular displacement, $\Delta\theta = ?$, when $\omega = 1.8$ rad/s
(b) Linear speed of blade tips, $v = ?$, at $t = 27$ s

SOLUTION

1. Use the torque and moment of inertia to calculate the angular acceleration of the turbine:

$$\alpha = \frac{\tau}{I} = \frac{1.6 \times 10^5 \text{ N} \cdot \text{m}}{8.4 \times 10^6 \text{ kg} \cdot \text{m}^2} = 0.019 \text{ rad/s}^2$$

Part (a)

2. Rearrange $\omega^2 = \omega_0^2 + 2\alpha\,\Delta\theta$ to calculate the angular displacement $\Delta\theta$:

$$\Delta\theta = \frac{(\omega^2 - \omega_0^2)}{2\alpha}$$
$$= \frac{(1.8 \text{ rad/s})^2 - (1.1 \text{ rad/s})^2}{2(0.019 \text{ rad/s}^2)} = 53 \text{ rad}$$

CONTINUED

3. Convert from radians to revolutions:

$$\Delta\theta = (53 \text{ rad})\left(\frac{1 \text{ rev}}{2\pi \text{ rad}}\right) = 8.4 \text{ rev}$$

Part (b)

4. Use $\omega = \omega_0 + \alpha t$ to find the final angular speed at $t = 27$ s:

$$\omega = \omega_0 + \alpha t$$
$$= (1.1 \text{ rad/s}) + (0.019 \text{ rad/s}^2)(27 \text{ s}) = 1.6 \text{ rad/s}$$

5. Calculate the corresponding linear speed with $v = r\omega$:

$$v = r\omega = (37 \text{ m})(1.6 \text{ rad/s}) = 59 \text{ m/s}$$

INSIGHT
The speed of the blade tips at $t = 27$ s is about 132 mph. This speed is an important consideration in wind turbine design because the greater the tip speed, the more noise the turbine makes.

PRACTICE PROBLEM
What is the linear acceleration of the blade tips at $t = 27$ s?   [**Answer:** $a = r\alpha = (37 \text{ m})(0.019 \text{ rad/s}^2) = 0.70 \text{ m/s}^2$. This result is independent of the time, as long as the angular acceleration remains constant.]

Some related homework problems: Problem 10, Problem 11, Problem 15

CONCEPTUAL EXAMPLE 11-5   WHICH BLOCK LANDS FIRST?   PREDICT/EXPLAIN

The rotating systems shown in the sketch differ only in that the two spherical movable masses are positioned either far from the axis of rotation (left) or near the axis of rotation (right). **(a)** If the hanging blocks are released simultaneously from rest, is it observed that the block on the left lands first, the block on the right lands first, or both blocks land at the same time? **(b)** Which of the following is the *best explanation* for your prediction?

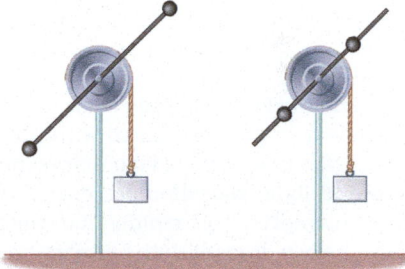

  I. The torque is the same in both cases, and therefore both blocks land at the same time.

 II. The moment of inertia is less for the system on the right, and hence it has the greater angular acceleration. As a result the block on the right lands first.

III. The moment of inertia is greater for the system on the left, and hence it has more kinetic energy as it rotates. As a result the block on the left lands first.

REASONING AND DISCUSSION
The net external torque, which is supplied by the hanging blocks, is the same for each of these systems. However, the moment of inertia of the system on the right is less than that of the system on the left, because the movable masses are closer to the axis of rotation. The angular acceleration is inversely proportional to the moment of inertia ($\alpha = \tau/I$), and hence the system on the right has the greater angular acceleration, and it wins the race.

ANSWER
(a) The block on the right lands first. **(b)** The best explanation is II.

EXAMPLE 11-6   DROP IT

A person holds his outstretched arm at rest in a horizontal position. The mass of the arm is m and its length is 0.740 m. When the person releases his arm, allowing it to drop freely, it begins to rotate about the shoulder joint. Find **(a)** the initial angular acceleration of the arm, and **(b)** the initial linear acceleration of the man's hand. (*Hint:* In calculating the torque, assume the mass of the arm is concentrated at its midpoint. In calculating the angular acceleration, use the moment of inertia of a uniform rod of length L about one end; $I = \frac{1}{3}mL^2$.)

PICTURE THE PROBLEM
The arm is initially horizontal and at rest. When released, it rotates downward about the shoulder joint. The force of gravity, mg, acts at a distance of $(0.740 \text{ m})/2 = 0.370$ m from the shoulder.

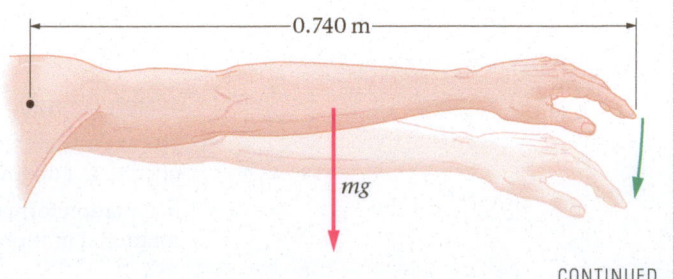

REASONING AND STRATEGY
The angular acceleration, α, can be found using $\tau = I\alpha$. In this case, the initial torque is $\tau = mg(L/2)$, where $L = 0.740$ m, and the moment of inertia is $I = \frac{1}{3}mL^2$.

CONTINUED

Once the initial angular acceleration is found, the corresponding linear acceleration is obtained from $a = r\alpha$.

Known Mass of arm, m; length of arm, $L = 0.740$ m.
Unknown (a) Initial angular acceleration of arm, $\alpha = ?$ (b) Initial linear acceleration, $a = ?$

SOLUTION

Part (a)

1. Use $\tau = I\alpha$ to find the angular acceleration, α:

$$\alpha = \frac{\tau}{I}$$

2. Write expressions for the initial torque, τ, and the moment of inertia, I:

$$\tau = mg\frac{L}{2}$$

$$I = \frac{1}{3}mL^2$$

3. Substitute τ and I into the expression for the angular acceleration. Notice that the mass of the arm cancels:

$$\alpha = \frac{\tau}{I} = \frac{mgL/2}{mL^2/3} = \frac{3g}{2L}$$

4. Substitute numerical values:

$$\alpha = \frac{3g}{2L} = \frac{3(9.81 \text{ m/s}^2)}{2(0.740 \text{ m})} = 19.9 \text{ rad/s}^2$$

Part (b)

5. Use $a = r\alpha$ to calculate the linear acceleration at the man's hand, a distance $r = L$ from the shoulder:

$$a = L\alpha = L\left(\frac{3g}{2L}\right) = \frac{3}{2}g = 14.7 \text{ m/s}^2$$

INSIGHT

The linear acceleration of the hand is 1.50 times greater than the acceleration of gravity, regardless of the mass of the arm. This can be demonstrated with the following simple experiment: Hold your arm straight out with a pen resting on your hand. Now, relax your deltoid muscles, and let your arm rotate freely downward about your shoulder joint. Notice that as your arm falls downward, your hand moves more rapidly than the pen, which appears to "lift off" your hand. The pen drops with the acceleration of gravity, which is clearly less than the acceleration of the hand. This effect can be seen in **FIGURE 11-9**.

PRACTICE PROBLEM

At what distance from the shoulder is the initial linear acceleration of the arm equal to the acceleration of gravity? [**Answer:** Set $a = r\alpha$ equal to g. This gives $r = 2L/3 = 0.493$ m.]

Some related homework problems: Problem 14, Problem 16, Problem 70

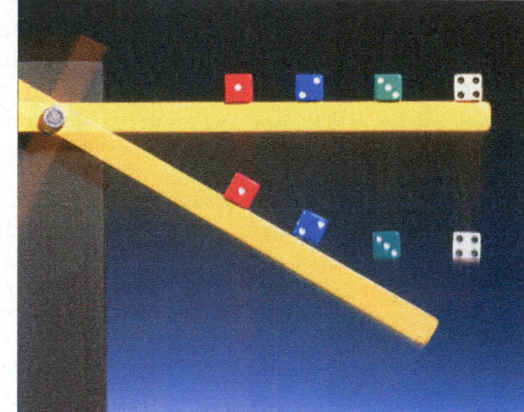

▶ **FIGURE 11-9** As a rod of length L rotates freely about one end, points farther from the axle than $2L/3$ have an acceleration greater than g (see the Practice Problem for Example 11-6). Thus, the rod falls out from under the last two dice.

Enhance Your Understanding   (Answers given at the end of the chapter)

2. Consider two objects with the following characteristics: Object 1 is a hoop with mass M and radius R; object 2 is a disk with mass $2M$ and radius $2R$. Suppose a torque τ is applied to object 1, and a torque 2τ is applied to object 2. Is the angular acceleration of object 1 greater than, less than, or equal to the angular acceleration of object 2? Explain.

Section Review

• The angular acceleration of an object is directly proportional to the net torque applied to it, and inversely proportional to the object's moment of inertia.

• In equation form that is similar to $F_{net} = ma$, we can write Newton's second law for rotational motion as $\tau_{net} = I\alpha$.

11-3 Zero Torque and Static Equilibrium

The parents of a young boy are supporting him on a long, lightweight plank, as illustrated in **FIGURE 11-10**. If the mass of the child is m, the upward forces exerted by the parents must sum to mg; that is,

$$F_1 + F_2 = mg$$

This condition ensures that the net force acting on the plank is zero. It *does not*, however, guarantee that the plank remains at rest.

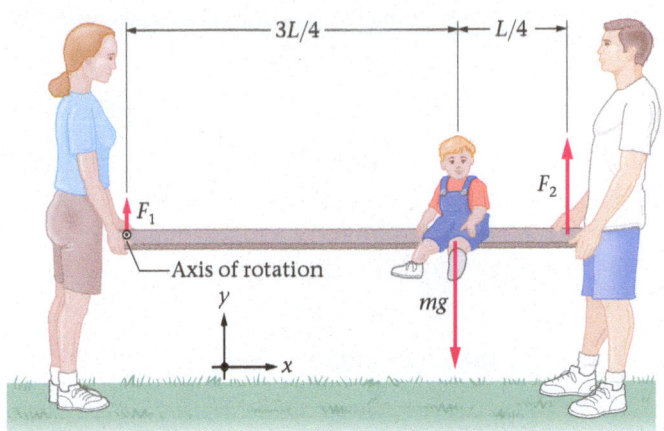

── **Axis of rotation**

── **FIGURE 11-10 Forces required for static equilibrium** Two parents support a child on a lightweight plank of length L. For the calculation described in the text, we choose the axis of rotation to be the left end of the plank.

To see why, imagine for a moment that the parent on the right lets go of the plank and that the parent on the left increases her force until it is equal to the weight of the child. In this case, $F_1 = mg$ and $F_2 = 0$, which clearly satisfies the force equation we have just written. The right end of the plank is no longer supported, however, and it drops toward the ground while the left end rises. In other words, the plank rotates in a clockwise sense. For the plank to remain completely at rest, with no translation or rotation, we must impose the following *two* conditions:

1. The net force acting on the plank must be zero, so that there is no translational acceleration.
2. The net torque acting on the plank must be zero, so that there is no rotational acceleration.

If both of these conditions are met, an extended object, like the plank, will remain at rest if it starts at rest. To summarize:

Conditions for Static Equilibrium
For an extended object to be in static equilibrium, the following two conditions must be met:

(i) The net force acting on the object must be zero:

$$\sum F_x = 0, \qquad \sum F_y = 0 \tag{11-5}$$

(ii) The net torque acting on the object must be zero:

$$\sum \tau = 0 \tag{11-6}$$

These conditions are independent of one another; that is, satisfying one does *not* guarantee that the other is satisfied.

Big Idea 3 A system is in static equilibrium if both the net force and the net torque acting on it are equal to zero.

Applying the Conditions for Equilibrium Let's apply these conditions to the plank that supports the child. First, we consider the forces acting on the plank, with upward chosen as the positive direction, as in Figure 11-10. Setting the net force equal to zero yields

$$F_1 + F_2 - mg = 0$$

Clearly, this agrees with the force equation we wrote earlier.

Next, we apply the torque condition. To do so, we must first choose an axis of rotation. For example, we might take the left end of the plank to be the axis, as in Figure 11-10. With this choice, we see that the force F_1 exerts zero torque, because it acts directly through the axis of rotation. On the other hand, F_2 acts at the far end of the plank, a distance L from the axis. In addition, F_2 would cause a counterclockwise (positive) rotation if it acted alone, as we can see in Figure 11-10. Therefore, the torque due to F_2 is positive and has the magnitude

$$\tau_2 = F_2 L$$

Finally, the weight of the child, mg, acts at a distance of $3L/4$ from the axis, and would cause a clockwise (negative) rotation if it acted alone. Hence, its torque is negative:

$$\tau_{mg} = -mg\left(\tfrac{3}{4}L\right)$$

Setting the net torque equal to zero, then, yields the following condition:

$$\tau_{net} = F_2 L - mg\left(\tfrac{3}{4}L\right) = 0$$

This torque condition, along with the force condition in $F_1 + F_2 - mg = 0$, can be used to determine the two unknowns, F_1 and F_2. For example, we can begin by canceling L in the torque equation to find F_2:

$$F_2 = \tfrac{3}{4}mg$$

Substituting this result into the force condition gives

$$F_1 + \tfrac{3}{4}mg - mg = 0$$

Therefore, F_1 is

$$F_1 = \tfrac{1}{4}mg$$

These two forces support the plank, *and* keep it from rotating. As one might expect, the force nearer the child is greater.

PROBLEM-SOLVING NOTE

Axis of Rotation

Any point in a system may be used as the axis of rotation when calculating torque. It is generally best, however, to choose an axis that gives zero torque for at least one of the unknown forces in the system. Such a choice simplifies the algebra needed to solve for the forces.

Any Location Can Be the Axis of Rotation   Our choice of the left end of the plank as the axis of rotation was completely arbitrary. In fact, if an object is in static equilibrium, the net torque acting on it is zero, regardless of the location of the axis of rotation. Hence, we are free to choose an axis of rotation that is most convenient for a given problem. In general, it is useful to pick the axis to be at the location of one of the unknown forces. This eliminates that force from the torque condition and simplifies the remaining algebra. In the following Example, we choose an axis of rotation that eliminates the unknown force F_2 from the torque equation.

EXAMPLE 11-7   FIND THE FORCES: AXIS ON THE RIGHT

A child of mass m is supported on a light plank by his parents, who exert the forces F_1 and F_2 as indicated in the sketch. Find the forces required to keep the plank in static equilibrium. Use the right end of the plank as the axis of rotation.

PICTURE THE PROBLEM
Our sketch shows the same physical system as in Figure 11-10, only this time the axis of rotation is at the right end of the plank.

REASONING AND STRATEGY
We set both the net force and the net torque equal to zero. The expression for the net torque is different this time, due to the new location for the axis of rotation, but the final result for the forces will be the same.

Known   Mass of child, m; distance for force 1, L; distance for force 2, 0; distance for weight (mg) of child, $L/4$.

Unknown $F_1 = ? F_2 = ?$

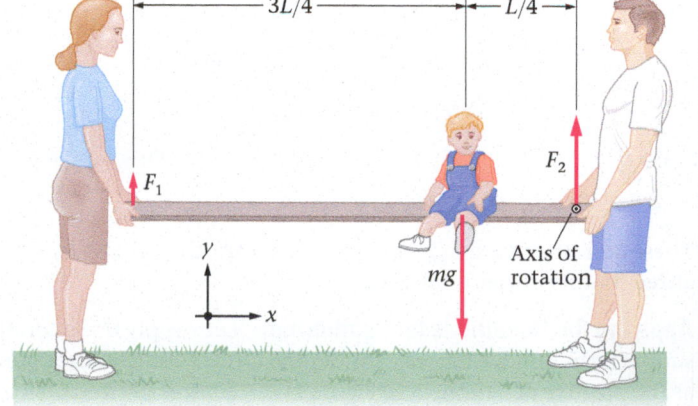

CONTINUED

SOLUTION

1. Set the net force acting on the plank equal to zero:

$$F_1 + F_2 - mg = 0$$

2. Set the net torque acting on the plank equal to zero. [Notice that the force F_1 rotates the plank clockwise (negative torque), and the weight mg rotates it counterclockwise (positive torque) about the axis.]:

$$-F_1(L) + mg\left(\tfrac{1}{4}L\right) = 0$$

3. Solve the torque condition for the force F_1:

$$F_1 = \tfrac{1}{4}mg$$

4. Substitute F_1 into the force condition to find F_2:

$$F_2 = mg - \tfrac{1}{4}mg = \tfrac{3}{4}mg$$

INSIGHT

As expected, the final results for the forces are identical to those obtained previously. Notice, however, that the torque produced by the child would cause a counterclockwise rotation in this case, and hence it is positive. Thus, the magnitude *and* sign of the torque produced by a given force depend on the location chosen for the axis of rotation.

PRACTICE PROBLEM — PREDICT/CALCULATE

Suppose the child moves to a new position, with the result that the force exerted by the father is reduced from $0.75mg$ to $0.60mg$. **(a)** Did the child move to the left or to the right? Explain. **(b)** Determine the distance and direction of the child's displacement. [**Answer: (a)** The child moved to the left, making for a more even distribution of the system's mass. **(b)** The new position of the child is $0.40L$ from the father. This means the child moved left a distance of $0.15L$.]

Some related homework problems: Problem 23, Problem 24

In the next Example, we show that the forces supporting a person or other object sometimes act in different directions. To emphasize the direction of the forces, we solve the Example in terms of the components of the relevant forces.

EXAMPLE 11-8   TAKING THE PLUNGE

A 5.00-m-long diving board of negligible mass is supported by two pillars. One pillar is at the left end of the diving board, as shown in the sketch; the other is 1.50 m away. Find the forces exerted by the pillars when a 90.0-kg diver stands at the far end of the board.

PICTURE THE PROBLEM

We choose upward to be the positive direction for the forces. When calculating torques, we use the left end of the diving board as the axis of rotation, which means that $\vec{F}_1$ causes zero torque. In addition, $\vec{F}_2$ would cause a counterclockwise rotation if it acted alone, so its torque is positive. On the other hand, $m\vec{g}$ would cause a clockwise rotation, so its torque is negative. Finally, $\vec{F}_2$ acts at a distance d from the axis of rotation, and $m\vec{g}$ acts at a distance L.

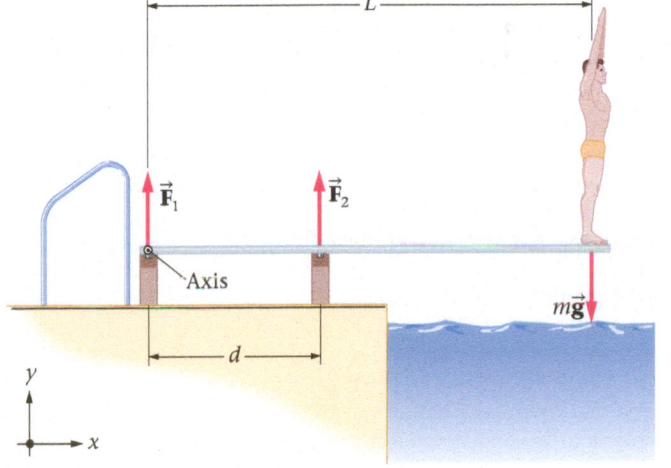

REASONING AND STRATEGY

As usual in static equilibrium problems, we use the conditions of (i) zero net force and (ii) zero net torque to determine the unknown forces, $\vec{F}_1$ and $\vec{F}_2$. In this system, all forces act in the positive or negative y direction; thus we need only set the net y component of force equal to zero.

Known   Length of diving board, $L = 5.00$ m; distance of second pillar, $d = 1.50$ m; mass of diver, $m = 90.0$ kg.

Unknown   Force exerted by pillar 1, $F_{1,y} = ?$; force exerted by pillar 2, $F_{2,y} = ?$

SOLUTION

1. Set the net y component of force acting on the diving board equal to zero:

$$\sum F_y = F_{1,y} + F_{2,y} - mg = 0$$

2. Calculate the torque due to each force, using the left end of the board as the axis of rotation. Notice that each force is at right angles to the radius and that $\vec{F}_1$ goes directly through the axis of rotation:

$$\tau_1 = F_{1,y}(0) = 0$$
$$\tau_2 = F_{2,y}(d)$$
$$\tau_3 = -mg(L)$$

CONTINUED

3. Set the net torque acting on the diving board equal to zero:

4. Solve the torque equation for the force $F_{2,y}$:

5. Use the force equation to determine $F_{1,y}$:

$$\sum \tau = F_{1,y}(0) + F_{2,y}(d) - mg(L) = 0$$

$$F_{2,y} = mg(L/d)$$
$$= (90.0 \text{ kg})(9.81 \text{ m/s}^2)(5.00 \text{ m}/1.50 \text{ m}) = 2940 \text{ N}$$

$$F_{1,y} = mg - F_{2,y}$$
$$= (90.0 \text{ kg})(9.81 \text{ m/s}^2) - 2940 \text{ N} = -2060 \text{ N}$$

INSIGHT

The first point to notice about our solution is that $F_{1,y}$ is negative, which means that $\vec{F}_1$ is actually directed *downward*, as shown to the right. To see why, imagine for a moment that the board is no longer connected to the first pillar. In this case, the board would rotate clockwise about the second pillar, and the left end of the board would move upward. Thus, a downward force is required on the left end of the board to hold it in place.

The second point is that both pillars exert forces with magnitudes that are considerably larger than the diver's weight, $mg = 883$ N. In particular, the first pillar must pull downward with a force of $2.33mg$, while the second pillar pushes upward with a force of $2.33mg + mg = 3.33mg$. This is not unusual. In fact, it is common for the forces in a structure, such as a bridge, a building, or the human body, to be much greater than the weight it supports.

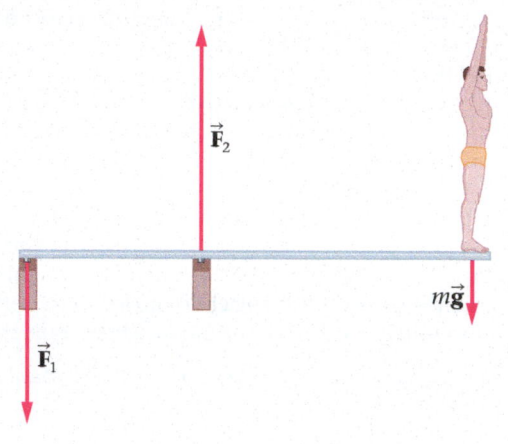

PRACTICE PROBLEM

Find the forces exerted by the pillars when the diver is 1.00 m to the left of the right end.   [**Answer:** $F_{1,y} = -1470$ N, $F_{2,y} = 2350$ N]

Some related homework problems: Problem 24, Problem 25, Problem 29

Walking the plank

RWP* To this point we have ignored the mass of the plank holding the child and the diving board holding the swimmer, since they were described as lightweight. If we want to consider the torque exerted by the force of gravity on an extended object of finite mass, however, we can simply treat it as if all its mass were concentrated at its center of mass, as was done in similar situations in Section 9-7. We consider such a system in the next Example.

EXAMPLE 11-9   WALKING THE PLANK: FIND THE MASS

A cat walks along a plank with mass $M = 6.00$ kg. The plank is supported by two sawhorses. The center of mass of the plank is a distance $d_1 = 0.850$ m to the left of sawhorse B. When the cat is a distance $d_2 = 1.11$ m to the right of sawhorse B, the plank just begins to tip. What is the mass of the cat, m?

PICTURE THE PROBLEM

In our sketch, the cat is at the point where the plank just begins to tip. As a result, the force exerted by sawhorse A is zero, $F_A = 0$. The distance from sawhorse B to the plank's center of mass is $d_1 = 0.850$ m, and the distance from sawhorse B to the cat is $d_2 = 1.11$ m.

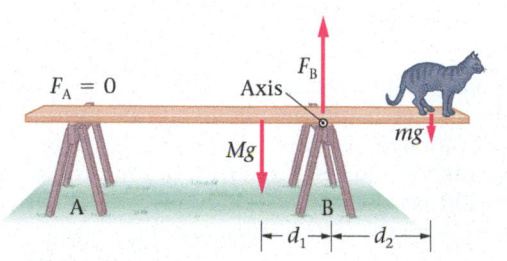

REASONING AND STRATEGY

We use sawhorse B as the axis of rotation, and we consider all of the mass of the plank to be concentrated at its center of mass. It follows that just two forces cause torques on the plank: (1) the weight of the cat, and (2) the weight of the plank. Setting the net torque caused by these two forces equal to zero determines the weight—and hence the mass—of the cat.

Known   Mass of plank, $M = 6.00$ kg; distance 1, $d_1 = 0.850$ m; distance 2, $d_2 = 1.11$ m.
Unknown   Mass of cat, $m = ?$

SOLUTION

1. Calculate the net torque about sawhorse B and set it equal to zero. Recall that clockwise rotation is negative, and counterclockwise rotation is positive:

$$\tau_{\text{net}} = -mg(d_2) + Mg(d_1) = 0$$

CONTINUED

---

*Real World Physics applications are denoted by the acronym RWP.

2. Solve for the mass of the cat and substitute numerical values:

$$mg(d_2) = Mg(d_1)$$
$$m = M\left(\frac{d_1}{d_2}\right) = (6.00 \text{ kg})\left(\frac{0.850 \text{ m}}{1.11 \text{ m}}\right) = 4.59 \text{ kg}$$

INSIGHT

We did not include a torque for sawhorse A because F_A is zero. The nonzero force F_B from sawhorse B also produces zero torque. This is because F_B goes through the axis of rotation. The magnitude of F_B is equal to $mg + Mg = 104$ N in order to make the net force acting on the system equal to zero.

PRACTICE PROBLEM

Write both the zero force and zero torque conditions for the case where the axis of rotation is at the center of mass of the plank.   [**Answer:** Zero force: $F_B - mg - Mg = 0$. Zero torque: $F_B(d_1) - mg(d_1 + d_2) = 0$. Combining these relationships gives $m = 4.59$ kg, as before.]

Some related homework problems: Problem 27, Problem 36

The same basic physics applies to a car or SUV that hits the brakes in a sudden stop, as illustrated in **FIGURE 11–11**. As the vehicle brakes, the frictional forces with the road tend to make it rotate in the clockwise sense. This is countered by a larger normal force on the front tires, which exerts a torque in the counterclockwise sense. The result is the vehicle goes into a "nose down" position during the braking.

Forces with Both Vertical and Horizontal Components

All of the previous examples dealt with forces that point either directly upward or directly downward. We now consider a more general situation, where forces may have both vertical and horizontal components. For example, consider the wall-mounted lamp (sconce) shown in **FIGURE 11-12**. The sconce consists of a light curved rod that is bolted to the wall at its lower end. Suspended from the upper end of the rod, a horizontal distance H from the wall, is the lamp of mass m. The rod is also connected to the wall by a horizontal wire a vertical distance V above the bottom of the rod.

Now, suppose we are designing this sconce to be placed in the lobby of a building on campus. To ensure its structural stability, we would like to know the tension T the wire must exert, as well as the vertical and horizontal components of the force $\vec{f}$ that must be exerted by the bolt on the rod. This information will be important in deciding on the type of wire and bolt to be used in the construction.

To find these forces, we apply the same conditions as before: The net force and the net torque must be zero. In this case, however, forces may have both horizontal and vertical components. Thus, the condition of zero net force is really two separate conditions: (i) zero net force in the horizontal direction, and (ii) zero net force in the vertical direction. These two conditions plus (iii), zero net torque, allow for a full solution of the problem.

Zero Net Torque   We begin with the torque condition. A convenient choice for the axis of rotation is the bottom end of the rod, because this eliminates one of the unknown forces ($\vec{f}$). With this choice we can readily calculate the net torque acting on the rod. We find

$$\sum \tau = T(V) - mg(H) = 0$$

This relationship can be solved immediately for the tension, giving

$$T = mg(H/V)$$

Notice that the tension is increased if the wire is connected closer to the bottom of the rod—that is, if V is reduced.

Zero Net Force   Next, we apply the force conditions. First, we sum the y components of all the forces and set the sum equal to zero:

$$\sum F_y = f_y - mg = 0$$

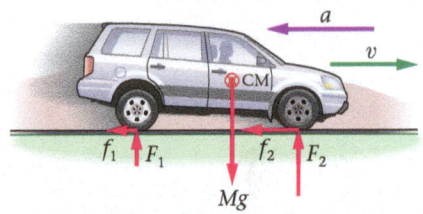

▲ **FIGURE 11-11** As the brakes are applied on this SUV, rotational equilibrium demands that the normal forces exerted on the front tires be greater than the normal forces exerted on the rear tires—which is why braking cars are "nose down" during a rapid stop. For this reason, many cars use disk brakes for the front wheels and less powerful drum brakes for the rear wheels. As the disk brakes wear, they tend to coat the front wheels with dust from the brake pads, which gives the front wheels a characteristic "dirty" look.

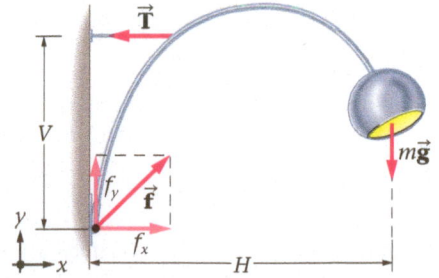

▲ **FIGURE 11-12 A lamp in static equilibrium** A wall-mounted lamp of mass m is suspended from a light curved rod. The bottom of the rod is bolted to the wall. The rod is also connected to the wall by a horizontal wire a vertical distance V above the bottom of the rod.

Thus, the vertical component of the force exerted by the bolt simply supports the weight of the lamp:

$$f_y = mg$$

Next, we sum the x components of the forces, and set that sum equal to zero as well:

$$\sum F_x = f_x - T = 0$$

Clearly, the x component of the force exerted by the bolt has the same magnitude as the tension, but it points in the opposite direction:

$$f_x = T = mg(H/V)$$

The bolt, then, pushes upward on the rod to support the lamp, and at the same time it pushes to the right to keep the rod from rotating.

For example, suppose the lamp in Figure 11-12 has a mass of 2.00 kg, and that $V = 12.0$ cm and $H = 15.0$ cm. In this case, we find the following forces:

$$T = mg(H/V) = (2.00 \text{ kg})(9.81 \text{ m/s}^2)(15.0 \text{ cm})/(12.0 \text{ cm}) = 24.5 \text{ N}$$
$$f_x = T = 24.5 \text{ N}$$
$$f_y = mg = (2.00 \text{ kg})(9.81 \text{ m/s}^2) = 19.6 \text{ N}$$

Notice that f_x and T are greater than the weight of the lamp, $mg = 19.6$ N. Just as we found with the diving board in Example 11-8, the forces required of structural elements can be greater than the weight of the object to be supported—an important consideration when designing a structure like a bridge, or a hanging sign like the one in **FIGURE 11-13**. The same effect occurs in the human body. We find in Problem 23, for example, that the force exerted by the biceps to support a baseball in the hand is several times larger than the baseball's weight. Similar conclusions apply to muscles throughout the body.

BIO In the next Example we consider another system in which forces have both vertical and horizontal components.

▲ **FIGURE 11-13** The chains that support this sign maintain it in a state of translational and rotational equilibrium. The forces in the chains are most easily analyzed by resolving them into vertical and horizontal components and applying the conditions for equilibrium. In particular, the net vertical force, the net horizontal force, and the net torque must all be zero.

EXAMPLE 11-10 ARM IN A SLING

A hiker who has broken his forearm rigs a temporary sling using a cord stretching from his shoulder to his hand. The cord holds the forearm level and makes an angle of 40.0° with the horizontal where it attaches to the hand. Considering the forearm and hand to be uniform, with a total mass of 1.30 kg and a length of 0.300 m, find **(a)** the tension in the cord and **(b)** the horizontal and vertical components of the force, $\vec{\mathbf{f}}$, exerted by the humerus (the bone of the upper arm) on the radius and ulna (the bones of the forearm).

PICTURE THE PROBLEM

In our sketch, we use the typical conventions for the positive x and y directions. In addition, the forearm and hand are assumed to be a uniform object, and hence we indicate the weight mg as acting at its center. The length of the forearm and hand is $L = 0.300$ m. Finally, two other forces act on the forearm: (i) the tension in the cord, $\vec{\mathbf{T}}$, at an angle of 40.0° above the negative x axis, and (ii) the force $\vec{\mathbf{f}}$ exerted at the elbow joint.

REASONING AND STRATEGY

In this system there are three unknowns: T, f_x, and f_y. These unknowns can be determined using the following three conditions: (i) net torque equals zero, (ii) net x component of force equals zero, and (iii) net y component of force equals zero.

We start with the torque condition, using the elbow joint as the axis of rotation. As we shall see, this choice of axis eliminates the force f, and gives a direct solution for the tension T. Next, we use T and the two force conditions to determine f_x and f_y.

CONTINUED

Known   Mass of arm and hand, $m = 1.30$ kg; length of arm and hand, $L = 0.300$ m; angle of cord, 40.0°.
Unknown (a) Tension in cord, $T = ?$ (b) Forces at elbow, $f_x = ?$ and $f_y = ?$

SOLUTION

Part (a)

1. Calculate the torque about the elbow joint. Notice that f causes zero torque, mg causes a negative torque, and the vertical component of T causes a positive torque. The horizontal component of T produces no torque, because it is on a line with the axis:

$$\sum \tau = (T \sin 40.0°)L - mg(L/2) = 0$$

2. Solve the torque condition for the tension, T:

$$T = \frac{mg}{2 \sin 40.0°} = \frac{(1.30 \text{ kg})(9.81 \text{ m/s}^2)}{2 \sin 40.0°} = 9.92 \text{ N}$$

Part (b)

3. Set the sum of the x components of force equal to zero, and solve for f_x:

$$\sum F_x = f_x - T \cos 40.0° = 0$$
$$f_x = T \cos 40.0° = (9.92 \text{ N}) \cos 40.0° = 7.60 \text{ N}$$

4. Set the sum of the y components of force equal to zero, and solve for f_y:

$$\sum F_y = f_y - mg + T \sin 40.0° = 0$$
$$f_y = mg - T \sin 40.0°$$
$$= (1.30 \text{ kg})(9.81 \text{ m/s}^2) - (9.92 \text{ N}) \sin 40.0° = 6.38 \text{ N}$$

INSIGHT

It is not necessary to determine T_x and T_y separately, because we know the direction of the cord. In particular, it is clear from our sketch that the components of $\vec{T}$ are $T_x = -T \cos 40.0° = -7.60$ N and $T_y = T \sin 40.0° = 6.38$ N.

Did you notice that $\vec{f}$ is at an angle of 40.0° with respect to the positive x axis, the same angle that $\vec{T}$ makes with the negative x axis? The reason for this symmetry, of course, is that mg acts at the center of the forearm. If mg were to act closer to the elbow, for example, $\vec{f}$ would make a larger angle with the horizontal, as we see in the following Practice Problem.

PRACTICE PROBLEM

Suppose that the forearm and hand are nonuniform, and that the center of mass is located at a distance of $L/4$ from the elbow joint. What are T, f_x, and f_y in this case?   [**Answer:** $T = 4.96$ N, $f_x = 3.80$ N, $f_y = 9.56$ N. In this case, $\vec{f}$ makes an angle of 68.3° with the horizontal.]

Some related homework problems: Problem 30, Problem 76, Problem 80

EXAMPLE 11-11   **DON'T WALK UNDER THE LADDER: FIND THE FORCES**

An 85-kg person stands on a lightweight ladder, as shown in the sketch. The floor is rough; hence, it exerts both a normal force, f_1, and a frictional force, f_2, on the ladder. The wall, on the other hand, is frictionless; it exerts only a normal force, f_3. Using the dimensions given in the figure, find the magnitudes of $f_1, f_2,$ and f_3.

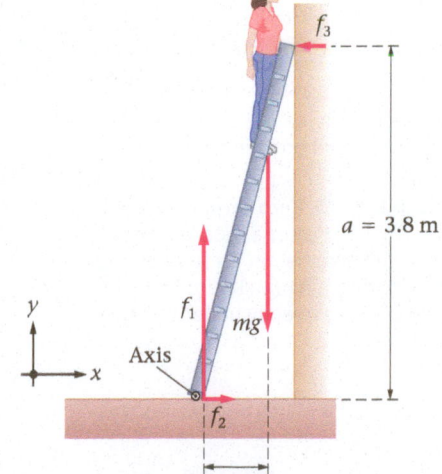

PICTURE THE PROBLEM

Our sketch shows the positive x and y directions, as well as the location of the axis of rotation, which is at the bottom of the ladder.

REASONING AND STRATEGY

As usual, we set both the net force and the net torque equal to zero. In this case, the net force must be set equal to zero in both the x and y directions.

Known   Mass of person, $m = 85$ kg; vertical distance, $a = 3.8$ m; horizontal distance, $b = 0.70$ m.
Unknown Forces exerted on ladder: $f_1 = ?$; $f_2 = ?$; $f_3 = ?$

SOLUTION

1. Set the net torque acting on the ladder equal to zero. Use the bottom of the ladder as the axis:

$$\tau_{net} = f_3(a) - mg(b) = 0$$

2. Solve for f_3:

$$f_3 = mg(b/a) = 150 \text{ N}$$

3. Sum the x components of force and set equal to zero:

$$f_2 - f_3 = 0$$

CONTINUED

4. Solve for f_2: $f_2 = f_3 = 150\,\text{N}$

5. Sum the y components of force and set equal to zero: $f_1 - mg = 0$

6. Solve for f_1: $f_1 = mg = 830\,\text{N}$

INSIGHT

If the floor is quite smooth, the ladder might slip—it depends on whether the coefficient of static friction is great enough to provide the needed force $f_2 = 150\,\text{N}$. In this system, the normal force exerted by the floor is $N = f_1 = 830\,\text{N}$. Therefore, if the coefficient of static friction is greater than 0.18 [noting that $0.18(830\,\text{N}) = 150\,\text{N}$], the ladder will stay put. Ladders often have rubberized pads on the bottom in order to increase the static friction, and hence increase the safety of the ladder.

PRACTICE PROBLEM

Write the zero-torque condition, and also the two zero-force conditions, for the case where the axis of rotation is at the top of the ladder.   [**Answer:** Zero horizontal force: $f_2 - f_3 = 0$; zero vertical force, $f_1 - mg = 0$; zero torque, $-f_1(b + c) + f_2(a) + mg(c) = 0$, where c is the horizontal distance from the wall to the line through the force mg. These relationships give the same results for f_1, f_2, and f_3, as expected, independent of the value of c.]

Some related homework problems: Problem 30, Problem 31

Enhance Your Understanding   (Answers given at the end of the chapter)

3. A "Physics" sign is supported symmetrically by two ropes, as shown in **FIGURE 11-14**. Is the magnitude of the tension in each rope greater than, less than, or equal to half the weight of the sign? Explain.

▲ **FIGURE 11-14** Enhance Your Understanding 3.

Section Review

- A system is in equilibrium when both the net force and the net torque acting on it are zero. The condition of zero net force applies equally to the x and y directions.

11-4 Center of Mass and Balance

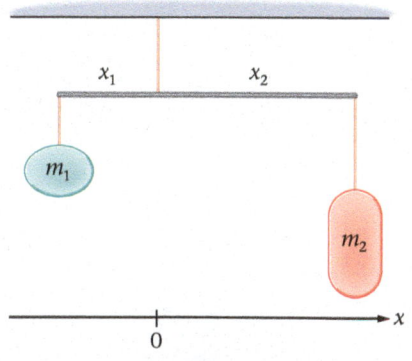

▲ **FIGURE 11-15 Zero torque and balance** One section of a mobile. The rod is balanced when the net torque acting on it is zero. This is equivalent to having the center of mass directly under the suspension point.

Suppose you decide to construct a mobile. To begin, you tie a thread to a light rod, as in **FIGURE 11-15**. Notice that the rod extends a distance x_1 to the left of the thread and a distance x_2 to the right. At the left end of the rod you attach an object of mass m_1. What mass, m_2, should be attached to the right end if the rod is to be balanced?

From the discussions in the previous sections, it is clear that if the rod is to be in static equilibrium (balanced), the net torque acting on it must be zero. Taking the point where the thread is tied to the rod as the axis of rotation, we can write this zero-torque condition as

$$m_1 g(x_1) - m_2 g(x_2) = 0$$

Canceling g and rearranging, we find

$$m_1 x_1 = m_2 x_2 \qquad\qquad 11\text{-}7$$

This gives the following result for m_2:

$$m_2 = m_1(x_1/x_2)$$

For example, if $x_2 = 2x_1$, it follows that m_2 should be one-half of m_1.

Let's now consider a slightly different question: Where is the center of mass of m_1 and m_2? Choosing the origin of the x axis to be at the location of the thread, as indicated in Figure 11-15, we can use the definition of the center of mass, Equation 9-14, to find x_{cm}:

$$x_{cm} = \frac{m_1(-x_1) + m_2(x_2)}{m_1 + m_2} = -\left(\frac{m_1 x_1 - m_2 x_2}{m_1 + m_2}\right)$$

Referring to the zero-torque condition in Equation 11-7, we see that $m_1x_1 - m_2x_2 = 0$; hence the center of mass is at the origin:

$$x_{cm} = 0$$

This is precisely where the string is attached. We conclude, then, that the rod balances when the center of mass is directly below the point from which the rod is suspended. This is a general result.

Let's apply this result to the case of the mobile shown in the next Example.

EXAMPLE 11-12   A WELL-BALANCED MEAL

As a grade-school project, students construct a mobile representing some of the major food groups. Their completed artwork is shown in the sketch. Find the masses m_1, m_2, and m_3 that are required for a perfectly balanced mobile. Assume the strings and the horizontal rods have negligible mass.

PICTURE THE PROBLEM
The dimensions of the horizontal rods and the values of the given masses are indicated in our sketch. Notice that each rod is balanced at its suspension point.

REASONING AND STRATEGY
We can find all three unknown masses by repeatedly applying the condition for balance, $m_1x_1 = m_2x_2$.

First, we apply the balance condition to m_1 and m_2, with the distances $x_1 = 12$ cm and $x_2 = 18$ cm. This gives a relationship between m_1 and m_2.

To get a second relationship between m_1 and m_2, we apply the balance condition again at the next higher level of the mobile. That is, the mass $(m_1 + m_2)$ at the distance 6.0 cm must balance the mass 0.30 kg at the distance 24 cm. These two conditions determine m_1 and m_2.

To find m_3 we again apply the balance condition, this time with the mass $(m_1 + m_2 + 0.30 \text{ kg} + 0.20 \text{ kg})$ at the distance 18 cm, and the mass m_3 at the distance 31 cm.

Known   Masses and distances are given in the sketch.
Unknown Masses $m_1 = ?$, $m_2 = ?$, and $m_3 = ?$

SOLUTION

1. Apply the balance condition to m_1 and m_2:

$m_1(12 \text{ cm}) = m_2(18 \text{ cm})$
$m_1 = (1.5)m_2$

2. Apply the balance condition to the next level up in the mobile. Solve for the sum, $m_1 + m_2$:

$(m_1 + m_2)(6.0 \text{ cm}) = (0.30 \text{ kg})(24 \text{ cm})$
$m_1 + m_2 = \dfrac{(0.30 \text{ kg})(24 \text{ cm})}{6.0 \text{ cm}} = 1.2 \text{ kg}$

3. Substitute $m_1 = (1.5)m_2$ into $m_1 + m_2 = 1.2$ kg to find m_2:

$(1.5)m_2 + m_2 = (2.5)m_2 = 1.2 \text{ kg}$
$m_2 = 1.2 \text{ kg}/2.5 = 0.48 \text{ kg}$

4. Use $m_1 = (1.5)m_2$ to find m_1:

$m_1 = (1.5)m_2 = (1.5)0.48 \text{ kg} = 0.72 \text{ kg}$

5. Apply the balance condition to the top level of the mobile:

$(0.72 \text{ kg} + 0.48 \text{ kg} + 0.30 \text{ kg} + 0.20 \text{ kg})(18 \text{ cm}) = m_3(31 \text{ cm})$

6. Solve for m_3:

$m_3 = \dfrac{(1.70 \text{ kg})(18 \text{ cm})}{31 \text{ cm}} = 0.99 \text{ kg}$

INSIGHT
With the values for m_1, m_2, and m_3 found above, the mobile balances at every level. In fact, the center of mass of the *entire* mobile is directly below the point where the uppermost string attaches to the ceiling.

PRACTICE PROBLEM
Find m_1, m_2, and m_3 if the 0.30-kg mass is replaced with a 0.40-kg mass.   [**Answer:** $m_1 = 0.96$ kg, $m_2 = 0.64$ kg, $m_3 = 1.3$ kg]

Some related homework problems: Problem 37, Problem 39

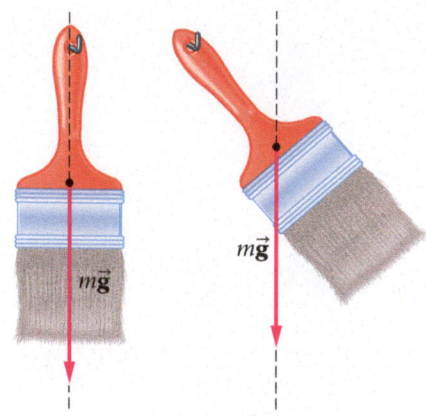

(a) Zero torque   **(b)** Nonzero torque

▲ **FIGURE 11-16 Equilibrium of a suspended object** **(a)** If an object's center of mass is directly below the suspension point, its weight creates zero torque and the object is in equilibrium. **(b)** When an object is rotated, so that the center of mass is no longer directly below the suspension point, the object's weight creates a torque. The torque tends to rotate the object to bring the center of mass under the suspension point.

Balancing a Meter Stick

Center of Mass for a Suspended Object   In general, if you allow an arbitrarily shaped object to hang freely, its center of mass is directly below the suspension point. To see why, notice that when the center of mass is directly below the suspension point, the torque due to gravity is zero, because the force of gravity extends right through the axis of rotation. This is shown in **FIGURE 11-16 (a)**. If the object is rotated slightly, as in **FIGURE 11-16 (b)**, the force of gravity is not in line with the axis of rotation—hence gravity produces a torque. This torque tends to rotate the object, bringing the center of mass back under the suspension point.

For example, suppose you cut a piece of wood into the shape of the continental United States, as shown in **FIGURE 11-17**. Next, drill a small hole in it, and hang it from the point A. The result is that the center of mass lies somewhere on the line aa'. Similarly, if a second hole is drilled at point B, we find that the center of mass lies somewhere on the line bb'. The only point that is on both the line aa' *and* the line bb' is the point CM, near Smith Center, Kansas, which marks the location of the center of mass.

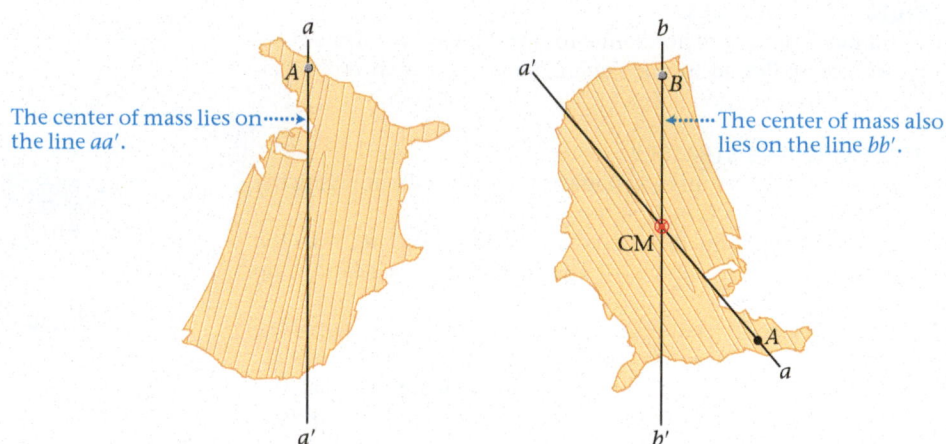

The center of mass lies on the line aa'.

The center of mass also lies on the line bb'.

▲ **FIGURE 11-17 The geometric center of the United States** To find the center of mass of an irregularly shaped object, such as a wooden model of the continental United States, suspend it from two or more points. The center of mass lies on a vertical line extending downward from the suspension point. The intersection of these vertical lines gives the precise location of the center of mass.

CONCEPTUAL EXAMPLE 11-13   COMPARE THE MASSES   PREDICT/EXPLAIN

A croquet mallet balances when suspended from its center of mass, as indicated in the left-hand drawing to the right. You now cut the mallet into two pieces at its center of mass, as in the right-hand drawing. **(a)** Is the mass of piece 1 greater than, less than, or equal to the mass of piece 2? **(b)** Which of the following is the *best explanation* for your prediction?

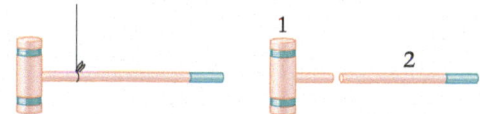

   I. Piece 1 has more mass because it balances piece 2, even though it extends a smaller distance from the balance point.

  II. Piece 1 has less mass because it is physically smaller than piece 2.

 III. The masses are equal because the mallet balanced at the cut point, and it would balance there only if there were equal mass on either side of that point.

REASONING AND DISCUSSION
The mallet balances because the *torques* due to the two pieces are of equal magnitude. The piece with the head of the mallet (piece 1) extends a smaller distance from the point of suspension than does the other piece, and hence its mass must be greater; that is, a large mass at a small distance creates the same torque as a small mass at a large distance.

ANSWER
(a) Piece 1 has the greater mass. **(b)** The best explanation is I.

▲ **FIGURE 11-18 Visualizing Concepts Balance and Center of Mass** (Left) Although her knowledge may be based more on practical experience than on physics, this woman knows exactly what she must do to keep from falling. By extending one leg backward as she leans forward, she keeps her center of mass safely positioned over the foot that supports her. (Center) A kangaroo has a long, heavy tail, which helps it to maintain its balance, both on solid ground and in flight between hops. (Right) Although it looks precarious, this rock in Arches National Park, Utah, has probably been balancing above the desert for many thousands of years. It will remain secure on its perch as long as its center of mass lies above its base of support.

Center of Mass for a Standing Object   Similar considerations apply to an object standing at rest on a surface, as opposed to being suspended from a point. In such a case, the object is in equilibrium as long as its center of mass is vertically above the base on which it is supported. For example, when you stand upright with normal posture, your feet provide a base of support, and your center of mass is above a point roughly halfway between your feet. If you lift your left foot from the floor—without changing your posture—you will begin to lose your balance and tip over. The reason is that your center of mass is no longer above the base of support, which is now your right foot. To balance on your right foot, you must lean slightly in that direction—and perhaps extend your left leg as well—so as to position your center of mass directly above the foot. This principle applies to everything from a person standing on a stool, to a kangaroo using its tail to balance, to one of the "balancing rocks" that are a familiar sight in the desert Southwest. Examples are shown in **FIGURE 11-18**. In Problem 38 we apply this condition for stability to a stack of books on the edge of a table.

Enhance Your Understanding   (Answers given at the end of the chapter)

4. A mobile made from three piggy banks (A, B, C) is shown in **FIGURE 11-19**. The banks are identical, except that they contain different numbers of coins and hence have different masses. Rank the banks in order of increasing mass. Indicate ties where appropriate. (Which bank would you rather have?)

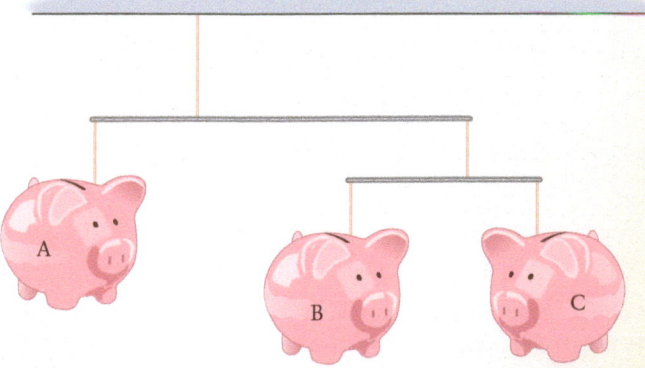

▲ **FIGURE 11-19** Enhance Your Understanding 4.

Section Review

• Extended objects balance when suspended from their center of mass.

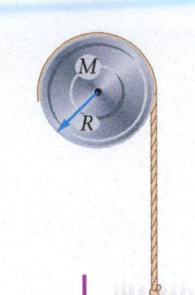

(a) Physical picture

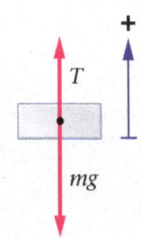

(b) Free-body diagram for mass

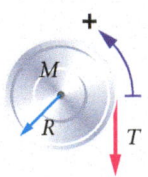

(c) Torque producing force acting on pulley

▲ **FIGURE 11-20 A mass suspended from a pulley** A mass m hangs from a string wrapped around the circumference of a disk-shaped pulley of radius R and mass M. When the mass is released, it accelerates downward. Positive directions of motion for the system are shown in parts (b) and (c). In part (c), the weight of the pulley acts downward at its center, and the axle exerts an upward force equal in magnitude to the weight of the pulley plus the tension in the string. Of the three forces acting on the pulley, only the tension in the string produces a torque about the axle.

11-5 Dynamic Applications of Torque

In this section we focus on applications of Newton's second law for rotational motion. To begin, consider a disk-shaped pulley of radius R and mass M with a string wrapped around its circumference, as in **FIGURE 11-20 (a)**. Hanging from the string is a mass m. When the mass is released, it accelerates downward and the pulley begins to rotate. If the pulley rotates without friction, and the string unwraps without slipping, what are the acceleration of the mass and the tension in the string?

At first it may seem that because the pulley rotates freely, the mass will simply fall with the acceleration of gravity. But remember, the pulley has a nonzero moment of inertia, $I > 0$, which means that it resists any change in its rotational motion. In order for the pulley to rotate, the string must pull downward on it with a tension T. This means that the string also pulls upward on the mass m with a tension T. As a result, the net downward force on m is less than mg, and hence its acceleration is less than g.

Applying Newton's Second Law to Both Linear and Rotational Motions To solve for the acceleration of the mass, we must apply Newton's second law to both the linear motion of the mass *and* the rotational motion of the pulley. The first step is to define a consistent choice of positive directions for the two motions. In Figure 11-20 (a) we note that when the pulley rotates counterclockwise, the mass moves upward. Thus, we choose counterclockwise to be positive for the pulley and upward to be positive for the mass.

With our positive directions established, we proceed to apply Newton's second law. Referring to the free-body diagram for the mass, shown in **FIGURE 11-20 (b)**, we see that

$$T - mg = ma \qquad 11\text{-}8$$

Similarly, the free-body diagram for the pulley is shown in **FIGURE 11-20 (c)**. Notice that the tension in the string, T, exerts a tangential force on the pulley at a distance R from the axis of rotation. This produces a torque of magnitude TR. The tension tends to cause a clockwise rotation, and hence it follows that the torque is negative; thus, $\tau = -TR$. As a result, Newton's second law for the pulley gives

$$-TR = I\alpha \qquad 11\text{-}9$$

Now, these two statements of Newton's second law are related by the fact that the string unwraps without slipping. As was discussed in Chapter 10, when a string unwraps without slipping, the angular and linear accelerations are related by

$$\alpha = \frac{a}{R}$$

Using this relationship in Equation 11-9, we have

$$-TR = I\frac{a}{R}$$

Dividing by R yields

$$T = -I\frac{a}{R^2}$$

Substituting this result into $T - mg = ma$, we find

$$-I\frac{a}{R^2} - mg = ma$$

Finally, dividing by m and rearranging yields the acceleration, a:

$$a = -\frac{g}{\left(1 + \dfrac{I}{mR^2}\right)} \qquad 11\text{-}10$$

Let's briefly check our solution for a. First, note that a is negative. This is to be expected because the mass accelerates downward, which is the negative direction. Second, if the moment of inertia were zero, $I = 0$, or if the mass m were infinite, $m \to \infty$, the mass would fall with the acceleration of gravity, $a = -g$. When I is greater than zero and m is finite, however, the acceleration of the mass has a magnitude less than g. In fact, in the limit of an infinite moment of inertia, $I \to \infty$, the acceleration vanishes— the mass is simply unable to cause the pulley to rotate in this case.

The next Example presents another system in which Newton's laws are used to relate linear and rotational motions.

EXAMPLE 11-14   THE PULLEY MATTERS

A 0.31-kg cart on a horizontal air track is attached to a string. The string passes over a disk-shaped pulley of mass 0.080 kg and radius 0.012 m and is pulled vertically downward with a constant force of 1.1 N. Find **(a)** the tension in the string between the pulley and the cart and **(b)** the acceleration of the cart.

PICTURE THE PROBLEM

The system is shown in the sketch. We label the mass of the cart with M, the mass of the pulley with m, and the radius of the pulley with r. The applied downward force creates a tension $T_1 = 1.1$ N in the vertical portion of the string. The horizontal portion of the string, from the pulley to the cart, has a tension T_2. If the pulley had zero mass, these two tensions would be equal. In this case, however, T_2 will have a different value than T_1.

We also show the relevant forces acting on the pulley and the cart. The positive direction of rotation is counterclockwise, and the corresponding positive direction of motion for the cart is to the left.

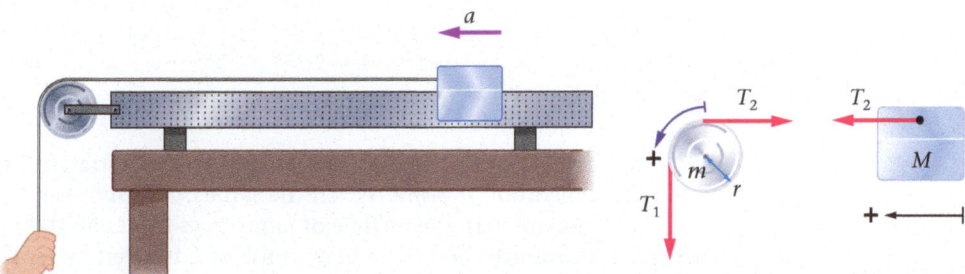

Physical picture   **Relevant forces**

REASONING AND STRATEGY

The two unknowns, T_2 and a, can be found by applying Newton's second law to both the pulley and the cart. This gives two equations for two unknowns.

In applying Newton's second law to the pulley, we note that the pulley is a disk and hence $I = \frac{1}{2}mr^2$. Also, because the string is not said to slip as it rotates the pulley, we can assume that the angular and linear accelerations are related by $\alpha = a/r$.

Known   Mass of cart, $M = 0.31$ kg; mass of pulley, $m = 0.080$ kg; radius of pulley, $r = 0.012$ m; downward force exerted on string, $T_1 = 1.1$ N.

Unknown   **(a)** Horizontal tension in string, $T_2 = ?$ **(b)** Acceleration of the cart, $a = ?$

SOLUTION

Part (a)

1. Apply Newton's second law to the cart:
$$T_2 = Ma$$

2. Apply Newton's second law to the pulley. Notice that T_1 causes a positive torque, and T_2 causes a negative torque. In addition, use the relationship $\alpha = a/r$:
$$\sum \tau = I\alpha$$
$$rT_1 - rT_2 = \left(\tfrac{1}{2}mr^2\right)\left(\frac{a}{r}\right) = \tfrac{1}{2}mra$$

3. Use the cart equation, $T_2 = Ma$, to eliminate the acceleration a in the pulley equation:
$$a = \frac{T_2}{M}$$
$$rT_1 - rT_2 = \tfrac{1}{2}mr\left(\frac{T_2}{M}\right)$$

4. Cancel r and solve for T_2:
$$T_2 = \frac{T_1}{1 + m/2M} = \frac{1.1 \text{ N}}{1 + 0.080 \text{ kg}/[2(0.31 \text{ kg})]} = 0.97 \text{ N}$$

Part (b)

5. Use $T_2 = Ma$ to find the acceleration of the cart:
$$a = \frac{T_2}{M} = \frac{0.97 \text{ N}}{0.31 \text{ kg}} = 3.1 \text{ m/s}^2$$

INSIGHT

We find that T_2 is less than T_1. As a result, the net torque acting on the pulley is in the counterclockwise direction, causing a rotation in that direction, as expected. If the mass of the pulley were zero ($m = 0$), the two tensions would be equal (as can be seen in Step 4), and the acceleration of the cart would be $T_1/M = 3.5$ m/s^2.

CONTINUED

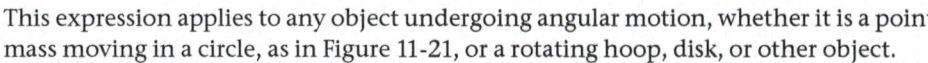

PRACTICE PROBLEM

What applied force is necessary to give the cart an acceleration of 2.2 m/s²?

[**Answer:** $T_1 = T_2(1 + m/2M) = (Ma)(1 + m/2M) = 0.77$ N. Only 0.68 N would be required if the pulley had zero mass.]

Some related homework problems: Problem 40, Problem 42

Enhance Your Understanding   (Answers given at the end of the chapter)

5. Refer to Figure 11-20 (a) for the following questions: **(a)** Is the tension in the rope greater than, less than, or equal to *mg*? Explain. **(b)** Is the tension in the rope greater than, less than, or equal to 0? Explain.

Section Review

- In systems containing both linear and rotational motions, the two must be analyzed together. This is done by applying Newton's second law to both motions, and taking advantage of any connections that exist between the motions.

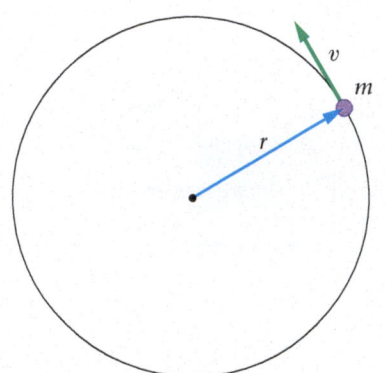

▲ **FIGURE 11-21 The angular momentum of circular motion** A particle of mass *m*, moving in a circle of radius *r* with a speed *v*. This particle has an angular momentum of magnitude $L = rmv$.

11-6 Angular Momentum

When an object of mass *m* moves with a speed *v* in a straight line, we say that it has a linear momentum, $p = mv$. When the same object moves with an angular speed ω along the circumference of a circle of radius *r*, as in **FIGURE 11-21**, we say that it has an **angular momentum**, *L*. The magnitude of *L* is given by replacing *m* and *v* in the expression for *p* with their angular analogs *I* and ω (Section 10-5). Thus, we define the angular momentum as follows:

Definition of Angular Momentum, L

$$L = I\omega \qquad\qquad 11\text{-}11$$

SI unit: $\text{kg} \cdot \text{m}^2/\text{s}$

This expression applies to any object undergoing angular motion, whether it is a point mass moving in a circle, as in Figure 11-21, or a rotating hoop, disk, or other object.

For a point mass *m* moving in a circle of radius *r*, the moment of inertia in this case is $I = mr^2$ (Equation 10-18). In addition, the linear speed of the mass is $v = r\omega$ (Equation 10-12). Combining these results, we find

$$L = I\omega = (mr^2)(v/r) = rmv$$

Noting that *mv* is the linear momentum *p*, we find that the angular momentum of a point mass can be written in the following form:

$$L = rmv = rp \qquad\qquad 11\text{-}12$$

It's important to recall that this expression applies specifically to a point particle moving along the circumference of a circle.

More generally, a point particle may be moving at an angle θ with respect to a radial line, as indicated in **FIGURE 11-22 (a)**. In this case, only the tangential component of the momentum, $p \sin\theta = mv \sin\theta$, contributes to the angular momentum, just as the tangential component of the force, $F \sin\theta$, is all that contributes to the torque. Thus, the magnitude of the angular momentum for a point particle is defined as:

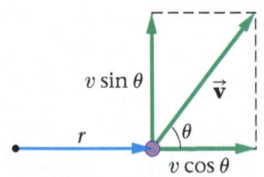

(a)

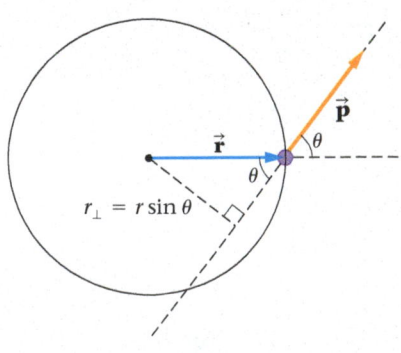

(b)

◀ **FIGURE 11-22 The angular momentum of nontangential motion (a)** When a particle moves at an angle θ with respect to the radial direction, only the tangential component of velocity, $v \sin\theta$, contributes to the angular momentum. In the case shown here, the particle's angular momentum has a magnitude given by $L = rmv \sin\theta$. **(b)** The angular momentum of an object can also be defined in terms of the moment arm, $r_\perp$. Since $r_\perp = r\sin\theta$, it follows that $L = rmv\sin\theta = r_\perp mv$. Note the similarity between this figure and Figure 11-5.

Angular Momentum, *L*, for a Point Particle

$$L = rp \sin\theta = rmv \sin\theta$$ 11-13

SI unit: $kg \cdot m^2/s$

If the particle moves in a circular path, the angle θ is 90° and the angular momentum is $L = rmv$, in agreement with Equation 11-12. On the other hand, if the object moves radially, so that $\theta = 0$, the angular momentum is zero; $L = rmv \sin 0 = 0$.

EXERCISE 11-15 ANGULAR MOMENTUM

Find the angular momentum of **(a)** a 16.0-g DVD (considered to be a uniform disk of radius 6.00 cm) spinning with an angular speed of 125 rad/s, and **(b)** a 125-kg motorcycle driving with a speed of 8.92 m/s on a circular track of radius 35.7 m.

REASONING AND SOLUTION

a. Recalling that $L = I\omega$, and that $I = \frac{1}{2}mR^2$ for a uniform disk (Table 10-1), we find

$$L = I\omega = \left(\tfrac{1}{2}mR^2\right)\omega = \tfrac{1}{2}(0.0160 \text{ kg})(0.0600 \text{ m})^2(125 \text{ rad/s}) = 0.00360 \text{ kg} \cdot m^2/s$$

b. Treating the motorcycle as a particle of mass m, we find

$$L = rmv = (35.7 \text{ m})(125 \text{ kg})(8.92 \text{ m/s}) = 39,800 \text{ kg} \cdot m^2/s$$

Angular Momentum and the Moment Arm An alternative definition of the angular momentum uses the moment arm, $r_\perp$, as was done for the torque in Equation 11-3. To apply this definition, start by extending a line through the momentum vector, $\vec{p}$, as in **FIGURE 11-22 (b)**. Next, draw a line from the axis of rotation perpendicular to the line through $\vec{p}$. The perpendicular distance from the axis of rotation to the line of $\vec{p}$ is the moment arm. From the figure we see that $r_\perp = r \sin\theta$. Hence, from Equation 11-13, the angular momentum is

$$L = r_\perp p = r_\perp mv$$

If an object moves in a circle of radius r, the moment arm is $r_\perp = r$ and the angular momentum reduces to our earlier result, $L = rp$.

CONCEPTUAL EXAMPLE 11-16   ANGULAR MOMENTUM?

Does an object moving in a straight line have nonzero angular momentum always, sometimes, or never?

REASONING AND DISCUSSION

The answer is sometimes, because it depends on the choice of the axis of rotation. If the axis of rotation is not on the line drawn through the momentum vector, as in the left-hand sketch, the moment arm is nonzero, and therefore $L = r_\perp p$ is also nonzero. If the axis of rotation is on the line of motion, as in the right-hand sketch, the moment arm is zero; hence the linear momentum is radial and L vanishes.

ANSWER

An object moving in a straight line may or may not have nonzero angular momentum, depending on the location of the axis of rotation.

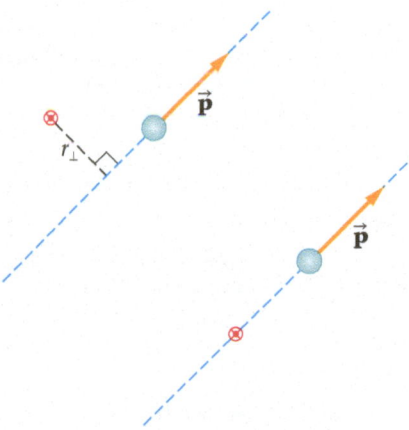

> **Big Idea ④** Angular momentum is the rotational analog of linear momentum. Angular momentum depends on the magnitude and direction of the linear momentum, and also on the distance from the axis of rotation.

PHYSICS IN CONTEXT
Looking Back

Angular momentum plays the same role in rotational systems as linear momentum does in linear systems in Chapter 9.

PHYSICS IN CONTEXT
Looking Ahead

Angular momentum is quantized (given discrete values) in the Bohr model of the hydrogen atom in Chapter 31.

An object moving with a momentum p in a straight line that does not go through the axis of rotation has an *angular* position that changes with time. This is illustrated in **FIGURE 11-23 (a)**. It is for this reason that such an object is said to have an *angular* momentum.

Sign of the Angular Momentum The sign of L is determined by whether the angle to a given object is increasing or decreasing with time. For example, the object moving counterclockwise in a circular path in **FIGURE 11-23 (b)** has a positive angular momentum because θ is increasing with time. Similarly, the object in Figure 11-23 (a) has an angle θ that increases with time, and hence its angular momentum is positive as well. On the other hand, if these objects were to have their direction of motion reversed, they would have angles that decrease with time and their angular momenta would be negative.

▶ **FIGURE 11-23 Angular momentum in linear and circular motions** An object moving in **(a)** a straight line and **(b)** a circular path. In both cases, the angular position increases with time; hence, the angular momentum is positive.

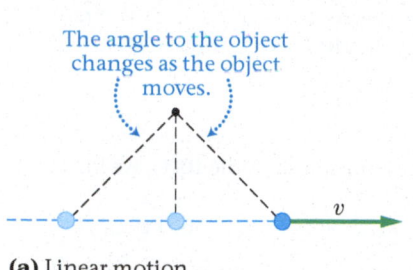

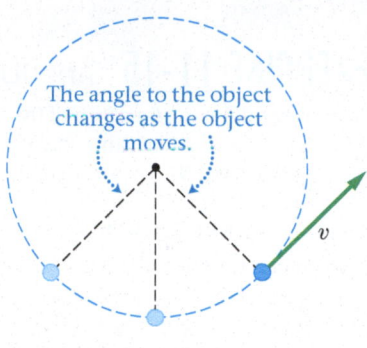

(a) Linear motion

(b) Circular motion

EXAMPLE 11-17 | HOCKEY SHOT

A 0.170-kg hockey puck moves straight toward the goalie with a speed of $v = 4.42$ m/s, as shown in the sketch. The puck is also observed by a defender. **(a)** Find the angular momentum as measured by the defender, given that $r = 1.45$ m and $\theta = 72.5°$. Use the expression $L = rmv \sin\theta$. **(b)** Find the moment arm, $r_\perp$, for the defender. **(c)** Check that the angular momentum is also given by $L = r_\perp mv$.

PICTURE THE PROBLEM

The hockey puck moves directly toward the goalie with a linear momentum $p = mv$. For the defender, the line of motion is displaced by a distance equal to the moment arm, $r_\perp = r \sin\theta$. Hence, the defender observes a nonzero angular momentum, while the goalie does not.

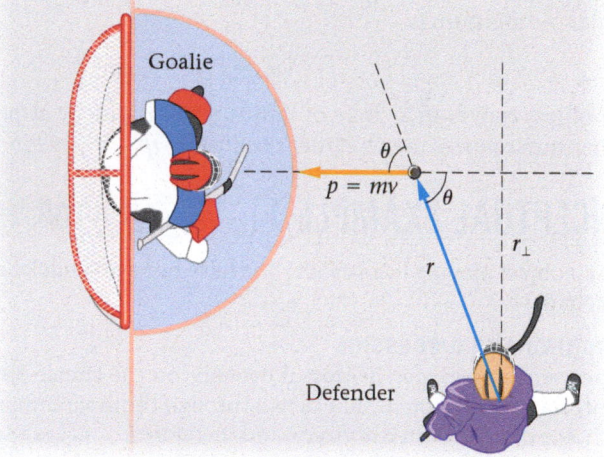

REASONING AND STRATEGY

a. The angular momentum of the puck can be found by applying $L = rmv \sin\theta$. In this case, the distance is $r = 1.45$ m and the angle between the line of sight and the puck's momentum is $\theta = 72.5°$. In addition, $m = 0.170$ kg and $v = 4.42$ m/s.

b. Notice that the moment arm, $r_\perp$, is the side of the right triangle opposite to the angle of $\theta = 72.5°$. It follows that $r_\perp = r \sin 72.5°$.

c. Use the moment arm from part **(b)** to calculate $L = r_\perp mv$.

Known Mass of puck, $m = 0.170$ kg; linear speed of puck, $v = 4.42$ m/s; distance to puck, $r = 1.45$ m; angle between $\vec{r}$ and $\vec{p}$, $\theta = 72.5°$.

Unknown **(a)** Angular momentum of puck, $L = ?$ **(b)** Moment arm, $r_\perp = ?$ **(c)** Angular momentum of puck, $L = ?$

SOLUTION

Part (a)

1. Evaluate $L = rmv \sin\theta$:

$$L = rmv \sin\theta$$
$$= (1.45 \text{ m})(0.170 \text{ kg})(4.42 \text{ m/s}) \sin 72.5° = 1.04 \text{ kg} \cdot \text{m}^2/\text{s}$$

Part (b)

2. Calculate the moment arm, $r_\perp$:

$$r_\perp = r \sin\theta = (1.45 \text{ m}) \sin 72.5° = 1.38 \text{ m}$$

Part (c)

3. Evaluate $L = r_\perp mv$:

$$L = r_\perp mv = (1.38 \text{ m})(0.170 \text{ kg})(4.42 \text{ m/s}) = 1.04 \text{ kg} \cdot \text{m}^2/\text{s}$$

CONTINUED

INSIGHT
The defender observes a nonzero angular momentum for the puck because she has to turn her head to follow its motion. The goalie always sees the puck at the same angle, and hence observes an angular momentum of zero.

PRACTICE PROBLEM — PREDICT/CALCULATE
As the puck gets closer to the goalie, the distance r from the defender increases and the angle θ decreases. **(a)** Does the puck's angular momentum, as observed by the defender, increase, decrease, or stay the same as it approaches the goalie? Explain. **(b)** Find the angular momentum relative to the defender when $r = 1.58$ m and $\theta = 61.0°$.   [**Answer:** **(a)** The distance r increases, and the angle θ decreases, but the moment arm $r_\perp = r \sin \theta$ stays the same. Hence the angular momentum also stays the same. **(b)** We find $r_\perp = 1.38$ m and $L = 1.04$ kg · m^2/s, as before.]

Some related homework problems: Problem 46, Problem 47

The Rate of Change of Angular Momentum   To conclude this section, we consider the rate of change of angular momentum with time. Assuming the moment of inertia is a constant—which is true as long as the mass and shape of the object remain unchanged—the change in L in a time interval Δt is

$$\frac{\Delta L}{\Delta t} = I \frac{\Delta \omega}{\Delta t}$$

Recall, however, that $\Delta \omega / \Delta t$ is the angular acceleration, α. Therefore, we have

$$\frac{\Delta L}{\Delta t} = I\alpha$$

Noting that $I\alpha$ is the torque, it follows that Newton's second law for rotational motion can be written as:

Newton's Second Law for Rotational Motion

$$\sum \tau = I\alpha = \frac{\Delta L}{\Delta t} \qquad \text{11-14}$$

Clearly, this is the rotational analog of $\sum F_x = ma_x = \Delta p_x / \Delta t$. Just as force can be expressed as the change in *linear* momentum in a given time interval, torque can be expressed as the change in *angular* momentum in a given time interval.

EXERCISE 11-18   TORQUE AND ANGULAR MOMENTUM

A ceiling fan experiences a constant torque of 1.9 N · m. If the fan is initially at rest, what is its angular momentum 2.6 s later?

REASONING AND SOLUTION
Solve $\sum \tau = \Delta L / \Delta t$ (Equation 11-14) for the change in angular momentum due to a single torque τ:

$$\Delta L = L_f - L_i = \left(\sum \tau \right) \Delta t = \tau \, \Delta t$$

Noting that the initial angular momentum of the fan is zero ($L_i = 0$), we find its final angular momentum is

$$L_f = \tau \, \Delta t = (1.9 \, \text{N·m})(2.6 \, \text{s}) = 4.9 \, \text{kg·m}^2\text{/s}$$

Enhance Your Understanding   (Answers given at the end of the chapter)

6. Consider two objects with the following characteristics: Object 1 is a hoop with mass $2M$ and radius R; object 2 is a disk with mass M and radius $2R$. If the angular speed of object 1 is ω, and the angular speed of object 2 is 2ω, which object has the greater angular momentum? Explain.

CONTINUED

▲ **FIGURE 11-24** Once she has launched herself into space, this diver is essentially a projectile. However, the principle of conservation of angular momentum allows her to control the rotational part of her motion. By curling her body up into a tight "tuck," she decreases her moment of inertia, thereby increasing the speed of her spin. To slow down for an elegant entry into the water, she will extend her body, thus increasing her moment of inertia.

   Spinning Person Drops Weights

Big Idea 5 In systems with zero net torque, the angular momentum is conserved. It follows that the total angular momentum of the universe is conserved, just like the total linear momentum and the total energy.

PHYSICS IN CONTEXT
Looking Ahead

The conservation of angular momentum plays a key role in the study of gravity. See, in particular, the discussion of Kepler's second law in Chapter 12.

Section Review

- Angular momentum plays the same role in rotational systems as linear momentum plays in linear systems.

- Angular momentum is given by $L = I\omega$ for a rotating system, or $L = rmv \sin\theta$ for a point particle.

- The rate of change of angular momentum in a system is equal to the net torque applied to the system.

11-7 Conservation of Angular Momentum

When an ice skater goes into a spin and pulls his arms inward to speed up, he probably doesn't think about angular momentum. Neither does a diver who springs into the air and folds her body to speed her rotation, as in **FIGURE 11-24**. Most people, in fact, are not aware that the actions of these athletes are governed by the same basic laws of physics that cause a collapsing star to spin faster as it becomes a rapidly rotating pulsar. Yet in all these cases, as we shall see, **conservation of angular momentum** is at work.

The Origin of Angular Momentum Conservation To see how angular momentum conservation comes about, consider an object with an initial angular momentum L_i acted on by a single torque τ. After a period of time, Δt, the object's angular momentum changes in accordance with Newton's second law:

$$\tau = \frac{\Delta L}{\Delta t}$$

Solving for ΔL, we find

$$\Delta L = L_f - L_i = \tau\,\Delta t$$

Thus, the final angular momentum of the object is

$$L_f = L_i + \tau\,\Delta t$$

If the torque acting on the object is zero ($\tau = 0$), it follows that the initial and final angular momenta are equal—that is, the angular momentum is conserved:

$$L_f = L_i \quad (\text{if } \tau = 0)$$

Angular momentum is also conserved in systems acted on by more than one torque, provided that the *net external torque* is zero. The reason internal torques can be ignored is that, just as internal forces come in equal and opposite pairs that cancel, so too do internal torques. As a result, the internal torques in a system sum to zero, and the net torque acting on it is simply the net external torque. Thus, for a general system, angular momentum is conserved if $\tau_{net, ext}$ is zero:

Conservation of Angular Momentum

$$L_f = L_i \quad (\text{if } \tau_{net, ext} = 0) \qquad\qquad 11\text{-}15$$

As an illustration of angular momentum conservation, we consider the case of a student rotating on a piano stool in the next Example. Notice how a change in the moment of inertia results in a change in angular speed.

EXAMPLE 11-19   GOING FOR A SPIN

For a classroom demonstration, a student sits on a piano stool holding a sizable mass in each hand. Initially, the student holds his arms outstretched and spins about the axis of the stool with an angular speed of 3.72 rad/s. The moment of inertia in this case is $5.33 \text{ kg} \cdot \text{m}^2$. While still spinning, the student pulls his arms in to his chest, reducing the moment of inertia to $1.60 \text{ kg} \cdot \text{m}^2$. **(a)** What is the student's angular speed now? **(b)** Find the initial and final angular momenta of the student.

CONTINUED

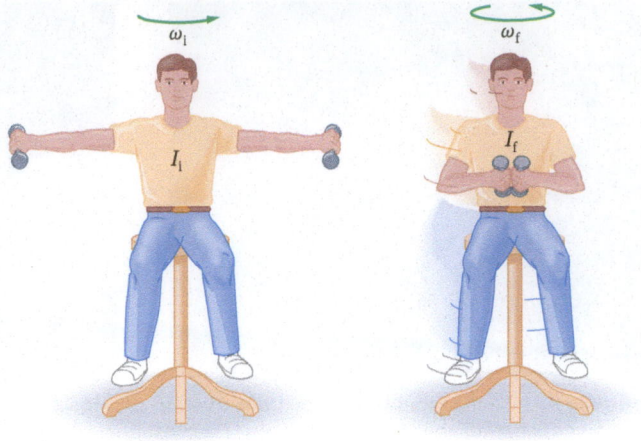

PICTURE THE PROBLEM

The initial and final configurations of the student are shown in our sketch. Clearly, the mass distribution in the final configuration, with the masses held closer to the axis of rotation, results in a smaller moment of inertia.

REASONING AND STRATEGY

Ignoring friction in the axis of the stool, we conclude that no external torques act on the system. As a result, the angular momentum is conserved. Therefore, setting the initial angular momentum, $L_i = I_i \omega_i$, equal to the final angular momentum, $L_f = I_f \omega_f$, yields the final angular speed.

Known Initial angular speed, $\omega_i = 3.72$ rad/s; initial moment of inertia, $I_i = 5.33$ kg · m²; final moment of inertia, $I_f = 1.60$ kg · m².

Unknown (a) Final angular speed, $\omega_f = ?$ (b) Initial and final angular momenta of student, $L_i = ?, L_f = ?$

SOLUTION

Part (a)

1. Apply angular momentum conservation to this system:

$$L_i = L_f$$
$$I_i \omega_i = I_f \omega_f$$

2. Solve for the final angular speed, ω_f:

$$\omega_f = \left(\frac{I_i}{I_f}\right)\omega_i$$

3. Substitute numerical values:

$$\omega_f = \left(\frac{5.33\ \text{kg} \cdot \text{m}^2}{1.60\ \text{kg} \cdot \text{m}^2}\right)(3.72\ \text{rad/s}) = 12.4\ \text{rad/s}$$

Part (b)

4. Use $L = I\omega$ to calculate the angular momentum. Substitute both initial and final values as a check:

$$L_i = I_i \omega_i = (5.33\ \text{kg} \cdot \text{m}^2)(3.72\ \text{rad/s}) = 19.8\ \text{kg} \cdot \text{m}^2/\text{s}$$
$$L_f = I_f \omega_f = (1.60\ \text{kg} \cdot \text{m}^2)(12.4\ \text{rad/s}) = 19.8\ \text{kg} \cdot \text{m}^2/\text{s}$$

INSIGHT

Initially the student completes one revolution roughly every two seconds. After pulling the weights in, the student's rotation rate has increased to almost two revolutions a second—quite a dizzying pace. The same physics applies to a rotating diver or a spinning ice skater.

PRACTICE PROBLEM

What moment of inertia would be required to give a final spin rate of 10.0 rad/s? [**Answer:** $I_f = (\omega_i/\omega_f)I_i = 1.98$ kg · m²]

Some related homework problems: Problem 54, Problem 58

Conservation of Angular Momentum in Nature An increasing angular speed, as experienced by the student in Example 11-19, can be observed in nature as well. Examples are shown in **FIGURE 11-25**. For example, a hurricane draws circulating air in at ground level toward its "eye," where it then rises to an altitude of 10 miles or more. As air moves inward toward the axis of rotation, its angular speed increases, just as the masses held by the student speed up when they are pulled inward. For example, if the wind has a speed of only 3.0 mph at a distance of 300 miles from the center of the hurricane, it would speed up to 150 mph when it comes to within 6.0 miles of the center. Of course, this analysis ignores friction, which would certainly decrease the wind speed. Still, the basic principle—that a decreasing distance from the axis of rotation implies an increasing speed—applies to both the student and the hurricane. Similar behavior is observed in tornadoes and waterspouts.

RWP Another example of conservation of angular momentum occurs in stellar explosions. On occasion a star explodes, sending a portion of its material out into space. After the explosion, the star collapses to a fraction of its original size, speeding up its rotation in the process. If the mass of the star is greater than 1.44 times the mass of

▲ **FIGURE 11-25 Visualizing Concepts Conservation of Angular Momentum** This 1992 satellite photo of Hurricane Andrew (left), one of the most powerful hurricanes in recent decades, clearly suggests the rotating structure of the storm. The violence of the hurricane winds can be attributed in large part to conservation of angular momentum: As air is pushed inward toward the low pressure near the eye of the storm, its rotational velocity increases. The same principle, operating on a smaller scale, explains the tremendous destructive power of tornadoes (right).

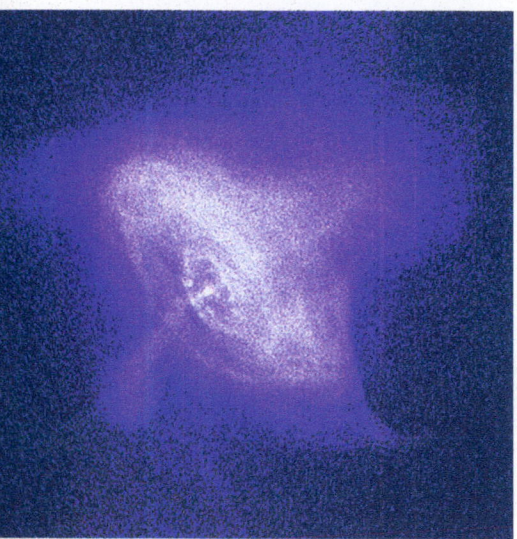

▲ **FIGURE 11-26** Among the fastest rotating objects known in nature are pulsars—stars that have collapsed to a tiny fraction of their original size. Since all the angular momentum of a star must be conserved when it collapses, the dramatic decrease in radius is accompanied by a correspondingly great increase in rotational speed. The Crab nebula pulsar spins at about 30 rev/s. This X-ray photograph shows rings and jets of high-energy particles flying outward from the whirling neutron star at the center.

the Sun, the collapse can continue until a *neutron star* is formed, with a radius of only about 10 to 20 km. Neutron stars have incredibly high densities; in fact, if you could bring a teaspoonful of neutron star material to the Earth, it would weigh about one billion tons! In addition, neutron stars produce powerful beams of X-rays and other radiation that sweep across the sky like a gigantic lighthouse beam as the star rotates. On the Earth we see pulses of radiation from these rotating beams, one for each revolution of the star. These "pulsating stars," or *pulsars*, typically have periods ranging from about 2 ms to nearly 1 s. The Crab nebula, shown in **FIGURE 11-26** (see Problem 6 in Chapter 10), is a famous example of such a system. The dependence of angular speed on radius for a collapsing star is considered in the next Quick Example.

QUICK EXAMPLE 11-20 A STELLAR PERFORMANCE: FIND THE ANGULAR SPEED

A star of radius $R = 2.3 \times 10^8$ m rotates with an angular speed $\omega = 2.4 \times 10^{-6}$ rad/s. If this star collapses to a radius of 20.0 km, find its final angular speed. (Treat the star as if it were a uniform sphere, and assume that no mass is lost as the star collapses.)

REASONING AND SOLUTION
Because the star is an isolated system, its angular momentum remains the same as it collapses. Thus, we set the initial angular momentum ($I_i \omega_i$) equal to the final angular momentum ($I_f \omega_f$) and solve for the final angular speed (ω_f).

1. Apply conservation of angular momentum:

$$I_i \omega_i = I_f \omega_f$$

2. Write expressions for the initial and final moments of inertia:

$$I_i = \tfrac{2}{5} M R_i^2 \quad \text{and} \quad I_f = \tfrac{2}{5} M R_f^2$$

3. Solve for the final angular speed:

$$\omega_f = (I_i/I_f)\omega_i = (R_i^2/R_f^2)\omega_i$$

4. Substitute numerical values:

$$\omega_f = \left(\frac{2.3 \times 10^8 \text{ m}}{20.0 \times 10^3 \text{ m}}\right)^2 (2.4 \times 10^{-6} \text{ rad/s})$$
$$= 320 \text{ rad/s}$$

RWP The final angular speed corresponds to a period of about 20 ms, a typical period for pulsars. Given that 320 rad/s is roughly 3000 rpm, it follows that a pulsar, which has the mass of a star, rotates as fast as the engine in a racing car.

Angular Momentum and the Length of the Day If the student in Example 11-19 were to stretch his arms back out again, the resulting *increase* in the moment of inertia would cause a *decrease* in his angular speed. The same effect might apply to the Earth one day. For example, a melting of the polar ice caps would lead to an increase in the Earth's moment of inertia (as we saw in Chapter 10) and thus, by angular momentum conservation, the angular speed of the Earth would decrease. This would mean that more time would be required for the Earth to complete a revolution about its axis of rotation; that is, the day would lengthen.

Angular Momentum and Kinetic Energy   Angular momentum is conserved in the systems we've studied so far, and hence it's natural to ask whether the energy is conserved as well. We consider this question in the next Conceptual Example.

CONCEPTUAL EXAMPLE 11-21   COMPARE KINETIC ENERGIES

A skater pulls in her arms, decreasing her moment of inertia by a factor of two, and doubling her angular speed. Is her final kinetic energy greater than, less than, or equal to her initial kinetic energy?

REASONING AND DISCUSSION
Let's calculate the initial and final kinetic energies, and compare them. The initial kinetic energy is

$$K_i = \tfrac{1}{2}I_i\omega_i^2$$

After pulling in her arms, the skater has half the moment of inertia and twice the angular speed. Hence, her final kinetic energy is

$$K_f = \tfrac{1}{2}I_f\omega_f^2 = \tfrac{1}{2}(I_i/2)(2\omega_i)^2 = 2\left(\tfrac{1}{2}I_i\omega_i^2\right) = 2K_i$$

Thus, the fact that K depends on the square of ω leads to an increase in the kinetic energy. The source of this additional energy is the work done by the muscles in the skater's arms as she pulls them in to her body.

ANSWER
The skater's kinetic energy increases.

Rotational Collisions

In the not-too-distant past, a person would play music by placing a record on a rotating turntable. Suppose, for example, that a turntable with a moment of inertia I_t is rotating freely with an initial angular speed ω_0. A record, with a moment of inertia I_r and initially at rest, is dropped straight down onto the rotating turntable, as in **FIGURE 11-27**. When the record lands, frictional forces between it and the turntable cause the record to speed up and the turntable to slow down, until they both have the same angular speed. Only internal forces are involved during this process, and hence it follows that the system's angular momentum is conserved. We can think of this event, then, as a "rotational collision."

Before the collision, the angular momentum of the system is

$$L_i = I_t\omega_0$$

After the collision, when both the record and the turntable are rotating with the angular speed ω_f, the system's angular momentum is

$$L_f = I_t\omega_f + I_r\omega_f$$

Setting $L_f = L_i$ yields the final angular speed:

$$\omega_f = \left(\frac{I_t}{I_t + I_r}\right)\omega_0 \qquad\qquad 11\text{-}16$$

Because this collision is completely inelastic, we expect the final kinetic energy to be less than the initial kinetic energy.

We conclude this section with a somewhat different example of a rotational collision. The physical principles involved are precisely the same, however.

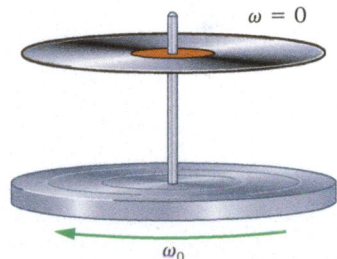
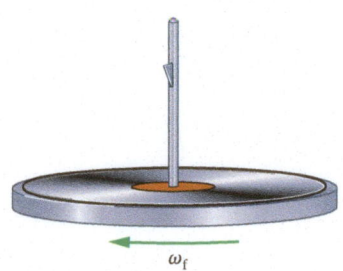

▲ **FIGURE 11-27 A rotational collision** A nonrotating record dropped onto a rotating turntable is an example of a "rotational collision." Since only internal forces are involved during the collision, the final angular momentum is equal to the initial angular momentum.

Off-Center Collision

QUICK EXAMPLE 11-22 CONSERVE ANGULAR MOMENTUM: FIND THE ANGULAR SPEED

A 34.0-kg child runs with a speed of 2.80 m/s tangential to the rim of a stationary merry-go-round. The merry-go-round has a moment of inertia of 512 kg·m² and a radius of 2.31 m. When the child jumps onto the merry-go-round, the entire system begins to rotate. What is the angular speed of the system? Ignore friction and any other type of external torque.

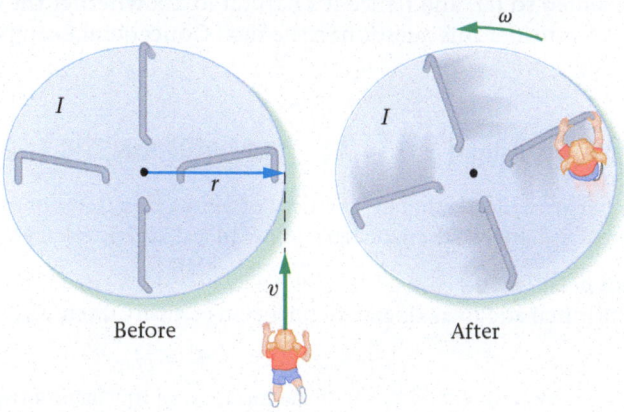

Before   After

Our sketch shows the system before and after the collision. Before the collision the merry-go-round is at rest and the child is running. After the collision the child is riding on the rotating merry-go-round.

REASONING AND SOLUTION
Angular momentum is conserved in this system, so we set the initial angular momentum equal to the final angular momentum to find the final angular speed. The initial angular momentum is the angular momentum of the running child. The final angular momentum is due to the rotation of the merry-go-round *and* the child about the axis of rotation.

1. Write the initial angular momentum of the child:

$$L_i = rmv$$

2. Write the final angular momentum of the system, noting that the moment of inertia of the merry-go-round is I and the moment of inertia of the child is mr^2:

$$L_f = (I + mr^2)\omega$$

3. Set $L_f = L_i$ and solve for the angular speed:

$$\omega = rmv/(I + mr^2)$$

4. Substitute numerical values:

$$\omega = \frac{(2.31\ \text{m})(34.0\ \text{kg})(2.80\ \text{m/s})}{512\ \text{kg}\cdot\text{m}^2 + (34.0\ \text{kg})(2.31\ \text{m})^2}$$
$$= 0.317\ \text{rad/s}$$

If the moment of inertia of the merry-go-round had been zero, $I = 0$, the angular speed of the merry-go-round would be $\omega = v/r$. This, in turn, would mean that the linear speed of the child ($v = r\omega$) is unchanged. If $I > 0$, however, the linear speed of the child is decreased. In this particular case, the child's linear speed after the collision is only $v = r\omega = 0.732$ m/s, compared with her initial linear speed of 2.80 m/s.

Enhance Your Understanding   (Answers given at the end of the chapter)

7. Consider the system described in Quick Example 11-22, where a child runs and then jumps onto a merry-go-round. Do you expect the initial kinetic energy of this system to be greater than, less than, or equal to the final kinetic energy? Explain.

Section Review

- Angular momentum is conserved in a system if the net torque exerted on it is zero.

11-8 Rotational Work and Power

Just as a force acting through a distance does work on an object, so too does a torque acting through an angular displacement. To see this, consider the case of fishing line pulled from a reel. If the line is pulled with a force F for a distance Δx, as in **FIGURE 11-28**, the work done on the reel is

$$W = F\Delta x$$

Now, the line is unwinding without slipping, and hence it follows that the linear displacement of the line, Δx, is related to the angular displacement of the reel, $\Delta\theta$, by the following relationship:

$$\Delta x = R\Delta\theta$$

In this equation, R is the radius of the reel, and $\Delta\theta$ is measured in radians. Thus, the work can be written as

$$W = F\Delta x = FR\Delta\theta$$

Finally, the torque exerted on the reel by the line is $\tau = RF$, from which it follows that the work done on the reel is simply torque times angular displacement:

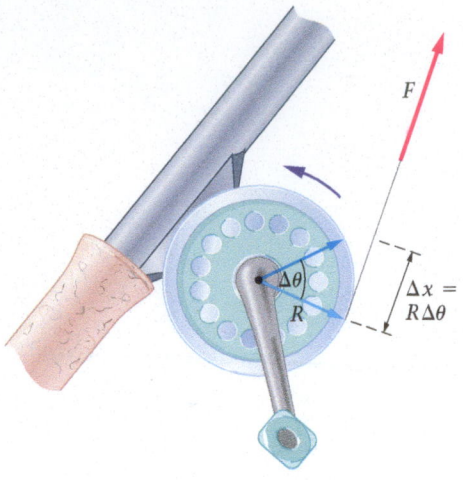

▲ **FIGURE 11-28 Rotational work** A force F pulling a length of line Δx from a fishing reel does the work $W = F\Delta x$. In terms of torque and angular displacement, the work can be expressed as $W = \tau\,\Delta\theta$.

Work Done by Torque

$$W = \tau\,\Delta\theta \qquad\qquad 11\text{-}17$$

Notice again the analogies between angular and linear quantities in $W = F\Delta x$ and $W = \tau\Delta\theta$. As usual, τ is the analog of F, and θ is the analog of x.

As we saw in Chapter 7, the net work done on an object is equal to the change in its kinetic energy. This is the work–energy theorem:

$$W = \Delta K = K_f - K_i \qquad\qquad 11\text{-}18$$

The work–energy theorem applies regardless of whether the work is done by a force acting through a distance or a torque acting through an angle.

Similarly, power is the amount of work done in a given time, regardless of whether the work is done by a force or a torque. In the case of a torque, we have $W = \tau\Delta\theta$, and hence:

Power Produced by a Torque

$$P = \frac{W}{\Delta t} = \tau\frac{\Delta\theta}{\Delta t} = \tau\omega \qquad\qquad 11\text{-}19$$

Again, the analogy is clear between $P = Fv$ for the linear case, and $P = \tau\omega$ for the rotational case.

EXERCISE 11-23   ROTATIONAL POWER

It takes a good deal of effort to make homemade ice cream. **(a)** If the torque required to turn the handle on an ice cream maker is 4.5 N·m, how much work is expended on each complete revolution of the handle? **(b)** How much power is required to turn the handle if each revolution is completed in 1.2 s?

REASONING AND SOLUTION

a. Applying $W = \tau\Delta\theta$ with $\tau = 4.5$ N·m and $\Delta\theta = 2\pi$ yields the work done in one revolution:

$$W = \tau\Delta\theta = (4.5\,\text{N·m})(2\pi\,\text{rad}) = 28\,\text{J}$$

b. Power is the work per time; that is,

$$P = W/\Delta t = (28\,\text{J})/(1.2\,\text{s}) = 23\,\text{W}$$

Equivalently, the angular speed of the handle is $\omega = (2\pi)/T = (2\pi)/(1.2\,\text{s}) = 5.2\,\text{rad/s}$, and therefore $P = \tau\omega$ (Equation 11-19) yields $P = \tau\omega = (4.5\,\text{N·m})(5.2\,\text{rad/s}) = 23\,\text{W}$, in agreement with our previous result.

8. In system 1, a torque of 20 N · m acts through an angular displacement of 45°. In system 2, a torque of 10 N · m acts through an angular displacement of $\pi/2$. Is the work done in system 1 greater than, less than, or equal to the work done in system 2? Explain.

Section Review

- The work done in a rotational system is equal to torque times angular displacement.

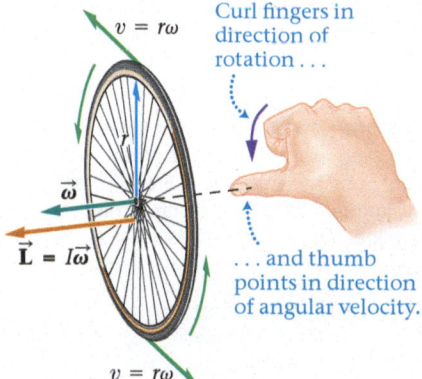

▲ **FIGURE 11-29 The right-hand rule for angular velocity** The angular velocity, $\vec{\omega}$, of a rotating wheel points along the axis of rotation. Its direction is given by the right-hand rule.

11-9 The Vector Nature of Rotational Motion

We've mentioned many times that the angular velocity is a vector, and that we must be careful to use the proper sign for ω. But if the angular velocity is a vector, what is its direction?

To address this question, consider the rotating wheel shown in **FIGURE 11-29**. Each point on the rim of this wheel has a velocity vector pointing in a different direction in the plane of rotation. But if different parts of the wheel move in different directions, how can we assign a single direction to the angular velocity vector, $\vec{\omega}$? The answer is that there is only one direction that remains fixed as the wheel rotates—the direction of the axis of rotation. By definition, then, the angular velocity vector, $\vec{\omega}$, is taken to point along the axis of rotation.

Given that $\vec{\omega}$ points along the axis of rotation, we must still decide whether it points to the left or to the right in Figure 11-29. The convention we use for assigning the direction of $\vec{\omega}$ is referred to as the right-hand rule:

Right-Hand Rule for Angular Velocity, $\vec{\omega}$

Curl the fingers of the right hand in the direction of rotation.

The thumb now points in the direction of the angular velocity, $\vec{\omega}$.

The right-hand rule (RHR) for $\vec{\omega}$ is illustrated in Figure 11-29. Additional examples of rotating objects are given in **FIGURE 11-30**.

▲ **FIGURE 11-30** Children have always been fascinated by tops—but not only children. The physicists in the photo at right, Wolfgang Pauli and Niels Bohr, seem as delighted by a spinning top as any child. Their contributions to modern physics, discussed in Chapter 30, helped to show that subatomic particles, the ultimate constituents of matter, have a property (now referred to as "spin") that is in some ways analogous to the rotation of a top or a gyroscope.

The same convention for direction applies to the angular momentum vector. First, recall that the angular momentum has a magnitude given by $L = I\omega$. Hence, we choose the direction of $\vec{L}$ to be the same as the direction of $\vec{\omega}$. That is,

$$\vec{L} = I\vec{\omega}$$

11-20

The angular momentum vector is also illustrated in Figure 11-29.

Similarly, torque is a vector, and it too is defined to point along the axis of rotation. The right-hand rule for torque is similar to that for angular velocity:

> **Right-Hand Rule for Torque, $\vec{\tau}$**
>
> Curl the fingers of the right hand in the direction of rotation that this torque would cause if it acted alone.
>
> The thumb now points in the direction of the torque vector, $\vec{\tau}$.

Two examples of torque vectors are shown in **FIGURE 11-31**.

As an example of torque and angular momentum vectors, consider the spinning bicycle wheel shown in **FIGURE 11-32**. The angular momentum vector for the wheel points to the left, along the axis of rotation. If a person pushes on the rim of the wheel in the direction indicated, the resulting torque is also to the left, as shown in the figure. If this torque lasts for a time Δt, the angular momentum changes by the amount

$$\Delta \vec{L} = \vec{\tau} \Delta t$$

Adding $\Delta \vec{L}$ to the original angular momentum $\vec{L}_i$ yields the final angular momentum $\vec{L}_f$, shown in Figure 11-32. Since $\vec{L}_f$ is in the same direction as $\vec{L}_i$, but with a greater magnitude, it follows that the wheel is spinning in the same direction as before, only faster. This is to be expected, considering the direction of the person's push on the wheel.

On the other hand, if a person pushes on the wheel in the opposite direction, the torque vector points to the right. As a result, $\Delta \vec{L}$ points to the right as well. When we add $\Delta \vec{L}$ and $\vec{L}_i$ to obtain the final angular momentum $\vec{L}_f$, we find that it has the same direction as $\vec{L}_i$, but a smaller magnitude. Hence, we conclude that the wheel spins more slowly, as one would expect.

Finally, a case of considerable interest is when the torque and angular momentum vectors are at right angles to one another. The classic example of such a system is the **gyroscope**. To begin, consider a gyroscope whose axis of rotation is horizontal, as in **FIGURE 11-33**. If the gyroscope were to be released with no spin it would simply fall, rotating counterclockwise downward about its point of support. Curling the fingers of the right hand in the counterclockwise direction, we see that the thumb, and hence the torque due to gravity, points out of the page.

Next, imagine the gyroscope to be spinning rapidly—as would normally be the case—with its angular momentum pointing to the left in Figure 11-33. If the gyroscope is released now, it doesn't fall as before, even though the torque is the same. To see what happens instead, consider the change in angular momentum, $\Delta \vec{L}$, caused by the torque, $\vec{\tau}$, acting for a small interval of time. As shown in **FIGURE 11-34**, the small

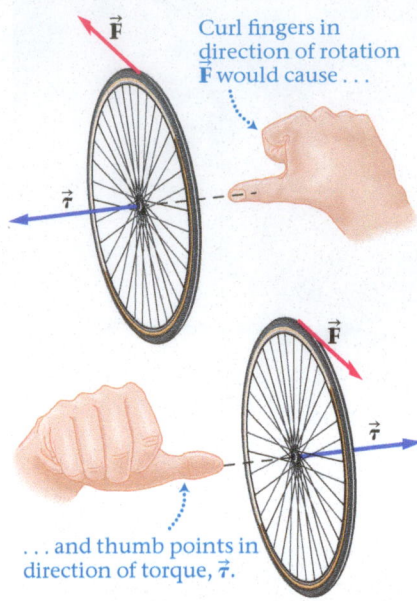

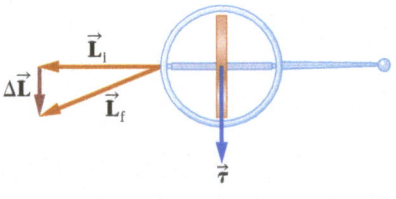

▲ **FIGURE 11-31 The right-hand rule for torque** Examples of torque vectors obtained using the right-hand rule.

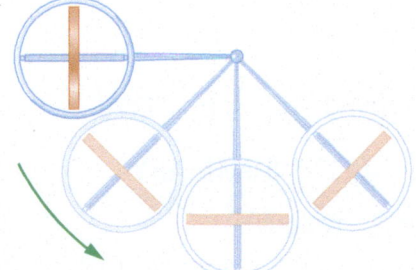

Top view

▲ **FIGURE 11-34 Precession of a gyroscope** The gyroscope as viewed from above. In a time Δt the angular momentum changes by the amount $\Delta \vec{L} = \vec{\tau} \Delta t$. This causes the angular momentum vector, and hence the gyroscope as a whole, to rotate in a counterclockwise direction.

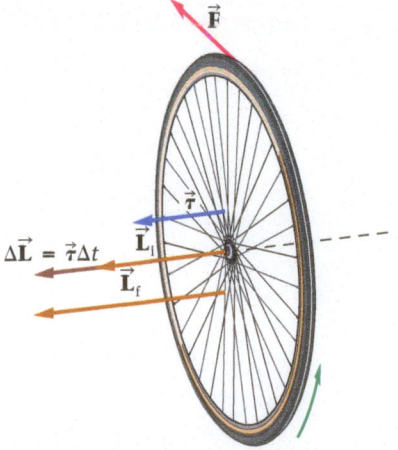

▲ **FIGURE 11-32 Torque and angular momentum vectors** A tangential push on the spinning wheel in the direction shown causes a torque to the left. As a result, the angular momentum increases. Hence, the wheel spins faster, as expected.

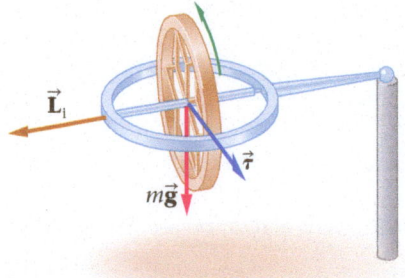

▲ **FIGURE 11-33 The torque exerted on a gyroscope** A spinning gyroscope has an initial angular momentum to the left. The torque due to gravity is out of the page.

change, $\Delta\vec{L}$, is at right angles to $\vec{L}_i$; hence the final angular momentum, $\vec{L}_f$, is essentially the same length as $\vec{L}_i$, but pointing in a direction slightly out of the page. With each small interval of time, the angular momentum vector continues to change in direction so that, viewed from above as in Figure 11-34, the gyroscope as a whole rotates in a counterclockwise sense around its support point. This type of motion, where the axis of rotation changes direction with time, is referred to as **precession**. The gyroscope in **FIGURE 11-35** is designed to use precession to test Einstein's theory of relativity.

RWP Because of its spinning motion about its rotational axis, the Earth may be considered as one rather large gyroscope. Gravitational forces exerted on the Earth by the Sun and the Moon subject it to external torques that cause its rotational axis to precess. At the moment, the rotational axis of the Earth points toward Polaris, the North Star, which remains almost fixed in position in time-lapse photographs while the other stars move in circular paths about it. In a few hundred years, however, Polaris will also move in a circular path in the sky because the Earth's axis of rotation will point in a different direction. After 26,000 years the Earth will complete one full cycle of precession, and Polaris will again be the pole star.

Enhance Your Understanding   (Answers given at the end of the chapter)

9. The angular velocity of the spinning bicycle wheel in **FIGURE 11-36** points out of the page, toward the reader. Does the wheel spin clockwise or counterclockwise?

▲ **FIGURE 11-36** Enhance Your Understanding 9.

Section Review

- The directions of the angular velocity, angular momentum, and torque are given by right-hand rules.

CHAPTER 11 REVIEW

CHAPTER SUMMARY

11-1   TORQUE

A force applied so as to cause an angular acceleration is said to exert a torque, τ.

Torque Due to a Tangential Force

A tangential force F applied at a distance r from the axis of rotation produces a torque

$$\tau = rF \qquad\qquad 11\text{-}1$$

Torque for a General Force

A force exerted at an angle θ with respect to the radial direction, and applied at a distance r from the axis of rotation, produces the torque

$$\tau = rF\sin\theta \qquad\qquad 11\text{-}2$$

11-2   TORQUE AND ANGULAR ACCELERATION

A single torque applied to an object gives it an angular acceleration.

Newton's Second Law for Rotation

The connection between torque and angular acceleration is

$$\sum \tau = I\alpha \qquad \qquad 11\text{-}4$$

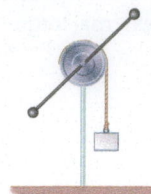

Rotational/Translational Analogies

Torque is analogous to force, the moment of inertia is analogous to mass, and angular acceleration is analogous to linear acceleration. Therefore, the rotational analog of $F = ma$ is $\tau = I\alpha$.

11-3   ZERO TORQUE AND STATIC EQUILIBRIUM

The conditions for an object to be in static equilibrium are that the total force and the total torque acting on the object must be zero:

$$\sum F_x = 0, \quad \sum F_y = 0, \quad \sum \tau = 0$$

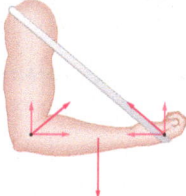

11-4   CENTER OF MASS AND BALANCE

An object balances when it is supported at its center of mass.

11-5   DYNAMIC APPLICATIONS OF TORQUE

Newton's second law can be applied to rotational systems in a way that is completely analogous to its application to linear systems.

11-6   ANGULAR MOMENTUM

A moving object has angular momentum as long as its direction of motion does not extend through the axis of rotation.

Angular Momentum and Angular Speed

Angular momentum can be expressed in terms of angular speed and the moment of inertia as follows:

$$L = I\omega \qquad \qquad 11\text{-}11$$

This is the rotational analog of $p = mv$.

Angular Momentum and Linear Speed

If an object of mass m is a distance r from the axis of rotation and moves with a linear speed v at an angle θ with respect to the radial direction, its angular momentum is

$$L = rmv \sin \theta \qquad \qquad 11\text{-}13$$

Newton's Second Law

Newton's second law can be expressed in terms of the rate of change of the angular momentum:

$$\sum \tau = I\alpha = \frac{\Delta L}{\Delta t} \qquad \qquad 11\text{-}14$$

This is the rotational analog of $\sum F = \Delta p / \Delta t$.

11-7 CONSERVATION OF ANGULAR MOMENTUM

If the net external torque acting on a system is zero, its angular momentum is conserved:

$$L_f = L_i$$

11-8 ROTATIONAL WORK AND POWER

A torque acting through an angle does work, just as a force acting through a distance does work.

Work Done by a Torque

A torque τ acting through an angle $\Delta\theta$ does work W given by

$$W = \tau\Delta\theta \qquad \text{11-17}$$

Work–Energy Theorem

The work–energy theorem is

$$W = \Delta K = K_f - K_i \qquad \text{11-18}$$

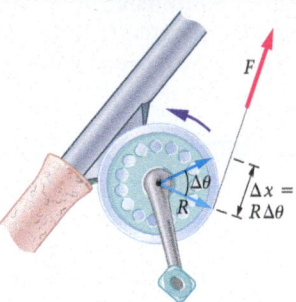

11–9 THE VECTOR NATURE OF ROTATIONAL MOTION

Rotational quantities have directions that point along the axis of rotation. The precise direction is given by the right-hand rule.

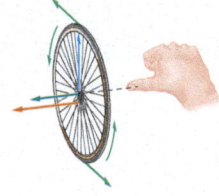

ANSWERS TO ENHANCE YOUR UNDERSTANDING QUESTIONS

1. Torque produced by F_3 < torque produced by F_2 = torque produced by F_4 < torque produced by F_1.

2. The angular acceleration of object 1 is greater (by a factor of 2) than the angular acceleration of object 2.

3. Greater than half the weight, because the ropes have both vertical and horizontal components.

4. Mass of B < mass of C < mass of A.

5. **(a)** The tension is less than mg, because the mass accelerates downward and hence the net force is downward. **(b)** The tension is greater than zero, because it is accelerating the pulley.

6. Object 2 has the greater angular momentum, because both objects have the same moment of inertia and object 2 has the greater angular speed.

7. The initial kinetic energy is greater than the final energy because the inelastic collision between the child and the merry-go-round dissipates energy.

8. The work done in system 1 is equal to the work done in system 2 because torque times angular displacement is the same for the two systems.

9. The wheel spins counterclockwise, as can be verified with the RHR for angular velocity.

CONCEPTUAL QUESTIONS

For instructor-assigned homework, go to www.masteringphysics.com.

(Answers to odd-numbered Conceptual Questions can be found in the back of the book.)

1. Two forces produce the same torque. Does it follow that they have the same magnitude? Explain.

2. A car pitches down in front when the brakes are applied sharply. Explain this observation in terms of torques.

3. A tightrope walker uses a long pole to aid in balancing. Why?

4. When a motorcycle accelerates rapidly from a stop it sometimes "pops a wheelie"; that is, its front wheel may lift off the ground. Explain this behavior in terms of torques.

5. Give an example of a system in which the net torque is zero but the net force is nonzero.

6. Give an example of a system in which the net force is zero but the net torque is nonzero.

7. Is the normal force exerted by the ground the same for all four tires on your car? Explain.

8. Give two everyday examples of objects that are not in static equilibrium.

9. Give two everyday examples of objects that are in static equilibrium.

10. Can an object have zero translational acceleration and, at the same time, have nonzero angular acceleration? If your answer is no, explain why not. If your answer is yes, give a specific example.

11. Stars form when a large rotating cloud of gas collapses. What happens to the angular speed of the gas cloud as it collapses?

12. What purpose does the tail rotor on a helicopter serve?

13. Is it possible to change the angular momentum of an object without changing its linear momentum? If your answer is no, explain why not. If your answer is yes, give a specific example.

14. Suppose a diver springs into the air with no initial angular velocity. Can the diver begin to rotate by folding into a tucked position? Explain.

PROBLEMS AND CONCEPTUAL EXERCISES

Answers to odd-numbered Problems and Conceptual Exercises can be found in the back of the book. **BIO** *identifies problems of biological or medical interest;* **CE** *indicates a conceptual exercise.* **Predict/Explain** *problems ask for two responses: (a) your prediction of a physical outcome, and (b) the best explanation among three provided; and* **Predict/Calculate** *problems ask for a prediction of a physical outcome, based on fundamental physics concepts, and follow that with a numerical calculation to verify the prediction. On all problems, bullets (•, ••, •••) indicate the level of difficulty.*

SECTION 11-1 TORQUE

1. • To tighten a spark plug, it is recommended that a torque of 15 N·m be applied. If a mechanic tightens the spark plug with a wrench that is 25 cm long, what is the minimum force necessary to create the desired torque?

2. • **Pulling a Weed** The gardening tool shown in **FIGURE 11-37** is used to pull weeds. If a 1.23-N·m torque is required to pull a given weed, what force did the weed exert on the tool?

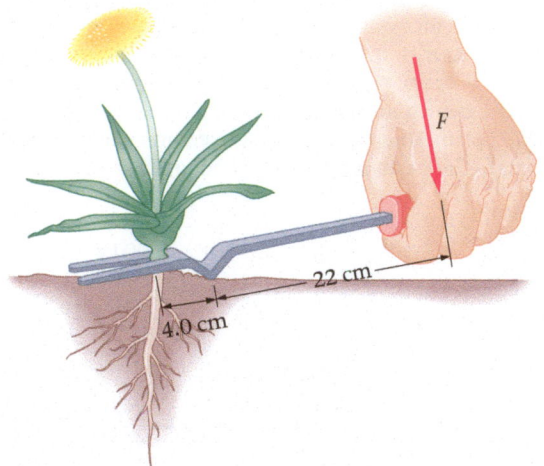

FIGURE 11-37 Problem 2

3. • A person slowly lowers a 3.6-kg crab trap over the side of a dock, as shown in **FIGURE 11-38**. What torque does the trap exert about the person's shoulder?

FIGURE 11-38 Problem 3

4. •• A squirrel-proof bird feeder has a lever that closes to protect the seeds when a 0.42-kg squirrel sits on it, but not when a 0.077-kg bird perches there. **(a)** If the lever radius is 9.6 cm, what torques do the squirrel and the bird each exert on it? **(b)** The torque produced by the animal is balanced by a spring that applies a perpendicular force a distance of 3.5 cm from the axis of rotation. If the squirrel must stretch the spring 2.5 cm in order to close the lever and protect the seeds, what should be the force constant of the spring?

5. •• At one position during its cycle, the foot pushes straight down with a 440-N force on a bicycle pedal arm that is rotated an angle $\phi = 65°$ from the vertical, as shown in **FIGURE 11-39**. If the force is applied $r = 0.17$ m from the axis of rotation, what torque is produced?

FIGURE 11-39 Problem 5

6. •• **BIO Predict/Calculate Force to Hold a Baseball** A person holds a 1.42-N baseball in his hand, a distance of 34.0 cm from the elbow joint, as shown in **FIGURE 11-40**. The biceps, attached at a distance of 2.75 cm from the elbow, exerts an upward force of 12.6 N on the forearm. Consider the forearm and hand to be a uniform rod with a mass of 1.20 kg. **(a)** Calculate the net torque acting on the forearm and hand. Use the elbow joint as the axis of rotation. **(b)** If the net torque obtained in part (a) is nonzero, in which direction will the forearm and hand rotate? **(c)** Would the torque exerted on the forearm by the biceps increase or decrease if the biceps were attached farther from the elbow joint?

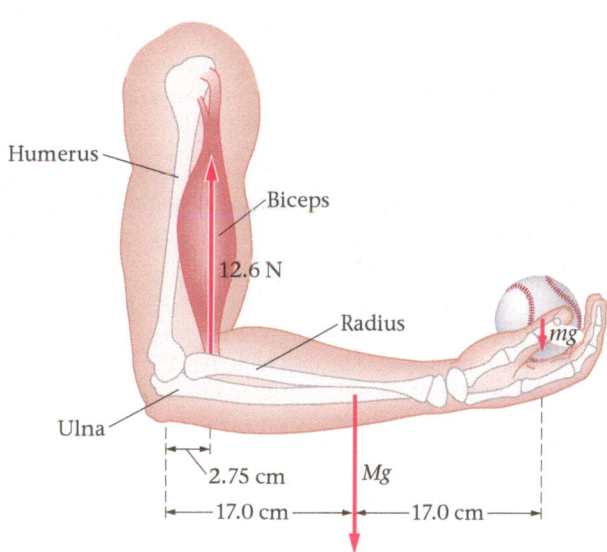

FIGURE 11-40 Problems 6 and 23

7. •• At the local playground, a 21-kg child sits on the right end of a horizontal teeter-totter, 1.8 m from the pivot point. On the left side of the pivot an adult pushes straight down on the teeter-totter with a force of 151 N. In which direction (clockwise or counterclockwise) does the teeter-totter rotate if the adult applies the force at a distance of **(a)** 3.0 m, **(b)** 2.5 m, or **(c)** 2.0 m from the pivot?

SECTION 11-2 TORQUE AND ANGULAR ACCELERATION

8. • **CE** Predict/Explain Consider the pulley–block systems shown in Conceptual Example 11-5. **(a)** Is the tension in the string on the left-hand rotating system (the one with the moveable masses at the ends of the rod) greater than, less than, or equal to the weight of the mass attached to that string? **(b)** Choose the *best explanation* from among the following:
   I. The mass is in free fall once it is released.
   II. The string rotates the pulley in addition to supporting the mass.
   III. The mass accelerates downward.

9. • **CE** Suppose a torque rotates your body about one of three different axes of rotation: case A, an axis through your spine; case B, an axis through your hips; and case C, an axis through your ankles. Rank these three axes of rotation in increasing order of the angular acceleration produced by the torque. Indicate ties where appropriate.

10. • A torque of 0.97 N · m is applied to a bicycle wheel of radius 35 cm and mass 0.75 kg. Treating the wheel as a hoop, find its angular acceleration.

11. • When a ceiling fan rotating with an angular speed of 3.66 rad/s is turned off, a frictional torque of 0.125 N · m slows it to a stop in 31.8 s. What is the moment of inertia of the fan?

12. • When the play button is pressed, a CD accelerates uniformly from rest to 450 rev/min in 3.0 revolutions. If the CD has a radius of 6.0 cm and a mass of 17 g, what is the torque exerted on it?

13. •• A person holds a ladder horizontally at its center. Treating the ladder as a uniform rod of length 3.66 m and mass 10.9 kg, find the torque the person must exert on the ladder to give it an angular acceleration of 0.333 rad/s^2.

14. •• A 0.180-kg wooden rod is 1.25 m long and pivots at one end. It is held horizontally and then released. **(a)** What is the angular acceleration of the rod after it is released? **(b)** What is the linear acceleration of a spot on the rod that is 0.95 m from the axis of rotation? **(c)** At what location along the rod should a die be placed (Figure 11-9) so that the die just begins to separate from the rod as it falls?

15. •• Predict/Calculate A wheel on a game show is given an initial angular speed of 1.22 rad/s. It comes to rest after rotating through 0.75 of a turn. **(a)** Find the average torque exerted on the wheel given that it is a disk of radius 0.71 m and mass 6.4 kg. **(b)** If the mass of the wheel is doubled and its radius is halved, will the angle through which it rotates before coming to rest increase, decrease, or stay the same? Explain. (Assume that the average torque exerted on the wheel is unchanged.)

16. •• The L-shaped object in **FIGURE 11-41** consists of three masses connected by light rods. What torque must be applied to this object to give it an angular acceleration of 1.20 rad/s^2 if it is rotated about **(a)** the x axis, **(b)** the y axis, or **(c)** the z axis (which is through the origin and perpendicular to the page)?

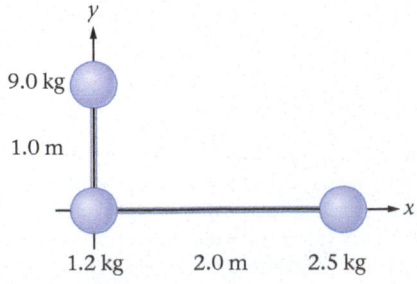

FIGURE 11-41 Problems 16, 17, and 66

17. •• **CE** The L-shaped object described in the previous problem can be rotated in one of the following three ways: case A, about the x axis; case B, about the y axis; and case C, about the z axis (which passes through the origin perpendicular to the plane of the figure). If the same torque τ is applied in each of these cases, rank them in increasing order of the resulting angular acceleration. Indicate ties where appropriate.

18. •• **CE** A motorcycle accelerates from rest, and both the front and rear tires roll without slipping. **(a)** Is the force exerted by the ground on the rear tire in the forward or in the backward direction? Explain. **(b)** Is the force exerted by the ground on the front tire in the forward or in the backward direction? Explain. **(c)** If the moment of inertia of the front tire is increased, will the motorcycle's acceleration increase, decrease, or stay the same? Explain.

19. •• Predict/Calculate A torque of 13 N · m is applied to the rectangular object shown in **FIGURE 11-42**. The torque can act about the x axis, the y axis, or the z axis, which passes through the origin and points out of the page. **(a)** In which case does the object experience the greatest angular acceleration? The least angular acceleration? Explain. Find the angular acceleration when the torque acts about **(b)** the x axis, **(c)** the y axis, and **(d)** the z axis.

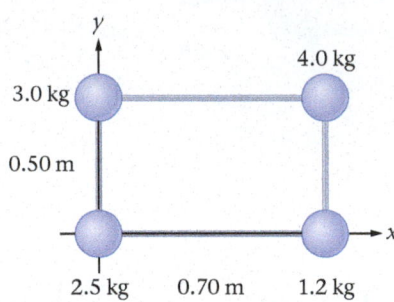

FIGURE 11-42 Problems 19 and 67

SECTION 11-3 ZERO TORQUE AND STATIC EQUILIBRIUM

20. • **CE** Predict/Explain Suppose the person in Example 11-11 climbs higher on the ladder. **(a)** As a result, is the ladder more likely, less likely, or equally likely to slip? **(b)** Choose the *best explanation* from among the following:
   I. The forces are the same regardless of the person's position.
   II. The magnitude of f_2 must increase as the person moves upward.
   III. When the person is higher, the ladder presses down harder on the floor.

21. • A string that passes over a pulley has a 0.321-kg mass attached to one end and a 0.635-kg mass attached to the other end. The pulley, which is a disk of radius 9.40 cm, has friction in its axle. What is the magnitude of the frictional torque that must be exerted by the axle if the system is to be in static equilibrium?

22. • To loosen the lid on a jar of jam 7.6 cm in diameter, a torque of 11 N · m must be applied to the circumference of the lid. If a jar wrench whose handle extends 15 cm from the center of the jar is attached to the lid, what is the minimum force required to open the jar?

23. •• **BIO** Predict/Calculate Referring to the person holding a baseball in Problem 6, suppose the biceps exert just enough upward force to keep the system in static equilibrium. **(a)** Is the force exerted by the biceps more than, less than, or equal to the combined weight of the forearm, hand, and baseball? Explain. **(b)** Determine the force exerted by the biceps.

24. •• **BIO** Predict/Calculate **A Person's Center of Mass** To determine the location of her center of mass, a physics student lies on a

lightweight plank supported by two scales 2.50 m apart, as indicated in **FIGURE 11-43**. If the left scale reads 435 N, and the right scale reads 183 N, find **(a)** the student's mass and **(b)** the distance from the student's head to her center of mass.

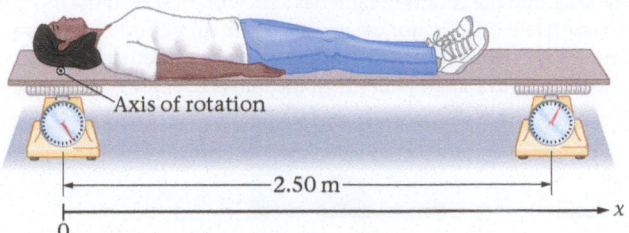

FIGURE 11-43 Problem 24

25. •• **Triceratops** A set of fossilized triceratops footprints discovered in Texas show that the front and rear feet were 3.2 m apart. The rear footprints were observed to be twice as deep as the front footprints. Assuming that the rear feet pressed down on the ground with twice the force exerted by the front feet, find the horizontal distance from the rear feet to the triceratops's center of mass.

26. •• **Predict/Calculate** A schoolyard teeter-totter with a total length of 6.4 m and a mass of 41 kg is pivoted at its center. A 21-kg child sits on one end of the teeter-totter. **(a)** Where should a parent push vertically downward with a force of 210 N in order to hold the teeter-totter level? **(b)** Where should the parent push with a force of 310 N? **(c)** How would your answers to parts (a) and (b) change if the mass of the teeter-totter were doubled? Explain.

27. •• A 0.122-kg remote control 23.0 cm long rests on a table, as shown in **FIGURE 11-44**, with a length L overhanging its edge. To operate the power button on this remote requires a force of 0.365 N. How far can the remote control extend beyond the edge of the table and still not tip over when you press the power button? Assume the mass of the remote is distributed uniformly, and that the power button is 1.41 cm from the overhanging end of the remote.

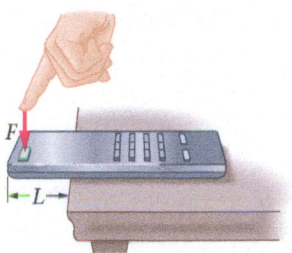

FIGURE 11-44 Problem 27

28. •• **Predict/Calculate** A 0.16-kg meterstick is held perpendicular to a vertical wall by a 2.5-m string going from the wall to the far end of the stick. **(a)** Find the tension in the string. **(b)** If a shorter string is used, will its tension be greater than, less than, or the same as that found in part (a)? **(c)** Find the tension in a 2.0-m string.

29. •• Babe Ruth steps to the plate and casually points to left center field to indicate the location of his next home run. The mighty Babe holds his bat across his shoulder, with one hand holding the small end of the bat. The bat is horizontal, and the distance from the small end of the bat to the shoulder is 22.5 cm. If the bat has a mass of 1.10 kg and has a center of mass that is 67.0 cm from the small end of the bat, find the magnitude and direction of the force exerted by **(a)** the hand and **(b)** the shoulder.

30. •• A uniform metal rod, with a mass of 2.0 kg and a length of 1.5 m, is attached to a wall by a hinge at its base. A horizontal wire bolted to the wall 0.70 m above the base of the rod holds the rod at an

angle of 28° above the horizontal. The wire is attached to the top of the rod. **(a)** Find the tension in the wire. Find **(b)** the horizontal and **(c)** the vertical components of the force exerted on the rod by the hinge.

31. •• A rigid, vertical rod of negligible mass is connected to the floor by a bolt through its lower end, as shown in **FIGURE 11-45**. The rod also has a wire connected between its top end and the floor. If a horizontal force F is applied at the midpoint of the rod, find **(a)** the tension in the wire, and **(b)** the horizontal and **(c)** the vertical components of force exerted by the bolt on the rod.

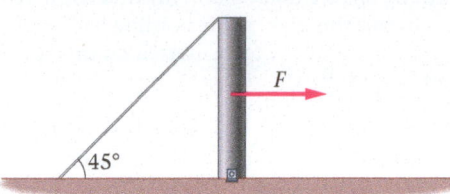

FIGURE 11-45 Problem 31

32. •• In **FIGURE 11-46** two acrobats perform a balancing maneuver. The arms of the top acrobat (in yellow), together with the weight of his head and shoulders, provide a downward force at a distance of 28 cm from the pivot point to balance the torque produced by his lower body. If the lower body of the top acrobat is modeled as a 54-kg mass at a distance of 39 cm from the pivot point, how much force must the arms and upper body produce?

FIGURE 11-46 Problem 32

33. ••• **BIO Forces in the Foot** In **FIGURE 11-47** we see the forces acting on a sprinter's foot just before she takes off at the start of the race. Find the magnitude of the force exerted on the heel by the Achilles tendon, F_H, and the magnitude of the force exerted on the foot at the ankle joint, F_J.

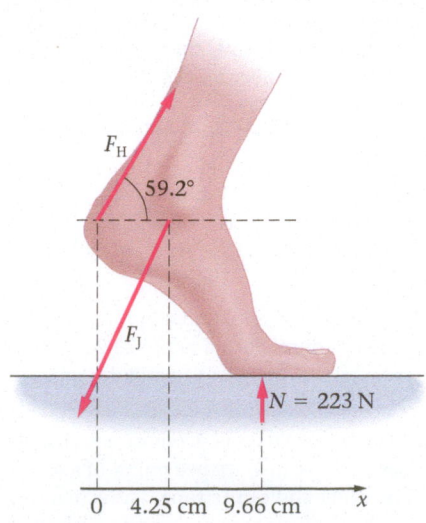

FIGURE 11-47 Problem 33

34. ••• A stick with a mass of 0.214 kg and a length of 0.436 m rests in contact with a bowling ball and a rough floor, as shown in **FIGURE 11-48**. The bowling ball has a diameter of 21.6 cm, and the angle the stick makes with the horizontal is 30.0°. You may assume there is no friction between the stick and the bowling ball, though friction with the floor must be taken into account. **(a)** Find the magnitude of the force exerted on the stick by the bowling ball. **(b)** Find the horizontal component of the force exerted on the stick by the floor. **(c)** Repeat part (b) for the vertical component of the force.

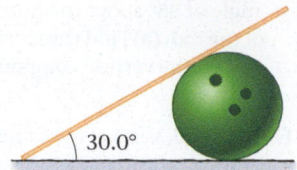

FIGURE 11-48 Problem 34

SECTION 11-4 CENTER OF MASS AND BALANCE

35. • A hand-held shopping basket 62.0 cm long has a 1.81-kg carton of milk at one end, and a 0.722-kg box of cereal at the other end. Where should a 1.80-kg container of orange juice be placed so that the basket balances at its center?

36. • If the cat in Example 11-9 has a mass of 3.9 kg, how far to the right of sawhorse B can it walk before the plank begins to tip?

37. •• Predict/Calculate A 0.34-kg meterstick balances at its center. If a necklace is suspended from one end of the stick, the balance point moves 9.5 cm toward that end. **(a)** Is the mass of the necklace more than, less than, or the same as that of the meterstick? Explain. **(b)** Find the mass of the necklace.

38. •• **Maximum Overhang** Three identical, uniform books of length L are stacked one on top the other. Find the maximum overhang distance d in **FIGURE 11-49** such that the books do not fall over.

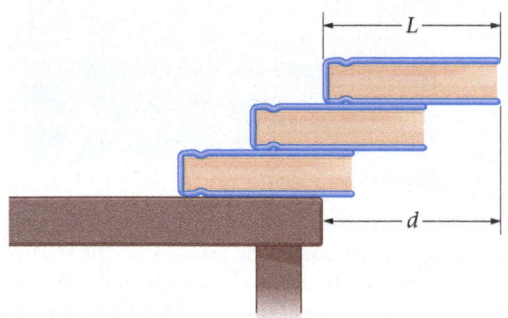

FIGURE 11-49 Problem 38

39. •• A baseball bat balances 71.1 cm from one end. If a 0.560-kg glove is attached to that end, the balance point moves 24.7 cm toward the glove. Find the mass of the bat.

SECTION 11-5 DYNAMIC APPLICATIONS OF TORQUE

40. •• A 2.85-kg bucket is attached to a rope wrapped around a disk-shaped pulley of radius 0.121 m and mass 0.742 kg. If the bucket is allowed to fall, **(a)** what is its linear acceleration? **(b)** What is the angular acceleration of the pulley? **(c)** How far does the bucket drop in 1.50 s?

41. •• A child exerts a tangential 53.4-N force on the rim of a disk-shaped merry-go-round with a radius of 2.29 m. If the merry-go-round starts at rest and acquires an angular speed of 0.125 rev/s in 3.50 s, what is its mass?

42. •• Predict/Calculate You pull downward with a force of 28 N on a rope that passes over a disk-shaped pulley of mass 1.2 kg and radius 0.075 m. The other end of the rope is attached to a 0.67-kg mass. **(a)** Is the tension in the rope the same on both sides of the pulley? If not, which side has the greater tension? **(b)** Find the tension in the rope on both sides of the pulley.

43. ••• One elevator arrangement includes the passenger car, a counterweight, and two large pulleys, as shown in **FIGURE 11-50**. Each pulley has a radius of 1.2 m and a moment of inertia of 380 kg · m². The top pulley is driven by a motor. The elevator car plus passengers has a mass of 3100 kg, and the counterweight has a mass of 2700 kg. If the motor is to accelerate the elevator car upward at 1.8 m/s², how much torque must it generate? *Hint:* The two pulleys move together, so you can model them as a single pulley with the sum of the moments of inertia.

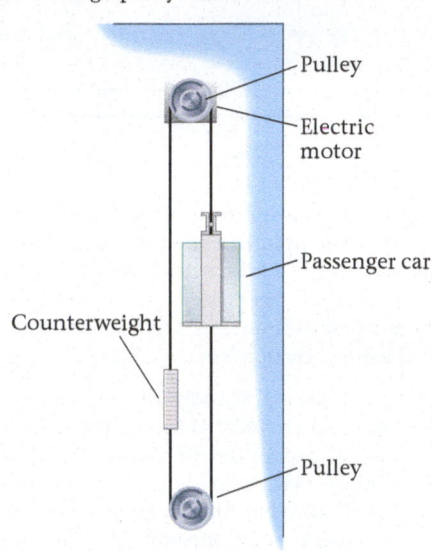

FIGURE 11-50 Problem 43

44. ••• **Atwood's Machine** An Atwood's machine consists of two masses, m_1 and m_2, connected by a string that passes over a pulley. If the pulley is a disk of radius R and mass M, find the acceleration of the masses.

SECTION 11-6 ANGULAR MOMENTUM

45. • A 1.4-kg bicycle tire with a radius of 33 cm rotates with an angular speed of 155 rpm. Find the angular momentum of the tire, assuming it can be modeled as a hoop.

46. • Jogger 1 in **FIGURE 11-51** has a mass of 65.3 kg and runs in a straight line with a speed of 3.35 m/s. **(a)** What is the magnitude of the jogger's linear momentum? **(b)** What is the magnitude of the jogger's angular momentum with respect to the origin, O?

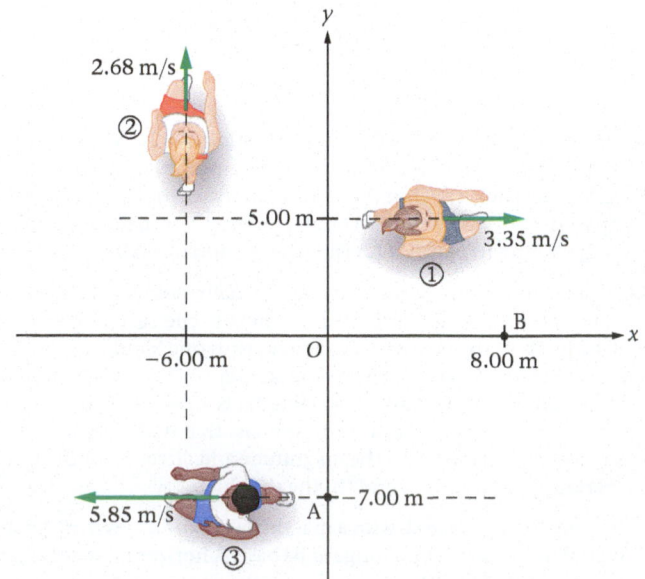

FIGURE 11-51 Problems 46 and 47

47. •• Predict/Calculate Suppose jogger 3 in Figure 11-51 has a mass of 62.2 kg and a speed of 5.85 m/s. **(a)** Is the magnitude of the jogger's angular momentum greater with respect to point A or point B? Explain. **(b)** Is the magnitude of the jogger's angular momentum with respect to point B greater than, less than, or the same as it is with respect to the origin, *O*? Explain. **(c)** Calculate the magnitude of the jogger's angular momentum with respect to points A, B, and *O*.

48. •• A torque of 0.12 N·m is applied to an egg beater. **(a)** If the egg beater starts at rest, what is its angular momentum after 0.65 s? **(b)** If the moment of inertia of the egg beater is 2.5×10^{-3} kg·m^2, what is its angular speed after 0.65 s?

49. •• A windmill has an initial angular momentum of 8500 kg·m^2/s. The wind picks up, and 5.86 s later the windmill's angular momentum is 9700 kg·m^2/s. What was the torque acting on the windmill, assuming it was constant during this time?

50. •• Two gerbils run in place with a linear speed of 0.55 m/s on an exercise wheel that is shaped like a hoop. Find the angular momentum of the system if each gerbil has a mass of 0.22 kg and the exercise wheel has a radius of 9.5 cm and a mass of 5.0 g.

SECTION 11-7 CONSERVATION OF ANGULAR MOMENTUM

51. • **CE** Predict/Explain A student rotates on a frictionless piano stool with his arms outstretched, a heavy weight in each hand. Suddenly he lets go of the weights, and they fall to the floor. As a result, does the student's angular speed increase, decrease, or stay the same? **(b)** Choose the *best explanation* from among the following:
  I. The loss of angular momentum when the weights are dropped causes the student to rotate more slowly.
  II. The student's moment of inertia is decreased by dropping the weights.
  III. Dropping the weights exerts no torque on the student, but the floor exerts a torque on the weights when they land.

52. • **CE** A puck on a horizontal, frictionless surface is attached to a string that passes through a hole in the surface, as shown in FIGURE 11-52. As the puck rotates about the hole, the string is pulled downward, bringing the puck closer to the hole. During this process, do the puck's **(a)** linear speed, **(b)** angular speed, and **(c)** angular momentum increase, decrease, or stay the same?

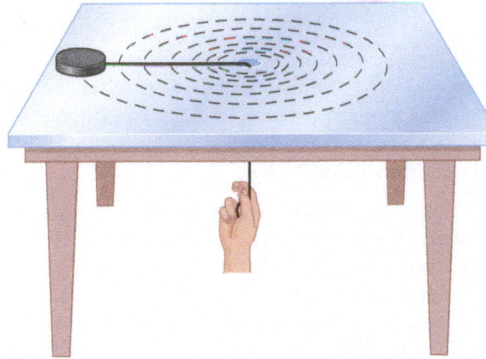

FIGURE 11-52 Problems 52 and 75

53. • **CE** A puck on a horizontal, frictionless surface is attached to a string that wraps around a pole of finite radius, as shown in FIGURE 11-53. **(a)** As the puck moves along the spiral path, does its speed increase, decrease, or stay the same? Explain. **(b)** Does its angular momentum increase, decrease, or stay the same? Explain.

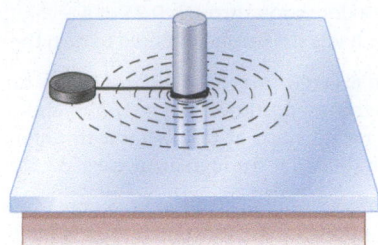

FIGURE 11-53 Problem 53

54. • As an ice skater begins a spin, his angular speed is 3.04 rad/s. After pulling in his arms, his angular speed increases to 5.24 rad/s. Find the ratio of the skater's final moment of inertia to his initial moment of inertia.

55. •• A disk-shaped merry-go-round of radius 2.63 m and mass 155 kg rotates freely with an angular speed of 0.641 rev/s. A 59.4-kg person running tangential to the rim of the merry-go-round at 3.41 m/s jumps onto its rim and holds on. Before jumping on the merry-go-round, the person was moving in the same direction as the merry-go-round's rim. What is the final angular speed of the merry-go-round?

56. •• A student sits at rest on a piano stool that can rotate without friction. The moment of inertia of the student-stool system is 4.1 kg·m^2. A second student tosses a 1.5-kg mass with a speed of 2.7 m/s to the student on the stool, who catches it at a distance of 0.40 m from the axis of rotation. What is the resulting angular speed of the student and the stool?

57. •• Predict/Calculate A turntable with a moment of inertia of 5.4×10^{-3} kg·m^2 rotates freely with an angular speed of $33\frac{1}{3}$ rpm. Riding on the rim of the turntable, 15 cm from the center, is a cute, 32-g mouse. **(a)** If the mouse walks to the center of the turntable, will the turntable rotate faster, slower, or at the same rate? Explain. **(b)** Calculate the angular speed of the turntable when the mouse reaches the center.

58. •• A student on a piano stool rotates freely with an angular speed of 2.95 rev/s. The student holds a 1.25-kg mass in each outstretched arm, 0.759 m from the axis of rotation. The combined moment of inertia of the student and the stool, ignoring the two masses, is 5.43 kg·m^2, a value that remains constant. **(a)** As the student pulls his arms inward, his angular speed increases to 3.54 rev/s. How far are the masses from the axis of rotation at this time, considering the masses to be points? **(b)** Calculate the initial and final kinetic energies of the system.

59. ••• **Walking on a Merry-Go-Round** A child of mass *m* stands at rest near the rim of a stationary merry-go-round of radius *R* and moment of inertia *I*. The child now begins to walk around the circumference of the merry-go-round with a tangential speed *v* with respect to the merry-go-round's surface. **(a)** What is the child's speed with respect to the ground? Check your result in the limits **(b)** $I \rightarrow 0$ and **(c)** $I \rightarrow \infty$.

SECTION 11-8 ROTATIONAL WORK AND POWER

60. • **CE** Predict/Explain Two spheres of equal mass and radius are rolling across the floor with the same speed. Sphere 1 is a uniform solid; sphere 2 is hollow. Is the work required to stop sphere 1 greater than, less than, or equal to the work required to stop sphere 2? **(b)** Choose the *best explanation* from among the following:
  I. Sphere 2 has the greater moment of inertia and hence the greater rotational kinetic energy.
  II. The spheres have equal mass and speed; therefore, they have the same kinetic energy.
  III. The hollow sphere has less kinetic energy.

61. • Turning a doorknob through 0.25 of a revolution requires 0.14 J of work. What is the torque required to turn the doorknob?

62. • A person exerts a tangential force of 36.1 N on the rim of a disk-shaped merry-go-round of radius 2.74 m and mass 167 kg. If the merry-go-round starts at rest, what is its angular speed after the person has rotated it through an angle of 32.5°?

63. • To prepare homemade ice cream, a crank must be turned with a torque of 3.95 N·m. How much work is required for each complete turn of the crank?

64. • **Power of a Dental Drill** A popular make of dental drill can operate at a speed of 42,500 rpm while producing a torque of 3.68 oz·in. What is the power output of this drill? Give your answer in watts.

65. •• For a home repair job you must turn the handle of a screwdriver 27 times. **(a)** If you apply an average force of 13 N tangentially to the 3.0-cm-diameter handle, how much work have you done? **(b)** If you complete the job in 24 seconds, what was your average power output?

66. •• The L-shaped object in Figure 11-40 consists of three masses connected by light rods. Find the work that must be done on this object to accelerate it from rest to an angular speed of 2.35 rad/s about **(a)** the x axis, **(b)** the y axis, and **(c)** the z axis (which is through the origin and perpendicular to the page).

67. •• The rectangular object in Figure 11-41 consists of four masses connected by light rods. What power must be applied to this object to accelerate it from rest to an angular speed of 2.5 rad/s in 6.4 s about **(a)** the x axis, **(b)** the y axis, and **(c)** the z axis (which is through the origin and perpendicular to the page)?

68. •• **Predict/Calculate** A circular saw blade accelerates from rest to an angular speed of 3620 rpm in 6.30 revolutions. **(a)** Find the torque exerted on the saw blade, assuming it is a disk of radius 15.2 cm and mass 0.755 kg. **(b)** Is the angular speed of the saw blade after 3.15 revolutions greater than, less than, or equal to 1810 rpm? Explain. **(c)** Find the angular speed of the blade after 3.15 revolutions.

GENERAL PROBLEMS

69. • **CE** A uniform disk stands upright on its edge, and rests on a sheet of paper placed on a tabletop. If the paper is pulled horizontally to the right, as in **FIGURE 11-54**, **(a)** does the disk rotate clockwise or counterclockwise about its center? Explain. **(b)** Does the center of the disk move to the right, move to the left, or stay in the same location? Explain.

FIGURE 11-54 Problem 69

70. • **CE** Consider the two rotating systems shown in **FIGURE 11-55**, each consisting of a mass m attached to a rod of negligible mass pivoted at one end. On the left, the mass is attached at the midpoint of the rod; to the right, it is attached to the free end of the rod. The rods are released from rest in the horizontal position at the same time. When the rod to the left reaches the vertical position, is the rod to the right not yet vertical (location A), vertical (location B), or past vertical (location C)? Explain.

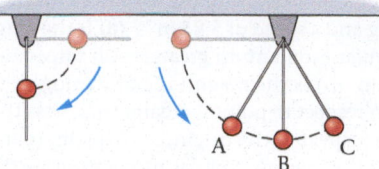

FIGURE 11-55 Problem 70

71. • **CE Predict/Explain** A disk and a hoop (bicycle wheel) of equal radius and mass each have a string wrapped around their circumferences. Hanging from the strings, halfway between the disk and the hoop, is a block of mass m, as shown in **FIGURE 11-56**. The disk and the hoop are free to rotate about their centers. When the block is allowed to fall, does it stay on the center line, move toward the right, or move toward the left? **(b)** Choose the *best explanation* from among the following:

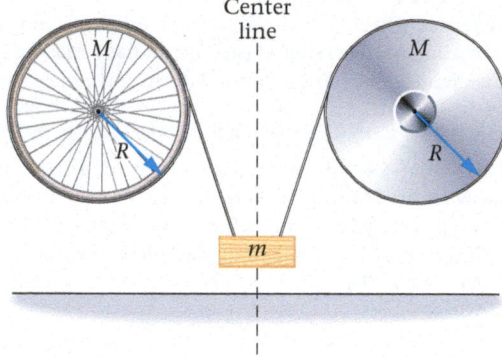

FIGURE 11-56 Problem 71

   I. The disk is harder to rotate, and hence its angular acceleration is less than that of the wheel.

  II. The wheel has the greater moment of inertia and unwinds more slowly than the disk.

 III. The system is symmetric, with equal mass and radius on either side.

72. • **CE** A beetle sits at the rim of a turntable that is at rest but is free to rotate about a vertical axis. Suppose the beetle now begins to walk around the perimeter of the turntable. Does the beetle move forward, backward, or does it remain in the same location relative to the ground? Answer for two different cases, **(a)** the turntable is much more massive than the beetle and **(b)** the turntable is massless.

73. • After getting a drink of water, a hamster jumps onto an exercise wheel for a run. A few seconds later the hamster is running in place with a speed of 1.3 m/s. Find the work done by the hamster to get the exercise wheel moving, assuming it is a hoop of radius 0.13 m and mass 6.5 g.

74. •• A 47.0-kg uniform rod 4.25 m long is attached to a wall with a hinge at one end. The rod is held in a horizontal position by a wire attached to its other end. The wire makes an angle of 30.0° with the horizontal, and is bolted to the wall directly above the hinge. If the wire can withstand a maximum tension of 1450 N before breaking, how far from the wall can a 68.0-kg person sit without breaking the wire?

75. •• **Predict/Calculate** A puck attached to a string moves in a circular path on a frictionless surface, as shown in Figure 11-52. Initially, the speed of the puck is v and the radius of the circle is r. If the string passes through a hole in the surface, and is pulled downward until the radius of the circular path is $r/2$, (a) does the speed of the puck increase, decrease, or stay the same? (b) Calculate the final speed of the puck.

76. •• **BIO** **The Masseter Muscle** The masseter muscle, the principal muscle for chewing, is one of the strongest muscles for its size in the human body. It originates on the lower edge of the zygomatic arch (cheekbone) and inserts in the angle of the mandible. Referring to the lower diagram in **FIGURE 11-57**, where $d = 7.60\,\text{cm}$, $h = 3.15\,\text{cm}$ and $D = 10.85\,\text{cm}$, (a) find the torque produced about the axis of rotation by the masseter muscle. The force exerted by the masseter muscle is $F_\text{M} = 455\,\text{N}$. (b) Find the biting force, F_B, exerted on the mandible by the upper teeth. Find (c) the horizontal and (d) the vertical component of the force F_J exerted on the mandible at the joint where it attaches to the skull. Assume that the mandible is in static equilibrium, and that upward is the positive vertical direction.

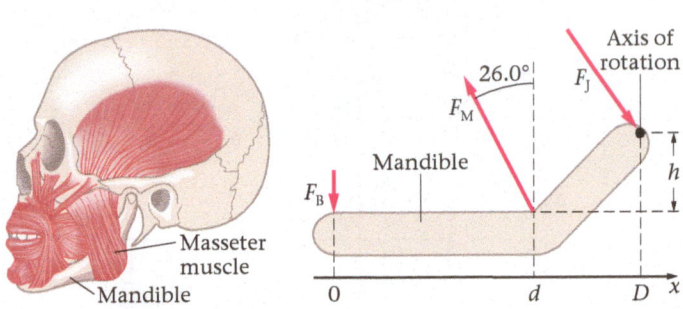

FIGURE 11-57 Problem 76

77. •• **Exercising the Biceps** You are designing exercise equipment to operate as shown in **FIGURE 11-58**, where a person pulls upward on an elastic cord. The cord behaves like an ideal spring and has an unstretched length of 31 cm. If you would like the torque about the elbow joint to be $81\,\text{N}\cdot\text{m}$ in the position shown, what force constant, k, is required for the cord?

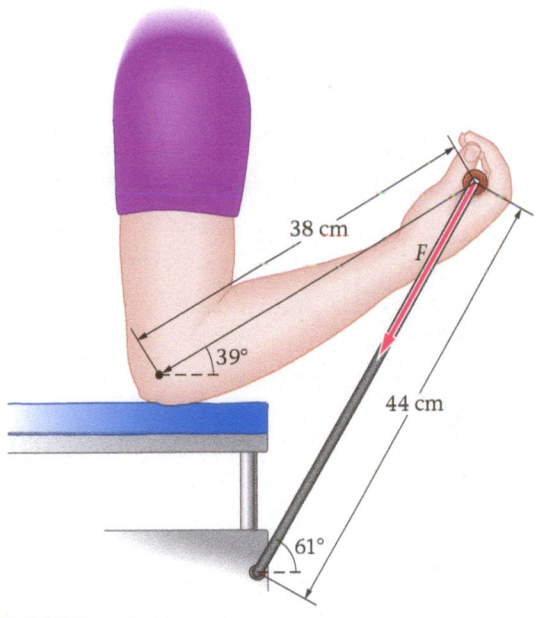

FIGURE 11-58 Problem 77

78. •• **Balancing a *T. rex*** Paleontologists believe that *Tyrannosaurus rex* stood and walked with its spine almost horizontal, as indicated in

FIGURE 11-59, and that its tail was held off the ground to balance its upper body about the hip joint. Given that the total mass of *T. rex* was 5400 kg, and that the placement of the center of mass of the tail and the upper body was as shown in Figure 11-59, find the mass of the tail required for balance.

FIGURE 11-59 Problem 78

79. •• In Example 11-11, suppose the ladder is uniform, 4.0 m long, and weighs 60.0 N. Find the forces exerted on the ladder when the person is (a) halfway up the ladder and (b) three-fourths of the way up the ladder.

80. •• When you arrive at Duke's Dude Ranch, you are greeted by the large wooden sign shown in **FIGURE 11-60**. The left end of the sign is held in place by a bolt, the right end is tied to a rope that makes an angle of $20.0°$ with the horizontal. If the sign is uniform, 3.20 m long, and has a mass of 16.0 kg, what are (a) the tension in the rope, and (b) the horizontal and vertical components of the force, $\vec{\textbf{F}}$, exerted by the bolt?

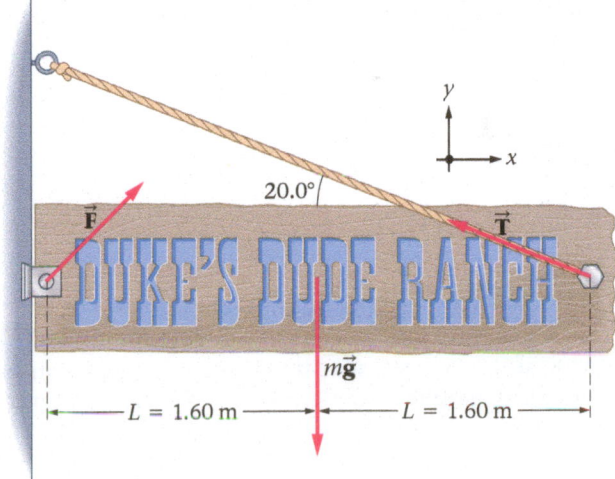

FIGURE 11-60 Problem 80

81. •• A 67.0-kg person stands on a lightweight diving board supported by two pillars, one at the end of the board, the other 1.10 m away. The pillar at the end of the board exerts a downward force of 828 N. (a) How far from that pillar is the person standing? (b) Find the force exerted by the second pillar.

82. •• **Flats Versus Heels** A woman might wear a pair of flat shoes to work during the day, as in **FIGURE 11-61 (a)**, but a pair of high heels, **FIGURE 11-61 (b)**, when going out for the evening. Assume that each foot supports half her weight, $w = W/2 = 279\,\text{N}$, and that the forces exerted by the floor on her feet occur at the points A and

B in both figures. Find the forces F_A (point A) and F_B (point B) for **(a)** flat shoes and **(b)** high heels. **(c)** How have the high heels changed the weight distribution between the woman's heels and toes?

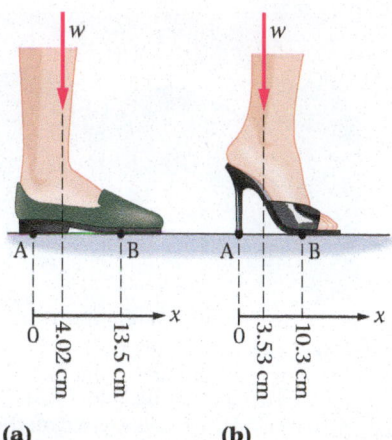

FIGURE 11-61 Problem 82

83. •• **BIO** A young girl sits at the edge of a dock by the bay, dipping her feet in the water. At the instant shown in **FIGURE 11-62**, she holds her lower leg stationary with her quadriceps muscle at an angle of 39° with respect to the horizontal. Use the information given in the figure, plus the fact that her lower leg has a mass of 3.4 kg, to determine the magnitude of the force, F_Q, exerted on the lower leg by the quadriceps.

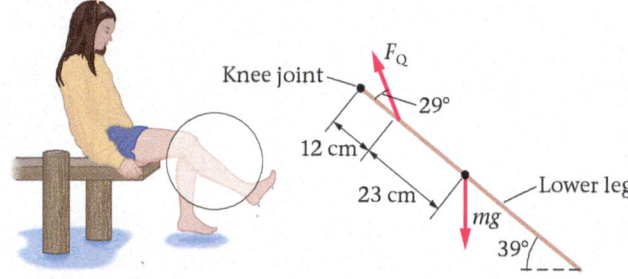

FIGURE 11-62 Problem 83

84. •• **BIO Deltoid Muscle** A crossing guard holds a STOP sign at arm's length, as shown in **FIGURE 11-63**. Her arm is horizontal, and we assume that the deltoid muscle is the only muscle supporting her arm. The weight of her upper arm is $W_u = 18$ N, the weight of her lower arm is $W_l = 11$ N, the weight of her hand is $W_h = 4.0$ N, and the weight of the sign is $W_s = 8.9$ N. The location where each of these forces acts on the arm is indicated in the figure. A force of magnitude f_d is exerted on the humerus by the deltoid, and the shoulder joint exerts a force on the humerus with horizontal and vertical components given by f_x and f_y, respectively. **(a)** Is the magnitude of f_d greater than, less than, or equal to the magnitude of f_x? Explain. Find **(b)** f_d, **(c)** f_x, and **(d)** f_y. (The weights in Figure 11-63 are drawn to scale; the unknown forces are to be determined. If a force is found to be negative, its direction is opposite to that shown.)

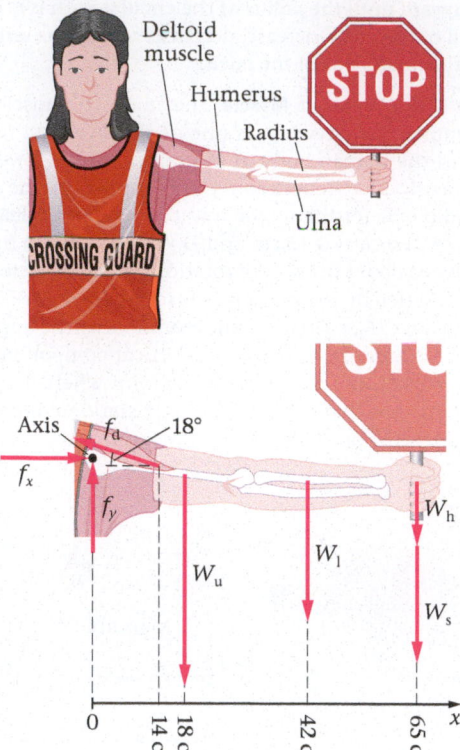

Free-Body Diagram of the Arm

FIGURE 11-63 Problem 84

85. •• **BIO Triceps** To determine the force a person's triceps muscle can exert, a doctor uses the procedure shown in **FIGURE 11-64**, where the patient pushes down with the palm of his hand on a force meter. Given that the weight of the lower arm is $Mg = 15.6$ N, and that the force meter reads $F = 89.0$ N, what is the force F_T exerted vertically upward by the triceps?

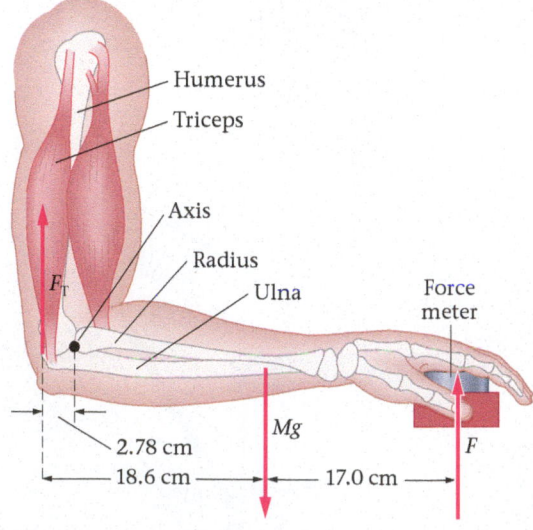

FIGURE 11-64 Problem 85

86. •• **Predict/Calculate** Suppose partial melting of the polar ice caps increases the moment of inertia of the Earth from $0.331\,M_E R_E^2$ to $0.332\,M_E R_E^2$. **(a)** Would the length of a day (the time required for the Earth to complete one revolution about its axis) increase or decrease? Explain. **(b)** Calculate the change in the length of a day. Give your answer in seconds.

87. ••• A bicycle wheel of radius R and mass M is at rest against a step of height $3R/4$, as illustrated in **FIGURE 11-65**. Find the minimum horizontal force F that must be applied to the axle to make the wheel start to rise up over the step.

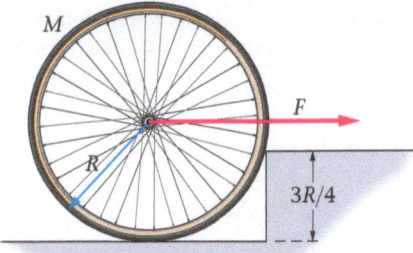

FIGURE 11-65 Problem 87

88. ••• A 0.101-kg yo-yo has an outer radius R that is 5.60 times greater than the radius r of its axle. The yo-yo is in equilibrium if a mass m is suspended from its outer edge, as shown in **FIGURE 11-66**. Find the tension in the two strings, T_1 and T_2, and the mass m.

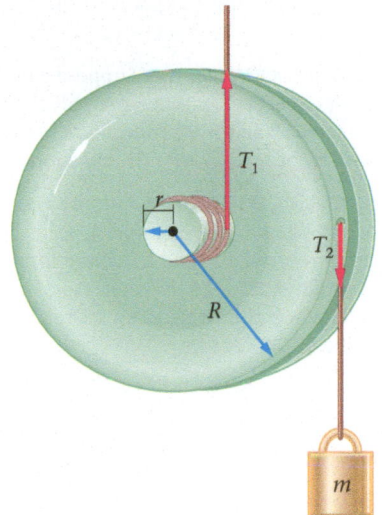

FIGURE 11-66 Problem 88

89. ••• **BIO Peak Pedaling Torque** The downward force produced by the quadriceps muscles during the power stroke of bicycle-pedaling motion is shown in **FIGURE 11-67** as a function of the crank angle ϕ (see Figure 11-39). The force from these muscles decreases linearly, but the torque depends on the crank angle according to $\tau = rF\sin(180° - \phi)$. Use the information in the graph, together with a computer spreadsheet, to find the angle ϕ at which the pedaling torque produced by the quadriceps muscle is a maximum. (Note that the actual torque applied to the crank is a result of the action of many muscles in addition to the quadriceps.)

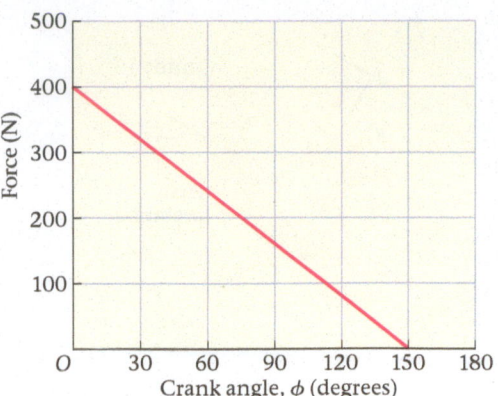

FIGURE 11-67 Problem 89

90. ••• A cylinder of mass m and radius r has a string wrapped around its circumference. The upper end of the string is held fixed, and the cylinder is allowed to fall. Show that its linear acceleration is $(2/3)g$.

91. ••• **Bricks in Equilibrium** Consider a system of four uniform bricks of length L stacked as shown in **FIGURE 11-68**. What is the maximum distance, x, that the middle bricks can be displaced outward before they begin to tip?

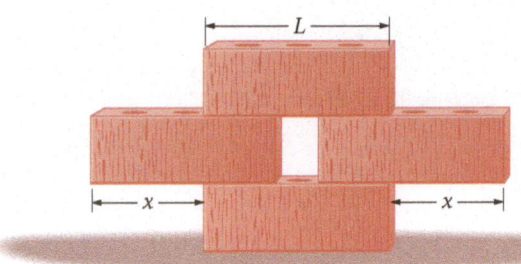

FIGURE 11-68 Problem 91

PASSAGE PROBLEMS

BIO Correcting Torsiversion

Torsiversion is a medical condition in which a tooth is rotated away from its normal position about the long axis of the root. Studies show that about 2 percent of the population suffer from this condition to some degree. For those who do, the improper alignment of the tooth can lead to tooth-to-tooth collisions during eating, as well as other problems. Typical patients display a rotation ranging from 20° to 60°, with an average around 30°.

An example is shown in **FIGURE 11-69 (a)**, where the first premolar is not only displaced slightly from its proper location in the negative y direction, but also rotated clockwise from its normal orientation. To correct this condition, an orthodontist might use an archwire and a bracket to apply both a force and a torque to the tooth. In the simplest case, two forces are applied to the tooth in different locations, as indicated by F_1 and F_2 in Figure 11-69 (a). These two forces, if chosen properly, can reposition the tooth by exerting a net force in the positive y direction, and also reorient it by applying a torque in the counterclockwise direction.

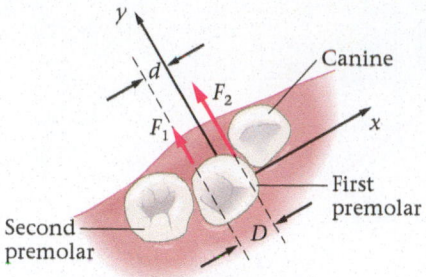

(a)

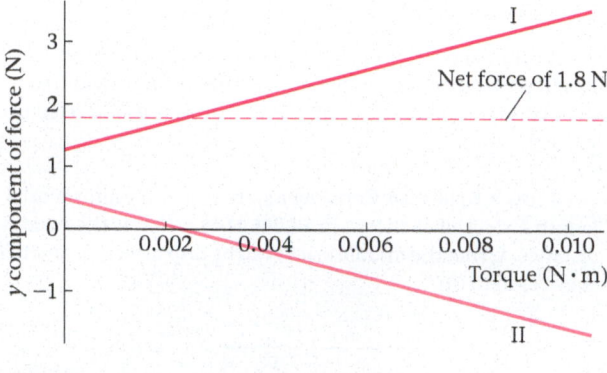

(b)

FIGURE 11-69 Problems 92, 93, 94, and 95

In a typical case, it may be desired to have a net force in the positive y direction of 1.8 N. In addition, the distances in Figure 11-69 (a) can be taken to be $d = 3.2$ mm and $D = 4.5$ mm. Given these conditions, a range of torques is possible for various values of the y components of the forces, F_{1y} and F_{2y}. For example, **FIGURE 11-69 (b)** shows the values of F_{1y} and F_{2y} necessary to produce a given torque, where the torque is measured about the center of the tooth (which is also the origin of the coordinate system). Notice that the two forces always add to 1.8 N in the positive y direction, though one of the forces changes sign as the torque is increased.

92. •• The two solid lines in Figure 11-69 (b) represent the two forces applied to the tooth. Which of the following statements about the two lines is correct?

A. Line I = F_{1y}, line II = F_{2y}.
B. The force indicated by line I changes sign.
C. Line I = F_{2y}, line II = F_{1y}.
D. The magnitude of the force indicated by line II is greater than the magnitude of the force indicated by line I.

93. • What is the value of the torque that corresponds to one of the forces being equal to zero?

A. 0.0023 N·m   B. 0.0058 N·m
C. 0.0081 N·m   D. 0.017 N·m

94. •• Find the values of F_{1y} and F_{2y} required to give zero net torque.

A. $F_{1y} = -1.2$ N, $F_{2y} = 3.0$ N
B. $F_{1y} = 1.1$ N, $F_{2y} = 0.75$ N
C. $F_{1y} = -0.73$ N, $F_{2y} = 2.5$ N
D. $F_{1y} = 0.52$ N, $F_{2y} = 1.3$ N

95. •• Find the values of F_{1y} and F_{2y} required to give a net torque of 0.0099 N·m. This is a torque that would be effective at rotating the tooth.

A. $F_{1y} = -1.7$ N, $F_{2y} = 3.5$ N
B. $F_{1y} = -3.8$ N, $F_{2y} = 5.6$ N
C. $F_{1y} = -0.23$ N, $F_{2y} = 2.0$ N
D. $F_{1y} = 4.0$ N, $F_{2y} = -2.2$ N

96. •• **REFERRING TO EXAMPLE 11-14** Suppose the mass of the pulley is doubled, to 0.160 kg, and that everything else in the system remains the same. **(a)** Do you expect the value of T_2 to increase, decrease, or stay the same? Explain. **(b)** Calculate the value of T_2 for this case.

97. •• **REFERRING TO EXAMPLE 11-14** Suppose the mass of the cart is doubled, to 0.62 kg, and that everything else in the system remains the same. **(a)** Do you expect the value of T_2 to increase, decrease, or stay the same? Explain. **(b)** Calculate the value of T_2 for this case.

98. •• **REFERRING TO QUICK EXAMPLE 11-22** Suppose the child runs with a different initial speed, but that everything else in the system remains the same. What initial speed does the child have if the angular speed of the system after the collision is 0.425 rad/s?

99. •• **REFERRING TO QUICK EXAMPLE 11-22** Suppose everything in the system is as described in Quick Example 11-22 except that the child approaches the merry-go-round in a direction that is not tangential. Find the angle θ between the direction of motion and the outward radial direction that is required if the final angular speed of the system is to be 0.272 rad/s.

Big Ideas

Gravity

1 Newton's law of universal gravitation states that all objects exert gravitational forces on all other objects.

2 The gravitational force of a spherical mass is the same as if all the mass were concentrated at the center of the sphere.

3 The orbits of planets, moons, comets, and asteroids are governed by Kepler's three laws.

4 Gravitational potential energy is a form of energy that must be taken into account when we apply energy conservation.

▲ Visitors from outer space have had significant impact on the Earth. For example, the Pingualuit crater in Ontario shows the result of a violent collision between the Earth and a rogue meteor or comet traveling many times faster than a bullet. But why do meteorites always seem to hit the Earth with such great speed? In this chapter we show that energy conservation, when applied to planetary bodies like the Earth, always implies a large impact speed, even when the alien object starts at rest far from the Earth.

The study of gravity has always been a central theme in physics—from Galileo's early experiments on free fall, to Einstein's general theory of relativity and Stephen Hawking's work on black holes. Perhaps the grandest milestone in this endeavor, however, was the discovery by Newton of the **law of universal gravitation.** With just one simple equation to describe the force of gravity, Newton was able to determine the orbits of planets, moons, and comets, and to explain such earthly phenomena as the tides and the fall of an apple.

12-1 Newton's Law of Universal Gravitation

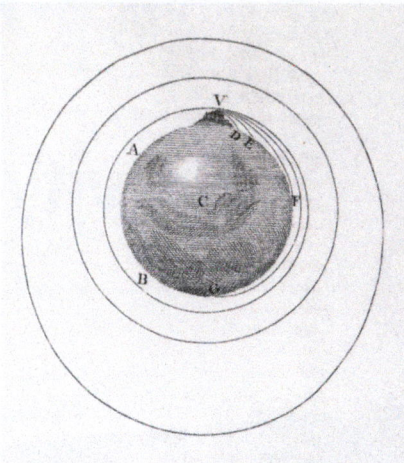

▲ **FIGURE 12-1** In this illustration from his great work, the *Principia*, published in 1687, Newton presents a "thought experiment" to show the connection between free fall and orbital motion. Imagine throwing a projectile horizontally from the top of a mountain. The greater the initial speed of the projectile, the farther it travels in free fall before it strikes the ground. In the absence of air resistance, a great enough initial speed could result in the projectile circling the Earth and returning to its starting point. Thus, an object orbiting the Earth is actually in free fall—it simply has a large horizontal speed.

Big Idea 1 Newton's law of universal gravitation states that all objects in the universe exert gravitational forces on all other objects. The magnitude of the force between two objects is proportional to the product of their masses, and inversely proportional to the square of the distance between them. The force is always attractive, and along the line connecting the two objects.

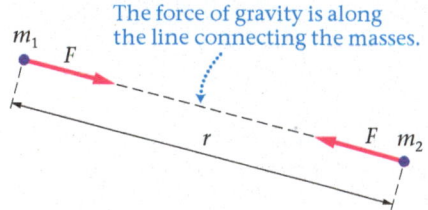

The force of gravity is along the line connecting the masses.

▲ **FIGURE 12-2 Gravitational force between point masses** Two point masses, m_1 and m_2, separated by a distance r exert equal and opposite attractive forces on one another. The magnitude of the forces, F, is given by Equation 12-1.

It's ironic, but the first fundamental force of nature to be recognized as such, **gravity**, is also the weakest of the fundamental forces. Still, it is the force most apparent to us in our everyday lives, and it is the force responsible for the motions of the Moon, the Earth, and the planets. Yet the connection between objects falling on Earth and planets moving in their orbits was not known before Newton.

Newton's Insight The flash of insight that came to Newton—whether it was due to seeing an apple fall to the ground or not—is simply this:

> The force that causes an apple to accelerate downward is the same force that causes the Moon to move in a circular path around the Earth.

To put it another way, Newton was the first to realize that the Moon is *constantly falling* toward the Earth, without getting closer to it, and that the Moon falls for the same reason that an apple falls. This is illustrated in a classic drawing by Newton, shown in **FIGURE 12-1**.

To be specific, in the case of the apple the motion is linear as it accelerates downward toward the center of the Earth. In the case of the Moon the motion is circular with constant speed. As discussed in Section 6-5, an object in uniform circular motion accelerates toward the center of the circle. It follows, therefore, that the Moon *also* accelerates toward the center of the Earth. In fact, the force responsible for the Moon's centripetal acceleration is the Earth's gravitational attraction, the same force responsible for the fall of the apple.

To describe the force of gravity, Newton proposed the following simple law:

Newton's Law of Universal Gravitation

The force of gravity between any two point objects of mass m_1 and m_2 is attractive and of magnitude

$$F = G\frac{m_1 m_2}{r^2} \qquad \text{12-1}$$

In this expression, r is the distance between the masses, and G is a constant referred to as the **universal gravitation constant.** Its value is

$$G = 6.67 \times 10^{-11} \, \text{N} \cdot \text{m}^2/\text{kg}^2 \qquad \text{12-2}$$

The force is directed along the line connecting the masses, as indicated in **FIGURE 12-2**.

Notice that each mass in Figure 12-2 experiences a force of the same magnitude, $F = Gm_1 m_2/r^2$, but acting in opposite directions. That is, the force of gravity between two objects forms an action-reaction pair.

Universal Gravitation According to Newton's law, all objects in the universe attract all other objects in the universe by way of the gravitational interaction. It is in this sense that the force law is called "universal." Thus, the net gravitational force acting on you is due not only to the planet on which you stand, which is certainly responsible for the majority of the net force, but also to people nearby, planets, and even stars in far-off galaxies. In short, everything in the universe "feels" everything else, thanks to gravity.

The fact that G is such a small number means that the force of gravity between everyday objects is imperceptibly small. This is shown in the following Exercise.

EXERCISE 12-1 THE FORCE OF GRAVITY

Two apples of mass $m_1 = 0.110$ kg and $m_2 = 0.220$ kg sit on a dining room table. Treating the apples as point objects, find the force of gravity between them when their separation is **(a)** 13.2 cm and **(b)** 132 cm.

REASONING AND SOLUTION

The force in each case can be found with Newton's law of universal gravitation (Equation 12-1): $F = Gm_1 m_2/r^2$.

CONTINUED

a. Substituting $m_1 = 0.110$ kg, $m_2 = 0.220$ kg, and $r = 0.132$ m into $F = Gm_1m_2/r^2$ yields

$$F = G\frac{m_1m_2}{r^2} = (6.67 \times 10^{-11}\,\text{N}\cdot\text{m}^2/\text{kg}^2)\frac{(0.110\,\text{kg})(0.220\,\text{kg})}{(0.132\,\text{m})^2} = 9.26 \times 10^{-11}\,\text{N}$$

b. Repeating the calculation for $r = 1.32$ m gives

$$F = G\frac{m_1m_2}{r^2} = (6.67 \times 10^{-11}\,\text{N}\cdot\text{m}^2/\text{kg}^2)\frac{(0.110\,\text{kg})(0.220\,\text{kg})}{(1.32\,\text{m})^2} = 9.26 \times 10^{-13}\,\text{N}$$

The forces found in Exercise 12-1 are imperceptibly small—hence we can pick up one apple and not feel a tug on it from the other apple. In general, gravitational forces are significant only when large masses, such as the Earth or the Moon, are involved. That's why the weight of each apple, which is the gravitational force between it and the Earth, is so much greater than the attractive force between the apples themselves. In addition, notice that the force experienced by each apple has exactly the same magnitude, even though one has twice the mass of the other.

Exercise 12-1 also illustrates how rapidly the force of gravity decreases with distance. In particular, because F varies as $1/r^2$, it is said to have an **inverse square dependence** on distance. Thus, for example, an increase in distance by a factor of 10 results in a decrease in the force by a factor of $10^2 = 100$, as the Example illustrates. A plot of the force of gravity versus distance is given in **FIGURE 12-3**. Notice that even though the force diminishes rapidly with distance, it never completely vanishes; thus, we say that gravity is a force of infinite range.

Vector Addition and Gravity If a given mass is acted on by gravitational interactions with a number of other masses, the net force acting on it is simply the vector sum of the individual forces. This property of gravity is referred to as **superposition.** As an example, superposition implies that the net gravitational force exerted on you at this moment is the vector sum of the force exerted by the Earth, plus the force exerted by the Moon, plus the force exerted by the Sun, and so on. The following Example illustrates superposition.

▶ **FIGURE 12-3 Dependence of the gravitational force on the separation distance, r** The $1/r^2$ dependence of the gravitational force means that it decreases rapidly with distance. Still, it never completely vanishes. For this reason, we say that gravity is a force of infinite range; that is, every mass in the universe experiences a nonzero force from every other mass in the universe, no matter how far away.

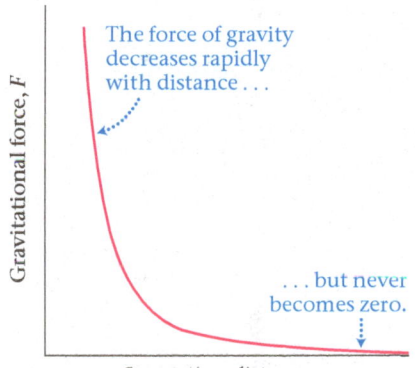

The force of gravity decreases rapidly with distance . . .

. . . but never becomes zero.

EXAMPLE 12-2   HOW MUCH FORCE IS WITH YOU?

As part of a daring rescue attempt, the *Millennium Eagle* passes between a pair of twin asteroids, as shown in the sketch. If the mass of the spaceship is 2.50×10^7 kg and the mass of each asteroid is 3.50×10^{11} kg, find the net gravitational force exerted on the *Millennium Eagle* **(a)** when it is at location A and **(b)** when it is at location B. Treat the spaceship and the asteroids as if they were point objects.

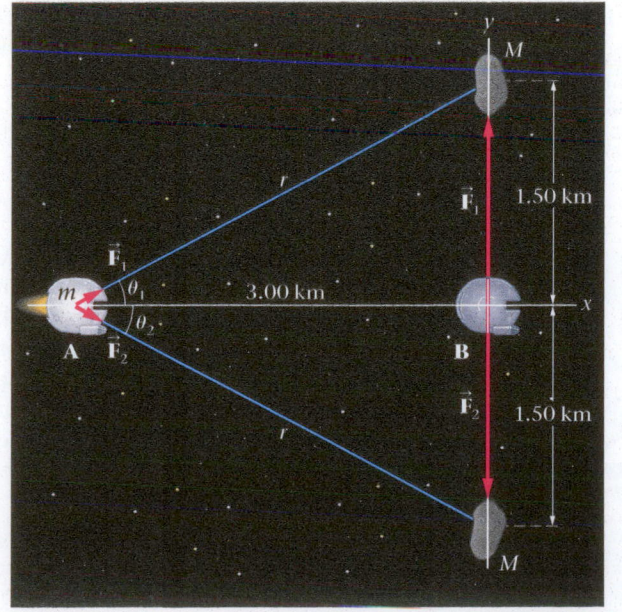

PICTURE THE PROBLEM
Our sketch shows the spaceship as it follows a path between the twin asteroids. The relevant distances and masses are indicated, as are the two points of interest, A and B. At location A the force $\vec{F}_1$ points above the x axis at the angle θ_1 (to be determined); the force $\vec{F}_2$ points below the x axis at the angle $\theta_2 = -\theta_1$, as can be seen by symmetry. At location B, the two forces act in opposite directions.

REASONING AND STRATEGY
To find the net gravitational force exerted on the spaceship, we first determine the magnitude of the force exerted on it by each asteroid. This is done by using $F = Gm_1m_2/r^2$ (Equation 12-1) and the distances given in our sketch. Next, we resolve these forces into x and y components. Finally, we sum the force components to find the net force.

CONTINUED

Known Mass of spaceship, $m = 2.50 \times 10^7$ kg; mass of each asteroid, $M = 3.50 \times 10^{11}$ kg; distances given in the sketch.

Unknown (a) Net force at location A, $\vec{F}_{net}$ = ? (b) Net force at location B, $\vec{F}_{net}$ = ?

SOLUTION

Part (a)

1. Use the Pythagorean theorem to find the distance r from point A to each asteroid. Also, refer to the sketch to find the angle between $\vec{F}_1$ and the x axis. The angle between $\vec{F}_2$ and the x axis has the same magnitude but the opposite sign:

$$r = \sqrt{(3.00 \times 10^3 \, \text{m})^2 + (1.50 \times 10^3 \, \text{m})^2} = 3350 \, \text{m}$$
$$\theta_1 = \tan^{-1}\left(\frac{1.50 \times 10^3 \, \text{m}}{3.00 \times 10^3 \, \text{m}}\right) = \tan^{-1}(0.500) = 26.6°$$
$$\theta_2 = -\theta_1 = -26.6°$$

2. Use r and $F = Gm_1m_2/r^2$ to calculate the magnitude of the forces $\vec{F}_1$ and $\vec{F}_2$ at point A:

$$F_1 = F_2 = G\frac{mM}{r^2}$$
$$= (6.67 \times 10^{-11} \, \text{N} \cdot \text{m}^2/\text{kg}^2)\frac{(2.50 \times 10^7 \, \text{kg})(3.50 \times 10^{11} \, \text{kg})}{(3350 \, \text{m})^2}$$
$$= 52.0 \, \text{N}$$

3. Use the values of θ_1 and θ_2 found in Step 1 to calculate the x and y components of $\vec{F}_1$ and $\vec{F}_2$:

$$F_{1,x} = F_1 \cos \theta_1 = (52.0 \, \text{N}) \cos 26.6° = 46.5 \, \text{N}$$
$$F_{1,y} = F_1 \sin \theta_1 = (52.0 \, \text{N}) \sin 26.6° = 23.3 \, \text{N}$$
$$F_{2,x} = F_2 \cos \theta_2 = (52.0 \, \text{N}) \cos(-26.6°) = 46.5 \, \text{N}$$
$$F_{2,y} = F_2 \sin \theta_2 = (52.0 \, \text{N}) \sin(-26.6°) = -23.3 \, \text{N}$$

4. Add the components of $\vec{F}_1$ and $\vec{F}_2$ to find the components of the net force. Write $\vec{F}_{net}$ in vector form:

$$F_x = F_{1,x} + F_{2,x} = 93.0 \, \text{N}$$
$$F_y = F_{1,y} + F_{2,y} = 0$$
$$\vec{F}_{net} = (93.0 \, \text{N})\hat{x}$$

Part (b)

5. Use $F = Gm_1m_2/r^2$ to find the magnitude of the forces exerted on the spaceship by the asteroids at location B:

$$F_1 = F_2 = G\frac{mM}{r^2}$$
$$= (6.67 \times 10^{-11} \, \text{N} \cdot \text{m}^2/\text{kg}^2)\frac{(2.50 \times 10^7 \, \text{kg})(3.50 \times 10^{11} \, \text{kg})}{(1.50 \times 10^3 \, \text{m})^2}$$
$$= 259 \, \text{N}$$

6. Use the fact that $\vec{F}_1$ and $\vec{F}_2$ have equal magnitudes and point in opposite directions to determine the net force, $\vec{F}_{net}$, acting on the spaceship:

$$\vec{F}_{net} = \vec{F}_1 + \vec{F}_2 = 0$$

INSIGHT

The net force at location A is in the positive x direction, as one would expect by symmetry. At location B, where the force exerted by each asteroid is about 5 times greater than it is at location A, the *net* force is zero because the attractive forces exerted by the two asteroids are equal and opposite, and thus cancel. Notice that the forces in our sketch have been drawn in correct proportion.

PRACTICE PROBLEM

Find the net gravitational force acting on the spaceship when it is at the location $x = 5.00 \times 10^3$ m and $y = 0$.
[**Answer:** The net force is 41.0 N in the negative x direction; that is, $\vec{F}_{net} = (-41.0 \, \text{N})\hat{x}$.]

Some related homework problems: Problem 8, Problem 10, Problem 11

BIO Although gravity is the weakest of the four fundamental forces of nature (the weak nuclear force, the electromagnetic force, and the strong nuclear force are discussed later in this book and summarized in Chapter 32), nearly every living organism on Earth has structures or features that help it adapt to an environment of uniform gravitation. For example, as plants grow, they tend to align their stems and roots along the direction of the gravitational field. This tendency, which is implemented by cellular elongation in response to differential concentrations of plant hormones known as auxins, is referred to as *gravitropism*. **FIGURE 12-4** illustrates the gravitropic response of a growing plant.

◄ **FIGURE 12-4** Plant stems grow vertically upward, opposite the force of gravity, because a concentration gradient of growth hormones called auxins develops in the stems when they are oriented away from the vertical.

Enhance Your Understanding

(Answers given at the end of the chapter)

1. Rank the four systems shown in **FIGURE 12-5** in order of increasing gravitational force exerted on each of the masses. Indicate ties where appropriate.

A m m

$\leftarrow r \rightarrow$

B m $2m$

$\leftarrow 2r \rightarrow$

C m $2m$

$\leftarrow r \rightarrow$

D $2m$ $2m$

$\leftarrow 2r \rightarrow$

▲ **FIGURE 12-5** Enhance Your Understanding 1.

Section Review

- The magnitude of the force of gravity between two masses m_1 and m_2 separated by a distance r is $F = Gm_1m_2/r^2$. In this expression, the universal gravitation constant has the value $G = 6.67 \times 10^{-11} \, \text{N} \cdot \text{m}^2/\text{kg}^2$.

- The force of gravity is always attractive, and points along the line connecting two masses. The net force due to multiple masses is given by vector summation.

PHYSICS IN CONTEXT
Looking Ahead

In Chapter 19, we shall see that the force between two electric charges has exactly the same mathematical form as the force of gravity between two masses. The equation describing the electric force is referred to as Coulomb's law.

12-2 Gravitational Attraction of Spherical Bodies

Newton's law of gravity applies to point objects. How, then, do we calculate the force of gravity for an object of finite size? In general, the approach is to divide the finite object into a collection of small mass elements, then use superposition and the methods of calculus to determine the net gravitational force. For an arbitrary shape, this calculation can be quite involved. For objects with a uniform spherical shape, however, the final result is remarkably simple, as was shown by Newton.

Uniform Sphere

Consider a uniform sphere of radius R and mass M, as in **FIGURE 12-6**. A point object of mass m is brought near the sphere, though still outside it at a distance r from its center. The object experiences a relatively strong attraction from mass near the point A, and a weaker attraction from mass near point B. In both cases the force is along the line connecting the mass m and the center of the sphere—that is, along the x axis. In addition, masses at the points C and D exert a net force that is also along the x axis—just as in the case of the twin asteroids in Example 12-2. Thus, the symmetry of the sphere guarantees that the net force it exerts on m is directed toward the sphere's center. The magnitude of the force exerted by the sphere must be calculated with the methods of calculus—which Newton invented and then applied to this problem. As a result of his calculations, Newton was able to show that **the net force exerted by the sphere on the mass m is the same as if all the mass of the sphere were concentrated at its center.** That is, the force between the mass m and the sphere of mass M has a magnitude F that is simply

$$F = G\frac{mM}{r^2}$$

12-3

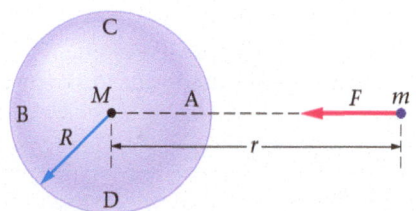

▲ **FIGURE 12-6 Gravitational force between a point mass and a sphere** The force is the same as if all the mass of the sphere were concentrated at its center.

The Force of Gravity on the Surface of the Earth   Let's apply this result to the case of a mass m on the surface of the Earth—assuming the Earth to be a uniform sphere of mass. If the mass of the Earth is M_E and its radius is R_E, it follows that the force exerted on m by the Earth is

$$F = G\frac{mM_E}{R_E{}^2} = m\left(\frac{GM_E}{R_E{}^2}\right)$$

We also know, however, that the gravitational force experienced by a mass m on the Earth's surface is simply $F = mg$, where g is the acceleration due to gravity. Therefore, we see that

$$m\left(\frac{GM_E}{R_E{}^2}\right) = mg$$

Big Idea 2   The gravitational force exerted by a spherical mass distribution is the same as if all the mass of the sphere were concentrated at its center.

Canceling the mass m, we find

$$g = \frac{GM_E}{R_E^2} = \frac{(6.67 \times 10^{-11}\,\text{N} \cdot \text{m}^2/\text{kg}^2)(5.97 \times 10^{24}\,\text{kg})}{(6.37 \times 10^6\,\text{m})^2} = 9.81\,\text{m/s}^2 \qquad 12\text{-}4$$

This result can be extended to objects above the Earth's surface, and hence farther from the center of the Earth, as we show in the next Example.

EXAMPLE 12-3   THE DEPENDENCE OF GRAVITY ON ALTITUDE

*RWP If you climb to the top of Mt. Everest, you will be about 5.50 mi above sea level. What is the acceleration due to gravity at this altitude?

PICTURE THE PROBLEM

At the top of the mountain, your distance from the center of the Earth is $r = R_E + h$, where $h = 5.50$ mi is the altitude.

REASONING AND STRATEGY

First, use $F = GmM_E/r^2$ to find the force due to gravity on the mountaintop. Then, set $F = mg_h$ to find the acceleration g_h at the height h.

Known   Height of Mt. Everest, $h = 5.50$ mi.
Unknown   Acceleration due to gravity at height h, $g_h = ?$

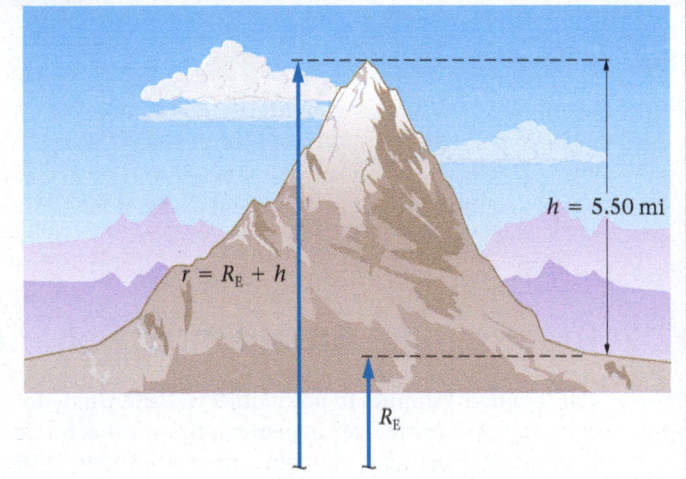

$h = 5.50$ mi

$r = R_E + h$

R_E

SOLUTION

1. Calculate the force F due to gravity at a height h above the Earth's surface:

$$F = G\frac{mM_E}{(R_E + h)^2}$$

2. Set F equal to mg_h and solve for g_h:

$$F = G\frac{mM_E}{(R_E + h)^2} = mg_h$$

$$g_h = G\frac{M_E}{(R_E + h)^2}$$

3. Factor out R_E^2 from the denominator, and use the fact that $GM_E/R_E^2 = g$:

$$g_h = \left(\frac{GM_E}{R_E^2}\right)\frac{1}{\left(1 + \dfrac{h}{R_E}\right)^2} = \frac{g}{\left(1 + \dfrac{h}{R_E}\right)^2}$$

4. Substitute numerical values, with $h = 5.50\,\text{mi} = (5.50\,\text{mi})(1609\,\text{m/mi}) = 8850\,\text{m}$ and $R_E = 6.37 \times 10^6\,\text{m}$:

$$g_h = \frac{g}{\left(1 + \dfrac{h}{R_E}\right)^2} = \frac{9.81\,\text{m/s}^2}{\left(1 + \dfrac{8850\,\text{m}}{6.37 \times 10^6\,\text{m}}\right)^2} = 9.78\,\text{m/s}^2$$

INSIGHT

As expected, the acceleration due to gravity is less as one moves farther from the center of the Earth. Thus, if you were to climb to the top of Mt. Everest, you would lose weight—not only because of the physical exertion required for the climb, but also because of the reduced gravity. In particular, a person with a mass of 60 kg (about 130 lb) would lose about half a pound of weight just by standing on the summit of the mountain.

A plot of g_h as a function of h is shown in **FIGURE 12-7 (a)**. The plot indicates the altitude of Mt. Everest and the orbit of the International Space Station. **FIGURE 12-7 (b)** shows g_h out to the orbits of communications and weather satellites, which are at an altitude of roughly 22,200 mi.

PRACTICE PROBLEM

Find the acceleration due to gravity at the altitude of the International Space Station's orbit, 322 km above the Earth's surface.   [**Answer:** $g_h = 8.89\,\text{m/s}^2$, a reduction of 9.38% compared to the acceleration of gravity on the surface of the Earth]

Some related homework problems: Problem 14, Problem 16

CONTINUED

_____

*Real World Physics applications are denoted by the acronym RWP.

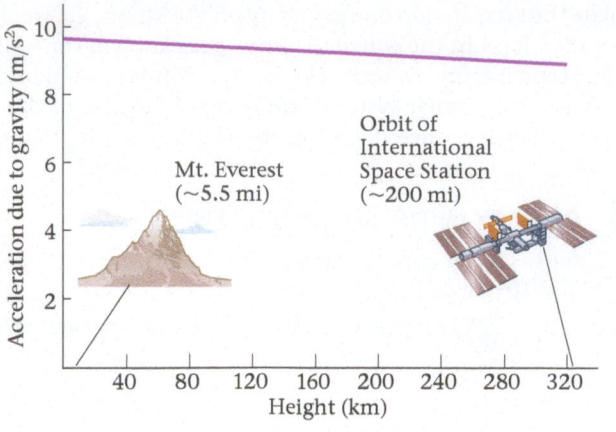
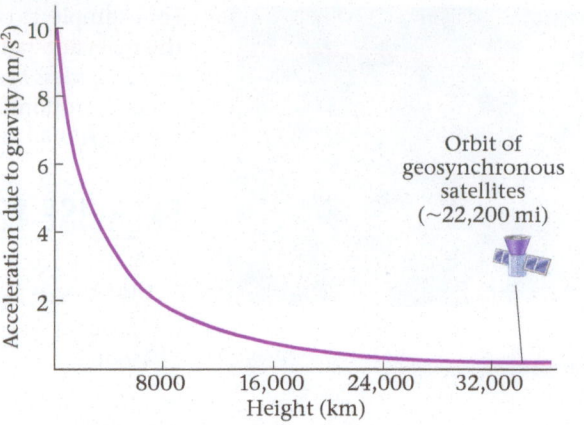

(a) Acceleration due to gravity near the Earth's surface

(b) Acceleration due to gravity far from the Earth

▲ **FIGURE 12-7** **The acceleration due to gravity at a height *h* above the Earth's surface** **(a)** In this plot, the peak of Mt. Everest is at about *h* = 5.50 mi, and the International Space Station orbit is at roughly *h* = 322 km. **(b)** This plot shows the decrease in the acceleration of gravity from the surface of the Earth to an altitude of about 25,000 mi. The orbit of geosynchronous satellites—ones that orbit above a fixed point on the Earth—is at roughly *h* = 22,200 mi.

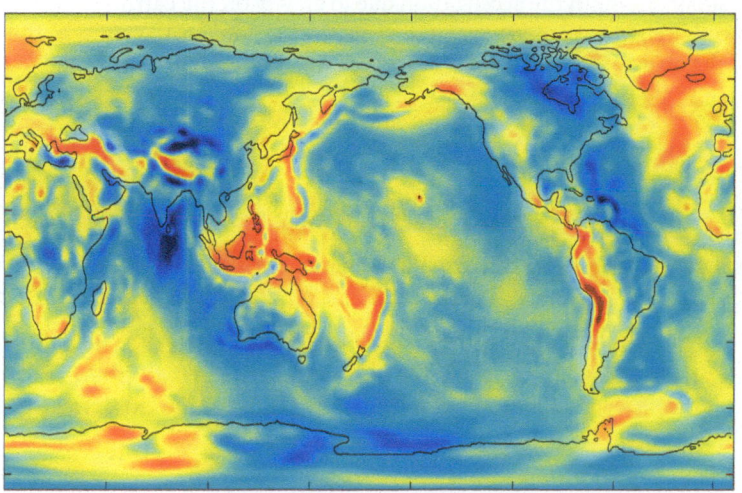

◄ **FIGURE 12-8** This global model of the Earth's gravitational strength was constructed from a combination of surface gravity measurements and satellite tracking data. It shows how the acceleration due to gravity varies from the value at an idealized "sea level" that takes into account the Earth's nonspherical shape. (The Earth is somewhat flattened at the poles—its radius is greatest at the equator.) Gravity is strongest in the red areas and weakest in the dark blue areas.

Of course, the Earth isn't actually a perfectly uniform sphere, though this is certainly a good first approximation. **FIGURE 12-8** shows the actual acceleration due to gravity on the Earth's surface. Variations are caused by the Earth's slightly nonspherical shape and nonuniform mass distribution.

RWP The local value of *g* can be measured with a *gravimeter*, the most sensitive of which can detect changes as small as $10^{-12}g$. A portable gravimeter is shown in **FIGURE 12-9 (a)**. Unusually dense materials nearby will slightly increase the value of *g* measured by a gravimeter, whereas a nearby deposit of low-density material will slightly decrease *g*. It follows that small variations in the acceleration of gravity can be used by those who search for buried oil and minerals, as well as those who study the motion of the Earth's crust. An orbiting gravimeter, the *Gravity Field and Steady-State Ocean Circulation Explorer* shown in **FIGURE 12-9 (b)**, recently completed a successful mission where it used small variations in *g* to map ocean currents, detect earthquakes, discover mantle plumes, and find remnants of ancient oceans and subduction zones.

Applications in the Solar System   The result $g = GM_E/R_E^2$ can be used to calculate the acceleration due to gravity on other objects in the solar system besides the Earth.

(a)

(b)

▶ **FIGURE 12-9** **(a)** A portable gravimeter. **(b)** An artist's conception of the European Space Agency's *Gravity Field and Steady-State Ocean Circulation Explorer* (GOCE) satellite, which has mapped Earth's gravitational field in unprecedented detail.

▲ **FIGURE 12-10 Visualizing Concepts**
Gravity in the Solar System (Top) The
weak lunar gravity permits astronauts,
even encumbered by their massive space
suits, to bound over the Moon's surface.
The low gravitational pull, only about one-
sixth that of Earth, is a consequence not
only of the Moon's smaller size, but also
of its lower average density. (Bottom) The
force of gravity on the surface of Mars is
only about 38% of its strength on Earth.
This was an important factor in designing
NASA's Phoenix Mars Lander, shown here
lifting a scoop of dirt on its 16th Martian
day after landing in May 2008.

For example, to calculate the acceleration due to gravity on the Moon, g_m, we simply use the mass and radius of the Moon in the equation. Once g_m is known, the weight of an object of mass m on the Moon is found by using $W_m = mg_m$. Similar comments apply to other astronomical bodies in the solar system, and everywhere else in the universe as well. **FIGURE 12-10** illustrates the concept for objects on both the Moon and Mars.

EXERCISE 12-4   GRAVITY IN THE SOLAR SYSTEM

a. Find the acceleration due to gravity on the surface of the Moon. (*Note:* The mass of the Moon is $M_m = 7.34 \times 10^{22}$ kg and its radius is $R_m = 1.74 \times 10^6$ m.)

b. The lunar rover had a mass of 225 kg. What was its weight on the Earth and on the Moon?

REASONING AND SOLUTION

a. To find the acceleration due to gravity on the surface of the Moon, we use $g = GM_E/R_E^2$ but with M_E replaced by $M_m = 7.34 \times 10^{22}$ kg, and R_E replaced with $R_m = 1.74 \times 10^6$ m. We find

$$g_m = \frac{GM_m}{R_m^2} = \frac{(6.67 \times 10^{-11}\,\text{N} \cdot \text{m}^2/\text{kg}^2)(7.34 \times 10^{22}\,\text{kg})}{(1.74 \times 10^6\,\text{m})^2} = 1.62\,\text{m/s}^2$$

This is about one-sixth the acceleration due to gravity on the Earth.

b. An object's weight is $W = mg$, where g is the acceleration due to gravity at the location of the object. Therefore, the lunar rover's weight on Earth was

$$W = mg = (225\,\text{kg})(9.81\,\text{m/s}^2) = 2210\,\text{N}$$

On the Moon, its weight was

$$W_m = mg_m = (225\,\text{kg})(1.62\,\text{m/s}^2) = 365\,\text{N}$$

As expected, this is roughly one-sixth its Earth weight.

The replacement of a sphere with a point mass at its center can be applied to many physical systems. For example, the force of gravity between two spheres of finite size is the same as if *both* were replaced by point masses. Thus, the gravitational force between the Earth, with mass M_E, and the Moon, with mass M_m, is

$$F = G\frac{M_E M_m}{r^2}$$

The distance r in this expression is the center-to-center distance between the Earth and the Moon, as shown in **FIGURE 12-11**. It follows, then, that in many calculations involving the solar system, moons and planets can be treated as point objects.

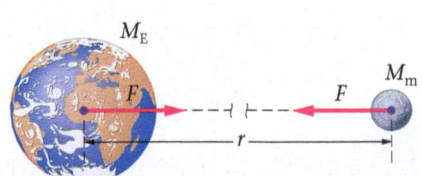

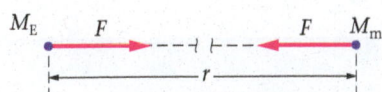

▲ **FIGURE 12-11 Gravitational force
between the Earth and the Moon** The
force is the same as if both the Earth and
the Moon were point masses. (The sizes
of the Earth and Moon are in correct proportion in this figure, but the separation
between the two should be much greater
than that shown here. In reality, it is about
30 times the diameter of the Earth, and so
would be about 2 ft on this scale.)

Weighing the Earth

The British physicist Henry Cavendish performed an experiment in 1798 that is often referred to as "weighing the Earth." What he did, in fact, was measure the value of the universal gravitation constant, G, that appears in Newton's law of gravity. As we've pointed out, G is a very small number; hence a sensitive experiment is needed for its measurement. It is because of this experimental difficulty that G was not measured until more than 100 years after Newton published the law of gravitation.

The Cavendish Experiment   In the Cavendish experiment, illustrated in **FIGURE 12-12**, two masses m are connected by a thin rod and suspended from a thin thread. Near each suspended mass is a large stationary mass M, as shown. Each suspended mass is attracted by the force of gravity toward the large mass near it; hence the rod holding the suspended masses tends to rotate and twist the thread. The angle through which the thread twists can be measured by bouncing a beam of light from a mirror attached to the thread. If the force required to twist the thread through a given angle is known (from previous experiments), a measurement of the twist angle gives the magnitude

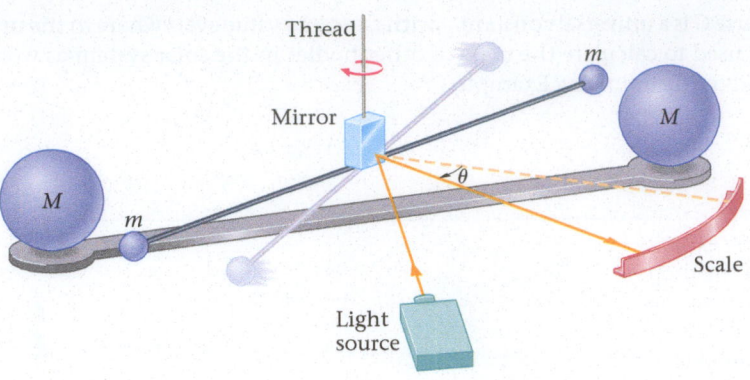

◀ **FIGURE 12-12 The Cavendish experiment** The gravitational attraction between the masses m and M causes the rod and the suspending thread to twist. Measurement of the twist angle allows for a direct measurement of the gravitational force.

of the force of gravity. Finally, knowing the masses, $m_1 = m$ and $m_2 = M$, and the distance between their centers, r, we can use $F = Gm_1m_2/r^2$ to solve for G. Cavendish found $6.754 \times 10^{-11} \, \text{N} \cdot \text{m}^2/\text{kg}^2$, in good agreement with the currently accepted value of $G = 6.67 \times 10^{-11} \, \text{N} \cdot \text{m}^2/\text{kg}^2$.

To see why Cavendish is said to have weighed the Earth, recall that the force of gravity on the surface of the Earth, mg, can be written as follows:

$$mg = G\frac{mM_E}{R_E{}^2}$$

Canceling m and solving for M_E yields

$$M_E = \frac{gR_E{}^2}{G} \qquad\qquad 12\text{-}5$$

Before the Cavendish experiment, the quantities g and R_E were known from direct measurement, but G had yet to be determined. When Cavendish measured G, he didn't actually "weigh" the Earth, of course. Instead, he calculated its mass, M_E.

EXERCISE 12-5   MASS OF THE EARTH

Use $M_E = gR_E{}^2/G$ to calculate the mass of the Earth.

REASONING AND SOLUTION
Substituting numerical values into $M_E = gR_E{}^2/G$, we find

$$M_E = \frac{gR_E{}^2}{G} = \frac{(9.81 \, \text{m/s}^2)(6.37 \times 10^6 \, \text{m})^2}{6.67 \times 10^{-11} \, \text{N} \cdot \text{m}^2/\text{kg}^2} = 5.97 \times 10^{24} \, \text{kg}$$

Connections with Geology   As soon as Cavendish determined the mass of the Earth, geologists were able to use the result to calculate its average density; that is, its average mass per volume. If we assume a spherical Earth of radius R_E, its total volume is

$$V_E = \tfrac{4}{3}\pi R_E{}^3 = \tfrac{4}{3}\pi(6.37 \times 10^6 \, \text{m})^3 = 1.08 \times 10^{21} \, \text{m}^3$$

Dividing the total volume into the total mass yields the average density, ρ:

$$\rho = \frac{M_E}{V_E} = \frac{5.97 \times 10^{24} \, \text{kg}}{1.08 \times 10^{21} \, \text{m}^3} = 5530 \, \text{kg/m}^3 = 5.53 \, \text{g/cm}^3$$

This is an interesting result because typical rocks found near the surface of the Earth, such as granite, have a density of only about $3.00 \, \text{g/cm}^3$. We conclude, then, that the interior of the Earth must have a greater density than its surface. In fact, by analyzing the propagation of seismic waves around the world, we now know that the Earth has a rather complex interior structure, including a solid inner core with a density of about $15.0 \, \text{g/cm}^3$ (see Section 10-5).

A similar calculation for the Moon yields an average density of about $3.33 \, \text{g/cm}^3$, essentially the same as the density of the lunar rocks brought back during the Apollo program. Hence, the Moon probably does not have an internal structure similar to that of the Earth.

Because G is a universal constant—with the same value everywhere in the universe—it can be used to calculate the mass of other bodies in the solar system as well. This is illustrated in the following Example.

EXAMPLE 12-6   MARS ATTRACTS!

After landing on Mars, an astronaut performs a simple experiment by dropping a rock. A quick calculation using the drop height and the time of fall yields a value of 3.73 m/s² for the rock's acceleration. (a) Find the mass of Mars, given that its radius is $R_M = 3.39 \times 10^6$ m. (b) What is the acceleration due to gravity at a distance $r = 2R_M$ from the center of Mars?

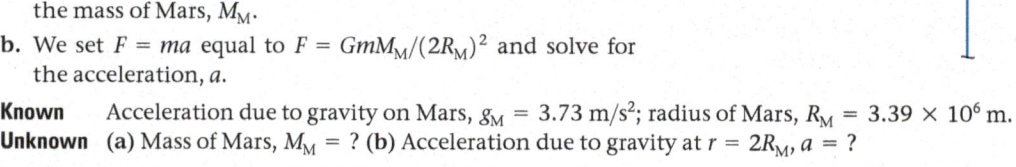

PICTURE THE PROBLEM

Our sketch shows an astronaut dropping a rock to the ground on the surface of Mars. When the acceleration of the rock is measured, we find $g_M = 3.73$ m/s², where the subscript M refers to Mars. In addition, we indicate the radius of Mars in our sketch, where $R_M = 3.39 \times 10^6$ m.

REASONING AND STRATEGY

a. The acceleration due to gravity is g_M on the surface of Mars, and hence it follows that the Martian force of gravity on an object of mass m is $F = mg_M$. This force is also given by Newton's law of gravity—that is, $F = GmM_M/R_M^2$. Setting these expressions for the force equal to one another yields the mass of Mars, M_M.

b. We set $F = ma$ equal to $F = GmM_M/(2R_M)^2$ and solve for the acceleration, a.

Known   Acceleration due to gravity on Mars, $g_M = 3.73$ m/s²; radius of Mars, $R_M = 3.39 \times 10^6$ m.
Unknown   (a) Mass of Mars, $M_M = $? (b) Acceleration due to gravity at $r = 2R_M$, $a = $?

SOLUTION

Part (a)

1. Set mg_M equal to GmM_M/R_M^2:

$$mg_M = G\frac{mM_M}{R_M^2}$$

2. Cancel m and solve for the mass of Mars:

$$M_M = \frac{g_M R_M^2}{G}$$

3. Substitute numerical values:

$$M_M = \frac{(3.73 \text{ m/s}^2)(3.39 \times 10^6 \text{ m})^2}{6.67 \times 10^{-11} \text{ N} \cdot \text{m}^2/\text{kg}^2} = 6.43 \times 10^{23} \text{ kg}$$

Part (b)

4. Apply Newton's law of gravity with $r = 2R_M$. Use the fact that $g_M = GM_M/R_M^2$ from Step 1, and the given numerical value, $g_M = 3.73$ m/s², to simplify the calculation:

$$ma = G\frac{mM_M}{(2R_M)^2} \quad \text{or}$$

$$a = G\frac{M_M}{(2R_M)^2} = \frac{1}{4}\left(G\frac{M_M}{R_M^2}\right) = \frac{1}{4}(g_M) = \frac{1}{4}(3.73 \text{ m/s}^2) = 0.933 \text{ m/s}^2$$

INSIGHT

The important point in this Example is that the universal gravitation constant, G, applies as well on Mars as on Earth, or any other object. Therefore, knowing the size and acceleration of gravity on the surface of an astronomical body is sufficient to determine its mass.

PRACTICE PROBLEM — PREDICT/CALCULATE

(a) If the radius of Mars were reduced to 3.00×10^6 m, with its mass remaining the same, would the acceleration due to gravity on the surface of Mars increase, decrease, or stay the same? Explain. (b) Check your answer by calculating the acceleration of gravity for this case.   [**Answer:** (a) The force of gravity increases as the radius is decreased, and hence the acceleration due to gravity would increase. (b) The acceleration due to gravity increases from 3.73 m/s² to 4.77 m/s².]

Some related homework problems: Problem 19, Problem 21

2. Suppose the Sun suddenly collapsed to half its current radius, with its mass remaining the same. Would the gravitational force exerted by the Sun on the Earth increase, decrease, or stay the same? Explain.

Section Review

• The force of gravity between a spherical mass M and a mass m outside the sphere is the same as if all the mass of the sphere were concentrated at its center.

12-3 Kepler's Laws of Orbital Motion

If you go outside each clear night and observe the position of Mars with respect to the stars, you will find that its apparent motion across the sky is rather complex. Instead of moving on a simple curved path, it occasionally reverses direction (this is known as *retrograde motion*). A few months later it reverses direction yet again and resumes its original direction of motion. Other planets exhibit similar odd behavior.

The Danish astronomer Tycho Brahe (1546–1601) followed the paths of the planets, and Mars in particular, for many years, even though the telescope had not yet been invented. He used, instead, an elaborate sighting device to plot the precise positions of the planets. Brahe was joined in his work by Johannes Kepler (1571–1630) in 1600, and after Brahe's death, Kepler inherited his astronomical observations.

Kepler made good use of Brahe's life work, extracting from his carefully collected data the three laws of orbital motion we know today as Kepler's laws. These laws make it clear that the Sun and the planets do not orbit the Earth, as Ptolemy—the ancient Greek astronomer—claimed, but rather that the Earth, along with the other planets, orbits the Sun, as proposed by Copernicus (1473–1543).

Why the planets obey Kepler's laws no one knew—not even Kepler—until Newton considered the problem decades after Kepler's death. Newton was able to show that each of Kepler's laws follows as a direct consequence of the law of universal gravitation. In the remainder of this section we consider Kepler's three laws one at a time, and point out the connection between them and the law of gravitation.

Kepler's First Law

Kepler tried long and hard to find a circular orbit around the Sun that would match Brahe's observations of Mars. After all, up to that time everyone from Ptolemy to Copernicus believed that celestial objects moved in circular paths of one sort or another. Though the orbit of Mars was exasperatingly close to being circular, the small differences between a circular path and the experimental observations just could not be ignored. Eventually, after a great deal of hard work and disappointment over the loss of circular orbits, Kepler discovered that Mars followed an orbit that was elliptical rather than circular. The same applied to the other planets. This observation became Kepler's first law:

> Planets follow elliptical orbits, with the Sun at one focus of the ellipse.

This is a fine example of the scientific method in action. Though Kepler expected, and wanted, to find circular orbits, he would not allow himself to ignore the data. If Brahe's observations had not been so accurate, Kepler probably would have chalked up the small differences between the data and a circular orbit to error. As it was, he had to discard a treasured—but incorrect—theory and move on to an unexpected, but ultimately correct, view of nature.

Kepler's first law is illustrated in **FIGURE 12-13**, along with a definition of an ellipse in terms of its two foci. In the case where the two foci merge, as in **FIGURE 12-14**, the ellipse reduces to a circle. Thus, a circular orbit *is* allowed by Kepler's first law, but only as a special case.

Newton was able to show that, because the force of gravity decreases with distance as $1/r^2$, closed orbits must have the form of ellipses or circles, as stated in Kepler's first

Big Idea **3** The orbits of planets, moons, comets, and asteroids are governed by Kepler's three laws, all of which can be derived directly from Newton's law of universal gravitation.

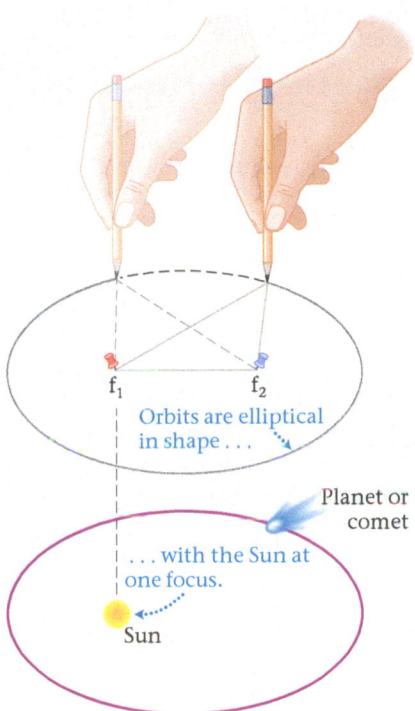

▲ **FIGURE 12-13 Drawing an ellipse** To draw an ellipse, put two tacks in a piece of cardboard. The tacks define the "foci" of the ellipse. Now connect a length of string to the two tacks, and use a pencil and the string to sketch out a smooth, closed curve, as shown. This closed curve is an ellipse. In a planetary orbit a planet follows an elliptical path, with the Sun at one focus. Nothing is at the other focus.

PHYSICS
IN CONTEXT
Looking Back

Conservation of angular momentum, in-troduced in Chapter 11, plays a key role in gravity. In fact, it is the basis for Kepler's second law.

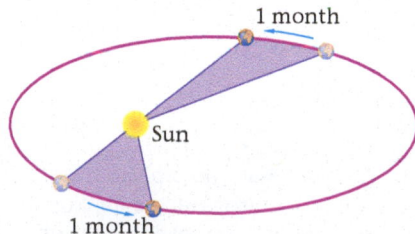

(a) Equal areas in equal times

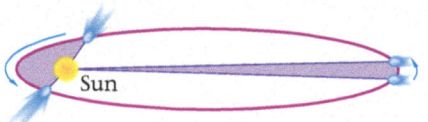

(b) Equal areas in equal times for highly elliptical orbit

▲ **FIGURE 12-15 Kepler's second law**
(a) The second law states that a planet sweeps out equal areas in equal times.
(b) In a highly elliptical orbit, the long, thin area is equal to the broad, fan-shaped area.

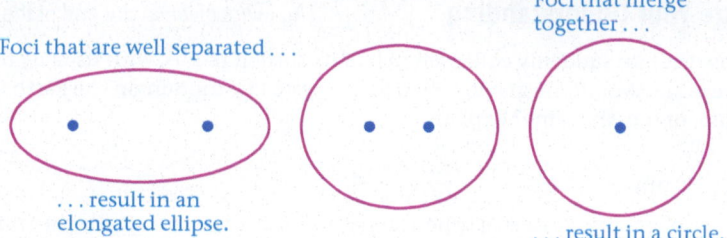

▲ **FIGURE 12-14 The circle as a special case of the ellipse** As the two foci of an ellipse approach one another, the ellipse becomes more circular. In the limit that the foci merge, the ellipse becomes a circle.

law. He also showed that orbits that are not closed—say, the orbit of a comet that passes by the Sun once and then leaves the solar system—are either parabolic or hyperbolic in shape.

Kepler's Second Law

When Kepler plotted the position of a planet on its elliptical orbit, indicating at each position the time the planet was there, he made an interesting observation. First, draw a line from the Sun to a planet at a given time. Then a certain time later—perhaps a month—draw a line again from the Sun to the new position of the planet. The result is that the planet has "swept out" a wedge-shaped area, as indicated in **FIGURE 12-15 (a)**. If this procedure is repeated when the planet is on a different part of its orbit, another wedge-shaped area is generated. Kepler's observation was that the areas of these two wedges are equal:

> As a planet moves in its orbit, it sweeps out an equal amount of area in an equal amount of time.

Kepler's second law follows from the fact that the force of gravity on a planet is directly toward the Sun. As a result, gravity exerts zero torque about the Sun, which means that the angular momentum of a planet in its orbit must be conserved. As Newton showed, conservation of angular momentum is equivalent to the equal-area law stated by Kepler.

CONCEPTUAL EXAMPLE 12-7　COMPARE SPEEDS　PREDICT/EXPLAIN

The Earth's orbit is slightly elliptical. In fact, the Earth is closer to the Sun during the northern hemisphere's winter than it is during the summer. **(a)** Is the orbital speed of the Earth during winter greater than, less than, or equal to its orbital speed during summer? **(b)** Which of the following is the *best explanation* for your prediction?

　I. The orbital speed of the Earth is conserved, and so its value is the same at all times.

　II. The Earth sweeps out equal areas in equal times, and hence it moves faster in winter when its distance from the Sun is smaller.

　III. The orbital speed of the Earth is smaller in winter, because a smaller distance from the Sun implies a smaller speed.

REASONING AND DISCUSSION
According to Kepler's second law, the area swept out by the Earth per month is the same in winter as it is in summer. In winter, however, the radius from the Sun to the Earth is smaller than it is in summer. Therefore, if this smaller radius is to sweep out the same area, the Earth must move more rapidly.

ANSWER
(a) The speed of the Earth is greater during the winter. **(b)** The best explanation is II.

Though we have stated the first two laws in terms of planets, they apply equally well to any object orbiting the Sun. For example, a comet might follow a highly ellipti-cal orbit, as in **FIGURE 12-15 (b)**. When it is near the Sun, it moves very quickly, for the reason discussed in Conceptual Example 12-7, sweeping out a broad, wedge-shaped area in a month's time. Later in its orbit, the comet is far from the Sun and moving slowly. In this case, the area it sweeps out in a month is a long, thin wedge. Still, the two wedges have equal areas.

Kepler's Third Law

Finally, Kepler studied the relationship between the mean distance of a planet from the Sun, r, and its period—that is, the time, T, it takes for the planet to complete one orbit. **FIGURE 12-16** shows a plot of period versus distance for the planets of the solar system. Kepler tried to "fit" these results to a simple dependence between T and r. If he tried a linear fit—that is, T proportional to r (the bottom curve in Figure 12-16)—he found that the period did not increase rapidly enough with distance. On the other hand, if he tried T proportional to r^2 (the top curve in Figure 12-16), the period increased too rapidly. Splitting the difference—trying T proportional to $r^{3/2}$—yields a good fit (the middle curve in Figure 12-16). This is Kepler's third law:

> The period, T, of a planet increases as its mean distance from the Sun, r, raised to the 3/2 power. That is,

$$T = (\text{constant})r^{3/2} \qquad \text{12-6}$$

Deriving Kepler's Third Law   It is straightforward to derive Kepler's third law for the special case of a circular orbit. Consider a planet orbiting the Sun at a distance r, as in **FIGURE 12-17**. The planet moves in a circular path, and hence a centripetal force must act on it, as we saw in Section 6-5. In addition, this force must be directed toward the center of the circle—that is, toward the Sun. It is as if you were to swing a ball on the end of a string in a circle above your head, as in Figure 6-22. In order for the ball to move in a circular path, you have to exert a force on the ball toward the center of the circular path. This force is exerted through the string. In the case of a planet orbiting the Sun, the centripetal force is provided by the force of gravity between the Sun and the planet.

If the planet has a mass m, and the Sun has a mass M_S, the force of gravity between them is

$$F = G\frac{mM_S}{r^2}$$

Now, this force creates the centripetal acceleration of the planet, a_{cp}, which, according to Equation 6-15, is

$$a_{cp} = \frac{v^2}{r}$$

Thus, the centripetal force necessary for the planet to orbit is ma_{cp}:

$$F = ma_{cp} = m\frac{v^2}{r}$$

Noting that the speed of the planet, v, is the circumference of the orbit, $2\pi r$, divided by the time to complete an orbit, T, we have

$$F = m\frac{v^2}{r} = m\frac{(2\pi r/T)^2}{r} = \frac{4\pi^2 rm}{T^2}$$

Setting the centripetal force equal to the force of gravity yields

$$\frac{4\pi^2 rm}{T^2} = G\frac{mM_S}{r^2}$$

Eliminating m and rearranging, we find

$$T^2 = \frac{4\pi^2}{GM_S}r^3$$

Taking the square root of both sides of the equation yields

$$T = \left(\frac{2\pi}{\sqrt{GM_S}}\right)r^{3/2} = (\text{constant})r^{3/2} \qquad \text{12-7}$$

As predicted by Kepler, T is proportional to $r^{3/2}$.

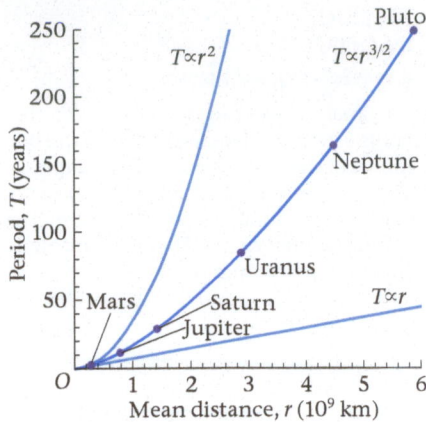

▲ **FIGURE 12-16 Kepler's third law and some near misses** These plots represent three possible mathematical relationships between period of revolution, T (in years), and mean distance from the Sun, r (in kilometers). The lower curve shows $T = (\text{constant})r$; the upper curve is $T = (\text{constant})r^2$. The middle curve, which fits the data, is $T = (\text{constant})r^{3/2}$. This is Kepler's third law.

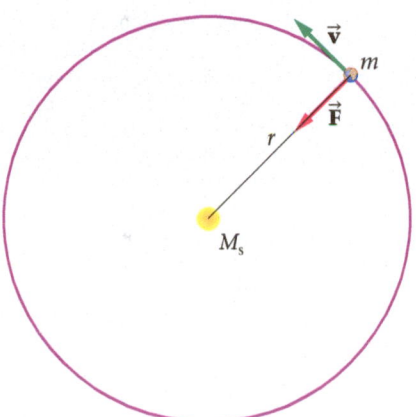

▲ **FIGURE 12-17 Centripetal force on a planet in orbit** As a planet revolves about the Sun in a circular orbit of radius r, the force of gravity between it and the Sun, $F = GmM_S/r^2$, provides the required centripetal force.

**PHYSICS
IN CONTEXT
Looking Ahead**

The analysis used to derive Kepler's third law is used again when we explore the Bohr model of the hydrogen atom in Chapter 31. The calculation is the same, but in hydrogen the Coulomb force between electric charges (Equation 19-5) is responsible for the orbital motion.

PROBLEM-SOLVING NOTE

The Mass in Kepler's Third Law

When applying Kepler's third law, recall that the mass in Equation 12-7, M_S, refers to the mass of the object being orbited. Thus, the third law can be applied to satellites of any object, as long as M_S is replaced by the orbited mass.

▶ **FIGURE 12-18** **Visualizing Concepts**
Orbital Motion Kepler's laws of orbital motion apply to planetary satellites as well as planets. Jupiter, the largest planet in the solar system, has at least 16 moons, all of which travel in elliptical orbits that obey Kepler's laws. (The moons in the photo at left, passing in front of Jupiter, are Europa and Io) Even some asteroids have been found to have their own satellites. The large cratered object in the photo at right is 243 Ida, an asteroid some 56 km long; its miniature companion at the top of the photo is Dactyl, about 1.5 km in diameter. Like all gravitationally bound bodies, Ida and Dactyl orbit their common center of mass.

Applying Kepler's Third Law Deriving Kepler's third law by using Newton's law of gravitation has allowed us to calculate the constant that multiplies $r^{3/2}$. Notice that the constant depends on the mass of the Sun; that is, *T depends on the mass being orbited.* It does not depend on the mass of the planet orbiting the Sun, however, as long as the planet's mass is much less than the mass of the Sun. As a result, Equation 12-7 applies equally to all the planets.

This result can also be applied to the case of a moon or a satellite (an artificial moon) orbiting a planet. To do so, we simply note that it is the planet that is being orbited, not the Sun. Hence, to apply Equation 12-7, we replace the mass of the Sun, M_S, with the mass of the appropriate planet. This concept is illustrated in **FIGURE 12-18**, where we show that it also applies to moons orbiting planets and asteroids.

As an example, let's calculate the mass of Jupiter. One of the four moons of Jupiter discovered by Galileo is Io, which completes one orbit every 42 h 27 min = 1.53×10^5 s. Given that the average distance from the center of Jupiter to Io is 4.22×10^8 m, we can find the mass of Jupiter as follows:

$$T = \left(\frac{2\pi}{\sqrt{GM_J}}\right)r^{3/2}$$

$$M_J = \frac{4\pi^2 r^3}{GT^2} = \frac{4\pi^2(4.22 \times 10^8 \text{ m})^3}{(6.67 \times 10^{-11} \text{ N}\cdot\text{m}^2/\text{kg}^2)(1.53 \times 10^5 \text{ s})^2} = 1.90 \times 10^{27} \text{ kg}$$

This is roughly 300 times the mass of the Earth.

 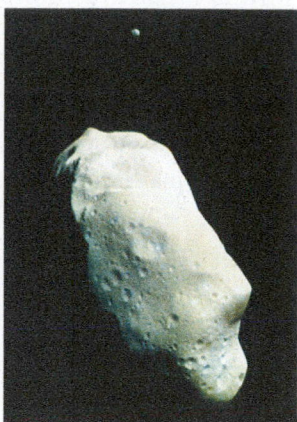

EXAMPLE 12-8 THE SUN AND MERCURY

The Earth revolves around the Sun once a year at an average distance of 1.50×10^{11} m. **(a)** Use this information to calculate the mass of the Sun. **(b)** Find the period of revolution for the planet Mercury, whose average distance from the Sun is 5.79×10^{10} m.

PICTURE THE PROBLEM
Our sketch shows the orbits of Mercury, Venus, and the Earth in correct proportion. In addition, each of these orbits is slightly elliptical, though the deviation from circularity is too small for the eye to see. Finally, we indicate that the orbital radius for Mercury is 5.79×10^{10} m and the orbital radius for the Earth is 1.50×10^{11} m.

REASONING AND STRATEGY

a. To find the mass of the Sun, we solve Equation 12-7 for M_S. Notice that the period $T = 1$ y must be converted to seconds before we evaluate the formula.

b. The period of Mercury is found by substituting $r = 5.79 \times 10^{10}$ m in Equation 12-7.

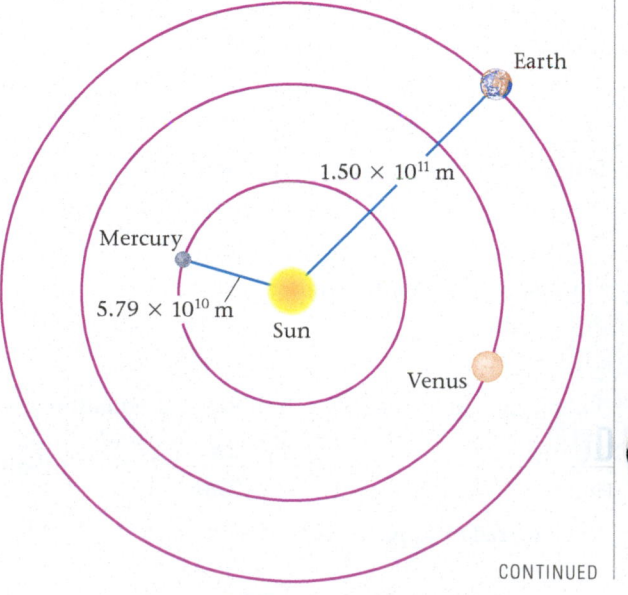

CONTINUED

Known Orbital radius of Earth, $r = 1.50 \times 10^{11}$ m; orbital radius of Mercury, $r = 5.79 \times 10^{10}$ m.
Unknown (a) Mass of the Sun, $M_S = ?$ (b) Period of Mercury, $T = ?$

SOLUTION

Part (a)

1. Solve Equation 12-7 for the mass of the Sun:

$$T = \left(\frac{2\pi}{\sqrt{GM_S}}\right)r^{3/2}$$

$$M_S = \frac{4\pi^2 r^3}{GT^2}$$

2. Calculate the period of the Earth in seconds:

$$T = 1\,\text{y}\left(\frac{365.24\,\text{days}}{1\,\text{y}}\right)\left(\frac{24\,\text{h}}{1\,\text{day}}\right)\left(\frac{3600\,\text{s}}{1\,\text{h}}\right) = 3.16 \times 10^7\,\text{s}$$

3. Substitute numerical values in the expression for the mass of the Sun obtained in Step 1:

$$M_S = \frac{4\pi^2 r^3}{GT^2}$$

$$= \frac{4\pi^2(1.50 \times 10^{11}\,\text{m})^3}{(6.67 \times 10^{-11}\,\text{N}\cdot\text{m}^2/\text{kg}^2)(3.16 \times 10^7\,\text{s})^2}$$

$$= 2.00 \times 10^{30}\,\text{kg}$$

Part (b)

4. Substitute $r = 5.79 \times 10^{10}$ m into Equation 12-7. In addition, use the mass of the Sun obtained in part (a):

$$T = \left(\frac{2\pi}{\sqrt{GM_S}}\right)r^{3/2}$$

$$= \left(\frac{2\pi}{\sqrt{(6.67 \times 10^{-11}\,\text{N}\cdot\text{m}^2/\text{kg}^2)(2.00 \times 10^{30}\,\text{kg})}}\right) \times (5.79 \times 10^{10}\,\text{m})^{3/2}$$

$$= 7.58 \times 10^6\,\text{s} = 0.240\,\text{y} = 87.7\,\text{days}$$

INSIGHT

In part (a), we find that the mass of the Sun is almost a million times greater than the mass of the Earth. In fact, the Sun accounts for 99.9% of all the mass in the solar system. In part (b) we see that Mercury, with its smaller orbital radius, has a shorter year than the Earth.

PRACTICE PROBLEM

Venus orbits the Sun with a period of 1.94×10^7 s. What is its average distance from the Sun?
[**Answer:** $r = 1.08 \times 10^{11}$ m]

Some related homework problems: Problem 25, Problem 28

RWP The 24 satellites of the Global Positioning System (GPS) are in relatively low orbits about the Earth. These satellites are used to provide a precise determination of an observer's position anywhere on Earth. The operating principle of the GPS is illustrated in **FIGURE 12-19.** Imagine, for example, that satellite 2 emits a radio signal at a particular time (all GPS satellites carry atomic clocks on board). This signal travels away from the satellite with the speed of light (see Chapter 25) and is detected a short time later by an observer's GPS receiver. Multiplying the time delay by the speed of light gives the distance of the receiver from satellite 2. Thus, in our example, the observer must lie somewhere on the red circle in Figure 12-19. Similar time-delay measurements for signals from satellite 11 show that the observer is also somewhere on the green circle; hence the observer is either at the point shown in Figure 12-19, or at the second intersection of the red and green circles on the other side of the planet. Measurements from satellite 6 can resolve the ambiguity and place the observer at the point shown in the figure. Measurements from additional satellites can even determine the observer's altitude. GPS receivers, which are used by hikers, boaters, and car navigation systems, typically use signals from as many as 12 satellites. As currently operated, the GPS gives positions with a typical accuracy of 2 m to 10 m. The altitude of GPS orbits is determined in the next Quick Example.

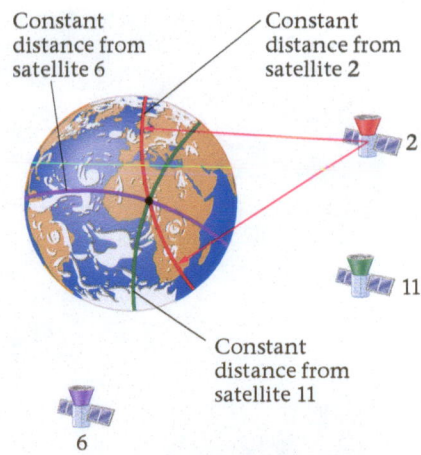

▲ **FIGURE 12-19 The Global Positioning System** A system of 24 satellites in orbit about the Earth makes it possible to determine a person's location with great accuracy. Measuring the distance of a person from satellite 2 places the person somewhere on the red circle. Similar measurements using satellite 11 place the person's position somewhere on the green circle, and further measurements can pinpoint the person's location.

QUICK EXAMPLE 12-9   FIND THE ALTITUDE OF A GPS SATELLITE

Find the altitude above the Earth's surface where a GPS satellite orbits with a period of 12 h. (Note that $R_E = 6.37 \times 10^6$ m, $M_E = 5.97 \times 10^{24}$ kg, and $T = 12\,\text{h} = 4.32 \times 10^4$ s.)

CONTINUED

REASONING AND SOLUTION

We find the period of an orbit around the Earth with $T = (2\pi/\sqrt{GM_E})r^{3/2}$, which is simply Equation 12-7 with the mass of the Earth, M_E, replacing the mass of the Sun, M_S.

1. Write the expression for the period of an object orbiting the Earth:

$$T = \left(\frac{2\pi}{\sqrt{GM_E}}\right)r^{3/2}$$

2. Solve for the radius, r:

$$r = \left(\frac{T}{2\pi}\right)^{2/3}(GM_E)^{1/3}$$

3. Substitute numerical values:

$$r = \left(\frac{4.32 \times 10^4\,\text{s}}{2\pi}\right)^{2/3}((6.67 \times 10^{-11}\,\text{N} \cdot \text{m}^2/\text{kg}^2)$$
$$\times (5.97 \times 10^{24}\,\text{kg}))^{1/3}$$
$$= 2.66 \times 10^7\,\text{m}$$

4. Subtract the radius of the Earth to find the altitude:

$$r - R_E = 2.66 \times 10^7\,\text{m} - 6.37 \times 10^6\,\text{m}$$
$$= 2.02 \times 10^7\,\text{m}$$

Thus, GPS satellites orbit at $2.02 \times 10^7\,\text{m} \sim 12{,}600$ mi above our heads.

Not all spacecraft are placed in GPS orbits, of course. The International Space Station, for example, orbits at an altitude of only about 320 km (200 mi). At that altitude, it takes about an hour and a half to complete one orbit. Some of the higher satellites in everyday use are considered next.

RWP A *geosynchronous satellite* is one that orbits above the equator with a period equal to one day. From the Earth, such a satellite appears to be in the same location in the sky at all times, making it particularly useful for applications such as communications and weather forecasting. From Kepler's third law, we know that a satellite has a period of one day only if its orbital radius has a particular value. This value can be found in the same way that the orbital altitude of GPS satellites was determined in Quick Example 12-9. The result is that geosynchronous satellites orbit at an altitude of 22,200 mi above the Earth. A satellite that was boosted from a low Earth orbit to a geosynchronous orbit is shown in **FIGURE 12-20**.

▲ **FIGURE 12-20 Visualizing Concepts Satellites** Many weather and communications satellites are placed in geosynchronous orbits that allow them to remain "stationary" in the sky—that is, fixed over one point on the Earth's equator. Because the Earth rotates, the period of such a satellite must exactly match that of the Earth. The altitude needed for such an orbit is about 36,000 km (22,200 mi). Other satellites, such as those used in the Global Positioning System (GPS), the Hubble Space Telescope, and the International Space Station, operate at much lower altitudes—typically just a few hundred miles. The photo at left shows the communications satellite Intelsat VI just prior to its capture by astronauts of the space shuttle *Endeavour*. A launch failure had left the satellite stranded in a low orbit. The astronauts snared the satellite (right) and fitted it with a new engine that boosted it to its geosynchronous orbit.

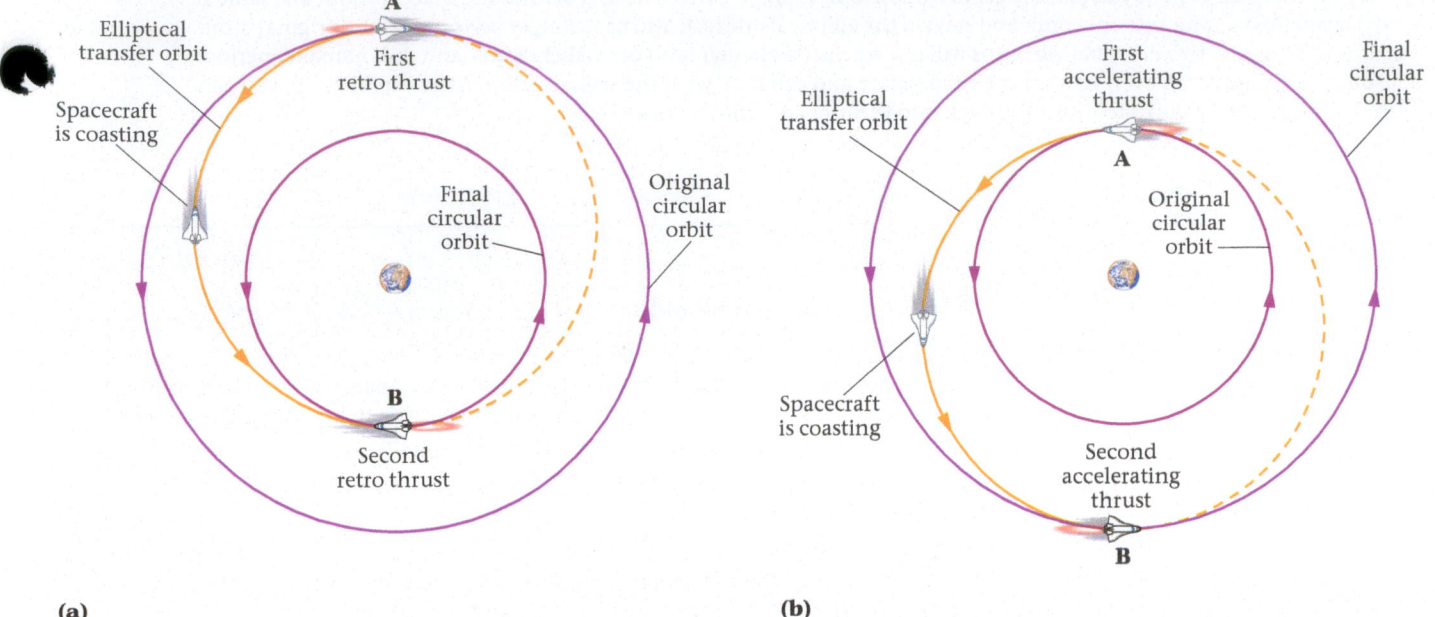

(a)   **(b)**

▲ **FIGURE 12-21 Orbital maneuvers (a)** The radius of a satellite's orbit can be decreased by firing the decelerating rockets once at point A and again at point B. Between firings the satellite follows an elliptical orbit. The satellite speeds up as it falls inward toward the Earth during this maneuver. For this reason its final speed in the new circular orbit is greater than its speed in the original orbit, even though the decelerating rockets have slowed it down twice. **(b)** The radius of a satellite's orbit can be increased by firing the accelerating rockets once at point A and again at point B. Between firings the satellite follows an elliptical orbit. The satellite slows down as it moves farther from the Earth during this maneuver. For this reason its final speed in the new circular orbit is less than its speed in the original orbit, even though the accelerating rockets have sped it up twice.

Orbital Maneuvers

RWP We now show how Kepler's laws can give insight into maneuvering a satellite in orbit. Suppose, for example, that you are piloting a spacecraft in a circular orbit, and you would like to move to a lower circular orbit. As you might expect, you should begin by using your rockets to decrease your speed—that is, fire the rockets that point in the forward direction so that their thrust (Section 9-8) is opposite to your direction of motion. The result of firing the decelerating rockets at a given point A in your original orbit is shown in **FIGURE 12-21 (a)**. Notice that your new orbit is not a circle, as desired, but rather an ellipse. To produce a circular orbit you can simply fire the decelerating rockets once again at point B, on the opposite side of the Earth from point A. The net result of these two firings is that you now move in a circular orbit of smaller radius.

Similarly, to move to a larger orbit, you must fire your accelerating rockets twice. The first firing puts you into an elliptical orbit that moves farther from the Earth, as **FIGURE 12-21 (b)** shows. After the second firing you are again in a circular orbit. This simplest type of orbital transfer, requiring just two rocket burns, is referred to as a *Hohmann transfer*. The Hohmann transfer is the basic maneuver used to send spacecraft such as the Mars lander from Earth's orbit about the Sun to the orbit of Mars.

CONCEPTUAL EXAMPLE 12-10   WHICH ROCKETS TO USE?

As you pilot your spacecraft in a circular orbit about the Earth, you notice the space station you want to dock with several miles ahead in the same orbit. To catch up with the space station, should you fire your accelerating rockets or fire your decelerating rockets?

REASONING AND DISCUSSION

You want to catch up with something miles ahead, and hence you must accelerate, right? Well, not in this case. Accelerating moves you into an elliptical orbit, as in Figure 12-21 (b), and with a second acceleration you can make your new orbit circular with a larger radius. Recall from Kepler's third law, however, that the larger the radius of an orbit, the greater the period.

CONTINUED

Thus, on your new higher path you take longer to complete an orbit, so you fall farther behind the space station. The same is true even if you fire your rockets only once and stay on the elliptical orbit—it also has a longer period than the original orbit.

On the other hand, two decelerating burns will put you into a circular orbit of smaller radius, and thus a smaller period. As a result, you complete an orbit in less time than before and catch up with the space station. After catching up, you can perform two accelerating burns to move you back into the original orbit to dock.

ANSWER
You should fire your decelerating rockets.

Enhance Your Understanding (Answers given at the end of the chapter)

3. Satellite A orbits the Earth at the radius r; satellite B orbits the Earth at the radius $2r$. **(a)** Is the orbital period of satellite A greater than, less than, or equal to the orbital period of satellite B? Explain. **(b)** Is the orbital speed of satellite A greater than, less than, or equal to the orbital speed of satellite B? Explain.

Section Review

- Kepler's three laws describe the orbital motions of planets, comets, satellites, and other objects in space. All three laws follow directly from Newton's law of universal gravitation.

12-4 Gravitational Potential Energy

In Chapter 8 we saw that the principle of conservation of energy can be used to solve a number of problems that would be difficult to handle with a straightforward application of Newton's laws of mechanics. Before we can apply energy conservation to astronomical situations, however, we must know the gravitational potential energy for a spherical object such as the Earth. Now you may be wondering: Don't we already know the potential energy of gravity? Well, in fact, in Chapter 8 we said that the gravitational potential energy a distance h above the Earth's surface is $U = mgh$. As was mentioned at the time, however, this result is valid only near the Earth's surface, where we can say that the acceleration due to gravity, g, is approximately constant.

General Result for Gravitational Potential Energy As the distance from the Earth increases, we know that g decreases, as was shown in Example 12-3. It follows that $U = mgh$ cannot be valid for arbitrary h. Indeed, it can be shown that the general result for the gravitational potential energy of a system consisting of a mass m a distance r from the center of the Earth is

$$U = -G\frac{mM_E}{r}$$

12-8

A plot of $U = -GmM_E/r$ is presented in **FIGURE 12-22**. Notice that U approaches zero as r approaches infinity. This is a common convention in astronomical systems. In fact, because only *differences* in potential energy matter, as was mentioned in Chapter 8, the

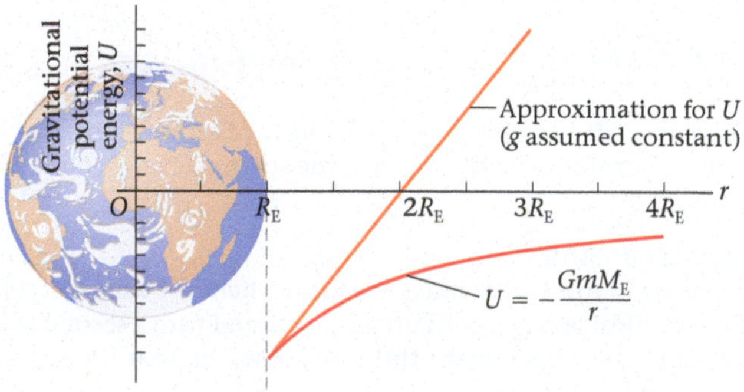

▶ **FIGURE 12-22 Gravitational potential energy as a function of distance r from the center of the Earth** The lower curve in this plot shows the gravitational potential energy, $U = -GmM_E/r$, for r greater than R_E. Near the Earth's surface, U is approximately linear, corresponding to the result $U = mgh$ given in Chapter 8.

choice of the reference point ($U = 0$) is completely arbitrary. When we considered systems that were near the Earth's surface, it was natural to let $U = 0$ at ground level. When we consider, instead, distances of astronomical scale, it is generally more convenient to choose the potential energy to be zero when objects are separated by an infinite distance.

EXERCISE 12-11   GRAVITATIONAL POTENTIAL ENERGY

Use $U = -GmM_E/r$ to find the gravitational potential energy of a 976-kg meteorite when it is **(a)** ten Earth radii from the center of the Earth and **(b)** on the surface of the Earth.

REASONING AND SOLUTION

a. In this case, the distance from the center of the Earth is $r = 10R_E$. It follows that the gravitational potential energy is

$$U = -G\frac{mM_E}{10R_E}$$

$$= -(6.67 \times 10^{-11}\,\text{N} \cdot \text{m}^2/\text{kg}^2)\frac{(976\,\text{kg})(5.97 \times 10^{24}\,\text{kg})}{10(6.37 \times 10^6\,\text{m})} = -6.10 \times 10^9\,\text{J}$$

b. Now, the distance from the center of the Earth is R_E, and therefore the gravitational potential energy is

$$U = -G\frac{mM_E}{R_E}$$

$$= -(6.67 \times 10^{-11}\,\text{N} \cdot \text{m}^2/\text{kg}^2)\frac{(976\,\text{kg})(5.97 \times 10^{24}\,\text{kg})}{(6.37 \times 10^6\,\text{m})} = -6.10 \times 10^{10}\,\text{J}$$

The potential energy in part (b) is ten times greater in magnitude than it is in part (a), because the distance from the center of the Earth to the meteorite is ten times smaller.

Connecting the General Gravitational Potential Energy with the Result Near the Surface of the Earth At first glance, $U = -GmM_E/r$ doesn't seem to bear any similarity to $U = mgh$, which we know to be valid near the surface of the Earth. Even so, there is a direct connection between these two expressions. Recall that when we say that the potential energy at a height h is mgh, what we mean is that when a mass m is raised from the ground to a height h, the potential energy of the system increases by the amount mgh. Let's calculate the corresponding difference in potential energy using $U = -GmM_E/r$.

First, at a height h above the surface of the Earth we have $r = R_E + h$; hence the potential energy there is

$$U = -G\frac{mM_E}{R_E + h}$$

On the surface of the Earth, where $r = R_E$, we have

$$U = -G\frac{mM_E}{R_E}$$

The corresponding difference in potential energy is

$$\Delta U = \left(-G\frac{mM_E}{R_E + h}\right) - \left(-G\frac{mM_E}{R_E}\right)$$

$$= \left(-G\frac{mM_E}{R_E}\right)\left(\frac{1}{1 + h/R_E}\right) - \left(-G\frac{mM_E}{R_E}\right)$$

If h is much smaller than the radius of the Earth, it follows that h/R_E is a small number. In this case, we can apply the useful approximation $1/(1 + x) \approx 1 - x$ [see Figure A-5 (b) in Appendix A] to write $1/(1 + h/R_E) \approx 1 - h/R_E$. As a result, we have

$$\Delta U = \left(-G\frac{mM_E}{R_E}\right)(1 - h/R_E) - \left(-G\frac{mM_E}{R_E}\right) = m\left[\frac{GM_E}{R_E^2}\right]h$$

PHYSICS IN CONTEXT
Looking Ahead

The electric potential energy (Chapter 20) has the same mathematical form as the gravitational potential energy introduced in this chapter.

The term in square brackets should look familiar—according to Equation 12-4 it is simply g. Hence, the increase in potential energy at the height h is

$$\Delta U = mgh$$

This result is valid for heights that are small compared with the radius of the Earth.

The straight line in Figure 12-22 corresponds to the potential energy mgh. Near the Earth's surface, it is clear that mgh and $-GmM_E/r$ are in close agreement. For larger r, however, the fact that gravity is getting weaker means that the potential energy does not continue rising as rapidly as it would if gravity were of constant strength.

Gravitational Potential Energy Is a Scalar An important distinction between the potential energy, U, and the gravitational force, $\vec{F}$, is that the force is a vector, whereas the potential energy is a scalar—that is, U is simply a number. As a result:

> The total gravitational potential energy of a system of objects is the sum of the gravitational potential energies of each pair of objects separately.

Because U is not a vector, there are no x or y components to consider, as would be the case with a vector. Finally, the potential energy $U = -GmM_E/r$ applies to a mass m a distance r from the center of the Earth, with mass M_E. More generally, if two point masses, m_1 and m_2, are separated by a distance r, their gravitational potential energy is:

PHYSICS IN CONTEXT
Looking Back

Just as the force of gravity is generalized in this chapter from the simple result $F = mg$ to $F = Gm_1m_2/r^2$, so too is the gravitational potential energy. Thus, the expression $U = mgh$ (Chapter 8) is generalized to $U = -Gm_1m_2/r$.

Gravitational Potential Energy, U

$$U = -G\frac{m_1 m_2}{r}$$

12-9

SI unit: joule, J

In the next Example we use this result, and the fact that U is a scalar, to find the total gravitational potential energy for a system of three point masses.

EXAMPLE 12-12   SIMPLE ADDITION

The sketch shows the spaceship *Millennium Condor* as it approaches a pair of twin asteroids. Assuming all three objects can be treated as point masses, find the total gravitational potential energy of the system, given that the mass of the spaceship is $m = 3.75 \times 10^7$ kg and the mass of each asteroid is $M = 1.54 \times 10^9$ kg. The relevant distances are given in the sketch.

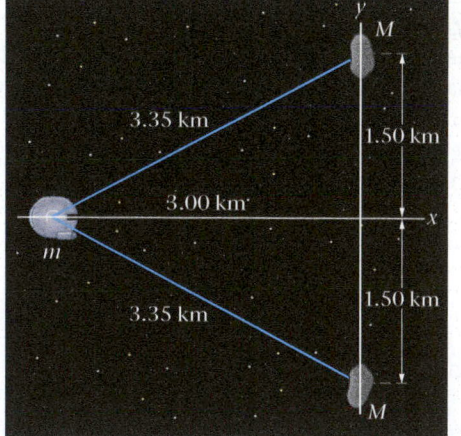

PICTURE THE PROBLEM
The positions of the spaceship and the two asteroids are shown in the sketch. Also shown are the distances between the various masses.

REASONING AND STRATEGY
The gravitational potential energy associated with each pair of masses is given by $U = -Gm_1m_2/r$ (Equation 12-9). The total gravitational potential energy of the system is the sum of the potential energies for each of the three pairs of masses.

Known   Mass of spaceship, $m = 3.75 \times 10^7$ kg; mass of asteroid, $M = 1.54 \times 10^9$ kg; distances given in the sketch.
Unknown   Total gravitational potential energy, $U_{total} = ?$

SOLUTION

1. Calculate the gravitational potential energy for the two asteroids, each with mass M:

$$U_{MM} = -G\frac{MM}{r_{MM}}$$

$$= -(6.67 \times 10^{-11}\,\text{N}\cdot\text{m}^2/\text{kg}^2)\frac{(1.54 \times 10^9\,\text{kg})^2}{2(1.50 \times 10^3\,\text{m})}$$

$$= -5.27 \times 10^4\,\text{J}$$

2. Calculate the gravitational potential energy for one of the asteroids (mass M) and the spaceship (mass m). The same calculation applies to the other asteroid as well:

$$U_{Mm} = -G\frac{Mm}{r_{Mm}}$$

$$= -(6.67 \times 10^{-11}\,\text{N}\cdot\text{m}^2/\text{kg}^2)\frac{(1.54 \times 10^9\,\text{kg})(3.75 \times 10^7\,\text{kg})}{(3.35 \times 10^3\,\text{m})}$$

$$= -1.15 \times 10^3\,\text{J}$$

CONTINUED

3. The total gravitational potential energy is the sum of the potential energies for the three pairs of objects:

$$U_{\text{total}} = U_{MM} + U_{Mm} + U_{Mm}$$
$$= -5.27 \times 10^4 \, \text{J} - 2(1.15 \times 10^3 \, \text{J}) = -5.50 \times 10^4 \, \text{J}$$

INSIGHT

The total gravitational potential energy of this system is negative, which is less than it would be if the separation of the three objects were to approach infinity, in which case $U_{\text{total}} = 0$. The implications of this change in potential energy are considered in the next section, when we explore the consequences of energy conservation.

PRACTICE PROBLEM — PREDICT/CALCULATE

(a) If the spaceship were moved to the origin of the coordinate system, halfway between the asteroids, would the potential energy of the system increase, decrease, or stay the same? Explain. (b) Verify your answer by calculating the potential energy in this case. [**Answer: (a)** The potential energy would decrease—that is, become more negative—because the distance from each asteroid to the spaceship would decrease. **(b)** $U_{\text{total}} = -5.78 \times 10^4 \, \text{J}$]

Some related homework problems: Problem 39, Problem 40

Enhance Your Understanding   (Answers given at the end of the chapter)

4. Consider the two systems shown in **FIGURE 12-23**. Is the total gravitational potential energy of system A greater than, less than, or equal to the total gravitational potential energy of system B? Explain.

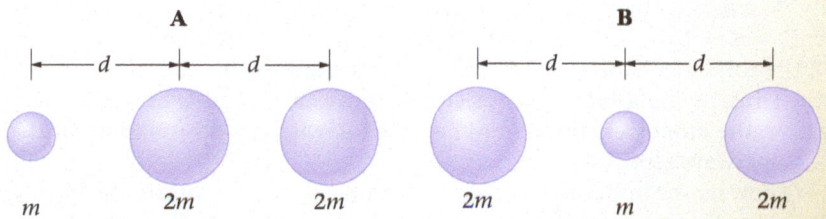

▲ **FIGURE 12-23** Enhance Your Understanding 4.

Section Review

- The gravitational potential energy of a system consisting of point masses m_1 and m_2 separated by a distance r is $U = -Gm_1m_2/r$.

- The gravitational potential energy of a system consisting of a mass m a distance r from the center of the Earth is $U = -GmM_E/r$. Near the surface of the Earth, this expression simplifies to $U = mgh$.

12-5 Energy Conservation

Now that we know the gravitational potential energy, U, at an arbitrary distance from a spherical object, we can apply energy conservation to astronomical situations in the same way we applied it to systems near the Earth's surface in Chapter 8. To be specific, the mechanical energy, E, of an object of mass m a distance r from the center of the Earth is

$$E = K + U = \tfrac{1}{2}mv^2 - G\frac{mM_E}{r} \qquad \text{12-10}$$

This more general expression replaces $E = K + U = \tfrac{1}{2}mv^2 + mgh$, which applies for objects near the Earth's surface.

Speed of an Impacting Asteroid If we set the initial mechanical energy of a system equal to its final mechanical energy, we can answer questions such as the following: Suppose that an asteroid has zero speed infinitely far from the Earth. If this asteroid were to fall directly toward the Earth, what speed would it have when it strikes the Earth's surface?

Big Idea **4** Gravitational potential energy is a form of energy that must be taken into account when we apply energy conservation. The general expression for gravitational potential energy is $U = -Gm_1m_2/r$, which simplifies to $U = mgh$ for objects near the Earth's surface.

PROBLEM-SOLVING NOTE

Energy Conservation in Astronomical Systems

To apply conservation of energy to an object that moves far from the surface of a planet, one must use $U = -GmM/r$, where r is the distance from the center of the planet.

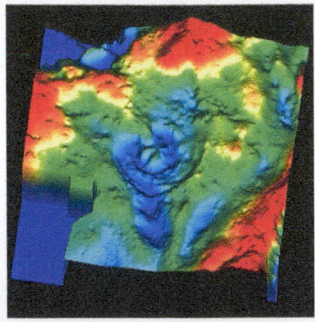

▲ FIGURE 12-24 Visualizing Concepts
Impacts with Earth Bodies from space have struck the Earth countless times in the past and continue to do so on a regular basis. Most such objects are relatively small, but larger objects, including the occasional comet or asteroid, also cross our path from time to time. The crater on top, in Rotorua, New Zealand, must be of relatively recent origin (thousands rather than millions of years old), because erosion has not yet erased this scar on the Earth's surface. The center image is a false-color gravity anomaly map of the Chicxulub impact crater in Mexico. The object that struck there some 65 million years ago may have produced such far-reaching climatic disruption that the dinosaurs and many other species became extinct as a result. The photo on the bottom shows the 12,500-metric-ton meteorite that blazed across the sky over Chelyabinsk, Russia, on February 15, 2013, injuring hundreds of people with its powerful shock wave.

RWP As you probably know, this is not an entirely academic question. Asteroids and comets, both large and small, have struck the Earth innumerable times during its history. Three such impacts are shown in **FIGURE 12-24**. A particularly large object appears to have struck the Earth on the Yucatan Peninsula in Mexico, near the town of Chicxulub, some 65 million years ago. Evidence suggests that this impact may have led to the mass extinctions at the end of the Cretaceous period, during which the dinosaurs disappeared from the Earth. Unfortunately, such events are not limited to the distant past. For example, as recently as 50,000 years ago, an iron asteroid tens of meters in diameter and shining 10,000 times brighter than the Sun (from atmospheric heating) slammed into the ground near Winslow, Arizona, forming the 1.2-km-wide Barringer Meteor Crater. More recently yet, at sunrise on June 30, 1908, a relatively small stony asteroid streaked through the atmosphere and exploded at an altitude of several kilometers near the Tunguska River in Siberia. The energy released by the explosion was comparable to that of an H-bomb, and it flattened the forest for kilometers in all directions. And on February 15, 2013, another meteorite struck in Siberia, causing extensive damage and injuring hundreds.

Applying Energy Conservation   Returning to our original question, we can use energy conservation to determine the speed such an asteroid or comet might have when it hits the Earth. To begin, we assume the asteroid starts at rest, and hence its initial kinetic energy is zero, $K_i = 0$. In addition, the initial potential energy of the system, U_i, is also zero, because $U = -GmM_E/r$ approaches zero as r approaches infinity. As a result, the total initial mechanical energy of the asteroid-Earth system is zero: $E_i = K_i + U_i = 0$. Because gravity is a conservative force (as discussed in Section 8-1), the total mechanical energy remains constant as the asteroid falls toward the Earth. Thus, as the asteroid moves closer to the Earth and U becomes increasingly negative, the kinetic energy K must become increasingly positive so that their sum, $U + K$, is always zero.

We now set the initial energy equal to the final energy to determine the final speed, v_f. Recalling that the final distance r is the radius of the Earth, R_E, we have

$$E_i = E_f$$

$$0 = \tfrac{1}{2}mv_f^2 - G\frac{mM_E}{R_E}$$

Solving for the final speed yields

$$v_f = \sqrt{\frac{2GM_E}{R_E}} \qquad\qquad 12\text{-}11$$

Substituting numerical values into this expression gives

$$v_f = \sqrt{\frac{2GM_E}{R_E}} = \sqrt{\frac{2(6.67\times10^{-11}\,\text{N}\cdot\text{m}^2/\text{kg}^2)(5.97\times10^{24}\,\text{kg})}{6.37\times10^6\,\text{m}}}$$

$$= 11{,}200\,\text{m/s}\;(25{,}000\,\text{mi/h}) \qquad 12\text{-}12$$

Thus, a typical asteroid hits the Earth moving at about 7.0 mi/s—about 16 times faster than a rifle bullet! This result is independent of the asteroid's mass.

Visualizing Energy Conservation   To help visualize energy conservation in this system, we plot the gravitational potential energy U in **FIGURE 12-25**. Also indicated in the plot is the total energy, $E = 0$. Noting that $U + K$ must always equal zero, we see that the value of K goes up as the value of U goes down. This is illustrated graphically in the figure with the help of histogram bars.

A somewhat more elaborate plot showing the same physics is presented in **FIGURE 12-26**. The two-dimensional surface in this case represents the potential energy function U as one moves away from the Earth in any direction. In particular, the dependence of U on distance r along any radial line in Figure 12-26 is the same as the shape of U versus r in Figure 12-25. Because the potential energy drops downward in such a plot, this type of situation is often referred to as a "potential well." If a marble is allowed to roll on such a surface, its motion is similar in many ways to the motion of an object near the Earth.

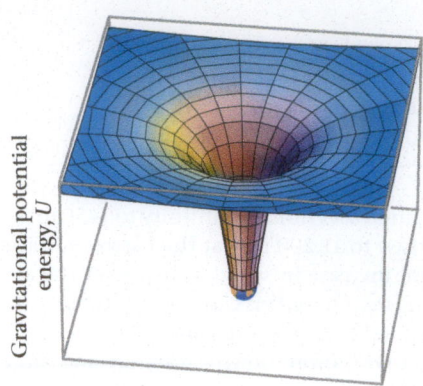

▲ **FIGURE 12-26 A gravitational potential "well"** A three-dimensional plot of the gravitational potential energy near an object such as the Earth. An object approaching the Earth speeds up as it "falls" into the gravitational potential well.

▲ **FIGURE 12-25 Potential and kinetic energies of an object falling toward Earth** As an object with zero total energy moves closer to the Earth, its gravitational potential energy, U, becomes increasingly negative. In order for the total energy to remain zero, $E = U + K = 0$, it is necessary for the kinetic energy to become increasingly positive.

In fact, if the marble is started with the right initial velocity, it will roll in a circular or elliptical "orbit" for a long time before falling into the center of the well. (Eventually, of course, the well does swallow up the marble. Though the retarding force of rolling friction is quite small, it still causes the marble to descend into a lower and lower orbit—just as air resistance causes a satellite to descend lower and lower into the Earth's atmosphere until it finally burns up.)

EXAMPLE 12-13   ARMAGEDDON RENDEZVOUS

In the movie *Armageddon*, a crew of hard-boiled oil drillers rendezvous with a menacing asteroid just as it passes the orbit of the Moon on its way toward Earth. Assuming the asteroid starts at rest infinitely far from the Earth, as in the previous discussion—and hence has zero initial energy—find its speed when it passes the Moon's orbit. Assume that the Moon orbits at a distance of $60R_E$ from the center of the Earth and that its gravitational influence may be ignored.

PICTURE THE PROBLEM
Our sketch shows the Earth, the Moon, and the asteroid. The initial position of the asteroid is at infinity, where its speed is zero. For the purposes of this problem, its final position is at the Moon's orbit, where its speed is v_f. At this point, the asteroid is heading directly for the Earth.

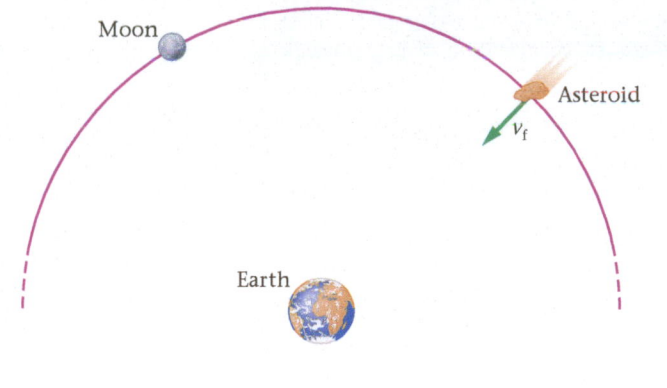

REASONING AND STRATEGY
The basic strategy is the same as that used above to obtain the speed of an asteroid when it hits the Earth; namely, we set the initial energy equal to the final energy and solve for the final speed v_f. In this case, the final radius is $r_f = 60R_E$. As before, the initial energy is zero; $E_i = 0$.

Known   Initial mechanical energy, $E_i = 0$; final radius, $r_f = 60R_E$.
Unknown Speed at Moon's orbit, $v_f = ?$

SOLUTION

1. Set the initial energy of the system equal to its final energy:

$$E_i = E_f$$
$$0 = \tfrac{1}{2}mv_f^2 - G\frac{mM_E}{60R_E}$$

CONTINUED

2. Solve for the final speed, v_f:

$$v_f = \sqrt{\frac{2GM_E}{60R_E}} = \frac{1}{\sqrt{60}}\left(\sqrt{\frac{2GM_E}{R_E}}\right)$$

3. Substitute the numerical value given in Equation 12-12 for the quantity in parentheses:

$$v_f = \frac{1}{\sqrt{60}}(11{,}200 \text{ m/s}) = 1450 \text{ m/s} \sim 3200 \text{ mi/h}$$

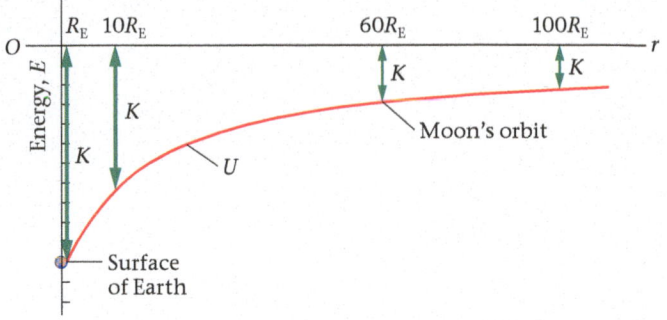

INSIGHT

The asteroid's speed increases from 0 at infinity to 1450 m/s at the Moon's orbit, and then to 11,200 m/s at the Earth's surface. The majority of this total increase in speed occurs after the asteroid passes the Moon. The reason for this can be seen in the accompanying plot of the gravitational potential energy, U.

This plot shows that U drops downward more and more rapidly as the Earth is approached. Thus, while there is relatively little increase in K from infinite distance to $r = 60R_E$, there is a substantially larger increase in K from $r = 60R_E$ to $r = R_E$.

PRACTICE PROBLEM

At what distance from the center of the Earth is the asteroid's speed equal to 3535 m/s?
[Answer: $r = 6.37 \times 10^7 \text{ m} = 10R_E$]

Some related homework problems: Problem 49, Problem 52

QUICK EXAMPLE 12-14   FIND THE DISTANCE TO A SATELLITE

A satellite in an elliptical orbit has a speed of 9.00 km/s when it is at its closest approach to the Earth (perigee). The satellite is 7.00×10^6 m from the center of the Earth at this time. When the satellite is at its greatest distance from the center of the Earth (apogee), its speed is 3.66 km/s. How far is the satellite from the center of the Earth at apogee? ($R_E = 6.37 \times 10^6$ m, $M_E = 5.97 \times 10^{24}$ kg)

REASONING AND SOLUTION

The mechanical energy (kinetic plus potential) of the satellite is conserved as it orbits the Earth. Setting the mechanical energy at perigee equal to the mechanical energy at apogee gives a relationship that can be solved for the distance at apogee.

1. Set the mechanical energy at perigee (speed v_1 at the distance r_1) equal to the energy at apogee (speed v_2 at the distance r_2):

$$\frac{1}{2}mv_1^2 - \frac{GmM_E}{r_1} = \frac{1}{2}mv_2^2 - \frac{GmM_E}{r_2}$$

2. Solve for $\frac{1}{r_2}$:

$$\frac{1}{r_2} = \frac{1}{r_1} + \frac{(v_2^2 - v_1^2)}{(2GM_E)}$$

3. Substitute numerical values:

$$\frac{1}{r_2} = 5.80 \times 10^{-8} \text{ m}^{-1}$$

4. Invert the fraction to obtain the distance at apogee, r_2:

$$r_2 = 1.72 \times 10^7 \text{ m}$$

In this case, apogee is about 2.5 times farther from the center of the Earth than is perigee.

▲ **FIGURE 12-27** Comet Hale-Bopp, one of the largest and brightest comets to visit our celestial neighborhood in recent decades, photographed in April 1997. While most of the planets and planetary satellites in the solar system have roughly circular orbits, the orbits of many comets are highly elliptical. In accordance with Kepler's second law, these objects spend most of their time moving slowly through cold, distant regions of the solar system (often far beyond the orbit of Pluto). Their visits to the inner solar system are infrequent and relatively brief.

The concepts in Quick Example 12-14 apply to comets as well. For example, the comet shown in **FIGURE 12-27** moves much faster when it passes close to the Sun than when its orbit takes it far from the Sun.

Escape Speed

Resisting the pull of Earth's gravity has always held a fascination for the human species, from Daedalus and Icarus with their wings of feathers and wax, to Leonardo da Vinci and his flying machine, to the Montgolfier brothers and their hot-air balloons. In his 1865 novel, *From the Earth to the Moon*, Jules Verne imagined launching a spacecraft to the Moon by firing it straight upward from a cannon. Not a bad idea—if you could survive the initial blast. Today, rockets are fired into space using the same basic idea, though they smooth out the initial blast by burning their engines over a period of several minutes.

Suppose, then, that you would like to launch a rocket of mass m with an initial speed sufficient not only to reach the Moon, but to allow it to escape the Earth altogether. If we refer to this speed as the **escape speed** for the Earth, v_e, the initial energy of the rocket is

$$E_i = K_i + U_i = \tfrac{1}{2}mv_e{}^2 - G\frac{mM_E}{R_E}$$

If the rocket just barely escapes the Earth, its speed will decrease to zero as its distance from the Earth approaches infinity. Therefore, the rocket's final kinetic energy is zero, as is the potential energy of the system, because $U = -GmM_E/r$ goes to zero as $r \to \infty$. It follows that

$$E_f = K_f + U_f = 0 - 0 = 0$$

Equating these energies yields

$$\tfrac{1}{2}mv_e{}^2 - G\frac{mM_E}{R_E} = 0$$

Therefore, the escape speed from the Earth is

$$v_e = \sqrt{\frac{2GM_E}{R_E}} = 11{,}200 \text{ m/s} \approx 25{,}000 \text{ mi/h} \qquad \text{12-13}$$

Notice that the escape speed is precisely the same as the speed of the asteroid calculated in Equation 12-12. This is not surprising when you consider that an object launched from the Earth to infinity is just the reverse of an object falling from infinity to the Earth.

The result given in Equation 12-13 can be applied to other astronomical objects as well by simply replacing M_E and R_E with the appropriate mass and radius for another object.

EXERCISE 12-15   ESCAPE SPEED

Calculate the escape speed for an object launched from the Moon.

REASONING AND SOLUTION

For the Moon we use $M_m = 7.34 \times 10^{22}$ kg and $R_m = 1.74 \times 10^6$ m. With these values, the escape speed is

$$v_e = \sqrt{\frac{2GM_m}{R_m}} = \sqrt{\frac{2(6.67 \times 10^{-11} \text{ N} \cdot \text{m}^2/\text{kg}^2)(7.34 \times 10^{22} \text{ kg})}{1.74 \times 10^6 \text{ m}}}$$

$$= 2370 \text{ m/s } (5300 \text{ mi/h})$$

The relatively low escape speed of the Moon means that it is much easier to launch a rocket into space from the Moon than from the Earth. For example, the tiny Lunar Module that blasted off from the Moon to return the Apollo astronauts to Earth could not have come close to escaping from the Earth.

RWP The Moon's low escape speed has another important consequence—it is the reason the Moon has no atmosphere. Even if you could magically supply the Moon with an atmosphere, it would soon evaporate into space because the individual molecules in the air move with speeds great enough to escape. On Earth, however, where the escape speed is much higher, gravity can prevent the rapidly moving molecules from moving off into space. Even so, light molecules, like hydrogen and helium, move faster at a given temperature than the heavier molecules like nitrogen and oxygen, as we shall see in Chapter 17. For this reason, the Earth's atmosphere contains virtually no hydrogen or helium. (In fact, helium was first discovered on the Sun, as we point out in Chapter 31—hence its name, which derives from the Greek word for the Sun, "helios.") Because a stable atmosphere is a likely requirement for the development of life, it follows that the escape speed is an important factor when considering the possibility of life on other planets.

CONCEPTUAL EXAMPLE 12-16 | COMPARE ESCAPE SPEEDS | PREDICT/EXPLAIN

(a) Is the escape speed for a 10,000-kg rocket greater than, less than, or equal to the escape speed for a 1000-kg rocket?
(b) Which of the following is the *best explanation* for your prediction?

I. The greater the mass of a rocket, the greater the energy needed to make it escape, and hence the greater the escape speed.

II. A rocket with more mass has more kinetic energy; hence less speed is needed to make it escape.

III. The escape speed depends on the mass of the Earth, but not on the mass of the rocket.

REASONING AND DISCUSSION

The derivation of the escape speed, $v_e = \sqrt{2GM_E/R_E}$ (Equation 12-13), shows that the mass of the rocket, m, cancels. Hence, the escape speed is the same for all objects, regardless of their mass. On the other hand, the kinetic energy required to give the 10,000-kg rocket the escape speed is 10 times greater than the kinetic energy required for the 1000-kg rocket.

ANSWER

(a) The escape speed is the same for both rockets. **(b)** The best explanation is III.

EXAMPLE 12-17 | HALF ESCAPE

Suppose Jules Verne's cannon launches a rocket straight upward with an initial speed equal to one-half the escape speed. How far from the center of the Earth does this rocket travel before momentarily coming to rest? (Ignore air resistance in the Earth's atmosphere.)

PICTURE THE PROBLEM

Our sketch shows the rocket launched vertically from the Earth's surface with an initial speed equal to half the escape speed, $v_0 = \frac{1}{2}v_e$. The rocket moves radially away from the Earth until it comes to rest, $v = 0$, at a distance r from the center of the Earth.

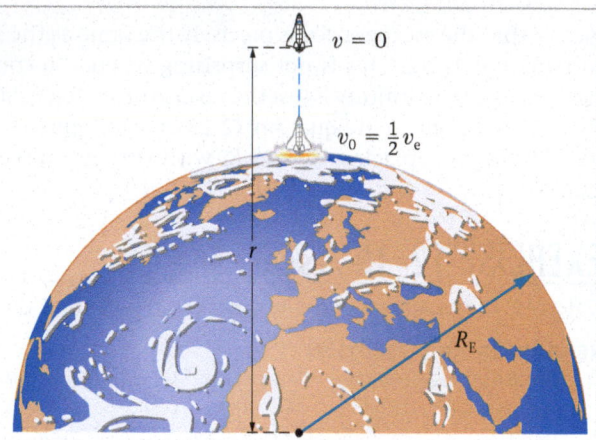

REASONING AND STRATEGY

We ignore air resistance in this system, and hence the final energy of the rocket, E_f, must be equal to its initial energy, E_0. Setting these energies equal determines the point where the rocket comes to rest.

Known   Initial speed, $v_0 = v_e/2$; initial radius, R_E.
Unknown   Final radius, $r = ?$

SOLUTION

1. The initial speed, v_0, is one-half the escape speed. Use $v_e = \sqrt{2GM_E/R_E}$ (Equation 12-13) to write an expression for v_0:

$$v_0 = \frac{1}{2}v_e = \frac{1}{2}\sqrt{2GM_E/R_E} = \sqrt{GM_E/2R_E}$$

2. Write the initial energy of the rocket, E_0. Substitute the result for v_0 from Step 1 and simplify:

$$E_0 = K_0 + U_0 = \frac{1}{2}mv_0^2 - \frac{GmM_E}{R_E}$$

$$= \frac{1}{2}m\left(\sqrt{\frac{GM_E}{2R_E}}\right)^2 - \frac{GmM_E}{R_E} = -\frac{3}{4}\frac{GmM_E}{R_E}$$

3. Write the final energy of the rocket. Recall that the rocket is a distance r from the center of the Earth when it comes to rest:

$$E_f = K_f + U_f = 0 - \frac{GmM_E}{r} = -\frac{GmM_E}{r}$$

4. Equate the initial and final energies:

$$-\frac{3}{4}\frac{GmM_E}{R_E} = -\frac{GmM_E}{r}$$

5. Solve for the final radius, r:

$$r = \frac{4}{3}R_E$$

INSIGHT

An initial speed of v_e allows the rocket to go to infinity before stopping. If the rocket is launched with half that initial speed, however, it can rise to a height of only $4R_E/3 - R_E = R_E/3$ above the Earth's surface. Quite a dramatic difference.

PRACTICE PROBLEM

Find the rocket's maximum distance from the center of the Earth, r, if its launch speed is $3v_e/4$.
[**Answer:** $r = 16R_E/7 = 2.29R_E$]

Some related homework problems: Problem 48, Problem 53, Problem 55

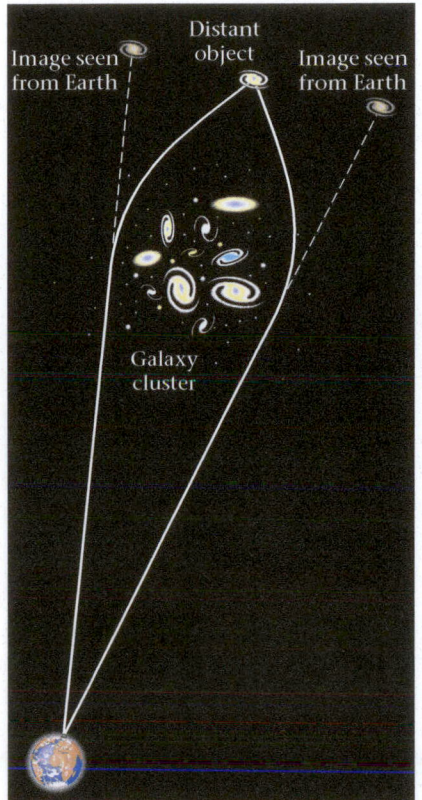

a function of distance from the center of the Earth, r, for various vertical launch speeds The lower two curves show launch speeds that are less than the escape speed, v_e. In these cases the rocket comes to rest momentarily at a finite height above the Earth. The next higher curve shows the speed of a rocket launched with the escape speed, v_e. In this case, the rocket slows to zero speed as the distance approaches infinity. The top curve corresponds to a launch speed greater than the escape speed—this rocket has a finite speed even at infinite distance.

A plot of the speed of a rocket as a function of its distance from the center of the Earth is presented in **FIGURE 12-28** for a variety of initial speeds. Notice that when the initial speed is less than the escape speed, the rocket comes to rest momentarily at a finite distance, r. In particular, if the launch speed is $0.75v_e$, as in the Practice Problem of Example 12-17, the rocket's maximum distance from the center of the Earth is $2.29R_E$.

Black Holes and Gravitational Lensing

As we can see from $v_e = \sqrt{2GM_E/R_E}$, the escape speed of an object increases with increasing mass and decreasing radius. Thus, for example, if a massive star were to collapse to a relatively small size, its escape speed would become very large. According to Einstein's theory of general relativity, the escape speed of a compressed, massive star could even exceed the speed of light. In this case nothing—not even light—could escape from the star. For this reason, such objects are referred to as *black holes*. Anything entering a black hole would be making a one-way trip to an unknown destiny.

RWP Just as a black hole can bend a beam of light back on itself and prevent it from escaping, any massive object can bend light—at least a little. For example, light from distant stars is deflected as it passes by the Sun by 1.75 seconds of arc (the angular size of a quarter at a distance of 1.8 miles). Light passing by an entire galaxy of stars, or even a cluster of galaxies, can be bent by more significant amounts, however, as **FIGURE 12-29** indicates. This effect is referred to as *gravitational lensing*, because the galaxies act much like the lenses we will study in Chapter 26. As a result of gravitational lensing, the images of very distant galaxies or quasars in deep-space astronomical photographs sometimes appear in duplicate, in quadruplicate, or even spread out into circular arcs.

Enhance Your Understanding   (Answers given at the end of the chapter)

5. Consider a distant planet that has twice the mass and twice the radius of the Earth. Is the escape speed on this planet greater than, less than, or equal to the escape speed on the Earth? Explain.

Section Review

- The total mechanical energy of a gravitational system is $E = K + U = \frac{1}{2}mv^2 - GmM/r$.

▲ **FIGURE 12-29 Gravitational lensing** Astronomers often find that very distant objects seem to produce multiple images in their photographs. The cause is the gravitational attraction of intervening galaxies or clusters of galaxies, which are so massive that they can significantly bend the light from remote objects as it passes by them on its way to Earth.

12-6 Tides

The reason for the ocean tides that rise and fall twice a day was a perplexing and enduring mystery until Newton introduced his law of universal gravitation. Even Galileo, who made so many advances in physics and astronomy, could not explain the tides. However, with the understanding that a force is required to cause an object to move in a circular path, and that the force of gravity becomes weaker with distance, it is possible to describe the tides in detail. In this section we show how it can be done. In addition, we extend the basic idea of tides to several related phenomena.

▶ **FIGURE 12-30** **The reason for two tides a day** **(a)** Tides are caused by a disparity between the gravitational force exerted at various points on a finite-sized object (dark red arrows) and the centripetal force needed for circular motion (light red arrows). Note that the gravitational force decreases with distance, as expected. On the other hand, the centripetal force required to keep an object moving in a circular path *increases* with distance. On the near side, therefore, the gravitational force is stronger than required, and the object is stretched inward. On the far side, the gravitational force is weaker than required, and the object stretches outward. **(b)** On the Earth, the water in the oceans responds more to the deforming effects of tides than do the solid rocks of the land. The result is two high tides and two low tides daily on opposite sides of the Earth.

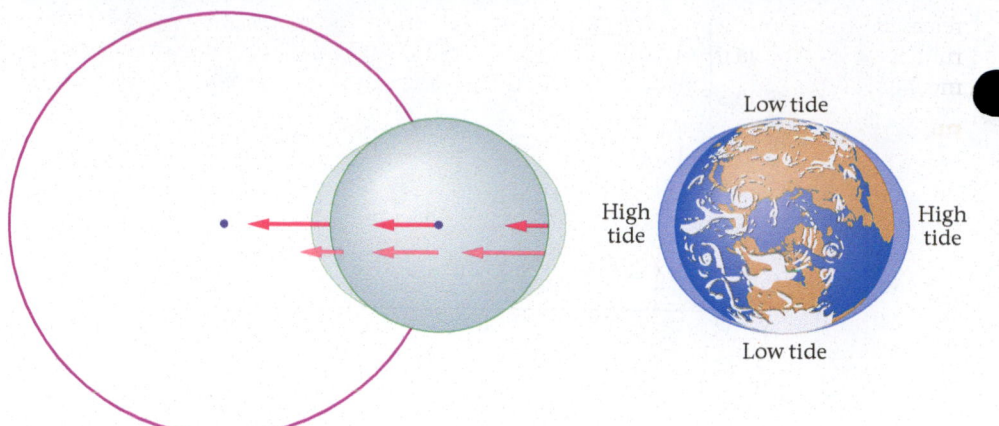

(a) The mechanism responsible for tides

(b) Tidal deformations on Earth

▲ **FIGURE 12-31** Tides on Earth are caused chiefly by the Moon's gravitational pull, though at full and new moon, when the Moon and Sun are aligned, the Sun's gravity can enhance the effect. In some places on Earth, such as the Bay of Fundy between Maine and Nova Scotia, local topographic conditions produce abnormally large tides.

Variation of Force To begin, consider the idealized situation shown in **FIGURE 12-30 (a)**. Here we see an object of finite size (a moon or a planet, for example) orbiting a point mass. If all the mass of the object were concentrated at its center, the gravitational force exerted on it by the central mass would be precisely the amount needed to cause it to move in its circular path. Because the object has a finite size, however, the force exerted on various parts of it has different magnitudes. For example, points closer to the central mass experience a greater force than points farther away.

To see the effect of this variation in force, we use a dark red vector in Figure 12-30 (a) to indicate the force exerted by the central mass at three different points on the object. In addition, we use a light red vector to show the force that is required at each of these three points to cause a mass at that distance to orbit the central mass. Comparing these vectors, we see that the forces are identical at the center of the object—as expected. On the near side of the object, however, the force exerted by the central mass is larger than the force needed to hold the object in orbit, and on the far side the force due to the central mass is less than the force needed to hold the object in orbit. The result is that the near side of the object is pulled closer to the central mass and the far side tends to move farther from the central mass. This causes an egg-shaped deformation of the object, as indicated in Figure 12-30 (a).

Any two objects orbiting one another cause deformations of this type. For example, the Earth causes a deformation in the Moon, and the Moon causes a similar deformation in the Earth. In **FIGURE 12-30 (b)** we show the Earth and the waters of its oceans deformed into an egg shape. Because the waters in the oceans can flow, they deform much more than the underlying rocky surface of the Earth. As a result, the water level relative to the surface of the Earth is higher at the *tidal bulges* shown in the figure. As the Earth rotates about its axis, a person at a given location will observe two high tides and two low tides each day. This is the basic mechanism of the tides on Earth, but the actual situation is complicated by the shape of the coastline at different locations and by the additional tidal effects due to the Sun. An example of a coastline that produces greater than normal tidal ranges is shown in **FIGURE 12-31**.

Tidal Locking The Moon has no oceans, of course, but the tidal bulges produced in it by the Earth are the reason we see only one side of the Moon. Specifically, the Earth exerts gravitational forces on the tidal bulges of the Moon, causing them to point directly toward the Earth. If the Moon were to rotate slightly away from this alignment, the forces exerted by the Earth would cause a torque that would return the Moon to the original alignment. The net result is that the Moon's period of rotation about its axis is equal to its period of revolution about the Earth. This effect, known as **tidal locking**, is common among the various moons in the solar system.

RWP A particularly interesting example of tidal locking is provided by Jupiter's moon Io, a site of intense volcanism. Io follows an elliptical orbit around Jupiter, and its tidal deformation is larger when it is closer to Jupiter than when it is farther away. As a result, each time Io orbits Jupiter it is squeezed into a greater deformation and then released. This continual flexing of Io causes its internal temperature to rise, just as a rubber ball gets warm if you squeeze and release it in your hand over and over. It is this mechanism that is largely responsible for Io's ongoing volcanic activity.

RWP In extreme cases, tidal deformation can become so large that an object is literally torn apart. Because tidal deformation increases as a moon moves closer to the planet it orbits, there is a limiting orbital radius—known as the **Roche limit**—inside of which the tidal force pulling the moon apart exceeds the gravitational attraction holding it together. A most spectacular example of this effect can be seen in the rings of Saturn, all of which exist well within the Roche limit. The small chunks of ice and other materials that make up the rings may be the remains of a moon that moved too close to Saturn and was destroyed by tidal forces. On the other hand, they may represent material that tidal forces prevented from aggregating to form a moon in the first place. In either case, this dramatic debris field will never coalesce to form a moon—tidal effects will not allow such a process to occur. Similar remarks apply to the smaller, much fainter rings that spacecraft have observed around Jupiter, Uranus, and Neptune.

Enhance Your Understanding   (Answers given at the end of the chapter)

6. If the radius of the Moon's orbit around the Earth were increased, would tides on the Earth be higher than, lower than, or the same as they are now? Explain. (The radius of the Moon's orbit is actually increasing at the rate of about 3.8 cm per year.)

Section Review

- Tides are caused by the difference in strength of the gravitational force from one side of an object to the other side.

CHAPTER 12 REVIEW

CHAPTER SUMMARY

12-1 NEWTON'S LAW OF UNIVERSAL GRAVITATION

The force of gravity between two point masses, m_1 and m_2, separated by a distance r is attractive and of magnitude

$$F = G\frac{m_1 m_2}{r^2}$$ 12-1

G is the universal gravitation constant:

$$G = 6.67 \times 10^{-11}\,\text{N}\cdot\text{m}^2/\text{kg}^2$$ 12-2

Inverse Square Dependence
The force of gravity decreases with distance, r, as $1/r^2$. This is referred to as an inverse square dependence.

Superposition
If more than one mass exerts a gravitational force on a given object, the net force is simply the vector sum of the individual forces.

Gravitational force, F vs *Separation distance, r*

12-2 GRAVITATIONAL ATTRACTION OF SPHERICAL BODIES

In calculations of gravitational forces, spherical objects can be replaced by point masses.

Uniform Sphere
If a mass m is outside a uniform sphere of mass M, the gravitational force between m and the sphere is equivalent to the force exerted by a point mass M located at the center of the sphere.

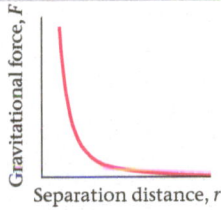

Acceleration of Gravity

Replacing the Earth with a point mass at its center, we find that the acceleration of gravity on the surface of the Earth is

$$g = \frac{GM_E}{R_E^2} \qquad \text{12-4}$$

12-3 KEPLER'S LAWS OF ORBITAL MOTION

Kepler determined three laws that describe the motion of the planets in our solar system. Newton showed that Kepler's laws are a direct consequence of his law of universal gravitation.

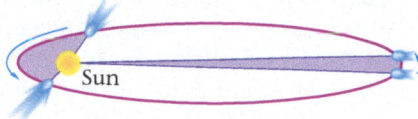

Kepler's First Law

The orbits of the planets are ellipses, with the Sun at one focus.

Kepler's Second Law

Planets sweep out equal areas in equal times.

Kepler's Third Law

The period of a planet's orbit, T, is proportional to the 3/2 power of its average distance from the Sun, r:

$$T = \left(\frac{2\pi}{\sqrt{GM_S}}\right)r^{3/2} = (\text{constant})r^{3/2} \qquad \text{12-7}$$

12-4 GRAVITATIONAL POTENTIAL ENERGY

The gravitational potential energy, U, between two point masses, m_1 and m_2, separated by a distance r is

$$U = -G\frac{m_1 m_2}{r} \qquad \text{12-9}$$

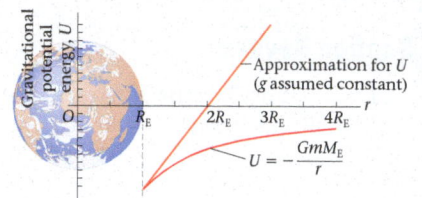

Zero Level

The zero level of the gravitational potential energy between two point masses is chosen to be at infinite separation of the two masses.

12-5 ENERGY CONSERVATION

With the gravitational potential energy given in Section 12-4, energy conservation can be applied to astronomical situations.

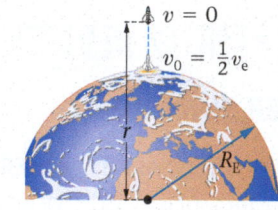

Total Mechanical Energy

An object with mass m, speed v, and at a distance r from the center of the Earth has a total mechanical energy given by

$$E = K + U = \frac{1}{2}mv^2 - \frac{GmM_E}{r} \qquad \text{12-10}$$

12-6 TIDES

Tides result from the variation of the gravitational force from one side of an astronomical object to the other side.

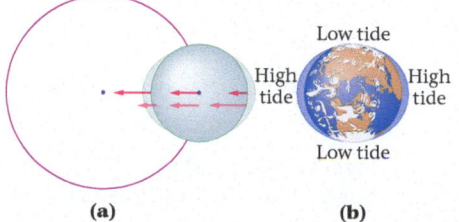

ANSWERS TO ENHANCE YOUR UNDERSTANDING QUESTIONS

1. $B < A = D < C$.

2. The gravitational force would stay the same because the masses and center-to-center distance did not change.

3. **(a)** The orbital period of satellite A is smaller because period increases with orbital radius. **(b)** The orbital speed of satellite A is greater. This follows because doubling the radius of satellite B doubles the circumference, which more than doubles the period, resulting in a lower speed (distance/time).

4. The total gravitational potential energy of system A is less (more negative) than that of system B. In general, a more "compact" system, like system A with the large masses close together, has a lower gravitational potential energy.

5. The escape speed on the distant planet is the same as the escape speed on Earth because the two factors of 2 (doubling the mass and doubling the radius) cancel.

6. The tides on Earth would be lower, because the gravitational forces exerted by the Moon on the Earth would be smaller than they are now.

CONCEPTUAL QUESTIONS

For instructor-assigned homework, go to www.masteringphysics.com.

(Answers to odd-numbered Conceptual Questions can be found in the back of the book.)

1. It is often said that astronauts in orbit experience weightlessness because they are beyond the pull of Earth's gravity. Is this statement correct? Explain.

2. When a person passes you on the street, you do not feel a gravitational tug. Explain.

3. Imagine bringing the tips of your index fingers together. Each finger contains a certain finite mass, and the distance between them goes to zero as they come into contact. From the force law $F = Gm_1m_2/r^2$ one might conclude that the attractive force between the fingers is infinite, and, therefore, that your fingers must remain forever stuck together. What is wrong with this argument?

4. Does the radius vector of Mars sweep out the same amount of area per time as that of the Earth? Why or why not?

5. When a communications satellite is placed in a geosynchronous orbit above the equator, it remains fixed over a given point on the ground. Is it possible to put a satellite into an orbit so that it remains fixed above the North Pole? Explain.

6. **The Mass of Pluto** On June 22, 1978, James Christy made the first observation of a moon orbiting Pluto. Until that time the mass of Pluto was not known, but with the discovery of its moon, Charon, its mass could be calculated with some accuracy. Explain.

7. Rockets are launched into space from Cape Canaveral in an easterly direction. Is there an advantage to launching to the east versus launching to the west? Explain.

8. One day in the future you may take a pleasure cruise to the Moon. While there you might climb a lunar mountain and throw a rock horizontally from its summit. If, in principle, you could throw the rock fast enough, it might end up hitting you in the back. Explain.

9. Apollo astronauts orbiting the Moon at low altitude noticed occasional changes in their orbit that they attributed to localized concentrations of mass below the lunar surface. Just what effect would such "mascons" have on their orbit?

10. If you light a candle on the International Space Station—which would not be a good idea—would it burn the same as on the Earth? Explain.

11. The force exerted by the Sun on the Moon is more than twice the force exerted by the Earth on the Moon. Should the Moon be thought of as orbiting the Earth or the Sun? Explain.

12. **The Path of the Moon** The Earth and Moon exert gravitational forces on one another as they orbit the Sun. As a result, the path they follow is not the simple circular orbit you would expect if either one orbited the Sun alone. Occasionally you will see a suggestion that the Moon follows a path like a sine wave centered on a circular path, as in the upper part of **FIGURE 12-32**. This is *incorrect*. The Moon's path is qualitatively like that shown in the lower part of Figure 12-32. Explain. (Refer to Conceptual Question 11.)

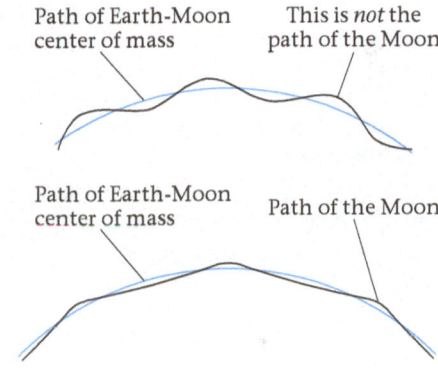

FIGURE 12-32 Conceptual Question 12

PROBLEMS AND CONCEPTUAL EXERCISES

Answers to odd-numbered Problems and Conceptual Exercises can be found in the back of the book. **BIO** *identifies problems of biological or medical interest;* **CE** *indicates a conceptual exercise.* **Predict/Explain** *problems ask for two responses: (a) your prediction of a physical outcome, and (b) the best explanation among three provided; and* **Predict/Calculate** *problems ask for a prediction of a physical outcome, based on fundamental physics concepts, and follow that with a numerical calculation to verify the prediction. On all problems, bullets (•, ••, •••) indicate the level of difficulty. Planetary data can be found in Appendix C.*

SECTION 12-1 NEWTON'S LAW OF UNIVERSAL GRAVITATION

1. • **CE** System A has masses m and m separated by a distance r; system B has masses m and $2m$ separated by a distance $2r$; system C has masses $2m$ and $3m$ separated by a distance $2r$; and system D has masses $4m$ and $5m$ separated by a distance $3r$. Rank these systems in order of increasing gravitational force. Indicate ties where appropriate.

2. • A 6.3-kg bowling ball and a 7.1-kg bowling ball rest on a rack 0.85 m apart. **(a)** What is the force of gravity exerted on each of the balls by the other ball? **(b)** At what separation is the force of gravity between the balls equal to 2.0×10^{-9} N?

3. • A communications satellite with a mass of 520 kg is in a circular orbit about the Earth. The radius of the orbit is 35,000 km as measured from the center of the Earth. Calculate **(a)** the weight of the satellite on the surface of the Earth and **(b)** the gravitational force exerted on the orbiting satellite by the Earth.

4. • **The Attraction of Ceres** Ceres, the largest asteroid known, has a mass of roughly 8.7×10^{20} kg. If Ceres passes within 14,000 km of the spaceship in which you are traveling, what force does it exert on you? (Use an approximate value for your mass, and treat yourself and the asteroid as point objects.)

5. • In one hand you hold a 0.13-kg apple, in the other hand a 0.22-kg orange. The apple and orange are separated by 0.75 m. What is the magnitude of the force of gravity that **(a)** the orange exerts on the apple and **(b)** the apple exerts on the orange?

6. •• **Predict/Calculate** A spaceship of mass m travels from the Earth to the Moon along a line that passes through the center of the Earth and the center of the Moon. **(a)** At what distance from the center of the Earth is the force due to the Earth twice the magnitude of the force due to the Moon? **(b)** How does your answer to part (a) depend on the mass of the spaceship? Explain.

7. •• At new moon, the Earth, Moon, and Sun are in a line, as indicated in **FIGURE 12-33**. Find the direction and magnitude of the net

gravitational force exerted on (a) the Earth, (b) the Moon, and (c) the Sun.

FIGURE 12-33 Problem 7

8. •• When the Earth, Moon, and Sun form a right triangle, with the Moon located at the right angle, as shown in **FIGURE 12-34**, the Moon is in its third-quarter phase. (The Earth is viewed here from above its North Pole.) Find the magnitude and direction of the net force exerted on the Moon. Give the direction relative to the line connecting the Moon and the Sun.

Sun   Moon
◉   ○

                                        Earth
                                          🌍

FIGURE 12-34 Problems 8, 9, and 69

9. •• Repeat the previous problem, this time finding the magnitude and direction of the net force acting on the Sun. Give the direction relative to the line connecting the Sun and the Moon.

10. •• Predict/Calculate Three 7.25-kg masses are at the corners of an equilateral triangle and located in space far from any other masses. (a) If the sides of the triangle are 0.610 m long, find the magnitude of the net force exerted on each of the three masses. (b) How does your answer to part (a) change if the sides of the triangle are doubled in length?

11. •• Predict/Calculate Four masses are positioned at the corners of a rectangle, as indicated in **FIGURE 12-35**. (a) Find the magnitude and direction of the net force acting on the 2.0-kg mass. (b) How do your answers to part (a) change (if at all) if all sides of the rectangle are doubled in length?

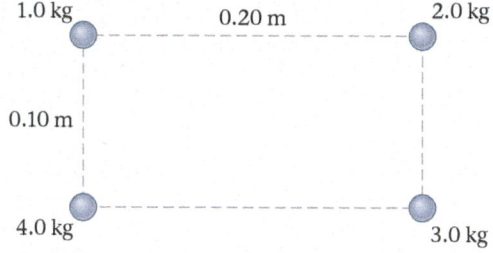

FIGURE 12-35 Problems 11 and 39

12. ••• Suppose that three astronomical objects (1, 2, and 3) are observed to lie on a line, and that the distance from object 1 to object 3 is D. Given that object 1 has four times the mass of object 3 and seven times the mass of object 2, find the distance between objects 1 and 2 for which the net force on object 2 is zero.

SECTION 12-2 GRAVITATIONAL ATTRACTION OF SPHERICAL BODIES

13. • Find the acceleration due to gravity on the surface of (a) Mercury and (b) Venus.

14. • At what altitude above the Earth's surface is the acceleration due to gravity equal to $g/4$?

15. • Two 6.4-kg bowling balls, each with a radius of 0.11 m, are in contact with one another. What is the gravitational attraction between the bowling balls?

16. • What is the acceleration due to Earth's gravity at a distance from the center of the Earth equal to the orbital radius of the Moon?

17. • **Extrasolar Planet Gravity** Kepler-62e is an exoplanet that orbits within the habitable zone around its parent star. The planet has a mass that is 3.57 times larger than Earth's and a radius that is 1.61 times larger than Earth's. Calculate the acceleration of gravity on the surface of Kepler-62e.

18. •• Predict/Calculate At a certain distance from the center of the Earth, a 5.0-kg object has a weight of 3.6 N. (a) Find this distance. (b) If the object is released at this location and allowed to fall toward the Earth, what is its initial acceleration? (c) If the object is now moved twice as far from the Earth, by what factor does its weight change? Explain. (d) By what factor does its initial acceleration change? Explain.

19. •• The acceleration due to gravity on the Moon's surface is known to be about one-sixth the acceleration due to gravity on the Earth. Given that the radius of the Moon is roughly one-quarter that of the Earth, find the mass of the Moon in terms of the mass of the Earth.

20. •• **Gravitational Tug of War** At some point along the direct path from the center of the Earth to the center of the Moon, the gravitational force of attraction on a spacecraft from the Moon becomes greater than the force from the Earth. (a) How far from the center of the Earth does this occur? (b) At this location, how far is the spacecraft from the surface of the Moon? How far is it from the surface of the Earth?

21. •• Predict/Calculate **An Extraterrestrial Volcano** Several volcanoes have been observed erupting on the surface of Jupiter's closest Galilean moon, Io. Suppose that material ejected from one of these volcanoes reaches a height of 5.00 km after being projected straight upward with an initial speed of 134 m/s. Given that the radius of Io is 1820 km, (a) outline a strategy that allows you to calculate the mass of Io. (b) Use your strategy to calculate Io's mass.

22. ••• Consider an asteroid with a radius of 19 km and a mass of 3.35×10^{15} kg. Assume the asteroid is roughly spherical. (a) What is the acceleration due to gravity on the surface of the asteroid? (b) Suppose the asteroid spins about an axis through its center, like the Earth, with a rotational period T. What is the smallest value T can have before loose rocks on the asteroid's equator begin to fly off the surface?

SECTION 12-3 KEPLER'S LAWS OF ORBITAL MOTION

23. • **CE** A satellite orbits the Earth in a circular orbit of radius r. At some point its rocket engine is fired so that its speed increases rapidly by a small amount. As a result, do the (a) apogee distance and (b) perigee distance increase, decrease, or stay the same?

24. • **CE** Predict/Explain **The Earth-Moon Distance Is Increasing** Laser reflectors left on the surface of the Moon by the Apollo astronauts show that the average distance from the Earth to the Moon is increasing at the rate of 3.8 cm per year. (a) As a result, will the length of the month increase, decrease, or remain the same? (b) Choose the *best explanation* from among the following:

   I. The greater the radius of an orbit, the greater the period, which implies a longer month.

  II. The length of the month will remain the same due to conservation of angular momentum.

  III. The speed of the Moon is greater with increasing radius; therefore, the length of the month will decrease.

25. • **Apollo Missions** On Apollo missions to the Moon, the command module orbited at an altitude of 110 km above the lunar surface. How much time did it take for the command module to complete one orbit?

26. • Find the orbital speed of a satellite in a geosynchronous circular orbit 3.58×10^7 m above the surface of the Earth.

27. • **An Extrasolar Planet** In July of 1999 a planet was reported to be orbiting the Sun-like star Iota Horologii with a period of 320 days. Find the radius of the planet's orbit, assuming that Iota Horologii has the same mass as the Sun. (This planet is presumably similar to Jupiter, but it may have large, rocky moons that enjoy a relatively pleasant climate.)

28. • Phobos, one of the moons of Mars, orbits at a distance of 9378 km from the center of the red planet. What is the orbital period of Phobos?

29. •• Predict/Calculate **An Asteroid with Its Own Moon** The asteroid 243 Ida has its own small moon, Dactyl. (See the photo on p. 394.) Another such system, asteroid 624 Hektor, has a mass of 7.9×10^{18} kg, with its moon orbiting at a radius of 623.5 km and a period of 2.965 days. (a) Given that the orbital radius of Dactyl is 108 km, and its period is 1.54 days, is the mass of 243 Ida greater than, less than, or equal to the mass of 624 Hektor? (b) Calculate the mass of 243 Ida.

30. •• **GPS Satellites** GPS (Global Positioning System) satellites orbit at an altitude of 2.0×10^7 m. Find (a) the orbital period, and (b) the orbital speed of such a satellite.

31. •• Predict/Calculate Two satellites orbit the Earth, with satellite 1 at a higher altitude than satellite 2. (a) Which satellite has the greater orbital speed? Explain. (b) Calculate the orbital speed of a satellite that orbits at an altitude of one Earth radius above the surface of the Earth. (c) Calculate the orbital speed of a satellite that orbits at an altitude of two Earth radii above the surface of the Earth.

32. •• Predict/Calculate Satellite A has a mass of 1000 kg and satellite B has a mass of 2000 kg. (a) When each satellite orbits at one Earth radius above the surface of the Earth, will the period of satellite B be longer, shorter, or the same as the period of satellite A? (b) Calculate the orbital periods of satellites A and B.

33. •• Predict/Calculate The Martian moon Deimos has an orbital period that is greater than the other Martian moon, Phobos. Both moons have approximately circular orbits. (a) Is Deimos closer to or farther from Mars than Phobos? Explain. (b) Calculate the distance from the center of Mars to Deimos given that its orbital period is 1.10×10^5 s.

34. •• Predict/Calculate (a) Calculate the orbital period of a satellite that orbits at two Earth radii above the surface of the Earth. (b) If the Earth suddenly became more massive while the satellite remains in circular orbit at the same altitude, will the orbital period increase, decrease, or remain the same? (c) Calculate the orbital period of the satellite if the Earth's mass were to double to 1.194×10^{25} kg and everything else were to remain the same.

35. ••• **Binary Stars** Alpha Centauri A and Alpha Centauri B are binary stars with a separation of 3.45×10^{12} m and an orbital period of 2.52×10^9 s. Assuming the two stars are equally massive (which is approximately the case), determine their mass.

SECTION 12-4 GRAVITATIONAL POTENTIAL ENERGY

36. • **Sputnik** The first artificial satellite to orbit the Earth was Sputnik I, launched October 4, 1957. The mass of Sputnik I was 83.5 kg, and its distances from the center of the Earth at apogee and perigee were 7330 km and 6610 km, respectively. Find the difference in gravitational potential energy for Sputnik I as it moved from apogee to perigee.

37. • How much gravitational potential energy is required to lift a 9380-kg *Progress* spacecraft to the altitude of the International Space Station, 422 km above the surface of the Earth?

38. •• CE Predict/Explain (a) Is the amount of energy required to get a spacecraft from the Earth to the Moon greater than, less than, or equal to the energy required to get the same spacecraft from the Moon to the Earth? (b) Choose the *best explanation* from among the following:
    I. The escape speed of the Moon is less than that of the Earth; therefore, less energy is required to leave the Moon.
    II. The situation is symmetric, and hence the same amount of energy is required to travel in either direction.
    III. It takes more energy to go from the Moon to the Earth because the Moon is orbiting the Earth.

39. •• Predict/Calculate Consider the four masses shown in Figure 12-35. (a) Find the total gravitational potential energy of this system. (b) How does your answer to part (a) change if all the masses in the system are doubled? (c) How does your answer to part (a) change if, instead, all the sides of the rectangle are halved in length?

40. •• Calculate the gravitational potential energy of a 9.50-kg mass (a) on the surface of the Earth and (b) at an altitude of 325 km. (c) Take the difference between the results for parts (b) and (a), and compare with mgh, where $h = 325$ km.

41. •• Three bowling balls form an equilateral triangle, as shown in **FIGURE 12-36**. Each ball has a radius of 10.8 cm and a mass of 7.26 kg. What is the total gravitational potential energy of this system?

42. •• Two 0.59-kg basketballs, each with a radius of 12 cm, are just touching. How much energy is required to change the separation between the centers of the basketballs to (a) 1.0 m and (b) 10.0 m? (Ignore any other gravitational interactions.)

FIGURE 12-36

43. •• Find the minimum kinetic energy needed for a 32,000-kg rocket to escape (a) the Moon or (b) the Earth.

SECTION 12-5 ENERGY CONSERVATION

44. • CE Predict/Explain Suppose the Earth were to suddenly shrink to half its current diameter, with its mass remaining constant. (a) Would the escape speed of the Earth increase, decrease, or stay the same? (b) Choose the best explanation from among the following:
    I. Since the radius of the Earth would be smaller, the escape speed would also be smaller.
    II. The Earth would have the same amount of mass, and hence its escape speed would be unchanged.
    III. The force of gravity would be much stronger on the surface of the compressed Earth, leading to a greater escape speed.

45. • CE Is the energy required to launch a rocket vertically to a height h greater than, less than, or equal to the energy required to put the same rocket into orbit at the height h? Explain.

46. • Suppose one of the Global Positioning System satellites has a speed of 4.46 km/s at perigee and a speed of 3.64 km/s at apogee. If the distance from the center of the Earth to the satellite at perigee is 2.00×10^4 km, what is the corresponding distance at apogee?

47. • **Meteorites from Mars** Several meteorites found in Antarctica are believed to have come from Mars, including the famous ALH84001

meteorite that was once thought to contain fossils of ancient life on Mars. Meteorites from Mars are thought to get to Earth by being blasted off the Martian surface when a large object (such as an asteroid or a comet) crashes into the planet. What speed must a rock have to escape Mars?

48. • What is the launch speed of a projectile that rises vertically above the Earth to an altitude equal to one Earth radius before coming to rest momentarily?

49. • A projectile launched vertically from the surface of the Moon rises to an altitude of 425 km. What was the projectile's initial speed?

50. •• (a) How much gravitational potential energy must a 3130-kg satellite acquire in order to attain a geosynchronous orbit? (b) How much kinetic energy must it gain? Note that because of the rotation of the Earth on its axis, the satellite had a velocity of 463 m/s relative to the center of the Earth just before launch.

51. •• Predict/Calculate **Halley's Comet** Halley's comet, which passes around the Sun every 76 years, has an elliptical orbit. When closest to the Sun (perihelion) it is at a distance of 8.823×10^{10} m and moves with a speed of 54.6 km/s. The greatest distance between Halley's comet and the Sun (aphelion) is 6.152×10^{12} m. (a) Is the speed of Halley's comet greater than or less than 54.6 km/s when it is at aphelion? Explain. (b) Calculate its speed at aphelion.

52. •• **The End of the Lunar Module** On Apollo Moon missions, the lunar module would blast off from the Moon's surface and dock with the command module in lunar orbit. After docking, the lunar module would be jettisoned and allowed to crash back onto the lunar surface. Seismometers placed on the Moon's surface by the astronauts would then pick up the resulting seismic waves. Find the impact speed of the lunar module, given that it is jettisoned from an orbit 110 km above the lunar surface moving with a speed of 1630 m/s.

53. •• If a projectile is launched vertically from the Earth with a speed equal to the escape speed, how high above the Earth's surface is it when its speed is one-third the escape speed?

54. •• Suppose a planet is discovered orbiting a distant star. If the mass of the planet is 10 times the mass of the Earth, and its radius is one-tenth the Earth's radius, how does the escape speed of this planet compare with that of the Earth?

55. •• A projectile is launched vertically from the surface of the Moon with an initial speed of 1010 m/s. At what altitude is the projectile's speed one-half its initial value?

56. •• To what radius would the Sun have to be contracted for its escape speed to equal the speed of light?

57. •• Predict/Calculate Two baseballs, each with a mass of 0.148 kg, are separated by a distance of 395 m in outer space, far from any other objects. (a) If the balls are released from rest, what speed do they have when their separation has decreased to 145 m? (b) Suppose the mass of the balls is doubled. Would the speed found in part (a) increase, decrease, or stay the same? Explain.

58. ••• On Earth, a person can jump vertically and rise to a height h. What is the radius of the largest spherical asteroid from which this person could escape by jumping straight upward? Assume that each cubic meter of the asteroid has a mass of 3500 kg.

SECTION 12-6   TIDES

59. •• The magnitude of the tidal force exerted on a linear object of mass m and length L is approximately $2GmML/r^3$. In this expression, M is the mass of the body causing the tidal force and r is the distance from the center of m to the center of M. Suppose you are 1 million miles (1.6×10^9 m) away from a black hole whose mass

is 1.99×10^{36} kg (one million times that of the Sun). (a) Estimate the tidal force exerted on your body ($L = 1.8$ m) by the black hole. (b) At what distance will the tidal force be approximately 10 times greater than your weight?

60. •• The magnitude of the tidal force between the International Space Station (ISS) and a nearby astronaut on a spacewalk is approximately $2GmMa/r^3$. In this expression, M is the mass of the Earth, $r = 6.79 \times 10^6$ m is the distance from the center of the Earth to the orbit of the ISS, $m = 125$ kg is the mass of the astronaut, and $a = 10$ m is the distance from the astronaut to the center of mass of the ISS. (a) Calculate the magnitude of the tidal force for this astronaut. This force tends to separate the astronaut from the ISS if the astronaut is located along the line that connects the center of the Earth with the center of mass of the ISS. (b) Calculate the force of gravitational attraction between the astronaut and the ISS if they are 10 m apart and the ISS has a mass of 420,000 kg. (c) Is the ISS orbit inside or outside the Roche limit?

61. ••• A dumbbell has a mass m on either end of a rod of length $2a$. The center of the dumbbell is a distance r from the center of the Earth, and the dumbbell is aligned radially. If $r \gg a$, show that the difference in the gravitational force exerted on the two masses by the Earth is approximately $4GmM_{\rm E}a/r^3$. (Note: The difference in force causes a tension in the rod connecting the masses. We refer to this as a *tidal force*.) [Hint: Use the fact that $1/(r - a)^2 - 1/(r + a)^2 \sim 4a/r^3$ for $r \gg a$.]

62. ••• Referring to the previous problem, suppose the rod connecting the two masses m is removed. In this case, the only force between the two masses is their mutual gravitational attraction. In addition, suppose the masses are spheres of radius a and mass $m = \frac{4}{3}\pi a^3\rho$ that touch each other. (The Greek letter ρ stands for the density of the masses.) (a) Write an expression for the gravitational force between the masses m. (b) Find the distance from the center of the Earth, r, for which the gravitational force found in part (a) is equal to the tidal force found in the previous problem. This distance is known as the *Roche limit*. (c) Calculate the Roche limit for Saturn, assuming $\rho = 3330$ kg/m³. (The famous rings of Saturn are within the Roche limit for that planet. Thus, the innumerable small objects, composed mostly of ice, that make up the rings will never coalesce to form a moon.)

GENERAL PROBLEMS

63. • CE You weigh yourself on a scale inside an airplane flying due east above the equator. If the airplane now turns around and heads due west with the same speed, will the reading on the scale increase, decrease, or stay the same? Explain.

64. • CE Rank objects A, B, and C in FIGURE 12-37 in order of increasing net gravitational force experienced by the object. Indicate ties where appropriate.

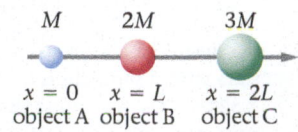

M $2M$ $3M$

$x = 0$ $x = L$ $x = 2L$
object A   object B   object C

FIGURE 12-37 Problems 64 and 65

65. • CE Referring to Figure 12-37, rank objects A, B, and C in order of increasing initial acceleration each would experience if it alone were allowed to move. Indicate ties where appropriate.

66. • CE **The Crash of Skylab** Skylab, the largest spacecraft ever to fall back to the Earth, met its fiery end on July 11, 1979, after flying directly over Everett, WA, on its last orbit. On the *CBS Evening News* the night before the crash, anchorman Walter Cronkite, in his rich baritone voice, made the following statement: "NASA says there is a little chance that Skylab will land in a populated area."

After the commercial, he immediately corrected himself by saying, "I meant to say '*there is little chance*' Skylab will hit a populated area." In fact, it landed primarily in the Indian Ocean off the west coast of Australia, though several pieces were recovered near the town of Esperance, Australia, which later sent the U.S. State Department a $400 bill for littering. The cause of Skylab's crash was the friction it experienced in the upper reaches of the Earth's atmosphere. As the radius of Skylab's orbit decreased, did its speed increase, decrease, or stay the same? Explain.

67. • Consider a system consisting of three masses on the *x* axis. Mass $m_1 = 1.00$ kg is at $x = 1.00$ m; mass $m_2 = 2.00$ kg is at $x = 2.00$ m; and mass $m_3 = 3.00$ kg is at $x = 3.00$ m. What is the total gravitational potential energy of this system?

68. •• An astronaut exploring a distant solar system lands on an unnamed planet with a radius of 3520 km. When the astronaut jumps upward with an initial speed of 2.91 m/s, she rises to a height of 0.525 m. What is the mass of the planet?

69. •• **Predict/Calculate** When the Moon is in its third-quarter phase, the Earth, Moon, and Sun form a right triangle, as shown in Figure 12-34. Calculate the magnitude of the force exerted on the Moon by **(a)** the Earth and **(b)** the Sun. **(c)** Does it make more sense to think of the Moon as orbiting the Sun, with a small effect due to the Earth, or as orbiting the Earth, with a small effect due to the Sun?

70. •• An equilateral triangle 10.0 m on a side has a 1.00-kg mass at one corner, a 2.00-kg mass at another corner, and a 3.00-kg mass at the third corner (**FIGURE 12-38**). Find the magnitude and direction of the net force acting on the 2.00-kg mass.

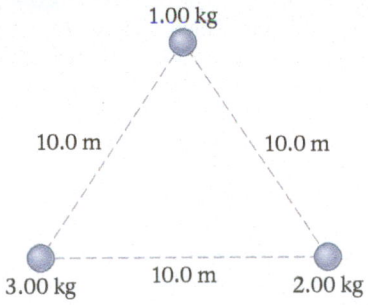

FIGURE 12-38 Problems 70 and 71

71. •• Suppose that each of the three masses in Figure 12-38 is replaced by a mass of 5.95 kg and radius 0.0714 m. If the masses are released from rest, what speed will they have when they collide at the center of the triangle? Ignore gravitational effects from any other objects.

72. •• **A Near Miss!** In the early morning hours of June 14, 2002, the Earth had a remarkably close encounter with an asteroid the size of a small city. The previously unknown asteroid, now designated 2002 MN, remained undetected until three days after it had passed the Earth. At its closest approach, the asteroid was 73,600 miles from the center of the Earth—about a third of the distance to the Moon. **(a)** Find the speed of the asteroid at closest approach, assuming its speed at infinite distance to be zero and considering only its interaction with the Earth. **(b)** Observations indicate the asteroid to have a diameter of about 0.730 km. Estimate the kinetic energy of the asteroid at closest approach, assuming it has an average density of 3.33 g/cm³. (For comparison, a 1-megaton nuclear weapon releases about 4.2×10^{15} J of energy.)

73. •• **Predict/Calculate** Suppose a planet is discovered that has the same total mass as the Earth, but half its radius. **(a)** Is the acceleration due to gravity on this planet more than, less than, or the same as the acceleration due to gravity on the Earth? Explain. **(b)** Calculate the acceleration due to gravity on this planet.

74. •• Show that the speed of a satellite in a circular orbit a height *h* above the surface of the Earth is

$$v = \sqrt{\frac{GM_E}{R_E + h}}$$

75. •• **Walking into Orbit** A spherical asteroid of average density would have a mass of 8.7×10^{13} kg if its radius were 2.0 km. **(a)** If you and your spacesuit had a mass of 125 kg, how much would you weigh when standing on the surface of this asteroid? **(b)** If you could walk on the surface of this asteroid, what minimum speed would you need to launch yourself into an orbit just above the surface of the asteroid?

76. •• In a binary star system, two stars orbit about their common center of mass, as shown in **FIGURE 12-39**. If $r_2 = 2r_1$, what is the ratio of the masses m_2/m_1 of the two stars?

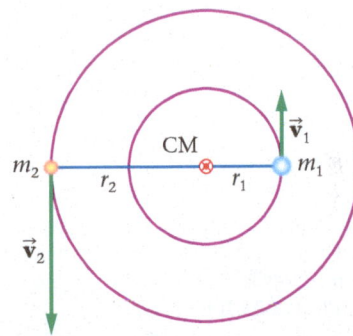

FIGURE 12-39 Problems 76 and 77

77. •• Find the orbital period of the binary star system described in the previous problem.

78. •• **Exploring Mars** In the not-too-distant future astronauts will travel to Mars to carry out scientific explorations. As part of their mission, it is likely that a "geosynchronous" satellite will be placed above a given point on the Martian equator to facilitate communications. At what altitude above the surface of Mars should such a satellite orbit? (*Note:* The Martian "day" is 24.6229 hours. Other relevant information can be found in Appendix C.)

79. •• **Comet Wild 2** In 2004, a NASA spacecraft named *Stardust* flew within 147 miles of Comet Wild 2 (pronounced "Vilt 2"), zooming by it at 6200 m/s, about six times the speed of a rifle bullet. Photos taken by *Stardust* show that the comet is roughly spherical, as shown in **FIGURE 12-40**, with a radius of 2.7 km. It has also been determined that the acceleration due to gravity on the surface of Wild 2 is 0.00010g. What is the minimum speed needed for an object to escape from the surface of Wild 2?

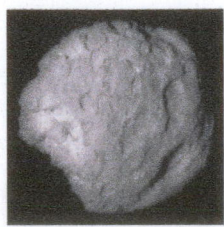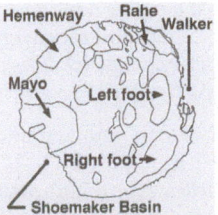

FIGURE 12-40 Comet Wild 2 and some of its surface features, including the Walker basin, the site of unusual jets of outward-flowing dust and rocks. (Problem 79)

80. •• **Predict/Calculate** (a) If you want to launch a probe that orbits the Sun with a period of 2.00 years, what should be its orbital radius? Give your answer in astronomical units. The astronomical unit AU is defined as the mean distance from the Sun to the Earth (1 AU = 1.50×10^{11} m). (b) Is the orbital speed of this probe greater than, less than, or equal to the orbital speed of the Earth? Explain. (c) What is the orbital speed of this probe?

81. •• **Predict/Calculate** A satellite is placed in Earth orbit at an altitude of 23,300 mi, which is 1100 miles higher than the altitude of a geosynchronous satellite. (a) Is the period of this satellite greater than or less than 24 hours? (b) As viewed from the surface of the Earth, does the satellite move eastward or westward? Explain. (c) Find the orbital period of this satellite.

82. •• Show that the force of gravity between the Moon and the Sun is always greater than the force of gravity between the Moon and the Earth.

83. •• The astronomical unit AU is defined as the mean distance from the Sun to the Earth (1 AU = 1.50×10^{11} m). Apply Kepler's third law (Equation 12-7) to the solar system, and show that it can be written as

$$T = Cr^{3/2}$$

In this expression, the period T is measured in years, the distance r is measured in astronomical units, and the constant C has a magnitude that you must determine.

84. •• (a) Find the kinetic energy of a 1940-kg satellite in a circular orbit about the Earth, given that the radius of the orbit is 12,600 miles. (b) How much energy is required to move this satellite to a circular orbit with a radius of 25,200 miles?

85. •• **Predict/Calculate Space Station Orbit** The International Space Station ($m = 4.5 \times 10^5$ kg) orbits at an altitude of 415 km above the Earth's surface. (a) Does the orbital speed of the station depend on its mass? Explain. (b) Find the speed of the station in its orbit. (c) How much time does it take for the station to complete one orbit of the Earth?

86. •• **Approaching the ISS** A Russian *Soyuz* module, with three astronauts and a full load of cargo, has a mass of 7500 kg. The International Space Station (ISS) has a mass of 420,000 kg. When *Soyuz* docks at the ISS, the two centers of mass are separated by 9.5 m. (a) Find the average of the gravitational force of attraction between the two spacecraft when *Soyuz* is docked and when it is 110 m from the ISS. (b) If *Soyuz* approached the ISS from rest at a distance of 110 m, a thruster would have to counteract the average force of gravitational attraction between it and the ISS. If this thruster has an exhaust velocity of 590 m/s, at what average rate must it burn fuel (in kg/s) to counteract the pull of the ISS? (See Section 9-8 for a discussion of thrust.) (c) If the approach requires 330 s, how much fuel will the thruster burn in order to counteract the gravitational attraction?

87. ••• **Predict/Calculate** Consider an object of mass m orbiting the Earth at a radius r. (a) Find the speed of the object. (b) Show that the total mechanical energy of this object is equal to (-1) times its kinetic energy. (c) Does the result of part (b) apply to an object orbiting the Sun? Explain.

88. ••• The Earth's orbit around the Sun is slightly elliptical (see Conceptual Example 12-7). At Earth's closest approach to the Sun (perihelion) the orbital radius is 1.471×10^{11} m, and at its farthest distance (aphelion) the orbital radius is 1.521×10^{11} m. Given

that the mass of the Earth is 5.972×10^{24} kg, and that of the Sun is 1.989×10^{30} kg, (a) find the difference in gravitational potential energy between when the Earth is at its aphelion and perihelion radii. (b) If the orbital speed of the Earth is 29,290 m/s at aphelion, what is its orbital speed at perihelion?

89. ••• Three identical stars, at the vertices of an equilateral triangle, orbit about their common center of mass (**FIGURE 12-41**). Find the period of this orbital motion in terms of the orbital radius, R, and the mass of each star, M.

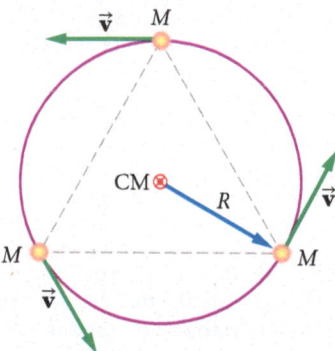

FIGURE 12-41 Problem 89

90. ••• Find an expression for the kinetic energy of a satellite of mass m in an orbit of radius r about a planet of mass M.

PASSAGE PROBLEMS

If life exists elsewhere in our solar system, it may not have developed independently from life on Earth. Instead, it's possible that microbes from Earth may have colonized other planets or moons by hitching a ride on a rock blasted from Earth's surface by a meteor impact. If the impact gives the rock enough energy to escape into space (while at the same time not raising its temperature so high as to "cook" the microbes), the rock may eventually reach another body in the solar system. In fact, rocks from Mars are known to have reached Earth in just this way, although none are currently known to have contained microbes. Computer modeling can be used to estimate the probability that a rock ejected from the surface of the Earth with a speed greater than the escape speed will reach another planet. These computer models indicate that under the influence of gravitational fields from the other objects in the solar system, an ejected rock can take millions of years to travel from one planet to another. During this time any life "aboard" is continually exposed to the high radiation levels of space. Some researchers have calculated that a 3.0-m-diameter rock at a typical rock density of 3.0 g/cm³ is sufficient to shield some types of microbes from the hostile environment of space for several million years of travel.

The accompanying plot shows the residual speed of an ejected object—that is, the speed the object would have when infinitely far from the Earth—as a function of its speed at the surface of the Earth (its original ejection speed). By simulating the motion of rocks ejected from the Earth with a variety of speeds, researchers conclude that 0.03% of the rocks ejected such that they have a residual speed of 2.5 km/s will have reached Mars 2.0 million years later. Although this doesn't seem like a high probability, there have been so many meteor impacts over the long history of the Earth that many ejected rocks must have reached Mars—though whether they carried microbes, and if they did, whether the microbes would have survived, are open questions.

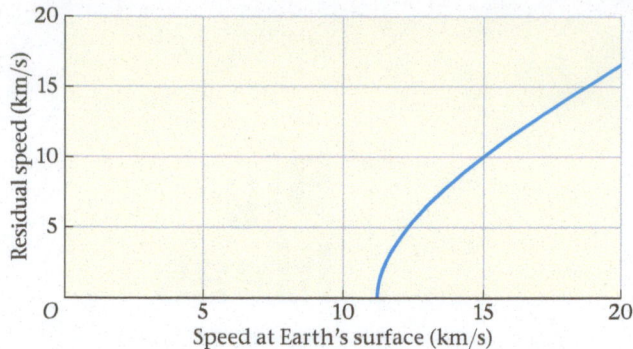

91. • What is the minimum kinetic energy an impact must give to a 3.0-m-diameter rock for the rock to be able to escape from the Earth?

    A. 5.3×10^3 J   **B.** 1.3×10^6 J
    C. 2.7×10^{12} J   **D.** 2.1×10^{13} J

92. • What is the speed a rock needs to be given at the surface of the Earth in order for it to have a residual speed of 2.5 km/s?

    A. 8.7 km/s   **B.** 10.9 km/s
    C. 11.5 km/s   **D.** 13.7 km/s

93. • Would increasing the ejection speed from 12 km/s to 13 km/s change the residual speed by more than, less than, or the same amount as increasing the ejection speed from 15 km/s to 16 km/s?

94. •• Consider a similar plot for rocks ejected from Mars. Where would this plot intercept the *x* axis?

A. The plot for Mars would intercept the *x* axis at 5.0 km/s.
B. The plot for Mars would intercept the *x* axis at 16.2 km/s.
C. The plot for Mars would intercept the *x* axis at 11.2 km/s.
D. The plot for Mars would not intercept the *x* axis.

95. •• **REFERRING TO EXAMPLE 12-8** Find the orbital radius that corresponds to a "year" of 150 days.

96. •• Predict/Calculate **REFERRING TO EXAMPLE 12-8** Suppose the mass of the Sun is suddenly doubled, but the Earth's orbital radius remains the same. **(a)** Would the length of an Earth year increase, decrease, or stay the same? **(b)** Find the length of a year for the case of a Sun with twice the mass. **(c)** Suppose the Sun retains its present mass, but the mass of the Earth is doubled instead. Would the length of the year increase, decrease, or stay the same?

97. •• Predict/Calculate **REFERRING TO EXAMPLE 12-17** **(a)** If the mass of the Earth were doubled, would the escape speed of a rocket increase, decrease, or stay the same? **(b)** Calculate the escape speed of a rocket for the case of an Earth with twice its present mass. **(c)** If the mass of the Earth retains its present value, but the mass of the rocket is doubled, does the escape speed increase, decrease, or stay the same?

98. •• Predict/Calculate **REFERRING TO EXAMPLE 12-17** Suppose the Earth is suddenly shrunk to half its present radius without losing any of its mass. **(a)** Would the escape speed of a rocket increase, decrease, or stay the same? **(b)** Find the escape speed for an Earth with half its present radius.

Oscillations About Equilibrium

1 Simple harmonic motion occurs when the restoring force is proportional to the displacement.

2 Simple harmonic motion is closely related to uniform circular motion.

3 The period of a mass on a spring depends on the mass and the force constant.

4 The period of a pendulum is proportional to the square root of its length.

5 An oscillator driven at its resonance frequency has a large amplitude of oscillation.

▲ A struck tuning fork and a plucked guitar string are just two examples of objects that oscillate about their equilibrium positions. In this chapter we explore the behavior of a variety of oscillating systems, including the classic case of a swinging pendulum, and lay the foundations for understanding many other natural phenomena as well.

A system in equilibrium is seldom left undisturbed for very long, it seems, before it's given a bump, a kick, or a nudge that displaces it from its equilibrium position. When this happens, the system **oscillates** back and forth from one side of the equilibrium position to the other. This chapter explores the basic physics of all oscillating systems, from a pendulum in a grandfather clock, to water molecules in a microwave oven, to the universe oscillating between "big bangs" and "big crunches."

13-1 Periodic Motion

A motion that repeats itself over and over is referred to as **periodic motion**. The beating of your heart, as illustrated in **FIGURE 13-1**, the ticking of a clock, and the movement of a child on a swing are familiar examples.

Period One of the key characteristics of a periodic system is the time required for the completion of one cycle of its repetitive motion. For example, the pendulum in a grandfather clock might take one second to swing from maximum displacement in one direction to maximum displacement in the opposite direction, and another second to swing back to its original position. Thus, the **period**, T, of this pendulum—the time required to complete one oscillation—is 2 s.

> **Definition of Period, T**
>
> T = time required for one cycle of a periodic motion
>
> SI unit: seconds/cycle = s

Notice that a cycle (that is, an oscillation) is dimensionless.

Frequency Closely related to the period is another common measure of periodic motion, the **frequency**, f. The frequency of an oscillation is simply the number of oscillations per unit of time. Thus, f tells us how frequently, or rapidly, an oscillation takes place—the higher the frequency, the more rapid the oscillations. By definition, the frequency is simply the inverse of the period, T:

> **Definition of Frequency, f**
>
> $$f = \frac{1}{T}$$ 13-1
>
> SI unit: cycle/second = 1/s = s^{-1}

If the period of an oscillation, T, is very small, resulting in rapid oscillations, the corresponding frequency, $f = 1/T$, is large, as expected.

A special unit has been introduced for the measurement of frequency. It is the **hertz** (Hz), named for the German physicist Heinrich Hertz (1857–1894), in honor of his pioneering studies of radio waves. By definition, one Hz is one cycle per second:

$$1 \text{ Hz} = 1 \text{ cycle/second}$$

High frequencies are often measured in kilohertz (kHz), where $1 \text{ kHz} = 10^3 \text{ Hz}$, or megahertz (MHz), where $1 \text{ MHz} = 10^6 \text{ Hz}$.

EXERCISE 13-1 COMPUTER PERIOD

The processing "speed" of a computer refers to the number of binary operations it can perform in 1 second, so it is really a frequency. If the processor of a certain computer operates at 3.70 GHz, how much time is required for one processing cycle?

REASONING AND SOLUTION
The frequency of the computer's processor is given as $f = 3.70 \text{ GHz}$. Therefore, we can rearrange $f = 1/T$ to solve for the processing period:

$$T = \frac{1}{f} = \frac{1}{3.70 \text{ GHz}} = \frac{1}{3.70 \times 10^9 \text{ cycles/s}} = 2.70 \times 10^{-10} \text{ s}$$

The high frequency of a processor corresponds to a very small period—less than a billionth of a second to complete one operation. We next consider a situation in which the frequency is considerably smaller.

EXERCISE 13-2 PLAYING CATCH

Two baseball players warm up with a game of catch. If it takes 6.6 s for the ball to go from one player to the other and then back to the first, what is the frequency of the ball's motion?

CONTINUED

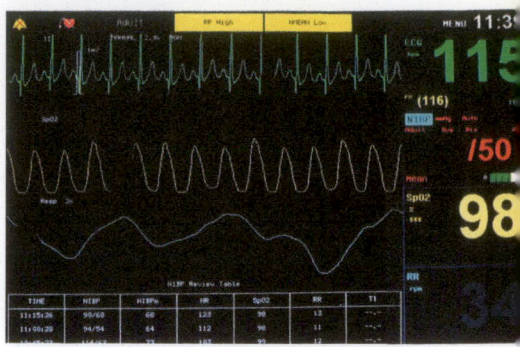

▲ **FIGURE 13-1** Periodic phenomena are found everywhere in nature, from the movements of the heavenly bodies to the vibration of individual atoms. The trace of an electrocardiogram (ECG or EKG), as shown here, records the rhythmic electrical activity that accompanies the beating of our hearts.

PHYSICS IN CONTEXT
Looking Ahead

Period, T, and frequency, f, play key roles in the study of sound waves, as shown in Chapter 14. They reappear when we study electromagnetic waves in Chapter 25.

PHYSICS IN CONTEXT
Looking Ahead

In Chapter 30 we study the beginnings of modern physics, and blackbody radiation in particular. As we shall see, the frequency f of light is directly related to the energy of a "particle of light," referred to as a photon.

REASONING AND SOLUTION

The period of this motion is the time for the ball to complete one round trip from one player to the other and back. Therefore, the period is $T = 6.6$ s, and the corresponding frequency is the inverse of the period:

$$f = \frac{1}{T} = \frac{1}{6.6\ \text{s}} = 0.15\ \text{Hz}$$

The period and frequency of periodic motion can vary over a remarkably large range. Table 13-1 gives a sampling of typical values of T and f.

TABLE 13-1 Typical Periods and Frequencies

System	Period (s)	Frequency (Hz)
Precession of the Earth	8.2×10^{11} (26,000 y)	1.2×10^{-12}
Hour hand of a clock	43,200 (12 h)	2.3×10^{-5}
Minute hand of a clock	3600	2.8×10^{-4}
Second hand of clock	60	0.017
Pendulum in grandfather clock	2.0	0.50
Human heartbeat	1.0	1.0
Lower range of human hearing	5.0×10^{-2}	20
Wing beat of housefly	5.0×10^{-3}	200
Upper range of human hearing	5.0×10^{-5}	20,000
Computer processor	5.6×10^{-10}	1.8×10^{9}

BIO An example of rapid periodic motion in nature is the drumming of a Hairy Woodpecker (*Picoides villosus*) (**FIGURE 13-2**), which can hammer its beak into a tree trunk as many as 25 times a second. It accomplishes this feat by using very strong thigh muscles to rotate its entire body, and then transferring the rotational motion to its head and beak. While the motion of its body can be roughly described by a sine or cosine function, the speed of its head is described by a more complex sawtooth function, achieving its maximum speed just before striking the tree. The rapid, periodic impacts allow the birds to drill for hidden bugs, construct hollowed-out nests, and make the characteristic woodpecker drumming sound, which itself is useful for declaring its territory and attracting a mate.

▲ **FIGURE 13-2** The Hairy Woodpecker can hammer a tree trunk as many as 25 times a second, and perform as many as 12,000 hits in a day. The impacts are severe enough (1000*g* or more) to render a person unconscious, but are just part of the daily routine for the woodpecker.

Enhance Your Understanding　　　(Answers given at the end of the chapter)

1. If the frequency of an oscillator is halved, by what factor does the period change?

Section Review

- Period, T, is the amount of time required to complete one cycle of an oscillation.
- Frequency, f, is the number of cycles per second. In terms of the period, $f = 1/T$.

13-2 Simple Harmonic Motion

Periodic motion can take many forms, such as a tennis ball going back and forth between players, or an expectant father pacing up and down a hospital hallway. There is one type of periodic motion, however, that is of particular importance. It is referred to as **simple harmonic motion**.

Oscillations of a Mass on a Spring　A classic example of simple harmonic motion is the oscillation of a mass attached to a spring. (See Section 6-2 for a discussion of ideal springs and the forces they exert.) To be specific, consider an air-track cart of mass m

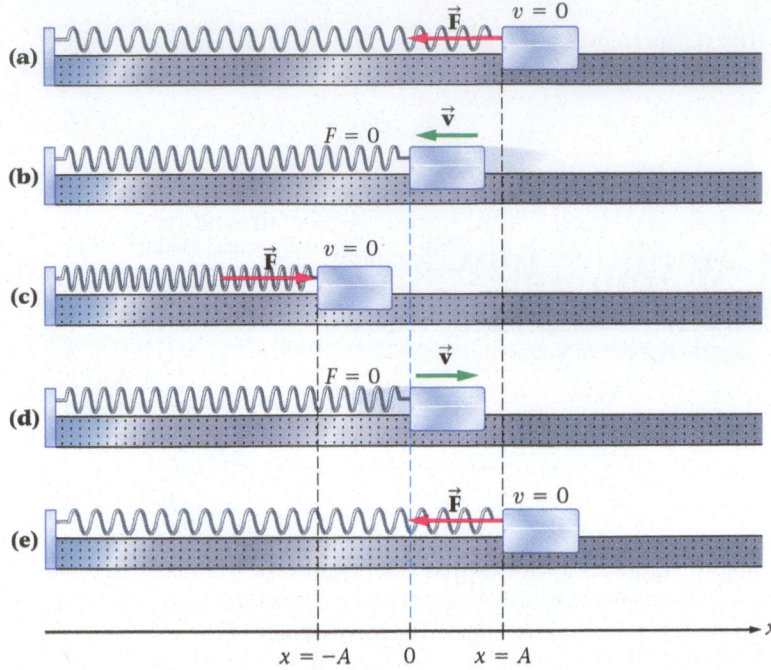

◄ **FIGURE 13-3 A mass attached to a
spring undergoes simple harmonic motion about x = 0** (**a**) The mass is at its
maximum positive value of x. Its velocity
is zero, and the force on it points to the
left with maximum magnitude. (**b**) The
mass is at the equilibrium position of the
spring. Here the speed has its maximum
value, and the force exerted by the spring
is zero. (**c**) The mass is at its maximum
displacement in the negative x direction.
The velocity is zero here, and the force
points to the right with maximum magnitude. (**d**) The mass is at the equilibrium
position of the spring, with zero force
acting on it and maximum speed. (**e**) The
mass has completed one cycle of its oscillation about $x = 0$.

attached to a spring of force constant k, as in **FIGURE 13-3**. When the spring is at its equilibrium length—neither stretched nor compressed—the cart is at the position $x = 0$,
where it will remain at rest if left undisturbed. If the cart is displaced from equilibrium by a distance x, however, the spring exerts a restoring force given by Hooke's law,
$F_x = -kx$. In words:

> A spring exerts a restoring force whose magnitude is proportional to the dis-
> tance it is displaced from equilibrium.

This direct proportionality between distance from equilibrium and force is the key feature of a mass–spring system that leads to simple harmonic motion. As for the direction of the spring force:

> The force exerted by a spring is opposite in direction to its displacement
> from equilibrium; this accounts for the minus sign in $F_x = -kx$. In general,
> a restoring force is one that *always* points toward the equilibrium position.

Big Idea 1 Simple harmonic
motion occurs when a system's
restoring force is proportional to the
displacement from equilibrium.

Now, suppose the cart is released from rest at the location $x = A$. As indicated in
Figure 13-3, the spring exerts a force on the cart to the left, causing the cart to accelerate
toward the equilibrium position. When the cart reaches $x = 0$, the net force acting on
it is zero. Its speed is not zero at this point, however, and so it continues to move to the
left. As the cart compresses the spring, it experiences a force to the right, causing it to
decelerate and finally come to rest at $x = -A$. The spring continues to exert a force to
the right; thus, the cart immediately begins to move to the right until it comes to rest
again at $x = A$, completing one oscillation in the time T.

Position as a Function of Time   If a pen is attached to the cart, it can trace its motion
on a strip of paper moving with constant speed, as indicated in **FIGURE 13-4**. On this
"strip chart" we obtain a record of the cart's motion as a function of time. As we see in
Figure 13-4, the motion of the cart looks like a sine or a cosine function.

Mathematical analysis, using the methods of calculus, shows that this is indeed
the case; that is, the position of the cart as a function of time can be represented by a
sine or a cosine function. The reason that either function works can be seen by considering **FIGURE 13-5**. If we take $t = 0$ to be at point 1, for example, the position as a function of time starts at zero, just like a sine function; if we choose $t = 0$ to be at point
2, however, the position versus time starts at its maximum value, just like a cosine
function. It's really the same mathematical function, differing only in the choice of
starting point.

▶ **FIGURE 13-4 Displaying position versus time for simple harmonic motion** As an air-track cart oscillates about its equilibrium position, a pen attached to it traces its motion onto a moving sheet of paper. This produces a "strip chart," showing that the cart's motion has the shape of a sine or a cosine function.

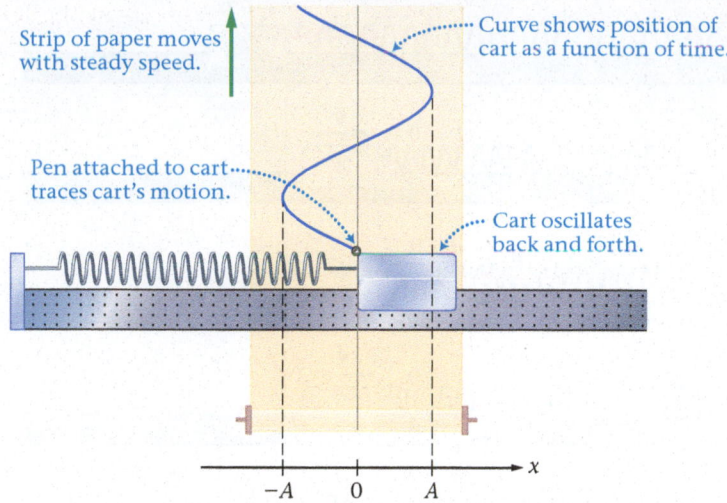

Strip of paper moves with steady speed.

Pen attached to cart traces cart's motion.

Curve shows position of cart as a function of time.

Cart oscillates back and forth.

Returning to Figure 13-3, notice that the position of the mass oscillates between $x = +A$ and $x = -A$. Because A represents the extreme displacement of the cart on either side of equilibrium, we refer to it as the **amplitude** of the motion. It follows that the amplitude is one-half the total range of motion. In addition, recall that the cart's motion repeats with a period T. As a result, the position of the cart is the same at the time $t + T$ as it is at the time t, as shown in Figure 13-5. Combining all these observations results in the following mathematical description of position versus time:

Position Versus Time in Simple Harmonic Motion

$$x = A \cos\left(\frac{2\pi}{T} t\right)$$

13-2

SI unit: m

This type of dependence on time—as a sine or a cosine—is characteristic of simple harmonic motion. With this particular choice, the position at $t = 0$ is $x = A \cos(0) = A$; thus, Equation 13-2 describes an object that has its maximum displacement at $t = 0$, like point 2 in Figure 13-5.

To see how Equation 13-2 works, recall that the cosine oscillates between $+1$ and -1. Therefore, $x = A \cos(2\pi t/T)$ will oscillate between $+A$ and $-A$, just as in the strip chart. Next, consider what happens if we replace the time t with the time $t + T$. This gives

$$x = A \cos\left(\frac{2\pi}{T}(t + T)\right)$$
$$= A \cos\left(\frac{2\pi}{T} t + \frac{2\pi}{T} T\right) = A \cos\left(\frac{2\pi}{T} t + 2\pi\right)$$

▶ **FIGURE 13-5 Simple harmonic motion as a sine or a cosine** The strip chart from Figure 13-4. The cart oscillates back and forth from $x = +A$ to $x = -A$, completing one cycle in the time T. The function traced by the pen can be represented as a sine function if $t = 0$ is taken to be at point 1, where the function is equal to zero. The function can be represented by a cosine if $t = 0$ is taken to be at point 2, where the function has its maximum value.

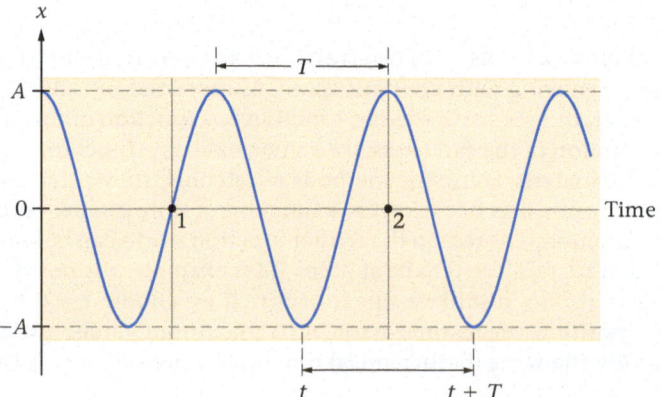

Using the fact that $\cos(\theta + 2\pi) = \cos\theta$ for any angle θ, we can rewrite the last expression as follows:

$$x = A\cos\left(\frac{2\pi}{T}t\right)$$

Therefore, as expected, the position at time $t + T$ is precisely the same as the position at time t.

EXAMPLE 13-3   SPRING TIME

An air-track cart attached to a spring completes one oscillation every 2.4 s. At $t = 0$ the cart is released from rest at a distance of 0.10 m from its equilibrium position. What is the position of the cart at **(a)** 0.30 s, **(b)** 0.60 s, **(c)** 2.7 s, and **(d)** 3.0 s?

PICTURE THE PROBLEM
In our sketch, we place the origin of the x-axis at the equilibrium position of the cart and choose the positive direction to point to the right. The cart is released from rest at $x = 0.10$ m, which means that its amplitude is $A = 0.10$ m. After it is released, the cart oscillates back and forth between $x = 0.10$ m and $x = -0.10$ m.

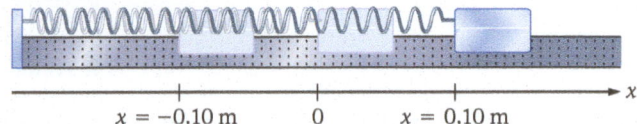

$x = -0.10$ m   0 $x = 0.10$ m

REASONING AND STRATEGY
Notice that the period of oscillation, $T = 2.4$ s, is given in the problem statement. Thus, we can find the position of the cart by evaluating $x = A\cos(2\pi t/T)$ at the desired times, using $A = 0.10$ m. (*Note:* Remember to have your calculator set to radian mode when evaluating these cosine functions.)

Known   Period of cart, $T = 2.4$ s; release point, $x = 0.10$ m.
Unknown   Position of cart at indicated times, $x = ?$

SOLUTION

Part (a)
1. Calculate x at the time $t = 0.30$ s:

$$x = A\cos\left(\frac{2\pi}{T}t\right) = (0.10\,\text{m})\cos\left[\left(\frac{2\pi}{2.4\,\text{s}}\right)(0.30\,\text{s})\right]$$
$$= (0.10\,\text{m})\cos(\pi/4) = 7.1\,\text{cm}$$

Part (b)
2. Now, substitute $t = 0.60$ s:

$$x = A\cos\left(\frac{2\pi}{T}t\right) = (0.10\,\text{m})\cos\left[\left(\frac{2\pi}{2.4\,\text{s}}\right)(0.60\,\text{s})\right]$$
$$= (0.10\,\text{m})\cos(\pi/2) = 0$$

Part (c)
3. Repeat with $t = 2.7$ s:

$$x = A\cos\left(\frac{2\pi}{T}t\right) = (0.10\,\text{m})\cos\left[\left(\frac{2\pi}{2.4\,\text{s}}\right)(2.7\,\text{s})\right]$$
$$= (0.10\,\text{m})\cos(9\pi/4) = 7.1\,\text{cm}$$

Part (d)
4. Repeat with $t = 3.0$ s:

$$x = A\cos\left(\frac{2\pi}{T}t\right) = (0.10\,\text{m})\cos\left[\left(\frac{2\pi}{2.4\,\text{s}}\right)(3.0\,\text{s})\right]$$
$$= (0.10\,\text{m})\cos(5\pi/2) = 0$$

INSIGHT
The results for parts (c) and (d) are the same as for parts (a) and (b), respectively. This is because the times in (c) and (d) are greater than the corresponding times in (a) and (b) by one period; that is, 2.7 s = 0.30 s + 2.4 s and 3.0 s = 0.60 s + 2.4 s.

PRACTICE PROBLEM
What is the first time the cart is at the position $x = -5.0$ cm?   [**Answer:** $t = T/3 = 0.80$ s]

Some related homework problems: Problem 10, Problem 17, Problem 18

Enhance Your Understanding   (Answers given at the end of the chapter)

2. Which of the following equations corresponds to the position-versus-time graph shown in **FIGURE 13-6**? In each equation, x is measured in meters and t is measured in

CONTINUED

seconds. **(a)** $x = (0.1 \text{ m}) \cos(2\pi t)$, **(b)** $x = (0.2 \text{ m}) \cos(2\pi t)$, **(c)** $x = (0.1 \text{ m}) \cos(\pi t)$, **(d)** $x = (0.2 \text{ m}) \cos(\pi t)$.

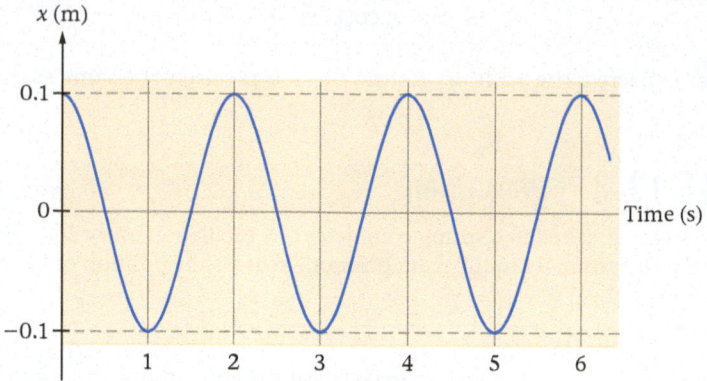

▲ **FIGURE 13-6** Enhance Your Understanding 2.

Section Review

- Simple harmonic motion can be represented mathematically by either a sine or a cosine function. In terms of the cosine function, we have $x = A \cos(2\pi t/T)$, where x is the position at time t, A is the amplitude (maximum displacement), and T is the period.

13-3 Connections Between Uniform Circular Motion and Simple Harmonic Motion

Though the motion of the air-track cart is strictly one-dimensional, it bears a close relationship to uniform circular motion. In this section, we explore this connection between simple harmonic motion and uniform circular motion in detail.

Uniform Circular Motion Imagine a turntable that rotates with a constant angular speed $\omega = 2\pi/T$, taking the time T to complete a revolution. At the rim of the turntable we place a small peg, as indicated in **FIGURE 13-7**. If we view the turntable from above, we see the peg undergoing uniform circular motion.

On the other hand, suppose we view the turntable from the side, so that the peg appears to move back and forth. Perhaps the easiest way to view this motion is to shine a light that casts a shadow of the peg on a screen, as shown in Figure 13-7. While the peg itself moves on a circular path, its shadow moves back and forth in a straight line.

To be specific, let the radius of the turntable be $r = A$, so that the shadow moves from $x = +A$ to $x = -A$. When the shadow is at $x = +A$, release a mass on a spring that is also at $x = +A$, so that the mass and the shadow start together. If we adjust the period of the turntable so that it completes one revolution in the same time T that the mass completes one oscillation, we find that the mass and the shadow move as one for all times. This is also illustrated in Figure 13-7. Recalling that the mass undergoes simple harmonic motion, it follows that the shadow does so as well.

We now use this connection, plus our knowledge of circular motion, to obtain detailed results for the position, velocity, and acceleration of a particle undergoing simple harmonic motion.

Position

In **FIGURE 13-8** we show the peg at the angular position θ, where θ is measured relative to the x axis. If the peg starts at $\theta = 0$ at $t = 0$, and the turntable rotates with a constant

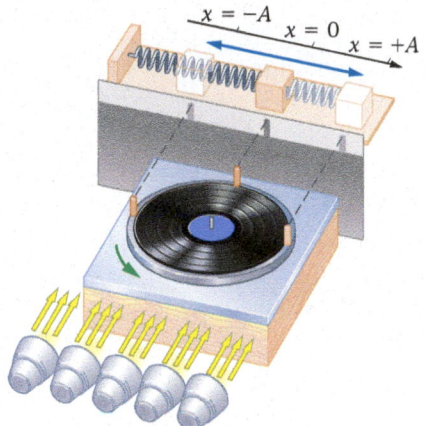

▲ **FIGURE 13-7 The relationship between uniform circular motion and simple harmonic motion** A peg is placed at the rim of a turntable that rotates with constant angular velocity. Viewed from above, the peg exhibits uniform circular motion. If the peg is viewed from the side, however, it appears to move back and forth in a straight line, as we can see by shining a light to cast a shadow of the peg onto a screen. The shadow moves with simple harmonic motion. If we compare this motion with the behavior of a mass on a spring, moving with the same period as the turntable and having an amplitude of motion equal to the radius of the turntable, we find that the mass and the shadow of the peg move together in simple harmonic motion.

angular speed ω, we know from Equation 10-10 that the angular position of the peg is simply

$$\theta = \omega t \qquad \text{13-3}$$

That is, the angular position increases linearly with time.

Now, imagine drawing a radius vector of length A to the position of the peg, as indicated in Figure 13-8. When we project the shadow of the peg onto the screen, the shadow is at the location $x = A \cos \theta$, which is the x component of the radius vector. Therefore, the position of the shadow as a function of time is:

Position of Shadow as a Function of Time

$$x = A \cos \theta = A \cos(\omega t) = A \cos\left(\frac{2\pi}{T}t\right) \qquad \text{13-4}$$

SI unit: m

Notice that we have used $\theta = \omega t$ (Equation 13-3) to express θ in terms of the time, t. Equations 13-4 and 13-2 are identical, so the shadow does indeed exhibit simple harmonic motion, just like a mass on a spring.

For notational simplicity, we often write the position of a mass on a spring in the form $x = A \cos(\omega t)$, which is more compact than $x = A \cos(2\pi t/T)$. When referring to a rotating turntable, ω is called the angular speed; when referring to simple harmonic motion, or other periodic motion, we have a slightly different name for ω. In these situations, ω is called the **angular frequency**:

Definition of Angular Frequency, ω

$$\omega = 2\pi f = \frac{2\pi}{T} \qquad \text{13-5}$$

SI unit: rad/s = s^{-1}

Velocity

We can find the velocity of the shadow in the same way that we determined its position; first find the velocity of the peg, then take its x component. The result of this calculation is the velocity as a function of time for simple harmonic motion.

To begin, recall that the velocity of an object in uniform circular motion of radius r has a magnitude equal to

$$v = r\omega$$

In addition, the velocity is tangential to the object's circular path, as indicated in **FIGURE 13-9 (a)**. Therefore, referring to the figure, we see that when the peg is at the angular position θ, the velocity vector makes an angle θ with the vertical. As a result, the x component of the velocity is $-v \sin \theta$. Combining these results, we find that the velocity of the peg, along the x axis, is

$$v_x = -v \sin \theta = -r\omega \sin \theta$$

In what follows we shall drop the x subscript, since we know that the shadow and a mass on a spring move only along the x axis. Recalling that $r = A$ and $\theta = \omega t$, we have:

Velocity in Simple Harmonic Motion

$$v = -A\omega \sin(\omega t) \qquad \text{13-6}$$

SI unit: m/s

We plot x and v for simple harmonic motion in **FIGURE 13-9 (b)**. Notice that when the displacement from equilibrium is a maximum, the velocity is zero. This is to be expected, because at $x = +A$ and $x = -A$ the object is momentarily at rest as it turns around. Appropriately, then, these points are referred to as **turning points** of the motion.

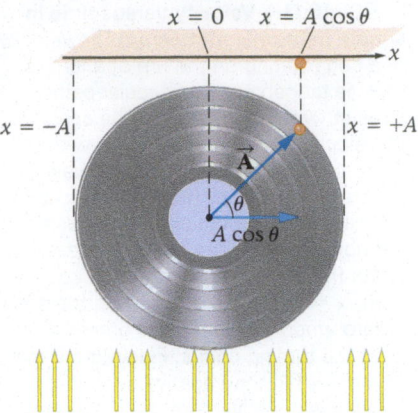

▲ **FIGURE 13-8 Position versus time in simple harmonic motion** A peg rotates on the rim of a turntable of radius A. When the peg is at the angular position θ, its shadow is at $x = A \cos \theta$. Note that $A \cos \theta$ is also the x component of the radius vector $\vec{A}$ from the center of the turntable to the peg.

Big Idea 2 Simple harmonic motion is closely related to uniform circular motion. In fact, the x or y component of an object in uniform circular motion has the same time dependence as an object in simple harmonic motion.

PHYSICS IN CONTEXT
Looking Back

Uniform circular motion (Chapter 10) is used to better understand simple harmonic motion.

◀ **FIGURE 13-9 Velocity versus time in simple harmonic motion** **(a)** The velocity of a peg rotating on the rim of a turntable is tangential to its circular path. As a result, when the peg is at the angle θ, its velocity makes an angle of θ with the vertical. The x component of the velocity, then, is $-v \sin \theta$. **(b)** Position, x, and velocity, v, as a function of time for simple harmonic motion. The speed is greatest when the object passes through equilibrium, $x = 0$. On the other hand, the speed is zero when the position is greatest—that is, at the turning points. Finally, note that as x moves in the negative direction, the velocity is negative. Similar remarks apply to the positive direction.

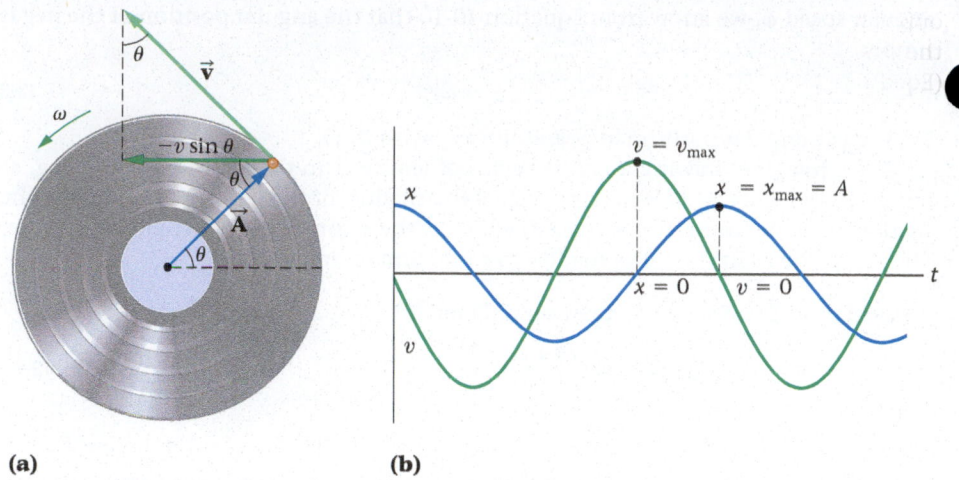

(a)   **(b)**

On the other hand, the speed is a maximum when the displacement from equilibrium is zero. Similarly, a mass on a spring is moving with its greatest speed as it goes through $x = 0$. From the expression $v = -A\omega \sin(\omega t)$, and the fact that the largest value of $\sin \theta$ is 1, we see that the maximum speed of the mass is

$$v_{max} = A\omega \qquad\qquad 13\text{-}7$$

After the mass passes $x = 0$ it begins either to compress or to stretch the spring, and hence it slows down.

Acceleration

The acceleration of an object in uniform circular motion has a magnitude given by

$$a_{cp} = r\omega^2$$

In addition, the direction of the acceleration is toward the center of the circular path, as indicated in **FIGURE 13-10 (a)**. Thus, when the angular position of the peg is θ, the acceleration vector is at an angle θ below the x axis, and its x component is $-a_{cp} \cos \theta$. Once again setting $r = A$ and $\theta = \omega t$, we find:

PROBLEM-SOLVING NOTE

Remember to Use Radians

Note that Equations 13-4, 13-6, and 13-8 must all be evaluated in terms of radians.

| **Acceleration in Simple Harmonic Motion** |
| $a = -A\omega^2 \cos(\omega t)$ 13-8 |
| SI unit: m/s² |

The Connection Between Acceleration and Position The position and acceleration for simple harmonic motion are plotted in **FIGURE 13-10 (b)**. If you look closely, you'll see that the acceleration and position vary with time in the same way but with opposite signs. That is, when the position has its maximum *positive* value, the acceleration has its maximum *negative* value, and so on. After all, the restoring force of the spring is

◀ **FIGURE 13-10 Acceleration versus time in simple harmonic motion** **(a)** The acceleration of a peg on the rim of a uniformly rotating turntable is directed toward the center of the turntable. Hence, when the peg is at the angle θ, the acceleration makes an angle θ with the horizontal. The x component of the acceleration is $-a \cos \theta$. **(b)** Position, x, and acceleration, a, as a function of time for simple harmonic motion. Note that when the position has its greatest positive value, the acceleration has its greatest negative value.

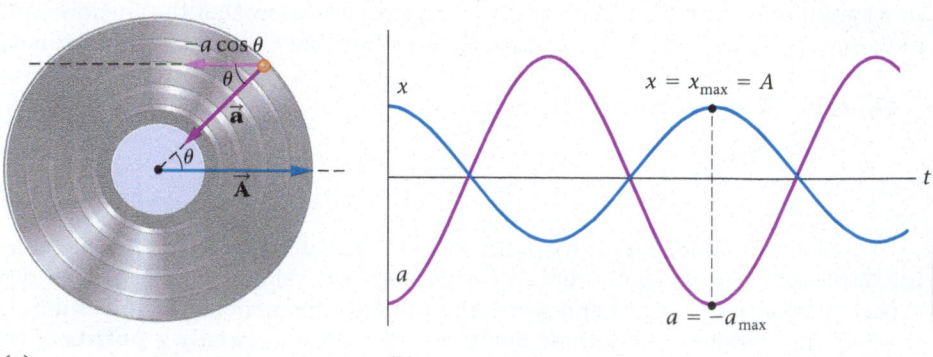

(a)   **(b)**

opposite to the position, and hence the acceleration, $a = F/m$, must also be opposite to the position. In fact, comparing $x = A\cos(\omega t)$ (Equation 13-4) and $a = -A\omega^2\cos(\omega t)$ (Equation 13-8), we see that the acceleration can be written as

$$a = -\omega^2 x$$

This expression is valid at all times. In addition, noting that the largest value of x is the amplitude A, we see that the maximum acceleration has the following magnitude:

$$a_{max} = A\omega^2 \qquad \text{13-9}$$

We conclude this section with a few examples using position, velocity, and acceleration in simple harmonic motion.

EXAMPLE 13-4   VELOCITY AND ACCELERATION

As in Example 13-3, an air-track cart attached to a spring completes one oscillation every 2.4 s. At $t = 0$ the cart is released from rest with the spring stretched 0.10 m from its equilibrium position. What are the velocity and acceleration of the cart at (a) 0.30 s and (b) 0.60 s?

PICTURE THE PROBLEM
Once again, we place the origin of the x axis at the equilibrium position of the cart, with the positive direction pointing to the right. In addition, the cart is released from rest at $x = 0.10$ m, which means that the amplitude of its motion is $A = 0.10$ m. It follows that the cart has zero speed at $x = 0.10$ m and $x = -0.10$ m. Its speed will be a maximum at $x = 0$, however, which is also the point where the acceleration is zero.

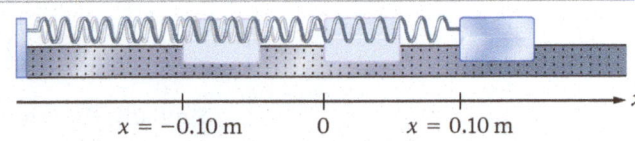

REASONING AND STRATEGY
After calculating the angular frequency, $\omega = 2\pi/T$, we simply substitute $t = 0.30$ s and $t = 0.60$ s into $v = -A\omega\sin(\omega t)$ and $a = -A\omega^2\cos(\omega t)$. Remember to set your calculators to radian mode.

Known   Period of cart, $T = 2.4$ s; release point, $x = 0.10$ m.
Unknown Velocity and acceleration ($v = ?$, $a = ?$) at indicated times

SOLUTION
1. Calculate the angular frequency for this motion:

$$\omega = \frac{2\pi}{T} = \frac{2\pi}{(2.4\text{ s})} = 2.6\text{ rad/s}$$

Part (a)

2. Calculate v at the time $t = 0.30$ s. Express ω in terms of π—that is, $\omega = 2\pi/(2.4\text{ s})$. This isn't necessary, but it makes it easier to evaluate the sine and cosine functions (see the discussion following the Example for additional details):

$$v = -A\omega\sin(\omega t)$$
$$= -(0.10\text{ m})(2.6\text{ rad/s})\sin\left[\left(\frac{2\pi}{2.4\text{ s}}\right)(0.30\text{ s})\right]$$
$$= -(26\text{ cm/s})\sin(\pi/4) = -18\text{ cm/s}$$

3. Similarly, calculate a at $t = 0.30$ s:

$$a = -A\omega^2\cos(\omega t)$$
$$= -(0.10\text{ m})(2.6\text{ rad/s})^2\cos\left[\left(\frac{2\pi}{2.4\text{ s}}\right)(0.30\text{ s})\right]$$
$$= -(68\text{ cm/s}^2)\cos\left(\frac{\pi}{4}\right) = -48\text{ cm/s}^2$$

Part (b)

4. Calculate v at the time $t = 0.60$ s:

$$v = -A\omega\sin(\omega t)$$
$$= -(0.10\text{ m})(2.6\text{ rad/s})\sin\left[\left(\frac{2\pi}{2.4\text{ s}}\right)(0.60\text{ s})\right]$$
$$= -(26\text{ cm/s})\sin\left(\frac{\pi}{2}\right) = -26\text{ cm/s}$$

5. Similarly, calculate a at $t = 0.60$ s:

$$a = -A\omega^2\cos(\omega t)$$
$$= -(0.10\text{ m})(2.6\text{ rad/s})^2\cos\left[\left(\frac{2\pi}{2.4\text{ s}}\right)(0.60\text{ s})\right]$$
$$= -(68\text{ cm/s}^2)\cos\left(\frac{\pi}{2}\right) = 0$$

CONTINUED

INSIGHT

Did you notice that the cart speeds up from $t = 0.30$ s to $t = 0.60$ s? In fact, the maximum speed of the cart, $v_{max} = A\omega = 26$ cm/s, occurs at $t = 0.60$ s $= T/4$. Referring to Example 13-3, we can see that this is precisely the time when the cart is at the equilibrium position, $x = 0$. As expected, the acceleration is zero at this time.

PRACTICE PROBLEM

What is the first time the velocity of the cart is $+26$ cm/s?   [**Answer:** $t = 3T/4 = 1.8$ s]

Some related homework problems: Problem 21, Problem 22

PROBLEM-SOLVING NOTE

Expressing Time in Terms of the Period

When evaluating the expression $x = A\cos(2\pi t/T)$ at the time t, it is often helpful to express t in terms of the period T.

In problems like the preceding Example, it's often useful to express the time t in terms of the period, T. For example, if the period is $T = 2.4$ s it follows that $t = 0.60$ s is $T/4$. Thus, the angular frequency multiplied by the time is

$$\omega t = \left(\frac{2\pi}{T}\right)\left(\frac{T}{4}\right) = \frac{\pi}{2}$$

This result was used in Steps 4 and 5 in Example 13-4. Using $\omega t = \pi/2$ we find that the position of the cart is

$$x = A\cos(\omega t) = A\cos\left(\frac{\pi}{2}\right) = 0$$

Similarly, the velocity of the cart is

$$v = -A\omega\sin(\omega t) = -A\omega\sin\left(\frac{\pi}{2}\right) = -A\omega$$

The acceleration of the cart is

$$a = -A\omega^2\cos(\omega t) = -A\omega^2\cos\left(\frac{\pi}{2}\right) = 0$$

When expressed in this way, it's clear why x and a are zero and why v has its maximum negative value.

EXAMPLE 13-5   TURBULENCE!

On December 29, 1997, a United Airlines flight from Tokyo to Honolulu was hit with severe turbulence 31 minutes after takeoff. Data from the airplane's "black box" indicated that the 747 moved up and down with an amplitude of 30.0 m and a maximum acceleration of 1.8g. Treating the up-and-down motion of the plane as simple harmonic motion, find **(a)** the time required for one complete oscillation and **(b)** the plane's maximum vertical speed.

PICTURE THE PROBLEM

Our sketch shows the 747 airliner moving up and down with an amplitude of $A = 30.0$ m relative to its normal horizontal flight path. The period of motion, T, is the time required for one complete cycle of this up-and-down motion.

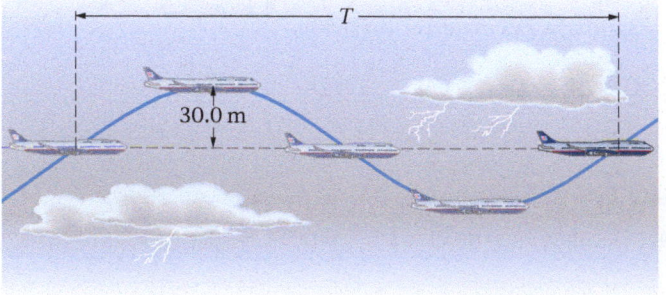

REASONING AND STRATEGY

We are given the maximum acceleration and the amplitude of motion. With these quantities, and the basic equations of simple harmonic motion, we can determine the other characteristics of the motion.

a. We know that the maximum acceleration of simple harmonic motion is $a_{max} = A\omega^2$ (Equation 13-9). This relationship can be solved for ω in terms of the known quantities a_{max} and A. We can then rearrange $\omega = 2\pi/T$ to solve for the period of motion, T.

b. The maximum vertical speed is found using $v_{max} = A\omega$ (Equation 13-7).

Known   Amplitude of motion, $A = 30.0$ m; maximum vertical acceleration, $a_{max} = 1.8g$.
Unknown **(a)** Period of motion, $T = ?$ **(b)** Maximum vertical speed, $v_{max} = ?$

CONTINUED

SOLUTION

Part (a)

1. Relate a_{max} to the angular frequency, ω:

$$a_{max} = A\omega^2$$

2. Solve for ω, and express in terms of T:

$$\omega = \sqrt{a_{max}/A} = 2\pi/T$$

3. Solve for T and substitute numerical values for g and A:

$$T = \frac{2\pi}{\sqrt{a_{max}/A}}$$

$$= \frac{2\pi}{\sqrt{1.8g/A}} = \frac{2\pi}{\sqrt{1.8(9.81 \text{ m/s}^2)/(30.0 \text{ m})}} = 8.2 \text{ s}$$

Part (b)

4. Calculate the maximum vertical speed using $v_{max} = A\omega = 2\pi A/T$:

$$v_{max} = A\omega = \frac{2\pi A}{T} = \frac{2\pi(30.0 \text{ m})}{8.2 \text{ s}} = 23 \text{ m/s}$$

INSIGHT

We don't expect the up-and-down motion of the plane to be exactly simple harmonic motion—after all, it's unlikely that the plane's path has precisely the shape of a cosine function. Still, the approximation is reasonable. In fact, many systems are well approximated by simple harmonic motion, and hence the equations derived in this section are used widely in physics.

Notice that the maximum vertical speed of the passengers (23 m/s) is roughly 50 mi/h, and that this speed is first in the upward direction and then 4.1 s later in the downward direction.

PRACTICE PROBLEM

What amplitude of motion would result in a maximum acceleration of 0.50g, with everything else remaining the same? [**Answer:** $A = 0.50g/\omega^2 = 0.50gT^2/4\pi^2 = 8.4$ m]

Some related homework problems: Problem 21, Problem 23

EXAMPLE 13-6   A BOBBING DRAGONFLY

When dragonflies hover, they tend to "bob" up and down with a motion that is a very good approximation to simple harmonic motion. Suppose that a dragonfly is at $x = A$ at time $t = 0$, and that its bobbing motion has an amplitude of $A = 0.40$ cm and a period of $T = 0.24$ s. Find (a) the position, (b) the velocity, and (c) the acceleration of the dragonfly at the time $t = 0.10$ s.

PICTURE THE PROBLEM

Our sketch shows a hovering dragonfly bobbing up and down along the x axis between $x = +A$ and $x = -A$. The positive x direction is chosen to be upward, and the dragonfly is at $x = +A$ at $t = 0$.

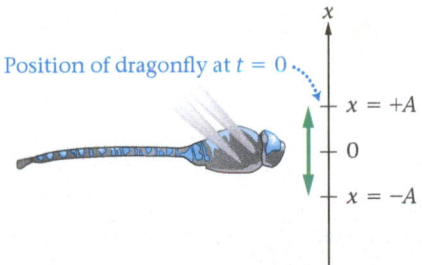

Position of dragonfly at $t = 0$

$x = +A$
0
$x = -A$

REASONING AND STRATEGY

The dragonfly bobs up and down in simple harmonic motion with an amplitude $A = 0.40$ cm and a period $T = 0.24$ s. It follows that the position of the dragonfly as a function of time is $x = A \cos(\omega t)$, where the angular frequency is $\omega = 2\pi/T$. (Notice that $x = A$ at $t = 0$, in agreement with the problem statement.) Similarly, the velocity is $v = -A\omega \sin(\omega t)$ and the acceleration is $a = -A\omega^2 \cos(\omega t)$. These expressions are to be evaluated at $t = 0.10$ s.

Known   Amplitude of motion, $A = 0.40$ cm; period of bobbing, $T = 0.24$ s; time, $t = 0.10$ s.
Unknown   (a) Position, $x = ?$ (b) Velocity, $v = ?$ (c) Acceleration, $a = ?$

SOLUTION

1. Evaluate ωt at the time $t = 0.10$ s. This quantity will be used in evaluating x, v, and a:

$$\omega t = \left(\frac{2\pi}{T}\right)t = \left(\frac{2\pi}{0.24 \text{ s}}\right)(0.10 \text{ s}) = 2.6 \text{ rad}$$

Part (a)

2. Evaluate $x = A \cos(\omega t)$ with $\omega t = 2.6$ rad, which corresponds to $t = 0.10$ s:

$$x = A \cos(\omega t)$$
$$= (0.40 \text{ cm}) \cos(2.6 \text{ rad}) = -0.34 \text{ cm}$$

Part (b)

3. Evaluate $v = -A\omega \sin(\omega t)$ with $\omega t = 2.6$ rad:

$$v = -A\omega \sin(\omega t)$$
$$= -(0.40 \text{ cm})\left(\frac{2\pi}{0.24 \text{ s}}\right) \sin(2.6 \text{ rad}) = -5.4 \text{ cm/s}$$

CONTINUED

Part (c)

4. Evaluate $a = -A\omega^2 \cos(\omega t)$ with $\omega t = 2.6$ rad:

$$a = -A\omega^2 \cos(\omega t)$$

$$= -(0.40 \text{ cm})\left(\frac{2\pi}{0.24 \text{ s}}\right)^2 \cos(2.6 \text{ rad}) = 230 \text{ cm/s}^2 = 2.3 \text{ m/s}^2$$

INSIGHT

At $t = 0.10$ s the dragonfly is almost at the bottom of its harmonic motion, which occurs at $t = T/2 = 0.12$ s. Hence, it makes sense that its position is negative, its velocity is small and negative, and its acceleration is positive.

PRACTICE PROBLEM

What are x, v, and a at the time $t = T/2 = 0.12$ s?   [**Answer:** $x = -0.40$ cm, $v = 0$, and $a = 2.7$ m/s^2]

Some related homework problems: Problem 21, Problem 23, Problem 27

Enhance Your Understanding   (Answers given at the end of the chapter)

3. An object moves with simple harmonic motion about $x = 0$. At one instant of time its position is $x = -0.1$ m and its velocity is 0. Is the acceleration of the object at this time positive, negative, or zero? Explain.

Section Review

- The position, velocity, and acceleration in simple harmonic motion are given by $x = A\cos(\omega t)$, $v = -A\omega \sin(\omega t)$, and $a = -A\omega^2 \cos(\omega t)$, respectively, where A is the amplitude and $\omega = 2\pi/T$.

13-4 The Period of a Mass on a Spring

In this section, we show how the period of a mass oscillating on a spring is related to the mass, m, the force constant of the spring, k, and the amplitude of motion, A. As a first step, we note that the net force acting on the mass at the position x is

$$F = -kx$$

Next, we apply Newton's second law, $F = ma$, to find the following:

$$ma = -kx$$

Substituting the time dependence of x and a, as given in Equations 13-4 and 13-8 in the previous section, we find

$$m[-A\omega^2 \cos(\omega t)] = -k[A \cos(\omega t)]$$

Canceling $-A \cos(\omega t)$ from each side of the equation yields

$$\omega^2 = \frac{k}{m}$$

Big Idea **3** The period of a mass oscillating on an ideal spring is proportional to the square root of the mass, and inversely protional to the square root of the force constant. The period is independent of the amplitude of motion.

Taking the square root on both sides of the equation yields

$$\omega = \sqrt{\frac{k}{m}} \qquad\qquad 13\text{-}10$$

Finally, recalling that $\omega = 2\pi/T$, it follows that the period of a mass oscillating on a spring is:

PHYSICS IN CONTEXT
Looking Back

Newton's second law, $F = ma$ (Chapter 5), played a key role in our discussion of a mass on a spring. We also used the force law for a spring, $F = -kx$ (Chapter 6).

Period of a Mass on a Spring

$$T = 2\pi\sqrt{\frac{m}{k}} \qquad\qquad 13\text{-}11$$

SI unit: s

Notice that the period is independent of the amplitude of motion, A. This is a key feature of simple harmonic motion.

EXERCISE 13-7   MASS, FORCE CONSTANT, AND PERIOD

A horizontal spring with a force constant of 11 N/m is connected to an air-track cart. When the cart is set in motion, the spring causes it to oscillate back and forth with a period of 1.2 s. What is the mass of the cart?

REASONING AND SOLUTION

The period of an object attached to a spring is related to the mass of the object, m, and the force constant of the spring, k, by the equation $T = 2\pi\sqrt{m/k}$. In this case we know k and T, and hence we can rearrange the equation to solve for the mass:

$$m = \frac{T^2 k}{4\pi^2} = \frac{(1.2\text{ s})^2(11\text{ N/m})}{4\pi^2} = 0.40\text{ kg}$$

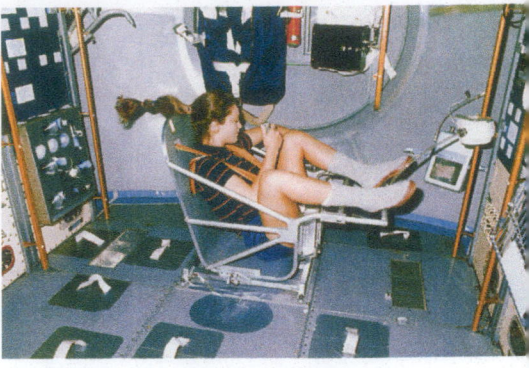

▲ **FIGURE 13-11** Because astronauts are in free fall when orbiting the Earth, they behave as if they were "weightless." It is nevertheless possible to determine the mass of an astronaut by exploiting the properties of oscillatory motion. The Body Mass Measurement Device (BMMD), into which astronaut Tamara Jernigan is strapped, is attached to a spring. If the force constant of the spring is known, her mass can be determined simply by measuring the period with which she rocks back and forth.

As one might expect, the period increases with the mass and decreases with the spring's force constant. For example, a larger mass has greater inertia, and hence it takes longer for the mass to move back and forth through an oscillation. On the other hand, a larger value of the force constant, k, indicates a stiffer spring. Clearly, a mass on a stiff spring completes an oscillation in less time than one on a soft, squishy spring.

RWP* The relationship between mass and period given by $T = 2\pi\sqrt{m/k}$ is used by NASA to measure the mass of astronauts in orbit. Recall that astronauts are in free fall as they orbit, as was discussed in Chapter 12, and therefore they are "weightless." As a result, they cannot simply step onto a bathroom scale to determine their mass. Thus, NASA has developed a device, known as the Body Mass Measurement Device (BMMD), to get around this problem. The BMMD is basically a spring attached to a chair, into which an astronaut is strapped, as shown in **FIGURE 13-11**. As the astronaut oscillates back and forth, the period of oscillation is measured. Since the force constant, k, and the period of oscillation, T, are known, the astronaut's mass can be determined in the same way we determined the mass of the air-track cart in Exercise 13-7.

Period and Amplitude   The period, $T = 2\pi\sqrt{m/k}$, is independent of the amplitude, A, which canceled in the derivation of this equation. This might seem counterintuitive at first: Shouldn't it take more time for a mass to cover the greater distance implied by a larger amplitude? While it's true that a mass will cover a greater distance when the amplitude is increased, it's also true that a larger amplitude implies a larger force exerted by the spring. With a greater force acting on it, the mass moves more rapidly; in fact, the speed of the mass is increased just enough that it covers the greater distance in precisely the same time.

PHYSICS IN CONTEXT
Looking Ahead

Though a mass on a spring may seem far removed from an electric circuit, we show in Chapter 24 that there is in fact a deep connection. In particular, the motion of a mass on a spring is directly analogous to the current in a specific type of circuit referred to as an *RLC* circuit.

EXAMPLE 13-8   SPRING INTO MOTION

When a 0.120-kg air-track cart is attached to a horizontal spring it oscillates with an amplitude of 0.0750 m and a maximum speed of 0.524 m/s. Find **(a)** the force constant and **(b)** the period of motion.

PICTURE THE PROBLEM

Our sketch shows a mass oscillating about the equilibrium position of a spring, which we place at $x = 0$. The amplitude of the oscillations is 0.0750 m, and therefore the mass moves back and forth between $x = 0.0750$ m and $x = -0.0750$ m. The maximum speed of the mass, which occurs at $x = 0$, is $v_{max} = 0.524$ m/s.

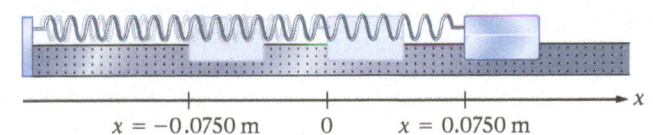

$x = -0.0750$ m   0 $x = 0.0750$ m

REASONING AND STRATEGY

a. To find the force constant, we first use the maximum speed, $v_{max} = A\omega$ (Equation 13-7), to determine the angular frequency ω. Once we know ω, we can obtain the force constant k with $\omega = \sqrt{k/m}$ (Equation 13-10).

b. We can find the period from the angular frequency, using $\omega = 2\pi/T$. Alternatively, we can use the force constant and the mass in $T = 2\pi\sqrt{m/k}$ (Equation 13-11).

Known   Mass of cart, $m = 0.120$ kg; amplitude of motion, $A = 0.0750$ m; maximum speed of cart, $v_{max} = 0.524$ m/s.
Unknown   **(a)** Force constant, $k = ?$ **(b)** Period of motion, $T = ?$

CONTINUED

---

*Real World Physics applications are denoted by the acronym RWP.

SOLUTION

Part (a)

1. Calculate the angular frequency in terms of the maximum speed:

$$v_{max} = A\omega$$
$$\omega = \frac{v_{max}}{A} = \frac{0.524 \text{ m/s}}{0.0750 \text{ m}} = 6.99 \text{ rad/s}$$

2. Solve $\omega = \sqrt{k/m}$ for the force constant:

$$\omega = \sqrt{\frac{k}{m}}$$
$$k = m\omega^2 = (0.120 \text{ kg})(6.99 \text{ rad/s})^2 = 5.86 \text{ N/m}$$

Part (b)

3. Use $\omega = 2\pi/T$ to find the period:

$$\omega = \frac{2\pi}{T}$$
$$T = \frac{2\pi}{\omega} = \frac{2\pi}{6.99 \text{ rad/s}} = 0.899 \text{ s}$$

4. Alternatively, use $T = 2\pi\sqrt{m/k}$ to find the period:

$$T = 2\pi\sqrt{\frac{m}{k}} = 2\pi\sqrt{\frac{0.120 \text{ kg}}{5.86 \text{ N/m}}} = 0.899 \text{ s}$$

INSIGHT

What would happen if we were to attach a greater mass to this same spring and release it with the same amplitude? The answer is that the period would increase. As a result, the angular frequency ω would decrease, and the maximum speed, $v_{max} = A\omega$, would decrease as well.

PRACTICE PROBLEM

What is the maximum acceleration of the mass described in this Example? [**Answer:** $a_{max} = A\omega^2 = 3.66 \text{ m/s}^2$]

Some related homework problems: Problem 36, Problem 39

These relationships between the motion of a mass on a spring and the mass, the force constant, and the amplitude are summarized in **FIGURE 13-12**.

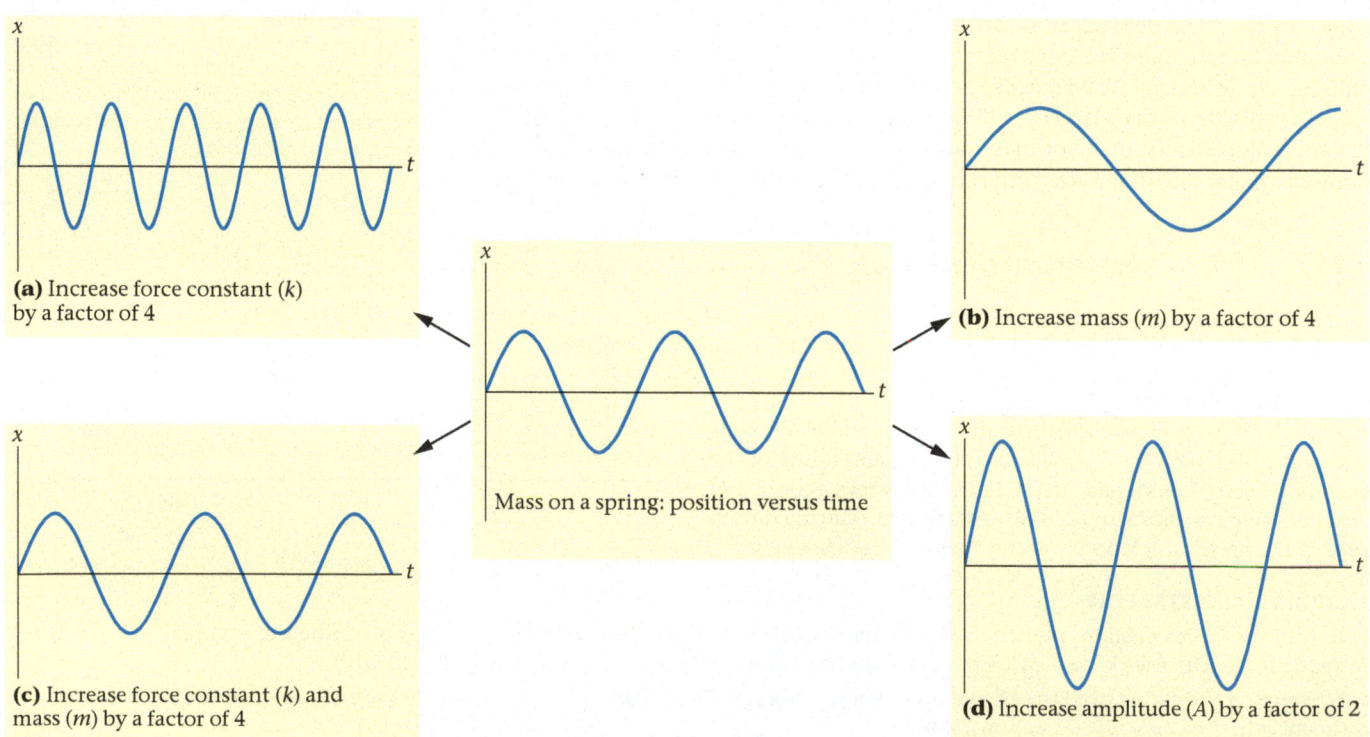

(a) Increase force constant (k) by a factor of 4

(b) Increase mass (m) by a factor of 4

Mass on a spring: position versus time

(c) Increase force constant (k) and mass (m) by a factor of 4

(d) Increase amplitude (A) by a factor of 2

▲ **FIGURE 13-12 Factors affecting the motion of a mass on a spring** The motion of a mass on a spring is determined by the force constant of the spring, k, the mass, m, and the amplitude, A. **(a)** Increasing the force constant causes the mass to oscillate with a greater frequency. **(b)** Increasing the mass lowers the frequency of oscillation. **(c)** If the force constant and the mass are both increased by the same factor, the effects described in parts (a) and (b) cancel, resulting in no change in the motion. **(d)** An increase in amplitude has no effect on the oscillation frequency. However, it increases the maximum speed and maximum acceleration of the mass.

QUICK EXAMPLE 13-9   MASS ON A SPRING: FIND THE FORCE CONSTANT AND THE MASS

When a 0.420-kg air-track cart is attached to a horizontal spring, it oscillates with a period of 0.350 s. If, instead, a different cart with mass m_2 is attached to the same spring, it oscillates with a period of 0.700 s. Find **(a)** the force constant of the spring and **(b)** the mass m_2.

REASONING AND STRATEGY

The information about the first cart allows us to determine the force constant of the spring. Once we know that, we can use the equation for the period of a mass on a spring to find the second mass.

Part (a)

1. Let the initial mass and period be m_1 and T_1, respectively:
$m_1 = 0.420$ kg
$T_1 = 0.350$ s

2. Use $T = 2\pi\sqrt{\dfrac{m}{k}}$ to write an expression for T_1:
$T_1 = 2\pi\sqrt{\dfrac{m_1}{k}}$

3. Solve this expression for the force constant, k:
$k = \dfrac{4\pi^2 m_1}{T_1^{\,2}} = 135$ N/m

Part (b)

4. Write an expression for T_2:
$T_2 = 2\pi\sqrt{\dfrac{m_2}{k}}$

5. Solve this expression for m_2:
$m_2 = \dfrac{kT_2^{\,2}}{4\pi^2} = 1.68$ kg

To double the period of a mass on a given spring—as in this case where it went from 0.350 s to 0.700 s—the mass must be increased by a factor of 4, in accordance with $T = 2\pi\sqrt{m/k}$. Therefore, an alternative way to find the second mass is $m_2 = 4(0.420\text{ kg}) = 1.68$ kg, in agreement with our result in Step 5.

A Vertical Spring

To this point we've considered only springs that are horizontal, and that are therefore unstretched at their equilibrium position. In many cases, however, we may wish to consider a vertical spring, as in **FIGURE 13-13**.

Now, when a mass m is attached to a vertical spring, it causes the spring to stretch. In fact, the vertical spring is in equilibrium when it exerts an upward force equal to the weight of the mass. That is, the spring stretches by an amount y_0 given by

$$ky_0 = mg$$

Rearranging, we find

$$y_0 = \frac{mg}{k} \qquad \text{13-12}$$

Thus, a mass on a vertical spring oscillates about the equilibrium point $y = -y_0$. In all other respects the oscillations are the same as for a horizontal spring. In particular, the motion is simple harmonic motion, and the period is given by $T = 2\pi\sqrt{m/k}$. Examples of a mass on a vertical spring are shown in **FIGURE 13-14**.

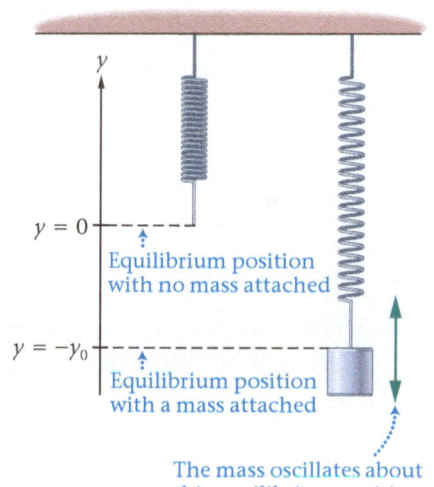

y

$y = 0$

Equilibrium position with no mass attached

$y = -y_0$

Equilibrium position with a mass attached

The mass oscillates about this equilibrium position.

▲ **FIGURE 13-13 A mass on a vertical spring** A mass stretches a vertical spring from its initial equilibrium at $y = 0$ to a new equilibrium at $y = -y_0 = -mg/k$. The mass executes simple harmonic motion about this new equilibrium.

◄ **FIGURE 13-14 Visualizing Concepts Mass on a Vertical Spring** A ball attached to a vertical spring oscillates up and down with simple harmonic motion. If successive images taken at equal time intervals are displaced laterally, as in the sequence of photos on the left, the ball appears to trace out a sinusoidal pattern (compare with Figure 13-5). The bungee jumper on the right will oscillate in a similar fashion, though friction and air resistance will reduce the amplitude of his bounces.

EXAMPLE 13-10 IT'S A STRETCH

A 0.260-kg mass is attached to a vertical spring. When the mass is put into motion, its period is 1.12 s. **(a)** How much does the mass stretch the spring when it is at rest in its equilibrium position? **(b)** Suppose this experiment is repeated on a planet where the acceleration due to gravity is twice what it is on Earth. By what multiplicative factors do the period and equilibrium stretch change?

PICTURE THE PROBLEM

We choose the vertical axis to have its origin at the unstretched position of the spring. Once the mass is attached, the spring stretches to the position $y = -y_0$. The mass oscillates about this point with a period of 1.12 s.

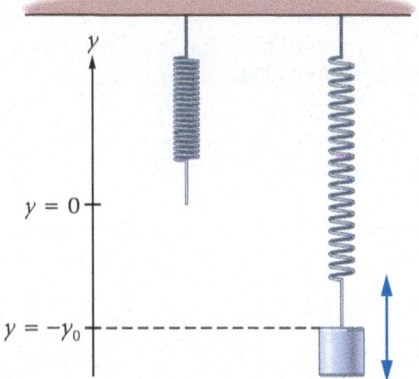

REASONING AND STRATEGY

a. In order to find the stretch of the spring, $y_0 = mg/k$, we need to know the force constant, k. We can find k from the period of oscillation—that is, from $T = 2\pi\sqrt{m/k}$.

b. We replace g with $2g$ in both $T = 2\pi\sqrt{m/k}$ and $y_0 = mg/k$. Notice that g does not occur in the expression for the period T.

Known Mass, $m = 0.260$ kg; period, $T = 1.12$ s.
Unknown **(a)** Equilibrium stretch, $y_0 = ?$ **(b)** Period and equilibrium stretch $(T = ?,\ y_0 = ?)$ when $g \rightarrow 2g$

SOLUTION

Part (a)

1. Use the period $T = 2\pi\sqrt{m/k}$ to solve for the force constant:

$$T = 2\pi\sqrt{\frac{m}{k}}$$

$$k = \frac{4\pi^2 m}{T^2} = \frac{4\pi^2(0.260\text{ kg})}{(1.12\text{ s})^2} = 8.18\text{ kg/s}^2 = 8.18\text{ N/m}$$

2. Set the magnitude of the spring force, ky_0, equal to mg to solve for y_0:

$$ky_0 = mg$$

$$y_0 = \frac{mg}{k} = \frac{(0.260\text{ kg})(9.81\text{ m/s}^2)}{8.18\text{ N/m}} = 0.312\text{ m}$$

Part (b)

3. Use $2g$ in place of g in the expressions for the period T and the magnitude of the equilibrium stretch, y_0:

$$T = 2\pi\sqrt{m/k} \xrightarrow{g \rightarrow 2g} T$$

$$y_0 = mg/k \xrightarrow{g \rightarrow 2g} 2y_0$$

INSIGHT

We see that doubling the force of gravity has no effect on the period (changes it by a factor of 1), but doubles the equilibrium stretch. Therefore, one would observe the same period of oscillation on the Moon or Mars—or even in orbit, as in the case of the Body Mass Measurement Device mentioned earlier in this section.

PRACTICE PROBLEM

A 0.170-kg mass stretches a vertical spring 0.250 m when at rest. What is its period when set into vertical motion?
[**Answer:** $T = 1.00$ s]

Some related homework problems: Problem 40, Problem 44

CONCEPTUAL EXAMPLE 13-11 COMPARE PERIODS   PREDICT/EXPLAIN

When a mass m is attached to a vertical spring with a force constant k, it oscillates with a period T. **(a)** If the spring is cut in half and the same mass is attached to it, is the period of oscillation greater than, less than, or equal to T? **(b)** Which of the following is the *best explanation* for your prediction?

  I. Cutting the spring in half effectively makes it stiffer, and hence the period is less than T.

  II. The spring is still made of the same material, and hence the period stays the same.

  III. A shorter spring is easier to stretch. As a result, the period is greater than T.

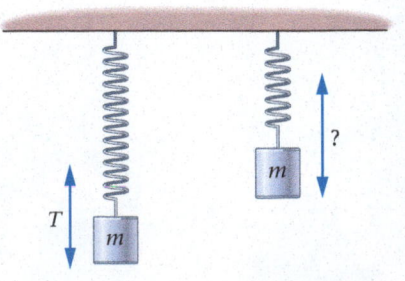

REASONING AND DISCUSSION

The downward force exerted by the mass is of magnitude mg, as is the upward force exerted by the spring. *Each coil* of the spring experiences the *same force*, just as each point in a string

CONTINUED

experiences the same tension. Therefore, each coil elongates by the same amount, regardless of how many coils there are in a given spring. It follows, then, that the total elongation of the spring with half the number of coils is half the total elongation of the longer spring. Half the elongation for the same applied force means a greater force constant, and hence the half-spring has a larger value of k—it is stiffer. As a result, its period of oscillation is less than the period of the full spring.

ANSWER
(a) The period for the half-spring is less than for the full spring. **(b)** The best explanation is I.

Enhance Your Understanding   (Answers given at the end of the chapter)

4. Rank the four mass–spring systems in **FIGURE 13-15** in order of increasing period. Indicate ties where appropriate.

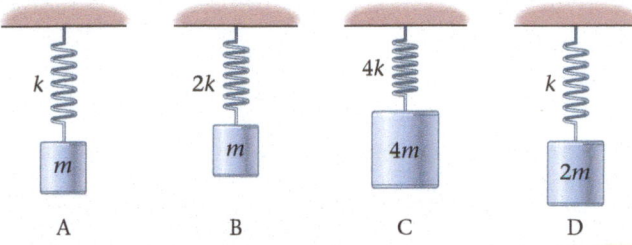

▲ **FIGURE 13-15** Enhance Your Understanding 4.

Section Review

*   The period of a mass on a spring is $T = 2\pi\sqrt{m/k}$, where m is the mass and k is the force constant.

13-5 Energy Conservation in Oscillatory Motion

In an ideal system, with no friction or other nonconservative forces, the total energy is conserved. For example, the total mechanical energy E of a mass on a horizontal spring is the sum of its kinetic energy, $K = \frac{1}{2}mv^2$, and its potential energy, $U = \frac{1}{2}kx^2$. Therefore, in the absence of nonconservative forces we have

$$E = K + U = \tfrac{1}{2}mv^2 + \tfrac{1}{2}kx^2 = \text{constant} \qquad \text{13-13}$$

Because E remains the same throughout the motion, it follows that there is a continual tradeoff between kinetic and potential energy.

This energy tradeoff is illustrated in **FIGURE 13-16**, where the horizontal line represents the total energy of the system, E, and the parabolic curve is the spring's potential

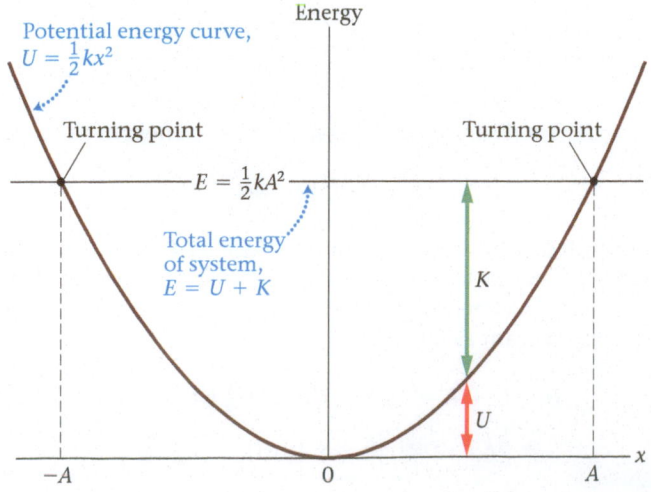

◀ **FIGURE 13-16 Energy as a function of position in simple harmonic motion**
The parabola represents the potential energy of the spring, $U = \frac{1}{2}kx^2$. The horizontal line shows the total energy of the system, $E = U + K$, which is constant. It follows that the distance from the parabola to the total-energy line is the kinetic energy, K. Note that the kinetic energy vanishes at the turning points, $x = A$ and $x = -A$. At these points the energy is purely potential, and thus the total energy of the system is $E = \frac{1}{2}kA^2$.

energy, U. At any given value of x, the sum of U and K must equal E; therefore, because U is the amount of energy from the axis to the parabola, K is the amount of energy from the parabola to the horizontal line. We can see, then, that the kinetic energy vanishes at the turning points, $x = +A$ and $x = -A$, as expected. On the other hand, the kinetic energy is greatest at $x = 0$, where the potential energy vanishes.

Potential and Kinetic Energy as a Function of Time   As a mass on a spring oscillates back and forth with time, its kinetic and potential energies also change with time. For example, the potential energy is

$$U = \tfrac{1}{2}kx^2$$

Letting $x = A\cos(\omega t)$, we have

$$U = \tfrac{1}{2}kA^2\cos^2(\omega t) \qquad\qquad \text{13-14}$$

Clearly, the maximum value of U is

$$U_{max} = \tfrac{1}{2}kA^2$$

From Figure 13-16, we see that the maximum value of U is simply the total energy of the system, E. Therefore

$$E = U_{max} = \tfrac{1}{2}kA^2 \qquad\qquad \text{13-15}$$

This result leads to the following conclusion:

> In simple harmonic motion, the total energy is proportional to the square of the amplitude of motion.

Similarly, the kinetic energy of the mass is

$$K = \tfrac{1}{2}mv^2$$

Letting $v = -A\omega\sin(\omega t)$ yields

$$K = \tfrac{1}{2}mA^2\omega^2\sin^2(\omega t)$$

It follows that the maximum kinetic energy is

$$K_{max} = \tfrac{1}{2}mA^2\omega^2 \qquad\qquad \text{13-16}$$

As noted, the maximum kinetic energy occurs when the potential energy is zero, and hence the maximum kinetic energy must equal the total energy, E. That is,

$$E = U + K = U_{max} + 0 = 0 + K_{max}$$

At first glance, Equation 13-16 $\left(E = K_{max} = \tfrac{1}{2}mA^2\omega^2\right)$ doesn't seem to be the same as Equation 13-15 $\left(E = U_{max} = \tfrac{1}{2}kA^2\right)$. However, if we recall that $\omega^2 = k/m$ (Equation 13-10), we see that

$$K_{max} = \tfrac{1}{2}mA^2\omega^2 = \tfrac{1}{2}mA^2\left(\frac{k}{m}\right) = \tfrac{1}{2}kA^2$$

Therefore, $K_{max} = U_{max} = E$, as expected, and the kinetic energy as a function of time is

$$K = \tfrac{1}{2}kA^2\sin^2(\omega t) \qquad\qquad \text{13-17}$$

K and U are plotted as functions of time in **FIGURE 13-17**. The horizontal line at the top is E, the sum of U and K at all times. This shows quite graphically the back-and-forth tradeoff of energy between kinetic and potential. Mathematically, we can see that the total energy E is constant as follows:

$$E = U + K = \tfrac{1}{2}kA^2\cos^2(\omega t) + \tfrac{1}{2}kA^2\sin^2(\omega t)$$
$$= \tfrac{1}{2}kA^2[\cos^2(\omega t) + \sin^2(\omega t)] = \tfrac{1}{2}kA^2$$

The last step follows from the fact that $\cos^2\theta + \sin^2\theta = 1$ for all θ.

PHYSICS IN CONTEXT
Looking Back

The concepts of kinetic energy (Chapter 7) and potential energy (Chapter 8) play key roles in understanding oscillatory motion.

PROBLEM-SOLVING NOTE

Maximum Potential and Kinetic Energy

The maximum potential energy of a mass–spring system is the same as the maximum kinetic energy of the mass. When the system has its maximum potential energy, the kinetic energy of the mass is zero; when the mass has its maximum kinetic energy, the potential energy of the system is zero.

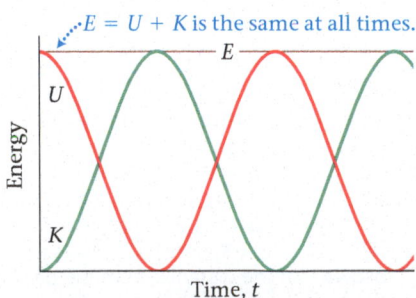

▲ **FIGURE 13-17 Energy as a function of time in simple harmonic motion** The sum of the potential energy, U, and the kinetic energy, K, is equal to the (constant) total energy E at all times. Note that when one energy (U or K) has its maximum value, the other energy is zero.

EXAMPLE 13-12 | STOP THE BLOCK

A 0.980-kg block slides on a frictionless, horizontal surface with a speed of 1.32 m/s. The block encounters an un-stretched spring with a force constant of 245 N/m, as shown in the sketch. **(a)** How far is the spring compressed before the block comes to rest? **(b)** For what amount of time is the block in contact with the spring before it comes to rest?

PICTURE THE PROBLEM

As our sketch shows, the initial energy of the system is entirely kinetic—namely, the kinetic energy of the block with mass $m = 0.980$ kg and speed $v_0 = 1.32$ m/s. When the block momentarily comes to rest after compressing the spring by the amount A, its kinetic energy has been converted into the potential energy of the spring.

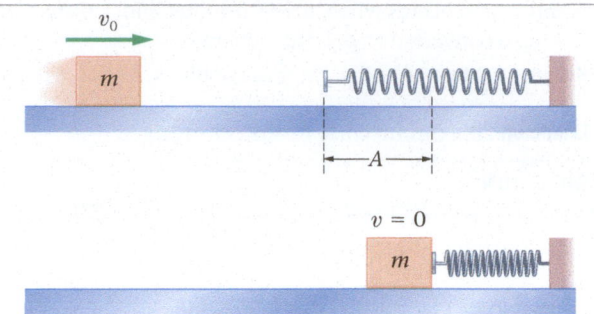

REASONING AND STRATEGY

a. We can find the compression, A, by using energy conservation. We set the initial kinetic energy of the block, $\frac{1}{2}mv_0^2$, equal to the spring potential energy, $\frac{1}{2}kA^2$, and solve for A.

b. If the mass were attached to the spring, it would complete one oscillation in the time $T = 2\pi\sqrt{m/k}$. In moving from the equilibrium position of the spring to maximum compression, the mass has undergone one-quarter of a cycle; thus the time is $T/4$.

Known Mass of block, $m = 0.980$ kg; initial speed of block, $v_0 = 1.32$ m/s; force constant, $k = 245$ N/m.
Unknown **(a)** Maximum compression, $A = ?$ **(b)** Time to come to rest, $t = ?$

SOLUTION

Part (a)

1. Set the initial kinetic energy of the block equal to the spring potential energy:

$$\tfrac{1}{2}mv_0^2 = \tfrac{1}{2}kA^2$$

2. Solve for A, the maximum compression:

$$A = v_0\sqrt{\frac{m}{k}} = (1.32 \text{ m/s})\sqrt{\frac{0.980 \text{ kg}}{245 \text{ N/m}}} = 0.0835 \text{ m}$$

Part (b)

3. Calculate the period of one oscillation:

$$T = 2\pi\sqrt{\frac{m}{k}} = 2\pi\sqrt{\frac{0.980 \text{ kg}}{245 \text{ N/m}}} = 0.397 \text{ s}$$

4. The block has been in contact with the spring for one-quarter of an oscillation, and hence $t = T/4$:

$$t = \tfrac{1}{4}T = \tfrac{1}{4}(0.397 \text{ s}) = 0.0993 \text{ s}$$

INSIGHT

If the horizontal surface had been rough, some of the block's initial kinetic energy would have been converted to thermal energy. In this case, the maximum compression of the spring would be less than that just calculated.

PRACTICE PROBLEM — PREDICT/CALCULATE

(a) If the initial speed of the mass in this Example is increased, does the time required to bring it to rest increase, decrease, or stay the same? Explain. **(b)** Check your answer by calculating the time for an initial speed of 1.50 m/s. [**Answer: (a)** Increasing v increases the amplitude, A. The period is independent of amplitude, however. Thus, the time is the same. **(b)** The time to bring the mass to rest is $t = 0.0993$ s, as before.]

Some related homework problems: Problem 53, Problem 83

In the next Example, we consider a bullet striking a block that is attached to a spring. As the bullet embeds itself in the block, the completely inelastic bullet–block collision dissipates some of the initial kinetic energy into thermal energy. Only the kinetic energy remaining after the collision is available for compressing the spring.

EXAMPLE 13-13 | BULLET–BLOCK COLLISION: FIND THE COMPRESSION

A bullet of mass m embeds itself in a block of mass M, which is attached to a spring of force constant k. If the initial speed of the bullet is v_0, find the maximum compression of the spring.

PICTURE THE PROBLEM

Our sketch shows the bullet with mass m and speed v_0 moving toward the block of mass M that is initially at rest. After the collision, the bullet and block move as if they were a single object until they reach the maximum compression of the spring, A, at which time the system is momentarily at rest before the spring expands again.

CONTINUED

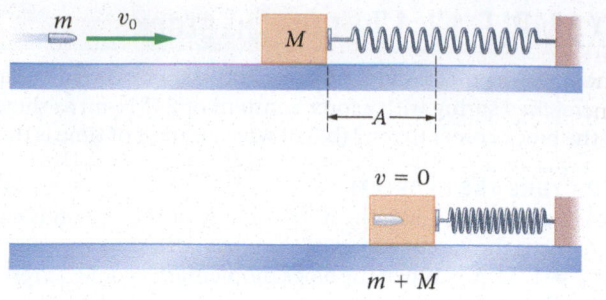

REASONING AND STRATEGY

We start by applying momentum conservation to the collision. This determines the speed of the bullet–block system after the collision. The kinetic energy of the bullet–block system after the collision is converted to the potential energy of the spring as it is compressed. The maximum compression occurs when all of the bullet–block system's kinetic energy has been converted to potential energy.

Known Mass of bullet, m; mass of block, M; force constant, k; initial speed of bullet, v_0.

Unknown Maximum compression of spring, $A = ?$

SOLUTION

1. Use momentum conservation to find the final speed, v, of the bullet–block system just after the collision:

$$mv_0 = (m + M)v$$
$$v = \frac{mv_0}{(m + M)}$$

2. Find the kinetic energy of the bullet–block system just after the collision:

$$\tfrac{1}{2}(m + M)v^2 = \tfrac{1}{2}\frac{m^2v_0^2}{(m + M)}$$

3. Set this kinetic energy equal to $\tfrac{1}{2}kA^2$ to find the maximum compression of the spring, A:

$$\tfrac{1}{2}\frac{m^2v_0^2}{(m + M)} = \tfrac{1}{2}kA^2$$
$$A = \frac{mv_0}{\sqrt{k(m + M)}}$$

INSIGHT

The maximum compression depends directly on the initial speed of the bullet. Thus, a measurement of the compression can be used to determine the bullet's speed.

PRACTICE PROBLEM — PREDICT/CALCULATE

Suppose the force constant of the spring is quadrupled. **(a)** Does the maximum compression, A, increase, decrease, or stay the same? Explain. **(b)** By what factor does the maximum compression, A, change? [**Answer: (a)** A stiffer spring compresses less for a given amount of energy, and hence the maximum compression decreases. **(b)** Substituting $4k$ for k in Step 3 shows that the maximum compression is halved.]

Some related homework problems: Problem 53, Problem 54

Enhance Your Understanding   (Answers given at the end of the chapter)

5. The total mechanical energy of an ideal oscillating mass–spring system is 15 J. **(a)** What is the potential energy of the system when the kinetic energy of the mass is 5 J? **(b)** What is the kinetic energy of the mass when the system's potential energy is 12 J?

Section Review

- The total mechanical energy of an oscillating mass–spring system is given by $E = K + U = \tfrac{1}{2}mv^2 + \tfrac{1}{2}kx^2$, assuming an absence of nonconservative forces in the system.

13-6 The Pendulum

One Sunday in 1583, as Galileo Galilei attended services in a cathedral in Pisa, Italy, he suddenly realized something interesting about the chandeliers hanging from the ceiling. Air currents circulating through the cathedral had set them in motion with small oscillations, and Galileo noticed that chandeliers of equal length oscillated with equal periods, even if their amplitudes were different. Indeed, as any given chandelier oscillated with decreasing amplitude, its period remained constant. He verified this observation by timing the oscillations with his pulse!

RWP Galileo was much struck by this observation, and after rushing home, he experimented with pendula constructed from different lengths of string and different weights. Continuing to use his pulse as a stopwatch, he observed that the period of a pendulum varies with its length, but is independent of the weight attached to the string. Thus, in one exhilarating afternoon, the young medical student discovered the key characteristics of a pendulum and launched himself on a new career in science. Later, he would go on to construct the first crude pendulum clock and a medical device, known as the pulsilogium, to measure a patient's pulse rate.

In modern terms, we would say that the chandeliers observed by Galileo were undergoing simple harmonic motion, as expected for small oscillations. As we know, the period of simple harmonic motion is independent of amplitude. The fact that the period is also independent of the mass is a special property of the pendulum, as we shall see next.

The Simple Pendulum

A simple pendulum consists of a mass m suspended by a light string or rod of length L. The pendulum has a stable equilibrium when the mass is directly below the suspension point, and oscillates about this position if displaced from it.

To understand the behavior of the pendulum, let's begin by considering the potential energy of the system. As shown in **FIGURE 13-18**, when the pendulum is at an angle θ with respect to the vertical, the mass m is above its lowest point by a vertical height $L(1 - \cos \theta)$. If we let the potential energy be zero at $\theta = 0$, the potential energy for general θ is

$$U = mgL(1 - \cos \theta) \qquad \text{13-18}$$

This function is plotted in **FIGURE 13-19**.

Notice that the stable equilibrium of the pendulum corresponds to a minimum of the potential energy, as expected. Near this minimum, the shape of the potential energy curve is approximately the same as for a mass on a spring, as indicated in Figure 13-19. As a result, when a pendulum oscillates with small displacements from the vertical, its motion is virtually the same as the motion of a mass on a spring; that is, the pendulum exhibits simple harmonic motion.

Next, we consider the forces acting on the mass m. In **FIGURE 13-20** we show the force of gravity, $m\vec{\mathbf{g}}$, and the tension force in the supporting string, $\vec{\mathbf{T}}$. The tension acts in the radial direction and supplies the force needed to keep the mass moving along its circular path. The net tangential force acting on m, then, is simply the tangential component of its weight:

$$F = mg \sin \theta$$

The direction of the net tangential force is always toward the equilibrium point. Thus F is a restoring force, as expected.

Now, for small angles θ (measured in *radians*), the sine of θ is approximately equal to the angle itself. That is,

$$\sin \theta \approx \theta$$

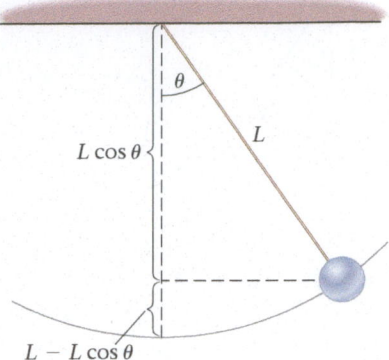

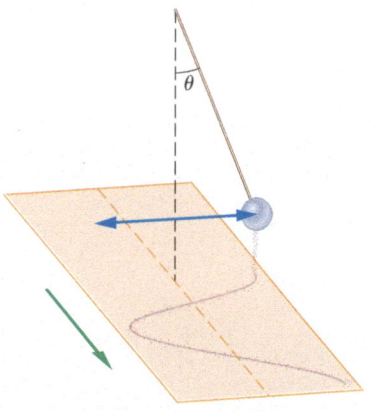

(a)

(b)

▲ **FIGURE 13-18 Motion of a pendulum** **(a)** As a pendulum swings away from its equilibrium position, it rises a vertical distance $L - L\cos \theta = L(1 - \cos \theta)$. **(b)** As a pendulum bob swings back and forth, it leaks a trail of sand onto a moving sheet of paper, creating a "strip chart" similar to that produced by a mass on a spring in Figure 13-4. In particular, the angle of the pendulum with the vertical varies with time like a sine or a cosine.

◀ **FIGURE 13-19 The potential energy of a simple pendulum** As a simple pendulum swings away from the vertical by an angle θ, its potential energy increases, as indicated by the solid curve. Near $\theta = 0$ the potential energy of the pendulum is essentially the same as that of a mass on a spring (dashed curve). Therefore, when a pendulum oscillates with small displacements from the vertical, it exhibits simple harmonic motion—the same as a mass on a spring.

Chin Basher?

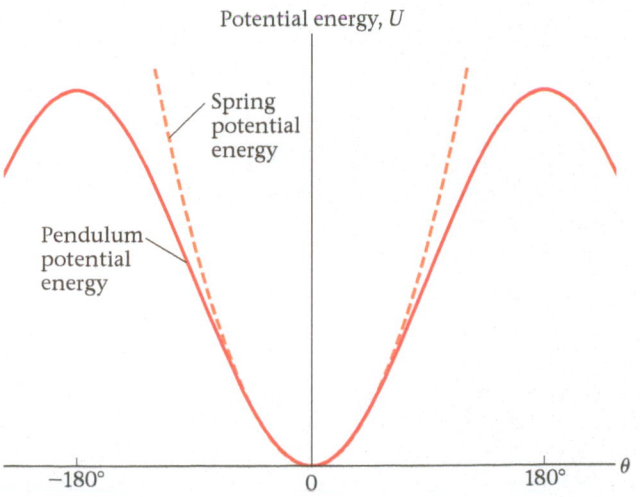

Potential energy, U

Spring potential energy

Pendulum potential energy

$-180°$ $\qquad$ 0 $\qquad$ $180°$ $\qquad$ θ

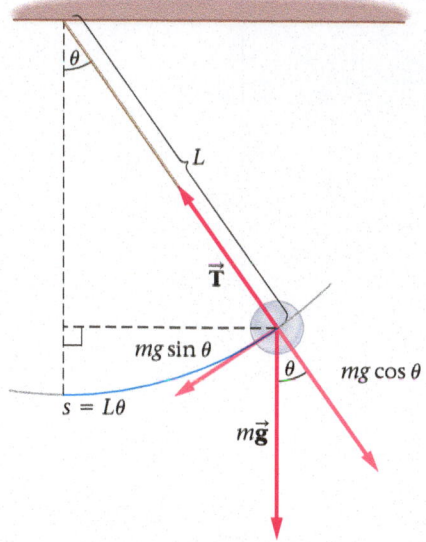

▲ **FIGURE 13-20 Forces acting on a pendulum bob** When the bob is displaced by an angle θ from the vertical, the restoring force is the tangential component of the weight, $mg \sin \theta$.

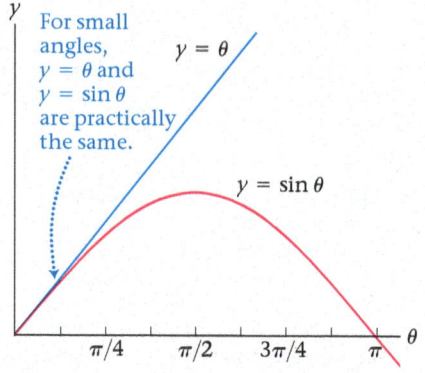

For small angles, $y = \theta$ and $y = \sin \theta$ are practically the same.

▲ **FIGURE 13-21 Relationship between sin θ and θ** For small angles measured in radians, $\sin \theta$ is approximately equal to θ. Thus, when considering small oscillations of a pendulum, we can replace $\sin \theta$ with θ. (See also Appendix A.)

Big Idea **4** The period of a pendulum is proportional to the square root of the length of the pendulum, and is independent of the mass of the pendulum. The period is also independent of the amplitude of oscillation, for small amplitudes.

This is illustrated in **FIGURE 13-21**. Notice that there is little difference between $\sin \theta$ and θ for angles smaller than about $\pi/8$ rad = 22.5 degrees. In addition, we see from Figure 13-20 that the arc length displacement of the mass from equilibrium is

$$s = L\theta$$

Equivalently,

$$\theta = \frac{s}{L}$$

Therefore, if the mass m is displaced from equilibrium by a small arc length s, the force it experiences is restoring and of magnitude

$$F = mg \sin \theta \approx mg\theta = \left(\frac{mg}{L}\right)s \qquad 13\text{-}19$$

The restoring force is proportional to the displacement, s, just as expected for simple harmonic motion.

Let's compare the pendulum to a mass on a spring. In the latter case, the restoring force has a magnitude given by

$$F = kx$$

The restoring force acting on the pendulum has precisely the same form if we let $x = s$ and

$$k = \frac{mg}{L}$$

Therefore, the period of a pendulum is simply the period of a mass on a spring, $T = 2\pi\sqrt{m/k}$, with k replaced by mg/L:

$$T = 2\pi\sqrt{\frac{m}{k}} = 2\pi\sqrt{\frac{m}{(mg/L)}}$$

Canceling the mass m, we find:

Period of a Pendulum (small amplitude)

$$T = 2\pi\sqrt{\frac{L}{g}} \qquad 13\text{-}20$$

SI unit: s

This is the classic formula for the period of a pendulum. Notice that T depends on the length of the pendulum, L, and on the acceleration of gravity, g. It is independent, however, of the mass, m, and the amplitude, A, as noted by Galileo.

The mass does not appear in the expression for the period of a pendulum, for the same reason that different masses free-fall with the same acceleration. In particular, a large mass tends to move more slowly because of its large inertia; on the other hand, the larger a mass, the greater the gravitational force acting on it. These two effects cancel in free fall, as well as in a pendulum.

EXERCISE 13-14 PERIOD AND LENGTH

The pendulum in a grandfather clock is designed to take 1.00 second to swing in each direction—that is, 2.00 seconds for a complete period. Find the length of a pendulum that has a period of 2.00 seconds.

REASONING AND SOLUTION

We can solve $T = 2\pi\sqrt{L/g}$ (Equation 13-20) for L and substitute numerical values. The result is

$$L = \frac{gT^2}{4\pi^2} = \frac{(9.81 \text{ m/s}^2)(2.00 \text{ s})^2}{4\pi^2} = 0.994 \text{ m}$$

CONCEPTUAL EXAMPLE 13-15   RAISE OR LOWER THE WEIGHT?

RWP If you look carefully at a grandfather clock, you will notice that the weight at the bottom of the pendulum can be moved up or down by turning a small screw. Suppose you have a grandfather clock at home that runs slow. Should you turn the adjusting screw so as to **(a)** raise the weight or **(b)** lower the weight?

REASONING AND DISCUSSION
To make the clock run faster, we want it to go from *tick* to *tick* more rapidly; in other words, we want the period of the pendulum to be decreased. From $T = 2\pi\sqrt{L/g}$ we can see that shortening the pendulum—that is, decreasing L—decreases the period. Hence, the weight should be raised, which effectively shortens the pendulum.

ANSWER
(a) The weight should be raised. This shortens the period and makes the clock run faster.

EXAMPLE 13-16   DROP TIME

A pendulum is constructed from a string 0.627 m long attached to a mass of 0.250 kg. When set in motion, the pendulum completes one oscillation every 1.59 s. If the pendulum is held at rest and the string is cut, how much time will it take for the mass to fall through a distance of 1.00 m?

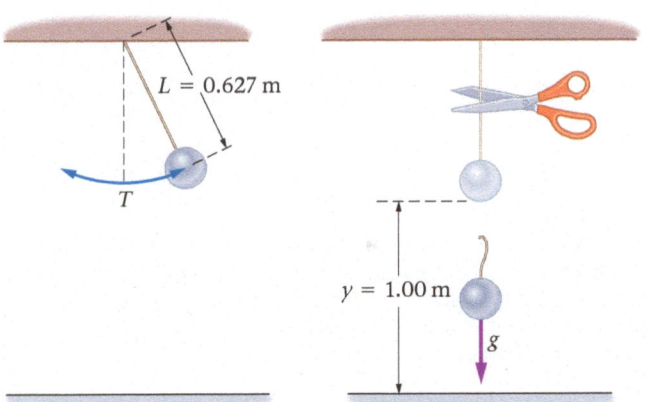

PICTURE THE PROBLEM
The pendulum has a length $L = 0.627$ m and a period of oscillation $T = 1.59$ s. When the pendulum is held at rest in a vertical position, a pair of scissors is used to cut the string. The mass then falls straight downward with an acceleration g through a distance $y = 1.00$ m.

REASONING AND STRATEGY
At first it might seem that the period of oscillation and the time of fall are unrelated. Recall, however, that the period of a pendulum depends on both its length *and* the acceleration of gravity g *at the location of the pendulum*.

    To solve this problem, then, we first use the period T to find the acceleration of gravity g. Once g is known, the time of fall is a straightforward kinematics problem (see Example 2-18).

Known   Length of pendulum, $L = 0.627$ m; mass of bob, $m = 0.250$ kg; period of oscillation, $T = 1.59$ s; distance of drop, $y = 1.00$ m.
Unknown Time to fall through drop distance, $t = ?$

SOLUTION

1. Use the formula for the period of a pendulum to solve for the acceleration of gravity:

$$T = 2\pi\sqrt{\frac{L}{g}} \quad \text{or} \quad g = \frac{4\pi^2 L}{T^2}$$

2. Substitute numerical values to find g:

$$g = \frac{4\pi^2 L}{T^2} = \frac{4\pi^2(0.627 \text{ m})}{(1.59 \text{ s})^2} = 9.79 \text{ m/s}^2$$

3. Use kinematics to solve for the time to drop from rest through a distance y:

$$y = \frac{1}{2}gt^2 \quad \text{or} \quad t = \sqrt{\frac{2y}{g}}$$

4. Substitute $y = 1.00$ m and the value of g just found to find the time:

$$t = \sqrt{\frac{2y}{g}} = \sqrt{\frac{2(1.00 \text{ m})}{9.79 \text{ m/s}^2}} = 0.452 \text{ s}$$

INSIGHT
The fact that the acceleration of gravity g varies from place to place on the Earth was mentioned in Chapter 2, and again in Chapter 12. In fact, "gravity maps" are valuable tools for geologists attempting to understand the underground properties of a given region. One instrument geologists use to make gravity maps is basically a very precise pendulum whose period can be accurately measured. Slight changes in the period from one location to another correspond to slight changes in g. More common in recent decades is an electronic "gravimeter" that uses a mass on a spring and relates the force of gravity to the stretch of the spring. These gravimeters can measure g with an accuracy of one part in 1000 million.

CONTINUED

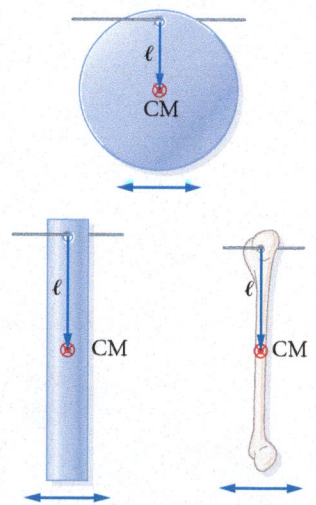

▲ **FIGURE 13-22 Examples of physical pendulums** In each case, an object of definite size and shape oscillates about a given pivot point. The period of oscillation depends in detail on the location of the pivot point as well as on the distance ℓ from it to the center of mass, CM.

The Physical Pendulum

In the ideal version of a simple pendulum, a bob of mass m swings back and forth a distance L from the suspension point. All of the pendulum's mass is assumed to be concentrated in the bob, which is treated as a point mass. On the other hand, a **physical pendulum** is one in which the mass is not concentrated at a point, but instead is distributed over a finite volume. Examples are shown in **FIGURE 13-22**. Detailed mathematical analysis shows that if the moment of inertia (Chapter 10) of a physical pendulum about its axis of rotation is I, and the distance from the axis to the center of mass is ℓ, the period of the pendulum is given by the following:

Period of a Physical Pendulum

$$T = 2\pi \sqrt{\frac{\ell}{g}} \left(\sqrt{\frac{I}{m\ell^2}} \right) \qquad\qquad 13\text{-}21$$

SI unit: s

Notice that the first part of the expression, $2\pi\sqrt{\ell/g}$, is the period of a simple pendulum with all its mass concentrated at the center of mass. The second factor, $\sqrt{I/m\ell^2}$, is a correction that takes into account the size and shape of the physical pendulum. Thus, writing the period in this form, rather than canceling one power of ℓ, makes for a convenient comparison with the simple pendulum.

To see how Equation 13-21 works in practice, we first apply it to a simple pendulum of mass m and length L. In this case, the moment of inertia is $I = mL^2$, and the distance to the center of mass is $\ell = L$. As a result, the period is

$$T = 2\pi \sqrt{\frac{\ell}{g}} \left(\sqrt{\frac{I}{m\ell^2}} \right) = 2\pi \sqrt{\frac{L}{g}} \left(\sqrt{\frac{mL^2}{mL^2}} \right) = 2\pi \sqrt{\frac{L}{g}}$$

Thus, as expected, Equation 13-21 also applies to a simple pendulum.

Walking Next, we apply Equation 13-21 to a nontrivial physical pendulum—namely, your leg. When you walk, your leg rotates about the hip joint much like a uniform rod pivoted about one end, as indicated in **FIGURE 13-23**. Thus, if we approximate your leg as a uniform rod of length L, its period can be found using Equation 13-21. Recall from Chapter 10 (see Table 10-1) that the moment of inertia of a rod of length L about one end is

$$I = \tfrac{1}{3} mL^2$$

Similarly, the center of mass of your leg is essentially at the center of your leg; thus,

$$\ell = \tfrac{1}{2} L$$

Combining these results, we find that the period of your leg is roughly

$$T = 2\pi \sqrt{\frac{\ell}{g}} \left(\sqrt{\frac{I}{m\ell^2}} \right) = 2\pi \sqrt{\frac{\frac{1}{2}L}{g}} \left(\sqrt{\frac{\frac{1}{3}mL^2}{m\left(\frac{1}{2}L\right)^2}} \right) = 2\pi \sqrt{\frac{L}{g}} \left(\sqrt{\frac{2}{3}} \right)$$

Given that a typical human leg is about a meter long ($L = 1.0$ m), we find that its period of oscillation about the hip is approximately $T = 1.6$ s.

RWP The significance of this result is that the natural walking pace of humans and other animals is largely controlled by the swinging motion of their legs as physical pendula. This is illustrated in **FIGURE 13-24**, where we compare the walking motions

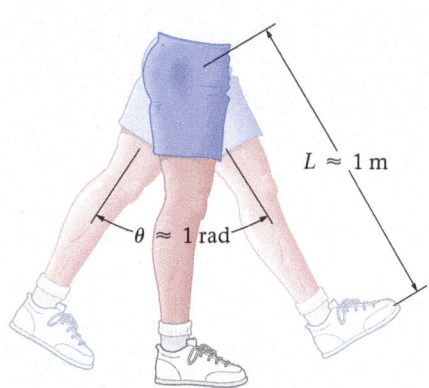

$L \approx 1$ m

$\theta \approx 1$ rad

▲ **FIGURE 13-23 The leg as a physical pendulum** As a person walks, each leg swings much like a physical pendulum. A reasonable approximation is to treat the leg as a uniform rod about 1 m in length.

PROBLEM-SOLVING NOTE

The Period of a Physical Pendulum

When finding the period of a physical pendulum, recall that I is the moment of inertia about the pivot point and ℓ is the distance from the pivot point to the center of mass.

of Beatles and elephants. In fact, a great deal of research in animal locomotion has focused on precisely this type of analysis. In the case just considered, suppose that the leg in Figure 13-23 swings through an angle of roughly 1.0 radian, as the foot moves from behind the hip to in front of the hip for the next step. The arc length through which the foot moves is $s = r\theta \approx (1.0\,\text{m})(1.0\,\text{rad}) = 1.0\,\text{m}$. Because it takes half a period ($T/2 = 0.80\,\text{s}$) for the foot to move from behind the hip to in front of it, the average speed of the foot is roughly

$$v = \frac{d}{t} \approx \frac{1.0\,\text{m}}{0.80\,\text{s}} \approx 1.3\,\text{m/s}$$

This is the typical speed of a person taking a brisk walk.

Returning to $T = 2\pi\sqrt{\ell/g}(\sqrt{I/m\ell^2})$, we can see that the smaller the moment of inertia I, the smaller the period. After all, an object with a small moment of inertia rotates quickly and easily. It therefore completes an oscillation in less time than an object with a larger moment of inertia. In the case of the human leg, the fact that its mass is distributed uniformly along its length—rather than concentrated at the far end—means that its moment of inertia and period are less than for a simple pendulum 1 m long.

QUICK EXAMPLE 13-17 PERIOD OF OSCILLATION

A Hula Hoop with mass $m = 0.68\,\text{kg}$ and radius $R = 0.51\,\text{m}$ hangs from a peg in a garage. When someone bumps the hoop, it begins to swing back and forth. What is the hoop's period of oscillation? (*Note:* The moment of inertia of a hoop about a point on its rim is $I = 2mR^2$.)

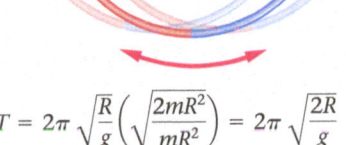

REASONING AND STRATEGY

The period is given by $T = 2\pi\sqrt{\ell/g}\left(\sqrt{I/(m\ell^2)}\right)$. In this case, the distance to the center of mass is $\ell = R$, and the moment of inertia is $I = 2mR^2$.

1. Substitute $I = 2mR^2$ and $\ell = R$ into $T = 2\pi\sqrt{\ell/g}\left(\sqrt{I/(m\ell^2)}\right)$:

$$T = 2\pi\sqrt{\frac{R}{g}}\left(\sqrt{\frac{2mR^2}{mR^2}}\right) = 2\pi\sqrt{\frac{2R}{g}}$$

2. Evaluate with $R = 0.51\,\text{m}$ and $g = 9.81\,\text{m/s}^2$:

$$T = 2\pi\sqrt{\frac{2(0.51\,\text{m})}{9.81\,\text{m/s}^2}} = 2.0\,\text{s}$$

This period is about 1.4 times greater than the period of a simple pendulum of length 0.51 m. This makes sense when you consider that much of the hoop's mass is farther from the axis of rotation than 0.51 m, resulting in a larger moment of inertia than is the case for a simple pendulum.

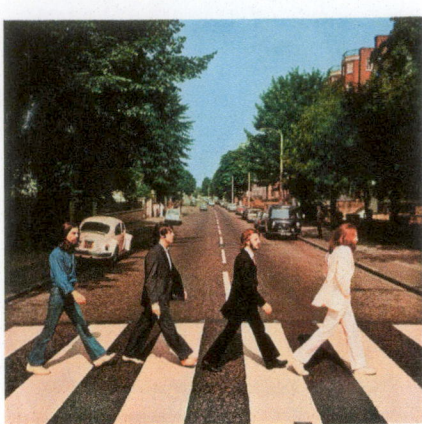

▲ **FIGURE 13-24 Visualizing Concepts**
Physical Pendulums and Walking When animals walk, the swinging movement of their legs can be approximated fairly well by treating the legs as physical pendula. Such analysis has proved useful in analyzing the gaits of various creatures, from Beatles to elephants.

Enhance Your Understanding (Answers given at the end of the chapter)

6. Rank the four pendulum systems in **FIGURE 13-25** in order of increasing period. Indicate ties where appropriate.

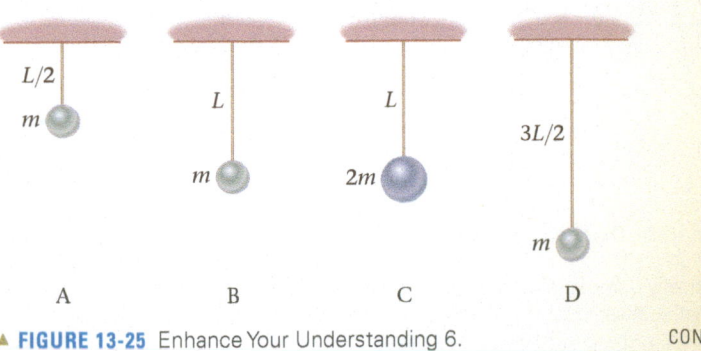

▲ **FIGURE 13-25** Enhance Your Understanding 6.

CONTINUED

13-7 Damped Oscillations

To this point we have restricted our considerations to oscillating systems in which no mechanical energy is gained or lost. In most physical systems, however, there is some loss of mechanical energy to friction, air resistance, or other nonconservative forces. As the mechanical energy of a system decreases, its amplitude of oscillation decreases as well, as expected from $E = \frac{1}{2}kA^2$ (Equation 13-15). This type of motion is referred to as a **damped oscillation**.

In a typical situation, an oscillating mass may lose its mechanical energy to a force such as air resistance that is proportional to the speed of the mass and opposite in direction. The force in such a case can be written as

$$\vec{\mathbf{F}} = -b\vec{\mathbf{v}}$$

The constant b is referred to as the **damping constant**; it is a measure of the strength of the damping force, and its SI units are kg/s.

If the damping constant is small, the system will continue to oscillate, but with a continuously decreasing amplitude. This type of motion, referred to as **underdamped**, is illustrated in **FIGURE 13-26 (a)**. In such cases, the amplitude decreases exponentially with time. Thus, if A_0 is the initial amplitude of an oscillating mass m, the amplitude at the time t is

$$A = A_0 e^{-bt/2m}$$

The exponential dependence of the amplitude is indicated by the dashed curve in the figure.

As the damping is increased, a point is reached where the system no longer oscillates, but instead simply relaxes back to the equilibrium position, as shown in **FIGURE 13-26 (b)**. A system with this type of behavior is said to be **critically damped**. If the damping is increased beyond this point, the system is said to be **overdamped**. In this case, the system still returns to equilibrium without oscillating, but the time required is greater. This is also illustrated in Figure 13-26 (b).

Some mechanical systems are designed to be near the condition for critical damping. If such a system is displaced from equilibrium, it will return to equilibrium, without oscillating, in the shortest possible time. For example, shock absorbers are designed so that a car that has just hit a bump will return to equilibrium quickly, without a lot of up-and-down oscillations.

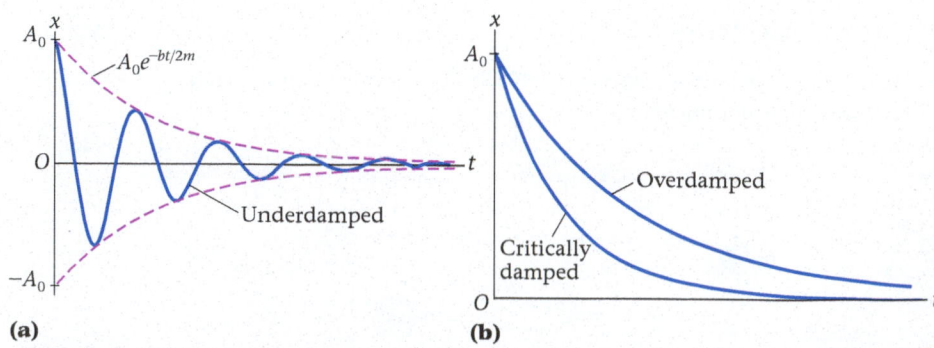

▲ **FIGURE 13-26 Damped oscillations (a)** In underdamped oscillations, the position continues to oscillate as a function of time, but the amplitude of oscillation decreases exponentially. **(b)** In critically damped and overdamped motion, no oscillations occur. Instead, an object simply settles back to its equilibrium position without overshooting. Equilibrium is reached most rapidly in the critically damped case.

Enhance Your Understanding   (Answers given at the end of the chapter)

7. The amplitude of a damped oscillation decreases from A at $t = 0$ to $\frac{3}{4}A$ at $t = T$. What is the amplitude of the system at $t = 2T$? Explain.

Section Review

• The amplitude of a damped oscillating system decreases with time as $A = A_0 e^{-bt/2m}$, where b is the damping constant.

13-8 Driven Oscillations and Resonance

In the previous section we considered the effects of removing energy from an oscillating system. It is also possible to increase the energy of a system, or to replace the energy lost to various forms of friction. This can be done by applying an external force that does positive work on the system.

Suppose, for example, that you hold the end of a string from which a small weight is suspended, as in **FIGURE 13-27**. If the weight is set in motion and you hold your hand still, it will soon stop oscillating. If you move your hand back and forth in a horizontal direction, however, you can keep the weight oscillating indefinitely. The motion of your hand is said to be "driving" the weight, leading to **driven oscillations**.

The response of the weight in this example depends on the frequency of your hand's back-and-forth motion, as you can readily verify for yourself. For instance, if you move your hand very slowly, the weight will simply track the motion of your hand. Similarly, if you oscillate your hand very rapidly, the weight will exhibit only small oscillations. Oscillating your hand at an intermediate frequency, however, can result in large amplitude oscillations for the weight.

Just what is an appropriate intermediate frequency? Well, to achieve a large response, your hand should drive the weight at the frequency at which it oscillates when not being driven—that is, when it oscillates by itself. This is referred to as the **natural frequency**, f_0, of the system. For example, the natural frequency of a pendulum of length L is simply the inverse of its period:

$$f_0 = \frac{1}{T} = \frac{1}{2\pi}\sqrt{\frac{g}{L}}$$

Similarly, the natural frequency of a mass on a spring is

$$f_0 = \frac{1}{T} = \frac{1}{2\pi}\sqrt{\frac{k}{m}}$$

In general, driving any system at a frequency near its natural frequency results in large oscillations.

As an example, **FIGURE 13-28** shows a plot of amplitude, A, versus driving frequency, f, for a mass on a spring. Notice the large amplitude for frequencies near f_0. This type of large response, due to frequency matching, is known as **resonance**, and the curves shown in Figure 13-28 are referred to as **resonance curves**. The five curves shown in Figure 13-28 correspond to different amounts of damping. As can be seen, systems with small damping have a high, narrow peak on their resonance curve. This means that resonance is a large effect in these systems, and that it is very sensitive to frequency. On the other hand, systems with large damping have resonance peaks that are broad and low.

RWP Resonance plays an important role in a variety of physical systems, from a pendulum to atoms in a laser to a tuner in a radio or TV. As we shall see in Chapter 24, for example, adjusting the tuning knob in a radio changes the resonance frequency of the electric circuit in the tuner. When its resonance frequency matches the frequency being broadcast by a station, that station is picked up. To change stations, we simply

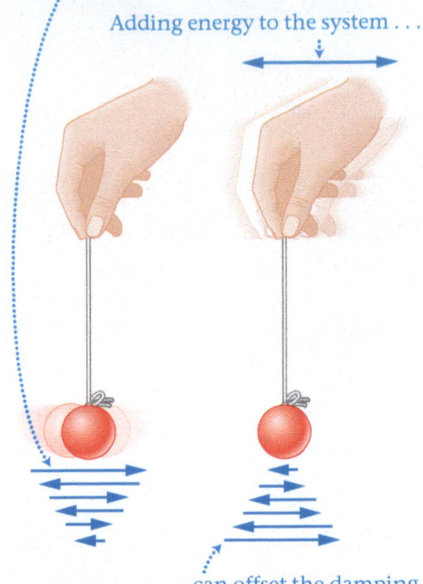

Damping reduces the amplitude of motion.

Adding energy to the system . . .

. . . can offset the damping.

▲ **FIGURE 13-27 Driven oscillations** If the support point of a pendulum is held still, its oscillations quickly die away due to damping. If the support point is oscillated back and forth, however, the pendulum will continue swinging. This is called "driving" the pendulum. If the driving frequency is close to the natural frequency of the pendulum—the frequency at which it oscillates when the support point is held still—its amplitude of motion can become quite large.

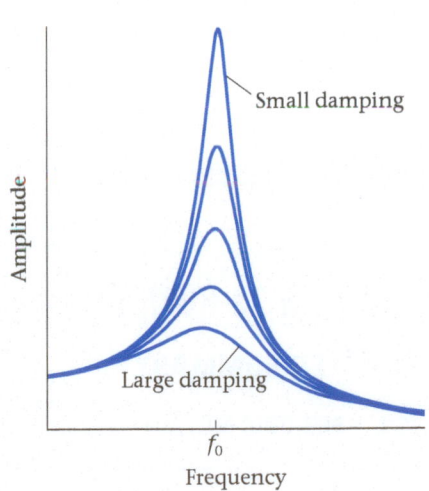

Small damping

Large damping

f_0

Amplitude

Frequency

▲ **FIGURE 13-28 Resonance curves for various amounts of damping** When the damping is small, the amplitude of oscillation can become very large for frequencies close to the natural frequency, f_0. When the damping is large, the amplitude has only a low, broad peak near the natural frequency.

(a)

(b)

Vibrating Rods

◄ **FIGURE 13-29 Visualizing Concepts Resonance (a)** Anyone who has ever pushed a child on a swing knows that the timing of the pushes is critical. If they are synchronized with the natural frequency of the swing, the amplitude can increase rapidly. **(b)** The phenomenon of resonance can have dangerous consequences. In 1940, the Tacoma Narrows Bridge, completed only four months earlier, collapsed when high winds set the bridge swaying at one of its resonant frequencies.

change the resonance frequency of the tuner to the frequency of another station. A good tuner will have little damping, so stations that are even slightly "off resonance" will have small response, and hence will not be heard.

Mechanical examples of resonance are all around us as well. In fact, you might want to try the following experiment the next time you notice a spider in its web. Move close to the web and hum, starting with a low pitch. Slowly increase the pitch of your humming, and soon you will notice the spider react excitedly. You have hit the resonance frequency of its web, causing it to vibrate and making the spider think he has snagged a lunch. If you continue to increase your humming frequency, the spider quiets down again, because you have gone past the resonance. Each individual spiderweb you encounter will resonate at a different, and specific, frequency.

RWP Man-made structures can show resonance effects as well. A playful, everyday example is shown in **FIGURE 13-29 (a)**. On a more serious note, one of the most dramatic and famous examples of mechanical resonance is the collapse of Washington's Tacoma Narrows Bridge in 1940. High winds through the narrows had often set the bridge into a gentle swaying motion, resulting in its being known by the affectionate nickname "Galloping Girdie." During one particular windstorm, however, the bridge experienced a resonance-like effect, and the amplitude of its swaying motion began to increase. Alarmed officials closed the bridge to traffic, and a short time later the swaying motion became so great that the bridge broke apart and fell into the waters below, as shown in **FIGURE 13-29 (b)**. Needless to say, bridges built since that time have been designed to prevent such catastrophic oscillations.

Big Idea **5** An oscillator driven at its resonance frequency responds with a large amplitude of oscillation. The smaller the damping, the larger the resonant amplitude.

Enhance Your Understanding (Answers given at the end of the chapter)

8. When you drive a pendulum at a frequency f_1, you observe an amplitude of 2 cm. When you gradually increase the driving frequency to $f_2 > f_1$, you observe that the amplitude decreases to 1 cm. Is the resonance frequency of the pendulum less than f_1, between f_1 and f_2, or greater than f_2? Explain.

Section Review

• The amplitude of a driven oscillating system is greatest when it is driven at its natural frequency.

CHAPTER 13 REVIEW

CHAPTER SUMMARY

13-1 PERIODIC MOTION

Periodic motion repeats after a definite length of time.

Period
The period, T, is the time required for a motion to repeat:

T = time required for one cycle of a periodic motion

Frequency
Frequency, f, is the number of oscillations per unit time. Equivalently, f is the inverse of the period:

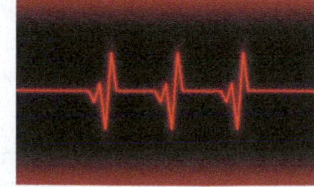

$$f = \frac{1}{T}$$

13-1

13-2   SIMPLE HARMONIC MOTION

A particular type of periodic motion is simple harmonic motion.

Restoring Force

Simple harmonic motion occurs when the restoring force is proportional to the displacement from equilibrium.

Amplitude

The maximum displacement from equilibrium is the amplitude, A.

Position Versus Time

The position, x, of an object in simple harmonic motion is

$$x = A \cos\left(\frac{2\pi}{T}t\right) = A \cos(\omega t) \qquad \text{13-2, 13-4}$$

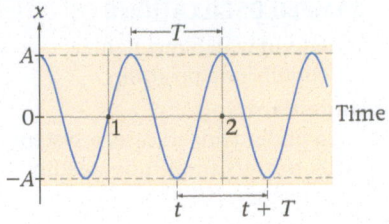

13-3   CONNECTIONS BETWEEN UNIFORM CIRCULAR MOTION AND SIMPLE HARMONIC MOTION

Circular motion viewed from the side—by projecting a shadow onto a screen, for example—is simple harmonic motion.

Angular Frequency

Angular frequency, ω, is 2π times the frequency:

$$\omega = 2\pi f = \frac{2\pi}{T} \qquad \text{13-5}$$

Velocity

Velocity as a function of time in simple harmonic motion is

$$v = -A\omega \sin(\omega t) \qquad \text{13-6}$$

Acceleration

Acceleration as a function of time in simple harmonic motion is

$$a = -A\omega^2 \cos(\omega t) \qquad \text{13-8}$$

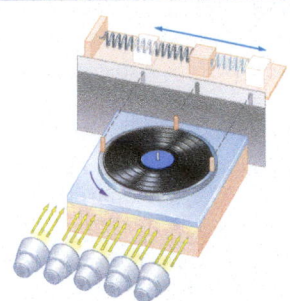

13-4   THE PERIOD OF A MASS ON A SPRING

An important special case of simple harmonic motion is a mass on a spring.

Period

The period of a mass m attached to a spring of force constant k is

$$T = 2\pi \sqrt{\frac{m}{k}} \qquad \text{13-11}$$

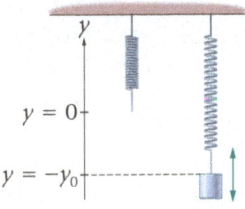

13-5   ENERGY CONSERVATION IN OSCILLATORY MOTION

In an ideal oscillatory system, the total energy remains constant.

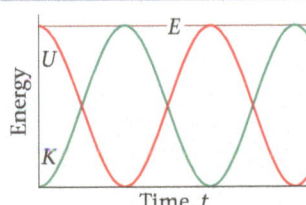

13-6   THE PENDULUM

A pendulum oscillating with small amplitude exhibits simple harmonic motion.

Period of a Simple Pendulum

The period of a simple pendulum of length L is

$$T = 2\pi \sqrt{\frac{L}{g}} \qquad \text{13-20}$$

Period of a Physical Pendulum

The period of a physical pendulum is

$$T = 2\pi \sqrt{\frac{\ell}{g}}\left(\sqrt{\frac{I}{m\ell^2}}\right) \qquad \text{13-21}$$

In this expression, I is the moment of inertia about the pivot point, and ℓ is the distance from the pivot point to the center of mass.

13-7 DAMPED OSCILLATIONS

Systems in which mechanical energy is lost to other forms eventually come to rest at the equilibrium position.

Underdamping

In an underdamped case, a system of mass m and damping constant b continues to oscillate as its amplitude steadily decreases with time. The decrease in amplitude is exponential:

$$A = A_0 e^{-bt/2m}$$

Critical Damping

In a system with critical damping, the system relaxes back to equilibrium with no oscillations in the least possible time.

Overdamping

An overdamped system also relaxes back to equilibrium with no oscillations, but more slowly than in critical damping.

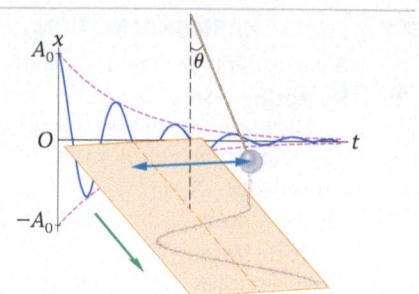

13-8 DRIVEN OSCILLATIONS AND RESONANCE

If an oscillating system is driven by an external force, it is possible for energy to be added to the system.

Resonance

Resonance is the response of an oscillating system to a driving force of the appropriate frequency.

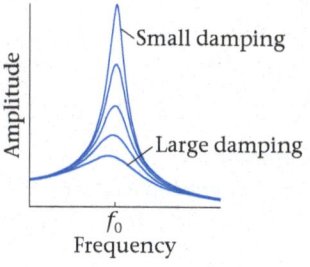

ANSWERS TO ENHANCE YOUR UNDERSTANDING QUESTIONS

1. The period, which is the inverse of the frequency, doubles.

2. **(c)**

3. The acceleration is positive because the object, which just came to rest momentarily, will begin to gain speed in the positive direction.

4. B < A = C < D

5. **(a)** 10 J. **(b)** 3 J.

6. A < B = C < D.

7. Each time interval of T reduces the amplitude by a factor of $\frac{3}{4}$. Thus, at $t = 2T$ the amplitude is $\left(\frac{3}{4}\right)^2 A = \left(\frac{9}{13}\right)A$.

8. The amplitude decreases as the frequency is increased beyond the value f_1. Therefore, the resonance frequency is less than f_1.

CONCEPTUAL QUESTIONS

For instructor-assigned homework, go to www.masteringphysics.com.

(Answers to odd-numbered Conceptual Questions can be found in the back of the book.)

1. A basketball player dribbles a ball with a steady period of T seconds. Is the motion of the ball periodic? Is it simple harmonic motion? Explain.

2. A person rides on a Ferris wheel that rotates with constant angular speed. If the Sun is directly overhead, does the person's shadow on the ground undergo periodic motion? Does it undergo simple harmonic motion? Explain.

3. An air-track cart bounces back and forth between the two ends of an air track. Is this motion periodic? Is it simple harmonic motion? Explain.

4. If a mass m and a mass $2m$ oscillate on identical springs with identical amplitudes, they both have the same maximum kinetic energy. How can this be? Shouldn't the larger mass have more kinetic energy? Explain.

5. An object oscillating with simple harmonic motion completes a cycle in a time T. If the object's amplitude is doubled, the time required for one cycle is still T, even though the object covers twice the distance. How can this be? Explain.

6. The position of an object undergoing simple harmonic motion is given by $x = A \cos(Bt)$. Explain the physical significance of the constants A and B. What is the frequency of this object's motion?

7. The pendulum bob in Figure 13-18 leaks sand onto the strip chart. What effect does this loss of sand have on the period of the pendulum? Explain.

PROBLEMS AND CONCEPTUAL EXERCISES

Answers to odd-numbered Problems and Conceptual Exercises can be found in the back of the book. **BIO** *identifies problems of biological or medical interest;* **CE** *indicates a conceptual exercise.* **Predict/Explain** *problems ask for two responses: (a) your prediction of a physical outcome, and (b) the best explanation among three provided; and* **Predict/Calculate** *problems ask for a prediction of a physical outcome, based on fundamental physics concepts, and follow that with a numerical calculation to verify the prediction. On all problems, bullets (•, ••, •••) indicate the level of difficulty.*

SECTION 13-1 PERIODIC MOTION

1. • A person in a rocking chair completes 12 cycles in 21 s. What are the period and frequency of the rocking?

2. • While fishing for catfish, a fisherman suddenly notices that the bobber (a floating device) attached to his line is bobbing up and down with a frequency of 1.8 Hz. What is the period of the bobber's motion?

3. • If you dribble a basketball with a frequency of 1.9 Hz, how much time does it take for you to complete 12 dribbles?

4. • You take your pulse and observe 74 heartbeats in a minute. What are the period and frequency of your heartbeat?

5. • **BIO** **Slow-Motion Dragonfly** A frame-by-frame analysis of a slow-motion video shows that a hovering dragonfly takes 7 frames to complete one wing beat. If the video is shot at 240 frames per second, **(a)** what is the period of the wing beat, and **(b)** how many wing beats does the dragonfly perform per second?

6. •• **Predict/Calculate** **(a)** Your heart beats with a frequency of 1.45 Hz. How many beats occur in a minute? **(b)** If the frequency of your heartbeat increases, will the number of beats in a minute increase, decrease, or stay the same? **(c)** How many beats occur in a minute if the frequency increases to 1.55 Hz?

7. •• You rev your car's engine to 3300 rpm (rev/min). **(a)** What are the period and frequency of the engine? **(b)** If you change the period of the engine to 0.033 s, how many rpms is it doing?

SECTION 13-2   SIMPLE HARMONIC MOTION

8. • **CE** A mass moves back and forth in simple harmonic motion with amplitude A and period T. **(a)** In terms of A, through what distance does the mass move in the time T? **(b)** Through what distance does it move in the time $5T/2$?

9. • **CE** A mass moves back and forth in simple harmonic motion with amplitude A and period T. **(a)** In terms of T, how much time does it take for the mass to move through a total distance of $2A$? **(b)** How much time does it take for the mass to move through a total distance of $3A$?

10. • The position of a mass oscillating on a spring is given by $x = (3.9\ \text{cm})\cos[2\pi t/(0.38\ \text{s})]$. **(a)** What is the period of this motion? **(b)** What is the first time the mass is at the position $x = 0$?

11. • The position of a mass oscillating on a spring is given by $x = (7.8\ \text{cm})\cos[2\pi t/(0.68\ \text{s})]$. **(a)** What is the frequency of this motion? **(b)** When is the mass first at the position $x = -7.8\ \text{cm}$?

12. •• **CE** A position-versus-time plot for an object undergoing simple harmonic motion is given in **FIGURE 13-30**. Rank the six points indicated in the figure in order of increasing **(a)** speed, **(b)** velocity, and **(c)** acceleration. Indicate ties where necessary.

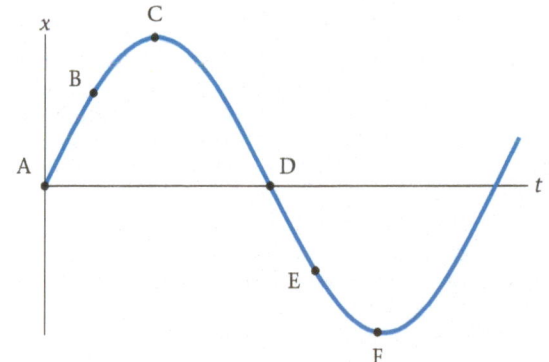

FIGURE 13-30 Problem 12

13. •• **CE** A mass on a spring oscillates with simple harmonic motion of amplitude A about the equilibrium position $x = 0$. Its maximum speed is v_{max} and its maximum acceleration is a_{max}. **(a)** What is the speed of the mass at $x = 0$? **(b)** What is the acceleration of the mass at $x = 0$? **(c)** What is the speed of the mass at $x = A$? **(d)** What is the acceleration of the mass at $x = A$?

14. •• A mass oscillates on a spring with a period of 0.63 s and an amplitude of 4.4 cm. Write an equation giving x as a function of time, assuming the mass starts at $x = A$ at time $t = 0$.

15. •• **Predict/Calculate** **Molecular Oscillations** An atom in a molecule oscillates about its equilibrium position with a frequency of $2.00 \times 10^{14}\ \text{Hz}$ and a maximum displacement of 3.50 nm. **(a)** Write an expression giving x as a function of time for this atom, assuming that $x = A$ at $t = 0$. **(b)** If, instead, we assume that $x = 0$ at $t = 0$, would your expression for position versus time use a sine function or a cosine function? Explain.

16. •• A mass oscillates on a spring with a period T and an amplitude 0.48 cm. The mass is at the equilibrium position $x = 0$ at $t = 0$, and is moving in the positive direction. Where is the mass at the times **(a)** $t = T/8$, **(b)** $t = T/4$, **(c)** $t = T/2$, and **(d)** $t = 3T/4$? **(e)** Plot your results for parts (a) through (d) with the vertical axis representing position and the horizontal axis representing time.

17. •• The position of a mass on a spring is given by $x = (3.8\ \text{cm})\cos[2\pi t/(0.88\ \text{s})]$. **(a)** What is the period, T, of this motion? **(b)** Where is the mass at $t = 0.55\ \text{s}$? **(c)** Show that the mass is at the same location at $0.55\ \text{s} + T$ seconds as it is at 0.55 s.

18. •• **Predict/Calculate** A mass attached to a spring oscillates with a period of 3.35 s. **(a)** If the mass starts from rest at $x = 0.0440\ \text{m}$ and time $t = 0$, where is it at time $t = 6.37\ \text{s}$? **(b)** Is the mass moving in the positive or negative x direction at $t = 6.37\ \text{s}$? Explain.

19. ••• A lawn sprinkler oscillates with simple harmonic motion of period $T = 52.0\ \text{s}$, and sprays water with an angle to the vertical $(\theta = 0)$ that varies from $\theta = -45°$ to $\theta = 45°$, as shown in **FIGURE 13-31**. Water from the sprinkler lands in a nearby garden patch when the angle the sprinkler makes with the vertical is greater than $\theta = 36°$. For what amount of time is the sprinkler watering the garden during one complete cycle of its oscillation?

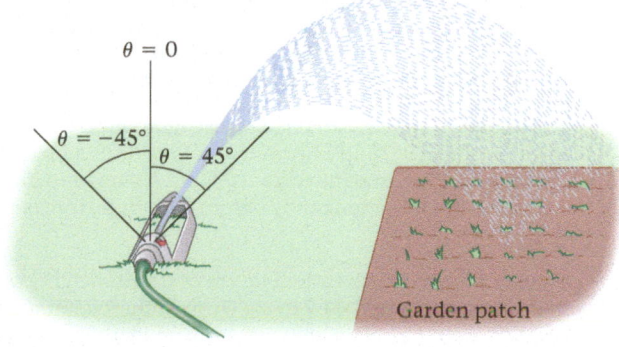

FIGURE 13-31 Problem 19

SECTION 13-3   CONNECTIONS BETWEEN UNIFORM CIRCULAR MOTION AND SIMPLE HARMONIC MOTION

20. • A ball rolls on a circular track of radius 0.62 m with a constant angular speed of 1.3 rad/s in the counterclockwise direction. If the angular position of the ball at $t = 0$ is $\theta = 0$, find the x component of the ball's position at the times 2.5 s, 5.0 s, and 7.5 s. Let $\theta = 0$ correspond to the positive x direction.

21. • An object executing simple harmonic motion has a maximum speed of 4.1 m/s and a maximum acceleration of 0.85 m/s². Find **(a)** the amplitude and **(b)** the period of this motion.

22. • A child rocks back and forth on a porch swing with an amplitude of 0.204 m and a period of 2.80 s. Assuming the motion is approximately simple harmonic, find the child's maximum speed.

23. •• **Predict/Calculate** A 30.0-g goldfinch lands on a slender branch, where it oscillates up and down with simple harmonic motion of amplitude 0.0335 m and period 1.65 s. **(a)** What is the maximum acceleration of the finch? Express your answer as a fraction of the acceleration of gravity, *g*. **(b)** What is the maximum speed of the goldfinch? **(c)** At the time when the goldfinch experiences its maximum acceleration, is its speed a maximum or a minimum? Explain.

24. •• **BIO Tuning Forks in Neurology** Tuning forks are used in the diagnosis of nervous afflictions known as large-fiber polyneuropathies, which are often manifested in the form of reduced sensitivity to vibrations. Disorders that can result in this type of pathology include diabetes and nerve damage from exposure to heavy metals. The tuning fork in **FIGURE 13-32** has a frequency of 165 Hz. If the tips of the fork move with an amplitude of 1.10 mm, find **(a)** their maximum speed and **(b)** their maximum acceleration. Give your answer to part (b) as a multiple of *g*.

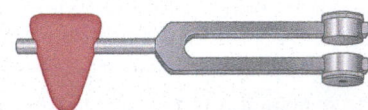

FIGURE 13-32 A Buck neurological hammer with tuning fork and Wartenburg pinwheel. (Problem 24)

25. •• A vibrating structural beam in a spacecraft can cause problems if the frequency of vibration is fairly high. Even if the amplitude of vibration is only a fraction of a millimeter, the acceleration of the beam can be several times greater than the acceleration due to gravity. As an example, find the maximum acceleration of a beam that vibrates with an amplitude of 0.25 mm at the rate of 110 vibrations per second. Give your answer as a multiple of *g*.

26. •• A peg on a turntable moves with a constant tangential speed of 0.82 m/s in a circle of radius 0.33 m. The peg casts a shadow on a wall. Find the following quantities related to the motion of the shadow: **(a)** the period, **(b)** the amplitude, **(c)** the maximum speed, and **(d)** the maximum magnitude of the acceleration.

27. •• The pistons in an internal combustion engine undergo a motion that is approximately simple harmonic motion. If the amplitude of the motion is 3.5 cm, and the engine runs at 1700 rev/min, find **(a)** the maximum acceleration of the pistons and **(b)** their maximum speed.

28. •• **Vomit Comet** NASA trains astronauts to deal with weightlessness (and its associated nausea) by flying them in the "vomit comet," a modified KC-135 airplane that flies in an oscillating path with a period of 72 seconds and a maximum acceleration of magnitude 9.81 m/s². What is the amplitude of the airplane's oscillation, assuming it undergoes simple harmonic motion?

29. •• A 0.84-kg air cart is attached to a spring and allowed to oscillate. If the displacement of the air cart from equilibrium is $x = (10.0 \text{ cm}) \cos[(2.00 \text{ s}^{-1})t + \pi]$, find **(a)** the maximum kinetic energy of the cart and **(b)** the maximum force exerted on it by the spring.

30. •• **Predict/Calculate** A person rides on a mechanical bucking horse (see **FIGURE 13-33**) that oscillates up and down with simple harmonic motion. The period of the bucking is 0.74 s and the amplitude is slowly increasing. At a certain amplitude the rider must hang on to prevent separating from the mechanical horse. **(a)** Give a strategy that will allow you to calculate this amplitude. **(b)** Carry out your strategy and find the desired amplitude.

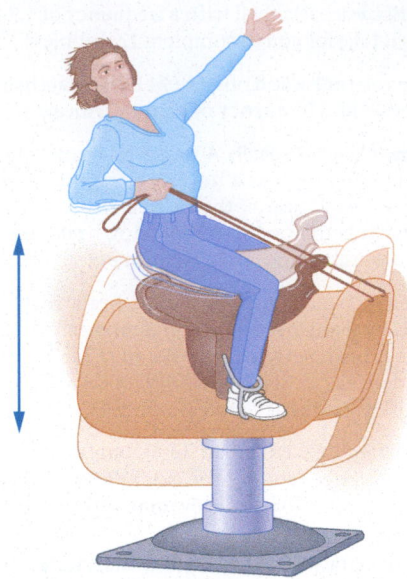

FIGURE 13-33 Problem 30

31. ••• An object moves with simple harmonic motion of period *T* and amplitude *A*. During one complete cycle, for what length of time is the speed of the object greater than $v_{max}/2$?

32. ••• An object executing simple harmonic motion has a maximum speed v_{max} and a maximum acceleration a_{max}. Find **(a)** the amplitude and **(b)** the period of this motion. Express your answers in terms of v_{max} and a_{max}.

SECTION 13-4 THE PERIOD OF A MASS ON A SPRING

33. • **CE Predict/Explain** If a mass *m* is attached to a given spring, its period of oscillation is *T*. If two such springs are connected end to end and the same mass *m* is attached, **(a)** is the resulting period of oscillation greater than, less than, or equal to *T*? **(b)** Choose the *best explanation* from among the following:

   I. Connecting two springs together makes the spring stiffer, which means that less time is required for an oscillation.

   II. The period of oscillation does not depend on the length of a spring, only on its force constant and the mass attached to it.

   III. The longer spring stretches more easily, and hence takes longer to complete an oscillation.

34. • **CE Predict/Explain** An old car with worn-out shock absorbers oscillates with a given frequency when it hits a speed bump. If the driver adds a couple of passengers to the car and hits another speed bump, **(a)** is the car's frequency of oscillation greater than, less than, or equal to what it was before? **(b)** Choose the *best explanation* from among the following:

   I. Increasing the mass on a spring increases its period, and hence decreases its frequency.

   II. The frequency depends on the force constant of the spring but is independent of the mass.

   III. Adding mass makes the spring oscillate more rapidly, which increases the frequency.

35. • **CE Predict/Explain** The two blocks in **FIGURE 13-34** have the same mass, *m*. All the springs have the same force constant, *k*, and are at their equilibrium length. When the blocks are set into oscillation, **(a)** is the period of block 1 greater than, less than, or equal to the period of block 2? **(b)** Choose the *best explanation* from among the following:

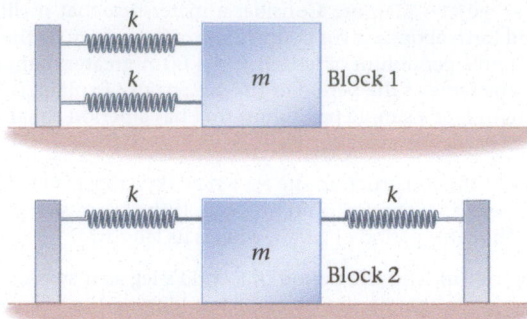

FIGURE 13-34 Problems 35 and 39

I. Springs in parallel are stiffer than springs in series; therefore the period of block 1 is smaller than the period of block 2.
II. The two blocks experience the same restoring force for a given displacement from equilibrium, and hence they have equal periods of oscillation.
III. The force of the two springs on block 2 partially cancel one another, leading to a longer period of oscillation.

36. • A 0.49-kg mass attached to a spring undergoes simple harmonic motion with a period of 0.67 s. What is the force constant of the spring?

37. • A freshly caught catfish is placed on a spring scale, and it oscillates up and down with a period of 0.204 s. If the spring constant of the scale is 2160 N/m, what is the mass of the catfish?

38. •• **CE** System A consists of a mass m attached to a spring with a force constant k; system B has a mass $2m$ attached to a spring with a force constant k; system C has a mass $3m$ attached to a spring with a force constant $6k$; and system D has a mass m attached to a spring with a force constant $4k$. Rank these systems in order of increasing period of oscillation.

39. •• Find the periods of block 1 and block 2 in Figure 13-34, given that $k = 49.2$ N/m and $m = 1.25$ kg.

40. •• When a 0.62-kg mass is attached to a vertical spring, the spring stretches by 12 cm. How much mass must be attached to the spring to result in a 0.67-s period of oscillation?

41. •• A spring with a force constant of 82 N/m is attached to a 0.47-kg mass. Assuming that the amplitude of motion is 3.4 cm, determine the following quantities for this system: (a) ω, (b) v_{max}, (c) T.

42. •• A bunch of grapes is placed in a spring scale at a supermarket. The grapes oscillate up and down with a period of 0.48 s, and the spring in the scale has a force constant of 650 N/m. What are (a) the mass and (b) the weight of the grapes?

43. •• Two people with a combined mass of 125 kg hop into an old car with worn-out shock absorbers. This causes the springs to compress by 8.00 cm. When the car hits a bump in the road, it oscillates up and down with a period of 1.65 s. Find (a) the total load supported by the springs and (b) the mass of the car.

44. •• A 0.95-kg mass attached to a vertical spring of force constant 130 N/m oscillates with a maximum speed of 0.45 m/s. Find the following quantities related to the motion of the mass: (a) the period, (b) the amplitude, (c) the maximum magnitude of the acceleration.

45. •• When a 0.184-kg mass is attached to a vertical spring, it causes the spring to stretch a distance d. If the mass is now displaced slightly from equilibrium, it is found to make 25.0 oscillations in 12.6 s. Find the stretch distance, d.

46. •• **Predict/Calculate** The springs of a 511-kg motorcycle have an effective force constant of 9130 N/m. (a) If a person sits on the motorcycle, does its period of oscillation increase, decrease, or stay the same? (b) By what percent and in what direction does the period of oscillation change when a 112-kg person rides the motorcycle?

47. ••• **Predict/Calculate** If a mass m is attached to a given spring, its period of oscillation is T. If two such springs are connected end to end, and the same mass m is attached, (a) is its period greater than, less than, or the same as with a single spring? (b) Verify your answer to part (a) by calculating the new period, T', in terms of the old period T.

SECTION 13-5 ENERGY CONSERVATION IN OSCILLATORY MOTION

48. • A 0.285-kg mass is attached to a spring with a force constant of 53.4 N/m. If the mass is displaced 0.180 m from equilibrium and released, what is its speed when it is 0.135 m from equilibrium?

49. • A 1.6-kg mass attached to a spring oscillates with an amplitude of 7.3 cm and a frequency of 2.8 Hz. What is its energy of motion?

50. •• **Predict/Calculate** A 0.40-kg mass is attached to a spring with a force constant of 26 N/m and released from rest a distance of 3.2 cm from the equilibrium position of the spring. (a) Give a strategy that allows you to find the speed of the mass when it is halfway to the equilibrium position. (b) Use your strategy to find this speed.

51. •• (a) What is the maximum speed of the mass in the previous problem? (b) How far is the mass from the equilibrium position when its speed is half the maximum speed?

52. •• **BIO** **Astronaut Mass** An astronaut uses a Body Mass Measurement Device to measure her mass. If the force constant of the spring is 2700 N/m, her mass is 75 kg, and the amplitude of her oscillation is 1.9 cm, what is her maximum speed during the measurement?

53. •• **Predict/Calculate** A 0.505-kg block slides on a frictionless horizontal surface with a speed of 1.18 m/s. The block encounters an unstretched spring and compresses it 23.2 cm before coming to rest. (a) What is the force constant of this spring? (b) For what length of time is the block in contact with the spring before it comes to rest? (c) If the force constant of the spring is increased, does the time required to stop the block increase, decrease, or stay the same? Explain.

54. •• A 3.55-g bullet embeds itself in a 1.47-kg block, which is attached to a spring of force constant 825 N/m. If the maximum compression of the spring is 5.88 cm, find (a) the initial speed of the bullet and (b) the time for the bullet–block system to come to rest.

SECTION 13-6 THE PENDULUM

55. • **CE** Metronomes, such as the penguin shown in **FIGURE 13-35**, are useful devices for music students. If it is desired to have the metronome tick with a greater frequency, should the penguin's bow tie be moved upward or downward?

56. • **CE** **Predict/Explain** A grandfather clock keeps correct time at sea level. If the clock is taken to the top of a nearby mountain, (a) would you expect it to keep correct time, run slow, or run fast? (b) Choose the *best explanation* from among the following:

FIGURE 13-35 How do you like my tie? (Problem 55)

I. Gravity is weaker at the top of the mountain, leading to a greater period of oscillation.
II. The length of the pendulum is unchanged, and therefore its period remains the same.
III. The extra gravity from the mountain causes the period to decrease.

57. • An observant fan at a baseball game notices that the radio commentators have lowered a microphone from their booth to just a few inches above the ground, as shown in **FIGURE 13-36**. The microphone is used to pick up sound from the field and from the fans. The fan also notices that the microphone is slowly swinging back and forth like a simple pendulum. Using her digital watch, she finds that 10 complete oscillations take 60.0 s. How high above the field is the radio booth? (Assume the microphone and its cord can be treated as a simple pendulum.)

FIGURE 13-36 Problem 57

58. • A simple pendulum of length 2.3 m makes 5.0 complete swings in 37 s. What is the acceleration of gravity at the location of the pendulum?

59. • **United Nations Pendulum** A large pendulum with a 200-lb gold-plated bob 12 inches in diameter is on display in the lobby of the United Nations building. The pendulum has a length of 75 ft. It is used to show the rotation of the Earth—for this reason it is referred to as a Foucault pendulum. What is the least amount of time it takes for the bob to swing from a position of maximum displacement to the equilibrium position of the pendulum? (Assume that the acceleration due to gravity is $g = 9.81$ m/s^2 at the UN building.)

60. •• **Predict/Calculate** If the pendulum in the previous problem were to be taken to the Moon, where the acceleration of gravity is $g/6$, **(a)** would its period increase, decrease, or stay the same? **(b)** Check your result in part (a) by calculating the period of the pendulum on the Moon.

61. •• A Hula Hoop hangs from a peg. Find the period of the hoop as it gently rocks back and forth on the peg. (For a hoop with axis at the rim $I = 2mR^2$, where R is the radius of the hoop.)

62. •• A fireman tosses his 0.98-kg hat onto a peg, where it oscillates as a physical pendulum (**FIGURE 13-37**). If the center of mass of the hat is 8.4 cm from the pivot point, and its period of oscillation is 0.73 s, what is the moment of inertia of the hat about the pivot point?

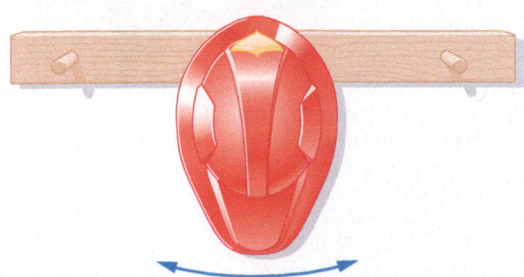

FIGURE 13-37 Problem 62

63. •• **Predict/Calculate** Consider a meterstick that oscillates back and forth about a pivot point at one of its ends. **(a)** Is the period of a simple pendulum of length $L = 1.00$ m greater than, less than, or the same as the period of the meterstick? Explain. **(b)** Find the length L of a simple pendulum that has a period equal to the period of the meterstick.

64. •• On the construction site for a new skyscraper, a uniform beam of steel is suspended from one end. If the beam swings back and forth with a period of 2.00 s, what is its length?

65. •• **BIO** **(a)** Find the period of a child's leg as it swings about the hip joint. Assume the leg is 0.55 m long and can be treated as a uniform rod. **(b)** Estimate the child's walking speed.

66. ••• Suspended from the ceiling of an elevator is a simple pendulum of length L. What is the period of this pendulum if the elevator **(a)** accelerates upward with an acceleration a, or **(b)** accelerates downward with an acceleration whose magnitude is greater than zero but less than g? Give your answer in terms of L, g, and a.

GENERAL PROBLEMS

67. • **CE** An object undergoes simple harmonic motion with a period T. In the time $3T/2$ the object moves through a total distance of $12D$. In terms of D, what is the object's amplitude of motion?

68. • **CE** If the amplitude of a simple harmonic oscillator is doubled, by what multiplicative factor do the following quantities change: **(a)** angular frequency, **(b)** frequency, **(c)** period, **(d)** maximum speed, **(e)** maximum acceleration, **(f)** total mechanical energy?

69. • **CE** A mass m is suspended from the ceiling of an elevator by a spring of force constant k. When the elevator is at rest, the period of the mass is T. Does the period increase, decrease, or remain the same when the elevator **(a)** moves upward with constant speed or **(b)** moves upward with constant acceleration?

70. • **CE** A pendulum of length L is suspended from the ceiling of an elevator. When the elevator is at rest, the period of the pendulum is T. Does the period increase, decrease, or remain the same when the elevator **(a)** moves upward with constant speed or **(b)** moves upward with constant acceleration?

71. • A 1.3-kg mass is attached to a spring with a force constant of 52 N/m. If the mass is released with a speed of 0.28 m/s at a distance of 8.1 cm from the equilibrium position of the spring, what is its speed when it is halfway to the equilibrium position?

72. • **BIO** **Measuring an Astronaut's Mass** An astronaut uses a Body Mass Measurement Device (BMMD) to determine her mass. What is the astronaut's mass, given that the force constant of the BMMD is 2600 N/m and the period of oscillation is 0.85 s? (See the discussion on page 431 for more details on the BMMD.)

73. • **Sunspot Observations** Sunspots vary in number as a function of time, exhibiting an approximately 11-year cycle. Galileo made the first European observations of sunspots in 1610, and daily observations were begun in Zurich in 1749. At the present time we are well into the 24th observed cycle. What is the frequency of the sunspot cycle? Give your answer in Hz.

74. • **BIO** **Weighing a Bacterium** Scientists are using tiny, nanoscale cantilevers 4 micrometers long and 500 nanometers wide—essentially miniature diving boards—as a sensitive way to measure mass. An example is shown in **FIGURE 13-38**. The cantilevers oscillate up and down with a frequency that depends on the mass placed near the tip, and a laser beam is used to measure the frequency. A single *E. coli* bacterium was measured to have a mass of 665 femtograms $= 6.65 \times 10^{-16}$ kg with this device, as the cantilever oscillated with a frequency of 14.5 MHz. Treating the cantilever as an ideal, massless spring, find its effective force constant.

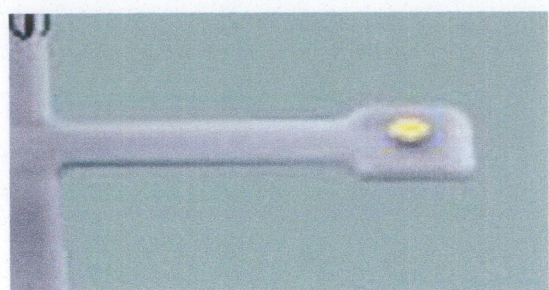

FIGURE 13-38 A silicon and silicon nitride cantilever with a 50-nanometer gold dot near its tip. (Problem 74)

75. •• **CE** An object undergoing simple harmonic motion with a period T is at the position $x = 0$ at the time $t = 0$. At the time $t = 0.25T$ the position of the object is positive. State whether x is positive, negative, or zero at the following times: **(a)** $t = 1.5T$, **(b)** $t = 2T$, **(c)** $t = 2.25T$, and **(d)** $t = 6.75T$.

76. •• The maximum speed of a 4.1-kg mass attached to a spring is 0.78 m/s, and the maximum force exerted on the mass is 13 N. **(a)** What is the amplitude of motion for this mass? **(b)** What is the force constant of the spring? **(c)** What is the frequency of this system?

77. •• The acceleration of a block attached to a spring is given by $a = -(0.302 \text{ m/s}^2) \cos([2.41 \text{ rad/s}]t)$. **(a)** What is the frequency of the block's motion? **(b)** What is the maximum speed of the block? **(c)** What is the amplitude of the block's motion?

78. •• **Helioseismology** In 1962, physicists at Cal Tech discovered that the surface of the Sun vibrates due to the violent nuclear reactions that roil within its core. This has led to a new field of solar science known as helioseismology. A typical vibration of the Sun is shown in **FIGURE 13-39**; it has a period of 5.7 minutes. The blue patches in Figure 13-39 are moving outward; the red patches are moving inward. **(a)** Find the angular frequency of this vibration. **(b)** The maximum speed at which a patch of the surface moves during a vibration is 4.5 m/s. What is the amplitude of the vibration, assuming it to be simple harmonic motion?

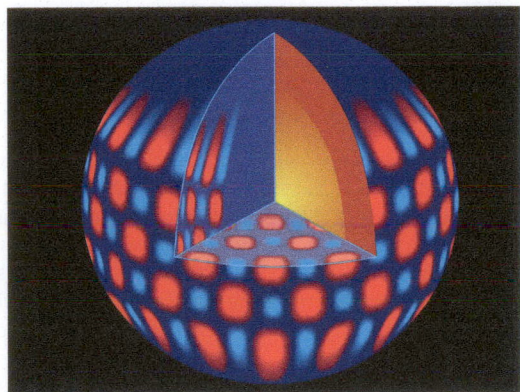

FIGURE 13-39 A typical vibration pattern of the Sun. (Problem 78)

79. •• **Predict/Calculate** A 9.50-g bullet, moving horizontally with an initial speed v_0, embeds itself in a 1.45-kg pendulum bob that is initially at rest. The length of the pendulum is $L = 0.745$ m. After the collision, the pendulum swings to one side and comes to rest when it has gained a vertical height of 12.4 cm. **(a)** Is the kinetic energy of the bullet–bob system immediately after the collision greater than, less than, or the same as the kinetic energy of the system just before the collision? Explain. **(b)** Find the initial speed of the bullet. **(c)** How much time does it take for the bullet–bob system to come to rest for the first time?

80. •• **BIO Spiderweb Oscillations** A 1.44-g spider oscillates on its web, which has a damping constant of 3.30×10^{-5} kg/s. How much time does it take for the spider's amplitude of oscillation to decrease by 10.0 percent?

81. •• A service dog tag (**FIGURE 13-40**) is a circular disk of radius 1.9 cm and mass 0.013 kg that can pivot about a small hole near its rim. Refer to Table 10-1 to determine the moment of inertia about the perpendicular axis that passes through the pivot hole, and find the period of oscillation of the dog tag.

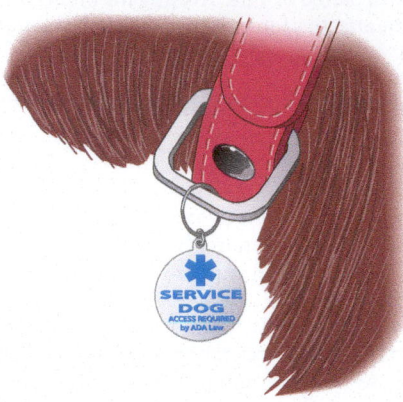

FIGURE 13-40 Problem 81

82. •• Calculate the ratio of the kinetic energy to the potential energy of a simple harmonic oscillator when its displacement is half its amplitude.

83. •• A 0.340-kg mass slides on a frictionless floor with a speed of 1.34 m/s. The mass strikes and compresses a spring with a force constant of 53.4 N/m. **(a)** How far does the mass travel after contacting the spring before it comes to rest? **(b)** How much time does it take for the spring to stop the mass?

84. •• A shock absorber is designed to quickly damp out the oscillations that a car would otherwise make because it is suspended on springs. **(a)** Find the period of oscillation of a 1610-kg car that is suspended by springs that make an effective force constant of 5.75×10^4 N/m. **(b)** Find the damping constant b that will reduce the amplitude of oscillations of this car by a factor of 5.00 within a time equal to half the period of oscillation.

85. •• **Predict/Calculate** **FIGURE 13-41** shows a displacement-versus-time graph of the periodic motion of a 3.8-kg mass on a spring. **(a)** Referring to the figure, do you expect the maximum speed of the mass to be greater than, less than, or equal to 0.50 m/s? Explain. **(b)** Calculate the maximum speed of the mass. **(c)** How much energy is stored in this system?

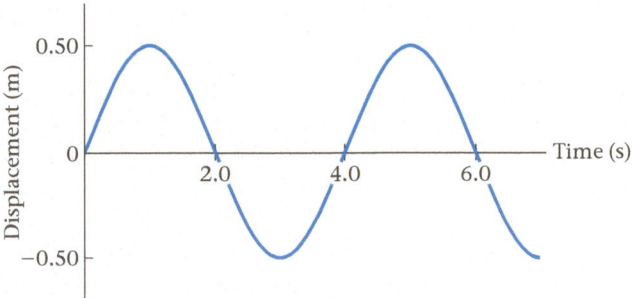

FIGURE 13-41 Problems 85 and 86

86. •• **Predict/Calculate** A 3.2-kg mass on a spring oscillates as shown in the displacement-versus-time graph in Figure 13-41.

(a) Referring to the graph, at what times between $t = 0$ and $t = 6.0$ s does the mass experience a force of maximum magnitude? Explain. **(b)** Calculate the magnitude of the maximum force exerted on the mass. **(c)** At what times shown in the graph does the mass experience zero force? Explain. **(d)** How much force is exerted on the mass at the time $t = 0.50$ s?

87. •• A 0.45-kg crow lands on a slender branch and bobs up and down with a period of 1.5 s. An eagle flies up to the same branch, scaring the crow away, and lands. The eagle now bobs up and down with a period of 4.8 s. Treating the branch as an ideal spring, find **(a)** the effective force constant of the branch and **(b)** the mass of the eagle.

88. ••• A mass m is connected to the bottom of a vertical spring whose force constant is k. Attached to the bottom of the mass is a string that is connected to a second mass m, as shown in **FIGURE 13-42**. Both masses are undergoing simple harmonic vertical motion of amplitude A. At the instant when the acceleration of the masses is a maximum in the upward direction the string breaks, allowing the lower mass to drop to the floor. Find the resulting amplitude of motion of the remaining mass.

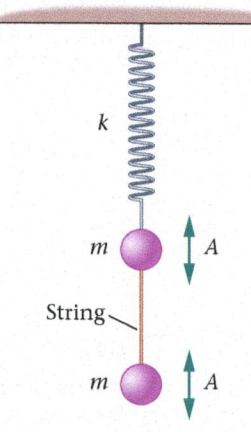

FIGURE 13-42 Problem 88

89. ••• **Predict/Calculate** Consider the pendulum shown in **FIGURE 13-43**. Note that the pendulum's string is stopped by a peg when the bob swings to the left, but moves freely when the bob swings to the right. **(a)** Is the period of this pendulum greater than, less than, or the same as the period of the same pendulum without the peg? **(b)** Calculate the period of this pendulum in terms of L and ℓ. **(c)** Evaluate your result for $L = 1.0$ m and $\ell = 0.25$ m.

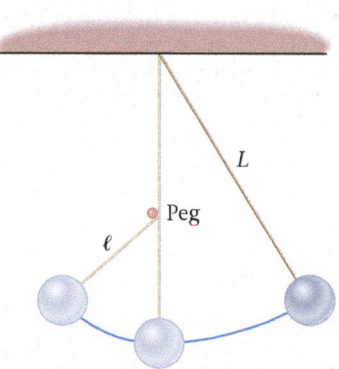

FIGURE 13-43 Problem 89

90. ••• An object undergoes simple harmonic motion of amplitude A and angular frequency ω about the equilibrium point $x = 0$. Use energy conservation to show that the speed of the object at the general position x is given by the following expression:

$$v = \omega\sqrt{A^2 - x^2}$$

91. ••• A physical pendulum consists of a light rod of length L suspended in the middle. A large mass m_1 is attached to one end of the rod, and a lighter mass m_2 is attached to the other end, as illustrated in **FIGURE 13-44**. Find the period of oscillation for this pendulum.

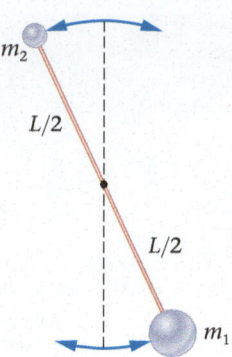

FIGURE 13-44 Problem 91

92. ••• **Predict/Calculate** A vertical hollow tube is connected to a speaker, which vibrates vertically with simple harmonic motion (**FIGURE 13-45**). The speaker operates with constant amplitude, A, but variable frequency, f. A slender pencil is placed inside the tube. **(a)** At low frequencies the pencil stays in contact with the speaker at all times; at higher frequencies the pencil begins to rattle. Explain the reason for this behavior. **(b)** Find an expression for the frequency at which rattling begins.

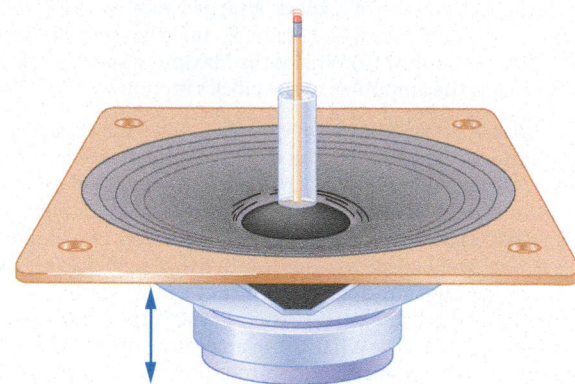

FIGURE 13-45 Problem 92

PASSAGE PROBLEMS

BIO A Cricket Thermometer, by Jiminy

Insects are ectothermic, which means their body temperature is largely determined by the temperature of their surroundings. This can have a number of interesting consequences. For example, the wing coloration in some butterfly species is determined by the ambient temperature, as is the body color of several species of dragonfly. In addition, the wing beat frequency of beetles taking flight varies with temperature due to changes in the resonant frequency of their thorax.

The origin of such temperature effects can be traced back to the fact that molecules have higher speeds and greater energy as temperature is increased (see Chapters 16 and 17). Thus, for example, molecules that collide and react as part of the metabolic process will do so more rapidly when the reactions are occurring at a higher temperature. As a result, development rates, heart rates, wing beats, and other processes all occur more rapidly.

One of the most interesting thermal effects is the temperature dependence of chirp rate in certain insects. This behavior has been observed in cone-headed grasshoppers, as well as several types of cricket. A particularly accurate connection between chirp rate and temperature is found in the snowy tree cricket (*Oecanthus fultoni* Walker), which chirps at a rate that follows the expression $N = T - 39$, where N is the number of chirps in 13 seconds, and T is the numerical value of the temperature in degrees Fahrenheit. This formula, which is known as Dolbear's law, is plotted in **FIGURE 13-46** (green line) along with data points (blue dots) for the snowy tree cricket.

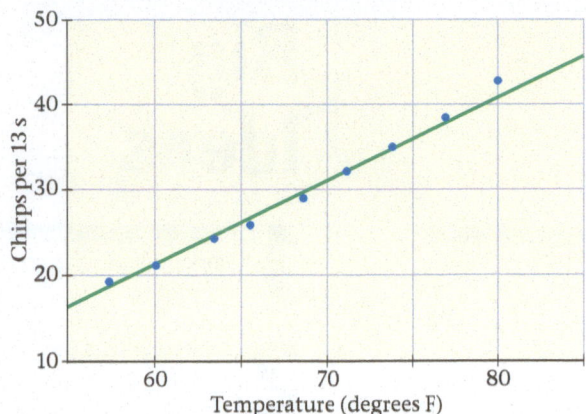

FIGURE 13-46 Problems 93, 94, 95, and 96

93. • If the temperature is increased by 10 degrees Fahrenheit, how many additional chirps are heard in a 13-s interval?

    A. 5   B. 10   C. 13   D. 39

94. • What is the temperature in degrees Fahrenheit if a cricket is observed to give 35 chirps in 13 s?

    A. 13 °F   B. 35 °F   C. 74 °F   D. 90 °F

95. • What is the frequency of the cricket's chirping (in Hz) when the temperature is 68 °F?

    A. 0.45 Hz   B. 2.2 Hz   C. 5.2 Hz   D. 29 Hz

96. •• Suppose the temperature decreases uniformly from 75 °F to 63 °F in 12 minutes. How many chirps does the cricket produce during this time?

    A. 28   B. 1700   C. 3800   D. 22,000

97. •• **Predict/Calculate** REFERRING TO EXAMPLE 13-5 Suppose we can change the plane's period of oscillation, while keeping its amplitude of motion equal to 30.0 m. **(a)** If we want to reduce the maximum acceleration of the plane, should we increase or decrease the period? Explain. **(b)** Find the period that results in a maximum acceleration of 1.0g.

98. •• **Predict/Calculate** REFERRING TO EXAMPLE 13-12 Suppose the force constant of the spring is doubled, but the mass and speed of the block are still 0.980 kg and 1.32 m/s, respectively. **(a)** By what multiplicative factor do you expect the maximum compression of the spring to change? Explain. **(b)** Find the new maximum compression of the spring. **(c)** Find the time required for the mass to come to rest after contacting the spring.

99. •• **Predict/Calculate** REFERRING TO EXAMPLE 13-12 **(a)** If the block's initial speed is increased, does the total time the block is in contact with the spring increase, decrease, or stay the same? **(b)** Find the total time of contact for $v_0 = 1.65$ m/s, $m = 0.980$ kg, and $k = 245$ N/m.

14

Waves and Sound

▲ Have you ever wondered why the instruments in a string quartet differ in size? It's not just tradition—there's also a physical reason, having to do with the way vibrating strings produce sound. But to understand this and other aspects of sound, it is first necessary to learn about waves in general. This chapter considers a variety of waves—including sound—and explores their characteristic properties and behavior.

I n this chapter, we extend the study of a single oscillator to the behavior of a series of oscillators that are connected to one another. Connecting the oscillators leads to an assortment of new phenomena, including waves on a string, water waves, and sound.

14-1 Types of Waves

Consider a group of swings in a playground swing set. We know that each swing by itself behaves like a simple pendulum—that is, like an oscillator. Now, let's connect the swings to one another. To be specific, suppose we tie a rope from the seat of the first swing to its neighbor, and then another rope from the second swing to the third swing, and so on. When the swings are at rest—in equilibrium—the connecting ropes have no effect. If you now sit in the first swing and begin oscillating—thus "disturbing" the equilibrium—the connecting ropes cause the other swings along the line to start oscillating as well. You have created a traveling disturbance.

In general, a disturbance that propagates from one place to another is referred to as a **wave**. Waves propagate with well-defined speeds determined by the properties of the material through which they travel. In addition, waves carry energy. For example, part of the energy you put into sound waves when you speak is carried to the ears of others, where some of the sound energy is converted into electrical energy carried by nerve impulses to the brain, which, in turn, creates the sensation of hearing.

It's important to distinguish between the motion of the wave itself and the motion of the individual particles that make up the wave. Common examples include the waves that propagate through a field of wheat. The individual wheat stalks sway back and forth as a wave passes, but they do not change their location. Similarly, surfers waiting for a big wave simply rise and fall in place as a small wave passes beneath them, as shown in FIGURE 14-1 (a). In addition, a "wave" at a ball game may propagate around the stadium more quickly than a person can run, as shown in FIGURE 14-1 (b), but the individual people making up the wave simply stand and sit in one place. From these simple examples it's clear that waves can come in a variety of types. We discuss some of the more common types in this section. In addition, we show how the speed of a wave is related to some of its basic properties.

Transverse Waves

Perhaps the easiest type of wave to visualize is a wave on a string, as illustrated in FIGURE 14-2. To generate such a wave, start by tying one end of a long string or rope to a wall. Pull on the free end with your hand, producing a tension in the string, and then move your hand up and down. As you do so, a wave will travel along the string toward the wall. In fact, if your hand moves up and down with simple harmonic motion, the wave on the string will have the shape of a sine or a cosine function; we refer to such a wave as a **harmonic wave.**

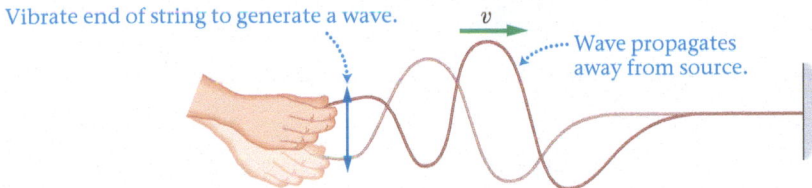

Vibrate end of string to generate a wave.

v

Wave propagates away from source.

Notice that the wave in Figure 14-2 travels in the horizontal direction, even though your hand oscillates vertically about one spot. In fact, if you look at any point on the string, it too moves vertically up and down, with no horizontal motion at all. This is shown in FIGURE 14-3, where we see that an individual point on a string oscillates up and down with an amplitude A as a wave travels past. Notice, in particular, that the displacement of particles in a string is at *right angles* to the direction of propagation of the wave. A wave with this property is called a **transverse wave:**

> In a transverse wave, the displacement of individual particles is at right angles to the direction of propagation of the wave.

Other examples of transverse waves include light and radio waves. These will be discussed in detail in Chapter 25.

(a)

(b)

▲ **FIGURE 14-1 Visualizing Concepts**
Waves A wave can be viewed as a disturbance that travels from one location to another. Although the wave itself moves steadily in one direction, the particles in the wave (people in the two cases shown here) do not share in this motion. Instead, they oscillate back and forth about their equilibrium positions as the wave moves past them.

◄ **FIGURE 14-2 A wave on a string**
Vibrating one end of a string with an up-and-down motion generates a wave that travels away from its point of origin.

PHYSICS IN CONTEXT
Looking Ahead

We next encounter waves when we study electricity and magnetism. In fact, we shall see in Chapter 25 that light is an electromagnetic wave, with both the electric and magnetic fields propagating much like a wave on a string.

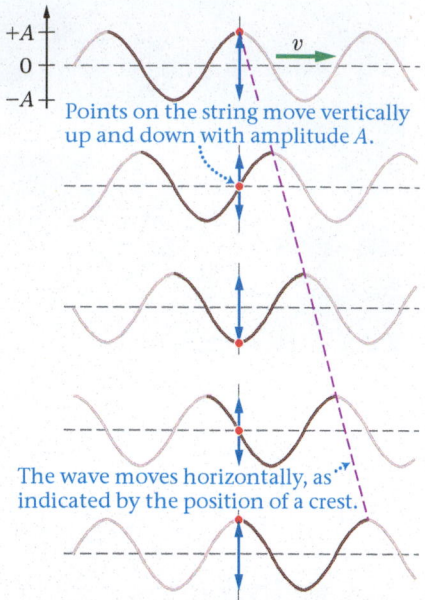

Points on the string move vertically up and down with amplitude A.

The wave moves horizontally, as indicated by the position of a crest.

▲ **FIGURE 14-3 The motion of a wave on a string** As a wave on a string moves horizontally, all points on the string vibrate in the vertical direction, as indicated by the blue arrow.

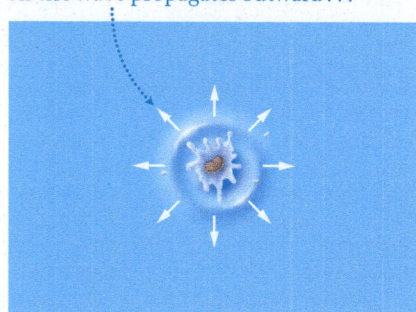

As the wave propagates outward . . .

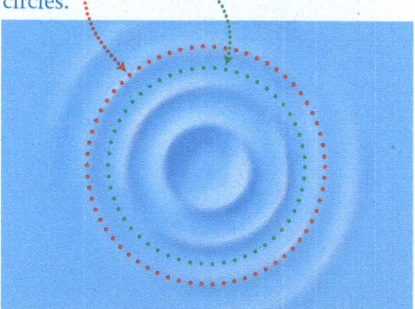

. . . the crests and troughs form concentric circles.

▲ **FIGURE 14-5 Water waves from a disturbance** An isolated disturbance in a pool of water, caused by a pebble dropped into the water, creates waves that propagate symmetrically away from the disturbance. The crests and troughs form concentric circles on the surface of the water as they move outward.

Big Idea **1** Waves can be transverse, longitudinal, or a combination of the two.

Longitudinal Waves

Longitudinal waves differ from transverse waves in the way that particles in the wave move. In particular, a longitudinal wave is defined as follows:

> In a longitudinal wave, the displacement of individual particles is *parallel* to the direction of propagation of the wave.

The classic example of a longitudinal wave is sound. When you speak, the vibrations in your vocal cords create a series of compressions and expansions (rarefactions) in the air. The same kind of situation occurs with a loudspeaker, as illustrated in **FIGURE 14-4**. Here we see a speaker diaphragm vibrating horizontally with simple harmonic motion. As it moves to the right it compresses the air momentarily; as it moves to the left it rarefies the air. A series of compressions and rarefactions then travel horizontally away from the loudspeaker with the speed of sound.

Figure 14-4 also indicates the motion of an individual particle in the air as a sound wave passes. Notice that the particle moves back and forth horizontally; that is, in the same direction as the propagation of the wave. The particle does not travel with the wave—each individual particle simply oscillates about a given position in space.

Water Waves

If a pebble is dropped into a pool of water, a series of concentric waves move away from the drop point. This is illustrated in **FIGURE 14-5**. To visualize the movement of the water as a wave travels by a given point, place a small piece of cork into the water. As a wave passes, the motion of the cork traces out the motion of the water itself, as indicated in **FIGURE 14-6**.

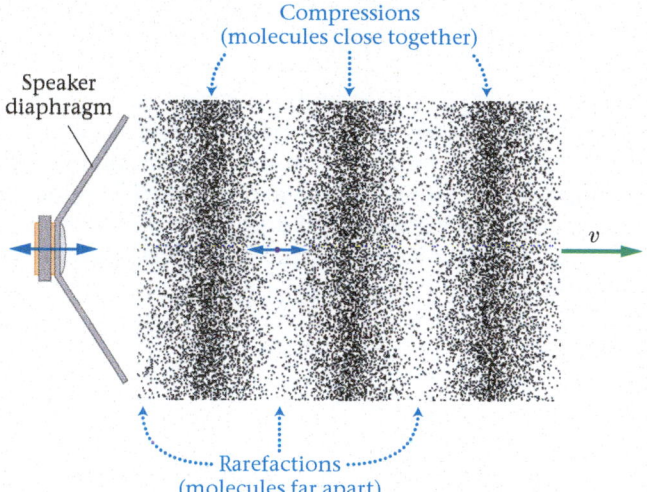

Compressions (molecules close together)

Speaker diaphragm

v

Rarefactions (molecules far apart)

▲ **FIGURE 14-4 Sound produced by a speaker** As the diaphragm of a speaker vibrates back and forth, it alternately compresses and rarefies the surrounding air. These regions of high and low density propagate away from the speaker with the speed of sound. Individual particles in the air oscillate back and forth about a given position, as indicated by the blue arrow.

Notice that the cork moves in a roughly circular path, returning to approximately its starting point. Thus, each element of water moves both vertically and horizontally as the wave propagates by in the horizontal direction. In this sense, a water wave is a combination of both transverse and longitudinal waves. This makes the water wave more difficult to analyze. Hence, most of our results will refer to the simpler cases of purely transverse and purely longitudinal waves.

Wavelength, Frequency, and Speed

A simple wave can be thought of as a regular, rhythmic disturbance that propagates from one point to another, repeating itself both in *space* and in *time*. We now show that the repeat length and the repeat time of a wave are directly related to its speed of propagation.

▶ **FIGURE 14-6 The motion of a water wave** As a water wave passes a given point, a molecule (or a small piece of cork) moves in a roughly circular path. This means that the water molecules move both vertically and horizontally. In this sense, the water wave has characteristics of both transverse and longitudinal waves.

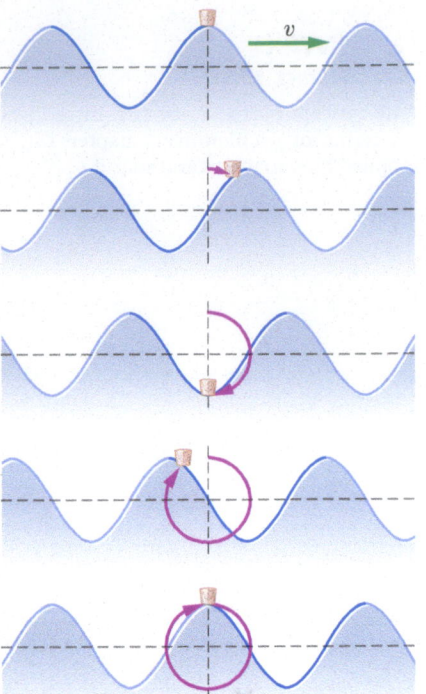

We begin by considering the snapshots of a wave shown in **FIGURE 14-7**. Points on the wave corresponding to maximum upward displacement are referred to as **crests**; points corresponding to maximum downward displacement are called **troughs**. The distance from one crest to the next, or from one trough to the next, is the repeat length—or **wavelength**, λ—of the wave.

Definition of Wavelength, λ

λ = distance over which a wave repeats

SI unit: m

Similarly, the repeat time—or **period**, T—of a wave is the time required for one wavelength to pass a given point, as illustrated in Figure 14-7. Closely related to the period of a wave is its **frequency**, f, which, as with oscillations, is defined as the inverse of the period: $f = 1/T$.

Combining these observations, we see that a wave travels a distance λ in the time T. If we apply the definition of speed—distance divided by time—it follows that the speed of a wave is:

Speed of a Wave

$$v = \frac{\text{distance}}{\text{time}} = \frac{\lambda}{T} = \lambda f$$

14-1

SI unit: m/s

This result applies to all waves.

EXERCISE 14-1 FREQUENCY AND WAVELENGTH

Sound waves travel in air with a speed of 343 m/s. **(a)** Middle C on a piano produces sound waves with a wavelength of 1.31 m. What is the frequency of middle C? **(b)** Dogs can hear sounds with frequencies as high as 45,000 Hz. What is the wavelength of such sounds?

REASONING AND SOLUTION

Wave speed, wavelength, and frequency are related by $v = \lambda f$. This relationship can be rearranged to solve for any one quantity in terms of the other two. **(a)** We solve $v = \lambda f$ for the frequency, and then substitute $v = 343$ m/s and $\lambda = 1.31$ m:

$$f = \frac{v}{\lambda} = \frac{343 \text{ m/s}}{1.31 \text{ m}} = 262 \text{ Hz}$$

(b) We solve $v = \lambda f$ for the wavelength, and then substitute $v = 343$ m/s and $f = 45,000$ Hz:

$$\lambda = \frac{v}{f} = \frac{343 \text{ m/s}}{45,000 \text{ Hz}} = 7.62 \text{ mm}$$

Enhance Your Understanding (Answers given at the end of the chapter)

1. Rank the following systems in order of increasing speed of the corresponding wave. Indicate ties where appropriate. **A**, $\lambda = 1$ m, $T = 1$ s; **B**, $\lambda = 4$ m, $T = 2$ s; **C**, $\lambda = 10$ m, $f = 1$ Hz; **D**, $\lambda = 1$ m, $f = 2$ Hz.

Section Review

- Transverse waves have particles that oscillate at right angles to the motion of the wave; longitudinal waves have particles that move parallel to the motion of the wave. Some waves combine characteristics of both transverse and longitudinal waves.

CONTINUED

$t = 0$

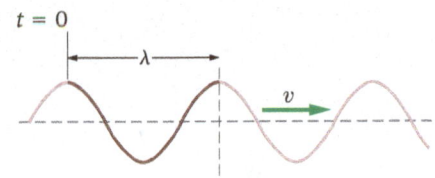

$t = T/4$ Crests

Troughs

$t = T/2$

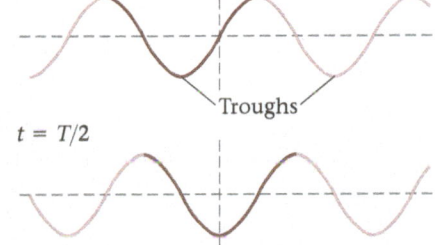

$t = 3T/4$

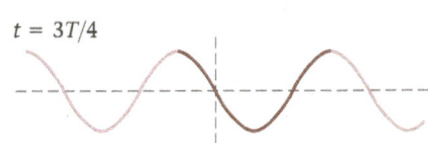

$t = T$

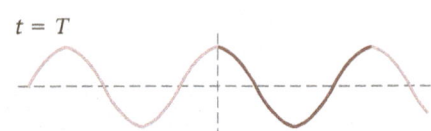

▲ **FIGURE 14-7 The speed of a wave** A wave repeats over a distance equal to the wavelength, λ. The time necessary for a wave to move one wavelength is the period, T. Thus, the speed of a wave is $v = \lambda/T = \lambda f$.

**PHYSICS
IN CONTEXT
Looking Back**

The concept of frequency, first introduced in terms of oscillations (Chapter 13), is applied here to waves and sound.

• The wavelength of a wave, λ, is the distance over which the wave repeats. The period of a wave is the time T for one wavelength to pass a given point, and the frequency f is the inverse of the period. The speed v of a wave is given by $v = \lambda f$.

14-2 Waves on a String

In this section, we consider some of the basic properties of waves traveling on a string. Our results apply equally well to waves on a rope, a flexible wire, or any similar linear medium.

The Speed of a Wave on a String

The speed of a wave is determined by the properties of the medium through which it propagates. In the case of a string of length L, there are two basic characteristics that determine the speed of a wave:

1. The tension in the string
2. The mass of the string

Let's begin with the tension, which is the force F transmitted through the string (we will use F for the tension rather than T, to avoid confusion between the tension and the period). Clearly, there must be a tension in a string in order for it to propagate a wave. Imagine, for example, that a string lies on a smooth floor with both ends free. If you shake one end of it, the portions of the string near your hand will oscillate slightly, but no wave will travel to the other end of the string. If someone else takes hold of the other end of the string and pulls enough to set up a tension, then any movement you make on your end will propagate to the other end. In fact, if the tension is increased, the waves will travel through the string more rapidly.

Next, let's consider the mass m of the string. A heavy string responds slowly to a given disturbance because of its inertia. Thus, if you try sending a wave through a kite string or a thick rope, both under the same tension, you will find that the wave in the rope travels more slowly. In general, the heavier a rope or string, the slower the speed of waves in it. Of course, the total mass of a string doesn't really matter; a longer string has more mass, but its other properties are basically the same. What is important is the mass of the string per length. We give this quantity the label μ:

Definition of Mass per Length, μ

$$\mu = \text{mass per length} = \frac{m}{L}$$

SI unit: kg/m

Applying Dimensional Analysis   From the above discussion, we expect the speed v to increase with the tension F and decrease with the mass per length, μ. Assuming these are the only factors determining the speed of a wave on a string, we can obtain the dependence of v on F and μ using dimensional analysis (see Section 1-3). First, we identify the dimensions of v, F, and μ:

$$[v] = \text{m/s}$$
$$[F] = \text{N} = \text{kg} \cdot \text{m/s}^2$$
$$[\mu] = \text{kg/m}$$

Next, we seek a combination of F and μ that has the dimensions of v; namely, m/s. Suppose, for example, that v depends on F to the power a and μ to the power b. Then, we have

$$v = F^a \mu^b$$

In terms of dimensions, this equation is

$$\text{m/s} = (\text{kg} \cdot \text{m/s}^2)^a (\text{kg/m})^b = \text{kg}^{a+b} \text{m}^{a-b} \text{s}^{-2a}$$

Comparing dimensions, we see that kg does not appear on the left side of the equation; therefore, we conclude that $a + b = 0$ so that kg does not appear on the right side of the equation. Hence, $a = -b$. Looking at the time dimension, s, we see that on the left we have s^{-1}; thus on the right side we must have $-2a = -1$, or $a = \frac{1}{2}$. It follows that $b = -a = -\frac{1}{2}$. This gives the following result:

Speed of a Wave on a String, v

$$v = \sqrt{\frac{F}{\mu}}$$

14-2

SI unit: m/s

As expected, the speed increases with F and decreases with μ.

Dimensional analysis does not guarantee that this is the complete, final result; there could be a dimensionless factor like $\frac{1}{2}$ or 2π left unaccounted for. It turns out, however, that a complete analysis based on Newton's laws gives precisely the same result.

PROBLEM-SOLVING NOTE

Mass Versus Mass-per-Length

To find the mass of a string, multiply its mass per length, μ, by its length L. That is, $m = \mu L$.

EXERCISE 14-2   FIND THE SPEED

A typical guitar string has a mass per length of 0.401×10^{-3} kg/m and is tightened to a tension of 71.4 N. What is the speed of waves on the string?

REASONING AND SOLUTION

The speed of waves on the guitar string is given by $v = \sqrt{F/\mu}$. We substitute $F = 71.4$ N and $\mu = 0.401 \times 10^{-3}$ kg/m to find the speed:

$$v = \sqrt{\frac{F}{\mu}} = \sqrt{\frac{71.4 \text{ N}}{0.401 \times 10^{-3} \text{ kg/m}}} = 422 \text{ m/s}$$

EXAMPLE 14-3   A WAVE ON A ROPE

A 12-m rope is pulled tight with a tension of 49 N. When one end is given a quick "flick," a wave is generated that takes 0.54 s to travel to the other end of the rope. What is the mass of the rope?

PICTURE THE PROBLEM

Our sketch shows a wave pulse traveling with a speed v from one end of the rope to the other, a distance of 12 m. The tension in the rope is 49 N, and the travel time of the pulse is 0.54 s.

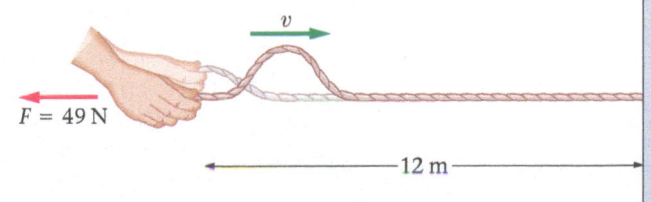

REASONING AND STRATEGY

We know that the speed of waves on a rope is determined by the tension and the mass per length. Thus, we first calculate the speed of the wave from the distance and time information given in the problem statement. Next, we solve for the mass per length in terms of speed and tension. Finally, we multiply by the length to get the mass.

Known   Length of rope, $L = 12$ m; force of tension in rope, $F = 49$ N; travel time, $t = 0.54$ s.
Unknown   Mass of rope, $m = ?$

SOLUTION

1. Calculate the speed of the wave:

$$v = \frac{d}{t} = \frac{12 \text{ m}}{0.54 \text{ s}} = 22 \text{ m/s}$$

2. Use $v = \sqrt{F/\mu}$ to solve for the mass per length:

$$\mu = \frac{F}{v^2}$$

3. Substitute numerical values for F and v:

$$\mu = \frac{F}{v^2} = \frac{49 \text{ N}}{(22 \text{ m/s})^2} = 0.10 \text{ kg/m}$$

4. Multiply μ by $L = 12$ m to find the mass:

$$m = \mu L = (0.10 \text{ kg/m})(12 \text{ m}) = 1.2 \text{ kg}$$

INSIGHT

The speed of a wave on this rope (almost 50 mi/h) is comparable to the speed of a car on a highway.

CONTINUED

In the following Conceptual Example, we consider the speed of a wave on a vertical rope of finite mass.

CONCEPTUAL EXAMPLE 14-4   SPEED OF A WAVE   PREDICT/EXPLAIN

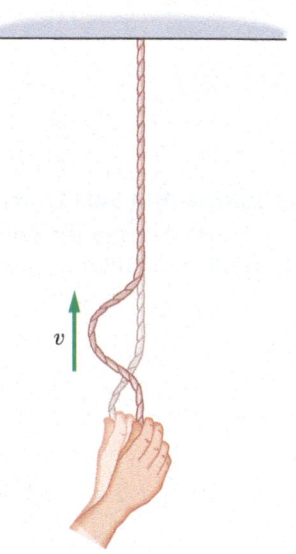

A rope of length L and mass M hangs from a ceiling. If the bottom of the rope is given a gentle wiggle, a wave will travel to the top of the rope. **(a)** As the wave travels upward, does its speed increase, decrease, or stay the same? **(b)** Which of the following is the *best explanation* for your prediction?

  I. The tension in the rope increases as the wave approaches the top, and hence the speed of the wave increases as well.

  II. The speed of the wave decreases as it moves upward because of the downward acceleration of gravity.

  III. The speed of the wave stays the same because it's the same rope all the way to the top.

REASONING AND DISCUSSION
The speed of the wave is determined by the tension in the rope and its mass per length. The mass per length is the same from bottom to top, but not the tension. In particular, the tension at any point in the rope is equal to the weight of the rope below that point. Thus, the tension increases from almost zero near the bottom to essentially Mg near the top. Because the tension increases with height, so too does the speed, according to the relationship $v = \sqrt{F/\mu}$.

ANSWER
(a) The speed increases. **(b)** The best explanation is I.

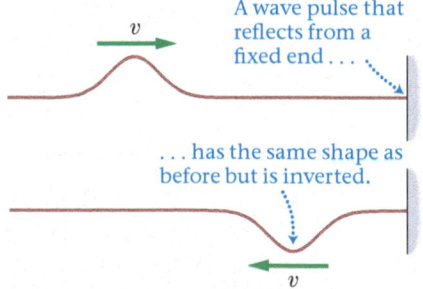

A wave pulse that reflects from a fixed end . . .

. . . has the same shape as before but is inverted.

▲ **FIGURE 14-8 A reflected wave pulse: fixed end** A wave pulse on a string is inverted when it reflects from an end that is tied down.

Reflections

Thus far we have discussed only the situation in which a wave travels along a string, but at some point the wave must reach the end of the string. What happens then? Clearly, we expect the wave to be reflected, but the precise way in which the reflection occurs needs to be considered.

    Suppose, for example, that the far end of a string is anchored firmly into a wall, as shown in **FIGURE 14-8**. If you give a flick to your end of the string, you set up a wave "pulse" that travels toward the far end. When it reaches the end, it exerts an upward force on the wall, trying to pull it up into the pulse. The end is tied down, however, and hence the wall exerts an equal and opposite downward force to keep the end at rest. Thus, the wall exerts a downward force on the string that is just the *opposite* of the upward force you exerted when you created the pulse. As a result, the reflection is an inverted, or upside-down, pulse, as indicated in Figure 14-8. We shall encounter this same type of inversion under reflection when we consider the reflection of light in Chapter 28.

    Another way to tie off the end of the string is shown in **FIGURE 14-9**. In this case, the string is tied to a small ring that slides vertically with little friction on a vertical pole. In this way, the string still has a tension in it, since it pulls on the ring, but it is also free to move up and down.

    Consider a pulse moving along such a string, as in Figure 14-9. When the pulse reaches the end, it lifts the ring upward and then lowers it back down. In fact, the pulse flicks the far end of the string in the *same* way that you flicked it when you created the pulse. Therefore, the far end of the string simply creates a new pulse, identical to the first, only traveling in the opposite direction. This is illustrated in the figure.

Thus, when waves reflect, they may or may not be inverted, depending on how the reflection occurs.

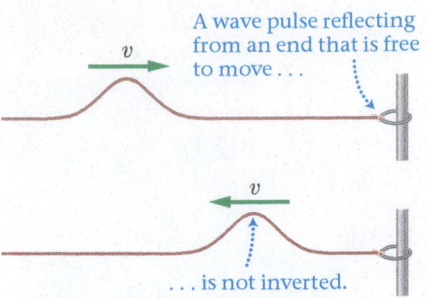

A wave pulse reflecting from an end that is free to move . . .

. . . is not inverted.

▲ **FIGURE 14-9 A reflected wave pulse: free end** A wave pulse on a string whose end is free to move is reflected without inversion.

Enhance Your Understanding

(Answers given at the end of the chapter)

2. Suppose the tension in a string is doubled, its mass remains the same, and its length is doubled. By what factor does the speed of waves on the string change?

Section Review

- The speed v of a wave on a string depends on the tension in the string, F, and the mass per unit length of the string, μ. The precise dependence is given by the equation $v = \sqrt{F/\mu}$.

- A wave pulse on a string is inverted when it reflects from a fixed end, but is not inverted when it reflects from an end that is free to move up and down.

14-3 Harmonic Wave Functions

If a wave is generated by oscillating one end of a string with simple harmonic motion, the waves will have the shape of a sine or a cosine function. This is shown in **FIGURE 14-10**, where the y direction denotes the vertical displacement of the string, and $y = 0$ corresponds to the flat string with no wave present. In what follows, we consider the mathematical formula that describes y as a function of time, t, and position, x, for such a harmonic wave.

First, notice that the harmonic wave in Figure 14-10 repeats when x increases by an amount equal to the wavelength, λ. Thus, the dependence of the wave on x must be of the form

$$y(x) = A \cos\left(\frac{2\pi}{\lambda} x\right) \qquad \text{14-3}$$

In this expression, A is the amplitude of the wave, and hence the vertical displacement ranges in value from $+A$ to $-A$.

To see that this is the correct dependence, notice that replacing x with $x + \lambda$ gives the same value for y:

$$y(x + \lambda) = A \cos\left[\frac{2\pi}{\lambda}(x + \lambda)\right] = A \cos\left(\frac{2\pi}{\lambda} x + 2\pi\right) = A \cos\left(\frac{2\pi}{\lambda} x\right) = y(x)$$

It follows that Equation 14-3 describes a vertical displacement that repeats with a wavelength λ, as desired for a wave.

This is only part of the "wave function," however, because we have not yet described the way the wave changes with time. This is illustrated in Figure 14-10, where we see a harmonic wave at times $t = 0$, $t = T/4$, $t = T/2$, $t = 3T/4$, and $t = T$. Notice that the peak in the wave that was originally at $x = 0$ at $t = 0$ moves to $x = \lambda/4$, $x = \lambda/2$, $x = 3\lambda/4$, and $x = \lambda$ for the times just given. Thus, the position x of this peak can be written as follows:

$$x = \lambda \frac{t}{T}$$

Equivalently, we can say that the peak that was at $x = 0$ is now at the location given by

$$x - \lambda \frac{t}{T} = 0$$

Similarly, the peak that was originally at $x = \lambda$ at $t = 0$ is at the following position at the general time t:

$$x - \lambda \frac{t}{T} = \lambda$$

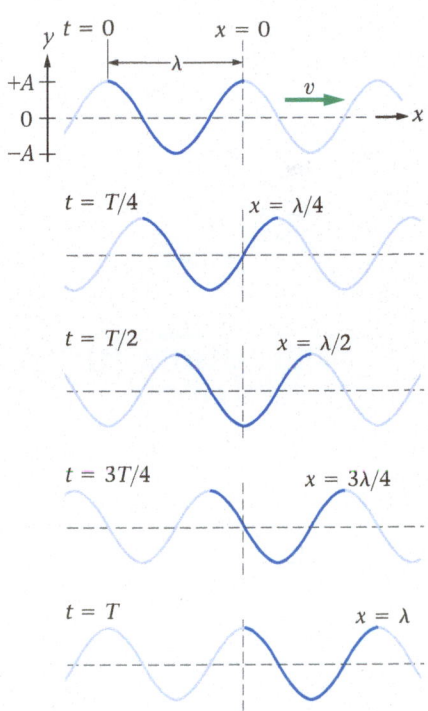

▲ **FIGURE 14-10 A harmonic wave moving to the right** As a wave moves, the peak that was at $x = 0$ at time $t = 0$ moves to the position $x = \lambda t/T$ at time t.

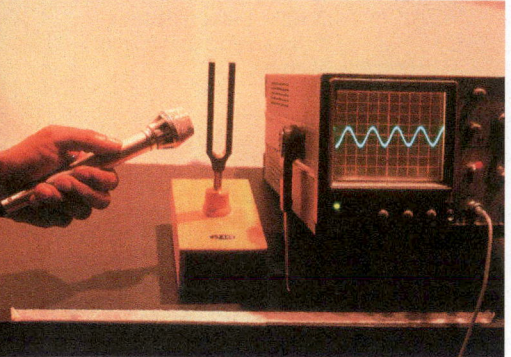

▲ **FIGURE 14-11** An oscilloscope connected to a microphone can be used to display the waveform of a pure tone, created here by a tuning fork. The trace on the screen shows that the wave has the shape of a sine or cosine function, as expected for a harmonic wave function.

In general, if the position of a given point on a wave at $t = 0$ is $x(0)$, and its position at the time t is $x(t)$, the relationship between these positions is $x(t) - \lambda t/T = x(0)$. Therefore, to take into account the time dependence of a wave, we replace $x = x(0)$ in Equation 14-3 with $x(0) = x - \lambda t/T$. This yields the harmonic wave function:

$$y(x, t) = A \cos\left[\frac{2\pi}{\lambda}\left(x - \lambda\frac{t}{T}\right)\right] = A \cos\left(\frac{2\pi}{\lambda}x - \frac{2\pi}{T}t\right) \qquad \text{14-4}$$

The wave function, $y(x, t)$, depends on both time and position, and repeats whenever position increases by the wavelength, λ, or time increases by the period, T, as desired. An example of a harmonic wave produced by a tuning fork is shown in **FIGURE 14-11**.

Enhance Your Understanding   (Answers given at the end of the chapter)

3. A particular harmonic wave is described by the following equation: $y(x, t) = (0.12\text{ m}) \cos(\pi x/4 - 8\pi t)$. In this expression, x is measured in meters, and t is measured in seconds. What are the **(a)** amplitude, **(b)** wavelength, and **(c)** period for this wave?

Section Review

- The mathematical function that describes a harmonic wave is $y(x, t) = A \cos\left(\frac{2\pi}{\lambda}x - \frac{2\pi}{T}t\right)$, where A is the amplitude of the wave, λ is the wavelength, and T is the period.

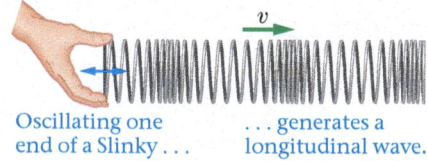

Oscillating one end of a Slinky . . .   . . . generates a longitudinal wave.

▲ **FIGURE 14-12 A wave on a Slinky** If one end of a Slinky is oscillated back and forth, a series of longitudinal waves are produced. These Slinky waves are analogous to sound waves.

14-4 Sound Waves

The first thing we do when we come into this world is make a sound. It's many years later before we realize that sound is a wave propagating through the air at a speed of about 770 mi/h. More years are required to gain an understanding of the physics of a sound wave.

A useful mechanical model of a sound wave is provided by a Slinky. If we oscillate one end of a Slinky back and forth horizontally, as in **FIGURE 14-12**, we send out a longitudinal wave that also travels in the horizontal direction. The wave consists of regions where the coils of the Slinky are compressed alternating with regions where the coils are more widely spaced.

In close analogy with the Slinky model, a loudspeaker produces sound waves by oscillating a diaphragm back and forth horizontally, as we saw in Figure 14-4. Just as with the Slinky, a wave travels away from the source horizontally. The wave consists of compressed regions alternating with rarefied regions.

At first glance, the sound wave seems very different from the wave on a string. In particular, the sound wave doesn't seem to have the nice sinusoidal shape of a wave. Certainly, Figure 14-4 gives no hint of such a wavelike shape.

If we plot the appropriate quantities, however, the classic wave shape emerges. For example, in **FIGURE 14-13 (a)** we plot the rarefactions and compressions of a typical sound wave, while in **FIGURE 14-13 (b)** we plot the fluctuations in the density of the air versus x. Clearly, the density oscillates in a wavelike fashion. Similarly, **FIGURE 14-13 (c)** shows a plot of the fluctuations in the pressure of the air as a function of x. In regions where the density is high, the pressure is also high; where the density is low, the pressure is low. Thus, pressure versus position again shows that a sound wave has the usual wavelike properties.

The Speed of Sound   The speed of sound is determined by the properties of the medium through which it propagates, just as with a wave on a string. In air, under normal atmospheric pressure and temperature, the speed of sound is approximately the following:

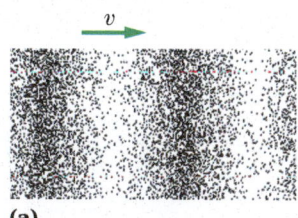

(a)

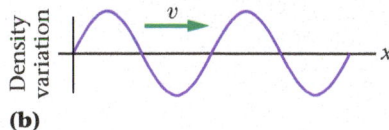

(b)

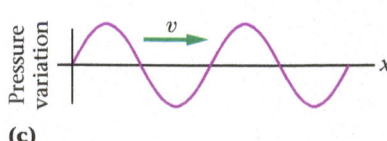

(c)

◄ **FIGURE 14-13 Wave properties of sound** A sound wave moving through the air **(a)** produces a wavelike disturbance in the **(b)** density and **(c)** pressure of the air.

Speed of Sound in Air (at room temperature, 20 °C)

$v = 343 \, \text{m/s} \approx 770 \, \text{mi/h}$

SI unit: m/s

When we refer to the speed of sound in this text we will always assume the value is 343 m/s, unless stated specifically otherwise.

As we shall see in Chapter 17, where we study the kinetic theory of gases, the speed of sound in air is directly related to the speed of the molecules themselves. Did you know, for example, that the air molecules colliding with your body at this moment have speeds that are essentially the speed of sound? As the air is heated the molecules will move faster, and hence the speed of sound also increases with temperature.

In a solid, the speed of sound is determined in part by how stiff the material is. The stiffer the material, the faster the sound wave, just as having more tension in a string causes a faster wave. Thus the speed of sound in plastic is rather high (2680 m/s), and in steel it is higher still (5960 m/s). Both speeds are much higher than the speed in air (343 m/s), which is certainly a "squishy" material in comparison. Table 14-1 gives a sampling of sound speed in a range of different materials.

TABLE 14-1 Speed of Sound in Various Materials

Material	Speed (m/s)
Aluminum	6420
Granite	6000
Steel	5960
Pyrex glass	5640
Copper	5010
Plastic	2680
Fresh water (20 °C)	1482
Fresh water (0 °C)	1402
Hydrogen (0 °C)	1284
Helium (0 °C)	965
Air (20 °C)	343
Air (0 °C)	331

CONCEPTUAL EXAMPLE 14-5 HOW FAR TO THE LIGHTNING?

Five seconds after a brilliant flash of lightning, thunder shakes the house. Was the lightning about a mile away, much closer than a mile, or much farther away than a mile?

REASONING AND DISCUSSION

As mentioned, the speed of sound is 343 m/s, which is just over 1000 ft/s. Thus, in five seconds sound travels slightly more than one mile. This gives rise to the following rule of thumb: The distance to a lightning strike (in miles) is the time for the thunder to arrive (in seconds) divided by 5.

Notice that we have ignored the travel time of light in our discussion. This is because light propagates with such a high speed (approximately 186,000 mi/s) that its travel time is about a million times less than that of sound.

ANSWER

The lightning was about a mile away.

EXAMPLE 14-6 WISHING WELL

You drop a stone from rest into a well that is 7.35 m deep. How much time elapses before you hear the splash?

PICTURE THE PROBLEM

Our sketch shows the well into which the stone is dropped. Notice that the depth of the well is $d = 7.35$ m. After the stone falls a distance d, the sound from the splash rises the same distance d before it is heard.

REASONING AND STRATEGY

The time until the splash is heard is the sum of (i) the time, t_1, for the stone to drop a distance d, and (ii) the time, t_2, for sound to travel a distance d.

For the time of the drop, we use one-dimensional kinematics with an initial velocity $v = 0$, since the stone is dropped from rest, and an acceleration g. Therefore, the relationship between distance and time for the stone is $d = \frac{1}{2}gt_1^2$, with $g = 9.81 \, \text{m/s}^2$.

For the sound wave, we use $d = vt_2$, with $v = 343$ m/s.

Known   Depth of well, $d = 7.35$ m.
Unknown   Time to hear splash, $t = ?$

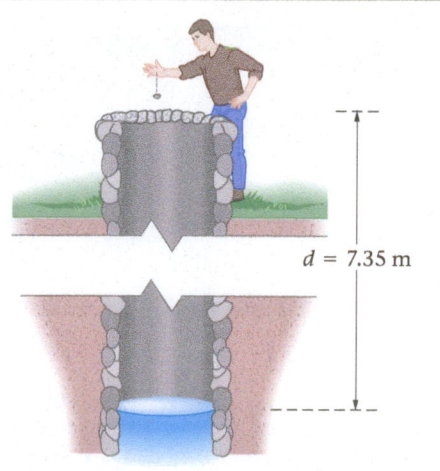

$d = 7.35$ m

CONTINUED

SOLUTION

1. Calculate the time for the stone to drop:

$$d = \tfrac{1}{2}gt_1^2$$

$$t_1 = \sqrt{\frac{2d}{g}} = \sqrt{\frac{2(7.35\ \text{m})}{9.81\ \text{m/s}^2}} = 1.22\ \text{s}$$

2. Calculate the time for the sound to travel a distance d:

$$d = vt_2$$

$$t_2 = \frac{d}{v} = \frac{7.35\ \text{m}}{343\ \text{m/s}} = 0.0214\ \text{s}$$

3. Sum the times found above:

$$t = t_1 + t_2 = 1.22\ \text{s} + 0.0214\ \text{s} = 1.24\ \text{s}$$

INSIGHT

The time of travel for the sound is quite small, only a couple hundredths of a second. It is still nonzero, however, and must be taken into account to obtain the correct total time.

In addition, notice that we use the same speed for a sound wave whether it is traveling horizontally or vertically—its speed is independent of its direction of motion. As a result, the waves emanating from a source of sound propagate outward in a spherical pattern, with the wave crests forming concentric spheres around the source.

PRACTICE PROBLEM

You drop a stone into a well and hear the splash 1.47 s later. How deep is the well?   [**Answer:** 10.2 m]

Some related homework problems: Problem 32, Problem 33

The Frequency of a Sound Wave

When we hear a sound, its frequency makes a great impression on us; in fact, the frequency determines the **pitch** of a sound. For example, the keys on a piano produce sounds with frequencies ranging from 55 Hz for the key farthest to the left to 4187 Hz for the rightmost key. Similarly, as you hum a song you change the shape and size of your vocal chords slightly to change the frequency of the sound you produce.

The frequency range of human hearing extends well beyond the range of a piano, however. As a rule of thumb, humans can hear sounds between 20 Hz on the low-frequency end and 20,000 Hz on the high-frequency end. Sounds with frequencies above this range are referred to as **ultrasonic**, while those with frequencies lower than 20 Hz are classified as **infrasonic**. Though we are unable to hear ultrasound and infrasound, these frequencies occur commonly in nature and are used in many technological applications as well.

BIO Bats and dolphins produce ultrasound almost continuously as they go about their daily lives. By listening to the echoes of their calls—that is, by using *echolocation*—they are able to navigate about their environment and detect their prey. As a defense mechanism, some of the insects that are preyed upon by bats have the ability to hear the ultrasonic frequency of a hunting bat and take evasive action. For instance, certain moths fold their wings in flight and drop into a precipitous dive toward the ground when they hear a bat on the prowl, as shown in **FIGURE 14-14 (a)**. In addition, the praying mantis has

Big Idea **2** Sound waves travel through the air with a speed of 343 m/s. The frequency of a sound wave is referred to as its pitch.

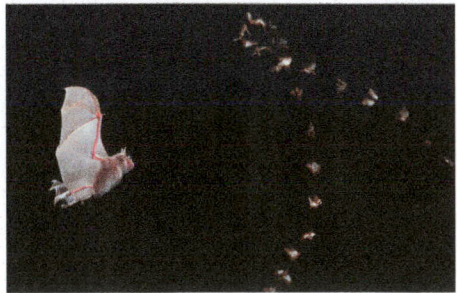

(a)

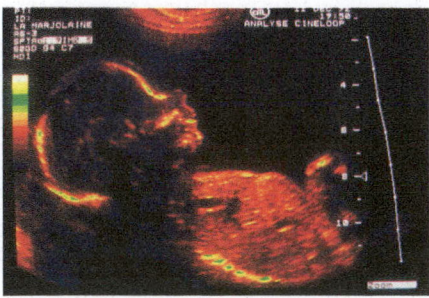

(b)

(c)

▲ **FIGURE 14-14 Visualizing Concepts Ultrasound** Sound waves with frequencies that are too high for human ears to detect are used by many animal species, and in technological applications as well. **(a)** Bats navigate in the dark and locate their prey by means of a system of biological sonar. They emit a continuous stream of ultrasonic sounds and detect the echoes from objects around them. **(b)** Ultrasound scans, or sonograms, are created by beaming ultrasonic pulses into the body and measuring the time required for the echoes to return. This technique is commonly used to evaluate heart function (echocardiograms) and to visualize the fetus in the uterus. **(c)** In shock wave lithotripsy, pulses of high-frequency sound waves are used to shatter kidney stones into fragments that can be passed in the urine.

a specialized ultrasound receptor on its abdomen that allows it to "duck and cover" in response to an approaching bat.

BIO Medical applications of ultrasound are also common. Perhaps the most familiar is the ultrasound scan that is used to image a fetus in the womb, as in **FIGURE 14-14 (b)**. By sending bursts of ultrasound into the body and measuring the time delay of the resulting echoes—the technological equivalent of echolocation—it is possible to map out the location of structures that lie hidden beneath the skin. In addition to imaging the interior of a body, ultrasound can also produce changes within the body that would otherwise require surgery. For example, in a technique called *shock wave lithotripsy* (SWL), an intense beam of ultrasound is concentrated onto a kidney stone that must be removed. Such a procedure is shown in **FIGURE 14-14 (c)**. After being hit with as many as 1000 to 3000 pulses of sound (at 23 joules per pulse), the stone is fractured into small pieces that the body can then eliminate on its own.

*RWP As for infrasound, it has been discovered in recent years that elephants (**FIGURE 14-15 (a)**) can communicate with one another using sounds with frequencies as low as 5 Hz. In fact, it may be that *most* elephant communication is infrasonic. These sounds, which humans feel as vibration rather than hear as sound, can carry over an area of about 30 square kilometers on the dry African savanna. And elephants are not alone in this ability. Whales, such as the blue whales shown in **FIGURE 14-15 (b)**, produce powerful infrasonic calls as well. Because sound generally travels farther in water than in air, the whale calls can be heard by others of their species over distances of thousands of kilometers.

RWP Not all sources of infrasound are biological, however. On February 15, 2013, a 12,000-ton asteroid crashed into the atmosphere near Chelyabinsk, Russia (**FIGURE 14-15 (c)**). It exploded with a flash of light bright enough to temporarily blind some people, and created a shock wave that shattered windows and injured hundreds. The event was observed not just visually, but with infrasound as well. A worldwide array of special microphones—originally designed to listen for clandestine nuclear weapons tests—heard the infrasonic boom created by the meteor. It was the largest infrasound wave ever recorded by the United Nations' monitoring system. In fact, the wave was strong enough to be detected after traveling 85,000 km, more than twice around the globe! Infrasound signals can be analyzed to discern the trajectories of meteors much smaller than the one that hit Chelyabinsk, as well as to study volcanic eruptions, earthquakes, and tsunamis.

It should be noted, in light of the wide range of frequencies observed in sound, that the speed of sound in air is the same for all frequencies. Thus, the speed v is a constant in the relationship

$$v = \lambda f$$

For example, if the frequency of a wave is doubled, its wavelength is halved, so that the speed v stays the same. The fact that different frequencies travel with the same speed is evident when we listen to an orchestra in a large room. Different instruments are producing sounds of different frequencies, but we hear the sounds at the same time. Otherwise, listening to music from a distance would be quite a different and inharmonious experience.

(a)

(b)

(c)

▲ **FIGURE 14-15 Visualizing Concepts Infrasound** Sound waves with frequencies that are too low to be heard with human ears occur in many natural settings. **(a)** Elephants use infrasound to communicate with one another with sounds that humans can't hear, but instead feel as "rumbles" in the ground. **(b)** Blue whales communicate over great distances by means of infrasonic sounds of about 10 Hz. **(c)** Even nonbiological systems, like a meteor streaking across the sky, can produce infrasonic sound waves.

Enhance Your Understanding   (Answers given at the end of the chapter)

4. Which is faster: wave 1 in medium 1 with a wavelength of 2 m and a frequency of 4 Hz, or wave 2 in medium 2 with a wavelength of 0.5 m and a frequency of 50 Hz?

Section Review

- Sound is a longitudinal wave that travels through the air at about 343 m/s. It also travels through liquids and solids, at speeds determined by the particular medium.

- The frequency of a sound is its pitch. Sounds with frequencies too high to be heard are referred to as ultrasonic; sounds with frequencies too low to be heard are referred to as infrasonic.

*Real World Physics applications are denoted by the acronym RWP.

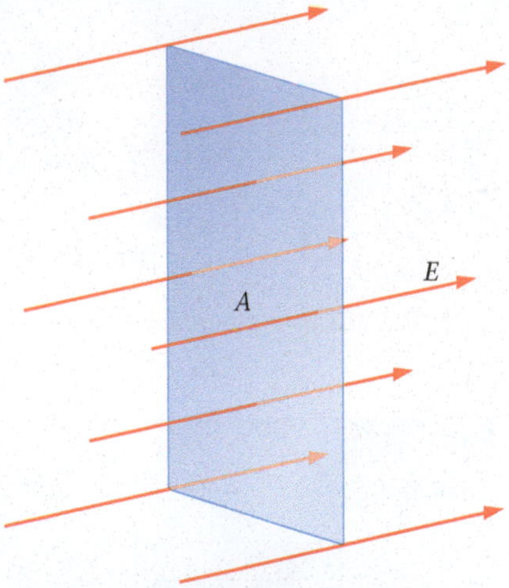

▲ **FIGURE 14-16 Intensity of a wave** If a wave carries an energy E through an area A in the time t, the corresponding intensity is $I = E/At = P/A$, where $P = E/t$ is the power.

PHYSICS IN CONTEXT
Looking Back

We use the concept of power, first developed in Chapter 7, in our definition of the intensity of a wave.

TABLE 14-2 Sound Intensities (W/m²)

Loudest sound produced in laboratory	10^9
Saturn V rocket at 50 m	10^8
Rupture of the eardrum	10^4
Jet engine at 50 m	10^2
Threshold of pain	1
Rock concert	10^{-1}
Jackhammer at 1 m	10^{-3}
Heavy street traffic	10^{-5}
Conversation at 1 m	10^{-6}
Classroom	10^{-7}
Whisper at 1 m	10^{-10}
Normal breathing	10^{-11}
Threshold of hearing	10^{-12}

PROBLEM-SOLVING NOTE

Intensity Variation with Distance

Suppose the intensity of a point source is I_1 at a distance r_1. This is enough information to find its intensity at any other distance. For example, to find the intensity I_2 at a distance r_2 we use the relationship $I_2 = (r_1/r_2)^2 I_1$.

14-5 Sound Intensity

The noise made by a jackhammer is much louder than the song of a sparrow. On this we can all agree. But how do we express such an observation physically? What physical quantity determines the loudness of a sound? We address these questions in this section, and we also present a quantitative scale by which loudness may be measured.

Intensity

Suppose you spread some jam on a piece of toast. The intensity of the taste is greater, the greater the amount of jam you use. On the other hand, the intensity of taste is smaller if you spread the same amount of jam over a larger area.

Similarly, the loudness of a sound is determined by its **intensity**—that is, by the amount of energy that passes through a given area in a given time. This is illustrated in **FIGURE 14-16**. If the energy E passes through the area A in the time t, the intensity I of the wave carrying the energy is

$$I = \frac{E}{At}$$

Recalling that power is energy per time, $P = E/t$, we can express the intensity as follows:

Definition of Intensity, I

$$I = \frac{P}{A}$$

14-5

SI unit: W/m²

The units are those of power (watts, W) divided by area (meters squared, m²).

Though we have introduced the concept of intensity in terms of sound, it applies to all types of waves. For example, the intensity of light from the Sun as it reaches the Earth's upper atmosphere is about 1380 W/m². If this intensity could be heard as sound, it would be painfully loud—roughly the equivalent of four jet airplanes taking off simultaneously. By comparison, the intensity of microwaves in a microwave oven is even greater, about 6000 W/m², whereas the intensity of a whisper is an incredibly tiny 10^{-10} W/m². A selection of representative intensities is given in Table 14-2.

EXERCISE 14-7 INTENSITY

Sounds from the street below come through an open window measuring 0.75 m by 0.88 m. If the power of the sound is 1.2×10^{-6} W, what is its intensity?

REASONING AND SOLUTION
Applying $I = P/A$ with $P = 1.2 \times 10^{-6}$ W and $A = (0.75 \text{ m})(0.88 \text{ m})$, we find

$$I = \frac{P}{A} = \frac{1.2 \times 10^{-6} \text{ W}}{(0.75 \text{ m})(0.88 \text{ m})} = 1.8 \times 10^{-6} \text{ W/m}^2$$

This is an intensity similar to that of normal conversation.

Dependence of Intensity on Distance   When we listen to a source of sound, such as a person speaking or a radio playing a song, we notice that the loudness of the sound decreases as we move away from the source. This means that the intensity of the sound is also decreasing. The reason for this reduction in intensity is simply that the energy emitted per time by the source spreads out over a larger area—just as spreading a certain amount of jam over a larger piece of toast reduces the intensity of the taste.

In **FIGURE 14-17** we show a source of sound (a bat) and two observers (moths) listening at the distances r_1 and r_2. The waves emanating from the bat propagate outward spherically, with the wave crests forming a series of concentric spheres. If we assume no reflections of sound, and a power output by the bat equal to P, the intensity detected by the first moth is

$$I_1 = \frac{P}{4\pi r_1^2}$$

In writing this expression, we have used the fact that the area of a sphere of radius r is $A = 4\pi r^2$. Similarly, the second moth hears the same sound with an intensity of

$$I_2 = \frac{P}{4\pi r_2{}^2}$$

The power P is the same in each case—it simply represents the amount of sound emitted by the bat. Solving for the intensity at moth 2 in terms of the intensity at moth 1, we find

$$I_2 = \left(\frac{r_1}{r_2}\right)^2 I_1 \qquad 14\text{-}6$$

In words, the intensity falls off with the square of the distance; doubling the distance reduces the intensity by a factor of four.

To summarize, the intensity a distance r from a point source of power P is:

Intensity with Distance from a Point Source

$$I = \frac{P}{4\pi r^2} \qquad 14\text{-}7$$

SI unit: W/m^2

This result assumes that no sound is reflected—which could increase the amount of energy passing through a given area—that no sound is absorbed, and that the sound propagates outward spherically. These assumptions are applied in the next Example.

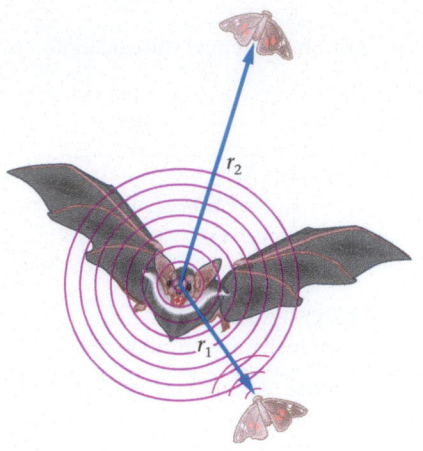

▲ **FIGURE 14-17 Echolocation** Two moths, at distances r_1 and r_2, hear the sonar signals sent out by a bat. The intensity of the signal decreases with the square of the distance from the bat. The bat, in turn, hears the echoes sent back by the moths. It can then use the direction and intensity of the returning echoes to locate its prey.

EXAMPLE 14-8   THE POWER OF SONG

Two people relaxing on a deck listen to a songbird sing. One person, only 1.66 m from the bird, hears the sound with an intensity of $2.86 \times 10^{-6}\ W/m^2$. **(a)** What intensity is heard by the second person, who is 6.71 m from the bird? Assume that no reflected sound is heard by either person. **(b)** What is the power output of the bird's song?

PICTURE THE PROBLEM
Our sketch shows the two observers, one at a distance of $r_1 = 1.66\ m$ from the bird, and the other at a distance of $r_2 = 6.71\ m$. The sound emitted by the bird is assumed to spread out spherically, with no reflections.

REASONING AND STRATEGY

a. The two intensities are related by $I_2 = (r_1/r_2)^2 I_1$ (Equation 14-6), with $r_1 = 1.66\ m$ and $r_2 = 6.71\ m$.

b. The power output can be obtained from the definition of intensity, $I = P/A$. We can calculate P for either observer, noting that $A = 4\pi r^2$.

Known   Distance to bird for person 1, $r_1 = 1.66$ m; distance for person 2, $r_2 = 6.71$ m; intensity for person 1, $I_1 = 2.86 \times 10^{-6}\ W/m^2$.
Unknown **(a)** Intensity for person 2, $I_2 = ?$ **(b)** Power output of bird, $P = ?$

SOLUTION

Part (a)

1. Substitute numerical values into $I_2 = (r_1/r_2)^2 I_1$:

$$I_2 = \left(\frac{r_1}{r_2}\right)^2 I_1 = \left(\frac{1.66\ m}{6.71\ m}\right)^2 (2.86 \times 10^{-6}\ W/m^2)$$
$$= 1.75 \times 10^{-7}\ W/m^2$$

Part (b)

2. Solve $I = P/A$ for the power, P, using data for observer 1:

$$I_1 = \frac{P}{A_1}$$
$$P = I_1 A_1 = (2.86 \times 10^{-6}\ W/m^2)[4\pi(1.66\ m)^2]$$
$$= 9.90 \times 10^{-5}\ W$$

CONTINUED

3. As a check, repeat the calculation for observer 2:

$$I_2 = \frac{P}{A_2}$$

$$P = I_2 A_2 = (1.75 \times 10^{-7}\,\text{W/m}^2)[4\pi(6.71\,\text{m})^2]$$
$$= 9.90 \times 10^{-5}\,\text{W}$$

INSIGHT

The intensity at observer 1 is about 16.3 times the intensity at observer 2. Even so, the bird doesn't *seem* to be 16.3 times *louder*, as perceived by observer 2; in fact, the bird seems to be only a little more than twice as loud. The connection between intensity and perceived (subjective) loudness is discussed in detail later in this section.

PRACTICE PROBLEM

If the intensity at a third observer is $7.43 \times 10^{-8}\,\text{W/m}^2$, how far is that observer from the bird? **[Answer:** $r_3 = 10.3\,\text{m}$**]**

Some related homework problems: Problem 38, Problem 39

Human Perception of Sound

BIO Hearing, like most of our senses, is incredibly versatile and sensitive. We can detect sounds that are about a million times fainter than a typical conversation, and listen to sounds that are a million times louder before experiencing pain. In addition, we are able to hear sounds over a wide range of frequencies, from 20 Hz to 20,000 Hz.

When detecting the faintest of sounds, our hearing is more sensitive than one would ever guess. For example, a faint sound, with an intensity of about $10^{-11}\,\text{W/m}^2$, causes a displacement of molecules in the air of about $10^{-10}\,\text{m}$. This displacement is roughly the diameter of an atom!

Equally interesting is the way we perceive the loudness of a sound. As an example, suppose you hear a sound of intensity I_1. Next, you listen to a second sound of intensity I_2, and this sound seems to be "twice as loud" as the first. If the two intensities are measured, it turns out that I_2 is about ten times greater than I_1. Similarly, a third sound, twice as loud as I_2, has an intensity I_3 that is ten times greater than I_2. Thus, $I_2 = 10\,I_1$ and $I_3 = 10\,I_2 = 100\,I_1$.

Our perception of sound, then, is such that a uniform increase in loudness corresponds to intensities that increase by multiplicative factors. For this reason, a convenient scale to measure loudness depends on the logarithm of intensity, as we discuss next.

Intensity Level and Decibels

In the study of sound, loudness is measured by the **intensity level** of a wave. Notice that we use the new term *intensity level*, which is different from *intensity*. In fact, the intensity level, designated by the symbol β, is defined as follows:

Definition of Intensity Level, β

$$\beta = (10\,\text{dB}) \log\!\left(\frac{I}{I_0}\right)$$

14-8

SI unit: decibel (dB), which is dimensionless

In this expression, log indicates the logarithm to the base 10, and I_0 is the intensity of the faintest sounds that can be heard. Experiments show this lowest detectable intensity to be

$$I_0 = 10^{-12}\,\text{W/m}^2$$

As is clear from its definition, the intensity level β is dimensionless; the only dimensions that enter into the definition are those of intensity, and they cancel in the logarithm. Still, just as with radians, it's convenient to label the values of intensity level with a name. The name we use—the *bel*—honors the work of Alexander Graham Bell (1847–1922), the inventor of the telephone. Because the bel is a fairly large unit, it's more common to measure β in units that are one-tenth of a bel. This unit is referred to as the **decibel**, and its abbreviation is dB.

PROBLEM-SOLVING NOTE

Intensity Versus Intensity Level

When reading a problem statement, be sure to note carefully whether it refers to the intensity, I, or to the intensity level, β. These two quantities have similar names but completely different meanings and units, as indicated in the following table:

Physical quantity	Physical meaning	Units
Intensity, I	Energy per time per area	W/m²
Intensity level, β	A measure of relative loudness	dB

PROBLEM-SOLVING NOTE

Calculating the Intensity Level

When determining the intensity level β, be sure to use the base-10 logarithm (log), as opposed to the "natural," or base e, logarithm (ln).

To get a feeling for the decibel scale, let's start with the faintest sounds. If a sound has an intensity $I = I_0$, the corresponding intensity level is

$$\beta = (10 \, \text{dB}) \log\left(\frac{I_0}{I_0}\right) = (10 \, \text{dB}) \log(1) = 0$$

Increasing the intensity by a factor of ten makes the sound seem twice as loud. In terms of decibels, we have

$$\beta = (10 \, \text{dB}) \log(10 I_0/I_0) = (10 \, \text{dB}) \log(10) = 10 \, \text{dB}$$

Going up in intensity by another factor of ten doubles the loudness of the sound again, and yields

$$\beta = (10 \, \text{dB}) \log(100 I_0/I_0) = (10 \, \text{dB}) \log(100) = 20 \, \text{dB}$$

Thus, *the loudness of a sound doubles with each increase in intensity level of 10 dB*. The *smallest* increase in intensity level that can be detected by the human ear is about 1 dB.

The intensity of a number of independent sound sources is simply the sum of the individual intensities. We use this fact in the following Example.

> **Big Idea** **3** The intensity of a sound wave is the power of the wave divided by the area through which it passes. The logarithm of a sound's intensity is related to its loudness, or intensity level. Increasing the intensity of a sound by a factor of ten doubles the perceived loudness.

EXAMPLE 14-9   PASS THE PACIFIER

A crying child emits sound with an intensity of $8.0 \times 10^{-6} \, \text{W/m}^2$. Find **(a)** the intensity level in decibels for the child's sounds, and **(b)** the intensity level for this child and its twin, both crying with identical intensities.

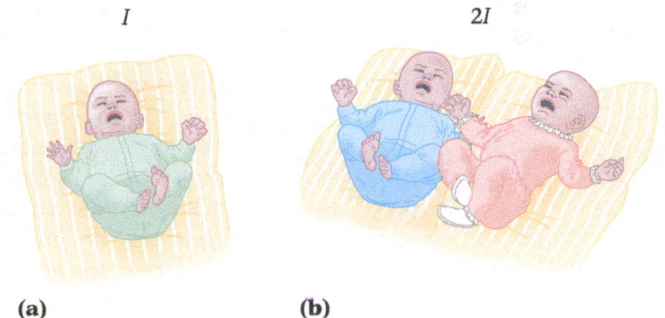

PICTURE THE PROBLEM
We consider the crying sounds of either one or two children. Each child emits sound with an intensity $I = 8.0 \times 10^{-6} \, \text{W/m}^2$. If two children are crying together, the total intensity of their sound is $2I$.

(a)   **(b)**

REASONING AND STRATEGY
The intensity level, β, is obtained by applying $\beta = (10 \, \text{dB}) \log(I/I_0)$ (Equation 14-8).

Known   Intensity of one crying child, $I = 8.0 \times 10^{-6} \, \text{W/m}^2$; intensity of two crying children, $2I$.
Unknown   **(a)** Intensity level for crying child, $\beta = ?$ **(b)** Intensity level for two crying children, $\beta = ?$

SOLUTION

Part (a)

1. Calculate β for $I = 8.0 \times 10^{-6} \, \text{W/m}^2$:

$$\beta = (10 \, \text{dB}) \log\left(\frac{I}{I_0}\right)$$

$$= (10 \, \text{dB}) \log\left(\frac{8.0 \times 10^{-6} \, \text{W/m}^2}{10^{-12} \, \text{W/m}^2}\right) = (10 \, \text{dB}) \log(8.0 \times 10^6)$$

$$= (10 \, \text{dB}) \log(8.0) + (10 \, \text{dB}) \log(10^6) = 69 \, \text{dB}$$

Part (b)

2. Repeat the calculation with I replaced by $2I$:

$$\beta = (10 \, \text{dB}) \log\left(\frac{2I}{I_0}\right)$$

$$= (10 \, \text{dB}) \log(2) + (10 \, \text{dB}) \log\left(\frac{I}{I_0}\right)$$

$$= 3.0 \, \text{dB} + 69 \, \text{dB} = 72 \, \text{dB}$$

INSIGHT
We see that when the intensity is doubled, as in this Example, the intensity level is increased by $(10 \, \text{dB}) \log(2) = 3 \, \text{dB}$. Similarly, when the intensity is *halved*, β decreases by 3 dB.

PRACTICE PROBLEM
What is the intensity level of four identically crying quadruplets? [**Answer:** $\beta = 75 \, \text{dB}$]

Some related homework problems: Problem 40, Problem 41

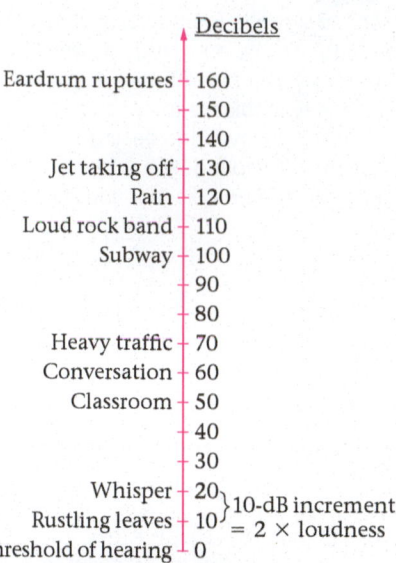

Decibels	
Eardrum ruptures	160
	150
	140
Jet taking off	130
Pain	120
Loud rock band	110
Subway	100
	90
	80
Heavy traffic	70
Conversation	60
Classroom	50
	40
	30
Whisper	20
Rustling leaves	10
Threshold of hearing	0

$\rbrace$ 10-dB increment = 2 × loudness

▲ **FIGURE 14-18 Representative intensity levels for common sounds**

Even though a change of 3 dB is relatively small—after all, a change of 10 dB is required to make a sound seem twice as loud—it still requires changing the intensity by a factor of two. For example, suppose a large nursery in a hospital has so many crying babies that the intensity level is 6 dB above the safe value, as determined by OSHA (Occupational Safety and Health Administration). To reduce the level by 6 dB it would be necessary to remove three-quarters of the children, leaving only one-quarter of the original number. To our ears, however, the nursery will *sound* only 40 percent quieter!

FIGURE 14-18 shows the decibel scale with representative values indicated for a variety of common sounds.

Enhance Your Understanding   (Answers given at the end of the chapter)

5. A sound has an intensity of I and a corresponding intensity level of β. If the intensity is increased to $10I$, what is the new intensity level? **(a)** 10β, **(b)** $\beta + 3$ dB, **(c)** $\beta + 10$ dB, **(d)** $\beta/2$.

Section Review

- The intensity of a sound, I, is given by $I = P/A$, where P is the power and A is the area through which the sound passes.

- The intensity level of a sound is $\beta = (10 \text{ dB}) \log(I/I_0)$, where $I_0 = 10^{-12} \text{ W/m}^2$.

14-6 The Doppler Effect

One of the most common physical phenomena involving sound is the change in pitch of a train whistle or a car horn as the vehicle moves past us. This change in pitch, due to the relative motion between a source of sound and the receiver, is called the **Doppler effect**, after the Austrian physicist Christian Doppler (1803–1853). If you listen carefully to the Doppler effect, you will notice that the pitch increases when the observer and the source are moving closer together, and decreases when the observer and the source are separating.

One of the more fascinating aspects of the Doppler effect is that it applies to all wave phenomena, not just to sound. In particular, the frequency of light is also Doppler-shifted when there is relative motion between the source and the receiver. For light, this change in frequency means a change in color. In fact, most distant galaxies are observed to be red-shifted in the color of their light, which means they are moving away from the Earth. Some galaxies, however, are moving toward us, and their light shows a blue shift.

In the remainder of this section, we focus on the Doppler effect in sound waves. We show that the effect is different depending on whether the observer or the source is moving. Finally, both observer and source may be in motion, and we present results for such cases as well.

Moving Observer

In **FIGURE 14-19** we see a stationary source of sound in still air. The radiated sound is represented by the circular patterns of compressions moving away from the source with a speed v. The distance between the compressions is the wavelength, λ, and the frequency of the sound is f. As for any wave, these quantities are related by

$$v = \lambda f$$

◀ **FIGURE 14-19 The Doppler effect: a moving observer** Sound waves from a stationary source form concentric circles moving outward with a speed v. To the observer, who moves toward the source with a speed u, the waves are moving with a speed $v + u$.

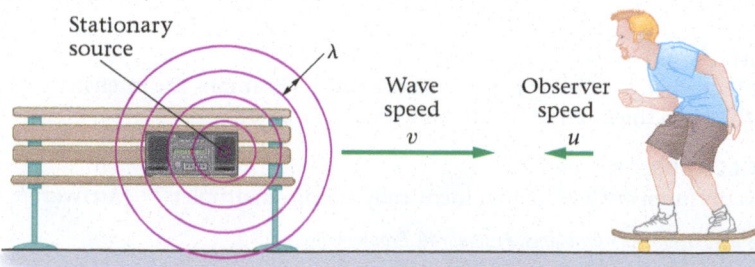

Stationary source

λ

Wave speed v

Observer speed u

For an observer moving toward the source with a speed u, as in Figure 14-19, the sound *appears* to have a higher speed, $v + u$ (though, of course, the speed of sound relative to the air is always the same). As a result, more compressions move past the observer in a given time than if the observer had been at rest. To the observer, then, the sound has a frequency, f', that is higher than the frequency of the source, f.

We can find the frequency f' by first noting that the wavelength of the sound does not change—it is still λ. The speed, however, has increased to $v' = v + u$. Thus, we can solve $v' = \lambda f'$ for the frequency, which yields

$$f' = \frac{v'}{\lambda} = \frac{v + u}{\lambda}$$

Finally, we recall that $v = \lambda f$ can be rearranged to give $\lambda = v/f$, and hence

$$f' = \frac{v + u}{(v/f)} = \left(\frac{v + u}{v}\right)f = (1 + u/v)f$$

Notice that f' is greater than f for this observer. This is the Doppler effect.

If the observer had been moving away from the source with a speed u, the sound would *appear* to the observer to have the reduced speed $v' = v - u$. Repeating the calculation just given, we find that

$$f' = \frac{v'}{\lambda} = \frac{v - u}{\lambda} = (1 - u/v)f$$

In this case the Doppler effect results in f' being less than f.

Combining these results, we have:

Doppler Effect for Moving Observer

$f' = (1 \pm u/v)f$ 14-9

SI unit: $1/s = s^{-1} = Hz$

In this expression u and v are speeds, with u referring to the observer and v referring to sound, and hence they are always positive. The appropriate signs are obtained by using the *plus* sign when the observer moves *toward* the source, and the *minus* sign when the observer moves *away from* the source.

EXAMPLE 14-10   A MOVING OBSERVER

A street musician sounds the A string of his violin, producing a tone of 440 Hz. What frequency does a bicyclist hear as she **(a)** approaches and **(b)** recedes from the musician with a speed of 11.0 m/s?

PICTURE THE PROBLEM
The sketch shows a stationary source of sound and a moving observer. In part (a) the observer approaches the source with a speed $u = 11.0$ m/s; in part (b) the observer has passed the source and is moving away with the same speed.

REASONING AND STRATEGY
The frequency heard by the observer is given by $f' = (1 \pm u/v)f$ (Equation 14-9), with the plus sign for part (a) and the minus sign for part (b).

Known   Frequency of sound, $f = 440$ Hz; speed of bicyclist, $u = 11.0$ m/s; speed of sound, $v = 343$ m/s.

Unknown   **(a)** Frequency heard by cyclist when approaching, $f' = ?$ **(b)** Frequency heard by cyclist when receding, $f' = ?$

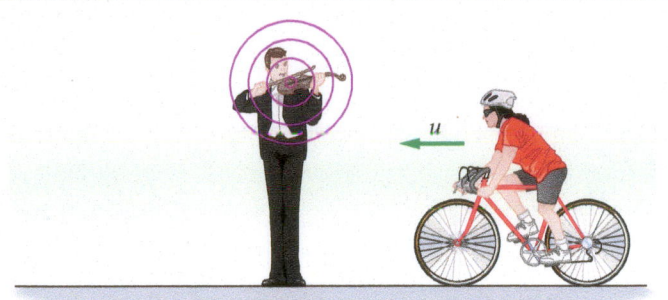

(a)

(b)

CONTINUED

SOLUTION

Part (a)

1. Apply $f' = (1 \pm u/v)f$ with the plus sign and $u = 11.0$ m/s: $f' = (1 + u/v)f = \left(1 + \dfrac{11.0 \text{ m/s}}{343 \text{ m/s}}\right)(440 \text{ Hz}) = 454 \text{ Hz}$

Part (b)

2. Now use the minus sign in $f' = (1 \pm u/v)f$: $f' = (1 - u/v)f = \left(1 - \dfrac{11.0 \text{ m/s}}{343 \text{ m/s}}\right)(440 \text{ Hz}) = 426 \text{ Hz}$

INSIGHT

As the bicyclist passes the musician, the observed frequency of sound decreases, giving a "wow" effect where the frequency of the sound changes rapidly. The difference in frequency is about 1 semitone, the frequency difference between adjacent notes on the piano. See Table 14-3 on page 485 for the frequencies of notes in the vicinity of middle C.

PRACTICE PROBLEM

If the bicyclist hears a frequency of 451 Hz when approaching the musician, what is her speed?   **[Answer: $u = 8.58$ m/s]**

Some related homework problems: Problem 45, Problem 46

Moving Source

In the case of a moving source, the Doppler effect is not due to the sound wave appearing to have a higher or lower speed, as when the observer moves. On the contrary, the speed of a wave is determined solely by the properties of the medium through which it propagates. Thus, once the source emits a sound wave, it travels through the medium with its characteristic speed v regardless of what the source is doing.

By way of analogy, consider a water wave. The speed of such waves is the same whether they are created by a rock dropped into the water or by a stick moved rapidly through the water. To take an extreme case, the waves coming to the beach from a slow-moving tugboat move with the same speed as the waves produced by a 100-mph speed boat. The same is true of sound waves.

▶ **FIGURE 14-20 The Doppler effect: a moving source** Sound waves from a moving source are bunched up in the forward direction, causing a shorter wavelength and a higher frequency.

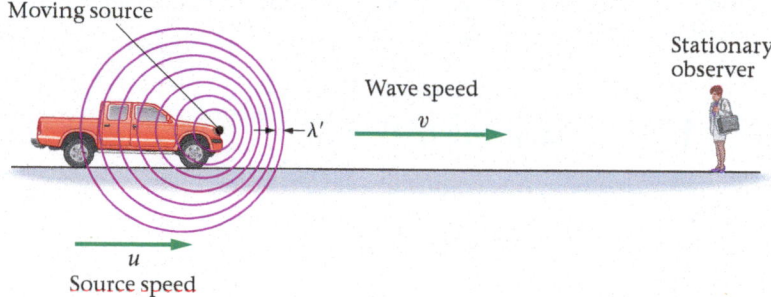

Consider, then, a source moving toward an observer with a speed u, as shown in **FIGURE 14-20**. If the frequency of the source is f, it emits one compression every T seconds, where $T = 1/f$. Therefore, during one cycle of the wave a compression travels a distance vT while the source moves a distance uT. As a result, the next compression is emitted a distance $vT - uT$ behind the previous compression, as illustrated in **FIGURE 14-21**. This means that the wavelength in the forward direction is

$$\lambda' = vT - uT = (v - u)T$$

Now, the speed of the wave is still v, as mentioned above, and hence

$$v = \lambda'f'$$

Solving for the new frequency, f', we find

$$f' = \frac{v}{\lambda'} = \frac{v}{(v - u)T}$$

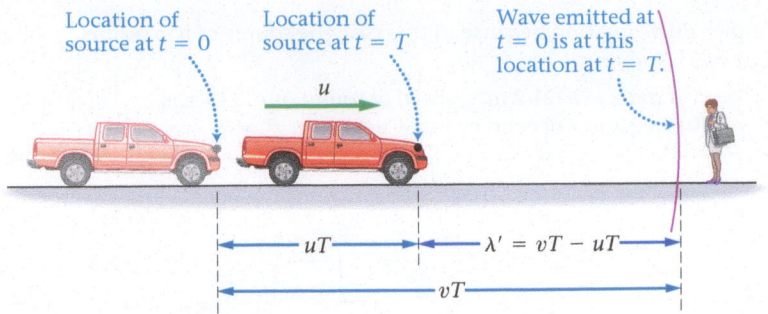

Location of source at $t = 0$

Location of source at $t = T$

Wave emitted at $t = 0$ is at this location at $t = T$.

u

uT — $\lambda' = vT - uT$

vT

◀ **FIGURE 14-21 The Doppler-shifted wavelength** During one period, T, the wave emitted at $t = 0$ moves through a distance vT. In the same time, the source moves toward the observer through the distance uT. At the time $t = T$ the next wave is emitted from the source; hence, the distance between the waves (the wavelength) is $\lambda' = vT - uT$.

Finally, recalling that $T = 1/f$, we have

$$f' = \frac{v}{(v - u)(1/f)} = \frac{v}{v - u}f = \left(\frac{1}{1 - u/v}\right)f$$

Notice that f' is greater than f, as expected.

In the reverse direction, the wavelength is increased by the amount uT. Thus,

$$\lambda' = vT + uT = (v + u)T$$

It follows that the Doppler-shifted frequency is

$$f' = \frac{v}{(v + u)T} = \left(\frac{1}{1 + u/v}\right)f$$

This is less than the source frequency, f.

Finally, we can combine these results to yield:

Doppler Effect for Moving Source

$$f' = \left(\frac{1}{1 \mp u/v}\right)f \qquad \text{14-10}$$

SI unit: $1/\text{s} = \text{s}^{-1} = \text{Hz}$

As before, u and v are positive quantities. The *minus* sign is used when the source moves *toward* the observer, and the *plus* sign when the source moves *away from* the observer. The speed v refers to the speed of sound, and the speed u refers to the speed of the source of the sound.

PHYSICS IN CONTEXT
Looking Back

Basic concepts from kinematics (Chapter 2) are used to derive the Doppler effect.

PROBLEM-SOLVING NOTE

Using Correct Signs

When a source approaches an observer, the frequency heard by the observer is greater than the frequency of the source; that is, $f' > f$. This means that we must use the minus sign in Equation 14-10, $f' = f/(1 - u/v)$, since this makes the denominator less than one. Similarly, we use the plus sign when the source moves away from the observer.

EXAMPLE 14-11   WHISTLE STOP

A train sounds its whistle as it approaches a tunnel in a cliff. The whistle produces a tone of 650.0 Hz, and the train travels with a speed of 21.2 m/s. **(a)** Find the frequency heard by an observer standing near the tunnel entrance. **(b)** The sound from the whistle reflects from the cliff back to the engineer in the train. What frequency does the engineer hear?

PICTURE THE PROBLEM
The train moves with a speed $u = 21.2$ m/s and emits sound with a frequency $f = 650.0$ Hz. The observer near the tunnel hears the Doppler-shifted frequency f', and the engineer on the train hears the reflected sound at an even higher frequency f''.

REASONING AND STRATEGY
Two Doppler shifts are involved in this problem. The first is due to the motion of the train toward the cliff. This shift causes the observer at the cliff to hear sound with a higher frequency f', given by $f' = (1/(1 \mp u/v))f$ (Equation 14-10) with the minus sign. The reflected sound has the same frequency, f'.

The second shift is due to the engineer moving toward the reflected sound. Thus, the engineer hears a frequency f'' that is greater than f'. We find f'' using $f' = (1 \pm u/v)f$

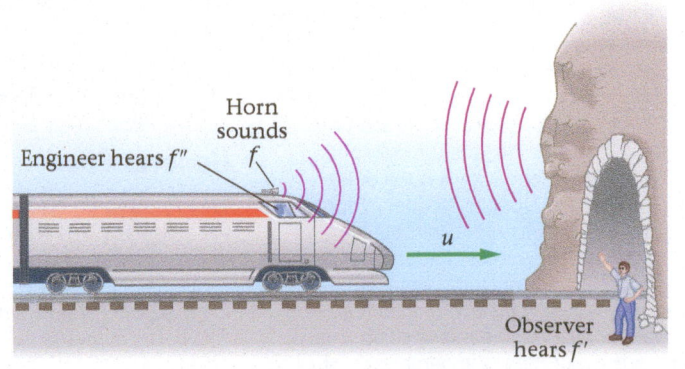

Horn sounds
f

Engineer hears f''

u

Observer hears f'

CONTINUED

(Equation 14-9) with the plus sign. We use this Doppler shift equation because in this case the source of the sound (the cliff) is stationary, while the observer (the engineer) is moving.

Known Frequency of train whistle, $f = 650.0$ Hz; speed of train, $u = 21.2$ m/s; speed of sound, $v = 343$ m/s.

Unknown (a) Frequency heard by observer at rest, $f' = ?$ (b) Frequency of echo heard by engineer, $f'' = ?$

SOLUTION

Part (a)

1. Use $f' = \left(\dfrac{1}{1 \mp u/v}\right) f$, with the minus sign, to Doppler-shift from f to f':

$$f' = \left(\frac{1}{1 - u/v}\right) f = \left(\frac{1}{1 - \dfrac{21.2\,\text{m/s}}{343\,\text{m/s}}}\right)(650.0\,\text{Hz})$$

$$= \left(\frac{1}{1 - 0.0618}\right)(650.0\,\text{Hz}) = 693\,\text{Hz}$$

Part (b)

2. Now use $f' = (1 \pm u/v) f$, with the plus sign, to Doppler-shift from f' to f'':

$$f'' = (1 + u/v) f' = \left(1 + \frac{21.2\,\text{m/s}}{343\,\text{m/s}}\right)(693\,\text{Hz})$$

$$= (1 + 0.0618)(693\,\text{Hz}) = 736\,\text{Hz}$$

INSIGHT

The reflected sound has the same frequency f' heard by the stationary observer, because all the cliff does is reverse the direction of motion of the sound heard at the cliff. Therefore, the cliff acts as a stationary source of sound at the frequency f'. The engineer's motion toward this stationary source results in the Doppler shift from f' to f''.

PRACTICE PROBLEM

If the stationary observer hears a frequency of 700.0 Hz, what are (a) the speed of the train and (b) the frequency heard by the engineer?   [**Answer:** (a) $u = 24.5$ m/s, (b) $f'' = 750$ Hz]

Some related homework problems: Problem 47, Problem 96

Comparing Doppler Shifts Now that we've obtained the Doppler effect for both moving observers and moving sources, it's interesting to compare the results. **FIGURE 14-22** shows the Doppler-shifted frequency versus speed for a 400-Hz source of sound. The upper curve corresponds to a moving source, the lower curve to a moving observer. Notice that while the two cases give similar results for low speed, the high-speed behavior is quite different. In fact, the Doppler frequency for the moving source grows without limit for speeds near the speed of sound, while the Doppler frequency for the moving observer is relatively small.

These results can be understood both in terms of mathematics—by simply comparing Equations 14-9 and 14-10—and physically. In physical terms, recall that a moving observer encounters wave crests separated by the wavelength, as indicated in Figure 14-19. Doubling the speed of the observer simply reduces the time required to move from one crest to the next by a factor of two, which doubles the frequency. Thus, in general, the frequency is proportional to the speed, as we see in the lower curve in Figure 14-22. In contrast, when the source moves, as in Figure 14-20, the wave crests become "bunched up" in the forward direction, because the source is almost keeping up with the propagating waves. As the speed of the source approaches the speed of sound, the separation between the crests approaches zero. Consequently, the frequency with which the crests pass a stationary observer approaches infinity, as indicated by the upper curve in Figure 14-22.

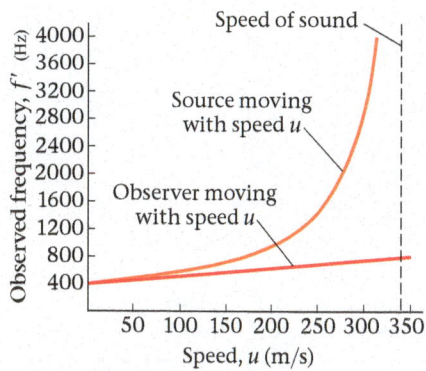

▲ **FIGURE 14-22 Doppler-shifted frequency versus speed for a 400-Hz sound source** The upper curve corresponds to a moving source, the lower curve to a moving observer. Notice that while the two cases give similar results for low speed, the high-speed behavior is quite different. In fact, the Doppler frequency for the moving source increases without limit for speeds near the speed of sound, while the Doppler frequency for the moving observer is relatively low. If a source moves faster than the speed of sound, the sound it produces is perceived not as a pure tone, but as a shock wave, commonly referred to as a sonic boom.

General Case

The results derived earlier in this section can be combined to give the Doppler effect for situations in which both observer and source move. Letting u_s be the speed of the source, and u_o be the speed of the observer, we have:

Doppler Effect for Moving Source and Observer

$$f' = \left(\frac{1 \pm u_o/v}{1 \mp u_s/v}\right) f \qquad\qquad 14\text{-}11$$

SI unit: $1/\text{s} = \text{s}^{-1} = \text{Hz}$

As in the previous cases, u_s, u_o, and v are positive quantities. In the numerator, the plus sign corresponds to the case in which the observer moves in the direction of the source, whereas the minus sign indicates motion in the opposite direction. In the denominator, the minus sign corresponds to the case in which the source moves in the direction of the observer, whereas the plus sign indicates motion in the opposite direction.

EXERCISE 14-12   GENERAL DOPPLER SHIFT

A car traveling north at 21 m/s sounds its 480-Hz horn. A bicyclist travels south on the same road with a speed of 6.7 m/s, heading directly toward the car—though at a safe distance. What frequency does the bicyclist hear?

REASONING AND SOLUTION

We use $f' = ((1 \pm u_o/v)/(1 \mp u_s/v))f$ (Equation 14-11) to find the Doppler-shifted frequency, because both the observer and the source are in motion. In this case, $u_s = 21$ m/s and $u_o = 6.7$ m/s. Because the source and the observer are approaching, we use the plus sign in the numerator (which raises the frequency) and the minus sign in the denominator (which also raises the frequency). The resulting frequency is

$$f' = \left(\frac{1 + u_o/v}{1 - u_s/v}\right)f = \left(\frac{1 + 6.7/343}{1 - 21/343}\right)(480\,\text{Hz}) = 520\,\text{Hz}$$

RWP The Doppler effect is used in an amazing variety of technological applications. Perhaps one of the most familiar is the "radar gun," which is used to measure the speed of a pitched baseball or a car that is breaking the speed limit. Although the radar gun uses radio waves rather than sound waves, the basic physical principle is the same: By measuring the Doppler-shifted frequency of waves reflected from an object, it is possible to determine its speed. Doppler radar, used in weather forecasting, applies the same technology to tracking the motion of precipitation inside storm clouds. An example showing the formation of a tornado is presented in **FIGURE 14-23 (a)**.

BIO In medicine, the Doppler shift is used to measure the speed of blood flow in an artery, as in **FIGURE 14-23 (b)**, or in the heart itself. In this application, a beam of ultrasound is directed toward an artery in a patient. Some of the sound is reflected back by red blood cells moving through the artery. The reflected sound is detected, and its frequency is used to determine the speed of blood flow. If this information is color coded, with different colors indicating different speeds and directions of flow, an impressive image of blood flow can be constructed.

RWP Finally, the Doppler effect applies to the light of distant galaxies as well, as in **FIGURE 14-23 (c)**. For example, if a galaxy moves away from us—as most do—the light we observe from that galaxy has a lower frequency than if the galaxy were at rest relative to us. Because red light has the lowest frequency of visible light, we refer to this reduction in frequency as a "red shift." Thus, by measuring the red shift of a galaxy, we can determine its speed relative to us. Such measurements show that the more distant a galaxy, the greater its speed relative to us—a result known as Hubble's law. This is just what one would expect from an explosion, or "Big Bang," in which rapidly moving pieces travel farther in a given amount of time. Thus the Doppler effect, and red-shift measurements, provide strong evidence for the Big Bang and an expanding universe.

Big Idea **4** The frequency of a sound wave depends on the motion of both the source of the sound and the observer. This is the Doppler effect. When the source and the observer approach one another, the observed frequency is higher; when the source and the observer move away from one another, the observed frequency decreases.

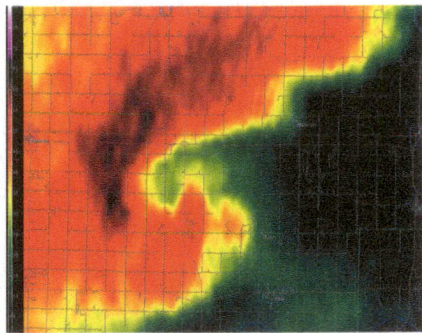

(a)

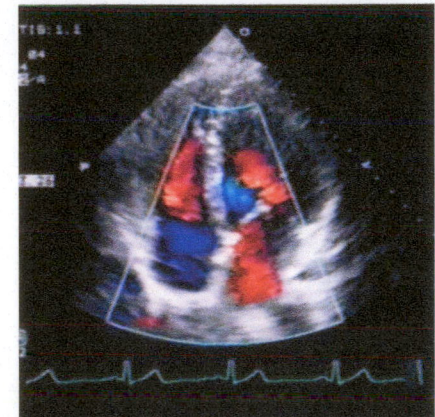

(b)

(c)

▶ **FIGURE 14-23 Visualizing Concepts The Doppler Effect** Many familiar and not-so-familiar devices utilize the Doppler effect. **(a)** Doppler radar, now widely used at airports and for weather forecasting, makes it possible for a forecaster to determine the speed and direction of winds in a distant storm by measuring the Doppler shift they produce. This image is the Doppler radar scan of a severe thunderstorm that struck Pawnee County, Oklahoma, on May 19, 2010. The hook-shaped echo is characteristic of tornadoes in the making. **(b)** A doppler echocardiogram of the human heart indicates the direction of blood flow with different colors. **(c)** Most distant galaxies are moving away from us, producing a "red shift" (lower frequency) in their light. The galaxies in this Hubble Deep Field photo have extreme red shifts—their velocities relative to us are near the speed of light.

6. Observer 1 approaches a stationary 1000-Hz source of sound with a speed of 25 m/s. Observer 2 is at rest and an identical 1000-Hz source approaches her with a speed of 25 m/s. Is the frequency heard by observer 1 greater than, less than, or equal to the frequency heard by observer 2?

Section Review

- The observed frequency of sound depends on the relative motion of the source and the observer. The observed frequency is higher when the source and observer approach one another, and lower when they separate from one another. This is referred to as the Doppler effect.

- The observed frequency f' from a source of sound with a frequency f is
  $$f' = \left(\frac{1 \pm u_o/v}{1 \mp u_s/v}\right)f,$$
  where u_o is the speed of the observer and u_s is the speed of the source. The plus and minus signs are chosen to give a higher frequency when the observer and the source are approaching one another, and to give a lower frequency when they are separating.

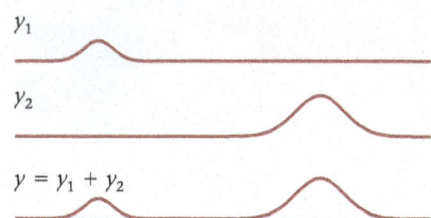

▲ **FIGURE 14-24 Wave superposition**
Waves of small amplitude superpose (that is, combine) by simple addition.

Big Idea 5 Waves that overlap one another add together algebraically, which can result in either an increase or decrease in amplitude. This is referred to as interference.

PHYSICS IN CONTEXT
Looking Ahead

We return to the wave properties of light in Chapter 28, where we see that superposition and interference play a similar role for light as they do for sound.

PHYSICS IN CONTEXT
Looking Ahead

Another type of wave behavior is known as diffraction. This concept is developed in Chapter 28.

14-7 Superposition and Interference

So far we've considered only a single wave at a time. In this section we turn our attention to the way waves combine when more than one is present. As we shall see, the behavior of waves is quite simple in this respect.

Superposition

The combination of two or more waves to form a resultant wave is referred to as **superposition**. When waves are of small amplitude, they superpose in the simplest of ways—they just add. For example, consider two waves on a string, as in **FIGURE 14-24**, described by the wave functions y_1 and y_2. If these two waves are present on the same string at the same time, the result is a wave given by

$$y = y_1 + y_2$$

The sum $y = y_1 + y_2$ is referred to as an algebraic sum, because y_1 and y_2 can each be either positive or negative.

To see how superposition works as a function of time, consider a string with two wave pulses on it, one traveling in each direction as shown in **FIGURE 14-25 (a)**. When the pulses arrive in the same region, they add, as stated. This is illustrated in Figure 14-25(a). The question is: What do the pulses look like after they have passed through one another? Does their interaction change them in any way?

The answer is that the waves are completely unaffected by their interaction. This is also shown in Figure 14-25 (a). After the wave pulses pass through one another they continue on as if nothing had happened. It's like listening to an orchestra, where many different instruments are playing simultaneously, and their sounds are combining throughout the concert hall. Even so, you can still hear individual instruments, each making its own sound as if the others were not present.

Interference

As simple as the principle of superposition is, it still leads to interesting consequences. For example, consider the wave pulses on a string shown in Figure 14-25 (a). When they combine, the resulting pulse has an amplitude equal to the sum of the amplitudes of the individual pulses. This is referred to as **constructive interference**.

On the other hand, two pulses like those in **FIGURE 14-25 (b)** may combine. When this happens, the positive displacement of one wave adds to the negative displacement

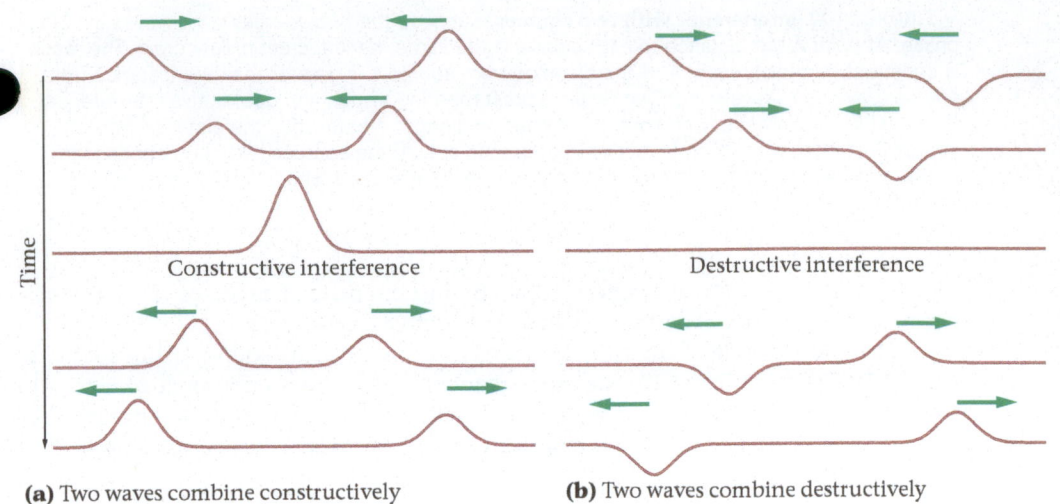

(a) Two waves combine constructively

(b) Two waves combine destructively

◄ **FIGURE 14-25 Interference** Wave pulses superpose as they pass through one another. Afterward, the pulses continue on unchanged. In **(a)**, an example of *constructive interference*, two positive pulses combine to give a larger amplitude. In **(b)**, an example of *destructive interference*, a positive pulse and a symmetrical negative pulse momentarily cancel each other.

of the other to create a net displacement of zero. That is, the pulses momentarily cancel one another. This is **destructive interference**.

It's important to note that the waves don't simply disappear when they experience destructive interference. For example, in Figure 14-25 (b) the wave pulses continue on unchanged after they interact. This makes sense from an energy point of view—after all, each wave carries energy, and hence the waves, along with their energy, cannot simply vanish. In fact, when the string is flat in Figure 14-25(b) it has its greatest transverse speed—just as a swing has its highest speed when it is in its equilibrium position. Therefore, the energy of the wave is still present at this instant of time—it is just in the form of the kinetic energy of the string.

Interference Patterns It should also be noted that interference is not limited to waves on a string; all waves exhibit interference effects. In fact, you could say that interference is one of the key characteristics that define waves. In general, when waves combine, they form **interference patterns** that include regions of both constructive and destructive interference. An example is shown in **FIGURE 14-26**, where two circular waves are interfering. Notice the regions of constructive interference separated by regions of destructive interference.

To understand the formation of such patterns, consider a system of two identical sources, as in **FIGURE 14-27**. Each source sends out waves consisting of alternating crests and troughs. We set up the system so that when one source emits a crest, the other emits a crest as well. Sources that are synchronized like this are said to be **in phase**.

Now, at a point like A, the distance to each source is the same. Thus, if the wave from one source produces a crest at point A, so too does the wave from the other source. As a result, with crest combining with crest, the interference at A is constructive.

Next consider point B. At this location the wave from source 1 must travel a greater distance than the wave from source 2. If the extra distance is half a wavelength, it follows that when the wave from source 2 produces a crest at B the wave from source 1 produces a trough. As a result, the waves combine to give destructive interference at B. At point C, on the other hand, the distance from source 1 is one wavelength greater than the distance from source 2. Hence the waves are in phase again at C, with crest meeting crest for constructive interference.

In general, then, we can say that constructive and destructive interference occur under the following conditions for two sources that are in phase:

Constructive interference occurs when the path length from two in-phase sources differs by $0, \lambda, 2\lambda, 3\lambda, \ldots$.

Destructive interference occurs when the path length from two in-phase sources differs by $\lambda/2, 3\lambda/2, 5\lambda/2, \ldots$.

Constructive interference along this line, where crest meets crest

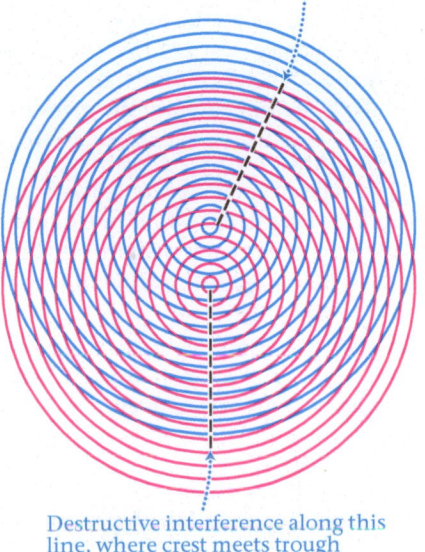

Destructive interference along this line, where crest meets trough

▲ **FIGURE 14-26 Interference of circular waves** Interference pattern formed by the superposition of two sets of circular waves. The light radial "rays" are regions where crest meets crest and trough meets trough (constructive interference). The dark areas in between the light rays are regions where the crest of one wave overlaps the trough of another wave (destructive interference).

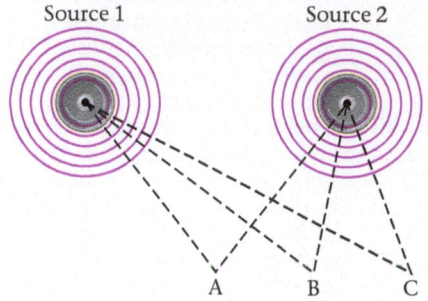

Source 1   Source 2

A   B   C

◀ **FIGURE 14-27 Interference with two sources** Suppose the two sources emit waves in phase. At point A the distance to each source is the same; hence, crest meets crest. The result is constructive interference. At B the distance from source 1 is greater than the distance from source 2 by half a wavelength. The result is crest meeting trough and destructive interference. Finally, at C the distance from source 1 is one wavelength greater than the distance from source 2. Hence, we find constructive interference at C. If the sources had been opposite in phase, then A and C would be points of destructive interference, and B would be a point of constructive interference.

Other path length differences result in intermediate degrees of interference, between the extremes of destructive and constructive interference.

A specific example of interference patterns is provided by sound, using speakers that emit sound in phase with the same frequency. This situation is analogous to the two sources in Figure 14-27. As a result, constructive or destructive interference is to be expected, depending on the path length from each speaker. This is illustrated in the next Example.

EXAMPLE 14-13   SOUND OFF

Two speakers separated by a distance of 4.30 m emit sound of frequency 221 Hz. The speakers are in phase with one another. A person listens from a location 2.80 m directly in front of one of the speakers. Does the person hear constructive or destructive interference?

PICTURE THE PROBLEM

Our sketch shows the two speakers emitting sound in phase at the frequency $f = 221$ Hz. The speakers are separated by the distance $D = 4.30$ m, and the observer is a distance $d_1 = 2.80$ m directly in front of one of the speakers.

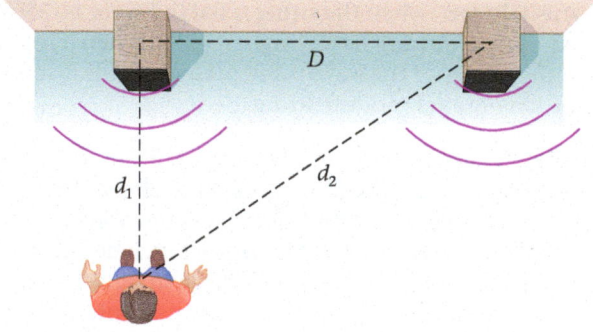

REASONING AND STRATEGY

The type of interference depends on whether the difference in path length, $d_2 - d_1$, is one or more wavelengths or an odd multiple of half a wavelength. Thus, we begin by calculating the wavelength, λ. Next, we find d_2, and compare the difference in path length to λ.

Known   Separation of speakers, $D = 4.30$ m; frequency of speakers, $f = 221$ Hz; distance from speaker 1 to person, $d_1 = 2.80$ m.

Unknown   Type of interference = ?

SOLUTION

1. Calculate the wavelength of this sound, using $v = \lambda f$. As usual, let $v = 343$ m/s be the speed of sound:

$$\lambda = \frac{v}{f} = \frac{343 \text{ m/s}}{221 \text{ Hz}} = 1.55 \text{ m}$$

2. Find the path length d_2:

$$d_2 = \sqrt{D^2 + d_1^2} = \sqrt{(4.30 \text{ m})^2 + (2.80 \text{ m})^2} = 5.13 \text{ m}$$

3. Determine the difference in path length, $d_2 - d_1$:

$$d_2 - d_1 = 5.13 \text{ m} - 2.80 \text{ m} = 2.33 \text{ m}$$

4. Divide λ into $d_2 - d_1$ to find the number of wavelengths that fit into the path difference:

$$\frac{d_2 - d_1}{\lambda} = \frac{2.33 \text{ m}}{1.55 \text{ m}} = 1.50$$

INSIGHT

The path difference is $3\lambda/2$, and hence we expect destructive interference. In the ideal case, the person would hear no sound. As a practical matter, some sound will be reflected from objects in the vicinity, resulting in a finite sound intensity.

PRACTICE PROBLEM

We know that 221 Hz gives destructive interference. What is the lowest frequency (corresponding to the largest wavelength) that gives constructive interference for the case described in this Example?   [**Answer:** Set $\lambda = d_2 - d_1 = 2.33$ m, the largest wavelength that gives constructive interference. This yields $f = 147$ Hz.]

Some related homework problems: Problem 57, Problem 59

RWP If the two speakers in Figure 14-27 have **opposite phase** (by connecting one of them with its wires reversed), the conditions for constructive and destructive interference are changed, as are the interference patterns. For example, at point A, where the distances from the two speakers are the same, the wave from one speaker is a compression when the wave from the other speaker is a rarefaction. Thus, point A is now a point of destructive interference rather than constructive interference. In general, then, the conditions for constructive and destructive interference are simply reversed—a path difference of $0, \lambda, 2\lambda, \ldots$ results in destructive interference, a path difference of $\lambda/2, 3\lambda/2, \ldots$ results in constructive interference.

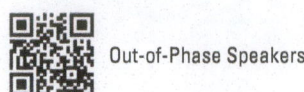

Out-of-Phase Speakers

QUICK EXAMPLE 14-14   OPPOSITE PHASE INTERFERENCE

The speakers shown in our sketch have opposite phase. They are separated by a distance of 5.20 m and emit sound with a frequency of 104 Hz. A person stands 3.00 m in front of the speakers and 1.30 m to one side of the center line between them. What type of interference occurs at the person's location?

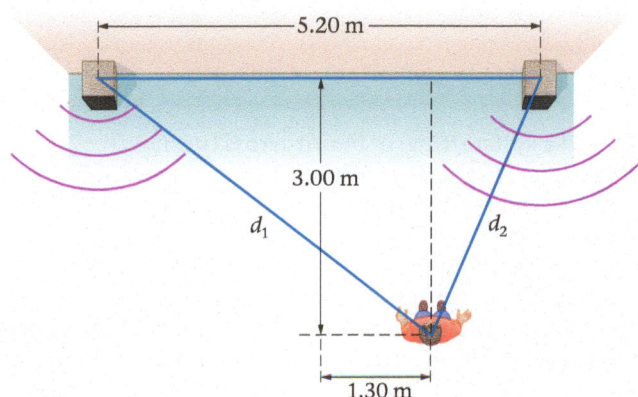

REASONING AND SOLUTION

We start by finding the wavelength of the sound and the distance to the person from each speaker. If the difference in distances is a multiple of the wavelength of the sound, the interference is destructive (because the speakers are opposite in phase); if the difference in distances is an odd multiple of half a wavelength, the interference is constructive.

1. Calculate the wavelength of the sound: $\lambda = 3.30 \text{ m}$

2. Find the path length d_1: $d_1 = 4.92 \text{ m}$

3. Find the path length d_2: $d_2 = 3.27 \text{ m}$

4. Calculate the path length difference, $d_1 - d_2$: $d_1 - d_2 = 1.65 \text{ m}$

5. Divide the path length difference by the wavelength: $\dfrac{d_1 - d_2}{\lambda} = 0.500$

The path length difference is half a wavelength, and the speakers have opposite phase. As a result, the person experiences constructive interference.

RWP Destructive interference can be used to reduce the intensity of noise in a variety of situations, such as a factory, a busy office, or even the cabin of an airplane. The process, referred to as active noise reduction (ANR), begins with a microphone that picks up the noise to be reduced. The signal from the microphone is then reversed in phase and sent to a speaker. As a result, the speaker emits sound that is opposite in phase to the incoming noise—in effect, the speaker produces "anti-noise." In this way, the noise is *actively* canceled by destructive interference, rather than simply reduced by absorption. The effect when wearing a pair of ANR headphones can be as much as a 30-dB reduction in the intensity level of noise.

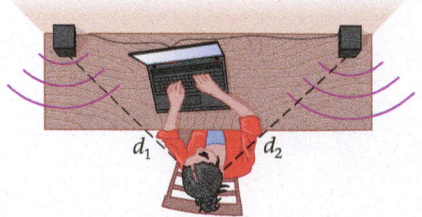

▲ **FIGURE 14-28** Enhance Your Understanding 7.

7. Two in-phase speakers emit sound with a wavelength of 2 m. An observer is a distance d_1 from one speaker and a distance d_2 from the other speaker, as shown schematically in **FIGURE 14-28**. For each of the following cases, state whether the observer experiences constructive or destructive interference: **(a)** $d_1 = 2$ m, $d_2 = 1$ m; **(b)** $d_1 = 4$ m, $d_2 = 2$ m; **(c)** $d_1 = 4$ m, $d_2 = 1$ m; **(d)** $d_1 = 6$ m, $d_2 = 1$ m.

Section Review

• Waves add by simple addition.

• If the crest of one wave meets the crest of another wave, the result is an increased amplitude—referred to as constructive interference. If the crest of one wave meets the trough of another wave, the result is a decreased amplitude—referred to as destructive interference.

14-8 Standing Waves

If you've ever plucked a guitar string, or blown across the mouth of a pop bottle to create a tone, you've generated **standing waves**. In general, a standing wave is one that oscillates with time, but remains fixed in its location. It's in this sense that the wave is said to be "standing."

In some respects, a standing wave can be considered as resulting from constructive interference of a wave with itself. As one might expect, then, standing waves occur only if specific conditions are satisfied. We explore these conditions in this section for two cases: (i) waves on a string and (ii) sound waves in a hollow, cylindrical structure.

Waves on a String

We begin by considering a string of length L that is tied down at both ends, as in **FIGURE 14-29 (a)**. If you pluck this string in the middle it vibrates as shown in **FIGURE 14-29 (b)**. This is referred to as the **fundamental mode** of vibration for this string or, also, as the **first harmonic**. Clearly, the string assumes a wavelike shape, but because of the boundary conditions—the ends tied down—the wave stays in place.

As is clear from **FIGURE 14-29 (c)**, the fundamental mode corresponds to half a wavelength of a usual wave on a string. One can think of the fundamental as being formed by this wave reflecting back and forth between the walls holding the string. If the frequency is just right, the reflections combine to give constructive interference and the fundamental is formed; if the frequency differs from the fundamental frequency, the reflections result in destructive interference and a standing wave does not result.

We can find the frequency of the fundamental, or first harmonic, as follows: First we use the fact that the wavelength of the first harmonic is twice the distance between the walls. Thus,

$$\lambda_1 = 2L$$

If the speed of waves on the string is v, it follows that the frequency of the first harmonic, f_1, is determined by the speed relationship $v = \lambda f$. In this case, $v = \lambda_1 f_1 = (2L)f_1$. Solving for the frequency yields

$$f_1 = \frac{v}{\lambda_1} = \frac{v}{2L}$$

Notice that the frequency of the first harmonic increases with the speed of the waves, and decreases as the string is lengthened.

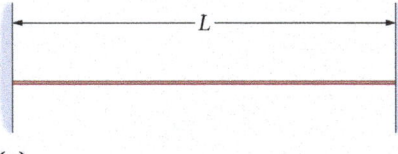

(a)

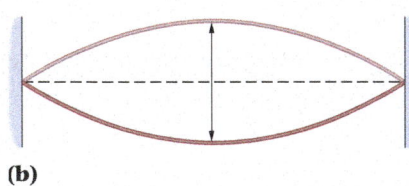

(b)

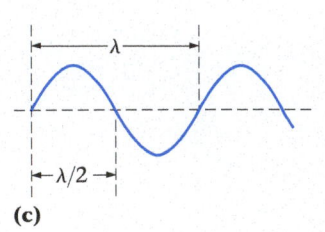

(c)

▲ **FIGURE 14-29 A standing wave (a)** A string is tied down at both ends. **(b)** If the string is plucked in the middle, a standing wave results. This is the fundamental mode of oscillation of the string. **(c)** The fundamental consists of one-half $\left(\frac{1}{2}\right)$ a wavelength between the two ends of the string. Hence, its wavelength is $2L$.

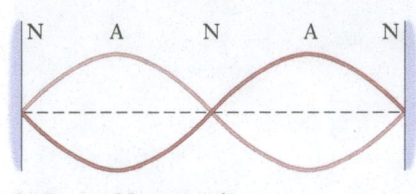

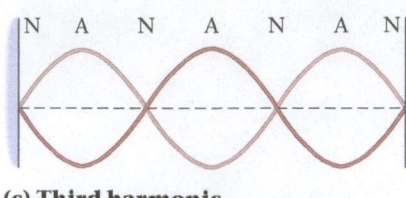

(a) First harmonic (fundamental)   **(b) Second harmonic**   **(c) Third harmonic**

▲ **FIGURE 14-30 Harmonics** The first three harmonics for a string tied down at both ends. Note that an extra half wavelength is added to go from one harmonic to the next. **(a)** $\lambda/2 = L$, $\lambda = 2L$; **(b)** $\lambda = L$; **(c)** $3\lambda/2 = L$, $\lambda = 2L/3$.

Higher Harmonics The first harmonic is not the only standing wave that can exist on a string. In fact, there are an infinite number of standing wave modes—or **harmonics**— for any given string, with frequencies that are integer multiples of the first harmonic. To find the higher harmonics, first note that the two ends of the string must remain fixed. Points on a standing wave that stay fixed are referred to as **nodes**. Halfway between any two nodes is a point on the wave that has a maximum displacement, as indicated in **FIGURE 14-30** and in the photo in **FIGURE 14-31**. Such a point is called an **antinode**. Referring to Figure 14-30 (a), then, we see that the first harmonic consists of two nodes (N) and one antinode (A); the sequence is N-A-N.

The *second* harmonic can be constructed by including one more half wavelength in the standing wave, as shown in Figure 14-30 (b). This mode has the sequence N-A-N-A-N, and has one complete wavelength between the walls; that is, $\lambda_2 = L$. Therefore, the frequency of the second harmonic, f_2, is

$$f_2 = \frac{v}{\lambda_2} = \frac{v}{L} = 2f_1$$

Similarly, the *third* harmonic again includes one more half wavelength, as in Figure 14-30 (c). Now there are one and a half wavelengths in the length L, and hence $(3/2)\lambda_3 = L$, or $\lambda_3 = 2L/3$. The corresponding third-harmonic frequency, f_3, is

$$f_3 = \frac{v}{\lambda_3} = \frac{v}{\frac{2}{3}L} = 3\frac{v}{2L} = 3f_1$$

Notice that the frequencies of the harmonics are increasing in integer steps; that is, each harmonic has a frequency that is an integer multiple of the first-harmonic frequency. Clearly, then, the sequence of standing waves can be characterized as follows:

Standing Waves on a String

The first harmonic (fundamental) has the following frequency and wavelength:

$$f_1 = \frac{v}{2L}$$

$$\lambda_1 = 2L \qquad\qquad 14\text{-}12$$

In general, the nth harmonic, with $n = 1, 2, 3, \ldots$, has the following frequency and wavelength:

$$f_n = nf_1 = n\frac{v}{2L} \qquad n = 1, 2, 3, \ldots$$

$$\lambda_n = \lambda_1/n = \frac{2L}{n} \qquad n = 1, 2, 3, \ldots \qquad 14\text{-}13$$

The integer n in these expressions represents the number of half wavelengths in the standing waves. In addition, notice that the difference in frequency between any two successive harmonics is equal to the first-harmonic frequency, f_1.

Waves on a string provide a one-dimensional example of standing waves. Standing waves can also exist in two- and three-dimensional systems. An example of a standing wave in two dimensions is shown in **FIGURE 14-32**.

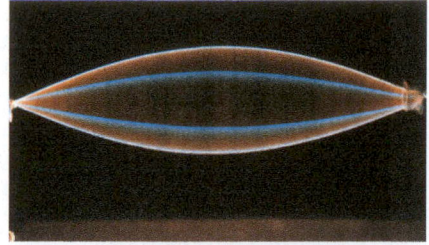

(a)

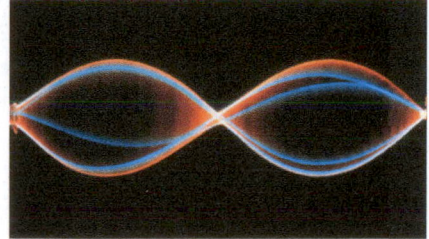

(b)

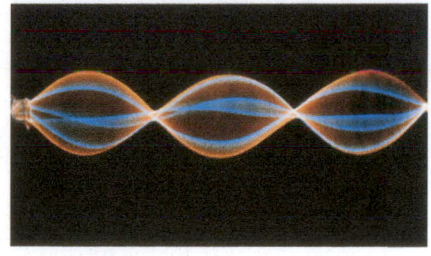

(c)

▲ **FIGURE 14-31** The string in these multiflash photographs vibrates in one of three different standing wave patterns, each with its own characteristic frequency. **(a)** The lowest frequency standing wave— also referred to as the fundamental, or first harmonic—is shown here. In this case, there is one antinode. **(b)** The second harmonic has two antinodes. **(c)** The third harmonic has three antinodes.

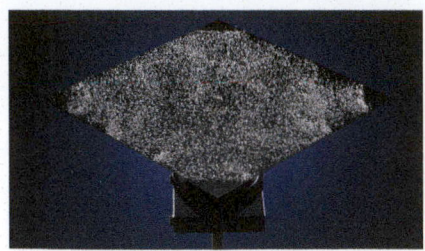 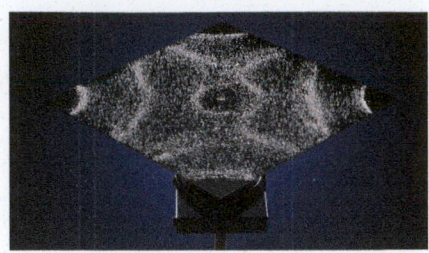 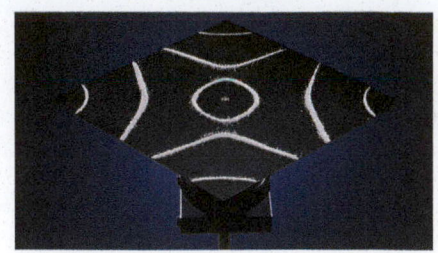

▲ **FIGURE 14-32** The photos show a time sequence (from left to right) as a square metal plate that is initially at rest is vibrated vertically about its center. Initially the plate is covered with a uniform coating of salt crystals. As the plate is vibrated, however, a standing wave develops. The salt makes the wave pattern visible by collecting at the nodes, where the plate is at rest. Clearly, standing waves on a two-dimensional plate can be much more complex than the standing waves on a string.

EXAMPLE 14-15 IT'S FUNDAMENTAL

One of the harmonics on a string 1.30 m long has a frequency of 15.60 Hz. The next higher harmonic has a frequency of 23.40 Hz. Find **(a)** the fundamental frequency and **(b)** the speed of waves on this string.

PICTURE THE PROBLEM

The problem statement does not tell us directly which two harmonics have the given frequencies. We do know, however, that they are *successive* harmonics of the string. For example, if one harmonic has one node between the two ends of the string, the next harmonic has two nodes. Our sketch illustrates this case, which turns out to be appropriate for this problem.

REASONING AND STRATEGY

a. We know from $f_n = nf_1 = n(v/2L)$ (Equation 14-13) that the frequencies of successive harmonics increase by f_1. That is, $f_2 = f_1 + f_1 = 2f_1$, $f_3 = f_2 + f_1 = 3f_1$, $f_4 = f_3 + f_1 = 4f_1, \ldots$. Therefore, we can find the fundamental frequency, f_1, by taking the difference between the given frequencies.

b. Once the fundamental frequency is determined, we can find the speed of waves in the string from the relationship $f_1 = v/2L$.

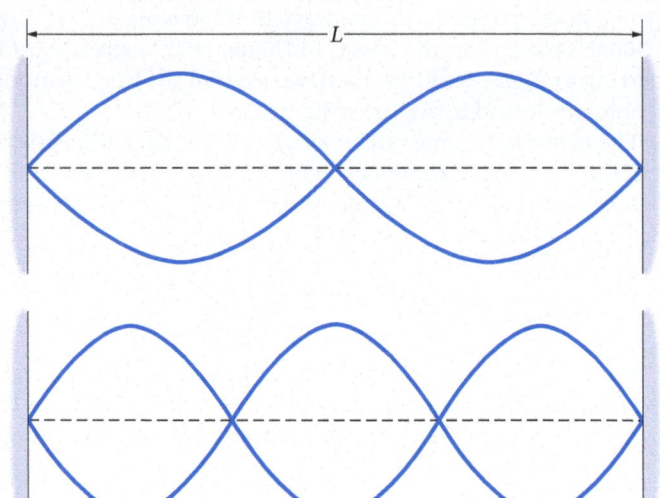

Known Length of string, $L = 1.30$ m; one harmonic of the string = 15.60 Hz; next higher harmonic of the string = 23.40 Hz.

Unknown **(a)** Fundamental frequency of the string, $f_1 = $? **(b)** Speed of waves on string, $v = $?

SOLUTION

Part (a)

1. The fundamental frequency is the difference between the two given frequencies:

$$f_1 = 23.40\,\text{Hz} - 15.60\,\text{Hz} = 7.80\,\text{Hz}$$

Part (b)

2. Solve $f_1 = v/2L$ for the speed, v:

$$f_1 = \frac{v}{2L}$$
$$v = 2Lf_1$$

3. Substitute numerical values:

$$v = 2Lf_1 = 2(1.30\,\text{m})(7.80\,\text{Hz}) = 20.3\,\text{m/s}$$

INSIGHT

Now that we know the fundamental frequency, we can identify the harmonics given in the problem statement. First, 15.6 Hz = 2(7.80 Hz), so this is the second harmonic. The next mode, 23.4 Hz = 3(7.80 Hz), is the third harmonic, as expected.

PRACTICE PROBLEM

Suppose the tension in this string is increased until the speed of the waves is 22.0 m/s. What are the frequencies of the first three harmonics in this case? [**Answer:** $f_1 = 8.46$ Hz, $f_2 = 16.9$ Hz, $f_3 = 25.4$ Hz]

Some related homework problems: Problem 69, Problem 71

Musical Instruments   When a guitar string is plucked or a piano string is struck, it vibrates primarily in its fundamental mode, with smaller contributions coming from the higher harmonics. It follows that notes of different pitch can be produced by using strings of different length. Recalling that the fundamental frequency for a string of length L is $f_1 = v/2L$, we see that long strings produce low frequencies and short strings produce high frequencies—all other variables remaining the same.

RWP This fact accounts for the general shape of a piano. Notice that the strings shorten toward the right side of the piano, where the notes are of higher frequency. Similarly, a double bass is a larger instrument with longer strings than a violin, as one would expect by the different frequencies the instruments produce. To tune a stringed instrument, the tension in the strings is adjusted—since changing the length of the instrument is impractical. This in turn varies the speed v of waves on the string, and hence the fundamental frequency $f_1 = v/2L$ can be adjusted as desired. Once the string is tuned, a performer can produce different notes by pressing the string against a fret to change its effective length—that is, the length that is free to vibrate.

RWP The human ear perceives frequencies as being equally spaced when they are related to one another by a multiplicative factor. For instance, middle C on the piano has a frequency of 261.7 Hz. If we move up one octave to the next C, the frequency is 523.3 Hz; up one more octave, the next C is 1047 Hz. Notice that with each octave the frequency doubles; that is, it goes up by a multiplicative factor of two. Since there are 12 semitones in one octave of the chromatic scale, it follows that the frequency increase from one semitone to the next is $(2)^{1/12}$. The frequencies for a full chromatic octave are given in Table 14-3.

TABLE 14-3 Chromatic Musical Scale

Note	Frequency (Hz)
Middle C	261.7
C♯ (C-sharp)	
D♭ (D-flat)	277.2
D	293.7
D♯, E♭	311.2
E	329.7
F	349.2
F♯, G♭	370.0
G	392.0
G♯, A♭	415.3
A	440.0
A♯, B♭	466.2
B	493.9
C	523.3

▲ **FIGURE 14-33** Three factors determine the pitch of a vibrating string: mass per unit length, μ; tension, F; and length, L. In a guitar, the first of these factors is fixed once the strings are put on the guitar. The strings vary in thickness; other things being equal, the heavier the string, the lower the pitch. The second factor, the tension, can be varied by twisting the pegs at the top of the neck (left), adjusting the pitch of each "open" string to its correct value. The third factor, the length of the string, is controlled by the musician pressing a string against one of the frets (right), changing its effective length—the length of string that is free to vibrate—and thus the note that is produced.

RWP On a guitar, a performer can produce two full octaves and more on a single string by pressing the string down against frets to change its effective length. The separation between frets is not uniform, however, as can be seen in **FIGURE 14-33**. In particular, suppose the unfretted string has a fundamental frequency of 250 Hz. Noting that one octave up on the scale would be twice the frequency, 500 Hz, the length of the string must be halved to produce that note. To go to the next octave, and double the frequency again to 1000 Hz, the string must be shortened by a factor of two again, to one-quarter of its original length. This is illustrated in **FIGURE 14-34**. Because the distance between successive octaves is decreasing—in this case from $L/2$ to $L/4$—but the number of notes in an octave stays the same, it follows that the spacing between frets must decrease as one goes to higher notes. As a result, the frets on a guitar are always more closely spaced as one moves toward the base of the neck.

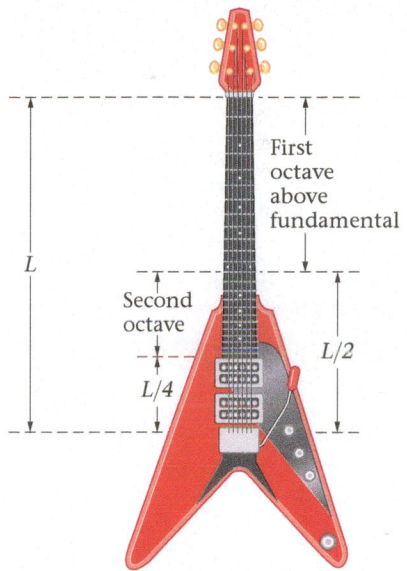

▲ **FIGURE 14-34 Frets on a guitar** To go up one octave from the fundamental, the effective length of a guitar string must be halved. To increase one more octave, it is necessary to halve the length of the string again. Thus, the distance between frets is not uniform; they are more closely spaced near the base of the neck.

Vibrating Columns of Air

If you blow across the open end of a pop bottle, as in **FIGURE 14-35**, you hear a tone of a certain frequency. If you pour some water into the bottle and repeat the experiment,

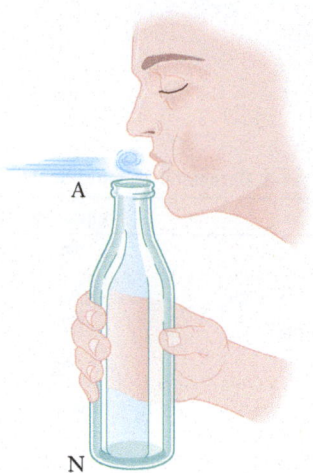

▲ FIGURE 14-35 Exciting a standing wave When air is blown across the open top of a soda pop bottle, the turbulent air flow can cause an audible standing wave. The standing wave will have an antinode, A, at the top (where the air is moving) and a node, N, at the bottom (where the air cannot move).

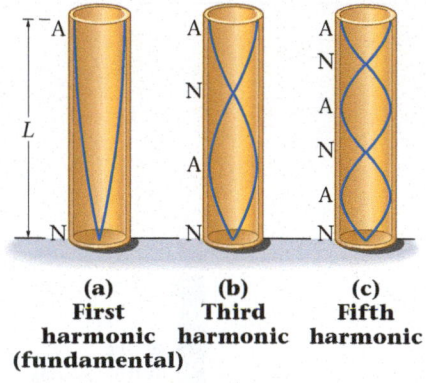

(a)	(b)	(c)
First	**Third**	**Fifth**
harmonic	**harmonic**	**harmonic**
(fundamental)		

▲ FIGURE 14-36 Standing waves in a pipe that is open at one end The first three harmonics for waves in a column of air of length L that is open at one end: **(a)** $\lambda/4 = L, \lambda = 4L$; **(b)** $3\lambda/4 = L, \lambda = 4L/3$; **(c)** $5\lambda/4 = L, \lambda = 4L/5$.

the sound you hear has a higher frequency. In both cases you have excited the fundamental mode of the column of air within the bottle. When water was added to the bottle, however, the column of air was shortened, leading to a higher frequency—in the same way that a shortened string has a higher frequency.

Columns Closed at One End Let's examine the situation of an empty bottle more carefully. When you blow across the opening in the bottle, the result is a swirling movement of air that excites rarefactions and compressions, as illustrated in Figure 14-35. For this reason, the opening is an antinode (A) for sound waves. In contrast, the bottom of the bottle is closed, preventing movement of the air; hence, it must be a node (N). Any standing wave in the bottle must have a node at the bottom and an antinode at the top.

The lowest frequency standing wave that is consistent with these conditions is shown in **FIGURE 14-36 (a)**. If we plot the density variation of the air for this wave, we see that one-quarter of a wavelength fits into the column of air in the bottle. Thus, if the length of the bottle is L, the first harmonic (fundamental) has a wavelength satisfying the following relationship:

$$\tfrac{1}{4}\lambda = L$$
$$\lambda = 4L$$

The first-harmonic frequency, f_1, is given by

$$v = \lambda f_1$$

Solving for f_1 we find

$$f_1 = \frac{v}{\lambda} = \frac{v}{4L}$$

This is half of the corresponding fundamental frequency for a wave on a string.

Higher Harmonics The next harmonic is produced by adding half a wavelength, just as in the case of the string. Thus, if the fundamental is represented by N-A, the second harmonic can be written as N-A-N-A. Because the distance from a node to an antinode is a quarter of a wavelength, we see that three-quarters $\left(\tfrac{3}{4}\right)$ of a wavelength fits into the bottle for this mode. This is shown in **FIGURE 14-36 (b)**. Therefore, $3\lambda/4 = L$, and hence

$$\lambda = \tfrac{4}{3}L$$

As a result, the frequency is

$$\frac{v}{\lambda} = \frac{v}{\tfrac{4}{3}L} = 3\frac{v}{4L} = 3f_1$$

Notice that this is the *third* harmonic of the pipe; that is, its frequency is three times f_1.

Similarly, the next-higher harmonic is written as N-A-N-A-N-A, as indicated in **FIGURE 14-36 (c)**. In this case, $5\lambda/4 = L$, and the frequency is

$$\frac{v}{\lambda} = \frac{v}{\tfrac{4}{5}L} = 5\frac{v}{4L} = 5f_1$$

This is the *fifth* harmonic of the pipe.

Clearly, the progression of harmonics for a column of air that is closed at one end and open at the other end is described by the following frequencies and wavelengths:

Standing Waves in a Column of Air Closed at One End

$$f_1 = \frac{v}{4L}$$
$$f_n = nf_1 = n\frac{v}{4L} \quad n = 1, 3, 5, \ldots$$
$$\lambda_n = \frac{\lambda_1}{n} = \frac{4L}{n} \quad n = 1, 3, 5, \ldots$$

14-14

Only odd harmonics are present in this case, as opposed to waves on a string, in which all integer harmonics occur.

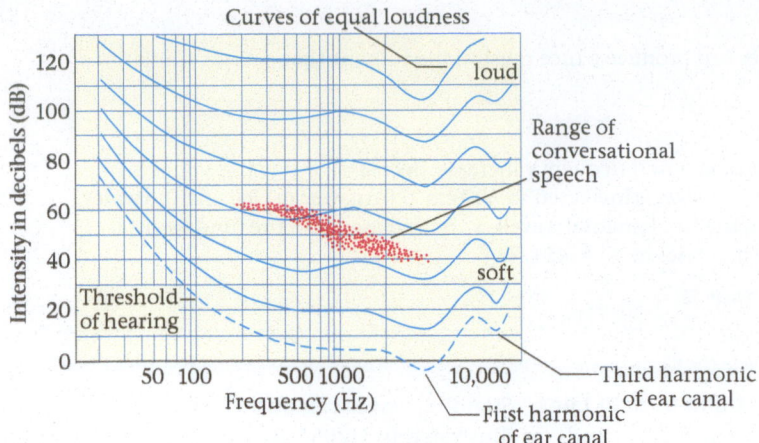

Curves of equal loudness

loud

Range of conversational speech

soft

Threshold of hearing

Third harmonic of ear canal

First harmonic of ear canal

◄ **FIGURE 14-37 Human response to sound** The human ear is more sensitive to some frequencies of sound than to others. For example, every point on a "curve of equal loudness" seems just as loud to us as any other point, even though the corresponding physical intensities may be quite different. To illustrate, note that the threshold of hearing is not equal to 0 dB for all frequencies. In fact, it is approximately 25 dB at 100 Hz, about 5 dB at 1000 Hz, and even slightly negative near 3500 Hz. Thus, regions where the curves dip downward correspond to increased sensitivity of the ear—in fact, near 3500 Hz we can hear sounds that are a thousand times less intense than sounds at 100 Hz. The two most prominent dips occur near 3500 Hz and 11,000 Hz, corresponding to standing waves in the ear canal analogous to those shown in Figure 14-36 (a) and (b), respectively. (See Problem 68.)

RWP The human ear canal is an example of a column of air that is closed at one end (the eardrum) and open at the other end. Standing waves in the ear canal can lead to an increased sensitivity of hearing. This is illustrated in **FIGURE 14-37**, which shows "curves of equal loudness" as a function of frequency. Where these curves dip downward, sounds of lower intensity seem just as loud as sounds of higher intensity at other frequencies. The two prominent dips near 3500 Hz and 11,000 Hz are due to standing waves in the ear canal corresponding to Figures 14-36 (a) and (b), respectively.

EXAMPLE 14-16 POP MUSIC

An empty soda pop bottle is to be used as a musical instrument in a band. In order to be tuned properly, the fundamental frequency of the bottle must be 440.0 Hz. **(a)** If the bottle is 26.0 cm tall, how high should it be filled with water to produce the desired frequency? Treat the bottle as a pipe that is closed at one end (the surface of the water) and open at the other end. **(b)** What is the frequency of the next higher harmonic for this bottle?

PICTURE THE PROBLEM
In our sketch, we label the height of the bottle with H = 26.0 cm, and the unknown height of water with h. Clearly, then, the length of the vibrating column of air is $L = H - h$.

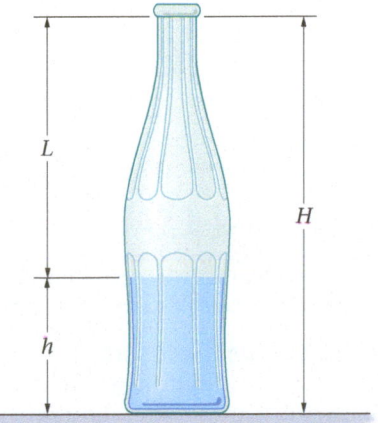

REASONING AND STRATEGY
a. Given the frequency of the fundamental (f_1 = 440.0 Hz) and the speed of sound in air (v = 343 m/s), we can use $f_1 = v/4L$ to solve for the length L of the air column. The height of water is then $h = H - L$.

b. The next higher harmonic for a pipe open at one end is the third harmonic ($n = 3$ in $f_n = n(v/4L)$). Thus, the next higher harmonic frequency for this bottle is $f_3 = 3f_1$.

Known Fundamental frequency of bottle, f_1 = 440.0 Hz; height of bottle, H = 0.260 m.
Unknown (a) Height of water in bottle, h = ? (b) Frequency of next higher harmonic, f_3 = ?

SOLUTION

Part (a)

1. Solve $f_1 = v/4L$ for the length L:

$$f_1 = \frac{v}{4L} \quad \text{or} \quad L = \frac{v}{4f_1}$$

2. Substitute numerical values:

$$L = \frac{v}{4f_1} = \frac{343 \text{ m/s}}{4(440.0 \text{ Hz})} = 0.195 \text{ m}$$

3. Use $h = H - L$ to find the height of the water:

$$h = H - L = 0.260 \text{ m} - 0.195 \text{ m} = 0.065 \text{ m} = 6.5 \text{ cm}$$

Part (b)

4. Calculate the frequency of the third harmonic (the next higher) with $f_3 = 3f_1$:

$$f_3 = 3f_1 = 3(440.0 \text{ Hz}) = 1320 \text{ Hz}$$

CONTINUED

INSIGHT
Filling a series of bottles with different levels of water can produce a nice musical instrument that sounds similar to a xylophone.

PRACTICE PROBLEM — PREDICT/CALCULATE
(a) If the water level in the bottle is raised, does the fundamental frequency increase, decrease, or stay the same? Explain. (b) Find the fundamental frequency if the water level is increased to 7.00 cm. [**Answer: (a)** Raising the water level reduces the length of the air column. As a result, the fundamental wavelength decreases, and the fundamental frequency increases. (b) The fundamental frequency increases to $f_1 = 451$ Hz.]

Some related homework problems: Problem 66, Problem 68, Problem 72

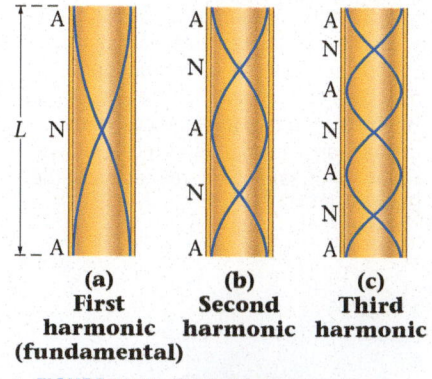

(a)
First harmonic (fundamental)

(b)
Second harmonic

(c)
Third harmonic

▲ **FIGURE 14-38 Standing waves in a pipe that is open at both ends** The first three harmonics for waves in a column of air of length L that is open at both ends: **(a)** $\lambda/2 = L$, $\lambda = 2L$; **(b)** $\lambda = L$; **(c)** $3\lambda/2 = L$, $\lambda = 2L/3$.

Columns Open at Both Ends Standing waves can also be produced in columns of air that are open at both ends, as illustrated in **FIGURE 14-38**. In this case there is an antinode at each end of the column. Hence, the first harmonic, or fundamental, is A-N-A, as shown in Figure 14-38 (a). In this case, half a wavelength fits into the pipe. As a result,

$$f_1 = \frac{v}{2L}$$

This is the same as the corresponding result for a wave on a string.
The next harmonic is A-N-A-N-A, which fits one complete wavelength in the pipe. This harmonic is shown in Figure 14-38 (b), and has the frequency

$$f_2 = \frac{v}{L} = 2f_1$$

This is the *second* harmonic of the pipe. The rest of the harmonics continue in exactly the same manner as for waves on a string, with all integer harmonics present. Thus, the frequencies and wavelengths in a column of air open at both ends are as follows:

Standing Waves in a Column of Air Open at Both Ends

$$f_1 = \frac{v}{2L}$$

$$f_n = nf_1 = n\frac{v}{2L} \quad n = 1, 2, 3, \ldots$$

$$\lambda_n = \frac{\lambda_1}{n} = \frac{2L}{n} \quad n = 1, 2, 3, \ldots$$

14-15

CONCEPTUAL EXAMPLE 14-17   TALKING WITH HELIUM

If you fill your lungs with helium and speak, you sound something like Donald Duck. From this observation, can we conclude that the speed of sound in helium is greater than, less than, or equal to the speed of sound in air?

REASONING AND DISCUSSION
When we speak with helium, our words are higher pitched. Looking at $f_n = n(v/2L)$ (Equation 14-15), we see that for the frequency to increase while the length of the vocal chords remains the same, the speed of sound must be higher.

ANSWER
The speed of sound is greater in helium than in air.

RWP A pipe organ uses a variety of pipes of different length, with some open at both ends and others open at one end only. When a key is pressed on the console of the organ, air is forced through a given pipe. By accurately adjusting the length of the pipe, we can create the desired tone. The principle is the same as playing a note on a pop bottle, as in **FIGURE 14-39**. In addition, because open and closed pipes have different harmonic frequencies, they sound distinctly different to the ear, even if they have the same fundamental frequency. Thus, by judiciously choosing both the length and the type of a pipe, an organist can play a range of different sounds, allowing the organ to mimic a trumpet, a trombone, a clarinet, and so on.

Standing waves have also been observed in the Sun. Like an enormous, low-frequency musical instrument, the Sun vibrates once roughly every five minutes, a result of the roiling nuclear reactions that take place within its core. One of the goals of SOHO, the Solar and Heliospheric Observatory, is to study these solar vibrations in detail. By observing the variety of standing waves produced in the Sun, we can learn more about its internal structure and dynamics.

Enhance Your Understanding   (Answers given at the end of the chapter)

8. When a string oscillates with the standing wave shown in **FIGURE 14-40 (a)** its frequency is 100 Hz. What is the frequency when the same string oscillates with the standing wave shown in **FIGURE 14-40 (b)**?

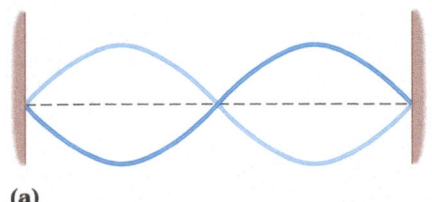 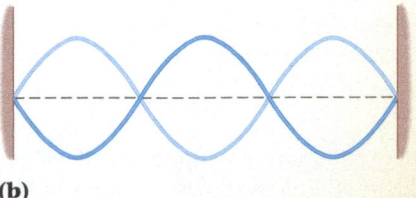

(a) **(b)**

▲ **FIGURE 14-40** Enhance Your Understanding 8.

Section Review

- Standing waves oscillate in place. The standing wave with the largest wavelength and lowest frequency is the fundamental, or first harmonic. Higher harmonics have higher frequencies.

- The standing waves on a string tied down at both ends have frequencies $f_n = n(v/2L)$, where n is the harmonic number, L is the length of the string, and v is the speed of waves on the string.

- The standing waves in a column of air of length L and closed at one end have frequencies $f_n = n(v/4L)$. In this expression, v is the speed of sound in air. The standing waves in a column of air of length L and open at both ends have frequencies $f_n = n(v/2L)$. This is the same expression as for the standing waves on a string.

▲ **FIGURE 14-39 Visualizing Concepts**
Waves in a Column of Air Blowing across the mouth of a bottle sets the air column within the bottle vibrating, producing a tone. This principle is put to use on a large scale in a pipe organ. A typical pipe organ may contain hundreds of pipes of different lengths, some open at both ends and some at only one, affording the performer great control over the tonal quality of the sound produced, as well as its pitch.

14-9 Beats

An interference pattern, such as that shown in Figure 14-26, is a snapshot at a given time, showing locations where constructive and destructive interference occur. It is an interference pattern in space. **Beats**, on the other hand, can be thought of as an interference pattern in time.

To be specific, imagine plucking two guitar strings that have slightly different frequencies. If you listen carefully, you notice that the sound produced by the strings is not constant in time. In fact, the intensity increases and decreases with a definite period. These fluctuations in intensity are the beats, and the frequency of successive maximum intensities is the **beat frequency**.

As an example, suppose two waves, with frequencies $f_1 = 1/T_1$ and $f_2 = 1/T_2$, interfere at a given, fixed location. At this location, each wave moves up and down with simple harmonic motion, as described by Equation 13-2. Applying this result to the vertical position, y, of each wave yields the following:

$$y_1 = A\cos\left(\frac{2\pi}{T_1}t\right) = A\cos(2\pi f_1 t)$$
$$y_2 = A\cos\left(\frac{2\pi}{T_2}t\right) = A\cos(2\pi f_2 t)$$

14-16

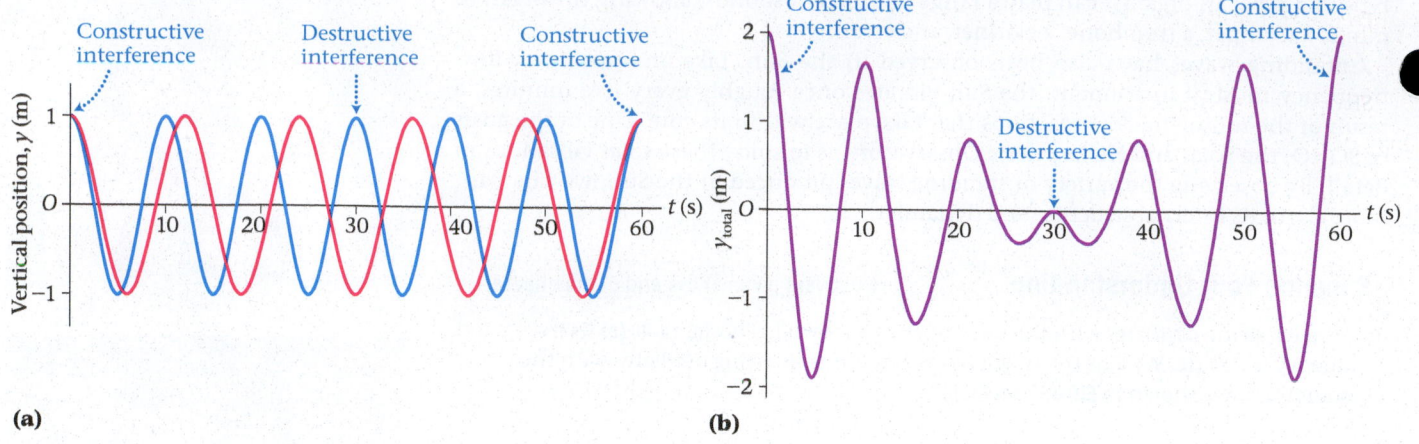

(a)

(b)

▲ **FIGURE 14-41 Interference of two waves with slightly different frequencies (a)** A plot of the two waves, y_1 (blue) and y_2 (red), given in Equations 14-16. **(b)** The resultant wave y_{total} for the two waves shown in part (a). Note the alternately constructive and destructive interference leading to beats.

These equations are plotted in **FIGURE 14-41 (a)**, with $A = 1\,\text{m}$, and their superposition, $y_{total} = y_1 + y_2$, is shown in **FIGURE 14-41 (b)**.

Notice that at the time $t = 0$, both y_1 and y_2 are equal to A; thus their superposition gives $2A$. Because the waves have different frequencies, however, they do not stay in phase. At a later time, t_1, we find that $y_1 = A$ and $y_2 = -A$; their superposition gives zero at this time. At a still later time, $t_2 = 2t_1$, the waves are again in phase and add to give $2A$. Thus, a person listening to these two waves hears a sound whose amplitude and loudness vary with time; that is, the person hears beats.

Superposing these waves mathematically, we find

$$
\begin{aligned}
y_{total} &= y_1 + y_2 \\
&= A\cos(2\pi f_1 t) + A\cos(2\pi f_2 t) \\
&= 2A\cos\left(2\pi\frac{f_1 - f_2}{2}t\right)\cos\left(2\pi\frac{f_1 + f_2}{2}t\right)
\end{aligned}
$$

14-17

Big Idea 6 When two sound waves of different frequency superpose, the result is sound with a loudness that varies with time. The frequency of the loudness variation is known as the beat frequency.

The final step in the expression follows from the trigonometric identities given in Appendix A. The first part of y_{total} is

$$
2A\cos\left(2\pi\frac{f_1 - f_2}{2}t\right)
$$

This gives the slowly varying amplitude of the beats, as indicated in **FIGURE 14-42**. A loud sound is heard whenever this term is equal to $-2A$ or $2A$—which happens twice during any given oscillation—and hence the beat frequency is:

▶ **FIGURE 14-42 Beats** Beats can be understood as oscillations at the average frequency, modulated by a slowly varying amplitude.

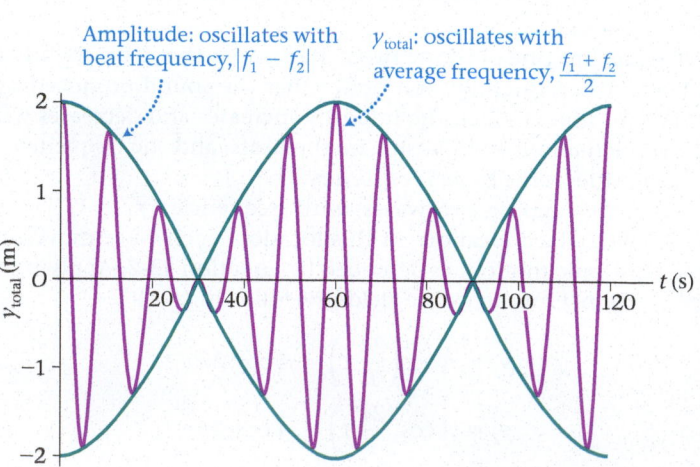

Definition of Beat Frequency

$$f_{\text{beat}} = |f_1 - f_2| \qquad \text{14-18}$$

SI unit: $1/s = s^{-1} = \text{Hz}$

Finally, the rapid oscillations within each beat are due to the second part of y_{total}:

$$\cos\left(2\pi\frac{f_1 + f_2}{2}t\right)$$

These oscillations have a frequency that is the average of the two input frequencies.

These results apply to any type of wave. In particular, if two sound waves produce beats, your ear will hear the average frequency with a loudness that varies with the beat frequency. For example, suppose the two guitar strings mentioned at the beginning of this section have the frequencies 438 Hz and 442 Hz. If you sound them simultaneously, you will hear the average frequency, 440 Hz, increasing and decreasing in loudness with a beat frequency of 4 Hz. This means that you will hear maximum loudness four times a second. If the frequencies are brought closer together, the beat frequency will be reduced, and fewer maxima will be heard each second.

Clearly, beats can be used to tune a musical instrument to a desired frequency. To tune a guitar string to 440 Hz, for example, the string can be played simultaneously with a 440-Hz tuning fork. Listening to the beats, the person tuning the guitar increases or decreases the tension in the string until the beat frequency becomes vanishingly small. This technique applies only to frequencies that are reasonably close to begin with, because the maximum beat frequencies the ear can detect are about 15 Hz to 20 Hz.

PROBLEM-SOLVING NOTE

Calculating the Beat Frequency

The beat frequency of two waves is the *magnitude* of the difference in their frequencies. Thus, the beat frequency is always positive.

EXAMPLE 14-18 GETTING A TUNE-UP

A guitarist tunes one of her strings to a 440-Hz tuning fork. Initially, a beat frequency of 5 Hz is heard when the string and tuning fork are sounded. When the guitarist tightens the string slowly and steadily, the beat frequency decreases to 4 Hz. What was the initial frequency of the string?

PICTURE THE PROBLEM

A guitarist is tightening a string at the same time that a tuning fork is giving off (emitting) sound with a frequency of 440 Hz. A beat frequency is heard, which decreases as the string is tightened.

Known   Frequency of tuning fork, $f = 440$ Hz; initial beat frequency = 5 Hz; final beat frequency = 4 Hz.

Unknown   Initial frequency of guitar string = ?

Beat frequency = 5 Hz   tightening string   Beat frequency = 4 Hz

440 Hz

REASONING AND SOLUTION

The fact that the initial beat frequency is 5 Hz means the initial frequency of the string is either 435 Hz or 445 Hz. Now, tightening the string increases its frequency. Why? Well, tightening the string means the tension is increased, which in turn means the speed of waves in the string is increased (recall $v = \sqrt{F/\mu}$ from Equation 14-2). As a result, the fundamental frequency is also increased (recall $f_1 = v/2L$ from Equation 14-12). The fact that increasing the frequency of the string decreased the beat frequency—bringing it closer to 440 Hz—means the initial frequency was 435 Hz.

INSIGHT

In this case, the initial frequency was too low (435 Hz). If tightening the string had increased the beat frequency, then the initial frequency would have been too high (445 Hz) and the guitarist would have to loosen the string.

CONTINUED

PRACTICE PROBLEM

Suppose that the initial beat frequency was 4 Hz, and that tightening the string caused the beat frequency to increase steadily to 6 Hz. What was the initial frequency of the string?   [**Answer: 444 Hz**]

Some related homework problems: Problem 78, Problem 79

Enhance Your Understanding   (Answers given at the end of the chapter)

9. Rank the following systems in order of increasing beat frequency. Indicate ties where appropriate. **A,** $f_1 = 100\,\text{Hz}$, $f_2 = 110\,\text{Hz}$; **B,** $f_1 = 90\,\text{Hz}$, $f_2 = 95\,\text{Hz}$; **C,** $f_1 = 110\,\text{Hz}$, $f_2 = 100\,\text{Hz}$; **D,** $f_1 = 303\,\text{Hz}$, $f_2 = 300\,\text{Hz}$.

Section Review

- Playing two sounds with different frequencies f_1 and f_2 simultaneously produces beats, which can be heard as an alternation in the loudness of the sound. The frequency of the beats is given by $f_{\text{beat}} = |f_1 - f_2|$.

CHAPTER 14 REVIEW

CHAPTER SUMMARY

14-1   TYPES OF WAVES

A wave is a propagating disturbance.

Transverse Waves and Longitudinal Waves

In a transverse wave individual particles move at right angles to the direction of wave propagation. In a longitudinal wave individual particles move in the same direction as the wave propagation.

Wavelength, Frequency, and Speed

The wavelength, λ, frequency, f, and speed, v, of a wave are related by

$$v = \lambda f \qquad \text{14-1}$$

14-2   WAVES ON A STRING

Transverse waves can propagate on a string held taut with a tension force, F.

Speed of a Wave on a String

The speed of a wave on a string with a tension force F and a mass per length μ is

$$v = \sqrt{\frac{F}{\mu}} \qquad \text{14-2}$$

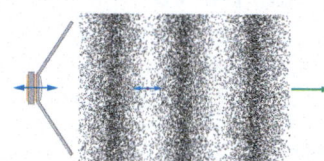

14-3   HARMONIC WAVE FUNCTIONS

A harmonic wave has the shape of a sine or a cosine function.

Wave Function

A harmonic wave of wavelength λ and period T is described by the following expression:

$$y(x, t) = A \cos\left(\frac{2\pi}{\lambda}x - \frac{2\pi}{T}t\right) \qquad \text{14-4}$$

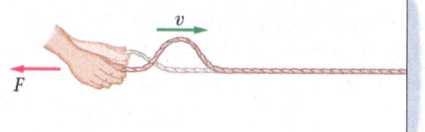
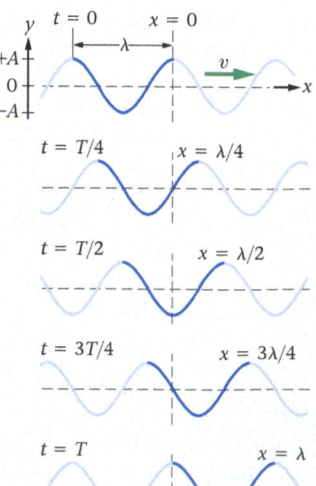

14-4   SOUND WAVES

A sound wave is a longitudinal wave of compressions and rarefactions that can travel through the air, as well as through other gases, liquids, and solids.

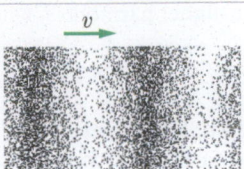

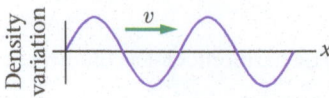

14-5   SOUND INTENSITY

The loudness of a sound is determined by its intensity.

Intensity
Intensity, I, is a measure of the amount of energy per time that passes through a given area. Since energy per time is power, P, the intensity of a wave is

$$I = \frac{P}{A} \qquad \text{14-5}$$

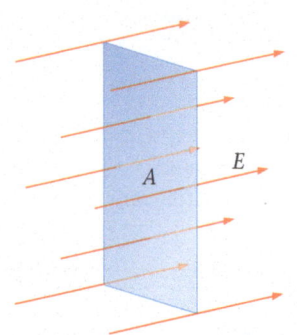

Intensity Level and Decibels
The intensity level, β, of a sound gives an indication of how loud it sounds to our ears. The intensity level is defined as

$$\beta = (10\,\text{dB}) \log\!\left(\frac{I}{I_0}\right) \qquad \text{14-8}$$

The value of β is given in decibels.

14-6   THE DOPPLER EFFECT

The change in frequency due to the relative motion between a source and a receiver is called the Doppler effect.

General Case
If the observer moves with a speed u_o and the source moves with a speed u_s, the Doppler effect gives

$$f' = \left(\frac{1 \pm u_o/v}{1 \mp u_s/v}\right) f \qquad \text{14-11}$$

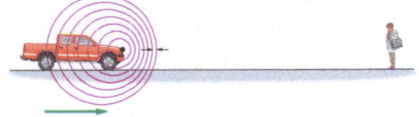

The plus and minus signs are chosen to give a higher frequency when the observer and the source are approaching one another, and to give a lower frequency when they are separating.

14-7   SUPERPOSITION AND INTERFERENCE

Waves can combine to give a variety of effects.

Constructive Interference
Waves that add to give a larger amplitude exhibit constructive interference.

Destructive Interference
Waves that add to give a smaller amplitude exhibit destructive interference.

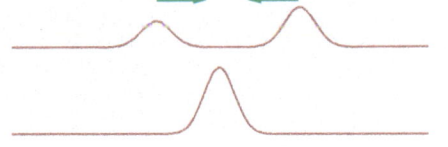

14-8   STANDING WAVES

Standing waves oscillate in a fixed location.

Waves on a String
The fundamental, or first harmonic, corresponds to half a wavelength fitting into the length of the string. The fundamental for waves of speed v on a string of length L is

$$f_1 = \frac{v}{2L}$$
$$\lambda_1 = 2L \qquad \text{14-12}$$

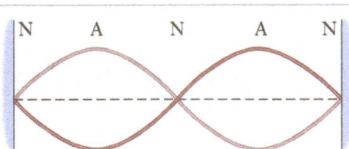

The higher harmonics, with $n = 1, 2, 3, \ldots$, are described by

$$f_n = nf_1 = n(v/2L)$$
$$\lambda_n = \lambda_1/n = 2L/n$$

14-13

Vibrating Columns of Air

The harmonics for a column of air closed at one end are

$$f_n = nf_1 = n(v/4L) \quad n = 1, 3, 5, \ldots$$
$$\lambda_n = \lambda_1/n = 4L/n$$

14-14

The harmonics for a column of air open at both ends are

$$f_n = nf_1 = n(v/2L) \quad n = 1, 2, 3, \ldots$$
$$\lambda_n = \lambda_1/n = 2L/n$$

14-15

In all four of these expressions the speed of sound is v and the length of the column is L.

14-9 BEATS

Beats occur when waves of slightly different frequencies interfere.

Beat Frequency

If waves of frequencies f_1 and f_2 interfere, the beat frequency is

$$f_{\text{beat}} = |f_1 - f_2|$$

14-18

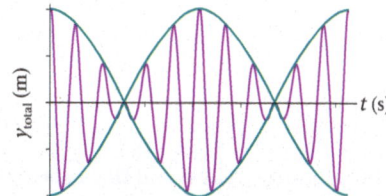

ANSWERS TO ENHANCE YOUR UNDERSTANDING QUESTIONS

1. $A < B = D < C$.

2. The speed doubles.

3. **(a)** $A = 0.12$ m. **(b)** $\lambda = 8$ m. **(c)** $T = 0.25$ s.

4. Wave 2 has the greater speed (25 m/s compared to 8 m/s).

5. **(c)**

6. Observer 1 hears a frequency that is less than the frequency heard by observer 2.

7. **(a)** Destructive. **(b)** Constructive. **(c)** Destructive. **(d)** Destructive.

8. 150 Hz.

9. $D < B < A = C$.

CONCEPTUAL QUESTIONS

For instructor-assigned homework, go to www.masteringphysics.com. **MP**

(Answers to odd-numbered Conceptual Questions can be found in the back of the book.)

1. A long nail has been driven halfway into the side of a barn. How should you hit the nail with a hammer to generate a longitudinal wave? How should you hit it to generate a transverse wave?

2. What type of wave is exhibited by "amber waves of grain"?

3. In a classic TV commercial, a group of cats feed from bowls of cat food that are lined up side by side. Initially there is one cat for each bowl. When an additional cat is added to the scene, it runs to a bowl at the end of the line and begins to eat. The cat that was there originally moves to the next bowl, displacing that cat, which moves to the next bowl, and so on down the line. What type of wave have the cats created? Explain.

4. Describe how the sound of a symphony played by an orchestra would be altered if the speed of sound depended on the frequency of sound.

5. A "radar gun" is often used to measure the speed of a major-league pitch by reflecting a beam of radio waves off a moving ball. Describe how the Doppler effect can give the speed of the ball from a measurement of the frequency of the reflected beam.

6. When you drive a nail into a piece of wood, you hear a tone with each blow of the hammer. In fact, the tone increases in pitch as the nail is driven farther into the wood. Explain.

7. Explain the function of the sliding part of a trombone.

8. On a guitar, some strings are single wires, others are wrapped with another wire to increase the mass per length. Which type of string would you expect to be used for a low-frequency note? Explain.

9. To play a C major chord on the piano, you hit the C, E, and G keys simultaneously. When you do so, you hear no beats. Why? (Refer to Table 14-3.)

PROBLEMS AND CONCEPTUAL EXERCISES

Answers to odd-numbered Problems and Conceptual Exercises can be found in the back of the book. **BIO** *identifies problems of biological or medical interest;* **CE** *indicates a conceptual exercise.* **Predict/Explain** *problems ask for two responses: (a) your prediction of a physical outcome, and (b) the best explanation among three provided; and* **Predict/Calculate** *problems ask for a prediction of a physical outcome, based on fundamental physics concepts, and follow that with a numerical calculation to verify the prediction. On all problems, bullets (•, ••, •••) indicate the level of difficulty. Planetary data can be found in Appendix C. Assume the speed of sound in air is 343 m/s unless stated otherwise.*

SECTION 14-1 TYPES OF WAVES

1. • A wave travels along a stretched horizontal rope. The vertical distance from crest to trough for this wave is 16 cm and the horizontal distance from crest to trough is 22 cm. What are **(a)** the wavelength and **(b)** the amplitude of this wave?

2. • A surfer floating beyond the breakers notes 17 waves per minute passing her position. If the wavelength of these waves is 38 m, what is their speed?

3. • The speed of surface waves in water decreases as the water becomes shallower. Suppose waves travel across the surface of a lake with a speed of 2.0 m/s and a wavelength of 1.5 m. When these waves move into a shallower part of the lake, their speed decreases to 1.6 m/s, though their frequency remains the same. Find the wavelength of the waves in the shallower water.

4. • **Tsunami** A tsunami traveling across deep water can have a speed of 750 km/h and a wavelength of 310 km. What is the frequency of such a wave?

5. •• A stationary boat bobs up and down with a period of 2.3 s when it encounters the waves from a moving boat. **(a)** What is the frequency of the waves? **(b)** If the crests of the waves are 8.5 m apart, what is their speed?

6. •• **Predict/Calculate** A 4.5-Hz wave with an amplitude of 12 cm and a wavelength of 27 cm travels along a stretched horizontal string. **(a)** How far does a given peak on the wave travel in a time interval of 0.50 s? **(b)** How far does a knot on the string travel in the same time interval? **(c)** How would your answers to parts (a) and (b) change if the amplitude of the wave were halved? Explain.

7. •• **Deepwater Waves** The speed of a deepwater wave with a wavelength λ is given approximately by $v = \sqrt{g\lambda/2\pi}$. Find **(a)** the speed and **(b)** the frequency of a deepwater wave with a wavelength of 4.5 m.

8. •• **Shallow-Water Waves** In shallow water of depth d the speed of waves is approximately $v = \sqrt{gd}$. Find **(a)** the speed and **(b)** the period of a wave with a wavelength of 2.6 cm in water that is 0.75 cm deep.

SECTION 14-2 WAVES ON A STRING

9. • **CE** Consider a wave on a string with constant tension. If the frequency of the wave is doubled, by what multiplicative factor do **(a)** the speed and **(b)** the wavelength change?

10. • **CE** Suppose you would like to double the speed of a wave on a string. By what multiplicative factor must you increase the tension?

11. • **CE Predict/Explain** Two strings are made of the same material and have equal tensions. String 1 is thick; string 2 is thin. **(a)** Is the speed of waves on string 1 greater than, less than, or equal to the speed of waves on string 2? **(b)** Choose the *best explanation* from among the following:

   I. Since the strings are made of the same material, the wave speeds will also be the same.

   II. A thick string implies a large mass per length and a slow wave speed.

   III. A thick string has a greater force constant, and therefore a greater wave speed.

12. • **CE Predict/Explain** Two strings are made of the same material and have waves of equal speed. String 1 is thick; string 2 is thin. **(a)** Is the tension in string 1 greater than, less than, or equal to the tension in string 2? **(b)** Choose the *best explanation* from among the following:

   I. String 1 must have a greater tension to compensate for its greater mass per length.

   II. String 2 will have a greater tension because it is thinner than string 1.

   III. Equal wave speeds implies equal tensions.

13. • **CE** The three waves, A, B and C, shown in **FIGURE 14-43** propagate on strings with equal tensions and equal mass per length. Rank the waves in order of increasing **(a)** frequency, **(b)** wavelength, and **(c)** speed. Indicate ties where appropriate.

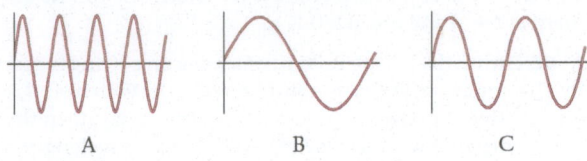

FIGURE 14-43 Problem 13

14. •• A brother and sister try to communicate with a string tied between two tin cans (**FIGURE 14-44**). If the string is 7.6 m long, has a mass of 29 g, and is pulled taut with a tension of 18 N, how much time does it take for a wave to travel from one end of the string to the other?

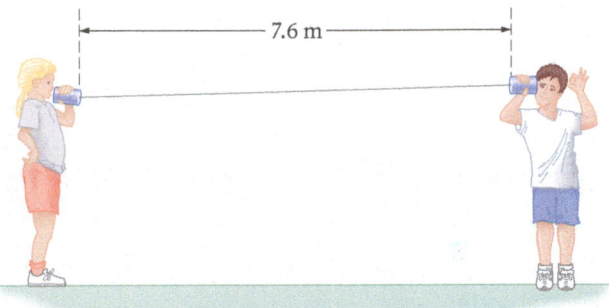

FIGURE 14-44 Problems 14 and 15

15. •• **Predict/Calculate** **(a)** Suppose the tension is increased in the previous problem. Does a wave take more, less, or the same time to travel from one end to the other? Calculate the time of travel for tensions of **(b)** 22 N and **(c)** 32 N.

16. •• **Predict/Calculate** A 5.7-m wire with a mass of 81 g is attached to the mast of a sailboat. If the wire is given a "thunk" at one end, it takes 0.084 s for the resulting wave to reach the other end. **(a)** What is the tension in the wire? **(b)** Would the tension found in part (a) be larger or smaller if the mass of the wire is greater than 81 g? **(c)** Calculate the tension for a 91-g wire.

17. •• A 4.5-m-long rope of mass 1.8 kg hangs from a ceiling. **(a)** What is the tension in the rope at the bottom end? **(b)** What is the wave speed in the rope at the bottom end? **(c)** What is the tension in the rope at the top end, where it is attached to the ceiling? **(d)** What is the wave speed in the rope at the top end? **(e)** It can be shown that the average wave speed in the rope is $\frac{1}{2}\sqrt{gL}$, where L is the length of the rope. Use the average wave speed to calculate the time required for a pulse produced at the bottom of the rope to travel to the top.

18. ••• Two steel guitar strings have the same length. String A has a diameter of 0.50 mm and is under 410.0 N of tension. String B has a diameter of 1.0 mm and is under a tension of 820.0 N. Find the ratio of the wave speeds, v_A/v_B, in these two strings.

19. ••• Use dimensional analysis to show how the speed v of a wave on a string of circular cross section depends on the tension in the string, F, the radius of the string, R, and its mass per volume, ρ.

SECTION 14-3 HARMONIC WAVE FUNCTIONS

20. • Write an expression for a harmonic wave with an amplitude of 0.19 m, a wavelength of 2.6 m, and a period of 1.2 s. The wave is transverse, travels to the right, and has a displacement of 0.19 m at $t = 0$ and $x = 0$.

21. • Write an expression for a transverse harmonic wave that has a wavelength of 2.6 m and propagates to the right with a speed of 14.3 m/s. The amplitude of the wave is 0.11 m, and its displacement at $t = 0$ and $x = 0$ is 0.11 m.

22. •• **CE** The vertical displacement of a wave on a string is described by the equation $y(x, t) = A \sin(Bx - Ct)$, in which A, B, and C are positive constants. (a) Does this wave propagate in the positive or negative x direction? (b) What is the wavelength of this wave? (c) What is the frequency of this wave? (d) What is the smallest positive value of x where the displacement of this wave is zero at $t = 0$?

23. •• As it travels through a crystal, a light wave is described by the function $E(x,t) = A \cos[(1.54 \times 10^7)x - (2.98 \times 10^{15})t]$. In this expression, x is measured in meters and t is measured in seconds. What is the speed of the light wave?

24. •• **Predict/Calculate** A wave on a string is described by the following equation:

$$y = (15\,\text{cm}) \cos\left(\frac{\pi}{5.0\,\text{cm}}x - \frac{\pi}{12\,\text{s}}t\right)$$

(a) What is the amplitude of this wave? (b) What is its wavelength? (c) What is its period? (d) What is its speed? (e) In which direction does the wave travel?

25. •• Consider a harmonic wave with the following wave function:

$$y = (12\,\text{cm}) \cos\left(\frac{\pi}{4.5\,\text{cm}}x - \frac{\pi}{16\,\text{s}}t\right)$$

Sketch this wave from $x = 0$ to $x = 9.0$ cm for the following times: (a) $t = 0$; (b) $t = 4.0$ s; (c) $t = 8.0$ s. (d) What is the least amount of time required for a given point on this wave to move from $y = 0$ to $y = 12$ cm?

26. •• **Predict/Calculate** Four waves are described by the following equations, in which all distances are measured in centimeters and all times are measured in seconds:

$$y_A = 10 \cos(3x - 4t)$$
$$y_B = 10 \cos(5x + 4t)$$
$$y_C = 20 \cos(-10x + 60t)$$
$$y_D = 20 \cos(-4x - 20t)$$

(a) Which of these waves travel in the +x direction? (b) Which of these waves travel in the −x direction? (c) Which wave has the highest frequency? (d) Which wave has the greatest wavelength? (e) Which wave has the greatest speed?

SECTION 14-4 SOUND WAVES

27. • At Zion National Park a loud shout produces an echo 1.55 s later from a colorful sandstone cliff. How far away is the cliff?

28. • **BIO Dolphin Ultrasound** Dolphins of the open ocean are classified as Type II Odontocetes (toothed whales). These animals use ultrasonic "clicks" with a frequency of about 55 kHz to navigate and find prey. (a) Suppose a dolphin sends out a series of clicks that are reflected back from the bottom of the ocean 75 m below. How much time elapses before the dolphin hears the echoes of the clicks? (The speed of sound in seawater is approximately 1530 m/s.) (b) What is the wavelength of 55-kHz sound in the ocean?

29. • The lowest note on a piano is A, four octaves below the A given in Table 14-3. The highest note on a piano is C, four octaves above middle C. Find the frequencies and wavelengths (in air) of these notes.

30. •• **Meteor Infrasound** A meteor that explodes in the atmosphere creates infrasound waves that can travel multiple times around the globe. On February 15, 2013, the Chelyabinsk meteor produced infrasound waves with a wavelength of 6.4 km that propagated 8840 km in 8.53 h. What was the frequency of these waves?

31. •• **Predict/Calculate** A sound wave in air has a frequency of 405 Hz. (a) What is its wavelength? (b) If the frequency of the sound is increased, does its wavelength increase, decrease, or stay the same? Explain. (c) Calculate the wavelength for a sound wave with a frequency of 485 Hz.

32. •• **Predict/Calculate** When you drop a rock into a well, you hear the splash 1.5 seconds later. (a) How deep is the well? (b) If the depth of the well were doubled, would the time required to hear the splash be greater than, less than, or equal to 3.0 seconds? Explain.

33. ••• A rock is thrown downward into a well that is 7.62 m deep. If the splash is heard 1.17 seconds later, what was the initial speed of the rock?

SECTION 14-5 SOUND INTENSITY

34. • **CE** If the distance to a point source of sound is doubled, by what multiplicative factor does the intensity change?

35. • The intensity level of sound in a truck is 88 dB. What is the intensity of this sound?

36. • The distance to a point source is decreased by a factor of three. (a) By what multiplicative factor does the intensity increase? (b) By what additive amount does the intensity level increase?

37. • Sound 1 has an intensity of 48.0 W/m². Sound 2 has an intensity level that is 3.5 dB greater than the intensity level of sound 1. What is the intensity of sound 2?

38. •• A bird-watcher is hoping to add the white-throated sparrow to her "life list" of species. How far could she be from the bird described in Example 14-8 and still be able to hear it? Assume no reflections or absorption of the sparrow's sound.

39. •• Residents of Hawaii are warned of the approach of a tsunami by sirens mounted on the tops of towers. Suppose a siren produces a sound that has an intensity level of 120 dB at a distance of 2.0 m. Treating the siren as a point source of sound, and ignoring reflections and absorption, find the intensity level heard by an observer at a distance of (a) 12 m and (b) 21 m from the siren. (c) How far away can the siren be heard?

40. •• In a pig-calling contest, a caller produces a sound with an intensity level of 107 dB. How many such callers would be required to reach the pain level of 120 dB?

41. •• **Predict/Calculate** Twenty violins playing simultaneously with the same intensity combine to give an intensity level of 82.5 dB. (a) What is the intensity level of each violin? (b) If the number of violins is increased to 40, will the combined intensity level be more than, less than, or equal to 165 dB? Explain.

42. •• **BIO The Human Eardrum** The radius of a typical human eardrum is about 4.0 mm. Find the energy per second received by an eardrum when it listens to sound that is (a) at the threshold of hearing and (b) at the threshold of pain.

SECTION 14-6 THE DOPPLER EFFECT

43. • **CE Predict/Explain** A horn produces sound with frequency f_0. Let the frequency you hear when you are at rest and the horn moves toward you with a speed u be f_1; let the frequency you hear when the horn is at rest and you move toward it with a speed u be

f_2. **(a)** Is f_1 greater than, less than, or equal to f_2? **(b)** Choose the *best explanation* from among the following:

   I. A moving observer encounters wave crests more often than a stationary observer, leading to a higher frequency.
  II. The relative speeds are the same in either case. Therefore, the frequencies will be the same as well.
 III. A moving source causes the wave crests to "bunch up," leading to a higher frequency than for a moving observer.

44. • **CE** You are heading toward an island in your speedboat when you see a friend standing on shore at the base of a cliff. You sound the boat's horn to get your friend's attention. Let the wavelength of the sound produced by the horn be λ_1, the wavelength as heard by your friend be λ_2, and the wavelength of the echo as heard on the boat be λ_3. Rank these wavelengths in order of increasing length. Indicate ties where appropriate.

45. • When the bell in a clock tower rings with a sound of 475 Hz, a pigeon roosting in the belfry flies directly away from the bell. If the pigeon hears a frequency of 448 Hz, what is its speed?

46. • A car approaches a train station with a speed of 24 m/s. A stationary train at the station sounds its 166-Hz horn. What frequency is heard by the driver of the car?

47. • **BIO** A bat moving with a speed of 3.25 m/s and emitting sound of 35.0 kHz approaches a moth at rest on a tree trunk. **(a)** What frequency is heard by the moth? **(b)** If the speed of the bat is increased, is the frequency heard by the moth higher or lower? **(c)** Calculate the frequency heard by the moth when the speed of the bat is 4.25 m/s.

48. • A motorcycle and a police car are moving toward one another. The police car emits sound with a frequency of 523 Hz and has a speed of 32.2 m/s. The motorcycle has a speed of 14.8 m/s. What frequency does the motorcyclist hear?

49. •• Hearing the siren of an approaching fire truck, you pull over to the side of the road and stop. As the truck approaches, you hear a tone of 460 Hz; as the truck recedes, you hear a tone of 410 Hz. How much time will it take for the truck to get from your position to the fire 5.0 km away, assuming it maintains a constant speed?

50. •• **Predict/Calculate** For a science experiment, you would like to use the Doppler effect to double the perceived frequency of a harmonica, from the 261.7-Hz tone (middle C) that it emits, to 523.3 Hz (C₅ note). **(a)** Which strategy will accomplish your goal more easily: move the harmonica, or move the observer? **(b)** Calculate the speed the harmonica must have in order for a stationary observer to hear 523.3 Hz. **(c)** Calculate the speed an observer must have in order to perceive a 523.3-Hz sound from a stationary harmonica that emits sound at 261.7 Hz.

51. •• **Predict/Calculate** Two bicycles approach one another, each traveling with a speed of 8.50 m/s. **(a)** If bicyclist A beeps a 315-Hz horn, what frequency is heard by bicyclist B? **(b)** Which of the following would cause the greater increase in the frequency heard by bicyclist B: (i) bicyclist A speeds up by 1.50 m/s, or (ii) bicyclist B speeds up by 1.50 m/s? Explain.

52. •• A train on one track moves in the same direction as a second train on the adjacent track. The first train, which is ahead of the second train and moves with a speed of 36.8 m/s, blows a horn whose frequency is 124 Hz. If the frequency heard on the second train is 135 Hz, what is its speed?

53. •• Two cars traveling with the same speed move directly away from one another. One car sounds a horn whose frequency is 422 Hz

and a person in the other car hears a frequency of 385 Hz. What is the speed of the cars?

54. ••• **The Bullet Train** The Shinkansen, the Japanese "bullet" train, runs at high speed from Tokyo to Nagoya. Riding on the Shinkansen, you notice that the frequency of a crossing signal changes markedly as you pass the crossing. As you approach the crossing, the frequency you hear is f; as you recede from the crossing the frequency you hear is $2f/3$. What is the speed of the train?

SECTION 14-7 SUPERPOSITION AND INTERFERENCE

55. • Pulses A, B, C, and D all travel at 10 m/s on the same string but in opposite directions. The string is depicted at time $t = 0$ in **FIGURE 14-45**. The small pulses have an amplitude of 2.0 cm, and the large pulses have an amplitude of 4.0 cm. Find the displacement of point P at times **(a)** $t = 0.10$ s and **(b)** $t = 0.20$ s.

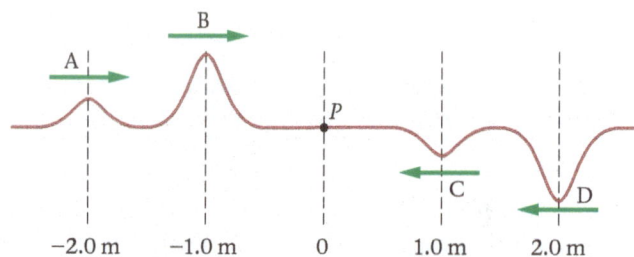

FIGURE 14-45 Problem 55

56. • Pulses A, B, C, and D all travel at 10 m/s on the same string but in opposite directions. The string is depicted at time $t = 0$ in **FIGURE 14-46**. The small pulses have an amplitude of 2.0 cm, and the large pulse has an amplitude of 4.0 cm. Find the displacement of point P at times **(a)** $t = 0.10$ s and **(b)** $t = 0.20$ s.

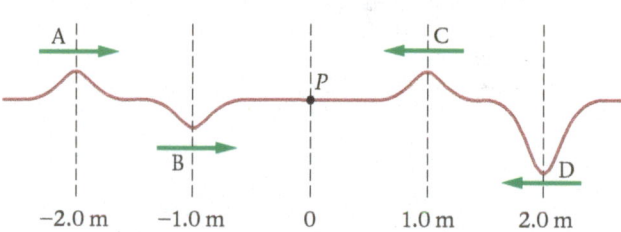

FIGURE 14-46 Problem 56

57. •• A pair of in-phase stereo speakers is placed side by side, 0.914 m apart. You stand directly in front of one of the speakers, 2.44 m from the speaker. What is the lowest frequency that will produce constructive interference at your location?

58. •• **Predict/Calculate** Two violinists, one directly behind the other, play for a listener directly in front of them. Both violinists sound concert A (440 Hz). **(a)** What is the smallest separation between the violinists that will produce destructive interference for the listener? **(b)** Does this smallest separation increase or decrease if the violinists produce a note with a higher frequency? **(c)** Repeat part (a) for violinists who produce sounds of 540 Hz.

59. •• Two loudspeakers are placed at either end of a gymnasium, both pointing toward the center of the gym and equidistant from it. The speakers emit 266-Hz sound that is in phase. An observer at the center of the gym experiences constructive interference. How far toward either speaker must the observer walk to first experience destructive interference?

60. •• Two speakers with opposite phase are positioned 3.5 m apart, both pointing toward a wall 5.0 m in front of them (**FIGURE 14-47**). An observer standing against the wall midway between the speakers hears destructive interference. If the observer hears constructive interference after moving 0.84 m to one side along the wall, what is the frequency of the sound emitted by the speakers?

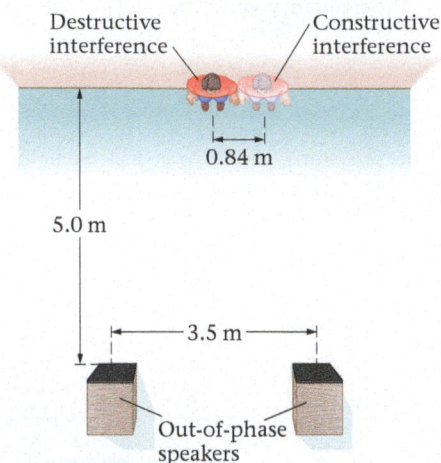

FIGURE 14-47 Problem 60

61. •• Suppose, in Example 14-13, that the speakers have opposite phase. What is the lowest frequency that gives destructive interference in this case?

SECTION 14-8 STANDING WAVES

62. • **CE** Predict/Explain When you blow across the opening of a partially filled 2-L soda pop bottle you hear a tone. (a) If you take a sip of the pop and blow across the opening again, does the tone you hear have a higher frequency, a lower frequency, or the same frequency as before? (b) Choose the *best explanation* from among the following:
  I. The same pop bottle will give the same frequency regardless of the amount of pop it contains.
  II. The greater distance from the top of the bottle to the level of the pop results in a higher frequency.
  III. A lower level of pop results in a longer column of air, and hence a lower frequency.

63. • An organ pipe that is open at both ends is 3.5 m long. What is its fundamental frequency?

64. • A string 2.5 m long with a mass of 3.6 g is stretched between two fixed points with a tension of 91 N. Find the frequency of the fundamental on this string.

65. • **BIO** Human Vocal Resonance The unique sound of a person's voice arises from the frequencies of the standing waves that occur in a person's chest, trachea, larynx, mouth, nose, and sinuses. Suppose a person's voice has a strong resonant frequency at 890 Hz when she is breathing air. What will be the frequency of the same resonance when the person breathes helium? (*Note:* The speed of sound in helium is 1007 m/s.)

66. •• The fundamental wavelength for standing sound waves in an empty pop bottle is 0.88 m. (a) What is the length of the bottle? (b) What is the frequency of the third harmonic of this bottle?

67. •• **CE** A string is tied down at both ends. Some of the standing waves on this string have the following frequencies: 100 Hz, 200 Hz, 250 Hz, and 300 Hz. It is also known that there are no standing waves with frequencies between 250 Hz and 300 Hz. (a) What is the fundamental frequency of this string? (b) What is the frequency of the third harmonic?

68. •• **BIO** Predict/Calculate **Standing Waves in the Human Ear** The human ear canal is much like an organ pipe that is closed at one end (at the tympanic membrane or eardrum) and open at the other (**FIGURE 14-48**). A typical ear canal has a length of about 2.4 cm. (a) What are the fundamental frequency and wavelength of the ear canal? (b) Find the frequency and wavelength of the ear canal's third harmonic. (Recall that the third harmonic in this case is the standing wave with the second-lowest frequency.) (c) Suppose a person has an ear canal that is shorter than 2.4 cm. Is the fundamental frequency of that person's ear canal greater than, less than, or the same as the value found in part (a)? Explain. [Note that the frequencies found in parts (a) and (b) correspond closely to the frequencies of enhanced sensitivity in Figure 14-37.]

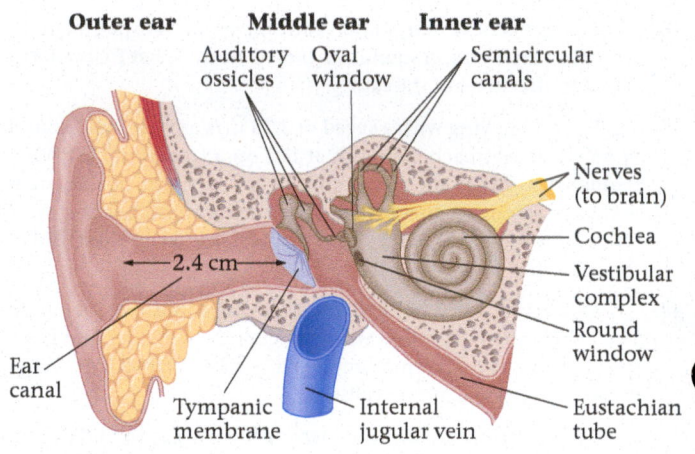

FIGURE 14-48 Problem 68

69. •• A guitar string 66 cm long vibrates with a standing wave that has three antinodes. (a) Which harmonic is this? (b) What is the wavelength of this wave?

70. •• Predict/Calculate A guitar string has a mass per length of 2.33×10^{-3} kg/m and a fundamental frequency of 146.8 Hz when it is under a tension of 82.4 N. The string breaks, and its owner has only a spare string of mass per length 6.79×10^{-3} kg/m. (a) In order to use the new string to play the same note, should the string tension be greater than, less than, or equal to 82.4 N? Explain. (b) What tension should the 6.79×10^{-3} kg/m string have so that its fundamental frequency is 146.8 Hz?

71. •• Predict/Calculate A 0.323-kg clothesline is stretched with a tension of 53.4 N between two poles 6.10 m apart. What is the frequency of (a) the fundamental and (b) the second harmonic? (c) If the tension in the clothesline is increased, do the frequencies in parts (a) and (b) increase, decrease, or stay the same? Explain.

72. •• The organ pipe in **FIGURE 14-49** is 2.75 m long. (a) What is the frequency of the standing wave shown in the pipe? (b) What is the fundamental frequency of this pipe?

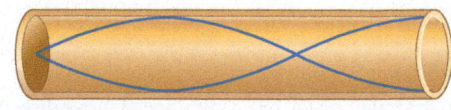

FIGURE 14-49 Problem 72

73. •• The frequency of the standing wave shown in **FIGURE 14-50** is 185 Hz. **(a)** What is the fundamental frequency of this pipe? **(b)** What is the length of the pipe?

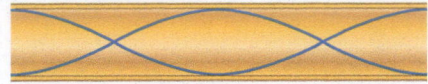

FIGURE 14-50 Problem 73

74. ••• An organ pipe open at both ends has a harmonic with a frequency of 440 Hz. The next higher harmonic in the pipe has a frequency of 495 Hz. Find **(a)** the frequency of the fundamental and **(b)** the length of the pipe.

SECTION 14-9 BEATS

75. • **CE** When guitar strings A and B are plucked at the same time, a beat frequency of 2 Hz is heard. If string A is tightened, the beat frequency increases to 3 Hz. Which of the two strings had the lower frequency initially?

76. • **CE** Predict/Explain **(a)** Is the beat frequency produced when a 245-Hz tone and a 240-Hz tone are played together greater than, less than, or equal to the beat frequency produced when a 140-Hz tone and a 145-Hz tone are played together? **(b)** Choose the *best explanation* from among the following:
   I. The beat frequency is determined by the difference in frequencies and is independent of their actual values.
   II. The higher frequencies will produce a higher beat frequency.
   III. The percentage change in frequency for 240 and 245 Hz is less than for 140 and 145 Hz, resulting in a lower beat frequency.

77. • You have three tuning forks with frequencies of 252 Hz, 256 Hz, and 259 Hz. What beat frequencies are possible with these tuning forks?

78. • **Tuning a Piano** To tune middle C on a piano, a tuner hits the key and at the same time sounds a 261-Hz tuning fork. If the tuner hears 3 beats per second, what are the possible frequencies of the piano key?

79. •• Two musicians are comparing their clarinets. The first clarinet produces a tone that is known to be 466 Hz. When the two clarinets play together they produce eight beats every 3.00 seconds. If the second clarinet produces a higher pitched tone than the first clarinet, what is the second clarinet's frequency?

80. •• Predict/Calculate Two strings that are fixed at each end are identical, except that one is 0.560 cm longer than the other. Waves on these strings propagate with a speed of 34.2 m/s, and the fundamental frequency of the shorter string is 212 Hz. **(a)** What beat frequency is produced if each string is vibrating with its fundamental frequency? **(b)** Does the beat frequency in part (a) increase or decrease if the longer string is increased in length? **(c)** Repeat part (a), assuming that the longer string is 0.761 cm longer than the shorter string.

81. •• Identical cellos are being tested. One is producing a fundamental frequency of 120.9 Hz on a string that is 1.25 m long and has a mass of 109 g. On the second cello an identical string is fingered to reduce the length that can vibrate. If the beat frequency produced by these two strings is 5.25 Hz, what is the vibrating length of the second string?

82. ••• A friend in another city tells you that she has two organ pipes of different lengths, one open at both ends, the other open at one end only. In addition, she has determined that the beat frequency caused by the second-lowest frequency of each pipe is equal to the beat frequency caused by the third-lowest frequency of each pipe. Her challenge to you is to calculate the length of the organ pipe that is open at both ends, given that the length of the other pipe is 1.00 m.

GENERAL PROBLEMS

83. • **CE FIGURE 14-51** shows a wave on a string moving to the right. For each of the points indicated on the string, A-F, state whether it is (I, moving upward; II, moving downward; or III, instantaneously at rest) at the instant pictured.

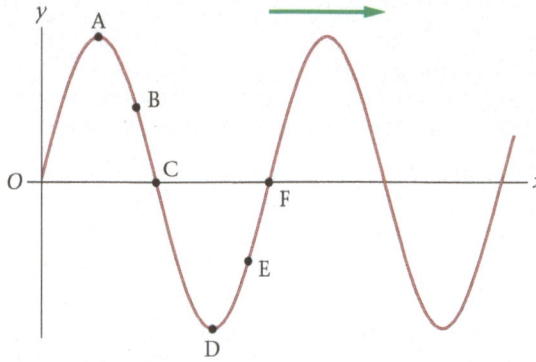

FIGURE 14-51 Problem 83

84. • The fundamental of an organ pipe that is closed at one end and open at the other end is 261.6 Hz (middle C). The second harmonic of an organ pipe that is open at both ends has the same frequency. What are the lengths of these two pipes?

85. • **The Loudest Animal** The loudest sound produced by a living organism on Earth is made by the bowhead whale (*Balaena mysticetus*). These whales can produce a sound that, if heard in air at a distance of 3.00 m, would have an intensity level of 127 dB. This is roughly the equivalent of 5000 trumpeting elephants. How far away can you be from a 127-dB sound and still just barely hear it? (Assume a point source, and ignore reflections and absorption.)

86. • **Hearing a Good Hit** Physicist Robert Adair, once appointed the "official physicist to the National League" by the commissioner of baseball, believes that the "crack of the bat" can tell an outfielder how well the ball has been hit. According to Adair, a good hit makes a sound of 510 Hz, while a poor hit produces a sound of 170 Hz. What is the difference in wavelength of these sounds?

87. • A standing wave of 603 Hz is produced on a string that is 1.33 m long and fixed on both ends. If the speed of waves on this string is 402 m/s, how many antinodes are there in the standing wave?

88. •• **Playing Harmonics** When a 63-cm-long guitar string is plucked, all possible standing waves simultaneously exist on the string. When you place your finger on the string, any standing waves with nonzero amplitude at that location are stopped, but any standing waves with a node at that location continue vibrating. If you wish to stop all standing waves with frequencies lower than the third harmonic on this string, how far from the end of the string should you place your finger?

89. •• **BIO** Measuring Hearing Loss To determine the amount of temporary hearing loss loud music can cause in humans, researchers studied a group of 20 adult females who were exposed to 110-dB music for 60 minutes. Eleven of the 20 subjects showed a 20.0-dB reduction in hearing sensitivity at 4000 Hz. What is the intensity corresponding to the threshold of hearing for these subjects?

90. •• **BIO** Hearing a Pin Drop The ability to hear a "pin drop" is the sign of sensitive hearing. Suppose a 0.48-g pin is dropped from a height of 31 cm, and that the pin emits sound for 1.2 s when it

lands. Assuming all of the mechanical energy of the pin is converted to sound energy, and that the sound radiates uniformly in all directions, find the maximum distance from which a person can hear the pin drop. (This is the ideal maximum distance, but atmospheric absorption and other factors will make the actual maximum distance considerably smaller.)

91. •• A cannon 105 m away from you shoots a cannonball straight up into the air with an initial speed of 46 m/s. What is the cannonball's speed when you hear the blast from the cannon? Ignore air resistance.

92. •• A machine shop has 120 equally noisy machines that together produce an intensity level of 92 dB. If the intensity level must be reduced to 82 dB, how many machines must be turned off?

93. •• **Predict/Calculate** A bottle has a standing wave with a frequency of 375 Hz. The next higher-frequency standing wave in the bottle is 625 Hz. **(a)** Is 375 Hz the fundamental frequency for the bottle? Explain. **(b)** What is the next-higher frequency standing wave after 625 Hz?

94. •• **Speed of a Tsunami** Tsunamis can have wavelengths between 100 and 400 km. Since this is much greater than the average depth of the oceans (about 4.3 km), the ocean can be considered as shallow water for these waves. Using the speed of waves in shallow water of depth *d* given in Problem 8, find the typical speed for a tsunami. (*Note*: In the open ocean, tsunamis generally have an amplitude of less than a meter, allowing them to pass ships unnoticed. As they approach shore, however, the water depth decreases and the waves slow down. This can result in an increase of amplitude to as much as 37 m or more.)

95. •• Two trains with 124-Hz horns approach one another. The slower of the two trains has a speed of 26 m/s. What is the speed of the fast train if an observer standing near the tracks between the trains hears a beat frequency of 4.4 Hz?

96. •• **Predict/Calculate** Jim is speeding toward James Island with a speed of 24 m/s when he sees Betsy standing on shore at the base of a cliff (**FIGURE 14-52**). Jim sounds his 330-Hz horn. **(a)** What frequency does Betsy hear? **(b)** Jim can hear the echo of his horn reflected back to him by the cliff. Is the frequency of this echo greater than or equal to the frequency heard by Betsy? Explain. **(c)** Calculate the frequency Jim hears in the echo from the cliff.

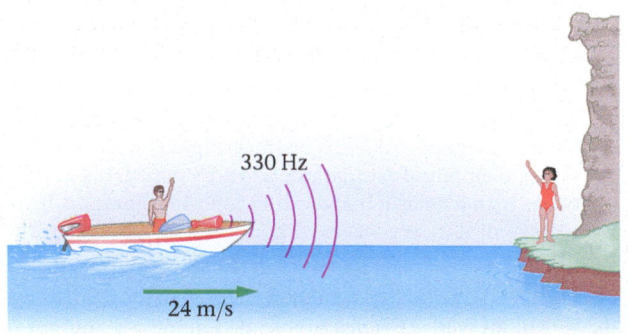

330 Hz

24 m/s

FIGURE 14-52 Problem 96

97. •• Two ships in a heavy fog are blowing their horns, both of which produce sound with a frequency of 175.0 Hz (**FIGURE 14-53**). One ship is at rest; the other moves on a straight line that passes through the one at rest. If people on the stationary ship hear a beat frequency of 3.5 Hz, what are the two possible speeds and directions of motion of the moving ship?

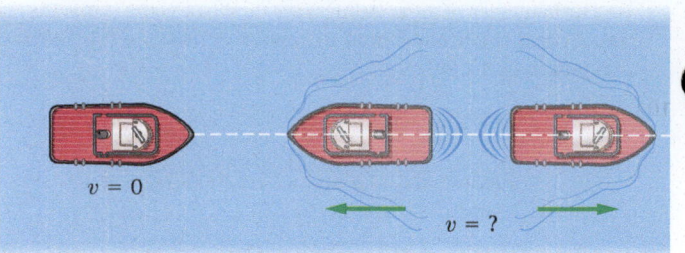

$v = 0$

$v = ?$

FIGURE 14-53 Problem 97

98. •• **BIO Cracking Your Knuckles** When you "crack" a knuckle, you cause the knuckle cavity to widen rapidly. This, in turn, allows the synovial fluid to expand into a larger volume. If this expansion is sufficiently rapid, it causes a gas bubble to form in the fluid in a process known as *cavitation*. This is the mechanism responsible for the cracking sound. (Cavitation can also cause pits in rapidly rotating ship's propellers.) If a "crack" produces a sound with an intensity level of 57 dB at your ear, which is 18 cm from the knuckle, how far from your knuckle can the "crack" be heard? Assume the sound propagates uniformly in all directions, with no reflections or absorption.

99. •• A steel guitar string has a tension *F*, length *L*, and diameter *D*. Give the multiplicative factor by which the fundamental frequency of the string changes under the following conditions: **(a)** The tension in the string is increased by a factor of 4. The diameter is *D* and the length is *L*. **(b)** The diameter of the string is increased by a factor of 3. The tension is *F* and the length is *L*. **(c)** The length of the string is halved. The tension is *F* and the diameter is *D*.

100. •• A Slinky has a mass of 0.28 kg and negligible length. When it is stretched 1.5 m, it is found that transverse waves travel the length of the Slinky in 0.75 s. **(a)** What is the force constant, *k*, of the Slinky? **(b)** If the Slinky is stretched farther, will the time required for a wave to travel the length of the Slinky increase, decrease, or stay the same? Explain. **(c)** If the Slinky is stretched 3.0 m, how much time does it take a wave to travel the length of the Slinky? (The Slinky stretches like an ideal spring, and propagates transverse waves like a rope with variable tension.)

101. •• **BIO Predict/Calculate OSHA Noise Standards** OSHA, the Occupational Safety and Health Administration, has established standards for workplace exposure to noise. According to OSHA's Hearing Conservation Standard, the permissible noise exposure per day is 95.0 dB for 4 hours or 105 dB for 1 hour. Assuming the eardrum is 9.5 mm in diameter, find the energy absorbed by the eardrum **(a)** with 95.0 dB for 4 hours and **(b)** with 105 dB for 1 hour. **(c)** Is OSHA's safety standard simply a measure of the amount of energy absorbed by the ear? Explain.

102. •• An organ pipe 3.4 m long is open at one end and closed at the other end. What is the linear distance between a node and the adjacent antinode for the third harmonic in this pipe?

103. •• Two identical strings with the same tension vibrate at 631 Hz. If the tension in one of the strings is increased by 2.25%, what is the resulting beat frequency?

104. •• **BIO The Love Song of the Midshipman Fish** When the plainfin midshipman fish (*Porichthys notatus*) migrates from deep Pacific waters to the west coast of North America each summer, the males begin to sing their "love song," which some describe as sounding like a low-pitched motorboat. Houseboat residents and shore dwellers are kept awake for nights on end by the amorous fish. The love song consists of a single note, the second A flat below middle C. **(a)** If the speed of sound in seawater is 1531 m/s,

what is the wavelength of the midshipman's song? **(b)** What is the wavelength of the sound after it emerges into the air? (Information on the musical scale is given in Table 14-3.)

105. ••• **Predict/Calculate** A rope of length L and mass M hangs vertically from a ceiling. The tension in the rope is only that due to its own weight. **(a)** Suppose a wave starts near the bottom of the rope and propagates upward. Does the speed of the wave increase, decrease, or stay the same as it moves up the rope? Explain. **(b)** Show that the speed of waves a height y above the bottom of the rope is $v = \sqrt{gy}$.

106. ••• **Beats and Standing Waves** In Problem 59, suppose the observer walks toward one speaker with a speed of 1.35 m/s. **(a)** What frequency does the observer hear from each speaker? **(b)** What beat frequency does the observer hear? **(c)** How far must the observer walk to go from one point of constructive interference to the next? **(d)** How many times per second does the observer hear maximum loudness from the speakers? Compare your result with the beat frequency from part (b).

PASSAGE PROBLEMS

BIO **The Sound of a Dinosaur**

Modern-day animals make extensive use of sounds in their interactions with others. Some sounds are meant primarily for members of the same species, like the cooing calls of a pair of doves, the long-range infrasound communication between elephants, or the songs of the humpback whale. Other sounds may be used as a threat to other species, such as the rattle of a rattlesnake or the roar of a lion.

There is little doubt that extinct animals used sounds in much the same ways. But how can we ever hear the call of a long-vanished animal like a dinosaur when sounds don't fossilize? In some cases, basic physics may have the answer.

Consider, for example, the long-crested, duck-billed dinosaur *Parasaurolophus walkeri*, which roamed the Earth 75 million years ago. This dinosaur, shown in **FIGURE 14-54 (a)**, possessed the largest crest of any duck bill—so long, in fact, that there was a notch in *P. walkeri*'s spine to make room for the crest when its head was tilted backward. Many paleontologists believe the air passages in the dinosaur's crest acted like bent organ pipes open at both ends, and that they produced sounds *P. walkeri* used to communicate with others of its kind. As air was forced through the passages, the predominant sound they produced would be the fundamental standing wave, with a small admixture of higher harmonics as well. The frequencies of these standing waves can be determined with basic physical principles. **FIGURE 14-54 (b)** presents a plot of the lowest ten harmonics of a pipe that is open at both ends as a function of the length of the pipe.

107. • Suppose the air passages in a certain *P. walkeri* crest produce a bent tube 2.7 m long. What is the fundamental frequency of this tube, assuming the bend has no effect on the frequency? (For comparison, a typical human hearing range is 20 Hz to 20 kHz.)

    A. 0.0039 Hz   **B.** 32 Hz   **C.** 64 Hz   **D.** 130 Hz

108. • Paleontologists believe the crest of a female *P. walkeri* was probably shorter than the crest of a male. If this was the case, would the fundamental frequency of a female be greater than, less than, or equal to the fundamental frequency of a male?

109. • Suppose the fundamental frequency of a particular female was 74 Hz. What was the length of the air passages in this female's crest?

    A. 1.2 m   **B.** 2.3 m   **C.** 2.7 m   **D.** 4.6 m

110. • As a young *P. walkeri* matured, the air passages in its crest might increase in length from 1.5 m to 2.7 m, causing a decrease in the standing wave frequencies. Referring to Figure 14-54, do you expect the change in the fundamental frequency to be greater than, less than, or equal to the change in the second harmonic frequency?

111. •• **Predict/Calculate** **REFERRING TO EXAMPLE 14-11** Suppose the engineer adjusts the speed of the train until the sound he hears reflected from the cliff is 775 Hz. The train's whistle still produces a tone of 650.0 Hz. **(a)** Is the new speed of the train greater than, less than, or equal to 21.2 m/s? Explain. **(b)** Find the new speed of the train.

112. •• **REFERRING TO EXAMPLE 14-11** Suppose the train is backing away from the cliff with a speed of 18.5 m/s and is sounding its 650.0-Hz whistle. **(a)** What is the frequency heard by the observer standing near the tunnel entrance? **(b)** What is the frequency heard by the engineer?

113. •• **Predict/Calculate** **REFERRING TO EXAMPLE 14-16** Suppose we add more water to the soda pop bottle. **(a)** Does the fundamental frequency increase, decrease, or stay the same? Explain. **(b)** Find the fundamental frequency if the height of water in the bottle is increased to 7.5 cm. The height of the bottle is still 26.0 cm.

114. •• **Predict/Calculate** **REFERRING TO EXAMPLE 14-16** The speed of sound increases slightly with temperature. **(a)** Does the fundamental frequency of the bottle increase, decrease, or stay the same as the air heats up on a warm day? Explain. **(b)** Find the fundamental frequency if the speed of sound in air increases to 348 m/s. Assume the bottle is 26.0 cm tall, and that it contains water to a depth of 6.5 cm.

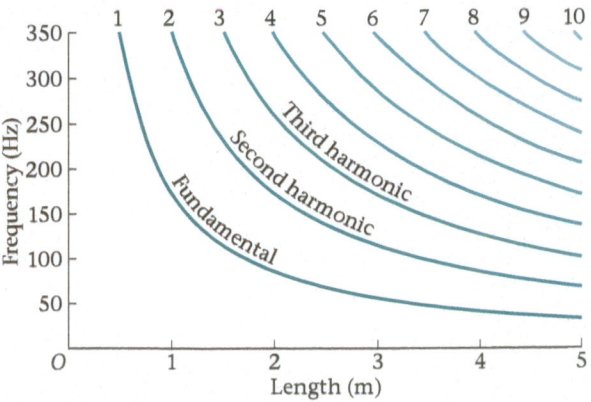

FIGURE 14-54 Standing wave frequencies as a function of length for a pipe open at both ends. The first ten harmonics ($n = 1 - 10$) are shown. (Problems 107, 108, 109, and 110)

15

Fluids

Big Ideas

1 Pressure is the force exerted by a fluid on an area.

2 Pressure increases with depth in a fluid.

3 Archimedes' principle states that the buoyant force is equal to the weight of the displaced fluid.

4 Bernoulli's equation is a statement of energy conservation applied to fluids.

5 Surface tension tends to make the surface of a fluid as small as possible.

▲ How do doughnuts cooking in oil illustrate physics in action? Well, did you notice that each one has a white ring around the middle? Why is that? When you think about it, you realize that as the doughnuts float in the cooking oil, more than half their volume is out of the oil—only about 45 percent is being cooked. When they're flipped over to cook the other side, another 45 percent is cooked, leaving 10 percent as a white uncooked ring around the middle. Archimedes' principle, one of the key topics in this chapter, has never had a tastier application.

When we speak of *fluids*, we are referring to substances like gases and liquids that can readily flow from place to place, and that take on the shape of a container rather than retain a shape of their own. In this chapter, we examine some of the fundamental principles that apply to fluids. As we shall see, all of these principles derive from the basic physics we have learned to this point.

15-1 Density

The properties of a fluid can be hard to pin down, given that it can flow, change shape, and either split into smaller portions or combine into a larger system. Thus, one of the best ways to quantify a fluid is in terms of its **density.** Specifically, the density, ρ, of a material (fluid or not) is defined as the mass, M, per volume, V:

TABLE 15-1 Densities of Common Substances

Substance	Density (kg/m^3)
Gold	19,300
Mercury	13,600
Lead	11,300
Silver	10,500
Iron	7860
Aluminum	2700
Ebony (wood)	1220
Ethylene glycol (antifreeze)	1114
Whole blood (37 °C)	1060
Seawater	1025
Freshwater	1000
Olive oil	920
Ice	917
Ethyl alcohol	806
Cherry (wood)	800
Balsa (wood)	120
Styrofoam	100
Oxygen	1.43
Air	1.29
Helium	0.179

Definition of Density, ρ

$$\rho = M/V \qquad \qquad 15\text{-}1$$

SI unit: kg/m^3

The denser a material, the more mass it has in any given volume.

To get a feel for densities in common substances, we start with water. For example, to fill a cubic container 1 meter on a side would take more than 2000 pounds of water. More precisely, water has the following density:

$$\rho_w = \text{density of water} = 1000 \text{ kg/m}^3$$

A gallon (1 gallon = 3.79 L = 3.79×10^{-3} m³) of water, then, has a mass of

$$M = \rho V = (1000 \text{ kg/m}^3)(3.79 \times 10^{-3} \text{ m}^3) = 3.79 \text{ kg}$$

As a rule of thumb, a gallon of water weighs just over 8 pounds.

In comparison, the density of the air in your room is roughly 1.29 kg/m³, and the helium in a helium-filled balloon has a density of only about 0.179 kg/m³. On the higher end of the density scale, solid gold "weighs in" with a hefty 19,300 kg/m³. The densities of some other common materials are listed in Table 15-1.

CONCEPTUAL EXAMPLE 15-1 THE WEIGHT OF AIR

One day you look in your refrigerator and find nothing but a dozen eggs (45 g each). A quick measurement shows that the inside of the refrigerator is 1.0 m by 0.61 m by 0.75 m. Is the weight of the *air* in your refrigerator much less than, about the same as, or much more than the weight of the *eggs*?

REASONING AND DISCUSSION

At first it might seem that the "thin air" in the refrigerator weighs practically nothing compared with a carton full of eggs. A brief calculation shows this is not the case. For the eggs, we have

$$m_{eggs} = 12(45 \text{ g}) = 0.54 \text{ kg}$$

For the air, with a density of 1.29 kg/m³, the mass is

$$m_{air} = \rho V = (1.29 \text{ kg/m}^3)(1.0 \text{ m} \times 0.61 \text{ m} \times 0.75 \text{ m}) = 0.59 \text{ kg}$$

Thus, the air, which has a mass of 0.59 kg (1.3 lb), actually weighs slightly more than the eggs, which have a mass of 0.54 kg (1.2 lb)!

ANSWER

The weight of the air is about the same as the weight of the eggs.

Enhance Your Understanding (Answers given at the end of the chapter)

1. System 1 consists of 10 kg of water; system 2 consists of 100 kg of water. **(a)** Is the volume of system 1 greater than, less than, or equal to the volume of system 2? Explain. **(b)** Is the density of system 1 greater than, less than, or equal to the density of system 2? Explain.

Section Review

• Density, ρ, is the amount of mass per volume: $\rho = M/V$.

PHYSICS IN CONTEXT
Looking Back

Pressure can be thought of as an application of the concept of force (Chapters 5 and 6) to fluids.

PHYSICS IN CONTEXT
Looking Ahead

The pressure of an ideal gas, and its connection with temperature and volume, will be explored in detail in Chapter 17 when we study thermodynamics.

15-2 Pressure

If you've ever pushed a button, or pressed a key on a keyboard, you've applied pressure. Now, you might object that you simply exerted a force on the button or the key, which is correct. That force is spread out over an area, however. For example, when you press a button, the tip of your finger contacts the button over a small but finite area. **Pressure**, P, is a measure of the amount of force, F, per area, A:

Definition of Pressure, P

$$P = F/A \qquad \qquad \text{15-2}$$

SI unit: N/m^2

The pressure is increased if the force applied to a given area is increased, or if a given force is applied to a smaller area. For example, if you press your finger against a balloon, not much happens—your finger causes a small indentation. On the other hand, if you push a needle against the balloon with the same force, you get an explosive pop. The difference is that the same force applied to the small area of a needle tip causes a large enough pressure to rupture the balloon.

EXAMPLE 15-2 POPPING A BALLOON

Find the pressure exerted on the surface of a balloon if you press with a force of 2.6 N using **(a)** your finger or **(b)** a needle. Assume the area of your fingertip is $1.5 \times 10^{-4} \, m^2$, and the area of the needle tip is $2.3 \times 10^{-7} \, m^2$. **(c)** Find the minimum force necessary to pop the balloon with the needle, given that the balloon pops with a pressure of $3.7 \times 10^5 \, N/m^2$.

PICTURE THE PROBLEM
Our sketch shows a balloon deformed by the press of a finger and of a needle. The difference is not the amount of force that is applied, but the area over which it is applied. In the case of the needle, the force may be sufficient to pop the balloon.

REASONING AND STRATEGY

a, b. The force F and area A are given in the problem statement. We can find the pressure by applying the definition: $P = F/A$.

c. Now we rearrange $P = F/A$ to find the force corresponding to a given pressure and area.

Known Force, $F = 2.6$ N; area of fingertip, $A_f = 1.5 \times 10^{-4} \, m^2$; area of needle tip, $A_n = 2.3 \times 10^{-7} \, m^2$; pressure to pop balloon, $P = 3.7 \times 10^5 \, N/m^2$.

Unknown **(a)** Pressure, $P = ?$ **(b)** Pressure, $P = ?$ **(c)** Force to pop balloon, $F = ?$

SOLUTION

Part (a)

1. Calculate the pressure exerted by the finger:
$$P = \frac{F}{A_f} = \frac{2.6 \, N}{1.5 \times 10^{-4} \, m^2} = 1.7 \times 10^4 \, N/m^2$$

Part (b)

2. Calculate the pressure exerted by the needle:
$$P = \frac{F}{A_n} = \frac{2.6 \, N}{2.3 \times 10^{-7} \, m^2} = 1.1 \times 10^7 \, N/m^2$$

Part (c)

3. Solve $P = F/A$ for the force:
$$F = PA_n$$

4. Substitute numerical values:
$$F = (3.7 \times 10^5 \, N/m^2)(2.3 \times 10^{-7} \, m^2) = 0.085 \, N$$

INSIGHT
The pressure exerted by the needle in part (b) is about 650 times greater than the pressure due to the finger in part (a). This increase in pressure with decreasing area accounts for the sharp tips found on a variety of objects, including nails, pens, pencils, and syringes.

PRACTICE PROBLEM
Find the area that a force of 2.6 N would have to act on to produce a pressure of $3.7 \times 10^5 \, N/m^2$.
[**Answer:** $A = 7.0 \times 10^{-6} \, m^2$]

Some related homework problems: Problem 7, Problem 8

RWP* An interesting example of force, area, and pressure in nature is provided by the family of small aquatic birds referred to as rails, and in particular by the gallinule. This bird has exceptionally long toes that are spread out over a large area, as can be seen in **FIGURE 15-1**. The result is that the weight of the bird causes only a relatively small pressure as it walks on the soft, muddy shorelines encountered in its habitat. In some species, the pressure exerted while walking is so small that the birds can actually walk across lily pads without sinking into the water.

Atmospheric Pressure and Gauge Pressure

Atmospheric Pressure   We're all used to working under pressure—about 14.7 pounds per square inch to be precise. This is **atmospheric pressure**, P_{at}, a direct result of the weight of the air above us. In SI units, atmospheric pressure has the following value:

> **Atmospheric Pressure, P_{at}**
>
> $P_{at} = 1.01 \times 10^5 \, \text{N/m}^2$ 15-3
>
> SI unit: N/m^2

A shorthand unit for N/m^2 is the **pascal** (Pa):

$$1 \, \text{Pa} = 1 \, \text{N/m}^2 \qquad \text{15-4}$$

The pascal honors the pioneering studies of fluids by the French scientist Blaise Pascal (1623–1662). Thus, atmospheric pressure can be written as

$$P_{at} = 101 \, \text{kPa}$$

In British units, pressure is measured in pounds per square inch, and

$$P_{at} = 14.7 \, \text{lb/in}^2$$

Finally, a common unit for atmospheric pressure in weather forecasting is the **bar**, defined as follows:

$$1 \, \text{bar} = 10^5 \, \text{Pa} \approx 1 \, P_{at}$$

EXERCISE 15-3   THE FORCE OF ATMOSPHERIC PRESSURE

Find the force exerted on the palm of your hand by atmospheric pressure. Assume your palm measures 0.077 m by 0.094 m.

REASONING AND SOLUTION
Rearranging $P = F/A$ to solve for the force, and using $P_{at} = 1.01 \times 10^5 \, \text{N/m}^2$, we find

$$F = P_{at}A = (1.01 \times 10^5 \, \text{Pa})(0.077 \, \text{m})(0.094 \, \text{m}) = 730 \, \text{N}$$

Thus, the atmosphere pushes on the palm of your hand with a force of approximately 160 pounds! Of course, it also pushes on the back of your hand with essentially the same force—but in the opposite direction.

FIGURE 15-2 illustrates the forces exerted on your hand by atmospheric pressure. If your hand is vertical, atmospheric pressure pushes to the right and to the left equally, so your hand feels zero net force. If your hand is horizontal, atmospheric pressure exerts upward and downward forces on your hand that are essentially the same in magnitude, again giving zero net force. This cancellation of forces occurs no matter what the orientation of your hand; thus, we can conclude the following:

> The pressure in a fluid acts equally in all directions, and acts at right angles to any surface.

A vivid illustration of the strength of atmospheric pressure is shown in **FIGURE 15-3 (a)**. When air is removed from the inside of the can, atmospheric pressure causes it to collapse. A method of overcoming atmospheric pressure to protect a broken arm is shown in **FIGURE 15-3 (b)**.

---

*Real World Physics applications are denoted by the acronym RWP.

PHYSICS IN CONTEXT
Looking Ahead

Pressure appears again in Chapter 18 when we consider thermal processes, and the back-and-forth conversion of thermal and mechanical energy.

PROBLEM-SOLVING NOTE

Pressure Is Force Per Area

Remember that pressure is proportional to the applied force and *inversely* proportional to the area over which it acts.

▲ **FIGURE 15-1** This bird exerts only a small pressure on the lily pad on which it walks because its weight is spread out over a large area by its long toes. Since the pressure is not enough to sink a lily pad, the bird can seemingly "walk on water."

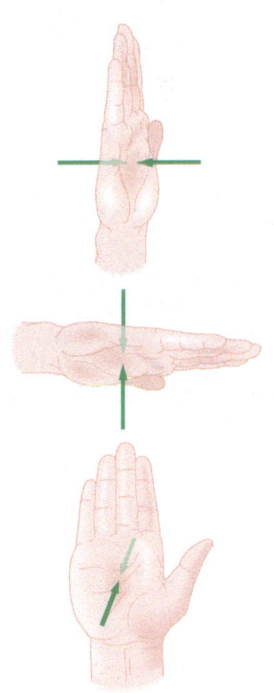

▲ **FIGURE 15-2** **Pressure is the same in all directions** The forces exerted on the two sides of a hand cancel, regardless of the hand's orientation. Hence, pressure acts equally in all directions.

▶ **FIGURE 15-3 Visualizing Concepts At-mospheric Pressure (a)** When the air is pumped out of a sealed can, atmospheric pressure produces an inward force that is unopposed. The resulting collapse of the can vividly illustrates the pressure that is all around us. **(b)** An air splint utilizes the same principle of unequal pressure. A plastic sleeve is placed around an injured limb and inflated to a pressure greater than that of the atmosphere—and thus of the body's internal pressure. The increased external pressure retards bleeding from the injured area, and also tends to immobilize the limb in case it might be fractured.

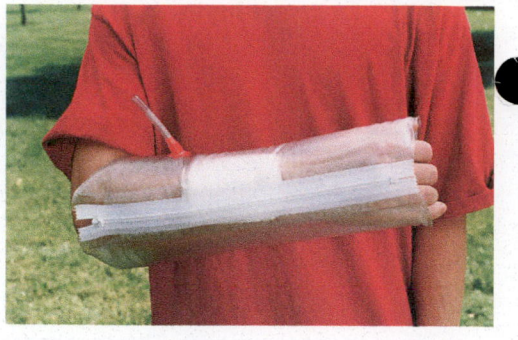

(a)   **(b)**

PROBLEM-SOLVING NOTE

Gauge Pressure

If a problem gives you the gauge pressure, recall that the actual pressure is the gauge pressure *plus* atmospheric pressure.

Gauge Pressure In many cases we're interested in the difference between a given pressure and atmospheric pressure. For example, a flat tire doesn't have zero pressure in it; the pressure in the tire is atmospheric pressure. To inflate the tire to 241 kPa (35 lb/in²), the pressure inside the tire must be greater than atmospheric pressure by this amount; that is, $P = 241 \text{ kPa} + P_{at} = 241 \text{ kPa} + 101 \text{ kPa} = 342 \text{ kPa}$.

To deal with such situations, we introduce the **gauge pressure,** P_g, defined as the amount of pressure that is in excess of atmospheric pressure:

$$P_g = P - P_{at} \qquad \text{15-5}$$

It is the gauge pressure that is determined by a tire gauge. Many problems in this chapter refer to the gauge pressure. Hence it must be remembered that the actual pressure in these cases is greater by the amount P_{at}; that is, $P = P_g + P_{at}$.

EXAMPLE 15-4   PRESSURING THE BALL: ESTIMATE THE GAUGE PRESSURE

Estimate the gauge pressure in a basketball by pushing down on it and noting the area of contact it makes with the surface on which it rests.

PICTURE THE PROBLEM

Our sketch shows the basketball both in its original state, and when a force F pushes downward on it. In the latter case, the ball flattens out on the bottom, forming a circular area of contact with the floor of diameter d.

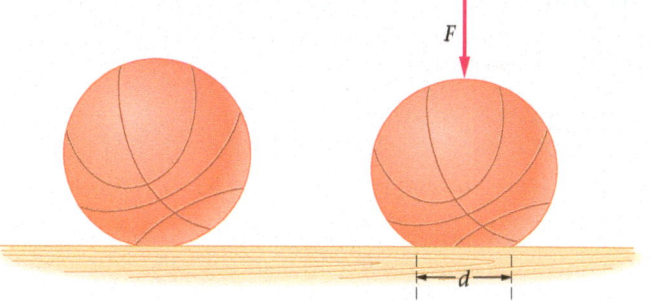

REASONING AND STRATEGY

To solve this problem, we have to make reasonable estimates of the force applied to the ball and the area of contact.

Suppose, for example, that we push down with a moderate force of 22 N (about 5 lb). The circular area of contact will probably have a diameter of about 2.0 centimeters. This can be verified by carrying out the experiment. Thus, given $F = 22$ N and $A = \pi(d/2)^2$, we can find the gauge pressure.

Known   Force of push, $F = 22$ N; diameter of contact with floor, $d = 2.0$ cm.

Unknown   Gauge pressure of basketball, $P_g = ?$

SOLUTION

1. Using these estimates, calculate the gauge pressure, P_g: $P_g = \dfrac{F}{A} = \dfrac{22 \text{ N}}{\pi\left(\dfrac{0.020 \text{ m}}{2}\right)^2} = 7.0 \times 10^4 \text{ Pa}$

INSIGHT

Given that a pressure of 1 atmosphere ($P_{at} = 101$ kPa $= 1.01 \times 10^5$ Pa) corresponds to 14.7 lb/in², it follows that the gauge pressure of the ball ($P_g = 7.0 \times 10^4$ Pa) corresponds to a pressure of roughly 10 lb/in². Thus, a basketball will typically have a gauge pressure in the neighborhood of 10 lb/in², and hence a total pressure inside the ball of about 25 lb/in².

CONTINUED

PRACTICE PROBLEM

What is the diameter of the circular area of contact if a basketball with a 12 lb/in^2 gauge pressure is pushed down with a force of 44 N (about 10 lb)?   [**Answer:** $d = 2.6$ cm]

Some related homework problems: Problem 9, Problem 11

Enhance Your Understanding   (Answers given at the end of the chapter)

2. A force F acts on a circular area of radius r. Rank the following systems in order of increasing pressure. Indicate ties where appropriate. **A,** $F = 10$ N, $r = 1$ cm; **B,** $F = 100$ N, $r = 10$ cm; **C,** $F = 20$ N, $r = 2$ cm; **D,** $F = 1$ N, $r = 0.1$ cm.

Section Review

- Pressure, P, is force per area: $P = F/A$.
- Atmospheric pressure is $P_{at} = 1.01 \times 10^5$ N/m^2 = 101 kPa, and gauge pressure is $P_g = P - P_{at}$.

15-3 Static Equilibrium in Fluids: Pressure and Depth

Pressure is exerted equally in all directions, but it is not equal at all *locations* within a fluid. Just think of the low air pressure on a mountaintop compared with the higher pressure at sea level. We explore this aspect of pressure in this section.

Pressure with Depth

Countless war movies have educated us on the perils of taking a submarine too deep. The hull creaks and groans, rivets start to pop, water begins to spray into the ship, and the captain keeps a close eye on the depth gauge. But what causes the pressure to increase as a submarine dives, and how much does it go up for a given increase in depth?

The answer to the first question is that the increased pressure is due to the added weight of water pressing on the submarine as it goes deeper. To see how this works, consider a cylindrical container filled to a height h with a fluid of density ρ, as in **FIGURE 15-4**. The top surface of the fluid is open to the atmosphere, with a pressure P_{at}. If the cross-sectional area of the container is A, the downward force exerted on the top surface by the atmosphere is

$$F_{top} = P_{at}A$$

Now, at the bottom of the container, the downward force is F_{top} *plus* the weight of the fluid. Recalling that $M = \rho V$, and that $V = hA$ for a cylinder of height h and area A, this weight is

$$W = Mg = \rho Vg = \rho(hA)g$$

Hence, we have

$$F_{bottom} = F_{top} + W = P_{at}A + \rho(hA)g$$

Finally, the pressure at the bottom is obtained by dividing F_{bottom} by the area A:

$$P_{bottom} = \frac{F_{bottom}}{A} = \frac{P_{at}A + \rho(hA)g}{A} = P_{at} + \rho gh$$

Of course, this relationship holds not only for the bottom of the container, but for *any depth h below the surface*. Thus, the answer to the second question is that if the depth increases by the amount h, the pressure increases by the amount ρgh. At a depth h below the surface of a fluid, then, the pressure P is given by

$$P = P_{at} + \rho gh \qquad\qquad \text{15-6}$$

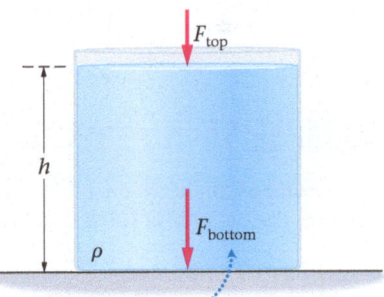

The force on the bottom is equal to the force on the top plus the weight of fluid in the flask.

▲ **FIGURE 15-4 Pressure and the weight of a fluid** The force pushing down on the bottom of the flask is greater than the force pushing down on the surface of the fluid. The difference in the force is the weight of the fluid in the flask.

Big Idea **2** Pressure increases with depth in a fluid. This is true whether the fluid is the air in the atmosphere or water in an ocean.

Pressure in Water and Alcohol

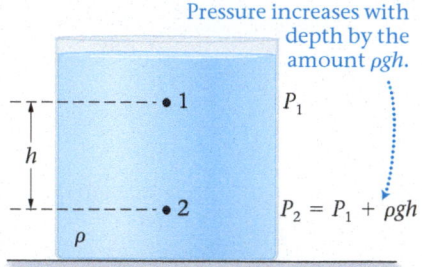

▲ **FIGURE 15-5 Pressure variation with depth** If point 2 is deeper than point 1 by the amount *h*, its pressure is greater by the amount *ρgh*.

This expression holds for any fluid with constant density ρ and a pressure P_{at} at its upper surface.

EXERCISE 15-5 TITANIC PRESSURE

RWP The *Titanic* was found in 1985 lying on the bottom of the North Atlantic at a depth of about 2.35 miles. What is the pressure at this depth?

REASONING AND SOLUTION

Applying $P = P_{at} + \rho gh$, with $\rho = 1025$ kg/m³ for seawater, we have

$$P = P_{at} + \rho gh = 1.01 \times 10^5 \text{ Pa} + (1025 \text{ kg/m}^3)(9.81 \text{ m/s}^2)\left(\frac{1609 \text{ m}}{1 \text{ mi}}\right)(2.35 \text{ mi})$$

$$= 3.81 \times 10^7 \text{ Pa}$$

This is about 377 atmospheres.

The relationship $P = P_{at} + \rho gh$ can be applied to any two points in a fluid. For example, if the pressure at one point is P_1, the pressure P_2 at a depth h below that point is the following:

Dependence of Pressure on Depth

$$P_2 = P_1 + \rho gh \tag{15-7}$$

This relationship is illustrated in **FIGURE 15-5** and utilized in the next Example.

EXAMPLE 15-6 PRESSURE AND DEPTH

A cubical box 20.00 cm on a side is completely immersed in a fluid. At the top of the box the pressure is 105.0 kPa; at the bottom the pressure is 106.8 kPa. What is the density of the fluid?

PICTURE THE PROBLEM

Our sketch shows the box at an unknown depth *d* below the surface of the fluid. The important dimension for this problem is the height of the box, which is 20.00 cm. We are also given that the pressures at the top and bottom of the box are $P_1 = 105.0$ kPa and $P_2 = 106.8$ kPa, respectively.

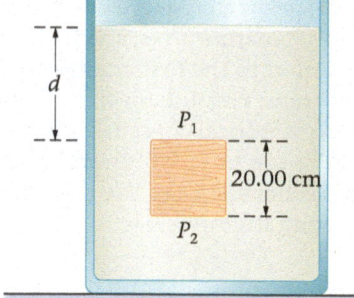

REASONING AND STRATEGY

The pressures at the top and bottom of the box are related by $P_2 = P_1 + \rho gh$. The pressures and the height of the box are given, and hence this relationship can be solved for the unknown density, ρ.

Known Vertical height of cubical box, $h = 20.00$ cm; pressure at top of box, $P_1 = 105.0$ kPa; pressure at bottom of box, $P_2 = 106.8$ kPa.

Unknown Density of fluid, $\rho = ?$

SOLUTION

1. Solve $P_2 = P_1 + \rho gh$ for the density:

$$\rho = \frac{P_2 - P_1}{gh}$$

2. Substitute numerical values:

$$\rho = \frac{(1.068 \times 10^5 \text{ Pa}) - (1.050 \times 10^5 \text{ Pa})}{(9.81 \text{ m/s}^2)(0.2000 \text{ m})} = 920 \text{ kg/m}^3$$

INSIGHT

Comparing this density with the values in Table 15-1, we see that the fluid in question is probably olive oil. If the box had been immersed in water instead, with its greater density, the difference in pressure between the top and bottom of the box would have been greater.

PRACTICE PROBLEM

Given the density obtained above, what is the depth of the fluid, *d*, at the top of the box? [**Answer:** $d = 0.44$ m]

Some related homework problems: Problem 14, Problem 19

CONCEPTUAL EXAMPLE 15-7   THE SIZE OF BUBBLES

One day, while swimming below the surface of the ocean, you let out a small bubble of air from your mouth. As the bubble rises toward the surface, does its diameter increase, decrease, or stay the same?

REASONING AND DISCUSSION

As the bubble rises, the pressure in the surrounding water decreases. This allows the air in the bubble to expand and occupy a larger volume.

ANSWER

The diameter of the bubble increases.

RWP An interesting application of the variation of pressure with depth is the **barometer,** which can be used to measure atmospheric pressure. We consider here the simplest type of barometer, first proposed by Evangelista Torricelli (1608–1647) in 1643. First, take a long glass tube—open at one end and closed at the other end—and fill it with a fluid of density ρ. Next, invert the tube and place its open end below the surface of the same fluid in a bowl, as shown in **FIGURE 15-6**. Some of the fluid in the tube will flow into the bowl, leaving an empty space (vacuum) at the top. Enough will remain, however, to create a difference in level, h, between the fluid in the bowl and the fluid in the tube.

The basic idea of the barometer is that this height difference is directly related to the atmospheric pressure that pushes down on the fluid in the bowl. To see how this works, first note that the pressure in the vacuum at the top of the tube is zero. Hence, the pressure in the tube at a depth h below the vacuum is $0 + \rho g h = \rho g h$. Now, at the level of the fluid in the bowl we know that the pressure is 1 atmosphere, P_{at}. Therefore, it follows that

$$P_{at} = \rho g h$$

If these pressures were not the same, there would be a pressure difference between the fluid in the tube and the fluid in the bowl, resulting in a net force and a flow of fluid. Thus, a measurement of h immediately gives atmospheric pressure.

A fluid that is often used in such a barometer is mercury (Hg), with a density of $\rho = 1.3595 \times 10^4 \ \mathrm{kg/m^3}$. Because of mercury's relatively high density, the corresponding height for a column of mercury is reasonably small. We find the following:

$$h = \frac{P_{at}}{\rho g} = \frac{1.013 \times 10^5 \ \mathrm{Pa}}{(1.3595 \times 10^4 \ \mathrm{kg/m^3})(9.81 \ \mathrm{m/s^2})} = 760 \ \mathrm{mm}$$

In fact, atmospheric pressure is *defined* in terms of millimeters of mercury (mmHg):

$$1 \text{ atmosphere} = P_{at} = 760 \ \mathrm{mmHg}$$

Table 15-2 lists the various expressions we have developed for atmospheric pressure.

Water Seeks Its Own Level

We're all familiar with the aphorism that water seeks its own level. In order for this to hold true, however, it's necessary that the pressure at the surface of the water (or other fluid) be the same everywhere on the surface. This was not the case for the barometer just discussed, where the pressure was P_{at} on one portion of the surface and zero on another. Let's take a moment, then, to consider the level assumed by a fluid with constant pressure on its surface. In doing so, we'll apply considerations involving force, pressure, and energy.

Balancing Forces   First, we consider the force-pressure point of view. In **FIGURE 15-7 (a)** we show a U-shaped tube containing a quantity of fluid of density ρ. The fluid rises to the same level in each arm of the U, where it is open to the atmosphere. Therefore, the

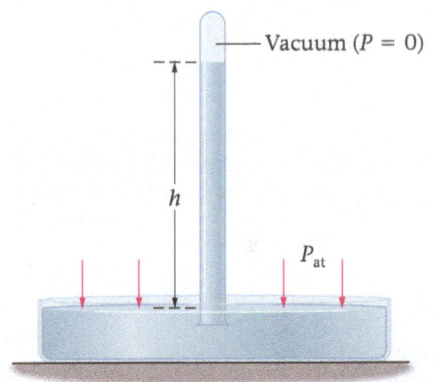

▲ **FIGURE 15-6 A simple barometer** Atmospheric pressure, P_{at}, is related to the height of fluid in the tube by $P_{at} = \rho g h$.

TABLE 15-2 Atmospheric Pressure

1 atmosphere	= P_{at}
	= 760 mmHg (definition)
	= 14.7 lb/in^2
	= 101 kPa
	= 101 kN/m^2
	~ 1 bar = 100 kPa

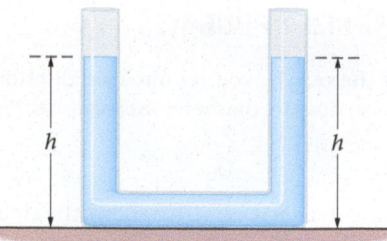

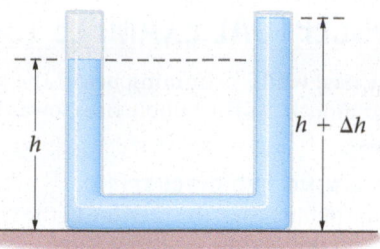

(a) A fluid is in equilibrium when the levels are equal

(b) Unequal levels result in unequal pressures and fluid flow

▲ **FIGURE 15-7 Fluids seek their own level (a)** When the levels are equal, the pressure is the same at the base of each arm of the U tube. As a result, the fluid in the horizontal section of the U is in equilibrium. **(b)** With unequal heights, the pressures are different. In this case, the pressure is greater at the base of the right arm; hence fluid will flow toward the left and the levels will equalize.

To create different levels, fluid must be moved from one side . . .

. . . to the other . . .

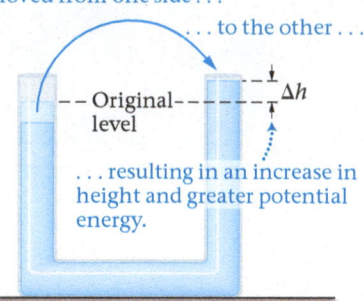

. . . resulting in an increase in height and greater potential energy.

▲ **FIGURE 15-8 Gravitational potential energy of a fluid** In order to create unequal levels in the two arms of the U tube, an element of fluid must be raised by the height Δh. This increases the gravitational potential energy of the system. The lowest potential energy corresponds to equal levels.

Water Level in Pascal's Vases

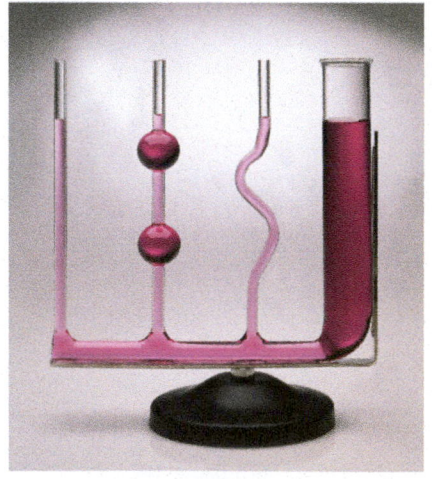

pressure at the base of each arm is the same: $P_{at} + \rho g h$. Thus, the fluid in the horizontal section of the U is pushed with equal force from each side, giving zero net force. As a result, the fluid remains at rest.

In contrast, in **FIGURE 15-7 (b)** the two arms of the U are filled to different levels. Therefore, the pressure at the base of the two arms is different, with the greater pressure at the base of the right arm. It follows that the fluid in the horizontal section experiences a net force to the left, causing it to move in that direction. This will tend to equalize the fluid levels of the two arms.

Minimizing Energy We can arrive at the same conclusion on the basis of energy minimization. Consider a U tube that is initially filled to the same level in both arms, as in Figure 15-7 (a). Now, consider moving a small element of fluid from one arm to the other, to create different levels, as in **FIGURE 15-8**. To move this fluid element to the other arm, it is necessary to lift it upward. This, in turn, causes an increase in the gravitational potential energy of the system. Nothing else in the system has changed its position, and therefore the only change in potential energy is the increase associated with the fluid element. Thus, we conclude that the system has a minimum energy when the fluid levels are the same and a higher energy when the levels are different. Just as a ball rolls to the bottom of a hill, where its energy is minimized, the fluid seeks its own level and a minimum energy. An illustration of a fluid seeking its own level is shown in **FIGURE 15-9**.

If two different liquids, with different densities, are combined in the same U tube, the levels in the arms are not the same. Still, the pressures at the base of each arm must be equal, as before. This is discussed in the next Example.

◄ **FIGURE 15-9** The containers shown here are connected at the bottom by a hollow tube, which allows fluid to flow freely between them. As a result the fluid level is the same in each container regardless of its shape and size.

EXAMPLE 15-8   OIL AND WATER DON'T MIX

A U-shaped tube is filled mostly with water, but a small amount of vegetable oil has been added to one side, as shown in the sketch. The density of the water is 1.00×10^3 kg/m³, and the density of the vegetable oil is 9.20×10^2 kg/m³. If the depth of the oil is 5.00 cm, what is the difference in level h between the top of the oil on one side of the U tube and the top of the water on the other side?

PICTURE THE PROBLEM

The U-shaped tube and the relevant dimensions are shown in our sketch. In particular, notice that the depth of the oil is 5.00 cm, and that the oil rises to a greater height on its side of the U tube than the water does on its side. This is due to the oil having the lower density. Both sides of the U tube are open to the atmosphere, so the pressure at the top surfaces is P_{at}.

CONTINUED

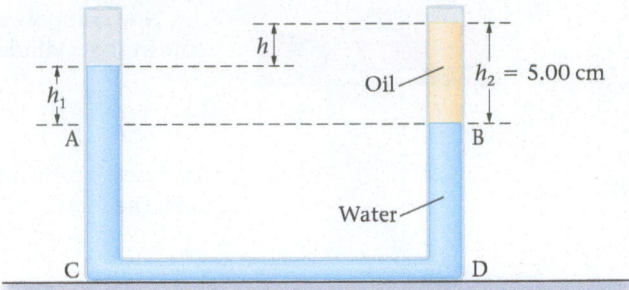

REASONING AND STRATEGY

For the system to be in equilibrium, it is necessary that the pressure be the same at the bottom of each side of the U tube—that is, at points C and D. If the pressure is the same at C and D, it will remain equal as one moves up through the water to points A and B. Above this point the pressures will differ because of the presence of the oil.

Therefore, setting the pressure at point A equal to the pressure at point B determines the depth h_1 in terms of the known depth h_2. It follows that the difference in level between the two sides of the U tube is $h = h_2 - h_1$.

Known   Density of water, $\rho_{water} = 1.00 \times 10^3 \, \text{kg/m}^3$; density of vegetable oil, $\rho_{oil} = 9.20 \times 10^2 \, \text{kg/m}^3$; depth of oil, $h_2 = 5.00 \, \text{cm}$.

Unknown   Height difference between the sides of the U tube, $h = ?$

SOLUTION

1. Find the pressure at point A, where the depth of the water is h_1:

$$P_A = P_{at} + \rho_{water} \, g h_1$$

2. Find the pressure at point B, where the depth of the oil is $h_2 = 5.00 \, \text{cm}$:

$$P_B = P_{at} + \rho_{oil} \, g h_2$$

3. Set P_A equal to P_B:

$$P_{at} + \rho_{water} \, g h_1 = P_{at} + \rho_{oil} \, g h_2$$

4. Solve the equation in Step 3 for the depth of the water, h_1, and substitute numerical values:

$$h_1 = h_2 \left(\frac{\rho_{oil}}{\rho_{water}} \right)$$

$$= (5.00 \, \text{cm}) \left(\frac{9.20 \times 10^2 \, \text{kg/m}^3}{1.00 \times 10^3 \, \text{kg/m}^3} \right) = 4.60 \, \text{cm}$$

5. Calculate the difference in levels between the water and oil sides of the U tube:

$$h = h_2 - h_1 = 5.00 \, \text{cm} - 4.60 \, \text{cm} = 0.40 \, \text{cm}$$

INSIGHT

The weight of the height h_1 of water is equal to the weight of the height h_2 of oil. This is a special case, however, and is due to the fact that our U tube has equal diameters on its two sides. If the oil side of the U tube had been wider, for example, the weight of the oil would have been greater, though the difference in height still would have been $h = 0.40 \, \text{cm}$.

PRACTICE PROBLEM

Find the pressure at points A and B.   [**Answer:** $P_A = P_B = P_{at} + 451 \, \text{Pa}$]

Some related homework problems: Problem 17, Problem 23, Problem 24

Pascal's Principle

Recall that if the surface of a fluid of density ρ is exposed to the atmosphere with a pressure P_{at}, the pressure at a depth h below the surface is

$$P = P_{at} + \rho g h$$

Suppose, now, that atmospheric pressure is increased from P_{at} to $P_{at} + \Delta P$. As a result, the pressure at the depth h is also increased:

$$P = P_{at} + \Delta P + \rho g h = (P_{at} + \rho g h) + \Delta P$$

Thus, by increasing the pressure at the top of the fluid by the amount ΔP, we have increased the pressure by the same amount everywhere in the fluid. This is **Pascal's principle:**

> An external pressure applied to an enclosed fluid is transmitted unchanged to every point within the fluid.

This principle applies no matter the size or shape of the container.

RWP A classic example of Pascal's principle at work is the **hydraulic lift,** which is shown schematically in **FIGURE 15-10**. Here we see two cylinders, one of cross-sectional area A_1, the other of cross-sectional area $A_2 > A_1$. The cylinders, each of which is fitted with a piston, are connected by a tube and filled with a fluid. Initially the pistons are at the same level and exposed to the atmosphere.

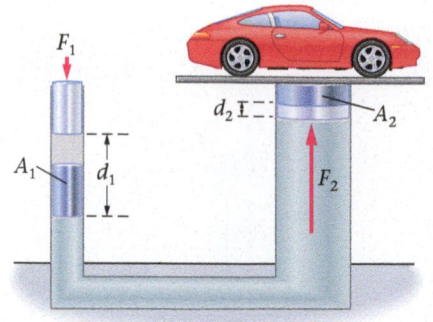

▲ **FIGURE 15-10 A hydraulic lift** A force F_1 exerted on the small piston causes a much larger force, F_2, to act on the large piston.

Now, suppose we push down on piston 1 with the force F_1. This increases the pressure in that cylinder by the amount

$$\Delta P = \frac{F_1}{A_1}$$

By Pascal's principle, the pressure in cylinder 2 increases by the *same* amount. Therefore, the increased upward force on piston 2 due to the fluid is

$$F_2 = (\Delta P)A_2$$

Substituting the increase in pressure, $\Delta P = F_1/A_1$, we find

$$F_2 = \left(\frac{F_1}{A_1}\right)A_2 = F_1\left(\frac{A_2}{A_1}\right) > F_1 \qquad \text{15-8}$$

To be specific, let's assume that A_2 is 100 times greater than A_1. Then, by pushing down on piston 1 with a force F_1, we push upward on piston 2 with a force of $100F_1$. Our force has been magnified 100 times!

If this sounds too good to be true, rest assured that we are not getting something for nothing. Just as with a lever, there is a tradeoff between the distance through which a force must be applied and the force magnification. This is illustrated in Figure 15-10, where we show piston 1 being pushed down through a distance d_1. This displaces a volume of fluid equal to A_1d_1. The same volume flows into cylinder 2, where it causes piston 2 to rise through a distance d_2. Equating the two volumes, we have

$$A_1d_1 = A_2d_2$$

Solving for d_2 gives

$$d_2 = d_1\left(\frac{A_1}{A_2}\right)$$

Thus, in the example just given, if we move piston 1 down a distance d_1, piston 2 rises a distance $d_2 = d_1/100$. Our force at piston 2 has been magnified 100 times, but the distance it moves has been reduced 100 times.

EXERCISE 15-9 HYDRAULIC LIFT

To inspect a 21,000-N truck, a mechanic raises it with a hydraulic lift. If the radius of the small piston in Figure 15-10 is 4.5 cm, and the radius of the large piston is 18 cm, find the force that must be exerted on the small piston to lift the truck.

REASONING AND SOLUTION

Solving $F_2 = (F_1/A_1)A_2$ for F_1, and noting that the area is πr^2, we find

$$F_1 = F_2\left(\frac{A_1}{A_2}\right) = (21{,}000\text{ N})\left[\frac{\pi(0.045\text{ m})^2}{\pi(0.18\text{ m})^2}\right] = 1300\text{ N}$$

Thus, the 4700-lb truck can be raised with a force of 290 lb.

Enhance Your Understanding (Answers given at the end of the chapter)

3. Is the increase in pressure from the surface of the ocean to a depth of 10 m greater than, less than, or equal to the increase in pressure from a depth of 100 m in the ocean to a depth of 110 m? Explain.

Section Review

- Pressure in a fluid with density ρ increases with depth h as follows: $P = P_{at} + \rho gh$.
- Pascal's principle states that an external pressure applied to a fluid is transmitted unchanged to every point within the fluid.

15-4 Archimedes' Principle and Buoyancy

The fact that a fluid's pressure increases with depth leads to many interesting consequences. Among them is the fact that a fluid exerts a net upward force on any object it surrounds. This is referred to as a **buoyant force.**

The Origin of Buoyancy   To see how buoyancy comes about, consider a cubical block immersed in a fluid of density ρ, as shown in **FIGURE 15-11**. The surrounding fluid exerts normal forces on all of its faces. Clearly, the horizontal forces pushing to the right and to the left are equal; hence they cancel and have no effect on the block.

The situation is quite different for the vertical forces, however. Notice, for example, that the downward force exerted on the top face is less than the upward force exerted on the lower face, because the pressure at the lower face is greater. This difference in forces gives rise to a net upward force—the buoyant force.

Let's calculate the buoyant force acting on the block. First, we assume that the cubical block is of length L on a side, and that the pressure on the top surface is P_1. The downward force on the block, then, is

$$F_1 = P_1 A = P_1 L^2$$

Note that we have used the fact that the area of a square face of side L is L^2. Next, we consider the bottom face. The pressure there is given by $P_2 = P_1 + \rho g h$, with a difference in depth of $h = L$:

$$P_2 = P_1 + \rho g L$$

Therefore, the upward force exerted on the bottom face of the cube is

$$F_2 = P_2 A = (P_1 + \rho g L)L^2 = P_1 L^2 + \rho g L^3 = F_1 + \rho g L^3$$

If we take upward as the positive direction, the net vertical force exerted by the fluid on the block—that is, the buoyant force, F_b—is

$$F_b = F_2 - F_1 = \rho g L^3$$

As expected, the block experiences a net upward force from the surrounding fluid.

Archimedes' Principle   The precise value of the buoyant force is of some significance, as we now show. First, recall that the volume of a cube is L^3. It follows that $\rho g L^3$ is the weight of *fluid* of density ρ that would occupy the same volume as the cube. Therefore, the buoyant force ($F_b = \rho g L^3$) is equal to the weight of fluid that is displaced by the cube. This is **Archimedes' principle:**

> An object immersed in a fluid experiences an upward buoyant force equal to the weight of the fluid displaced by the object.

This principle applies to an object of any size and shape, and to any type of fluid. Stated mathematically, if a volume V of an object is immersed in a fluid of density ρ_{fluid}, the buoyant force F_b is the following:

Buoyant Force When a Volume V Is Submerged in a Fluid of Density ρ_{fluid}

$$F_b = \rho_{fluid} g V \qquad\qquad 15\text{-}9$$

SI unit: N

The volume V may be the total volume of the object, or any fraction of the total volume that is immersed in the fluid.

To verify that Archimedes' principle is completely general, consider the submerged object shown in **FIGURE 15-12 (a)**. If we were to replace this object with an equivalent volume of fluid, as in **FIGURE 15-12 (b)**, the container would hold nothing but fluid and

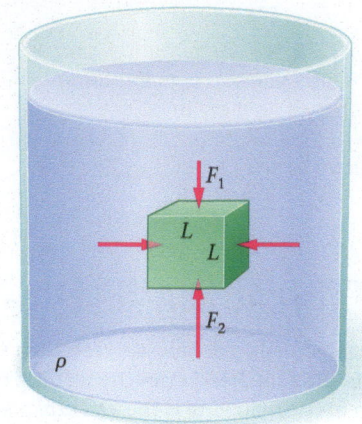

▲ **FIGURE 15-11 Buoyant force due to a fluid** A fluid surrounding an object exerts a buoyant force in the upward direction. This is due to the fact that pressure increases with depth, and hence the upward force on the object, F_2, is greater than the downward force, F_1. Forces acting to the left and to the right cancel.

PROBLEM-SOLVING NOTE

The Buoyant Force

Note that the buoyant force is equal to the weight of displaced fluid. It does not depend on the weight of the object that displaces the fluid.

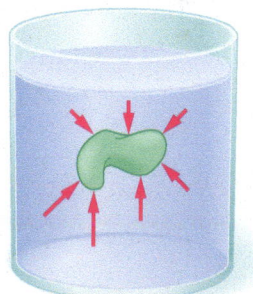

(a) The forces acting on an object surrounded by fluid

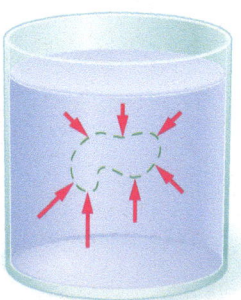

(b) The same forces act when the object is replaced by fluid

▶ **FIGURE 15-12 Buoyant force equals the weight of displaced fluid** The buoyant force acting on the object in **(a)** is equal to the weight of the "fluid object" (with the same size and shape) in **(b)**. This is because the fluid in (b) is at rest; hence the buoyant force acting on the fluid object must cancel its weight. The same forces act on the original object in (a), however, and therefore the buoyant force it experiences is also equal to the weight of the fluid object.

Big Idea 3 Archimedes' principle states that the buoyant force exerted by a fluid on an object is equal to the weight of the *fluid* that is displaced by the object.

would be in static equilibrium. As a result, we conclude that the net buoyant force acting on this "fluid object" must be upward and equal in magnitude to its weight. Now here's the key idea: Because the original object and the "fluid object" occupy the same position, the forces acting on their surfaces are identical, and hence the net buoyant force is the same for both objects. Therefore, the original object experiences a buoyant force equal to the weight of fluid that it displaces—that is, equal to the weight of the "fluid object."

CONCEPTUAL EXAMPLE 15-10   HOW IS THE SCALE READING AFFECTED?   PREDICT/EXPLAIN

A partially filled flask of water rests on a scale. **(a)** If you dip your finger into the water, without touching the flask, does the reading on the scale increase, decrease, or stay the same? **(b)** Which of the following is the *best explanation* for your prediction?

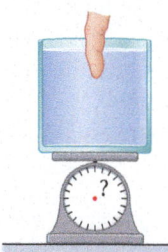

  I. The upward buoyant force on your finger results in a downward force on the flask, which increases the reading on the scale.

 II. The reading on the scale decreases because your finger has displaced some water.

III. There is no change in the reading of the scale because your finger doesn't actually touch the flask, and hence you can't exert a downward force on it.

REASONING AND DISCUSSION
Your finger experiences an upward buoyant force when it is dipped into the water. By Newton's third law, the water experiences an equal and opposite reaction force acting downward. This downward force is transmitted to the scale, which in turn gives a higher reading.

    Another way to look at this result is to note that when you dip your finger into the water, its depth increases. This results in a greater pressure at the bottom of the flask, and hence a greater downward force on the flask. The scale reads this increased downward force.

ANSWER
(a) The reading on the scale increases. **(b)** The best explanation is I.

Enhance Your Understanding   (Answers given at the end of the chapter)

  4. Is the buoyant force exerted on a cubical block of aluminum 10 cm on a side greater than, less than, or equal to the buoyant force exerted on a cubical block of lead 10 cm on a side, when both are completely submerged in water? Explain.

Section Review

•   Archimedes' principle is as follows: An object immersed in a fluid experiences an upward buoyant force equal to the weight of the fluid displaced by the object.

•   The mathematical statement of Archimedes' principle is $F_b = \rho_{\text{fluid}} g V$.

15-5 Applications of Archimedes' Principle

In this section we consider a variety of applications of Archimedes' principle. We begin with situations in which an object is fully immersed. Later we consider systems in which an object floats.

Complete Submersion

An interesting application of complete submersion can be found in an apparatus commonly used in determining a person's percentage of body fat. We consider the basic physics of the apparatus and the measurement procedure in the next Example. After the Example, we derive the relationship between overall body density and the body-fat percentage.

Weighing Weights in Water

EXAMPLE 15-11   MEASURING THE BODY'S DENSITY

RWP A person who weighs 720.0 N in air is lowered into a tank of water to about chin level. He sits in a harness of negligible mass suspended from a scale that reads his apparent weight. He now exhales as much air as possible and dunks his head underwater, submerging his entire body. If his apparent weight while submerged is 34.3 N, find **(a)** his volume and **(b)** his density.

PICTURE THE PROBLEM

The scale, the tank, and the person are shown in our sketch. We also show the free-body diagram for the person. The weight of the person in air is designated by $W = mg = 720.0$ N, and the apparent weight in water, which is the upward force exerted by the scale on the person, is designated by $W_a = 34.3$ N. The buoyant force exerted by the water on the person is F_b.

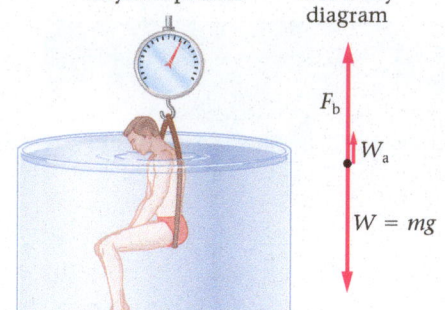
Physical picture   Free-body diagram

REASONING AND STRATEGY

To find the volume, V_p, and density, ρ_p, of the person, we must use two separate conditions. They are as follows:

a. When the person is submerged, the surrounding water exerts an upward buoyant force given by Archimedes' principle: $F_b = \rho_{water}V_pg$. This relationship, and Newton's second law, can be used to determine V_p.

b. The weight of the person in air is $W = mg = \rho_pV_pg$. Combining this relationship with the volume, V_p, found in part (a) allows us to determine the density, ρ_p.

Known   Weight of person, $W = 720.0$ N; apparent weight, $W_a = 34.3$ N.
Unknown   **(a)** Volume of person, $V_p = ?$ **(b)** Density of person, $\rho_p = ?$

SOLUTION

Part (a)

1. Apply Newton's second law to the person. The person remains at rest, and therefore the net force acting on him is zero:

$$W_a + F_b - W = 0$$

2. Substitute $F_b = \rho_{water}V_pg$ for the buoyant force, and solve for the volume of the person, V_p:

$$W_a + \rho_{water}V_pg - W = 0$$
$$V_p = \frac{W - W_a}{\rho_{water}g} = \frac{720.0\text{ N} - 34.3\text{ N}}{(1.00 \times 10^3\text{ kg/m}^3)(9.81\text{ m/s}^2)}$$
$$= 6.99 \times 10^{-2}\text{ m}^3$$

Part (b)

3. Use $W = \rho_pV_pg$ to solve for the density of the person, ρ_p:

$$W = \rho_pV_pg$$
$$\rho_p = \frac{W}{V_pg} = \frac{720.0\text{ N}}{(6.99 \times 10^{-2}\text{ m}^3)(9.81\text{ m/s}^2)}$$
$$= 1050\text{ kg/m}^3$$

INSIGHT

As in Conceptual Example 15-10, the water exerts an upward buoyant force on the person, and an equal and opposite reaction force acts downward on the tank and water, making it press against the floor with a greater force.

In addition, notice that the density of the person (1050 kg/m³) is only slightly greater than the density of seawater (1025 kg/m³), as given in Table 15-1.

PRACTICE PROBLEM

The person can float in water if his lungs are partially filled with air, which increases the volume of his body. What volume must his body have to just float?   [**Answer:** $V_p = 7.34 \times 10^{-2}$ m³]

Some related homework problems: Problem 43, Problem 44

BIO Once the overall density of the body is determined, the percentage of body fat can be obtained by noting that body fat has a density of $\rho_f = 9.00 \times 10^2$ kg/m³, whereas the lean body mass (muscles and bone) has a density of $\rho_l = 1.10 \times 10^3$ kg/m³. Suppose, for example, that a fraction x_f of the total body mass M is fat mass, and a fraction $(1 - x_f)$ is lean mass; that is, the fat mass is $m_f = x_fM$ and the lean mass is $m_l = (1 - x_f)M$. The total volume of the body is $V = V_f + V_l$. Using the fact that $V = m/\rho$, we can write the total volume as $V = m_f/\rho_f + m_l/\rho_l = x_fM/\rho_f + (1 - x_f)M/\rho_l$. Combining these results, we find that the overall density of a person's body is

$$\rho_p = \frac{M}{V} = \frac{1}{\dfrac{x_f}{\rho_f} + \dfrac{(1 - x_f)}{\rho_l}}$$

Solving for the body-fat fraction, x_f, yields

$$x_f = \frac{1}{\rho_p}\left(\frac{\rho_l\rho_f}{\rho_l - \rho_f}\right) - \frac{\rho_f}{\rho_l - \rho_f}$$

Finally, substituting the values for ρ_f and ρ_l, we find

$$x_f = \frac{(4950 \text{ kg/m}^3)}{\rho_p} - 4.50$$

This result is known in medical circles as *Siri's formula*. For example, if $\rho_p = 900$ kg/m^3 (all fat), we find $x_f = 1$; if $\rho_p = 1100$ kg/m^3 (no fat), we find $x_f = 0$. In the case of Example 15-11, where $\rho_p = 1050$ kg/m^3, we find that this person's body-fat fraction is $x_f = 0.214$, for a percentage of 21.4%. This is a reasonable value for a healthy adult male.

A recent refinement to the measurement of body-fat percentage is the Bod Pod (FIGURE 15-13), an egg-shaped, airtight chamber in which a person sits comfortably—high and dry, surrounded only by air. This device works on the same physical principle as submerging a person in water, only it uses air instead of water. Since air is about a thousand times less dense than water, the measurements of apparent weight must be roughly a thousand times more sensitive. Fortunately, this is possible with today's technology, allowing for a much more convenient means of measurement.

Next we consider a low-density object, such as a piece of wood, held down below the surface of a denser fluid.

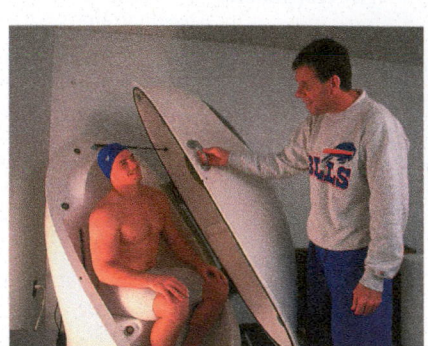

▲ **FIGURE 15-13** This device, known as the Bod Pod, measures the body-fat percentage of a person inside it by varying the air pressure in the chamber and measuring the corresponding changes in the person's apparent weight. Archimedes' principle is at work here, just as it is in Example 15-11.

EXAMPLE 15-12 FIND THE TENSION IN THE STRING

A piece of wood with a density $\rho_{wood} = 706$ kg/m^3 is tied with a string to the bottom of a water-filled flask. The wood is completely immersed and has a volume $V_{wood} = 8.00 \times 10^{-6}$ m^3. What is the tension in the string? (Recall that the density of water is $\rho_{water} = 1.00 \times 10^3$ kg/m^3.)

PICTURE THE PROBLEM
Our sketch shows the piece of wood held in place underwater by a string. Also shown is the free-body diagram for the wood, indicating the three forces acting on it.

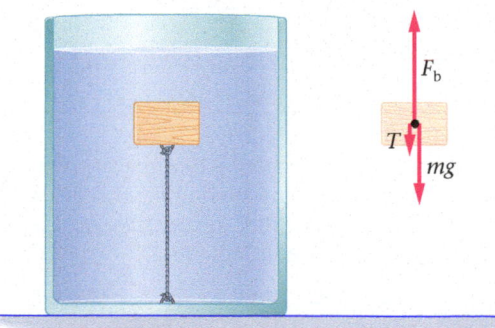

REASONING AND STRATEGY
To find the tension in the string we apply Newton's second law to the piece of wood, noting that the wood is at rest and hence has zero acceleration. The weight of the wood has a magnitude of $\rho_{wood}gV_{wood}$, and the buoyant force has a magnitude of $\rho_{water}gV_{wood}$.

Known Density of wood, $\rho_{wood} = 706$ kg/m^3; volume of wood, $V_{wood} = 8.00 \times 10^{-6}$ m^3; density of water, $\rho_{water} = 1.00 \times 10^3$ kg/m^3.

Unknown Tension in the string, $T = ?$

SOLUTION

1. Apply Newton's second law to the piece of wood: $\quad F_b - T - mg = 0$

2. Solve for the tension, T: $\quad T = F_b - mg$

3. Calculate the buoyant force:
$$F_b = \rho_{water}gV_{wood}$$
$$= (1.00 \times 10^3 \text{ kg/m}^3)(9.81 \text{ m/s}^2)(8.00 \times 10^{-6} \text{ m}^3)$$
$$= 0.0785 \text{ N}$$

4. Calculate the weight of the wood:
$$mg = \rho_{wood}gV_{wood} = (706 \text{ kg/m}^3)(9.81 \text{ m/s}^2)(8.00 \times 10^{-6} \text{ m}^3)$$
$$= 0.0554 \text{ N}$$

5. Subtract to obtain the tension:
$$T = F_b - mg$$
$$= 0.0785 \text{ N} - 0.0554 \text{ N} = 0.0231 \text{ N}$$

CONTINUED

INSIGHT

The wood is less dense than water, and hence its buoyant force when completely immersed (0.0785 N) is greater than its weight (0.0554 N).

PRACTICE PROBLEM — PREDICT/CALCULATE

(a) If the density of the wood is increased, does the tension in the string increase, decrease, or stay the same? Explain.
(b) What is the tension in the string if the piece of wood has a density of 822 kg/m^3? [**Answer: (a)** The tension decreases because the weight of the wood is closer to the value of the buoyant force. **(b)** The tension is 0.0140 N, which is less than that found above.]

Some related homework problems: Problem 40, Problem 93

▶ **FIGURE 15-14 Floatation (a)** The block of wood displaces some water, but not enough to equal its weight. Thus, the block would not float at this position. **(b)** The weight of displaced water equals the weight of the block in this case. The block floats now.

Floatation

When an object floats, the buoyant force acting on it equals its weight. For example, suppose we slowly lower a block of wood into a flask of water. At first, as in **FIGURE 15-14 (a)**, only a small amount of water is displaced and the buoyant force is a fraction of the block's weight. If we were to release the block now, it would drop farther into the water. As we continue to lower the block, more water is displaced, increasing the buoyant force.

Eventually, we reach the situation pictured in **FIGURE 15-14 (b)**, where the block begins to float. In this case, the buoyant force equals the weight of the wood. This, in turn, means that the weight of the displaced water is equal to the weight of the wood. In general:

> An object floats when it displaces an amount of fluid whose weight is equal to the weight of the object.

This is illustrated in **FIGURE 15-15 (a)**, where we show the volume of water equal to the weight of a block of wood. The wood displaces just this amount of water when it floats. A similar case involving human floatation is shown in **FIGURE 15-16**.

On the other hand, in **FIGURE 15-15 (b)** we show the amount of water necessary to have the same weight as a block of metal. Clearly, if the metal is completely submerged, the buoyant force is only a fraction of its weight, and so it sinks. On the other hand, if the metal is formed into the shape of a bowl, as in **FIGURE 15-15 (c)**, it can displace a

(a) Some water is displaced, but not enough to float the wood

(b) More water is displaced, and now the wood floats

(a) Block of wood and water of equal weight

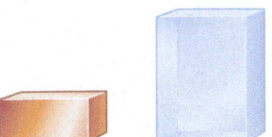

(b) Block of metal and water of equal weight

(c) Same block of metal as in (b), but shaped like a bowl to displace more water

▲ **FIGURE 15-15 Floating an object that is more dense than water (a)** A wood block and the volume of water that has the same weight. Because the wood has a larger volume than this, it floats. **(b)** A metal block and the volume of water that has the same weight. Since the metal displaces less water than this, it sinks. **(c)** If the metal in (b) is shaped like a bowl, it can displace more water than the volume of the metal itself. In fact, it can displace enough water to float.

▲ **FIGURE 15-16** The water of the Dead Sea is unusually dense because of its high salt content. As a result, swimmers can float higher in the water than they are accustomed and engage in recreational activities that we don't ordinarily associate with a dip in the ocean.

▶ **FIGURE 15-17** **Visualizing Concepts**
Floatation **(a)** Although steel is denser than water, a ship's bowl-like hull displaces enough water to allow it to float. (The boundary between the red and black areas of the hull is the Plimsoll line mentioned in Conceptual Example 15-14.) **(b)** For hot-air balloons, the key to buoyancy is the low density of the heated air in its envelope.

(a)　　　　　　　　　　　　　　　　　　　　　**(b)**

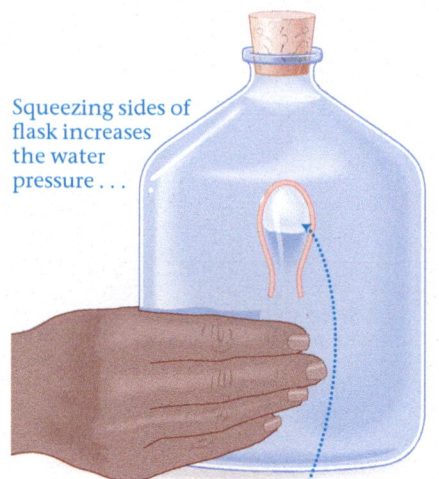

Squeezing sides of flask increases the water pressure . . .

. . . and shrinks the air bubble, causing the diver to drop.

▲ **FIGURE 15-18** **A Cartesian diver** A Cartesian diver floats because of the bubble of air trapped within it. When the bottle is squeezed, which increases the pressure in the water, the bubble is reduced in size and the diver sinks.

▲ **FIGURE 15-19** Felix Baumgartner prepares to jump from an altitude of 38,969 m (24.2 mi). He was carried to that altitude by a helium-filled balloon that operates on Archimedes' principle.

volume of water equal to its weight and float. A similar example is shown in **FIGURE 15-17 (a)**.

Another way to change the buoyancy of an object is to alter its overall density. A good example is provided by the hot-air balloons in **FIGURE 15-17 (b)**, which use low-density heated air to produce an overall density less than that of air. Similar considerations apply to the *Cartesian diver* shown in **FIGURE 15-18**. As illustrated, the diver is simply a small glass tube, similar in shape to an eyedropper, with an air bubble trapped inside. Initially, the overall density of the tube and the air bubble is less than the density of water, and the diver floats. When the bottle containing the diver is squeezed, however, the pressure in the water rises, and the air bubble is compressed to a smaller volume. Now, the overall density of the tube and air bubble is greater than that of water, and the diver descends. By adjusting the pressure on the bottle, the diver can be made to float at any depth in the bottle.

BIO The principle behind the Cartesian diver also applies to the swim bladder of ray-finned bony fishes. The swim bladder is basically an air sac whose volume can be controlled by the fish. By adjusting the size of the swim bladder, the fish can give itself "neutral buoyancy"—that is, the fish can float without effort at a given depth, just like the Cartesian diver. Similar considerations apply to certain diving sea mammals, such as the bottlenose dolphin, Weddell seal, elephant seal, and blue whale. All of these animals are capable of diving to great depths—in fact, some of the seals have been observed at depths of nearly 400 m. In order to conserve energy on their long dives, they take advantage of the fact that the pressure of the surrounding water compresses their bodies and flattens the air sacs in their lungs. Just as with the Cartesian diver, this decreases their buoyancy to the point where they begin to sink. As a result, they can glide effortlessly to the desired depth, saving energy for the swim back to the surface.

RWP Just as fish and dolphins are surrounded by an ocean of liquid water, balloons are surrounded by an ocean of air. The warm air inside a hot-air balloon is less dense than the surrounding cold air, so that the buoyant force on the balloon (from the displacement of the cold air) is larger than its weight, and the balloon rises into the sky. Hydrogen or helium-filled balloons work on the same principle, and are frequently used to make weather measurements. The behemoth high-altitude balloons, some as large as 30 m in diameter, rise into the atmosphere until the average density of the balloon matches the density of the surrounding atmosphere, often at an altitude of 30 km (100,000 ft) or higher. On October 14, 2012, Austrian skydiver Felix Baumgartner rode a helium-filled balloon to an altitude of 38,969 m (127,852 ft, or 24.2 mi) and jumped (**FIGURE 15-19**), setting a record for the highest skydive ever, and becoming the first human to break the sound barrier without any form of engine power.

QUICK EXAMPLE 15-13　FLOATING A BLOCK OF WOOD

How much water (density 1000 kg/m³) must be displaced to float a cubical block of wood (density 634 kg/m³) that is 13.3 cm on a side?

REASONING AND SOLUTION
We start by finding the volume of the wood block, V_{wood}, from its dimensions. Once we have this, we can find its mass ($m_{wood} = \rho_{wood} V_{wood}$) and weight ($W_{wood} = m_{wood} g$). Next,

CONTINUED

we set the weight of a volume of water ($W_{water} = \rho_{water} V_{water} g$) equal to the weight of the wood. This expression can be solved for the only unknown, the volume of the water.

1. Calculate the volume of the wood: $V_{wood} = (0.133 \text{ m})^3 = 0.00235 \text{ m}^3$

2. Find the weight of the wood:
$$W_{wood} = m_{wood}g = \rho_{wood}V_{wood}g$$
$$= (634 \text{ kg/m}^3)(0.00235 \text{ m}^3)(9.81 \text{ m/s}^2)$$
$$= 14.6 \text{ N}$$

3. Write an expression for the weight of an arbitrary volume of water, V_{water}:
$$W_{water} = m_{water}g = \rho_{water}V_{water}g$$

4. Set the weight of water equal to the weight of the wood:
$$W_{water} = \rho_{water}V_{water}g$$
$$= W_{wood} = 14.6 \text{ N}$$

5. Solve for the volume of water:
$$V_{water} = \frac{W_{wood}}{\rho_{water}g} = \frac{14.6 \text{ N}}{(1000 \text{ kg/m}^3)(9.81 \text{ m/s}^2)}$$
$$= 0.00149 \text{ m}^3 = 0.634 V_{wood}$$

As expected, only a fraction of the wood (63.4%) must be submerged for it to float.

CONCEPTUAL EXAMPLE 15-14 THE PLIMSOLL MARK

On the side of a cargo ship you may see a horizontal line indicating "maximum load." (The line is sometimes known as the "Plimsoll mark," after the nineteenth-century British legislator who caused it to be adopted.) When a ship is loaded to capacity, the maximum load line is at water level. The ship shown here has two maximum load lines, one for freshwater and one for salt water. Which line should be marked "maximum load for salt water": **(a)** the top line or **(b)** the bottom line?

REASONING AND DISCUSSION
If a ship sails from freshwater into salt water it floats higher, just as it is easier for you to float in an ocean than in a lake. The reason is that salt water is denser than freshwater; hence less of it needs to be displaced to provide a given buoyant force. Because the ship floats higher in salt water, the bottom line should be used to indicate maximum load.

ANSWER
(b) The bottom line should be used in salt water.

Tip of the Iceberg

As we've seen, an object floats when its weight is equal to the weight of the fluid it displaces. Let's use this condition to determine just how much of a floating object is submerged. We will then apply our result to the classic case of an iceberg, as shown in **FIGURE 15-20**.

Consider a solid of density ρ_s floating in a fluid of density ρ_f, as in **FIGURE 15-21**. If the solid has a volume V_s, its total weight is

$$W_s = \rho_s V_s g$$

Similarly, the weight of a volume V_f of displaced fluid is

$$W_f = \rho_f V_f g$$

Equating these weights, we find the following:

$$W_s = W_f$$
$$\rho_s V_s g = \rho_f V_f g$$

Canceling g, and solving for the volume of displaced fluid, we have

$$V_f = V_s(\rho_s/\rho_f)$$

Noting that the volume of displaced fluid, V_f, is the same as the volume of the solid that is submerged, V_{sub}, we find:

▲ **FIGURE 15-20** Most people know that the bulk of an iceberg lies below the surface of the water. But, as with ships and swimmers, the actual proportion that is submerged depends on whether the water is fresh or salt (see Example 15-15).

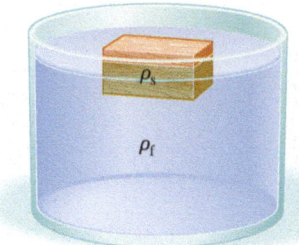

◄ **FIGURE 15-21 Submerged volume of a floating object** A solid, of volume V_s and density ρ_s, floats in a fluid of density ρ_f. The volume of the solid that is submerged is $V_{sub} = V_s(\rho_s/\rho_f)$.

Submerged Volume V_{sub} for a Solid of Volume V_s and Density ρ_s Floating in a Fluid of Density ρ_f

$$V_{sub} = V_s(\rho_s/\rho_f) \qquad\qquad\qquad 15\text{-}10$$

SI unit: m^3

As a check, notice that this result agrees with our calculations in Quick Example 15-13. We now apply this result to the classic case of ice floating in water.

EXAMPLE 15-15   THE TIP OF THE ICEBERG

RWP A chunk of ice floats in fresh water. What percentage of the ice projects above the level of the water? Assume a density of 917 kg/m^3 for the ice and 1.00×10^3 kg/m^3 for the water.

PICTURE THE PROBLEM

Our sketch shows a chunk of ice floating in a glass of water. In this case the solid object is ice, with a density $\rho_s = \rho_{ice} = 917$ kg/m^3, and the fluid is water, with a density $\rho_f = \rho_{water} = 1.00 \times 10^3$ kg/m^3. Thus, in this system, we make the following identifications: s = solid → ice, f = fluid → water.

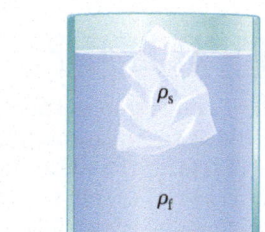

REASONING AND STRATEGY

We can use Equation 15-10, $V_{sub} = V_s(\rho_s/\rho_f)$, to find the volume of ice that is below the water (that is, *submerged*). It follows that the fraction of the total volume of the ice, V_s, that is submerged is $V_{sub}/V_s = \rho_s/\rho_f$. The fraction of the ice that is above the water, then, is $1 - V_{sub}/V_s$. Multiplying this fraction by 100 yields the percentage p above water.

Known   Density of ice, $\rho_s = \rho_{ice} = 917$ kg/m^3; density of water, $\rho_f = \rho_{water} = 1.00 \times 10^3$ kg/m^3.

Unknown   Percentage of ice above water level, $p = 100(1 - V_{sub}/V_s) = ?$

SOLUTION

1. Calculate the fraction of the total volume of the ice that is submerged:

$$\frac{V_{sub}}{V_s} = \frac{\rho_s}{\rho_f} = \frac{917 \text{ kg/m}^3}{1.00 \times 10^3 \text{ kg/m}^3} = 0.917$$

2. Calculate the fraction of the ice that is above the water level:

$$1 - \frac{V_{sub}}{V_s} = 1 - 0.917 = 0.083$$

3. Multiply by 100 to obtain a percentage:

$$p = 100\left(1 - \frac{V_{sub}}{V_s}\right) = 100(0.083) = 8.3\%$$

INSIGHT

Because we seek a percentage, it's not necessary to know the total volume of the ice. Thus, our result that 8.3% of the ice is above the water level applies whether we are talking about an ice cube in a drinking glass, or an iceberg floating in a freshwater lake. If an iceberg floats in the ocean, however, it will float higher due to the higher density of seawater. We consider this case in the following Practice Problem.

PRACTICE PROBLEM

What percentage of an ice chunk is above the water level if it floats in seawater? (The density of seawater can be found in Table 15-1.)   [**Answer: 10.5%**]

Some related homework problems: Problem 46, Problem 47, Problem 87

CONCEPTUAL EXAMPLE 15-16   THE NEW WATER LEVEL I

A cup is filled to the brim with water and a floating ice cube. When the ice melts, which of the following occurs: **(a)** water overflows the cup, **(b)** the water level is lower, or **(c)** the water level remains the same?

CONTINUED

REASONING AND DISCUSSION

The ice cube floats, and hence it displaces a volume of water equal to its weight. But when the ice cube melts, it becomes water, and its weight is the same. Hence, the melted water fills exactly the same volume that the ice cube displaced when floating. As a result, the water level is unchanged.

ANSWER

(c) The water level remains the same.

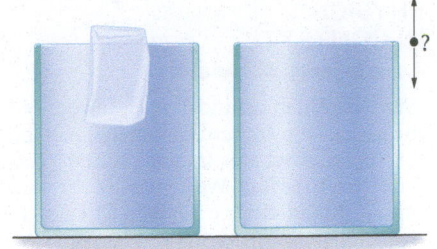

CONCEPTUAL EXAMPLE 15-17   THE NEW WATER LEVEL II

A cup is filled to the brim with water and a floating ice cube. Resting on top of the ice cube is a small pebble. When the ice melts, which of the following occurs: **(a)** water overflows the cup, **(b)** the water level is lower, or **(c)** the water level remains the same?

REASONING AND DISCUSSION

We know from the previous Conceptual Example that the ice itself makes no difference to the water level. As for the pebble, when it floats on the ice it displaces an amount of water equal to its *weight*. When the ice melts, the pebble drops to the bottom of the cup, where it displaces a volume of water equal to its own *volume*. Because the volume of the pebble is less than the volume of water with the same weight, we conclude that less water is displaced after the ice melts. Hence, the water level is lower.

ANSWER

(b) The water level is lower.

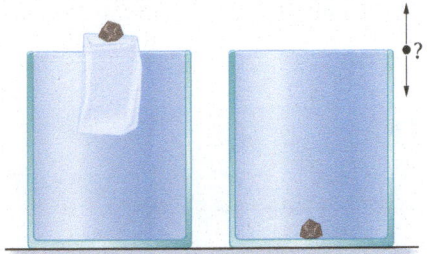

Enhance Your Understanding   (Answers given at the end of the chapter)

5. A cup is filled to the brim with water. Floating in the water is an ice cube, on top of which rests a small block of wood, as shown in **FIGURE 15-22**. When the ice melts, and the block of wood floats in the water, is the level of the water higher (causing the cup to overflow), is it lower, or does it stay the same? Explain.

Section Review

- The condition for an object to float in a fluid is as follows: An object floats when it displaces an amount of fluid whose weight is equal to the weight of the object.

▲ **FIGURE 15-22** Enhance Your Understanding 5.

15-6 Fluid Flow and Continuity

Suppose you want to water the yard, but you don't have a spray nozzle for the end of the hose. Without a nozzle the water flows rather slowly from the hose and hits the ground within half a meter. But if you place your thumb over the end of the hose, narrowing the opening to a fraction of its original size, the water sprays out with a high speed and a large range. Why does decreasing the size of the opening have this effect?

Equation of Continuity   To answer this question, we begin by considering a simple system that shows the same behavior. Imagine a fluid that flows with a speed v_1 through a cylindrical pipe of cross-sectional area A_1, as in the left-hand portion of **FIGURE 15-23**. If the pipe narrows to a cross-sectional area A_2, as in the right-hand portion of Figure 15-23, the fluid will flow with a new speed, v_2.

We can find the speed in the narrow section of the pipe by assuming that any amount of fluid that passes point 1 in a given time, Δt, must also flow past point 2 in the same time. If this were not the case, the system would be gaining or losing fluid. To

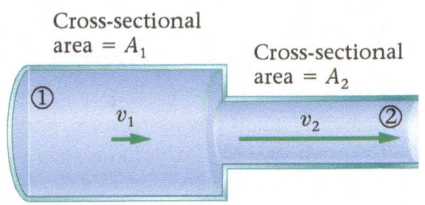

▲ **FIGURE 15-23 Fluid flow through a pipe of varying diameter** As a fluid flows from a large pipe to a small pipe, the same mass of fluid passes a given point in a given amount of time. Thus, the speed in the small pipe is greater than it is in the large pipe.

PROBLEM-SOLVING NOTE

Continuity of Flow

The speed of an incompressible fluid is inversely proportional to the area through which it flows.

▲ **FIGURE 15-24** Narrowing the opening in a hose with a nozzle (or thumb) increases the velocity of flow, as one would expect from the equation of continuity.

find the mass of fluid passing point 1 in the time Δt, note that the fluid moves through a distance $v_1 \Delta t$ in this time. As a result, the volume of fluid going past point 1 is

$$\Delta V_1 = A_1 v_1 \Delta t$$

Hence, the mass of fluid passing point 1 is

$$\Delta m_1 = \rho_1 \Delta V_1 = \rho_1 A_1 v_1 \Delta t$$

Similarly, the mass passing point 2 in the time Δt is

$$\Delta m_2 = \rho_2 \Delta V_2 = \rho_2 A_2 v_2 \Delta t$$

Notice that we have allowed for the possibility of the fluid having different densities at points 1 and 2.

Finally, equating these two masses yields the relationship between v_1 and v_2:

$$\Delta m_1 = \Delta m_2$$
$$\rho_1 A_1 v_1 \Delta t = \rho_2 A_2 v_2 \Delta t$$

Canceling Δt we find:

Equation of Continuity

$$\rho_1 A_1 v_1 = \rho_2 A_2 v_2 \qquad \text{15-11}$$

This relationship is referred to as the **equation of continuity.**

Incompressible Fluids   Most gases are readily compressed, which means that their densities can change. In contrast, most liquids are practically incompressible, so their densities are essentially constant. Unless stated otherwise, we will assume that all liquids discussed in this text are perfectly incompressible. Thus, for liquids, ρ_1 and ρ_2 are the same in Equation 15-11, and the equation of continuity reduces to the following:

Equation of Continuity for an Incompressible Fluid

$$A_1 v_1 = A_2 v_2 \qquad \text{15-12}$$

Reducing the area of flow increases the speed, as we see in **FIGURE 15-24**.

We next apply this relationship to the case of water flowing through the nozzle of a fire hose.

EXAMPLE 15-18   SPRAY I

Water travels through a 9.8-cm-diameter fire hose with a speed of 1.5 m/s. At the end of the hose, the water flows out through a nozzle whose diameter is 2.6 cm. What is the speed of the water coming out of the nozzle?

PICTURE THE PROBLEM
In our sketch, we label the speed of the water in the hose with v_1 and the speed of the water coming out the nozzle with v_2. We are given that $v_1 = 1.5$ m/s. We also know that the diameter of the hose is $d_1 = 9.8$ cm and the diameter of the nozzle is $d_2 = 2.6$ cm.

REASONING AND STRATEGY
We can find the water speed in the nozzle by applying $A_1 v_1 = A_2 v_2$
(Equation 15-12). In addition, we assume that the hose and nozzle are circular in cross section; hence, their areas are given by $A = \pi d^2/4$, where d is the diameter.

Known   Diameter of fire hose, $d_1 = 9.8$ cm; speed of water in fire hose, $v_1 = 1.5$ m/s; diameter of nozzle, $d_2 = 2.6$ cm.
Unknown   Speed of water in nozzle, $v_2 = ?$

SOLUTION

1. Solve $A_1 v_1 = A_2 v_2$ for v_2, the speed of water in the nozzle: $v_2 = v_1(A_1/A_2)$

2. Replace the areas with $A = \pi d^2/4$: $v_2 = v_1\left(\dfrac{\pi d_1^2/4}{\pi d_2^2/4}\right) = v_1\left(\dfrac{d_1^2}{d_2^2}\right)$

3. Substitute numerical values: $v_2 = v_1\left(\dfrac{d_1^2}{d_2^2}\right) = (1.5 \text{ m/s})\left(\dfrac{9.8 \text{ cm}}{2.6 \text{ cm}}\right)^2 = 21 \text{ m/s}$

CONTINUED

INSIGHT
Clearly, a small-diameter nozzle can give very high speeds. In fact, the speed depends inversely on the diameter squared.

PRACTICE PROBLEM
What diameter would be required to give the water a speed of 14 m/s in the nozzle?   [**Answer:** $d_2 = 3.2$ cm]

Some related homework problems: Problem 51, Problem 56, Problem 57

Enhance Your Understanding   (Answers given at the end of the chapter)

6. Water flows with a speed v through a pipe. If the diameter of the pipe is decreased to half its initial value, what is the new speed of the water? Explain.

Section Review

• The equation of continuity for a flowing fluid is $\rho_1 A_1 v_1 = \rho_2 A_2 v_2$. For an incompressible fluid, the density is constant, and hence the equation of continuity simplifies to $A_1 v_1 = A_2 v_2$.

15-7 Bernoulli's Equation

In this section, we apply the work-energy theorem to fluids. The result is a relationship among the pressure of a fluid, its speed, and its height. This relationship is known as **Bernoulli's equation.**

Change in Speed

We begin by considering a system in which the speed of the fluid changes, but its height remains the same. To be specific, the system of interest is the same as that shown in Figure 15-23. We have already shown that the speed of the fluid increases as it flows from region 1 to region 2; we now investigate the corresponding change in pressure.

Our plan of attack is to first calculate the total work done on the fluid as it moves from one region to the next. This result will depend on the pressure in the fluid. Once the total work is obtained, the work-energy theorem allows us to equate it to the change in kinetic energy of the fluid. This will give the pressure-speed relationship we desire.

Consider an element of fluid of length Δx_1, as shown in **FIGURE 15-25**. This element is pushed in the direction of motion by the pressure P_1. Thus, P_1 does positive work, ΔW_1, on the fluid element. Since the force exerted on the element is $F_1 = P_1 A_1$, and work is force times distance, the work done on the element by P_1 is

$$\Delta W_1 = F_1 \Delta x_1 = P_1 A_1 \Delta x_1$$

The volume of the fluid element is $\Delta V_1 = A_1 \Delta x_1$, so the work done by P_1 is

$$\Delta W_1 = P_1 \Delta V_1$$

Next, when the fluid element emerges into region 2, it experiences a force opposite to its direction of motion due to the pressure P_2. Thus, P_2 does negative work on the element. Following the same steps given previously, we can write the work done by P_2 as

$$\Delta W_2 = -P_2 \Delta V_2$$

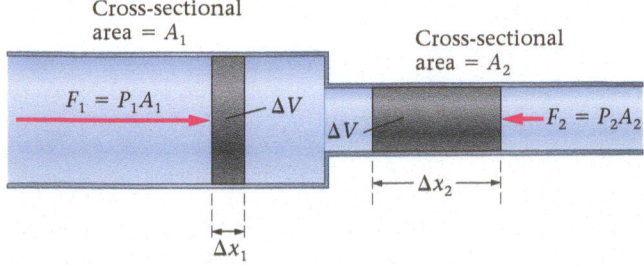

◄ **FIGURE 15-25 Work done on a fluid element** As an incompressible fluid element of volume ΔV moves from pipe 1 to pipe 2, the pressure P_1 does a positive work $P_1 \Delta V$ and the pressure P_2 does a negative work $P_2 \Delta V$. Since P_1 is greater than P_2, the net result is that positive work is done, and the fluid element speeds up.

Now, for an incompressible fluid, the volume of the element does not change as it goes from region 1 to region 2. Therefore,

$$\Delta V_1 = \Delta V_2 = \Delta V$$

Using this result, we can write the total work done on the fluid element as follows:

$$\Delta W_{total} = \Delta W_1 + \Delta W_2 = P_1 \Delta V - P_2 \Delta V = (P_1 - P_2)\Delta V$$

The final step is to equate the total work to the change in kinetic energy:

$$\Delta W_{total} = (P_1 - P_2)\Delta V = K_{final} - K_{initial} = K_2 - K_1 \qquad \text{15-13}$$

What is the kinetic energy of the fluid element? Well, the mass of the element is

$$\Delta m = \rho \Delta V$$

Thus, its kinetic energy is simply

$$K = \tfrac{1}{2}(\Delta m)v^2 = \tfrac{1}{2}(\rho \Delta V)v^2$$

Using this expression in Equation 15-13, we have

$$\Delta W_{total} = (P_1 - P_2)\Delta V = \left(\tfrac{1}{2}\rho v_2{}^2 - \tfrac{1}{2}\rho v_1{}^2\right)\Delta V$$

Canceling the common factor ΔV and rearranging, we find the following:

**PHYSICS
IN CONTEXT**
Looking Back

The connection between a force acting over a distance and kinetic energy (Chapter 7) is applied to a fluid in the derivation of Bernoulli's equation.

> **Bernoulli's Equation** (constant height)
>
> $$P_1 + \tfrac{1}{2}\rho v_1{}^2 = P_2 + \tfrac{1}{2}\rho v_2{}^2 \qquad \text{15-14}$$

Notice that $P_1 + \tfrac{1}{2}\rho v_1{}^2 = P_2 + \tfrac{1}{2}\rho v_2{}^2$ is equivalent to saying that $P + \tfrac{1}{2}\rho v^2$ is constant. Thus, there is a tradeoff between the pressure in a fluid and its speed—as the fluid speeds up, its pressure decreases. If this seems odd, recall that P_1 acts to increase the speed of the fluid element and P_2 acts to decrease its speed. The element will speed up, then, only if P_2 is less than P_1. The fluid speeds up as it enters a region of lower pressure.

EXAMPLE 15-19 SPRAY II

Predict/Calculate

Suppose the pressure in the fire hose in Example 15-18 is 350 kPa. (a) Is the pressure in the nozzle greater than, less than, or equal to 350 kPa? (b) Find the pressure in the nozzle.

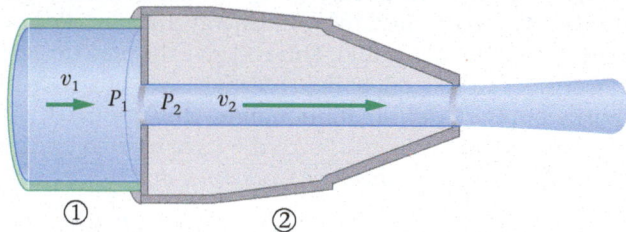

PICTURE THE PROBLEM
In our sketch, we use the same numbering system as in Example 15-18; that is, 1 refers to the hose, 2 refers to the nozzle. Therefore, $P_1 = 350$ kPa, and P_2 is to be determined.

REASONING AND STRATEGY

a. The pressure and speed of the fluid at constant height are related by $P_1 + \tfrac{1}{2}\rho v_1{}^2 = P_2 + \tfrac{1}{2}\rho v_2{}^2$. This result can be used to compare P_1 to P_2.

b. In Example 15-18, we used the equation of continuity ($A_1 v_1 = A_2 v_2$) to determine the speed of water in the nozzle, v_2. We now use this result, along with $P_1 + \tfrac{1}{2}\rho v_1{}^2 = P_2 + \tfrac{1}{2}\rho v_2{}^2$, to determine P_2.

Known   Pressure in fire hose, $P_1 = 350$ kPa; speed in fire hose, $v_1 = 1.5$ m/s; speed in nozzle, $v_2 = 21$ m/s.
Unknown   Pressure in nozzle, $P_2 = ?$

SOLUTION

Part (a)

1. Apply $P_1 + \tfrac{1}{2}\rho v_1{}^2 = P_2 + \tfrac{1}{2}\rho v_2{}^2$, noting that v_2 is greater than v_1, as determined in Example 15-18:

   The pressure and speed are related by $P + \tfrac{1}{2}\rho v^2 = $ constant. Thus, an increase in speed (v) can occur only when there is a corresponding decrease in pressure (P). **Answer:** P_2 (nozzle) is less than P_1 (hose); that is, $P_2 < 350$ kPa.

Part (b)

2. Solve $P_1 + \tfrac{1}{2}\rho v_1{}^2 = P_2 + \tfrac{1}{2}\rho v_2{}^2$ for the pressure in the nozzle, P_2:

   $$P_2 = P_1 + \tfrac{1}{2}\rho(v_1{}^2 - v_2{}^2)$$

CONTINUED

3. Substitute numerical values, including $v_1 = 1.5$ m/s and $v_2 = 21$ m/s from Example 15-18:

$$P_2 = 350 \text{ kPa} + \tfrac{1}{2}(1.00 \times 10^3 \text{ kg/m}^3)[(1.5 \text{ m/s})^2 - (21 \text{ m/s})^2]$$
$$= 131 \text{ kPa}$$

INSIGHT
The pressure in the nozzle is less than the pressure in the hose by roughly a factor of 2.7. This is because part of the energy associated with the high pressure in the hose has been converted to kinetic energy as the water passes into the nozzle. The connection between pressure and energy will be explored in more detail later in this section.

PRACTICE PROBLEM
What nozzle speed would be required to give a nozzle pressure of 161 kPa? [**Answer:** $v_2 = 19.5$ m/s]

Some related homework problems: Problem 60, Problem 62

Change in Height

If a fluid flows through the pipe shown in **FIGURE 15-26**, its height increases from y_1 to y_2 as it goes from one section of the pipe to the next. Since the cross-sectional area of the pipe is constant, however, the speed of the fluid is unchanged, according to $A_1v_1 = A_2v_2$. Thus, the change in kinetic energy of the fluid element shown in Figure 15-26 is zero.

The total work done on the fluid element is the sum of the works done by the pressure in each of the two regions, plus the work done by gravity. As before, the work done by pressure is

$$\Delta W_{\text{pressure}} = \Delta W_1 + \Delta W_2 = (P_1 - P_2)\Delta V$$

As the fluid element rises, gravity does negative work on it. Recall that the mass of the element is

$$\Delta m = \rho \Delta V$$

Therefore, the work done by gravity is

$$\Delta W_{\text{gravity}} = -\Delta mg(y_2 - y_1) = -\rho \Delta V g(y_2 - y_1)$$

Setting the total work equal to zero (since $\Delta K = 0$) yields

$$\Delta W_{\text{total}} = \Delta W_{\text{pressure}} + \Delta W_{\text{gravity}}$$
$$= (P_1 - P_2)\Delta V - \rho g(y_2 - y_1)\Delta V = 0$$

Canceling ΔV and rearranging gives:

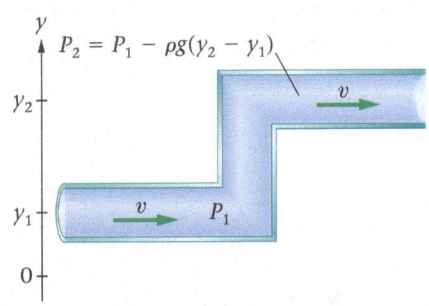

▲ **FIGURE 15-26 Fluid pressure in a pipe of varying elevation** Fluid of density ρ flows in a pipe of uniform cross-sectional area from height y_1 to height y_2. As it does so, its pressure decreases by the amount $\rho g(y_2 - y_1)$.

Bernoulli's Equation (constant speed)

$$P_1 + \rho gy_1 = P_2 + \rho gy_2 \qquad \qquad 15\text{-}15$$

In this case, it is $P + \rho gy$ that is constant—hence, pressure decreases as the height within a fluid increases. In fact, $P_1 + \rho gy_1 = P_2 + \rho gy_2$ (Equation 15-15) is precisely the same as the pressure–depth relationship, Equation 15-7 ($P_2 = P_1 + \rho gh$), which was obtained using force considerations. Here we obtained the same result using energy considerations and the work-energy theorem.

EXERCISE 15-20 PRESSURE AT THE TOP

Water flows with constant speed through a garden hose that goes up a step 27.5 cm high, as in Figure 15-26. If the water pressure is 132 kPa at the bottom of the step, what is its pressure at the top of the step?

REASONING AND SOLUTION
We apply $P_1 + \rho gy_1 = P_2 + \rho gy_2$, letting subscript 1 refer to the bottom of the step and subscript 2 refer to the top of the step. Rearranging and solving for P_2, we have

$$P_2 = P_1 + \rho g(y_1 - y_2)$$
$$= 132 \text{ kPa} + (1.00 \times 10^3 \text{ kg/m}^3)(9.81 \text{ m/s}^2)(0 - 0.275 \text{ m}) = 129 \text{ kPa}$$

This is precisely the pressure difference that would be observed if the water had been at rest.

General Case

In a more general case, both the height of a fluid and its speed may change. Combining the results obtained in Equations 15-14 and 15-15 yields the general form of Bernoulli's equation:

Bernoulli's Equation (general case)

$$P_1 + \tfrac{1}{2}\rho v_1^2 + \rho g y_1 = P_2 + \tfrac{1}{2}\rho v_2^2 + \rho g y_2 \qquad\qquad \text{15-16}$$

Thus, in general, the quantity $P + \tfrac{1}{2}\rho v^2 + \rho g y$ is a constant within a fluid. This is basically a statement of energy conservation. For example, recalling the definition of density in Equation 15-1, we find that $\tfrac{1}{2}\rho v^2$ is $\tfrac{1}{2}(M/V)v^2 = \left(\tfrac{1}{2}Mv^2\right)/V$. Clearly, this term represents the kinetic energy per volume of the fluid. Similarly, the term $\rho g y$ can be written as $(M/V)gy = (Mgy)/V$, which is the gravitational potential energy per volume.

Finally, the first term in Bernoulli's equation—the pressure—can also be thought of as an energy per volume. Recall that $P = F/A$. If we multiply numerator and denominator by a distance, d, we have $P = Fd/Ad$. But Fd is the work done by the force F as it acts through the distance d, and Ad is the volume swept out by an area A moved through a distance d. Therefore, the pressure can be thought of as work (energy) per volume: $P = W/V$.

As a result, Bernoulli's equation is simply a restatement of the work-energy theorem in terms of energy per volume. Of course, this relationship holds only as long as we can ignore frictional losses, which would lead to heating. We will consider the energy aspects of heat in Chapter 16.

QUICK EXAMPLE 15-21   FIND THE PRESSURE

A hose goes up a step of height 26.4 cm. The speed of water in the hose at the bottom of the step is 1.31 m/s, and the cross-sectional area of the hose on top of the step is half the cross-sectional area at the bottom of the step. If the pressure in the water at the bottom of the step is 147 kPa, what is the pressure in the water on top of the step?

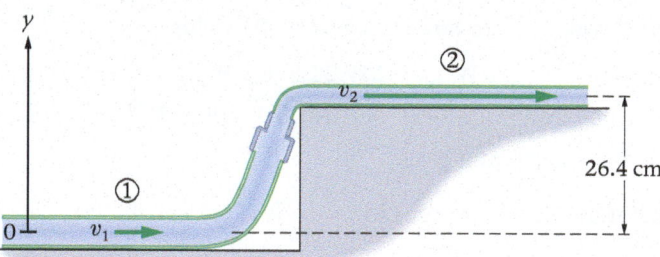

REASONING AND SOLUTION

First, we use the equation of continuity to find the speed of water on the top of the step. Combining this with the known height at the top of the step allows us to find the pressure there. (Let subscript 1 refer to the bottom of the step, and subscript 2 refer to the top of the step.)

1. Use the continuity equation to find the water's speed on the top of the step:

$$v_2 = \left(\frac{A_1}{A_2}\right)v_1 = 2v_1 = 2(1.31 \text{ m/s}) = 2.62 \text{ m/s}$$

2. Use Bernoulli's equation for the general case to solve for P_2:

$$P_2 = P_1 - \rho g(y_2 - y_1) - \tfrac{1}{2}\rho(v_2^2 - v_1^2)$$

3. Substitute numerical values:

$$P_2 = 147 \text{ kPa} - (1000 \text{ kg/m}^3)(9.81 \text{ m/s}^2)(0.264 \text{ m})$$
$$\quad -\tfrac{1}{2}(1000 \text{ kg/m}^3)[(2.62 \text{ m/s})^2 - (1.31 \text{ m/s})^2]$$
$$= 142 \text{ kPa}$$

The pressure on the top of the step is less than the pressure at the bottom of the step for two reasons: (i) The top of the step is higher, and (ii) the water has a greater speed at the top of the step.

Enhance Your Understanding   (Answers given at the end of the chapter)

7. Water flows through a pipe with a varying diameter, as shown in **FIGURE 15-27**. Small vertical tubes open to the air indicate the pressure in each segment of the pipe. (**a**) Is the height of water in tube B greater than, less than, or equal to the height of water in tube A? Explain. (**b**) Is the height of water in tube C greater than, less than, or equal to the height of water in tube A? Explain. (**c**) Is the height of water in tube C greater than, less than, or equal to the height of water in tube B? Explain.

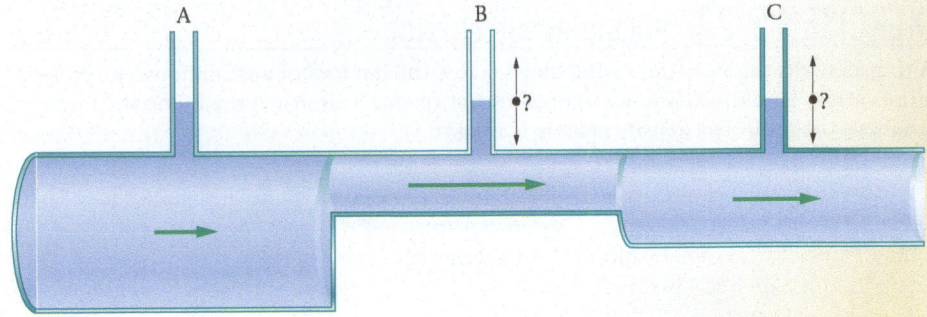

▲ **FIGURE 15-27** Enhance Your Understanding 7.

Section Review

- Bernoulli's equation for the general case is $P_1 + \frac{1}{2}\rho v_1^2 + \rho g y_1 = P_2 + \frac{1}{2}\rho v_2^2 + \rho g y_2$. This is basically a statement of energy conservation per volume for a fluid.

15-8 Applications of Bernoulli's Equation

We now consider a variety of real-world examples that illustrate various applications of Bernoulli's equation.

Pressure and Speed

As mentioned, it often seems counterintuitive that a fast-moving fluid should have less pressure than a slow-moving one. Remember, however, that pressure can be thought of as a form of energy. From this point of view, there is an energy tradeoff between pressure and kinetic energy.

Perhaps the easiest way to demonstrate the dependence of pressure on speed is to blow across the top of a piece of paper. If you hold the paper as shown in **FIGURE 15-28**, then blow over the top surface, the paper will lift upward. The reason is that there is a difference in air speed between the top and the bottom of the paper, with the higher speed on top. As a result, the pressure above the paper is lower. This pressure difference, in turn, results in a net upward force, referred to as **lift**, and the paper rises.

A similar example of pressure and speed is provided by the airplane wing. A cross section of a typical wing is shown in **FIGURE 15-29**. The shape of the wing is designed so that air flows more rapidly over the top surface than the lower surface. As a result, the pressure is less on top. As with the piece of paper, the pressure difference results in a net upward force (lift) on the wing. (The wing also deflects air downward, which results in an upward force on the wing due to Newton's third law.) It follows that aerodynamic lift is a dynamic effect; it requires a flow of air. The greater the speed difference, the greater the upward force on the wing.

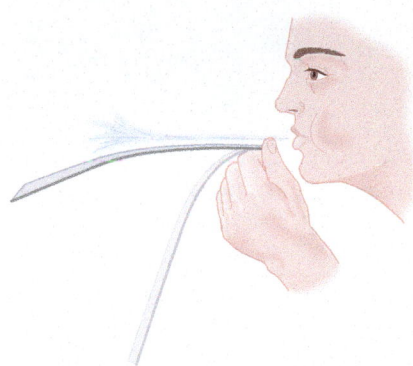

▲ **FIGURE 15-28 The bernoulli effect on a sheet of paper** If you hold a piece of paper by its end, it will bend downward. Blowing across the top of the paper reduces the pressure there, resulting in a net upward force that lifts the paper to a nearly horizontal position.

Air Jet Blows between Bowling Balls

▶ **FIGURE 15-29 Air flow and lift in an airplane wing** This figure shows the cross section of an airplane wing with air flowing past it. The wing is shaped so that air flows more rapidly over the top of the wing than along the bottom. As a result, the pressure on the top of the wing is reduced, and a net upward force (lift) is generated. The wing also deflects moving air downward, which also results in an upward force on the wing.

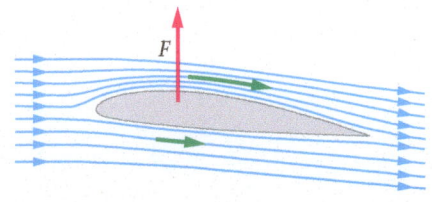

(a)

(b)

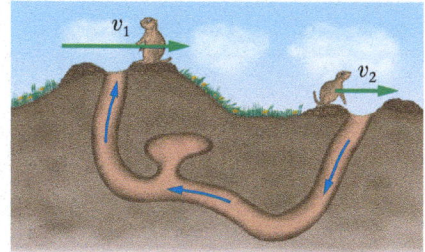

▲ **FIGURE 15-31 Air circulation in a prairie dog burrow** A prairie dog burrow typically has a high mound on one end and a low mound on the other. Since the wind speed increases with height above the ground, the pressure is smaller at the high-mound end of the burrow. The result is a very convenient circulation of fresh air through the burrow.

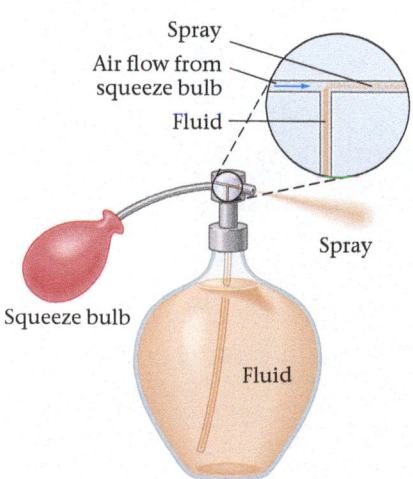

◀ **FIGURE 15-30 Visualizing Concepts Bernoulli's Equation (a)** We often say that a hurricane or tornado "blew the roof off a house." Actually, the high-speed wind passing over the roof creates a region of reduced pressure, and normal atmospheric pressure inside the house then "blows" the roof off. **(b)** Prairie dogs use the Bernoulli effect to ventilate their burrows. One end of a burrow is always at a greater height than the other. This means that the wind speed is different at the two ends, and the resulting pressure difference produces a flow of air through the burrow.

EXERCISE 15-22   PRESSURE DIFFERENCE

During a hurricane, a 42.6 m/s wind blows across the flat roof of a small house, as shown in the sketch. Find the difference in pressure between the air inside the home and the air just above the roof. (The density of air is 1.29 kg/m³.)

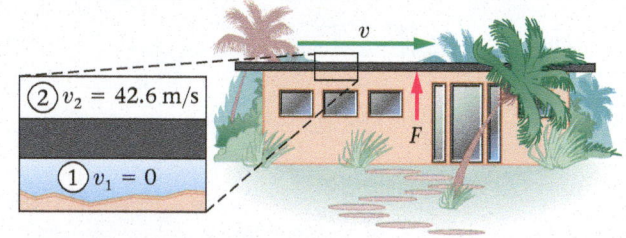

REASONING AND SOLUTION

Use Bernoulli's equation with point 1 just under the roof and point 2 just above the roof. The air speed is zero at point 1, but is equal to the wind speed at point 2. There is little difference in elevation between these points, however, and hence $y_1 = y_2 = y$. Thus, Bernoulli's general equation reduces to

$$P_1 + 0 + \rho g y = P_2 + \tfrac{1}{2}\rho v_2^2 + \rho g y$$

Solving this expression for the pressure difference, $P_1 - P_2$, yields the following:

$$P_1 - P_2 = \tfrac{1}{2}\rho v_2^2 = \tfrac{1}{2}(1.29 \text{ kg/m}^3)(42.6 \text{ m/s})^2 = 1.17 \text{ kPa}$$

A difference in pressure of 1.17 kPa might seem rather small, considering that atmospheric pressure is 101 kPa. However, it can still cause a significant force on a relatively large area, such as a roof. If a typical roof has an area of about 130 m², for example, a pressure difference of 1.17 kPa results in an upward force of over 34,000 pounds! This is why roofs are often torn from houses during severe windstorms, as illustrated in **FIGURE 15-30 (a)**.

On a lighter note, the prairie dogs in **FIGURE 15-30 (b)** *benefit* from the effects of Bernoulli's equation. A schematic prairie dog burrow is shown in **FIGURE 15-31**. Notice that one of the entrance/exit mounds is higher than the other. This is significant because the speed of air flow due to the incessant prairie winds varies with height; the speed goes to zero right at ground level, and increases to its maximum value within a few feet above the surface. As a result, the speed of air over the higher mound is greater than that over the lower mound. This causes the pressure to be less over the higher mound. With a pressure difference between the two mounds, air is drawn through the burrow, giving a form of natural air conditioning.

Similar effects are seen in an atomizer, which sprays perfume in a fine mist. As the bulb shoots a gust of air, as in **FIGURE 15-32**, it passes through a narrow orifice, which causes the air speed to increase. The pressure decreases as a result, and perfume is drawn up by the pressure difference into the stream of air.

◀ **FIGURE 15-32 An atomizer** The operation of an "atomizer" can be understood in terms of Bernoulli's equation. The high-speed jet of air made by squeezing the bulb creates a low pressure at the top of the vertical tube. This causes fluid to be drawn up the tube and expelled with the air jet as a fine spray.

CONCEPTUAL EXAMPLE 15-23   A RAGTOP ROOF

A small ranger vehicle has a soft, ragtop roof. When the car is at rest, the roof is flat. When the car is cruising at highway speeds with its windows rolled up, does the roof bow upward, remain flat, or bow downward?

REASONING AND DISCUSSION

When the car is in motion, air flows over the top of the roof, while the air inside the car is at rest—since the windows are closed. Thus, there is less pressure over the roof than under it. As a result, the roof bows upward.

ANSWER

The roof bows upward.

Torricelli's Law

Our final example of Bernoulli's equation deals with the speed of a fluid as it flows through a hole in a container. Consider, for example, the tank of water shown in **FIGURE 15-33**. If a hole is poked through the side of the tank at a depth h below the surface, what is the speed of the water as it emerges?

To answer this question, we apply Bernoulli's equation to the two points shown in the figure. First, at point 1 we notice that the water is open to the atmosphere; thus $P_1 = P_{at}$. Next, with the origin at the level of the hole, the height of the water surface is $y_1 = h$. Finally, if the hole is relatively small and the tank is large, the top surface of the water will have essentially zero speed; thus we can set $v_1 = 0$. Collecting these results in Bernoulli's general equation, we have the following for point 1:

$$P_1 + \tfrac{1}{2}\rho v_1^2 + \rho g y_1 = P_{at} + 0 + \rho g h$$

Now for point 2. At this point the height is $y_2 = 0$, by definition of the origin, and the speed of the escaping water is the unknown, v_2. *Now here is the key step*: The pressure P_2 is *atmospheric pressure*, because the hole opens the water to the atmosphere. Thus, for point 2 we have the following:

$$P_2 + \tfrac{1}{2}\rho v_2^2 + \rho g y_2 = P_{at} + \tfrac{1}{2}\rho v_2^2 + 0$$

Equating the results for points 1 and 2 yields

$$P_1 + \tfrac{1}{2}\rho v_1^2 + \rho g y_1 = P_2 + \tfrac{1}{2}\rho v_2^2 + \rho g y_2$$
$$P_{at} + \rho g h = P_{at} + \tfrac{1}{2}\rho v_2^2$$

Eliminating P_{at} and ρ we find

$$v_2 = \sqrt{2gh} \qquad\qquad 15\text{-}17$$

This result is known as **Torricelli's law.**

This expression for v_2 should look familiar; it is the speed of an object that falls freely through a distance h. That is, the water emerges from the tank with the same speed as if it had fallen from the surface of the water to the hole. Similarly, if the emerging stream of water were to be directed upward, as in **FIGURE 15-34**, it would have just enough speed to rise through a height h—right back to the water's surface. This is precisely what one would expect on the basis of energy conservation.

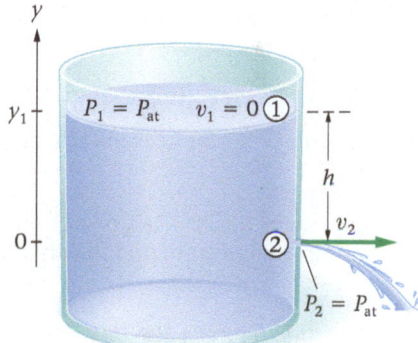

▲ **FIGURE 15-33 Fluid emerging from a hole in a container** Since the fluid exiting the hole is in contact with the atmosphere, the pressure there is just as it is on the top surface of the fluid.

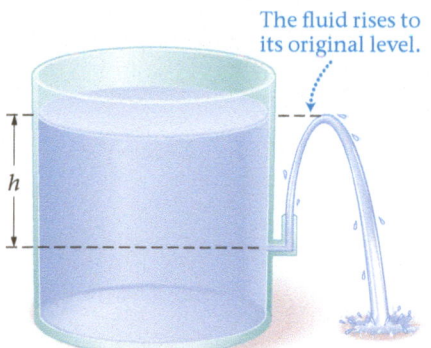

▲ **FIGURE 15-34 Maximum height of a stream of water** If the fluid emerging from a hole in a container is directed upward, it has just enough speed to reach the surface level of the fluid. This is an example of energy conservation.

EXAMPLE 15-24   A WATER FOUNTAIN

In designing a backyard water fountain, a gardener wants a stream of water to exit from the bottom of one tub (tub 1) and land in a second one (tub 2), as shown in the sketch. The top of tub 2 is 0.775 m below the hole in tub 1. In addition, tub 1 is filled with water to a depth of 0.218 m. How far to the right of tub 1 must the center of tub 2 be placed to catch the stream of water?

PICTURE THE PROBLEM

Our sketch labels the pertinent quantities for this problem. We know that $h = 0.218$ m and $H = 0.775$ m. The distance D is to be found. We have also chosen an appropriate coordinate system, with the y direction vertical and the x direction horizontal. The $y = 0$ level is at the top of tub 2.

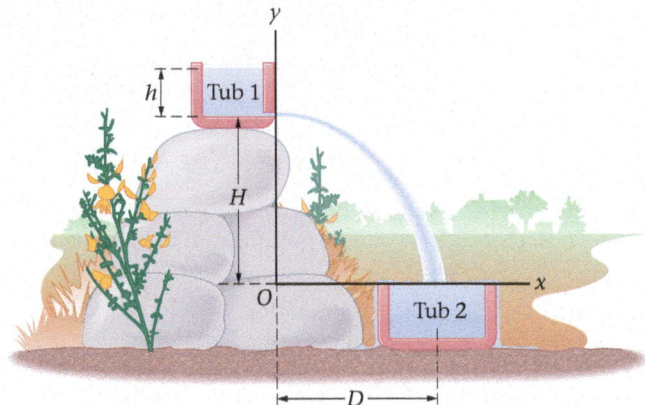

REASONING AND STRATEGY

This problem combines Torricelli's law and kinematics. First, we find the speed v of the stream of water as it leaves tub 1 using Equation 15-17 ($v = \sqrt{2gh}$). Next, we find the time t required for the stream to fall freely through a distance H to tub 2 using kinematics (Chapter 2). Finally, because the stream moves with constant speed in the x direction, the distance D is given by $D = vt$.

Known   Water depth in tub 1, $h = 0.218$ m; vertical drop to tub 2, $H = 0.775$ m.

Unknown   Horizontal distance to tub 2, $D = ?$

SOLUTION

1. Find the speed v of the stream when it leaves tub 1:

$$v = \sqrt{2gh} = \sqrt{2(9.81 \text{ m/s}^2)(0.218 \text{ m})} = 2.07 \text{ m/s}$$

2. Find the time t for free fall through a height H:

$$y = H - \tfrac{1}{2}gt^2 = 0$$

$$t = \sqrt{\frac{2H}{g}} = \sqrt{\frac{2(0.775 \text{ m})}{9.81 \text{ m/s}^2}} = 0.397 \text{ s}$$

3. Multiply v times t to find the distance D:

$$x = vt = (2.07 \text{ m/s})(0.397 \text{ s}) = 0.822 \text{ m} = D$$

INSIGHT

Our solution for x can also be written symbolically as $x = vt = \sqrt{2gh}(\sqrt{2H/g}) = 2\sqrt{hH}$. Thus, if the values of h and H are interchanged, the distance $x = D$ remains the same. Also, notice that x is independent of the acceleration due to gravity; therefore, the fountain would work just as well on the Moon with its reduced gravity.

PRACTICE PROBLEM

Find the distance D for $h = 0.775$ m and $H = 0.218$ m.   [**Answer:** $D = 2\sqrt{hH} = 0.822$ m, as expected.]

Some related homework problems: Problem 65, Problem 106

Enhance Your Understanding   (Answers given at the end of the chapter)

8. A rock is dropped into the container shown in Figure 15-33 before the hole is made at the $y = 0$ level. When the hole is made, does the water emerge with a speed that is greater than, less than, or equal to the speed without the rock? Explain.

Section Review

- When fluid emerges from a hole in a container at a depth h below the surface of the fluid, its speed is $v = \sqrt{2gh}$. This is known as Torricelli's law.

15-9 Viscosity and Surface Tension

To this point we have considered only "ideal" fluids. In particular, we've assumed that fluids flow without frictional losses and that the molecules in a fluid have no interaction with one another. In this section, we consider the consequences that follow from relaxing these assumptions.

▶ **FIGURE 15-35 Fluid flow through a tube (a)** An ideal fluid flows through a tube with a speed that is the same everywhere in the fluid. **(b)** In a fluid with finite viscosity, the speed of the fluid goes to zero on the walls of the tube and reaches its maximum value in the center of the tube. The average speed of the fluid depends on the pressure difference between the ends of the tube, $P_1 - P_2$; the length of the tube, L; the cross-sectional area of the tube, A; and the coefficient of viscosity of the liquid, η.

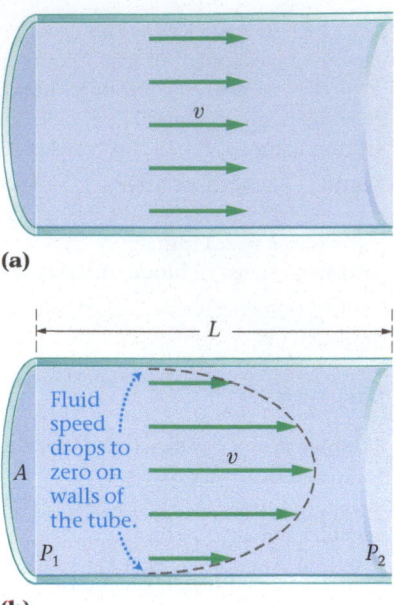

(a)

(b)

Viscosity

When a block slides across a rough floor, it experiences a frictional force opposing the motion. Similarly, a fluid flowing past a stationary surface experiences a force opposing the flow. This tendency to resist flow is referred to as the **viscosity** of a fluid. Fluids like air have low viscosities, thicker fluids like water are more viscous, and fluids like honey and motor oil are characterized by high viscosity.

To be specific, consider a situation of great practical importance—the flow of a fluid through a tube. Examples of this type of system include water flowing through a metal pipe in a house and blood flowing through an artery or a vein. If the fluid were ideal, with zero viscosity, it would flow through the tube with a speed that is the same throughout the fluid, as indicated in **FIGURE 15-35 (a)**. Real fluids with finite viscosity are found to have flow patterns like the one shown in **FIGURE 15-35 (b)**. In this case, the fluid is at rest next to the walls of the tube and flows with its greatest speed in the center of the tube. Because adjacent portions of the fluid flow past one another with differ-ent speeds, a force must be exerted on the fluid to maintain the flow, just as a force is required to keep a block sliding across a rough surface.

The force causing a viscous fluid to flow is provided by the pressure difference, $P_1 - P_2$, across a given length, L, of tube. Experiments show that the required pressure difference is proportional to the length of the tube and to the average speed, v, of the fluid. In addition, it is inversely proportional to the cross-sectional area, A, of the tube. Combining these observations, we can write the pressure difference in the following form:

$$P_1 - P_2 \propto \frac{vL}{A}$$

The constant of proportionality between the pressure difference and vL/A is related to the **coefficient of viscosity, η,** of a fluid. In fact, the viscosity is *defined* in such a way that the pressure difference is given by the following expression:

$$P_1 - P_2 = 8\pi\eta\frac{vL}{A} \qquad \text{15-18}$$

From this equation we can see that the dimensions of the coefficient of viscosity are $N \cdot s/m^2$. A common unit in the study of viscous fluids is the **poise,** named for the French physiologist Jean Leonard Marie Poiseuille (1797–1869) and defined as

$$1 \text{ poise} = 1 \text{ dyne} \cdot s/cm^2 = 0.1\,N \cdot s/m^2$$

For example, the viscosity of water at room temperature is $1.0016 \times 10^{-3}\,N \cdot s/m^2$ and the viscosity of blood at 37 °C is $2.72 \times 10^{-3}\,N \cdot s/m^2$. Some additional viscosities are given in Table 15-3.

TABLE 15-3 Viscosities (η) of Various Fluids ($N \cdot s/m^2$)

Honey	10
Glycerine (20 °C)	1.50
10-wt motor oil (30 °C)	0.250
Whole blood (37 °C)	2.72×10^{-3}
Water (0 °C)	1.79×10^{-3}
Water (20 °C)	1.0055×10^{-3}
Water (100 °C)	2.82×10^{-4}
Air (20 °C)	1.82×10^{-5}

EXAMPLE 15-25 BLOOD SPEED IN THE PULMONARY ARTERY

BIO The pulmonary artery, which connects the heart to the lungs, is 8.9 cm long and has a pressure difference over this length of 450 Pa. If the inside radius of the artery is 2.3 mm, what is the average speed of blood in the pulmonary artery?

PICTURE THE PROBLEM

Our sketch shows a schematic representation of the pulmonary artery, not drawn to scale. The length (8.9 cm) and radius (2.3 mm) of the artery are indicated. In addition, we note that the pressure difference between the two ends of the artery is 450 Pa.

CONTINUED

REASONING AND STRATEGY

The average speed of the blood can be found by using $P_1 - P_2 = 8\pi\eta(vL/A)$. Notice that the pressure difference, $P_1 - P_2$, is given as 450 Pa = 450 N/m², and that the cross-sectional area of the blood vessel is $A = \pi r^2$.

Known   Length of artery, $L = 8.9$ cm; pressure difference, $P_1 - P_2 = 450$ N/m²; inside radius of artery, $r = 2.3$ mm.

Unknown   Speed of blood in artery, $v = ?$

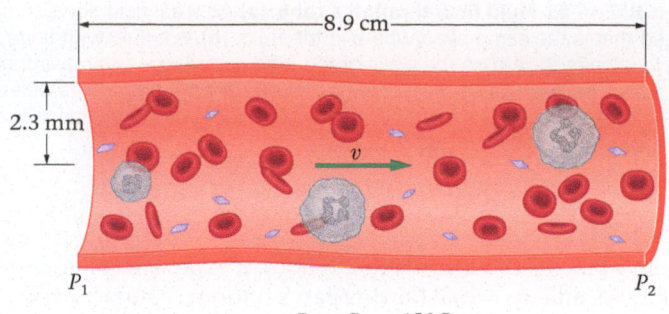

$P_1 - P_2 = 450$ Pa

SOLUTION

1. Solve $P_1 - P_2 = 8\pi\eta\,(vL/A)$ for the average speed, v:

$$v = \frac{(P_1 - P_2)A}{8\pi\eta L}$$

2. Replace the cross-sectional area A with πr^2 and cancel π from the numerator and denominator:

$$v = \frac{(P_1 - P_2)r^2}{8\eta L}$$

3. Substitute numerical values:

$$v = \frac{(450\ \text{Pa})(0.0023\ \text{m})^2}{8(0.00272\ \text{N}\cdot\text{s/m}^2)(0.089\ \text{m})} = 1.2\ \text{m/s}$$

INSIGHT

The viscosity of blood increases rapidly with its hematocrit value—that is, with the concentration of red blood cells in the whole blood (see Chapter 10 for more information on the hematocrit value of blood). Thus, thick blood, with a high hematocrit value, requires a significantly larger pressure difference for a given rate of blood flow. This higher pressure must be provided by the heart, which consequently works harder with each beat.

PRACTICE PROBLEM

What pressure difference is required to give the blood in this pulmonary artery an average speed of 1.5 m/s?
[**Answer:** 550 Pa]

Some related homework problems: Problem 99, Problem 100

Volume Flow Rate   A convenient way to characterize the flow of a fluid is in terms of its volume flow rate—the volume of fluid that passes a given point in a given amount of time. Referring to Section 15-6, we see that the volume flow rate of a fluid is simply vA, where v is the average speed of the fluid and A is the cross-sectional area of the tube through which it flows. Solving $P_1 - P_2 = 8\pi\eta(vL/A)$ (Equation 15-18) for the average speed gives $v = (P_1 - P_2)A/8\pi\eta L$. Multiplying this result by the cross-sectional area of the tube yields the volume flow rate:

$$\text{volume flow rate} = \frac{\Delta V}{\Delta t} = vA = \frac{(P_1 - P_2)A^2}{8\pi\eta L}$$

Using the fact that the cross-sectional area of a cylindrical tube is $A = \pi r^2$, where r is its radius, we obtain the result known as **Poiseuille's equation:**

$$\frac{\Delta V}{\Delta t} = \frac{(P_1 - P_2)\pi r^4}{8\eta L} \qquad \text{15-19}$$

Notice that the volume flow rate varies with the *fourth power* of the tube's radius; thus a small change in radius corresponds to a large change in volume flow rate.

To see the significance of the r^4 dependence, consider an artery that branches into an arteriole with half the artery's radius. Letting r go to $r/2$ in Poiseuille's equation, and solving for the pressure difference, we find

$$P_1 - P_2 = \frac{8\eta L}{\pi(r/2)^4}\left(\frac{\Delta V}{\Delta t}\right) = 16\left[\frac{8\eta L}{\pi r^4}\left(\frac{\Delta V}{\Delta t}\right)\right]$$

Thus, the pressure difference across a given length of arteriole is 16 times what it is across the same length of artery. In fact, in the human body the pressure drop along an artery is small compared to the rather large pressure drop observed in the arterioles. This is a direct consequence of the increased viscous drag of the blood as it flows through narrower blood vessels.

Similarly, a narrowing, or *stenosis*, of an artery can produce significant increases in blood pressure. For example, a reduction in radius of only 20%, from r to $0.8r$, causes an increase in pressure by a factor of $(1/0.8)^4 \sim 2.4$. Thus, even a small narrowing of an artery can lead to an increased risk for heart disease and stroke.

Surface Tension

A small insect or spider resting on the surface of a pond or a lake is a common sight in the summertime. If you look carefully, you can see that the insect creates tiny dimples in the water's surface (**FIGURE 15-36**), almost as if it were supported by a thin sheet of rubber. In fact, the surface of water and other fluids behaves in many respects as if it were an elastic membrane. This effect is known as **surface tension.**

To understand the origin of surface tension, we start by noting that the molecules in a fluid exert attractive forces on one another. Thus, a molecule deep within a fluid experiences forces in all directions, as indicated in **FIGURE 15-37**, due to the molecules that surround it on all sides. The net force on such a molecule is zero. As a molecule nears the surface, however, it experiences a net force away from the surface, because there are no fluid molecules on the other side of the surface to attract it in that direction. It follows that work must be done on a molecule to move it from within a fluid to the surface, and that the *energy* of a fluid is increased for every molecule on its surface.

In general, physical systems tend toward configurations of minimum energy. For example, a ball on a slope rolls downhill, lowering its gravitational potential energy. If the energy of a droplet of liquid is to be minimized, it must have the smallest surface area possible for a given volume; that is, it must have the shape of a sphere. This is the reason small drops of dew are always spherical. Larger drops of water may be distorted by the downward pull of Earth's gravity, but in orbit drops of all sizes are spherical.

Energy is required to increase the area of a liquid surface, and hence the situation is similar to the energy required to stretch a spring or to stretch a sheet of rubber. Thus, the surface of a liquid behaves as if it were elastic, resisting tendencies to increase its area. For example, if a drop of dew is distorted into an ellipsoid, it quickly returns to its original spherical shape. Similarly, when an insect alights on the surface of a pond, it creates dimples that increase the surface area. The water resists this distortion with a force sufficient to support the weight of the insect—if the insect is not too large. In fact, even a needle or a razor blade can be supported on the surface of water if it is put into place gently, even though it has a density significantly greater than the density of water.

BIO Surface tension is important in many biological systems, and it plays a particularly important role in human breathing. For example, the crucial exchange of oxygen and carbon dioxide between inspired air and the blood occurs across the membranes of small balloonlike structures called *alveoli*. During inhalation, the alveoli expand from a radius of about 0.050 mm to 0.10 mm as they draw in fresh air. The walls of the alveoli are coated with a thin film of water, however, and in order to expand they must push outward against the water's surface tension—like trying to inflate a balloon against the surface tension of the rubber. To reduce the rather large surface tension of water, and make breathing easier, the lungs produce a substance called a *surfactant* that mixes with the water. This surfactant is produced rather late in the development of a fetus, however, and therefore premature infants may experience respiratory distress and even death as a result of too much surface tension in their alveoli.

▲ **FIGURE 15-36** Surface tension causes the surface of a liquid to behave like an elastic skin or membrane. When a small force is applied to the liquid surface, it tends to stretch, resisting penetration. This phenomenon enables a fishing spider to launch itself vertically upward as high as 4 cm from the water surface, as if from a trampoline. Such an acrobatic maneuver can save the spider from the attack of a predatory fish.

Big Idea **5** Surface tension tends to make the surface area of a fluid as small as possible. Because of this tendency, the surface of a fluid acts much like an elastic membrane, resisting forces that try to stretch it and increase its area.

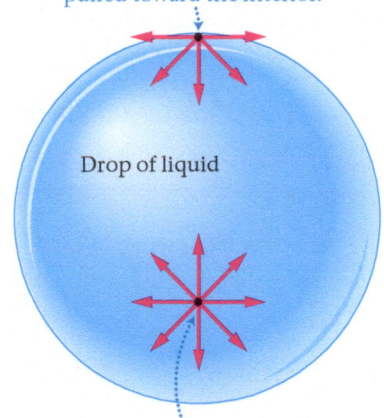

A molecule on the surface is pulled toward the interior.

Drop of liquid

An interior molecule is pulled by its neighbors equally in all directions.

▲ **FIGURE 15-37** **The origin of surface tension** A molecule in the interior of a fluid experiences attractive forces of equal magnitude in all directions, giving a net force of zero. A molecule near the surface of the fluid experiences a net attractive force toward the interior of the fluid. This causes the surface to be pulled inward, resulting in a surface of minimum area.

Enhance Your Understanding   (Answers given at the end of the chapter)

9. Which pipe requires a greater pressure difference to make a given fluid flow through it with a speed v: pipe 1 with $L = 1$ m and a diameter of 0.1 m, or pipe 2 with $L = 2$ m and a diameter of 0.2 m?

Section Review

- The pressure difference required to push a fluid with a speed v through a pipe of length L and cross-sectional area A is $P_1 - P_2 = 8\pi\eta(vL/A)$. In this expression, η is the coefficient of viscosity.

CHAPTER 15 REVIEW

CHAPTER SUMMARY

15-1 DENSITY

The density, ρ, of a material is its mass M per volume V:

$$\rho = M/V \qquad \text{15-1}$$

15-2 PRESSURE

Pressure, P, is force F per area A:

$$P = F/A \qquad \text{15-2}$$

Atmospheric Pressure

The pressure exerted by the atmosphere is $P_{at} = 1.01 \times 10^5 \text{ N/m}^2 \approx 14.7 \text{ lb/in}^2$.

Pascals

Pressure is often given in terms of the pascal (Pa), where 1 Pa = 1 N/m^2.

Gauge Pressure

Gauge pressure is the difference between the actual pressure and atmospheric pressure:

$$P_g = P - P_{at} \qquad \text{15-5}$$

15-3 STATIC EQUILIBRIUM IN FLUIDS: PRESSURE AND DEPTH

The pressure of a fluid in static equilibrium increases with depth.

Pressure with Depth

If the pressure at one point in a fluid is P_1, the pressure at a depth h below that point is

$$P_2 = P_1 + \rho g h \qquad \text{15-7}$$

Pascal's Principle

An external pressure applied to an enclosed fluid is transmitted unchanged to every point within the fluid.

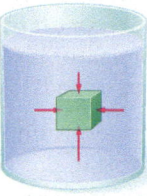

15-4 ARCHIMEDES' PRINCIPLE AND BUOYANCY

The fact that the pressure in a fluid increases with depth leads to a net upward force on any object that is immersed in the fluid. This upward force is referred to as a buoyant force. The magnitude of the buoyant force is equal in magnitude to the weight of fluid displaced by the object.

$$F_b = \rho_{fluid} g V \qquad \text{15-9}$$

15-5 APPLICATIONS OF ARCHIMEDES' PRINCIPLE

Archimedes' principle applies equally well to objects that are completely immersed, partially immersed, or floating.

Floatation

An object floats when it displaces an amount of fluid equal to its weight.

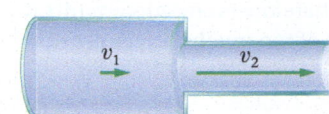

15-6 FLUID FLOW AND CONTINUITY

The speed of a fluid changes as the cross-sectional area of the pipe through which it flows changes.

Equation of Continuity for an Incompressible Fluid

$$A_1 v_1 = A_2 v_2 \qquad \text{15-12}$$

15-7 BERNOULLI'S EQUATION

Bernoulli's equation can be thought of as energy conservation per volume for a fluid.

General Case

The pressure, speed, and height of a fluid at two different points are related by

$$P_1 + \tfrac{1}{2}\rho v_1^2 + \rho g y_1 = P_2 + \tfrac{1}{2}\rho v_2^2 + \rho g y_2 \qquad \text{15-16}$$

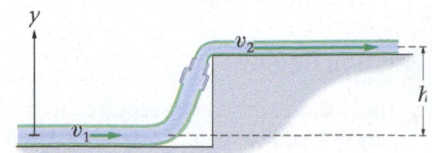

15-8   APPLICATIONS OF BERNOULLI'S EQUATION

Bernoulli's equation applies to a wide range of everyday situations, including airplane wings and prairie dog burrows.

Torricelli's Law

According to Torricelli's law, if a hole is poked in a container at a depth h below the surface, the fluid exits with the speed

$$v = \sqrt{2gh} \qquad \text{15-17}$$

15-9   VISCOSITY AND SURFACE TENSION

Viscosity in a fluid is similar to friction between two solid surfaces.

$$P_1 - P_2 = 8\pi\eta\frac{vL}{A} \qquad \text{15-18}$$

Surface Tension

A fluid tends to pull inward on its surface, resulting in a surface of minimum area. The surface of the fluid behaves much like an elastic membrane enclosing the fluid.

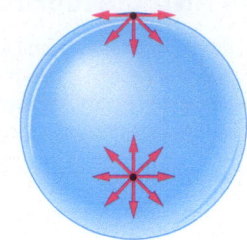

ANSWERS TO ENHANCE YOUR UNDERSTANDING QUESTIONS

1. **(a)** The volume of system 1 is less than the volume of system 2 because it contains less of the same substance (water). **(b)** The density of system 1 is equal to the density of system 2 because they consist of the same substance.

2. B < C < A < D.

3. The increase in pressure from 0 to 10 m is the same as the increase in pressure from 100 m to 110 m because the density of water is the same at 0 and 100 m.

4. The buoyant force exerted on the aluminum block is the same as the buoyant force on the lead block because they displace the same amount of water.

5. The water level stays the same because the wood displaces the same amount of water whether it rests on the ice cube or floats by itself.

6. The new speed is $4v$ because the area has decreased by a factor of $2^2 = 4$.

7. **(a)** The height in tube B is less than in tube A because the speed is greater at B. **(b)** The height in tube C is less than in tube A because the speed is greater at C. **(c)** The height in tube C is greater than in tube B because the speed is smaller at C.

8. The speed is greater because the rock raises the water level.

9. A greater pressure difference is required for pipe 1.

CONCEPTUAL QUESTIONS

For instructor-assigned homework, go to www.masteringphysics.com. **MP**

(Answers to odd-numbered Conceptual Questions can be found in the back of the book.)

1. Suppose you drink a liquid through a straw. Explain why the liquid moves upward, against gravity, into your mouth.

2. Considering your answer to the previous question, is it possible to sip liquid through a straw on the surface of the Moon? Explain.

3. Water towers on the roofs of buildings have metal bands wrapped around them for support. The spacing between bands is smaller near the base of a tower than near its top. Explain.

4. What holds a suction cup in place?

5. Suppose a force of 400 N is required to push the top off a wine barrel. In a famous experiment, Blaise Pascal attached a tall, thin tube to the top of a filled wine barrel, as shown in **FIGURE 15-38**. Water was slowly added to the tube until the barrel burst. The puzzling result found by Pascal was that the barrel broke when the weight of water in the tube was much less than 400 N. Explain Pascal's observation.

FIGURE 15-38 Conceptual Question 5 and Problem 18

6. Why is it more practical to use mercury in the barometer shown in Figure 15-6 than water?

7. An object's density can be determined by first weighing it in air, then in water (provided the density of the object is greater than the density of water, so that it is totally submerged when placed in water). Explain how these two measurements can give the desired result.

8. How does a balloonist control the vertical motion of a hot-air balloon?

9. Why is it possible for people to float without effort in Utah's Great Salt Lake?

10. **Physics in the Movies** In the movie *Voyage to the Bottom of the Sea*, the Earth is experiencing a rapid warming. In one scene, large icebergs break up into small, car-size chunks that drop downward through the water and bounce off the hull of the submarine *Seaview*. Is this an example of good, bad, or ugly physics? Explain.

11. One day, while snorkeling near the surface of a crystal-clear ocean, it occurs to you that you could go considerably deeper by simply lengthening the snorkel tube. Unfortunately, this does not work well at all. Why?

12. Since metal is more dense than water, how is it possible for a metal boat to float?

13. A sheet of water passing over a waterfall is thicker near the top than near the bottom. Similarly, a stream of water emerging from a water faucet becomes narrower as it falls. Explain.

14. It is a common observation that smoke rises more rapidly through a chimney when there is a wind blowing outside. Explain.

15. Is it best for an airplane to take off against the wind or with the wind? Explain.

16. If you have a hair dryer and a Ping Pong ball at home, try this demonstration. Direct the air from the dryer in a direction just above horizontal. Next, place the Ping Pong ball in the stream of air. If done just right, the ball will remain suspended in midair. Use the Bernoulli effect to explain this behavior.

17. Suppose a pitcher wants to throw a baseball so that it rises as it approaches the batter. How should the ball be spinning to accomplish this feat? Explain.

PROBLEMS AND CONCEPTUAL EXERCISES

Answers to odd-numbered Problems and Conceptual Exercises can be found in the back of the book. **BIO** *identifies problems of biological or medical interest;* **CE** *indicates a conceptual exercise.* **Predict/Explain** *problems ask for two responses: (a) your prediction of a physical outcome, and (b) the best explanation among three provided; and* **Predict/Calculate** *problems ask for a prediction of a physical outcome, based on fundamental physics concepts, and follow that with a numerical calculation to verify the prediction. On all problems, bullets (•, ••, •••) indicate the level of difficulty.*

SECTION 15-1 DENSITY

1. • Estimate the weight of the air in your physics classroom.

2. • What weight of water is required to fill a 25-gallon aquarium?

3. • You buy a "gold" ring at a pawn shop. The ring has a mass of 4.7 g and a volume of 0.55 cm³. Is the ring solid gold?

4. • A cube of metal has a mass of 0.347 kg and measures 3.21 cm on a side. Calculate the density and identify the metal.

SECTION 15-2 PRESSURE

5. • What is the downward force exerted by the atmosphere on a soccer field whose dimensions are 105 m by 55 m?

6. • **BIO Bioluminescence** Some species of dinoflagellate (a type of unicellular plankton) can produce light as the result of biochemical reactions within the cell. This light is an example of bioluminescence. It is found that bioluminescence in dinoflagellates can be triggered by deformation of the cell surface with a pressure as low as one dyne (10^{-5} N) per square centimeter. What is this pressure in **(a)** pascals and **(b)** atmospheres?

7. • A 71-kg person sits on a 3.9-kg chair. Each leg of the chair makes contact with the floor in a circle that is 1.1 cm in diameter. Find the pressure exerted on the floor by each leg of the chair, assuming the weight is evenly distributed.

8. • To prevent damage to floors (and to increase friction), a crutch will often have a rubber tip attached to its end. If the end of the crutch is a circle of radius 0.95 cm without the tip, and the tip is a circle of radius 2.0 cm, by what factor does the tip reduce the pressure exerted by the crutch?

9. •• Suppose that when you ride on your 7.85-kg bike the weight of you and the bike is supported equally by the two tires. If the gauge pressure in the tires is 70.2 lb/in² and the area of contact between each tire and the road is 7.03 cm², what is your weight?

10. •• **Shock Wave Pressure** On February 15, 2013, a 12,000,000-kg asteroid exploded in the atmosphere above Chelyabinsk, Russia, generating a shock wave with a pressure amplitude of 750 Pa. What force (in N) would this pressure difference exert on a window with dimensions of 0.86 m × 1.17 m? (The shock wave shattered many glass windows, resulting in a number of injuries.)

11. •• **Predict/Calculate** The weight of your 1420-kg car is supported equally by its four tires, each inflated to a gauge pressure of 35.0 lb/in². **(a)** What is the area of contact each tire makes with the road? **(b)** If the gauge pressure is increased, does the area of contact increase, decrease, or stay the same? **(c)** What gauge pressure is required to give an area of contact of 116 cm² for each tire?

SECTION 15-3 STATIC EQUILIBRIUM IN FLUIDS: PRESSURE AND DEPTH

12. • **CE** Two drinking glasses, 1 and 2, are filled with water to the same depth. Glass 1 has twice the diameter of glass 2. **(a)** Is the weight of the water in glass 1 greater than, less than, or equal to the weight of the water in glass 2? **(b)** Is the pressure at the bottom of glass 1 greater than, less than, or equal to the pressure at the bottom of glass 2?

13. • **CE FIGURE 15-39** shows four containers, each filled with water to the same level. Rank the containers in order of increasing pressure at the depth *h*. Indicate ties where appropriate.

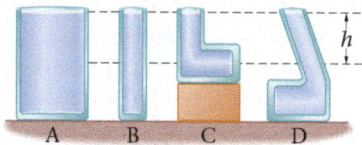

FIGURE 15-39 Problem 13

14. • Water in the lake behind Hoover Dam is 221 m deep. What is the water pressure at the base of the dam?

15. • In a classroom demonstration, the pressure inside a soft drink can is suddenly reduced to essentially zero. Assuming the can to be a cylinder with a height of 12 cm and a diameter of 6.5 cm, find the net inward force exerted on the vertical sides of the can due to atmospheric pressure.

16. •• As a storm front moves in, you notice that the column of mercury in a barometer rises to only 736 mm. **(a)** What is the air pressure? **(b)** If the mercury in this barometer is replaced with water, to what height does the column of water rise? Assume the same air pressure found in part (a).

17. •• In the hydraulic system shown in **FIGURE 15-40**, the piston on the left has a diameter of 5.1 cm and a mass of 2.3 kg. The piston on the right has a diameter of 13 cm and a mass of 3.6 kg. If the density of the fluid is 750 kg/m³, what is the height difference *h* between the two pistons?

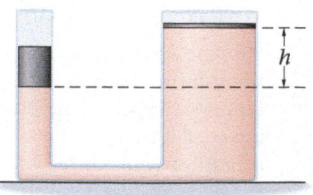

FIGURE 15-40 Problem 17

18. •• A circular wine barrel 75 cm in diameter will burst if the net upward force exerted on the top of the barrel is 643 N. A tube 1.0 cm in diameter extends into the barrel through a hole in the top, as indicated in Figure 15-38. Initially, the barrel is filled to the top and the tube is empty above that level. What weight of water must be poured into the tube in order to burst the barrel?

19. •• A cylindrical container with a cross-sectional area of 65.2 cm² holds a fluid of density 806 kg/m³. At the bottom of the container the pressure is 126 kPa. **(a)** What is the depth of the fluid? **(b)** Find the pressure at the bottom of the container after an additional $2.05 \times 10^{-3}\,\text{m}^3$ of this fluid is added to the container. Assume that no fluid spills out of the container.

20. •• **Predict/Calculate Tourist Submarine** A submarine called the *Deep View 66* is being developed to take 66 tourists at a time on sightseeing trips to tropical coral reefs. According to guidelines of the American Society of Mechanical Engineers (ASME), to be safe for human occupancy the *Deep View 66* must be able to withstand a pressure of 10.0 N per square millimeter. **(a)** To what depth can the *Deep View 66* safely descend in seawater? **(b)** If the submarine is used in freshwater instead, is its maximum safe depth greater than, less than, or the same as in seawater? Explain.

21. •• **Predict/Calculate** A water storage tower is filled with freshwater to a depth of 6.4 m. What is the pressure at **(a)** 4.5 m and **(b)** 5.5 m below the surface of the water? **(c)** Why are the metal bands on such towers more closely spaced near the base of the tower?

22. •• **Predict/Calculate** You step into an elevator holding a glass of water filled to a depth of 7.9 cm. After a moment, the elevator moves upward with constant acceleration, increasing its speed from 0 to 2.4 m/s in 2.9 s. **(a)** During the period of acceleration, is the pressure exerted on the bottom of the glass greater than, less than, or the same as before the elevator began to move? Explain. **(b)** Find the change in the pressure exerted on the bottom of the glass as the elevator accelerates.

23. •• Suppose you pour water into a container until it reaches a depth of 14 cm. Next, you carefully pour in a 7.5-cm thickness of olive oil so that it floats on top of the water. What is the pressure at the bottom of the container?

24. •• Referring to Example 15-8, suppose that some vegetable oil has been added to both sides of the U tube. On the right side of the tube, the depth of oil is 5.00 cm, as before. On the left side of the tube, the depth of the oil is 3.00 cm. Find the difference in fluid level between the two sides of the tube.

25. •• **Predict/Calculate** As a stunt, you want to sip some water through a very long, vertical straw. **(a)** First, explain why the liquid moves upward, against gravity, into your mouth when you sip. **(b)** What is the tallest straw that you could, in principle, drink from in this way?

26. •• **BIO Predict/Calculate** The patient in **FIGURE 15-41** is to receive an intravenous injection of medication. In order to work properly, the pressure of fluid containing the medication must be 109 kPa at the injection point. **(a)** If the fluid has a density of 1020 kg/m³, find the height at which the bag of fluid must be suspended above the patient. Assume that the pressure inside the bag is one atmosphere. **(b)** If a less dense fluid is used instead, must the height of suspension be increased or decreased? Explain.

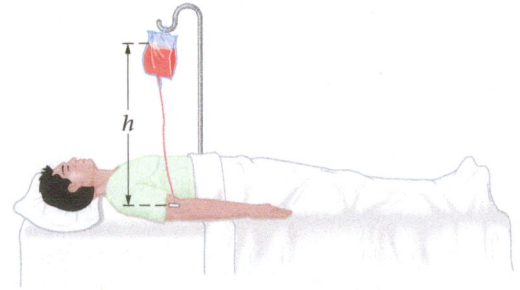

FIGURE 15-41 Problem 26

27. ••• A cylindrical container 1.0 m tall contains mercury to a certain depth, *d*. The rest of the cylinder is filled with water. If the pressure at the bottom of the cylinder is two atmospheres, what is the depth *d*?

SECTION 15-4 ARCHIMEDES' PRINCIPLE AND BUOYANCY

28. • **CE Predict/Explain Beebe and Barton** On Wednesday, August 15, 1934, William Beebe and Otis Barton made history by descending in the Bathysphere—basically a steel sphere 4.75 ft in diameter—3028 ft below the surface of the ocean, deeper than anyone had been before. **(a)** As the Bathysphere was lowered, was the buoyant force exerted on it at a depth of 10 ft greater than, less than, or equal to the buoyant force exerted on it at a depth of 50 ft? **(b)** Choose the *best explanation* from among the following:
 I. The buoyant force depends on the density of the water, which is essentially the same at 10 ft and 50 ft.
 II. The pressure increases with depth, and this increases the buoyant force.
 III. The buoyant force decreases as an object sinks below the surface of the water.

29. • **CE** Lead is more dense than aluminum. **(a)** Is the buoyant force on a solid lead sphere greater than, less than, or equal to the buoyant force on a solid aluminum sphere of the same diameter? **(b)** Does your answer to part (a) depend on the fluid that is causing the buoyant force?

30. • A fish adjusts its buoyancy to hover in one place in a small bowl as it drops a pebble it was holding in its mouth. When the pebble is released, does the water level in the bowl rise, fall, or stay the same?

31. • A raft is 3.7 m wide and 6.1 m long. When a horse is loaded onto the raft, it sinks 3.8 cm deeper into the water. What is the weight of the horse?

32. •• **BIO Predict/Calculate Swim Bladder** Many ray-finned fish adjust their buoyancy by changing the volume of gas in a *swim bladder* located just below their spine. **(a)** If the volume of the fish increases while the mass remains the same, does the buoyant force on the fish increase, decrease, or remain the same? **(b)** Calculate the change in the buoyant force on a fish if its volume increases by 11 cm³, assuming it swims in seawater.

33. •• **Predict/Calculate Submarine Buoyancy** Submarines adjust their buoyancy by changing the amount of water held in rigid containers called *ballast tanks*. **(a)** If the mass of a submerged submarine increases while the volume remains the same, does the buoyant force on the submarine increase, decrease, or remain the same? **(b)** Calculate the change in the buoyant force on a submerged submarine if it pumps 0.85 m³ of seawater into its ballast tank.

34. •• A 3.2-kg balloon is filled with helium (density = 0.179 kg/m³). If the balloon is a sphere with a radius of 4.9 m, what is the maximum weight it can lift?

35. •• A hot-air balloon plus cargo has a mass of 312 kg and a volume of 2210 m³ on a day when the outside air density is 1.22 kg/m³. The balloon is floating at a constant height of 9.14 m above the ground. What is the density of the hot air in the balloon?

36. ••• In the lab you place a beaker that is half full of water (density ρ_w) on a scale. You now use a light string to suspend a piece of metal of volume *V* in the water. The metal is completely submerged, and none of the water spills out of the beaker. Give a symbolic expression for the change in reading of the scale.

SECTION 15-5 APPLICATIONS OF ARCHIMEDES' PRINCIPLE

37. • **CE Predict/Explain** A block of wood has a steel ball glued to one surface. The block can be floated with the ball "high and dry" on its top surface. **(a)** When the block is inverted, and the ball is immersed in water, does the volume of wood that is submerged increase, decrease, or stay the same? **(b)** Choose the *best explanation* from among the following:

    I. When the block is inverted the ball pulls it downward, causing more of the block to be submerged.

    II. The same amount of mass is supported in either case, therefore the amount of the block that is submerged is the same.

    III. When the block is inverted the ball experiences a buoyant force, which reduces the buoyant force that must be provided by the wood.

38. • **CE Predict/Explain** In the preceding problem, suppose the block of wood with the ball "high and dry" is floating in a tank of water. **(a)** When the block is inverted, does the water level in the tank increase, decrease, or stay the same? **(b)** Choose the *best explanation* from among the following:

    I. Inverting the block makes the block float higher in the water, which lowers the water level in the tank.

    II. The same mass is supported by the water in either case, and therefore the amount of displaced water is the same.

    III. The inverted block floats lower in the water, which displaces more water and raises the level in the tank.

39. • **CE Measuring Density with a Hydrometer** A hydrometer, a device for measuring fluid density, is constructed as shown in **FIGURE 15-42**. If the hydrometer samples fluid 1, the small float inside the tube is submerged to level 1. When fluid 2 is sampled, the float is submerged to level 2. Is the density of fluid 1 greater than, less than, or equal to the density of fluid 2? (This is how mechanics test your antifreeze level. Since antifreeze [ethylene glycol] is more dense than water, the higher the density of coolant in your radiator the more antifreeze protection you have.)

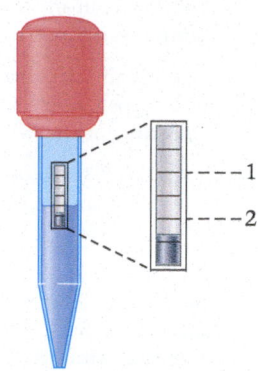

FIGURE 15-42 Problem 39

40. • **CE Predict/Explain** Referring to Example 15-12, suppose the flask with the wood tied to the bottom is placed on a scale. At some point the string breaks and the wood rises to the surface where it floats. **(a)** When the wood is floating, is the reading on the scale greater than, less than, or equal to its previous reading? **(b)** Choose the *best explanation* from among the following:

    I. The same mass is supported by the scale before and after the string breaks, and therefore the reading on the scale remains the same.

    II. When the block is floating the water level drops, and this reduces the reading on the scale.

    III. When the block is floating it no longer pulls upward on the flask; therefore, the reading on the scale increases.

41. • **CE** On a planet in a different solar system the acceleration of gravity is greater than it is on Earth. If you float in a pool of water on this planet, do you float higher than, lower than, or at the same level as when you float in water on Earth?

42. • An air mattress is 2.3 m long, 0.66 m wide, and 14 cm deep. If the air mattress itself has a mass of 0.22 kg, what is the maximum mass it can support in freshwater?

43. •• A solid block is attached to a spring scale. When the block is suspended in air, the scale reads 21.2 N; when it is completely immersed in water, the scale reads 18.2 N. What are **(a)** the volume and **(b)** the density of the block?

44. •• As in the previous problem, a solid block is suspended from a spring scale. If the reading on the scale when the block is completely immersed in water is 25.0 N, and the reading when it is completely immersed in alcohol of density 806 kg/m³ is 25.7 N, what are **(a)** the block's volume and **(b)** its density?

45. •• **BIO** A person weighs 756 N in air and has a body-fat percentage of 28.1%. **(a)** What is the overall density of this person's body? **(b)** What is the volume of this person's body? **(c)** Find the apparent weight of this person when completely submerged in water.

46. •• **Predict/Calculate** A log floats in a river with one-fourth of its volume above the water. **(a)** What is the density of the log? **(b)** If the river carries the log into the ocean, does the portion of the log above the water increase, decrease, or stay the same? Explain.

47. •• A person with a mass of 78 kg and a volume of 0.086 m³ floats quietly in water. **(a)** What is the volume of the person that is above water? **(b)** If an upward force *F* is applied to the person by a friend, the volume of the person above water increases by 0.0032 m³. Find the force *F*.

48. •• **Predict/Calculate** A block of wood floats on water. A layer of oil is now poured on top of the water to a depth that more than covers the block, as shown in **FIGURE 15-43**. **(a)** Is the volume of wood submerged in water greater than, less than, or the same as before? **(b)** If 90% of the wood is submerged in water before the oil is added, find the fraction submerged when oil with a density of 875 kg/m³ covers the block.

FIGURE 15-43 Problem 48

49. •• A piece of lead has the shape of a hockey puck, with a diameter of 7.5 cm and a height of 2.5 cm. If the puck is placed in a mercury bath, it floats. How deep below the surface of the mercury is the bottom of the lead puck?

50. ••• **Predict/Calculate** A lead weight with a volume of 0.82×10^{-5} m³ is lowered on a fishing line into a lake to a depth of 1.0 m. **(a)** What tension is required in the fishing line to give the weight an upward acceleration of 2.1 m/s²? **(b)** If the initial depth of the weight is increased to 2.0 m, does the tension found in part (a) increase, decrease, or stay the same? Explain. **(c)** What acceleration will the weight have if the tension in the fishing line is 1.2 N? Give both direction and magnitude.

SECTION 15-6 FLUID FLOW AND CONTINUITY

51. • To water the yard, you use a hose with a diameter of 3.6 cm. Water flows from the hose with a speed of 1.3 m/s. If you partially block the end of the hose so the effective diameter is now 0.52 cm, with what speed does water spray from the hose?

52. • Water flows through a pipe with a speed of 2.4 m/s. Find the flow rate in kg/s if the diameter of the pipe is 3.1 cm.

53. • To fill a child's inflatable wading pool, you use a garden hose with a diameter of 2.9 cm. Water flows from this hose with a speed of 1.3 m/s. How much time will it take to fill the pool to a depth of 26 cm if the pool is circular and has a diameter of 2.0 m?

54. • **BIO Heart Pump Rate** When at rest, your heart pumps blood at the rate of 5.00 liters per minute (L/min). What are the volume and mass of blood pumped by your heart in one day?

55. •• **BIO Blood Speed in an Arteriole** A typical arteriole has a diameter of 0.080 mm and carries blood at the rate of 9.6×10^{-5} cm³/s. **(a)** What is the speed of the blood in an arteriole? **(b)** Suppose an arteriole branches into 8800 capillaries, each with a diameter of

6.0×10^{-6} m. What is the blood speed in the capillaries? (The low speed in capillaries is beneficial; it promotes the diffusion of materials to and from the blood.)

56. •• **Predict/Calculate** Water flows at the rate of 3.11 kg/s through a hose with a diameter of 3.22 cm. **(a)** What is the speed of water in this hose? **(b)** If the hose is attached to a nozzle with a diameter of 0.732 cm, what is the speed of water in the nozzle? **(c)** Is the number of kilograms per second flowing through the nozzle greater than, less than, or equal to 3.11 kg/s? Explain.

57. •• A river narrows at a rapids from a width of 12 m to a width of only 5.8 m. The depth of the river before the rapids is 2.7 m; the depth in the rapids is 0.85 m. Find the speed of water flowing in the rapids, given that its speed before the rapids is 1.2 m/s. Assume the river has a rectangular cross section.

58. •• **BIO How Many Capillaries?** The aorta has an inside diameter of approximately 2.1 cm, compared to that of a capillary, which is about 1.0×10^{-5} m (10 μm). In addition, the average speed of flow is approximately 1.0 m/s in the aorta and 0.30 mm/s in a capillary. Assuming that all the blood that flows through the aorta also flows through the capillaries, how many capillaries does the circulatory system have?

SECTION 15-7 BERNOULLI'S EQUATION

59. • **BIO Plaque in an Artery** The buildup of plaque on the walls of an artery may decrease its diameter from 1.1 cm to 0.75 cm. If the speed of blood flow was 15 cm/s before reaching the region of plaque buildup, find **(a)** the speed of blood flow and **(b)** the pressure drop within the plaque region.

60. • A horizontal pipe contains water at a pressure of 120 kPa flowing with a speed of 1.9 m/s. When the pipe narrows to one-half its original diameter, what are **(a)** the speed and **(b)** the pressure of the water?

61. •• Unfiltered olive oil must flow at a minimum speed of 3.0 m/s to prevent settling of debris in a pipe. The oil leaves a pump at a pressure of 88 kPa through a pipe of radius 9.5 mm. It then enters a horizontal pipe at atmospheric pressure. Ignore the effects of viscosity. **(a)** What is the speed of the oil as it leaves the pump if it flows at 3.0 m/s in the horizontal pipe? **(b)** What is the radius of the horizontal pipe?

62. •• **BIO A Blowhard** Tests of lung capacity show that adults are able to exhale 1.5 liters of air through their mouths in as little as 1.0 second. **(a)** If a person blows air at this rate through a drinking straw with a diameter of 0.60 cm, what is the speed of air in the straw? **(b)** If the air from the straw in part (a) is directed horizontally across the upper end of a second straw that is vertical, as shown in **FIGURE 15-44**, to what height does water rise in the vertical straw?

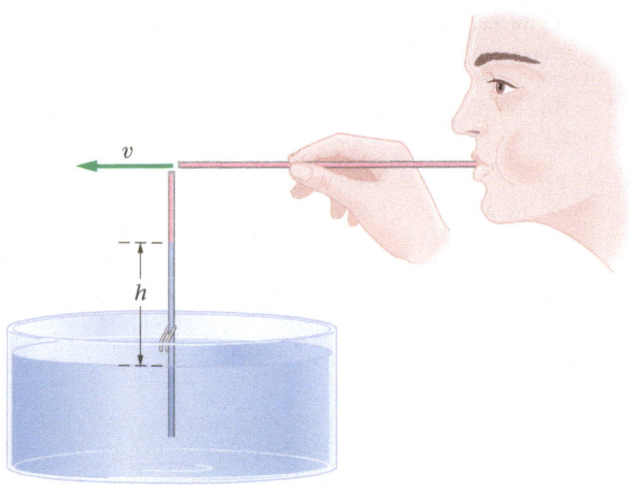

FIGURE 15-44 Problem 62

63. •• **Predict/Calculate** Water flows through a horizontal tube of diameter 2.5 cm that is joined to a second horizontal tube of diameter 1.2 cm. The pressure difference between the tubes is 7.3 kPa. **(a)** Which tube has the higher pressure? **(b)** Which tube has the higher speed of flow? **(c)** Find the speed of flow in the first tube.

64. •• A garden hose is attached to a water faucet on one end and a spray nozzle on the other end. The water faucet is turned on, but the nozzle is turned off so that no water flows through the hose. The hose lies horizontally on the ground, and a stream of water sprays vertically out of a small leak to a height of 0.68 m. What is the pressure inside the hose?

SECTION 15-8 APPLICATIONS OF BERNOULLI'S EQUATION

65. • A water tank springs a leak. Find the speed of water emerging from the hole if the leak is 2.9 m below the surface of the water, which is open to the atmosphere.

66. •• **(a)** Find the pressure difference on an airplane wing if air flows over the upper surface with a speed of 125 m/s, and along the bottom surface with a speed of 109 m/s. **(b)** If the area of the wing is 32 m^2, what is the net upward force exerted on the wing?

67. •• On a vacation flight, you look out the window of the jet and wonder about the forces exerted on the window. Suppose the air outside the window moves with a speed of approximately 170 m/s shortly after takeoff, and that the air inside the plane is at atmospheric pressure. **(a)** Find the pressure difference between the inside and outside of the window. **(b)** If the window is 25 cm by 42 cm, find the force exerted on the window by air pressure.

68. •• **Pitot Tube** The pitot tube is commonly used to measure the air speed of an aircraft. Air flows into a small opening at the end of a tube that is closed at the other end, bringing the air to rest and allowing the measurement of the pressure difference between air at rest inside the tube and air moving rapidly just outside the tube. If the high-altitude air density is 0.364 kg/m^3, and the pressure difference between inside and outside the tube is 9460 Pa, what is the airplane's speed relative to the air?

69. •• **Predict/Calculate** During a thunderstorm, winds with a speed of 47.7 m/s blow across a flat roof with an area of 668 m^2. **(a)** Find the magnitude of the force exerted on the roof as a result of this wind. **(b)** Is the force exerted on the roof in the upward or downward direction? Explain.

70. •• A garden hose with a diameter of 1.6 cm has water flowing in it with a speed of 1.3 m/s and a pressure of 1.5 atmospheres. At the end of the hose is a nozzle with a diameter of 0.64 cm. Find **(a)** the speed of water in the nozzle and **(b)** the pressure in the nozzle.

71. ••• **Predict/Calculate** Water flows in a cylindrical, horizontal pipe. As the pipe narrows to half its initial diameter, the pressure in the pipe changes. **(a)** Is the pressure in the narrow region greater than, less than, or the same as the initial pressure? Explain. **(b)** Calculate the change in pressure between the wide and narrow regions of the pipe. Give your answer symbolically in terms of the density of the water, ρ, and its initial speed v.

SECTION 15-9 VISCOSITY AND SURFACE TENSION

72. • **BIO Vasodilation** When the body requires an increased blood flow rate in a particular organ or muscle, it can accomplish this by increasing the diameter of arterioles in that area. This is referred to as vasodilation. What percentage increase in the diameter of an arteriole is required to double the volume flow rate of blood, all other factors remaining the same?

73. • **BIO (a)** Find the volume of blood that flows per second through the pulmonary artery described in Example 15-25. **(b)** If the radius

of the artery is reduced by 18%, by what factor is the blood flow rate reduced? Assume that all other properties of the artery remain unchanged.

74. • **BIO** **An Occlusion in an Artery** Suppose an occlusion in an artery reduces its diameter by 15%, but the volume flow rate of blood in the artery remains the same. By what factor has the pressure drop across the length of this artery increased?

75. •• **Motor Oil** The viscosity of 5W-30 motor oil changes from 6.4 N·s/m² at –30 °C to 0.0090 N·s/m² at 100 °C. Model the oil circulation system of a car engine as a tube of radius 5.0 mm and length 2.1 m, driven by a pump that produces a pressure difference of 110 kPa across the ends of the tube. (a) What volume flow rate does the pump produce when the oil is at –30 °C? (b) What is the volume flow rate when the oil is at the operating temperature of 100 °C?

76. •• **Predict/Calculate** Water at 20 °C flows through a horizontal garden hose at the rate of 5.0×10^{-4} m³/s. The diameter of the garden hose is 2.5 cm. (a) What is the water speed in the hose? (b) What is the pressure drop across a 15-m length of hose? Suppose the cross-sectional area of the hose is halved, but the length and pressure drop remain the same. (c) By what factor does the water speed change? (d) By what factor does the volume flow rate change? Explain.

GENERAL PROBLEMS

77. • **CE Reading a Weather Glass** A weather glass, as shown in **FIGURE 15-45**, is used to give an indication of a change in the weather. Does the water level in the neck of the weather glass move up or move down when a low-pressure system approaches?

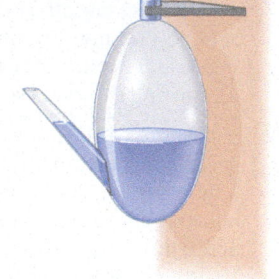

FIGURE 15-45 Problem 77

78. • **CE Predict/Explain** A person floats in a boat in a small backyard swimming pool. Inside the boat with the person are some bricks. (a) If the person drops the bricks overboard to the bottom of the pool, does the water level in the pool increase, decrease, or stay the same? (b) Choose the *best explanation* from among the following:

   I. When the bricks sink they displace less water than when they were floating in the boat; hence, the water level decreases.

   II. The same mass (boat + bricks + person) is in the pool in either case, and therefore the water level remains the same.

   III. The bricks displace more water when they sink to the bottom than they did when they were above the water in the boat; therefore the water level increases.

79. • **CE** A person floats in a boat in a small backyard swimming pool. Inside the boat with the person are several blocks of wood. Suppose the person now throws the blocks of wood into the pool, where they float. (a) Does the boat float higher, lower, or at the same level relative to the water? (b) Does the water level in the pool increase, decrease, or stay the same?

80. • **CE** The three identical containers in **FIGURE 15-46** are open to the air and filled with water to the same level. A block of wood floats in container A; an identical block of wood floats in container B, supporting a small lead weight; container C holds only water. (a) Rank the three containers in order of increasing weight of water they contain. Indicate ties where appropriate. (b) Rank the three containers in order of increasing weight of the container plus its contents. Indicate ties where appropriate.

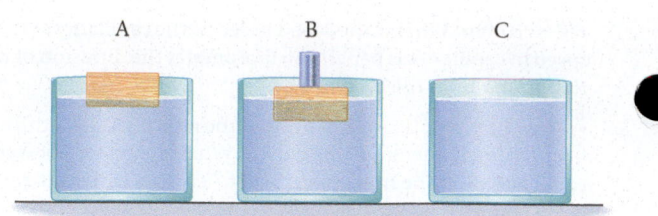

FIGURE 15-46 Problem 80

81. • **CE BIO Sphygmomanometer** When a person's blood pressure is taken with a device known as a sphygmomanometer, it is measured on the arm, at approximately the same level as the heart. If the measurement were to be taken on a standing patient's leg instead, would the reading on the sphygmomanometer be greater than, less than, or the same as when the measurement is made on the arm?

82. • A water main broke on Lake Shore Drive in Chicago on November 8, 2002, shooting water straight upward to a height of 8.0 ft. What was the pressure in the pipe?

83. • **BIO Measuring Pain Threshold** A useful instrument for evaluating fibromyalgia and trigger-point tenderness is the doloriometer or algorimeter. This device consists of a force meter attached to a circular probe that is pressed against the skin until pain is experienced. If the reading of the force meter is 3.25 lb, and the diameter of the circular probe is 1.39 cm, what is the pressure applied to the skin? Give your answer in pascals.

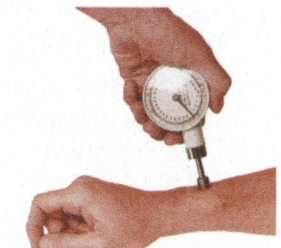

A doloriometer/algorimeter used to evaluate pain threshold. (Problem 83)

84. •• **BIO Power Output of the Heart** The power output of the heart is given by the product of the average blood pressure, 1.33 N/cm², and the flow rate, 105 cm³/s. (a) Find the power of the heart. Give your answer in watts. (b) How much energy does the heart expend in a day? (c) Suppose the energy found in part (b) is used to lift a 72-kg person vertically to a height h. Find h, in meters.

85. •• A solid block is suspended from a spring scale. When the block is in air, the scale reads 35.0 N, when immersed in water the scale reads 31.1 N, and when immersed in oil the scale reads 31.8 N. (a) What is the density of the block? (b) What is the density of the oil?

86. •• A wooden block with a density of 710 kg/m³ and a volume of 0.012 m³ is attached to the top of a vertical spring whose force constant is $k = 540$ N/m. Find the amount by which the spring is stretched or compressed if it and the wooden block are (a) in air or (b) completely immersed in water. [The density of air may be neglected in part (a).]

87. •• **Predict/Calculate Floating a Ball and Block** A 1.35-kg wooden block has an iron ball of radius 1.22 cm glued to one side. (a) If the block floats in water with the iron ball "high and dry," what is the volume of wood that is submerged? (b) If the block is now inverted, so that the iron ball is completely immersed, does the volume of wood that is submerged in water increase, decrease, or remain the same? Explain. (c) Calculate the volume of wood that is submerged when the block is in the inverted position.

88. •• **The Depth of the Atmosphere** Evangelista Torricelli (1608–1647) was the first to put forward the idea that we live at the bottom of

an ocean of air. **(a)** Given the value of atmospheric pressure at the surface of the Earth, and the fact that there is zero pressure in the vacuum of space, determine the depth of the atmosphere, assuming that the density of air and the acceleration of gravity are constant. **(b)** According to this model, what is the atmospheric pressure at the summit of Mt. Everest, 29,029 ft above sea level. (In fact, the density of air and the acceleration of gravity decrease with altitude, so the result obtained here is less than the actual depth of the atmosphere. Still, this is a reasonable first estimate.)

89. •• **The Hydrostatic Paradox I** Consider the lightweight containers shown in **FIGURE 15-47**. Both containers have bases of area $A_{base} = 24 \text{ cm}^2$ and depths of water equal to 18 cm. As a result, the downward force on the base of container 1 is equal to the downward force on container 2, even though the containers clearly hold different weights of water. This is referred to as the hydrostatic paradox. **(a)** Given that container 2 has an annular (ring-shaped) region of area $A_{ring} = 72 \text{ cm}^2$, determine the *net* downward force exerted on the container by the water. **(b)** Show that your result from part (a) is equal to the weight of the water in container 2.

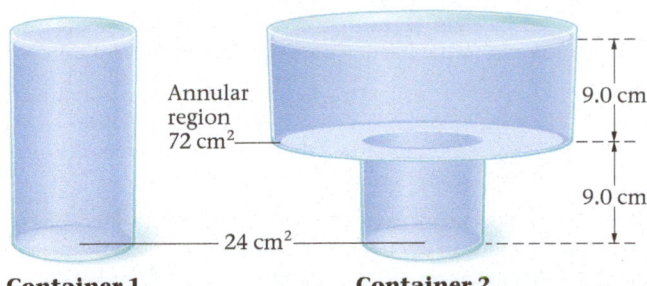

Container 1   **Container 2**

Annular region 72 cm²   9.0 cm
                           9.0 cm
——— 24 cm² ———

FIGURE 15-47 Problem 89

90. •• **The Hydrostatic Paradox II** Consider the two lightweight containers shown in **FIGURE 15-48**. As in the previous problem, these containers have equal forces on their bases but contain different weights of water. This is another version of the hydrostatic paradox. **(a)** Determine the *net* downward force exerted by the water on container 2. Note that the bases of the containers have an area $A_{base} = 24 \text{ cm}^2$, the annular region has an area $A_{ring} = 18 \text{ cm}^2$, and the depth of the water is 18 cm. **(b)** Show that your result from part (a) is equal to the weight of the water in container 2. **(c)** If a hole is poked in the annular region of container 2, how fast will water exit the hole? **(d)** How high above the hole will the stream of water rise?

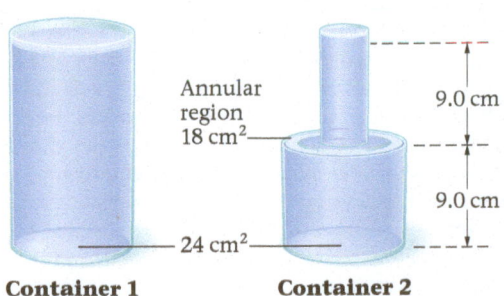

Annular region 18 cm²   9.0 cm
                           9.0 cm
——— 24 cm² ———

Container 1   **Container 2**

FIGURE 15-48 Problem 90

91. •• **Predict/Calculate** A backyard swimming pool is circular in shape and contains water to a uniform depth of 42 cm. It is 2.4 m in diameter and is not completely filled. **(a)** What is the pressure at the bottom of the pool? **(b)** If a person gets into the pool and floats peacefully, does the pressure at the bottom of the pool increase, decrease, or stay the same? **(c)** Calculate the pressure due to the

water and the person at the bottom of the pool if the floating person has a mass of 73 kg.

92. •• A prospector finds a solid rock composed of granite ($\rho = 2650 \text{ kg/m}^3$) and gold. If the volume of the rock is $3.84 \times 10^{-4} \text{ m}^3$, and its mass is 3.40 kg, **(a)** what mass of gold is contained in the rock? What percentage of the rock is gold by **(b)** volume and **(c)** mass?

93. •• **Predict/Calculate (a)** If the tension in the string in Example 15-12 is 0.89 N, what is the volume of the wood? Assume that everything else remains the same. **(b)** If the string breaks and the wood floats on the surface, does the water level in the flask rise, drop, or stay the same? Explain. **(c)** Assuming the flask is cylindrical with a cross-sectional area of 62 cm², find the change in water level after the string breaks.

94. •• **Predict/Calculate A Siphon for Irrigation** A siphon is a device that allows water to flow from one level to another. The siphon shown in **FIGURE 15-49** delivers water from an irrigation canal to a field of crops. To operate the siphon, water is first drawn through the length of the tube. After the flow is started in this way it continues on its own. **(a)** Using points 1 and 3 in Figure 15-49, find the speed v of the water leaving the siphon at its lower end. Give a symbolic answer. **(b)** Is the speed of the water at point 2 greater than, less than, or the same as its speed at point 3? Explain.

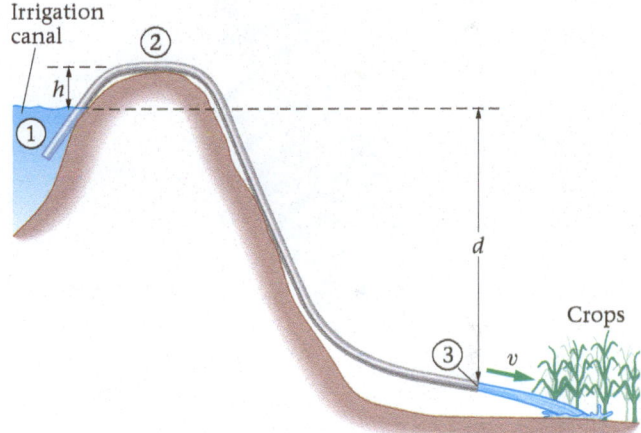

Irrigation canal
②
h
①
d
Crops
③ v

FIGURE 15-49 Problem 94

95. •• **Milking Cows** The density of raw milk at body temperature is 1023 kg/m³. After being extracted from the cow, the milk is pumped through a pipe to the bottom of a cooling tank filled to a depth of $h = 1.8$ m (**FIGURE 15-50**). **(a)** If the pressure at location 2 in Figure 15-50 is atmospheric pressure, what minimum speed v_1 must the milk have in order to enter the cooling tank? Assume the milk at location 2 is essentially at rest. **(b)** What is the pressure at the bottom of the cooling tank?

$P_2 = P_{at}$ $v_2 = 0$ ② h
① v_1 0

FIGURE 15-50 Problem 95

96. •• A tin can is filled with water to a depth of 39 cm. A hole 11 cm above the bottom of the can produces a stream of water that is directed at an angle of 36° above the horizontal. Find **(a)** the range and **(b)** the maximum height of this stream of water.

97. •• **BIO** A person weighs 685 N in air but only 497 N when standing in water up to the hips. Find **(a)** the volume of each of the person's legs and **(b)** the mass of each leg, assuming they have a density that is 1.05 times the density of water.

98. •• **Thunderstorm Outflow** Rain-cooled air near the core of a thunderstorm sinks and then spreads out in front of the storm in a *forward flank downdraft gust front* (**FIGURE 15-51**). These gusts can vary from a cool breeze to a violent and damaging wind. Thunderstorms are extremely complex, but modeling air as an incompressible fluid can offer some insight. **(a)** Suppose 1.0 m³ of rain-cooled air has a density of 0.835 kg/m³, while the warmer air surrounding it has a density of 0.819 kg/m³. Taking into account the buoyant force on the parcel of air, find its downward acceleration. **(b)** If the parcel maintains the same acceleration from rest at an altitude of 4000 m, what is its speed when it arrives at the surface? **(c)** Now model the air as an incompressible fluid of constant density 1.02 kg/m³ that is at rest and has a pressure of 61.6 kPa at $h = 4000$ m altitude, but is moving and has a pressure of 101.3 kPa at $h = 0$. What is the speed of the air at the surface?

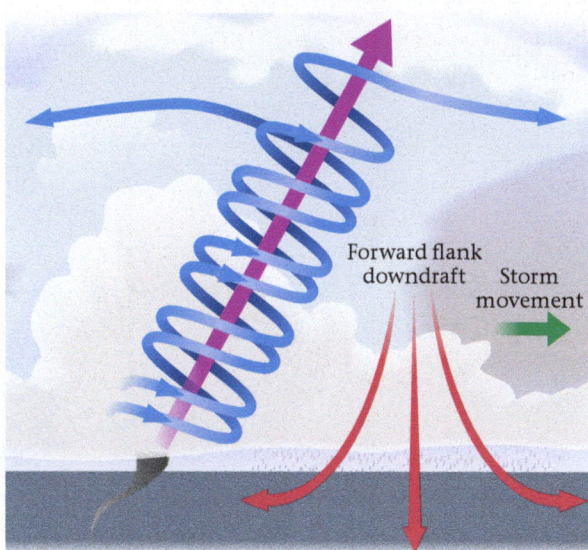

FIGURE 15-51 Problem 98

99. •• A horizontal pipe carries oil whose coefficient of viscosity is 0.00012 N·s/m². The diameter of the pipe is 5.2 cm, and its length is 55 m. **(a)** What pressure difference is required between the ends of this pipe if the oil is to flow with an average speed of 1.2 m/s? **(b)** What is the volume flow rate in this case?

100. •• **BIO** A patient is given an injection with a hypodermic needle 3.3 cm long and 0.26 mm in diameter. Assuming the solution being injected has the same density and viscosity as water at 20 °C, find the pressure difference needed to inject the solution at the rate of 1.5 g/s.

101. •• **Going Over Like a *Mythbuster* Lead Balloon** On one episode of *Mythbusters*, Jamie and Adam try to make a lead balloon that will float when filled with helium. The balloon they constructed was approximately cubical in shape, and 10 feet on a side. They used a thin lead foil, which gave the finished balloon a mass of 11 kg. **(a)** What was the thickness of the foil? **(b)** Would the lead balloon float if filled with helium? **(c)** If the balloon does float, what would be the most mass it could lift in addition to its own mass?

102. ••• A round wooden log with a diameter of 73 cm floats with one-half of its radius out of the water. What is the log's density?

103. ••• The hollow, spherical glass shell shown in **FIGURE 15-52** has an inner radius R and an outer radius $1.2R$. The density of the glass is ρ_g. What fraction of the shell is submerged when it floats in a liquid of density $\rho = 1.5\rho_g$? (Assume the interior of the shell is a vacuum.)

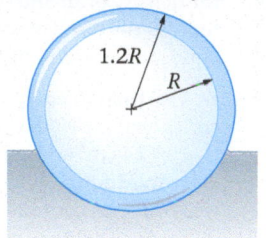

FIGURE 15-52 Problem 103

104. ••• A geode is a hollow rock with a solid shell and an air-filled interior. Suppose a particular geode weighs twice as much in air as it does when completely submerged in water. If the density of the solid part of the geode is 2500 kg/m³, what fraction of the geode's volume is hollow?

105. ••• A tank of water filled to a depth d has a hole in its side a height h above the table on which it rests. Show that water emerging from the hole hits the table at a horizontal distance of $2\sqrt{(d-h)h}$ from the base of the tank.

106. ••• The water tank in **FIGURE 15-53** is open to the atmosphere and has two holes in it, one 0.80 m and one 3.6 m above the floor on which the tank rests. If the two streams of water strike the floor in the same place, what is the depth of water in the tank?

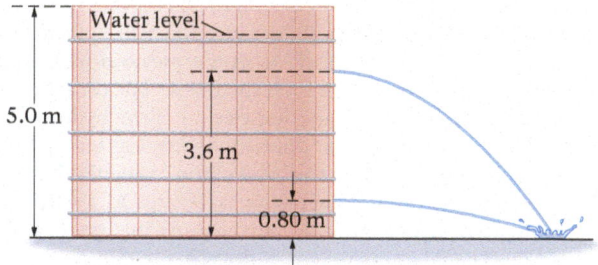

FIGURE 15-53 Problem 106

PASSAGE PROBLEMS

Cooking Doughnuts

Doughnuts are cooked by dropping the dough into hot vegetable oil until it changes from white to a rich, golden brown. One popular brand of doughnut automates this process; the doughnuts are made on an assembly line that customers can view in operation as they wait to order. Watching the doughnuts cook gives the customers time to develop an appetite as they ponder the physics of the process.

First, the uncooked doughnut is dropped into hot vegetable oil, whose density is $\rho = 919$ kg/m³. There it browns on one side as it floats on the oil. After cooking for the proper amount of time, a mechanical lever flips the doughnut over so it can cook on the other side. The doughnut floats fairly high in the oil, with less than half of its volume submerged. As a result, the final product has a characteristic white stripe around the middle where the dough is always out of the oil, as shown in **FIGURE 15-54**.

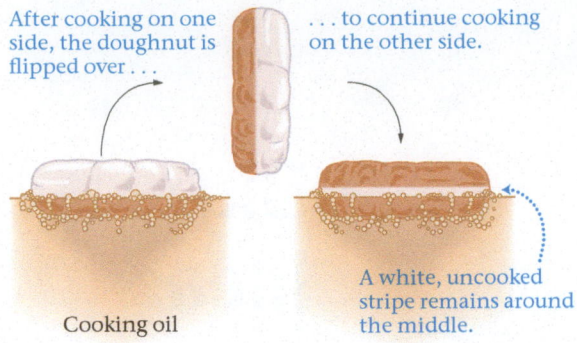

After cooking on one side, the doughnut is flipped over . . .

. . . to continue cooking on the other side.

Cooking oil

A white, uncooked stripe remains around the middle.

FIGURE 15-54 Steps in cooking a doughnut

The relationship between the density of the doughnut and the height of the white stripe is graphed in **FIGURE 15-55**. On the *x* axis we plot the density of the doughnut as a fraction of the density of the vegetable cooking oil; the *y* axis shows the height of the white stripe as a fraction of the total height of the doughnut. Notice that the height of the white stripe is plotted for both positive and negative values.

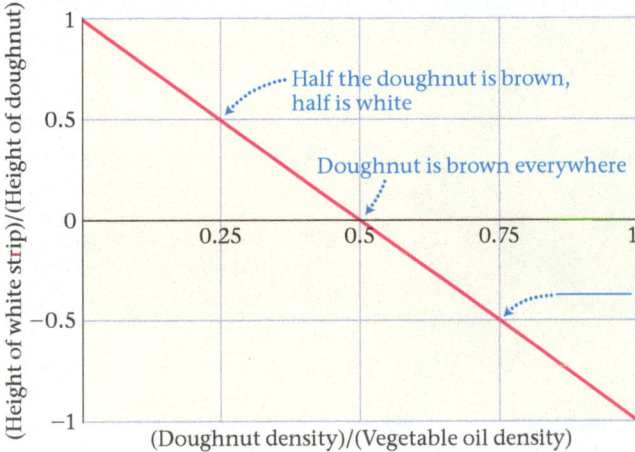

Half the doughnut is brown, half is white

Doughnut is brown everywhere

(Height of white strip)/(Height of doughnut)

(Doughnut density)/(Vegetable oil density)

FIGURE 15-55 Cooking a floating doughnut A finished doughnut has a stripe around the middle whose height is related to the density of the doughnut by the straight line shown here. (Problems 107, 108, 109, and 110)

107. • Assuming the doughnut has a cylindrical shape of height *H* and diameter *D*, and that the height of the white stripe is 0.22*H*, what is the density of the doughnut?

A. 260 kg/m³   B. 360 kg/m³
C. 720 kg/m³   D. 820 kg/m³

108. • Figure 15-57 has comments where the "height" of the white stripe is 0.5*H* and 0. The comment for −0.5*H* has been left blank, however. Which of the following comments is most appropriate for this case?

A. The doughnut sinks.
B. The white stripe is still present, but has a negative height.
C. Half the doughnut is light brown, half is dark brown.
D. The top and bottom of the doughnut are white, the middle one-half is brown.

109. •• A new doughnut is being planned whose density will be 330 kg/m³. If the height of the doughnut is *H*, what will be the height of the white stripe?

A. 0.14*H*   B. 0.24*H*
C. 0.28*H*   D. 0.64*H*

110. •• Suppose the density of a doughnut is 550 kg/m³. In terms of the height of the doughnut, *H*, what is the height of the portion of the doughnut that is out of the oil?

A. 0.20*H*   B. 0.40*H*
C. 0.67*H*   D. 0.80*H*

111. •• **Predict/Calculate REFERRING TO EXAMPLE 15-8** Suppose we use a different vegetable oil that has a higher density than the one in Example 15-8. (a) If everything else remains the same, will the height difference, *h*, increase, decrease, or remain the same? Explain. (b) Find the height difference for an oil that has a density of 9.60 × 10² kg/m³.

112. •• **REFERRING TO EXAMPLE 15-8** Find the height difference, *h*, if the depth of the oil is increased to 7.50 cm. Assume everything else in the problem remains the same.

113. •• **REFERRING TO EXAMPLE 15-24** (a) Find the height *H* required to make *D* = 0.655 m. Assume everything else in the problem remains the same. (b) Find the depth *h* required to make *D* = 0.455 m. Assume everything else in the problem remains the same.

114. •• **REFERRING TO EXAMPLE 15-24** Suppose both *h* and *H* are increased by a factor of two. By what factor is the distance *D* increased?

16

Temperature and Heat

Big Ideas

1 Heat is energy that is transferred due to a difference in temperature.

2 Most substances expand when their temperature is increased.

3 Mechanical work can be converted into an equivalent amount of heat.

4 Different substances require different amounts of heat for a given increase in temperature.

5 Thermal energy can be transferred by conduction, convection, or radiation.

▲ Penguins are gregarious animals, but it's more than a desire to be close to one another that drives these birds to huddle together in tight groups. By packing in closely, the penguins reduce their surface area in contact with the frigid temperatures of the Antarctic. As a result, the transfer of thermal energy from their warm bodies to the surrounding air is minimized. This chapter explores these and other issues involving temperature, heat, and thermal energy.

In our study of physics so far, every calculation has been in terms of mass, length, and time, or some combination of the three. We now add a fourth physical quantity—*temperature*.

With the introduction of temperature, we broaden the scope of physics, allowing the study of a wide variety of new physical situations that mechanics alone cannot address.

16-1 Temperature and the Zeroth Law of Thermodynamics

Even as children we learn to avoid objects that are "hot." We also discover early on that if we forget to wear our coats outside we can become "cold." Later, we associate high values of something called "temperature" with hot objects, and low values with cold objects. These basic notions about temperature carry over into physics, though with a bit more precision.

Heat and the Flow of Energy   When we put a pan of cool water on a hot stove burner, we say that "heat flows" from the hot burner to the cool water. To be more precise, we define **heat** as follows:

> Heat is the energy that is *transferred* between objects because of a temperature difference.

Therefore, when we say there is a "transfer of heat" or a "heat flow" from object A to object B, what we mean is that the total energy of object A decreases and the total energy of object B increases. The total amount of energy in a substance—the sum of all of its kinetic and potential energy—is referred to as its *internal energy*, or **thermal energy**. Thus, an object's thermal energy refers both to the random motion of its particles (kinetic energy) and to the separation and orientation of the particles relative to one another (potential energy). Adding thermal energy to a system is known as *heating* and removing thermal energy is known as *cooling*. To summarize, an object does not "contain" heat—it has a certain energy content, and the energy it *exchanges* with other objects due to temperature differences is called heat.

Now, objects are said to be in **thermal contact** if heat can flow between them. In general, when a hot object is brought into thermal contact with a cold object, heat will be exchanged. The result is that the hot object cools off—and its molecules move more slowly—while the cold object warms up—and its molecules move more rapidly. After some time in thermal contact, the transfer of heat ceases. At this point, we say that the objects are in **thermal equilibrium**. The study of physical processes involving the transfer of heat is the area of physics referred to as **thermodynamics**. Because thermodynamics deals with the flow of energy within and between objects, it has great practical importance in all areas of the life sciences, physical sciences, and engineering.

Zeroth Law of Thermodynamics

We now introduce perhaps the most fundamental law obeyed by thermodynamic systems—referred to, appropriately enough, as the zeroth law of thermodynamics. The zeroth law spells out the basic properties of temperature. Its name reflects not only its basic importance, but also its subtlety, in that it was not recognized as a separate law until after the other three laws of thermodynamics had been accepted. Later, in Chapter 18, we introduce the remaining three laws of thermodynamics. These laws enable us to analyze the behaviors of engines and refrigerators, and to show, among other things, that perpetual motion machines are not possible.

The basic idea of the zeroth law of thermodynamics is that thermal equilibrium is determined by a single physical quantity—the **temperature**. For example, two objects in thermal contact are in equilibrium when they have the *same* temperature. If one object has a higher temperature, heat flows from that object to the other object until their temperatures are equal.

This may seem almost too obvious to mention, at least until you give it more thought. Suppose, for example, that you have a piece of metal and a pool of water, and you want to know if heat will flow between them when you put the metal in the pool. You measure the temperature of each, and if they are the same, you can conclude that no heat will flow. If the temperatures are different, however, it follows that there will be a flow of heat. Nothing else matters—not the type of metal, its mass, its shape, the amount of water, whether the water is fresh or salt, and so on—all that matters is one number, the temperature.

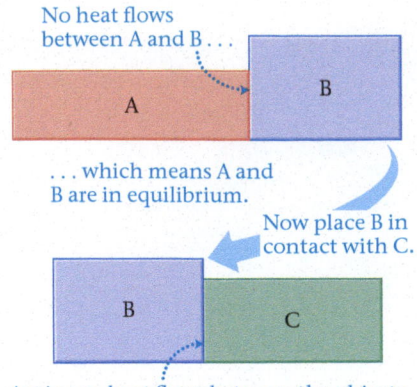

No heat flows between A and B . . .

. . . which means A and B are in equilibrium.

Now place B in contact with C.

Again, no heat flows between the objects. Therefore, C is also in equilibrium with B.

If A and C are now placed in contact, we find that no heat flows, so they too are in equilibrium.

▲ **FIGURE 16-1 An illustration of the zeroth law of thermodynamics** If A and C are each in thermal equilibrium with B, then A will be in thermal equilibrium with C when they are brought into thermal contact.

Big Idea ❶ Heat is energy that is transferred from one physical system to another due to a difference in temperature. If two systems have the same temperature, there will be no transfer of energy when they are brought into thermal contact.

PHYSICS IN CONTEXT Looking Ahead

We make extensive use of the concept of temperature throughout Chapters 17 and 18.

To summarize, the **zeroth law of thermodynamics** can be stated as follows:

> If object A is in thermal equilibrium with object B, and object C is also in thermal equilibrium with object B, then objects A and C will be in thermal equilibrium if brought into thermal contact.

This law is illustrated in **FIGURE 16-1**. We begin with objects A and B in thermal contact and in equilibrium. Next, object B is separated from A, and placed in contact with C. Objects C and B are also found to be in equilibrium. Hence, by the zeroth law, we are assured that when A and C are placed in contact they also will be in equilibrium.

To apply this principle to our example, let object A be the piece of metal and object C be the pool of water. Object B, then, can be a thermometer, used to measure the temperature of the metal and the water. If A and C are each separately in equilibrium with B—which means that they have the same temperature—they will be in equilibrium with one another.

Enhance Your Understanding   (Answers given at the end of the chapter)

1. When bricks A and B are placed in thermal contact, energy flows from brick B to brick A. Which brick had the higher initial temperature?

Section Review

• Heat is the energy transferred between objects due to a difference in temperature. If objects have the same temperature, they are said to be in thermal equilibrium.

• Two objects in thermal equilibrium with a third object will be in thermal equilibrium with one another if placed in thermal contact.

16-2 Temperature Scales

A variety of temperature scales are commonly used both in everyday situations and in physics. Some are related to familiar reference points, such as the temperature of boiling or freezing water. Others have more complex, historical rationales for their values. Here we consider three of the more frequently used temperature scales. We also examine the connections between them.

The Celsius Scale

Perhaps the easiest temperature scale to remember is the Celsius scale, named in honor of the Swedish astronomer Anders Celsius (1701–1744). Originally, Celsius assigned zero degrees to boiling water and 100 degrees to freezing water. These values were later reversed by the biologist Carolus Linnaeus (1707–1778). Thus, today we say that water freezes at zero degrees Celsius, which we abbreviate as 0 °C, and boils at 100 °C.

The choice of zero level for a temperature scale is quite arbitrary, as is the number of degrees between any two reference points. In the Celsius scale, as in others, there is no upper limit to the value a temperature may have. There is a lower limit, however. For the Celsius scale, the lowest possible temperature is −273.15 °C, as we shall see later in this section.

One bit of notation should be pointed out before we continue. When we write a Celsius temperature, we give a numerical value followed by the degree symbol, °, and the capital letter C. For example, 5 °C is the temperature five degrees Celsius. On the other hand, if the temperature of an object is *changed* by a given amount, we use the notation C°. Thus, if we increase the temperature by five degrees on the Celsius scale, we say that the change in temperature is 5 C°—that is, five Celsius degrees. This is summarized below:

A temperature T of five degrees is 5 °C (five degrees Celsius).

A temperature change ΔT of five degrees is 5 C° (five Celsius degrees).

The Fahrenheit Scale

The Fahrenheit scale was developed by Daniel Gabriel Fahrenheit (1686–1736), who chose zero to be the lowest temperature he was able to achieve in his laboratory. He also chose 96 degrees to be body temperature, though why he made this choice is not known. In the modern version of the Fahrenheit scale, body temperature is 98.6 °F; in addition, water freezes at 32 °F and boils at 212 °F. Lastly, using the same convention as for °C and C°, we say that an increase of 180 F° is required to bring water from freezing to boiling.

The Fahrenheit scale not only has a different zero than the Celsius scale, it also has a different "size" for its degree. As just noted, 180 Fahrenheit degrees are required for the same change in temperature as 100 Celsius degrees. Hence, the Fahrenheit degrees are smaller by a factor of $100/180 = 5/9$.

To convert between a Fahrenheit temperature, T_F, and a Celsius temperature, T_C, we start by writing a linear relationship between them. Thus, let

$$T_F = aT_C + b$$

We would like to determine the constants a and b. This requires two independent pieces of information, which we have in the freezing and boiling points of water. Using the freezing point, we find

$$32 \,°F = a(0 \,°C) + b = b$$

Thus, b is 32 °F. Next, the boiling point gives

$$212 \,°F = a(100 \,°C) + 32 \,°F$$

Solving for the constant a, we find

$$a = \frac{(212 \,°F - 32 \,°F)}{(100 \,°C)} = \frac{180 \,F°}{100 \,C°} = \tfrac{9}{5} F°/C°$$

Combining our results gives the following conversion relationship:

Conversion Between Degrees Celsius and Degrees Fahrenheit

$$T_F = \left(\tfrac{9}{5} F°/C°\right)T_C + 32 \,°F \qquad\qquad \text{16-1}$$

Similarly, this relationship can be rearranged to convert from Fahrenheit to Celsius:

Conversion Between Degrees Fahrenheit and Degrees Celsius

$$T_C = \left(\tfrac{5}{9} C°/F°\right)(T_F - 32 \,°F) \qquad\qquad \text{16-2}$$

Because conversion factors like $\tfrac{9}{5} F°/C°$ are a bit clumsy and tend to clutter up an equation, we will generally drop the degree symbols until the final result. For example, to convert 10 °C to degrees Fahrenheit, we write

$$T_F = \tfrac{9}{5} T_C + 32 = \tfrac{9}{5}(10) + 32 = 50 \,°F$$

Similar remarks apply in converting from Fahrenheit to Celsius.

EXAMPLE 16-1 TEMPERATURE CONVERSIONS

(a) On a fine spring day you notice that the temperature is 75 °F. What is the corresponding temperature on the Celsius scale? **(b)** If the temperature on a brisk winter morning is −2.0 °C, what is the corresponding Fahrenheit temperature?

PICTURE THE PROBLEM

Our sketch shows a circular thermometer with both a Fahrenheit and a Celsius scale. The range of the scales extends well beyond the temperatures 75 °F and −2.0 °C needed for the problem.

CONTINUED

REASONING AND STRATEGY

The conversions asked for in this problem are straightforward applications of the relationships between T_F and T_C. In particular, for (a) we use $T_C = \frac{5}{9}(T_F - 32)$, and for (b) we use $T_F = \frac{9}{5}T_C + 32$.

Known (a) Temperature is $T_F = 75\,°F$. (b) Temperature is $T_C = -2.0\,°C$.
Unknown (a) Temperature in Celsius, $T_C = ?$ (b) Temperature in Fahrenheit, $T_F = ?$

SOLUTION

Part (a)

1. Substitute $T_F = 75\,°F$ in $T_C = \frac{5}{9}(T_F - 32)$: $T_C = \frac{5}{9}(75 - 32) = 24\,°C$

Part (b)

2. Substitute $T_C = -2.0\,°C$ in $T_F = \frac{9}{5}T_C + 32$: $T_F = \frac{9}{5}(-2.0) + 32 = 28\,°F$

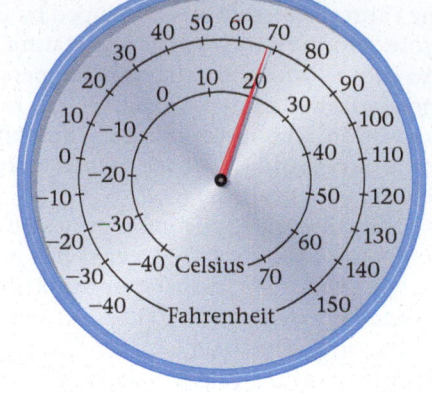

INSIGHT

The results given here agree with the scales shown in the sketch.

PRACTICE PROBLEM

Find the Celsius temperature that corresponds to 110 °F.   [**Answer:** $T_C = 43\,°C$]

Some related homework problems: Problem 2, Problem 3

QUICK EXAMPLE 16-2   SAME TEMPERATURE

What temperature is the same on both the Celsius and Fahrenheit scales?

REASONING AND SOLUTION

We can use either $T_F = \left(\frac{9}{5}\,F°/C°\right)T_C + 32\,°F$ or $T_C = \left(\frac{5}{9}\,C°/F°\right)(T_F - 32\,°F)$ to set the two temperatures equal to one another: $T_F = T_C = t$. The algebra is a bit simpler with $T_F = \left(\frac{9}{5}\,F°/C°\right)T_C + 32\,°F$, so we'll use that relationship.

1. Set $T_F = T_C = t$ in $T_F = \frac{9}{5}T_C + 32$: $t = 9t/5 + 32$

2. Move all terms involving t to the left side of the equation: $-4t/5 = 32$

3. Solve for t: $t = -40$

4. As a check, substitute $T_F = -40\,°F$ in $T_C = \frac{5}{9}(T_F - 32)$: $T_C = \frac{5}{9}(-40 - 32) = -40\,°C$

Thus, $-40\,°F$ is the same as $-40\,°C$. This is consistent with the sketch shown in Example 16-1.

Absolute Zero

Experiments show conclusively that there is a lowest temperature below which it is impossible to cool an object. This is referred to as **absolute zero**. Though absolute zero can be approached from above arbitrarily closely, it can never be attained.

To give an idea of just where absolute zero is on the Celsius scale, we start with the following observation: If a given volume V of air—say, the air in a balloon—is cooled from 100 °C to 0 °C, its volume decreases by roughly $V/4$. Imagine this trend continuing uninterrupted. In this case, cooling from 0 °C to -100 °C would reduce the volume by another $V/4$, from -100 °C to -200 °C by another $V/4$, and finally, from -200 °C to -300 °C by another $V/4$, which brings the volume down to zero. Clearly, it doesn't make sense for the volume to be less than zero, and hence absolute zero must be roughly -300 °C.

This result, though crude, is in the right ballpark. A precise determination of absolute zero can be made with a device known as a **constant-volume gas thermometer** (FIGURE 16-2). The basic idea is that by adjusting the level of mercury in the right-hand tube, we can set the level of mercury in the left-hand tube to a fixed reference level. With the mercury so adjusted, the gas occupies a constant volume and its pressure is simply $P_{gas} = P_{at} + \rho g h$ (Equation 15-7), where ρ is the density of mercury.

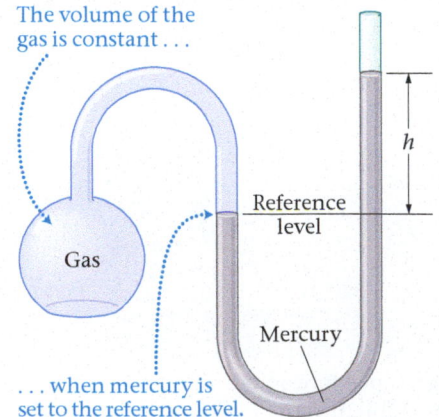

The volume of the gas is constant . . .

. . . when mercury is set to the reference level.

▲ **FIGURE 16-2 A constant-volume gas thermometer** By adjusting the height of mercury in the right-hand tube, we can set the level in the left-hand tube at the reference level. This assures that the gas occupies a constant volume.

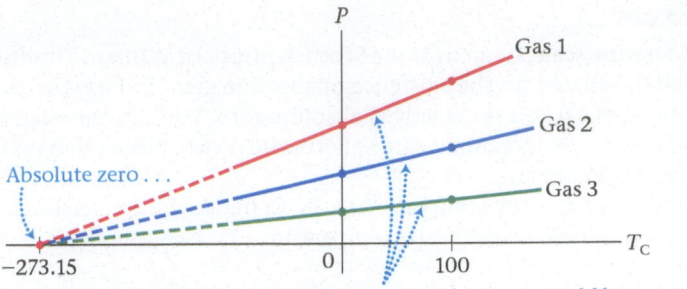

As the temperature of the gas is changed, the mercury level in the right-hand tube can be readjusted as described. The gas pressure can be determined again, and the process repeated. The results of a series of such measurements are shown in **FIGURE 16-3**.

Notice that as a gas is cooled, its pressure decreases, as one would expect. In fact, the decrease in pressure is approximately linear. At low enough temperatures the gas eventually liquefies, and its behavior changes, but if we extrapolate the straight line obtained before liquefaction, we see that it reaches zero pressure (the lowest pressure possible) at −273.15 °C.

What is remarkable about this result is that it is independent of the gas we use in the thermometer. For example, gas 2 and gas 3 in Figure 16-3 have pressures that are different from one another, and from gas 1. Yet all three gases extrapolate to zero pressure at precisely the same temperature. Thus, we conclude that there is indeed a *unique* value of absolute zero, below which further cooling is not possible.

EXAMPLE 16-3   IT'S A GAS

The gas in a constant-volume gas thermometer has a pressure of 97.6 kPa at 0.00 °C. Assuming ideal behavior, as in Figure 16-3, what is the pressure of this gas at 105 °C?

PICTURE THE PROBLEM
Our sketch plots the pressure of the gas as a function of temperature. At $T_C = 0.00$ °C the pressure is 97.6 kPa, and at $T_C = -273.15$ °C the pressure extrapolates linearly to zero. We also indicate the point corresponding to 105 °C on the graph.

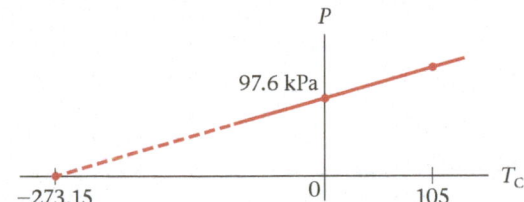

REASONING AND STRATEGY
We assume that the pressure lies on a straight line, as in Figure 16-3. To find the pressure at $T_C = 105$ °C we simply extend the straight line.

Thus, we start with the information that the pressure increases linearly from 0 to 97.6 kPa when the temperature increases from −273.15 °C to 0.00 °C. This rate of increase in pressure must also apply to an increase in temperature from −273.15 °C to 105 °C. Using this rate we can find the desired pressure.

Known $P = 0$ at $T_C = -273.15$, $P = 97.6$ kPa at $T_C = 0.00$ °C.
Unknown $P = ?$ at $T_C = 105$ °C.

SOLUTION

1. Calculate the rate at which pressure increases with temperature for this gas:

$$\text{rate} = \frac{97.6 \text{ kPa}}{273.15 \text{ C}°} = 0.357 \text{ kPa/C}°$$

2. Multiply this rate by the temperature change from −273.15 °C to 105 °C (a change of 378.15 C°):

$$(0.357 \text{ kPa/C}°)(378.15 \text{ C}°) = 135 \text{ kPa}$$

INSIGHT
The pressure of this gas increases from slightly less than one atmosphere at 0.00 °C to roughly 34% more than one atmosphere at 105 °C.

PRACTICE PROBLEM
Find the temperature at which the pressure of the gas is 74 kPa.   [**Answer:** $T_C = -66$ °C]

Some related homework problems: Problem 7, Problem 8

The Kelvin Scale

The Kelvin temperature scale, named for the Scottish physicist William Thomson, Lord Kelvin (1824–1907), is based on the existence of absolute zero. In fact, the zero of the Kelvin scale, abbreviated 0 K, is set exactly at absolute zero. Thus, in this scale there are no negative equilibrium temperatures. The Kelvin scale is also chosen to have the same degree size as the Celsius scale.

As mentioned, absolute zero occurs at −273.15 °C; hence the conversion between a Kelvin-scale temperature, T, and a Celsius temperature, T_C, is as follows:

> **Conversion Between a Celsius Temperature and a Kelvin Temperature**
>
> $$T = T_C + 273.15$$ 16-3

The difference between the Celsius and Kelvin scales is simply a difference in the zero level.

The notation for the Kelvin scale differs somewhat from that for the Celsius and Fahrenheit scales. In particular, by international agreement, the degree terminology and the degree symbol, °, are not used in the Kelvin scale. Instead, a temperature of 5 K is read simply as 5 kelvin. In addition, a change in temperature of 5 kelvin is written 5 K, the same as for a temperature of 5 kelvin.

Though the Celsius and Fahrenheit scales are the ones most commonly used in everyday situations, the Kelvin scale is used more than any other in physics. This stems from the fact that the Kelvin scale incorporates the significant concept of absolute zero. As a result, the thermal energy of a system depends in a very simple way on the Kelvin temperature. This will be discussed in detail in the next chapter.

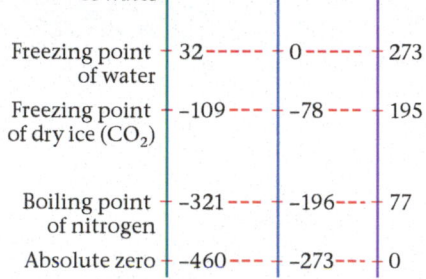

▲ **FIGURE 16-4 Temperature scales** A comparison of the Fahrenheit, Celsius, and Kelvin temperature scales. Physically significant temperatures, such as the freezing and boiling points of water, are indicated for each scale.

EXERCISE 16-4   KELVIN SCALE

You visit Palm Springs on a nice warm day, when the temperature is 113 °F. Convert this temperature to the Kelvin temperature scale.

REASONING AND SOLUTION

We will do two temperature conversions to go from Fahrenheit to Kelvin. First, we convert from °F to °C using $T_C = \left(\frac{5}{9}\,C°/F°\right)\left(T_F - 32\,°F\right)$:

$$T_C = \frac{5}{9}(113 - 32) = 45\,°C$$

Next, we convert °C to K with $T = T_C + 273.15$:

$$T = 45 + 273.15 = 318\,K$$

The three temperature scales presented in this section are shown side by side in **FIGURE 16-4**. Temperatures of particular interest are indicated as well. This permits a useful visual comparison between the scales.

Enhance Your Understanding   (Answers given at the end of the chapter)

2. Is the "size" of a degree in the Fahrenheit scale greater than, less than, or equal to the size of a degree in the Kelvin scale?

Section Review

- Three temperature scales are in common usage. The conversions between these scales are as follows: Celsius to Fahrenheit, $T_F = \left(\frac{9}{5}\,F°/C°\right)T_C + 32\,°F$; Fahrenheit to Celsius, $T_C = \left(\frac{5}{9}\,C°/F°\right)(T_F - 32\,°F)$; Celsius to Kelvin, $T = T_C + 273.15$.

16-3 Thermal Expansion

Most substances expand when heated. For example, power lines on a hot summer day hang low compared to their appearance on a cold day in winter. In fact, thermal expansion is the basis for many thermometers, including the familiar thermometers used for measuring a fever. The expansion of a liquid, such as mercury or alcohol, results in a column of liquid of variable height within the glass neck of the thermometer. The height is read against markings on the glass, which gives the temperature.

You may wonder what bizarre substance could possibly be an exception to this common response to heating. The most important exception occurs in a substance you drink every day—water. This is just one of the many special properties that sets water apart from most other substances.

In this section we consider the physics of thermal expansion, including linear, area, and volume expansion. We also discuss briefly the unusual thermal behavior of water, and some of its more significant consequences.

Big Idea 2 Most substances expand when their temperature is increased. A notable exception is water, which expands as its temperature is lowered from 4 °C to 0 °C.

Linear Expansion

Consider a rod whose length is L_0 at the temperature T_0. Experiments show that when the rod is heated or cooled, its length changes in direct proportion to the temperature change. Thus, if the change in temperature is ΔT, the change in length of the rod, ΔL, is

$$\Delta L = (\text{constant})\Delta T$$

The constant of proportionality depends, among other things, on the substance from which the rod is made.

CONCEPTUAL EXAMPLE 16-5   COMPARE EXPANSIONS

When rod 1 is heated by an amount ΔT, its length increases by ΔL. If rod 2, which is twice as long as rod 1 and made of the same material, is heated by the same amount, does its length increase by **(a)** ΔL, **(b)** $2\Delta L$, or **(c)** $\Delta L/2$?

REASONING AND DISCUSSION
We can imagine rod 2 to be composed of two copies of rod 1 placed end to end, as shown.

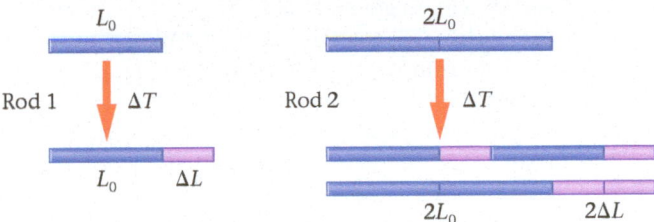

When the temperature is increased by ΔT, each copy of rod 1 expands by ΔL. Hence, the total expansion of the two copies is $2\Delta L$, as is the total expansion of rod 2.

ANSWER
(b) The rod that is twice as long expands twice as much, by $2\Delta L$.

We conclude, then, on the basis of the preceding Conceptual Example, that the change in length is proportional to *both* the initial length, L_0, and the temperature change, ΔT. The constant of proportionality is referred to as α, the **coefficient of linear expansion**, and is defined as follows:

Definition of Coefficient of Linear Expansion, α

$$\Delta L = \alpha L_0 \Delta T \qquad \text{16-4}$$

SI unit for α: $K^{-1} = (C°)^{-1}$

Table 16-1 gives values of α for a variety of substances.

EXERCISE 16-6   THERMAL EXPANSION

The Eiffel Tower (**FIGURE 16-5**), constructed in 1889 by Alexandre Eiffel, is an impressive latticework structure made of iron. If the tower is 324 m high on a 0.0 °C day, by how much does its height increase on a day when the temperature is 29 °C?   CONTINUED

▲ **FIGURE 16-5** The Eiffel Tower in Paris gains about a sixth of an inch in height for each Celsius degree that the temperature rises.

REASONING AND SOLUTION

We can calculate the change in height with $\Delta L = \alpha L_0 \Delta T$. Table 16-1 shows that the coefficient of linear expansion for iron is $\alpha = 12 \times 10^{-6} \, \text{K}^{-1}$, and we know that the initial length (height in this case) is $L_0 = 324 \, \text{m}$. The change in temperature is $\Delta T = 29 \, \text{C}° = 29 \, \text{K}$, and the corresponding change in height is

$$\Delta L = \alpha L_0 \Delta T = (12 \times 10^{-6} \, \text{K}^{-1})(324 \, \text{m})(29 \, \text{K}) = 11 \, \text{cm}$$

TABLE 16-1 Coefficients of Thermal Expansion near 20°C

Substance	Coefficient of linear expansion, $\alpha(\text{K}^{-1})$	Substance	Coefficient of volume expansion, $\beta(\text{K}^{-1})$
Lead	29×10^{-6}	Ether	1.51×10^{-3}
Aluminum	24×10^{-6}	Carbon tetrachloride	1.18×10^{-3}
Brass	19×10^{-6}		
Copper	17×10^{-6}	Alcohol	1.01×10^{-3}
Iron (steel)	12×10^{-6}	Gasoline	0.95×10^{-3}
Concrete	12×10^{-6}	Olive oil	0.68×10^{-3}
Window glass	11×10^{-6}	Water	0.21×10^{-3}
Pyrex glass	3.3×10^{-6}	Mercury	0.18×10^{-3}
Quartz	0.50×10^{-6}		

RWP* An interesting application of thermal expansion is in the behavior of a bimetallic strip. As the name suggests, a bimetallic strip consists of two metals bonded together to form a linear strip of metal. This is illustrated in **FIGURE 16-6**. Because two different metals will, in general, have different coefficients of linear expansion, α, the two sides of the strip will change lengths by different amounts when heated or cooled.

For example, suppose metal B in Figure 16-6 (a) has the larger coefficient of linear expansion. This means that its length will change by greater amounts than metal A for the same temperature change. Hence, if this strip is cooled, the B side will shrink more than the A side, resulting in the strip bending toward the B side, as in Figure 16-6 (b). On the other hand, if the strip is heated, the B side expands by a greater amount than the A side, and the strip curves toward the A side, as in Figure 16-6 (c). Thus, the shape of the bimetallic strip depends sensitively on temperature.

Because of this property, bimetallic strips are used in a variety of thermal applications. For example, a bimetallic strip can be used as a thermometer; as the strip

If B has a greater coefficient of linear expansion than A . . .

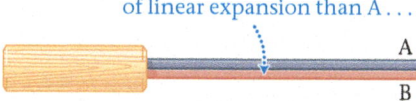

(a) A bimetallic strip

. . . then it shrinks more when cooled . . .

(b) Chilling the strip

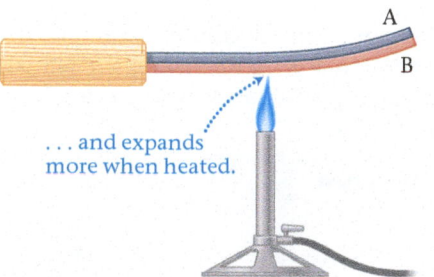

. . . and expands more when heated.

(c) Heating the strip

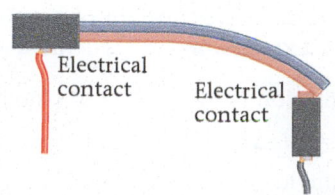

Electrical contact

Electrical contact

(d) A bimetallic strip in a thermostat

▲ **FIGURE 16-6 A bimetallic strip** **(a)** A bimetallic strip composed of metals A and B. **(b)** If metal B has a larger coefficient of linear expansion than metal A, it will shrink more when cooled and **(c)** expand more when heated. **(d)** A bimetallic strip can be used to construct a thermostat. If the temperature falls, the strip bends downward and closes the electric circuit, which then operates a heater. When the temperature rises, the strip deflects in the opposite direction, breaking the circuit and turning off the heater.

*Real World Physics applications are denoted by the acronym RWP.

changes its shape, it can move a needle to indicate the temperature. Similarly, many thermostats have a bimetallic strip to turn on or shut off a heater. This is shown in Figure 16-6 (d). As the temperature of the room changes, the bimetallic strip deflects in one direction or the other, which either closes or breaks the electric circuit connected to the heater.

RWP Another common use of thermal expansion is the *antiscalding device* for water faucets. An antiscalding device is simply a valve inside a water faucet that is attached to a spring. When the water temperature is at a safe level, the valve permits water to flow freely through the faucet. If the temperature of the water reaches a dangerous level, however, the thermal expansion of the spring is enough to close the valve and stop the flow of water—thus preventing inadvertent scalding. When the water cools down again, the valve reopens and the flow of water resumes.

RWP Finally, thermal expansion can have undesirable effects in some cases. For example, you may have noticed that bridges often have gaps between different sections of the structure, as seen in **FIGURE 16-7 (a)**. When the air temperature rises in the summer, the sections of the bridge can expand freely into these gaps. If the gaps were not present, the expansion of the different sections could cause the bridge to buckle and warp. Thus, these gaps, referred to as *expansion joints*, are a way of avoiding this type of heat-related damage. Expansion joints can also be found in railroad tracks and oil pipelines (**FIGURE 16-7 (b)**), to name just two other examples.

Area Expansion

The length of an object changes with temperature, and hence it follows that its area changes as well. To see precisely how the area changes, consider a square piece of metal of length L on a side. The initial area of the square is $A = L^2$. If the temperature of the square is increased by ΔT, the length of each side increases from L to $L + \Delta L = L + \alpha L \Delta T$. As a result, the square has an increased area, A', given by

$$A' = (L + \Delta L)^2 = (L + \alpha L \Delta T)^2$$
$$= L^2 + 2\alpha L^2 \Delta T + \alpha^2 L^2 \Delta T^2$$

Now, if $\alpha \Delta T$ is much less than one—which is certainly the case for typical changes in temperature, ΔT—then $\alpha^2 \Delta T^2$ is even smaller. Hence, if we ignore this small contribution, we find

$$A' \approx L^2 + 2\alpha L^2 \Delta T = A + 2\alpha A \Delta T$$

As a result, the change in area, ΔA, is

$$\Delta A = A' - A \approx 2\alpha A \Delta T \qquad \text{16-5}$$

Notice the similarity between this relationship and Equation 16-4 ($\Delta L = \alpha L_0 \Delta T$); the length L has been replaced with the area A, and the expansion coefficient has been doubled to 2α.

Though this calculation was done for the simple case of a square, the result applies to an area of any shape. For example, the area of a circular disk of radius r and area $A = \pi r^2$ will increase by the amount $2\alpha A \Delta T$ with an increase in temperature of ΔT. What about a washer, however, which is a disk of metal with a circular hole cut out of its center? What happens to the area of the hole as the washer is heated? Does it expand along with everything else, or does the expanding washer "expand into the hole" to make it smaller? We consider this question in the next Conceptual Example.

(a)

(b)

▲ **FIGURE 16-7 Visualizing Concepts**
Thermal Expansion Thermal expansion, though small, is far from negligible in many everyday situations. This is especially true when long objects such as railroad tracks, bridges, or pipelines are involved. **(a)** Bridges and elevated highways must include expansion joints to prevent the roadway from buckling when it expands in hot weather. **(b)** Pipelines typically include loops that allow for expansion and contraction when the temperature changes.

PROBLEM-SOLVING NOTE

Expansion of a Hole

A hole in a material expands the same as if it were made of the material itself. Thus, to find the expansion of a hole in a steel plate, we use the coefficient of linear expansion for steel.

CONCEPTUAL EXAMPLE 16-7   HEATING A HOLE   PREDICT/EXPLAIN

A washer has a hole in the middle. **(a)** As the washer is heated, does the hole expand, shrink, or stay the same size? **(b)** Which of the following is the *best explanation* for you prediction?

  I. The hole shrinks because the metal of the washer expands inward into the hole.

  II. The hole expands along with everything else—every dimension of the washer expands.

 III. The size of the hole remains the same because it is empty space, rather than metal like the washer.

CONTINUED

REASONING AND DISCUSSION

To make a washer from a disk of metal, we can cut along a circular shown in the sketch, and remove the inner disk. If we now heat the both the washer and the inner disk expand. On the other hand, left the inner disk in place and heated the original disk, it would pand. Removing the *heated* inner disk would create an expanded with an expanded hole in the middle. We obtain the same result we remove the inner disk and then heat, or heat first and then re- the inner disk.

Thus heating the washer causes both it and its hole to expand, and they both expand with the same coefficient of linear expan- sically, the system behaves the same as if we had produced a photo- enlargement—everything expands.

ANSWER

(a) The hole expands along with everything else. (b) The best ex- tion is II.

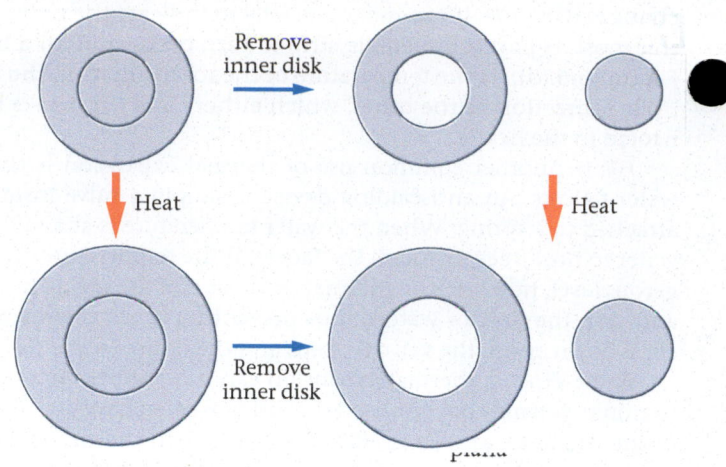

Remove inner disk

Heat   Heat

Remove inner disk

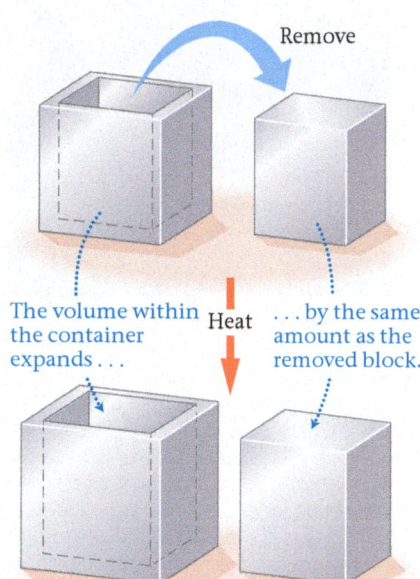

Remove

The volume within the container expands... Heat ...by the same amount as the removed block.

◀ **FIGURE 16-8 Volume expansion** A portion of a cube is removed to create a container. When heated, the removed portion expands, just as the volume within the container expands.

Volume Expansion

Just as the hole in a washer increases in area with heating, so does the empty volume within a cup or other container. This is illustrated in **FIGURE 16-8**, where we show a block of material with a volume removed to convert it into a container. As the system is heated, there will be an expansion of the container, of the volume within it, and of the volume that was removed. As with the area of a hole, the volume within a con- tainer expands with the same coefficient of expansion as the container itself.

To calculate the change in volume, consider a cube of length L on a side. The ini- tial volume of the cube is $V = L^3$. Increasing the temperature results in an increased volume given by

$$V' = (L + \Delta L)^3 = (L + \alpha L \Delta T)^3$$
$$= L^3 + 3\alpha L^3 \Delta T + 3\alpha^2 L^3 \Delta T^2 + \alpha^3 L^3 \Delta T^3$$

Neglecting the smaller contributions, as we did with the area, we find

$$V' \approx L^3 + 3\alpha L^3 \Delta T = V + 3\alpha V \Delta T$$

Therefore, the change in volume, ΔV, is

$$\Delta V = V' - V \approx 3\alpha V \Delta T$$

This expression, though calculated for a cube, applies to any volume.

In general, volume expansion is described in the same way as linear expansion, but with a **coefficient of volume expansion**, β, defined as follows:

Definition of Coefficient of Volume Expansion, β

$$\Delta V = \beta V \Delta T \qquad \text{16-6}$$

SI unit for β: $K^{-1} = (C°)^{-1}$

Values of β are given in Table 16-1 for a number of different liquids. If Table 16-1 lists a value of α for a given substance, but not for β, its change in volume is calculated as follows:

$$\Delta V = \beta V \Delta T \approx 3\alpha V \Delta T \qquad \text{16-7}$$

That is, we simply make the identification $\beta = 3\alpha$. For substances that have a specific value for β listed in Table 16-1, we simply use Equation 16-6. This is illustrated in the next Example.

EXAMPLE 16-8   OIL SPILL

A copper flask with a volume of 201 cm³ is filled to the brim with olive oil. If the temperature of the system is increased from 11 °C to 36 °C, how much oil spills from the flask?

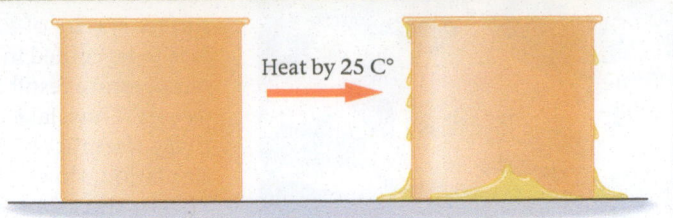

Heat by 25 C°

PICTURE THE PROBLEM
Our sketch shows the flask filled to the top initially, then spilling over when heated by 25 C°. (*Note:* Because degrees have the same size on the Celsius and Kelvin scales, it follows that $\Delta T = 25\ C° = 25\ K$.)

REASONING AND STRATEGY
As the system is heated, both the flask and the olive oil expand. Thus, we start by calculating the expansion of the oil and the flask separately. A quick glance at Table 16-1 shows that olive oil, with $\beta_{oil} = 0.68 \times 10^{-3}\ K^{-1}$, expands much more with temperature than does the copper flask, with $\beta_{copper} = 3\alpha_{copper} = 3(17 \times 10^{-6}\ K^{-1})$. The difference in expansion volumes is what spills out.

Known   Volume of copper flask, $V_{flask} = 201\ cm^3$; change in temperature, $\Delta T = 25\ K$; coefficients of thermal expansion, $\beta_{oil} = 0.68 \times 10^{-3}\ K^{-1}$ and $\beta_{copper} = 3\alpha_{copper} = 3(17 \times 10^{-6}\ K^{-1})$.

Unknown   Amount of oil that spills out of flask $= \Delta V_{oil} - \Delta V_{flask} = ?$

SOLUTION

1. Calculate the change in volume of the olive oil:

$$\Delta V_{oil} = \beta_{oil} V \Delta T$$
$$= (0.68 \times 10^{-3}\ K^{-1})(201\ cm^3)(25\ K) = 3.4\ cm^3$$

2. Calculate the change in volume of the copper flask:

$$\Delta V_{flask} = \beta_{copper} V \Delta T = 3\alpha_{copper} V \Delta T$$
$$= 3(17 \times 10^{-6}\ K^{-1})(201\ cm^3)(25\ K) = 0.26\ cm^3$$

3. Find the difference in volume expansions. This is the volume of oil that spills out:

$$\Delta V_{oil} - \Delta V_{flask} = 3.4\ cm^3 - 0.26\ cm^3 = 3.1\ cm^3$$

INSIGHT
If the system were cooled, the oil would lose volume more rapidly than the flask. This would result in a drop in the level of the oil.

This Example illustrates why fuel tanks on cars are designed to stop filling before the gas reaches the top—otherwise, the tank could overflow on a hot day.

PRACTICE PROBLEM — PREDICT/CALCULATE
Suppose the copper flask is initially filled to the brim with gasoline rather than olive oil. **(a)** Referring to Table 16-1, do you expect the amount of liquid that spills with heating by 25 °C to be greater than, less than, or the same as in the case of olive oil? **(b)** Find the volume of spilled gasoline in this case.   [**Answer: (a)** More liquid will spill because gasoline has a greater coefficient of thermal expansion. **(b)** $\Delta V = 4.5\ cm^3$]

Some related homework problems: Problem 22, Problem 27

Special Properties of Water

RWP As we've mentioned, water is a substance rich with unusual behavior. For example, in the last chapter we discussed the fact that the solid form of water (ice) is less dense than the liquid form. Hence, icebergs float. What is remarkable about icebergs floating is not that 90% is submerged, but that 10%, or any amount at all, is above the water. The solids of most substances are denser than their liquids; hence when they freeze, their solids immediately sink.

Here we consider the unusual *thermal* behavior of water. **FIGURE 16-9** shows the density of water over a wide range of temperatures. Notice that the density is a maximum at about 4 °C. Thus, when you *heat* water from 0 °C to 4 °C, it actually *shrinks*, rather than expands, and becomes *more dense*. The reason is that water molecules that were once part of the rather open crystal structure of ice are now able to pack more closely together in the liquid. Above 4 °C water expands with heating, just like other substances. The implications of this expansion are quite evident in the El Niño effect, shown in **FIGURE 16-10**.

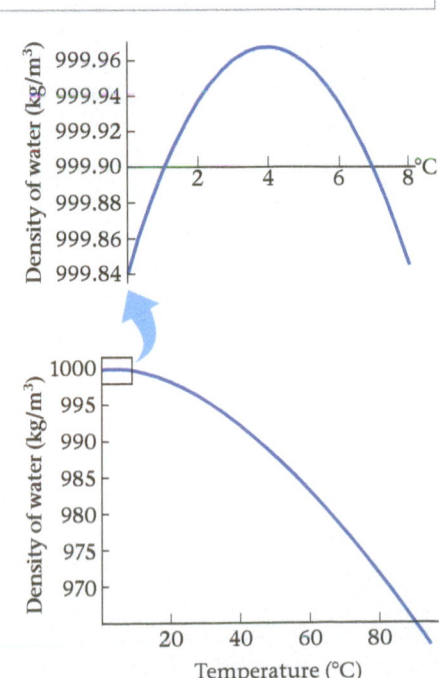

▶ **FIGURE 16-9   The unusual behavior of water near 4 °C** The density of water actually *increases* as the water is heated between 0 °C and 4 °C. Maximum density occurs near 4 °C.

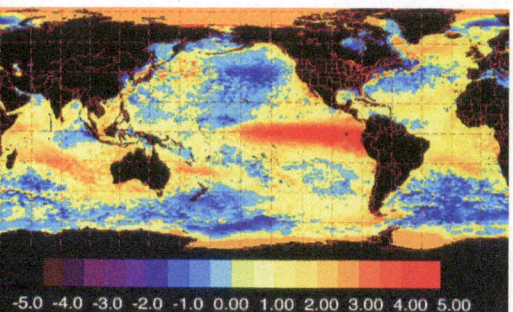

▲ **FIGURE 16-10** Above 4 °C water expands with increasing temperature like any normal liquid. This photo shows the expansion of water on a grand scale. It depicts an El Niño event: the appearance of a mass of unusually warm water in the equatorial region of the eastern Pacific Ocean every few years. An El Niño causes worldwide changes in climate and weather patterns. In this satellite image, the warmest water temperatures are represented by red, the coolest by violet. The El Niño is the red spike west of the South American coast. In this region, sea levels are as much as 20 cm higher than their average values, due to thermal expansion of the water.

▲ **FIGURE 16-11** The Arctic permafrost landscape can be significantly altered by frost heaving, which often results in polygonal patterns.

RWP The expansion of water from 0 °C to 4 °C has significant consequences for the ecology of lakes in northern latitudes. When temperatures drop in the winter, the surface waters of a lake cool first and sink, allowing warmer water to rise to the surface to be cooled in turn. Eventually, a lake can fill with water at 4 °C. Further drops in temperature result in cooler, less dense water near the surface, where it floats until it freezes. Thus, lakes freeze on the top surface first, with the bottom water staying relatively warm at about 4 °C. In addition, the ice and snow on top of a lake act as thermal insulation, slowing the continued growth of ice.

On the other hand, if water had more ordinary behavior—like shrinking when cooled, and a solid form that is more dense than the liquid—a lake would freeze from the bottom up. There would be no insulating layer of ice on top, and if the winter were long enough, and cold enough, the lake could freeze solid. This, of course, would be disastrous for fish and other creatures that live in the water.

RWP The same physics that is responsible for floating icebergs and ice-covered lakes is to blame for water pipes that burst in the winter. Even a water pipe made of steel is not strong enough to keep from rupturing when the ice forming within it expands outward. Later, when the temperature rises above freezing again, the burst pipe will make itself known by springing a leak.

Freezing water can also transform the landscape. In Arctic regions, for instance, the thawing and freezing cycles of water-saturated soils leads to *frost heaving*, a phenomenon that can create fantastic shapes in the landscape, including a pattern of polygons, as shown in **FIGURE 16-11**.

Enhance Your Understanding   (Answers given at the end of the chapter)

3. The following systems consist of a metal rod with a given initial length. The rods are subjected to the indicated increases in temperature. Rank the systems in order of increasing change in length. Indicate ties where appropriate. (Refer to Table 16-1 for the coefficients of thermal expansion.)

System	A	B	C	D
Metal	aluminum	steel	steel	aluminum
Initial length	2 m	2 m	1 m	1 m
Temperature increase	40 C°	20 C°	30 C°	10 C°

Section Review

- Most substances expand with an increase in temperature, as described by the equation $\Delta L = \alpha L_0 \Delta T$. Water from 0 °C to 4 °C is an important exception.

16-4 Heat and Mechanical Work

As mentioned previously, heat is the energy *transferred* from one object to another. At one time it was thought—erroneously—that an object contained a certain amount of "fluid," or caloric, that could flow from one place to another. This idea was overturned by the observations of Benjamin Thompson (1753–1814), also known as Count Rumford, the American-born physicist, spy, and social reformer who at one point in his eclectic career supervised the boring of cannon barrels by large drills. He observed that as long as mechanical work was done to turn the drill bits, they continued to produce heat in unlimited quantities. Clearly, the unlimited heat observed in boring the cannons was not present initially in the metal, but instead was produced by continually turning the drill bit.

With this observation, it became clear that heat was simply another form of energy transfer that must be taken into account when applying conservation of energy. For example, if you rub sandpaper back and forth over a piece of wood, you do work. The energy associated with that work is not lost; instead, it produces an increase in temperature. Taking into account the energy associated with this temperature change, we find that energy is indeed conserved. In fact, no observation has ever indicated a situation in which energy is not conserved.

The Equivalence Between Work and Heat   The equivalence between work and heat was first explored quantitatively by James Prescott Joule (1818–1889), the British physicist. In one of his experiments, Joule observed the increase in temperature in a device similar to that shown in **FIGURE 16-12**. Here, a total mass $2m$ falls through a certain distance h, during which gravity does the mechanical work, $2mgh$. As the mass falls, it turns the paddles in the water, which results in a slight warming of the water. By measuring the mechanical work, $2mgh$, and the increase in the water's temperature, ΔT, Joule was able to show that energy was indeed conserved—it had been converted from gravitational potential energy into an increased thermal energy of the water, as indicated by a higher temperature. Joule's experiments established the precise amount of mechanical work that has the same effect as a given transfer of heat.

Before Joule's work, heat was measured in a unit called the calorie (cal). In particular, one kilocalorie (kcal) was defined as the amount of heat needed to raise the temperature of 1 kg of water from 14.5 °C to 15.5 °C. With his experiments, Joule was able to show that 1 kcal = 4186 J, or equivalently, that one calorie of heat transfer is the equivalent of 4.186 J of mechanical work. This is referred to as the **mechanical equivalent of heat**:

The Mechanical Equivalent of Heat

1 cal = 4.186 J   16-8

SI unit: J

Other Units of Heat   In studies of nutrition a different calorie is used. It is the Calorie, with a capital C, and it is simply a kilocalorie; that is, 1 C = 1 kcal. Perhaps this helps people to feel a little better about their calorie intake. After all, a 250-C candy bar sounds a lot better than a 250,000-cal candy bar. They are equivalent, however.

Another common unit for measuring heat is the British thermal unit (Btu). By definition, a Btu is the energy required to heat 1 lb of water from 63 °F to 64 °F. In terms of calories and joules, a Btu is as follows:

$$1 \text{ Btu} = 0.252 \text{ kcal} = 1055 \text{ J} \qquad 16\text{-}9$$

Finally, we shall use the symbol Q to denote heat:

Heat, Q

Q = heat = energy transferred due to temperature differences   16-10

SI unit: J

Using the mechanical equivalent of heat as the conversion factor, we will typically give heat in either calories or joules.

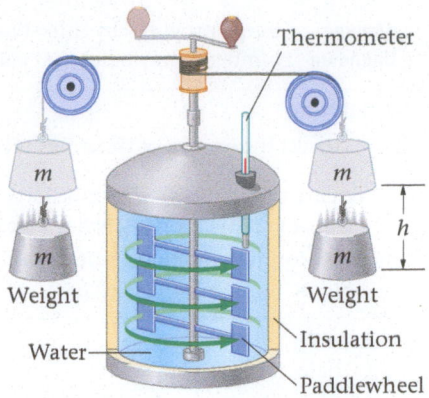

▲ **FIGURE 16-12 The mechanical equivalent of heat** A device of this type was used by James Joule to measure the mechanical equivalent of heat.

Big Idea **3** Mechanical work can be converted into an equivalent amount of heat. For example, the work done by friction on a sliding object increases the temperature of the object.

PHYSICS IN CONTEXT
Looking Back

The concept of work (Chapter 7) plays a key role in this section, where we relate mechanical work to its equivalent in the form of heat.

PHYSICS IN CONTEXT
Looking Ahead

Heat plays an important role in Chapter 18, where we consider the heat and work exchanged during different types of thermal processes.

EXAMPLE 16-9   LIFT THAT BALE!

A farmhand lifts a 23.5-kg bale of hay to a height $H = 1.1$ m to get it into the back of a flatbed truck. How many bales of hay can the farmhand lift using the energy provided by a thick, rich, 485-C chocolate milkshake?

PICTURE THE PROBLEM
Our sketch shows the farmhand lifting a bale of hay of mass m from the ground to the height H to get it into the truck. Work must be done against gravity to lift the bale, and the energy for that work is provided by a milkshake.

REASONING AND STRATEGY
We know that the energy provided by the milkshake is equivalent to the heat $Q = 485,000$ cal. This energy can be converted to joules by using the relationship 1 cal = 4.186 J. Finally, we set the energy of the milkshake equal to the work done against gravity in lifting N bales of hay to a height H; that is, $Q = N(mgH)$. We can solve this for N.

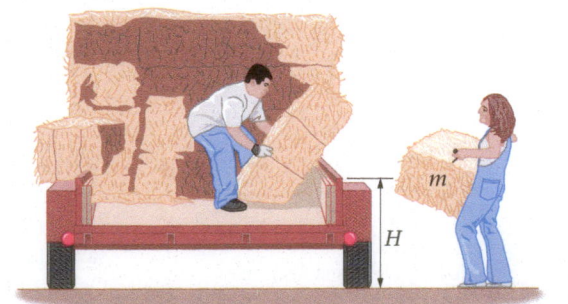

CONTINUED

Known   Energy provided by milkshake, $Q = 485,000$ cal; mass of hay bale, $m = 23.5$ kg; height of lift, $H = 1.1$ m.
Unknown   Number of hay bales that can be lifted, $N = ?$

SOLUTION

1. Convert the energy of the milkshake to joules:

$$Q = 485,000 \text{ cal} = 485,000 \text{ cal}\left(\frac{4.186 \text{ J}}{1 \text{ cal}}\right) = 2.03 \times 10^6 \text{ J}$$

2. Equate the energy of the milkshake to the work done against gravity in lifting N hay bales to a height H:

$$Q = N(mgH)$$

3. Solve for the number of bales N and substitute numerical values:

$$N = \frac{Q}{mgH} = \frac{2.03 \times 10^6 \text{ J}}{(23.5 \text{ kg})(9.81 \text{ m/s}^2)(1.1 \text{ m})} = 8000$$

INSIGHT

8000 bales is a lot of hay! Even assuming a metabolic efficiency of 25%, the milkshake could give the energy to lift 2000 bales of hay. Of course, energy is also expended in walking from bale to bale, in bending over and standing up, and other physical activities. The point, however, is that the chemical energy stored in food is large compared with the mechanical energy you expend while doing everyday tasks.

PRACTICE PROBLEM

If the farmhand lifts 100 bales of hay into the truck, how many Calories are burned? Assume 25% metabolic efficiency. [**Answer: 24 C**]

Some related homework problems: Problem 29, Problem 31

▲ **FIGURE 16-13** Enhance Your Understanding 4.

Enhance Your Understanding   (Answers given at the end of the chapter)

4. Your friend claims that stirring a glass of water with a spoon, as in **FIGURE 16-13**, increases the temperature of the water. Is she correct? Explain.

Section Review

- Doing 4.186 J of mechanical work on a system, and converting that work to an increase in temperature, is equivalent to heating the system by adding 1 cal of thermal energy.

16-5 Specific Heats

In the previous section, we mentioned that it takes 4186 J of heat to raise the temperature of 1 kg of water by 1 C°. We have to be clear about the fact that we are heating water, however, because the heat required for a 1 C° increase varies considerably from one substance to another. For example, it takes only 128 J of heat to increase the temperature of 1 kg of lead by 1 C°. In general, the heat required for a given increase in temperature is given by the **heat capacity** of a substance.

Heat Capacity

Suppose we add the heat Q to a given object, and its temperature increases by the amount ΔT. The heat capacity, C, of this object is defined as follows:

Definition of Heat Capacity, *C*

$$C = \frac{Q}{\Delta T} \qquad\qquad 16\text{-}11$$

SI unit: J/K = J/C°

The units of heat capacity are joules per kelvin (J/K). Because the degree size is the same for the Kelvin and Celsius scales, C can also be expressed in units of joules per Celsius degree (J/C°). (Note that the heat capacity is written with an italic capital C, as opposed to the roman capital C used for degrees Celsius.)

The name "heat capacity" is perhaps a bit unfortunate. It derives from the mistaken idea of a "heat fluid," mentioned in the previous section. Objects were imagined to "contain" a certain amount of this nonexistent fluid. Today, we know that an object can readily gain or release heat when it is in thermal contact with other objects—objects cannot be thought of as holding a certain amount of heat.

Instead, the heat capacity should be viewed as the amount of heat necessary for a given temperature change. An object with a large heat capacity, like water, requires a large amount of heat for each increment in temperature. Just the opposite is true for an object with a small heat capacity, like a piece of lead.

To find the heat required for a given ΔT, we simply rearrange $C = Q/\Delta T$ to solve for Q. This yields

$$Q = C\Delta T \qquad\qquad\qquad 16\text{-}12$$

It should be noted that the *heat capacity is always positive*—just like a speed. Thus, the relationship $Q = C\Delta T$ implies that the heat Q and the temperature change ΔT have the same sign. This observation leads to the following sign conventions for Q:

Q is positive if ΔT is positive—that is, if heat is *added* to a system.

Q is negative if ΔT is negative—that is, if heat is *removed* from a system.

EXERCISE 16-10 TEMPERATURE CHANGE

The heat capacity of 1.00 kg of iron is 448 J/K. **(a)** What is the temperature change of the iron if 493 J of heat is added to the system? **(b)** How much heat must be removed from the iron for its temperature to drop by 2.68 K?

REASONING AND SOLUTION

a. Heat, heat capacity, and temperature change are related by $Q = C\Delta T$. Solving for the change in temperature, we find

$$\Delta T = \frac{Q}{C} = \frac{493\,\text{J}}{448\,\text{J/K}} = 1.10\,\text{K}$$

b. Heat is removed in this case, and hence Q is negative. Substituting $\Delta T = -2.68$ K in $Q = C\Delta T$, we find

$$Q = C\Delta T = (448\,\text{J/K})(-2.68\,\text{K}) = -1200\,\text{J}$$

Big Idea 4 Different substances are characterized by their specific heats; that is, they require different amounts of heat for a given increase in temperature.

Specific Heat

It takes 4186 J to increase the temperature of one kilogram of water by one degree Celsius, and hence it takes twice that much to make the same temperature change in two kilograms of water, and so on. Thus, the heat capacity varies not only with the type of substance, but also with the mass of the substance.

It is useful, then, to define a new quantity—the **specific heat**, c—that depends only on the substance, and not on its mass. This definition is as follows:

Definition of Specific Heat, c

$$c = \frac{Q}{m\Delta T} \qquad\qquad 16\text{-}13$$

SI unit: $J/(kg \cdot K) = J/(kg \cdot C°)$

Thus, for example, the specific heat of water is

$$c_{\text{water}} = 4186\,\text{J/(kg}\cdot\text{K)}$$

Specific heats for common substances are listed in Table 16-2. Notice that the specific heat of water is by far the largest of any common material. This is just another of the many unusual properties of water. Having such a large specific heat means that water can give off or take in large quantities of heat with little change in temperature. It is for this reason that if you take a bite of a pie that is just out of the oven, you are much

TABLE 16-2 Specific Heats at Atmospheric Pressure

Substance	Specific heat, c [J/(kg · K)]
Water	4186
Ice	2090
Steam	2010
Beryllium	1820
Air	1004
Aluminum	900
Glass	837
Silicon	703
Iron (steel)	448
Copper	387
Silver	234
Gold	129
Lead	128

▲ **FIGURE 16-14** These two blocks of metal (aluminum on the left and lead on the right) have equal volumes. In addition, they were heated to equal temperatures before being placed on the block of paraffin wax. Notice, however, that the aluminum melted more wax—and hence gave off more heat—even though the lead block is about four times heavier than the aluminum block. The reason is that lead has a very small specific heat; in fact, lead's specific heat is about seven times smaller than the specific heat of aluminum, as we see in Table 16-2. As a result, even this relatively large mass of lead melts considerably less wax per degree of temperature change than the lightweight aluminum—a direct visual illustration of lead's low specific heat.

PROBLEM-SOLVING NOTE

Heat Flow and Thermal Equilibrium

To find the temperature of thermal equilibrium when two objects with different temperatures are brought into thermal contact, we simply use the idea that the heat that flows *out* of one of the objects flows *into* the other object.

Water Balloon Held over Candle Flame

PHYSICS IN CONTEXT
Looking Back

Energy and energy conservation (Chapter 8) are important in our understanding of specific heats.

more likely to burn your tongue on the fruit filling (which has a high water content) than on the much drier crust. An illustration of the difference in specific heats of aluminum and lead is shown in **FIGURE 16-14**.

RWP Water's unusually large specific heat also accounts for the moderate climates in regions near great bodies of water. In particular, the enormous volume and large heat capacity of an ocean serve to maintain a nearly constant temperature in the water, which in turn acts to even out the temperature of adjacent coastal areas. For example, the West Coast of the United States benefits from the moderating effect of the Pacific Ocean, aided by the prevailing breezes that come from the ocean onto the coastal regions. In the Midwest, on the other hand, temperature variations can be considerably greater as the land (with a relatively small specific heat) quickly heats up in the summer and cools off in the winter.

Calorimetry

Let's use the specific heat to solve a practical problem. Suppose a block of mass m_b, specific heat c_b, and initial temperature T_b is dropped into a **calorimeter** (basically, a lightweight, insulated flask) containing water. If the water has a mass m_w, a specific heat c_w, and an initial temperature T_w, find the final temperature of the block and the water. Assume that the calorimeter is light enough that it can be ignored, and that no heat is transferred from the calorimeter to its surroundings.

There are two basic ideas involved in solving this problem:

1. The total energy of the system is conserved.

2. The system is in thermal equilibrium at the end, and hence the final temperature of the block is equal to the final temperature of the water.

In particular, the first condition means that the amount of heat lost by the block is equal to the heat gained by the water—or vice versa if the water's initial temperature is higher.

Mathematically, we can write condition 1 as

$$Q_b + Q_w = 0 \qquad \text{16-14}$$

This means that the heat flow from the block is equal and opposite to the heat flow from the water; in other words, energy is conserved. If we write the heats Q in terms of the specific heats and temperatures, and let the common final temperature be T, to satisfy condition 2, we have

$$m_b c_b (T - T_b) + m_w c_w (T - T_w) = 0$$

Notice that for each heat we write the change in temperature as T_{final} minus $T_{initial}$, as it should be. Solving for the final temperature, T, we find

$$T = \frac{m_b c_b T_b + m_w c_w T_w}{m_b c_b + m_w c_w} \qquad \text{16-15}$$

This result need not be memorized, of course. Whenever solving a problem of this sort, one simply writes down energy conservation and solves for the desired unknown.

In some cases, we may wish to consider the influence of the container itself. This is illustrated in the following Quick Example.

QUICK EXAMPLE 16-11 FIND THE FINAL TEMPERATURE

Suppose 54 g of water at 35 °C are poured into a 65-g aluminum cup with an initial temperature of 11 °C. Find the final temperature of the system, assuming no heat is exchanged with the surroundings.

REASONING AND SOLUTION
We apply energy conservation—setting the heat flow out of the water equal to the heat flow into the aluminum—and solve for the common final temperature.

1. Write an expression for the heat flow out of the water:

$$Q_w = m_w c_w (T - T_w)$$

CONTINUED

2. Write an expression for the heat flow into the aluminum:

$$Q_a = m_a c_a (T - T_a)$$

3. Apply energy conservation:

$$Q_w + Q_a = 0$$

4. Solve for the final temperature, T:

$$T = \frac{m_a c_a T_a + m_w c_w T_w}{m_a c_a + m_w c_w} = 30\,°C$$

As one might expect from water's large specific heat (five times that of aluminum), the final common temperature T is much closer to the initial temperature of the water (T_w) than to that of the aluminum (T_a), even though the mass of the aluminum cup is greater than the mass of the water.

EXAMPLE 16-12   COOLING OFF

A 0.50-kg block of metal with an initial temperature of 54.5 °C is dropped into a container holding 1.1 kg of water at 20.0 °C. If the final temperature of the block–water system is 21.4 °C, what is the specific heat of the metal block? Assume the container can be ignored, and that no heat is exchanged with the surroundings.

PICTURE THE PROBLEM

Initially, when the block is first dropped into the water, the temperatures of the block and water are T_b = 54.5 °C and T_w = 20.0 °C, respectively. When thermal equilibrium is established, both the block and the water have the same temperature, T = 21.4 °C.

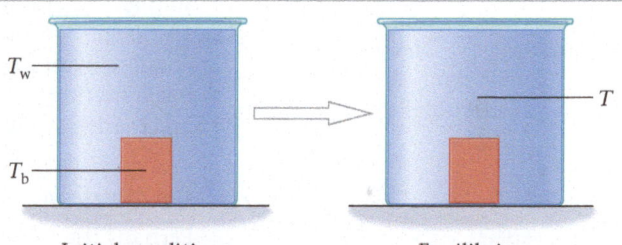

Initial conditions   Equilibrium

REASONING AND STRATEGY

Heat flows from the block to the water. Setting the heat flow *out of the block* plus the heat flow *into the water* equal to zero (conservation of energy) yields the block's specific heat.

Known   Mass of metal block, m_b = 0.50 kg; initial temperature of block, T_b = 54.5 °C; mass of water, m_w = 1.1 kg; initial temperature of water, T_w = 20.0 °C; final temperature of system, T = 21.4 °C.

Unknown   Specific heat of metal block, c_b = ?

SOLUTION

1. Write an expression for the heat flow out of the block. Notice that Q_{block} is negative, since T is less than T_b:

$$Q_{block} = m_b c_b (T - T_b)$$

2. Write an expression for the heat flow into the water. Notice that Q_{water} is positive, since T is greater than T_w:

$$Q_{water} = m_w c_w (T - T_w)$$

3. Set the sum of the heats equal to zero:

$$Q_{block} + Q_{water} = m_b c_b (T - T_b) + m_w c_w (T - T_w) = 0$$

4. Solve for the specific heat of the block, c_b

$$c_b = \frac{m_w c_w (T - T_w)}{m_b (T_b - T)}$$

5. Substitute numerical values:

$$c_b = \frac{(1.1\,kg)\,[4186\,J/(kg \cdot K)]\,(21.4\,°C - 20.0\,°C)}{(0.50\,kg)(54.5\,°C - 21.4\,°C)}$$

$$= 390\,J/(kg \cdot K)$$

INSIGHT

From Table 16-2 we can see that the block is probably made of copper.

In addition, notice that the final temperature is much closer to the initial temperature of the water than to the initial temperature of the block. This is due in part to the fact that the mass of the water is about twice that of the block; more important, however, is the fact that the water's specific heat is more than ten times greater than that of the block. In the following Practice Problem we set the mass of the water equal to the mass of the block, so that we can see clearly the effect of the different specific heats.

PRACTICE PROBLEM

If the mass of the water is also 0.50 kg, what is the equilibrium temperature?   [**Answer:** T = 23 °C—still much closer to the water's initial temperature than to the block's]

Some related homework problems: Problem 40, Problem 41

5. Objects A and B have the same mass. When they receive the same amount of thermal energy, the temperature of object A increases more than the temperature of object B. Is the specific heat of object A greater than, less than, or equal to the specific heat of object B? Explain.

Section Review

*   Heat capacity, $C = Q/\Delta T$, is the amount of heat per temperature change. Specific heat, $c = Q/(m\Delta T)$, is the amount of heat per mass per temperature change.

*   Calorimetry uses energy conservation to determine the final temperature of a system.

16-6 Conduction, Convection, and Radiation

Heat can be exchanged in a variety of ways. The Sun, for example, warms the Earth from across 93 million miles of empty space by a process known as radiation. As the sunlight strikes the ground and raises its temperature, the ground-level air gets warmer and begins to rise, producing a further exchange of heat by means of convection. Finally, if you walk across the ground in bare feet, you will feel the warming effect of heat entering your body by conduction. In this section we consider each of these three mechanisms of heat exchange in detail.

Conduction

Perhaps the most familiar form of heat exchange is **conduction**, which is the flow of heat directly through a physical material. For example, if you hold one end of a metal rod and put the other end into a fire, it doesn't take long before you begin to feel warmth on your end. The heat you feel is transported along the rod by conduction.

Let's consider this observation from a microscopic point of view. To begin, when you placed one end of the rod into the fire, the high temperature at that location caused the molecules to vibrate with an increased amplitude. These molecules in turn jostle their neighbors and cause them to vibrate with greater amplitude as well. Eventually, the effect travels from molecule to molecule across the length of the rod, resulting in the macroscopic phenomenon of conduction.

If you were to repeat the experiment, this time with a wooden rod, the hot end of the rod would heat up so much that it might even catch on fire, but your end would still be comfortably cool. Thus, conduction depends on the type of material involved. Some materials, called **conductors**, conduct heat very well, whereas other materials, called **insulators**, conduct heat poorly.

Just how much heat flows as a result of conduction? To answer this question we consider the simple system shown in **FIGURE 16-15**. Here we show a rod of length L and cross-sectional area A, with one end at the temperature T_1 and the other at the temperature $T_2 > T_1$. Experiments show that the amount of heat Q that flows through the rod has these properties:

* Q increases in proportion to the rod's cross-sectional area, A.

* Q increases in proportion to the temperature difference, $\Delta T = T_2 - T_1$.

* Q increases steadily with time, t.

* Q decreases with the length of the rod, L.

Combining these observations in a mathematical expression gives the following equation:

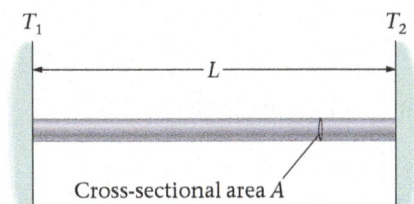

T_1 T_2

L

Cross-sectional area A

▲ **FIGURE 16-15 Heat conduction through a rod** The amount of heat that flows through a rod of length L and cross-sectional area A per time is proportional to $A(T_2 - T_1)L$.

Heat Flow by Conduction

$$Q = kA\left(\frac{\Delta T}{L}\right)t$$ 16-16

TABLE 16-3 Thermal Conductivities

Substance	Thermal conductivity, $k[\text{W}/(\text{m}\cdot\text{K})]$
Silver	417
Copper	395
Gold	291
Aluminum	217
Steel, low carbon	66.9
Lead	34.3
Stainless steel— alloy 302	16.3
Ice	1.6
Concrete	1.3
Glass	0.84
Water	0.60
Asbestos	0.25
Wood	0.10
Wool	0.040
Air	0.0234

◄ **FIGURE 16-16 Visualizing Concepts**
Conduction Maintaining proper body temperature in an environment that is often too hot or too cold is a problem for many animals. When the sand is blazing hot, this lizard (top) keeps its contact with the ground to a minimum. By standing on two legs instead of four, it reduces conduction of heat from the ground to its body. Polar bears (bottom) have the opposite problem. The loss of precious body heat to their surroundings is retarded by their thick fur, which is actually made up of hollow fibers. Air trapped within these fibers provides enhanced insulation, just as it does in our thermal blankets and double-paned windows.

PHYSICS IN CONTEXT
Looking Ahead

The flux of heat transferred between two objects in thermal contact is a useful analogy in understanding the flux of electric fields in Chapter 19 and magnetic fields in Chapter 23.

The constant k is referred to as the **thermal conductivity** of the rod. It varies from material to material, as indicated in Table 16-3. An illustration of animals that deal with thermal conduction in different ways is shown in **FIGURE 16-16**.

CONCEPTUAL EXAMPLE 16-13 THE FEEL OF TILE IN THE MORNING

You get up in the morning and walk barefoot from the bedroom to the bathroom. In the bedroom you walk on carpet, but in the bathroom the floor is tile. Does the tile feel warmer, cooler, or the same temperature as the carpet?

REASONING AND DISCUSSION
Everything in the house is at the same temperature, so it might seem that the carpet and the tile would feel the same. As you probably know from experience, however, the tile feels cooler. The reason is that tile has a much larger thermal conductivity than carpet, which is actually a fairly good insulator. As a result, more heat flows from your skin to the tile than from your skin to the carpet. To your feet, then, it is as if the tile were much cooler than the carpet.

To get an idea of the thermal conductivities that would apply in this case, let's examine Table 16-3. For the tile, we might expect a thermal conductivity of roughly 0.84, the value for glass. For the carpet, the thermal conductivity might be as low as 0.04, the thermal conductivity of wool. Thus, the tile could have a thermal conductivity that is 20 times larger than that of the carpet.

ANSWER
The tile feels cooler.

Thermal conductivity is an important consideration when insulating a home. We consider some of these issues in the next Example.

EXAMPLE 16-14 WHAT A PANE!

One of the windows in a house has the shape of a square 1.0 m on a side. The glass in the window is 0.50 cm thick, and has a coefficient of thermal conductivity of $k = 0.84 \text{ W}/(\text{m}\cdot\text{K})$. **(a)** How much heat is lost through this window in one day if the temperature difference between the inside and outside of the window is 21 C°? **(b)** Suppose all the dimensions of the window—height, width, and thickness—are doubled. If everything else remains the same, by what factor does the heat flow change?

CONTINUED

PICTURE THE PROBLEM

The glass from the window is shown in our sketch, along with its relevant dimensions. Heat flows across a temperature difference of 21 C° through a length equal to the thickness of the glass, $L = 0.50$ cm.

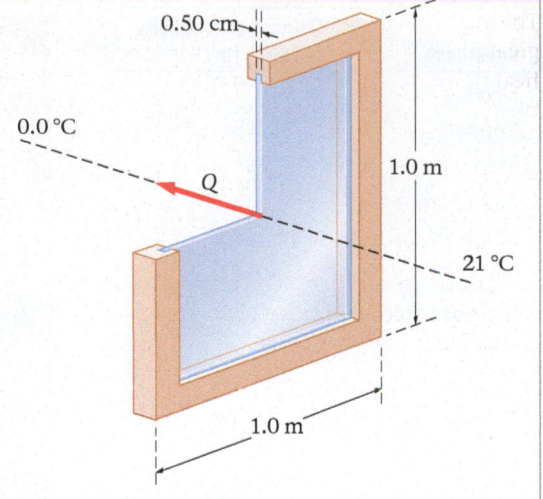

REASONING AND STRATEGY

a. The heat flow is given by $Q = kA(\Delta T/L)t$ (Equation 16-16). The area is $A = (1.0 \text{ m})^2$ and the length over which heat is conducted—in this case the thickness of the glass—is $L = 0.0050$ m. The temperature difference is $\Delta T = 21 \text{ C}° = 21$ K. Also, recall from Section 7-4 that 1 W = 1 J/s.

b. Doubling all dimensions increases the thickness by a factor of two and increases the area by a factor of four; that is, $L \to 2L$ and $A \to (2 \times \text{height}) \times (2 \times \text{width}) = 4A$. We can use these results in $Q = kA(\Delta T/L)t$.

Known   Area of window, $A = (1.0 \text{ m})^2$; thickness of glass, $L = 0.50$ cm; temperature difference across window, $\Delta T = 21$ K; thermal conductivity of glass, $k = 0.84 \text{ W}/(\text{m} \cdot \text{K})$.

Unknown   (a) Heat lost during one day, $Q = ?$ (b) Heat lost when dimensions are doubled, $Q = ?$

SOLUTION

Part (a)

1. Calculate the heat flow for a given time, t:

$$Q = kA\left(\frac{\Delta T}{L}\right)t$$

$$= [0.84 \text{ W}/(\text{m} \cdot \text{K})](1.0 \text{ m})^2\left(\frac{21 \text{ K}}{0.0050 \text{ m}}\right)t = (3500 \text{ W})t$$

2. Substitute the number of seconds in a day (86,400 s) for the time t:

$$Q = (3500 \text{ W})t = (3500 \text{ W})(86,400 \text{ s}) = 3.0 \times 10^8 \text{ J}$$

Part (b)

3. Replace L with $2L$ and A with $4A$ in Step 1. The result is a doubling of the heat flow, Q:

$$Q = kA\left(\frac{\Delta T}{L}\right)t \to k(4A)\left[\frac{\Delta T}{(2L)}\right]t \to 2\left[kA\left(\frac{\Delta T}{L}\right)t\right] = 2Q$$

INSIGHT

Q is a sizable amount of heat, roughly equivalent to the energy released in burning a gallon of gasoline. A considerable reduction in heat loss can be obtained by using a double-paned window, which has an insulating layer of air (actually argon or krypton) sandwiched between the two panes of glass. This is discussed in more detail later in this section, and is explored in Homework Problems 56 and 90.

PRACTICE PROBLEM

Suppose the window is replaced with a plate of solid silver. How thick must this plate be to have the same heat flow in a day as the glass?   **[Answer: $L = 2.5$ m]**

Some related homework problems: Problem 53, Problem 55

Next, we consider the heat flow through a combination of two different materials with different thermal conductivities.

CONCEPTUAL EXAMPLE 16-15   COMPARE THE HEAT FLOW   PREDICT/EXPLAIN

Two metal rods are to be used to conduct heat from a region at 100 °C to a region at 0 °C. The rods can be placed in parallel, as shown on the left, or in series, as on the right. (a) Is the heat conducted in the parallel arrangement greater than, less than, or the same as the heat conducted with the rods in series? (b) Which of the following is the *best explanation* for your prediction?

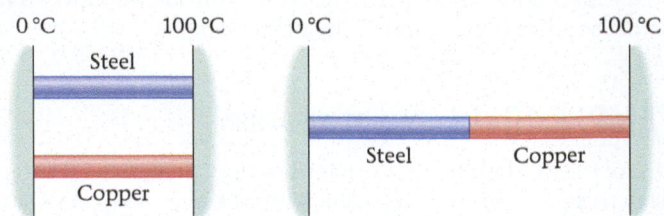

   I. The parallel arrangement conducts more heat because the cross-sectional area through which heat flows is greater.

  II. More heat is conducted in series because the heat that flows through one rod also flows through the other rod.

III. The heat conducted is the same in both cases because the heat has to flow through the same two rods.

CONTINUED

REASONING AND DISCUSSION

The parallel arrangement conducts more heat for two reasons. First, the cross-sectional area available for heat flow is twice as great for the parallel rods. A greater cross-sectional area gives a greater heat flow—everything else being equal. Second, more heat flows through each rod in the parallel configuration because they both have the full temperature difference of 100 C° between their ends. In the series configuration, each rod has a smaller temperature difference between its ends, so less heat flows.

ANSWER

(a) More heat is conducted when the rods are in parallel. (b) The best explanation is I.

In the next Example, we consider the case of heat conduction when two rods are placed in parallel. In the homework we shall consider the corresponding case of the same two rods conducting heat in series.

EXAMPLE 16-16   PARALLEL RODS

Two 0.525-m rods, one lead and the other copper, are connected between metal plates held at 2.00 °C and 106 °C. The rods have a square cross section, 1.50 cm on a side. How much heat flows through the two rods in 1.00 s? Assume that no heat is exchanged between the rods and the surroundings, except at the ends.

PICTURE THE PROBLEM

The lead and the copper rods, each 0.525 m long, are shown in the sketch. Both rods have a temperature difference of 104 C° = 104 K between their ends.

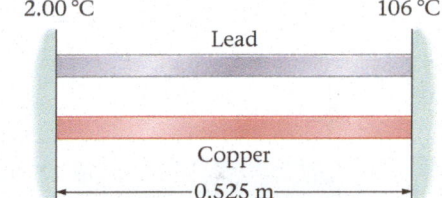

REASONING AND STRATEGY

The heat flowing through each rod can be calculated using $Q = kA(\Delta T/L)t$ and the value of k given in Table 16-3. The total heat flow is simply the sum of that calculated for each rod.

Known   Length of rods, $L = 0.525$ m; metal in top rod = lead; metal in bottom rod = copper; temperature difference across rods, $\Delta T = 104$ K; cross-sectional area of each rod, $A = (0.0150 \text{ m})^2$; time during which heat flows, $t = 1.00$ s.

Unknown   Total amount of heat that flows, $Q = ?$

SOLUTION

1. Calculate the heat flow in 1.00 second through the lead rod:

$$Q_{lead} = k_{lead}A\left(\frac{\Delta T}{L}\right)t$$

$$= [34.3 \text{ W}/(\text{m} \cdot \text{K})](0.0150 \text{ m})^2\left(\frac{104 \text{ K}}{0.525 \text{ m}}\right)(1.00 \text{ s}) = 1.53 \text{ J}$$

2. Calculate the heat flow in 1.00 second through the copper rod:

$$Q_{copper} = k_{copper}A\left(\frac{\Delta T}{L}\right)t$$

$$= [395 \text{ W}/(\text{m} \cdot \text{K})](0.0150 \text{ m})^2\left(\frac{104 \text{ K}}{0.525 \text{ m}}\right)(1.00 \text{ s}) = 17.6 \text{ J}$$

3. Sum the heats found in Steps 1 and 2 to get the total heat:

$$Q_{total} = Q_{lead} + Q_{copper} = 1.53 \text{ J} + 17.6 \text{ J} = 19.1 \text{ J}$$

INSIGHT

As our results show, the copper rod is by far the better conductor of heat. It is also a very good conductor of electricity, as we shall see in Chapter 21.

PRACTICE PROBLEM

What temperature difference would be required for the total heat flow in 1.00 second to be 15.0 J?
[**Answer:** $\Delta T = 81.5$ C° = 81.5 K]

Some related homework problems: Problem 57, Problem 58

As we shall see in Homework Problem 59, the heat conducted by the same two rods connected in series is only 1.41 J, which is less than either of the two rods conducts when placed in parallel. This verifies the conclusion reached in Conceptual Example 16-15.

▲ **FIGURE 16-17** Many wading birds use countercurrent exchange in their circulatory systems. This mechanism allows them to keep the temperature of their legs well below that of their bodies. In this way they reduce the conductive loss of body heat to the water.

Candle Chimneys

▲ **FIGURE 16-19** When you light a candle, it heats the air near the flame as it burns. Because hot air is less dense—and hence more buoyant—than cool air, a circulation pattern is established with hot air rising and being replaced from below by cool, oxygenated air. Thus, convection is necessary for a candle to continue burning. When the burning candle in this jar is dropped, it suddenly finds itself in free fall—an essentially "weightless" environment where buoyancy has no effect. As a result, convection ceases and the flame is quickly extinguished as it consumes all the oxygen in its immediate vicinity.

RWP An application of thermal conductivity in series is the *insulated window*. Most homes today have insulated windows as a means of increasing their energy efficiency. If you look closely at one of these windows, you will see that it is actually constructed from two panes of glass separated by a gas-filled gap—usually argon or krypton gas. Thus, heat flows through three different materials in series as it passes into or out of a home. The fact that the thermal conductivity of a gas is about 40 times smaller than that of glass means that the insulated window results in significantly less heat flow than would be experienced with a single pane of glass.

BIO As a final example of conduction, we note that many biological systems transfer heat by a mechanism known as *countercurrent exchange*. Consider, for example, a wading bird (**FIGURE 16-17**) standing in cool water all day. As warm blood flows through the arteries on its way to the legs and feet of the bird, it passes through constricted regions where the legs join to the body. Here, where the arteries and veins are packed closely together, the body-temperature arterial blood flowing into the legs transfers heat to the much cooler venous blood returning to the body. Thus, the counter-flowing streams of blood serve to maintain the core body temperature of the bird, while at the same time keeping the legs and feet at much cooler temperatures. The feet still receive the oxygen and nutrients carried by the blood, but they stay at a relatively low temperature to reduce the amount of heat lost to the water.

BIO Similar effects occur in humans. It is common, for example, to hear complaints that a person's hands or feet are cold. There is good reason for this, because they are in fact much cooler than the core body temperature. Just as with the wading birds, the warm arterial blood flowing to the hands and feet exchanges heat with the cool venous blood flowing in the opposite direction (**FIGURE 16-18**). This helps to reduce the heat loss to our surroundings, and to maintain the desired temperature in the core of the body.

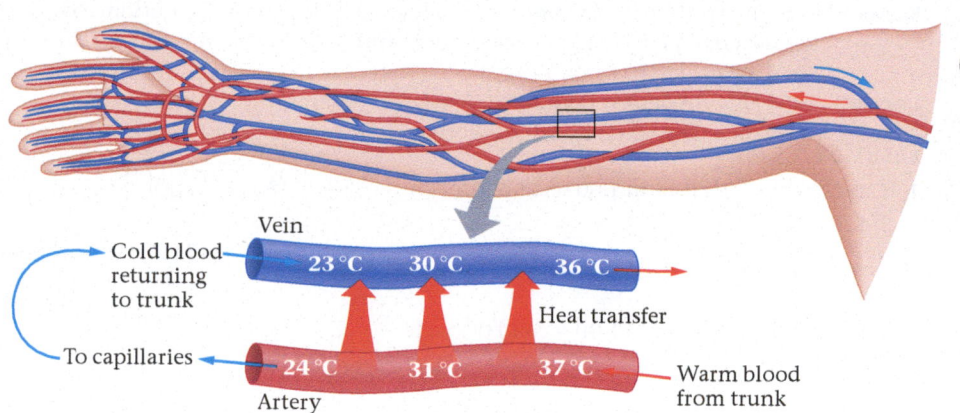

▶ **FIGURE 16-18 Countercurrent heat exchange in the human arm** Arteries bringing warm blood to the limbs lie close to veins returning cooler blood to the body. This arrangement assures that a temperature difference (gradient) is maintained over the entire length that the vessels run parallel to one another, maximizing heat exchange between the warm arterial blood and the cooler venous blood.

Convection

Suppose you want to heat a small room. To do so, you bring a portable electric heater into the room and turn it on. As the heating coils get red hot, they heat the air in their vicinity; as this air warms, it expands, becoming less dense. Because of its lower density, the warm air rises, to be replaced by cold, dense air descending from overhead. This sets up a circulating flow of air that transports heat from the heating coils to the air throughout the room. Heat exchange of this type is referred to as **convection**. Convection is also important in the burning of a candle, as illustrated in **FIGURE 16-19**, where we also show what happens when convection is "shut off."

In general, convection occurs when a fluid is unevenly heated. As with the room heater, the warm portions of the fluid rise because of their lower density and the cool portions sink because of their higher density. Thus, in convection, temperature differences result in a flow of fluid. It is this physical flow of matter that carries heat throughout the system.

Convection occurs on an enormous range of length scales. For example, the same type of uneven heating produced by an electric heater in a room can occur in the atmosphere of the Earth as well. The common seashore occurrence of sea breezes during the day and land breezes in the evening is one such example. (See **FIGURE 16-20** for an illustration of this effect.) On a larger scale, the Sun causes greater warming near the equator of the Earth than near the poles; as a result, warm equatorial air rises, cool polar air descends, and global convection patterns are established. Similar convection patterns occur in ocean waters, and plate tectonics is believed to be caused, at least in part, by convection currents in the Earth's mantle. The Sun also has convection currents, due to the intense heating that occurs in its interior, and disturbances in these currents are often visible as sunspots.

Radiation

Convection and conduction occur primarily in specific situations, but *all* objects give off energy as a result of **radiation**. The energy radiated by an object is in the form of electromagnetic waves (Chapter 25), which include visible light as well as infrared and ultraviolet radiation. Thus, unlike convection and conduction, radiation has no need for a physical material to mediate the energy transfer, because electromagnetic waves can propagate through empty space—that is, through a vacuum. Therefore, the heat you feel radiated from a hot furnace would reach you even if the air were suddenly removed—just as radiant energy from the Sun reaches the Earth across 150 million kilometers of vacuum. Examples of infrared radiation from objects as diverse a house, a baby, and an oil pipeline are shown in **FIGURE 16-21**.

RWP Because radiation often includes visible light, it's sometimes possible to "see" the temperature of an object. This is the physical basis of the **optical pyrometer**, invented by Josiah Wedgwood (1730–1795), the renowned English potter and relative of Charles Darwin. For instance, when objects are about 800 °C, as in an oven's heating coil or in molten lava, they appear to be a dull "red hot," as shown in **FIGURE 16-22 (a)**. Molten steel at 1500 °C glows yellow, and the 6000 °C plasma created by arc welding is bluish-white hot. Even stars have colors that indicate their surface temperature. For example, red stars like Betelgeuse in **FIGURE 16-22 (b)** (upper left) are cooler than blue stars like the nearby Rigel (lower right).

The energy radiated per time by an object—that is, the radiated power, *P*—is proportional to the surface area, *A*, over which the radiation occurs. It also depends on

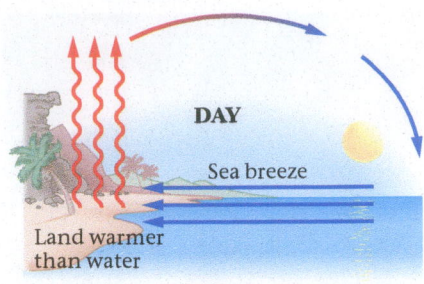

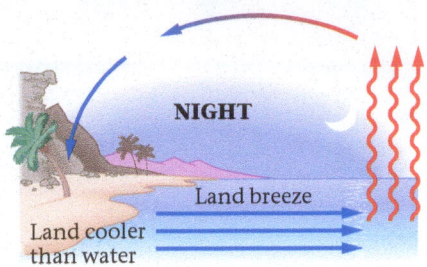

▲ **FIGURE 16-20 Alternating land and sea breezes (a)** During the day, the sun warms the land more rapidly than the water. This is because the land, which is mostly rocks, has a lower specific heat than the water. The warm land heats the air above it, which becomes less dense and rises. Cooler air from over the water flows in to take its place, producing a "sea breeze." **(b)** At night, the land cools off more rapidly than the water—again because of its lower specific heat. Now it is the air above the relatively warm water that rises and is replaced by cooler air from over the land, producing a "land breeze."

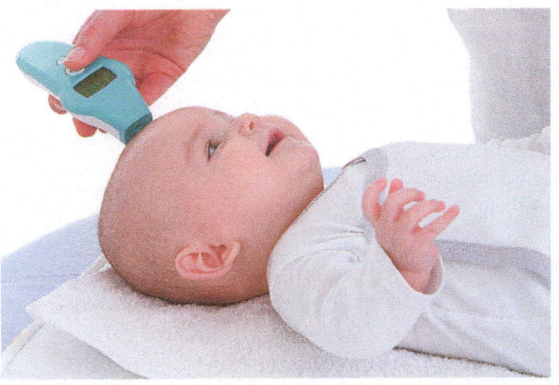

(a)   (b)   (c)

▲ **FIGURE 16-21 Visualizing Concepts Infrared Radiation** Many everyday objects radiate energy in the infrared part of the electromagnetic spectrum (see Chapter 25), which is invisible to our eyes. **(a)** Careful measurements of the infrared radiation emitted by a house (depicted here in false color) can be used to locate thermal leaks. **(b)** A baby's temperature can be taken easily and unobtrusively with an infrared sensor. **(c)** Infrared radiation is given off by the finned "radiators" attached to the supports of the Alaska pipeline. These fins are designed to prevent melting of the environmentally sensitive permafrost over which the pipeline runs.

(a)

(b)

◀ **FIGURE 16-22** **Visualizing Concepts** **Visible Radiation** **(a)** Red-hot volcanic lava is just hot enough to radiate in the visible range. Even when it cools enough to stop glowing, it still emits energy, but most of it is in the form of invisible infrared radiation. **(b)** Stars have colors that are indicative of their surface temperatures. Thus, the red supergiant Betelgeuse (upper left) in the constellation Orion is much cooler than the blue supergiant Rigel (lower right).

the temperature of the object. In fact, the dependence is on the fourth power of the temperature, T^4, where T is the Kelvin-scale temperature. Thus, for instance, if T is doubled, the radiated power increases by a factor of 16. All this behavior is contained in the **Stefan–Boltzmann law:**

Stefan–Boltzmann Law for Radiated Power, P

$$P = e\sigma A T^4 \qquad \text{16-17}$$

SI unit: W

The constant σ in this expression is a fundamental physical constant, the **Stefan–Boltzmann constant:**

$$\sigma = 5.67 \times 10^{-8} \, \text{W/(m}^2 \cdot \text{K}^4) \qquad \text{16-18}$$

The other constant in the Stefan–Boltzmann law is the **emissivity**, e. The emissivity is a dimensionless number between 0 and 1 that indicates how effective an object is in radiating energy. A value of 1 means that the object is a perfect radiator. In general, a dark-colored object will have an emissivity near 1, and a light-colored object will have an emissivity closer to 0. People wearing garments with an emissivity near 1 are shown in **FIGURE 16-23**.

EXERCISE 16-17 RADIATED POWER

Calculate the power radiated by a cube with a side length of 7.70 cm at the temperature 339 K (about 151 °F). Assume the emissivity is unity.

REASONING AND SOLUTION
A cube has 6 faces, each with an area of L^2, where L is the length of a side. The total surface area of this cube, then, is $A = 6L^2$, with $L = 0.0770$ m. Substituting A and $T = 339$ K in $P = e\sigma A T^4$ yields

$$P = e\sigma A T^4 = (1)[5.67 \times 10^{-8} \, \text{W/(m}^2 \cdot \text{K}^4)]6(0.0770 \, \text{m})^2(339 \, \text{K})^4 = 26.6 \, \text{W}$$

Experiments show that objects absorb radiation from their surroundings according to the same law, the Stefan–Boltzmann law, by which they emit radiation. Thus, if the temperature of an object is T, and its surroundings are at the temperature T_s, the *net* power radiated by the object is:

Net Radiated Power, P_{net}

$$P_{net} = e\sigma A(T^4 - T_s^4) \qquad \text{16-19}$$

SI unit: W

If the object's temperature is greater than its surroundings, it radiates more energy than it absorbs and P_{net} is positive. On the other hand, if its temperature is lower than the surroundings, it absorbs more energy than it radiates and P_{net} is negative. When the object has the same temperature as its surroundings, it is in equilibrium and the net power is zero.

PROBLEM-SOLVING NOTE

Radiated Power

To correctly calculate the radiated power, Equations 16-17 and 16-19 the temperatures must be expressed in the Kelvin scale.

◀ **FIGURE 16-23** Wear black in the desert? It actually helps—by radiating heat away from the wearer more efficiently.

EXAMPLE 16-18   HUMAN POLAR BEARS

On New Year's Day, several human "polar bears" prepare for their annual dip into the icy waters of Narragansett Bay. One of these hardy souls has a surface area of 1.15 m² and a surface temperature of 303 K (~30 °C). Find the net radiated power from this person **(a)** in a dressing room, where the temperature is 293 K (~20 °C), and **(b)** outside, where the temperature is 273 K (~0 °C). Assume an emissivity of 0.900 for the person's skin.

PICTURE THE PROBLEM

Our sketch shows the person radiating power in a room where the surroundings are at 293 K, and outside where the temperature is 273 K. The person also absorbs radiation from the surroundings; hence the net radiated power is greater when the surroundings are cooler.

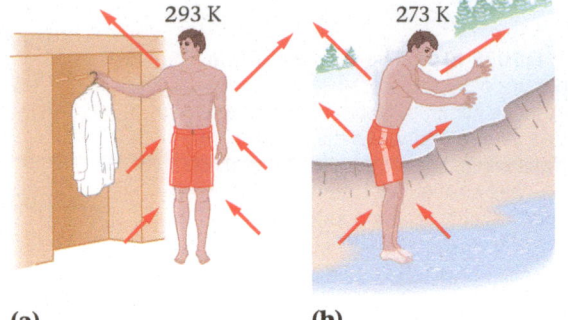

(a)   **(b)**

REASONING AND STRATEGY

A straightforward application of $P_{net} = e\sigma A(T^4 - T_s^4)$ can be used to find the net power for both parts (a) and (b). The only difference is the temperature of the surroundings, T_s.

Known   Surface area of person, $A = 1.15$ m²; surface temperature of person, $T = 303$ K; temperature in room, $T_s = 293$ K; temperature outside, $T_s = 273$ K; emissivity of person's skin, $e = 0.900$.

Unknown   **(a)** Net radiated power in room, $P_{net} = ?$ **(b)** Net radiated power outside, $P_{net} = ?$

SOLUTION

Part (a)

1. Calculate the net power using $P_{net} = e\sigma A(T^4 - T_s^4)$ (Equation 16-19) and $T_s = 293$ K:

$$\begin{aligned} P_{net} &= e\sigma A(T^4 - T_s^4) \\ &= (0.900)[5.67 \times 10^{-8} \text{ W/(m}^2 \cdot \text{K}^4)](1.15 \text{ m}^2) \\ &\quad \times [(303 \text{ K})^4 - (293 \text{ K})^4] \\ &= 62.1 \text{ W} \end{aligned}$$

Part (b)

2. Calculate the net power using $P_{net} = e\sigma A(T^4 - T_s^4)$ and $T_s = 273$ K:

$$\begin{aligned} P_{net} &= e\sigma A(T^4 - T_s^4) \\ &= (0.900)[5.67 \times 10^{-8} \text{ W/(m}^2 \cdot \text{K}^4)](1.15 \text{ m}^2) \\ &\quad \times [(303 \text{ K})^4 - (273 \text{ K})^4] \\ &= 169 \text{ W} \end{aligned}$$

INSIGHT

In the warm room the net radiated power is roughly that of a small lightbulb (about 60 W); outdoors, the net radiated power has more than doubled, and is comparable to that of a 150-W lightbulb.

PRACTICE PROBLEM

What temperature must the surroundings have if the net radiated power is to be 155 W? [**Answer:** $T_s = 276$ K ≈ 3 °C]

Some related homework problems: Problem 54, Problem 62

The same emissivity e applies to both the emission and absorption of energy. Thus, a perfect emitter ($e = 1$) is also a perfect absorber. Such an object is referred to as a **blackbody**. As we shall see later, in Chapter 30, the study of blackbody radiation near the turn of the twentieth century ultimately led to one of the most fundamental revolutions in science—the introduction of quantum physics.

RWP The opposite of a blackbody is an ideal reflector, which absorbs *no* radiation ($e = 0$). It follows that an ideal reflector also radiates no energy. This is why the inside of a Thermos bottle is highly reflective. As an almost ideal reflector, the inside of the bottle radiates very little of the energy contained in the hot liquid that it holds. In addition to its shiny interior, a Thermos bottle has a vacuum between its inner and outer walls, as shown in **FIGURE 16-24**. Because convection and conduction cannot occur in a vacuum, this design vastly reduces heat exchange between the contents of the bottle and the surroundings. This type of double-walled insulating container was invented by Sir James Dewar (1842–1923), a Scottish physicist and chemist.

PHYSICS IN CONTEXT
Looking Back

Notice that we have used the concept of power (Chapter 7) in talking about the amount of energy radiated by an object in a given amount of time.

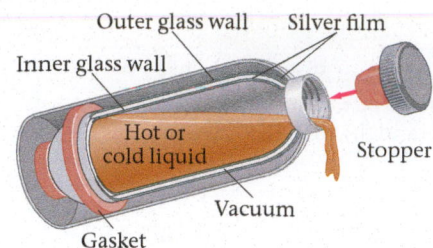

Outer glass wall   Silver film

Inner glass wall

Hot or cold liquid

Stopper

Vacuum

Gasket

▲ **FIGURE 16-24 The Thermos bottle** The hot or cold liquid stored in a Thermos bottle is separated from the outside world by a vacuum between the inner and outer glass walls. In addition, the inner glass wall has a reflective coating so that it is a good reflector and a poor radiator.

Enhance Your Understanding   (Answers given at the end of the chapter)

6. The following systems consist of a cylindrical metal rod with a given length and radius. Rank the systems in order of increasing heat conducted in 1 s, assuming each rod has the same temperature difference between its ends. Indicate ties where appropriate. (Refer to Table 16-3 for thermal conductivities.)

System	A	B	C	D
Metal	lead	copper	silver	aluminum
Length	20 cm	20 cm	10 cm	10 cm
Radius	3 cm	2 cm	1 cm	10 cm

Section Review

- The amount of heat conducted through a substance of length L and cross-sectional area A in the time t is $Q = kA(\Delta T/L)t$. The constant k is the thermal conductivity of the substance, and ΔT is the temperature difference across the substance.

- The net power radiated by an object with surface area A and temperature T in surroundings of temperature T_s is $P_{net} = e\sigma A(T^4 - T_s^4)$. The constant e is the emissivity (a number between 0 and 1), and σ is the Stefan–Boltzmann constant, $\sigma = 5.67 \times 10^{-8} \, \text{W/(m}^2 \cdot \text{K}^4)$.

CHAPTER 16 REVIEW

CHAPTER SUMMARY

16-1 TEMPERATURE AND THE ZEROTH LAW OF THERMODYNAMICS

This section defines several new terms dealing with heat and temperature.

Heat

Heat is the energy transferred between objects because of a temperature difference.

Thermal Contact

Objects are in thermal contact if heat can flow between them.

Thermal Equilibrium

Objects that are in thermal contact, but have no heat exchange between them, are said to be in thermal equilibrium.

Zeroth Law of Thermodynamics

If objects A and C are in thermal equilibrium with object B, then they are in thermal equilibrium with each other.

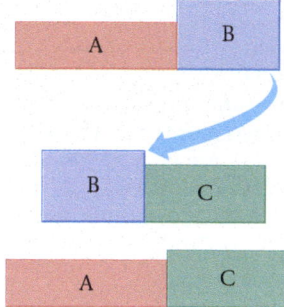

16-2 TEMPERATURE SCALES

Temperature is commonly measured in terms of three different scales.

Celsius Scale

In the Celsius scale, water freezes at 0 °C and boils at 100 °C.

Fahrenheit Scale

In the Fahrenheit scale, water freezes at 32 °F and boils at 212 °F.

Absolute Zero

The lowest temperature attainable is referred to as absolute zero. It is impossible to cool an object to a temperature lower than absolute zero, which is −273.15 °C.

Kelvin Scale

In the Kelvin scale, absolute zero is 0 K. In addition, water freezes at 273.15 K and boils at 373.15 K. The degree size is the same for the Kelvin and Celsius scales.

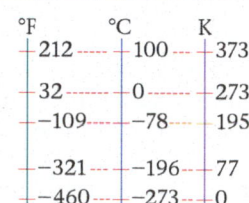

°F	°C	K
212	100	373
32	0	273
−109	−78	195
−321	−196	77
−460	−273	0

16-3 THERMAL EXPANSION

Most, though not all, substances expand when heated.

Linear Expansion

When an object of length L_0 is heated by the amount ΔT, its length increases by ΔL:

$$\Delta L = \alpha L_0 \Delta T \qquad \text{16-4}$$

The constant α is the coefficient of linear expansion (Table 16-1).

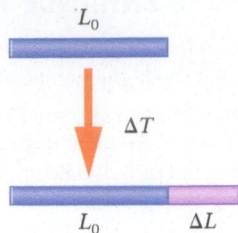

16-4 HEAT AND MECHANICAL WORK

An important step forward in the understanding of heat was the recognition that it is a form of energy.

Mechanical Equivalent of Heat

$$1 \text{ cal} = 4.186 \text{ J} \qquad \text{16-8}$$

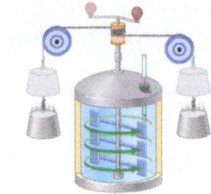

16-5 SPECIFIC HEATS

Heat Capacity

The heat capacity of an object is the heat Q divided by the associated temperature change, ΔT:

$$C = \frac{Q}{\Delta T} \qquad \text{16-11}$$

Specific Heat

The specific heat is the heat capacity per unit mass. Thus, the specific heat is independent of the quantity of a material in a given object:

$$c = \frac{Q}{m \Delta T} \qquad \text{16-13}$$

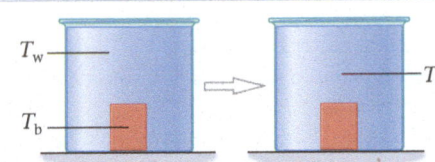

16-6 CONDUCTION, CONVECTION, AND RADIATION

This section considers three common mechanisms of heat exchange.

Conduction

In conduction, heat flows through a material with no bulk motion. The heat flows as a result of the interactions of individual atoms with their neighbors.

If the thermal conductivity of a material is k, its cross-sectional area is A, and its length L, the heat exchanged in the time t is

$$Q = kA\left(\frac{\Delta T}{L}\right)t \qquad \text{16-16}$$

Convection

Convection is heat exchange due to the bulk motion of an unevenly heated fluid.

Radiation

Radiation is the heat exchange due to electromagnetic radiation, such as infrared rays and light.

The energy per time, or power P, radiated by an object with a surface area A at the Kelvin temperature T is

$$P = e\sigma A T^4 \qquad \text{16-17}$$

where e is the emissivity (a constant between 0 and 1) and σ is the Stefan–Boltzmann constant: $\sigma = 5.67 \times 10^{-8} \text{ W/(m}^2 \cdot \text{K}^4)$.

Since an object will also absorb radiation from its surroundings at the temperature T_s, the net radiated power is

$$P_{\text{net}} = e\sigma A(T^4 - T_s^4) \qquad \text{16-19}$$

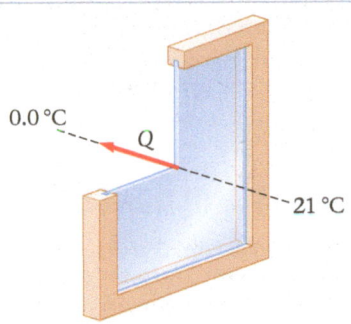

ANSWERS TO ENHANCE YOUR UNDERSTANDING QUESTIONS

1. Heat flows from high temperature to low temperature, so brick B has the higher temperature.

2. Fahrenheit degrees are "smaller" than Kelvin degrees because there are 180 Fahrenheit degrees between freezing and boiling water, but only 100 Kelvin degrees.

3. D < C < B < A.

4. Yes, part of the mechanical work done in stirring the water goes into increasing its temperature.

5. The temperature increase is inversely proportional to the specific heat; thus the specific heat of object A is less than that of object B.

6. A < C < B < D.

CONCEPTUAL QUESTIONS

For instructor-assigned homework, go to www.masteringphysics.com **MP**

(Answers to odd-numbered Conceptual Questions can be found in the back of the book.)

1. A cup of hot coffee is placed on the table. Is it in thermal equilibrium? What condition determines when the coffee is in equilibrium?

2. To find the temperature at the core of the Sun, you consult some Web sites on the Internet. One site says the temperature is about 15 million °C, another says it is 15 million kelvin. Is this a serious discrepancy? Explain.

3. Is it valid to say that a hot object contains more heat than a cold object?

4. If the glass in a glass thermometer had the same coefficient of volume expansion as mercury, the thermometer would not be very useful. Explain.

5. When a mercury-in-glass thermometer is inserted into a hot liquid the mercury column first drops and then rises. Explain this behavior.

6. Sometimes the metal lid on a glass jar has been screwed on so tightly that it is very difficult to open. Explain why holding the lid under hot running water often loosens it enough for easy opening.

7. Two different objects receive the same amount of heat. Give at least two reasons why their temperature changes may not be the same.

8. The specific heat of concrete is greater than that of soil. Given this fact, would you expect a major-league baseball field or the parking lot that surrounds it to cool off more in the evening following a sunny day?

9. When you touch a piece of metal and a piece of wood that are both at room temperature the metal feels cooler. Why?

10. The rate of heat flow through a slab does *not* depend on which of the following? (a) The temperature difference between opposite faces of the slab. (b) The thermal conductivity of the slab. (c) The thickness of the slab. (d) The cross-sectional area of the slab. (e) The specific heat of the slab.

11. If a lighted match is held beneath a balloon inflated with air, the balloon quickly bursts. If, instead, the lighted match is held beneath a balloon filled with water, the balloon remains intact, even if the flame comes in contact with the balloon. Explain.

12. Updrafts of air allow hawks and eagles to glide effortlessly, all the while gaining altitude. What causes the updrafts?

13. **BIO** The fur of polar bears consists of hollow fibers. (Sometimes algae will grow in the hollow regions, giving the fur a green cast.) Explain why hollow hairs can be beneficial to the polar bears.

14. Object 2 has twice the emissivity of object 1, though they have the same size and shape. If the two objects radiate the same power, what is the ratio of their Kelvin temperatures?

PROBLEMS AND CONCEPTUAL EXERCISES

Answers to odd-numbered Problems and Conceptual Exercises can be found in the back of the book. **BIO** *identifies problems of biological or medical interest;* **CE** *indicates a conceptual exercise.* **Predict/Explain** *problems ask for two responses: (a) your prediction of a physical outcome, and (b) the best explanation among three provided; and* **Predict/Calculate** *problems ask for a prediction of a physical outcome, based on fundamental physics concepts, and follow that with a numerical calculation to verify the prediction. On all problems, bullets (•, ••, •••) indicate the level of difficulty.*

SECTION 16-2 TEMPERATURE SCALES

1. • **CE** **Predict/Explain** The temperature inside a freezer is 22 °F and the temperature outside is 42 °F. The temperature difference is 20 F°. (a) Is the temperature difference ΔT in degrees Celsius greater than, less than, or equal to 20 C°? (b) Choose the *best explanation* from among the following:
   I. The temperature difference is equal to 20 C° because temperature differences are the same in all temperature scales.
   II. The temperature difference is less than 20 C° because $\Delta T_C = \frac{5}{9}(20 \,°F) = 11 \,°C$.
   III. The temperature difference is greater than 20 C° because $\Delta T_C = \frac{9}{5}(20 \,°F) + 32 = 68 \,°C$.

2. • **Lowest Temperature on Earth** The official record for the lowest temperature ever recorded on Earth was set at Vostok, Antarctica, on July 21, 1983. The temperature on that day fell to −89.2 °C, well below the temperature of dry ice. What is this temperature in degrees Fahrenheit?

3. • Incandescent lightbulbs heat a tungsten filament to a temperature of about 4580 °F, which is almost half as hot as the surface of the Sun. What is this temperature in degrees Celsius?

4. • Normal body temperature for humans is 98.6 °F. What is the corresponding temperature in (a) degrees Celsius and (b) kelvins?

5. • The temperature at the surface of the Sun is about 6000 K. Convert this temperature to the (a) Celsius and (b) Fahrenheit scales.

6. • One day you notice that the outside temperature increased by 17 F° between your early morning jog and your lunch at noon. What is the corresponding change in temperature in the (a) Celsius and (b) Kelvin scales?

7. • The gas in a constant-volume gas thermometer has a pressure of 93.5 kPa at 105 °C. (a) What is the pressure of the gas at 47.5 °C? (b) At what temperature does the gas have a pressure of 105 kPa?

8. •• **Predict/Calculate** A constant-volume gas thermometer has a pressure of 80.3 kPa at −10.0 °C and a pressure of 86.4 kPa at

10.0 °C. **(a)** At what temperature does the pressure of this system extrapolate to zero? **(b)** What are the pressures of the gas at the freezing and boiling points of water? **(c)** In general terms, how would your answers to parts (a) and (b) change if a different constant-volume gas thermometer is used? Explain.

9. •• **Greatest Change in Temperature** A world record for the greatest change in temperature was set in Spearfish, SD, on January 22, 1943. At 7:30 A.M. the temperature was −4.0 °F; two minutes later the temperature was 45 °F. Find the average rate of temperature change during those two minutes in kelvins per second.

10. •• We know that −40 °C corresponds to −40 °F. What temperature has the same value in both the Fahrenheit and Kelvin scales?

11. •• At what temperature is the reading on a Fahrenheit scale twice the reading on a Celsius scale?

12. ••• When the bulb of a constant-volume gas thermometer is placed in a beaker of boiling water at 100 °C, the pressure of the gas is 227 mmHg. When the bulb is moved to an ice–salt mixture, the pressure of the gas drops to 162 mmHg. Assuming ideal behavior, as in Figure 16-3, what is the Celsius temperature of the ice–salt mixture?

SECTION 16-3 THERMAL EXPANSION

13. • **CE** Bimetallic strip A is made of copper and steel; bimetallic strip B is made of aluminum and steel. **(a)** Referring to Table 16-1, which strip bends more for a given change in temperature? **(b)** Which of the metals listed in Table 16-1 would give the greatest amount of bend when combined with steel in a bimetallic strip?

14. • **CE** Referring to Table 16-1, which would be more accurate for all-season outdoor use: a tape measure made of steel or one made of aluminum?

15. • **CE** Predict/Explain A brass plate has a circular hole whose diameter is slightly smaller than the diameter of an aluminum ball. If the ball and the plate are always kept at the same temperature, **(a)** should the temperature of the system be increased or decreased in order for the ball to fit through the hole? **(b)** Choose the *best explanation* from among the following:
   I. The aluminum ball changes its diameter more with temperature than the brass plate, and therefore the temperature should be decreased.
   II. Changing the temperature won't change the fact that the ball is larger than the hole.
   III. Heating the brass plate makes its hole larger, and that will allow the ball to pass through.

16. • **CE** FIGURE 16-25 shows five metal plates, all at the same temperature and all made from the same material. They are all placed in an oven and heated by the same amount. Rank the plates in order of increasing expansion in **(a)** the vertical and **(b)** the horizontal direction. Indicate ties where appropriate.

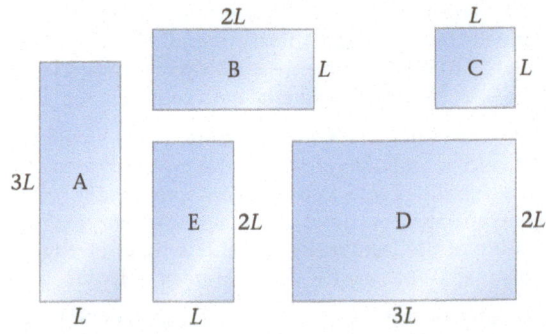

FIGURE 16-25 Problem 16

17. • **Longest Suspension Bridge** The world's longest suspension bridge is the Akashi Kaikyo Bridge in Japan. The bridge is 3910 m long and is constructed of steel. How much longer is the bridge on a warm summer day (30.0 °C) than on a cold winter day (−5.00 °C)?

18. •• A vinyl siding panel for a house is installed on a day when the temperature is 15.6 °C. If the coefficient of thermal expansion for vinyl siding is 55.8×10^{-6} K^{-1}, how much room (in mm) should the installer leave for expansion of a 3.66-m length if the sunlit temperature of the siding could reach 48.9 °C?

19. •• A cylinder bore in an aluminum engine block has a diameter of 96.00 mm at 20.00 °C. **(a)** What is the diameter of the bore when the engine operates at 119.0 °C? **(b)** At what temperature is the diameter of the hole equal to 95.85 mm?

20. •• Predict/Calculate It is desired to slip an aluminum ring over a steel bar (FIGURE 16-26). At 10.00 °C the inside diameter of the ring is 4.000 cm and the diameter of the rod is 4.040 cm. **(a)** In order for the ring to slip over the bar, should the ring be heated or cooled? Explain. **(b)** Find the temperature of the ring at which it fits over the bar. The bar remains at 10.0 °C.

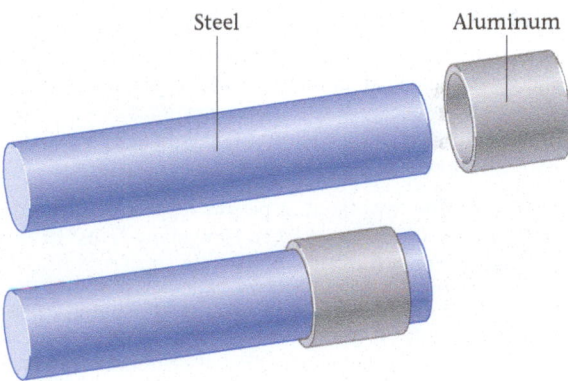

Steel   Aluminum

FIGURE 16-26 Problems 20 and 77

21. •• At 18.75 °C a brass sleeve has an inside diameter of 2.21988 cm and a steel shaft has a diameter of 2.22258 cm. It is desired to shrink-fit the sleeve over the steel shaft. **(a)** To what temperature must the sleeve be heated in order for it to slip over the shaft? **(b)** Alternatively, to what temperature must the shaft be cooled before it is able to slip through the sleeve?

22. •• Early in the morning, when the temperature is 5.5 °C, gasoline is pumped into a car's 53-L steel gas tank until it is filled to the top. Later in the day the temperature rises to 27 °C. Since the volume of gasoline increases more for a given temperature increase than the volume of the steel tank, gasoline will spill out of the tank. How much gasoline spills out in this case?

23. •• Some cookware has a stainless steel interior ($\alpha = 17.3 \times 10^{-6}$ K^{-1}) and a copper bottom ($\alpha = 17.0 \times 10^{-6}$ K^{-1}) for better heat distribution. Suppose an 8.0-in. pot of this construction is heated to 610 °C on the stove. If the initial temperature of the pot is 22 °C, what is the difference in diameter change for the copper and the steel?

24. •• Predict/Calculate You construct two wire-frame cubes, one using copper wire, the other using aluminum wire. At 23 °C the cubes enclose equal volumes of 0.016 m³. **(a)** If the temperature of the cubes is increased, which cube encloses the greater volume? **(b)** Find the difference in volume between the cubes when their temperature is 97 °C.

25. •• A metal ball that is 1.2 m in diameter expands by 2.2 mm when the temperature is increased by 95 °C. **(a)** What is the coefficient

of thermal expansion for the ball? **(b)** What substance is this ball likely made from? (Refer to Table 16-1.)

26. •• A copper ball with a radius of 1.7 cm is heated until its diameter has increased by 0.16 mm. Assuming an initial temperature of 21 °C, find the final temperature of the ball.

27. •• Predict/Calculate An aluminum saucepan with a diameter of 23 cm and a height of 6.0 cm is filled to the brim with water. The initial temperature of the pan and water is 19 °C. The pan is now placed on a stove burner and heated to 88 °C. **(a)** Will water overflow from the pan, or will the water level in the pan decrease? Explain. **(b)** Calculate the volume of water that overflows or the drop in water level in the pan, whichever is appropriate.

SECTION 16-4   HEAT AND MECHANICAL WORK

28. • **BIO** Sleeping Metabolic Rate When people sleep, their metabolic rate is about 2.6×10^{-4} C/(s·kg). How many Calories does an 81-kg person metabolize while getting a good night's sleep of 8.5 h?

29. •• **BIO** An exercise machine indicates that you have worked off 2.1 Calories in a minute-and-a-half of running in place. What was your power output during this time? Give your answer in both watts and horsepower.

30. •• **BIO** A certain sandwich cookie contains 53 C of nutritional energy. **(a)** For what amount of time must you swim in order to work off three cookies if swimming consumes 540 C/h? **(b)** Instead of swimming you decide to run for 45 min at a pace that works off 850 C/h. How many cookies have you worked off?

31. •• **BIO** During a workout, a person repeatedly lifts a 16-lb barbell through a distance of 1.1 ft. How many "reps" of this lift are required to work off 150 C?

32. •• Predict/Calculate Consider the apparatus that Joule used in his experiments on the mechanical equivalent of heat, shown in Figure 16-12. Suppose both blocks have a mass of 0.95 kg and that they fall through a distance of 0.48 m. **(a)** Find the expected rise in temperature of the water, given that 6200 J are needed for every 1.0 C° increase. Give your answer in Celsius degrees. **(b)** Would the temperature rise in Fahrenheit degrees be greater than or less than the result in part (a)? Explain. **(c)** Find the rise in temperature in Fahrenheit degrees.

33. •• **BIO** It was shown in Example 16-18 that a typical person radiates about 62 W of power at room temperature. Given this result, how much time does it take for a person to radiate away the energy acquired by consuming a 230-C doughnut?

SECTION 16-5   SPECIFIC HEATS

34. • **CE** Predict/Explain Two objects are made of the same material but have different temperatures. Object 1 has a mass m, and object 2 has a mass $2m$. If the objects are brought into thermal contact, **(a)** is the temperature change of object 1 greater than, less than, or equal to the temperature change of object 2? **(b)** Choose the *best explanation* from among the following:
   I. The larger object gives up more heat, and therefore its temperature change is greatest.
   II. The heat given up by one object is taken up by the other object. Since the objects have the same heat capacity, the temperature changes are the same.
   III. One object loses heat of magnitude Q, the other gains heat of magnitude Q. With the same magnitude of heat involved, the smaller object has the greater temperature change.

35. • **CE** Predict/Explain A certain amount of heat is transferred to 2 kg of aluminum, and the same amount of heat is transferred to 1 kg of ice. Referring to Table 16-2, **(a)** is the increase in temperature of the aluminum greater than, less than, or equal to the increase in temperature of the ice? **(b)** Choose the *best explanation* from among the following:
   I. Twice the specific heat of aluminum is less than the specific heat of ice, and hence the aluminum has the greater temperature change.
   II. The aluminum has the smaller temperature change since its mass is less than that of the ice.
   III. The same heat will cause the same change in temperature.

36. • Suppose 72.3 J of heat are added to a 101-g piece of aluminum at 20.5 °C. What is the final temperature of the aluminum?

37. • Estimate the heat required to heat a 0.15-kg apple from 12 °C to 36 °C. (Assume the apple is mostly water.)

38. • A 9.7-g lead bullet is fired into a fence post. The initial speed of the bullet is 720 m/s, and when it comes to rest, half its kinetic energy goes into heating the bullet. How much does the bullet's temperature increase?

39. •• Thermal energy is added to 150 g of water at the rate of 55 J/s for 2.5 min. How much does the temperature of the water increase?

40. •• Predict/Calculate Silver pellets with a mass of 1.0 g and a temperature of 85 °C are added to 220 g of water at 14 °C. **(a)** How many pellets must be added to increase the equilibrium temperature of the system to 25 °C? Assume no heat is exchanged with the surroundings. **(b)** If copper pellets are used instead, does the required number of pellets increase, decrease, or stay the same? Explain. **(c)** Find the number of copper pellets that are required.

41. •• A 225-g lead ball at a temperature of 81.2 °C is placed in a light calorimeter containing 155 g of water at 20.3 °C. Find the equilibrium temperature of the system.

42. •• If 2200 J of heat are added to a 190-g object, its temperature increases by 12 C°. **(a)** What is the heat capacity of this object? **(b)** What is the object's specific heat?

43. •• **Chips by the Ton** Tortilla chips are manufactured by submerging baked and partially cooled *masa* at 41 °C into corn oil at 182 °C. **(a)** If the specific heat of masa is 1850 J/(kg·K), how much thermal energy is required to warm 0.454 kg of masa to the temperature of the oil? **(b)** If the production line makes 227 kg of chips every hour, at what rate (in kW) must thermal energy be supplied to the corn oil, assuming that all of the energy goes into warming the masa?

44. •• A 117-g lead ball is dropped from rest from a height of 7.62 m. The collision between the ball and the ground is totally inelastic. Assuming all the ball's kinetic energy goes into heating the ball, find its change in temperature.

45. •• To determine the specific heat of an object, a student heats it to 100 °C in boiling water. She then places the 34.5-g object in a 151-g aluminum calorimeter containing 114 g of water. The aluminum and water are initially at a temperature of 20.0 °C, and are thermally insulated from their surroundings. If the final temperature is 23.6 °C, what is the specific heat of the object? Referring to Table 16-2, identify the material in the object.

46. •• Predict/Calculate A student drops a 0.33-kg piece of steel at 42 °C into a container of water at 22 °C. The student also drops a 0.51-kg chunk of lead into the same container at the same time. The temperature of the water remains the same. **(a)** Was the temperature of the lead greater than, less than, or equal to 22 °C? Explain. **(b)** What was the temperature of the lead?

47. ••• A ceramic coffee cup, with $m = 116$ g and $c = 1090$ J/(kg·K), is initially at room temperature (24.0 °C). If 225 g of 80.3 °C

coffee and 12.2 g of 5.00 °C cream are added to the cup, what is the equilibrium temperature of the system? Assume that no heat is exchanged with the surroundings, and that the specific heat of coffee and cream are the same as the specific heat of water.

SECTION 16-6   CONDUCTION, CONVECTION, AND RADIATION

48. • **CE Predict/Explain** In a popular lecture demonstration, a sheet of paper is wrapped around a rod that is made from wood on one half and metal on the other half. If held over a flame, the paper on one half of the rod is burned while the paper on the other half is unaffected. **(a)** Is the burned paper on the wooden half of the rod, or on the metal half of the rod? **(b)** Choose the *best explanation* from among the following:
    I. The metal will be hotter to the touch than the wood; therefore the metal side will be burnt.
    II. The metal conducts heat better than the wood, and hence the paper on the metal half is unaffected.
    III. The metal has the smaller specific heat; hence it heats up more and burns the paper on its half of the rod.

49. • **CE FIGURE 16-27** shows a composite slab of three different materials with equal thickness but different thermal conductivities. The opposite sides of the composite slab are held at the fixed temperatures T_1 and T_2. Given that $k_B > k_A > k_C$, rank the materials in order of the temperature difference across them, starting with the smallest. Indicate ties where appropriate.

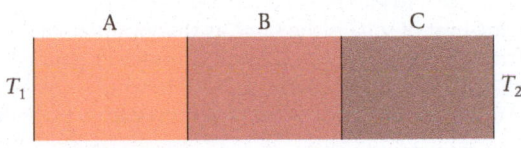

FIGURE 16-27 Problem 49

50. • **CE** Heat is transferred from an area where the temperature is 20 °C to an area where the temperature is 0 °C through a composite slab consisting of four different materials, each with the same thickness. The temperatures at the interface between each of the materials are given in **FIGURE 16-28**. Rank the four materials in order of increasing thermal conductivity. Indicate ties where appropriate.

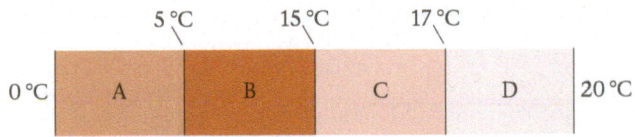

FIGURE 16-28 Problem 50

51. • **CE Predict/Explain** Two identical bowls of casserole need to be kept warm. The cook covers bowl 1 with transparent plastic wrap and bowl 2 with shiny aluminum foil, in each case trapping a layer of air between the covering and the casserole. **(a)** After a few minutes, will the temperature of casserole 1 be greater than, less than, or equal to the temperature of casserole 2? **(b)** Choose the *best explanation* from among the following:
    I. The lower thermal conductivity of the plastic wrap over casserole 1 will suppress heat loss by conduction better than aluminum foil.
    II. In each case heat loss by convection is suppressed, so the final temperatures will be the same.
    III. The high reflectivity of the aluminum foil over casserole 2 will suppress heat loss by radiation, while the heat losses by conduction and convection for casseroles 1 and 2 will be similar.

52. • **CE** Two bowls of soup with identical temperatures are placed on a table. Bowl 1 has a metal spoon in it; bowl 2 does not. After a few minutes, is the temperature of the soup in bowl 1 greater than, less than, or equal to the temperature of the soup in bowl 2?

53. • A glass window 0.33 cm thick measures 81 cm by 39 cm. How much heat flows through this window per minute if the inside and outside temperatures differ by 13 C°?

54. • **BIO** Assuming your skin temperature is 37.2 °C and the temperature of your surroundings is 20.0 °C, determine the length of time required for you to radiate away the energy gained by eating a 278-C ice cream cone. Let the emissivity of your skin be 0.915 and its area be 1.78 m².

55. • Find the heat that flows in 1.0 s through a lead brick 25 cm long if the temperature difference between the ends of the brick is 8.5 C°. The cross-sectional area of the brick is 14 cm².

56. •• Consider a double-paned window consisting of two panes of glass, each with a thickness of 0.500 cm and an area of 0.725 m², separated by a layer of air with a thickness of 1.75 cm. The temperature on one side of the window is 0.00 °C; the temperature on the other side is 20.0 °C. In addition, note that the thermal conductivity of glass is roughly 36 times greater than that of air. **(a)** Approximate the heat transfer through this window by ignoring the glass. That is, calculate the heat flow per second through 1.75 cm of air with a temperature difference of 20.0 C°. (The exact result for the complete window is 19.1 J/s.) **(b)** Use the approximate heat flow found in part (a) to find an approximate temperature difference across each pane of glass. (The exact result is 0.157 C°.)

57. •• **Predict/Calculate** Two metal rods of equal length—one aluminum, the other stainless steel—are connected in parallel with a temperature of 20.0 °C at one end and 118 °C at the other end. Both rods have a circular cross section with a diameter of 3.50 cm. **(a)** Determine the length the rods must have if the combined rate of heat flow through them is to be 27.5 J per second. **(b)** If the length of the rods is doubled, by what factor does the rate of heat flow change?

58. •• Two cylindrical metal rods—one copper, the other lead—are connected in parallel with a temperature of 21.0 °C at one end and 112 °C at the other end. Both rods are 0.650 m in length, and the lead rod is 2.76 cm in diameter. If the combined rate of heat flow through the two rods is 33.2 J/s, what is the diameter of the copper rod?

59. •• **Predict/Calculate** Two metal rods—one lead, the other copper—are connected in series, as shown in **FIGURE 16-29**. These are the same two rods that were connected in parallel in Example 16-16. Note that each rod is 0.525 m in length and has a square cross section 1.50 cm on a side. The temperature at the lead end of the rods is 2.00 °C; the temperature at the copper end is 106 °C. **(a)** The average temperature of the two ends is 54.0 °C. Is the temperature in the middle, at the lead-copper interface, greater than, less than, or equal to 54.0 °C? Explain. **(b)** Given that the heat flow through each of these rods in 1.00 s is 1.41 J, find the temperature at the lead-copper interface.

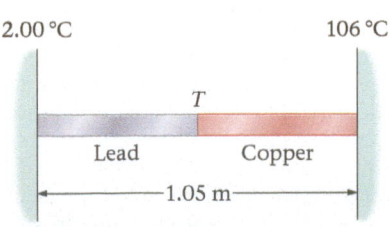

FIGURE 16-29 Problem 59

60. •• **Predict/Calculate** Consider two cylindrical metal rods with equal cross section—one lead, the other aluminum—connected in series. The temperature at the lead end of the rods is 22.5 °C; the temperature at the aluminum end is 85.0 °C. **(a)** Given that the temperature at the lead-aluminum interface is 52.5 °C, and that the lead rod is 14.5 cm long, what condition can you use to find the length of the aluminum rod? **(b)** Find the length of the aluminum rod.

61. •• A copper rod 85 cm long is used to poke a fire. The hot end of the rod is maintained at 115 °C and the cool end has a constant temperature of 22 °C. What is the temperature of the rod 25 cm from the cool end?

62. •• Two identical objects are placed in a room at 24 °C. Object 1 has a temperature of 99 °C, and object 2 has a temperature of 27 °C. What is the ratio of the net power emitted by object 1 to that radiated by object 2?

63. ••• A block has the dimensions L, $2L$, and $3L$. When one of the $L \times 2L$ faces is maintained at the temperature T_1 and the other $L \times 2L$ face is held at the temperature T_2, the rate of heat conduction through the block is P. Answer the following questions in terms of P. **(a)** What is the rate of heat conduction in this block if one of the $L \times 3L$ faces is held at the temperature T_1 and the other $L \times 3L$ face is held at the temperature T_2? **(b)** What is the rate of heat conduction in this block if one of the $2L \times 3L$ faces is held at the temperature T_1 and the other $2L \times 3L$ face is held at the temperature T_2?

GENERAL PROBLEMS

64. • **CE Predict/Explain** A pendulum is made from an aluminum rod with a mass attached to its free end. If the pendulum is cooled, **(a)** does the pendulum's period increase, decrease, or stay the same? **(b)** Choose the *best explanation* from among the following:
   I. The period of a pendulum depends only on its length and the acceleration of gravity. It is independent of mass and temperature.
   II. Cooling makes everything move more slowly, and hence the period of the pendulum increases.
   III. Cooling shortens the aluminum rod, which decreases the period of the pendulum.

65. • **CE** A copper ring stands on edge with a metal rod placed inside it, as shown in **FIGURE 16-30**. As this system is heated, will the rod ever touch the top of the ring? Answer yes or no for the case of a rod that is made of **(a)** copper, **(b)** aluminum, and **(c)** steel.

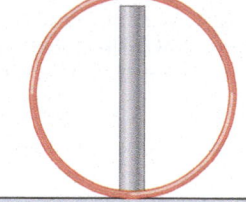

FIGURE 16-30 Problems 65 and 66

66. • **CE** Referring to the copper ring in the previous problem, imagine that initially the ring is hotter than room temperature, and that an aluminum rod that is colder than room temperature fits snugly inside the ring. When this system reaches thermal equilibrium at room temperature, is the rod firmly wedged in the ring **(A)** or can it be removed easily **(B)**?

67. • **CE Predict/Explain** The specific heat of alcohol is about half that of water. Suppose you have 0.5 kg of alcohol at the temperature 20 °C in one container, and 0.5 kg of water at the temperature 30 °C in a second container. When these fluids are poured into the same container and allowed to come to thermal equilibrium, **(a)** is the final temperature greater than, less than, or equal to 25 °C? **(b)** Choose the *best explanation* from among the following:
   I. The low specific heat of alcohol pulls in more heat, giving a final temperature that is less than 25 °C.

   II. More heat is required to change the temperature of water than to change the temperature of alcohol. Therefore, the final temperature will be greater than 25 °C.
   III. Equal masses are mixed together; therefore, the final temperature will be 25 °C, the average of the two initial temperatures.

68. • **Making Steel Sheets** In the continuous-caster process, steel sheets 25.4 cm thick, 2.03 m wide, and 10.0 m long are produced at a temperature of 872 °C. What are the dimensions of a steel sheet once it has cooled to 20.0 °C?

69. • **The Coldest Place in the Universe** The Boomerang nebula holds the distinction of having the lowest recorded temperature in the universe, a frigid −272 °C. What is this temperature in kelvins?

70. • **BIO The Hottest Living Things** From the surreal realm of deep-sea hydrothermal vents 200 miles offshore from Puget Sound, comes a newly discovered hyperthermophilic—or extreme heat-loving—microbe that holds the record for the hottest existence known to science. This microbe is tentatively known as Strain 121 for the temperature at which it thrives: 121 °C. (At sea level, water at this temperature would boil vigorously, but the extreme pressures at the ocean floor prevent boiling from occurring.) What is this temperature in degrees Fahrenheit?

71. •• Thermal energy is added to 180 g of water at a constant rate for 3.5 min, resulting in an increase in temperature of 12 C°. What is the heating rate, in joules per second?

72. •• You wish to chill 12 cans of soda at 25.0 °C down to 5.0 °C before serving them to guests. Each can has a mass of 0.354 kg, and the specific heat of soda is the same as that of water. **(a)** How much thermal energy must be removed in order to chill all 12 cans? **(b)** If the chilling process requires 4.0 h, what is the average rate (in W) at which thermal energy is removed from the soda?

73. •• **BIO Brain Power** As you read this problem, your brain is consuming about 22 W of power. **(a)** If your mass is 65 kg, how many steps with a height of 21 cm must you climb to expend a mechanical energy equivalent to one hour of brain operation? **(b)** A typical human brain, which is 77% water, has a mass of 1.4 kg. Assuming that the 22 W of brain power is converted to heat, what temperature rise would you estimate for the brain in one hour of operation? Ignore the significant heat transfer that occurs between a human head and its surroundings, as well as the 23% of the brain that is not water.

74. •• **BIO Brain Food** Your brain consumes about 22 W of power, and avocados have been shown to promote brain health. If each avocado contains 310 C, and your brain were powered entirely by the energy from the avocados, how many must you eat each day?

75. •• **BIO The Cricket Thermometer** The rate of chirping of the snowy tree cricket (*Oecanthus fultoni* Walker) varies with temperature in a predictable way. A linear relationship provides a good match to the chirp rate, but an even more accurate relationship is the following:

$$N = (5.63 \times 10^{10})e^{-(6290\text{ K})/T}$$

In this expression, N is the number of chirps in 13.0 s and T is the temperature in kelvins. If a cricket is observed to chirp 185 times in 60.0 s, what is the temperature in degrees Fahrenheit?

76. •• **Predict/Calculate** A pendulum consists of a large weight suspended by a steel wire that is 0.9500 m long. **(a)** If the temperature increases, does the period of the pendulum increase, decrease, or stay the same? Explain. **(b)** Calculate the change in length of the pendulum if the temperature increase is 150.0 C°. **(c)** Calculate

the period of the pendulum before and after the temperature increase. (Assume that the coefficient of linear expansion for the wire is $12.00 \times 10^{-6}\,\text{K}^{-1}$, and that $g = 9.810\,\text{m/s}^2$ at the location of the pendulum.)

77. •• **Predict/Calculate** Once the aluminum ring in Problem 20 is slipped over the bar, the ring and bar are allowed to equilibrate at a temperature of 22 °C. The ring is now stuck on the bar. **(a)** If the temperatures of both the ring and the bar are changed together, should the system be heated or cooled to remove the ring? **(b)** Find the temperature at which the ring can be removed.

78. •• A 256-kg rock sits in full sunlight on the edge of a cliff 5.75 m high. The temperature of the rock is 33.2 °C. If the rock falls from the cliff into a pool containing 6.00 m³ of water at 15.5 °C, what is the final temperature of the rock–water system? Assume that the specific heat of the rock is 1010 J/(kg·K).

79. •• Water going over Iguaçu Falls on the border of Argentina and Brazil drops through a height of about 72 m. Suppose that all the gravitational potential energy of the water goes into raising its temperature. Find the increase in water temperature at the bottom of the falls as compared with the top.

80. •• **Thermal Storage** Solar heating of a house is much more efficient if there is a way to store the thermal energy collected during the day to warm the house at night. Suppose one solar-heated home utilizes a concrete slab of area 12 m² and 25 cm thick. **(a)** If the density of concrete is 2400 kg/m³, what is the mass of the slab? **(b)** The slab is exposed to sunlight and absorbs energy at a rate of 1.4×10^7 J/h for 10 h. If it begins the day at 22 °C and has a specific heat of 750 J/(kg·K), what is its temperature at sunset? **(c)** Model the concrete slab as being surrounded on both sides (contact area 24 m²) with a 2.0-m-thick layer of air in contact with a surface that is 5.0 °C cooler than the concrete. At sunset, what is the rate at which the concrete loses thermal energy by conduction through the air layer? **(d)** Model the concrete slab as having a surface area of 24 m² and surrounded by an environment 5.0 °C cooler than the concrete. If its emissivity is 0.94, what is the rate at which the concrete loses thermal energy by radiation at sunset?

81. •• **Pave It Over** Suppose city 1 leaves an entire block (100 m × 100 m) as a park with trees and grass (emissivity 0.96) while city 2 paves the same area over with asphalt (emissivity 1.0). Sunlight heats each surface to 40.0 °C by sunset, and then the surface radiates its heat into a cube of air 100 m on a side and at 30.0 °C. **(a)** At what rate does the park in city 1 deliver energy to the air at sunset? **(b)** At what rate does the asphalt in city 2 deliver energy to the air at sunset? **(c)** If each city block maintains the same radiated power for 2.0 h and there are no other energy losses, what are the final temperatures of the cubes of air above each city block? The density of air at 30.0 °C is 1.16 kg/m³. Although this example is oversimplified, a more sophisticated analysis recently showed that a city park can cool the air that passes through it by more than 4 °C.

82. •• **BIO** Suppose you could convert the 495 C in the cheeseburger you ate for lunch into mechanical energy with 100% efficiency. **(a)** How high could you throw a 0.141-kg baseball with the energy contained in the cheeseburger? **(b)** How fast would the ball be moving at the moment of release?

83. •• You turn a crank on a device similar to that shown in Figure 16-12 and produce a power of 0.16 hp. If the paddles are immersed in 0.75 kg of water, for what length of time must you turn the crank to increase the temperature of the water by 5.5 C°?

84. •• **Predict/Calculate BIO Heat Transport in the Human Body** The core temperature of the human body is 37.0 °C, and the skin, with a

surface area of 1.40 m², has a temperature of 34.0 °C. **(a)** Find the rate of heat transfer out of the body under the following assumptions: (i) The average thickness of tissue between the core and the skin is 1.20 cm; (ii) the thermal conductivity of the tissue is that of water. **(b)** Without repeating the calculation of part (a), what rate of heat transfer would you expect if the skin temperature were to fall to 31.0 °C? Explain.

85. •• **The Solar Constant** The surface of the Sun has a temperature of 5500 °C. **(a)** Treating the Sun as a perfect blackbody, with an emissivity of 1.0, find the power that it radiates into space. The radius of the Sun is 7.0×10^8 m, and the temperature of space can be taken to be 3.0 K. **(b)** The *solar constant* is the number of watts of sunlight power falling on a square meter of the Earth's upper atmosphere. Use your result from part (a) to calculate the solar constant, given that the distance from the Sun to the Earth is 1.5×10^{11} m.

86. ••• Bars of two different metals are bolted together, as shown in **FIGURE 16-31**. Show that the distance D does not change with temperature if the lengths of the two bars have the following ratio: $L_A/L_B = \alpha_B/\alpha_A$.

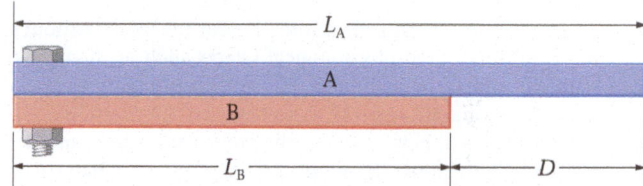

FIGURE 16-31 Problem 86

87. ••• A grandfather clock has a simple brass pendulum of length L. One night, the temperature in the house is 25.0 °C and the period of the pendulum is 1.00 s. The clock keeps correct time at this temperature. If the temperature in the house quickly drops to 17.1 °C just after 10 P.M., and stays at that value, what is the actual time when the clock indicates that it is 10 A.M. the next morning?

88. ••• **Predict/Calculate** A sheet of aluminum has a circular hole with a diameter of 10.0 cm. A 9.99-cm-long steel rod is placed inside the hole, along a diameter of the circle, as shown in **FIGURE 16-32**. It is desired to change the temperature of this system until the steel rod just touches both sides of the circle. **(a)** Should the temperature of the system be increased or decreased? Explain. **(b)** By how much should the temperature be changed?

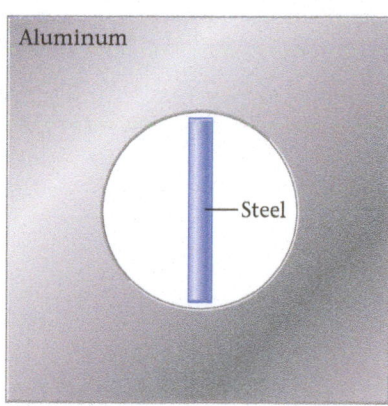

FIGURE 16-32 Problem 88

89. ••• A layer of ice has formed on a small pond. The air just above the ice is at −5.4 °C, the water-ice interface is at 0 °C, and the water at the bottom of the pond is at 4.0 °C. If the total depth from the top of the ice to the bottom of the pond is 1.4 m, how thick is the layer of ice? *Note*: The thermal conductivity of ice is 1.6 W/(m·C°) and that of water is 0.60 W/(m·C°).

90. ••• **A Double-Paned Window** An energy-efficient double-paned window consists of two panes of glass, each with thickness L_1 and thermal conductivity k_1, separated by a layer of air of thickness L_2 and thermal conductivity k_2. Show that the equilibrium rate of heat flow through this window per unit area, A, is

$$\frac{Q}{At} = \frac{(T_2 - T_1)}{2L_1/k_1 + L_2/k_2}$$

In this expression, T_1 and T_2 are the temperatures on either side of the window.

PASSAGE PROBLEMS

Cool Medicine In situations in which the brain is deprived of oxygen, such as in a heart attack or traumatic brain injury, irreversible damage to the brain can happen very quickly. It's been proposed that cooling the body of a patient after cardiac arrest or traumatic brain injury may help protect the brain against long-term damage, perhaps by reducing the metabolic rate and thus the oxygen demand of the brain.

In order to achieve some protection for the brain without causing other undesired physiological effects such as disruptions to heart rhythms, the goal is to reduce the core body temperature from its normal value of 37 °C to between 30 °C and 35 °C. Animal experiments have shown some promising results, and human trials are being done. In one trial, 95% of patients had their core temperature reduced to 34 °C in 2.0 hours. To cool the body this quickly, each patient was placed between two cooled blankets, one above the patient and one below. Water and alcohol were sprayed on the patient, and any skin that wasn't covered by the top blanket was exposed to the environment.

91. • Which method of cooling the patient in this study relies primarily on conduction?

   A. Exposing skin to the environment
   B. Spraying water and alcohol on the patient
   C. Using cooled blankets above and below
   D. None of these

92. •• What total amount of heat must be removed to drop the whole body temperature of a typical 65-kg patient in this study by 1.0 °C?

Because the human body is mostly water, the average specific heat of the human body is relatively high, 3.5×10^3 J/(kg·K).

   A. 3.5×10^3 J   B. 2.3×10^5 J
   C. 5.4×10^6 J   D. 6.2×10^7 J

93. •• During the cooling of a typical 65-kg patient in this study, what is the average rate of heat loss, assuming the whole body's temperature changes by the same amount as the core temperature does?

   A. 95 W   B. 200 W
   C. 1100 W   D. 8700 W

94. •• For a patient in a typical hospital setting, the most important mechanism of heat loss is radiation. How does the power radiated by a patient change when the skin temperature drops from 34 °C to 33 °C, assuming no other changes?

   A. It's now 99% of what it originally was.
   B. It's now 97% of what it originally was.
   C. It's now 89% of what it originally was.
   D. It doesn't change.

95. •• **REFERRING TO EXAMPLE 16-12** Suppose the mass of the block is to be increased enough to make the final temperature of the system equal to 22.5 °C. What is the required mass? Everything else in Example 16-12 remains the same.

96. •• **REFERRING TO EXAMPLE 16-12** Suppose the initial temperature of the block is to be increased enough to make the final temperature of the system equal to 22.5 °C. What is the required initial temperature? Everything else remains the same as in Example 16-12.

97. •• **Predict/Calculate REFERRING TO EXAMPLE 16-16** Suppose the lead rod is replaced with a second copper rod. **(a)** Will the heat that flows in 1.00 s increase, decrease, or stay the same? Explain. **(b)** Find the heat that flows in 1.00 s with two copper rods. Everything else remains the same as in Example 16-16.

98. •• **Predict/Calculate REFERRING TO EXAMPLE 16-16** Suppose the temperature of the hot plate is to be changed to give a total heat flow of 25.2 J in 1.00 s. **(a)** Should the new temperature of the hot plate be greater than or less than 106 °C? Explain. **(b)** Find the required temperature of the hot plate. Everything else is the same as in Example 16-16.

Big Ideas

Phases and Phase Changes

1 The pressure of an ideal gas depends on the number of molecules, the temperature, and the volume.

2 The pressure of a gas is due to molecules constantly hitting the wall of a container.

3 The average kinetic energy of a gas is proportional to its absolute temperature.

4 When two phases are in equilibrium, adding heat does not change the temperature.

5 The amount of heat needed to change one phase to another is the latent heat.

▲ Though many people are surprised to learn that a liquid can boil and freeze at the same time, it is in fact true. At a unique combination of temperature and pressure called the triple point, all three phases of matter—solid, liquid, and gas—coexist in equilibrium. In this chapter we'll learn more about the three phases of matter, their differences and similarities, and what happens when a substance changes from one phase to another.

The matter we encounter in everyday life is generally a solid, a liquid, or a gas. The air we breathe is a gas, the water we drink is a liquid, and the salt on our popcorn is a solid. In this chapter we consider these three *phases* of matter in detail, as well as changes from one phase to another.

The number of molecules and the volume of the gas are the same . . .

Reference level

Gas

Mercury

h

. . . as long as the mercury is kept at the reference level.

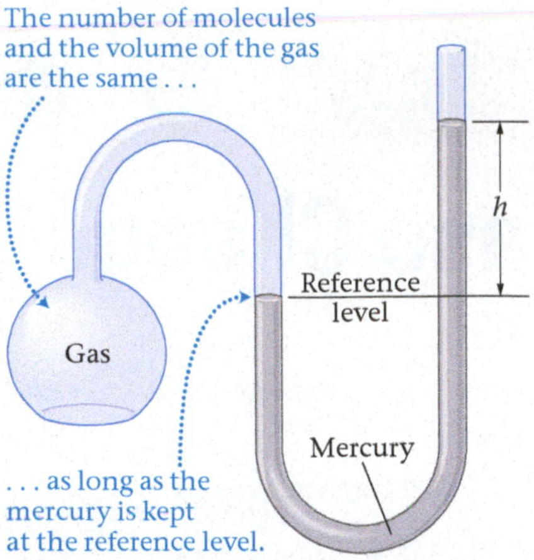

▲ **FIGURE 17-1** **A constant-volume gas thermometer** A device like this can be used as a thermometer. Note that both the volume of the gas and the number of molecules in the gas remain constant.

The pump forces more molecules into the ball . . .

. . . which increases the pressure.

▲ **FIGURE 17-2** **Inflating a basketball** A hand pump can be used to increase the number of molecules inside a basketball. This raises the pressure of the gas in the ball, causing it to inflate.

PHYSICS IN CONTEXT
Looking Back

We use the concept of temperature (Chapter 16) throughout this chapter. We also make use of pressure (Chapter 15) in our discussion of ideal gases.

▲ **FIGURE 17-3** **Increasing pressure by decreasing volume** Sitting on a basketball reduces its volume and increases the pressure of the gas it contains.

17-1 Ideal Gases

In Chapter 16 we discussed the thermal behavior of gases. In particular, we saw that the pressure of a constant volume of gas decreases linearly with decreasing temperature over a wide range of temperatures. Eventually, however, when the temperature is low enough, real gases change to liquid and then solid form, and their behavior changes. These changes are due to interactions between the molecules in a gas. Hence, the weaker these interactions, the wider the range of temperatures over which the simple, linear gas behavior persists. We now consider a simplified model of a gas, the **ideal gas**, in which intermolecular interactions are vanishingly small.

Though ideal gases do not actually exist in nature, the behavior of real gases is generally quite well approximated by that of an ideal gas, especially when the gas is dilute. By studying the simple "ideal" case, we can gain considerable insight into the workings of a real gas. This is similar to the kind of idealizations we made in mechanics. For example, we often considered a surface to be perfectly frictionless, when real surfaces have at least some friction, and we imagined springs to obey Hooke's law exactly, though real springs show some deviations. The ideal gas plays an analogous role in our study of thermodynamics.

Equation of State

We can describe the way the pressure, P, of an ideal gas depends on temperature, T, number of molecules, N, and volume, V, from just a few simple observations. First, imagine holding the number of molecules and the volume of a gas constant, as in the constant-volume gas thermometer shown in **FIGURE 17-1**. As we have already noted, the pressure of a gas under these conditions varies linearly with temperature. Therefore, we conclude the following:

$$P = (\text{constant})T$$
(fixed volume, V; fixed number of molecules, N)

In this expression, the constant multiplying the temperature depends on the number of molecules in the gas and its volume; the temperature T is measured on the Kelvin scale.

Second, imagine you have a basketball that is just slightly underinflated, as in **FIGURE 17-2**. The ball has the size and shape of a basketball, but is a bit too squishy. To increase the pressure you "pump" more molecules from the atmosphere into the ball. Thus, while the temperature and volume of the gas in the ball remain constant, its pressure increases as the number of molecules increases:

$$P = (\text{constant})N$$
(fixed volume, V; fixed temperature, T)

Our third and final observation concerns the volume dependence of pressure. Returning to the basketball, suppose that instead of pumping it up, you sit on it, as pictured in **FIGURE 17-3**. This deforms it a bit, *reducing* its volume. At the same time, the pressure of the gas in the ball *increases*. Thus, when the number of molecules and temperature remain constant, the pressure varies *inversely* with volume:

$$P = \frac{(\text{constant})}{V}$$
(fixed number of molecules, N; fixed temperature, T)

Combining these observations, we arrive at the following mathematical expression for the pressure of a gas:

$$P = k\frac{NT}{V}$$

The constant k in this expression is a fundamental constant of nature, the **Boltzmann constant,** named for the Austrian physicist Ludwig Boltzmann (1844-1906):

Boltzmann Constant, k

$$k = 1.38 \times 10^{-23} \text{ J/K} \qquad \text{17-1}$$

SI unit: J/K

In general, a relationship between the thermal properties of a substance is referred to as an **equation of state.** Rearranging slightly, we can write the equation of state for an ideal gas:

PROBLEM-SOLVING NOTE

Ideal-Gas Law

Equation of State for an Ideal Gas

$$PV = NkT \qquad \text{17-2}$$

When using the equation of state for an ideal gas, remember that the temperature must always be given in terms of the Kelvin scale.

We apply this result to the gas contained in a person's lungs in the next Example.

EXAMPLE 17-1   TAKE A DEEP BREATH

Predict/Calculate

BIO A person's lungs can hold 6.0 L ($1 \text{ L} = 10^{-3} \text{ m}^3$) of air at body temperature (310 K) and atmospheric pressure (101 kPa). **(a)** Given that air is 21% oxygen, find the number of oxygen molecules in the lungs. **(b)** If the person now climbs to the top of a mountain, where the air pressure is considerably less than 101 kPa, does the number of molecules in the lungs increase, decrease, or stay the same?

PICTURE THE PROBLEM
Our sketch shows a person's lungs, with their combined volume of $V = 6.0$ L. In addition, we indicate that the pressure in the lungs is $P = 101$ kPa and the temperature is $T = 310$ K.

REASONING AND STRATEGY

a. We will treat the air in the lungs as an ideal gas. Given the volume, temperature, and pressure of the gas, we can use the equation of state, $PV = NkT$, to solve for the number, N.
    Finally, only 21% of the molecules in the air are oxygen. Therefore, we multiply N by 0.21 to find the number of oxygen molecules.

b. We apply $PV = NkT$ again, but this time with a reduced pressure P. Notice that V and T remain the same, however, because they are determined by the shape and size of person's body.

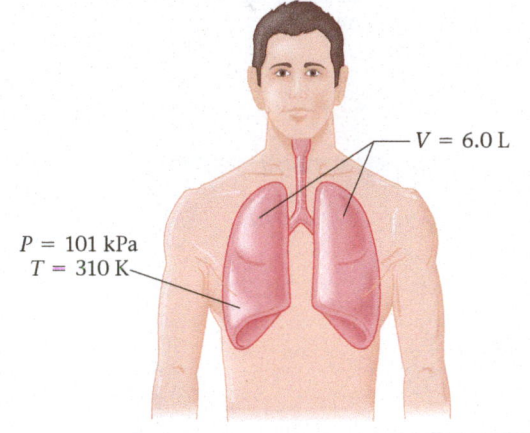

$V = 6.0$ L

$P = 101$ kPa
$T = 310$ K

Known   Volume of lungs, $V = 6.0 \times 10^{-3} \text{ m}^3$; temperature, $T = 310$ K; pressure, $P = 101$ kPa; fraction of molecules that are oxygen, $f = 0.21$.
Unknown   **(a)** Number of oxygen molecules, $N_{\text{oxy}} = ?$ **(b)** Change in number of molecules?

SOLUTION

Part (a)

1. Solve the equation of state for the number of molecules:

$$PV = NkT \quad \text{or} \quad N = PV/kT$$

2. Substitute numerical values to find the number of molecules in the lungs:

$$N = \frac{PV}{kT} = \frac{(1.01 \times 10^5 \text{ Pa})(6.0 \times 10^{-3} \text{ m}^3)}{(1.38 \times 10^{-23} \text{ J/K})(310 \text{ K})} = 1.4 \times 10^{23}$$

3. Multiply N by $f = 0.21$ to find the number of molecules that are oxygen, O_2:

$$N_{\text{oxy}} = fN = 0.21N = 0.21(1.4 \times 10^{23}) = 2.9 \times 10^{22}$$

Part (b)
From Step 1 we see that the number of molecules is $N = PV/kT$. As a result, if the volume V of a person's lungs remains the same, along with the body temperature T, it follows that the number of molecules N will decrease linearly with pressure. For example, at an altitude of 10,000 ft the air pressure is about 70% of its value at sea level, meaning that the number of oxygen molecules in the lungs is only 70% of its usual value. No wonder people have a hard time "catching their breath" at high altitudes.

INSIGHT
As one might expect, the number of molecules in even an average-size pair of lungs is enormous. In fact, a number this large is difficult to compare to anything we are familiar with. For example, the number of stars in the Milky Way galaxy is estimated to be "only" about 10^{11}. Thus, if each star in the Milky Way were itself a galaxy of 10^{11} stars, then all of these stars combined would just begin to approach the number of oxygen molecules in your lungs at this very moment!

CONTINUED

PRACTICE PROBLEM

If the person takes a particularly deep breath, so that the lungs hold a total of 1.5×10^{23} molecules, what is the new volume of the lungs?   [**Answer:** $V = 6.4 \times 10^{-3}\,\text{m}^3 = 6.4\,\text{L}$]

Some related homework problems: Problem 4, Problem 5

CONCEPTUAL EXAMPLE 17-2   AIR PRESSURE

Feeling a bit cool, you turn up the thermostat in your living room. A short time later the air is warmer. Assuming the room is well sealed, is the pressure of the air greater than, less than, or the same as before you turned up the heat?

REASONING AND DISCUSSION

Because the room is well sealed, the number of molecules and the volume they occupy are approximately constant. Thus, increasing T, while holding N and V fixed, leads to an increased pressure, P.

ANSWER

The air pressure is greater.

PHYSICS IN CONTEXT
Looking Ahead

The equation of state for an ideal gas plays an important role in Chapter 18, especially when we consider thermal processes as well as specific heats at constant pressure and constant volume.

Describing a Gas in Terms of Moles   Another common way to write the ideal-gas equation of state is in terms of the number of **moles** in a gas—as opposed to using the number of molecules, N, as in Equation 17-2. In the SI system of units, the mole (mol) is defined in terms of the most abundant isotope of carbon, which is referred to as carbon-12. The definition is as follows:

> A mole is the amount of a substance that contains as many elementary entities as there are atoms in 12 g of carbon-12.

The phrase "elementary entities" refers to molecules, which may contain more than one atom, as in water (H_2O), or just a single atom, as in carbon (C) and helium (He). Experiments show that the number of atoms in a mole of carbon-12 is 6.022×10^{23}. This number is known as **Avogadro's number,** N_A, named for the Italian physicist and chemist Amedeo Avogadro (1776–1856):

Avogadro's Number, N_A

$$N_A = 6.022 \times 10^{23}\ \text{molecules/mol}$$ 17-3

SI unit: mol^{-1}; the number of molecules is dimensionless

Examples of one mole of various substances are shown in **FIGURE 17-4**.

Now, if we let n denote the number of moles in a gas, the number of molecules it contains is

$$N = nN_A$$

Substituting this into the ideal-gas equation of state yields

$$PV = nN_A kT$$

The constants N_A and k are combined to form the **universal gas constant,** R, defined as follows:

Universal Gas Constant, R

$$R = N_A k = (6.022 \times 10^{23}\ \text{molecules/mol})(1.38 \times 10^{-23}\ \text{J/K})$$
$$= 8.31\ \text{J/(mol} \cdot \text{K)}$$ 17-4

SI unit: $\text{J/(mol} \cdot \text{K)}$

Thus, an alternative form of the equation of state is:

Equation of State for an Ideal Gas

$$PV = nRT$$ 17-5

This relationship is used in the following Quick Example.

▲ **FIGURE 17-4 Moles of various substances** A mole of any substance contains 6.022×10^{23} entities (atoms or molecules). This number is referred to as Avogadro's number (N_A). At standard temperature and pressure, N_A molecules of a gas occupy a volume of 22.4 L. One mole of a solid or liquid substance, on the other hand, is a mass in grams that is equal to its atomic mass. Thus, the mole bridges the realm of atoms and molecules and the macroscopic world of observable masses and volumes. This photo shows molar amounts of four different substances. From left to right they are sugar, salt, carbon, hydrogen, and copper.

QUICK EXAMPLE 17-3 THE AMOUNT OF AIR IN A BASKETBALL

How many moles of air are in an inflated basketball? Assume the pressure in the ball is 159 kPa, the temperature is 293 K, and the diameter of the ball is 23.0 cm.

REASONING AND SOLUTION

First, we solve the equation of state, $PV = nRT$, for the number of moles, n. Next, we substitute the numerical values $P = 159$ kPa and $T = 293$ K. The volume of the (spherical) basketball is $V = 4\pi r^3/3$, with $r = \frac{1}{2}(0.230 \text{ m}) = 0.115$ m.

1. Solve $PV = nRT$ for the number of moles, n:

$$n = \frac{PV}{RT}$$

2. Calculate the volume of the ball:

$$V = \frac{4\pi r^3}{3} = \frac{4\pi(0.115 \text{ m})^3}{3} = 0.00637 \text{ m}^3$$

3. Substitute the numerical values in Step 1:

$$n = \frac{PV}{RT} = \frac{(159 \times 10^3 \text{ Pa})(0.00637 \text{ m}^3)}{(8.31 \text{ J/(mol} \cdot \text{K)})(293 \text{ K})} = 0.416 \text{ mol}$$

Thus, a typical basketball contains slightly less than one-half mole of air.

The Mass of a Mole As mentioned before, a mole of anything has precisely the same number (N_A) of particles. What differs from substance to substance is the mass of one mole. For example, one mole of helium atoms has a mass of 4.00260 g, and one mole of copper atoms has a mass of 63.546 g. In general, we define the **atomic or molecular mass**, M, of a substance to be the mass in grams of one mole of that substance. Thus, the atomic mass for helium is $M = 4.00260$ g/mol and for copper it is $M = 63.546$ g/mol. The periodic table in Appendix E gives the atomic masses for all elements.

The atomic mass provides a convenient bridge between the macroscopic world, where we measure the mass of a substance in grams, and the microscopic world, where the number of molecules is typically 10^{23} or greater. As we have seen, if you measure out a mass of copper equal to 63.546 g, you have, in effect, counted out $N_A = 6.022 \times 10^{23}$ atoms of copper. It follows that the mass of an individual copper atom, m, is the atomic mass of copper divided by Avogadro's number; that is,

$$m = \frac{M}{N_A} \qquad \text{17-6}$$

We use this relationship to find the masses of an iron atom and a nitrogen molecule in the following Exercise.

EXERCISE 17-4 FIND THE MASS

Find the masses of **(a)** an atom of iron, Fe, and **(b)** a molecule of nitrogen, N_2. Atomic masses are listed in Appendix E.

REASONING AND SOLUTION

a. The atomic mass of iron is 55.85 g/mol (Appendix E). Thus, an iron atom has the mass

$$m_{Fe} = \frac{M}{N_A} = \frac{55.85 \text{ g/mol}}{6.022 \times 10^{23} \text{ atoms/mol}} = 9.274 \times 10^{-23} \text{ g/atom} = 9.274 \times 10^{-26} \text{ kg/atom}$$

b. The atomic mass of nitrogen is 14.01 g/mol (Appendix E), and hence the molecular mass of N_2 is $2(14.01 \text{ g/mol}) = 28.02$ g/mol. Therefore, the mass of an N_2 molecule is

$$m_{N_2} = \frac{M}{N_A} = \frac{28.02 \text{ g/mol}}{6.022 \times 10^{23} \text{ molecules/mol}} = 4.653 \times 10^{-23} \text{ g/molecule}$$
$$= 4.653 \times 10^{-26} \text{ kg/molecule}$$

Isotherms

Historically, the ideal-gas equation of state was arrived at piece by piece, as a result of the combined efforts of a number of researchers. For example, the English scientist Robert Boyle (1627–1691) established the fact that the pressure of a gas varies inversely

PROBLEM-SOLVING NOTE

Pressure and Volume in an Isotherm

If the temperature of an ideal gas is constant, the pressure and volume change in such a way that their product has a constant value.

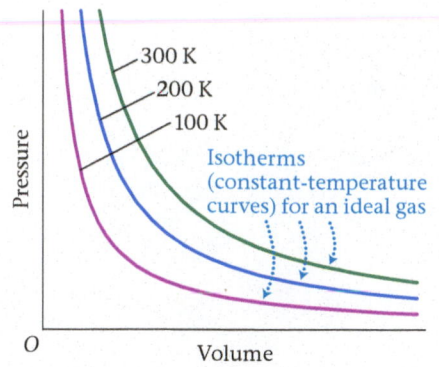

▲ FIGURE 17-5 Ideal-gas isotherms
Pressure-versus-volume isotherms for an ideal gas. Each isotherm is of the form $P = NkT/V = $ constant$/V$. The three isotherms shown are for the temperatures 100 K, 200 K, and 300 K.

with volume—as long as temperature and the number of molecules are held constant. This is known as **Boyle's law:**

$$P_i V_i = P_f V_f \qquad \text{17-7}$$

(fixed number of molecules, N; fixed temperature, T)

To see that Boyle's law is consistent with $PV = NkT$, notice that if N and T are constant, then so too is PV. Another way of saying that PV is a constant is to say that its initial value must be equal to its final value. This is Boyle's law.

In addition, when N and T are constant, the equation of state $PV = NkT$ implies that

$$P = \frac{NkT}{V} = \frac{\text{constant}}{V}$$

This result is plotted in **FIGURE 17-5**, where we show P as a function of V. The greater the temperature, T, the larger the constant in the numerator. Therefore, the curves farther from the origin correspond to higher temperatures, as indicated.

Each of the curves in Figure 17-5 corresponds to a different, fixed temperature. As a result, these curves are known as **isotherms,** which means, literally, "constant temperature." We shall return to the topic of isotherms in the next chapter, when we consider various thermal processes.

EXAMPLE 17-5 | UNDER PRESSURE

A cylindrical flask of cross-sectional area A is fitted with an airtight piston that is free to slide up and down. Contained within the flask is an ideal gas. Initially the pressure applied by the piston is 140 kPa and the height of the piston above the base of the flask is 22 cm. When additional mass is added to the piston, the pressure increases to 180 kPa. Assuming the system is always at the temperature 290 K, find the new height of the piston.

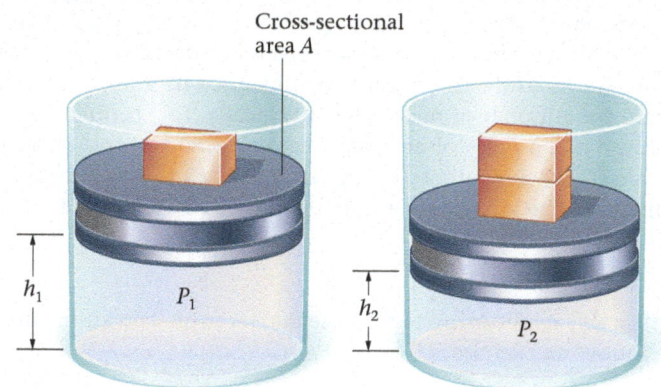

PICTURE THE PROBLEM
The physical system is shown in our sketch. The initial pressure in the flask is $P_1 = 140$ kPa, and the initial height of the piston is $h_1 = 22$ cm. When additional mass is placed on the piston, the pressure increases to $P_2 = 180$ kPa and the height decreases to h_2. The temperature remains constant.

REASONING AND STRATEGY
The temperature is held constant in this system, and hence it follows that $PV = NkT$ is also constant. As a result, $P_1 V_1 = P_2 V_2$, as in Boyle's law. This relationship can be used to find V_2 (since we are given P_1, P_2, and V_1).

Next, the cylindrical volume is related to the height by $V = Ah$. The area A is the same in both cases (and hence it will cancel), allowing us to solve for the height, h_2.

Known Initial pressure, $P_1 = 140$ kPa; initial height of piston, $h_1 = 22$ cm; final pressure, $P_2 = 180$ kPa; temperature, $T = 290$ K.
Unknown Final height of piston, $h_2 = $?

SOLUTION

1. Set the initial and final values of PV equal to one another, then solve for V_2:

$$P_1 V_1 = P_2 V_2$$
$$V_2 = V_1 \left(\frac{P_1}{P_2} \right)$$

2. Substitute $V_1 = Ah_1$ and $V_2 = Ah_2$, and solve for h_2:

$$Ah_2 = Ah_1 \left(\frac{P_1}{P_2} \right)$$
$$h_2 = h_1 \left(\frac{P_1}{P_2} \right)$$

3. Substitute numerical values:

$$h_2 = h_1 \frac{P_1}{P_2} = (22 \text{ cm}) \frac{140 \text{ kPa}}{180 \text{ kPa}} = 17 \text{ cm}$$

CONTINUED

► **FIGURE 17-6** Charles's law states that the volume of a gas at constant pressure is proportional to its Kelvin temperature. For example, this balloon was fully inflated at room temperature (293 K) and atmospheric pressure. When the balloon was cooled by vapors from liquid nitrogen (77 K), however, the volume of the air inside the balloon decreased markedly, causing it to shrivel.

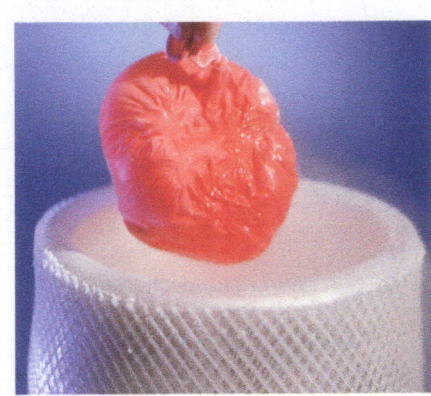

Finally, recall that in Conceptual Example 15-7 we considered the behavior of an air bubble rising from a swimmer in shallow water. At the time, we said that the diameter of the bubble increases as it rises because the pressure of the surrounding water is decreasing. This is certainly the case, but now we can be more precise. If we assume that the water temperature is constant, and that no gas molecules are added to or removed from the bubble, then the volume of the bubble has the following dependence:

$$V = \frac{NkT}{P} = \frac{\text{constant}}{P}$$

This is just our isotherm, again, and we see that the volume indeed increases as pressure decreases.

Constant Pressure

Another aspect of ideal-gas behavior was discovered by the French scientist Jacques Charles (1746–1823) and later studied in greater detail by fellow Frenchman Joseph Gay-Lussac (1778–1850). Known today as **Charles's law,** their result is that the volume of a gas divided by its temperature is constant, as long as the pressure and number of molecules are constant:

$$\frac{V_i}{T_i} = \frac{V_f}{T_f}$$ 17-8

(fixed number of molecules, N; fixed pressure, P)

As with Boyle's law, this result follows immediately from the ideal-gas equation of state. Solving $PV = NkT$ for V/T, we have

$$\frac{V}{T} = \frac{Nk}{P}$$

If N and P are constant, then so too is the quantity V/T.

Charles's law can be rewritten as a linear relationship between volume and temperature:

$$V = (\text{constant})T$$

The constant in this expression is Nk/P. This result is illustrated by the shrunken balloon in **FIGURE 17-6** and also in **FIGURE 17-7**, where we see that the volume of an ideal gas vanishes as the temperature approaches absolute zero.

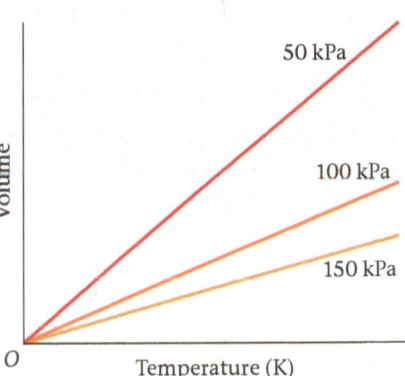

▲ **FIGURE 17-7 Volume versus temperature for an ideal gas at constant pressure** The temperature in this plot is the Kelvin temperature, T; hence $T = 0$ corresponds to absolute zero. At absolute zero the volume attains its lowest possible value—zero.

PROBLEM-SOLVING NOTE

Volume and Temperature in an Ideal Gas

If the pressure of an ideal gas is constant, the ratio of the volume and temperature remains constant. This is true no matter what the value of the pressure.

EXAMPLE 17-6   FIND THE VOLUME

A weather balloon is being prepared for launch. Initially it holds 6.86 m³ of hydrogen at 277 K. As pre-launch preparations continue in the morning sun, the temperature of the hydrogen rises to 288 K. What is the volume of the hydrogen at this new temperature? (The skin of the balloon is loose at launch, and has negligible effect on the pressure of the gas it contains.)

CONTINUED

PICTURE THE PROBLEM

Our sketch shows the weather balloon partially inflated before launch. (At high altitude, the balloon expands and eventually bursts.) The initial temperature of the gas in the balloon is T_1, and its initial volume is V_1.

REASONING AND STRATEGY

The number of molecules and the pressure in the enclosed gas remain the same, and hence the volume changes linearly with the absolute temperature. Another way of saying this is that V/T is constant. Therefore, we set the initial value of V/T equal to its final value. This allows us to solve for the final volume of the hydrogen in the balloon.

Known Initial temperature, $T_1 = 277$ K; initial volume, $V_1 = 6.86$ m³; final temperature, $T_2 = 288$ K.

Unknown Final volume, $V_2 = ?$

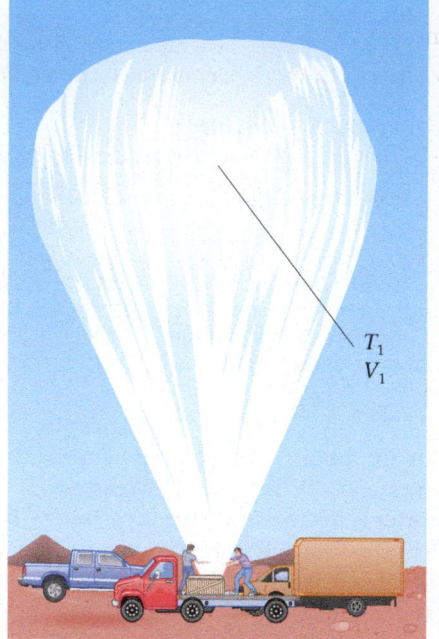

T_1
V_1

SOLUTION

1. Set the initial value of V/T equal to the final value of V/T:

$$\frac{V_1}{T_1} = \frac{V_2}{T_2}$$

2. Solve for the final volume V_2:

$$V_2 = V_1\left(\frac{T_2}{T_1}\right)$$

3. Substitute numerical values in Step 2:

$$V_2 = V_1\left(\frac{T_2}{T_1}\right) = (6.86 \text{ m}^3)\left(\frac{288 \text{ K}}{277 \text{ K}}\right) = 7.13 \text{ m}^3$$

INSIGHT

The volume of the gas increased by the same factor that the absolute temperature increased.

PRACTICE PROBLEM

At what temperature is the volume of the gas equal to 6.97 m³? **[Answer: 281 K]**

Some related homework problems: Problem 8, Problem 10

Enhance Your Understanding   (Answers given at the end of the chapter)

1. Rank the following ideal-gas systems in order of increasing pressure. Indicate ties where appropriate.

System	A	B	C	D
n (moles)	1	2	10	4
T (K)	300	350	300	350
V (m³)	1	5	5	10

Section Review

- The equation of state for an ideal gas is $PV = NkT$, where N is the number of molecules, T is the Kelvin temperature, and $k = 1.38 \times 10^{-23}$ J/K is Boltzmann's constant. Similarly, the equation of state can be written as $PV = nRT$, where n is the number of moles and $R = 8.31$ J/(mol·K) is the universal gas constant.

Big Idea 2 The pressure of a gas is due to the continuous bombardment of molecules against the wall of a container.

17-2 The Kinetic Theory of Gases

We can readily measure the pressure and temperature of a gas using a pressure gauge and a thermometer. These are *macroscopic* quantities that apply to the gas as a whole. It is not so easy, however, to measure *microscopic* quantities, such as the position or velocity of an individual molecule. Still, there must be some connection between what happens on the microscopic level and what we observe on the macroscopic level. This connection is described by the **kinetic theory of gases.**

In the kinetic theory, we imagine a gas to be made up of a collection of molecules moving about inside a container of volume V. In particular, we assume the following:

- The container holds a very large number N of identical molecules, each of mass m. The molecules themselves can be thought of as "point particles"—that is, the volume of the molecules is negligible compared to the volume of the container, and the size of the molecules is negligible compared to the distance between them. (This is why dilute real gases are the best approximation to the ideal case.)

- The molecules move about the container in a random manner. They obey Newton's laws of motion at all times.

- When molecules hit the walls of the container or collide with one another, they bounce elastically. Other than these collisions, the molecules have no interactions.

With these basic assumptions we can relate the pressure of a gas to the behavior of the molecules themselves.

The Origin of Pressure

As we shall see, the pressure exerted by a gas is due to the innumerable collisions between gas molecules and the walls of their container. Each collision results in a change of momentum for a given molecule, just like throwing a ball at a wall and having it bounce back. The total change in momentum of the molecules in a given time, divided by the time, is equal to the force a wall must exert on the gas to contain it (see Section 9-2). The average of this force over time, and over the area of a wall, is the pressure of the gas.

To be specific, imagine a container that is a cube of length L on a side. Its volume, then, is $V = L^3$. In addition, consider a given molecule that happens to be moving in the negative x direction toward a wall, as in **FIGURE 17-8**. If its speed is v_x, its initial momentum is $p_{i,x} = -mv_x$. After bouncing from the wall, it moves in the positive x direction with the same speed (since the collision is elastic), and hence its final momentum is $p_{f,x} = +mv_x$. As a result, the molecule's change in momentum is

$$\Delta p_x = p_{f,x} - p_{i,x} = mv_x - (-mv_x) = 2mv_x$$

The wall exerts a force on the molecule to cause this momentum change.

After the bounce, the molecule travels to the other side of the container and back before bouncing off the same wall again. The time required for this round trip of length $2L$ is

$$\Delta t = \frac{2L}{v_x}$$

Thus, by Newton's second law, the average force exerted by the wall on the molecule is

$$F = \frac{\Delta p_x}{\Delta t} = \frac{2mv_x}{2L/v_x} = \frac{mv_x^2}{L}$$

The average pressure exerted by this wall on a single molecule, then, is simply the force divided by the area. Because the area of the wall is $A = L^2$, we have

$$P_{molecule} = \frac{F}{A} = \frac{(mv_x^2/L)}{L^2} = \frac{mv_x^2}{L^3} = \frac{mv_x^2}{V} \qquad 17\text{-}9$$

Notice that we have used the fact that the volume of the container is $V = L^3$.

In this calculation we assumed that the molecule moves in the x direction. This was merely to simplify the derivation. If, instead, the molecule moves at some angle to the x axis, the calculation applies to its x component of motion. The final conclusions are unchanged.

Speed Distribution of Molecules

In deriving Equation 17-9 ($P_{molecule} = mv_x^2/V$), we considered a single molecule with a particular speed. Other molecules will have different speeds, of course. In addition,

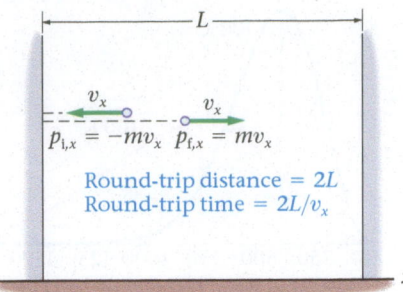

▲ **FIGURE 17-8 Force exerted by a molecule on the wall of a container**
A molecule bounces off a wall of a container, changing its momentum from $-mv_x$ to $+mv_x$; the change in momentum is $2mv_x$. A round trip will be completed in the time $\Delta t = 2L/v_x$, so the average force exerted on the molecule by the wall is $F = \Delta p/\Delta t = 2mv_x/(2L/v_x) = mv_x^2/L$.

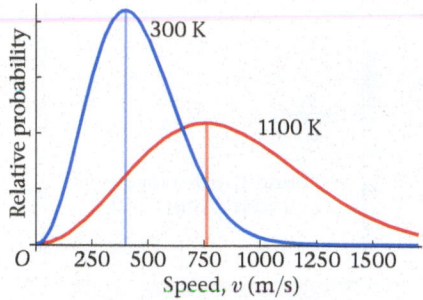

▲ **FIGURE 17-9 The Maxwell speed distribution** The Maxwell distribution of molecular speeds for O_2 at the temperatures $T = 300$ K and $T = 1100$ K. Note that the most probable speed increases with increasing temperature.

the speed of any given molecule changes with time as it collides with other molecules in the gas. What remains constant, however, is the overall **distribution of speeds.**

This is illustrated in **FIGURE 17-9**, where we present the results obtained by the Scottish physicist James Clerk Maxwell (1831–1879). This plot shows the relative probability that an O_2 molecule will have a given speed. For example, on the curve labeled 300 K, the most probable speed is about 390 m/s. When the temperature is increased to 1100 K, the most probable speed increases to roughly 750 m/s. Other speeds occur as well, from speeds near zero to those that are very large, but these have much lower probabilities.

Thus, in the expression $P_{molecule} = mv_x^2/V$, the term v_x^2 should be replaced with the average of v_x^2 over all the molecules in the gas, to take into the account the different speeds of different molecules. Writing this average as $(v_x^2)_{av}$, we have

$$P_{molecule} = \frac{m(v_x^2)_{av}}{V}$$

But all N molecules in the gas follow the same distribution, and hence the pressure exerted by the gas as a whole is N times the average for a single molecule:

$$P = NP_{molecule} = N\frac{m(v_x^2)_{av}}{V} = \left(\frac{N}{V}\right)m(v_x^2)_{av} \qquad 17\text{-}10$$

Of course, there's nothing special about the x direction—the result $P = (N/V)m(v_x^2)_{av}$ applies equally well with $(v_y^2)_{av}$ or $(v_z^2)_{av}$ in place of $(v_x^2)_{av}$. Thus, it would be preferable to express the pressure of the gas in terms of the overall speed of the molecules, rather than in terms of a single component. The speed squared of a molecule is

$$v^2 = v_x^2 + v_y^2 + v_z^2$$

Hence, the average of v^2 is

$$(v^2)_{av} = (v_x^2)_{av} + (v_y^2)_{av} + (v_z^2)_{av}$$

Noting that the x, y, and z directions are equivalent, it follows that

$$(v_x^2)_{av} = (v_y^2)_{av} = (v_z^2)_{av}$$

As a result,

$$(v^2)_{av} = (v_x^2)_{av} + (v_y^2)_{av} + (v_z^2)_{av} = 3(v_x^2)_{av}$$

Equivalently,

$$(v_x^2)_{av} = \tfrac{1}{3}(v^2)_{av}$$

Using this replacement in $P = (N/V)m(v_x^2)_{av}$ yields

$$P = \tfrac{1}{3}\left(\frac{N}{V}\right)m(v^2)_{av}$$

The last part of this expression, $m(v^2)_{av}$, is simply twice the average kinetic energy of a molecule. Thus, using K for the kinetic energy, as in Chapters 7 and 8, we can write the pressure as follows:

Pressure in the Kinetic Theory of Gases

$$P = \tfrac{1}{3}\left(\frac{N}{V}\right)2K_{av} = \tfrac{2}{3}\left(\frac{N}{V}\right)K_{av} = \tfrac{2}{3}\left(\frac{N}{V}\right)\left(\tfrac{1}{2}mv^2\right)_{av} \qquad 17\text{-}11$$

To summarize, using the kinetic theory, we have shown that the pressure of a gas is proportional to the number of molecules and inversely proportional to the volume. We discussed both of these dependences before when considering the ideal gas. In addition, we can make the following observation:

> The pressure of an ideal gas is directly proportional to the average kinetic energy of its individual molecules.

This is the key connection between microscopic behavior and macroscopic observables.

Kinetic Energy and Temperature

If we compare the ideal-gas equation of state, $PV = NkT$, with the result from kinetic theory, $P = \frac{2}{3}(N/V)\left(\frac{1}{2}mv^2\right)_{av}$, we find

$$PV = NkT = \frac{2}{3}N\left(\frac{1}{2}mv^2\right)_{av}$$

As a result, it follows that

$$\frac{2}{3}\left(\frac{1}{2}mv^2\right)_{av} = kT$$

Equivalently:

Kinetic Energy and Temperature

$$\left(\tfrac{1}{2}mv^2\right)_{av} = K_{av} = \tfrac{3}{2}kT \qquad \text{17-12}$$

This is one of the most important results of kinetic theory. It says that the average kinetic energy of a gas molecule is directly proportional to the Kelvin temperature, T. Thus, when we heat a gas, what happens on the microscopic level is that the molecules move with speeds that are, on average, greater. Similarly, cooling a gas causes the molecules to slow. This result also shows the fundamental significance of the Kelvin temperature scale.

EXERCISE 17-7 AVERAGE KINETIC ENERGY

Find the average kinetic energy of oxygen molecules in the air. Assume the air is at a temperature of 23 °C.

REASONING AND SOLUTION

Using $K_{av} = \frac{3}{2}kT$, and the fact that 23 °C = 296 K, we find

$$K_{av} = \tfrac{3}{2}kT = \tfrac{3}{2}(1.38 \times 10^{-23}\,\text{J/K})(296\,\text{K}) = 6.13 \times 10^{-21}\,\text{J}$$

This is also the average kinetic energy of *nitrogen* molecules in the air. In fact, the type of molecule is not important in determining the average kinetic energy; all that matters is the temperature of the gas.

Returning to Equation 17-12, $\left(\frac{1}{2}mv^2\right)_{av} = \frac{3}{2}kT$, we find with a slight rearrangement that

$$\left(v^2\right)_{av} = \frac{3kT}{m}$$

Now, the square root of $\left(v^2\right)_{av}$ is given a special name—it is called the **root mean square** (rms) speed. Specifically, the root mean square speed is the square *root* of the *mean* of the *square* of the speed. For gas molecules, then, the rms speed, v_{rms}, is the following:

RMS Speed of a Gas Molecule

$$v_{rms} = \sqrt{\left(v^2\right)_{av}} = \sqrt{\frac{3kT}{m}} \qquad \text{17-13}$$

SI unit: m/s

Rewriting this in terms of the molecular mass, M, we have

$$v_{rms} = \sqrt{\frac{3kT}{m}} = \sqrt{\frac{3kT}{(M/N_A)}} = \sqrt{\frac{3N_A kT}{M}}$$

Finally, using $N_A k = R$ yields

$$v_{rms} = \sqrt{\frac{3RT}{M}} \qquad \text{17-14}$$

The rms speed is one of the characteristic speeds of the Maxwell speed distribution. As shown in **FIGURE 17-10**, v_{rms} is slightly greater than the most probable speed, v_{mp}, and the average speed, v_{av}.

PHYSICS IN CONTEXT
Looking Back

Kinetic energy (Chapter 7) plays a central role in our understanding of an ideal gas. In particular, we can make a clear and specific connection between the mechanical concept of kinetic energy and the thermodynamic concept of temperature.

Big Idea **3** The average kinetic energy of a gas is proportional to the absolute temperature of the gas.

PHYSICS IN CONTEXT
Looking Ahead

The rms value of a quantity is an important consideration in Chapter 24, when we study alternating-current circuits.

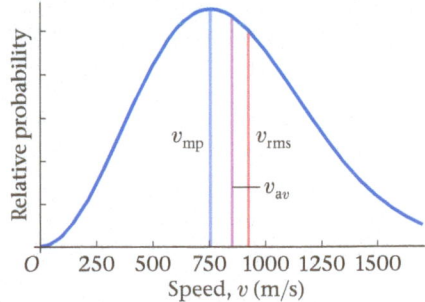

▲ **FIGURE 17-10 Most probable, average, and rms speeds** Characteristic speeds for O_2 at the temperature $T = 1100$ K. From left to right, the indicated speeds are the most probable speed, v_{mp}, the average speed, v_{av}, and the rms speed, v_{rms}.

EXAMPLE 17-8   FRESH AIR

Predict/Calculate

The atmosphere is composed primarily of nitrogen N_2 (78%) and oxygen O_2 (21%). **(a)** Is the rms speed of N_2 (28.0 g/mol) greater than, less than, or the same as the rms speed of O_2 (32.0 g/mol)? **(b)** Find the rms speed of N_2 and O_2 at 293 K.

PICTURE THE PROBLEM

Our sketch shows molecules in the air bouncing off the walls of a room—they also bounce off a person in the room. Notice that the molecules move in random directions with a variety of speeds. Most of the molecules are N_2, but about 21% are O_2.

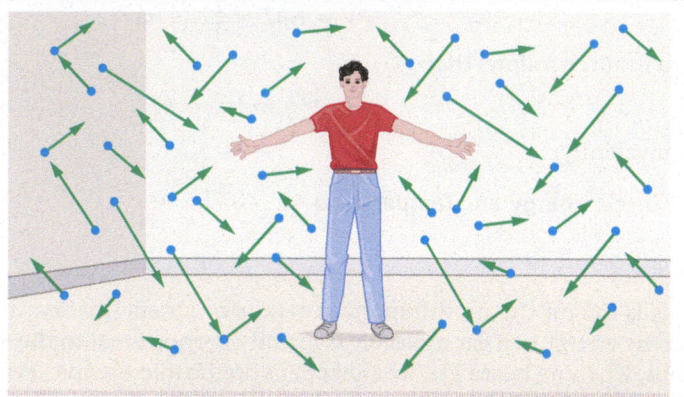

REASONING AND STRATEGY

The rms speeds are calculated by straightforward substitution into the equation $v_{rms} = \sqrt{3RT/M}$. The only point to be careful about is the molecular mass of each molecule. In particular, notice that for nitrogen the molecular mass is 28.0 g/mol = 0.0280 kg/mol, and that for oxygen the molecular mass is 32.0 g/mol = 0.0320 kg/mol.

Known   Molecular mass of N_2, M = 0.0280 kg/mol; molecular mass of O_2, M = 0.0320 kg/mol; temperature, T = 293 K.
Unknown The rms speed of N_2, v_{rms} = ? The rms speed of O_2, v_{rms} = ?

SOLUTION

Part (a)
Both molecules are at the same temperature, and hence they have the same kinetic energy. The nitrogen molecule has less mass, however; thus, if it is to have the same kinetic energy, it must have a higher speed. This is also what one would expect on the basis of $v_{rms} = \sqrt{3RT/M}$ because M is smaller for the nitrogen molecule.

Part (b)

1. To find the rms speed of N_2, substitute M = 0.0280 kg/mol in $v_{rms} = \sqrt{3RT/M}$:

$$v_{rms} = \sqrt{\frac{3[8.31 \text{ J/(mol} \cdot \text{K)}](293 \text{ K})}{0.0280 \text{ kg/mol}}} = 511 \text{ m/s}$$

2. To find the rms speed of O_2, substitute M = 0.0320 kg/mol in $v_{rms} = \sqrt{3RT/M}$:

$$v_{rms} = \sqrt{\frac{3[8.31 \text{ J/(mol} \cdot \text{K)}](293 \text{ K})}{0.0320 \text{ kg/mol}}} = 478 \text{ m/s}$$

INSIGHT

As expected, N_2 has the higher rms speed. For comparison, the speed of sound at 293 K is 343 m/s, which is just over 750 mi/h. Thus, the molecules in the air are bouncing off your skin with the speed of a supersonic jet.

The N_2 and O_2 molecules each contribute a *partial pressure* on your skin because of these collisions. The sum of these partial pressures, about 78 kPa for N_2 and 21 kPa for O_2, is the total pressure of the gas. This relationship is called *Dalton's law*. (See also Problems 43 and 44.)

PRACTICE PROBLEM

One of the substances found in air is monatomic and has an rms speed of 428 m/s, at 293 K. What is this substance? [**Answer:** M = 0.0399 kg/mol; thus the substance is argon, Ar, which comprises about 0.94% of the atmosphere.]

Some related homework problems: Problem 21, Problem 22

RWP* The speed of molecules in the air determines the average rate at which one molecule will collide with another molecule. For air at room temperature, the collision rate is roughly 1.6 billion collisions per second. The speed of molecules also limits the speed of sound in a gas, because sound is a longitudinal wave that depends on those collisions in order to propagate (see Section 14-4). In fact, the speed of sound is proportional to the square root of the temperature, and thus sound travels about 10% faster in air on a hot summer day (33°C) than it does on a bitter cold winter day (−18°C). In addition, the rate at which molecules collide with each other determines the rate at which chemical reactions can occur. This is why a cricket chirps at a faster rate with increasing temperature, making it a novel type of thermometer. (See the Passage Problems in Chapters 1 and 13, and Problem 75 in Chapter 16.)

---

*Real World Physics applications are denoted by the acronym RWP.

CONCEPTUAL EXAMPLE 17-9   COMPARE SPEEDS

Each of two containers of equal volume holds an ideal gas. Container A has twice as many molecules as container B. If the gas pressure is the same in the two containers, is the rms speed of the molecules in container A greater than, less than, or the same as the rms speed of the molecules in container B?

REASONING AND DISCUSSION

Because P and V are the same, you might think the rms speeds are the same as well. Recall, however, that *the pressure of a gas is caused by the collisions of gas molecules with the walls of the container*. If more molecules occupy a given volume, and they bounce off the walls with the same speed as in a container with fewer molecules, the pressure will be greater. Therefore, in order for the pressure in the two containers to be the same, it is necessary that the rms speed of the molecules in container A be less than the rms speed in container B.

Mathematically, we note that $P_A = P_B$ and $V_A = V_B$; hence $P_A V_A = P_B V_B$. Using the ideal-gas equation, $PV = NkT$, we can write this condition as $N_A k T_A = N_B k T_B$. Thus, if $N_A = 2N_B$, it follows that $T_A = T_B/2$. Because the temperature is lower in container A, its molecules have the lower rms speed.

ANSWER

The rms speed is less in container A.

The Internal Energy of an Ideal Gas

The internal energy of a substance is the sum of all its potential and kinetic energies. In an ideal gas there are no interactions between molecules, other than perfectly elastic collisions; hence, there is no potential energy. As a result, the total energy of the system is the sum of the kinetic energy of each of its molecules. Thus, for an ideal gas of N pointlike molecules—that is, a monatomic gas—the internal energy is simply:

Internal Energy of a Monatomic Ideal Gas

$$U = \tfrac{3}{2} NkT \qquad\qquad\qquad \text{17-15}$$

SI unit: J

In terms of moles, this result is

$$U = \tfrac{3}{2} nRT \qquad\qquad\qquad \text{17-16}$$

We shall return to this result in the next chapter.

EXERCISE 17-10   INTERNAL ENERGY

A basketball at 290 K holds 0.95 mol of air molecules. What is the internal energy of the air in the ball?

REASONING AND SOLUTION

Applying the relationship $U = \tfrac{3}{2} nRT$, we find

$$U = \tfrac{3}{2} nRT = \tfrac{3}{2}(0.95 \text{ mol})[8.31 \text{ J/(mol} \cdot \text{K)}](290 \text{ K}) = 3400 \text{ J}$$

This is roughly the kinetic energy a basketball would have if you dropped it from a height of 700 m.

If an ideal gas is diatomic, there are additional contributions to the internal energy. For example, a diatomic molecule, which is shaped somewhat like a dumbbell, can have rotational kinetic energy. Diatomic molecules can also vibrate along the line joining the two atoms, which is yet another contribution to the total energy. Thus, the result $U = \tfrac{3}{2} nRT$ applies only to the simplest case, the ideal monatomic gas.

Enhance Your Understanding   (Answers given at the end of the chapter)

2. If the Kelvin temperature of a gas is doubled, by what factor does the root mean square speed of the molecules increase?

CONTINUED

Section Review

- In the kinetic theory of gases, the pressure of a gas is viewed as the result of innumerable collisions of molecules bouncing off the walls of a container. The average kinetic energy of a monatomic molecule in an ideal gas is $\frac{3}{2}kT$.

17-3 Solids and Elastic Deformation

The defining characteristic of a solid object is that it has a particular shape. For example, when Michelangelo carved the statue of David from a solid block of marble, he was confident it would retain its shape long after his work was done. Even so, the shape of a solid *can* be changed—though usually only slightly—if it is acted on by a force. In this section we consider various types of deformations that can occur in solids and the way these deformations are related to the forces that cause them.

Changing the Length of a Solid

A useful model for a solid is a lattice of small balls representing molecules connected to one another by springs representing the intermolecular forces. Pulling on a solid rod with a force F, for example, causes each "intermolecular spring" in the direction of the force to expand by an amount proportional to F. The net result is that the entire solid increases its length by an amount proportional to the force, $\Delta L \propto F$, as shown in **FIGURE 17-11**.

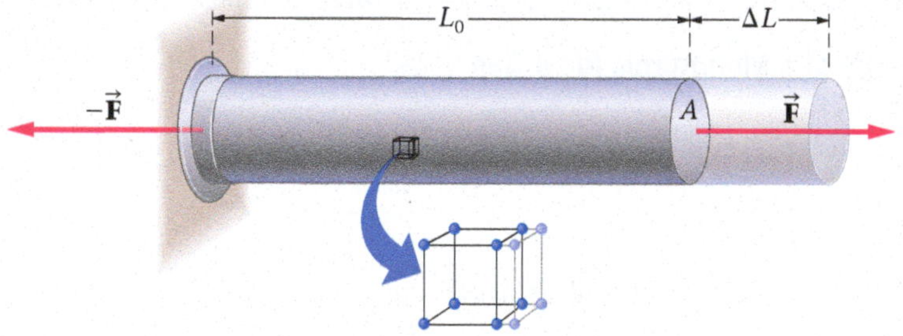

▸ **FIGURE 17-11 Stretching a rod** Equal and opposite forces applied to the ends of a rod cause it to stretch. On the atomic level, the forces stretch the "intermolecular springs" in the solid, resulting in an overall increase in length. The stretch is proportional to the force F and the initial length L_0, and inversely proportional to the cross-sectional area A.

TABLE 17-1 Young's Modulus for Various Materials

Material	Young's modulus, Y (N/m^2)
Tungsten	36×10^{10}
Steel	20×10^{10}
Copper	11×10^{10}
Brass	9.0×10^{10}
Aluminum	6.9×10^{10}
Pyrex glass	6.2×10^{10}
Lead	1.6×10^{10}
Bone	
Tension	1.6×10^{10}
Compression	0.94×10^{10}
Nylon	0.37×10^{10}

PROBLEM-SOLVING NOTE

Area and Young's Modulus

When using Young's modulus, remember that A is the area that is at right angles to the applied force.

The stretch ΔL is also proportional to the initial length of the rod, L_0. To see why, we first note that each intermolecular spring expands by the same amount, giving a total stretch that is proportional to the number of such springs. But the number of intermolecular springs is proportional to the total initial length of the rod. Thus, it follows that $\Delta L \propto FL_0$.

Finally, the amount of stretch for a given force F is inversely proportional to the cross-sectional area A of the rod. For example, a rod with a cross-sectional area $2A$ is like two rods of cross-sectional area A placed side by side. Thus, applying a force F to a rod of area $2A$ is equivalent to applying a force $F/2$ to two rods of area A. The result is half the stretch when the area is doubled; that is, $\Delta L \propto FL_0/A$. Solving this relationship for the amount of force F required to produce a given stretch ΔL, we find $F \propto (\Delta L/L_0)A$ or, as an equality,

$$F = Y\left(\frac{\Delta L}{L_0}\right)A \qquad 17\text{-}17$$

The proportionality constant Y in this expression is **Young's modulus,** named for the English physicist Thomas Young (1773–1829). Comparing the two sides of Equation 17-17, we see that Young's modulus has the units of force per area (N/m^2).

Typical values for Young's modulus are given in Table 17-1. Notice that the values vary from material to material, but are all rather large. This means that a large force is required to cause even a small stretch in a solid. Of course, Equation 17-17 applies equally well to a compression or a stretch. However, some materials have a slightly different Young's modulus for compression and stretching. For example, human bones under tension (stretching) have a Young's modulus of 1.6×10^{10} N/m^2, while bones under compression have a slightly smaller Young's modulus of 9.4×10^{9} N/m^2.

EXAMPLE 17-11 | STRETCHING A BONE

BIO At the local airport, a person carries a 21-kg duffel bag in one hand. Assuming the humerus (the upper arm bone) supports the entire weight of the bag, determine the amount by which the bone stretches. (The humerus may be assumed to be 33 cm in length and to have an effective cross-sectional area of 5.2×10^{-4} m^2.)

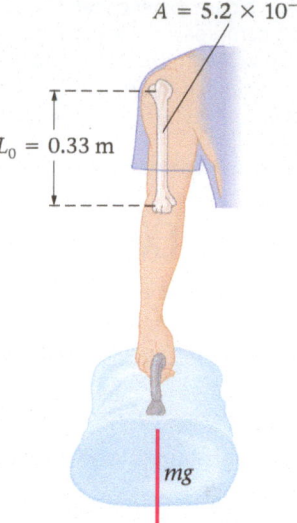

$A = 5.2 \times 10^{-4}$ m^2

$L_0 = 0.33$ m

mg

PICTURE THE PROBLEM
The humerus is oriented vertically, and hence the weight of the duffel bag applies tension to the bone. The Young's modulus in this case is 1.6×10^{10} N/m^2. In addition, we note that the initial length of the bone is $L_0 = 0.33$ m and its cross-sectional area is $A = 5.2 \times 10^{-4}$ m^2.

REASONING AND STRATEGY
We can find the amount of stretch by solving the relationship $F = Y(\Delta L/L_0)A$ for the quantity ΔL. The force applied to the bone is simply the weight of the duffel bag, $F = mg$, with $m = 21$ kg. (We ignore the relatively small weight of the forearm and hand.)

Known Mass of duffel bag, $m = 21$ kg; length of humerus, $L_0 = 0.33$ m; cross-sectional area of humerus, $A = 5.2 \times 10^{-4}$ m^2.
Unknown Change in length of humerus, $\Delta L = ?$

SOLUTION

1. Solve $F = Y(\Delta L/L_0)A$ for the amount of stretch, ΔL:

$$\Delta L = \frac{FL_0}{YA}$$

2. Calculate the force applied to the humerus:

$$F = mg = (21 \text{ kg})(9.81 \text{ m/s}^2) = 210 \text{ N}$$

3. Substitute numerical values into the expression for ΔL:

$$\Delta L = \frac{FL_0}{YA} = \frac{(210 \text{ N})(0.33 \text{ m})}{(1.6 \times 10^{10} \text{ N/m}^2)(5.2 \times 10^{-4} \text{ m}^2)}$$
$$= 8.3 \times 10^{-6} \text{ m}$$

INSIGHT
We find that the amount of stretch is imperceptibly small. The reason for this, of course, is that Young's modulus is such a large number. If the bone had been compressed rather than stretched, its change in length, though still minuscule, would have been greater by a factor of 16/9.4.

PRACTICE PROBLEM
Suppose the humerus is reduced uniformly in size by a factor of two. This means that both the length and diameter are halved. What is the amount of stretch in this case? [**Answer:** With $L_0 \rightarrow L_0/2$ and $A \rightarrow A/4$, it follows that the stretch is doubled. Thus, $\Delta L \rightarrow 2 \Delta L = 1.7 \times 10^{-5}$ m.]

Some related homework problems: Problem 31, Problem 32

There is a straightforward connection between $F = Y(\Delta L/L_0)A$ (Equation 17-17) and Hooke's law for a spring, $F = kx$ (Equation 6-5). To see the connection, we rewrite Equation 17-17 as

$$F = \left(\frac{YA}{L_0}\right)\Delta L$$

Notice that the force required to cause a certain stretch is proportional to the stretch—just as in Hooke's law. In fact, if we identify ΔL with the displacement x of a spring from equilibrium and YA/L_0 with the force constant k, we have Hooke's law:

$$F = \left(\frac{YA}{L_0}\right)\Delta L = kx$$

Thus, we see that the force constant of a spring depends on the Young's modulus, Y, of the material from which it is made, the cross-sectional area A of the wire, and the length of the wire, L_0. Two structures with changes in length due to applied forces are shown in **FIGURE 17-12**.

**PHYSICS
IN CONTEXT
Looking Back**

We gain a deeper understanding of Hooke's law (Chapter 6) in terms of the elastic deformation of solids.

▶ **FIGURE 17-12 Visualizing Concepts**
Changing the Length of a Solid (a) The cables in this suspension bridge have significant forces pulling on them from either end. As a result, they are under tension, just like the rod in Figure 17-11. It follows that the length of each cable increases by an amount that is proportional to its initial length. **(b)** This dam experiences forces that tend to compress it. The magnitude of its change in length can be described in terms of $F = Y(\Delta L/L_0)A$, with the appropriate value of Young's modulus.

(a)

(b)

CONCEPTUAL EXAMPLE 17-12   COMPARE FORCE CONSTANTS   PREDICT/EXPLAIN

Two identical springs are connected end to end. **(a)** Is the force constant of the resulting compound spring greater than, less than, or equal to the force constant of a single spring? **(b)** Which of the following is the *best explanation* for your prediction?

 I. The resulting force constant is greater because two springs are acting together, doubling the force.

 II. The two springs produce a longer compound spring—but the material is the same, and therefore the force constant is the same as well.

 III. The force constant is less for the compound spring because it stretches twice as far for the same applied force.

REASONING AND DISCUSSION
It might seem that the force constant would be the same, because we simply have twice the length of the same spring. On the other hand, it might seem that the force constant is greater, because two springs are exerting a force rather than just one. In fact, the force constant decreases.

The reason for the reduced force constant is that if we apply a force F to the compound spring, each individual spring stretches by a certain amount. The compound spring, then, stretches by twice this amount. By Hooke's law, $F = kx$, a spring that stretches twice as far for the same applied force has half the force constant.

We can also obtain this result by recalling that the force constant is $k = YA/L_0$. Thus, if the length of a spring is doubled—as in the compound spring—the force constant is halved.

ANSWER
(a) The force constant of the compound spring is less, by a factor of two. **(b)** The best explanation is III.

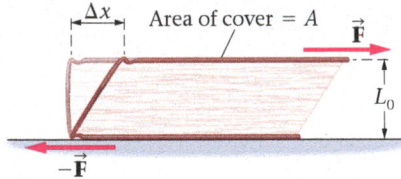

TABLE 17-2 Shear Modulus for Various Materials

Material	Shear modulus, S (N/m²)
Tungsten	15×10^{10}
Steel	8.1×10^{10}
Bone	8.0×10^{10}
Copper	4.2×10^{10}
Brass	3.5×10^{10}
Aluminum	2.4×10^{10}
Lead	0.54×10^{10}

◀ **FIGURE 17-13 Shear deformation** Equal and opposite forces applied to the top and bottom of a book result in a shear deformation. The amount of deformation is proportional to the force F and the thickness of the book L_0, and inversely proportional to the area A.

Changing the Shape of a Solid

Another type of deformation, referred to as a **shear deformation,** changes the *shape* of a solid. Consider a book of thickness L_0 resting on a table, as shown in **FIGURE 17-13**. A force F is applied to the right on the top cover of the book, and static friction applies a force F to the left on the bottom cover of the book. The result is that the book remains at rest but becomes slanted by the amount Δx. The force required to cause a given amount of slant is proportional to Δx, inversely proportional to the thickness of the book L_0, and proportional to the surface area A of the book's cover; that is, $F \propto A\Delta x/L_0$. Writing this as an equality, we have

$$F = S\left(\frac{\Delta x}{L_0}\right)A \qquad\qquad 17\text{-}18$$

The constant of proportionality in this case is the **shear modulus,** S. Like Young's modulus, the shear modulus has the units N/m². Typical values of the shear modulus are collected in Table 17-2. As with Young's modulus, the shear modulus is large in magnitude, meaning that most solids require a large force to cause even a small amount of shear.

The expressions $F = Y(\Delta L/L_0)A$ (Equation 17-17) and $F = S(\Delta x/L_0)A$ (Equation 17-18) are similar in structure, but it's important to be aware of their differences as well. For example, the term L_0 in the Young's modulus equation refers to the length of a solid measured in the direction of the applied force. In contrast, L_0 in the shear modulus equation refers to the thickness of the solid as measured in a direction perpendicular to the applied force. Similarly, the area A in Equation 17-17 is the cross-sectional area of the solid, and this area is perpendicular to the applied force. In contrast, the area A in Equation 17-18 is the area of the solid in the plane of the applied force.

PROBLEM-SOLVING NOTE

Area and the Shear Modulus

When using the shear modulus, remember that A is the area that is parallel to the applied force.

EXAMPLE 17-13   DEFORMING A STACK OF PANCAKES: FIND THE SHEAR MODULUS

A horizontal force of 1.3 N is applied to the top of a stack of pancakes 16 cm in diameter and 8.5 cm high. The result is a shear deformation of 2.4 cm. What is the shear modulus for these pancakes?

PICTURE THE PROBLEM

In our sketch, we see the stack of pancakes deformed by a force of magnitude $F = 1.3$ N to the right at the top of the stack and a force of equal magnitude to the left at the bottom of the stack. The result is a shear deformation of 2.4 cm.

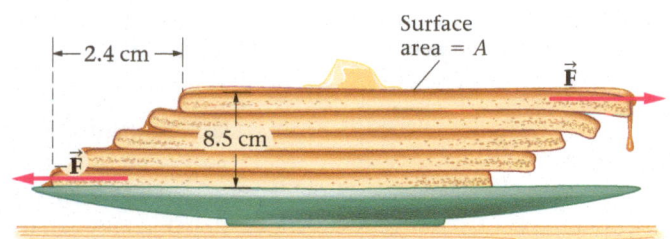

REASONING AND STRATEGY

We can solve $F = S(\Delta x/L_0)A$ for the shear modulus, S. The other quantities in the equation are $F = 1.3$ N, $\Delta x = 2.4$ cm, $A = \pi d^2/4$ with $d = 16$ cm, and $L_0 = 8.5$ cm.

Known   Horizontal force, $F = 1.3$ N; diameter of pancakes, $d = 0.16$ m; height of pancakes, $L_0 = 0.085$ m; shear of pancakes, $\Delta x = 0.024$ m.

Unknown   Shear modulus of pancakes, $S = ?$

SOLUTION

1. Solve $F = S(\Delta x/L_0)A$ for the shear modulus:

$$S = \frac{FL_0}{A\Delta x}$$

2. Calculate the area of the pancakes:

$$A = \frac{\pi d^2}{4} = \frac{\pi(0.16\,\text{m})^2}{4} = 0.020\,\text{m}^2$$

3. Substitute numerical values in Step 1:

$$S = \frac{FL_0}{A\Delta x} = \frac{(1.3\,\text{N})(0.085\,\text{m})}{(0.020\,\text{m}^2)(0.024\,\text{m})} = 230\,\text{N/m}^2$$

INSIGHT

Notice the small value of the pancakes' shear modulus, especially when compared to the shear modulus of a typical solid. This is a reflection of the fact that the pancake stack is easily deformed.

PRACTICE PROBLEM

Suppose the stack of pancakes is doubled in height, but everything else in the system remains the same. By what factor does the shear deformation change?   [**Answer:** The shear deformation doubles.]

Some related homework problems: Problem 33, Problem 81

Changing the Volume of a Solid

If a piece of Styrofoam is taken deep into the ocean, the tremendous pressure of the water causes it to shrink to a fraction of its original volume, as can be seen in **FIGURE 17-14**. This is an extreme example of the volume change that occurs in all solids when the pressure of their surroundings is changed. The general situation is illustrated in **FIGURE 17-15**, where we show a spherical solid whose volume decreases by the amount

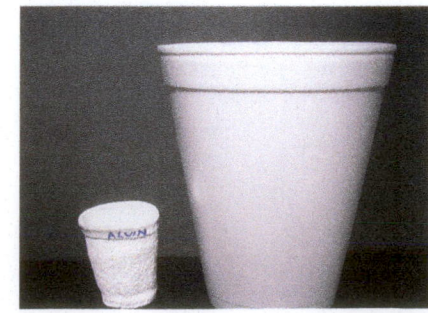

▶ **FIGURE 17-14** Styrofoam has a very small bulk modulus, which means that even a relatively small increase in pressure can cause a large decrease in volume. The Styrofoam cup at left was immersed in water to a depth of 1955 m, where the pressure was nearly 200 atm.

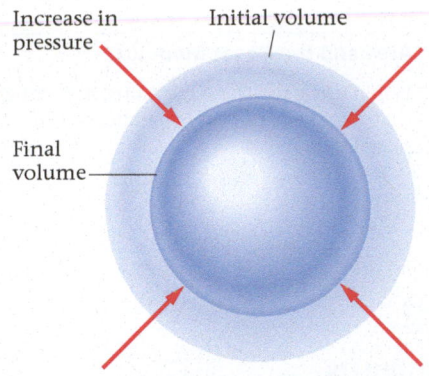

Increase in pressure

Initial volume

Final volume

◀ **FIGURE 17-15 Changing the volume of a solid** As the pressure surrounding an object increases, its volume decreases. The amount of volume change is proportional to the initial volume and to the change in pressure.

ΔV when the pressure acting on it increases by the amount ΔP. Experiments show that the pressure difference required to cause a given change in volume, ΔV, is proportional to ΔV and inversely proportional to the initial volume of the object, V_0. Therefore, we can write ΔP as follows:

$$\Delta P = -B\left(\frac{\Delta V}{V_0}\right) \qquad 17\text{-}19$$

The constant of proportionality in this case, B, is called the **bulk modulus.** Like Young's modulus and the shear modulus, the bulk modulus is a positive quantity with units of N/m^2.

Table 17-3 gives some representative values of the bulk modulus. Notice the large magnitudes of B, indicating that even small volume changes require large changes in pressure.

TABLE 17-3 Bulk Modulus for Various Materials

Material	Bulk modulus, B (N/m²)
Gold	22×10^{10}
Tungsten	20×10^{10}
Steel	16×10^{10}
Copper	14×10^{10}
Aluminum	7.0×10^{10}
Brass	6.1×10^{10}
Ice	0.80×10^{10}
Water	0.22×10^{10}
Oil	0.17×10^{10}

Stress and Strain

In general, we refer to an applied force per area as a **stress** and the resulting deformation as a **strain.** If the stress applied to an object is not too large, a proportional relationship between the strain and stress is found to hold, and a plot of strain versus stress gives a straight line, as indicated in **FIGURE 17-16.** This straight-line relationship is simply a generalization of Hooke's law—extending the result for a spring to any solid object. As the stress becomes larger, the strain eventually begins to increase at a rate that is greater than the straight line.

For stresses less than the elastic limit, the deformation of an object is reversible; that is, the deformation vanishes as the stress that caused it is reduced to zero. This is just like a spring returning to its equilibrium position when the applied force vanishes. When a deformation is reversible, we say that it is an **elastic deformation.** For stresses greater than the *elastic limit*, on the other hand, the object becomes permanently deformed, and it will not return to its original size and shape even when the stress is removed. The elastic limit is indicated in Figure 17-16. Exceeding the elastic limit is like a spring that has been stretched too far, or a car fender that has been dented. If the stress on an object is increased even further, the object eventually tears apart or fractures.

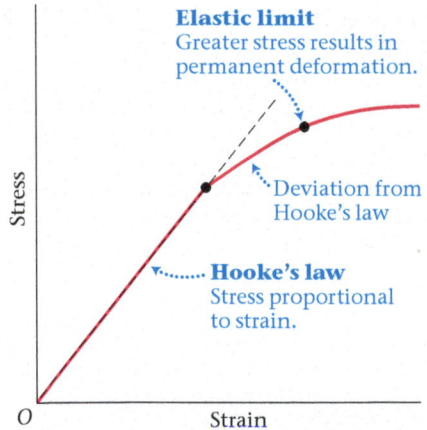

Elastic limit
Greater stress results in permanent deformation.

Deviation from Hooke's law

Hooke's law
Stress proportional to strain.

Stress

Strain

▲ **FIGURE 17-16 Stress Versus Strain** When stress and strain are small, they are proportional, as given by Hooke's law. Larger stresses can result in deviations from Hooke's law. Stresses greater than the elastic limit cause permanent deformation of the material.

Enhance Your Understanding   (Answers given at the end of the chapter)

3. A metal rod of a given initial length and cross-sectional area stretches when a force is applied to it. Rank the following rods in order of increasing amount of stretch. (All the rods are made of the same material.) Indicate ties where appropriate.

Rod	A	B	C	D
F (N)	5	10	50	100
L_0 (m)	1	0.5	2	1
A (10^{-4} m²)	4	10	30	5

Section Review

- When a force is applied to a solid, the result is a change in its length, shape, or volume.

- A force applied to a solid causes a change in length given by $F = Y(\Delta L/L_0)A$. Similar expressions apply to changes in shape and volume.

17-4 Phase Equilibrium and Evaporation

To this point, we have studied the behavior of substances when they are in a single phase of matter. For example, we considered the properties of liquids in Chapter 15, and in this chapter we turned our attention to the characteristics of gases and solids. It's common, however, for a substance to exist in more than one phase at a time. In particular, if a substance has two or more phases that coexist in a stable fashion, we say that the substance exhibits **phase equilibrium.**

Equilibrium Vapor Pressure   To see what phase equilibrium means in a specific case, consider a closed container that is partially filled with a liquid, as in **FIGURE 17-17**. The container is kept at the constant temperature T_0, and initially the volume above the liquid is empty of molecules—it is a vacuum. Soon, however, some of the faster molecules in the liquid begin to escape the relatively strong intermolecular forces of their neighbors in the liquid and start to form a low-density gas. Occasionally a gas molecule collides with the liquid and reenters it, but initially, more molecules are entering the gas than returning to the liquid.

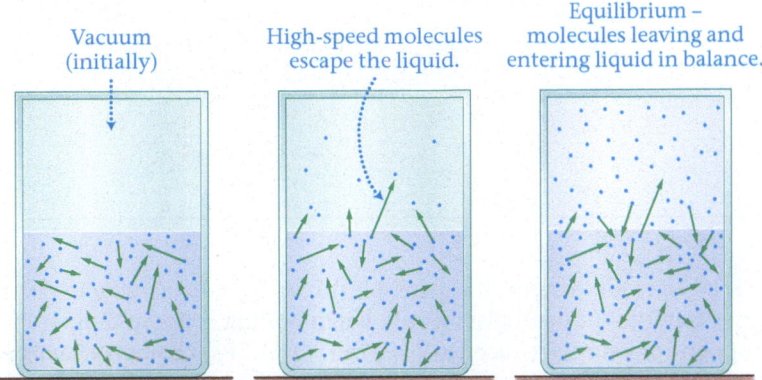

◀ **FIGURE 17-17 A liquid in equilibrium with its vapor** Initially, a liquid is placed in a container, and the volume above it is a vacuum. High-speed molecules in the liquid are able to escape into the upper region of the container, forming a low-density gas. As the gas becomes more dense, the number of molecules leaving the liquid is balanced by the number returning to the liquid. At this point the system is in equilibrium.

This process continues until the gas is dense enough that the number of molecules returning to the liquid equals the number entering the gas. There is a constant "flow" of molecules in both directions, but when *phase equilibrium* is reached, these flows cancel, and the number of molecules in each phase remains constant. The pressure of the gas when equilibrium is established is referred to as the **equilibrium vapor pressure.** In **FIGURE 17-18** we plot the equilibrium vapor pressure of water.

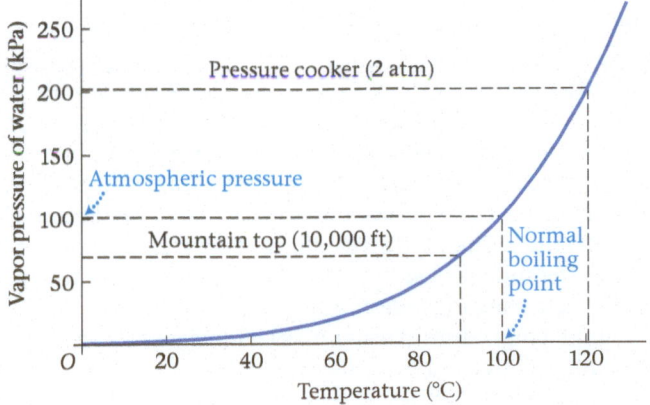

◀ **FIGURE 17-18 The vapor-pressure curve for water** The vapor pressure of water increases with increasing temperature. In particular, at $T = 100\,°C$ the vapor pressure is one atmosphere, 101 kPa.

What happens when we change the temperature? Well, if we increase the temperature, there will be more high-speed molecules in the liquid that can escape into the gas. Thus, to have an equal number of gas molecules returning to the liquid, it will be necessary for the pressure of the vapor to be greater. It follows that the equilibrium vapor pressure increases with temperature. This is also illustrated in Figure 17-18.

For each temperature there is just one equilibrium vapor pressure—that is, just one pressure where the precise balance between the phases is established. Thus, when we plot the equilibrium vapor pressure versus temperature, as in Figure 17-18, the result is a curve—the **vapor-pressure curve.** The significance of this curve is that it determines the boiling point of a liquid. In particular:

A liquid boils at the temperature at which its vapor pressure equals the external pressure.

In Figure 17-18, we can see that the vapor pressure is equal to atmospheric pressure, P_{at} = 101 kPa, when the temperature of the water is 100 °C, as expected.

CONCEPTUAL EXAMPLE 17-14   BOILING TEMPERATURE

When water boils at the top of a mountain, is its temperature greater than, less than, or equal to 100 °C?

REASONING AND DISCUSSION

At the top of a mountain, air pressure is less than it is at sea level. Therefore, according to Figure 17-18, the boiling temperature will be lower as well.

For example, the atmospheric pressure at the top of a 10,000-ft mountain is roughly three-quarters what it is at sea level. The corresponding boiling point, as shown in Figure 17-18, is about 90 °C. Thus, cooking food in boiling water may be very difficult at such altitudes, because the water really isn't that hot.

ANSWER

The temperature of the water is less than 100 °C.

▲ **FIGURE 17-19** Old Faithful geyser in Yellowstone National Park erupts on schedule.

BIO If taken to an extreme, the principle discussed in Conceptual Example 17-14 implies that water in an open glass would begin boiling at room temperature when the atmospheric pressure is reduced to 2350 Pa (2.3% of P_{at}). You may have heard that this effect would cause an astronaut's blood to boil if he or she were exposed to the vacuum of space, but that would not occur because the heart can maintain a high pressure inside a person's closed circulatory system. Humans who have been exposed to very low pressures lose consciousness in less than 30 s due to a lack of oxygen, but they recover quickly with no ill effects once air pressure is restored.

If, instead, the atmospheric pressure were greatly increased, the boiling temperature of water would greatly increase as well. An *autoclave* (which is basically an elaborate pressure cooker) sterilizes surgical tools at temperatures well above 100 °C by using this principle. In a kitchen pressure cooker, this elevated temperature results in a reduced cooking time.

This principle is closely linked to how *geysers* work. A geyser, like the one shown in **FIGURE 17-19**, is a naturally heated spring of water that erupts periodically because of its underground structure. Below the geyser lies a tall column of water (with an average depth of 2000 m) that has at least one constriction. The narrow part suppresses thermal convection and allows the deepest water to become superheated by nearby magma. When the deepest water finally begins to boil, the bubbling action reduces the pressure, which causes the surrounding water to suddenly become far above the boiling-point temperature. The explosive boiling event that follows can hurl water as high as 60 m above the surface opening. After the eruption, cool water again fills the water column and begins heating to restart the cycle.

RWP The boiling temperature of water may be increased by adding a pinch of salt. The salt dissolves into sodium and chloride ions in the water. These ions have a strong interaction with the water molecules, making it harder for them to break free of the liquid and enter the vapor phase. As a result, the boiling temperature rises. This is why salt is often added to water when boiling eggs; the result is a higher temperature of boiling, which helps to solidify any material that might leak out of small cracks in the eggs.

Phase Diagram The vapor-pressure curve separates areas of liquid and gas on a graph of pressure versus temperature, as shown in **FIGURE 17-20**. A curve similar to the vapor-pressure curve indicates where the solid and liquid phases are in equilibrium.

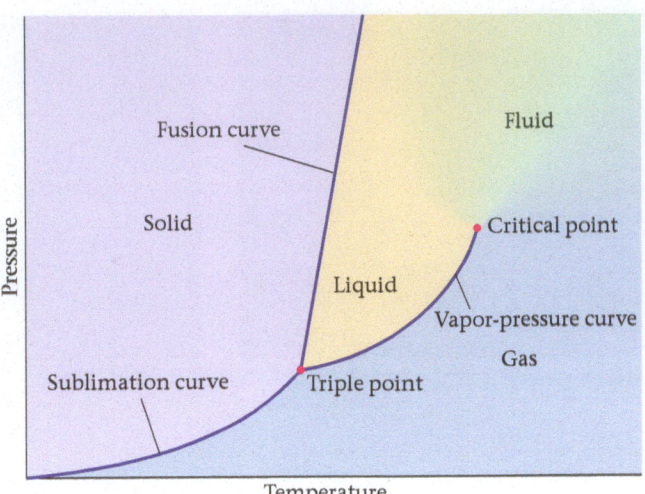

◄ **FIGURE 17-20 A typical phase diagram**
This phase diagram shows the regions corresponding to each of the three common phases of matter. Note that all three phases are in equilibrium at the triple point. In addition, the liquid and gas phases become indistinguishable beyond the critical point, where they are referred to as a fluid.

PHYSICS IN CONTEXT
Looking Ahead

The concept of a phase change is used in Chapter 21 when we discuss the fact that certain materials become superconducting (have zero electrical resistance) below a critical temperature.

PHYSICS IN CONTEXT
Looking Ahead

Phase changes appear again in Chapter 32 when we discuss the fundamental forces of nature and how they evolved as the universe cooled over time.

It is referred to as the **fusion curve.** Similarly, equilibrium between the solid and gas phases occurs along the **sublimation curve.** All three equilibrium curves are shown in Figure 17-20, on a plot that is referred to as a **phase diagram.**

Notice that the vapor-pressure curve comes to an end at a finite temperature and pressure. This end point is called the **critical point.** Beyond the critical point there is no longer a distinction between liquid and gas. They are one and the same, and are referred to, simply, as a fluid. Thus, as mentioned before, the liquid and gas phases are very similar—the liquid is just more dense than the gas and its molecules are somewhat less mobile.

The phase diagram also indicates that there is one particular temperature and pressure where all three phases are in equilibrium. This point is called the **triple point.** In water, the triple point occurs at the temperature $T = 273.16$ K and the pressure $P = 611.2$ Pa. At this temperature and pressure, ice, water, and steam are all in equilibrium with one another.

Finally, there is one feature of the fusion line that is of particular interest. In most substances, this line has a positive slope, as in Figure 17-20. This means that as the pressure is increased, the melting temperature of the substance also increases. This makes sense, because a solid is generally more dense than the corresponding liquid. Hence, if you apply pressure to a liquid—with the temperature held constant—the system will tend to become more dense and eventually solidify. This is indicated in **FIGURE 17-21 (a)**, where we see that an increase in pressure at constant temperature (as from A to B) results in crossing the fusion curve into the solid region.

RWP There are exceptions to this rule, however, and it will probably come as no surprise that water is one such exception. This is related to the fact that ice is less dense than water. As a result, the fusion curve for water has a negative slope coming out of the triple point. This is illustrated in **FIGURE 17-21 (b)**. What this means is that if the temperature is held constant, ice will melt when the pressure applied to it is increased by a large amount. This effect plays a role in how glaciers move: The tremendous pressure at the bottom of the glacier can melt the bottom layer of ice, providing a low-friction layer that helps the glacier to flow downhill. Similarly, the pressure exerted by an ice skate's blade can cause the ice beneath it to melt.

RWP The expansion of water as it freezes has important implications in geology as well. Suppose, for example, that water finds its way into cracks and fissures on a rocky cliff. If the temperature dips below freezing, the water in these cracks will begin to freeze and expand. This tends to split the rock farther apart in a process known as *frost wedging*. Over time, the repeated freezing and melting of water can break a "solid" rock cliff into a pile of debris, forming a talus slope at the base of the cliff. Similarly, repeated thermal expansion and contraction can result in fractured and cracked *patterned ground*. Examples on both Earth and Mars are shown in **FIGURE 17-22**.

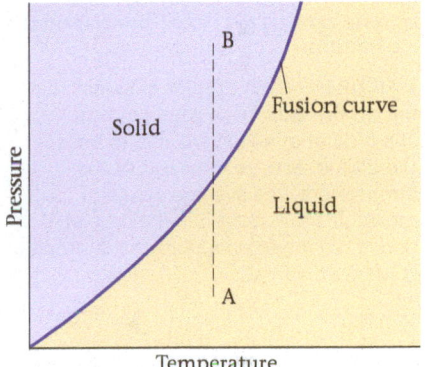

(a)

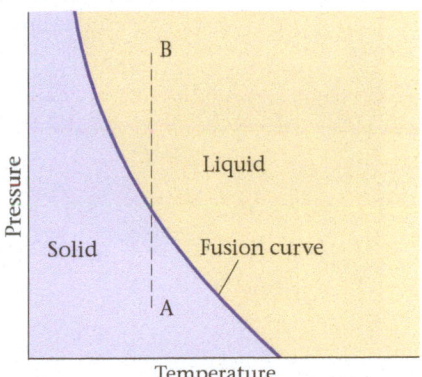

(b)

▲ **FIGURE 17-21 The solid-liquid phase boundary** Fusion curves for **(a)** a typical substance and **(b)** water. For the typical substance, we note that an increase in pressure at constant temperature results in a liquid being converted to a solid. For water, increasing the pressure exerted on ice at constant temperature can result in the ice melting.

▶ **FIGURE 17-22 Visualizing Concepts**
Thermal Expansion The repeated expansion and contraction of water as it freezes and thaws over long periods of time can cause the ground to crack and buckle. Eventually, the resulting cracks may combine to form polygonal networks referred to as patterned ground (left), a common feature in Arctic regions of the Earth. The same type of patterned ground has been observed on the surface of Mars (right). This, along with recent observations of dark streaking on hillsides, gives strong evidence that water exists just below the surface of Mars in certain parts of the planet.

High-speed molecules leave the droplet . . . . . . reducing its temperature . . .

. . . and causing heat to be drawn from the warm skin.

▲ **FIGURE 17-23 A droplet of sweat resting on the skin** High-speed molecules in a droplet of sweat are able to escape the droplet and become part of the atmosphere. The average speed of the molecules that remain in the droplet is reduced, so the temperature of the droplet is reduced as well.

▲ **FIGURE 17-24** Although the body temperature of this athlete is well below the boiling point of water, the water contained in his sweat is evaporating rapidly from his skin. (Since water vapor is invisible, the "steam" we see in this photo, though it indicates the presence of evaporation, is not the water vapor itself. Rather, it is a cloud of tiny droplets that form when the water vapor loses heat to the cold air around it and condenses back to the liquid state.)

BIO Finally, the expansion of ice can have devastating effects in living systems. If the blood cells of an organism were to freeze, for example, the resulting expansion could rupture the cells. To combat such effects, the crocodile icefish (*Chionodraco hamatus*), a small white-blooded fish found in southern polar seas, has a high concentration of glycoproteins in its bloodstream that serve as a biological antifreeze, inhibiting the formation and growth of ice crystals.

Evaporation

If you think about it for a moment, you realize that evaporation is a bit odd. After all, when you're hot and sweaty from physical exertion, steam rises from your skin. But how can this be, when water boils at 100 °C? How can a skin temperature of only 30 °C or 35 °C result in steam?

Recall that if a liquid is placed in a closed container with a vacuum above it, some of its fastest molecules will break loose and form a gas. When the gas becomes dense enough, its pressure rises to the equilibrium vapor pressure of the liquid, and the system attains equilibrium. But what if you open the container and let a breeze blow across it, removing much of the gas? In that case, the release of molecules from the liquid into the gas—that is, **evaporation**—continues without reaching equilibrium. As the molecules are continually removed, the liquid progressively evaporates until none is left. This is the basic mechanism of evaporation.

BIO Let's investigate how evaporation helps to cool us when we exercise or work up a sweat. First, consider a droplet of sweat on the skin, as illustrated in **FIGURE 17-23**. As mentioned before, the high-speed molecules in the drop are the ones that will escape from the liquid into the surrounding air. The breeze takes these molecules away as soon as they escape; hence, the chance of their reentering the drop of sweat is very small.

Now, what does this mean for the molecules that are left behind in the sweat droplet? Well, the droplet is preferentially losing high-speed molecules; therefore, the average kinetic energy of the remaining molecules must decrease. As we know from kinetic theory, this means that the temperature of the droplet must also decrease. Because the droplet is now cooler than its surroundings, including the skin on which it rests, it draws in heat from the body. This warms the droplet, increasing the speed of its molecules and continuing the evaporation process at more or less the same rate. Thus, sweat droplets are an effective means of drawing heat from the body and releasing it into the surrounding air in the form of high-speed water molecules. A dramatic example of evaporative cooling is shown in **FIGURE 17-24**, and two more examples are presented in **FIGURE 17-25**.

To look at this a bit more quantitatively, consider the Maxwell speed distribution for water molecules in a sweat droplet at 30 °C. This is shown in **FIGURE 17-26**. The rms speed at this temperature is 648 m/s. Also shown in Figure 17-26 is the speed distribution for water molecules at 100 °C. In this case the rms speed is 719 m/s, only slightly greater than that at 30 °C. Thus, it's clear that if water molecules at 100 °C have enough speed to escape into the gas phase, many molecules at 30 °C will be able to escape as well.

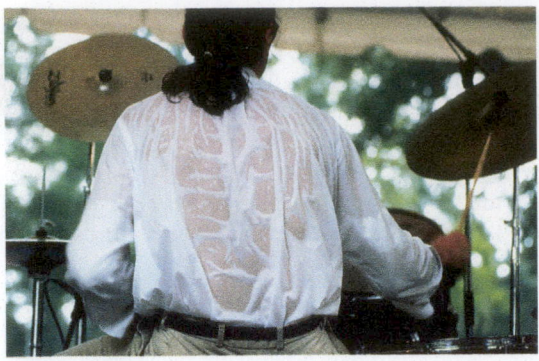

(a)   (b)

◄ **FIGURE 17-25 Visualizing Concepts**
Evaporative Cooling (a) Human beings keep cool by sweating. Because the most energetic molecules are the ones most likely to escape by evaporation, significant quantities of heat are removed from the body when perspiration evaporates from the skin. **(b)** Dogs, lacking sweat glands, nevertheless take advantage of the same mechanism to help regulate body temperature. In hot weather they pant to promote evaporation from the tongue. Of course, sitting in a cool pond can also help. (What mechanism is involved here?)

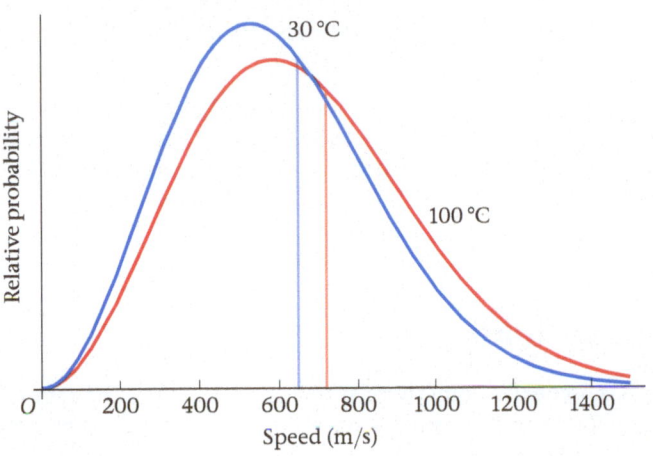

◄ **FIGURE 17-26 Speed distribution for water** The blue curve shows the speed distribution for water at 30 °C, and the red curve corresponds to 100 °C. Note that the rms speed for 100 °C is only slightly greater than that for 30 °C.

The Evaporating Atmosphere

RWP The atmosphere of a planet or a moon can evaporate in much the same way a drop of sweat evaporates from your forehead. In the case of an atmosphere, however, it is the force of gravity that must be overcome. If an astronomical body has a weak gravitational field, some molecules in its atmosphere may be moving rapidly enough to escape; that is, some molecules may have speeds in excess of the escape speed for that body.

Consider the Earth, for example. If a rocket is to escape the Earth, it must have a speed of at least 11,200 m/s. Recall from Chapter 12, however, that the escape speed is independent of the mass of the rocket. Thus, it applies equally well to molecules and rockets. As a result, molecules moving faster than 11,200 m/s may escape the Earth.

Now, let's compare the escape speed to the speeds of some of the molecules in the Earth's atmosphere. We have already seen in this chapter that the speeds of nitrogen and oxygen are on the order of the speed of sound—that is, several hundred meters per second. This is far below the Earth's escape speed. Thus, the odds against a nitrogen or oxygen molecule having enough speed to escape the Earth are truly astronomical. Good thing, too, because this is why these molecules have persisted in our atmosphere for billions of years.

On the other hand, consider a lightweight molecule like hydrogen, H_2. The fact that its average kinetic energy is the same as that of all the other molecules in the air (see Equation 17-12) means that its speed will be much greater than the speed of, say, oxygen or nitrogen. This is illustrated in **FIGURE 17-27**, where we show the speed distributions for both O_2 and H_2 at 20 °C. Clearly, it's very likely to find an H_2 molecule with a speed on the order of a couple thousand meters per second. Because of the higher speeds for H_2, the probability that an H_2 molecule will have enough speed to escape the Earth is about

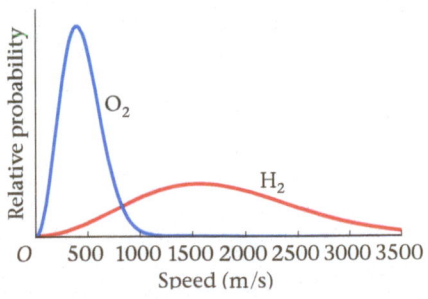

▲ **FIGURE 17-27 Speed distribution for O_2 and H_2 at 20 °C** The typical speeds of an H_2 molecule (red curve) are much greater than those for an O_2 molecule (blue curve). In fact, some H_2 molecules move fast enough to escape from the Earth's atmosphere.

300 orders of magnitude greater than the corresponding probability for O_2! It is no surprise, then, that Earth's atmosphere contains essentially no hydrogen.

On Jupiter, however, gravity is more intense than on Earth and the temperature is lower. As a result, not even hydrogen moves fast enough to escape. In fact, Jupiter's atmosphere is composed of mostly H_2 and He.

At the other extreme, the Moon has a rather weak gravitational field. In fact, it is unable to maintain any atmosphere at all. Whatever atmosphere it may have had early in its history has long since evaporated. You might say that the Moon's atmosphere is "lost in space."

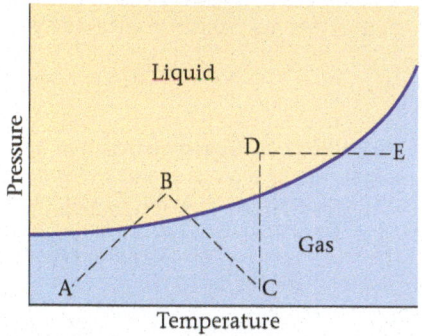

▲ **FIGURE 17-28** Enhance Your Understanding 4.

Enhance Your Understanding   (Answers given at the end of the chapter)

4. A portion of a substance's phase diagram is shown in **FIGURE 17-28**. **(a)** Is the curve in the figure the fusion curve, the vapor-pressure curve, or the sublimation curve? **(b)** Which of the processes (AB, BC, CD, DE) corresponds to heating the substance at constant pressure? **(c)** Which of the processes (AB, BC, CD, DE) corresponds to increasing the pressure of the substance at constant temperature?

Section Review

- When two phases of a substance are in equilibrium, the number of molecules in each phase remains constant.

17-5 Latent Heats

When two phases coexist, something surprising happens—the temperature remains the same even when you add a small amount of heat. How can that be?

To understand this behavior, let's start by considering an ice cube initially at a temperature of $-10\,°C$. Adding an amount of heat Q results in a temperature increase ΔT given by the relationship $Q = mc_{ice}\,\Delta T$, as discussed in Chapter 16 . When the ice cube's temperature reaches $0\,°C$, however, adding more heat does not cause an additional increase in temperature. Instead, the heat goes into converting some of the ice into water. On a microscopic level, the added heat causes some of the molecules in the solid ice to vibrate with enough energy to break loose from neighboring molecules and become part of the liquid.

Thus, as long as any ice remains in a cup of water, *and* the water and ice are in equilibrium, you can be sure that both the ice and the water are at $0\,°C$. If heat is added to the system, the amount of ice decreases; if heat is removed, the amount of ice increases. The amount of heat required to completely convert 1 kg of ice to water is referred to as the **latent heat,** L. In general, the latent heat is defined as follows:

> The latent heat, L, is the heat that must be added to or removed from one kilogram of a substance to convert it from one phase to another.
>
> During the conversion process, the temperature of the system remains constant.

Just as with the specific heat, the latent heat is always a positive quantity.

In mathematical form, we can say that the heat Q required to convert a mass m from one phase to another is $Q = mL$. This gives the following definition for the latent heat:

Definition of Latent Heat, L

$$L = \frac{Q}{m}$$

17-20

SI unit: J/kg

A Variety of Latent Heats   The value of the latent heat depends not only on which substance we're considering, but also on which phases are involved. For example, the latent heat to melt (or fuse) a substance is referred to as the **latent heat of fusion,**

Big Idea 4 When two phases of a substance are in equilibrium, the addition or removal of heat does not change the temperature of the substance. Instead, the amount of each phase in the system changes.

Big Idea 5 The amount of heat necessary to change one phase of a substance to another at constant temperature is the latent heat. The latent heat depends on the substance, and also on the two phases in question.

L_f. Similarly, the latent heat required to convert a liquid to a gas is the **latent heat of vaporization,** L_v, and the latent heat needed to convert a solid directly to a gas is the **latent heat of sublimation,** L_s. A novel example of a sublimating system on Mars is shown in **FIGURE 17-29.**

Typical latent heats for a variety of substances are given in Table 17-4.

TABLE 17-4 Latent Heats for Various Materials

Material	Latent heat of fusion, L_f (J/kg)	Latent heat of vaporization, L_v (J/kg)
Water	33.5×10^4	22.6×10^5
Ammonia	33.2×10^4	13.7×10^5
Copper	20.7×10^4	47.3×10^5
Benzene	12.6×10^4	3.94×10^5
Ethyl alcohol	10.8×10^4	8.55×10^5
Gold	6.28×10^4	17.2×10^5
Nitrogen	2.57×10^4	2.00×10^5
Lead	2.32×10^4	8.59×10^5
Oxygen	1.39×10^4	2.13×10^5

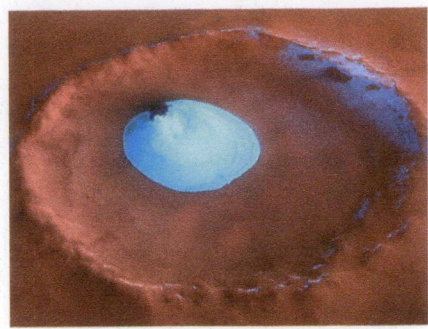

▲ **FIGURE 17-29** This recently discovered ice lake on the surface of Mars lies on the floor of an impact crater. It grows or shrinks with the Martian seasons. Because atmospheric pressure is so low on Mars, however, the ice does not melt during the Martian summer—instead, it sublimates directly to the vapor phase. On Mars, water ice would have behavior similar to that of dry ice here on Earth.

The relationship between the temperature of a substance (in this case water) and the heat added to it is illustrated in **FIGURE 17-30.** Initially 1 kg of H_2O is in the form of ice at $-20\,°C$. As heat is added to the ice, its temperature rises until it begins to melt at $0\,°C$. The temperature then remains constant until the latent heat of fusion is supplied to the system. When all the ice has melted to water at $0\,°C$, continued heating results in a renewed increase in temperature. When the temperature of the water rises to $100\,°C$, boiling begins and the temperature again remains constant—this time until an amount of heat equal to the latent heat of vaporization is added to the system. Finally, with the entire system converted to steam, continued heating again produces an increasing temperature.

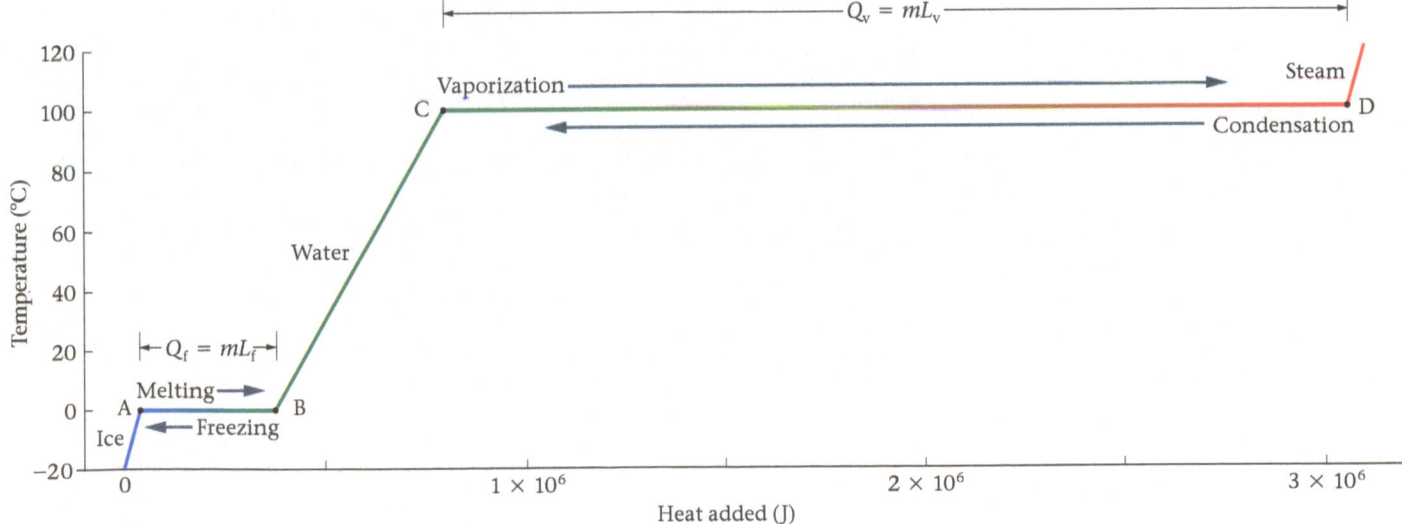

▲ **FIGURE 17-30 Temperature versus heat added or removed** The temperature of $m = 1.000$ kg of water as heat is added to or removed from the system. Note that the temperature stays the same—even as heat is added—when the system is changing from one phase to another. The points A, B, C, and D are referred to in Homework Problems 56 and 57.

CONCEPTUAL EXAMPLE 17-15   SEVERITY OF A BURN

Both water at 100 °C and steam at 100 °C can cause serious burns. Is a burn produced by steam likely to be more severe, less severe, or the same severity as a burn produced by water?

REASONING AND DISCUSSION

As the water or steam comes into contact with the skin, it cools from 100 °C to a skin temperature of something like 35 °C. For the case of water, this means that a certain amount of heat is transferred to the skin, which can cause a burn. The steam, on the other hand, must first give off the heat required for it to condense to water at 100 °C. After that, the condensed water cools to body temperature, as before. Thus, the heat transferred to the skin will be greater in the case of steam, resulting in a more serious burn.

ANSWER

The steam burn is more severe.

▶ **FIGURE 17-31 Heat required for a given change in temperature** The amount of heat required to raise 0.550 kg of H_2O from ice at −20.0 °C to water at 20.0 °C is the heat difference between points A and D. To calculate this heat we sum the following three heats: (1) the heat to warm the ice from −20.0 °C to 0 °C; (2) the heat to melt all the ice; (3) the heat to warm the water from 0 °C to 20.0 °C.

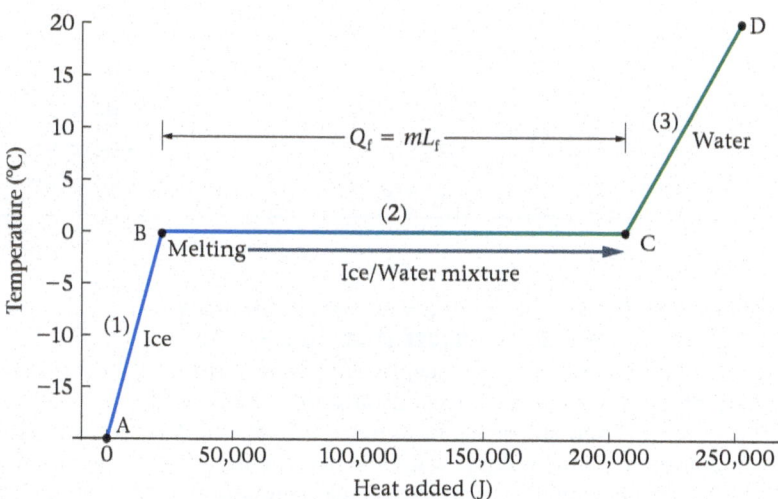

Practical Applications of Latent Heats and Specific Heats

As a numerical example of using latent heat, let's calculate the heat energy required to raise the temperature of 0.550 kg of ice from −20.0 °C to water at 20.0 °C—that is, from point A to point D in **FIGURE 17-31**. The way to approach a problem like this is to take it one step at a time—that is, each phase, and each conversion from one phase to another, should be treated separately.

Thus, the first step (A to B in Figure 17-31) is to find the heat necessary to warm the ice from −20.0 °C to 0 °C. Using the specific heat of ice, $c_{ice} = 2090 \, J/(kg \cdot C°)$, we find

$$Q_1 = mc_{ice} \, \Delta T = (0.550 \text{ kg})[2090 \, J/(\text{kg} \cdot C°)](20.0 \, C°) = 23,000 \, J$$

The second step in the process (B to C) is to melt the ice at 0 °C. The heat required for this is found using the latent heat of fusion for water ($L_f = 33.5 \times 10^4 \, J/kg$):

$$Q_2 = mL_f = (0.550 \text{ kg})(33.5 \times 10^4 \, J/kg) = 184,000 \, J$$

Finally, the third step (C to D) is to heat the water at 0 °C to 20.0 °C. This time we use the specific heat for water, $c_{water} = 4186 \, J/(kg \cdot C°)$:

$$Q_3 = mc_{water} \, \Delta T = (0.550 \text{ kg})[4186 \, J/(\text{kg} \cdot C°)](20.0 \, C°) = 46,000 \, J$$

The total heat required for this process, then, is

$$Q_{total} = Q_1 + Q_2 + Q_3$$
$$= 23,000 \, J + 184,000 \, J + 46,000 \, J = 253,000 \, J$$

We consider a similar problem in the following Example.

EXAMPLE 17-16 | STEAM HEAT

To make steam, you add 5.60×10^5 J of heat to 0.220 kg of water at an initial temperature of 50.0 °C. Find the final temperature of the steam.

PICTURE THE PROBLEM

Our sketch shows the temperature versus the heat added for water. The initial point for this system, placed at the origin, is at 50.0 °C. As we shall see, adding the given amount of heat, $Q = 5.60 \times 10^5$ J, raises the temperature to the point labeled "Final" in the plot.

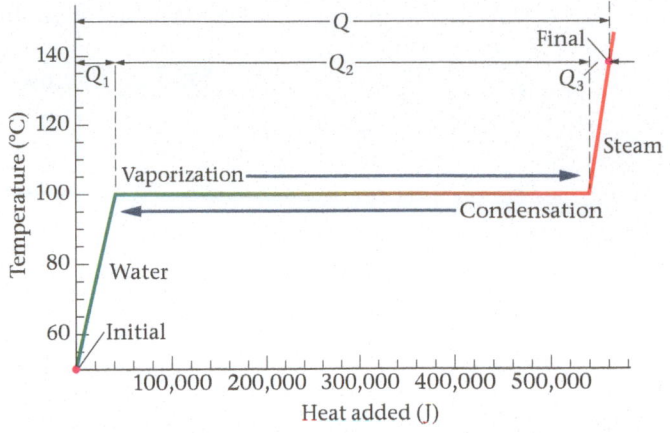

REASONING AND STRATEGY

To find the final temperature, we first calculate the amount of heat that must be added to heat the water to 100 °C. If this is less than the total heat added, we continue by calculating the amount of heat needed to vaporize all the water. If the sum of these two heats is still less than the total heat added to the water, we calculate the increase in temperature when the remaining heat is added to the steam.

Known Heat added, $Q = 5.60 \times 10^5$ J; mass of water, $m = 0.220$ kg; initial temperature of water, $T_i = 50.0$ °C.
Unknown Final temperature of water, $T_f = ?$

SOLUTION

1. Calculate the heat that must be added to the water to heat it to 100 °C. Call the result Q_1:

$$Q_1 = mc_{water} \, \Delta T$$
$$= (0.220 \text{ kg})[4186 \text{ J/(kg} \cdot \text{C°)}](50.0 \text{ C°})$$
$$= 4.60 \times 10^4 \text{ J}$$

2. Next, calculate the heat that must be added to the water to convert it to steam. Let this result be Q_2:

$$Q_2 = mL_f = (0.220 \text{ kg})(22.6 \times 10^5 \text{ J/kg})$$
$$= 4.97 \times 10^5 \text{ J}$$

3. Determine the heat that is still to be added to the system. Let this remaining heat be Q_3:

$$Q_3 = 5.60 \times 10^5 \text{ J} - Q_1 - Q_2$$
$$= 5.60 \times 10^5 \text{ J} - 4.60 \times 10^4 \text{ J} - 4.97 \times 10^5 \text{ J}$$
$$= 17,000 \text{ J}$$

4. Use Q_3 to find the increase in temperature of the steam:

$$Q_3 = mc_{steam} \, \Delta T$$
$$\Delta T = \frac{Q_3}{mc_{steam}} = \frac{17,000 \text{ J}}{(0.220 \text{ kg})[(2010 \text{ J})/(\text{kg} \cdot \text{C°})]} = 38 \text{ C°}$$

5. The final temperature is 100 °C plus the increase in temperature found in Step 4:

$$T_f = 100 \text{ °C} + 38 \text{ C°} = 138 \text{ °C}$$

INSIGHT

Thus, the system ends up completely converted to steam at a temperature of 138 °C. If the amount of heat added to the system had been greater than Q_1 but less than $Q_1 + Q_2$, the final temperature of the system would have been 100 °C. In this case, the final state of the system is a mixture of liquid water and steam.

PRACTICE PROBLEM

Find the final temperature if the amount of heat added to the system is 3.40×10^5 J. [**Answer:** Only 0.130 kg of water vaporizes into steam. Hence the final temperature is 100 °C, with water and steam in equilibrium.]

Some related homework problems: Problem 49, Problem 54

RWP A pleasant application of latent heat is found in the making of homemade ice cream. As you may know, it is necessary to add salt to the ice-water mixture surrounding the container holding the ingredients for the ice cream. The dissolved salt molecules interact with water molecules in the liquid, impairing their ability to interact with one another and freeze. This means that ice and water are no longer in equilibrium at 0 °C; a lower temperature is required. The result is that ice begins to melt in the ice-water mixture, and, in the process of melting, the ice draws the required latent heat from its surroundings—which includes the ice cream. Thus, the salt together with the ice produces a temperature lower than the melting temperature of ice alone.

Enhance Your Understanding   (Answers given at the end of the chapter)

5.   Which requires more heat: melting 100 kg of copper, or boiling 10 kg of water?

Section Review

* Latent heat is the amount of heat needed to change one phase of a substance to another phase at constant temperature. Latent heat depends on the substance, and also on the phases in question.

17-6 Phase Changes and Energy Conservation

In the last section, we considered problems in which a given amount of heat is simply added to or removed from a system. We now turn to a more interesting type of problem involving energy conservation. In these problems, heat is exchanged within a system—that is, between its parts—but not with the external world. As a result, the total energy of the system is constant. Still, the heat flow within the system can cause changes in temperatures and phases.

The basic idea in solving energy-conservation problems is the following:

> Set the magnitude of the heat *lost* by one part of the system equal to the magnitude of the heat *gained* by another part of the system.

For example, consider the following problem: A 0.0420-kg ice cube at $-10.0\,°C$ is placed in a Styrofoam cup containing 0.350 kg of water at $0\,°C$. Assuming that the cup can be ignored, and that no heat is exchanged with the surroundings, find the mass of ice in the system when the equilibrium temperature of $0\,°C$ is reached.

Applying Energy Conservation   The initial setup for this problem is illustrated in **FIGURE 17-32 (a)**. Because the water is warmer than the ice, it follows that heat flows into the ice from the water, as indicated in **FIGURE 17-32 (b)**. The amount of heat will be just enough to raise the temperature of the ice from $-10.0\,°C$ to $0\,°C$. Thus,

$$Q = \text{heat gained by the ice}$$
$$= mc_{\text{ice}}\,\Delta T = (0.0420\ \text{kg})[2090\ \text{J/(kg}\cdot\text{C°)}](10.0\ \text{C°}) = 878\ \text{J}$$

By energy conservation, we can say that the heat lost by the water has the same magnitude as the heat gained by the ice. (In general, in these problems it is simplest to calculate the magnitude of each heat—so that all the heats are positive—and then set the "lost" and "gained" magnitudes equal.) In this case, then, we have

$$\text{heat lost by the water} = \text{heat gained by the ice} = 878\ \text{J}$$

Thus, 878 J of heat are removed from the water.

PROBLEM-SOLVING NOTE

Determining Equilibrium

In problems involving two different phases—like solid and liquid—it may not be clear in advance whether both phases or only one is present in the equilibrium state. It may be necessary to assume one case or the other and proceed with the calculation based on that assumption. If the result obtained in this way is not physical, the assumption must be changed.

The heat Q that leaves the water . . .

. . . causes some of it to freeze, enlarging the ice cube.

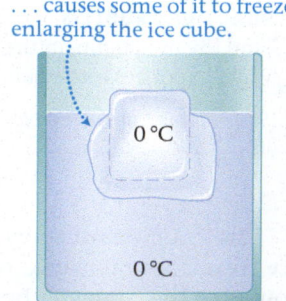

▶ **FIGURE 17-32 Water freezing to form ice** An ice cube at $-10.0\,°C$ is placed in water at $0\,°C$. Since the ice is colder than the water, heat flows *from* the water *into* the ice. Heat that leaves the water results in the formation of additional ice in the system.

(a) Initial

(b) Heat goes from the warmer water into the colder ice

(c) Final

Now, noting that the water is already at 0 °C, removing heat from it does not lower its temperature—instead, it merely converts some of the water to ice. How much is converted? The amount is determined by the latent heat of fusion. In particular, we have

$$Q = \text{heat lost by the water}$$
$$= mL_f = 878 \text{ J}$$

In this expression, m is the mass of water that has been converted to ice. Solving for the mass yields

$$m = \frac{Q}{L_f} = \frac{878 \text{ J}}{33.5 \times 10^4 \text{ J/kg}} = 0.00262 \text{ kg}$$

Thus, the final amount of ice in the system is 0.0420 kg + 0.00262 kg = 0.0446 kg. This is illustrated in **FIGURE 17-32 (c)**.

Finally, in the next Example we consider a system in which a quantity of ice completely melts in warm water. The result is a container of liquid water at an intermediate temperature. Still, the basic idea is to apply energy conservation.

EXAMPLE 17-17   WARM PUNCH

A large punch bowl holds 3.95 kg of lemonade (which is essentially just water) at 20.0 °C. A 0.0450-kg ice cube at 0 °C is placed in the lemonade. What is the final temperature of the system, and how much ice (if any) remains when the system reaches equilibrium? Ignore any heat exchange with the bowl or the surroundings.

PICTURE THE PROBLEM
Our sketch shows the various stages imagined for this problem, starting with the initial condition in which the ice is at 0 °C and the lemonade (water) is at 20.0 °C. As one might guess from the large amount of water and the small amount of ice, all of the ice will melt. Therefore, the final condition is a container of liquid water at a final temperature, T_f.

0 °C
20.0 °C
Initial condition

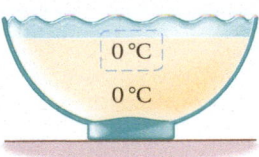
0 °C
Calculate heat Q_w lost by water, 20 °C → 0 °C

0 °C
0 °C
Calculate heat Q_{ice} needed to melt ice cube

0 °C
Calculate remaining heat $Q = Q_w - Q_{ice}$

T_f
Calculate final temperature when heat Q is added to water and melted ice

REASONING AND STRATEGY
To apply energy conservation to this problem, we first calculate the heat that would be lost by the water, Q_w, if we cooled it to 0 °C. We then imagine using part of this heat, Q_{ice}, to melt the ice.

As we shall see, a great deal of heat, $Q = Q_w - Q_{ice}$, is left after the ice is melted. We imagine adding this heat back into the system, which now contains 3.95 kg + 0.0450 kg of water at 0 °C. Calculating the increase in temperature caused by the heat Q gives us the final temperature of the system.

Known   Initial mass of water, m_{water} = 3.95 kg; initial temperature of water, 20.0 °C; mass of ice cube, m_{ice} = 0.0450 kg; initial temperature of ice, 0 °C.
Unknown   Final temperature of system, T_f = ?

SOLUTION

1. Find the heat lost by the water, Q_w, if it is cooled to 0 °C:

$$Q_w = m_{water} c_{water} \Delta T$$
$$= (3.95 \text{ kg})[4186 \text{ J/(kg} \cdot \text{C°})](20.0 \text{ C°})$$
$$= 3.31 \times 10^5 \text{ J}$$

2. Calculate the amount of heat, Q_{ice}, needed to melt all the ice:

$$Q_{ice} = m_{ice} L_f$$
$$= (0.0450 \text{ kg})(33.5 \times 10^4 \text{ J/kg}) = 1.51 \times 10^4 \text{ J}$$

3. Determine the amount of heat, Q, that is left:

$$Q = Q_w - Q_{ice}$$
$$= 3.31 \times 10^5 \text{ J} - 1.51 \times 10^4 \text{ J} = 3.16 \times 10^5 \text{ J}$$

CONTINUED

4. Use the remaining heat Q to warm the 3.95 kg + 0.0450 kg of water at 0 °C to its final temperature:

$$Q = (m_{water} + m_{ice})c_{water}\,\Delta T$$

$$\Delta T = \frac{Q}{(m_{water} + m_{ice})c_{water}}$$

$$= \frac{3.16 \times 10^5\,\text{J}}{(3.95\,\text{kg} + 0.0450\,\text{kg})[4186\,\text{J}/(\text{kg}\cdot\text{C}°)]}$$

$$= 18.9\,\text{C}°$$

5. The final temperature is 0 °C plus the change in temperature found in Step 4:

$$T_f = 0\,°\text{C} + 18.9\,\text{C}° = 18.9\,°\text{C}$$

INSIGHT

As expected, the relatively small ice cube did not lower the temperature of the system very much. In fact, the temperature of the lemonade (water) decreased only from 20.0 °C to 18.9 °C.

PRACTICE PROBLEM

What would the final temperature of the system be if the ice cube's mass were 0.0750 kg? [**Answer:** $T_f = 18.2\,°\text{C}$]

Some related homework problems: Problem 60, Problem 64

Enhance Your Understanding   (Answers given at the end of the chapter)

6. An ice cube is placed in a cup of water. A few minutes later half the ice has melted, and the remainder is in equilibrium with the water. **(a)** Was the initial temperature of the water greater than, less than, or equal to 0 °C? Explain. **(b)** Is the final temperature of the water greater than, less than, or equal to 0 °C? Explain.

Section Review

• When heat is exchanged between various parts of a closed system, the amount of each phase may change, but the total energy of the system remains constant.

CHAPTER 17 REVIEW

CHAPTER SUMMARY

17-1 IDEAL GASES

An ideal gas is a simplified model of a real gas in which interactions between molecules are ignored.

Equation of State

The equation of state for an ideal gas is

$$PV = NkT \qquad\qquad 17\text{-}2$$

In this expression, N is the number of molecules, T is the Kelvin temperature, and $k = 1.38 \times 10^{-23}\,\text{J/K}$ is Boltzmann's constant.

In terms of the universal gas constant, $R = 8.31\,\text{J}/(\text{mol}\cdot\text{K})$, and the number of moles in the gas, n, the ideal-gas equation of state is

$$PV = nRT \qquad\qquad 17\text{-}5$$

Avogadro's Number and Moles

The number of molecules in a mole (mol) is Avogadro's number, $N_A = 6.022 \times 10^{23}$.

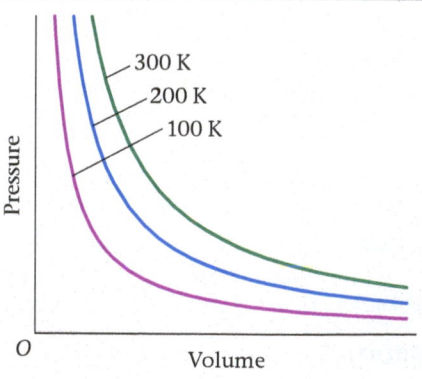

17-2 THE KINETIC THEORY OF GASES

In kinetic theory, a gas is imagined to be composed of a large number of pointlike molecules bouncing off the walls of a container.

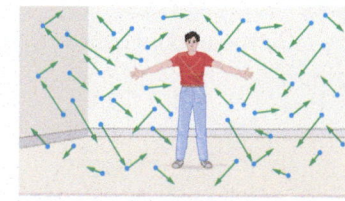

Kinetic Energy and Temperature

Kinetic theory relates the average kinetic energy of the molecules in a gas to the Kelvin temperature of the gas, T:

$$\left(\tfrac{1}{2}mv^2\right)_{av} = K_{av} = \tfrac{3}{2}kT \qquad \text{17-12}$$

RMS Speed

The rms (root mean square) speed of the molecules in a gas at the Kelvin temperature T is

$$v_{rms} = \sqrt{\frac{3kT}{m}} = \sqrt{\frac{3RT}{M}} \qquad \text{17-13}$$

Internal Energy of an Ideal Gas

The internal energy of a monatomic ideal gas is

$$U = \tfrac{3}{2}NkT = \tfrac{3}{2}nRT \qquad \text{17-15}$$

17-3 SOLIDS AND ELASTIC DEFORMATION

When a force is applied to a solid, its size and shape may change.

Changing the Length of a Solid

The force required to change the length of a solid by the amount ΔL is

$$F = Y\left(\frac{\Delta L}{L_0}\right)A \qquad \text{17-17}$$

In this expression, Y is Young's modulus, L_0 is the initial length parallel to the applied force, and A is the cross-sectional area perpendicular to the applied force.

Shear Deformation

The force required to shear, or deform, a solid by the amount Δx is

$$F = S\left(\frac{\Delta x}{L_0}\right)A \qquad \text{17-18}$$

In this expression, S is the shear modulus, L_0 is the initial length perpendicular to the applied force, and A is the cross-sectional area parallel to the applied force.

Changing the Volume of a Solid

The change in pressure required to change the volume of a solid by the amount ΔV is

$$\Delta P = -B\left(\frac{\Delta V}{V_0}\right) \qquad \text{17-19}$$

In this expression, B is the bulk modulus and V_0 is the initial volume.

17-4 PHASE EQUILIBRIUM AND EVAPORATION

The three most common phases of matter are solid, liquid, and gas. Solids maintain a definite shape, whereas gases and liquids flow to take on the shape of their container.

Equilibrium Between Phases

When phases are in equilibrium, the number of molecules in each phase remains constant.

17-5 LATENT HEATS

The latent heat, L, is the amount of heat per unit mass that must be added to or removed from a substance to convert it from one phase to another.

Latent Heat of Fusion

The heat required for melting or freezing is called the latent heat of fusion, L_f.

Latent Heat of Vaporization

The heat required for vaporizing or condensing is the latent heat of vaporization, L_v.

Latent Heat of Sublimation

The heat required to sublime a solid directly to a gas, or to condense a gas to a solid, is the latent heat of sublimation, L_s.

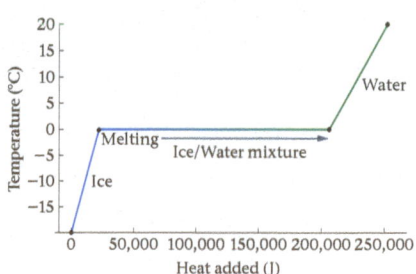

17-6 PHASE CHANGES AND ENERGY CONSERVATION

When heat is exchanged within a system, with no exchanges with the surroundings, the energy of the system is conserved.

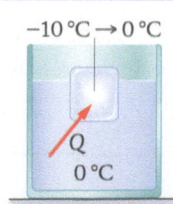

ANSWERS TO ENHANCE YOUR UNDERSTANDING QUESTIONS

1. B = D < A < C.

2. The root mean square speed increases by a factor of the square root of 2.

3. B < A < C < D.

4. **(a)** The plot shows the vapor-pressure curve. **(b)** DE corresponds to heating at constant pressure. **(c)** CD corresponds to increasing pressure at constant temperature.

5. More heat is required to boil the water.

6. **(a)** The initial temperature of the water was greater than 0 °C because heat flowed from the water to the ice as half of the ice melted. **(b)** The final temperature of the water is 0 °C because the system is in equilibrium.

CONCEPTUAL QUESTIONS

For instructor-assigned homework, go to www.masteringphysics.com.

(Answers to odd-numbered Conceptual Questions can be found in the back of the book.)

1. How is the air pressure in a tightly sealed house affected by operating the furnace? Explain.

2. The average speed of air molecules in your room is on the order of the speed of sound. What is their average velocity?

3. Is it possible to change both the pressure and the volume of an ideal gas without changing the average kinetic energy of its molecules? If your answer is no, explain why not. If your answer is yes, give a specific example.

4. **An Airport at Great Elevation** One of the highest airports in the world is located in La Paz, Bolivia. Pilots prefer to take off from this airport in the morning or the evening, when the air is quite cold. Explain.

5. A camping stove just barely boils water on a mountaintop. When the stove is used at sea level, will it be able to boil water? Explain your answer.

6. An autoclave is a device used to sterilize medical instruments. It is essentially a pressure cooker that heats the instruments in water under high pressure. This ensures that the sterilization process occurs at temperatures greater than the normal boiling point of water. Explain why the autoclave produces such high temperatures.

7. As the temperature of ice is increased, it changes first into a liquid and then into a vapor. On the other hand, dry ice, which is solid carbon dioxide, changes directly from a solid to a vapor as its temperature is increased. How might one produce liquid carbon dioxide?

8. **BIO** Isopropyl alcohol is sometimes rubbed onto a patient's arms and legs to lower their body temperature. Why is this effective?

9. A drop of water on a kitchen counter evaporates in a matter of minutes. However, only a relatively small fraction of the molecules in the drop move rapidly enough to escape through the drop's surface. Why, then, does the entire drop evaporate rather than just a small fraction of it?

PROBLEMS AND CONCEPTUAL EXERCISES

Answers to odd-numbered Problems and Conceptual Exercises can be found in the back of the book. **BIO** *identifies problems of biological or medical interest;* **CE** *indicates a conceptual exercise.* **Predict/Explain** *problems ask for two responses: (a) your prediction of a physical outcome, and (b) the best explanation among three provided; and* **Predict/Calculate** *problems ask for a prediction of a physical outcome, based on fundamental physics concepts, and follow that with a numerical calculation to verify the prediction. On all problems, bullets (•, ••, •••) indicate the level of difficulty*

SECTION 17-1 IDEAL GASES

1. • **CE (a)** Is the number of molecules in one mole of N_2 greater than, less than, or equal to the number of molecules in one mole of O_2? **(b)** Is the mass of one mole of N_2 greater than, less than, or equal to the mass of one mole of O_2?

2. • **CE Predict/Explain** If you put a helium-filled balloon in the refrigerator, **(a)** will its volume increase, decrease, or stay the same? **(b)** Choose the *best explanation* from among the following:
   I. Lowering the temperature of an ideal gas at constant pressure results in a reduced volume.
   II. The same amount of gas is in the balloon; therefore, its volume remains the same.
   III. The balloon can expand more in the cool air of the refrigerator, giving an increased volume.

3. • **CE** Two containers hold ideal gases at the same temperature. Container A has twice the volume and half the number of molecules as container B. What is the ratio P_A/P_B, where P_A is the pressure in container A and P_B is the pressure in container B?

4. • Standard temperature and pressure (STP) is defined as a temperature of 0 °C and a pressure of 101.3 kPa. What is the volume occupied by one mole of an ideal gas at STP?

5. • **BIO** After emptying her lungs, a person inhales 4.3 L of air at 0.0 °C and holds her breath. How much does the volume of the air increase as it warms to her body temperature of 35 °C?

6. • An automobile tire has a volume of 0.0185 m³. At a temperature of 294 K the absolute pressure in the tire is 212 kPa. How many moles of air must be pumped into the tire to increase its pressure to 252 kPa, given that the temperature and volume of the tire remain constant?

7. • **Amount of Helium in a Blimp** The Goodyear blimp *Spirit of Akron* is 62.6 m long and contains 7023 m³ of helium. When the temperature of the helium is 285 K, its absolute pressure is 112 kPa. Find the mass of the helium in the blimp.

8. • A compressed-air tank holds 0.500 m³ of air at a temperature of 295 K and a pressure of 820 kPa. What volume would the air occupy if it were released into the atmosphere, where the pressure is 101 kPa and the temperature is 303 K?

9. •• **CE** Four ideal gases have the following pressures, *P*, volumes, *V*, and mole numbers, *n*: gas A, *P* = 100 kPa, *V* = 1 m³, *n* = 10 mol; gas B, *P* = 200 kPa, *V* = 2 m³, *n* = 20 mol; gas C, *P* = 50 kPa, *V* = 1 m³, *n* = 50 mol; gas D, *P* = 50 kPa, *V* = 4 m³, *n* = 5 mol. Rank these gases in order of increasing temperature. Indicate ties where appropriate.

10. •• A balloon contains 3.9 liters of nitrogen gas at a temperature of 84 K and a pressure of 101 kPa. If the temperature of the gas is allowed to increase to 28 °C and the pressure remains constant, what volume will the gas occupy?

11. •• **Predict/Calculate** A balloon is filled with helium at a pressure of 2.4×10^5 Pa. The balloon is at a temperature of 18 °C and has a radius of 0.25 m. **(a)** How many helium atoms are contained in the balloon? **(b)** Suppose we double the number of helium atoms in the balloon, keeping the pressure and the temperature fixed. By what factor does the radius of the balloon increase? Explain.

12. •• **Predict/Calculate** A bicycle tire with a volume of 0.00212 m^3 is filled to its recommended absolute pressure of 495 kPa on a cold winter day when the tire's temperature is −15 °C. The cyclist then brings his bicycle into a hot laundry room at 32 °C. **(a)** If the tire warms up while its volume remains constant, will the pressure increase be greater than, less than, or equal to the manufacturer's stated 10% overpressure limit? **(b)** Find the absolute pressure in the tire when it warms to 32 °C at constant volume.

13. •• A 515-cm^3 flask contains 0.460 g of a gas at a pressure of 153 kPa and a temperature of 322 K. What is the molecular mass of this gas?

14. •• **Predict/Calculate** **The Atmosphere of Mars** On Mars, the average temperature is −64 °F and the average atmospheric pressure is 0.92 kPa. **(a)** What is the number of molecules per volume in the Martian atmosphere? **(b)** Is the number of molecules per volume on the Earth greater than, less than, or equal to the number per volume on Mars? Explain your reasoning. **(c)** Estimate the number of molecules per volume in Earth's atmosphere.

15. •• The air inside a hot-air balloon has an average temperature of 79.2 °C. The outside air has a temperature of 20.3 °C. What is the ratio of the density of air in the balloon to the density of air in the surrounding atmosphere?

16. •• A cylindrical flask is fitted with an airtight piston that is free to slide up and down, as shown in **FIGURE 17-33**. A mass rests on top of the piston. The initial temperature of the system is 313 K and the pressure of the gas is held constant at 137 kPa. The temperature is now increased until the height of the piston rises from 23.4 cm to 26.0 cm. What is the final temperature of the gas?

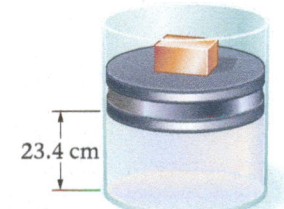

23.4 cm

FIGURE 17-33 Problems 16 and 17

17. •• Consider the system described in the previous problem. Contained within the flask is an ideal gas at a constant temperature of 313 K. Initially the pressure applied by the piston and the mass is 137 kPa and the height of the piston above the base of the flask is 23.4 cm. When additional mass is added to the piston, the height of the piston decreases to 20.0 cm. Find the new pressure applied by the piston.

18. ••• **Predict/Calculate** A gas has a temperature of 310 K and a pressure of 101 kPa. **(a)** Find the volume occupied by 1.25 mol of this gas, assuming it is ideal. **(b)** Assuming the gas molecules can be approximated as small spheres of diameter 2.5×10^{-10} m, determine the fraction of the volume found in part (a) that is occupied by the molecules. **(c)** In determining the properties of an ideal gas, we assume that molecules are points of zero volume. Discuss the validity of this assumption for the case considered here.

SECTION 17-2 THE KINETIC THEORY OF GASES

19. • **CE** **Predict/Explain** The air in your room is composed mostly of oxygen (O_2) and nitrogen (N_2) molecules. The oxygen molecules are more massive than the nitrogen molecules. **(a)** Is the rms speed of the O_2 molecules greater than, less than, or equal to the rms speed of the N_2 molecules? **(b)** Choose the *best explanation* from among the following:
  I. The more massive oxygen molecules have greater momentum and therefore greater speed.
  II. Equal temperatures for the oxygen and nitrogen molecules imply they have equal rms speeds.
  III. The temperature is the same for both molecules, and hence their average kinetic energies are equal. As a result, the more massive oxygen molecules have lower speeds.

20. • **CE** If the translational speed of molecules in an ideal gas is doubled, by what factor does the Kelvin temperature change? Explain.

21. • At what temperature is the rms speed of H_2 equal to the rms speed that O_2 has at 303 K?

22. • Suppose a planet has an atmosphere of pure ammonia at 5.5 °C. What is the rms speed of the ammonia molecules? (The molecular weight of ammonia, NH_3, is 17.03 g/mol.)

23. •• **Predict/Calculate** Three moles of oxygen gas (that is, 3.0 mol of O_2) are placed in a portable container with a volume of 0.0035 m^3. If the temperature of the gas is 295 °C, find **(a)** the pressure of the gas and **(b)** the average kinetic energy of an oxygen molecule. **(c)** Suppose the volume of the gas is doubled, while the temperature and number of moles are held constant. By what factor do your answers to parts (a) and (b) change? Explain.

24. •• **Predict/Calculate** The rms speed of O_2 is 1550 m/s at a given temperature. **(a)** Is the rms speed of H_2O at this temperature greater than, less than, or equal to 1550 m/s? Explain. **(b)** Find the rms speed of H_2O at this temperature.

25. •• **Predict/Calculate** An ideal gas is kept in a container of constant volume. The pressure of the gas is also kept constant. **(a)** If the number of molecules in the gas is doubled, does the rms speed increase, decrease, or stay the same? Explain. **(b)** If the initial rms speed is 1300 m/s, what is the final rms speed?

26. •• What is the temperature of a gas of CO_2 molecules whose rms speed is 309 m/s?

27. •• The rms speed of a sample of gas is increased by 1%. **(a)** What is the percent change in the temperature of the gas? **(b)** What is the percent change in the pressure of the gas, assuming its volume is held constant?

28. ••• **Enriching Uranium** In naturally occurring uranium atoms, 99.3% are ^{238}U (atomic mass = 238 u, where u = 1.6605×10^{-27} kg) and only 0.7% are ^{235}U (atomic mass = 235 u). Uranium-fueled reactors require an enhanced proportion of ^{235}U. Since both isotopes of uranium have identical chemical properties, they can be separated only by methods that depend on their differing masses. One such method is gaseous diffusion, in which uranium hexafluoride (UF_6) gas diffuses through a series of porous barriers. The lighter $^{235}UF_6$ molecules have a slightly higher rms speed at a given temperature than the heavier $^{238}UF_6$ molecules, and this allows the two isotopes to be separated. Find the ratio of the rms speeds of the two isotopes at 23.0 °C.

29. ••• A 380-mL spherical flask contains 0.065 mol of an ideal gas at a temperature of 283 K. What is the average force exerted on the walls of the flask by a single molecule?

SECTION 17-3 SOLIDS AND ELASTIC DEFORMATION

30. • **CE** Predict/Explain A hollow cylindrical rod (rod 1) and a solid cylindrical rod (rod 2) are made of the same material. The two rods have the same length and the same outer radius. If the same compressional force is applied to each rod, (a) is the change in length of rod 1 greater than, less than, or equal to the change in length of rod 2? (b) Choose the *best explanation* from among the following:

   I. The solid rod has the larger effective cross-sectional area, since the empty part of the hollow rod doesn't resist compression. Therefore, the solid rod has the smaller change in length.
   II. The rods have the same outer radius and hence the same cross-sectional area. As a result, their change in length is the same.
   III. The walls of the hollow rod are hard and resist compression more than the uniform material in the solid rod. Therefore the hollow rod has the smaller change in length.

31. • A rock climber hangs freely from a nylon rope that is 16 m long and has a diameter of 8.1 mm. If the rope stretches 4.2 cm, what is the mass of the climber?

32. • **BIO** To stretch a relaxed biceps muscle 2.5 cm requires a force of 25 N. Find the Young's modulus for the muscle tissue, assuming it to be a uniform cylinder of length 0.24 m and cross-sectional area 47 cm².

33. • A 22-kg chimpanzee hangs from the end of a horizontal, broken branch 1.1 m long, as shown in **FIGURE 17-34**. The branch is a uniform cylinder 4.6 cm in diameter, and the end of the branch supporting the chimp sags downward through a vertical distance of 13 cm. What is the shear modulus for this branch?

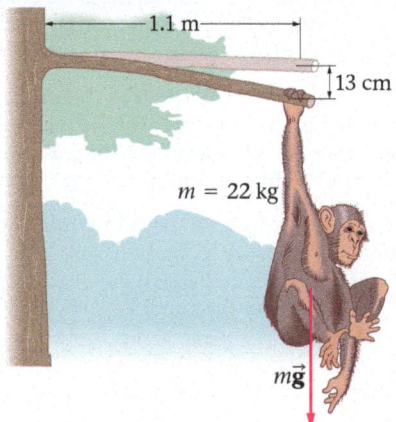

FIGURE 17-34 Problem 33

34. • **The Marianas Trench** The deepest place in all the oceans is the Marianas Trench, where the depth is 10.9 km and the pressure is 1.10×10^8 Pa. If a copper ball 15.0 cm in diameter is taken to the bottom of the trench, by how much does its volume decrease?

35. •• **CE** Four cylindrical rods with various cross-sectional areas and initial lengths are stretched by an applied force, as in Figure 17-11. The resulting change in length of each rod is given in the following table. Rank these rods in order of increasing Young's modulus. Indicate ties where appropriate.

Rod	Applied Force	Cross-Sectional Area	Initial Length	Change in Length
1	F	A	L	ΔL
2	F	$2A$	$2L$	ΔL
3	$2F$	$2A$	L	$2\Delta L$
4	$3F$	A	$L/2$	ΔL

36. •• Predict/Calculate A steel wire 4.1 m long stretches 0.13 cm when it is given a tension of 380 N. (a) What is the diameter of the wire? (b) If it is desired that the stretch be less than 0.13 cm, should its diameter be increased or decreased? Explain.

37. ••• **BIO** **Spiderweb** An orb weaver spider with a mass of 0.26 g hangs vertically by one of its threads. The thread has a Young's modulus of 4.7×10^9 N/m² and a radius of 9.8×10^{-6} m. (a) What is the fractional increase in the thread's length caused by the spider? (b) Suppose a 76-kg person hangs vertically from a nylon rope. What radius must the rope have if its fractional increase in length is to be the same as that of the spider's thread?

38. •• Predict/Calculate Two rods of equal length (0.55 m) and diameter (1.7 cm) are placed end to end. One rod is aluminum, the other is brass. If a compressive force of 8400 N is applied to the rods, (a) how much does their combined length decrease? (b) Which of the rods changes its length by the greatest amount? Explain.

39. •• A piano wire 0.82 m long and 0.93 mm in diameter is fixed on one end. The other end is wrapped around a tuning peg 3.5 mm in diameter. Initially the wire, whose Young's modulus is 2.4×10^{10} N/m², has a tension of 14 N. Find the tension in the wire after the tuning peg has been turned through one complete revolution.

SECTION 17-4 PHASE EQUILIBRIUM AND EVAPORATION

40. • **CE** The formation of ice from water is accompanied by which of the following: (a) an absorption of heat by the water; (b) an increase in temperature; (c) a decrease in volume; (d) a removal of heat from the water; (e) a decrease in temperature?

41. • **Vapor Pressure for Water** FIGURE 17-35 shows a portion of the vapor-pressure curve for water. Referring to the figure, estimate the pressure that would be required for water to boil at 30 °C.

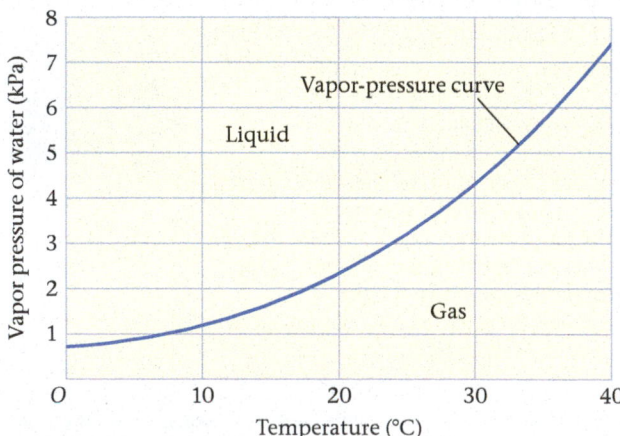

FIGURE 17-35 Problems 41, 42, 43, and 44

42. • Using the vapor-pressure curve given in Figure 17-35, find the temperature at which water boils when the pressure is 1.5 kPa.

43. •• **Dew-Point Temperature** One measure of the amount of water in the atmosphere is the *dew point*, the temperature at which the atmosphere's water content would be in equilibrium with liquid water and droplets of dew would begin to form. (a) On a warm summer day when the dew-point temperature is 68 °F, what is the pressure of water in the atmosphere? Refer to Figure 17-35 and note that the pressure of the water vapor (called the *partial pressure* of water) is equal to the vapor pressure at the dew-point temperature. (b) If the weather changes and the partial pressure of water is reduced to 1.7 kPa, what is the new dew-point temperature?

44. •• **Relative Humidity** One measure of the amount of water in the atmosphere is the *relative humidity*, the ratio of the actual partial pressure of water vapor to the maximum possible partial pressure of water. If the partial pressure of water vapor exceeds the vapor pressure (Figure 17-35) at that temperature, the vapor condenses to liquid and either fog or dew begins to form. In that case the relative humidity would be 100%. Referring to Figure 17-35, if the partial pressure of water in the atmosphere is 1.7 kPa when the air temperature is 25 °C, what is the relative humidity?

45. •• **Predict/Calculate The Vapor Pressure of CO_2** A portion of the vapor-pressure curve for carbon dioxide is given in **FIGURE 17-36**. **(a)** Estimate the pressure at which CO_2 boils at 0 °C. **(b)** If the temperature is increased, does the boiling pressure increase, decrease, or stay the same? Explain.

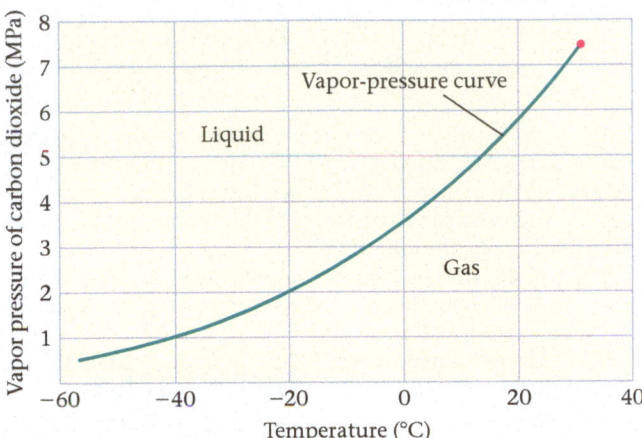

FIGURE 17-36 Problem 45

46. •• **Phase Diagram for Water** The phase diagram for water is shown in **FIGURE 17-37**. **(a)** What is the temperature T_1 on the phase diagram? **(b)** What is the temperature T_2 on the phase diagram? **(c)** What happens to the melting/freezing temperature of water if atmospheric pressure is *decreased*? Justify your answer by referring to the phase diagram. **(d)** What happens to the boiling/condensation temperature of water if atmospheric pressure is *increased*? Justify your answer by referring to the phase diagram.

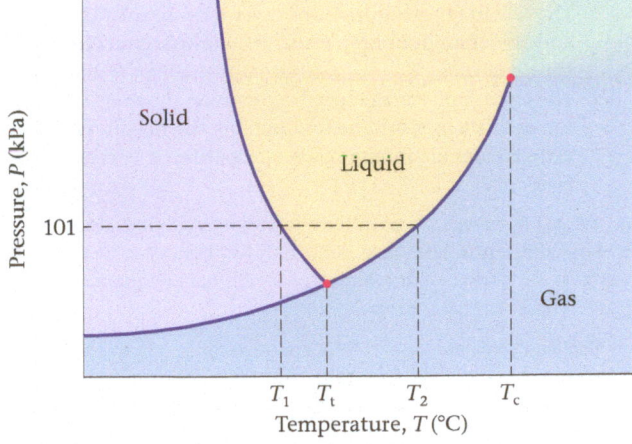

FIGURE 17-37 Problems 46 and 48

47. •• **Phase Diagram for CO_2** The phase diagram for CO_2 is shown in **FIGURE 17-38**. **(a)** What is the phase of CO_2 at $T = 20$°C and $P = 500$ kPa? **(b)** What is the phase of CO_2 at $T = -80$°C and $P = 120$ kPa? **(c)** For reasons of economy and convenience, bulk CO_2 is often transported in liquid form in pressurized tanks. Using

the phase diagram, determine the minimum pressure required to keep CO_2 in the liquid phase at 20 °C.

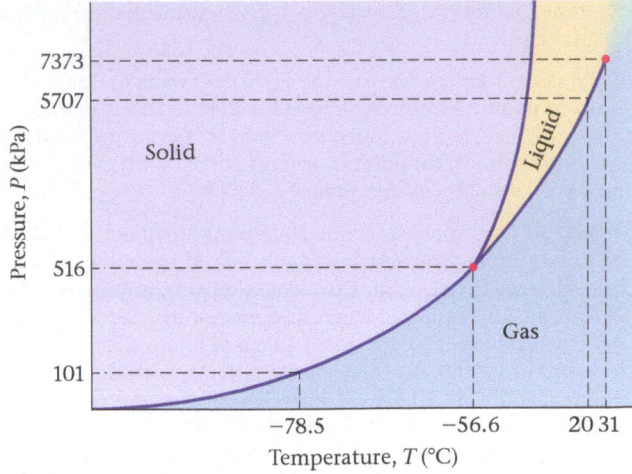

FIGURE 17-38 Problem 47

48. •• A sample of liquid water at atmospheric pressure has a temperature just above the freezing point. Refer to Figure 17-37 to answer the following questions. **(a)** What phase changes occur if the temperature of the system is increased while the pressure is held constant? **(b)** Suppose, instead, that the temperature of the system is held constant just above the freezing point while the pressure is decreased. What phase changes occur now?

SECTION 17-5 LATENT HEATS

49. • How much heat must be removed from 1.96 kg of water at 0 °C to make ice cubes at 0 °C?

50. • A heat transfer of 9.5×10^5 J is required to convert a block of ice at -15 °C to water at 15 °C. What was the mass of the block of ice?

51. • How much heat must be added to 2.55 kg of copper to change it from a solid at 1358 K to a liquid at 1358 K?

52. • An ammonia refrigeration cycle involves the conversion of 0.85 kg of liquid ammonia into vapor every minute at the boiling-point temperature. At what rate does the ammonia absorb energy?

53. •• **CE** Four liquids are at their freezing temperature. Heat is now removed from each of the liquids until it becomes completely solidified. The amount of heat that must be removed, Q, and the mass, m, of each of the liquids are as follows: liquid A, $Q = 33,500$ J, $m = 0.100$ kg; liquid B, $Q = 166,000$ J, $m = 0.500$ kg; liquid C, $Q = 31,500$ J, $m = 0.250$ kg; liquid D, $Q = 5400$ J, $m = 0.0500$ kg. Rank these liquids in order of increasing latent heat of fusion. Indicate ties where appropriate.

54. •• **Predict/Calculate** A 1.1-kg block of ice is initially at a temperature of -5.0 °C. **(a)** If 5.2×10^5 J of heat are added to the ice, what is the final temperature of the system? Find the amount of ice, if any, that remains. **(b)** Suppose the amount of heat added to the ice block is doubled. By what factor must the mass of the ice be increased if the system is to have the same final temperature? Explain.

55. •• **Predict/Calculate** Referring to the previous problem, suppose the amount of heat added to the block of ice is reduced by a factor of 2 to 2.6×10^5 J. Note that this amount of heat is still sufficient to melt at least some of the ice. **(a)** Do you expect the temperature increase in this case to be one-half that found in the previous problem? Explain. **(b)** What is the final temperature of the system in this case? Find the amount of ice, if any, that remains.

56. •• Figure 17-30 shows a temperature-versus-heat plot for 1.000 kg of water. **(a)** Calculate the heat corresponding to the points A, B, C, and D. **(b)** Calculate the slope of the line from point B to point C. Show that this slope is equal to $1/c$, where c is the specific heat of liquid water.

57. •• **Predict/Calculate** Suppose the 1.000 kg of water in Figure 17-30 starts at point A at time zero. Heat is added to this system at the rate of 12,250 J/s. How much time does it take for the system to reach **(a)** point B, **(b)** point C, and **(c)** point D? **(d)** Describe the physical state of the system at time $t = 63.00$ s.

58. •• **BIO** In Conceptual Example 17-15 we pointed out that steam can cause more serious burns than water at the same temperature. Here we examine this effect quantitatively, noting that flesh becomes badly damaged when its temperature reaches 50.0 °C. **(a)** Calculate the heat released as 12.5 g of liquid water at 100 °C is cooled to 50.0 °C. **(b)** Calculate the heat released when 12.5 g of steam at 100 °C is condensed and cooled to 50.0 °C. **(c)** Find the mass of flesh that can be heated from 37.0 °C (normal body temperature) to 50.0 °C for the cases considered in parts (a) and (b). (The average specific heat of flesh is 3500 J/kg · K.)

59. •• When you go out to your car one cold winter morning you discover a 0.58-cm-thick layer of ice on the windshield, which has an area of 1.6 m². If the temperature of the ice is −2.0 °C, and its density is 917 kg/m³, find the heat required to melt all the ice.

SECTION 17-6 PHASE CHANGES AND ENERGY CONSERVATION

60. •• A large punch bowl holds 3.99 kg of lemonade (which is essentially water) at 20.5 °C. A 0.0550-kg ice cube at −10.2 °C is placed in the lemonade. What are the final temperature of the system, and the amount of ice (if any) remaining? Ignore any heat exchange with the bowl or the surroundings.

61. •• A 155-g aluminum cylinder is removed from a liquid nitrogen bath, where it has been cooled to −196 °C. The cylinder is immediately placed in an insulated cup containing 80.0 g of water at 15.0 °C. What is the equilibrium temperature of this system? If your answer is 0 °C, determine the amount of water that has frozen. The average specific heat of aluminum over this temperature range is 653 J/(kg · K).

62. •• An 825-g iron block is heated to 352 °C and placed in an insulated container (of negligible heat capacity) containing 40.0 g of water at 20.0 °C. What is the equilibrium temperature of this system? If your answer is 100 °C, determine the amount of water that has vaporized. The average specific heat of iron over this temperature range is 560 J/(kg · K).

63. •• **Party Planning** You are expecting to serve 32 cups of soft drinks to your guests tonight. Each cup will hold 285 g of a soft drink that has a specific heat of 4186 J/(kg · K) and an initial temperature of 24 °C. If each guest would like to enjoy the drink at 3.0 °C, how much ice (in kg) should you buy? Assume the initial temperature of the ice is 0 °C, and ignore the heat exchange with the plastic cups and the surroundings.

64. •• **Predict/Calculate** A 35-g ice cube at 0.0 °C is added to 110 g of water in a 62-g aluminum cup. The cup and the water have an initial temperature of 23 °C. **(a)** Find the equilibrium temperature of the cup and its contents. **(b)** Suppose the aluminum cup is replaced with one of equal mass made from silver. Is the equilibrium temperature with the silver cup greater than, less than, or the same as with the aluminum cup? Explain.

65. •• A 48-g block of copper at −12 °C is added to 110 g of water in a 75-g aluminum cup. The cup and the water have an initial temperature of 4.1 °C. **(a)** Find the equilibrium temperature of the cup and its contents. **(b)** What mass of ice, if any, is present when the system reaches equilibrium?

66. •• A 0.075-kg ice cube at 0.0 °C is dropped into a Styrofoam cup holding 0.33 kg of water at 14 °C. **(a)** Find the final temperature of the system and the amount of ice (if any) remaining. Assume the cup and the surroundings can be ignored. **(b)** Find the initial temperature of the water that would be enough to *just barely* melt all of the ice.

67. •• To help keep her barn warm on cold days, a farmer stores 865 kg of warm water in the barn. How many hours would a 2.00-kilowatt electric heater have to operate to provide the same amount of heat as is given off by the water as it cools from 20.0 °C to 0 °C and then freezes at 0 °C?

GENERAL PROBLEMS

68. • **CE** As you go up in altitude, do you expect the ratio of oxygen to nitrogen in the atmosphere to increase, decrease, or stay the same? Explain.

69. • **CE** Predict/Explain Suppose the Celsius temperature of an ideal gas is doubled from 100 °C to 200 °C. **(a)** Does the average kinetic energy of the molecules in this gas increase by a factor that is greater than, less than, or equal to 2? **(b)** Choose the *best explanation* from among the following:

   I. Changing the temperature from 100 °C to 200 °C goes beyond the boiling point, which will increase the kinetic energy by more than a factor of 2.

   II. The average kinetic energy is directly proportional to the temperature, so doubling the temperature doubles the kinetic energy.

   III. Doubling the Celsius temperature from 100 °C to 200 °C changes the Kelvin temperature from 373.15 K to 473.15 K, which is an increase of less than a factor of 2.

70. • **CE** Predict/Explain Suppose the absolute temperature of an ideal gas is doubled from 100 K to 200 K. **(a)** Does the average speed of the molecules in this gas increase by a factor that is greater than, less than, or equal to 2? **(b)** Choose the *best explanation* from among the following:

   I. Doubling the Kelvin temperature doubles the average kinetic energy, but this implies an increase in the average speed by a factor of $\sqrt{2} = 1.414\ldots$, which is less than 2.

   II. The Kelvin temperature is the one we use in the ideal-gas law, and therefore doubling it also doubles the average speed of the molecules.

   III. The change in average speed depends on the mass of the molecules in the gas, and hence doubling the Kelvin temperature generally results in an increase in speed that is greater than a factor of 2.

71. • **Largest Raindrops** Atmospheric scientists studying clouds in the Marshall Islands have observed what they believe to be the world's largest raindrops, with a radius of 0.52 cm. How many molecules are in these monster drops?

72. • **Cooling Computers** Researchers are developing "heat exchangers" for laptop computers that take heat from the laptop—to keep it from being damaged by overheating—and use it to vaporize methanol. Given that 5100 J of heat is removed from the laptop when 4.6 g of methanol is vaporized, what is the latent heat of vaporization for methanol?

73. •• A 1.10-kg block of ice is initially at a temperature of −5.0 °C. **(a)** If 3.1×10^5 J of thermal energy are added to the ice, what is the amount of ice that remains? **(b)** How much additional thermal energy must be added to this system to convert it to 1.10 kg of water at 5.0 °C?

74. •• **Scuba Tanks** In scuba diving circles, "an 80" refers to a scuba tank that holds 80 cubic feet of air, a standard amount for recreational diving. Given that a scuba tank is a cylinder 2 feet long and half a foot in diameter, determine **(a)** the volume of a tank and **(b)** the pressure in a tank when 80 cubic feet of air is compressed into its relatively small volume. **(c)** What is the mass of air in a tank that holds 80 cubic feet of air? Assume the temperature is 21 °C and that the walls of the tank are of negligible thickness.

75. •• **Evaporating Atmosphere** Hydrogen gas evaporates into space even though its rms speed is less than one-fifth of the gravitational escape speed. This is because the distribution of molecular speeds at equilibrium (see Figure 17-27) shows that some of the molecules do have speeds that exceed the escape speed. To what temperature must the atmosphere be heated for the air around us to have an rms speed that is one-fifth of the gravitational escape speed? Treat the air as having a molecular mass of 0.029 kg/mol.

76. •• To make steam, you add 5.60×10^5 J of thermal energy to 0.220 kg of water at an initial temperature of 50.0 °C. Find the final temperature of the steam.

77. •• **A Boiling Geyser** **(a)** The column of water that forms as a geyser erupts is 2100 m tall. What is the pressure at the bottom of the column of water? **(b)** The water vapor-pressure curve (Figure 17-18) near the bottom of the column can be approximated by $P = (1.86 \times 10^5)T - 4.80 \times 10^7$, where P is the pressure in Pa and T is the Celsius temperature. What is the boiling-point temperature for water at the bottom of the column?

78. •• **A Melting Glacier** **(a)** A glacier is made of ice of density 850 kg/m^3 and is 92 m thick. Treating the glacial ice as if it were a liquid, what is the pressure at the bottom of the ice? **(b)** The water solid-liquid curve (Figure 17-21 (b)) near the bottom of the glacier can be approximated by $P = (-1.31 \times 10^7)T + 1.01 \times 10^5$, where P is the pressure in Pa and T is the Celsius temperature. What is the melting-point temperature for water at the bottom of the glacier?

79. •• Peter catches a 4.2-kg striped bass on a fishing line 0.55 mm in diameter and begins to reel it in. He fishes from a pier well above the water, and his fish hangs vertically from the line out of the water. The fishing line has a Young's modulus of 5.1×10^9 N/m^2. **(a)** What is the fractional increase in length of the fishing line if the fish is at rest? **(b)** What is the fractional increase in the fishing line's length when the fish is pulled upward with a constant speed of 2.5 m/s? **(c)** What is the fractional increase in the fishing line's length when the fish is pulled upward with a constant acceleration of 2.5 m/s^2?

80. •• A steel ball (density = 7860 kg/m^3) with a diameter of 6.4 cm is tied to an aluminum wire 82 cm long and 2.5 mm in diameter. The ball is whirled about in a vertical circle with a tangential speed of 7.8 m/s at the top of the circle and 9.3 m/s at the bottom of the circle. Find the amount of stretch in the wire **(a)** at the top and **(b)** at the bottom of the circle.

81. •• A lead brick with the dimensions shown in **FIGURE 17-39** rests on a rough solid surface. A force of 2400 N is applied as indicated. Find **(a)** the change in height of the brick and **(b)** the amount of shear deformation.

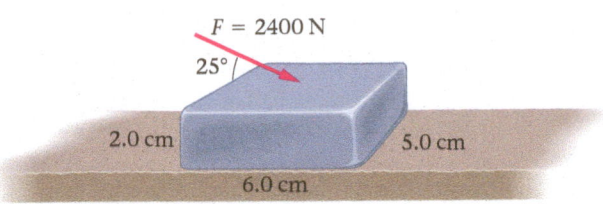

FIGURE 17-39 Problem 81

82. •• **(a)** Find the amount of heat that must be extracted from 1.3 kg of steam at 120 °C to convert it to ice at 0.0 °C. **(b)** What speed would this 1.3-kg block of ice have if its translational kinetic energy were equal to the thermal energy calculated in part (a)?

83. ••• **Mighty Ice Lift** A tremendous force is generated when water freezes into ice and expands in volume by 9.0%. Suppose a 1.000-m^3 cube of liquid water freezes into ice that is 1.000 m on a side by 1.090 m tall. How many 68-kg students would the ice be able to lift? Determine this by calculating the amount of force on the top 1.000-m^2 face that would be required to squeeze 1.090 m^3 of ice back to 1.000 m^3, assuming all of the volume change occurs along the vertical direction.

84. ••• **Orthopedic Implants** Metals such as titanium and stainless steel are frequently used for orthopedic implants such as artificial hip and knee joints. As with most metals, though, their elastic properties are significantly different from those of bone. Recently, metal "foams" made from aluminum and steel have been shown to have promising properties for use as implants. **(a)** A 0.5-m-long piece of bone with a certain cross-sectional area shortens by 0.10 mm under a given compressive force. By how much does a piece of steel with the same length and cross-sectional area shorten if the same force is applied? **(b)** In order to determine whether a material such as the aluminum-steel foam behaves similarly to bone, plots of measured stress (force per area) versus strain ($\Delta L/L_0$) such as the ones shown in **FIGURE 17-40** may be used. Which one of these plots corresponds to a material with elastic properties equal to those of bone in compression? **(c)** What is the value of Young's modulus for the material represented by curve D?

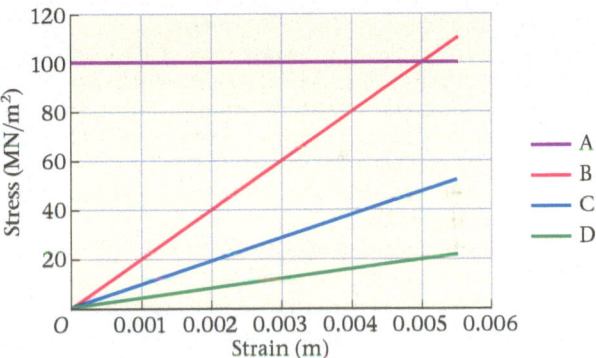

FIGURE 17-40 Problem 84

85. ••• Students on a spring break picnic bring a cooler that contains 5.1 kg of ice at 0.0 °C. The cooler has walls that are 3.8 cm thick and are made of Styrofoam, which has a thermal conductivity of 0.030 W/(m·C°). The surface area of the cooler is 1.5 m^2, and it rests in the shade where the air temperature is 21 °C. **(a)** Find the rate at which heat flows into the cooler. **(b)** How much time does it take for the ice in the cooler to melt?

86. ••• A 5.9-kg block of ice at −1.5 °C slides on a horizontal surface with a coefficient of kinetic friction equal to 0.069. The initial speed of the block is 7.1 m/s and its final speed is 5.3 m/s. Assuming that all the energy dissipated by kinetic friction goes into melting a small mass m of the ice, and that the rest of the ice block remains at −1.5 °C, determine the value of m.

87. ••• A cylindrical copper rod 37 cm long and 7.5 cm in diameter is placed upright on a hot plate held at a constant temperature of 120 °C, as indicated in **FIGURE 17-41**. A small depression on top of the rod holds a 25-g ice cube at an initial temperature of 0.0 °C. How much time does it take for the ice cube to melt? Assume there

is no heat loss through the vertical surface of the rod, and that the thermal conductivity of copper is 390 W/(m · C°).

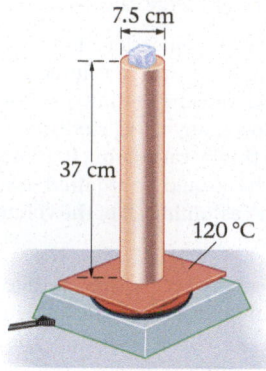

7.5 cm

37 cm

120 °C

FIGURE 17-41 Problem 87

PASSAGE PROBLEMS

Diving in the Bathysphere

The American naturalist Charles William Beebe (1877–1962) set a world record in 1934 when he and Otis Barton (1899–1992) made a dive to a depth of 923 m below the surface of the ocean. The dive was made just 10 miles from Nonsuch Island, off the coast of Bermuda, in a device known as the bathysphere, designed and built by Barton. The bathysphere was basically a steel sphere 4.75 ft in diameter with three small ports made of fused quartz. Lowered into the ocean on a steel cable whose radius was 1.85 cm, the bathysphere also carried bottles of oxygen, chemicals to absorb carbon dioxide, and a telephone line to the surface.

Beebe was fascinated by the new forms of life he and Barton encountered on their numerous dives. At one point he saw a "creature, several feet long, dart toward the window, turn sideways and—explode. At the flash, which was so strong that it illumined my face . . . I saw the great red shrimp and the outpouring fluid of flame." No wonder he considered the ocean depths to be "a world as strange as that of Mars."

The dives were not without their risks, however. It was not uncommon, for example, to have the bathysphere return to the surface partially filled with water after a window seal failed. On one deep dive, water began to stream rapidly into the sphere. Beebe quickly called to the surface and asked—not to be raised quickly—but to be lowered more rapidly, in the hope that increasing water pressure would force the leaking window into its seals to stop the leak. It worked, showing that Beebe was not only an exceptional naturalist, but also a cool-headed scientist with a good knowledge of basic physics!

88. • What pressure did the bathysphere experience at its record depth?

    A. 9.37 atm   B. 89.6 atm
    C. 91.9 atm   D. 92.9 atm

89. • How many moles of air did the bathysphere contain when it was sealed at the surface, assuming a temperature of 297 K and ignoring the thickness of the metal shell? (*Note*: A resting person breathes roughly 0.5 mol of air per minute.)

    A. 65.2 mol   B. 270 mol
    C. 392 mol   D. 523 mol

90. •• How much did the volume of the bathysphere decrease as it was lowered to its record depth? (For simplicity, treat the bathysphere as a solid metal sphere.)

    A. 9.0×10^{-5} m^3 B. 9.2×10^{-5} m^3
    C. 1.1×10^{-4} m^3 D. 3.8×10^{-4} m^3

91. •• Suppose the bathysphere and its occupants had a combined mass of 12,700 kg. How much did the cable stretch when the bathysphere was at a depth of 923 m? (Neglect the weight of the cable itself, but include the effects of the bathysphere's buoyancy.)

    A. 47 cm   B. 48 cm
    C. 52 cm   D. 53 cm

92. •• **REFERRING TO EXAMPLE 17-17 (a)** Find the final temperature of the system if *two* 0.0450-kg ice cubes are added to the warm lemonade. The temperature of the ice is 0 °C; the temperature and mass of the warm lemonade are 20.0 °C and 3.95 kg, respectively. **(b)** How many 0.0450-kg ice cubes at 0 °C must be added to the original warm lemonade if the final temperature of the system is to be at least as cold as 15.0 °C?

93. •• **REFERRING TO EXAMPLE 17-17 (a)** Find the final temperature of the system if a single 0.045-kg ice cube at 0 °C is added to 2.00 kg of lemonade at 1.00 °C. **(b)** What initial temperature of the lemonade will be just high enough to melt all of the ice in a single ice cube and result in an equilibrium temperature of 0 °C? The mass of the lemonade is 2.00 kg and the temperature of the ice cube is 0 °C.

Big Ideas

1 Energy conservation applies to thermodynamic systems, just as it does to mechanical systems.

2 Heat engines use a difference in temperature to convert thermal energy to mechanical energy.

3 Entropy is a measure of disorder in a physical system.

4 The lowest temperature possible is absolute zero.

The Laws of Thermodynamics

▲ Did this plane just fly through a cloud? No, it created its own cloud. As the plane moves forward at high speed it creates a low-pressure region near its tail. When the surrounding air expands into this region, it cools by the process of adiabatic cooling. As a result, water vapor in the air condenses, creating the plane's own personal cloud that follows it around. This chapter explores various thermodynamic processes, including adiabatic cooling, and shows how these processes can be used to understand the laws of thermodynamics and the operation of heat engines.

In this chapter we discuss the fundamental laws of nature that govern thermodynamic processes. Of particular interest is the second law of thermodynamics, which introduces the idea of directionality in the behavior of nature. Melting ice, cooling lava, and the crumbling ruins of the Parthenon all illustrate the second law in action.

18-1 The Zeroth Law of Thermodynamics

Although the zeroth law of thermodynamics has already been presented in Chapter 16, we repeat it here so that all the laws of thermodynamics can be collected together in one chapter. As you recall, the zeroth law states the conditions for thermal equilibrium between objects. To be precise:

> **Zeroth Law of Thermodynamics**
> If object A is in thermal equilibrium with object C, and object B is separately in thermal equilibrium with object C, then objects A and B will be in thermal equilibrium if they are placed in thermal contact.

The physical quantity that is equal when two objects are in thermal equilibrium is the temperature. In particular, if two objects have the same temperature, then we can be assured that *no* energy in the form of heat will flow when they are placed in thermal contact. On the other hand, if heat does flow between two objects, it follows that they are not in thermal equilibrium, and they do not have the same temperature.

Any temperature scale can be used to determine whether objects will be in thermal equilibrium. As we saw in the previous chapter, however, the Kelvin scale is particularly significant in physics. For example, the average kinetic energy of a gas molecule is directly proportional to the Kelvin temperature, as is the volume of an ideal gas. In this chapter we present additional illustrations of the special significance of the Kelvin scale.

Enhance Your Understanding   (Answers given at the end of the chapter)

1. System 1 is at 0 °C and system 2 is at 0 °F. If these systems are placed in thermal contact, which of the following statements is correct? **(A)** Heat flows from system 1 to system 2. **(B)** Heat flows from system 2 to system 1. **(C)** No heat flows.

Section Review

- Objects are in thermal equilibrium when their temperatures are the same.

18-2 The First Law of Thermodynamics

The first law of thermodynamics is a statement of energy conservation that specifically includes energy transferred in the form of heat. For example, consider the system shown in **FIGURE 18-1**. The internal energy of this system—that is, the sum of all its potential and kinetic energies—has the initial value U_i. If an amount of heat Q flows into the system, its internal energy increases to the final value $U_f = U_i + Q$. Thus,

$$\Delta U = U_f - U_i = Q \qquad 18\text{-}1$$

Of course, if heat is removed from the system, its internal energy decreases. We can take this into account by giving Q a *positive* value when the system *gains* heat, and a *negative* value when it *loses* heat. (In the following, for the sake of brevity, we will generally just say "heat" as opposed to a more complete statement like "energy flow in the form of heat.")

Similarly, suppose the system under consideration does a work W on the external world, as in **FIGURE 18-2**. If the system is insulated so that no heat can flow in or out, the energy to do the work must come from the internal energy of the system. Thus, if the initial internal energy is U_i, the final internal energy is $U_f = U_i - W$. Therefore,

$$\Delta U = U_f - U_i = -W \qquad 18\text{-}2$$

On the other hand, if work is done *on* the system, its internal energy increases. Thus, we use the following sign conventions for the work, W:

> W is *positive* when a system *does work on the external world*.
>
> W is *negative* when *work is done on a system*.

These sign conventions are summarized in Table 18-1.

Adding the heat Q to the system . . .   . . . increases the system's energy from U_i to $U_i + Q$.

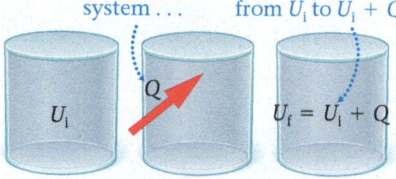

▲ **FIGURE 18-1 The internal energy of a system** A system initially has the internal energy U_i (left). After the heat Q is added, the system's new internal energy is $U_f = U_i + Q$ (right). The system has rigid walls; hence, it can do no work on the external world.

PHYSICS IN CONTEXT
Looking Back

The concepts of heat, work, and temperature, from Chapters 16 and 17, are used throughout this chapter.

Big Idea **1** Energy conservation applies to thermodynamic systems, just as it does to mechanical systems. When we apply energy conservation, we must consider not only mechanical work, but energy transferred in the form of heat as well.

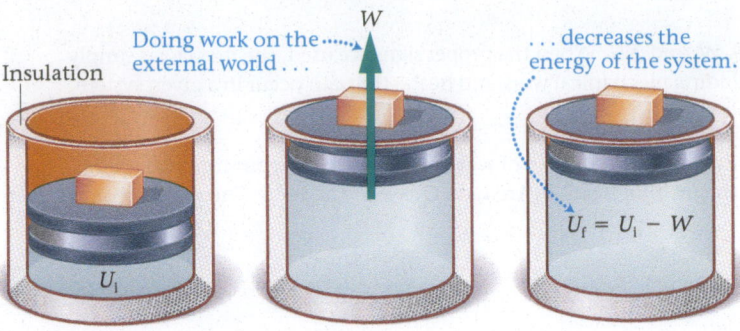

◀ **FIGURE 18-2 Work and internal energy**
A system initially has the internal energy U_i (left). After the system does the work W on the external world, its remaining internal energy is $U_f = U_i - W$ (right). Note that the insulation guarantees that no heat is gained or lost by the system.

Combining the results in Equations 18-1 and 18-2 yields the first law of thermodynamics:

First Law of Thermodynamics

The change in a system's internal energy, ΔU, is related to the heat Q and the work W as follows:

$$\Delta U = Q - W \qquad \qquad \text{18-3}$$

If you apply the sign conventions given here, it is straightforward to verify that adding heat to a system, and/or doing work on it, increases the internal energy. On the other hand, if the system does work, and/or heat is removed, its internal energy decreases. Example 18-1 gives a specific numerical application of the first law.

TABLE 18-1 Signs of Q and W

Q positive	System *gains* heat
Q negative	System *loses* heat
W positive	Work done *by* system
W negative	Work done *on* system

PROBLEM-SOLVING NOTE

Proper Signs for Q and W

When applying the first law of thermodynamics, it is important to determine the proper signs for Q, W, and ΔU.

EXAMPLE 18-1   SHELL GAME

(a) Rowing your shell to work one day, you give off 4.8×10^5 J of heat and do 6.2×10^5 J of work. What is the change in your internal energy? **(b)** On your return trip, you give off 3.9×10^5 J of heat and your internal energy decreases by 5.1×10^5 J. How much work did you do on the return trip?

PICTURE THE PROBLEM

Our sketch shows a person rowing a racing shell. The fact that the person does work on the external world means that W is positive. As for the heat, the fact that heat is given off by the person means that Q is negative.

REASONING AND STRATEGY

The signs of W and Q have been determined in our sketch, and the magnitudes are given in the problem statement. To find ΔU for part (a), we simply use the first law of thermodynamics, $\Delta U = Q - W$. To find the work W for part (b), we solve the first law for W, which yields $W = Q - \Delta U$.

Known (a) Heat, $Q = -4.8 \times 10^5$ J; work, $W = 6.2 \times 10^5$ J.
(b) Heat, $Q = -3.9 \times 10^5$ J; change in internal energy, $\Delta U = -5.1 \times 10^5$ J.

Unknown (a) Change in internal energy, $\Delta U = ?$ (b) Work, $W = ?$

SOLUTION

Part (a)

1. Calculate ΔU, using $Q = -4.8 \times 10^5$ J and $W = 6.2 \times 10^5$ J:

$$\Delta U = Q - W$$
$$= (-4.8 \times 10^5 \text{ J}) - 6.2 \times 10^5 \text{ J} = -1.1 \times 10^6 \text{ J}$$

Part (b)

2. Solve $\Delta U = Q - W$ for the work, W:

$$W = Q - \Delta U$$

3. Substitute $Q = -3.9 \times 10^5$ J and $\Delta U = -5.1 \times 10^5$ J to find the work:

$$W = -3.9 \times 10^5 \text{ J} - (-5.1 \times 10^5 \text{ J}) = 1.2 \times 10^5 \text{ J}$$

CONTINUED

INSIGHT
Notice how important it is to use the correct signs for Q, W, and ΔU. When the proper signs are used, the first law is simply a way of keeping track of all the energy exchanges—including mechanical work and heat—that can occur in a given system.

PRACTICE PROBLEM
After coasting for a few minutes you begin to row again, doing 5.7×10^5 J of work and decreasing your internal energy by 7.6×10^5 J. How much energy did you give off in the form of heat?   [**Answer:** $Q = -1.9 \times 10^5$ J. The negative sign means you have given off heat, as expected.]

Some related homework problems: Problem 3, Problem 5

Comparing Heat, Work, and Internal Energy   Just looking at the first law, $\Delta U = Q - W$, we can easily get the false impression that U, Q, and W are basically the same type of physical quantity. Certainly, they are all measured in the same units (J). In other respects, however, they are quite different. For example, the heat Q represents energy that flows through thermal contact. In contrast, the work W indicates a transfer of energy by the action of a force through a distance.

The most important distinction between these quantities, however, is the way they depend on the **state of a system,** which is determined by its temperature, pressure, and volume. The internal energy, for example, depends only on the state of a system, and not on how the system is brought to that state. A simple example is the ideal gas, where U depends only on the temperature T, and not on any previous values T may have had. Because U depends only on the *state* of a system—whether the system is an ideal gas or something more complicated—it is referred to as a **state function.**

On the other hand, Q and W are *not* state functions; they depend on the precise way—that is, on the process—by which a system is changed from one state, A, to another state, B. For example, one process connecting the states A and B may result in a heat $Q_1 = 19$ J and a work $W_1 = 32$ J. With a different process connecting the *same two states*, we may find a heat $Q_2 = -24$ J $\neq Q_1$ and a work $W_2 = -11$ J $\neq W_1$. Still, the difference in internal energy—which depends only on the initial and final states—must be the same for all processes connecting A and B. It follows that

$$\Delta U_{AB} = Q_1 - W_1 = Q_2 - W_2 = -13\,\text{J}$$

Therefore, the internal energy of this system always decreases by 13 J when it changes from state A to state B.

Enhance Your Understanding   (Answers given at the end of the chapter)

2. The following systems exchange energy with the external world in the form of heat Q and work W. Rank the systems in order of increasing change in internal energy, from most negative to most positive. Indicate ties where appropriate.

System	A	B	C	D
Q	10 J	−10 J	50 J	−50 J
W	10 J	10 J	−20 J	−20 J

Section Review

*   The first law of thermodynamics is a statement of energy conservation that combines both work and heat. It can be stated as $\Delta U = Q - W$, where U is the internal energy, Q is the heat, and W is the work. Q is positive when heat energy is added to the system, and W is positive when work is done on the system.

18-3 Thermal Processes

In this section we consider a variety of thermodynamic processes that can be used to change the state of a system. Among these are ones that take place at constant pressure, constant volume, or constant temperature. We also consider processes in which no heat is allowed to flow into or out of the system.

All of the processes discussed in this section are assumed to be **quasi-static,** which is a fancy way of saying they occur so slowly that at any given time the system and its surroundings are essentially in equilibrium. Thus, in a quasi-static process, the pressure and temperature are always uniform throughout the system. Furthermore, we assume that the system in question is free from friction or other dissipative forces.

These assumptions can be summarized by saying that the processes we consider are **reversible.** To be precise, a reversible process is defined by the following condition:

> For a process to be reversible, it must be possible to return both the system and its surroundings to exactly the same states they were in before the process began.

The fact that both the system and its surroundings must be returned to their initial states is the key element of this definition. For example, if there is friction between a piston and the cylinder in which it slides, a reversible process is not possible. Even if the piston is returned to its original location, the heat generated by friction will warm the cylinder and eventually flow into the surrounding air. Thus, it is not possible to "undo" the effects of friction, and such a process is said to be **irreversible.** Even without friction, a process can be irreversible if it occurs rapidly enough to cause effects such as turbulence, or if the system is far from equilibrium.

PHYSICS IN CONTEXT
Looking Ahead

Chapter 21 shows how an electric current can generate heat, much like the heating caused by friction as one object (like a piston) slides against another (like a cylinder wall).

Constant-temperature heat bath

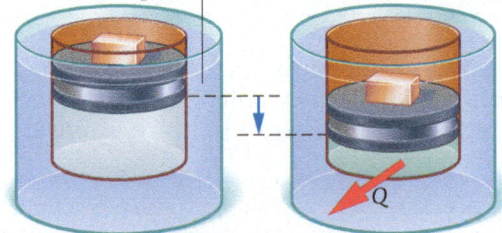

(a) Compressing at constant temperature expels heat.

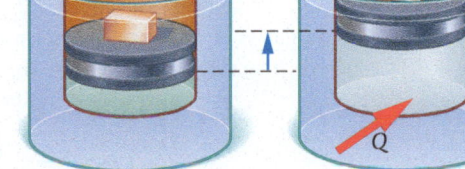

(b) Expanding at constant temperature draws in heat.

▲ **FIGURE 18-3 An idealized reversible process (a)** A piston is slowly moved downward, compressing a gas. In order for its temperature to remain constant, a heat Q goes from the gas into the constant-temperature heat bath, which may be nothing more than a large volume of water. **(b)** As the piston is allowed to slowly rise back to its initial position, it draws the same amount of heat Q from the heat bath that it gave to the bath in (a). Hence, both the system (gas) and the surroundings (heat bath) return to their initial states.

In practice, then, all real processes are irreversible to some extent. It is still possible, however, to have a process that closely approximates a perfectly reversible process, just as we can have systems that are practically free of friction. For example, in **FIGURE 18-3 (a)** we consider a "frictionless" piston that is slowly forced downward, while the gas in the cylinder is kept at constant temperature. An amount of heat, Q, goes from the gas to its surroundings in order to keep the temperature from rising. In **FIGURE 18-3 (b)** the piston is slowly moved back upward, drawing in the same heat Q from its surroundings to keep the temperature from dropping. In an "ideal" case like this, the process is reversible, because the system and its surroundings are left unchanged. We shall assume that all the processes described in this section are reversible.

Constant-Pressure Processes

We begin by considering a process that occurs at **constant pressure.** To be specific, suppose a gas with the pressure P_0 is held in a cylinder of cross-sectional area A, as in **FIGURE 18-4.** If the piston moves outward, so that the volume of the gas increases from an initial value V_i to a final value V_f, the process can be represented graphically as shown in **FIGURE 18-5.** Here we plot pressure P versus volume V in a PV plot; the process just described is the bold horizontal line segment with an arrow on it.

As the gas expands, it does work on the piston. First, the gas exerts a force on the piston equal to the pressure times the area:

$$F = P_0 A$$

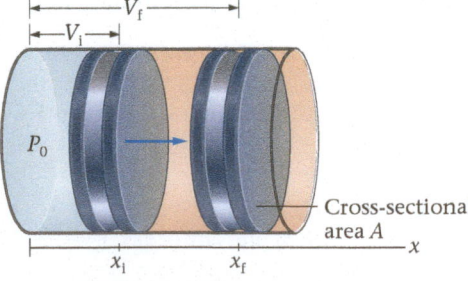

▲ **FIGURE 18-4 Work done by an expanding gas** A gas in a cylinder of cross-sectional area A expands with a constant pressure of P_0 from an initial volume $V_i = Ax_i$ to a final volume $V_f = Ax_f$. As it expands, it does the work $W = P_0(V_f - V_i)$.

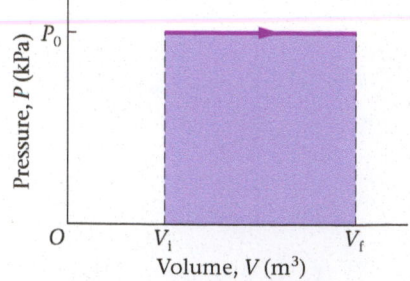

▲ **FIGURE 18-5 A constant-pressure process** A PV plot representing the constant-pressure process shown in Figure 18-4. The area of the shaded region, $P_0(V_f - V_i)$, is equal to the work done by the expanding gas in Figure 18-4.

PROBLEM-SOLVING NOTE

Work and the *PV* Diagram

When finding the work done by calculating the area on a PV diagram, recall that a pressure of 1 Pa times a volume of 1 m³ gives an energy equal to 1 J.

Second, the gas moves the piston from the position x_i to the position x_f, where $V_i = Ax_i$ and $V_f = Ax_f$, as indicated in Figure 18-4. Thus, the work done by the gas is the force times the distance through which the piston moves:

$$W = F(x_f - x_i) = P_0A(x_f - x_i) = P_0(Ax_f - Ax_i) = P_0(V_f - V_i)$$

In general, if the pressure, P, is constant, and the volume changes by the amount ΔV, the work done by the gas is

$$W = P\Delta V \qquad (constant\ pressure) \qquad 18\text{-}4$$

The gas does positive work if the volume increases, because ΔV is positive.

EXERCISE 18-2 WORK DONE BY A GAS

A gas with a constant pressure of 126 kPa expands from a volume of 0.982 m³ to a volume of 1.19 m³. How much work does the gas do on its surroundings?

REASONING AND SOLUTION
The work done by the gas can be found with $W = P\Delta V$. Substituting numerical values, we find

$$W = P\Delta V = (126\ \text{kPa})(1.19\ \text{m}^3 - 0.982\ \text{m}^3) = 26{,}200\ \text{J}$$

For comparison, this is the energy required to raise the temperature of 1.0 kg of water by 6.26 C°.

Looking closely at the PV plot in Figure 18-5, we can see that the work done by the gas is equal to the area under the horizontal line representing the constant-pressure process. In particular, the shaded region is a rectangle of height P_0 and width $\Delta V = V_f - V_i$, and therefore its area is

$$\text{area} = P_0(V_f - V_i) = W$$

Though this result was obtained for the special case of constant pressure, it applies to any process:

> The work done by an expanding gas is equal to the area under the curve representing the process in a PV plot.

This result is applied in the next Example.

EXAMPLE 18-3 WORK AREA

A gas expands from an initial volume of 0.40 m³ to a final volume of 0.62 m³ as the pressure increases linearly from 110 kPa to 230 kPa. Find the work done by the gas.

PICTURE THE PROBLEM
The sketch shows the process for this problem. As the volume increases from $V_i = 0.40$ m³ to $V_f = 0.62$ m³, the pressure increases linearly from $P_i = 110$ kPa to $P_f = 230$ kPa.

REASONING AND STRATEGY
The work done by this gas is equal to the shaded area in the sketch. We can calculate this area as the sum of the area of a rectangle plus the area of a triangle.

  In particular, the rectangle has a height P_i and a width $(V_f - V_i)$. Similarly, the triangle has a height $(P_f - P_i)$ and a base of $(V_f - V_i)$.

Known   Initial volume of gas, $V_i = 0.40$ m³; final volume of gas, $V_f = 0.62$ m³; initial pressure of gas, $P_i = 110$ kPa; final pressure of gas, $P_f = 230$ kPa.
Unknown Work done by gas, $W = ?$

SOLUTION

1. Calculate the area of the rectangular portion of the total area:

$$A_{\text{rectangle}} = P_i(V_f - V_i)$$
$$= (110\ \text{kPa})(0.62\ \text{m}^3 - 0.40\ \text{m}^3) = 2.4 \times 10^4\ \text{J}$$

CONTINUED

2. Next, calculate the area of the triangular portion of the total area:

$$A_{triangle} = \tfrac{1}{2}(P_f - P_i)(V_f - V_i)$$
$$= \tfrac{1}{2}(230\,\text{kPa} - 110\,\text{kPa})(0.62\,\text{m}^3 - 0.40\,\text{m}^3)$$
$$= 1.3 \times 10^4\,\text{J}$$

3. Sum these areas to find the work done by the gas:

$$W = A_{rectangle} + A_{triangle}$$
$$= 2.4 \times 10^4\,\text{J} + 1.3 \times 10^4\,\text{J} = 3.7 \times 10^4\,\text{J}$$

INSIGHT

We could also have solved this problem by noting that the pressure varies linearly, and hence its average value is simply $P_{av} = \tfrac{1}{2}(P_f + P_i) = 170\,\text{kPa}$. The work done by the gas, then, is $W = P_{av}\,\Delta V = (170\,\text{kPa})(0.22\,\text{m}^3) = 3.7 \times 10^4\,\text{J}$, as before.

PRACTICE PROBLEM

Suppose the pressure varies linearly from 110 kPa to 290 kPa. How much work does the gas do in this case?
[**Answer:** $W = 4.4 \times 10^4\,\text{J}$]

Some related homework problems: Problem 18, Problem 19

Constant-Volume Processes

Next, we consider a **constant-volume** process. Suppose, for example, that heat is added to a gas in a container of fixed volume, as in **FIGURE 18-6**, causing the pressure to increase. Noting that there is no displacement of any of the walls, we conclude that the force exerted by the gas does no work. Thus, for *any* constant-volume process, we have

$$W = 0 \qquad\qquad (constant\ volume)$$

From the first law of thermodynamics, $\Delta U = Q - W$, it follows that $\Delta U = Q$; that is, the change in internal energy is equal to the amount of heat that is added to or removed from the system. In addition, the conclusion that zero work is done in a constant-volume process is consistent with our earlier statement that work is equal to the area under the curve representing the process. For example, in **FIGURE 18-7** we show a constant-volume process in which the pressure is increased from P_i to P_f. This line is vertical, and hence the area under it is zero; that is, $W = 0$, as expected.

▲ **FIGURE 18-6 Adding heat to a system of constant volume** Heat is added to a system of constant volume, increasing its pressure from P_i to P_f. Since there is no displacement of the walls, there is no work done in this process. Therefore, $W = 0$ and the change in internal energy is simply equal to the heat added to or removed from the system, $\Delta U = Q$.

QUICK EXAMPLE 18-4   FIND THE TOTAL WORK

A gas undergoes the three-part process shown in the sketch, connecting the states A and B. The three parts of the process are labeled 1, 2, and 3. Find the total work done by the gas during this process.

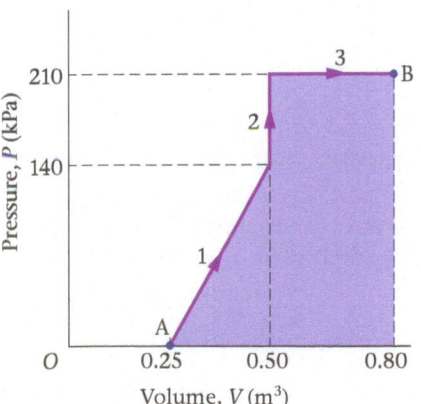

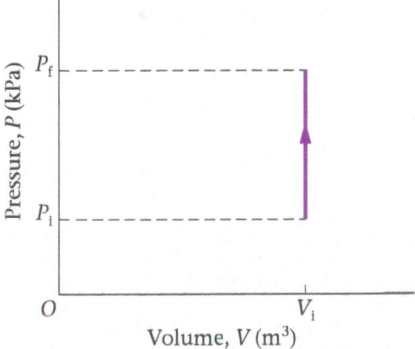

▲ **FIGURE 18-7 A constant-volume process** The pressure increases from P_i to P_f, just as in Figure 18-6, while the volume remains constant at its initial value, V_i. The area under this process is zero, as is the work.

REASONING AND SOLUTION

As before, the work done by the gas is equal to the shaded area in the sketch. We can calculate this area as the sum of the areas of each part of the total process.

1. Calculate the work done during part 1 of the total process:

$$W_1 = \tfrac{1}{2}(140\,\text{kPa})(0.25\,\text{m}^3) = 17{,}500\,\text{J}$$

2. Calculate the work done during part 2 of the total process:

$$W_2 = 0$$

CONTINUED

3. Calculate the work done during part 3 of the total process: $W_3 = (210 \text{ kPa})(0.30 \text{ m}^3) = 63{,}000 \text{ J}$

4. Sum the results to find the total work: $W = W_1 + W_2 + W_3 = 80{,}500 \text{ J}$

The work done by the gas would be different if the process had been a straight-line (uniform) expansion from point A to point B. In that case, the total work done would be 57,750 J. Thus, the work W—which is *not* a state function—depends on the process, as expected.

PHYSICS IN CONTEXT
Looking Back

Our understanding of ideal gases from Chapter 17 is put to good use here, in the study of common thermal processes.

Isothermal Processes

Another common process is one that takes place at constant temperature—that is, an **isothermal process.** For an ideal gas the isotherm has a relatively simple form. In particular, if T is constant, it follows that PV is a constant as well:

$$PV = NkT = \text{constant}$$

Thus, ideal-gas isotherms have the following pressure-volume relationship:

$$P = \frac{NkT}{V} = \frac{\text{constant}}{V}$$

This is illustrated in **FIGURE 18-8**, where we show several isotherms corresponding to different temperatures. (Recall that we've seen isotherms before, in Figure 17-5.)

▶ **FIGURE 18-8 Isotherms on a PV plot** These isotherms are for one mole of an ideal gas at the temperatures 300 K, 500 K, 700 K, and 900 K. Notice that each isotherm has the shape of a hyperbola. As the temperature is increased, however, the isotherms move farther from the origin. Thus, the pressure corresponding to a given volume increases with temperature, as one would expect.

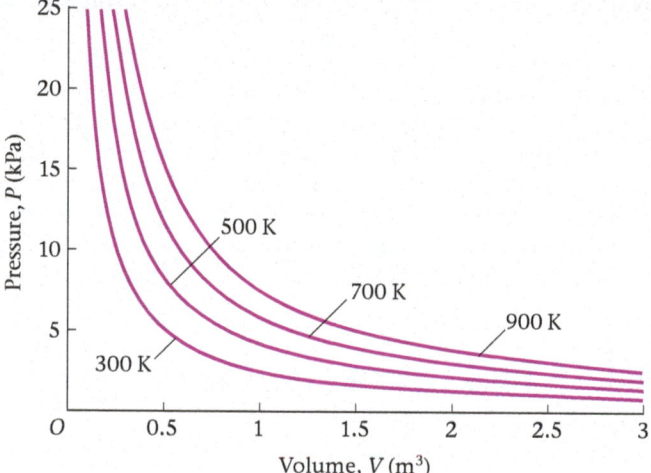

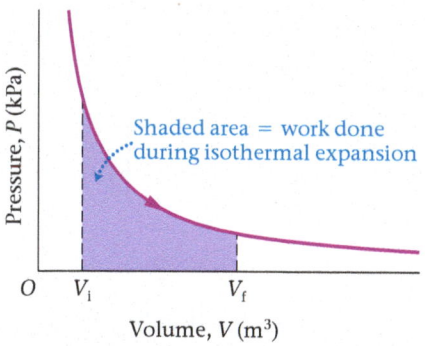

▲ **FIGURE 18-9 An isothermal expansion** In an isothermal expansion from the volume V_i to the volume V_f, the work done is equal to the shaded area. For n moles of an ideal gas at the temperature T, the work done by the gas is $W = nRT \ln(V_f/V_i)$.

For any given isotherm, such as the one shown in **FIGURE 18-9**, the work done by an expanding gas is equal to the area under the curve, as usual. In particular, the work done in expanding from V_i to V_f in Figure 18-9 is equal to the shaded area. This area may be derived by using the methods of calculus. The result is found to be

$$W = NkT \ln\left(\frac{V_f}{V_i}\right) = nRT \ln\left(\frac{V_f}{V_i}\right) \qquad (\textit{constant temperature}) \qquad \text{18-5}$$

The symbol "ln" stands for the natural logarithm; that is, log to the base e. This result is utilized in the next Example.

EXAMPLE 18-5   ISOTHERMAL HEAT FLOW

A cylinder holds 0.54 mol of a monatomic ideal gas at a temperature of 310 K. As the gas expands isothermally from an initial volume of $4.2 \times 10^{-3} \text{ m}^3$ to a final volume of $6.8 \times 10^{-3} \text{ m}^3$, determine the amount of heat that must be added to the gas to maintain a constant temperature.

CONTINUED

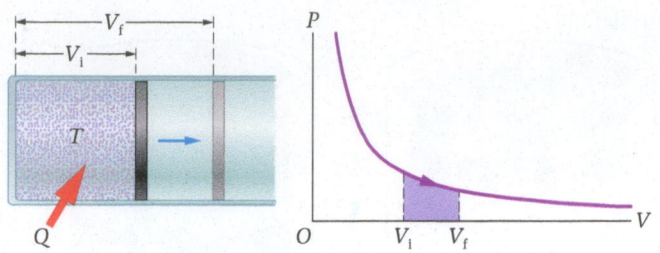

PICTURE THE PROBLEM

The physical process is illustrated on the left in the sketch. Notice that heat flows into the gas as it expands in order to keep its temperature constant at $T = 310$ K. The graph on the right is a PV plot showing the same process. The work done by the expanding gas is equal to the shaded area from $V_i = 4.2 \times 10^{-3}$ m³ to $V_f = 6.8 \times 10^{-3}$ m³.

REASONING AND STRATEGY

We can use the first law of thermodynamics, $\Delta U = Q - W$, to find the heat Q in terms of W and ΔU; that is, $Q = \Delta U + W$. We can find W and ΔU as follows: (i) The work W is found using $W = nRT \ln(V_f/V_i)$ (Equation 18-5). (ii) To determine ΔU, recall that the internal energy of an ideal gas depends only on the temperature. Specifically, the internal energy of a monatomic ideal gas is $U = \frac{3}{2} nRT$. Because the temperature is constant in this process, there is no change in internal energy; that is, $\Delta U = 0$.

Known   Moles of gas, $n = 0.54$ mol; constant temperature, $T = 310$ K; initial volume, $V_i = 4.2 \times 10^{-3}$ m³; final volume, $V_f = 6.8 \times 10^{-3}$ m³.

Unknown   Heat added to gas, $Q = ?$

SOLUTION

1. Solve the first law of thermodynamics for the heat, Q:

$$\Delta U = Q - W$$
$$Q = \Delta U + W$$

2. Calculate the change in the gas's internal energy:

$$\Delta U = \frac{3}{2} nR(T_f - T_i) = 0$$

3. Calculate the work done by the expanding gas:

$$W = nRT \ln\left(\frac{V_f}{V_i}\right)$$
$$= (0.54 \text{ mol})[8.31 \text{ J/(mol} \cdot \text{K)}](310 \text{ K}) \ln\left(\frac{6.8 \times 10^{-3} \text{ m}^3}{4.2 \times 10^{-3} \text{ m}^3}\right)$$
$$= 670 \text{ J}$$

4. Substitute numerical values to find Q:

$$Q = \Delta U + W = 0 + 670 \text{ J} = 670 \text{ J}$$

INSIGHT

Thus, when an ideal gas undergoes an isothermal expansion, the heat gained by the gas is equal to the work it does; that is, $Q = W$. This is a direct consequence of the first law of thermodynamics, $\Delta U = Q - W$, and the fact that $\Delta U = 0$ for an ideal gas at constant temperature.

PRACTICE PROBLEM

Find the final volume of this gas when it has expanded enough to do 730 J of work.
[**Answer:** $V_f = 7.1 \times 10^{-3}$ m³]

Some related homework problems: Problem 20, Problem 21

Adiabatic Processes

The final process we consider is one in which no heat flows into or out of a system. Such a process, in which $Q = 0$, is said to be **adiabatic.** One way to produce an adiabatic process is illustrated in **FIGURE 18-10 (a)**. Here we see a cylinder that is insulated well enough that no heat can pass through the insulation (*adiabatic* means, literally, "not passable"). When the piston is pushed downward in the cylinder—decreasing the volume and *compressing* the gas—the gas heats up and its pressure increases. Similarly, in **FIGURE 18-10 (b)**, an adiabatic *expansion* causes the temperature of the gas to decrease, as does the pressure.

What does an adiabatic process look like on a PV plot? Certainly, its general shape must be similar to that of an isotherm; in particular, as the volume is decreased, the pressure increases. However, it can't be identical to an isotherm because, as we have pointed out, the temperature changes during an adiabatic process. The comparison between an adiabatic curve and an isothermal curve is the subject of the following Conceptual Example.

Insulation

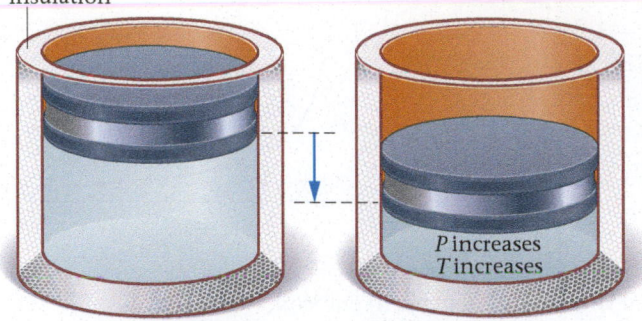

(a) Compressing with no heat flow increases P and T.

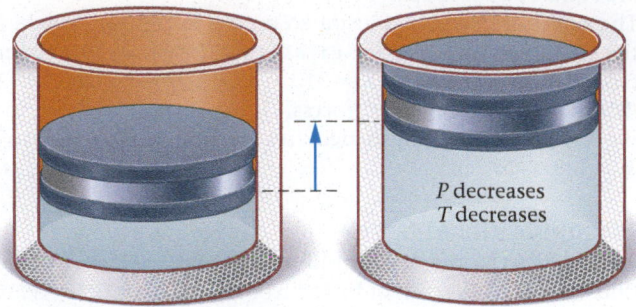

(b) Expanding with no heat flow decreases P and T.

▲ **FIGURE 18-10 An adiabatic process** In adiabatic processes, no heat flows into or out of the system. In the cases shown in this figure, heat flow is prevented by insulation. **(a)** An adiabatic compression increases both the pressure and the temperature. **(b)** An adiabatic expansion results in a decrease in pressure and temperature.

CONCEPTUAL EXAMPLE 18-6 PRESSURE VERSUS VOLUME

A certain gas has an initial volume and pressure given by point A in the PV plot. If the gas is compressed isothermally, its pressure rises as indicated by the curve labeled "Isotherm." If, instead, the gas is compressed adiabatically from point A, does its pressure follow curve i, curve ii, or curve iii?

REASONING AND DISCUSSION
In an isothermal compression, some heat flows out of the system in order for its temperature to remain constant. No heat flows out of the system in the adiabatic process, however, and therefore its temperature rises. As a result, the pressure is greater for any given volume when the compression is adiabatic, as compared to isothermal. It follows that the adiabatic curve, or adiabat, is represented by curve iii.

ANSWER
Curve iii is the adiabat.

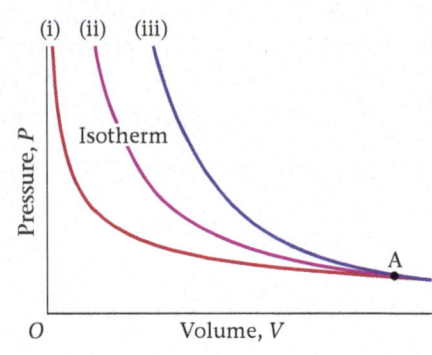

As we've seen, an adiabatic curve is similar to an isotherm, only steeper. The precise mathematical relationship describing an adiabat is presented in the next section.

EXAMPLE 18-7 WORK INTO ENERGY

When a certain gas is compressed adiabatically, the amount of work done on it is 730 J. Find the change in internal energy of the gas.

PICTURE THE PROBLEM
Our sketch shows a piston being pushed downward, compressing a gas in an insulated cylinder. The insulation ensures that no heat can flow—as required in an adiabatic process.

REASONING AND STRATEGY
We know that 730 J of work is done on the gas, and we know that no heat is exchanged ($Q = 0$) in an adiabatic process. Thus, we can find ΔU by substituting Q and W into the first law of thermodynamics. One note of caution: Be careful to use the correct sign for the work. In particular, recall that the work done *on* a system is *negative*.

Known   Work, $W = -730\,\text{J}$; process = adiabatic.
Unknown Change in internal energy, $\Delta U = ?$

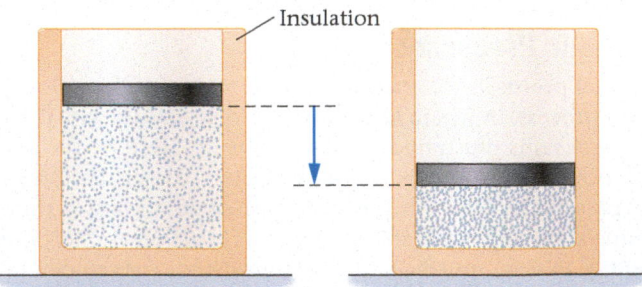

Insulation

SOLUTION
1. Identify the work and heat for this process:

$W = -730\,\text{J}$
$Q = 0$

CONTINUED

2. Substitute Q and W into the first law of thermodynamics to find the change in internal energy, ΔU:

$$\Delta U = Q - W = 0 - (-730\,\text{J}) = 730\,\text{J}$$

INSIGHT

Because no energy can enter or leave the system in the form of heat, all the work done on the system goes into increasing its internal energy. This follows from the first law of thermodynamics, $\Delta U = Q - W$, which for $Q = 0$ reduces to $\Delta U = -W$. As a result, the temperature of the gas increases.

A familiar example of this type of effect is the heating that occurs when you pump air into a tire or a ball—the work done on the pump appears as an increased temperature. The effect occurs in the reverse direction as well. When air is let out of a tire, for example, it does work on the atmosphere as it expands, producing a cooling effect that can be quite noticeable. In extreme cases, the cooling can be great enough to create frost on the valve stem of the tire.

PRACTICE PROBLEM

If a system's internal energy decreases by 550 J in an adiabatic process, how much work was done by the system? [**Answer:** $W = +550\,\text{J}$]

Some related homework problems: Problem 17, Problem 24

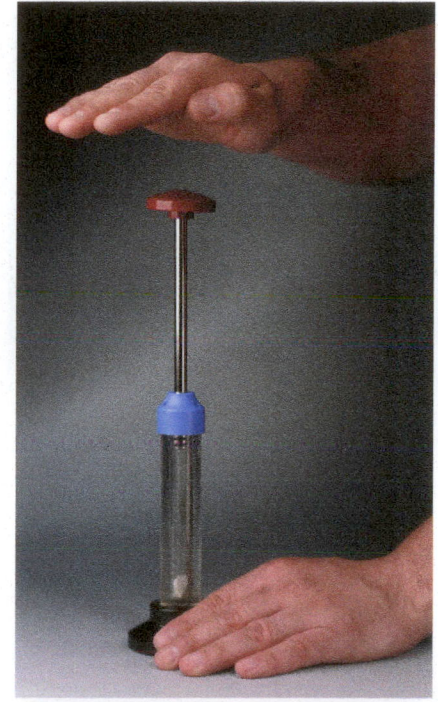

▶ **FIGURE 18-11 Adiabatic heating** When a piston that fits snugly inside a cylinder is pushed downward rapidly, the temperature of the gas within the cylinder increases before there is time for heat to flow out of the system. Thus, the process is essentially adiabatic. As a result, the temperature of the gas can increase enough to ignite bits of paper in the cylinder. In a diesel engine, the same principle is used to ignite an air-gasoline mixture without a spark plug.

Examples of Adiabatic Processes An adiabatic process can occur when the system is thermally insulated, as in Figure 18-10, or in a system where the change in volume occurs rapidly. For example, if an expansion or compression happens quickly enough, there is no time for heat flow to occur. As a result, the process is adiabatic, even if there is no insulation.

RWP* An example of a rapid process is shown in **FIGURE 18-11**. Here, a piston is fitted into a cylinder that contains a certain volume of gas and a small piece of tissue paper. If the piston is driven downward rapidly—by a sharp impulsive blow, for example—the gas is compressed before heat has a chance to flow. As a result, the temperature of the gas rises rapidly. In fact, the rise in temperature can be enough for the paper to burst into flames.

The same principle applies to the operation of a diesel engine. As you may know, a diesel differs from a standard internal combustion engine in that it has no spark plugs. It doesn't need them. Instead of using a spark to ignite the fuel in a cylinder, it uses adiabatic heating. Fuel and air are admitted into the cylinder; then the piston rapidly compresses the air-fuel mixture. Just as with the piece of paper in Figure 18-11, the rising temperature is sufficient to ignite the fuel and run the engine.

The characteristics of constant-pressure, constant-volume, isothermal, and adiabatic processes are summarized in Table 18-2.

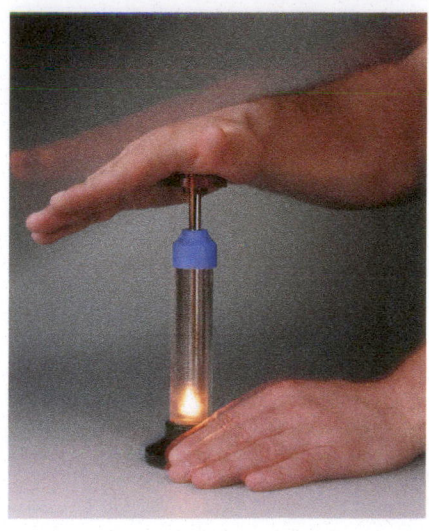

TABLE 18-2 Thermodynamic Processes and Their Characteristics

Constant pressure	$W = P\Delta V$	$Q = \Delta U + P\Delta V$
Constant volume	$W = 0$	$Q = \Delta U$
Isothermal (constant temperature)	$W = Q$	$\Delta U = 0$
Adiabatic (no heat flow)	$W = -\Delta U$	$Q = 0$

Enhance Your Understanding   (Answers given at the end of the chapter)

3. The pressure of a system doubles during an isothermal process. What happens to its volume?

Section Review

- An isothermal process is one that occurs with no change in temperature. An adiabatic process is one that occurs with no energy transfer in the form of heat.

---

*Real World Physics applications are denoted by the acronym RWP.

18-4 Specific Heats for an Ideal Gas: Constant Pressure, Constant Volume

Recall that the specific heat of a substance is the amount of heat needed to raise the temperature of 1 kg of the substance by 1 Celsius degree. As we know, however, the amount of heat depends on the type of process used to raise the temperature. Thus, we should specify, for example, whether a specific heat applies to a process at constant pressure or constant volume.

Constant Volume A constant-volume process is illustrated in **FIGURE 18-12**. Here we see an ideal gas of mass m in a container of fixed volume V. A heat Q flows into the container. As a result of this added heat, the temperature of the gas rises by the amount ΔT, and its pressure increases as well. Now, the specific heat at constant volume, c_v, is defined by the following relationship:

$$Q_v = mc_v\Delta T$$

In what follows, it will be more convenient to use the **molar specific heat,** denoted by a capital letter C, which is defined in terms of the number of moles rather than the mass of the substance. Thus, if a gas contains n moles, its molar specific heat at constant volume is given by

$$Q_v = nC_v\Delta T$$

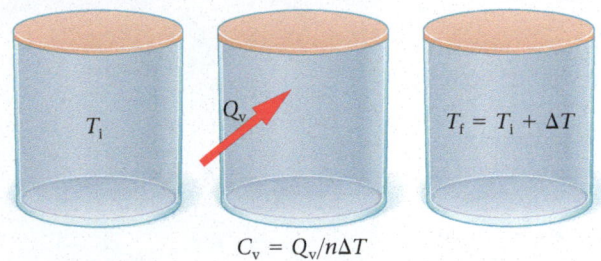

$$C_v = Q_v/n\Delta T$$

▲ **FIGURE 18-12 Heating at constant volume** If the heat Q_v is added to n moles of gas, and the temperature rises by ΔT, the molar specific heat at constant volume is $C_v = Q_v/n\Delta T$. No mechanical work is done at constant volume.

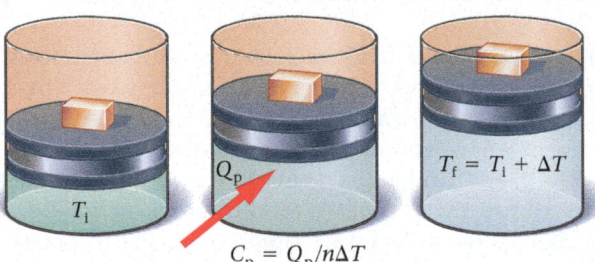

$$C_p = Q_p/n\Delta T$$

▲ **FIGURE 18-13 Heating at constant pressure** If the heat Q_p is added to n moles of gas, and the temperature rises by ΔT, the molar specific heat at constant pressure is $C_p = Q_p/n\Delta T$. Note that the heat Q_p must increase the temperature *and* do mechanical work by lifting the piston.

Constant Pressure Similarly, a constant-pressure process is illustrated in **FIGURE 18-13**. In this case, the gas is held in a container with a movable piston that applies a constant pressure P. As a heat Q is added to the gas, its temperature increases, which causes the piston to rise—after all, if the piston didn't rise, the pressure of the gas would increase. If the temperature of the gas increases by the amount ΔT, the molar specific heat at constant pressure is given by

$$Q_p = nC_p\Delta T$$

We would now like to obtain a relationship between C_p and C_v.

Before we carry out the calculation, let's consider the qualitative relationship between these specific heats. This is addressed in the following Conceptual Example.

CONCEPTUAL EXAMPLE 18-8 COMPARING SPECIFIC HEATS

How does the molar specific heat at constant pressure, C_p, compare with the molar specific heat at constant volume, C_v, for a given system? **(a)** $C_p > C_v$; **(b)** $C_p = C_v$; **(c)** $C_p < C_v$.

REASONING AND DISCUSSION

In a constant-volume process, as in Figure 18-12, the heat that is added to a system goes entirely into increasing the temperature, since no work is done. On the other hand, at constant pressure, the heat added to a system increases the temperature

CONTINUED

and does mechanical work. This is illustrated in Figure 18-13, where we see that the heat must not only raise the temperature, but also supply enough energy to lift the piston. Thus, more heat is required in the constant-pressure process, and hence that specific heat is greater.

ANSWER

(a) $C_p > C_v$; the specific heat at constant pressure is greater than the specific heat at constant volume.

Comparing C_v and C_p We turn now to a detailed calculation of both C_v and C_p for a monatomic ideal gas. To begin, we rearrange the first law of thermodynamics, $\Delta U = Q - W$, to solve for the heat, Q:

$$Q = \Delta U + W$$

Recall from the previous section, however, that the work is zero, $W = 0$, for any constant-volume process. Hence, for constant volume we have

$$Q_v = \Delta U$$

Recalling that $U = \frac{3}{2}NkT = \frac{3}{2}nRT$ for a monatomic ideal gas, we find

$$Q_v = \Delta U = \frac{3}{2}nR\Delta T$$

Comparing with the definition of the molar specific heat, it follows that:

> **Molar Specific Heat for a Monatomic Ideal Gas at Constant Volume**
>
> $C_v = \frac{3}{2}R$ 18-6

Let's perform a similar calculation for constant pressure. In this case, we know that the work done by a gas is $W = P\Delta V$. In an ideal gas, for which $PV = nRT$, it follows that

$$W = P\Delta V = nR\Delta T$$

Combining this with the first law of thermodynamics yields

$$Q_p = \Delta U + W$$
$$= \frac{3}{2}nR\Delta T + nR\Delta T = \frac{5}{2}nR\Delta T$$

Applying the definition of molar specific heat yields:

> **Molar Specific Heat for a Monatomic Ideal Gas at Constant Pressure**
>
> $C_p = \frac{5}{2}R$ 18-7

As expected, the specific heat at constant pressure is larger than the specific heat at constant volume, and the difference is precisely the extra contribution due to the work done in lifting the piston in the constant-pressure case. In particular, we see that

$$C_p - C_v = R \qquad \text{18-8}$$

Though this relationship was derived for a monatomic ideal gas, it holds for all ideal gases, regardless of the structure of their molecules. It is also a good approximation for most real gases, as can be seen in Table 18-3.

EXERCISE 18-9 FIND THE HEAT

Find the heat required to raise the temperature of 0.209 mol of a monatomic ideal gas by 7.45 C° at (a) constant volume and (b) constant pressure.

REASONING AND SOLUTION

In this system, we know that $n = 0.209$ mol and $\Delta T = 7.45$ C°. The value of the specific heat depends on the type of process.

(a) To find the heat required at constant volume we use $Q_v = nC_v\Delta T$ with $C_v = \frac{3}{2}R$. This gives

$$Q_v = nC_v\Delta T = \frac{3}{2}nR\Delta T = \frac{3}{2}(0.209\text{ mol})[8.31\text{ J}/(\text{mol}\cdot\text{K})](7.45\text{ K}) = 19.4\text{ J}$$

CONTINUED

PHYSICS IN CONTEXT
Looking Back

The specific heat, defined in Chapter 16, is revisited in this section. In this case, we've extended the concept to processes that occur at constant pressure or constant volume.

TABLE 18-3 $C_p - C_v$ **for Various Gases**

Helium	$0.995R$
Nitrogen	$1.00R$
Oxygen	$1.00R$
Argon	$1.01R$
Carbon dioxide	$1.01R$
Methane	$1.01R$

PROBLEM-SOLVING NOTE

Constant Volume Versus Constant Pressure

The heat required to increase the temperature of an ideal gas depends on whether the process is at constant pressure or constant volume. More heat is required when the process occurs at constant pressure.

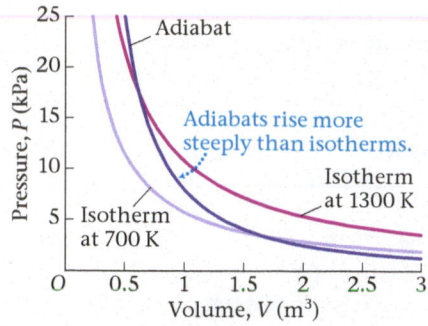

▲ FIGURE 18-14 A comparison between isotherms and adiabats Two isotherms are shown, one for 700 K and one for 1300 K. An adiabat is also shown. Note that the adiabat is a steeper curve than the isotherms.

PHYSICS IN CONTEXT
Looking Ahead

The expansion of the universe is discussed in Chapter 32. During the expansion, the temperature of the universe has decreased, very much as it does for an ideal gas undergoing an adiabatic expansion.

(b) For constant pressure we use $Q_p = nC_p\Delta T$ with $C_p = \frac{5}{2}R$. This gives

$$Q_p = nC_p\Delta T = \frac{5}{2}nR\Delta T = \frac{5}{2}(0.209 \text{ mol})[8.31 \text{ J/(mol·K)}](7.45 \text{ K}) = 32.3 \text{ J}$$

As expected, the required heat is greater at constant pressure.

Adiabatic Processes

We return now briefly to a consideration of adiabatic processes. As we shall see, the relationship between C_p and C_v is important in determining the behavior of a system undergoing an adiabatic process.

FIGURE 18-14 shows an adiabatic curve and two isotherms. As mentioned before, the adiabatic curve is steeper and it cuts across the isotherms. For the isotherms, we recall that the curves are described by the equation

$$PV = \text{constant} \qquad (isothermal)$$

A similar equation applies to adiabats. In this case, calculus may be used to show that the appropriate equation is

$$PV^\gamma = \text{constant} \qquad (adiabatic) \qquad 18\text{-}9$$

In this expression, the constant γ is the ratio C_p/C_v:

$$\gamma = \frac{C_p}{C_v} = \frac{\frac{5}{2}R}{\frac{3}{2}R} = \frac{5}{3}$$

This value of γ applies to monatomic ideal gases—and is a good approximation for monatomic real gases as well. The value of γ is different, however, for gases that are diatomic, triatomic, and so on.

EXAMPLE 18-10 HOT AIR

A container with an initial volume of 0.0607 m³ holds 2.45 moles of a monatomic ideal gas at a temperature of 325 K. The gas is now compressed adiabatically to a volume of 0.0377 m³. Find **(a)** the final pressure and **(b)** the final temperature of the gas.

PICTURE THE PROBLEM
Our sketch shows a gas being compressed in an insulated container. The gas starts with a temperature of 325 K, but because no heat can flow outward through the insulation (adiabatic process), the work done on the gas results in an increase in temperature.

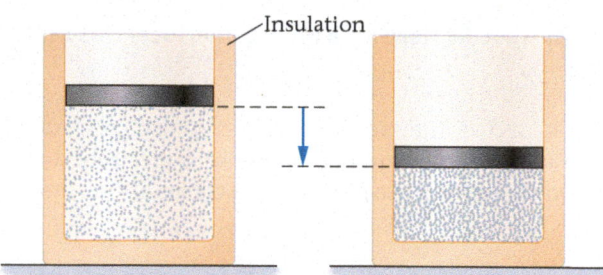

REASONING AND STRATEGY

a. We can find the final pressure as follows: (i) Find the initial pressure using the ideal-gas equation of state, $P_iV_i = nRT_i$. (ii) Next, let $P_iV_i^\gamma = P_fV_f^\gamma$, since this is an adiabatic process. Solve this equation for the final pressure.

b. Use the final pressure and volume to find the final temperature, using the ideal-gas equation: $P_fV_f = nRT_f$.

Known Initial volume, $V_i = 0.0607 \text{ m}^3$; final volume, $V_f = 0.0377 \text{ m}^3$; number of moles, $n = 2.45 \text{ mol}$; initial temperature, $T_i = 325 \text{ K}$; process = adiabatic.
Unknown **(a)** Final pressure, $P_f = ?$ **(b)** Final temperature, $T_f = ?$

SOLUTION

Part (a)

1. Find the initial pressure, using $PV = nRT$:

$$P_i = \frac{nRT_i}{V_i}$$

$$= \frac{(2.45 \text{ mol})[8.31 \text{ J/(mol·K)}](325 \text{ K})}{0.0607 \text{ m}^3} = 109 \text{ kPa}$$

2. Use $PV^\gamma = \text{constant}$ to find a relationship for P_f:

$$P_iV_i^\gamma = P_fV_f^\gamma$$

$$P_f = P_i\left(\frac{V_i}{V_f}\right)^\gamma$$

CONTINUED

3. Substitute numerical values:

$$P_f = (109 \text{ kPa})\left(\frac{0.0607 \text{ m}^3}{0.0377 \text{ m}^3}\right)^{5/3} = 241 \text{ kPa}$$

Part (b)

4. Use $PV = nRT$ to solve for the final temperature:

$$T_f = \frac{P_f V_f}{nR}$$

$$= \frac{(241 \text{ kPa})(0.0377 \text{ m}^3)}{(2.45 \text{ mol})[8.31 \text{ J/(mol} \cdot \text{K)}]} = 446 \text{ K}$$

INSIGHT

Decreasing the volume of the gas increased its pressure from 109 kPa to 241 kPa, and at the same time increased its temperature from 325 K to 446 K. This is a specific example of adiabatic heating. Adiabatic cooling is the reverse effect, where the temperature of a gas decreases as its volume increases.

PRACTICE PROBLEM

In part (a), to what volume must the gas be compressed to yield a final pressure of 278 kPa?   [**Answer:** $V_f = 0.0346 \text{ m}^3$]

Some related homework problems: Problem 39, Problem 40

RWP Adiabatic heating and cooling can have important effects on the climate of a given region. For example, moisture-laden winds blowing from the Pacific Ocean into western Oregon are deflected upward when they encounter the Cascade Mountains. As the air rises, the atmospheric pressure decreases (see Chapter 15), allowing the air to expand and undergo adiabatic cooling. The result is that the moisture in the air condenses to form clouds and precipitation on the west side of the mountains. In some cases, where the air holds relatively little moisture, this mechanism may result in isolated, lens-shaped (lenticular) clouds just above the peak of a mountain. An example is shown in **FIGURE 18-15**. When the winds continue on to the east side of the mountains, they have little moisture remaining; thus, eastern Oregon is in the *rain shadow* of the Cascade Mountains. In addition, as the air descends on the east side of the mountains, it undergoes adiabatic heating. These are the primary reasons the summers in western Oregon are moist and mild, while the summers in eastern Oregon are hot and dry.

▲ **FIGURE 18-15** A spectacular lenticular (lens-shaped) cloud floats above a mountain in Tierra del Fuego, at the southern tip of Chile. Lenticular clouds are often seen "parked" above and just downwind of high mountain peaks, even when there are no other clouds in the sky. The reason is that as moisture-laden winds are deflected upward by the mountain, the moisture they contain cools due to adiabatic expansion and condenses to form a cloud.

Enhance Your Understanding   (Answers given at the end of the chapter)

4. Rank the following systems in order of the amount of heat needed to produce the indicated temperature change, from least to greatest. Indicate ties where appropriate.

System	A	B	C	D
Moles, n	1	2	4	10
ΔT (K)	2	10	10	1
Type of process	constant pressure	constant volume	constant pressure	constant volume

Section Review

- The molar specific heat for a monatomic ideal gas at constant volume is $C_v = \frac{3}{2}R$. The corresponding molar specific heat for constant pressure is $C_p = \frac{5}{2}R$.

18-5 The Second Law of Thermodynamics

Have you ever warmed your hands by pressing them against a block of ice? Probably not. But if you think about it, you might wonder why it doesn't work. After all, the first law of thermodynamics would be satisfied if energy simply flowed from the ice to your hands. The ice would get colder while your hands got warmer, and the energy of the universe would remain the same.

As we know, however, this sort of thing just doesn't happen—the spontaneous flow of heat is *always* from warmer objects to cooler objects, and never in the reverse direction. This simple observation, in fact, is one of many ways of expressing the **second law of thermodynamics:**

> **Second Law of Thermodynamics: Heat Flow**
> When objects of different temperatures are brought into thermal contact, the spontaneous flow of heat that results is always from the high-temperature object to the low-temperature object. Spontaneous heat flow never proceeds in the reverse direction.

Thus, the second law of thermodynamics is more restrictive than the first law; it says that of all the processes that conserve energy, only those that proceed in a certain direction actually occur. In a sense, the second law implies a definite "directionality" to the behavior of nature. For this reason, the second law is sometimes referred to as the "arrow of time."

For example, suppose you saw a movie that showed a snowflake landing on a person's hand and melting to a small drop of water. Nothing would seem particularly noteworthy about the scene from a physics point of view. But if the movie showed a drop of water on a person's hand suddenly freeze into the shape of a snowflake, and then lift off the person's hand into the air, it wouldn't take long to realize the film was running backward. It's clear in which direction time should "flow."

We study further consequences of the second law in the next few sections. As we do so, we'll find other more precise, but equivalent, ways of stating the second law.

Enhance Your Understanding (Answers given at the end of the chapter)

5. Is any process that conserves energy allowed in a physical system? Give an example to support your answer.

Section Review

- The second law of thermodynamics states that heat always flows from a hot object to a cool object, and never in the reverse direction.

(a)

(b)

▲ **FIGURE 18-16 Visualizing Concepts**
Heat Engines (a) A modern version of Hero's engine, invented by the Greek mathematician and engineer Hero of Alexandria. In this simple heat engine, the steam that escapes from a heated vessel of water is directed tangentially, causing the vessel to rotate. This converts the thermal energy supplied to the water into mechanical energy, in the form of rotational motion. **(b)** A steam engine of slightly more recent design hauls passengers up and down Mt. Washington in New Hampshire. Note that the locomotive is belching two clouds, one black and one white. Can you explain their origin?

18-6 Heat Engines and the Carnot Cycle

RWP A **heat engine,** simply put, is a device that converts heat into work. A simple, easy-to-understand example is shown in **FIGURE 18-16 (a)**, but the classic example is the steam engine, as shown in **FIGURE 18-16 (b)**. The basic elements of a steam engine are illustrated in **FIGURE 18-17**. First, some form of fuel (oil, wood, coal, etc.) is used to vaporize water in the boiler. The resulting steam is then allowed to enter the engine itself, where it expands against a piston, doing mechanical work. As the piston moves, it causes gears or wheels to rotate, which delivers the mechanical work to the external world. After leaving the engine, the steam proceeds to the condenser, where it gives off heat to the cool air in the atmosphere and condenses to liquid form.

All heat engines have the following features in common:

A high-temperature reservoir that supplies heat to the engine (the boiler in the steam engine)

A low-temperature reservoir where "waste" heat is released (the condenser in the steam engine)

An engine that converts thermal energy to mechanical work in a cyclic fashion

These features are illustrated schematically in **FIGURE 18-18**. In addition, though not shown in the figure, heat engines have a working substance (steam in the steam engine) that causes the engine to operate.

To begin our analysis of heat engines, we note that a certain amount of heat, Q_h, is supplied to the engine from the high temperature or "hot" reservoir during each cycle.

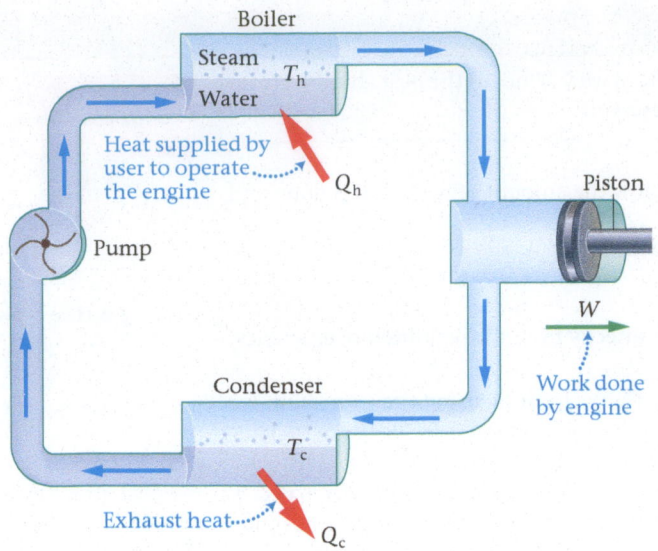

▲ **FIGURE 18-17 A schematic steam engine** The basic elements of a steam engine are a boiler—in which heat converts water to steam—and a piston that can be displaced by the expanding steam. In some engines, the steam is simply exhausted into the atmosphere after it has expanded against the piston. More sophisticated engines send the exhaust steam to a condenser, where it is cooled and condensed back to liquid water, then recycled to the boiler.

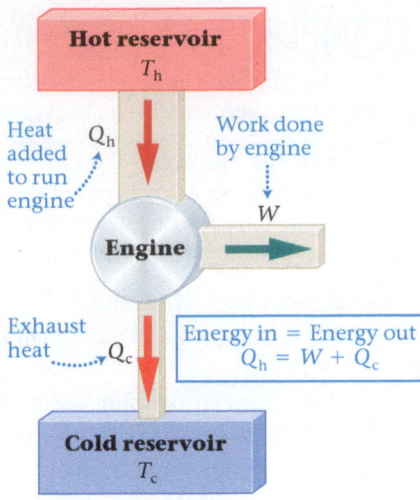

▲ **FIGURE 18-18 A schematic heat engine** The engine absorbs a heat Q_h from the hot reservoir, performs the work W, and gives off the heat Q_c to the cold reservoir. Energy conservation gives $Q_h = W + Q_c$, where Q_h and Q_c are the magnitudes of the hot- and cold-temperature heats.

Of this heat, a fraction appears as work, W, during each cycle, and the rest is given off as waste heat, Q_c, at a relatively low temperature to the "cold" reservoir. There is no change in energy for the engine, because it returns to its initial state at the completion of each cycle. To simplify notation, we shall let Q_h and Q_c denote magnitudes for the remainder of this chapter, so that both quantities are positive. With this convention, energy conservation can be written as follows:

$$\text{energy in } (Q_h) = \text{energy out } (W + Q_c)$$

Rearranging this relationship, we find that the work done during a cycle is

$$W = Q_h - Q_c \qquad \text{18-10}$$

As we shall see, the second law of thermodynamics requires that heat engines must *always* exhaust a finite amount of heat to a cold reservoir.

The Efficiency of a Heat Engine   Of particular interest for any engine is its **efficiency**, *e*, which is simply the fraction of the heat supplied to the engine that appears as work. Thus, we define the efficiency to be

$$e = \frac{W}{Q_h} \qquad \text{18-11}$$

Using the energy-conservation result just derived in Equation 18-10 ($W = Q_h - Q_c$), we find:

Efficiency of a Heat Engine, *e*

$$e = \frac{W}{Q_h} = \frac{Q_h - Q_c}{Q_h} = 1 - \frac{Q_c}{Q_h} \qquad \text{18-12}$$

SI unit: dimensionless

For example, if $e = 0.20$, we say that the engine is 20% efficient. In this case, 20% of the input heat is converted to work, $W = 0.20Q_h$, and 80% goes to waste heat, $Q_c = 0.80Q_h$. Efficiency can be thought of as the ratio of how much you receive (work) to how much you have to pay to run the engine (input heat).

PHYSICS IN CONTEXT
Looking Ahead

In Chapter 23 we shall see how the mechanical work produced by a heat engine can be converted to electrical energy by a generator, in the form of an electric current.

EXAMPLE 18-11 HEAT INTO WORK

A heat engine with an efficiency of 24.0% performs 1250 J of work. Find **(a)** the heat absorbed from the hot reservoir and **(b)** the heat given off to the cold reservoir.

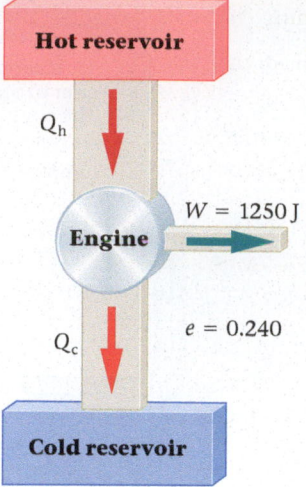

PICTURE THE PROBLEM

Our sketch shows a schematic of the heat engine. We know the amount of work that is done and the efficiency of the engine. We seek the heats Q_h and Q_c.

Note that an efficiency of 24.0% means that $e = 0.240$.

REASONING AND STRATEGY

a. We can find the heat absorbed from the hot reservoir directly from the definition of efficiency, $e = W/Q_h$.

b. We can find Q_c by using energy conservation, $W = Q_h - Q_c$, or by using the expression for efficiency in terms of the heats, $e = 1 - Q_c/Q_h$.

Known Efficiency, $e = 0.240$; work done, $W = 1250$ J.
Unknown **(a)** Heat absorbed from hot reservoir, $Q_h = ?$ **(b)** Heat given off to cold reservoir, $Q_c = ?$

SOLUTION

Part (a)

1. Use $e = W/Q_h$ to solve for the heat Q_h:

$$e = \frac{W}{Q_h}$$
$$Q_h = \frac{W}{e} = \frac{1250\,\text{J}}{0.240} = 5210\,\text{J}$$

Part (b)

2. Use energy conservation to solve for Q_c:

$$W = Q_h - Q_c$$
$$Q_c = Q_h - W = 5210\,\text{J} - 1250\,\text{J} = 3960\,\text{J}$$

3. As a double check, use the efficiency $(e = 1 - Q_c/Q_h)$ to find Q_c:

$$e = 1 - \frac{Q_c}{Q_h}$$
$$Q_c = (1 - e)Q_h = (1 - 0.240)(5210\,\text{J}) = 3960\,\text{J}$$

INSIGHT

When the efficiency of a heat engine is less than one-half (50%), as in this case, the amount of heat given off as waste to the cold reservoir is greater than the amount of heat converted to work.

PRACTICE PROBLEM

What is the efficiency of a heat engine that does 1250 J of work and gives off 5250 J of heat to the cold reservoir? **[Answer:** $e = 0.192$**]**

Some related homework problems: Problem 43, Problem 44

A temperature difference is essential to the operation of a heat engine. As heat flows from the hot to the cold reservoir in Figure 18-18, for example, the heat engine is able to tap into that flow and convert part of it to work—the greater the efficiency of the engine, the more heat that is converted to work. The second law of thermodynamics imposes limits, however, on the maximum efficiency a heat engine can have. We explore these limits next.

The Carnot Cycle and Maximum Efficiency

In 1824, the French engineer Sadi Carnot (1796–1832) published a book entitled *Reflections on the Motive Power of Fire* in which he considered the following question: Under what conditions will a heat engine have maximum efficiency? To address this question, let's consider a heat engine that operates between a single hot reservoir at the fixed temperature T_h and a single cold reservoir at the fixed temperature T_c. Carnot's result, known today as **Carnot's theorem,** can be expressed as follows:

Carnot's Theorem

If an engine operating between two constant-temperature reservoirs is to have maximum efficiency, it must be an engine in which all processes are reversible.

In addition, all reversible engines operating between the same two temperatures, T_c and T_h, have the same efficiency.

We should point out that no real engine can ever be perfectly reversible, just as no surface can be perfectly frictionless. Nonetheless, the concept of a reversible engine is a useful idealization.

Carnot's theorem is remarkable for a number of reasons. First, consider what the theorem says: No engine, no matter how sophisticated or technologically advanced, can exceed the efficiency of a reversible engine. We can strive to improve the technology of heat engines, but there is an upper limit to the efficiency that can never be exceeded. Second, the theorem is just as remarkable for what it does not say. It says nothing, for example, about the working substance that is used in the engine—it is as valid for a liquid or solid working substance as for one that is gaseous. Furthermore, it says nothing about the type of reversible engine that is used, what the engine is made of, or how it is constructed. Diesel engine, jet engine, rocket engine—none of these things matter. In fact, all that *does* matter are the two temperatures, T_c and T_h.

Recall that the efficiency of a heat engine can be written as follows:

$$e = 1 - \frac{Q_c}{Q_h}$$

Because the efficiency e depends only on the temperatures T_c and T_h, according to Carnot's theorem, it follows that Q_c/Q_h must also depend only on T_c and T_h. In fact, Lord Kelvin used this observation to propose that, instead of using a thermometer to measure temperature, we measure the efficiency of a heat engine and from this determine the temperature. Thus, he suggested that we *define* the ratio of the temperatures of two reservoirs, T_c/T_h, to be equal to the ratio of the heats Q_c/Q_h:

$$\frac{Q_c}{Q_h} = \frac{T_c}{T_h}$$

If we choose the size of a degree in this temperature scale to be equal to 1 C°, then we have, in fact, defined the Kelvin temperature scale discussed in Chapter 16. Thus, if T_h and T_c are given in kelvins, the maximum efficiency of a heat engine is:

Maximum Efficiency of a Heat Engine

$$e_{max} = 1 - \frac{T_c}{T_h} \qquad \text{18-13}$$

Limits to Efficiency Suppose for a moment that we *could* construct an ideal engine, perfectly reversible and free from all forms of friction. Would this ideal engine have 100% efficiency? No, it would not. From Equation 18-13 $[e_{max} = 1 - (T_c/T_h)]$ we can see that the only way the efficiency of a heat engine could be 100% (that is, $e_{max} = 1$) would be for T_c to be 0 K. As we shall see in the last section of this chapter, this is ruled out by the third law of thermodynamics. Hence, the maximum efficiency will always be less than 100%. No matter how perfect the engine, some of the input heat will always be wasted—given off as Q_c—rather than converted to work.

Since the efficiency is defined to be $e = W/Q_h$, it follows that the maximum work a heat engine can do with the input heat Q_h is

$$W_{max} = e_{max}Q_h = \left(1 - \frac{T_c}{T_h}\right)Q_h \qquad \text{18-14}$$

If the hot and cold reservoirs have the same temperature, so that $T_c = T_h$, it follows that the maximum efficiency is zero. As a result, the amount of work that such an engine can do is also zero. As mentioned before, a heat engine requires different temperatures

PROBLEM-SOLVING NOTE

Maximum Efficiency

The maximum efficiency a heat engine can have is determined solely by the temperature of the hot (T_h) and cold (T_c) reservoirs. The numerical value of this efficiency is $e = 1 - T_c/T_h$. Remember, however, that the temperatures must be expressed in the Kelvin scale for this expression to be valid.

Big Idea **2** Heat engines use a difference in temperature to convert thermal energy to mechanical energy. The maximum efficiency of the conversion is determined solely by the operating temperatures of the engine.

in order to operate. For example, for a fixed T_c, the higher the temperature of T_h, the greater the efficiency.

Finally, even though Carnot's theorem may seem quite different from the second law of thermodynamics, they are, in fact, equivalent. It can be shown, for example, that if Carnot's theorem were violated, it would be possible for heat to flow spontaneously from a cold object to a hot object.

CONCEPTUAL EXAMPLE 18-12 COMPARING EFFICIENCIES

Suppose you have a heat engine that can operate in one of two different modes. In mode 1, the temperatures of the two reservoirs are $T_c = 200$ K and $T_h = 400$ K; in mode 2, the temperatures are $T_c = 400$ K and $T_h = 600$ K. Is the maximum efficiency of mode 1 greater than, less than, or equal to the maximum efficiency of mode 2?

REASONING AND DISCUSSION

At first, you might think that because the temperature difference is the same in the two modes, the maximum efficiency is the same as well. This is not the case, however, because the maximum efficiency depends on the *ratio* of the two temperatures ($e = 1 - T_c/T_h$) rather than on their difference. In this case, the maximum efficiency of mode 1 is $e_1 = 1 - (200/400) = 1/2$ and the maximum efficiency of mode 2 is $e_2 = 1 - (400/600) = 1/3$. Thus, mode 1, even though it operates at the lower temperatures, is more efficient.

ANSWER

The efficiency of mode 1 is greater than the efficiency of mode 2.

QUICK EXAMPLE 18-13 FIND THE TEMPERATURE

If the heat engine in Example 18-11 is operating at its maximum efficiency of 24.0%, and its cold reservoir is at a temperature of 295 K, what is the temperature of the hot reservoir?

REASONING AND SOLUTION

Our sketch shows the schematic for this heat engine. We know that the maximum efficiency of the engine is 0.240, and that the temperature of the cold reservoir is $T_c = 295$ K.

We can solve the efficiency equation ($e = 1 - T_c/T_h$) for the temperature of the hot reservoir, T_h.

1. Write the efficiency, e, in terms of the hot and cold temperatures:

$$e = 1 - \frac{T_c}{T_h}$$

2. Solve for T_h:

$$T_h = T_c/(1 - e)$$

3. Substitute numerical values for T_c and e to find T_h:

$$T_h = \frac{295 \text{ K}}{(1 - 0.240)} = 388 \text{ K}$$

Although an efficiency of 24.0% may seem low, it is typical of many real engines.

We can summarize the conclusions of this section as follows: The first law of thermodynamics states that you cannot get something for nothing. To be specific, you cannot get more work out of a heat engine than the amount of heat you put in. The best you can do is break even. The second law of thermodynamics is more restrictive than the first law; it says that *you can't even break even*—some of the input heat must be wasted. It's a law of nature.

Enhance Your Understanding   (Answers given at the end of the chapter)

6. Heat engines A and B operate with different hot and cold reservoirs, as indicated in **FIGURE 18-19**. Is the maximum efficiency of engine A greater than, less than, or equal to the maximum efficiency of engine B? Explain.

CONTINUED

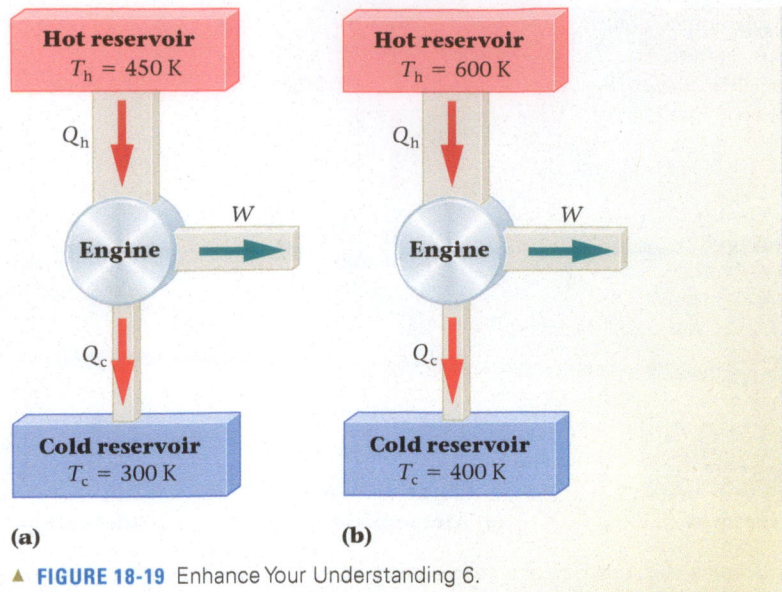

(a)   **(b)**

▲ **FIGURE 18-19** Enhance Your Understanding 6.

Section Review

- The maximum efficiency of a heat engine is $e_{max} = 1 - T_c/T_h$, where the temperatures are measured on the Kelvin scale.

18-7 Refrigerators, Air Conditioners, and Heat Pumps

When we stated the second law of thermodynamics in Section 18-5, we said that the spontaneous flow of heat is always from high temperature to low temperature. The key word here is *spontaneous*. It *is* possible for heat to flow "uphill," from a cold object to a hot one, but it doesn't happen spontaneously—work must be done on the system to make it happen, just as work must be done to pump water up from a well. Refrigerators, air conditioners, and heat pumps are devices that use work to transfer heat from a cold object to a hot object.

RWP Let's begin by comparing the operations of a heat engine, **FIGURE 18-20 (a)**, and a **refrigerator, FIGURE 18-20 (b)**. Notice that all the arrows are reversed in the refrigerator; in effect, *a refrigerator is a heat engine running backward*. In particular, the refrigerator uses a work W to remove a certain amount of heat, Q_c, from the cold reservoir (the interior of the refrigerator). It then exhausts an even larger amount of heat, Q_h, to the hot reservoir (the air in the kitchen). By energy conservation, it follows that

$$Q_h = Q_c + W$$

Thus, as a refrigerator operates, it warms the kitchen at the same time that it cools the food stored within it.

To design an effective refrigerator, you would like it to remove as much heat from its interior as possible for the smallest amount of work. After all, the work is supplied by electrical energy that must be paid for each month. Thus, we define the **coefficient of performance (COP) for a refrigerator** as an indicator of its effectiveness:

Coefficient of Performance for a Refrigerator, COP

$$COP = \frac{Q_c}{W} \qquad\qquad 18\text{-}15$$

SI unit: dimensionless

Typical values for the coefficient of performance range from 2 to 6.

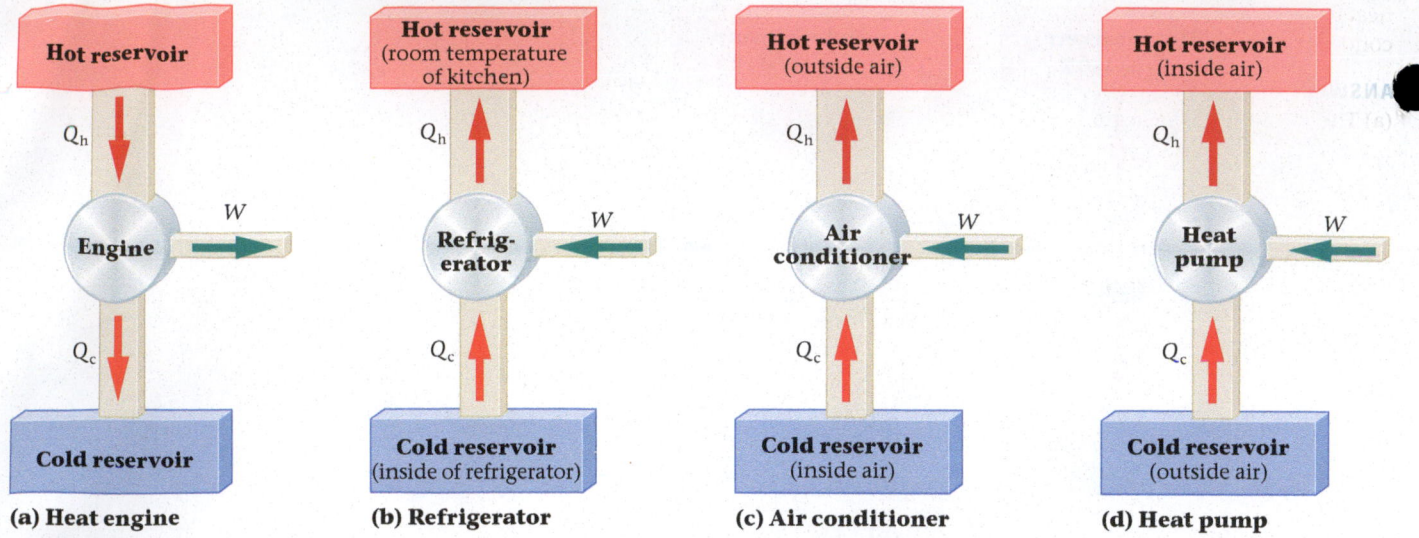

▲ **FIGURE 18-20 Cooling and heating devices** Schematic comparison of a generalized heat engine **(a)** with a refrigerator, an air conditioner, and a heat pump. **(b)** In the refrigerator, a work W is done to remove a heat Q_c from the cold reservoir (the inside of the refrigerator). By energy conservation, the heat given off to the hot reservoir (the kitchen) is $Q_h = W + Q_c$. **(c)** An air conditioner removes heat from the cool air inside a house and exhausts it into the hot air of the atmosphere. **(d)** A heat pump moves heat from a cold reservoir to a hot reservoir, just like an air conditioner. The difference is that the reservoirs are switched, so that heat is pumped into the house rather than out to the atmosphere.

EXERCISE 18-14   WORK SUPPLIED

A refrigerator has a coefficient of performance of 3.25. How much work must be supplied to this refrigerator in order to remove 261 J of heat from its interior?

REASONING AND SOLUTION
Rearranging COP $= Q_c/W$ to solve for the work yields

$$W = \frac{Q_c}{\text{COP}} = \frac{261\,\text{J}}{3.25} = 80.3\,\text{J}$$

Thus, 80.3 J of work removes 261 J of heat from the refrigerator, and exhausts 80.3 J + 261 J = 341 J of heat into the kitchen.

RWP An **air conditioner,** as illustrated in **FIGURE 18-20 (c)**, is basically a refrigerator in which the cold reservoir is the room that is being cooled. To be specific, the air conditioner uses electrical energy to "pump" heat from the cool room to the warmer air outside. As with the refrigerator, more heat is exhausted to the hot reservoir than is removed from the cold reservoir; that is, $Q_h = Q_c + W$, as before.

CONCEPTUAL EXAMPLE 18-15   ROOM TEMPERATURE   PREDICT/EXPLAIN

You haven't had time to install your new air conditioner in the window yet, so as a short-term measure you decide to place it on the dining-room table and turn it on to cool things off a bit. **(a)** As a result, does the air in the dining room get warmer, get cooler, or stay at the same temperature? **(b)** Which of the following is the *best explanation* for your prediction?

  I. The room gets warmer due to the combined effect of the heat exhausted by the air conditioner and the work done by its motor.

 II. The room gets cooler because the air conditioner cools the air.

III. The room temperature stays the same because the warm air in the exhaust of the air conditioner offsets the cool air it produces.

REASONING AND DISCUSSION
You might think the room stays at the same temperature, because the air conditioner draws heat from the room as usual, but then exhausts heat back into the room that would normally be sent outside. However, the motor of the air conditioner is doing work in order to draw heat from the room, and the heat that would normally be exhausted outdoors is equal to the

CONTINUED

heat drawn from the room *plus* the work done by the motor: $Q_h = Q_c + W$. Thus, the net effect is that the motor of the air conditioner is continually adding heat to the room, causing it to get warmer.

ANSWER
(a) The air in the dining room gets warmer. **(b)** The best explanation is I.

RWP A **heat pump** can be thought of as an air conditioner with the reservoirs switched. As we see in **FIGURE 18-20 (d)**, a heat pump does a work W to remove an amount of heat Q_c from the cold reservoir of outdoor air, then exhausts a heat Q_h into the hot reservoir of air in the room. Just as with the refrigerator and the air conditioner, the heat going to the hot reservoir is $Q_h = Q_c + W$.

In an **ideal**, reversible heat pump with only two operating temperatures, T_c and T_h, the Carnot relationship $Q_c/Q_h = T_c/T_h$ holds, just as it does for a heat engine. Thus, if you want to add a heat Q_h to a room, the work that must be done to accomplish this is

$$W = Q_h - Q_c = Q_h\left(1 - \frac{Q_c}{Q_h}\right) = Q_h\left(1 - \frac{T_c}{T_h}\right) \qquad \text{18-16}$$

We use this result in the next Example.

EXAMPLE 18-16   PUMPING HEAT

An ideal heat pump, one that satisfies the Carnot relationship $W = Q_h(1 - T_c/T_h)$, is used to heat a room that is at 293 K. If the pump does 275 J of work, how much heat does it supply to the room if the outdoor temperature is **(a)** 273 K or **(b)** 263 K?

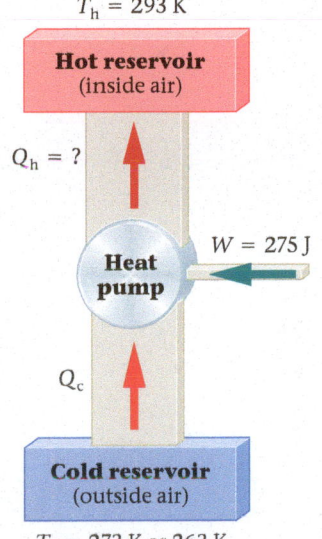

$T_h = 293$ K

Hot reservoir
(inside air)

$Q_h = ?$

$W = 275$ J

Heat pump

Q_c

Cold reservoir
(outside air)

$T_c = 273$ K or 263 K

PICTURE THE PROBLEM
Our sketch shows the heat pump doing 275 J of mechanical work to transfer the heat Q_h to the hot reservoir at the temperature 293 K. The temperature of the cold reservoir is 273 K for part (a) and 263 K for part (b). We wish to determine Q_h for each of these cases.

REASONING AND STRATEGY
For an ideal heat pump, we know that $W = Q_h(1 - T_c/T_h)$. Therefore, given the hot and cold temperatures, as well as the mechanical work, it is straightforward to determine the heat delivered to the hot reservoir, Q_h.

Known   Temperature of room (hot reservoir), $T_h = 293$ K; work done by pump, $W = 275$ J; temperature of outdoors (cold reservoir), **(a)** $T_c = 273$ K, **(b)** $T_c = 263$ K.
Unknown   Heat supplied to room (hot reservoir), $Q_h = ?$

SOLUTION

1. Solve $W = Q_h(1 - T_c/T_h)$ for the heat Q_h: $Q_h = W/(1 - T_c/T_h)$

Part (a)

2. Substitute $W = 275$ J, $T_c = 273$ K, and $T_h = 293$ K into the expression for Q_h:

$$Q_h = \frac{W}{1 - \dfrac{T_c}{T_h}} = \frac{275\,\text{J}}{1 - \dfrac{273\,\text{K}}{293\,\text{K}}} = 4030\,\text{J}$$

Part (b)

3. Substitute $W = 275$ J, $T_c = 263$ K, and $T_h = 293$ K into the expression for Q_h:

$$Q_h = \frac{W}{1 - \dfrac{T_c}{T_h}} = \frac{275\,\text{J}}{1 - \dfrac{263\,\text{K}}{293\,\text{K}}} = 2690\,\text{J}$$

INSIGHT
As one might expect, the same amount of work provides less heat when the outside temperature is lower. That is, more work must be done on a colder day to provide the same heat to the inside air.

In addition, notice that if 275 J of heat is supplied to an electric heater, then 275 J of heat is given to the air in the room. When that same energy is used to run a heat pump, a good deal more than 275 J of heat is added to the room.

CONTINUED

PRACTICE PROBLEM

How much work must be done by this heat pump to supply 2550 J of heat on a day when the outside temperature is 253 K?
[**Answer:** $W = 348$ J]

Some related homework problems: Problem 58, Problem 59

The purpose of a heat pump is to add heat to a room; in fact, we would like to add as much heat (Q_h) as possible for the least work (W). As a result, the **coefficient of performance (COP) for a heat pump** is defined as follows:

Coefficient of Performance for a Heat Pump, COP

$$\text{COP} = \frac{Q_h}{W} \qquad \qquad \text{18-17}$$

SI unit: dimensionless

The COP for a heat pump, which usually ranges from 3 to 4, depends on the inside and outside temperatures. We show how to use the COP in the next Exercise.

EXERCISE 18-17   WORK REQUIRED

A heat pump with a coefficient of performance equal to 3.65 supplies 3220 J of heat to a room. How much work is required?

REASONING AND SOLUTION

Work is related to heat and the coefficient of performance by $\text{COP} = Q_h/W$. Rearranging this equation, and solving for the work, W, we find

$$W = \frac{Q_h}{\text{COP}} = \frac{3220 \, \text{J}}{3.65} = 882 \, \text{J}$$

Enhance Your Understanding   (Answers given at the end of the chapter)

7. If the coefficient of performance (COP) of a particular refrigerator is increased, does the work required to remove a certain amount of heat from the cold reservoir increase, decrease, or remain the same? Explain.

Section Review

- A heat engine can be run in reverse to serve as a refrigerator, a device that uses work to "pump" heat from a cold reservoir to a hot reservoir.

18-8 Entropy

In this section we introduce a new quantity that is as fundamental to physics as energy or temperature. This quantity is referred to as the *entropy*, and it is related to the amount of disorder in a system. For example, a messy room has more entropy than a neat one, a pile of bricks has more entropy than a building constructed from the bricks, a freshly laid egg has more entropy than one that is just about to hatch, and a puddle of water has more entropy than the block of ice from which it melted. We begin by considering the connection between entropy and heat, and later develop more fully the connection between disorder and entropy.

When discussing heat engines, we saw that if an engine is reversible, it satisfies the following relationship:

$$\frac{Q_c}{Q_h} = \frac{T_c}{T_h}$$

Rearranging slightly, we can rewrite this as

$$\frac{Q_c}{T_c} = \frac{Q_h}{T_h}$$

Notice that the quantity Q/T is the same for both the hot and the cold reservoirs. This relationship prompted the German physicist Rudolf Clausius (1822–1888) to propose the following definition: The **entropy**, S, is a quantity whose change is given by the heat Q divided by the absolute temperature T:

Definition of Entropy Change, ΔS

$$\Delta S = \frac{Q}{T} \qquad \text{18-18}$$

SI unit: J/K

For this definition to be valid, the heat Q must be transferred reversibly at the fixed Kelvin temperature T. Notice that if heat is added to a system ($Q > 0$), the entropy of the system increases; if heat is removed from a system ($Q < 0$), its entropy decreases.

Entropy is a state function, just like the internal energy, U. This means that the value of S depends only on the state of a system, and not on how the system gets to that state. It follows, then, that the *change* in entropy, ΔS, depends only on the initial and final states of a system. Thus, if a process is irreversible—so that Equation 18-18 ($\Delta S = Q/T$) *does not* hold—we can still calculate ΔS by using a reversible process to connect the same initial and final states.

PROBLEM-SOLVING NOTE

Calculating Entropy Change

When calculating the entropy change, $\Delta S = Q/T$, be sure to convert the temperature T to kelvins. It is *only* when this is done that the correct value for ΔS can be obtained.

EXAMPLE 18-18   MELTS IN YOUR HAND

(a) Calculate the change in entropy when a 0.125-kg chunk of ice melts at 0 °C. Assume the melting occurs reversibly.
(b) Suppose heat is now drawn reversibly from the 0 °C meltwater, causing a decrease in entropy of 112 J/K. How much ice freezes in the process?

PICTURE THE PROBLEM

In our sketch we show a 0.125-kg chunk of ice at the temperature 0 °C. As the ice absorbs the heat Q from its surroundings, it melts to water at 0 °C. Because the system absorbs heat, its entropy increases. When heat is drawn out of the meltwater, the entropy will decrease.

REASONING AND STRATEGY

a. The entropy change is $\Delta S = Q/T$, where $T = 0\,°C = 273$ K. To find the heat Q, we note that to melt the ice we must add to it the latent heat of fusion, L_f. Thus, the heat is $Q = mL_f$, where $L_f = 33.5 \times 10^4$ J/kg.

b. The magnitude of heat withdrawn from the meltwater is $Q = T|\Delta S|$, where $\Delta S = -112$ J/K. Using this heat, we can calculate the mass of re-frozen ice using $m = Q/L_f$.

Known   **(a)** Mass of ice, $m = 0.125$ kg; temperature, $T = 0\,°C = 273$ K. **(b)** Change in entropy, $\Delta S = -112$ J/K.
Unknown   **(a)** Change in entropy, $\Delta S = ?$ **(b)** Mass of ice that freezes, $m = ?$

SOLUTION

Part (a)

1. Find the heat that must be absorbed by the ice for it to melt:
$$Q = mL_f = (0.125\text{ kg})(33.5 \times 10^4\text{ J/kg}) = 4.19 \times 10^4\text{ J}$$

2. Calculate the change in entropy:
$$\Delta S = \frac{Q}{T} = \frac{4.19 \times 10^4\text{ J}}{273\text{ K}} = 153\text{ J/K}$$

Part (b)

3. Find the magnitude of the heat that is removed from the meltwater:
$$|Q| = T|\Delta S| = (273\text{ K})(112\text{ J/K}) = 3.06 \times 10^4\text{ J}$$

4. Use this heat to determine the amount of water that is re-frozen:
$$m = \frac{|Q|}{L_f} = \frac{(3.06 \times 10^4\text{ J})}{(33.5 \times 10^4\text{ J/kg})} = 0.0913\text{ kg}$$

INSIGHT

Notice that we were careful to convert the temperature of the system from 0 °C to 273 K before we applied $\Delta S = Q/T$. This must always be done when calculating the entropy change. If we had neglected to do the conversion in this case, we would have found an infinite increase in entropy—which is clearly not possible physically.

In addition, notice that the entropy change is positive in part (a) and negative in part (b). This illustrates the general rule that entropy increases when heat is added to a system, and decreases when it is removed.

CONTINUED

PRACTICE PROBLEM
Find the mass of ice that would be required to give an entropy change of 264 J/K. [**Answer:** $m = 0.215$ kg]

Some related homework problems: Problem 65, Problem 66

Entropy and Heat Engines Let's apply the definition of entropy change to the case of a reversible heat engine. First, a heat Q_h leaves the hot reservoir at the temperature T_h. Thus, the entropy of this reservoir decreases by the amount Q_h/T_h:

$$\Delta S_h = -\frac{Q_h}{T_h}$$

Recall that Q_h is the magnitude of the heat leaving the hot reservoir; hence, the minus sign is used to indicate a decrease in entropy. Similarly, heat is added to the cold reservoir; hence, its entropy increases by the amount Q_c/T_c:

$$\Delta S_c = \frac{Q_c}{T_c}$$

The total entropy change for this system is

$$\Delta S_{total} = \Delta S_h + \Delta S_c = -\frac{Q_h}{T_h} + \frac{Q_c}{T_c}$$

Recalling that $Q_h/T_h = Q_c/T_c$, as pointed out at the beginning of this section, we find that the total entropy change vanishes:

$$\Delta S_{total} = -\frac{Q_h}{T_h} + \frac{Q_c}{T_c} = 0$$

What is special about a reversible engine, then, is the fact that its entropy does not change.

On the other hand, a real engine will always have a lower efficiency than a reversible engine operating between the same temperatures. This means that in a real engine less of the heat from the hot reservoir is converted to work; hence, more heat is given off as waste heat to the cold reservoir. Thus, for a given value of Q_h, the heat Q_c is greater in an irreversible engine than in a reversible one. As a result, instead of $Q_c/T_c = Q_h/T_h$, we have

$$\frac{Q_c}{T_c} > \frac{Q_h}{T_h}$$

Therefore, if an engine is irreversible, the total entropy change is positive:

$$\Delta S_{total} = -\frac{Q_h}{T_h} + \frac{Q_c}{T_c} > 0$$

In general, *any irreversible process results in an increase of entropy*.

These results can be summarized in the following general statements:

Entropy in the Universe
The total entropy of the universe *increases* whenever an *irreversible* process occurs.

The total entropy of the universe is *unchanged* whenever a *reversible* process occurs.

Because all *real* processes are irreversible (with reversible processes being a useful idealization), the total entropy of the universe continually increases. Thus, in terms of the entropy, the universe moves in only one direction—toward an ever-increasing entropy. This is quite different from the behavior with regard to energy, which remains constant no matter what type of process occurs.

In fact, this statement about entropy in the universe is yet another way of expressing the second law of thermodynamics. Recall, for example, that our original statement of the second law said that heat flows spontaneously from a hot object to a cold object. During this flow of heat the entropy of the universe increases, as we show in the next Example. Hence, the direction in which heat flows is seen to be the result of the

PROBLEM-SOLVING NOTE

Entropy Change

Though it is tempting to treat entropy like energy, setting the final value equal to the initial value, this is not the case in general. Only in a reversible process is the entropy unchanged—otherwise, it increases. Still, the entropy of part of a system can decrease, as long as the entropy of other parts increases by the same amount or more.

Big Idea **3** Entropy is a measure of disorder in a physical system. All thermodynamic processes proceed in a direction that either increases entropy or leaves it unchanged. The entropy of the universe never decreases.

general principle of entropy increase in the universe. Again, we see a directionality in nature—the ever-present "arrow of time."

EXAMPLE 18-19   ENTROPY IS NOT CONSERVED!

A hot reservoir at the temperature 576 K transfers 1050 J of heat irreversibly to a cold reservoir at the temperature 305 K. Find the change in entropy of the universe.

PICTURE THE PROBLEM

The relevant physical situation is shown in our sketch. Notice that the heat $Q = 1050$ J is transferred from the hot reservoir at the temperature $T_h = 576$ K directly to the cold reservoir at the temperature $T_c = 305$ K.

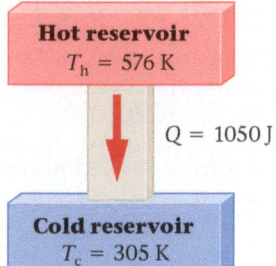

REASONING AND STRATEGY

As the heat Q leaves the hot reservoir, its entropy *decreases* by Q/T_h. When the same heat Q enters the cold reservoir, its entropy *increases* by the amount Q/T_c. Summing these two contributions gives the entropy change of the universe.

Known   Temperature of hot reservoir, $T_h = 576$ K; temperature of cold reservoir, $T_c = 305$ K; heat, $Q = 1050$ J.

Unknown   Change in entropy of the universe, $\Delta S_{universe} = ?$

SOLUTION

1. Calculate the entropy change of the hot reservoir. This entropy change is negative:

$$\Delta S_h = -\frac{Q}{T_h} = -\frac{1050 \text{ J}}{576 \text{ K}} = -1.82 \text{ J/K}$$

2. Calculate the entropy change of the cold reservoir. This entropy change is positive:

$$\Delta S_c = \frac{Q}{T_c} = \frac{1050 \text{ J}}{305 \text{ K}} = 3.44 \text{ J/K}$$

3. Sum these contributions to obtain the entropy change of the universe:

$$\Delta S_{universe} = \Delta S_h + \Delta S_c = -\frac{Q}{T_h} + \frac{Q}{T_c}$$
$$= -1.82 \text{ J/K} + 3.44 \text{ J/K} = 1.62 \text{ J/K}$$

INSIGHT

The decrease in entropy of the hot reservoir is more than made up for by the increase in entropy of the cold reservoir. This is a general result.

PRACTICE PROBLEM

What amount of heat must be transferred between these reservoirs for the entropy of the universe to increase by 1.50 J/K? [**Answer:** $Q = 972$ J]

Some related homework problems: Problem 67, Problem 71

When certain processes occur, it sometimes appears as if the entropy of the universe has decreased. On closer examination, however, it can always be shown that there is a larger increase in entropy elsewhere that results in an overall increase. This issue is addressed in the next Conceptual Example.

CONCEPTUAL EXAMPLE 18-20   ENTROPY CHANGE

You put a tray of water in the kitchen freezer, and some time later it has turned to ice. Has the entropy of the universe increased, decreased, or stayed the same?

REASONING AND DISCUSSION

It might seem that the entropy of the universe has decreased. After all, heat is removed from the water to freeze it, and, as we know, removing heat from an object lowers its entropy. On the other hand, we also know that the freezer does work to draw heat from the water; hence, it exhausts more heat into the kitchen than it absorbs from the water. As a result, the entropy increase of the heated air in the kitchen is greater than the magnitude of the entropy decrease of the water. Thus, the entropy of the universe increases—as it must for *any* real process.

ANSWER

The entropy of the universe has increased.

Useful Thermal Energy As the entropy in the universe increases, the amount of work that can be done is diminished. For example, the heat flow in Example 18-19 resulted in an increase in the entropy of the universe by the amount 1.62 J/K. If this same heat had been used in a reversible engine, however, it could have done work, and because the engine is reversible, the entropy of the universe would stay the same. In the next Example, we calculate the work that could be done with a reversible engine in this situation.

EXAMPLE 18-21 FIND THE WORK

Suppose a reversible heat engine operates between the two heat reservoirs described in Example 18-19. Find the amount of work done by such an engine when 1050 J of heat is drawn from the hot reservoir.

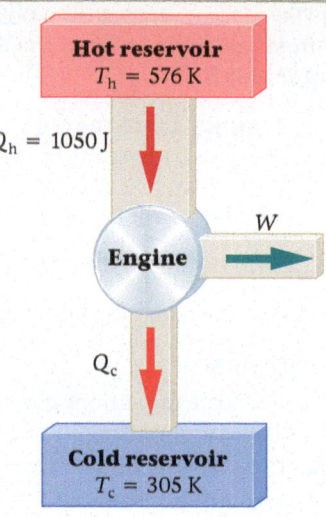

PICTURE THE PROBLEM
Our sketch shows a reversible heat engine operating between a hot reservoir at 576 K and a cold reservoir at 305 K.

REASONING AND STRATEGY
We first calculate the efficiency of this engine using $e = 1 - T_c/T_h$. We then use $W = eQ_h$ to find the work done.

Known Temperature of hot reservoir, $T_h = 576$ K; temperature of cold reservoir, $T_c = 305$ K; heat, $Q = 1050$ J.
Unknown Work done, $W = ?$

SOLUTION

1. Calculate the efficiency of this engine:
$$e = 1 - \frac{T_c}{T_h} = 1 - \left(\frac{305\text{ K}}{576\text{ K}}\right) = 0.470$$

2. Multiply the efficiency by Q_h to find the work done:
$$W = eQ_h = (0.470)(1050\text{ J}) = 494\text{ J}$$

INSIGHT
This engine is reversible, and hence its total entropy change must be zero. The decrease in entropy of the hot reservoir is $-(1050\text{ J})/576\text{ K} = -1.82$ J/K. It follows that the increase in entropy of the cold reservoir must have the same magnitude. The amount of heat that flows into the cold reservoir is $Q_c = Q_h - W = 1050\text{ J} - 494\text{ J} = 556\text{ J}$. This heat causes an entropy increase equal to $(556\text{ J})/(305\text{ K}) = +1.82$ J/K, as expected. If the engine were irreversible, it would exhaust more than 556 J of heat, and this would create a net increase in entropy and a reduction in the amount of work done. Clearly, then, a reversible engine produces the maximum amount of work.

PRACTICE PROBLEM
Suppose this engine is not reversible, and that only 457 J of work is done when 1050 J of heat is drawn from the hot reservoir. What is the entropy increase of the universe in this case? **[Answer: 0.12 J/K]**

Some related homework problems: Problem 71, Problem 86

The Quality of Energy When 1050 J of heat is transferred directly from the hot reservoir to the cold reservoir in Example 18-19, the entropy of the universe increases by 1.62 J/K. When this same heat is transferred reversibly, with an ideal engine, the entropy of the universe stays the same, but 494 J of work is done. The connection between the work done by a reversible engine, and the entropy increase of the irreversible process, is very simple:

$$W = T_c \Delta S_{\text{universe}} = (305\text{ K})(1.62\text{ J/K}) = 494\text{ J}$$

Thus, the flow of thermal energy can increase the entropy of the universe, or it can be tapped to produce useful work.

To see that this expression is valid in general, recall that in Example 18-19 the total change in entropy is $\Delta S_{\text{universe}} = Q/T_c - Q/T_h$. That is, the heat $Q_h = Q$ is withdrawn from the hot reservoir (lowering the entropy by the amount Q/T_h), and the same heat

$Q_c = Q$ is added to the cold reservoir (increasing the entropy by the larger amount, Q/T_c). If we multiply this increase in entropy by the temperature of the cold reservoir, T_c, we have $T_c \Delta S_{universe} = Q - QT_c/T_h = Q(1 - T_c/T_h)$. Recalling that the efficiency of an ideal engine is $e = 1 - T_c/T_h$, we see that $T_c \Delta S_{universe} = Qe$. Finally, the work done by an ideal engine is $W = eQ_h$, or in this case, $W = eQ$, since $Q_h = Q$. Therefore, we see that $W = T_c \Delta S_{universe}$, as expected.

In general, a process in which the entropy of the universe increases is one in which less work is done than if the process had been reversible. Thus, we lose forever the ability for that work to be done, because to restore the universe to its former state would mean lowering its entropy, which cannot be done. We conclude that with every increase in entropy, there is that much less useful work that can be done by the universe.

For this reason, entropy is sometimes referred to as a measure of the "quality" of energy. When an irreversible process occurs, and the entropy of the universe increases, we say that the energy of the universe has been "degraded" because less of it is available to do work. This process of increasing degradation of energy and increasing entropy in the universe is a continuing aspect of nature.

Enhance Your Understanding   (Answers given at the end of the chapter)

8. In system A, the heat Q is added to a cold reservoir with a temperature of 300 K. In system B, the same heat is added to a cold reservoir with a temperature of 350 K. Is the increase in entropy for system A greater than, less than, or equal to the increase in entropy for system B? Explain.

Section Review

-   The change in entropy is defined as $\Delta S = Q/T$. Entropy is a state function, and depends only on the state of a system and not on how the state is produced.

-   The total entropy of the universe *increases* whenever an *irreversible* process occurs. The total entropy of the universe is *unchanged* whenever a *reversible* process occurs.

18-9 Order, Disorder, and Entropy

In the previous section we considered entropy from the point of view of thermodynamics. We saw that as heat flows from a hot object to a cold object, the entropy of the universe increases. In this section we show that entropy can also be thought of as a measure of the amount of **disorder** in the universe.

We begin with the situation of heat flow from a hot to a cold object. In **FIGURE 18-21 (a)** we show two bricks, one hot and the other cold. As we know from kinetic theory, the molecules in the hot brick have more kinetic energy than the molecules in the cold brick. This means that the system is rather orderly, in that all the high-kinetic-energy molecules are grouped together in the hot brick, and all the low-kinetic-energy molecules are grouped together in the cold brick. There is a definite regularity, or order, to the distribution of the molecular speeds.

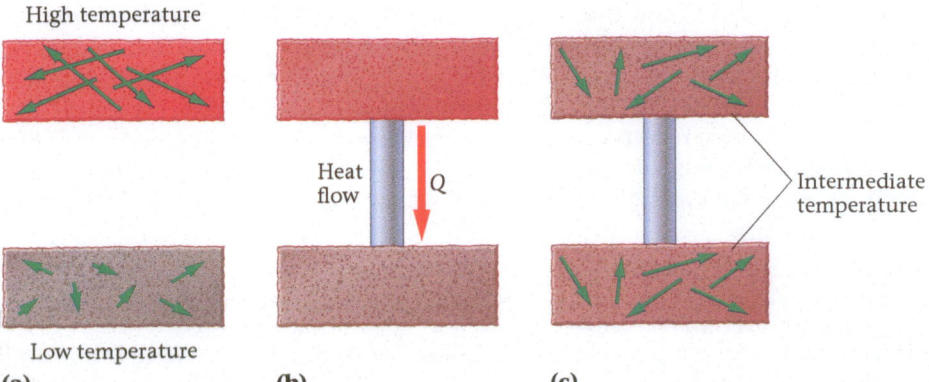

High temperature

Heat flow Q

Intermediate temperature

Low temperature

(a)   **(b)**   **(c)**

◄ **FIGURE 18-21 Heat flow and disorder**
(a) Initially, two bricks have different temperatures, and hence different average kinetic energies. **(b)** Heat flows from the hot brick to the cold brick. **(c)** The final result is that both bricks have the same intermediate temperature, and all the molecules have the same average kinetic energy. Thus, the initial orderly segregation of molecules by kinetic energy has been lost.

▲ **FIGURE 18-22** All processes that occur spontaneously increase the entropy of the universe. In this case, the movements of the water molecules become more random and chaotic when they reach the tumultuous, swirling pool at the bottom of the falls. In addition, some of their kinetic energy is converted to thermal energy—the most disordered and degraded form of energy.

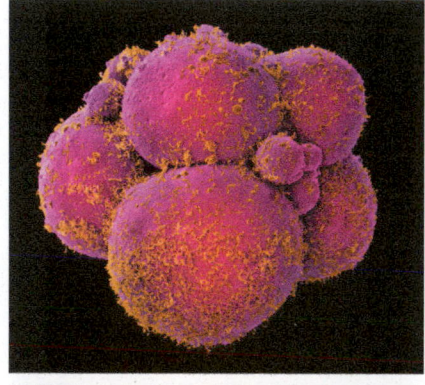

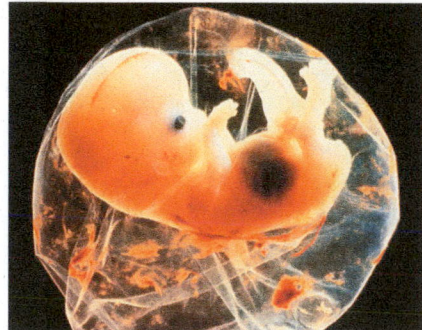

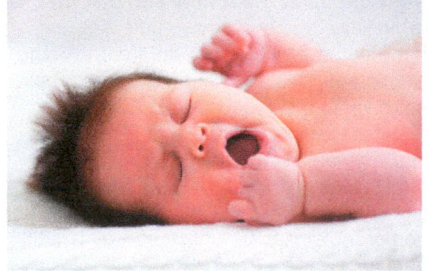

◄ **FIGURE 18-23** Many species, including humans, develop from a single fertilized egg into a complex multicellular organism. In the process they create large, intricately ordered molecules such as proteins and DNA from smaller, simpler precursors. In the metabolic processes of living things, however, heat is produced, which increases the entropy of the universe as a whole. Thus, the second law is not violated.

Now bring the bricks into thermal contact, as in **FIGURE 18-21 (b)**. The result is a flow of heat from the hot brick to the cold brick until the temperatures become equal. The final result is indicated in **FIGURE 18-21 (c)**. During the heat transfer the entropy of the universe increases, as we know, and the system loses the nice orderly distribution it had in the beginning. In the final state, all the molecules have the same average kinetic energy; hence, the system is randomized, or disordered. We are led to the following conclusion:

As the entropy of a system increases, its disorder increases as well; that is, an *increase* in entropy is the same as a *decrease* in order.

If heat had flowed in the opposite direction—from the cold brick to the hot brick—the ordered distribution of molecules would have been reinforced, rather than lost.

To take another example, consider the 0.125-kg chunk of ice discussed in Example 18-18. As we saw there, the entropy of the universe increases as the ice melts. Now let's consider what happens on the molecular level. To begin, the molecules are well ordered in their crystalline positions. As heat is absorbed, however, the molecules begin to free themselves from the ice and move about randomly in the growing puddle of water. Thus, the regular order of the solid is lost. Again, we see that as entropy increases, so too does the disorder of the molecules.

Thus, the second law of thermodynamics can be stated as the principle that the disorder of the universe is continually increasing. Everything that happens in the universe, such as the water pouring over a falls in **FIGURE 18-22**, is simply making it a more disorderly place. And there is nothing you can do to prevent it—nothing you can do will result in the universe being more ordered. Just as freezing a tray of water to make ice actually results in more entropy—and more disorder—in the universe, so does any action you take.

Heat Death

If one carries the previous discussion to its logical conclusion, it seems that the universe is "running down." That is, the disorder of the universe constantly increases, and as it does, the amount of energy available to do work decreases. If this process continues, might there come a day when no more work can be done? And, if that day does come, what then?

This is one possible scenario for the fate of the universe, and it is referred to as the "heat death" of the universe. In this scenario, heat continues to flow from hotter regions in space (like stars) to cooler regions (like planets) until, after many billions of years, all objects in the universe have the same temperature. With no temperature differences left, there can be no work done, and the universe would cease to do anything of particular interest. Not a pretty picture, but certainly a possibility. The universe may simply continue with its present expansion until the stars burn out and the galaxies fade away like the dying embers of a scattered campfire.

Living Systems

BIO So far we have focused on the rather gloomy prospect of the universe constantly evolving toward greater disorder. Is it possible, however, that life is an exception to this rule? After all, we know that an embryo utilizes simple raw materials to produce a complex, highly ordered living organism, as illustrated in **FIGURE 18-23**. Similarly, the well-known biological aphorism "ontogeny recapitulates phylogeny," while not strictly correct, expresses the fact that the development of an individual organism from embryo to adult often reflects certain aspects of the evolutionary development of the species as a whole. Thus, over time, species often evolve toward more complex forms. Finally, living systems are able to use disordered raw materials in the environment to produce orderly structures in which to live. It seems, then, that there are many ways in which living systems produce increasing order.

This conclusion is flawed, however, because it fails to take into account the entropy of the environment in which the organism lives. It is similar to the conclusion that a freezer violates the second law of thermodynamics because it reduces the entropy of water as it freezes it into ice. This analysis neglects the fact that the freezer exhausts heat into the room, increasing the entropy of the air by an amount that is greater than the entropy decrease of the water. In the same way, living organisms constantly give off heat to the atmosphere as a by-product of their metabolism, increasing its entropy. Thus, if we build a house from a pile of bricks—decreasing the entropy of the bricks—the heat we give off during our exertions increases the entropy of the atmosphere more than enough to give a net increase in entropy.

Finally, all living organisms can be thought of as heat engines, tapping into the flow of energy from one place to another to produce mechanical work. Plants, for example, tap into the flow of energy from the high temperature of the Sun to the cold of deep space and use a small fraction of this energy to sustain themselves and reproduce. Animals consume plants and generate heat within their bodies as they metabolize their food. A fraction of the energy released by the metabolism is in turn converted to mechanical work. Living systems, then, obey the same laws of physics as steam engines and refrigerators—they simply produce different results as they move the universe toward greater disorder.

Enhance Your Understanding   (Answers given at the end of the chapter)

9. Is the entropy of a bag of popcorn greater than, less than, or equal to the entropy of the kernels of corn before popping? Explain.

Section Review

- Entropy can be thought of as a measure of the disorder of a system. When the entropy of a system increases, its disorder increases as well.

18-10 The Third Law of Thermodynamics

Finally, we consider the **third law of thermodynamics,** which states that there is no temperature lower than absolute zero, and that absolute zero is unattainable. It is possible to cool an object to temperatures arbitrarily close to absolute zero—experiments have reached temperatures as low as 4.5×10^{-10} K—but no object can ever be cooled to precisely 0 K.

As an analogy to cooling toward absolute zero, imagine walking toward a wall, with each step half the distance between you and the wall. Even if you take an infinite number of steps, you will still not reach the wall. You can get arbitrarily close, of course, but you never get all the way there.

The same sort of thing happens when cooling. To cool an object, you can place it in thermal contact with an object that is colder. Heat transfer will occur, with your object ending up cooler and the other object ending up warmer. In particular, suppose you had a collection of objects at 0 K to use for cooling. You put your object in contact with one of the 0 K objects and your object cools, while the 0 K object warms slightly. You continue this process, each time throwing away the "warmed up" 0 K object and using a new one. Each time you do this your object gets closer to 0 K, but it ever actually gets there.

In light of this discussion, we can express the third law of thermodynamics as follows:

Third Law of Thermodynamics
It is impossible to lower the temperature of an object to absolute zero in a finite number of steps.

Big Idea 4 The lowest temperature possible is absolute zero. Systems can be cooled arbitrarily close to absolute zero, but can never attain it.

As with the second law of thermodynamics, this law can be expressed in a number of different but equivalent ways. The essential idea, however, is always the same: Absolute zero is the limiting temperature and, though it can be approached arbitrarily closely, it can never be attained.

Enhance Your Understanding   (Answers given at the end of the chapter)

10. Explain the difference between 0 °C, 0 °F, and 0 K.

Section Review

- No equilibrium temperature can be less than absolute zero, 0 K. A system can be cooled arbitrarily close to 0 K, but can never actually reach it.

CHAPTER 18 REVIEW

CHAPTER SUMMARY

18-1 THE ZEROTH LAW OF THERMODYNAMICS

When two objects have the same temperature, they are in thermal equilibrium.

18-2 THE FIRST LAW OF THERMODYNAMICS

The first law of thermodynamics is a statement of energy conservation that includes heat.

If U is the internal energy of an object, Q is the heat added to it, and W is the work done by the object, the first law of thermodynamics can be written as follows:

$$\Delta U = Q - W \qquad \text{18-3}$$

18-3 THERMAL PROCESSES

Reversible
In a reversible process it is possible to return the system and its surroundings to their initial states.

Irreversible
Irreversible processes cannot be "undone." When the system is returned to its initial state, the surroundings have been altered.

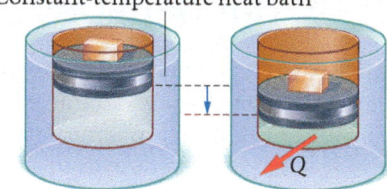
Constant-temperature heat bath

Work
In general, the work done during a process is equal to the area under the process curve in a PV plot.

Constant Pressure
In a PV plot, a constant-pressure process is represented by a horizontal line. The work done at constant pressure is $W = P\Delta V$.

Constant Volume
In a PV plot, a constant-volume process is represented by a vertical line. The work done at constant volume is zero; $W = 0$.

Isothermal Process
In a PV plot, an isothermal process is represented by PV = constant. The work done in an isothermal expansion from V_i to V_f is $W = nRT \ln(V_f/V_i)$.

Adiabatic Process
An adiabatic process occurs with no heat transfer; that is, $Q = 0$.

18-4 SPECIFIC HEATS FOR AN IDEAL GAS: CONSTANT PRESSURE, CONSTANT VOLUME

Specific heats have different values depending on whether they apply to a process at constant pressure or a process at constant volume.

Molar Specific Heat

The molar specific heat, C, is defined by $Q = nC\Delta T$, where n is the number of moles.

Constant Volume

The molar specific heat for an ideal monatomic gas at constant volume is

$$C_v = \frac{3}{2}R \qquad \text{18-6}$$

Constant Pressure

The molar specific heat for an ideal monatomic gas at constant pressure is

$$C_p = \frac{5}{2}R \qquad \text{18-7}$$

Adiabatic Process

In a PV plot, an adiabatic process is represented by $PV^\gamma = $ constant, where γ is the ratio C_p/C_v. For a monatomic, ideal gas $\gamma = \frac{5}{3}$.

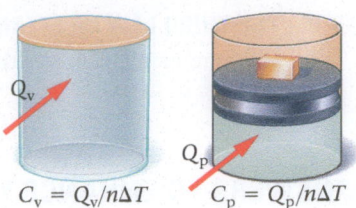

$$C_v = Q_v/n\Delta T \qquad C_p = Q_p/n\Delta T$$

18-5 THE SECOND LAW OF THERMODYNAMICS

When objects of different temperatures are brought into thermal contact, the spontaneous flow of heat that results is always from the high-temperature object to the low-temperature object. Spontaneous heat flow never proceeds in the reverse direction.

18-6 HEAT ENGINES AND THE CARNOT CYCLE

A heat engine is a device that converts heat into work—for example, a steam engine.

Carnot's Theorem

If an engine operating between two constant-temperature reservoirs is to have maximum efficiency, it must be an engine in which all processes are reversible. In addition, all reversible engines operating between the same two temperatures have the same efficiency.

Maximum Efficiency

The maximum efficiency of a heat engine operating between the Kelvin temperatures T_h and T_c is

$$e_{max} = 1 - \frac{T_c}{T_h} \qquad \text{18-13}$$

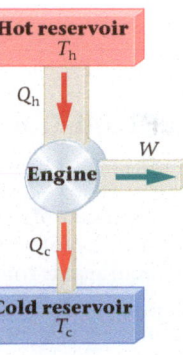

18-7 REFRIGERATORS, AIR CONDITIONERS, AND HEAT PUMPS

Refrigerators, air conditioners, and heat pumps are devices that use work to make heat flow from a cold region to a hot region.

Coefficient of Performance

The coefficient of performance of a refrigerator or air conditioner doing the work W to remove a heat Q_c from a cold reservoir is

$$\text{COP} = \frac{Q_c}{W} \qquad \text{18-15}$$

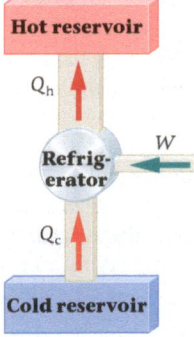

18-8 ENTROPY

Like the internal energy U, the entropy S is a state function.

Change in Entropy

The change in entropy during a reversible exchange of the heat Q at the Kelvin temperature T is

$$\Delta S = \frac{Q}{T} \qquad \text{18-18}$$

18-9 ORDER, DISORDER, AND ENTROPY

Entropy is a measure of the disorder of a system. As entropy increases, a system becomes more disordered.

Heat Death

A possible fate of the universe is heat death, in which everything is at the same temperature and no more work can be done.

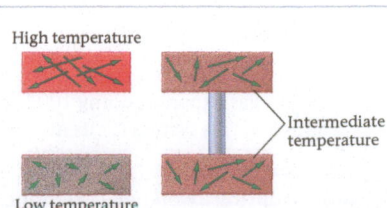

18-10 THE THIRD LAW OF THERMODYNAMICS

It is impossible to lower the temperature of an object to absolute zero in a finite number of steps.

ANSWERS TO ENHANCE YOUR UNDERSTANDING QUESTIONS

1. **A.** Heat flows from system 1 (the warmer system) to system 2.

2. D < B < A < C.

3. The volume is halved, because pressure and volume are inversely related in an isotherm.

4. A < D < B < C.

5. No. A broken dish coming back together and flying up into your hand would conserve energy, but would not be allowed in a physical system.

6. The efficiency of heat engine A is equal to the efficiency of heat engine B because both have the same ratio of temperatures between the cold and hot reservoirs.

7. The required work decreases, because a higher COP implies that less heat is needed to remove a given amount of energy.

8. The increase in entropy is greater for system A, because it has the lower temperature.

9. The entropy of the popped popcorn is greater, because popping the corn is an irreversible process.

10. The temperature 0 K (absolute zero) is the lowest temperature possible. The temperatures 0 °F and 0 °C are arbitrary assignments of zero in these everyday temperature scales—they have no deeper significance.

CONCEPTUAL QUESTIONS

For instructor-assigned homework, go to www.masteringphysics.com.

(Answers to odd-numbered Conceptual Questions can be found in the back of the book.)

1. The temperature of a substance is held fixed. Is it possible for heat to flow **(a)** into or **(b)** out of this system? For each case, give an explanation if your answer is no. If your answer is yes, give a specific example.

2. Heat is added to a substance. Is it safe to conclude that the temperature of the substance will rise? Give an explanation if your answer is no. If your answer is yes, give a specific example.

3. Are there thermodynamic processes in which all the heat absorbed by an ideal gas goes completely into mechanical work? If so, give an example.

4. An ideal gas is held in an insulated container at the temperature T. All the gas is initially in one-half of the container, with a partition separating the gas from the other half of the container, which is a vacuum. If the partition ruptures, and the gas expands to fill the entire container, what is its final temperature?

5. Which of the following processes are approximately reversible? **(a)** Lighting a match. **(b)** Pushing a block up a frictionless inclined plane. **(c)** Frying an egg. **(d)** Swimming from one end of a pool to the other. **(e)** Stretching a spring by a small amount. **(f)** Writing a report for class.

6. Which law of thermodynamics would be violated if heat were to spontaneously flow between two objects of equal temperature?

7. Heat engines always give off a certain amount of heat to a low-temperature reservoir. Would it be possible to use this "waste" heat as the heat input to a second heat engine, and then use the "waste" heat of the second engine to run a third engine, and so on?

8. A heat pump uses 100 J of energy as it operates for a given time. Is it possible for the heat pump to deliver more than 100 J of heat to the inside of the house in this same time? Explain.

9. Which law of thermodynamics is most pertinent to the statement that "all the king's horses and all the king's men couldn't put Humpty Dumpty back together again"?

10. Which has more entropy: **(a)** popcorn kernels, or the resulting popcorn; **(b)** two eggs in a carton, or an omelet made from the eggs; **(c)** a pile of bricks, or the resulting house; **(d)** a piece of paper, or the piece of paper after it has been burned?

PROBLEMS AND CONCEPTUAL EXERCISES

Answers to odd-numbered Problems and Conceptual Exercises can be found in the back of the book. **BIO** *identifies problems of biological or medical interest;* **CE** *indicates a conceptual exercise.* Predict/Explain *problems ask for two responses: (a) your prediction of a physical outcome, and (b) the best explanation among three provided; and* Predict/Calculate *problems ask for a prediction of a physical outcome, based on fundamental physics concepts, and follow that with a numerical calculation to verify the prediction. On all problems, bullets (•, ••, •••) indicate the level of difficulty.*

SECTION 18-2 THE FIRST LAW OF THERMODYNAMICS

1. • **CE** Give the change in internal energy of a system if **(a)** $W = 50$ J, $Q = 50$ J; **(b)** $W = -50$ J, $Q = -50$ J; or **(c)** $W = 50$ J, $Q = -50$ J.

2. • **CE** A gas expands, doing 100 J of work. How much heat must be added to this system for its internal energy to increase by 200 J?

3. • A swimmer does 7.7×10^5 J of work and gives off 3.9×10^5 J of heat during a workout. Determine ΔU, W, and Q for the swimmer.

4. • When 1310 J of heat are added to one mole of an ideal monatomic gas, its temperature increases from 272 K to 276 K. Find the work done by the gas during this process.

5. • Three different processes act on a system. **(a)** In process A, 42 J of work are done on the system and 77 J of heat are added to the system. Find the change in the system's internal energy. **(b)** In process B, the system does 42 J of work and 77 J of heat are added to the system. What is the change in the system's internal energy? **(c)** In process C, the system's internal energy decreases by 120 J

while the system performs 120 J of work on its surroundings. How much heat was added to the system?

6. • An ideal gas is taken through the four processes shown in **FIGURE 18-24**. The changes in internal energy for three of these processes are as follows: $\Delta U_{AB} = +82$ J; $\Delta U_{BC} = +15$ J; $\Delta U_{DA} = -56$ J. Find the change in internal energy for the process from C to D.

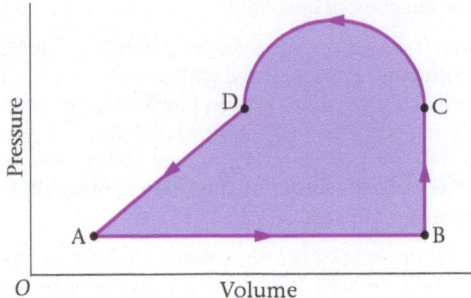

FIGURE 18-24 Problem 6

7. •• A basketball player does 4.81×10^5 J of work during her time in the game, and evaporates 0.151 kg of water. Assuming a latent heat of 2.26×10^6 J/kg for the perspiration (the same as for water), determine **(a)** the change in the player's internal energy and **(b)** the number of nutritional calories the player has converted to work and heat.

8. •• **Predict/Calculate** One mole of an ideal monatomic gas is initially at a temperature of 274 K. **(a)** Find the final temperature of the gas if 3260 J of heat are added to it and it does 712 J of work. **(b)** Suppose the amount of gas is doubled to two moles. Does the final temperature found in part (a) increase, decrease, or stay the same? Explain.

9. •• **Predict/Calculate** **Energy from Gasoline** Burning a gallon of gasoline releases 1.19×10^8 J of internal energy. If a certain car requires 5.20×10^5 J of work to drive one mile, **(a)** how much heat is given off to the atmosphere each mile, assuming the car gets 25.0 miles to the gallon? **(b)** If the miles per gallon of the car is increased, does the amount of heat released to the atmosphere increase, decrease, or stay the same? Explain.

10. •• A cylinder contains 4.0 moles of a monatomic gas at an initial temperature of 27 °C. The gas is compressed by doing 560 J of work on it, and its temperature increases by 130 °C. How much heat flows into or out of the gas?

11. •• An ideal gas is taken through the three processes shown in **FIGURE 18-25**. Fill in the missing entries in the following table:

	Q	W	ΔU
A → B	− 136 J	(a)	(b)
B → C	− 341 J	− 136 J	(c)
C → A	(e)	153 J	(d)

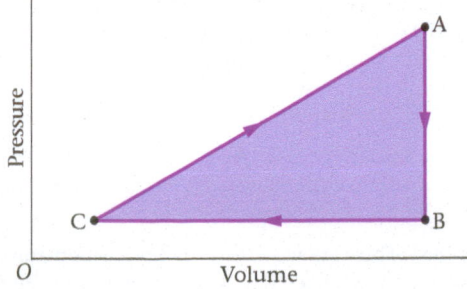

FIGURE 18-25 Problems 11 and 88

SECTION 18-3 THERMAL PROCESSES

12. • **CE** **FIGURE 18-26** shows three different multistep processes, labeled A, B, and C. Rank these processes in order of increasing work done by a gas that undergoes the process. Indicate ties where appropriate.

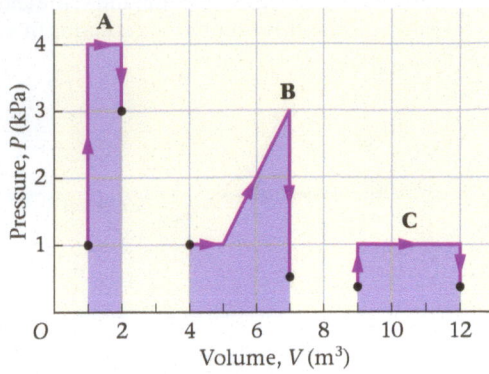

FIGURE 18-26 Problem 12

13. • The volume of a monatomic ideal gas triples in an isothermal expansion. By what factor does its pressure change?

14. • An ideal gas is compressed at constant pressure to one-half its initial volume. If the pressure of the gas is 120 kPa, and 790 J of work is done on it, find the initial volume of the gas.

15. • As an ideal gas expands at constant pressure from a volume of 0.74 m³ to a volume of 2.3 m³ it does 93 J of work. What is the gas pressure during this process?

16. •• A system consisting of an ideal gas at the constant pressure of 120 kPa gains 960 J of heat. Find the change in volume of the system if the internal energy of the gas increases by **(a)** 920 J or **(b)** 360 J.

17. •• **Predict/Calculate** **(a)** If the internal energy of a system increases as the result of an adiabatic process, is work done on the system or by the system? **(b)** Calculate the work done on or by the system in part (a) if its internal energy increases by 670 J.

18. •• **(a)** Find the work done by a monatomic ideal gas as it expands from point A to point C along the path shown in **FIGURE 18-27**. **(b)** If the temperature of the gas is 267 K at point A, what is its temperature at point C? **(c)** How much heat has been added to or removed from the gas during this process?

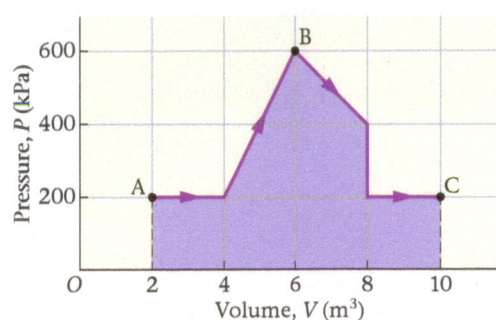

FIGURE 18-27 Problems 18 and 19

19. •• **Predict/Calculate** A fluid expands from point A to point B along the path shown in Figure 18-27. **(a)** How much work is done by the fluid during this expansion? **(b)** Does your answer to part (a) depend on whether the fluid is an ideal gas? Explain.

20. •• **Predict/Calculate** If 9.50 moles of a monatomic ideal gas at a temperature of 235 K are expanded isothermally from a volume of 1.12 L to a volume of 4.33 L, calculate **(a)** the work done and

(b) the heat flow into or out of the gas. **(c)** If the number of moles is doubled, by what factors do your answers to parts (a) and (b) change? Explain.

21. •• Suppose 118 moles of a monatomic ideal gas undergo an isothermal expansion from 1.00 m³ to 4.00 m³, as shown in **FIGURE 18-28**. **(a)** What is the temperature at the beginning and at the end of this process? **(b)** How much work is done by the gas during this expansion?

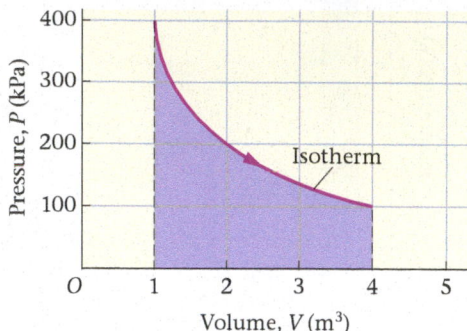

Volume, V (m³)

FIGURE 18-28 Problems 21 and 76

22. •• A weather balloon contains an ideal gas and has a volume of 2.5 m³ at launch, with a pressure of 101 kPa. As it rises slowly through the atmosphere, the gas temperature remains constant at 273 K. **(a)** High in the atmosphere the pressure drops to 19.3 kPa. What is the volume of the balloon at that point? **(b)** How much work is done by the gas in the balloon as it expands?

23. •• Predict/Calculate **(a)** A monatomic ideal gas expands at constant pressure. Is heat added to the system or taken from the system during this process? **(b)** Find the heat added to or taken from the gas in part (a) if it expands at a pressure of 130 kPa from a volume of 0.76 m³ to a volume of 0.93 m³.

24. •• During an adiabatic process, the temperature of 3.22 moles of a monatomic ideal gas drops from 475 °C to 215 °C. For this gas, find **(a)** the work it does, **(b)** the heat it exchanges with its surroundings, and **(c)** the change in its internal energy.

25. •• An ideal gas follows the three-part process shown in **FIGURE 18-29**. At the completion of one full cycle, find **(a)** the net work done by the system, **(b)** the net change in internal energy of the system, and **(c)** the net heat absorbed by the system.

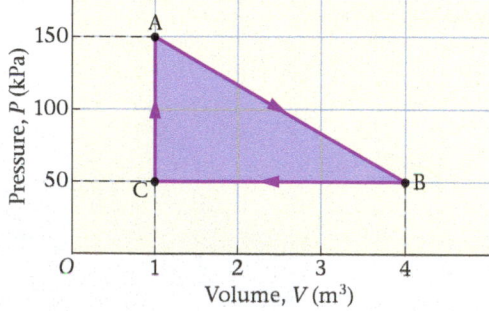

Volume, V (m³)

FIGURE 18-29 Problems 25 and 27

26. •• With the pressure held constant at 260 kPa, 43 mol of a monatomic ideal gas expands from an initial volume of 0.75 m³ to a final volume of 1.9 m³. **(a)** How much work was done by the gas during the expansion? **(b)** What were the initial and final temperatures of the gas? **(c)** What was the change in the internal energy of the gas? **(d)** How much heat was added to the gas?

27. •• Predict/Calculate Suppose 67.5 moles of an ideal monatomic gas undergo the series of processes shown in Figure 18-29. **(a)** Calculate the temperature at the points A, B, and C. **(b)** For each

process, A → B, B → C, and C → A, state whether heat enters or leaves the system. Explain in each case. **(c)** Calculate the heat exchanged with the gas during each of the three processes.

28. •• A system expands by 0.75 m³ at a constant pressure of 125 kPa. Find the heat that flows into or out of the system if its internal energy **(a)** increases by 65 kJ or **(b)** decreases by 33 kJ. In each case, give the direction of heat flow.

29. •• Predict/Calculate An ideal monatomic gas is held in a perfectly insulated cylinder fitted with a movable piston. The initial pressure of the gas is 110 kPa, and its initial temperature is 280 K. By pushing down on the piston, you are able to increase the pressure to 140 kPa. **(a)** During this process, did the temperature of the gas increase, decrease, or stay the same? Explain. **(b)** Find the final temperature of the gas.

30. ••• A certain amount of a monatomic ideal gas undergoes the process shown in **FIGURE 18-30**, in which its pressure doubles and its volume triples. In terms of the number of moles, n, the initial pressure, P_i, and the initial volume, V_i, determine **(a)** the work done by the gas W, **(b)** the change in internal energy of the gas U, and **(c)** the heat added to the gas Q.

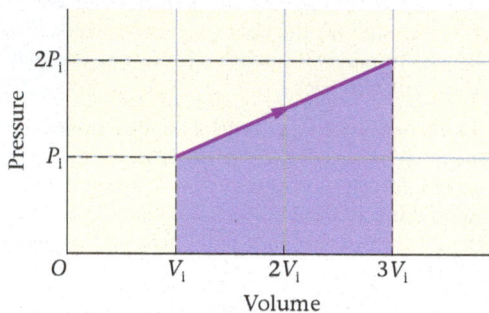

Volume

FIGURE 18-30 Problem 30

31. ••• An ideal gas doubles its volume in one of three different ways: (i) at constant pressure; (ii) at constant temperature; (iii) adiabatically. Explain your answers to each of the following questions: **(a)** In which expansion does the gas do the most work? **(b)** In which expansion does the gas do the least work? **(c)** Which expansion results in the highest final temperature? **(d)** Which expansion results in the lowest final temperature?

SECTION 18-4 SPECIFIC HEATS FOR AN IDEAL GAS: CONSTANT PRESSURE, CONSTANT VOLUME

32. * CE Predict/Explain You plan to add a certain amount of heat to a gas in order to raise its temperature. **(a)** If you add the heat at constant volume, is the increase in temperature greater than, less than, or equal to the increase in temperature if you add the heat at constant pressure? **(b)** Choose the *best explanation* from among the following:
   I. The same amount of heat increases the temperature by the same amount, regardless of whether the volume or the pressure is held constant.
   II. All the heat goes into raising the temperature when added at constant volume; none goes into mechanical work.
   III. Holding the pressure constant will cause a greater increase in temperature than simply having a fixed volume.

33. • Find the amount of heat needed to increase the temperature of 5.5 mol of an ideal monatomic gas by 27 K if **(a)** the pressure or **(b)** the volume is held constant.

34. • **(a)** If 585 J of heat are added to 49 moles of a monatomic gas at constant volume, how much does the temperature of the gas increase? **(b)** Repeat part (a), this time for a constant-pressure process.

35. • A system consists of 3.5 mol of an ideal monatomic gas at 315 K. How much heat must be added to the system to double its internal energy at **(a)** constant pressure or **(b)** constant volume?

36. • Find the change in temperature if 170 J of heat are added to 2.8 mol of an ideal monatomic gas at **(a)** constant pressure or **(b)** constant volume.

37. •• **Gasoline Ignition** Consider a short time span just before and after the spark plug in a gasoline engine ignites the fuel-air mixture and releases 1970 J of thermal energy into a volume of 47.6 cm³ that is at a gas pressure of 1.17×10^6 Pa. In this short period of time the volume of the gas can be considered constant. **(a)** Treating the fuel-air mixture as a monatomic ideal gas at an initial temperature of 325 K, what is its temperature after ignition? **(b)** What is the pressure in the gas after ignition?

38. •• **Predict/Calculate** A cylinder contains 18 moles of a monatomic ideal gas at a constant pressure of 160 kPa. **(a)** How much work does the gas do as it expands 3200 cm³, from 5400 cm³ to 8600 cm³? **(b)** If the gas expands by 3200 cm³ again, this time from 2200 cm³ to 5400 cm³, is the work it does greater than, less than, or equal to the work found in part (a)? Explain. **(c)** Calculate the work done as the gas expands from 2200 cm³ to 5400 cm³.

39. •• **Predict/Calculate** The volume of a monatomic ideal gas doubles in an adiabatic expansion. By what factor do **(a)** the pressure and **(b)** the temperature of the gas change? **(c)** Verify your answers to parts (a) and (b) by considering 135 moles of gas with an initial pressure of 330 kPa and an initial volume of 1.2 m³. Find the pressure and temperature of the gas after it expands adiabatically to a volume of 2.4 m³.

40. •• A monatomic ideal gas is held in a thermally insulated container with a volume of 0.0750 m³. The pressure of the gas is 101 kPa, and its temperature is 325 K. **(a)** To what volume must the gas be compressed to increase its pressure to 145 kPa? **(b)** At what volume will the gas have a temperature of 295 K?

41. ••• Consider the expansion of 60.0 moles of a monatomic ideal gas along processes 1 and 2 in **FIGURE 18-31**. In process 1 the gas is heated at constant volume from an initial pressure of 106 kPa to a final pressure of 212 kPa. In process 2 the gas expands at constant pressure from an initial volume of 1.00 m³ to a final volume of 3.00 m³. **(a)** How much heat is added to the gas during these two processes? **(b)** How much work does the gas do during this expansion? **(c)** What is the change in the internal energy of the gas?

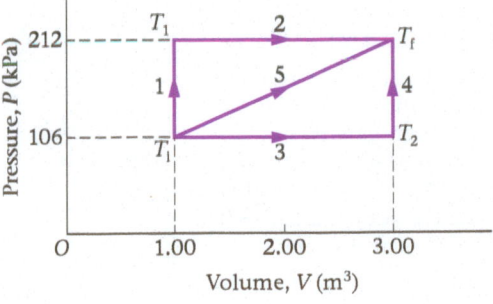

FIGURE 18-31 Problems 41 and 78

SECTION 18-6 HEAT ENGINES AND THE CARNOT CYCLE

42. • **CE** A Carnot engine can be operated with one of the following four sets of reservoir temperatures: A, 400 K and 800 K; B, 400 K and 600 K; C, 800 K and 1200 K; and D, 800 K and 1000 K. Rank these reservoir temperatures in order of increasing efficiency of the Carnot engine. Indicate ties where appropriate.

43. • What is the efficiency of an engine that exhausts 840 J of heat in the process of doing 360 J of work?

44. • An engine receives 660 J of heat from a hot reservoir and gives off 420 J of heat to a cold reservoir. What are **(a)** the work done and **(b)** the efficiency of this engine?

45. • A Carnot engine operates between the temperatures 410 K and 290 K. **(a)** How much heat must be given to the engine to produce 2500 J of work? **(b)** How much heat is discarded to the cold reservoir as this work is done?

46. • A nuclear power plant has a reactor that produces heat at the rate of 838 MW. This heat is used to produce 253 MW of mechanical power to drive an electrical generator. **(a)** At what rate is heat discarded to the environment by this power plant? **(b)** What is the thermal efficiency of the plant?

47. • At a coal-burning power plant a steam turbine is operated with a power output of 548 MW. The thermal efficiency of the power plant is 32.0%. **(a)** At what rate is heat discarded to the environment by this power plant? **(b)** At what rate must heat be supplied to the power plant by burning coal?

48. •• **Predict/Calculate** A portable generator produces 2.5 kW of mechanical work with an efficiency of 0.20 to supply electricity at a work site. **(a)** If the efficiency is increased but the work output remains the same, will the rate of fuel consumption by the engine increase, decrease, or stay the same? Explain. **(b)** Will the rate of waste heat emission increase, decrease, or stay the same? Explain. **(c)** Calculate the rates of fuel consumption and waste heat emission for engine efficiencies of 0.20 and 0.30.

49. •• **Predict/Calculate** The efficiency of a particular Carnot engine is 0.280. **(a)** If the high-temperature reservoir is at a temperature of 565 K, what is the temperature of the low-temperature reservoir? **(b)** To increase the efficiency of this engine to 40.0%, must the temperature of the low-temperature reservoir be increased or decreased? Explain. **(c)** Find the temperature of the low-temperature reservoir that gives an efficiency of 0.400.

50. •• During each cycle a reversible engine absorbs 3100 J of heat from a high-temperature reservoir and performs 1900 J of work. **(a)** What is the efficiency of this engine? **(b)** How much heat is exhausted to the low-temperature reservoir during each cycle? **(c)** What is the ratio, T_h/T_c, of the two reservoir temperatures?

51. •• **Solar Thermal Power** The Ivanpah Solar Electric Generating System is the largest solar thermal power-tower facility in the world. It collects 1321 MW of solar thermal energy to heat special collectors to 566 °C. The thermal energy is then used to make steam to operate an electric generator. **(a)** What is the maximum efficiency of the power plant, taking the cold reservoir to be the surrounding environment at 12 °C. **(b)** How much mechanical work (electricity) could the facility generate if it operated at its maximum efficiency? **(c)** If the facility generates 392 MW of electricity, what is its actual efficiency?

52. ••• The operating temperatures for a Carnot engine are T_c and $T_h = T_c + 55$ K. The efficiency of the engine is 11%. Find T_c and T_h.

53. ••• A certain Carnot engine takes in the heat Q_h and exhausts the heat $Q_c = 2Q_h/3$, as indicated in **FIGURE 18-32**. **(a)** What is the efficiency of this engine? **(b)** Using the Kelvin temperature scale, find the ratio T_c/T_h.

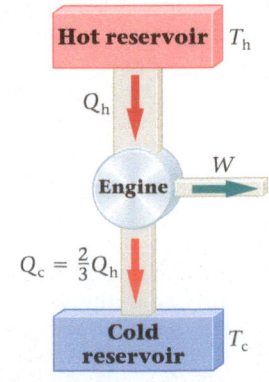

FIGURE 18-32 Problem 53

SECTION 18-7   REFRIGERATORS, AIR CONDITIONERS, AND HEAT PUMPS

54. • **CE** Predict/Explain **(a)** If the temperature in the kitchen is decreased, is the cost (work needed) to freeze a dozen ice cubes greater than, less than, or equal to what it was before the kitchen was cooled? **(b)** Choose the *best explanation* from among the following:
 I. The difference in temperature between the inside and the outside of the refrigerator is decreased, and hence less work is required to freeze the ice.
 II. The same amount of ice is frozen in either case, which requires the same amount of heat to be removed and hence the same amount of work.
 III. Cooling the kitchen means that the refrigerator must do more work, both to freeze the ice cubes and to warm the kitchen.

55. • The refrigerator in your kitchen does 490 J of work to remove 95 J of heat from its interior. **(a)** How much heat does the refrigerator exhaust into the kitchen? **(b)** What is the refrigerator's coefficient of performance?

56. • A refrigerator with a coefficient of performance of 1.75 absorbs 3.45×10^4 J of heat from the low-temperature reservoir during each cycle. **(a)** How much mechanical work is required to operate the refrigerator for a cycle? **(b)** How much heat does the refrigerator discard to the high-temperature reservoir during each cycle?

57. •• Suppose you must remove an average of 3.4×10^8 J of thermal energy per day to keep your house cool during the summer. **(a)** If you upgrade from an old air conditioner with a COP of 2.0 to a new air conditioner with a COP of 6.0, by how many joules is the required mechanical work reduced each day? **(b)** If the cost of supplying your air conditioner with mechanical work is $0.12 per 3.6×10^6 J, how much money will you save during a 92-day cooling season?

58. •• To keep a room at a comfortable 20.5 °C, a Carnot heat pump does 315 J of work and supplies the room with 4180 J of heat. **(a)** How much heat is removed from the outside air by the heat pump? **(b)** What is the temperature of the outside air?

59. •• An air conditioner is used to keep the interior of a house at a temperature of 21 °C while the outside temperature is 32 °C. If heat leaks into the house at the rate of 11 kW, and the air conditioner has the efficiency of a Carnot engine, what is the mechanical power required to keep the house cool?

60. •• A reversible refrigerator has a coefficient of performance equal to 9.50. What is its efficiency?

61. •• A freezer has a coefficient of performance equal to 4.0. How much electrical energy must this freezer use to produce 1.5 kg of ice at −5.0 °C from water at 15 °C?

SECTION 18-8   ENTROPY

62. • **CE** Predict/Explain **(a)** If you rub your hands together, does the entropy of the universe increase, decrease, or stay the same? **(b)** Choose the *best explanation* from among the following:
 I. Rubbing hands together draws heat from the surroundings, and therefore lowers the entropy.
 II. No mechanical work is done by the rubbing, and hence the entropy does not change.
 III. The heat produced by rubbing raises the temperature of your hands and the air, which increases the entropy.

63. • **CE** Predict/Explain **(a)** An ideal gas is expanded slowly and isothermally. Does its entropy increase, decrease, or stay the same? **(b)** Choose the *best explanation* from among the following:
 I. Heat must be added to the gas to maintain a constant temperature, and this increases the entropy of the gas.
 II. The temperature of the gas remains constant, which means its entropy also remains constant.
 III. As the gas is expanded its temperature and entropy will decrease.

64. • **CE** Predict/Explain **(a)** A gas is expanded reversibly and adiabatically. Does its entropy increase, decrease, or stay the same? **(b)** Choose the *best explanation* from among the following:
 I. The process is reversible, and no heat is added to the gas. Therefore, the entropy of the gas remains the same.
 II. Expanding the gas gives it more volume to occupy, and this increases its entropy.
 III. The gas is expanded with no heat added to it, and hence its temperature will decrease. This, in turn, will lower its entropy.

65. • Find the change in entropy when 1.85 kg of water at 100 °C is boiled away to steam at 100 °C.

66. • Determine the change in entropy that occurs when 4.3 kg of water freezes at 0 °C.

67. •• **CE** You heat a pan of water on the stove. Rank the following temperature increases in order of increasing entropy change. Indicate ties where appropriate: **A,** 25 °C to 35 °C; **B,** 35 °C to 45 °C; **C,** 45 °C to 50 °C; and **D,** 50 °C to 55 °C.

68. •• On a cold winter's day heat leaks slowly out of a house at the rate of 20.0 kW. If the inside temperature is 22 °C, and the outside temperature is −14.5 °C, find the rate of entropy increase.

69. •• An 88-kg parachutist descends through a vertical height of 380 m with constant speed. Find the increase in entropy produced by the parachutist, assuming the air temperature is 21 °C.

70. •• Predict/Calculate Consider the air-conditioning system described in Problem 59. **(a)** Does the entropy of the universe increase, decrease, or stay the same as the air conditioner keeps the imperfectly insulated house cool? Explain. **(b)** What is the rate at which the entropy of the universe changes during this process?

71. •• A heat engine operates between a high-temperature reservoir at 610 K and a low-temperature reservoir at 320 K. In one cycle, the engine absorbs 6400 J of heat from the high-temperature reservoir and does 2200 J of work. What is the net change in entropy as a result of this cycle?

72. ••• It can be shown that as a mass m with specific heat c changes temperature from T_i to T_f its change in entropy is $\Delta S = mc \ln (T_f/T_i)$ if the temperatures are expressed in kelvin. Suppose you put 79 g of milk at 278 K into an insulated cup containing 296 g of coffee at 355 K, and that each has the specific heat of water. The system comes to an equilibrium temperature of 339 K. **(a)** What is the entropy change of the milk? **(b)** What is the entropy change of the coffee? **(c)** What is the entropy change of the universe due to adding the milk to the coffee?

GENERAL PROBLEMS

73. • **CE** Consider the three-process cycle shown in **FIGURE 18-33.** For each process in the cycle, **(a)** A→B, **(b)** B→C, and **(c)** C→A, state whether the work done by the system is positive, negative, or zero.

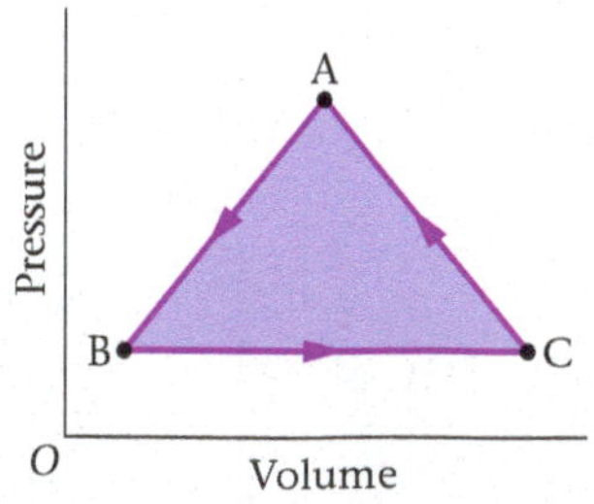

FIGURE 18-33 Problem 73

74. • **CE** An ideal gas has the pressure and volume indicated by point I in **FIGURE 18-34.** At this point its temperature is T_1. The temperature of the gas can be increased to T_2 by using the constant-volume process, I→II , or the constant-pressure process, I→III. Is the entropy change for the process I→II greater than, less than, or equal to the entropy change on the process I→III? Explain.

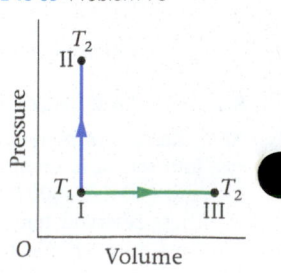

FIGURE 18-34 Problem 74

75. •• The heat that goes into a particular Carnot engine is 4.00 times greater than the work it performs. What is the engine's efficiency?

76. •• **Predict/Calculate** Consider 132 moles of a monatomic gas undergoing the isothermal expansion shown in Figure 18-28. (a) What is the temperature T of this expansion? (b) Does the entropy of the gas increase, decrease, or stay the same during the process? Explain. (c) Calculate the change in entropy for the gas, ΔS, if it is nonzero. (d) Calculate the work done by the gas during this process, and compare to $T\Delta S$.

77. •• **Inflating a Hot-Air Balloon** A hot-air balloon is inflated by heating air at constant pressure. Assume the number of air molecules inside the balloon remains constant and that it expands at atmospheric pressure. (a) If the initial temperature and volume of the air inside the balloon are 18 °C and 1600 m³, respectively, what is the final volume of the balloon when its temperature is increased to 65 °C? (b) How much work does the balloon do to expand against atmospheric pressure?

78. •• **Predict/Calculate** Referring to Figure 18-31, suppose 60.0 moles of a monatomic ideal gas are expanded along process 5. (a) How much work does the gas do during this expansion? (b) What is the change in the internal energy of the gas? (c) How much heat is added to the gas during this process?

79. •• **Predict/Calculate** Engine A has an efficiency of 66%. Engine B absorbs the same amount of heat from the hot reservoir and exhausts twice as much heat to the cold reservoir. (a) Which engine has the greater efficiency? Explain. (b) What is the efficiency of engine B?

80. •• **Nuclear Versus Natural Gas Energy** Because of safety concerns and other constraints, nuclear power plants must be run at lower efficiencies than natural gas-fired power plants. The efficiency of a particular gas-fired power plant is 44.8% and that of a nuclear plant is 32.6%. (a) For every 1.00 kJ of mechanical work (to generate electricity) produced by the gas-fired power plant, how much energy is exhausted as waste heat? (b) How much waste heat is generated for every 1.00 kJ of work produced by the nuclear power plant? (c) What is the percentage increase of waste heat generated by the nuclear plant over the natural gas plant as a result of the lower efficiency?

81. •• A freezer with a coefficient of performance of 3.88 is used to convert 1.75 kg of water to ice in one hour. The water starts at a temperature of 20.0 °C, and the ice that is produced is cooled to a temperature of −5.00 °C. (a) How much heat must be removed from the water for this process to occur? (b) How much electrical energy does the freezer use during this hour of operation? (c) How much heat is discarded into the room that houses the freezer?

82. •• **Entropy and the Sun** The surface of the Sun has a temperature of 5500 °C and the temperature of deep space is 3.0 K. (a) Find the entropy increase produced by the Sun in one day, given that it radiates heat at the rate of 3.80×10^{26} W. (b) How much work could have been done if this heat had been used to run an ideal heat engine?

83. •• The following table lists results for various processes involving n moles of a monatomic ideal gas. Fill in the missing entries.

	Q	W	ΔU
Constant pressure	$\frac{5}{2}nR\Delta T$	(a)	$\frac{3}{2}nR\Delta T$
Adiabatic	(b)	$-\frac{3}{2}nR\Delta T$	(c)
Constant volume	$\frac{3}{2}nR\Delta T$	(d)	(e)
Isothermal	(f)	$nRT\ln(V_f/V_i)$	(g)

84. •• A cylinder with a movable piston holds 2.95 mol of argon at a constant temperature of 235 K. As the gas is compressed isothermally, its pressure increases from 101 kPa to 121 kPa. Find (a) the final volume of the gas, (b) the work done by the gas, and (c) the heat added to the gas.

85. •• **Making Ice** You place 0.410 kg of cold water inside a freezer that has a constant temperature of 0 °C. The water eventually freezes and becomes ice at 0 °C. (a) What is the change in entropy of the water as it turns into ice? (See Table 17-4 for the appropriate latent heat.) (b) If the coefficient of performance of the freezer is 4.50, how much heat does the freezer exhaust into your kitchen as a result of freezing the water? (c) If your kitchen temperature is 22 °C, what is the increase in entropy of your kitchen? (d) What is the change in entropy of the universe during the freezing process?

86. •• An inventor claims a new cyclic engine that uses organic grape juice as its working material. According to the claims, the engine absorbs 1250 J of heat from a 1010 K reservoir and performs 1120 J of work each cycle. The waste heat is exhausted to the atmosphere at a temperature of 302 K. (a) What is the efficiency that is implied by these claims? (b) What is the efficiency of a reversible engine operating between the same high and low temperatures used by this engine? (Should you invest in this invention?)

87. •• **Predict/Calculate** A small dish containing 530 g of water is placed outside for the birds. During the night the outside temperature drops to −5.0 °C and stays at that value for several hours. (a) When the water in the dish freezes at 0 °C, does its entropy increase, decrease, or stay the same? Explain. (b) Calculate the change in entropy that occurs as the water freezes. (c) When the water freezes, by how much does the entropy of the environment increase? (d) What is the change in entropy of the universe as a result of this process?

88. •• **Predict/Calculate** An ideal gas is taken through the three processes shown in Figure 18-25. Fill in the missing entries in the following table.

	Q	W	ΔU
A → B	(a)	(b)	−95.3 J
B → C	(c)	−90.8 J	−136 J
C → A	331 J	(d)	(e)

89. ••• One mole of an ideal monatomic gas follows the three-part cycle shown in **FIGURE 18-35**. (a) Fill in the table below. (b) What is the efficiency of this cycle?

	Q	W	ΔU
A → B			
B → C			
C → A			

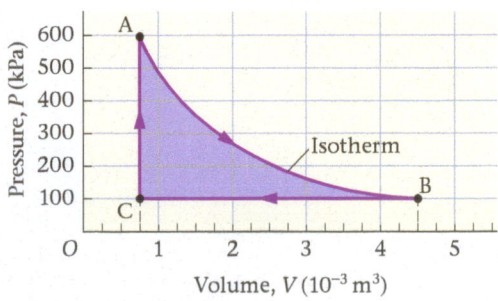

FIGURE 18-35 Problem 89

90. ••• When a heat Q is added to a monatomic ideal gas at constant pressure, the gas does a work W. Find the ratio, W/Q.

91. ••• **The Carnot Cycle FIGURE 18-36** shows an example of a Carnot cycle. The cycle consists of the following four processes: (1) an isothermal expansion from V_1 to V_2 at the temperature T_h; (2) an adiabatic expansion from V_2 to V_3 during which the temperature drops from T_h to T_c; (3) an isothermal compression from V_3 to V_4 at the temperature T_c; and (4) an adiabatic compression from V_4 to V_1 during which the temperature increases from T_c to T_h. Show that the efficiency of this cycle is $e = 1 - T_c/T_h$, as expected.

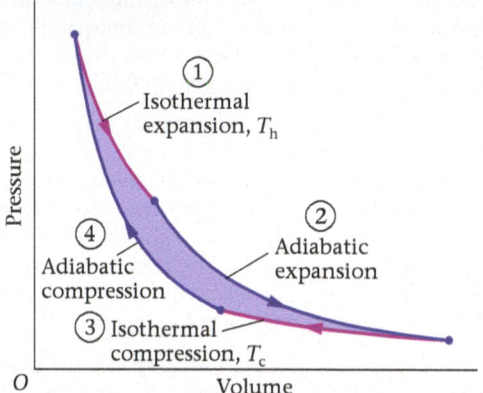

FIGURE 18-36 Problem 91

92. ••• A Carnot engine and a Carnot refrigerator operate between the same two temperatures. Show that the coefficient of performance, COP, for the refrigerator is related to the efficiency, e, of the engine by the following expression: COP $= (1 - e)/e$.

PASSAGE PROBLEMS

Energy from the Ocean

Whenever two objects are at different temperatures, thermal energy can be extracted with a heat engine. A case in point is the ocean, where one "object" is the warm water near the surface, and the other is the cold water at considerable depth. Tropical seas, in particular, can have significant temperature differences between the sun-warmed surface waters, and the cold, dark water 1000 m or more below the surface. A typical oceanic "temperature profile" is shown in **FIGURE 18-37**, where we see a rapid change in temperature—a thermocline—between depths of approximately 400 m and 900 m.

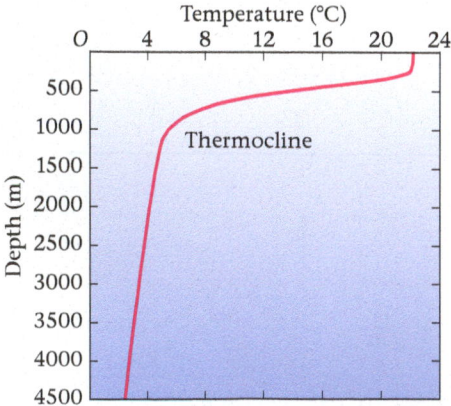

FIGURE 18-37 Temperature versus depth for ocean waters in the tropics (Problems 93, 94, 95, and 96)

The idea of tapping this potential source of energy has been around for a long time. In 1870, for example, Captain Nemo in Jules Verne's *Twenty Thousand Leagues Under the Sea*, said, "I owe all to the ocean; it produces electricity, and electricity gives heat, light, motion, and, in a word, life to the Nautilus." Just 11 years later, the French physicist Jacques Arsene d'Arsonval proposed a practical system referred to as Ocean Thermal Energy Conversion (OTEC), and in 1930 Georges Claude, one of d'Arsonval's students, built and operated the first experimental OTEC system off the coast of Cuba.

OTEC systems, which are potentially low-cost and carbon neutral, can provide not only electricity, but also desalinated water as part of the process. In fact, an OTEC plant generating 2 MW of electricity is expected to produce over 14,000 cubic feet of desalinated water a day. The governments of Hawaii, Japan, and Australia are actively pursuing plans for OTEC systems.

93. • Suppose an OTEC system operates with surface water at 22 °C and deep water at 4.0 °C. What is the maximum efficiency this system could have?

    A. 6.10%   **B.** 8.20%   **C.** 9.40%   **D.** 18.0%

94. • If we could go deeper for colder water, say to a depth of 4500 m, what is the maximum efficiency that can be achieved?

    A. 6.64%   **B.** 7.52%   **C.** 14.0%   **D.** 25.5%

95. • After they are used in the OTEC plant, the warm surface water and cold deep water (which is richer in inorganic nutrients) may be mixed together and discharged back into the ocean. In this process, an environmental concern is to not affect the local temperature profile so as to reduce impacts on sea life. If 22 °C and 5.5 °C water are mixed in equal volumes, at about what depth should the mixture be discharged so that the discharged mixture is at the same temperature as the surroundings?

    A. 0 m   **B.** 100 m   **C.** 500 m   **D.** 1000 m

96. •• A commercial OTEC system may take in 1500 kg of water per second at 22 °C and cool it to 4.0 °C. How much energy is released in one second by this system? (For comparison, the energy released in burning a gallon of gasoline is 1.3×10^8 J.)

    A. 2.5×10^7 J   **B.** 1.1×10^8 J
    C. 1.4×10^8 J   **C.** 1.6×10^8 J

97. •• **Predict/Calculate REFERRING TO EXAMPLE 18-21** Suppose we lower the temperature of the cold reservoir to 295 K; the temperature of the hot reservoir is still 576 K. **(a)** Is the new efficiency of the engine greater than, less than, or equal to 0.470? Explain. **(b)** What is the new efficiency? **(c)** Find the work done by this engine when 1050 J of heat is drawn from the hot reservoir.

98. •• **Predict/Calculate REFERRING TO EXAMPLE 18-21** Suppose the temperature of the hot reservoir is increased by 16 K, from 576 K to 592 K, *and* that the temperature of the cold reservoir is also increased by 16 K, from 305 K to 321 K. **(a)** Is the new efficiency greater than, less than, or equal to 0.470? Explain. **(b)** What is the new efficiency? **(c)** What is the change in entropy of the hot reservoir when 1050 J of heat is drawn from it? **(d)** What is the change in entropy of the cold reservoir?

Big Ideas

1 Electric charge comes in discrete units of two types—positive and negative.

2 Like charges repel one another; unlike charges attract.

3 The electric field points in the direction of the force that a positive charge would experience.

4 The electric flux through a closed surface is proportional to the amount of charge it encloses.

Electric Charges, Forces, and Fields

▲ Amber, a form of fossilized tree resin long used to make beautiful beads and other ornaments, has also made contributions to two different sciences. Pieces of amber have preserved prehistoric insects and pollen grains for modern students of evolution. And, more than 2500 years ago, amber provided Greek scientists with their first opportunity to study electric forces—the subject of this chapter.

We are all made of electric charges. Every atom in our body contains both positive and negative charges held together by an attractive force that is similar to gravity—only vastly stronger. In this chapter, we discuss the basic properties of electric charges and the forces that act between them.

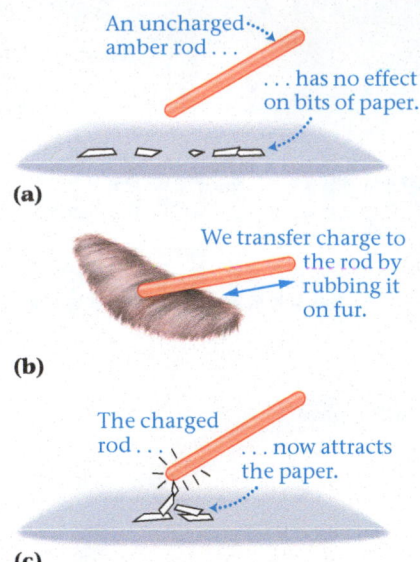

An uncharged
amber rod . . .

. . . has no effect
on bits of paper.

(a)

We transfer charge to
the rod by
rubbing it
on fur.

(b)

The charged
rod . . .

. . . now attracts
the paper.

(c)

▲ **FIGURE 19-1 Charging an amber rod** An uncharged amber rod **(a)** exerts no force on scraps of paper. When the rod is rubbed against a piece of fur **(b)**, it becomes charged and then attracts the paper **(c)**.

▶ **FIGURE 19-2 Likes repel, opposites attract** A charged amber rod is suspended by a thread. The charge on the amber is designated as negative. **(a)** When another charged amber rod is brought near the suspended rod, it rotates away, indicating a repulsive force between like charges. **(b)** When a charged glass rod is brought close to the suspended amber rod, the amber rotates toward the glass, indicating an attractive force and the existence of a second type of charge, which we designate as positive.

19-1 Electric Charge

The effects of electric charge have been known since at least 600 B.C. About that time, the Greeks noticed that amber—a solid, translucent material formed from the fossilized resin of extinct coniferous trees—has a peculiar property. If a piece of amber is rubbed with animal fur, it attracts small, lightweight objects. This phenomenon is illustrated in **FIGURE 19-1**.

For some time, it was thought that amber was unique in its ability to become "charged." Much later, it was discovered that other materials behave in this way as well. For example, if glass is rubbed with a piece of silk, it too can attract small objects. In this respect, glass and amber seem to be the same. It turns out, however, that these two materials have different types of charge.

To see this, imagine suspending a small, charged rod of amber from a thread, as in **FIGURE 19-2**. If a second piece of charged amber is brought near the rod, as shown in Figure 19-2 (a), the rod rotates away, indicating a repulsive force between the two pieces of amber. Thus, "like" charges repel. On the other hand, if a piece of charged glass is brought near the amber rod, the amber rotates toward the glass, indicating an attractive force. This is illustrated in Figure 19-2 (b). Clearly, then, the *different* charges on the glass and amber attract one another. We refer to different charges as being the "opposite" of one another, as in the familiar expression "opposites attract." Examples of electrostatic forces in action are shown in **FIGURE 19-3**.

Positive and Negative Charges We know today that the two types of charge found on amber and glass are, in fact, the only types that exist, and we still use the purely arbitrary names—**positive** (+) charge and **negative** (−) charge—proposed by

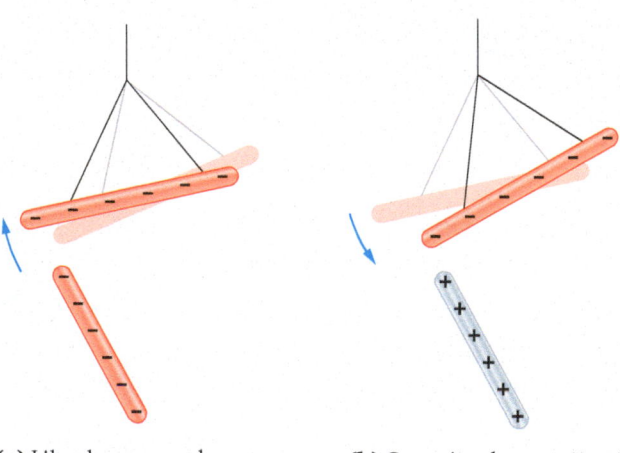

(a) Like charges repel

(b) Opposite charges attract

(a)

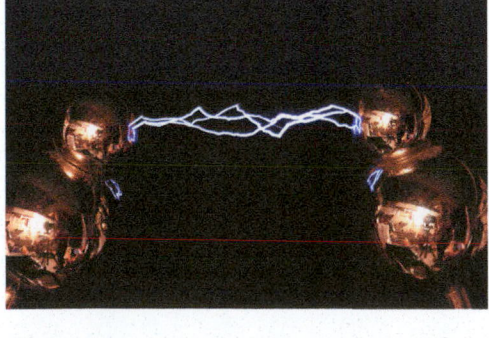

(b)

(c)

▲ **FIGURE 19-3 Visualizing Concepts Electrostatic Force (a)** The Van de Graaff generator these children are touching can produce very large charges of static electricity. The children are clearly not frightened, so why is their hair standing on end? Could it be because like charges repel? **(b)** The two metal spheres in this photo have opposite charges, resulting in a spark between them. **(c)** Lightning bolts are electrical discharges (large-scale sparks) between oppositely charged clouds and the ground.

Benjamin Franklin (1706–1790) in 1747. In accordance with Franklin's original suggestion, the charge of amber is negative, and the charge of glass is positive (the opposite of negative). Calling the different charges + and − is actually quite useful mathematically; for example, an object that contains an equal amount of positive and negative charge has a net charge that adds up to zero. Objects with zero net charge are said to be electrically **neutral**.

A familiar example of an electrically neutral object is the atom. Atoms have a small, dense nucleus with a positive charge surrounded by a cloud of negatively charged electrons (from the Greek word for amber, *elektron*) with an equal but opposite negative charge. Two types of particles are found in the nucleus; one is positively charged (proton), and the other is electrically neutral (neutron). A simplified pictorial representation of an atom is shown in **FIGURE 19-4**.

Charge Is Quantized All electrons have exactly the same electric charge. This charge is very small and is defined to have a magnitude, *e*, given by the following:

Magnitude of an Electron's Charge, *e*

$$e = 1.60 \times 10^{-19}\,\text{C} \qquad\qquad 19\text{-}1$$

SI unit: coulomb, C

In this expression, C is a unit of charge referred to as the **coulomb**, named for the French physicist Charles-Augustin de Coulomb (1736–1806). (The precise definition of the coulomb is in terms of electric current, which we shall discuss in Chapter 21.) Clearly, the charge on an electron, which is negative, is −*e*. This is one of the defining, or *intrinsic*, properties of the electron. Another intrinsic property of the electron is its mass, m_e:

$$m_e = 9.11 \times 10^{-31}\,\text{kg} \qquad\qquad 19\text{-}2$$

In contrast, the charge on a proton—one of the main constituents of nuclei—is *exactly* +*e*. Therefore, the net charge on atoms, which have equal numbers of electrons and protons, is precisely zero. The mass of the proton is

$$m_p = 1.673 \times 10^{-27}\,\text{kg} \qquad\qquad 19\text{-}3$$

Notice that this is about 2000 times larger than the mass of the electron. The other main constituent of the nucleus is the neutron, which, as its name implies, has zero charge. Its mass is slightly larger than that of the proton:

$$m_n = 1.675 \times 10^{-27}\,\text{kg} \qquad\qquad 19\text{-}4$$

Because the magnitude of the charge per electron is so small, 1.60×10^{-19} C/electron, it follows that the number of electrons in 1 C of charge is enormous:

$$\frac{1\,\text{C}}{1.60 \times 10^{-19}\,\text{C/electron}} = 6.25 \times 10^{18}\,\text{electrons}$$

As we shall see when we consider the force between charges, a coulomb is a significant amount of charge; even a powerful lightning bolt delivers only 20 to 30 C. A more common unit of charge is the microcoulomb, μC, where $1\,\mu\text{C} = 10^{-6}$ C. Still, the amount of charge contained in everyday objects is very large, even in units of the coulomb, as we show in the following Exercise.

EXERCISE 19-1 FIND THE CHARGE

Find the amount of positive electric charge in one mole of sodium atoms. (*Note*: The nucleus of a sodium atom contains 11 protons.)

REASONING AND SOLUTION
Each sodium atom contains 11 positive charges of magnitude *e*. Therefore, the total positive charge in a mole of sodium atoms is

$$N_A(11e) = (6.02 \times 10^{23})(11)(1.60 \times 10^{-19}\,\text{C}) = 1.06 \times 10^{6}\,\text{C}$$

Thus, a mere 23 g of sodium (the mass of one mole) contains over a million coulombs of positive charge—and the same amount of negative charge as well.

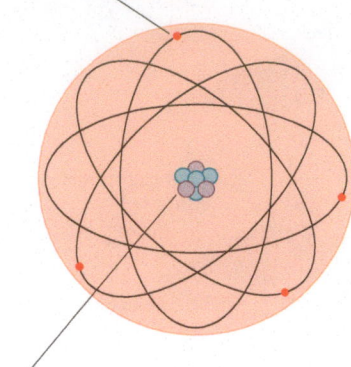

Negatively charged electron

Positively charged nucleus

(a) Simplified picture of an atom

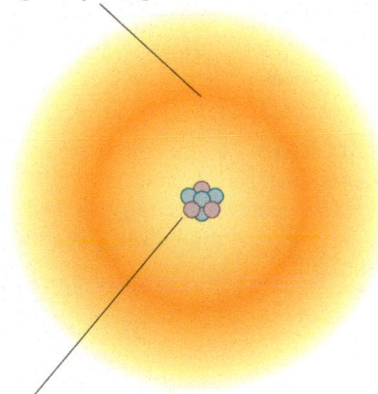

Negatively charged electron cloud

Positively charged nucleus

(b) More accurate picture of an atom

▲ **FIGURE 19-4 The structure of an atom** **(a)** A crude representation of an atom, showing the positively charged nucleus at its center and the negatively charged electrons orbiting about it. **(b)** More accurately, the electrons should be thought of as forming a "cloud" of negative charge surrounding the nucleus. The density of the cloud at a given location is proportional to the probability that an electron will be found there.

PHYSICS IN CONTEXT
Looking Ahead

When electric charge flows from one location to another, it produces an electric current. We consider electric circuits with a one-way, direct current (dc) in Chapter 21, and circuits with an alternating current (ac) in Chapter 24.

Uncharged rod
and fur

(a)

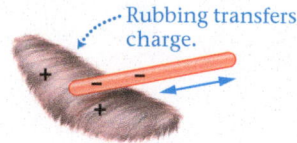

Rubbing transfers
charge.

(b)

The net charge
is still zero . . .

. . . but the rod and fur
are now oppositely
charged.

(c)

◄ **FIGURE 19-5 Charge transfer (a)** Initially, an amber rod and a piece of fur are electrically neutral; that is, they each contain equal quantities of positive and negative charges. **(b)** As they are rubbed together, charge is transferred from one to the other. **(c)** In the end, the fur and the rod have charges of equal magnitude but opposite sign.

Charge Separation

How is it that rubbing a piece of amber with fur gives the amber a charge? Originally, it was thought that the friction of rubbing *created* the observed charge. We now know, however, that rubbing the fur across the amber simply results in a *transfer* of charge from the fur to the amber—with the total amount of charge remaining unchanged. This is indicated in **FIGURE 19-5**. Before charging, the fur and the amber are both neutral. During the rubbing process some electrons are transferred from the fur to the amber, giving the amber a net negative charge, and leaving the fur with a net positive charge. At no time during this process is charge ever created or destroyed. This, in fact, is an example of one of the fundamental conservation laws of physics:

> **Conservation of Electric Charge**
>
> The total electric charge of the universe is constant. No physical process can result in an increase or decrease in the total amount of electric charge in the universe.

Big Idea ❶ Electric charge comes in discrete units of two types—positive and negative. The total charge in the universe is conserved.

When charge is transferred from one object to another, it is generally due to the movement of electrons. In a typical solid, the nuclei of the atoms are fixed in position. The outer electrons of these atoms, however, are often weakly bound and fairly easily separated. As a piece of fur rubs across amber, for example, some of the electrons that were originally part of the fur are separated from their atoms and deposited onto the amber. The atom that loses an electron is now a **positive ion**, and the atom that receives an extra electron becomes a **negative ion**. This is charging by separation.

**PHYSICS
IN CONTEXT
Looking Back**

The concept of a conserved quantity, like energy (Chapter 8) or momentum (Chapter 9), appears again here in terms of the electric charge.

Triboelectric Charging In general, when two materials are rubbed together, the magnitude *and* sign of the charge that each material acquires depends on how strongly it holds onto its electrons. For example, if silk is rubbed against glass, the glass acquires a positive charge, as was mentioned earlier in this section. It follows that electrons have moved from the glass to the silk, giving the silk a *negative* charge. If silk is rubbed against amber, however, the silk becomes *positively* charged, as electrons in this case pass from the silk to the amber.

These results can be understood by referring to Table 19-1, which presents the relative charging due to rubbing—also known as **triboelectric charging**—for a variety of materials. The more plus signs associated with a material, the more readily it gives up electrons and becomes positively charged. Similarly, the more minus signs for a material, the more readily it acquires electrons. For example, we know that amber becomes negatively charged when rubbed against fur, but a greater negative charge is obtained if rubber, PVC, or Teflon is rubbed with fur instead. In general, when two materials in Table 19-1 are rubbed together, the one higher in the list becomes positively charged, and the one lower in the list becomes negatively charged. The greater the separation on the list, the greater the magnitude of the charge.

Charge separation occurs not only when one object is rubbed against another, but also when objects collide. For example, colliding crystals of ice in a rain cloud cause charge separation that may ultimately result in bolts of lightning to bring the charges together again. Similarly, particles in the rings of Saturn are constantly undergoing collisions and becoming charged as a result. In fact, when the *Voyager* spacecraft examined the rings of Saturn, it observed electrostatic discharges, similar to lightning bolts on Earth. In addition, ghostly radial "spokes" that extend across the rings of Saturn— which cannot be explained by gravitational forces alone—are the result of electrostatic interactions.

TABLE 19-1 Triboelectric Charging

Material	Relative charging with rubbing
Rabbit fur	++++++
Glass	+++++
Human hair	++++
Nylon	+++
Silk	++
Paper	+
Cotton	−
Wood	−−
Amber	−−−
Rubber	−−−−
PVC	−−−−−
Teflon	−−−−−−

CONCEPTUAL EXAMPLE 19-2   COMPARE THE MASS

Is the mass of an amber rod after it is charged with fur greater than, less than, or the same as its mass before the charging?

REASONING AND DISCUSSION

The amber rod becomes negatively charged, and hence it acquires electrons from the fur. Each electron has a small, but nonzero, mass. Therefore, the mass of the rod increases ever so slightly as it is charged.

ANSWER

The mass of the amber rod is greater after the charging.

Because electrons always have the charge $-e$, and protons always have the charge $+e$, it follows that all objects must have a net charge that is an integral multiple of e. This conclusion was confirmed early in the twentieth century by the American physicist Robert A. Millikan (1868–1953) in a classic series of experiments. He found that the charge on a drop of oil is $\pm e$, $\pm 2e$, $\pm 3e$, and so on, but never $1.5e$ or $-9.3847e$, for example. We describe this restriction by saying that electric charge is **quantized**.

Charged Rod and Aluminum Can

Polarization   We know that charges of opposite sign attract, but it is also possible for a charged rod to attract small objects that have zero net charge. The mechanism responsible for this attraction is called **polarization**.

To see how polarization works, consider **FIGURE 19-6 (a)**. Here we show a positively charged rod held close to an enlarged view of a neutral object. An atom near the surface of the neutral object will become elongated because the negative electrons in it are attracted to the rod while the positive protons are repelled. As a result, a net negative charge develops on the surface near the rod—the so-called polarization charge. The attractive force between the rod and this *induced* polarization charge leads to a net attraction between the rod and the entire neutral object.

Of course, the same conclusion is reached if we consider a negative rod held near a neutral object—except in that case the polarization charge is positive. Thus, the effect of polarization is to give rise to an attractive force regardless of the sign of the charged object. It is for this reason that both charged amber and charged glass attract neutral objects—even though their charges are opposite.

RWP* If you've ever rubbed a balloon on your hair and observed it carefully, you've seen electric polarization at work. Your hair donates electrons to the rubber balloon (Table 19-1), and the negatively charged balloon can polarize nearby neutral molecules, such as those in the uncharged stream of water in **FIGURE 19-6 (b)**, leading to an

▼ **FIGURE 19-6 Visualizing Concepts**
Electric Polarization (a) When a charged rod is far from a neutral object, the atoms in the object are undistorted, as in Figure 19-4. As the rod is brought closer, however, the atoms distort, producing an excess of one type of charge on the surface of the object (in this case a negative charge). This induced charge is referred to as a polarization charge. Because the sign of the polarization charge is the opposite of the sign of the charge on the rod, there is an attractive force between the rod and the object. **(b)** A charged balloon induces a polarization charge on a stream of water, and deflects it toward the balloon. **(c)** A charged balloon induces a polarization charge on the paper figures, and the electrostatic attraction holds the figures against the balloon.

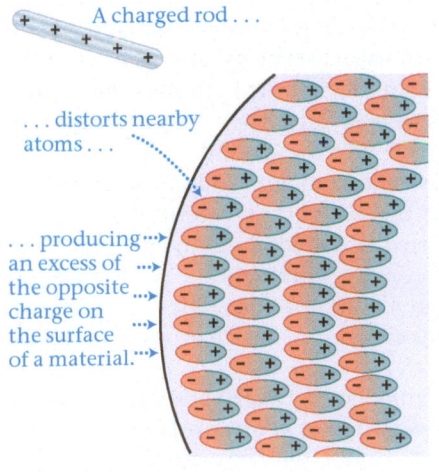

A charged rod . . .

. . . distorts nearby atoms . . .

. . . producing an excess of the opposite charge on the surface of a material.

(a)

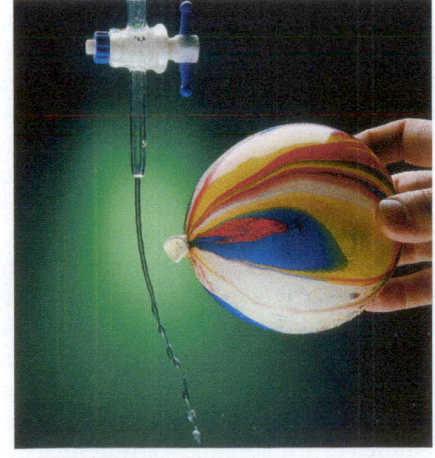

(b)

(c)

---

*Real World Physics applications are denoted by the acronym RWP.

attractive force between the water and the balloon. Similarly, the charged balloon can polarize the molecules in a wall, as shown in **FIGURE 19-6 (c)**. The small attractive force that results is sufficient to support the weight of a balloon.

BIO A potentially dangerous, and initially unsuspected, medical application of polarization occurs in endoscopic surgery. In these procedures, a tube carrying a small video camera is inserted into the body. The resulting video image is shown on a computer screen, which is often electrically charged when dusted with a cloth. Minute airborne particles in the operating room—including dust, lint, and skin cells—are polarized by the charge on the screen, and are attracted to its exterior surface. The problem comes when a surgeon touches the screen to point out an important feature to others in the medical staff. Even the slightest touch can transfer particles—many of which carry bacteria—from the screen to the surgeon's finger and from there to the patient.

▲ **FIGURE 19-7** People who work with electricity must be careful to use gloves made of nonconducting materials. Rubber, an excellent insulator, is often used for this purpose.

Enhance Your Understanding   (Answers given at the end of the chapter)

1. When rabbit fur is rubbed on silk, both materials become charged. Referring to Table 19-1, determine the sign of the charge on each material.

Section Review

*   Electric charge comes in two types, referred to as positive and negative. Charge comes in discrete, quantized units of magnitude $e = 1.60 \times 10^{-19}$ C. The charge of an electron is $-e$, and the charge of a proton is $+e$.

*   The total charge in the universe is conserved.

19-2 Insulators and Conductors

Suppose you rub one end of an amber rod with fur, being careful not to touch the other end. The result is that the rubbed portion becomes charged, whereas the other end remains neutral. In particular, the negative charge transferred to the rubbed end stays put; it does not move about from one end of the rod to the other. Materials like amber, in which charges are not free to move, are referred to as **insulators**. Most insulators are nonmetallic substances, and most are also good thermal insulators. An example is given in **FIGURE 19-7**.

In contrast, most metals are good **conductors** of electricity, in the sense that they allow charges to move about more or less freely. For example, suppose an uncharged metal sphere is placed on an insulating base. If a charged rod is brought into contact with the sphere, as in **FIGURE 19-8 (a)**, some charge will be transferred to the sphere at the point of contact. The charge does not stay put, however. Because the metal is a good conductor of electricity, the charges are free to move about the sphere, which they do because of their mutual repulsion. The result is a uniform distribution of charge over the surface of the sphere, as shown in **FIGURE 19-8 (b)**. Notice that the insulating base prevents charge from flowing away from the sphere into the ground.

On a microscopic level, the difference between conductors and insulators is that the atoms in conductors allow one or more of their outermost electrons to become detached. These detached electrons, often referred to as "conduction electrons," can move freely throughout the conductor. In a sense, the conduction electrons behave almost like gas molecules moving about within a container. Insulators, in contrast, have very few, if any, free electrons; the electrons are bound to their atoms and cannot move from place to place within the material.

Some materials have properties that are intermediate between those of a good conductor and a good insulator. These materials, referred to as **semiconductors**, can be fine-tuned to display almost any desired degree of conductivity by controlling the concentration of the various components from which they are made. The great versatility of semiconductors is one reason they have found such wide areas of application in electronics and computers.

RWP Exposure to light can sometimes determine whether a given material is an insulator or a conductor. An example of such a **photoconductive** material is

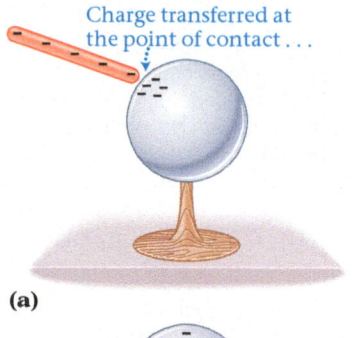

Charge transferred at the point of contact . . .

(a)

. . . spreads over the entire surface of a conductor.

(b)

▲ **FIGURE 19-8 Charging a conductor**
(a) When an uncharged metal sphere is touched by a charged rod, some charge is transferred at the point of contact.
(b) Because like charges repel, and charges move freely in a conductor, the transferred charge quickly spreads out and covers the entire surface of the sphere.

selenium, which conducts electricity when light shines on it but is an insulator when in the dark. Because of this special property, selenium plays a key role in the production of photocopies. To see how, we first note that at the heart of most photocopiers is a selenium-coated aluminum drum. Initially, the selenium is given a positive charge and kept in the dark—which causes it to retain its charge. When flash lamps illuminate a document to be copied, an image of the document falls on the drum. Where the document is light, the selenium is illuminated and becomes a conductor, and the positive charge flows away into the aluminum drum, leaving the selenium uncharged. Where the document is dark, the selenium is not illuminated, meaning that it is an insulator, and its charge remains in place. At this point, a negatively charged "toner" powder is wiped across the drum, where it sticks to those positively charged portions of the drum that were not illuminated. Next, the drum is brought into contact with paper, transferring the toner to it. Finally, the toner is fused into the paper fibers with heat, the drum is cleaned of excess toner, and the cycle repeats. Thus, a slight variation in electrical properties due to illumination is the basis of an entire technology.

Enhance Your Understanding   (Answers given at the end of the chapter)

2. Are metals generally good conductors or good insulators? Explain.

Section Review

- Electrons move freely in conductors, but are not free to move in insulators.

19-3 Coulomb's Law

We've already discussed the fact that electric charges exert forces on one another. The precise law describing these forces was first determined by Coulomb in the late 1780s. His result is remarkably simple. Suppose, for example, that an idealized point charge q_1 is separated by a distance r from another point charge q_2. Both charges are at rest; that is, the system is **electrostatic**. According to Coulomb's law, the magnitude of the electrostatic force between these charges is proportional to the product of the magnitude of the charges, $|q_1||q_2|$, and inversely proportional to the square of the distance, r^2, between them:

Coulomb's Law for the Magnitude of the Electrostatic Force Between Point Charges

$$F = k\frac{|q_1||q_2|}{r^2} \qquad \text{19-5}$$

SI unit: newton, N

In this expression, the proportionality constant k has the value

$$k = 8.99 \times 10^9 \text{ N} \cdot \text{m}^2/\text{C}^2 \qquad \text{19-6}$$

Notice that the units of k are simply those required for the force F to have the units of newtons.

The direction of the force in Coulomb's law is along the line connecting the two charges. In addition, we know from the observations described in Section 19-1 that like charges repel and opposite charges attract. These properties are illustrated in **FIGURE 19-9**, where force vectors are shown for charges of various signs. Thus, when applying Coulomb's law, we first calculate the magnitude of the force using Equation 19-5 ($F = k|q_1||q_2|/r^2$), and then determine its direction with the "likes repel, opposites attract" rule.

Finally, notice how Newton's third law applies to each of the cases shown in Figure 19-9. For example, the force exerted on charge 1 by charge 2, $\vec{F}_{12}$, is always equal in magnitude and opposite in direction to the force exerted on charge 2 by charge 1, $\vec{F}_{21}$; that is, $\vec{F}_{21} = -\vec{F}_{12}$.

Big Idea ❷ The force between two electric charges is proportional to the product of the charges and inversely proportional to the square of the distance separating them. Like charges repel one another, and unlike charges attract.

PHYSICS IN CONTEXT
Looking Back

Coulomb's law for the electrostatic force between two charges (Equation 19-5) is virtually identical to Newton's law of universal gravitation between two masses (Chapter 12), but with electric charge replacing mass.

PHYSICS IN CONTEXT
Looking Ahead

The electric force is conservative, and hence it has an associated electric potential energy. We will determine this potential energy in Chapter 20. We will also point out the close analogy between the electric potential energy and the gravitational potential energy (Chapter 12).

PHYSICS IN CONTEXT
Looking Ahead

Coulomb's law comes up again in atomic physics, where it plays a key role in the Bohr model of the hydrogen atom in Chapter 31.

▶ **FIGURE 19-9 Forces between point charges** The electrostatic forces exerted by two point charges on one another are always along the line connecting the charges. If the charges have the same sign, as in **(a)** and **(c)**, the forces are repulsive; that is, each charge experiences a force that points away from the other charge. Charges of opposite sign, as in **(b)**, experience attractive forces. Notice that in all cases the forces exerted on the two charges form an action-reaction pair; that is, $\vec{F}_{21} = -\vec{F}_{12}$.

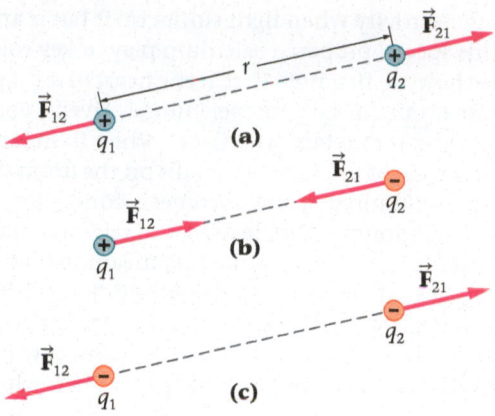

CONCEPTUAL EXAMPLE 19-3 WHERE DO THEY COLLIDE?

An electron and a proton, initially separated by a distance d, are released from rest simultaneously. The two particles are free to move. When they collide, are they at the midpoint of their initial separation, closer to the initial position of the proton, or closer to the initial position of the electron?

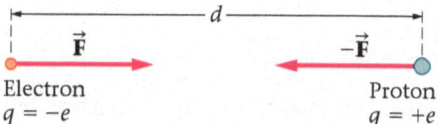

Electron
$q = -e$

Proton
$q = +e$

REASONING AND DISCUSSION

Because of Newton's third law, the forces exerted on the electron and proton are equal in magnitude and opposite in direction. For this reason, it might seem that the particles meet at the midpoint. The masses of the particles, however, are quite different. In fact, as mentioned in Section 19-1, the mass of the proton is about 2000 times greater than the mass of the electron; therefore, the proton's acceleration ($a = F/m$) is about 2000 times less than the electron's acceleration. As a result, the particles collide near the initial position of the proton. More specifically, they collide at the location of the center of mass of the system, which remains at rest throughout the process.

ANSWER

The particles collide closer to the initial position of the proton.

Comparing the Electric and Gravitational Forces Before we continue, let's point out some of the similarities and differences between Coulomb's law, $F = k|q_1||q_2|/r^2$, and Newton's law of gravity, $F = Gm_1m_2/r^2$. As for similarities, the force in each case decreases as the square of the distance between the two objects. In addition, both forces depend on a product of intrinsic quantities: In the case of the electric force, the intrinsic quantity is the charge; in the case of gravity, it is the mass.

Equally significant, however, are the differences. For example, the force of gravity is always attractive, whereas the electric force can be attractive or repulsive. As a result, the net electric force between neutral objects, such as the Earth and the Moon, is essentially zero because attractive and repulsive forces cancel one another. Because gravity is always attractive, however, the net gravitational force between the Earth and the Moon is nonzero. Thus, in astronomy, gravity rules, and electric forces play hardly any role.

Just the opposite is true in atomic systems. To see this, let's compare the electric and gravitational forces between a proton and an electron in a hydrogen atom. Taking the distance between the two particles to be the radius of hydrogen, $r = 5.29 \times 10^{-11}$ m, we find that the gravitational force has a magnitude

$$F_g = G\frac{m_e m_p}{r^2}$$

$$= (6.67 \times 10^{-11}\,\text{N} \cdot \text{m}^2/\text{kg}^2)\frac{(9.11 \times 10^{-31}\,\text{kg})(1.673 \times 10^{-27}\,\text{kg})}{(5.29 \times 10^{-11}\,\text{m})^2}$$

$$= 3.63 \times 10^{-47}\,\text{N}$$

Similarly, the magnitude of the electric force between the electron and the proton is

$$F_e = k \frac{|q_1||q_2|}{r^2}$$

$$= (8.99 \times 10^9 \, \text{N} \cdot \text{m}^2/\text{C}^2) \frac{|-1.60 \times 10^{-19} \, \text{C}||1.60 \times 10^{-19} \, \text{C}|}{(5.29 \times 10^{-11} \, \text{m})^2}$$

$$= 8.22 \times 10^{-8} \, \text{N}$$

Taking the ratio, we find that the electric force is greater than the gravitational force by a factor of

$$\frac{F_e}{F_g} = \frac{8.22 \times 10^{-8} \, \text{N}}{3.63 \times 10^{-47} \, \text{N}} = 2.26 \times 10^{39}$$

$$= 2,260,000,000,000,000,000,000,000,000,000,000,000,000$$

This huge factor explains why a small piece of charged amber can lift bits of paper off the ground, even though the entire mass of the Earth is pulling downward on the paper. It also shows why gravity generally plays no role in atomic systems.

Next, we use the electric force to get an idea of the speed of an electron in a hydrogen atom and the frequency of its orbital motion.

PROBLEM-SOLVING NOTE

Distance Dependence of the Coulomb Force

The Coulomb force has an inverse-square dependence on distance. Be sure to divide the product of the charges, $k|q_1||q_2|$, by r^2 when calculating the force.

EXAMPLE 19-4 THE BOHR ORBIT

In an effort to better understand the behavior of atomic systems, the Danish physicist Niels Bohr (1885–1962) introduced a simple model for the hydrogen atom. In the Bohr model, as it is known today, the electron is imagined to move in circular orbits about a stationary proton. The force responsible for the electron's circular motion is the electric force of attraction between the electron and the proton. **(a)** The Bohr orbit with the smallest radius is called the first Bohr orbit. Its radius is 5.29×10^{-11} m. Given that the mass of the electron is $m_e = 9.11 \times 10^{-31}$ kg, find the electron's speed in this orbit. **(b)** What is the frequency of the electron's orbital motion?

PICTURE THE PROBLEM

Our sketch shows the electron moving with a speed v in the first Bohr orbit, with a radius r. Because the proton is so much more massive than the electron, it is essentially stationary at the center of the orbit. Notice that the orbiting electron has a charge $q_1 = -e$ and the proton has a charge $q_2 = +e$.

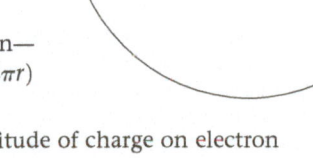

REASONING AND STRATEGY

a. The idea behind this model is that a force is required to make the electron move in a circular path, and this force is provided by the electric force of attraction between the electron and the proton, $F = ke^2/r^2$. Thus, as with any circular motion, we set the force acting on the electron equal to its mass times its centripetal acceleration. This allows us to solve for the centripetal acceleration, $a_{cp} = v^2/r$ (Equation 6-15), which in turn gives us the speed v.

b. The frequency of the electron's orbital motion is $f = 1/T$, where T is the period of the motion—that is, the time for one complete orbit. The time for an orbit, in turn, is the circumference $(2\pi r)$ divided by the speed, or $T = 2\pi r/v$. Taking the inverse immediately yields the frequency.

Known Bohr radius, $r = 5.29 \times 10^{-11}$ m; mass of electron, $m_e = 9.11 \times 10^{-31}$ kg; magnitude of charge on electron and proton, $e = 1.60 \times 10^{-19}$ C.

Unknown **(a)** Speed of electron, $v = ?$ **(b)** Frequency of orbital motion, $f = ?$

SOLUTION

Part (a)

1. Set the Coulomb force between the electron and proton equal to the centripetal force required for the electron's circular orbit. Note that $|q_1| = |q_2| = e$:

$$k \frac{|q_1||q_2|}{r^2} = m_e a_{cp}$$

$$k \frac{e^2}{r^2} = m_e \frac{v^2}{r}$$

2. Solve for the speed of the electron, v:

$$v = e\sqrt{\frac{k}{m_e r}}$$

3. Substitute numerical values:

$$v = (1.60 \times 10^{-19} \, \text{C}) \sqrt{\frac{8.99 \times 10^9 \, \text{N} \cdot \text{m}^2/\text{C}^2}{(9.11 \times 10^{-31} \, \text{kg})(5.29 \times 10^{-11} \, \text{m})}}$$

$$= 2.19 \times 10^6 \, \text{m/s}$$

CONTINUED

Part (b)

4. Calculate the time for one orbit, T, which is the distance (circumference = $2\pi r$) divided by the speed (v):

$$T = \frac{2\pi r}{v} = \frac{2\pi(5.29 \times 10^{-11}\,\text{m})}{2.19 \times 10^6\,\text{m/s}} = 1.52 \times 10^{-16}\,\text{s}$$

5. Take the inverse of T to find the frequency:

$$f = \frac{1}{T} = \frac{1}{1.52 \times 10^{-16}\,\text{s}} = 6.58 \times 10^{15}\,\text{Hz}$$

INSIGHT

If you could travel around the world at the speed of an electron in the first Bohr orbit, your trip would take only about 18 s, but your centripetal acceleration would be a more-than-lethal 76,000 times the acceleration of gravity!

The frequency of the electron in the first Bohr orbit is incredibly large. We won't encounter frequencies this high again until we study light waves in Chapter 25.

PRACTICE PROBLEM — PREDICT/CALCULATE

The second Bohr orbit has a radius that is four times the radius of the first Bohr orbit. **(a)** Is the speed of the electron in the second Bohr orbit greater than, less than, or equal to its speed in the first Bohr orbit? Explain. **(b)** What is the speed of an electron in the second Bohr orbit?   **[Answer: (a)** As we can see from Step 2, the speed of the electron decreases with the square root of the radius of the orbit. Thus, the electron's speed in the second Bohr orbit is less than in the first Bohr orbit by a factor of two. **(b)** $v = 1.09 \times 10^6\,\text{m/s}$]

Some related homework problems: Problem 14, Problem 21, Problem 33

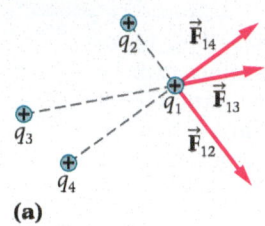

(a)

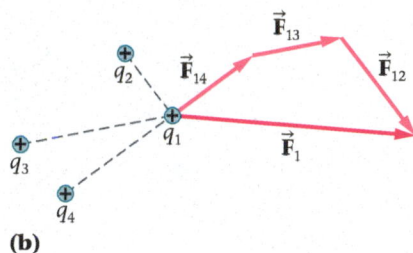

(b)

▲ **FIGURE 19-10 Superposition of forces** **(a)** Forces are exerted on q_1 by the charges q_2, q_3, and q_4. These forces are $\vec{F}_{12}$, $\vec{F}_{13}$, and $\vec{F}_{14}$, respectively. **(b)** The net force acting on q_1, which we label $\vec{F}_1$, is the vector sum of $\vec{F}_{12}$, $\vec{F}_{13}$, and $\vec{F}_{14}$.

Another indication of the strength of the electric force is given in the following Exercise.

EXERCISE 19-5   ELECTRIC FORCE

Find the electric force between two 5.00-C charges separated by 1.00 m.

REASONING AND SOLUTION

Substituting $q_1 = q_2 = 5.00\,\text{C}$ and $r = 1.00\,\text{m}$ in Coulomb's law, we find

$$F = k\frac{|q_1||q_2|}{r^2} = (8.99 \times 10^9\,\text{N}\cdot\text{m}^2/\text{C}^2)\frac{(5.00\,\text{C})(5.00\,\text{C})}{(1.00\,\text{m})^2} = 2.25 \times 10^{11}\,\text{N}$$

This is an enormous force in everyday terms, equivalent to about 23 million metric tons of force!

Exercise 19-5 shows that electric charges of five coulombs exert a force of about 23 million tons on one another when separated by a distance of a meter. Similarly, if the charge in your body could be separated into a pile of positive charges on one side of the room and a pile of negative charges on the other side, the force needed to hold them apart would be roughly 10^{24} tons! Thus, everyday objects are never far from electrical neutrality, because disturbing neutrality requires such tremendous forces.

Superposition of Forces

The electric force, like all forces, is a vector quantity. Hence, when a charge experiences forces due to two or more other charges, the net force on it is simply the *vector* sum of the forces taken individually. For example, in **FIGURE 19-10**, the total force on charge 1, $\vec{F}_1$, is the vector sum of the forces due to charges 2, 3, and 4:

$$\vec{F}_1 = \vec{F}_{12} + \vec{F}_{13} + \vec{F}_{14}$$

This is referred to as the **superposition** of forces.

Notice that the total force acting on a given charge is the sum of interactions involving just *two* charges at a time, with the force between each pair of charges given by Coulomb's law. For example, the total force acting on charge 1 in Figure 19-10 is the sum of the forces between q_1 and q_2, q_1 and q_3, and q_1 and q_4. Therefore, superposition of forces can be thought of as the generalization of Coulomb's law to systems containing more than two charges. In our first numerical Example of superposition, we consider three charges in a line.

PHYSICS
IN CONTEXT
Looking Back

The electric force is a vector quantity. Therefore, the material on vector addition and vector components (Chapter 3) finds extensive use in this chapter.

EXAMPLE 19-6   NET FORCE

A charge $q_1 = -6.5\,\mu C$ is at the origin, and a charge $q_2 = -2.42\,\mu C$ is on the x axis at $x = 1.00$ m. Find the net force acting on a charge $q_3 = +1.4\,\mu C$ located at $x = 0.55$ m.

PICTURE THE PROBLEM

The physical situation is shown in our sketch, with each charge at its appropriate location. Notice that the forces exerted on charge q_3 by the charges q_1 and q_2 are in opposite directions. We give the force on q_3 due to q_1 the label $\vec{F}_{31}$, and the force on q_3 due to q_2 the label $\vec{F}_{32}$. (The location of q_3 and the length of the force vectors are drawn to scale.)

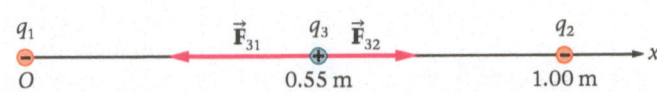

REASONING AND STRATEGY

The net force on q_3 is the vector sum of the forces due to q_1 and q_2. In particular, notice that $\vec{F}_{31}$ points in the negative x direction $(-\hat{x})$, whereas $\vec{F}_{32}$ points in the positive x direction $(+\hat{x})$. The magnitude of $\vec{F}_{31}$ is $k|q_1||q_3|/r^2$, with $r = 0.55$ m. Similarly, the magnitude of $\vec{F}_{32}$ is $k|q_2||q_3|/r^2$, with $r = 0.45$ m.

Known   At the origin, charge $q_1 = -6.5\,\mu C$; at $x = 1.00$ m, charge $q_2 = -2.42\,\mu C$; at $x = 0.55$ m, charge $q_3 = +1.4\,\mu C$.
Unknown   Net force on charge 3, $\vec{F}_3 = ?$

SOLUTION

1. Find the force acting on q_3 due to q_1. This force is in the negative x direction, as indicated in the sketch, and hence we give it a negative sign:

$$\vec{F}_{31} = -k\frac{|q_1||q_3|}{r^2}\hat{x}$$

$$= -(8.99 \times 10^9\,\text{N}\cdot\text{m}^2/\text{C}^2)\frac{(6.5 \times 10^{-6}\,\text{C})(1.4 \times 10^{-6}\,\text{C})}{(0.55\,\text{m})^2}\hat{x}$$

$$= -0.27\,\text{N}\,\hat{x}$$

2. Find the force acting on q_3 due to q_2. This force is in the positive x direction, as indicated in the sketch, and hence we give it a positive sign:

$$\vec{F}_{32} = k\frac{|q_2||q_3|}{r^2}\hat{x}$$

$$= (8.99 \times 10^9\,\text{N}\cdot\text{m}^2/\text{C}^2)\frac{(2.42 \times 10^{-6}\,\text{C})(1.4 \times 10^{-6}\,\text{C})}{(0.45\,\text{m})^2}\hat{x}$$

$$= 0.15\,\text{N}\,\hat{x}$$

3. Superpose (add) these forces to find the total force, $\vec{F}_3$, acting on q_3:

$$\vec{F}_3 = \vec{F}_{31} + \vec{F}_{32} = -0.27\,\text{N}\,\hat{x} + 0.15\,\text{N}\,\hat{x}$$
$$= -0.12\,\text{N}\,\hat{x}$$

INSIGHT

The net force acting on q_3 has a magnitude of 0.12 N, and points in the negative x direction.

PRACTICE PROBLEM

Find the net force on q_3 if it is at the location $x = 0.71$ m.   [**Answer:** $\vec{F}_3 = 0.20\,\text{N}\,\hat{x}$. The net force is now in the positive x direction.]

Some related homework problems: Problem 17, Problem 23, Problem 24

QUICK EXAMPLE 19-7   FIND THE LOCATION OF ZERO NET FORCE

In Example 19-6, the net force acting on the charge q_3 at $x = 0.55$ m is to the left (negative x direction). In the Practice Problem to that Example, the net force at $x = 0.71$ m is to the right (positive x direction). To what value of x should q_3 be moved for the net force on it to be zero?

REASONING AND SOLUTION

The net force on q_3 is zero when the force to the left and the force to the right are equal in magnitude, as indicated in our sketch. This condition can be used to solve for the one unknown, x. (The location of q_3 and the length of the force vectors are drawn to scale.)

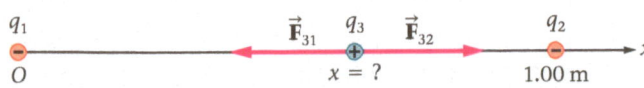

1. Write the magnitude of the force due to q_1:

$$F_{31} = k\frac{|q_1||q_3|}{x^2}$$

CONTINUED

2. Write the magnitude of the force due to q_2:

$$F_{32} = k\frac{|q_2||q_3|}{(1.00\text{ m} - x)^2}$$

3. Set these magnitudes equal to one another, and cancel common terms:

$$\frac{|q_1|}{x^2} = \frac{|q_2|}{(1.00\text{ m} - x)^2}$$

4. Take the square root of both sides, rearrange, and solve for x:

$$x = (1.00\text{ m})\left(1 + \sqrt{\frac{|q_2|}{|q_1|}}\right)^{-1}$$

5. Substitute numerical values:

$$x = 0.62\text{ m}$$

If q_3 is placed between $x = 0$ and $x = 0.62$ m, the net force acting on it is to the left. On the other hand, if q_3 is placed between $x = 0.62$ m and $x = 1.00$ m, the net force acting on it is to the right. These results agree with Example 19-6.

PROBLEM-SOLVING NOTE

Determining the Direction of the Electric Force

When determining the total force acting on a charge, begin by calculating the magnitude of each of the individual forces acting on it. Next, assign appropriate directions to the forces based on the principle "opposites attract, likes repel" and perform a vector sum.

Next, we consider systems in which the individual forces are not along the same line. In such cases, it is often useful to resolve the individual force vectors into components and then perform the required vector sum component by component. This technique is illustrated in the following Example and Conceptual Example.

EXAMPLE 19-8   SUPERPOSITION

Three charges, each equal to $+2.90\ \mu$C, are placed at three corners of a square 0.500 m on a side, as shown in the sketch. Find the magnitude and direction of the net force on charge 3.

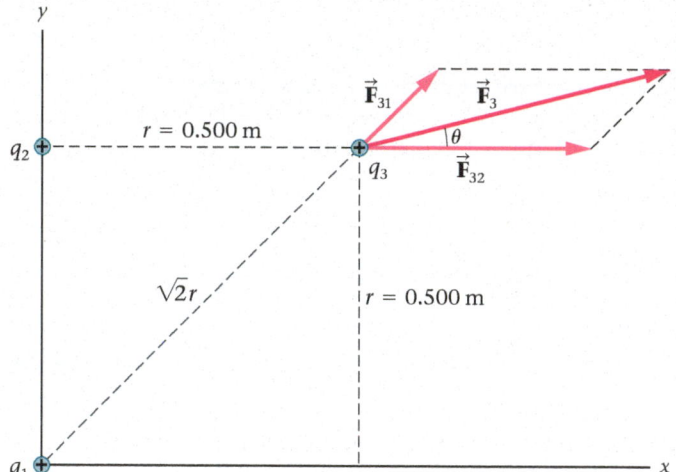

PICTURE THE PROBLEM

The positions of the three charges are shown in the sketch. We also show the force produced by charge 1, $\vec{F}_{31}$, and the force produced by charge 2, $\vec{F}_{32}$. Notice that $\vec{F}_{31}$ is 45.0° above the x axis and that $\vec{F}_{32}$ is in the positive x direction. Also, the distance from charge 2 to charge 3 is $r = 0.500$ m, and the distance from charge 1 to charge 3 is $\sqrt{2}r$.

REASONING AND STRATEGY

To find the net force, we first calculate the magnitudes of $\vec{F}_{31}$ and $\vec{F}_{32}$ and then their components. Summing these components yields the components of the net force, $\vec{F}_3$. Once we know the components of $\vec{F}_3$, we can calculate its magnitude and direction in the same way as for any other vector.

Known   Three charges, $q_1 = q_2 = q_3 = +2.90\ \mu$C; side length of square, $r = 0.500$ m.
Unknown   Magnitude of net force on charge 3, $F_3 = $? Direction of net force on charge 3, $\theta = $?

SOLUTION

1. Find the magnitude of $\vec{F}_{31}$:

$$F_{31} = k\frac{|q_1||q_3|}{(\sqrt{2}r)^2} = (8.99 \times 10^9\text{ N}\cdot\text{m}^2/\text{C}^2)\frac{(2.90 \times 10^{-6}\text{ C})^2}{2(0.500\text{ m})^2}$$
$$= 0.151\text{ N}$$

2. Find the magnitude of $\vec{F}_{32}$:

$$F_{32} = k\frac{|q_2||q_3|}{r^2} = (8.99 \times 10^9\text{ N}\cdot\text{m}^2/\text{C}^2)\frac{(2.90 \times 10^{-6}\text{ C})^2}{(0.500\text{ m})^2}$$
$$= 0.302\text{ N}$$

3. Calculate the components of $\vec{F}_{31}$ and $\vec{F}_{32}$:

$$F_{31,x} = F_{31}\cos 45.0° = (0.151\text{ N})(0.707) = 0.107\text{ N}$$
$$F_{31,y} = F_{31}\sin 45.0° = (0.151\text{ N})(0.707) = 0.107\text{ N}$$
$$F_{32,x} = F_{32}\cos 0° = (0.302\text{ N})(1) = 0.302\text{ N}$$
$$F_{32,y} = F_{32}\sin 0° = (0.302\text{ N})(0) = 0$$

4. Find the components of $\vec{F}_3$:

$$F_{3,x} = F_{31,x} + F_{32,x} = 0.107\text{ N} + 0.302\text{ N} = 0.409\text{ N}$$
$$F_{3,y} = F_{31,y} + F_{32,y} = 0.107\text{ N} + 0 = 0.107\text{ N}$$

CONTINUED

5. Find the magnitude of $\vec{F}_3$:

$$F_3 = \sqrt{F_{3,x}^2 + F_{3,y}^2} = \sqrt{(0.409 \text{ N})^2 + (0.107 \text{ N})^2} = 0.423 \text{ N}$$

6. Find the direction of $\vec{F}_3$:

$$\theta = \tan^{-1}\left(\frac{F_{3,y}}{F_{3,x}}\right) = \tan^{-1}\left(\frac{0.107 \text{ N}}{0.409 \text{ N}}\right) = 14.7°$$

INSIGHT

Thus, the net force on charge 3 has a magnitude of 0.423 N and points in a direction 14.7° above the x axis. Notice that charge 1, which is $\sqrt{2}$ times farther away from charge 3 than is charge 2, produces only half as much force as charge 2.

PRACTICE PROBLEM

Find the magnitude and direction of the net force on charge 3 if its magnitude is doubled to 5.80 μC. Assume that charge 1 and charge 2 are unchanged.   [**Answer:** $F_3 = 2(0.423 \text{ N}) = 0.846 \text{ N}, \theta = 14.7°$. The angle is unchanged.]

Some related homework problems: Problem 27, Problem 28

CONCEPTUAL EXAMPLE 19-9 COMPARE THE FORCES PREDICT/EXPLAIN

A charge $-q$ is to be placed at either point A or point B in our sketch. Assume points A and B lie on a line that is midway between the two positive charges. (a) Is the net force experienced at point A greater than, less than, or equal to the net force experienced at point B? (b) Which of the following is the *best explanation* for your prediction?

I. The net force at point A is greater because the charges are closer together there.

II. The net force is less at point A because the forces due to the positive charges cancel.

III. The force is the same at points A and B because the same charges are involved in both cases.

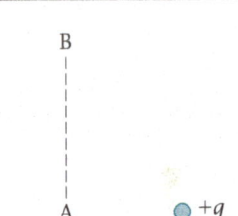

REASONING AND SOLUTION

Point A is closer to the two positive charges than is point B. As a result, the force exerted by each positive charge is greater when the charge $-q$ is placed at A. The *net* force, however, is zero at point A, because the equal attractive forces due to the two positive charges cancel.

At point B, on the other hand, the attractive forces combine to give a net downward force. Hence, the charge $-q$ experiences a greater net force at point B.

ANSWER

(a) The net force at point A is less than the net force at point B. (b) The best explanation is II.

Spherical Charge Distributions

Although Coulomb's law is stated in terms of point charges, it can be applied to any type of charge distribution by using the appropriate mathematics. For example, suppose a sphere has a charge Q distributed uniformly over its surface. If a point charge q is outside the sphere, a distance r from its center, the methods of calculus show that the magnitude of the force between the point charge and the sphere is simply

$$F = k\frac{|q||Q|}{r^2}$$

In situations like this, the spherical charge distribution behaves the same as if all its charge were concentrated in a point at its center. For point charges inside a charged spherical shell, the net force exerted by the shell is zero. In general, the electrical behavior of spherical *charge* distributions is analogous to the gravitational behavior of spherical *mass* distributions.

In the next Example, we consider a system in which a charge Q is distributed uniformly over the surface of a sphere. In such a case it is often convenient to specify the amount of *charge per area* on the sphere. This is referred to as the **surface charge density**, σ. If a sphere has an area A and a surface charge density σ, its total charge is

$$Q = \sigma A \qquad \text{19-7}$$

Notice that the SI unit of σ is C/m^2. If the radius of the sphere is R, then $A = 4\pi R^2$ and $Q = \sigma(4\pi R^2)$.

PROBLEM-SOLVING NOTE

Spherical Charge Distributions

Remember that a uniform spherical charge distribution can be replaced with a point charge only when considering points outside the charge distribution.

QUICK EXAMPLE 19-10   FIND THE FORCE EXERTED BY A SPHERE

An insulating sphere of radius $R = 0.12$ m has a uniform surface charge density equal to $5.5\,\mu\text{C/m}^2$. A point charge $q = 0.83\,\mu\text{C}$ is a distance $r = 0.44$ m from the center of the sphere. Find the magnitude of the force exerted by the sphere on the point charge.

REASONING AND SOLUTION

We can multiply the surface charge density by the surface area of the sphere to calculate the total charge on the sphere. The force on the point charge is the same as if all the charge on the sphere is concentrated at its center.

1. Find the area of the sphere:

$$A = 4\pi R^2 = 0.18\text{ m}^2$$

2. Calculate the total charge on the sphere:

$$Q = (5.5\,\mu\text{C/m}^2)(0.18\text{ m}^2) = 0.99\,\mu\text{C}$$

3. Use Coulomb's law to calculate the magnitude of the force between the sphere and the point charge:

$$F = k\frac{qQ}{r^2} = 0.038\text{ N}$$

As long as the point charge is outside the sphere, and the charge distribution remains spherically uniform, the sphere may be treated as a point charge.

Enhance Your Understanding   (Answers given at the end of the chapter)

3. Positive and negative charges of equal magnitude are placed at the upper left and lower right corners of a square, as shown in **FIGURE 19-11**. **(a)** If a positive charge of the same magnitude is placed at the upper right corner of the square, which of the vectors (1, 2, 3, 4) represents the net force exerted on the charge? **(b)** If the charge at the upper right corner is negative, but with the same magnitude as the other charges, which of the vectors (1, 2, 3, 4) represents the net force exerted on the charge?

Section Review

- The force between two point charges, q_1 and q_2, separated by a distance r, is given by Coulomb's law, $F = k\,|q_1|\,|q_2|/r^2$. The constant in Coulomb's law is $k = 8.99 \times 10^9\,\text{N}\cdot\text{m}^2/\text{C}^2$.

- The total force produced by multiple charges is given by superposition—that is, by vector addition.

- A spherical charge distribution exerts a force on objects outside it that is the same as if all its charge were concentrated at its center.

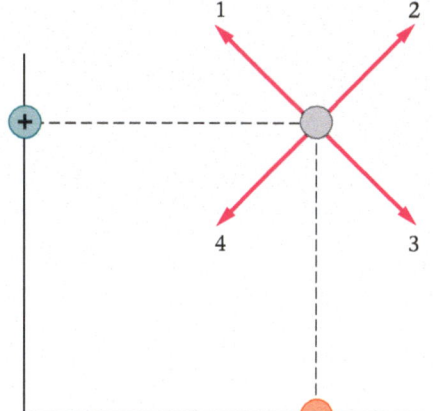

▲ **FIGURE 19-11** Enhance Your Understanding 3.

19-4 The Electric Field

You've probably encountered the notion of a "force field" in science fiction movies. A concrete example of a force field is the force between electric charges. Consider, for example, a positive point charge q at the origin of a coordinate system, as in **FIGURE 19-12**. If a positive "test charge," q_0, is placed at point A, the force exerted on it by q is indicated by the vector $\vec{\mathbf{F}}_A$. If the test charge is placed at point B, the force it experiences there is $\vec{\mathbf{F}}_B$. At every point in space there is a corresponding force. In this sense, Figure 19-12 allows us to visualize the "force field" associated with the charge q.

Because the magnitude of the force at every point in Figure 19-12 is proportional to q_0 (due to Coulomb's law), it's convenient to divide by q_0 and define a *force per charge* at every point in space that is independent of q_0. We refer to the force per charge as the electric field, $\vec{\mathbf{E}}$. Its precise definition is as follows:

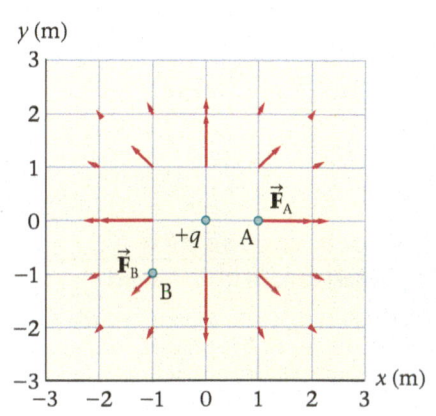

▲ **FIGURE 19-12 An electrostatic force field** The positive charge $+q$ at the origin of this coordinate system exerts a different force on a given charge at every point in space. Here we show the force vectors associated with q for a grid of points.

Definition of the Electric Field, $\vec{\mathbf{E}}$

If a test charge q_0 experiences a force $\vec{\mathbf{F}}$ at a given location, the electric field $\vec{\mathbf{E}}$ at that location is

$$\vec{\mathbf{E}} = \frac{\vec{\mathbf{F}}}{q_0}$$

19-8

SI unit: N/C

It should be noted that this definition applies whether the force $\vec{\mathbf{F}}$ is due to a single charge, as in Figure 19-12, or to a group of charges. In addition, it is assumed that the test charge is small enough that it does not disturb the position of any other charges in the system.

To summarize, *the electric field is the force per charge at a given location*. Therefore, if we know the electric field vector $\vec{\mathbf{E}}$ at a given point, the force that a charge q experiences at that point is

$$\vec{\mathbf{F}} = q\vec{\mathbf{E}} \qquad \qquad 19\text{-}9$$

Notice that the direction of the force depends on the sign of the charge. In particular,

- A positive charge experiences a force *in the direction* of $\vec{\mathbf{E}}$.
- A negative charge experiences a force *in the opposite direction* of $\vec{\mathbf{E}}$.

Finally, the magnitude of the force is the product of the magnitudes of q and $\vec{\mathbf{E}}$:

- The magnitude of the force acting on a charge q is $F = |q|E$.

As we continue in this chapter, we will determine the electric field for a variety of different charge distributions. In some cases $\vec{\mathbf{E}}$ will decrease with distance as $1/r^2$ (a point charge), in other cases as $1/r$ (a line of charge), and in others $\vec{\mathbf{E}}$ will be a constant (a charged plane). Before we calculate the electric field itself, however, we first consider the force exerted on charges by a constant electric field.

PROBLEM-SOLVING NOTE

The Force Exerted by an Electric Field

The force exerted on a charge by an electric field can point in only one of two directions—parallel or antiparallel to the direction of the field.

EXAMPLE 19-11   FORCE FIELD

In a certain region of space, a uniform electric field has a magnitude of 5.30×10^4 N/C and points in the positive x direction. Find the magnitude and direction of the force this field exerts on a charge of **(a)** $+2.80\,\mu$C and **(b)** $-9.30\,\mu$C.

PICTURE THE PROBLEM
In our sketch we indicate the uniform electric field and the two charges mentioned in the problem. The positive charge experiences a force in the positive x direction (the direction of $\vec{\mathbf{E}}$), and the negative charge experiences a force in the negative x direction (opposite to $\vec{\mathbf{E}}$).

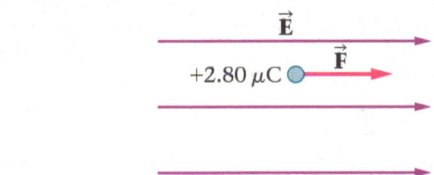

REASONING AND STRATEGY
To find the magnitude of each force, we use $F = |q|E$. The direction has already been indicated in our sketch.

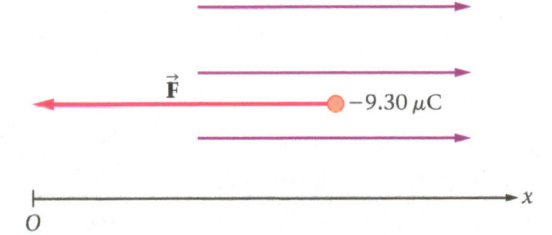

Known	Magnitude of electric field, $E = 5.30 \times 10^4$ N/C; direction of electric field $= +x$ direction;.
	(a) Charge, $q = +2.80\,\mu$C; **(b)** Charge, $q = -9.30\,\mu$C.
Unknown	Magnitude of force on charge, $F = $? Direction of force $= $?

SOLUTION

Part (a)

1. Find the magnitude of the force on the $+2.80$-μC charge: $F = |q|E = (2.80 \times 10^{-6}\,\text{C})(5.30 \times 10^4\,\text{N/C}) = 0.148$ N

Part (b)

2. Find the magnitude of the force on the -9.30-μC charge: $F = |q|E = (9.30 \times 10^{-6}\,\text{C})(5.30 \times 10^4\,\text{N/C}) = 0.493$ N

INSIGHT
To summarize, the force on the $+2.80$-μC charge is of magnitude 0.148 N and points in the positive x direction; the force on the -9.30-μC charge is of magnitude 0.493 N and points in the negative x direction.

PRACTICE PROBLEM
If the $+2.80$-μC charge experiences a force of 0.21 N, what is the magnitude of the electric field?
[**Answer:** $E = 7.5 \times 10^4$ N/C]

Some related homework problems: Problem 40, Problem 44

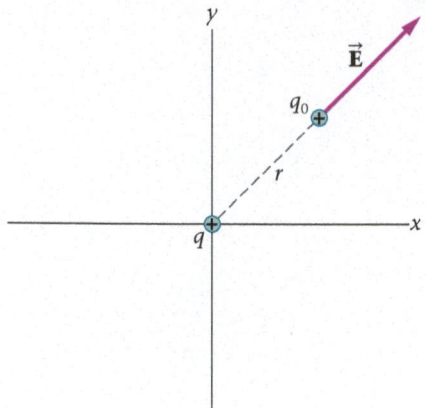

▲ **FIGURE 19-13 The electric field of a point charge** The electric field $\vec{E}$ due to a positive charge q at the origin is directed radially outward. Its magnitude is $E = k|q|/r^2$.

RWP The fact that charges of opposite sign experience forces in opposite directions in an electric field is used to purify water in the process known as **electrodialysis**. This process depends on the fact that most minerals that dissolve in water dissociate into positive and negative ions. Probably the most common example is table salt (NaCl), which dissociates into positive sodium ions (Na^+) and negative chlorine ions (Cl^-). When brackish water is passed through a strong electric field in an electrodialysis machine, the mineral ions move in opposite directions and pass through two different types of semipermeable membrane—one that allows only positive ions to pass through, the other only negative ions. This process leaves water that is purified of dissolved minerals and suitable for drinking.

The Electric Field of a Point Charge

Perhaps the simplest example of an electric field is the field produced by an idealized point charge. To be specific, suppose a positive point charge q is at the origin in **FIGURE 19-13**. If a positive test charge q_0 is placed a distance r from the origin, the force it experiences is directed away from the origin and is of magnitude

$$F = k\frac{|q||q_0|}{r^2}$$

Applying our definition of the electric field in Equation 19-8 ($\vec{E} = \vec{F}/q_0$), we find that the magnitude of the field is

$$E = \frac{F}{q_0} = \frac{\left(k\dfrac{|q||q_0|}{r^2}\right)}{q_0} = k\frac{|q|}{r^2}$$

Because a positive charge experiences a force that is directed radially outward, that too is the direction of $\vec{E}$.

In general, then, we can say that the electric field a distance r from a point charge q has the following magnitude:

Magnitude of the Electric Field Due to a Point Charge

$$E = k\frac{|q|}{r^2} \qquad\qquad 19\text{-}10$$

If the charge q is positive, the field points radially outward from the charge; if it is negative, the field points radially inward. This is illustrated in **FIGURE 19-14**. Thus, to determine the electric field due to a point charge, we first use $E = k|q|/r^2$ to find its magnitude, and then use the rule illustrated in Figure 19-14 to find its direction.

EXERCISE 19-12   ELECTRIC FIELD

Find the electric field produced by a 2.9-μC point charge at a distance of **(a)** 1.0 m and **(b)** 2.0 m.

REASONING AND SOLUTION

a. Applying Equation 19-10 ($E = k|q|/r^2$) with $q = 2.9\,\mu$C and $r = 1.0$ m yields

$$E = k\frac{|q|}{r^2} = (8.99 \times 10^9\,\mathrm{N\cdot m^2/C^2})\frac{(2.9 \times 10^{-6}\,\mathrm{C})}{(1.0\,\mathrm{m})^2} = 2.6 \times 10^4\,\mathrm{N/C}$$

b. Noting that E depends on $1/r^2$, we see that doubling the distance from 1.0 m to 2.0 m results in a reduction in the electric field by a factor of four:

$$E = \tfrac{1}{4}(2.6 \times 10^4\,\mathrm{N/C}) = 0.65 \times 10^4\,\mathrm{N/C}$$

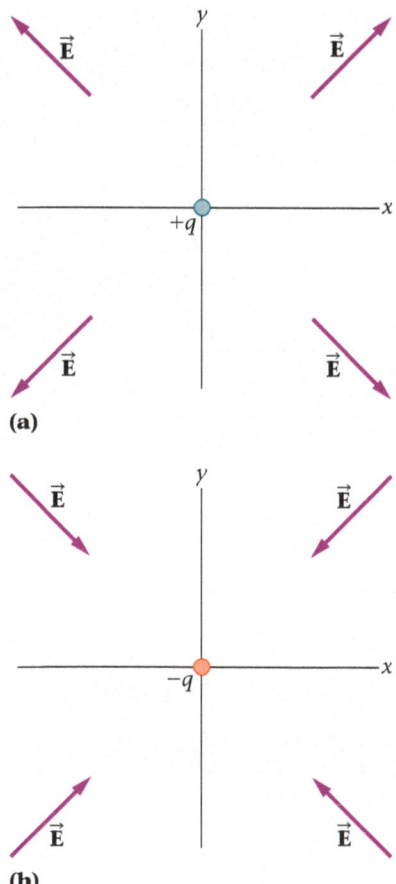

(a)

(b)

▲ **FIGURE 19-14 The direction of the electric field (a)** The electric field due to a positive charge at the origin points radially outward. **(b)** If the charge at the origin is negative, the electric field points radially inward.

Superposition of Fields

Many electrical systems consist of more than two charges. In such cases, the total electric field can be found by using superposition—just as when we find the total force due to a system of charges. In particular, the total electric field is found by calculating the vector sum—often using components—of the electric fields due to each charge separately.

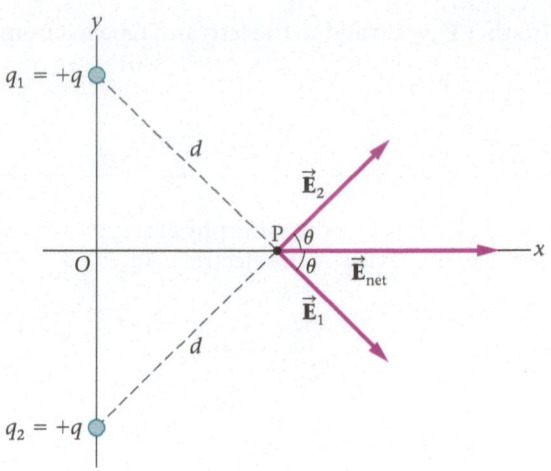

◀ **FIGURE 19-15 Superposition of the electric field** The net electric field at the point P is the vector sum of the fields due to the charges q_1 and q_2. Note that $\vec{E}_1$ and $\vec{E}_2$ point away from the charges q_1 and q_2, respectively. This is as expected, since both of these charges are positive.

For example, let's calculate the total electric field at point P in **FIGURE 19-15**. First we sketch the directions of the fields $\vec{E}_1$ and $\vec{E}_2$ due to the charges $q_1 = +q$ and $q_2 = +q$, respectively. In particular, if a positive test charge is at point P, the force due to q_1 is down and to the right, whereas the force due to q_2 is up and to the right. From the geometry of the figure we see that $\vec{E}_1$ is at an angle θ below the x axis, and—by symmetry—$\vec{E}_2$ is at the same angle θ above the axis. Because the two charges have the same magnitude, and the distances from P to the charges are the same, it follows that $\vec{E}_1$ and $\vec{E}_2$ have the same magnitude:

$$E_1 = E_2 = E = k\frac{|q|}{d^2}$$

To find the net electric field $\vec{E}_{net}$, we use components. First, consider the y direction. In this case, we have $E_{1,y} = -E \sin \theta$ and $E_{2,y} = +E \sin \theta$. Hence, the y component of the net electric field is zero:

$$E_{net,y} = E_{1,y} + E_{2,y} = -E \sin \theta + E \sin \theta = 0$$

From Figure 19-15 it is apparent that this result could have been anticipated by symmetry considerations. Finally, we determine the x component of E_{net}:

$$E_{net,x} = E_{1,x} + E_{2,x} = E \cos \theta + E \cos \theta = 2E \cos \theta$$

Thus, the net electric field at P is in the positive x direction, as shown in Figure 19-15, and has a magnitude equal to $2E \cos \theta$.

CONCEPTUAL EXAMPLE 19-13 THE SIGN OF THE CHARGES

Two charges, q_1 and q_2, have equal magnitudes q and are placed as shown in the accompanying sketch. The net electric field at point P is directed vertically upward. Can we conclude that q_1 is positive and q_2 is negative, q_1 is negative and q_2 is positive, or q_1 and q_2 have the same sign?

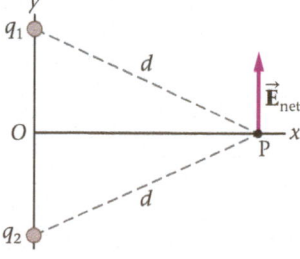

REASONING AND DISCUSSION

If the net electric field at P is directed vertically upward, the x components of $\vec{E}_1$ and $\vec{E}_2$ must cancel, and the y components must both be in the positive y direction. The only way for this to happen is to have q_1 negative and q_2 positive, as shown in the following diagram.

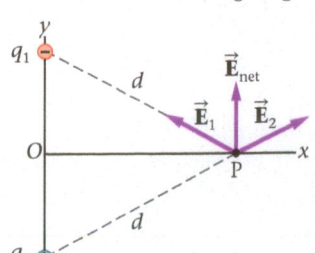

With this choice, a positive test charge at P is attracted to q_1 (so that $\vec{\mathbf{E}}_1$ is up and to the left) and repelled from q_2 (so that $\vec{\mathbf{E}}_2$ is up and to the right).

ANSWER
Charge q_1 is negative and q_2 is positive.

In the next Example, we consider the same physical system presented in Example 19-8, this time from the point of view of the electric field.

EXAMPLE 19-14   SUPERPOSITION IN THE FIELD

Two charges, each equal to $+2.90\,\mu\text{C}$, are placed at two corners of a square 0.500 m on a side, as shown in the sketch. Find the magnitude and direction of the net electric field at a third corner of the square, the point labeled 3 in the sketch.

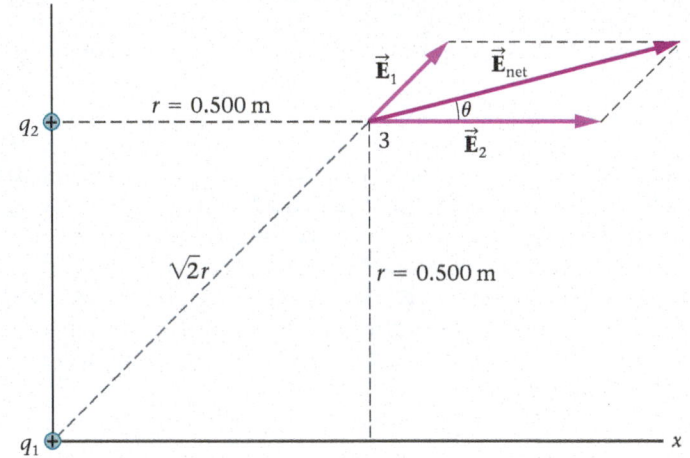

PICTURE THE PROBLEM
The positions of the two charges are shown in the sketch. We also show the electric field produced by each charge. The key difference between this sketch and the one in Example 19-8 is that in this case there is no charge at point 3; the electric field still exists there, even though it has no charge on which to exert a force.

REASONING AND STRATEGY
In analogy with Example 19-8, we first calculate the magnitudes of $\vec{\mathbf{E}}_1$ and $\vec{\mathbf{E}}_2$ and then their components. Summing these components yields the components of the net electric field, $\vec{\mathbf{E}}_{\text{net}}$. Once we know the components of $\vec{\mathbf{E}}_{\text{net}}$, we find its magnitude and direction in the same way as for $\vec{\mathbf{F}}_{\text{net}}$ in Example 19-8.

Known   Charges, $q_1 = q_2 = +2.90\,\mu\text{C}$; side length of square, $r = 0.500\,\text{m}$.
Unknown   Magnitude of net electric field at point 3, $E_{\text{net}} = ?$ Direction of net electric field at point 3, $\theta = ?$

SOLUTION

1. Find the magnitude of $\vec{\mathbf{E}}_1$:

$$E_1 = k\frac{|q_1|}{(\sqrt{2}r)^2} = (8.99 \times 10^9\,\text{N}\cdot\text{m}^2/\text{C}^2)\frac{(2.90 \times 10^{-6}\,\text{C})}{2(0.500\,\text{m})^2}$$
$$= 5.21 \times 10^4\,\text{N/C}$$

2. Find the magnitude of $\vec{\mathbf{E}}_2$:

$$E_2 = k\frac{|q_2|}{r^2} = (8.99 \times 10^9\,\text{N}\cdot\text{m}^2/\text{C}^2)\frac{(2.90 \times 10^{-6}\,\text{C})}{(0.500\,\text{m})^2}$$
$$= 1.04 \times 10^5\,\text{N/C}$$

3. Calculate the components of $\vec{\mathbf{E}}_1$ and $\vec{\mathbf{E}}_2$:

$$E_{1,x} = E_1\cos 45.0° = (5.21 \times 10^4\,\text{N/C})(0.707) = 3.68 \times 10^4\,\text{N/C}$$
$$E_{1,y} = E_1\sin 45.0° = (5.21 \times 10^4\,\text{N/C})(0.707) = 3.68 \times 10^4\,\text{N/C}$$
$$E_{2,x} = E_2\cos 0° = (1.04 \times 10^5\,\text{N/C})(1) = 1.04 \times 10^5\,\text{N/C}$$
$$E_{2,y} = E_2\sin 0° = (1.04 \times 10^5\,\text{N/C})(0) = 0$$

4. Find the components of $\vec{\mathbf{E}}_{\text{net}}$:

$$E_{\text{net},x} = E_{1,x} + E_{2,x} = 3.68 \times 10^4\,\text{N/C} + 1.04 \times 10^5\,\text{N/C}$$
$$= 1.41 \times 10^5\,\text{N/C}$$
$$E_{\text{net},y} = E_{1,y} + E_{2,y} = 3.68 \times 10^4\,\text{N/C} + 0 = 3.68 \times 10^4\,\text{N/C}$$

5. Find the magnitude of $\vec{\mathbf{E}}_{\text{net}}$:

$$E_{\text{net}} = \sqrt{E_{\text{net},x}^2 + E_{\text{net},y}^2} = \sqrt{(1.41 \times 10^5\,\text{N/C})^2 + (3.68 \times 10^4\,\text{N/C})^2}$$
$$= 1.46 \times 10^5\,\text{N/C}$$

6. Find the direction of $\vec{\mathbf{E}}_{\text{net}}$:

$$\theta = \tan^{-1}\left(\frac{E_{\text{net},y}}{E_{\text{net},x}}\right) = \tan^{-1}\left(\frac{3.68 \times 10^4\,\text{N/C}}{1.41 \times 10^5\,\text{N/C}}\right) = 14.6°$$

INSIGHT
As one would expect, the direction of the net electric field is the same as the direction of the net force in Example 19-8 (except for a small discrepancy in the last decimal place due to rounding off in the calculations). In addition, the magnitude of the force exerted by the electric field on a charge of $2.90\,\mu\text{C}$ is $F = qE_{\text{net}} = (2.90\,\mu\text{C})(1.46 \times 10^5\,\text{N/C}) = 0.423\,\text{N}$, the same as was found in Example 19-8.

CONTINUED

PRACTICE PROBLEM

Find the magnitude and direction of the net electric field at the bottom right corner of the square. [**Answer:** $E_{\text{net}} = 1.46 \times 10^5\ \text{N/C}, \theta = -14.6°$]

Some related homework problems: Problem 47, Problem 48

BIO Many aquatic creatures are capable of producing electric fields. For example, African freshwater fishes in the family Mormyridae can generate weak electric fields from modified tail muscles and are able to detect variations in this field as they move through their environment. With this capability, these nocturnal feeders have an electrical guidance system that assists them in locating obstacles, enemies, and food. Much stronger electric fields are produced by electric eels and electric skates. In particular, the electric eel (*Electrophorus electricus*) generates electric fields strong enough to kill small animals and stun larger animals, including humans.

BIO Sharks are well known for their sensitivity to weak electric fields in their surroundings. In fact, they possess specialized organs for this purpose, known as the ampullae of Lorenzini, which assist in the detection of prey. Recently, this sensitivity has been put to use as a method of repelling sharks in order to protect swimmers and divers. A device called the SharkPOD (Protective Oceanic Device) consists of two metal electrodes, one attached to a diver's air tank, the other to one of the diver's fins. These electrodes produce a strong electric field that completely surrounds the diver and causes sharks to turn away out to a distance as far as 7 m.

The SharkPOD was used in the 2000 Summer Olympic Games in Sydney to protect swimmers competing in the triathlon. The swimming part of the event was held in Sydney Harbour, where great white sharks are a common sight. To protect the swimmers, divers wearing the SharkPOD swam along the course, a couple meters below the athletes. The race was completed without incident.

▲ **FIGURE 19-16** Enhance Your Understanding 4.

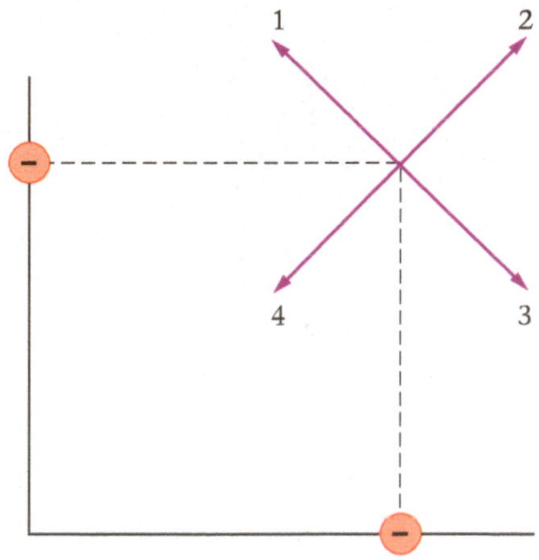

Enhance Your Understanding (Answers given at the end of the chapter)

4. Negative charges of equal magnitude are placed at the upper left and lower right corners of a square, as shown in **FIGURE 19-16**. Which of the four vectors (1, 2, 3, 4) represents the electric field at the upper right corner of the square?

Section Review

- The electric field $\vec{\mathbf{E}}$ is the electric force per charge, $\vec{\mathbf{E}} = \vec{\mathbf{F}}/q_0$. Given the electric field, the force on a charge q is $\vec{\mathbf{F}} = q\vec{\mathbf{E}}$.
- The electric field points away from positive charges and toward negative charges.
- The electric field for a point charge is $E = k|q|/r^2$.

19-5 Electric Field Lines

When you look at plots like those in Figures 19-12 and 19-14, it's tempting to imagine a pictorial representation of the electric field. This thought is reinforced when one considers a photograph like **FIGURE 19-17**, which shows grass seeds suspended in oil. Because of polarization effects, the grass seeds tend to align in the direction of the electric field, much like the elongated atoms shown in Figure 19-6 (a). In this case, the seeds are aligned radially, due to the electric field of the charged rod seen end-on in the middle of the photograph. Clearly, a set of radial lines would seem to represent the electric field in this case.

In fact, an entirely consistent method of drawing electric field lines is obtained by using the following set of rules:

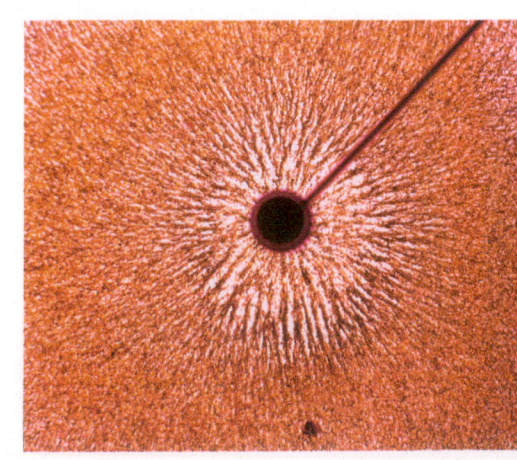

► **FIGURE 19-17 Grass seeds in an electric field** Grass seeds aligning with electric field lines.

Big Idea **3** At every point in space, the electric field points in the direction of the force that a positive charge would experience. Electric field lines go from positive charges to negative charges, and the stronger the field, the more closely spaced the lines.

Rules for Drawing Electric Field Lines

Electric field lines have the following properties:

1. They point in the *direction* of the electric field vector $\vec{E}$ at every point.
2. They *start* at positive (+) charges or at infinity.
3. They *end* at negative (−) charges or at infinity.
4. They are more *dense* where $\vec{E}$ has a greater magnitude. In particular, the number of lines entering or leaving a charge is proportional to the magnitude of the charge.

We now show how these rules are applied.

For example, the electric field lines for two different point charges are presented in **FIGURE 19-18**. First, we know that the electric field points directly away from the charge (+q) in Figure 19-18 (a); hence from rule 1 the field lines are radial. In agreement with rule 2 the field lines start on a + charge, and in agreement with rule 3 they end at infinity. Finally, as anticipated from rule 4, the field lines are closer together near the charge, where the field is more intense. Similar considerations apply to Figure 19-18 (b), where the charge is −2q. In this case, however, the direction of the field lines is reversed, because the charge is negative, and the number of lines is doubled, because the magnitude of the charge has been doubled.

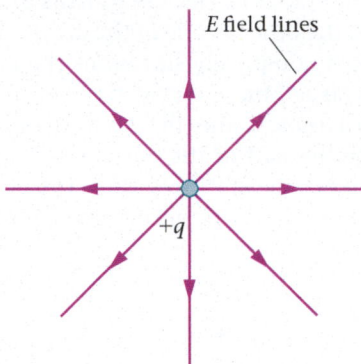

(a) E field lines point away from positive charges

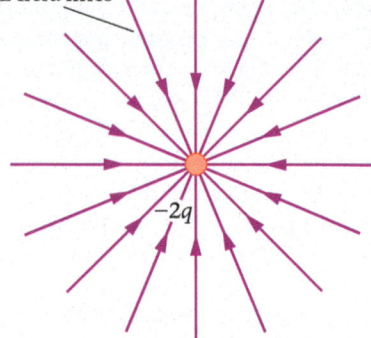

(b) E field lines point toward negative charges

▲ **FIGURE 19-18 Electric field lines for a point charge (a)** Near a positive charge the field lines point radially away from the charge. The lines start on the positive charge and end at infinity. **(b)** Near a negative charge the field lines point radially inward. They start at infinity and end on a negative charge and are more dense where the field is more intense. Notice that the number of lines drawn for part (b) is twice the number drawn for part (a), a reflection of the relative magnitudes of the charges.

CONCEPTUAL EXAMPLE 19-15 INTERSECT OR NOT?

Which of the following statements is correct: Electric field lines **(a)** can or **(b)** cannot intersect?

REASONING AND DISCUSSION

By definition, electric field lines are always tangent to the electric field. Because the electric force, and hence the electric field, can point in only one direction at any given location, it follows that field lines cannot intersect. If they did, the field at the intersection point would have two conflicting directions.

ANSWER

(b) Electric field lines cannot intersect.

Electric Field Lines for Multiple Charges In **FIGURE 19-19** we see examples of electric field lines for various combinations of charges. In systems like these, we draw a set of curved field lines that are tangent to the electric field vector, $\vec{E}$, at every point. This is illustrated for a variety of points in Figure 19-19 (a), and similar considerations apply to all such field diagrams. In addition, recall that the magnitude of $\vec{E}$ is greater in those regions of Figure 19-19 where the field lines are more closely packed together. Clearly,

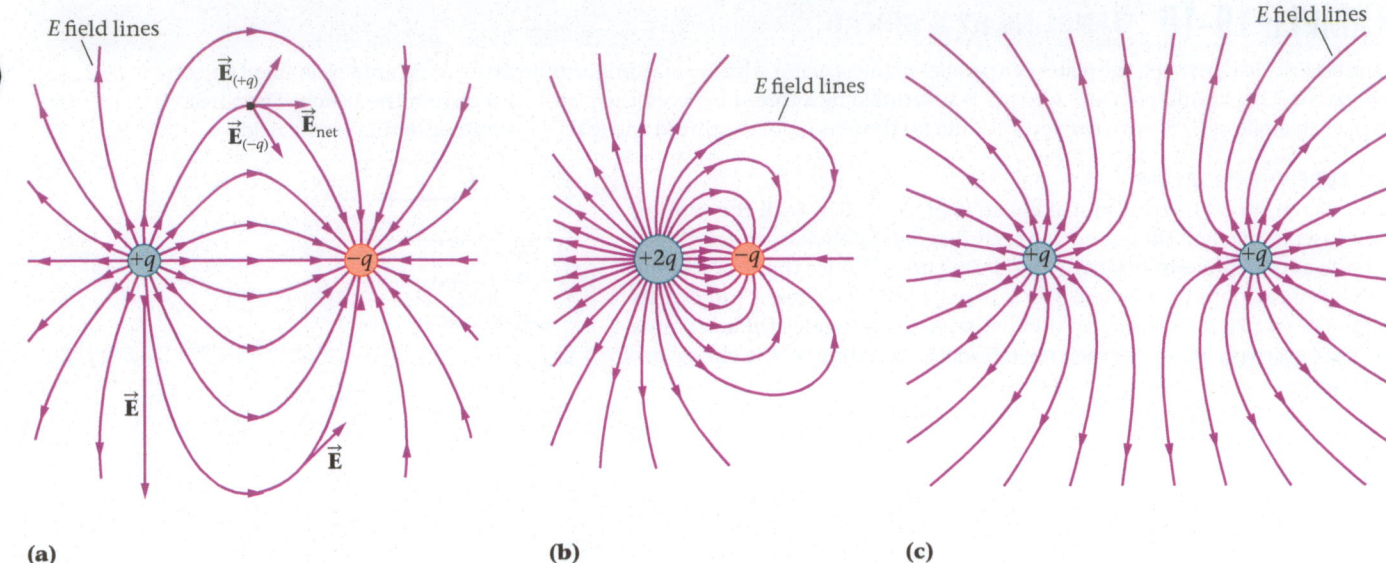

(a)   **(b)**   **(c)**

▲ **FIGURE 19-19 Electric field lines for systems of charges (a)** The electric field lines for a dipole form closed loops that become more widely spaced with distance from the charges. Note that at each point in space, the electric field vector $\vec{\mathbf{E}}$ is tangent to the field lines. **(b)** In a system with a net charge, some field lines extend to infinity. If the charges have opposite signs, some field lines start on one charge and terminate on the other charge. **(c)** All of the field lines in a system with charges of the same sign extend to infinity.

then, we expect an intense electric field between the charges in Figure 19-19 (b) and a vanishing field between the charges in Figure 19-19 (c).

Of particular interest is the $+q$ and $-q$ charge combination in Figure 19-19 (a). In general, a system of equal and opposite charges separated by a nonzero distance is known as an **electric dipole**. The total charge of the dipole is zero, but because the charges are separated, the electric field does not vanish. Instead, the field lines form "loops" that are characteristic of dipoles.

Many molecules are polar—water is a common example—which means they have an excess of positive charge near one end and a corresponding excess of negative charge near the other end. As a result, they produce an electric dipole field. Similarly, a typical bar magnet produces a *magnetic* dipole field, as we shall see in Chapter 22.

Finally, the electric field representations in Figures 19-18 and 19-19 are two-dimensional "slices" through the full field, which is three-dimensional. Therefore, one should imagine a similar set of field lines in these figures coming out of the page and going into the page.

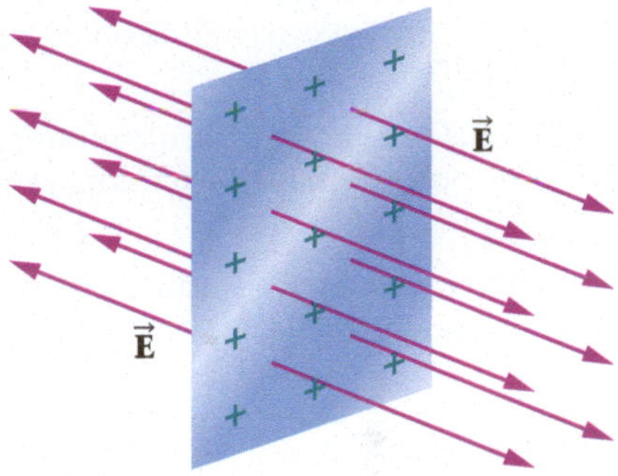

Parallel-Plate Capacitor

A particularly simple and important field picture results when charge is spread uniformly over a large plate, as illustrated in **FIGURE 19-20**. At points that are not near the edge of the plate, the electric field is uniform in both direction and magnitude. That is, the field points in a single direction—perpendicular to the plate—and its magnitude is independent of the distance from the plate. This result can be proved using Gauss's law, as we show in Section 19-7.

If two such conducting plates with opposite charge are placed parallel to one another and separated by a distance d, as in **FIGURE 19-21**, the result is referred to as a **parallel-plate capacitor**. The field for such a system is uniform between the plates, and zero outside the plates. This is the ideal case, which is exactly true for an infinite plate and a good approximation for real plates. The field lines are illustrated in Figure 19-21. Parallel-plate capacitors are discussed further in the next chapter and will be of particular interest in Chapters 21 and 24, when we consider electric circuits.

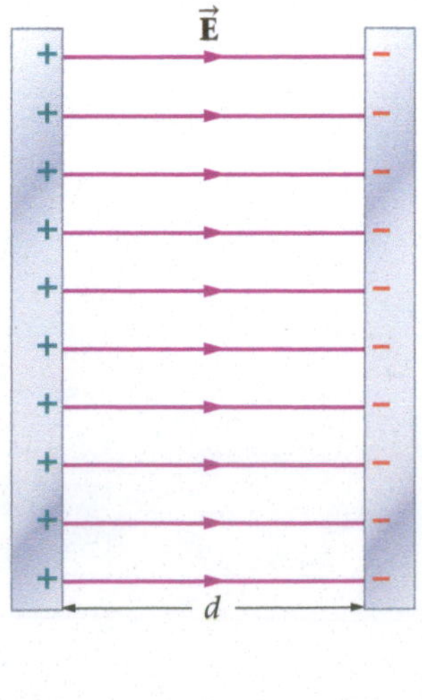

▲ **FIGURE 19-20 The electric field of a charged plate** The electric field near a large charged plate is uniform in direction and magnitude.

▶ **FIGURE 19-21 A parallel-plate capacitor** In the ideal case, the electric field is uniform between the plates and zero outside.

EXAMPLE 19-16   DANGLING BY A THREAD

The electric field between the plates of a parallel-plate capacitor is horizontal, uniform, and has a magnitude E. A small object of mass 0.0250 kg and charge $-3.10\ \mu C$ is suspended by a thread between the plates, as shown in the sketch. The thread makes an angle of 10.5° with the vertical. Find **(a)** the tension in the thread and **(b)** the magnitude of the electric field.

PICTURE THE PROBLEM

Our sketch shows the thread making an angle $\theta = 10.5°$ with the vertical. The inset to the right shows the free-body diagram for the suspended object, as well as our choice of positive x and y directions. Notice that we label the charge of the object $-q$, where $q = 3.10\ \mu C$, in order to clearly indicate its sign. The electric force acting on the charge has a magnitude of qE and points in the negative x direction; the gravitational force acting on the charge has a magnitude of mg and points in the negative y direction. The tension in the thread has an x component of $T\sin\theta$ and a y component of $T\cos\theta$.

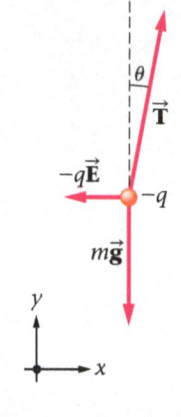

REASONING AND STRATEGY

The relevant physical principle in this problem is that because the object is at rest, the net force acting on it must vanish. Thus, setting the x and y components of the net force equal to zero yields two conditions, which can be used to solve for the two unknowns, T and E.

Known   Mass of object, $m = 0.0250$ kg; charge of object, $-q = -3.10\ \mu C$; angle of thread with vertical, $\theta = 10.5°$.
Unknown   **(a)** Tension in the thread, $T = ?$ **(b)** Magnitude of electric field, $E = ?$

SOLUTION

1. Set the net force in the x direction equal to zero:

   $$-qE + T\sin\theta = 0$$

2. Set the net force in the y direction equal to zero:

   $$T\cos\theta - mg = 0$$

Part (a)

3. Because we know all the quantities in the y force equation except for the tension, we use it to solve for T:

   $$T = \frac{mg}{\cos\theta} = \frac{(0.0250\text{ kg})(9.81\text{ m/s}^2)}{\cos(10.5°)} = 0.249\text{ N}$$

Part (b)

4. Now use the x force equation to find the magnitude of the electric field, E:

   $$E = \frac{T\sin\theta}{q} = \frac{(0.249\text{ N})\sin(10.5°)}{3.10 \times 10^{-6}\text{ C}} = 1.46 \times 10^4\text{ N/C}$$

INSIGHT

As expected, the negatively charged object is attracted to the positively charged plate. This means that the electric force exerted on it is opposite in direction to the electric field.

PRACTICE PROBLEM

Suppose the electric field between the plates is 2.50×10^4 N/C, but the charge of the object and the angle of the thread with the vertical are the same as before. Find the tension in the thread and the mass of the object.
[**Answer:** $T = 0.425$ N, $m = 0.0426$ kg]

Some related homework problems: Problem 56, Problem 86, Problem 91

RWP A technology that makes use of a capacitor-like arrangement is the photovoltaic solar panel used on many homes today (see **FIGURE 19-22**). Inside the panel is an interface between two different semiconductors that results in a thin layer of positive charge on one surface and a thin layer of negative charge on the other surface. Under most circumstances the charges are not free to move, and the interface behaves like a charged parallel-plate capacitor, with a uniform electric field pointing from the positive charges to the negative charges. When light strikes the interface, it gives some electrons sufficient energy to move freely, and the electric field pulls them toward the positive layer. The electrons can now flow through the semiconductor and into an external circuit; that is, the photovoltaic panel acts like a light-activated electric battery.

◀ **FIGURE 19-22 Solar energy** Solar photovoltaic panels are an increasingly popular way to harness renewable energy from the Sun.

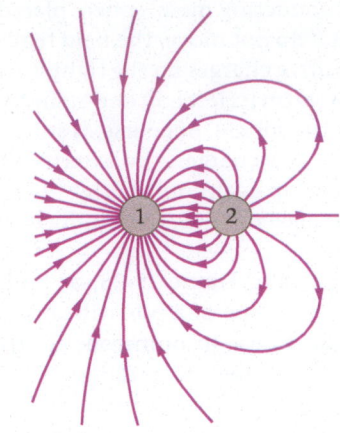

▲ **FIGURE 19-23** Enhance Your Understanding 5.

Electroscope in Conducting Shell

(a)

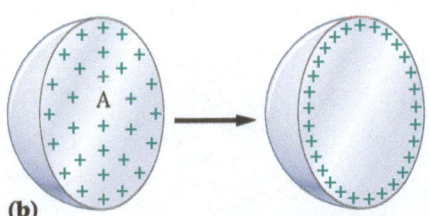

(b)

▲ **FIGURE 19-24 Charge distribution on a conducting spher (a)** A charge placed on a conducting sphere distributes itself uniformly on the surface of the sphere; none of the charge is within the volume of the sphere. **(b)** If the charge were distributed uniformly throughout the volume of a sphere, any given charge, like that at point A, would experience forces due to other charges. As a result, all charges would move as far from one another as possible—that is, to the surface of the conductor.

19-6 Shielding and Charging by Induction

In a perfect conductor there are enormous numbers of electrons completely free to move about within the conductor. This simple fact has some rather interesting consequences. Consider, for example, a solid metal sphere attached to an insulating base as in **FIGURE 19-24**. Suppose a positive charge Q is placed on the sphere. The question is: How does this charge distribute itself on the sphere when it is in equilibrium—that is, when all the charges are at rest? In particular, does the charge spread itself uniformly throughout the volume of the sphere, or does it concentrate on the surface?

The answer is that the charge concentrates on the surface, as shown in Figure 19-24 (a), but let's investigate why this should be the case. First, assume the opposite—that the charge is spread uniformly throughout the sphere's volume, as indicated in Figure 19-24 (b). If this were the case, a charge at location A would experience an outward force due to the spherical distribution of charge between it and the center of the sphere. Because charges are free to move, the charge at A would respond to this force by moving toward the surface. Clearly, then, a uniform distribution of charge within the sphere's volume is not in equilibrium. In fact, the argument that a charge at point A will move toward the surface can be applied to any charge within the sphere. Thus, the net result is that *all* the excess charge Q moves onto the surface of the sphere, which allows the individual charges to be spread out as far from one another as possible.

The preceding result holds no matter what the shape of the conductor. In general:

Excess Charge on a Conductor
Excess charge placed on a conductor, whether positive or negative, moves to the exterior surface of the conductor.

We specify the exterior surface in this statement because a conductor may contain one or more cavities. When an excess charge is applied to such a conductor, all the charge ends up on the exterior surface, with none on the interior surfaces.

Electrostatic Shielding
The ability of electrons to move freely within a conductor has another important consequence; namely, the electric field within a conductor vanishes.

Zero Field within a Conductor
When electric charges are at rest, the electric field within a conductor is zero: $E = 0$.

By *within* a conductor, we mean a location in the actual material of the conductor, as opposed to a location in a cavity within the material.

The best way to see the validity of this statement is to again consider the opposite. If there were a nonzero field within a conductor, electrons would move in response to the field. They would continue to move until finally the field was reduced to zero, at which point the system would be in equilibrium and no more charges would move. Thus, equilibrium and $E = 0$ within a conductor go hand in hand.

RWP A straightforward extension of this idea explains the phenomenon of **shielding**, in which a conductor "shields" its interior from external electric fields. For

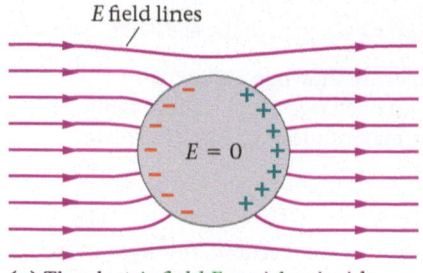

(a) The electric field *E* vanishes inside a conductor

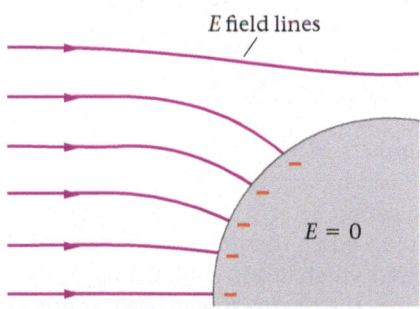

(b) *E* field lines meet a conducting surface at right angles

▲ **FIGURE 19-25 Electric field near a conducting surface (a)** When an uncharged conductor is placed in an electric field, the field induces opposite charges on opposite sides of the conductor. The net charge on the conductor is still zero, however. The induced charges produce a field within the conductor that *exactly* cancels the external field, leading to *E* = 0 inside the conductor. This is an example of electrical shielding. **(b)** Electric field lines meet the surface of a conductor at right angles.

Charged Conductor with Teardrop Shape

example, in **FIGURE 19-25 (a)** we show an uncharged, conducting metal sphere placed in an electric field. Because the positive ions in the metal do not move, the field tends to move negative charges to the left and leave excess positive charges on the right; hence, it causes the sphere to have an **induced** negative charge on its left half and an induced positive charge on its right half. The total charge on the sphere, of course, is still zero. Because field lines end on (−) charges and begin on (+) charges, the external electric field ends on the left half of the sphere and starts up again on the right half. In between, within the conductor, the field is zero, as expected. Thus, the conductor has shielded its interior from the applied electric field.

Shielding occurs whether the conductor is solid, as in Figure 19-25, or hollow. In fact, even a thin sheet of metal foil formed into the shape of a box will shield its interior from external electric fields. This effect is put to use in numerous electrical devices, which often have a metal foil or wire mesh enclosure surrounding the sensitive electric circuits. In this way, a given device can be isolated from the effects of other nearby devices that might otherwise interfere with its operation. A dramatic example of shielding is shown in **FIGURE 19-26 (a)**. Such shielding can even spare you from injury if your car or airplane is struck by lightning, as dramatically illustrated in **FIGURE 19-26 (b)**.

Notice also in Figure 19-25 that the field lines bend slightly near the surface of the sphere. In fact, on closer examination, as in **FIGURE 19-25 (b)**, we see that the field lines always contact the surface at right angles. This is true for any conductor:

Electric Fields at Conductor Surfaces
Electric field lines contact conductor surfaces at right angles.

If an electric field contacted a conducting surface at an angle other than 90°, the result would be a component of force parallel to the surface. This would result in a movement of electrons and, hence, would not correspond to equilibrium. Instead, electrons would move until the parallel component of the electric field was canceled.

RWP Another example of field lines near conducting surfaces is shown in **FIGURE 19-27**. Notice that the field lines are more densely packed near a sharp point, indicating that the field is more intense in such regions. This effect illustrates the basic principle behind the operation of lightning rods. If you look closely, you will notice that all lightning rods have a sharply pointed tip. During an electrical storm, the electric field at the tip becomes so intense that electric charge is given off into the atmosphere. In this way, a lightning rod acts to discharge the area near a house—by giving off a steady stream of charge—thus preventing a bolt of lightning, which transfers charge in one sudden blast. If a lightning bolt does occur, a lightning rod can conduct it safely to the ground, as in **FIGURE 19-26 (c)**. Sharp points on the rigging of a ship at sea can also give off streams of charge during a storm, often producing glowing lights referred to as Saint Elmo's Fire.

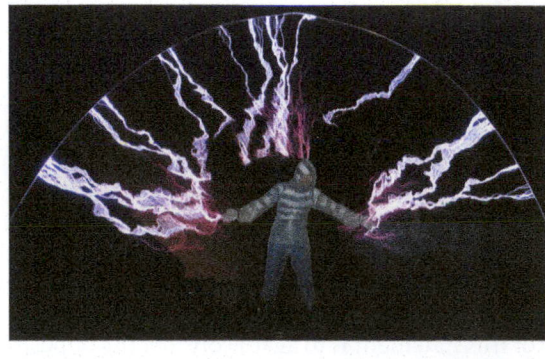

(a)

(b)

(c)

▲ **FIGURE 19-26 Visualizing Concepts Electric Discharge and Shielding (a)** A member of the band Thunderbolt Craziness produces artificial lightning bolts in time with the music. The metal suit he wears protects him from harm. **(b)** The passengers inside an airplane are unharmed when it was struck by a lightning bolt, as in this test with a model, thanks to the shielding effect of the airplane's metal body. **(c)** Lightning rods have sharp points, because that is where the electric field of a conductor is most intense. If a bolt strikes, the lightning rod safely conducts the flow of charge to the ground, as in this lightning strike to the CN Tower in Toronto.

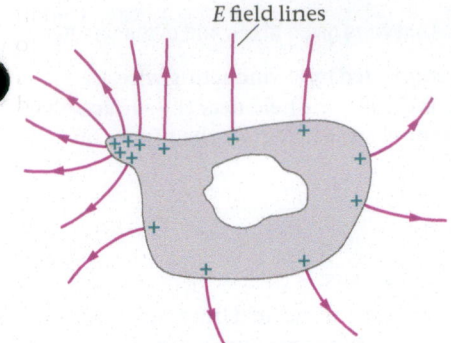

▲ **FIGURE 19-27 Intense electric field near a sharp point** Electric charges and field lines are more densely packed near a sharp point. This means that the electric field is more intense in such regions as well. (Note that there are no electric charges on the interior surface surrounding the cavity.)

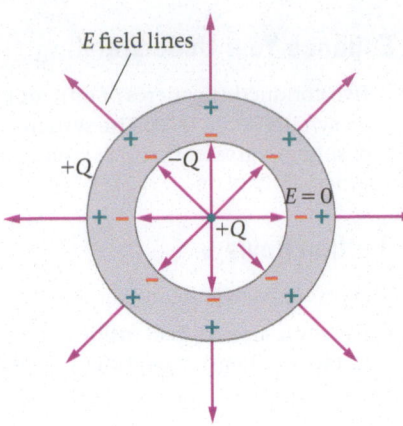

▲ **FIGURE 19-28 Shielding works in only one direction** A conductor does not shield the external world from charges it encloses. Still, the electric field is zero within the conductor itself.

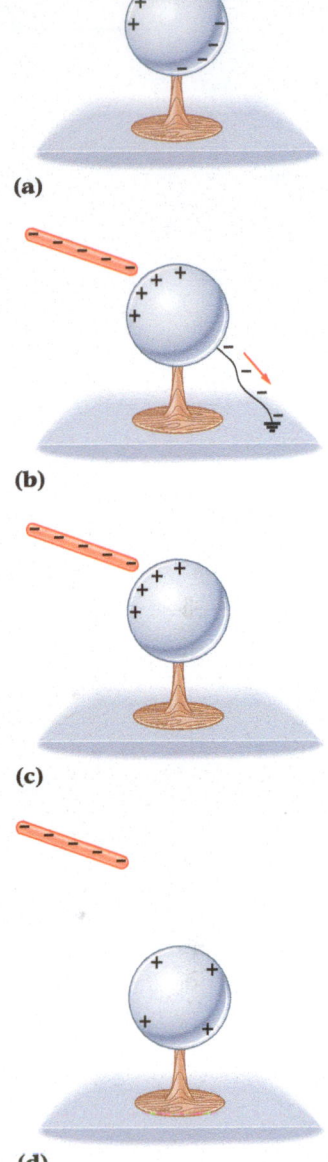

(a)

(b)

(c)

(d)

▲ **FIGURE 19-29 Charging by induction (a)** A charged rod induces + and − charges on opposite sides of the conductor. **(b)** When the conductor is grounded, charges that are repelled by the rod enter the ground. There is now a net charge on the conductor. **(c)** Removing the grounding wire, with the charged rod still in place, traps the net charge on the conductor. **(d)** The charged rod can now be removed, and the conductor retains a charge that is opposite in sign to that on the charged rod.

RWP The same principle is used to clean the air we breathe, in devices known as *electrostatic precipitators*. In an electrostatic precipitator, smoke and other airborne particles in a smokestack are given a charge as they pass by sharply pointed electrodes—like lightning rods—within the stack. Once the particles are charged, they are removed from the air by charged plates that exert electrostatic forces on them. The resultant emission from the smokestack contains drastically reduced amounts of potentially harmful particulates.

One final note regarding shielding is that it works in one direction only: A conductor shields its interior from external fields, but it does not shield the external world from fields within it. This phenomenon is illustrated in **FIGURE 19-28** for the case of an uncharged conductor. First, the charge $+Q$ in the cavity induces a charge $-Q$ on the interior surface, in order for the field in the conductor to be zero. Because the conductor is uncharged, a charge $+Q$ will be induced on its exterior surface. As a result, the external world will experience a field due, ultimately, to the charge $+Q$ within the cavity of the conductor.

Charging by Induction

One way to charge an object is to touch it with a charged rod, but since electric forces can act at a distance, it is also possible to charge an object without making direct physical contact. This type of charging is referred to as **charging by induction.**

To see how this type of charging works, consider an uncharged metal sphere on an insulating base. If a negatively charged rod is brought close to the sphere without touching it, as in **FIGURE 19-29 (a)**, electrons in the sphere are repelled. An induced positive charge is produced on the near side of the sphere, and an induced negative charge on the far side. At this point the sphere is still electrically neutral, however.

The key step in this charging process, shown in **FIGURE 19-29 (b)**, is to connect the sphere to the ground using a conducting wire. As one might expect, this is referred to as **grounding** the sphere, and is indicated by the symbol ⏚. (A table of electrical symbols can be found in Appendix D.) Because the ground is a fairly good conductor of electricity, and Earth can receive or give up practically unlimited numbers of electrons, the effect of grounding the sphere is that the electrons repelled by the charged rod enter the ground. Now the sphere has a net positive charge. With the rod kept in place, the grounding wire is removed, as in **FIGURE 19-29 (c)**, trapping the net positive charge on the sphere. The rod can now be pulled away, as in **FIGURE 19-29 (d)**.

Notice that the *induced* charge on the sphere is opposite in sign to the charge on the rod. In contrast, when the sphere is charged by *touch* as in Figure 19-8, it acquires a charge with the same sign as the charge on the rod.

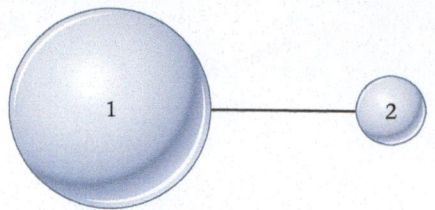

▲ **FIGURE 19-30** Enhance Your Understanding 6.

6. Two conducting spheres of different radii are connected by a conducting wire, as shown in **FIGURE 19-30**. The system is charged. Is the electric field near the surface of sphere 1 greater than, less than, or equal to the electric field near the surface of sphere 2? Explain.

Section Review

• Excess charge on a conductor moves to its exterior surface.

• The static electric field inside a conductor is zero, and electric field lines contact the surface of a conductor at right angles. The field is intense near sharp points.

19-7 Electric Flux and Gauss's Law

In this section, we introduce the idea of an electric flux and show that it can be used to calculate the electric field. The precise connection between electric flux and the charges that produce the electric field is provided by Gauss's law.

Electric Flux

Consider a uniform electric field $\vec{E}$, as in **FIGURE 19-31 (a)**, passing through an area A that is perpendicular to the field. Looking at the electric field lines with their arrows, we can easily imagine a "flow" of electric field through the area. Though there is no actual flow, of course, the analogy is a useful one. It is with this in mind that we define an **electric flux,** Φ, for this case as follows:

$$\Phi = EA$$

On the other hand, if the area A is parallel to the field lines, as in **FIGURE 19-31 (b)**, none of the $\vec{E}$ lines pierce the area, and hence there is no flux of electric field:

$$\Phi = 0$$

In an intermediate case, as shown in **FIGURE 19-31 (c)**, the $\vec{E}$ lines pierce the area A at an angle θ away from the perpendicular. As a result, the component of $\vec{E}$ perpendicular to the surface is $E \cos \theta$, and the component parallel to the surface is $E \sin \theta$. Because only the perpendicular component of $\vec{E}$ causes a flux (the parallel component does not pierce the area), the flux in the general case is the following:

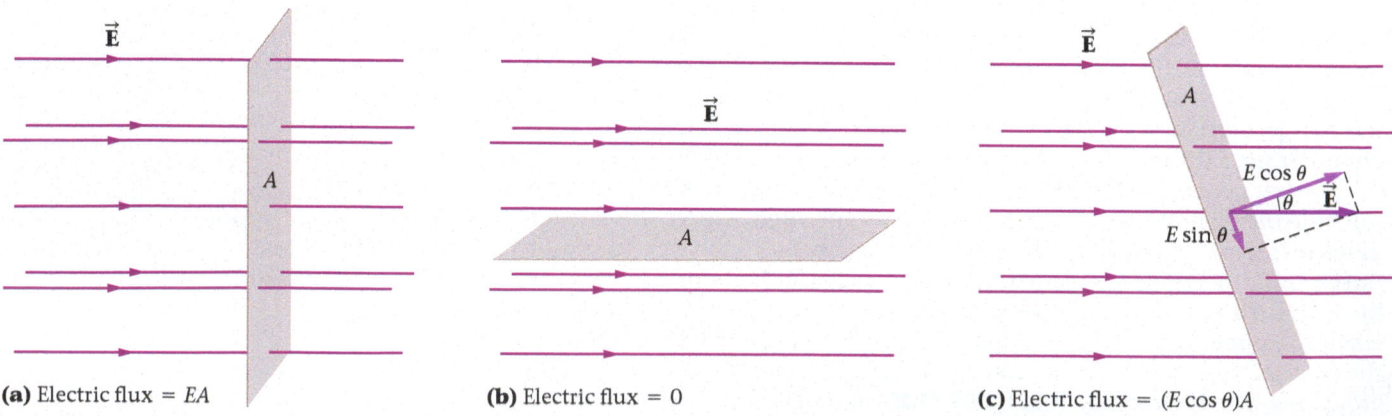

(a) Electric flux = EA (b) Electric flux = 0   (c) Electric flux = $(E \cos \theta)A$

▲ **FIGURE 19-31 Electric flux (a)** When an electric field $\vec{E}$ passes perpendicularly through the plane of an area A, the electric flux is $\Phi = EA$. **(b)** When the plane of an area is parallel to $\vec{E}$, so that no field lines "pierce" the area, the electric flux is zero, $\Phi = 0$. **(c)** When the normal to the plane of an area is tilted at an angle θ away from the electric field $\vec{E}$, only the perpendicular component of $\vec{E}$, $E \cos \theta$, contributes to the electric flux. Thus, the flux is $\Phi = (E \cos \theta)A$.

Definition of Electric Flux, Φ

$$\Phi = EA \cos \theta \qquad \text{19-11}$$

SI unit: $N \cdot m^2/C$

Finally, if the surface through which the flux is calculated is *closed*, the sign of the flux is as follows:

- The flux is *positive* for field lines that *leave* the enclosed volume of the surface.
- The flux is *negative* for field lines that *enter* the enclosed volume of the surface.

Gauss's Law

As a simple example of electric flux, consider a positive point charge q and a spherical surface of radius r centered on the charge, as in **FIGURE 19-32 (a)**. The electric field on the surface of the sphere has the constant magnitude

$$E = k\frac{q}{r^2}$$

Because the electric field is everywhere perpendicular to the spherical surface, it follows that the electric flux is simply E times the area $A = 4\pi r^2$ of the sphere:

$$\Phi = EA = \left(k\frac{q}{r^2}\right)(4\pi r^2) = 4\pi kq$$

We will often find it convenient to express k in terms of another constant, ϵ_0, as follows: $k = 1/(4\pi\epsilon_0)$. This new constant, which we call the **permittivity of free space**, is

$$\epsilon_0 = \frac{1}{4\pi k} = 8.85 \times 10^{-12} \, C^2/N \cdot m^2 \qquad \text{19-12}$$

In terms of ϵ_0, the flux through the spherical surface reduces to

$$\Phi = 4\pi kq = \frac{q}{\epsilon_0}$$

Thus, we find the very simple result that the electric flux through a sphere that encloses a charge q is the charge divided by the permittivity of free space, ϵ_0. This is a nice result, but what makes it truly remarkable is that it is equally true for *any* surface that encloses the charge q. For example, if one were to calculate the electric flux through the closed arbitrary surface shown in **FIGURE 19-32 (b)**—which would be a difficult task—the result, nonetheless, would still be simply q/ϵ_0. This, in fact, is a special case of Gauss's law:

Gauss's Law

If a charge q is enclosed by an arbitrary surface, the total electric flux through the surface, Φ, is

$$\Phi = \frac{q}{\epsilon_0} \qquad \text{19-13}$$

SI unit: $N \cdot m^2/C$

Notice that we use q rather than $|q|$ in Gauss's law. This is because the electric flux can be positive or negative, depending on the sign of the charge. In particular, if the charge q is positive, the field lines leave the enclosed volume and the flux is positive; if the charge is negative, the field lines enter the enclosed volume and the flux is negative.

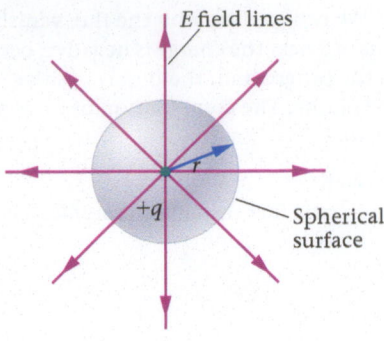

(a)

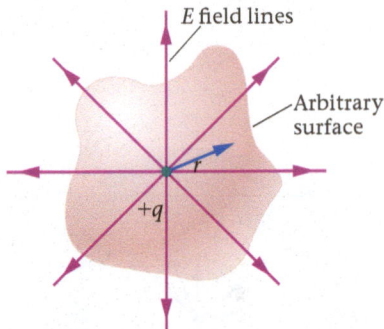

(b)

▲ **FIGURE 19-32 Electric flux for a point charge** The electric flux through the spherical surface surrounding a positive point charge q is $\Phi = EA = (kq/r^2)(4\pi r^2) = q/\epsilon_0$. The electric flux through an arbitrary surface is the same as for the sphere, but the calculation of the flux would be much more difficult due to the lack of symmetry.

Big Idea **4** The flux of the electric field is the product of the normal component of the electric field and the area through which it passes. The electric flux through a closed surface is proportional to the amount of charge it encloses.

CONCEPTUAL EXAMPLE 19-17 SIGN OF THE ELECTRIC FLUX

Consider the surface S shown in the sketch. Is the electric flux through this surface negative, positive, or zero?

REASONING AND DISCUSSION

Because the surface S encloses no charge, the net electric flux through it must be zero, by Gauss's law. The fact that a charge $+q$ is nearby is irrelevant, because it is outside the volume enclosed by the surface.

CONTINUED

We can explain why the flux vanishes in another way. Notice that the flux on portions of S near the charge is negative, because field lines enter the enclosed volume there. On the other hand, the flux is positive on the outer portions of S, where field lines exit the volume. The combination of these positive and negative contributions is a net flux of zero.

ANSWER

The electric flux through the surface S is zero.

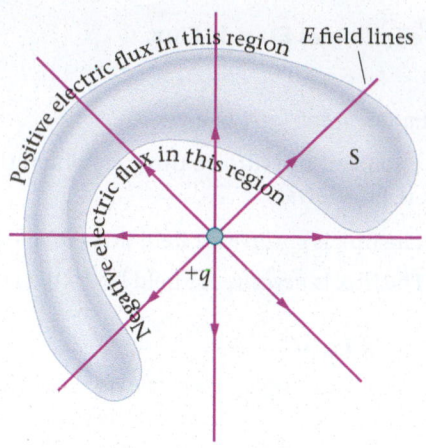

Using a Gaussian Surface Although Gauss's law holds for an arbitrary surface, its real utility is seen when the system in question has a simple symmetry. For example, consider a point charge $+Q$ in the center of a hollow, conducting spherical shell, as illustrated in **FIGURE 19-33**. The shell has an inside radius R_A and an outside radius R_B, and is uncharged. To calculate the field inside the cavity, where $r < R_A$, we consider an imaginary spherical surface—a so-called **Gaussian surface**—with radius r_1, and centered on the charge $+Q$, as indicated in Figure 19-33. The electric flux through this Gaussian surface is $\Phi = E(4\pi r_1{}^2) = Q/\epsilon_0$. Therefore, the magnitude of the electric field, as expected, is

$$E = \frac{Q}{4\pi\epsilon_0 r_1{}^2} = k\frac{Q}{r_1{}^2}$$

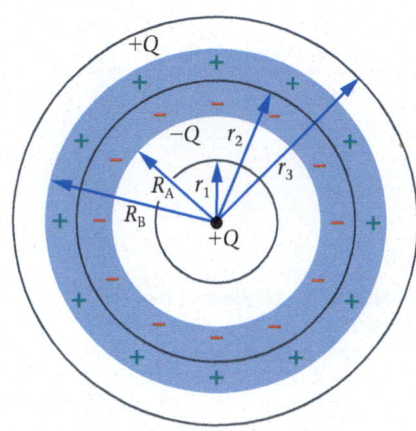

▲ **FIGURE 19-33 Gauss's law applied to a spherical shell** A simple system with three different Gaussian surfaces.

Notice, in particular, that the charges on the spherical shell do not affect the electric flux through this Gaussian surface, because they are not contained within the surface.

Next, consider a Gaussian surface within the shell, with $R_A < r_2 < R_B$, as in Figure 19-33. Because the field within a conductor is zero, $E = 0$, it follows that the electric flux for this surface is zero: $\Phi = EA = 0$. This means that the net charge within the Gaussian surface is also zero; that is, the induced charge on the inner surface of the shell is $-Q$.

Finally, consider a spherical Gaussian surface that encloses the entire spherical shell, with a radius $r_3 > R_B$. In this case, also shown in Figure 19-33, the flux is $\Phi = E(4\pi r_3{}^2) = $ (enclosed charge)$/\epsilon_0$. What is the enclosed charge? Well, we know that the spherical shell is uncharged—the induced charges of $+Q$ and $-Q$ on its outer and inner surfaces sum to zero—hence the total enclosed charge is simply $+Q$ from the charge at the center of the shell. Therefore, $\Phi = E(4\pi r_3{}^2) = Q/\epsilon_0$, and

$$E = \frac{Q}{4\pi\epsilon_0 r_3{}^2} = k\frac{Q}{r_3{}^2}$$

Thus, the field outside the shell is the same as if the shell were not present, showing that the conducting shell does not shield the external world from charges within it, in agreement with our conclusions in the previous section.

Gaussian surfaces do not need to be spherical, however. Consider, for example, a thin sheet of charge that extends to infinity, as in **FIGURE 19-34**. We expect the field to be at right angles to the sheet because, by symmetry, there is no reason for it to tilt in one direction more than in any other direction. Hence, we choose our Gaussian surface to be a cylinder, as in Figure 19-34. With this choice, no field lines pierce the curved surface of the cylinder. The electric flux through this Gaussian surface, then, is due solely to the contributions of the two end caps, each of area A. Hence, the flux is $\Phi = E(2A)$. If the charge per area on the sheet is σ, the enclosed charge is σA, and hence Gauss's law states that

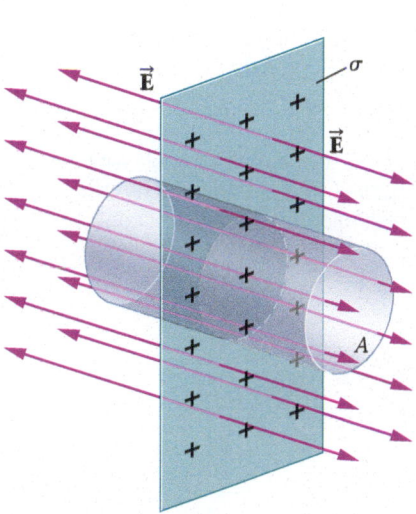

▲ **FIGURE 19-34 Gauss's law applied to a sheet of charge** A charged sheet of infinite extent and the Gaussian surface used to calculate the electric field.

$$\Phi = E(2A) = \frac{(\sigma A)}{\epsilon_0}$$

Canceling the area, we find

$$E = \frac{\sigma}{2\epsilon_0}$$

Notice that E does not depend in any way on the distance from the sheet, as was mentioned in Section 19-5.

   We conclude this chapter with an additional example of Gauss's law in action.

EXAMPLE 19-18   FIND THE ELECTRIC FIELD

Use the cylindrical Gaussian surface shown in the sketch to calculate the electric field between the metal plates of a parallel-plate capacitor. Each plate has a charge per area of magnitude σ.

PICTURE THE PROBLEM

Our sketch shows the two plates of the parallel-plate capacitor, and the cylindrical Gaussian surface to be used in calculating the electric field. The left end cap of the Gaussian surface is within the metal of the left plate, where the electric field is zero, and the right end cap of the surface is in the region between the plates, where the electric field has a magnitude E. The cross-sectional area of the cylindrical Gaussian surface is A, and the charge per area on the left plate is σ.

REASONING AND STRATEGY

To calculate the flux through the Gaussian surface, we first note that there is zero flux through the curved surface of the cylindrical surface, because it is parallel to the electric field. There is also zero flux through the left end cap of the cylinder, which is in the metal of the conducting plate where the electric field is zero. The only part of the cylindrical surface that has electric flux is the right end cap, which has the electric field E exiting through the area A. This gives a positive flux of EA. Setting this flux equal to the enclosed charge (σA) divided by ϵ_0 allows us to determine E.

Known   Charge per area $= \sigma$.
Unknown Electric field between plates, $E = ?$

SOLUTION

1. Calculate the electric flux through the curved surface of the cylinder: 0

2. Calculate the electric flux through the left and right end caps of the cylinder: $0 + EA$

3. Determine the charge enclosed by the cylinder: σA

4. Apply Gauss's law, and cancel the area A, to find the electric field, E:

$$EA = \frac{(\sigma A)}{\epsilon_0}$$

$$E = \frac{\sigma}{\epsilon_0}$$

INSIGHT

Only one surface of our Gaussian surface had a nonzero electric flux, but it was all we needed to apply Gauss's law and determine E.

PRACTICE PROBLEM

Suppose we extend the Gaussian surface so that its right end cap is within the metal of the right plate. The left end of the Gaussian surface remains in its original location. What is the electric flux through this new Gaussian surface? Explain.   [**Answer:** The flux is zero, because the net charge enclosed by the Gaussian surface is zero.]

Some related homework problems: Problem 57, Problem 59, Problem 65, Problem 68

Enhance Your Understanding   (Answers given at the end of the chapter)

7. Four Gaussian surfaces (A, B, C, D) are shown in **FIGURE 19-35**, along with four charges of equal magnitude. The two charges on the left are positive; the two charges on the right are negative. Rank the four Gaussian surfaces in order of increasing electric flux, from most negative to most positive. Indicate ties where appropriate.

CONTINUED

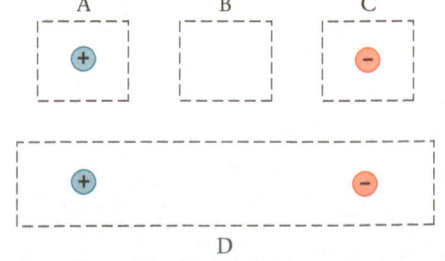

▲ **FIGURE 19-35** Enhance Your Understanding 7.

> **Section Review**
>
> - Electric flux Φ is equal to the normal component of the electric field, $E \cos \theta$, times the area A through which it passes, $\Phi = EA \cos \theta$. Electric flux is positive when field lines exit an enclosed volume, and negative when they enter a volume.
> - Gauss's law states that the electric flux Φ is equal to the enclosed charge q divided by ϵ_0: $\Phi = q/\epsilon_0$. In symmetric systems, Gauss's law can be used to calculate the electric field.

CHAPTER 19 REVIEW

CHAPTER SUMMARY

19-1 ELECTRIC CHARGE

Electric charge is one of the fundamental properties of matter. Electrons have a negative charge, $-e$, and protons have a positive charge, $+e$.

Magnitude of an Electron's Charge
The charge on an electron has the following magnitude:

$$e = 1.60 \times 10^{-19}\,\text{C} \qquad \text{19-1}$$

The SI unit of charge is the coulomb, C.

Charge Conservation
The total charge in the universe is constant.

19-2 INSULATORS AND CONDUCTORS

An insulator does not allow electrons within it to move from atom to atom. In conductors, each atom gives up one or more electrons that are then free to move throughout the material. Semiconductors have properties that are intermediate between those of insulators and conductors.

19-3 COULOMB'S LAW

Electric charges exert forces on one another along the line connecting them: Like charges repel, opposite charges attract.

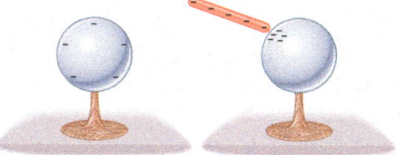

Coulomb's Law
The magnitude of the force between two point charges, q_1 and q_2, separated by a distance r is

$$F = k\frac{|q_1||q_2|}{r^2} \qquad \text{19-5}$$

The constant k in this expression is

$$k = 8.99 \times 10^9\,\text{N}\cdot\text{m}^2/\text{C}^2 \qquad \text{19-6}$$

Superposition
The electric force on one charge due to two or more other charges is the vector sum of each individual force.

19-4 THE ELECTRIC FIELD

The electric field is the force per charge at a given location in space.

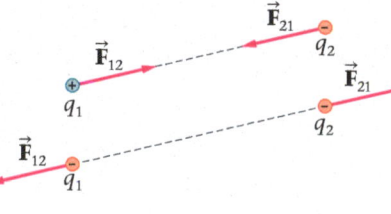

Direction of $\vec{E}$
$\vec{E}$ points in the direction of the force experienced by a *positive* test charge.

Point Charge
The electric field a distance r from a point charge q has a magnitude given by

$$E = k\frac{|q|}{r^2} \qquad \text{19-10}$$

Superposition
The total electric field due to two or more charges is given by the vector sum of the fields due to each charge individually.

19-5 ELECTRIC FIELD LINES

The electric field can be visualized by drawing lines according to a given set of rules.

Rules for Drawing Electric Field Lines

Electric field lines (1) point in the direction of the electric field vector $\vec{E}$ at all points; (2) start at + charges or infinity; (3) end at − charges or infinity; and (4) are more dense the greater the magnitude of $\vec{E}$.

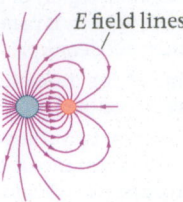

E field lines

19-6 SHIELDING AND CHARGING BY INDUCTION

Any excess charge placed on a conductor moves to its exterior surface.

The electric field within a conductor in equilibrium is zero.

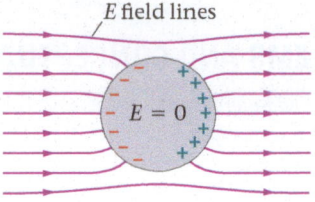

E field lines

19-7 ELECTRIC FLUX AND GAUSS'S LAW

Gauss's law relates the charge enclosed by a given surface to the electric flux through the surface.

Electric Flux

If an area A is tilted at an angle θ to an electric field $\vec{E}$, the electric flux through the area is

$$\Phi = EA \cos \theta \qquad \text{19-11}$$

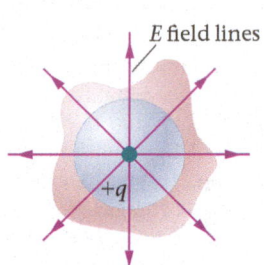

E field lines

$+q$

Gauss's Law

Gauss's law states that if a charge q is enclosed by a surface, the electric flux through the surface is

$$\Phi = \frac{q}{\epsilon_0} \qquad \text{19-13}$$

The constant appearing in this equation is the permittivity of free space, ϵ_0:

$$\epsilon_0 = \frac{1}{4\pi k} = 8.85 \times 10^{-12} \, \text{C}^2/\text{N} \cdot \text{m}^2 \qquad \text{19-12}$$

Gauss's law is used to calculate the electric field in highly symmetric systems.

ANSWERS TO ENHANCE YOUR UNDERSTANDING QUESTIONS

1. Rabbit fur has a positive charge, and silk has a negative charge.

2. Electrons move freely in a conductor.

3. **(a)** The net force is given by vector 3. **(b)** The net force is given by vector 1.

4. The electric field is given by vector 4.

5. **(a)** Charge 1 is negative. **(b)** Charge 2 is positive. **(c)** The magnitude of charge 1 is greater than the magnitude of charge 2.

6. The electric field at the surface of sphere 1 is less than the electric field at the surface of sphere 2, because sphere 2 has a smaller radius of curvature (is "sharper" than sphere 1).

7. C < B = D < A.

CONCEPTUAL QUESTIONS

For instructor-assigned homework, go to www.masteringphysics.com.

(Answers to odd-numbered Conceptual Questions can be found in the back of the book.)

1. The fact that the electron has a negative charge and the proton has a positive charge is due to a convention established by Benjamin Franklin. Would there have been any significant consequences if Franklin had chosen the opposite convention? Is there any advantage to naming charges plus and minus as opposed to, say, A and B?

2. Explain why a comb that has been rubbed through your hair attracts small bits of paper, even though the paper is uncharged.

3. Small bits of paper are attracted to an electrically charged comb, but as soon as they touch the comb they are strongly repelled. Explain this behavior.

4. A charged rod is brought near a suspended object, which is repelled by the rod. Can we conclude that the suspended object is charged? Explain.

5. A charged rod is brought near a suspended object, which is attracted to the rod. Can we conclude that the suspended object is charged? Explain.

6. A point charge $+Q$ is fixed at a height H above the ground. Directly below this charge is a small ball with a charge $-q$ and a mass m. When the ball is at a height h above the ground, the net force (gravitational plus electrical) acting on it is zero. Is this a stable equilibrium for the object? Explain.

7. A proton moves in a region of constant electric field. Does it follow that the proton's velocity is parallel to the electric field? Does it follow that the proton's acceleration is parallel to the electric field? Explain.

8. Describe some of the differences between charging by induction and charging by contact.

9. A system consists of two charges of equal magnitude and opposite sign separated by a distance d. Since the total electric charge of this system is zero, can we conclude that the electric field produced by the system is also zero? Does your answer depend on the separation d? Explain.

10. The force experienced by charge 1 at point A is different in direction and magnitude from the force experienced by charge 2 at point B. Can we conclude that the electric fields at points A and B are different? Explain.

11. Can an electric field exist in a vacuum? Explain.

12. Gauss's law can tell us how much charge is contained within a Gaussian surface. Can it tell us where inside the surface it is located? Explain.

PROBLEMS AND CONCEPTUAL EXERCISES

Answers to odd-numbered Problems and Conceptual Exercises can be found in the back of the book. **BIO** *identifies problems of biological or medical interest;* **CE** *indicates a conceptual exercise.* **Predict/Explain** *problems ask for two responses: (a) your prediction of a physical outcome, and (b) the best explanation among three provided; and* **Predict/Calculate** *problems ask for a prediction of a physical outcome, based on fundamental physics concepts, and follow that with a numerical calculation to verify the prediction. On all problems, bullets (•, ••, •••) indicate the level of difficulty.*

SECTION 19-1 ELECTRIC CHARGE

1. • **CE Predict/Explain** An electrically neutral object is given a positive charge. **(a)** In principle, does the object's mass increase, decrease, or stay the same as a result of being charged? **(b)** Choose the *best explanation* from among the following:
   I. To give the object a positive charge we must remove some of its electrons; this will reduce its mass.
   II. Since electric charges have mass, giving the object a positive charge will increase its mass.
   III. Charge is conserved, and therefore the mass of the object will remain the same.

2. • **CE (a)** Based on the materials listed in Table 19-1, is the charge of a rubber balloon more likely to be positive or negative? Explain. **(b)** If the charge on the balloon is reversed, will the stream of water deflect toward or away from the balloon? Explain.

3. • **CE** This problem refers to the information given in Table 19-1. **(a)** If rabbit fur is rubbed against glass, what is the sign of the charge each acquires? Explain. **(b)** Repeat part (a) for the case of glass and rubber. **(c)** Comparing the situations described in parts (a) and (b), in which case is the magnitude of the triboelectric charge greater? Explain.

4. • Find the net charge of a system consisting of **(a)** 6.15×10^6 electrons and 7.44×10^6 protons or **(b)** 212 electrons and 165 protons.

5. • Find the total electric charge of 2.5 kg of **(a)** electrons and **(b)** protons.

6. • A container holds a gas consisting of 2.85 moles of oxygen molecules. One in a million of these molecules has lost a single electron. What is the net charge of the gas?

7. • **The Charge on Adhesive Tape** When adhesive tape is pulled from a dispenser, the detached tape acquires a positive charge and the remaining tape in the dispenser acquires a negative charge. If the tape pulled from the dispenser has $0.14 \, \mu$C of charge per centimeter, what length of tape must be pulled to transfer 1.8×10^{13} electrons to the remaining tape?

8. •• **CE** Four pairs of conducting spheres, all with the same radius, are shown in **FIGURE 19-36**, along with the net charge placed on them initially. The spheres in each pair are now brought into contact, allowing charge to transfer between them. Rank

FIGURE 19-36 Problem 8

the pairs of spheres in order of increasing magnitude of the charge transferred. Indicate ties where appropriate.

9. ••• A system of 1525 particles, each of which is either an electron or a proton, has a net charge of -5.456×10^{-17} C. **(a)** How many electrons are in this system? **(b)** What is the mass of this system?

SECTION 19-3 COULOMB'S LAW

10. • **CE** A charge $+q$ and a charge $-q$ are placed at opposite corners of a square. Will a third point charge experience a greater force if it is placed at one of the empty corners of the square, or at the center of the square? Explain.

11. • **CE** Consider the three electric charges, A, B, and C, shown in **FIGURE 19-37**. Rank the charges in order of increasing magnitude of the net force they experience. Indicate ties where appropriate.

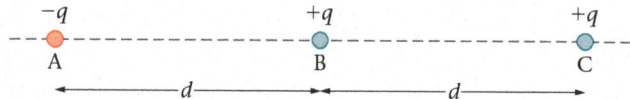

FIGURE 19-37 Problems 11 and 43

12. • **CE Predict/Explain** Suppose the charged sphere in Quick Example 19-10 is made from a conductor, rather than an insulator. **(a)** Do you expect the magnitude of the force between the point charge and the conducting sphere to be greater than, less than, or equal to the force between the point charge and an insulating sphere? **(b)** Choose the *best explanation* from among the following:
   I. The conducting sphere will allow the charges to move, resulting in a greater force.
   II. The charge of the sphere is the same whether it is conducting or insulating, and therefore the force is the same.
   III. The charge on a conducting sphere will move as far away as possible from the point charge. This results in a reduced force.

13. • At what separation is the electrostatic force between a $+13.1$-μC point charge and a $+26.1$-μC point charge equal in magnitude to 1.77 N?

14. • How much equal charge should be placed on the Earth and the Moon so that the electrical repulsion balances the gravitational force of 1.98×10^{20} N? Treat the Earth and Moon as point charges a distance 3.84×10^8 m apart.

15. • **Predict/Calculate** Two point charges, the first with a charge of $+3.13 \times 10^{-6}$ C and the second with a charge of -4.47×10^{-6} C, are separated by 25.5 cm. **(a)** Find the magnitude of the electrostatic force experienced by the positive charge. **(b)** Is the magnitude of the

force experienced by the negative charge greater than, less than, or the same as that experienced by the positive charge? Explain.

16. • When two identical ions are separated by a distance of 6.2×10^{-10} m, the electrostatic force each exerts on the other is 5.4×10^{-9} N. How many electrons are missing from each ion?

17. • Given that $q = +18 \, \mu C$ and $d = 21$ cm, find the direction and magnitude of the net electrostatic force exerted on the point charge q_1 in **FIGURE 19-38**.

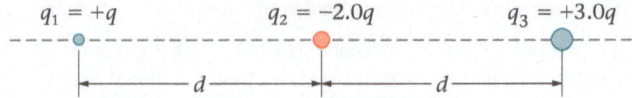

$q_1 = +q$ $q_2 = -2.0q$ $q_3 = +3.0q$

FIGURE 19-38 Problems 17, 23, 24, and 54

18. •• **CE** Five point charges, $q_1 = +q$, $q_2 = +2q$ $q_3 = -3q$, $q_4 = -4q$, and $q_5 = -5q$, are placed in the vicinity of an insulating spherical shell with a charge $+Q$ distributed uniformly over its surface, as indicated in **FIGURE 19-39**. Rank the point charges in order of increasing magnitude of the force exerted on them by the sphere. Indicate ties where appropriate.

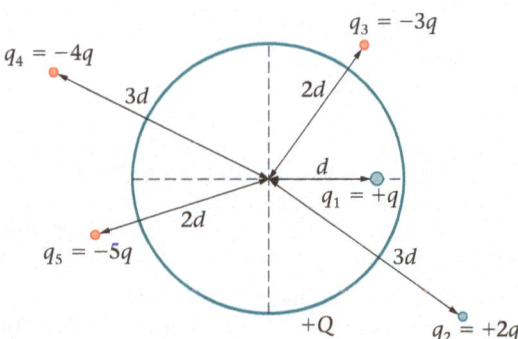

FIGURE 19-39 Problem 18

19. •• **CE** Three charges, $q_1 = +q$, $q_2 = -q$, and $q_3 = +q$, are at the vertices of an equilateral triangle, as shown in **FIGURE 19-40**. (a) Rank the three charges in order of increasing magnitude of the electric force they experience. Indicate ties where appropriate. (b) Give the direction angle, θ, of the net electric force experienced by charge 1. Note that θ is measured counterclockwise from the positive x axis. (c) Repeat part (b) for charge 2. (d) Repeat part (b) for charge 3.

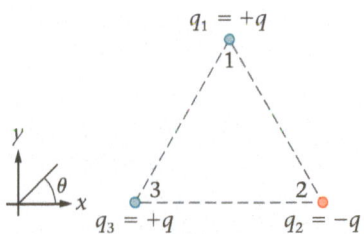

FIGURE 19-40 Problems 19 and 80

20. •• The attractive electrostatic force between the point charges $+8.44 \times 10^{-6}$ C and Q has a magnitude of 0.975 N when the separation between the charges is 1.31 m. Find the sign and magnitude of the charge Q.

21. •• If the speed of the electron in Example 19-4 were 7.3×10^5 m/s, what would be the corresponding orbital radius?

22. •• A sphere of radius 4.22 cm and uniform surface charge density $+12.1 \, \mu C/m^2$ exerts an electrostatic force of magnitude 46.9×10^{-3} N on a point charge of $+1.95 \, \mu C$. Find the separation between the point charge and the center of the sphere.

23. •• **Predict/Calculate** Given that $q = +12 \, \mu C$ and $d = 19$ cm, (a) find the direction and magnitude of the net electrostatic force exerted on the point charge q_2 in Figure 19-38. (b) How would your answers to part (a) change if the distance d were tripled?

24. •• Suppose the charge q_2 in Figure 19-38 can be moved left or right along the line connecting the charges q_1 and q_3. Given that $q = +12 \, \mu C$, find the distance from q_1 where q_2 experiences a net electrostatic force of zero. (The charges q_1 and q_3 are separated by a fixed distance of 32 cm.)

25. •• A point charge $q = -0.55$ nC is fixed at the origin. Where must a proton be placed in order for the electric force acting on it to be exactly opposite to its weight? (Let the y axis be vertical and the x axis be horizontal.)

26. •• A point charge $q = -0.55$ nC is fixed at the origin. Where must an electron be placed in order for the electric force acting on it to be exactly opposite to its weight? (Let the y axis be vertical and the x axis be horizontal.)

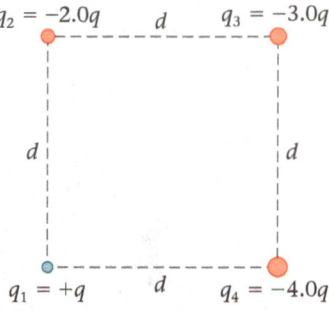

FIGURE 19-41 Problems 27 and 28

27. •• Find the direction and magnitude of the net electrostatic force exerted on the point charge q_2 in **FIGURE 19-41**. Let $q = +1.8 \, \mu C$ and $d = 42$ cm.

28. •• **Predict/Calculate** (a) Find the direction and magnitude of the net electrostatic force exerted on the point charge q_3 in Figure 19-41. Let $q = +2.4 \, \mu C$ and $d = 27$ cm. (b) How would your answers to part (a) change if the distance d were doubled?

29. •• **Predict/Calculate** Two point charges lie on the x axis. A charge of $+8.9 \, \mu C$ is at the origin, and a charge of $-6.1 \, \mu C$ is at $x = 12$ cm. (a) At what position x would a third charge q_3 be in equilibrium? (b) Does your answer to part (a) depend on whether q_3 is positive or negative? Explain.

30. •• A system consists of two positive point charges, q_1 and $q_2 > q_1$. The total charge of the system is $+62.0 \, \mu C$, and each charge experiences an electrostatic force of magnitude 85.0 N when the separation between them is 0.270 m. Find q_1 and q_2.

31. •• **Predict/Calculate** The point charges in **FIGURE 19-42** have the following values: $q_1 = +2.1 \, \mu C$, $q_2 = +6.3 \, \mu C$, $q_3 = -0.89 \, \mu C$. (a) Given that the distance d in Figure 19-42 is 4.35 cm, find the direction and magnitude of the net electrostatic force exerted on the point charge q_1. (b) How would your answers to part (a) change if the distance d were doubled? Explain.

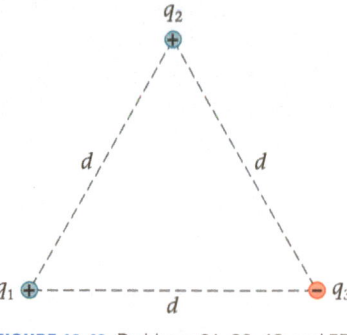

FIGURE 19-42 Problems 31, 32, 46, and 55

32. •• Referring to the previous problem, suppose that the magnitude of the net electrostatic force exerted on the point charge q_2 in Figure 19-42 is 0.65 N. (a) Find the distance d. (b) What is the direction of the net force exerted on q_2?

33. •• **Predict/Calculate** (a) If the nucleus in Example 19-4 had a charge of $+2e$ (as would be the case for a nucleus of helium),

would the speed of the electron be greater than, less than, or the same as that found in the Example? Explain. (Assume the radius of the electron's orbit is the same.) **(b)** Find the speed of the electron for a nucleus of charge $+2e$.

34. •• Four point charges are located at the corners of a square with sides of length a. Two of the charges are $+q$, and two are $-q$. Find the magnitude and direction of the net electric force exerted on a charge $+Q$, located at the center of the square, for each of the following two arrangements of charge: **(a)** The charges alternate in sign $(+q, -q, +q, -q)$ as you go around the square; **(b)** the two positive charges are on the top corners, and the two negative charges are on the bottom corners.

35. ••• **Predict/Calculate** Two identical point charges in free space are connected by a string 7.6 cm long. The tension in the string is 0.21 N. **(a)** Find the magnitude of the charge on each of the point charges. **(b)** Using the information given in the problem statement, is it possible to determine the sign of the charges? Explain. **(c)** Find the tension in the string if $+1.0\,\mu$C of charge is transferred from one point charge to the other. Compare with your result from part (a).

36. ••• Two spheres with uniform surface charge density, one with a radius of 7.8 cm and the other with a radius of 4.1 cm, are separated by a center-to-center distance of 38 cm. The spheres have a combined charge of $+55\,\mu$C and repel one another with a force of 0.75 N. What is the surface charge density on each sphere?

37. ••• Point charges, q_1 and q_2, are placed on the x axis, with q_1 at $x = 0$ and q_2 at $x = d$. A third point charge, $+Q$, is placed at $x = 3d/4$. If the net electrostatic force experienced by the charge $+Q$ is zero, how are q_1 and q_2 related?

SECTION 19-4 THE ELECTRIC FIELD

38. • **CE** Two electric charges are separated by a finite distance. Somewhere between the charges, on the line connecting them, the net electric field they produce is zero. **(a)** Do the charges have the same or opposite signs? Explain. **(b)** If the point of zero field is closer to charge 1, is the magnitude of charge 1 greater than or less than the magnitude of charge 2? Explain.

39. • What is the magnitude of the electric field produced by a charge of magnitude $4.50\,\mu$C at a distance of **(a)** 1.00 m and **(b)** 3.00 m?

40. • A $+5.0$-μC charge experiences a 0.64-N force in the positive y direction. If this charge is replaced with a -2.7-μC charge, what force will it experience?

41. • Two point charges lie on the x axis. A charge of $+6.2\,\mu$C is at the origin, and a charge of $-9.5\,\mu$C is at $x = 10.0$ cm. What is the net electric field at **(a)** $x = -4.0$ cm and **(b)** $x = +4.0$ cm?

42. •• Two point charges lie on the x axis. A charge of $+2.40$ pC is at the origin, and a charge of -4.80 pC is at $x = -10.0$ cm. What third charge should be placed at $x = +20.0$ cm so that the total electric field at $x = +10.0$ cm is zero?

43. •• **CE** The electric field on the dashed line in Figure 19-37 vanishes at infinity, but also at two different points a finite distance from the charges. Identify the regions in which you can find $E = 0$ at a finite distance from the charges: region 1, to the left of point A; region 2, between points A and B; region 3, between points B and C; region 4, to the right of point C.

44. •• An object with a charge of $-2.1\,\mu$C and a mass of 0.0044 kg experiences an upward electric force, due to a uniform electric field, equal in magnitude to its weight. **(a)** Find the direction and magnitude of the electric field. **(b)** If the electric charge on the

object is doubled while its mass remains the same, find the direction and magnitude of its acceleration.

45. •• **Predict/Calculate** Figure 19-42 shows a system consisting of three charges, $q_1 = +5.00\,\mu$C, $q_2 = +5.00\,\mu$C, and $q_3 = -5.00\,\mu$C, at the vertices of an equilateral triangle of side $d = 2.95$ cm. **(a)** Find the magnitude of the electric field at a point halfway between the charges q_1 and q_2. **(b)** Is the magnitude of the electric field halfway between the charges q_2 and q_3 greater than, less than, or the same as the electric field found in part (a)? Explain. **(c)** Find the magnitude of the electric field at the point specified in part (b).

46. •• Two point charges of equal magnitude are 8.3 cm apart. At the midpoint of the line connecting them, their combined electric field has a magnitude of 51 N/C. Find the magnitude of the charges.

47. •• **Predict/Calculate** A point charge $q = +4.7\,\mu$C is placed at each corner of an equilateral triangle with sides 0.21 m in length. **(a)** What is the magnitude of the electric field at the midpoint of any of the three sides of the triangle? **(b)** Is the magnitude of the electric field at the center of the triangle greater than, less than, or the same as the magnitude at the midpoint of a side? Explain.

48. ••• **Predict/Calculate** Four point charges, each of magnitude q, are located at the corners of a square with sides of length a. Two of the charges are $+q$, and two are $-q$. The charges are arranged in one of the following two ways: (1) The charges alternate in sign $(+q, -q, +q, -q)$ as you go around the square; (2) the top two corners of the square have positive charges $(+q, +q)$, and the bottom two corners have negative charges $(-q, -q)$. **(a)** In which case will the electric field at the center of the square have the greatest magnitude? Explain. **(b)** Calculate the electric field at the center of the square for each of these two cases.

49. ••• The electric field at the point $x = 5.00$ cm and $y = 0$ points in the positive x direction with a magnitude of 10.0 N/C. At the point $x = 10.0$ cm and $y = 0$ the electric field points in the positive x direction with a magnitude of 15.0 N/C. Assuming this electric field is produced by a single point charge, find **(a)** its location and **(b)** the sign and magnitude of its charge.

SECTION 19-5 ELECTRIC FIELD LINES

50. • **Predict/Calculate** The electric field lines surrounding three charges are shown in **FIGURE 19-43**. The center charge is $q_2 = -22.8\,\mu$C. **(a)** What are the signs of q_1 and q_3? **(b)** Find q_1. **(c)** Find q_3.

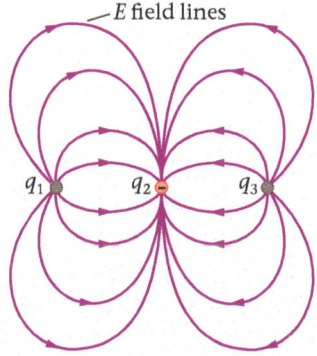

FIGURE 19-43 Problems 50 and 51

51. •• Referring to Figure 19-43, suppose q_2 is not known. Instead, it is given that $q_1 + q_2 = -1.8\,\mu$C. Find q_1, q_2, and q_3.

52. •• **CE** The electric field lines surrounding three charges are shown in **FIGURE 19-44**. **(a)** Which of the charges q_A, q_B, and q_C are positively charged, and which are negatively charged? **(b)** Rank the charges in order of increasing magnitude.

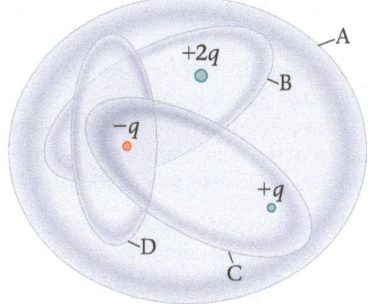

FIGURE 19-44 Problem 52

53. •• Make a qualitative sketch of the electric field lines produced by the four charges, $+q$, $-q$, $+q$, and $-q$, arranged clockwise on the four corners of a square with sides of length d.

54. •• Sketch the electric field lines for the system of charges shown in Figure 19-38.

55. •• Sketch the electric field lines for the system of charges described in Problem 31.

56. •• Suppose the magnitude of the electric field between the plates in Example 19-16 is changed, and a new object with a charge of $-2.05 \,\mu\text{C}$ is attached to the thread. If the tension in the thread is 0.450 N, and the angle it makes with the vertical is 16°, what are **(a)** the mass of the object and **(b)** the magnitude of the electric field?

SECTION 19-7 ELECTRIC FLUX AND GAUSS'S LAW

57. • **CE Predict/Explain** Gaussian surface 1 has twice the area of Gaussian surface 2. Both surfaces enclose the same charge Q. **(a)** Is the electric flux through surface 1 greater than, less than, or the same as the electric flux through surface 2? **(b)** Choose the *best explanation* from among the following:
   I. Gaussian surface 2 is closer to the charge, since it has the smaller area. It follows that it has the greater electric flux.
   II. The two surfaces enclose the same charge, and hence they have the same electric flux.
   III. Electric flux is proportional to area. As a result, Gaussian surface 1 has the greater electric flux.

58. • **CE** Suppose the conducting shell in Figure 19-33—which has a point charge $+Q$ at its center—has a nonzero net charge. How much charge is on the inner and outer surface of the shell when the net charge of the shell is **(a)** $-2Q$, **(b)** $-Q$, and **(c)** $+Q$?

59. • **CE** Rank the Gaussian surfaces shown in **FIGURE 19-45** in order of increasing electric flux, starting with the most negative. Indicate ties where appropriate.

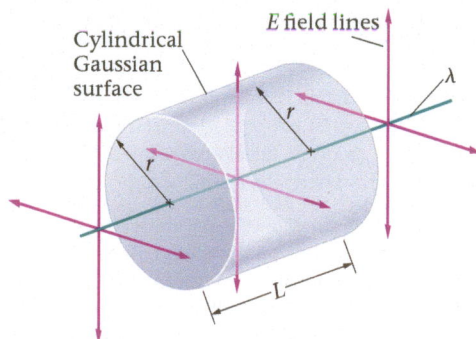

FIGURE 19-45 Problems 59 and 72

60. • A uniform electric field of magnitude 35,000 N/C makes an angle of 47° with a plane surface of area 0.0153 m². What is the electric flux through this surface?

61. • In order to work well, a square antenna must intercept a flux of at least $0.045 \,\text{N} \cdot \text{m}^2/\text{C}$ when it is perpendicular to a uniform electric field of magnitude 5.0 N/C. What is the minimum side length of the antenna?

62. • A surface encloses the charges $q_1 = 3.2 \,\mu\text{C}$, $q_2 = 6.9 \,\mu\text{C}$, and $q_3 = -4.1 \,\mu\text{C}$. Find the electric flux through this surface.

63. •• **BIO Nerve Cells** Nerve cells are long, thin cylinders along which electrical disturbances (nerve impulses) travel. The cell membrane of a typical nerve cell consists of an inner and an outer wall separated by a distance of $0.10 \,\mu\text{m}$. The electric field within the cell membrane is $7.0 \times 10^5 \,\text{N/C}$. Approximating the cell membrane as a parallel-plate capacitor, determine the magnitude of the charge density on the inner and outer cell walls.

64. •• The electric flux through each of the six sides of a rectangular box are as follows:

$$\Phi_1 = +150.0 \,\text{N} \cdot \text{m}^2/\text{C}; \Phi_2 = +250.0 \,\text{N} \cdot \text{m}^2/\text{C};$$
$$\Phi_3 = -350.0 \,\text{N} \cdot \text{m}^2/\text{C}; \Phi_4 = +175.0 \,\text{N} \cdot \text{m}^2/\text{C};$$
$$\Phi_5 = -100.0 \,\text{N} \cdot \text{m}^2/\text{C}; \Phi_6 = +450.0 \,\text{N} \cdot \text{m}^2/\text{C}.$$

How much charge is in this box?

65. •• Consider a spherical Gaussian surface and three charges: $q_1 = 2.03 \,\mu\text{C}$, $q_2 = -3.28 \,\mu\text{C}$, and $q_3 = 4.89 \,\mu\text{C}$. Find the electric flux through the Gaussian surface if it completely encloses **(a)** only charges q_1 and q_2, **(b)** only charges q_2 and q_3, and **(c)** all three charges. **(d)** Suppose a fourth charge, Q, is added to the situation described in part (c). Find the sign and magnitude of Q required to give zero electric flux through the surface.

66. •• The surface charge per area on the outside of a conducting spherical shell (Figure 19-33) of outer radius 2.5 cm is measured to be $-3.6 \,\mu\text{C/m}^2$. What charge is enclosed by the shell?

67. •• **Photovoltaic Field** Suppose the field in the interface region of a photovoltaic panel is $1.1 \times 10^6 \,\text{N/C}$. Modeling the interface as a parallel-plate capacitor, what is the charge density σ on either side of the interface?

68. ••• A thin wire of infinite extent has a charge per unit length of λ. Using the cylindrical Gaussian surface shown in **FIGURE 19-46**, show that the electric field produced by this wire at a radial distance r has a magnitude given by

$$E = \frac{\lambda}{2\pi\epsilon_0 r}$$

Note that the direction of the electric field is always radially away from the wire.

Cylindrical Gaussian surface

E field lines

λ

r

L

FIGURE 19-46 Problems 68 and 82

GENERAL PROBLEMS

69. • **CE Predict/Explain** An electron and a proton are released from rest in space, far from any other objects. The particles move toward each other, due to their mutual electrical attraction. **(a)** When they meet, is the kinetic energy of the electron greater than, less than, or equal to the kinetic energy of the proton? **(b)** Choose the *best explanation* from among the following:

I. The proton has the greater mass. Since kinetic energy is proportional to mass, it follows that the proton will have the greater kinetic energy.

II. The two particles experience the same force, but the light electron moves farther than the massive proton. Therefore, the work done on the electron, and hence its kinetic energy, is greater.

III. The same force acts on the two particles. Therefore, they will have the same kinetic energy and energy will be conserved.

70. • **CE Predict/Explain** In Conceptual Example 19-9, suppose the charge to be placed at either point A or point B is $+q$ rather than $-q$. **(a)** Is the magnitude of the net force experienced by the movable charge at point A greater than, less than, or equal to the magnitude of the net force at point B? **(b)** Choose the *best explanation* from among the following:

I. Point B is farther from the two fixed charges. As a result, the net force at point B is less than at point A.

II. The net force at point A cancels, just as it does in Conceptual Example 19-9. Therefore, the nonzero net force at point B is greater in magnitude than the zero net force at point A.

III. The net force is greater in magnitude at point A because at that location the movable charge experiences a net repulsion from each of the fixed charges.

71. • **CE** Under normal conditions, the electric field at the surface of the Earth points downward, into the ground. What is the sign of the electric charge on the ground?

72. • **CE** A Gaussian surface for the charges shown in Figure 19-45 has an electric flux equal to $+3q/\epsilon_0$. Which charges are contained within this Gaussian surface?

73. • A proton is released from rest in a uniform electric field of magnitude 2.18×10^5 N/C. Find the speed of the proton after it has traveled **(a)** 1.00 cm and **(b)** 10.0 cm.

74. • **BIO Ventricular Fibrillation** If a charge of 0.30 C passes through a person's chest in 1.0 s, the heart can go into ventricular fibrillation—a nonrhythmic "fluttering" of the ventricles that results in little or no blood being pumped to the body. If this rate of charge transfer persists for 4.5 s, how many electrons pass through the chest?

75. • A point charge at the origin of a coordinate system produces the electric field $\vec{E} = (56,000 \text{ N/C})\,\hat{x}$ on the x axis at the location $x = -0.65$ m. Determine the sign and magnitude of the charge.

76. •• Find the orbital radius for which the kinetic energy of the electron in Example 19-4 is 1.51 eV. (*Note*: 1 eV = 1 electronvolt = 1.6×10^{-19} J.)

77. •• **The Balloon and Your Hair** Suppose 7.5×10^{10} electrons are transferred to a balloon by rubbing it on your hair. Treating the balloon and your hair as point charges separated by 0.15 m, what is the force of attraction between them?

78. •• **The Balloon and the Wall** When a charged balloon sticks to a wall, the downward gravitational force is balanced by an upward static friction force. The normal force is provided by the electrical attraction between the charged balloon and the equal but oppositely charged polarization induced in the wall's molecules. If the mass of a balloon is 1.9 g, its coefficient of static friction with the wall is 0.74, and the average distance between the opposite charges is 0.45 mm, what minimum amount of charge must be placed on the balloon in order for it to stick to the wall?

79. •• **CE** Four lightweight, plastic spheres, labeled A, B, C, and D, are suspended from threads in various combinations, as illustrated in **FIGURE 19-47**. It is given that the net charge on sphere D is $+Q$, and that the other spheres have net charges of $+Q$, $-Q$, or 0. From the results of the four experiments shown in Figure 19-47, and the fact that the spheres have equal masses, determine the net charge of **(a)** sphere A, **(b)** sphere B, and **(c)** sphere C.

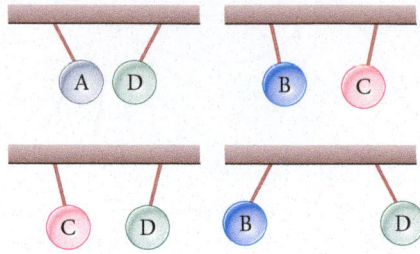

FIGURE 19-47 Problem 79

80. •• Find **(a)** the direction and **(b)** the magnitude of the net electric field at the center of the equilateral triangle in Figure 19-40. Give your answers in terms of the angle θ, as defined in Figure 19-40, and E, the magnitude of the electric field produced by any *one* of the charges at the center of the triangle.

81. •• A small object of mass 0.0150 kg and charge 3.1 μC hangs from the ceiling by a thread. A second small object, with a charge of 4.2 μC, is placed 1.2 m vertically below the first charge. Find **(a)** the electric field at the position of the upper charge due to the lower charge and **(b)** the tension in the thread.

82. •• The electric field at a radial distance of 47.7 cm from the thin charged wire shown in Figure 19-46 has a magnitude of 35,400 N/C. **(a)** Using the result given in Problem 68, what is the magnitude of the charge per length on this wire? **(b)** At what distance from the wire is the magnitude of the electric field equal to $\frac{1}{2}(35,400 \text{ N/C})$?

83. •• **Predict/Calculate** Three charges are placed at the vertices of an equilateral triangle of side $a = 0.93$ m, as shown in **FIGURE 19-48**. Charges 1 and 3 are $+7.3$ μC; charge 2 is -7.3 μC. **(a)** Find the magnitude and direction of the net force acting on charge 3. **(b)** If charge 3 is moved to the origin, will the net force acting on it there be greater than, less than, or equal to the net force found in part (a)? Explain. **(c)** Find the net force on charge 3 when it is at the origin.

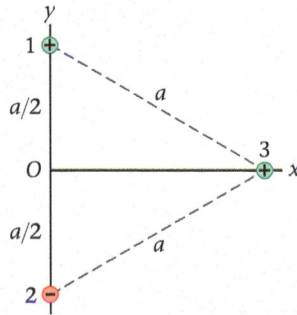

FIGURE 19-48 Problem 83

84. •• **Predict/Calculate BIO Cell Membranes** The cell membrane in a nerve cell has a thickness of 0.12 μm. **(a)** Approximating the cell membrane as a parallel-plate capacitor with a surface charge density of 5.9×10^{-6} C/m², find the electric field within the membrane. **(b)** If the thickness of the membrane were doubled, would your answer to part (a) increase, decrease, or stay the same? Explain.

85. •• A square with sides of length L has a point charge at each of its four corners. Two corners that are diagonally opposite have charges equal to $+3.25$ μC; the other two diagonal corners have charges Q. Find the magnitude and sign of the charges Q such that each of the $+3.25$-μC charges experiences zero net force.

86. •• Two small plastic balls hang from threads of negligible mass. Each ball has a mass of 0.22 g and a charge of magnitude q. The balls are attracted to each other, and the threads attached to the balls

make an angle of 20.0° with the vertical, as shown in **FIGURE 19-49**. Find **(a)** the magnitude of the electric force acting on each ball, **(b)** the tension in each of the threads, and **(c)** the magnitude of the charge on the balls.

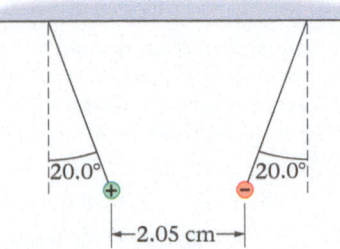

FIGURE 19-49 Problem 86

87. •• A small sphere with a charge of $+2.44\,\mu C$ is attached to a relaxed horizontal spring whose force constant is 89.2 N/m. The spring extends along the x axis, and the sphere rests on a frictionless surface with its center at the origin. A point charge $Q = -8.55\,\mu C$ is now moved slowly from infinity to a point $x = d > 0$ on the x axis. This causes the small sphere to move to the position $x = 0.124$ m. Find d.

88. •• Twelve identical point charges q are equally spaced around the circumference of a circle of radius R. The circle is centered at the origin. One of the twelve charges, which happens to be on the positive x axis, is now moved to the center of the circle. Find **(a)** the direction and **(b)** the magnitude of the net electric force exerted on this charge.

89. •• **BIO Nerve Impulses** When a nerve impulse propagates along a nerve cell, the electric field within the cell membrane changes from 7.0×10^5 N/C in one direction to 3.0×10^5 N/C in the other direction. Approximating the cell membrane as a parallel-plate capacitor, find the magnitude of the change in charge density on the walls of the cell membrane.

90. •• Predict/Calculate **The Electric Field of the Earth** The Earth produces an approximately uniform electric field at ground level. This electric field has a magnitude of 110 N/C and points radially inward, toward the center of the Earth. **(a)** Find the surface charge density (sign and magnitude) on the surface of the Earth. **(b)** Given that the radius of the Earth is 6.38×10^6 m, find the total electric charge on the Earth. **(c)** If the Moon had the same amount of electric charge distributed uniformly over its surface, would its electric field at the surface be greater than, less than, or equal to 110 N/C? Explain.

91. •• An object of mass $m = 2.5$ g and charge $Q = +42\,\mu C$ is attached to a string and placed in a uniform electric field that is inclined at an angle of 30.0° with the horizontal (**FIGURE 19-50**). The object is in static equilibrium when the string is horizontal. Find **(a)** the magnitude of the electric field and **(b)** the tension in the string.

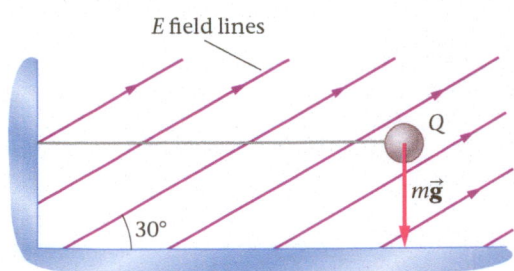

FIGURE 19-50 Problem 91

92. •• Four identical charges, $+Q$, occupy the corners of a square with sides of length a. A fifth charge, q, can be placed at any desired location. Find the location of the fifth charge, and the value of q, such that the net electric force acting on each of the original four charges, $+Q$, is zero.

93. •• Two charges, $+q$ and $-q$, occupy two corners of an equilateral triangle, as shown in **FIGURE 19-51**. **(a)** If $q = 1.8\,\mu C$ and $r = 0.50$ m, find the magnitude and direction of the electric field at point A, the other vertex of the triangle. Let the direction angle be measured counterclockwise from the positive x axis. **(b)** What is the total electric flux through the surface indicated in the figure? **(c)** Explain why Gauss's law cannot easily be used to find the magnitude of the electric field at point A.

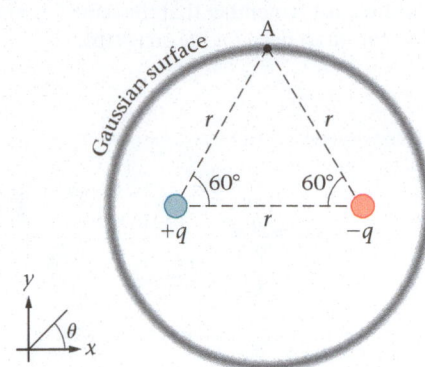

FIGURE 19-51 Problem 93

94. •• **FIGURE 19-52** shows an electron entering a parallel-plate capacitor with a speed of 5.45×10^6 m/s. The electric field of the capacitor has deflected the electron downward by a distance of 0.618 cm at the point where the electron exits the capacitor. Find **(a)** the magnitude of the electric field in the capacitor and **(b)** the speed of the electron when it exits the capacitor.

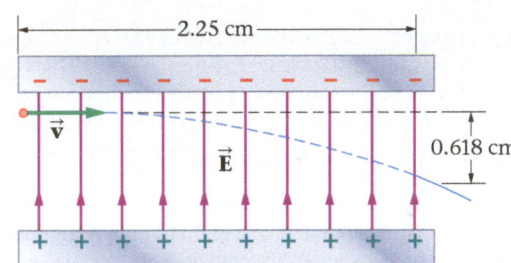

FIGURE 19-52 Problem 94

95. ••• Two identical conducting spheres are separated by a fixed center-to-center distance of 35 cm and have different charges. Initially, the spheres attract each other with a force of 0.099 N. The spheres are now connected by a thin conducting wire. After the wire is removed, the spheres are positively charged and repel one another with a force of 0.031 N. Find **(a)** the final and **(b)** the initial charges on the spheres.

PASSAGE PROBLEMS

Bumblebees and Static Cling

Have you ever pulled clothes from a dryer only to have them "cling" together? Have you ever walked across a carpet and had a "shocking" experience when you touched a doorknob? If so, you already know a lot about static electricity.

Ben Franklin showed that the same kind of spark we experience on a carpet, when scaled up in size, is responsible for bolts of lightning. His insight led to the invention of lightning rods to conduct electricity safely away from a building into the ground. Today, we employ static electricity in many technological applications, ranging from photocopiers to electrostatic precipitators that clean emissions from smokestacks. We even use electrostatic salting machines to give potato chips the salty taste we enjoy!

Living organisms also use static electricity—in fact, static electricity plays an important role in the pollination process. Imagine a bee busily flitting from flower to flower. As air rushes over its body and wings it acquires an electric charge—just as you do when your feet rub against a carpet. A bee might have only 93.0 pC of charge, but that's more than enough to attract grains of pollen from a distance, like a charged comb attracting bits of paper. The result is a bee covered with grains of pollen, as illustrated in the accompanying photo, unwittingly transporting pollen from one flower to another. So, the next time you experience annoying static cling in your clothes, just remember that the same force helps pollinate the plants that we all need for life on Earth.

A white-tailed bumblebee with static cling.
(Problems 96, 97, 98, and 99)

96. • How many electrons must be transferred away from a bee to produce a charge of +93.0 pC?

    A. 1.72×10^{-9} B. 5.81×10^{8}
    C. 1.02×10^{20} D. 1.49×10^{29}

97. • Suppose two bees, each with a charge of 93.0 pC, are separated by a distance of 1.20 cm. Treating the bees as point charges, what is the magnitude of the electrostatic force experienced by the bees? (In comparison, the weight of a 0.140-g bee is 1.37×10^{-3} N.)

    A. 6.01×10^{-17} N   B. 6.48×10^{-9} N
    C. 5.40×10^{-7} N   D. 5.81×10^{-3} N

98. •• The force required to detach a grain of pollen from an avocado stamen is approximately 4.0×10^{-8} N. What is the maximum distance at which the electrostatic force between a bee and a grain of pollen is sufficient to detach the pollen? Treat the bee and pollen as point charges, and assume the pollen has a charge opposite in sign and equal in magnitude to the bee.

    A. 4.7×10^{-7} m   B. 1.9 mm
    C. 4.4 cm   D. 220 m

99. •• Pollen of the lisianthus plant requires a force 10 times that of the avocado pollen to be detached. If the lisianthus pollen has the same electric charge as does the avocado pollen, at what maximum distance can the same bee detach lisianthus pollen, assuming point charges?

    A. 0.12 cm   B. 1.4 cm   C. 4.0 cm   D. 12 cm

100. •• **Predict/Calculate** REFERRING TO EXAMPLE 19-14 Suppose $q_1 = +2.90 \, \mu C$ is no longer at the origin, but is now on the y axis between $y = 0$ and $y = 0.500$ m. The charge $q_2 = +2.90 \, \mu C$ is at $x = 0$ and $y = 0.500$ m, and point 3 is at $x = y = 0.500$ m. (a) Is the magnitude of the net electric field at point 3, which we call E_{net}, greater than, less than, or equal to its previous value? Explain. (b) Is the angle θ that E_{net} makes with the x axis greater than, less than, or equal to its previous value? Explain. Find the new values of (c) E_{net} and (d) θ if q_1 is at $y = 0.250$ m.

101. •• **Predict/Calculate** REFERRING TO EXAMPLE 19-14 In this system, the charge q_1 is at the origin, the charge q_2 is at $x = 0$ and $y = 0.500$ m, and point 3 is at $x = y = 0.500$ m. Suppose that $q_1 = +2.90 \, \mu C$, but that q_2 is increased to a value greater than $+2.90 \, \mu C$. As a result, do (a) E_{net} and (b) θ increase, decrease, or stay the same? Explain. If $E_{net} = 1.66 \times 10^5$ N/C, find (c) q_2 and (d) θ.

102. •• **Predict/Calculate** REFERRING TO EXAMPLE 19-16 The magnitude of the charge is changed until the angle the thread makes with the vertical is $\theta = 15.0°$. The electric field is 1.46×10^4 N/C and the mass of the object is 0.0250 kg. (a) Is the new magnitude of q greater than or less than its previous value? Explain. (b) Find the new value of q.

103. •• REFERRING TO EXAMPLE 19-16 Suppose the magnitude of the electric field is adjusted to give a tension of 0.253 N in the thread. This will also change the angle the thread makes with the vertical. (a) Find the new value of E. (b) Find the new angle between the thread and the vertical.

Big Ideas

1 Electric potential energy is similar to gravitational potential energy.

2 Energy conservation can be extended to include electrical systems.

3 The electric potential of a point charge is kq/r.

4 The electric potential of a system of charges is the sum of the potentials due to each charge.

5 Equipotential surfaces are surfaces on which the electric potential has a constant value.

6 Capacitors store both electric charge and electric potential energy.

Electric Potential and Electric Potential Energy

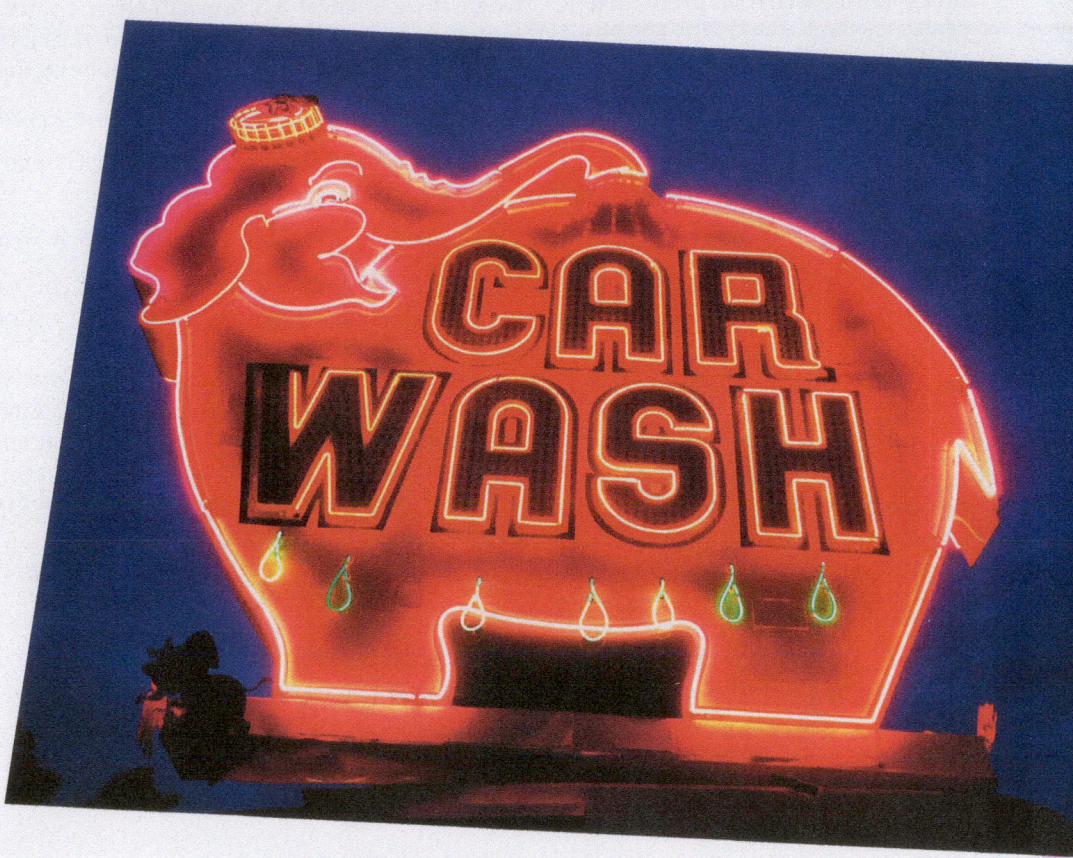

▲ Neon signs do in fact contain neon, which gives the red color, as well as noble gases such as argon, helium, krypton, and xenon to produce a wide range of brilliant colors. A neon sign operates with a difference in electric potential of about 10,000 volts. In contrast, muscle contractions in the human heart produce electric potentials of only about a thousandth of a volt. The physics underlying electric potential—and its application to everything from neon signs to EKGs—is the subject of this chapter.

Before paramedics can use a defibrillator on a heart-attack victim, they must wait a few seconds for it to be "charged up." During the charging process, electric charge builds up on a capacitor, storing electrical energy. When the defibrillator is activated, the energy is released in a sudden burst that can save a person's life. In this chapter we develop the concept of electrical energy, and discuss how both energy and charge are stored in a capacitor.

20-1 Electric Potential Energy and the Electric Potential

Quantity	Symbol	Units	Meaning
Electric potential energy	U	joules	energy associated with electric force
Electric potential	V	volts	electric potential energy per charge
Potential (shorthand for electric potential)	V	volts	same as electric potential

Big Idea **1** Electric potential energy is analogous to gravitational potential energy. Both can store energy for later release. Electric potential is defined as the electric potential energy per charge.

Electric and gravitational forces have many similarities. One of the most important of these is that both forces are conservative. As a result, there must be an **electric potential energy**, U, associated with the electric force, just as there is a gravitational potential energy, $U = mgy$, associated with the force of gravity. (Conservative forces and potential energies are discussed in Chapter 8.)

To illustrate the concept of electric potential energy, consider a uniform electric field, $\vec{E}$, as shown in **FIGURE 20-1 (a)**. A positive test charge q_0 is placed in this field, where it experiences a downward electric force of magnitude $F = q_0 E$. If the charge is moved upward through a distance d, the electric force and the displacement are in opposite directions; hence, the work done by the electric force is negative:

$$W = -q_0 E d$$

Using the definition of potential energy change given in Equation 8-1, $\Delta U = -W$, we find that the potential energy of the charge is changed by the amount

$$\Delta U = -W = q_0 E d \qquad \text{20-1}$$

The electric potential energy increases in this case, just as the gravitational potential energy of a ball increases when it is raised against the force of gravity to a higher altitude, as indicated in **FIGURE 20-1 (b)**.

On the other hand, if the charge q_0 is negative, the electric force acting on it will be upward. In this case, the electric force does *positive* work as the charge is raised through the distance d, and the change in potential energy is *negative*. Thus, the change in potential energy depends on the sign of a charge as well as on its magnitude.

In the last chapter we found it convenient to define an electric field, $\vec{E}$, as the force per charge:

$$\vec{E} = \frac{\vec{F}}{q_0}$$

Similarly, it is useful to define a quantity that is equal to the change in electric potential energy per charge, $\Delta U / q_0$ This quantity is referred to as the **electric potential**, V:

Definition of Electric Potential, V

$$\Delta V = \frac{\Delta U}{q_0} = \frac{-W}{q_0} \qquad \text{20-2}$$

SI unit: joule/coulomb = volt, V

In common usage, the electric potential is often referred to simply as the "potential."

▶ **FIGURE 20-1 Change in electric potential energy (a)** A positive test charge q_0 experiences a downward force due to the electric field $\vec{E}$. If the charge is moved upward a distance d, the work done by the electric field is $-q_0 Ed$. At the same time, the electric potential energy of the system increases by $q_0 Ed$. The situation is analogous to that of an object in a gravitational field. **(b)** If a ball is lifted against the force exerted by gravity, the gravitational potential energy of the system increases.

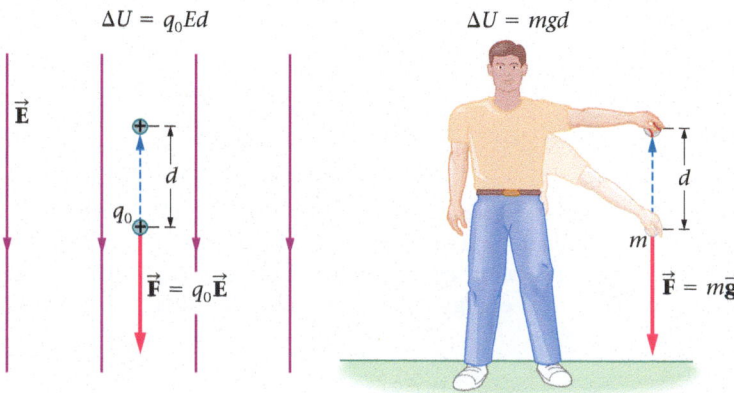

(a) Moving a charge in an electric field

(b) Moving a mass in a gravitational field

Notice that our definition of V gives only the *change* in electric potential, ΔV. As with gravitational potential energy, the electric potential can be set to zero at any desired location—only changes in electric potential are measurable. In addition, just as the potential energy U is a scalar (that is, simply a number) so too is the potential V.

Units for Electric Potential We know that potential energy is measured in joules, and charge is measured in coulombs; hence, the SI units of electric potential are joules per coulomb. This combination of units is referred to as the volt, in honor of Alessandro Volta (1745–1827), who invented a predecessor to the modern battery. To be specific, the volt (V) is defined as follows:

$$1\,\text{V} = 1\,\text{J/C} \qquad \qquad 20\text{-}3$$

Rearranging slightly, we see that energy can be expressed as charge times voltage: $1\,\text{J} = (1\,\text{C})(1\,\text{V})$.

A convenient and commonly used unit of energy in atomic systems is the **electron volt** (eV), defined as the product of the electron charge and a potential difference of 1 volt:

$$1\,\text{eV} = (1.60 \times 10^{-19}\,\text{C})(1\,\text{V}) = 1.60 \times 10^{-19}\,\text{J}$$

It follows that the electron volt is the energy change an electron experiences when it moves through a potential difference of 1 V. As we shall see in later chapters, typical atomic energies are in the eV range, whereas typical energies in nuclear systems are in the MeV (10^6 eV) range. Even the MeV is a minuscule energy, however, when compared to the joule.

EXERCISE 20-1 ELECTRIC POTENTIAL AND ELECTRIC POTENTIAL ENERGY

a. Find the change in electric potential energy, ΔU, as a charge of 3.65×10^{-6} C moves from a point A to a point B, given that the change in electric potential between these points is $\Delta V = V_B - V_A = 12.0$ V.

b. Find the change in electric potential when the electric potential energy of a charge of 1.35×10^{-6} C decreases by 2.43×10^{-5} J.

REASONING AND SOLUTION

These questions can be answered by applying the definition, $\Delta V = \Delta U / q_0$, that relates electric potential, V, electric potential energy, U, and charge, q_0.

a. Solving $\Delta V = \Delta U / q_0$ for ΔU, we find

$$\Delta U = q_0\,\Delta V = (3.65 \times 10^{-6}\,\text{C})(12.0\,\text{V}) = 4.38 \times 10^{-5}\,\text{J}$$

b. Applying $\Delta V = \Delta U / q_0$ directly, we find

$$\Delta V = \frac{\Delta U}{q_0} = \frac{-2.43 \times 10^{-5}\,\text{J}}{1.35 \times 10^{-6}\,\text{C}} = -18.0\,\text{V}$$

Electric Field and the Rate of Change of Electric Potential

There is a connection between the electric field and the electric potential that is both straightforward and useful. To obtain this relationship, let's apply the definition $\Delta V = -W/q_0$ to the case of a test charge that moves through a distance Δs in the direction of the electric field, as in **FIGURE 20-2**. The work done by the electric field in this case is simply the magnitude of the electric force, $F = q_0 E$, times the distance, Δs:

$$W = q_0 E \Delta s$$

Therefore, the change in electric potential is

$$\Delta V = \frac{-W}{q_0} = \frac{-(q_0 E \Delta s)}{q_0} = -E \Delta s$$

PHYSICS IN CONTEXT
Looking Back

The development of the concepts of electric potential energy and electric potential parallels that of gravitational potential energy for a uniform gravitational field (Chapter 8).

PROBLEM-SOLVING NOTE

Electric Potential and Its Unit of Measurement

Be careful to distinguish between the electric potential V and the corresponding unit of measurement, the volt V, since both are designated by the same symbol. For example, in the expression $\Delta V = 12$ V, the first V refers to the electric potential, and the second V refers to the units of the volt. Similarly, when someone says he has a "12-volt battery," what he really means is that the battery produces a difference in electric potential, ΔV, of 12 volts.

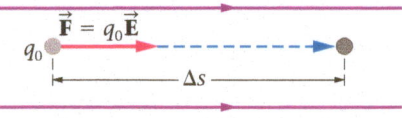

▲ **FIGURE 20-2 Electric field and electric potential** As a charge q_0 moves in the direction of the electric field, $\vec{\mathbf{E}}$, the electric potential, V, decreases. In particular, if the charge moves a distance Δs, the electric potential decreases by the amount $\Delta V = -E \Delta s$.

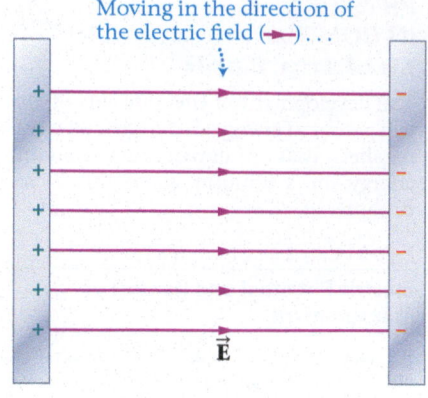

Moving in the direction of the electric field (→) ...

$\vec{E}$

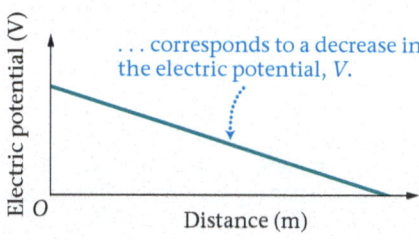

Electric potential (V)

... corresponds to a decrease in the electric potential, V.

O Distance (m)

◄ **FIGURE 20-3 The electric potential for a constant electric field** As a general rule, the electric potential, V, decreases as one moves in the direction of the electric field. In the case shown here, the electric field is constant; as a result, the electric potential decreases uniformly with distance. We have arbitrarily set the potential equal to zero at the right-hand plate.

Solving for the electric field, we find:

Connection Between the Electric Field and the Electric Potential

$$E = -\frac{\Delta V}{\Delta s}$$

20-4

SI unit: volts/meter, V/m

This relationship shows that the electric field, which can be expressed in units of N/C, also has the units of volts per meter. That is,

$$1\,\text{N/C} = 1\,\text{V/m}$$

20-5

To summarize:

> The electric field depends on the rate of change of the electric potential with position.

In terms of our gravitational analogy, you can think of the potential V as the height of a hill and the electric field E as the slope of the hill.

In addition, the change in potential, $\Delta V = -E\Delta s$, is negative when E and Δs are in the same direction (both positive or both negative). In other words:

> The electric potential decreases in the direction of the electric field.

For example, in cases where the electric field is *constant*, as between the plates of a parallel-plate capacitor (Section 19-5), the electric potential decreases *linearly* in the direction of the field. These observations are illustrated in **FIGURE 20-3**.

EXAMPLE 20-2   PLATES AT DIFFERENT POTENTIALS

A uniform electric field is established by connecting the plates of a parallel-plate capacitor to a 12-V battery. **(a)** If the plates are separated by 0.63 cm, what is the magnitude of the electric field in the capacitor? **(b)** A charge of $+4.6 \times 10^{-6}$ C moves from the positive plate to the negative plate. Find the change in electric potential energy for the system. (In electrical systems we shall assume that gravity can be ignored, unless specifically instructed otherwise.)

PICTURE THE PROBLEM

Our sketch shows the parallel-plate capacitor connected to a 12-V battery. The battery guarantees that the potential difference between the plates is 12 V, with the positive plate at the higher potential. The separation of the plates is $d = 0.63$ cm $= 0.0063$ m, and the charge that moves from the positive to the negative plate is $q = +4.6 \times 10^{-6}$ C.

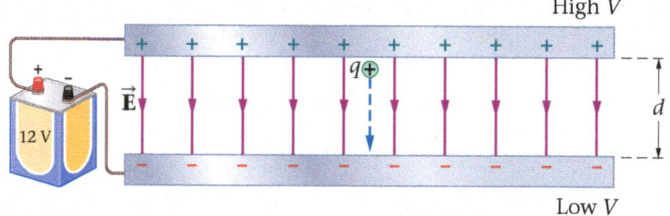

High V

$\vec{E}$ q ⊕ d

Low V

REASONING AND STRATEGY

a. The electric field can be calculated using $E = -\Delta V/\Delta s$. Thus, if one moves in the direction of the electric field from the positive plate to the negative plate ($\Delta s = 0.63$ cm), the electric potential decreases by 12 V; that is, $\Delta V = -12$ V.

b. The change in electric potential energy is $\Delta U = q\,\Delta V$.

Known   Voltage difference, $\Delta V = -12$ V; plate separation, $\Delta s = 0.0063$ m; charge, $q = +4.6 \times 10^{-6}$ C.
Unknown   **(a)** Magnitude of electric field, $E = ?$ **(b)** Change in electric potential energy, $\Delta U = ?$

SOLUTION

Part (a)

1. Substitute $\Delta s = 0.0063$ m and $\Delta V = -12$ V in $E = -\Delta V/\Delta s$:

$$E = -\frac{\Delta V}{\Delta s} = -\frac{(-12\,\text{V})}{0.0063\,\text{m}} = 1900\,\text{V/m}$$

Part (b)

2. Evaluate $\Delta U = q\Delta V$:

$$\Delta U = q\Delta V$$
$$= (4.6 \times 10^{-6}\,\text{C})(-12\,\text{V}) = -5.5 \times 10^{-5}\,\text{J} = -55\,\mu\text{J}$$

CONTINUED

INSIGHT
The electric potential energy of the system decreases as the positive charge moves in the direction of the electric field, just as the gravitational potential energy of a system decreases when an object moves downward (like a falling ball). In the next section, we show how this decrease in electric potential energy shows up as an increase in the charge's kinetic energy, just as the kinetic energy of a falling ball increases.

PRACTICE PROBLEM
Find the separation of the plates that results in an electric field of 2300 V/m.   [**Answer:** 0.52 cm. Notice that *decreasing* the separation *increases* the field.]

Some related homework problems: Problem 6, Problem 9

A similar situation is considered in the following Quick Example.

QUICK EXAMPLE 20-3   FIND THE ELECTRIC FIELD AND POTENTIAL DIFFERENCE

The electric potential at point B in the parallel-plate capacitor shown in the sketch is less than the electric potential at point A by 4.63 V. The separation between points A and B is 0.670 cm, and the separation between the plates is 2.01 cm. Find **(a)** the electric field within the capacitor and **(b)** the potential difference between the plates.

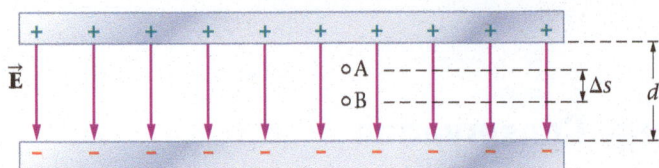

REASONING AND SOLUTION
The electric field is given by $E = -\Delta V/\Delta s$. We apply this relationship directly in part (a) for points A and B, noting that a displacement from A to B is in the direction of the electric field (and hence $\Delta x = 0.670$ cm), and that the corresponding change in electric potential is $\Delta V = -4.63$ V. For part (b) we solve $E = -\Delta V/\Delta s$ for the change in electric potential between the plates, and use the electric field E found in part (a). This is valid because E is uniform within a parallel-plate capacitor.

Part (a)
1. Calculate the electric field, $E = -\Delta V/\Delta s$, using the information for points A and B; namely, $\Delta V = V_B - V_A = -4.63$ V and $\Delta s = 0.670$ cm:

$$E = -\frac{\Delta V}{\Delta s} = -\frac{(-4.63\ \text{V})}{0.00670\ \text{m}} = 691\ \text{V/m}$$

Part (b)
2. Calculate the change in electric potential between the plates, $\Delta V = V_{(-)} - V_{(+)}$, using $\Delta V = -E\,\Delta s$, with $E = 691$ V/m and $\Delta s = 2.01$ cm:

$$\Delta V = -E\,\Delta s$$
$$= -(691\ \text{V})(0.0201\ \text{m}) = -13.9\ \text{V}$$

Notice that displacements in the direction of the electric field E correspond to positive values for Δs.

CONCEPTUAL EXAMPLE 20-4   CONSTANT ELECTRIC POTENTIAL

In a certain region of space the electric potential V is known to be constant. Is the electric field in this region positive, negative, or zero?

REASONING AND DISCUSSION
The electric field is related to the *rate of change* of the electric potential with position, not to the value of the potential. Because the rate of change of a constant potential is zero, so too is the electric field.

In particular, if one moves a distance Δs from one point to another in this region, the change in potential is zero; $\Delta V = 0$. Hence, the electric field vanishes; $E = -\Delta V/\Delta s = 0$.

ANSWER
The electric field is zero.

In our calculations of the electric potential difference $\Delta V = -E\Delta s$ to this point, we have always taken Δs to be a displacement in the direction of the electric field, which means that Δs is positive. More generally, Δs can be a displacement in any direction, as long as we replace E with the component of $\vec{E}$ in the direction of Δs. Thus, for example, the change in electric potential as we move in the x direction is $\Delta V = -E_x \Delta x$, and the change in electric potential as we move in the y direction is $\Delta V = -E_y \Delta y$. We shall point out the utility of these expressions in Section 20-4.

Enhance Your Understanding   (Answers given at the end of the chapter)

1. The electric potential in system A changes uniformly by 100 V over a distance of 1 m; in system B the electric potential changes uniformly by 1 V over a distance of 1 cm. Is the magnitude of the electric field in system A greater than, less than, or equal to the magnitude of the electric field in system B? Explain.

Section Review

- The change in electric potential, V, is defined to be the change in electric potential energy, U, per charge, $\Delta V = \Delta U/q_0$.

- The electric field E is equal to minus the rate of change of electric potential with distance, $E = -(\Delta V/\Delta s.)$

20-2 Energy Conservation

When a ball is dropped in a gravitational field, its gravitational potential energy decreases as it falls. At the same time, its kinetic energy increases. If nonconservative forces such as air resistance can be ignored, we know that the decrease in gravitational potential energy is equal to the increase in kinetic energy—in other words, the total energy of the ball is conserved.

Because the electric force is also conservative, the same considerations apply to a charged object in an electric field. Therefore, ignoring other forces, we can say that the total energy of an electric charge must be conserved. As a result, the sum of its kinetic and electric potential energies must be the same at any two points, say A and B:

$$K_A + U_A = K_B + U_B$$

Expressing the kinetic energy as $\frac{1}{2}mv^2$, we can write energy conservation as

$$\tfrac{1}{2}mv_A{}^2 + U_A = \tfrac{1}{2}mv_B{}^2 + U_B$$

This expression applies to any conservative force. In the case of a uniform gravitational field the potential energy is $U = mgy$; for an ideal spring it is $U = \frac{1}{2}kx^2$; in the previous section we showed that the electric potential energy is

$$U = qV$$

Therefore, energy conservation for an electrical system can be written as follows:

$$\tfrac{1}{2}mv_A{}^2 + qV_A = \tfrac{1}{2}mv_B{}^2 + qV_B$$

For example, suppose a particle of mass $m = 1.75 \times 10^{-5}$ kg and charge $q = 5.20 \times 10^{-5}$ C is released from rest at a point A. As the particle moves to another point, B, the electric potential decreases by 60.0 V; that is, $V_A - V_B = 60.0$ V. The particle's speed at point B can be found using energy conservation. To do so, we first solve for the kinetic energy at point B:

$$\tfrac{1}{2}mv_B{}^2 = \tfrac{1}{2}mv_A{}^2 + q(V_A - V_B)$$

Next we set $v_A = 0$, because the particle starts at rest, and solve for v_B:

$$\tfrac{1}{2}mv_B{}^2 = q(V_A - V_B)$$

$$v_B = \sqrt{\frac{2q(V_A - V_B)}{m}} = \sqrt{\frac{2(5.20 \times 10^{-5} \text{ C})(60.0 \text{ V})}{1.75 \times 10^{-5} \text{ kg}}} = 18.9 \text{ m/s} \qquad \text{20-6}$$

PHYSICS IN CONTEXT
Looking Back

Energy conservation (Chapter 8) is used here, this time with the electric potential energy.

PROBLEM-SOLVING NOTE

Energy Conservation and Electric Potential Energy

Energy conservation applies to charges in an electric field. Therefore, to relate the speed or kinetic energy of a charge to its location simply requires setting the initial energy equal to the final energy. Note that the potential energy for a charge q is $U = qV$.

Thus, the decrease in electric potential energy appears as an increase in kinetic energy—and a corresponding increase in speed.

QUICK EXAMPLE 20-5 FIND THE SPEED WITH A RUNNING START

Suppose the particle described in the preceding discussion has an initial speed of 8.00 m/s at point A. What is its speed at point B?

REASONING AND SOLUTION
The method of solution is the same as in the preceding discussion; the only difference is that $v_A = 8.00$ m/s.

1. Write the equation for energy conservation:

$$\tfrac{1}{2}mv_A{}^2 + qV_A = \tfrac{1}{2}mv_B{}^2 + qV_B$$

2. Solve for the kinetic energy at point B:

$$\tfrac{1}{2}mv_B{}^2 = \tfrac{1}{2}mv_A{}^2 + q(V_A - V_B)$$

3. Solve for the speed at point B, v_B:

$$v_B = \sqrt{v_A{}^2 + \frac{2q(V_A - V_B)}{m}}$$

4. Substitute numerical values:

$$v_B = \sqrt{(8.00\ \text{m/s})^2 + \frac{2(5.20 \times 10^{-5}\ \text{C})(60.0\ \text{V})}{1.75 \times 10^{-5}\ \text{kg}}}$$

$$= 20.5\ \text{m/s}$$

The final speed is not 8.00 m/s greater than 18.9 m/s; in fact, it is only 1.6 m/s greater. As we have seen in previous chapters, this is due to the fact that the kinetic energy depends on v^2 rather than on v.

> **Big Idea** ❷ Energy conservation can be extended to electrical systems by including the electric potential energy. Electric potential energy can be converted to other forms of energy, like kinetic energy or other forms of potential energy.

PHYSICS IN CONTEXT
Looking Ahead

The concept of electrical energy is generalized yet again in Chapter 23, where we show how an electric motor can convert electrical energy to mechanical energy. We also show that the reverse process is possible, with a generator converting mechanical energy to electrical energy.

In the preceding discussions, a positive charge moves to a region where the electric potential is less, and its speed increases. As one might expect, the situation is just the opposite for a negative charge—when a negative charge moves to a region with a larger electric potential, its speed increases. In general:

> *Positive* charges accelerate in the direction of *decreasing* electric potential.

> *Negative* charges accelerate in the direction of *increasing* electric potential.

In both cases, the charge moves to a region of lower electric potential energy.

EXAMPLE 20-6 FROM PLATE TO PLATE

Suppose the charge in Example 20-2 is released from rest at the positive plate and that it reaches the negative plate with a speed of 3.5 m/s. What are (a) the mass of the charge and (b) its final kinetic energy?

PICTURE THE PROBLEM
The physical situation is the same as in Example 20-2. In this case, however, we know that the charge starts at rest at the positive plate, $v_A = 0$, and hits the negative plate with a speed of $v_B = 3.5$ m/s. Notice that the electric potential of the positive plate is 12 V greater than that of the negative plate. Therefore $V_A - V_B = 12$ V.

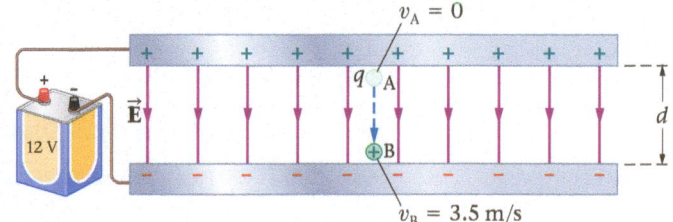

REASONING AND STRATEGY
a. The energy of the charge is conserved as it moves from one plate to the other. Setting the initial energy equal to the final energy gives us an equation in which there is only one unknown, the mass of the charge.

b. The final kinetic energy is simply $\tfrac{1}{2}mv_B{}^2$.

Known Charge, $q = 4.6 \times 10^{-6}$ C; initial speed, $v_A = 0$; final speed, $v_B = 3.5$ m/s; change in electric potential, $V_A - V_B = 12$ V.

Unknown (a) Mass of charge, $m = ?$ (b) Final kinetic energy of charge, $K = ?$

CONTINUED

SOLUTION

Part (a)

1. Apply energy conservation to this system:

$$\tfrac{1}{2}mv_A^2 + qV_A = \tfrac{1}{2}mv_B^2 + qV_B$$

2. Solve for the mass, m:

$$m = \frac{2q(V_A - V_B)}{v_B^2 - v_A^2}$$

3. Substitute numerical values:

$$m = \frac{2(4.6 \times 10^{-6}\,C)(12\,V)}{(3.5\,m/s)^2 - 0} = 9.0 \times 10^{-6}\,kg$$

Part (b)

4. Calculate the final kinetic energy:

$$K_B = \tfrac{1}{2}mv_B^2$$
$$= \tfrac{1}{2}(9.0 \times 10^{-6}\,kg)(3.5\,m/s)^2 = 5.5 \times 10^{-5}\,J$$

INSIGHT

We see that the final kinetic energy is precisely equal to the decrease in electric potential energy calculated in Example 20-2, as expected by energy conservation.

PRACTICE PROBLEM

If the mass of the charge had been 6.5×10^{-5} kg, what would be its **(a)** final speed and **(b)** final kinetic energy? [**Answer: (a)** $v_B = 1.3$ m/s, **(b)** $K_B = 5.5 \times 10^{-5}$ J. The final kinetic energy is the same, regardless of the mass, because energy is conserved.]

Some related homework problems: Problem 17, Problem 18

CONCEPTUAL EXAMPLE 20-7 FINAL SPEED
PREDICT/EXPLAIN

An electron, with a charge of -1.60×10^{-19} C, accelerates from rest through a potential difference V. A proton, with a charge of $+1.60 \times 10^{-19}$ C, accelerates from rest through a potential difference $-V$. **(a)** Is the final speed of the electron greater than, less than, or the same as the final speed of the proton? **(b)** Which of the following is the *best explanation* for your prediction?

  I. The electron has the greater speed because it has the same kinetic energy as the proton but less mass.

 II. The electron has less speed because it has less mass than the proton.

III. Both particles have the same speed because they accelerate through a potential difference of the same magnitude.

REASONING AND DISCUSSION

The electron and proton have charges of equal magnitude, and therefore they have equal changes in electric potential energy. As a result, their final kinetic energies are equal. The electron has less mass than the proton, however, and hence its speed is greater.

ANSWER

(a) The electron is moving faster than the proton. **(b)** The best explanation is I.

Enhance Your Understanding   (Answers given at the end of the chapter)

2. Particle A accelerates from rest through a potential difference of 1 V and acquires a speed of 1 m/s. Particle B accelerates from rest through a potential difference of 10 V and acquires a speed of 2 m/s. If the particles have the same mass, is the magnitude of the charge of particle A greater than, less than, or equal to the magnitude of the charge of particle B? Explain.

Section Review

- For an electrical system, conservation of energy, $\tfrac{1}{2}mv_A^2 + U_A = \tfrac{1}{2}mv_B^2 + U_B$, can be applied with the electric potential energy, $U = qV$.

20-3 The Electric Potential of Point Charges

Consider a point charge, $+q$, that is fixed at the origin of a coordinate system, as in **FIGURE 20-4**. Suppose, in addition, that a positive test charge, $+q_0$, is held at rest at point A, a distance r_A from the origin. At this location the test charge experiences a repulsive force with a magnitude given by Coulomb's law, $F = k|q_0||q|/r_A^2$.

If the test charge is now released, the repulsive force between it and the charge $+q$ will cause it to accelerate away from the origin. When it reaches an arbitrary point B, its kinetic energy will have increased by the same amount that its electric potential energy has decreased. Thus, we conclude that the electric potential energy is greater at point A than at point B. In fact, the methods of integral calculus can be used to show that the difference in electric potential energy between points A and B is

$$U_A - U_B = \frac{kq_0 q}{r_A} - \frac{kq_0 q}{r_B}$$

Notice that the electric potential energy for point charges depends inversely on their separation, the same as the distance dependence of the gravitational potential energy for point masses (Equation 12–9).

The corresponding change in electric potential is found by dividing the electric potential energy by the test charge, q_0:

$$V_A - V_B = \frac{1}{q_0}(U_A - U_B) = \frac{kq}{r_A} - \frac{kq}{r_B}$$

If the test charge is moved infinitely far away from the origin, so that $r_B \to \infty$, the term kq/r_B vanishes, and the difference in electric potential becomes

$$V_A - V_B = \frac{kq}{r_A}$$

Because the potential can be set to zero at any convenient location, we choose to set V_B equal to zero; in other words:

> We choose the electric potential to be zero infinitely far from a given charge.

With this choice, the potential of a point charge is $V_A = kq/r_A$. Dropping the subscript, we can write the electric potential at an arbitrary distance r as follows:

Electric Potential for a Point Charge

$$V = \frac{kq}{r}$$ (20-7)

SI unit: volt, V

Recall that this expression for V actually represents a *change* in potential; in particular, V is the change in potential from a distance of infinity to a distance r. The corresponding difference in electric potential energy for the test charge q_0 is simply $U = q_0 V$; that is:

Electric Potential Energy for Point Charges q and q_0 Separated by a Distance r

$$U = q_0 V = \frac{kq_0 q}{r}$$ (20-8)

SI unit: joule, J

Notice that *the electric potential energy of two charges separated by an infinite distance is zero.*

One final point in regard to $V = kq/r$ and $U = kq_0 q/r$ is that r is a *distance* and hence is always a positive quantity. For example, suppose a charge q is at the origin of a coordinate system. It follows that the potential at the point $x = 1$ m is the same as the potential at $x = -1$ m, because in both cases the distance is $r = 1$ m.

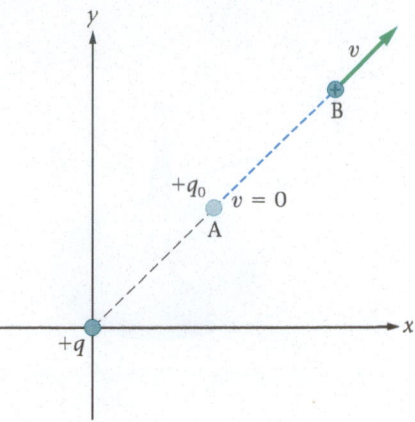

▲ **FIGURE 20-4 Energy conservation in an electrical system** A test charge, $+q_0$, is released from rest at point A. When it reaches point B, its kinetic energy will have increased by the same amount that its electric potential energy has decreased.

PHYSICS IN CONTEXT
Looking Back

The electric potential of a point charge is developed in this section. The results are similar to those obtained for the gravitational potential energy of a point mass in Chapter 12.

Big Idea 3 The electric potential of a point charge is proportional to the charge, and inversely proportional to the distance from the charge.

PHYSICS IN CONTEXT
Looking Ahead

The electric potential energy for a point charge is applied to Bohr's model of the hydrogen atom in Chapter 31. With this energy we can determine the colors of light that hydrogen atoms emit.

Big Idea 4 The electric potential due to a system of charges is the algebraic sum of the electric potentials produced by each charge individually.

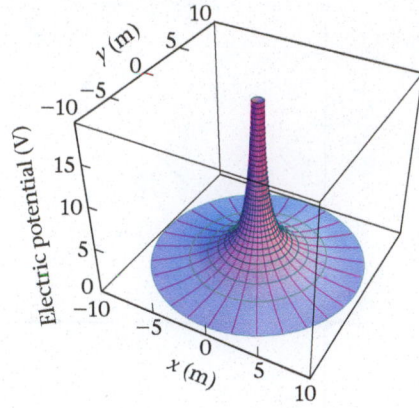

(a) Electric potential near a positive charge

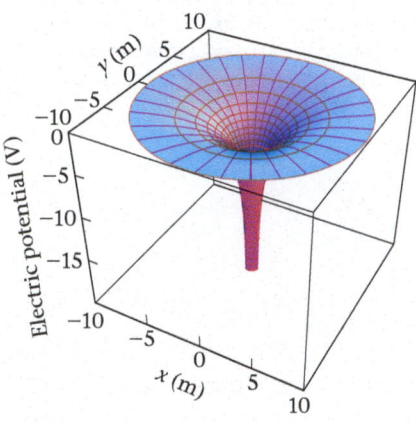

(b) Electric potential near a negative charge

▲ **FIGURE 20-5 The electric potential of a point charge** Electric potential near **(a)** a positive and **(b)** a negative charge at the origin. In the case of the positive charge, the electric potential forms a "potential hill" near the charge. Near the negative charge we observe a "potential well."

PROBLEM-SOLVING NOTE

Superposition for the Electric Potential

Recall that to find the total electric potential due to a system of charges, you need only sum the potentials due to each charge separately—being careful to take into account the appropriate sign. Note that there are no components to deal with, as there are when we superpose electric fields.

EXERCISE 20-8   POINT CHARGE

(a) Find the electric potential produced by a point charge of 7.22×10^{-7} C at a distance of 2.75 m. **(b)** Find the point charge that produces an electric potential of -5120 V at a distance of 3.55 m from the charge.

REASONING AND SOLUTION

For a point charge, the electric potential, electric charge, and distance are related by $V = kq/r$. This relationship applies to both parts of this Example.

(a) Substitute $q = 7.22 \times 10^{-7}$ C and $r = 2.75$ m in $V = kq/r$ to find

$$V = \frac{kq}{r} = \frac{(8.99 \times 10^9 \,\text{N} \cdot \text{m}^2/\text{C}^2)(7.22 \times 10^{-7} \,\text{C})}{2.75 \,\text{m}} = 2360 \,\text{V}$$

Thus, the potential at $r = 2.75$ m due to this point charge is 2360 V greater than the potential at infinity.

(b) This time, we rearrange $V = kq/r$ to solve for the charge q, and substitute $V = -5120$ V and $r = 3.55$ m to find

$$q = \frac{rV}{k} = \frac{(3.55 \,\text{m})(-5120 \,\text{V})}{(8.99 \times 10^9 \,\text{N} \cdot \text{m}^2/\text{C}^2)} = -2.02 \times 10^{-6} \,\text{C} = -2.02 \,\mu\text{C}$$

Dependence on the Sign of q   Notice that V depends on the sign of the charge in question. For example, in **FIGURE 20-5 (A)** we show the electric potential for a positive charge at the origin. The potential increases to positive infinity near the origin and decreases to zero far away, forming a "potential hill." On the other hand, the potential for a negative charge, shown in **FIGURE 20-5 (B)**, approaches negative infinity near the origin, forming a "potential well."

Thus, if the charge at the origin is positive, a positive test charge will move away from the origin, as if sliding "downhill" on the electric potential surface. Similarly, if the charge at the origin is negative, a positive test charge will move toward the origin, which again means that it slides downhill, this time into a potential well. Negative test charges, in contrast, always tend to slide "uphill" on electric potential curves, like bubbles rising in water.

Superposition of the Electric Potential

Like many physical quantities, the electric potential obeys a simple superposition principle. In particular:

> The total electric potential due to two or more charges is equal to the algebraic sum of the potentials due to each charge separately.

By *algebraic sum* we mean that the potential of a given charge may be positive or negative, and hence the *algebraic sign* of each potential must be taken into account when calculating the total potential. In particular, positive and negative contributions may cancel to give zero potential at a given location.

EXAMPLE 20-9   TWO POINT CHARGES

A charge $q = 4.11 \times 10^{-9}$ C is placed at the origin, $x = 0$, and a second charge equal to $-2q$ is placed on the x axis at the location $x = 1.00$ m. **(a)** Find the electric potential at $x = 0.500$ m. **(b)** The electric potential vanishes at some point between the charges—that is, for a value of x between 0 and 1.00 m. Find this value of x.

PICTURE THE PROBLEM

Our sketch shows the two charges placed on the x axis as described in the problem statement. Clearly, the electric potential is large and positive near the origin, and large and negative near $x = 1.00$ m. Thus, it follows that V must vanish at some point between $x = 0$ and $x = 1.00$ m.

CONTINUED

A plot of V as a function of x is shown as an aid in visualizing the calculations given in this Example.

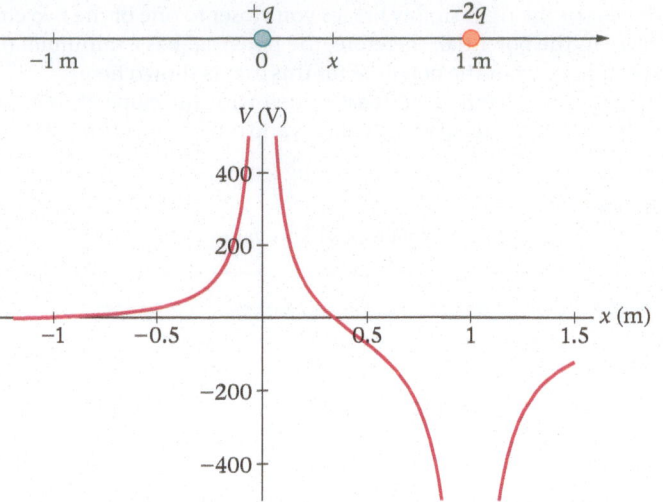

REASONING AND STRATEGY

a. As indicated in the sketch, an arbitrary point between $x = 0$ and $x = 1.00$ m is a distance x from the charge $+q$ and a distance $(1.00 \text{ m} - x)$ from the charge $-2q$. Thus, by superposition, the total electric potential at a point x is $V = kq/x + k(-2q)/(1.00 \text{ m} - x)$.

b. Setting $V = 0$ allows us to solve for the unknown, x.

Known   Charge, $q = 4.11 \times 10^{-9}$ C; location of first charge, $x = 0$; location of second charge, $x = 1.00$ m.

Unknown   (a) Electric potential at $x = 0.500$ m, $V = ?$ (b) Location where $V = 0$, $x = ?$

SOLUTION

Part (a)

1. Use superposition to write an expression for V at an arbitrary point x between $x = 0$ and $x = 1.00$ m:

$$V = \frac{kq}{x} + \frac{k(-2q)}{1.00 \text{ m} - x}$$

2. Substitute numerical values into the expression for V, including $x = 0.500$ m:

$$V = \frac{(8.99 \times 10^9 \, \text{N} \cdot \text{m}^2/\text{C}^2)(4.11 \times 10^{-9} \, \text{C})}{0.500 \text{ m}}$$

$$+ \frac{(8.99 \times 10^9 \, \text{N} \cdot \text{m}^2/\text{C}^2)(-2)(4.11 \times 10^{-9} \, \text{C})}{1.00 \text{ m} - 0.500 \text{ m}}$$

$$= -73.9 \, \text{N} \cdot \text{m/C} = -73.9 \, \text{V}$$

Part (b)

3. Set the expression for V in Step 1 equal to zero, and simplify by canceling the common factor, kq:

$$V = \frac{kq}{x} + \frac{k(-2q)}{1.00 \text{ m} - x} = 0$$

$$\frac{1}{x} = \frac{2}{1.00 \text{ m} - x}$$

4. Solve for x:

$$1.00 \text{ m} - x = 2x$$

$$x = \tfrac{1}{3}(1.00 \text{ m}) = 0.333 \text{ m}$$

INSIGHT

Suppose a small positive test charge is released from rest at $x = 0.500$ m. In which direction will it move? From the point of view of the Coulomb force, we know it will move to the right—repelled by the positive charge at the origin and attracted to the negative charge at $x = 1.00$ m. We come to the same conclusion when considering the electric potential, because we know that a positive test charge will "slide downhill" on the potential curve.

PRACTICE PROBLEM

The electric potential in this system also vanishes at a point on the negative x axis. Find this point.

[**Answer:** $V = 0$ at $x = -1.00$ m. Notice that the point $x = -1.00$ m is 1.00 m from the charge $+q$, and 2.00 m from the charge $-2q$. Hence, at $x = -1.00$ m we have $V = kq/(1.00 \text{ m}) + k(-2q)/(2.00 \text{ m}) = 0$.]

Some related homework problems: Problem 32, Problem 33

CONCEPTUAL EXAMPLE 20-10   A PEAK OR A VALLEY?

Two point charges, each equal to $+q$, are placed on the x axis at $x = -1$ m and $x = +1$ m. Does the electric potential near the origin look like a peak or a valley?

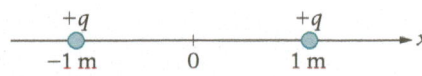

REASONING AND DISCUSSION

We know that the potential is large and positive near each of the charges. As you move away from the charges, the potential tends to decrease. In particular, at very large positive or negative values of x the potential approaches zero.

CONTINUED

Between $x = -1$ m and $x = +1$ m the potential has its lowest value when you are as far away from the two charges as possible. This occurs at the origin. Moving slightly to the left or the right simply brings you closer to one of the two charges, resulting in an increase in the potential; therefore, the potential has a minimum (bottom of a valley) at the origin. A plot of the potential for this case is shown here.

Finally, the electric field is associated with the slope of this curve. Clearly, the field has a large magnitude near the charges and is zero at the origin.

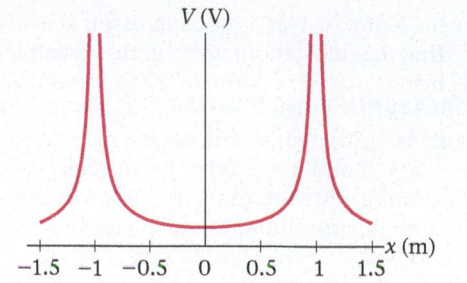

ANSWER

Near the origin, the potential looks like a valley.

Superposition applies equally well to the electric potential energy. The following Example illustrates superposition for both the electric potential and the electric potential energy.

EXAMPLE 20-11 FLY AWAY

Two charges, $+q$ and $+2q$, are held in place on the x axis at the locations $x = -d$ and $x = +d$, respectively. A third charge, $+3q$, is released from rest on the y axis at $y = d$. **(a)** Find the electric potential due to the first two charges at the initial location of the third charge. **(b)** Find the initial electric potential energy of the third charge. **(c)** What is the kinetic energy of the third charge when it has moved infinitely far away from the other two charges?

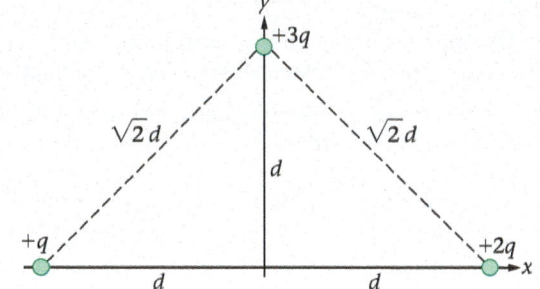

PICTURE THE PROBLEM

As indicated in the sketch, the third charge, $+3q$, is separated from the other two charges by the initial distance, $\sqrt{2}d$. When the third charge is released, the repulsive forces due to $+q$ and $+2q$ will cause it to move away to an infinite distance.

REASONING AND STRATEGY

a. The electric potential at the initial position of the third charge is the sum of the potentials due to $+q$ and $+2q$, each at a distance of $\sqrt{2}d$.

b. The initial electric potential energy of the third charge, U_i, is simply its charge, $+3q$, times the potential, V, found in part (a).

c. We can find the final kinetic energy using energy conservation: $U_i + K_i = U_f + K_f$. Because $K_i = 0$ (third charge starts at rest) and $U_f = 0$ (third charge infinitely far away), we find that $K_f = U_i$.

Known Charge $+q$ at $x = -d$; charge $+2q$ at $x = +d$; charge $+3q$ at $y = d$.
Unknown **(a)** Initial electric potential at third charge, $V_i = ?$ **(b)** Initial electric potential energy of system, $U_i = ?$
(c) Final kinetic energy of third charge, $K_f = ?$

SOLUTION

Part (a)

1. Calculate the net electric potential at the initial position of the third charge:

$$V_i = \frac{k(+q)}{\sqrt{2}d} + \frac{k(+2q)}{\sqrt{2}d} = \frac{3kq}{\sqrt{2}d}$$

Part (b)

2. Multiply V by $(+3q)$ to find the initial electric potential energy, U_i, of the third charge:

$$U_i = (+3q)V_i = (+3q)\frac{3kq}{\sqrt{2}d} = \frac{9kq^2}{\sqrt{2}d}$$

Part (c)

3. Use energy conservation to find the final (infinite separation) kinetic energy:

$$U_i + K_i = U_f + K_f$$
$$\frac{9kq^2}{\sqrt{2}d} + 0 = 0 + K_f$$
$$K_f = \frac{9kq^2}{\sqrt{2}d}$$

CONTINUED

INSIGHT
If the third charge had started closer to the other two charges, it would have been higher up on the "potential hill," and hence its kinetic energy at infinity would have been greater, as shown in the following Practice Problem.

PRACTICE PROBLEM
Suppose the third charge is released from rest just above the origin. What is its final kinetic energy in this case?
[**Answer:** $K_f = U_i = 9kq^2/d$. As expected, this is greater than $9kq^2/\sqrt{2}d$.]

Some related homework problems: Problem 35, Problem 40

The electric potential energy for a pair of charges, q_1 and q_2, separated by a distance r is $U = kq_1q_2/r$. In systems that contain more than two charges, the total electric potential energy is the sum of terms like $U = kq_1q_2/r$ for each pair of charges in the system.

Enhance Your Understanding   (Answers given at the end of the chapter)

3. The following systems consist of a point charge at a distance r from the measurement location. Rank the systems in order of increasing electric potential, from most negative to most positive. Indicate ties where appropriate.

System	A	B	C	D
q (C)	1	−1	10	−10
r (m)	1	2	3	10

Section Review

- The electric potential a distance r from a point charge q is $V = kq/r$.

- The total electric potential due to a system of point charges is equal to the algebraic sum of the potentials due to each charge separately.

- The electric potential energy for the point charges q_0 and q separated by a distance r is $U = kq_0q/r$.

20-4 Equipotential Surfaces and the Electric Field

A contour map is a useful tool for serious hikers and backpackers. The first thing you notice when looking at a map such as the one shown in Figure 8-21 is a series of closed curves—the contours—each denoting a different altitude. When the contours are closely spaced, the altitude changes rapidly, indicating steep terrain. Conversely, widely spaced contours indicate a fairly flat surface.

A similar device can help us visualize the electric potential due to one or more electric charges. Consider, for example, a single positive charge located at the origin. As we saw in the previous section, the electric potential due to this charge approaches zero far from the charge and rises to form an infinitely high "potential hill" near the charge. A three-dimensional representation of the potential is plotted in Figure 20-5 (a). The same potential is shown as a contour map in **FIGURE 20-6**. In this case, the contours, rather than representing altitude, indicate the value of the potential. Because the value of the potential at any point on a given contour is equal to the value at any other point on the same contour, we refer to the contours as **equipotential surfaces**, or simply, **equipotentials**.

Equipotentials and the Electric Field   An equipotential plot also contains important information about the magnitude and direction of the electric field. For example, in Figure 20-6 we know that the electric field is more intense near the charge, where

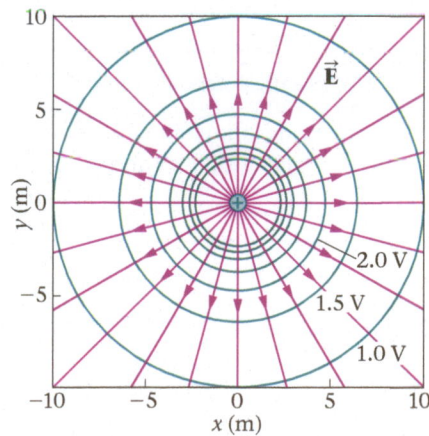

▲ **FIGURE 20-6 Equipotentials for a point charge** Equipotential surfaces for a positive point charge located at the origin. Near the origin, where the equipotentials are closely spaced, the potential varies rapidly with distance and the electric field is large. This is a top view of Figure 20-5 (a).

Big Idea 5 Equipotential surfaces are defined as surfaces on which the electric potential has a constant value. Electric field lines and equipotential surfaces always intersect at right angles.

**PHYSICS
IN CONTEXT
Looking Back**

Notice that we have returned to the idea of equipotential curves—curves along which the potential energy is constant—in direct analogy to the contour maps discussed at the end of Chapter 8.

the equipotentials are closely spaced, than it is far from the charge, where the equipotentials are widely spaced. This simply illustrates that the electric field, $E = -\Delta V/\Delta s$, depends on the rate of change of the potential with position—the greater the change in potential, ΔV, over a given distance, Δs, the larger the magnitude of E.

The direction of the electric field is given by the minus sign in $E = -\Delta V/\Delta s$. For example, if the electric potential decreases over a distance Δs, it follows that ΔV is negative; $\Delta V < 0$. Thus, $-\Delta V$ is positive, and hence $E > 0$. To summarize:

> The electric field points in the direction of decreasing electric potential.

This is also illustrated in Figure 20-6, where we see that the electric field points away from the charge $+q$, in the direction of decreasing electric potential.

Not only does the field point in the direction of decreasing potential, it is, in fact, *perpendicular* to the equipotential surfaces:

> The electric field is always perpendicular to the equipotential surfaces.

To see why this is the case, we note that zero work is done when a charge is moved perpendicular to an electric field. That is, the work $W = Fd\cos\theta$ is zero when the angle θ is 90°. If zero work is done, it follows from $\Delta V = -W/q_0$ (Equation 20-2) that there is no change in potential. Therefore, the potential is constant (equipotential) in a direction perpendicular to the electric field.

CONCEPTUAL EXAMPLE 20-12   EQUIPOTENTIAL SURFACES

Is it possible for equipotential surfaces to intersect?

REASONING AND DISCUSSION

To answer this question, it is useful to consider the analogous case of contours on a contour map (Figure 8-21). As we know, each contour corresponds to a different altitude. Because each point on the map has only a single value of altitude, it follows that it is impossible for contours to intersect.

Precisely the same reasoning applies to the electric potential, and hence equipotential surfaces never intersect.

ANSWER

No. Equipotential surfaces cannot intersect.

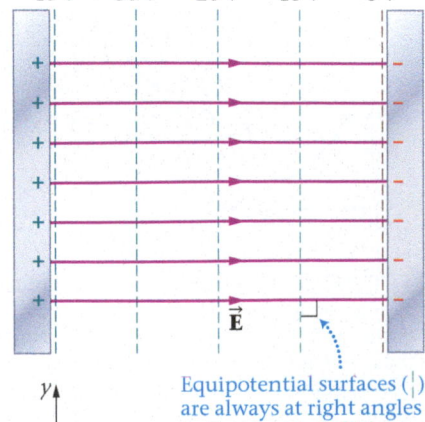

The electric potential decreases ($40\,V \rightarrow 30\,V \rightarrow 20\,V \rightarrow \cdots$) in the direction of the electric field (→).

40 V   30 V   20 V   10 V   0 V

$\vec{E}$

Equipotential surfaces (|) are always at right angles to the electric field (→).

Plotting Equipotentials and the Electric Field   The graphical relationship between equipotentials and electric fields can be illustrated with a few examples. We begin with the simple case of a uniform electric field, as between the plates of a parallel-plate capacitor. In **FIGURE 20-7**, we plot both the electric field lines, which point in the positive x direction, and the corresponding equipotential surfaces. As expected, the field lines are perpendicular to the equipotentials, and $\vec{E}$ points in the direction of decreasing V.

To see this last result more clearly, note that $\vec{E} = E\hat{x}$ in Figure 20-7, from which it follows that $E_x = E > 0$ and $E_y = 0$. Referring to Section 20-1, we see that if we move in the positive x direction, the change in electric potential is negative, $\Delta V = -E_x \Delta x = -E\Delta x < 0$, as expected. On the other hand, if we move in the y direction, the change in potential is zero, $\Delta V = -E_y \Delta y = 0$. This is why the equipotentials, which are always perpendicular to $\vec{E}$, are parallel to the y axis in Figure 20-7.

FIGURE 20-8 (a) shows a similar plot for the case of two positive charges of equal magnitude. Notice that the electric field lines always cross the equipotentials at right angles. In addition, the electric field is more intense where the equipotential surfaces are closely spaced. In the region midway between the two charges, where the electric field is essentially zero, the potential is practically constant.

◀ **FIGURE 20-7 Equipotential surfaces for a uniform electric field** The electric field is always perpendicular to equipotential surfaces, and points in the direction of decreasing electric potential. In this case the electric field is (i) uniform and (ii) horizontal. As a result, (i) the electric potential decreases at a uniform rate, and (ii) the equipotential surfaces are vertical.

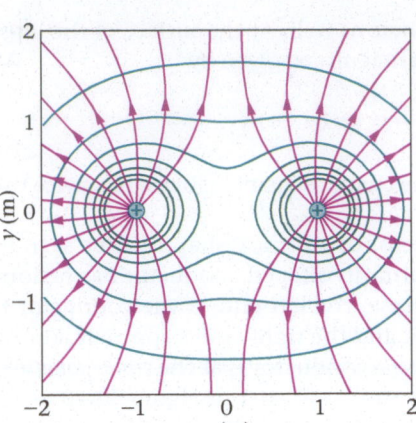

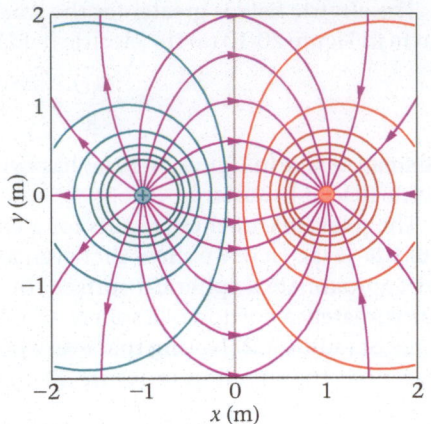

(a) Equipotentials for two positive charges   (b) Equipotentials for a dipole

◀ **FIGURE 20-8 Equipotential surfaces for two point charges** **(a)** In the case of two equal positive charges, the electric field between them is weak because the field produced by one charge effectively cancels the field produced by the other. As a result, the electric potential is practically constant between the charges. **(b)** For equal charges of opposite sign (a dipole), the electric field is strong between the charges, and the potential changes rapidly.

Finally, in **FIGURE 20-8 (b)** we show the equipotentials for two charges of opposite sign, one $+q$ (at $x = -1$ m) and the other $-q$ (at $x = +1$ m), forming an electric dipole. In this case, the amber-color equipotentials denote negative values of V. Teal-color equipotentials correspond to positive V, and the tan-color equipotential has the value $V = 0$. The electric field is nonzero between the charges, even though the potential is zero there—recall that $E = -\Delta V/\Delta s$ is related to the *rate of change of V*, not to its value. The relatively large number of equipotential surfaces between the charges shows that V is indeed *changing rapidly* in that region.

Ideal Conductors

A charge placed on an ideal conductor is free to move. As a result, a charge can be moved from one location on a conductor to another with no work being done. Because the work is zero, the change in potential is also zero; therefore, every point on or within an ideal conductor is at the same potential.

> Ideal conductors are equipotential surfaces; every point on or within such a conductor is at the same potential.

For example, when a charge Q is placed on a conductor, as in **FIGURE 20-9**, it distributes itself over the surface in such a way that the potential is the same everywhere. If the conductor has the shape of a sphere, the charge spreads uniformly over its surface, as in Figure 20-9 (a). If, however, the conductor has a sharp end and a blunt end, as in Figure 20-9 (b), the charge is more concentrated near the sharp end, which results in a large electric field. We pointed out this effect earlier, in Section 19-6.

To see why this should be the case, consider a conducting sphere of radius R with a charge Q distributed uniformly over its surface, as in **FIGURE 20-10 (a)**. The charge density on this sphere is $\sigma = Q/4\pi R^2$, and the potential at its surface is the same as for a point charge Q at the center of the sphere; that is, $V = kQ/R$. Writing this in terms of the surface charge density, we have

$$V = \frac{kQ}{R} = \frac{k\sigma(4\pi R^2)}{R} = 4\pi k\sigma R$$

Now, consider a sphere of radius $R/2$. For this sphere to have the *same potential* as the large sphere, it must have twice the charge density, 2σ. In particular, letting σ go to 2σ and R go to $R/2$ in the expression $V = 4\pi k\sigma R$ yields the same result as for the large sphere:

$$V = 4\pi k(2\sigma)\left(\frac{R}{2}\right) = 4\pi k\sigma R$$

Clearly, then, the smaller the radius of a sphere with potential V, the greater its charge density.

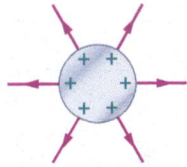

(a) Uniform sphere, uniform charge distribution

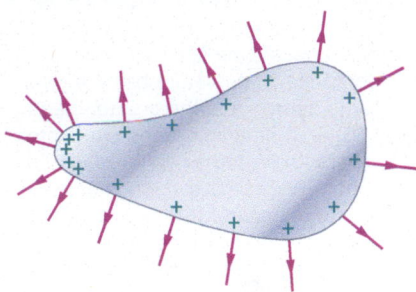

(b) Charge is concentrated at pointed end

▲ **FIGURE 20-9 Electric charges on the surface of ideal conductors** **(a)** On a spherical conductor, the charge is distributed uniformly over the surface. **(b)** On a conductor of arbitrary shape, the charge is more concentrated, and the electric field is more intense, where the conductor is more sharply curved. Note that in all cases the electric field is perpendicular to the surface of the conductor.

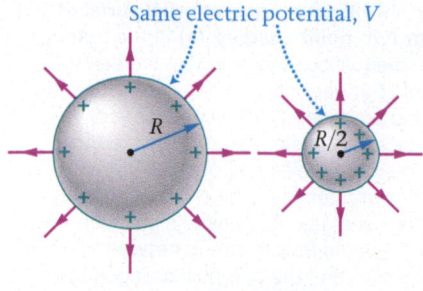

(a)

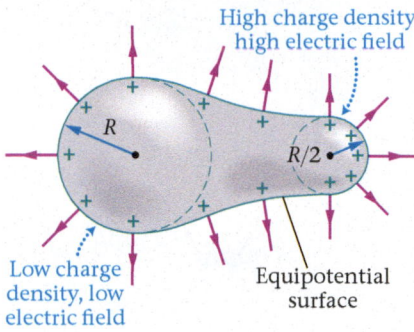

Same electric potential, V

High charge density, high electric field

R

$R/2$

Low charge density, low electric field

Equipotential surface

(b)

▲ **FIGURE 20-10 Charge concentration near points (a)** If two spheres of different radii have the same electric potential at their surfaces, the sphere with the smaller radius of curvature has the greater charge density and the greater electric field. **(b)** An arbitrarily shaped conductor can be approximated by spheres with the same potential at the surface and varying radii of curvature. It follows that the more sharply curved end of a conductor has a greater charge density and a more intense field.

The electric field is greater for the small sphere as well. At the surface of the large sphere in Figure 20-10 (a) the electric field has a magnitude given by

$$E = \frac{kQ}{R^2} = \frac{k\sigma(4\pi R^2)}{R^2} = 4\pi k\sigma$$

The small sphere in Figure 20-10 (a) has twice the charge density; hence it has twice the electric field at its surface.

The relevance of these results for a conductor of arbitrary shape can be seen in **FIGURE 20-10 (b)**. Here we show that even an arbitrarily shaped conductor has regions that approximate a spherical surface. In this case, the left end of the conductor is approximately a portion of a sphere of radius R, and the right end is approximately a sphere of radius $R/2$. Noting that every point on a conductor is at the *same potential*, we see that the situation in Figure 20-10 (b) is similar to that in Figure 20-10 (a). It follows that the sharper end of the conductor has the greater charge density and the greater electric field. If a conductor has a sharp point, as in a lightning rod, the corresponding radius of curvature is very small, which can result in an enormous electric field. We consider this possibility in more detail in the next section.

Finally, we note that because the surface of an ideal conductor is an equipotential, the electric field must meet the surface at right angles. This is also shown in Figures 20-9 and 20-10.

Electric Potential and the Human Body

The human body is a relatively good conductor of electricity, but it is not an ideal conductor. If it were, the entire body would be one large equipotential surface. Instead, muscle activity, the beating of the heart, and nerve impulses in the brain all lead to slight differences in electric potential from one point on the skin to another.

BIO In the case of the heart, the powerful waves of electrical activity as the heart muscles contract result in potential differences that are typically in the range of 1 mV. These potential differences can be detected and displayed with an instrument known as an **electrocardiograph**, abbreviated ECG or EKG. (The K derives from the original Dutch name for the device.) A typical EKG signal is displayed in **FIGURE 20-11 (a)**.

To understand the various features of an EKG signal, we note that a heartbeat begins when the heart's natural "pacemaker," the sinoatrial (SA) node in the right atrium, triggers a wave of muscular contraction across both atria that pumps blood into the ventricles. This activity gives rise to the pulse known as the *P wave* in Figure 20-11 (a). Following this contraction, the atrioventricular (AV) node, located between the ventricles, initiates a more powerful wave of contraction in the ventricles, sending oxygen-poor blood to the lungs and oxygenated blood to the rest of the body. These events are reflected in the series of pulses known as the *QRS complex*. Other features in the EKG signal can be interpreted as well. For example, the *T wave* is associated with repolarization of the ventricles in preparation for the next contraction cycle. Even

▶ **FIGURE 20-11 The electrocardiograph (a)** A typical electrocardiograph tracing, with the major features labeled. The EKG records the electrical activity that accompanies the rhythmic contraction and relaxation of heart muscle tissue. The main pumping action of the heart is associated with the *QRS* complex: Contraction of the ventricles begins just after the *R* peak. **(b)** The simplest arrangement of electrodes, or "leads," for an EKG. More precise information can be obtained by the use of additional leads.

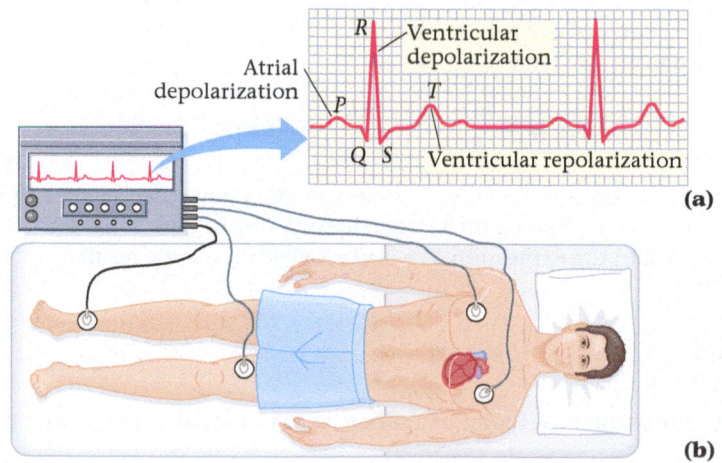

Atrial depolarization

R Ventricular depolarization

P T

Q S Ventricular repolarization

(a)

(b)

small irregularities in the magnitude, shape, sequence, or timing of these features can provide an experienced physician with essential clues for the diagnosis of many cardiac abnormalities and pathologies. For example, an inverted *T* wave often indicates inadequate blood supply to the heart muscle (ischemia), possibly caused by a heart attack (myocardial infarction).

To record these signals, electrodes are attached to the body in a variety of locations. In the simplest case, illustrated in **FIGURE 20-11 (b)**, three electrodes are connected in an arrangement known as *Einthoven's triangle*, after the Dutch pioneer of these techniques. Two of these three electrodes are attached to the shoulders, the third to the left groin. For convenience, the two shoulder electrodes are often connected to the wrists, and the groin electrode is connected to the left ankle—the signal at these locations is practically the same as for the locations in Figure 20-11 (b) because of the high conductivity of the human limbs. By convention, the electrode on the right leg is used as the ground. In current applications of the EKG, it is typical to use 12 electrodes to gain more detailed information about the heart's activity (see **FIGURE 20-12**).

BIO Electrical activity in the brain can be detected and displayed with an **electroencephalograph** or EEG. In a typical application of this technique, a regular array of 8 to 16 electrodes is placed around the head. The potential differences in this case—in the range of 0.1 mV—are much smaller than those produced by the heart, and much more complex to interpret. One of the key characteristics of an EEG signal, such as the one shown in **FIGURE 20-13**, is the frequency of the waves. For example, waves at 0.5 to 3.5 Hz, referred to as *D waves*, are common during sleep. A relaxed brain produces *a waves*, with frequencies in the range of 8 to 13 Hz, and an alert brain generates *b waves*, with frequencies greater than 13 Hz. Finally, *q waves*—with frequencies of 5 to 8 Hz—are common in newborns but indicate severe stress in adults.

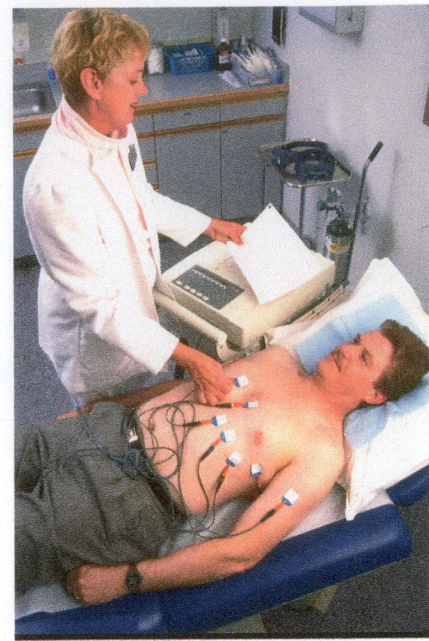

▲ **FIGURE 20-12** An electrocardiograph can be made with as few as three electrodes and a ground. However, more detailed information about the heart's electrical activity can be obtained by using additional "leads" at precise anatomical locations. Modern EKGs often use a 12-lead array.

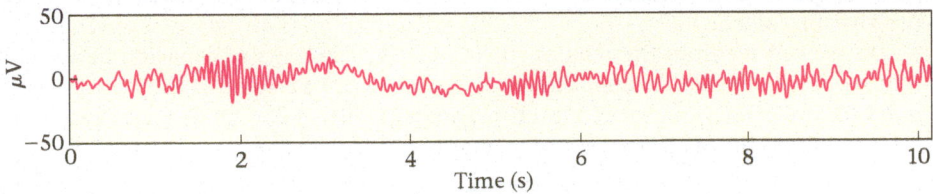

▲ **FIGURE 20-13** **The electroencephalograph** A typical EEG signal.

Enhance Your Understanding   (Answers given at the end of the chapter)

4. **FIGURE 20-14** shows a series of equipotential surfaces for a certain system that are equally spaced in terms of the value of their electric potential. Is the magnitude of the electric field near point A greater than, less than, or equal to the magnitude of the electric field near point B? Explain.

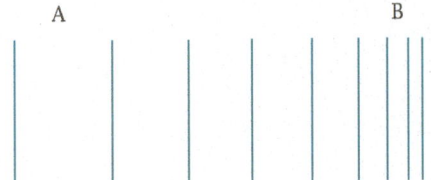

▲ **FIGURE 20-14** Enhance Your Understanding 4.

Section Review

- Equipotential surfaces are defined as surfaces with a constant value of the electric potential. When equipotential surfaces are closely spaced, the corresponding electric field is large.

- The electric field always intersects equipotential surfaces at right angles.

20-5 Capacitors and Dielectrics

A **capacitor** gets its name from the fact that it has a *capacity* to store both electric charge and energy. Capacitors are a common and important element in modern electronic devices. They can provide large bursts of energy to a circuit, or protect delicate circuitry from excess charge originating elsewhere.

In general, a capacitor is nothing more than two conductors, referred to as *plates*, separated by a finite distance. When the plates of a capacitor are connected to the terminals of a battery, they become charged—one plate acquires a charge $+Q$, the other an equal and opposite charge, $-Q$. The greater the charge Q for a given voltage V, the greater the **capacitance**, C, of the capacitor.

To be specific, suppose a certain battery produces a potential difference of V volts between its terminals. When this battery is connected to a capacitor, a charge of magnitude Q appears on each plate. If a different battery, with a voltage of $2V$, is connected to the same capacitor, the charge on the plates doubles in magnitude to $2Q$. Thus, the charge Q is proportional to the applied voltage V. We define the constant of proportionality to be the capacitance C:

$$Q = CV$$

Solving for the capacitance, we have:

Definition of Capacitance, C

$$C = \frac{Q}{V} \qquad \text{20-9}$$

SI unit: coulomb/volt = farad, F

In this expression, Q is the magnitude of the charge on either plate, and V is the magnitude of the voltage difference between the plates. By definition, the capacitance is a positive quantity.

As we can see from the relationship $C = Q/V$, the units of capacitance are coulombs per volt. In the SI system this combination of units is referred to as the farad (F), in honor of the English physicist Michael Faraday (1791–1867), a pioneering researcher into the properties of electricity and magnetism. In particular,

$$1\,\text{F} = 1\,\text{C/V} \qquad \text{20-10}$$

Just as the coulomb is a rather large unit of charge, so too is the farad a rather large unit of capacitance. More typical values for capacitance are in the picofarad ($1\,\text{pF} = 10^{-12}\,\text{F}$) to microfarad ($1\,\mu\text{F} = 10^{-6}\,\text{F}$) range.

EXERCISE 20-13   CHARGE ON A CAPACITOR

(a) A capacitor of $0.55\,\mu\text{F}$ is charged to a voltage of 12 V. What is the magnitude of the charge on each plate of the capacitor? **(b)** What is the capacitance of a capacitor that has a charge of 1.9×10^{-6} C when connected to a 9.0-V battery?

REASONING AND SOLUTION

We use the defining equation for capacitance, $C = Q/V$, to relate the charge, capacitance, and voltage of a given capacitor.

(a) Rearranging $C = Q/V$ to solve for Q, we find

$$Q = CV = (0.55 \times 10^{-6}\,\text{F})(12\,\text{V}) = 6.6 \times 10^{-6}\,\text{C} = 6.6\,\mu\text{C}$$

(b) Applying $C = Q/V$ directly yields

$$C = \frac{Q}{V} = \frac{1.9 \times 10^{-6}\,\text{C}}{9.0\,\text{V}} = 2.1 \times 10^{-7}\,\text{F} = 0.21\,\mu\text{F}$$

A bucket of water provides a useful analogy when thinking about capacitors. In this analogy we make the following identifications: (i) The cross-sectional area of the bucket is the capacitance C; (ii) the amount of water in the bucket is the charge Q; and

(iii) the depth of the water is the voltage difference V between the plates. Therefore, just as a wide bucket can hold more water than a narrow bucket—when filled to the same level—a large capacitor can hold more charge than a small capacitor when they both have the same potential difference. Charging a capacitor, then, is like pouring water into a bucket: If the capacitance is large, a large amount of charge can be placed on the plates with little potential difference between them.

Parallel-Plate Capacitor

A particularly simple capacitor is the parallel-plate capacitor, first introduced in Section 19-5 . In this device, two parallel plates of area A are separated by a distance d, as indicated in **FIGURE 20-15**. We would like to determine the capacitance of such a capacitor. As we shall see, the capacitance is related in a simple way to the parameters A and d.

To begin, we notice that when a charge $+Q$ is placed on one plate, and a charge $-Q$ is placed on the other, a uniform electric field is produced between the plates. The magnitude of this field is given by Gauss's law, and was determined in Example 19-18. There we found that $E = \sigma/\epsilon_0$. Observing that the charge per area is $\sigma = Q/A$, we have

$$E = \frac{Q}{\epsilon_0 A} \qquad \text{20-11}$$

The corresponding potential difference between the plates is $\Delta V = -E\Delta s = -(Q/\epsilon_0 A)d$. The magnitude of this potential difference is simply $V = (Q/\epsilon_0 A)d$.

Now that we have determined the potential difference for a parallel-plate capacitor with a charge Q on its plates, we can apply $C = Q/V$ to find an expression for the capacitance:

$$C = \frac{Q}{V} = \frac{Q}{(Q/\epsilon_0 A)d}$$

Canceling the charge Q and rearranging slightly gives the final result:

Capacitance of a Parallel-Plate Capacitor

$$C = \frac{\epsilon_0 A}{d} \qquad \text{20-12}$$

As mentioned previously, the capacitance depends in a straightforward way on the area of the plates, A, and their separation, d. We use this expression for capacitance in the next Example.

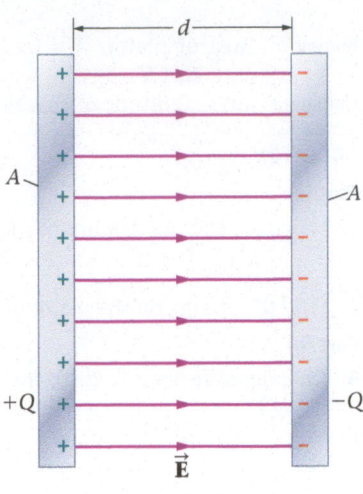

▲ **FIGURE 20-15 A parallel-plate capacitor** A parallel-plate capacitor, with plates of area A, separation d, and charges of magnitude Q. The capacitance of such a capacitor is $c = \epsilon_0 A/d$.

PROBLEM-SOLVING NOTE

Using Magnitudes with Capacitance

Note that the capacitance C is always positive. Therefore, when applying a relationship like $C = Q/V$, we always use magnitudes for the charge and the electric potential.

EXAMPLE 20-14 ALL CHARGED UP

A parallel-plate capacitor is constructed with plates of area 0.0280 m^2 and separation 0.550 mm. **(a)** Find the magnitude of the charge on each plate of this capacitor when the potential difference between the plates is 20.1 V. **(b)** What is the magnitude of the electric field between the plates of the capacitor?

PICTURE THE PROBLEM

The capacitor is shown in our sketch. Notice that the plates of the capacitor have an area $A = 0.0280 \text{ m}^2$, a separation $d = 0.550 \text{ mm}$, and a potential difference $V = 20.1 \text{ V}$. The charge on each plate has a magnitude Q. The electric field is uniform, and points from the positive plate to the negative plate.

REASONING AND STRATEGY

a. The charge on the plates is given by $Q = CV$. We know the potential difference, V, but we must determine the capacitance, C. We can do this using the relationship $C = \epsilon_0 A/d$, along with the given information for A and d.

b. We can find the magnitude of the electric field with Equation 20-11; that is, $E = Q/\epsilon_0 A$. As in part (a), the charge on the plates is $Q = CV$. In addition, using the symbolic expression $C = \epsilon_0 A/d$ will allow us to simplify the final result for E.

CONTINUED

Known Area of plates, $A = 0.0280 \text{ m}^2$; separation of plates, $d = 0.550 \text{ mm}$; potential difference between plates, $V = 20.1 \text{ V}$.

Unknown (a) Magnitude of charge on each plate, $Q = ?$ (b) Magnitude of electric field, $E = ?$

SOLUTION

Part (a)

1. Calculate the capacitance of the capacitor:

$$C = \frac{\epsilon_0 A}{d} = \frac{(8.85 \times 10^{-12} \text{ C}^2/\text{N} \cdot \text{m}^2)(0.0280 \text{ m}^2)}{0.550 \times 10^{-3} \text{ m}}$$
$$= 4.51 \times 10^{-10} \text{ F}$$

2. Find the charge on the plates of the capacitor:

$$Q = CV = (4.51 \times 10^{-10} \text{ F})(20.1 \text{ V}) = 9.06 \times 10^{-9} \text{ C}$$

Part (b)

3. Calculate the magnitude of the electric field using $E = Q/\epsilon_0 A$. Use a symbolic expression for $Q = CV$ to simplify the final expression:

$$E = \frac{Q}{\epsilon_0 A} = \frac{CV}{\epsilon_0 A} = \frac{(\epsilon_0 A/d)V}{\epsilon_0 A} = \frac{V}{d}$$

4. Substitute numerical values:

$$E = \frac{V}{d} = \frac{20.1 \text{ V}}{0.550 \times 10^{-3} \text{ m}} = 36,500 \text{ V/m}$$

INSIGHT

The total amount of charge in the capacitor is zero, $+Q + (-Q) = 0$, but the fact that there is a charge separation—with one plate positive and the other negative—means that the capacitor stores energy, as we shall see in the next section.

If a capacitor is connected to a battery, the battery maintains a constant voltage between the plates of the capacitor. It follows from $Q = CV$ that any change in the capacitance—by changing the plate area, A, or separation, d, for example—results in a different amount of charge on the plates.

PRACTICE PROBLEM

What separation d is necessary to give an increased charge of magnitude 2.00×10^{-8} C on each plate of this capacitor? Assume all other quantities in the system remain the same. [**Answer:** $d = 0.249 \text{ mm}$. Notice that a *smaller* separation results in a *greater* amount of stored charge.]

Some related homework problems: Problem 50, Problem 51

We see from Equation 20-12 that the capacitance of a parallel-plate capacitor increases with the area A of its plates—basically, the greater the area, the more room there is in the capacitor to hold charge. Again, this is like pouring water into a bucket with a large cross-sectional area. On the other hand, the capacitance decreases with increasing separation d. The reason for this dependence is that with a constant electric field between the plates (as we saw in Section 19-5), the potential difference between the plates is proportional to their separation: $V = Ed$. Thus, the capacitance, which is inversely proportional to the potential difference ($C = Q/V$), is also inversely proportional to the plate separation—the greater the separation, the greater the potential difference required to store a given amount of charge. All capacitors, regardless of their design, share these general features. Thus, to produce a large capacitance, one would like to have plates of large area close together. This is often accomplished by inserting a thin piece of paper between two large sheets of metal foil. The foil is then rolled up tightly to form a compact, large-capacity capacitor. Examples of common capacitors are shown in **FIGURE 20-16**.

▶ **FIGURE 20-16 Visualizing Concepts**
Capacitors Capacitors come in a variety of physical forms (left), with many different types of dielectric (insulating material) occupying the space between their plates. Capacitors are also rated for the maximum voltage that can be applied to them before the dielectric breaks down, allowing a spark to jump across the gap between the plates. A variable air capacitor (right), often used in early radios, has interleaved plates that can be rotated to vary their area of overlap. This changes the effective area of the capacitor, and hence its capacitance.

CONCEPTUAL EXAMPLE 20-15   CHARGE ON THE PLATES

A parallel-plate capacitor is connected to a battery that maintains a constant potential difference V between the plates. If the plates are pulled away from each other, increasing their separation, does the magnitude of the charge on the plates increase, decrease, or remain the same?

REASONING AND DISCUSSION

The capacitance of a parallel-plate capacitor is $C = \epsilon_0 A/d$, and hence increasing the separation, d, decreases the capacitance. With a smaller value of C, and a constant value for V, the charge $Q = CV$ will decrease. The same general behavior can be expected with any capacitor.

ANSWER

The charge on the plates decreases.

Sometimes a capacitor is first connected to a battery to be charged and is then disconnected. In this case, the charge on the plates is "trapped"—it has no place to go—and hence Q must remain constant. If the capacitance is changed now, say by changing the plate separation, the result is a different potential difference, $V = Q/C$, between the plates.

Dielectrics

One way to increase the capacitance of a capacitor is to insert an insulating material, referred to as a **dielectric**, between its plates. With a dielectric in place, a capacitor can store more charge or energy in the same volume. Thus, dielectrics play an important role in miniaturizing electronic devices.

To see how this works, consider the parallel-plate capacitor shown in **FIGURE 20-17 (a)**. Initially the plates are separated by a vacuum and connected to a battery, giving the plates the charges $+Q$ and $-Q$. The battery is now removed, and the charge on the plates remains constant. The electric field between the plates is uniform and has a magnitude E_0. If the distance between the plates is d, the corresponding potential difference is $V_0 = E_0 d$, and the capacitance is

$$C_0 = \frac{Q}{V_0}$$

Now, insert a dielectric slab, as illustrated in **FIGURES 20-17 (b)** and **(c)**. If the molecules in the dielectric have a permanent dipole moment, they will align with the field, as shown in Figure 20-17 (b). Even without a permanent dipole moment, however, the

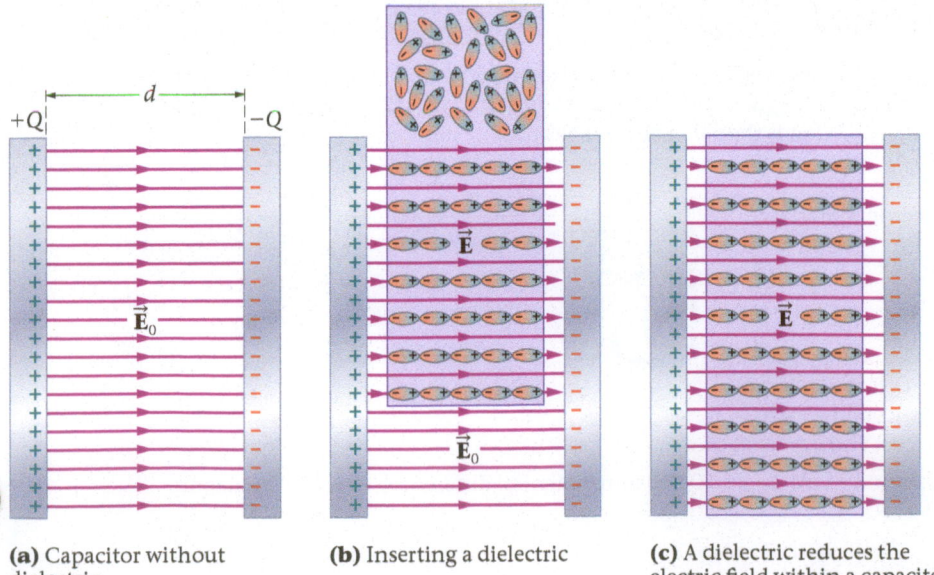

(a) Capacitor without dielectric

(b) Inserting a dielectric

(c) A dielectric reduces the electric field within a capacitor

◄ **FIGURE 20-17 The effect of a dielectric on the electric field of a capacitor** When a dielectric is placed in the electric field between the plates of a capacitor, the molecules of the dielectric tend to become oriented with their positive ends pointing toward the negatively charged plate and their negative ends pointing toward the positively charged plate. The result is a buildup of positive charge on one surface of the dielectric and of negative charge on the other. Since field lines start on positive charges and end on negative charges, we see that the number of field lines within the dielectric is reduced. Thus, within the dielectric the applied electric field $\vec{E}_0$ is partially canceled. Because the strength of the electric field is less, the voltage between the plates is less as well. Since V is smaller while Q remains the same, the capacitance, $C = Q/V$, is increased by the dielectric.

TABLE 20-1 Dielectric Constants

Substance	Dielectric constant, κ
Water	80.4
Neoprene rubber	6.7
Pyrex glass	5.6
Mica	5.4
Paper	3.7
Mylar	3.1
Teflon	2.1
Air	1.00059
Vacuum	1

molecules will become polarized by the field (see Section 19-1). This polarization leads to the same type of alignment, although the effect is weaker. The result of this alignment is a positive charge on the surface of the slab near the negative plate and a negative charge on the surface near the positive plate.

Recalling that electric field lines terminate on negative charges and start on positive charges, we can see from Figure 20-17 (c) that fewer field lines exist within the dielectric. Consequently, there is a reduced field, E, in a dielectric, which we characterize with a dimensionless **dielectric constant**, κ, as follows:

$$E = \frac{E_0}{\kappa} \qquad \text{20-13}$$

In the case of a vacuum, $\kappa = 1$ and $E = E_0$, as before. For an insulating material, however, the value of κ is greater than one. For example, paper has a dielectric constant of roughly 4, which means that the electric field within paper is about one-quarter what it would be in a vacuum. Typical values of κ are listed in Table 20-1.

Thus, a dielectric reduces the field between the plates of a capacitor by a factor of κ. This, in turn, decreases the *potential difference* between the plates by the same factor:

$$V = Ed = \left(\frac{E_0}{\kappa}\right)d = \frac{E_0 d}{\kappa} = \frac{V_0}{\kappa}$$

Finally, because the potential difference is smaller, the capacitance must be larger:

$$C = \frac{Q}{V} = \frac{Q}{(V_0/\kappa)} = \kappa\frac{Q}{V_0} = \kappa C_0 \qquad \text{20-14}$$

PROBLEM-SOLVING NOTE

The Effects of a Dielectric

Dielectrics reduce the electric field in a capacitor, which results in a reduced potential difference between the plates. As a result, a dielectric always increases the capacitance.

In effect, the dielectric partially shields one plate from the other, making it easier to build up a charge on the plates. If the space between the plates of a capacitor is filled with paper, for example, the capacitance will be about four times larger than if the space had been a vacuum.

The relationship $C = \kappa C_0$ applies to any capacitor. For the special case of a parallel-plate capacitor filled with a dielectric, we have:

PHYSICS IN CONTEXT
Looking Ahead

Capacitors play a key role in direct-current (dc) circuits in Chapter 21, and in alternating-current (ac) circuits in Chapter 24.

Capacitance of a Parallel-Plate Capacitor Filled with a Dielectric

$$C = \frac{\kappa \epsilon_0 A}{d} \qquad \text{20-15}$$

We apply this equation in the next Example.

EXAMPLE 20-16 | EVEN MORE CHARGED UP

A parallel-plate capacitor is constructed with plates of area 0.0280 m^2 and separation 0.550 mm. The space between the plates is filled with a dielectric with dielectric constant κ. When the capacitor is connected to a 12.0-V battery, each of the plates has a charge of magnitude 3.62×10^{-8} C. What is the value of the dielectric constant, κ?

PICTURE THE PROBLEM
Our sketch shows the capacitor with a dielectric material inserted between the plates. In other respects, the capacitor is the same as the one considered in Example 20-14.

REASONING AND STRATEGY
We are given the potential difference V and the charge Q, and hence we can find the capacitance using $C = Q/V$. Next, we relate the capacitance to the physical characteristics of the capacitor with $C = \kappa \mathcal{E}_0 A/d$. Using the given values for A and d, we solve for κ.

Known Area of plates, $A = 0.0280$ m^2; separation of plates, $d = 0.550$ mm; potential difference, $V = 12.0$ V; magnitude of charge on plates, $Q = 3.62 \times 10^{-8}$ C.

Unknown Dielectric constant, $\kappa = ?$

CONTINUED

SOLUTION

1. Determine the value of the capacitance:

$$C = \frac{Q}{V} = \frac{(3.62 \times 10^{-8}\,\text{C})}{12.0\,\text{V}} = 3.02 \times 10^{-9}\,\text{F}$$

2. Solve $C = \kappa\epsilon_0 A/d$ for the dielectric constant, κ:

$$C = \frac{\kappa\epsilon_0 A}{d}$$

$$\kappa = \frac{Cd}{\epsilon_0 A}$$

3. Substitute numerical values to find κ:

$$\kappa = \frac{Cd}{\epsilon_0 A}$$

$$= \frac{(3.02 \times 10^{-9}\,\text{F})(0.550 \times 10^{-3}\,\text{m})}{(8.85 \times 10^{-12}\,\text{C}^2/\text{N}\cdot\text{m}^2)(0.0280\,\text{m}^2)} = 6.70$$

INSIGHT

Comparing our result with the dielectric constants given in Table 20-1, we see that the dielectric may be neoprene rubber.

PRACTICE PROBLEM — PREDICT/CALCULATE

(a) If a different dielectric with a smaller dielectric constant is inserted into the capacitor, which is still connected to a 12.0-V battery, does the charge on the plates increase, decrease, or remain the same? Explain. (b) Find the charge on the plates for $\kappa = 3.5$.   [**Answer: (a)** Decreasing κ decreases the capacitance, and hence the charge $Q = CV$ decreases as well (recall that V remains the same). **(b)** $Q = 1.89 \times 10^{-8}\,\text{C}$]

Some related homework problems: Problem 54, Problem 56

RWP* The fact that the capacitance of a capacitor depends on the separation of its plates finds a number of interesting applications. For example, if you have ever typed on a computer keyboard, you have probably been utilizing the phenomenon of capacitance without realizing it. Many computer keyboards are designed in such a way that each key is connected to the upper plate of a parallel-plate capacitor, as illustrated in **FIGURE 20-18**. When you depress a given key, the separation between the plates of that capacitor decreases, and the corresponding capacitance increases. The circuitry of the computer detects this change in capacitance, thereby determining which key you have pressed.

RWP Many of the ubiquitous touchscreens (**FIGURE 20-19**) on smartphones, tablets, and other devices also work by the measurement of capacitance. In a common arrangement, the bottom surface of a thin plate of glass is coated with thin strips of a transparent conductor such as indium tin oxide (ITO), with a dielectric material sandwiched in between. Because human skin is also a conductor, touching the glass means bringing some electric charge nearby, distorting the local electrostatic field and changing the capacitance at that spot. By forming a grid with rows and columns of ITO and continuously measuring the capacitance of each intersection in the grid, a touchscreen can determine where it has been touched, even if by multiple fingers simultaneously.

RWP Another, less well-known application of capacitance is the theremin (**FIGURE 20-20**), a musical instrument that you play without touching! Two antennas on the theremin are used to control the sound it makes; one antenna adjusts the volume, the other adjusts the pitch. When a person places a hand near one of the antennas, the effect is similar to that of a parallel-plate capacitor, with the hand playing the role of one plate and the antenna playing the role of the other plate. Changing the separation between hand and antenna changes the capacitance, which the theremin's circuitry then converts into a corresponding change of volume or pitch. Theremins have been used to provide "ethereal" music for a number of science fiction films, and some popular bands use theremins in their musical arrangements.

Dielectric Breakdown

If the electric field applied to a dielectric is large enough, it can literally tear the atoms apart, allowing the dielectric to conduct electricity. This condition is referred to as **dielectric breakdown**. The maximum field a dielectric can withstand before breakdown is called the **dielectric strength**. Typical values are given in Table 20-2.

---
*Real World Physics applications are denoted by the acronym RWP.

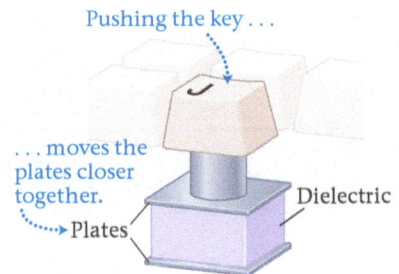

Pushing the key . . .

. . . moves the plates closer together.

Plates

Dielectric

▲ **FIGURE 20-18 Capacitance and the computer keyboard** The keys on many computer keyboards form part of a parallel-plate capacitor. Depressing the key changes the plate separation. The corresponding change in capacitance can be detected by the computer's circuitry.

▲ **FIGURE 20-19** Touchscreen technology measures the capacitance at the intersections of a grid of transparent conductors to determine the locations where the screen is touched.

▲ **FIGURE 20-20** A musician plays a theremin at an outdoor concert.

TABLE 20-2 Dielectric Strengths

Substance	Dielectric Strength (V / m)
Mica	100×10^6
Teflon	60×10^6
Paper	16×10^6
Pyrex glass	14×10^6
Neoprene rubber	12×10^6
Air	3.0×10^6

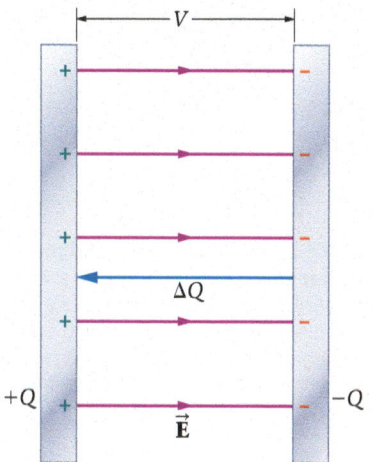

▲ **FIGURE 20-22 The energy required to charge a capacitor** A capacitor has a charge of magnitude Q on its plates and a potential difference V between the plates. Transferring a small charge increment, $+\Delta Q$, from the negative plate to the positive plate increases the electric potential energy of the capacitor by the amount $\Delta U = (\Delta Q)V$.

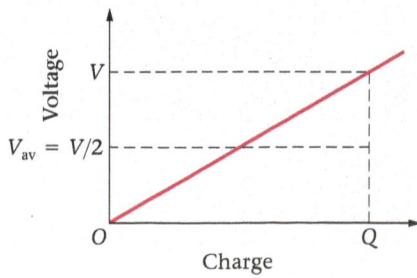

For example, if the electric field in air exceeds about 3,000,000 V/m, dielectric breakdown will occur, leading to a spark on a small scale or a bolt of lightning on a larger scale. Next time you walk across a carpet and get a shock when reaching for the doorknob, think about the fact that you have just produced an electric field of roughly *3 million volts per meter!* The sharp tip of a lightning rod, which has a high electric field in its vicinity, helps to initiate and guide lightning to the ground, or to dissipate charge harmlessly so that no lightning occurs at all. St. Elmo's Fire—the glow of light around the rigging of a ship in a storm—is another example of dielectric breakdown in air.

Enhance Your Understanding   (Answers given at the end of the chapter)

5. Two parallel-plate capacitors are identical, except that capacitor B has a solid piece of metal inserted between its plates, as indicated in **FIGURE 20-21**. Is the capacitance of capacitor A greater than, less than, or equal to the capacitance of capacitor B? Explain.

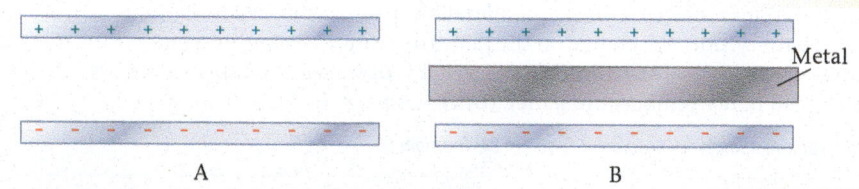

Metal

A   B

▲ **FIGURE 20-21** Enhance Your Understanding 5.

Section Review

- Capacitance is defined as the amount of charge stored in a capacitor for a given voltage, $C = Q/V$.

- The capacitance of a parallel-plate capacitor with a plate area A and a separation d is $C = \epsilon_0 A/d$.

- A dielectric reduces the voltage between the plates of a capacitor for a given amount of charge, and this increases the capacitance. A dielectric with dielectric constant κ increases the capacitance of a parallel-plate capacitor to $C = \kappa \epsilon_0 A/d$.

20-6 Electrical Energy Storage

Capacitors store more than just charge—they also store energy. To see how, consider a capacitor that has charges of magnitude Q on its plates, and a potential difference of V. Now, imagine transferring a small amount of charge, ΔQ, from one plate to the other, as in **FIGURE 20-22**. This charge must be moved across a potential difference of V, and hence the change in electric potential energy is $\Delta U = (\Delta Q)V$. It follows that the potential energy of the capacitor increases by $(\Delta Q)V$ when the magnitude of the charge on its plates is increased from Q to $Q + \Delta Q$. As more charge is transferred from one plate to the other, more electric potential energy is stored in the capacitor.

To find the total electric energy stored in a capacitor, we must take into account the fact that the potential difference between the plates increases as the charge on the plates increases. In fact, recalling that the potential difference is given by $V = Q/C$, we see that V increases linearly with the charge, as illustrated in **FIGURE 20-23**. In particular, if the final potential difference is V, the average potential during charging is $\frac{1}{2}V$. Therefore, the total energy U stored in a capacitor with charge Q and potential difference V can be written as follows:

$$U = QV_{av} = \tfrac{1}{2}QV \qquad \text{20-16}$$

◀ **FIGURE 20-23 The voltage of a capacitor being charged** The voltage V between the plates of a capacitor increases linearly with the charge Q on the plates, $V = Q/C$. Therefore, if a capacitor is charged to a final voltage of V, the average voltage during charging is $V_{av} = \frac{1}{2}V$.

▶ **FIGURE 20-24** An electronic flash unit like the one at left includes a capacitor that can store a large amount of charge. When the charge is released, the resulting flash can be as short as a millisecond or less, allowing photographers to "freeze" motion, as in the photo at right. Even faster strobe units can be used to photograph explosions, shock waves, or speeding bullets.

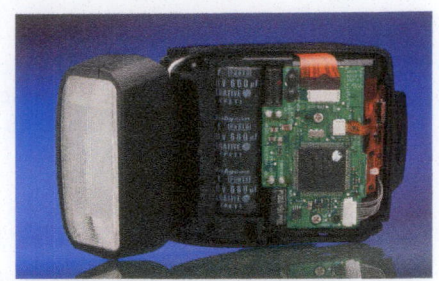

Equivalently, noting that $Q = CV$, the energy stored in a capacitor of capacitance C and voltage V is

$$U = \tfrac{1}{2}CV^2 \qquad\qquad \textbf{20-17}$$

Finally, using $V = Q/C$, we find that the energy stored in a capacitor of charge Q and capacitance C is

$$U = \frac{Q^2}{2C} \qquad\qquad \textbf{20-18}$$

All these expressions are equivalent; they simply give the energy in terms of different variables.

RWP The energy stored in a capacitor can be put to a number of practical uses. Any time you take a flash photograph, for example, you are triggering the rapid release of energy from a capacitor, as indicated in **FIGURE 20-24**. The flash unit typically contains a capacitor with a capacitance of 100 to 400 μF. When fully charged to a voltage of about 300 V, the capacitor contains roughly 15 J of energy. Activating the flash causes the stored energy, which took several seconds to accumulate, to be released in less than a millisecond. Because of the rapid release of energy, the power output of a flash unit is impressively large—about 10 to 20 kW. This is far in excess of the power provided by the battery that operates the unit. Similar considerations apply to the defibrillator used in the treatment of heart-attack victims, as we show in the next Example.

EXAMPLE 20-17 | THE DEFIBRILLATOR: DELIVERING A SHOCK TO THE SYSTEM

BIO When a person's heart undergoes ventricular fibrillation—a rapid, uncoordinated twitching of the heart muscles—it often takes a strong jolt of electrical energy to restore the heart's regular beating and save the person's life. The device that delivers this jolt of energy is known as a defibrillator (**FIGURE 20-25**), and it uses a capacitor to store the necessary energy. In a typical defibrillator, a 175-μF capacitor is charged until the potential difference between the plates is 2240 V. **(a)** What is the magnitude of the charge on each plate of the fully charged capacitor? **(b)** Find the energy stored in the charged-up defibrillator.

PICTURE THE PROBLEM
Our sketch shows a simplified representation of a capacitor. The values of the capacitance and the potential difference are indicated.

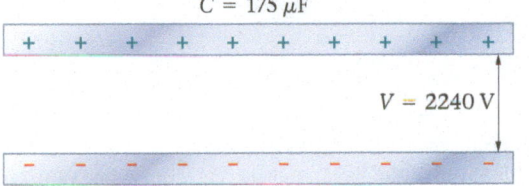

$C = 175\ \mu\text{F}$

$V = 2240\ \text{V}$

REASONING AND STRATEGY

a. We can find the charge stored on the capacitor plates using $Q = CV$.

b. The energy stored in the capacitor can be determined immediately using $U = \tfrac{1}{2}CV^2$. In addition, now that we know the charge on each plate of the capacitor, the energy can be found with the relationships $U = \tfrac{1}{2}QV$ and $U = Q^2/2C$, as an additional check on our result.

Known Capacitance, $C = 175 \times 10^{-6}$ F; potential difference, $V = 2240$ V.
Unknown (a) Magnitude of charge on plates, $Q = ?$ (b) Energy stored in capacitor, $U = ?$

SOLUTION

Part (a)

1. Use $Q = CV$ to find the charge on the plates:

$$Q = CV = (175 \times 10^{-6}\ \text{F})(2240\ \text{V}) = 0.392\ \text{C}$$

Part (b)

2. Find the stored energy using $U = \tfrac{1}{2}CV^2$:

$$U = \tfrac{1}{2}CV^2 = \tfrac{1}{2}(175 \times 10^{-6}\ \text{F})(2240\ \text{V})^2 = 439\ \text{J}$$

3. As a check, use $U = \tfrac{1}{2}QV$:

$$U = \tfrac{1}{2}QV = \tfrac{1}{2}(0.392\ \text{C})(2240\ \text{V}) = 439\ \text{J}$$

4. Finally, use the relationship $U = Q^2/2C$:

$$U = \frac{Q^2}{2C} = \frac{(0.392\ \text{C})^2}{2(175 \times 10^{-6}\ \text{F})} = 439\ \text{J}$$

CONTINUED

INSIGHT
Of the 439 J stored in the defibrillator's capacitor, typically about 200 J will actually pass through the person's body in a pulse lasting about 2 ms. The power delivered by the pulse is approximately $P = U/t = (200\,\text{J})/(0.002\,\text{s}) = 100\,\text{kW}$. This is significantly larger than the power delivered by the battery, which can take up to 30 s to fully charge the capacitor.

PRACTICE PROBLEM
Suppose the defibrillator is "fired" when the voltage is only half its maximum value of 2240 V. How much energy is stored in this case?   [**Answer:** $E = (439\,\text{J})/4 = 110\,\text{J}$]

Some related homework problems: Problem 66, Problem 71

Big Idea 6 Capacitors store both electric charge and electric potential energy. Work is done to charge a capacitor—that is, to produce a charge separation—and that work can be recovered later.

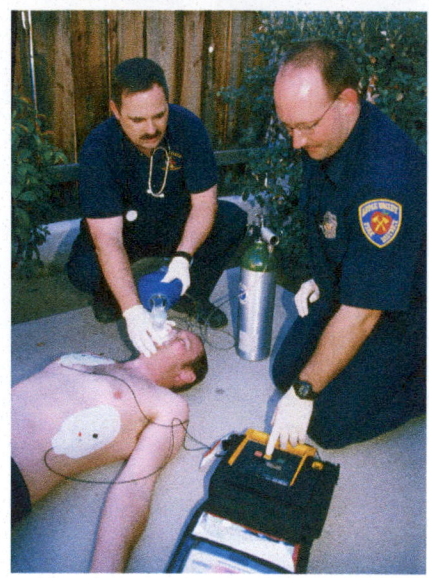

▲ **FIGURE 20-25** A jolt of electric current from a defibrillator can restore a normal heartbeat when the heart muscle has begun to twitch irregularly or has stopped beating altogether. A capacitor is used to store electricity, discharging it in a burst lasting only a couple of milliseconds.

Energy and the Electric Field   We've discussed many examples of energy stored in a capacitor, but where exactly is the energy located? The answer is that the energy can be thought of as stored in the electric field, E, between the plates. To be specific, consider the relationship

$$\text{energy} = \tfrac{1}{2}QV$$

In the case of a parallel-plate capacitor of area A and separation d, we know that $Q = \epsilon_0 EA$ (Equation 20-11) and $V = Ed$. Thus, the energy stored in the capacitor can be written as

$$U = \text{energy} = \tfrac{1}{2}(\epsilon_0 EA)(Ed) = \tfrac{1}{2}\epsilon_0 E^2(Ad)$$

We have grouped A and d together because the product Ad is simply the total volume between the plates. Therefore, the **energy density** (energy per volume) is given by the following:

$$u_E = \text{electric energy density} = \frac{\text{electric energy}}{\text{volume}} = \tfrac{1}{2}\epsilon_0 E^2 \qquad \text{20-19}$$

This result, though derived for a capacitor, is valid for any electric field, whether it occurs within a capacitor or anyplace else.

Enhance Your Understanding   (Answers given at the end of the chapter)

6. The following systems consist of a capacitor for which two of the three quantities Q, V, and C are known. Rank the systems in order of increasing energy stored in the capacitors. Indicate ties where appropriate.

System	A	B	C	D
Q (C)	—	1	0.3	—
V (V)	5	—	20	10
C (μF)	100	200	—	25

Section Review

- The energy stored in a capacitor can be written as $U = \tfrac{1}{2}QV = \tfrac{1}{2}CV^2 = Q^2/2C$. These expressions are equivalent; they simply express the energy in terms of different variables.

CHAPTER 20 REVIEW

CHAPTER SUMMARY

20-1 ELECTRIC POTENTIAL ENERGY AND THE ELECTRIC POTENTIAL
The electric force is conservative, just like the force of gravity. As a result, there is a potential energy U associated with the electric force.

Electric Potential Energy, U
The change in electric potential energy is defined by $\Delta U = -W$, where W is the work done by the electric field.

Electric Potential, V

The change in electric potential is defined to be $\Delta V = \Delta U/q_0$.

Relation Between the Electric Field and the Electric Potential

The electric field is related to the rate of change of the electric potential. In particular, if the electric potential changes by the amount ΔV with a displacement Δs, the electric field in the direction of the displacement is

$$E = -\frac{\Delta V}{\Delta s} \qquad \text{20-4}$$

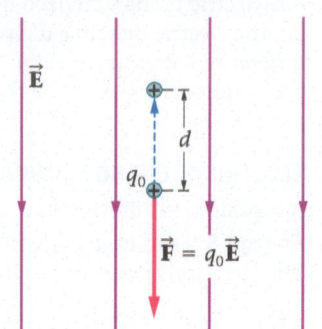

$$\Delta U = q_0 E d$$

20-2 ENERGY CONSERVATION

Energy Conservation

As usual, energy conservation can be expressed as follows:

$$\tfrac{1}{2}mv_A^2 + U_A = \tfrac{1}{2}mv_B^2 + U_B$$

In the case of the electric force, the potential energy is $U = q_0 V$.

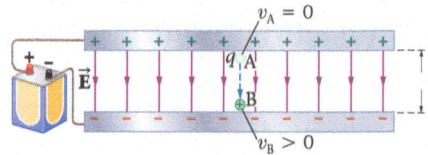

20-3 THE ELECTRIC POTENTIAL OF POINT CHARGES

If we define the electric potential of a point charge q to be zero at an infinite distance from the charge, the electric potential at a distance r is

$$V = \frac{kq}{r} \qquad \text{20-7}$$

Electric Potential Energy

We define the electric potential energy of two charges, q_0 and q, to be zero when the separation between them is infinite. When the charges are separated by a distance r, the potential energy of the system is

$$U = \frac{kq_0 q}{r} \qquad \text{20-8}$$

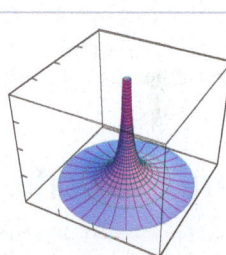

Electric potential of
a positive point charge

Superposition

The electric potential of two or more point charges is simply the algebraic sum of the potentials due to each charge separately.

20-4 EQUIPOTENTIAL SURFACES AND THE ELECTRIC FIELD

Equipotential surfaces are defined as surfaces on which the electric potential is constant. Different equipotential surfaces correspond to different values of the potential.

Electric Field

The electric field is always perpendicular to the equipotential surfaces, and it points in the direction of decreasing electric potential.

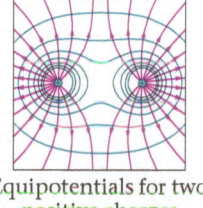

Equipotentials for two
positive charges

20-5 CAPACITORS AND DIELECTRICS

A capacitor is a device that stores electric charge.

Capacitance

Capacitance is defined as the amount of charge Q stored in a capacitor per volt of potential difference V between the plates of the capacitor. Thus,

$$C = \frac{Q}{V} \qquad \text{20-9}$$

Parallel-Plate Capacitor

The capacitance of a parallel-plate capacitor, with plates of area A and separation d, is

$$C = \frac{\epsilon_0 A}{d} \qquad \text{20-12}$$

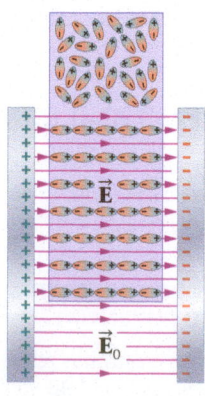

Dielectrics

A dielectric is an insulating material that increases the capacitance of a capacitor.

Dielectric Constant

A dielectric is characterized by the dimensionless dielectric constant, κ. In particular, the electric field in a dielectric is reduced by the factor κ, $E = E_0/\kappa$; the potential difference between capacitor plates is decreased by the factor κ, $V = V_0/\kappa$; and the capacitance is increased by the factor κ:

$$C = \kappa C_0 \qquad\qquad\qquad 20\text{-}14$$

20-6 ELECTRICAL ENERGY STORAGE

A capacitor, in addition to storing charge, stores electrical energy.

Energy Stored in a Capacitor

The electrical energy stored in a capacitor can be expressed as follows:

$$U = \tfrac{1}{2}QV = \tfrac{1}{2}CV^2 = Q^2/2C \qquad\qquad 20\text{-}16, 17, 18$$

ANSWERS TO ENHANCE YOUR UNDERSTANDING QUESTIONS

1. The magnitude of the electric field is the same for the two systems, because the rate of change of the electric potential is the same.

2. The magnitude of the charge of particle B is less than the magnitude of the charge of particle A because its kinetic energy is only 4 times greater than that of particle A, even though it accelerated through a potential difference that was 10 times greater.

3. D < B < A < C.

4. The magnitude of the electric field near point A is less than it is near point B because the rate of change of the electric potential is greater near point B—that is, the equipotential surfaces near point B are more "bunched up."

5. The capacitance of capacitor A is less than the capacitance of capacitor B because the metal bar (which has zero electric field within it) reduces the potential difference between the plates.

6. A = D < C < B.

CONCEPTUAL QUESTIONS

For instructor-assigned homework, go to www.masteringphysics.com. **MP**

(Answers to odd-numbered Conceptual Questions can be found in the back of the book.)

1. In one region of space the electric potential has a positive constant value. In another region of space the potential has a negative constant value. What can be said about the electric field within each of these two regions of space?

2. If the electric field is zero in some region of space is the electric potential zero there as well? Explain.

3. Sketch the equipotential surface that goes through point 1 in **FIGURE 20-26**. Repeat for point 2 and for point 3.

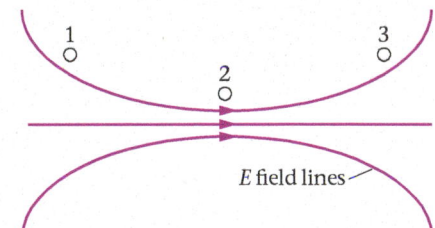

FIGURE 20-26 Conceptual Question 3 and Problems 43 and 44

4. How much work is required to move a charge from one location on an equipotential to another point on the same equipotential? Explain.

5. It is known that the electric potential is constant on a given two-dimensional surface. What can be said about the electric field on this surface?

6. Explain why equipotentials are always perpendicular to the electric field.

7. Two charges are at locations that have the same value of the electric potential. Is the electric potential energy the same for these charges? Explain.

8. A capacitor is connected to a battery and fully charged. What becomes of the charge on the capacitor when it is disconnected from the battery? What becomes of the charge when the two terminals of the capacitor are connected to one another?

9. On which of the following quantities does the capacitance of a capacitor depend: **(a)** the charge on the plates; **(b)** the separation of the plates; **(c)** the voltage difference between the plates; **(d)** the electric field between the plates; or **(e)** the area of the plates?

10. We say that a capacitor stores charge, yet the total charge in a capacitor is zero; that is, $Q + (-Q) = 0$. In what sense does a capacitor store charge if the net charge within it is zero?

PROBLEMS AND CONCEPTUAL EXERCISES

Answers to odd-numbered Problems and Conceptual Exercises can be found in the back of the book. **BIO** *identifies problems of biological or medical interest;* **CE** *indicates a conceptual exercise.* **Predict/Explain** *problems ask for two responses: (a) your prediction of a physical outcome, and (b) the best explanation among three provided; and* **Predict/Calculate** *problems ask for a prediction of a physical outcome, based on fundamental physics concepts, and follow that with a numerical calculation to verify the prediction. On all problems, bullets (•, ••, •••) indicate the level of difficulty.*

SECTION 20-1 ELECTRIC POTENTIAL ENERGY AND THE ELECTRIC POTENTIAL

1. • **CE** An electron is released from rest in a region of space with a nonzero electric field. As the electron moves, does it experience an increasing or decreasing electric potential? Explain.

2. • A uniform electric field of magnitude 3.8×10^5 N/C points in the positive x direction. Find the change in electric potential energy of a 7.5-μC charge as it moves from the origin to the points **(a)** $(0, 6.0\,\text{m})$; **(b)** $(6.0\,\text{m}, 0)$; and **(c)** $(6.0\,\text{m}, 6.0\,\text{m})$.

3. • A uniform electric field of magnitude 6.8×10^5 N/C points in the positive x direction. Find the change in electric potential between the origin and the points **(a)** $(0, 6.0 \, \text{m})$; **(b)** $(6.0 \, \text{m}, 0)$; and **(c)** $(6.0 \, \text{m}, 6.0 \, \text{m})$.

4. • **BIO Electric Potential Across a Cell Membrane** In a typical living cell, the electric potential inside the cell is 0.070 V lower than the electric potential outside the cell. The thickness of the cell membrane is 0.10 μm. What are the magnitude and direction of the electric field within the cell membrane?

5. • An old-fashioned computer monitor accelerates electrons and directs them to the screen in order to create an image. If the accelerating plates are 0.953 cm apart, and have a potential difference of 22,500 V, what is the magnitude of the uniform electric field between them?

6. • A parallel-plate capacitor has plates separated by 0.95 mm. If the electric field between the plates has a magnitude of **(a)** 1.2×10^5 V/m or **(b)** 2.4×10^4 N/C, what is the potential difference between the plates?

7. • When an ion accelerates through a potential difference of -1880 V, its electric potential energy increases by 6.02×10^{-16} J. What is the charge on the ion?

8. • **The Electric Potential of the Earth** The Earth has a vertical electric field with a magnitude of approximately 100 V/m near its surface. What is the magnitude of the potential difference between a point on the ground and a point on the same level as the top of the Washington Monument (555 ft high)?

9. •• A uniform electric field with a magnitude of 6860 N/C points in the positive x direction. Find the change in electric potential energy when a $+14.5$-μC charge is moved 5.75 cm in **(a)** the positive x direction, **(b)** the negative x direction, and **(c)** the positive y direction.

10. •• **Predict/Calculate** A spark plug in a car has electrodes separated by a gap of 0.025 in. To create a spark and ignite the air-fuel mixture in the engine, an electric field of 3.0×10^6 V/m is required in the gap. **(a)** What potential difference must be applied to the spark plug to initiate a spark? **(b)** If the separation between electrodes is increased, does the required potential difference increase, decrease, or stay the same? Explain. **(c)** Find the potential difference for a separation of 0.050 in.

11. •• A uniform electric field with a magnitude of 1200 N/C points in the negative x direction, as shown in **FIGURE 20-27**. **(a)** What is the difference in electric potential, $\Delta V = V_B - V_A$, between points A and B? **(b)** What is the difference in electric potential, $\Delta V = V_B - V_C$, between points B and C? **(c)** What is the difference in electric potential, $\Delta V = V_C - V_A$, between points C and A? **(d)** From the information given in this problem, is it possible to determine the value of the electric potential at point A? If so, determine V_A; if not, explain why.

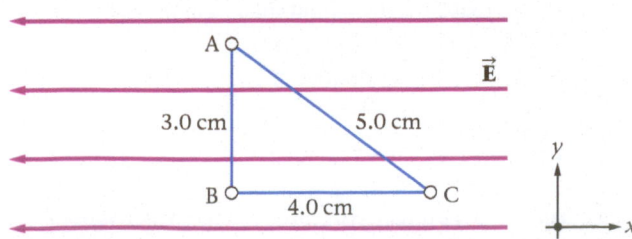

FIGURE 20-27 Problems 11 and 20

12. •• **A Charged Battery** A typical 12-V car battery can deliver 2.0×10^5 C of charge. If the energy supplied by the battery could

be converted entirely to kinetic energy, what speed would it give to a 1600-kg car that is initially at rest?

13. •• **BIO Predict/Calculate The Sodium Pump** Living cells actively "pump" positive sodium ions (Na^+) from inside the cell to outside the cell. This process is referred to as pumping because work must be done on the ions to move them from the negatively charged inner surface of the membrane to the positively charged outer surface. Given that the electric potential is 0.070 V higher outside the cell than inside the cell, and that the cell membrane is 0.10 μm thick, **(a)** calculate the work that must be done (in joules) to move one sodium ion from inside the cell to outside. **(b)** If the thickness of the cell membrane is increased, does your answer to part (a) increase, decrease, or stay the same? Explain. (It is estimated that as much as 20% of the energy we consume in a resting state is used in operating this "sodium pump.")

14. •• **Predict/Calculate** The electric potential of a system as a function of position along the x axis is given in **FIGURE 20-28**. **(a)** In which of the regions, 1, 2, 3, or 4, do you expect E_x to be greatest? In which region does E_x have its greatest magnitude? Explain. **(b)** Calculate the value of E_x in each of the regions, 1, 2, 3, and 4.

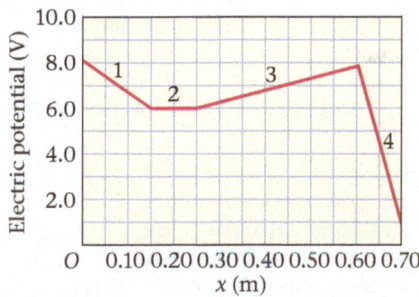

FIGURE 20-28 Problem 14

15. ••• Points A and B have electric potentials of 332 V and 149 V, respectively. When an electron released from rest at point A arrives at point C, its kinetic energy is K_A. When the electron is released from rest at point B, however, its kinetic energy when it reaches point C is $K_B = 2K_A$. What are **(a)** the electric potential at point C and **(b)** the kinetic energy K_A?

SECTION 20-2 ENERGY CONSERVATION

16. • **CE Predict/Explain** An electron is released from rest in a region of space with a nonzero electric field. **(a)** As the electron moves, does the electric potential energy of the system increase, decrease, or stay the same? **(b)** Choose the *best explanation* from among the following:

    I. Because the electron has a negative charge its electric potential energy doesn't decrease, as one might expect, but increases instead.

    II. As the electron begins to move, its kinetic energy increases. The increase in kinetic energy is equal to the decrease in the electric potential energy of the system.

    III. The electron will move perpendicular to the electric field, and hence its electric potential energy will remain the same.

17. • Calculate the speed of **(a)** a proton and **(b)** an electron after each particle accelerates from rest through a potential difference of 355 V.

18. • **Ion Thrusters** NASA's *Deep Space 1* and *Dawn* spacecraft employed ion thrusters in which xenon ions with a charge of $+1.60 \times 10^{-19}$ C and a mass of 2.18×10^{-25} kg are accelerated through a potential difference of 1315 V. What is the speed of the xenon ions as they exit the thruster?

19. • Find the potential difference required to accelerate protons from rest to 10% of the speed of light. (Above this speed, relativistic effects start to become significant.)

20. •• **Predict/Calculate** A particle with a mass of 3.8 g and a charge of +0.045 μC is released from rest at point A in Figure 20-27. **(a)** In which direction will this charge move? **(b)** What speed will it have after moving through a distance of 5.0 cm? The electric field has a magnitude of 1200 N/C. **(c)** Suppose the particle continues moving for another 5.0 cm. Will its increase in speed for the second 5.0 cm be greater than, less than, or equal to its increase in speed in the first 5.0 cm? Explain.

21. •• **Conduction Electrons** In the microscopic view of electrical conduction in a copper wire, electrons are accelerated by an electric field and then collide with metal atoms after traveling about 3.9×10^{-8} m. If an electron begins from rest and is accelerated by a field of 0.065 N/C, what is its speed when it collides with a metal atom?

22. •• A proton has an initial speed of 5.5×10^5 m/s. **(a)** What potential difference is required to bring the proton to rest? **(b)** What potential difference is required to reduce the initial speed of the proton by a factor of 2? **(c)** What potential difference is required to reduce the initial kinetic energy of the proton by a factor of 2?

SECTION 20-3 THE ELECTRIC POTENTIAL OF POINT CHARGES

23. • In **FIGURE 20-29**, $q_1 = +1.8$ nC and $q_2 = -2.1$ nC, and the side length of the square is 0.10 m. **(a)** What is the electric potential at point A? **(b)** What is the electric potential at point B?

24. • In Figure 20-29, it is given that $q_1 = +Q$. **(a)** What value must q_2 have if the electric potential at point A is to be zero? **(b)** With the value for q_2 found in part (a), is the electric potential at point B positive, negative, or zero? Explain.

FIGURE 20-29 Problems 23, 24, 25, and 26

25. • **CE** The charge q_1 in Figure 20-29 has the value $+Q$. **(a)** What value must q_2 have if the electric potential at point B is to be zero? **(b)** With the value for q_2 found in part (a), is the electric potential at point A positive, negative, or zero? Explain.

26. • **CE** It is given that the electric potential is zero at the center of the square in Figure 20-29. **(a)** If $q_1 = +Q$, what is the value of the charge q_2? **(b)** Is the electric potential at point A positive, negative, or zero? Explain. **(c)** Is the electric potential at point B positive, negative, or zero? Explain.

27. • The electric potential 1.6 m from a point charge q is 3.8×10^4 V. What is the value of q?

28. • A point charge of 9.2 μC is at the origin. What is the electric potential at **(a)** (3.0 m, 0); **(b)** (−3.0 m, 0); and **(c)** (3.0 m, −3.0 m)?

29. • **The Bohr Atom** The hydrogen atom consists of one electron and one proton. In the Bohr model of the hydrogen atom, the electron orbits the proton in a circular orbit of radius 0.529×10^{-10} m. What is the electric potential due to the proton at the electron's orbit?

30. •• How far must the point charges $q_1 = +6.22$ μC and $q_2 = -22.1$ μC be separated for the electric potential energy of the system to be −106 J?

31. •• **CE** Four different arrangements of point charges are shown in **FIGURE 20-30**. In each case the charges are the same distance from the origin. Rank the four arrangements in order of increasing electric potential at the origin, taking the potential at infinity to be zero. Indicate ties where appropriate.

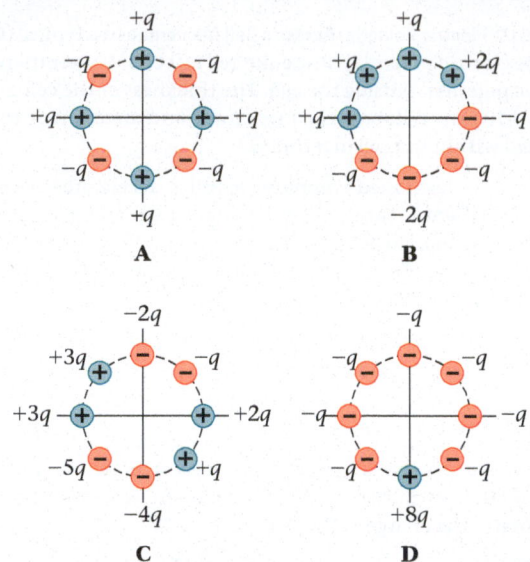

FIGURE 20-30 Problem 31

32. •• **Predict/Calculate** Point charges +4.1 μC and −2.2 μC are placed on the x axis at (11 m, 0) and (−11 m, 0), respectively. **(a)** Sketch the electric potential on the x axis for this system. **(b)** Your sketch should show one point on the x axis between the two charges where the potential vanishes. Is this point closer to the +4.1-μC charge or closer to the −2.2-μC charge? Explain. **(c)** Find the point referred to in part (b).

33. •• In **FIGURE 20-31**, the charge $q = 4.11 \times 10^{-9}$ C. **(a)** Find the electric potential at $x = 0.500$ m. **(b)** Find the value of x between 0 and 1.00 m where the electric potential is zero.

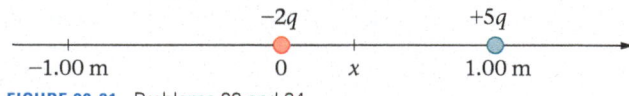

FIGURE 20-31 Problems 33 and 34

34. •• **Predict/Calculate** In Figure 20-31, the charge $q = 3.67 \times 10^{-9}$ C. **(a)** Find the value of x between −1.00 m and 0 where the electric potential is zero. **(b)** At values of x to the left of −1.00 m, do you expect the electric potential to be greater than, less than, or equal to zero? **(c)** Calculate the electric potential at $x = -2.00$ m.

35. •• A charge of 4.07 μC is held fixed at the origin. A second charge of 3.45 μC is released from rest at the position (1.25 m, 0.570 m). **(a)** If the mass of the second charge is 2.36 g, what is its speed when it moves infinitely far from the origin? **(b)** At what distance from the origin does the second charge attain half the speed it will have at infinity?

36. •• **Predict/Calculate** A charge of 20.2 μC is held fixed at the origin. **(a)** If a −5.25-μC charge with a mass of 3.20 g is released from rest at the position (0.925 m, 1.17 m), what is its speed when it is halfway to the origin? **(b)** Suppose the −5.25-μC charge is released from rest at the point $x = \frac{1}{2}(0.925$ m) and $y = \frac{1}{2}(1.17$ m). When it is halfway to the origin, is its speed greater than, less than, or equal to the speed found in part (a)? Explain. **(c)** Find the speed of the charge for the situation described in part (b).

37. •• A charge of −2.505 μC is located at (3.055 m, 4.501 m), and a charge of 1.875 μC is located at (−2.533 m, 0). **(a)** Find the

electric potential at the origin. **(b)** At one point on the line connecting these two charges the potential is zero. Find this point.

38. •• **Predict/Calculate** **FIGURE 20-32** shows three charges at the corners of a rectangle. **(a)** How much work must be done to move the $+2.7\text{-}\mu\text{C}$ charge to infinity? **(b)** Suppose, instead, that we move the $-6.1\text{-}\mu\text{C}$ charge to infinity. Is the work required in this case greater than, less than, or the same as when we moved the $+2.7\text{-}\mu\text{C}$ charge to infinity? Explain. **(c)** Calculate the work needed to move the $-6.1\text{-}\mu\text{C}$ charge to infinity.

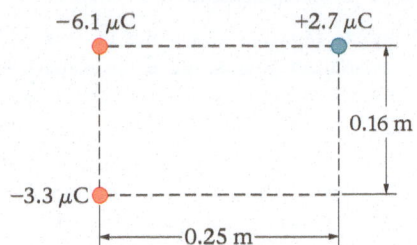

FIGURE 20-32 Problems 38 and 39

39. •• How much work must be done to move the three charges in Figure 20-32 infinitely far from one another?

40. •• **(a)** Find the electric potential at point P in **FIGURE 20-33**. **(b)** Suppose the three charges shown in Figure 20-33 are held in place. A fourth charge, with a charge of $+4.82\,\mu\text{C}$ and a mass of 2.33 g, is released from rest at point P. What is the speed of the fourth charge when it has moved infinitely far away from the other three charges?

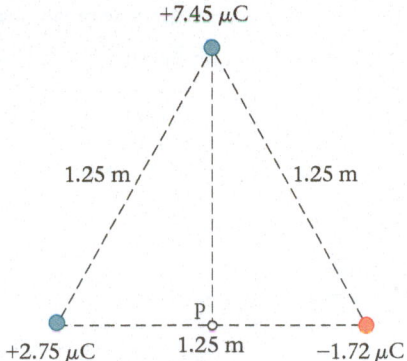

FIGURE 20-33 Problems 40 and 88

41. ••• A square of side a has a charge $+Q$ at each corner. What is the electric potential energy of this system of charges?

42. ••• A square of side a has charges $+Q$ and $-Q$ alternating from one corner to the next, as shown in **FIGURE 20-34**. Find the electric potential energy for this system of charges.

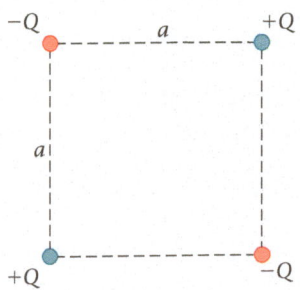

FIGURE 20-34 Problems 42 and 98

SECTION 20-4 EQUIPOTENTIAL SURFACES AND THE ELECTRIC FIELD

43. • **CE Predict/Explain** **(a)** Is the electric potential at point 1 in Figure 20-26 greater than, less than, or equal to the electric

potential at point 3? **(b)** Choose the *best explanation* from among the following:

  I. The electric field lines point to the right, indicating that the electric potential is greater at point 3 than at point 1.
  II. The electric potential is large where the electric field lines are close together and small where they are widely spaced. Therefore, the electric potential is the same at points 1 and 3.
  III. The electric potential decreases as we move in the direction of the electric field, as shown in Figure 20-3. Therefore, the electric potential is greater at point 1 than at point 3.

44. • **CE Predict/Explain** Imagine sketching a large number of equipotential surfaces in Figure 20-26, with a constant difference in electric potential between adjacent surfaces. **(a)** Would the equipotentials at point 2 be more closely spaced, be less closely spaced, or have the same spacing as equipotentials at point 1? **(b)** Choose the *best explanation* from among the following:

  I. When electric field lines are close together, the corresponding equipotentials are far apart.
  II. Equipotential surfaces, by definition, always have equal spacing between them.
  III. The electric field is more intense at point 2 than at point 1, which means the equipotential surfaces are more closely spaced in that region.

45. • Two point charges are on the x axis. Charge 1 is $+q$ and is located at $x = -1.0\text{ m}$; charge 2 is $-2q$ and is located at $x = 1.0\text{ m}$. Make sketches of the equipotential surfaces for this system **(a)** out to a distance of about 2.0 m from the origin and **(b)** far from the origin. In each case, indicate the direction in which the potential increases.

46. •• **CE** **FIGURE 20-35** shows a series of equipotentials in a particular region of space, and five different paths along which an electron is moved. **(a)** Does the electric field in this region point to the right, to the left, up, or down? Explain. **(b)** For each path, indicate whether the work done on the electron by the electric field is positive, negative, or zero. **(c)** Rank the paths in order of increasing amount of work done on the electron by the electric field. Indicate ties where appropriate. **(d)** Is the electric field near path A greater than, less than, or equal to the electric field near path E? Explain.

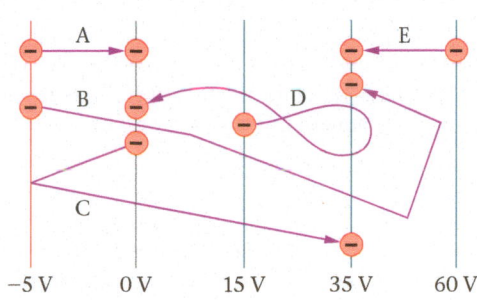

FIGURE 20-35 Problem 46

47. •• **Predict/Calculate** Consider a region in space where a uniform electric field $E = 6500\text{ N/C}$ points in the negative x direction. **(a)** What is the orientation of the equipotential surfaces? Explain. **(b)** If you move in the positive x direction, does the electric potential increase or decrease? Explain. **(c)** What is the distance between the $+14\text{-V}$ and the $+16\text{-V}$ equipotentials?

48. •• A given system has the equipotential surfaces shown in **FIGURE 20-36**. **(a)** What are the magnitude and direction of the electric field? **(b)** What is the shortest distance one can move to undergo a change in potential of 10.0 V?

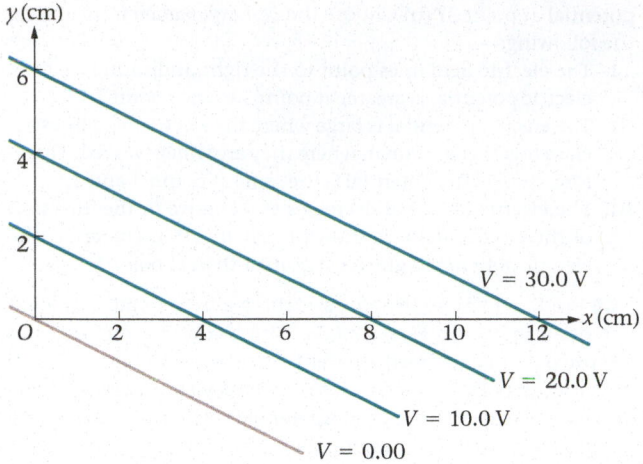

FIGURE 20-36 Problems 48, 49, and 89

49. •• A given system has the equipotential surfaces shown in Figure 20-36. (a) What is the electric potential at the point (2.0 cm, 2.0 cm)? (b) How much work is done by the electric force when a 0.10-C charge is moved from the point (2.0 cm, 2.0 cm) to the point (4.0 cm, 4.0 cm)?

SECTION 20-5 CAPACITORS AND DIELECTRICS

50. • A 0.75-μF capacitor is connected to a 9.0-V battery. How much charge is on each plate of the capacitor?

51. • It is desired that 7.7 μC of charge be stored on each plate of a 5.2-μF capacitor. What potential difference is required between the plates?

52. • To operate a given flash lamp requires a charge of 38 μC. What capacitance is needed to store this much charge in a capacitor with a potential difference between its plates of 9.0 V?

53. • **Planet Capacitor** It can be shown that the capacitance of a conducting sphere of radius R is R/k, where k is the Coulomb's law constant, $k = 8.99 \times 10^9 \, \text{N} \cdot \text{m}^2/\text{C}^2$. In this arrangement the second conductor of the capacitor is taken to be a concentric conducting spherical shell of infinite radius. Treating the Earth as a conducting sphere of radius 6.37×10^6 m, what is its capacitance?

54. •• A parallel-plate capacitor is made from two aluminum-foil sheets, each 3.8 cm wide and 6.1 m long. Between the sheets is a Teflon strip of the same width and length that is 0.025 mm thick. What is the capacitance of this capacitor? (The dielectric constant of Teflon is 2.1.)

55. •• A parallel-plate capacitor is constructed with circular plates of radius 0.051 m. The plates are separated by 0.25 mm, and the space between the plates is filled with a dielectric with dielectric constant κ. When the charge on the capacitor is 1.8 μC the potential difference between the plates is 1150 V. Find the value of κ.

56. •• Predict/Calculate A parallel-plate capacitor has plates with an area of 0.012 m² and a separation of 0.88 mm. The space between the plates is filled with a dielectric whose dielectric constant is 2.0. (a) What is the potential difference between the plates when the charge on the capacitor plates is 4.7 μC? (b) Will your answer to part (a) increase, decrease, or stay the same if the dielectric constant is increased? Explain. (c) Calculate the potential difference for the case where the dielectric constant is 4.0.

57. •• Predict/Calculate A 72-nF parallel-plate capacitor has plates that sandwich a dielectric layer that is 75 μm thick. (a) In order to reduce the capacitance, should the thickness of the dielectric layer be increased or decreased? (b) Calculate the thickness of the dielectric layer that would reduce the capacitance to 64 nF.

58. •• Predict/Calculate Consider a parallel-plate capacitor constructed from two circular metal plates of radius R. The plates are separated by a distance of 1.5 mm. (a) What radius must the plates have if the capacitance of this capacitor is to be 1.0 μF? (b) If the separation between the plates is increased, should the radius of the plates be increased or decreased to maintain a capacitance of 1.0 μF? Explain. (c) Find the radius of the plates that gives a capacitance of 1.0 μF for a plate separation of 3.0 mm.

59. •• A parallel-plate capacitor has plates of area 3.75×10^{-4} m². What plate separation is required if the capacitance is to be 1580 pF? Assume that the space between the plates is filled with (a) air or (b) paper.

60. •• Predict/Calculate A parallel-plate capacitor filled with air has plates of area 0.0066 m² and a separation of 0.45 mm. (a) Find the magnitude of the charge on each plate when the capacitor is connected to a 12-V battery. (b) Will your answer to part (a) increase, decrease, or stay the same if the separation between the plates is increased? Explain. (c) Calculate the magnitude of the charge on the plates if the separation is 0.90 mm.

61. •• Suppose that after walking across a carpeted floor you reach for a doorknob and just before you touch it a spark jumps 0.50 cm from your finger to the knob. Find the minimum voltage needed between your finger and the doorknob to generate this spark.

62. •• (a) What plate area is required for an air-filled, parallel-plate capacitor with a plate separation of 2.6 mm to have a capacitance of 22 pF? (b) What maximum voltage can be applied to this capacitor without causing dielectric breakdown?

63. •• **Lightning** As a crude model for lightning, consider the ground to be one plate of a parallel-plate capacitor and a cloud at an altitude of 550 m to be the other plate. Assume the surface area of the cloud to be the same as the area of a square that is 0.50 km on a side. (a) What is the capacitance of this capacitor? (b) How much charge can the cloud hold before the dielectric strength of the air is exceeded and a spark (lightning) results?

64. ••• A parallel-plate capacitor is made from two aluminum-foil sheets, each 3.00 cm wide and 10.0 m long. Between the sheets is a mica strip of the same width and length that is 0.0225 mm thick. What is the maximum charge that can be stored in this capacitor?

SECTION 20-6 ELECTRICAL ENERGY STORAGE

65. • Calculate the work done by a 9.0-V battery as it charges a 7.2-μF capacitor in the flash unit of a camera.

66. • **BIO Defibrillator** An automatic external defibrillator (AED) delivers 135 J of energy at a voltage of 725 V. What is the capacitance of this device?

67. •• **BIO Predict/Calculate Cell Membranes** The membrane of a living cell can be approximated by a parallel-plate capacitor with plates of area 4.75×10^{-9} m², a plate separation of 8.5×10^{-9} m, and a dielectric with a dielectric constant of 4.5. (a) What is the energy stored in such a cell membrane if the potential difference across it is 0.0725 V? (b) Would your answer to part (a) increase, decrease, or stay the same if the thickness of the cell membrane is increased? Explain.

68. •• A capacitor with plate area 0.0440 m² and plate separation 85.0 μm is to be charged to 12.0 V and store 7.50 μJ of energy. What should be the dielectric constant of the material between the plates?

69. •• Find the electric energy density between the plates of a 225-μF parallel-plate capacitor. The potential difference between the plates is 345 V, and the plate separation is 0.223 mm.

70. •• What electric field strength would store 17.5 J of energy in every 1.00 mm^3 of space?

71. •• An electronic flash unit for a camera contains a capacitor with a capacitance of 750 μF. When the unit is fully charged and ready for operation, the potential difference between the capacitor plates is 360 V. (a) What is the magnitude of the charge on each plate of the fully charged capacitor? (b) Find the energy stored in the "charged-up" flash unit.

72. ••• A parallel-plate capacitor has plates with an area of 405 cm^2 and an air-filled gap between the plates that is 2.25 mm thick. The capacitor is charged by a battery to 575 V and then is disconnected from the battery. (a) How much energy is stored in the capacitor? (b) The separation between the plates is now increased to 4.50 mm. How much energy is stored in the capacitor now? (c) How much work is required to increase the separation of the plates from 2.25 mm to 4.50 mm? Explain your reasoning.

GENERAL PROBLEMS

73. • **CE Predict/Explain** A proton is released from rest in a region of space with a nonzero electric field. (a) As the proton moves, does the electric potential energy of the system increase, decrease, or stay the same? (b) Choose the *best explanation*:
    I. As the proton begins to move, its kinetic energy increases. The increase in kinetic energy is equal to the decrease in the electric potential energy of the system.
    II. Because the proton has a positive charge, its electric potential energy will always increase.
    III. The proton will move perpendicular to the electric field, and hence its electric potential energy will remain the same.

74. • **CE** The plates of a parallel-plate capacitor have constant charges of $+Q$ and $-Q$. Do the following quantities increase, decrease, or remain the same as the separation of the plates is increased? (a) The electric field between the plates; (b) the potential difference between the plates; (c) the capacitance; (d) the energy stored in the capacitor.

75. • **CE** A parallel-plate capacitor is connected to a battery that maintains a constant potential difference V between the plates. If the plates of the capacitor are pulled farther apart, do the following quantities increase, decrease, or remain the same? (a) The electric field between the plates; (b) the charge on the plates; (c) the capacitance; (d) the energy stored in the capacitor.

76. • **CE** The plates of a parallel-plate capacitor have constant charges of $+Q$ and $-Q$. Do the following quantities increase, decrease, or remain the same as a dielectric is inserted between the plates? (a) The electric field between the plates; (b) the potential difference between the plates; (c) the capacitance; (d) the energy stored in the capacitor.

77. • **CE** A parallel-plate capacitor is connected to a battery that maintains a constant potential difference V between the plates. If a dielectric is inserted between the plates of the capacitor, do the following quantities increase, decrease, or remain the same? (a) The electric field between the plates; (b) the charge on the plates; (c) the capacitance; (d) the energy stored in the capacitor.

78. • Find the difference in electric potential, $\Delta V = V_B - V_A$, between the points A and B for the following cases: (a) The electric field does 0.052 J of work as you move a +5.7-μC charge from A to B.

(b) The electric field does −0.052 J of work as you move a −5.7-μC charge from A to B. (c) You perform 0.052 J of work as you slowly move a +5.7-μC charge from A to B.

79. •• A 0.32-μF capacitor is charged by a 1.5-V battery. After being charged, the capacitor is connected to a small electric motor. Assuming 100% efficiency, (a) to what height can the motor lift a 7.5-g mass? (b) What initial voltage must the capacitor have if it is to lift a 7.5-g mass through a height of 1.0 cm?

80. •• A charge of 22.5 μC is located at (4.40 m, 6.22 m), and a charge of −14.2 μC is located at (−4.50 m, 6.75 m). What charge must be located at (2.23 m, −3.31 m) if the electric potential is to be zero at the origin?

81. •• **The Bohr Model** In the Bohr model of the hydrogen atom (see Problem 29) what is the smallest amount of work that must be done on the electron to move it from its circular orbit, with a radius of 0.529×10^{-10} m, to an infinite distance from the proton? This value is referred to as the ionization energy of hydrogen.

82. •• **Predict/Calculate** A +1.2-μC charge and a −1.2-μC charge are placed at (0.50 m, 0) and (−0.50 m, 0), respectively. (a) In **FIGURE 20-37**, at which of the points A, B, C, or D is the electric potential smallest in value? At which of these points does it have its greatest value? Explain. (b) Calculate the electric potential at points A, B, C, and D.

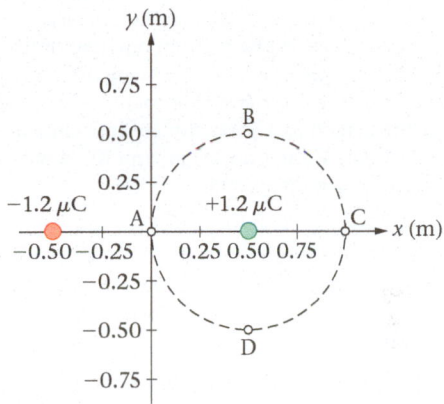

FIGURE 20-37   Problem 82

83. •• How much work is required to bring three protons, initially infinitely far apart, to a configuration where each proton is 1.5×10^{-15} m from the other two? (This is a typical separation for protons in a nucleus.)

84. •• A point charge $Q = +87.1\,\mu$C is held fixed at the origin. A second point charge, with mass $m = 0.0576$ kg and charge $q = -2.87\,\mu$C, is placed at the location (0.323 m, 0). (a) Find the electric potential energy of this system of charges. (b) If the second charge is released from rest, what is its speed when it reaches the point (0.121 m, 0)?

85. •• **Electron Escape Speed** An electron is at rest just above the surface of a sphere with a radius of 2.7 mm and a uniformly distributed positive charge of 1.8×10^{-15} C. Like a rocket blasting off from the Earth, the electron is given an initial speed v_e radially outward from the sphere. If the electron coasts to infinity, where its kinetic energy drops to zero, what is the escape speed, v_e?

86. •• **Quark Model of the Neutron** According to the quark model of fundamental particles, neutrons—the neutral particles in an atom's nucleus—are composed of three quarks. Two of these quarks are "down" quarks, each with a charge of $-e/3$; the third quark is an "up" quark, with a charge of $+2e/3$. This gives the neutron a net charge of zero. What is the electric potential energy of these three quarks,

assuming they are equidistant from one another, with a separation distance of 1.3×10^{-15} m? (Quarks are discussed in Chapter 32.)

87. •• A parallel-plate capacitor is charged to an electric potential of 305 V by moving 3.75×10^{16} electrons from one plate to the other. How much work is done in charging the capacitor?

88. •• **Predict/Calculate** The three charges shown in Figure 20-33 are held in place as a fourth charge, q, is brought from infinity to the point P. The charge q starts at rest at infinity and is also at rest when it is placed at the point P. **(a)** If q is a positive charge, is the work required to bring it to the point P positive, negative, or zero? Explain. **(b)** Find the value of q if the work needed to bring it to point P is -1.3×10^{-11} J.

89. •• **(a)** In Figure 20-36 we see that the electric potential increases by 10.0 V as one moves 4.00 cm in the positive x direction. Use this information to calculate the x component of the electric field. (Ignore the y direction for the moment.) **(b)** Apply the same reasoning as in part (a) to calculate the y component of the electric field. **(c)** Combine the results from parts (a) and (b) to find the magnitude and direction of the electric field for this system.

90. •• **BIO** **Predict/Calculate** **Electric Catfish** The electric catfish (*Malapterurus electricus*) is an aggressive fish, 1.0 m in length, found today in tropical Africa. The catfish is capable of generating jolts of electricity up to 350 V by producing a positively charged region of muscle near the head and a negatively charged region near the tail. **(a)** For the same amount of charge, can the catfish generate a higher voltage by separating the charge from one end of its body to the other, as it does, or from one side of the body to the other? Explain. **(b)** Estimate the charge generated at each end of a catfish as follows: Treat the catfish as a parallel-plate capacitor with plates of area 1.8×10^{-2} m², separation 1.0 m, and filled with a dielectric with a dielectric constant $\kappa = 95$.

91. •• **Regenerative Braking** Many electric cars can capture the kinetic energy of the car and store it in a capacitor or a battery by means of an electric generator. If a 1380-kg electric car brakes from 11.0 m/s to rest and charges a capacitor to 202 V, what capacitance is needed to store all of the car's kinetic energy?

92. •• **Predict/Calculate** **Computer Keyboards** Many computer keyboards operate on the principle of capacitance. As shown in Figure 20-18, each key forms a small parallel-plate capacitor whose separation is reduced when the key is depressed. **(a)** Does depressing a key increase or decrease its capacitance? Explain. **(b)** Suppose the plates for each key have an area of 47.5 mm² and an initial separation of 0.550 mm. In addition, let the dielectric have a dielectric constant of 3.75. If the circuitry of the computer can detect a change in capacitance of 0.425 pF, what is the minimum distance a key must be depressed to be detected?

93. •• **Predict/Calculate** A point charge of mass 0.081 kg and charge $+6.77\,\mu$C is suspended by a thread between the vertical parallel plates of a parallel-plate capacitor, as shown in **FIGURE 20-38**. **(a)** If the charge deflects to the right of vertical, as indicated in the figure, which of the two plates is at the higher electric potential? **(b)** If the angle of deflection is 22°, and the separation between the plates is 0.025 m, what is the potential difference between the plates?

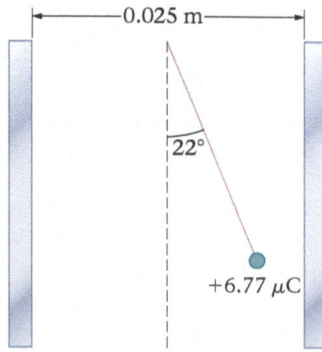

FIGURE 20-38 Problems 93 and 99

94. •• **BIO** **Cell Membranes and Dielectrics** Many cells in the body have a cell membrane whose inner and outer surfaces carry opposite charges, just like the plates of a parallel-plate capacitor. Suppose a typical cell membrane has a thickness of 8.1×10^{-9} m, and its inner and outer surfaces carry charge densities of -0.58×10^{-3} C/m² and $+0.58 \times 10^{-3}$ C/m², respectively. In addition, assume that the material in the cell membrane has a dielectric constant of 5.5. **(a)** Find the direction and magnitude of the electric field within the cell membrane. **(b)** Calculate the potential difference between the inner and outer walls of the membrane, and indicate which wall of the membrane has the higher potential.

95. •• **BIO** **Mitochondrial Membrane** Every cell in the body has organelles called *mitochondria* that can generate a voltage difference between their interior and exterior. **(a)** If the capacitance of a mitochondrion is 4.3×10^{-11} F and the potential difference between the interior and exterior is 0.18 V, how much electrical energy does it store? **(b)** If a mitochondrion were to use all of its stored electrical energy to produce ATP molecules, and each ATP molecule requires 9.5×10^{-20} J, how many molecules could it produce? (In reality, ATP molecules are produced by the flow of protons caused by the *proton motive force*, and not by the direct conversion of electrical energy stored by the capacitance of mitochondria.)

96. •• Long, long ago, on a planet far, far away, a physics experiment was carried out. First, a 0.250-kg ball with zero net charge was dropped from rest at a height of 1.00 m. The ball landed 0.552 s later. Next, the ball was given a net charge of 7.75 μC and dropped in the same way from the same height. This time the ball fell for 0.680 s before landing. What is the electric potential at a height of 1.00 m above the ground on this planet, given that the electric potential at ground level is zero? (Air resistance can be ignored.)

97. •• **Rutherford's Planetary Model of the Atom** In 1911, Ernest Rutherford developed a planetary model of the atom, in which a small positively charged nucleus is orbited by electrons. The model was motivated by an experiment carried out by Rutherford and his graduate students, Geiger and Marsden. In this experiment, they fired alpha particles with an initial speed of 1.75×10^7 m/s at a thin sheet of gold. (Alpha particles are obtained from certain radioactive decays. They have a charge of $+2e$ and a mass of 6.64×10^{-27} kg.) How close can the alpha particles get to a gold nucleus (charge $= +79e$), assuming the nucleus remains stationary? (This calculation sets an upper limit on the size of the gold nucleus. See Chapter 31 for further details.)

98. ••• **Predict/Calculate** **(a)** One of the $-Q$ charges in Figure 20-34 is given an outward "kick" that sends it off with an initial speed v_0 while the other three charges are held at rest. If the moving charge has a mass m, what is its speed when it is infinitely far from the other charges? **(b)** Suppose the remaining $-Q$ charge, which also has a mass m, is now given the same initial speed, v_0. When it is infinitely far away from the two $+Q$ charges, is its speed greater than, less than, or the same as the speed found in part (a)? Explain.

99. ••• Figure 20-38 shows a charge $q = +6.77\,\mu$C with a mass $m = 0.081$ kg suspended by a thread of length $L = 0.022$ m between the plates of a capacitor. **(a)** Plot the electric potential energy of the system as a function of the angle θ the thread makes with the vertical. (The electric field between the plates has a magnitude $E = 4.16 \times 10^4$ V/m.) **(b)** Repeat part (a) for the case of the gravitational potential energy of the system. **(c)** Show that the total potential energy of the system (electric plus gravitational) is a minimum when the angle θ satisfies the equilibrium condition for the charge, $\tan \theta = qE/mg$. This relation implies that $\theta = 22°$.

100. ••• The electric potential a distance r from a point charge q is 2.70×10^4 V. One meter farther away from the charge the potential is 6140 V. Find the charge q and the initial distance r.

101. ••• When the potential difference between the plates of a capacitor is increased by 3.25 V, the magnitude of the charge on each plate increases by 13.5 μC. What is the capacitance of this capacitor?

102. ••• The electric potential a distance r from a point charge q is 155 V, and the magnitude of the electric field is 2240 N/C. Find the values of q and r.

PASSAGE PROBLEMS

BIO **The Electric Eel**

Of the many unique and unusual animals that inhabit the rainforests of South America, one stands out because of its mastery of electricity. The electric eel (*Electrophorus electricus*), one of the few creatures on Earth able to generate, store, and discharge electricity, can deliver a powerful series of high-voltage discharges reaching 650 V. These jolts of electricity are so strong, in fact, that electric eels have been known to topple a horse crossing a stream 20 feet away, and to cause respiratory paralysis, cardiac arrhythmia, and even death in humans.

Though similar in appearance to an eel, the electric "eel" is actually more closely related to catfish. They are found primarily in the Amazon and Orinoco river basins, where they navigate the slow-moving, muddy water with low-voltage electric organ discharges (EOD), saving the high-voltage EODs for stunning prey and defending against predators. Obligate air breathers, electric eels obtain about 80% of their oxygen by gulping air at the water's surface. Even so, they are able to attain lengths of 2.5 m and a mass of 20 kg.

The organs that produce the eel's electricity take up most of its body, and consist of thousands of modified muscle cells—called electroplaques—stacked together like the cells in a battery. Each electroplaque is capable of generating a voltage of 0.15 V, and together they produce a positive charge near the head of the eel and a negative charge near its tail.

103. • Electric eels produce an electric field within their body. In which direction does the electric field point?

    A. toward the head   **B.** toward the tail
    C. upward   **D.** downward

104. • As a rough approximation, consider an electric eel to be a parallel-plate capacitor with plates of area 1.8×10^{-2} m^2 separated by 2.0 m and filled with a dielectric whose dielectric constant is $\kappa = 95$. What is the capacitance of the eel in this model?

    A. 8.0×10^{-14} F   **B.** 7.6×10^{-12} F
    C. 1.5×10^{-11} F   **D.** 9.3×10^{-8} F

105. • In terms of the parallel-plate model of the previous problem, how much charge does an electric eel generate at each end of its body when it produces a voltage of 650 V?

    A. 1.2×10^{-14} C   **B.** 5.2×10^{-11} C
    C. 4.9×10^{-9} C   **D.** 6.1×10^{-5} C

106. • How much energy is stored by an electric eel when it is charged up to 650 V? Use the same parallel-plate model discussed in the previous two problems.

    A. 1.8×10^{-17} J   **B.** 1.7×10^{-8} J
    C. 1.6×10^{-6} J   **D.** 2.0×10^{-2} J

107. •• **Predict/Calculate** **REFERRING TO EXAMPLE 20-9** Suppose the charge $-2q$ at $x = 1.00$ m is replaced with a charge $-3q$, where $q = 4.11 \times 10^{-9}$ C. The charge $+q$ is at the origin. **(a)** Is the electric potential positive, negative, or zero at the point $x = 0.333$ m? Explain. **(b)** Find the point between $x = 0$ and $x = 1.00$ m where the electric potential vanishes. **(c)** Is there a point in the region $x < 0$ where the electric potential passes through zero?

108. •• **REFERRING TO EXAMPLE 20-9** Suppose we can change the location of the charge $-2q$ on the x axis. The charge $+q$ (where $q = 4.11 \times 10^{-9}$ C) is still at the origin. **(a)** Where should the charge $-2q$ be placed to ensure that the electric potential vanishes at $x = 0.500$ m? **(b)** With the location of $-2q$ found in part (a), where does the electric potential pass through zero in the region $x < 0$?

109. •• **Predict/Calculate** **REFERRING TO EXAMPLE 20-9** Suppose the charge $+q$ at the origin is replaced with a charge $+5q$, where $q = 4.11 \times 10^{-9}$ C. The charge $-2q$ is still at $x = 1.00$ m. **(a)** Is there a point in the region $x < 0$ where the electric potential passes through zero? **(b)** Find the location between $x = 0$ and $x = 1.00$ m where the electric potential passes through zero. **(c)** Find the location in the region $x > 1.00$ m where the electric potential passes through zero.

21

Electric Current and Direct-Current Circuits

▲ A battery is a device that uses chemical energy to separate positive and negative charges, producing a potential difference between its terminals. In this case, the chemical energy comes from reactions that take place between the metal electrodes and the acid in the lemon juice. The potential difference causes a current to flow in the wires, as indicated by the glowing light. This chapter explores simple electric circuits, like the one seen here, and shows how to analyze more complex ones as well.

Big Ideas

1 Electric current is the flow of electric charge from one location to another.

2 A potential difference is needed for electric current to flow through a resistor.

3 Resistors dissipate energy in electric circuits.

4 Kirchhoff's rules apply charge conservation and energy conservation to electric circuits.

5 In an *RC* circuit, a finite time is required for charge to build up on a capacitor.

As you read this paragraph, your heart is pumping blood through the arteries and veins in your body, much like a battery "pumping" electric charge through an electric circuit.

In this chapter we consider some of the basic properties of moving electric charges and electric circuits.

21-1 Electric Current

A flow of electric charge from one place to another is called an **electric current**. Often, the charge is carried by electrons moving through a metal wire. Though the analogy should not be pushed too far, the electrons flowing through a wire are much like water molecules flowing through a garden hose or blood cells flowing through an artery.

To be specific, suppose a charge ΔQ flows past a given point in a wire in a time Δt. In such a case, we say that the electric current, I, in the wire is:

Definition of Electric Current, I

$$I = \frac{\Delta Q}{\Delta t}$$ 21-1

SI unit: coulomb per second, C/s = ampere, A

The unit of current, the ampere (A) or *amp* for short, is named for the French physicist André-Marie Ampère (1775–1836) and is defined simply as 1 coulomb per second:

$$1\,A = 1\,C/s$$

The following Example shows that the number of electrons involved in typical electric circuits is extremely large—not unlike the large number of water molecules flowing through a garden hose.

EXAMPLE 21-1   A BATTERY OF ELECTRONS

A cell phone contains a battery rated at 2600 mA·h = 2.60 A·h. How many electrons can this battery supply to the phone? (*Note:* 1 A·h = 1 ampere·hour = charge delivered by a current of 1 A in a time of 1 hour.)

PICTURE THE PROBLEM

The sketch shows a cell phone with a 2.60-A·h battery. As the cell phone operates, it draws electrons from the battery until the battery delivers 2.60 A·h of electrons.

2.60-A·h battery

REASONING AND STRATEGY

First, we calculate the charge ΔQ that is the equivalent of 2.60 A·h. Once we know the charge, the number of electrons, N, is simply ΔQ divided by the magnitude of the electron's charge: $N = \Delta Q/e$.

Known   Rating of battery = charge the battery can deliver = 2.60 A·h.
Unknown   Number of electrons, N = ?

SOLUTION

1. Calculate the charge, ΔQ, that corresponds to 2.60 A·h. Recall that 1 h = 3600 s:

   $\Delta Q = I\,\Delta t = (2.60\,A)(3600\,s) = 9360\,C$

2. Divide by the magnitude of the electron's charge, e, to find the number of electrons:

   $N = \dfrac{\Delta Q}{e} = \dfrac{9360\,C}{1.60 \times 10^{-19}\,C/electron}$

   $= 5.85 \times 10^{22}$ electrons

INSIGHT

Thus, even a typical cell phone battery can deliver an extremely large number of electrons.

PRACTICE PROBLEM

What A·h rating must a D-cell battery have to deliver 2.95×10^{23} electrons?   [**Answer:** 13.1 A·h]

Some related homework problems: Problem 1, Problem 2, Problem 3

When charge flows through a closed path and returns to its starting point, we refer to the closed path as an *electric circuit*. In this chapter we consider **direct-current circuits,** also known as dc circuits, in which the current always flows in the same direction. Two examples from nature are shown in **FIGURE 21-1**. Circuits with currents that periodically reverse their direction are referred to as **alternating-current circuits.** These ac circuits are considered in detail in Chapter 24.

PHYSICS IN CONTEXT
Looking Ahead

In Chapter 24 we extend our discussion of electric circuits from those in which the current flows in only one direction (dc) to circuits in which the current alternates in direction (ac, or alternating current).

► **FIGURE 21-1 Visualizing Concepts**
Electric Current Electric currents are not confined to the wires in our houses and machines, but occur in nature as well. A lightning bolt is simply an enormous, brief current. It flows when the difference in electric potential between cloud and ground (or cloud and cloud) becomes so great that it exceeds the breakdown strength of air. An enormous quantity of charge leaps across the gap in a fraction of a second. Some organisms, such as this electric torpedo ray, have internal organic "batteries" that can produce significant electric potentials. The resulting current is used to stun their prey.

Batteries and Electromotive Force

Although electrons move rather freely in metal wires, they do not flow unless the wires are connected to a source of electrical energy. A close analogy is water in a garden hose. Imagine that you and a friend each hold one end of a garden hose filled with water. If the two ends are held at the same level, as in **FIGURE 21-2 (a)**, the water does not flow. If, however, one end is raised above the other, as in **FIGURE 21-2 (b)**, water flows from the high end—where the gravitational potential energy is high—to the low end.

► **FIGURE 21-2 Water flow as an analogy for electric current** Water can flow quite freely through a garden hose, but if both ends are at the same level **(a)**, there is no flow. If the ends are held at different levels **(b)**, the water flows from the region where the gravitational potential energy is high to the region where it is low.

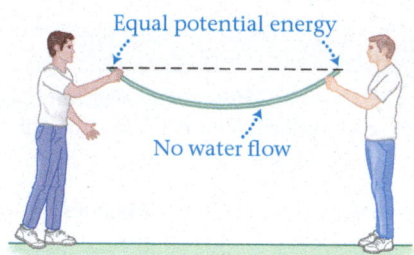
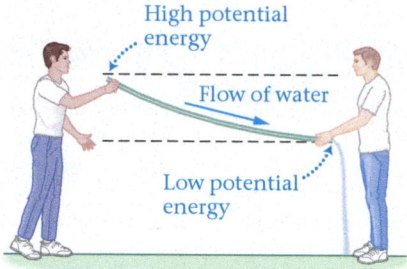

(a) No water flows when potential energies are equal.

(b) Water flows from high to low potential energy.

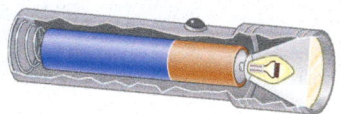

(a) A simple flashlight

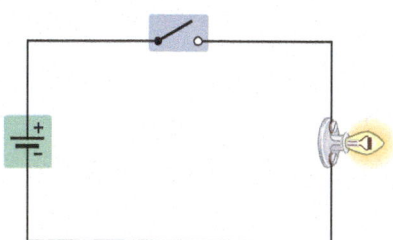

(b) Circuit diagram for flashlight

▲ **FIGURE 21-3 The flashlight: A simple electric circuit (a)** A simple flashlight, consisting of a battery, a switch, and a lightbulb. **(b)** When the switch is in the open position, the circuit is "broken" and no charge can flow. When the switch is closed, electrons flow through the circuit and the light glows.

A **battery** performs a similar function in an electric circuit. To put it simply, a battery uses chemical reactions to produce a difference in electric potential between its two ends, or **terminals.** The symbol for a battery is ⊣⊢. The terminal corresponding to a high electric potential is denoted by a +, and the terminal corresponding to a low electric potential is denoted by a −. When the battery is connected to a circuit, electrons move in a closed path from the negative terminal of the battery, through the circuit, and back to the positive terminal.

A simple example of an electrical system is shown in **FIGURE 21-3 (a)**, where we show a battery, a switch, and a lightbulb as they might be connected in a flashlight. In the schematic circuit shown in **FIGURE 21-3 (b)**, the switch is "open"—creating an **open circuit**—which means there is no closed path through which the electrons can flow. As a result, the light is off. When the switch is closed—which "closes" the circuit—charge flows around the circuit, causing the light to glow.

A mechanical analog to the flashlight circuit is shown in **FIGURE 21-4**. In this system, the person raising the water from a low to a high level is analogous to the battery, the paddle wheel is analogous to the lightbulb, and the water is analogous to the electric charge. Notice that the person does work in raising the water; later, as the water falls back to its original level, it does work on the external world by turning the paddle wheel.

When a battery is disconnected from a circuit and carries no current, the difference in electric potential between its terminals is referred to as its *electromotive force*, or *emf* ($\mathcal{E}$). It follows that the units of emf are the same as those of electric potential—namely, volts. Clearly, then, the electromotive force is not really a force at all. Instead, the emf determines the amount of work a battery does to move a certain amount of charge around a circuit (like the person lifting water in Figure 21-4). To be specific, the magnitude of the work done by a battery of emf $\mathcal{E}$ as a charge ΔQ is moved from one of its terminals to the other is given by Equation 20-2:

$$W = \Delta Q \mathcal{E}$$

We apply this relationship to a flashlight circuit in the following Quick Example.

QUICK EXAMPLE 21-2   OPERATING A FLASHLIGHT: FIND THE CHARGE AND THE WORK

A battery with an emf of 1.5 V delivers a current of 0.42 A to a flashlight bulb for 77 s (see Figure 21-3). Find **(a)** the charge that passes through the circuit and **(b)** the work done by the battery.

REASONING AND SOLUTION

We can find the charge delivered, ΔQ, by rearranging the definition of current, $I = \Delta Q / \Delta t$. Once the charge is determined, the work done is charge times emf: $W = \Delta Q \mathcal{E}$.

Part (a)

1. Use the definition of current, $I = \Delta Q / \Delta t$, to find the charge that flows through the circuit:

$$\Delta Q = I \Delta t$$
$$= (0.42 \text{ A})(77 \text{ s}) = 32 \text{ C}$$

Part (b)

2. Once we know ΔQ, we can use $W = \Delta Q \mathcal{E}$ to find the work:

$$W = \Delta Q \mathcal{E}$$
$$= (32 \text{ C})(1.5 \text{ V}) = 48 \text{ J}$$

The more charge a battery moves through a circuit, the more work it does. Similarly, the greater the emf, the greater the work. We can see, then, that a car battery that operates at 12 volts and delivers several amps of current does much more work than a flashlight battery—as expected.

Internal Resistance in a Battery The emf of a battery is the potential difference it can produce between its terminals under ideal conditions. In real batteries, however, there is always some internal loss, leading to a potential difference that is less than the ideal value. In fact, the greater the current flowing through a battery, the greater the reduction in potential difference between its terminals, as we shall see in Section 21-4. Only when the current is zero can a real battery produce its full emf. Because most batteries have relatively small internal losses, we shall treat batteries as ideal—always producing a potential difference precisely equal to $\mathcal{E}$—unless specifically stated otherwise.

Direction of Current Flow When we draw an electric circuit, it will be useful to draw an arrow indicating the flow of current. By convention, the direction of the current arrow is given in terms of a positive test charge, in much the same way that the direction of the electric field is determined:

> The direction of the current in an electric circuit is the direction in which a *positive* test charge would move.

Notice that a positive charge would flow from a region of high electric potential, near the positive terminal of the battery, to a region of low electric potential, near the negative terminal, as indicated in **FIGURE 21-5**. Of course, in a typical circuit the charges that flow are actually *negatively* charged electrons. As a result, the flow of electrons and the current arrow point in opposite directions, as shown in Figure 21-5.

Speed of Conduction Electrons It may seem surprising at first, but electrons move rather slowly through a typical wire. They suffer numerous collisions with the atoms in the wire, and hence their path is rather tortuous and roundabout, as indicated in **FIGURE 21-6**. Like a car contending with a series of speed bumps, the electron's average speed, or **drift speed** as it is often called, is limited by the repeated collisions—in fact, their average speed is commonly about 10^{-4} m/s. Thus, if you switch on the headlights of a car, for example, an electron leaving the battery will take about an hour to reach the lightbulb, yet the lights seem to shine from the instant the switch is turned on. How is this possible?

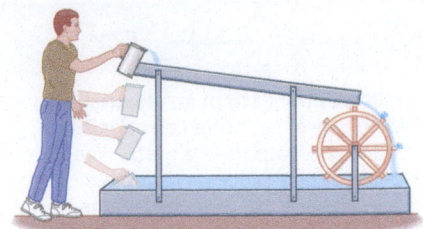

▲ **FIGURE 21-4 A mechanical analog to the flashlight circuit** The person lifting the water corresponds to the battery in Figure 21-3, and the paddle wheel corresponds to the lightbulb.

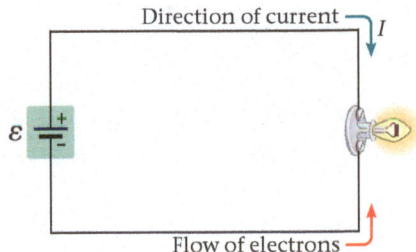

▲ **FIGURE 21-5 Direction of current and electron flow** In the flashlight circuit, electrons flow from the negative terminal of the battery to the positive terminal. The direction of the current, I, is just the opposite: from the positive terminal to the negative terminal.

Big Idea 1 Electric current is the flow of electric charge from one location to another. By convention, electric current points in the direction that positive charge would flow in a circuit, even though most electric currents are caused by the flow of negative charges.

PHYSICS IN CONTEXT
Looking Ahead

A dc circuit with a current flowing through it will play an important role in our discussion of magnetism in Chapter 22. We will also consider the magnetic force exerted on a current-carrying wire in Chapter 22.

▲ FIGURE 21-6 Path of an electron in a wire Typical path of an electron as it bounces off atoms in a metal wire. Because of the tortuous path the electron follows, its average velocity is rather small.

The answer is that as an electron begins to move away from the battery, it exerts a force on its neighbors, causing them to move in the same general direction and, in turn, to exert a force on their neighbors, and so on. This process generates a propagating influence that travels through the wire at nearly the speed of light. The phenomenon is analogous to a bowling ball hitting one end of a line of balls; the effect of the colliding ball travels through the line at roughly the speed of sound, although the individual balls have very little displacement. Similarly, the electrons in a wire move with a rather small average velocity as they collide with and bounce off the atoms making up the wire, whereas the influence they have on one another races ahead and causes the light to shine.

Enhance Your Understanding   (Answers given at the end of the chapter)

1. The following systems consist of a battery supplying current to a circuit. Rank the systems in order of increasing work done by the battery. Indicate ties where appropriate.

System	A	B	C	D
Battery emf (V)	3	9	12	15
Current (A)	1	2	5	10
Time (s)	40	10	2	1

Section Review

*   Current I is the amount of charge that flows in a given amount of time, $I = \Delta Q/\Delta t$.
*   The work done by a battery with an emf $\mathcal{E}$ as it supplies a charge ΔQ to a circuit is $W = \Delta Q \mathcal{E}$.

21-2 Resistance and Ohm's Law

Electrons flow through metal wires with relative ease. In the ideal case, they move freely. In real wires, however, electrons experience a **resistance** to their motion, in much the same way that friction slows a box sliding across the floor.

Ohm's Law   In order to move electrons against the resistance of a wire, it's necessary to apply a potential difference between its ends. For a wire with constant resistance R, the potential difference V necessary to create a current I is given by Ohm's law:

Ohm's Law

$$V = IR$$ 21-2

SI unit: volt, V

Ohm's law is named for the German physicist Georg Simon Ohm (1789–1854).
    Solving Ohm's law for the resistance, we find

$$R = \frac{V}{I}$$

From this expression it's clear that the units of resistance are volts per amp. In particular, we define 1 volt per amp to be 1 **ohm.** Letting the Greek letter omega, Ω, designate the ohm, we have

$$1\ \Omega = 1\ \text{V/A}$$

A device for measuring resistance is called an ohmmeter.
    In an electric circuit, we indicate a resistor with a zigzag line: ⌇⌇⌇. In contrast, a straight line in a circuit, ——, indicates an ideal wire of zero resistance.

Big Idea 2 A potential difference is needed to cause electric current to flow through a resistor. The greater the resistance, the greater the required potential difference.

▶ **FIGURE 21-7** A light-emitting diode (LED) is a relatively small, nonohmic device (top), but groups of LEDs can be used to form displays of practically any size (bottom). Because LEDs are extremely durable, and predicted to last 20 years or more, they are becoming the illumination of choice in high-reliability applications such as traffic lights, emergency exit signs, and brake lights. You'll probably see several on your way home today.

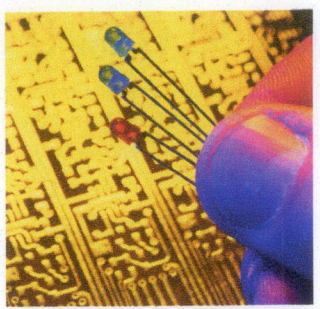

EXERCISE 21-3 USING OHM'S LAW

(a) A potential difference of 12 V is applied to a 250-Ω resistor. How much current flows through the resistor? (b) What is the voltage across a 150-Ω resistor that carries a current of 0.026 A?

REASONING AND SOLUTION
We can use Ohm's law directly, or rearrange it slightly, to solve both of these problems.

(a) Rearrange $V = IR$ to solve for the current, I:

$$I = \frac{V}{R} = \frac{12\,\text{V}}{250\,\Omega} = \frac{12\,\text{V}}{250\,\text{V/A}} = 0.048\,\text{A}$$

(b) Use Ohm's law in its standard form to solve for the voltage V:

$$V = IR = (0.026\,\text{A})(150\,\Omega) = 3.9\,\text{V}$$

It should be noted that Ohm's law is not a law of nature. Instead, it's a useful rule of thumb—like Hooke's law for springs or the ideal-gas laws for gases. Materials that are well approximated by Ohm's law are said to be "ohmic" in their behavior; they show a simple linear relationship between the voltage applied to them and the current that results. In particular, if one plots current versus voltage for an ohmic material, the result is a straight line, with a constant slope equal to $1/R$. We shall assume all materials considered in this text obey Ohm's law, unless specifically stated otherwise.

Nonohmic materials, on the other hand, have more complex relationships between voltage and current. A plot of current versus voltage for a nonohmic material is nonlinear, and hence the material does not have a constant resistance. (As an example, see Problem 7.) It is precisely these "nonlinearities," however, that can make such materials so useful in the construction of electronic devices, like the light-emitting diodes (LEDs) shown in **FIGURE 21-7**.

Resistivity

Suppose you have a piece of wire of length L and cross-sectional area A. The resistance of this wire depends on the particular material from which it is constructed. If the wire is made of copper, for instance, its resistance will be less than if it is made of iron. The quantity that characterizes the resistance of a given material is its **resistivity**, ρ. For a wire of given dimensions, the greater the resistivity, the greater the resistance.

The resistance of a wire also depends on its length and area. To understand the dependence on L and A, consider again the analogy of water flowing through a hose. If the hose is very long, the resistance it presents to the water will be correspondingly large, whereas a wider hose—one with a larger cross-sectional area—will offer less resistance to the water. After all, water flows more easily through a short fire hose than through a long soda straw; hence, the resistance of a hose—and similarly a piece of wire—should be proportional to L and inversely proportional to A; that is, proportional to (L/A).

Combining these observations, we can write the resistance of a wire of length L, area A, and resistivity ρ in the following way:

Definition of Resistivity, ρ

$$R = \rho\left(\frac{L}{A}\right) \qquad \text{21-3}$$

Because the units of L are m and the units of A are m², it follows that the units of resistivity are $(\Omega \cdot \text{m})$. Typical values for ρ are given in Table 21-1. Notice the enormous range in the values of ρ, with the resistivity of an insulator like rubber about 10^{21} times greater than the resistivity of a good conductor like silver!

TABLE 21-1 Resistivities

Substance	Resistivity, ρ ($\Omega \cdot$ m)
Insulators	
Quartz (fused)	7.5×10^{17}
Rubber	1 to 100×10^{13}
Glass	1 to $10{,}000 \times 10^{9}$
Semiconductors	
Silicon*	0.10 to 60
Germanium*	0.001 to 0.5
Conductors	
Lead	22×10^{-8}
Iron	9.71×10^{-8}
Tungsten	5.6×10^{-8}
Aluminum	2.65×10^{-8}
Gold	2.20×10^{-8}
Copper	1.68×10^{-8}
Copper, annealed	1.72×10^{-8}
Silver	1.59×10^{-8}

*The resistivity of a semiconductor varies greatly with the type and amount of impurities it contains. This property makes them particularly useful in electronic applications.

CONCEPTUAL EXAMPLE 21-4   COMPARE THE RESISTANCE

Wire 1 has a length L and a circular cross section of diameter D. Wire 2 is constructed from the same material as wire 1 and has the same shape, but its length is $2L$ and its diameter is $2D$. Is the resistance of wire 2 the same as that of wire 1, twice that of wire 1, or half that of wire 1?

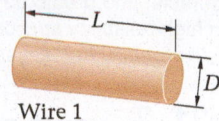

Wire 1

REASONING AND DISCUSSION

First, the resistance of wire 1 is

$$R_1 = \rho\left(\frac{L}{A}\right) = \rho\frac{L}{(\pi D^2/4)}$$

Notice that we have used the fact that the area of a circle of diameter D is $\pi D^2/4$. For wire 2 we replace L with $2L$ and D with $2D$:

$$R_2 = \rho\frac{2L}{[\pi(2D)^2/4]} = \left(\frac{1}{2}\right)\rho\frac{L}{(\pi D^2/4)} = \frac{1}{2}R_1$$

Wire 2

Thus, increasing the length by a factor of 2 increases the resistance by a factor of 2; on the other hand, increasing the diameter by a factor of 2 increases the area, and decreases the resistance, by a factor of 4. Overall, then, the resistance of wire 2 is half that of wire 1.

ANSWER

The resistance of wire 2 is half that of wire 1: $R_2 = R_1/2$.

EXAMPLE 21-5   A CURRENT-CARRYING WIRE

A current of 1.97 A flows through an annealed copper wire 2.25 m long and 1.47 mm in diameter. Find the potential difference between the ends of the wire. (The value of ρ for annealed copper can be found in Table 21-1.)

PICTURE THE PROBLEM

The wire carries a current $I = 1.97$ A, and its total length L is 2.25 m. We assume that the wire has a circular cross section, with a diameter $D = 1.47$ mm $= 0.00147$ m.

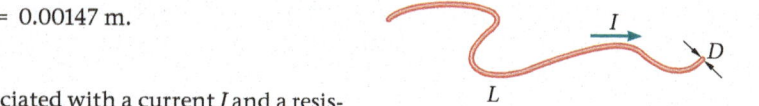

REASONING AND STRATEGY

We know from Ohm's law that the potential difference associated with a current I and a resistance R is $V = IR$. We are given the current in the wire, but not the resistance. The resistance is easily determined, however, using $R = \rho(L/A)$ with $A = \pi D^2/4$. Thus, we first calculate R and then substitute the result into $V = IR$ to obtain the potential difference.

Known   Current, $I = 1.97$ A; length of wire, $L = 2.25$ m; diameter of wire, $D = 0.00147$ m; resistivity of annealed copper, $\rho = 1.72 \times 10^{-8}$ Ω·m.
Unknown Potential difference, $V = ?$

SOLUTION

1. Calculate the resistance of the wire:

$$R = \rho\left(\frac{L}{A}\right) = \rho\left(\frac{L}{\pi D^2/4}\right)$$

$$= (1.72 \times 10^{-8}\,\Omega\cdot\text{m})\left[\frac{2.25\,\text{m}}{\pi(0.00147\,\text{m})^2/4}\right] = 0.0228\,\Omega$$

2. Multiply R by the current, I, to find the potential difference: $V = IR = (1.97\,\text{A})(0.0228\,\Omega) = 0.0449$ V

INSIGHT

Copper is an excellent conductor; therefore, both the resistance and the potential difference are quite small.

PRACTICE PROBLEM — PREDICT/CALCULATE

Suppose a second annealed copper wire has the same length as the one in this Example, but requires a greater potential difference to carry the same current. **(a)** Is the diameter of the second wire greater than or less than the diameter of the first wire? Explain. **(b)** Find the diameter of an annealed copper wire that carries a current of 1.97 A when the potential difference between its ends is 0.0760 V?   **[Answer: (a)** The diameter is smaller because this will produce a greater resistance. **(b)** 1.13 mm]

Some related homework problems: Problem 15, Problem 16

Temperature Dependence and Superconductivity

We know from everyday experience that a wire carrying an electric current can become warm—even quite hot, as in the case of a burner on a stove or the filament in an incandescent lightbulb. This follows from our earlier discussion of the fact that electrons collide with the atoms in a wire as they flow through an electric circuit. These collisions cause the atoms to jiggle with greater kinetic energy about their equilibrium positions. As a result, the temperature of the wire increases (see Section 17-2, and Equation 17-12 in particular). For example, the wire filament in an incandescent lightbulb can reach temperatures of roughly 2800 °C (in comparison, the surface of the Sun has a temperature of about 5500 °C), and the heating coil on a stove has a temperature of about 750 °C.

As a wire is heated, its resistivity tends to increase. This is because atoms that are jiggling more rapidly are more likely to collide with electrons and slow their progress through the wire. In fact, many metals show an approximately linear increase of ρ over a wide range of temperatures. Once the dependence of ρ on T is known for a given material, the change in resistivity can be used as a means of measuring temperature.

The first practical application of this principle was in a device known as the **bolometer.** Invented in 1880, the bolometer is an extremely sensitive thermometer that uses the temperature variation in the resistivity of platinum, nickel, or bismuth as a means of detecting temperature changes as small as 0.0001 C°. Soon after its invention, a bolometer was used to detect infrared radiation from the stars.

Some materials, like semiconductors, actually show a drop in resistivity as temperature is increased. This is because the resistivity of a semiconductor is strongly dependent on the number of electrons that are free to move about and conduct a current. As the temperature is increased in a semiconductor, more electrons are able to break free from their atoms, leading to an increased current and a reduced resistivity. Electronic devices incorporating such temperature-dependent semiconductors are known as **thermistors.** The digital fever thermometer so common in today's hospitals uses a thermistor to provide accurate measurements of a patient's temperature.

Superconductors   Resistivity typically increases with temperature; hence, it follows that if a wire is cooled below room temperature, its resistivity will *decrease*. Quite surprising, however, was a discovery made in the laboratory of Heike Kamerlingh Onnes in 1911. Measuring the resistance of a sample of mercury at temperatures just a few degrees above absolute zero, researchers found that at about 4.2 K the resistance of the mercury suddenly dropped to zero—not just to a very small value, but to *zero*. At this temperature, we say that the mercury becomes **superconducting,** a hitherto unknown phase of matter. Since that time many different superconducting materials have been discovered, with various different **critical temperatures,** T_c, at which superconductivity begins. Today we know that superconductivity is a result of quantum effects (Chapter 30).

When a material becomes superconducting, a current can flow through it with absolutely no resistance. In fact, if a current is initiated in a superconducting ring of material, it will flow undiminished for as long as the ring is kept cool enough. In some cases, circulating currents have been maintained for years, with absolutely no sign of diminishing.

In 1986, a new class of ceramic-based superconductors was discovered that has zero resistance at temperatures significantly higher than those of previously known superconducting materials. An example is shown in **FIGURE 21-8**. I remember attending the "Woodstock of physics" in March 1987, where thousands of physicists jammed a cavernous meeting room at the New York Hilton to hear the latest news about these new high-temperature (high-T_c) superconductors. We all knew a Nobel Prize in Physics would soon follow these discoveries, and indeed the prize was awarded for high-T_c superconductors that very fall. At the moment, the highest temperature at which superconductivity has been observed is about 133 K. Since the discovery of high-T_c superconductors, hopes have been raised that it may one day be possible to produce room-temperature superconductors. The practical benefits of such a breakthrough, including power transmission with no losses, improved MRI

▲ **FIGURE 21-8** When cooled below their critical temperature, superconductors not only lose their resistance to current flow but also exhibit new magnetic properties, such as repelling an external magnetic field. Here, a superconductor (bottom) levitates a small magnet.

scanners, and magnetically levitated trains, could be immense. Research continues along these lines.

Enhance Your Understanding   (Answers given at the end of the chapter)

2. If the voltage and resistance are doubled in a simple battery-resistor circuit, does the current increase, decrease, or stay the same? Explain.

Section Review

- Ohm's law, $V = IR$, is a rule of thumb relating the voltage V, resistance R, and current I in a resistor. The greater the voltage, the greater the current.

21-3 Energy and Power in Electric Circuits

When a charge ΔQ moves across a potential difference V, its electric potential energy, U, changes by the amount

$$\Delta U = (\Delta Q)V$$

Recalling that power is the rate at which energy changes, $P = \Delta U / \Delta t$, we can write the electrical power as follows:

$$P = \frac{\Delta U}{\Delta t} = \frac{(\Delta Q)V}{\Delta t}$$

Recalling that the electric current is given by $I = \Delta Q / \Delta t$, we have:

Electrical Power

$$P = IV \qquad\qquad\qquad \text{21-4}$$

SI unit: watt, W

Thus, a current of 1 amp flowing across a potential difference of 1 V produces a power of 1 W.

EXERCISE 21-6   BATTERY CURRENT

A handheld electric fan operates on a 3.00-V battery. If the power consumed by the fan is 1.77 W, what is the current supplied by the battery?

REASONING AND SOLUTION
The power of an electrical device is the current it draws times the voltage; that is, $P = IV$. Solving this relationship for the current, we find

$$I = \frac{P}{V} = \frac{1.77\ \text{W}}{3.00\ \text{V}} = 0.590\ \text{A}$$

The expression $P = IV$ applies to any electrical system. In the special case of a resistor, the electrical power is dissipated in the form of heat. Applying Ohm's law ($V = IR$) to this case, we can write the power dissipated in a resistor as

$$P = IV = I(IR) = I^2 R \qquad\qquad \text{21-5}$$

Similarly, using Ohm's law to solve for the current, $I = V/R$, we have

$$P = IV = \left(\frac{V}{R}\right)V = \frac{V^2}{R} \qquad\qquad \text{21-6}$$

These three power relationships ($P = IV = I^2 R = V^2/R$) are equivalent for resistors that obey Ohm's law, and which one we use depends on the quantities that are known in a given situation. These relationships also apply to incandescent lightbulbs, which are basically resistors that become hot enough to glow, and to heating elements, such as the one shown in **FIGURE 21-9.**

PHYSICS IN CONTEXT
Looking Back

The concept of electric potential energy (Chapter 20) is used here, where we talk about the energy associated with an electric circuit.

PHYSICS IN CONTEXT
Looking Back

We discuss the power of an electric circuit in this section. For this we refer back to mechanics, where power was originally introduced in Chapter 7.

▲ **FIGURE 21-9** The heating element of an electric space heater is nothing more than a length of resistive wire coiled up for compactness. As electric current flows through the wire, the power it dissipates ($P = I^2R$) is converted to heat and light. The coils near the center are the hottest, and hence they glow with a higher-frequency, yellowish light.

CONCEPTUAL EXAMPLE 21-7 COMPARE LIGHTBULBS

A battery that produces a potential difference V is connected to a 5-W lightbulb. Later, the 5-W lightbulb is replaced with a 10-W lightbulb. **(a)** In which case does the battery supply more current? **(b)** Which lightbulb has the greater resistance?

REASONING AND DISCUSSION

a. To compare the currents, we need consider only the relationship $P = IV$. Solving for the current yields $I = P/V$. When the voltage V is the same, it follows that the greater the power, the greater the current. In this case, then, the current in the 10-W bulb is twice the current in the 5-W bulb.

b. We now consider the relationship $P = V^2/R$, which gives resistance in terms of voltage and power. Rearranging, we find $R = V^2/P$. Again, with V the same, it follows that the smaller the power, the greater the resistance. Thus, the resistance of the 5-W bulb is twice that of the 10-W bulb.

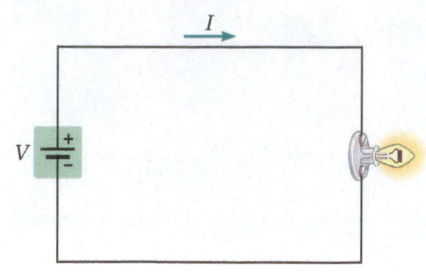

ANSWER

(a) When the battery is connected to the 10-W bulb, it delivers twice as much current as when it is connected to the 5-W bulb. **(b)** The 5-W bulb has twice as much resistance as the 10-W bulb.

On a microscopic level, the power dissipated by a resistor is the result of incessant collisions between electrons moving through the circuit and atoms making up the resistor. Specifically, the electric potential difference produced by the battery causes electrons to accelerate until they bounce off an atom of the resistor. At this point the electrons transfer energy to the atoms, causing them to jiggle more rapidly. The increased kinetic energy of the atoms is reflected in an increased temperature of the resistor (see Section 17-2). After each collision, the potential difference accelerates the electrons again and the process repeats—like a car bouncing through a series of speed bumps—resulting in a continuous transfer of energy from the electrons to the atoms.

Big Idea 3 Resistors dissipate energy in electric circuits in much the same way that friction acting on a sliding block dissipates mechanical energy.

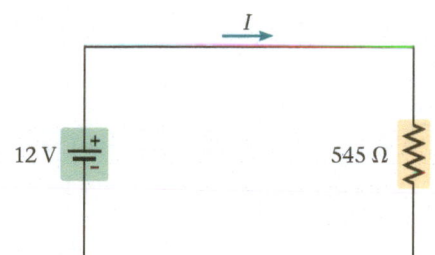

Resistance in Copper and Nichrome

EXAMPLE 21-8 HEATED RESISTANCE

A battery with an emf of 12 V is connected to a 545-Ω resistor. How much energy is dissipated in the resistor in 65 s?

PICTURE THE PROBLEM

The circuit, consisting of a battery and a resistor, is shown in our sketch. We show the current flowing from the positive terminal of the 12-V battery, through the 545-Ω resistor, and into the negative terminal of the battery.

REASONING AND STRATEGY

We know that a current flowing through a resistor dissipates power (energy per time), which means that the energy it dissipates in a given time is simply the power multiplied by the time: $\Delta U = P \Delta t$. The time is given ($\Delta t = 65$ s), and the power can be found using $P = IV$, $P = I^2R$, or $P = V^2/R$. The last expression is most convenient in this case, because the problem statement gives us the voltage and resistance.

To summarize, we first calculate the power, then multiply by the time.

Known Battery emf, $V = 12$ V; resistor, $R = 545\ \Omega$; time, $\Delta t = 65$ s.
Unknown Energy dissipated, $\Delta U = ?$

SOLUTION

1. Calculate the power dissipated in the resistor: $\qquad P = \dfrac{V^2}{R} = \dfrac{(12\ \mathrm{V})^2}{(545\ \Omega)} = 0.26\ \mathrm{W}$

2. Multiply the power by the time to find the energy dissipated: $\qquad \Delta U = P \Delta t = (0.26\ \mathrm{W})(65\ \mathrm{s}) = 17\ \mathrm{J}$

INSIGHT

The current in this circuit is $I = V/R = 0.022$ A. Using this result, we find that the same power is given by the other two power relationships; that is, $P = I^2R = IV = 0.26$ W, as expected.

PRACTICE PROBLEM

How much energy is dissipated in the resistor if the voltage is doubled to 24 V? [**Answer:** 4(17 J) = 68 J]

Some related homework problems: Problem 27, Problem 30

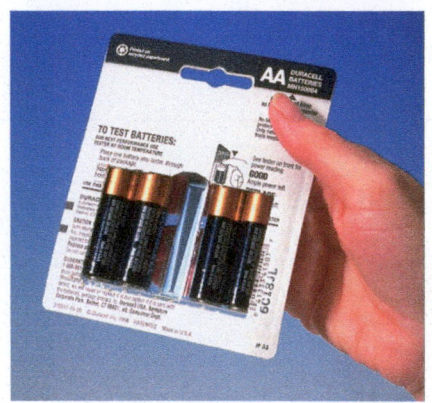

▲ **FIGURE 21-10** The battery testers now often built into battery packages (top photo) employ a tapered graphite strip. The narrow end (bottom of the lower photo) has the highest resistance, and thus produces the most heat when a current flows through the strip. The heat is used to produce the display on the front that indicates the strength of the battery—if the current is sufficient to warm even the top of the strip, where the resistance is lowest, the battery is fresh.

Battery Check Meters A commonly encountered application of resistance heating is found in the "battery check" meters often included with packs of batteries, as shown in **FIGURE 21-10**. To operate one of these meters, you simply press the contacts on either end of the meter against the corresponding terminals of the battery to be checked. This allows a current to flow through the main working element of the meter—a tapered strip of graphite.

The reason the strip is tapered is to provide a variation in resistance. According to the relationship given in Equation 21-3, $R = \rho(L/A)$, the smaller the cross-sectional area A of the strip, the larger the resistance R. It follows that the narrow end has a higher resistance than the wide end. Because the same current I flows through all parts of the strip, the power dissipated is expressed most conveniently in the form $P = I^2R$. It follows that at the narrow end of the strip, where R is largest, the heating due to the current will be the greatest. Pressing the meter against the terminals of the battery, then, results in an overall warming of the graphite strip, with the narrow end warmer than the wide end.

The final element in the meter is a thin layer of liquid crystal (similar to the material used in liquid-crystal displays) that responds to small increases in temperature. In particular, this liquid crystal is black and opaque at room temperature but transparent when heated slightly. The liquid crystal is placed in front of a colored background, which can be seen in those regions where the graphite strip is warm enough to make the liquid crystal transparent. If the battery is weak, only the narrow portion of the strip becomes warm enough, and the meter shows only a small stripe of color. A strong battery, on the other hand, heats the entire strip enough to make the liquid crystal transparent, resulting in a colored stripe the full length of the meter.

Energy Usage

When you get a bill from the local electric company, you will find the number of kilowatt-hours of electricity that you have used. Notice that a kilowatt-hour (kWh) has the units of energy:

$$1 \text{ kilowatt-hour} = (1000 \text{ W})(3600 \text{ s}) = (1000 \text{ J/s})(3600 \text{ s}) = 3.6 \times 10^6 \text{ J}$$

Thus, the electric company is charging for the amount of energy you use—as one would expect—and not for the rate at which you use it. The following Example considers the energy and monetary cost for a typical everyday situation.

EXAMPLE 21-9 | YOUR GOOSE IS COOKED

A holiday goose is cooked in the kitchen oven for 4.00 h. Assume that the stove draws a current of 20.0 A, operates at a voltage of 220.0 V, and uses electrical energy that costs $0.068 per kWh. How much does it cost to cook your goose?

PICTURE THE PROBLEM

We show a schematic representation of the stove cooking the goose in our sketch. The current in the circuit is 20.0 A, and the voltage difference across the heating coils is 220 V.

REASONING AND STRATEGY

The cost is simply the energy usage (in kWh) times the cost per kilowatt-hour ($0.068). To find the energy used, we note that energy is power multiplied by time. The time is given, and the power associated with a current I and a voltage V is $P = IV$.

Thus, we find the power, multiply by the time, and then multiply by $0.068 to find the cost.

Known Time of cooking, $\Delta t = 4.00$ h; current, $I = 20.0$ A; voltage, $V = 220.0$ V; cost of electricity = $0.068 per kWh.

Unknown Cost to cook your goose = ?

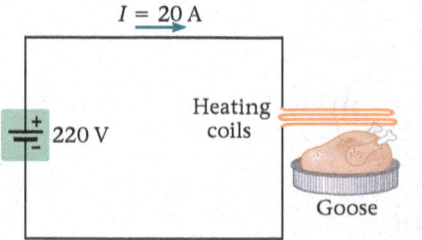

CONTINUED

SOLUTION

1. Calculate the power delivered to the stove:

$P = IV = (20.0\,\text{A})(220.0\,\text{V}) = 4.40\,\text{kW}$

2. Multiply power by time to determine the total energy supplied to the stove during the 4.00 h of cooking:

$\Delta U = P\,\Delta t = (4.40\,\text{kW})(4.00\,\text{h}) = 17.6\,\text{kWh}$

3. Multiply by the cost per kilowatt-hour to find the total cost of cooking:

$\text{cost} = (17.6\,\text{kWh})(\$0.068/\text{kWh}) = \$1.20$

INSIGHT

Thus, your goose can be cooked for just over a dollar.

PRACTICE PROBLEM

If the voltage and current are reduced by a factor of 2 each, for how much time must the goose be cooked to use the same amount of energy?   [**Answer:** 4(4.00 h) = 16.0 h. *Note:* You should be able to answer a question like this by referring to your previous solution, without repeating the calculation in detail.]

Some related homework problems: Problem 27, Problem 28

Enhance Your Understanding   (Answers given at the end of the chapter)

3. In the following systems, two of the quantities I, V, and R are known for a given resistor. Rank the systems in order of increasing power dissipated in the resistor. Indicate ties where appropriate.

System	A	B	C	D
I (A)	1	—	2	5
V (V)	5	12	—	9
R (Ω)	—	100	300	—

Section Review

- The power P dissipated in an electric circuit is the product of the current and voltage: $P = IV$.

- In the specific case of a resistor that obeys Ohm's law, the power dissipated can be written as $P = I^2R$ or $P = V^2/R$. These expressions are equivalent, and either one can be used depending on the quantities that are known.

21-4 Resistors in Series and Parallel

Electric circuits often contain a number of resistors connected in various ways. In this section we consider simple circuits that contain only resistors and batteries. For each type of circuit considered, we calculate the **equivalent resistance** produced by a group of individual resistors.

Resistors in Series

When resistors are connected one after the other, like dancers in a conga line, we say that they are in *series*. **FIGURE 21-11 (a)** shows three resistors, R_1, R_2, and R_3, connected in series. The three resistors acting together have the same effect—that is, they draw the same current—as a single resistor, referred to as the equivalent resistor, R_{eq}. This equivalence is illustrated in **FIGURE 21-11 (b)**. We now calculate the value of the equivalent resistance.

The first thing to notice about the circuit in Figure 21-11 (a) is that the same current I must flow through each of the resistors—there is no other place for the current to go. As a result, the potential differences across the three resistors are $V_1 = IR_1$, $V_2 = IR_2$, and $V_3 = IR_3$. The total potential difference from point A to point B must be the emf of the battery, $\mathcal{E}$, and hence it follows that

$$\mathcal{E} = V_1 + V_2 + V_3$$

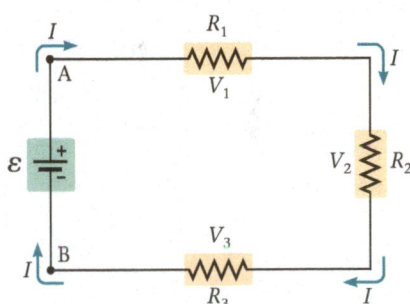

(a) Three resistors in series

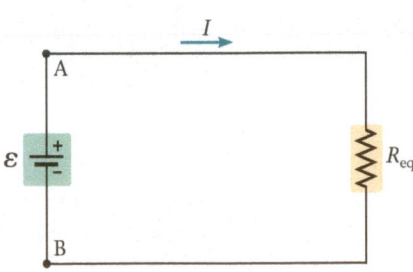

(b) Equivalent resistance has the same current

▲ **FIGURE 21-11 Resistors in series**
(a) Three resistors, R_1, R_2, and R_3, connected in series. Note that the same current I flows through each resistor. **(b)** The equivalent resistance, $R_{eq} = R_1 + R_2 + R_3$, has the same current I flowing through it as the current I in the original circuit.

Writing each of the potentials in terms of the current and resistance, we find

$$\mathcal{E} = IR_1 + IR_2 + IR_3 = I(R_1 + R_2 + R_3)$$

Now, let's compare this expression with the result we obtain for the equivalent circuit in Figure 21-11 (b). In this circuit, the potential difference across the battery is $V = IR_{eq}$. This potential must also be the same as the emf of the battery, and thus

$$\mathcal{E} = IR_{eq}$$

Comparing this expression with $\mathcal{E} = I(R_1 + R_2 + R_3)$, we see that the equivalent resistance is simply the sum of the individual resistances:

$$R_{eq} = R_1 + R_2 + R_3$$

In general, for any number of resistors in series, the equivalent resistance is the sum of the individual resistances:

Equivalent Resistance for Resistors in Series

$$R_{eq} = R_1 + R_2 + R_3 + \cdots = \sum R \qquad \text{21-7}$$

SI unit: ohm, Ω

Notice that the equivalent resistance is greater than the greatest resistance of any of the individual resistors. Connecting the resistors in series is like making a single resistor increasingly longer; as its length increases, so too does its resistance.

EXAMPLE 21-10   THREE RESISTORS IN SERIES

A circuit consists of three resistors connected in series to a 24.0-V battery. The current in the circuit is 0.0466 A. Given that $R_1 = 225\ \Omega$ and $R_2 = 155\ \Omega$, find **(a)** the value of R_3 and **(b)** the potential difference across each resistor.

PICTURE THE PROBLEM
The circuit is shown in our sketch. Notice that the 24.0-V battery delivers the same current, $I = 0.0466$ A, to each of the three resistors. This is the key characteristic of a series circuit.

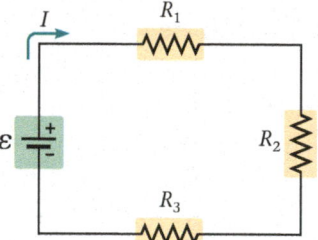

REASONING AND STRATEGY

a. First, we can obtain the equivalent resistance of the circuit using Ohm's law (as in Equation 21-2): $R_{eq} = \mathcal{E}/I$. Because the resistors are in series, we also know that $R_{eq} = R_1 + R_2 + R_3$. We can solve this relationship for the only unknown, R_3.

b. We can now calculate the potential difference across each resistor using Ohm's law, $V = IR$.

Known   Emf of battery, $\mathcal{E} = 24.0$ V; current, $I = 0.0466$ A; $R_1 = 225\ \Omega$; $R_2 = 155\ \Omega$.
Unknown   **(a)** $R_3 = ?$ **(b)** $V_1 = ?$; $V_2 = ?$; $V_3 = ?$

SOLUTION

Part (a)

1. Use Ohm's law to find the equivalent resistance of the circuit:
$$R_{eq} = \frac{\mathcal{E}}{I} = \frac{24.0\ \text{V}}{0.0466\ \text{A}} = 515\ \Omega$$

2. Set R_{eq} equal to the sum of the individual resistances, and solve for R_3:
$$R_{eq} = R_1 + R_2 + R_3$$
$$R_3 = R_{eq} - R_1 - R_2$$
$$= 515\ \Omega - 225\ \Omega - 155\ \Omega = 135\ \Omega$$

Part (b)

3. Use Ohm's law to determine the potential difference across R_1:
$$V_1 = IR_1 = (0.0466\ \text{A})(225\ \Omega) = 10.5\ \text{V}$$

4. Find the potential difference across R_2:
$$V_2 = IR_2 = (0.0466\ \text{A})(155\ \Omega) = 7.22\ \text{V}$$

5. Find the potential difference across R_3:
$$V_3 = IR_3 = (0.0466\ \text{A})(135\ \Omega) = 6.29\ \text{V}$$

INSIGHT
Notice that the sum of the individual potential differences is 10.5 V + 7.22 V + 6.29 V = 24.0 V, as expected.

PRACTICE PROBLEM
Find the power dissipated in each resistor.   [**Answer:** Using $P = I^2R$, we find $P_1 = 0.489$ W, $P_2 = 0.337$ W, $P_3 = 0.293$ W.]

Some related homework problems: Problem 40, Problem 41

Internal Resistance An everyday example of resistors in series is the **internal resistance**, r, of a battery. As was mentioned in Section 21-1, real batteries have internal losses that cause the potential difference between their terminals to be less than ε and to depend on the current in the battery. The simplest way to model a real battery is to imagine it consists of an ideal battery of emf ε in series with an internal resistance r, as shown in **FIGURE 21-12**. If this battery is then connected to an external resistance, R, the equivalent resistance of the circuit is $r + R$. As a result, the current flowing through the circuit is $I = \varepsilon/(r + R)$, and the potential difference between the terminals of the battery is $\varepsilon - Ir$. Thus, we see that the potential difference produced by the battery is less than ε by an amount that is proportional to the current I. Only in the limit of zero current, or zero internal resistance, will the battery produce its full emf. (See Problems 48 and 51 for examples of batteries with internal resistance.)

Resistors in Parallel

Resistors are in *parallel* when they are connected across the same potential difference, as in **FIGURE 21-13 (a)**. In a case like this, which is similar to parallel lanes of traffic, the current has parallel paths through which it can flow. As a result, the total current in the circuit, I, is equal to the sum of the currents through each of the three resistors:

$$I = I_1 + I_2 + I_3$$

Because the potential difference is the same for each of the resistors, it follows that the currents flowing through them are as follows:

$$I_1 = \frac{\varepsilon}{R_1}, \quad I_2 = \frac{\varepsilon}{R_2}, \quad I_3 = \frac{\varepsilon}{R_3}$$

Summing these three currents, we find

$$I = \frac{\varepsilon}{R_1} + \frac{\varepsilon}{R_2} + \frac{\varepsilon}{R_3} = \varepsilon\left(\frac{1}{R_1} + \frac{1}{R_2} + \frac{1}{R_3}\right) \qquad \text{21-8}$$

Now, in the equivalent circuit shown in **FIGURE 21-13 (b)**, Ohm's law gives $\varepsilon = IR_{eq}$, or

$$I = \varepsilon\left(\frac{1}{R_{eq}}\right) \qquad \text{21-9}$$

Comparing Equations 21-8 and 21-9, we find that the equivalent resistance for three resistors in parallel is

$$\frac{1}{R_{eq}} = \frac{1}{R_1} + \frac{1}{R_2} + \frac{1}{R_3}$$

In general, for any number of resistors in parallel, we have:

Equivalent Resistance for Resistors in Parallel

$$\frac{1}{R_{eq}} = \frac{1}{R_1} + \frac{1}{R_2} + \frac{1}{R_3} + \cdots = \sum \frac{1}{R} \qquad \text{21-10}$$

SI unit: ohm, Ω

As a simple example, consider a circuit with two identical resistors R connected in parallel. The equivalent resistance in this case is given by

$$\frac{1}{R_{eq}} = \frac{1}{R} + \frac{1}{R} = \frac{2}{R}$$

Solving for R_{eq}, we find $R_{eq} = \frac{1}{2}R$. If we connect three such resistors in parallel, the corresponding result is

$$\frac{1}{R_{eq}} = \frac{1}{R} + \frac{1}{R} + \frac{1}{R} = \frac{3}{R}$$

In this case, $R_{eq} = \frac{1}{3}R$. Thus, the more resistors we connect in parallel, the smaller the equivalent resistance. Each time we add a new resistor in parallel with the others,

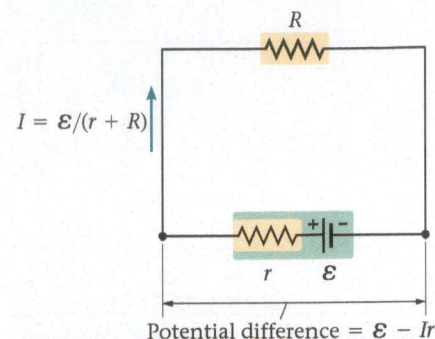

$I = \varepsilon/(r + R)$

Potential difference $= \varepsilon - Ir$

▲ **FIGURE 21-12 The internal resistance of a battery** Real batteries always dissipate some energy in the form of heat. These losses can be modeled by a small "internal" resistance, r, within the battery. As a result, the potential difference between the terminals of a real battery is less than its ideal emf, ε. For example, if a battery produces a current I, the potential difference between its terminals is $\varepsilon - Ir$. In the case shown here, a battery is connected in series with the resistor R. Instead of producing the current $I = \varepsilon/R$, as in the ideal case, it produces the current $I = \varepsilon/(r + R)$.

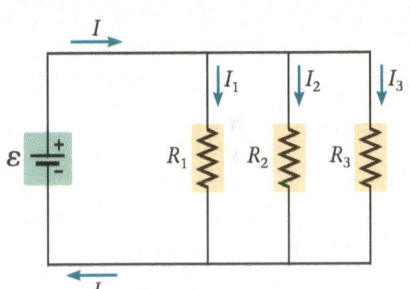

(a) Three resistors in parallel

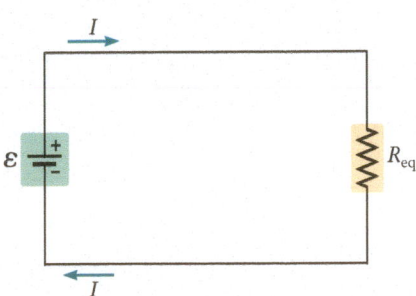

(b) Equivalent resistance has the same current

▲ **FIGURE 21-13 Resistors in parallel (a)** Three resistors, R_1, R_2, and R_3, connected in parallel. Note that each resistor is connected across the same potential difference ε. **(b)** The equivalent resistance, $1/R_{eq} = 1/R_1 + 1/R_2 + 1/R_3$, has the same current flowing through it as the total current I in the original circuit.

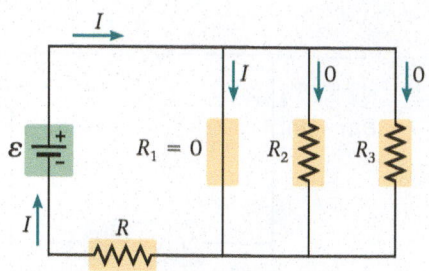

we give the battery a new path through which current can flow—analogous to opening an additional lane of traffic on a busy highway. Stated another way, giving the current multiple paths through which it can flow is equivalent to using a wire with a greater cross-sectional area. From either point of view, the fact that more current flows with the same potential difference means that the equivalent resistance has been reduced.

Finally, if any one of the resistors in a parallel connection is equal to zero, the equivalent resistance is also zero. This situation is referred to as a **short circuit,** and is illustrated in **FIGURE 21-14**. In this case, all of the current flows through the path of zero resistance. In general, the equivalent resistance of resistors in parallel is less than or equal to the resistance of the smallest resistor.

PROBLEM-SOLVING NOTE

The Equivalent Resistance of Resistors in Parallel

After summing the inverses of resistors in parallel, remember to take one more inverse at the end of your calculation to find the equivalent resistance.

EXAMPLE 21-11 | THREE RESISTORS IN PARALLEL

Consider a circuit with three resistors, $R_1 = 225\ \Omega$, $R_2 = 155\ \Omega$, and $R_3 = 135\ \Omega$, connected in parallel with a 24.0-V battery. (These are the same resistances and battery as in the series circuit in Example 21-10.) Find **(a)** the total current supplied by the battery and **(b)** the current through each resistor.

PICTURE THE PROBLEM
The sketch indicates the parallel connection of the resistors with the battery. Each of the resistors experiences precisely the same potential difference—namely, the 24.0 V produced by the battery. This is the feature that characterizes parallel connections.

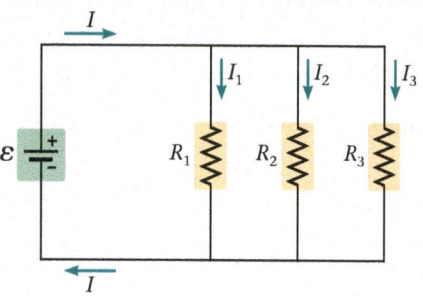

REASONING AND STRATEGY

a. We can find the total current from $I = \varepsilon/R_{eq}$, where $1/R_{eq} = 1/R_1 + 1/R_2 + 1/R_3$.

b. For each resistor, the current is given by Ohm's law, $I = \varepsilon/R$.

Known Emf of battery, $\varepsilon = 24.0\ \text{V}$; $R_1 = 225\ \Omega$; $R_2 = 155\ \Omega$; $R_3 = 135\ \Omega$.
Unknown **(a)** Total current, $I = ?$ **(b)** $I_1 = ?$; $I_2 = ?$; $I_3 = ?$

SOLUTION

Part (a)

1. Find the equivalent resistance of the circuit:

$$\frac{1}{R_{eq}} = \frac{1}{R_1} + \frac{1}{R_2} + \frac{1}{R_3}$$

$$= \frac{1}{225\ \Omega} + \frac{1}{155\ \Omega} + \frac{1}{135\ \Omega} = 0.0183\ \Omega^{-1}$$

$$R_{eq} = (0.0183\ \Omega^{-1})^{-1} = 54.6\ \Omega$$

2. Use Ohm's law to find the total current:

$$I = \frac{\varepsilon}{R_{eq}} = \frac{24.0\ \text{V}}{54.6\ \Omega} = 0.440\ \text{A}$$

Part (b)

3. Calculate I_1 using $I_1 = \varepsilon/R_1$ with $\varepsilon = 24.0\ \text{V}$:

$$I_1 = \frac{\varepsilon}{R_1} = \frac{24.0\ \text{V}}{225\ \Omega} = 0.107\ \text{A}$$

4. Repeat the preceding calculation for resistors 2 and 3:

$$I_2 = \frac{\varepsilon}{R_2} = \frac{24.0\ \text{V}}{155\ \Omega} = 0.155\ \text{A}$$

$$I_3 = \frac{\varepsilon}{R_3} = \frac{24.0\ \text{V}}{135\ \Omega} = 0.178\ \text{A}$$

INSIGHT
As expected, the equivalent resistance, $R_{eq} = 54.6\ \Omega$, is less than the smallest resistance, $R_3 = 135\ \Omega$. In addition, the smallest resistor carries the greatest current, and the three currents combined yield the total current; that is, $I_1 + I_2 + I_3 = 0.107\ \text{A} + 0.155\ \text{A} + 0.178\ \text{A} = 0.440\ \text{A} = I$. Comparing with the same three resistors connected in series (Example 21-10), we see that the current in the parallel circuit (0.440 A) is 9.4 times greater than the current in the series circuit (0.0466 A). In general, parallel circuits tend to have more current because their equivalent resistances are small.

PRACTICE PROBLEM
Find the power dissipated in each resistor. [**Answer:** Using $P = V^2/R$, we find $P_1 = 2.56\ \text{W}$, $P_2 = 3.72\ \text{W}$, $P_3 = 4.27\ \text{W}$.]

Some related homework problems: Problem 42, Problem 43

Power in Series and Parallel   In comparing Examples 21-10 and 21-11, notice the differences in the power dissipated in each circuit. First, the total power dissipated in the parallel circuit (10.6 W) is much greater than that dissipated in the series circuit (1.12 W). This is due to the fact that the equivalent resistance of the parallel circuit is smaller than the equivalent resistance of the series circuit, and the power delivered by a voltage V to a resistance R is inversely proportional to the resistance ($P = V^2/R$). In addition, notice that the smallest resistor, R_3, has the smallest power in the series circuit (0.293 W), but it has the largest power in the parallel circuit (4.27 W). These issues are explored further in the following Conceptual Example.

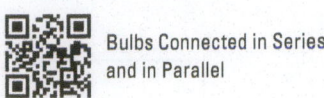

Bulbs Connected in Series and in Parallel

CONCEPTUAL EXAMPLE 21-12   SERIES VERSUS PARALLEL   PREDICT/EXPLAIN

Two identical lightbulbs are connected to a battery, either in series or in parallel. **(a)** When the bulbs are connected in series are they brighter than, dimmer than, or the same brightness as when they are connected in parallel? **(b)** Which of the following is the *best explanation* for your prediction?

  I. The bulbs are brighter when connected in series because the same current passes through each of the bulbs, as compared to the parallel case where the current is split between the two bulbs.

 II. The bulbs are dimmer when connected in series because the voltage across each bulb is less in series than it is when they are connected in parallel.

III. The bulbs have the same brightness in either case because the same bulbs are connected to the same battery.

REASONING AND DISCUSSION

When the bulbs are connected in parallel, the voltage across each one is V, and hence the power dissipated by each bulb is $P = V^2/R$. When the bulbs are connected in series, the voltage across each one is only $V/2$, and hence each bulb dissipates the power $P = (V/2)^2/R = \frac{1}{4}(V^2/R)$—only one-quarter of the power dissipated in parallel. As a result, the bulbs are dimmer in series than in parallel.

ANSWER

(a) The bulbs connected in series are dimmer than the bulbs connected in parallel. **(b)** The best explanation is II.

RWP* Notice from Equation 21-7 that series circuits tend to have high resistance and low current, and from Equation 21-10 that parallel circuits tend to have low resistance and high current. This has a bearing on the electrical design of holiday lights. Ideally, the lights should be wired in parallel, so that if one light burns out, the other lights remain lit. However, a parallel connection requires a very large resistance for each bulb in order to limit the current and the power consumed, a resistance that is difficult to achieve with a standard wire filament or LED bulb. The clever solution featured by many holiday lights is to wire the lights in series, but equip each bulb with a shunt resistor that activates if the bulb burns out, ensuring that current can continue to flow through the other lights. The drawback is that the equivalent resistance of the circuit decreases whenever a light burns out, which increases the current through the circuit and accelerates the failure rate of the other lights. This is why holiday lights come with the warning, "To reduce the risk of overheating, replace burned out lamps promptly."

Combination Circuits

The rules we've developed for series and parallel resistors can be applied to more complex circuits as well. For example, consider the circuit shown in **FIGURE 21-15 (a)**, where four resistors, each equal to R, are connected in a way that combines series and parallel features. To analyze this circuit, we first note that the two vertically oriented resistors are connected in parallel with one another. Therefore, the equivalent resistance of this unit is given by $1/R_{eq} = 1/R + 1/R$, or $R_{eq} = R/2$. Replacing these two resistors with $R/2$ yields the circuit shown in **FIGURE 21-15 (b)**, which consists of three resistors in series. As a result, the equivalent resistance of the entire circuit is $R + R/2 + R = 2.5R$, as indicated in **FIGURE 21-15 (c)**. Similar methods can be applied to a wide variety of circuits.

PROBLEM-SOLVING NOTE

Analyzing a Complex Circuit

When considering an electric circuit with resistors in series and parallel, work from the smallest units of the circuit outward to ever larger units.

---

*Real World Physics applications are denoted by the acronym RWP.

▲ **FIGURE 21-15 Analyzing a complex circuit of resistors** (a) The two vertical resistors are in parallel with one another; hence, they can be replaced with their equivalent resistance, $R/2$. (b) Now the circuit consists of three resistors in series. The equivalent resistance of these three resistors is 2.5 R. (c) The original circuit reduced to a single equivalent resistance.

(a) Replace parallel resistors   (b) Replace series resistors   (c) Final equivalent resistance

EXAMPLE 21-13   COMBINATION SPECIAL

In the circuit shown in the sketch, the emf of the battery is 12.0 V, and each resistor has a resistance of 200.0 Ω. Find **(a)** the current supplied by the battery to this circuit and **(b)** the current through the lower two resistors.

PICTURE THE PROBLEM

The circuit for this problem has three resistors connected to a battery. The lower two resistors are in series with one another, and in parallel with the upper resistor. The battery has an emf of 12.0 V.

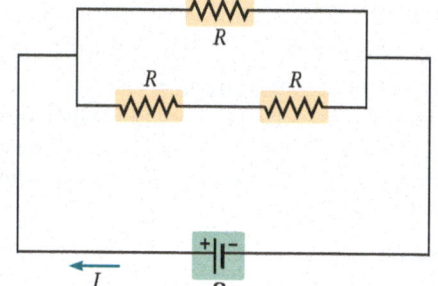

REASONING AND STRATEGY

a. The current supplied by the battery, I, is given by Ohm's law, $I = \mathcal{E}/R_{eq}$, where R_{eq} is the equivalent resistance of the three resistors. To find R_{eq}, we first note that the lower two resistors are in series, giving a net resistance of $2R$. Next, the upper resistor, R, is in parallel with $2R$. Calculating this equivalent resistance yields the desired R_{eq}.

b. Because the voltage across the lower two resistors is $\mathcal{E}$, the current through them is $I_{lower} = \mathcal{E}/R_{eq,lower} = \mathcal{E}/2R$.

Known   Emf of battery, $\mathcal{E} = 12.0$ V; resistance of each resistor, $R = 200.0\,\Omega$.
Unknown   (a) Current supplied by battery, $I = ?$ (b) Current through lower resistors, $I_{lower} = ?$

SOLUTION

Part (a)

1. Calculate the equivalent resistance of the lower two resistors:

$$R_{eq,lower} = R + R = 2R$$

2. Calculate the equivalent resistance of R in parallel with $2R$:

$$\frac{1}{R_{eq}} = \frac{1}{R} + \frac{1}{2R} = \frac{3}{2R}$$

$$R_{eq} = \tfrac{2}{3}R = \tfrac{2}{3}(200.0\,\Omega) = 133.3\,\Omega$$

3. Find the current supplied by the battery, I:

$$I = \frac{\mathcal{E}}{R_{eq}} = \frac{12.0\text{ V}}{133.3\,\Omega} = 0.0900\text{ A}$$

Part (b)

4. Use $\mathcal{E}$ and $R_{eq,lower}$ to find the current in the lower two resistors:

$$I_{lower} = \frac{\mathcal{E}}{R_{eq,lower}} = \frac{12.0\text{ V}}{2(200.0\,\Omega)} = 0.0300\text{ A}$$

INSIGHT

The total resistance of the three 200.0-Ω resistors is less than 200.0 Ω—in fact, it is only 133.3 Ω. Notice also that 0.0300 A flows through the lower two resistors, and therefore twice that much—0.0600 A—flows through the upper resistor.

PRACTICE PROBLEM — PREDICT/CALCULATE

Suppose the upper resistor is changed from R to $2R$, and the lower two resistors remain the same. **(a)** Will the current supplied by the battery increase, decrease, or stay the same? Explain. **(b)** Find the new current.   [**Answer:** (a) The current will decrease because there is greater resistance to its flow. (b) 0.0600 A]

Some related homework problems: Problem 45, Problem 46, Problem 48

A similar combination circuit is shown in **FIGURE 21-16**. Originally (left), identical bulbs 1 and 2 are in series with a battery and are equally bright. Next (right), a third identical bulb (3) is added to the circuit in parallel with bulb 2. How does this affect the brightness of bulbs 1 and 2? Well, adding bulb 3 creates a new path for the current and increases the total current in the circuit by a factor of $\frac{4}{3}$ (check this yourself). This increases the brightness of bulb 1. The current passing through bulb 1 is equally split between bulbs 2 and 3, and hence the new current in bulb 2 is $\frac{1}{2}\left(\frac{4}{3}\right) = \frac{2}{3}$ of its original value. Thus, bulb 2 becomes dimmer. Bulb 3 is in parallel with bulb 2, and has the same brightness.

Enhance Your Understanding

(Answers given at the end of the chapter)

4. The two circuits shown in **FIGURE 21-17** have identical batteries and resistors, but the arrangement of the resistors is different. Is the current in circuit 1 greater than, less than, or equal to the current in circuit 2? Explain.

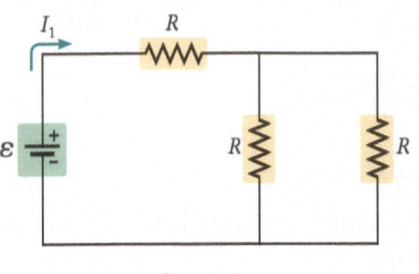

Circuit 1

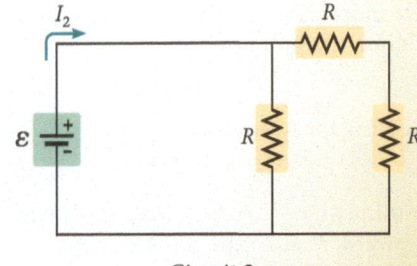

Circuit 2

▲ **FIGURE 21-17** Enhance Your Understanding 4.

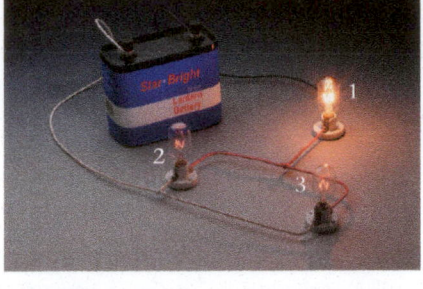

▲ **FIGURE 21-16** Adding a third bulb to this circuit changes the brightness of each of the bulbs. The changes can be understood in terms of the properties of resistors in series and parallel.

Section Review

- The equivalent resistance of resistors connected in series is the sum of the individual resistances: $R_{eq} = R_1 + R_2 + R_3 + \cdots = \sum R$.

- The equivalent resistance of resistors connected in parallel is the inverse of the sum of the inverses of the individual resistances: $\frac{1}{R_{eq}} = \frac{1}{R_1} + \frac{1}{R_2} + \frac{1}{R_3} + \cdots = \sum \frac{1}{R}$.

21-5 Kirchhoff's Rules

To find the currents and voltages in a general electric circuit—even one that can't be broken up into parts with resistors in parallel and other parts with resistors in series—we use two rules first introduced by the German physicist Gustav Kirchhoff (1824–1887). The *Kirchhoff rules* are simply ways of expressing charge conservation (the junction rule) and energy conservation (the loop rule) in a closed circuit. These conservation laws are always obeyed in nature, and hence the Kirchhoff rules are completely general.

The Junction Rule

The junction rule follows from the observation that the current entering any point in a circuit must equal the current leaving that point. If this were not the case, charge would either build up or disappear from a circuit.

As an example, consider the circuit shown in **FIGURE 21-18**. At point A, three wires join to form a **junction.** (In general, a *junction* is any point in a circuit where three or more wires meet.) The current carried by each of the three wires is indicated in the figure. Notice that the current entering the junction is I_1; the current leaving the junction is $I_2 + I_3$. Setting the incoming and outgoing currents equal, we have $I_1 = I_2 + I_3$, or equivalently

$$I_1 - I_2 - I_3 = 0$$

This is Kirchhoff's junction rule applied to the junction at point A.

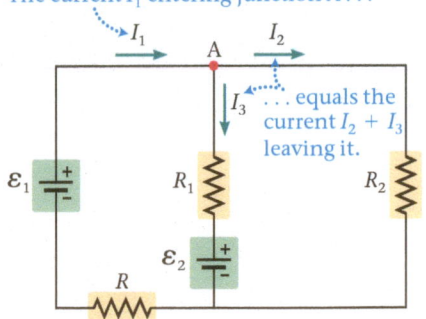

The current I_1 entering junction A . . .

. . . equals the current $I_2 + I_3$ leaving it.

▲ **FIGURE 21-18 Kirchhoff's junction rule** Kirchhoff's junction rule states that the sum of the currents entering a junction must equal the sum of the currents leaving the junction. In this case, for the junction labeled A, $I_1 = I_2 + I_3$ or $I_1 - I_2 - I_3 = 0$.

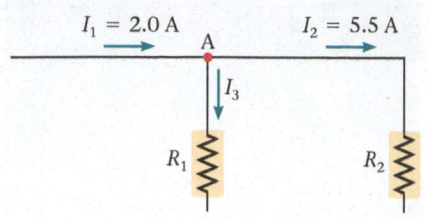

▲ **FIGURE 21-19 A specific application of Kirchhoff's junction rule** Applying Kirchhoff's junction rule to the junction A, $I_1 - I_2 - I_3 = 0$, yields the result $I_3 = -3.5$ A. The minus sign indicates that I_3 flows opposite to the direction shown; that is, I_3 is actually upward.

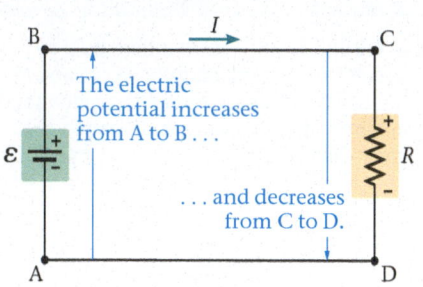

▲ **FIGURE 21-20 Kirchhoff's loop rule** Kirchhoff's loop rule states that as one moves around a closed loop in a circuit, the algebraic sum of all potential differences must be zero. The electric potential increases as one moves from the − to the + plate of a battery; it decreases as one moves through a resistor in the direction of the current.

In general, we associate a + sign with currents entering a junction and a − sign with currents leaving a junction, and state Kirchhoff's junction rule as follows:

> The algebraic sum of all currents meeting at any junction in a circuit must equal zero.

In the example just discussed, I_1 enters the junction (+), I_2 and I_3 leave the junction (−); hence, the algebraic sum of currents at the junction is $I_1 - I_2 - I_3$. Setting this sum equal to zero recovers our previous result.

In some cases we may not know the direction of all the currents meeting at a junction in advance. When this happens, we simply choose a direction for the unknown currents, apply the junction rule, and continue as usual. If the value we obtain for a given current is negative, it simply means that the direction we chose was wrong; the current actually flows in the opposite direction.

For example, suppose we know both the direction and magnitude of the currents I_1 and I_2 in **FIGURE 21-19**. To find the third current, we apply the junction rule—but first we must choose a direction for I_3. If we choose I_3 to point downward, as shown in the figure, the junction rule gives

$$I_1 - I_2 - I_3 = 0$$

Solving for I_3, we have

$$I_3 = I_1 - I_2 = 2.0 \text{ A} - 5.5 \text{ A} = -3.5 \text{ A}$$

Noting that I_3 is negative, we conclude that the actual direction of this current is upward; that is, the 2.0-A current and the 3.5-A current enter the junction and combine to yield the 5.5-A current that leaves the junction.

The Loop Rule

Imagine taking a day hike on a mountain path. First, you gain altitude to reach a scenic viewpoint; later, you descend below your starting point into a valley; finally, you gain altitude again and return to the trailhead. During the hike you sometimes increase your gravitational potential energy, and sometimes you decrease it, but the net change at the end of the hike is zero—after all, you return to the same altitude from which you started. Kirchhoff's loop rule is an application of the same idea to an electric circuit.

For example, consider the simple circuit shown in **FIGURE 21-20**. The electric potential increases by the amount $\mathcal{E}$ in going from point A to point B, because we move from the low-potential (−) terminal of the battery to the high-potential (+) terminal. This is like gaining altitude in the hiking analogy. Next, there is no potential change as we go from point B to point C, because these points are connected by an ideal wire. As we move from point C to point D, however, the potential does change—recall that a potential difference is required to force a current through a resistor. We label the potential difference across the resistor ΔV_{CD}. Finally, there is no change in potential between points D and A, because they too are connected by an ideal wire.

We can now apply the idea that the net change in electric potential (the analog to gravitational potential energy in the hike) must be zero around any closed loop. In this case, we have

$$\mathcal{E} + \Delta V_{CD} = 0$$

Thus, we find that $\Delta V_{CD} = -\mathcal{E}$; that is, the electric potential *decreases* as one moves across the resistor *in the direction of the current*. To indicate this drop in potential, we label the side where the current enters the resistor with a + (indicating high potential) and the side where the current leaves the resistor with a − (indicating low potential). Finally, we can use Ohm's law to set the magnitude of the potential drop equal to IR and find the current in the circuit:

$$|\Delta V_{CD}| = \mathcal{E} = IR$$
$$I = \frac{\mathcal{E}}{R}$$

This, of course, is the expected result.

In general, Kirchhoff's loop rule can be stated as follows:

> The algebraic sum of all potential differences around any closed loop in a circuit is zero.

We now consider a variety of applications in which both the junction rule and the loop rule are used to find the various currents and potentials in a circuit.

Applications of Kirchhoff's Rules

We begin by considering the relatively simple circuit shown in **FIGURE 21-21**. The currents and voltages in this circuit can be found by considering various parallel and series combinations of the resistors, as we did in the previous section. Thus, Kirchhoff's rules are not strictly needed in this case. Still, applying the rules to this circuit illustrates many of the techniques that can be used when studying more complex circuits.

Let's suppose that all the resistors have the value $R = 100.0\ \Omega$, and that the emf of the battery is $\mathcal{E} = 15.0\ \text{V}$. The equivalent resistance of the resistors can be obtained by noting that the vertical resistors are connected in parallel with one another and in series with the horizontal resistor. The vertical resistors combine to give a resistance of $R/2$, which, when added to the horizontal resistor, gives an equivalent resistance of $R_{\text{eq}} = 3R/2 = 150.0\ \Omega$. The current in the circuit, then, is $I = \mathcal{E}/R_{\text{eq}} = 15.0\ \text{V}/150.0\ \Omega = 0.100\ \text{A}$.

Now we approach the same problem from the point of view of Kirchhoff's rules. First, we apply the junction rule to point A:

$$I_1 - I_2 - I_3 = 0 \qquad \text{(junction A)} \qquad 21\text{-}11$$

Notice that current I_1 splits at point A into currents I_2 and I_3, which combine again at point B to give I_1 flowing through the horizontal resistor. We can apply the junction rule to point B, which gives $-I_1 + I_2 + I_3 = 0$, but this differs from Equation 21-11 by only a minus sign, and hence no new information is gained.

Next, we apply the loop rule. Because there are three unknowns, I_1, I_2, and I_3, we need three independent equations for a full solution. One has already been given by the junction rule; thus, we expect that two loop equations will be required to complete the solution. To begin, we consider loop 1, which is shown in **FIGURE 21-22 (a)**. We choose to move around this loop in the clockwise direction. (If we were to choose the counterclockwise direction instead, the same information would be obtained but with a minus sign.) For loop 1, then, we have an increase in potential as we move across the battery, a drop in potential across the vertical resistor of $I_3 R$, and another drop in potential across the horizontal resistor, this time of magnitude $I_1 R$. Applying the loop rule, we find the following:

$$\mathcal{E} - I_3 R - I_1 R = 0 \qquad \text{(loop 1)} \qquad 21\text{-}12$$

Similarly, we can apply the loop rule to loop 2, shown in **FIGURE 21-22 (b)**. In this case we cross the right-hand vertical resistor in the direction of the current, implying a drop in potential, and we cross the left-hand vertical resistor against the current, implying an increase in potential. Therefore, the loop rule gives

$$I_3 R - I_2 R = 0 \qquad \text{(loop 2)} \qquad 21\text{-}13$$

There is a third possible loop, shown in **FIGURE 21-22 (c)**, but the information it gives is not different from that already obtained. In fact, *any two of the three loops* is sufficient to complete our solution.

Notice that R cancels in Equation 21-13; hence, we see that $I_3 - I_2 = 0$, or $I_3 = I_2$. Substituting this result into the junction rule (Equation 21-11), we obtain

$$I_1 - I_2 - I_3 = I_1 - I_2 - I_2$$
$$= I_1 - 2I_2 = 0$$

Solving this equation for I_2 gives us $I_2 = I_1/2 = I_3$. Finally, using the first loop equation (Equation 21-12), we find

$$\mathcal{E} - \left(\frac{I_1}{2}\right)R - I_1 R = \mathcal{E} - \tfrac{3}{2}I_1 R = 0$$

The only unknown in this equation is the current I_1. Solving for this current, we find

$$I_1 = \frac{\mathcal{E}}{\tfrac{3}{2}R} = \frac{15.0\ \text{V}}{\tfrac{3}{2}(100.0\ \Omega)} = 0.100\ \text{A}$$

As expected, our result using Kirchhoff's rules agrees with the result obtained previously. Finally, the other two currents in the circuit are $I_2 = I_3 = I_1/2 = 0.0500\ \text{A}$.

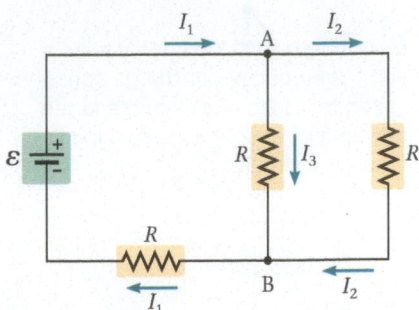

▲ **FIGURE 21-21 Analyzing a simple circuit** A simple circuit that can be studied using either equivalent resistance or Kirchhoff's rules.

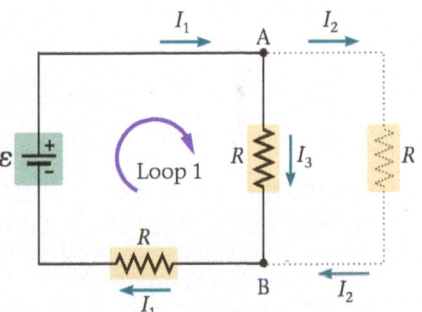

(a)

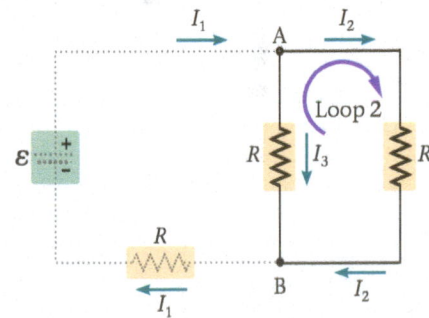

(b)

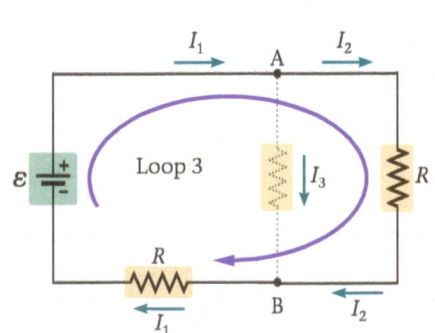

(c)

▲ **FIGURE 21-22 Using loops to analyze a circuit** Three loops associated with the circuit in Figure 21-21.

Big Idea 4 Kirchhoff's rules apply the principle of charge conservation to nodes in a circuit, and the principle of energy conservation to loops in a circuit.

EXERCISE 21-14   LOOP EQUATION

Write the loop equation for loop 3 in Figure 21-22 (c).

REASONING AND SOLUTION

Proceeding in a clockwise direction, as indicated in the figure, we find an increase in potential across the battery, and a drop in potential across each of the two resistors. Therefore, we have

$$+\mathcal{E} - I_2R - I_1R = 0$$

If we proceed in a counterclockwise direction around loop 3, we find

$$-\mathcal{E} + I_2R + I_1R = 0$$

This result is the same as the clockwise result except for an overall minus sign, and therefore it contains no new information. In general, it does not matter in which direction we choose to go around a loop. In addition, these loop equations agree with the results obtained above from the other two loops.

PROBLEM-SOLVING NOTE

Applying Kirchhoff's Rules

When applying Kirchhoff's rules, be sure to use the appropriate signs for currents and potential differences.

Clearly, the Kirchhoff approach is more involved than the equivalent-resistance method. However, it is not possible to analyze all circuits in terms of equivalent resistances. In such cases, Kirchhoff's rules are the only option, as illustrated in the next Example.

EXAMPLE 21-15   TWO LOOPS, TWO BATTERIES: FIND THE CURRENTS

Find the currents in the circuit shown in the sketch.

PICTURE THE PROBLEM

Our sketch shows the circuit in question, which includes two batteries and three resistors. It isn't possible to analyze this circuit using equivalent resistances, but it can be solved by applying Kirchhoff's rules to junction A (or B) and loops 1 and 2.

REASONING AND STRATEGY

There are three unknown currents in this circuit, I_1, I_2, and I_3, and hence three independent conditions are required to find them. These three conditions can be obtained by applying the junction rule to point A, and the loop rule to loops 1 and 2. (*Hint:* Use the junction rule at point A to solve for I_3 in terms of I_1 and I_2. Substitute this result in the two loop equations. This gives two equations in two unknowns, which can be solved for I_1 and I_2. Return to the junction rule to determine I_3.)

Known   Battery in loop 1 = 15 V; battery in loop 2 = 9.0 V; resistors, $R = 100.0\ \Omega$.
Unknown   Currents, $I_1 = ?, I_2 = ?, I_3 = ?$

SOLUTION

1. Apply the junction rule to point A: $I_1 - I_2 - I_3 = 0$

2. Apply the loop rule to loop 1 (let $R = 100.0\ \Omega$): $15\ \text{V} - I_3R - I_1R = 0$

3. Apply the loop rule to loop 2 (let $R = 100.0\ \Omega$): $-9.0\ \text{V} - I_2R + I_3R = 0$

4. Solve for I_1, I_2, and I_3: $I_1 = 0.070\ \text{A}$
$I_2 = -0.010\ \text{A}$
$I_3 = 0.080\ \text{A}$

INSIGHT

Notice that I_2 is negative. This means that the direction of that current is opposite to the direction shown in the circuit diagram.

PRACTICE PROBLEM

Suppose the polarity of the 9.0-V battery is reversed. What are the currents in this case?   [**Answer:** $I_1 = 0.13$ A, $I_2 = 0.11$ A, $I_3 = 0.020$ A]

Some related homework problems: Problem 54, Problem 56

Enhance Your Understanding   (Answers given at the end of the chapter)

5. Which of the following is the correct statement of Kirchhoff's rules for loop 1 in **FIGURE 21-23**? (a) $12\,\text{V} + (200\,\Omega)I_2 + (300\,\Omega)I_1 = 0$, (b) $12\,\text{V} + (200\,\Omega)I_3 - (300\,\Omega)I_1 = 0$, (c) $12\,\text{V} - (200\,\Omega)I_3 + (300\,\Omega)I_2 = 0$, (d) $12\,\text{V} - (200\,\Omega)I_3 - (300\,\Omega)I_1 = 0$.

Section Review

• Kirchhoff's rules can solve for currents in circuits that are too complex for equivalent resistances. The junction rule is a statement of charge conservation, and the loop rule is a statement of energy conservation.

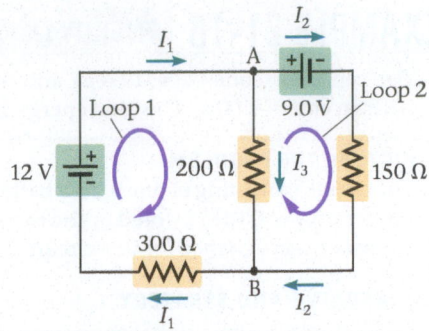

▲ **FIGURE 21-23** Enhance Your Understanding 5.

21-6 Circuits Containing Capacitors

To this point we have considered only resistors and batteries in electric circuits. Capacitors, which can also play an important role, are represented by a set of parallel lines (reminiscent of a parallel-plate capacitor): ╪. We now investigate simple circuits involving batteries and capacitors, leaving for the next section circuits that combine all three circuit elements.

Capacitors in Parallel

The simplest way to combine capacitors, as we shall see, is by connecting them in parallel. For example, **FIGURE 21-24 (a)** shows three capacitors connected in parallel with a battery of emf $\mathcal{E}$. As a result, each capacitor has the same potential difference, $\mathcal{E}$, between its plates. The magnitudes of the charges on each capacitor are as follows:

$$Q_1 = C_1\mathcal{E}, \quad Q_2 = C_2\mathcal{E}, \quad Q_3 = C_3\mathcal{E}$$

As a result, the total charge on the three capacitors is

$$Q = Q_1 + Q_2 + Q_3 = \mathcal{E}C_1 + \mathcal{E}C_2 + \mathcal{E}C_3 = (C_1 + C_2 + C_3)\mathcal{E}$$

If an equivalent capacitor is used to replace the three in parallel, as in **FIGURE 21-24 (b)**, the charge on its plates must be the same as the total charge on the individual capacitors:

$$Q = C_{eq}\mathcal{E}$$

Comparing $Q = C_{eq}\mathcal{E}$ with $Q = (C_1 + C_2 + C_3)\mathcal{E}$, we see that the equivalent capacitance is simply

$$C_{eq} = C_1 + C_2 + C_3$$

In general, the equivalent capacitance of capacitors connected in parallel is the sum of the individual capacitances:

Equivalent Capacitance for Capacitors in Parallel	
$C_{eq} = C_1 + C_2 + C_3 + \cdots = \sum C$	21-14
SI unit: farad, F	

Thus, connecting capacitors in parallel produces an equivalent capacitance greater than the greatest individual capacitance. It is as if the plates of the individual capacitors are connected together to give one large set of plates, with a correspondingly large capacitance.

PHYSICS IN CONTEXT
Looking Back

Capacitors were first introduced in Chapter 20, and are used now in dc circuits.

PHYSICS IN CONTEXT
Looking Ahead

Capacitors appear again in ac circuits in Chapter 24, and are especially important in *RLC* circuits.

PROBLEM-SOLVING NOTE

Finding the Equivalent Capacitance of a Circuit

When calculating the equivalent capacitance of capacitors in series, be sure to take one final inverse at the end of your calculation to find C_{eq}. Also, when considering circuits with capacitors in both series and parallel, start with the smallest units of the circuit and work your way out to the larger units.

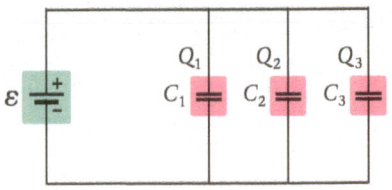

(a) Three capacitors in parallel

(b) Equivalent capacitance with same total charge

▶ **FIGURE 21-24 Capacitors in parallel** (a) Three capacitors, C_1, C_2, and C_3, connected in parallel. Note that each capacitor is connected across the same potential difference, $\mathcal{E}$. (b) The equivalent capacitance, $C_{eq} = C_1 + C_2 + C_3$, has the same charge on its plates as the total charge on the three original capacitors.

EXAMPLE 21-16 ENERGY IN PARALLEL

Two capacitors, one $C_1 = 12.0\ \mu F$ and the other of unknown capacitance C_2, are connected in parallel across a battery with an emf of 9.00 V. The total energy stored in the two capacitors is 0.0115 J. What is the value of the capacitance C_2?

PICTURE THE PROBLEM

The circuit, consisting of one 9.00-V battery and two capacitors, is shown in the sketch. The total energy of 0.0115 J stored in the two capacitors is the same as the energy stored in the equivalent capacitance for this circuit.

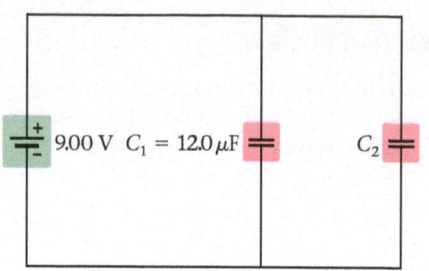

REASONING AND STRATEGY

Recall from Chapter 20 that the energy stored in a capacitor can be written as $U = \frac{1}{2}CV^2$.

It follows, then, that for an equivalent capacitance, C_{eq}, the energy is $U = \frac{1}{2}C_{eq}V^2$. We know the energy and voltage, and hence we can solve for the equivalent capacitance. Finally, the equivalent capacitance is the sum of the individual capacitances, $C_{eq} = C_1 + C_2 = 12.0\ \mu F + C_2$. We use this relationship to solve for C_2.

Known $C_1 = 12.0\ \mu F$; emf of battery, $\mathcal{E} = 9.00$ V; energy stored, $U = 0.0115$ J.
Unknown $C_2 = ?$

SOLUTION

1. Solve $U = \frac{1}{2}C_{eq}V^2$ for the equivalent capacitance:

$$U = \frac{1}{2}C_{eq}V^2$$
$$C_{eq} = \frac{2U}{V^2}$$

2. Substitute numerical values to find C_{eq}:

$$C_{eq} = \frac{2U}{V^2} = \frac{2(0.0115\ J)}{(9.00\ V)^2} = 284\ \mu F$$

3. Solve for C_2 in terms of the equivalent capacitance:

$$C_{eq} = C_1 + C_2 = 12.0\ \mu F + C_2$$
$$C_2 = C_{eq} - 12.0\ \mu F = 284\ \mu F - 12.0\ \mu F = 272\ \mu F$$

INSIGHT

The energy stored in the $C_1 = 12.0$-μF capacitor is $U = \frac{1}{2}C_1V^2 = 0.000486$ J. In comparison, the $C_2 = 272$-μF capacitor stores an energy equal to 0.0110 J. Thus, the larger capacitor stores the greater amount of energy. Though this may seem only natural, one needs to be careful. When we examine capacitors in *series* later in this section, we shall find exactly the opposite result.

PRACTICE PROBLEM

What is the total charge stored on the two capacitors?   [**Answer:** $Q = C_{eq}\mathcal{E} = 2.56 \times 10^{-3}$ C]

Some related homework problems: Problem 69, Problem 71

Although you probably haven't thought about it, when you turn on a "touch-sensitive" lamp, you are part of a circuit with capacitors in parallel. In fact, you're one of the capacitors! When you touch such a lamp, a small amount of charge moves onto your body—your body is like the plate of a capacitor. Because you have effectively increased the plate area—as always happens when capacitors are connected in parallel—the capacitance of the circuit increases. The electronic circuitry in the lamp senses this increase in capacitance and triggers the switch to turn the light on or off. A similar strategy is employed by the touchscreen technology used in smartphones, tablets, and computers, as described in Chapter 20.

Capacitors in Series

You have probably noticed from Equation 21-14 that capacitors connected in *parallel* combine in the same way as resistors connected in *series*. Similarly, capacitors connected in *series* obey the same rules as resistors connected in *parallel*, as we now show.

Consider three capacitors—initially uncharged—connected in series with a battery, as in **FIGURE 21-25 (a)**. The battery causes the left plate of C_1 to acquire a positive charge, $+Q$. This charge, in turn, attracts a negative charge $-Q$ onto the right plate of the capacitor. Because the capacitors start out uncharged, there is zero net charge between C_1 and C_2. As a result, the negative charge $-Q$ on the right plate of C_1 leaves a corresponding positive charge $+Q$ on the upper plate of C_2. The charge $+Q$ on the upper plate of C_2 attracts a negative charge $-Q$ onto its lower plate, leaving a corresponding positive charge $+Q$ on the right plate of C_3. Finally, the positive charge on the right plate of C_3 attracts a negative charge $-Q$ onto its left plate. The result is that all three capacitors have charge of the same magnitude on their plates.

With the same charge Q on all the capacitors, the potential difference for each is as follows:

$$V_1 = \frac{Q}{C_1}, \quad V_2 = \frac{Q}{C_2}, \quad V_3 = \frac{Q}{C_3}$$

Noting that the total potential difference across the three capacitors must equal the emf of the battery, we have

$$\mathcal{E} = V_1 + V_2 + V_3 = \frac{Q}{C_1} + \frac{Q}{C_2} + \frac{Q}{C_3} = Q\left(\frac{1}{C_1} + \frac{1}{C_2} + \frac{1}{C_3}\right) \quad 21\text{-}15$$

An equivalent capacitor connected to the same battery, as in **FIGURE 21-25 (b)**, will satisfy the relationship $Q = C_{eq}\mathcal{E}$, or

$$\mathcal{E} = Q\left(\frac{1}{C_{eq}}\right) \quad 21\text{-}16$$

A comparison of Equations 21-15 and 21-16 yields the result

$$\frac{1}{C_{eq}} = \frac{1}{C_1} + \frac{1}{C_2} + \frac{1}{C_3}$$

Thus, in general, we have the following rule for combining capacitors in series:

Equivalent Capacitance for Capacitors in Series

$$\frac{1}{C_{eq}} = \frac{1}{C_1} + \frac{1}{C_2} + \frac{1}{C_3} + \cdots = \sum \frac{1}{C} \quad 21\text{-}17$$

SI unit: farad, F

It follows that the equivalent capacitance of a group of capacitors connected in series is less than the smallest individual capacitance. In this case, it is as if the plate separations of the individual capacitors add to give a larger effective separation, and a correspondingly smaller capacitance.

More complex circuits, with some capacitors in series and others in parallel, can be handled in the same way as was done earlier with resistors. This is illustrated in the following Example.

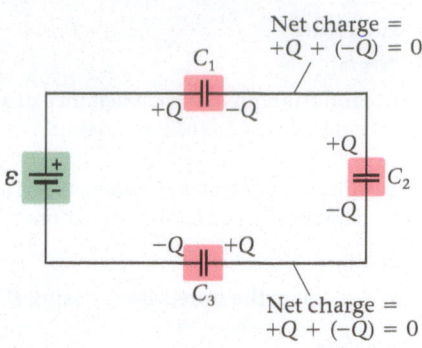

(a) Three capacitors in series

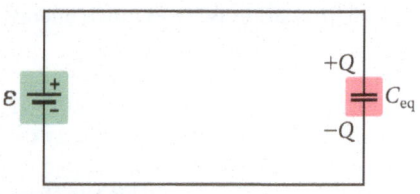

(b) Equivalent capacitance with same total charge

▲ **FIGURE 21-25 Capacitors in series**
(a) Three capacitors, C_1, C_2, and C_3, connected in series. Note that each capacitor has the same magnitude of charge on its plates. **(b)** The equivalent capacitance, $1/C_{eq} = 1/C_1 + 1/C_2 + 1/C_3$, has the same charge as the original capacitors.

EXAMPLE 21-17 FIND THE EQUIVALENT CAPACITANCE AND THE STORED ENERGY

Consider the electric circuit shown in the sketch, consisting of a 12.0-V battery and three capacitors connected partly in series and partly in parallel. Find **(a)** the equivalent capacitance of this circuit and **(b)** the total energy stored in the circuit.

PICTURE THE PROBLEM
The circuit for this Example contains a battery and three capacitors. The capacitances are indicated in the circuit. Notice that the 10.0-μF and 5.00-μF capacitors are in series with one another, and as a group they are in parallel with the 20.0-μF capacitor.

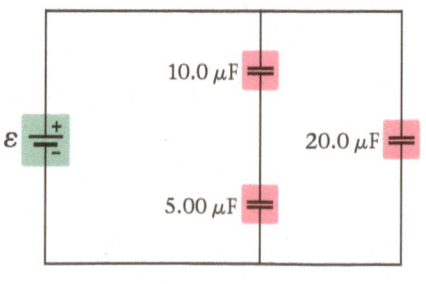

REASONING AND STRATEGY

a. We start by finding the equivalent capacitance of the 10.0-μF and 5.00-μF capacitors in series by taking the inverse of the sum of the inverses. Call this capacitance C_{series}. Next, we combine C_{series} in parallel with the 20.0-μF capacitor by simple addition to find the equivalent capacitance of the circuit, $C_{circuit}$.

b. Knowing the voltage of the battery and the equivalent capacitance of the circuit, we can find the stored energy with $U = \frac{1}{2}C_{circuit}V^2$.

Known Voltage of battery = 12.0 V; capacitors 10.0 μF, 5.00 μF, and 20.0 μF.
Unknown (a) Equivalent capacitance of circuit, $C_{circuit}$ = ? (b) Energy stored in circuit, U = ?

CONTINUED

SOLUTION

Part (a)

1. Find the equivalent capacitance of a 10.0-μF capacitor in series with a 5.00-μF capacitor:

$$\frac{1}{C_{series}} = \frac{1}{10.0\,\mu F} + \frac{1}{5.00\,\mu F} = \frac{3}{10.0\,\mu F}$$
$$C_{series} = 3.33\,\mu F$$

2. Find the equivalent capacitance of a 3.33-μF capacitor in parallel with a 20.0-μF capacitor:

$$C_{circuit} = 3.33\,\mu F + 20.0\,\mu F = 23.3\,\mu F$$

Part (b)

3. Calculate the stored energy using $U = \frac{1}{2}C_{circuit}V^2$:

$$U = \frac{1}{2}C_{circuit}V^2 = \frac{1}{2}(23.3\,\mu F)(12.0\,V)^2 = 1.68 \times 10^{-3}\,J$$

INSIGHT

The 10.0-μF capacitor and the 5.00-μF capacitor are connected in series in this circuit. As you might expect, one of these capacitors stores twice as much energy as the other. Which is it? Check the Practice Problem for the answer.

PRACTICE PROBLEM — PREDICT/CALCULATE

(a) Is the energy stored in the 10.0-μF capacitor greater than or less than the energy stored in the 5.00-μF capacitor? Explain. (b) Check your answer by calculating the energy stored in each of the two capacitors. [**Answer:** (a) Capacitors in series have the same charge, and hence we can compare their stored energy with $U = Q^2/2C$. Clearly, then, the 5.00-μF capacitor has twice the stored energy of the 10.0-μF capacitor. (b) The charge on the 10.0-μF and 5.00-μF capacitors is $Q = 40\,\mu C$. Hence, for the 10.0-μF capacitor we have $U = 8.00 \times 10^{-5}\,J$; for the 5.00-$\mu$F capacitor we have $U = 16.0 \times 10^{-5}\,J$—twice the energy.]

Some related homework problems: Problem 69, Problem 71

Enhance Your Understanding   (Answers given at the end of the chapter)

6. Do two capacitors give a larger equivalent capacitance if they are connected in parallel or in series? Explain.

Section Review

- Capacitors in parallel combine in the same way as resistors in series; thus $C_{eq} = C_1 + C_2 + C_3 + \cdots = \sum C$ (parallel).

- Capacitors in series combine in the same way as resistors in parallel; thus
$$\frac{1}{C_{eq}} = \frac{1}{C_1} + \frac{1}{C_2} + \frac{1}{C_3} + \cdots = \sum \frac{1}{C}\ \text{(series)}.$$

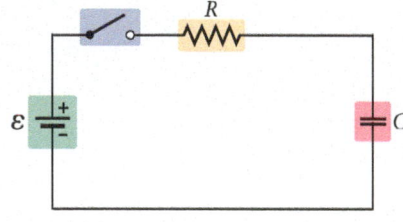

(a) $t < 0$

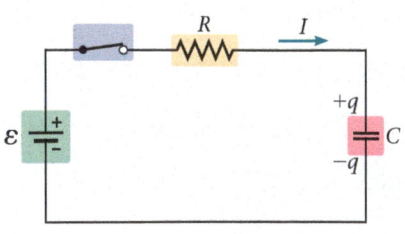

(b) $t > 0$

▲ **FIGURE 21-26 A typical *RC* circuit**
(a) Before the switch is closed ($t < 0$), there is no current in the circuit and no charge on the capacitor. **(b)** After the switch is closed ($t > 0$), current flows and the charge on the capacitor builds up over a finite time. As $t \to \infty$, the charge on the capacitor approaches $Q = C\mathcal{E}$.

21-7 *RC* Circuits

When the switch is closed on a circuit that contains only batteries and capacitors, the charge on the capacitor plates appears almost instantaneously—essentially at the speed of light. This is not the case, however, in circuits that also contain resistors. In these situations, the resistors limit the rate at which charge can flow, and an appreciable amount of time may be required before the capacitors acquire a significant charge. A useful analogy is the amount of time needed to fill a bucket with water. If you use a fire hose, which has little resistance to the flow of water, the bucket fills almost instantly. If you use a garden hose, which presents a much greater resistance to the water, filling the bucket may take a minute or more.

Simple *RC* Circuit The simplest example of such a circuit, a so-called ***RC* circuit**, is shown in **FIGURE 21-26**. Initially (before $t = 0$) the switch is open, and there is no current in the resistor or charge on the capacitor. At $t = 0$ the switch is closed and current begins to flow. Without a resistor, the capacitor would immediately take on the charge $Q = C\mathcal{E}$. The effect of the resistor, however, is to slow the charging process—in fact, the larger the resistance, the longer it takes for the capacitor to charge. One way to think of this is to note that as long as a current flows in the circuit, as in Figure 21-26 (b), there is a potential drop across the resistor; hence, the potential difference between the plates of the capacitor is less than the emf of the battery. With less voltage across the capacitor there will be less charge on its plates compared with the charge that would result if the plates were connected directly to the battery.

The methods of calculus can be used to show that the charge on the capacitor in Figure 21-26 varies with time in an exponential fashion, as follows:

$$q(t) = C\mathcal{E}(1 - e^{-t/\tau}) \qquad 21\text{-}18$$

In this expression, e is Euler's number ($e = 2.718 \ldots$) or, more precisely, the base of natural logarithms (see Appendix A). The quantity τ is referred to as the **time constant** of the circuit. The time constant is related to the resistance and capacitance of a circuit by the following simple relationship: $\tau = RC$. As we shall see, τ can be thought of as a characteristic time for the behavior of an RC circuit.

For example, at time $t = 0$ the exponential term is $e^{-0/\tau} = e^0 = 1$; therefore, the charge on the capacitor is zero at $t = 0$, as expected:

$$q(0) = C\mathcal{E}(1 - 1) = 0$$

In the opposite limit, $t \to \infty$, the exponential vanishes: $e^{-\infty/\tau} = 0$. Thus the charge in this limit is $C\mathcal{E}$:

$$q(t \to \infty) = C\mathcal{E}(1 - 0) = C\mathcal{E}$$

This is just the charge Q the capacitor would have had from $t = 0$ on if there had been no resistor in the circuit. Finally, at time $t = \tau$ the charge on the capacitor is $q = C\mathcal{E}(1 - e^{-1}) = C\mathcal{E}(1 - 0.368) = 0.632C\mathcal{E}$, which is 63.2% of its final charge. The charge on the capacitor as a function of time is plotted in **FIGURE 21-27**. The tick marks on the horizontal axis indicate the times $\tau, 2\tau, 3\tau$, and 4τ. Notice that the capacitor is almost completely charged by the time $t = 4\tau$.

Before we continue, let's check to see that the quantity $\tau = RC$ is in fact a time. Suppose, for example, that the resistor and capacitor in an RC circuit have the values $R = 120 \ \Omega$ and $C = 3.5 \ \mu\text{F}$, respectively. Multiplying R and C, we find

$$\tau = RC = (120 \text{ ohm})(3.5 \times 10^{-6} \text{ farad})$$
$$= \left(\frac{120 \text{ volt}}{\text{coulomb/second}}\right)\left(\frac{3.5 \times 10^{-6} \text{ coulomb}}{\text{volt}}\right) = 4.2 \times 10^{-4} \text{ second}$$

As we can see, RC is indeed a time, and its units are seconds.

Figure 21-27 also shows that the charge on the capacitor increases rapidly initially, indicating a large current in the circuit. Eventually, the charging slows down, because the greater the charge on the capacitor, the harder it is to transfer additional charge against the repulsive electric force. Later, the charge barely changes with time, which means that the current is essentially zero. In fact, the mathematical expression for the current—again derived from calculus—is the following:

$$I(t) = \left(\frac{\mathcal{E}}{R}\right)e^{-t/\tau} \qquad 21\text{-}19$$

This expression is plotted in **FIGURE 21-28**, where we see that significant variation in the current occurs over times ranging from $t = 0$ to $t \sim 4\tau$. At time $t = 0$ the current is $I(0) = \mathcal{E}/R$, which is the value it would have if the capacitor were replaced by an ideal wire. As $t \to \infty$, the current approaches zero, as expected: $I(t \to \infty) \to 0$. In this limit, the capacitor is essentially fully charged, so that no more charge can flow onto its plates. Thus, in this limit, the capacitor behaves like an open switch.

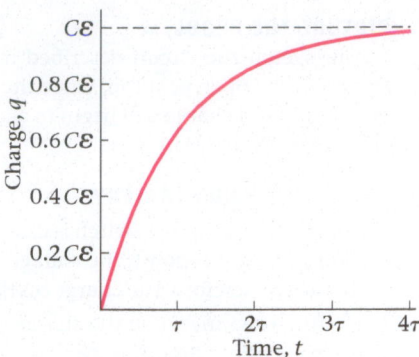

▲ **FIGURE 21-27 Charge versus time for the _RC_ circuit in Figure 21-26** The horizontal axis shows time in units of the characteristic time, $\tau = RC$. The vertical axis shows the magnitude of the charge on the capacitor in units of $C\mathcal{E}$.

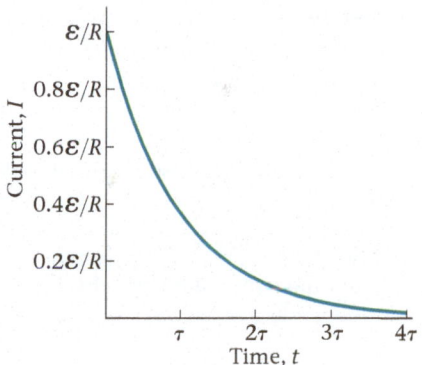

▲ **FIGURE 21-28 Current versus time for the _RC_ circuit in Figure 21-26** Initially the current is $\mathcal{E}/R$, the same as if the capacitor were not present. The current approaches zero after a period equal to several time constants, $\tau = RC$.

PROBLEM-SOLVING NOTE

The Limiting Behavior of Capacitors

Capacitors in RC circuits act like short circuits at $t = 0$ and open circuits as $t \to \infty$.

EXAMPLE 21-18 CHARGING A CAPACITOR

A circuit consists of a resistor $R_1 = 126 \ \Omega$, a resistor $R_2 = 275 \ \Omega$, a capacitor $C = 182 \ \mu\text{F}$, a switch, and an $\mathcal{E} = 3.00\text{-V}$ battery, all connected in series. Initially the capacitor is uncharged and the switch is open. At time $t = 0$ the switch is closed. **(a)** What charge will the capacitor have a long time after the switch is closed? **(b)** At what time will the charge on the capacitor be 80.0% of the value found in part (a)?

CONTINUED

PICTURE THE PROBLEM

In our sketch the circuit described in the problem statement is shown with the switch in the open position. Once the switch is closed at $t = 0$, current will flow in the circuit and charge will begin to accumulate on the capacitor plates. Notice that the two resistors are in series.

REASONING AND STRATEGY

a. A long time after the switch is closed, the current stops and the capacitor is fully charged. At this point, the voltage across the capacitor is equal to the emf of the battery. Therefore, the charge on the capacitor is $Q = C\mathcal{E}$.

b. To find the time when the charge will be 80.0% of the full charge, we set $q(t) = C\mathcal{E}(1 - e^{-t/\tau}) = 0.800C\mathcal{E}$ and solve for the desired time, t. Notice that τ is equal to RC, where in this circuit R is the equivalent resistance of two resistors in series; thus $R = R_1 + R_2 = 126\,\Omega + 275\,\Omega$.

Known   Emf of battery, $\mathcal{E} = 3.00\,\text{V}; R_1 = 126\,\Omega; R_2 = 275\,\Omega; C = 182\,\mu\text{F}$.
Unknown   (a) Final charge on capacitor, $Q = ?$ (b) Time for 80% charge on capacitor, $t = ?$

SOLUTION

Part (a)

1. Evaluate $Q = C\mathcal{E}$ for this circuit:

$$Q = C\mathcal{E} = (182\,\mu\text{F})(3.00\,\text{V}) = 546\,\mu\text{C}$$

Part (b)

2. Set $q(t) = 0.800C\mathcal{E}$ in $q(t) = C\mathcal{E}(1 - e^{-t/\tau})$ and cancel $C\mathcal{E}$:

$$q(t) = 0.800C\mathcal{E} = C\mathcal{E}(1 - e^{-t/\tau})$$
$$0.800 = 1 - e^{-t/\tau}$$

3. Solve for t in terms of the time constant τ:

$$e^{-t/\tau} = 1 - 0.800 = 0.200$$
$$t = -\tau \ln(0.200)$$

4. Calculate τ and use the result to find the time t:

$$\tau = RC = (126\,\Omega + 275\,\Omega)(182\,\mu\text{F}) = 73.0\,\text{ms}$$
$$t = -(73.0\,\text{ms}) \ln(0.200) = -(73.0\,\text{ms})(-1.61) = 118\,\text{ms}$$

INSIGHT

The time required for the charge on a capacitor to reach 80.0% of its final value is 1.61 time constants. This result is independent of the values of R and C in an RC circuit.

PRACTICE PROBLEM

What is the current in this circuit at the time found in part (b)?
[**Answer:** $I(t) = (\mathcal{E}/R)e^{-t/\tau} = [(3.00\,\text{V})/(126\,\Omega + 275\,\Omega)](0.200) = (7.48\,\text{mA})(0.200) = 1.50\,\text{mA}$]

Some related homework problems: Problem 75, Problem 78

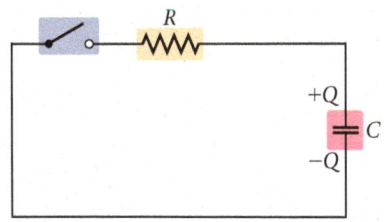

(a) $t < 0$

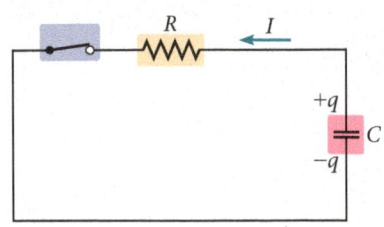

(b) $t > 0$

Discharge Speed for Series and Parallel Capacitors

◀ **FIGURE 21-29   Discharging a capacitor   (a)** A charged capacitor is connected to a resistor. Initially the circuit is open, and no current can flow. **(b)** When the switch is closed, current flows from the + plate of the capacitor to the − plate. The charge remaining on the capacitor approaches zero after several time constants, $\tau = RC$.

Similar behavior occurs when a charged capacitor is allowed to discharge in an RC circuit, as in **FIGURE 21-29**. In this case, the initial charge on the capacitor is Q. If the switch is closed at $t = 0$, the charge for later times is

$$q(t) = Qe^{-t/\tau} \qquad \text{21-20}$$

Like charging, the discharging of a capacitor occurs with a characteristic time $\tau = RC$.

To summarize, circuits with resistors and capacitors have the following general characteristics:

- Charging and discharging occur over a finite, characteristic time given by the time constant, $\tau = RC$.

- At $t = 0$ current flows freely through a capacitor being charged; it behaves like a short circuit.

- As $t \to \infty$ the current flowing into a capacitor approaches zero. In this limit, a capacitor behaves like an open switch.

We explore these features further in the following Conceptual Example.

CONCEPTUAL EXAMPLE 21-19   CURRENT IN AN *RC* CIRCUIT

What current flows through the battery in the circuit in the sketch **(a)** immediately after the switch is closed and **(b)** a long time after the switch is closed?

REASONING AND DISCUSSION

a. Immediately after the switch is closed, the capacitor acts like a short circuit—that is, as if the battery were connected to two resistors R in parallel. The equivalent resistance in this case is $R/2$; therefore, the current is $I = \mathcal{E}/(R/2) = 2\mathcal{E}/R$.

b. After current has been flowing in the circuit for a long time, the capacitor acts like an open switch. Now current can flow only through the one resistor, R; hence, the current is $I = \mathcal{E}/R$, half of its initial value.

ANSWER

(a) The current is $2\mathcal{E}/R$; **(b)** the current is $\mathcal{E}/R$.

Applications of *RC* Circuits   An example of a circuit that contains many resistors and capacitors is shown in **FIGURE 21-30**. In general, the fact that *RC* circuits have a characteristic time makes them useful in a variety of applications. On a rather mundane level, *RC* circuits are used to determine the time delay on windshield wipers. When you adjust the delay knob in your car, you change a resistance or a capacitance, which in turn changes the time constant of the circuit. This results in a greater or a smaller delay. The blinking rate of turn signals is also determined by the time constant of an *RC* circuit.

A more critical and interesting application of *RC* circuits is the heart pacemaker. In the simplest case, these devices use an *RC* circuit to deliver precisely timed pulses directly to the heart. The more sophisticated pacemakers available today can even "sense" when a patient's heart rate falls below a predetermined value. The pacemaker then begins sending appropriate pulses to the heart to increase its rate. Many pacemakers can even be reprogrammed after they are surgically implanted to respond to changes in a patient's condition.

Normally, the heart's rate of beating is determined by its own natural pacemaker, the sinoatrial or SA node, located in the upper right chamber of the heart. If the SA node is not functioning properly, it may cause the heart to beat slowly or irregularly. To correct the problem, a pacemaker is implanted just under the collarbone, as shown in **FIGURE 21-31**, and an electrode is introduced intravenously via the cephalic vein. The distal end of the electrode is positioned, with fluoroscopic guidance, in the right ventricular apex. From that point on, the operation of the pacemaker follows the basic principles of electric circuits, as described in this chapter.

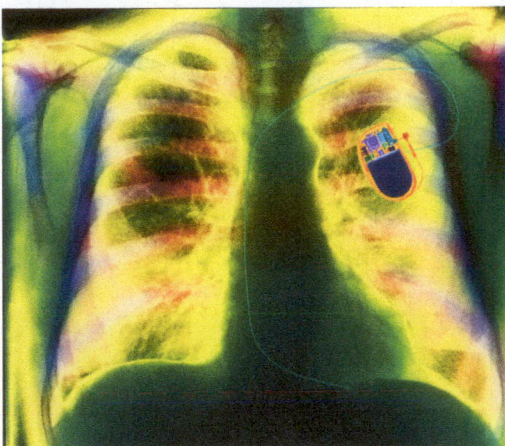

▲ **FIGURE 21-30** A modern-day circuit board incorporates numerous resistors (cylinders with colored bands) and capacitors (cylinders without bands).

▲ **FIGURE 21-31** An X-ray showing a pacemaker installed in a person's chest. The timing of the electrical pulses that keep the heart beating regularly is determined by an *RC* circuit powered by a small, long-lived battery.

Enhance Your Understanding   (Answers given at the end of the chapter)

7. Give a symbolic expression for the current that flows through the circuit in **FIGURE 21-32** **(a)** immediately after the switch is closed, and **(b)** a long time after the switch is closed.

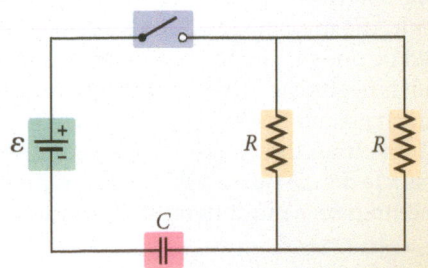

▲ **FIGURE 21-32** Enhance Your Understanding 7.

Section Review

- The charge on a capacitor in an *RC* circuit varies with time as follows:
  $q(t) = C\mathcal{E}(1 - e^{-t/\tau})$. The characteristic time in this expression is $\tau = RC$.
- The current in an *RC* circuit varies with time as follows: $I(t) = (\mathcal{E}/R)e^{-t/\tau}$.

Big Idea **5** In an *RC* circuit, a finite amount of time is required for charge to build up on a capacitor. Similarly, a finite amount of time is required for a capacitor to discharge. The characteristic time is determined by the product of the resistance R and capacitance C.

21-8 Ammeters and Voltmeters

Devices for measuring currents and voltages in a circuit are referred to as **ammeters** and **voltmeters,** respectively. In each case, the ideal situation is for the meter to measure the desired quantity without altering the characteristics of the circuit being studied. This is accomplished in different ways for these two types of meters, as we shall see.

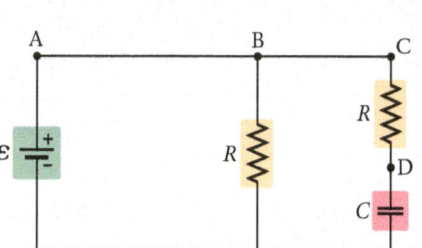 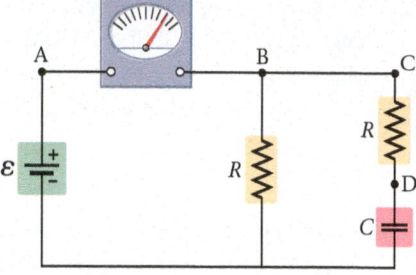

(a) Typical electric circuit

(b) Measuring the current between A and B

▲ **FIGURE 21-33 Measuring the current in a circuit** To measure the current flowing between points A and B in **(a)**, an ammeter is inserted into the circuit, as shown in **(b)**. An ideal ammeter would have zero resistance.

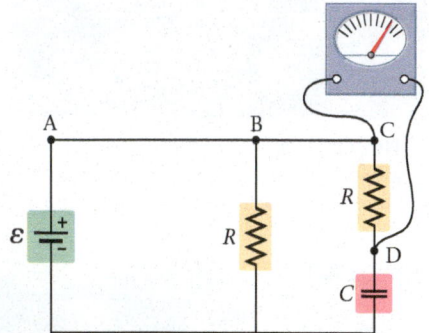

▲ **FIGURE 21-34 Measuring the voltage in a circuit** The voltage difference between points C and D can be measured by connecting a voltmeter in parallel to the original circuit. An ideal voltmeter would have infinite resistance.

▲ **FIGURE 21-35** A typical digital multimeter, which can measure resistance (teal settings), current (yellow settings), or voltage (red settings). This meter is measuring the voltage of a "9-volt" battery.

First, the ammeter is designed to measure the flow of current through a particular portion of a circuit. For example, we may want to know the current flowing between points A and B in the circuit shown in **FIGURE 21-33 (a)**. To measure this current, we insert the ammeter into the circuit in such a way that all the current flowing from A to B must also flow through the meter. This is done by connecting the meter "in series" with the other circuit elements between A and B, as indicated in **FIGURE 21-33 (b)**.

If the ammeter has a finite resistance—which must be the case for real meters—the presence of the meter in the circuit will alter the current it is intended to measure. Thus, an *ideal* ammeter would be one with zero resistance. In practice, if the resistance of the ammeter is much less than the other resistances in the circuit, its reading will be reasonably accurate.

Second, a voltmeter measures the potential drop between any two points in a circuit. Referring again to the circuit in Figure 21-33 (a), we may wish to know the difference in potential between points C and D. To measure this voltage, we connect the voltmeter "in parallel" to the circuit at the appropriate points, as in **FIGURE 21-34**.

A real voltmeter always allows some current to flow through it, which means that the current flowing through the circuit is less than before the meter was connected. As a result, the measured voltage is altered from its ideal value. An *ideal* voltmeter, then, would be one in which the resistance is infinite, so that the current it draws from the circuit is negligible. In practical situations it is sufficient that the resistance of the meter be much greater than the resistances in the circuit.

Sometimes the functions of an ammeter, voltmeter, and ohmmeter are combined in a single device called a **multimeter.** An example is shown in **FIGURE 21-35**. Adjusting the settings on a multimeter allows a variety of circuit properties to be measured.

Section Review

- Ammeters measure the current through a given part of a circuit, and are connected in series.

- Voltmeters measures the voltage between two points in a circuit, and are connected in parallel.

CHAPTER 21 REVIEW

CHAPTER SUMMARY

21-1 ELECTRIC CURRENT

Electric current is the flow of electric charge.

Definition

If a charge ΔQ passes a given point in the time Δt, the corresponding electric current is

$$I = \frac{\Delta Q}{\Delta t} \qquad \text{21-1}$$

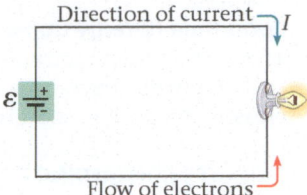

Direction of current

Flow of electrons

Ampere

The unit of current is the ampere, or amp for short. By definition, 1 amp is 1 coulomb per second: $1\,\text{A} = 1\,\text{C/s}$.

Electromotive Force

The electromotive force, or emf, $\mathcal{E}$, is the potential difference between the terminals of a battery under ideal conditions.

Work Done by a Battery

As a battery moves a charge ΔQ around a circuit, it does the work $W = (\Delta Q)\mathcal{E}$.

Direction of Current

By definition, the direction of the current I in a circuit is the direction in which *positive* charges would move.

21-2 RESISTANCE AND OHM'S LAW

In order to move electrons against the resistance of a wire, it is necessary to apply a potential difference between the ends of the wire.

Ohm's Law

To produce a current I through a wire with resistance R, the following potential difference, V, is required:

$$V = IR \qquad \text{21-2}$$

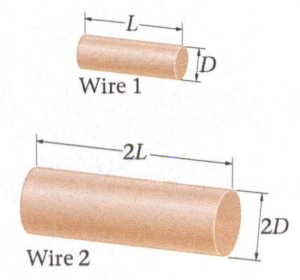

Wire 1

Wire 2

21-3 ENERGY AND POWER IN ELECTRIC CIRCUITS

Electrical Power

If a current I flows across a potential difference V, the corresponding electrical power is

$$P = IV \qquad \text{21-4}$$

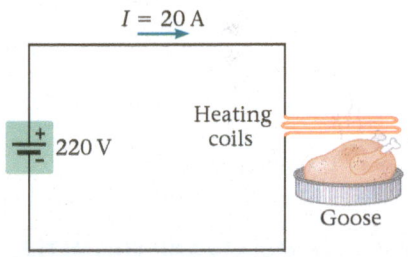

$I = 20\,\text{A}$

220 V

Heating coils

Goose

Power Dissipation in a Resistor

If a potential difference V produces a current I in a resistor R, the electrical power converted to heat is

$$P = I^2 R = V^2/R \qquad \text{21-5, 21-6}$$

Energy Usage and the Kilowatt-Hour

The energy equivalent of 1 kilowatt-hour (kWh) is

$$1\,\text{kWh} = 3.6 \times 10^6\,\text{J}$$

21-4 RESISTORS IN SERIES AND PARALLEL

Resistors connected end to end are said to be in series. Resistors connected across the same potential difference are said to be connected in parallel.

Series

The equivalent resistance, R_{eq}, of resistors connected in series is equal to the sum of the individual resistances:

$$R_{eq} = R_1 + R_2 + R_3 + \cdots = \sum R \qquad \text{21-7}$$

Parallel

The equivalent resistance, R_{eq}, of resistors connected in parallel is given by the following:

$$\frac{1}{R_{eq}} = \frac{1}{R_1} + \frac{1}{R_2} + \frac{1}{R_3} + \cdots = \sum \frac{1}{R} \qquad \text{21-10}$$

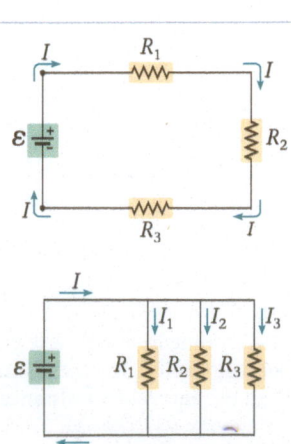

21-5 KIRCHHOFF'S RULES

Junction Rule (Charge Conservation)

The algebraic sum of all currents meeting at a junction must equal zero. Currents entering the junction are taken to be positive; currents leaving are taken to be negative.

Loop Rule (Energy Conservation)

The algebraic sum of all potential differences around a closed loop is zero. The potential increases in going from the − to the + terminal of a battery and decreases when crossing a resistor in the direction of the current.

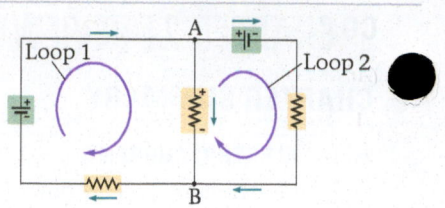

21-6 CIRCUITS CONTAINING CAPACITORS

Capacitors connected end to end are said to be in series. Capacitors connected across the same potential difference are said to be connected in parallel.

Parallel

The equivalent capacitance, C_{eq}, of capacitors connected in parallel is equal to the sum of the individual capacitances:

$$C_{eq} = C_1 + C_2 + C_3 + \cdots = \sum C \qquad 21\text{-}14$$

Series

The equivalent capacitance, C_{eq}, of capacitors connected in series is given by

$$\frac{1}{C_{eq}} = \frac{1}{C_1} + \frac{1}{C_2} + \frac{1}{C_3} + \cdots = \sum \frac{1}{C} \qquad 21\text{-}17$$

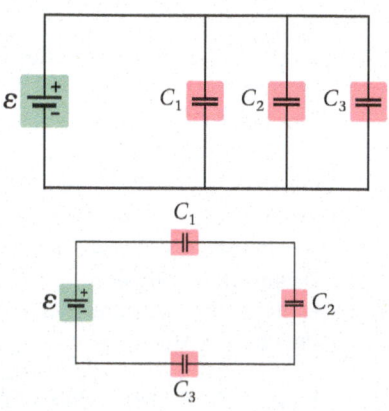

21-7 *RC* CIRCUITS

In circuits that contain both resistors and capacitors, there is a characteristic time, $\tau = RC$, during which significant changes occur.

Charging a Capacitor

The charge on a capacitor in an *RC* circuit varies with time as follows:

$$q(t) = C\mathcal{E}(1 - e^{-t/\tau}) \qquad 21\text{-}18$$

The corresponding current is given by

$$I(t) = \left(\frac{\mathcal{E}}{R}\right)e^{-t/\tau} \qquad 21\text{-}19$$

Behavior near t = 0

Just after the switch is closed in an *RC* circuit, capacitors offer no resistance to the flow of current.

Behavior as t → ∞

Long after the switch is closed in an *RC* circuit, capacitors behave like open circuits.

21-8 AMMETERS AND VOLTMETERS

Ammeters and voltmeters are devices for measuring currents and voltages, respectively, in electric circuits.

ANSWERS TO ENHANCE YOUR UNDERSTANDING QUESTIONS

1. A = C < D < B.
2. The current stays the same because current is voltage divided by resistance.
3. B < A < D < C.
4. The current in circuit 2 is greater than the current in circuit 1 because the equivalent resistance of circuit 1 ($3R/2$) is greater than the equivalent resistance of circuit 2 ($2R/3$).
5. (d).
6. The equivalent capacitance is larger when capacitors are connected in parallel because it is as if the areas of their plates were combined into a single larger capacitor.
7. (a) $I = 2\mathcal{E}/R$. (b) $I = 0$.

CONCEPTUAL QUESTIONS

For instructor-assigned homework, go to www.masteringphysics.com. **(MP)**

(Answers to odd-numbered Conceptual Questions can be found in the back of the book.)

1. Your body is composed of electric charges. Does it follow, then, that you produce an electric current when you walk?

2. Suppose you charge a comb by rubbing it through the fur on your dog's back. Do you produce a current when you walk across the room carrying the comb?

3. An electron moving through a wire has an average drift speed that is very small. Does this mean that its instantaneous velocity is also very small?

4. Are car headlights connected in series or parallel? Give an every-day observation that supports your answer.

5. Is it possible to connect a group of resistors of value *R* in such a way that the equivalent resistance is less than *R*? If so, give a specific example.

6. What physical quantity do resistors connected in series have in common?

7. What physical quantity do resistors connected in parallel have in common?

8. Explain how electrical devices can begin operating almost imme-diately after you throw a switch, even though individual electrons in the wire may take hours to reach the device.

9. Explain the difference between resistivity and resistance.

10. Explain why birds can roost on high-voltage wires without being electrocuted.

11. Consider the circuit shown in Figure 21-36, in which a light of resistance R and a capacitor of capacitance *C* are connected in series. The capacitor has a large capacitance, and is initially uncharged. The battery provides enough power to light the bulb when connected to the battery directly. Describe the behavior of the light after the switch is closed.

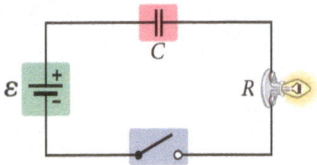

FIGURE 21-36 Conceptual Question 11

PROBLEMS AND CONCEPTUAL EXERCISES

Answers to odd-numbered Problems and Conceptual Exercises can be found in the back of the book. **BIO** *identifies problems of biological or medical interest;* **CE** *indicates a conceptual exercise.* **Predict/Explain** *problems ask for two responses: (a) your prediction of a physical outcome, and (b) the best explanation among three provided; and* **Predict/Calculate** *problems ask for a prediction of a physical outcome, based on fundamental physics concepts, and follow that with a numerical calculation to verify the prediction. On all problems, bullets (•, ••, •••) indicate the level of difficulty.*

SECTION 21-1   ELECTRIC CURRENT

1. • A flashlight bulb carries a current of 0.38 A for 98 s. How much charge flows through the bulb in this time? How many electrons?

2. • **Predict/Calculate** A car battery does 360 J of work on the charge passing through it as it starts an engine. **(a)** If the emf of the bat-tery is 12 V, how much charge passes through the battery during the start? **(b)** If the emf is doubled to 24 V while the work remains the same, does the amount of charge passing through the battery increase or decrease? By what factor?

3. • Highly sensitive ammeters can measure currents as small as 10.0 fA. How many electrons per second flow through a wire with a 10.0-fA current?

4. •• A television set connected to a 120-V outlet consumes 21 W of power. **(a)** How much current flows through the television? **(b)** How much time does it take for 10 million electrons to pass through the TV?

5. •• **BIO** **Pacemaker Batteries** Pacemakers designed for long-term use commonly employ a lithium-iodine battery capable of sup-plying 0.42 A·h of charge. **(a)** How many coulombs of charge can such a battery supply? **(b)** If the average current produced by the pacemaker is 5.6 μA, what is the expected lifetime of the device?

SECTION 21-2   RESISTANCE AND OHM'S LAW

6. • **CE** A conducting wire is quadrupled in length and tripled in diameter. **(a)** Does its resistance increase, decrease, or stay the same? Explain. **(b)** By what factor does its resistance change?

7. • **CE** **FIGURE 21-37** shows a plot of current versus voltage for two different materials, A and B. Which of these materials satisfies Ohm's law? Explain.

8. • **CE** **Predict/Explain** Current-versus-voltage plots for two materials, A and B, are shown in Figure 21-37. **(a)** Is the resis-tance of material A greater than, less than, or equal to the resistance of material B at the voltage V_1? **(b)** Choose the *best explanation* from among the following:
   I. Curve B is higher in value than curve A.
   II. A larger slope means a larger value of I/V, and hence a smaller value of *R*.
   III. Curve B has the larger slope at the voltage V_1 and hence the larger resistance.

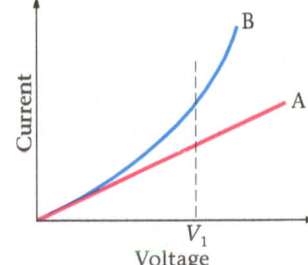

FIGURE 21-37 Problems 7 and 8

9. • A silver wire is 6.9 m long and 0.79 mm in diameter. What is its resistance?

10. • When a potential difference of 12 V is applied to a given wire, it conducts 0.45 A of current. What is the resistance of the wire?

11. • The tungsten filament of a lightbulb has a resistance of 0.07 Ω. If the filament is 27 cm long, what is its diameter?

12. • A wire is made out of aluminum with a cross-sectional area of 5.25×10^{-4} m^2 and a length of 65 km. What is its resistance?

13. • **Transcranial Direct-Current Stimulation** In a tDCS treatment proce-dure, 0.85 mA of current flows through a patient's brain when

there is a 7.4-V potential difference between the electrodes. What is the equivalent resistance between the electrodes?

14. •• **CE** The four conducting cylinders shown in **FIGURE 21-38** are all made of the same material, though they differ in length and/or diameter. They are connected to four different batteries, which supply the necessary voltages to give the circuits the same current, I. Rank the four voltages, V_1, V_2, V_3, and V_4, in order of increasing value. Indicate ties where appropriate.

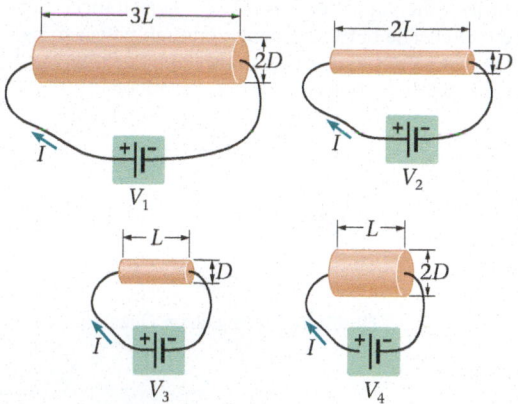

FIGURE 21-38 Problem 14

15. •• **Predict/Calculate** A bird lands on a bare copper wire carrying a current of 32 A. The wire is 6 gauge, which means that its cross-sectional area is 0.13 cm². **(a)** Find the difference in potential between the bird's feet, assuming they are separated by a distance of 6.0 cm. **(b)** Will your answer to part (a) increase or decrease if the separation between the bird's feet increases? Explain.

16. •• A current of 0.86 A flows through a copper wire 0.74 mm in diameter when it is connected to a potential difference of 15 V. How long is the wire?

17. •• **Predict/Calculate BIO Current Through a Cell Membrane** A typical cell membrane is 8.0 nm thick and has an electrical resistivity of $1.3 \times 10^7 \ \Omega \cdot m$. **(a)** If the potential difference between the inner and outer surfaces of a cell membrane is 75 mV, how much current flows through a square area of membrane $1.0 \ \mu m$ on a side? **(b)** Suppose the thickness of the membrane is doubled, but the resistivity and potential difference remain the same. Does the current increase or decrease? By what factor?

18. •• When a potential difference of 12 V is applied to a wire 6.9 m long and 0.33 mm in diameter, the result is an electric current of 2.1 A. What is the resistivity of the wire?

19. •• **Predict/Calculate (a)** What is the resistance per meter of a gold wire with a cross-sectional area of $1.6 \times 10^{-7} \ m^2$? **(b)** Would your answer to part (a) increase, decrease, or stay the same if the diameter of the wire were increased? Explain. **(c)** Repeat part (a) for a wire with a cross-sectional area of $2.2 \times 10^{-7} \ m^2$.

20. •• **BIO Resistance and Current in the Human Finger** The interior of the human body has an electrical resistivity of $0.15 \ \Omega \cdot m$. **(a)** Estimate the resistance for current flowing the length of your index finger. (For this calculation, ignore the much higher resistivity of your skin.) **(b)** Your muscles will contract when they carry a current greater than 15 mA. What voltage is required to produce this current through your finger?

21. ••• Consider a rectangular block of metal of height A, width B, and length C, as shown in **FIGURE 21-39**. If a potential difference V is maintained between the two $A \times B$ faces of the block, a current I_{AB} is observed to flow. Find the current that flows if the same potential difference V is applied between the two $B \times C$ faces of the block. Give your answer in terms of I_{AB}.

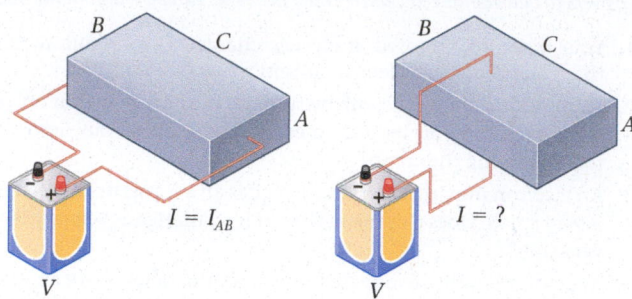

FIGURE 21-39 Problem 21

SECTION 21-3   ENERGY AND POWER IN ELECTRIC CIRCUITS

22. • **CE** Light A has four times the power rating of light B when operated at the same voltage. **(a)** Is the resistance of light A greater than, less than, or equal to the resistance of light B? Explain. **(b)** What is the ratio of the resistance of light A to the resistance of light B?

23. • **CE** Two lightbulbs operate on the same potential difference. Bulb A has four times the power output of bulb B. **(a)** Which bulb has the greater current passing through it? Explain. **(b)** What is the ratio of the current in bulb A to the current in bulb B?

24. • **CE** Two lightbulbs operate on the same current. Bulb A has four times the power output of bulb B. **(a)** Is the potential difference across bulb A greater than or less than the potential difference across bulb B? Explain. **(b)** What is the ratio of the potential difference across bulb A to that across bulb B?

25. • A 65-V generator supplies 4.8 kW of power. How much current does the generator produce?

26. • A portable CD player operates with a current of 18 mA at a potential difference of 4.3 V. What is the power usage of the player?

27. • Find the power dissipated in a 22-Ω electric heater connected to a 120-V outlet.

28. • The current in a 120-V reading lamp is 2.6 A. If the cost of electrical energy is $0.075 per kilowatt-hour, how much does it cost to operate the light for an hour?

29. •• Circuit A in a house has a voltage of 208 V and is limited by a 40.0-A circuit breaker. Circuit B is at 120.0 V and has a 20.0-A circuit breaker. What is the ratio of the maximum power delivered by circuit A to that delivered by circuit B?

30. •• **Predict/Calculate** A 65-W lightbulb operates on a potential difference of 95 V. Find **(a)** the current in the bulb and **(b)** the resistance of the bulb. **(c)** If this bulb is replaced with one whose resistance is half the value found in part (b), is its power rating greater than or less than 65 W? By what factor?

31. •• **Rating Car Batteries** Car batteries are rated by the following two numbers: **(1)** cranking amps = current the battery can produce for 30.0 seconds while maintaining a terminal voltage of at least 7.2 V and **(2)** reserve capacity = number of minutes the battery can produce a 25-A current while maintaining a terminal voltage of at least 10.5 V. One particular battery is advertised as having 905 cranking amps and a 155-minute reserve capacity. Which of these two ratings represents the greater amount of energy delivered by the battery?

SECTION 21-4   RESISTORS IN SERIES AND PARALLEL

32. • **CE Predict/Explain** A dozen identical lightbulbs are connected to a given emf. **(a)** Will the lights be brighter if they are connected

in series or in parallel? **(b)** Choose the *best explanation* from among the following:

    I. When connected in parallel each bulb experiences the maximum emf and dissipates the maximum power.

    II. Resistors in series have a larger equivalent resistance and dissipate more power.

    III. Resistors in parallel have a smaller equivalent resistance and dissipate less power.

33. • **CE** A circuit consists of three resistors, $R_1 < R_2 < R_3$, connected in series to a battery. Rank these resistors in order of increasing **(a)** current through them and **(b)** potential difference across them. Indicate ties where appropriate.

34. • **CE** Predict/Explain Two resistors are connected in parallel. **(a)** If a third resistor is now connected in parallel with the original two, does the equivalent resistance of the circuit increase, decrease, or remain the same? **(b)** Choose the *best explanation* from among the following:

    I. Adding a resistor generally tends to increase the resistance, but putting it in parallel tends to decrease the resistance; therefore the effects offset and the resistance stays the same.

    II. Adding more resistance to the circuit will increase the equivalent resistance.

    III. The third resistor gives yet another path for current to flow in the circuit, which means that the equivalent resistance is less.

35. • What is the minimum number of 88-Ω resistors that must be connected in parallel to produce an equivalent resistance of 12 Ω or less?

36. • Find the equivalent resistance between points A and B for the group of resistors shown in **FIGURE 21-40**.

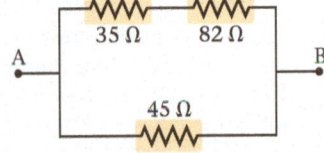

FIGURE 21-40 Problems 36, 37, and 97

37. •• A 9.00-V battery is connected across the terminals A and B for the group of resistors shown in Figure 21-40. What is the potential difference across each resistor?

38. •• **Holiday Lights** In a string of holiday lights, 50 lights—with a resistance of 13.5 Ω each—are connected in series. **(a)** If the string of lights is plugged into a 120.0-V outlet, what current flows through it? **(b)** Three of the lights burn out and activate their shunt resistors, effectively reducing the resistance of each one to 0.15 Ω. How much current now flows through the entire string of lights?

39. •• Your toaster has a power cord with a resistance of 0.020 Ω connected in series with a 9.6-Ω nichrome heating element. If the potential difference between the terminals of the toaster is 120 V, how much power is dissipated in **(a)** the power cord and **(b)** the heating element?

40. •• A circuit consists of a 12.0-V battery connected to three resistors (37 Ω, 22 Ω, and 141 Ω) in series. Find **(a)** the current that flows through the battery and **(b)** the potential difference across each resistor.

41. •• Predict/Calculate Three resistors, 11 Ω, 53 Ω, and R, are connected in series with a 24.0-V battery. The total current flowing through the battery is 0.16 A. **(a)** Find the value of resistance R. **(b)** Find the potential difference across each resistor. **(c)** If the voltage of the battery had been greater than 24.0 V, would your answer to part (a) have been larger or smaller? Explain.

42. •• A circuit consists of a battery connected to three resistors (65 Ω, 25 Ω, and 170 Ω) in parallel. The total current through the resistors is 2.8 A. Find **(a)** the emf of the battery and **(b)** the current through each resistor.

43. •• Predict/Calculate Three resistors, 22 Ω, 67 Ω, and R, are connected in parallel with a 12.0-V battery. The total current flowing through the battery is 0.88 A. **(a)** Find the value of resistance R. **(b)** Find the current through each resistor. **(c)** If the total current in the battery had been greater than 0.88 A, would your answer to part (a) have been larger or smaller? Explain.

44. •• A 99-Ω resistor has a current of 0.82 A and is connected in series with a 110-Ω resistor. What is the emf of the battery to which the resistors are connected?

45. •• The equivalent resistance between points A and B of the resistors shown in **FIGURE 21-41** is 33 Ω. Find the value of resistance R.

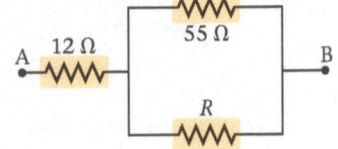

FIGURE 21-41 Problems 45, 49, and 87

46. •• Find the equivalent resistance between points A and B shown in **FIGURE 21-42**.

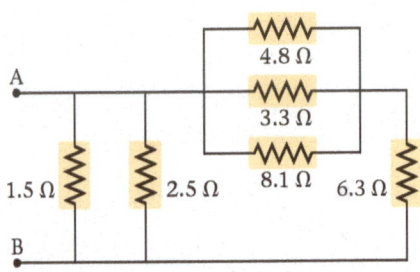

FIGURE 21-42 Problems 46 and 50

47. •• How many 23-W lightbulbs can be connected in parallel across a potential difference of 92 V before the total current in the circuit exceeds 3.0 A?

48. •• The circuit in **FIGURE 21-43** includes a battery with a finite internal resistance, $r = 0.50$ Ω. **(a)** Find the current flowing through the 7.1-Ω and the 3.2-Ω resistors. **(b)** How much current flows through the battery? **(c)** What is the potential difference between the terminals of the battery?

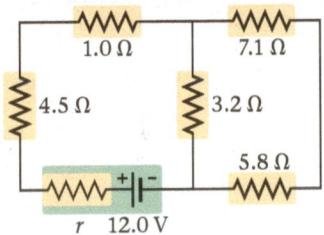

FIGURE 21-43 Problems 48 and 51

49. •• Predict/Calculate A 12-V battery is connected to terminals A and B in Figure 21-41. **(a)** Given that $R = 85$ Ω, find the current in each resistor. **(b)** Suppose the value of R is increased. For each resistor in turn, state whether the current flowing through it increases or decreases. Explain.

50. •• Predict/Calculate The terminals A and B in Figure 21-42 are connected to a 9.0-V battery. **(a)** Find the current flowing through each resistor. **(b)** Is the potential difference across the 6.3-Ω resistor greater than, less than, or the same as the potential difference across the 1.5-Ω resistor? Explain.

51. •• Predict/Calculate Suppose the battery in Figure 21-43 has an internal resistance $r = 0.25$ Ω. **(a)** How much current flows through the battery? **(b)** What is the potential difference between the terminals of the battery? **(c)** If the 3.2-Ω resistor is increased in value, will the current in the battery increase or decrease? Explain.

52. ••• Predict/Calculate The current flowing through the 8.45-Ω resistor in **FIGURE 21-44** is 1.52 A. **(a)** What is the voltage of the battery? **(b)** If the 17.2-Ω resistor is increased in value, will the current provided by the battery increase, decrease, or stay the same? Explain.

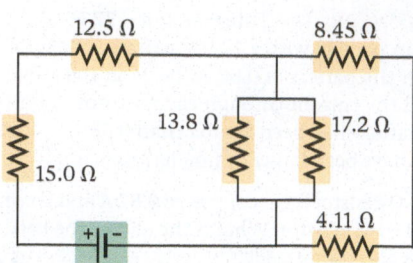

FIGURE 21-44 Problem 52

53. ••• **Predict/Calculate** Four identical resistors are connected to a battery as shown in **FIGURE 21-45**. When the switch is open, the current through the battery is I_0. **(a)** When the switch is closed, will the current through the battery increase, decrease, or stay the same? Explain. **(b)** Calculate the current that flows through the battery when the switch is closed. Give your answer in terms of I_0.

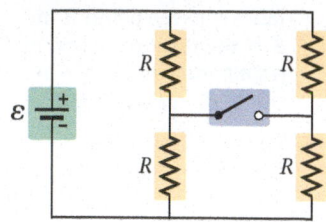

FIGURE 21-45 Problem 53

SECTION 21-5 KIRCHHOFF'S RULES

54. • Find the magnitude and direction (clockwise or counterclockwise) of the current in **FIGURE 21-46**.

55. • **Predict/Calculate** Suppose the polarity of the 11.5-V battery in Figure 21-46 is reversed. **(a)** Do you expect this to increase or decrease the amount of current flowing in the circuit? Explain. **(b)** Calculate the magnitude and direction (clockwise or counterclockwise) of the current in this case.

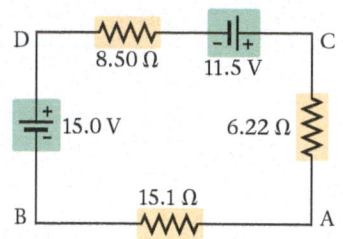

FIGURE 21-46 Problems 54, 55, and 56

56. •• **Predict/Calculate** It is given that point A in Figure 21-46 is grounded ($V = 0$). **(a)** Is the potential at point B greater than or less than zero? Explain. **(b)** Is the potential at point C greater than or less than zero? Explain. **(c)** Calculate the potential at point D.

57. •• Consider the circuit shown in **FIGURE 21-47**. Find the current through each resistor using **(a)** the rules for series and parallel resistors and **(b)** Kirchhoff's rules.

58. •• Suppose point A is grounded ($V = 0$) in Figure 21-47. Find the potential at points B and C.

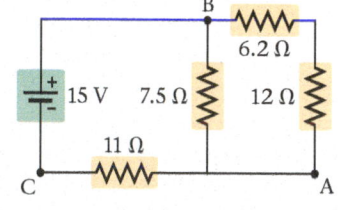

FIGURE 21-47 Problems 57 and 58

59. •• **Predict/Calculate** **(a)** Find the current in each resistor in **FIGURE 21-48**. **(b)** Is the potential at point A greater than, less than, or equal to the potential at point B? Explain. **(c)** Determine the potential difference between the points A and B.

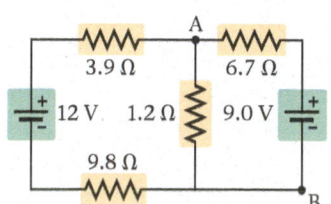

FIGURE 21-48 Problem 59

60. ••• Two batteries and three resistors are connected as shown in **FIGURE 21-49**. How much current flows through each battery when the switch is **(a)** closed and **(b)** open?

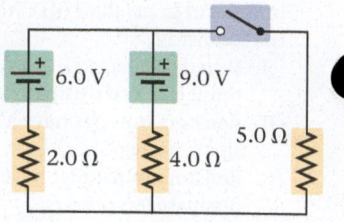

FIGURE 21-49 Problem 60

SECTION 21-6 CIRCUITS CONTAINING CAPACITORS

61. • **CE** Two capacitors, $C_1 = C$ and $C_2 = 2C$, are connected to a battery. **(a)** Which capacitor stores more energy when they are connected to the battery in series? Explain. **(b)** Which capacitor stores more energy when they are connected in parallel? Explain.

62. • **CE** **Predict/Explain** Two capacitors are connected in series. **(a)** If a third capacitor is now connected in series with the original two, does the equivalent capacitance increase, decrease, or remain the same? **(b)** Choose the *best explanation* from among the following:

  I. Adding a capacitor generally tends to increase the capacitance, but putting it in series tends to decrease the capacitance; therefore, the net result is no change.

  II. Adding a capacitor in series will increase the total amount of charge stored, and hence increase the equivalent capacitance.

  III. Adding a capacitor in series decreases the equivalent capacitance since each capacitor now has less voltage across it, and hence stores less charge.

63. • **CE** **Predict/Explain** Two capacitors are connected in parallel. **(a)** If a third capacitor is now connected in parallel with the original two, does the equivalent capacitance increase, decrease, or remain the same? **(b)** Choose the *best explanation* from among the following:

  I. Adding a capacitor tends to increase the capacitance, but putting it in parallel tends to decrease the capacitance; therefore, the net result is no change.

  II. Adding a capacitor in parallel will increase the total amount of charge stored, and hence increase the equivalent capacitance.

  III. Adding a capacitor in parallel decreases the equivalent capacitance since each capacitor now has less voltage across it, and hence stores less charge.

64. • A 252-μF capacitor is connected in series with a 126-μF capacitor. What is the equivalent capacitance of the pair?

65. • A 36-μF capacitor is connected in parallel with an 18-μF capacitor. What is the equivalent capacitance of the pair?

66. • Find the equivalent capacitance between points A and B for the group of capacitors shown in **FIGURE 21-50**.

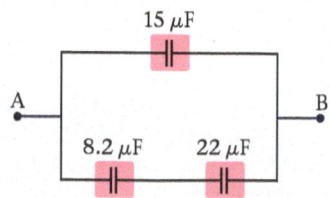

FIGURE 21-50 Problems 66 and 69

67. •• A 15-V battery is connected to three capacitors in series. The capacitors have the following capacitances: 4.5 μF, 12 μF, and 32 μF. Find the voltage across the 32-μF capacitor.

68. •• **CE** Three different circuits, each containing a switch and two capacitors, are shown in **FIGURE 21-51**. Initially, the plates of the capacitors are charged as shown. The switches are then closed, allowing charge to move freely between the capacitors. Rank the circuits in order of increasing final charge on the left plate of **(a)** the upper capacitor and **(b)** the lower capacitor. Indicate ties where appropriate.

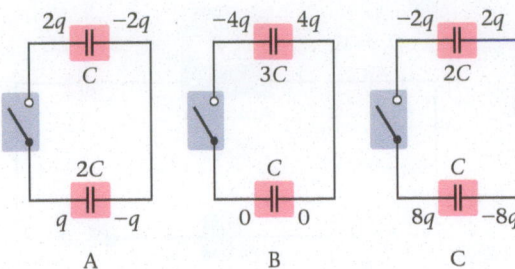

FIGURE 21-51 Problem 68

69. •• Terminals A and B in Figure 21-50 are connected to an 18-V battery. Find the energy stored in each capacitor.

70. •• **Predict/Calculate** You would like to add a second capacitor to a 24-μF capacitor to obtain an equivalent capacitance of 15 μF. (a) Should you connect the second capacitor in series or in parallel with the 24-μF capacitor? (b) Find the value of the second capacitance you will need.

71. •• Two capacitors, one 7.5 μF and the other 15 μF, are connected in parallel across a 15-V battery. (a) Find the equivalent capacitance of the two capacitors. (b) Find the charge stored in each capacitor. (c) Find the energy stored in each capacitor.

72. •• The equivalent capacitance of the capacitors shown in **FIGURE 21-52** is 12.4 μF. Find the value of capacitance C.

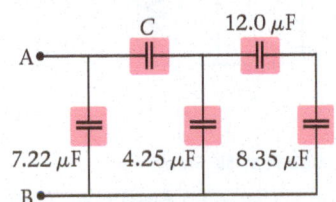

FIGURE 21-52 Problems 72 and 100

73. ••• With the switch in position A, the 11.2-μF capacitor in **FIGURE 21-53** is fully charged by the 12.0-V battery, and the 9.50-μF capacitor is uncharged. The switch is now moved to position B. As a result, charge flows between the capacitors until they have the same voltage across their plates. Find this voltage.

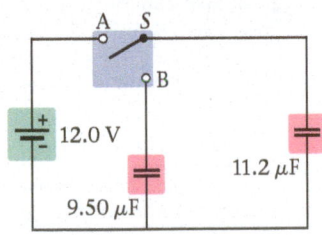

FIGURE 21-53 Problem 73

SECTION 21-7 RC CIRCUITS

74. • The switch on an RC circuit is closed at $t = 0$. Given that $\mathcal{E} = 6.0\,\text{V}, R = 92\,\Omega$, and $C = 28\,\mu\text{F}$, how much charge is on the capacitor at time $t = 4.0\,\text{ms}$?

75. • The capacitor in an RC circuit ($R = 120\,\Omega, C = 45\,\mu\text{F}$) is initially uncharged. Find (a) the charge on the capacitor and (b) the current in the circuit one time constant ($\tau = RC$) after the circuit is connected to a 9.0-V battery.

76. • **CE** Three RC circuits have the emf, resistance, and capacitance given in the accompanying table. Initially, the switch on the circuit is open and the capacitor is uncharged. Rank these circuits in order of increasing (a) initial current (immediately after the switch is closed) and (b) time for the capacitor to acquire half its final charge. Indicate ties where appropriate.

	$\mathcal{E}$ (V)	$R\,(\Omega)$	$C\,(\mu\text{F})$
Circuit A	12	4	3
Circuit B	9	3	1
Circuit C	9	9	2

77. •• Consider an RC circuit with $\mathcal{E} = 12.0\,\text{V}, R = 195\,\Omega$, and $C = 45.7\,\mu\text{F}$. Find (a) the time constant for the circuit, (b) the maximum charge on the capacitor, and (c) the initial current in the circuit.

78. •• The resistor in an RC circuit has a resistance of 125 Ω. (a) What capacitance must be used in this circuit if the time constant is to be 4.5 ms? (b) Using the capacitance determined in part (a), calculate the current in the circuit 9.0 ms after the switch is closed. Assume that the capacitor is uncharged initially and that the emf of the battery is 9.0 V.

79. •• A flash unit for a camera has a capacitance of 1500 μF. What resistance is needed in this RC circuit if the flash is to charge to 90% of its full charge in 21 s?

80. •• **FIGURE 21-54** shows a simplified circuit for a photographic flash unit. This circuit consists of a 9.0-V battery, a 50.0-kΩ resistor, a 140-μF capacitor, a flashbulb, and two switches. Initially, the capacitor is uncharged and the two switches are open. To charge the unit, switch S_1 is closed; to fire the flash, switch S_2 (which is connected to the camera's shutter) is closed. How much time does it take to charge the capacitor to 5.0 V?

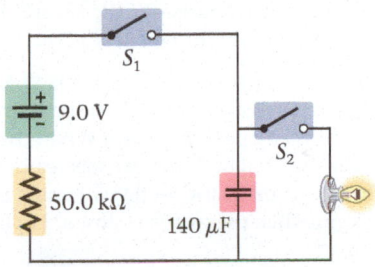

FIGURE 21-54 Problem 80

81. •• **Nerve Impulse Propagation** The speed with which nerve impulses travel is determined in large part by the characteristic time constant $\tau = RC$ of the circuit formed by the resistivity of the axon and the capacitance of its wall. The resistance of a 1.00-mm-long segment of an axon is 25.5 MΩ. (a) For nerve axons with no protective myelin sheath, the wall capacitance is about 3.14×10^{-10} F for each segment of length $L = 1.00$ mm. Find the speed of the nerve impulses given by $v = L/\tau$. (b) Many axons are surrounded by a myelin sheath that decreases the wall capacitance to 1.57×10^{-12} F. What is the speed of nerve impulses along such myelinated axons?

82. •• **Predict/Calculate** Consider the RC circuit shown in **FIGURE 21-55**. Find (a) the time constant and (b) the initial current for this circuit. (c) It is desired to increase the time constant of this circuit by adjusting the value of the 6.5-Ω resistor. Should the resistance of this resistor be increased or decreased to have the desired effect? Explain.

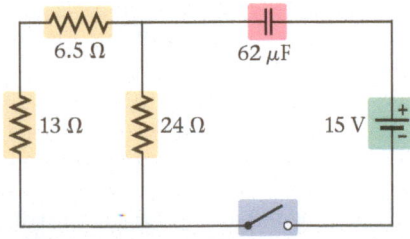

FIGURE 21-55 Problem 82

GENERAL PROBLEMS

83. • **CE** Consider the circuit shown in **FIGURE 21-56**, in which three lights, each with a resistance R, are connected in series. The circuit also contains an open switch. **(a)** When the switch is closed, does the intensity of light 2 increase, decrease, or stay the same? Explain.

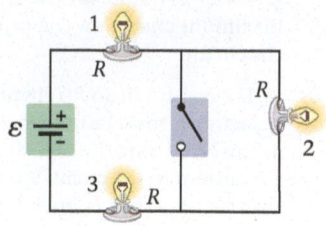

FIGURE 21-56 Problems 83 and 84

(b) Do the intensities of lights 1 and 3 increase, decrease, or stay the same when the switch is closed? Explain.

84. • **CE** Predict/Explain **(a)** Referring to Problem 83 and the circuit in Figure 21-56, does the current supplied by the battery increase, decrease, or remain the same when the switch is closed? **(b)** Choose the *best explanation* from among the following:

 I. The current decreases because only two resistors can draw current from the battery when the switch is closed.

 II. Closing the switch makes no difference to the current since the second resistor is still connected to the battery as before.

 III. Closing the switch shorts out the second resistor, decreases the total resistance of the circuit, and increases the current.

85. • **CE** Consider the circuit shown in **FIGURE 21-57**, in which three lights, each with a resistance R, are connected in parallel. The circuit also contains an open switch. **(a)** When the switch is closed, does the intensity of light 3 increase, decrease, or stay the same? Explain. **(b)** Do the intensities of lights 1 and 2 increase, decrease, or stay the same when the switch is closed? Explain.

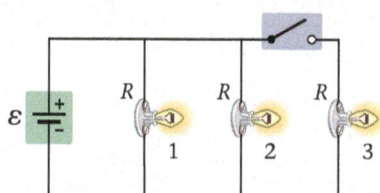

FIGURE 21-57 Problems 85 and 86

86. • **CE** Predict/Explain **(a)** When the switch is closed in the circuit shown in Figure 21-57, does the total power dissipated in the circuit increase, decrease, or stay the same? **(b)** Choose the *best explanation* from among the following:

 I. Closing the switch adds one more resistor to the circuit. This makes it harder for the battery to supply current, which decreases the power dissipated.

 II. The equivalent resistance of the circuit is reduced by closing the switch, but the voltage remains the same. Therefore, from $P = V^2/R$ we see that the power dissipated increases.

 III. The power dissipated remains the same because power, $P = IV$, is independent of resistance.

87. • Suppose that points A and B in Figure 21-41 are connected to a 12-V battery. Find the power dissipated in each of the resistors assuming that $R = 65\ \Omega$.

88. •• **CE** The circuit shown in **FIGURE 21-58** shows a resistor and two capacitors connected in series with a battery of voltage V. The circuit also has an ammeter and a switch. Initially, the switch is open and both capacitors are uncharged. The following questions refer to a time long after the switch is closed and current has ceased to flow. **(a)** In terms of V, what is the voltage across the capacitor C_1? **(b)** In terms of CV, what is the charge on the right plate of C_2? **(c)** What is the net charge that flowed through the ammeter during charging? Give your answer in terms of CV.

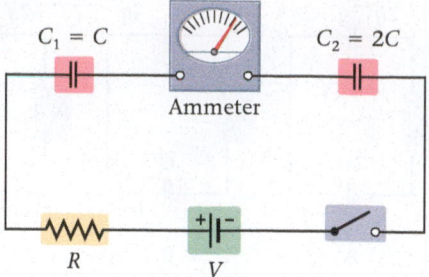

FIGURE 21-58 Problem 88

89. •• **CE** The three circuits shown in **FIGURE 21-59** have identical batteries, resistors, and capacitors. Initially, the switches are open and the capacitors are uncharged. Rank the circuits in order of increasing **(a)** final charge on the capacitor and **(b)** time for the current to drop to 90% of its initial value. Indicate ties where appropriate.

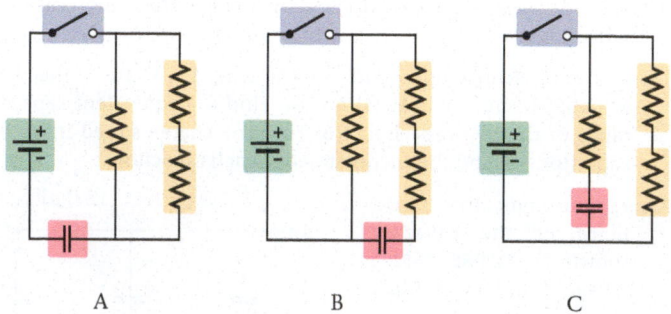

FIGURE 21-59 Problem 89

90. •• **Electrical Safety Codes** For safety reasons, electrical codes have been established that limit the amount of current a wire of a given size can carry. For example, an 18-gauge (cross-sectional area = 0.823 mm²), rubber-insulated extension cord with copper wires can carry a maximum current of 5.0 A. Find the voltage drop in a 12-ft, 18-gauge extension cord carrying a current of 5.0 A. (*Note:* In an extension cord, the current must flow through two lengths—down and back.)

91. •• A portable CD player uses a current of 7.5 mA at a potential difference of 3.5 V. **(a)** How much energy does the player use in 35 s? **(b)** Suppose the player has a mass of 0.65 kg. For what length of time could the player operate on the energy required to lift it through a height of 1.0 m?

92. •• An electrical heating coil is immersed in 6.6 kg of water at 22 °C. The coil, which has a resistance of 250 Ω, warms the water to 32 °C in 15 min. What is the potential difference at which the coil operates?

93. •• Predict/Calculate Consider the circuit shown in **FIGURE 21-60**. **(a)** Is the current flowing through the battery immediately after the switch is closed greater than, less than, or the same as the current flowing through the battery long after the switch is closed? Explain. **(b)** Find the current flowing through the battery immediately after the switch is closed. **(c)** Find the current in the battery long after the switch is closed.

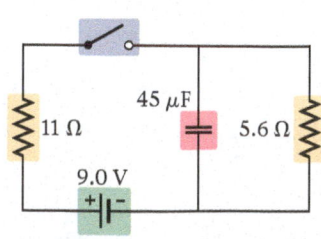

FIGURE 21-60 Problem 93

94. •• Two resistors are connected in series to a battery with an emf of 12 V. The voltage across the first resistor is 2.7 V and the current through the second resistor is 0.15 A. Find the resistance of the two resistors.

95. •• BIO Pacemaker Pulses A pacemaker sends a pulse to a patient's heart every time the capacitor in the pacemaker charges to a voltage of 0.25 V. It is desired that the patient receive 75 pulses per minute. Given that the capacitance of the pacemaker is $110 \, \mu F$ and that the battery has a voltage of 9.0 V, what value should the resistance have?

96. •• Three resistors $(R, \frac{1}{2}R, 2R)$ are connected to a battery. **(a)** If the resistors are connected in series, which one has the greatest rate of energy dissipation? **(b)** Repeat part (a), this time assuming that the resistors are connected in parallel.

97. •• Predict/Calculate Suppose we connect a 12.0-V battery to terminals A and B in Figure 21-40. **(a)** Is the current in the 45-Ω resistor greater than, less than, or the same as the current in the 35-Ω resistor? Explain. **(b)** Calculate the current flowing through each of the three resistors in this circuit.

98. •• National Electric Code In the United States, the National Electric Code sets standards for maximum safe currents in insulated copper wires of various diameters. The accompanying table gives a portion of the code. Notice that wire diameters are identified by the *gauge* of the wire, and that 1 mil = 10^{-3} in. Find the maximum power dissipated per length in **(a)** an 8-gauge wire and **(b)** a 10-gauge wire.

Gauge	Diameter (mils)	Safe current (A)
8	129	35
10	102	25

99. ••• Solar Panel Power The current-versus-voltage plot for a solar panel is shown in **FIGURE 21-61**. **(a)** The short-circuit current, I_{sc}, of a solar panel is the current it can generate when a wire connects its output terminals, making the load resistance $R = 0$. What is I_{sc} for this panel? **(b)** The open-circuit potential, V_{oc}, of a solar panel is the voltage it can generate when it is disconnected from any circuit (infinite load resistance R). What is V_{oc} for this panel? **(c)** Calculate the power output of the solar panel at points A, B, C, D, and E in Figure 21-61.

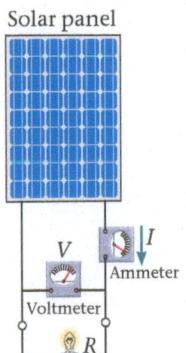
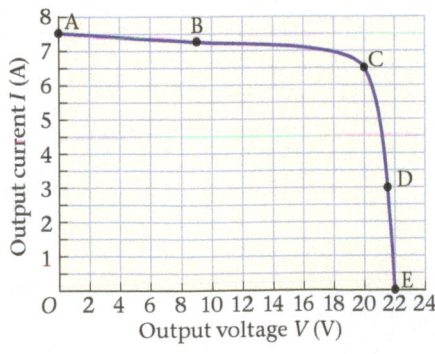

(a)   **(b)**

FIGURE 21-61 Problem 99

100. ••• Predict/Calculate A 15.0-V battery is connected to terminals A and B in Figure 21-52. **(a)** Given that $C = 15.0 \, \mu F$, find the charge on each of the capacitors. **(b)** Find the total energy stored in this system. **(c)** If the 7.22-μF capacitor is increased in value, will the total energy stored in the circuit increase or decrease? Explain.

101. ••• When two resistors, R_1 and R_2, are connected in series across a 6.0-V battery, the potential difference across R_1 is 4.0 V. When R_1 and R_2 are connected in parallel to the same battery, the current through R_2 is 0.45 A. Find the values of R_1 and R_2.

102. ••• The circuit shown in **FIGURE 21-62** is known as a Wheatstone bridge. Find the value of the resistor R such that the current through the 85.0-Ω resistor is zero.

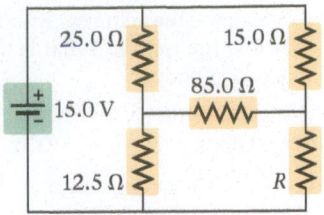

FIGURE 21-62 Problem 102

PASSAGE PROBLEMS

BIO Footwear Safety

The American National Standards Institute (ANSI) specifies safety standards for a number of potential workplace hazards. For example, ANSI requires that footwear provide protection against the effects of compression from a static weight, impact from a dropped object, puncture from a sharp tool, and cuts from saws. In addition, to protect against the potentially lethal effects of an electrical shock, ANSI provides standards for the electrical resistance that a person and footwear must offer to the flow of electric current.

Specifically, regulation ANSI Z41-1999 states that the resistance of a person and his or her footwear must be tested with the circuit shown in **FIGURE 21-63**. In this circuit, the voltage supplied by the battery is $\varepsilon = 50.0$ V and the resistance in the circuit is $R = 1.00 \, M\Omega$. Initially the circuit is open and no current flows. When a person touches the metal sphere attached to the battery, however, the circuit is closed and a small current flows through the person, the shoes, and back to the battery. The amount of current flowing through the person can be determined by using a voltmeter to measure the voltage drop V across the resistor R. To be safe, the current should not exceed $150 \, \mu A$.

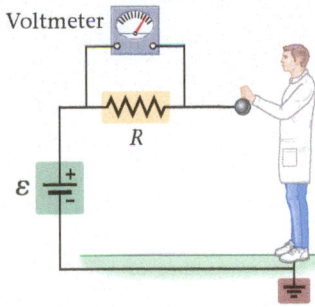

FIGURE 21-63 Problems 103, 104, 105, and 106

Notice that the experimental setup in Figure 21-63 is a dc circuit with two resistors in series—the resistance R and the resistance of the person and footwear, R_{pf}. It follows that the current in the circuit is $I = \varepsilon/(R + R_{pf})$. We also know that the current is $I = V/R$, where V is the reading of the voltmeter. These relationships can be combined to relate the voltage V to the resistance R_{pf}, with the result shown in **FIGURE 21-64**. According to ANSI regulations, Type II footwear must give a resistance R_{pf} in the range of $0.1 \times 10^7 \, \Omega$ to $100 \times 10^7 \, \Omega$.

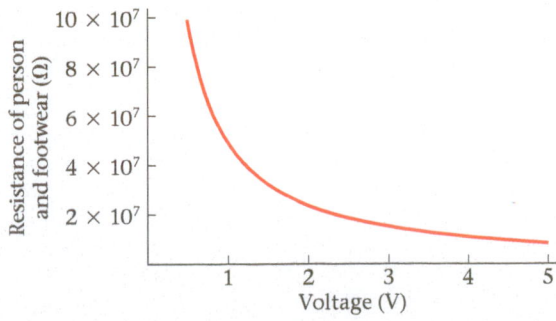

FIGURE 21-64 Problems 103, 104, 105, and 106

103. • Suppose the voltmeter measures a potential difference of 3.70 V across the resistor. What is the current that flows through the person's body?

A. 3.70×10^{-6} A   B. 5.00×10^{-5} A
C. 0.0740 A   D. 3.70 A

104. • What is the resistance of the person and footwear when the voltmeter reads 3.70 V?

A. $1.25 \times 10^7 \, \Omega$ B. $1.35 \times 10^7 \, \Omega$
C. $4.63 \times 10^7 \, \Omega$ D. $1.71 \times 10^8 \, \Omega$

105. • The resistance of a given person and footwear is $4.00 \times 10^7 \, \Omega$. What is the reading on the voltmeter when this person is tested?

A. 0.976 V   B. 1.22 V   C. 1.25 V   D. 50.0 V

106. • The standard specifies that footwear should be tested three times: with the person standing on only the left shoe; with the person standing on only the right shoe; and with the person standing on the pair. Do you expect the resistance of the person standing on the left shoe alone to be larger than, smaller than, or the same as the resistance of the person standing on the pair, when measured under the test conditions?

107. •• **REFERRING TO EXAMPLE 21-13** Suppose the three resistors in this circuit have the values $R_1 = 100.0 \, \Omega$, $R_2 = 200.0 \, \Omega$, and $R_3 = 300.0 \, \Omega$, and the emf of the battery is 12.0 V. (The resistor numbers are given in the Interactive Figure.) (a) Find the potential difference across each resistor. (b) Find the current that flows through each resistor.

108. •• **REFERRING TO EXAMPLE 21-13** Suppose $R_1 = R_2 = 225 \, \Omega$ and $R_3 = R$. The emf of the battery is 12.0 V. (The resistor numbers are given in the Interactive Figure.) (a) Find the value of R such that the current supplied by the battery is 0.0750 A. (b) Find the value of R that gives a potential difference of 2.65 V across resistor 2.

109. •• Predict/Calculate **REFERRING TO EXAMPLE 21-18** Suppose the resistance of the 126-Ω resistor is reduced by a factor of 2. The other resistor is 275 Ω, the capacitor is 182 μF, and the battery has an emf of 3.00 V. (a) Does the final value of the charge on the capacitor increase, decrease, or stay the same? Explain. (b) Does the time for the capacitor to charge to 80.0% of its final value increase, decrease, or stay the same? Explain. (c) Find the time referred to in part (b).

110. •• Predict/Calculate **REFERRING TO EXAMPLE 21-18** Suppose the capacitance of the 182-μF capacitor is reduced by a factor of 2. The two resistors are 126 Ω and 275 Ω, and the battery has an emf of 3.00 V. (a) Find the final value of the charge on the capacitor. (b) Does the time for the capacitor to charge to 80.0% of its final value increase, decrease, or stay the same? Explain. (c) Find the time referred to in part (b).

Magnetism

Big Ideas

1 Magnetic fields are produced by magnets.

2 The magnetic force on a particle is proportional to its charge and speed.

3 Magnetic fields exert forces on current-carrying wires.

4 A loop of current in a magnetic field experiences a torque.

5 Current-carrying wires produce magnetic fields that circulate around the wire.

6 Current loops produce magnetic fields that are similar to those produced by bar magnets.

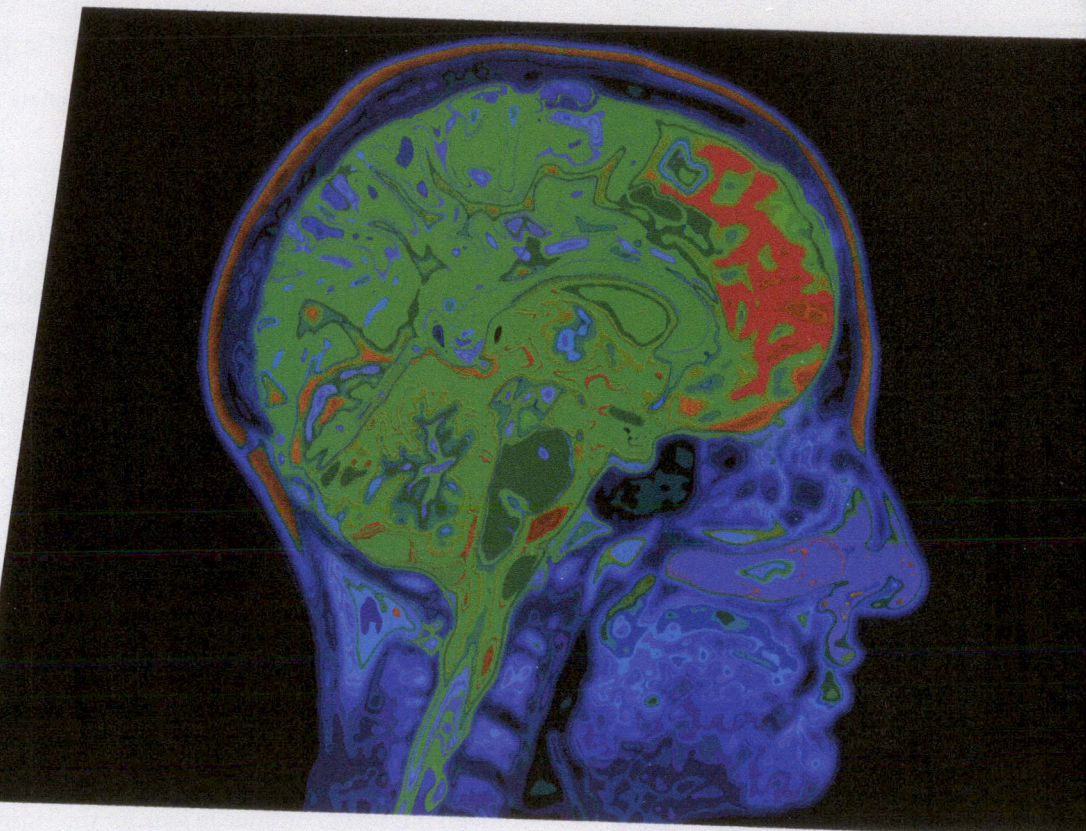

▲ One of the special abilities of a well-known superhero is X-ray vision. You might argue, however, that it would be better to have "magnetic vision." For example, magnetic resonance imaging (MRI) can produce images like the one shown here, where intricate details are seen within the body. In this chapter, we concentrate on the magnetic field, how it is produced, and how it exerts a force on charged particles. As we shall see, the magnetic field is a close cousin of the electric field.

The effects of magnetism have been known since antiquity. This chapter focuses on the basic aspects of magnetism, and on some of the important connections between magnetism and electricity.

Newton's third law (Chapter 5) applies to the force between bar magnets. In fact, Newton's laws apply to all types of forces, including gravitational, electric, and magnetic forces.

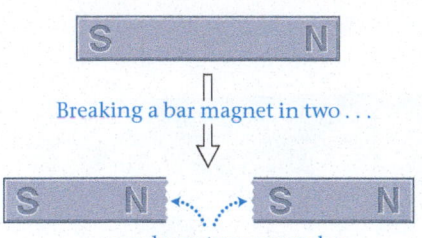

Breaking a bar magnet in two . . .

. . . produces two new poles.

▲ FIGURE 22-2 **Magnets always have two poles** When a bar magnet is broken in half, two new poles appear. Each half has both a north pole and a south pole, just like any other bar magnet.

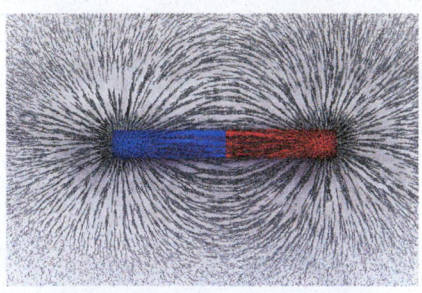

(a)

(b)

▲ FIGURE 22-3 **Magnetic field lines** The field of a bar magnet **(a)** or horseshoe magnet **(b)** can be visualized using iron filings on a sheet of glass or paper. The filings orient themselves with the field lines, creating a "snapshot" of the magnetic field.

22-1 The Magnetic Field

We begin our study of magnetism with a few general observations regarding magnets and the fields they produce. These observations apply over a wide range of scales—from the behavior of small, handheld bar magnets to the global effects associated with the magnetic field of the Earth.

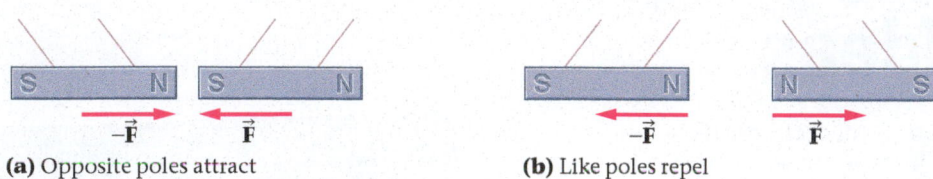

(a) Opposite poles attract   **(b)** Like poles repel

▲ FIGURE 22-1 **The force between two bar magnets**

Permanent Magnets

Our first direct experience with magnets is often a playful exploration of the way small permanent magnets, called bar magnets, affect one another. As we know, a bar magnet can attract another magnet or repel it, depending on which ends of the magnets are brought together. One end of a magnet is referred to as its *north pole*; the other end is its *south pole*. The poles are defined as follows: Imagine suspending a bar magnet from a string so that it is free to rotate in a horizontal plane—much like the needle of a compass. The end of the magnet that points toward the north geographic pole of the Earth we refer to as the "north-seeking pole," or simply the north pole. The opposite end of the magnet is the "south-seeking pole," or the south pole.

The rule for whether two bar magnets attract or repel each other is learned at an early age: Opposites attract; likes repel. Thus, if two magnets are brought together so that opposite poles approach each other, as in **FIGURE 22-1 (a)**, the force they experience is attractive. Like poles brought close together, as in **FIGURE 22-1 (b)**, experience a repulsive force.

An interesting aspect of magnets is that they *always* have two poles. You might think that if you break a bar magnet in two, each of the halves will have just one pole. Instead, breaking a magnet in half results in the appearance of two new poles on either side of the break, as illustrated in **FIGURE 22-2**. This behavior is fundamentally different from that found in electricity, where the two types of charge can exist separately. Though physicists continue to look for the elusive "magnetic monopole," and speculate as to its possible properties, none has been found.

Magnetic Field Lines

Just as an electric charge creates an electric "field" in its vicinity, so too does a magnet create a magnetic field. As in the electric case, the magnetic field represents the effect a magnet has on its surroundings. For example, in Figure 19-17 we saw a visual indication of the electric field $\vec{E}$ of a point charge using grass seeds suspended in oil. Similarly, the magnetic field, which we represent with the symbol $\vec{B}$, can be visualized using small iron filings sprinkled onto a smooth surface. In **FIGURE 22-3 (a)**, for example, a sheet of glass or plastic is placed on top of a bar magnet. When iron filings are dropped onto the sheet, they align with the magnetic field in their vicinity, giving a good representation of the overall field produced by the magnet. Similar effects are seen in **FIGURE 22-3 (b)**, with a magnet that has been bent to bring its poles close together. Because of its shape, this type of magnet is referred to as a horseshoe magnet.

In both cases shown in Figure 22-3, notice that the filings are bunched together near the poles of the magnets. This is where the magnetic field is most intense. We illustrate this by drawing field lines that are densely packed near the poles, as in **FIGURE 22-4**. As one moves away from a magnet in any direction, the field weakens, as indicated by the increasingly wider separations between the field lines. Thus, we indicate the magnitude of the vector $\vec{B}$ by the spacing of the field lines.

We can also assign a direction to the magnetic field. In particular, the direction of $\vec{\mathbf{B}}$ at a given location is defined as follows:

> The direction of the magnetic field, $\vec{\mathbf{B}}$, at a given location is the direction in which the north-seeking arrow of a compass needle points when placed at that location.

In Figure 22-4 the north-seeking arrow of the compass is the red arrow.

Let's apply this definition to the bar magnet in Figure 22-4. Imagine, for example, placing a compass near the south pole of the magnet. Because opposites attract, the north-seeking pole of the compass needle—the red arrow—will point toward the south pole of the magnet, as shown in the figure. Hence, according to our definition, the direction of the magnetic field at that location is toward the bar magnet's south pole. Similarly, one can see that the magnetic field must point away from the north pole of the bar magnet. In general, we can make the following observation:

> Magnetic field lines exit from the north pole of a magnet and enter at the south pole.

The field lines continue even within the body of a magnet. In fact, magnetic field lines always form closed loops; they never start or stop anywhere, in contrast to electric field lines. Again, this is related to the fact that there are no magnetic monopoles, where a magnetic field line could start or stop, whereas electric field lines start on positive charges and stop on negative charges.

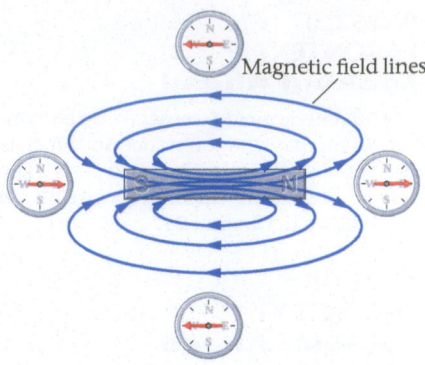

▲ **FIGURE 22-4 Magnetic field lines for a bar magnet** The field lines are closely spaced near the poles, where the magnetic field $\vec{\mathbf{B}}$ is most intense. In addition, the lines form closed loops that leave at the north pole of the magnet and enter at the south pole.

Big Idea 1 Magnetic fields are produced by magnets in much the same way that electric fields are produced by electric charges.

CONCEPTUAL EXAMPLE 22-1   MAGNETIC FIELD LINES

Can magnetic field lines cross one another?

REASONING AND DISCUSSION

Recall that the direction in which a compass points at any given location is the direction of the magnetic field at that point. Because a compass can point in only one direction, there must be only one direction for the field $\vec{\mathbf{B}}$. If field lines were to cross, however, there would be two directions for $\vec{\mathbf{B}}$ at the crossing point, and this is not allowed.

ANSWER

No. Magnetic field lines never cross.

RWP* An interesting example of magnetic field lines is provided by the humble refrigerator magnet. The flexible variety of these popular magnets has the unusual property that one side sticks quite strongly to the refrigerator, whereas the other side (the printed side) does not stick at all. Clearly, the magnetic field produced by such a magnet is not like that of a simple bar magnet, which generates a symmetrical field. Instead, these refrigerator magnets are composed of multiple magnetic stripes of opposite polarity, as indicated in **FIGURE 22-5**. The net effect is a magnetic field similar to the field that would be produced by a large number of tiny horseshoe magnets placed side by side, pointing in the same direction. In this way, the field is intense on the side containing the poles of the tiny magnets, as in Figure 22-3 (b), and very weak on the other side.

Geomagnetism

RWP The Earth, like many planets, produces its own magnetic field. In many respects, the Earth's magnetic field is like that of a giant bar magnet, as illustrated in **FIGURE 22-6**, with a pole near each geographic pole of the Earth. The magnetic poles are not perfectly aligned with the rotational axis of the Earth, however, but are inclined at an angle that varies slowly with time. Presently, the magnetic poles are tilted away from the rotational axis by an angle of about 11.5°. The current location of the north magnetic pole is just west of Ellef Ringnes Island, one of the Queen Elizabeth Islands of extreme northern Canada.

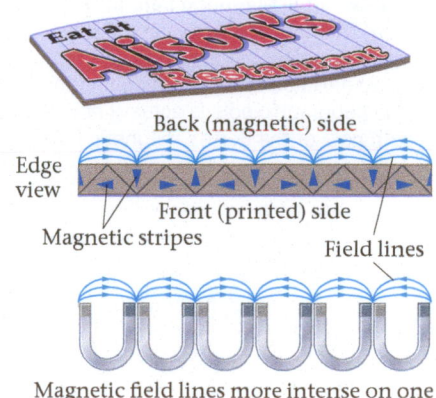

Magnetic field lines more intense on one side than the other

▲ **FIGURE 22-5 Refrigerator magnet** A flexible refrigerator magnet is made from a large number of narrow magnetic stripes with magnetic fields in different directions. The net effect is a field similar to that of a series of parallel horseshoe magnets placed side by side.

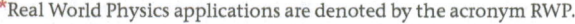

*Real World Physics applications are denoted by the acronym RWP.

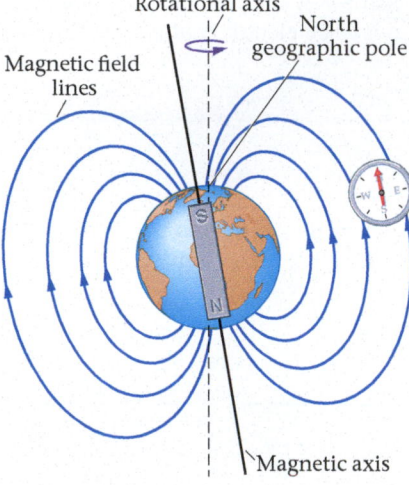

▲ **FIGURE 22-6 Magnetic field of the Earth** The Earth's magnetic field is similar to that of a giant bar magnet tilted slightly from the rotational axis. Notice that the north geographic pole is actually near the south pole of the Earth's magnetic field.

Because the north-seeking pole of a compass needle points toward the north magnetic pole of the Earth, and because opposites attract, we arrive at the following conclusion:

> The *north* geographic pole of the Earth is actually near the *south* pole of the Earth's magnetic field.

This is indicated in Figure 22-6. The figure also shows that the field lines are essentially horizontal (parallel to the Earth's surface) near the equator but enter or leave the Earth vertically near the poles. Thus, for example, if you were to stand near the north geographic pole, your compass would try to point straight down.

Although the Earth's magnetic field is similar in many respects to that of a huge bar magnet, it is far more complex in both its shape and behavior—in fact, even the mechanism by which the field is produced is still not completely understood. It seems likely, however, that flowing currents of molten material in the Earth's core are the primary cause of the field, as expressed in the **dynamo theory** of magnetism.

One of the reasons for lingering uncertainty about the Earth's magnetic field is that its behavior over time is rather complicated. For example, we've already mentioned that the magnetic poles drift slowly with time. But this is just the beginning of the field's interesting behavior. We've learned in the last century that the Earth's field has actually *reversed* direction many times over the ages. The last reversal occurred about 780,000 years ago—in fact, from 980,000 years ago to 780,000 years ago your compass would have pointed opposite to its direction today.

These ancient field reversals have left a permanent record in the rocks of the ocean floors, among other places. By analyzing the evidence of these reversals, geologists have found strong support for the theories of continental drift and plate tectonics and have developed a whole new branch of geology referred to as **paleomagnetism.** We shall return to these topics again briefly near the end of this chapter.

Enhance Your Understanding   (Answers given at the end of the chapter)

1. Is pole 1 in the bar magnet shown in **FIGURE 22-7** a north magnetic pole (N) or a south magnetic pole (S)?

▲ **FIGURE 22-7** Enhance Your Understanding 1.

Section Review

- Magnets always have two "poles," which are referred to as the "north pole" and the "south pole." Opposite poles attract; like poles repel.

- Magnetic field lines exit from the north pole of a magnet and enter at the south pole. In general, the direction of the magnetic field, $\vec{\mathbf{B}}$, is the direction the north-seeking arrow of a compass points when placed at a given location.

- The *north geographic pole* of the Earth is actually near the *south pole* of the Earth's magnetic field.

22-2 The Magnetic Force on Moving Charges

We've discussed briefly the familiar forces that act between one magnet and another. In this section we consider the force a magnetic field exerts on a moving electric charge. As we shall see, both the magnitude and the direction of this force have rather unusual characteristics. We begin with the magnitude.

Magnitude of the Magnetic Force

Consider a magnetic field $\vec{\mathbf{B}}$ that points from left to right in the plane of the page, as indicated in **FIGURE 22-8**. Three particles of charge q move through this region with velocities in different directions. The angle between $\vec{\mathbf{v}}$ and $\vec{\mathbf{B}}$ for the particle on the right has a magnitude denoted by the symbol θ. Experiment shows that the magnitude of the force $\vec{F}$ experienced by this particle is given by the following:

Magnitude of the Magnetic Force, F

$$F = |q|vB \sin \theta \qquad\qquad 22\text{-}1$$

SI unit: newton, N

Thus, the magnetic force depends on several factors. Two of these are the same as for the electric force:

    (i) the charge of the particle, q;

    (ii) the magnitude of the field—in this case, the magnetic field, B.

However, the magnetic force also depends on two factors that do not affect the strength of the electric force:

    (iii) the speed of the particle, v;

    (iv) the magnitude of the angle between the velocity vector and the magnetic field vector, θ.

It follows that the behavior of particles in magnetic fields is significantly different from their behavior in electric fields.

    In particular, a particle must have a charge and must be moving if the magnetic field is to exert a force on it. Even then the force vanishes if the particle moves in the direction of the field (that is, if $\theta = 0$, as for the bottom charge in Figure 22-8) or in the direction opposite to the field ($\theta = 180°$). The maximum force is experienced when the particle moves at right angles to the field (top charge in Figure 22-8), so that $\theta = 90°$ and $\sin \theta = 1$.

    Finally, $F = |q|vB \sin \theta$ gives only the magnitude of the force, and hence depends on the magnitude of the charge, $|q|$. As we shall see later in this section, the *sign* of q is important in determining the *direction* of the magnetic force. In addition, the angle θ in Equation 22-1 is the magnitude of the angle between $\vec{v}$ and $\vec{B}$, and always has a value in the range $0 \le \theta \le 180°$.

    In practice, the magnitude—or strength—of a magnetic field is *defined* by the relationship $F = |q|vB \sin \theta$. Thus, B is given by the following:

Definition of the Magnitude of the Magnetic Field, B

$$B = \frac{F}{|q|v \sin \theta} \qquad\qquad 22\text{-}2$$

SI unit: 1 tesla = 1 T = 1 N/(A·m)

The units of B are those of force divided by the product of charge and speed—that is, N/[C·(m/s)]. Rearranging slightly, we can write these units as N/[(C/s)·m]. Finally, noting that 1 A = 1 C/s, we find that 1 N/[C·(m/s)] = 1 N/(A·m). The latter combination of units is named the **tesla,** in recognition of the pioneering electrical and magnetic studies of the Croatian-born American engineer Nikola Tesla (1856–1943). In particular,

$$1 \text{ tesla} = 1 \text{ T} = 1 \text{ N/(A·m)}$$

We now give an example of the application of the magnetic force.

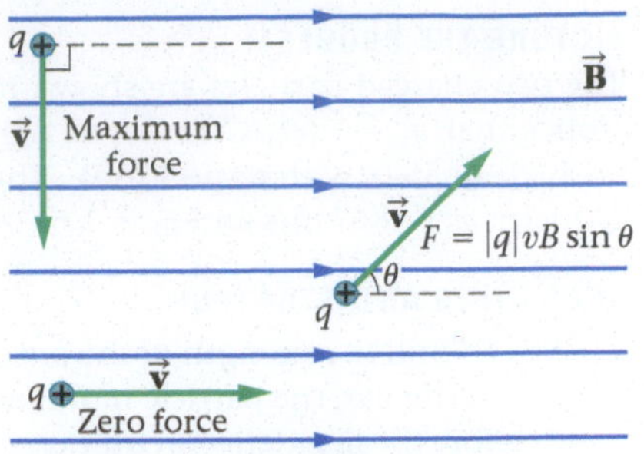

▲ **FIGURE 22-8 The magnetic force on a moving charged particle** A particle of charge q moves through a region of magnetic field $\vec{B}$ with a velocity $\vec{v}$. The magnitude of the force experienced by the charge is $F = |q|vB \sin \theta$. Note that the force is a maximum when the velocity is perpendicular to the field and is zero when the velocity is parallel to the field.

PROBLEM-SOLVING NOTE

The Properties of Magnetic Forces

Remember that a magnetic force occurs only when a charged particle moves in a direction that is not along the line defined by the magnetic field. Stationary particles experience no magnetic force.

Big Idea 2 The force on a particle due to a magnetic field is proportional to the charge of the particle and to its speed. In addition, the force has its maximum value when the velocity of the particle is at right angles to the magnetic field, and is zero when the velocity and magnetic field are parallel to one another.

EXAMPLE 22-2   A TALE OF TWO CHARGES

Particle 1, with a charge $q_1 = 3.60 \, \mu\text{C}$ and a speed $v_1 = 862$ m/s, travels at right angles to a uniform magnetic field. The magnetic force it experiences is 3.17×10^{-3} N. Particle 2, with a charge $q_2 = 73.0 \, \mu\text{C}$ and a speed $v_2 = 1.61 \times 10^3$ m/s, moves at an angle of 55.0° relative to the same magnetic field. Find **(a)** the strength of the magnetic field and **(b)** the magnitude of the magnetic force exerted on particle 2.

CONTINUED

PICTURE THE PROBLEM

The two charged particles are shown in our sketch, along with the magnetic field lines. Notice that $q_1 = 3.60\,\mu C$ moves at right angles to $\vec{B}$; the charge $q_2 = 73.0\,\mu C$ moves at an angle of 55.0° with respect to $\vec{B}$. The magnetic force also depends on the speeds of the particles, $v_1 = 862$ m/s and $v_2 = 1.61 \times 10^3$ m/s.

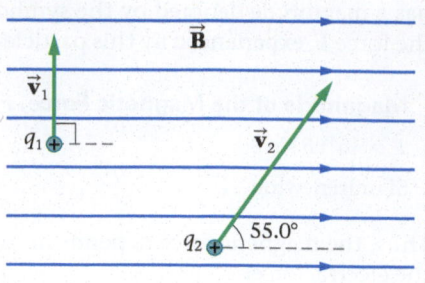

REASONING AND STRATEGY

a. We can find the strength of the magnetic field using the information given for particle 1. In particular, the particle moves at right angles to the magnetic field ($\theta_1 = 90°$), and therefore the magnetic force it experiences is $F_1 = |q_1|v_1B \sin \theta_1 = q_1v_1B$. Given that we know F_1, q_1, and v_1, we can solve for B.

b. Once we have determined B in part (a), it is straightforward to calculate the magnetic force on particle 2 using $F_2 = |q_2|v_2B \sin \theta_2$.

Known   Charge of particle 1, $q_1 = 3.60\,\mu C$; speed of particle 1, $v_1 = 862$ m/s; charge of particle 2, $q_2 = 73.0\,\mu C$; speed of particle 2, $v_2 = 1.61 \times 10^3$ m/s; magnetic force on particle 1, $F_1 = 3.17 \times 10^{-3}$ N.

Unknown   (a) Magnitude of magnetic field, $B = ?$ (b) Magnitude of magnetic force on particle 2, $F_2 = ?$

SOLUTION

Part (a)

1. Set $\theta_1 = 90°$ in $F_1 = |q_1|v_1B \sin \theta_1$, then solve for B:

$$F_1 = |q_1|v_1B \sin 90° = |q_1|v_1B$$
$$B = \frac{F_1}{|q_1|v_1}$$

2. Substitute numerical values:

$$B = \frac{F_1}{|q_1|v_1} = \frac{3.17 \times 10^{-3}\,\text{N}}{(3.60 \times 10^{-6}\,\text{C})(862\,\text{m/s})} = 1.02\,\text{T}$$

Part (b)

3. Use $B = 1.02$ T and $\theta_2 = 55.0°$ in $F_2 = |q_2|v_2B \sin \theta_2$ to find the magnetic force exerted on particle 2:

$$F_2 = |q_2|v_2B \sin \theta_2$$
$$= (73.0 \times 10^{-6}\,\text{C})(1.61 \times 10^3\,\text{m/s})(1.02\,\text{T}) \sin 55.0°$$
$$= 0.0982\,\text{N}$$

INSIGHT

The charge and speed of a particle are not enough to determine the magnetic force acting on it—its direction of motion relative to the magnetic field is needed as well.

PRACTICE PROBLEM

At what angle relative to the magnetic field must particle 2 move if the magnetic force it experiences is to be 0.0813 N? [**Answer:** $\theta = 42.7°$]

Some related homework problems: Problem 6, Problem 9

Units for the Magnetic Field

The tesla is a fairly large unit of magnetic strength, especially when compared with the magnetic field at the surface of the Earth, which is roughly 5.0×10^{-5} T. Thus, another commonly used unit of magnetism is the gauss (G), defined as follows:

$$1\,\text{gauss} = 1\,\text{G} = 10^{-4}\,\text{T}$$

In terms of the gauss, the Earth's magnetic field on the surface of the Earth is approximately 0.5 G. It should be noted, however, that the gauss is not an SI unit of magnetic field. Even so, it finds wide usage because of its convenient magnitude. The magnitudes of some typical magnetic fields are given in Table 22-1.

TABLE 22-1 Typical Magnetic Fields

Physical system	Magnetic field (G)
Magnetar (a magnetic neutron star formed in a supernova explosion)	10^{15}
Strongest man-made magnetic field	6×10^5
High-field MRI	15,000
Low-field MRI	2000
Sunspots	1000
Bar magnet	100
Earth	0.50

Magnetic Force Right-Hand Rule (RHR)

We now consider the *direction* of the magnetic force, which is rather interesting and unexpected. Instead of pointing in the direction of the magnetic field, $\vec{B}$, or in the direction of the velocity, $\vec{v}$, as one might expect, the following behavior is observed:

The magnetic force $\vec{F}$ points in a direction that is perpendicular to both $\vec{B}$ and $\vec{v}$.

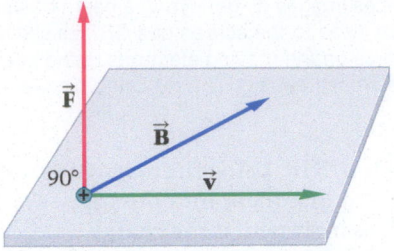

(a) $\vec{F}$ is perpendicular to both $\vec{v}$ and $\vec{B}$

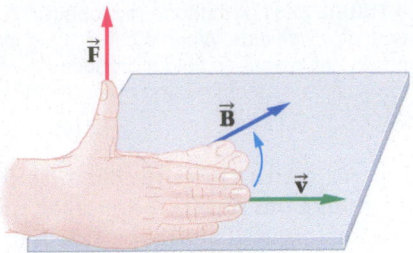

(b) Curl fingers from $\vec{v}$ to $\vec{B}$; thumb points in direction of $\vec{F}$

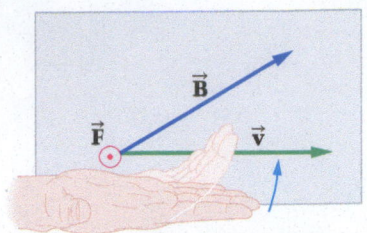

(c) Top view, looking down on $\vec{F}$

▲ **FIGURE 22-9 The magnetic force right-hand rule (a)** The magnetic force, $\vec{F}$, is perpendicular to both the velocity, $\vec{v}$, and the magnetic field, $\vec{B}$. (The force vectors shown in this figure are for the case of a positive charge. The force on a negative charge would be in the opposite direction.) **(b)** As the fingers of the right hand are curled from $\vec{v}$ to $\vec{B}$, the thumb points in the direction of $\vec{F}$. **(c)** An overhead view, looking down on the plane defined by the vectors $\vec{v}$ and $\vec{B}$. In this two-dimensional representation, the force vector comes out of the page and is indicated by a circle with a dot inside. If the charge was negative, the force would point into the page, and the symbol indicating $\vec{F}$ would be a circle with an X inside.

As an example, consider the vectors $\vec{B}$ and $\vec{v}$ in **FIGURE 22-9 (a)**, which lie in the indicated plane. The force on a positive charge, $\vec{F}$, as shown in the figure, is perpendicular to this plane and hence to both $\vec{B}$ and $\vec{v}$.

Notice that $\vec{F}$ could equally well point downward in Figure 22-9 (a) and still be perpendicular to the plane. The way we determine the precise direction of $\vec{F}$ is with a right-hand rule—similar to the right-hand rules used in calculating torques and angular momentum in Chapter 11. To be specific, the direction of $\vec{F}$ is found using the *magnetic force* right-hand rule:

> **Magnetic Force Right-Hand Rule (RHR)**
>
> To find the direction of the magnetic force on a positive charge, start by pointing the fingers of your right hand in the direction of the velocity, $\vec{v}$. Now curl your fingers toward the direction of $\vec{B}$, as illustrated in **FIGURE 22-9 (b)**. Your thumb points in the direction of $\vec{F}$. If the charge is negative, the force points opposite to the direction of your thumb.

Applying this rule to Figure 22-9 (a), we see that $\vec{F}$ does indeed point upward, as indicated.

A mathematical way to write the magnetic force that includes both its magnitude and direction is in terms of the vector **cross product.** Details on the cross product are given in Appendix A, but basically we can write the magnetic force $\vec{F}$ as follows:

$$\vec{F} = q\vec{v} \times \vec{B}$$

In this expression, the term $\vec{v} \times \vec{B}$ is referred to as the cross product of $\vec{v}$ and $\vec{B}$. From the definition of the cross product, the magnitude of the force is $F = |q|vB \sin\theta$, where θ is the magnitude of the angle between $\vec{v}$ and $\vec{B}$, precisely as in Equation 22-1. In addition, the direction of $\vec{F}$ is given by the magnetic force RHR. Thus, Equation 22-1 plus the magnetic force RHR are all we need to calculate magnetic forces—the cross product simply provides a more compact way of saying the same thing.

Representing Vectors Perpendicular to the Page Three-dimensional plots like the one in Figure 22-9 (b) are somewhat difficult to sketch in your notebook or on the blackboard, and hence we often use a shorthand for indicating vectors that point into or out of the page. (See Appendix A for a fuller discussion and examples.) First, imagine a vector as an arrow, with a pointed tip at one end and crossed feathers at the other end. If a vector points out of the page—toward us—we see its tip. This situation is indicated by drawing a circle with a dot inside it, as in **FIGURE 22-9 (c)**.

On the other hand, if a vector points into the page, we see the crossed feathers at its base; hence, we draw a circle with an X inside it to symbolize the feathers. As an example, **FIGURE 22-10** indicates a uniform magnetic field $\vec{B}$ that points into the page.

PHYSICS IN CONTEXT
Looking Back

The direction of the force exerted on a moving charge by a magnetic field is given by the right-hand rule, which was discussed in Chapter 11 in relation to torques and angular momentum.

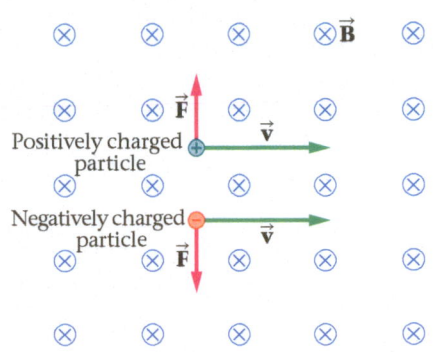

▲ **FIGURE 22-10 The magnetic force for positive and negative charges** The direction of the magnetic force depends on the sign of the charge. Specifically, the force exerted on a negatively charged particle is opposite in direction to the force exerted on a positively charged particle.

◄ **FIGURE 22-11** A cathode-ray tube (CRT) produces images by deflecting a beam of electrons with electric fields. When a bar magnet is moved close to the screen of a CRT, as shown in this photo, the magnetic field it produces exerts a force on the moving electrons in the beam. This force changes the direction of motion of the electrons, resulting in a distorted picture.

A particle with a positive charge q moves to the right. Using the magnetic force RHR—extending our fingers to the right and then curling them into the page—we see that the force exerted on this particle is upward, as indicated. If the charge is negative, the direction of $\vec{F}$ is reversed, as is also illustrated in Figure 22-10. An example of the effect of magnetic forces acting on moving electrons is shown in **FIGURE 22-11**.

CONCEPTUAL EXAMPLE 22-3   CHARGE OF A PARTICLE

Three particles travel through a region of space where the magnetic field is out of the page, as shown in the sketch on the left. For each of the three particles, state whether the particle's charge is positive, negative, or zero.

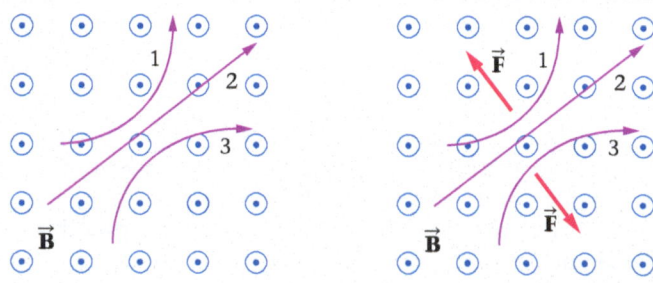

REASONING AND DISCUSSION

In the sketch on the right, we indicate the general direction of the force required to cause the observed motion. The force indicated for particle 3 is in the direction given by the magnetic force RHR; hence, particle 3 must have a positive charge. The force acting on particle 1 is in the opposite direction; hence, that particle must be negatively charged. Finally, particle 2 is undeflected; hence, its charge, and the force acting on it, must be zero.

ANSWER

Particle 1, negative; particle 2, zero; particle 3, positive.

We conclude this section with an Example illustrating the magnetic force RHR.

EXAMPLE 22-4   ELECTRIC AND MAGNETIC FIELDS

A particle with a charge of 5.75 μC and a speed of 351 m/s is acted on by both an electric and a magnetic field. The particle moves along the x axis in the positive direction, the magnetic field has a strength of 2.26 T and points in the positive y direction, and the electric field points in the positive z direction with a magnitude of 1.16×10^3 N/C. Find the magnitude of the net force acting on the particle.

PICTURE THE PROBLEM

The physical situation is shown in our sketch. In particular, we show a three-dimensional coordinate system with each of the three relevant vectors, $\vec{v}$, $\vec{B}$, and $\vec{E}$, indicated. The charge is positive, and therefore $|q| = q$.

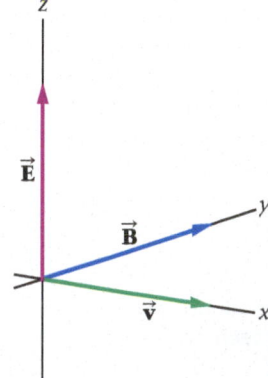

REASONING AND STRATEGY

First, the force due to the electric field is in the direction of $\vec{E}$ (positive z direction), and it has a magnitude qE. Second, the direction of the magnetic force is given by the RHR. In this case curling the fingers of a right hand from $\vec{v}$ to $\vec{B}$ results in the thumb (the direction of $\vec{F}$) pointing in the positive z direction. Thus, in this system the forces due to the electric and magnetic fields are in the same direction. Finally, because the angle between $\vec{v}$ and $\vec{B}$ is 90°, the magnitude of the magnetic force is qvB.

Known   Charge of particle, $q = 5.75$ μC; speed of particle, $v = 351$ m/s; magnetic field, $B = 2.26$ T; electric field, $E = 1.16 \times 10^3$ N/C.

Unknown   Magnitude of net force, $F_{net} = ?$

SOLUTION

1. Calculate the magnitude of the electric force exerted on the particle. This force is in the positive z direction:

$$F_E = qE$$
$$= (5.75 \times 10^{-6}\,\text{C})(1.16 \times 10^3\,\text{N/C}) = 6.67 \times 10^{-3}\,\text{N}$$

CONTINUED

2. Calculate the magnitude of the magnetic force exerted on the particle. This force is also in the positive z direction:

$F_B = qvB$
$= (5.75 \times 10^{-6}\,\text{C})(351\,\text{m/s})(2.26\,\text{T}) = 4.56 \times 10^{-3}\,\text{N}$

3. Because both forces are in the positive z direction, we simply add them to obtain the magnitude of the net force:

$F_{net} = F_E + F_B = (6.67 \times 10^{-3}\,\text{N}) + (4.56 \times 10^{-3}\,\text{N})$
$= 1.12 \times 10^{-2}\,\text{N}$

INSIGHT

The net force on the particle is in the positive z direction, and it will be in that direction if the velocity vector, $\vec{v}$, points in *any* direction in the x-y plane. The magnitude of the net force, however, will depend on the direction of $\vec{v}$, as we see in the following Practice Problem.

PRACTICE PROBLEM

Find the net force on the particle if its velocity is reversed and it moves in the negative x direction.   [**Answer:** In this case, the magnetic force is in the negative z direction. Therefore, the net force is $F_{net} = F_E + F_B = (6.67 \times 10^{-3}\,\text{N}) + (-4.56 \times 10^{-3}\,\text{N}) = 2.11 \times 10^{-3}\,\text{N}$. This is in the positive z direction.]

Some related homework problems: Problem 12, Problem 14, Problem 15

Enhance Your Understanding   (Answers given at the end of the chapter)

2. A positively charged particle moves with a velocity $\vec{v}$ at an angle θ relative to a magnetic field $\vec{B}$, as shown in **FIGURE 22-12**. Does the magnetic force exerted on the particle point into the page or out of the page?

Section Review

- The magnitude of the magnetic force on a particle of charge q moving with a speed v at an angle θ relative to a magnetic field of strength B is $F = |q|vB \sin \theta$.

- To find the direction of the magnetic force on a positive charge, point the fingers of your right hand in the direction of $\vec{v}$ and curl your fingers toward $\vec{B}$. Your thumb now points in the direction of $\vec{F}$. If the charge is negative, the force is in the opposite direction to the force on a positive charge.

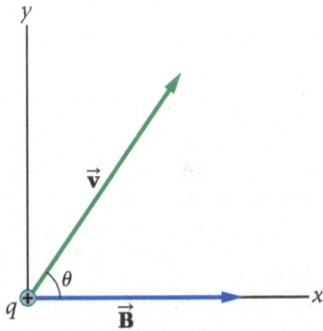

▲ **FIGURE 22-12** Enhance Your Understanding 2.

22-3 The Motion of Charged Particles in a Magnetic Field

As we've seen, the magnetic force has characteristics that set it apart from the force exerted by electric or gravitational fields. In particular, the magnetic force depends not only on the speed of a particle but on its direction of motion as well. We now explore some of the consequences that follow from these characteristics and relate them to the type of motion that occurs in magnetic fields.

Electric Versus Magnetic Forces

We begin by investigating the motion of a charged particle as it is projected into a region with either an electric or a magnetic field. For example, in **FIGURE 22-13 (a)** we consider a uniform electric field pointing downward. A positively charged particle moving horizontally into this region experiences a constant downward force—much like a mass in a uniform gravitational field. As a result, the particle begins to accelerate downward and follow a parabolic path.

If the same particle, moving to the right, encounters a magnetic field instead, as in **FIGURE 22-13 (b)**, the resulting motion is quite different. We again assume that the field is uniform and pointing downward. In this case, however, the magnetic force RHR shows

▶ **FIGURE 22-13 Differences between motion in electric and magnetic fields** (a) A positively charged particle moving into a region with an electric field experiences a downward force that causes it to accelerate. (b) A charged particle entering a magnetic field experiences a horizontal force at right angles to its direction of motion. In this case, the speed of the particle remains constant.

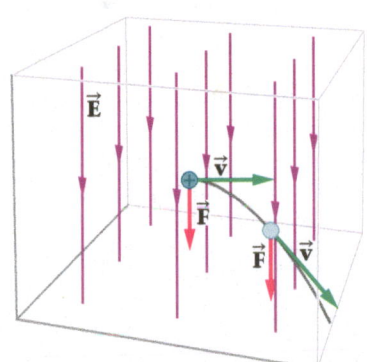

(a) Motion in an electric field

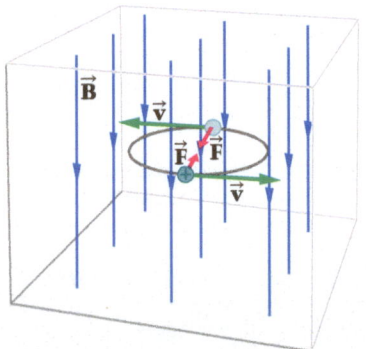

(b) Motion in a magnetic field

that $\vec{F}$ points into the page. Our particle now begins to follow a horizontal path into the page. As we shall see, this path is circular.

Perhaps even more significant than these differences in motion is the fact that an electric field can do work on a charged particle, whereas a constant magnetic field cannot. In Figure 22-13 (a), for example, as soon as the particle begins to move downward, a component of its velocity is in the direction of the electric force. This means that the electric force does work on the particle, and its speed increases—again, just like a mass falling in a gravitational field. If the particle moves through a magnetic field, however, no work is done on it because the magnetic force is *always* at right angles to the direction of motion. Thus, the speed of the particle in the magnetic field remains constant.

CONCEPTUAL EXAMPLE 22-5   DIRECTION OF THE MAGNETIC FIELD

In a device called a **velocity selector,** charged particles move through a region of space with both an electric and a magnetic field. If the speed of the particle has a particular value, the net force acting on it is zero. Assume that a positively charged particle moves in the positive x direction, as shown in the sketch, and the electric field is in the positive y direction. Should the magnetic field be in **(a)** the positive z direction, **(b)** the negative y direction, or **(c)** the negative z direction in order to give zero net force?

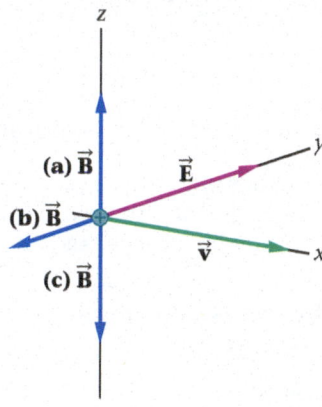

REASONING AND DISCUSSION

The force exerted by the electric field is in the positive y direction; hence, the magnetic force must be in the negative y direction if it is to cancel the electric force. If we simply try the three possible directions for $\vec{B}$ one at a time, applying the magnetic force RHR in each case, we find that only a magnetic field along the positive z axis gives rise to a force in the negative y direction, as desired.

ANSWER

(a) $\vec{B}$ should point in the positive z direction.

Magnet and Electron Beam

Velocity Selector   To follow up on Conceptual Example 22-5, let's determine the speed for which the net force is zero. First, recall that the electric force has a magnitude qE. Second, the magnitude of the magnetic force is qvB. With $\vec{E}$ in the positive y direction and $\vec{B}$ in the positive z direction, these forces are in opposite directions. Setting the magnitudes of the forces equal yields $qE = qvB$. Canceling the charge q and solving for v, we find

$$v = \frac{E}{B}$$

(velocity selector)

A particle with this speed, regardless of its charge, passes through the velocity selector with zero net force and hence no deflection.

RWP The physical principle illustrated in the velocity selector can be used to measure the speed of blood with a device known as an *electromagnetic flowmeter*. Suppose an artery passes between the poles of a magnet, as shown in **FIGURE 22-14**. Charged ions in the blood will be deflected at right angles to the artery by the magnetic field. The resulting charge separation produces an electric field that opposes the magnetic deflection.

▶ **FIGURE 22-14 The electromagnetic flowmeter** As blood flows through a magnetic field, the charged ions it contains are deflected. The deflection of charged particles results in an electric field opposing the deflection. If the speed of the blood is v, the electric field generated by ions moving through the magnetic field satisfies the equation $v = E/B$.

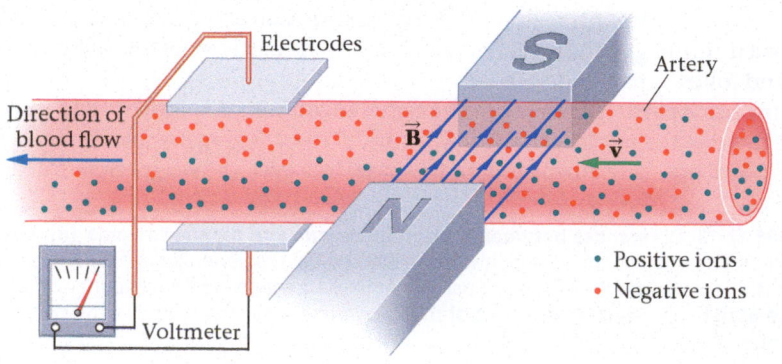

When the electric field is strong enough, the deflection ceases and the blood flows normally through the artery. If the electric field is measured, and the magnetic field is known, the speed of the blood flow is simply $v = E/B$, as in a standard velocity selector.

Constant-Velocity, Straight-Line Motion

Recall that the magnetic force is zero if a particle's velocity $\vec{v}$ is parallel (or antiparallel) to the magnetic field $\vec{B}$. In such a case the particle's acceleration is zero; therefore, its velocity remains constant. Thus, the simplest type of motion in a magnetic field is a constant-velocity, straight-line drift along the magnetic field lines.

Circular Motion

The next simplest case is a particle with a velocity that is perpendicular to the magnetic field. Consider, for example, the situation shown in **FIGURE 22-15**. Here a particle of mass m, charge $+q$, and speed v moves in a region with a constant magnetic field $\vec{B}$ pointing out of the page. Because $\vec{v}$ is at right angles to $\vec{B}$, the magnitude of the magnetic force is $F = |q|vB \sin 90° = |q|vB$.

Now we consider the particle's motion. At point 1 the particle is moving to the right; hence, the magnetic force is downward, causing the particle to accelerate in the downward direction. When the particle reaches point 2, it is moving downward, and now the magnetic force is to the left. This causes the particle to accelerate to the left. At point 3, the force exerted on it is upward, and so on, as the particle continues moving.

Thus, at every point on the particle's path the magnetic force is at right angles to the velocity, pointing toward a common center—but this is just the condition required for circular motion. As we saw in Section 6-5, in circular motion the acceleration is toward the center of the circle; for this reason, a centripetal force is required to cause the motion. In this case the centripetal force is supplied by the magnetic force—in the same way that a string exerts a centripetal force on a ball being whirled about in a circle.

Recall that the centripetal acceleration of a particle moving with a speed v in a circle of radius r is

$$a_{cp} = \frac{v^2}{r}$$

Therefore, setting ma_{cp} equal to the magnitude of the magnetic force, $|q|vB$, yields the following condition:

$$m\frac{v^2}{r} = |q|vB$$

Canceling one power of the speed, v, we find that the radius of the circular orbit is

$$r = \frac{mv}{|q|B} \qquad \text{22-3}$$

We can see, therefore, that the faster and more massive the particle, the larger the circle. Conversely, the stronger the magnetic field and the greater the charge, the smaller the circle.

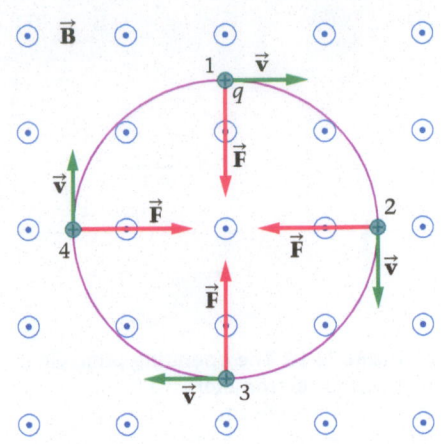

▲ **FIGURE 22-15 Circular motion in a magnetic field** A charged particle moves in a direction perpendicular to the magnetic field. At each point on the particle's path the magnetic force is at right angles to the velocity and hence toward the center of a circle.

EXERCISE 22-6 SPEED OF AN ELECTRON

An electron moving perpendicular to a magnetic field of 4.31×10^{-3} T follows a circular path of radius 3.70 mm. What is the electron's speed?

REASONING AND SOLUTION

Solving $r = mv/|q|B$ for the speed v, and using $m = 9.11 \times 10^{-31}$ kg and $|q| = 1.60 \times 10^{-19}$ C for an electron, we find

$$v = \frac{r|q|B}{m} = \frac{(3.70 \times 10^{-3}\,\text{m})(1.60 \times 10^{-19}\,\text{C})(4.31 \times 10^{-3}\,\text{T})}{9.11 \times 10^{-31}\,\text{kg}} = 2.80 \times 10^6\,\text{m/s}$$

The speed of this electron is about 1% of the speed of light.

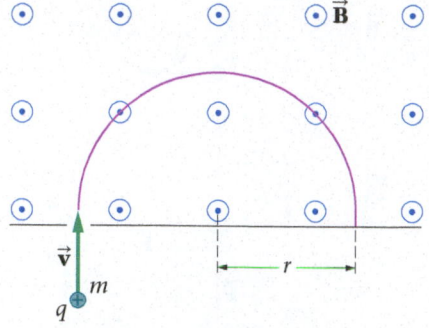

▲ **FIGURE 22-16 The operating principle of a mass spectrometer** In a mass spectrometer, a beam of charged particles enters a region with a magnetic field perpendicular to the velocity. The particles then follow a circular orbit of radius $r = mv/|q|B$. Particles of different mass will follow different paths.

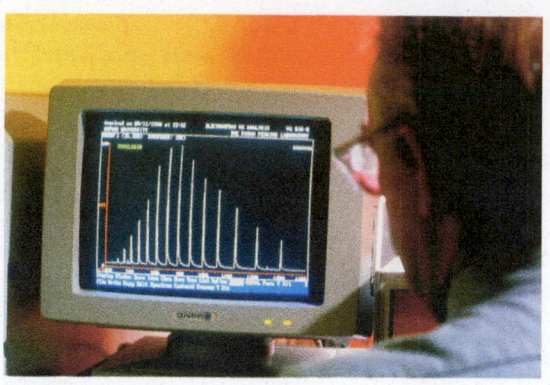

▲ **FIGURE 22-17** In addition to separating and "weighing" isotopes, mass spectrometers can be used to study the chemical composition and structure of large, biologically important molecules. Here, a researcher studies the spectrum of myoglobin, an oxygen-carrying protein found in muscle tissue. Each peak represents a fragment of the molecule with its own characteristic combination of mass and charge.

Mass Spectrometer One of the applications of circular motion in a magnetic field is in a device known as a **mass spectrometer.** A mass spectrometer can be used to separate isotopes (atoms of the same element with different masses) and to measure atomic masses. It finds many uses in medicine (anesthesiologists use it to measure respiratory gases), biology (to determine reaction mechanisms in photosynthesis), geology (to date fossils), space science (to determine the atmospheric composition of Mars), and a variety of other fields.

RWP The basic principles of a mass spectrometer are illustrated in **FIGURE 22-16**. Here we see a beam of ions of mass m and charge $+q$ entering a region of constant magnetic field with a speed v. The field causes the ions to move along a circular path, with a radius that depends on the mass and charge of the ion. Thus, different isotopes follow different paths and hence can be separated and identified, as indicated in **FIGURE 22-17**. A specific situation is considered in the next Example.

EXAMPLE 22-7   URANIUM SEPARATION

Two isotopes of uranium, ^{235}U ($m_{235} = 3.90 \times 10^{-25}$ kg) and ^{238}U ($m_{238} = 3.95 \times 10^{-25}$ kg), are sent into a mass spectrometer with a speed of 1.05×10^5 m/s, as indicated in the sketch. Given that each isotope is singly ionized, and that the strength of the magnetic field is 0.750 T, what is the separation distance d between the two isotopes after they complete half a circular orbit?

PICTURE THE PROBLEM
The relevant features of the mass spectrometer are indicated in our sketch. Notice that both isotopes enter at the same location with the same speed. Because they have different masses, however, they follow circular paths of different radii. This difference results in the separation d after half an orbit.

REASONING AND STRATEGY
We begin by calculating the radius of each isotope's circular path using $r = mv/|q|B$. The masses, speeds, and magnetic field are given. We can infer the charge from the fact that the isotopes are "singly ionized," which means that a single electron has been removed from each atom. Because the atoms were electrically neutral before the electron was removed, it follows that the charge of the isotopes is $q = e = 1.60 \times 10^{-19}$ C.

Finally, once we know the radii, the separation between the isotopes is given by $d = 2r_{238} - 2r_{235}$, as indicated in the sketch.

Known   Mass of ^{235}U atom, $m_{235} = 3.90 \times 10^{-25}$ kg; mass of ^{238}U atom, $m_{238} = 3.95 \times 10^{-25}$ kg; speed of atoms, $v = 1.05 \times 10^5$ m/s; charge of atoms, $q = 1.60 \times 10^{-19}$ C; magnetic field, $B = 0.750$ T.
Unknown   Separation distance between isotopes, $d = ?$

CONTINUED

SOLUTION

1. Determine the radius of the circular path of ^{235}U atoms:

$$r_{235} = \frac{m_{235}v}{|q|B}$$

$$= \frac{(3.90 \times 10^{-25}\,\text{kg})(1.05 \times 10^5\,\text{m/s})}{(1.60 \times 10^{-19}\,\text{C})(0.750\,\text{T})} = 34.1\,\text{cm}$$

2. Determine the radius of the circular path of ^{238}U atoms:

$$r_{238} = \frac{m_{238}v}{|q|B}$$

$$= \frac{(3.95 \times 10^{-25}\,\text{kg})(1.05 \times 10^5\,\text{m/s})}{(1.60 \times 10^{-19}\,\text{C})(0.750\,\text{T})} = 34.6\,\text{cm}$$

3. Calculate the separation distance between the isotopes:

$$d = 2r_{238} - 2r_{235} = 2(34.6\,\text{cm} - 34.1\,\text{cm}) = 1\,\text{cm}$$

INSIGHT

The difference in masses is quite small, and would be difficult to measure directly, but the mass spectrometer converts the tiny difference to an easily measured distance of separation.

PRACTICE PROBLEM — PREDICT/CALCULATE

(a) Does the separation d increase or decrease if the magnetic field is increased? Explain. (b) Check your answer by calculating the separation for $B = 1.00\,\text{T}$, with everything else remaining the same.   [**Answer: (a)** A stronger field decreases the radii and hence decreases the separation d. **(b)** If B changes by a factor x, the separation changes by the factor $1/x$. In this case, the factor is $x = \frac{4}{3}$; that is, $B = \left(\frac{4}{3}\right)(0.750\,\text{T}) = 1.00\,\text{T}$. Therefore, the new separation distance is $d = \left(\frac{3}{4}\right)(1\,\text{cm}) = \frac{3}{4}\,\text{cm}$.]

Some related homework problems: Problem 19, Problem 24

QUICK EXAMPLE 22-8   FIND THE TIME FOR ONE ORBIT

Calculate the time T required for a particle of mass m and charge q to complete a circular orbit in a magnetic field of strength B.

REASONING AND SOLUTION

As the particle moves in a circular orbit, its speed v is equal to the circumference of the orbit, $2\pi r$, divided by the time for an orbit, T. In addition, the radius of the orbit is related to the speed by $r = mv/|q|B$. Combining these relationships yields an expression for the time T in terms of m, $|q|$, and B.

1. Write the speed, v, of a particle in terms of the time T to complete an orbit of radius r:

$$v = \frac{2\pi r}{T}$$

2. Substitute this expression for v into the condition for a circular orbit, $r = mv/|q|B$:

$$r = \frac{m(2\pi r/T)}{|q|B}$$

3. Cancel r and rearrange to solve for T:

$$T = \frac{2\pi m}{|q|B}$$

It's interesting to note that the time required for an orbit is independent of the speed of the particle.

RWP A particularly useful application of the magnetic force on a moving charge is the *synchrotron undulator*, a device that is capable of producing very bright X-rays (part of the electromagnetic spectrum that we will explore in Chapter 25). In a synchrotron undulator, a beam of electrons moving at extremely high velocities passes through a series of closely spaced magnets with opposite poles, as illustrated in **FIGURE 22-18**. The magnetic forces accelerate the electrons first in one direction, and then in the opposite direction, so that the beam undulates or wiggles back and forth. This undulating action is responsible for the production of the X-rays. Synchrotrons with undulators can produce very bright and stable beams of X-rays that are useful for analyzing biological molecules, exploring the properties of materials, and manufacturing microscopic structures using lithography, among many other uses.

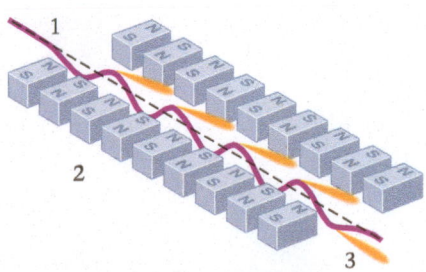

▲ **FIGURE 22-18** In a synchrotron undulator, a beam of high-speed electrons (1) passes between the poles of a series of alternating magnets (2). This undulates (or wiggles) the beam, causing the electrons to emit X-rays (3).

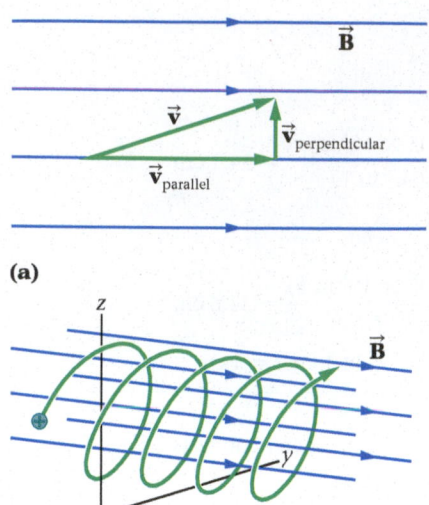

(a)

(b)

◀ **FIGURE 22-19 Helical motion in a magnetic field (a)** A velocity at an angle to a magnetic field $\vec{B}$ can be resolved into parallel and perpendicular components. The parallel component gives a constant-velocity drift in the direction of the field. The perpendicular component gives circular motion perpendicular to the field.
(b) Helical motion is a combination of linear motion and circular motion.

Helical Motion

The final type of motion we consider is a combination of the two motions discussed already. Suppose, for example, that a particle has an initial velocity at an angle to the magnetic field, as in **FIGURE 22-19 (a)**. In this case there is a component of velocity parallel to $\vec{B}$ and a component perpendicular to $\vec{B}$. The parallel component of the velocity remains constant with time (zero force in this direction), whereas the perpendicular component results in a circular motion, as just discussed. Combining the two motions, we can see that the particle follows a helical path, as shown in **FIGURE 22-19 (b)**.

RWP If a magnetic field is curved, as in the case of a bar magnet or the magnetic field of the Earth, the helical motion of charged particles will be curved as well. Specifically, the axis of the helical motion will follow the direction of $\vec{B}$. For example, electrons and protons emitted by the "solar wind" frequently encounter the Earth's magnetic field and begin to move in helical paths following the field lines. Near the poles, where the field lines are concentrated as they approach the Earth's surface, the circulating electrons begin to collide with atoms and molecules in the atmosphere. These collisions can excite and ionize the atmospheric atoms and molecules, resulting in the emission of light known as the **aurora borealis** (northern lights) in the northern hemisphere and the **aurora australis** (southern lights) in the southern hemisphere. The yellowish-green light in an aurora is given off by oxygen atoms, and the blue or purplish-red light is generally due to nitrogen molecules. **FIGURE 22-20** shows an eruption on the Sun that contributes to the solar wind, as well as images of the aurora borealis produced when the particles in the solar wind interact with the Earth's magnetic field.

Enhance Your Understanding   (Answers given at the end of the chapter)

3. A particle orbits in a magnetic field with a radius of 1 m. If the mass of the particle is doubled, while everything else remains the same, what is the new radius of the orbit? **(a)** 4 m, **(b)** 2 m, **(c)** 1 m, **(d)** $\frac{1}{2}$ m.

Section Review

- The radius of a particle's orbit in a magnetic field of strength B is $r = mv/|q|B$. In this expression, m is the mass of the particle, v is its speed, and q is its charge.

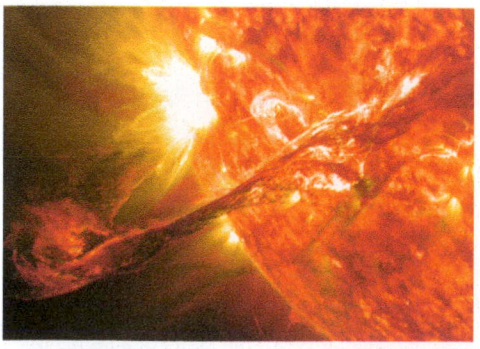

▲ **FIGURE 22-20 Visualizing Concepts Magnetic Fields** (Left) An enormous eruption of matter from the surface of the Sun, large enough to encompass the entire Earth many times over. The loop structure is created by the Sun's own complex magnetic field. (Center) A glowing aurora borealis surrounds the Earth's north geographic pole in this photograph from space. (Right) An auroral display seen from Earth. The characteristic red and green colors are produced by ionized nitrogen molecules and oxygen atoms, respectively.

▶ **FIGURE 22-21 The magnetic force on a current-carrying wire** A current-carrying wire in a magnetic field experiences a force, unless the current is parallel or antiparallel to the field. **(a)** For a wire segment of length L the magnitude of the force is $F = ILB \sin \theta$. **(b)** The direction of the force is given by the magnetic force RHR; the only difference is that you start by pointing the fingers of your right hand in the direction of the current I. In this case the force points out of the page.

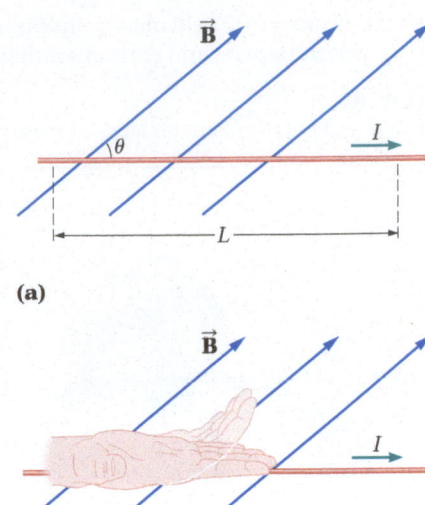

(a)

(b)

22-4 The Magnetic Force Exerted on a Current-Carrying Wire

A charged particle experiences a force when it moves across magnetic field lines. This is true whether the particle travels in a vacuum or in a current-carrying wire. Thus, a wire with a current will experience a force that is simply the resultant of all the forces experienced by the individual moving charges responsible for the current.

Specifically, consider a straight wire segment of length L with a current I flowing from left to right, as in **FIGURE 22-21 (a)**. Also present in this region of space is a magnetic field $\vec{\mathbf{B}}$ at an angle θ to the length of the wire. If the conducting charges move through the wire with an average drift speed v, the time required for them to move from one end of the wire segment to the other is $\Delta t = L/v$. The amount of charge that flows through the wire in this time is $q = I \Delta t = IL/v$. Therefore, the force exerted on the wire is

$$F = qvB \sin \theta = \left(\frac{IL}{v}\right)vB \sin \theta$$

Canceling v, we find that the force on a wire segment of length L with a current I at an angle θ to the magnetic field $\vec{\mathbf{B}}$ is:

Magnetic Force on a Current-Carrying Wire	
$F = ILB \sin \theta$	22-4
SI unit: newton, N	

As with single charges, the maximum force occurs when the current is perpendicular to the magnetic field ($\theta = 90°$) and is zero if the current is in the same direction as $\vec{\mathbf{B}}$ ($\theta = 0$).

The direction of the magnetic force on a wire is given by the same right-hand rule used earlier for single charges. Thus, to find the direction of the force in **FIGURE 22-21 (b)**, start by pointing the fingers of your right hand in the direction of the current I. This assumes that positive charges are flowing in the direction of I, consistent with our convention from Chapter 21. Now, curl your fingers toward the direction of $\vec{\mathbf{B}}$. Your thumb, which points out of the page, indicates the direction of $\vec{\mathbf{F}}$.

Of course, the current is actually caused by negatively charged electrons flowing in the opposite direction to the current. The magnetic force on these negatively charged particles moving in the opposite direction is the same as the force on positively charged particles moving in the direction of I. Thus, in all cases pertaining to current-carrying wires, we can simply think of the current as the direction in which positively charged particles move.

PROBLEM-SOLVING NOTE

The Magnetic Force on a Current-Carrying Wire

Note that a current-carrying wire experiences no force when it is in the same direction as the magnetic field. The maximum force occurs when the wire is perpendicular to the magnetic field.

Big Idea **3** Magnetic fields exert forces on current-carrying wires that depend on the magnitude of the current and the direction of the current relative to the magnetic field.

Current-Carrying Wire in Magnetic Field

CONCEPTUAL EXAMPLE 22-9   MAGNETIC POLES

When the switch is closed in the circuit shown in the sketch on the left, the wire between the poles of the horseshoe magnet deflects downward. Is the left end of the magnet a north magnetic pole or a south magnetic pole?

REASONING AND DISCUSSION

Once the switch is closed, the current in the wire is into the page, as shown in the sketch on the right. Applying the magnetic force RHR, we

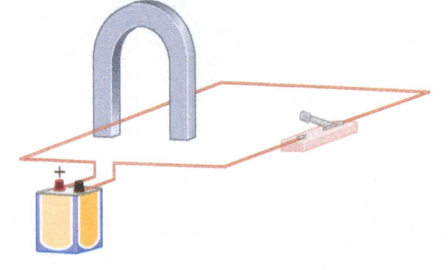

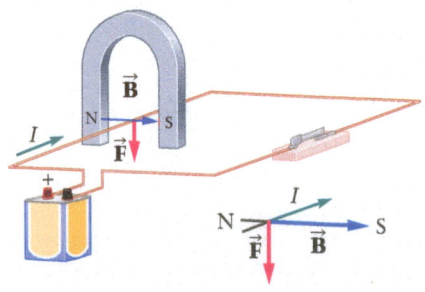

CONTINUED

see that the magnetic field must point from left to right in order for the force to be downward. Because magnetic field lines leave from north poles and enter at south poles, it follows that the left end of the magnet must be a north magnetic pole.

ANSWER
The left end of the magnet is a north magnetic pole.

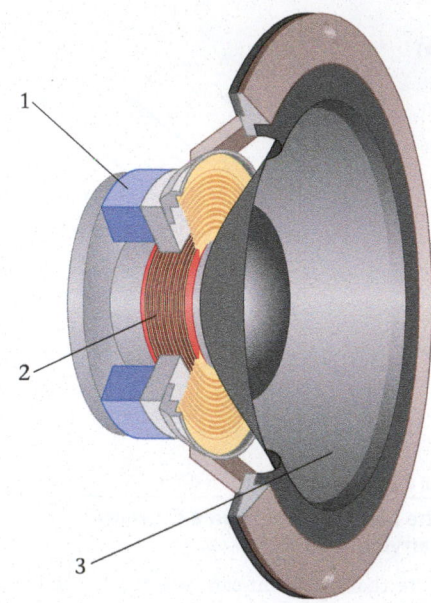

◄ **FIGURE 22-22** In a loudspeaker, a permanent magnet (1) creates a magnetic field around a coil of wire (2) that is attached to the base of a conical diaphragm (3). When a current flows through the coil, the magnetic field produces a force on the coil that moves the diaphragm. This in turn moves air molecules and produces sound.

RWP The magnetic force on a conductor is the fundamental principle behind the operation of a *loudspeaker*. A permanent magnet creates a strong field near a fine coil of wire that is attached to a flexible diaphragm, as shown in **FIGURE 22-22**. When the alternating electric current produced by the audio source flows through the coil, the mechanical force on the coil produced by the magnetic field moves the coil and the diaphragm rapidly. The back-and-forth motion of the diaphragm then generates the sound wave that we hear. The same idea forms the basis of nearly every loudspeaker, from ear buds and headphones to the giant speakers used at outdoor concerts.

The magnetic force exerted on a current-carrying wire can be quite substantial. In the following Example, we consider the current necessary to levitate a copper rod.

EXAMPLE 22-10   MAGNETIC LEVITY

A copper rod 0.175 m long and with a mass of 0.0529 kg is suspended from two thin, flexible wires, as shown in the sketch. At right angles to the rod is a uniform magnetic field of 0.615 T pointing into the page. Find **(a)** the direction and **(b)** the magnitude of the electric current needed to levitate the copper rod.

PICTURE THE PROBLEM
Our sketch shows the physical situation with the relevant quantities labeled. The direction of the current has been indicated as well. With the current in this direction, the RHR gives an upward magnetic force, as required for the rod to be levitated.

REASONING AND STRATEGY
To find the magnitude of the current, we set the magnetic force equal in magnitude to the force of gravity. For the magnetic force we have $F = ILB \sin \theta$. In this case the field is at right angles to the current; hence, $\theta = 90°$, and the force simplifies to $F = ILB$. Thus, setting ILB equal to mg determines the current.

Known   Length of rod, $L = 0.175$ m; mass of rod, $m = 0.0529$ kg; magnetic field, $B = 0.615$ T.
Unknown   **(a)** Direction of current = ? **(b)** Magnitude of current, I = ?

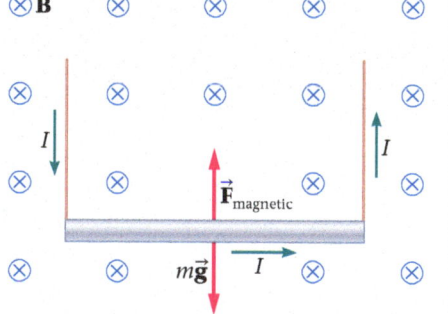

SOLUTION

Part (a)

1. Determine the direction of I:

The current, I, points to the right. To verify, point the fingers of your right hand to the right, curl them into the page, and your thumb will point upward, as desired.

Part (b)

2. Set the magnitude of the magnetic force equal to the magnitude of the force of gravity:

$$ILB = mg$$

3. Solve for the current, I:

$$I = \frac{mg}{LB} = \frac{(0.0529 \text{ kg})(9.81 \text{ m/s}^2)}{(0.175 \text{ m})(0.615 \text{ T})} = 4.82 \text{ A}$$

CONTINUED

INSIGHT

Magnetic forces, like electric forces, can easily exceed the force of gravity. In fact, when we consider atomic systems in Chapter 31 we shall see that gravity plays no role in the behavior of an atom—only electric and magnetic forces are important on the atomic level.

PRACTICE PROBLEM

Suppose the rod is doubled in length, which also doubles its mass. Does the current needed to levitate the rod increase, decrease, or stay the same? [**Answer:** The levitation current is given by $I = mg/LB$; hence, it's clear that doubling both m and L has no effect on I; the current needed to levitate the rod remains the same.]

Some related homework problems: Problem 28, Problem 31

Enhance Your Understanding   (Answers given at the end of the chapter)

4. The following systems consist of a current-carrying wire in a magnetic field of magnitude B. Rank the systems in order of increasing magnetic force. Indicate ties where appropriate.

System	A	B	C	D
I (A)	1	2	0.5	0.1
L (m)	0.5	1	2	4
θ (°)	0	30	60	90

Section Review

*   The force on a wire of length L that carries a current I at an angle θ relative to a magnetic field of strength B is $F = ILB \sin \theta$.

22-5 Loops of Current and Magnetic Torque

The fact that a current-carrying wire experiences a force when placed in a magnetic field is one of the fundamental discoveries that makes modern applications of electric power possible. In most of these applications, including electric motors and generators, the wire is shaped into a current-carrying loop. We will examine some of these applications further in Chapter 23; in this section we lay the groundwork by considering what happens when a simple current loop is placed in a magnetic field.

Rectangular Current Loops

Consider a rectangular loop of height h and width w carrying a current I, as shown in **FIGURE 22-23**. The loop is placed in a region of space with a uniform magnetic field $\vec{B}$ that is parallel to the plane of the loop. From Figure 22-23, it is clear that the horizontal segments of the loop experience zero force, since they are parallel to the field. The vertical segments, on the other hand, are perpendicular to the field; hence, they experience forces of magnitude $F = IhB$. One of these forces is into the page (left side); the other is out of the page (right side).

Perhaps the best way to visualize the torque caused by these forces is to use a top view and look directly down on the loop, as in **FIGURE 22-24 (a)**. Here we can see more clearly that the forces on the vertical segments are equal in magnitude and opposite in direction. If we imagine an axis of rotation through the center of the loop, at the point O, it's clear that the forces exert a torque that tends to rotate the loop clockwise. The magnitude of this torque for each vertical segment is the force ($F = IhB$) times the moment arm ($w/2$). Noting that both vertical segments exert a torque in the same direction, we see that the total torque is simply the sum of the torques produced by each segment:

$$\tau = (IhB)\left(\frac{w}{2}\right) + (IhB)\left(\frac{w}{2}\right) = IB(hw)$$

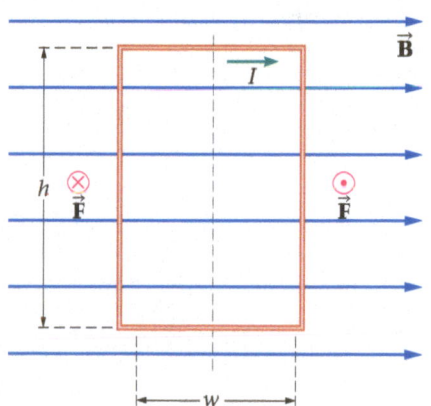

▲ **FIGURE 22-23 Magnetic forces on a current loop** A rectangular current loop in a magnetic field. Only the vertical segments of the loop experience forces, and they tend to rotate the loop about a vertical axis.

PHYSICS IN CONTEXT
Looking Back

A current-carrying loop in a magnetic field can experience a torque. Recall, that torque was defined and studied in Chapter 11.

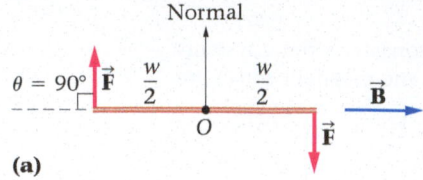

(a)

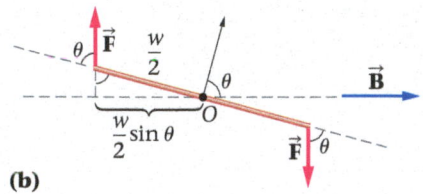

(b)

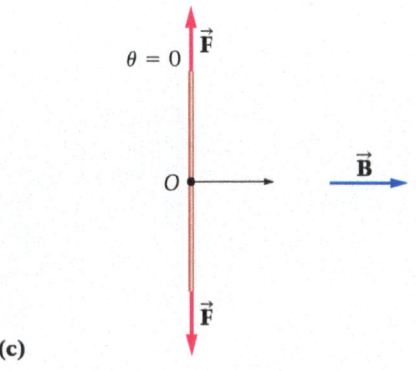

(c)

◄ FIGURE 22-24 Magnetic torque on a current loop A current loop placed in a magnetic field produces a torque. **(a)** The torque is greatest when the plane of the loop is parallel to the magnetic field—that is, when the normal to the loop is perpendicular to the magnetic field. **(b)** As the loop rotates, the torque decreases by a factor of sin θ. **(c)** The torque vanishes when the plane of the loop is perpendicular to the magnetic field.

Finally, observing that the area of the rectangular loop is $A = hw$, we can express the torque as follows:

$$\tau = IAB$$

As the loop begins to rotate, the situation will be like that shown in **FIGURE 22-24 (b)**. Here we see that the forces still have the same magnitude, IhB, but now the moment arms are $(w/2)\sin\theta$ rather than $w/2$. Thus, for a general angle, the torque must include the factor $\sin\theta$:

Torque Exerted on a Rectangular Loop of Area A

$$\tau = IAB \sin\theta \qquad\qquad\qquad 22\text{-}5$$

SI unit: N · m

Notice that the angle θ is the angle between the plane of the loop and the magnetic force exerted on each side of the loop. Equivalently, θ is the angle between the normal to the loop and the magnetic field.

In the case shown in Figure 22-24 (a), the angle θ is 90°, and the normal to the plane of the loop is perpendicular to the magnetic field. In this orientation the torque attains its maximum value, $\tau = IBA$. When θ is zero, as in **FIGURE 22-24 (c)**, the torque vanishes because the moment arm of the magnetic forces is zero. Thus, there is no torque when the magnetic field is parallel to the normal of a loop.

General Loops

We have shown that the torque exerted on a rectangular loop of area A is $\tau = IBA \sin\theta$. A more detailed derivation shows that the same relationship applies to *any* planar loop—that is, the torque is proportional to the area of the loop, no matter what its shape. For example, a circular loop of radius r and area πr^2 experiences a torque given by the expression $\tau = IB\pi r^2 \sin\theta$.

In many applications it's desirable to produce as large a torque as possible. A simple way to increase the torque is to wrap a long wire around a loop N times, creating a coil of N "turns." Each of the N turns produces the same torque as a single loop; hence, the total torque is increased by a factor of N. In general, then, the torque produced by a loop with N turns is:

Torque Exerted on a General Loop of Area A and N Turns

$$\tau = NIAB \sin\theta \qquad\qquad\qquad 22\text{-}6$$

SI unit: N · m

Big Idea 4 Loops of current in a magnetic field experience zero net force. However, a current loop does experience a torque in a magnetic field, which is the basis for many practical applications, including electric motors.

Notice that the torque depends on a number of factors in the system. First, it depends on the strength of the magnetic field, B, and on its orientation, θ, with respect to the normal of the loop. In addition, the torque depends on the current in the loop, I, the area of the loop, A, and the number of turns in the loop, N. The product of these "loop factors," NIA, is referred to as the **magnetic moment** of the loop. The magnetic moment, which has units of A · m², is proportional to the amount of torque a given loop can exert.

EXAMPLE 22-11   TORQUE ON A COIL

A rectangular coil with 225 turns is 5.0 cm high and 4.0 cm wide. When the coil is placed in a magnetic field of 0.25 T, its maximum torque is 0.36 N · m. What is the current in the coil?

CONTINUED

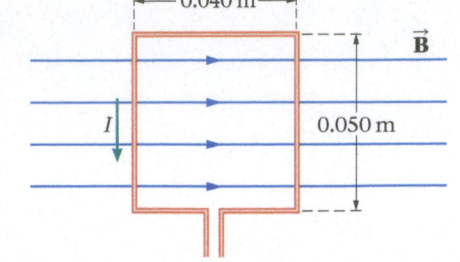

PICTURE THE PROBLEM

The coil, along with its dimensions, is shown in the sketch. The sketch also reflects the fact that the maximum torque is produced when the magnetic field is in the plane of the coil—that is, when the angle θ between the magnetic field $\vec{B}$ and the normal to the coil is 90°.

REASONING AND STRATEGY

The torque is given by the expression $\tau = NIAB \sin \theta$. Clearly, the maximum torque occurs when $\sin \theta = 1$; that is, $\tau_{max} = NIAB$. Solving this relationship for I yields the current in the coil.

Known   Number of turns, $N = 225$; area of coil, $A = (0.050 \text{ m})(0.040 \text{ m})$; magnetic field, $B = 0.25$ T; maximum torque, $\tau = 0.36 \text{ N} \cdot \text{m}$.

Unknown   Current in coil, $I = ?$

SOLUTION

1. Write an expression for the maximum torque:

$$\tau_{max} = NIAB$$

2. Solve for the current, I:

$$I = \frac{\tau_{max}}{NAB}$$

3. Substitute numerical values:

$$I = \frac{0.36 \text{ N} \cdot \text{m}}{(225)(0.050 \text{ m})(0.040 \text{ m})(0.25 \text{ T})} = 3.2 \text{ A}$$

INSIGHT

This calculation gives the magnitude of the current, but not its direction. The direction of the current will determine whether the torque on the coil is clockwise or counterclockwise.

PRACTICE PROBLEM

If the shape of this coil were changed to circular, keeping a constant perimeter, would the maximum torque increase, decrease, or stay the same?   [**Answer:** In general, a circle has the greatest area for a given perimeter. Therefore, the maximum torque would increase if the coil were made circular, because its area would be larger.]

Some related homework problems: Problem 38, Problem 39

Applications of Torque

RWP The torque exerted by a magnetic field finds a number of useful applications. For example, if a needle is attached to a coil, as in **FIGURE 22-25**, it can be used as part of a meter. When the coil is connected to an electric circuit, it experiences a torque, and a corresponding deflection of the needle, that is proportional to the current. The result is that the current in a circuit is indicated by the reading on the meter. A simple device of this type is referred to as a **galvanometer.**

Of even greater practical importance is the fact that magnetic torque can be used to power a motor. For example, an electric current passing through the coils of a motor causes a torque that rotates the axle of the motor. As the coils rotate, a device known as the commutator reverses the direction of the current as the orientation of the coil reverses, which ensures that the torque is always in the same direction and that the coils continue to turn in the same direction. Electric motors, which will be discussed in greater detail in Chapter 23, are used in everything from DVD players to electric razors to electric cars.

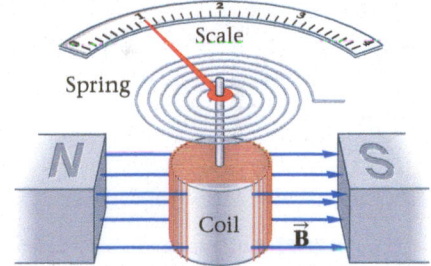

▲ **FIGURE 22-25** **The galvanometer** The basic elements of a galvanometer are a coil, a magnetic field, a spring, and a needle attached to the coil. As current passes through the coil, a torque acts on it, causing it to rotate. The spring ensures that the angle of rotation is proportional to the current in the coil.

Enhance Your Understanding   (Answers given at the end of the chapter)

5. Two current-carrying loops are identical, except for the direction of the magnetic field in their vicinity. The magnetic field at loop 1 is parallel to the plane of the loop, and the magnetic field at loop 2 is perpendicular to the plane of the loop. Is the magnetic torque exerted on loop 1 greater than, less than, or equal to the magnetic torque exerted on loop 2? Explain.

Section Review

- The torque exerted on a current-carrying coil is $\tau = NIAB \sin \theta$. In this expression, N is the number of turns in the coil, I is the current, A is the area of the coil, B is the strength of the magnetic field, and θ is the angle between the normal to the coil and the magnetic field.

22-6 Electric Currents, Magnetic Fields, and Ampère's Law

Electricity and magnetism were once thought to be completely unrelated phenomena. The first connection between them was discovered accidentally by Hans Christian Oersted in 1820. In fact, Oersted was giving a public lecture on various aspects of science when, at one point, he closed a switch and allowed a current to flow through a wire. What he noticed was that a nearby compass needle deflected from its usual orientation when the switch was closed—Oersted had just discovered that electric currents create magnetic fields. In this section we focus on the connection between electric currents and magnetic fields. In so doing, our attention shifts from the effects of magnetic fields—the subject of the previous sections—to their production.

▶ **FIGURE 22-26 The magnetic field of a current-carrying wire (a)** An electric current flowing through a wire produces a magnetic field. In the case of a long, straight wire, the field circulates around the wire. **(b)** Compass needles point along the circumference of a circle centered on the wire.

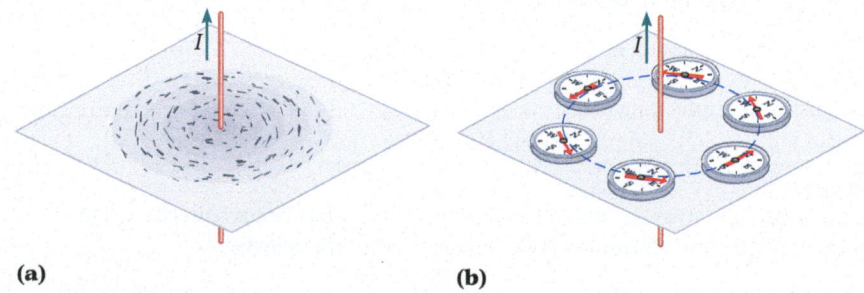

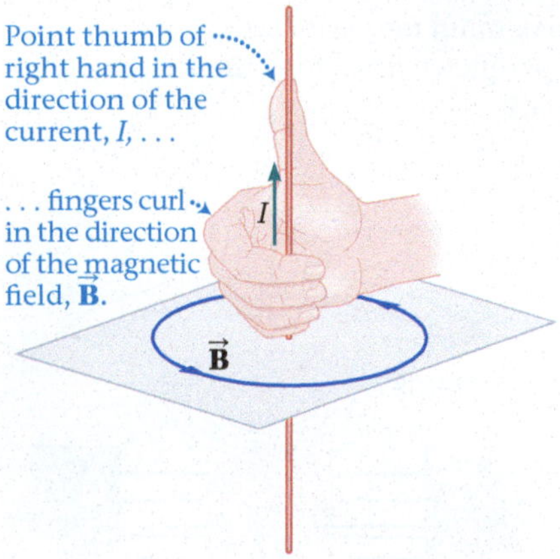

(a)   **(b)**

A Long, Straight Wire

We start with the simplest possible case—a straight, infinitely long wire that carries a current I. To visualize the magnetic field such a wire produces, we shake iron filings onto a sheet of paper that is pierced by the wire, as indicated in **FIGURE 22-26 (a)**. The result is that the filings form into circular patterns centered on the wire—evidently, the magnetic field "circulates" around the wire.

We can gain additional information about the field by placing a number of small compasses around the wire, as in **FIGURE 22-26 (b)**. In addition to confirming the circular shape of the field lines, the compass needles show the field's direction. To understand this direction, we must again utilize a right-hand rule—this time, we refer to the rule as the *magnetic field* right-hand rule:

Point thumb of right hand in the direction of the current, I, . . .

. . . fingers curl in the direction of the magnetic field, **B**.

▲ **FIGURE 22-27 The magnetic field right-hand rule** The magnetic field right-hand rule determines the direction of the magnetic field produced by a current-carrying wire. With the thumb of the right hand pointing in the direction of the current, the fingers curl in the direction of the field.

Magnetic Field Right-Hand Rule

To find the direction of the magnetic field due to a current-carrying wire, point the thumb of your right hand along the wire in the direction of the current I. Your fingers are now curling around the wire in the direction of the magnetic field.

This rule is illustrated in **FIGURE 22-27**, where we see that it predicts the same direction as that indicated by the compass needles in Figure 22-26 (b).

CONCEPTUAL EXAMPLE 22-12   DIRECTION OF THE CURRENT

The magnetic field shown in the sketch is due to the horizontal, current-carrying wire. Does the current in the wire flow to the left or to the right?

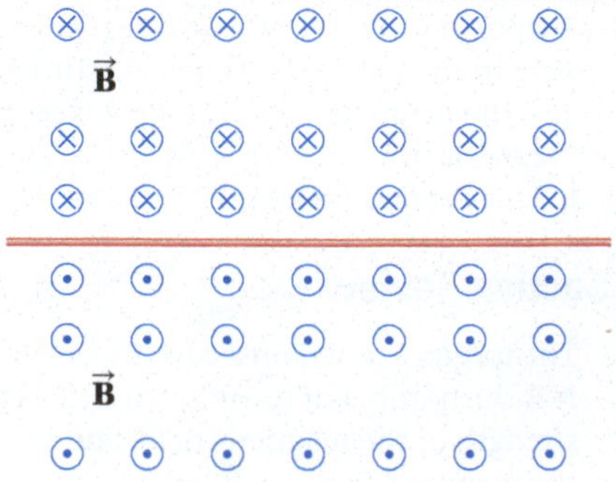

REASONING AND DISCUSSION
If you point the thumb of your right hand along the wire to the left, your fingers curl into the page above the wire and out of the page below the wire, as shown in the figure. Thus, the current flows to the left.

ANSWER
The current in the wire flows to the left.

Experiment shows that the field produced by a current-carrying wire doubles if the current I is doubled. In addition, the field decreases by a factor of two if the distance from the wire, r, is doubled. Hence, we conclude that the magnetic field B must be proportional to I/r; that is,

$$B = (\text{constant})\frac{I}{r}$$

The precise expression for B will now be derived using a law of nature known as Ampère's law.

Ampère's Law

Ampère's law relates the magnetic field along a closed path to the electric current enclosed by the path. Specifically, consider the current-carrying wires shown in **FIGURE 22-28**. These wires are enclosed by the closed path P, which can be divided into many small, straight-line segments of length ΔL. On each of these segments, the magnetic field $\vec{B}$ can be resolved into a component parallel to the segment, $B_\parallel$, and a component perpendicular to the segment, $B_\perp$. Of particular interest is the product $B_\parallel \Delta L$. According to Ampère's law, the sum of $B_\parallel \Delta L$ over all segments of a closed path is equal to a constant times the current enclosed by the path. This relationship can be written symbolically as follows:

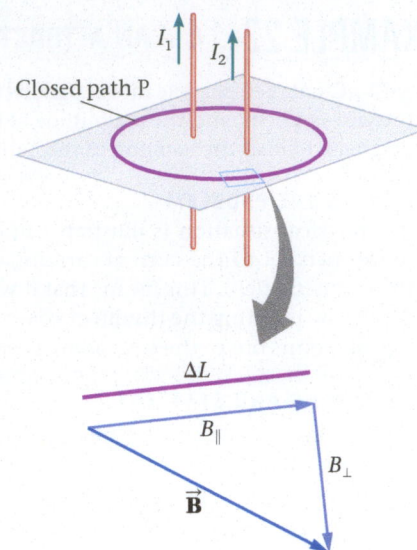

▲ **FIGURE 22-28 Illustrating Ampère's law** A closed path P encloses the currents I_1 and I_2. According to Ampère's law, the sum of $B_\parallel \Delta L$ around the path P is equal to $\mu_0 I_{\text{enclosed}}$. In this case, $I_{\text{enclosed}} = I_1 + I_2$.

Ampère's Law

$$\sum B_\parallel \Delta L = \mu_0 I_{\text{enclosed}} \qquad\qquad \text{22-7}$$

In this expression, μ_0 is a constant called the **permeability of free space.** Its value is

$$\mu_0 = 4\pi \times 10^{-7}\,\text{T}\cdot\text{m/A} \qquad\qquad \text{22-8}$$

We should emphasize that Ampère's law is a law of nature—it is valid for all magnetic fields and currents that are constant in time.

Let's apply Ampère's law to the case of a long, straight wire carrying a current I. We already know that the field circulates around the wire, as illustrated in **FIGURE 22-29**. It is reasonable, then, to choose a circular path of radius r (and circumference $2\pi r$) to enclose the wire. Because the magnetic field is parallel to the circular path at every point, and all points on the path are the same distance from the wire, it follows that $B_\parallel = B = \text{constant}$. Therefore, the sum of $B_\parallel \Delta L$ around the closed path gives

$$\sum B_\parallel \Delta L = B \sum \Delta L = B(2\pi r)$$

According to Ampère's law, this sum must equal $\mu_0 I_{\text{enclosed}} = \mu_0 I$. Therefore, we have

$$B(2\pi r) = \mu_0 I$$

Solving for B, we obtain:

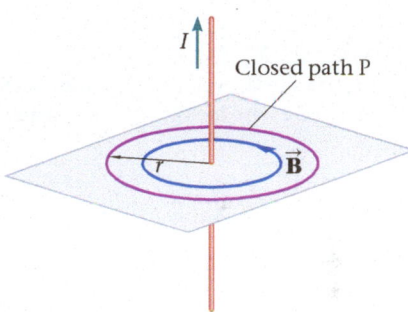

▲ **FIGURE 22-29 Applying Ampère's law** To apply Ampère's law to a long, straight wire, we consider a circular path centered on the wire. Since the magnetic field is everywhere parallel to this path, and has constant magnitude B at all points on it, the sum of $B_\parallel \Delta L$ over the path is $B(2\pi r)$. Setting this equal to $\mu_0 I$ yields the magnetic field of the wire: $B = \mu_0 I/2\pi r$.

Magnetic Field for a Long, Straight Wire

$$B = \frac{\mu_0 I}{2\pi r} \qquad\qquad \text{22-9}$$

SI unit: tesla, T

As expected, the field is equal to a constant $(\mu_0/2\pi)$ times I/r.

Big Idea **5** Current-carrying wires produce magnetic fields that circulate around the wire.

EXERCISE 22-13 MAGNETIC FIELD

Find the magnitude of the magnetic field 2 m from a long, straight wire carrying a current of 10 A.

REASONING AND SOLUTION
Straightforward substitution in the expression for the magnetic field produced by a wire, $B = \mu_0 I/2\pi r$, yields

$$B = \frac{\mu_0 I}{2\pi r} = \frac{(4\pi \times 10^{-7}\,\text{T}\cdot\text{m/A})(10\,\text{A})}{2\pi(2\,\text{m})} = 1 \times 10^{-6}\,\text{T}$$

This is a weak field, only about 2% of the strength of the Earth's magnetic field.

EXAMPLE 22-14 | AN ATTRACTIVE WIRE

A 52-μC charged particle moves parallel to a long wire with a speed of 720 m/s. The separation between the particle and the wire is 13 cm, and the magnitude of the force exerted on the particle is 1.4×10^{-7} N. Find **(a)** the magnitude of the magnetic field at the location of the particle and **(b)** the current in the wire.

PICTURE THE PROBLEM

The physical situation is illustrated in our sketch. Notice that the charged particle moves parallel to the current-carrying wire; hence, its velocity is at right angles to the magnetic field. This means that it will experience the maximum magnetic force. Finally, by pointing the thumb of your right hand in the direction of I, you can verify that $\vec{\mathbf{B}}$ points out of the page above the wire and into the page below it.

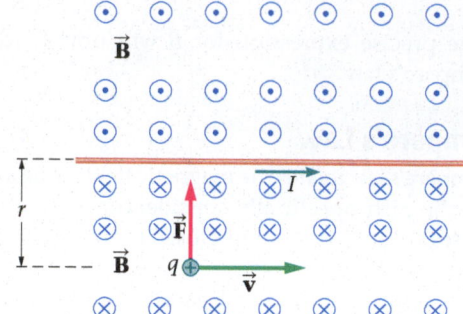

REASONING AND STRATEGY

a. The maximum magnetic force on the charged particle is $F = qvB$. This relationship can be solved for B.

b. The magnetic field is produced by the current in the wire. Therefore, $B = \mu_0 I / 2\pi r$. Using B from part (a), we can solve for the current I.

Known Charge of particle, $q = 52\ \mu$C; speed of particle, $v = 720$ m/s; distance to wire, $r = 0.13$ m; magnitude of force, $F = 1.4 \times 10^{-7}$ N.

Unknown **(a)** Magnetic field, $B = ?$ **(b)** Current in wire, $I = ?$

SOLUTION

Part (a)

1. Use $F = qvB$ to solve for the magnetic field:

$$F = qvB$$
$$B = \frac{F}{qv} = \frac{1.4 \times 10^{-7}\,\text{N}}{(5.2 \times 10^{-5}\,\text{C})(720\,\text{m/s})} = 3.7 \times 10^{-6}\,\text{T}$$

Part (b)

2. Rearrange the relationship for the magnetic field of a wire to solve for the current:

$$B = \frac{\mu_0 I}{2\pi r}$$
$$I = \frac{2\pi r B}{\mu_0}$$

3. Substitute the value for B found in part (a) and evaluate I:

$$I = \frac{2\pi r B}{\mu_0} = \frac{2\pi(0.13\,\text{m})(3.7 \times 10^{-6}\,\text{T})}{4\pi \times 10^{-7}\,\text{T} \cdot \text{m/A}} = 2.4\,\text{A}$$

INSIGHT

Using the magnetic force RHR, we can see that the direction of the force exerted on the charged particle is toward the wire, as illustrated in our sketch. If the particle had been negatively charged, the force would have been away from the wire.

PRACTICE PROBLEM

Suppose a particle with a charge of 52 μC is 13 cm above the wire and moving with a speed of 720 m/s to the left. Find the magnitude and direction of the force acting on this particle.   [**Answer:** $F = 1.4 \times 10^{-7}$ N, away from wire]

Some related homework problems: Problem 46, Problem 49

QUICK EXAMPLE 22-15 | FIND THE MAGNETIC FIELD

Two wires separated by a distance of 22 cm carry currents in the same direction. The current in wire 1 is $I_1 = 1.5$ A, and the current in wire 2 is $I_2 = 4.5$ A. Find the magnitude of the magnetic field at the point P, halfway between the wires.

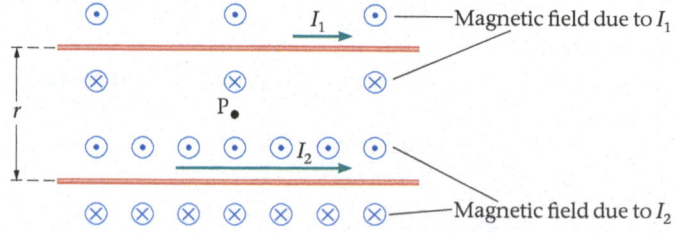

CONTINUED

REASONING AND SOLUTION

Each of the wires carries a current in the same direction. As a result, they produce fields in opposite directions between the wires, as indicated in the sketch. The magnitude of each field is given by $B = \mu_0 I/(2\pi r)$, with $r = 0.11$ m for each wire, and $I_1 = 1.5$ A for the top wire and $I_2 = 4.5$ A for the bottom wire. We will choose *out of the page* to be the *positive* direction (though the opposite choice would work just as well).

1. Find the magnitude and direction of the magnetic field produced by wire 1:

$$B_1 = \frac{\mu_0(1.5\text{ A})}{2\pi(0.11\text{ m})} = 2.7 \times 10^{-6}\text{ T, into page}$$

2. Find the magnitude and direction of the magnetic field produced by wire 2:

$$B_2 = \frac{\mu_0(4.5\text{ A})}{2\pi(0.11\text{ m})} = 8.2 \times 10^{-6}\text{ T, out of page}$$

3. Calculate the magnitude of the net field:

$$B = B_2 - B_1 = 5.5 \times 10^{-6}\text{ T, out of page}$$

Because the field produced by I_2 has the greater magnitude, the net field is out of the page. If the currents were equal, the net field midway between the wires would be zero.

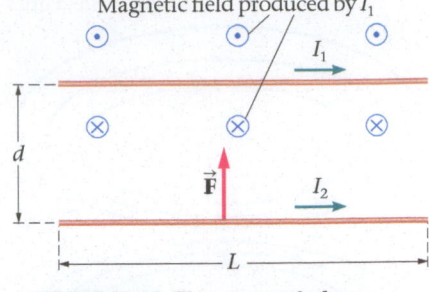

▲ **FIGURE 22-30 The magnetic force between current-carrying wires** A current in wire 1 produces a magnetic field, $B_1 = \mu_0 I_1/2\pi d$, at the location of wire 2. The result is a force exerted on a length L of wire 2 of magnitude $F = \mu_0 I_1 I_2 L/2\pi d$.

Forces Between Current-Carrying Wires

We know that a current-carrying wire in a magnetic field experiences a force. We also know that a current-carrying wire produces a magnetic field. It follows, then, that one current-carrying wire will exert a force on another.

To work out this relationship in detail, consider the two wires with parallel currents and separation d shown in **FIGURE 22-30**. The magnetic field produced by wire 1 circulates around it, coming out of the page above the wire and entering the page below the wire. Thus, wire 2 experiences a magnetic field pointing into the page with a magnitude given by Equation 22-9: $B = \mu_0 I_1/2\pi d$. The force experienced by wire 2, therefore, has a magnitude given by $F = ILB \sin\theta$ with $\theta = 90°$:

$$F = I_2 LB = I_2 L\left(\frac{\mu_0 I_1}{2\pi d}\right) = \frac{\mu_0 I_1 I_2}{2\pi d} L \qquad 22\text{-}10$$

The direction of the force acting on wire 2, as given by the magnetic force RHR, is upward—that is, toward wire 1. A similar calculation starting with the field produced by wire 2 gives a force of the same magnitude acting downward on wire 1. This is to be expected, because the forces acting on wires 1 and 2 form an action-reaction pair, as indicated in **FIGURE 22-31 (a)**. Hence, wires with parallel currents attract one another.

If the currents in wires 1 and 2 are in opposite directions, as in **FIGURE 22-31 (b)**, the situation is similar to that discussed in the preceding paragraph, except that the direction of the forces is reversed. Thus, wires with opposite currents repel one another.

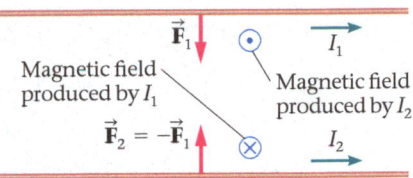

(a) Currents in same direction

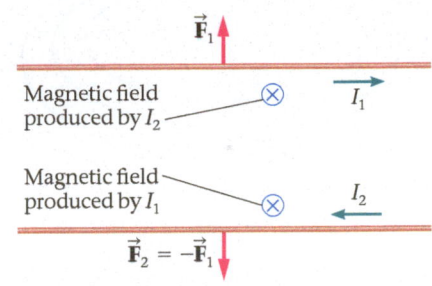

(b) Currents in opposite directions

▲ **FIGURE 22-31 The direction of the magnetic force between current-carrying wires** The forces between current-carrying wires depend on the relative direction of their currents. **(a)** If the currents are in the same direction, the force is attractive. **(b)** Wires with oppositely directed currents experience repulsive forces.

Enhance Your Understanding   (Answers given at the end of the chapter)

6. Two current-carrying wires cross at right angles, as shown in **FIGURE 22-32**. The wires have currents of equal magnitude I in the directions indicated. For each of the four points, 1, 2, 3, and 4, state whether the total magnetic field there is strong (large magnitude) or weak (small magnitude).

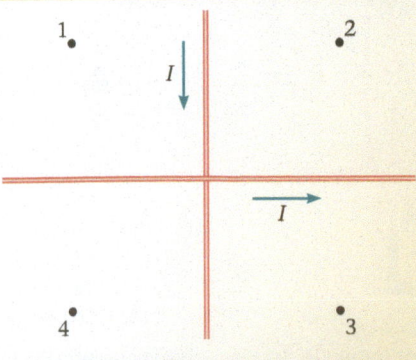

▲ **FIGURE 22-32** Enhance Your Understanding 6.

CONTINUED

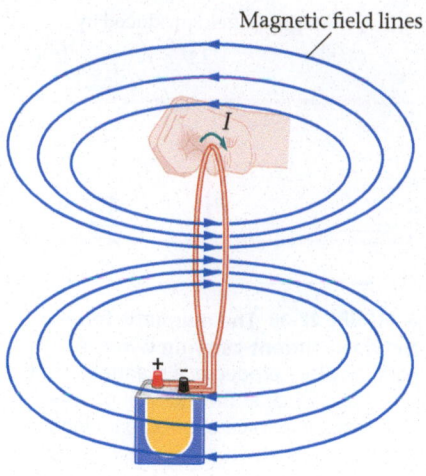

Magnetic field lines

I

(a) Magnetic field of a current loop

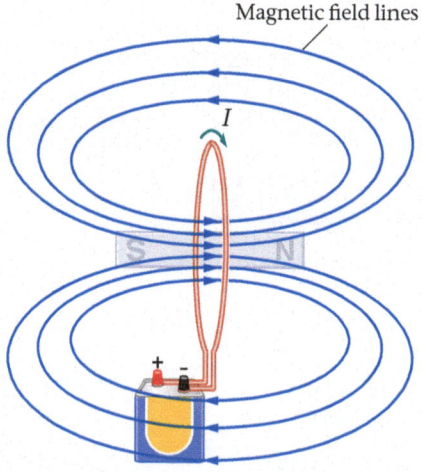

Magnetic field lines

I

S   N

(b) Magnetic field of bar magnet is similar

▲ **FIGURE 22-33 The magnetic field of a current loop** **(a)** The magnetic field produced by a current loop is relatively intense within the loop and falls off rapidly outside the loop. **(b)** A permanent magnet produces a field that is very similar to the field of a current loop.

22-7 Current Loops and Solenoids

We now consider the magnetic fields produced when a current-carrying wire has a circular or a helical geometry—as opposed to the straight wires considered in the previous section. As we shall see, there are many practical applications for such geometries.

Circular Current Loop

We begin by considering the magnetic field produced by a current-carrying wire that is formed into the shape of a circular loop. In **FIGURE 22-33 (a)** we show a wire loop connected to a battery producing a current in the direction indicated. Using the magnetic field RHR, as shown in the figure, we see that $\vec{\mathbf{B}}$ points from left to right as it passes through the loop. Notice also that the field lines are bunched together within the loop, indicating an intense field there, but are more widely spaced outside the loop.

The most interesting aspect of the field produced by the loop is its close resemblance to the field of a bar magnet. This similarity is illustrated in **FIGURE 22-33 (b)**, where we see that one side of the loop behaves like a north magnetic pole and the other side like a south magnetic pole. Thus, if two loops with identical currents are placed near each other, as in **FIGURE 22-34 (a)**, the force between them will be similar to the force between two bar magnets pointing in the same direction—that is, they will attract each other because opposite poles are brought close together. If the loops have oppositely directed currents, as in **FIGURE 22-34 (b)**, the force between them is repulsive because like poles are brought together. Notice the similarity of the results for straight wires, Figure 22-31, and for loops, Figure 22-34.

The magnitude of the magnetic field produced by a circular loop of N turns, radius R, and current I varies from point to point; however, it can be shown that at the center of the loop, the field is given by the following simple expression:

$$B = \frac{N\mu_0 I}{2R} \quad \text{(center of circular loop of radius } R\text{)} \qquad 22\text{-}11$$

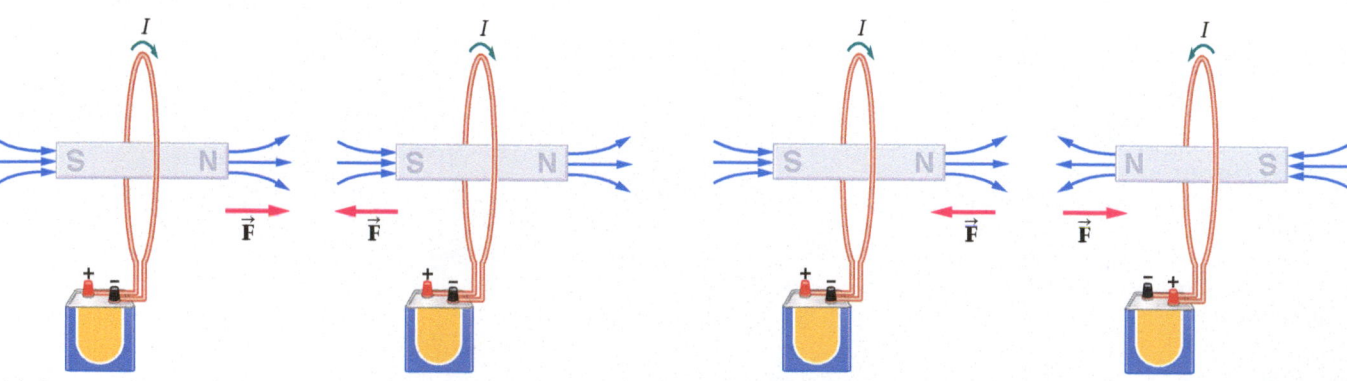

I I I I

S   N   S   N   S   N   N   S

$\vec{\mathbf{F}}$ $\vec{\mathbf{F}}$ $\vec{\mathbf{F}}$ $\vec{\mathbf{F}}$

(a) Currents in same direction   **(b)** Currents in opposite directions

▲ **FIGURE 22-34 Magnetic forces between current loops** To decide whether current loops will experience an attractive or repulsive force, it is useful to think in terms of the corresponding permanent magnets. **(a)** Current loops with currents in the same direction are like two bar magnets lined up in the same direction; they attract each other. **(b)** Current loops with opposite currents act like bar magnets with opposite orientations; they repel each other.

▶ **FIGURE 22-35 The solenoid** A solenoid is formed from a long wire that is wound into a succession of current loops. The magnetic field inside the solenoid is relatively strong and uniform. Outside the solenoid the field is weak. In the ideal case, we consider the field outside the solenoid to be zero and the field inside to be uniform and parallel to the solenoid's axis.

Notice that the field is proportional to the current in the loop and inversely proportional to its radius.

Helical Solenoid

A **solenoid** is an electrical device in which a long wire has been wound into a succession of closely spaced loops with the geometry of a helix. Also referred to as an **electromagnet,** a solenoid carrying a current produces an intense, nearly uniform, magnetic field inside the loops, as indicated in **FIGURE 22-35**. Notice that each loop of the solenoid carries a current in the same direction; therefore, the magnetic force between loops is attractive and serves to hold the loops tightly together.

The magnetic field lines in Figure 22-35 are tightly packed inside the solenoid but are widely spaced outside. In the ideal case of a very long, tightly packed solenoid, the magnetic field outside is practically zero—especially when compared with the intense field inside the solenoid. We can use this idealization, in combination with Ampère's law, to calculate the magnitude of the field inside the solenoid.

To do so, consider the rectangular path of width L and height h shown in **FIGURE 22-36**. Notice that the parallel component of the field on side 1 is simply $\vec{B}$. On sides 2 and 4 the parallel component is zero, because $\vec{B}$ is perpendicular to those sides. Finally, on side 3 (which is outside the solenoid) the magnetic field is essentially zero. Using these results, we obtain the sum of $B_\parallel \Delta L$ over the rectangular loops:

$$\sum B_\parallel \Delta L = \sum_{\text{side 1}} B_\parallel \Delta L + \sum_{\text{side 2}} B_\parallel \Delta L + \sum_{\text{side 3}} B_\parallel \Delta L + \sum_{\text{side 4}} B_\parallel \Delta L$$

$$= BL + 0 + 0 + 0 = BL$$

Next, the current enclosed by the rectangular circuit is NI, where N is the number of loops in the length L. Therefore, Ampère's law gives

$$BL = \mu_0 NI$$

Solving for B and letting the number of loops per length be $n = N/L$, we find

Magnetic Field of a Solenoid

$$B = \mu_0 \left(\frac{N}{L}\right)I = \mu_0 nI \qquad \text{22-12}$$

SI unit: tesla, T

This result is independent of the cross-sectional area of the solenoid.

When used as an electromagnet, a solenoid has many useful properties. First and foremost, it produces a strong magnetic field that can be turned on or off at the flip of a switch—unlike the field of a permanent magnet. In addition, the magnetic field can be further intensified by filling the core of the solenoid with an iron bar. In such a case, the magnetic field of the solenoid magnetizes the iron bar, which then adds its field to the overall field of the system. These properties and others make solenoids useful devices in a variety of electric circuits. Further examples are given in the next chapter.

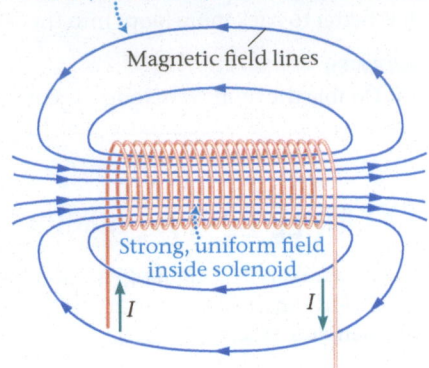

Weak (almost zero) field outside solenoid

Magnetic field lines

Strong, uniform field inside solenoid

I I

PROBLEM-SOLVING NOTE

The Magnetic Field Inside a Solenoid

In an ideal solenoid, the magnetic field inside the solenoid points along the axis and is uniform. Outside the solenoid the field is zero. Therefore, when calculating the field produced by a solenoid, it is not necessary to specify a particular point inside the solenoid; all inside points have the same field.

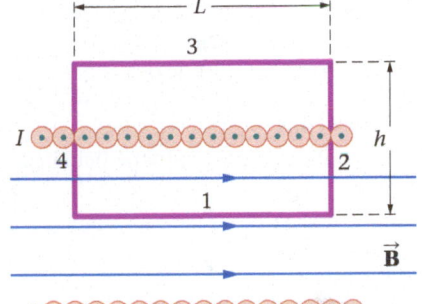

▲ **FIGURE 22-36 Ampère's law and the magnetic field in a solenoid** To calculate the magnetic field inside a solenoid, we apply Ampère's law to the rectangular path shown here. The only side of the rectangle that has a nonzero, parallel component of $\vec{B}$ is side 1.

PHYSICS IN CONTEXT
Looking Ahead

In Chapter 23 we introduce the idea of an inductor, which is basically a solenoid. In particular, we will show that an inductor can store energy in the form of a magnetic field, just as a capacitor (Chapter 20) can store energy in an electric field.

CONCEPTUAL EXAMPLE 22-16   MAGNETIC FIELD IN A SOLENOID

If you want to increase the strength of the magnetic field inside a solenoid, is it better to **(a)** double the number of loops, keeping the length the same, or **(b)** double the length, keeping the number of loops the same?

REASONING AND DISCUSSION
Referring to the expression $B = \mu_0(N/L)I$, we see that doubling the number of loops ($N \rightarrow 2N$) while keeping the length the same ($L \rightarrow L$) results in a doubled magnetic field ($B \rightarrow 2B$). On the other hand, doubling the length ($L \rightarrow 2L$) while keeping

CONTINUED

the number of loops the same ($N \rightarrow N$) reduces the magnetic field by a factor of two ($B \rightarrow B/2$). Hence, to increase the field, it is better to pack more loops into the same length.

ANSWER
(a) Double the number of loops, keeping the same length.

EXAMPLE 22-17   THROUGH THE CORE OF A SOLENOID

A solenoid is 21.5 cm long, has 275 loops, and carries a current of 2.95 A. Find the magnitude of the force exerted on a 12.0-μC charged particle moving at 1050 m/s through the interior of the solenoid, at an angle of 11.5° relative to the solenoid's axis.

PICTURE THE PROBLEM
Our sketch shows the solenoid, along with the uniform magnetic field it produces parallel to its axis. Inside the solenoid, a positively charged particle moves at an angle of $\theta = 11.5°$ relative to the magnetic field.

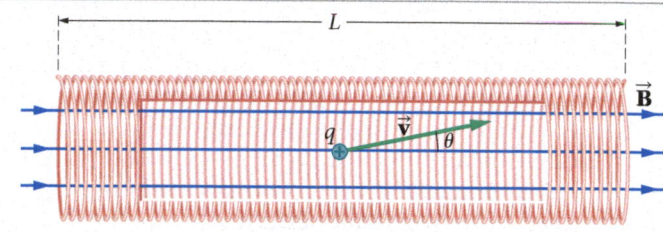

REASONING AND STRATEGY
The force exerted on the charged particle is magnetic; hence, we start by calculating the magnetic field produced by the solenoid: $B = \mu_0(N/L)I$. Next, we note that the magnetic field in a solenoid is parallel to its axis. It follows that the magnitude of the force exerted on the charge is given by $F = |q|vB \sin \theta$, with $\theta = 11.5°$.

Known Length of solenoid, $L = 0.215$ m; number of loops, $N = 275$; current, $I = 2.95$ A; charge of particle, $q = 12.0\,\mu$C; speed of particle, $v = 1050$ m/s; angle of motion, $\theta = 11.5°$.
Unknown Magnitude of force, $F = ?$

SOLUTION

1. Calculate the magnetic field inside the solenoid:

$$B = \mu_0\left(\frac{N}{L}\right)I$$
$$= (4\pi \times 10^{-7}\,\text{T}\cdot\text{m/A})\left(\frac{275}{0.215\,\text{m}}\right)(2.95\,\text{A}) = 4.74 \times 10^{-3}\,\text{T}$$

2. Use B to find the force exerted on the charged particle:

$$F = |q|vB \sin \theta$$
$$= (12.0 \times 10^{-6}\,\text{C})(1050\,\text{m/s})(4.74 \times 10^{-3}\,\text{T}) \sin 11.5°$$
$$= 1.19 \times 10^{-5}\,\text{N}$$

INSIGHT
The magnetic field strength inside this modest solenoid is approximately 100 times greater than the magnetic field at the surface of the Earth.

PRACTICE PROBLEM
What current would be required to make the force acting on the particle equal to 1.45×10^{-5} N?   [**Answer:** $I = 3.59$ A]

Some related homework problems: Problem 56, Problem 57

Big Idea **6** Current loops produce magnetic fields that are similar to the fields produced by bar magnets. Solenoids produce magnetic fields that are intense and uniform inside the solenoid, but are practically zero outside of it.

RWP Magnetic resonance imaging (MRI) instruments utilize solenoids large enough to accommodate a person within their coils. Not only are these solenoids large, they are also capable of producing extremely powerful magnetic fields on the order of 1 or 2 tesla (see Table 22-1). So powerful are these fields, in fact, that metallic objects such as mop buckets and stretchers have been known to be pulled from across the room into the bore of the magnet. A metal oxygen bottle in the same room can be turned into a dangerous, high-speed projectile.

Enhance Your Understanding   (Answers given at the end of the chapter)

7. Rank the following solenoids in order of increasing magnetic field within the core of the solenoid. Indicate ties where appropriate.

CONTINUED

Solenoid	A	B	C	D
Number of loops, N	100	200	500	300
Length, L (m)	0.5	0.5	1.5	1.5
Current, I (A)	0.1	0.2	0.3	0.5

Section Review

- A solenoid produces a magnetic field that is parallel to its axis and has a magnitude given by $B = \mu_0 nI$. In this expression, n is the number of turns per length; that is, $n = N/L$. The magnetic field is practically zero outside the solenoid.

22-8 Magnetism in Matter

Some materials have strong magnetic fields; others do not. To understand these differences, we must consider the behavior of matter on the atomic level.

Microscopic Origin of Magnetic Fields   To begin, recall that circulating electric currents produce magnetic fields much like those produced by bar magnets. This is significant because atoms have electrons orbiting their nucleus, which means they have circulating electric currents. The fields produced by these orbiting electrons can be sizable—on the order of 10 T or so at the nucleus of the atom. In most atoms, though, the fields produced by the various individual electrons tend to cancel one another, resulting in a very small or zero net magnetic field.

Another type of circulating electric current present in atoms produces fields that occasionally do not cancel. These currents are associated with the *spin* that all electrons have. A simple model of the electron—which should *not* be taken literally—is a spinning sphere of charge. The circulating charge associated with the spinning motion gives rise to a magnetic field. Electrons tend to "pair up" in atoms in such a way that their spins are opposite to one another, again resulting in zero net field. However, in some atoms—such as iron, nickel, and cobalt—the net field due to the spinning electrons is nonzero, and strong magnetic effects can occur as the magnetic field of one atom tends to align with the magnetic field of another. A full understanding of these types of effects requires the methods of quantum mechanics, which will be discussed in Chapters 30 and 31.

Ferromagnetism

If the tendency of magnetic atoms to self-align is strong enough, as it is in materials such as iron and nickel, the result can be an intense magnetic field—as in a bar magnet. Materials with this type of behavior are called **ferromagnets.** Counteracting the tendency of atoms to align, however, is the disorder caused by increasing temperature. In fact, all ferromagnets lose their magnetic field if the temperature is high enough to cause their atoms to orient in random directions. For example, the magnetic field of a bar magnet made of iron vanishes if its temperature exceeds 770 °C.

This type of temperature behavior shows that the simple picture of a large bar magnet within the Earth cannot be correct. As we know, temperature increases with depth below the Earth's surface. In fact, at depths of about 15 miles the temperature is already above 770 °C, so the magnetism of an iron magnet would be lost due to thermal effects. Of course, at even greater depths an iron magnet would melt to form a liquid. In fact, it appears that the magnetic field of the Earth is caused by circulating currents of molten iron, nickel, and other metals. These circulating currents create a magnetic field in much the same way as the circulating current in a solenoid.

Temperature also plays a key role in the magnetization that is observed in rocks on the ocean floor. As molten rock is extruded from mid-ocean ridges, it has no net magnetization because of its high temperature. When the rock cools, however, it becomes magnetized in the direction of the Earth's magnetic field (**FIGURE 22-37**). In effect, the Earth's magnetic field becomes "frozen" in the solidified rock. As the seafloor spreads,

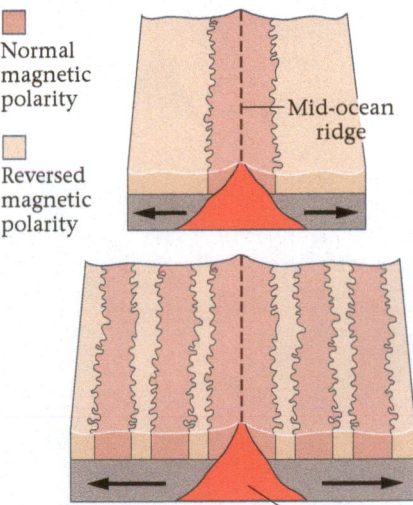

Normal
magnetic
polarity

Reversed
magnetic
polarity

Mid-ocean
ridge

Molten rock

▲ **FIGURE 22-37 Mid-Ocean Ridge** Molten rock extruded at a mid-ocean ridge magnetizes in the direction of the Earth's magnetic field as it cools. After cooling, the direction of the Earth's field remains "frozen" in the rocks. Newly extruded material near the ridge undergoes the same process. As a result, the magnetization of the rocks on either side of a mid-ocean ridge produces a geological record of the polarity of the Earth's magnetic field over time, as well as convincing confirmation of seafloor spreading.

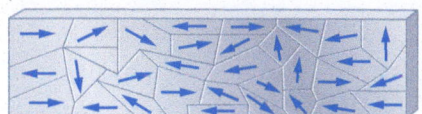

(a) Magnetic domains in zero external field

$\vec{B}$

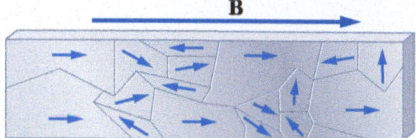

(b) Domains in direction of external magnetic field grow in size

▲ **FIGURE 22-38 Magnetic domains (a)** A ferromagnetic material tends to form into numerous domains with magnetization pointing in different directions. **(b)** When an external field is applied, the domains pointing in the direction of the field often grow in size at the expense of domains pointing in other directions.

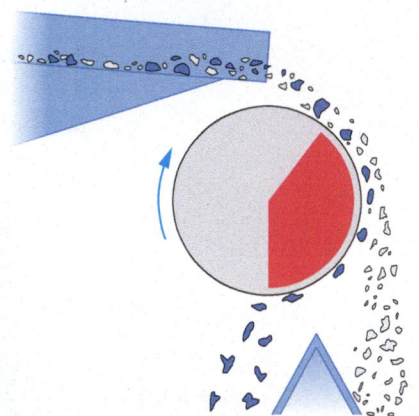

▲ **FIGURE 22-39** A drum made of nonmagnetic material rotates around a stationary core that is partially filled with a strong permanent magnet (red area in the diagram). Iron and other ferromagnetic materials (blue) stick to the drum until they rotate past the stationary magnet, at which point they separate from the drum and fall onto a different conveyer belt from the one that catches the nonmagnetic materials (white).

and more material is formed along the ridge, a continuous record of the Earth's magnetic field is formed. In particular, if the Earth's field reverses at some point in time, the field in the solidified rocks will record that fact.

Another important feature of ferromagnets is that their magnetism is characterized by magnetic **domains** within the material, as illustrated in **FIGURE 22-38 (a)**. Each domain has a strong magnetic field, but different domains are oriented differently, so that the net effect may be small. The typical size of these domains is on the order of 10^{-4} cm to 10^{-1} cm. When an external field is applied to such a material, the magnetic domains that are pointing in the direction of the applied field often grow in size at the expense of domains with different orientations, as indicated in **FIGURE 22-38 (b)**. The result is that the applied external field produces a net magnetization in the material.

RWP The fact that not all metals are ferromagnetic comes in handy at the processing facility for recyclable materials, where the materials are passed over a rotating drum surrounding a stationary magnet (see **FIGURE 22-39**). Ferromagnetic materials like iron and steel stick to the drum, but nonferromagnetic metals like aluminum and copper do not. Because these metals appear frequently in waste streams and must be recycled differently from each other, a rapid and efficient sorting mechanism is essential for reclaiming these valuable materials so that they do not languish in a landfill.

BIO Many living organisms are known to incorporate small ferromagnetic crystals, consisting of *magnetite*, in their bodies. For example, some species of bacteria use magnetite crystals to help orient themselves with respect to the Earth's magnetic field. Magnetite has also been found in the brains of bees and pigeons, where it is suspected to play a role in navigation. It is even found in human brains, though its possible function there is unclear.

Whatever the role of magnetite in people, observations show that a magnetic field can affect the way the human brain operates. In recent experiments involving people viewing optical illusions, it has been found that if the parietal lobe on one side of the brain is exposed to a highly focused magnetic field of about 1 T, a temporary interruption in much of the neural activity in that hemisphere results. Fortunately, fields of 1 T are generally not encountered in everyday life.

Paramagnetism and Diamagnetism

Not all magnetic materials are ferromagnetic, however. In some cases a ferromagnet has zero magnetic field simply because it is at too high a temperature. In other cases, the tendency for the self-alignment of individual atoms in a given material is too weak to produce a net magnetic field, even at low temperatures. In either case, a strong external magnetic field applied to the material can make the material magnetic by forcing the atoms into alignment. Magnetic effects of this type are referred to as **paramagnetism.**

Finally, all materials display a further magnetic effect referred to as **diamagnetism.** In the diamagnetic effect, an applied magnetic field on a material produces an oppositely directed field in response. The resulting repulsive force is usually too weak to be noticed, except in superconductors. The basic mechanism responsible for diamagnetism will be discussed in greater detail in the next chapter.

RWP If a magnetic field is strong enough, however, even the relatively weak repulsion of diamagnetism can lead to significant effects. For example, researchers at the Nijmegen High Field Magnet Laboratory have used a field of 16 T to levitate a strawberry, a cricket, and even a frog, as shown in **FIGURE 22-40 (a)**. The diamagnetic repulsion of the water in these organisms is great enough, in a field that strong, to counteract the gravitational force of the Earth. The researchers reported that the living frog showed no visible signs of discomfort during its levitation, and that it hopped away normally after the experiment.

A similar diamagnetic effect can be used to levitate a small magnet between a person's fingertips, as in **FIGURE 22-40 (b)**. In the case shown in the photograph, a powerful magnet about 8 ft above the person's hand applies a magnetic field strong enough to counteract the weight of the small magnet. By itself, this system of two magnets is not stable—if the small magnet is displaced upward slightly, the increased attractive force

▶ **FIGURE 22-40 (a)** The repulsive diamagnetic forces produced by water molecules in the body of a frog are strong enough to levitate it in an intense magnetic field of 16 T. (Animals levitated in this way appear to suffer no harm or discomfort.) **(b)** This small magnet is suspended in the strong magnetic field produced by a larger magnet above it (outside the photo). Ordinarily, such an arrangement would be highly unstable. The addition of small, repulsive diamagnetic forces due to a person's fingers, however, is sufficient to convert it into a stable equilibrium.

of the upper magnet raises it farther; if it is displaced downward, the force of gravity pulls it down even farther. With the diamagnetic effect of the fingers, however, the small magnet is stabilized. If it is displaced upward now, the repulsive diamagnetic effect of the finger pushes it back down; if it is displaced downward, the diamagnetic repulsion of the thumb pushes it back up. The result is a stable magnetic levitation—with no smoke or mirrors.

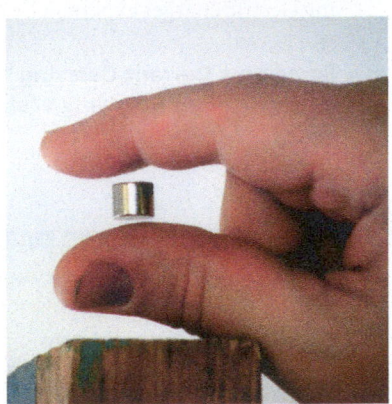

Section Review

- In ferromagnetism, a material develops a strong permanent magnetic field if the temperature is not too high.

- In paramagnetism, a material produces a magnetic field in response to an applied magnetic field.

- All materials show a small diamagnetic effect, in which the material produces a magnetic field that opposes an applied magnetic field.

CHAPTER 22 REVIEW

CHAPTER SUMMARY

22-1 THE MAGNETIC FIELD

The magnetic field gives an indication of the effect a magnet will have in a given region.

Magnetic Poles

A magnet is characterized by two poles, referred to as the north pole and the south pole.

Magnetic Field Lines

Magnetic field lines point away from north poles and toward south poles, and always form closed loops.

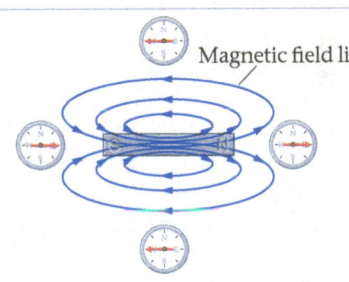

Magnetic field lines

22-2 THE MAGNETIC FORCE ON MOVING CHARGES

In order for a magnetic field to exert a force on a particle, the particle must have charge and must be moving.

Magnitude of the Magnetic Force

The magnitude of the magnetic force is

$$F = |q|vB \sin \theta \qquad \text{22-1}$$

In this expression, q is the charge of the particle, v is its speed, B is the magnitude of the magnetic field, and θ is the angle between the velocity vector $\vec{\mathbf{v}}$ and the magnetic field vector $\vec{\mathbf{B}}$.

Magnetic Force Right-Hand Rule (RHR)

For a positive charge, point the fingers of your right hand in the direction of $\vec{\mathbf{v}}$ and curl them toward the direction of $\vec{\mathbf{B}}$. Your thumb then points in the direction of the force $\vec{\mathbf{F}}$. The force on a negative charge is in the opposite direction to that on a positive charge.

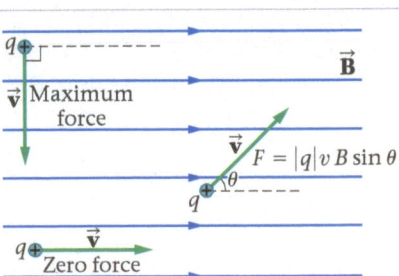

22-3   THE MOTION OF CHARGED PARTICLES IN A MAGNETIC FIELD

The motion of a charged particle in a magnetic field is quite different from that in an electric field.

Circular Motion

If a charged particle moves perpendicular to a magnetic field, it will orbit with constant speed in a circle of radius $r = mv/|q|B$.

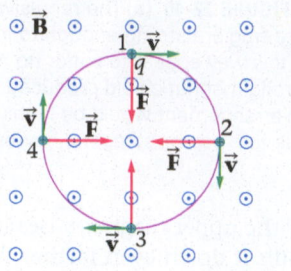

22-4   THE MAGNETIC FORCE EXERTED ON A CURRENT-CARRYING WIRE

An electric current in a wire is caused by the movement of electric charges. Since moving electric charges experience magnetic forces, it follows that a current-carrying wire will as well.

Force on a Current-Carrying Wire

A wire of length L carrying a current I at an angle θ to a magnetic field B experiences a force given by

$$F = ILB \sin \theta \qquad\qquad 22\text{-}4$$

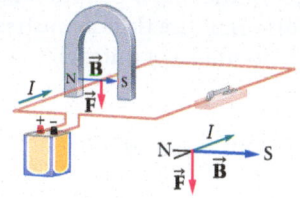

22-5   LOOPS OF CURRENT AND MAGNETIC TORQUE

A current loop placed in a magnetic field experiences a torque that depends on the relative orientation of the plane of the loop and the magnetic field.

Torque on a General Loop

The magnetic torque exerted on a current loop is given by

$$\tau = NIAB \sin \theta \qquad\qquad 22\text{-}6$$

In this expression, N is the number of turns around the loop, I is the current, A is the area of the loop, B is the strength of the magnetic field, and θ is the angle between the plane of the loop and the magnetic force.

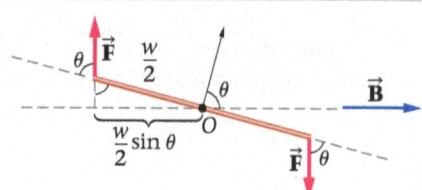

22-6   ELECTRIC CURRENTS, MAGNETIC FIELDS, AND AMPÈRE'S LAW

The key observation that serves to unify electricity and magnetism is that electric currents cause magnetic fields.

Magnetic Field Right-Hand Rule

The direction of the magnetic field produced by a current is found by pointing the thumb of the right hand in the direction of the current. Then the fingers of the right hand curl in the direction of the field.

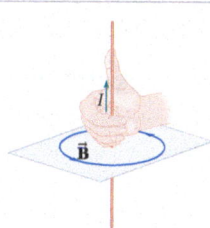

Ampère's Law

Ampère's law can be expressed as follows:

$$\sum B_{\parallel} \, \Delta L = \mu_0 I_{\text{enclosed}} \qquad\qquad 22\text{-}7$$

In this expression, $B_{\parallel}$ is the component of the magnetic field parallel to a segment of a closed path of length ΔL, I is the current enclosed by the path, and $\mu_0 = 4\pi \times 10^{-7} \, \text{T} \cdot \text{m/A}$ is a constant called the permeability of free space.

Magnetic Field of a Long, Straight Wire

A long, straight wire carrying a current I produces a magnetic field of magnitude B given by

$$B = \frac{\mu_0 I}{2\pi r} \qquad\qquad 22\text{-}9$$

In this expression, r is the radial distance from the wire.

Forces Between Current-Carrying Wires

Two wires, carrying the currents I_1 and I_2, exert forces on each other. If the wires are separated by a distance d, the force exerted on a length L is

$$F = \frac{\mu_0 I_1 I_2}{2\pi d} L \qquad\qquad 22\text{-}10$$

Wires that carry current in the same direction attract one another; wires with opposite-directed currents repel one another.

22-7 CURRENT LOOPS AND SOLENOIDS

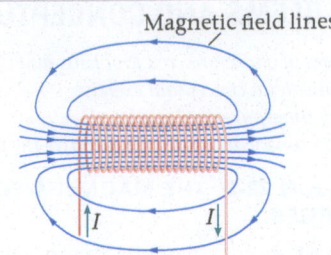

Magnetic field lines

Current Loop

The magnetic field at the center of a current loop of N turns, radius R, and current I is

$$B = \frac{N\mu_0 I}{2R} \qquad\qquad 22\text{-}11$$

Magnetic Field of a Solenoid

The magnetic field inside a solenoid is nearly uniform and aligned along the solenoid's axis. If the solenoid has N loops in a length L and carries a current I, its magnetic field is

$$B = \mu_0\left(\frac{N}{L}\right)I = \mu_0 n I \qquad\qquad 22\text{-}12$$

In the second expression, n is the number of loops per length; $n = N/L$. The magnetic field outside a solenoid is small, and in the ideal case can be considered to be zero.

22-8 MAGNETISM IN MATTER

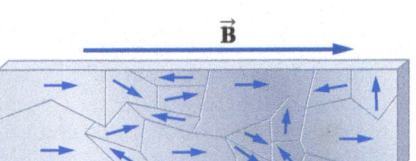

The ultimate origin of the magnetic fields we observe around us is circulating electric currents on the atomic level.

Paramagnetism

A paramagnetic material has no magnetic field unless an external magnetic field is applied to it.

Ferromagnetism

A ferromagnetic material produces a magnetic field even in the absence of an external magnetic field.

Diamagnetism

Diamagnetism is an induced magnetic field in the opposite direction to an applied external magnetic field.

ANSWERS TO ENHANCE YOUR UNDERSTANDING QUESTIONS

1. Pole 1 is a north (N) magnetic pole.

2. The magnetic force points into the page.

3. **(b)** 2 m.

4. A < D < C < B.

5. The magnetic torque is greater on loop 1, with the magnetic field parallel to the plane of the loop, as illustrated in Figure 22-24.

6. Point 1, weak field; point 2, strong field; point 3, weak field; point 4, strong field.

7. A < B < C = D.

CONCEPTUAL QUESTIONS

For instructor-assigned homework, go to www.masteringphysics.com.

(Answers to odd-numbered Conceptual Questions can be found in the back of the book.)

1. Two charged particles move at right angles to a magnetic field and deflect in opposite directions. Can one conclude that the particles have opposite charges?

2. An electron moves with constant velocity through a region of space that is free of magnetic fields. Can one conclude that the electric field is zero in this region? Explain.

3. An electron moves with constant velocity through a region of space that is free of electric fields. Can one conclude that the magnetic field is zero in this region? Explain.

4. Describe how the motion of a charged particle can be used to distinguish between an electric and a magnetic field.

5. Explain how a charged particle moving in a circle of small radius can take the same amount of time to complete an orbit as an identical particle orbiting in a circle of large radius.

6. A current-carrying wire is placed in a region with a uniform magnetic field. The wire experiences zero magnetic force. Explain.

PROBLEMS AND CONCEPTUAL EXERCISES

Answers to odd-numbered Problems and Conceptual Exercises can be found in the back of the book. **BIO** *identifies problems of biological or medical interest;* **CE** *indicates a conceptual exercise.* **Predict/Explain** *problems ask for two responses: (a) your prediction of a physical outcome, and (b) the best explanation among three provided; and* **Predict/Calculate** *problems ask for a prediction of a physical outcome, based on fundamental physics concepts, and follow that with a numerical calculation to verify the prediction. On all problems, bullets (•, ••, •••) indicate the level of difficulty.*

SECTION 22-2 THE MAGNETIC FORCE ON MOVING CHARGES

1. • **CE Predict/Explain** Proton 1 moves with a speed v from the east coast to the west coast in the continental United States; proton 2 moves with the same speed from the southern United States toward Canada. **(a)** Is the magnitude of the magnetic force experienced by proton 2 greater than, less than, or equal to the force experienced by proton 1? **(b)** Choose the *best explanation* from among the following:

    I. The protons experience the same force because the magnetic field is the same and their speeds are the same.

    II. Proton 1 experiences the greater force because it moves at right angles to the magnetic field.

    III. Proton 2 experiences the greater force because it moves in the same direction as the magnetic field.

2. • **CE** An electron moving in the positive x direction, at right angles to a magnetic field, experiences a magnetic force in the positive y direction. What is the direction of the magnetic field?

3. • **CE** Suppose particles A, B, and C in **FIGURE 22-41** have identical masses and charges of the same magnitude. Rank the particles in order of increasing speed. Indicate ties where appropriate.

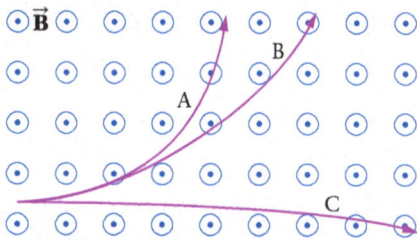

FIGURE 22-41 Problems 3 and 4

4. • **CE** Referring to Figure 22-41, what is the sign of the charge for each of the three particles? Explain.

5. • What is the acceleration of a proton moving with a speed of 7.5 m/s at right angles to a magnetic field of 1.4 T?

6. • An electron moves at right angles to a magnetic field of 0.23 T. What is its speed if the force exerted on it is 8.9×10^{-15} N?

7. • A negatively charged ion moves due north with a speed of 1.5×10^6 m/s at the Earth's equator. What is the magnetic force exerted on this ion?

8. •• A proton high above the equator approaches the Earth moving straight downward with a speed of 375 m/s. Find the acceleration of the proton, given that the magnetic field at its altitude is 4.05×10^{-5} T.

9. •• A 0.32-μC particle moves with a speed of 16 m/s through a region where the magnetic field has a strength of 0.95 T. At what angle to the field is the particle moving if the force exerted on it is **(a)** 4.8×10^{-6} N, **(b)** 3.0×10^{-6} N, or **(c)** 1.0×10^{-7} N?

10. •• A particle with a charge of 18 μC experiences a force of 2.8×10^{-4} N when it moves at right angles to a magnetic field with a speed of 24 m/s. What force does this particle experience when it moves with a speed of 6.3 m/s at an angle of 25° relative to the magnetic field?

11. •• An ion experiences a magnetic force of 6.2×10^{-16} N when moving in the positive x direction but no magnetic force when moving in the positive y direction. What is the magnitude of the magnetic force exerted on the ion when it moves in the x-y plane along the line $x = y$? Assume that the ion's speed is the same in all cases.

12. •• An electron moving with a speed of 4.0×10^5 m/s in the positive x direction experiences zero magnetic force. When it moves in the positive y direction, it experiences a force of 3.2×10^{-13} N that points in the positive z direction. What are the direction and magnitude of the magnetic field?

13. •• **Predict/Calculate** Two charged particles with different speeds move one at a time through a region of uniform magnetic field. The particles move in the same direction and experience equal magnetic forces. **(a)** If particle 1 has four times the charge of particle 2, which particle has the greater speed? Explain. **(b)** Find the ratio of the speeds, v_1/v_2.

14. •• A 6.60-μC particle moves through a region of space where an electric field of magnitude 1450 N/C points in the positive x direction, and a magnetic field of magnitude 1.22 T points in the positive z direction. If the net force acting on the particle is 6.23×10^{-3} N in the positive x direction, find the magnitude and direction of the particle's velocity. Assume the particle's velocity is in the x-y plane.

15. ••• When at rest, a proton experiences a net electromagnetic force of magnitude 8.0×10^{-13} N pointing in the positive x direction. When the proton moves with a speed of 1.5×10^6 m/s in the positive y direction, the net electromagnetic force on it decreases in magnitude to 7.5×10^{-13} N, still pointing in the positive x direction. Find the magnitude and direction of **(a)** the electric field and **(b)** the magnetic field.

SECTION 22-3 THE MOTION OF CHARGED PARTICLES IN A MAGNETIC FIELD

16. • **CE** A velocity selector is to be constructed using a magnetic field in the positive y direction. If positively charged particles move through the selector in the positive z direction, **(a)** what must be the direction of the electric field? **(b)** Repeat part (a) for the case of negatively charged particles.

17. • Charged particles pass through a velocity selector with electric and magnetic fields at right angles to each other, as shown in **FIGURE 22-42**. If the electric field has a magnitude of 450 N/C and the magnetic field has a magnitude of 0.18 T, what speed must the particles have to pass through the selector undeflected?

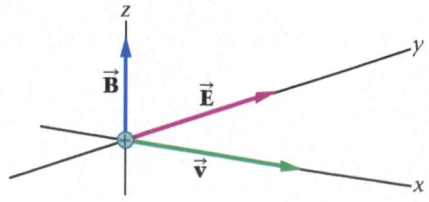

FIGURE 22-42 Problem 17

18. • The velocity selector in **FIG-URE 22-43** is designed to allow charged particles with a speed of 4.5×10^3 m/s to pass through undeflected. Find the direction and magnitude of the required electric field, given that the magnetic field has a magnitude of 0.96 T.

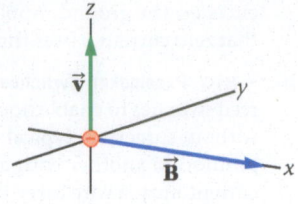

FIGURE 22-43 Problem 18

19. •• Find the radius of the orbit when (a) an electron or (b) a proton moves perpendicular to a magnetic field of 0.86 T with a speed of 6.47×10^5 m/s.

20. •• **BIO** Predict/Calculate The artery in Figure 22-14 has an inside diameter of 2.75 mm and passes through a region where the magnetic field is 0.065 T. (a) If the voltage difference between the electrodes is 195 μV, what is the speed of the blood? (b) Which electrode is at the higher potential? Does your answer depend on the sign of the ions in the blood? Explain.

21. •• An electron accelerated from rest through a voltage of 750 V enters a region of constant magnetic field. If the electron follows a circular path with a radius of 27 cm, what is the magnitude of the magnetic field?

22. •• A 10.2-μC particle with a mass of 2.80×10^{-5} kg moves perpendicular to a 0.850-T magnetic field in a circular path of radius 29.3 m. (a) How fast is the particle moving? (b) How much time will it take for the particle to complete one orbit?

23. •• Predict/Calculate When a charged particle enters a region of uniform magnetic field, it follows a circular path, as indicated in **FIGURE 22-44**. (a) Is this particle positively or negatively charged? Explain. (b) Suppose that the magnetic field has a magnitude of 0.180 T, the particle's speed is 6.0×10^6 m/s, and the radius of its path is 52.0 cm. Find the mass of the particle, given that its charge has a magnitude of 1.60×10^{-19} C. Give your result in atomic mass units, u, where 1 u = 1.67×10^{-27} kg.

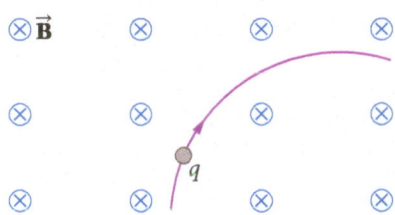

FIGURE 22-44 Problem 23

24. •• A proton with a kinetic energy of 4.6×10^{-16} J moves perpendicular to a magnetic field of 0.36 T. What is the radius of its circular path?

25. •• Predict/Calculate An alpha particle (the nucleus of a helium atom) consists of two protons and two neutrons, and has a mass of 6.64×10^{-27} kg. A horizontal beam of alpha particles is injected with a speed of 1.3×10^5 m/s into a region with a vertical magnetic field of magnitude 0.155 T. (a) How much time does it take for an alpha particle to move halfway through a complete circle? (b) If the speed of the alpha particle is doubled, does the time found in part (a) increase, decrease, or stay the same? Explain. (c) Repeat part (a) for alpha particles with a speed of 2.6×10^5 m/s.

26. ••• An electron and a proton move in circular orbits in a plane perpendicular to a uniform magnetic field $\vec{B}$. Find the ratio of the radii of their circular orbits when the electron and the proton have (a) the same momentum and (b) the same kinetic energy.

27. ••• **Helical Motion** As a model of the physics of the aurora, consider a proton emitted by the Sun that encounters the magnetic field of the Earth while traveling at 4.3×10^5 m/s. (a) The proton arrives at an angle of 33° from the direction of $\vec{B}$ (refer to Figure 22-19). What is the radius of the circular portion of its path if $B = 3.5 \times 10^{-5}$ T? (b) Calculate the time required for the proton to complete one circular orbit in the magnetic field. (c) How far parallel to the magnetic field does the proton travel during the time to complete a circular orbit? This is called the *pitch* of its helical motion.

SECTION 22-4 THE MAGNETIC FORCE EXERTED ON A CURRENT-CARRYING WIRE

28. • What is the magnetic force exerted on a 2.35-m length of wire carrying a current of 0.819 A perpendicular to a magnetic field of 0.920 T?

29. • A wire with a current of 2.1 A is at an angle of 38.0° relative to a magnetic field of 0.78 T. Find the force exerted on a 2.25-m length of the wire.

30. • The magnetic force exerted on a 1.2-m segment of straight wire is 1.6 N. The wire carries a current of 3.0 A in a region with a constant magnetic field of 0.50 T. What is the angle between the wire and the magnetic field?

31. •• A 0.61-m copper rod with a mass of 0.043 kg carries a current of 15 A in the positive x direction. What are the magnitude and direction of the minimum magnetic field needed to levitate the rod?

32. •• The long, thin wire shown in **FIGURE 22-45** is in a region of constant magnetic field $\vec{B}$. The wire carries a current of 6.2 A and is oriented at an angle of 7.5° to the direction of the magnetic field. (a) If the magnetic force exerted on this wire per meter is 0.038 N, what is the magnitude of the magnetic field? (b) At what angle will the force exerted on the wire per meter be equal to 0.016 N?

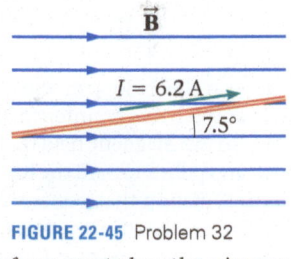

FIGURE 22-45 Problem 32

33. •• A wire with a length of 3.8 m and a mass of 0.65 kg is in a region of space with a magnetic field of 0.74 T. What is the minimum current needed to levitate the wire?

34. •• **Loudspeaker Force** The coil in a loudspeaker has 50 turns and a radius of 3.8 cm. The magnetic field is perpendicular to the wires in the coil and has a magnitude of 0.33 T. If the current in the coil is 280 mA, what is the total force on the coil?

35. •• A high-voltage power line carries a current of 110 A at a location where the Earth's magnetic field has a magnitude of 0.59 G and points to the north, 72° below the horizontal. Find the direction and magnitude of the magnetic force exerted on a 250-m length of wire if the current in the wire flows (a) horizontally toward the east or (b) horizontally toward the south.

36. ••• A metal bar of mass m and length L is suspended from two conducting wires, as shown in **FIGURE 22-46**. A uniform magnetic field of magnitude B points vertically downward. Find the angle θ the suspending wires make with the vertical when the bar carries a current I.

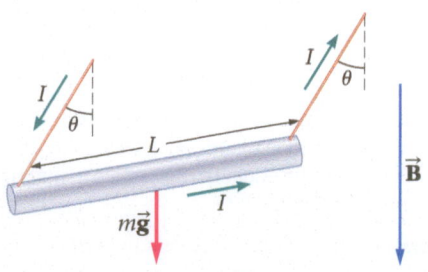

FIGURE 22-46 Problem 36

SECTION 22-5 LOOPS OF CURRENT AND MAGNETIC TORQUE

37. • CE For each of the three situations shown in **FIGURE 22-47**, indicate whether there will be a tendency for the square current loop to rotate clockwise, counterclockwise, or not at all, when viewed from above the loop along the indicated axis.

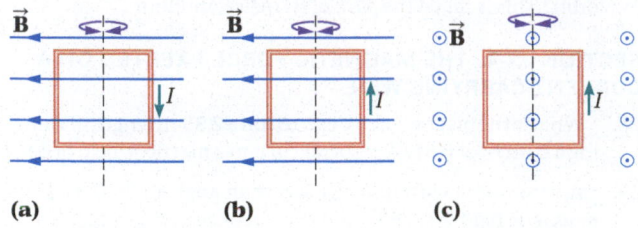

(a)   **(b)**   **(c)**

FIGURE 22-47 Problem 37

38. • A rectangular loop of 280 turns is 35 cm wide and 19 cm high. What is the current in this loop if the maximum torque in a field of 0.48 T is 23 N·m?

39. • A single circular loop of radius 0.15 m carries a current of 3.1 A in a magnetic field of 0.91 T. What is the maximum torque exerted on this loop?

40. •• In the previous problem, find the angle the plane of the loop must make with the field if the torque is to be half its maximum value.

41. •• A square loop of wire 0.15 m on a side lies on a horizontal table and carries a counterclockwise current of 4.2 A. The component of Earth's magnetic field that is in the plane of the loop is 2.5×10^{-5} T and points toward the top of the loop. **(a)** Consider a horizontal axis of rotation that passes through the center of the loop from left to right. Does the top wire of the loop want to rotate toward you (up from the table) or away from you (down into the table)? **(b)** Calculate the magnitude of the torque about the axis described in (a).

42. ••• Predict/Calculate Each of the 10 turns of wire in a vertical, rectangular loop carries a current of 0.22 A. The loop has a height of 8.0 cm and a width of 15 cm. A horizontal magnetic field of magnitude 0.050 T is oriented at an angle of $\theta = 65°$ relative to the normal to the plane of the loop, as indicated in **FIGURE 22-48**.

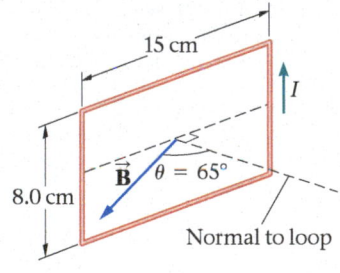

FIGURE 22-48 Problem 42

Find **(a)** the magnetic force on each side of the loop, **(b)** the net magnetic force on the loop, and **(c)** the magnetic torque on the loop. **(d)** If the loop can rotate about a vertical axis with only a small amount of friction, will it end up with an orientation given by $\theta = 0, \theta = 90°$, or $\theta = 180°$? Explain.

SECTION 22-6 ELECTRIC CURRENTS, MAGNETIC FIELDS, AND AMPÈRE'S LAW

43. • Find the magnetic field 7.25 cm from a long, straight wire that carries a current of 6.81 A.

44. • How much current must pass through a horizontal power transmission cable in order for the magnetic field at a location 11 m directly below it to be equal to the Earth's magnetic field, which is approximately 5.0×10^{-5} T?

45. • You travel to the north magnetic pole of the Earth, where the magnetic field points vertically downward. There, you draw a circle on the ground. Applying Ampère's law to this circle, show that zero current passes through its area.

46. • BIO Pacemaker Switches Some pacemakers employ magnetic reed switches to enable doctors to change their mode of operation without surgery. A typical reed switch can be switched from one position to another with a magnetic field of 5.0×10^{-4} T. What current must a wire carry if it is to produce a 5.0×10^{-4} T field at a distance of 0.50 m?

47. •• Two power lines, each 290 m in length, run parallel to each other with a separation of 23 cm. If the lines carry parallel currents of 120 A, what are the magnitude and direction of the magnetic force each exerts on the other?

48. •• Predict/Calculate Consider the long, straight, current-carrying wires shown in **FIGURE 22-49**. One wire carries a current of 6.2 A in the positive y direction; the other wire carries a current of 4.5 A in the positive x direction. **(a)** At which of the two points, A or B, do you expect the magnitude of the net magnetic field to be greater? Explain. **(b)** Calculate the magnitude of the net magnetic field at points A and B.

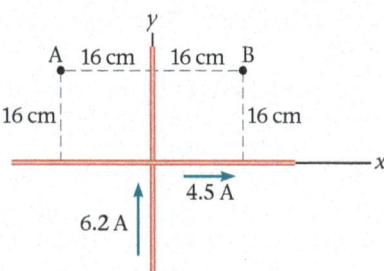

FIGURE 22-49 Problem 48

49. •• In Oersted's experiment, suppose that the compass was 0.15 m from the current-carrying wire. If a magnetic field of one-third the Earth's magnetic field of 5.0×10^{-5} T was required to give a noticeable deflection of the compass needle, what current must the wire have carried?

50. •• Predict/Calculate Two long, straight wires are separated by a distance of 9.25 cm. One wire carries a current of 2.75 A, the other carries a current of 4.33 A. **(a)** Find the force per meter exerted on the 2.75-A wire. **(b)** Is the force per meter exerted on the 4.33-A wire greater than, less than, or the same as the force per meter exerted on the 2.75-A wire? Explain.

51. ••• Two long, straight wires are oriented perpendicular to the page, as shown in **FIGURE 22-50**. The current in one wire is $I_1 = 3.0$ A, pointing into the page, and the current in the other wire is $I_2 = 4.0$ A, pointing out of the page. Find the magnitude and direction of the net magnetic field at point P.

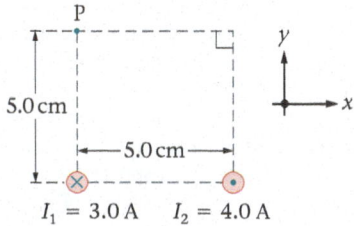

FIGURE 22-50 Problem 51

SECTION 22-7 CURRENT LOOPS AND SOLENOIDS

52. • CE A loop of wire is connected to the terminals of a battery, as indicated in **FIGURE 22-51**. If the loop is to attract the bar magnet, which of the terminals, A or B, should be the positive terminal of the battery? Explain.

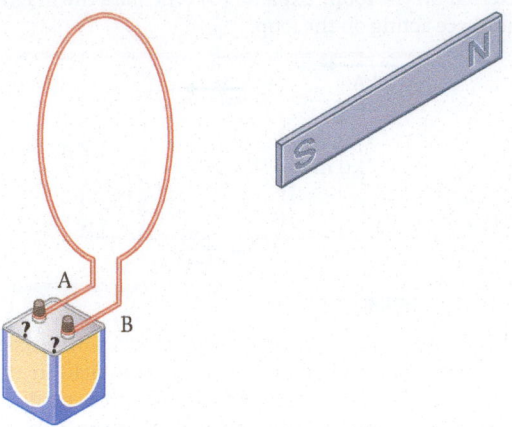

FIGURE 22-51 Problem 52

53. • **CE** Predict/Explain The number of turns in a solenoid is doubled, and at the same time its length is doubled. (a) Does the magnetic field within the solenoid increase, decrease, or stay the same? (b) Choose the *best explanation* from among the following:
I. Doubling the number of turns in a solenoid doubles its magnetic field, and hence the field increases.
II. Making a solenoid longer decreases its magnetic field, and therefore the field decreases.
III. The magnetic field remains the same because the number of turns per length is unchanged.

54. • A circular coil of wire has a radius of 7.5 cm and has 125 turns of wire that carries 6.6 A of current. What is the magnetic field at the center of the coil?

55. • The solenoid for an automobile power door lock is 2.5 cm long and has 190 turns of wire that carry 1.6 A of current. What is the magnitude of the magnetic field that it produces?

56. • It is desired that a solenoid 25 cm long and with 350 turns produce a magnetic field within it equal to the Earth's magnetic field $(5.0 \times 10^{-5} \text{ T})$. What current is required?

57. • A solenoid that is 72 cm long produces a magnetic field of 1.8 T within its core when it carries a current of 8.1 A. How many turns of wire are contained in this solenoid?

58. • The maximum current in a superconducting solenoid can be as large as 3.75 kA. If the number of turns per meter in such a solenoid is 3650, what is the magnitude of the magnetic field it produces?

59. •• To construct a solenoid, you wrap insulated wire uniformly around a plastic tube 7.6 cm in diameter and 33 cm in length. You would like a 3.0-A current to produce a 2.5-kG magnetic field inside your solenoid. What is the total length of wire you will need to meet these specifications?

GENERAL PROBLEMS

60. • **CE** A proton is to orbit the Earth at the equator using the Earth's magnetic field to supply part of the necessary centripetal force. Should the proton move eastward or westward? Explain.

61. • **CE** FIGURE 22-52 shows an electron beam whose initial direction of motion is horizontal, from right to left. A magnetic field deflects the beam downward. What is the direction of the magnetic field?

FIGURE 22-52 A horizontal electron beam is deflected downward by a magnetic field. (Problem 61)

62. • **CE** The three wires shown in FIGURE 22-53 are long and straight, and they each carry a current of the same magnitude, I. The currents in wires 1 and 3 are out of the page; the current in wire 2 is into the page. What is the direction of the magnetic force experienced by wire 3?

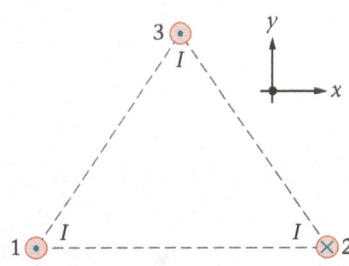

FIGURE 22-53 Problems 62 and 63

63. • **CE** Each of the current-carrying wires in Figure 22-53 is long and straight, and carries the current I either into or out of the page, as shown. What is the direction of the net magnetic field produced by these three wires at the center of the triangle?

64. • **CE** The four wires shown in FIGURE 22-54 are long and straight, and they each carry a current of the same magnitude, I. The currents in wires 1, 2, and 3 are out of the page; the current in wire 4 is into the page. What is the direction of the magnetic force experienced by wire 2?

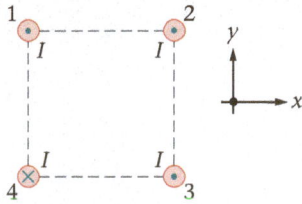

FIGURE 22-54 Problems 64 and 65

65. • **CE** Each of the current-carrying wires in Figure 22-54 is long and straight, and carries the current I either into or out of the page, as shown. What is the direction of the net magnetic field produced by these four wires at the center of the square?

66. • **BIO** Brain Function and Magnetic Fields Experiments have shown that thought processes in the brain can be affected if the parietal lobe is exposed to a magnetic field with a strength of 1.0 T. How much current must a long, straight wire carry if it is to produce a 1.0-T magnetic field at a distance of 0.50 m? (For comparison, a typical lightning bolt carries a current of about 20,000 A, which would melt most wires.)

67. • **Credit-Card Magnetic Strips** Experiments carried out on the television show *Mythbusters* determined that a magnetic field of 1000 gauss is needed to corrupt the information on a credit card's magnetic strip. (They also busted the myth that a credit card can be demagnetized by an electric eel or an eelskin wallet.) Suppose a long, straight wire carries a current of 3.5 A. How close can a credit card be held to this wire without damaging its magnetic strip?

68. • **Superconducting Solenoid** Cryomagnetics, Inc., advertises a high-field, superconducting solenoid that produces a magnetic field of 17 T with a current of 105 A. What is the number of turns per meter in this solenoid?

69. •• Consider a current loop immersed in a magnetic field, as in Figure 22-47 (a). It is given that $B = 0.42$ T and $I = 8.8$ A. In addition, the loop is a square 0.38 m on a side. Find the magnitude of the magnetic force exerted on each side of the loop.

70. •• **CE** A positively charged particle moves through a region with a uniform electric field pointing toward the top of the page and a uniform magnetic field pointing into the page. The particle can have one of the four velocities shown in **FIGURE 22-55**. **(a)** Rank the four possibilities in order of increasing magnitude of the net force (F_1, F_2, F_3, and F_4) the particle experiences. Indicate ties where appropriate. **(b)** Which of the four velocities could potentially result in zero net force?

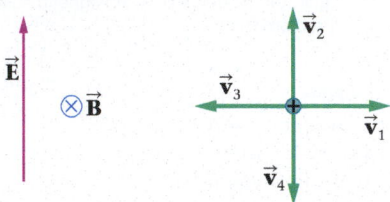

FIGURE 22-55 Problem 70

71. •• **CE** A proton follows the path shown in **FIGURE 22-56** as it moves through three regions with different uniform magnetic fields, B_1, B_2, and B_3. In each region the proton completes a half-circle, and the magnetic field is perpendicular to the page. **(a)** Rank the three fields in order of increasing magnitude. Indicate ties where appropriate. **(b)** Give the direction (into or out of the page) for each of the fields.

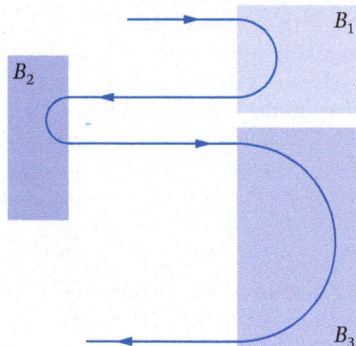

FIGURE 22-56 Problems 71 and 72

72. •• **CE Predict/Explain** Suppose the initial speed of the proton in Figure 22-56 is increased. **(a)** Does the radius of each half-circular path segment increase, decrease, or stay the same? **(b)** Choose the *best explanation* from among the following:
   I. The radius of a circular orbit in a magnetic field is proportional to the speed of the proton; therefore, the radius of the half-circular path will increase.
   II. A greater speed means the proton will experience more force from the magnetic field, resulting in a decrease of the radius.
   III. The increase in speed offsets the increase in magnetic force, resulting in no change of the radius.

73. •• **BIO Magnetic Resonance Imaging** An MRI (magnetic resonance imaging) solenoid produces a magnetic field of 1.5 T. The solenoid is 2.5 m long, 1.0 m in diameter, and wound with insulated wires 2.2 mm in diameter. Find the current that flows in the solenoid. (Your answer should be rather large.)

74. •• **Predict/Calculate** A long, straight wire carries a current of 14 A. Next to the wire is a square loop with sides 1.0 m in length, as

shown in **FIGURE 22-57**. The loop carries a current of 2.5 A in the direction indicated. **(a)** What is the direction of the net force exerted on the loop? Explain. **(b)** Calculate the magnitude of the net force acting on the loop.

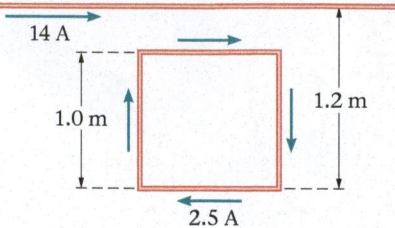

FIGURE 22-57 Problem 74

75. •• A particle with a charge of 38 μC moves with a speed of 77 m/s in the positive x direction. The magnetic field in this region of space has a component of 0.45 T in the positive y direction, and a component of 0.85 T in the positive z direction. What are the magnitude and direction of the magnetic force on the particle?

76. •• **Predict/Calculate** A beam of protons with various speeds is directed in the positive x direction. The beam enters a region with a uniform magnetic field of magnitude 0.52 T pointing in the negative z direction, as indicated in **FIGURE 22-58**. It is desired to use a uniform electric field (in addition to the magnetic field) to select from this beam only those protons with a speed of 1.42×10^5 m/s— that is, only these protons should be undeflected by the two fields. **(a)** Determine the magnitude and direction of the electric field that yields the desired result. **(b)** Suppose the electric field is to be produced by a parallel-plate capacitor with a plate separation of 2.5 cm. What potential difference is required between the plates? **(c)** Which plate in Figure 22-58 (top or bottom) should be positively charged? Explain.

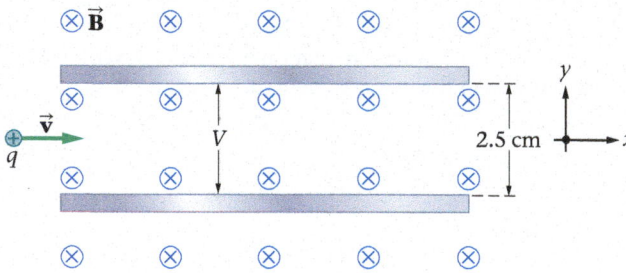

FIGURE 22-58 Problem 76

77. •• Two parallel wires, each carrying a current of 2.2 A in the same direction, are shown in **FIGURE 22-59**. Find the direction and magnitude of the net magnetic field at points A, B, and C.

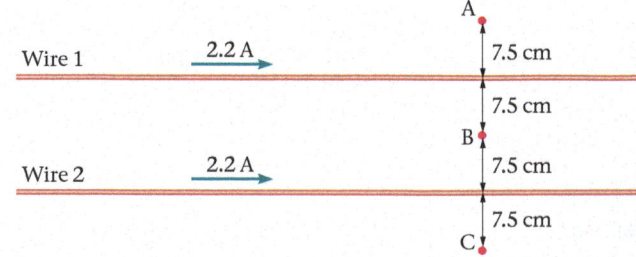

FIGURE 22-59 Problems 77 and 78

78. •• Repeat Problem 77 for the case where the current in wire 1 is reversed in direction.

79. •• **Electric Motor** A current of 2.4 A flows through a circular coil of wire with 52 turns and a radius of 0.64 cm. The coil rotates in a uniform 0.12-T magnetic field. **(a)** What is the maximum torque exerted on the loop of wire? **(b)** The time-averaged torque on the loop is half its maximum value. If the moment of inertia of the loop is 2.5×10^{-5} kg·m^2 and the loop undergoes constant angular acceleration, in what time will it have reached its maximum angular speed of 2500 rev/min?

80. •• **Balanced Coil** A student wishes to suspend a square loop of wire at a 45° angle to the vertical using the Earth's magnetic field. The geometry is depicted in **FIGURE 22-60**. At the student's location the Earth's magnetic field has a magnitude of 5.2×10^{-5} T and is oriented 60.0° below horizontal. All of the counterclockwise torque acting on the loop arises from the magnetic force on the bottom side of the square. **(a)** Is the current in the loop clockwise or counterclockwise, as viewed from above? **(b)** The loop is a square with mass $m = 0.0089$ kg and sides of length $L = 0.10$ m. How much current must flow through the loop to suspend it as shown? (Note: The angle between the plane of the loop and the magnetic force is 105°.)

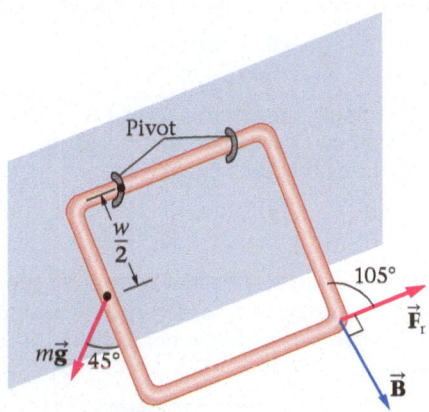

FIGURE 22-60 Problem 80

81. •• **Lightning Bolts** A powerful bolt of lightning can carry a current of 225 kA. **(a)** Treating a lightning bolt as a long, thin wire, calculate the magnitude of the magnetic field produced by such a bolt of lightning at a distance of 35 m. **(b)** If two such bolts strike simultaneously at a distance of 35 m from each other, what is the magnetic force per meter exerted by one bolt on the other?

82. •• **Predict/Calculate** Consider the two current-carrying wires shown in **FIGURE 22-61**. The current in wire 1 is 3.7 A; the current in wire 2 is adjusted to make the net magnetic field at point A equal to zero. **(a)** Is the magnitude of the current in wire 2 greater than, less than, or the same as that in wire 1? Explain. **(b)** Find the magnitude and direction of the current in wire 2.

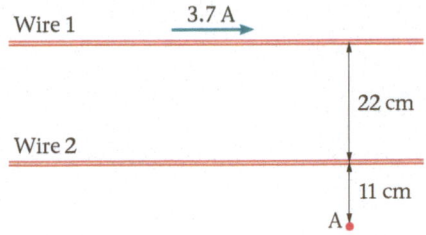

FIGURE 22-61 Problem 82

83. •• **Magnetars** The astronomical object 4U014 + 61 has the distinction of creating the most powerful magnetic field ever observed. This object is referred to as a "magnetar" (a subclass of pulsars), and its magnetic field is 1.3×10^{15} times greater than the Earth's magnetic field. **(a)** Suppose a 2.5-m straight wire carrying a current of 1.1 A is placed in this magnetic field at an angle of 65° to the

field lines. What force does this wire experience? **(b)** A field this strong can significantly change the behavior of an atom. To see this, consider an electron moving with a speed of 2.2×10^6 m/s. Compare the maximum magnetic force exerted on the electron to the electric force a proton exerts on an electron in a hydrogen atom. The radius of the hydrogen atom is 5.29×10^{-11} m.

84. •• Consider a system consisting of two concentric solenoids, as illustrated in **FIGURE 22-62**. The current in the outer solenoid is $I_1 = 1.25$ A, and the current in the inner solenoid is $I_2 = 2.17$ A. Given that the number of turns per centimeter is 105 for the outer solenoid and 125 for the inner solenoid, find the magnitude and direction of the magnetic field **(a)** between the solenoids and **(b)** inside the inner solenoid.

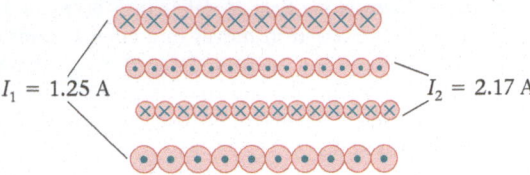

FIGURE 22-62 Problem 84

85. •• Solenoids produce magnetic fields that are relatively intense for the amount of current they carry. To make a direct comparison, consider a solenoid with 48.0 turns per centimeter, a radius of 1.27 cm, and a current of 0.776 A. **(a)** Find the magnetic field at the center of the solenoid. **(b)** What current must a long, straight wire carry to have the same magnetic field as that found in part (a)? Let the distance from the wire be the same as the radius of the solenoid, 1.27 cm.

86. •• The current in a solenoid with 28 turns per centimeter is 0.55 A. The solenoid has a radius of 1.5 cm. A long, straight wire runs along the axis of the solenoid, carrying a current of 16 A. Find the magnitude of the net magnetic field a radial distance of 0.75 cm from the straight wire.

87. •• **BIO** **Predict/Calculate** **Transcranial Magnetic Stimulation** One way to study brain function is to produce a rapidly changing magnetic field within the brain. When this technique, known as transcranial magnetic stimulation (TMS), is applied to the prefrontal cortex, for example, it can reduce a person's ability to conjugate verbs, though other thought processes are unaffected. The rapidly varying magnetic field is produced with a circular coil of 21 turns and a radius of 6.0 cm placed directly on the head. The current in this loop increases at the rate of 1.2×10^7 A/s (by discharging a capacitor). **(a)** At what rate does the magnetic field at the center of the coil increase? **(b)** Suppose a second coil with half the area of the first coil is used instead. Would your answer to part (a) increase, decrease, or stay the same? By what factor?

88. •• **Synchrotron Undulator** In one portion of a synchrotron undulator, electrons traveling at 2.99×10^8 m/s enter a region of uniform magnetic field with a strength of 0.844 T. **(a)** What is the acceleration of an electron in this region? **(b)** The total power of X-rays emitted by these electrons is given by $P = (1.07 \times 10^{-45})a^2$ W, where a is the acceleration in m/s^2. What power is emitted by the electrons in this portion of the undulator?

89. •• **Predict/Calculate** A single current-carrying circular loop of radius R is placed next to a long, straight wire, as shown in **FIGURE 22-63**. The current in the wire points to the right and is of

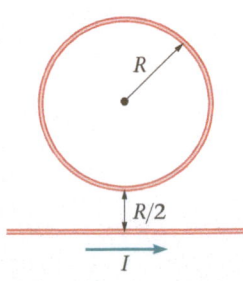

FIGURE 22-63 Problem 89

magnitude I. **(a)** In which direction must current flow in the loop to produce zero magnetic field at its center? Explain. **(b)** Calculate the magnitude of the current in part (a).

90. ••• A thin ring of radius R and charge per length λ rotates with an angular speed ω about an axis perpendicular to its plane and passing through its center. Find the magnitude of the magnetic field at the center of the ring.

91. ••• A solenoid is made from a 25-m length of wire of resistivity $2.3 \times 10^{-8} \, \Omega \cdot m$. The wire, whose radius is 2.1 mm, is wrapped uniformly onto a plastic tube 4.5 cm in diameter and 1.65 m long. Find the emf to which the ends of the wire must be connected to produce a magnetic field of 0.015 T within the solenoid.

92. ••• **Magnetic Fields in the Bohr Model** In the Bohr model of the hydrogen atom, the electron moves in a circular orbit of radius 5.29×10^{-11} m about the nucleus. Given that the charge on the electron is -1.60×10^{-19} C, and that its speed is 2.2×10^{6} m/s, find the magnitude of the magnetic field the electron produces at the nucleus of the atom.

93. ••• A single-turn square loop carries a current of 18 A. The loop is 15 cm on a side and has a mass of 0.035 kg. Initially the loop lies flat on a horizontal tabletop. When a horizontal magnetic field is turned on, it is found that only one side of the loop experiences an upward force. Find the minimum magnetic field, B_{min}, necessary to start tipping the loop up from the table.

PASSAGE PROBLEMS

BIO Magnetoencephalography

To read and understand this sentence your brain must process visual input from your eyes and translate it into words and thoughts. As you do so, minute electric currents flow through the neurons in your visual cortex. These currents, like any electric current, produce magnetic fields. In fact, even your innermost thoughts and dreams produce magnetic fields that can be detected outside your head.

Magnetoencephalography (MEG) is the study of magnetic fields produced by electrical activity in the brain. Though completely noninvasive, MEG can provide detailed information on spontaneous brain function—like alpha waves and pathological epileptic spikes—as well as brain activity that is evoked by visual, auditory, and tactile stimuli.

The magnetic fields produced by brain activity are incredibly weak—roughly 100 million times smaller than the Earth's magnetic field. Even so, sensitive detectors called SQUIDS (superconducting quantum interference devices), which were invented by physicists as a research tool, can detect fields as small as 1.0×10^{-15} T. Coupled with sophisticated electronics and software, and operating at liquid helium temperatures ($-269 \, ^\circ C$), SQUIDS can localize

the source of brain activity to within millimeters. When the information from MEG is overlaid with the anatomical data from an MRI scan, the result is a richly detailed "map" of the electrical activity within the brain.

94. • Approximating a neuron by a straight wire, what electric current is needed to produce a magnetic field of 1.0×10^{-15} T at a distance of 5.0 cm?

    A. 4.0×10^{-22} A   **B.** 7.9×10^{-11} A
    C. 2.5×10^{-10} A   **D.** 1.0×10^{-7} A

95. • Suppose a neuron in the brain carries a current of 5.0×10^{-8} A. Treating the neuron as a straight wire, what is the magnetic field it produces at a distance of 7.5 cm?

    A. 1.3×10^{-13} T   **B.** 4.2×10^{-12} T
    C. 1.1×10^{-10} T   **D.** 3.3×10^{-7} T

96. • A given neuron in the brain carries a current of 3.1×10^{-8} A. If the SQUID detects a magnetic field of 2.8×10^{-14} T, how far away is the neuron? Treat the neuron as a straight wire.

    A. 22 cm   **B.** 70 cm
    C. 140 cm   **D.** 176 cm

97. •• A SQUID detects a magnetic field of 1.8×10^{-14} T at a distance of 13 cm. How many electrons flow through the neuron per second? Treat the neuron as a straight wire.

    A. 1.2×10^{10} **B.** 2.3×10^{10}
    C. 7.3×10^{10} **D.** 9.2×10^{10}

98. •• **Predict/Calculate** **REFERRING TO EXAMPLE 22-7** Suppose the speed of the isotopes is doubled. **(a)** Does the separation distance, d, increase, decrease, or stay the same? Explain. **(b)** Find the separation distance for this case.

99. •• **Predict/Calculate** **REFERRING TO EXAMPLE 22-7** Suppose we change the initial speed of ^{238}U, leaving everything else the same. **(a)** If we want the separation distance to be zero, should the initial speed of ^{238}U be increased or decreased? Explain. **(b)** Find the required initial speed.

100. •• **REFERRING TO QUICK EXAMPLE 22-15** The current I_1 is adjusted until the magnetic field halfway between the wires has a magnitude of 7.5×10^{-7} T and points *into* the page. Everything else in the system remains the same as in Quick Example 22-15. Find the magnitude and direction of I_1.

101. •• **REFERRING TO QUICK EXAMPLE 22-15** The current I_2 is adjusted until the magnetic field 5.5 cm below wire 2 has a magnitude of 2.5×10^{-6} T and points *out* of the page. Everything else in the system remains the same as in Quick Example 22-15. Find the magnitude and direction of I_2.

Big Ideas

Magnetic Flux and Faraday's Law of Induction

1 A changing magnetic field produces an electromotive force just like the emf of a battery.

2 The induced emf is directly proportional to the rate of change of the magnetic flux.

3 A changing magnetic flux produces an electric current that opposes the change.

4 A changing magnetic flux is the basis for the operation of electric generators and motors.

5 An inductor opposes changes in the current in an electric circuit.

6 Inductors produce magnetic fields that store energy.

7 Transformers use a change in magnetic flux to produce a change in voltage.

▲ Have you used Faraday's law of induction today? Well, if you've "swiped" a credit card through a card reader you have. Credit cards have magnetic strips on the back that record three tracks of information encoded in a magnetic field that has many reversals in polarity—like a series of dots and dashes in Morse code. As you swipe a card through a reader, the changing magnetic field induces electric currents in wire coils—as described by Faraday's law—which the reader uses to identify your account. This is just one of many examples of Faraday's law covered in this chapter.

In the last chapter we saw a number of connections between electricity and static magnetic fields. We now explore some of the interesting new phenomena that occur when magnetic fields change with time.

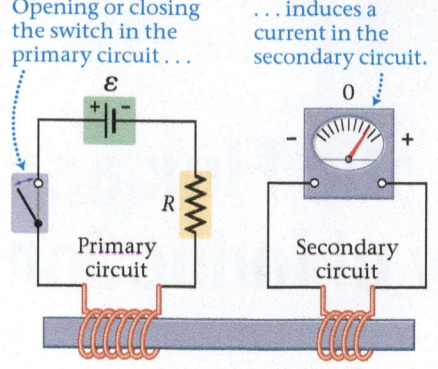

Opening or closing
the switch in the
primary circuit...

...induces a
current in the
secondary circuit.

Primary
circuit

Secondary
circuit

▲ FIGURE 23-1 **Magnetic induction** Basic setup of Faraday's experiment on magnetic induction. When the position of the switch on the primary circuit is changed from open to closed or from closed to open, an electromotive force (emf) is induced in the secondary circuit. The induced emf causes a current in the secondary circuit, and the current is detected by the ammeter. If the current in the primary circuit does not change, no matter how large it may be, there is no induced current in the secondary circuit.

23-1 Induced Electromotive Force

When Hans Christian Oersted observed in 1820 that an electric current produces a magnetic field, it was pure serendipity. In contrast, Michael Faraday (1791–1867), an English chemist and physicist who was aware of Oersted's results, purposefully set out to determine whether a similar effect operates in the reverse direction—that is, whether a magnetic field can produce an electric field. His careful experimentation showed that such a connection does indeed exist.

Inducing an emf   A simplified version of the type of experiment performed by Faraday is shown in **FIGURE 23-1**. Two electric circuits are involved. The first, called the **primary circuit,** consists of a battery, a switch, a resistor to control the current, and a coil of several turns around an iron bar. When the switch is closed on the primary circuit, a current flows through the coil, producing a magnetic field that is particularly intense within the iron bar.

The **secondary circuit** also has a coil wrapped around the iron bar, and this coil is connected to an ammeter to detect any current in the circuit. Notice, however, that there is no battery in the secondary circuit, and no direct physical contact between the two circuits. What does link the circuits, instead, is the magnetic field in the iron bar— it ensures that the field experienced by the secondary coil is approximately the same as that produced by the primary coil.

Now, let's look at the experimental results. When the switch is closed on the primary circuit, the magnetic field in the iron bar rises from zero to some finite amount, and the ammeter in the secondary coil deflects to one side briefly, and then returns to zero. As long as the current in the primary circuit is maintained at a constant value, the ammeter in the secondary circuit gives zero reading. If the switch on the primary circuit is now opened, so that the magnetic field decreases again to zero, the ammeter in the secondary circuit deflects briefly in the opposite direction, and then returns to zero. We can summarize these observations as follows:

- The current in the secondary circuit is zero as long as the current in the primary circuit is constant—which means, in turn, that the magnetic field in the iron bar is constant. It does not matter whether the constant value of the magnetic field is zero or nonzero.

- When the magnetic field passing through the secondary coil increases, a current is observed to flow in one direction in the secondary circuit; when the magnetic field decreases, a current is observed in the opposite direction.

Recall that the current in the secondary circuit appears without any direct contact between the primary and secondary circuits. For this reason we refer to the secondary current as an **induced current.** Because the induced current behaves the same as a current produced by a battery with an electromotive force (emf), we say that the changing magnetic field creates an **induced emf** in the secondary circuit. This leads to our final experimental observation:

- The magnitudes of the induced current and induced emf are found to be proportional to the rate of change of the magnetic field—the more rapidly the magnetic field changes, the greater the induced emf.

In Section 23-3 (Faraday's Law of Induction) we return to these observations and state them in the form of a mathematical law.

In Faraday's experiment, the changing magnetic field is caused by a changing current in the primary circuit. Any means of altering the magnetic field is just as effective, however. For example, in **FIGURES 23-2** and **23-3** we show a common classroom demonstration of induced emf. In this case, there is no primary circuit; instead, the magnetic field is changed by simply moving a permanent magnet toward or away from a coil connected to an ammeter. When the magnet is moved toward the coil, the meter deflects in one direction; when it is pulled away from the coil, the meter deflects in the opposite direction. There is no induced emf, however, when the magnet is held still.

Big Idea **1** A changing magnetic field produces an induced electromotive force that is just like the emf produced by a battery. The induced emf is in one direction when the magnetic field increases, and in the opposite direction when the magnetic field decreases.

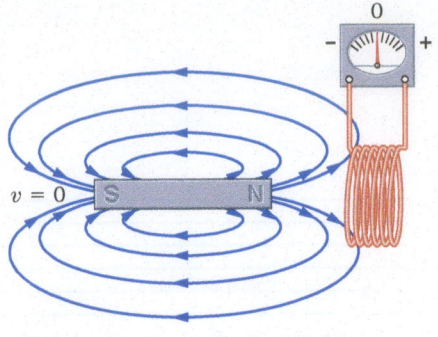

(a) Moving magnet toward coil induces current in one direction

(b) No motion, no induced current

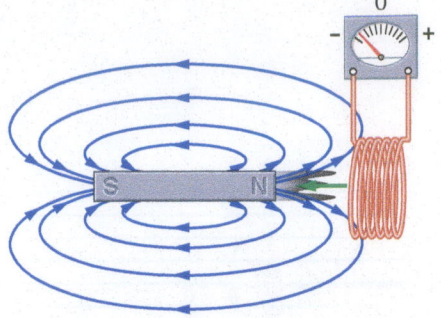

(c) Moving magnet away from coil induces current in opposite direction

▲ **FIGURE 23-2 Induced current produced by a moving magnet** A coil experiences an induced current when the magnetic field passing through it varies. **(a)** When the magnet moves toward the coil, the current is in one direction. **(b)** No current is induced while the magnet is held still. **(c)** When the magnet is pulled away from the coil, the current is in the other direction.

Enhance Your Understanding   (Answers given at the end of the chapter)

1. Which of the following situations results in an induced emf? **(a)** A large, constant magnetic field; **(b)** a decreasing magnetic field; **(c)** an increasing magnetic field; **(d)** an oscillating magnetic field.

Section Review

- A changing magnetic field induces an emf, similar to the action of a battery. In this case, however, the emf is in one direction when the magnetic field increases, and in the opposite direction when the magnetic field decreases. A constant magnetic field produces zero induced emf.

23-2 Magnetic Flux

In the previous section, we saw that an emf can be induced in a coil of wire by simply changing the strength of the magnetic field that passes through the coil. This isn't the only way to induce an emf, however. More generally, we can change the direction of the magnetic field rather than its magnitude. We can even change the orientation of the coil in a constant magnetic field, or change the cross-sectional area of the coil. All of these situations can be described in terms of the change of a single physical quantity, the magnetic flux.

Basically, **magnetic flux** is a measure of the number of magnetic field lines that cross a given area, in complete analogy with the electric flux discussed in Section 19-7. Suppose, for example, that a magnetic field $\vec{B}$ crosses a surface area A at right angles, as in **FIGURE 23-4 (a)**. The magnetic flux, Φ, in this case is simply the magnitude of the magnetic field times the area:

$$\Phi = BA$$

If, on the other hand, the magnetic field is parallel to the surface, as in **FIGURE 23-4 (b)**, we see that *no field lines cross the surface*. In a case like this the magnetic flux is zero:

$$\Phi = 0$$

In general, only the component of $\vec{B}$ that is *perpendicular* to a surface contributes to the magnetic flux. The magnetic field in **FIGURE 23-4 (c)**, for example, crosses the surface at an angle θ relative to the normal, and hence its perpendicular component is $B \cos \theta$. The magnetic flux, then, is simply $B \cos \theta$ times the area A:

▲ **FIGURE 23-3** When a magnet moves toward or away from a coil, the magnetic field within the coil changes. This, in turn, induces an electric current in the coil, which is detected by the meter.

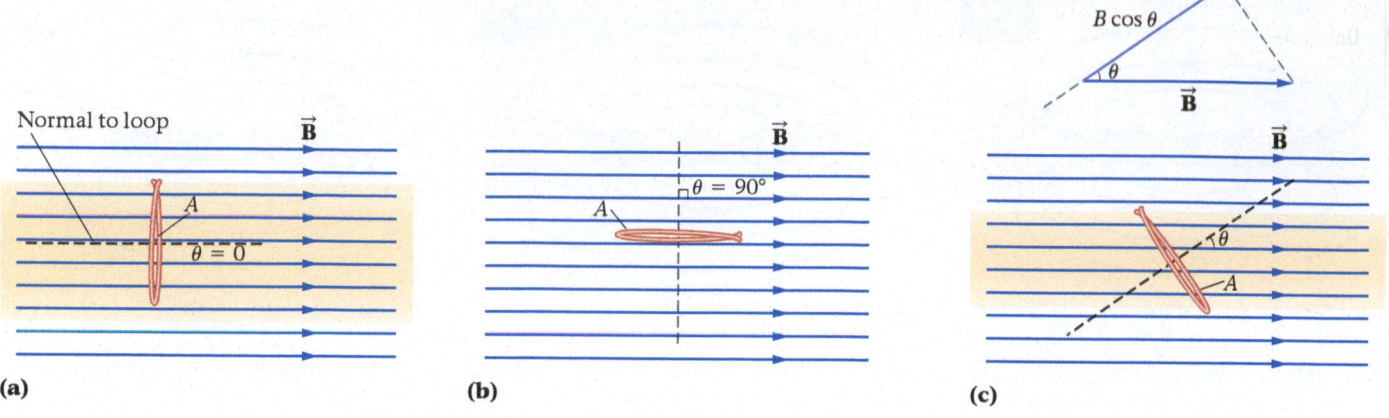

▲ **FIGURE 23-4 The magnetic flux through a loop** The magnetic flux through a loop of area A is $\Phi = BA \cos \theta$, where θ is the angle between the normal to the loop and the magnetic field. **(a)** The loop is perpendicular to the field; hence, $\theta = 0$ and $\Phi = BA$. **(b)** The loop is parallel to the field; therefore, $\theta = 90°$ and $\Phi = 0$. **(c)** For a general angle θ, the component of the field that is perpendicular to the loop is $B \cos \theta$; hence, the flux is $\Phi = BA \cos \theta$.

PHYSICS
IN CONTEXT
Looking Back

The concept of electric flux, so useful in Gauss's law in Chapter 19, is extended to magnetic flux in this chapter, where it plays a key role in Faraday's law.

Definition of Magnetic Flux, Φ

$$\Phi = (B \cos \theta)A = BA \cos \theta \qquad 23\text{-}1$$

SI unit: $1\ \text{T} \cdot \text{m}^2 = 1\ \text{weber} = 1\ \text{Wb}$

Notice that $\Phi = BA \cos \theta$ gives $\Phi = BA$ when $\vec{B}$ is perpendicular to the surface ($\theta = 0$), and $\Phi = 0$ when $\vec{B}$ is parallel to the surface ($\theta = 90°$), as expected. Finally, the unit of magnetic flux is the **weber (Wb),** named after the German physicist Wilhelm Weber (1804–1891). It is defined as follows:

$$1\ \text{Wb} = 1\ \text{T} \cdot \text{m}^2 \qquad 23\text{-}2$$

As we have seen, magnetic flux depends on the magnitude of the magnetic field, B, its orientation with respect to a surface, θ, and the area of the surface, A. A change in any of these variables results in a change in the flux. For example, in the case of the permanent magnet moved toward or away from the coil in the previous section, it is the change in *magnitude* of the field, B, that results in a change in flux. In the following Example, we consider the effect of changing the *orientation* of a wire loop in a region of constant magnetic field.

EXAMPLE 23-1   A SYSTEM IN FLUX

Consider a circular loop with a 2.50-cm radius in a constant magnetic field of 0.625 T. Find the magnetic flux through this loop when its normal makes an angle of **(a)** 0°, **(b)** 30.0°, **(c)** 60.0°, and **(d)** 90.0° with the direction of the magnetic field $\vec{B}$.

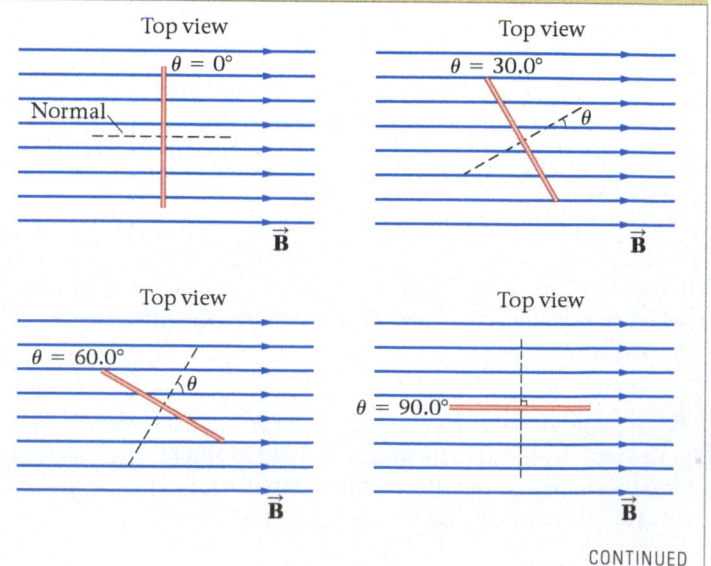

PICTURE THE PROBLEM
Our sketch shows four top views of the system, looking directly down on the top edge of the circular loop. The orientation of the loop and its normal are shown in each of the four panels. Notice that the specified angles between the normal and the magnetic field direction are $\theta = 0°$, 30.0°, 60.0°, and 90.0°.

REASONING AND STRATEGY
To find the magnetic flux, we use the following relationship: $\Phi = BA \cos \theta$. In this expression, B is the magnitude of the magnetic field and $A = \pi r^2$ is the area of a circular loop of radius r. The values of B, r, and θ are given in the problem statement.

CONTINUED

Known	Radius of loop, $r = 0.0250$ m; magnetic field, $B = 0.625$ T; angle between normal to loop and magnetic field, $\theta = 0°, 30.0°, 60.0°, 90.0°$.
Unknown	Magnetic flux, $\Phi = ?$

SOLUTION

Part (a)

1. Substitute $\theta = 0°$ in $\Phi = BA \cos \theta$, with $A = \pi r^2$:

$$\Phi = BA \cos \theta$$
$$= (0.625 \text{ T})\pi(0.0250 \text{ m})^2 \cos 0° = 1.23 \times 10^{-3} \text{ T} \cdot \text{m}^2$$

Part (b)

2. Substitute $\theta = 30.0°$ in $\Phi = BA \cos \theta$:

$$\Phi = BA \cos \theta$$
$$= (0.625 \text{ T})\pi(0.0250 \text{ m})^2 \cos 30.0° = 1.06 \times 10^{-3} \text{ T} \cdot \text{m}^2$$

Part (c)

3. Substitute $\theta = 60.0°$ in $\Phi = BA \cos \theta$:

$$\Phi = BA \cos \theta$$
$$= (0.625 \text{ T})\pi(0.0250 \text{ m})^2 \cos 60.0° = 6.14 \times 10^{-4} \text{ T} \cdot \text{m}^2$$

Part (d)

4. Substitute $\theta = 90.0°$ in $\Phi = BA \cos \theta$:

$$\Phi = BA \cos \theta$$
$$= (0.625 \text{ T})\pi(0.0250 \text{ m})^2 \cos 90.0° = 0$$

INSIGHT

Thus, even if a magnetic field is uniform in space and constant in time, the magnetic flux through a given area will change if the orientation of the area changes. This is particularly relevant to the case of a coil that rotates in a field and constantly changes its orientation. As we shall see in the next section, a changing magnetic flux can create an electric current, and in Section 23-6 we show how this applies to generators and motors.

PRACTICE PROBLEM

At what angle is the flux in this system equal to $1.00 \times 10^{-4} \text{ T} \cdot \text{m}^2$?   [**Answer:** $\theta = 85.3°$]

Some related homework problems: Problem 1, Problem 3

CONCEPTUAL EXAMPLE 23-2   MAGNETIC FLUX

The three loops of wire shown in the sketch are all in a region of space with a uniform, constant magnetic field. Loop 1 swings back and forth like the bob on a pendulum; loop 2 rotates about a vertical axis; and loop 3 oscillates vertically on the end of a spring. Which loop or loops have a magnetic flux that changes with time?

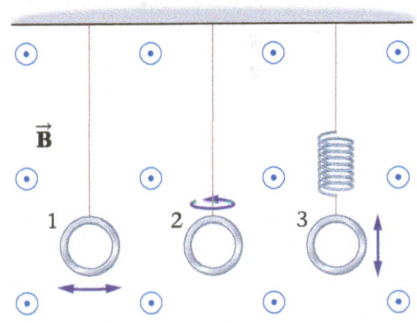

REASONING AND DISCUSSION

Loop 1 moves back and forth, and loop 3 moves up and down, but the magnetic field is uniform, and hence the flux doesn't depend on the loop's position. Loop 2, on the other hand, changes its orientation relative to the field as it rotates; hence, its flux does change with time.

ANSWER

Only loop 2 has a changing magnetic flux.

Enhance Your Understanding   (Answers given at the end of the chapter)

2. What is the angle θ in the definition of magnetic flux $\Phi = BA \cos \theta$ through a surface? **(a)** The angle between $\vec{B}$ and the plane of the surface; **(b)** the angle between $\vec{B}$ and the x axis; **(c)** the angle between $\vec{B}$ and the normal to the surface; **(d)** the angle between the surface and the x axis.

Section Review

• Magnetic flux is defined as $\Phi = BA \cos \theta$. The magnetic flux through a loop depends on the magnetic field strength, B, the area of the loop, A, and the loop's orientation relative to the magnetic field, θ.

23-3 Faraday's Law of Induction

Now that the magnetic flux is defined, we can be more precise about the experimental observations described in Section 23-1. In particular, Faraday found that the secondary coil in Figure 23-1 experiences an induced emf only when the *magnetic flux* through it changes with time. Furthermore, the induced emf for a given loop is found to be proportional to the *rate* at which the flux changes with time, $\Delta\Phi/\Delta t$. If there are N loops in a coil, each with the same magnetic flux, Faraday found that the induced emf is given by the following relationship:

Faraday's Law of Induction

$$\varepsilon = -N\frac{\Delta\Phi}{\Delta t} = -N\frac{\Phi_{final} - \Phi_{initial}}{t_{final} - t_{initial}} \qquad \text{23-3}$$

In this expression, known as **Faraday's law of induction,** the minus sign in front of N indicates that the induced emf opposes the change in magnetic flux, as we show in the next section. When we are concerned only with magnitudes, as is often the case, we use the form of Faraday's law given next:

Magnitude of Induced emf

$$|\varepsilon| = N\left|\frac{\Delta\Phi}{\Delta t}\right| = N\left|\frac{\Phi_{final} - \Phi_{initial}}{t_{final} - t_{initial}}\right| \qquad \text{23-4}$$

Faraday's law gives the *emf* that is induced in a circuit or a loop of wire. The current that is induced as a result of the emf depends on the characteristics of the circuit itself—for example, how much resistance it contains.

EXERCISE 23-3   TIME FOR A CHANGE

The induced emf in a single loop of wire has a magnitude of 1.35 V when the magnetic flux is changed from $0.99\ \text{T}\cdot\text{m}^2$ to $0.26\ \text{T}\cdot\text{m}^2$. How much time is required for this change in flux?

REASONING AND SOLUTION
A "single loop of wire" is one that has a single "turn"; that is, $N = 1$. Using this value in $|\varepsilon| = N|\Delta\Phi/\Delta t|$, and rearranging to solve for the elapsed time, we find

$$|\Delta t| = N\frac{|\Delta\Phi|}{|\varepsilon|} = (1)\frac{|0.26\ \text{T}\cdot\text{m}^2 - 0.99\ \text{T}\cdot\text{m}^2|}{1.35\ \text{V}} = 0.54\ \text{s}$$

In the next Example, we consider a system in which the magnetic field varies in both magnitude and direction over the area of a coil. We can still calculate the magnetic flux in such a case if we simply replace the perpendicular component of the magnetic field ($B\cos\theta$) with its average value, $(B\cos\theta)_{av}$. With this substitution, the magnetic flux, $\Phi = BA\cos\theta$, becomes $\Phi = (B\cos\theta)_{av}A$.

Big Idea 2 The induced emf is directly proportional to the rate of change of the magnetic flux.

PHYSICS IN CONTEXT
Looking Ahead

The detailed connections between electric and magnetic fields developed in Chapters 22 and 23 are central to understanding electromagnetic waves (radio waves, microwaves, visible light, X-rays, etc.), as described in Chapter 25.

EXAMPLE 23-4   BAR MAGNET INDUCTION

A bar magnet is moved rapidly toward a 40-turn circular coil of wire. As the magnet moves, the average value of $B\cos\theta$ over the area of the coil increases from 0.0125 T to 0.450 T in 0.250 s. If the radius of the coil is 3.05 cm, and the resistance of its wire is 3.55 Ω, find the magnitude of **(a)** the induced emf and **(b)** the induced current.

PICTURE THE PROBLEM
The motion of the bar magnet relative to the coil is shown in our sketch. As the magnet approaches the coil, the average value of the perpendicular component of the magnetic field increases. As a result, the magnetic flux through the coil increases with time.

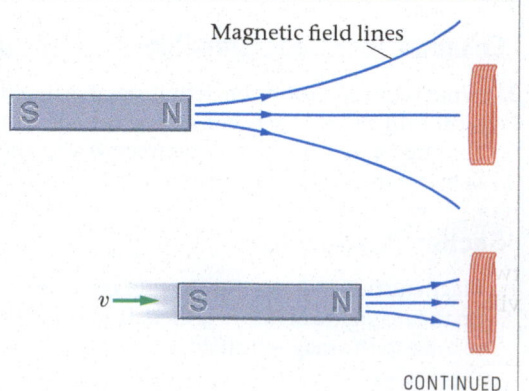

Magnetic field lines

CONTINUED

REASONING AND STRATEGY
To find the induced emf and the induced current we must first calculate the rate of change of the magnetic flux. To find the flux, we use $\Phi = B_{av}A$, where $A = \pi r^2$ and B_{av} is the average value of $B\cos\theta$. Once we have found the induced emf using Faraday's law, $|\varepsilon| = N|\Delta\Phi/\Delta t|$, we obtain the induced current using Ohm's law, $I = V/R$, with $V = |\varepsilon|$.

Known   Number of turns in coil, $N = 40$; area of coil, $A = \pi(0.0305\text{ m})^2$; resistance of coil, $R = 3.55\ \Omega$; average initial value of $B\cos\theta$, $(B\cos\theta)_{av,i} = 0.0125\ T$; average final value of $B\cos\theta$, $(B\cos\theta)_{av,f} = 0.450\ T$; elapsed time, $\Delta t = 0.250$ s.

Unknown   (a) Induced emf $= ?$ (b) Induced current, $I = ?$

SOLUTION

Part (a)

1. Calculate the initial and final values of the magnetic flux through the coil:

$$\Phi_i = (B\cos\theta)_{av,i}A$$
$$= (0.0125\text{ T})\pi(0.0305\text{ m})^2 = 3.65 \times 10^{-5}\text{ T}\cdot\text{m}^2$$
$$\Phi_f = (B\cos\theta)_{av,f}A$$
$$= (0.450\text{ T})\pi(0.0305\text{ m})^2 = 1.32 \times 10^{-3}\text{ T}\cdot\text{m}^2$$

2. Use Faraday's law to find the magnitude of the induced emf:

$$|\varepsilon| = N\left|\frac{\Phi_f - \Phi_i}{\Delta t}\right|$$
$$= (40)\left|\frac{1.32 \times 10^{-3}\text{ T}\cdot\text{m}^2 - 3.65 \times 10^{-5}\text{ T}\cdot\text{m}^2}{0.250\text{ s}}\right|$$
$$= 0.205\text{ V}$$

Part (b)

3. Use Ohm's law to calculate the induced current:

$$I = \frac{V}{R} = \frac{0.205\text{ V}}{3.55\ \Omega} = 0.0577\text{ A}$$

INSIGHT
If the magnet is now pulled back to its original position in the same amount of time, the induced emf and current will have the same magnitudes; their directions will be reversed, however.

PRACTICE PROBLEM
How many turns of wire would be required in this coil to give an induced current of at least 0.100 A?
[**Answer:** $N = 70$]

Some related homework problems: Problem 10, Problem 20, Problem 21

RWP* An everyday application of Faraday's law of induction can be found in a type of microphone known as a *dynamic microphone*. These devices employ a stationary permanent magnet and a wire coil attached to a movable diaphragm, as illustrated in **FIGURE 23-5**. When a sound wave strikes the microphone, it causes the diaphragm to oscillate, moving the coil alternately closer to and farther from the magnet. This movement changes the magnetic flux through the coil, which in turn produces an induced emf. Connecting the coil to an amplifier increases the magnitude of the induced emf enough that it can power a set of large speakers. The same principle is employed in the *seismograph*, shown in **FIGURE 23-6 (a)**, except in this case the oscillations that produce the induced emf are generated by earthquakes and the vibrations they send through the ground.

   RWP The American pop-jazz guitarist Les Paul (1915–2009) applied the same basic physics to musical instruments when he made the first solid-body *electric guitar* in 1941. A typical electric guitar is shown in **FIGURE 23-6 (b)**. The pickup is simply a small permanent magnet with a coil wrapped around it, as shown in **FIGURE 23-7**. This magnet produces a field that is strong enough to produce a magnetization in the steel guitar string, which is the moving part in the system. When the string is plucked, the oscillating string changes the magnetic flux in the coil, inducing an emf that can be amplified. A typical electric guitar (or should we say, magnetic guitar) will have two or three sets of pickups, each positioned to amplify a different harmonic of the vibrating strings.

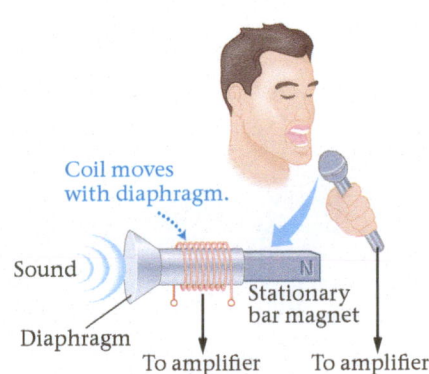

Coil moves with diaphragm.

Sound

Diaphragm

Stationary bar magnet

To amplifier   To amplifier

▲ **FIGURE 23-5 A dynamic microphone**

---

*Real World Physics applications are denoted by the acronym RWP.

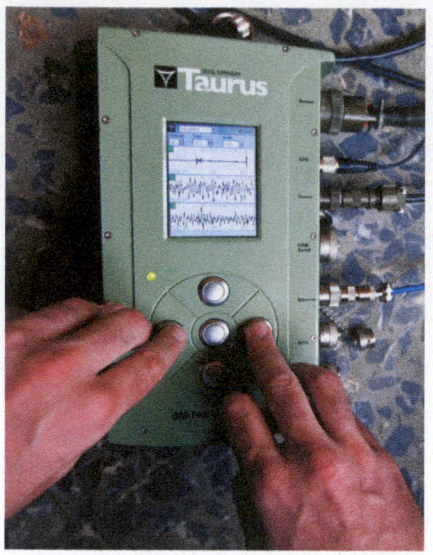

(a)

(b)

▲ **FIGURE 23-6 Visualizing Concepts Faraday's Law** Many devices take advantage of Faraday's law to translate physical oscillations into electrical impulses. In all such devices, the oscillations are used to change the relative position of a coil and a magnet, thus causing the electric current in the coil to vary. **(a)** The vibrations produced by an earthquake or underground explosion produce an oscillating current in a seismograph that can be amplified and used to drive a plotting pen. **(b)** The magnetic pickups located under the strings of an electric guitar function in much the same way. In this case, plucking a string (which is magnetized by the pickup) causes it to vibrate and generate an oscillating electrical signal in a pickup. This signal is amplified and fed to speakers, which produce most of the sound we hear.

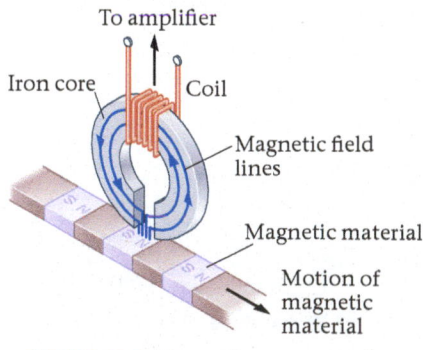

▲ **FIGURE 23-7 The pickup on an electric guitar**

RWP Faraday's law of induction also applies to the operation of computer disk drives, credit card readers at grocery stores, audio tape players, and ATM machines. In these devices, information is recorded on a magnetic material by using an electromagnet to produce regions of opposite magnetic polarity, as indicated in **FIGURE 23-8**. When the information is to be "read out," the magnetic material is moved past the gap in an iron core that is wrapped with many turns of a wire coil. As magnetic regions of different polarity pass between the poles of the magnet, its field is alternately increased or decreased—resulting in a changing magnetic flux through the coil. The corresponding induced emf is sent to electronic circuitry to be translated into information a computer can process.

▲ **FIGURE 23-8 Magnetic tape recording**

Enhance Your Understanding   (Answers given at the end of the chapter)

3. In system 1 the magnetic flux through a coil with 2 turns changes from $1 \text{ T} \cdot \text{m}^2$ to $11 \text{ T} \cdot \text{m}^2$ in 5 s; in system 2 the magnetic flux through a coil with 3 turns changes from $1 \text{ T} \cdot \text{m}^2$ to $2 \text{ T} \cdot \text{m}^2$ in 0.5 s. Is the magnitude of the induced emf in system 1 greater than, less than, or equal to the magnitude of the induced emf in system 2? Explain.

Section Review

*   Faraday's law relates the rate of change of magnetic flux to the induced emf as follows: $\mathcal{E} = -N(\Delta\Phi/\Delta t)$. The minus sign is the subject of the next section. In terms of magnitudes we can write Faraday's law as $|\mathcal{E}| = N|\Delta\Phi/\Delta t|$.

23-4 Lenz's Law

In this section we discuss **Lenz's law,** a physical way of expressing the meaning of the minus sign in Faraday's law of induction. The basic idea of Lenz's law, first stated by the Estonian physicist Heinrich Lenz (1804–1865), is the following:

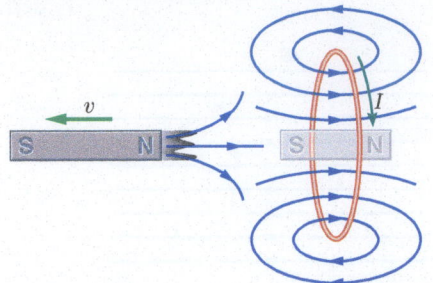

Magnetic field lines

(a) Moving magnet toward coil induces a field that repels the magnet

(b) Moving magnet away from coil induces a field that attracts the magnet

◀ **FIGURE 23-9 Applying Lenz's law to a magnet moving toward and away from a current loop** **(a)** If the north pole of a magnet is moved toward a conducting loop, the induced current produces a north pole pointing toward the magnet's north pole. This creates a repulsive force opposing the change that caused the current. **(b)** If the north pole of a magnet is pulled away from a conducting loop, the induced current produces a south magnetic pole near the magnet's north pole. The result is an attractive force opposing the motion of the magnet.

Lenz's Law
An induced current always flows in a direction that *opposes* the change that caused it.

This general physical principle is closely related to energy conservation, as we shall see at several points in our discussion. The remainder of this section, however, is devoted to specific applications of Lenz's law.

As a first example, consider a bar magnet that is moved toward a conducting ring, as in **FIGURE 23-9 (a)**. If the north pole of the magnet approaches the ring, a current is induced that tends to oppose the motion of the magnet. To be specific, the current in the ring creates a magnetic field that has a north pole (that is, diverging field lines) facing the north pole of the magnet, as indicated in the figure. There is now a repulsive force acting on the magnet, opposing its motion.

On the other hand, if the magnet is pulled away from the ring, as in **FIGURE 23-9 (b)**, the induced current creates a south pole facing the north pole of the magnet. The resulting attractive force again opposes the magnet's motion. We can obtain the same results using Faraday's law directly, but Lenz's law gives a simpler, more physical way to approach the problem.

Big Idea **3** A changing magnetic flux produces an electric current that opposes the change.

CONCEPTUAL EXAMPLE 23-5   FALLING MAGNETS   PREDICT/EXPLAIN

The magnets shown in the sketch are dropped from rest through the middle of conducting rings. Notice that the ring on the right has a small break in it, whereas the ring on the left forms a closed loop. **(a)** As the magnets drop toward the rings, do you predict that the magnet on the left has an acceleration that is greater than, less than, or the same as that of the magnet on the right? **(b)** Which of the following is the *best explanation* for your prediction?

  I. Both magnets fall with the acceleration due to gravity. The rings have no effect.

  II. The magnet on the left has a greater downward acceleration because the ring exerts an attractive force on it.

  III. The magnet on the right has the greater downward acceleration because the gap prevents a current that would exert a repulsive force on the magnet.

REASONING AND DISCUSSION

As the magnet on the left approaches the ring, it induces a circulating current. According to Lenz's law, this current produces a magnetic field that exerts a repulsive force on the magnet—to oppose its motion. In contrast, the ring on the right has a break, so it cannot have a circulating current. As a result, it exerts no force on its magnet. Therefore, the magnet on the right falls with the acceleration of gravity; the magnet on the left falls with a smaller acceleration.

ANSWER

(a) The magnet on the left has an acceleration that is less than that of the magnet on the right. **(b)** The best explanation is III.

In the examples given so far in this section, the "change" involved the motion of a bar magnet, but Lenz's law applies no matter how the magnetic flux is changed. Suppose, for example, that a magnetic field is decreased with time, as illustrated in **FIGURE 23-10**. In this case the change is a decrease of magnetic flux through the ring.

▶ **FIGURE 23-10 Lenz's law applied to a decreasing magnetic field** As the magnetic field is decreased, in going from **(a)** to **(b)**, the induced current produces a magnetic field that passes through the ring in the same direction as $\vec{B}$.

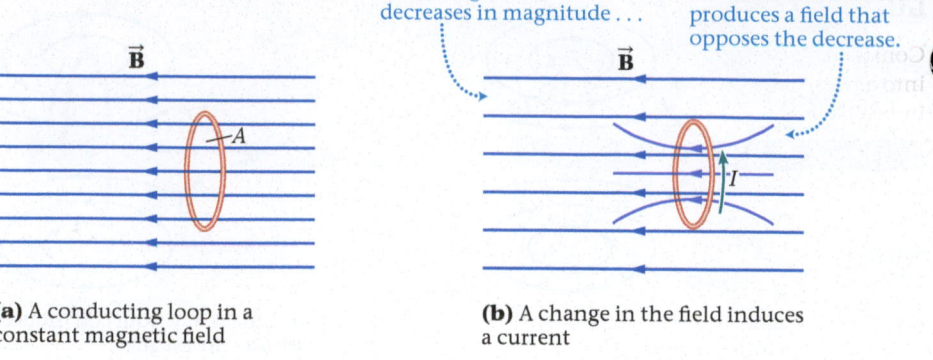

If the magnetic field decreases in magnitude . . .   . . . the induced current produces a field that opposes the decrease.

(a) A conducting loop in a constant magnetic field

(b) A change in the field induces a current

The induced current flows in a direction that tends to oppose this change; that is, the induced current must produce a field within the ring that is in the same direction as the decreasing $\vec{B}$ field, as we see in Figure 23-10 (b).

Motional emf: Qualitative

Next, we consider the physical situation shown in **FIGURE 23-11**. Here we see a metal rod that is free to move in the vertical direction, and a constant magnetic field that points out of the page. The rod is in frictionless contact with two vertical wires; hence, a current can flow in a loop through the rod, the wires, and the lightbulb. It is the motion of the rod that produces an emf in this system, and hence it is referred to as **motional emf.** The question to be answered at this point is this: What is the direction of the current when the rod is released from rest and allowed to fall?

In this system the magnetic field is constant, but the magnetic flux through the current loop, $\Phi = BA$, decreases anyway. The reason is that as the rod falls, the area A enclosed by the loop decreases. To oppose this decrease in flux, Lenz's law implies that the induced current must flow in a direction that strengthens the magnetic field $\vec{B}$ within the loop. This direction is counterclockwise, as indicated in **FIGURE 23-12**.

Notice that the current in the rod flows from right to left. This direction is also in agreement with Lenz's law, because it opposes the change that caused it. To be specific, the change for the rod is that it begins to fall *downward*. In order to oppose this change, the current through the rod must produce an *upward* magnetic force. As we see in Figure 23-12, using the right-hand rule, a current from right to left does just that.

When the rod is first released, there is no current and its downward acceleration is the acceleration of gravity—the rod speeds up as its gravitational potential energy is converted to kinetic energy. As the rod falls, the induced current begins to flow, however, and the rod is acted on by the upward magnetic force. This causes its acceleration to decrease. Ultimately, the magnetic force equals the force of gravity in magnitude, and the rod's acceleration vanishes. From this point on it falls with constant speed and hence constant kinetic energy. Now you may wonder what happens to the rod's gravitational potential energy. In fact, all the rod's gravitational potential energy is now being converted to heat and light in the lightbulb—none of it goes to increasing the kinetic energy of the rod.

Imagine for a moment the consequences that would follow if Lenz's law were not obeyed. If the current in the rod were from left to right, for example, the magnetic force would be downward, causing the rod to increase its downward acceleration. As the rod sped up, the induced current would become even larger and the downward force would increase as well. Thus, the rod would simply move faster and faster, without limit. Clearly, energy would not be conserved in such a case. The connections between mechanical and electrical energy are explored in greater detail in the next section.

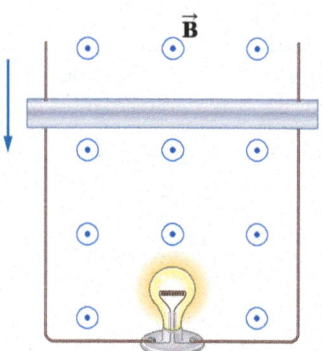

▲ **FIGURE 23-11 Motional emf** Motional emf is created in this system as the rod falls. The result is an induced current, which causes the light to shine.

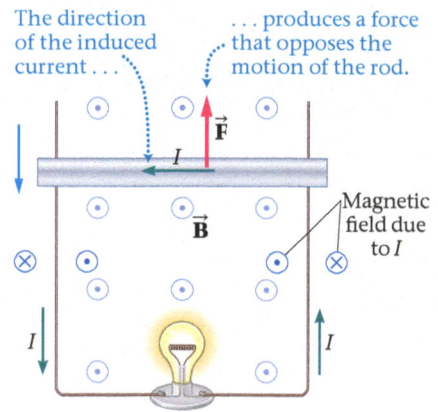

The direction of the induced current . . .   . . . produces a force that opposes the motion of the rod.

Magnetic field due to I

◀ **FIGURE 23-12 Determining the direction of an induced current** The direction of the current induced by the rod's downward motion is counterclockwise, because this direction produces a magnetic field within the loop that points out of the page—in the same direction as the original field. Notice that a current flowing in this direction through the rod interacts with the original magnetic field to give an upward force, opposing the downward motion of the rod.

CONCEPTUAL EXAMPLE 23-6   THE DIRECTION OF INDUCED CURRENT

Consider a system in which a metal ring is falling out of a region with a magnetic field and into a field-free region, as shown in our sketch to the right. According to Lenz's law, is the induced current in the ring as it exits the field region clockwise or counterclockwise?

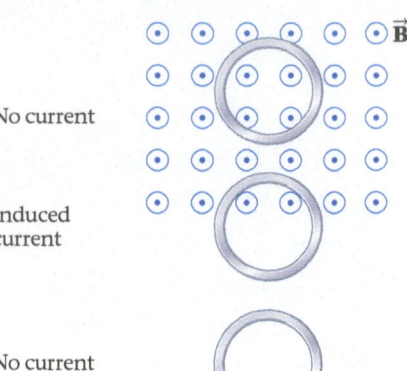

REASONING AND DISCUSSION

The induced current must be in a direction that opposes the change in the system. In this case, the change is that fewer magnetic field lines are piercing the area of the loop and pointing out of the page. The induced current can oppose this change by generating more field lines out of the page *within the loop*. As shown on the left in the figure below, the induced current must be counterclockwise to accomplish this.

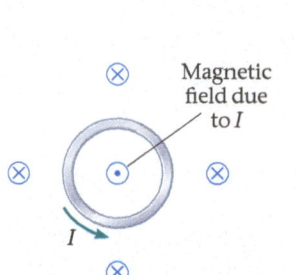

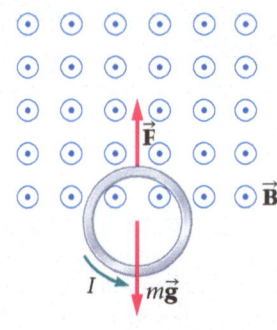

Finally, on the right in the same figure, notice that the induced current generates an upward magnetic force at the top of the ring, but no magnetic force on the bottom, where the magnetic field is zero. Hence, the motion of the ring is retarded as it drops out of the field—again, an example of Lenz's law.

ANSWER

The induced current is counterclockwise.

Eddy Currents and Magnetic Braking   The retarding effect on a ring leaving a magnetic field, as just described in Conceptual Example 23-6, allows us to understand the behavior of **eddy currents.** Suppose, for example, that a sheet of metal falls from a region with a magnetic field to a region with no field, as in **FIGURE 23-13**. In the portion of the sheet that is just leaving the field, a circulating current—an eddy current—is induced in the metal, just like the circulating current set up in the ring the Conceptual Example. As in the case of the ring, the current will retard the motion of the sheet of metal, somewhat like a frictional force.

The frictionlike effect of eddy currents is the basis for the phenomenon known as *magnetic braking*. An important advantage of this type of braking is that no direct physical contact is needed, thus eliminating frictional wear. In addition, the force of magnetic braking is stronger, the greater the speed of the metal with respect to the magnetic field. In contrast, kinetic friction is independent of the relative speed of the surfaces. Magnetic braking is used in everything from fishing reels to exercise bicycles to roller coasters.

RWP Magnetic braking is also the operating principle behind the common analog *speedometer*. In these devices, a cable is connected at one end to a gear in the car's transmission and at the other end to a small permanent magnet. When the car is under way, the magnet rotates within a metal cup. The resulting "magnetic friction" between the rotating magnet and the metal cup—which is proportional to the speed of rotation—causes the cup to rotate in the same direction. A hairspring attached to the cup stops the rotation when the force of the spring equals the force of magnetic friction. Thus, the cup—and the needle attached to it—rotate through an angle that is proportional to the speed of the car.

RWP Finally, eddy currents are also used in the kitchen. In an *induction stove*, a metal coil is placed just beneath the cooking surface. When an alternating current is sent through the coil, it sets up an alternating magnetic field that, in turn, induces eddy currents in

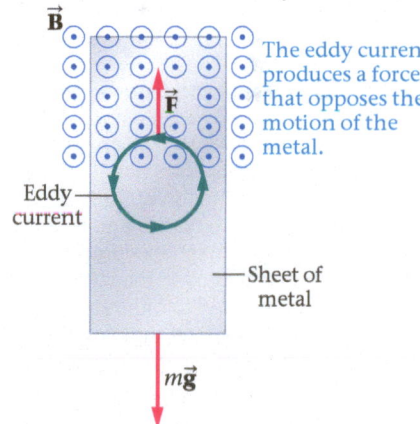

The eddy current produces a force that opposes the motion of the metal.

▲ **FIGURE 23-13 Eddy currents** A circulating current is induced in a sheet of metal near where it leaves a region with a magnetic field. This "eddy current" is similar to the current induced in the ring described in Conceptual Example 23-6. In both cases, the induced current exerts a retarding force on the moving object, opposing its motion.

Eddy Currents in Different Metals

nearby metal objects. The finite resistance of a metal pan, for instance, will result in heating as the eddy current dissipates power according to the relationship $P = I^2R$. Nonmetallic objects, like the surface of the stove and glassware, remain cool to the touch because they are poor electrical conductors and thus have negligible eddy currents.

Enhance Your Understanding (Answers given at the end of the chapter)

4. A metal ring moves to the right from a field-free region into a region with a magnetic field pointing into the page, as shown in **FIGURE 23-14**. As the ring begins to enter the region with the magnetic field, is the induced current in the ring clockwise, counterclockwise, or zero? Explain.

Section Review

- Lenz's law states that induced currents always flow in a direction that *opposes* the change that caused them.

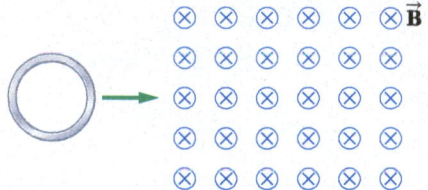

▲ **FIGURE 23-14** Enhance Your Understanding 4.

23-5 Mechanical Work and Electrical Energy

To see how mechanical work can be converted to electrical energy, we turn to a detailed calculation of the forces and currents involved in a simple physical system. This analysis will form the basis for our understanding of electric generators and motors in the next section.

Motional emf: Quantitative

We again consider motional emf, though this time in a somewhat simpler system where gravity plays no role. As shown in **FIGURE 23-15**, our setup consists of a rod that slides horizontally without friction on a U-shaped wire connected to a lightbulb of resistance R. A constant magnetic field points out of the page, and the rod is pushed by an external agent—your hand, for instance—so that it moves to the right with a constant speed v.

▶ **FIGURE 23-15 Force and induced current** A conducting rod slides without friction on horizontal wires in a region where the magnetic field is uniform and pointing out of the page. The motion of the rod to the right induces a clockwise current and a corresponding magnetic force to the left. An external force of magnitude $F = B^2v\ell^2/R$ is required to offset the magnetic force and to keep the rod moving with a constant speed v.

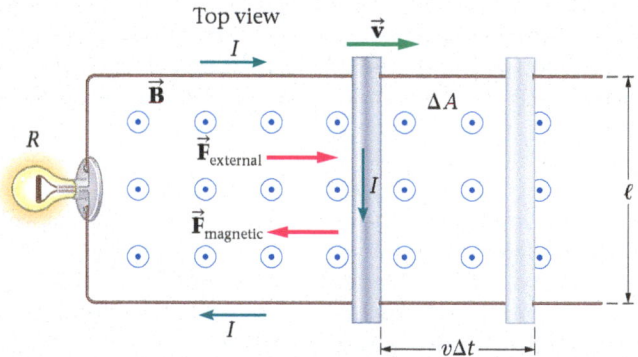

Let's begin by calculating the rate at which the magnetic flux changes with time. Because $\vec{B}$ is perpendicular to the area of the loop, it follows that $\Phi = BA$. Now the area increases with time because of the motion of the rod. In fact, the rod moves a horizontal distance $v\Delta t$ in the time Δt; therefore, the area of the loop increases by the amount $\Delta A = (v\Delta t)\ell$, as indicated in Figure 23-15. As a result, the increase in magnetic flux is

$$\Delta\Phi = B\,\Delta A = Bv\ell\,\Delta t$$

With this result in hand, we can use Faraday's law to calculate the magnitude of the induced emf. We find that for this single-turn loop,

$$|\varepsilon| = N\left|\frac{\Delta\Phi}{\Delta t}\right| = (1)\frac{Bv\ell\,\Delta t}{\Delta t} = Bv\ell \qquad \text{23-5}$$

As one might expect, the induced emf is proportional to the strength of the magnetic field and to the speed of the rod.

Now, let's calculate the electric field caused by the motion of the rod. First, we know that the emf from one end of the rod to the other is $\mathcal{E} = B v \ell$, where $\mathcal{E}$ is an electric potential difference measured in volts. Next, recall that the difference in electric potential caused by an electric field E over a length ℓ is $V = E\ell$. Applying these results to the emf in the rod, we have $E\ell = B v \ell$, or

$$E = Bv \qquad\qquad 23\text{-}6$$

Thus, a changing magnetic flux does indeed create an electric field, as Faraday suspected. In particular, moving through a magnetic field B at a speed v causes an electric field of magnitude $E = Bv$ in a direction that is at right angles to both the velocity and the magnetic field.

EXERCISE 23-7 FIND THE SPEED

As the rod in Figure 23-15 moves through a 0.321-T magnetic field, it experiences an induced electric field of 0.899 V/m. How fast is the rod moving?

REASONING AND SOLUTION
Solving $E = Bv$ for the speed v, we find

$$v = \frac{E}{B} = \frac{0.899 \text{ V/m}}{0.321 \text{ T}} = 2.80 \text{ m/s}$$

If we assume that the only resistance in the system is the lightbulb, R, it follows that the current in the circuit has the following magnitude:

$$I = \frac{|\mathcal{E}|}{R} = \frac{B v \ell}{R} \qquad\qquad 23\text{-}7$$

Using Lenz's law, we can determine that the direction of the current is clockwise, as indicated in Figure 23-15. Next, we show how both the magnitude and direction of the current play a key role in the energy aspects of this system.

Mechanical and Electrical Power

We begin with a calculation of the force and power required to keep the rod moving with a constant speed v. First, notice that the rod carries a current of magnitude $I = B v \ell / R$ at right angles to a magnetic field of strength B. The length of the rod is ℓ, and thus it follows that the magnetic force exerted on it has a magnitude

$$F = I\ell B = \left(\frac{B v \ell}{R}\right)(\ell)B = \frac{B^2 v \ell^2}{R} \qquad\qquad 23\text{-}8$$

The direction of the current in the rod is downward on the page in Figure 23-15; hence, the magnetic force *RHR* shows that $\vec{F}_{\text{magnetic}}$ is directed to the left. As a result, an external force, $\vec{F}_{\text{external}}$, must act to the right with equal magnitude to maintain the rod's constant speed. The mechanical power delivered by the external force is simply the force times the speed:

$$P_{\text{mechanical}} = Fv = \left(\frac{B^2 v \ell^2}{R}\right)v = \frac{B^2 v^2 \ell^2}{R} \qquad\qquad 23\text{-}9$$

Let's compare this mechanical power with the electrical power converted to light and heat in the lightbulb. Recalling that the power dissipated in a resistor with a resistance R is $P = I^2 R$, we have

$$P_{\text{electrical}} = I^2 R = \left(\frac{B v \ell}{R}\right)^2 R = \frac{B^2 v^2 \ell^2}{R} \qquad\qquad 23\text{-}10$$

This is *exactly* the same as the mechanical power. Thus, with the aid of a magnetic field, we can convert *mechanical* power directly to *electrical* power. This simple example of motional emf illustrates the basic principle behind the generation of virtually all the world's electrical energy.

**PHYSICS
IN CONTEXT
Looking Back**

Work, energy, and power (Chapters 7 and 8) are discussed in relation to electrical energy in this section. These concepts are also important in understanding the operation of electric motors and generators in Section 23-6.

EXAMPLE 23-8 LIGHT POWER

The lightbulb in the circuit shown in the sketch has a resistance of 22 Ω and consumes 5.0 W of power; the rod is 1.25 m long and moves to the left with a constant speed of 3.1 m/s. **(a)** What is the strength of the magnetic field? **(b)** What external force is required to maintain the rod's constant speed?

PICTURE THE PROBLEM

Our sketch shows the rod being pushed to the left by the force $\vec{F}_{external}$. The motion of the rod, in turn, generates an electric current whose direction, according to Lenz's law, produces a magnetic force, $\vec{F}_{magnetic}$, that opposes the motion of the rod. We know that the rod moves with constant speed, and therefore the applied force $\vec{F}_{external}$ exactly cancels the magnetic force $\vec{F}_{magnetic}$.

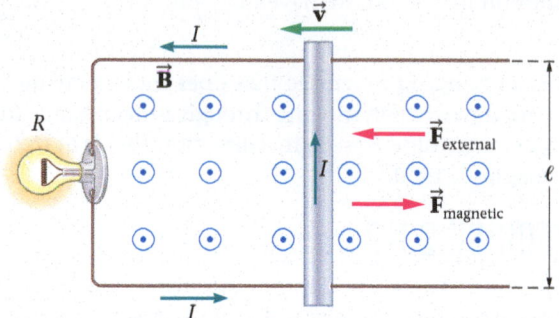

REASONING AND STRATEGY

a. The power consumed by the lightbulb is given by the expression $P = B^2 v^2 \ell^2 / R$. This relationship can be solved to give the strength of the magnetic field, B.

b. The magnitude of the external force is given by $F = B^2 v \ell^2 / R$, where B has the value obtained in part (a). Alternatively, we can solve for the force using the power relationship $P = Fv$.

Known   Resistance, $R = 22\ \Omega$; power, $P = 5.0$ W; length of rod, $\ell = 1.25$ m; speed of rod, $v = 3.1$ m/s.
Unknown **(a)** Magnetic field, $B = ?$ **(b)** External force, $F_{external} = ?$

SOLUTION

Part (a)

1. Write the power expression for this case, then rearrange to solve for the magnetic field, B:

$$P = \frac{B^2 v^2 \ell^2}{R}$$

$$B = \frac{\sqrt{PR}}{v\ell}$$

2. Substitute numerical values:

$$B = \frac{\sqrt{PR}}{v\ell} = \frac{\sqrt{(5.0\ \text{W})(22\ \Omega)}}{(3.1\ \text{m/s})(1.25\ \text{m})} = 2.7\ \text{T}$$

Part (b)

3. Calculate the magnitude of the external force by substituting numerical values into $F = B^2 v \ell^2 / R$:

$$F_{external} = \frac{B^2 v \ell^2}{R}$$

$$= \frac{(2.7\ \text{T})^2 (3.1\ \text{m/s})(1.25\ \text{m})^2}{22\ \Omega} = 1.6\ \text{N}$$

4. As a double check, we can find the magnitude of the external force by rearranging the power relationship $P = Fv$:

$$F_{external} = \frac{P}{v} = \frac{5.0\ \text{W}}{3.1\ \text{m/s}} = 1.6\ \text{N}$$

INSIGHT

A constant external *force* is required to move the rod with constant *speed*, and the magnitude of the external force is proportional to the speed. This is similar to the behavior of a "drag" force like air resistance. In addition, notice that the *net* force acting on the rod is zero—as expected for an object that moves with constant speed.

PRACTICE PROBLEM

How fast will the rod move if the external force exerted on it is 1.4 N?   **[Answer:** $v = 2.7$ m/s**]**

Some related homework problems: Problem 38, Problem 39, Problem 40

QUICK EXAMPLE 23-9   FIND THE WORK CONVERTED TO LIGHT ENERGY

Referring to Example 23-8, find **(a)** the work done by the external force in 0.58 s and **(b)** the energy consumed by the lightbulb in the same time.

REASONING AND SOLUTION

(a) In general, work is equal to force times distance, $W = Fd$. We know the force in this case ($F = 1.6$ N), and we can find the distance from velocity times the time, $d = vt$. **(b)** Recall that power is energy per time. Thus, we can find the energy by multiplying power times the time; that is, energy $= Pt$.

CONTINUED

Part (a)

1. Find the distance covered by the rod in 0.58 s:

$$d = vt = (3.1 \text{ m/s})(0.58 \text{ s}) = 1.8 \text{ m}$$

2. Find the work done using $W = Fd$:

$$W = Fd = (1.6 \text{ N})(1.8 \text{ m}) = 2.9 \text{ J}$$

Part (b)

3. Calculate the energy used in the lightbulb by multiplying the power by the time:

$$\text{energy} = Pt = (5.0 \text{ W})(0.58 \text{ s}) = 2.9 \text{ J}$$

As expected, the energy used by the lightbulb is equal to the work done by the external force as it pushes the rod against the magnetic force.

RWP Motional emfs are generated whenever a magnet moves relative to a conductor or when a conductor moves relative to a magnetic field. For example, *magnetic antitheft devices* in libraries and stores detect the motional emf produced by a small magnetic strip placed in a book or other item. If this strip is not demagnetized, it will trigger an alarm as it moves past a sensitive set of conductors designed to pick up the induced emf.

RWP A similar effect describes the operating principle of a bicycle speedometer. A magnet mounted on a wheel spoke passes near a coil of wire mounted on the bike's frame, which induces a current in the coil. By measuring the time between current pulses, and knowing the circumference of the bicycle wheel (the user must supply this information), the speedometer can determine the speed of the bicycle.

BIO In numerous biomedical applications of the same principle, it is the conductor that moves while the magnetic field remains stationary. For example, researchers have tracked the head and body movements of several flying insects, including blowflies, hover flies, and honeybees. They attach lightweight, flexible wires to a small metal coil on the insect's head, and another on its thorax, and then allow the insect to fly in a stationary magnetic field. As the coils move through the field, they experience induced emfs that can be analyzed by a computer to determine the corresponding orientation of the head and thorax.

BIO The same technique is used by researchers investigating the movements of the human eye. You may not be aware of it, but your eyes are jiggling back and forth rather rapidly at this very moment. This constant motion, referred to as **saccadic motion,** prevents an image from appearing at the same position on the retina for a period of time. Without saccadic motion, our brains would tend to ignore stationary images that we are focused on in favor of moving images of little interest. To study the dynamics of saccadic motion, a small coil of wire is attached to a modified contact lens. As the eye moves the coil through a stationary magnetic field, the induced motional emf gives the eye's orientation and rotational speed.

Enhance Your Understanding   (Answers given at the end of the chapter)

5. Suppose the speed of the rod in Example 23-8 is doubled. **(a)** By what factor does the magnitude of the external force change? **(b)** By what factor does the electrical power change?

Section Review

- Work must be done to move a conductor through a magnetic field. This work can be converted to other forms of energy, such as heat in a resistor or light in a lightbulb.

23-6 Generators and Motors

We've just seen that a changing magnetic flux can serve as a means of converting mechanical work to electrical energy. In this section we discuss this concept in greater detail as we explore the workings of an electric generator. We also point out that the energy conversion can run in the other direction as well, with electrical energy being converted to mechanical work in an electric motor.

Big Idea **4** A changing magnetic flux is the basis for the operation of electric generators and motors.

PROBLEM-SOLVING NOTE

Maximum and Minimum Values of Induced emf

Note that the induced emf of a rotating coil has a maximum value of $NBA\omega$ and a minimum value of $-NBA\omega$. This follows from Equation 23-11 and the fact that the maximum and minimum values of $\sin \omega t$ are $+1$ and -1, respectively.

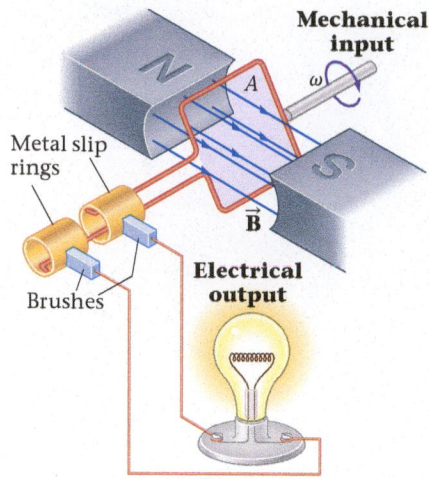

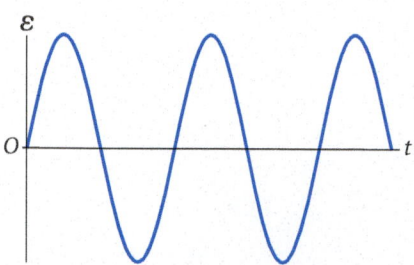

▲ **FIGURE 23-17 Induced emf of a rotating coil** Induced emf produced by an electric generator like the one shown in Figure 23-16. The emf alternates in sign, so the current changes sign as well. We say that a generator of this type produces an "alternating current."

◄ **FIGURE 23-16 An electric generator** The basic operating elements of an electric generator are shown in a schematic representation. As the coil is rotated by an external source of mechanical work, it produces an emf that can be used to power an electric circuit.

Electric Generators

RWP An electric **generator** is a device designed to efficiently convert mechanical energy to electrical energy. The mechanical energy can be supplied by any of a number of sources, such as falling water in a hydroelectric dam, expanding steam in a coal-fired power plant, or the output of a small gasoline motor that powers a portable generator. In all cases the basic operating principle is the same—mechanical energy is used to move a conductor through a magnetic field to produce a motional emf.

The linear motion of a metal rod through a magnetic field, as in Figure 23-15, results in motional emf, but to continue producing an electric current in this way the rod must move through ever greater distances. A practical way to employ the same effect is to use a coil of wire that can be *rotated* in a magnetic field. By rotating the coil, which in turn changes the magnetic flux, we can continue the process indefinitely.

Imagine, then, a coil of area A located in the magnetic field between the poles of a magnet, as illustrated in **FIGURE 23-16**. Metal rings are attached to either end of the wire that makes up the coil, and carbon brushes are in contact with the rings to allow the induced emf to be communicated to the outside world. As the coil rotates with the angular speed ω, the emf produced in it is given by Faraday's law, $\mathcal{E} = -N(\Delta\Phi/\Delta t)$. Referring to the similar case of uniform circular motion in Chapter 13, and Equation 13-6 in particular, we can show that the explicit expression for the emf of a rotating coil is as follows:

$$\mathcal{E} = NBA\omega \sin \omega t \qquad \text{23-11}$$

This result is plotted in **FIGURE 23-17**. Notice that the induced emf in the coil alternates in sign, which means that the current in the coil alternates in direction. For this reason, this type of generator is referred to as an **alternating-current generator** or, simply, an ac generator. Electrical energy production with large-scale ac generators is shown in **FIGURE 23-18**.

▲ **FIGURE 23-18 Visualizing Concepts Generating Electricity** Hydroelectric power provides an excellent example of energy transformations. First, the gravitational potential energy of water held behind a dam is converted to kinetic energy as the water descends through pipes in the generating station (left). Next, the kinetic energy of the falling water is used to spin turbines (center). Each turbine rotates the coil of a generator in a strong magnetic field, thus converting mechanical energy into an electric current in the coil (right). Eventually the electrical energy is delivered to homes, factories, and cities, where it is converted to light (in incandescent bulbs and fluorescent tubes), heat (in devices such as toasters and broilers), and work (in various machines).

EXAMPLE 23-10 GENERATOR NEXT

The coil in an electric generator has 125 turns and an area of 2.42×10^{-3} m^2. It is desired that the maximum emf of the coil be 169 V when it rotates at the rate of 60.0 cycles per second. **(a)** What is the angular speed of the generator? **(b)** Find the strength of the magnetic field B that is required for this generator to operate at the desired voltage.

PICTURE THE PROBLEM

In our sketch, a square conducting coil of area A and angular speed ω rotates between the poles of a magnet, where the magnetic field has a strength B. As the coil rotates, the magnetic flux through it changes continuously. The result is an induced emf that varies sinusoidally with time.

REASONING AND STRATEGY

a. We need to find the angular speed, ω, since it is not given directly in the problem statement. What we are given, however, is that the frequency of the coil's rotation is $f = 60.0$ cycles per second $= 60.0$ Hz. The corresponding angular speed is $\omega = 2\pi f$. (See Chapter 13.)

b. The induced emf of the generator is given by $\mathcal{E} = NBA\omega \sin \omega t$. Therefore, the maximum emf occurs when $\sin \omega t$ has its maximum value of 1. It follows that $\mathcal{E}_{max} = NBA\omega$. This relationship can be solved for the magnetic field, B.

Known   Number of turns, $N = 125$; area of coil, $A = 2.42 \times 10^{-3}$ m^2; rotation frequency, $f = 60.0$ Hz; maximum emf, $\mathcal{E}_{max} = 169$ V.

Unknown   **(a)** Angular speed of generator, $\omega = $? **(b)** Magnetic field, $B = $?

SOLUTION

Part (a)

1. Calculate the angular speed of the coil:

$$\omega = 2\pi f = 2\pi(60.0 \text{ Hz}) = 377 \text{ rad/s}$$

Part (b)

2. Solve for the magnetic field in terms of the maximum emf:

$$\mathcal{E}_{max} = NBA\omega \quad \text{or} \quad B = \frac{\mathcal{E}_{max}}{NA\omega}$$

3. Substitute numerical values into the expression for B:

$$B = \frac{\mathcal{E}_{max}}{NA\omega} = \frac{169 \text{ V}}{(125)(2.42 \times 10^{-3} \text{ m}^2)(377 \text{ rad/s})} = 1.48 \text{ T}$$

INSIGHT

The emf of a generator coil depends on its area A but not on its shape. In addition, each turn of the coil generates the same induced emf; therefore, the total emf of the coil is directly proportional to the number of turns. Finally, the more rapidly the coil rotates, the more rapidly the magnetic flux through it changes. As a result, the induced emf is also proportional to the angular speed of the generator.

PRACTICE PROBLEM

What is the maximum emf of this generator if its rotation frequency is reduced to 50.5 Hz, while keeping the magnetic field the same?   [**Answer: 142 V**]

Some related homework problems: Problem 41, Problem 42

RWP All electric generators convert mechanical energy to electrical energy. Most generators operate as indicated in Figure 23-16, but other types are possible as well. For example, some generators use the mechanical energy of a shaking motion to produce electrical energy by making a magnet move back and forth inside a coil of wire. The basic situation is similar to that depicted in Figure 23-2. As the magnet moves inside the coil, an electric current is produced because of the changing magnetic flux. Circuitry then captures the electrical energy to charge a capacitor or recharge a battery. This type of generator has been marketed in "shake" flashlights (**FIGURE 23-19 (a)**), which are great for emergency situations because they never need batteries. Similarly, some wristwatches never need replacement batteries because they can use the mechanical energy of everyday motion as a source of electrical energy (**FIGURE 23-19 (b)**).

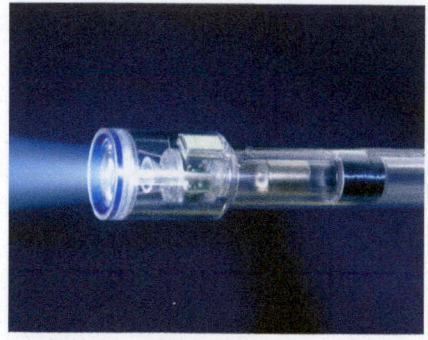

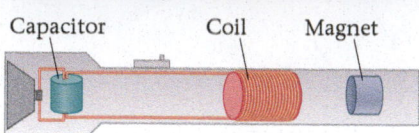

Capacitor   Coil   Magnet

(a)

◄ **FIGURE 23-19 Motion electric generators** Some devices convert the kinetic energy of a user's motion into electrical energy by causing a magnet to move relative to a nearby coil of wire. **(a)** A "shake" flashlight (also known as a Faraday flashlight) uses a shaking motion to power the light. **(b)** Some watches use the shaking motion of everyday movements to keep their battery charged.

Electric Motors

RWP The basic physical principle behind the operation of electric motors has already been discussed. In particular, we saw in Section 22-5 that a current-carrying loop in a magnetic field experiences a torque that tends to make it rotate. If such a loop is mounted on an axle, as in **FIGURE 23-20**, it is possible for the magnetic torque to be applied to the external world.

To see just how this works in practice, notice that the torque exerted on the loop at the moment shown in Figure 23-20 causes it to rotate clockwise *toward* the vertical position. Once it reaches this orientation, and continues past it due to its angular momentum, the alternating current from the electrical input reverses direction. This reverses the torque on the loop and causes the loop to rotate *away* from the vertical—which means it is still rotating in the clockwise sense. The next time the loop becomes vertical, the current again reverses, causing the loop to continue rotating clockwise. The result is an axle continually turning in the same direction.

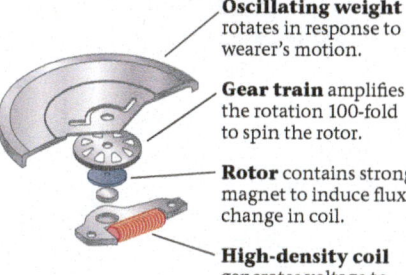

Oscillating weight rotates in response to wearer's motion.

Gear train amplifies the rotation 100-fold to spin the rotor.

Rotor contains strong magnet to induce flux change in coil.

High-density coil generates voltage to recharge the watch battery.

(b)

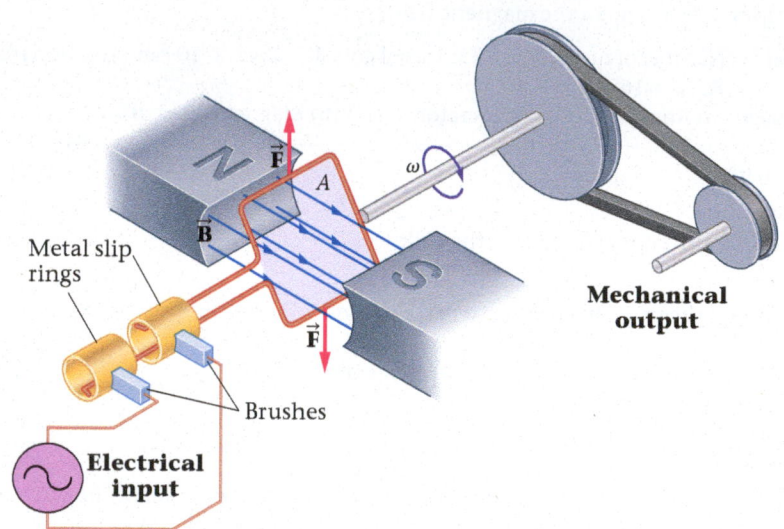

Metal slip rings

Brushes

Electrical input

Mechanical output

▲ **FIGURE 23-20 A simple electric motor** An electric current causes the coil of a motor to rotate and deliver mechanical work to the outside world.

What we have just described is precisely the reverse of a generator—instead of doing work to turn a coil and produce an electric current, as in a generator, a motor uses an electric current to produce rotation in a coil, which then does work. Thus, a motor, in some sense, is a generator run in reverse. In fact, it is possible to have one electric motor turn the axle of another identical motor and thereby produce electricity from the "motor"-turned generator.

RWP Similarly, if a car is powered by an electric motor, its motor can double as a generator when it helps to slow the car during braking. This results in a more efficient mode of transportation, because some of the car's kinetic energy that is normally converted to heat during braking can instead be used to recharge the batteries and extend the range of the car. Many of the newer "hybrid" cars make use of this "energy recovery" technology.

Enhance Your Understanding   (Answers given at the end of the chapter)

6. Consider the electric generator shown in Figure 23-16. **(a)** Is the magnetic flux through the loop a maximum or a minimum when the plane of the loop is perpendicular to the

CONTINUED

magnetic field? **(b)** Is the induced emf produced by the generator a maximum or a minimum when the plane of the loop is perpendicular to the magnetic field?

Section Review

- The induced emf of a rotating coil is given by $\mathcal{E} = NBA\omega \sin \omega t$. In this expression, N is the number of turns in the coil, A is its area, B is the strength of the magnetic field, and ω is the angular frequency of the coil.

23-7 Inductance

We began this chapter with a description of Faraday's induction experiment, in which a changing current in one coil induces a current in another coil. This type of interaction between coils is referred to as **mutual inductance.** It is also possible, however, for a single, isolated coil to have a similar effect—that is, a coil with a changing current can induce an emf, and hence a current, in *itself*. This type of process is known as **self-inductance.**

To be more precise, consider a coil of wire in a region free of magnetic fields, as in **FIGURE 23-21 (a).** Initially the current in the coil is zero, so the magnetic flux through the coil is zero as well. The current in the coil is now steadily increased from zero, as indicated in **FIGURE 23-21 (b).** As a result, the magnetic flux through the coil increases with time, which, according to Faraday's law, induces an emf in the coil.

The direction of the induced emf is given by Lenz's law, which says that the induced current opposes the change that caused it. In this case, the change is the increasing current in the coil; hence, the induced emf is in a direction opposite to the current. This situation is illustrated in **FIGURE 23-22,** where we schematically show the effect of the induced emf (often referred to as a "back" emf) in the circuit. Notice that in order to increase the current in the coil, it is necessary to *force* the current against an opposing emf.

This is a rather interesting result. It says that even if the wires in a coil are ideal, with no resistance whatsoever, work must still be done to increase the current. Conversely, if a current already exists in a coil and it decreases with time, Lenz's law ensures that the self-induced current will oppose this change as well, in an attempt to keep the current constant. Hence, a coil tends to resist changes in its current, regardless of whether the change is an increase or a decrease.

Defining Inductance To make a quantitative connection between a changing current and the associated self-induced emf, we start with Faraday's law in magnitude form:

$$|\mathcal{E}| = N\left|\frac{\Delta\Phi}{\Delta t}\right|$$

Now, since the magnetic field is proportional to the current in a coil, it follows that the magnetic flux Φ is also proportional to I. As a result, the time rate of change of the magnetic flux, $\Delta\Phi/\Delta t$, must be proportional to the time rate of change of the current, $\Delta I/\Delta t$. Therefore, the induced emf can be written as follows:

Definition of Inductance, L

$$|\mathcal{E}| = N\left|\frac{\Delta\Phi}{\Delta t}\right| = L\left|\frac{\Delta I}{\Delta t}\right| \qquad \text{23-12}$$

SI unit: $1 \text{ V} \cdot \text{s/A} = 1 \text{ henry} = 1 \text{ H}$

This expression defines the constant of proportionality, L, which is referred to as the **inductance** of a coil. Rearranging Equation 23-12, we see that

$$L = N\left|\frac{\Delta\Phi}{\Delta I}\right|$$

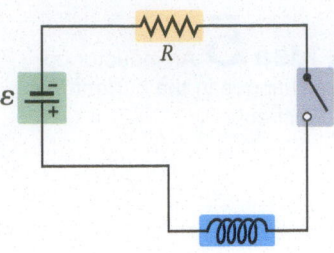

(a) No current, no field in coil

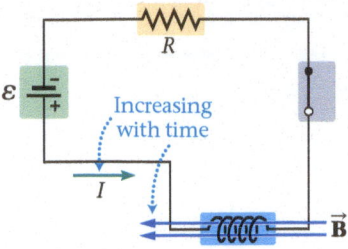

(b) Increasing current produces increasing magnetic field in coil

▲ **FIGURE 23-21 A changing current in an inductor (a)** A coil of wire with no current and no magnetic flux. **(b)** The current is now increasing with time, which produces a magnetic flux that also increases with time.

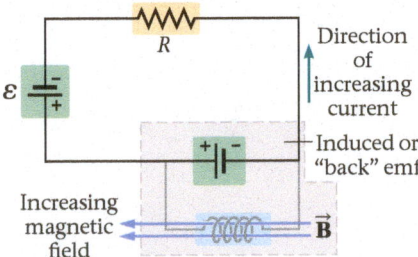

▲ **FIGURE 23-22 The back emf of an inductor** The effect of an increasing current in a coil is an induced emf that opposes the increase. This is indicated schematically by replacing the coil with the opposing, or "back," emf.

Big Idea **5** An inductor opposes changes in the current in an electric circuit.

The SI unit of inductance is the **henry (H)**, named to honor the work of Joseph Henry (1797–1878), an American physicist. As can be seen from Equation 23-12, the henry is one volt-second per amp:

$$1\,H = 1\,V \cdot s/A \qquad \text{23-13}$$

Common inductances are generally in the millihenry (mH) range.

EXERCISE 23-11   FIND THE INDUCTANCE

The coil in an electromagnet carries a constant direct current of 3.90 A. When a switch is suddenly opened, the current drops to zero over a time of 2.50×10^{-3} s = 2.50 ms. If the magnitude of the average induced emf during this time is 8.10 V, what is the inductance of the coil?

REASONING AND SOLUTION

Solving $|\mathcal{E}| = L|\Delta I/\Delta t|$ for the inductance L, and using only magnitudes, we find

$$L = |\mathcal{E}|\left|\frac{\Delta t}{\Delta I}\right| = (8.10\,V)\frac{(2.50 \times 10^{-3}\,s)}{3.90\,A} = 5.19 \times 10^{-3}\,H = 5.19\,mH$$

Inductance of a Solenoid   Of particular interest is the inductance of an ideal solenoid with N turns and length ℓ. For example, suppose the initial current in a solenoid is zero and the final current is I. It follows that the initial magnetic flux is zero. The final flux is simply $\Phi_f = BA\cos\theta$, where, according to Equation 22-12, the magnetic field in the solenoid is $B = \mu_0(N/\ell)I$. Finally, we know that the field in a solenoid is perpendicular to its cross-sectional area, A; thus $\theta = 0$ and $\Phi_f = BA$. Combining these results, we obtain

$$L = \frac{N(BA - 0)}{(I - 0)} = \frac{N[\mu_0(N/\ell)I]A}{I} = \mu_0\left(\frac{N^2}{\ell}\right)A$$

Denoting the number of turns per length as $n = N/\ell$, we can express the inductance of a solenoid in the following forms:

Inductance of a Solenoid

$$L = \mu_0\left(\frac{N^2}{\ell}\right)A = \mu_0 n^2 A\ell \qquad \text{23-14}$$

Notice that doubling the number of turns per length quadruples the inductance. In addition, the inductance is proportional to the volume of a solenoid, $V = A\ell$.

In the following Examples we consider problems involving solenoids and their inductance.

EXAMPLE 23-12   SOLENOID SELF-INDUCTION

A 500-turn solenoid is 8.0 cm long. When the current in this solenoid is increased from 0 to 2.5 A in 0.35 s, the magnitude of the induced emf is 0.012 V. Find **(a)** the inductance and **(b)** the cross-sectional area of the solenoid.

PICTURE THE PROBLEM

The solenoid described in the problem statement is shown in our sketch. Notice that it is 8.0 cm in length and contains 500 turns.

500 turns

8.0 cm

REASONING AND STRATEGY

a. We can use the definition of inductance, $|\mathcal{E}| = L|\Delta I/\Delta t|$, to solve for the inductance.

b. Once we know L from part (a), we can find the cross-sectional area from $L = \mu_0 n^2 A\ell$, where $n = N/\ell$.

Known   Number of turns, $N = 500$; change in current, $\Delta I = 2.5$ A; elapsed time, $\Delta t = 0.35$ s; induced emf, $\mathcal{E} = 0.012$ V; length of solenoid, $\ell = 8.0$ cm.

Unknown   **(a)** Inductance, $L = ?$ **(b)** Cross-sectional area of solenoid, $A = ?$

CONTINUED

SOLUTION

Part (a)

1. Using magnitudes, solve $|\varepsilon| = L|\Delta I/\Delta t|$ for the inductance:

$$L = \frac{|\varepsilon|}{|\Delta I/\Delta t|} = \frac{0.012 \text{ V}}{(2.5 \text{ A}/0.35 \text{ s})} = 1.7 \text{ mH}$$

Part (b)

2. Solve $L = \mu_0 n^2 A \ell$ for the cross-sectional area of the solenoid:

$$A = \frac{L}{\mu_0 n^2 \ell} = \frac{L\ell}{\mu_0 N^2}$$

3. Substitute numerical values:

$$A = \frac{(1.7 \times 10^{-3} \text{ H})(0.080 \text{ m})}{(4\pi \times 10^{-7} \text{ T} \cdot \text{m/A})(500)^2} = 4.3 \times 10^{-4} \text{ m}^2$$

INSIGHT

This cross-sectional area corresponds to a radius of about 1.2 cm. Therefore, it would take about 37 m of wire to construct this inductor.

PRACTICE PROBLEM

Find the inductance and induced emf for a solenoid with a cross-sectional area of $1.3 \times 10^{-3} \text{ m}^2$. The solenoid, which is 9.0 cm long, is made from a piece of wire that is 44 m in length. Assume that the rate of change of current remains the same. [**Answer:** $L = 2.2 \text{ mH}$, $|\varepsilon| = 0.016 \text{ V}$]

Some related homework problems: Problem 49, Problem 50

QUICK EXAMPLE 23-13 FIND THE NUMBER OF TURNS

The inductance of a solenoid 5.00 cm long is 0.157 mH. If its cross-sectional area is $1.00 \times 10^{-4} \text{ m}^2$, how many turns does the solenoid have?

REASONING AND SOLUTION

We can use the basic relationship for the inductance of a solenoid, $L = \mu_0(N^2/\ell)A$, to solve for the one unknown, the number of turns, N.

1. Rearrange $L = \mu_0(N^2/\ell)A$ to solve for the number of turns, N:

$$N = (L\ell/\mu_0 A)^{1/2}$$

2. Substitute numerical values:

$$N = \left(\frac{(0.157 \times 10^{-3} \text{ H})(0.0500 \text{ m})}{(4\pi \times 10^{-7} \text{ T} \cdot \text{m/A})(1.00 \times 10^{-4} \text{ m}^2)} \right)^{1/2}$$
$$= 250$$

A coil with a finite inductance—an **inductor** for short—resists changes in its electric current. Similarly, a particle of finite mass has the property of inertia, by which it resists changes in its velocity. These analogies, between inductance and mass on the one hand and current and velocity on the other, are useful ones, as we shall see in greater detail when we consider the behavior of ac circuits in the next chapter.

Enhance Your Understanding   (Answers given at the end of the chapter)

7. Rank the following solenoids in order of increasing inductance. Indicate ties where appropriate.

Solenoid	A	B	C	D
N	100	200	300	400
A (m^{-4})	2	4	1	10
ℓ (cm)	5	10	15	20

Section Review

- Inductance L is defined by the following relationship: $|\varepsilon| = L|\Delta I/\Delta t|$. An inductance resists changes in the current in an electric circuit.

- The inductance of a solenoid with a cross-sectional area A, length ℓ, and number of turns per length n, is $L = \mu_0 n^2 A \ell$. Equivalently, we can write $L = \mu_0(N^2/\ell)A$, where N is the total number of turns in the solenoid.

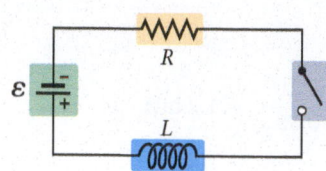

▲ **FIGURE 23-23** **An *RL* circuit** The characteristic time for an *RL* circuit is $\tau = L/R$. After the switch has been closed for a time much greater than τ, the current in the circuit is simply $I = \mathcal{E}/R$. That is, when the current is steady, the inductor behaves like a zero-resistance wire.

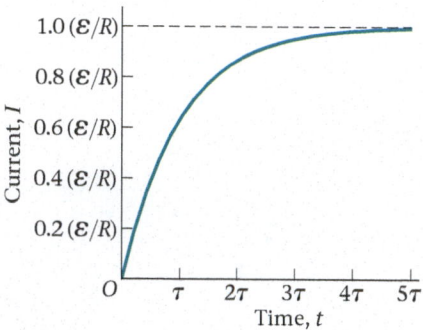

▲ **FIGURE 23-24** **Current as a function of time in an *RL* circuit** Notice that the current approaches $\mathcal{E}/R$ after several characteristic time intervals, $\tau = L/R$, have elapsed.

PROBLEM-SOLVING NOTE

Opening or Closing a Switch in an *RL* Circuit

When the switch is first closed in an *RL* circuit, the inductor has a back emf equal to that of the battery. Long after the switch is closed, current flows freely through the inductor, with no resistance at all.

PHYSICS
IN CONTEXT
Looking Ahead

Inductors are important elements in any electric circuit, but especially in alternating current (ac) circuits. This is discussed in detail in Chapter 24.

23-8 *RL* Circuits

The circuit shown in **FIGURE 23-23** is referred to as an **RL circuit**; it consists of a resistor, *R*, and an inductor, *L*, in series with a battery of emf $\mathcal{E}$. Notice that the inductor is represented by a symbol reminiscent of a coil with several loops, as shown here:

We assume the inductor is ideal, which means that the wires forming its loops have no resistance—the only resistance in the circuit is the resistor *R*. It follows, then, that after the switch has been closed for a long time, the current in the circuit is simply $I = \mathcal{E}/R$.

You may have noticed that we said the current would be $\mathcal{E}/R$ after the switch had been closed "for a long time." Why did we make this qualification? The point is that because of the inductor, and its tendency to resist changes in the current, it is not possible for the current to rise immediately to its final value. The inductor slows the process, causing the current to rise over a finite period of time, just as the charge on a capacitor builds up over the characteristic time $\tau = RC$ in an *RC* circuit.

The corresponding characteristic time for an *RL* circuit is given by

$$\tau = \frac{L}{R} \qquad 23\text{-}15$$

As expected, the larger the inductance, the longer the time required for the current to build up. In addition, the current approaches its final value of $\mathcal{E}/R$ with an exponential time dependence, much like the case of an *RC* circuit. With the aid of calculus, it can be shown that the current as a function of time is given by

$$I = \frac{\mathcal{E}}{R}\left(1 - e^{-t/\tau}\right) = \frac{\mathcal{E}}{R}\left(1 - e^{-tR/L}\right) \qquad 23\text{-}16$$

Notice the similarity between this expression, which is plotted in **FIGURE 23-24**, and the expression given in Equation 21-18 for an *RC* circuit.

EXERCISE 23-14   *RL* CIRCUIT

Find the resistance in the *RL* circuit of Figure 23-23, given that the inductance is 0.15 H and the characteristic time is 6.2 ms.

REASONING AND SOLUTION
Solving $\tau = L/R$ for the resistance, we find

$$R = \frac{L}{\tau} = \frac{0.15\ H}{6.2 \times 10^{-3}\ s} = 24\ \Omega$$

In the next Example we consider a circuit with one inductor and two resistors. By replacing the resistors with their equivalent resistance, however, we can still treat the circuit as a simple *RL* circuit.

EXAMPLE 23-15   *RL* IN PARALLEL

The circuit shown in the sketch consists of a 12-V battery, a 56-mH inductor, and two 150-Ω resistors in parallel. **(a)** Find the characteristic time for this circuit. What is the current in this circuit **(b)** one characteristic time interval after the switch is closed and **(c)** a long time after the switch is closed?

CONTINUED

PICTURE THE PROBLEM
The circuit in question is shown in diagram (a). Noting that two resistors R connected in parallel have an equivalent resistance equal to $R/2$, we can replace the original circuit with the equivalent RL circuit shown in diagram (b).

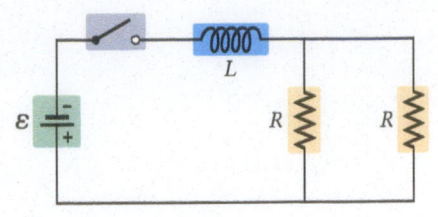

(a)

REASONING AND STRATEGY
a. We can find the characteristic time using $\tau = L/R_{eq}$, where $R_{eq} = R/2 = 75\ \Omega$.

b. To find the current after one characteristic time interval, we substitute $t = \tau$ in $I = (\mathcal{E}/R)(1 - e^{-t/\tau})$.

c. A long time after the switch has been closed $(t \to \infty)$, the exponential term in $I = (\mathcal{E}/R)(1 - e^{-t/\tau})$ is essentially zero; hence, the current in this case is simply $I = \mathcal{E}/R_{eq}$.

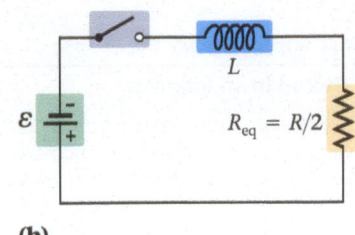

(b)

Known Emf, $\mathcal{E} = 12$ V; inductance, $L = 56 \times 10^{-3}$ H; resistance, $R = 150\ \Omega$.
Unknown (a) Characteristic time, $\tau = ?$ (b) Current after the time τ, $I = ?$, (c) Current after a long time, $I = ?$

SOLUTION

Part (a)

1. Calculate the characteristic time, $\tau = L/R_{eq}$, using 75 Ω for the equivalent resistance:

$$\tau = \frac{L}{R_{eq}} = \frac{56 \times 10^{-3}\,\text{H}}{75\ \Omega} = 7.5 \times 10^{-4}\,\text{s}$$

Part (b)

2. Substitute $t = \tau$ in $I = \dfrac{\mathcal{E}}{R}(1 - e^{-t/\tau})$:

$$I = \frac{\mathcal{E}}{R_{eq}}(1 - e^{-t/\tau}) = \left(\frac{12\,\text{V}}{75\ \Omega}\right)(1 - e^{-1}) = 0.10\,\text{A}$$

Part (c)

3. Substitute $t \to \infty$ in $I = \dfrac{\mathcal{E}}{R}(1 - e^{-t/\tau})$:

$$I = \frac{\mathcal{E}}{R_{eq}}(1 - e^{-t/\tau}) = \frac{\mathcal{E}}{R_{eq}}(1 - e^{-\infty}) = \frac{\mathcal{E}}{R_{eq}} = \frac{12\,\text{V}}{75\ \Omega} = 0.16\,\text{A}$$

INSIGHT
The current rises to almost two-thirds of its final value after just one characteristic time interval. It takes only a few more time intervals for the current to essentially "saturate" at its final value, $\mathcal{E}/R_{eq}$.

PRACTICE PROBLEM — PREDICT/CALCULATE
(a) If L is doubled to $2(56\ \text{mH}) = 112\ \text{mH}$, will the current after one characteristic time interval be greater than, less than, or the same as that found in part (b) of Example 23-15? Explain. (b) Find the current at $t = \tau$ in this case.
[**Answer:** (a) The current is the same when $t = \tau$ because $\mathcal{E}$ and R are unchanged. Notice that the value of τ is doubled, however, so it actually takes twice as much time for the current to rise to 0.10 A. (b) $I = 0.10$ A]

Some related homework problems: Problem 53, Problem 56

Enhance Your Understanding
(Answers given at the end of the chapter)

8. Consider the circuit shown in **FIGURE 23-25**. (a) Is the current supplied by the battery immediately after the switch is closed greater than, less than, or equal to the current it supplies a long time after the switch is closed? Explain. (b) What is the current supplied by the battery a long time after the switch is closed?

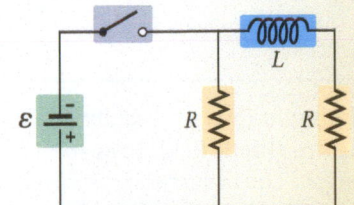

▲ **FIGURE 23-25** Enhance Your Understanding 8.

Section Review
- An inductor in a RL circuit slows the increase in current to its final value of $\mathcal{E}/R$. The current as a function of time is given by $I = (\mathcal{E}/R)(1 - e^{-t/\tau})$, where the characteristic time is $\tau = L/R$.

PHYSICS IN CONTEXT
Looking Ahead

A particular type of circuit containing an inductor—the *RLC* circuit—plays an important role in ac circuits in Chapter 24.

PHYSICS IN CONTEXT
Looking Back

We use our experience with *RC* circuits (Chapter 21) to gain a better understanding of *RL* circuits.

23-9 Energy Stored in a Magnetic Field

Work is required to establish a current in an inductor. For example, consider a circuit that consists of just two elements—a battery and an inductor. Even though there are no resistors in the circuit, the battery must still do electrical work to force charge to flow through the inductor, in opposition to its self-induced back emf. What happens to the energy expended to increase the current? We know that it can't be lost; after all, energy is always conserved. We also know that it hasn't been converted to thermal energy (dissipated to heat), because there is no resistance in the circuit. So where *did* the energy go? It turns out that the energy is stored in the magnetic field, just as the energy in a capacitor is stored in its electric field.

To see how this energy storage works, imagine increasing the current in an inductor L from $I_i = 0$ to $I_f = I$ in the time $\Delta t = T$. The magnitude of the emf required to accomplish this increase in current is

$$\mathcal{E} = L\frac{\Delta I}{\Delta t} = L\frac{I - 0}{T} = \frac{LI}{T} \qquad 23\text{-}17$$

During this time the average current is $I/2$, as indicated in **FIGURE 23-26**. Thus, the average power supplied by the emf, $P_{av} = I_{av}V$, is

$$P_{av} = \frac{1}{2}I\mathcal{E} = \frac{1}{2}\frac{LI^2}{T} \qquad 23\text{-}18$$

The total *energy* delivered by the emf is simply the average power times the total time; hence, the energy U required to increase the current in the inductor from 0 to I is

$$U = P_{av}T$$

Using the result given in Equation 23-18, we find:

Energy Stored in an Inductor

$$U = \frac{1}{2}LI^2 \qquad 23\text{-}19$$

SI unit: joule, J

Note the similarity of Equation 23-19 to $U = \frac{1}{2}CV^2$ for the energy stored in a capacitor.

EXERCISE 23-16   FIND THE INDUCTANCE

An inductor carrying a current of 3.9 A stores 0.45 J of energy. What is the inductance of this inductor?

REASONING AND SOLUTION

Solving $U = \frac{1}{2}LI^2$ for the inductance, we find

$$L = \frac{2U}{I^2} = \frac{2(0.45\text{ J})}{(3.9\text{ A})^2} = 0.059\text{ H}$$

Magnetic Energy Density   As mentioned earlier, the energy of an inductor is stored in its magnetic field. To be specific, consider a solenoid of length ℓ with n turns per unit length. The inductance of this solenoid, as determined in Section 23-7, is

$$L = \mu_0 n^2 A\ell$$

Therefore, the energy stored in the solenoid's magnetic field when it carries a current I is

$$U = \frac{1}{2}LI^2 = \frac{1}{2}(\mu_0 n^2 A\ell)I^2$$

Recalling that the magnetic field inside a solenoid has a magnitude given by $B = \mu_0 nI$, we can write the energy as follows:

$$U = \frac{1}{2\mu_0}B^2 A\ell$$

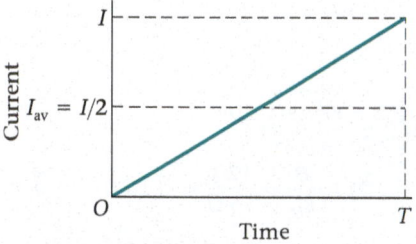

▲ **FIGURE 23-26 Current in an *RL* circuit with *R* = 0** In a circuit with only a battery and an inductor, the current rises linearly with time, because the characteristic time ($\tau = L/R$) with $R = 0$ is infinite. If the current after a time T is I, the average current during this time is $I_{av} = I/2$.

Big Idea 6 Inductors produce magnetic fields that store energy, in much the same way that capacitors produce electric fields that store energy.

PHYSICS IN CONTEXT
Looking Back

In Chapter 20 we introduced the concept of a capacitor (C), and showed how a capacitor stores energy in its electric field. Here we generalize this concept to inductors (L), which store energy in their magnetic fields.

Finally, we note that the volume where the magnetic field exists—that is, the volume inside the solenoid—is area times length, $A\ell$. Therefore, the magnetic energy per volume is

$$u_B = \frac{\text{magnetic energy}}{\text{volume}} = \frac{B^2}{2\mu_0} \qquad \text{23-20}$$

Although derived for the case of an ideal solenoid, this expression applies to any magnetic field no matter how it is generated.

Finally, recall that in the case of an electric field we found that the energy density was $u_E = \varepsilon_0 E^2/2$. Notice, in both cases, that the energy density depends on the magnitude of the field squared. We shall return to these expressions in the next chapter and again when we study electromagnetic waves, such as radio waves and light, in Chapter 25.

EXAMPLE 23-17   AN UNKNOWN RESISTANCE

After the switch in the circuit shown in the sketch has been closed for a long time, the energy stored in the inductor is 3.11×10^{-3} J. What is the value of the resistance R?

PICTURE THE PROBLEM
Our sketch shows the circuit for this system, with a 36.0-V battery, a 92.5-Ω resistor, and a 75.0-mH inductor all connected in series with an unknown resistor, R. Initially, the switch is open, but after it has been closed for a long time the current settles down to a constant value, as does the amount of energy stored in the inductor.

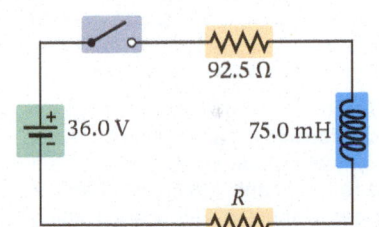

REASONING AND STRATEGY
First, the energy stored in the inductor, $U = \frac{1}{2}LI^2$, gives us the current in the circuit.

Second, we note that after the switch has been closed for a long time the current has a constant value. Thus, there is no opposing emf produced by the inductor. As a result, the final current is simply the emf divided by the equivalent resistance of two resistors in series: $I = \varepsilon/(92.5\ \Omega + R)$. This relationship can be used to solve for R.

Known   Energy in inductor, $U = 3.11 \times 10^{-3}$ J; inductance, $L = 75.0 \times 10^{-3}$ H; top resistor, 92.5 Ω; emf of battery, $\varepsilon = 36.0$ V.

Unknown   Unknown resistance, $R = ?$

SOLUTION

1. Use the energy stored in the inductor to solve for the current in the circuit:

$$U = \frac{1}{2}LI^2$$
$$I = \sqrt{\frac{2U}{L}}$$

2. Substitute numerical values:

$$I = \sqrt{\frac{2U}{L}} = \sqrt{\frac{2(3.11 \times 10^{-3}\ \text{J})}{75.0 \times 10^{-3}\ \text{H}}} = 0.288\ \text{A}$$

3. Use Ohm's law to solve for the resistance R:

$$I = \frac{\varepsilon}{92.5\ \Omega + R}$$
$$R = \frac{\varepsilon}{I} - 92.5\ \Omega$$

4. Substitute numerical values:

$$R = \frac{36.0\ \text{V}}{0.288\ \text{A}} - 92.5\ \Omega = 32.5\ \Omega$$

INSIGHT
As long as the switch is closed, electrical energy is dissipated in the resistors (converted to thermal energy) at a constant rate. On the other hand, *no* energy is dissipated in the inductor—it simply stores a constant amount of energy in the magnetic field.

PRACTICE PROBLEM — PREDICT/CALCULATE
(a) If R is increased, does the energy stored in the inductor increase, decrease, or stay the same? Explain. (b) Find the stored energy with $R = 106\ \Omega$.   [**Answer: (a)** With a larger R, the current is less; hence U is less. **(b)** With $R = 106\ \Omega$ we find $U = 1.23 \times 10^{-3}$ J.]

Some related homework problems: Problem 60, Problem 64

23-10 Transformers

In electrical applications it is often useful to be able to change the voltage from one value to another. For example, high-voltage power lines may operate at voltages as high as 750,000 V, but before the electrical power can be used in the home it must be "stepped down" to 120 V. Similarly, the 120 V from a wall socket may be stepped down again to 9 V or 12 V to power a portable CD player, or stepped up to give the 15,000 V needed in a television tube. The electrical device that performs these voltage conversions is called a **transformer.**

The basic idea of a transformer has already been described in connection with Faraday's induction experiment in Section 23-1. In fact, the device shown in Figure 23-1 can be thought of as a crude transformer. Basically, the magnetic flux created by a primary coil induces a voltage in a secondary coil. If the number of turns is different in the two coils, however, the voltages will be different as well.

To see how this works, consider the system shown in **FIGURE 23-27**. Here an ac generator produces an alternating current in the primary circuit of voltage V_p. The primary circuit then loops around an iron core with N_p turns. The iron core intensifies and concentrates the magnetic flux and ensures, at least to a good approximation, that the secondary coil experiences the same magnetic flux as the primary coil. The secondary coil has N_s turns around its iron core and is connected to a secondary circuit that may operate a CD player, a lightbulb, or some other device. We would now like to determine the voltage, V_s, in the secondary circuit.

To find this voltage, we apply Faraday's law of induction to each of the coils. For the primary coil we have

$$\mathcal{E}_p = -N_p \frac{\Delta\Phi_p}{\Delta t}$$

Similarly, the emf of the secondary coil is

$$\mathcal{E}_s = -N_s \frac{\Delta\Phi_s}{\Delta t}$$

Now, recall that the magnetic fluxes are equal, $\Delta\Phi_p = \Delta\Phi_s$, because of the iron core. As a result, we can divide these two equations (which cancels the flux) to find

$$\frac{\mathcal{E}_p}{\mathcal{E}_s} = \frac{N_p}{N_s}$$

Assuming that the resistance in the coils is negligible, so that the emf in each is essentially the same as its voltage, we obtain the following result:

Transformer Equation

$$\frac{V_p}{V_s} = \frac{N_p}{N_s}$$

23-21

Equation 23-21 is known as the **transformer equation.**

It follows that the voltage in the secondary circuit is simply

$$V_s = V_p \left(\frac{N_s}{N_p}\right)$$

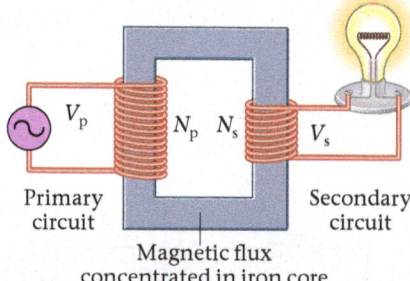

▲ **FIGURE 23-27 The basic elements of a transformer** An alternating current in the primary circuit creates an alternating magnetic flux and, hence, an alternating induced emf in the secondary circuit. The ratio of the emfs in the two circuits, V_s/V_p, is equal to the ratio of the number of turns in each circuit, N_s/N_p.

Big Idea **7** Transformers use a change in magnetic flux to produce a change in voltage in an electric circuit.

Therefore, if the number of turns in the secondary coil is less than the number of turns in the primary coil, the voltage is stepped down, and $V_s < V_p$. Conversely, if the number of turns in the secondary coil is higher, the voltage is stepped up, and $V_s > V_p$.

Now, before you begin to think that transformers give us something for nothing, we note that there is more to the story. Because energy must always be conserved, the average power in the primary circuit must be the same as the average power in the secondary circuit. Noting that power can be written as $P = IV$, it follows that

$$I_p V_p = I_s V_s$$

Therefore, the ratio of the current in the secondary coil to the current in the primary coil is:

Transformer Equation (Current and Voltage)

$$\frac{I_s}{I_p} = \frac{V_p}{V_s} = \frac{N_p}{N_s} \qquad 23\text{-}22$$

Thus, if the voltage is stepped up, the current is stepped down.

Suppose, for example, that the number of turns in the secondary coil is twice the number of turns in the primary coil. As a result, the transformer doubles the voltage in the secondary circuit, $V_s = 2 V_p$, and at the same time halves the current, $I_s = I_p/2$. In general, there is always a tradeoff between voltage and current in a transformer:

> If a transformer increases the voltage by a given factor, it decreases the current by the same factor. Similarly, if it decreases the voltage, it increases the current.

This behavior is similar to that of a lever, in which there is a tradeoff between the force that can be exerted and the distance through which it is exerted. In the case of a lever, as in the case of a transformer, the tradeoff is due to energy conservation: The work done on one end of the lever, $F_1 d_1$, must be equal to the work done on the other end of the lever, $F_2 d_2$.

PROBLEM-SOLVING NOTE

Energy Conservation in Transformers

Energy conservation requires that the product of I and V be the same for each coil of a transformer. Thus, as the voltage of the secondary goes up with the number of turns, the current decreases by the same factor.

QUICK EXAMPLE 23-18 FIND THE NUMBER OF TURNS ON THE SECONDARY

A common summertime sound in many backyards is the *zap* heard when an unfortunate insect flies into a high-voltage "bug zapper." Typically, such devices operate with voltages of about 4000 V obtained from a transformer plugged into a standard 120-V outlet. How many turns are on the secondary coil of the transformer if the primary coil has 27 turns?

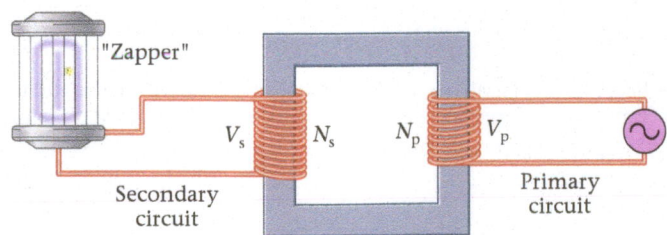

REASONING AND SOLUTION

The transformer in this case steps the voltage up from 120 V in the primary coil to 4000 V in the secondary coil. Given the number of turns in the primary coil ($N_p = 27$), we can use the transformer equation, $V_p/V_s = N_p/N_s$, to find the number of turns in the secondary coil, N_s.

1. Solve the transformer equation for the number of turns in the secondary coil:

$$N_s = N_p\left(\frac{V_s}{V_p}\right)$$

2. Substitute numerical values:

$$N_s = (27)\left(\frac{4000\text{ V}}{120\text{ V}}\right)$$

$$= 900$$

Of course, any transformer with a ratio of turns given by $N_s/N_p = 900/27$ will produce the same secondary voltage.

▶ **FIGURE 23-28 Visualizing Concepts**
Transmitting Electricity The transmission of electric power over long distances would not be feasible without transformers. A step-up transformer near the power plant (left) boosts the plant's output voltage from 12,000 V to the 240,000 V carried by high-voltage lines that crisscross the country (center). A series of step-down transformers near the destination reduce the voltage, first to 2400 V at local substations for distribution to neighborhoods, and then to the 240 V supplied to most houses. The familiar gray cylinders commonly seen on utility poles (right) are the transformers responsible for this last voltage reduction.

Notice that the operation of a transformer depends on a *changing* magnetic flux to create an induced emf in the secondary coil. If the current is constant in direction and magnitude, there is simply no induced emf. This is an important advantage that ac circuits have over dc circuits and is one reason that most electrical power systems today operate with alternating currents.

RWP Finally, transformers play an important role in the *transmission* of electrical energy from the power plants that produce it to the communities and businesses where it is used, as illustrated in **FIGURE 23-28**. As was pointed out in Equation 21-3, the resistance of a wire is directly proportional to its length. Therefore, when electrical energy is transmitted over a large distance, the finite resistivity of the wires that carry the current becomes significant. If the wire carries a current I and has a resistance R, the power dissipated as waste heat is $P = I^2 R$. One way to reduce this energy loss in a given wire is to reduce the current, I. A transformer that steps up the voltage of a power plant by a factor of 15 (from 10,000 V to 150,000 V, for example) at the same time reduces the current by a factor of 15 and reduces the energy dissipation by a factor of $15^2 = 225$. When the electrical energy reaches the location where it is to be used, step-down transformers can convert the voltage to levels (such as 120 V or 240 V) typically used in the home or workplace.

Enhance Your Understanding   (Answers given at the end of the chapter)

10. If a transformer doubled both the voltage and the current in a circuit, what physical principle would it violate?

Section Review

- The voltage and number of turns in the primary circuit are related to the voltage and number of turns in the secondary circuit by $V_p/V_s = N_p/N_s$. When a transformer increases the voltage by a given factor, it decreases the current by the same factor.

CHAPTER 23 REVIEW

CHAPTER SUMMARY

23-1 INDUCED ELECTROMOTIVE FORCE

A changing magnetic field can induce a current in a circuit. The driving force behind the current is referred to as the induced electromotive force.

Basic Features of Magnetic Induction
An induced current occurs when there is a *change* in the magnetic field. The magnitude of the induced current depends on the *rate* of change of the magnetic field.

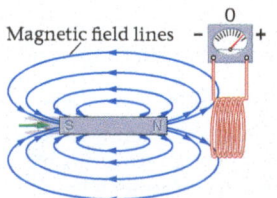

23-2 MAGNETIC FLUX

Magnetic Flux, Defined

If a magnetic field of strength B crosses an area A at an angle θ relative to the area's normal, the magnetic flux is

$$\Phi = (B \cos \theta)A = BA \cos \theta \qquad \text{23-1}$$

Units of Magnetic Flux

Magnetic flux is measured in webers (Wb), defined as follows:

$$1 \, \text{Wb} = 1 \, \text{T} \cdot \text{m}^2 \qquad \text{23-2}$$

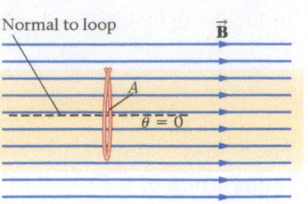

23-3 FARADAY'S LAW OF INDUCTION

Faraday's law of induction gives precise mathematical form to the basic features of magnetic induction discussed in Section 23-1.

Faraday's Law

If the magnetic flux in a coil of N turns changes by the amount $\Delta\Phi$ in the time Δt, the induced emf is

$$\varepsilon = -N \frac{\Delta\Phi}{\Delta t} = -N \frac{\Phi_{\text{final}} - \Phi_{\text{initial}}}{t_{\text{final}} - t_{\text{initial}}} \qquad \text{23-3}$$

Magnitude of the Induced emf

In many cases we are concerned only with the magnitude of the magnetic flux and the induced current. In such cases we use the following form of Faraday's law:

$$|\varepsilon| = N \left| \frac{\Delta\Phi}{\Delta t} \right| = N \left| \frac{\Phi_{\text{final}} - \Phi_{\text{initial}}}{t_{\text{final}} - t_{\text{initial}}} \right| \qquad \text{23-4}$$

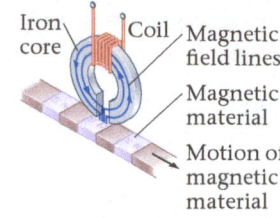

23-4 LENZ'S LAW

Lenz's law states that an induced current flows in the direction that opposes the change that caused the current.

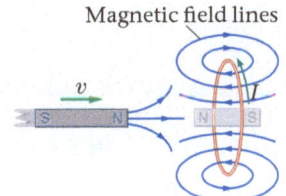

23-5 MECHANICAL WORK AND ELECTRICAL ENERGY

With the help of a magnetic field, mechanical work can be converted into electrical energy in the form of currents flowing through a circuit.

Motional emf

If a conductor of length ℓ is moved at right angles to a magnetic field of strength B with a speed v, the magnitude of the induced emf in the conductor is

$$|\varepsilon| = Bv\ell \qquad \text{23-5}$$

The corresponding electric field is

$$E = Bv \qquad \text{23-6}$$

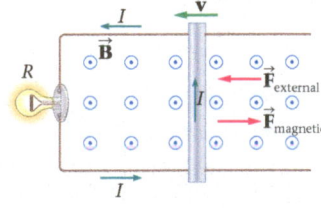

23-6 GENERATORS AND MOTORS

Generators and motors are practical devices for converting between electrical and mechanical energy.

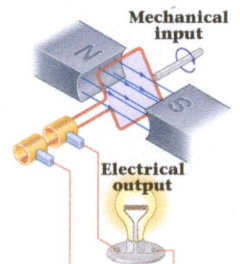

23-7 INDUCTANCE

Self-inductance, or inductance for short, is the effect produced when a coil with a changing current induces an emf in itself.

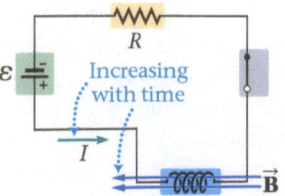

Inductance

Inductance, L, is defined as follows:

$$|\mathcal{E}| = L\left|\frac{\Delta I}{\Delta t}\right|$$ 23-12

The SI unit of inductance is the henry (H): $1\ H = 1\ V \cdot s/A$.

Inductance of a Solenoid

The inductance of a solenoid of length ℓ with N turns of wire and a cross-sectional area A is

$$L = \mu_0 n^2 A\ell = \mu_0\left(\frac{N^2}{\ell}\right)A$$ 23-14

Note that the number of turns per length is $n = N/\ell$.

23-8 *RL* CIRCUITS

A simple *RL* circuit consists of a battery, a switch, a resistor R, and an inductor L connected in series.

Characteristic Time

The characteristic time over which the current changes in an *RL* circuit is

$$\tau = \frac{L}{R}$$ 23-15

Increasing Current Versus Time

When the switch is closed in an *RL* circuit at time $t = 0$, the current increases with time as follows:

$$I = \frac{\mathcal{E}}{R}(1 - e^{-t/\tau}) = \frac{\mathcal{E}}{R}(1 - e^{-tR/L})$$ 23-16

In this expression $\mathcal{E}$ is the emf of the battery.

23-9 ENERGY STORED IN A MAGNETIC FIELD

Energy can be stored in a magnetic field, just as it is in an electric field.

Energy Stored in an Inductor

An inductor of inductance L carrying a current I stores the energy

$$U = \tfrac{1}{2}LI^2$$ 23-19

23-10 TRANSFORMERS

Transformer Equation

The equation relating the voltage V, current I, and number of turns N in the primary (p) and secondary (s) circuits of a transformer is

$$\frac{I_s}{I_p} = \frac{V_p}{V_s} = \frac{N_p}{N_s}$$ 23-22

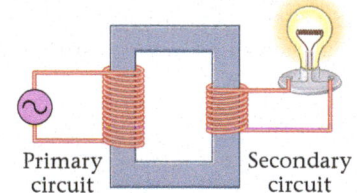

Primary   Secondary
circuit   circuit

ANSWERS TO ENHANCE YOUR UNDERSTANDING QUESTIONS

1. An induced emf is produced in (b), (c), and (d).

2. The angle θ in $\Phi = BA \cos \theta$ is described by (c).

3. The induced emf is greater in system 2 because the rate of change of the magnetic flux is the same in both systems, but system 2 has more turns in the coil.

4. The induced current in the ring is counterclockwise, which produces a magnetic field within the area of the ring that is out of the page, opposing the change caused by entering a field that points into the page.

5. (a) The magnitude of the force doubles. (b) The electrical power quadruples.

6. (a) The flux is a maximum. (b) The induced emf is a minimum.

7. $A < C < B < D$.

8. (a) The current immediately after the switch is closed is less than the current after a long time, because no current flows through the inductor initially. (b) The current supplied a long time after the switch is closed is $I = 2(\mathcal{E}/R)$.

9. Doubling the current will store more energy than doubling the inductance.

10. Doubling both current and voltage would violate energy conservation.

CONCEPTUAL QUESTIONS

For instructor-assigned homework, go to www.masteringphysics.com.

(Answers to odd-numbered Conceptual Questions can be found in the back of the book.)

1. Explain the difference between a magnetic field and a magnetic flux.

2. A metal ring with a break in its perimeter is dropped from a field-free region of space into a region with a magnetic field. What effect does the magnetic field have on the ring?

3. Many equal-arm balances have a small metal plate attached to one of the two arms. The plate passes between the poles of a magnet mounted in the base of the balance. Explain the purpose of this arrangement.

4. **FIGURE 23-29** shows a vertical iron rod with a wire coil of many turns wrapped around its base. A metal ring slides over the rod and rests on the wire coil. Initially the switch connecting the coil to a battery is open, but when it is closed, the ring flies into the air. Explain why this happens.

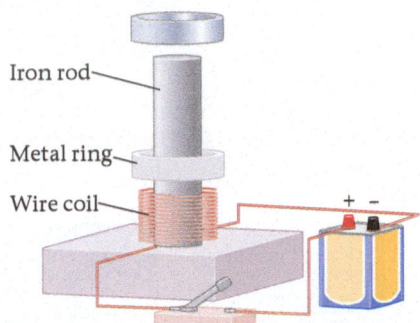

Iron rod

Metal ring

Wire coil

FIGURE 23-29 Conceptual Question 4

5. A metal rod of resistance R can slide without friction on two zero-resistance rails, as shown in **FIGURE 23-30**. The rod and the rails are immersed in a region of constant magnetic field pointing out of the page. Describe the motion of the rod when the switch is closed. Your discussion should include the effects of motional emf.

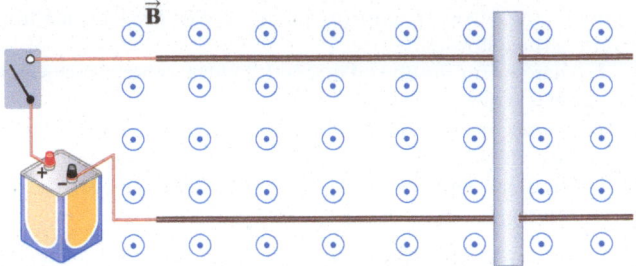

FIGURE 23-30 Conceptual Question 5

6. Recently, NASA tested a power generation system that involves connecting a small satellite to the space shuttle with a conducting wire several miles long. Explain how such a system can generate electrical power.

7. Explain what happens when the angular speed of the coil in an electric generator is increased.

PROBLEMS AND CONCEPTUAL EXERCISES

Answers to odd-numbered Problems and Conceptual Exercises can be found in the back of the book. **BIO** *identifies problems of biological or medical interest;* **CE** *indicates a conceptual exercise.* **Predict/Explain** *problems ask for two responses: (a) your prediction of a physical outcome, and (b) the best explanation among three provided; and* **Predict/Calculate** *problems ask for a prediction of a physical outcome, based on fundamental physics concepts, and follow that with a numerical calculation to verify the prediction. On all problems, bullets (•, ••, •••) indicate the level of difficulty.*

SECTION 23-2 MAGNETIC FLUX

1. • A 0.085-T magnetic field passes through a circular ring of radius 4.1 cm at an angle of 22° with the normal. Find the magnitude of the magnetic flux through the ring.

2. • A uniform magnetic field of 0.0250 T points vertically upward. Find the magnitude of the magnetic flux through each of the five sides of the open-topped rectangular box shown in **FIGURE 23-31**, given that the dimensions of the box are $L = 32.5$ cm, $W = 12.0$ cm, and $H = 10.0$ cm.

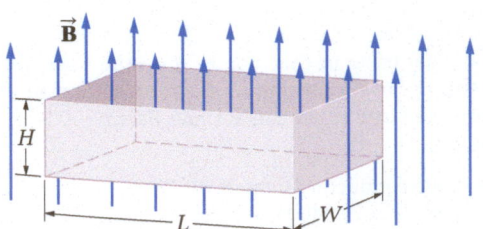

FIGURE 23-31 Problem 2

3. • A magnetic field is oriented at an angle of 67° to the normal of a rectangular area 5.9 cm by 7.8 cm. If the magnetic flux through

this surface has a magnitude of 4.8×10^{-5} T·m², what is the strength of the magnetic field?

4. • **MRI Solenoid** The magnetic field produced by an MRI solenoid 2.5 m long and 1.2 m in diameter is 1.7 T. Find the magnitude of the magnetic flux through the core of this solenoid.

5. •• Find the magnitude of the magnetic flux through the floor of a house that measures 22 m by 18 m. Assume that the Earth's magnetic field at the location of the house has a horizontal component of 2.6×10^{-5} T pointing north, and a downward vertical component of 4.2×10^{-5} T.

6. •• At a certain location, the Earth's magnetic field has a magnitude of 5.4×10^{-5} T and points in a direction that is 68° below the horizontal. Find the magnitude of the magnetic flux through the top of a desk at this location that measures 76 cm by 45 cm.

7. •• **Predict/Calculate** A solenoid with 385 turns per meter and a diameter of 17.0 cm has a magnetic flux through its core of magnitude 1.28×10^{-4} T·m². **(a)** Find the current in this solenoid. **(b)** How would your answer to part (a) change if the diameter of the solenoid were doubled? Explain.

8. ••• A single-turn square loop of side L is centered on the axis of a long solenoid. In addition, the plane of the square loop is

perpendicular to the axis of the solenoid. The solenoid has 1250 turns per meter and a diameter of 6.00 cm, and carries a current of 2.50 A. Find the magnetic flux through the loop when **(a)** $L = 3.00$ cm, **(b)** $L = 6.00$ cm, and **(c)** $L = 12.0$ cm.

SECTION 23-3 FARADAY'S LAW OF INDUCTION

9. • **CE Predict/Explain** A bar magnet is inside a closed cubical box. Magnetic field lines pointing outward from the box produce positive flux, and those pointing inward produce negative magnetic flux. **(a)** Is the total magnetic flux through the surface of the box positive, negative, or zero? **(b)** Choose the *best explanation* from among the following:
   I. There are more outward magnetic field lines because the magnet is inside the box.
   II. The magnet draws field lines inward toward itself, producing more negative flux.
   III. The same number of magnetic field lines leave the box as enter it.

10. • A 0.65-T magnetic field is perpendicular to a circular loop of wire with 73 turns and a radius of 18 cm. If the magnetic field is reduced to zero in 0.12 s, what is the magnitude of the induced emf?

11. • **FIGURE 23-32** shows the magnetic flux through a coil as a function of time. At what times shown in this plot do **(a)** the magnetic flux and **(b)** the induced emf have the greatest magnitude?

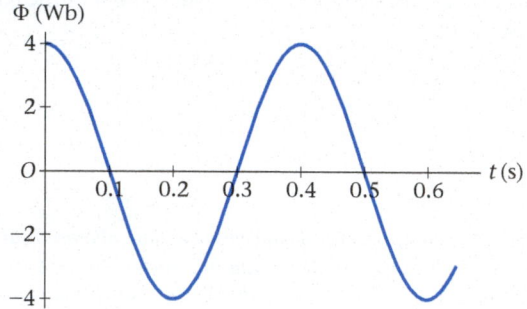

FIGURE 23-32 Problems 11 and 17

12. • **FIGURE 23-33** shows the magnetic flux through a single-loop coil as a function of time. What is the induced emf in the coil at **(a)** $t = 0.050$ s, **(b)** $t = 0.15$ s, and **(c)** $t = 0.50$ s?

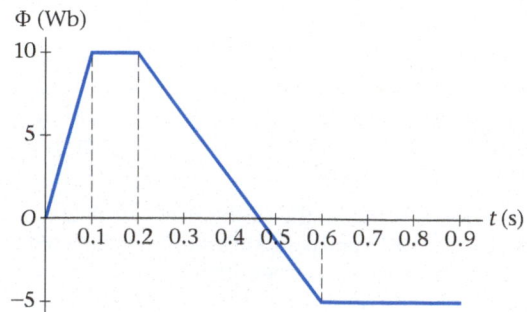

FIGURE 23-33 Problems 12 and 16

13. • One type of antenna for receiving AM radio signals is a square loop of wire, 0.15 m on a side, that has 25 turns. If the magnetic field from the radio waves changes at a rate of 9.2×10^{-4} T/s and is perpendicular to the loop, what is the magnitude of the induced emf in the loop?

14. •• **CE** A wire loop is placed in a magnetic field that is perpendicular to its plane. The field varies with time as shown in **FIGURE 23-34**.

Rank the six regions of time in order of increasing magnitude of the induced emf. Indicate ties where appropriate.

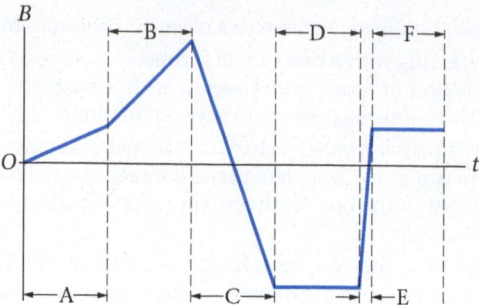

FIGURE 23-34 Problem 14

15. •• **CE FIGURE 23-35** shows four different situations in which a metal ring moves to the right with constant speed through a region with a varying magnetic field. The intensity of the color indicates the intensity of the field, and in each case the field either increases or decreases at a uniform rate from the left edge of the colored region to the right edge. The direction of the field in each region is indicated. For each of the four cases, state whether the induced emf is clockwise, counterclockwise, or zero.

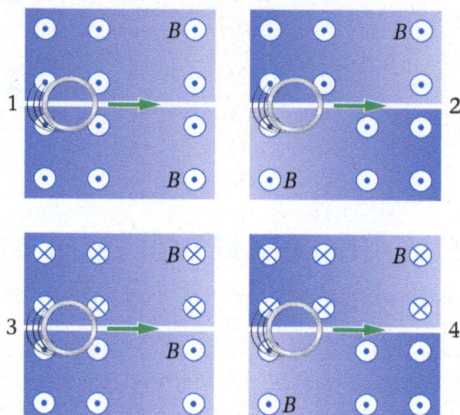

FIGURE 23-35 Problem 15

16. •• **Predict/Calculate** The magnetic flux through a single-loop coil is given by Figure 23-33. **(a)** Is the magnetic flux at $t = 0.25$ s greater than, less than, or the same as the magnetic flux at $t = 0.55$ s? Explain. **(b)** Is the induced emf at $t = 0.25$ s greater than, less than, or the same as the induced emf at $t = 0.55$ s? Explain. **(c)** Calculate the induced emf at the times $t = 0.25$ s and $t = 0.55$ s.

17. •• **Predict/Calculate** Consider a single-loop coil whose magnetic flux is given by Figure 23-32. **(a)** Is the magnitude of the induced emf in this coil greater near $t = 0.4$ s or near $t = 0.5$ s? Explain. **(b)** At what times in this plot do you expect the induced emf in the coil to have a maximum magnitude? Explain. **(c)** Estimate the induced emf in the coil at times near $t = 0.3$ s, $t = 0.4$ s, and $t = 0.5$ s.

18. •• A single conducting loop of wire has an area of 7.2×10^{-2} m² and a resistance of 110 Ω. Perpendicular to the plane of the loop is a magnetic field of strength 0.88 T. At what rate (in T/s) must this field change if the induced current in the loop is to be 0.43 A?

19. •• The area of a 120-turn coil oriented with its plane perpendicular to a 0.33-T magnetic field is 0.050 m². Find the average induced emf in this coil if the magnetic field reverses its direction in 0.11 s.

20. •• An emf is induced in a conducting loop of wire 1.40 m long as its shape is changed from square to circular. Find the average magnitude of the induced emf if the change in shape occurs in 0.125 s and the local 0.455-T magnetic field is perpendicular to the plane of the loop.

21. •• A magnetic field increases from 0 to 0.55 T in 1.6 s. How many turns of wire are needed in a circular coil 12 cm in diameter to produce an induced emf of 9.0 V?

SECTION 23-4 LENZ'S LAW

22. • **CE** Predict/Explain A metal ring is dropped into a localized region of constant magnetic field, as indicated in **FIGURE 23-36**. The magnetic field is zero above and below the region where it is finite. **(a)** For each of the three indicated locations (1, 2, and 3), is the induced current clockwise,

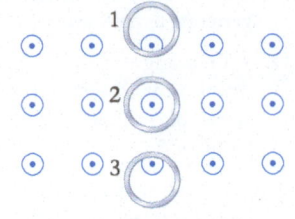

FIGURE 23-36 Problems 22 and 23

counterclockwise, or zero? **(b)** Choose the *best explanation* from among the following:
  I. Clockwise at 1 to oppose the field; zero at 2 because the field is uniform; counterclockwise at 3 to try to maintain the field.
  II. Counterclockwise at 1 to oppose the field; zero at 2 because the field is uniform; clockwise at 3 to try to maintain the field.
  III. Clockwise at 1 to oppose the field; clockwise at 2 to maintain the field; clockwise at 3 to oppose the field.

23. • **CE** Predict/Explain A metal ring is dropped into a localized region of constant magnetic field, as indicated in Figure 23-36. The magnetic field is zero above and below the region where it is finite. **(a)** For each of the three indicated locations (1, 2, and 3), is the magnetic force exerted on the ring upward, downward, or zero? **(b)** Choose the *best explanation* from among the following:
  I. Upward at 1 to oppose entering the field; zero at 2 because the field is uniform; downward at 3 to help leaving the field.
  II. Upward at 1 to oppose entering the field; upward at 2 where the field is strongest; upward at 3 to oppose leaving the field.
  III. Upward at 1 to oppose entering the field; zero at 2 because the field is uniform; upward at 3 to oppose leaving the field.

24. • **CE** Predict/Explain **FIGURE 23-37** shows two metal disks of the same size and material oscillating in and out of a region with a magnetic field. One disk is solid; the other has a series of slots. **(a)** Is the damping effect of eddy currents on the solid disk greater than, less than, or equal to the damping effect on the slotted disk? **(b)** Choose the *best explanation* from among the following:
  I. The solid disk experiences a greater damping effect because eddy currents in it flow freely and are not interrupted by the slots.
  II. The slotted disk experiences the greater damping effect because the slots allow more magnetic field to penetrate the disk.
  III. The disks are the same size and made of the same material; therefore, they experience the same damping effect.

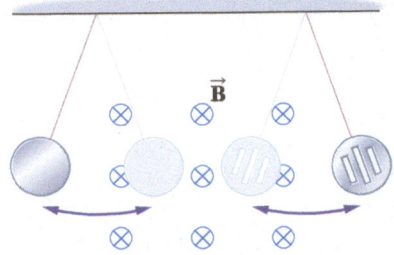

FIGURE 23-37 Problems 24 and 25

25. • **CE** Predict/Explain **(a)** As the solid metal disk in Figure 23-37 swings to the right, from the region with no field into the region with a finite magnetic field, is the induced current in the disk clockwise, counterclockwise, or zero? **(b)** Choose the *best explanation* from among the following:
  I. The induced current is clockwise, since this produces a field within the disk in the same direction as the magnetic field that produced the current.
  II. The induced current is counterclockwise to generate a field within the disk that points out of the page.
  III. The induced current is zero because the disk enters a region where the magnetic field is uniform.

26. • **CE** A bar magnet with its north pole pointing downward is falling toward the center of a horizontal conducting ring. As viewed from above, is the direction of the induced current in the ring clockwise or counterclockwise? Explain.

27. • **CE A Wire Loop and a Magnet** A loop of wire is dropped and allowed to fall between the poles of a horseshoe magnet, as shown in **FIGURE 23-38**. State whether the induced current in the loop is clockwise or counterclockwise when **(a)** the loop is above the magnet and **(b)** the loop is below the magnet.

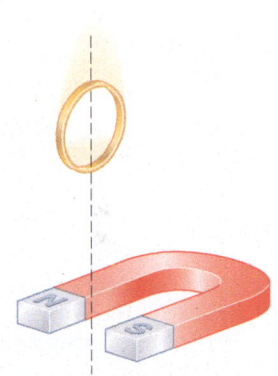

FIGURE 23-38 Problems 27 and 28

28. •• **CE** Suppose we change the situation shown in Figure 23-38 as follows: Instead of allowing the loop to fall on its own, we attach a string to it and lower it with constant speed along the path indicated by the dashed line. Is the tension in the string greater than, less than, or equal to the weight of the loop? Give specific answers for times when **(a)** the loop is above the magnet and **(b)** the loop is below the magnet. Explain in each case.

29. •• **CE FIGURE 23-39** shows a current-carrying wire and a circuit containing a resistor R. **(a)** If the current in the wire is constant, is the induced current in the circuit clockwise, counterclockwise, or zero? Explain. **(b)** If the current in the wire increases, is the induced current in the circuit clockwise, counter-

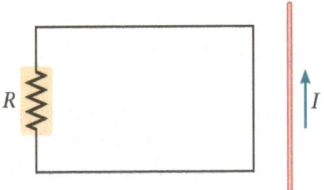

FIGURE 23-39 Problems 29 and 30

clockwise, or zero? Explain.

30. •• **CE** Consider the physical system shown in Figure 23-39. If the current in the wire changes direction, is the induced current in the circuit clockwise, counterclockwise, or zero? Explain.

31. •• **CE** A long, straight wire carries a current I, as indicated in **FIGURE 23-40**. Three small metal rings are placed near the current-carrying wire (A and C) or directly on top of it (B). If the current in the wire is increasing with

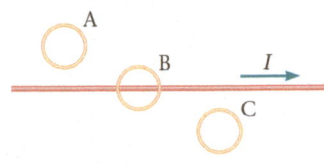

FIGURE 23-40 Problem 31

time, indicate whether the induced emf in each of the rings is clockwise, counterclockwise, or zero. Explain your answer for each ring.

32. •• **FIGURE 23-41** shows a circuit with an area of 0.060 m² containing a R = 1.0-Ω resistor and a C = 250-μF uncharged capacitor.

Pointing into the plane of the circuit is a uniform magnetic field of magnitude 0.25 T. In 0.010 s the magnetic field reverses direction at a constant rate to become 0.25 T pointing out of the plane. What maximum charge (sign and magnitude) accumulates on the upper plate of the capacitor in the diagram?

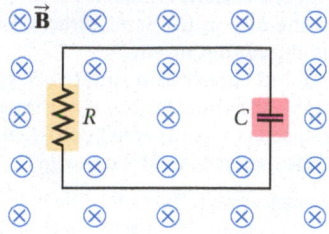

FIGURE 23-41 Problems 32 and 33

33. •• Figure 23-41 shows a circuit with an area of 0.060 m² containing a $R = 1.0\text{-}\Omega$ resistor and a $C = 250\text{-}\mu F$ uncharged capacitor. Pointing into the plane of the circuit is a uniform magnetic field of magnitude 0.15 T. In 0.010 s the magnetic field strengthens at a constant rate to become 0.80 T pointing into the plane. What maximum charge (sign and magnitude) accumulates on the upper plate of the capacitor in the diagram?

SECTION 23-5 MECHANICAL WORK AND ELECTRICAL ENERGY

34. • **CE** A conducting rod slides on two wires in a region with a magnetic field. The two wires are not connected. Is a force required to keep the rod moving with constant speed? Explain.

35. • A metal rod 0.95 m long moves with a speed of 2.4 m/s perpendicular to a magnetic field. If the induced emf between the ends of the rod is 0.65 V, what is the strength of the magnetic field?

36. •• **Airplane emf** A Boeing KC-135A airplane has a wingspan of 39.9 m and flies at constant altitude in a northerly direction with a speed of 850 km/h. If the vertical component of the Earth's magnetic field is 5.0×10^{-5} T, and its horizontal component is 1.4×10^{-5} T, what is the induced emf between the wing tips?

37. •• **Predict/Calculate FIGURE 23-42** shows a zero-resistance rod sliding to the right on two zero-resistance rails separated by the distance $L = 0.500$ m. The rails are connected by a 10.0-Ω resistor, and the entire system is in a uniform magnetic field with a magnitude of 0.750 T. (a) Find the speed at which the bar must be moved to produce a current of 0.175 A in the resistor. (b) Would your answer to part (a) change if the bar was moving to the left instead of to the right? Explain.

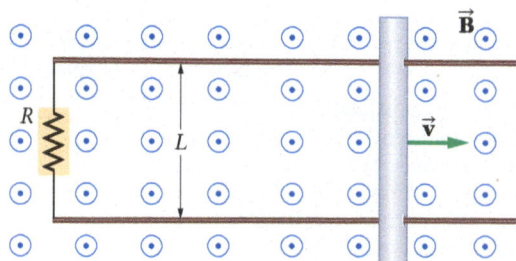

FIGURE 23-42 Problems 37 and 38

38. •• Referring to part (a) of Problem 37, (a) find the force that must be exerted on the rod to maintain a constant current of 0.175 A in the resistor. (b) What is the rate of energy dissipation in the resistor? (c) What is the mechanical power delivered to the rod?

39. •• (a) Find the current that flows in the circuit shown in Example 23-8. (b) What speed must the rod have if the current in the circuit is to be 1.0 A?

40. •• Suppose the mechanical power delivered to the rod in Example 23-8 is 8.9 W. Find (a) the current in the circuit and (b) the speed of the rod.

SECTION 23-6 GENERATORS AND MOTORS

41. • The maximum induced emf in a generator rotating at 210 rpm is 35 V. How fast must the rotor of the generator rotate if it is to generate a maximum induced emf of 55 V?

42. • A rectangular coil 25 cm by 45 cm has 150 turns. This coil produces a maximum emf of 75 V when it rotates with an angular speed of 190 rad/s in a magnetic field of strength B. Find the value of B.

43. • A 1.6-m wire is wound into a coil with a radius of 3.2 cm. If this coil is rotated at 85 rpm in a 0.075-T magnetic field, what is its maximum emf?

44. •• **Shake Flashlight** A shake flashlight uses the mechanical energy of the shaking action to charge a capacitor that operates the LED light. (a) If you shake the flashlight with an average force of 2.0 N at a speed of 0.85 m/s, what mechanical power do you produce? (b) If you shake the flashlight for 30 seconds, and 75% of the mechanical energy is converted to electrical energy, how much energy is stored in the capacitor? (c) If the LED lamp draws 0.032 A at 3.1 V, for how much time will the capacitor operate the LED?

45. •• **Predict/Calculate** A circular coil with a diameter of 22.0 cm and 155 turns rotates about a vertical axis with an angular speed of 1250 rpm. The only magnetic field in this system is that of the Earth. At the location of the coil, the horizontal component of the magnetic field is 3.80×10^{-5} T, and the vertical component is 2.85×10^{-5} T. (a) Which component of the magnetic field is important when calculating the induced emf in this coil? Explain. (b) Find the maximum emf induced in the coil.

46. •• A generator is designed to produce a maximum emf of 190 V while rotating with an angular speed of 3800 rpm. Each coil of the generator has an area of 0.016 m². If the magnetic field used in the generator has a magnitude of 0.052 T, how many turns of wire are needed?

SECTION 23-7 INDUCTANCE

47. • Find the induced emf when the current in a 48.0-mH inductor increases from 0 to 535 mA in 15.5 ms.

48. • How many turns should a solenoid of cross-sectional area 0.035 m² and length 0.32 m have if its inductance is to be 65 mH?

49. •• The inductance of a solenoid with 450 turns and a length of 24 cm is 7.3 mH. (a) What is the cross-sectional area of the solenoid? (b) What is the induced emf in the solenoid if its current drops from 3.2 A to 0 in 55 ms?

50. •• Determine the inductance of a solenoid with 660 turns in a length of 34 cm. The circular cross section of the solenoid has a radius of 4.6 cm.

51. •• A solenoid with a cross-sectional area of 1.14×10^{-3} m² is 0.610 m long and has 985 turns per meter. Find the induced emf in this solenoid if the current in it is increased from 0 to 2.00 A in 33.3 ms.

52. ••• **Predict/Calculate** A solenoid has N turns of area A distributed uniformly along its length, ℓ. When the current in this

solenoid increases at the rate of 2.0 A/s, an induced emf of 75 mV is observed. **(a)** What is the inductance of this solenoid? **(b)** Suppose the spacing between coils is doubled. The result is a solenoid that is twice as long but with the same area and number of turns. Will the induced emf in this new solenoid be greater than, less than, or equal to 75 mV when the current changes at the rate of 2.0 A/s? Explain. **(c)** Calculate the induced emf for part **(b)**.

SECTION 23-8 *RL* CIRCUITS

53. • How much time does it take for the current in an *RL* circuit with $R = 130\ \Omega$ and $L = 68$ mH to reach half its final value?

54. • A simple *RL* circuit includes a 0.125-H inductor. If the circuit is to have a characteristic time of 0.500 ms, what should be the value of the resistance?

55. •• **CE** The four electric circuits shown in **FIGURE 23-43** have identical batteries, resistors, and inductors. Rank the circuits in order of increasing current supplied by the battery long after the switch is closed. Indicate ties where appropriate.

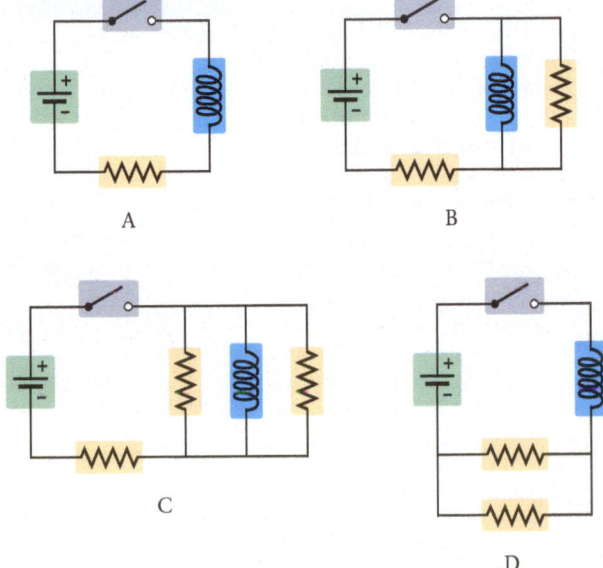

A   B

C

D

FIGURE 23-43 Problem 55

56. •• The circuit shown in **FIGURE 23-44** consists of a 6.0-V battery, a 31-mH inductor, and four 61-Ω resistors. **(a)** Find the characteristic time for this circuit. What is the current supplied by this battery **(b)** two characteristic time intervals after closing the switch and **(c)** a long time after the switch is closed?

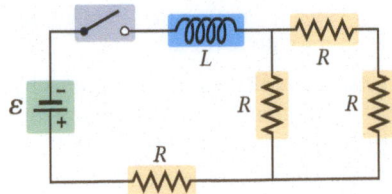

FIGURE 23-44 Problem 56

57. •• The current in an *RL* circuit increases to 95% of its final value 2.24 s after the switch is closed. **(a)** What is the characteristic time for this circuit? **(b)** If the inductance in the circuit is 0.275 H, what is the resistance?

58. ••• **Predict/Calculate** An AWG 24 copper wire with a diameter of 0.511 mm is tightly wrapped (one turn per wire diameter) around a 2.54-cm-diameter tube. As the number of turns increases, the inductance *L* of the resulting solenoid increases, and so does the resistance *R* of the wire because its length increases. **(a)** Write an expression for the characteristic time $\tau = L/R$ of the solenoid in terms of the wire diameter *d*, the number of turns *N*, the solenoid diameter *D*, and the resistivity ρ of copper (Table 21-1). **(b)** If *N* increases, will τ increase, decrease, or remain the same? **(c)** Find τ for $N = 325$ turns.

59. ••• Consider the *RL* circuit shown in **FIGURE 23-45**. When the switch is closed, the current in the circuit is observed to increase from 0 to 0.32 A in 0.15 s. **(a)** What is the inductance *L*? **(b)** At what time after the switch is closed $(t = 0)$ does the current have the value 0.50 A? **(c)** What is the maximum current that flows in this circuit?

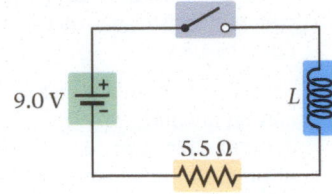

FIGURE 23-45 Problems 59 and 60

SECTION 23-9 ENERGY STORED IN A MAGNETIC FIELD

60. • Consider the circuit shown in Figure 23-45. Assuming the inductor in this circuit has the value $L = 9.5$ mH, how much energy is stored in the inductor after the switch has been closed for a long time?

61. • A solenoid is 1.8 m long and has 450 turns per meter. What is the cross-sectional area of this solenoid if it stores 0.39 J of energy when it carries a current of 12 A?

62. •• **Alcator Fusion Experiment** In the Alcator fusion experiment at MIT, a magnetic field of 50.0 T is produced. **(a)** What is the magnetic energy density in this field? **(b)** Find the magnitude of the electric field that would have the same energy density found in part **(a)**.

63. •• **Superconductor Energy Storage** An engineer proposes to store 75 kJ of energy by flowing 1.5 kA of current through a superconducting solenoid that forms an inductor. **(a)** What is the required inductance *L*? **(b)** If the solenoid is to have a radius of 1.2 m and be wound with no more than 5 turns per meter, what length must the solenoid have?

64. •• **Predict/Calculate** After the switch in **FIGURE 23-46** has been closed for a long time, the energy stored in the inductor is 0.110 J. **(a)** What is the value of the resistance *R*? **(b)** If it is desired that more energy be stored in the inductor, should the resistance *R* be greater than or less than the value found in part **(a)**? Explain.

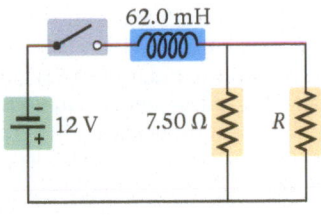

FIGURE 23-46 Problems 64 and 65

65. •• **Predict/Calculate** Suppose the resistor in Figure 23-46 has the value $R = 14\ \Omega$ and that the switch is closed at time $t = 0$. **(a)** How much energy is stored in the inductor at the time $t = \tau$? **(b)** How much energy is stored in the inductor at the time $t = 2\tau$? **(c)** If the value of *R* is increased, does the characteristic time, τ, increase or decrease? Explain.

66. ••• You would like to store 9.9 J of energy in the magnetic field of a solenoid. The solenoid has 580 circular turns of diameter 7.2 cm distributed uniformly along its 28-cm length. **(a)** How much current is needed? **(b)** What is the magnitude of the magnetic field inside the solenoid? **(c)** What is the energy density (energy/volume) inside the solenoid?

SECTION 23-10 TRANSFORMERS

67. • **CE** Transformer 1 has a primary voltage V_p and a secondary voltage V_s. Transformer 2 has twice the number of turns on both its primary and secondary coils compared with transformer 1. If the primary voltage on transformer 2 is $2V_p$, what is its secondary voltage? Explain.

68. • The electric motor in a toy train requires a voltage of 4.5 V. Find the ratio of turns on the primary coil to turns on the secondary coil in a transformer that will step the 120-V household voltage down to 4.5 V.

69. • **Predict/Calculate** A disk drive plugged into a 120-V outlet operates on a voltage of 9.0 V. The transformer that powers the disk drive has 125 turns on its primary coil. **(a)** Should the number of turns on the secondary coil be greater than or less than 125? Explain. **(b)** Find the number of turns on the secondary coil.

70. • A transformer with a turns ratio (secondary/primary) of 1:18 is used to step down the voltage from a 120-V wall socket to be used in a battery recharging unit. What is the voltage supplied to the recharger?

71. • A neon sign that requires a voltage of 11,000 V is plugged into a 120-V wall outlet. What turns ratio (secondary/primary) must a transformer have to power the sign?

72. •• A step-down transformer produces a voltage of 6.0 V across the secondary coil when the voltage across the primary coil is 120 V. What voltage appears across the primary coil of this transformer if 120 V is applied to the secondary coil?

73. •• A step-up transformer has 30 turns on the primary coil and 750 turns on the secondary coil. If this transformer is to produce an output of 4800 V with a 15-mA current, what input current and voltage are needed?

GENERAL PROBLEMS

74. • **CE Predict/Explain** An airplane flies level to the ground toward the north pole. **(a)** Is the induced emf from wing tip to wing tip when the plane is at the equator greater than, less than, or equal to the wing-tip-to-wing-tip emf when it is at the latitude of New York? **(b)** Choose the *best explanation* from among the following:
    I. The induced emf is the same because the strength of the Earth's magnetic field is the same at the equator and at New York.
    II. The induced emf is greater at New York because the vertical component of the Earth's magnetic field is greater there than at the equator.
    III. The induced emf is less at New York because at the equator the plane is flying parallel to the magnetic field lines.

75. • **CE** You hold a circular loop of wire at the north magnetic pole of the Earth. Consider the magnetic flux through this loop due to the Earth's magnetic field. Is the flux when the normal to the loop points horizontally greater than, less than, or equal to the flux when the normal points vertically downward? Explain.

76. • **CE** The inductor shown in **FIGURE 23-47** is connected to an electric circuit with a changing current. At the moment in question, the inductor has an induced emf with the indicated direction. Is the current in the circuit at this time increasing and to the right, increasing and to the left, decreasing and to the right, or decreasing and to the left?

FIGURE 23-47 Problem 76

77. • **Interstellar Magnetic Field** The *Voyager I* spacecraft moves through interstellar space with a speed of 8.0×10^3 m/s. The magnetic field

in this region of space has a magnitude of 2.0×10^{-10} T. Assuming that the 5.0-m-long antenna on the spacecraft is at right angles to the magnetic field, find the induced emf between its ends.

78. • **BIO Blowfly Flight** The coils used to measure the movements of a blowfly, as described in Section 23-5, have a diameter of 2.0 mm. In addition, the fly is immersed in a magnetic field of magnitude 0.15 mT. Find the maximum magnetic flux experienced by one of these coils.

79. • **BIO Electrognathography** Computerized jaw tracking, or electrognathography (EGN), is an important tool for diagnosing and treating temporomandibular disorders (TMDs) that affect a person's ability to bite effectively. The first step in applying EGN is to attach a small permanent magnet to the patient's gum below the lower incisors. Then, as the jaw undergoes a biting motion, the resulting change in magnetic flux is picked up by wire coils placed on either side of the mouth, as

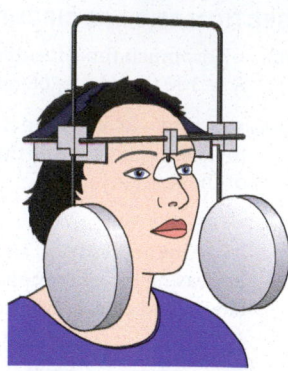

FIGURE 23-48 Problem 79

shown in **FIGURE 23-48**. Suppose this person's jaw moves to her right and that the north pole of the permanent magnet also points to her right. From her point of view, is the induced current in the coil to **(a)** her right and **(b)** her left clockwise or counterclockwise? Explain.

80. •• A rectangular loop of wire 24 cm by 72 cm is bent into an L shape, as shown in **FIGURE 23-49**. The magnetic field in the vicinity of the loop has a magnitude of 0.035 T and points in a direction 25° below the y axis. The magnetic field has no x component. Find the magnitude of the magnetic flux through the loop.

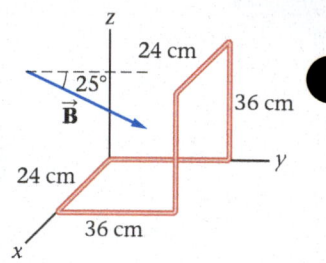

FIGURE 23-49 Problem 80

81. •• Consider a rectangular loop of wire 6.8 cm by 9.2 cm in a uniform magnetic field of magnitude 1.7 T. The loop is rotated from a position of zero magnetic flux to a position of maximum flux in 21 ms. What is the average induced emf in the loop?

82. •• **Predict/Calculate** A car with a vertical radio antenna 85 cm long drives due east at 25 m/s. The Earth's magnetic field at this location has a magnitude of 5.9×10^{-5} T and points northward, 72° below the horizontal. **(a)** Is the top or the bottom of the antenna at the higher potential? Explain. **(b)** Find the induced emf between the ends of the antenna.

83. •• The rectangular coils in a 355-turn generator are 13 cm by 19 cm. What is the maximum emf produced by this generator when it rotates with an angular speed of 505 rpm in a magnetic field of 0.55 T?

84. •• A cubical box 22 cm on a side is placed in a uniform 0.35-T magnetic field. Find the net magnetic flux through the box.

85. •• **BIO MRI Scanner** An MRI scanner is based on a solenoid magnet that produces a large magnetic field. The magnetic field doesn't stop at the solenoid's edge, however, but extends into the area around the magnet. Suppose a technician walks toward the scanner at 0.80 m/s from a region 1.0 m from the scanner where the magnetic field is negligible, into a region next to the scanner

where the field is 4.0 T and points horizontally. **(a)** As a result of this motion, what is the maximum magnitude of the change in flux through a loop defined by the outside of the technician's head? Assume the loop is vertical and has a circular cross section with a diameter of 18 cm. **(b)** What is the magnitude of the average induced emf around the outside of the technician's head during the time she's moving toward the scanner? (The biological effects due to motion in a strong magnetic field include nausea and dizziness.)

86. •• **BIO Transcranial Magnetic Stimulation** Transcranial magnetic stimulation (TMS) is a noninvasive method for studying brain function, and possibly for treatment as well. In this technique, a conducting loop is held near a person's head, as shown in **FIGURE 23-50**. When the current in the loop is changed rapidly, the magnetic field it creates can change at the rate of $3.00 \times 10^4 \, \text{T/s}$. This rapidly changing magnetic field induces an electric current in a restricted region of the brain that can cause a finger to twitch, bright spots to appear in the visual field (magnetophosphenes), or a feeling of complete happiness to overwhelm a person. If the magnetic field changes at the previously mentioned rate over an area of $1.13 \times 10^{-2} \, \text{m}^2$, what is the induced emf?

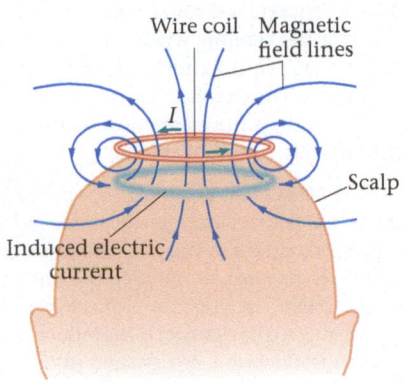

FIGURE 23-50 Transcranial magnetic stimulation (Problem 86)

87. •• A magnetic field with the time dependence shown in **FIGURE 23-51** is at right angles to a 155-turn circular coil with a diameter of 3.75 cm. What is the induced emf in the coil at **(a)** $t = 2.50 \, \text{ms}$, **(b)** $t = 7.50 \, \text{ms}$, **(c)** $t = 15.0 \, \text{ms}$, and **(d)** $t = 25.0 \, \text{ms}$?

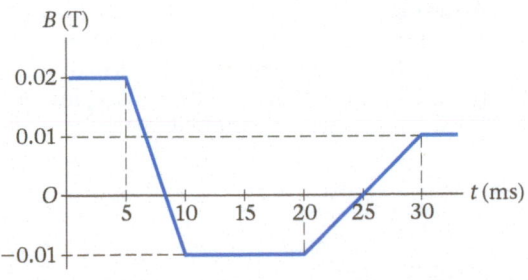

FIGURE 23-51 Problem 87

88. •• The characteristic time of an RL circuit with $L = 25 \, \text{mH}$ is twice the characteristic time of an RC circuit with $C = 45 \, \mu\text{F}$. Both circuits have the same resistance R. Find **(a)** the value of R and **(b)** the characteristic time of the RL circuit.

89. •• A 6.0-V battery is connected in series with a 49-mH inductor, a 150-Ω resistor, and an open switch. **(a)** At what time after the switch is closed ($t = 0$) will the current in the circuit be equal to 12 mA? **(b)** How much energy is stored in the inductor when the current reaches its maximum value?

90. •• A 9.0-V battery is connected in series with a 31-mH inductor, a 180-Ω resistor, and an open switch. **(a)** What is the current in the circuit 0.120 ms after the switch is closed? **(b)** How much energy is stored in the inductor at this time?

91. •• **BIO Blowfly Maneuvers** Suppose the fly described in Problem 78 turns through an angle of 90° in 37 ms. If the magnetic flux through one of the coils on the insect goes from a maximum to zero during this maneuver, and the coil has 85 turns of wire, find the magnitude of the induced emf.

92. ••• **Predict/Calculate** A conducting rod of mass m is in contact with two vertical conducting rails separated by a distance L, as shown in **FIGURE 23-52**. The entire system is immersed in a magnetic field of magnitude B pointing out of the page. Assuming the rod slides without friction, **(a)** describe the motion of the rod after it is released from rest. **(b)** What is the direction of the induced current (clockwise or counterclockwise) in the circuit? **(c)** Find the speed of the rod after it has fallen for a long time.

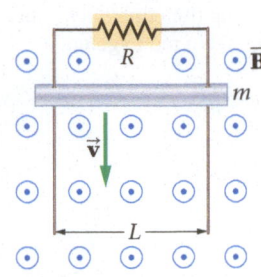

FIGURE 23-52 Problem 92

93. ••• **Predict/Calculate** A single-turn rectangular loop of width W and length L moves parallel to its length with a speed v. The loop moves from a region with a magnetic field $\vec{B}$ perpendicular to the plane of the loop to a region where the magnetic field is zero, as shown in **FIGURE 23-53**. Find the rate of change in the magnetic flux through the loop **(a)** before it enters the region of zero field, **(b)** just after it enters the region of zero field, and **(c)** once it is fully within the region of zero field. **(d)** For each of the cases considered in parts (a), (b), and (c), state whether the induced current in the loop is clockwise, counterclockwise, or zero. Explain in each case.

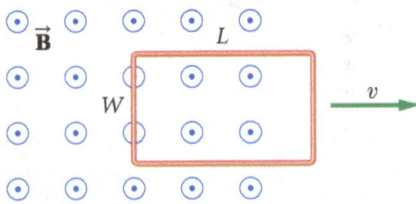

FIGURE 23-53 Problem 93

94. ••• **Predict/Calculate** The switch in the circuit shown in **FIGURE 23-54** is open initially. **(a)** Find the current in the circuit a long time after the switch is closed. **(b)** Describe the behavior of the light-bulb from the time the switch is closed until the current reaches the value found in part (a). **(c)** Now, suppose the switch is opened after having been closed for a long time. If the inductor is large, it is observed that the light flashes brightly and then burns out. Explain this behavior. **(d)** Find the voltage across the lightbulb just before and just after the switch is opened.

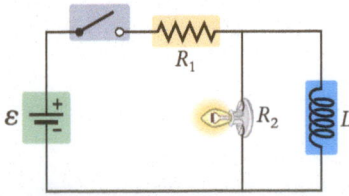

FIGURE 23-54 Problem 94

95. ••• **Energy Density in E and B Fields** An electric field E and a magnetic field B have the same energy density. **(a)** Express the ratio E/B in

terms of the fundamental constants ε_0 and μ_0. **(b)** Evaluate E/B numerically, and compare your result with the speed of light.

PASSAGE PROBLEMS

Loop Detectors on Roadways

"Smart" traffic lights are controlled by loops of wire embedded in the road (**FIGURE 23-55**). These "loop detectors" sense the change in magnetic field as a large metal object—such as a car or a truck—moves over the loop. Once the object is detected, electric circuits in the controller check for cross traffic, and then turn the light from red to green.

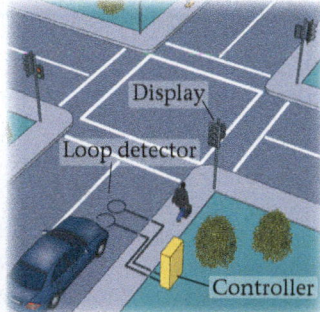

FIGURE 23-55 Problems 96, 97, 98, and 99

A typical loop detector consists of three or four loops of 14-gauge wire buried 3 in. below the pavement. You can see the marks on the road where the pavement has been cut to allow for installation of the wires. There may be more than one loop detector at a given intersection; this allows the system to recognize that an object is moving as it activates first one detector and then another over a short period of time. If the system determines that a car has entered the intersection while the light is red, it can activate one camera to take a picture of the car from the front—to see the driver's face—and then a second camera to take a picture of the car and its license plate from behind. This red-light camera system was used to good effect during an exciting chase scene through the streets of London in the movie *National Treasure: Book of Secrets*.

Motorcycles are small enough that they often fail to activate the detectors, leaving the cyclist waiting and waiting for a green light. Some companies have begun selling powerful neodymium magnets to mount on the bottom of a motorcycle to ensure that they are "seen" by the detectors.

96. • Suppose the downward vertical component of the magnetic field increases as a car drives over a loop detector. As viewed from above, is the induced current in the loop clockwise, counterclockwise, or zero?

97. •• A car drives onto a loop detector and increases the downward component of the magnetic field within the loop from 1.2×10^{-5} T to 2.6×10^{-5} T in 0.38 s. What is the induced emf in the detector if it is circular, has a radius of 0.67 m, and consists of four loops of wire?

A. 0.66×10^{-4} V **B.** 1.5×10^{-4} V
C. 2.1×10^{-4} V **D.** 6.2×10^{-4} V

98. •• A truck drives onto a loop detector and increases the downward component of the magnetic field within the loop from 1.2×10^{-5} T to the larger value B in 0.38 s. The detector is circular, has a radius of 0.67 m, and consists of three loops of wire. What is B, given that the induced emf is 8.1×10^{-4} V?

A. 3.6×10^{-5} T **B.** 7.3×10^{-5} T
C. 8.5×10^{-5} T **D.** 24×10^{-5} T

99. •• Suppose a motorcycle increases the downward component of the magnetic field within a loop only from 1.2×10^{-5} T to 1.9×10^{-5} T. The detector is square, is 0.75 m on a side, and has four loops of wire. Over what period of time must the magnetic field increase if it is to induce an emf of 1.4×10^{-4} V?

A. 0.028 s **B.** 0.11 s **C.** 0.35 s **D.** 0.60 s

100. •• **REFERRING TO CONCEPTUAL EXAMPLE 23-6** Suppose the ring is initially to the left of the field region, where there is no field, and is moving to the right. When the ring is partway into the field region, **(a)** is the induced current in the ring clockwise, counterclockwise, or zero, and **(b)** is the magnetic force exerted on the ring to the right, to the left, or zero? Explain.

101. •• **REFERRING TO CONCEPTUAL EXAMPLE 23-6** Suppose the ring is completely inside the field region initially and is moving to the right. **(a)** Is the induced current in the ring clockwise, counterclockwise, or zero, and **(b)** is the magnetic force on the ring to the right, to the left, or zero? Explain. The ring now begins to emerge from the field region, still moving to the right. **(c)** Is the induced current in the ring clockwise, counterclockwise, or zero, and **(d)** is the magnetic force on the ring to the right, to the left, or zero? Explain.

102. •• **REFERRING TO EXAMPLE 23-8** **(a)** What external force is required to give the rod a speed of 3.49 m/s, everything else remaining the same? **(b)** What is the current in the circuit in this case?

103. •• **Predict/Calculate REFERRING TO EXAMPLE 23-8** Suppose the direction of the magnetic field is reversed. Everything else in the system remains the same. **(a)** Is the magnetic force exerted on the rod to the right, to the left, or zero? Explain. **(b)** Is the direction of the induced current clockwise, counterclockwise, or zero? Explain. **(c)** Suppose we now adjust the strength of the magnetic field until the speed of the rod is 2.49 m/s, keeping the force equal to 1.60 N. What is the new magnitude of the magnetic field?

Alternating-Current Circuits

Big Ideas

1 The voltage across a resistor in an ac circuit is in phase with the current.

2 Root mean square (rms) values play a key role in analyzing ac circuits.

3 The voltage across a capacitor in an ac circuit lags behind the current by 90°.

4 The voltage in an ac *RC* circuit can be found by adding the voltage phasors.

5 The voltage across an inductor in an ac circuit leads the current by 90°.

6 Capacitors and inductors have simple behavior at high and low frequencies.

▲ The familiar 60-Hz hum of high-voltage power lines makes it clear they are part of an alternating current (ac) circuit. This chapter explores some of the key features of ac circuits, including phasor analysis and resonance.

We conclude our study of electric circuits with the ones that are most common in everyday life—circuits with an alternating current (ac). As we shall see, ac circuits have a rich variety of behaviors that make them useful in practical applications, as well as interesting from a physics point of view.

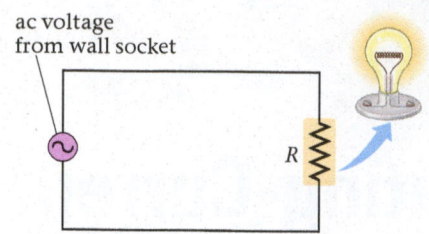

ac voltage from wall socket

R

▲ **FIGURE 24-1 An ac generator connected to a lamp** Simplified alternating-current circuit diagram for a lamp plugged into a wall socket. The lightbulb is replaced in the circuit with its equivalent resistance, R.

**PHYSICS
IN CONTEXT
Looking Ahead**

An ac generator appears again in Chapter 25, where we show how the alternating current in an electric circuit can produce electromagnetic waves.

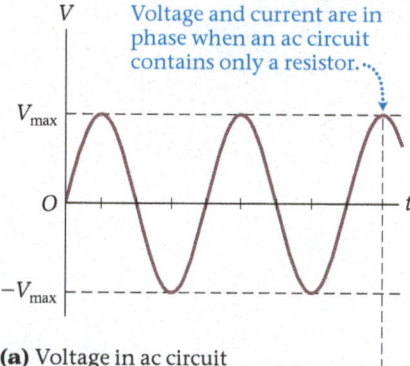

Voltage and current are in phase when an ac circuit contains only a resistor.

V_{max}

O

$-V_{max}$

(a) Voltage in ac circuit

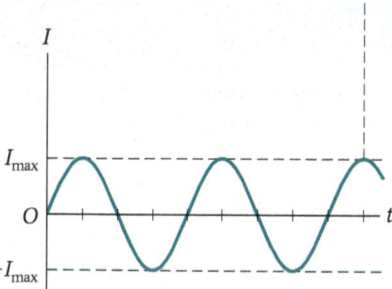

I_{max}

O

$-I_{max}$

(b) Current in ac circuit with resistance only

▲ **FIGURE 24-2 Voltage and current in an ac resistor circuit (a)** An ac voltage described by $V = V_{max} \sin \omega t$. **(b)** The alternating current, $I = I_{max} \sin \omega t = (V_{max}/R) \sin \omega t$, corresponding to the ac voltage in (a). Note that the voltage of the generator and the current in the resistor are in phase with each other; that is, their maxima and minima occur at precisely the same times.

24-1 Alternating Voltages and Currents

When you plug a lamp into a wall socket, the voltage and current supplied to the light-bulb change sinusoidally with a steady frequency, completing 60 cycles each second. Because the current periodically reverses direction, we say that the wall socket provides an **alternating current.** A simplified alternating-current (ac) circuit diagram for a lamp is shown in **FIGURE 24-1.** In this circuit, we indicate the bulb by its equivalent resistance, R, and the wall socket by an **ac generator,** represented as a circle enclosing one cycle of a sine wave.

The voltage delivered by an ac generator, which is plotted in **FIGURE 24-2 (a),** can be represented mathematically as follows:

$$V = V_{max} \sin \omega t \qquad \text{24-1}$$

In this expression, V_{max} is the largest value attained by the voltage during a cycle, and the angular frequency is $\omega = 2\pi f$, where $f = 60$ Hz. (Notice the similarity between the sinusoidal time dependence in an ac circuit and the time dependence given in Section 13-2 for simple harmonic motion.) From Ohm's law, $I = V/R$, it follows that the current in the lightbulb is

$$I = \frac{V}{R} = \left(\frac{V_{max}}{R}\right) \sin \omega t = I_{max} \sin \omega t \qquad \text{24-2}$$

This result is plotted in **FIGURE 24-2 (b).**

Notice that the voltage and current plots have the same time variation. In particular, the voltage reaches its maximum value at precisely the same time as the current. We express this relationship between the voltage and current in a resistor by saying that they are *in phase* with one another. As we shall see later in this chapter, other circuit elements, like capacitors and inductors, have different phase relationships between the voltage and current. For these elements, the voltage and current reach maximum values at different times.

Phasors

A convenient way to represent an alternating voltage and the corresponding current is with counterclockwise-rotating vectors referred to as **phasors.** Phasors allow us to take advantage of the connection between uniform circular motion and linear, sinusoidal motion—just as we did when we studied oscillations in Chapter 13. In that case, we related the motion of a peg on a rotating turntable to the oscillating motion of a mass on a spring by projecting the position of the peg onto a screen. We use a similar projection with phasors.

For example, **FIGURE 24-3** shows a vector of magnitude V_{max} rotating counterclockwise about the origin with an angular speed ω. This rotating vector is the voltage phasor. If the voltage phasor makes an angle $\theta = \omega t$ with the x axis, it follows that its y component is $V_{max} \sin \omega t$; that is, if we project the voltage phasor onto the y axis, the projection gives the instantaneous value of the voltage, $V = V_{max} \sin \omega t$, in agreement with Equation 24-1. In general, we can make the following identification:

> The instantaneous value of a quantity represented by a phasor is the projection of the phasor onto the y axis.

Figure 24-3 also shows the current phasor, represented by a rotating vector of magnitude $I_{max} = V_{max}/R$. The current phasor points in the same direction as the voltage phasor; hence, the instantaneous current is $I = I_{max} \sin \omega t$, as in Equation 24-2.

The fact that the voltage and current phasors always point in the same direction for a resistor is an equivalent way of saying that the voltage and current are in phase:

> The voltage phasor for a resistor points in the same direction as the current phasor.

In circuits that also contain capacitors or inductors, the current and voltage phasors will usually point in different directions.

Root Mean Square (rms) Values

Both the voltage and the current in Figure 24-2 have average values that are zero. Thus, V_{av} and I_{av} give very little information about the actual behavior of V and I. A more useful type of average, or mean, is the **root mean square**, or **rms** for short.

To see the significance of a *root mean square*, consider the current as a function of time, as given in Equation 24-2. First, we *square* the current to obtain

$$I^2 = I_{max}^2 \sin^2 \omega t$$

Clearly, I^2 is always positive, as we can see in **FIGURE 24-4**, and hence its average value will not vanish. Next, we calculate the *mean* value of I^2. This can be done by inspecting Figure 24-4, where we see that I^2 varies *symmetrically* between 0 and I_{max}^2; that is, it spends equal amounts of time above and below the value $\frac{1}{2} I_{max}^2$. Hence, the mean value of I^2 is one-half of I_{max}^2:

$$(I^2)_{av} = \frac{1}{2} I_{max}^2$$

Finally, we take the square *root* of this average so that our final result is a current rather than a current squared. This calculation yields the rms value of the current:

$$I_{rms} = \sqrt{(I^2)_{av}} = \frac{1}{\sqrt{2}} I_{max} \qquad 24\text{-}3$$

In general, any quantity x that varies with time as $x = x_{max} \sin \omega t$ or $x = x_{max} \cos \omega t$ obeys the same relationships among its average, maximum, and rms values:

RMS Value of a Quantity with Sinusoidal Time Dependence

$$(x^2)_{av} = \frac{1}{2} x_{max}^2$$
$$x_{rms} = \frac{1}{\sqrt{2}} x_{max} \qquad 24\text{-}4$$

As an example, the rms value of the voltage in an ac circuit is

$$V_{rms} = \frac{1}{\sqrt{2}} V_{max} \qquad 24\text{-}5$$

This result is applied to standard household voltages in the following Exercise.

EXERCISE 24-1   MAXIMUM VOLTAGE

Typical household circuits operate with an rms voltage of 120 V. What is the maximum, or peak, value of the voltage in these circuits?

REASONING AND SOLUTION

Solving $V_{rms} = \frac{1}{\sqrt{2}} V_{max}$ for the maximum voltage, V_{max}, we find

$$V_{max} = \sqrt{2} V_{rms} = \sqrt{2}(120 \text{ V}) = 170 \text{ V}$$

Whenever we refer to an ac generator in this chapter, we shall assume that the time variation is sinusoidal, so that the rms relationships given in Equation 24-4 hold. If a different form of time variation is to be considered, it will be specified explicitly. Because the rms and maximum values of a sinusoidally varying quantity are proportional, it follows that *any* relationship between rms values, like $I_{rms} = V_{rms}/R$, is equally valid as a relationship between maximum values, $I_{max} = V_{max}/R$.

Average Power   We now show how rms values are related to the average power consumed by a circuit. Referring again to the lamp circuit shown in Figure 24-1, we see that the instantaneous power dissipated in the resistor is $P = I^2R$. Using the time dependence $I = I_{max} \sin \omega t$, we find

$$P = I^2 R = I_{max}^2 R \sin^2 \omega t$$

This equation shows that P is always positive, as one might expect; after all, a current always dissipates energy as it passes through a resistor, regardless of its direction. To find the average power dissipated in the resistor, we note again that the average of $\sin^2 \omega t$ is $\frac{1}{2}$; thus

$$P_{av} = I_{max}^2 R (\sin^2 \omega t)_{av} = \frac{1}{2} I_{max}^2 R$$

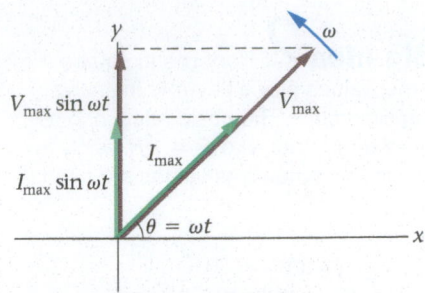

▲ **FIGURE 24-3 Phasor diagram for an ac resistor circuit** Since the current and voltage are in phase in a resistor, the corresponding phasors point in the same direction at all times. Both phasors rotate about the origin with an angular speed ω, and the vertical component of each is the instantaneous value of that quantity.

Big Idea 1 The voltage across a resistor in an ac circuit is in phase with the current. This means that when the current in the circuit is a maximum, so too is the voltage across the resistor.

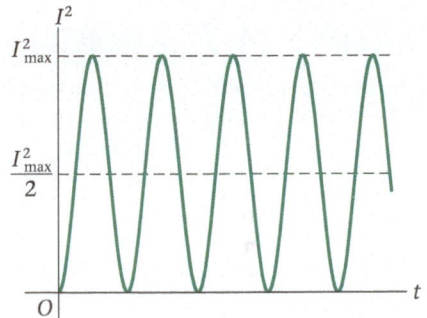

▲ **FIGURE 24-4 The square of a sinusoidally varying current** Note that I^2 varies symmetrically about the value $\frac{1}{2} I_{max}^2$. The average of I^2 over time, then, is $\frac{1}{2} I_{max}^2$.

Big Idea 2 Root mean square (rms) values play a key role in analyzing ac circuits. The basic relationships from dc circuits, like $P = I^2R$, apply in ac circuits when rms values are used.

PHYSICS IN CONTEXT
Looking Back

The concepts of current, resistance, capacitance, and inductance (Chapters 21 and 23) are used throughout this chapter. We also make extensive use of the concept of electrical power (Chapter 21).

PROBLEM-SOLVING NOTE

Maximum Versus RMS Values

When reading a problem statement, be sure to determine whether a given voltage or current is a maximum value or an rms value. If an rms current is given, for example, it follows that the maximum current is $I_{max} = \sqrt{2}I_{rms}$.

In terms of the rms current, $I_{rms} = I_{max}/\sqrt{2}$, it follows that

$$P_{av} = I_{rms}^2 R$$

Therefore, we arrive at the following conclusion:

> $P = I^2R$ gives the instantaneous power consumption in both ac and dc circuits. To find the *average* power in an ac circuit, we simply replace I with I_{rms}.

Similar conclusions apply to many other dc and instantaneous formulas as well. For example, the power can also be written as $P = V^2/R$. Using the time-dependent voltage of an ac circuit, we have

$$P = \frac{V^2}{R} = \left(\frac{V_{max}^2}{R}\right) \sin^2 \omega t$$

Clearly, the average power is

$$P_{av} = \left(\frac{V_{max}^2}{R}\right)\left(\frac{1}{2}\right) = \frac{V_{rms}^2}{R} \qquad \text{24-6}$$

As before, we convert the dc power, $P = V^2/R$, to an average ac power by using the rms value of the voltage.

The close similarity between dc expressions and the corresponding ac expressions with rms values is one of the advantages of working with rms values. Another is that electrical meters, such as ammeters and voltmeters, generally give readings of rms values, rather than peak values, when used in ac circuits. In the remainder of this chapter, we shall make extensive use of rms quantities.

EXAMPLE 24-2   A RESISTOR CIRCUIT

An ac generator with a maximum voltage of 24.0 V and a frequency of 60.0 Hz is connected to a resistor with a resistance $R = 265\ \Omega$. Find **(a)** the rms voltage and **(b)** the rms current in the circuit. In addition, determine **(c)** the average power and **(d)** the maximum power dissipated in the resistor.

PICTURE THE PROBLEM
The circuit in this case consists of a 60.0-Hz ac generator connected directly to a 265-Ω resistor. The maximum voltage of the generator is 24.0 V.

REASONING AND STRATEGY
a. The rms voltage is simply $V_{rms} = V_{max}/\sqrt{2}$.
b. Ohm's law gives the rms current: $I_{rms} = V_{rms}/R$.
c. The average power can be found using $P_{av} = I_{rms}^2 R$ or $P_{av} = V_{rms}^2/R$.
d. The maximum power, $P_{max} = V_{max}^2/R$, is twice the average power.

Known   Maximum voltage, $V_{max} = 24.0$ V; frequency, $f = 60.0$ Hz; resistance, $R = 265\ \Omega$.

Unknown **(a)** rms voltage, $V_{rms} = ?$ **(b)** rms current, $I_{rms} = ?$ **(c)** Average power, $P_{av} = ?$
(d) Maximum power, $P_{max} = ?$

SOLUTION

Part (a)

1. Use $V_{rms} = V_{max}/\sqrt{2}$ to find the rms voltage:

$$V_{rms} = \frac{V_{max}}{\sqrt{2}} = \frac{24.0\ \text{V}}{\sqrt{2}} = 17.0\ \text{V}$$

Part (b)

2. Divide the rms voltage by the resistance, R, to find the rms current:

$$I_{rms} = \frac{V_{rms}}{R} = \frac{17.0\ \text{V}}{265\ \Omega} = 0.0642\ \text{A}$$

Part (c)

3. Use $I_{rms}^2 R$ to find the average power:

$$P_{av} = I_{rms}^2 R = (0.0642\ \text{A})^2(265\ \Omega) = 1.09\ \text{W}$$

4. As a double check, use V_{rms}^2/R to find the average power:

$$P_{av} = \frac{V_{rms}^2}{R} = \frac{(17.0\ \text{V})^2}{265\ \Omega} = 1.09\ \text{W}$$

CONTINUED

Part (d)

5. Use $P_{max} = V^2_{max}/R$ to find the maximum power:

$$P_{max} = \frac{V^2_{max}}{R} = 2\left(\frac{V^2_{rms}}{R}\right) = 2(1.09 \text{ W}) = 2.18 \text{ W}$$

INSIGHT

The instantaneous power in the resistor, $P = (V^2_{max}/R)\sin^2 \omega t$, oscillates symmetrically between 0 and twice the average power. This is another example of the type of averaging illustrated in Figure 24-4.

PRACTICE PROBLEM — PREDICT/CALCULATE

Suppose we would like the average power dissipated in the resistor to be 5.00 W. **(a)** Should the resistance be increased or decreased, assuming the same ac generator? Explain. **(b)** Find the required value of R.   [**Answer: (a)** The resistance should be decreased. This follows because $P_{av} = V^2_{rms}/R$, and hence a smaller resistance (at constant rms voltage V_{rms}) corresponds to a larger dissipated power. **(b)** The required resistance is $R = 57.8 \, \Omega$, which is less than the original 265 Ω.]

Some related homework problems: Problem 3, Problem 4

CONCEPTUAL EXAMPLE 24-3   COMPARE AVERAGE POWER   PREDICT/EXPLAIN

(a) If the frequency of the ac generator in Example 24-2 is increased, while everything else remains the same, does the average power dissipated in the resistor increase, decrease, or stay the same? **(b)** Which of the following is the *best explanation* for your prediction?

  I. The greater the frequency, the greater the power dissipated.

 II. As the frequency is increased there is less time for energy to be dissipated, so the power decreases.

III. The resistance of a resistor doesn't depend on frequency, so the power it dissipates stays the same.

REASONING AND DISCUSSION

None of the results in Example 24-2 depend on the frequency of the generator. For example, the relationship $V_{rms} = V_{max}/\sqrt{2}$ depends only on the fact that the voltage varies sinusoidally with time and not at all on the frequency of the oscillations. The same frequency independence applies to the rms current and the average power. In addition, we note that resistance itself is independent of frequency—something that is not true of capacitors and inductors, as we shall see later in this chapter.

ANSWER

(a) The average power remains the same. **(b)** The best explanation is III.

   RWP* Household electricity is ac because the transformers that allow for the efficient transmission of electricity, as described in Section 23-10, work only with alternating current. However, many electronic devices require dc power, and *rectifying circuits* are responsible for converting the ac power into dc. These circuits are contained in the small plastic boxes (power supplies) that you plug into the wall outlet. On the other hand, sometimes it is necessary to convert the dc power of a solar panel or a battery into ac power, and this is accomplished with an *inverter circuit*. The ability to convert between ac and dc power is a key ingredient in our ability to harness electrical power.

Safety Features in Household Electric Circuits

With electrical devices as common these days as a horse and buggy in an earlier era, we sometimes forget that household electric circuits pose potential dangers to homes and their occupants. For example, if several electrical devices are plugged into a single outlet, the current in the wires connected to that outlet may become quite large. The corresponding power dissipation in the wires ($P = I^2R$) can turn them red hot and lead to a fire.

   To protect against this type of danger, household circuits use fuses and circuit breakers. Enclosed within a fuse is a thin metal strip that melts and breaks the circuit if the current flowing through it exceeds its designated limit (typically 15 A). Thus, when a fuse "burns out," it is an indication that too many devices are operating on that circuit.

---

*Real World Physics applications are denoted by the acronym RWP.

(a) (b) (c)

▲ **FIGURE 24-5 Visualizing Concepts Electrical Safety Devices** Many common safety devices help to minimize the risks associated with electricity. **(a)** Most houses are protected by circuit breaker panels. If too much current is being drawn (perhaps as the result of a short circuit), the heat "trips" the circuit breaker (by bending a bimetallic strip), interrupting the circuit and cutting off the current before the wires can become hot enough to start a fire. **(b)** The danger of electric shock is reduced by the use of polarized plugs (top) or three-prong, grounded plugs (bottom) on appliances. Each of these provides a low-resistance path to ground that can be wired to the case of an appliance. In the event that a "hot" wire touches the case, most of the current will flow through the grounded wire rather than through the user. **(c)** Even more protection is afforded by a ground fault circuit interrupter, or GFCI. This device, which is much faster and more sensitive than an ordinary circuit breaker, utilizes magnetic induction to interrupt the current in a circuit that has developed a short.

Circuit breakers, as shown in **FIGURE 24-5 (a)**, provide similar protection with a switch that incorporates a bimetallic strip (Figure 16-6). When the bimetallic strip is cool, it closes the switch, allowing current to flow. When the strip is heated by a large current, however, it bends enough to open the switch and stop the current. Unlike the fuse, which cannot be used after it burns out, the circuit breaker can be reset when the bimetallic strip cools and returns to its original shape.

BIO Household circuits also pose a shock hazard to the occupants of a home, and at much lower current levels than the 15 A it takes to trigger a fuse or a circuit breaker. For example, it takes a current of only about 0.001 A to give a person a mild tingling sensation. Currents in the range of 0.01 A to 0.02 A produce muscle spasms that make it difficult to let go of the wire delivering the current, or may even result in respiratory arrest. When currents reach 0.1 A to 0.2 A, the heartbeat is interrupted by an uncoordinated twitching referred to as ventricular fibrillation. As a result, currents in this range can prove to be fatal in a matter of seconds.

RWP Several strategies are employed to reduce the danger of electric shock. The first line of defense is the *polarized plug*, shown at the top in **FIGURE 24-5 (b)**, in which one prong of the plug is wider than the other prong. The corresponding wall socket will accept the plug in only one orientation, with the wide prong in the wide receptacle. The narrow receptacle of the outlet is wired to the high-potential side of the circuit; the wide receptacle is connected to the low-potential side, which is essentially at "ground" potential. A polarized plug provides protection by ensuring that the outer enclosure of an electrical appliance, which manufacturers design to be connected to the wide prong, is at low potential. Furthermore, when an electrical device with a polarized plug is turned off, the high potential extends only from the wall outlet to the switch, leaving the rest of the device at zero potential.

Another line of defense against accidental shock is the three-prong grounded plug, shown at the bottom in Figure 24-5 (b). In this plug, the rounded third prong is connected directly to ground when it is plugged into a three-prong receptacle. In addition, the third prong is wired to the outer enclosure of an electrical appliance. If something goes wrong within the appliance, and a high-potential wire comes into contact with the outer enclosure, the resulting current flows through the third prong, rather than through the body of a person who happens to touch the case.

RWP An even greater level of protection is provided by a device known as a *ground fault circuit interrupter* (GFCI), shown in **FIGURE 24-5 (c)**. The basic operating principle of a GFCI is illustrated in **FIGURE 24-6**. Notice that the wires carrying an ac current to the protected appliance pass through a small iron ring. When the appliance operates normally, the two wires carry equal currents in opposite directions—in one wire the current goes to the appliance, in the other the same current returns from the appliance. Each of these wires produces a magnetic field (Equation 22-9), but because their currents are in opposite directions, the magnetic fields are in opposite directions as well. As a result, the magnetic fields of the two wires cancel. If a malfunction occurs in the appliance—say, a wire frays and contacts the case of the appliance—current that would ordinarily return through the power cord may pass through the user's body instead and into the ground. In such a situation, the wire carrying current to the appliance now produces a net magnetic field within the iron ring that varies with the frequency of the ac generator. The changing magnetic field in the ring induces a current in the sensing coil wrapped around the ring, and the induced current triggers a circuit breaker in the GFCI. This cuts the flow of current to the appliance within a millisecond, protecting the user.

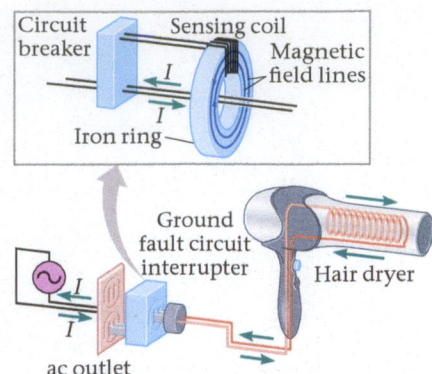

▲ **FIGURE 24-6 The ground fault circuit interrupter** A ground fault circuit interrupter, can cut off the current in a shorted circuit in less than a millisecond, before any harm can be done.

Enhance Your Understanding (Answers given at the end of the chapter)

1. A circuit has a sinusoidally varying current whose maximum value is 1 A. Is the rms current in the circuit greater than, less than, or equal to the average current in the circuit? Explain.

Section Review

- The root mean square (rms) voltage in an ac circuit is related to the maximum voltage as follows: $V_{rms} = \frac{1}{\sqrt{2}} V_{max}$. A similar equation applies to the current.

- The average power in an ac circuit can be written as $P_{av} = I_{rms}^2 R$. This expression is basically the same as the power in a dc circuit ($P = I^2 R$), except we are now calculating the average power, and the current is replaced by the rms value of the current. Similar comments apply to many equations originally derived for dc circuits.

24-2 Capacitors in ac Circuits

In an ac circuit that contains only a generator and a capacitor, the relationship between current and voltage is different in many important respects from the behavior seen with a resistor. In this section we explore these differences and show that the average power consumed by a capacitor is zero.

Capacitive Reactance

Consider a simple circuit consisting of an ac generator and a capacitor, as in **FIGURE 24-7**. The generator supplies an rms voltage, V_{rms}, to the capacitor. We would like to answer the following question: How is the rms current in this circuit related to the capacitance of the capacitor, C, and the frequency of the generator, ω?

To answer this question requires the methods of calculus, but the final result is quite straightforward. In fact, the rms current is simply

$$I_{rms} = \omega C V_{rms} \qquad \text{24-7}$$

In analogy with the expression $I_{rms} = V_{rms}/R$ for a resistor, we will find it convenient to rewrite this result in the following form:

$$I_{rms} = \frac{V_{rms}}{X_C} \qquad \text{24-8}$$

In this expression, X_C is referred to as the **capacitive reactance;** it plays the same role for a capacitor that resistance does for a resistor. In particular, to find the rms

▲ **FIGURE 24-7 An ac generator connected to a capacitor** The rms current in an ac capacitor circuit is $I_{rms} = V_{rms}/X_C$, where the capacitive reactance is $X_C = 1/\omega C$. Therefore, the rms current in this circuit is proportional to the frequency of the generator.

current in a resistor, we divide V_{rms} by R; to find the rms current that flows into one side of a capacitor and out the other side, we divide V_{rms} by X_C. Comparing $I_{rms} = \omega C V_{rms}$ and $I_{rms} = V_{rms}/X_C$, we see that the capacitive reactance can be written as follows:

Capacitive Reactance, X_C

$$X_C = \frac{1}{\omega C}$$

24-9

SI unit: ohm, Ω

It's straightforward to show that the unit of $X_C = 1/\omega C$ is the ohm, the same unit as for resistance.

EXERCISE 24-4   CAPACITIVE REACTANCE

Find the capacitive reactance of a 45.8-μF capacitor in a 60.0-Hz circuit.

REASONING AND SOLUTION

The capacitive reactance is defined as $X_C = 1/\omega C$. In this case, the capacitance is $C = 45.8 \times 10^{-6}$ F and the frequency is $f = 60.0$ Hz. Using these values, and recalling that the angular frequency is $\omega = 2\pi f$, we find

$$X_C = \frac{1}{\omega C} = \frac{1}{2\pi(60.0 \text{ s}^{-1})(45.8 \times 10^{-6} \text{ F})} = 57.9 \ \Omega$$

Unlike resistance, the capacitive reactance of a capacitor depends on the frequency of the ac generator. For example, at low frequencies the capacitive reactance becomes very large, and hence the rms current, $I_{rms} = V_{rms}/X_C$, is very small. This is to be expected—after all, in the limit of low frequency, the current in the circuit becomes constant; that is, the ac generator becomes a dc battery. In this case we know that the capacitor becomes fully charged, and the current ceases to flow.

In the limit of high frequency, the capacitive reactance is small and the current is large. The reason is that at high frequency the current changes direction so rapidly that there is never enough time to fully charge the capacitor. As a result, the charge on the capacitor is never very large, and therefore it offers essentially no resistance to the flow of charge—it's like a zero-resistance wire.

The behavior of a capacitor as a function of frequency is considered in the next Exercise.

EXERCISE 24-5   RMS CURRENT

Suppose the capacitance in Figure 24-7 is 6.25 μF and the rms voltage of the generator is 120 V. Find the rms current in the circuit when the frequency of the generator is **(a)** 60.0 Hz and **(b)** 30.0 Hz.

REASONING AND SOLUTION

(a) The rms current is $I_{rms} = V_{rms}/X_C$. Noting that the capacitive reactance is $X_C = 1/\omega C$, and the angular frequency is $\omega = 2\pi f$ with $f = 60.0$ Hz, we find

$$I_{rms} = \frac{V_{rms}}{X_C} = \omega C V_{rms} = (2\pi)(60.0 \text{ s}^{-1})(6.25 \times 10^{-6} \text{ F})(120 \text{ V}) = 0.28 \text{ A}$$

(b) Similarly, using $f = 30.0$ Hz for the frequency, we find

$$I_{rms} = \frac{V_{rms}}{X_C} = \omega C V_{rms} = (2\pi)(30.0 \text{ s}^{-1})(6.25 \times 10^{-6} \text{ F})(120 \text{ V}) = 0.14 \text{ A}$$

When the frequency is reduced, the capacitive reactance becomes larger. As a result, the rms current decreases.

PROBLEM-SOLVING NOTE

Drawing a Capacitor Phasor

When drawing a voltage phasor for a capacitor, always draw it at right angles to the current phasor. In addition, be sure the capacitor's voltage phasor is 90° *clockwise* from the current phasor.

The rms current in a capacitor is proportional to its capacitance, C, for all frequencies. This stands to reason, because a capacitor with a large capacitance can store and

▶ **FIGURE 24-8 Voltage, current, and power in an ac capacitor circuit (a)** Note that the time dependences are different for the voltage and the current. In particular, the voltage reaches its maximum value $\pi/2$ rad, or 90°, *after* the current. Thus we say that the voltage *lags* behind the current by 90°. **(b)** The power consumed by a capacitor in an ac circuit, $P = IV$, has an average value of zero.

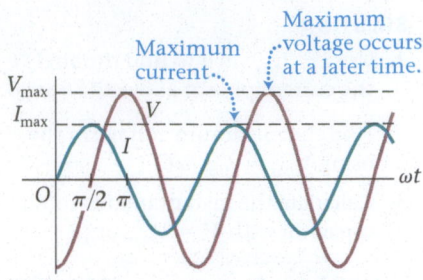

(a) V and I in an ac capacitor circuit

release large amounts of charge, which results in a large current. Finally, we note that rms expressions like $I_{rms} = V_{rms}/X_C$ (Equation 24-8) are equally valid in terms of maximum quantities; that is, $I_{max} = V_{max}/X_C$.

Phasor Diagrams: Capacitor Circuits

FIGURE 24-8 (a) shows the time dependence of the voltage and current in an ac capacitor circuit. Notice that the voltage and current are not in phase. For example, the current has its maximum value at the time when $\omega t = \pi/2 = 90°$, whereas the voltage does not reach its maximum value until a later time, when $\omega t = \pi = 180°$. Thus, we say that the maximum voltage *lags behind* the maximum current—that is, the maximum voltage occurs at a *larger value of the time* than the maximum current. This relationship applies to more than just maximum values. In general:

> The voltage across a capacitor *lags behind* the current by 90°.

This 90° difference between current and voltage can probably best be seen in a phasor diagram. For example, in **FIGURE 24-9** we show both the current and the voltage phasors for a capacitor. The current phasor, with a magnitude I_{max}, is shown at the angle $\theta = \omega t$; it follows that the instantaneous current is $I = I_{max} \sin \omega t$. (For consistency, all phasor diagrams in this chapter will show the current phasor at the angle $\theta = \omega t$.) The voltage phasor, with a magnitude $V_{max} = I_{max}X_C$, is at right angles to the current phasor, pointing in the direction $\theta = \omega t - 90°$. The instantaneous value of the voltage is $V = V_{max} \sin(\omega t - 90°)$. Because the phasors rotate counterclockwise, we see that the voltage phasor lags behind the current phasor.

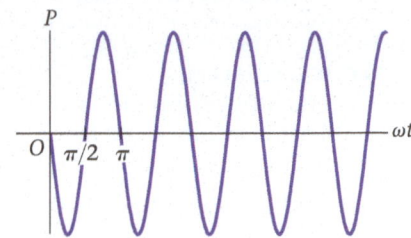

(b) Power in an ac capacitor circuit

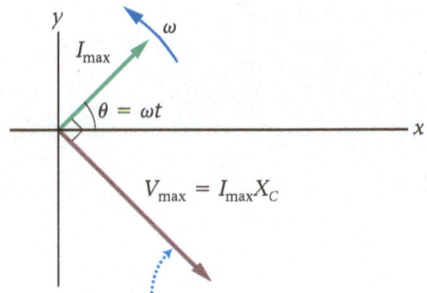

Voltage phasor lags current phasor by 90° in an ac capacitor circuit.

▶ **FIGURE 24-9 Phasor diagram for an ac capacitor circuit** Both the current and voltage phasors rotate counterclockwise about the origin. The voltage phasor lags behind the current phasor by 90°. This means that the voltage points in a direction that is 90° clockwise from the direction of the current.

EXAMPLE 24-6 INSTANTANEOUS VOLTAGE

Consider a capacitor circuit, as in Figure 24-7, where the capacitance is $C = 34.0\,\mu F$, the maximum current is 1.45 A, and the frequency of the generator is 60.0 Hz. At a given time, the current flowing through the capacitor is 0.950 A and increasing. What is the voltage across the capacitor at this time?

PICTURE THE PROBLEM

The appropriate phasor diagram for this system is shown in our sketch. To see how this diagram was drawn, we first recall that the current has a positive value and is increasing. The fact that it is positive means it must point in a direction θ between 0 and 180°. The additional fact that it is increasing means θ must be between 0 and 90°; if it were between 90° and 180°, it would be decreasing with time. Finally, the voltage phasor lags behind the current phasor by 90° in a capacitor.

REASONING AND STRATEGY

There are several pieces that we must put together to complete this solution. First, we know the maximum current, I_{max}, and the instantaneous current, $I = I_{max} \sin \omega t = I_{max} \sin \theta$. We can solve this relationship for the angle θ.

Second, the maximum voltage is given by $V_{max} = I_{max}X_C = I_{max}/\omega C$.

Finally, the instantaneous voltage is $V = V_{max} \sin(\theta - 90°)$, as indicated in the sketch.

Known Capacitance, $C = 34.0\,\mu F$; maximum current, $I_{max} = 1.45$ A; instantaneous current, $I = 0.950$ A; frequency, $f = 60.0$ Hz.

Unknown Voltage across capacitor, $V = ?$

CONTINUED

SOLUTION

1. Use $I = I_{max} \sin\theta$ to find the angle θ of the current phasor above the x axis:

$$\theta = \sin^{-1}\left(\frac{I}{I_{max}}\right) = \sin^{-1}\left(\frac{0.950 \text{ A}}{1.45 \text{ A}}\right) = 40.9°$$

2. Find the maximum voltage in the circuit using $V_{max} = I_{max}/\omega C$:

$$V_{max} = \frac{I_{max}}{\omega C} = \frac{1.45 \text{ A}}{2\pi(60.0 \text{ s}^{-1})(34.0 \times 10^{-6} \text{ F})} = 113 \text{ V}$$

3. Calculate the instantaneous voltage across the capacitor with $V = V_{max}\sin(\theta - 90°)$:

$$V = V_{max}\sin(\theta - 90°)$$
$$= (113 \text{ V})\sin(40.9° - 90°) = -85.4 \text{ V}$$

INSIGHT

We can see from the phasor diagram that the current is positive and increasing in magnitude, whereas the voltage is negative and decreasing in magnitude. It follows that both the current and the voltage are becoming more positive with time.

PRACTICE PROBLEM

What is the voltage across the capacitor when the current in the circuit has its maximum value of 1.45 A? Is the capacitor's voltage increasing or decreasing at this time?   [**Answer:** $V = 0$, increasing. At this time, the current phasor points in the positive y direction, and the voltage phasor points in the positive x direction.]

Some related homework problems: Problem 11, Problem 14

Big Idea **3** The voltage across a capacitor in an ac circuit lags behind the current by 90°. This means that when the current passes through a maximum and begins to decrease, the voltage on the capacitor passes through zero and begins to increase.

To gain a qualitative understanding of the phase relationship between the voltage and current in a capacitor, notice that at the time when $\omega t = \pi/2$, the voltage across the capacitor in Figure 24-8 (a) is zero. Because the capacitor offers no resistance to the flow of current at this time, the current in the circuit is now a maximum. As the current continues to flow, charge builds up on the capacitor, and its voltage increases. This causes the current to decrease. At the time when $\omega t = \pi$, the capacitor voltage reaches a maximum and the current vanishes. As the current begins to flow in the opposite direction, charge flows out of the capacitor and its voltage decreases. When the voltage goes to zero, the current is once again a maximum, though this time in the opposite direction. It follows, then, that the variations of current and voltage are 90° out of phase—that is, when one is a maximum or a minimum, the other is zero—just like position and velocity in simple harmonic motion.

Power

As a final observation on the behavior of a capacitor in an ac circuit, we consider the power it consumes. Recall that the instantaneous power for any circuit can be written as $P = IV$. In **FIGURE 24-8 (b)** we plot the power $P = IV$ corresponding to the current and voltage shown in Figure 24-8 (a). The result is a power that changes sign with time.

In particular, notice that the power is negative when the current and voltage have opposite signs, as between $\omega t = 0$ and $\omega t = \pi/2$, but is positive when they have the same sign, as between $\omega t = \pi/2$ and $\omega t = \pi$. This means that between $\omega t = 0$ and $\omega t = \pi/2$ the capacitor delivers energy to the generator, but between $\omega t = \pi/2$ and $\omega t = \pi$ it draws energy from the generator. As a result, the average power as a function of time is zero, as can be seen by the symmetry about zero power in Figure 24-8 (b). Thus, *a capacitor in an ac circuit consumes zero net energy.*

Enhance Your Understanding   (Answers given at the end of the chapter)

2. Circuit 1 has a capacitor with a capacitance C and an ac generator with an angular frequency ω. Circuit 2 is similar, but its capacitor has a capacitance of $C/2$, and its generator has an angular frequency of 3ω. Is the capacitive reactance in circuit 1 greater than, less than, or equal to the capacitive reactance in circuit 2? Explain.

Section Review

- The rms current in a capacitor in an ac circuit is $I_{rms} = V_{rms}/X_C$, where the capacitive reactance is $X_C = 1/\omega C$.

- The voltage across a capacitor in an ac circuit lags behind the current by 90°.

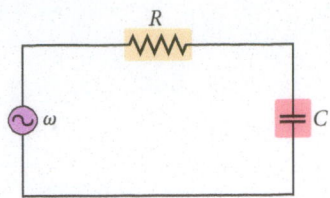

24-3 *RC* Circuits

We now consider a resistor and a capacitor in series in the same ac circuit. This leads to a useful generalization of resistance known as the *impedance*.

Impedance

The circuit shown in **FIGURE 24-10** consists of an ac generator, a resistor, R, and a capacitor, C, connected in series. It is assumed that the values of R and C are known, as well as the maximum voltage, V_{max}, and angular frequency, ω, of the generator. In terms of these quantities we would like to determine the maximum current in the circuit and the maximum voltages across both the resistor and the capacitor.

To begin, we note that the magnitudes of the voltages are readily determined. For instance, the magnitude of the maximum voltage across the resistor is $V_{max,R} = I_{max}R$, and for the capacitor it is $V_{max,C} = I_{max}X_C = I_{max}/\omega C$. The total voltage in this circuit is *not* the sum of these two voltages, however, because they are not in phase—they do not attain their maximum values at the same time. To take these phase differences into account we turn to a phasor diagram.

FIGURE 24-11 shows the phasor diagram for a simple *RC* circuit. To construct this diagram, we start by drawing the current phasor with a length I_{max} at the angle $\theta = \omega t$, as indicated in the figure. Next, we draw the voltage phasor associated with the resistor. This has a magnitude of $I_{max}R$ and points in the same direction as the current phasor. Finally, we draw the voltage phasor for the capacitor. This phasor has a magnitude of $I_{max}X_C$ and points in a direction that is rotated 90° clockwise from the current phasor.

Now, to obtain the total voltage in the circuit in terms of the individual voltages, we perform a vector sum of the resistor-voltage phasor and the capacitor-voltage phasor, as indicated in Figure 24-11. The total voltage, then, is the hypotenuse of the right triangle formed by these two phasors. Its magnitude, using the Pythagorean theorem, is simply

$$V_{max} = \sqrt{V_{max,R}^2 + V_{max,C}^2} \qquad \text{24-10}$$

This equation applies equally well to rms voltages. Substituting the preceding expressions for the voltages across R and C, we find

$$V_{max} = \sqrt{(I_{max}R)^2 + (I_{max}X_C)^2} = I_{max}\sqrt{R^2 + X_C^2}$$

The last quantity in the preceding expression is given a special name; it is called the **impedance**, Z:

Impedance in an *RC* Circuit

$$Z = \sqrt{R^2 + X_C^2} = \sqrt{R^2 + \left(\frac{1}{\omega C}\right)^2} \qquad \text{24-11}$$

SI unit: ohm, Ω

Clearly, the impedance has units of ohms, the same as those of resistance and reactance.

EXERCISE 24-7   IMPEDANCE

A given *RC* circuit has $R = 225\ \Omega$, $C = 32.8\ \mu F$, and $f = 60.0$ Hz. What is the impedance of the circuit?

REASONING AND SOLUTION
The impedance can be found by applying $Z = \sqrt{R^2 + (1/\omega C)^2}$ with $\omega = 2\pi f$. The result in this case is

$$Z = \sqrt{R^2 + \left(\frac{1}{\omega C}\right)^2} = \sqrt{(225\ \Omega)^2 + \left(\frac{1}{2\pi(60.0\ \text{s}^{-1})(32.8 \times 10^{-6}\ \text{F})}\right)^2} = 239\ \Omega$$

We are now in a position to calculate the maximum current in an *RC* circuit. First, given R, C, and $\omega = 2\pi f$, we can determine the value of the impedance Z. Next, we solve for the maximum current by rearranging the equation

$$V_{max} = I_{max}\sqrt{R^2 + X_C^2} = I_{max}Z$$

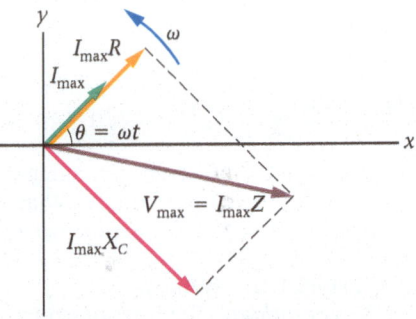

▲ **FIGURE 24-10 An alternating-current *RC* circuit** An ac *RC* circuit consists of a generator, a resistor, and a capacitor connected in series.

▲ **FIGURE 24-11 Phasor diagram for an *RC* circuit** The maximum current in an *RC* circuit is $I_{max} = V_{max}/Z$, where V_{max} is the maximum voltage of the ac generator and $Z = \sqrt{R^2 + X_C^2}$ is the impedance of the circuit. The maximum voltage across the resistor is $V_{max,R} = I_{max}R$ and the maximum voltage across the capacitor is $V_{max,C} = I_{max}X_C$.

Big Idea **4** The voltage in an ac *RC* circuit can be found by adding the voltage phasors of the resistor and capacitor.

This yields

$$I_{max} = \frac{V_{max}}{Z}$$

In a circuit that contains only a resistor, the maximum current is $I_{max} = V_{max}/R$; thus, we see that the impedance Z is indeed a generalization of resistance that can be applied to more complex circuits.

To check some limits of Z, recall that in the capacitor circuit discussed in the previous section (Figure 24-7), the resistance is zero. Hence, in that case the impedance is

$$Z = \sqrt{0 + X_C{}^2} = X_C$$

The maximum voltage across the capacitor, then, is $V_{max} = I_{max}Z = I_{max}X_C$, as expected. Similarly, in a resistor circuit with no capacitor, as in Figure 24-1, the capacitive reactance is zero; hence:

$$Z = \sqrt{R^2 + 0} = R$$

Thus, Z includes X_C and R as special cases.

EXAMPLE 24-8   FIND THE FREQUENCY

An ac generator with an rms voltage of 110 V is connected in series with a 35-Ω resistor and a 11-μF capacitor. The rms current in the circuit is 1.2 A. What are (a) the impedance and (b) the capacitive reactance of this circuit? (c) What is the frequency, f, of the generator?

PICTURE THE PROBLEM

The appropriate circuit is shown in our sketch. We are given the rms voltage of the generator, $V_{rms} = 110$ V; the resistance, $R = 35\ \Omega$; and the capacitance, $C = 11\ \mu$F. The only remaining variable that affects the current is the frequency of the generator, f.

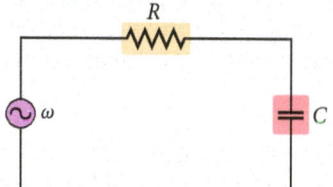

REASONING AND STRATEGY

a. The impedance, Z, can be solved directly from $V_{rms} = I_{rms}Z$.

b. Once the impedance is known, we can use $Z = \sqrt{R^2 + X_C{}^2}$ to find the capacitive reactance, X_C.

c. Now that the reactance is known, we can use the relationships $X_C = 1/\omega C = 1/2\pi f C$ to solve for the frequency, f.

Known   rms voltage, $V_{rms} = 110$ V; rms current, $I_{rms} = 1.2$ A; resistance, $R = 35\ \Omega$; capacitance, $C = 11\ \mu$F.
Unknown   (a) Impedance, $Z = $? (b) Capacitive reactance, $X_C = $? (c) Frequency, $f = $?

SOLUTION

Part (a)

1. Use $V_{rms} = I_{rms}Z$ to find the impedance, Z:

$$Z = \frac{V_{rms}}{I_{rms}} = \frac{110\ \text{V}}{1.2\ \text{A}} = 92\ \Omega$$

Part (b)

2. Solve for X_C in terms of Z and R:

$$Z = \sqrt{R^2 + X_C{}^2}$$
$$X_C = \sqrt{Z^2 - R^2} = \sqrt{(92\ \Omega)^2 - (35\ \Omega)^2} = 85\ \Omega$$

Part (c)

3. Use the value of X_C from part (b) to find the frequency, f:

$$X_C = \frac{1}{\omega C} = \frac{1}{2\pi f C}$$
$$f = \frac{1}{2\pi X_C C} = \frac{1}{2\pi (85\ \Omega)(11 \times 10^{-6}\ \text{F})} = 170\ \text{Hz}$$

INSIGHT

The rms voltage across the resistor is $V_{rms,R} = I_{rms}R = 42$ V, and the rms voltage across the capacitor is $V_{rms,C} = I_{rms}X_C = 102$ V. As expected, these voltages *do not* add up to the generator voltage of 110 V—in fact, they add up to a considerably larger voltage. The point is, however, that the sum of 42 V and 102 V is physically meaningless because the voltages are 90° out of phase, and hence their maximum values do not occur at the same time. If the voltages are combined in the following way, $V_{rms} = \sqrt{V_{rms,R}^2 + V_{rms,C}^2}$, which takes into account the 90° phase difference between the voltages, we find $V_{rms} = \sqrt{V_{rms,R}^2 + V_{rms,C}^2} = \sqrt{(42\ \text{V})^2 + (102\ \text{V})^2} = 110$ V, as expected.

CONTINUED

QUICK EXAMPLE 24-9   FIND THE RESISTANCE

An ac generator with an rms voltage of 120 V and a frequency of 60.0 Hz is connected in series with a 340-μF capacitor and a resistor of resistance R. What value must R have if the rms current in this circuit is to be 2.5 A?

REASONING AND SOLUTION
The resistance can be determined with $Z = \sqrt{R^2 + X_C^2}$ once we know the impedance and capacitive reactance. The impedance can be found with $Z = V_{rms}/I_{rms}$, and the capacitive reactance with $X_C = 1/\omega C = 1/(2\pi fC)$.

1. Rearrange $Z = \sqrt{R^2 + X_C^2}$ to solve for the resistance in terms of the impedance and capacitive reactance:

$$R = \sqrt{Z^2 - X_C^2}$$

2. Calculate the impedance of the circuit:

$$Z = \frac{V_{rms}}{I_{rms}} = \frac{120\ \text{V}}{2.5\ \text{A}} = 48\ \Omega$$

3. Calculate the capacitive reactance:

$$X_C = \frac{1}{2\pi fC} = \frac{1}{2\pi(60.0\ \text{s}^{-1})(340 \times 10^{-6}\ \text{F})} = 7.8\ \Omega$$

4. Use the above results to calculate the resistance:

$$R = \sqrt{Z^2 - X_C^2} = \sqrt{(48\ \Omega)^2 - (7.8\ \Omega)^2} = 47\ \Omega$$

At this frequency the capacitor has relatively little effect in the circuit, as can be seen by comparing X_C, R, and Z.

We can also gain considerable insight about an ac circuit with qualitative reasoning—that is, without going through detailed calculations like those just given. This is illustrated in the following Conceptual Example.

CONCEPTUAL EXAMPLE 24-10   COMPARE BRIGHTNESS: CAPACITOR CIRCUIT

Shown in the sketch are two circuits with identical ac generators and lightbulbs. Circuit 2 differs from circuit 1 by the addition of a capacitor in series with the lightbulb. Does the lightbulb in circuit 2 shine more brightly, less brightly, or with the same intensity as that in circuit 1?

REASONING AND DISCUSSION
Because circuit 2 has both a resistance (in the lightbulb) and a capacitive reactance, its impedance, $Z = \sqrt{R^2 + X_C^2}$, is greater than that of circuit 1, which has only the resistance R. Therefore, the current in circuit 2, $I_{rms} = V_{rms}/Z$, is less than the current in circuit 1. As a result, the average power dissipated in the lightbulb, $P_{av} = I_{rms}^2 R$, is less in circuit 2, so its bulb shines less brightly.

Notice that the bulb dims, even though no power is consumed by the capacitor.

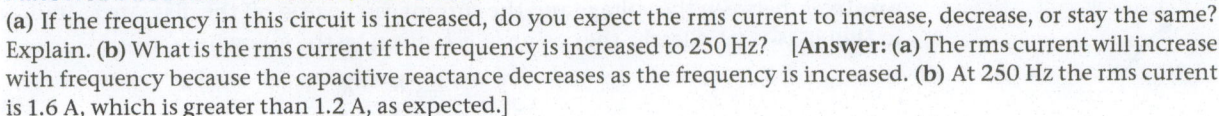

Circuit 1   Circuit 2

ANSWER
The lightbulb in circuit 2 shines less brightly.

Phase Angle and Power Factor

We have seen how to calculate the current in an *RC* circuit and how to find the voltages across each element. Next we consider the phase relationship between the total voltage in the circuit and the current. As we shall see, there is a direct connection between this phase relationship and the power consumed by a circuit.

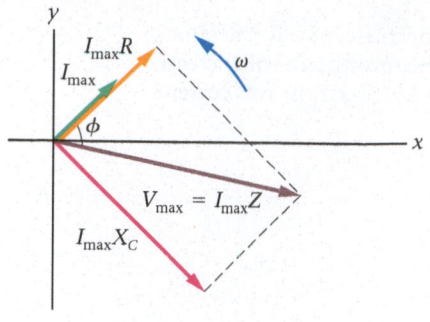

(a) Phase angle ϕ

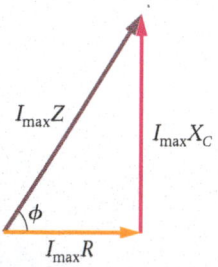

(b) $\cos \phi = R/Z$

▲ **FIGURE 24-12 Phase angle for an RC circuit** **(a)** The phase angle, ϕ, is the angle between the voltage phasor, $V_{max} = I_{max}Z$, and the current phasor, I_{max}. Because I_{max} and $I_{max}R$ are in the same direction, we can say that ϕ is the angle between $I_{max}Z$ and $I_{max}R$. **(b)** From this triangle we can see that $\cos \phi = R/Z$, a result that is valid for any circuit.

To do this, consider the phasor diagram shown in **FIGURE 24-12 (a)**. The **phase angle**, ϕ, between the voltage and the current can be read off the diagram as indicated in **FIGURE 24-12 (b)**. Clearly, the cosine of ϕ is given by the following:

$$\cos \phi = \frac{I_{max}R}{I_{max}Z} = \frac{R}{Z} \qquad 24\text{-}12$$

As we shall see, both the magnitude of the voltage and its phase angle relative to the current play important roles in the behavior of a circuit.

Consider, for example, the power consumed by the circuit shown in Figure 24-10. One way to obtain this result is to recall that no power is consumed by the capacitor at all, as was shown in the previous section. Thus the total power of the circuit is simply the power dissipated by the resistor. This power can be written as

$$P_{av} = I_{rms}^2 R$$

An equivalent expression for the power is obtained by replacing one power of I_{rms} with the expression $I_{rms} = V_{rms}/Z$. This replacement yields

$$P_{av} = I_{rms}^2 R = I_{rms}I_{rms}R = I_{rms}\left(\frac{V_{rms}}{Z}\right)R$$

Finally, recalling that $R/Z = \cos \phi$, we can write the power as follows:

$$P_{av} = I_{rms}V_{rms}\cos \phi \qquad 24\text{-}13$$

Thus, a knowledge of the current and voltage in a circuit, along with the value of $\cos \phi$, gives the power. The multiplicative factor $\cos \phi$ is referred to as the **power factor**.

Writing the power in terms of ϕ allows one to get a feel for the power in a circuit simply by inspecting the phasor diagram. For example, in the case of a circuit with only a capacitor, the phasor diagram (Figure 24-9) shows that the angle between the current and voltage has a magnitude of 90°, so the power factor is zero; $\cos \phi = \cos 90° = 0$. Thus no power is consumed in this circuit, as expected. In contrast, in a purely resistive circuit the phasor diagram shows that $\phi = 0$, as in Figure 24-3, giving a power factor of 1. In this case, the power is simply $P_{av} = I_{rms}V_{rms}$. Therefore, the angle between the current and the voltage in a phasor diagram gives an indication of the power being used by the circuit. We explore this feature in greater detail in the next Example and in **FIGURE 24-13**.

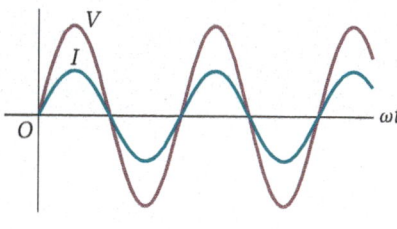

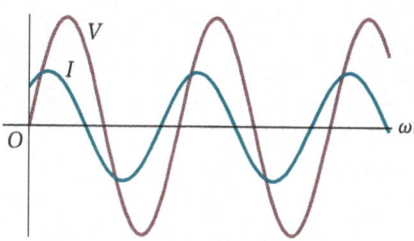

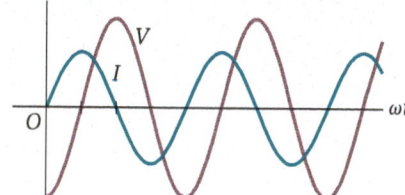

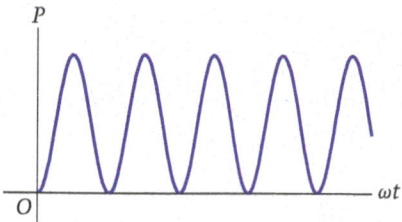

(a) V and I in phase ($\phi = 0$)

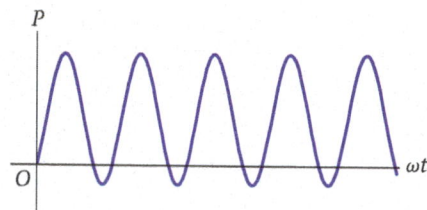

(b) V lags I by 45° ($\phi = 45°$)

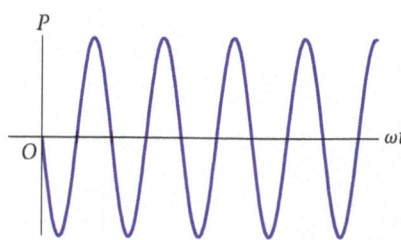

(c) V lags I by 90° ($\phi = 90°$)

▲ **FIGURE 24-13 Voltage, current, and power for three phase angles, ϕ** **(a)** $\phi = 0$. In this case the power is always positive. **(b)** $\phi = 45°$. The power oscillates between positive and negative values, with an average that is positive. **(c)** $\phi = 90°$. The power oscillates symmetrically about its average value of zero.

EXAMPLE 24-11 | POWER AND RESISTANCE

An *RC* circuit has an ac generator with an rms voltage of 240 V. The rms voltage in the circuit lags behind the rms current by 56°, and the rms current is 2.5 A. Find **(a)** the value of the resistance, *R*, and **(b)** the average power consumed by the circuit.

PICTURE THE PROBLEM

The phasor diagram appropriate to this circuit is shown in the sketch. Notice that the voltage lags behind the current by the phase angle $\phi = 56°$.

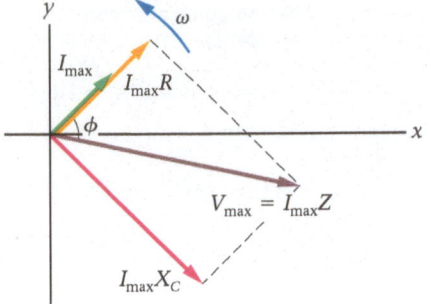

REASONING AND STRATEGY

a. To find the resistance in the circuit, recall that $\cos\phi = R/Z$; thus, $R = Z\cos\phi$. We can find the impedance, *Z*, from $V_{rms} = I_{rms}Z$, since we know both V_{rms} and I_{rms}.

b. The average power consumed by the circuit is simply $P_{av} = I_{rms}V_{rms}\cos\phi$.

Known   rms voltage, V_{rms} = 240 V; rms current, I_{rms} = 2.5 A; phase angle, ϕ = 56°.
Unknown   **(a)** Resistance, *R* = ? **(b)** Average power, P_{av} = ?

SOLUTION

Part (a)

1. Use $V_{rms} = I_{rms}Z$ to find the impedance of the circuit:

$$Z = \frac{V_{rms}}{I_{rms}} = \frac{240\ V}{2.5\ A} = 96\ \Omega$$

2. Now use $\cos\phi = R/Z$ to find the resistance:

$$R = Z\cos\phi = (96\ \Omega)\cos 56° = 54\ \Omega$$

Part (b)

3. Calculate the average power consumed by the circuit with the expression $P_{av} = I_{rms}V_{rms}\cos\phi$:

$$P_{av} = I_{rms}V_{rms}\cos\phi = (2.5\ A)(240\ V)\cos 56° = 340\ W$$

INSIGHT

Of course, the average power consumed by this circuit is simply the average power dissipated in the resistor: $P_{av} = I_{rms}^2 R = (2.5\ A)^2(54\ \Omega) = 340\ W$. Notice that care must be taken if the average power dissipated in the resistor is calculated using $P_{av} = V_{rms}^2/R$. The potential pitfall is that one might use 240 V for the rms voltage; but this is the rms voltage of the generator, *not* the rms voltage across the resistor (which is 135 V). This problem didn't arise with $P_{av} = I_{rms}^2 R$ because the same current flows through both the generator and the resistor.

PRACTICE PROBLEM

What is the capacitive reactance in this circuit?   [**Answer:** X_C = 79 Ω]

Some related homework problems: Problem 24, Problem 25

Enhance Your Understanding   (Answers given at the end of the chapter)

3. The two circuits shown in **FIGURE 24-14** have identical lightbulbs and generators. Circuit 1 has a capacitor with a capacitance of *C*, and circuit 2 has two capacitors, each with a capacitance of *C*, connected in series. Does the light in circuit 1 shine brighter than, dimmer than, or with the same brightness as the light in circuit 2? Explain.

Section Review

- The impedance of an *RC* circuit is $Z = \sqrt{R^2 + X_C^2} = \sqrt{R^2 + (1/\omega C)^2}$.

- The average power consumed in an *RC* circuit is $P_{av} = I_{rms}V_{rms}\cos\phi$, where $\cos\phi = R/Z$ is referred to as the power factor.

Circuit 1   Circuit 2

▲ **FIGURE 24-14** Enhance Your Understanding 3.

PROBLEM-SOLVING NOTE

The Angle in the Power Factor

The angle in the power factor, $\cos\phi$, is the angle between the voltage phasor and the current phasor. Be careful not to identify ϕ with the angle between the voltage phasor and the *x* or *y* axis.

24-4 Inductors in ac Circuits

We turn now to the case of ac circuits that contain inductors. As we shall see, the behavior of inductors is, in many respects, just the opposite of that of capacitors.

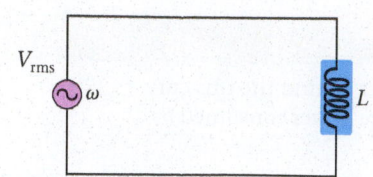

▲ **FIGURE 24-15 An ac generator connected to an inductor** The rms current in an ac inductor circuit is $I_{rms} = V_{rms}/X_L$, where the inductive reactance is $X_L = \omega L$.

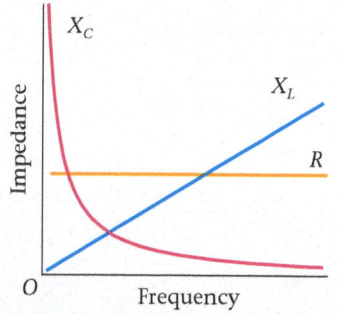

▲ **FIGURE 24-16 Frequency variation of the inductive reactance, X_L, the capacitive reactance, X_C, and the resistance, R** The resistance is independent of frequency. In contrast, the capacitive reactance becomes large with decreasing frequency, and the inductive reactance becomes large with increasing frequency.

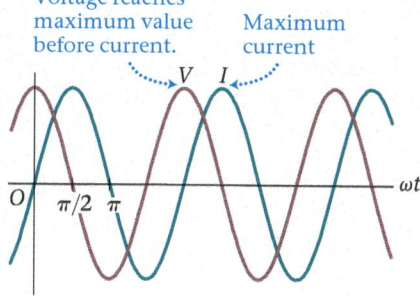

(a) V and I in an ac inductor circuit

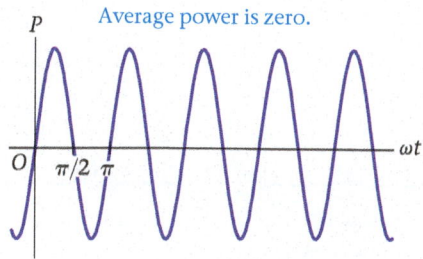

(b) Power in an ac inductor circuit

▲ **FIGURE 24-17 Voltage, current, and power in an ac inductor circuit (a)** Voltage V and current I in an inductor. Note that V reaches a maximum $\pi/2$ rad = 90° *before* the current. Thus, the voltage *leads* the current by 90°. **(b)** The power consumed by an inductor. Note that the average power is zero.

Inductive Reactance

The voltage across a capacitor is given by $V = IX_C$, where $X_C = 1/\omega C$ is the capacitive reactance. Similarly, the voltage across an inductor connected to an ac generator, as in **FIGURE 24-15**, can be written as $V = IX_L$, which defines the **inductive reactance, X_L**. The precise expression for X_L, in terms of frequency and inductance, can be derived using the methods of calculus. The result of such a calculation is the following:

Inductive Reactance, X_L

$$X_L = \omega L \qquad\qquad 24\text{-}14$$

SI unit: ohm, Ω

A plot of X_L versus frequency is given in **FIGURE 24-16**, where it is compared with X_C and R. The rms current in an ac inductor circuit is $I_{rms} = V_{rms}/X_L = V_{rms}/\omega L$.

Notice that X_L increases with frequency, in contrast to the behavior of X_C. This is easily understood when one recalls that the voltage across an inductor has a magnitude given by $\mathcal{E} = L\,\Delta I/\Delta t$. Thus the higher the frequency, the more rapidly the current changes with time, and hence the greater the voltage across the inductor.

EXERCISE 24-12   CURRENT IN AN INDUCTOR

A 35-mH inductor is connected to an ac generator with an rms voltage of 24 V and a frequency of 60.0 Hz. What is the rms current in the inductor?

REASONING AND SOLUTION

The rms current is $I_{rms} = V_{rms}/X_L = V_{rms}/\omega L$. Substituting numerical values, we find

$$I_{rms} = \frac{V_{rms}}{\omega L} = \frac{24\text{ V}}{2\pi(60.0\text{ s}^{-1})(35 \times 10^{-3}\text{ H})} = 1.8\text{ A}$$

Phasor Diagrams: Inductor Circuits

Now that we can calculate the magnitude of the voltage across an inductor, we turn to the question of the phase of the inductor's voltage relative to the current in the circuit. With both phase and magnitude determined, we can construct the appropriate phasor diagram for an inductor circuit.

Suppose the current in an inductor circuit is as shown in **FIGURE 24-17 (a)**. At time zero the current is zero, but increasing at its maximum rate. Because the voltage across an inductor depends on the rate of change of current, it follows that the inductor's voltage is a maximum at $t = 0$. When the current reaches a maximum value, at the time $\omega t = \pi/2$, its rate of change becomes zero; hence, the voltage across the inductor falls to zero at that point, as is also indicated in Figure 24-17 (a). Thus we see that the current and the inductor's voltage are a quarter of a cycle (90°) out of phase. More specifically, because the voltage reaches its maximum *before* the current reaches its maximum—that is, at a smaller value of the time—we say that the voltage *leads* the current:

The voltage across an inductor *leads* the current by 90°.

This is just the opposite of the behavior in a capacitor.

The phase relationship between current and voltage in an inductor is shown with a phasor diagram in **FIGURE 24-18**. Here we see the current, I_{max}, and the corresponding inductor voltage, $V_{max} = I_{max}X_L$. As you can see, the voltage is rotated counterclockwise (that is, ahead) of the current by 90°.

Because of the 90° angle between the current and voltage, the power factor for an inductor is zero; $\cos 90° = 0$. Thus, an ideal inductor—like an ideal capacitor—consumes zero average power; that is,

$$P_{av} = I_{rms}V_{rms}\cos\phi = I_{rms}V_{rms}\cos 90° = 0$$

The instantaneous power in an inductor alternates in sign, as shown in **FIGURE 24-17 (b)**. Thus energy enters the inductor at one time, only to be given up at a later time, for a net gain (on average) of zero energy.

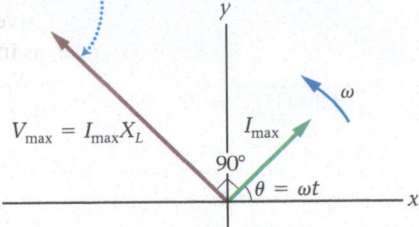

Voltage phasor leads current phasor by 90°.

▲ **FIGURE 24-18 Phasor diagram for an ac inductor circuit** The maximum value of the inductor's voltage is $I_{max}X_L$, and its angle is 90° ahead (counterclockwise) of the current.

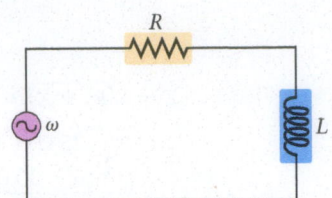

▲ **FIGURE 24-19 An alternating-current RL circuit** An ac RL circuit consists of a generator, a resistor, and an inductor connected in series.

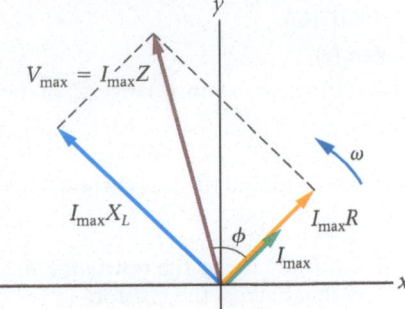

▲ **FIGURE 24-20 Phasor diagram for an RL circuit** The maximum current in an RL circuit is $I_{max} = V_{max}/Z$, where V_{max} is the maximum voltage of the ac generator and $Z = \sqrt{R^2 + X_L^2}$ is the impedance of the circuit. The angle between the maximum voltage phasor and the current phasor is ϕ. Note that the voltage leads the current.

RL Circuits

Next, we consider an ac circuit that contains both a resistor and an inductor, as shown in **FIGURE 24-19**. The corresponding phasor diagram is drawn in **FIGURE 24-20**, where we see the resistor voltage in phase with the current, and the inductor voltage 90° ahead. The total voltage, of course, is given by the vector sum of these two phasors. Therefore, the magnitude of the total voltage is

$$V_{max} = \sqrt{(I_{max}R)^2 + (I_{max}X_L)^2} = I_{max}\sqrt{R^2 + X_L^2} = I_{max}Z$$

This expression defines the impedance, just as with an RC circuit. In this case, the impedance is:

Impedance in an RL Circuit

$$Z = \sqrt{R^2 + X_L^2} = \sqrt{R^2 + (\omega L)^2} \qquad \text{24-15}$$

SI unit: ohm, Ω

The impedance for an RL circuit has the same form as for an RC circuit, except that X_L replaces X_C. Similarly, the power factor for an RL circuit can be written as follows:

$$\cos\phi = \frac{R}{Z} = \frac{R}{\sqrt{R^2 + (\omega L)^2}}$$

We consider an RL circuit in the following Example.

PROBLEM-SOLVING NOTE

Drawing an Inductor Phasor

When drawing a voltage phasor for an inductor, always draw it at right angles to the current phasor. In addition, be sure the inductor's voltage phasor is 90° *counterclockwise* from the current phasor.

Big Idea 5 The voltage across an inductor in an ac circuit leads the current by 90°. This means that when the current passes through a maximum and begins to decrease, the voltage across the inductor passes through zero and begins to decrease.

EXAMPLE 24-13 RL CIRCUIT

A 0.380-H inductor and a 225-Ω resistor are connected in series to an ac generator with an rms voltage of 30.0 V and a frequency of 60.0 Hz. Find **(a)** the rms current in the circuit, **(b)** the rms voltage across the resistor, and **(c)** the rms voltage across the inductor.

PICTURE THE PROBLEM
Our sketch shows a 60.0-Hz generator connected in series with a 225-Ω resistor and a 0.380-H inductor. Because of the series connection, the same current flows through each of the circuit elements.

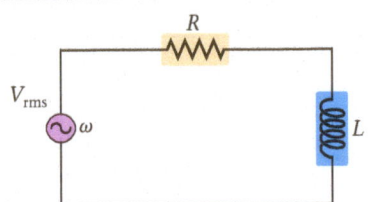

REASONING AND STRATEGY
a. The rms current in the circuit is $I_{rms} = V_{rms}/Z$, where the impedance is $Z = \sqrt{R^2 + (\omega L)^2}$.
b. The rms voltage across the resistor is $V_{rms,R} = I_{rms}R$.
c. The rms voltage across the inductor is $V_{rms,L} = I_{rms}X_L = I_{rms}\omega L$.

Known Inductance, $L = 0.380$ H; resistance, $R = 225$ Ω; rms voltage, $V_{rms} = 30.0$ V; frequency, $f = 60.0$ Hz.
Unknown **(a)** rms current, $I_{rms} = ?$ **(b)** rms voltage across resistor, $V_{rms,R} = ?$ **(c)** rms voltage across inductor, $V_{rms,L} = ?$

CONTINUED

SOLUTION

Part (a)

1. Calculate the impedance, Z, of the circuit:

$$Z = \sqrt{R^2 + X_L^2} = \sqrt{R^2 + (\omega L)^2}$$
$$= \sqrt{(225\ \Omega)^2 + [2\pi(60.0\ \text{s}^{-1})(0.380\ \text{H})]^2} = 267\ \Omega$$

2. Use Z to find the rms current:

$$I_{\text{rms}} = \frac{V_{\text{rms}}}{Z} = \frac{30.0\ \text{V}}{267\ \Omega} = 0.112\ \text{A}$$

Part (b)

3. Multiply I_{rms} by the resistance, R, to find the rms voltage across the resistor:

$$V_{\text{rms},R} = I_{\text{rms}}R = (0.112\ \text{A})(225\ \Omega) = 25.2\ \text{V}$$

Part (c)

4. Multiply I_{rms} by the inductive reactance, X_L, to find the rms voltage across the inductor:

$$V_{\text{rms},L} = I_{\text{rms}}X_L = I_{\text{rms}}\omega L$$
$$= (0.112\ \text{A})2\pi(60.0\ \text{s}^{-1})(0.380\ \text{H}) = 16.0\ \text{V}$$

INSIGHT

As with the RC circuit, the individual voltages *do not* add up to the generator voltage. However, the generator rms voltage *is* equal to $\sqrt{V_{\text{rms},R}^2 + V_{\text{rms},L}^2}$.

PRACTICE PROBLEM — PREDICT/CALCULATE

(a) If the frequency in this circuit is increased, do you expect the rms current to increase, decrease, or stay the same? Explain. (b) What is the current if the frequency is increased to 125 Hz?   [**Answer: (a)** Increasing the frequency increases the inductive reactance, which means the current will decrease. **(b)** At 125 Hz, the current is reduced from 0.112 A to 0.0802 A.]

Some related homework problems: Problem 35, Problem 36

The preceding Practice Problem shows that the current in an RL circuit decreases with increasing frequency. Conversely, as we saw in Example 24-8, an RC circuit has the opposite behavior. These results are summarized and extended in **FIGURE 24-21**, where we show the currents in RL and RC circuits as a function of frequency. It is assumed in this plot that both circuits have the same resistance, R. The horizontal line at the top of the plot indicates the current that would flow if the circuits contained *only* the resistor.

CONCEPTUAL EXAMPLE 24-14   COMPARE BRIGHTNESS: INDUCTOR CIRCUIT

Shown in the sketch are two circuits with identical ac generators and lightbulbs. Circuit 2 differs from circuit 1 by the addition of an inductor in series with the light. Does the lightbulb in circuit 2 shine more brightly, less brightly, or with the same intensity as that in circuit 1?

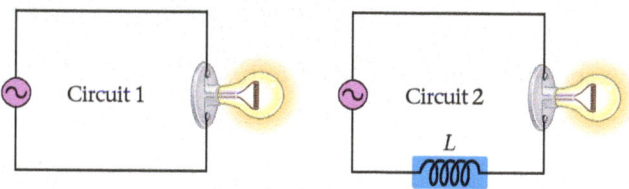

REASONING AND DISCUSSION

Circuit 2 has both a resistance and an inductive reactance, and hence its impedance, $Z = \sqrt{R^2 + X_L^2}$, is greater than that of circuit 1. Therefore, the current in circuit 2, $I_{\text{rms}} = V_{\text{rms}}/Z$, is less than the current in circuit 1, and the average power dissipated in lightbulb 2 is less.

ANSWER

The lightbulb in circuit 2 shines less brightly.

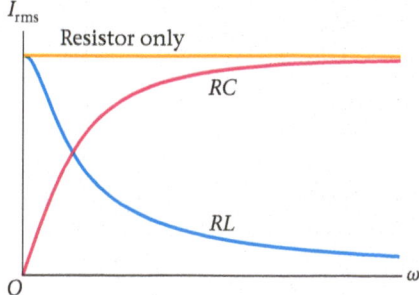

RWP The basic principle illustrated in Conceptual Example 24-14 is used commercially in light dimmers. When you rotate the knob on a light dimmer in one direction or the other, adjusting the light's intensity to the desired level, what you are actually doing is moving an iron rod into or out of the coils of an inductor. This changes both the inductance of the inductor and the intensity of the light. For example, if the

◄ **FIGURE 24-21 RMS currents in RL and RC circuits as a function of frequency** A circuit with only a resistor has the same rms current regardless of frequency. In an RC circuit the current is low where the capacitive reactance is high (low frequency), and in the RL circuit the current is low where the inductive reactance is high (high frequency).

inductance is increased—by moving the iron rod deeper within the inductor's coil—the current in the circuit is decreased and the light dims. The advantage of dimming a light in this way is that no energy is dissipated by the inductor. In contrast, if one were to dim a light by placing a resistor in the circuit, the reduction in light would occur at the expense of wasted energy in the form of heat in the resistor. This needless dissipation of energy is avoided with the inductive light dimmer.

Enhance Your Understanding   (Answers given at the end of the chapter)

4. A circuit consists of an ac generator and an inductor with an inductive reactance X_L. If the period of the generator is doubled, what is the inductive reactance in the circuit? **(a)** $4X_L$, **(b)** $2X_L$, **(c)** X_L, **(d)** $X_L/2$, **(e)** $X_L/4$.

Section Review

*   The inductive reactance is given by $X_L = \omega L$.
*   The impedance of an *RL* circuit is $Z = \sqrt{R^2 + X_L^2} = \sqrt{R^2 + (\omega L)^2}$, and the power factor is $\cos \phi = R/Z = R/\sqrt{R^2 + (\omega L)^2}$.

24-5 *RLC* Circuits

After having considered *RC* and *RL* circuits separately, we now consider circuits with all three elements, *R*, *L*, and *C*. In particular, if *R*, *L*, and *C* are connected in series, as in **FIGURE 24-22**, the resulting circuit is referred to as an **RLC circuit.** We now consider the behavior of such a circuit, using our previous results as a guide.

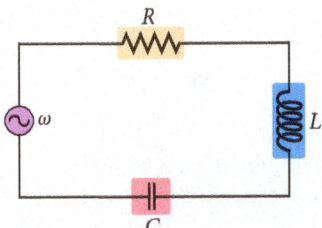

▲ **FIGURE 24-22 An alternating-current *RLC* circuit** An ac *RLC* circuit consists of a generator, a resistor, an inductor, and a capacitor connected in series. The voltage may lead or lag the current, depending on the frequency of the generator and the values of *L* and *C*.

Phasor Diagram

As one might expect, a useful way to analyze the behavior of an *RLC* circuit is with the assistance of a phasor diagram. Thus, the phasor diagram for the circuit in Figure 24-22 is shown in **FIGURE 24-23**. Notice that in addition to the current phasor, we show three separate voltage phasors corresponding to the resistor, inductor, and capacitor. The phasor diagram also shows that the voltage of the resistor is in phase with the current, the voltage of the inductor is 90° ahead of the current, and the voltage of the capacitor is 90° behind the current.

To find the total voltage in the system we must, as usual, perform a vector sum of the three voltage phasors. This process can be simplified if we first sum the inductor and capacitor phasors, which point in opposite directions along the same line. In the case shown in Figure 24-23 we can see that X_L is greater than X_C, so that the sum of these two voltage phasors is $I_{max}X_L - I_{max}X_C$. Combining this result with the phasor for the resistor voltage, and applying the Pythagorean theorem, we obtain the total maximum voltage:

$$V_{max} = \sqrt{(I_{max}R)^2 + (I_{max}X_L - I_{max}X_C)^2} = I_{max}\sqrt{R^2 + (X_L - X_C)^2} = I_{max}Z$$

The impedance of the circuit is thus defined as follows:

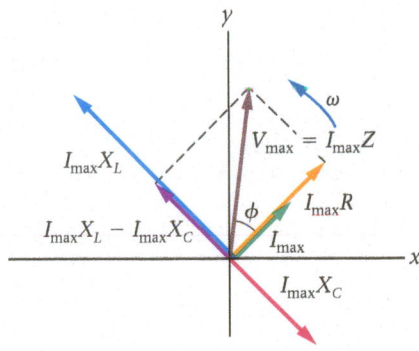

▲ **FIGURE 24-23 Phasor diagram for a typical *RLC* circuit** In the case shown here, we assume that X_L is greater than X_C. All the results are the same for the opposite case, X_C greater than X_L, except that the phase angle ϕ changes sign, as can be seen in Equation 24-17.

Impedance of an *RLC* Circuit

$$Z = \sqrt{R^2 + (X_L - X_C)^2} = \sqrt{R^2 + \left(\omega L - \frac{1}{\omega C}\right)^2} \qquad \text{24-16}$$

SI unit: ohm, Ω

This result for the impedance contains the expressions given for the *RC* and *RL* circuits as special cases. For example, in the *RC* circuit the inductance is zero; hence, $X_L = 0$, and the impedance becomes

$$Z = \sqrt{R^2 + X_C^2}$$

This relationship is identical with the result given for the *RC* circuit in Equation 24-11.

In addition to the magnitude of the total voltage, we are also interested in the phase angle ϕ between it and the current. From the phasor diagram it is clear that this angle is given by

$$\tan \phi = \frac{I_{max}(X_L - X_C)}{I_{max}R} = \frac{X_L - X_C}{R} \qquad \text{24-17}$$

In particular, if X_L is greater than X_C, as in Figure 24-23, then ϕ is positive and the voltage leads the current. On the other hand, if X_C is greater than X_L, it follows that ϕ is negative, so the voltage lags behind the current. In the special case $X_C = X_L$ the phase angle is zero, and the current and voltage are in phase. There is special significance to this last case, as we shall see in greater detail in the next section.

Finally, the power factor, $\cos \phi$, can also be obtained from the phasor diagram. Referring again to Figure 24-23, we see that

$$\cos \phi = \frac{R}{Z}$$

This is precisely the result given earlier for *RC* and *RL* circuits.

EXAMPLE 24-15 DRAW YOUR PHASORS!

An ac generator with a frequency of 60.0 Hz and an rms voltage of 120.0 V is connected in series with a 175-Ω resistor, a 90.0-mH inductor, and a 15.0-μF capacitor. Draw the appropriate phasor diagram for this system and calculate the phase angle, ϕ.

PICTURE THE PROBLEM
Our sketch shows the *RLC* circuit described in the problem statement. In particular, the 60.0-Hz generator is connected in series with a 175-Ω resistor, a 90.0-mH inductor, and a 15.0-μF capacitor.

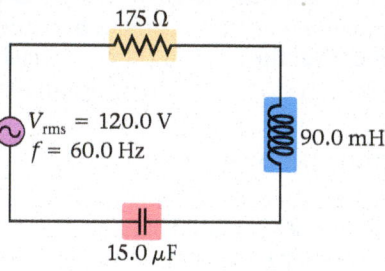

REASONING AND STRATEGY
To draw an appropriate phasor diagram we must first determine the values of X_C and X_L. These values determine immediately whether the voltage leads the current or lags behind it.

The precise value of the phase angle between the current and voltage is given by $\tan \phi = (X_L - X_C)/R$.

Known Frequency, $f = 60.0$ Hz; rms voltage, $V_{rms} = 120.0$ V; resistance, $R = 175\ \Omega$; inductance, $L = 90.0$ mH; capacitance, $C = 15.0\ \mu$F.

Unknown Phase angle, $\phi = ?$

SOLUTION

1. Calculate the capacitive and inductive reactances:

$$X_C = \frac{1}{\omega C} = \frac{1}{2\pi(60.0\ \text{s}^{-1})(15.0 \times 10^{-6}\ \text{F})} = 177\ \Omega$$

$$X_L = \omega L = 2\pi(60.0\ \text{s}^{-1})(90.0 \times 10^{-3}\ \text{H}) = 33.9\ \Omega$$

2. Use $\tan \phi = (X_L - X_C)/R$ to find the phase angle, ϕ:

$$\phi = \tan^{-1}\left(\frac{X_L - X_C}{R}\right)$$

$$= \tan^{-1}\left(\frac{33.9\ \Omega - 177\ \Omega}{175\ \Omega}\right) = -39.3°$$

INSIGHT
We can now draw the phasor diagram for this circuit. First, the fact that X_C is greater than X_L means that the voltage of the generator lags behind the current and that the phase angle is negative. In fact, we know that the phase angle has a magnitude of 39.3°. It follows that the phasor diagram for this circuit is as shown in the accompanying diagram.

Notice that the lengths of the phasors $I_{max}R$, $I_{max}X_C$, and $I_{max}X_L$ in this diagram are drawn in proportion to the values of R, X_C, and X_L, respectively.

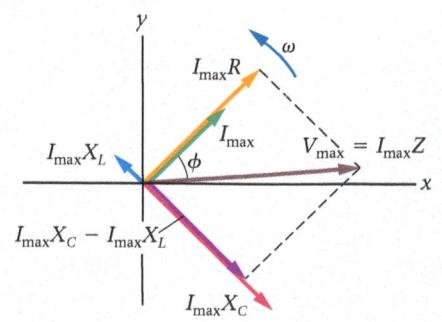

PRACTICE PROBLEM
Find the impedance and rms current for this circuit.
[**Answer:** $Z = 226\ \Omega$, $I_{rms} = V_{rms}/Z = 0.531$ A]

Some related homework problems: Problem 52, Problem 55

The phase angle can also be obtained from the power factor: $\cos\phi = R/Z = (175\,\Omega)/(226\,\Omega) = 0.774$. Notice, however, that $\cos^{-1}(0.774) = \pm39.3°$. Thus, because of the symmetry of the cosine, the power factor determines only the magnitude of ϕ, not its sign. The sign can be obtained from the phasor diagram, of course, but using $\phi = \tan^{-1}[(X_L - X_C)/R]$ yields both the magnitude and sign of ϕ.

QUICK EXAMPLE 24-16 FIND THE INDUCTOR VOLTAGE

The circuit elements shown in the sketch are connected to an ac generator at points A and B. If the rms voltage of the generator is 41 V and its frequency is 75 Hz, what is the rms voltage across the inductor?

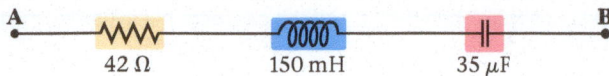

A — 42 Ω — 150 mH — 35 μF — B

REASONING AND SOLUTION

We start by finding the capacitive reactance, the inductive reactance, and the impedance, Z. Once Z is known, we can determine the rms current in the circuit with $I_{rms} = V_{rms}/Z$. With this result, we can find the rms voltage across the inductor with $V_{rms,L} = I_{rms}X_L$.

1. Calculate the capacitive reactance: $X_C = \dfrac{1}{\omega C} = \dfrac{1}{2\pi(75\text{ s}^{-1})(35\times 10^{-6}\text{ F})} = 61\ \Omega$

2. Calculate the inductive reactance: $X_L = \omega L = 2\pi(75\text{ s}^{-1})(0.150\text{ H}) = 71\ \Omega$

3. Use $Z = \sqrt{R^2 + (X_L - X_C)^2}$ to determine the impedance of the circuit:
$$Z = \sqrt{R^2 + (X_L - X_C)^2}$$
$$= \sqrt{(42\ \Omega)^2 + (71\ \Omega - 61\ \Omega)^2} = 43\ \Omega$$

4. Divide the rms voltage, V_{rms}, by the impedance, Z, to find the rms current:
$$I_{rms} = \frac{V_{rms}}{Z} = \frac{41\text{ V}}{43\ \Omega} = 0.95\text{ A}$$

5. Multiply the rms current, I_{rms}, by the inductive reactance, X_L, to find the rms voltage across the inductor:
$$V_{rms,L} = I_{rms}X_L = (0.95\text{ A})(71\ \Omega) = 67\text{ V}$$

Notice that the voltage across individual components in a circuit can be larger than the applied voltage. In this case the voltage across the inductor is 67 V, whereas the applied voltage is only 41 V.

Table 24-1 summarizes the various results for ac circuit elements (R, L, and C) and their combinations (RC, RL, and RLC).

TABLE 24-1 Properties of AC Circuit Elements and Their Combinations

Circuit element	Impedance, Z	Average power, P_{av}	Phase angle, ϕ	Phasor diagram
⟋⟍⟋⟍	$Z = R$	$P_{av} = I_{rms}^2 R = V_{rms}^2/R$	$\phi = 0°$	Figure 24-3
⊣⊢	$Z = X_C = \dfrac{1}{\omega C}$	$P_{av} = 0$	$\phi = -90°$	Figure 24-9
⟋⟍⟋⟍ (coil)	$Z = X_L = \omega L$	$P_{av} = 0$	$\phi = +90°$	Figure 24-18
⟋⟍ ⊣⊢	$Z = \sqrt{R^2 + X_C^2} = \sqrt{R^2 + \left(\dfrac{1}{\omega C}\right)^2}$	$P_{av} = I_{rms}V_{rms}\cos\phi$	$-90° < \phi < 0°$	Figure 24-11
⟋⟍ (coil)	$Z = \sqrt{R^2 + X_L^2} = \sqrt{R^2 + (\omega L)^2}$	$P_{av} = I_{rms}V_{rms}\cos\phi$	$0° < \phi < 90°$	Figure 24-20
⟋⟍ (coil) ⊣⊢	$Z = \sqrt{R^2 + (X_L - X_C)^2}$	$P_{av} = I_{rms}V_{rms}\cos\phi$	$-90° < \phi < 0°$ $(X_C > X_L)$	Figure 24-23
	$= \sqrt{R^2 + \left(\omega L - \dfrac{1}{\omega C}\right)^2}$		$0° < \phi < 90°$ $(X_C < X_L)$	

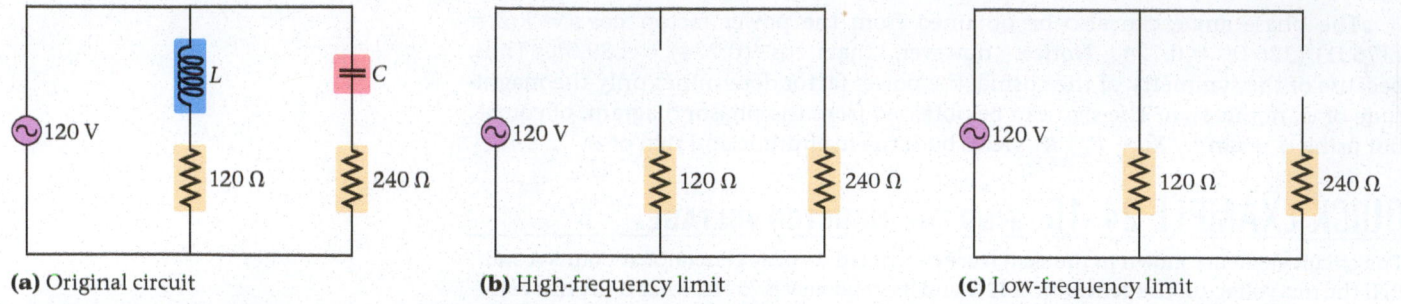

(a) Original circuit **(b)** High-frequency limit **(c)** Low-frequency limit

▲ **FIGURE 24-24 High-frequency and low-frequency limits of an ac circuit (a)** A complex circuit containing resistance, inductance, and capacitance. Although this is not a simple *RLC* circuit, we can still obtain useful results about the circuit in the limits of high and low frequencies. **(b)** The high-frequency limit, in which the inductor is essentially an open circuit, and the capacitor behaves like an ideal wire. **(c)** The low-frequency limit. In this case, the inductor is like an ideal wire, and the capacitor acts like an open circuit.

Big Idea 6 Capacitors act like zero-resistance wires at high frequency, and like open circuits at low frequency. Inductors act like open circuits at high frequency, and like zero-resistance wires at low frequency.

Large and Small Frequencies

In the limit of very large or small frequencies, the behavior of capacitors and inductors is quite simple, which allows us to investigate more complex circuits that contain *R*, *L*, and *C*. For example, in the limit of large frequency the reactance of an inductor becomes very large, whereas that of a capacitor becomes vanishingly small. This means that a capacitor acts like a segment of ideal wire, with no resistance, and an inductor behaves like a very large resistor with practically no current—essentially, an open circuit.

By applying these observations to the circuit in **FIGURE 24-24 (a)**, for example, we can predict the current supplied by the generator at high frequencies. First, we replace the capacitor with an ideal wire and the inductor with an open circuit. This results in the simplified circuit shown in **FIGURE 24-24 (b)**. Clearly, the current in this circuit is $I_{rms} = V_{rms}/R = (120\text{ V})/(240\ \Omega) = 0.50\text{ A}$; hence, we expect the current in the original circuit to approach this value as the frequency is increased.

In the opposite extreme of very small frequency, we obtain the behavior that would be expected if the ac generator were replaced with a battery. Specifically, the reactance of an inductor vanishes as the frequency goes to zero, whereas that of a capacitor becomes extremely large. Thus, the roles of the inductor and capacitor are now reversed; it is the inductor that acts like an ideal wire, and the capacitor that behaves like an open circuit. For small frequencies, then, the circuit in Figure 24-24 (a) will behave the same as the circuit shown in **FIGURE 24-24 (c)**. The current in this circuit, and in the original circuit at low frequency, is $I_{rms} = V_{rms}/R = (120\text{ V})/(120\ \Omega) = 1.0\text{ A}$.

EXAMPLE 24-17 FIND *R*

The circuit shown in the sketch is connected to an ac generator with an rms voltage of 120 V. What value must *R* have if the rms current in this circuit is to approach 1.0 A at high frequency?

PICTURE THE PROBLEM
The top diagram shows the original circuit with its various elements. The high-frequency behavior of this circuit is indicated in the bottom diagram, where the inductor has been replaced with an open circuit and the capacitors have been replaced with ideal wires.

REASONING AND STRATEGY
The high-frequency circuit has only one remaining path through which current can flow. On this path the resistors with resistance *R* and 100 Ω are in series. Therefore, the total resistance of the circuit is $R_{total} = R + 100\ \Omega$. Finally, the rms current is $I_{rms} = V_{rms}/R_{total}$. Setting I_{rms} equal to 1.0 A gives us the value of *R*.

Known rms voltage, $V_{rms} = 120$ V; rms current at high frequency, $I_{rms} = 1.0$ A.
Unknown Resistance, $R = ?$

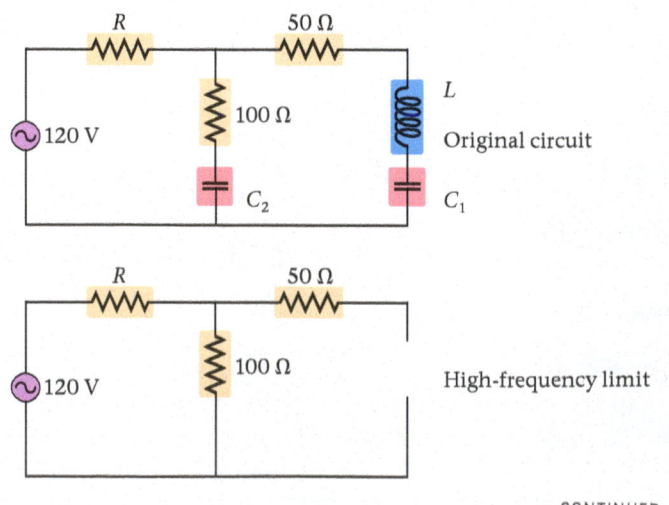

Original circuit

High-frequency limit

CONTINUED

SOLUTION

1. Calculate the total resistance of the high-frequency circuit: $R_{\text{total}} = R + 100 \ \Omega$

2. Write an expression for the rms current in the circuit: $I_{\text{rms}} = \dfrac{V_{\text{rms}}}{R_{\text{total}}} = \dfrac{V_{\text{rms}}}{R + 100 \ \Omega}$

3. Solve for the resistance R: $R = \dfrac{V_{\text{rms}}}{I_{\text{rms}}} - 100 \ \Omega = \dfrac{120 \ \text{V}}{1.0 \ \text{A}} - 100 \ \Omega = 20 \ \Omega$

INSIGHT

Notice that no values are given for the capacitances and the inductance. At high enough frequencies the precise values of these quantities are unimportant.

PRACTICE PROBLEM

What is the rms current in this circuit in the limit of small frequency?   [**Answer:** The current approaches zero.]

Some related homework problems: Problem 46, Problem 51

Enhance Your Understanding   (Answers given at the end of the chapter)

5. Two ac circuits are shown in **FIGURE 24-25**. At high frequencies, is the current in circuit 1 greater than, less than, or equal to the current in circuit 2? Explain.

Section Review

- The impedance of an *RLC* circuit is given by $Z = \sqrt{R^2 + (X_L - X_C)^2} = \sqrt{R^2 + (\omega L - 1/\omega C)^2}$.
- At high frequency a capacitor acts like an ideal wire, and at low frequency it acts like an open circuit. Just the opposite is the case for an inductor.

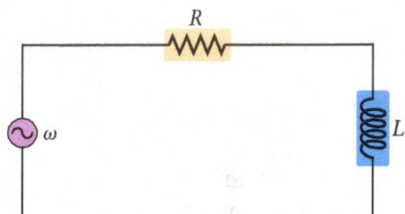

Circuit 1

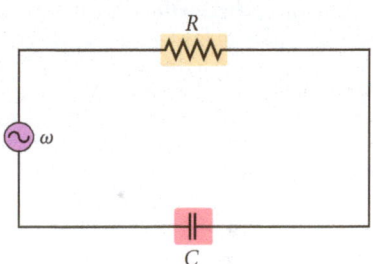

Circuit 2

▲ **FIGURE 24-25** Enhance Your Understanding 5.

24-6 Resonance in Electric Circuits

Many physical systems have natural frequencies of oscillation. For example, we saw in Chapter 13 that a child on a swing oscillates about the vertical with a definite, characteristic frequency determined by the length of the swing and the acceleration due to gravity. Similarly, an object attached to a spring oscillates about its equilibrium position with a frequency determined by the "stiffness" of the spring and the mass of the object. Certain electric circuits have analogous behavior—their electric currents oscillate with characteristic frequencies. In this section we consider some examples of "oscillating" electric circuits.

LC Circuits

Perhaps the simplest circuit that displays an oscillating electric current is an *LC* **circuit** with no generator—that is, one that consists of nothing more than an inductor and a capacitor. Suppose, for example, that at $t = 0$ a charged capacitor has just been connected to an inductor and there is no current in the circuit, as shown in **FIGURE 24-26 (a)**. The capacitor is charged, and hence it has a voltage, $V = Q/C$, which causes a current to begin flowing through the inductor, as in **FIGURE 24-26 (b)**. Soon all the charge drains from the capacitor and its voltage drops to zero, but the current continues to flow because an inductor resists changes in its current. In fact, the current continues flowing until the capacitor becomes charged enough in the *opposite* direction to stop the current, as in **FIGURE 24-26 (c)**. At this point, the current begins to flow back the way it came, and the same sequence of events occurs again, leading to a *series of oscillations* in the current.

In the ideal case, the oscillations can continue forever, since neither an inductor nor a capacitor dissipates energy. The situation is completely analogous to a mass oscillating on a spring with no friction, as we indicate in Figure 24-26. At $t = 0$ the capacitor has a charge of magnitude Q on its plates, which means it stores the energy

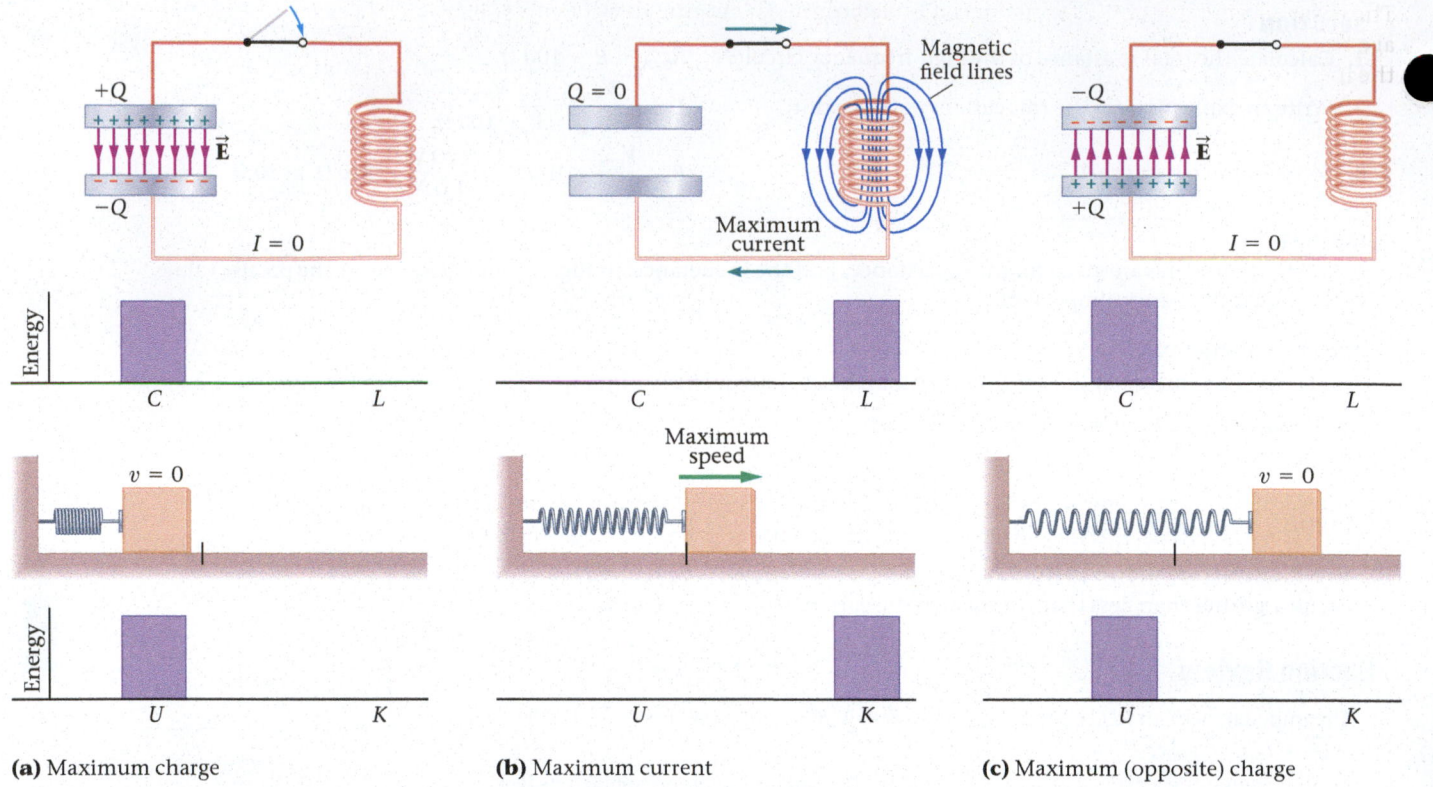

(a) Maximum charge   **(b)** Maximum current   **(c)** Maximum (opposite) charge

▲ **FIGURE 24-26 Oscillations in an *LC* circuit with no generator** The current oscillations in an *LC* circuit are analogous to the oscillations of a mass on a spring. **(a)** At the moment the switch is closed, all the energy in the circuit is stored in the electric field of the charged capacitor. This is analogous to a mass at rest against a compressed spring, where all the energy of the system is stored in the spring. **(b)** A quarter of the way through a cycle, the capacitor is uncharged and the current in the inductor is a maximum. At this time, all the energy in the circuit is stored in the magnetic field of the inductor. In the mass–spring analog, the spring is at equilibrium and the mass has its maximum speed. All the system's energy is now in the form of kinetic energy. **(c)** After half a cycle, the capacitor is fully charged in the opposite direction and holds all the system's energy. This corresponds to a fully extended spring (with all the system's energy) and the mass at rest.

$U_C = Q^2/2C$. This situation is analogous to a spring being compressed a distance x and storing the potential energy $U = \frac{1}{2}kx^2$. At a later time the charge on the capacitor is zero, so it no longer stores any energy. The energy is not lost, however. Instead, it is now in the inductor, which carries a current I and stores the energy $U_L = \frac{1}{2}LI^2 = U_C$. In the mass–spring system, this corresponds to the mass being at the equilibrium position of the spring. At this time all the system's energy is the kinetic energy of the mass, $K = \frac{1}{2}mv^2 = U$, and none is stored in the spring. As the current continues to flow, it charges the capacitor with the opposite polarity until it reaches the magnitude Q and stores the same energy, U_C, as at $t = 0$. In the mass–spring system, this corresponds to the spring being stretched by the same distance x that it was originally compressed, which again stores all the initial energy as potential energy.

Thus, we see that a close analogy exists between a capacitor and a spring, and between an inductor and a mass. In addition, the charge on the capacitor is analogous to the displacement of the spring, and the current in the inductor is analogous to the speed of the mass. Thus, for example, the energy stored in the inductor, $\frac{1}{2}LI^2$, corresponds precisely to the kinetic energy of the mass, $\frac{1}{2}mv^2$. Comparing the potential energy of a spring, $\frac{1}{2}kx^2$, and the energy stored in a capacitor, $Q^2/2C$, we see that the stiffness of a spring is analogous to $1/C$. This makes sense, because a capacitor with a large capacitance, C, can store large quantities of charge with ease, just as a spring with a small force constant (if C is large, then $k = 1/C$ is small) can be stretched quite easily.

In the mass–spring system the natural angular frequency of oscillation is determined by the characteristics of the system. In particular, recall from Section 13-8 that

$$f_0 = \frac{1}{2\pi}\sqrt{\frac{k}{m}} \quad \text{or} \quad \omega_0 = 2\pi f_0 = \sqrt{\frac{k}{m}}$$

The natural frequency of the *LC* circuit can be determined by noting that the rms voltage across the capacitor *C* in Figure 24-26 must be the same as the rms voltage across the inductor *L*. This condition can be written as follows:

$$V_{rms,C} = V_{rms,L}$$

$$I_{rms}X_C = I_{rms}X_L$$

$$I_{rms}\left(\frac{1}{\omega_0 C}\right) = I_{rms}(\omega_0 L)$$

Solving for ω_0, we find:

Natural Frequency of an *LC* Circuit

$$\omega_0 = \frac{1}{\sqrt{LC}} = 2\pi f_0 \qquad\qquad 24\text{-}18$$

SI unit: s^{-1}

Note again the analogy between this result and that for a mass on a spring: If we make the following replacements, $m \rightarrow L$ and $k \rightarrow 1/C$, in the mass–spring result, we find the expected *LC* result:

$$\omega_0 = \sqrt{\frac{k}{m}} = \sqrt{\frac{1}{LC}}$$

We summarize the mass–spring/*LC* circuit analogies in Table 24-2.

TABLE 24-2 Analogies Between a Mass on a Spring and an *LC* Circuit

Mass–spring system		*LC* circuit	
Position	x	Charge	q
Velocity	$v = \Delta x/\Delta t$	Current	$I = \Delta q/\Delta t$
Mass	m	Inductance	L
Force constant	k	Inverse capacitance	$1/C$
Natural frequency	$\omega_0 = \sqrt{k/m}$	Natural frequency	$\omega_0 = \sqrt{1/LC}$

EXERCISE 24-18   FIND THE CAPACITANCE

It is desired to tune the natural frequency of an *LC* circuit to match the 91.3-MHz broadcast signal of an FM radio station. If a capacitor of 2.45×10^{-12} F is to be used in the circuit, what inductance is required?

REASONING AND SOLUTION

The natural frequency is given by $\omega_0 = 1/\sqrt{LC}$. Solving this relationship for the inductance, we find

$$L = \frac{1}{\omega_0^2 C} = \frac{1}{[2\pi(91.3 \times 10^6 \text{ s}^{-1})]^2 (2.45 \times 10^{-12} \text{ F})} = 1.24 \times 10^{-6} \text{ H}$$

Resonance

Whenever a physical system has a natural frequency, we can expect to find resonance when it is driven near that frequency. In a mass–spring system, for example, if we move the top end of the spring up and down with a frequency near the natural frequency, the displacement of the mass can become quite large. Similarly, if we push a person on a swing at the right frequency, the amplitude of motion will increase. These are examples of resonance in mechanical systems.

To drive an electric circuit, we can connect it to an ac generator. As we adjust the frequency of the generator, the current in the circuit will be a maximum at the natural frequency of the circuit. For example, consider the circuit shown in Figure 24-22. Here

PHYSICS IN CONTEXT
Looking Back

Alternating-current circuits have many similarities with oscillating mechanical systems. In particular, this chapter presents detailed connections between the behavior of a mass on a spring (Chapter 13) and the behavior of an *RLC* circuit.

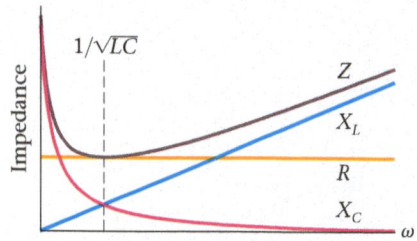

(a) Impedance is a minimum ($Z = R$) at resonance

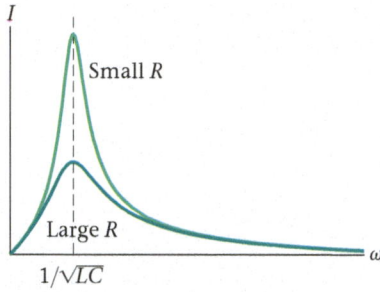

(b) Resonance curves for the current

▲ FIGURE 24-28 **Airport metal detector**

PROBLEM-SOLVING NOTE

A Circuit at Resonance

The resonance frequency of an *RLC* circuit depends only on the inductance, L, and the capacitance, C; that is, $\omega_0 = 1/\sqrt{LC}$. On the other hand, the impedance at resonance depends only on the resistance, R; that is, $Z = R$.

◄ FIGURE 24-27 **Resonance in an *RLC* circuit (a)** The impedance, Z, of an *RLC* circuit varies as a function of frequency. The minimum value of Z—which corresponds to the largest current—occurs at the resonance frequency $\omega_0 = 1/\sqrt{LC}$, where $X_C = X_L$. At this frequency $Z = R$. **(b)** Typical resonance peaks for an *RLC* circuit. The location of the peak is independent of resistance, but the resonance effect becomes smaller and more spread out with increasing R.

an ac generator drives a circuit containing an inductor, a capacitor, and a resistor. As we have already seen, the inductor and capacitor together establish a natural frequency $\omega_0 = 1/\sqrt{LC}$, and the resistor provides for energy dissipation.

The phasor diagram for this circuit is like the one shown in Figure 24-23. Recall that the maximum current in this circuit is

$$I_{max} = \frac{V_{max}}{Z}$$

Thus, the smaller the impedance, the larger the current. Hence, to obtain the *largest* possible current, we must have the *smallest Z*. Recall that the impedance is given by

$$Z = \sqrt{R^2 + (X_L - X_C)^2}$$

Writing Z in terms of frequency, we have

$$Z = \sqrt{R^2 + \left(\omega L - \frac{1}{\omega C}\right)^2} \qquad \text{24-19}$$

This expression for Z is plotted in **FIGURE 24-27 (a)**. Notice that the smallest value of the impedance is $Z = R$, and that this value is attained precisely at the frequency where $X_L = X_C$. We can see this mathematically by setting $X_L = X_C$ in the expression $Z = \sqrt{R^2 + (X_L - X_C)^2}$, which yields $Z = \sqrt{R^2 + 0} = R$, as expected. The frequency at which $X_L = X_C$ is the frequency for which $\omega L = 1/\omega C$. This frequency is $\omega = 1/\sqrt{LC} = \omega_0$, the *natural frequency* found in Equation 24-18 for LC circuits.

FIGURE 24-27 (b) shows typical plots of the current in an *RLC* circuit. Notice that the current peaks at the resonance frequency. Note also that increasing the resistance, which reduces the maximum current in the circuit, does not change the resonance frequency. It does, however, make the resonance peak flatter and broader. As a result, the resonance effect occurs over a wide range of frequencies and gives only a small increase in current. If the resistance is small, however, the peak is high and sharp. In this case, the resonance effect yields a large current that is restricted to a very small range of frequencies.

RWP Radio and television tuners use low-resistance *RLC* circuits so that they can pick up one station at a frequency f_1 without also picking up a second station at a nearby frequency f_2. In a typical radio tuner, for example, turning the knob of the tuner rotates one set of capacitor plates between a second set of plates, effectively changing both the plate separation and the plate area. This changes the capacitance in the circuit and the frequency at which it resonates. If the resonance peak is high and sharp—as occurs with low resistance—only the station broadcasting at the resonance frequency will be picked up and amplified. Other stations at nearby frequencies will produce such small currents in the tuning circuit that they will be undetectable.

RWP Metal detectors also use resonance in *RLC* circuits, although in this case it is the inductance that changes rather than the capacitance. For example, when you walk through a metal detector at an airport, as in **FIGURE 24-28**, you are actually walking through a large coil of wire—an inductor. Metal objects on your person cause the inductance of the coil to increase slightly. This increase in inductance results in a small decrease in the resonance frequency of the *RLC* circuit to which the coil is connected. If the resonance peak is sharp and high, even a slight change in frequency results in a large change in current. It is this change in current that sets off the detector, indicating the presence of metal.

CONCEPTUAL EXAMPLE 24-19 PHASE OF THE VOLTAGE

An *RLC* circuit is driven at its resonance frequency. Is its voltage ahead of, behind, or in phase with the current?

REASONING AND DISCUSSION
At resonance the capacitive and inductive reactances are equal, which means that the voltage across the capacitor is equal in magnitude and opposite in direction to the voltage across the inductor. As a result, the net voltage in the system is simply the voltage across the resistor, which is in phase with the current.

ANSWER
The voltage and current are in phase.

EXAMPLE 24-20 A CIRCUIT IN RESONANCE

An ac generator with an rms voltage of 25 V is connected in series to a 10.0-Ω resistor, a 53-mH inductor, and a 65-μF capacitor. Find (a) the resonance frequency of the circuit and (b) the rms current at resonance. In addition, sketch the phasor diagram at resonance.

PICTURE THE PROBLEM
As mentioned in Conceptual Example 24-19, the magnitude of the voltage across an inductor at resonance is equal to the magnitude of the voltage across the capacitor. Because the phasors corresponding to these voltages point in opposite directions, however, they cancel. This leads to a net voltage phasor that is simply $I_{max}R$ in phase with the current, as indicated in our sketch. Notice that the lengths of the phasors $I_{max}R$, $I_{max}X_C$, and $I_{max}X_L$ are drawn in proportion to the values of R, X_C, and X_L, respectively.

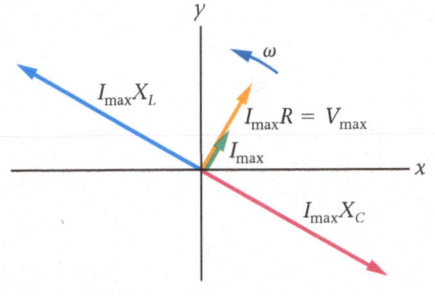

REASONING AND STRATEGY
a. We find the resonance frequency by substituting numerical values into $\omega_0 = 1/\sqrt{LC} = 2\pi f_0$.

b. At resonance the impedance of the circuit is simply the resistance; that is, $Z = R$. Thus, the rms current in the circuit is $I_{rms} = V_{rms}/Z = V_{rms}/R$.

Known rms voltage, $V_{rms} = 25$ V; resistance, $R = 10.0\ \Omega$; inductance, $L = 53$ mH; capacitance, $C = 65\ \mu$F.
Unknown (a) Resonance frequency, $f_0 = ?$ (b) rms current at resonance, $I_{rms} = ?$

SOLUTION

Part (a)

1. Calculate the resonance frequency:

$$\omega_0 = \frac{1}{\sqrt{LC}} = 2\pi f_0$$

$$f_0 = \frac{1}{2\pi\sqrt{LC}}$$

$$= \frac{1}{2\pi\sqrt{(53 \times 10^{-3}\ \text{H})(65 \times 10^{-6}\ \text{F})}} = 86\ \text{Hz}$$

Part (b)

2. Determine the impedance at resonance:

$$Z = R = 10.0\ \Omega$$

3. Divide the rms voltage by the impedance to find the rms current:

$$I_{rms} = \frac{V_{rms}}{Z} = \frac{25\ \text{V}}{10.0\ \Omega} = 2.5\ \text{A}$$

INSIGHT
If the frequency of this generator is increased above resonance, the inductive reactance, X_L, will be larger than the capacitive reactance, X_C, and hence the voltage will lead the current. If the frequency is lowered below resonance, the voltage will lag behind the current.

PRACTICE PROBLEM
What is the magnitude of the rms voltage across the capacitor?   [**Answer:** 71 V. Notice again that the voltage across a given circuit element can be much larger than the applied voltage.]

Some related homework problems: Problem 66, Problem 67

**PHYSICS
IN CONTEXT
Looking Ahead**

Electromagnetic waves can be detected with an *RLC* circuit whose frequency is matched to the frequency of the electromagnetic wave. This is discussed in Chapter 25.

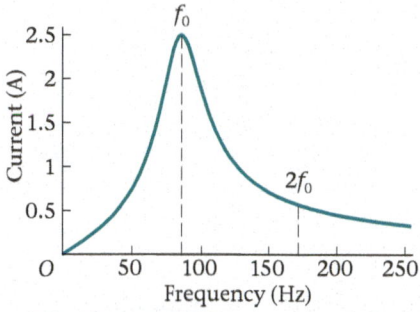

▲ **FIGURE 24-29 RMS current in an *RLC* circuit** This plot shows the rms current versus frequency for the circuit considered in Example 24-20 and Quick Example 24-21. The currents determined in these Examples occur at the frequencies indicated by the vertical dashed lines.

QUICK EXAMPLE 24-21   FIND THE OFF-RESONANCE CURRENT

Referring to the system in the previous Example, find the rms current when the generator operates at a frequency that is twice the resonance frequency.

REASONING AND SOLUTION
We now use the frequency $f = 2f_0 = 2(86\ \text{Hz})$ to find the inductive reactance, the capacitive reactance, and the impedance, Z. The rms current is then given by $I_{rms} = V_{rms}/Z$.

1. Calculate the inductive reactance at the frequency $f = 2(86\ \text{Hz})$:

$$X_L = \omega L = 2\pi[2(86\ \text{Hz})](53 \times 10^{-3}\ \text{H})$$
$$= 57\ \Omega$$

2. Calculate the capacitive reactance at the frequency $f = 2(86\ \text{Hz})$:

$$X_C = \frac{1}{\omega C} = \frac{1}{2\pi[2(86\ \text{Hz})](65 \times 10^{-6}\ \text{F})} = 14\ \Omega$$

3. Use these results, along with the resistance R, to find the impedance, Z:

$$Z = \sqrt{R^2 + (X_L - X_C)^2}$$
$$= \sqrt{(10.0\ \Omega)^2 + (57\ \Omega - 14\ \Omega)^2} = 44\ \Omega$$

4. Divide the rms voltage by the impedance to find the rms current:

$$I_{rms} = \frac{V_{rms}}{Z} = \frac{25\ \text{V}}{44\ \Omega} = 0.57\ \text{A}$$

As expected, the inductive reactance is greater than the capacitive reactance at this frequency, and the current is less than the current at resonance.

24-20 and Quick Example 24-21 is shown in **FIGURE 24-29**. Notice that the two currents just calculated, at $f_0 = 86\ \text{Hz}$ and $f = 2f_0 = 2(86\ \text{Hz}) = 172\ \text{Hz}$, are indicated on the graph.

Enhance Your Understanding   (Answers given at the end of the chapter)

6. An *RLC* circuit has a resonance frequency of 60 Hz. When this circuit is operated at 30 Hz, does the maximum voltage of the generator lead or lag behind the current in the circuit? Explain.

Section Review

- The resonance of an *RLC* circuit occurs at the frequency $\omega_0 = 1/\sqrt{LC}$.

CHAPTER 24 REVIEW

CHAPTER SUMMARY

24-1   ALTERNATING VOLTAGES AND CURRENTS

An ac generator produces a voltage that varies with time as

$$V = V_{max} \sin \omega t \qquad \text{24-1}$$

Phasors
A phasor is a rotating vector representing a voltage or a current in an ac circuit.

RMS Values
The rms, or root mean square, of a quantity x is the square root of the average value of x^2.

Standard dc formulas like $P = I^2R$ can be converted to ac average formulas by using rms values; for example, $P_{av} = I_{max}^2 R$.

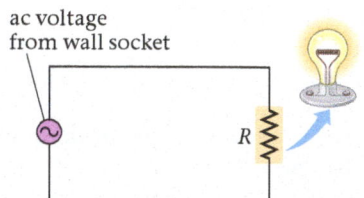

ac voltage
from wall socket

R

24-2   CAPACITORS IN AC CIRCUITS

A capacitor in an ac circuit has a current that depends on the frequency and is out of phase with the voltage.

Capacitive Reactance

The rms current in a capacitor is

$$I_{rms} = \frac{V_{rms}}{X_C} \qquad \text{24-8}$$

In this expression, X_C is the capacitive reactance:

$$X_C = \frac{1}{\omega C} \qquad \text{24-9}$$

Phase Relationship Between Voltage and Current in a Capacitor

The voltage across a capacitor lags behind the current by 90°.

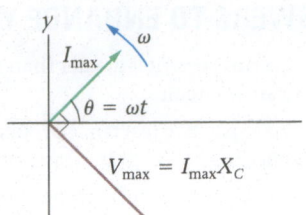

24-3 *RC* CIRCUITS

Impedance

The impedance, Z, of an RC circuit is

$$Z = \sqrt{R^2 + X_C{}^2} = \sqrt{R^2 + \left(\frac{1}{\omega C}\right)^2} \qquad \text{24-11}$$

Voltage and Current

The rms voltage and current in an RC circuit are related by

$$V_{rms} = I_{rms}\sqrt{R^2 + X_C{}^2} = I_{rms}Z$$

Power Factor

If the phase angle between the current and voltage in an RC circuit is ϕ, the average power consumed by the circuit is

$$P_{av} = I_{rms}V_{rms}\cos\phi \qquad \text{24-13}$$

The term $\cos\phi$ is referred to as the power factor.

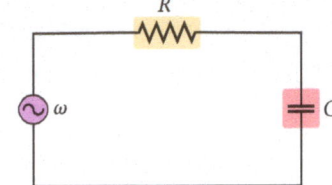

24-4 INDUCTORS IN AC CIRCUITS

Inductive Reactance

The inductive reactance is

$$X_L = \omega L \qquad \text{24-14}$$

The rms current in an inductor is

$$I_{rms} = \frac{V_{rms}}{X_L}$$

RL Circuits

The impedance of an RL circuit is

$$Z = \sqrt{R^2 + X_L{}^2} = \sqrt{R^2 + (\omega L)^2} \qquad \text{24-15}$$

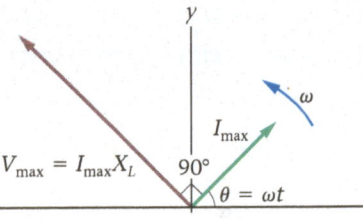

24-5 *RLC* CIRCUITS

Impedance

The impedance of an RLC circuit is

$$Z = \sqrt{R^2 + (X_L - X_C)^2} = \sqrt{R^2 + \left(\omega L - \frac{1}{\omega C}\right)^2} \qquad \text{24-16}$$

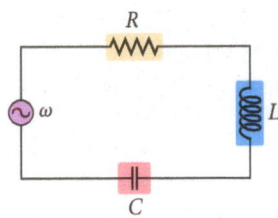

Large and Small Frequencies

In the limit of large frequencies, inductors behave like open circuits and capacitors are like ideal wires. For very small frequencies, the behaviors are reversed; inductors are like ideal wires, and capacitors act like open circuits.

24-6 RESONANCE IN ELECTRIC CIRCUITS

Electric circuits can have natural frequencies of oscillation, just like a pendulum or a mass on a spring.

LC Circuits

Circuits containing only an inductor and a capacitor have a natural frequency given by

$$\omega_0 = \frac{1}{\sqrt{LC}} = 2\pi f_0 \qquad \text{24-18}$$

Resonance

An RLC circuit connected to an ac generator has maximum current at the frequency $\omega = 1/\sqrt{LC}$. This effect is referred to as resonance.

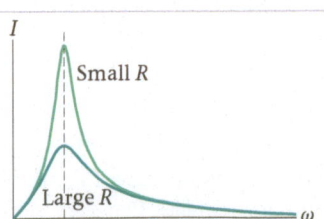

ANSWERS TO ENHANCE YOUR UNDERSTANDING QUESTIONS

1. The rms current is greater than the average current, which is zero in an ac circuit.

2. The capacitive reactance of circuit 1 is greater than that of circuit 2, because $3\omega(C/2)$ is greater than ωC.

3. The light in circuit 1 shines brighter than the light in circuit 2, because circuit 1 has the smaller impedance.

4. (d) $X_L/2$.

5. At high frequency the current in circuit 1 is less than that in circuit 2, because the inductor in circuit 1 acts like an open circuit at high frequency.

6. The voltage lags the current because at frequencies less than the resonance frequency, the capacitor dominates the circuit (that is, $X_C > X_L$), and voltage lags current in capacitor circuits.

CONCEPTUAL QUESTIONS

For instructor-assigned homework, go to www.masteringphysics.com. (MP)

(Answers to odd-numbered Conceptual Questions can be found in the back of the book.)

1. Why is the current in an ac circuit not always in phase with its voltage?

2. An *LC* circuit is driven at a frequency higher than its resonance frequency. What can be said about the phase angle, ϕ, for this circuit?

3. An *LC* circuit is driven at a frequency lower than its resonance frequency. What can be said about the phase angle, ϕ, for this circuit?

4. In Conceptual Example 24-14 we considered an ac circuit consisting of a lightbulb in series with an inductor. The effect of the

inductor was to cause the bulb to shine less brightly. Would the same be true in a direct-current (dc) circuit? Explain.

5. How do the resistance, capacitive reactance, and inductive reactance change when the frequency in a circuit is increased?

6. Two *RLC* circuits have different values of *L* and *C*. Is it possible for these two circuits to have the same resonance frequency? Explain.

7. Can an *RLC* circuit have the same impedance at two different frequencies? Explain.

PROBLEMS AND CONCEPTUAL EXERCISES

Answers to odd-numbered Problems and Conceptual Exercises can be found in the back of the book. **BIO** *identifies problems of biological or medical interest;* **CE** *indicates a conceptual exercise.* **Predict/Explain** *problems ask for two responses: (a) your prediction of a physical outcome, and (b) the best explanation among three provided; and* **Predict/Calculate** *problems ask for a prediction of a physical outcome, based on fundamental physics concepts, and follow that with a numerical calculation to verify the prediction. On all problems, bullets (•, ••, •••) indicate the level of difficulty.*

SECTION 24-1 ALTERNATING VOLTAGES AND CURRENTS

1. • An ac generator produces a peak voltage of 75 V. What is the rms voltage of this generator?

2. • **European Electricity** In many European homes the rms voltage available from a wall socket is 240 V. What is the maximum voltage in this case?

3. • An rms voltage of 120 V produces a maximum current of 4.1 A in a certain resistor. Find the resistance of this resistor.

4. • The rms current in an ac circuit with a resistance of 150 Ω is 0.15 A. What are (a) the average and (b) the maximum power consumed by this circuit?

5. •• A 3.33-kΩ resistor is connected to a generator with a maximum voltage of 241 V. Find (a) the average and (b) the maximum power delivered to this circuit.

6. •• A "75-watt" lightbulb uses an average power of 75 W when connected to an rms voltage of 120 V. (a) What is the resistance of the lightbulb? (b) What is the maximum current in the bulb? (c) What is the maximum power used by the bulb at any given instant of time?

7. •• **Inverter Efficiency** An array of solar panels produces 9.45 A of direct current at a potential difference of 192 V. The current flows into an inverter that produces a 60-Hz alternating current with $V_{max} = 171$ V and $I_{max} = 19.5$ A. (a) What rms power is produced by the inverter? (b) Use the rms values to find the power efficiency P_{out}/P_{in} of the inverter.

8. ••• **Square-Wave Voltage I** The relationship $V_{rms} = V_{max}/\sqrt{2}$ is valid only for voltages that vary sinusoidally. Find the relationship between V_{rms} and V_{max} for the "square-wave" voltage shown in **FIGURE 24-30**.

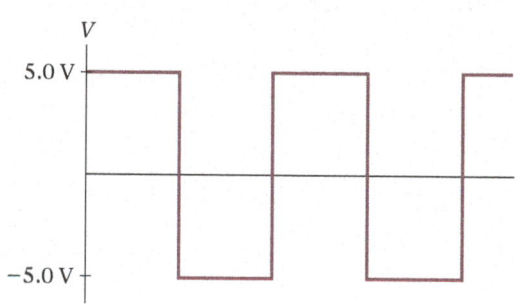

FIGURE 24-30 Problem 8

SECTION 24-2 CAPACITORS IN AC CIRCUITS

9. • The reactance of a capacitor is 65 Ω at a frequency of 87 Hz. What is its capacitance?

10. • The capacitive reactance of a capacitor at 60.0 Hz is 105 Ω. At what frequency is its capacitive reactance 72.5 Ω?

11. • A 105-μF capacitor is connected to an ac generator with an rms voltage of 32.5 V and a frequency of 125 Hz. What is the rms current in this circuit?

12. •• The rms voltage across a 0.010-μF capacitor is 2.8 V at a frequency of 75 Hz. What are (a) the rms and (b) the maximum current through the capacitor?

13. •• **Predict/Calculate** The rms current through a 55.5-μF capacitor is 1.51 A when it is connected to an ac generator. (a) If a second capacitor is connected in series with the first, will the rms current increase or decrease? (b) Calculate the rms current when a 38.7-μF capacitor is connected in series with the 55.5-μF capacitor.

14. •• The maximum current in a 22-μF capacitor connected to an ac generator with a frequency of 120 Hz is 0.15 A. (a) What is the

maximum voltage of the generator? (b) What is the voltage across the capacitor when the current in the circuit is 0.10 A and increasing? (c) What is the voltage across the capacitor when the current in the circuit is 0.10 A and decreasing?

15. •• **Predict/Calculate** An rms voltage of 20.5 V with a frequency of 1.00 kHz is applied to a 0.395-μF capacitor. (a) What is the rms current in this circuit? (b) By what factor does the current change if the frequency of the voltage is doubled? (c) Calculate the current for a frequency of 2.00 kHz.

16. •• A circuit consists of a 2.00-kHz generator and a capacitor. When the rms voltage of the generator is 0.800 V, the rms current in the circuit is 0.515 mA. (a) What is the reactance of the capacitor at 2.00 kHz? (b) What is the capacitance of the capacitor? (c) If the rms voltage is maintained at 0.800 V, what is the rms current at 4.00 kHz? At 20.0 kHz?

17. •• A 0.22-μF capacitor is connected to an ac generator with an rms voltage of 12 V. For what range of frequencies will the rms current in the circuit be less than 1.0 mA?

18. •• At what frequency will a generator with an rms voltage of 504 V produce an rms current of 7.50 mA in a 0.0150-μF capacitor?

19. ••• **Predict/Calculate** A 22.0-μF capacitor is connected to an ac generator with an rms voltage of 118 V and a frequency of 60.0 Hz. (a) What is the rms current in the circuit? (b) If you wish to increase the rms current, should you add a second capacitor in series or in parallel? (c) Find the value of the capacitance that should be added to increase the rms current to 1.50 A.

SECTION 24-3   *RC* CIRCUITS

20. • Find the impedance of a 60.0-Hz circuit with a 65.5-Ω resistor connected in series with a 85.0-μF capacitor.

21. • An ac generator with a frequency of 125 Hz and an rms voltage of 42.5 V is connected in series with a 10.0-kΩ resistor and a 0.250-μF capacitor. What is the rms current in this circuit?

22. • The rms current in an *RC* circuit is 0.72 A. The capacitor in this circuit has a capacitance of 13 μF and the ac generator has a frequency of 150 Hz and an rms voltage of 95 V. What is the resistance in this circuit?

23. • When an ac generator with a frequency of 180 Hz and an rms voltage of 36 V is connected to an *RC* circuit, the rms current is 0.28 A. If the resistance has a value of 54 Ω, what is the value of the capacitance?

24. •• A 50.0-Hz generator with an rms voltage of 115 V is connected in series to a 3.12-kΩ resistor and a 1.65-μF capacitor. Find (a) the rms current in the circuit and (b) the phase angle, ϕ, between the current and the voltage.

25. •• (a) At what frequency must the circuit in Problem 24 be operated for the current to lead the voltage by 25.0°? (b) Using the frequency found in part (a), find the average power consumed by this circuit.

26. •• Find the power factor for an *RC* circuit connected to a 60.0-Hz generator with an rms voltage of 195 V. The values of *R* and *C* in this circuit are 105 Ω and 82.4 μF, respectively.

27. •• **Predict/Calculate** (a) Determine the power factor for an *RC* circuit with $R = 4.0$ kΩ and $C = 0.35$ μF that is connected to an ac generator with an rms voltage of 24 V and a frequency of 150 Hz. (b) Will the power factor for this circuit increase, decrease, or stay the same if the frequency of the generator is increased? Explain.

28. ••• **Square-Wave Voltage II** The "square-wave" voltage shown in **FIG-URE 24-31** is applied to an *RC* circuit. Sketch the shape of the instantaneous voltage across the capacitor, assuming the time constant of the circuit is equal to the period of the applied voltage.

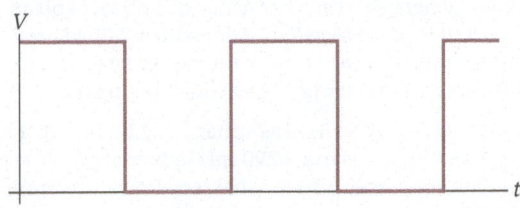

FIGURE 24-31 Problems 28, 43, and 91

SECTION 24-4   INDUCTORS IN AC CIRCUITS

29. • **CE Predict/Explain** When a long copper wire of finite resistance is connected to an ac generator, as shown in **FIGURE 24-32 (a)**, a certain amount of current flows through the wire. The wire is now wound into a coil of many loops and reconnected to the generator, as indicated in **FIGURE 24-32 (b)**. (a) Is the current supplied to the coil greater than, less than, or the same as the current supplied to the uncoiled wire? (b) Choose the *best explanation* from among the following:

 I. More current flows in the circuit because the coiled wire is an inductor, and inductors tend to keep the current flowing in an ac circuit.

 II. The current supplied to the circuit is the same because the wire is the same. Simply wrapping the wire in a coil changes nothing.

 III. Less current is supplied to the circuit because the coiled wire acts as an inductor, which increases the impedance of the circuit.

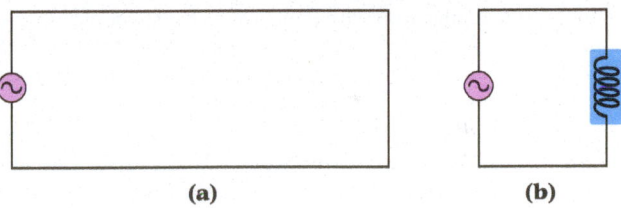

FIGURE 24-32 Problem 29

30. • An inductor has a reactance of 56.5 Ω at 85.0 Hz. What is its reactance at 60.0 Hz?

31. • What is the rms current in a 97.5-mH inductor when it is connected to a 60.0-Hz generator with an rms voltage of 135 V?

32. • What rms voltage is required to produce an rms current of 2.1 A in a 66-mH inductor at a frequency of 25 Hz?

33. • **Loudspeaker Impedance** A loudspeaker's impedance is listed as 8.0 Ω at 2.0 kHz. Assuming its resistance and capacitance are negligibly small, what is the inductance of the speaker coil?

34. •• **Fluorescent Lamp Ballast** An inductor is used to limit the current through a fluorescent lamp that is operated with an rms voltage of 120 V at $f = 60$ Hz. If the maximum rms current should be 0.33 A, and assuming the lamp has zero resistance, what value of inductance *L* should be used as the ballast?

35. •• A 525-Ω resistor and a 295-mH inductor are connected in series with an ac generator with an rms voltage of 20.0 V and a frequency of 60.0 Hz. What is the rms current in this circuit?

36. •• The rms current in an *RL* circuit is 0.26 A when it is connected to an ac generator with a frequency of 60.0 Hz and an rms voltage of 25 V. (a) Given that the inductor has an inductance of

145 mH, what is the resistance of the resistor? **(b)** Find the rms voltage across the resistor. **(c)** Find the rms voltage across the inductor. **(d)** Use your results from parts (b) and (c) to show that $\sqrt{V_{rms,R}^2 + V_{rms,L}^2}$ is equal to 25 V.

37. •• An ac generator with a frequency of 1.55 kHz and an rms voltage of 20.8 V is connected in series with a 2.00-kΩ resistor and a 292-mH inductor. **(a)** What is the power factor for this circuit? **(b)** What is the average power consumed by this circuit?

38. •• **Predict/Calculate** An rms voltage of 22.2 V with a frequency of 1.00 kHz is applied to a 0.290-mH inductor. **(a)** What is the rms current in this circuit? **(b)** By what factor does the current change if the frequency of the voltage is doubled? **(c)** Calculate the current for a frequency of 2.00 kHz.

39. •• A 0.22-μH inductor is connected to an ac generator with an rms voltage of 12 V. For what range of frequencies will the rms current in the circuit be less than 1.0 mA?

40. •• The phase angle in a certain *RL* circuit is 68° at a frequency of 60.0 Hz. If $R = 2.1 \, \Omega$ for this circuit, what is the value of the inductance, *L*?

41. •• **(a)** Sketch the phasor diagram for an ac circuit with a 105-Ω resistor in series with a 22.5-mH inductor. The frequency of the generator is 60.0 Hz. **(b)** If the rms voltage of the generator is 120 V, what is the average power consumed by the circuit?

42. •• A large air conditioner has a resistance of 7.0 Ω and an inductive reactance of 15 Ω. If the air conditioner is powered by a 60.0-Hz generator with an rms voltage of 240 V, find **(a)** the impedance of the air conditioner, **(b)** its rms current, and **(c)** the average power consumed by the air conditioner.

43. ••• **Square-Wave Voltage III** The "square-wave" voltage shown in Figure 24-31 is applied to an *RL* circuit. Sketch the shape of the instantaneous voltage across the inductor, assuming the time constant of the circuit is much less than the period of the applied voltage.

SECTION 24-5 *RLC* CIRCUITS

44. • **CE** An inductor and a capacitor are to be connected to a generator. Will the generator supply more current at *high* frequency if the inductor and capacitor are connected in series or in parallel? Explain.

45. • **CE** An inductor and a capacitor are to be connected to a generator. Will the generator supply more current at *low* frequency if the inductor and capacitor are connected in series or in parallel? Explain.

46. • **CE** Predict/Explain **(a)** When the ac generator in **FIGURE 24-33** operates at high frequency, is the rms current in the circuit greater than, less than, or the same as when the generator operates at low frequency? **(b)** Choose the *best explanation* from among the following:
    I. The current is the same because at high frequency the inductor is like an open circuit, and at low frequency the capacitor is like an open circuit. In either case the resistance of the circuit is *R*.
    II. Less current flows at high frequency because in that limit the inductor acts like an open circuit, allowing no current to flow.
    III. More current flows at high frequency because in that limit the capacitor acts like an ideal wire of zero resistance.

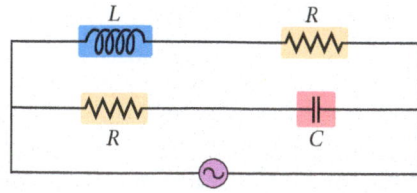

FIGURE 24-33 Problems 46 and 72

47. • **CE** Predict/Explain **(a)** When the ac generator in **FIGURE 24-34** operates at high frequency, is the rms current in the circuit greater than, less than, or the same as when the generator operates at low frequency? **(b)** Choose the *best explanation* from among the following:
    I. The current at high frequency is greater because the higher the frequency the more charge that flows through a circuit.
    II. Less current flows at high frequency because in that limit the inductor is like an open circuit and current has only one path to flow through.
    III. The inductor has zero resistance, and therefore the resistance of the circuit is the same at all frequencies. As a result the current is the same at all frequencies.

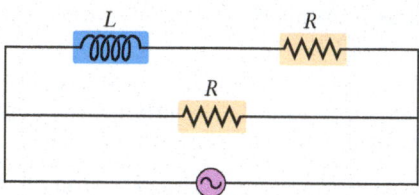

FIGURE 24-34 Problems 47 and 51

48. • **CE** Predict/Explain **(a)** When the ac generator in **FIGURE 24-35** operates at high frequency, is the rms current in the circuit greater than, less than, or the same as when the generator operates at low frequency? **(b)** Choose the *best explanation* from among the following:
    I. The capacitor has no resistance, and therefore the resistance of the circuit is the same at all frequencies. As a result the current is the same at all frequencies.
    II. Less current flows at high frequency because in that limit the capacitor is like an open circuit and current has only one path to flow through.
    III. More current flows at high frequency because in that limit the capacitor is like a short circuit and current has two parallel paths to flow through.

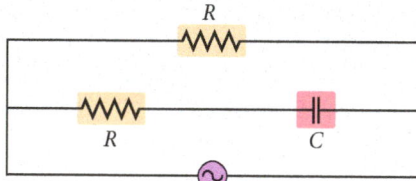

FIGURE 24-35 Problems 48 and 72

49. • Find the rms voltage across each element in an *RLC* circuit with $R = 8.8 \, k\Omega$, $C = 0.12 \, \mu F$, and $L = 28$ mH. The generator supplies an rms voltage of 115 V at a frequency of 60.0 Hz.

50. • What is the impedance of a 1.50-kΩ resistor, a 135-mH inductor, and a 22.8-μF capacitor connected in series with a 60.0-Hz ac generator?

51. •• Consider the circuit shown in Figure 24-34. The ac generator in this circuit has an rms voltage of 65 V. Given that $R = 15 \, \Omega$ and $L = 0.22$ mH, find the rms current in this circuit in the limit of **(a)** high frequency and **(b)** low frequency.

52. •• What is the phase angle in an *RLC* circuit with $R = 9.9 \, k\Omega$, $C = 1.5 \, \mu F$, and $L = 250$ mH? The generator supplies an rms voltage of 115 V at a frequency of 60.0 Hz.

53. •• An ac voltmeter, which displays the rms voltage between the two points touched by its leads, is used to measure voltages in the

circuit shown in **FIGURE 24-36**. In this circuit, the ac generator has an rms voltage of 9.00 V and a frequency of 25.0 kHz. The inductance in the circuit is 0.250 mH, the capacitance is 0.150 μF, and the resistance is 2.50 Ω. What is the reading on a voltmeter when it is connected to points **(a)** A and B, **(b)** B and C, **(c)** A and C, and **(d)** A and D?

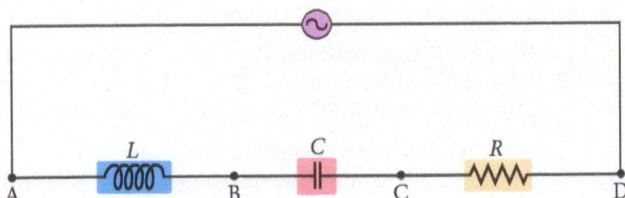

FIGURE 24-36 Problems 53 and 54

54. •• **Predict/Calculate** Consider the ac circuit shown in Figure 24-36, where we assume that the values of R, L, and C are the same as in the previous problem, and that the rms voltage of the generator is still 9.00 V. The frequency of the generator, however, is doubled to 50.0 kHz. Calculate the rms voltage across **(a)** the resistor, R, **(b)** the inductor, L, and **(c)** the capacitor, C. **(d)** Do you expect the sum of the rms voltages in parts (a), (b), and (c) to be greater than, less than, or equal to 9.00 V? Explain.

55. •• **(a)** Sketch the phasor diagram for an ac circuit with a 105-Ω resistor in series with a 22.5-mH inductor and a 32.2-μF capacitor. The frequency of the generator is 60.0 Hz. **(b)** If the rms voltage of the generator is 120 V, what is the average power consumed by the circuit?

56. •• A generator connected to an RLC circuit has an rms voltage of 150 V and an rms current of 39 mA. If the resistance in the circuit is 3.3 kΩ and the capacitive reactance is 6.6 kΩ, what is the inductive reactance of the circuit?

57. ••• **Manufacturing Plant Power** A manufacturing plant uses 2.22 kW of electric power provided by a 60.0-Hz ac generator with an rms voltage of 485 V. The plant uses this power to run a number of high-inductance electric motors. The plant's total resistance is $R = 25.0\ \Omega$ and its inductive reactance is $X_L = 45.0\ \Omega$. **(a)** What is the total impedance of the plant? **(b)** What is the plant's power factor? **(c)** What is the rms current used by the plant? **(d)** What capacitance, connected in series with the power line, will increase the plant's power factor to unity? **(e)** If the power factor is unity, how much current is needed to provide the 2.22 kW of power needed by the plant? Compare your answer with the current found in part (c). (Because power-line losses are proportional to the square of the current, a utility company will charge an industrial user with a low power factor a higher rate per kWh than a company with a power factor close to unity.)

SECTION 24-6 RESONANCE IN ELECTRIC CIRCUITS

58. • **CE Predict/Explain** In an RLC circuit a second capacitor is added in *series* to the capacitor already present. **(a)** Does the resonance frequency increase, decrease, or stay the same? **(b)** Choose the *best explanation* from among the following:
    I. The resonance frequency stays the same because it depends only on the resistance in the circuit.
    II. Adding a capacitor in series increases the equivalent capacitance, and this decreases the resonance frequency.
    III. Adding a capacitor in series decreases the equivalent capacitance, and this increases the resonance frequency.

59. • An RLC circuit has a resonance frequency of 1.9 kHz. If the inductance is 0.13 mH, what is the capacitance?

60. • **The Magnetron** A magnetron in a kitchen microwave oven resonates at 2.45 GHz in a manner analogous to an RLC circuit. If the capacitance of its resonator is 4.4×10^{-13} F, what is the value of its inductance?

61. • At resonance, the rms current in an RLC circuit is 5.8 A. If the rms voltage of the generator is 120 V, what is the resistance, R?

62. •• **CE** The resistance in an RLC circuit is doubled. **(a)** Does the resonance frequency increase, decrease, or stay the same? Explain. **(b)** Does the maximum current in the circuit increase, decrease, or stay the same? Explain.

63. •• **Predict/Calculate** The capacitive reactance in an RLC circuit is determined to be 45 Ω when the inductive reactance is 65 Ω. **(a)** Is the generator frequency ω higher or lower than the natural frequency ω_0? **(b)** Is the phase angle ϕ positive or negative? **(c)** Calculate the phase angle ϕ if $R = 33\ \Omega$.

64. •• **Predict/Calculate** The capacitive reactance in an RLC circuit is determined to be 31 Ω when the inductive reactance is 16 Ω. **(a)** Is the generator frequency ω higher or lower than the natural frequency ω_0? **(b)** Is the phase angle ϕ positive or negative? **(c)** Calculate the phase angle ϕ if $R = 12\ \Omega$.

65. •• A 115-Ω resistor, a 67.6-mH inductor, and a 189-μF capacitor are connected in series to an ac generator. **(a)** At what frequency will the current in the circuit be a maximum? **(b)** At what frequency will the impedance of the circuit be a minimum?

66. •• **(a)** Find the frequency at which an 18-μF capacitor has the same reactance as an 18-mH inductor. **(b)** What is the resonance frequency of an LC circuit made with this inductor and capacitor?

67. •• Consider an RLC circuit with $R = 105\ \Omega, L = 518$ mH, and $C = 0.200\ \mu$F. **(a)** At what frequency is this circuit in resonance? **(b)** Find the impedance of this circuit if the frequency has the value found in part (a), but the capacitance is increased to $0.220\ \mu$F. **(c)** What is the power factor for the situation described in part (b)?

68. •• **Predict/Calculate** An RLC circuit has a resonance frequency of 155 Hz. **(a)** If both L and C are doubled, does the resonance frequency increase, decrease, or stay the same? Explain. **(b)** Find the resonance frequency when L and C are doubled.

69. •• An RLC circuit has a capacitance of $0.29\ \mu$F. **(a)** What inductance will produce a resonance frequency of 95 MHz? **(b)** It is desired that the impedance at resonance be one-fifth the impedance at 11 kHz. What value of R should be used to obtain this result?

GENERAL PROBLEMS

70. • **CE BIO Persistence of Vision** Although an incandescent lightbulb appears to shine with constant intensity, this is an artifact of the eye's persistence of vision. In fact, the intensity of a bulb's light rises and falls with time due to the alternating current used in household circuits. If you could perceive these oscillations, would you see the light attain maximum brightness 60 or 120 times per second? Explain.

71. • **CE** An RLC circuit is driven at its resonance frequency. Is its impedance greater than, less than, or equal to R? Explain.

72. • **CE Predict/Explain** Suppose the circuits shown in Figures 24-33 and 24-35 are connected to identical batteries, rather than to ac generators. **(a)** Assuming the value of R is the same in the two circuits, is the current in Figure 24-33 greater than, less than, or the

same as the current in Figure 24-35? **(b)** Choose the *best explanation* from among the following:

I. The circuits have the same current because the capacitor acts like an open circuit and the inductor acts like a short circuit.

II. The current in Figure 24-33 is larger because it has more circuit elements, each of which can carry current.

III. The current in Figure 24-35 is larger because it has fewer circuit elements, meaning less resistance to current flow.

73. • **CE Predict/Explain** Consider a circuit consisting of a lightbulb and a capacitor, as shown in circuit 2 of Conceptual Example 24-10. **(a)** If the frequency of the generator is increased, does the intensity of the lightbulb increase, decrease, or stay the same? **(b)** Choose the *best explanation* from among the following:

I. As the frequency increases it becomes harder to force current through the capacitor, and therefore the intensity of the lightbulb decreases.

II. The intensity of the lightbulb increases because as the frequency becomes higher the capacitor acts more like a short circuit, allowing more current to flow.

III. The intensity of the lightbulb is independent of frequency because the circuit contains a capacitor but not an inductor.

74. • **CE** Consider a circuit consisting of a lightbulb and an inductor, as shown in Conceptual Example 24-14. If the frequency of the generator is increased, does the intensity of the lightbulb increase, decrease, or stay the same? Explain.

75. • A 4.40-μF and an 8.80-μF capacitor are connected in *parallel* to a 60.0-Hz generator operating with an rms voltage of 115 V. What is the rms current supplied by the generator?

76. • A 4.40-μF and an 8.80-μF capacitor are connected in *series* to a 60.0-Hz generator operating with an rms voltage of 115 V. What is the rms current supplied by the generator?

77. •• A 12.5-μF capacitor and a 47.5-μF capacitor are connected to an ac generator with a frequency of 60.0 Hz. What is the capacitive reactance of this pair of capacitors if they are connected **(a)** in parallel or **(b)** in series?

78. •• **CE** A generator drives an *RLC* circuit with the voltage *V* shown in **FIGURE 24-37**. The corresponding current *I* is also shown in the figure. **(a)** Is the inductive reactance of this circuit greater than, less than, or equal to its capacitive reactance? Explain. **(b)** Is the frequency of this generator greater than, less than, or equal to the resonance frequency of the circuit? Explain.

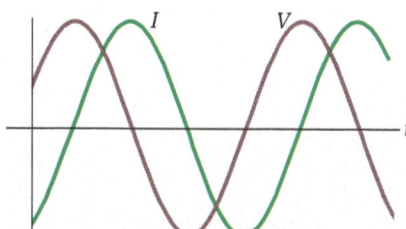

FIGURE 24-37 Problem 78

79. •• Trapped on a deserted island, you salvage some copper wire and a generator to build a makeshift radio transmitter. **(a)** If you wrap the wire into an 8-turn coil of radius 3.8 cm and length 25 cm, what is its inductance? **(b)** What capacitance do you need to make an *RLC* circuit with the coil that oscillates at the emergency beacon frequency of 406 MHz? **(c)** If you fashion the parallel-plate capacitor out of two aluminum squares, 5.0 cm on a side, what should be the separation distance between the two plates?

80. •• **Predict/Calculate** When a certain resistor is connected to an ac generator with a maximum voltage of 15 V, the average power dissipated in the resistor is 22 W. **(a)** What is the resistance of the resistor? **(b)** What is the rms current in the circuit? **(c)** We know that $P_{av} = I_{rms}^2 R$, and hence it seems that reducing the resistance should reduce the average power. On the other hand, we also know that $P_{av} = V_{rms}^2/R$, which suggests that reducing R increases P_{av}. Which conclusion is correct? Explain.

81. •• Find the average power consumed by an *RC* circuit connected to a 60.0-Hz generator with an rms voltage of 172 V. The values of *R* and *C* in this circuit are 6.30 kΩ and 2.05 μF, respectively.

82. •• A 1.15-kΩ resistor and a 505-mH inductor are connected in series to a 1250-Hz generator with an rms voltage of 14.2 V. **(a)** What is the rms current in the circuit? **(b)** What capacitance must be inserted in series with the resistor and inductor to reduce the rms current to half the value found in part (a)?

83. •• **Predict/Calculate RLC Phasor** The phasor diagram for an *RLC* circuit is shown in **FIGURE 24-38**. **(a)** If the resistance in this circuit is 525 Ω, what is the impedance? **(b)** If the frequency in this circuit is increased, will the impedance increase, decrease, or stay the same? Explain.

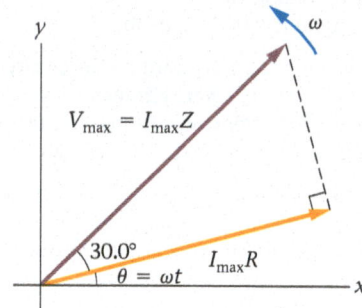

FIGURE 24-38 Problems 83, 84, and 85

84. •• **Predict/Calculate** Figure 24-39 shows the phasor diagram for an *RLC* circuit in which the impedance is 337 Ω. **(a)** What is the resistance, *R*, in this circuit? **(b)** Is this circuit driven at a frequency that is greater than, less than, or equal to the resonance frequency of the circuit? Explain.

85. •• **Predict/Calculate** An *RLC* circuit has a resistance $R = 25\,\Omega$ and an inductance $L = 160$ mH, and is connected to an ac generator with a frequency of 55 Hz. The phasor diagram for this circuit is shown in Figure 24-39. Find **(a)** the impedance, *Z*, and **(b)** the capacitance, *C*, for this circuit. **(c)** If the value of *C* is decreased, will the impedance of the circuit increase, decrease, or stay the same? Explain.

86. •• **Predict/Calculate Black-Box Experiment** You are given a sealed box with two electrical terminals. The box contains a 5.00-Ω resistor in series with either an inductor or a capacitor. When you attach an ac generator with an rms voltage of 0.750 V to the terminals of the box, you find that the current increases with increasing frequency. **(a)** Does the box contain an inductor or a capacitor? Explain. **(b)** When the frequency of the generator is 25.0 kHz, the rms current is 87.2 mA. What is the capacitance or inductance of the unknown component in the box?

87. •• An *RLC* circuit with $R = 20.0\,\Omega$, $L = 295$ mH, and $C = 49.7\,\mu$F is connected to an ac generator with an rms voltage of 21.5 V. Determine the average power delivered to this circuit when the frequency of the generator is **(a)** equal to the resonance frequency, **(b)** twice the resonance frequency, and **(c)** half the resonance frequency.

88. •• **A Light-Dimmer Circuit** The intensity of a lightbulb with a resistance of 120 Ω is controlled by connecting it in series with an inductor whose inductance can be varied from $L = 0$ to $L = L_{max}$. This "light dimmer" circuit is connected to an ac generator with a frequency of 60.0 Hz and an rms voltage of 110 V. **(a)** What is the average power dissipated in the lightbulb when $L = 0$? **(b)** The inductor is now adjusted so that $L = L_{max}$. In this case, the average power dissipated in the lightbulb is one-fourth the value found in part (a). What is the value of L_{max}?

89. ••• An electric motor with a resistance of 15 Ω and an inductance of 53 mH is connected to a 60.0-Hz ac generator. **(a)** What is the power factor for this circuit? **(b)** In order to increase the power factor of this circuit to 0.80, a capacitor is connected in series with the motor and inductor. Find the required value of the capacitance.

90. ••• **Predict/Calculate** **Tuning a Radio** A radio tuning circuit contains an RLC circuit with $R = 5.0\,\Omega$ and $L = 2.8\,\mu H$. **(a)** What capacitance is needed to produce a resonance frequency of 85 MHz? **(b)** If the capacitance is increased above the value found in part (a), will the impedance increase, decrease, or stay the same? Explain. **(c)** Find the impedance of the circuit at resonance. **(d)** Find the impedance of the circuit when the capacitance is 1% higher than the value found in part (a).

91. ••• If the maximum voltage in the square wave shown in Figure 24-31 is V_{max}, what are **(a)** the average voltage, V_{av}, and **(b)** the rms voltage, V_{rms}?

92. ••• An ac generator supplies an rms voltage of 5.00 V to an RL circuit. At a frequency of 20.0 kHz the rms current in the circuit is 45.0 mA; at a frequency of 25.0 kHz the rms current is 40.0 mA. What are the values of R and L in this circuit?

93. ••• An RC circuit consists of a resistor $R = 32\,\Omega$, a capacitor $C = 25\,\mu F$, and an ac generator with an rms voltage of 120 V. **(a)** At what frequency will the rms current in this circuit be 2.9 A? For this frequency, what are **(b)** the rms voltage across the resistor, $V_{rms,R}$, and **(c)** the rms voltage across the capacitor, $V_{rms,C}$? **(d)** Show that $V_{rms,R} + V_{rms,C} > 120\,V$, but that $\sqrt{V_{rms,R}^2 + V_{rms,C}^2} = 120\,V$.

PASSAGE PROBLEMS

Human Impedance

BIO Human tissue is made up of, among other things, cells with cell membranes that act as capacitors and ionic fluids that have resistance. So it's not surprising that tissue has electrical properties—including an electrical impedance that may vary with frequency and with the type of tissue. Some researchers have been exploring these electrical properties in order to see whether measurements of them can be used to screen for breast cancer or skin cancer, or to distinguish between different types of tissue disorders.

One research group has proposed that electrical impedance measurements might aid in the diagnosis of two different skin conditions: clavus ("corns," callused skin) and verruca (plantar warts). Although clavus and verruca are similar in appearance, they have different causes (repeated friction on the skin in the case of clavus, and the human papilloma virus in the case of verruca) and are treated differently. In an investigation of this proposal, samples of tissue were collected from patients with one of the two conditions, and a controlled area of each of the samples was tested at different frequencies with an RLC meter that measures resistance, impedance, capacitance, and phase angle. It was found that at 80 Hz, the typical (median) clavus sample had a capacitance of 1.2 nF and a resistance of 1100 kΩ. The corresponding values for a typical sample of verruca were 5.85 nF and 10.20 kΩ.

94. • At the measured frequency, what is the ratio of the capacitive reactance of a typical clavus sample to that of verruca?

   A. 0.21   B. 4.9   C. 16   D. 110

95. • What is the value of the capacitive reactance of the typical clavus sample at 80 Hz?

   A. 96 Ω   B. 10 kΩ   C. 550 kΩ   D. 1.7 MΩ

96. •• If the sample behaves in a way that can be modeled by a series RC circuit, what is the expected value of the impedance for the typical clavus sample at this frequency?

   A. 96 Ω   B. 2000 Ω   C. 2.0 MΩ   D. 10 MΩ

97. •• If the sample behaves in a way that can be modeled by a series RC circuit, what is the expected value of the phase angle for the typical clavus sample at this frequency?

   A. −6.0°   B. −57°   C. −84°   D. +33°

98. •• **Predict/Calculate** **REFERRING TO EXAMPLE 24-15** Suppose we would like to change the phase angle for this circuit to $\phi = -25.0°$, and we would like to accomplish this by changing the resistor to a value other than 175 Ω. The inductor is still 90.0 mH, the capacitor is 15.0 μF, the rms voltage is 120.0 V, and the ac frequency is 60.0 Hz. **(a)** Should the resistance be increased or decreased? Explain. **(b)** Find the resistance that gives the desired phase angle. **(c)** What is the rms current in the circuit with the resistance found in part (b)?

99. •• **Predict/Calculate** **REFERRING TO EXAMPLE 24-15** You plan to change the frequency of the generator in this circuit to produce a phase angle of smaller magnitude. The resistor is still 175 Ω, the inductor is 90.0 mH, the capacitor is 15.0 μF, and the rms voltage is 120.0 V. **(a)** Should you increase or decrease the frequency? Explain. **(b)** Find the frequency that gives a phase angle of −22.5°. **(c)** What is the rms current in the circuit at the frequency found in part (b)?

25

Electromagnetic Waves

Big Ideas

1 An electromagnetic wave consists of oscillating electric and magnetic fields.

2 An accelerating electric charge produces an electromagnetic wave.

3 Light waves show a Doppler effect similar to that of sound waves.

4 Electromagnetic waves have frequencies that range from zero to infinity.

5 Electromagnetic waves carry both energy and momentum.

6 Electromagnetic waves can be polarized.

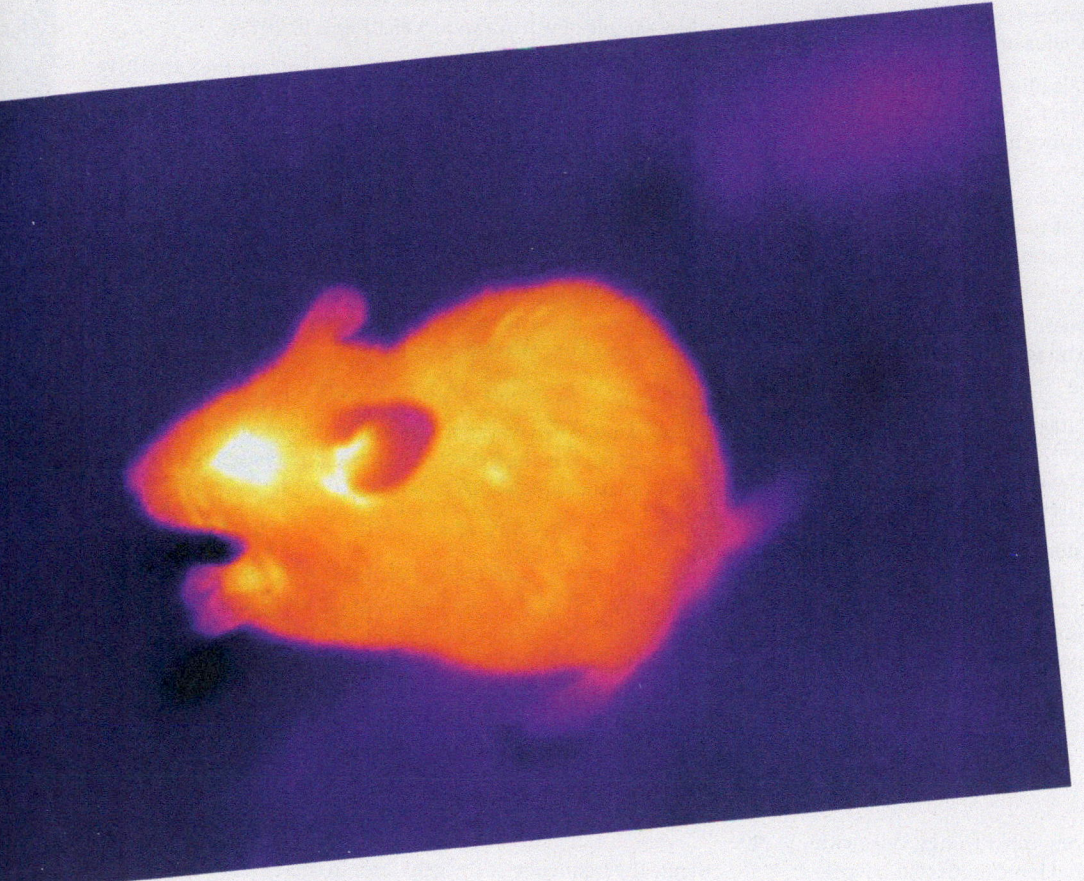

▲ It's funny when you think about it, but we see very little of the world around us. Our eyes respond to only a small range of electromagnetic waves—waves formed by oscillating electric and magnetic fields. Electromagnetic waves with wavelengths greater than those of visible light (infrared) can be used to produce thermographs like this one of a small mouse. Electromagnetic waves with shorter wavelengths than visible light (ultraviolet) produce suntans. The full spectrum of electromagnetic waves is the subject of this chapter.

n this chapter, we show that electric and magnetic fields can work together to create traveling waves called **electromagnetic waves.** These waves are responsible for everything from the visible light we see all around us, to the X-rays that reveal our internal structure, and much more.

25-1 The Production of Electromagnetic Waves

Electromagnetic waves were predicted, and their properties were studied theoretically, decades before they were first produced with electric circuits in the lab. The prediction came from Scottish physicist James Clerk Maxwell (1831–1879), who, in 1864, hypothesized that because a changing magnetic field produces an electric field (Faraday's law), a changing electric field should similarly produce a magnetic field. In effect, Maxwell suggested a sort of "symmetry" that links electric and magnetic fields.

Maxwell followed up on his suggestion by working out its mathematical consequences. Among these was that electric and magnetic fields, acting together, could produce an **electromagnetic wave** that travels with the speed of light. As a result, Maxwell proposed that visible light—which had previously been thought of as a phenomenon completely separate from electricity and magnetism—was, in fact, an electromagnetic wave. His theory also implied that electromagnetic waves would not be limited to visible light, and that it should be possible to produce them with oscillating electric circuits similar to those studied in the previous chapter.

Producing Electromagnetic Waves in the Lab   The first person to produce and observe electromagnetic waves in the lab was the German physicist Heinrich Hertz (1857–1894) in 1887. Hertz used what was basically an *LC* circuit to generate an alternating current, and he found that energy could be transferred from one such circuit to a similar circuit several meters away—a precursor to the Wi-Fi circuits so familiar today. He was able to show, in addition, that the energy transfer exhibited such standard wave phenomena as reflection, refraction, interference, diffraction, and polarization. There could be no doubt that waves were produced by the first circuit and that they propagated across the room to the second circuit. Even more significantly, Hertz was able to show that the speed of the waves was roughly the speed of light, as predicted by Maxwell.

It took only a few years for Hertz's experimental apparatus to be refined and improved to the point where it could be used in practical applications. The first to do so was Guglielmo Marconi (1874–1937), who immediately recognized the implications of the electromagnetic-wave experiments—namely, that waves could be used for communications, eliminating the wires necessary for telegraphy. He patented his first system in 1896 and gained worldwide attention when, in 1901, he received a radio signal in St. John's, Newfoundland, that had been sent from Cornwall, England. When Maxwell died, electromagnetic waves were still just a theory; twenty years later, they were revolutionizing communications. Today, they are a significant part of everyday life, as evidenced in **FIGURE 25-1**.

To gain an understanding of electromagnetic waves, consider the simple electric circuit shown in **FIGURE 25-2**. Here we show an ac generator of period *T* connected to the center of an antenna, which is basically a long, straight wire with a break in the middle. Suppose at time $t = 0$ the generator gives the upper segment of the antenna a maximum positive charge and the lower segment a maximum negative charge, as shown in Figure 25-2 (a). A positive test charge placed on the *x* axis at point P will experience a downward force; hence, the electric field there is downward. A short time later, when the charge on the antenna is reduced in magnitude, the electric field at P also has a smaller magnitude. We show this result in Figure 25-2 (b).

More important, Figure 25-2 (b) also shows that the electric field produced at time $t = 0$ has not vanished, nor has it simply been replaced with the new, reduced-magnitude field. Instead, the original field has *moved farther away from the antenna,* to point Q. The reason that the reduction in charge on the antenna is felt at point P *before* it is felt at point Q is simply that it takes a finite time for this change in charge to be felt at a distance. This is analogous to the fact that a person near a lightning strike hears the thunder before a person who is half a mile away, or that a wave pulse on a string takes a finite time to move from one end of the string to the other.

After the generator has completed one-quarter of a cycle, at time $t = \frac{1}{4}T$, the antenna is uncharged and the field vanishes, as in Figure 25-2 (c). Still later the charges

▲ **FIGURE 25-1** Electromagnetic waves are produced by (and detected as) oscillating electric currents in a wire or similar conducting element. The actual antenna is often much smaller than is commonly imagined—the bowl-shaped structures that we tend to think of as antennas, such as these microwave relay dishes, serve to focus the transmitted beam in a particular direction or concentrate the received signal on the actual detector.

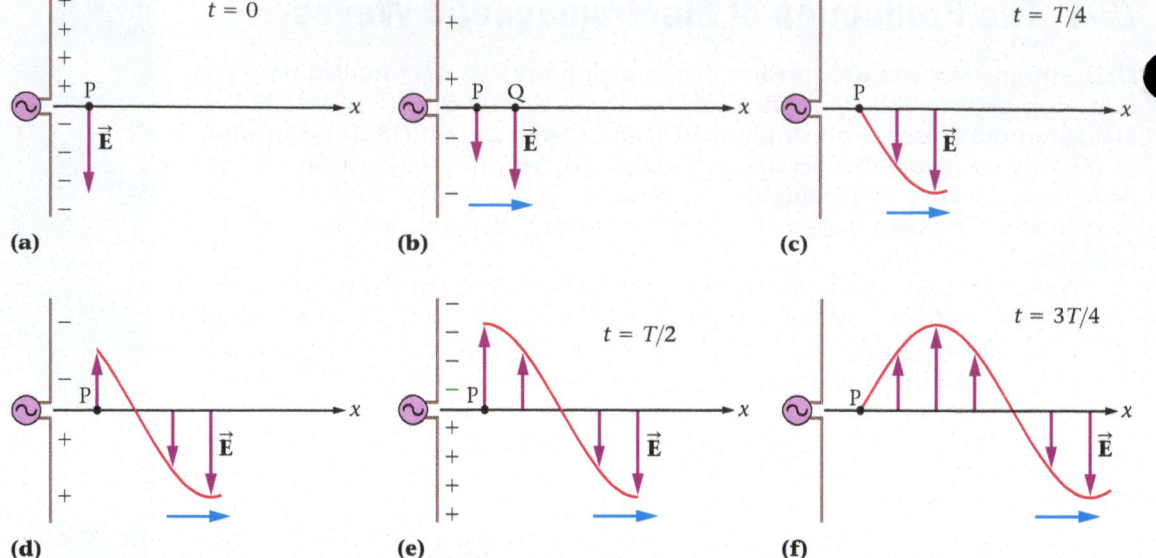

▲ **FIGURE 25-2 Producing an electromagnetic wave** A traveling electromagnetic wave produced by an ac genera-tor attached to an antenna. **(a)** At $t = 0$, the electric field at point P is downward. **(b)** A short time later, the electric field at P is still downward, but now with a reduced magnitude. Note that the field created at $t = 0$ has moved to point Q. **(c)** After one-quarter of a cycle, at $t = \frac{1}{4}T$, the electric field at P vanishes. **(d)** The charge on the antenna has reversed polarity now, and the electric field at P points upward. **(e)** When the oscillator has completed half a cycle, $t = \frac{1}{2}T$, the field at point P is upward and of maximum magnitude. **(f)** At $t = \frac{3}{4}T$, the field at P vanishes again. The fields produced at earlier times continue to move away from the antenna.

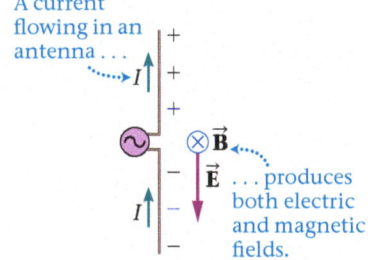

▲ **FIGURE 25-3 Field directions in an electromagnetic wave** At a time when the electric field produced by the antenna points downward, the magnetic field points into the page. In general, the electric and magnetic fields in an electromagnetic wave are always at right angles to each other.

Big Idea 1 An oscillating elec-tric field produces an oscillating magnetic field, and the two travel together as a wave called an electro-magnetic wave.

on the antenna segments change sign, giving rise to an electric field that points upward, as we see in Figures 25-2 (d) and (e). The field vanishes again after three-quarters of a cycle, at $t = \frac{3}{4}T$, as shown in Figure 25-2 (f). Immediately after this time, the electric field begins to point downward once more. The net result is a wavelike electric field moving steadily away from the antenna. To summarize:

> The electric field produced by an antenna connected to an ac generator propagates away from the antenna, analogous to a wave on a string moving away from your hand as you wiggle it up and down.

This is really only half of the electromagnetic wave, however; the other half is a similar wave in the magnetic field. To see this, consider **FIGURE 25-3**, where we show the current in the antenna flowing upward at a time when the upper segment is positive. Pointing the thumb of the right hand in the direction of the current, and curling the fingers around the wire, as specified in the magnetic field right-hand rule (RHR), we see that $\vec{B}$ points into the page at the same time that $\vec{E}$ points downward. It follows, then, that $\vec{E}$ and $\vec{B}$ are at right angles to each other. A more detailed analysis for distances far from the antenna shows that $\vec{E}$ and $\vec{B}$ in an electromagnetic wave are perpendicu-lar to each other at all times, and that they are also in phase; that is, when the magni-tude of $\vec{E}$ is at its maximum, so is the magnitude of $\vec{B}$.

Combining the preceding results, we can represent the electric and magnetic fields in an electromagnetic wave as shown in **FIGURE 25-4**. Notice that not only are $\vec{E}$ and $\vec{B}$ perpendicular to each other, they are also perpendicular to the direction of propa-gation; hence, electromagnetic waves are **transverse** waves. (See Section 14-1 for a comparison of various types of waves.) The direction of propagation is given by a right-hand rule:

Direction of Propagation for Electromagnetic Waves
Point the fingers of your right hand in the direction of $\vec{E}$, and curl your fingers toward $\vec{B}$. Your thumb now points in the direction of propagation.

This rule is consistent with the direction of propagation shown in Figure 25-4.

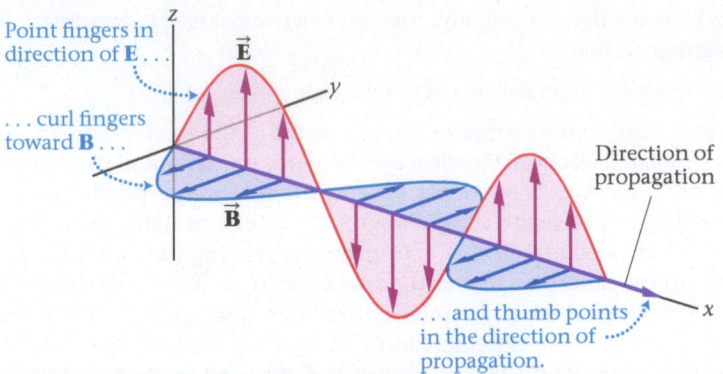

▲ **FIGURE 25-4 The right-hand rule applied to an electromagnetic wave** An electromagnetic wave propagating in the positive x direction. Note that $\vec{\mathbf{E}}$ and $\vec{\mathbf{B}}$ are perpendicular to each other and in phase. The direction of propagation is given by the thumb of the right hand, after pointing the fingers in the direction of $\vec{\mathbf{E}}$ and curling them toward $\vec{\mathbf{B}}$.

PHYSICS IN CONTEXT
Looking Back

Electric and magnetic fields (see Chapters 19 and 22, respectively) are used throughout this chapter. In fact, we show that $\vec{\mathbf{E}}$ and $\vec{\mathbf{B}}$ fields are intimately linked, and that together they can form traveling waves (Chapter 14).

CONCEPTUAL EXAMPLE 25-1   DIRECTION OF THE MAGNETIC FIELD

An electromagnetic wave propagates in the positive y direction, as shown in the sketch. If the electric field at the origin is in the positive z direction, is the magnetic field at the origin in the positive x direction, the negative x direction, or the negative y direction?

REASONING AND DISCUSSION
Pointing the fingers of the right hand in the positive z direction (the direction of $\vec{\mathbf{E}}$), we see that in order for the thumb to point in the direction of propagation (the positive y direction) the fingers must be curled toward the positive x direction. Therefore, $\vec{\mathbf{B}}$ points in the positive x direction.

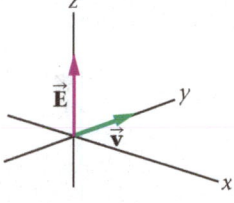

ANSWER
$\vec{\mathbf{B}}$ is in the positive x direction.

RWP* Electromagnetic waves can be detected in much the same way they are generated. Suppose, for instance, that an electromagnetic wave moves to the right, as in **FIGURE 25-5**. As the wave continues to move, its electric field exerts a force on electrons in the antenna that is alternately up and down, resulting in an alternating current. Thus the electromagnetic field makes the antenna behave much like an ac generator. If the antenna is connected to an LC circuit, as indicated in the figure, the resulting current can be relatively large if the resonant frequency of the circuit matches the frequency of the wave. This is the basic principle behind radio and television tuners. In fact, when you adjust the tuning knob on a radio, you are actually changing the capacitance or the inductance in an LC circuit and, therefore, changing the resonance frequency.

Finally, though we have discussed the production of electromagnetic waves by means of an electric circuit and an antenna, this is certainly not the only way electromagnetic

Big Idea **2** Any time an electric charge accelerates, it produces an electromagnetic wave.

PHYSICS IN CONTEXT
Looking Ahead

The straight-line propagation of light presented in this section is the basis for the geometrical optics to be studied in Chapters 26 and 27.

◄ **FIGURE 25-5 Receiving radio waves** Basic elements of a tuning circuit used to receive radio waves. First, an incoming wave sets up an alternating current in the antenna. Next, the resonance frequency of the LC circuit is adjusted to match the frequency of the radio wave, resulting in a relatively large current in the circuit. This current is then fed into an amplifier to further increase the signal.

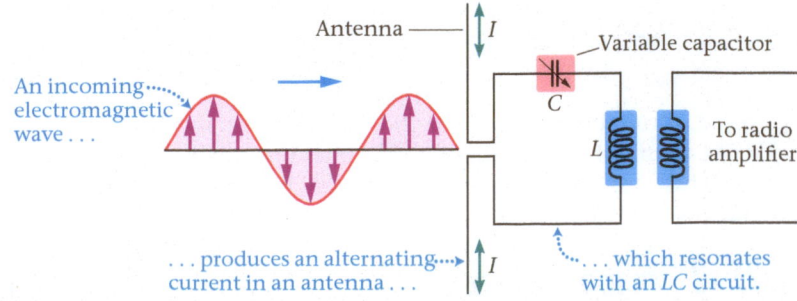

*Real World Physics applications are denoted by the acronym RWP.

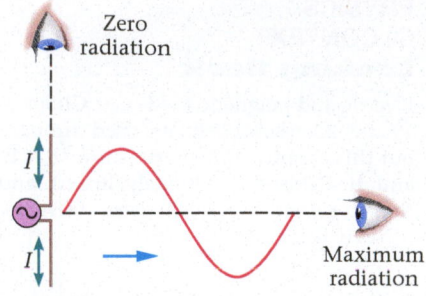

▲ **FIGURE 25-6 Electromagnetic waves and the line of sight** Electromagnetic radiation is greatest when charges accelerate at right angles to the line of sight. Zero radiation is observed when the charges accelerate along the line of sight. These observations apply to electromagnetic waves of all frequencies.

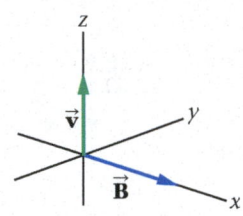

▲ **FIGURE 25-7** Enhance Your Understanding 1.

waves can be generated. In fact, any time an electric charge is accelerated, it produces electromagnetic waves:

> Accelerated charges radiate electromagnetic waves.

This condition applies no matter what the cause of the acceleration. In addition, the intensity of radiated electromagnetic waves depends on the orientation of the acceleration relative to the viewer. For example, viewing the antenna perpendicular to its length, so that the charges accelerate at right angles to the line of sight, results in maximum intensity, as illustrated in **FIGURE 25-6**. Conversely, viewing the antenna straight down from above, in the same direction as the acceleration, results in zero intensity.

RWP A device called a *synchrotron* takes electromagnetic wave production by charge acceleration to the extreme. Electrons moving at nearly the speed of light are caused to wiggle (accelerate) by the application of strong magnetic fields in opposing directions (see Figure 22-18). Due to their acceleration, the electrons produce electromagnetic waves. Synchrotrons produce very bright and stable beams of ultraviolet and X-ray light (discussed in Section 25-3) that are especially useful for biological research and materials analysis.

Enhance Your Understanding   (Answers given at the end of the chapter)

1. An electromagnetic wave propagates in the positive z direction, as shown in **FIGURE 25-7**. The magnetic field at the origin points in the positive x direction. Does the electric field at the origin point in the negative x direction, the negative y direction, the positive y direction, or the negative z direction?

Section Review

- Electromagnetic waves consist of oscillating electric and magnetic fields that travel at the speed of light.

- Accelerating electric charges radiate electromagnetic waves.

25-2 The Propagation of Electromagnetic Waves

A sound wave or a wave on a string requires a medium through which it can propagate. For example, when the air is pumped out of a jar containing a ringing bell, its sound can no longer be heard. In contrast, we can still *see* that the bell is ringing. Thus, light can propagate through a vacuum, as can all other types of electromagnetic waves, such as radio waves and microwaves. In fact, electromagnetic waves travel through a vacuum with the maximum speed that *any* form of energy can have, as we discuss in detail in Chapter 29.

The Speed of Light

All electromagnetic waves travel through a vacuum with precisely the same speed, c. Because light is the form of electromagnetic wave most familiar to us, we refer to c as the *speed of light in a vacuum*. The approximate value of this speed is as follows:

Speed of Light in a Vacuum

$$c = 3.00 \times 10^8 \text{ m/s} \qquad 25\text{-}1$$

This is a large speed, corresponding to about 186,000 mi/s. Put another way, a beam of light could travel around the world about seven times in a single second. In air the speed of light is slightly less than it is in a vacuum, and in denser materials, such as glass or water, the speed of light is reduced to about two-thirds of its vacuum value.

EXERCISE 25-2   TIME OF TRAVEL

The distance between Earth and the Sun is 1.50×10^{11} m. How much time does it take for light to cover this distance?

CONTINUED

REASONING AND SOLUTION

Given that speed is distance divided by time ($v = d/t$), it follows that the time t to cover a distance d is $t = d/v$. Using $v = c$ for the speed, we find

$$t = \frac{d}{c} = \frac{1.50 \times 10^{11} \text{ m}}{3.00 \times 10^8 \text{ m/s}} = 500 \text{ s}$$

Noting that 500 s is $8\frac{1}{3}$ min, we can say that Earth is a distance of just over 8 light-minutes from the Sun.

Measuring the Speed of Light   Because the speed of light is so large, its value is somewhat difficult to determine. The first scientific attempt to measure the speed of light was made by Galileo (1564–1642), who used two lanterns for the experiment. Galileo opened the shutters of one lantern, and an assistant a considerable distance away was instructed to open the shutter on the second lantern as soon as he observed the light from Galileo's lantern. Galileo then attempted to measure the time that elapsed before he saw the light from his assistant's lantern. There was no perceptible time lag, beyond the normal human reaction time, and hence Galileo could conclude only that the speed of light must be very great indeed.

The first to give a finite, numerical value to the speed of light was the Danish astronomer Ole Romer (1644–1710), though he did not set out to measure the speed of light at all. Romer was measuring the times at which the moons of Jupiter disappeared behind the planet, and he noticed that these eclipses occurred earlier when Earth was closer to Jupiter and later when Earth was farther away from Jupiter. This difference is illustrated in **FIGURE 25-8**. From the results of Exercise 25-2, we know that light requires about 16 minutes to travel from one side of Earth's orbit to the other, and this is roughly the discrepancy in eclipse times observed by Romer. In 1676 he announced the first crude value for the speed of light, 2.25×10^8 m/s.

The first laboratory measurement of the speed of light was performed by the French scientist Armand Fizeau (1819–1896). The basic elements of his experiment, shown in **FIGURE 25-9**, are a mirror and a rotating, notched wheel. Light passing through one notch travels to a mirror a considerable distance away, is reflected back, and then, if the rotational speed of the wheel is adjusted properly, passes through the *next notch* in the wheel. By measuring the rotational speed of the wheel and the distance from the wheel to the mirror, Fizeau was able to obtain a value of 3.13×10^8 m/s for the speed of light.

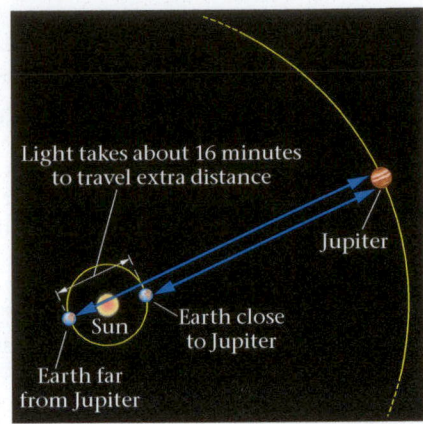

▲ **FIGURE 25-8 Using Jupiter to determine the speed of light** When the Earth is at its greatest distance from Jupiter, light takes about 16 minutes longer to travel between them. This time lag allowed Ole Romer to estimate the speed of light.

◀ **FIGURE 25-9 Fizeau's experiment to measure the speed of light** If the time required for light to travel to the far mirror and back is equal to the time it takes the wheel to rotate from one notch to the next, light will pass through the wheel and on to the observer.

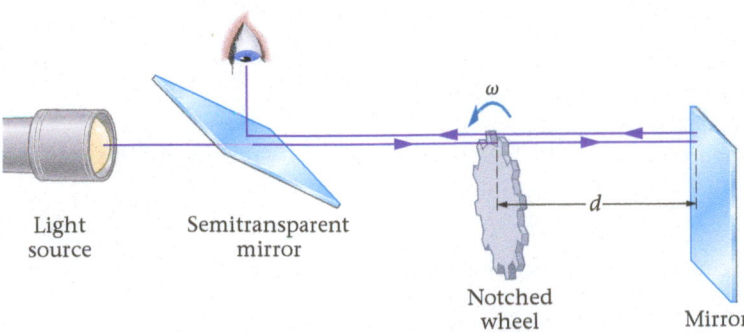

Today, experiments to measure the speed of light have been refined to such a degree that we now use it to *define* the meter, as was mentioned in Chapter 1. Thus, by definition, the speed of light in a vacuum is

$$c = 299\,792\,458 \text{ m/s}$$

For most routine calculations, however, the value $c = 3.00 \times 10^8$ m/s is adequate.

Maxwell's theoretical description of electromagnetic waves allowed him to obtain a simple expression for c in terms of previously known physical quantities. In particular, he found that c could be written as follows:

$$c = \frac{1}{\sqrt{\epsilon_0 \mu_0}}$$

25-2

Recall that $\epsilon_0 = 8.85 \times 10^{-12}\,\mathrm{C^2/(N\cdot m^2)}$ occurs in the expression for the electric field due to a point charge; in fact, ϵ_0 determines the strength of the electric field. The constant $\mu_0 = 4\pi \times 10^{-7}\,\mathrm{T\cdot m/A}$ plays an equivalent role for the magnetic field. Thus, Maxwell was able to show that these two constants, which were determined by electrostatic and magnetostatic measurements, also combine to yield the speed of light—again demonstrating the symmetrical role that electric and magnetic fields play in electromagnetic waves. Substituting the values for ϵ_0 and μ_0, we find

$$c = \frac{1}{\sqrt{(8.85 \times 10^{-12}\,\mathrm{C^2/(N\cdot m^2)})(4\pi \times 10^{-7}\,\mathrm{T\cdot m/A})}} = 3.00 \times 10^8\,\mathrm{m/s}$$

Clearly, Maxwell's theoretical expression agrees with experiment.

EXAMPLE 25-3   FIZEAU'S RESULTS

Consider a Fizeau experiment in which the wheel has 450 notches and rotates with a speed of 35 rev/s. Light passing through one notch travels to the mirror and back just in time to pass through the next notch. If the distance from the wheel to the mirror is 9500 m, what is the speed of light obtained by this measurement?

PICTURE THE PROBLEM
Our sketch shows an experimental setup similar to Fizeau's. The notched wheel is 9500 m from a mirror and spins with an angular speed of 35 rev/s. We show a few notches in our sketch to represent the 450 notches on the actual wheel.

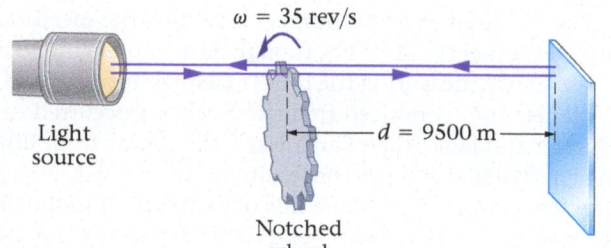

REASONING AND STRATEGY
The speed of light is the distance traveled, $2d$, divided by the time required, Δt. To find the time, we note that the wheel rotates from one notch to the next during this time; that is, it rotates through an angle $\Delta\theta = (1/450)$ rev. Knowing the rotational speed, ω, of the wheel, we can find the time using the relationship $\Delta\theta = \omega\,\Delta t$ (Section 10-1).

Known   Distance from wheel to mirror, $d = 9500$ m; number of notches in wheel, 450; angular separation of notches, $\Delta\theta = (1/450)$ rev; angular speed, $\omega = 35$ rev/s.

Unknown   Speed of light, $c = ?$

SOLUTION

1. Find the time required for the wheel to rotate from one notch to the next:

$$\Delta t = \frac{\Delta\theta}{\omega} = \frac{(1/450)\ \mathrm{rev}}{35\ \mathrm{rev/s}} = 6.3 \times 10^{-5}\,\mathrm{s}$$

2. Divide the time into the distance to find the speed of light:

$$c = \frac{2d}{\Delta t} = \frac{2(9500\ \mathrm{m})}{6.3 \times 10^{-5}\,\mathrm{s}} = 3.0 \times 10^8\,\mathrm{m/s}$$

INSIGHT
Even with a rather large distance for the round trip, the travel time of the light is small, only 0.063 millisecond. This illustrates the great difficulty experimentalists face in attempting to make an accurate measurement of c.

PRACTICE PROBLEM
If the wheel has 430 notches, what rotational speed is required for the return beam of light to pass through the next notch? [**Answer:** $\omega = 37$ rev/s]

Some related homework problems: Problem 15, Problem 16

Although the speed of light is enormous by earthly standards, it is useful to look at it from an astronomical perspective. Imagine, for example, that you could shrink the solar system to fit onto a football field, with the Sun at one end zone and Pluto at the other. On this scale, Earth would be a grain of sand located at the 2.5-yard line from the Sun, and light would take 8 min to cover that distance. To travel to Pluto, at the other end of the field, light would require about 5.5 h. Thus, on this scale, the speed of light is like the crawl of a small caterpillar. When one recalls that the solar system is but a speck on the outskirts of the Milky Way galaxy, and that the nearest major galaxy to our own—the Andromeda galaxy, shown in **FIGURE 25-10**—is about 2.2 million light-years away, the speed of light doesn't appear so great after all.

The Doppler Effect

In Section 14-6 we discussed the Doppler effect for sound waves—the familiar increase or decrease in frequency as a source of sound approaches or recedes. A similar Doppler effect applies to electromagnetic waves. There are two fundamental differences, however. First, sound waves require a medium through which to travel, whereas light can propagate across a vacuum. Second, the speed of sound can be different for different observers. For example, an observer approaching a source of sound measures an increased speed of sound, whereas an observer detecting sound from a moving source measures the usual speed of sound. For this reason, the Doppler effect with sound is different for a moving observer than it is for a moving source (see Figure 14-22 for a direct comparison). In contrast, the speed of electromagnetic waves is *independent* of the motion of the source and observer, as we shall see in Chapter 29. Therefore, there is just one Doppler effect for electromagnetic waves, and it depends only on the *relative speed* between the observer and the source.

For relative speeds u that are small compared with the speed of light, the observed frequency f' from a source with frequency f is

$$f' = f\left(1 \pm \frac{u}{c}\right) \qquad \text{25-3}$$

Notice that u in this expression is a speed and hence is always positive. The appropriate sign in front of the term u/c is chosen for a given situation—the plus sign applies to a source that is approaching the observer, the minus sign to a receding source. In addition, u is a *relative* speed between the source and the observer, both of which may be moving. For example, if an observer is moving in the positive x direction with a speed of 5 m/s, and a source ahead of the observer is moving in the positive x direction with a speed of 12 m/s, the relative speed is $u = 12$ m/s $- 5$ m/s $= 7$ m/s. Because the distance between the observer and the source is increasing with time in this case, we would choose the minus sign in $f' = f(1 \pm u/c)$.

EXERCISE 25-4   OBSERVED FREQUENCY

An FM radio station broadcasts at a frequency of 91.5 MHz. If you drive your car toward the station at 26.1 m/s, what change in frequency do you observe?

REASONING AND SOLUTION

We can find the change in frequency, $f' - f$, by rearranging the equation $f' = f(1 \pm u/c)$. In addition, we use the plus sign because the car is approaching the source. Thus,

$$f' - f = f\frac{u}{c} = (91.5 \times 10^6 \text{ Hz})\frac{26.1 \text{ m/s}}{3.00 \times 10^8 \text{ m/s}} = 7.96 \text{ Hz}$$

Thus, the frequency changes by only 7.96 Hz = 0.00000796 MHz.

RWP Common applications of the Doppler effect include the radar units used to measure the speed of automobiles and the Doppler radar that is used to monitor the weather. In Doppler radar, electromagnetic waves are sent out into the atmosphere and are reflected back to the receiver. The change in frequency of the reflected beam relative to the outgoing beam provides a way of measuring the speed of the clouds and precipitation that reflected the beam. Thus, Doppler radar gives more information than just where a rainstorm is located; it also tells how it is moving. Measurements of this type are particularly important for airports, where information regarding areas of possible wind shear can be crucial for safety. The Doppler effect is also used to determine the speed of distant galaxies relative to the Earth, as shown in **FIGURE 25-11**.

▶ **FIGURE 25-11** This long-exposure photograph, taken from orbit by the Hubble Space Telescope, shows more than 1600 galaxies when examined closely. Most of them exhibit a Doppler red shift—that is, their light is shifted to lower frequencies by the Doppler effect, indicating that they are receding from Earth as the universe expands. The red shifts marked on the photo correspond to distances ranging from about 1.3 billion light-years to over 13 billion light-years (nearly 10^{23} miles).

▲ **FIGURE 25-10** Even traveling at 300 million meters per second, light from the Andromeda galaxy takes more than 2 million years to reach us. Yet Andromeda is one of our nearest cosmic neighbors.

Big Idea 3 The Doppler effect, where the observed frequency of a sound wave depends on the speed of the source, applies in a simplified way to light waves.

PHYSICS IN CONTEXT
Looking Back

The Doppler effect, which we studied for the case of sound in Chapter 14, is applied to light in this section.

PROBLEM-SOLVING NOTE

Evaluating the Doppler Shift

Since everyday objects generally move with speeds much lower than the speed of light, the Doppler-shifted frequency differs little from the original frequency. To see the Doppler effect more clearly, it is often useful to calculate the difference in frequency, $f' - f$, rather than the Doppler-shifted frequency itself.

EXAMPLE 25-5 NEXRAD

RWP The Doppler weather radar used by the National Weather Service is referred to as Nexrad, which stands for next-generation radar. Nexrad commonly operates at a frequency of 2.7 GHz. If a Nexrad wave reflects from an approaching weather system that is moving with a speed of 28 m/s, find the difference in frequency between the outgoing and returning waves. (Assume the speed of the radar waves is c.)

PICTURE THE PROBLEM
Our sketch shows the outgoing radar wave, the incoming weather system, and the returning radar wave. The speed of the weather system relative to the radar station is $u = 28$ m/s, and the frequency of the outgoing wave is $f = 2.7$ GHz.

REASONING AND STRATEGY
Two Doppler effects are involved in this problem. First, the outgoing wave is seen to have a frequency $f' = f(1 + u/c)$ by the weather system—we use the plus sign because the weather system moves toward the source. Second, the weather system acts like a source of radar with frequency f' moving toward the receiver, and hence an observer at the radar facility detects a frequency $f'' = f'(1 + u/c)$. Thus, given u and f, we can calculate the difference, $f'' - f$.

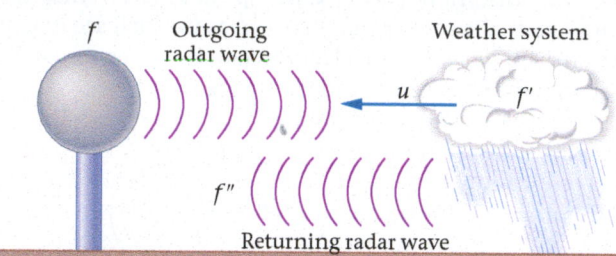

Known Frequency of radar, $f = 2.7 \times 10^9$ Hz; speed of weather system, $u = 28$ m/s.

Unknown Difference in frequency, $f'' - f = ?$

SOLUTION

1. Use $f' = f(1 + u/c)$ to calculate the difference in frequency, $f' - f$, observed by the weather system:

$$f' - f = f\frac{u}{c} = \frac{(2.7 \times 10^9 \text{ Hz})(28 \text{ m/s})}{3.00 \times 10^8 \text{ m/s}} = 250 \text{ Hz}$$

2. Now, use $f'' = f'(1 + u/c)$ to find the difference in frequency between the returning signal, f'', and the frequency observed by the weather system, f':

$$f'' - f' = f'\frac{u}{c}$$

3. Use the results of Step 1 to replace f' with $f + 250$ Hz:

$$f'' - f' = f'\frac{u}{c}$$
$$f'' - (f + 250 \text{ Hz}) = (f + 250 \text{ Hz})\frac{u}{c}$$

4. Solve for the frequency difference, $f'' - f$. Recall that $f = 2.7 \times 10^9$ Hz:

$$f'' - f = (f + 250 \text{ Hz})\frac{u}{c} + 250 \text{ Hz}$$
$$= \frac{(2.7 \times 10^9 \text{ Hz} + 250 \text{ Hz})(28 \text{ m/s})}{3.00 \times 10^8 \text{ m/s}} + 250 \text{ Hz}$$
$$= 500 \text{ Hz}$$

INSIGHT
Notice that we focus on the *difference* in frequency between the very large numbers $f = 2,700,000,000$ Hz and $f'' = 2,700,000,500$ Hz. Clearly, it is more convenient to simply write $f'' - f = 500$ Hz.

In addition, notice that the two Doppler shifts in this problem are analogous to the two Doppler shifts we found for the case of a train approaching a tunnel in Example 14-11.

PRACTICE PROBLEM
Find the difference in frequency if the weather system is receding with a speed of 21 m/s.
[**Answer:** $f'' - f = -380$ Hz]

Some related homework problems: Problem 22, Problem 24

Enhance Your Understanding   (Answers given at the end of the chapter)

2. A distant galaxy is moving away from the Earth. **(a)** Is the speed of the light we see from the galaxy greater than, less than, or equal to c? **(b)** Is the frequency of the light we receive from the galaxy Doppler shifted to a higher frequency, Doppler shifted to a lower frequency, or at the same frequency at which it was emitted?

Section Review

- All electromagnetic waves travel with the speed of light, regardless of their frequency. In a vacuum, the speed of light is $c = 3.00 \times 10^8$ m/s.

- Electromagnetic waves experience a simplified version of the Doppler effect in sound waves. For speeds u small compared with the speed of light, the Doppler effect for electromagnetic waves is $f' = f(1 \pm u/c)$.

25-3 The Electromagnetic Spectrum

When white light passes through a prism it spreads out into a rainbow of colors, with red on one end and violet on the other. All these various colors of light are electromagnetic waves, of course; they differ only in their frequency and, hence, their wavelength. The relationship between frequency and wavelength for any wave with a speed v is simply $v = f\lambda$, as was shown in Section 14-1. Because all electromagnetic waves in a vacuum have the same speed, c, it follows that f and λ are related as follows:

$$c = f\lambda \qquad \text{25-4}$$

Thus, as the frequency of an electromagnetic wave increases, its wavelength decreases.

In the following Example we calculate the frequency of red and violet light, given the corresponding wavelengths. The wavelengths are given in units of nanometers (nm), where 1 nm = 10^{-9} m.

EXAMPLE 25-6 ROSES ARE RED, VIOLETS ARE VIOLET

Find the frequency of **(a)** red light, with a wavelength of 700.0 nm, and **(b)** violet light, with a wavelength of 400.0 nm.

PICTURE THE PROBLEM

The visible electromagnetic spectrum, along with representative wavelengths, is shown in our diagram. In addition to the wavelengths of 700.0 nm for red light and 400.0 nm for violet light, we include 600.0 nm for yellowish-orange light and 500.0 nm for greenish-blue light.

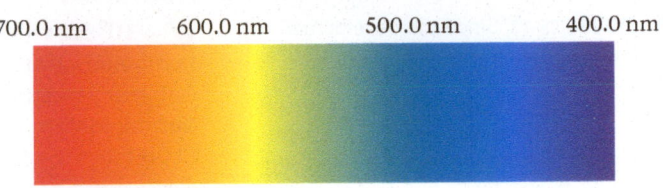

700.0 nm 600.0 nm 500.0 nm 400.0 nm

REASONING AND STRATEGY

We obtain the frequency by rearranging $c = f\lambda$ to yield $f = c/\lambda$.

Known Wavelength of red light, $\lambda_{red} = 700.0$ nm; wavelength of violet light, $\lambda_{violet} = 400.0$ nm.

Unknown **(a)** Frequency of red light, $f_{red} = ?$ **(b)** Frequency of violet light, $f_{violet} = ?$

SOLUTION

1. Substitute $\lambda_{red} = 700.0$ nm for red light:

$$f_{red} = \frac{c}{\lambda_{red}} = \frac{3.00 \times 10^8 \text{ m/s}}{700.0 \times 10^{-9} \text{ m}} = 4.29 \times 10^{14} \text{ Hz}$$

2. Substitute $\lambda_{violet} = 400.0$ nm for violet light:

$$f_{violet} = \frac{c}{\lambda_{violet}} = \frac{3.00 \times 10^8 \text{ m/s}}{400.0 \times 10^{-9} \text{ m}} = 7.50 \times 10^{14} \text{ Hz}$$

INSIGHT

The frequency of visible light is extremely high. In fact, even for the relatively low frequency of red light, it takes only 2.33×10^{-15} s to complete one cycle. The *range* of visible frequencies is relatively small, however, when compared with other portions of the electromagnetic spectrum.

PRACTICE PROBLEM

What is the wavelength of light that has a frequency of 5.25×10^{14} Hz? **[Answer: 571 nm]**

Some related homework problems: Problem 26, Problem 27

In principle, the frequency of an electromagnetic wave can have any positive value, and this full range of frequencies is known as the **electromagnetic spectrum.** Certain bands of the spectrum are given special names, as indicated in **FIGURE 25-12.**

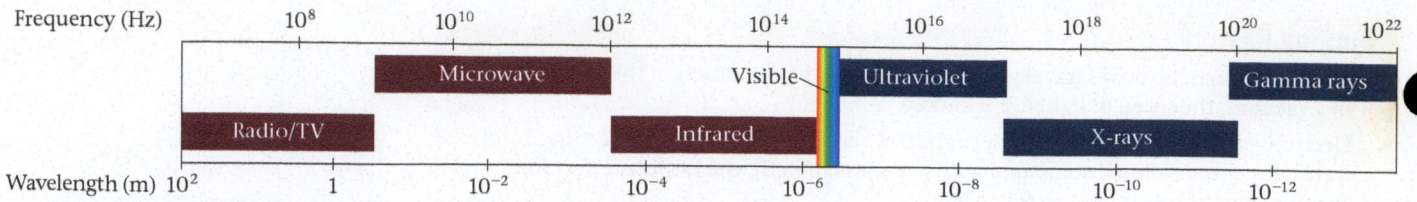

▲ **FIGURE 25-12 The electromagnetic spectrum** Note that the visible portion of the spectrum is relatively narrow. The boundaries between various bands of the spectrum are not sharp but, instead, are somewhat arbitrary.

Big Idea **4** Electromagnetic waves have frequencies that range from zero to infinity. Waves in different parts of this spectrum of frequencies—like radio waves, infrared waves, light waves, and X-rays—have different characteristics.

For example, we have just seen that visible light occupies a relatively narrow band of frequencies from 4.29×10^{14} Hz to 7.50×10^{14} Hz. In what follows, we discuss the various regions of the electromagnetic spectrum that have the greatest relevance to humans and our technology, in order of increasing frequency.

Radio Waves

$(f \sim 10^6$ **Hz to** 10^9 **Hz,** $\lambda \sim 300$ **m to 0.3 m)** The lowest-frequency electromagnetic waves of practical importance are *radio* and *television waves* in the frequency range of roughly 10^6 Hz to 10^9 Hz. Waves in this frequency range are produced in a variety of ways. For example, molecules and accelerated electrons in space give off radio waves, which radio astronomers detect with large dish receivers. Radio waves are also produced when a piece of adhesive tape is slowly peeled from a surface, as you can confirm by holding a portable radio near the tape and listening for pops and snaps coming from the speaker. Most commonly, the radio waves we pick up with our radios and televisions are produced by alternating currents in metal antennas. Radio waves are also emitted by many astronomical objects, and can be detected by radio telescopes, like those shown in **FIGURE 25-13**.

▲ **FIGURE 25-13** Since the development of the first radio telescopes in the 1950s, the radio portion of the electromagnetic spectrum has provided astronomers with a valuable new window on the universe. These antennas are part of the Very Large Array (VLA), located in San Augustin, New Mexico.

Microwaves

$(f \sim 10^9$ **Hz to** 10^{12} **Hz,** $\lambda \sim 300$ **mm to 0.3 mm)** Electromagnetic radiation with frequencies from 10^9 Hz to about 10^{12} Hz is referred to as *microwaves*. Waves in this frequency range are used to carry telephone conversations, to operate the Global Positioning System, and to cook our food. Microwaves, with wavelengths of about 1 mm to 30 cm, are the highest-frequency electromagnetic waves that can be produced by electronic circuitry.

Infrared Waves

BIO $(f \sim 10^{12}$ **Hz to** 4.3×10^{14} **Hz,** $\lambda \sim 0.3$ **mm to 700 nm)** Electromagnetic waves with frequencies just below that of red light—roughly 10^{12} Hz to 4.3×10^{14} Hz—are known as *infrared* rays. These waves can be felt as heat on our skin but cannot be seen with our eyes. Many creatures, including various types of pit vipers, have specialized infrared receptors that allow them to "see" the infrared rays given off by a warm-blooded prey animal, even in total darkness. Infrared rays are often generated by the rotations and vibrations of molecules. In turn, when infrared rays are absorbed by an object, its molecules rotate and vibrate more vigorously, resulting in an increase in the object's temperature. Finally, many remote controls—for items ranging from TVs to DVD players to gas fireplaces—operate on a beam of infrared light with a wavelength of about 1000 nm. This infrared light is so close to the visible spectrum and so low in intensity that it cannot be felt as heat. Examples of systems involving infrared radiation are shown in **FIGURE 25-14**, and thermograms taken with infrared radiation are shown in **FIGURE 25-15**.

Visible Light

$(f \sim 4.3 \times 10^{14}$ **Hz, to** 7.5×10^{14} **Hz,** $\lambda \sim 700$ **nm to 400 nm)** The portion of the electromagnetic spectrum most familiar to us is the spectrum of visible light, represented by the full range of colors seen in a rainbow. Each of the different colors, as perceived by our eyes and nervous system, is nothing more than an electromagnetic wave

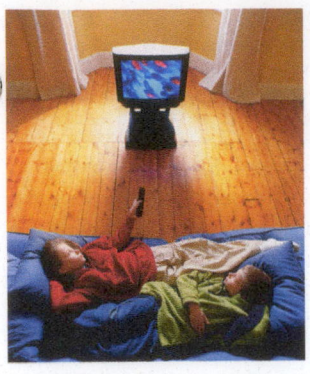

◄ **FIGURE 25-14** Visualizing Concepts
Infrared Radiation We use infrared rays all the time, even though they are invisible to us. If you change the channel with a remote control, your signal is sent by an infrared ray; if you move your hand in front of a no-touch water faucet, an infrared ray detects the motion. In contrast, snakes called pit vipers can actually "see" infrared rays with the "pit" organs located just in front of their eyes. What must a remote control look like to one of these creatures?

with a different frequency. Waves in this frequency range (4.3×10^{14} to 7.5×10^{14} Hz) are produced primarily by electrons changing their positions within an atom, as we discuss in detail in Chapter 31.

Ultraviolet Light

BIO ($f \sim 7.5 \times 10^{14}$ **Hz to** 10^{17} **Hz,** $\lambda \sim$ **400 nm to 3 nm**) When electromagnetic waves have frequencies just above that of violet light—from about 7.5×10^{14} Hz to 10^{17} Hz—they are called *ultraviolet* or *UV rays*. Although these rays are invisible, they often make their presence known by causing suntans with moderate exposure. More prolonged or intense exposure to UV rays can have harmful consequences, including an increased probability of developing a skin cancer. Fortunately, most of the UV radiation that reaches Earth from the Sun is absorbed in the upper atmosphere by ozone (O_3) and other molecules. A significant reduction in the ozone concentration in the stratosphere could result in an unwelcome increase of UV radiation on Earth's surface. When UV light is intense enough, it can be used to kill bacteria on tools and surfaces for medical sanitation, as shown in **FIGURE 25-16**, as well as to sterilize drinking water and wastewater.

X-Rays

($f \sim 10^{17}$ **Hz to** 10^{20} **Hz,** $\lambda \sim$ **3 nm to 0.003 nm**) As the frequency of electromagnetic waves is raised even higher, into the range between about 10^{17} Hz and 10^{20} Hz, we reach the part of the spectrum known as *X-rays*. Typically, the X-rays used in medicine are generated by the rapid deceleration of high-speed electrons projected against a metal target, as we show in Section 31-7. These energetic rays, which are only weakly absorbed by the skin and soft tissues, pass through our bodies rather freely, except when they encounter bones, teeth, or other relatively dense material. This property makes X-rays most valuable for medical diagnosis, research, and treatment. Still, X-rays can cause damage to human tissue, and it is desirable to reduce unnecessary exposure to these rays as much as possible.

Gamma Rays

BIO ($f \sim 10^{20}$ **Hz and higher,** $\lambda \sim$ **0.003 nm and smaller**) Finally, electromagnetic waves with frequencies above about 10^{20} Hz are generally referred to as *gamma* (γ) *rays*. These rays, which are even more energetic than X-rays, are often produced as neutrons and protons rearrange themselves within a nucleus, or when a particle collides with its antiparticle, and the two annihilate each other. These processes are discussed in detail in Chapter 32. Gamma rays are also highly penetrating and destructive to living cells. It is for this reason that they are used to kill cancer cells and, more recently, microorganisms in food. Irradiated food, however, is a concept that has yet to become popular with the general public, even though NASA has irradiated astronauts' food since the 1960s. If you happen to see irradiated food in the grocery store, you will know that it has been exposed to γ rays from cobalt-60 for 20 to 30 minutes. An illustration of the benefits of irradiation for food preservation is shown in **FIGURE 25-17**.

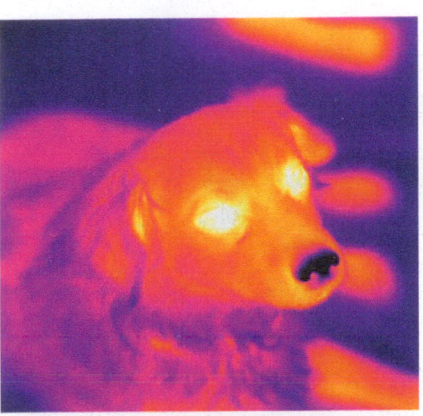

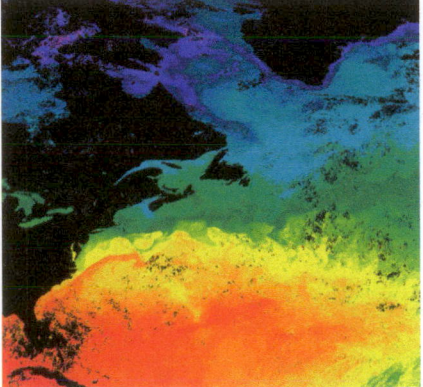

▲ **FIGURE 25-15** Visualizing Concepts
Thermograms Photographs made with infrared radiation are often called thermograms, since most infrared wavelengths can be felt as heat by the human skin. Thermograms provide a useful remote sensing technique for measuring temperature. In the photo on the top, the areas of the dog's head that are warmest (pink) and coolest (blue) can be clearly identified. In the photo on the bottom, an infrared satellite image of the Atlantic Ocean off the coast of North America, warmer colors indicate higher sea surface temperatures. The swirling red streak running from the lower left toward the upper right is the Gulf Stream.

◄ **FIGURE 25-16** The safety glasses in this cabinet are sterilized by the ultraviolet light from the germicidal lamp. The lamp produces a sufficient amount of light at a wavelength of 254 nm to severely damage the DNA of any bacteria and prevent them from reproducing.

Notice that the visible part of the electromagnetic spectrum, so important to life on Earth, is actually the smallest of the frequency bands we have named. This accounts for the fact that a rainbow produces only a narrow band of color in the sky—if the visible band were wider, the rainbow would be wider as well. It should be remembered, however, that there is nothing particularly special about the visible band; in fact, it is even species-dependent. For example, some bees and butterflies can see ultraviolet light, and, as mentioned previously, certain snakes can form images from infrared radiation.

One of the main factors in determining the visible range of frequencies is Earth's atmosphere. For example, if one examines the transparency of the atmosphere as a function of frequency, it is found that there is a relatively narrow range of frequencies for which the atmosphere is highly transparent. As eyes evolved in living systems on Earth, they could have evolved to be sensitive to various different frequency ranges. It so happens, however, that the range of frequencies that most animal eyes can detect matches nicely with the range of frequencies that the atmosphere allows to reach Earth's surface. This is a nice example of natural adaptation.

▲ **FIGURE 25-17** The use of radiation to preserve food can be quite effective, but the technique is still controversial. Both boxes of strawberries shown here were stored for about 2 weeks at refrigerator temperature. Before storage, the box at right was irradiated to kill microorganisms and mold spores.

Enhance Your Understanding (Answers given at the end of the chapter)

3. If the frequency of an electromagnetic wave is increased, **(a)** does its speed increase, decrease, or stay the same? **(b)** Does its wavelength increase, decrease, or stay the same?

Section Review

- Electromagnetic waves can have any frequency. Electromagnetic waves with different frequencies can have quite different properties.

25-4 Energy and Momentum in Electromagnetic Waves

All waves transmit energy, and electromagnetic waves are no exception. When you walk outside on a sunny day, for example, you feel warmth where the sunlight strikes your body. The energy creating this warm sensation originated in the Sun, and it has just completed a 93-million-mile trip when it reaches you. In fact, the energy necessary for most of the life on Earth is transported here across the vacuum of space by electromagnetic waves traveling at the speed of light.

PHYSICS IN CONTEXT
Looking Back

In this section, we study the energy, momentum, and intensity associated with electromagnetic waves. These are the same concepts introduced previously in Chapter 8 (energy), Chapter 9 (momentum), and Chapter 14 (intensity), respectively.

The Energy of an Electromagnetic Waves The fact that electromagnetic waves carry energy is no surprise when you recall that they are composed of electric and magnetic fields, each of which has an associated energy density. For example, in Section 20-6 we showed that the energy density of an electric field of magnitude E is

$$u_E = \tfrac{1}{2}\epsilon_0 E^2$$

Similarly, we showed in Section 23-9 that the energy density of a magnetic field of magnitude B is

$$u_B = \frac{1}{2\,\mu_0} B^2$$

It follows that the total energy density, u, of an electromagnetic wave is simply

$$u = u_E + u_B = \tfrac{1}{2}\varepsilon_0 E^2 + \frac{1}{2\,\mu_0} B^2$$

25-5

As expected, both E and B contribute to the total energy carried by a wave. Not only that, but it can be shown that the electric and magnetic energy densities in an electromagnetic wave are, in fact, equal to each other—again demonstrating the symmetrical role played by the electric and magnetic fields. Thus, the total energy density of an electromagnetic field can be written in the following equivalent forms:

$$u = \tfrac{1}{2}\epsilon_0 E^2 + \frac{1}{2\mu_0}B^2 = \epsilon_0 E^2 = \frac{1}{\mu_0}B^2 \qquad \text{25-6}$$

Because E and B vary sinusoidally with time, as indicated in Figure 25-4, it follows that their average values are zero—just like the current or voltage in an ac circuit. Therefore, to find the average energy density of an electromagnetic wave, we must use the rms values of E and B:

$$u_{av} = \tfrac{1}{2}\epsilon_0 E_{rms}^2 + \frac{1}{2\mu_0}B_{rms}^2 = \epsilon_0 E_{rms}^2 = \frac{1}{\mu_0}B_{rms}^2 \qquad \text{25-7}$$

Recall that the rms value of a sinusoidally varying quantity x is related to its maximum value as $x_{rms} = x_{max}/\sqrt{2}$. Thus, for the electric and magnetic fields in an electromagnetic wave we have

$$E_{rms} = \frac{E_{max}}{\sqrt{2}}$$
$$\text{25-8}$$
$$B_{rms} = \frac{B_{max}}{\sqrt{2}}$$

The fact that the electric and magnetic energy densities are equal in an electromagnetic wave has a further interesting consequence. Setting u_E equal to u_B, we obtain

$$\tfrac{1}{2}\epsilon_0 E^2 = \frac{1}{2\mu_0}B^2$$

Rearranging slightly and taking the square root, we find

$$E = \frac{1}{\sqrt{\epsilon_0\mu_0}}B$$

Finally, from the fact that the speed of light is given by the relationship $c = 1/\sqrt{\epsilon_0\mu_0}$, it follows that

$$E = cB \qquad \text{25-9}$$

Thus, not only does an electromagnetic wave have both an electric and a magnetic field, the fields must also have the specific ratio $E/B = c$.

EXERCISE 25-7   MAGNETIC FIELD

At a given instant of time the electric field in a beam of sunlight has a magnitude of 429 N/C. What is the magnitude of the magnetic field at this instant?

REASONING AND SOLUTION
In any electromagnetic wave, the magnitudes of the E and B fields must satisfy the relationship $E = cB$. Rearranging this equation to solve for B, we find

$$B = \frac{E}{c} = \frac{429 \text{ N/C}}{3.00 \times 10^8 \text{ m/s}} = 1.43 \times 10^{-6}\text{ T}$$

Notice that the units are consistent, because $1\text{ T} = 1\text{ N/(C}\cdot\text{m/s)}$.

The Intensity of an Electromagnetic Wave   The amount of energy a wave delivers to a unit area in a unit time is referred to as its **intensity**, I. (Equivalently, noting that power is energy per time, the intensity of a wave is the power per unit area.) Imagine, for example, an electromagnetic wave of area A moving in the positive x direction, as in **FIGURE 25-18**. In the time Δt the wave moves through a distance $c\Delta t$; hence, all the energy in the volume $\Delta V = A(c\Delta t)$ is deposited on the area A in this time. Because energy is equal to the energy density times the volume, it follows that the energy in

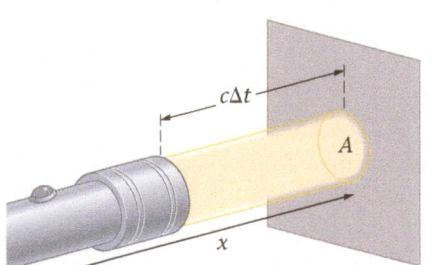

▲ **FIGURE 25-18 The energy in a beam of light**   A beam of light of cross-sectional area A shines on a surface. All the light energy contained in the volume $\Delta V = A(c\Delta t)$ strikes the surface in the time Δt.

the volume ΔV is $\Delta U = u\Delta V$. Therefore, the intensity of the wave (energy per area per time) is

$$I = \frac{\Delta U}{A\Delta t} = \frac{u(Ac\Delta t)}{A\Delta t} = uc$$

In terms of the electric and magnetic fields, we have

Point of Equal Brightness between
Two Light Sources

$$I = uc = \frac{1}{2}c\epsilon_0 E^2 + \frac{1}{2\mu_0}cB^2 = c\epsilon_0 E^2 = \frac{c}{\mu_0}B^2 \qquad \text{25-10}$$

As before, to calculate an average intensity, we must replace E and B with their rms values. Notice that the intensity is proportional to the square of the fields. This is analogous to the case of simple harmonic motion, where the energy of oscillation is proportional to the square of the amplitude, as we found in Section 13-5.

EXAMPLE 25-8   LIGHTBULB FIELDS

A garage is illuminated by a single incandescent lightbulb dangling from a wire. If the bulb radiates light uniformly in all directions, and consumes an average electrical power of 50.0 W, what are **(a)** the average intensity of the light and **(b)** the rms values of E and B at a distance of 1.00 m from the bulb? (Assume that 5.00% of the electrical power consumed by the bulb is converted to light.)

PICTURE THE PROBLEM
The physical situation is shown in the sketch. We assume that all the power radiated by the bulb passes uniformly through the area of a sphere of radius $r = 1.00$ m centered on the bulb.

REASONING AND STRATEGY

a. Recall that intensity is power per unit area; therefore, $I_{av} = P_{av}/A$. In this case the area is $A = 4\pi r^2$, the surface area of a sphere of radius r. The average power of the light is 5.00% of 50.0 W.

b. Once we know the intensity, we can obtain the E and B fields. The intensity is an average value, I_{av}, and hence the corresponding fields are rms values: $I_{av} = c\epsilon_0 E_{rms}^2 = cB_{rms}^2/\mu_0$.

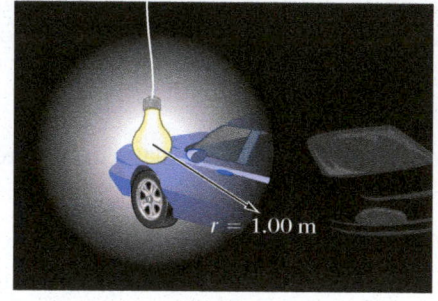

$r = 1.00$ m

Known   Average power of lightbulb, $P_{av} = 50.0$ W; efficiency of lightbulb, 0.0500; distance to lightbulb, $r = 1.00$ m.
Unknown   **(a)** Average intensity of light, $I_{av} = ?$ **(b)** rms electric field, $E_{rms} = ?$ rms magnetic field, $B_{rms} = ?$

SOLUTION

Part (a)

1. Calculate the average intensity at the surface of the sphere of radius $r = 1.00$ m:

$$I_{av} = \frac{P_{av}}{A} = \frac{(50.0 \text{ W})(0.0500)}{4\pi(1.00 \text{ m})^2} = 0.199 \text{ W/m}^2$$

Part (b)

2. Use $I_{av} = c\epsilon_0 E_{rms}^2$ to find E_{rms}:

$$I_{av} = c\epsilon_0 E_{rms}^2$$

$$E_{rms} = \sqrt{\frac{I_{av}}{c\epsilon_0}} = \sqrt{\frac{0.199 \text{ W/m}^2}{(3.00 \times 10^8 \text{ m/s})(8.85 \times 10^{-12} \text{ C}^2/\text{N}\cdot\text{m}^2)}}$$
$$= 8.66 \text{ N/C}$$

3. Use $I_{av} = cB_{rms}^2/\mu_0$ to find B_{rms}:

$$I_{av} = \frac{c}{\mu_0}B_{rms}^2$$

$$B_{rms} = \sqrt{\frac{\mu_0 I_{av}}{c}} = \sqrt{\frac{(4\pi \times 10^{-7} \text{ N}\cdot\text{s}^2/\text{C}^2)(0.199 \text{ W/m}^2)}{(3.00 \times 10^8 \text{ m/s})}}$$
$$= 2.89 \times 10^{-8} \text{ T}$$

INSIGHT
The electric field in the light from the bulb has a magnitude (8.66 N/C = 8.66 V/m) that is relatively easy to measure. The magnetic field, however, is so small that a direct measurement would be difficult. This is common in electromagnetic waves because $B = E/c$, and c is such a large number.

CONTINUED

Momentum in Electromagnetic Waves Electromagnetic waves also carry momentum. In fact, it can be shown that if a total energy U is absorbed by a given area, the momentum, p, that it receives is

$$p = \frac{U}{c} \qquad\qquad 25\text{-}11$$

For an electromagnetic wave absorbed by an area A, the total energy received in the time Δt is $U = u_{av}Ac\Delta t$; hence, the momentum, Δp, received in this time is

$$\Delta p = \frac{u_{av}Ac\Delta t}{c} = \frac{I_{av}A\Delta t}{c}$$

Since the average force is $F_{av} = \Delta p/\Delta t = u_{av}A = I_{av}A/c$, it follows that the average pressure (force per area) is simply

$$\text{pressure}_{av} = \frac{I_{av}}{c} \qquad\qquad 25\text{-}12$$

The pressure exerted by light is commonly referred to as **radiation pressure.**

> **Big Idea 5** Electromagnetic waves carry both energy and momentum.

CONCEPTUAL EXAMPLE 25-9   MOMENTUM TRANSFER

When an electromagnetic wave carrying an energy U is absorbed by an object, the momentum the object receives is $p = U/c$. If, instead, the object reflects the wave, is the momentum the object receives greater than, less than, or the same as when it absorbs the wave?

REASONING AND DISCUSSION
When the object reflects the wave, it must supply not only enough momentum to stop the wave, $p = U/c$, but also an equivalent amount of momentum to send the wave back in the opposite direction. Thus, the momentum the object receives is $p = 2U/c$. This situation is completely analogous to the momentum transfer of a ball that either sticks to a wall or bounces back.

ANSWER
The object receives twice as much momentum.

Radiation pressure is a very real effect, though in everyday situations it is too small to be noticed. In principle, turning on a flashlight should give the user a "kick," like the recoil from firing a gun. In practice, the effect is much too small to be felt. In the following Quick Example, we calculate the radiation pressure due to sunlight to get a feel for the magnitudes involved.

QUICK EXAMPLE 25-10   RADIATION PRESSURE ON THE KEPLER TELESCOPE

NASA's Kepler Space Telescope is an orbiting observatory that searches for Earth-like planets orbiting distant stars. When two of the reaction wheels used to stabilize the telescope failed, the mission almost came to an end. Fortunately, a plan emerged to use radiation pressure from sunlight shining on Kepler's solar panels to help point the telescope. If the average intensity of sunlight at Kepler's orbit is 1350 W/m², what is the average force it exerts on the solar panels? Assume the panels have an area of 10.2 m², and that they absorb all the light that falls on them.

CONTINUED

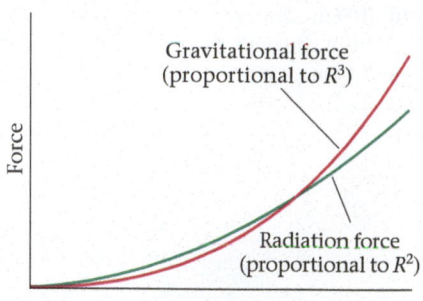

Force

Gravitational force
(proportional to R^3)

Radiation force
(proportional to R^2)

Radius, R

▲ **FIGURE 25-19 Radiation versus gravitational forces** The forces exerted on a small particle of radius R by gravity and radiation pressure. For large R the gravitational force dominates; for small R the radiation force is larger.

PHYSICS IN CONTEXT
Looking Ahead

Light is a fascinating phenomenon, with characteristics of both waves and particles. In this chapter we focused on the wavelike properties of light, and we will expand on this topic in Chapter 28. The particle-like aspects of light will be presented in Chapter 30, when we study quantum physics.

▲ **FIGURE 25-20** Like many comets, comet Hale-Bopp, which passed relatively close to Earth in 1997, developed two tails as it approached the Sun. The straighter, blue tail in this photograph is gas boiling off the comet's head as volatile material is vaporized by the Sun's heat. The curved, whiter tail consists mostly of dust particles, blown away by the pressure of sunlight.

Big Idea 6 Electromagnetic waves have a property not found in sound waves—they can be polarized. A linearly polarized electromagnetic wave has an electric field that oscillates along a fixed direction.

REASONING AND SOLUTION

First, we find the average radiation pressure exerted on the solar panels. This is given by the average intensity of the sunlight divided by the speed of light; that is, $\text{pressure}_{av} = I_{av}/c$. Next, we multiply the average pressure by the area of the panels to find the average force they experience.

1. Calculate the average radiation pressure acting on the solar panels:

$$\text{pressure}_{av} = \frac{I_{av}}{c}$$

$$= \frac{1350 \text{ W/m}^2}{3.00 \times 10^8 \text{ m/s}} = 4.50 \times 10^{-6} \text{ N/m}^2$$

2. Multiply the average pressure by the area to find the average force exerted on the panels:

$$F_{av} = (\text{pressure}_{av})A$$
$$= (4.50 \times 10^{-6} \text{ N/m}^2)(10.2 \text{ m}^2) = 4.59 \times 10^{-5} \text{ N}$$

This is a very small force, but because it acts for days on end in a frictionless environment, it can have significant effects.

Even though the radiation force on a beach towel is negligible, the effects of radiation pressure can be significant on particles that are very small. Consider a small object of radius R drifting through space somewhere in our solar system. The gravitational attraction it feels from the Sun depends on its mass. Because mass is proportional to volume, and volume is proportional to the cube of the particle's radius, the gravitational force on the particle varies as R^3. On the other hand, the radiation pressure is exerted over the area of the object, and the area is proportional to R^2. As R becomes smaller, the radiation pressure, with its R^2 dependence, decreases less rapidly than the gravitational force, with its R^3 dependence, as illustrated in **FIGURE 25-19**. Thus, for small enough particles, the radiation pressure from sunlight is actually more important than the gravitational force. It is for this reason that dust particles given off by a comet are "blown" away by sunlight, giving the comet a long tail that streams away from the Sun, as shown in **FIGURE 25-20**. Some imaginative thinkers have envisaged a "sailing" ship designed to travel through the cosmos using this light-pressure "wind."

Finally, it has been discovered that the energy carried by light comes in discrete units, as if there were "particles" of energy in a light beam. So, in many ways, light behaves just like any other wave, showing the effects of diffraction and interference, whereas in other ways it acts like a particle. Therefore, light, and all electromagnetic waves, have a "dual" nature, exhibiting both wave and particle properties. As we shall see when we consider quantum physics in Chapter 30, this *wave–particle duality* plays a fundamental role in our understanding of modern physics.

Enhance Your Understanding   (Answers given at the end of the chapter)

4. If the electric field in an electromagnetic field is doubled, by what factor does the magnetic field change? Explain.

Section Review

- The average energy density carried by an electromagnetic wave is $u_{av} = \epsilon_0 E_{rms}^2 = B_{rms}^2/\mu_0$.
- The average intensity of an electromagnetic wave is $I_{av} = c\epsilon_0 E_{rms}^2 = c_0 B_{rms}^2/\mu_0$,
- Electromagnetic waves carry an average momentum given by $\text{pressure}_{av} = I_{av}/c$.

25-5 Polarization

When looking into the blue sky on a crystal-clear day, we see light that is fairly uniform—as long as we refrain from looking too close to the Sun. However, for some animals, like the common honeybee, the light in the sky is far from uniform. The reason is that honeybees are sensitive to the **polarization** of light, an ability that aids in their navigation from hive to flower and back.

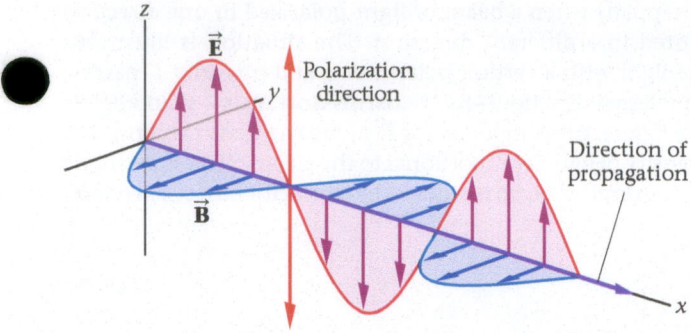

(a) Polarization in *z* direction

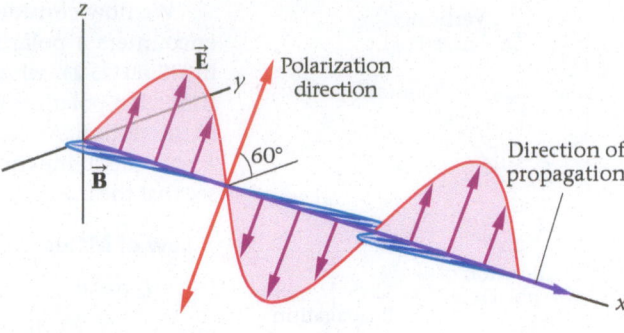

(b) Polarization 60° to *y* direction

The Meaning of Polarization

To understand what is meant by the polarization of light, or any other electromagnetic wave, consider the electromagnetic waves pictured in **FIGURE 25-21**. Each of these waves has an electric field that points along a single line. For example, the electric field in Figure 25-21 (a) points in either the positive or negative *z* direction. We say, then, that this wave is **linearly polarized** in the *z* direction. A wave of this sort might be produced by a straight-wire antenna oriented along the *z* axis. Similarly, the direction of polarization for the wave in Figure 25-21 (b) is in the *y-z* plane at an angle of 60° relative to the *y* direction. In general, we can make the following statement:

> The polarization of an electromagnetic wave is the direction along which its electric field oscillates.

▲ **FIGURE 25-21 Polarization of electromagnetic waves** The polarization of an electromagnetic wave is the direction along which its electric field vector, $\vec{E}$, points. The cases shown illustrate **(a)** polarization in the *z* direction and **(b)** polarization in the *y-z* plane, at an angle of 60° with respect to the *y* axis.

(a) Vertically polarized light

(b) Unpolarized light

A convenient way to represent the polarization of a beam of light is shown in **FIGURE 25-22**. In part (a) of this figure, we indicate light that is polarized in the vertical direction. Not all light is polarized, however. In part (b) we show light that is a combination of many waves with polarizations in different, random directions. Such light is said to be **unpolarized.** A common incandescent lightbulb produces unpolarized light because each atom in the heated filament sends out light of random polarization. Similarly, the light from the Sun is unpolarized.

◄ **FIGURE 25-22 Polarized versus unpolarized light** A beam of light that is **(a)** polarized in the vertical direction and **(b)** unpolarized.

Parallel-Wire Polarizer for Microwaves

Passing Light Through Polarizers

A beam of unpolarized light can be polarized in a number of ways, including by passing it through a **polarizer.** To be specific, a polarizer is a material that is composed of long, thin, electrically conductive molecules oriented in a specific direction. When a beam of light strikes a polarizer, it is readily absorbed if its electric field is parallel to the molecules; light whose electric field is perpendicular to the molecules passes through the material with little absorption. As a result, the light that passes through a polarizer is preferentially polarized along a specific direction. Common examples of a polarizer are the well-known Polaroid sheets used to make Polaroid sunglasses.

A simple mechanical analog of a polarizer is shown in **FIGURE 25-23**. Here we see a wave that displaces a string in the vertical direction as it propagates toward a slit cut into a block of wood. If the slit is oriented vertically, as in Figure 25-23 (a), the wave passes through unhindered. Conversely, when the slit is oriented horizontally, it stops the wave, as indicated in Figure 25-23 (b). A polarizer performs a similar function on a beam of light.

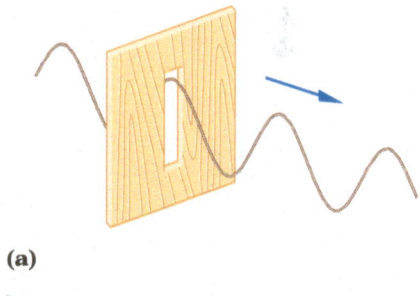

(a)

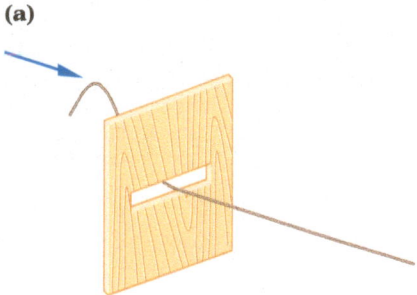

(b)

▲ **FIGURE 25-23 A mechanical analog of a polarizer** **(a)** The polarization of the wave is in the same direction as the polarizer; hence, the wave passes through unaffected. **(b)** The polarization of the wave is at right angles to the direction of the polarizer. In this case the wave is absorbed.

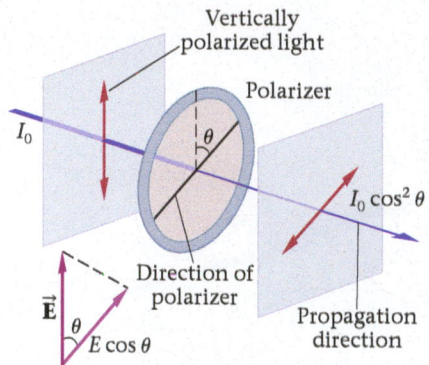

▲ **FIGURE 25-24 Transmission of polarized light through a polarizer** A polarized beam of light, with intensity I_0, encounters a polarizer oriented at an angle θ relative to the polarization direction. The intensity of light transmitted through the polarizer is $I = I_0 \cos^2 \theta$. After passing through the polarizer, the light is polarized in the same direction as the polarizer.

PROBLEM-SOLVING NOTE

Transmission Through Polarizing Filters

The intensity of light transmitted through a pair of polarizing filters depends only on the relative angle θ between the filters; it is independent of whether the filters are rotated clockwise or counterclockwise relative to each other.

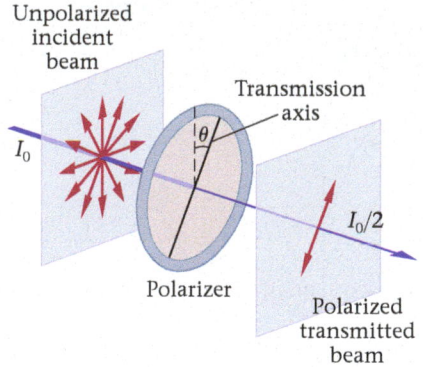

▲ **FIGURE 25-25 Transmission of unpolarized light through a polarizer** When an unpolarized beam of intensity I_0 passes through a polarizer, the transmitted beam has an intensity of $\frac{1}{2} I_0$ and is polarized in the direction of the polarizer.

▶ **FIGURE 25-26 A polarizer and an analyzer** An unpolarized beam of intensity I_0 is polarized in the vertical direction by a polarizer with a vertical transmission axis. Next, it passes through another polarizer, the analyzer, whose transmission axis is at an angle θ relative to the transmission axis of the polarizer. The final intensity of the beam is $I = \frac{1}{2} I_0 \cos^2 \theta$.

We now consider what happens when a beam of light polarized in one direction encounters a polarizer oriented in a different direction. The situation is illustrated in **FIGURE 25-24**, where we see light with a vertical polarization and intensity I_0 passing through a polarizer with its preferred direction—its **transmission axis**—at an angle θ to the vertical. As shown in the figure, the component of $\vec{\mathbf{E}}$ along the transmission axis is $E \cos \theta$. Recalling that the intensity of light is proportional to the electric field squared, we see that the intensity, I, of the transmitted beam is reduced by the factor $\cos^2 \theta$. Therefore:

Law of Malus

$$I = I_0 \cos^2 \theta \qquad\qquad 25\text{-}13$$

This result is known as the **law of Malus,** after the French engineer Etienne-Louis Malus (1775–1812). Notice that the intensity is unchanged if $\theta = 0$, and is zero if $\theta = 90°$.

EXERCISE 25-11 TRANSMITTED INTENSITY

Vertically polarized light with an intensity of 428 W/m^2 passes through a polarizer oriented at an angle θ to the vertical. Find the transmitted intensity of the light for **(a)** $\theta = 15.0°$, **(b)** $\theta = 45.0°$, and **(c)** $\theta = 90.0°$.

REASONING AND SOLUTION

Applying $I = I_0 \cos^2 \theta$ to each case, we obtain

   a. $I = (428 \text{ W/m}^2) \cos^2 15.0° = 399 \text{ W/m}^2$
   b. $I = (428 \text{ W/m}^2) \cos^2 45.0° = 214 \text{ W/m}^2$
   c. $I = (428 \text{ W/m}^2) \cos^2 90.0° = 0$

Change in the Direction of Polarization Just as important as the change in intensity is what happens to the *polarization* of the transmitted light:

> The transmitted beam of light is no longer polarized in its original direction; it is now polarized in the direction of the polarizer.

This effect also is illustrated in Figure 25-24. Thus, a polarizer changes both the intensity *and* the polarization of a beam of light.

FIGURE 25-25 shows an unpolarized beam of light with an intensity I_0 encountering a polarizer. In this case, there is no single angle θ; instead, to obtain the transmitted intensity, we must average $\cos^2 \theta$ over all angles. This has already been done in Section 24-1, where we considered rms values in ac circuits. As was shown there, the average of $\cos^2 \theta$ is one-half; thus, the transmitted intensity for an unpolarized beam with an intensity of I_0 is:

Transmitted Intensity for an Unpolarized Beam

$$I = \frac{1}{2} I_0 \qquad\qquad 25\text{-}14$$

A common type of polarization experiment is shown in **FIGURE 25-26**. An unpolarized beam is first passed through a polarizer to give the light a specified polarization.

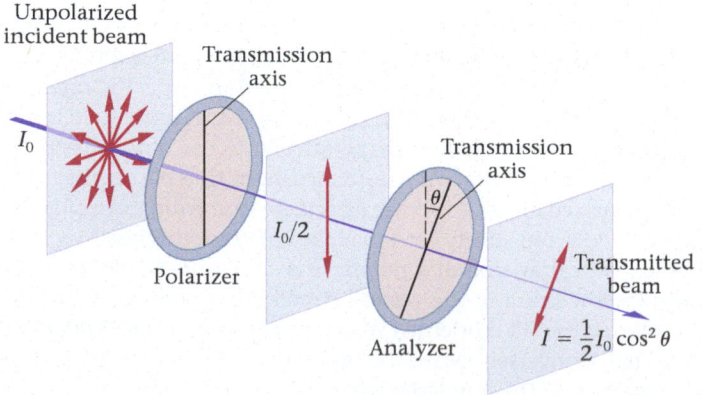

The light next passes through a second polarizer, referred to as the **analyzer,** whose transmission axis is at an angle θ relative to the polarizer. The orientation of the analyzer can be adjusted to give a beam of light of variable intensity and polarization. We consider a situation of this type in the next Example.

EXAMPLE 25-12   ANALYZE THIS

In the polarization experiment shown in our sketch, the final intensity of the beam is $0.200\,I_0$. What is the angle θ between the transmission axes of the analyzer and polarizer?

PICTURE THE PROBLEM
The experimental setup is shown in the sketch. As indicated, the intensity of the unpolarized incident beam, I_0, is reduced to $\frac{1}{2}I_0$ after passing through the first polarizer. The second polarizer reduces the intensity further, to $0.200\,I_0$.

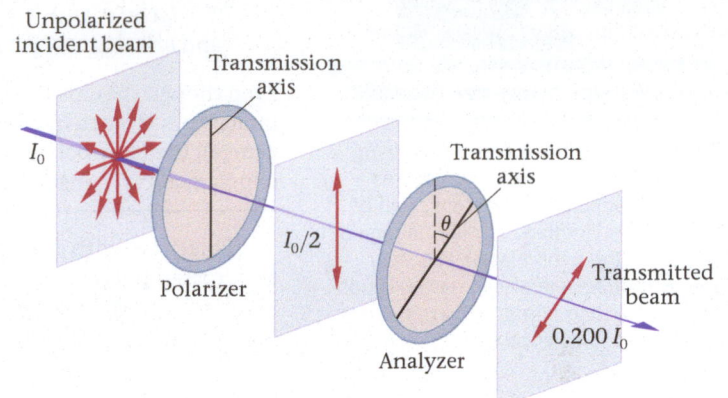

REASONING AND STRATEGY
It is clear from the sketch that the analyzer reduces the intensity of the light by a factor of 1/2.50; that is, the intensity is reduced from $I_0/2$ to $(1/2.50)(I_0/2) = I_0/5.00 = 0.200\,I_0$. Thus we must find the angle θ that gives this reduction. Recalling that an analyzer reduces intensity according to the relationship $I = I_0 \cos^2\theta$, we set $\cos^2\theta = 1/2.50$ and solve for θ.

Known   Final intensity, $I = 0.200 I_0$.
Unknown   Angle between polarizing axes, $\theta = ?$

SOLUTION

1. Set $\cos^2\theta$ equal to 1/2.50 and solve for $\cos\theta$:

$$\cos^2\theta = \frac{1}{2.50}$$

$$\cos\theta = \frac{1}{\sqrt{2.50}}$$

2. Solve for the angle θ:

$$\theta = \cos^{-1}\!\left(\frac{1}{\sqrt{2.50}}\right) = 50.8°$$

INSIGHT
The analyzer absorbs part of the light as the beam passes through, and it also absorbs energy. Therefore, in principle, the analyzer would experience a slight heating. As always, energy must be conserved.

PRACTICE PROBLEM — PREDICT/CALCULATE
(a) If the angle θ is increased slightly, does the final intensity increase, decrease, or stay the same? Explain. **(b)** Check your answer by finding the final intensity for 60.0°.   [**Answer: (a)** The final intensity decreases because the transmission axis of the analyzer is at a larger angle to the transmission axis of the polarizer—that is, at an angle closer to 90°. **(b)** For $\theta = 60.0°$, the final intensity is $I = 0.125\,I_0$.]

Some related homework problems: Problem 73, Problem 79

QUICK EXAMPLE 25-13   FIND THE TRANSMITTED INTENSITY

Calculate the transmitted intensity for the following two cases: **(a)** A vertically polarized beam of intensity I_0 passes through a polarizer with its transmission axis at 60° to the vertical. **(b)** A vertically polarized beam of intensity I_0 passes through two polarizers, the first with its transmission axis at 30° to the vertical, and the second with its transmission axis rotated an additional 30° to the vertical. (*Note:* In both cases, the final beam is polarized at 60° to the vertical.)

REASONING AND SOLUTION
In part (a), we apply $I = I_0 \cos^2\theta$ directly, with $\theta = 60°$. In part (b), we apply $I = I_0 \cos^2\theta$ twice, once for each polarizer, and each time with $\theta = 30°$. The intensity after the first

CONTINUED

polarizer—we'll call it the intermediate intensity—is the initial intensity for the second polarizer.

Part (a)

1. Calculate the final intensity using $I = I_0 \cos^2 \theta$, with $\theta = 60°$:

$$I = I_0 \cos^2 60°$$
$$= I_0(0.5)^2 = \tfrac{1}{4}I_0 = 0.25\,I_0$$

Part (b)

2. Calculate the intermediate intensity using $I = I_0 \cos^2 \theta$, with $\theta = 30°$:

$$I_{intermediate} = I_0 \cos^2 30° = I_0\left(\sqrt{3}/2\right)^2 = \tfrac{3}{4}I_0$$

3. Calculate the final intensity using $I = I_0 \cos^2 \theta$, with $\theta = 30°$ and $I_0 = I_{intermediate}$:

$$I_{final} = I_{intermediate} \cos^2 30°$$
$$= \left(\tfrac{3}{4}I_0\right)\left(\sqrt{3}/2\right)^2 = \tfrac{9}{16}I_0 = 0.56\,I_0$$

Even though the direction of polarization is rotated by a total of 60° in each case, the final intensity is more than twice as great when two polarizers are used instead of just one. In general, the more polarizers that are used—and, hence, the more smoothly the direction of polarization changes—the greater the final intensity.

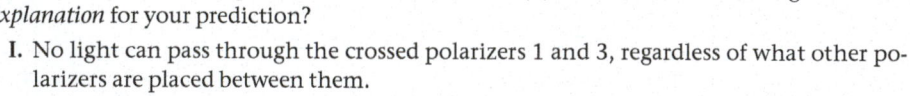

▲ **FIGURE 25-27 Overlapping polarizing sheets** When unpolarized light strikes a single layer of polarizing material, half of the light is transmitted. This is true regardless of how the transmission axis is oriented. But no light at all can pass through a pair of polarizing filters with axes at right angles (crossed polarizers).

Polarizers with transmission axes at right angles to one another are referred to as "crossed polarizers." The transmission through a pair of crossed polarizers is zero, because $\theta = 90°$ in Malus's law. Crossed polarizers are illustrated in **FIGURE 25-27** and are referred to in the following Conceptual Example.

CONCEPTUAL EXAMPLE 25-14   LIGHT TRANSMISSION   PREDICT/EXPLAIN

Consider a set of three polarizers. Polarizer 1 has a vertical transmission axis, and polarizer 3 has a horizontal transmission axis. Taken together, polarizers 1 and 3 are a pair of crossed polarizers, so that when a beam of unpolarized light shines on polarizer 1 from the left, no light emerges from polarizer 3. Polarizer 2, with a transmission angle at 45° to the vertical, is now placed between polarizers 1 and 3, as shown in the sketch. **(a)** Is the transmission of light through the three polarizers zero or nonzero? **(b)** Which of the following is the *best explanation* for your prediction?

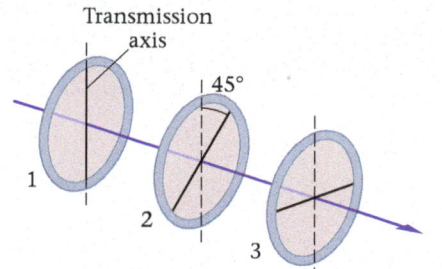

I. No light can pass through the crossed polarizers 1 and 3, regardless of what other polarizers are placed between them.

II. Polarizer 2 changes the polarization direction by 45°, and light can now pass through polarizer 3.

REASONING AND DISCUSSION

Polarizers 1 and 3 are still crossed, and hence it might seem that no light can be transmitted. When one recalls, however, that a polarizer causes a beam to have a polarization in the direction of its transmission axis, it becomes clear that transmission is indeed possible.

To be specific, some of the light that passes through polarizer 1 will also pass through polarizer 2, because the angle between their transmission axes is less than 90°. After passing through polarizer 2, the light is polarized at 45° to the vertical. As a result, some of this light can pass through polarizer 3 because, again, the angle between the polarization direction and the transmission axis is less than 90°.

In fact, the intensity of the unpolarized incident light is reduced by a factor of two when it passes through polarizer 1, by a factor of two when it passes through polarizer 2 (because $\cos^2 45° = \tfrac{1}{2}$), and again by a factor of two when it passes through polarizer 3. The final intensity, then, is one-eighth the incident intensity.

ANSWER

(a) The transmission is nonzero. **(b)** The best explanation is II.

RWP There are many practical uses for crossed polarizers. For example, engineers often construct a plastic replica of a building, bridge, or similar structure to study the relative amounts of stress in its various parts in a technique known as *photoelastic stress analysis*. Dentists use the same technique to study stresses in teeth, and doctors use stress analysis when they design prosthetic joints, as in **FIGURE 25-28**. In this application, the plastic replica plays the role of polarizer 2 in Conceptual Example 25-14. In particular, in those regions of the structure where the stress is high, the plastic acts to rotate the plane of polarization and—like polarizer 2 in the Conceptual Example—allows light to pass through the system. By examining such models with crossed polarizers, engineers can gain valuable insight into the safety of the structures they plan to build, dentists can determine where a tooth is likely to break, and doctors can see where an artificial hip joint needs to be strengthened.

RWP Another use of crossed polarizers is in the operation of a liquid-crystal display (LCD). There are basically three essential elements to each active area of an LCD—two crossed polarizers and a thin cell that holds a fluid composed of long, thin molecules known as a **liquid crystal.** The liquid crystal is selected for its ability to rotate the direction of polarization, and the thickness of the liquid-crystal cell is adjusted to give a rotation of 90°. Thus, in its "off" state, the liquid crystal rotates the direction of polarization and light passes through the crossed polarizers, as illustrated in **FIGURE 25-29 (a)**. In this state, the LCD is transparent—it allows ambient light to enter the display, reflect off the back, and exit the display, giving it the characteristic light background. When a voltage is applied to the liquid crystal, it no longer rotates the direction of polarization, and light is no longer transmitted through that area of the display. Thus, in the "on" state, shown in **FIGURE 25-29 (b)**, an area of the LCD appears black, which is how the black characters are formed on the light background, as shown in **FIGURE 25-29 (c)**. Because very little energy is required to give the voltage necessary to turn a liquid-crystal cell "on," the LCD is very energy efficient. In addition, the LCD uses light already present in the environment; it does not need to produce its own light, as do some displays.

Finally, there are many organic compounds that are capable of rotating the polarization direction of a beam of light. Such compounds, which include sugar, turpentine, and tartaric acid, are said to be **optically active.** When a solution containing optically active molecules is placed between crossed polarizers, the amount of light that passes through the system gives a direct measure of the concentration of the active molecules in the solution.

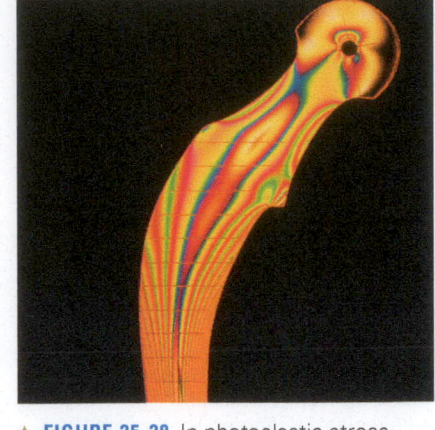

▲ **FIGURE 25-28** In photoelastic stress analysis, a plastic model of an object being studied is placed between crossed polarizers. In this case, the object is a prosthetic hip joint. If the polarization of the light is unchanged by the plastic, it will not pass through the second polarizer. In areas where the plastic is stressed, however, it rotates the plane of polarization, allowing some of the light to pass through.

Polarization by Scattering and Reflection

When unpolarized light is scattered from atoms or molecules, as in the atmosphere, or reflected by a solid or liquid surface, the light can acquire some degree of polarization. The basic reason for this is illustrated in **FIGURE 25-30**. In this case, we consider a vertically polarized beam of light that encounters a molecule. The light causes electrons in the molecule to oscillate in the vertical direction—that is, in the direction of the light's

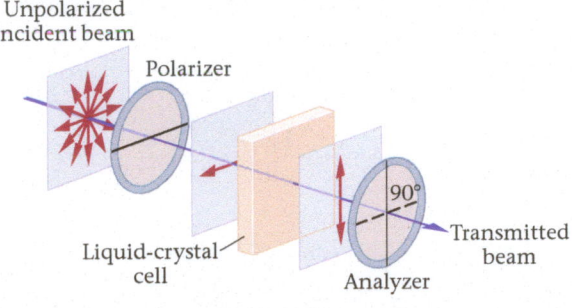

(a) Off (transmitted light gives bright background)

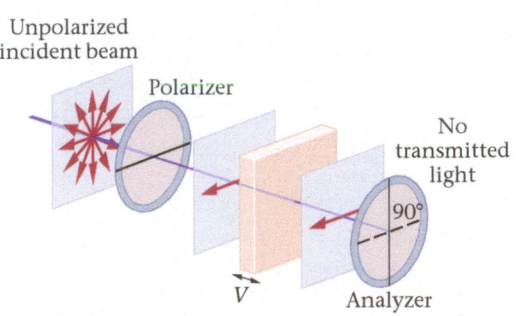

(b) On (dark characters formed where no light is transmitted)

(c) Resulting LCD display

▲ **FIGURE 25-29 Basic operation of a liquid-crystal display (LCD) (a)** In the "off" state, the liquid crystal in the cell rotates the polarization of the light by 90°, allowing light to pass through the display. This produces the light background of the display. **(b)** In the "on" state, a voltage V is applied to the liquid-crystal cell, which means that the polarization of the light is no longer rotated. As a result, no light is transmitted and this element of the display appears black. **(c)** The final result is a display where the dark characters are caused by a voltage applied to the appropriate liquid-crystal cell.

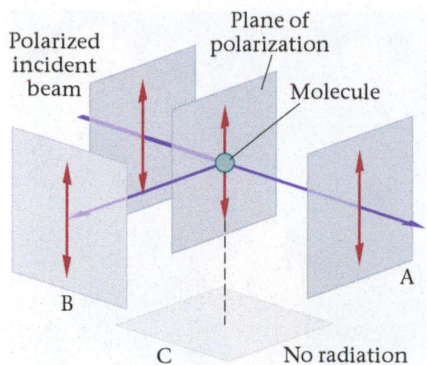

▲ **FIGURE 25-30 Vertically polarized light scattering from a molecule** The molecule radiates scattered light, much like a small antenna. At points A and B strong scattered rays are observed; at point C no radiation is seen.

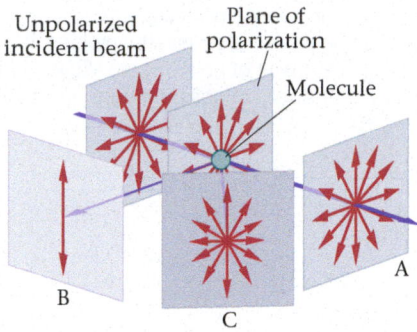

▲ **FIGURE 25-31 Unpolarized light scattering from a molecule** In the forward direction, A, the scattered light is unpolarized. At right angles to the initial beam of light, B, the scattered light is polarized. Along other directions the light is only partially polarized.

electric field. As we know, an accelerating charge radiates; thus, the molecule radiates as if it were a small vertical antenna, and this radiation is what we observe as scattered light. We also know, however, that an antenna gives off maximum radiation in the direction perpendicular to its length, and no radiation at all along its axis. Therefore, an observer at point A or B in Figure 25-30 sees maximum scattered radiation, whereas an observer at point C sees no radiation at all.

If we apply this same idea to an initially unpolarized beam of light scattering from a molecule, we find the situation pictured in **FIGURE 25-31**. In this case, electrons in the molecule oscillate in all directions within the plane of polarization; hence, an observer in the forward direction, at point A, sees scattered light of all polarizations—that is, unpolarized light. An observer at point B, however, sees no radiation from electrons oscillating horizontally, only from those oscillating vertically. Hence this observer sees vertically polarized light. An observer at an intermediate angle sees an intermediate amount of polarization.

BIO This mechanism produces polarization in the light coming from the sky. In particular, maximum polarization is observed in a direction at right angles to the Sun, as can be seen in **FIGURE 25-32**. Thus, to a creature that is sensitive to the polarization of light, like a bee or certain species of birds, the light from the sky varies with the direction relative to the Sun. Experiments show that such creatures can use polarized light as an aid in navigation.

RWP Another point of interest follows from the general observation that the amount of light scattered from a molecule is greatest when the wavelength of the light is comparable to the size of the molecule. The molecules in the atmosphere are generally much smaller than the wavelength of visible light, but blue light, with its relatively short wavelength, scatters more effectively than red light, with its longer wavelength. Similarly, microscopic particles of dust in the upper atmosphere scatter the short-wavelength blue light most effectively. This basic mechanism is the answer to the age-old question, Why is the sky blue? Similarly, a sunset appears red because you are viewing the Sun through the atmosphere, and most of the Sun's blue light has been scattered off in other directions. In fact, the blue light that is missing from your sunset—giving it a red color—is the blue light of someone else's blue sky, as indicated in Figure 25-32.

RWP Polarization also occurs when light reflects from a smooth surface, like the top of a table or the surface of a calm lake. A typical reflection situation, with unpolarized light from the Sun reflecting from the surface of a lake, is shown in **FIGURE 25-33**. When the light encounters molecules in the water, their electrons oscillate in the plane of polarization. For an observer at point A, however, only oscillations at right angles to the line of sight give rise to radiation. As a result, the reflected light from the lake is polarized horizontally. Polaroid sunglasses take advantage of this effect by using sheets of Polaroid material with a vertical transmission axis. With this orientation, the horizontally polarized reflected light—the glare—is not transmitted. A variety of effects due to scattering and polarization are shown in **FIGURE 25-34**.

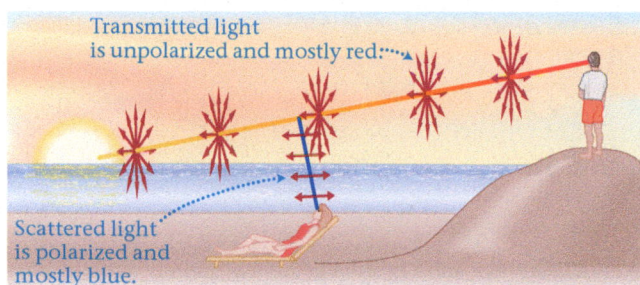

▲ **FIGURE 25-32 The effects of scattering on sunlight** The scattering of sunlight by the atmosphere produces polarized light for an observer looking at right angles to the direction of the Sun. This observer also sees more blue light than red. An observer looking toward the Sun sees unpolarized light that contains more red light than blue.

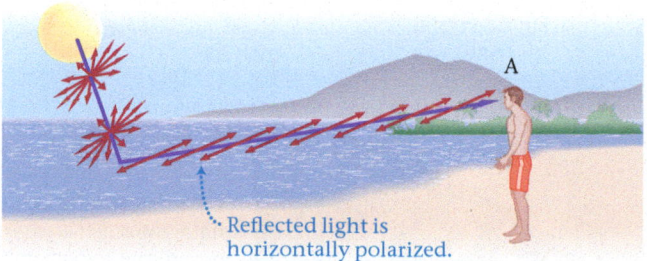

▲ **FIGURE 25-33 Polarization by reflection** Because the observer sees no radiation from electrons moving along the line of sight, the radiation that is observed is polarized horizontally. Polaroid sunglasses with a vertical transmission axis reduce this kind of reflected glare.

▲ **FIGURE 25-34** Visualizing Concepts **Scattering and Polarization** As light is scattered by air molecules and dust particles in the atmosphere, it becomes polarized. The shorter wavelengths at the blue end of the spectrum are scattered most effectively, creating our familiar blue skies. Sunsets are red (left) because much of the blue light has been scattered as it passes through the atmosphere on the way to our eyes. The polarization of light from the sky can be demonstrated with a pair of photographs like those at center—one taken without a polarizing filter (top center), the other with a polarizer (bottom center). A smooth surface such as a lake (right) can also act as a polarizing filter. That is why the sky is darker and the clouds are more easily seen in the reflected image than in the direct view.

Polarization presents a potential problem for the makers of digital watches and electronic devices with LCD displays. Suppose the person using such a device is wearing Polaroid sunglasses—not an unusual circumstance. As we saw in Figure 25-29, the light emerging from an LCD display is linearly polarized. If the polarization direction is vertical, the light will pass through the sunglasses and the display can be read as usual. On the other hand, if the polarization direction of the display is horizontal, it will appear completely black through a pair of Polaroid sunglasses—no light will pass through at all. The same effect can be seen by observing an LCD display through a pair of Polaroid sunglasses and slowly rotating the glasses or the display through 90°. Try this experiment sometime with a computer display or a digital watch, and you will see the result shown in **FIGURE 25-35**. Clearly, it would not be wise to make an LCD display with a horizontal polarization direction—a person wearing Polaroid sunglasses would think the display was broken.

Enhance Your Understanding

(Answers given at the end of the chapter)

5. In the system shown in **FIGURE 25-36**, a vertically polarized incident beam of light encounters two polarizers. If polarizer 1 is removed from the system, does the intensity of the transmitted beam increase, decrease, or stay the same? Explain.

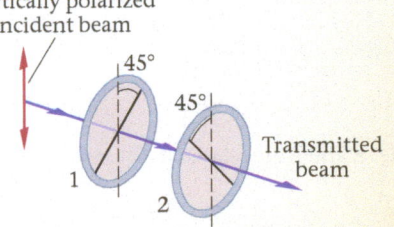

Vertically polarized incident beam

45°
45°

Transmitted beam

1
2

▲ **FIGURE 25-36** Enhance Your Understanding 5.

▲ **FIGURE 25-35** To eliminate the glare from reflected light, which is horizontally polarized, the lenses of Polaroid sunglasses have a vertical transmission axis. LCD displays are usually constructed so that the polarized light that comes from them is also vertically polarized. Thus a person wearing Polaroid glasses can view an LCD display (top). If the pair of glasses or the LCD device is rotated 90°, the display becomes invisible (bottom).

Section Review

- The transmitted intensity of light passing through a polarizer whose transmission axis is at an angle θ to the polarization direction of the light is given by the law of Malus, $I = I_0 \cos^2 \theta$. In this expression, I_0 is the initial intensity of the light and I is the transmitted intensity.

- When light passes through a polarizer, its direction of polarization is changed to the direction of the polarizer's transmission axis.

CHAPTER 25 REVIEW

CHAPTER SUMMARY

25-1 THE PRODUCTION OF ELECTROMAGNETIC WAVES

Electromagnetic waves are traveling waves of oscillating electric and magnetic fields.

Direction of Propagation

To find the direction of propagation of an electromagnetic wave, point the fingers of your right hand in the direction of $\vec{E}$, then curl them toward $\vec{B}$. Your thumb will be pointing in the direction of propagation.

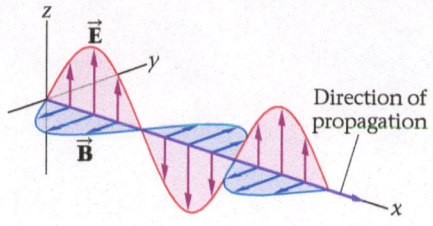

25-2 THE PROPAGATION OF ELECTROMAGNETIC WAVES

Electromagnetic waves can travel through a vacuum, and all electromagnetic waves in a vacuum have precisely the same speed, c = 3.00×10^8 m/s. This is referred to as the *speed of light in a vacuum*.

Doppler Effect

For relative speeds, u, that are small compared with the speed of light, c, the Doppler effect shifts the frequency of an electromagnetic wave as follows:

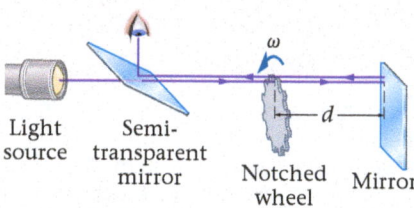

$$f' = f\left(1 \pm \frac{u}{c}\right) \qquad \text{25-3}$$

In this expression, the plus sign is used when the source and receiver are approaching, the minus sign when they are receding.

25-3 THE ELECTROMAGNETIC SPECTRUM

Electromagnetic waves can have any frequency from zero to infinity. The entire range of waves with different frequencies is referred to as the electromagnetic spectrum.

Frequency–Wavelength Relationship

The frequency, f, and wavelength, λ, of all electromagnetic waves in a vacuum are related as follows:

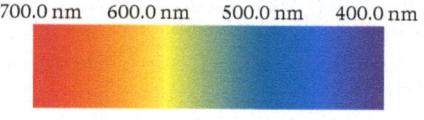

$$c = f\lambda \qquad \text{25-4}$$

25-4 ENERGY AND MOMENTUM IN ELECTROMAGNETIC WAVES

Electromagnetic waves carry both energy and momentum, shared equally between the electric and magnetic fields.

Energy Density

The energy density, u, of an electromagnetic wave can be written in several equivalent forms:

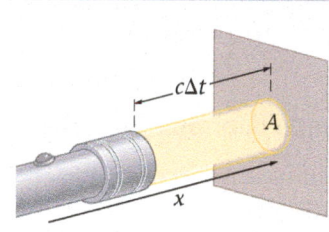

$$u = \tfrac{1}{2}\epsilon_0 E^2 + \frac{1}{2\mu_0}B^2 = \epsilon_0 E^2 = \frac{1}{\mu_0}B^2 \qquad \text{25-6}$$

Ratio of Electric and Magnetic Fields

In an electromagnetic field, the magnitudes E and B are related as follows:

$$E = cB \qquad \text{25-9}$$

Intensity

The intensity, I, of an electromagnetic field is the power per area. It can be expressed in the following forms:

$$I = uc = \tfrac{1}{2}c\epsilon_0 E^2 + \frac{1}{2\mu_0}cB^2 = c\epsilon_0 E^2 = \frac{c}{\mu_0}B^2 \qquad \text{25-10}$$

Radiation Pressure

If an electromagnetic wave shines on an object with an average intensity I_{av}, the average pressure exerted on the object by the radiation is

$$\text{pressure}_{av} = \frac{I_{av}}{c} \qquad \text{25-12}$$

25-5 POLARIZATION

The polarization of a beam of light is the direction along which its electric field points. An unpolarized beam has components with polarization in random directions.

Polarizer

A polarizer transmits only light whose electric field has a component in the direction of the polarizer's transmission axis.

Law of Malus

If light with intensity I_0 encounters a polarizer with a transmission axis at a direction θ relative to its polarization, the transmitted intensity, I, is

$$I = I_0 \cos^2 \theta \qquad\qquad 25\text{-}13$$

Unpolarized Light Passing Through a Polarizer

When an unpolarized beam of light of intensity I_0 passes through a polarizer, the transmitted intensity, I, is reduced by half:

$$I = \tfrac{1}{2} I_0 \qquad\qquad 25\text{-}14$$

Polarization by Scattering

Light scattered by the atmosphere is polarized when viewed at right angles to the Sun.

Polarization by Reflection

When light reflects from a horizontal surface, like a tabletop or the surface of a lake, it is partially polarized in the horizontal direction.

ANSWERS TO ENHANCE YOUR UNDERSTANDING QUESTIONS

1. The electric field points in the negative y direction.

2. **(a)** The speed of light is equal to c, independent of the speed of the source. **(b)** The frequency of light from the galaxy is Doppler shifted to lower frequency.

3. **(a)** The speed stays the same. **(b)** The wavelength decreases.

4. The magnetic field is doubled as well, because the electric and magnetic fields are proportional to one another in an electromagnetic wave.

5. The intensity increases if polarizer 1 is removed, because the light then passes through a single polarizer at an angle of 45°. When polarizer 1 is present, no light passes through the system, because polarizers 1 and 2 have polarizer directions that are perpendicular to one another.

CONCEPTUAL QUESTIONS

For instructor-assigned homework, go to www.masteringphysics.com. (MP)

(Answers to odd-numbered Conceptual Questions can be found in the back of the book.)

1. Explain why the "invisible man" would be unable to see.

2. While wearing your Polaroid sunglasses at the beach, you notice that they reduce the glare from the water better when you are sitting upright than when you are lying on your side. Explain.

3. You want to check the time while wearing your Polaroid sunglasses. If you hold your forearm horizontally, you can read the time easily. If you hold your forearm vertically, however, so that you are looking at your watch sideways, you notice that the display is black. Explain.

4. **BIO Polarization and the Ground Spider** The ground spider *Drassodes cupreus*, like many spiders, has several pairs of eyes. It has been discovered that one of these pairs of eyes acts as a set of polarization filters, with one eye's polarization direction oriented at 90° to the other eye's polarization direction. In addition, experiments show that the spider uses these eyes to aid in navigating to and from its burrow. Explain how such eyes might aid navigation.

5. You are given a sheet of Polaroid material. Describe how to determine the direction of its transmission axis if none is indicated on the sheet.

6. Can sound waves be polarized? Explain.

7. At a garage sale you find a pair of "Polaroid" sunglasses priced to sell. You are not sure, however, if the glasses are truly Polaroid, or if they simply have tinted lenses. How can you tell which is the case? Explain.

8. **3-D Movies** Modern-day 3-D movies are produced by projecting two different images onto the screen, with polarization directions that are at 90° relative to one another. Viewers must wear headsets with polarizing filters to experience the 3-D effect. Explain how this works.

PROBLEMS AND CONCEPTUAL EXERCISES

Answers to odd-numbered Problems and Conceptual Exercises can be found in the back of the book. **BIO** *identifies problems of biological or medical interest;* **CE** *indicates a conceptual exercise.* Predict/Explain *problems ask for two responses: (a) your prediction of a physical outcome, and (b) the best explanation among three provided; and* Predict/Calculate *problems ask for a prediction of a physical outcome, based on fundamental physics concepts, and follow that with a numerical calculation to verify the prediction. On all problems, bullets (•, ••, •••) indicate the level of difficulty.*

SECTION 25-1 THE PRODUCTION OF ELECTROMAGNETIC WAVES

1. • **CE** If the electric field in an electromagnetic wave is increasing in magnitude at a particular time, is the magnitude of the magnetic field at the same time increasing or decreasing? Explain.

2. • The electric field of an electromagnetic wave points in the positive y direction. At the same time, the magnetic field of this wave points in the positive z direction. In what direction is the wave traveling?

3. • An electric charge on the x axis oscillates sinusoidally about the origin. A distant observer is located at a point on the $+z$ axis. **(a)** In what direction will the electric field oscillate at the observer's location? **(b)** In what direction will the magnetic field oscillate at the observer's location? **(c)** In what direction will the electromagnetic wave propagate at the observer's location?

4. • An electric charge on the y axis oscillates sinusoidally about the origin. A distant observer is located at a point on the $+x$ axis. **(a)** In what direction will the electric field oscillate at the observer's location? **(b)** In what direction will the magnetic field oscillate at the observer's location? **(c)** In what direction will the electromagnetic wave propagate at the observer's location?

5. •• Give the direction (N, S, E, W, up, or down) of the missing quantity for each of the four electromagnetic waves listed in Table 25-1.

TABLE 25-1 Problem 5

Direction of propagation	Direction of electric field	Direction of magnetic field
N	W	**(a)**
N	**(b)**	W
up	S	**(c)**
(d)	down	S

6. •• Give the direction ($\pm x$, $\pm y$, $\pm z$) of the missing quantity for each of the four electromagnetic waves listed in Table 25-2.

TABLE 25-2 Problem 6

Direction of propagation	Direction of electric field	Direction of magnetic field
$+x$	$+y$	**(a)**
$+x$	**(b)**	$+y$
$-y$	$+z$	**(c)**
(d)	$+z$	$+y$

SECTION 25-2 THE PROPAGATION OF ELECTROMAGNETIC WAVES

7. • **CE** Three electromagnetic waves have electric and magnetic fields pointing in the directions shown in **FIGURE 25-37**. For each of the three cases, state whether the wave propagates in the $+x$, $-x$, $+y$, $-y$, $+z$, or $-z$ direction.

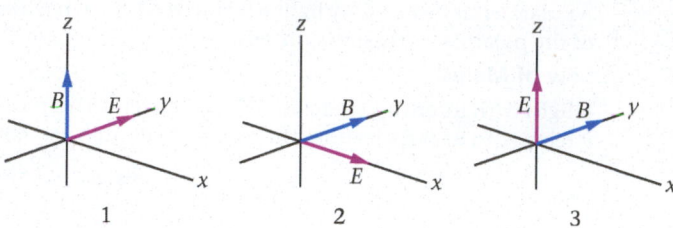

FIGURE 25-37 Problem 7

8. • The light-year (ly) is a unit of distance commonly used in astronomy. It is defined as the distance traveled by light in a vacuum in one year. **(a)** Express 1 ly in km. **(b)** Express the speed of light, c, in units of ly per year. **(c)** Express the speed of light in feet per nanosecond.

9. • Alpha Centauri, the closest star to the Sun, is 4.3 ly away. How far is this in meters?

10. • **Mars Rover** When the Mars rover *Sojourner* was deployed on the surface of Mars in July 1997, radio signals took about 12 min to travel from Earth to the rover. How far was Mars from Earth at that time?

11. • A fighter jet is traveling at 515 m/s directly away from a communication antenna that broadcasts at 406 MHz. What change in frequency does the fighter jet observe?

12. •• A distant star is traveling directly away from Earth with a speed of 49,500 km/s. By what factor are the wavelengths in this star's spectrum changed?

13. •• Predict/Calculate The frequency of light reaching Earth from a particular galaxy is 15% lower than the frequency the light had when it was emitted. **(a)** Is this galaxy moving toward or away from Earth? Explain. **(b)** What is the speed of this galaxy relative to the Earth? Give your answer as a fraction of the speed of light.

14. •• Predict/Calculate When an airplane communicates with a satellite using a frequency of 1.535×10^9 Hz, the signal received by the satellite is shifted higher by 207 Hz. **(a)** Is the airplane moving toward or away from the satellite? Explain. **(b)** What is the component of the airplane's velocity toward or away from the satellite?

15. •• **Measuring the Speed of Light** Galileo attempted to measure the speed of light by measuring the time elapsed between his opening a lantern and his seeing the light return from his assistant's lantern. The experiment is illustrated in **FIGURE 25-38**. What distance, d, must separate Galileo and his assistant in order for the human reaction time, $\Delta t = 0.2$ s, to introduce no more than a 15% error in the speed of light?

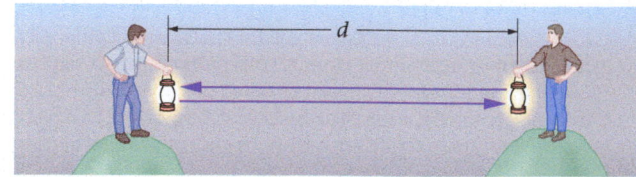

FIGURE 25-38 Problem 15

16. •• **Measuring the Speed of Light: Michelson** In 1926, Albert Michelson measured the speed of light with a technique similar to that used by Fizeau. Michelson used an eight-sided mirror rotating at 528 rev/s in place of the toothed wheel, as illustrated in **FIGURE 25-39**. The distance from the rotating mirror to a distant reflector was 35.5 km. If the light completed the 71.0-km round trip in the time it took the mirror to complete one-eighth of a revolution, what is the speed of light?

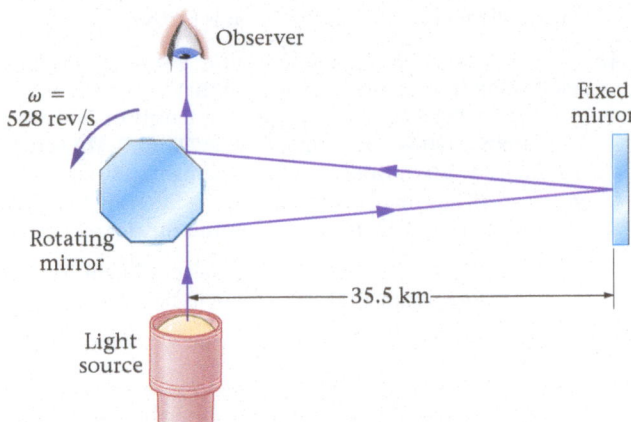

FIGURE 25-39 Problems 16 and 91

17. •• **Communicating with the *Voyager* Spacecraft** The *Voyager 1* spacecraft has traveled farther than any other man-made object, and in August 2012 it entered into interstellar space when it was a distance of 1.8×10^{13} m from Earth. How many hours elapsed between the time a command was sent from Earth and the time the command was received by *Voyager* when it entered interstellar space?

18. •• A father and his daughter are interested in the same baseball game. The father sits next to his radio at home and listens to the game; his daughter attends the game and sits in the outfield bleachers. In the bottom of the ninth inning a home run is hit. If the father's radio is 132 km from the radio station, and the daughter is 115 m from home plate, who hears the home run first? (Assume that there is no time delay between the baseball being hit and its sound being broadcast by the radio station. In addition, let the speed of sound in the stadium be 343 m/s.)

19. •• **Predict/Calculate** (a) How fast would a motorist have to be traveling for a yellow ($\lambda = 590$ nm) traffic light to appear green ($\lambda = 550$ nm) because of the Doppler shift? (b) Should the motorist be traveling toward or away from the traffic light to see this effect? Explain.

20. •• Most of the galaxies in the universe are observed to be moving away from Earth. Suppose a particular galaxy emits orange light with a frequency of 5.000×10^{14} Hz. If the galaxy is receding from Earth with a speed of 4375 km/s, what is the frequency of the light when it reaches Earth?

21. •• Two starships, the *Enterprise* and the *Constitution,* are approaching each other head-on from a great distance. The separation between them is decreasing at a rate of 782.5 km/s. The *Enterprise* sends a laser signal toward the *Constitution*. If the *Constitution* observes a wavelength $\lambda = 670.3$ nm, what wavelength was emitted by the *Enterprise*?

22. •• Baseball scouts often use a radar gun to measure the speed of a pitch. One particular model of radar gun emits a microwave signal at a frequency of 10.525 GHz. What will be the increase in frequency if these waves are reflected from a 95.0-mi/h fastball headed straight toward the gun? (*Note:* 1 mi/h = 0.447 m/s)

23. •• A state highway patrol car radar unit uses a frequency of 8.00×10^9 Hz. What frequency difference will the unit detect from a car receding at a speed of 64.5 m/s from a stationary patrol car?

24. •• Consider a spiral galaxy that is moving directly away from Earth with a speed $V = 3.240 \times 10^5$ m/s at its center, as shown in **FIGURE 25-40**. The galaxy is also rotating about its center, such that points in its spiral arms are moving with a speed $v = 5.750 \times 10^5$ m/s relative to the center. If light with a frequency of 7.308×10^{14} Hz is emitted in both arms of the galaxy, what frequency is detected by astronomers observing the arm that is moving (a) toward and (b) away from Earth? (Measurements of this type are used to map out the speed of various regions in distant, rotating galaxies.)

FIGURE 25-40 Problem 24

25. ••• **Predict/Calculate** A highway patrolman sends a 28.250-GHz radar beam toward a speeding car. The reflected wave is lower in frequency by 2.97 kHz. (a) Is the car moving toward or away from the radar gun? Explain. (b) What is the speed of the car? [*Hint:* For small values of x, the following approximation may be used: $(1 + x)^2 \approx 1 + 2x$.]

SECTION 25-3   THE ELECTROMAGNETIC SPECTRUM

26. • **BIO Dental X-rays** The X-rays produced in the dentist's office typically have a wavelength of 0.30 nm. What is the frequency of these rays?

27. • Find the frequency of green light with a wavelength of 555 nm.

28. • Yellow light has a wavelength $\lambda = 590$ nm. How many of these waves would span the 1.35-mm thickness of a dime?

29. • How many red wavelengths ($\lambda = 705$ nm) tall are you?

30. • A cell phone transmits at a frequency of 1.94×10^9 Hz. What is the wavelength of the electromagnetic wave used by this phone?

31. • **Microwave Oven** If a microwave oven produces electromagnetic waves with a frequency of 2.45 GHz, what is their wavelength?

32. • **BIO Human Radiation** Under normal conditions, humans radiate electromagnetic waves with a wavelength of about 9.0 microns. (a) What is the frequency of these waves? (b) To what portion of the electromagnetic spectrum do these waves belong?

33. • **BIO UV Radiation** Ultraviolet light is typically divided into three categories. UV-A, with wavelengths between 400 nm and 320 nm, has been linked with malignant melanomas. UV-B radiation, which is the primary cause of sunburn and other skin cancers, has wavelengths between 320 nm and 280 nm. Finally, the region known as UV-C extends to wavelengths of 100 nm. (a) Find the range of frequencies for UV-B radiation. (b) In which of these three categories does radiation with a frequency of 7.9×10^{14} Hz belong?

34. • **Communicating with a Submarine** Normal radiofrequency waves cannot penetrate more than a few meters below the surface of the ocean. One method of communicating with submerged submarines uses very low frequency (VLF) radio waves. What is the wavelength (in air) of a 10.0-kHz VLF radio wave?

35. •• **Predict/Calculate** When an electromagnetic wave travels from one medium to another with a different speed of propagation, the frequency of the wave remains the same. Its wavelength, however, changes. (a) If the wave speed decreases, does the wavelength increase

or decrease? Explain. **(b)** Consider a case where the wave speed decreases from c to $\frac{3}{4}c$. By what factor does the wavelength change?

36. •• **Predict/Calculate (a)** Which color of light has the higher frequency, red or violet? **(b)** Calculate the frequency of blue light with a wavelength of 470 nm, and red light with a wavelength of 680 nm.

37. •• ULF (ultra low frequency) electromagnetic waves, produced in the depths of outer space, have been observed with wavelengths in excess of 29 million kilometers. What is the period of such a wave?

38. •• A television is tuned to a station broadcasting at a frequency of 2.04×10^8 Hz. For best reception, the rabbit-ear antenna used by the TV should be adjusted to have a tip-to-tip length equal to half a wavelength of the broadcast signal. Find the optimum length of the antenna.

39. •• An AM radio station's antenna is constructed to be $\lambda/4$ tall, where λ is the wavelength of the radio waves. How tall should the antenna be for a station broadcasting at a frequency of 810 kHz?

40. •• As you drive by an AM radio station, you notice a sign saying that its antenna is 112 m high. If this height represents one quarter-wavelength of its signal, what is the frequency of the station?

41. •• Find the difference in wavelength $(\lambda_1 - \lambda_2)$ for each of the following pairs of radio waves: **(a)** $f_1 = 50$ kHz and $f_2 = 52$ kHz, **(b)** $f_1 = 500$ kHz and $f_2 = 502$ kHz.

42. ••• **Synchrotron Frequency** In one portion of a synchrotron, electrons traveling at 2.99×10^8 m/s enter a region of uniform magnetic field with a strength of 0.599 T. **(a)** What is the acceleration of an electron in this region? (Ignore the effects of relativity.) **(b)** The largest amount of light is emitted by the synchrotron at a frequency given by $f = (0.0433)\,a$ Hz, where a is the acceleration in m/s². What are this frequency and its corresponding wavelength? In what portion of the electromagnetic spectrum do these waves belong?

SECTION 25-4 ENERGY AND MOMENTUM IN ELECTROMAGNETIC WAVES

43. • **CE** If the rms value of the electric field in an electromagnetic wave is doubled, **(a)** by what factor does the rms value of the magnetic field change? **(b)** By what factor does the average intensity of the wave change?

44. • **CE** The radiation pressure exerted by beam of light 1 is half the radiation pressure of beam of light 2. If the rms electric field of beam 1 has the value E_0, what is the rms electric field in beam 2?

45. • The maximum magnitude of the electric field in an electromagnetic wave is 0.0675 V/m. What is the maximum magnitude of the magnetic field in this wave?

46. • What is the rms value of the electric field in a sinusoidal electromagnetic wave that has a maximum electric field of 99 V/m?

47. •• The magnetic field in an electromagnetic wave has a peak value given by $B = 2.9\,\mu$T. For this wave, find **(a)** the peak electric field strength, **(b)** the peak intensity, and **(c)** the average intensity.

48. •• What is the maximum value of the electric field in an electromagnetic wave whose maximum intensity is 7.55 W/m²?

49. •• What is the maximum value of the electric field in an electromagnetic wave whose average intensity is 7.55 W/m²?

50. •• **Predict/Calculate** Electromagnetic wave 1 has a maximum electric field of $E_0 = 52$ V/m, and electromagnetic wave 2 has a maximum magnetic field of $B_0 = 1.5\,\mu$T. **(a)** Which wave has the greater intensity? **(b)** Calculate the intensity of each wave.

51. •• A 75-kW radio station broadcasts its signal uniformly in all directions. **(a)** What is the average intensity of its signal at a distance of

250 m from the antenna? **(b)** What is the average intensity of its signal at a distance of 2500 m from the antenna?

52. •• At what distance will a 45-W lightbulb have the same apparent brightness as a 120-W bulb viewed from a distance of 25 m? (Assume that both bulbs convert electrical power to light with the same efficiency, and radiate light uniformly in all directions.)

53. •• What is the ratio of the sunlight intensity reaching Pluto compared with the sunlight intensity reaching Earth? (On average, Pluto is 39 times as far from the Sun as is Earth.)

54. •• **Predict/Calculate** In the following, assume that lightbulbs radiate uniformly in all directions and that 5.0% of their power is converted to light. **(a)** Find the average intensity of light at a point 2.0 m from a 120-W red lightbulb ($\lambda = 710$ nm). **(b)** Is the average intensity 2.0 m from a 120-W blue lightbulb ($\lambda = 480$ nm) greater than, less than, or the same as the intensity found in part (a)? Explain. **(c)** Calculate the average intensity for part (b).

55. •• A 7.7-mW laser produces a narrow beam of light. How much energy is contained in a 1.0-m length of its beam?

56. •• What length of a 7.7-mW laser's beam will contain 9.5 mJ of energy?

57. •• **Sunlight Intensity** After filtering through the atmosphere, the Sun's radiation illuminates Earth's surface with an average intensity of 1.0 kW/m². Assuming this radiation strikes the 15-m × 45-m black, flat roof of a building at normal incidence, calculate the average force the radiation exerts on the roof.

58. •• **Predict/Calculate (a)** Find the electric and magnetic field amplitudes in an electromagnetic wave that has an average energy density of 1.0 J/m³. **(b)** By what factor must the field amplitudes be increased if the average energy density is to be doubled to 2.0 J/m³?

59. •• **Lasers for Fusion** Some of the most powerful lasers in the world are used in nuclear fusion experiments. The NOVA laser produced 40.0 kJ of energy in a pulse that lasted 2.50 ns, and the NIF laser produces a 20.0-ns pulse with 4.20 MJ of energy. **(a)** Which laser produces more energy in each pulse? **(b)** Which laser produces the greater average power during each pulse? **(c)** If the beam diameters are the same, which laser produces the greater average intensity?

60. •• **BIO** You are standing 2.5 m from a 150-W lightbulb. **(a)** If the pupil of your eye is a circle 5.0 mm in diameter, how much energy enters your eye per second? (Assume that 5.0% of the lightbulb's power is converted to light.) **(b)** Repeat part (a) for the case of a 1.0-mm-diameter laser beam with a power of 0.50 mW.

61. •• **BIO Laser Safety**
A 0.75-mW laser emits a narrow beam of light that enters the eye, as shown in **FIGURE 25-41**. **(a)** How much energy is absorbed

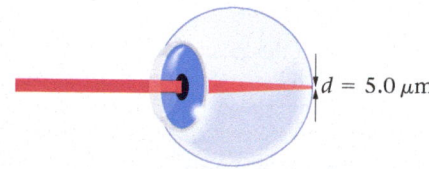

FIGURE 25-41 Problems 61 and 92

by the eye in 0.2 s? **(b)** The eye focuses this beam to a tiny spot on the retina, perhaps 5.0 μm in diameter. What is the average intensity of light (in W/cm²) at this spot? **(c)** Damage to the retina can occur if the average intensity of light exceeds 1.0×10^{-2} W/cm². By what factor has the intensity of this laser beam exceeded the safe value?

62. •• Find the rms electric and magnetic fields at a point 2.50 m from a lightbulb that radiates 115 W of light uniformly in all directions.

63. •• A 0.50-mW laser produces a beam of light with a diameter of 1.5 mm. **(a)** What is the average intensity of this beam? **(b)** At what

distance does a 150-W lightbulb have the same average intensity as that found for the laser beam in part (a)? (Assume that 5.0% of the bulb's power is converted to light.)

64. •• A laser emits a cylindrical beam of light 3.4 mm in diameter. If the average power of the laser is 2.5 mW, what is the rms value of the electric field in the laser beam?

65. •• (a) If the laser in Problem 64 shines its light on a perfectly absorbing surface, how much energy does the surface receive in 12 s? (b) What is the radiation pressure exerted by the beam?

66. ••• **BIO** **Laser Surgery** Each pulse produced by an argon–fluoride excimer laser used in PRK and LASIK ophthalmic surgery lasts only 10.0 ns but delivers an energy of 2.50 mJ. (a) What is the power produced during each pulse? (b) If the beam has a diameter of 0.850 mm, what is the average intensity of the beam during each pulse? (c) If the laser emits 55 pulses per second, what is the average power it generates?

67. ••• A pulsed laser produces brief bursts of light. One such laser emits pulses that carry 0.350 J of energy but last only 225 fs. (a) What is the average power during one of these pulses? (b) Assuming the energy is emitted in a cylindrical beam of light 2.00 mm in diameter, calculate the average intensity of this laser beam. (c) What is the rms electric field in this wave?

SECTION 25-5   POLARIZATION

68. • **CE** Predict/Explain Consider the two polarization experiments shown in **FIGURE 25-42**. (a) If the incident light is unpolarized, is the transmitted intensity in case A greater than, less than, or the same as the transmitted intensity in case B? (b) Choose the *best explanation* from among the following:

  I. The transmitted intensity is the same in either case; the first polarizer lets through one-half the incident intensity, and the second polarizer is at an angle θ relative to the first.
  II. Case A has a smaller transmitted intensity than case B because the first polarizer is at an angle θ relative to the incident beam.
  III. Case B has a smaller transmitted intensity than case A because the direction of polarization is rotated by an angle θ in the clockwise direction in case B.

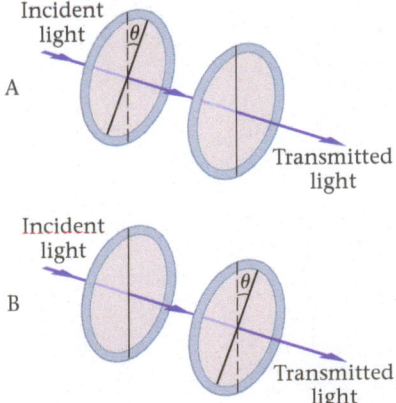

FIGURE 25-42   Problems 68 and 69

69. • **CE** Predict/Explain Consider the two polarization experiments shown in Figure 25-42. (a) If the incident light is polarized in the horizontal direction, is the transmitted intensity in case A greater than, less than, or the same as the transmitted intensity in case B? (b) Choose the *best explanation* from among the following:

  I. The two cases have the same transmitted intensity because the angle between the polarizers is θ in each case.
  II. The transmitted intensity is greater in case B because all of the initial beam gets through the first polarizer.

III. The transmitted intensity in case B is smaller than in case A; in fact, the transmitted intensity in case B is zero because the first polarizer is oriented vertically.

70. • **CE** An incident beam of light with an intensity I_0 passes through a polarizing filter whose transmission axis is at an angle θ to the vertical. As the angle is changed from $\theta = 0$ to $\theta = 90°$, the intensity as a function of angle is given by one of the curves in **FIG-URE 25-43**. Give the color of the curve corresponding to an incident beam that is (a) unpolarized, (b) vertically polarized, and (c) horizontally polarized.

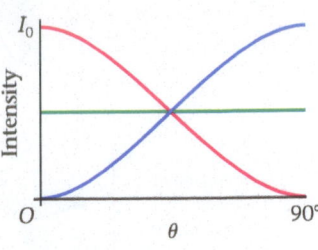

FIGURE 25-43   Problem 70

71. • Vertically polarized light with an intensity of 0.55 W/m² passes through a polarizer whose transmission axis is at an angle of 85.0° with the vertical. What is the intensity of the transmitted light?

72. • A person riding in a boat observes that the sunlight reflected by the water is polarized parallel to the surface of the water. The person is wearing polarized sunglasses with the polarization axis vertical. If the wearer leans at an angle of 21.5° to the vertical, what fraction of the reflected light intensity will pass through the sunglasses?

73. •• Unpolarized light passes through two polarizers whose transmission axes are at an angle of 30.0° with respect to each other. What fraction of the incident intensity is transmitted through the polarizers?

74. •• In Problem 73, what should be the angle between the transmission axes of the polarizers if it is desired that one-tenth of the incident intensity be transmitted?

75. •• **CE** Unpolarized light is incident with intensity I_0 on a polarizer whose transmission axis is vertical, as in Figure 25-26. It then falls on an analyzer whose transmission axis is at an angle θ to the vertical. Which of the graphs in **FIGURE 25-44** depicts the transmitted intensity as θ is changed from 0° to 360°?

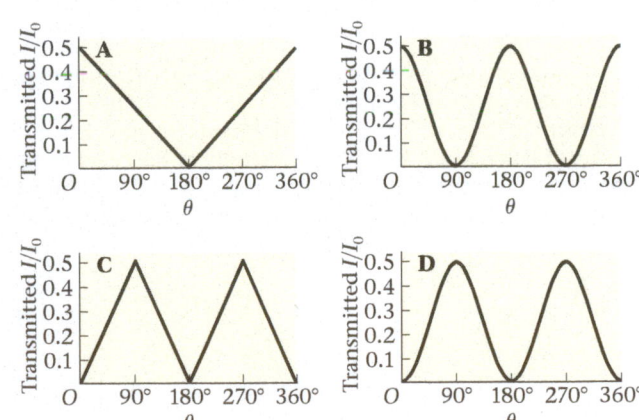

FIGURE 25-44   Problem 75

76. •• Predict/Calculate A beam of vertically polarized light encounters two polarizing filters, as shown in **FIGURE 25-45**. (a) Rank the three cases, A, B, and C, in order of increasing transmitted intensity. Indicate ties where appropriate. (b) Calculate the transmitted intensity for each of the cases in Figure 25-45, assuming that the incident intensity is 55.5 W/m². Verify that your numerical results agree with the rankings in part (a).

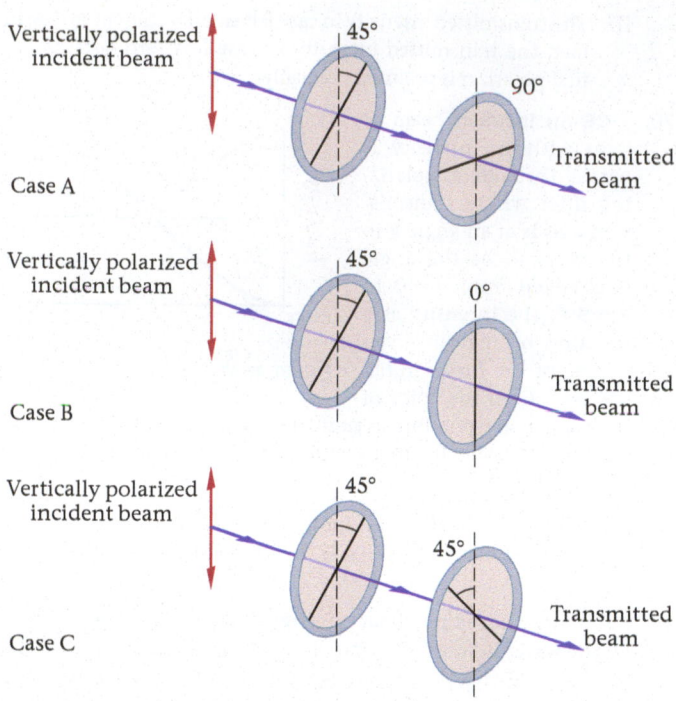

FIGURE 25-45 Problems 76 and 77

77. •• **Predict/Calculate** Repeat Problem 76, this time assuming that the polarizers to the left in Figure 25-45 are at an angle of 22.5° to the vertical rather than 45°. The incident intensity is again 55.5 W/m².

78. •• **BIO Predict/Calculate Optical Activity** Optically active molecules have the property of rotating the direction of polarization of linearly polarized light. Many biologically important molecules have this property, some causing a counterclockwise rotation (negative rotation angle), others causing a clockwise rotation (positive rotation angle). For example, a 5.00 gram per 100 mL solution of *l*-leucine causes a rotation of −0.550°; the same concentration of *d*-glutamic acid causes a rotation of 0.620°. **(a)** If placed between crossed polarizers, which of these solutions transmits the greater intensity? Explain. **(b)** Find the transmitted intensity for each of these solutions when placed between crossed polarizers. The incident beam is unpolarized and has an intensity of 12.5 W/m².

79. •• A helium–neon laser emits a beam of unpolarized light that passes through three Polaroid filters, as shown in **FIGURE 25-46**. The intensity of the laser beam is I_0. **(a)** What is the intensity of the beam at point A? **(b)** What is the intensity of the beam at point B? **(c)** What is the intensity of the beam at point C? **(d)** If filter 2 is removed, what is the intensity of the beam at point C?

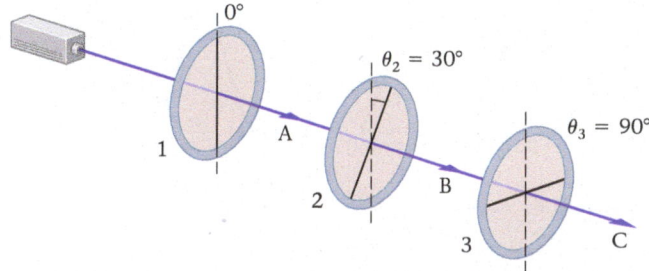

FIGURE 25-46 Problems 79, 80, and 98

80. ••• Referring to Figure 25-46, suppose that filter 3 is at a general angle θ with the vertical, rather than the angle 90°. **(a)** Find an expression for the transmitted intensity as a function of θ. **(b)** Plot your result from part (a), and determine the maximum transmitted intensity. **(c)** At what angle θ does maximum transmission occur?

GENERAL PROBLEMS

81. • **CE** Suppose the magnitude of the electric field in an electromagnetic wave is doubled. **(a)** By what factor does the magnitude of the magnetic field change? **(b)** By what factor does the maximum intensity of the wave change?

82. • **CE** If "sailors" of the future use radiation pressure to propel their ships, should the surfaces of their sails be absorbing or reflecting? Explain.

83. • **BIO** A typical medical X-ray has a frequency of 1.50×10^{19} Hz. What is the wavelength of such an X-ray?

84. • **BIO Radiofrequency Ablation** In radiofrequency (RF) ablation, a small needle is inserted into a cancerous tumor. When radiofrequency oscillating currents are sent into the needle, ions in the neighboring tissue respond by vibrating rapidly, causing local heating to temperatures as high as 100 °C. This kills the cancerous cells and, because of the small size of the needle, relatively few of the surrounding healthy cells. A typical RF ablation treatment uses a frequency of 750 kHz. What is the wavelength that such radio waves would have in a vacuum?

85. •• **Predict/Calculate** At a particular instant of time, a light beam traveling in the positive z direction has an electric field given by $\vec{E} = (6.22 \text{ N/C})\hat{x} + (2.87 \text{ N/C})\hat{y}$. The magnetic field in the beam has a magnitude of 2.28×10^{-8} T at the same time. **(a)** Does the magnetic field at this time have a z component that is positive, negative, or zero? Explain. **(b)** Write $\vec{B}$ in terms of unit vectors.

86. •• **Predict/Calculate** A light beam traveling in the negative z direction has a magnetic field $\vec{B} = (3.02 \times 10^{-9} \text{ T})\hat{x} + (-5.28 \times 10^{-9} \text{ T})\hat{y}$ at a given instant of time. The electric field in the beam has a magnitude of 1.82 N/C at the same time. **(a)** Does the electric field at this time have a z component that is positive, negative, or zero? Explain. **(b)** Write $\vec{E}$ in terms of unit vectors.

87. •• **CE FIGURE 25-47** shows four polarization experiments in which unpolarized incident light passes through two polarizing filters with different orientations. Rank the four cases in order of increasing amount of transmitted light. Indicate ties where appropriate.

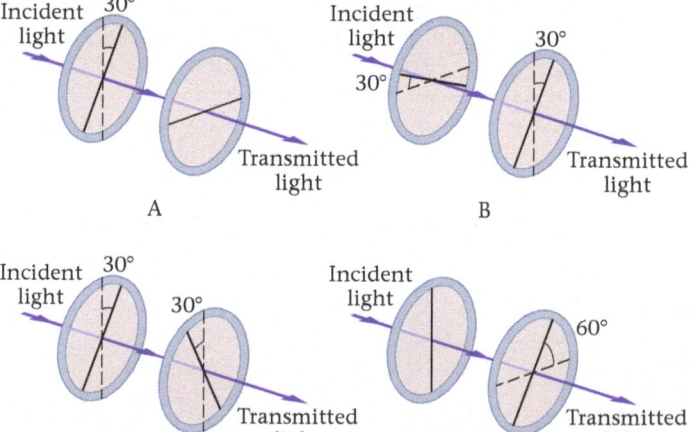

FIGURE 25-47 Problem 87

88. •• **Lightning and Thunder** During a thunderstorm a bolt of lightning strikes 2.41 km away from you. **(a)** How much time elapses between when the lightning strikes and when the light reaches your eyes? **(b)** If the speed of sound is 343 m/s, how much time elapses before the sound of thunder reaches your ears?

89. •• **The Apollo 11 Reflector** One of the experiments placed on the Moon's surface by Apollo 11 astronauts was a reflector that is

used to measure the Earth–Moon distance with high accuracy. A laser beam on Earth is bounced off the reflector, and its round-trip travel time is recorded. If the travel time can be measured to within an accuracy of 0.030 ns, what is the uncertainty in the Earth–Moon distance?

90. •• **International Space Station** The International Space Station (ISS) orbits Earth at an altitude of 422 km. (a) What is its orbital speed? (b) What is the speed of a receiving antenna on Earth's equator? (c) Suppose the ISS is moving directly toward the antenna, and the antenna is moving directly away from the ISS. If the ISS transmits a radio signal at 145.8000 MHz, what frequency is received by the antenna?

91. •• **Predict/Calculate** Suppose the distance to the fixed mirror in Figure 25-39 is decreased to 20.5 km. (a) Should the angular speed of the rotating mirror be increased or decreased to ensure that the experiment works as described in Problem 16? (b) Find the required angular speed, assuming the speed of light is 3.00×10^8 m/s.

92. •• **BIO** **Predict/Calculate** Consider the physical situation illustrated in Figure 25-41. (a) Is E_{rms} in the incident laser beam greater than, less than, or the same as E_{rms} where the beam hits the retina? Explain. (b) If the intensity of the beam at the retina is equal to the damage threshold, 1.0×10^{-2} W/cm², what is the value of E_{rms} at that location? (c) If the diameter of the spot on the retina is reduced by a factor of 2, by what factor does the intensity increase? By what factor does E_{rms} increase?

93. •• **BIO** **Polaroid Vision in a Spider** Experiments show that the ground spider *Drassodes cupreus* uses one of its several pairs of eyes as a polarization detector. In fact, the two eyes in this pair have polarization directions that are at right angles to one another. Suppose linearly polarized light with an intensity of 775 W/m² shines from the sky onto the spider, and that the intensity transmitted by one of the polarizing eyes is 274 W/m². (a) For this eye, what is the angle between the polarization direction of the eye and the polarization direction of the incident light? (b) What is the intensity transmitted by the other polarizing eye?

94. •• A state highway patrol car radar unit uses a frequency of 9.00×10^9 Hz. What frequency difference will the unit detect from a car approaching a parked patrol car with a speed of 28.6 m/s?

95. •• What is the ratio of the sunlight intensity reaching Mercury compared with the sunlight intensity reaching Earth? (On average, Mercury's distance from the Sun is 0.39 that of Earth's.)

96. •• What area is needed for a solar collector to absorb 45.0 kW of power from the Sun's radiation if the collector is 75.0% efficient? (At the surface of Earth, sunlight has an average intensity of 1.00×10^3 W/m².)

97. •• **BIO** **Near-Infrared Brain Scans** Light in the near-infrared (close to visible red) can penetrate surprisingly far through human tissue, a fact that is being used to "illuminate" the interior of the brain in a noninvasive technique known as near-infrared spectroscopy (NIRS). In this procedure, illustrated in **FIGURE 25-48**, an optical fiber carrying a beam of infrared laser light with a power of 1.5 mW and a cross-sectional diameter of 1.2 mm is placed against the skull. Some of the light enters the brain, where it scatters from hemoglobin in the blood. The scattered light is picked up by a detector and analyzed by a computer. (a) According to the **Beer–Lambert law,** the intensity of light, I, decreases with penetration distance, d, as $I = I_0 e^{-\mu d}$, where I_0 is the initial intensity of the beam and $\mu = 4.7$ cm⁻¹ for a typical case. Find the intensity of the laser beam after it penetrates through 3.0 cm of tissue. (b) Find the electric field of the initial light beam.

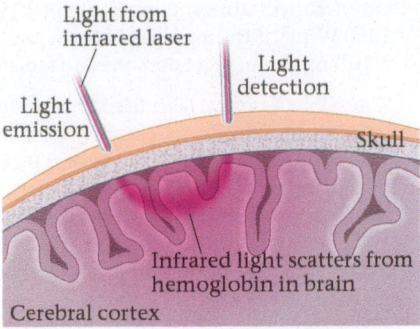

FIGURE 25-48 The basic elements of a near-infrared brain scan. (Problem 97)

98. •• Three polarizers are arranged as shown in Figure 25-46. If the incident beam of light is unpolarized and has an intensity of 1.60 W/m², find the transmitted intensity (a) when $\theta_2 = 25.0°$ and $\theta_3 = 50.0°$, and (b) when $\theta_2 = 50.0°$ and $\theta_3 = 25.0°$.

99. •• **Glare Reduction** The light that reaches a person's eyes is a combination of 625 W/m² of unpolarized light and 282 W/m² of horizontally polarized light. The person puts on sunglasses that are ideal polarizers with a vertical transmission axis, but also darkening lenses that absorb 33.0% of all light. What intensity is transmitted through these sunglasses?

100. •• **Orbital Drift** The radiation pressure exerted by the Sun on the Earth counteracts the gravitational attraction and puts the Earth into an orbit that is farther from the Sun than if there were no radiation pressure. It can be shown that the distance added to Earth's orbital radius is given by $(F_{rad}/F_g)r$, where F_{rad} is the radiation force exerted on Earth by the Sun, F_g is the gravitational force between the Earth and Sun, and r is the average orbital radius. (a) If the intensity of sunlight that strikes Earth is 1360 W/m², what is F_{rad}, assuming the Earth is a perfect absorber? (b) Assuming $r = 1.50 \times 10^{11}$ m, what is F_g? (c) What is the additional distance added to Earth's orbital radius by the radiation pressure?

101. •• A lightbulb emits light uniformly in all directions. If the rms electric field of this light is 16.0 N/C at a distance of 1.35 m from the bulb, what is the average total power radiated by the bulb?

102. •• **Radio Reception** A 125-kW radio station broadcasts its signal uniformly in all directions. (a) What is the average intensity of its signal at a distance of 8.05 km from the station? (b) What is the rms value of the electric field in this radio wave? (c) If a 1.22-m receiving antenna is oriented parallel to the electric field of the radio wave, what rms voltage appears between its ends?

103. ••• **Light Rocket** Stranded 12 m from your spacecraft, you realize that your flashlight makes a directed beam of intensity 950 W/m² and can be used to propel you back to safety. (a) If the radius of the beam is 3.8 cm, how much force does it exert on you? (b) If your mass (with spacesuit and gear) is 95 kg, what acceleration does the flashlight give you? (c) How much time will it take for you to reach the spacecraft? (d) Frustrated by the slow pace, you throw the 0.82-kg flashlight at 22 m/s directly away from the spacecraft. In how much time will you reach the spacecraft after you throw the flashlight?

104. ••• A typical home may require a total of 2.00×10^3 kWh of energy per month. Suppose you would like to obtain this energy from sunlight, which has an average daylight intensity of 1.00×10^3 W/m². Assuming that sunlight is available 8.0 h per day, 25 d per month (accounting for cloudy days), and that you have a way to store energy from your collector when the Sun isn't shining, determine the smallest collector size that will provide the needed energy, given a conversion efficiency of 25%.

105. ••• At the top of Earth's atmosphere, sunlight has an average intensity of 1360 W/m². If the average distance from Earth to the Sun is 1.50×10^{11} m, at what rate does the Sun radiate energy?

106. ••• **Predict/Calculate** A typical laser used in introductory physics laboratories produces a continuous beam of light about 1.0 mm in diameter. The average power of such a laser is 0.75 mW. What are **(a)** the average intensity, **(b)** the peak intensity, and **(c)** the average energy density of this beam? **(d)** If the beam is reflected from a mirror, what is the maximum force the laser beam can exert on it? **(e)** Describe the orientation of the laser beam relative to the mirror for the case of maximum force.

107. ••• Four polarizers are set up so that the transmission axis of each successive polarizer is rotated clockwise by an angle θ relative to the previous polarizer. Find the angle θ for which unpolarized light is transmitted through these four polarizers with its intensity reduced by a factor of 25.

108. ••• **BIO Optical Activity of Sugar** The sugar concentration in a solution (e.g., in a urine specimen) can be measured conveniently by using the *optical activity* of sugar and other asymmetric molecules. In general, an optically active molecule, like sugar, will rotate the plane of polarization through an angle that is proportional to the thickness of the sample and to the concentration of the molecule. To measure the concentration of a given solution, a sample of known thickness is placed between two polarizing filters that are at right angles to each other, as shown in **FIGURE 25-49**. The intensity of light transmitted through the two filters can be compared with a calibration chart to determine the concentration. **(a)** What percentage of the incident (unpolarized) light will pass through the first filter? **(b)** If no sample is present, what percentage of the initial light will pass through the second filter? **(c)** When a particular sample is placed between the two filters, the intensity of light emerging from the second filter is 40.0% of the incident intensity. Through what angle did the sample rotate the plane of polarization? **(d)** A second sample has half the sugar concentration of the first sample. Find the intensity of light emerging from the second filter in this case.

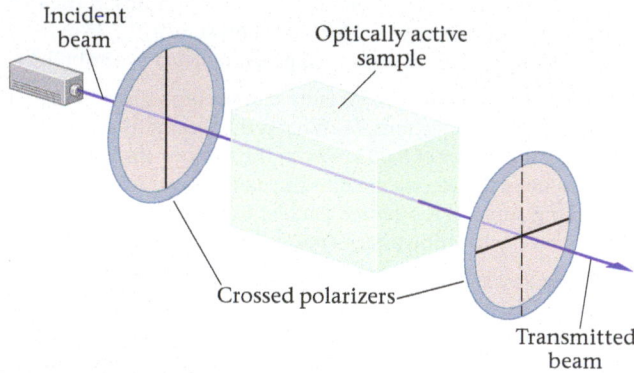

FIGURE 25-49 Problem 108

PASSAGE PROBLEMS

Visible-Light Curing in Dentistry

An essential part of modern dentistry is visible-light curing (VLC), a procedure that hardens the restorative materials used in fillings, veneers, and other applications. These "curing lights" work by activating molecules known as photoinitiators within the restorative materials. The photoinitiators, in turn, start a process of polymerization that causes monomers to link together to form a tough, solid polymer network. Thus, with VLC a dentist can apply and shape soft restorative materials as desired, shine a bright light

on the result as shown in **FIGURE 25-50 (a)**, and in 20 seconds have a completely hardened—or cured—final product.

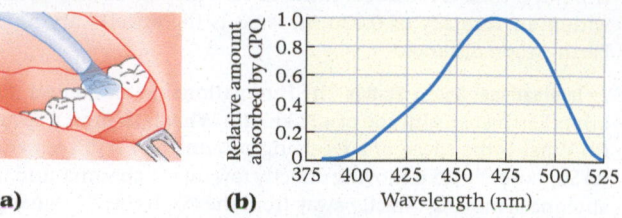

(a)　　　　　**(b)**

FIGURE 25-50 An intense beam of light cures, or hardens, the restorative material used to fill a cavity. (Problems 109, 110, 111, and 112)

The most common photoinitiator is camphoroquinone (CPQ). To cure CPQ in the least time, one should illuminate it with light having a wavelength that matches the wavelength at which CPQ absorbs the most light. A rough plot of the relative absorption of CPQ for light of different wavelengths is shown in **FIGURE 25-50 (b)**. Many VLC units use a halogen light, but there are some drawbacks to this approach. First, the filament in a halogen light is heated to a temperature of about 3000 K, which can cause heat degradation of components in the curing unit itself. Second, less than 1% of the energy given off by a halogen bulb is visible light, so a halogen bulb must have a high power rating to produce the desired light intensity at the wavelengths that CPQ will actually absorb.

More recently, VLC units have begun to use LEDs as their light source. These lights stay cool, emit nearly all of their energy output as visible light at the desired wavelength, and provide light with an intensity as high as 1000 mW/cm², which is about 10 times the intensity of sunlight on the surface of the Earth.

109. • What is the color of the light that is most effective at activating the photoinitiator CPQ?

　A. red　　　　　　　B. green
　C. blue　　　　　　 D. ultraviolet

110. • Suppose a VLC unit uses an LED that produces light with an average intensity of 400 mW/cm². What is the rms value of the electric field in this beam of light?

　A. 1230 N/C　　　　 B. 390 N/C
　C. 1700 N/C　　　　 D. 2.1×10^5 N/C

111. • How much radiation pressure does the beam of light in Problem 110 exert on a tooth, assuming the tooth absorbs all the light?

　A. 6.67×10^{-8} N/m²　B. 2.00×10^{-6} N/m²
　C. 1.33×10^{-5} N/m²　D. 4.00×10^{3} N/m²

112. • Assuming the light from the VLC unit has a beam 0.50 cm in diameter, how much energy does the light deliver in 20 seconds?

　A. 0.025 J　　　　　 B. 3.9 J
　C. 63 J　　　　　　 D. 5000 J

113. •• **Predict/Calculate REFERRING TO EXAMPLE 25-12** Suppose the incident beam of light is linearly polarized in the same direction θ as the transmission axis of the analyzer. The transmission axis of the polarizer remains vertical. **(a)** What value must θ have if the transmitted intensity is to be $0.200\, I_0$? **(b)** If θ is increased from the value found in part (a), does the transmitted intensity increase, decrease, or stay the same? Explain.

114. •• **REFERRING TO EXAMPLE 25-12** Suppose the incident beam of light is linearly polarized in the vertical direction. In addition, the transmission axis of the analyzer is at an angle of 80.0° to the vertical. What angle should the transmission axis of the polarizer make with the vertical if the transmitted intensity is to be a maximum?

Geometrical Optics

Big Ideas

1 The angle of reflection is equal to the angle of incidence.

2 A plane mirror forms an upright image that is reversed right to left.

3 Spherical mirrors can form images with various sizes and orientations.

4 When light passes from one medium into another, its direction of propagation changes—it refracts.

5 When light passes into a medium where its speed is reduced, it bends closer to the normal.

6 Lenses can form images with various sizes and orientations.

7 Light of different frequency refracts, or bends, by different amounts.

▲ Examples of optics are all around us. Take a close look at a dew-covered lawn and you'll see thousands of water droplets. Look even closer, and you'll see that each droplet is basically a small convex lens forming an upside-down image of the objects behind it. In this chapter we explore the laws of geometrical optics, and use them to understand how lenses, and mirrors, form images.

When you look into a mirror, or through a lens, you see images of objects in your surroundings. In the case of mirrors, light is *reflected* onto a new path. In lenses the speed of light is reduced, resulting in a change of direction referred to as *refraction*. By changing the direction in which light travels, both mirrors and lenses can be used to create images of various sizes and orientations.

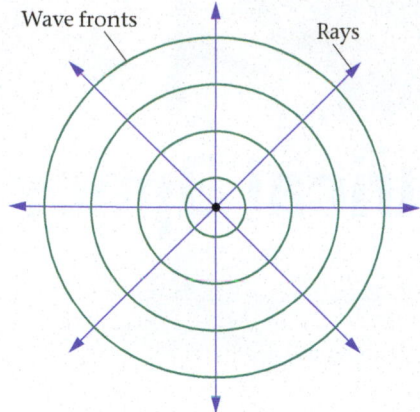

▲ **FIGURE 26-1 Wave fronts and rays** In this case, the wave fronts indicate the crests of water waves. The rays indicate the local direction of motion.

26-1 The Reflection of Light

The basis of geometrical optics is changing the direction in which light travels. Perhaps the simplest way to change the direction of light is by reflection from a shiny surface. To understand this process in detail, it is convenient to describe light in terms of "wave fronts" and "rays." As we shall see later in this chapter, these concepts are equally useful in understanding the behavior of lenses.

Wave Fronts and Rays

Consider the waves created by a rock dropped into a still pool of water. As we know, these waves form concentric outward-moving circles. A simplified representation of this system is given in **FIGURE 26-1**, where the circles indicate the crests of the waves. We refer to these circles as **wave fronts.** In addition, the radial motion of the waves is indicated by the outward-pointing arrows, referred to as **rays.** Notice that the rays are always at right angles to the wave fronts.

A similar situation applies to electromagnetic waves radiated by a small source, as illustrated in **FIGURE 26-2 (a)**. In this case, however, the waves move outward in three dimensions, giving rise to spherical wave fronts. As expected, **spherical waves** have rays that point radially outward.

**PHYSICS
IN CONTEXT
Looking Back**

The concepts of light rays and wave fronts relate directly to the discussion of light propagation presented in Chapter 25.

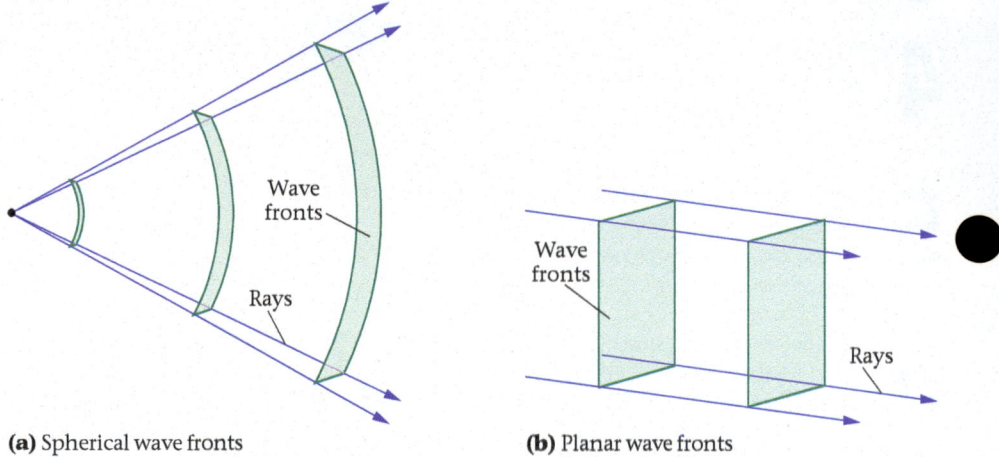

(a) Spherical wave fronts   **(b)** Planar wave fronts

▲ **FIGURE 26-2 Spherical and planar wave fronts (a)** As spherical waves move farther from their source, the wave fronts become increasingly flat. **(b)** In the limit of large distances the wave fronts are planes, and the corresponding rays are parallel to one another.

In **FIGURE 26-2 (b)** we show that as one moves farther from the source of spherical waves, the wave fronts become increasingly flat (like the surface of the Earth), and the rays are more nearly parallel. In the limit of increasing distance, the wave fronts approach perfectly flat planes. Such **plane waves,** with their planar wave fronts and parallel rays, are useful idealizations for investigating the properties of mirrors and lenses.

We will usually simplify our representation of light beams by omitting the wave fronts and plotting only one, or a few, rays. For example, in **FIGURE 26-3** both the incident plane wave and the reflected plane wave are shown as single rays pointing in the direction of propagation. The direction of these rays is considered next.

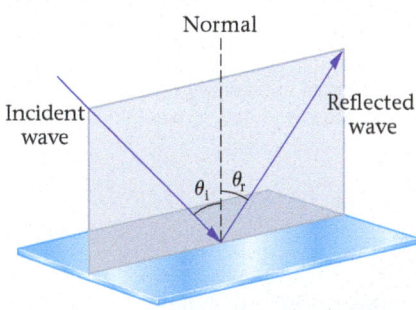

▲ **FIGURE 26-3 Reflection from a smooth surface** In this simplified representation, the incident and reflected waves are indicated by single rays pointing in the direction of propagation. Notice that the angle of reflection, θ_r, is equal to the angle of incidence, θ_i. In addition, the incident ray, the reflected ray, and the normal all lie in the same plane.

The Law of Reflection

To characterize the behavior of light as it reflects from a mirror or other shiny object, we begin by drawing the normal, which is simply a line perpendicular to the reflecting surface at the point of incidence. Relative to the normal, the incident ray strikes the surface at the angle θ_i, the *angle of incidence*, as shown in Figure 26-3. Similarly, the *angle*

(a) Specular reflection

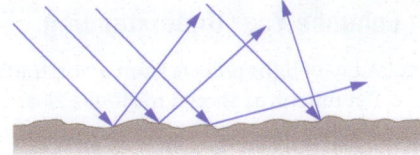

(b) Diffuse reflection

◄ **FIGURE 26-4 Reflection from smooth and rough surfaces (a)** A smooth surface gives specular reflection, in which all the reflected light propagates in a single direction. **(b)** A rough surface gives rise to reflected waves propagating over a range of directions. This process is referred to as diffuse reflection.

of reflection, θ_r, is the angle the reflected ray makes with the normal. The relationship between these two angles is very simple—they are equal:

Law of Reflection

$$\theta_r = \theta_i \hspace{6cm} 26\text{-}1$$

Notice, in addition, that the incident ray, the normal, and the reflected ray all lie in the same plane, as is also clear from Figure 26-3.

The reflection of light from a smooth shiny surface, as in **FIGURE 26-4 (a)** and **FIGURE 26-5**, is referred to as **specular reflection.** Notice that all the reflected light moves in the same direction. In contrast, if a surface is rough, as in **FIGURE 26-4 (b)**, the reflected light is sent out in a variety of directions, giving rise to **diffuse reflection.** For example, when the surface of a road is wet, the water creates a smooth surface, and headlights reflecting from the road undergo specular reflection. As a result, the reflected light goes in a single direction, giving rise to an intense glare. When the same road is dry, its surface is microscopically rough; hence, light is reflected in many directions and glare is not observed. The law of reflection is obeyed in either case, of course; it is the surface that is different, not the underlying physics.

RWP*A novel variation on specular versus diffuse reflection occurs in a new type of electronic chip known as a Digital Micromirror Device (DMD), an example of which is shown in **FIGURE 26-6**. These small devices consist of as many as 1.3 million microscopic flat mirrors, each of which, though smaller than the diameter of a human hair, can be oriented independently in response to electrical signals. The reflection from each micromirror is specular, and if all 1.3 million mirrors are oriented in the same direction, the DMD acts like a small flat mirror. Conversely, if the mirrors are oriented randomly, the reflection from the DMD is diffuse. When a DMD is used to project a movie, each micromirror plays the role of a single pixel in the projected image. In such a system, the light directed onto the DMD cycles rapidly from red to green to blue, and each of the mirrors reflects only the appropriate colors for that pixel onto the screen, as illustrated in **FIGURE 26-7**. The result is a projected image of great brilliance and vividness that eliminates the need for film.

Big Idea **1** The angle of reflection of light reflecting from a mirror is equal to the angle of incidence.

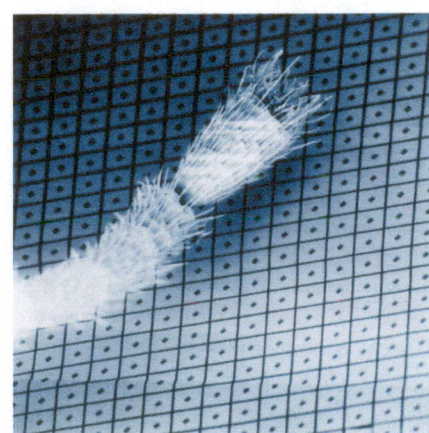

▲ **FIGURE 26-5** When rays of light are reflected from a smooth, flat surface, both the incident and reflected rays make the same angles with the normal. Consequently, parallel rays are reflected in the same direction, as shown in Figure 26-4 (a). The result is specular reflection, which produces a "mirror image" of the source of the rays.

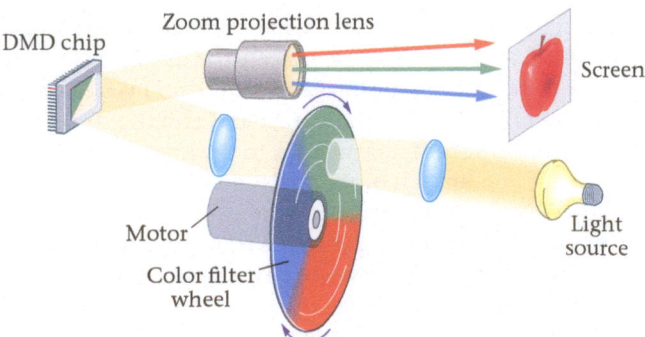

▲ **FIGURE 26-7 A Digital Micromirror Projection System** A digital projection system based on micromirrors reflects incoming light onto a distant screen. Each micromirror produces one pixel of the final image. The color and intensity of a given pixel are determined by the amount of red, green, and blue light the corresponding micromirror reflects to the screen.

▲ **FIGURE 26-6** An ant's leg provides a sense of scale in this photo of an array of Digital Micromirror Device (DMD) mirrors. Each mirror has an area of 16 μm^2 and can pivot 10° in either direction about a diagonal axis.

---

*Real World Physics applications are denoted by the acronym RWP.

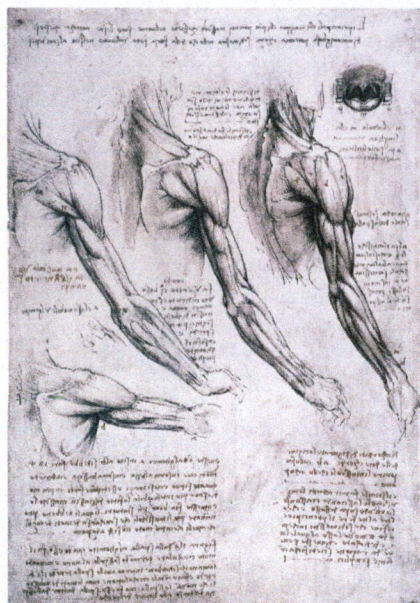

▲ **FIGURE 26-9** Leonardo da Vinci (1452–1519), the quintessential Renaissance man, tried to keep the contents of his notebooks private by setting down his observations in "mirror writing."

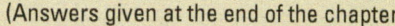
26-2 Forming Images with a Plane Mirror

We are all familiar with looking at ourselves in a mirror. If the mirror is perfectly flat—that is, a **plane mirror**—we see an upright image of ourselves as far behind the mirror as we are in front. In addition, the image is reversed right to left; for example, if we raise our right hand, the mirror image raises its left hand. Text read in a mirror is reversed right to left as well, as in the famous example of Leonardo da Vinci's notebooks, a sample of which is shown in **FIGURE 26-9**. This is the reason ambulances and other emergency vehicles use mirror-image writing on the front of their vehicles—when viewed in the rearview mirror of a car, the writing looks normal. In this section we use the law of reflection to derive these and other results.

Before we delve into the details of mirror images, however, let's pause for a moment to consider the process by which physical objects produce images in our eyes. First, any object that is near you at this moment is bathed in light coming at it from all directions. As this object reflects the incoming light back into the room, every point on it acts like a source of light. When we view the object, the light coming from any given point enters our eyes and is focused to a point on our retina. This is the case *for every point* that we can see on the object—each point on the object is detected by a different point on the retina. This results in a one-to-one mapping between the physical object and the image on the retina.

The formation of a mirror image occurs in a similar manner, except that the light from an object reflects from a mirror before it enters our eyes. This is illustrated in **FIGURE 26-10 (a)**, where we show an object—a small flower—placed in front of a plane mirror. Rays of light leaving the top of the flower at point P are shown reflecting from the mirror and entering the eye of an observer. To the observer, it appears that the rays are emanating from point P' *behind the mirror*.

We can show that the image is the same distance behind the mirror as the object is in front. Consider **FIGURE 26-10 (b)**, where we indicate the distance of the object from the mirror by d_o, and the distance from the image to the mirror by d_i. One ray from the top of the flower is shown reflecting from the mirror and entering the observer's eye. We also show the extension of the reflected ray back to the image. By the law of reflection, if the angle of incidence at point A is θ, the angle of reflection is also θ. Therefore, the straight line from the observer to the image cuts across the normal line with an angle θ on either side, as shown. Clearly, then, the angles indicated by ϕ must also be equal to one another. Combining these results, we see that triangle PAQ shares a side and two adjacent angles with triangle P'AQ; hence, the two triangles are equal. It follows that the distance d_o is equal to the distance d_i, as expected.

In the following Example, we once again apply the law of reflection to the flower and its mirror image.

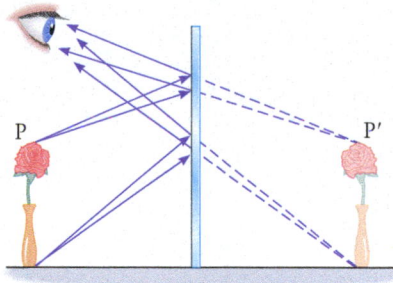

(a) Image formed by a plane mirror

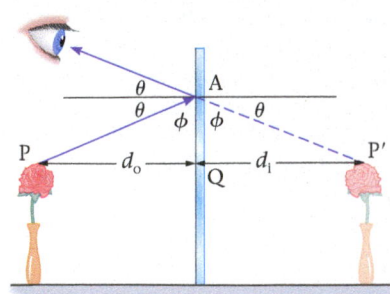

(b) Object distance (d_o) equals image distance (d_i)

◀ **FIGURE 26-10 Locating a mirror image (a)** Rays of light from point P at the top of the flower appear to originate from point P' behind the mirror. **(b)** Construction showing that the distance from the object to the mirror is the same as the distance from the image to the mirror.

EXAMPLE 26-1   REFLECTING ON A FLOWER

An observer is at table level, a distance d to the left of a flower of height h. The flower itself is a distance d to the left of a mirror, as shown in the sketch. A ray of light propagating from the top of the flower to the observer's eye reflects from the mirror at a height y above the table. Find y in terms of the height of the flower, h.

PICTURE THE PROBLEM

The physical situation is shown in the sketch, along with a ray from the top of the flower to the eye of the observer. The point where the ray hits the mirror is a height y above the table. The flower is a distance d to the left of the mirror, and its image is a distance d to the right of the mirror. Finally, the observer's eye is a distance $2d$ to the left of the mirror.

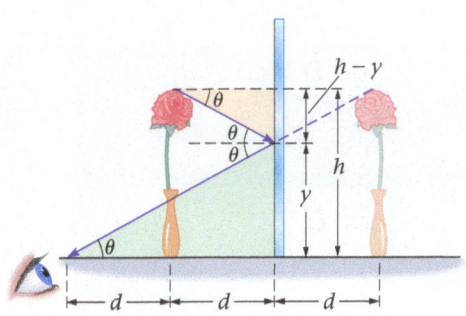

REASONING AND STRATEGY

To find y, we must use the fact that the angle of reflection is equal to the angle of incidence. In the sketch we indicate these two angles by the symbol θ. It follows, then, that $\tan \theta$ obtained from the small yellow triangle must be equal to $\tan \theta$ obtained from the larger green triangle. Setting these expressions for $\tan \theta$ equal to one another yields a relationship that can be solved for y.

Known   Distance from observer to flower = d; distance from flower to mirror = d; height of flower = h.
Unknown   Height where light from top of flower reflects from the mirror, y = ?

SOLUTION

1. Write an expression for $\tan \theta$ using the yellow triangle:

$$\tan \theta = \frac{h - y}{d}$$

2. Write a similar expression for $\tan \theta$ using the green triangle:

$$\tan \theta = \frac{y}{2d}$$

3. Set the two expressions for $\tan \theta$ equal to one another:

$$\frac{h - y}{d} = \frac{y}{2d}$$

4. Rearrange the preceding equation and solve for y:

$$h - y = \frac{y}{2}$$

$$y = \tfrac{2}{3} h$$

INSIGHT

This result means that the observer can see the entire image of the flower in a section of mirror that is only two-thirds the height of the flower.

PRACTICE PROBLEM — PREDICT/CALCULATE

(a) If the observer moves farther to the left of the base of the flower, does the point of reflection move upward, downward, or stay at the same location? Explain. **(b)** As a check on your answer, calculate the point of reflection for the case where the distance between the observer and the base of the flower is $2d$.   **[Answer: (a)** The point of reflection moves upward, because θ is smaller as one moves farther from the mirror image of the flower. **(b)** $y = \frac{3}{4} h$, which is greater than $\frac{2}{3} h$, as expected.**]**

Some related homework problems: Problem 6, Problem 8

To summarize, the basic features of reflection by a plane mirror are as follows:

Properties of Mirror Images Produced by Plane Mirrors

- A mirror image is upright, but appears reversed right to left.

- A mirror image appears to be the same distance behind the mirror that the object is in front of the mirror.

- A mirror image is the same size as the object.

Big Idea **2** A plane mirror forms an upright image the same distance behind the mirror that the object is in front of the mirror. The image is the same size as the object, but reversed right to left.

CONCEPTUAL EXAMPLE 26-2   HEIGHT OF MIRROR

To save expenses, you would like to buy the shortest mirror that will allow you to see your entire body. Should the mirror be half your height, two-thirds your height, or equal to your height?

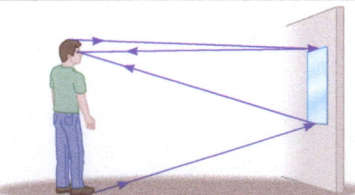

REASONING AND DISCUSSION

First, to see your feet, the mirror must extend from your eyes downward to a point halfway between your eyes and feet, as shown in the sketch. Similarly, the mirror must extend upward from your eyes half the distance to the top of your head. Altogether, then, the mirror must have a height equal to half the distance from your eyes to your feet plus half the distance from your eyes to the top of your head—that is, half your total height.

ANSWER
The mirror needs to be only half your height.

▲ **FIGURE 26-11** In this heads-up display in an airplane cockpit, important flight information is reflected on a transparent screen in the windshield, enabling the pilot to view the data without diverting attention from the scene ahead.

RWP An application of mirror images that has been used in military aircraft for many years, and is now beginning to appear in commercial automobiles, is the *heads-up display*. An example is shown in **FIGURE 26-11**. In the case of a car, a small illuminated display screen is recessed in the dashboard, out of direct sight of the driver. The screen shows important information, like the speed of the car, in mirror image. The driver sees the information not by looking directly at the screen—which is hidden from view—but by looking at its reflection in the windshield. Thus, while still looking straight ahead (heads up), the driver can see both the road and the reading of the speedometer.

A similar device is used in theaters to provide subtitles for the hearing impaired. In this case, a transparent plastic screen is mounted on the arm of a person's chair. This screen is adjusted in such a way that the person can look through it to see the movie, and at the same time see the reflection of a screen in the back of the theater that gives the subtitles in mirror-image form.

EXAMPLE 26-3   TWO-DIMENSIONAL CORNER REFLECTOR

Two mirrors are placed at right angles, as shown in the sketch. An incident ray of light makes an angle of 30° with the x axis and reflects from the lower mirror. Find the angle the outgoing ray makes with the y axis after it reflects once from each mirror.

PICTURE THE PROBLEM
The physical system is shown in the sketch. Notice that the normal for the first reflection is vertical, whereas the normal is horizontal for the second reflection. It follows that the angle of incidence for the first reflection is $\theta = 90° - 30° = 60°$. Similarly, the angle of incidence for the second reflection is $90° - \theta = 30°$.

Known   Angle of incidence for first mirror = 60°; angle of incidence for second mirror = 30°.
Unknown   Angle of outgoing light relative to y axis, θ = ?

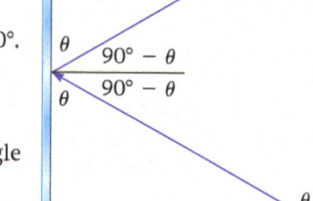

REASONING AND SOLUTION
We apply the law of reflection to each of the two reflections. For the first reflection, the angle of incidence is $\theta = 60°$; hence, the reflected ray also makes a 60° angle with the vertical.

At the second reflection, the normal is horizontal; hence, the angle of incidence is $90° - \theta = 30°$. After the second reflection, the outgoing ray is 30° above the horizontal and hence it makes an angle $\theta = 60°$ with respect to the y axis, as shown.

INSIGHT
Notice that the outgoing ray travels parallel to the incoming ray, but in the opposite direction. This is true regardless of the value of the initial angle θ.

PRACTICE PROBLEM
If the incoming ray hits the horizontal mirror farther to the right, is the angle the outgoing ray makes with the vertical greater than θ, equal to θ, or less than θ?   [**Answer:** The angle of the outgoing ray is still equal to θ. The distance between the incoming and outgoing rays is increased, however.]

Some related homework problems: Problem 9, Problem 15

Corner Reflectors   If three plane mirrors are joined at right angles, as in **FIGURE 26-12**, the result is a **corner reflector.** A corner reflector behaves in three dimensions the same as two mirrors at right angles in two dimensions; namely, a ray incident on the corner reflector is sent back in the same direction from which it came. This type of behavior has led to many useful applications for corner reflectors.

RWP One of the more interesting applications of corner reflectors involves the only Apollo experiments still returning data from the Moon. On *Apollos 11, 14,* and *15* the astronauts placed retroreflector arrays consisting of 100 corner reflectors on the lunar surface. Scientists at observatories on Earth can send a laser beam to the appropriate location on the Moon, where the retroreflector sends it back to its source. By measuring the round-trip travel time of the light, scientists can determine the Earth–Moon distance to an accuracy of about 3 cm out of a total distance of roughly 385,000 km! Over the years, these measurements have revealed that the Moon is slowly moving away from Earth at the rate of 3.8 cm per year.

RWP Corner reflectors, or retroreflectors more generally, also appear in many ordinary items, like bicycle reflectors, highway signs, and reflective clothing. A close inspection of a bicycle reflector, like the one in **FIGURE 26-13 (a)**, shows that the plastic has been molded into dozens of corner reflectors that send light from a car's headlights directly back to the driver's eyes. Highway signs and the reflective stripes on clothing become highly visible at night by a similar effect, as can be seen in **FIGURE 26-13 (b)**. These items are coated with millions of tiny, clear plastic spheres that behave in much the same way as corner reflectors. Spherical water droplets produce a similar retroreflector effect that can be seen as the *heiligenschein* around the shadow of a camera. An example of heiligenschein is shown in **FIGURE 26-13 (c)**.

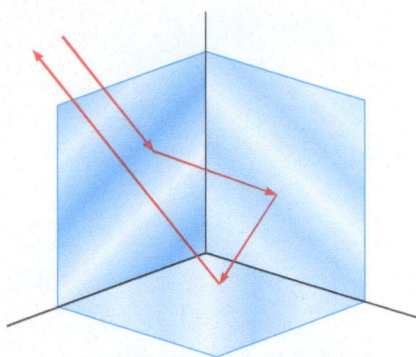

▲ **FIGURE 26-12 A corner reflector** A three-dimensional corner reflector constructed from three plane mirrors at right angles to one another. A ray entering the reflector is sent back in the direction from which it came.

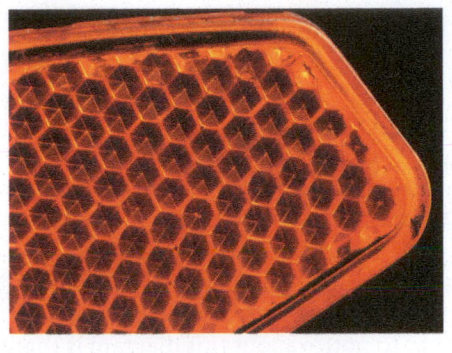

(a)

(b)

(c)

▲ **FIGURE 26-13 Visualizing Concepts Retroreflectors** Retroreflectors, which reflect light back to the source, occur in a variety of everyday situations. **(a)** A magnified view of a bicycle reflector shows that it is covered with tiny corner reflectors (retroreflectors) that are molded into the plastic. **(b)** Reflective strips on clothing contain millions of tiny clear plastic beads, each of which acts as a retroreflector. **(c)** *Heiligenschein* is the term for the bright area on the ground where water droplets are acting like retroreflectors.

Enhance Your Understanding   (Answers given at the end of the chapter)

2. A meterstick is placed 40 cm in front of a plane mirror, as shown in **FIGURE 26-14**. What is the perceived distance between the 50 cm mark on the meterstick and the mirror image of the 25 cm mark?

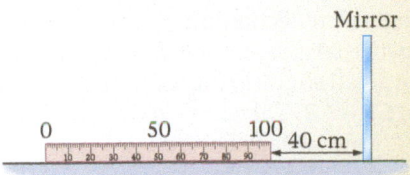

▲ **FIGURE 26-14** Enhance Your Understanding 2.

Section Review

*   Images formed by plane mirrors are upright, the same size as the object, reversed from right to left, and the same distance behind the mirror that the object is in front of the mirror.

26-3 Spherical Mirrors

We now consider another type of mirror that is encountered frequently in everyday life—the spherical mirror. These mirrors get their name from the fact that they have the same shape as a section of a sphere, as is shown in **FIGURE 26-15**, where a portion of a spherical shell of radius *R* is cut away from the rest of the shell. If the outside of this spherical section is a reflecting surface, the result is a **convex** mirror; if the inside surface is reflecting, we have a **concave** mirror.

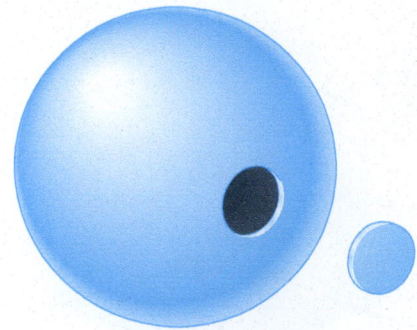

▲ **FIGURE 26-15 Spherical mirrors** A spherical mirror has the same shape as a section of a sphere. If the outside surface of this section is reflecting, the mirror is convex. If the inside surface reflects, the mirror is concave.

Convex and concave spherical mirrors are illustrated in **FIGURE 26-16**, where we also indicate the **center of curvature,** C, and the **principal axis.** The center of curvature is the center of the sphere with radius R of which the mirror is a section, and the principal axis is a straight line drawn through the center of curvature and the midpoint of the mirror. Notice that the principal axis intersects the mirror at right angles.

Focal Point and Focal Length To investigate the behavior of spherical mirrors, suppose a beam of light is directed toward either a convex or a concave mirror along its principal axis. For example, several parallel rays of light are approaching a convex mirror from the left in **FIGURE 26-17**. After reflecting from the mirror, the rays diverge as if they originated from a single point *behind* the mirror called the **focal point,** F. (This is strictly true only for light rays close to the principal axis, as we point out in greater detail later in this section. All of our results for spherical mirrors assume rays close to the principal axis.)

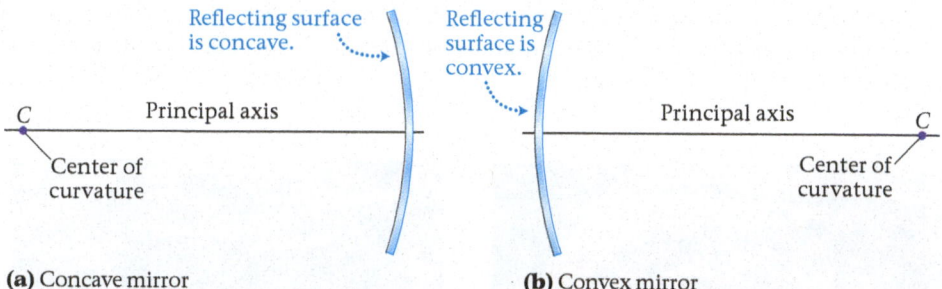

(a) Concave mirror   **(b)** Convex mirror

▲ **FIGURE 26-16 Concave and convex mirrors (a)** A concave mirror and its center of curvature, C, which is on the same side as the reflecting surface. **(b)** A convex mirror and its center of curvature, C. In this case C is on the opposite side of the mirror from its reflecting surface.

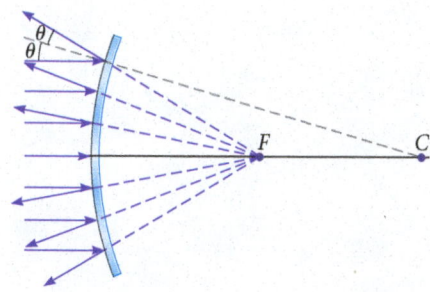

▲ **FIGURE 26-17 Parallel rays on a convex mirror** When parallel rays of light reflect from a convex mirror, they diverge as if originating from a focal point halfway between the surface of the mirror and its center of curvature.

Now, let's find the **focal length**—that is, the distance from the surface of the mirror to the focal point. This can be done with the aid of **FIGURE 26-18**, which shows a single ray reflecting from the mirror. The first thing to notice about this diagram is that a straight line drawn through the center of curvature always intersects the mirror at right angles; hence, the line through C is the normal to the surface at the point of incidence, A. The incoming ray is parallel to the principal axis, and hence it follows that the angle of incidence, θ, is equal to the angle FCA. Next, the law of reflection states that the angle of reflection must equal θ, which means that the angle CAF is also θ. We see, then, that CAF is an isosceles triangle, with the sides CF and FA of equal length. Finally, for small angles θ, the length CF is approximately equal to half the length CA = R; that is, $CF \sim \frac{1}{2}R$. Therefore, to this same approximation, the distance FB is also $\frac{1}{2}R$.

Thus, when considering a convex mirror of radius R, we will always use the following result for the focal length:

Focal Length for a Convex Mirror of Radius R

$$f = -\tfrac{1}{2}R$$

26-2

SI unit: m

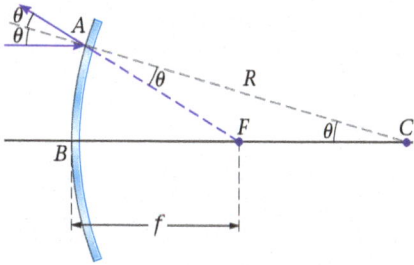

▲ **FIGURE 26-18 Ray diagram for a convex mirror** Ray diagram used to locate the focal point, F.

The minus sign in this expression is used to indicate that the focal point lies behind the mirror. This is part of a general sign convention for mirrors that will be discussed in detail in the next section.

The situation is similar for a concave mirror. First, rays parallel to the principal axis are reflected by the mirror and brought together at a focal point, F, as shown in **FIGURE 26-19**. The same type of analysis used for the convex mirror can be applied to the concave mirror as well, with the result that the focal point is a distance $\frac{1}{2}R$ in front of the mirror. Thus, for a concave mirror:

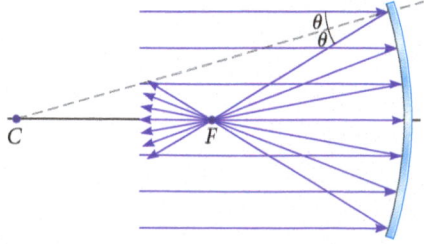

◀ **FIGURE 26-19 Parallel rays on a concave mirror** Parallel rays reflecting from a concave mirror pass through the focal point F, which is halfway between the surface of the mirror and the center of curvature.

Focal Length for a Concave Mirror of Radius *R*

$$f = \tfrac{1}{2}R \qquad\qquad 26\text{-}3$$

SI unit: m

This time f is positive, indicating that the focal point is in front of the mirror. In this case, the rays of light actually pass through the focal point, in contrast to the behavior of the convex mirror.

CONCEPTUAL EXAMPLE 26-4   STARTING A FIRE

Suppose you would like to use the Sun to start a fire in the wilderness. Which type of mirror, concave or convex, would work best?

REASONING AND DISCUSSION

The rays from the Sun are essentially parallel, due to the fact that it is at such a great distance from Earth. Therefore, if a convex mirror is used, the situation is like that shown in Figure 26-17, in which the rays are spread out after reflection. On the other hand, a concave mirror will bring the rays together at a point, as in Figure 26-19. Clearly, the concave mirror should be used to start a fire.

ANSWER

The concave mirror is the one to use.

Aberrations A final point regarding the approximation made earlier in this section—namely, that the angle θ is small. This is equivalent to saying that the distance between the principal axis of the mirror and the incoming rays is much less than the radius of curvature of the mirror, R. When rays are displaced from the axis by distances comparable to the radius R, as in **FIGURE 26-20 (a)**, the result is that they do not all pass through the focal point: The farther a ray is from the axis, the more it misses the focal point. In a case like this, the mirror will produce a blurred image, an effect known as **spherical aberration.** This effect can be reduced to undetectable levels by restricting the incoming rays to only those near the axis. These axis-hugging rays are referred to as **paraxial rays.**

   RWP Another way to eliminate spherical aberration is to construct a mirror with a *parabolic* cross section, as shown in **FIGURE 26-20 (b)**. One of the key properties of a parabola is that rays parallel to its axis are reflected through the same point F regardless of their distance from the axis. Thus a parabolic mirror produces a sharp image from all the rays coming into it. For this reason, mirrors in astronomical telescopes, like that of the Hale telescope on Mount Palomar in California, are polished to a parabolic shape to give the greatest possible light-gathering ability and the sharpest possible images. A slight change in the shape of a mirror, as happened with the Hubble Space Telescope shown in **FIGURE 26-21**, can lead to distorted images. Satellite dishes are also parabolic to maximize the intensity of the received signal.

   The same principle works in reverse as well. For example, if a source of light is placed at the focal point of a parabolic mirror, as at point F in Figure 26-20 (b), the mirror will redirect the light into an intense, unidirectional beam that can be aimed in a precise direction. Applications of this effect include flashlights, car headlights, and the giant arc lights that sweep across the sky to announce a grand opening.

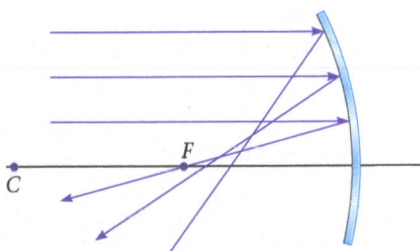

(a) A spherical mirror blurs the focus

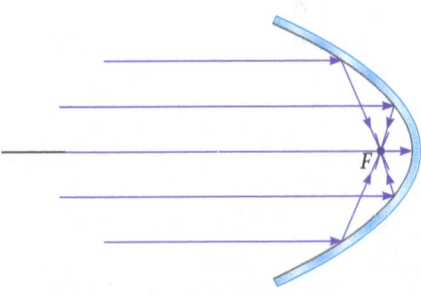

(b) A parabolic mirror has a single focal point

▲ **FIGURE 26-20 Spherical aberration and the parabolic mirror (a)** Rays far from the axis of a spherical mirror are not reflected through its focal point. The result is a blurred image referred to as spherical aberration. **(b)** A mirror with a parabolic cross section brings all parallel rays to a focus at one point, regardless of how far they are from the axis of the mirror.

Enhance Your Understanding   (Answers given at the end of the chapter)

3. Rank the following spherical mirrors in order of increasing focal length, from most negative to most positive. Indicate ties where appropriate.

System	A	B	C	D
Mirror type	concave	convex	concave	convex
Radius of curvature (cm)	20	50	60	80

CONTINUED

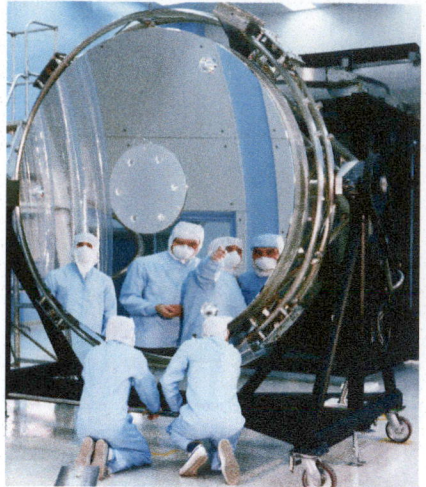

▲ **FIGURE 26-21** The mirror of the Hubble Space Telescope (HST) was to have a perfectly parabolic shape. Unfortunately, owing to a mistake in the grinding of the mirror, the images produced by the HST were marred by spherical aberration. This necessitated a repair mission to the orbiting telescope, during which astronauts installed corrective optics to compensate for the defect.

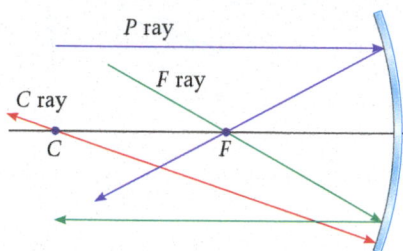

▲ **FIGURE 26-22 Principal rays used in ray tracing for a concave mirror** The parallel ray (P ray) reflects through the focal point, the focal-point ray (F ray) reflects parallel to the axis, and the center-of-curvature ray (C ray) reflects back along its incoming path.

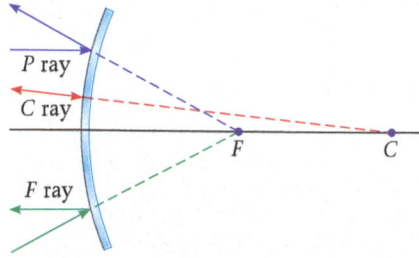

▲ **FIGURE 26-23 Principal rays used in ray tracing for a convex mirror** The parallel ray (P ray) reflects along a direction that extends back to the focal point. Similarly, the focal-point ray (F ray) moves toward the focal point until it reflects from the mirror, after which it moves parallel to the axis. The center-of-curvature ray (C ray) is directed toward the center of curvature. It reflects from the mirror back along its incoming path.

26-4 Ray Tracing and the Mirror Equation

The images formed by a spherical mirror can be more varied than those produced by a plane mirror. In a plane mirror the image is always upright, the same size as the object, and the same distance from the mirror as the object. In the case of spherical mirrors, the image can be either upright or inverted, larger or smaller than the object, and closer to or farther away from the mirror than the object.

To find the orientation, size, and location of an image in a spherical mirror, we use two techniques. The first, referred to as **ray tracing,** gives the orientation of the image as well as qualitative information on its location and size. If drawn carefully to scale, a ray diagram can also give quantitative results. The second method, using a relationship referred to as the **mirror equation,** provides precise quantitative information without the need for accurate scale drawings. Both methods are presented in this section.

Ray Tracing

The basic idea behind ray tracing is to follow the path of representative rays of light as they reflect from a mirror and form an image. This was done for the plane mirror in Section 26-2. Ray tracing for spherical mirrors is a straightforward extension of the same basic techniques.

There are three rays with simple behavior that are used most often in ray tracing with spherical mirrors. These rays are illustrated in **FIGURE 26-22** for a concave mirror, and in **FIGURE 26-23** for a convex mirror. We start with the parallel ray (P ray), which, as its name implies, is parallel to the principal axis of the mirror. As we know from the previous section, a parallel ray is reflected through the focal point of a concave mirror, as shown by the purple ray in Figure 26-22. Similarly, a P ray reflects from a convex mirror along a line that *extends back* through the focal point, like the purple ray in Figure 26-23.

Next, a ray that passes through the focal point of a concave mirror is reflected parallel to the axis, as indicated by the green ray in Figure 26-22. Thus, in a sense, a focal-point ray (F ray) is the reverse of a P ray. In general, any ray is equally valid in either the forward or reverse direction. The corresponding F ray for a convex mirror is shown in green in Figure 26-23.

Finally, notice that any straight line extending from the center of curvature intersects the mirror at right angles. Thus, a ray moving along such a path is reflected back along the same path. Center-of-curvature rays (C rays) are illustrated in red in Figures 26-22 and 26-23.

Using Rays to Form an Image To see how these rays can be used to obtain an image, consider the convex mirror shown in **FIGURE 26-24**. In front of the mirror is an object,

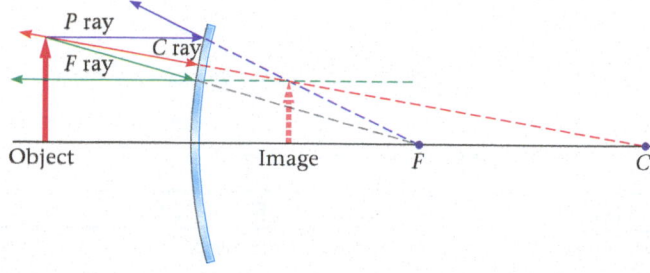

▲ **FIGURE 26-24 Image formation with a convex mirror** Ray diagram showing an image formed by a convex mirror. The three outgoing rays (P, F, and C) extend back to a single point at the top of the image.

represented symbolically by a red arrow. Also indicated in the figure are the three rays described above. Notice that these rays diverge from the mirror as if they had originated from the tip of the orange dashed arrow behind the mirror. This arrow is the image of the object; in fact, because no light passes through the image, we call it a **virtual image.** As we can see from the diagram, the virtual image is upright, smaller than the object, and located between the mirror and the focal point F.

Even though we drew three rays in Figure 26-24, any two would have given the intersection point at the tip of the virtual image. This is commonly the case with ray diagrams. When possible, it is useful to draw all three rays as a check on your results.

In the limit that the object is very close to a convex mirror, the mirror is essentially flat and behaves like a plane mirror. Thus the virtual image will be about the same distance behind the mirror that the image is in front, and about the same size as the object. Conversely, if the object is far from the mirror, the image is very small and practically at the focal point. These limits are illustrated in **FIGURE 26-25**.

Next, we consider the image formed by a concave mirror. In particular, we examine the following three cases: (i) The object is farther from the mirror than the center of curvature; (ii) the object is between the center of curvature and the focal point; and (iii) the object is between the mirror and the focal point. The F and P rays for case (i) are drawn in **FIGURE 26-26 (a)**, showing that the image is inverted, closer to the mirror, and smaller than the object. In addition, because light rays pass through the image, we call it a **real image.** The corresponding ray diagram for case (ii) is shown in **FIGURE 26-26 (b)**. In this case, the image is again real and inverted, but it is now farther from the mirror and larger than the object—as the object moves closer to the focal point, an observer will see a large inverted image of the object far in front of the mirror. The C ray is not useful in these cases and therefore is omitted from the diagrams.

The final case, in which the object is between the mirror and the focal point, is considered in the next Example.

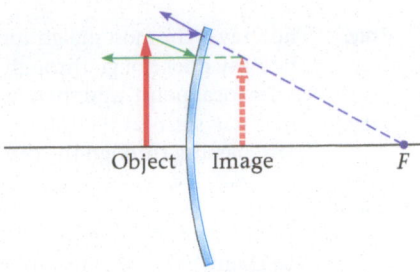

(a) An object close to a convex mirror

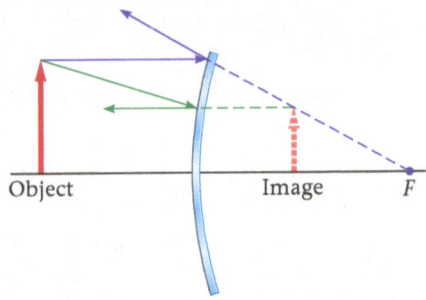

(b) An object far from a convex mirror

▲ **FIGURE 26-25 Image size and location in a convex mirror (a)** When an object is close to a convex mirror, the image is practically the same size and distance from the mirror. **(b)** In the limit that the object is very far from the mirror, the image is small and close to the focal point.

PROBLEM-SOLVING NOTE

Using Rays to Locate the Image of a Spherical Mirror

To find the top of an image, draw the three rays, P, F, and C, from the top of the object. The rays will either intersect at the top of the image (real image) or extend backward to the top of the image (virtual image). When drawing the rays, remember that the P ray is *parallel* to the axis of the mirror, the C ray goes either through or toward the mirror's *center* of curvature, and the F ray goes either through or toward the *focal* point of the mirror.

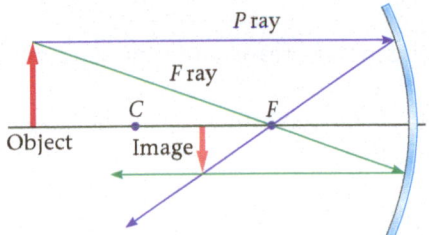

(a) An object beyond C

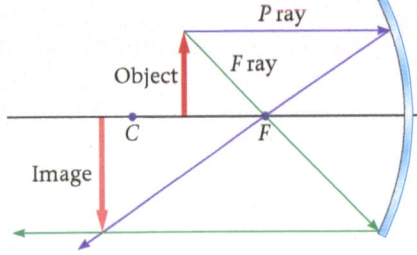

(b) An object between C and F

▲ **FIGURE 26-26 Image formation with a concave mirror** Ray diagrams showing the image formed by a concave mirror when the object is **(a)** beyond the center of curvature and **(b)** between the center of curvature and the focal point.

EXAMPLE 26-5   IMAGE FORMATION

Use a ray diagram to find the location, size, and orientation of the image formed by a concave mirror when the object is between the mirror and the focal point.

PICTURE THE PROBLEM
The physical system is shown in the sketch, along with the three principal rays. Notice that after reflection these three rays diverge from one another, just as if they had originated at a point behind the mirror.

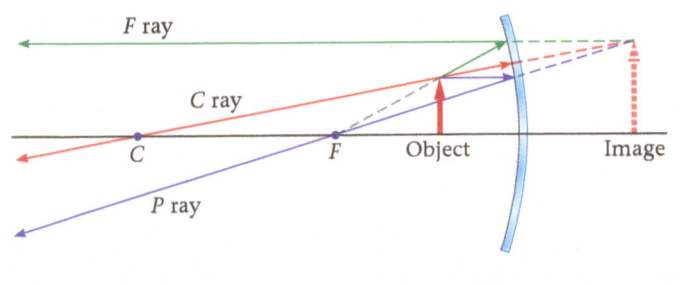

REASONING AND SOLUTION
The three outgoing rays (P, F, and C) extend back to form an image behind the mirror that is upright (same orientation as the object) and enlarged. We now discuss these three rays one at a time:

CONTINUED

P ray: The *P* ray is the most straightforward of the three. It starts parallel to the axis, and then reflects through the focal point.

F ray: The *F* ray does not go through the focal point, as is usually the case. Instead, it starts on a line that *extends back* to the focal point, and then reflects parallel to the axis.

C ray: The *C* ray starts at the top of the object, contacts the mirror at right angles, and reflects back along its initial path, passing through the center of curvature.

INSIGHT

Makeup mirrors are concave mirrors with fairly large focal lengths. The person applying makeup is between the mirror and its focal point, as is the object in this Example, and therefore sees an upright and enlarged image, as desired.

PRACTICE PROBLEM

If the object in the diagram is moved closer to the mirror, does the image increase or decrease in size? [**Answer:** As the object moves closer, the mirror behaves more like a plane mirror. Thus, the image becomes smaller, so that it is closer in size to the object.]

Some related homework problems: Problem 25, Problem 26

CONCEPTUAL EXAMPLE 26-6   REARVIEW MIRRORS

The passenger-side rearview mirrors in cars often have warning labels that read: OBJECTS IN MIRROR ARE CLOSER THAN THEY APPEAR. Are these rearview mirrors concave or convex?

REASONING AND DISCUSSION

Objects in the mirror are closer than they appear because the mirror produces an image that is reduced in size, which makes the object look as if it is farther away. In addition, we know that the rearview mirror always gives an upright image, no matter how close or far away the object. The mirror that always produces upright and reduced images is the convex mirror.

ANSWER

The mirrors are convex.

The imaging characteristics of convex and concave mirrors are summarized in Table 26-1.

We now show how to obtain results about the images formed by mirrors in a quantitative manner.

TABLE 26-1 Imaging Characteristics of Convex and Concave Spherical Mirrors

CONVEX MIRROR Object location	Image orientation	Image size	Image type
Arbitrary	Upright	Reduced	Virtual

CONCAVE MIRROR Object location	Image orientation	Image size	Image type
Beyond *C*	Inverted	Reduced	Real
C	Inverted	Same as object	Real
Between *F* and *C*	Inverted	Enlarged	Real
Just beyond *F*	Inverted	Approaching infinity	Real
Just inside *F*	Upright	Approaching infinity	Virtual
Between mirror and *F*	Upright	Enlarged	Virtual

The Mirror Equation

The mirror equation is a precise mathematical relationship between the object distance and the image distance for a given mirror. To obtain this relationship, we use the ray diagrams shown in **FIGURE 26-27**. Notice that the distance from the mirror to the object is d_o, the distance from the mirror to the image is d_i, and the distance from the mirror to the center of curvature is R. In addition, the height of the object is h_o, and the height of the image is h_i. Because the image is inverted in this case, its height is negative; thus $-h_i$ is positive.

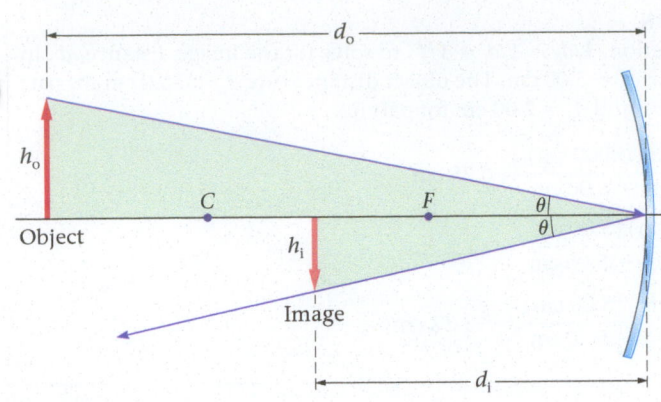

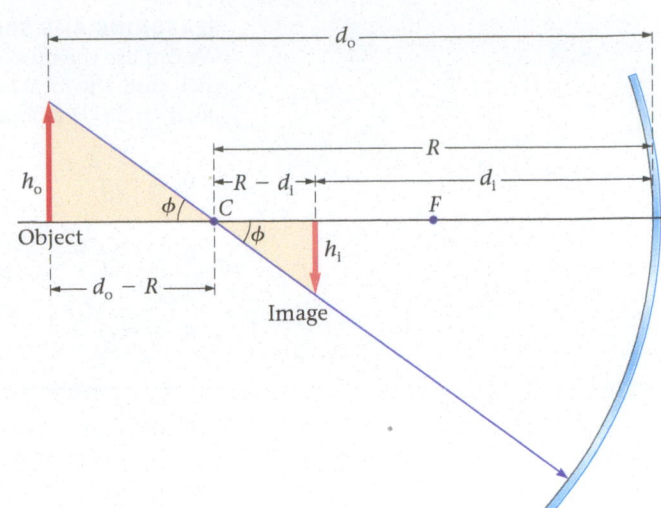

(a) Triangles to derive Equation 26-4

(b) Triangles to derive Equation 26-5

▲ **FIGURE 26-27 Ray diagrams used to derive the mirror equation** (a) The two similar triangles in this case are used to obtain Equation 26-4. (b) These similar triangles yield Equation 26-5.

The ray in Figure 26-27 (a) hits the mirror at its midpoint, where the principal axis is the normal to the mirror. As a result, the ray reflects at an angle θ below the principal axis that is equal to its incident angle θ above the axis. Therefore, the two green triangles in this diagram are similar, from which it follows that $h_o/d_o = (-h_i)/d_i$, or

$$\frac{h_o}{-h_i} = \frac{d_o}{d_i} \qquad \text{26-4}$$

Next, Figure 26-27 (b) shows the C ray for this mirror. From the figure it's clear that the two yellow triangles in this diagram are also similar, because they are both right triangles and they share the common angle ϕ. Thus, $h_o/(d_o - R) = (-h_i)/(R - d_i)$, or

$$\frac{h_o}{-h_i} = \frac{d_o - R}{R - d_i} \qquad \text{26-5}$$

Setting these two expressions for $h_o/(-h_i)$ equal gives

$$\frac{d_o}{d_i} = \frac{d_o - R}{R - d_i} \quad \text{or} \quad 1 = \frac{1 - \dfrac{R}{d_o}}{\dfrac{R}{d_i} - 1}$$

Rearranging, we find $R/d_o + R/d_i = 2$. Finally, dividing by R and recalling that $f = \frac{1}{2}R$ for a concave mirror, we obtain the **mirror equation**:

The Mirror Equation

$$\frac{1}{d_o} + \frac{1}{d_i} = \frac{1}{f} \qquad \text{26-6}$$

We apply the mirror equation to a simple system in the following Exercise.

EXERCISE 26-7 SPOON OPTICS

The concave side of a spoon (the side that forms a "bowl") has a focal length given by $f = 5.00$ cm. Find the image distance for this "mirror" when the object distance is (a) 25.0 cm, (b) 9.00 cm, and (c) 2.00 cm. These three cases correspond to Figure 26-26 (a), Figure 26-26 (b), and Example 26-5, respectively.

CONTINUED

REASONING AND SOLUTION

We can use the mirror equation, $1/d_o + 1/d_i = 1/f$, to solve for the image distance, d_i. In each case, the focal length is $f = 5.00$ cm. The object distances are $d_o = 25.0$ cm for part (a), $d_o = 9.00$ cm for part (b), and $d_o = 2.00$ cm for part (c).

a. $d_i = \dfrac{d_o f}{d_o - f} = \dfrac{(25.0 \text{ cm})(5.00 \text{ cm})}{25.0 \text{ cm} - 5.00 \text{ cm}} = 6.25$ cm

b. $d_i = \dfrac{d_o f}{d_o - f} = \dfrac{(9.00 \text{ cm})(5.00 \text{ cm})}{9.00 \text{ cm} - 5.00 \text{ cm}} = 11.3$ cm

c. $d_i = \dfrac{d_o f}{d_o - f} = \dfrac{(2.00 \text{ cm})(5.00 \text{ cm})}{2.00 \text{ cm} - 5.00 \text{ cm}} = -3.33$ cm

The image distance in Exercise 26-7 is negative when the object is closer to the mirror than the focal point [$d_o < f$, as in part (c)]. We know from Example 26-5 that this is also the case where the image is *behind* the mirror. Thus, the *sign* of the image distance indicates the *side* of the mirror on which the image is located. As long as we are discussing signs, it should be noted that the mirror equation applies equally well to a *convex* mirror, as long as we recall that the focal length in this case is *negative*. The following Exercise calculates the image distance for a convex mirror.

EXERCISE 26-8 CONVEX MIRROR

The convex mirror in Figure 26-24 has a 20.0-cm radius of curvature. Find the image distance for this mirror when the object distance is 6.33 cm, as it is in Figure 26-24.

REASONING AND SOLUTION

Recalling that $f = -\frac{1}{2}R$ for a convex mirror, we find $f = -10.0$ cm. Using this and the mirror equation, $1/d_o + 1/d_i = 1/f$, we can solve for the image distance, d_i. We find

$$d_i = \frac{d_o f}{d_o - f} = \frac{(6.33 \text{ cm})(-10.0 \text{ cm})}{6.33 \text{ cm} - (-10.0 \text{ cm})} = -3.88 \text{ cm}$$

This result agrees with the image location shown in Figure 26-24.

Magnification Next, we consider the height of an image, which is given by the relationship in Equation 26-4 ($h_o/-h_i = d_o/d_i$). Solving this equation for h_i, we find

$$h_i = -\left(\frac{d_i}{d_o}\right)h_o \qquad\qquad 26\text{-}7$$

The ratio of the height of the image to the height of the object is defined as the **magnification,** *m*; that is, $m = h_i/h_o$. From the above result, we see that:

Magnification, m

$$m = \frac{h_i}{h_o} = -\frac{d_i}{d_o} \qquad\qquad 26\text{-}8$$

The sign of the magnification gives the orientation of the image. For example, if both d_o and d_i are positive, as in Figure 26-26, the magnification is negative and the image is inverted. Conversely, if the image is behind the mirror, so that d_i is negative, the magnification is positive, and the image is upright. An example of this case is shown in Example 26-5 and Figure 26-24. Finally, the magnitude of the magnification gives the factor by which the size of the image is increased or decreased compared with the object. In the special case of an image with the same size and orientation as the object, as in a plane mirror, the magnification is 1.

The sign conventions for mirrors are summarized below:

Focal Length

f is positive for concave mirrors.
f is negative for convex mirrors.

PROBLEM-SOLVING NOTE

Applying the Mirror Equation

To use the mirror equation correctly, be careful to use the appropriate signs for all the known quantities. The final answer will also have a sign, which gives additional information about the system.

Big Idea 3 Spherical mirrors can form images that are upright or inverted, behind the mirror or in front of the mirror, and either enlarged or reduced in size. The specific characteristics of the image can be determined by ray tracing or the mirror equation.

Magnification

> m is positive for upright images.
> m is negative for inverted images.

Image Distance

> d_i is positive for images in front of a mirror (real images).
> d_i is negative for images behind a mirror (virtual images).

Object Distance

> d_o is positive for objects in front of a mirror (real objects).
> d_o is negative for objects behind a mirror (virtual objects).

The case of a negative object distance—that is, a virtual object—can occur when the image from one mirror serves as the object for another mirror or lens. For example, if mirror 1 produces an image that is behind mirror 2, we say that the image of mirror 1 is a virtual object for mirror 2. Such situations will be considered in the next chapter.

We now apply the mirror and magnification equations to specific Examples.

EXAMPLE 26-9   CHECKING IT TWICE

After leaving some presents under the tree, Santa notices his image in a shiny, spherical Christmas ornament. The ornament has a radius of 4.52 cm and is 0.866 m away from Santa. Curious to know the location and size of his image, Santa consults a book on physics. Knowing that Santa likes to "check it twice," what results should he obtain, assuming his height is 1.76 m?

PICTURE THE PROBLEM

The physical situation is illustrated in the sketch, along with the image Santa sees in the ornament. Because the spherical ornament is a convex mirror, the image it forms is upright and reduced.

REASONING AND STRATEGY

The ornament is a convex mirror with a radius $R = 0.0452$ m. We can find the location of the image, then, by using the mirror equation with the focal length given by $f = -\frac{1}{2}R$.

Once we have determined the location of the image, we can find its size using $h_i = -(d_i/d_o)h_o = mh_o$, where $h_o = 1.76$ m.

Known Radius of ornament, $R = 0.0452$ m; object distance, $d_o = 0.866$ m; object height, $h_o = 1.76$ m.
Unknown Image location, $d_i = ?$ Image height, $h_i = ?$

SOLUTION

1. Calculate the focal length for the ornament:

$$f = -\tfrac{1}{2}R = -\tfrac{1}{2}(4.52 \text{ cm}) = -0.0226 \text{ m}$$

2. Use the mirror equation to find the image distance, d_i:

$$\frac{1}{d_i} = \frac{1}{f} - \frac{1}{d_o} = \frac{1}{(-0.0226 \text{ m})} - \frac{1}{0.866 \text{ m}} = -45.4 \text{ m}^{-1}$$

$$d_i = \frac{1}{(-45.4 \text{ m}^{-1})} = -0.0220 \text{ m}$$

3. Determine the magnification of the image using $m = -(d_i/d_o)$:

$$m = -\left(\frac{d_i}{d_o}\right) = -\left(\frac{-0.0220 \text{ m}}{0.866 \text{ m}}\right) = 0.0254$$

4. Find the image height with $h_i = mh_o$:

$$h_i = mh_o = (0.0254)(1.76 \text{ m}) = 0.0447 \text{ m}$$

INSIGHT

Thus, Santa's image is 2.20 cm behind the surface of the ornament—about halfway between the surface and the center—and 4.47 cm high. His image fits on the surface of the ornament with room to spare, and therefore Santa can see the reflection of his entire body.

PRACTICE PROBLEM

How far from the ornament must Santa stand if his image is to be 2.15 cm behind its surface? **[Answer: $d_o = 44.2$ cm]**

Some related homework problems: Problem 29, Problem 33, Problem 38

QUICK EXAMPLE 26-10   SAY AHHH: FIND THE MAGNIFICATION OF THE TOOTH

A dentist uses a small mirror attached to a thin rod to examine one of your teeth. When the tooth is 1.38 cm in front of the mirror, the image it forms is 9.98 cm behind the mirror. Find (a) the focal length of the mirror and (b) the magnification of the image.

REASONING AND SOLUTION

Given the image and object distances, we can use the mirror equation, $1/d_o + 1/d_i = 1/f$, to find the focal length of the mirror.

1. Use the information given to identify the object and image distances: $d_o = 1.38 \text{ cm}, d_i = -9.98 \text{ cm}$

Part (a)

2. Use the mirror equation, $1/d_o + 1/d_i = 1/f$, to calculate the focal length:

$$\frac{1}{f} = \frac{1}{d_o} + \frac{1}{d_i} = \frac{1}{1.38 \text{ cm}} + \frac{1}{-9.98 \text{ cm}} = 0.624 \text{ cm}^{-1}$$

$$f = \frac{1}{0.624 \text{ cm}^{-1}} = 1.60 \text{ cm}$$

Part (b)

3. Use $m = -(d_i/d_o)$ to find the magnification:

$$m = -\frac{d_i}{d_o} = -\left(\frac{-9.98 \text{ cm}}{1.38 \text{ cm}}\right) = 7.23$$

Because the focal length is positive, and the magnification is greater than 1, we conclude that the dentist's mirror is concave. This mirror will make the tooth look 7.23 times larger than life. A convex mirror, in contrast, always produces a magnification less than 1 and would not be useful in this situation.

Finally, recall that a spherical mirror behaves like a plane mirror when the object is close to the mirror. But just what exactly do we mean by *close*? Well, in this case, *close* means that the object distance should be small in comparison with the radius of curvature. For example, if the radius of curvature were to go to infinity, the mirror would behave like a plane mirror for all object distances. Simply put, a sphere with $R \to \infty$ has a surface that is essentially as flat as a plane. Letting $f = \frac{1}{2}R$ go to infinity in the mirror equation yields

$$\frac{1}{d_o} + \frac{1}{d_i} = \frac{1}{f} \to 0$$

Therefore, in this limit $d_i = -d_o$, as expected for a plane mirror.

Enhance Your Understanding   (Answers given at the end of the chapter)

4. A spherical mirror with an object at the distance $d_o = 15$ cm produces an image at the distance $d_i = 25$ cm. Is the image (a) upright or inverted and (b) larger than or smaller than the object?

Section Review

- Ray tracing can determine the approximate location, orientation, and size of the image formed by a mirror.

- The precise location of the image formed by a mirror is given by the mirror equation: $1/d_o + 1/d_i = 1/f$. In this expression, f is the focal length of the mirror, d_o is the object distance, and d_i is the image distance.

- The magnification, m, of an image in a mirror is given by $m = -d_i/d_o$.

26-5 The Refraction of Light

When a wave propagates from a medium in which its speed is v_1 to another medium in which it has a different speed, $v_2 \neq v_1$, it will, in general, change its direction of motion. This phenomenon is called **refraction.**

Understanding Refraction   To see the origin of refraction, consider the behavior of a marching band as it moves from a solid section of ground into an area where the ground is muddy, as illustrated in **FIGURE 26-28**. In this analogy, the rows of the marching band correspond to the wave fronts of a traveling wave, and the solid and muddy sections of the ground represent media in which the wave speed is different.

If the speed of the marchers on solid ground is v_1, then after a time Δt they have advanced a distance $v_1 \Delta t$. On the other hand, if the marchers in the mud move at the reduced speed v_2, they advance only a distance $v_2 \Delta t$ in the same time. This causes a bend in the line of marchers, as can be seen in Figure 26-28, and therefore a change in the direction of motion.

FIGURE 26-29 shows a simplified version of Figure 26-28, with three "wave fronts" taking the place of the marchers. Also shown is a ray drawn at right angles to the wave fronts, and a normal to the interface between the two types of ground. The angle of incidence is θ_1, and from the figure we can see that this is also the angle between the incoming wave front and the interface. The outgoing wave front and its ray are characterized by a different angle, θ_2. From the geometry of the figure, we see that the green and brown triangles share a common side, AB. Therefore,

$$\sin \theta_1 = \frac{v_1 \Delta t}{AB} \quad \text{and} \quad \sin \theta_2 = \frac{v_2 \Delta t}{AB}$$

Eliminating the common factor $\Delta t / AB$ yields

$$\frac{\sin \theta_1}{v_1} = \frac{\sin \theta_2}{v_2} \qquad \text{26-9}$$

Thus, we see that the direction of propagation is directly related to the speed of propagation.

Index of Refraction   In general, the speed of light depends on the medium through which it travels. For example, we know that in a vacuum the speed of light is $c = 3.00 \times 10^8$ m/s. When light propagates through water, however, its speed is reduced by a factor of 1.33. In general, the speed of light in a given medium, v, is determined by the medium's **index of refraction,** n, defined as follows:

Definition of the Index of Refraction, n

$$v = \frac{c}{n} \qquad \text{26-10}$$

Values of the index of refraction for a variety of media are given in Table 26-2.

EXERCISE 26-11   TIME IN WATER

How much time does it take for light to travel 1.20 m in water?

REASONING AND SOLUTION
The speed of light in water is c/n, where $n = 1.33$. Therefore, the time required to cover 1.20 m is

$$t = \frac{d}{v} = \frac{d}{(c/n)} = \frac{1.20 \text{ m}}{\left(\dfrac{3.00 \times 10^8 \text{ m/s}}{1.33}\right)} = 5.32 \times 10^{-9} \text{ s}$$

Snell's Law   Returning to the direction of propagation, let's suppose light has the speed $v_1 = c/n_1$ in one medium and the speed $v_2 = c/n_2$ in a second medium. The direction of propagation in these two media is related by Equation 26-9:

$$\frac{\sin \theta_1}{(c/n_1)} = \frac{\sin \theta_2}{(c/n_2)}$$

Elimination of the common factor c yields the following relationship, known as Snell's law:

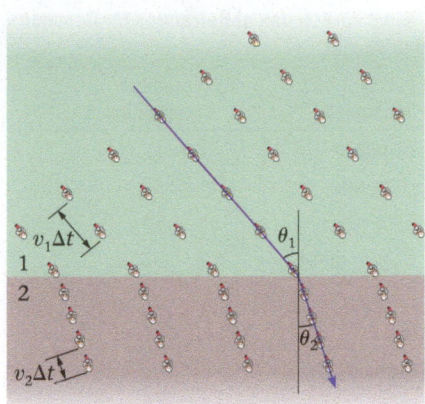

▲ **FIGURE 26-28 An analogy for refraction**   As a marching band moves from an area where the ground is solid into one where it is soft and muddy, the direction of motion changes.

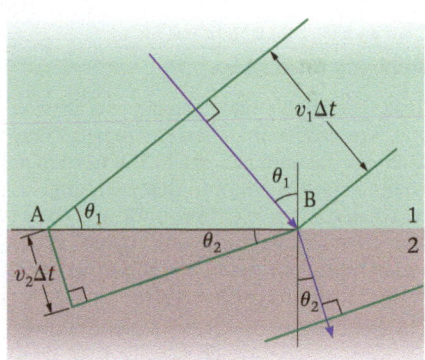

▲ **FIGURE 26-29 The basic mechanism of refraction**   Refraction is the bending of wave fronts and a change in the direction of propagation due to a change in speed.

Big Idea **4** When light changes speed on passing from one medium into another, its direction of propagation changes as well. This is referred to as refraction.

PHYSICS IN CONTEXT
Looking Ahead

The index of refraction plays an important role in Chapter 28, when we study interference in thin films. As we shall see, the wavelength of light depends on the index of refraction of the medium in which it propagates. The wavelength, in turn, is needed to determine whether the light experiences constructive or destructive interference.

TABLE 26-2 Index of Refraction for Common Substances

Substance	Index of refraction, n
SOLIDS	
Diamond	2.42
Flint glass	1.66
Crown glass	1.52
Fused quartz (glass)	1.46
Ice	1.31
LIQUIDS	
Benzene	1.50
Ethyl alcohol	1.36
Water	1.33
GASES	
Carbon dioxide	1.00045
Air	1.000293

PROBLEM-SOLVING NOTE

Applying Snell's Law

To apply Snell's law correctly, recall that the two angles in $n_1 \sin \theta_1 = n_2 \sin \theta_2$ are always measured relative to the normal at the interface. Also, note that each angle is associated with its corresponding index of refraction. For example, the angle θ_1 is the angle to the normal in the substance with an index of refraction equal to n_1.

Big Idea 5 When light passes into a medium where its speed is reduced, it bends closer to the normal between the two media.

PHYSICS IN CONTEXT
Looking Back

We make extensive use of trigonometric functions in the study of refraction. Trigonometric functions were first introduced in Chapter 3 when we discussed vector components.

Snell's Law

$$n_1 \sin \theta_1 = n_2 \sin \theta_2 \qquad 26\text{-}11$$

A typical application of Snell's law is given in the following Exercise.

EXERCISE 26-12 ANGLE OF REFRACTION

A beam of light in air enters **(a)** water ($n = 1.33$) or **(b)** diamond ($n = 2.42$) at an angle of 60.0° relative to the normal. Find the angle of refraction for each case.

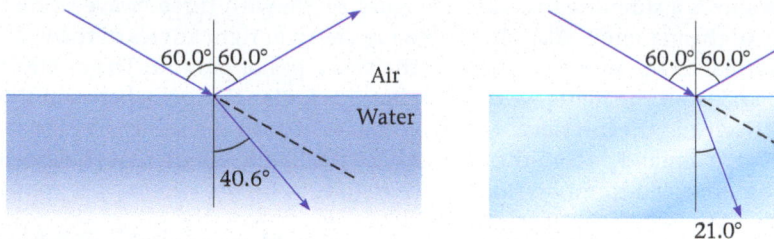

REASONING AND SOLUTION

Because the beam starts in air, we refer to Table 26-2 and set $n_1 = 1.000293$, or simply $n_1 = 1.00$ to three significant figures. We also solve Snell's law for θ_2, giving $\theta_2 = \sin^{-1}\left(\dfrac{n_1}{n_2} \sin \theta_1\right)$.

a. With $n_2 = 1.33$ we find

$$\theta_2 = \sin^{-1}\left(\frac{1.00}{n_2} \sin \theta_1\right) = \sin^{-1}\left(\frac{1.00}{1.33} \sin 60.0°\right) = 40.6°$$

b. Setting $n_2 = 2.42$, we find

$$\theta_2 = \sin^{-1}\left(\frac{1.00}{n_2} \sin \theta_1\right) = \sin^{-1}\left(\frac{1.00}{2.42} \sin 60.0°\right) = 21.0°$$

From the preceding Exercise we can see that the greater the difference in the index of refraction between two different materials, the greater the difference in the direction of propagation. In addition, light is bent closer to the normal in the medium where its speed is reduced. Of course, the opposite is true when light passes into a medium in which its speed is greater, as can be seen by reversing the incident and refracted rays. The qualitative features of refraction are as follows:

- When a ray of light enters a medium where the index of refraction is increased, and hence the speed of the light is *decreased*, the ray is bent *toward* the normal. A good example is shown in **FIGURE 26-30**.

- When a ray of light enters a medium where the index of refraction is decreased, and hence the speed of the light is *increased*, the ray is bent *away* from the normal.

- The greater the change in the index of refraction, the greater the change in the propagation direction. If there is no change in the index of refraction, there is no change in the direction of propagation.

- If a ray of light goes from one medium into another along the normal, it is undeflected, regardless of the index of refraction of each medium.

The last property listed follows directly from Snell's law: If θ_1 is zero, then $0 = n_2 \sin \theta_2$, which means that $\theta_2 = 0$. Refraction is explored further in the following Example.

◄ **FIGURE 26-30** The water in this tank contains a fluorescent dye, making it easier to see the refraction of the beam as it passes from the air into the water.

EXAMPLE 26-13 — SITTING ON A DOCK OF THE BAY

One night while on vacation in the Caribbean you walk to the end of a dock and, for no particular reason, shine your laser pointer into the water. When you shine the beam of light on the water a horizontal distance of 2.4 m from the dock, you see a glint of light from a shiny object on the sandy bottom—perhaps a gold doubloon. If the pointer is 1.8 m above the surface of the water, and the water is 5.5 m deep, what is the horizontal distance from the end of the dock to the shiny object?

PICTURE THE PROBLEM

The person standing at the end of the dock and the shiny object on the bottom are shown in the sketch. All of the known distances are indicated in the sketch, along with the angle of incidence, θ_1, and the angle of refraction, θ_2. Finally, the appropriate indices of refraction from Table 26-2 are given as well.

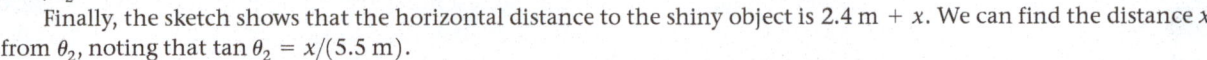

REASONING AND STRATEGY

We can use Snell's law and basic trigonometry to find the horizontal distance to the shiny object.

First, the information given in the problem determines the angle of incidence, θ_1. In particular, we can see from the sketch that $\tan \theta_1 = (2.4\ \text{m})/(1.8\ \text{m})$.

Second, Snell's law, ($n_1 \sin \theta_1 = n_2 \sin \theta_2$), gives the angle of refraction, θ_2.

Finally, the sketch shows that the horizontal distance to the shiny object is 2.4 m + x. We can find the distance x from θ_2, noting that $\tan \theta_2 = x/(5.5\ \text{m})$.

Known   Horizontal distance to spot on the water = 2.4 m; height of laser pointer above the water = 1.8 m; depth of water = 5.5 m.

Unknown   Horizontal distance to object = ?

SOLUTION

1. Find the angle of incidence from the information given in the problem:

$$\theta_1 = \tan^{-1}\left(\frac{2.4\ \text{m}}{1.8\ \text{m}}\right) = 53°$$

2. Use Snell's law to calculate the angle of refraction:

$$\theta_2 = \sin^{-1}\left(\frac{n_1}{n_2}\sin \theta_1\right) = \sin^{-1}\left[\left(\frac{1.00}{1.33}\right)\sin 53°\right] = 37°$$

3. Calculate x using $\tan \theta_2 = x/(5.5\ \text{m})$:

$$\tan \theta_2 = \frac{x}{5.5\ \text{m}}$$
$$x = (5.5\ \text{m})\tan \theta_2 = (5.5\ \text{m}) \tan 37° = 4.1\ \text{m}$$

4. Add 2.4 m to x to find the total horizontal distance to the shiny object:

$$\text{distance} = 2.4\ \text{m} + x = 2.4\ \text{m} + 4.1\ \text{m} = 6.5\ \text{m}$$

INSIGHT

If the water were to be removed, the incident beam of light would continue along its original path, characterized by the angle θ_1, and overshoot the hoped-for doubloon by a significant distance. Similarly, if you were to simply stand at the end of the dock and look out into the water at a glint of gold on the bottom, you would be looking in a direction (again characterized by θ_1) that is too high—the gold would be below your line of sight and closer to you than you think.

PRACTICE PROBLEM — PREDICT/CALCULATE

(a) If the index of refraction for water were 1.35 instead of 1.33, would the doubloon be farther from the dock, nearer the dock, or the same distance calculated above? Explain. (b) Check your answer by calculating the distance with $n_2 = 1.35$.   [**Answer: (a)** The doubloon would be closer to the dock, because a larger index of refraction causes a greater change in the direction of the light beam. **(b)** The horizontal distance in this case is 6.4 m, which is less than 6.5 m as expected.]

Some related homework problems: Problem 49, Problem 50, Problem 54

RWP Refraction is responsible for a number of common "optical illusions." For example, we all know that a pencil placed in a glass of water appears to be bent, though it is still perfectly straight. The cause of this illusion is shown in **FIGURE 26-31**, where we see that rays leaving the water bend away from the normal and make the pencil appear to be above its actual position. This is an example of what is known as *apparent depth*, in

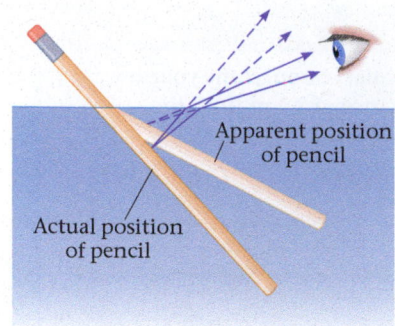

▲ **FIGURE 26-31 Refraction and the "bent" pencil** Refraction causes a pencil to appear bent when placed in water. Note that rays leaving the water are bent away from the normal and hence extend back to a point that is higher than the actual position of the pencil.

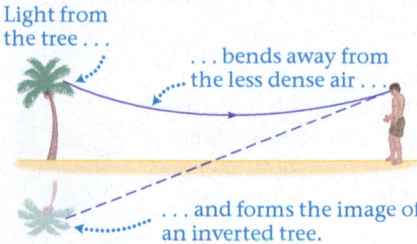

▲ **FIGURE 26-32 A mirage** A mirage is produced when light bends upward due to the low index of refraction of heated air near the ground.

▲ **FIGURE 26-33** One of the most common mirages, often seen in hot weather, makes a stretch of road look like the surface of a lake. The blue color that so resembles water to our eyes is actually an image of the sky, refracted by the hot, low-density air above the road.

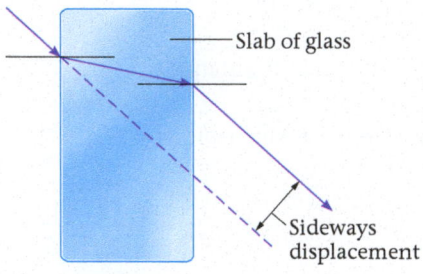

▲ **FIGURE 26-34 Light propagating through a glass slab** When a ray of light passes through a glass slab, it first refracts toward the normal, then away from the normal. The net result is that the ray continues in its original direction but is displaced sideways by a finite distance.

which an object appears to be closer to the water's surface than it really is. The relationship between the true depth and apparent depth is considered in Problem 55.

RWP Similarly, refraction can cause a mirage, as when hot, dry ground in the distance appears to be covered with water. Basically, hot air near the surface is less dense—and hence has a smaller index of refraction—than the cooler, denser air higher up. Thus, as light propagates toward the ground, it bends away from the normal until, eventually, it travels upward and enters the eye of an observer, as indicated in **FIGURE 26-32**. What appears to be a reflecting pool of water in the distance, as shown in **FIGURE 26-33**, is actually an image of the sky.

Finally, if light passes through a refracting slab, like the sheet of glass in **FIGURE 26-34**, it undergoes two refractions—one at each surface of the slab. The first refraction bends the light rays closer to the normal, and the second refraction bends the rays away from the normal. As can be seen in the figure, the two changes in direction cancel, so that the final direction of the light is the same as its original direction. The light has been displaced sideways, however, by an amount proportional to the thickness of the slab. The displacement distance is calculated in Problem 103.

CONCEPTUAL EXAMPLE 26-14 　REFRACTION IN A PRISM

A horizontal ray of light encounters a prism, as shown in the first sketch. After passing through the prism, is the ray deflected upward, still horizontal, or deflected downward?

REASONING AND DISCUSSION
When the ray enters the prism, it is bent toward the normal, which deflects it *downward*, as shown in the second sketch. When it leaves through the opposite side of the prism, it is bent away from the normal. Because the sides of a prism are angled in opposite directions, however, bending away from the normal in the second refraction also causes a *downward* deflection.

The net result, then, is a downward deflection of the ray.

ANSWER
The ray is deflected downward.

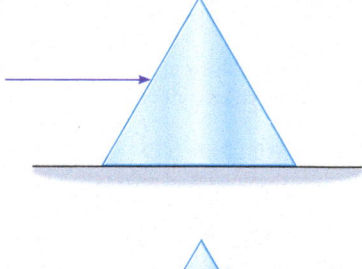

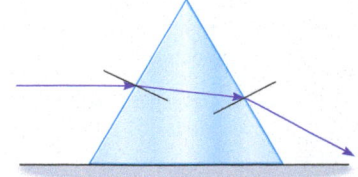

We will use the results of this Conceptual Example in the next section when we investigate the behavior of lenses. For now, we turn to two additional phenomena associated with refraction.

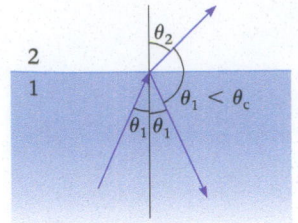

(a) Small angle of incidence

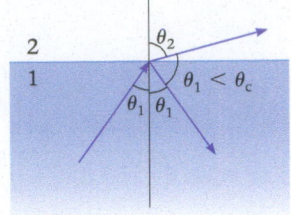

(b) Larger angle of incidence

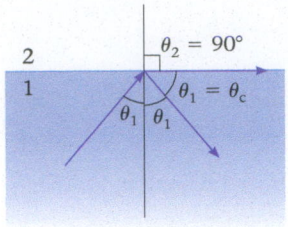

(c) Refracted beam parallel to interface

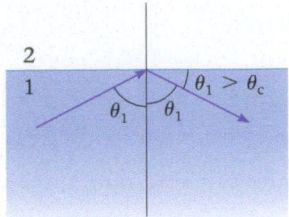

(d) Total internal reflection

Total Internal Reflection

Sometimes refraction can "trap" a ray of light and prevent it from leaving a material. For example, **FIGURE 26-35 (a)** shows a ray of light in water encountering a water–air interface. In such a case, it is observed that part of the light is reflected back into the water at the interface—as from the surface of a mirror—while the rest emerges into the air along a direction that is bent away from the normal according to Snell's law. If the angle of incidence is increased, as in **FIGURE 26-35 (b)**, the angle of refraction increases as well. At some critical angle of incidence, θ_c, the refracted beam no longer enters the air but instead is directed parallel to the water–air interface (**FIGURE 26-35 (c)**). In this case, the angle of refraction is 90°. For angles of incidence greater than the critical angle, as in **FIGURE 26-35 (d)**, it is observed that all of the light is reflected back into the water—it's effectively trapped in the water. This phenomenon is referred to as **total internal reflection.**

We can find the critical angle for total internal reflection by setting $\theta_2 = 90°$ and applying Snell's law:

$$n_1 \sin \theta_c = n_2 \sin 90° = n_2$$

Therefore, the critical angle is given by the following relationship:

Critical Angle for Total Internal Reflection, θ_c

$$\sin \theta_c = \frac{n_2}{n_1} \qquad \text{26-12}$$

Because $\sin \theta$ is always less than or equal to 1, the index of refraction, n_1, must be larger than the index of refraction, n_2, if Equation 26-12 is to give a physical solution. Thus, total internal reflection can occur only when light in one medium encounters an interface with another medium in which the speed of light is greater. For example, light moving from water to air can undergo total internal reflection, as shown in Figure 26-35, but light moving from air to water cannot.

▲ **FIGURE 26-35 Total internal reflection** Total internal reflection can occur when light propagates from a region with a high index of refraction to one with a lower index of refraction. Part **(a)** shows an incident ray in medium 1 encountering the interface with medium 2, which has a lower index of refraction. A portion of the ray is transmitted to region 2 at the angle θ_2, given by Snell's law, and a portion of the ray is reflected back into medium 1 at the angle of incidence, θ_1. The sum of the intensities of the refracted and reflected rays equals the intensity of the incident ray. In part **(b)** the angle of incidence has been increased, and the refracted beam makes a smaller angle with the interface. When the angle of incidence equals the critical angle, θ_c, as in part **(c)**, the refracted ray propagates parallel to the interface. For incident angles greater than θ_c there is no refracted ray at all, as shown in part **(d)**, and all of the incident intensity goes into the reflected ray.

PROBLEM-SOLVING NOTE

Total Internal Reflection

Remember that total internal reflection can occur only on the side of an interface between two different substances that has the greater index of refraction.

EXAMPLE 26-15   LIGHT TOTALLY REFLECTED

Consider a sample of glass whose index of refraction is $n = 1.65$. Find the critical angle for total internal reflection for light traveling from this glass to **(a)** air ($n = 1.00$) and **(b)** water ($n = 1.33$).

PICTURE THE PROBLEM

Our sketch shows the two cases considered in the problem. In each case the incident medium is glass; therefore $n_1 = 1.65$. For part (a) it follows that $n_2 = 1.00$, and for part (b) $n_2 = 1.33$.

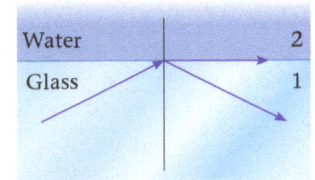

REASONING AND STRATEGY

The critical angle, θ_c, is defined by $\sin \theta_c = n_2/n_1$. It is straightforward to obtain θ_c by simply substituting the appropriate indices of refraction for each case in the relationship $\theta_c = \sin^{-1}(n_2/n_1)$.

Known   Index of refraction of glass, $n = 1.65$; **(a)** index of refraction of air, $n = 1.00$; **(b)** index of refraction of water, $n = 1.33$.

Unknown Critical angle, $\theta_c = ?$

CONTINUED

SOLUTION

Part (a)

1. Solve $\theta_c = \sin^{-1}(n_2/n_1)$ for θ_c, using $n_1 = 1.65$ and $n_2 = 1.00$:

$$\theta_c = \sin^{-1}\left(\frac{n_2}{n_1}\right) = \sin^{-1}\left(\frac{1.00}{1.65}\right) = 37.3°$$

Part (b)

2. Solve $\theta_c = \sin^{-1}(n_2/n_1)$ for θ_c, using $n_1 = 1.65$ and $n_2 = 1.33$:

$$\theta_c = \sin^{-1}\left(\frac{n_2}{n_1}\right) = \sin^{-1}\left(\frac{1.33}{1.65}\right) = 53.7°$$

INSIGHT

In the case of water and glass, the two indices of refraction are close in value; hence, light escapes from glass to water over a wider range of incident angles (0° to 53.7°) than from glass to air (0° to 37.3°). In general, a *large* change in the index of refraction, such as from diamond to air, means that very little light escapes (the majority reflects), which is why diamonds sparkle more than glass.

PRACTICE PROBLEM — PREDICT/CALCULATE

Suppose the incident ray is in a different type of glass, with a glass–air critical angle of 40.0°. **(a)** Is the index of refraction of this glass greater than or less than 1.65? Explain. **(b)** Verify your answer with a calculation.   [**Answer: (a)** The critical angle is larger for this glass, which means that its index of refraction must be less than 1.65. **(b)** The calculated value is 1.56.]

Some related homework problems: Problem 56, Problem 57

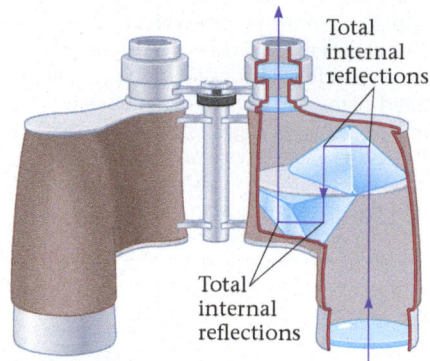

▲ FIGURE 26-36 Porro prisms Prisms are used to "fold" the light path within a pair of binoculars. This makes the binoculars easier to handle.

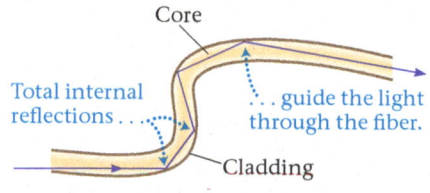

▲ FIGURE 26-37 An optical fiber An optical fiber channels light along its core by a series of total internal reflections between the core and the cladding.

RWP Total internal reflection is frequently put to practical use. For example, many binoculars contain a set of prisms—referred to as *Porro prisms*—that use total internal reflection to "fold" a relatively long light path into the short length of the binoculars, as shown in **FIGURE 26-36**. This allows the user to hold with ease a relatively short device that has the same optical behavior as a set of long, unwieldy telescopes. Thus, the characteristic zigzag shape of a binocular is not a fashion statement, but a reflection of its internal optical construction.

RWP *Optical fibers* are another important and common application of total internal reflection. These thin fibers are generally composed of a glass or plastic core with a high index of refraction surrounded by an outer coating, or cladding, with a low index of refraction. Light is introduced into the core of the fiber at one end. It then propagates along the fiber in a zigzag path, undergoing one total internal reflection after another, as indicated in **FIGURE 26-37**. The core is so transparent that even after light propagates through a 1-km length of fiber, the amount of absorption is roughly the same as if the light had simply passed through a glass window. In addition, the total internal reflections allow the fiber to go around corners, and even to be tied into knots, and still deliver the light to the other end. Examples of total internal reflection are shown in **FIGURE 26-38**.

The ability of optical fibers to convey light along curved paths has been put to good use in various fields of medicine. In particular, devices known as *endoscopes* allow physicians to examine the interior of the body by snaking a flexible tube containing optical fibers into the part of the body to be examined. For example, a type of endoscope called the bronchoscope can be inserted into the nose or throat, threaded through the bronchial tubes, and eventually placed in the lungs. There, the bronchoscope delivers light through one set of fibers and returns an image to the physician through another set of fibers. In some cases, the bronchoscope can even be used to retrieve small samples from the lung for further analysis. Similarly, the colonoscope can be used to examine the colon, making it one of the most important weapons in the fight against colon cancer.

Finally, optical fibers are important components of telecommunication systems. Not only are they small, light, and flexible, but they are also immune to the type of electrical interference that can degrade information carried on copper wires. Even more important, however, is that optical fibers can carry thousands of times more information than an electric current in a wire. For example, it takes only a single optical fiber to transmit several television programs and tens of thousands of telephone conversations, all at the same time.

Total Polarization

As explained in Section 25-5, light reflected from a nonmetallic surface is generally polarized to some degree. For example, the light reflected from the surface of a lake

▶ **FIGURE 26-38 Visualizing Concepts Total Internal Reflection** At right (top), a beam of light enters a tank of water from above and is reflected by mirrors oriented at different angles. Most of the light in the first two beams passes through the water–air interface, undergoing refraction as it leaves the water. Only a small portion of the light is reflected back down into the tank. (The weak beams are hard to see, but the spots they make on the bottom of the tank are clearly visible.) The third beam, however, strikes the interface at an angle of incidence greater than the critical angle. As a result, all of the beam is reflected, as if the surface of the water were a mirror. This phenomenon of total internal reflection makes it possible to "pipe" light through tiny optical fibers such as the one shown at right (bottom).

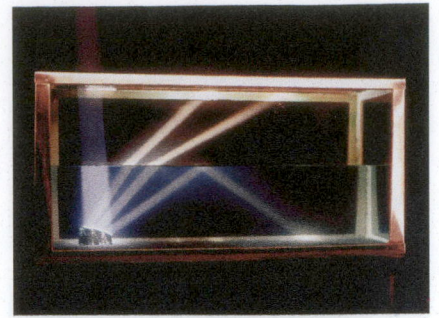

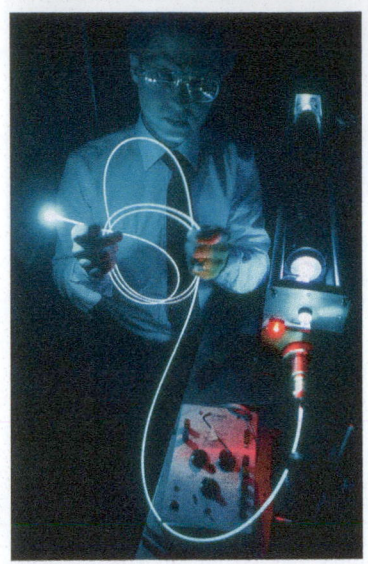

is preferentially polarized in the horizontal direction. The polarization of reflected light is complete for one special angle of incidence, **Brewster's angle**, θ_B, defined as follows:

> Reflected light is completely polarized when the reflected and refracted beams are at right angles to one another. The direction of polarization is parallel to the reflecting surface.

This situation is illustrated in **FIGURE 26-39**. Brewster's angle is named for its discoverer, and the inventor of the kaleidoscope, the Scottish physicist Sir David Brewster (1781–1868).

To calculate Brewster's angle, we begin by applying Snell's law with an incident angle equal to θ_B:

$$n_1 \sin \theta_B = n_2 \sin \theta_2$$

Next, we note from Figure 26-39 that the angles on the right side of the normal line add to 180°; that is, $\theta_B + 90° + \theta_2 = 180°$. Rearranging slightly, we have $\theta_B + \theta_2 = 90°$, or $\theta_2 = 90° - \theta_B$. Therefore, it follows that $\sin \theta_2 = \sin(90° - \theta_B)$. A standard trigonometric identity [$\sin (90° - \theta) = \cos \theta$, see Appendix A] leads to the following relationship: $\sin \theta_2 = \sin(90° - \theta_B) = \cos \theta_B$. Combining this result with the preceding equation, and noting that $\tan \theta_B = \sin \theta_B / \cos \theta_B$, we obtain:

Brewster's Angle, θ_B

$$\tan \theta_B = \frac{n_2}{n_1} \qquad \qquad 26\text{-}13$$

We calculate Brewster's angle for a typical situation in the next Exercise.

EXERCISE 26-16   BREWSTER'S ANGLE

Find Brewster's angle for light reflected from the top of a soda-lime glass ($n = 1.52$) coffee table.

REASONING AND SOLUTION

Using $\tan \theta_B = n_2/n_1$, and letting $n_1 = 1.00$ and $n_2 = 1.52$, we find

$$\theta_B = \tan^{-1}\left(\frac{n_2}{n_1}\right) = \tan^{-1}\left(\frac{1.52}{1.00}\right) = 56.7°$$

At this angle of incidence, the reflected beam is completely polarized in the horizontal direction.

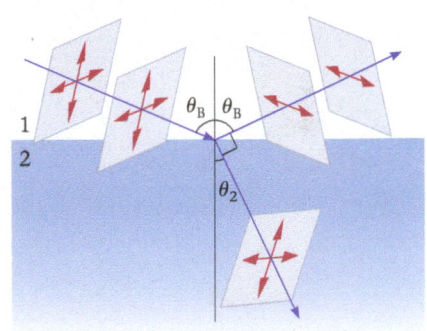

▲ **FIGURE 26-39 Brewster's angle** When light is incident at Brewster's angle, θ_B, the reflected and refracted rays are perpendicular to each other. In addition, the reflected light is completely polarized parallel to the reflecting surface.

Enhance Your Understanding   (Answers given at the end of the chapter)

5. **(a)** As a beam of light passes from flint glass to diamond, does it bend closer to or farther from the normal? **(b)** As a beam of light passes from flint glass to crown glass, does it bend closer to or farther from the normal?

Section Review

- The index of refraction n is the factor by which the speed of light is reduced from its vacuum value c to its value v in a given medium; that is, $v = c/n$.

CONTINUED

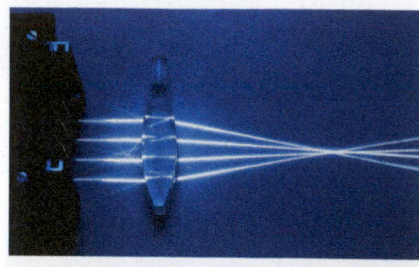

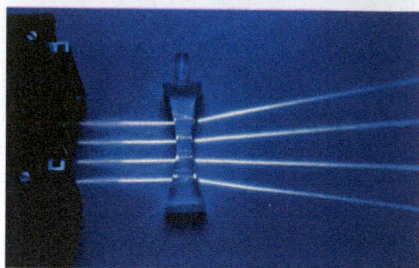

▲ **FIGURE 26-40** The paths of light rays through a convex (converging) lens (top) and a concave (diverging) lens (bottom).

▶ **FIGURE 26-41 A variety of converging and diverging lenses** Converging and diverging lenses come in a variety of shapes. Generally speaking, converging lenses are thicker in the middle than at the edges, and diverging lenses are thinner in the middle. We will use double convex and double concave lenses when we present examples of converging and diverging lenses, respectively.

- Snell's law, $n_1 \sin \theta_1 = n_2 \sin \theta_2$, relates the direction of the propagation of light in two different media that have different indices of refraction.
- The critical angle for total reflection at the boundary between two different media is $\sin \theta_c = n_2/n_1$. Brewster's angle for total polarization is given by $\tan \theta_B = n_2/n_1$.

26-6 Ray Tracing for Lenses

As we've seen, a ray of light can be redirected as it passes from one medium to another. A device that takes advantage of this effect, and uses it to focus light and form images, is a **lens**. Typically, a lens is a thin piece of glass or other transparent substance that can be characterized by the effect it has on light. In particular, converging lenses take parallel rays of light and bring them together at a focus; diverging lenses cause parallel rays to spread out as if diverging from a point source. Examples are shown in **FIGURE 26-40**. A variety of converging and diverging lenses are illustrated in **FIGURE 26-41**, though we consider only the most basic types here—namely, the double concave (or simply concave) and the double convex (or simply convex). Notice, in general, that converging lenses are thicker in the middle, and diverging lenses are thinner in the middle.

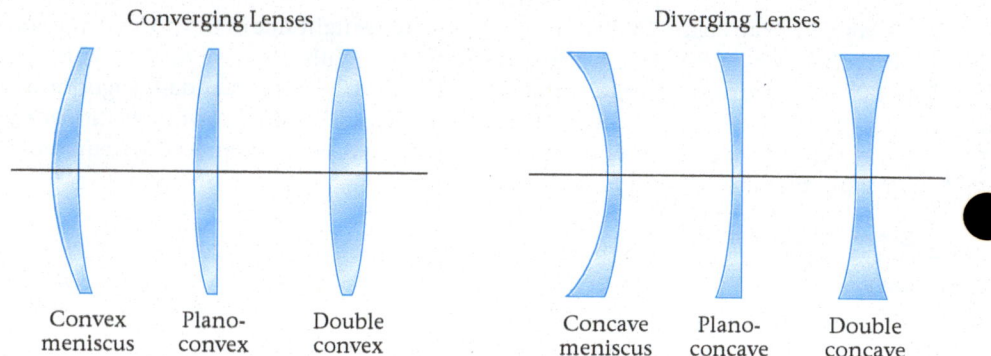

Converging Lenses			Diverging Lenses		
Convex meniscus	Plano-convex	Double convex	Concave meniscus	Plano-concave	Double concave

Let's start by considering a convex lens, as shown in **FIGURE 26-42**. To see qualitatively why such a lens is converging, we note that it is similar in shape to two prisms placed back to back. Recalling the bending of light described for a prism in Conceptual Example 26-14, we expect parallel rays of light to be brought together. In fact, convex lenses are shaped so that they bring parallel light to a focus at a **focal point**, F, along the center line, or axis, of the lens, as indicated in the figure.

▶ **FIGURE 26-42 A convex lens compared with a pair of prisms** The behavior of a convex lens is similar to that of two prisms placed back to back. In both cases, light parallel to the axis is made to converge. Note that the lens, because of its curved shape, brings light to a focus at the focal point, F.

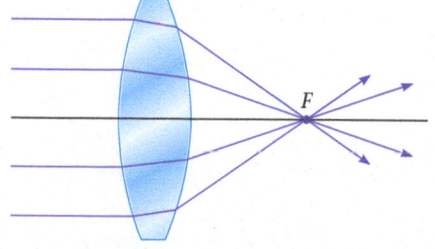

A double convex lens

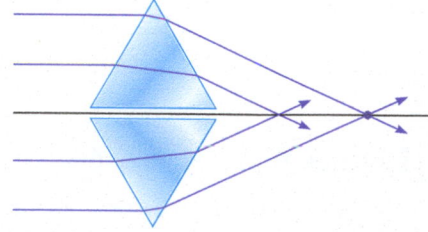

Two prisms stacked "back to back"

Similarly, a concave lens is similar in shape to two prisms brought together point to point, as we see in **FIGURE 26-43**. In this case, parallel rays are bent away from the axis of the lens. When the diverging rays from a concave lens are extended back, they appear to originate at a focal point F on the axis of the lens.

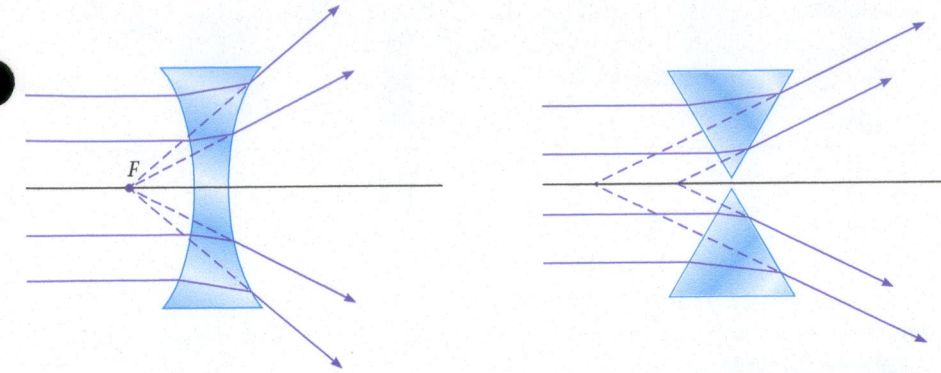

A double concave lens   Two prisms stacked "tip to tip"

▲ **FIGURE 26-43 A concave lens compared with a pair of prisms** A concave lens and two prisms placed point to point have similar behavior. In both cases, parallel light is made to diverge.

Ray Tracing To determine the type of image formed by a convex or concave lens, we can use ray tracing, as we did with mirrors. The three principal rays for lenses are shown in **FIGURES 26-44** and **26-45**. Their properties are as follows:

- The *P* ray—or parallel ray—approaches the lens parallel to its axis. The *P* ray is bent so that it passes through the focal point of a convex lens, or extrapolates back to the focal point on the same side of a concave lens.

- The *F* ray (focal-point ray) on a convex lens is drawn through the focal point and on to the lens, as pictured in Figure 26-44. The lens bends the ray parallel to the axis— basically the reverse of a *P* ray. For a concave lens, the *F* ray is drawn toward the focal point *on the other side* of the lens, as in Figure 26-45. Before it gets there, however, it passes through the lens and is bent parallel to the axis.

- The midpoint ray (*M* ray) goes through the middle of the lens, which is basically like a thin slab of glass. For ideal lenses, which are infinitely thin, the *M* ray continues in its original direction with negligible displacement after passing through the lens.

To illustrate the use of ray tracing, we start with the concave lens shown in **FIGURE 26-46**. Notice that the three rays originating from the top of the object extend backward to a single point on the left side of the lens—to an observer on the right side of the lens this point is the top of the image. Our ray diagram also shows that the image is upright and reduced in size. In addition, the image is virtual, because it is on the same side of the lens as the object. These are general features of the image formed by a concave lens.

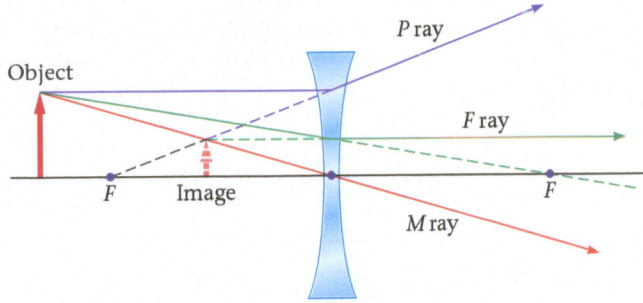

▲ **FIGURE 26-46 The image formed by a concave lens** Ray tracing can be used to find the image produced by a concave lens. Note that the *P, F,* and *M* rays all extend back to the top of the virtual image, which is upright and reduced in size.

The behavior of a convex lens is more interesting in that the type of image it forms depends on the location of the object. For example, if the object is placed beyond the focal point, as in **FIGURE 26-47 (a)**, the image is inverted, on the opposite side of the lens, and light passes through it—it is a real image. An example of images formed by convex lenses (dew drops in this case) is shown in **FIGURE 26-48**. If the object is placed between the lens and the focal point, as in **FIGURE 26-47 (b)**, the result is an image that is virtual (on the same side as the object) and upright.

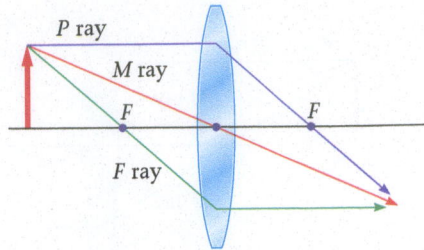

▲ **FIGURE 26-44 The three principal rays used for ray tracing with convex lenses** The image formed by a convex lens can be found by using the rays shown here. The *P* ray propagates parallel to the principal axis until it encounters the lens, where it is refracted to pass through the focal point on the far side of the lens. The *F* ray passes through the focal point on the near side of the lens, then leaves the lens parallel to the principal axis. The *M* ray passes through the middle of the lens with no deflection. Note that in each case we consider the lens to be of negligible thickness (thin-lens approximation). Therefore, the *P* and *F* rays are depicted as undergoing refraction at the center of the lens, rather than at its front and back surfaces, and the *M* ray has zero displacement. This convention is adopted in all subsequent figures.

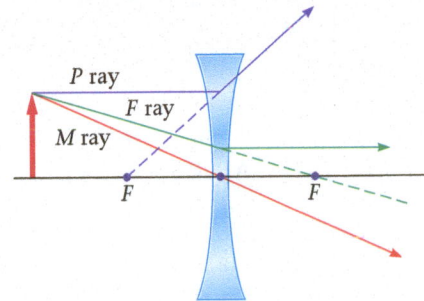

▲ **FIGURE 26-45 The three principal rays used for ray tracing with concave lenses**

 Partially Covering a Lens

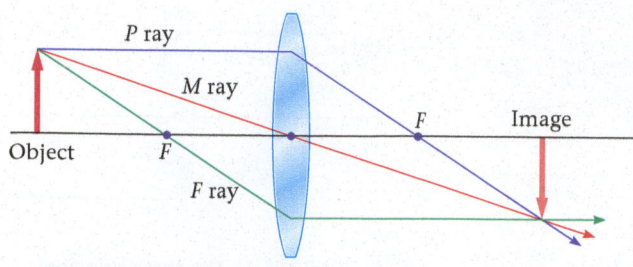

(a) Object beyond focal point *F*

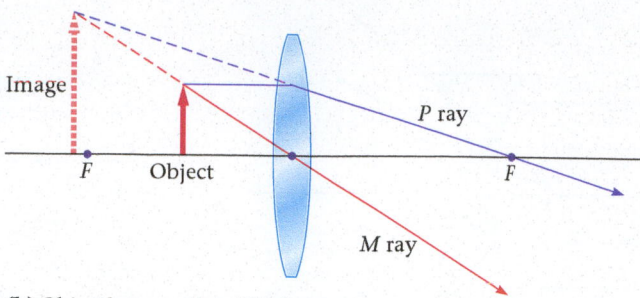

(b) Object between *F* and the lens

▲ **FIGURE 26-47 Ray tracing for a convex lens (a)** The object is beyond the focal point. The image in this case is real and inverted. **(b)** The object is between the lens and the focal point. In this case the image is virtual, upright, and enlarged.

▲ **FIGURE 26-48** Drops of dew can serve as double convex lenses, producing tiny, inverted images of objects beyond their focal points.

The imaging characteristics of concave and convex lenses are summarized in Table 26-3. Compare with Table 26-1, which presents the same information for mirrors.

The location of the focal point depends on the index of refraction of the lens, as well as that of the surrounding medium. The effect of the surrounding medium is considered in the next Conceptual Example.

TABLE 26-3 Imaging Characteristics of Concave and Convex Lenses

CONCAVE LENS Object location	Image orientation	Image size	Image type
Arbitrary	Upright	Reduced	Virtual
CONVEX LENS Object location	Image orientation	Image size	Image type
Beyond *F*	Inverted	Reduced or enlarged	Real
Just beyond *F*	Inverted	Approaching infinity	Real
Just inside *F*	Upright	Approaching infinity	Virtual
Between lens and *F*	Upright	Enlarged	Virtual

CONCEPTUAL EXAMPLE 26-17   A LENS IN WATER   PREDICT/EXPLAIN

The lens shown in the sketch on the left is generally used in air. **(a)** If it is placed in water instead, does its focal length increase, decrease, or stay the same? **(b)** Which of the following is the *best explanation* for your prediction?

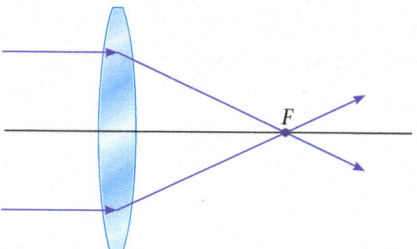

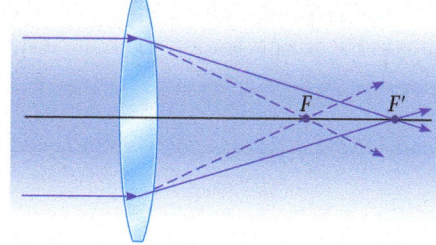

I. The focal length increases because the lens has an index of refraction close to that of water, and hence it bends light less in water than it does in air.

II. The focal length decreases because the index of refraction of water is greater than the index of refraction of air, and this results in more refraction and a smaller focal length.

III. The lens is the same, so the focal length is the same as well.

REASONING AND DISCUSSION

In water, the difference in index of refraction between the lens and its surroundings is less than when it is in air. Therefore, recalling the discussion following Exercise 26-12, we conclude that light is bent less by the lens when it is in water, as illustrated in the sketch on the right.

As a result, the focal length of the lens increases.

BIO This explains why our vision is so affected when we immerse our eyes in water—the focusing ability of the eye is greatly altered by the water, as we can see from the sketch. On the other hand, if we wear goggles, so that our eyes are still in contact with air, our vision is normal.

ANSWER

(a) The focal length increases. **(b)** The best explanation is I.

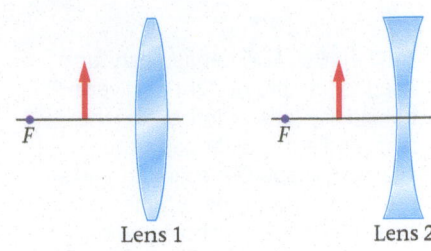

Enhance Your Understanding   (Answers given at the end of the chapter)

6. The lenses shown in **FIGURE 26-49** have objects that are identical in size and location relative to the lens. Is the image produced by lens 1 larger than, smaller than, or the same size as the image formed by lens 2? Explain.

Section Review

- The approximate location, orientation, and size of the image formed by a lens can be determined by ray tracing, similar to the ray tracing used for mirrors.

▲ **FIGURE 26-49** Enhance Your Understanding 6.

26-7 The Thin-Lens Equation

To calculate the precise location and size of the image formed by a lens, we use an equation that is analogous to the mirror equation. This equation can be derived by referring to **FIGURE 26-50**, which shows the image produced by a convex lens, along with the P and M rays that locate the image.

First, notice that the P ray creates two similar green-shaded triangles on the right side of the lens in Figure 26-50 (a). Because the triangles are similar, it follows that

$$\frac{h_o}{f} = \frac{-h_i}{d_i - f} \qquad 26\text{-}14$$

In this expression, f is the **focal length**—that is, the distance from the lens to the focal point, F—and we use $-h_i$ on the right side of the equation, because h_i is negative for an inverted image. Next, the M ray forms another pair of similar triangles, shown with yellow shading in Figure 26-50 (b), from which we obtain the following:

$$\frac{h_o}{d_o} = \frac{-h_i}{d_i} \qquad 26\text{-}15$$

Combining these two relationships, we obtain a result known as the thin-lens equation:

Thin-Lens Equation
$$\frac{1}{d_o} + \frac{1}{d_i} = \frac{1}{f} \qquad 26\text{-}16$$

Finally, the magnification, m, of the image is defined in the same way as for a mirror:

$$h_i = mh_o \qquad 26\text{-}17$$

Rearranging Equation 26-15 ($h_o/d_o = -h_i/d_i$), we find that $h_i = -(d_i/d_o)h_o$. Therefore, the magnification for a lens, just as for a mirror, is:

Magnification, m
$$m = -\frac{d_i}{d_o} \qquad 26\text{-}18$$

As before, the sign of the magnification indicates the orientation of the image, and the magnitude gives the amount by which its size is enlarged or reduced compared with the object.

The sign conventions for lenses can be summarized as follows:

Focal Length

   f is positive for converging (convex) lenses.
   f is negative for diverging (concave) lenses.

Magnification

   m is positive for upright images (same orientation as object).
   m is negative for inverted images (opposite orientation of object).

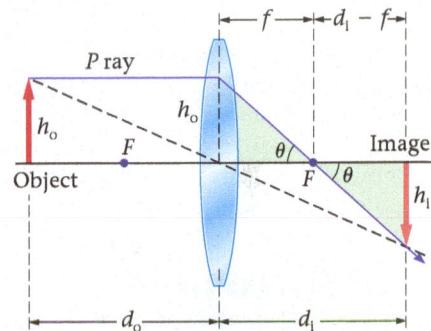

(a) Triangles to derive Equation 26-14

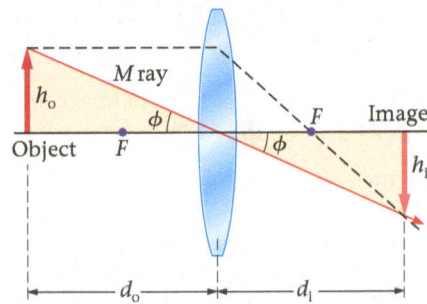

(b) Triangles to derive Equation 26-15

▲ **FIGURE 26-50 Ray diagrams used to derive the thin-lens equation (a)** The two similar triangles in this case are used to obtain Equation 26-14. **(b)** These similar triangles yield Equation 26-15.

PROBLEM-SOLVING NOTE

Applying the Thin-Lens Equation

To use the thin-lens equation correctly, be careful to use the appropriate signs for all the known quantities. The final answer will also have a sign, which gives additional information about the system.

Big Idea 6 Lenses can form images that are upright or inverted, behind the lens or in front of the lens, and either enlarged or reduced in size. The specific characteristics of the image can be determined by ray tracing or the thin-lens equation.

Image Distance

d_i is positive for real images (images on the opposite side of the lens from the object).

d_i is negative for virtual images (images on the same side of the lens as the object).

Object Distance

d_o is positive for real objects (from which light diverges).

d_o is negative for virtual objects (toward which light converges).

Examples of virtual objects will be presented in Chapter 27.

We now apply the thin-lens equation and the definition of magnification to typical lens systems.

EXAMPLE 26-18 — OBJECT DISTANCE AND FOCAL LENGTH

A lens produces a real image that is twice as large as the object, inverted, and located 15 cm from the lens. Find **(a)** the object distance and **(b)** the focal length of the lens.

PICTURE THE PROBLEM

Because the image is *real*, the lens must be convex, and the object must be outside the focal point, as we indicate in our sketch. [Compare with Figure 26-47 (a).] In addition, the image is inverted (negative magnification), which means the magnification is $m = -2$. Finally, the distance to the real image is given as $d_i = 15$ cm.

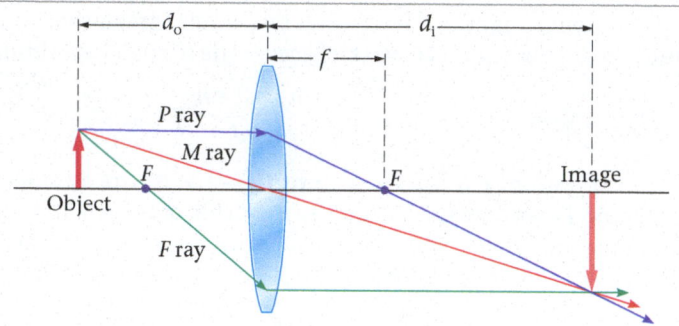

REASONING AND STRATEGY

To find both d_o and f requires two independent relationships. One is provided by the magnification, the other by the thin-lens equation.

a. We can use the magnification, $m = -d_i/d_o$, to find the object distance, d_o. As noted before, $m = -2$ in this case.

b. We now use the values of d_i and d_o in the thin-lens equation, $1/d_o + 1/d_i = 1/f$, to find the focal length, f.

Known   Magnification of image, $m = -2$; image distance, $d_i = 15$ cm.
Unknown   **(a)** Object distance, $d_o = ?$ **(b)** Focal length, $f = ?$

SOLUTION

Part (a)

1. Use $m = -d_i/d_o$ to find the object distance, d_o:

$$m = -\frac{d_i}{d_o} = -2$$

$$d_o = -\frac{d_i}{m} = -\left(\frac{15 \text{ cm}}{-2}\right) = 7.5 \text{ cm}$$

Part (b)

2. Use the thin-lens equation to find $1/f$:

$$\frac{1}{f} = \frac{1}{d_o} + \frac{1}{d_i} = \frac{1}{7.5 \text{ cm}} + \frac{1}{15 \text{ cm}} = \frac{1}{5.0 \text{ cm}}$$

3. Invert $1/f$ to find the focal length, f:

$$f = \left(\frac{1}{5.0 \text{ cm}}\right)^{-1} = 5.0 \text{ cm}$$

INSIGHT

As expected for a convex lens, the focal length is positive. In addition, notice that the object distance is greater than the focal length, in agreement with both Figure 26-47 (a) and our sketch for this Example. Finally, the magnification produced by this lens is not always −2. In fact, it depends on the precise location of the object, as we see in the following Practice Problem.

PRACTICE PROBLEM — PREDICT/CALCULATE

Suppose we would like to have a magnification of −3 using this same lens. **(a)** Should the object be moved closer to the lens or farther from it? Explain. **(b)** Find the object and image distances for this case.   [**Answer: (a)** The object should be moved closer to the lens. This moves the image farther from the lens and makes it larger. **(b)** We find $d_o = 6.67$ cm (which is less than 7.5 cm, as expected) and $d_i = 3d_o = 20.0$ cm.]

Some related homework problems: problem 70, problem 73, problem 99

QUICK EXAMPLE 26-19   FIND THE MAGNIFICATION

An object is placed 12 cm in front of a diverging lens with a focal length of −7.9 cm. Find (a) the image distance and (b) the magnification.

REASONING AND SOLUTION

(a) Given the focal length and object distance, we can use the thin-lens equation to find the image distance. (b) Once the image and object distances are known, we can use them to find the magnification.

a. Use the thin-lens equation to find the image distance, d_i:

$$\frac{1}{d_i} = \frac{1}{f} - \frac{1}{d_o} = \frac{1}{(-7.9 \text{ cm})} - \frac{1}{12 \text{ cm}} = -0.21 \text{ cm}^{-1}$$

$$d_i = \frac{1}{-0.21 \text{ cm}^{-1}} = -4.8 \text{ cm}$$

b. Use $m = -d_i/d_o$ to find the magnification:

$$m = -\frac{d_i}{d_o} = -\frac{(-4.8 \text{ cm})}{12 \text{ cm}} = 0.40$$

Because the image distance is negative, it follows that the image is virtual and on the same side of the lens as the object, as expected for a concave (diverging) lens. In addition, the fact that the magnification is positive means that the image is upright. These numerical values correspond to the system illustrated in Figure 26-46.

Enhance Your Understanding   (Answers given at the end of the chapter)

7. An object at the distance d_o = 15 cm from a lens produces an inverted image. Is the focal length of the lens greater than, less than, or equal to 15 cm? Explain.

Section Review

- The thin-lens equation relates the image distance, object distance, and focal length for a thin lens. The relationship is the same as that for a mirror—namely, $1/d_o + 1/d_i = 1/f$.

- The magnification m for a thin lens obeys the same relationship as for a mirror—that is, $m = -d_i/d_o$.

26-8 Dispersion and the Rainbow

As discussed earlier in this chapter, different materials—such as air, water, and glass—have different indices of refraction. There is more to the story, however. The index of refraction for a given material also depends on the frequency of the light; in general, the higher the frequency, the higher the index of refraction. This means, for example, that violet light—with its high frequency—has a larger index of refraction, and bends more when refracted, than does red light. The result is that white light, with its mixture of frequencies, is spread out by refraction, so that different colors travel in different directions. This "spreading out" of light according to color is known as **dispersion.**

Big Idea 7 Light of different frequency refracts, or bends, by different amounts. This disperses the colors contained in the light, and is responsible for the spectrum of colors seen in a rainbow.

EXAMPLE 26-20   PRISMATICS

A flint-glass prism is made in the shape of a 30°-60°-90° triangle, as shown in the sketch. Red and violet light are incident on the prism at right angles to its vertical side. Given that the index of refraction of flint glass is 1.66 for red light and 1.70 for violet light, find the angle each ray makes with the horizontal when it emerges from the prism.

PICTURE THE PROBLEM

The prism and the red and violet rays are shown in our sketch. The angle of incidence on the vertical side of the prism is 0°; hence, the angle of refraction is also 0° for both rays. On the slanted side of the prism, the rays have an angle of incidence equal to 30.0°. Their angles of refraction are different, however.

CONTINUED

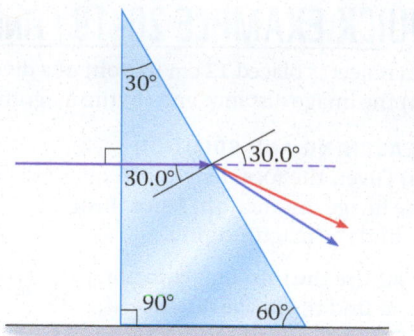

REASONING AND STRATEGY

To find the final angle for each ray, we apply Snell's law with the appropriate index of refraction. Notice, however, that the angle of refraction is measured relative to the normal, which itself is 30.0° above the horizontal. Therefore, the angle each ray makes with the horizontal is the angle of refraction minus 30.0°.

Known   Index of refraction for red light, $n = 1.66$; index of refraction for violet light, $n = 1.70$.

Unknown   Deflection angles for red light and violet light = ?

SOLUTION

1. Apply Snell's law with $n_1 = 1.66$, $\theta_1 = 30.0°$, and $n_2 = 1.00$ to find the angle of refraction, θ_2, for red light:

$$n_1 \sin \theta_1 = n_2 \sin \theta_2$$
$$\theta_2 = \sin^{-1}\left(\frac{n_1}{n_2} \sin \theta_1\right) = \sin^{-1}\left(\frac{1.66}{1.00} \sin 30.0°\right) = 56.1°$$

2. Subtract 30.0° to find the angle relative to the horizontal:

$$56.1° - 30.0° = 26.1°$$

3. Next, apply Snell's law with $n_1 = 1.70$, $\theta_1 = 30.0°$, and $n_2 = 1.00$ to find the angle of refraction, θ_2, for violet light:

$$\theta_2 = \sin^{-1}\left(\frac{n_1}{n_2} \sin \theta_1\right) = \sin^{-1}\left(\frac{1.70}{1.00} \sin 30.0°\right) = 58.2°$$

4. Subtract 30.0° to find the angle relative to the horizontal:

$$58.2° - 30.0° = 28.2°$$

INSIGHT

This difference in the angles (26.1° for red light and 28.2° for violet light) is the reason for the familiar "rainbow" of colors seen with a prism. It is also the cause of a common defect of simple lenses—referred to as chromatic aberration—in which different colors focus at different points. In the next chapter, we show how two or more lenses in combination can be used to correct this problem.

PRACTICE PROBLEM

If green light emerges from the prism at an angle of 27.0° below the horizontal, what is the index of refraction for this color of light?   [**Answer:** $n = 1.68$]

Some related homework problems: Problem 84, Problem 85

▲ **FIGURE 26-51** A rainbow over Isaac Newton's childhood home in the manor house of Woolsthorpe, near Grantham, Lincolnshire, England. (Note the apple tree near the right side of the house.)

PHYSICS IN CONTEXT
Looking Ahead

The discussion of dispersion and the rainbow relates directly to chromatic aberration, one of the lens aberrations discussed in Chapter 27.

RWP Perhaps the most famous and striking example of dispersion is provided by the rainbow, which is caused by the dispersion of light in droplets of rain (see **FIGURE 26-51**). The physical situation is illustrated in **FIGURE 26-52**, which shows a single drop of rain and an incident beam of sunlight. When sunlight enters the drop, it is separated into its red and violet components by dispersion, as shown. The light then reflects from the back of the drop and finally refracts and undergoes additional dispersion as it leaves the drop.

Notice that the final direction of the light is almost opposite to its incident direction, falling short by only 40° to 42°, depending on the color of the light. To be specific, violet light—which is bent the most by refraction—changes its direction of propagation by 320°, so it is only 40° away from moving in the direction of the Sun. Red light, which is not bent as much as violet light, changes its direction by 318° and hence moves in a direction that is 42° away from the Sun.

To see how the rainbow is formed in the sky, imagine standing with your back to the setting Sun, looking toward an area where rain is falling. Consider a single drop as it falls toward the ground. This drop, like all the other drops in the area, is sending out light of all colors in different directions. When the drop is at an angle of 42° above the horizontal, we see the red light coming from it, as indicated in **FIGURE 26-53**. As the drop continues to fall, its angle above the horizontal decreases. Eventually it reaches a height where its angle with the horizontal is 40°, at which point the violet light from the drop reaches our eye. In between, the drop has sent all the other colors of the rainbow to us for our enjoyment. Two reflections within a raindrop can cause a double rainbow, as seen in Figure 26-51 over Isaac Newton's boyhood home.

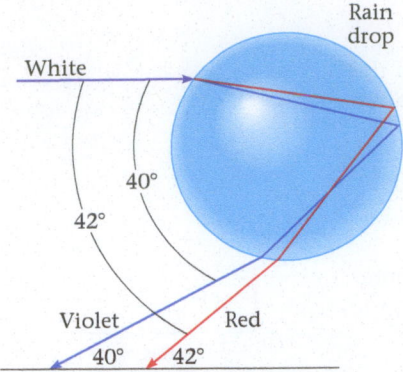

▲ **FIGURE 26-52 Dispersion in a raindrop** White light entering a raindrop is spread out by dispersion into its various color components—like red and violet. (The angles shown in this figure are exaggerated for clarity.) Rays of light that reflect twice inside the raindrop before exiting produce the "secondary" rainbow, which is above the normal, or "primary," rainbow. The sequence of colors in a secondary rainbow is reversed from that in the primary rainbow.

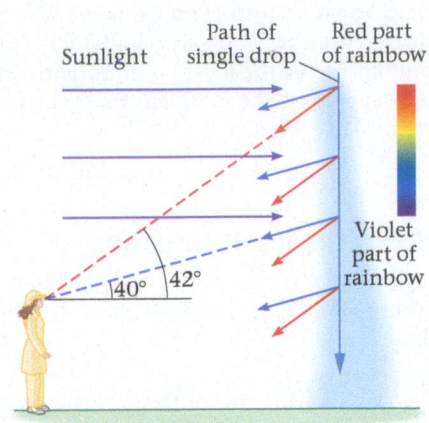

▲ **FIGURE 26-53 How rainbows are produced** As a single drop of rain falls toward the ground, it sends all the colors of the rainbow to an observer. Note that the top of the rainbow is red; the bottom is violet. (The angles in this figure have been exaggerated for clarity.)

Enhance Your Understanding   (Answers given at the end of the chapter)

8. Referring to Figure 26-53, do you expect the index of refraction in water for yellow light to be greater than, less than, or equal to that of green light? Explain.

Section Review

- Dispersion is the splitting up of white light into its component colors. A rainbow is a dramatic example of dispersion.

CHAPTER 26 REVIEW

CHAPTER SUMMARY

26-1   THE REFLECTION OF LIGHT

The direction of light can be changed by reflecting it from a shiny surface.

Law of Reflection
The law of reflection states that the angle of reflection, θ_r, is equal to the angle of incidence, θ_i; that is, $\theta_r = \theta_i$.

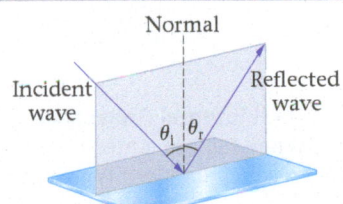

26-2   FORMING IMAGES WITH A PLANE MIRROR

The image formed by a plane mirror has the following characteristics:

- The image is upright, but appears reversed right to left.
- The image appears to be the same distance behind the mirror that the object is in front of the mirror.
- The image is the same size as the object.

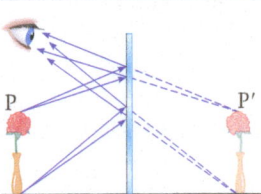

26-3   SPHERICAL MIRRORS

Focal Point and Focal Length for a Convex Mirror
A convex mirror reflects rays that are parallel to its principal axis so that they diverge, as if they had originated from a focal point, F, behind the mirror. The focal length of a convex mirror is

$$f = -\frac{1}{2}R \qquad\qquad 26\text{-}2$$

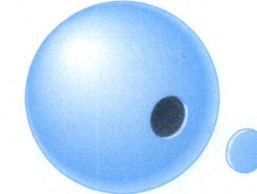

Focal Point and Focal Length for a Concave Mirror

A concave mirror reflects rays that are parallel to its principal axis so that they pass through a point known as the focal point, F. The distance from the surface of the mirror to the focal point is the focal length, f. The focal length for a mirror with a radius of curvature R is

$$f = \tfrac{1}{2} R \qquad \text{26-3}$$

26-4   RAY TRACING AND THE MIRROR EQUATION

The location, size, and orientation of an image produced by a mirror can be found qualitatively using a ray diagram or quantitatively using a relationship known as the mirror equation.

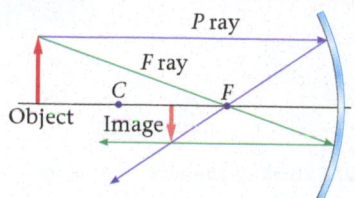

Ray Tracing

Ray tracing involves drawing two or three of the rays that have particularly simple behavior. These rays originate at a point on the object and intersect at the corresponding point on the image.

Real/Virtual Images

An image is said to be real if light passes through the apparent position of the image itself; it is virtual if light does not pass through the image.

Mirror Equation

The mirror equation relates the object distance, d_o, image distance, d_i, and focal length, f:

$$\frac{1}{d_o} + \frac{1}{d_i} = \frac{1}{f} \qquad \text{26-6}$$

Magnification

The magnification of an image is

$$m = -\frac{d_i}{d_o} \qquad \text{26-8}$$

26-5   THE REFRACTION OF LIGHT

Refraction is the change in the direction of light due to a change in its speed.

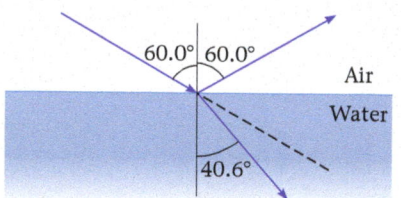

Index of Refraction

The index of refraction, n, quantifies how much a medium slows the speed of light. In particular, the speed of light in a medium is

$$v = \frac{c}{n} \qquad \text{26-10}$$

Snell's Law

Snell's law relates the index of refraction and angle of incidence in one medium (n_1, θ_1) to the index of refraction and angle of refraction in another medium (n_2, θ_2):

$$n_1 \sin \theta_1 = n_2 \sin \theta_2 \qquad \text{26-11}$$

Total Internal Reflection

When light in a medium in which its speed is relatively low encounters a medium in which its speed is greater, the light will be totally reflected back into its original medium if its angle of incidence exceeds the critical angle, θ_c, given by

$$\sin \theta_c = \frac{n_2}{n_1} \qquad \text{26-12}$$

Total Polarization

Reflected light is totally polarized parallel to the surface when the reflected and refracted rays are at right angles. This condition occurs at Brewster's angle, θ_B, given by

$$\tan \theta_B = \frac{n_2}{n_1} \qquad \text{26-13}$$

26-6   RAY TRACING FOR LENSES

As with mirrors, ray tracing is a convenient way to determine the qualitative features of an image formed by a lens.

Lens

A lens is an object that uses refraction to bend light and form images.

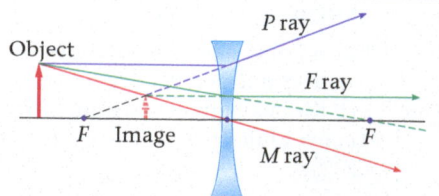

26-7 THE THIN-LENS EQUATION

The precise location of an image formed by a lens can be obtained using the thin-lens equation, which has the same form as the mirror equation.

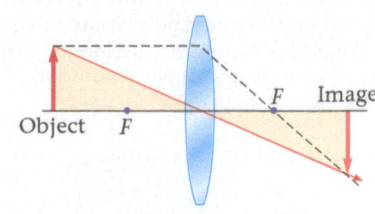

Thin-Lens Equation

The thin-lens equation relates the object distance, d_o, the image distance, d_i, and the focal length, f, for a lens:

$$\frac{1}{d_o} + \frac{1}{d_i} = \frac{1}{f} \qquad\qquad 26\text{-}16$$

Magnification

The magnification, m, of an image formed by a lens is given by the same expression used for the images produced by mirrors:

$$m = -\frac{d_i}{d_o} \qquad\qquad 26\text{-}18$$

26-8 DISPERSION AND THE RAINBOW

The index of refraction depends on frequency—generally, the higher the frequency, the higher the index of refraction. This difference in index of refraction causes light of different colors to be refracted in different directions (dispersion). Rainbows are caused by dispersion within raindrops.

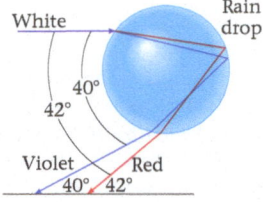

ANSWERS TO ENHANCE YOUR UNDERSTANDING QUESTIONS

1. The angle of incidence is the angle from the ray to the normal to the surface. In this case, the angle of incidence is $90° − 25° = 65°$.

2. The mirror image of the 25 cm mark is $75\text{ cm} + 40\text{ cm} = 115\text{ cm}$ behind the mirror. The 50 cm mark on the meterstick is $50\text{ cm} + 40\text{ cm} = 90\text{ cm}$ in front of the mirror. Therefore, the distance between these locations is $90\text{ cm} + 115\text{ cm} = 205\text{ cm}$.

3. $D < B < A < C$.

4. The situation is like that pictured in Figure 26-26 (b). **(a)** The image is inverted. **(b)** The image is larger than the object.

5. **(a)** The light bends closer to the normal, because diamond has the larger index of refraction. **(b)** The light bends away from the normal, because crown glass has the smaller index of refraction.

6. The image produced by lens 1 is larger than the image produced by lens 2, because concave lenses always produce reduced images, but a convex lens produces an enlarged image when the object is between the focal point and the lens, as in this case.

7. Only a convex lens produces an inverted image, and this occurs when the object is farther from the lens than the focal point. Therefore, the focal length of the lens is less than 15 cm.

8. The index of refraction for yellow light is less than that for green light, because yellow light is closer to red light in the visible spectrum (see Figure 26-52), and red light has the smallest index of refraction of any visible light.

CONCEPTUAL QUESTIONS

For instructor-assigned homework, go to www.masteringphysics.com.

(Answers to odd-numbered Conceptual Questions can be found in the back of the book.)

1. Two plane mirrors meet at right angles at the origin, as indicated in **FIGURE 26-54**. Suppose an L-shaped object has the position and orientation labeled A. Draw the location and orientation of *all* the images of object A formed by the two mirrors.

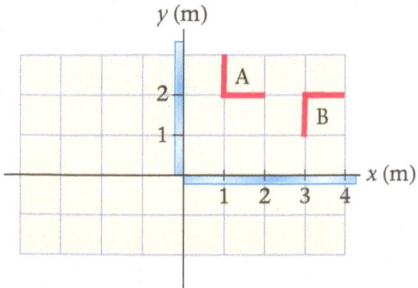

FIGURE 26-54 Conceptual Questions 1 and 2

2. Two plane mirrors meet at right angles at the origin, as indicated in Figure 26-54. Suppose an L-shaped object has the position and orientation labeled B. Draw the location and orientation of *all* the images of object B formed by the two mirrors.

3. What is the radius of curvature of a plane mirror? What is its focal length? Explain.

4. Dish receivers for satellite TV always use the concave side of the dish, never the convex side. Explain.

5. Suppose you would like to start a fire by focusing sunlight onto a piece of paper. In Conceptual Example 26-4 we saw that a concave mirror would be better than a convex mirror for this purpose. At what distance from the mirror should the paper be held for best results?

6. When light propagates from one medium to another, does it always bend toward the normal? Explain.

7. A swimmer at point B in **FIGURE 26-55** needs help. Two lifeguards depart simultaneously from their tower at point A, but they

follow different paths. Although both lifeguards run with equal speed on the sand and swim with equal speed in the water, the lifeguard who follows the longer path, ACB, arrives at point B before the lifeguard who follows the shorter, straight-line path from A to B. Explain.

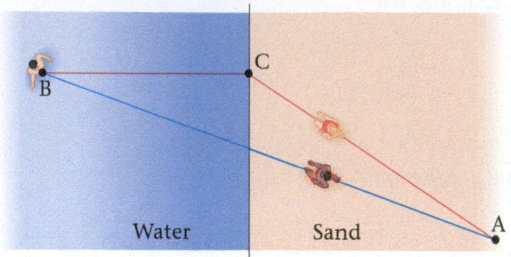

FIGURE 26-55 Conceptual Question 7

8. When you observe a mirage on a hot day, what are you actually seeing when you gaze at the "pool of water" in the distance?

9. Sitting on a deserted beach one evening, you watch as the last bit of the Sun approaches the horizon. Just before the Sun disappears from sight, is the top of the Sun actually above or below the horizon? That is, if Earth's atmosphere could be instantly removed just before the Sun disappeared, would the Sun still be visible, or would it be below the horizon? Explain.

10. **The Disappearing Eyedropper** The photograph in **FIGURE 26-56** shows eyedroppers partially immersed in oil (left) and water (right).

Explain why the dropper is invisible in the oil.

11. **The Invisible Man** In the H. G. Wells novel *The Invisible Man*, a person becomes invisible by altering his index of refraction to match that of air. This is the idea behind the disappearing eyedropper in Conceptual Question 10. If the invisible man could actually do this, would he be able to see? Explain.

12. **What's the Secret?** The top of **FIGURE 26-57** shows the words SECRET CODE written in different colors. If you place a cylindrical rod of glass or plastic just above the words, you find that SECRET appears inverted, but CODE does not. Explain.

FIGURE 26-56 What happened to the dropper? (Conceptual Questions 10 and 11)

FIGURE 26-57 Conceptual Question 12

PROBLEMS AND CONCEPTUAL EXERCISES

Answers to odd-numbered Problems and Conceptual Exercises can be found in the back of the book. **BIO** *identifies problems of biological or medical interest;* **CE** *indicates a conceptual exercise.* **Predict/Explain** *problems ask for two responses: (a) your prediction of a physical outcome, and (b) the best explanation among three provided; and* **Predict/Calculate** *problems ask for a prediction of a physical outcome, based on fundamental physics concepts, and follow that with a numerical calculation to verify the prediction. On all problems, bullets (•, ••, •••) indicate the level of difficulty.*

SECTION 26-1 THE REFLECTION OF LIGHT

1. • A laser beam is reflected by a plane mirror. It is observed that the angle between the incident and reflected beams is 28°. If the mirror is now rotated so that the angle of incidence increases by 5.0°, what is the new angle between the incident and reflected beams?

2. • The angle between the Sun and a rescue aircraft is 54°. What should be the angle of incidence for sunlight on a plane mirror so that the rescue pilot sees the reflected light?

3. •• The reflecting surfaces of two mirrors form a vertex with an angle of 110°. If a ray of light strikes mirror 1 with an angle of incidence of 45°, find the angle of reflection of the ray when it leaves mirror 2.

4. •• A ray of light reflects from a plane mirror with an angle of incidence of 27°. If the mirror is rotated by an angle θ, through what angle is the reflected ray rotated?

5. •• **Predict/Calculate** A small vertical mirror hangs on the wall, 1.40 m above the floor. Sunlight strikes the mirror, and the reflected beam forms a spot on the floor 2.50 m from the wall. Later in the day, you notice that the spot has moved to a point 3.75 m from the wall. **(a)** Were your two observations made in the morning or in the afternoon? Explain. **(b)** What was the change in the Sun's angle of elevation between your two observations?

6. •• Sunlight enters a room at an angle of 32° above the horizontal and reflects from a small mirror lying flat on the floor. The reflected light forms a spot on a wall that is 2.0 m behind the mirror, as shown in **FIGURE 26-58**. If you now place a pencil under the edge of the mirror nearer the wall, tilting it upward by 5.0°, how much higher on the wall (Δy) is the spot?

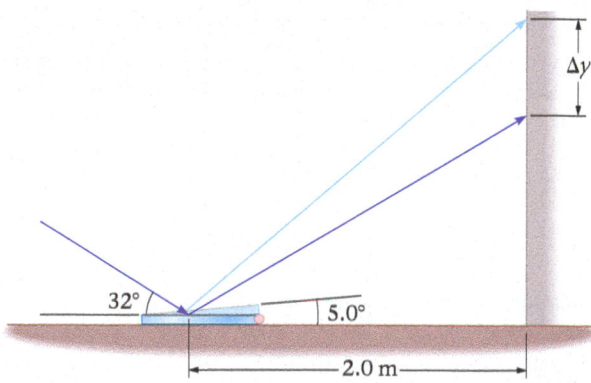

FIGURE 26-58 Problem 6

7. •• You stand 1.50 m in front of a wall and gaze downward at a small vertical mirror mounted on it. In this mirror you can see the reflection of your shoes. If your eyes are 1.85 m above your feet, through what angle should the mirror be tilted for you to see your eyes reflected in the mirror? (The location of the mirror remains the same, only its angle to the vertical is changed.)

8. •• **Predict/Calculate** Standing 2.3 m in front of a small vertical mirror, you see the reflection of your belt buckle, which is 0.72 m below your eyes. **(a)** What is the vertical location of the mirror relative to the level of your eyes? **(b)** What angle do your eyes make

with the horizontal when you look at the buckle? **(c)** If you now move backward until you are 6.0 m from the mirror, will you still see the buckle, or will you see a point on your body that is above or below the buckle? Explain.

9. •• How many times does the light beam shown in **FIGURE 26-59** reflect from **(a)** the top and **(b)** the bottom mirror?

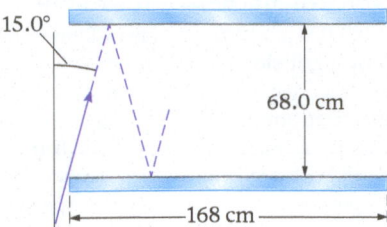

FIGURE 26-59 Problem 9

SECTION 26-2 FORMING IMAGES WITH A PLANE MIRROR

10. • **CE** If you view a clock in a mirror, as in **FIGURE 26-60**, do the hands rotate clockwise or counterclockwise?

FIGURE 26-60 Which way do the hands go? (Problem 10)

11. • A 13.5-foot-long, nearsighted python is stretched out perpendicular to a plane mirror, admiring its reflected image. If the greatest distance to which the snake can see clearly is 28.0 ft, how close must its head be to the mirror for it to see a clear image of its tail?

12. •• **(a)** How rapidly does the distance between you and your mirror image decrease if you walk directly toward a mirror with a speed of 2.9 m/s? **(b)** Repeat part (a) for the case in which you walk toward a mirror but at an angle of 32° to its normal.

13. •• You are 1.8 m tall and stand 2.8 m from a plane mirror that extends vertically upward from the floor. On the floor 1.0 m in front of the mirror is a small table, 0.80 m high. What is the minimum height the mirror must have for you to be able to see the top of the table in the mirror?

14. •• The rear window in a car is approximately a rectangle, 1.7 m wide and 0.45 m high. The inside rearview mirror is 0.50 m from the driver's eyes, and 1.90 m from the rear window. What are the minimum dimensions for the rearview mirror if the driver is to be able to see the entire width and height of the rear window in the mirror without moving her head?

15. •• **Predict/Calculate** You hold a small plane mirror 0.50 m in front of your eyes, as shown in **FIGURE 26-61** (not to scale). The mirror is 0.32 m high, and in it you see the image of a tall building behind you. **(a)** If the building is 95 m behind you, what vertical height of the building, *H*, can be seen in the mirror at any one time? **(b)** If you move the mirror closer to your eyes, does your answer to part (a) increase, decrease, or stay the same? Explain.

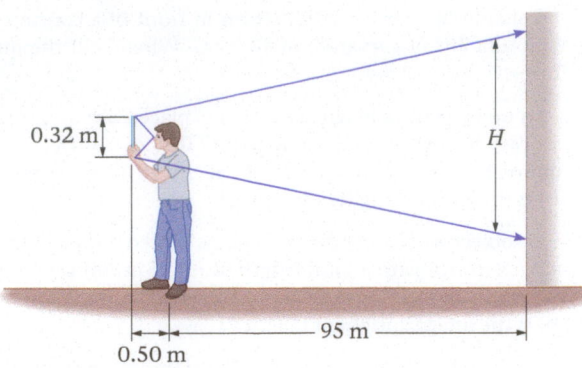

FIGURE 26-61 Problem 15

16. •• Two rays of light converge toward each other, as shown in **FIGURE 26-62**, forming an angle of 27°. Before they intersect, however, they are reflected from a circular plane mirror with a diameter of 11 cm. If the mirror can be moved horizontally to the left or right, what is the greatest possible distance *d* from the mirror to the point where the reflected rays meet?

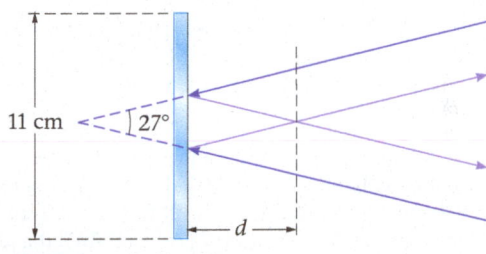

FIGURE 26-62 Problem 16

SECTION 26-3 SPHERICAL MIRRORS

17. • **CE** Astronomers often use large mirrors in their telescopes to gather as much light as possible from faint distant objects. Should the mirror in their telescopes be concave or convex? Explain.

18. • A section of a sphere has a radius of curvature of 0.72 m. If this section is painted with a reflective coating on both sides, what is the focal length of **(a)** the convex side and **(b)** the concave side?

19. • A mirrored-glass gazing globe in a garden is 31.9 cm in diameter. What is the focal length of the globe?

20. • Sunlight reflects from a concave piece of broken glass, converging to a point 18 cm from the glass. What is the radius of curvature of the glass?

SECTION 26-4 RAY TRACING AND THE MIRROR EQUATION

21. • **CE** You hold a shiny tablespoon at arm's length and look at the back side of the spoon. **(a)** Is the image you see of yourself upright or inverted? **(b)** Is the image enlarged or reduced? **(c)** Is the image real or virtual?

22. • **CE** You hold a shiny tablespoon at arm's length and look at the front side of the spoon. **(a)** Is the image you see of yourself upright or inverted? **(b)** Is the image enlarged or reduced? **(c)** Is the image real or virtual?

23. • **CE** An object is placed to the left of a concave mirror, beyond its focal point. In which direction will the image move when the object is moved farther to the left?

24. • **CE** An object is placed to the left of a convex mirror. In which direction will the image move when the object is moved farther to the left?

25. • A small object is located 36.0 cm in front of a concave mirror with a radius of curvature of 48.0 cm. Where will the image be formed?

26. • An object with a height of 33 cm is placed 2.0 m in front of a concave mirror with a focal length of 0.75 m. **(a)** Determine the approximate location and size of the image using a ray diagram. **(b)** Is the image upright or inverted?

27. • An object with a height of 33 cm is placed 2.0 m in front of a convex mirror with a focal length of −0.75 m. **(a)** Determine the approximate location and size of the image using a ray diagram. **(b)** Is the image upright or inverted?

28. •• Find the location and magnification of the image produced by the mirror in Problem 26 using the mirror and magnification equations.

29. •• Find the location and magnification of the image produced by the mirror in Problem 27 using the mirror and magnification equations.

30. •• During a daytime football game you notice that a player's reflective helmet forms an image of the Sun 4.8 cm behind the surface of the helmet. What is the radius of curvature of the helmet, assuming it to be roughly spherical?

31. •• A convex mirror on the passenger side of a car produces an image of a vehicle that is 18.3 m from the mirror. If the image is located 55.4 cm behind the mirror, what is the radius of curvature of the mirror?

32. •• **Predict/Calculate** A magician wishes to create the illusion of a 2.74-m-tall elephant. He plans to do this by forming a virtual image of a 50.0-cm-tall model elephant with the help of a spherical mirror. **(a)** Should the mirror be concave or convex? **(b)** If the model must be placed 3.00 m from the mirror, what radius of curvature is needed? **(c)** How far from the mirror will the image be formed?

33. •• A person 1.8 m tall stands 0.86 m from a reflecting globe in a garden. **(a)** If the diameter of the globe is 18 cm, where is the image of the person, relative to the surface of the globe? **(b)** How large is the person's image?

34. •• Shaving/makeup mirrors typically have one flat and one concave (magnifying) surface. You find that you can project a magnified image of a lightbulb onto the wall of your bathroom if you hold the mirror 1.8 m from the bulb and 3.5 m from the wall. **(a)** What is the magnification of the image? **(b)** Is the image erect or inverted? **(c)** What is the focal length of the mirror?

35. •• **The Hale Telescope** The 200-inch-diameter concave mirror of the Hale telescope on Mount Palomar has a focal length of 16.9 m. An astronomer stands 20.0 m in front of this mirror. **(a)** Where is her image located? Is it in front of or behind the mirror? **(b)** Is her image real or virtual? How do you know? **(c)** What is the magnification of her image?

36. •• A concave mirror produces a virtual image that is three times as tall as the object. **(a)** If the object is 16 cm in front of the mirror, what is the image distance? **(b)** What is the focal length of this mirror?

37. •• A concave mirror produces a real image that is three times as large as the object. **(a)** If the object is 22 cm in front of the mirror, what is the image distance? **(b)** What is the focal length of this mirror?

38. •• The virtual image produced by a convex mirror is one-third the size of the object. **(a)** If the object is 27 cm in front of the mirror, what is the image distance? **(b)** What is the focal length of this mirror?

39. •• You view a nearby tree in a concave mirror. The inverted image of the tree is 4.8 cm high and is located 7.0 cm in front of the mirror. If the tree is 28 m from the mirror, what is its height?

40. ••• A shaving/makeup mirror produces an erect image that is magnified by a factor of 2.2 when your face is 25 cm from the mirror. What is the mirror's radius of curvature?

41. ••• A concave mirror with a focal length of 36 cm produces an image whose distance from the mirror is one-third the object distance. Find the object and image distances.

SECTION 26-5 THE REFRACTION OF LIGHT

42. • **CE Predict/Explain** When a ray of light enters a glass lens surrounded by air, it slows down. **(a)** As it leaves the glass, does its speed increase, decrease, or stay the same? **(b)** Choose the *best explanation* from among the following:
  I. Its speed increases because the ray is now propagating in a medium with a smaller index of refraction.
  II. The speed decreases because the speed of light decreases whenever light moves from one medium to another.
  III. The speed will stay the same because the speed of light is a universal constant.

43. • **CE Samurai Fishing** A humorous scene in Akira Kurosawa's classic film *The Seven Samurai* shows the young samurai Kikuchiyo wading into a small stream and plucking a fish from it for his dinner. **(a)** As Kikuchiyo looks through the water to the fish, does he see it in the general direction of point 1 or point 2 in **FIGURE 26-63**? **(b)** If the fish looks up at Kikuchiyo, does it see Kikuchiyo's head in the general direction of point 3 or point 4?

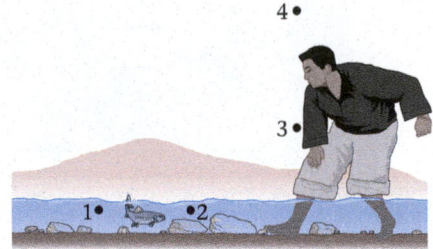

FIGURE 26-63 Problem 43

44. • **CE Day Versus Night** **(a)** Imagine for a moment that the Earth has no atmosphere. Over the period of a year, is the number of daylight hours at your home greater than, less than, or equal to the number of nighttime hours? **(b)** Repeat part (a), only this time take into account the Earth's atmosphere.

45. • **CE Predict/Explain** A kitchen has twin side-by-side sinks. One sink is filled with water, the other is empty. **(a)** Does the sink with water appear to be deeper, shallower, or the same depth as the empty sink? **(b)** Choose the *best explanation* from among the following:
  I. The sink with water appears deeper because you have to look through the water to see the bottom.
  II. Water bends the light, making an object under the water appear to be closer to the surface. Thus the water-filled sink appears shallower.
  III. The sinks are identical, and therefore have the same depth. This doesn't change by putting water in one of them.

46. • Light travels a distance of 0.902 m in 4.00 ns in a given substance. What is the index of refraction of this substance?

47. • Light enters a container of benzene at an angle of 43° to the normal; the refracted beam makes an angle of 27° with the normal. Calculate the index of refraction of benzene.

48. • The angle of refraction of a ray of light traveling into an ice cube from air is 46°. Find the angle of incidence.

49. •• **Ptolemy's Optics** One of the many works published by the Greek astronomer Ptolemy (A.D. ca. 100–170) was *Optics*. In this book

Ptolemy reports the results of refraction experiments he conducted by observing light passing from air into water. His results are as follows: angle of incidence = 10.0°, angle of refraction = 8.00°; angle of incidence = 20.0°, angle of refraction = 15.5°. Find the percentage error in the calculated index of refraction of water for each of Ptolemy's measurements.

50. •• A submerged scuba diver looks up toward the calm surface of a freshwater lake and notes that the Sun appears to be 35° from the vertical. The diver's friend is standing on the shore of the lake. At what angle above the horizon does the friend see the sun?

51. •• A pond with a total depth (ice + water) of 3.25 m is covered by a transparent layer of ice, with a thickness of 0.38 m. Find the time required for light to travel vertically from the surface of the ice to the bottom of the pond.

52. •• Light is refracted as it travels from a point A in medium 1 to a point B in medium 2. If the index of refraction is 1.33 in medium 1 and 1.51 in medium 2, how much time does it take for light to go from A to B, assuming it travels 331 cm in medium 1 and 151 cm in medium 2?

53. •• You have a semicircular disk of glass with an index of refraction of $n = 1.52$. Find the incident angle θ for which the beam of light in **FIGURE 26-64** will hit the indicated point on the screen.

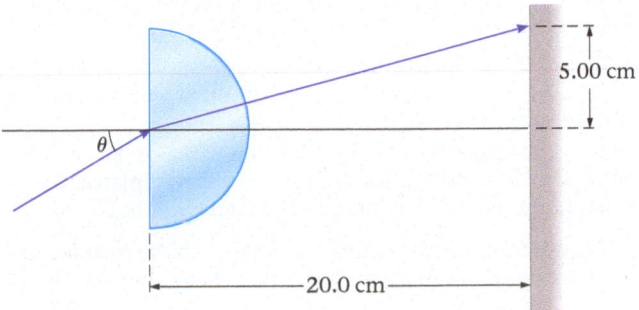

FIGURE 26-64 Problem 53x

54. •• The observer in **FIGURE 26-65** is positioned so that the far edge of the bottom of the empty glass (not to scale) is just visible. When the glass is filled to the top with water, the center of the bottom of the glass is just visible to the observer. Find the height, H, of the glass, given that its width is $W = 5.7$ cm.

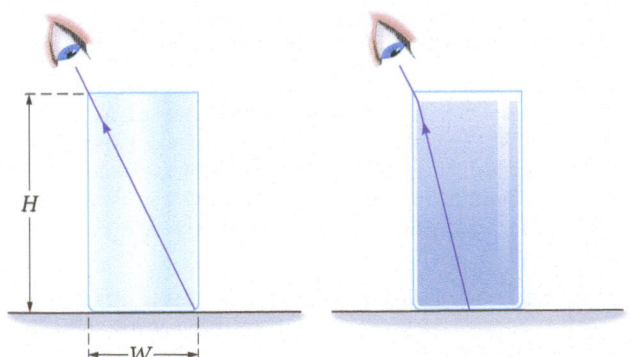

FIGURE 26-65 Problem 54

55. •• A coin is lying at the bottom of a pool of water that is 5.5 feet deep. Viewed from directly above the coin, how far below the surface of the water does the coin appear to be? (The coin is assumed to be small in diameter; therefore, we can use the small-angle approximations $\sin \theta \cong \tan \theta \cong \theta$.)

56. •• A ray of light enters the long side of a 45°-90°-45° prism and undergoes two total internal reflections, as indicated in **FIGURE 26-66**.

The result is a reversal in the ray's direction of propagation. Find the minimum value of the prism's index of refraction, n, for these internal reflections to be total.

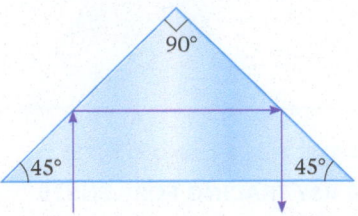

FIGURE 26-66 Problem 56

57. •• **Predict/Calculate** A glass paperweight with an index of refraction n rests on a desk, as shown in **FIGURE 26-67**. An incident ray of light enters the horizontal top surface of the paperweight at an angle $\theta = 77°$ to the vertical. **(a)** Find the minimum value of n for which the internal reflection on the vertical surface of the paperweight is total. **(b)** If θ is decreased, is the minimum value of n increased or decreased? Explain.

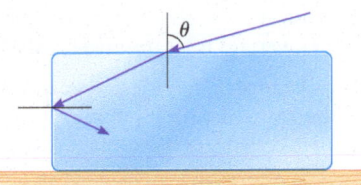

FIGURE 26-67 Problems 57, 58, and 82

58. •• **Predict/Calculate** Suppose the glass paperweight in Figure 26-67 has an index of refraction $n = 1.38$. **(a)** Find the value of θ for which the reflection on the vertical surface of the paperweight exactly satisfies the condition for total internal reflection. **(b)** If θ is increased, is the reflection at the vertical surface still total? Explain.

59. •• While studying physics at the library late one night, you notice the image of the desk lamp reflected from the varnished tabletop. When you turn your Polaroid sunglasses sideways, the reflected image disappears. If this occurs when the angle between the incident and reflected rays is 110°, what is the index of refraction of the varnish?

60. ••• A horizontal beam of light enters a 45°-90°-45° prism at the center of its long side, as shown in **FIGURE 26-68**. The emerging ray moves in a direction that is 34° below the horizontal. What is the index of refraction of this prism?

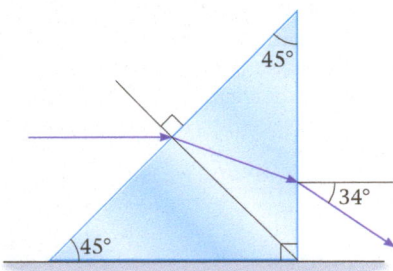

FIGURE 26-68 Problems 60 and 106

61. ••• A laser beam enters one of the sloping faces of the equilateral glass prism ($n = 1.42$) in **FIGURE 26-69** and refracts through the prism. Within the prism the light travels horizontally. What is the angle θ between the direction of the incident ray and the direction of the outgoing ray?

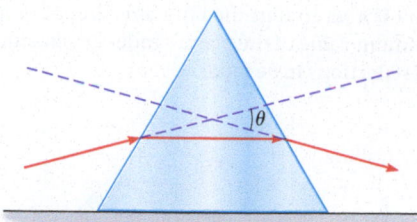

FIGURE 26-69 Problems 61 and 85

SECTION 26-6 RAY TRACING FOR LENSES

62. • **(a)** Use a ray diagram to determine the approximate location of the image produced by a concave lens when the object is at a distance $\frac{1}{2}|f|$ from the lens. **(b)** Is the image upright or inverted? **(c)** Is the image real or virtual? Explain.

63. • **(a)** Use a ray diagram to determine the approximate location of the image produced by a concave lens when the object is at a distance $2|f|$ from the lens. **(b)** Is the image upright or inverted? **(c)** Is the image real or virtual? Explain.

64. • An object is a distance $f/2$ from a convex lens. **(a)** Use a ray diagram to find the approximate location of the image. **(b)** Is the image upright or inverted? **(c)** Is the image real or virtual? Explain.

65. • An object is a distance $2f$ from a convex lens. **(a)** Use a ray diagram to find the approximate location of the image. **(b)** Is the image upright or inverted? **(c)** Is the image real or virtual? Explain.

66. •• Two lenses that are 35 cm apart are used to form an image, as shown in **FIGURE 26-70**. Lens 1 is converging and has a focal length $f_1 = 14$ cm; lens 2 is diverging and has a focal length $f_2 = -7.0$ cm. The object is placed 24 cm to the left of lens 1. **(a)** Use a ray diagram to find the approximate location of the image. **(b)** Is the image upright or inverted? **(c)** Is the image real or virtual? Explain.

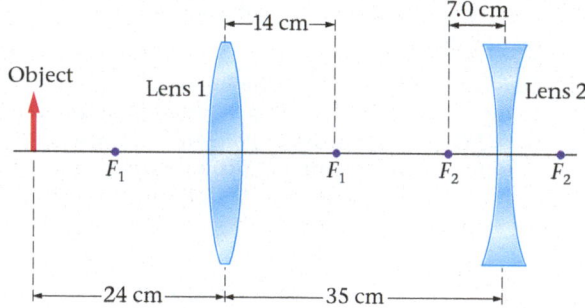

FIGURE 26-70 Problems 66 and 75

67. •• Two lenses that are 35 cm apart are used to form an image, as shown in **FIGURE 26-71**. Lens 1 is diverging and has a focal length $f_1 = -7.0$ cm; lens 2 is converging and has a focal length $f_2 = 14$ cm. The object is placed 24 cm to the left of lens 1. **(a)** Use a ray diagram to find the approximate location of the image. **(b)** Is the image upright or inverted? **(c)** Is the image real or virtual? Explain.

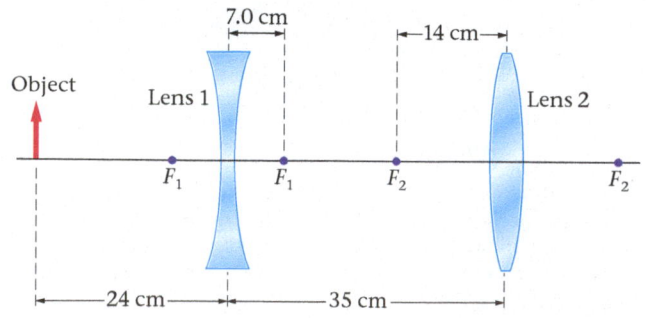

FIGURE 26-71 Problems 67 and 76

SECTION 26-7 THE THIN-LENS EQUATION

68. • A convex lens is held over a piece of paper outdoors on a sunny day. When the paper is held 26 cm below the lens, the sunlight is focused on the paper and the paper ignites. What is the focal length of the lens?

69. • A concave lens has a focal length of −39 cm. Find the image distance and magnification that result when an object is placed 31 cm in front of the lens.

70. • When an object is located 38 cm to the left of a lens, the image is formed 22 cm to the right of the lens. What is the focal length of the lens?

71. •• An object with a height of 2.54 cm is placed 36.3 mm to the left of a lens with a focal length of 35.0 mm. **(a)** Where is the image located? **(b)** What is the height of the image?

72. •• A lens for a digital camera has a focal length given by $f = 45.0$ mm. **(a)** How close to the image sensor should the lens be placed to form a sharp image of an object that is 5.00 m away? **(b)** What is the magnification of the image on the sensor?

73. •• **Predict/Calculate** An object is located to the left of a convex lens whose focal length is 34 cm. The magnification produced by the lens is $m = 3.0$. **(a)** To increase the magnification to 4.0, should the object be moved closer to the lens or farther away? Explain. **(b)** Calculate the distance through which the object should be moved.

74. •• **Predict/Calculate** You have two lenses at your disposal, one with a focal length $f_1 = +40.0$ cm, the other with a focal length $f_2 = -40.0$ cm. **(a)** Which of these two lenses would you use to project an image of a lightbulb onto a wall that is far away? **(b)** If you want to produce an image of the bulb that is enlarged by a factor of 2.00, how far from the wall should the lens be placed?

75. •• **(a)** Determine the distance from lens 1 to the final image for the system shown in Figure 26-70. **(b)** What is the magnification of this image?

76. •• **(a)** Determine the distance from lens 1 to the final image for the system shown in Figure 26-71. **(b)** What is the magnification of this image?

77. •• **Predict/Calculate** An object is located to the left of a concave lens whose focal length is −34 cm. The magnification produced by the lens is $m = \frac{1}{3}$. **(a)** To decrease the magnification to $m = \frac{1}{4}$, should the object be moved closer to the lens or farther away? **(b)** Calculate the distance through which the object should be moved.

78. •• **BIO Predict/Calculate** Albert is nearsighted, and without his eyeglasses he can focus only on objects less than 2.2 m away. **(a)** Are Albert's eyeglasses concave or convex? Explain. **(b)** To correct Albert's nearsightedness, his eyeglasses must produce a virtual, upright image at a distance of 2.2 m when viewing an infinitely distant object. What is the focal length of Albert's eyeglasses?

79. •• A small insect viewed through a convex lens is 1.8 cm from the lens and appears 2.5 times larger than its actual size. What is the focal length of the lens?

80. ••• **Predict/Calculate** A friend tells you that when he takes off his eyeglasses and holds them 23 cm above a printed page the image of the print is erect but reduced to 0.67 of its actual size. **(a)** Is the image real or virtual? How do you know? **(b)** What is the focal length of your friend's glasses? **(c)** Are the lenses in the glasses concave or convex? Explain.

81. ••• **Predict/Calculate** A friend tells you that when she takes off her eyeglasses and holds them 23 cm above a printed page the image of the print is erect but enlarged to 1.5 times its actual size.

(a) Is the image real or virtual? How do you know? (b) What is the focal length of your friend's glasses? (c) Are the lenses in the glasses concave or convex? Explain.

SECTION 26-8 DISPERSION AND THE RAINBOW

82. • **CE** Referring to Figure 26-52, which color, violet or red, travels more rapidly through water? Explain.

83. • **CE** Predict/Explain You take a picture of a rainbow with an infrared camera, and your friend takes a picture at the same time with visible light. (a) Is the height of the rainbow in the infrared picture greater than, less than, or the same as the height of the rainbow in the visible-light picture? (b) Choose the *best explanation* from among the following:
   I. The height will be greater because the top of a rainbow is red, and so infrared light would be even higher.
   II. The height will be less because infrared light is below the visible spectrum.
   III. A rainbow is the same whether seen in visible light or infrared; therefore the height is the same.

84. •• The index of refraction for red light in a certain liquid is 1.305; the index of refraction for violet light in the same liquid is 1.336. Find the dispersion $(\theta_v - \theta_r)$ for red and violet light when both are incident from air onto the flat surface of the liquid at an angle of 45.00° to the normal.

85. •• A horizontal incident beam consisting of white light passes through an equilateral prism, like the one shown in Figure 26-69. What is the dispersion $(\theta_v - \theta_r)$ of the outgoing beam if the prism's index of refraction is $n_v = 1.505$ for violet light and $n_r = 1.421$ for red light?

86. •• The focal length of a lens is inversely proportional to the quantity $(n - 1)$, where n is the index of refraction of the lens material. The value of n, however, depends on the wavelength of the light that passes through the lens. For example, one type of flint glass has an index of refraction of $n_r = 1.572$ for red light and $n_v = 1.605$ in violet light. Now, suppose a white object is placed 24.00 cm in front of a lens made from this type of glass. If the red light reflected from this object produces a sharp image 55.00 cm from the lens, where will the violet image be found?

GENERAL PROBLEMS

87. • **CE** Jurassic Park A *T. rex* chases the heroes of Steven Spielberg's *Jurassic Park* as they desperately try to escape in their Jeep. The *T. rex* is closing in fast, as they can see in the outside rearview mirror. Near the bottom of the mirror they also see the following helpful message: OBJECTS IN THE MIRROR ARE CLOSER THAN THEY APPEAR. Is this mirror concave or convex? Explain.

88. • **CE** Predict/Explain If a lens is immersed in water, its focal length changes, as discussed in Conceptual Example 26-17. (a) If a spherical mirror is immersed in water, does its focal length increase, decrease, or stay the same? (b) Choose the *best explanation* from among the following:
   I. The focal length will increase because the water will cause more bending of light.
   II. Water will refract the light. This, combined with the reflection due to the mirror, will result in a decreased focal length.
   III. The focal length stays the same because it depends on the fact that the angle of incidence is equal to the angle of reflection for a mirror. This is unaffected by the presence of water.

89. • **CE** Predict/Explain A glass slab surrounded by air causes a sideways displacement in a beam of light. (a) If the slab is now placed in water, does the displacement it causes increase, decrease, or stay the same? (b) Choose the *best explanation* from among the following:
   I. The displacement of the beam increases because of the increased refraction due to the water.
   II. The displacement of the beam is decreased because with water surrounding the slab there is a smaller difference in index of refraction between the slab and its surroundings.
   III. The displacement stays the same because it is determined only by the properties of the slab; in particular, the material it is made of and its thickness.

90. •• **CE** Inverse Lenses Suppose we mold a hollow piece of plastic into the shape of a double concave lens. The "lens" is watertight, and its interior is filled with air. We now place this lens in water and shine a beam of light on it. (a) Does the lens converge or diverge the beam of light? Explain. (b) If our hollow lens is double convex instead, does it converge or diverge a beam of light when immersed in water? Explain.

91. •• Standing 2.5 m in front of a small vertical mirror you see the reflection of your belt buckle, which is 0.70 m below your eyes. If you remain 2.5 m from the mirror but climb onto a stool, how high must the stool be to allow you to see your knees in the mirror? Assume that your knees are 1.2 m below your eyes.

92. •• Predict/Calculate Apparent Size of Floats in a Termometro Lentos The Galileo thermometer, or Termometro Lentos (slow thermometer in Italian), consists of a vertical, cylindrical flask containing a fluid and several glass floats of different color. The floats all have the same dimensions, but they appear to differ in size depending on their location within the cylinder. (a) Does a float near the front surface of the cylinder (the surface closest to you) appear to be larger or smaller than a float near the back surface? (b) **FIGURE 26-72** shows a ray diagram for a float near the front surface of the cylinder. Draw a ray diagram for a float at the center of the cylinder, and show that the change in apparent size agrees with your answer to part (a).

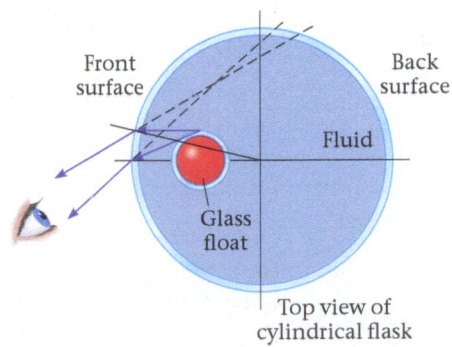

FIGURE 26-72 Problem 92

93. •• (a) Find the two locations where an object can be placed in front of a concave mirror with a radius of curvature of 46.0 cm such that its image is twice its size. (b) In each of these cases, state whether the image is real or virtual, upright or inverted.

94. •• A convex mirror with a focal length of −85 cm is used to give a truck driver a view behind the vehicle. (a) If a person who is 1.7 m tall stands 2.2 m from the mirror, where is the person's image located? (b) Is the image upright or inverted? (c) What is the size of the image?

95. •• Predict/Calculate The three laser beams shown in **FIGURE 26-73** meet at a point at the back of a solid, transparent sphere. (a) What is the index of refraction of the sphere? (b) Is there a finite index of refraction that will make the three beams come to a focus at the center of the sphere? If your answer is yes, give the required index of refraction; if your answer is no, explain why not.

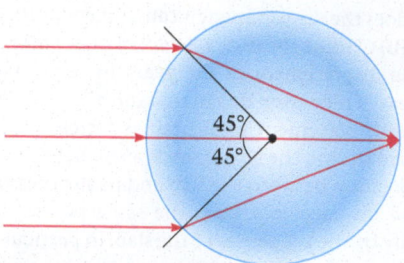

FIGURE 26-73 Problem 95

96. •• **Predict/Calculate** A film of oil, with an index of refraction of 1.48 and a thickness of 1.50 cm, floats on a pool of water, as shown in **FIGURE 26-74**. A beam of light is incident on the oil at an angle of 60.0° to the vertical. **(a)** Find the angle θ the light beam makes with the vertical as it travels through the water. **(b)** How does your answer to part (a) depend on the thickness of the oil film? Explain.

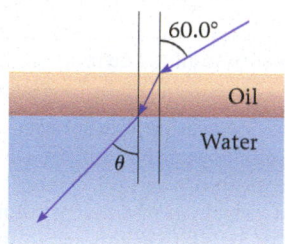

FIGURE 26-74 Problem 96

97. •• **FIGURE 26-75** shows a ray of light entering one end of an optical fiber at an angle of incidence $\theta_i = 50.0°$. The index of refraction of the fiber is 1.62. **(a)** Find the angle θ the ray makes with the normal when it reaches the curved surface of the fiber. **(b)** Show that the internal reflection from the curved surface is total.

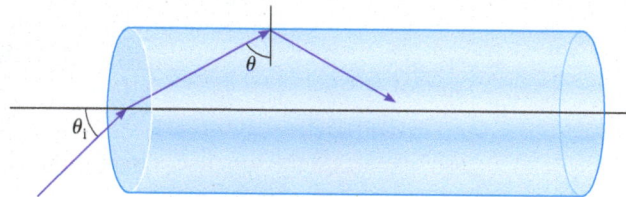

FIGURE 26-75 Problem 97 and 98

98. •• Suppose the fiber depicted in Figure 26-75 has an index of refraction of 1.62 and its curved surface is surrounded by a cladding material with an index of refraction of 1.52. **(a)** What is the critical angle for total internal reflection of rays at the fiber-cladding boundary? **(b)** What is the maximum angle θ_i at which rays can enter the fiber from air as shown and undergo total internal reflection? **(c)** If the diameter of the fiber core is 9.0 μm, how many reflections per meter does the ray undergo when it arrives at the boundary at the critical angle?

99. •• An arrow 2.00 cm long is located 75.0 cm from a lens that has a focal length $f = 30.0$ cm. **(a)** If the arrow is perpendicular to the principal axis of the lens, as in **FIGURE 26-76 (a)**, what is its lateral magnification, defined as h_i/h_o? **(b)** Suppose, instead, that the arrow lies *along* the principal axis, extending from 74.0 cm to 76.0 cm from the lens, as indicated in **FIGURE 26-76 (b)**. What is the longitudinal

magnification of the arrow, defined as L_i/L_o? (*Hint:* Use the thin-lens equation to locate the image of each end of the arrow.)

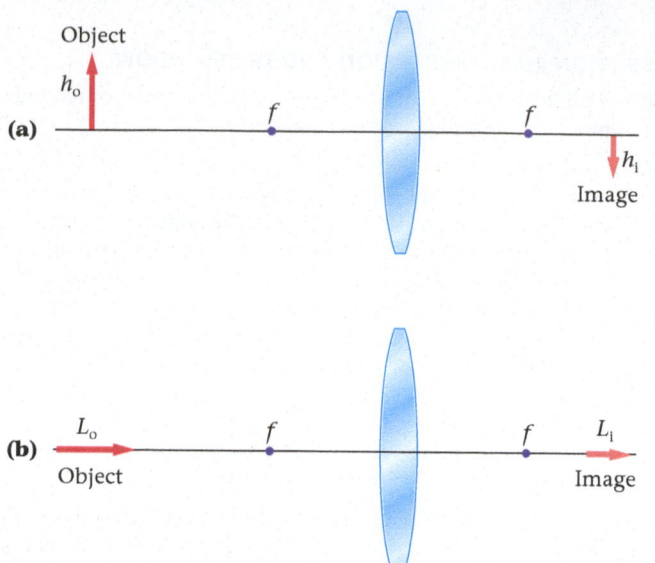

FIGURE 26-76 Problem 99

100. •• A convex lens with $f_1 = 20.0$ cm is mounted 40.0 cm to the left of a concave lens. When an object is placed 30.0 cm to the left of the convex lens, a real image is formed 60.0 cm to the right of the concave lens. What is the focal length f_2 of the concave lens?

101. ••• Two thin lenses, with focal lengths f_1 and f_2, are placed in contact. What is the effective focal length of the double lens?

102. ••• When an object is placed a distance d_o in front of a curved mirror, the resulting image has a magnification m. Find an expression for the focal length of the mirror, f, in terms of d_o and m.

103. ••• **A Slab of Glass** Give a symbolic expression for the sideways displacement d of a light ray passing through the slab of glass shown in **FIGURE 26-77**. The thickness of the glass is t, its index of refraction is n, and the angle of incidence is θ.

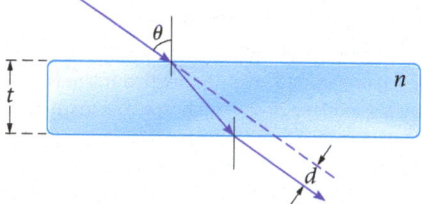

FIGURE 26-77 Problem 103

104. ••• **Least Time** A beam of light propagates from point A in medium 1 to point B in medium 2, as shown in **FIGURE 26-78**. The index of refraction is different in these two media; therefore, the light follows a refracted path that obeys Snell's law. **(a)** Calculate the time required for light to travel from A to B along the refracted path. **(b)** Compare the time found in part (a) with the time it takes for light to travel from A to B along a straight-line path. (Note that the time on the straight-line path is longer than the time on the refracted path. In general, the shortest time between two points in different media is along the path given by Snell's law.)

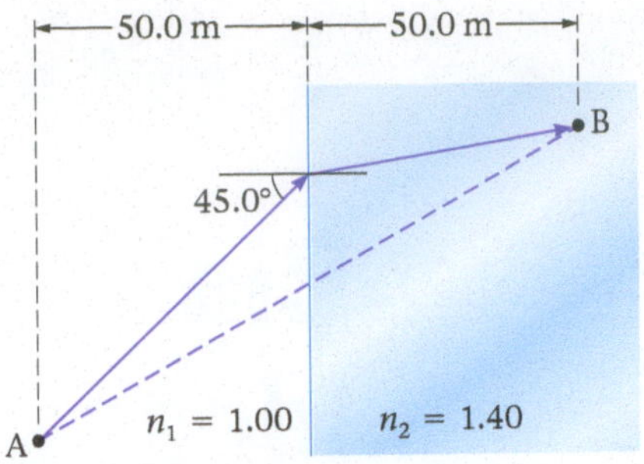

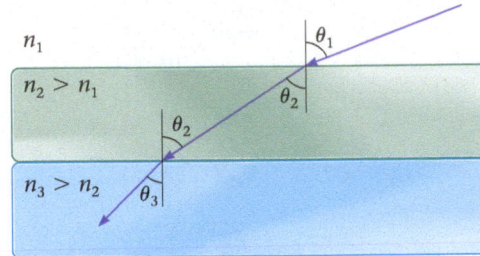

FIGURE 26-78 Problem 104

105. ••• The ray of light shown in **FIGURE 26-79** passes from medium 1 to medium 2 to medium 3. The index of refraction in medium 1 is n_1, in medium 2 it is $n_2 > n_1$, and in medium 3 it is $n_3 > n_2$. Show that medium 2 can be ignored when calculating the angle of refraction in medium 3; that is, show that $n_1 \sin \theta_1 = n_3 \sin \theta_3$.

FIGURE 26-79 Problem 105

106. ••• **Predict/Calculate** A beam of light enters the sloping side of a 45°-90°-45° glass prism with an index of refraction $n = 1.66$. The situation is similar to that shown in Figure 26-68, except that the angle of incidence of the incoming beam can be varied. **(a)** Find the angle of incidence for which the reflection on the vertical side of the prism exactly satisfies the condition for total internal reflection. **(b)** If the angle of incidence is increased, is the reflection at the vertical surface still total? Explain. **(c)** What is the minimum value of n such that a horizontal beam like that in Figure 26-68 undergoes total internal reflection at the vertical side of the prism?

PASSAGE PROBLEMS

The Focal Length of a Lens

A number of factors play a role in determining the focal length of a lens. First and foremost is the shape of the lens. As a general rule, a lens that is thicker in the middle will converge light, a lens that is thinner in the middle will diverge light.

Another important factor is the index of refraction of the lens material, n_{lens}. For example, imagine comparing two lenses with identical shapes but made of different materials. The lens with the larger index of refraction bends light more, bringing it to a focus in a shorter distance. As a result, a larger index of refraction implies a focal length with a smaller magnitude. In fact, the focal length of a lens surrounded by air ($n = 1$) is given by the **lens maker's formula:**

$$\frac{1}{f_{\text{in air}}} = (n_{lens} - 1)\left(\frac{1}{R_1} - \frac{1}{R_2}\right)$$

In this expression R_1 and R_2 are the radii of curvature of the front and back surfaces of the lens, respectively. For given values of R_1 and R_2—that is, for a given shape—the focal length of the lens becomes smaller in magnitude as the index of refraction increases.

A lens is not always surrounded by air, however. More generally, the fluid in which the lens is immersed may have an index of refraction given by n_{fluid}. In this case, the focal length is given by

$$\frac{1}{f_{\text{in fluid}}} = \left(\frac{n_{lens} - n_{fluid}}{n_{fluid}}\right)\left(\frac{1}{R_1} - \frac{1}{R_2}\right)$$

It follows, then, that the focal lengths of a lens surrounded by air or by a general fluid are related by

$$f_{\text{in fluid}} = \left[\frac{(n_{lens} - 1)n_{fluid}}{n_{lens} - n_{fluid}}\right]f_{\text{in air}}$$

This relation shows that the surrounding fluid can change the magnitude of the focal length, or even cause it to become infinite. The fluid can also change the sign of the focal length, which determines whether the lens is diverging or converging.

107. • A converging lens with a focal length in air of $f = +5.25$ cm is made from ice. What is the focal length of this lens if it is immersed in benzene? (Refer to Table 26-2.)

A. −20.7 cm
B. −18.1 cm
C. −12.8 cm
D. −11.2 cm

108. • A diverging lens with $f = -12.5$ cm is made from ice. What is the focal length of this lens if it is immersed in ethyl alcohol? (Refer to Table 26-2.)

A. 102 cm
B. 105 cm
C. 118 cm
D. 122 cm

109. • Calculate the focal length of a lens in water, given that the index of refraction of the lens is $n_{lens} = 1.52$ and its focal length in air is 25.0 cm. (Refer to Table 26-2.)

A. 57.8 cm
B. 66.0 cm
C. 91.0 cm
D. 104 cm

110. • Suppose a lens is made from fused quartz (glass), and that its focal length in air is −7.75 cm. What is the focal length of this lens if it is immersed in benzene? (Refer to Table 26-2.)

A. −130 cm
B. 134 cm
C. 141 cm
D. −145 cm

111. •• **REFERRING TO EXAMPLE 26-5** Suppose the radius of curvature of the mirror is 5.0 cm. **(a)** Find the object distance that gives an upright image with a magnification of 1.5. **(b)** Find the object distance that gives an inverted image with a magnification of −1.5.

112. •• **Predict/Calculate REFERRING TO EXAMPLE 26-5** An object is 4.5 cm in front of the mirror. **(a)** What radius of curvature must the mirror have if the image is to be 2.2 cm in front of the mirror? **(b)** What is the magnification of the image? **(c)** If the object is moved closer to the mirror, does the magnification of the image increase in magnitude, decrease in magnitude, or stay the same?

113. •• **REFERRING TO EXAMPLE 26-18 (a)** What object distance is required to give an image with a magnification of +2.0? Assume that the focal length of the lens is +5.0 cm. **(b)** What is the location of the image in this case?

114. •• **Predict/Calculate REFERRING TO EXAMPLE 26-18** Suppose the convex lens is replaced with a concave lens with a focal length of −5.0 cm. **(a)** Where must the object be placed to form an image with a magnification of 0.50? **(b)** What is the location of the image in this case? **(c)** If we now move the object closer to the lens, does the magnification of the image increase, decrease, or stay the same?

27

Optical Instruments

Big Ideas

1 The human eye uses the cornea and lens to form an image on the retina.

2 In a group of lenses, the image formed by one lens acts as the object for the next lens.

3 A nearsighted eye converges light more than a normal eye.

4 A farsighted eye converges light less than a normal eye.

5 With a magnifying glass, you can view objects closer than with your eyes alone.

6 Common aberrations (defects) of lenses include a distorted shape and dispersion of colors.

▲ Optical instruments play an important role in our lives, whether they are used to produce a big laugh, or to correct a person's vision. In this chapter, we consider a range of optical instruments, from eyeglasses and contact lenses to microscopes and telescopes. All of these applications can be understood by applying the basic principles of geometrical optics presented in the previous chapter.

Human vision can be aided in a number of ways by the practical application of optics. In this chapter, we consider a variety of optical instruments designed to either correct or extend our abilities to see the natural world.

27-1 The Human Eye and the Camera

The Human Eye

The eye is a marvelously sensitive and versatile optical instrument, allowing us to observe objects as distant as the stars or as close as the book in our hands. Perhaps more amazing is the fact that the eye can accomplish all this even though its basic structure is that of a spherical bag of water 2.5 cm in diameter. The slight differences that set the eye apart from a bag of water make all the difference in terms of optical performance, however.

BIO The fundamental elements of a human eye are illustrated in **FIGURE 27-1**. Basically, light enters the eye through the transparent outer coating of the eye, the *cornea*, and then passes through the *aqueous humor*, the adjustable *lens*, and the jellylike *vitreous humor* before reaching the light-sensitive *retina* at the back of the eye, as illustrated in **FIGURE 27-2**. The retina is covered with millions of small structures known as *rods* and *cones*, which, when stimulated by light, send electrical impulses along the *optic nerve* to the brain. How the nerve impulses are processed, so that we interpret the upside-down image on the retina as a right-side-up object, is another story altogether; here we concentrate on the optical properties of the eye.

Most of the refraction needed to produce an image occurs at the cornea, as light first enters the eye. The reason is that the difference in the index of refraction is greater at the air–cornea interface than at any interface within the eye. Specifically, the index of refraction of air is $n = 1.00$, whereas that of the cornea is about $n = 1.38$, just slightly greater than the index of refraction of water ($n = 1.33$). When light passes from the cornea to the aqueous humor, the index of refraction changes from $n = 1.38$ to $n = 1.33$. Next, light encounters the lens ($n = 1.40$) and then the vitreous humor ($n = 1.34$) before arriving at the retina.

In the end, the lens accounts for only about a quarter of the total refraction produced by the eye—but it is a crucial contribution nonetheless. By altering the shape of the lens with the *ciliary muscles*, we are able to change the precise amount of refraction the lens produces, which, in turn, changes its focal length. Specifically, when we view a distant object, our ciliary muscles are *relaxed*, as shown in **FIGURE 27-3 (a)**, allowing the lens to be relatively flat. As a result, it causes little refraction and its focal length is at its greatest. When we view a nearby object, the lens must shorten its focal length and cause more refraction, as shown in **FIGURE 27-3 (b)**. Thus, the ciliary muscles *tense* to give the lens a greater curvature. The process of changing the shape of the lens, and hence adjusting its focal length, is referred to as **accommodation.** Producing the proper accommodation is no easy feat for a newborn, but is automatic for an adult.

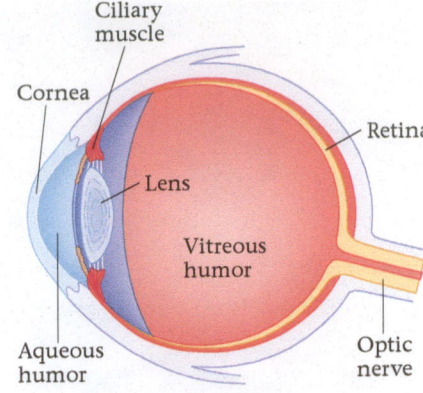

▲ **FIGURE 27-1 Basic elements of the human eye** Light enters the eye through the cornea and the lens. It is focused onto the retina by the ciliary muscles, which change the shape of the lens.

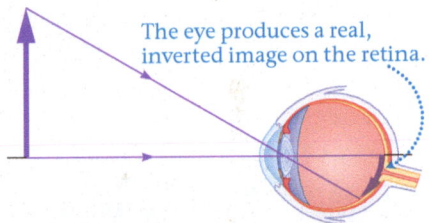

The eye produces a real, inverted image on the retina.

▲ **FIGURE 27-2 Image production in the eye**

Big Idea **1** The human eye uses the cornea and lens to form an image on the retina, in much the same way that a camera uses a lens to focus an image on film or an image sensor.

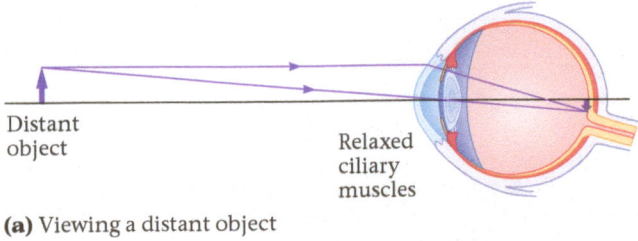

(a) Viewing a distant object

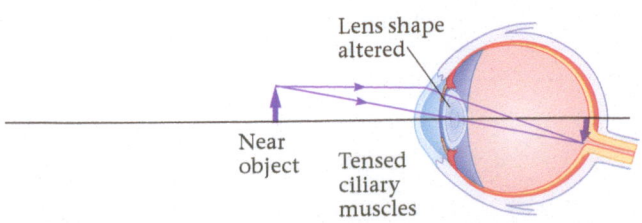

(b) Viewing a near object

◄ **FIGURE 27-3 Accommodation in the human eye (a)** When the eye is viewing a distant object, the ciliary muscles are relaxed and the focal length of the lens is at its greatest. **(b)** When the eye is focusing on a near object, the ciliary muscles are tensed, changing the shape and reducing the focal length of the lens.

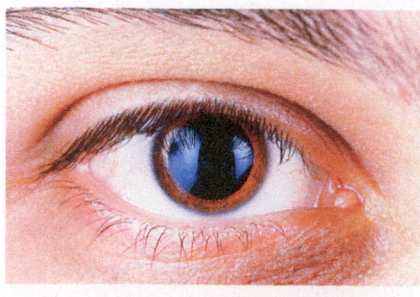

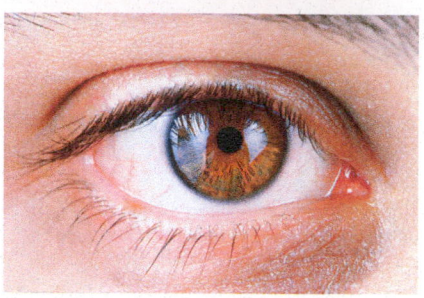

▲ **FIGURE 27-4** The pigmented iris of the human eye responds automatically to changing levels of illumination, dilating the pupil in dim light and contracting it in bright light.

The fact that the ciliary muscles must be tensed to focus on nearby objects means that our eyes can "tire" from muscular strain. That is why it is beneficial to pause occasionally from reading and to look off into the distance. Viewing distant objects allows the ciliary muscles to relax, thus reducing the strain on our eyes.

The amount of light that reaches the retina is controlled by a colored diaphragm called the *iris*. As the iris expands or contracts, it adjusts the size of the *pupil*, the opening through which light enters the eye, as shown in **FIGURE 27-4**. In bright light the pupil closes down to about 1 mm in diameter. On the darkest nights, the dark-adapted pupil can open up to a diameter of about 7 mm.

The Near and Far Points The lens can be distorted only so much, and hence there is a limit to how close the eye can focus. The shortest distance at which a sharp focus can be obtained is the **near point**—anything closer will appear fuzzy no matter how hard we try to focus on it. For young people the near-point distance, N, is typically about 25 cm, but it increases with age. Persons 40 years of age may experience a near point that is 40 cm from the eye, and in later years the near point may move to 500 cm or more. The extension of the near point to greater distances with age is referred to as *presbyopia* (or "short arm" syndrome) and is due to the lens becoming less flexible. Thus, as one ages it is not uncommon to have to move a piece of paper away from the eyes in order to focus, and eventually reading glasses may be necessary.

At the other end of the scale, the **far point** is the greatest distance an object can be from the eye and still be in focus. Since we can focus on the Moon and stars, it is clear that the normal far point is essentially infinity.

EXAMPLE 27-1 | JOURNEY TO NEAR POINT

BIO The near-point distance of a given eye is $N = 25$ cm. Treating the eye as if it were a single thin lens a distance 2.3 cm from the retina, find the focal length of the lens when it is focused on an object **(a)** at the near point and **(b)** at infinity. (Typical values for the effective lens–retina distance range from 1.7 cm to 2.5 cm.)

PICTURE THE PROBLEM
The eye, and the simplified thin-lens equivalent, are shown in the sketch. Notice that the horizontal axis is broken in order to display the object and eye in the same sketch. The object and image distances are indicated as well.

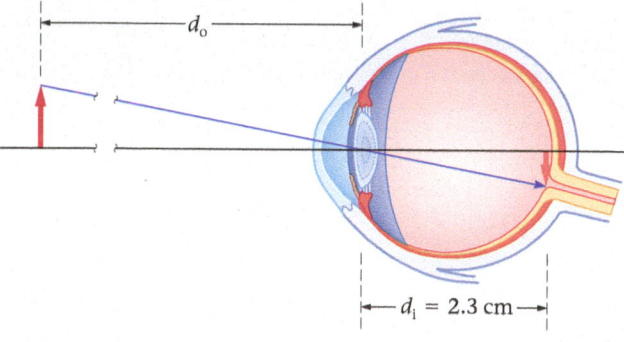

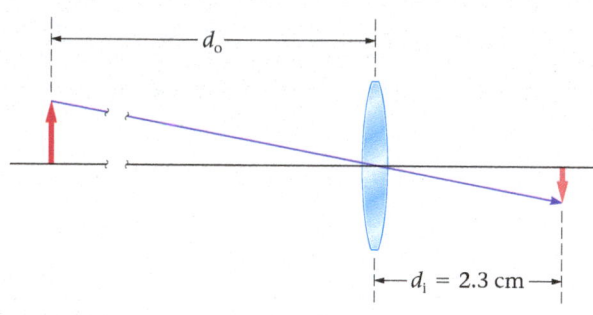

REASONING AND STRATEGY
The focal length can be found using the thin-lens equation, $1/d_o + 1/d_i = 1/f$. The image distance is $d_i = 2.3$ cm for both (a) and (b). For part (a) the object distance is $d_o = 25$ cm; for part (b) the object distance is $d_o = \infty$.

Known Near-point distance, $N = 25$ cm; distance from lens to retina, $d_i = 2.3$ cm.
Unknown **(a)** Focal length, $f = ?$, when $d_o = 25$ cm. **(b)** Focal length, $f = ?$, when $d_o = \infty$.

SOLUTION

Part (a)

1. Substitute $d_o = 25$ cm and $d_i = 2.3$ cm into the thin-lens equation and solve for f:

$$\frac{1}{f} = \frac{1}{d_o} + \frac{1}{d_i} = \frac{1}{25 \text{ cm}} + \frac{1}{2.3 \text{ cm}} = 0.47 \text{ cm}^{-1}$$

$$f = \frac{1}{0.47 \text{ cm}^{-1}} = 2.1 \text{ cm}$$

CONTINUED

Part (b)

2. Substitute $d_o = \infty$ and $d_i = 2.3$ cm into the thin-lens equation and solve for f:

$$\frac{1}{f} = \frac{1}{d_o} + \frac{1}{d_i} = \frac{1}{\infty} + \frac{1}{2.3 \text{ cm}} = \frac{1}{2.3 \text{ cm}}$$

$$f = 2.3 \text{ cm}$$

INSIGHT

The effective focal length of the eye changes by only about 2 mm in changing focus from the near point to infinity. Thus, if the shape of the eye is changed even slightly—so that the distance from the lens to the retina is increased or decreased by a couple millimeters—the eye will no longer be able to function properly. We return to this point in the next section when we discuss near- and farsightedness.

PRACTICE PROBLEM

At what object distance is the effective focal length of the eye equal to 2.0 cm? [**Answer:** $d_o = 15$ cm]

Some related homework problems: Problem 4, Problem 5

Floaters Perhaps you've noticed that when you look at a light-colored background, or a clear sky, one or more "spots" may float across your field of vision. Many people have these "floaters." In fact, some people are occasionally fooled by one of these spots into thinking a fly is buzzing about their head. For this reason, these floaters are called *muscae volitantes*, which means, literally, "flying flies." Muscae volitantes are caused by cells, cell fragments, or other small impurities suspended either in the vitreous humor of the eye or in the lens itself. These are usually harmless, but can be symptoms of a detached retina in extreme cases.

The Camera

A simple camera, such as the one illustrated in **FIGURE 27-5**, operates in much the same way as the eye. In particular, the lens of the camera forms a real, inverted image on a light-sensitive material—which in this case is usually a charge-coupled device (CCD) in a digital camera. The focusing mechanism is different, however. To focus a camera at different distances the lens is moved either toward or away from the CCD. Thus, the eye focuses by changing the shape of a stationary lens; the camera focuses by moving a lens of fixed shape.

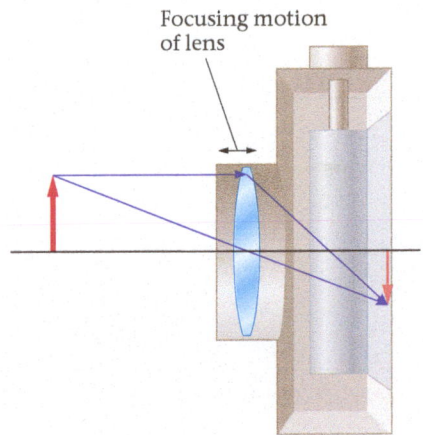

Focusing motion of lens

▲ **FIGURE 27-5 Basic elements of a camera** A camera forms a real, inverted image on photographic film or an electronic sensor, and is focused by moving the lens back and forth. Unlike the adjustable shape of the human eye, the shape of the camera lens does not change.

QUICK EXAMPLE 27-2 FIND THE DISPLACEMENT OF THE LENS

A simple camera uses a thin lens with a focal length of 6.50 cm. How far, and in what direction, must the lens be moved to change the focus of the camera from a person 21.1 m away to a person only 2.85 m away?

REASONING AND SOLUTION

We can use the thin-lens equation ($1/f = 1/d_o + 1/d_i$) to find the image distance for each case. The difference between these image distances is the distance the lens must be moved.

1. Calculate the image distance for an object distance of 21.1 m = 2110 cm:

$$\frac{1}{d_{i1}} = \frac{1}{f} - \frac{1}{d_{o1}} = \frac{1}{6.50 \text{ cm}} - \frac{1}{2110 \text{ cm}}$$

$$d_{i1} = \frac{(6.50 \text{ cm})(2110 \text{ cm})}{(2110 \text{ cm} - 6.50 \text{ cm})} = 6.52 \text{ cm}$$

2. Calculate the image distance for an object distance of 2.85 m = 285 cm:

$$\frac{1}{d_{i2}} = \frac{1}{f} - \frac{1}{d_{o2}} = \frac{1}{6.50 \text{ cm}} - \frac{1}{285 \text{ cm}}$$

$$d_{i2} = \frac{(6.50 \text{ cm})(285 \text{ cm})}{(285 \text{ cm} - 6.50 \text{ cm})} = 6.65 \text{ cm}$$

3. Find the difference in image distance:

$$d_{i2} - d_{i1} = 6.65 \text{ cm} - 6.52 \text{ cm}$$
$$= 0.13 \text{ cm} = 1.3 \text{ mm}$$

The image distance is greater for the person at 2.85 m, and hence the lens must be moved *away* from the film by 1.3 mm to change the focus the desired amount.

RWP* The aperture of a camera is analogous to the pupil of an eye, and like the pupil, its size can be adjusted—the greater the size, the more light that is available to make an image. Photographers often characterize the size of the aperture with a dimensionless quantity called the **f-number**, which is defined as follows:

$$f\text{-number} = \frac{\text{focal length}}{\text{diameter of aperture}} = \frac{f}{D} \qquad \text{27-1}$$

Notice that the larger the diameter of the aperture, the smaller the f-number; in fact, $D = f/(f\text{-number})$.

Cameras typically have aperture settings indicated by a sequence of f-numbers such as the following:

$$1.4, \quad 2, \quad 2.8, \quad 4, \quad 5.6, \quad 8, \quad 11, \quad 16$$

For example, a camera lens with a focal length of 50.0 mm and an aperture setting of 4 has an aperture diameter of $D = (50.0\text{ mm})/4 = 12.5$ mm. This is often referred to by photographers as an $f/4$ setting for the aperture. Similarly, changing the aperture setting to $f/2$ opens the aperture to a diameter of $(50.0\text{ mm})/2 = 25.0$ mm. Because the aperture is an approximately circular opening, the area through which light enters the camera, $A = \pi D^2/4$, varies as the square of the aperture diameter D and thus inversely as the square of the f-number.

The amount of light that falls on the photographic film or CCD in a camera is also controlled by the shutter speed. A shutter speed of 1/500, for example, means that the shutter is open for only 1/500 of a second—very effective for "freezing" the motion of a high-speed object. Changing the shutter speed to 1/250 doubles the time the shutter is open, which doubles the light received by the film. Typical camera shutter speeds are 1/1000, 1/500, 1/250, 1/125, and so on, usually indicated on a dial as 1000, 500, 250, 125, respectively.

Suppose, for example, that a photograph receives the proper amount of light when the shutter speed is 1/500 and the aperture f-number is 4. If the photographer decides to take a second shot with a shutter speed of 1/250, what f-number is required to maintain the correct exposure? Well, the slower shutter speed doubles the light entering the camera, and hence the area of the aperture must be halved to compensate. To halve the area, the diameter must be *reduced* by a factor of $\sqrt{2} \approx 1.4$, which means the f-number must be *increased* by a factor of 1.4. Therefore, the photographer must change the aperture setting to $4 \times (1.4) = 5.6$. Notice that each of the f-numbers listed earlier is larger than the previous one by a factor of roughly 1.4, leading to a factor of 2 difference in the light received by the film. This is the reason for the otherwise unusual progression of f-numbers.

Enhance Your Understanding   (Answers given at the end of the chapter)

1. If the f-number on a camera is increased, does the diameter of the aperture increase, decrease, or stay the same? Explain.

Section Review

- The human eye and a camera operate in the same general way, using a lens to project an inverted image on a light-sensitive material such as the retina or a CCD chip.

27-2 Lenses in Combination and Corrective Optics

A normal eye can provide sharp images for objects over a wide range of distances, but some eyes have difficulty focusing on distant objects, and others are unable to focus as close as the normal near point. Problems such as these can be corrected with glasses or contact lenses used in combination with the eye's own lens. In general, a combination of lenses can have beneficial properties not possible with a single lens, as we shall see

---

*Real World Physics applications are denoted by the acronym RWP.

throughout the rest of this chapter. First, we discuss how to analyze a system with more than one lens; we then apply these results to correcting near- and farsightedness.

Lenses in Combination The basic operating principle for a system consisting of more than one lens is the following:

> The *image* produced by one lens acts as the *object* for the next lens. This is true regardless of whether the image produced by the first lens is real or virtual, or whether it is in front of or behind the second lens.

This principle finds many applications. In fact, most optical instruments—like cameras, microscopes, and telescopes—use a number of lenses to produce the desired results.

A Two-Lens System As an example of a lens system, consider the two lenses shown in **FIGURE 27-6 (a)**. An object is 20.0 cm to the left of the convex lens (lens 1), which is 50.0 cm from a concave lens (lens 2) to its right. Given that the focal lengths of the convex and concave lenses are 10.0 cm and −12.5 cm, respectively, we would like to find the location and orientation of the image produced by the two lenses acting together.

The first step is illustrated in **FIGURE 27-6 (b)**, where we show the image formed by the convex lens with a ray diagram. The precise location of image 1 is given by the thin-lens equation:

$$\frac{1}{d_o} + \frac{1}{d_i} = \frac{1}{f} \quad \text{or} \quad \frac{1}{20.0 \text{ cm}} + \frac{1}{d_i} = \frac{1}{10.0 \text{ cm}}$$

$$d_i = 20.0 \text{ cm}$$

As indicated in the ray diagram, image 1 is inverted and 20.0 cm to the right of the convex lens.

The next step is to note that image 1 is 50.0 cm − 20.0 cm = 30.0 cm to the left of the concave lens. Considering image 1 to be the object for the concave lens, we obtain the ray diagram shown in **FIGURE 27-6 (c)**. The thin-lens equation, with $d_o = 30.0$ cm and $f = -12.5$ cm, yields the following image distance:

$$\frac{1}{d_o} + \frac{1}{d_i} = \frac{1}{f} \quad \text{or} \quad \frac{1}{30.0 \text{ cm}} + \frac{1}{d_i} = \frac{1}{-12.5 \text{ cm}}$$

$$d_i = -8.82 \text{ cm}$$

Thus, image 2, the final image of the two-lens system, is 8.82 cm to the left of the concave lens.

The orientation and size of the final image can be found as follows:

> The total magnification produced by a lens system is equal to the *product* of the magnifications produced by each lens individually.

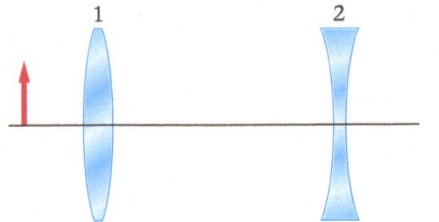

(a) Original object

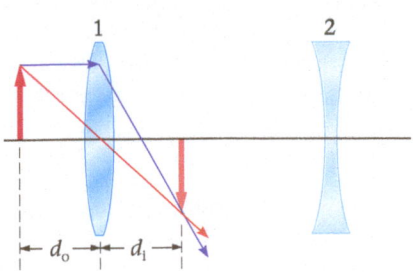

(b) Image from first lens

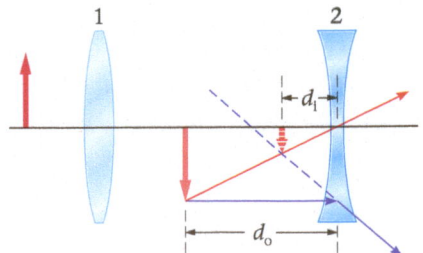

(c) Image from second lens

▲ **FIGURE 27-6** **A two-lens system** In this system, a convex and a concave lens are separated by 50.0 cm. **(a)** An object is placed 20.0 cm to the left of the convex lens, whose focal length is 10.0 cm. **(b)** The image formed by the convex lens is 20.0 cm to its right. This image is the object for the concave lens. **(c)** The object for this lens is 30.0 cm to its left. Because the focal length of this lens is −12.5 cm, it forms an image 8.82 cm to its left. This is the final image of the system.

PHYSICS IN CONTEXT
Looking Ahead

Ray diagrams showing light reflected from a mirror are used in Einstein's theory of relativity, in Chapter 29. There we analyze a "light clock," consisting of two mirrors and a beam of light, and show that a moving clock runs at a slower rate than a clock at rest.

Using Equation 26-18 ($m = -d_i/d_o$), we find that lens 1 produces the magnification $m_1 = -(20.0 \text{ cm})/(20.0 \text{ cm}) = -1$. Similarly, lens 2 produces the magnification $m_2 = -(-8.82 \text{ cm})/(30.0 \text{ cm}) = 0.294$. The total magnification of the system, then, is $m_{total} = m_1 m_2 = -0.294$. As a result, the final image is inverted and reduced in size by a factor of 0.294 compared with the original object. This is in agreement with the results shown in the ray diagrams of Figure 27-6.

The following Example considers the effect of reversing the order of the two lenses.

EXAMPLE 27-3 FIND THE FINAL IMAGE

Find the location and total magnification of the final image produced by the lens system shown in Figure 27-6 if the positions of the two lenses are reversed.

PICTURE THE PROBLEM

The situation is shown to scale in our sketch (except for the enlargement). Notice that the first lens is concave, and the second lens is convex.

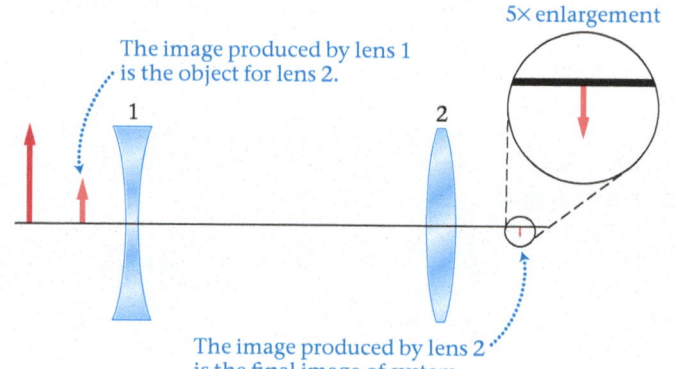

The image produced by lens 1 is the object for lens 2.

5× enlargement

The image produced by lens 2 is the final image of system.

REASONING AND STRATEGY

Lens 1 produces a reduced, upright image to its left. This image is the object for lens 2. The final image of the system, produced by lens 2, is tiny and inverted. We show a 5× enlargement for clarity. The thin-lens equation can be used to find both images in this system.

Known Focal length of lens 1, $f_1 = -12.5$ cm; object distance for lens 1, $d_{o1} = 20.0$ cm; separation between lenses, 50.0 cm; focal length of lens 2, $f_2 = 10.0$ cm.

Unknown Location of final image, $d_{i2} = ?$; total magnification, $m_{total} = ?$

SOLUTION

1. Use the thin-lens equation to find the image distance for the concave lens (lens 1), with $d_{o1} = 20.0$ cm:

$$\frac{1}{d_{o1}} + \frac{1}{d_{i1}} = \frac{1}{f_1} \quad \text{or} \quad \frac{1}{20.0 \text{ cm}} + \frac{1}{d_{i1}} = \frac{1}{-12.5 \text{ cm}}$$

$$d_{i1} = -7.69 \text{ cm}$$

2. Find the object distance for the convex lens (lens 2):

$$d_{o2} = 50.0 \text{ cm} + 7.69 \text{ cm} = 57.7 \text{ cm}$$

3. Use this object distance and the thin-lens equation to find the location of the final image:

$$\frac{1}{d_{o2}} + \frac{1}{d_{i2}} = \frac{1}{f_2} \quad \text{or} \quad \frac{1}{57.7 \text{ cm}} + \frac{1}{d_{i2}} = \frac{1}{10.0 \text{ cm}}$$

$$d_{i2} = 12.1 \text{ cm}$$

INSIGHT

The final image is 12.1 cm to the right of lens 2. The magnification produced by lens 1 is $m_1 = -d_{i1}/d_{o1} = 0.385$, and the magnification produced by lens 2 is $m_2 = -d_{i2}/d_{o2} = -0.210$. Hence, the total magnification is $m_{total} = m_1 m_2 = -0.0809$, showing that the final image is reduced in size by a factor of 0.0809 and inverted. Notice how different the results are when we simply reverse the order of the lenses—it's just like the different result you get when you look through the wrong end of a pair of binoculars.

PRACTICE PROBLEM

What is the location of the final image if both lenses are concave, each with a focal length of -12.5 cm? Everything else remains the same. **[Answer:** $d_{i2} = -10.3$ cm—that is, 10.3 cm to the left of the right lens**]**

Some related homework problems: Problem 25, Problem 28

We now show how a two-lens system (the eye plus an external lens) can correct abnormal vision.

Nearsightedness

When a person with normal vision relaxes the ciliary muscles of the eye, an object at infinity is in focus. In a nearsighted (myopic) person, however, a totally relaxed eye focuses only out to a finite distance from the eye—the far point. Thus a person with this

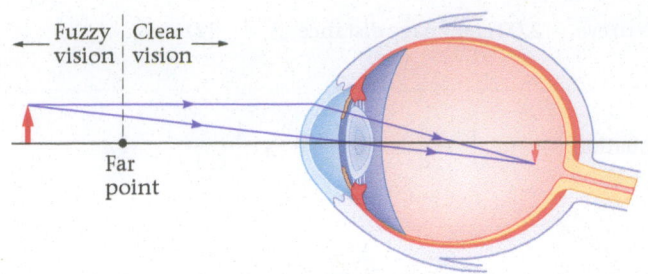

► **FIGURE 27-7 Eye shape and nearsightedness** An eye that is elongated can cause nearsightedness. In this case, an object at infinity comes to a focus in front of the retina.

condition is said to be nearsighted because objects near the eye can be focused—the person's vision is good there—whereas objects beyond the far point appear fuzzy.

The problem in this situation is that the eye converges the light coming into it in too short a distance; in other words, the focal length of the eye is less than the distance from the lens to the retina. This condition is illustrated in **FIGURE 27-7**, where we see that an object at infinity forms an image in front of the retina, because of the elongation of the eye. The effect need not be large; as we saw in the previous section, an elongation of only a millimeter or two is enough to cause a problem.

To correct this condition, we need to "undo" some of the excess convergence produced by the eye, so that images again fall on the retina. This can be done by placing a *diverging* lens in front of the eye. Specifically, consider an object at infinity—which would ordinarily appear blurry to a nearsighted person. If a concave lens with the proper focal length produces a virtual image of this object at the nearsighted person's far point, as in **FIGURE 27-8**, the person's relaxed eye can now focus on the object. We consider this situation in the next Example.

PROBLEM-SOLVING NOTE

Correcting Nearsightedness

To correct for nearsightedness, one must use a lens that produces an image at the person's far-point distance when the object is at infinity. Note that the far point is closer to the lens in a pair of glasses than it is to the eye, since the glasses are a finite distance in front of the eyes.

Big Idea ③ A nearsighted eye converges light more than a normal eye, allowing the eye to focus on nearby objects but not on distant objects. Nearsightedness can be corrected by placing a diverging lens in front of the eye.

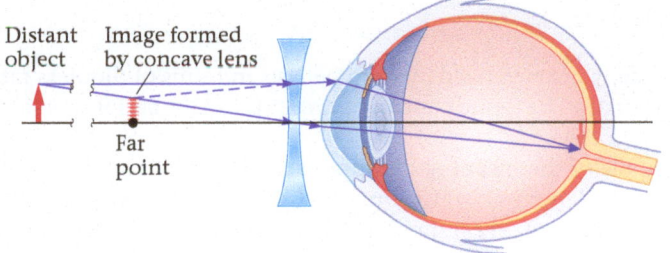

◄ **FIGURE 27-8 Correcting nearsightedness** A diverging lens in front of the eye can correct for nearsightedness. The concave lens focuses light from an object beyond the far point to produce an image that is at the far point. The eye can now focus on the image of the object.

EXAMPLE 27-4   EXTENDED VISION

BIO A nearsighted person has a far point that is 323 cm from her eye. If the lens in a pair of glasses is 2.00 cm in front of this person's eye, what focal length must it have to allow her to focus on distant objects?

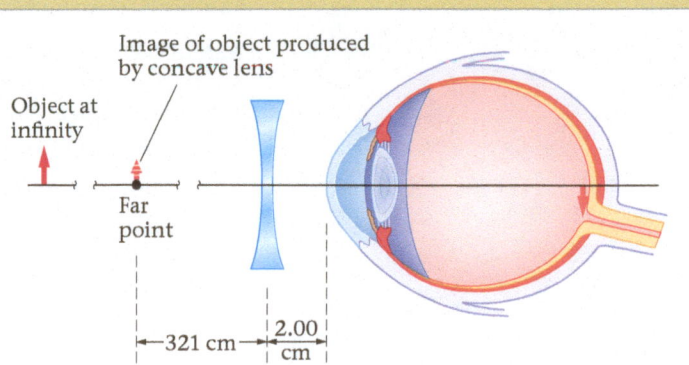

PICTURE THE PROBLEM
In our sketch, we show an object at infinity and the corresponding image produced by the concave lens. As is usual with a concave lens, the image is upright, reduced, and on the same side of the lens as the object. In addition, the image is placed at the person's far point, where the eye can focus on it.

REASONING AND STRATEGY
We can find the focal length of the lens using the thin-lens equation, $1/d_o + 1/d_i = 1/f$. In this case, $d_o = \infty$, because the object is infinitely far away. As for the image, it must be 323 cm from the eye, which is 323 cm − 2.00 cm = 321 cm in front of the lens. Thus, the image distance is $d_i = -321$ cm, where the minus sign is required because the image is on the same side of the lens as the object. With these values for d_o and d_i, it is straightforward to find the focal length, f.

CONTINUED

Known Far-point distance = 323 cm; distance from lens to eye = 2.00 cm; image distance, d_i = −321 cm.
Unknown Focal length, f = ?, for d_o = ∞.

SOLUTION

1. Substitute d_o = ∞ and d_i = −321 cm in the thin-lens equation:

$$\frac{1}{d_o} + \frac{1}{d_i} = \frac{1}{f} = \frac{1}{\infty} + \frac{1}{-321 \text{ cm}}$$

2. Solve for the focal length, f:

$$f = -321 \text{ cm}$$

INSIGHT

With a lens of this focal length, the person can focus on distant objects with a relaxed eye. In addition, notice that the focal length is equal to the image distance, $f = d_i$. This is always the case when the object distance is infinite.

PRACTICE PROBLEM

If these glasses are used to view an object 525 cm from the eye, how far from the eye is the image produced by the concave lens? [**Answer:** The image is 201 cm in front of the eye.]

Some related homework problems: Problem 37, Problem 38

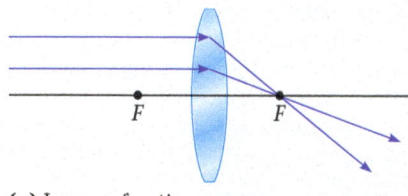

(a) Large refractive power

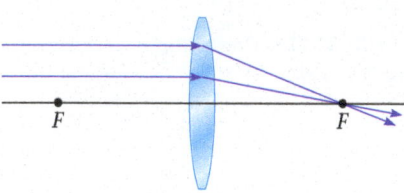

(b) Small refractive power

▲ **FIGURE 27-9 Refractive power** Light is bent (refracted) more by a lens with a short focal length than one with a long focal length. Therefore, lens **(a)** has a greater refractive power than lens **(b)**.

Refractive Power The ability of a lens to refract light—its **refractive power**—is related to its focal length. For example, the shorter the focal length, the more strongly a lens refracts light, as indicated in **FIGURE 27-9**. Thus refractive power depends inversely on the focal length. By definition, then, we say that the refractive power of a lens is $1/f$, where f is measured in meters:

Refractive Power

$$\text{refractive power} = \frac{1}{f} \qquad 27\text{-}2$$

SI unit: diopter = m⁻¹

Lenses are typically characterized by optometrists in terms of diopters rather than in terms of focal length. The type of test an optometrist might perform on a person's eyes is shown in **FIGURE 27-10**.

As an example of the meaning of diopters, a lens with a refractive power of 12.5 diopters has a focal length of $1/(12.5 \text{ m}^{-1})$ = 8.00 cm (a converging lens), and a lens with a refractive power of −12.5 diopters has a focal length of −8.00 cm (a diverging lens). In Example 27-4, the lens required to correct nearsightedness had a refractive power of $1/(-3.21 \text{ m})$ = −0.312 diopter.

QUICK EXAMPLE 27-5   FIND THE REFRACTIVE POWER

BIO A person has a far point that is 5.70 m from his eyes. If this person is to wear glasses that are 2.00 cm from his eyes, what refractive power, in diopters, must his lenses have?

REASONING AND SOLUTION

We start by using the thin-lens equation to find the required focal length, f. The inverse of the focal length, in meters, is the refractive power, in diopters.

1. Identify the object distance: $d_o = \infty$

2. Identify the image distance: $d_i = -(570 \text{ cm} - 2.00 \text{ cm}) = -568 \text{ cm}$

3. Use the thin-lens equation to calculate the focal length of the lenses in the glasses:

$$\frac{1}{d_o} + \frac{1}{d_i} = \frac{1}{\infty} + \frac{1}{-568 \text{ cm}} = \frac{1}{f}$$
$$f = -568 \text{ cm}$$

4. Convert f to meters and invert to find the refractive power:

$$\text{refractive power} = \frac{1}{f} = \frac{1}{-5.68 \text{ m}}$$
$$= -0.176 \text{ diopter}$$

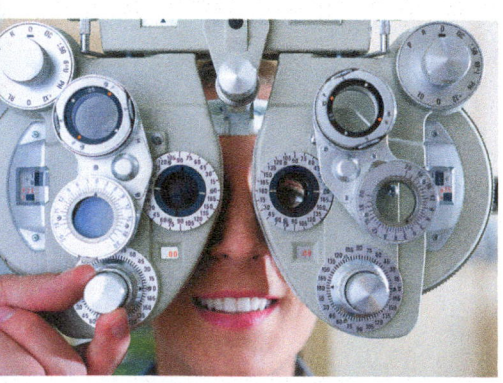

▲ **FIGURE 27-10** In order to prescribe the right corrective lenses, an optometrist needs to measure the refractive properties of the patient's eyes. This device, known as a *phoropter*, allows the optometrist to see how lenses with various optical characteristics affect the patient's vision.

The fact that this person's far point is farther away (closer to infinity) than the far point in Example 27-4 means that the lenses do not have to be as "strong" to correct the vision. As a result, the magnitude of the refractive power in this case (0.176 diopter) is less than the corresponding magnitude (0.312 diopter) in Example 27-4.

Medical Procedures to Correct Nearsightedness   Today, in addition to glasses and contact lenses, a number of medical procedures are available to correct nearsightedness. Some of these procedures involve using laser beams to reshape the cornea of the eye. These techniques are discussed in Chapter 31, where we consider lasers in detail. Here we present two alternative procedures that change the shape of a cornea by mechanical means.

BIO   Perhaps the simplest such technique is the implantation of either an *intracorneal ring* or an *Intact* in the cornea of an eye. An Intact consists of two small, clear crescents made of the same material used in contact lenses. These crescents are slipped into "tunnels" that are cut into the cornea. When the crescents are in place they tend to stretch the cornea outward and flatten its surface. This flattening increases the focal length of the eye and corrects for nearsightedness by −1.0 to −3.0 diopters. If necessary, the Intacts can be removed and replaced with others that cause more or less flattening of the cornea. Intracorneal rings are similar, except that they consist of a single ring rather than the two crescents used in Intacts.

BIO   A more complex method involves making radial incisions in the cornea with a highly precise diamond blade that cuts to a specified depth. This method is referred to as *radial keratotomy*, or RK, and is illustrated in **FIGURE 27-11**. The cuts allow the peripheral parts of the cornea to bulge outward, which, in turn, causes the central portion to flatten. As with an Intact, the flattening of the cornea corrects for nearsightedness.

Farsightedness

A person who is farsighted (hyperopic) can see clearly beyond a certain distance—the near point—but cannot focus on closer objects. Basically, the vision of a farsighted person differs from that of a person with normal vision by having a near point that is much farther from the eye than the usual 25 cm. As a result, a farsighted person is typically unable to read clearly, because a book would have to be held at such a great distance to come into focus.

Farsightedness can be caused by an eyeball that is shorter than normal, as illustrated in **FIGURE 27-12**, or by a lens that becomes sufficiently stiff with age that it can no longer take on the shape required to focus on nearby objects. In such cases, rays from an object inside the near point are brought to a focus behind the retina. Thus the focal length of the farsighted eye is too large—stated another way, the farsighted eye does not converge the incoming light strongly enough to focus on the retina.

This problem can be corrected by "preconverging" the light—that is, by using a converging lens in front of the eye to add to its insufficient convergence. For example, suppose an object is inside a person's near point, as in **FIGURE 27-13**. If a converging lens placed in front of the eye can produce an image that is far away—that is, beyond the near point—the farsighted person can view the object with ease. Such a system is considered in the next Example.

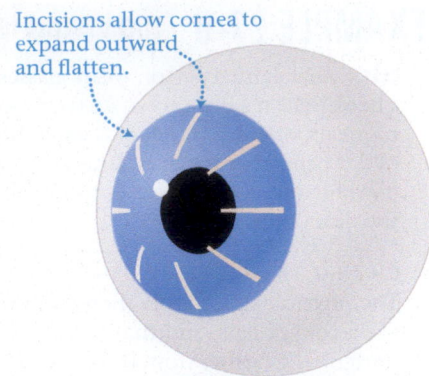

▲ **FIGURE 27-11 Radial keratotomy** In radial keratotomy, a series of radial incisions is made around the periphery of the cornea. This allows the cornea to expand outward, resulting in a flattening of the cornea's central region. The reduced curvature of the cornea increases the focal length of the eye, allowing it to focus on distant objects.

Incisions allow cornea to expand outward and flatten.

Big Idea 4 A farsighted eye converges light less than a normal eye, allowing the eye to focus on distant objects but not on nearby objects. Farsightedness can be corrected by placing a converging lens in front of the eye.

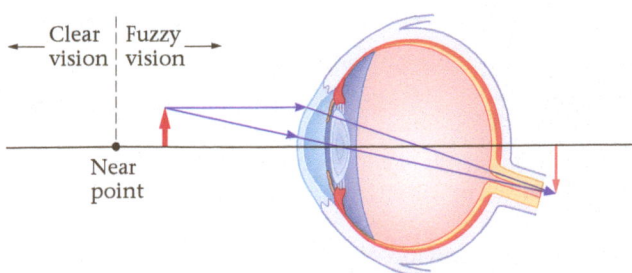

▲ **FIGURE 27-12 Eye shape and farsightedness** An eye that is shorter than normal can cause farsightedness. Note that an object inside the near point comes to a focus behind the retina.

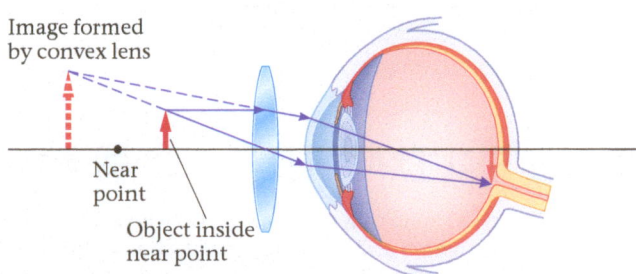

▲ **FIGURE 27-13 Correcting farsightedness** A converging lens in front of the eye can correct for farsightedness. The convex lens focuses light from an object inside the near point to produce an image that is beyond the near point. The eye can now focus on the image of the object.

EXAMPLE 27-6 HIS VISION IS A FAR SIGHT BETTER

BIO A farsighted person wears glasses that enable him to read a book held at a distance of 25.0 cm from his eyes, even though his near-point distance is 57.0 cm. If his glasses are at a distance of 2.00 cm from his eyes, find the focal length and refractive power required of his lenses to place the image of the book at the near point.

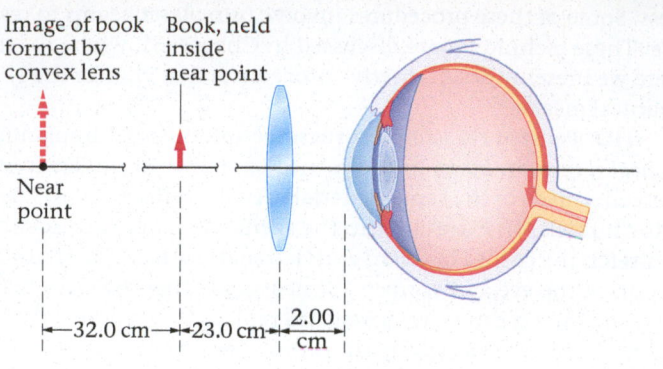

PICTURE THE PROBLEM

The physical situation is shown in our sketch. Notice that we use a convex lens, and that the image produced by the lens is upright and farther from the eye than the book. These features are in agreement with the qualitative characteristics shown in Figure 27-13. In this case, however, the image of the book is exactly at the near point.

REASONING AND STRATEGY

The focal length of the lens can be found using the thin-lens equation. First, the object distance is $d_o = 23.0$ cm, taking into account the 2.00 cm separation between the glasses and the eye. Similarly, the desired image distance is $d_i = -55.0$ cm. The minus sign is required on the image distance because the image is on the same side of the lens as the object.

Known   Near-point distance = 57.0 cm; object distance, $d_o = 23.0$ cm; image distance, $d_i = -55.0$ cm.
Unknown Focal length, $f = ?$ Refractive power = ?

SOLUTION

1. Use the thin-lens equation to find the focal length, f:

$$\frac{1}{f} = \frac{1}{d_o} + \frac{1}{d_i} = \frac{1}{23.0 \text{ cm}} + \frac{1}{-55.0 \text{ cm}} = 0.0253 \text{ cm}^{-1}$$

$$f = \frac{1}{0.0253 \text{ cm}^{-1}} = 39.5 \text{ cm}$$

2. The refractive power is $1/f$, with f measured in meters:

$$\text{refractive power} = \frac{1}{f} = \frac{1}{0.395 \text{ m}} = 2.53 \text{ diopters}$$

INSIGHT

The object (book) in this Example is between the lens and its focal point. As a result, the image is upright [see Figure 26-47 (b)], as desired for reading glasses. In addition, the image is virtual, as was also the case for the concave lens in Example 27-4. Thus, even though a "virtual image" can sound like one that isn't real or of practical importance, you are viewing a virtual image every time you look through a pair of glasses.

PRACTICE PROBLEM — PREDICT/CALCULATE

Suppose a second person has a near-point distance that is greater than 57.0 cm. **(a)** Is the refractive power of the lenses required for this person greater than or less than 2.53 diopters? Explain. **(b)** To check your answer, calculate the refractive power for a near-point distance of 67.0 cm.   [**Answer: (a)** The refractive power must be greater than 2.53 diopters because more preconvergence is required to correct for the greater near-point distance. **(b)** We find 2.81 diopters for a near-point distance of 67.0 cm, which is greater than 2.53 diopters, as expected.]

Some related homework problems: Problem 33, Problem 36

CONCEPTUAL EXAMPLE 27-7 EYEGLASSES TO START A FIRE   PREDICT/EXPLAIN

Bill and Ted are on an excellent camping trip when they decide to start a fire by focusing sunlight with a pair of eyeglasses. **(a)** If Bill is nearsighted and Ted is farsighted, should they use Bill's glasses or Ted's glasses, or will either pair of glasses work just as well? **(b)** Which of the following is the *best explanation* for your prediction?

  I. Use Bill's glasses because they are concave, which means they will focus light from the Sun to a point.

 II. Use Ted's glasses because they are convex, which is the correct type to use to focus light from the Sun to a point.

III. Either pair of glasses will work just as well, because they both work by focusing light to a point.

CONTINUED

REASONING AND DISCUSSION

To focus the parallel rays of light from the Sun to a point requires a converging (convex) lens. As we have seen, nearsightedness is corrected with a diverging lens; farsightedness is corrected with a converging lens. Therefore, the eyeglasses of farsighted Ted are converging, and they should be used to start the fire.

ANSWER

(a) Ted's glasses are the more excellent choice. (b) The best explanation is II.

To consider the effect of a pair of contact lenses, we simply take into account that contacts are placed directly against the eye. This means that the eye–object distance is the same as the lens–object distance. We make use of this fact in the following Quick Example.

PROBLEM-SOLVING NOTE

Finding the Focal Length of Contact Lenses

QUICK EXAMPLE 27-8   FIND THE FOCAL LENGTH OF CONTACT LENSES

Find the focal length of a contact lens that will allow a person with a near-point distance of 156 cm to read a newspaper held 24.5 cm from the eyes.

Contact lenses are so named because they are in direct contact with the eye. Therefore, the object distance for a contact lens is the same as the distance from the eye to the object. Similarly, the near point is the same distance from both the contact lens and the eye.

REASONING AND SOLUTION

Once we identify the object distance and the image distance, the thin-lens formula can be used to find the focal length of the contact lenses. The object distance is the distance to the newspaper (24.5 cm); the image distance is minus the distance to the near point (–156 cm) because the image is on the same side of the lens as the object.

1. Identify the object distance: $d_o = 24.5 \text{ cm}$

2. Identify the image distance: $d_i = -156 \text{ cm}$

3. Use the thin-lens equation to calculate the focal length:
$$\frac{1}{f} = \frac{1}{d_o} + \frac{1}{d_i} = \frac{1}{24.5 \text{ cm}} + \frac{1}{-156 \text{ cm}} = 0.0344 \text{ cm}^{-1}$$
$$f = \frac{1}{0.0344 \text{ cm}^{-1}} = 29.1 \text{ cm}$$

The refractive power of these contact lenses is 3.44 diopters.

BIO An important step in assuring that contact lenses fit a patient's eye properly is to measure the radius of curvature of the cornea. This is accomplished with a device known as a *keratometer*. To begin the procedure, a brightly lit object is brought near the eye. The light from the object reflects from the front surface of the cornea, just as light reflects from a convex, spherical mirror. In the next step of the procedure, the keratometer measures the magnification of the mirror image produced by the cornea. Finally, a straightforward application of the mirror equation determines the radius of curvature. A specific example of a keratometer in use is given in Problem 88.

Another eye condition that requires corrective optics is **astigmatism.** In most cases, astigmatism is due to an irregular curvature of the cornea, with a greater curvature in one direction than in another. For example, the curvature in the vertical plane may be greater than the curvature in the horizontal plane. As a result, if the eye is focused for light coming into it from one direction, it will be out of focus for light arriving from a different direction. Almost everyone has some degree of astigmatism, usually quite mild, but in serious cases it can cause distorted or blurry vision at all distances.

Enhance Your Understanding   (Answers given at the end of the chapter)

2. When a nearsighted eye focuses on an object at infinity, does it form an image that is in front of the retina, on the retina, or behind the retina?

CONTINUED

Section Review

- Nearsightedness can be corrected with a diverging lens; farsightedness can be corrected with a converging lens.

- Refractive power is defined as follows: refractive power = $1/f$. If the focal length is expressed in meters, the refractive power is given in diopters. The larger the magnitude of the refractive power, the more the corresponding lens bends light.

27-3 The Magnifying Glass

Big Idea 5 A magnifying glass allows a person to view an object from a closer distance than would be possible with the unaided eye. This gives an enlarged view of the object.

A **magnifying glass** is nothing more than a simple convex lens. Working together with the eye as part of a two-lens system, a *magnifier* can make objects appear to be many times larger than their actual size. As we shall see, the magnifying glass works by moving the near point closer to the eye—much as a converging lens corrects farsightedness. Basically, the magnifier allows an object to be viewed from a reduced distance, and this is what makes it appear larger.

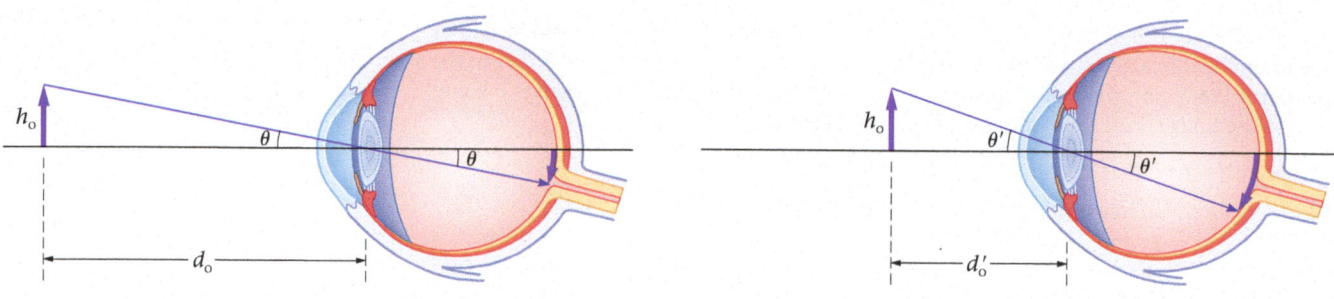

(a) Small angular size

(b) Same object, larger angular size

▲ **FIGURE 27-14 Angular size and distance** The angular size of an object depends on its distance from the eye, even though its height, h_o, remains the same.

Apparent Size To be specific about the apparent size of an object, we consider the angle it subtends on the retina—after all, the more area an image takes up on the retina, the larger it will seem to us. For example, suppose an object of height h_o is a distance d_o from the eye, as in **FIGURE 27-14 (a)**. The image formed by this object subtends an angle θ on the retina of the eye, as shown in the figure. If the angle is small, as is often the case, it can be approximated by the tangent of the angle, because $\tan \theta \approx \theta$ for small angles. Referring to the figure, we see that $\tan \theta = h_o/d_o$, and hence the angle θ is approximately

$$\theta \approx \frac{h_o}{d_o}$$

If the object is now moved closer to the eye, to a distance d_o', the angle subtended by the image is larger. Using the small-angle approximation again, and referring to **FIGURE 27-14 (b)**, we find that the new angle is

$$\theta' \approx \frac{h_o}{d_o'} > \theta$$

Thus, moving an object closer to the eye increases its apparent size, because its image covers a larger portion of the retina. There is a limit, however, to how close an object can come to the unaided eye and still be in focus—the near point. This is where a magnifier comes into play.

Suppose, then, that you would like to see as much detail as possible on a small object—a feather, perhaps, or a flower petal. With the unaided eye you can bring the object to the near point, a distance N from the eye, as in **FIGURE 27-15 (a)**. If the height of the object is h_o, the angular size of the object on the retina is approximately

$$\theta = \frac{h_o}{N}$$

Now, consider placing a convex lens of focal length $f < N$ just in front of the eye, as shown in **FIGURE 27-15 (b)**. If the object is brought to the focal point of this lens, its image will be infinitely far from the eye, where it can be viewed in focus with ease. As we can see from the figure, the angular size of the image is approximately

$$\theta' = \frac{h_o}{f}$$

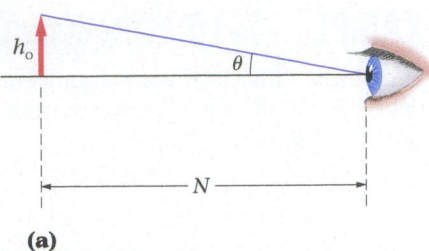

(a)

Note that θ' is greater than θ—because f is less than N—and hence the object appears larger. The factor by which the object is enlarged, referred to as the **angular magnification**, M, is defined as follows:

Angular Magnification, M

$$M = \frac{\theta'}{\theta} \qquad \text{27-3}$$

SI unit: dimensionless

Using the angles $\theta' = h_o/f$ and $\theta = h_o/N$ obtained previously, we find that $M = (h_o/f)/(h_o/N)$, which simplifies to:

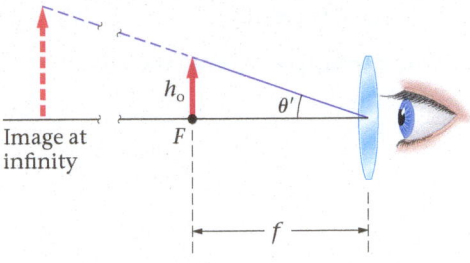

(b)

▲ **FIGURE 27-15 How a simple magnifier works** **(a)** An object viewed with the unaided eye at the near point of the eye subtends an angle $\theta \approx h_o/N$. **(b)** With a magnifier, the object can be viewed from the distance f, which is less than N. As a result, the angular size of the object is $\theta' \approx h_o/f$, which is greater than θ.

Angular Magnification for a Magnifying Glass, M

$$M = \frac{N}{f} \qquad \text{27-4}$$

SI unit: dimensionless

A typical situation is considered in the following Exercise.

EXERCISE 27-9   MAGNIFYING GLASS FOCAL LENGTH

A person with a near-point distance of 26 cm examines a stamp with a magnifying glass. If the magnifying glass produces an angular magnification of 4.0, what is its focal length?

REASONING AND SOLUTION
Rearranging the equation for angular magnification, $M = N/f$, to solve for the focal length, we find

$$f = \frac{N}{M} = \frac{26 \text{ cm}}{4.0} = 6.5 \text{ cm}$$

An object examined at close range with a magnifier appears larger precisely because it is closer to the eye. In fact, the angular size of the object in Figure 27-15 (b) is h_o/f regardless of whether the magnifier is present or not. If the magnifier is not present, however, the object is out of focus, and then its increased angular size is of no practical value. The magnifier is beneficial in that it brings the object into focus at this close distance, as illustrated in **FIGURE 27-16**.

In addition, because the image produced by the magnifier is at infinity, the rays entering the eye are parallel. This means that a person can view the image with a completely relaxed eye. If the same object is viewed with the unaided eye at the near point, the ciliary muscles are fully tensed, causing eye strain. Thus, not only does the magnifier enlarge the object, it also makes it more comfortable to view. As we shall see in the next two sections, the magnifier also plays an important role in both the microscope and the telescope.

It is possible to obtain a magnification that is slightly greater than the value N/f given in Equation 27-4. After all, this magnification is for an image formed at infinity; if the image were closer to the eye, it would appear larger. The closest the image can be—and still be in focus—is the near point. In the next Example we compare the magnifications that result when the image is at infinity and at the near point.

▲ **FIGURE 27-16** A magnifying glass produces an enlarged image when held near an object. The image is upright and virtual.

EXAMPLE 27-10 COMPARING MAGNIFICATIONS

A biologist with a near-point distance of $N = 26$ cm examines an insect wing through a magnifying glass whose focal length is 4.3 cm. Find the angular magnification when the image produced by the magnifier is **(a)** at infinity and **(b)** at the near point.

PICTURE THE PROBLEM

Our sketch shows the situation in which the image formed by the magnifying glass is at the near point. To produce an image at this point, the object must be closer to the magnifier than its focal point. Therefore, the object appears larger in this case than it does in the case where the object is at the focal length and the image is at infinity.

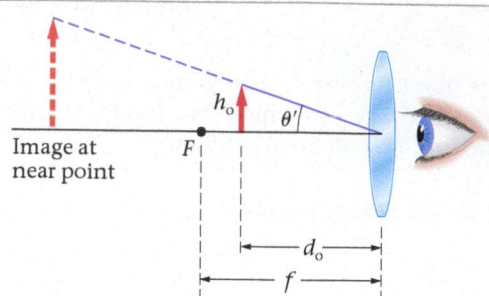

REASONING AND STRATEGY

a. The magnification when the image is at infinity is $M = N/f$, as given in Equation 27-4.

b. The angular size of the object in this case is $\theta' \approx h_o/d_o$, where d_o is the object distance that places the image at the near point. Thus the magnification is $M = \theta'/\theta = (h_o/d_o)/(h_o/N) = N/d_o$. This result differs from the magnification in part (a) only in that f has been replaced with d_o. All that remains is to calculate d_o using the thin-lens equation.

Known Near-point distance, $N = 26$ cm; focal length of magnifying glass, $f = 4.3$ cm.
Unknown Angular magnification, $M = ?$, for **(a)** $d_i = \infty$ and **(b)** $d_i = -26$ cm.

SOLUTION

Part (a)

1. Substitute $N = 26$ cm and $f = 4.3$ cm into $M = N/f$:

$$M = \frac{N}{f} = \frac{26 \text{ cm}}{4.3 \text{ cm}} = 6.0$$

Part (b)

2. Use the thin-lens equation to find d_o, given $f = 4.3$ cm and $d_i = -26$ cm:

$$\frac{1}{d_o} = \frac{1}{f} - \frac{1}{d_i} = \frac{1}{4.3 \text{ cm}} - \frac{1}{-26 \text{ cm}} = 0.27 \text{ cm}^{-1}$$

$$d_o = \frac{1}{0.27 \text{ cm}^{-1}} = 3.7 \text{ cm}$$

3. Calculate the magnification using $M = N/d_o$:

$$M = \frac{N}{d_o} = \frac{26 \text{ cm}}{3.7 \text{ cm}} = 7.0$$

INSIGHT

We see that moving the object closer to the lens, which brings the image in from infinity to the near point, results in an increase in magnification of 1.0—from 6.0 to 7.0.

PRACTICE PROBLEM

What is the magnification if the object is placed 4.0 cm in front of the magnifying glass? [**Answer:** $M = N/d_o = 6.5$]

Some related homework problems: Problem 50, Problem 53

We see from Example 27-10 that the magnification of a magnifying glass with the object at a distance d_o is $M = N/d_o$, where N is the near-point distance of the observer. In the special case of an image at infinity, the object distance is the focal length, and $M = N/f$. It can be shown (see Problem 99) that the greatest magnification—which occurs when the image is at the near point—is $M = 1 + N/f$. The magnification results are summarized here:

$$M = \frac{N}{f} \quad \text{(image at infinity)}$$

$$M = 1 + \frac{N}{f} \quad \text{(image at near point)}$$

27-5

Example 27-10 is simply a special case of these general expressions.

The early microscopes produced by Antonie van Leeuwenhoek (1632–1723) were, in fact, simply powerful magnifying glasses using a single lens mounted in a hole in a flat metal plate. The object to be examined with the microscope was placed on the head of a small, movable pin placed just below the lens. The observer would then hold the top surface of the lens close to the eye for the most stable, wide-angle view. Leeuwenhoek's instruments were capable of magnifying objects by as much as 275 times, enough that he could make the first detailed microscopic descriptions of single-celled animals, red blood cells, plant cells, and much more. Microscopes in common use today have more than one lens, as we describe in the next section.

▲ **FIGURE 27-17** Enhance Your Understanding 3.

Enhance Your Understanding (Answers given at the end of the chapter)

3. A magnifying glass is held over a ruled piece of paper, as shown in **FIGURE 27-17**. What is the magnification of this magnifying glass? Explain.

Section Review

- The magnification M produced by a magnifying glass is $M = N/f$. In this expression, N is the near-point distance, and f is the focal length of the magnifying glass.

27-4 The Compound Microscope

Although a magnifying glass is a useful device, higher magnifications and improved optical quality can be obtained with a **microscope.** The simplest microscope consists of two converging lenses fixed at either end of a tube. Such an instrument, illustrated in **FIGURE 27-18**, is sometimes referred to as a *compound microscope.*

The basic optical elements of a microscope are the **objective** and the **eyepiece.** The objective is a converging lens with a relatively short focal length that is placed near the object to be viewed. It forms a real, inverted, and enlarged image, as shown in **FIGURE 27-19**. The precise location of the image is adjusted when the microscope is focused by moving the objective up or down. This image serves as the object for the second lens in the microscope—the eyepiece. In fact, the eyepiece is simply a magnifier that views the image of the objective, giving it an additional enlargement.

The final magnification of the microscope, then, is simply the product of the magnification of the objective and the magnification of the eyepiece. For example, a microscope might have a 10×eyepiece (meaning it magnifies 10 times) and a 50× objective. When these two lenses are used together, the magnification of the microscope is $10 \cdot 50 = 500$, or 500×.

In a typical situation, the object to be examined is placed only a small distance beyond the focal point of the objective, which means that $d_o \approx f_{objective}$. The magnification produced by the objective is given by Equation 26-18:

$$m_{objective} = -\frac{d_i}{d_o} \approx -\frac{d_i}{f_{objective}}$$

Next, the image formed by the objective is essentially at the focal point of the eyepiece. This means that the eyepiece forms a virtual image at infinity that the observer can view with a relaxed eye. The angular magnification of the eyepiece is given by Equation 27-4 ($M = N/f$):

$$M_{eyepiece} = \frac{N}{f_{eyepiece}}$$

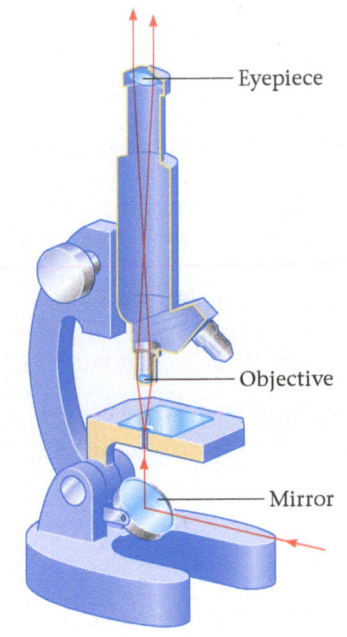

▲ **FIGURE 27-18 Basic rlements of a compound microscope** A compound microscope consists of two lenses—an objective and an eyepiece—fixed at either end of a movable tube.

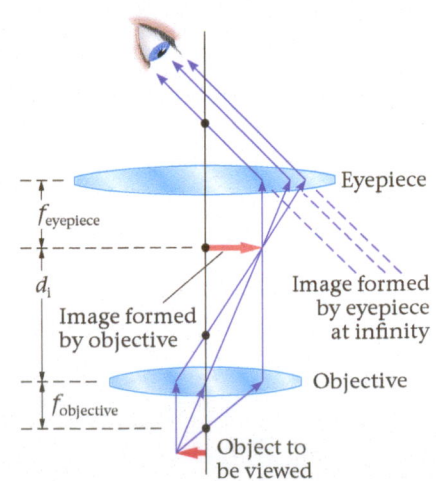

▶ **FIGURE 27-19 The operation of a compound microscope** In a compound microscope the object is placed just outside the focal point of the objective. The resulting enlarged image is then enlarged further by the eyepiece, which is basically a magnifying glass.

PROBLEM-SOLVING NOTE

Lens Placement in a Microscope

In a working microscope, the distance between the lenses is greater than the sum of their focal lengths.

Multiplying these magnifications, we find the total magnification of the microscope:

$$M_{total} = m_{objective} M_{eyepiece} = \left(-\frac{d_i}{f_{objective}}\right)\left(\frac{N}{f_{eyepiece}}\right)$$

$$= -\frac{d_i N}{f_{objective} f_{eyepiece}}$$

27-6

The minus sign indicates that the image is inverted.
Typical numerical values are considered in the following Example.

EXAMPLE 27-11　A MICROSCOPIC VIEW

In biology class, a student with a near-point distance of N = 25 cm uses a microscope to view an amoeba. If the objective has a focal length of 1.0 cm, the eyepiece has a focal length of 2.5 cm, and the amoeba is 1.1 cm from the objective, what is the magnification produced by the microscope?

PICTURE THE PROBLEM
The lenses of the microscope are shown in the sketch. Notice that the distance from the objective to the object is d_o and the distance from the objective to its image is d_i.

REASONING AND STRATEGY
The magnification of the microscope is given by Equation 27-6 ($M_{total} = -(d_i N/f_{objective} f_{eyepiece})$). The only unknown in this expression is the image distance, d_i. We can find this by using the thin-lens equation.

Known Near-point distance, N = 25 cm; objective focal length, $f_{objective}$ = 1.0 cm; eyepiece focal length, $f_{eyepiece}$ = 2.5 cm; object distance, d_o = 1.1 cm.
Unknown Total magnification, M_{total} = ?

SOLUTION

1. Use the thin-lens equation to find the image distance, d_i:

$$\frac{1}{d_i} = \frac{1}{f} - \frac{1}{d_o} = \frac{1}{1.0 \text{ cm}} - \frac{1}{1.1 \text{ cm}} = 0.091 \text{ cm}^{-1}$$

$$d_i = \frac{1}{0.091 \text{ cm}^{-1}} = 11 \text{ cm}$$

2. Use ($M_{total} = -(d_i N/f_{objective} f_{eyepiece})$) to find the magnification of the microscope:

$$M_{total} = -\frac{d_i N}{f_{objective} f_{eyepiece}} = -\frac{(11 \text{ cm})(25 \text{ cm})}{(1.0 \text{ cm})(2.5 \text{ cm})} = -110$$

INSIGHT
The image of the amoeba is inverted and 110 times larger than life size.

PRACTICE PROBLEM — PREDICT/CALCULATE
(a) If the focal length of the eyepiece is increased, does the magnitude of the magnification increase or decrease? Explain. (b) Check your response by calculating the magnification when the focal length of the eyepiece is 3.5 cm. [**Answer: (a)** The magnification decreases in magnitude because it is inversely proportional to the focal length of the eyepiece. **(b)** The new value of the magnification is −79.]

Some related homework problems: Problem 60, Problem 64

Enhance Your Understanding　　　(Answers given at the end of the chapter)

4. Rank the following microscopes in order of increasing magnitude of the total magnification. Indicate ties where appropriate.

Microscope	A	B	C	D
d_i (cm)	10	15	10	15
$f_{objective}$ (cm)	2	1	1	2
$f_{eyepiece}$ (cm)	2	3	4	3

CONTINUED

Section Review

- The total magnification M_{total} produced by a microscope is $(M_{total} = -(d_i N / f_{objective} f_{eyepiece}))$. In this expression, N is the near-point distance, d_i is the image distance, $f_{objective}$ is the focal length of the objective, and $f_{eyepiece}$ is the focal length of the eyepiece.

27-5 Telescopes

A telescope is similar in many respects to a microscope—both use two converging lenses to give a magnified image of an object with small angular size. In the case of a microscope, the object itself is small and close at hand; in the case of the telescope, the object may be as large as a galaxy, but its angular size can be very small because of its great distance. The major difference between these instruments is that the telescope must deal with an object that is essentially infinitely far away.

For this reason, it is clear that the light entering the objective of a telescope from a distant object is focused at the focal point of the objective, as shown in **FIGURE 27-20**. If the image formed by the objective has a height h_i, the angular size of the object is approximately

$$\theta = -\frac{h_i}{f_{objective}}$$

The minus sign is included because h_i is negative for the inverted image formed by the objective.

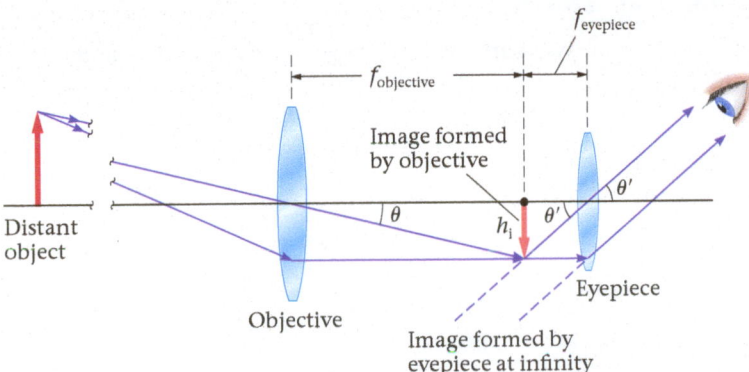

Distant object · Objective · $f_{objective}$ · Image formed by objective · h_i · θ · θ' · $f_{eyepiece}$ · Eyepiece · Image formed by eyepiece at infinity

◄ **FIGURE 27-20 Basic elements of a telescope** A telescope focuses light from a distant object at its focal point. The image of the objective is placed at the focal point of the eyepiece to produce an enlarged image that can be viewed with a relaxed eye.

As in the microscope, the image of the objective is the object for the eyepiece, which is basically a magnifier. Thus, if the image of the objective is placed at the focal point of the eyepiece, it will form an image that is at infinity, as indicated in Figure 27-20. In this configuration, the observer can view the final image of the telescope with a completely relaxed eye. The angular size of the image formed by the eyepiece, θ', is shown in Figure 27-20. This angle is approximately

$$\theta' = -\frac{h_i}{f_{eyepiece}}$$

To find the total angular magnification of the telescope, we take the ratio of θ' to θ:

$$M_{total} = \frac{\theta'}{\theta} = \frac{f_{objective}}{f_{eyepiece}} \qquad \text{27-7}$$

Thus, for example, a telescope with an objective whose focal length is 1500 mm and an eyepiece whose focal length is 10.0 mm produces an angular magnification of 150.

PROBLEM-SOLVING NOTE

Lens Placement in a Telescope

In a working telescope, the distance between the lenses is approximately equal to the sum of their focal lengths. In addition, the focal length of the eyepiece is significantly less than that of the objective.

CONCEPTUAL EXAMPLE 27-12 COMPARING TELESCOPES

Two telescopes have identical eyepieces, but telescope A is twice as long as telescope B. Is the magnification of telescope A greater than the magnification of telescope B, is it less than the magnification of telescope B, or is there no way to tell?

REASONING AND DISCUSSION

Notice in Figure 27-20 that the total length of the telescope is $f_{objective}$ plus $f_{eyepiece}$. Thus, because the scopes have identical eyepieces, it follows that telescope A must have a greater objective focal length than telescope B. We know, then, from $M_{total} = f_{objective}/f_{eyepiece}$, that the magnification of telescope A is greater than the magnification of telescope B.

ANSWER

Telescope A has a greater magnification than telescope B.

Telescopes that use two or more lenses, as in Figure 27-20, are referred to as *refractors*. In fact, the first telescopes constructed for astronomical purposes, made by Galileo starting in 1609, were refractors. Galileo's telescopes differed from the telescopes in common use today, however, in that they used a diverging lens for the eyepiece. We consider this type of telescope in Problems 76, 84, and 91. By the end of 1609, Galileo had produced a telescope whose angular magnification was 20. This was more than enough to enable him to see—for the first time in human history—mountains on the Moon, stars in the Milky Way, the phases of Venus, and moons orbiting Jupiter. As a result of his telescopic observations, Galileo became a firm believer in the Copernican model of the solar system.

In the next Example we consider a standard refractor with two converging lenses. In particular, we show how the length of such a telescope is related to the focal lengths of its objective and eyepiece.

EXAMPLE 27-13 CONSIDERING THE TELESCOPE AT LENGTH

A telescope has a magnification of 40.0 and a length of 1230 mm. What are the focal lengths of the objective and eyepiece?

PICTURE THE PROBLEM

Our sketch shows that the overall length of the telescope, L, is equal to the sum of the focal lengths of the objective and the eyepiece. In addition, we see that the objective forms an inverted image of the distant object, which the eyepiece enlarges. The final image seen by the observer, then, is inverted.

REASONING AND STRATEGY

We can determine the two unknowns, $f_{objective}$ and $f_{eyepiece}$, from the two independent pieces of information given in the problem statement: (1) The magnification, $M_{total} = f_{objective}/f_{eyepiece}$, is equal to 40.0. (2) The total length, $L = f_{objective} + f_{eyepiece}$, is equal to 1230 mm.

Combining these statements gives us the two focal lengths.

Known Total magnification, $M_{total} = 40.0$; length of telescope, $L = 1230$ mm.
Unknown Objective focal length, $f_{objective} = ?$ Eyepiece focal length, $f_{eyepiece} = ?$

SOLUTION

1. Use the magnification equation to write $f_{objective}$ in terms of $f_{eyepiece}$:

$$M_{total} = \frac{f_{objective}}{f_{eyepiece}} = 40.0$$

$$f_{objective} = 40.0 f_{eyepiece}$$

2. Substitute $f_{objective} = 40.0 f_{eyepiece}$ into the expression for the total length of the telescope:

$$L = f_{objective} + f_{eyepiece}$$
$$= 40.0 f_{eyepiece} + f_{eyepiece} = 41.0 f_{eyepiece} = 1230 \text{ mm}$$

3. Divide 1230 mm by 41.0 to find the focal length of the eyepiece:

$$f_{eyepiece} = \frac{1230 \text{ mm}}{41.0} = 30.0 \text{ mm}$$

4. Multiply $f_{eyepiece}$ by 40.0 to find the focal length of the objective:

$$f_{objective} = 40.0 f_{eyepiece} = 40.0(30.0 \text{ mm}) = 1200 \text{ mm}$$

CONTINUED

INSIGHT
The focal lengths found in this example are typical of those used in many popular amateur telescopes. A telescope with a magnification of 40.0 can easily show the moons of Jupiter and the rings of Saturn.

PRACTICE PROBLEM
A telescope 1820 mm long has an objective with a focal length of 1780 mm. Find the focal length of the eyepiece and the magnification of the telescope.   [**Answer:** $f_{eyepiece} = 40.0$ mm, $M_{total} = 44.5$]

Some related homework problems: Problem 69, Problem 70, Problem 78

Astronomical Telescopes   Because telescopes are typically used to view objects that are very dim, it is desirable to have an objective with as large a diameter, or aperture, as possible. If the diameter is doubled, for example, the light gathered by a telescope increases by the same factor as the area of its objective: $2^2 = 4$. In addition, a larger aperture results in a higher-resolution image, as we shall see in the next chapter. Examples of telescopes with large apertures are shown in **FIGURE 27-21**.

▲ **FIGURE 27-21 Visualizing Concepts Astronomical Telescopes** At left, the 1-m objective lens of the Yerkes refractor—the largest refracting telescope ever constructed. At right, the twin 10-m Keck reflecting telescopes atop Mauna Kea on the Big Island of Hawaii, where the air is cold, thin, and dry—ideal conditions for astronomical observation.

For a refractor, a large aperture means a very large and heavy piece of glass. In fact, the world's largest refractor, the Yerkes refractor at Williams Bay, Wisconsin, has an objective that is 1 m across. If refractors were made much larger, the objective lens would sag and distort under its own weight.

The telescopes with the largest apertures are reflectors, one example of which is illustrated in **FIGURE 27-22**. Invented by Isaac Newton in 1671, and referred to as a Newtonian reflector, this type of telescope uses a mirror in place of an objective lens. As with the refractor, the mirror forms an image that the eyepiece then magnifies. Because a mirror can be much thinner and lighter than a lens, and can be supported all over its back surface instead of just around the edges as with a lens, it has many advantages for a large scope. For example, the twin Keck reflecting telescopes atop Hawaii's Mauna Kea have hexagonal objective mirrors that are 10 m across, which means that each gathers 100 times as much light as the Yerkes refractor. When used together, the Keck telescopes constitute the world's largest pair of binoculars!

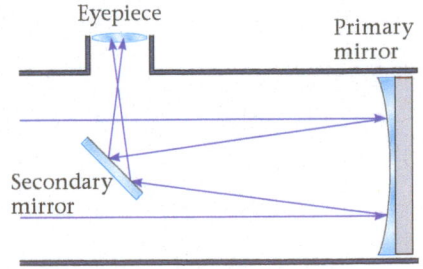

▲ **FIGURE 27-22 A Newtonian reflecting telescope** The large primary mirror collects incoming light and reflects it off a small secondary mirror into the eyepiece.

Enhance Your Understanding   (Answers given at the end of the chapter)

5. In a typical telescope, is $f_{eyepiece}$ greater than, less than, or equal to $f_{objective}$? Explain.

CONTINUED

Big Idea 6 Common defects (aberrations) of lenses include a distorted shape, which focuses rays at different points, and dispersion, which focuses different colors at different locations.

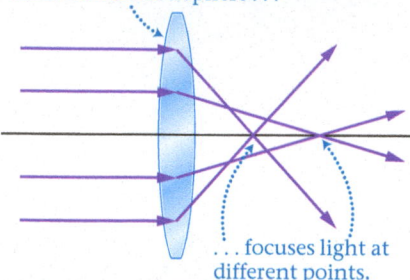

A lens whose surface is shaped like a section of a sphere . . .

. . . focuses light at different points.

▲ **FIGURE 27-23 Spherical aberration** In a lens with spherical aberration, light striking the lens at different locations comes together at different focal points.

PHYSICS IN CONTEXT
Looking Back

Dispersion, which was discussed in Chapter 26 in relation to rainbows, appears again in this chapter as the cause of chromatic aberration.

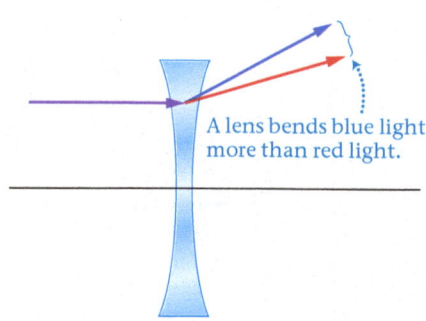

A lens bends blue light more than red light.

▲ **FIGURE 27-26 Chromatic aberration in a diverging lens** A diverging lens also bends blue light more than red light, though the rays are bent in the opposite direction compared with the bending by a converging lens.

27-6 Lens Aberrations

An ideal lens brings all parallel rays of light that strike it together at a single focal point. Real lenses, however, never quite live up to the ideal. Instead, a real lens blurs the focal point into a small but finite region of space. This, in turn, blurs the image it forms. The deviation of a lens from ideal behavior is referred to as an **aberration.**

One example is **spherical aberration,** which is related to the shape of a lens. FIGURE 27-23 shows parallel rays of light passing through a lens with spherical aberration; notice that the rays do not meet at a single focal point. This is completely analogous to the spherical aberration present in mirrors that have a spherical cross section, as shown in Figure 26-20. To prevent spherical aberration, a lens must be ground and polished to a very precise nonspherical shape.

Another common aberration in lenses is **chromatic aberration,** which is due to the unavoidable dispersion present in any refracting material. Just as a prism splits white light into a spectrum of colors, each color bent by a different amount, so too does a lens bend light of different colors by different amounts as shown in FIGURE 27-24. The result is that white light passing through a lens does not focus to a single point. This is why you sometimes see a fringe of color around an image seen through a simple lens, as shown in FIGURE 27-25.

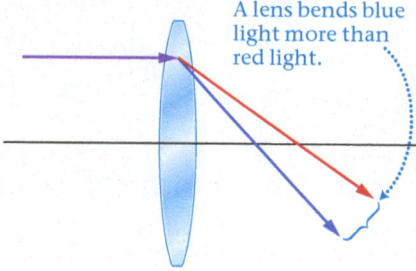

A lens bends blue light more than red light.

▲ **FIGURE 27-24 Chromatic aberration in a converging lens** Chromatic aberration is caused by the fact that blue light bends more than red light as it passes through a lens. As a result, different colors have different focal points.

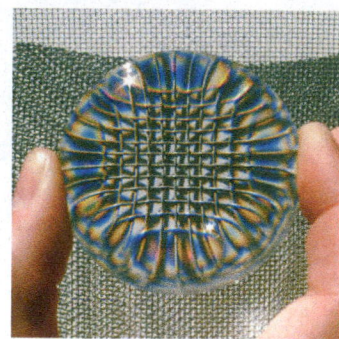

▲ **FIGURE 27-25** Chromatic aberration typical of a simple lens.

Chromatic aberration can be corrected by combining two or more lenses to form a compound lens. For example, suppose a convex lens made from glass A bends red and blue light as shown in Figure 27-24. Notice that the blue light is bent more, as expected. A second lens, concave this time, is made of glass B. This lens also bends blue light more than red light, but since it is a diverging lens, it bends both the red and blue light in the opposite direction compared with the convex lens, as we see in FIGURE 27-26. If the amount of bending of blue light compared with red light is different for glasses A and B, it is possible to construct a convex and a concave lens in such a way that the opposite directions in which they bend red and blue light can be made to cancel while still causing the light to come to a focus. This combination is illustrated in FIGURE 27-27. Such a lens is said to be **achromatic.** Three lenses connected similarly can produce even better correction for chromatic aberration. Lenses of this type are referred to as **apochromatic.**

Many lenses in optical instruments are not single lenses at all but, instead, are compound achromatic lenses. For example, the "lens" in a 35-mm camera may

▶ **FIGURE 27-27 Correcting for chromatic aberration** An achromatic doublet is a combination of a converging and a diverging lens made of different types of glass. In the converging lens, the blue light is bent toward the axis of the lens more than the red light. In the diverging lens, the opposite is the case. The net result is that the red and blue light pass through the same focal point.

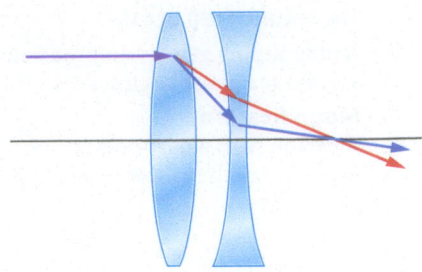

Two lenses canceling chromatic aberration

actually contain five or more individual lenses—some of them converging, some diverging, some made of one type of glass, some made of another. A zoom lens for a camera must not only correct for chromatic aberration, but must also be able to change its focal length. Thus, it is not uncommon for such a lens to contain as many as 12 individual lenses.

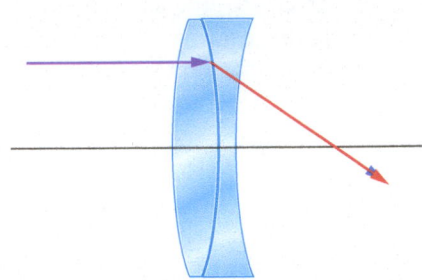

An achromatic doublet

Enhance Your Understanding   (Answers given at the end of the chapter)

6. One advantage of reflecting telescopes over refracting telescopes is that they are practically immune to which of these defects: **(a)** spherical aberration or **(b)** chromatic aberration?

Section Review

- Spherical aberration is caused by a lens with a spherical cross section. The result is that parallel rays of light are not focused at the same point.

- Chromatic aberration is caused by the dispersion (or spreading out) of colors in a lens. The result is that different colors come to a focus at different locations.

CHAPTER 27 REVIEW

CHAPTER SUMMARY

27-1   THE HUMAN EYE AND THE CAMERA

The human eye forms a real, inverted image on the retina; a camera forms a real, inverted image on light-sensitive material.

Focusing the Eye
The eye is focused by the ciliary muscles, which change the shape of the lens. This process is referred to as accommodation.

Near Point
The near point is the closest distance to which the eye can focus. A typical value for the near-point distance, N, is 25 cm.

Far Point
The far point is the greatest distance at which the eye can focus. In a normal eye the far point is infinity.

f-Number
The f-number of a lens relates the diameter of the aperture, D, to the focal length, f, as follows:

$$f\text{-number} = \frac{\text{focal length}}{\text{diameter of aperture}} = \frac{f}{D} \qquad 27\text{-}1$$

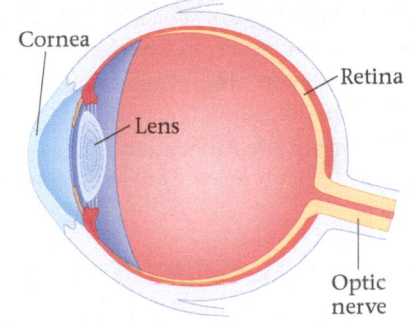

27-2   LENSES IN COMBINATION AND CORRECTIVE OPTICS

The basic idea used in analyzing systems that have more than one lens is the following: The *image* produced by one lens acts as the *object* for the next lens. The total magnification of a system of lenses is given by the product of the magnifications produced by each lens individually.

Nearsightedness
Nearsightedness is a condition in which clear vision is restricted to a region relatively close to the eye. This condition can be corrected with diverging lenses placed in front of the eyes.

Farsightedness
A person who is farsighted can see clearly only at relatively large distances from the eye. This condition can be corrected by placing converging lenses in front of the eyes.

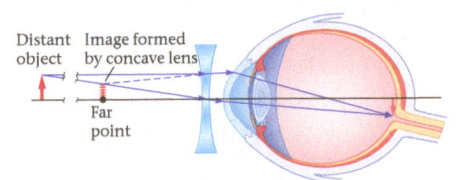

27-3 THE MAGNIFYING GLASS

A magnifying glass is simply a converging lens. It works by allowing an object to be viewed at a distance closer than the near-point distance.

Magnification

A person with a near-point distance N using a magnifying glass with a focal length f experiences the following magnification, M:

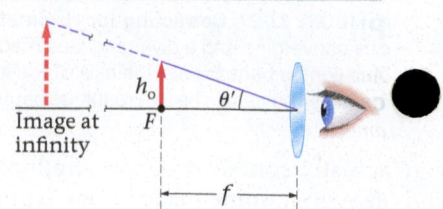

$$M = \frac{N}{f} \quad \text{(image at infinity)}$$

27-4 THE COMPOUND MICROSCOPE

A compound microscope uses two lenses in combination—an objective and an eyepiece—to produce a magnified image.

Magnification

The magnification produced by a compound microscope is

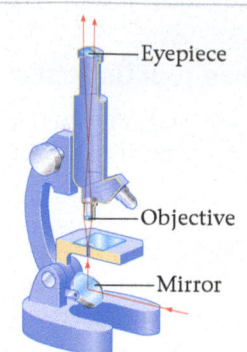

$$M_{\text{total}} = -\frac{d_i N}{f_{\text{objective}} f_{\text{eyepiece}}} \qquad \text{27-6}$$

In this expression, N is the near-point distance and d_i is the image distance for the objective.

27-5 TELESCOPES

A telescope provides magnified views of distant objects.

Magnification

The total magnification produced by a telescope is

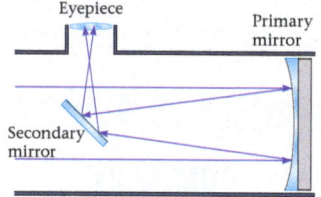

$$M_{\text{total}} = \frac{f_{\text{objective}}}{f_{\text{eyepiece}}} \qquad \text{27-7}$$

27-6 LENS ABERRATIONS

Any deviation of a lens from ideal behavior is referred to as an aberration.

Spherical Aberration

Spherical aberration is related to the shape of a lens.

Chromatic Aberration

Chromatic aberration results from dispersion within a refracting material.

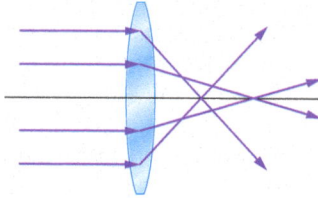

ANSWERS TO ENHANCE YOUR UNDERSTANDING QUESTIONS

1. The f-number is inversely related to the diameter of the aperture. Therefore, increasing the f-number decreases the aperture.

2. When viewing an object at infinity, a nearsighted eye forms an image in front of the retina.

3. The spacing of the lines on the paper is doubled by the magnifying glass; therefore the magnification is 2.

4. A = C = D < B.

5. A typical telescope has a total magnification that is greater than 1. It follows that f_{eyepiece} is less than $f_{\text{objective}}$.

6. **(b)** Reflecting telescopes reflect all colors of light at the same angle, and hence they do not have chromatic aberration like refracting telescopes.

CONCEPTUAL QUESTIONS

For instructor-assigned homework, go to www.masteringphysics.com.

(Answers to odd-numbered Conceptual Questions can be found in the back of the book.)

1. Why is it restful to your eyes to gaze off into the distance?

2. If a lens is cut in half through a plane perpendicular to its surface, does it show only half an image?

3. If your near-point distance is N, how close can you stand to a mirror and still be able to focus on your image?

4. When you open your eyes underwater, everything looks blurry. Can this be thought of as an extreme case of nearsightedness or farsightedness? Explain.

5. When you use a simple magnifying glass, does it matter whether you hold the object to be examined closer to the lens than its focal length or farther away? Explain.

6. Does chromatic aberration occur in mirrors? Explain.

PROBLEMS AND CONCEPTUAL EXERCISES

Answers to odd-numbered Problems and Conceptual Exercises can be found in the back of the book. **BIO** *identifies problems of biological or medical interest;* **CE** *indicates a conceptual exercise.* Predict/Explain *problems ask for two responses: (a) your prediction of a physical outcome, and (b) the best explanation among those provided; and* Predict/Calculate *problems ask for a prediction of a physical outcome, based on fundamental physics concepts, and follow that with a numerical calculation to verify the prediction. On all problems, bullets (•, ••, •••) indicate the level of difficulty.*

SECTION 27-1 THE HUMAN EYE AND THE CAMERA

1. • **CE BIO** Predict/Explain **Octopus Eyes** To focus its eyes, an octopus does not change the shape of its lens, as is the case in humans. Instead, an octopus moves its rigid lens back and forth, as in a camera. This changes the distance from the lens to the retina and brings an object into focus. **(a)** If an object moves closer to an octopus, must the octopus move its lens closer to or farther from its retina to keep the object in focus? **(b)** Choose the *best explanation* from among the following:

    I. The lens must move closer to the retina—that is, farther away from the object—to compensate for the object moving closer to the eye.

    II. When the object moves closer to the eye, the image produced by the lens will be farther behind the lens; therefore, the lens must move farther from the retina.

2. • Your friend is 1.7 m tall. **(a)** When she stands 3.2 m from you, what is the height of her image formed on the retina of your eye? (Consider the eye to consist of a thin lens 2.5 cm from the retina.) **(b)** What is the height of her image when she is 4.2 m from you?

3. • Which forms the larger image on the retina of your eye: a 43-ft tree seen from a distance of 210 ft, or a 12-in. flower viewed from a distance of 2.0 ft?

4. • Approximating the eye as a single thin lens 2.70 cm from the retina, find the eye's near-point distance if the smallest focal length the eye can produce is 2.10 cm.

5. • Approximating the eye as a single thin lens 2.70 cm from the retina, find the focal length of the eye when it is focused on an object at a distance of **(a)** 255 cm and **(b)** 25.5 cm?

6. • Find the far-point distance of a person whose relaxed eye has a focal length of 2.68 cm, treating the eye as if it were a single thin lens 2.71 cm from the retina.

7. • Four camera lenses have the following focal lengths and f-numbers:

Lens	Focal length (mm)	f-number
A	150	$f/1.2$
B	150	$f/5.6$
C	35	$f/1.2$
D	35	$f/5.6$

    Rank these lenses in order of increasing aperture diameter. Indicate ties where appropriate.

8. • **BIO** The focal length of the human eye is approximately 1.7 cm. **(a)** What is the f-number for the human eye in bright light, when the pupil diameter is 2.0 mm? **(b)** What is the f-number in dim light, when the pupil diameter has expanded to 7.0 mm?

9. • Predict/Calculate A camera with a 65-mm-focal-length lens has aperture f-numbers of 2.8, 4, 8, 11, and 16. **(a)** Which setting has the largest aperture diameter? **(b)** Calculate the five possible aperture diameters for this camera.

10. •• The actual light sensor size of a digital camera is 14.9 mm × 22.3 mm. You want to take a photo of your friend, who is 1.9 m tall. Your camera has a 55-mm-focal-length lens. How far from the camera should your friend stand in order to produce a 22-mm-tall image on the light sensor?

11. •• **(a)** Find the f-number of a telescope with an objective diameter of 8.3 cm and a focal length of 91 cm. **(b)** Find the aperture diameter of an $f/1.2$ all-sky meteor camera lens with a focal length of 2.5 mm.

12. •• You are taking a photo of a poster on the wall of your dorm room, so you can't back away any farther than 3.0 m to take the shot. The poster is 0.80 m wide and 1.2 m tall, and you want the image to fit on the 15-mm × 22-mm image sensor in your camera. What is the longest focal length lens that will work?

13. •• You are taking pictures of the beach at sunset. Just before the Sun sets, a shutter speed of 1/100 and an aperture of $f/11$ produces a properly exposed picture. Shortly after the Sun sets, however, your light meter indicates that the scene is only one-quarter as bright as before. **(a)** If you don't change the aperture, what approximate shutter speed is needed for your second shot? **(b)** If, instead, you keep the shutter speed at 1/100 s, what approximate f-stop will be needed for the second shot?

14. •• Predict/Calculate You are taking a photograph of a horse race. A shutter speed of 125 at $f/5.6$ produces a properly exposed image, but the running horses give a blurred image. Your camera has f-stops of 2, 2.8, 4, 5.6, 8, 11, and 16. **(a)** To use the shortest possible exposure time (i.e., highest shutter speed), which f-stop should you use? **(b)** What is the shortest exposure time you can use and *still* get a properly exposed image?

15. •• **The Hale Telescope** The 200-in. (5.08-m) diameter mirror of the Hale telescope on Mount Palomar has a focal length $f = 16.9$ m. **(a)** When the detector is placed at the focal point of the mirror (the "prime focus"), what is the f-number for this telescope? **(b)** The coudé focus arrangement uses additional mirrors to bend the light path and increase the effective focal length to 155.4 m. What is the f-number of the telescope when the coudé focus is being used? (*Coudé* is French for "elbow," since the light path is "bent like an elbow." This arrangement is useful when the light needs to be focused onto a distant instrument.)

SECTION 27-2 LENSES IN COMBINATION AND CORRECTIVE OPTICS

16. • **CE** Predict/Explain Two professors are stranded on a deserted island. Both wear glasses, though one is nearsighted and the other is farsighted. **(a)** Which person's glasses should be used to focus the rays of the Sun and start a fire? **(b)** Choose the *best explanation* from among the following:

    I. A nearsighted person can focus close, so that person's glasses should be used to focus the sunlight on a piece of moss at a distance of a couple inches.

    II. A farsighted person can't focus close, so the glasses to correct that person's vision are converging. A converging lens is what you need to concentrate the rays of the Sun.

17. • **CE** A clerk at the local grocery store wears glasses that make her eyes look larger than they actually are. Is the clerk nearsighted or farsighted? Explain.

18. • **CE** The umpire at a baseball game wears glasses that make his eyes look smaller than they actually are. Is the umpire nearsighted or farsighted? Explain.

19. • **CE** A police detective discovers eyeglasses with a focal length of −80.0 cm at a crime scene. **(a)** If the eyeglasses belong to the suspect, is the suspect nearsighted or farsighted? Explain. **(b)** A search of a person's home reveals an eyeglass prescription of −0.80 diopter. Is the person the suspect? Explain.

20. • **BIO** The cornea of a normal human eye has an optical power of +44.0 diopters. What is its focal length?

21. • A myopic student is shaving without his glasses. If his eyes have a far point of 1.9 m, what is the greatest distance he can stand from the mirror and still see his image clearly?

22. • An eyeglass prescription calls for a lens with an optical power of +2.9 diopters. What is the focal length of this lens?

23. • An optometrist prescribes contact lenses with a power of −0.75 diopter for you. What is your far-point distance?

24. •• Two thin lenses, with $f_1 = +25.0$ cm and $f_2 = -42.5$ cm, are placed in contact. What is the focal length of this combination?

25. •• Two concave lenses, each with $f = -15$ cm, are separated by 7.5 cm. An object is placed 25 cm in front of one of the lenses. Find **(a)** the location and **(b)** the magnification of the final image produced by this lens combination.

26. •• **BIO** Predict/Calculate The focal length of a relaxed human eye is approximately 1.7 cm. When we focus our eyes on a close-up object, we can change the refractive power of the eye by about 16 diopters. **(a)** Does the refractive power of our eyes increase or decrease by 16 diopters when we focus closely? Explain. **(b)** Calculate the focal length of the eye when we focus closely.

27. •• **BIO** Predict/Calculate **Diopter Change in Diving Cormorants** Double-crested cormorants (*Phalacrocorax auritus*) are extraordinary birds—they can focus on objects in the air, just like we can, but they can also focus underwater as they pursue their prey. To do so, they have one of the largest *accommodation ranges* in nature—that is, they can change the focal length of their eyes by amounts that are greater than is possible in other animals. When a cormorant plunges into the ocean to catch a fish, it can change the refractive power of its eyes by about 45 diopters, as compared to only 16 diopters of change possible in the human eye. **(a)** Should this change of 45 diopters be an increase or a decrease? Explain. **(b)** If the focal length of the cormorant's eyes is 4.2 mm before it enters the water, what is the focal length after the refractive power changes by 45 diopters?

28. •• A converging lens of focal length 9.000 cm is 18.0 cm to the left of a diverging lens of focal length −6.00 cm. A coin is placed 12.0 cm to the left of the converging lens. Find **(a)** the location and **(b)** the magnification of the coin's final image.

29. •• Repeat Problem 28, this time with the coin placed 18.0 cm to the right of the diverging lens.

30. •• Find the focal length of contact lenses that would allow a farsighted person with a near-point distance of 166 cm to read a book at a distance of 10.1 cm.

31. •• Find the focal length of contact lenses that would allow a nearsighted person with a 125-cm far point to focus on the stars at night.

32. •• What focal length should a pair of contact lenses have if they are to correct the vision of a person with a near point of 66 cm?

33. •• Reading glasses with a power of +1.50 diopters make reading a book comfortable for you when you wear them 2.2 cm from your eye. If you hold the book 28.0 cm from your eye, what is your near-point distance?

34. •• A nearsighted person wears contacts with a focal length of −8.5 cm. If this person's far-point distance with her contacts is 8.5 m, what is her uncorrected far-point distance?

35. •• Without his glasses, Isaac can see objects clearly only if they are less than 3.8 m from his eyes. What focal length glasses worn 2.0 cm from his eyes will allow Isaac to see distant objects clearly?

36. •• A person whose near-point distance is 42.5 cm wears a pair of glasses that are 2.1 cm from her eyes. With the aid of these glasses, she can now focus on objects 25 cm away from her eyes. Find the focal length and refractive power of her glasses.

37. •• A pair of eyeglasses is designed to allow a person with a far-point distance of 2.50 m to read a road sign at a distance of 25.0 m. Find the focal length required of these glasses if they are to be worn **(a)** 2.00 cm or **(b)** 1.00 cm from the eyes.

38. •• Predict/Calculate Your favorite aunt can read a newspaper only if it is within 15.0 cm of her eyes. **(a)** Is your aunt nearsighted or farsighted? Explain. **(b)** Should your aunt wear glasses that are converging or diverging to improve her vision? Explain. **(c)** How many diopters of refractive power must her glasses have if they are worn 2.00 cm from the eyes and allow her to read a newspaper at a distance of 25.0 cm?

39. •• Predict/Calculate The relaxed eyes of a patient have a refractive power of 48.5 diopters. **(a)** Is this patient nearsighted or farsighted? Explain. **(b)** If this patient is nearsighted, find the far point. If this person is farsighted, find the near point. (For the purposes of this problem, treat the eye as a single-lens system, with the retina 2.40 cm from the lens.)

40. •• Without glasses, your Uncle Albert can see things clearly only if they are between 35 cm and 160 cm from his eyes. **(a)** What power eyeglass lens will correct your uncle's myopia? Assume the lenses will sit 2.0 cm from his eyes. **(b)** What is your uncle's near point when wearing these glasses?

41. •• A 2.05-cm-tall object is placed 30.0 cm to the left of a converging lens with a focal length $f_1 = 20.5$ cm. A diverging lens, with a focal length $f_2 = -42.5$ cm, is placed 30.0 cm to the right of the first lens. How tall is the final image of the object?

42. •• A simple camera telephoto lens consists of two lenses. The objective lens has a focal length $f_1 = +39.0$ cm. Precisely 36.0 cm behind this lens is a concave lens with a focal length $f_2 = -10.0$ cm. The object to be photographed is 4.00 m in front of the objective lens. **(a)** How far behind the concave lens should the film be placed? **(b)** What is the linear magnification of this lens combination?

43. •• Predict/Calculate With unaided vision, a librarian can focus only on objects that lie at distances between 5.0 m and 0.50 m. **(a)** Which type of lens (converging or diverging) is needed to correct his nearsightedness? Explain. **(b)** Which type of lens will correct his farsightedness? Explain. **(c)** Find the refractive power needed for each part of the bifocal eyeglass lenses that will give the librarian normal visual acuity from 25 cm out to infinity. (Assume the lenses rest 2.0 cm from his eyes.)

44. •• A person's prescription for her new bifocal glasses calls for a refractive power of −0.445 diopter in the distance-vision part, and a power of +1.85 diopters in the close-vision part. What are the near and far points of this person's uncorrected vision? Assume the glasses are 2.00 cm from the person's eyes, and that the person's near-point distance is 25.0 cm when wearing the glasses.

45. •• A person's prescription for his new bifocal eyeglasses calls for a refractive power of −0.0625 diopter in the distance-vision part and a power of +1.05 diopters in the close-vision part. Assuming the glasses rest 2.00 cm from his eyes and that the corrected near-point distance is 25.0 cm, determine the near and far points of this person's uncorrected vision.

46. ••• Two lenses, with $f_1 = +20.0$ cm and $f_2 = +30.0$ cm, are placed on the x axis, as shown in **FIGURE 27-28**. An object is fixed 50.0 cm to the left of lens 1, and lens 2 is a variable distance x to the right of lens 1. Find the lateral magnification and location of the final image relative to lens 2 for the following cases: **(a)** $x = 115$ cm; **(b)** $x = 30.0$ cm; **(c)** $x = 0$. **(d)** Show that your result for part (c) agrees with the relation for the effective focal length of two lenses in contact, $1/f_{\text{eff}} = 1/f_1 + 1/f_2$.

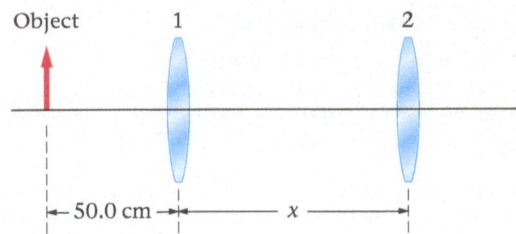

FIGURE 27-28 Problem 46

47. ••• A converging lens with a focal length of 4.0 cm is to the left of a second identical lens. When a feather is placed 12 cm to the left of the first lens, the final image is the same size and orientation as the feather itself. What is the separation between the lenses?

SECTION 27-3 THE MAGNIFYING GLASS

48. • **CE** Two magnifying glasses are for sale at a store. Magnifying glass 1 has a 4-in. diameter with a long focal length, and glass 2 has a 1-in. diameter with a short focal length. **(a)** Which magnifying glass should you purchase if you wish to examine tiny insects? Explain. **(b)** Which glass should you purchase if you wish to start a campfire using sunlight? Explain.

49. • The Moon is 3476 km in diameter and orbits the Earth at an average distance of 384,400 km. **(a)** What is the angular size of the Moon as seen from Earth? **(b)** A penny is 19 mm in diameter. How far from your eye should the penny be held to produce the same angular diameter as the Moon?

50. •• A magnifying glass is a single convex lens with a focal length of $f = +14.0$ cm. **(a)** What is the angular magnification when this lens forms a (virtual) image at $-\infty$? How far from the object should the lens be held? **(b)** What is the angular magnification when this lens forms a (virtual) image at the person's near point (assumed to be 25 cm)? How far from the object should the lens be held in this case?

51. •• Calculate the focal length of a magnifying lens designed to produce an angular magnification of 8.50 while producing the image at the standard near-point distance of 25.0 cm.

52. •• **Predict/Calculate** A student has two lenses, one of focal length $f_1 = 5.0$ cm and the other with focal length $f_2 = 13$ cm. **(a)** When used as a simple magnifier, which of these lenses can produce the greater magnification? Explain. **(b)** Find the maximum magnification produced by each of these lenses.

53. •• A beetle 4.93 mm long is examined with a simple magnifier of focal length $f = 10.1$ cm. If the observer's eye is relaxed while using the magnifier, and has a near-point distance of 26.0 cm, what is the apparent length of the beetle?

54. •• To engrave wishes of good luck on a watch, an engraver uses a magnifier whose focal length is 8.75 cm. If the image formed by the magnifier is at the engraver's near point of 24.6 cm, find **(a)** the distance between the watch and the magnifier and **(b)** the angular magnification of the engraving. Assume the magnifying glass is directly in front of the engraver's eyes.

55. •• A jeweler examines a diamond with a magnifying glass. If the near-point distance of the jeweler is 22.8 cm, and the focal length of the magnifying glass is 7.70 cm, find the angular magnification when the diamond is held at the focal point of the magnifier. Assume the magnifying glass is directly in front of the jeweler's eyes.

56. •• In Problem 55, find the angular magnification when the diamond is held 5.59 cm from the magnifying glass.

57. ••• A person with a near-point distance of 25 cm finds that a magnifying glass gives an angular magnification that is 1.5 times larger when the image of the magnifier is at the near point than when the image is at infinity. What is the focal length of the magnifying glass?

SECTION 27-4 THE COMPOUND MICROSCOPE

58. • **CE** You have two lenses: lens 1 with a focal length of 0.45 cm and lens 2 with a focal length of 1.9 cm. If you construct a microscope with these lenses, which one should you use as the objective? Explain.

59. • **Predict/Calculate** Microscope objective A is labeled 15× and objective B is labeled 25×. **(a)** Which objective has the longer focal length? Explain. If the image formed by the objective is designed to be 16.3 cm from the lens, calculate the focal length of the **(b)** 15× and **(c)** 25× objectives.

60. • A compound microscope has an objective lens with a focal length of 2.2 cm and an eyepiece with a focal length of 5.4 cm. If the image produced by the objective is 12 cm from the objective, what magnification does this microscope produce?

61. • **BIO** A typical red blood cell subtends an angle of only 1.9×10^{-5} rad when viewed at a person's near-point distance of 25 cm. Suppose a red blood cell is examined with a compound microscope in which the objective and eyepiece are separated by a distance of 12.0 cm. Given that the focal length of the eyepiece is 2.7 cm, and the focal length of the objective is 0.49 cm, find the magnitude of the angle subtended by the red blood cell when viewed through this microscope.

62. •• **(a)** If you treat a 10× eyepiece of a microscope as a magnifying glass that produces an image at infinity, what is its focal length? **(b)** The microscope is designed for an image distance of 163 mm from the objective, and the objective is marked 25×. What is its focal length?

63. •• The medium-power objective lens in a laboratory microscope has a focal length $f_{\text{objective}} = 3.75$ mm. **(a)** If this lens produces a magnification of -50.0, what is its "working distance"; that is, what is the distance from the object to the objective lens? **(b)** What is the focal length of an eyepiece lens that will provide an overall magnification of -125?

64. •• A compound microscope has the objective and eyepiece mounted in a tube that is 18.0 cm long. The focal length of the eyepiece is 2.62 cm, and the near-point distance of the person using the microscope is 25.0 cm. If the person can view the image produced by the microscope with a completely relaxed eye, and the magnification is -4525, what is the focal length of the objective?

65. •• The barrel of a compound microscope is 15 cm in length. The specimen will be mounted 1.0 cm from the objective, and the eyepiece has a 5.0-cm focal length. Determine the focal length of the objective lens.

66. •• A compound microscope uses a 75.0-mm lens as the objective and a 2.0-cm lens as the eyepiece. The specimen will be mounted 122 mm from the objective. Determine **(a)** the barrel length and **(b)** the total magnification produced by the microscope.

67. ••• The "tube length" of a microscope is defined to be the difference between the (objective) image distance and objective focal length: $L = d_i - f_{objective}$. Many microscopes are standardized to a tube length of $L = 160$ mm. Consider such a microscope whose objective lens has a focal length $f_{objective} = 7.50$ mm. **(a)** How far from the object should this lens be placed? **(b)** What focal length eyepiece would give an overall magnification of -55? **(c)** What focal length eyepiece would give an overall magnification of -110?

SECTION 27-5   TELESCOPES

68. • **CE** Two telescopes of different lengths produce the same angular magnification. Is the focal length of the long telescope's eyepiece greater than or less than the focal length of the short telescope's eyepiece? Explain.

69. • A grade school student plans to build a 35-power telescope as a science fair project. She starts with a magnifying glass with a focal length of 7.5 cm as the eyepiece. What focal length is needed for her objective lens?

70. • A 75-power refracting telescope has an eyepiece with a focal length of 5.0 cm. How long is the telescope?

71. • An amateur astronomer wants to build a small refracting telescope. The only lenses available to him have focal lengths of 5.00 cm, 10.0 cm, 20.0 cm, and 30.0 cm. **(a)** What is the greatest magnification that can be obtained using two of these lenses? **(b)** How long is the telescope with the greatest magnification?

72. • A pirate sights a distant ship with a spyglass that gives an angular magnification of 22. If the focal length of the eyepiece is 11 mm, what is the focal length of the objective?

73. •• A telescope has lenses with focal lengths $f_1 = +30.0$ cm and $f_2 = +5.0$ cm. **(a)** What distance between the two lenses will allow the telescope to focus on an infinitely distant object and produce an infinitely distant image? **(b)** What distance between the lenses will allow the telescope to focus on an object that is 5.0 m away and to produce an infinitely distant image?

74. •• Jason has a 25-power telescope whose objective lens has a focal length of 120 cm. To make his sister appear smaller than normal, he turns the telescope around and looks through the objective lens. What is the angular magnification of his sister when viewed through the "wrong" end of the telescope?

75. •• **Roughing It with Science** A professor shipwrecked on Hooligan's Island decides to build a telescope from his eyeglasses and some coconut shells. Fortunately, the professor's eyes require different prescriptions, with the left lens having a power of $+5.0$ diopters and the right lens having a power of $+2.0$ diopters. **(a)** Which lens should he use as the objective? **(b)** What is the angular magnification of the professor's telescope?

76. •• **Galileo's Telescope** Galileo's first telescope used a convex objective lens and a concave eyepiece, as shown in **FIGURE 27-29**. When this telescope is focused on an infinitely distant object, it produces an infinitely distant image. **(a)** What is the focal length of the eyepiece of a Galilean telescope that has an objective focal length of 1.25 m and a magnification of $+4.00$? **(b)** How far apart are the two lenses?

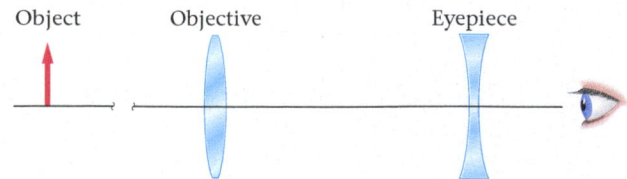

FIGURE 27-29 Problems 76, 84, and 91

77. •• The Moon has an angular size of $0.50°$ when viewed with unaided vision from Earth. Suppose the Moon is viewed through a telescope with an objective whose focal length is 68 cm and an eyepiece whose focal length is 17 mm. What is the angular size of the Moon as seen through this telescope?

78. •• A telescope is 275 mm long and has an objective lens with a focal length of 257 mm. **(a)** What is the focal length of the eyepiece? **(b)** What is the magnification of this telescope?

SECTION 27-6   LENS ABERRATIONS

79. • **CE** The focal length for light that strikes near the center of a spherical convex lens is 15 cm. Referring to Figure 27-23, will the focal length for light that strikes near the edge of the lens be greater than, less than, or equal to 15 cm?

80. • **CE** The focal length for red light that strikes a spherical concave lens is -15 cm. Referring to Figure 27-26, will the magnitude of the (negative) focal length for blue light be greater than, less than, or equal to 15 cm?

GENERAL PROBLEMS

81. • **CE BIO** Predict/Explain **Intracorneal Ring** An intracorneal ring is a small plastic device implanted in a person's cornea to change its curvature. By changing the shape of the cornea, the intracorneal ring can correct a person's vision. **(a)** If a person is nearsighted, should the ring increase or decrease the cornea's curvature? **(b)** Choose the *best explanation* from among the following:
  I. The intracorneal ring should increase the curvature of the cornea so that it bends light more. This will allow it to focus on light coming from far away.
  II. The intracorneal ring should decrease the curvature of the cornea so it's flatter and bends light less. This will allow parallel rays from far away to be focused properly on the retina.

82. • **CE BIO** The lens in a normal human eye, with aqueous humor on one side and vitreous humor on the other side, has a refractive power of 15 diopters. Suppose a lens is removed from an eye and surrounded by air. In this case, is its refractive power greater than, less than, or equal to 15 diopters? Explain.

83. • **CE BIO** Predict/Explain **Treating Cataracts** When the lens in a person's eye becomes clouded by a cataract, the lens can be removed with a process called phacoemulsification and replaced with a man-made intraocular lens. The intraocular lens restores clear vision, but its focal length cannot be changed to allow the user to focus at different distances. In most cases, the intraocular lens is adjusted for viewing of distant objects, and corrective glasses are worn when viewing nearby objects. **(a)** Should the refractive power of the corrective glasses be positive or negative? **(b)** Choose the *best explanation* from among the following:
  I. The refractive power should be positive—converging—because the intraocular lens will make the person farsighted.
  II. A negative refractive power is required to bring the focal point of the intraocular lens in from infinity to a finite value.

84. •• Galileo's original telescope (Figure 27-29) used a convex objective and a concave eyepiece. Use a ray diagram to show that this telescope produces an upright image when a distant object is being viewed. Assume that the eyepiece is to the right of the object and that the right-hand focal point of the eyepiece is just to the left of the objective's right-hand focal point. In addition, assume that the focal length of the eyepiece has a magnitude that is about one-quarter the focal length of the objective.

85. •• Predict/Calculate For each of the following cases, use a ray diagram to show that the angular sizes of the image and the object are identical if both angles are measured *from the center of the lens*. **(a)** A convex lens with the object outside the focal length. **(b)** A convex lens with the object inside the focal length. **(c)** A concave lens with the object outside the focal length. **(d)** Given that the angular size does not change, how does a simple magnifier work? Explain.

86. •• **Predict/Calculate** You have two lenses, with focal lengths $f_1 = +2.60$ cm and $f_2 = +20.4$ cm. **(a)** How would you arrange these lenses to form a magnified image of the Moon? **(b)** What is the maximum angular magnification these lenses could produce? **(c)** How would you arrange the same two lenses to form a magnified image of an insect? **(d)** If you use the magnifier of part (c) to view an insect, what is the angular magnification when the insect is held 2.90 cm from the objective lens?

87. •• **BIO** The eye is actually a multiple-lens system, but we can approximate it with a single-lens system for most of our purposes. When the eye is focused on a distant object, the optical power of the equivalent single lens is +41.4 diopters. **(a)** What is the effective focal length of the eye? **(b)** How far in front of the retina is this "equivalent lens" located?

88. •• **BIO** **Fitting Contact Lenses with a Keratometer** When a patient is being fitted with contact lenses, the curvature of the patient's cornea is measured with an instrument known as a keratometer. A lighted object is held near the eye, and the keratometer measures the magnification of the image formed by reflection from the front of the cornea. If an object is held 10.0 cm in front of a patient's eye, and the reflected image is magnified by a factor of 0.035, what is the radius of curvature of the patient's cornea?

89. •• **Pricey Stamp** A rare 1918 "Jenny" stamp, depicting a misprinted, upside-down Curtiss JN-4 "Jenny" airplane, sold at auction for $525,000. A collector uses a simple magnifying glass to examine the "Jenny," obtaining a linear magnification of 2.5 when the stamp is held 2.76 cm from the lens. What is the focal length of the magnifying glass?

90. •• **BIO** **Predict/Calculate** **A Big Eye** The largest eye ever to exist on Earth belonged to an extinct species of ichthyosaur, *Temnodontosaurus platyodon*. This creature had an eye that was 26.4 cm in diameter. It is estimated that this ichthyosaur also had a relatively large pupil, giving it an effective aperture setting of about $f/1.1$. **(a)** Assuming its pupil was one-third the diameter of the eye, what was the approximate focal length of the ichthyosaur's eye? **(b)** When the ichthyosaur narrowed its pupil in bright light, did its f-number increase or decrease? Explain.

91. •• Consider a Galilean telescope, as illustrated in Figure 27-29, constructed from two lenses with focal lengths of 75.6 cm and −18.0 mm. **(a)** What is the distance between these lenses if an infinitely distant object is to produce an infinitely distant image? **(b)** What is the angular magnification when the lenses are separated by the distance calculated in part (a)?

92. •• A farsighted person uses glasses with a refractive power of 3.6 diopters. The glasses are worn 2.5 cm from his eyes. What is this person's near point when not wearing glasses?

93. ••• **Landing on an Aircraft Carrier** The Fresnel Lens Optical Landing System (FLOLS) used to ensure safe landings on aircraft carriers consists of a series of Fresnel lenses of different colors. Each lens focuses light in a different, specific direction, and hence which light a pilot sees on approach determines whether the plane is above, below, or on the proper landing path. The basic idea behind a Fresnel lens, which has the same optical properties as an ordinary lens, is shown in **FIGURE 27-30**. Suppose an object is 17.1 cm behind a Fresnel lens, and that the corresponding image is a distance $d_i = d$ in front of the lens. If the object is moved to a distance of 12.0 cm behind the lens, the image distance doubles to $d_i = 2d$. In the FLOLS, it is desired to have the image of the lightbulb at infinity. What object distance will give this result for this particular lens?

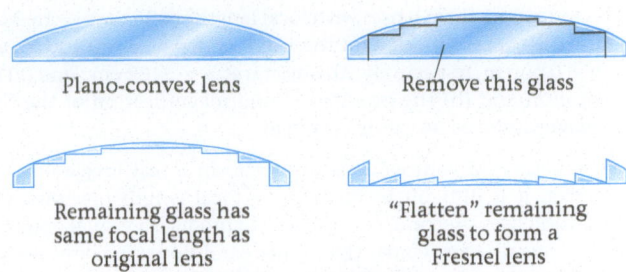

Plano-convex lens   Remove this glass

Remaining glass has same focal length as original lens   "Flatten" remaining glass to form a Fresnel lens

FIGURE 27-30 Fresnel Lenses and the FLOLS A lens causes light to refract at its surface; therefore, the interior glass can be removed without changing its optical properties. This produces a Fresnel lens, which is much lighter than the original lens. (Problem 93)

94. ••• A Cassegrain astronomical telescope uses two mirrors to form the image. The larger (concave) objective mirror has a focal length $f_1 = +50.0$ cm. A small convex secondary mirror is mounted 43.0 cm in front of the primary. As shown in **FIGURE 27-31**, light is reflected from the secondary through a hole in the center of the primary, thereby forming a real image 8.00 cm behind the primary mirror. What is the radius of curvature of the secondary mirror?

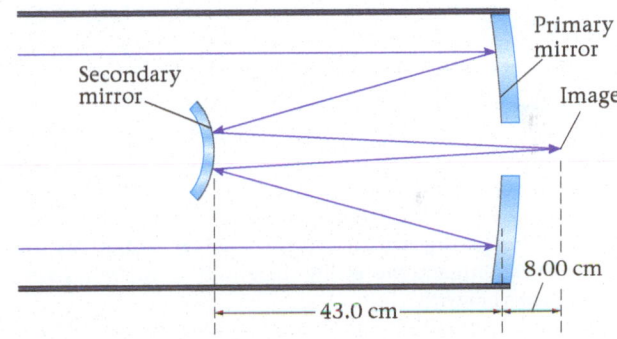

Secondary mirror

Primary mirror

Image

8.00 cm

43.0 cm

FIGURE 27-31 Problem 94

95. ••• **Predict/Calculate** A convex lens ($f = 20.0$ cm) is placed 10.0 cm in front of a plane mirror. A matchstick is placed 25.0 cm in front of the lens, as shown in **FIGURE 27-32**. **(a)** If you look through the lens toward the mirror, where will you see the image of the matchstick? **(b)** Is the image real or virtual? Explain. **(c)** What is the magnification of the image? **(d)** Is the image upright or inverted?

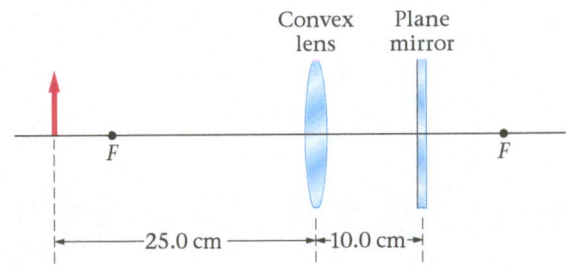

Convex lens   Plane mirror

F F

25.0 cm   10.0 cm

FIGURE 27-32 Problem 95

96. ••• The diameter of a collimated laser beam can be expanded or reduced by using two converging lenses, with focal lengths f_1 and f_2, mounted a distance $f_1 + f_2$ from each other, as shown in **FIGURE 27-33**. What is the ratio of the two beam diameters, (d_1/d_2), expressed in terms of the focal lengths?

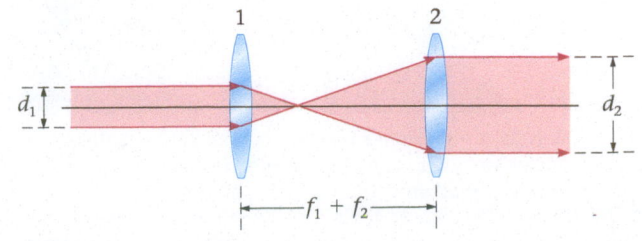

1   2

d_1 d_2

$f_1 + f_2$

FIGURE 27-33 Problem 96

97. ••• Consider three lenses with focal lengths of 25.0 cm, −15.0 cm, and 11.0 cm positioned on the x axis at $x = 0$, $x = 0.400$ m, and $x = 0.500$ m, respectively. An object is at $x = -122$ cm. Find **(a)** the location and **(b)** the orientation and magnification of the final image produced by this lens system.

98. ••• Because a concave lens cannot form a real image of a real object, it is difficult to measure its focal length precisely. One method uses a second, convex, lens to produce a virtual object for the concave lens. Under the proper conditions, the concave lens will form a real image of the virtual object! A student conducting a laboratory project on concave lenses makes the following observations: When a lamp is placed 42.0 cm to the left of a particular convex lens, a real (inverted) image is formed 37.5 cm to the right of the lens. The lamp and convex lens are kept in place while a concave lens is mounted 15.0 cm to the right of the convex lens. A real image of the lamp is now formed 35.0 cm to the right of the concave lens. What is the focal length of each lens?

99. ••• A person with a near-point distance N uses a magnifying glass with a focal length f. Show that the greatest magnification that can be achieved with this magnifier is $M = 1 + N/f$.

PASSAGE PROBLEMS

BIO Cataracts and Intraocular Lenses

A cataract is an opacity or "cloudiness" that develops in the lens of an eye. The result can be serious degradation of vision, or even blindness. Cataracts can be caused by prolonged exposure to electromagnetic radiation of almost any form. For example, cataracts are unusually common among airline pilots, who encounter intense UV exposure at high altitude. To delay cataract formation, doctors recommend protecting the eyes from sunlight with sunglasses or a hat with a brim.

Cataracts are generally treated by removing the affected lens with a technique referred to as phacoemulsification. After the natural lens is removed, it is replaced with a man-made, intraocular lens, or IOL. In many cases, the IOL is rigid; neither its focal length nor location can be changed. In most cases these lenses are designed to allow the eye to see clearly at infinity, but corrective glasses or contacts must be worn for close vision. In extreme cases, a "piggyback" IOL may need to be installed along with the original IOL to fine-tune the optics.

More recently, adaptive IOLs have been developed that flex when the focusing muscles of the eye contract, thus allowing a degree of accommodation. This is illustrated in **FIGURE 27-34**, where we see the IOL move forward to focus on a close object. Notice that the focal length of the adaptive IOL is fixed, just as with a normal IOL, but the eye muscles can change its location—the same as in a camera when it focuses.

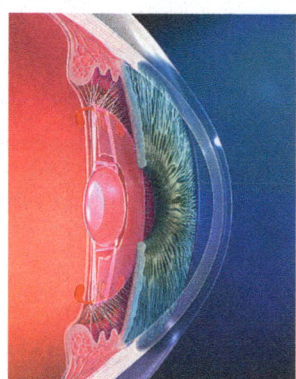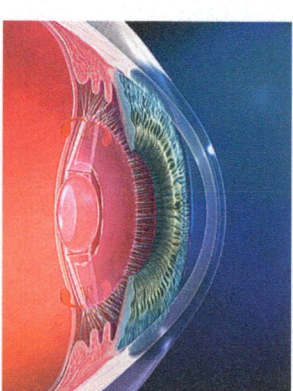

FIGURE 27-34 Intraocular Lenses In cases of severe cataract, an intraocular lens (IOL) can be implanted in place of the eye's natural lens. (Problems 100, 101, 102, and 103)

100. • A patient receives a rigid IOL whose focus cannot be changed—it is designed to provide clear vision of objects at infinity. The patient will use corrective contacts to allow for close vision. Should the refractive power of the corrective contacts be positive or negative?

101. • Referring to the previous problem, find the refractive power of contacts that will allow the patient to focus on a book at a distance of 23.0 cm.

    A. 0.0435 diopter   **B.** 0.230 diopter
    C. 4.35 diopters   **D.** 8.70 diopters

102. • After a fixed IOL is installed, it is found to have the wrong focal length, so a piggyback IOL is installed as well. The piggyback IOL has a power of −1.50 diopters. What is its focal length and type?

    A. converging, 1.50 m   **B.** converging, 66.7 cm
    C. diverging, 1.50 m   **D.** diverging, 66.7 cm

103. • Suppose a flexible, adaptive IOL has a focal length of 3.00 cm. How far forward must the IOL move to change the focus of the eye from an object at infinity to an object at a distance of 50.0 cm?

    A. 1.9 mm   **B.** 2.8 mm
    C. 3.1 mm   **D.** 3.2 mm

104. •• **Predict/Calculate** REFERRING TO EXAMPLE 27-4 Suppose a person's eyeglasses have a focal length of −301 cm, are 2.00 cm in front of the eyes, and allow the person to focus on distant objects. **(a)** Is this person's far point greater than or less than 323 cm, which is the far point for glasses the same distance from the eyes and with a focal length of −321 cm? Explain. **(b)** Find the far point for this person.

105. •• **Predict/Calculate** REFERRING TO EXAMPLE 27-4 In Example 27-4, a person has a far-point distance of 323 cm. If this person wears glasses 2.00 cm in front of the eyes with a focal length of −321 cm, distant objects can be brought into focus. Suppose a second person's far point is 353 cm. **(a)** Is the magnitude of the focal length of the eyeglasses that allow this person to focus on distant objects greater than or less than 321 cm? Assume the glasses are 2.00 cm in front of the eyes. **(b)** Find the required focal length for the second person's eyeglasses.

106. •• **Predict/Calculate** REFERRING TO EXAMPLE 27-6 Suppose a person's eyeglasses have a refractive power of 2.75 diopters and that they allow the person to focus on an object that is just 25.0 cm from the eye. The glasses are 2.00 cm in front of the eyes. **(a)** Is this person's near point greater than or less than 57.0 cm, which is the near-point distance when the glasses have a refractive power of 2.53 diopters? Explain. **(b)** Find the near point for this person.

107. •• **Predict/Calculate** REFERRING TO EXAMPLE 27-6 Suppose a person's near-point distance is 67.0 cm. **(a)** Is the refractive power of the eyeglasses that allow this person to focus on an object just 25.0 cm from the eye greater than or less than 2.53 diopters, which is the refractive power when the near-point distance is 57.0 cm? The glasses are worn 2.00 cm in front of the eyes. **(b)** Find the required refractive power for this person's eyeglasses.

Physical Optics: Interference and Diffraction

Big Ideas

1 Light waves can show the effects of constructive and destructive interference.

2 Monochromatic light that passes through two slits produces bright and dark fringes.

3 Light reflecting from a medium with a larger index of refraction changes phase by half a wavelength.

4 When light encounters an obstacle, or passes through an opening, it diffracts (bends).

5 The ability to separate different sources of light depends on the wavelength of the light.

6 Diffraction gratings spread light out to form a wide spectrum of colors, similar to a prism.

▲ Most of the colors we see every day can be understood in terms of absorption and reflection—an old-fashioned fire engine absorbs green light and reflects red light, a grassy lawn absorbs red light and reflects green light, and so on. Yet the distinctive iridescent hues of this rose chafer beetle are created by different physical mechanisms. These mechanisms, known as interference and diffraction, cannot be understood solely in terms of geometrical optics. Instead, they have their origin in the wave nature of light, the subject of this chapter.

Our discussion of optics to this point has been in terms of "rays" that propagate in straight lines. In this chapter we show that the wave properties of light are important as well; in fact, they are responsible for such phenomena as the operation of a DVD, the appearance of images on a television screen, and the brilliant iridescent colors of a butterfly's wing.

▲ **FIGURE 28-1** Two sets of water waves, radiating outward in circular patterns from point sources, create an interference pattern where they overlap.

28-1 Superposition and Interference

One of the fundamental aspects of wave behavior is **superposition,** in which the net displacement caused by a combination of waves is the algebraic sum of the displacements caused by each wave individually. If waves add to cause a larger displacement, we say that they interfere **constructively;** if the net displacement is reduced, the interference is **destructive.** Examples of wave superposition on a string are given in Figures 14-24 and 14-25, and an example with water waves is shown in **FIGURE 28-1.** Because light also exhibits wave behavior, with propagating electric and magnetic fields, it too can show interference effects and a resulting increase or decrease in brightness.

Observing Interference   Interference is noticeable, however, only if certain conditions are met. In the case of light, for example, the light should be **monochromatic;** that is, it should have a single color and hence a single frequency. In addition, if two or more sources of light are to show interference, they must maintain a constant phase relationship with one another—that is, they must be **coherent.** Sources whose relative phases vary randomly with time show no discernible interference patterns and are said to be **incoherent.** Incoherent light sources include incandescent and fluorescent lights. In contrast, lasers emit light that is both monochromatic and coherent.

The basic principle that determines whether waves will interfere constructively or destructively is their phase relative to one another. For example, if two waves have zero phase difference—that is, the waves are in phase—they add constructively, and the net result is an increased amplitude, as **FIGURE 28-2 (a)** indicates. If waves of equal amplitude are 180° out of phase, however, the net result is zero amplitude and destructive interference, as shown in **FIGURE 28-2 (b).** Notice that a *180° difference in phase* corresponds to waves being out of step by *half a wavelength.* Similarly, if the phase difference between two waves is 360°—which corresponds to one full wavelength—the interference is again constructive, as in **FIGURE 28-2 (c).**

In Figure 28-2 the lower wave is shifted to the right, meaning it is *ahead* of the upper wave by half a wavelength in part (b), and *ahead* by a full wavelength in part (c). The same results are obtained, however, if the lower wave is *behind* the upper wave. Thus, for example, a phase difference of −180° produces destructive interference, the same as in Figure 28-2 (b), and a phase difference of −360° produces constructive interference, just as in Figure 28-2 (c).

Let's apply the preceding observations to a system that consists of two radio antennas radiating electromagnetic waves of frequency f and wavelength λ, as in **FIGURE 28-3.** If the antennas are connected to the same transmitter, they emit waves that are in phase and coherent. When waves from each antenna reach point P_0, they have

Big Idea **1** Light waves can show the effects of constructive and destructive interference, just like sound waves. Waves that are in phase, or an integer number of wavelengths out of phase, interfere constructively, resulting in light that is more intense. Waves that are an odd number of half wavelengths out of phase interfere destructively, resulting in an absence of light.

▶ **FIGURE 28-2 Constructive and destructive interference (a)** Waves that are in phase add to give a larger amplitude. This is constructive interference. **(b)** Waves that are half a wavelength out of phase interfere destructively. If the individual waves have equal amplitudes, as here, their sum will have zero amplitude. **(c)** When waves are one wavelength out of phase, the result is again constructive interference, exactly the same as when the waves are in phase.

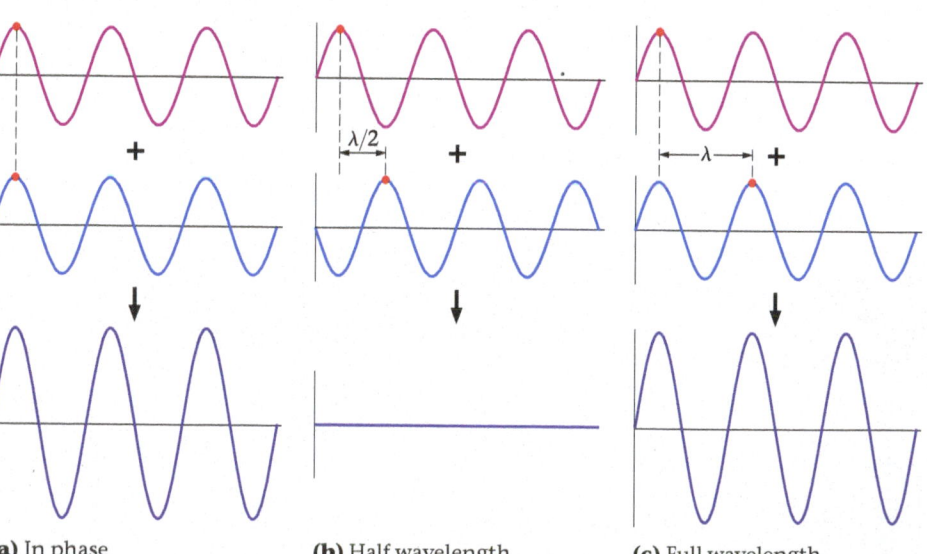

(a) In phase   **(b)** Half wavelength out of phase   **(c)** Full wavelength out of phase

▶ **FIGURE 28-3 Two radio antennas transmitting the same signal** At point P_0, midway between the antennas, the waves travel the same distance, and hence they interfere constructively. At point P_1 the distance ℓ_2 is greater than the distance ℓ_1 by one wavelength; thus, P_1 is also a point of constructive interference. At Q_1 the distance ℓ_2 is greater than the distance ℓ_1 by half a wavelength, and the waves interfere destructively at that point.

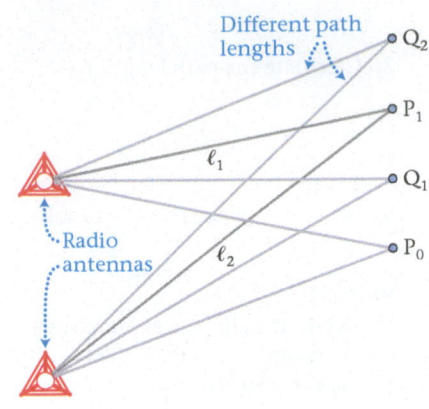

traveled the same distance—that is, the same number of wavelengths—and hence they are still in phase. As a result, point P_0 experiences constructive interference and the radio signal is strong. This location corresponds to the situation shown in Figure 28-2 (a).

To reach point P_1, the waves from the two antennas must travel different distances, ℓ_1 and ℓ_2, as indicated in Figure 28-3. If the difference in these distances is one wavelength, $\ell_2 - \ell_1 = \lambda$, the waves are 360° out of phase, and point P_1 also has constructive interference. This location corresponds to Figure 28-2 (c). Similarly, if the difference in path lengths to a point P_m is an integer $m = 0, 1, 2, \ldots$ times the wavelength, $\ell_2 - \ell_1 = m\lambda$, that point will also be a location of constructive interference.

On the other hand, the difference in path lengths to point Q_1 is half a wavelength, $\ell_2 - \ell_1 = \lambda/2$, and hence the waves cancel there, just as in Figure 28-2 (b). Destructive interference also occurs at Q_2, where the difference in path lengths is one and a half wavelengths, $\ell_2 - \ell_1 = 3\lambda/2$. In general, for $m = 1, 2, 3, \ldots$, destructive interference occurs at points Q_m where the difference in path lengths is $\ell_2 - \ell_1 = \left(m - \frac{1}{2}\right)\lambda$.

To summarize, *constructive interference* satisfies the conditions

$$\ell_2 - \ell_1 = m\lambda \qquad m = 0, 1, 2, \ldots \qquad \textit{(constructive interference)}$$

Similarly, *destructive interference* satisfies the conditions

$$\ell_2 - \ell_1 = \left(m - \frac{1}{2}\right)\lambda \quad m = 1, 2, 3, \ldots \qquad \textit{(destructive interference)}$$

These conditions are applied to a specific physical system in the following Example.

PHYSICS IN CONTEXT
Looking Back

Superposition and interference in light waves are completely analogous to superposition and interference in sound waves, as discussed in Chapter 14.

EXAMPLE 28-1   TWO MAY NOT BE BETTER THAN ONE

Two friends tune their radios to the same frequency and pick up a signal transmitted simultaneously by a pair of antennas. The friend who is equidistant from the antennas, at P_0, receives a strong signal (a maximum). The friend at point Q_1 receives a very weak signal (a minimum). Find the wavelength of the radio waves if $d = 7.50$ km, $L = 14.0$ km, and $y = 1.88$ km. Assume that Q_1 is the first point of minimum signal as one moves away from P_0 in the y direction.

PICTURE THE PROBLEM
Our sketch shows the radio antennas and the two locations mentioned in the problem statement. Notice that the radio antennas are separated by a distance $d = 7.50$ km in the y direction, and that the points P_0 and Q_1 have a y-direction separation of $y = 1.88$ km. The distance to P_0 and Q_1 in the x direction is $L = 14.0$ km.

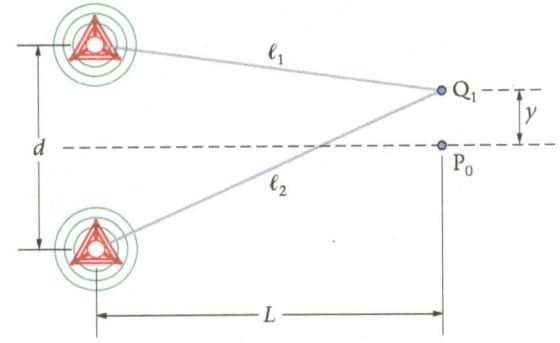

REASONING AND STRATEGY
Because point Q_1 is the first *minimum* in the y direction from the *maximum* at point P_0, we know that the path difference, $\ell_2 - \ell_1$, is one half a wavelength. Thus we can determine λ by calculating the lengths ℓ_2 and ℓ_1 and setting their difference equal to $\lambda/2$.

Known   Separation between radio sources, $d = 7.50$ km; distance to right of sources, $L = 14.0$ km; distance between observers, $y = 1.88$ km.
Unknown   Wavelength of radio waves, $\lambda = ?$

SOLUTION

1. Calculate the path length ℓ_1:

$$\ell_1 = \sqrt{L^2 + \left(\frac{d}{2} - y\right)^2}$$

$$= \sqrt{(14.0 \text{ km})^2 + \left(\frac{7.50 \text{ km}}{2} - 1.88 \text{ km}\right)^2} = 14.1 \text{ km}$$

CONTINUED

2. Calculate the path length ℓ_2:

$$\ell_2 = \sqrt{L^2 + \left(\frac{d}{2} + y\right)^2}$$

$$= \sqrt{(14.0 \text{ km})^2 + \left(\frac{7.50 \text{ km}}{2} + 1.88 \text{ km}\right)^2} = 15.1 \text{ km}$$

3. Set $\ell_2 - \ell_1$ equal to $\lambda/2$ and solve for the wavelength:

$$\ell_2 - \ell_1 = \tfrac{1}{2}\lambda$$
$$\lambda = 2(\ell_2 - \ell_1) = 2(15.1 \text{ km} - 14.1 \text{ km}) = 2.0 \text{ km}$$

INSIGHT

We see that radio waves are rather large. In fact, the distance from one crest to the next (the wavelength) for these waves is about 1.2 miles. Recalling that radio waves travel at the speed of light, the corresponding frequency is $f = c/\lambda = 150$ kHz.

PRACTICE PROBLEM — PREDICT/CALCULATE

Suppose the wavelength broadcast by these antennas is changed, and that the y distance between P_0 and Q_1 (the first minimum) increases as a result. **(a)** Is the new wavelength greater than or less than 2.0 km? Explain. **(b)** Find the wavelength for the case $y = 2.91$ km.   [**Answer: (a)** The new wavelength is greater than 2.0 km, because the wavelength increases with increasing separation between the points P_0 and Q_1. **(b)** $\lambda = 3.0$ km]

Some related homework problems: Problem 1, Problem 4, Problem 5

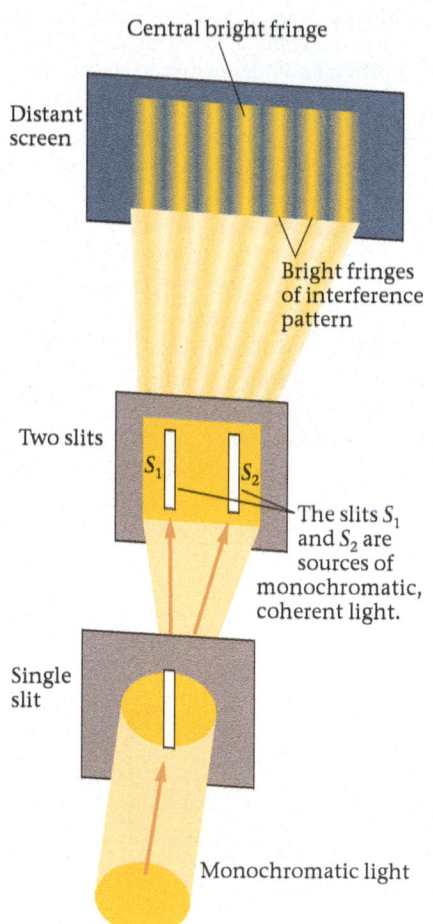

▲ FIGURE 28-4 Young's two-slit experiment The first screen produces a small source of light that illuminates the two slits, S_1 and S_2. After passing through these slits, the light spreads out into an interference pattern of alternating bright and dark fringes on a distant screen.

Enhance Your Understanding   (Answers given at the end of the chapter)

1. Two beams of light that have the same phase are said to be **(a)** monochromatic, **(b)** incoherent, **(c)** coherent, **(d)** random.

Section Review

• Constructive interference results when the path difference between waves is a multiple of a wavelength: $\ell_2 - \ell_1 = m\lambda$, where $m = 0, 1, 2, \ldots$.

• Destructive interference results when the path difference between waves is half a wavelength, one and a half wavelengths, and so on. In this case, the path difference satisfies the relationship $\ell_2 - \ell_1 = \left(m - \frac{1}{2}\right)\lambda$, where $m = 1, 2, 3, \ldots$.

28-2 Young's Two-Slit Experiment

We now consider a classic physics experiment that not only demonstrates the wave nature of light, but also allows one to determine the wavelength of a beam of light, just as we did for the radio waves in Example 28-1. The experiment was first performed in 1801 by the English physician and physicist Thomas Young (1773–1829), whose medical background contributed to his studies of vision, and whose love of languages made him a key figure in deciphering the Rosetta Stone. The experiment, in simplest form, consists of a beam of monochromatic light that passes through a small slit in a screen and then illuminates two slits, S_1 and S_2, in a second screen. After the light passes through the two slits, it shines on a distant screen, as **FIGURE 28-4** shows, where an interference pattern of bright and dark "fringes" is observed. An example of this kind of interference pattern, using a green laser as the light source, is shown in **FIGURE 28-5**.

In this "two-slit experiment" the slit in the first screen serves only to produce a small source of light that prevents the interference pattern on the distant screen from becoming smeared out. The key elements in the experiment are the two slits in the second screen. Because they are equidistant from the single slit, as shown in Figure 28-4, the light passing through them has the same phase. Thus the two slits act as monochromatic, coherent sources of light—analogous to the two radio antennas in Example 28-1—as needed to produce interference.

If light were composed of small particles, or "corpuscles" as Newton referred to them, they would simply pass straight through the two slits and illuminate the distant screen directly behind each slit. If light is a wave, on the other hand, each slit acts as the source of new waves, analogous to water waves passing through two small openings. This is referred to as **Huygens's principle** and is illustrated in **FIGURE 28-6**. Notice that light radiates away from the slits in all forward directions—not just in the direction of the incoming light. The result is that light is spread out over a large area on the distant screen; it is not localized in small regions directly behind the slits. Thus, an experiment like this can readily distinguish between the two models of light.

Constructive and Destructive Interference Of key importance is the fact that the illumination of the distant screen is not only spread out, but also consists of alternating bright and dark fringes, as indicated in Figure 28-4. These fringes are the direct result of constructive and destructive interference. For example, the central bright fringe is midway between the two slits; hence, the path lengths for light from the slits are equal. Because light coming from the slits is in phase, it follows that the light is also in phase at the midpoint, giving rise to constructive interference—just like at point P_0 in Figure 28-3.

The next bright fringe occurs when the difference in path lengths from the two slits is equal to one wavelength of light, as with point P_1 in Figure 28-3. In most experimental situations the distance to the screen is much greater than the separation d of the slits; hence, the light travels to a point on the screen along approximately parallel paths, as indicated in **FIGURE 28-7**. Therefore, the path difference, $\Delta \ell$, is

$$\Delta \ell = d \sin \theta$$

As a result, the bright fringe closest to the midpoint occurs at the angle θ given by the condition $d \sin \theta = \lambda$, or $\sin \theta = \lambda/d$. In general, a bright fringe occurs whenever the path difference, $\Delta \ell = d \sin \theta$, is equal to $m\lambda$, where $m = 0, \pm 1, \pm 2, \ldots$. Therefore, we find that bright fringes satisfy the following conditions:

Conditions for Bright Fringes (Constructive Interference) in a Two-Slit Experiment

$d \sin \theta = m\lambda \qquad m = 0, \pm 1, \pm 2, \ldots$ 　　　　　　　　　　　　　　28-1

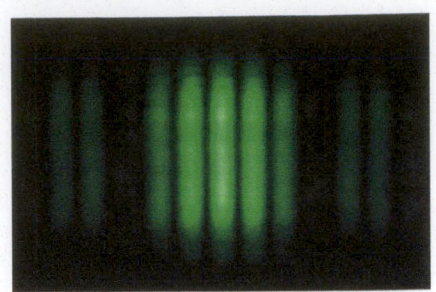

▲ **FIGURE 28-5** An interference pattern created by monochromatic laser light passing through two slits.

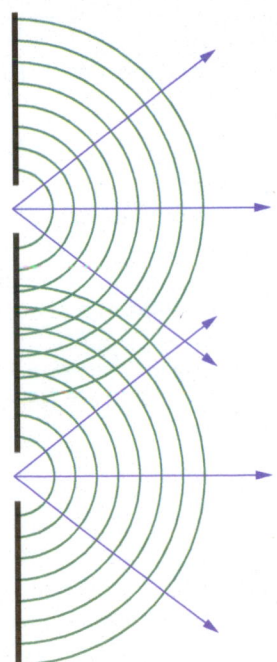

▲ **FIGURE 28-6 Huygens's principle** According to Huygens's principle, each of the two slits in Young's experiment acts as a source of light waves propagating outward in all forward directions. It follows that light from the two sources can overlap, resulting in an interference pattern.

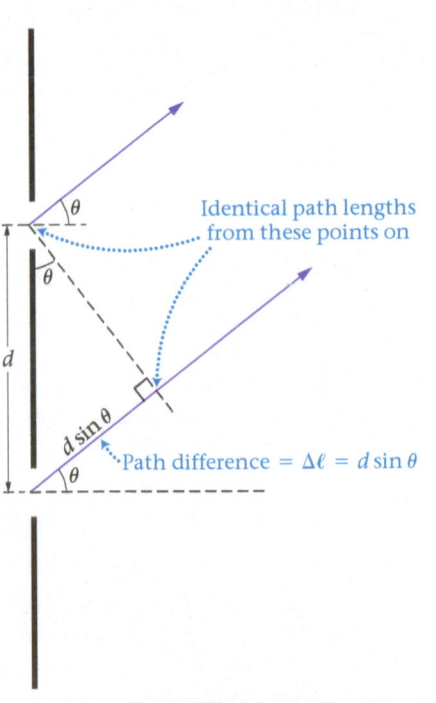

Identical path lengths from these points on

Path difference $= \Delta \ell = d \sin \theta$

▲ **FIGURE 28-7 Path difference in the two-slit experiment** Light propagating from two slits to a distant screen along parallel paths; note that the paths make an angle θ relative to the normal to the slits. The difference in path lengths is $\Delta \ell = d \sin \theta$, where d is the slit separation.

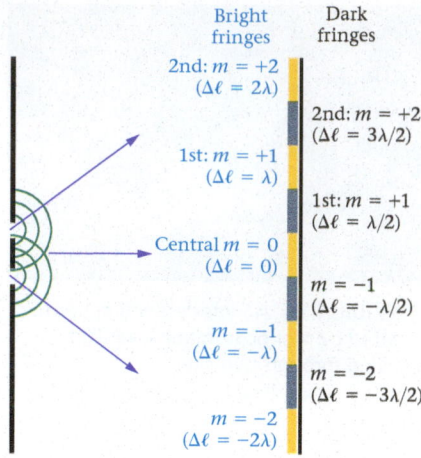

▲ **FIGURE 28-8 The two-slit pattern** Numbering systems for bright and dark fringes.

Big Idea 2 Monochromatic

light that passes through two slits produces an interference pattern of bright and dark fringes. The separation between the fringes is directly related to the wavelength of the light.

PHYSICS IN CONTEXT
Looking Ahead

One of the key experiments related to relativity was the measurement of the speed of light in different directions by Michelson and Morley (Chapter 29). Their experiment was based on observing interference fringes, just like those seen in Young's two-slit experiment.

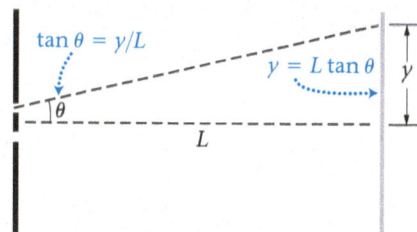

▲ **FIGURE 28-9 Linear distance in an interference pattern** If light propagates at an angle θ relative to the normal to the slits, it is displaced a linear distance $y = L \tan \theta$ on the distant screen.

PROBLEM-SOLVING NOTE

Angular Versus Linear Position of Fringes

The angle at which a bright or dark fringe occurs is determined by the wavelength of the light and the separation of the slits. The linear position of a fringe on a screen is determined by the distance from the slits to the screen.

Notice that the $m = 0$ fringe occurs at $\theta = 0$, which corresponds to the central fringe. In addition, positive values of m indicate fringes above the central bright fringe; negative values indicate fringes below the central bright fringe.

Between the bright fringes we find dark fringes, where destructive interference occurs. The conditions for a dark fringe are that the difference in path lengths be $\pm\lambda/2, \pm3\lambda/2, \pm5\lambda/2, \ldots$. The dark fringes are analogous to the points $Q_1, Q_2, \ldots$ in Figure 28-3. It follows that the angles corresponding to the dark fringes are given by the following conditions:

Conditions for Dark Fringes (Destructive Interference) in a Two-Slit Experiment

$d \sin \theta = \left(m - \frac{1}{2}\right)\lambda \quad m = 1, 2, 3, \ldots \quad$ *(above central bright fringe)*

$d \sin \theta = \left(m + \frac{1}{2}\right)\lambda \quad m = -1, -2, -3, \ldots$ *(below central bright fringe)*

28-2

Clearly, $m = +1$ corresponds to a path difference of $\Delta\ell = \lambda/2$, the value $m = +2$ corresponds to a path difference of $\Delta\ell = 3\lambda/2$, and so on. These dark fringes are above the central bright fringe. Similarly, $m = -1$ corresponds to a path difference of $-\lambda/2$, and $m = -2$ corresponds to a path difference of $-3\lambda/2$. These dark fringes are below the central bright fringe. **FIGURE 28-8** shows the numbering systems for both bright and dark fringes. (All problems in this text refer to fringes above the central bright fringe, so only the first set of conditions in Equations 28-2 will be needed.)

Now, because $\sin \theta$ is less than or equal to 1, it follows from Equation 28-1 ($d \sin \theta = m\lambda$) that d must be greater than or equal to λ to satisfy the conditions for the $m = \pm 1$ bright fringes. In a typical situation, d may be 100 times larger than λ, which means that the angle to the first dark or bright fringe will be roughly half a degree. If d is too much larger than λ, however, the angle between successive minima and maxima is so small that they tend to merge together, making the interference pattern difficult to discern.

EXERCISE 28-2 ANGLE TO FRINGES

Red light ($\lambda = 734$ nm) passes through a pair of slits with a separation of 6.45×10^{-5} m. Find the angles corresponding to **(a)** the first bright fringe and **(b)** the second dark fringe above the central bright fringe.

REASONING AND SOLUTION

a. The condition for bright fringes is $d \sin \theta = m\lambda$. Referring to Figure 28-8 we find that $m = +1$ corresponds to the first bright fringe above the central bright fringe. Solving $d \sin \theta = m\lambda$ for the angle θ, we find

$$\theta = \sin^{-1}\left(m\frac{\lambda}{d}\right) = \sin^{-1}\left[(1)\frac{7.34 \times 10^{-7}\ \text{m}}{6.45 \times 10^{-5}\ \text{m}}\right] = 0.652°$$

b. The condition for dark fringes is $d \sin \theta = \left(m - \frac{1}{2}\right)\lambda$. Referring to Figure 28-8, we see that the second dark fringe corresponds to $m = +2$. Solving $d \sin \theta = \left(m - \frac{1}{2}\right)\lambda$ for the angle θ yields

$$\theta = \sin^{-1}\left[\left(m - \frac{1}{2}\right)\frac{\lambda}{d}\right] = \sin^{-1}\left[\left(2 - \frac{1}{2}\right)\frac{7.34 \times 10^{-7}\ \text{m}}{6.45 \times 10^{-5}\ \text{m}}\right] = 0.978°$$

A convenient way to characterize the location of interference fringes is in terms of their linear distance from the central fringe, as indicated in **FIGURE 28-9**. If the distance to the screen is L—and L is much greater than the slit separation d—it follows that the linear distance y is given by the following expression:

Linear Distance from Central Fringe

$y = L \tan \theta$

28-3

In the next Example we show how a measurement of the linear distance between fringes can determine the wavelength of light.

EXAMPLE 28-3   BLUE LIGHT SPECIAL

Two slits with a separation of 8.5×10^{-5} m create an interference pattern on a screen 2.3 m away. **(a)** If the tenth bright fringe above the central fringe is a linear distance of 12 cm above the central fringe, what is the wavelength of light used in the experiment? **(b)** What is the linear distance from the central bright fringe to the tenth dark fringe above it?

PICTURE THE PROBLEM

Our sketch shows that the first bright fringe above the central fringe corresponds to $m = +1$ in $d \sin \theta = m\lambda$, the second bright fringe corresponds to $m = +2$, and so on. Therefore, $m = +10$ gives the position of the tenth bright fringe. Similarly, the first dark fringe corresponds to $m = +1$, and so the tenth dark fringe is given by $m = +10$ in $d \sin \theta = \left(m - \frac{1}{2}\right)\lambda$.

Finally, we note that the separation of the slits is $d = 8.5 \times 10^{-5}$ m, the distance to the screen is $L = 2.3$ m, and the vertical distance to the $m = +10$ bright fringe is $y = 12$ cm $= 0.12$ m.

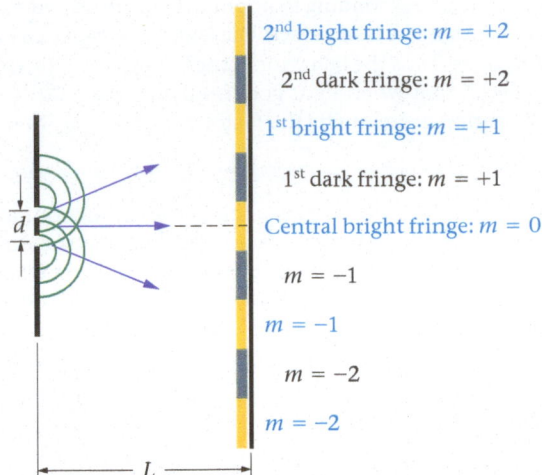

2nd bright fringe: $m = +2$

2nd dark fringe: $m = +2$

1st bright fringe: $m = +1$

1st dark fringe: $m = +1$

Central bright fringe: $m = 0$

$m = -1$

$m = -1$

$m = -2$

$m = -2$

REASONING AND STRATEGY

a. To find the wavelength, we first determine the angle to the tenth fringe using $y = L \tan \theta$ (Equation 28-3). Once we know θ, we use the condition for bright fringes to determine the wavelength. That is, we set $m = +10$ in Equation 28-1 ($d \sin \theta = m\lambda$) and solve for λ.

b. We use $d \sin \theta = \left(m - \frac{1}{2}\right)\lambda$ (Equations 28-2) with $m = +10$ and λ from part (a) to determine the angle θ. Next, we substitute θ in $y = L \tan \theta$ to find the linear distance.

Known   Slit separation, $d = 8.5 \times 10^{-5}$ m; distance to screen, $L = 2.3$ m; distance to tenth bright fringe, $y = 0.12$ m.
Unknown   **(a)** Wavelength of light, $\lambda = ?$ **(b)** Distance to tenth dark fringe, $y = ?$

SOLUTION

Part (a)

1. Calculate the angle to the tenth bright fringe:

$$y = L \tan \theta$$
$$\theta = \tan^{-1}\left(\frac{y}{L}\right) = \tan^{-1}\left(\frac{0.12 \text{ m}}{2.3 \text{ m}}\right) = 3.0°$$

2. Use $\sin \theta = m\lambda/d$ to find the wavelength:

$$\lambda = \frac{d}{m}\sin \theta = \left(\frac{8.5 \times 10^{-5} \text{ m}}{10}\right)\sin(3.0°)$$
$$= 4.4 \times 10^{-7} \text{ m} = 440 \text{ nm}$$

Part (b)

3. Find the angle corresponding to the tenth dark fringe:

$$d \sin \theta = \left(m - \frac{1}{2}\right)\lambda$$
$$\theta = \sin^{-1}\left[\left(m - \frac{1}{2}\right)\lambda/d\right]$$
$$= \sin^{-1}\left[\left(10 - \frac{1}{2}\right)(4.4 \times 10^{-7} \text{ m})/(8.5 \times 10^{-5} \text{ m})\right] = 2.8°$$

4. Use $y = L \tan \theta$ to find the linear distance:

$$y = L \tan \theta = (2.3 \text{ m}) \tan (2.8°) = 0.11 \text{ m}$$

INSIGHT

We've expressed the wavelength of the light in nanometers, a common unit of measure for light waves. Referring to the electromagnetic spectrum shown in Example 25-6, we see that light with a wavelength of 440 nm is dark blue.

PRACTICE PROBLEM — PREDICT/CALCULATE

(a) If the wavelength of light used in this experiment is increased, does the linear distance to the tenth bright fringe above the central fringe increase, decrease, or stay the same? Explain. **(b)** Check your reasoning by calculating the linear distance to the tenth bright fringe for a wavelength of 550 nm.   [**Answer: (a)** The linear distance is directly related to the wavelength, and hence it increases. **(b)** The linear distance in this case is $y = 0.15$ m, which is greater than 0.12 m, as expected.]

Some related homework problems: Problem 17, Problem 25

Finally, we consider the effect of changing the medium through which the light propagates in a two-slit experiment.

CONCEPTUAL EXAMPLE 28-4 FRINGE SPACING

A two-slit experiment is performed in the air. Later, the same apparatus is immersed in water and the experiment is repeated. When the apparatus is in water, are the interference fringes more closely spaced, more widely spaced, or spaced the same as when the apparatus was in air?

REASONING AND DISCUSSION

The angles corresponding to bright fringes are related to the wavelength by the equation $d \sin \theta = m\lambda$. From this relationship it is clear that if λ is increased, the angle θ (and hence the spacing between fringes) also increases; if λ is decreased, the angle θ decreases. Thus the behavior of the two-slit experiment in water depends on how the wavelength of light changes in water.

Recall that when light goes from air ($n = 1.00$) to a medium in which the index of refraction is $n > 1$, the speed of propagation decreases by the factor n:

$$v = \frac{c}{n}$$

The frequency of light, f, is unchanged throughout as it goes from one medium to another. Therefore, the fact that the speed $v = \lambda f$ decreases by a factor n means that the wavelength λ decreases by the same factor. Hence, if the wavelength of light is λ when $n = 1$, its wavelength in a medium with an index of refraction $n > 1$ is

$$\lambda_n = \frac{\lambda}{n} \qquad\qquad 28\text{-}4$$

As a result, the wavelength of light is smaller in water than in air, and therefore the interference fringes are more closely spaced when the experiment is performed in water.

ANSWER

The fringes are more closely spaced.

The relationship $\lambda_n = \lambda/n$ plays an important role in the interference observed in thin films, as we shall see in the next section.

Enhance Your Understanding   (Answers given at the end of the chapter)

2. If the wavelength in a two-slit experiment is increased, does the angle to the first bright fringe above the central bright fringe increase, decrease, or stay the same?

Section Review

- Bright fringes in a two-slit experiment satisfy the following conditions: $d \sin \theta = m\lambda$, where $m = 0, \pm 1, \pm 2, \ldots$. Dark fringes above the central bright fringe satisfy these conditions: $d \sin \theta = \left(m - \frac{1}{2}\right)\lambda$, where $m = 1, 2, 3, \ldots$.

28-3 Interference in Reflected Waves

Waves that reflect from objects at different locations can interfere with one another, just like the light from two different slits. In fact, interference due to reflected waves is observed in many everyday circumstances, as we show next. Before we can understand the physics behind this type of interference, however, we must note that reflected waves change their phase in two completely different ways. First, the phase changes in proportion to the distance the waves travel—just as with light in the two-slit experiment. For example, the phase of a wave that travels half a wavelength changes by 180° and the phase of a wave that travels one wavelength changes by 360°. Second, the phase of a reflected wave can change as a result of the reflection process itself. We begin by considering the latter type of phase change.

Phase Changes Due to Reflection

PHYSICS IN CONTEXT
Looking Back

Phase changes due to the reflection of light from an interface are just like the phase changes seen when a wave on a string reflects from an end that is either tied down or free to move. See in particular Section 14-2.

Phase changes due to reflection have been discussed before in connection with waves on a string in Chapter 14. In particular, we observed at that time that a wave on a string reflects differently depending on whether the end of the string is tied to a solid support, as in Figure 14-8, or is free to move up and down, as in Figure 14-9. The wave with

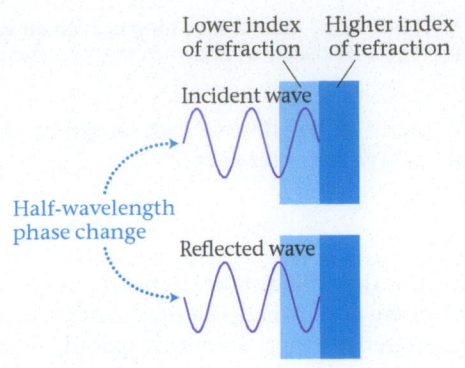

Higher index of refraction | Lower index of refraction
Incident wave
No phase change
Reflected wave

(a) Reflection from lower index of refraction

Lower index of refraction | Higher index of refraction
Incident wave
Half-wavelength phase change
Reflected wave

(b) Reflection from higher index of refraction

▲ **FIGURE 28-10 Phase change with reflection (a)** An electromagnetic wave reflects with no phase change when it encounters a medium with a lower index of refraction. **(b)** An electromagnetic wave reflects with a half-wavelength (180°) phase change when it encounters a medium with a larger index of refraction.

a loose end is reflected back exactly as it approached the end; that is, there is no phase change. Conversely, a wave on a string that is tied down is inverted when reflected. This is equivalent to changing the phase of the wave by 180°, or half a wavelength.

Because light is a wave, it undergoes an analogous phase change on reflection. As indicated in **FIGURE 28-10 (a)**, a light wave that encounters a region with a lower index of refraction is reflected with no phase change, like a wave on a string whose end is free to move. In contrast, when light encounters a region with a larger index of refraction, as in **FIGURE 28-10 (b)**, it is reflected with a phase change of half a wavelength, like a wave on a string whose end is fixed. This half-wavelength phase change also applies to light reflected from a solid surface, such as a mirror.

To summarize:

> There is no phase change when light reflects from a region with a lower index of refraction.

> There is a half-wavelength phase change when light reflects from a region with a higher index of refraction, or from a solid surface.

We now apply these observations to the case of an air wedge.

Air Wedge

An interesting example of reflection interference is provided by two plates of glass that touch at one end and have a small separation at the other, as shown in **FIGURE 28-11**. Notice that the air between the plates occupies a thin, wedge-shaped region; hence, this type of arrangement is referred to as an **air wedge.**

The predominant interference effect in this system is between light reflected from the bottom surface of the top glass plate and light reflected from the upper surface of the lower plate, because these surfaces are physically so close together. Consider, for example, the two rays illustrated in Figure 28-11. Ray 1 reflects at the glass-air interface; it experiences no phase change. Ray 2 travels a distance d through the air ($n = 1.00$), reflects from the air-glass interface, then travels essentially the same distance d in the opposite direction before rejoining ray 1 (the rays in Figure 28-11 are shown widely separated for clarity). The reflection from the air-glass interface results in a 180° phase change—the same phase change as if the wave had traveled half a wavelength—and hence the *effective* path length of ray 2 is

$$\Delta \ell_{eff} = d + \tfrac{1}{2}\lambda + d = \tfrac{1}{2}\lambda + 2d$$

If the effective path length is an integer number of wavelengths, $\lambda/2 + 2d = m\lambda$, rays 1 and 2 will interfere constructively. Dividing by the wavelength, we have the following condition for **constructive interference:**

$$\frac{1}{2} + \frac{2d}{\lambda} = m \qquad m = 1, 2, 3, \ldots \qquad \text{28-5}$$

Big Idea 3 When light reflects from a medium with a lower index of refraction, there is no change in phase. When light reflects from a medium with a larger index of refraction, its phase changes by half a wavelength.

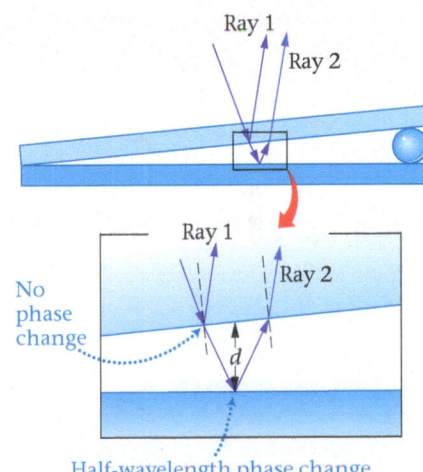

Ray 1
Ray 2

Ray 1
No phase change
Ray 2
d
Half-wavelength phase change

▲ **FIGURE 28-11 An air wedge** In an air wedge, interference occurs between the light reflected from the bottom surface of the top plate of glass (ray 1) and the light reflected from the top surface of the bottom plate of glass (ray 2). (The two rays are shown widely separated for clarity; in reality, they are almost on top of one another.)

PROBLEM-SOLVING NOTE

Interference of Reflected Light

When determining whether reflected light interferes constructively or destructively, it is essential to take into account the phase changes that can occur under reflection.

Side view

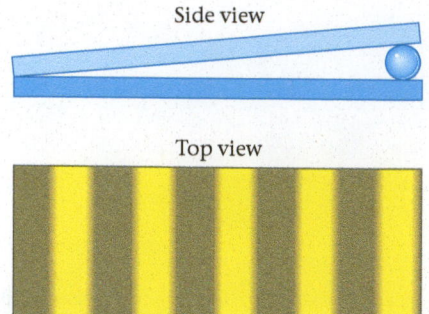

Top view

◄ **FIGURE 28-12 Interference fringes in an air wedge** The interference fringes in an air wedge are regularly spaced, as shown in the top view of the wedge.

Similarly, if the effective path length of ray 2 is an odd half integer, there will be **destructive interference:**

$$\frac{1}{2} + \frac{2d}{\lambda} = m + \frac{1}{2} \qquad m = 0, 1, 2, \ldots \qquad \text{28-6}$$

Noting that the distance between the plates, d, increases linearly with the distance from the point where the glass plates touch, we can conclude that the dark and bright interference fringes are evenly spaced, as shown in **FIGURE 28-12**.

CONCEPTUAL EXAMPLE 28-5   DARK OR BRIGHT FRINGE?

Is the point where the glass plates touch in an air wedge a dark fringe or a bright fringe?

REASONING AND DISCUSSION
At the point where the glass plates touch, the separation d is zero. Setting d equal to zero in Equation 28-5 $\left(\frac{1}{2} + 2d/\lambda = m\right)$ gives $\frac{1}{2} = m$, which can never be satisfied with an integer value of m. As a result, we conclude that the point of contact is not a bright fringe.

Considering Equation 28-6 $\left(\frac{1}{2} + 2d/\lambda = m + \frac{1}{2}\right)$ in the limit of $d = 0$ yields $\frac{1}{2} = m + \frac{1}{2}$, which is satisfied by $m = 0$. Therefore, the point of contact of the glass plates is the first dark fringe in the system.

We can understand the origin of the dark fringe at the point of contact by recalling that ray 1 undergoes no phase change on reflection, whereas ray 2 experiences a 180° phase change due to reflection. It follows, then, that when the path difference, $2d$, approaches zero, rays 1 and 2 will cancel with destructive interference.

ANSWER
A dark fringe occurs where the two glass plates touch.

The next Example uses the number of fringes observed in an air wedge to calculate the thickness of a human hair.

EXAMPLE 28-6   SPLITTING HAIRS

An air wedge is formed by placing a human hair between two glass plates on one end, and allowing them to touch on the other end. When this wedge is illuminated with red light ($\lambda = 771$ nm), it is observed to have 179 bright fringes. How thick is the hair?

PICTURE THE PROBLEM
The air wedge used in this experiment is shown in the sketch. The separation of the plates on the end with the hair is equal to the thickness of the hair, t. It is at this point that the 179th bright fringe is observed.

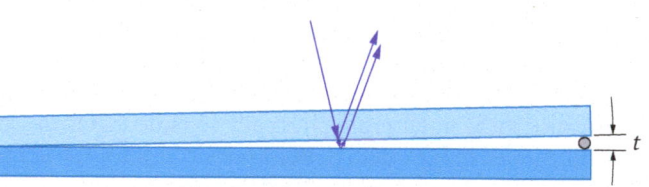

REASONING AND STRATEGY
Recall that the condition for bright fringes is $\frac{1}{2} + 2d/\lambda = m$, and that $m = 1$ corresponds to the first bright fringe, $m = 2$ corresponds to the second bright fringe, and so on. Clearly, then, the 179th bright fringe is given by $m = 179$. Substituting this value for m in $\frac{1}{2} + 2d/\lambda = m$, and setting the plate separation, d, equal to the thickness of the hair, t, allows us to solve for t.

Known   Wavelength of light, $\lambda = 771$ nm; number of bright fringes, $m = 179$.
Unknown   Thickness of hair, $t = ?$

SOLUTION

1. Solve the bright-fringe condition, $\frac{1}{2} + 2d/\lambda = m$, for the plate separation, d:

$$\frac{1}{2} + \frac{2d}{\lambda} = m$$

$$d = \frac{\lambda}{2}\left(m - \frac{1}{2}\right)$$

CONTINUED

2. Set $m = 179$, and solve for the hair thickness, $t = d$:

$$t = \frac{\lambda}{2}\left(m - \frac{1}{2}\right)$$

$$= \frac{(771 \times 10^{-9}\,\text{m})}{2}\left(179 - \frac{1}{2}\right)$$

$$= 6.88 \times 10^{-5}\,\text{m} = 68.8\,\mu\text{m}$$

INSIGHT

Thus, a human hair has a diameter of roughly 70 micrometers. To measure the thickness of a hair, we have used a "ruler" with units that are comparable to the distance to be measured. In this case, the hair has a thickness about 100 times larger than the wavelength of the light that we used.

PRACTICE PROBLEM — PREDICT/CALCULATE

(a) If a thicker hair is used in this experiment, will the number of bright fringes increase, decrease, or stay the same? Explain. **(b)** How many bright fringes will be observed if the hair has a thickness of 80.0 μm? [**Answer: (a)** A larger value for d means that m must be larger in order to satisfy $\frac{1}{2} + 2d/\lambda = m$. Therefore, the number of bright fringes will increase. **(b)** In this case, there are 208 bright fringes.]

Some related homework problems: Problem 37, Problem 42

Clearly, the location of interference fringes is very sensitive to even extremely small changes in the plate separation. This can be illustrated dramatically by simply pressing down lightly on the upper plate with a finger. Even though the finger causes no visible change in the glass plate, the fringes are observed to move. By measuring the displacement of the fringes it is possible to calculate the tiny deflection, or bend, the finger caused in the plate. Devices using this type of mechanism are frequently used to show small displacements that would be completely invisible to the naked eye.

Newton's Rings

RWP* A system similar to an air wedge, but with a slightly different geometry, is shown in **FIGURE 28-13 (a)**. In this case, the upper glass plate is replaced by a curved piece of glass with a spherical cross section. Still, the mechanism producing interference is the same as before. The result is a series of circular interference fringes, as shown in **FIGURE 28-13 (b)**, referred to as **Newton's rings.**

Notice that the fringes become more closely spaced as one moves farther from the center of the pattern. The reason is that the curved surface of the upper piece of glass moves away from the lower plate at a progressively faster rate. As a result, the horizontal distance required to go from one fringe to the next becomes smaller as one moves away from the center.

Newton's rings can be used to test the shape of a lens. Imperfections in the ring pattern indicate slight distortions in the lens. As in the case of an air wedge, even a very small change in shape can cause a significant displacement of the interference fringes.

Thin Films

Perhaps the most commonly observed examples of interference in light are provided by thin films, such as those found in soap bubbles and oil slicks. We are all familiar, for example, with the swirling patterns of color that are seen on the surface of a bubble, as shown in **FIGURE 28-14**. These colors are the result of the constructive and destructive interference that can occur when white light reflects from a thin film. In particular, some colors undergo destructive interference and are *eliminated* from the incident light, while others colors are *enhanced* by constructive interference.

To determine the conditions for constructive and destructive interference in a thin film, consider **FIGURE 28-15**. Here we show a thin film of thickness t and index of refraction $n > 1$, with air ($n = 1.00$) on either side. To analyze this system, we proceed in much the same way as we did for the air wedge earlier in this section. Specifically, we

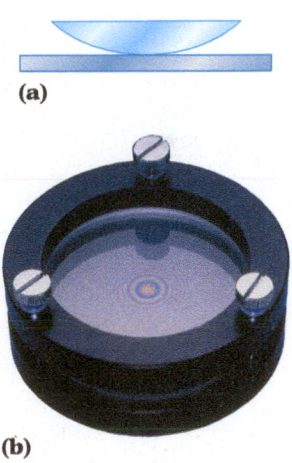

(a)

(b)

▲ **FIGURE 28-13** **A system for generating Newton's rings** **(a)** A variation on the air wedge is produced by placing a piece of glass with a spherical cross section on top of a plate of glass. **(b)** A top view of the system shown in part (a). The circular interference fringes are referred to as Newton's rings.

▲ **FIGURE 28-14** The swirling colors typical of soap bubbles are created by interference, both destructive and constructive, which eliminates certain wavelengths from the reflected light while enhancing others. Which colors are removed or enhanced at a given point depends on the precise thickness of the film in that region.

---

*Real World Physics applications are denoted by the acronym RWP.

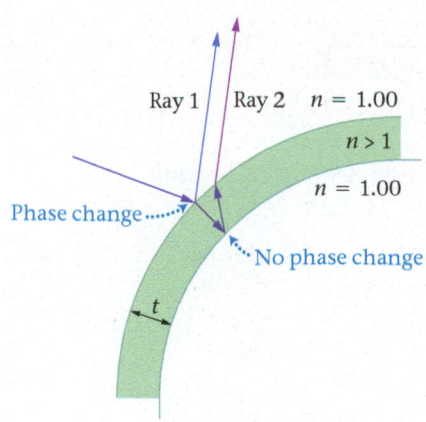

▲ **FIGURE 28-15 Interference in thin films** The phase of ray 1 changes by half a wavelength due to reflection; the phase of ray 2 changes by $2t/\lambda_n$, where λ_n is the wavelength of light within the thin film.

focus on the phase change of rays 1 and 2, taking into account phase changes due to both reflection and path difference.

First, notice that ray 1 reflects from the air-film interface; hence, its phase changes by half a wavelength. The effective path length for ray 1, then, is

$$\ell_{eff,1} = \tfrac{1}{2}\lambda$$

Dividing by the wavelength, λ, gives the phase change of ray 1 in terms of the wavelength:

$$\frac{\ell_{eff,1}}{\lambda} = \frac{1}{2} \qquad \text{28-7}$$

Recall that if two rays of light differ in phase by half a wavelength, the result is destructive interference.

Next, ray 2 reflects from the film-air interface; hence, it experiences no phase change due to reflection. It does, however, have a phase change as a result of traveling an extra distance $2t$ through the film. Thus the effective path length for ray 2 is

$$\ell_{eff,2} = 2t$$

To put this path length in terms of the wavelength, we must recall that if the wavelength of light in a vacuum is λ_{vacuum}, its wavelength in a medium with an index of refraction n is

$$\lambda_n = \frac{\lambda_{vacuum}}{n}$$

Dividing the path length by λ_n gives us the phase change of ray 2 in terms of the wavelength within the film:

$$\frac{\ell_{eff,2}}{\lambda_n} = \frac{2t}{\lambda_n} = \frac{2nt}{\lambda_{vacuum}} \qquad \text{28-8}$$

Finally, we can calculate the difference in phase changes for rays 1 and 2 using the preceding results:

$$\text{difference in phase changes} = \frac{2nt}{\lambda_{vacuum}} - \frac{1}{2} \qquad \text{28-9}$$

In the limit of zero film thickness, $t = 0$, the phase difference is $-\tfrac{1}{2}$. This corresponds to destructive interference, because moving a wave back half a wavelength is equivalent to moving it ahead half a wavelength. In general, then, destructive interference occurs if any of the following conditions are satisfied:

$$\frac{2nt}{\lambda_{vacuum}} - \frac{1}{2} = -\frac{1}{2}, \frac{1}{2}, \frac{3}{2}, \cdots$$

Adding $\tfrac{1}{2}$ to each side yields our final result for **destructive interference**:

$$\frac{2nt}{\lambda_{vacuum}} = m \qquad m = 0, 1, 2, \ldots \qquad \text{28-10}$$

Similarly, if the phase difference between the rays is equal to an integer, m, the result is **constructive interference**:

$$\frac{2nt}{\lambda_{vacuum}} - \frac{1}{2} = m \qquad m = 0, 1, 2, \ldots \qquad \text{28-11}$$

These conditions are applied in the next Example.

EXAMPLE 28-7   RED LIGHT SPECIAL

A beam consisting of red light ($\lambda_{vacuum} = 662\ \text{nm}$) and blue light ($\lambda_{vacuum} = 465\ \text{nm}$) is directed at right angles onto a thin soap film. If the film has an index of refraction $n = 1.33$ and is suspended in air ($n = 1.00$), find the smallest nonzero thickness for which it appears red in reflected light.

PICTURE THE PROBLEM

Our sketch shows a soap film of thickness t and index of refraction $n = 1.33$ suspended in air ($n = 1.00$). We consider each of the colors in the incoming beam of light separately, with one blue ray and one red ray. If the film is to look red in reflected light, it follows that the reflected blue light must cancel due to destructive interference, as indicated.

CONTINUED

REASONING AND STRATEGY

As mentioned above, the desired thickness of the film is such that blue light satisfies one of the conditions for destructive interference given in Equation 28-10 $(2nt/\lambda_{vacuum} = m)$. Because the case $m = 0$ corresponds to zero thickness, $t = 0$, it follows that the smallest nonzero thickness is given by $m = 1$.

Known Wavelength of blue light, $\lambda_{vacuum} = 465$ nm; index of refraction of film, $n = 1.33$; index of refraction of air, $n = 1.00$.

Unknown Smallest thickness for destructive interference, $t = $?

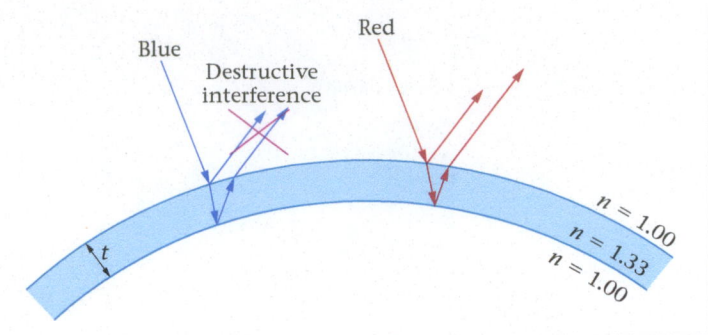

SOLUTION

1. Solve $2nt/\lambda_{vacuum} = m$ for the thickness t:

$$\frac{2nt}{\lambda_{vacuum}} = m \quad \text{or} \quad t = \frac{m\lambda_{vacuum}}{2n}$$

2. Calculate the thickness using $m = 1$:

$$t = \frac{m\lambda_{vacuum}}{2n} = \frac{(1)(465 \text{ nm})}{2(1.33)} = 175 \text{ nm}$$

INSIGHT

Although blue light is canceled at this thickness, red light is not. In fact, repeating the preceding calculation shows that red light is not canceled until the thickness of the film is increased to 249 nm.

PRACTICE PROBLEM

What is the smallest thickness of film that gives *constructive* interference for the blue light in this system?
[**Answer:** $t = 87.4$ nm]

Some related homework problems: Problem 31, Problem 32

The connection between the thickness of a film and the color it shows in reflected light is illustrated in **FIGURE 28-16**. In thicker regions of the film, the longer wavelengths of light interfere destructively, resulting in reflected light that is bluish. In the limit of zero thickness, the condition for destructive interference, $2nt/\lambda_{vacuum} = m$, is satisfied for all wavelengths with m set equal to zero. Hence, an extremely thin film appears dark in reflected light—it is essentially one large dark fringe.

Another example of a thin-film interference is provided by a film floating on a liquid or coating a solid surface. For example, **FIGURE 28-17** shows a thin film floating on water. If the index of refraction of the film is greater than that of water, the situation in terms of interference is essentially the same as for a thin film suspended in air. In particular, the ray reflected from the top surface of the film undergoes a phase change of half a wavelength; the ray reflected from the bottom of the film has no phase change due to reflection. Thus the interference conditions given in Equation 28-10 $(2nt/\lambda_{vacuum} = m)$ and Equation 28-11 $\left(2nt/\lambda_{vacuum} - \frac{1}{2} = m\right)$ still apply.

On the other hand, suppose a thin film has an index of refraction that is greater than 1.00 but less than the index of refraction of the material on which it is supported, as in **FIGURE 28-18**. In a case like this, there is a reflection phase change at both the upper and the lower surfaces of the film. The condition for destructive interference for this type of system is discussed in the following Example.

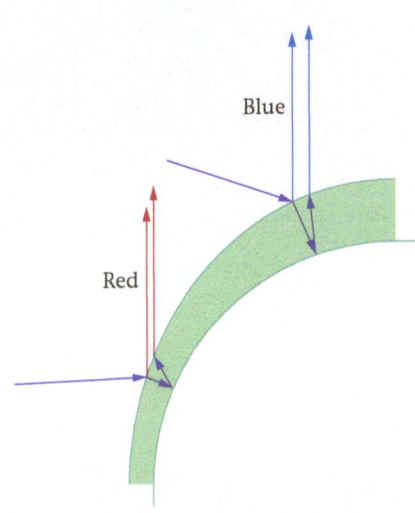

▲ **FIGURE 28-16 Thickness and color in a thin film** Thicker portions of thin film appear blue, since the long-wavelength red light experiences destructive interference. Thinner regions appear red because the short-wavelength blue light interferes destructively.

QUICK EXAMPLE 28-8 NONREFLECTIVE COATING: FIND THE THICKNESS

Camera lenses $(n = 1.52)$ are often coated with a thin film of magnesium fluoride $(n = 1.38)$. These "nonreflective coatings" use destructive interference to reduce unwanted reflections. Find the condition for destructive interference in this case, and calculate the minimum thickness required to give destructive interference for light in the middle of the visible spectrum (yellow-green light, $\lambda_{vacuum} = 569$ nm).

REASONING AND SOLUTION

The coating on the lens has an index of refraction less than that of the glass in the lens; thus the situation is like the one shown in Figure 28-18. We calculate the change in phase

CONTINUED

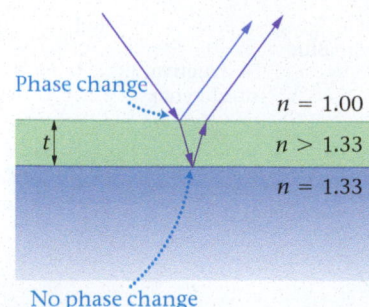

▲ FIGURE 28-17 A thin film with one phase change If the index of refraction of the film is greater than that of the water, the situation in terms of phase changes is essentially the same as for a thin film suspended in air.

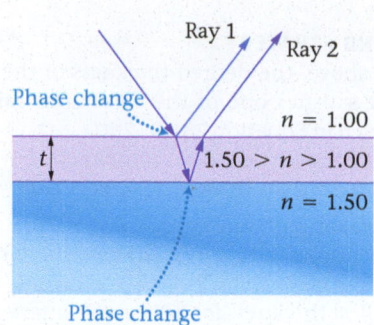

▲ FIGURE 28-18 A thin film with two phase changes A thin film is applied to a material with a relatively large index of refraction. If the index of refraction of the film is less than that of the material that supports it, there will be a phase change for reflections from both surfaces of the film. Films of this type are often used in nonreflective coatings.

▲ FIGURE 28-19 The lenses of binoculars and cameras often have a blue, purple, or amber tint, the product of their antireflection coating. The coating is a thin film that reduces reflection from the lens surfaces by destructive interference.

for both ray 1 and ray 2 shown in Figure 28-18, and set the difference between these phase changes equal to half a wavelength. This will result in destructive interference between the rays.

1. The phase change (in units of the wavelength) for ray 1 is $\frac{1}{2}$, due to reflection:

$$\frac{1}{2}$$

2. The phase change (in units of the wavelength) for ray 2 is $\frac{1}{2}$ (reflection) plus $2t/\lambda_n$ (path-length difference):

$$\frac{1}{2} + \frac{2t}{\lambda_n}$$

3. Set the difference in phase changes equal to $\frac{1}{2}$:

$$\left(\frac{1}{2} + \frac{2t}{\lambda_n}\right) - \left(\frac{1}{2}\right) = \frac{1}{2}$$

$$\frac{2t}{\lambda_n} = \frac{1}{2}$$

4. Solve for the thickness:

$$t = \frac{\lambda_n}{4}$$

5. Substitute numerical values:

$$t = \frac{\lambda_n}{4} = \frac{\lambda_{\text{vacuum}}}{4n} = \frac{569 \text{ nm}}{4(1.38)} = 103 \text{ nm}$$

RWP Because the thickness of the film should be a quarter of a wavelength, these nonreflective films are often referred to as quarter-wave coatings. An example of a nonreflective coating on a pair of binoculars is shown in **FIGURE 28-19**.

Interference in CDs and DVDs

RWP Destructive interference plays a crucial role in the operation of a CD or DVD. The basic idea behind these devices is that information is encoded in a series of "bumps" on an otherwise smooth reflecting surface, as shown in **FIGURE 28-20 (a)**. A laser beam directed onto the surface is reflected back to a detector, and as the intensity of the reflected

◄ FIGURE 28-20 (a) This microscopic view of the surface of a DVD reveals the pattern of tiny bumps that encodes the information. **(b)** A DVD uses red light to read the stored information. A Blu-ray uses blue light whose shorter wavelength allows for smaller "bumps" and hence more tightly packed information.

(a)

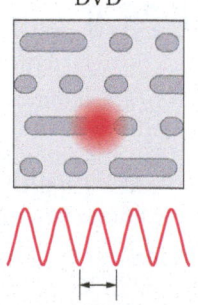

(b)

DVD

Blu-ray

$\lambda = 650$ nm $\lambda = 405$ nm

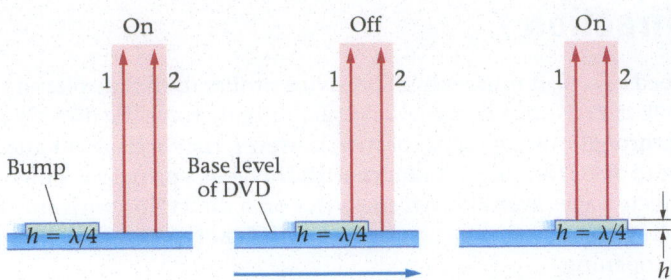

beam varies due to the bumps, the information on the DVD is decoded—similar to using dots and dashes to send information in Morse code. The laser that reads a regular DVD is red and has a wavelength of 650 nm. On the other hand, a Blu-ray DVD uses light that is deep blue (hence the name), with a wavelength of only 405 nm. The shorter the wavelength of light, the smaller the bumps that encode the information; therefore, a Blu-ray can hold much more information than a regular DVD, as indicated in **FIGURE 28-20 (b)**.

Imagine the laser beam illuminating a small area on the surface of a DVD. As a bump moves into the beam, as in **FIGURE 28-21**, there are two components to the reflected beam—one from the top of the bump, the other from the bottom. If these two beams are out of phase by half a wavelength, there will be destructive interference and the detector will receive a weak signal. When the beam is reflected solely from the top of the bump, the detector again receives a strong signal, because there is no longer an interference effect. As the bump moves out of the beam we again have the condition for destructive interference, and the detector signal again weakens. Thus the bumps give the detector a series of "on" and "off" signals that can be converted to sound, pictures, or other types of information.

Let's find the necessary height h of a bump if the light of a laser is to have destructive interference. Figure 28-21 shows the situation, in which ray 1 reflects from the top of a bump, and ray 2 reflects from the base level of the DVD. The path difference between the rays is $2h$; thus, to have destructive interference, $2h$ must be equal to half a wavelength, $2h = \lambda/2$, or

$$h = \frac{\lambda}{4}$$

In the case of a DVD, the required height of a bump is about 163 nm.

Enhance Your Understanding   (Answers given at the end of the chapter)

3. For each of the cases shown in **FIGURE 28-22**, state whether the phase change for reflection 1 is greater than, less than, or equal to the phase change for reflection 2. (a) $n_1 = 1.2$, $n_2 = 1.4$; (b) $n_1 = 1.4$, $n_2 = 1.2$; (c) $n_1 = 1.2$, $n_2 = 1.1$; (d) $n_1 = 1.5$, $n_2 = 1.6$.

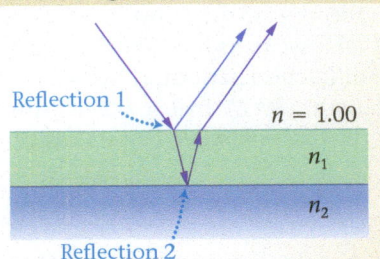

▲ **FIGURE 28-22** Enhance Your Understanding 3.

Section Review

- Constructive interference in an air wedge satisfies the following conditions: $\frac{1}{2} + 2d/\lambda = m$, where $m = 1, 2, 3, \ldots$. Destructive interference satisfies $\frac{1}{2} + 2d/\lambda = m + \frac{1}{2}$, where $m = 0, 1, 2, \ldots$.

- Constructive interference for a thin film in air, like a soap bubble, satisfies the following conditions: $2nt/\lambda_{vacuum} - \frac{1}{2} = m$, where $m = 0, 1, 2, \ldots$. Destructive interference satisfies $2nt/\lambda_{vacuum} = m$, where $m = 0, 1, 2, \ldots$.

- Conditions for constructive and destructive interference in other systems can be derived in the same way as was done for the air wedge and the thin film.

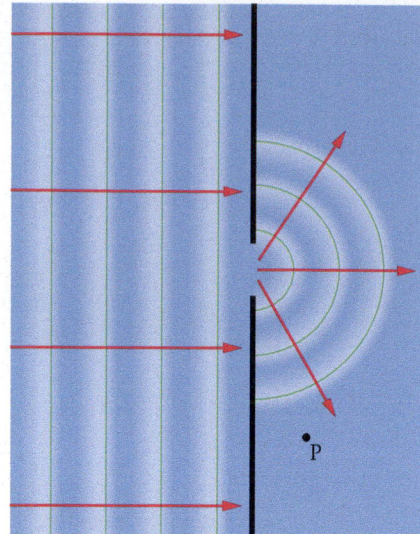

▲ FIGURE 28-23 Diffraction of water waves As water waves pass through an opening, they diffract, or change direction. Thus an observer at point P detects waves even though this point is not on a line with the original direction of the waves and the opening. All waves exhibit similar diffraction behavior.

Big Idea 4 When light encounters an obstacle, or passes through an opening, it diffracts—which means it bends away from its original direction of propagation.

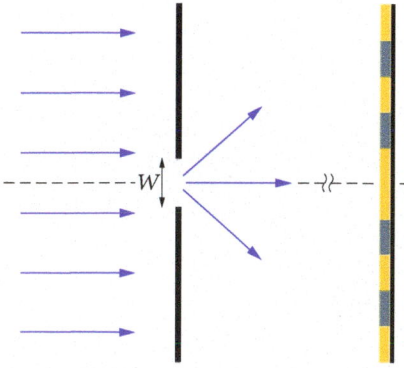

▲ FIGURE 28-24 Single-slit diffraction When light of wavelength λ passes through a slit of width W, a "diffraction pattern" of bright and dark fringes is formed.

28-4 Diffraction

If light is indeed a wave, it must exhibit behavior similar to that displayed by the water waves in **FIGURE 28-23**. Notice that the waves are initially traveling directly to the right. After passing through the gap in the barrier, however, they spread out and travel in all possible forward directions, in accordance with Huygens's principle. Thus, an observer at point P detects waves even though she is not on a direct line with the initial waves and the gap. In general, waves always bend, or **diffract**, when they pass by a barrier or through an opening.

A familiar example of diffraction is the observation that you can hear a person talking even when that person is out of sight around a corner. The sound waves from the person bend around the corner, just like the water waves in Figure 28-23. It might seem, then, that light cannot be a wave, because it does not bend around a corner along with the sound. There is a significant difference between sound and light waves, however; namely, their wavelengths, which differ by many orders of magnitude—from a meter or so for sound to about 10^{-7} m for light. As we shall see, the angle through which a wave bends is greater, the larger the wavelength of the wave; hence, diffraction effects in light are typically small compared with those in sound waves and water waves.

To investigate the diffraction of light we start by considering the behavior of a beam of light as it passes through a single, narrow slit in a screen.

Single-Slit Diffraction

Consider monochromatic light of wavelength λ passing through a narrow slit of width W, as shown in **FIGURE 28-24**. After passing through the slit, the light shines on a distant screen, which we assume to be much farther from the slit than the width W. According to Huygens's principle, each point within the slit can be considered as a source of new waves that radiate toward the screen. The interference of these sources with one another generates a diffraction pattern.

We can understand the origin of a single-slit diffraction pattern by referring to **FIGURE 28-25**, where we show light propagating to the screen from various points in the slit. For example, consider waves from points 1 and 1' propagating to the screen at an angle θ relative to the initial direction of the light. Because the screen is distant, the waves from 1 and 1' travel on approximately parallel paths to reach the screen. From the figure, then, it is clear that the path difference for these waves is $(W/2) \sin \theta$. Similarly, the same path difference applies to the "wave pairs" from points 2 and 2', the wave pairs from points 3 and 3', and so on through all points in the slit.

In the forward direction, $\theta = 0°$, the path difference is zero: $(W/2) \sin 0° = 0$. As a result, all wave pairs interfere constructively, giving maximum intensity at the center of the diffraction pattern. However, if θ is increased until the path difference is half a wavelength, $(W/2) \sin \theta = \lambda/2$, each wave pair experiences destructive interference. Thus, the *first minimum*, or dark fringe, in the diffraction pattern occurs at the angle given by

$$W \sin \theta = \lambda$$

To find the second dark fringe, imagine dividing the slit into four regions, as illustrated in **FIGURE 28-26**. Within the upper two regions, we perform the same wave-pair construction described above; the same is done with the lower two regions. In this case, the path difference between wave pairs 1 and 1' is $(W/4) \sin \theta$. As before, destructive interference first occurs when the path difference is $\lambda/2$. Solving for sin θ in this case gives us the condition for the *second dark fringe*:

$$W \sin \theta = 2\lambda$$

The next dark fringe can be found by dividing the slit into six regions, with a path difference between wave pairs of $(W/6) \sin \theta$. In this case, the condition for destructive interference is $W \sin \theta = 3\lambda$. In general, then, dark fringes satisfy the following conditions:

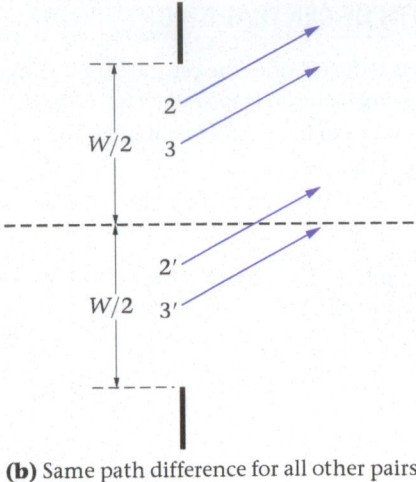

(a) Path difference for rays 1 and 1′

(b) Same path difference for all other pairs of rays

▲ **FIGURE 28-25 Locating the first dark fringe in single-slit diffraction** The location of the first dark fringe in single-slit diffraction can be determined by considering pairs of waves radiating from the top and bottom half of a slit. **(a)** A wave pair originating at points 1 and 1′ has a path difference of $(W/2) \sin \theta$. These waves interfere destructively if the path difference is equal to half a wavelength. **(b)** The rest of the light coming from the slit can be considered to consist of additional wave pairs, like 2 and 2′, 3 and 3′, and so on. Each wave pair has the same path difference.

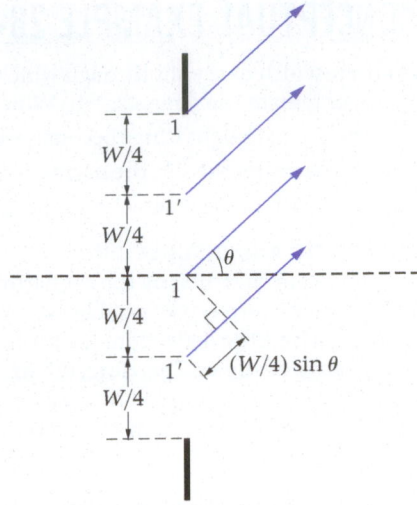

▲ **FIGURE 28-26 Locating the second dark fringe in single-slit diffraction** To find the second dark fringe in a single-slit diffraction pattern, we divide the slit into four regions and consider wave pairs that originate from points separated by a distance $W/4$. Destructive interference occurs when the path difference, $(W/4) \sin \theta$, is half a wavelength.

Conditions for Dark Fringes in Single-Slit Interference

$$W \sin \theta = m\lambda \qquad m = \pm 1, \pm 2, \pm 3, \ldots \qquad \text{28-12}$$

By including both positive and negative values for m, we have taken into account the symmetry of the diffraction pattern about its midpoint.

PHYSICS IN CONTEXT
Looking Ahead

The Heisenberg uncertainty principle that is so fundamental to quantum physics can be understood as analogous to the diffraction of light, as we shall see in Section 30-6.

EXERCISE 28-9 WAVELENGTH OF LIGHT

Monochromatic light passes through a slit of width 1.75×10^{-5} m. If the first dark fringe of the resulting diffraction pattern is at angle $\theta = 2.28°$, what is the wavelength of the light?

REASONING AND SOLUTION
Solving Equation 28-12 ($W \sin \theta = m\lambda$) for the wavelength gives

$$\lambda = \frac{W \sin \theta}{m} = \frac{(1.75 \times 10^{-5} \text{ m}) \sin(2.28°)}{1} = 696 \text{ nm}$$

We used $m = 1$ because this corresponds to the *first* dark fringe.

The bright fringes in a diffraction pattern consist of the central fringe plus additional bright fringes approximately halfway between successive dark fringes. In addition, the central fringe is about twice as wide as the other bright fringes. In the small-angle approximation ($\sin \theta \sim \theta$), the central fringe extends from $\theta = \lambda/W$ to $\theta = -\lambda/W$, and hence its width is given by the following:

$$\text{approximate angular width of central fringe} = 2\frac{\lambda}{W} \qquad \text{28-13}$$

Thus we see that the wavelength λ plays a crucial role in diffraction patterns and that λ/W gives the characteristic angle of "bending" produced by diffraction. In the following Conceptual Example we consider the role played by the width of the slit.

CONCEPTUAL EXAMPLE 28-10   WIDTH OF CENTRAL BRIGHT FRINGE   PREDICT/EXPLAIN

(a) If the width of the slit through which light passes is reduced, does the central bright fringe become wider, become narrower, or remain the same size? (b) Which of the following is the *best explanation* for your prediction?

  I. The central bright fringe becomes wider because a beam of light spreads out more when it passes through a smaller slit.

  II. The narrower the slit, the narrower the central bright fringe.

  III. The width of the central bright fringe depends only on the wavelength of light, and not on the width of the slit.

REASONING AND DISCUSSION

It might seem that making the slit narrower causes the diffraction pattern to be narrower as well. Recall, however, that the diffraction pattern is produced by waves propagating from all parts of the slit. If the slit is wide, the incoming wave passes through with little deflection. If the slit is small, on the other hand, it acts like a point source, and light is radiated over a broad range of angles. Therefore, the smaller slit produces a wider central fringe.

    This result is confirmed by considering Equation 28-13 (angular width of central fringe $= 2\lambda/W$), where we see that a smaller value of W results in a wider central fringe.

ANSWER

(a) The central bright fringe becomes wider. (b) The best explanation is I.

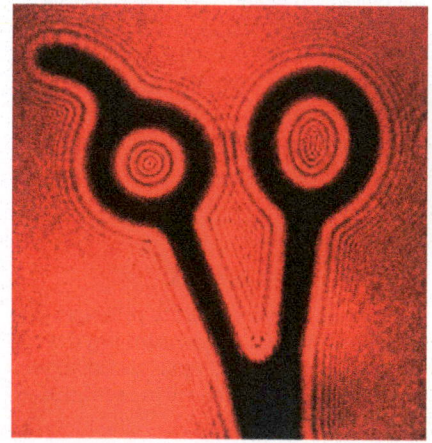

◄ **FIGURE 28-27** On close inspection, the shadow of a sharp edge is seen to consist of numerous fringes produced by diffraction.

    As mentioned earlier, waves diffract whenever they encounter some sort of barrier or opening. It follows, then, that the shadow cast by an object is really not as sharp as it may seem. On closer examination, as shown in **FIGURE 28-27**, the shadow of an object such as a pair of scissors actually consists of a tightly spaced series of diffraction fringes. Thus, shadows are not the sharp boundaries implied by geometrical optics but, instead, are smeared out on a small length scale. Under ordinary conditions the diffraction pattern in a shadow is not readily visible. However, a similar diffraction pattern can be observed by simply holding two fingers close together before your eyes—try it.

EXAMPLE 28-11   EXPLORING THE DARK SIDE

Light with a wavelength of 511 nm forms a diffraction pattern after passing through a single slit of width 2.20×10^{-6} m. Find the angle associated with (a) the first and (b) the second dark fringe above the central bright fringe.

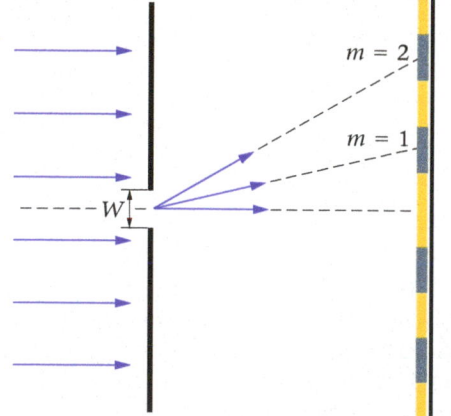

PICTURE THE PROBLEM

In our sketch we identify the first and second dark fringes above the central bright fringe. The first dark fringe corresponds to $m = 1$, and the second corresponds to $m = 2$. Finally, the width of the slit is $W = 2.20 \times 10^{-6}$ m.

REASONING AND STRATEGY

We can find the desired angles by using the condition for dark fringes, $W \sin \theta = m\lambda$ (Equation 28-12). As mentioned above, we use $m = 1$ for part (a), and $m = 2$ for part (b). The values of λ and W are given in the problem statement.

Known   Wavelength of light, $\lambda = 511$ nm; width of slit, $W = 2.20 \times 10^{-6}$ m.
Unknown   Angle, $\theta = $?, of (a) the first dark fringe and (b) the second dark fringe.

SOLUTION

Part (a)

1. Solve for θ using $m = 1$: $\theta = \sin^{-1}\left(\dfrac{m\lambda}{W}\right) = \sin^{-1}\left[\dfrac{(1)(511 \times 10^{-9}\,\text{m})}{2.20 \times 10^{-6}\,\text{m}}\right] = 13.4°$

Part (b)

1. Solve for θ using $m = 2$: $\theta = \sin^{-1}\left(\dfrac{m\lambda}{W}\right) = \sin^{-1}\left[\dfrac{(2)(511 \times 10^{-9}\,\text{m})}{2.20 \times 10^{-6}\,\text{m}}\right] = 27.7°$

CONTINUED

INSIGHT
The angle to the second dark fringe is *not* simply twice the angle to the first dark fringe. This is because the angle θ depends on the sine function, which is not linear. If you look at a plot of the sine function, as in Figure 13-21, you will see that $\sin \theta$ is slightly less than θ for the angles considered here. Therefore, to double the value of $\sin \theta$, you must increase θ by slightly more than a factor of 2.

PRACTICE PROBLEM — PREDICT/CALCULATE
Suppose the wavelength of light in this experiment is changed to give the first dark fringe at an angle greater than 13.4°. **(a)** Is the required wavelength greater than or less than 511 nm? Explain. **(b)** Check your answer by calculating the wavelength required to give the first dark fringe at $\theta = 15.0°$.   [**Answer: (a)** A larger angle in $W \sin \theta = m\lambda$ implies that the wavelength must be greater. **(b)** $\lambda = 569$ nm]

Some related homework problems: Problem 44, Problem 45, Problem 49

QUICK EXAMPLE 28-12   FIND THE LINEAR DISTANCE

In a single-slit experiment, light passes through the slit and forms a diffraction pattern on a screen 2.54 m away. If the wavelength of light is 655 nm, and the width of the slit is 4.41×10^{-5} m, find the linear distance on the screen from the center of the diffraction pattern to the first dark fringe.

REASONING AND SOLUTION
We can find the angle to the first dark fringe with $W \sin \theta = m\lambda$. Once the angle is known, the linear distance is given by the distance to the screen multiplied by the tangent of the angle.

1. Find the angle to the first dark fringe:

$$\theta = \sin^{-1}\left(\frac{m\lambda}{W}\right) = \sin^{-1}\left[\frac{(1)(655 \times 10^{-9}\ \text{m})}{4.41 \times 10^{-5}\ \text{m}}\right] = 0.851°$$

2. Use $y = L \tan \theta$ to find the linear distance:

$$y = L \tan \theta = (2.54\ \text{m}) \tan (0.851°) = 3.77\ \text{cm}$$

The linear distance is found using Equation 28-3, just as for the two-slit experiment.

Poisson's Bright Spot   The photo in **FIGURE 28-28** shows a particularly interesting diffraction phenomenon: the shadow produced by a penny. As expected, the edges of the shadow show diffraction fringes, but of even greater interest is the small bright point of light seen in the center of the shadow. This is referred to as "Poisson's bright spot," after the French scientist Siméon D. Poisson (1781–1840) who predicted its existence. It should be noted that Poisson did not believe that light is a wave. In fact, he used his prediction of a bright spot to show that the wave model of light was absurd and must be rejected—after all, how could a bright spot occur in the darkest part of a shadow? When experiments soon after his prediction showed conclusively that the bright spot does exist, however, the wave model of light gained almost universal acceptance. We know today that light has both wave and particle properties, a fact referred to as *wave-particle duality*. This will be discussed in detail in Chapter 30.

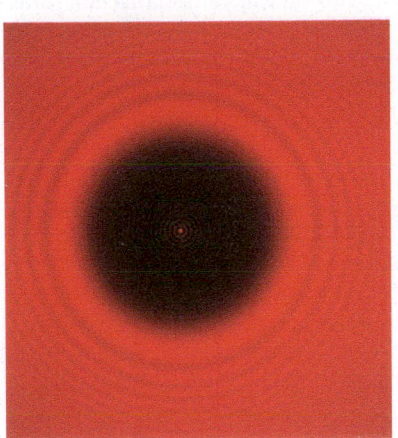

▲ **FIGURE 28-28** The paradoxical bright dot at the center of a circular shadow, known as Poisson's bright spot, is convincing proof of the wave nature of light.

Enhance Your Understanding   (Answers given at the end of the chapter)

4. If the wavelength of light passing through a single slit is decreased, does the width of the central bright fringe increase, decrease, or stay the same? Assume everything else in the system is the same.

Section Review

- The condition for bright fringes in single-slit diffraction is $W \sin \theta = m\lambda$, where $m = \pm 1, \pm 2, \pm 3, \dots$.

- The angular width of the central bright fringe in single-slit diffraction is given approximately by $2\lambda/W$. This angle is measured in radians.

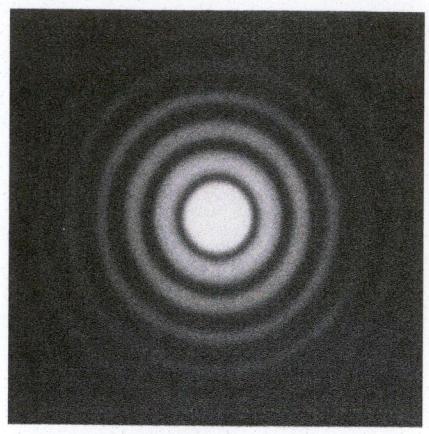

▲ **FIGURE 28-29 Diffraction from a circular opening** Light passing through a circular aperture creates a circular diffraction pattern of alternating bright and dark regions.

PROBLEM-SOLVING NOTE

Angular Resolution and the Index of Refraction

When applying the condition $\theta_{min} = 1.22\lambda/D$, it is important to remember that λ refers to the wavelength in the region between the aperture and the screen (retina, film, etc.) on which the diffraction pattern is observed. If the index of refraction in this region is n, the wavelength is reduced from λ to λ/n. (See Conceptual Example 28-4.)

Big Idea 5 The ability to resolve, or distinguish between, different sources of light depends on the wavelength of the light. The greater the wavelength, the less the ability to resolve.

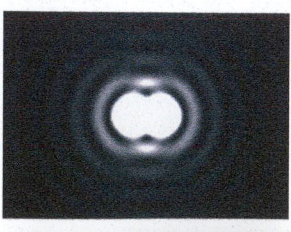

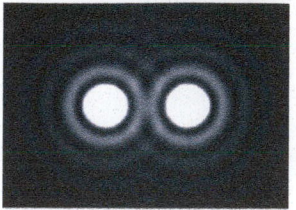

28-5 Resolution

Diffraction affects, and ultimately limits, the way we see the world. There is a difference, for example, in how sharply we can see a scene depending on the size of our pupils—in general, the larger the pupils, the sharper the vision. For example, an eagle, which has pupils that are even larger than ours, can see a small creature on the ground with greater acuity than is possible for a person. Similarly, a camera or a telescope with a large aperture can "see" with greater detail than the human eye. The sharpness of vision—in particular, the ability to visually separate closely spaced objects—is referred to as **resolution.**

Circular Aperture The way in which diffraction affects resolution can be seen by considering the diffraction pattern created by a circular aperture—such as the pupil of an eye. Just as with the diffraction pattern of a slit, a circular aperture creates a pattern of alternating bright and dark regions. The difference, as one might expect, is that the diffraction pattern of a circular opening is circular in shape, as we see in **FIGURE 28-29**.

In a slit pattern, the first dark fringe occurs at the angle given by $\sin\theta = \lambda/W$, where W is the width of the slit. A circular aperture of diameter D produces a central bright spot and a dark fringe at an angle θ from the center line given by the following expression:

First Dark Fringe for the Diffraction Pattern of a Circular Opening

$$\sin\theta = 1.22\frac{\lambda}{D}$$

28-14

So we see that the change in geometry from a slit of width W to a circular opening of diameter D is reflected in the replacement of $(1)\lambda/W$ with $(1.22)\lambda/D$.

The physical significance of this result is that even if you focus perfectly on a point source of light, it will form a circular image of finite size on the retina. This blurs the image, replacing a point with a small circle. Thus, the diffraction of light through the pupil limits the resolution of your eye. In addition, it should be noted that the wavelength in Equation 28-14 ($\sin\theta = 1.22\lambda/D$) refers to the wavelength in the medium in which the diffraction pattern is observed. For example, the wavelength that should be used in considering the eye is $\lambda_n = \lambda/n$, where n is the eye's average index of refraction (approximately $n = 1.36$), whereas the wavelength in the air-filled interior of a telescope is simply λ.

Diffraction-induced smearing also makes it difficult to visually separate objects that are close to one another. In particular, if two closely spaced sources of light are smeared by diffraction, the circles they produce may overlap, making it difficult to tell if there are two sources or only one. The condition that is used to determine whether two sources can be visually separated is called **Rayleigh's criterion,** which can be stated as follows: *If the first dark fringe of one circular diffraction pattern passes through the center of a second diffraction pattern, the two sources responsible for the patterns will appear to be a single source.* This condition is illustrated in **FIGURE 28-30**.

To put Rayleigh's criterion in quantitative terms, notice that for small angles (as is usually the case) the location of the first dark fringe is given by $\theta = 1.22\lambda/D$, where θ is measured in radians. Therefore, Rayleigh's criterion gives the following condition:

Rayleigh's Criterion

Two objects can be seen as separate only if their angular separation is greater than the minimum angle θ_{min} defined as follows:

$$\theta_{min} = 1.22\frac{\lambda}{D}$$

28-15

The larger the diameter D of the aperture, the smaller the angular separation that can be resolved and, hence, the greater the resolution.

◄ **FIGURE 28-30 Resolving two point sources: Rayleigh's criterion** If the angular separation between two sources is not great enough (top), their diffraction patterns overlap to the point where they appear as a single elongated source. With greater angular separation (bottom), the individual sources can be distinguished as separate.

EXERCISE 28-13 MINIMUM ANGLE

Find θ_{min} for green light (535 nm) and an aperture diameter of 5.77 mm.

REASONING AND SOLUTION

Substituting into Equation 28-15 ($\theta_{min} = 1.22\lambda/D$), we find

$$\theta_{min} = 1.22\frac{\lambda}{D} = 1.22\left(\frac{535 \times 10^{-9}\text{ m}}{5.77 \times 10^{-3}\text{ m}}\right) = 0.000113\text{ rad} = 0.00648°$$

Because our pupils have diameters of about 5.00 mm, the small value of θ_{min} indicates that human vision has the potential for relatively high resolution.

An example of diffraction-limited resolution is illustrated in the photos in **FIGURE 28-31**. In the first photo we see a brilliant light in the distance that may be the single headlight of an approaching motorcycle or the unresolved image of two headlights on a car. If we are seeing the headlights of a car, the angular separation between them will increase as the car approaches. When the angular separation exceeds $1.22\lambda/D$, as in the second photo, we are able to distinguish the two headlights as separate sources of light. As the car continues to approach, its individual headlights become increasingly distinct, as shown in the third photo.

▲ **FIGURE 28-31 Resolving the headlights of an approaching car.** If the headlights were true point sources and the atmosphere perfectly transparent (or absent), the individual headlights could be distinguished at a much greater distance, as Example 28-14 shows.

The distance at which the two headlights can be resolved increases as the size of the aperture increases. This dependence is considered in detail in the following Example.

EXAMPLE 28-14 MOTORCYCLE OR CAR?

The linear distance separating the headlights of a car is 1.1 m. Assuming light of 460 nm, a pupil diameter of 5.0 mm, and an average index of refraction for the eye of 1.36, find the maximum distance at which the headlights can be distinguished as two separate sources of light.

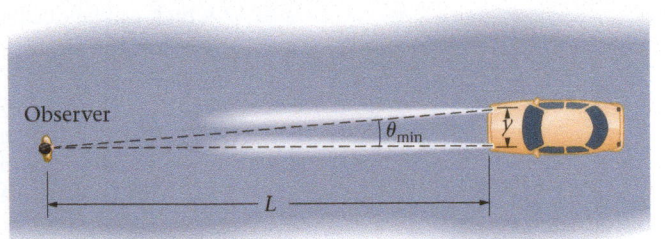

PICTURE THE PROBLEM

Our sketch shows the car a distance L from the observer, with the headlights separated by a linear distance $y = 1.1$ m. The condition for the headlights to be resolved is that their angular separation is at least θ_{min}.

REASONING AND SOLUTION

To find the maximum distance, we must first determine the minimum angular separation, which is given by $\theta_{min} = 1.22\lambda_n/D$. In this expression, $\lambda_n = \lambda/1.36$.

Once we know θ_{min}, we can find the distance L using the trigonometric relationship $\tan\theta_{min} = y/L$, which follows directly from our sketch.

CONTINUED

Known   Separation of headlights, $y = 1.1$ m; wavelength of light, $\lambda = 460$ nm; pupil diameter, $D = 5.0 \times 10^{-3}$ m; index of refraction of eye, $n = 1.36$.

Unknown   Maximum distance to separate light sources, $L = ?$

SOLUTION

1. Find the minimum angular separation for the headlights to be resolved:

$$\theta_{min} = 1.22\frac{\lambda/n}{D}$$

$$= 1.22\left[\frac{(460 \times 10^{-9}\text{ m})/1.36}{5.0 \times 10^{-3}\text{ m}}\right] = 8.3 \times 10^{-5}\text{ rad}$$

2. Solve the relationship $\tan\theta_{min} = y/L$ for the distance L:

$$L = \frac{y}{\tan\theta_{min}}$$

3. Substitute numerical values to find L:

$$L = \frac{y}{\tan\theta_{min}} = \frac{1.1\text{ m}}{\tan(8.3 \times 10^{-5}\text{ rad})} = 13{,}000\text{ m}$$

INSIGHT

Thus, the car must be about 8 mi away before the headlights appear to merge. This, of course, is the ideal case. In the real world, the finite size of the headlights and the blurring effects of the atmosphere greatly reduce the maximum distance.

PRACTICE PROBLEM

If the pupil diameter for an eagle is 6.2 mm, from what distance can it resolve the car's headlights under ideal conditions?   [**Answer:** In this case, the distance is 16,000 m, or about 9.9 mi.]

Some related homework problems: Problem 60, Problem 77

QUICK EXAMPLE 28-15   RESOLVING IDA AND DACTYL

The asteroid Ida is orbited by its own small "moon" called Dactyl. If the separation between these two asteroids is 2.5 km, what is the maximum distance at which the Hubble Space Telescope (aperture diameter = 2.4 m) can still resolve them with 550-nm light?

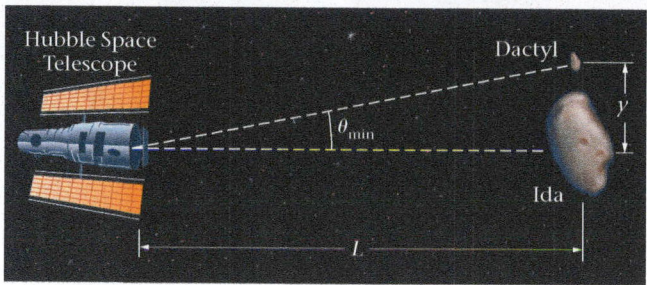

REASONING AND SOLUTION

The minimum angular separation for Ida and Dactyl to be resolved is $\theta_{min} = 1.22\lambda/D$, which can be calculated with the information given in the problem statement. Knowing θ_{min} and the linear separation y, we can find the distance L with the trigonometric relationship $\tan(\theta_{min}) = y/L$.

1. Calculate the minimum angular separation for the asteroids to be resolved, with $\lambda = 550$ nm and $D = 2.4$ m:

$$\theta_{min} = 1.22\frac{\lambda}{D} = 1.22\left(\frac{550 \times 10^{-9}\text{ m}}{2.4\text{ m}}\right)$$
$$= 2.8 \times 10^{-7}\text{ rad}$$

2. Express L in terms of y and θ_{min}:

$$\tan(\theta_{min}) = \frac{y}{L} \quad \text{or} \quad L = \frac{y}{\tan(\theta_{min})}$$

3. Substitute $y = 2.5$ km and $\theta_{min} = 2.8 \times 10^{-7}$ rad to find L:

$$L = \frac{y}{\tan(\theta_{min})} = \frac{2500\text{ m}}{\tan(2.8 \times 10^{-7}\text{ rad})}$$
$$= 8.9 \times 10^{9}\text{ m} = 5.5 \times 10^{6}\text{ mi}$$

If the asteroids are farther away than this distance, the Hubble Space Telescope will not be able to image them as two separate objects.

RWP We conclude this section by considering two interesting examples of diffraction-limited resolution. First, in the artistic style known as *pointillism*, the artist applies paint to a canvas in the form of small dots of color. When viewed from a distance, the individual dots cannot be resolved and the painting appears to be painted with continuous colors. An example is shown in FIGURE 28-32. If this painting is viewed from a distance of more than a few meters, the color dots blend into one another.

RWP The second example is the formation of a picture on a television screen. Although a television can show all the colors of the rainbow, it in fact produces only three colors—red, green, and blue, the so-called additive primaries. These three colors are grouped together tightly to form the "pixels" on the screen, as FIGURE 28-33 shows. From a distance the three individual color spots can no longer be distinguished, and the eye sees the net effect of the three colors combined. Because any color can be created with the proper amounts of the three primary colors, the television screen can reproduce any picture.

To see this effect, try the following: Look for a region on a television screen or a computer monitor where the picture is yellow. Because yellow is created by mixing red and green light equally, you will see on close examination (perhaps with the aid of a magnifying glass) that pixels in the yellow region of the screen have both the red and green dots illuminated, but the blue dots are dark. As you slowly move away from the screen, notice that the red and green dots merge, leaving the brain with the sensation of a yellow light, even though there are no yellow dots on the screen.

RWP Diffraction also plays a role in observational astronomy. As described in Quick Example 28-15, the resolution of a telescope is limited by the diffraction of the light that passes through its aperture. One way astronomers avoid this problem is by using astronomical interferometry, a process by which two or more telescopes are linked together and their signals are combined by a computer. The superposition of the signals can be analyzed to reconstruct an image of the source with better resolution than the diffraction limit imposed by the aperture of a single telescope. Such interferometry is something like Young's two-slit experiment in reverse; instead of light from two sources producing an interference pattern on a distant screen, a distant source produces an interference pattern when it is detected by two telescopes.

▲ **FIGURE 28-32 Pointillism** Paul Signac's *The Mills at Owerschie* (1905), an example of pointillism.

▲ **FIGURE 28-33 Pixels on a television screen** A typical pixel on the screen of a color television consists of three closely spaced color spots: one red, one blue, and one green. These are the only colors the television actually produces.

Enhance Your Understanding   (Answers given at the end of the chapter)

5. If you view the world with blue light, is your resolution greater than, less than, or equal to your resolution when viewing the world with red light?

Section Review

- For two objects to be seen as separate, they must have an angular separation that is greater than the following minimum angle: $\theta_{min} = 1.22\lambda/D$, where θ_{min} is measured in radians. This is known as Rayleigh's criterion.

28-6 Diffraction Gratings

As we saw earlier in this chapter, a screen with one or two slits can produce striking patterns of interference fringes. It is natural to wonder what interference effects may be produced if the number of slits is increased significantly. In general, we refer to a system with a large number of slits as a **diffraction grating.** There are many ways of producing a grating; for example, one might use a diamond stylus to cut thousands of slits in the aluminum coating on a sheet of glass. Alternatively, one might photoreduce an image of parallel lines onto a strip of film. In some cases it is possible to produce gratings with as many as 40,000 slits—or "lines," as they are often called—per centimeter.

The interference pattern formed by a diffraction grating consists of a series of sharp, widely spaced bright fringes—called *principal maxima*—separated by relatively dark regions with a number of weak secondary maxima, as indicated in FIGURE 28-34 for the case of five slits. In the limit of a large number of slits, the principal maxima become more sharply peaked, and the secondary maxima become insignificant. As one might

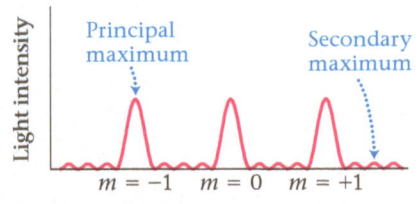

▲ **FIGURE 28-34 Diffraction pattern for five slits** The interference pattern produced by a diffraction grating with five slits. The large "principal" maxima are sharper than the maxima in the two-slit apparatus. The small "secondary" maxima are negligible compared with the principal maxima.

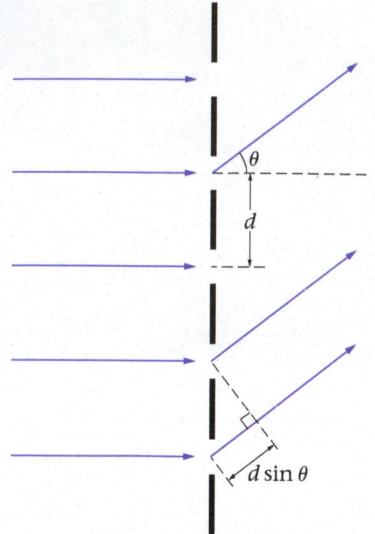

▲ **FIGURE 28-35 Path-length difference in a diffraction grating** A simple diffraction grating consists of a number of slits with a spacing d. The difference in path lengths for rays from neighboring slits is $d \sin \theta$.

Big Idea ❻ Diffraction gratings use the interference effects of a large number of closely spaced sources to disperse (spread out) light into a wide spectrum of colors, similar to the behavior of a prism.

expect, the angle at which a principal maximum occurs depends on the wavelength of light that passes through the grating. In this way, a grating acts much like a prism—sending the various colors of white light off in different directions. A grating, however, can spread the light out over a wider range of angles than a prism.

To determine the angles at which principal maxima are found, consider a grating with a large number of slits, each separated from the next by a distance d, as shown in **FIGURE 28-35**. A beam of light with wavelength λ is incident on the grating from the left and is diffracted onto a distant screen. At an angle θ to the incident direction, the path difference between successive slits is $d \sin \theta$, as Figure 28-35 shows. Therefore, constructive interference—and a principal maximum—occurs when the path difference is an integral number of wavelengths, λ:

> **Constructive Interference in a Diffraction Grating**
>
> $d \sin \theta = m\lambda \qquad m = 0, \pm 1, \pm 2, \ldots$ 　　　　　　 28-16

Notice that the angle θ becomes larger as d is made smaller. In particular, if a grating has more lines per centimeter (smaller d), light passing through it will be spread out over a larger range of angles.

EXERCISE 28-16 　SLIT SPACING

Find the slit spacing necessary for 450-nm light to have a first-order ($m = 1$) principal maximum at 15°.

REASONING AND SOLUTION
Solving Equation 28-16 ($d \sin \theta = m\lambda$) for d, we obtain

$$d = \frac{m\lambda}{\sin \theta} = \frac{(1)(450 \times 10^{-9}\,\text{m})}{\sin 15°} = 1.7 \times 10^{-6}\,\text{m}$$

A grating is often characterized in terms of its number of lines per unit length, N. For example, a particular grating might have 2250 lines per centimeter. The corresponding slit separation, d, is simply the inverse of the number of lines per length. In this case the slit separation is $d = 1/N = 1/(2250\,\text{cm}^{-1}) = 4.44 \times 10^{-4}\,\text{cm} = 4.44 \times 10^{-6}\,\text{m}$.

EXAMPLE 28-17 　A SECOND-ORDER MAXIMUM

When 546-nm light passes through a particular diffraction grating, a second-order principal maximum is observed at an angle of 16.0°. How many lines per centimeter does this grating have?

PICTURE THE PROBLEM
Our sketch shows the first few principal maxima on either side of the center of the diffraction pattern. The second-order maximum ($m = 2$) is the second maximum above the central peak, and it occurs at an angle of 16.0°.

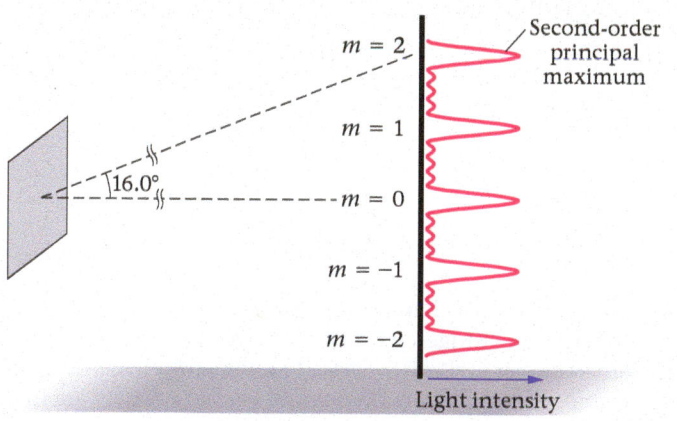

REASONING AND STRATEGY
First, we can use $\sin \theta = m\lambda/d$ to find the necessary spacing d, given that m, λ, and θ are specified in the problem statement. Next, the number of lines per centimeter, N, is simply the inverse of the spacing d; that is, $N = 1/d$.

Known　Wavelength of light, $\lambda = 546$ nm; angle to second-order principal maximum, $\theta = 16.0°$.
Unknown　Lines per centimeter on grating, $N = ?$

SOLUTION

1. Calculate the distance d between slits:
$$d = \frac{m\lambda}{\sin \theta} = \frac{(2)(546 \times 10^{-9}\,\text{m})}{\sin 16.0°} = 3.96 \times 10^{-6}\,\text{m}$$

2. Take the inverse of d to find the number of lines per meter:
$$N = \frac{1}{d} = \frac{1}{3.96 \times 10^{-6}\,\text{m}} = 2.53 \times 10^{5}\,\text{m}^{-1}$$

CONTINUED

3. Convert to lines per centimeter: $\qquad N = 2.53 \times 10^5 \text{ m}^{-1}\left(\dfrac{1 \text{ m}}{100 \text{ cm}}\right) = 2530 \text{ cm}^{-1}$

INSIGHT

Thus, this diffraction grating has 2530 lines per centimeter. Though this is a lot of lines to pack into a distance of 1 cm, it is common to find this many lines or more in a typical school-laboratory diffraction grating.

PRACTICE PROBLEM — PREDICT/CALCULATE

(a) If the grating is ruled with more lines per centimeter, does the angle to the second-order maximum increase, decrease, or stay the same? Explain. **(b)** Check your answer by calculating the angle for a grating with 3530 lines per centimeter.   [**Answer: (a)** More lines per centimeter means that d is smaller, which implies that θ increases. **(b)** $\theta = 22.7°$]

Some related homework problems: Problem 61, Problem 62, Problem 72

RWP Of the many ways to produce a diffraction grating, one of the more novel is by *acousto-optic modulation* (AOM). In this technique, light is diffracted not by a series of slits but by a series of high-density wave fronts produced by a sound wave propagating through a solid or a liquid. For example, in the AOM device shown in **FIGURE 28-36**, sound waves propagate through a quartz crystal, producing a series of closely spaced, parallel wave fronts. An incoming beam of light diffracts from these wave fronts, giving rise to an intense outgoing beam. If the sound is turned off, however, the incoming light passes through the crystal without being deflected. Thus, simply turning the sound on or off causes the diffracted beam to be switched on or off, whereas changing the frequency of the sound can change the angle of the diffracted beam. Many laser printers use AOMs to control the laser beam responsible for "drawing" the desired image on a light-sensitive surface.

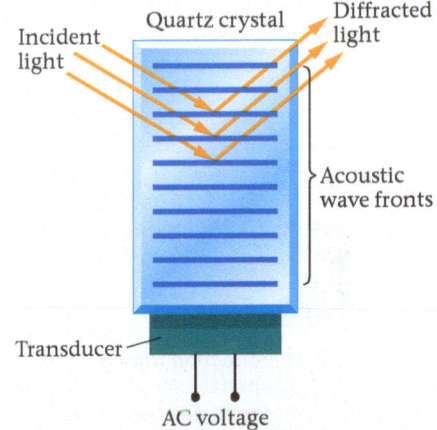

▲ **FIGURE 28-36 An acousto-optic modulator** Acousto-optic modulators use sound waves, and the density variations they produce, to diffract light. The angle at which the light diffracts can be controlled by the frequency of the sound.

X-Ray Diffraction

There is another type of diffraction grating that is not made by clever applications of technology but occurs naturally—the crystal. The key characteristic of a crystal is that it has a regular, repeating structure. In particular, crystals generally consist of regularly spaced planes of atoms or ions; these planes, just like the wave fronts in an AOM, can diffract an incoming beam of electromagnetic radiation.

For a diffraction grating to be effective, however, the wavelength of the radiation, λ, must be comparable to the spacing, d, in Equation 28-16. In a typical crystal, the spacing between atomic planes is roughly an angstrom; that is, $d \sim 10^{-10}$ m ~ 0.1 nm. Notice that this distance is much less than the wavelength of visible light, which is roughly 400 nm to 700 nm. Therefore, visible light will not produce useful diffraction effects from a crystal. However, if we consider the full electromagnetic spectrum, as presented in Figure 25-12, we see that wavelengths of 0.1 nm fall within the X-ray portion of the spectrum. Indeed, X-rays produce vivid diffraction patterns when sent through crystals. An example is shown in **FIGURE 28-37**.

Today, *X-ray diffraction* is a valuable scientific tool. First, the angle at which principal maxima occur in an X-ray diffraction pattern can determine the precise distance between various planes of atoms in a particular crystal. Second, the symmetry of the pattern determines the type of crystal structure. More sophisticated analysis of X-ray diffraction patterns—in particular, the angles and intensities of diffraction maxima—can be used to help determine the structures of even large organic molecules. In fact, it was in part through examination of X-ray diffraction patterns that J. D. Watson and F. H. C. Crick were able to deduce the double-helix structure of DNA in 1953.

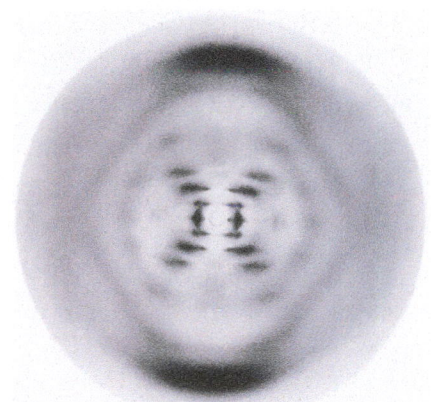

▲ **FIGURE 28-37 X-ray diffraction** X-ray diffraction pattern produced by DNA. A photo like this one helped Watson and Crick deduce the double-helix structure of the DNA molecule in 1953.

Gratings

RWP As noted earlier in this section, a grating can produce a wide separation in the various colors contained in a beam of light. This phenomenon is used as a means of measuring the corresponding wavelengths with an instrument known as a *grating spectroscope*. As we see in **FIGURE 28-38**, light entering a grating spectroscope is diffracted as

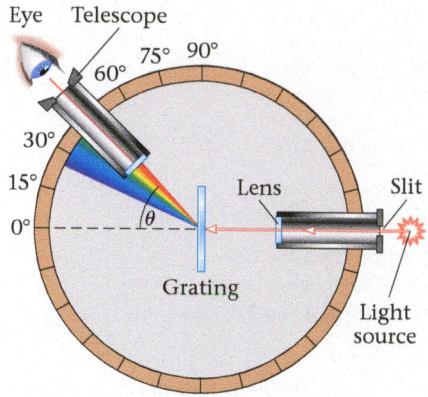

▲ **FIGURE 28-38 A grating spectroscope** Because gratings can spread light out over a wider angle than prisms, they are used in most modern spectroscopes.

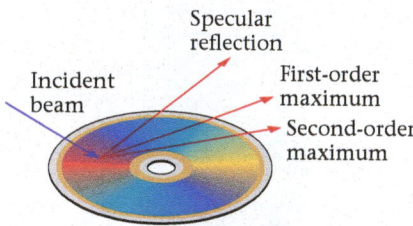

▲ **FIGURE 28-39 Reflection from a DVD** When a laser beam shines on a DVD, a number of reflected beams are observed. The most intense of these is the specular beam, whose angle of reflection is equal to its angle of incidence—the same as if the DVD were a plane mirror. Additional reflected beams are observed at angles corresponding to the principal maxima of the grating.

it passes through a grating. Next, the angle of diffraction of a given color is determined by a small telescope mounted on a rotating base. Finally, application of Equation 28-16 ($d \sin \theta = m\lambda$) allows one to determine the wavelength of the light to great precision. Devices of this type have played a key role in elucidating the expanding nature of the universe by carefully measuring the red shift (see Chapter 25) of light emitted by distant stars and galaxies.

Yet another way to produce a diffraction grating is to inscribe lines on a reflecting surface, with the regions between the lines acting as coherent sources of light. An everyday example of this type of *reflection grating* is a CD or a DVD. As described earlier in this chapter, the information on a DVD is encoded in a series of bumps, and these bumps spiral around the DVD, creating a tightly spaced set of lines. When a beam of monochromatic light is incident on a DVD, as in **FIGURE 28-39**, a series of reflected beams is created. One beam reflects at an angle equal to the incident angle—this is referred to as the *specular beam*, because it is the beam that would be expected from a smooth plane mirror. In addition, a number of other reflected beams are observed, each corresponding to a different principal maximum of the grating.

BIO If white light is reflected from a CD or DVD, different colors in the incident light are reflected at different angles. This is why light reflected from a DVD shows the colors of the rainbow, as in **FIGURE 28-40 (a)**. A similar effect occurs in light reflected from feathers, or from the wings of a butterfly. For example, a microscopic examination of a butterfly wing shows that it consists of a multitude of plates, much like shingles on a roof. On each of these plates is a series of closely spaced ridges. These ridges act like the grooves on a DVD, producing reflected light of different colors in different directions. This type of coloration is referred to as **iridescence.** The next time you examine an iridescent object, notice how the color changes as you change the angle from which it is viewed. A biological example of iridescence is provided by the fly's eye shown in **FIGURE 28-40 (b)**.

(a)

(b)

▲ **FIGURE 28-40 Visualizing Concepts Iridescence** Diffraction can occur when light falls on a surface that has grooves or lines with a spacing comparable to the wavelength of light. **(a)** Many common surfaces can act as diffraction gratings, including artificially created ones, such as DVDs. **(b)** Natural objects can also show iridescence, as illustrated by the rainbow colors in a fly's eye.

Enhance Your Understanding   (Answers given at the end of the chapter)

6. Suppose a diffraction grating has slits separated by 6 times the wavelength of light used to illuminate the grating. **(a)** What is the angle corresponding to the $m = 3$ principal maximum? **(b)** What is the angle corresponding to the $m = 6$ principal maximum? **(c)** Are there principal maxima for $m > 6$?

Section Review

- The condition for constructive interference in a diffraction grating is $d \sin \theta = m\lambda$, where $m = 0, \pm 1, \pm 2, \ldots$. In this expression, d is the spacing between lines on the grating, λ is the wavelength of light, θ is the angle relative to the normal, and m is the order of the diffraction maximum.

CHAPTER 28 REVIEW

CHAPTER SUMMARY

28-1 SUPERPOSITION AND INTERFERENCE

The simple addition of two or more waves to give a resultant wave is referred to as superposition. When waves are superposed, the result may be a wave of greater amplitude (constructive interference) or of reduced amplitude (destructive interference).

28-2 YOUNG'S TWO-SLIT EXPERIMENT

Interference effects in light are shown clearly in Young's two-slit experiment, in which light passing through two slits forms bright and dark interference "fringes."

Conditions for Bright Fringes

Bright fringes in a two-slit experiment occur at angles θ given by the following relationship:

$$d \sin \theta = m\lambda \quad m = 0, \pm 1, \pm 2, \ldots \qquad 28\text{-}1$$

In this expression, λ is the wavelength of the light and d is the separation of the slits. The various values of the integer m correspond to different bright fringes.

Conditions for Dark Fringes

The locations of dark fringes in a two-slit experiment are given by the following:

$$d \sin \theta = \left(m - \tfrac{1}{2}\right)\lambda \quad m = 1, 2, 3, \ldots \ \text{(above central bright fringe)}$$
$$d \sin \theta = \left(m + \tfrac{1}{2}\right)\lambda \quad m = -1, -2, -3, \ldots \ \text{(below central bright fringe)} \qquad 28\text{-}2$$

28-3 INTERFERENCE IN REFLECTED WAVES

Phase Changes Due to Reflection

No phase change occurs when light is reflected from a region with a lower index of refraction, whereas a 180° (half wavelength) phase change occurs when light reflects from a region with a higher index of refraction, or from a solid surface.

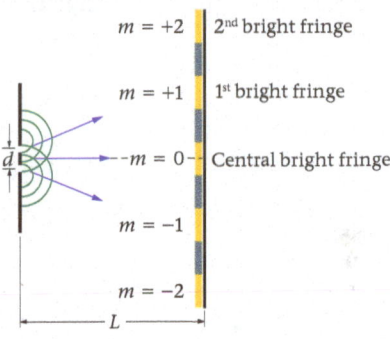

Side view

Top view

Air Wedge

Two plates of glass that touch on one end and have a small separation on the other end form an air wedge. When light of wavelength λ shines on an air wedge, bright fringes occur when the separation between the plates, d, is such that

$$\frac{1}{2} + \frac{2d}{\lambda} = m \quad m = 1, 2, 3, \ldots \qquad 28\text{-}5$$

Similarly, dark fringes occur when the following conditions are satisfied:

$$\frac{1}{2} + \frac{2d}{\lambda} = m + \frac{1}{2} \quad m = 0, 1, 2, \ldots \qquad 28\text{-}6$$

Thin Films

Constructive interference for a thin film in air satisfies $2nt/\lambda_{\text{vacuum}} - 1/2 = m$; destructive interference satisfies $2nt/\lambda_{\text{vacuum}} = m$. In both cases, $m = 0, 1, 2, \ldots$.

28-4 DIFFRACTION

When a wave encounters an obstacle, or passes through an opening, it changes direction. This phenomenon is referred to as diffraction.

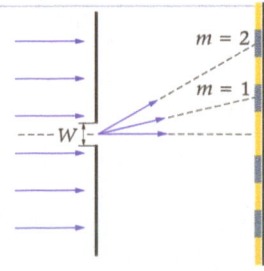

Single-Slit Diffraction

When monochromatic light of wavelength λ passes through a single slit of width W, it forms a diffraction pattern of alternating bright and dark fringes.

Condition for Dark Fringes

The condition that determines the location of dark fringes in single-slit diffraction is

$$W \sin \theta = m\lambda \quad m = \pm 1, \pm 2, \pm 3, \ldots \qquad 28\text{-}12$$

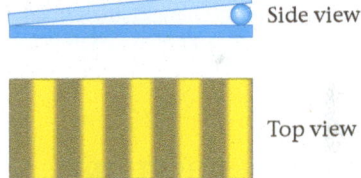

28-5   RESOLUTION

Resolution refers to the ability of a visual system, like the eye or a camera, to distinguish closely spaced objects.

First Dark Fringe

A circular aperture of diameter D produces a circular diffraction pattern in which the first dark fringe occurs at the angle θ given by the following condition:

$$\sin \theta = 1.22 \frac{\lambda}{D} \qquad \text{28-14}$$

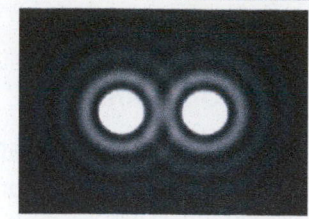

Rayleigh's Criterion: Quantitative Statement

In quantitative terms, Rayleigh's criterion states that if the angular separation between two objects is less than a certain minimum, $\theta_{min} = 1.22\lambda/D$, they will appear to be a single object.

28-6   DIFFRACTION GRATINGS

A diffraction grating is a large number of slits through which a beam of light can pass.

Principal Maxima

The principal maxima produced by a diffraction grating occur at the angles given by the following conditions:

$$d \sin \theta = m\lambda \qquad m = 0, \pm 1, \pm 2, \dots \qquad \text{28-16}$$

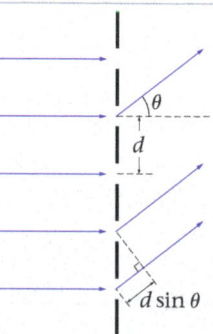

In this expression, d is the distance between successive slits and λ is the wavelength of light.

ANSWERS TO ENHANCE YOUR UNDERSTANDING QUESTIONS

1. **(c)** coherent.

2. Increasing the wavelength increases the angle to the first bright fringe.

3. **(a)** Equal phase changes. **(b)** A greater phase change for reflection 1. **(c)** A greater phase change for reflection 1. **(d)** Equal phase changes.

4. The width of the central bright fringe decreases.

5. Your resolution is greater with blue light.

6. **(a)** For $m = 3$, we have $\sin \theta = 1/2$, or $\theta = 30°$. **(b)** For $m = 6$, we have $\sin \theta = 1$, or $\theta = 90°$. **(c)** No, because that would correspond to $\sin \theta > 1$.

CONCEPTUAL QUESTIONS

For instructor-assigned homework, go to www.masteringphysics.com.

(Answers to odd-numbered Conceptual Questions can be found in the back of the book.)

1. When two light waves interfere destructively, what happens to their energy?

2. What happens to the two-slit interference pattern if the separation between the slits is less than the wavelength of light?

3. If a radio station broadcasts its signal through two different antennas simultaneously, does this guarantee that the signal you receive will be stronger than from a single antenna? Explain.

4. How would you expect the interference pattern of a two-slit experiment to change if white light is used instead of monochromatic light?

5. Describe the changes that would be observed in the two-slit interference pattern if the entire experiment were to be submerged in water.

6. Two identical sheets of glass are coated with films of different materials but equal thickness. The colors seen in reflected light from the two films are different. Give a reason that can account for this observation.

7. A cat's eye has a pupil that is elongated in the vertical direction. How does the resolution of a cat's eye differ in the horizontal and vertical directions?

8. Which portion of the soap film in the accompanying photograph is thinnest? Explain.

Conceptual Question 8

9. The color of an iridescent object, like a butterfly wing or a feather, appears to be different when viewed from different directions. The color of a painted surface appears the same from all viewing angles. Explain the difference.

PROBLEMS AND CONCEPTUAL EXERCISES

Answers to odd-numbered Problems and Conceptual Exercises can be found in the back of the book. **BIO** *identifies problems of biological or medical interest;* **CE** *indicates a conceptual exercise.* **Predict/Explain** *problems ask for two responses: (a) your prediction of a physical outcome, and (b) the best explanation among three provided; and* **Predict/Calculate** *problems ask for a prediction of a physical outcome, based on fundamental physics concepts, and follow that with a numerical calculation to verify the prediction. On all problems, bullets (•, ••, •••) indicate the level of difficulty.* (In all problems involving sound waves, take the speed of sound to be 343 m/s.)

SECTION 28-1 SUPERPOSITION AND INTERFERENCE

1. • Two sources emit waves that are coherent, in phase, and have wavelengths of 26.0 m. Do the waves interfere constructively or destructively at an observation point 78.0 m from one source and 143 m from the other source?

2. • In an experiment to demonstrate interference, you connect two antennas to a single radio receiver. When the two antennas are adjacent to each other, the received signal is strong. You leave one antenna in place and move the other one directly away from the radio transmission tower. How far should the second antenna be moved in order to receive a minimum signal from a station that broadcasts at 97.9 MHz?

3. • A theme park creates a new kind of water wave pool with large waves caused by constructive interference. There are two wave generators in phase with each other along either side of a pool that is 24.0 m wide. A swimmer that is 8.0 m from one generator and 16.0 m from the other notices that she is in a region with almost no wave amplitude, but there are large-amplitude waves on either side of her. What is the longest wavelength that will produce this interference pattern?

4. •• Two sources emit waves that are in phase with each other. What is the longest wavelength that will give constructive interference at an observation point 161 m from one source and 295 m from the other source?

5. •• A person driving at 17 m/s crosses the line connecting two radio transmitters at right angles, as shown in **FIGURE 28-41**. The transmitters emit identical signals in phase with each other, which the driver receives on the car radio. When the car is at point A, the radio picks up a maximum net signal. **(a)** What is the longest possible wavelength of the radio waves? **(b)** At what time after the car passes point A does the radio experience a minimum in the net signal? Assume that the wavelength has the value found in part (a).

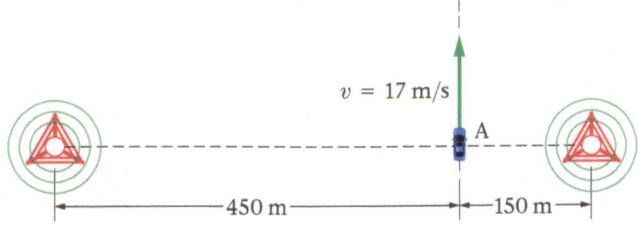

FIGURE 28-41 Problem 5

6. •• Two students in a dorm room listen to a pure tone produced by two loudspeakers that are in phase. Students A and B in **FIGURE 28-42** hear a maximum sound. What is the lowest possible frequency of the loudspeakers?

7. •• If the loudspeakers in Problem 6 are 180° out of phase, determine whether a 185-Hz tone heard at location B is a maximum or a minimum.

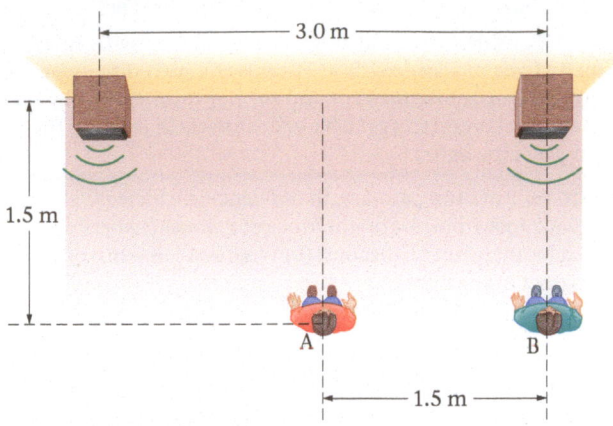

FIGURE 28-42 Problems 6 and 7

8. •• A microphone is located on the line connecting two speakers that are 0.938 m apart and oscillating in phase. The microphone is 2.83 m from the midpoint of the two speakers. What are the lowest two frequencies that produce an interference maximum at the microphone's location?

9. •• A microphone is located on the line connecting two speakers that are 0.845 m apart and oscillating 180° out of phase. The microphone is 2.25 m from the midpoint of the two speakers. What are the lowest two frequencies that produce an interference maximum at the microphone's location?

10. •• **Predict/Calculate** Radio waves of frequency 1.427 GHz arrive at two telescopes that are connected by a computer to perform interferometry. One portion of the same wave front travels 1.051 m farther than the other before the two signals are combined. **(a)** Will the two waves combine constructively or destructively? **(b)** Calculate the value of m for the path difference between the two signals.

11. •• Moe, Larry, and Curly stand in a line with a spacing of 1.00 m. Larry is 3.00 m in front of a pair of stereo speakers 0.800 m apart, as shown in **FIGURE 28-43**. The speakers produce a single-frequency tone, vibrating in phase with each other. What are the two lowest frequencies that allow Larry to hear a loud tone while Moe and Curly hear very little?

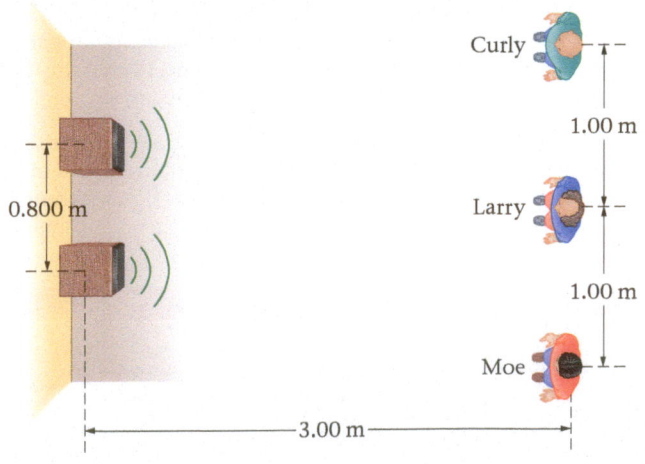

FIGURE 28-43 Problems 11 and 12

12. •• **Predict/Calculate** In Figure 28-43 the two speakers emit sound that is 180° out of phase and of a single frequency, f. **(a)** Does Larry hear a sound intensity that is a maximum or a minimum? Does your answer depend on the frequency of the sound? Explain. **(b)** Find the lowest two frequencies that produce a maximum sound intensity at the positions of Moe and Curly.

SECTION 28-2 YOUNG'S TWO-SLIT EXPERIMENT

13. • **CE** Consider a two-slit interference pattern, with monochromatic light of wavelength λ. What is the path difference $\Delta\ell$ for **(a)** the fourth bright fringe and **(b)** the third dark fringe above the central bright fringe? Give your answers in terms of the wavelength of the light.

14. • **CE (a)** Does the path-length difference $\Delta\ell$ increase or decrease as you move from one bright fringe of a two-slit experiment to the next bright fringe farther out? **(b)** What is $\Delta\ell$ in terms of the wavelength λ of the light?

15. • **CE Predict/Explain** A two-slit experiment with red light produces a set of bright fringes. **(a)** Will the spacing between the fringes increase, decrease, or stay the same if the color of the light is changed to blue? **(b)** Choose the *best explanation* from among the following:
   I. The spacing between the fringes will increase because blue light has a greater frequency than red light.
   II. The fringe spacing decreases because blue light has a shorter wavelength than red light.
   III. Only the wave property of light is important in producing the fringes, not the color of the light. Therefore the spacing stays the same.

16. • Laser light with a wavelength $\lambda = 690$ nm illuminates a pair of slits at normal incidence. What slit separation will produce first-order maxima at angles of $\pm 25°$ from the incident direction?

17. • Monochromatic light passes through two slits separated by a distance of 0.0374 mm. If the angle to the third maximum above the central fringe is 3.61°, what is the wavelength of the light?

18. • In Young's two-slit experiment, the first dark fringe above the central bright fringe occurs at an angle of 0.44°. What is the ratio of the slit separation, d, to the wavelength of the light, λ?

19. •• **Predict/Calculate** A two-slit experiment with slits separated by 48.0×10^{-5} m produces a second-order maximum at an angle of 0.0990°. **(a)** Find the wavelength of the light used in this experiment. **(b)** If the slit separation is increased but the second-order maximum stays at the same angle, does the wavelength increase, decrease, or stay the same? Explain. **(c)** Calculate the wavelength for a slit separation of 68.0×10^{-5} m.

20. •• A two-slit pattern is viewed on a screen 1.00 m from the slits. If the two third-order minima are 22.0 cm apart, what is the width (in cm) of the central bright fringe?

21. •• Light from a He–Ne laser ($\lambda = 632.8$ nm) strikes a pair of slits at normal incidence, forming a double-slit interference pattern on a screen located 1.16 m from the slits. **FIGURE 28-44** shows the interference pattern observed on the screen. What is the slit separation?

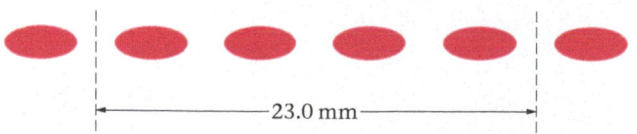

←————————23.0 mm————————→

FIGURE 28-44 Problems 21, 22, and 24

22. •• For a science fair demonstration you would like to create a diffraction pattern similar to that shown in Figure 28-44. If you plan to shine a laser pointer ($\lambda = 652$ nm) through a pair of slits that are separated by 175 μm, how far away from the slits should you place the screen?

23. •• Light with a wavelength of 576 nm passes through two slits and forms an interference pattern on a screen 8.85 m away. If the linear distance on the screen from the central fringe to the first bright fringe above it is 5.36 cm, what is the separation of the slits?

24. •• **Predict/Calculate** Suppose the interference pattern shown in Figure 28-44 is produced by monochromatic light passing through two slits, with a separation of 135 μm, and onto a screen 1.20 m away. **(a)** What is the wavelength of the light? **(b)** If the frequency of this light is increased, will the bright spots of the pattern move closer together or farther apart? Explain.

25. •• A physics instructor wants to produce a double-slit interference pattern large enough for her class to see. For the size of the room, she decides that the distance between successive bright fringes on the screen should be at least 2.50 cm. If the slits have a separation $d = 0.0175$ mm, what is the minimum distance from the slits to the screen when 632.8-nm light from a He–Ne laser is used?

26. •• **Predict/Calculate** When green light ($\lambda = 505$ nm) passes through a pair of double slits, the interference pattern shown in **FIGURE 28-45 (a)** is observed. When light of a different color passes through the same pair of slits, the pattern shown in **FIGURE 28-45 (b)** is observed. **(a)** Is the wavelength of the second color longer or shorter than 505 nm? Explain. **(b)** Find the wavelength of the second color. (Assume that the angles involved are small enough to set $\sin\theta \approx \tan\theta$.)

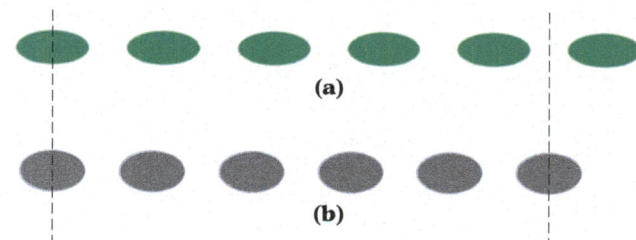

(a)

(b)

FIGURE 28-45 Problems 26 and 27

27. •• **Predict/Calculate** The interference pattern shown in Figure 28-45 (a) is produced by green light with a wavelength of $\lambda = 505$ nm passing through two slits with a separation of 127 μm. After passing through the slits, the light forms a pattern of bright and dark spots on a screen located 1.25 m from the slits. **(a)** What is the distance between the two vertical, dashed lines in Figure 28-45 (a)? **(b)** If it is desired to produce a more tightly packed interference pattern, like the one shown in Figure 28-45 (b), should the frequency of the light be increased or decreased? Explain.

SECTION 28-3 INTERFERENCE IN REFLECTED WAVES

28. • **CE FIGURE 28-46** shows four different cases where light of wavelength λ reflects from both the top and the bottom of a thin film of thickness d. The indices of refraction of the film and the media above and below it are indicated in the figure. For which of the cases will the two reflected rays undergo constructive interference if **(a)** $d = \lambda/4$ or **(b)** $d = \lambda/2$?

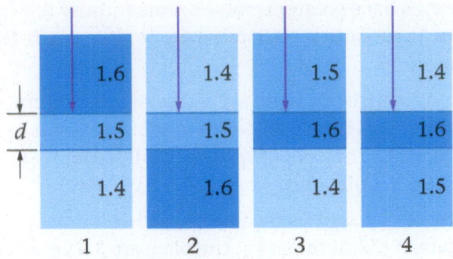

FIGURE 28-46 Problem 28

29. • **CE** The oil film floating on water in the accompanying photo appears dark near the edges, where it is thinnest. Is the index of refraction of the oil greater than or less than that of the water? Explain.

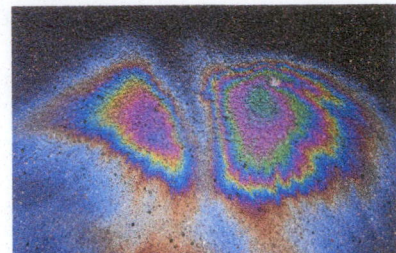

Light reflected from a film of oil. (Problem 29)

30. • A soap bubble with walls 418 nm thick floats in air. If this bubble is illuminated perpendicularly with sunlight, what wavelength (and color) will be absent in the reflected light? Assume that the index of refraction of the soap film is 1.33. (Refer to Example 25-6 for the connection between wavelength and color.)

31. • A soap film (n = 1.33) is 825 nm thick. White light strikes the film at normal incidence. What visible wavelengths will be constructively reflected if the film is surrounded by air on both sides? (Refer to Example 25-6 for the range of visible wavelengths.)

32. • White light is incident on a soap film (n = 1.30) in air. The reflected light looks bluish because the red light (λ = 670 nm) is absent in the reflection. What is the minimum thickness of the soap film?

33. • A 742-nm-thick soap film (n_{film} = 1.33) rests on a glass plate (n_{glass} = 1.52). White light strikes the film at normal incidence. What visible wavelengths will be constructively reflected from the film? (Refer to Example 25-6 for the range of visible wavelengths.)

34. • An oil film (n = 1.46) floats on a water puddle. You notice that green light (λ = 538 nm) is absent in the reflection. What is the minimum thickness of the oil film?

35. •• A radio broadcast antenna is 36.00 km from your house. Suppose an airplane is flying 2.230 km above the line connecting the broadcast antenna and your radio, and that waves reflected from the airplane travel 88.00 wavelengths farther than waves that travel directly from the antenna to your house. (a) Do you observe constructive or destructive interference between the direct and reflected waves? (Hint: Does a phase change occur when the waves are reflected?) (b) The situation just described occurs when the plane is above a point on the ground that is two-thirds of the way from the antenna to your house. What is the wavelength of the radio waves?

36. •• **Predict/Calculate Newton's Rings** Monochromatic light with λ = 648 nm shines down on a plano-convex lens lying on a piece of plate glass, as shown in **FIGURE 28-47**. When viewing from above, one sees a set of concentric dark and bright fringes, referred to as Newton's rings. (See Figure 28-13 (b) for a photo of Newton's rings.)

(a) If the radius of the twelfth dark ring from the center is measured to be 1.56 cm, what is the radius of curvature, R, of the lens? (b) If light with a longer wavelength is used with this system, will the radius of the twelfth dark ring be greater than or less than 1.56 cm? Explain.

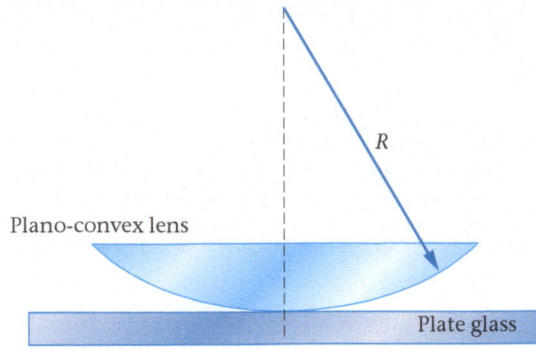

FIGURE 28-47 Problems 36 and 84

37. •• Light is incident from above on two plates of glass, separated on both ends by small wires of diameter d = 0.600 μm. Considering only interference between light reflected from the bottom surface of the upper plate and light reflected from the upper surface of the lower plate, state whether the following wavelengths give constructive or destructive interference: (a) λ = 600.0 nm; (b) λ = 800.0 nm; (c) λ = 343.0 nm.

38. •• **Submarine Saver** A naval engineer is testing an nonreflective coating for submarines that would help them avoid detection by producing destructive interference for sonar waves that have a frequency of 512 Hz and travel at 1480 m/s. If there is a phase change for waves reflected from both the top and bottom surfaces of the coating, what is the minimum thickness of the coating? For sonar waves let n_{water} = 1.00 and $n_{coating}$ = 13.5.

39. •• **Predict/Calculate** A thin layer of magnesium fluoride (n = 1.38) is used to coat a flint-glass lens (n = 1.61). (a) What thickness should the magnesium fluoride film have if the reflection of 565-nm light is to be suppressed? Assume that the light is incident at right angles to the film. (b) If it is desired to suppress the reflection of light with a higher frequency, should the coating of magnesium fluoride be made thinner or thicker? Explain.

40. •• **Holographic Interferometry** An interferometric hologram is produced by exposing the film twice, once when the object is unstressed and once when it is stressed, causing it to flex slightly. The resulting hologram has light and dark bands that are analogous to the interference fringes of an air wedge. If 22 bright fringes are observed between one side of an object and the other when the hologram is produced by light of wavelength 514 nm, by what distance did one side of the object flex with respect to the other?

41. ••• White light is incident normally on a thin soap film (n = 1.33) suspended in air. (a) What are the two minimum thicknesses that will constructively reflect yellow (λ = 590 nm) light? (b) What are the two minimum thicknesses that will destructively reflect yellow (λ = 590 nm) light?

42. ••• Two glass plates are separated by fine wires with diameters d_1 = 0.0500 mm and d_2 = 0.0520 mm, as indicated in **FIGURE 28-48**. The wires are parallel and separated by a distance of 7.00 cm. If monochromatic light with λ = 589 nm is incident from above, what is the distance (in cm) between adjacent dark bands in the reflected light? (Consider interference only between light reflected from the bottom surface of the upper plate and light reflected from the upper surface of the lower plate.)

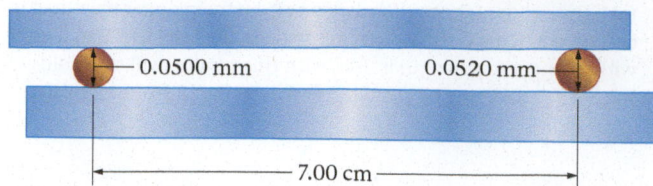

FIGURE 28-48 Problem 42

SECTION 28-4   DIFFRACTION

43. • **CE** A single-slit diffraction pattern is formed on a distant screen. Assuming the angles involved are small, by what factor will the width of the central bright spot on the screen change if (a) the wavelength is doubled, (b) the slit width is doubled, or (c) the distance from the slit to the screen is doubled?

44. • What width single slit will produce first-order diffraction minima at angles of $\pm 28°$ from the central maximum with 710-nm light?

45. • Diffraction also occurs with sound waves. Consider 1400-Hz sound waves diffracted by a door that is 94 cm wide. What is the angle between the two first-order diffraction minima?

46. •• Green light ($\lambda = 546$ nm) strikes a single slit at normal incidence. What width slit will produce a central maximum that is 2.75 cm wide on a screen 1.80 m from the slit?

47. •• Light with a wavelength of 696 nm passes through a slit 7.64 μm wide and falls on a screen 1.95 m away. Find the linear distance on the screen from the central bright fringe to the first bright fringe above it.

48. •• **Predict/Calculate** A single slit is illuminated with 610-nm light, and the resulting diffraction pattern is viewed on a screen 2.3 m away. (a) If the linear distance between the first and second dark fringes of the pattern is 12 cm, what is the width of the slit? (b) If the slit is made wider, will the distance between the first and second dark fringes increase or decrease? Explain.

49. •• How many dark fringes will be produced on either side of the central maximum if green light ($\lambda = 553$ nm) is incident on a slit that is 8.00 μm wide?

50. •• **Predict/Calculate** The diffraction pattern shown in **FIGURE 28-49** is produced by passing He–Ne laser light ($\lambda = 632.8$ nm) through a single slit and viewing the pattern on a screen 1.50 m behind the slit. (a) What is the width of the slit? (b) If monochromatic yellow light with a wavelength of 591 nm is used with this slit instead, will the distance indicated in Figure 28-49 be greater than or less than 15.2 cm? Explain.

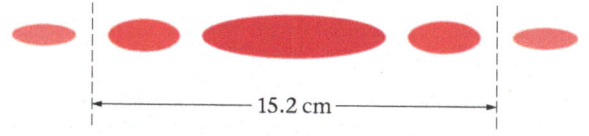

FIGURE 28-49 Problems 50 and 86

51. •• A screen is placed 1.50 m behind a single slit. The central maximum in the resulting diffraction pattern on the screen is 1.60 cm wide—that is, the two first-order diffraction minima are separated by 1.60 cm. What is the distance between the two second-order minima?

SECTION 28-5   RESOLUTION

52. • **CE** Predict/Explain (a) In principle, do your eyes have greater resolution on a dark cloudy day or on a bright sunny day? (b) Choose the *best explanation* from among the following:
  I. Your eyes have greater resolution on a cloudy day because your pupils are open wider to allow more light to enter the eye.

  II. Your eyes have greater resolution on a sunny day because the bright light causes your pupil to narrow down to a smaller opening.

53. • Two point sources of light are separated by 5.5 cm. As viewed through a 12-μm-diameter pinhole, what is the maximum distance from which they can be resolved (a) if red light ($\lambda = 690$ nm) is used, or (b) if violet light ($\lambda = 420$ nm) is used?

54. • A spy camera is said to be able to read the numbers on a car's license plate. If the numbers on the plate are 5.5 cm apart, and the spy satellite is at an altitude of 180 km, what must be the diameter of the camera's aperture? (Assume light with a wavelength of 550 nm.)

55. • **Splitting Binary Stars** As seen from Earth, the red dwarfs Krüger 60A and Krüger 60B form a binary star system with an angular separation of 2.5 arc seconds. What is the smallest diameter telescope that could theoretically resolve these stars using 550-nm light? (*Note:* 1 arc sec = 1/3600°.)

56. • **Very Large Telescope Interferometer** A series of optical telescopes produced an image that has a resolution of about 0.00350 arc second. What is the smallest diameter telescope that could theoretically resolve these features using light with a wavelength of 2.30 μm? (*Note:* 1 arc sec = 1/3600°)

57. • Find the minimum aperture diameter of a camera that can resolve detail on the ground the size of a person (2.0 m) from an SR-71 Blackbird airplane flying at an altitude of 24 km. (Assume light with a wavelength of 480 nm.)

58. •• **The Resolution of Hubble** The Hubble Space Telescope (HST) orbits Earth at an altitude of 613 km. It has an objective mirror that is 2.4 m in diameter. If the HST were to look down on Earth's surface (rather than up at the stars), what is the minimum separation of two objects on the Earth's surface that could be resolved using 550-nm light?

59. •• A lens that is "optically perfect" is still limited by diffraction effects. Suppose a lens has a diameter of 120 mm and a focal length of 640 mm. (a) Find the angular width (that is, the angle from the bottom to the top) of the central maximum in the diffraction pattern formed by this lens when illuminated with 540-nm light. (b) What is the linear width (diameter) of the central maximum at the focal distance of the lens?

60. •• Early cameras were little more than a box with a pinhole on the side opposite the film. (a) What angular resolution would you expect from a pinhole with a 0.50-mm diameter? (b) What is the greatest distance from the camera at which two point objects 15 cm apart can be resolved? (Assume light with a wavelength of 520 nm.)

SECTION 28-6   DIFFRACTION GRATINGS

61. • A grating has 797 lines per centimeter. Find the angles of the first three principal maxima above the central fringe when this grating is illuminated with 645-nm light.

62. • Suppose you want to produce a diffraction pattern with X-rays whose wavelength is 0.030 nm. If you use a diffraction grating, what separation between lines is needed to generate a pattern with the first maximum at an angle of 14°? (For comparison, a typical atom is a few tenths of a nanometer in diameter.)

63. • A diffraction grating has 2500 lines/cm. What is the angle between the first-order maximum for red light ($\lambda = 680$ nm) and the first-order maximum for blue light ($\lambda = 410$ nm)?

64. • The yellow light from a helium discharge tube has a wavelength of 587.5 nm. When this light illuminates a certain diffraction grating it produces a first-order principal maximum at an angle of 1.450°. Calculate the number of lines per centimeter on the grating.

65. •• A diffraction grating with 365 lines/mm is 1.25 m in front of a screen. What is the wavelength of light whose first-order maxima will be 16.4 cm from the central maximum on the screen?

66. •• **Protein Structure** X-rays with a wavelength of 0.0711 nm create a diffraction pattern when they pass through a protein crystal. If the angular spacing between adjacent bright spots ($m = 0$ and $m = 1$) in the diffraction pattern is 28.9°, what is the average spacing between atoms in the crystal?

67. •• White light strikes a grating with 7600 lines/cm at normal incidence. How many complete visible spectra will be formed on either side of the central maximum? (Refer to Example 25-6 for the range of visible wavelengths.)

68. •• White light strikes a diffraction grating (615 lines/mm) at normal incidence. What is the highest-order visible maximum that is formed? (Refer to Example 25-6 for the range of visible wavelengths.)

69. •• **CD Reflection** The rows of bumps on a CD form lines that are separated by 1.60 µm. When white light reflects from its surface, at what angle from the specular reflection ray would you expect to see the first-order maximum for 555-nm green light?

70. •• A light source emits two distinct wavelengths [$\lambda_1 = 430$ nm (violet); $\lambda_2 = 630$ nm (orange)]. The light strikes a diffraction grating with 450 lines/mm at normal incidence. Identify the colors of the first eight interference maxima on either side of the central maximum.

71. •• A laser emits two wavelengths ($\lambda_1 = 420$ nm; $\lambda_2 = 630$ nm). When these two wavelengths strike a grating with 450 lines/mm, they produce maxima (of different orders) that coincide. **(a)** What is the order (m) of each of the two overlapping lines? **(b)** At what angle does this overlap occur?

72. •• **Predict/Calculate** When blue light with a wavelength of 465 nm illuminates a diffraction grating, it produces a first-order principal maximum but no second-order maximum. **(a)** Explain the absence of higher-order principal maxima. **(b)** What is the maximum spacing between lines on this grating?

73. •• Monochromatic light strikes a diffraction grating at normal incidence before illuminating a screen 2.50 m away. If the first-order maxima are separated by 1.43 m on the screen, what is the distance between the two second-order maxima?

74. ••• A diffraction grating with a slit separation d is illuminated by a beam of monochromatic light of wavelength λ. The diffracted beam is observed at an angle ϕ relative to the incident direction. If the plane of the grating bisects the angle between the incident and diffracted beams, show that the mth maximum will be observed at an angle that satisfies the relation $m\lambda = 2d \sin(\phi/2)$, with $m = 0, \pm 1, \pm 2, \ldots$.

GENERAL PROBLEMS

75. • **CE Predict/Explain (a)** If a thin liquid film floating on water has an index of refraction less than that of water, will the film appear bright or dark in reflected light as its thickness goes to zero? **(b)** Choose the *best explanation* from among the following:
    - I. The film will appear bright because as the thickness of the film goes to zero the phase difference for reflected rays goes to zero.
    - II. The film will appear dark because there is a phase change at both interfaces, and this will cause destructive interference of the reflected rays.

76. • **CE** If the index of refraction of an eye could be magically reduced, would the eye's resolution increase or decrease? Explain.

77. •• When reading the printout from a laser printer, you are actually looking at an array of tiny dots. If the pupil of your eye is 4.3 mm in diameter when reading a page held 28 cm from your eye, what is the minimum separation of adjacent dots that can be resolved? (Assume light with a wavelength of 540 nm, and use 1.36 as the index of refraction for the interior of the eye.)

78. •• The headlights of a pickup truck are 1.36 m apart. What is the greatest distance at which these headlights can be resolved

as separate points of light on a photograph taken with a camera whose aperture has a diameter of 15.5 mm? (Take $\lambda = 555$ nm.)

79. •• **Antireflection Coating** A glass lens ($n_{glass} = 1.52$) has an antireflection coating of MgF$_2$ ($n = 1.38$). **(a)** For 545-nm light, what minimum thickness of MgF$_2$ will cause the reflected rays R_2 and R_4 in **FIGURE 28-50** to interfere destructively, assuming normal incidence? **(b)** Interference will also occur between the forward-moving rays R_1 and R_3 in Figure 28-50. What minimum thickness of MgF$_2$ will cause these two rays to interfere constructively?

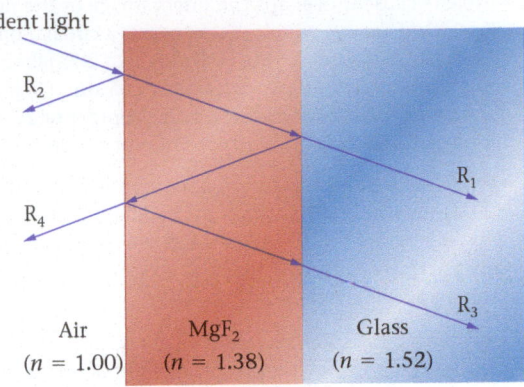

Incident light

Air
($n = 1.00$) | MgF$_2$
($n = 1.38$) | Glass
($n = 1.52$)

FIGURE 28-50 Problem 79

80. •• A thin film of oil ($n = 1.30$) floats on water ($n = 1.33$). When sunlight is incident at right angles to this film, the only colors that are enhanced by reflection are blue (458 nm) and red (687 nm). Estimate the thickness of the oil film.

81. •• The yellow light of sodium, with wavelengths of 588.99 nm and 589.59 nm, is normally incident on a grating with 494 lines/cm. Find the linear distance between the first-order maxima for these two wavelengths on a screen 2.55 m from the grating.

82. •• **Predict/Calculate** A thin soap film ($n = 1.33$) suspended in air has a uniform thickness. When white light strikes the film at normal incidence, violet light ($\lambda_V = 420$ nm) is constructively reflected. **(a)** If we would like green light ($\lambda_G = 560$ nm) to be constructively reflected, instead, should the film's thickness be increased or decreased? **(b)** Find the new thickness of the film. (Assume the film has the minimum thickness that can produce these reflections.)

83. •• **Predict/Calculate** A thin film of oil ($n = 1.40$) floats on water ($n = 1.33$). When sunlight is incident at right angles to this film, the only colors that are absent from the reflected light are blue (458 nm) and red (687 nm). Estimate the thickness of the oil film.

84. •• **Predict/Calculate** Sodium light, with a wavelength of $\lambda = 589$ nm, shines downward onto the system shown in Figure 28-47. When viewed from above, you see a series of concentric circles known as Newton's rings. **(a)** Do you expect a bright or a dark spot at the center of the pattern? Explain. **(b)** If the radius of curvature of the plano-convex lens is $R = 26.1$ m, what is the radius of the tenth-largest dark ring? (Only rings of nonzero radius will be counted as "rings.")

85. •• **BIO The Largest Eye** The colossal squid (*Mesonychoteuthis hamiltoni*) has the largest eyes documented in the animal kingdom, at least 27 cm in diameter. The eyes are large in order to see in dim light, but they also have terrific resolution. If the squid's eyes are free of defects, find the maximum distance at which the squid could distinguish a pair of lights on a ship that are separated by 1.8 m. Assume 555-nm light, a pupil diameter of 11 cm, and an average index of refraction for the eye of 1.36.

86. •• **Predict/Calculate** Figure 28-49 shows a single-slit diffraction pattern formed by light passing through a slit of width $W = 11.2$ µm and illuminating a screen 0.855 m behind the slit.

(a) What is the wavelength of the light? **(b)** If the width of the slit is decreased, will the distance indicated in Figure 28-49 be greater than or less than 15.2 cm? Explain.

87. •• **BIO** **Entoptic Halos** Images produced by structures within the eye (like lens fibers or cell fragments) are referred to as entoptic images. These images can sometimes take the form of "halos" around a bright light seen against a dark background. The halo in such a case is actually the bright outer rings of a circular diffraction pattern, like Figure 28-29, with the central bright spot not visible because it overlaps the direct image of the light. Find the diameter of the eye structure that causes a circular diffraction pattern with the first dark ring at an angle of 2.7° when viewed with monochromatic light of wavelength 605 nm. (Typical eye structures of this type have diameters on the order of $10\,\mu m$. Also, the index of refraction of the vitreous humor is 1.36.)

88. ••• White light is incident on a soap film ($n = 1.33$, thickness = 800.0 nm) suspended in air. If the incident light makes a 45° angle with the normal to the film, what visible wavelength(s) will be constructively reflected? (Refer to Example 25-6 for the range of visible wavelengths.)

89. ••• **Predict/Calculate** A system like that shown in Figure 28-35 consists of N slits, each transmitting light of intensity I_0. The light from each slit has the same phase and the same wavelength. The net intensity I observed at an angle θ due to all N slits is

$$I = I_0 \left[\frac{\sin(N\phi/2)}{\sin(\phi/2)} \right]^2$$

In this expression, $\phi = (2\pi d/\lambda)\sin\theta$, where λ is the wavelength of the light. **(a)** Show that the intensity in the limit $\theta \to 0$ is $I = N^2 I_0$. This is the maximum intensity of the interference pattern. **(b)** Show that the first points of zero intensity on either side of $\theta = 0$ occur at $\phi = 2\pi/N$ and $\phi = -2\pi/N$. **(c)** Does the central maximum ($\theta = 0$) of this pattern become narrower or broader as the number of slits is increased? Explain.

90. ••• A curved piece of glass with a radius of curvature R rests on a flat plate of glass. Light of wavelength λ is incident normally on this system. Considering only interference between waves reflected from the curved (lower) surface of glass and the top surface of the plate, show that the radius of the nth dark ring is

$$r_n = \sqrt{n\lambda R - n^2\lambda^2/4}$$

91. ••• **BIO** **The Resolution of the Eye** The resolution of the eye is ultimately limited by the pupil diameter. What is the smallest diameter spot the eye can produce on the retina if the pupil diameter is 3.55 mm? Assume light with a wavelength of $\lambda = 525$ nm. (*Note:* The distance from the pupil to the retina is 25.4 mm. In addition, the space between the pupil and the retina is filled with a fluid whose index of refraction is $n = 1.36$.)

PASSAGE PROBLEMS

Resolving Lines on an HDTV

The American Television Systems Committee (ATSC) sets the standards for high-definition television (HDTV). One of the approved HDTV formats is 1080p, which means 1080 horizontal lines scanned progressively (p)—that is, one line after another in sequence from top to bottom. Another standard is 1080i, which stands for 1080 lines interlaced (i). In this system it takes two scans of the screen to show a complete picture: the first scan shows the "even" horizontal lines, the second scan shows the "odd" horizontal lines. Interlacing was the norm for television displays until the 1970s, and is still used in most standard-definition TVs today. Progressive scanning

became more popular with the advent of computer monitors, and is used today in LCD, DLP, and plasma HDTVs.

In addition, the ATSC sets the standard for the shape of displays. For example, it defines a "wide screen" to be one with a 16:9 ratio; that is, the width of the display is greater than the height by the factor 16/9. This ratio is just a little larger than the golden ratio, $\phi = (1 + \sqrt{5})/2 = 1.618\ldots$, which is generally believed to be especially pleasing to the eye. Whatever the shape or definition of a TV, the ATSC specifies that it project 30 frames per second on a progressive display, or 60 fields per second on an interlace display, where each field is half the horizontal lines.

For the following problems, assume that 1080 horizontal lines are displayed on a television with a screen that is 15.7 inches high (32-inch diagonal), and that the light coming from the screen has a wavelength of 645 nm. Also, assume that the pupil of your eye has a diameter of 5.50 mm, and that the index of refraction of the interior of the eye is 1.36.

92. • What is the minimum angle your eye can resolve, according to the Rayleigh criterion and the above assumptions?

  A. 0.862×10^{-4} rad   B. 1.05×10^{-4} rad
  C. 1.43×10^{-4} rad   D. 1.95×10^{-4} rad

93. • What is the linear separation between horizontal lines on the screen?

  A. 0.0235 mm   B. 0.145 mm
  C. 0.369 mm   D. 0.926 mm

94. • What is the angular separation of the horizontal lines as viewed from a distance of 12.0 feet?

  A. 1.01×10^{-4} rad   B. 2.53×10^{-4} rad
  C. 2.56×10^{-4} rad   D. 12.1×10^{-4} rad

95. • According to the Rayleigh criterion, what is the closest you can be to the TV screen before resolving the individual horizontal lines? (In practice you can be considerably closer than this distance before resolving the lines.)

  A. 3.51 ft   B. 4.53 ft   C. 11.5 ft   D. 14.0 ft

96. •• **Predict/Calculate** **REFERRING TO EXAMPLE 28-3** Suppose we change the slit separation to a value other than 8.5×10^{-5} m, with the result that the linear distance to the tenth bright fringe above the central bright fringe increases from 12 cm to 18 cm. The screen is still 2.3 m from the slits, and the wavelength of the light is 440 nm **(a)** Did we increase or decrease the slit separation? Explain. **(b)** Find the new slit separation.

97. •• **Predict/Calculate** **REFERRING TO EXAMPLE 28-3** The wavelength of the light is changed to a value other than 440 nm, with the result that the linear distance to the *seventh* bright fringe above the central bright fringe is 12 cm. The screen is still 2.3 m from the slits, and the slit separation is 8.5×10^{-5} m. **(a)** Is the new wavelength longer or shorter than 440 nm? Explain. **(b)** Find the new wavelength.

98. •• **Predict/Calculate** **REFERRING TO EXAMPLE 28-11** The light used in this experiment has a wavelength of 511 nm. **(a)** If the width of the slit is decreased, will the angle to the first dark fringe above the central bright fringe increase or decrease? Explain. **(b)** Find the angle to the first dark fringe if the reduced slit width is 1.50×10^{-6} m.

99. •• **Predict/Calculate** **REFERRING TO EXAMPLE 28-11** The width of the slit in this experiment is 2.20×10^{-6} m. **(a)** If the frequency of the light is decreased, will the angle to the first dark fringe above the central bright fringe increase or decrease? Explain. **(b)** Find the angle to the first dark fringe if the reduced frequency is 5.22×10^{14} Hz.

Relativity

Big Ideas

1 The speed of light is the same in all inertial frames of reference.

2 Light travels in a vacuum with a speed c that is greater than the speed of any object of finite mass.

3 Moving clocks run slower than clocks at rest.

4 Moving objects are shorter in the direction of motion than their counterparts at rest.

5 Energy and mass are two forms of the same thing, and can be converted from one to the other.

6 Experiments in an accelerating frame of reference give the same results as in a gravitational field.

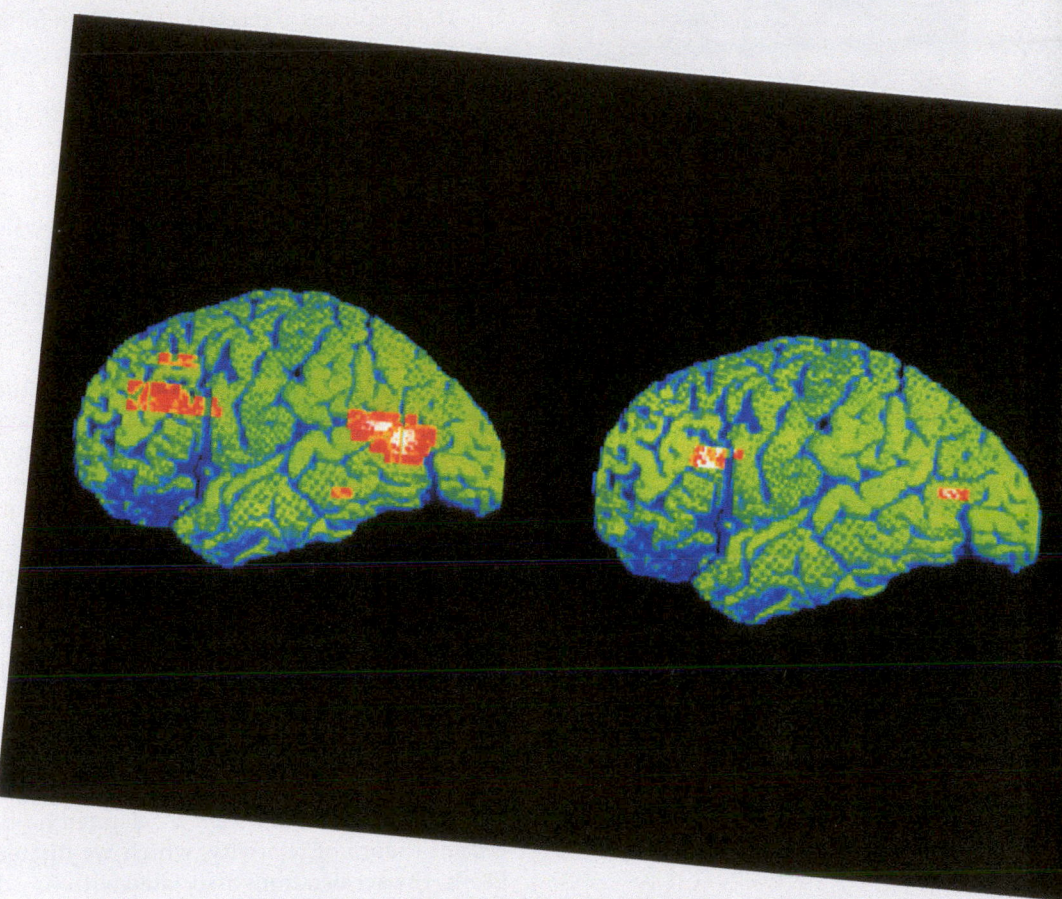

▲ The images shown here are the result of positron emission tomography (PET) scans of the brain, in which a radioactive tracer is injected into the bloodstream. The bright spots show areas of the brain that are activated when various mental tasks are performed. What is really amazing about these images, however, is that they are produced by matter (electrons) annihilating with antimatter (positrons emitted by the radioactive tracer) inside a person's brain. The energy released in these annihilations is given by the familiar equation $E = mc^2$, developed by a 26-year-old patent clerk who would soon become the most famous scientist of our time. This chapter explores some of the major contributions of Albert Einstein, including his special and general theories of relativity, and shows how they have radically altered our views of space, time, matter, and energy.

Modern physics started around the beginning of the twentieth century, when two fundamentally new ways of looking at nature were introduced. One of these was Albert Einstein's theory of relativity, the subject of this chapter, and the other was quantum physics, the subject of the next chapter. As we shall see, relativity reveals many unsuspected aspects of nature, including the slowing of moving clocks and the conversion between mass and energy.

▲ **FIGURE 29-1** Albert Einstein.

Big Idea **1** The laws of nature are the same in all inertial frames of reference. In addition, the speed of light is the same in all inertial frames of reference, independent of the motion of the source or the receiver.

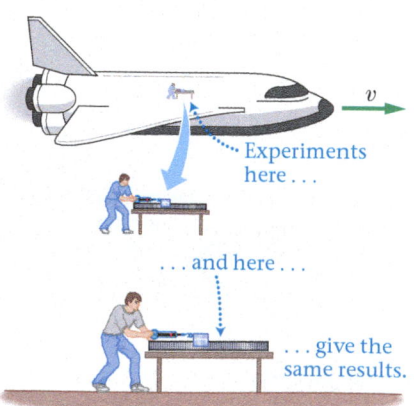

▲ **FIGURE 29-2 Inertial frames of reference** The two observers shown are in different inertial frames of reference. According to the first postulate of relativity, physical experiments will give identical laws of nature in the two frames.

29-1 The Postulates of Special Relativity

At a time when some scientists thought physics was almost completely understood, with only minor details to be straightened out, physics was changed forever with the introduction of the **special theory of relativity.** Published in 1905 by Albert Einstein (1879–1955), a 26-year-old patent clerk (third class) in Berne, Switzerland, the theory fundamentally altered our understanding of such basic physical concepts as time, length, mass, and energy. Einstein at the time of his development of relativity is shown in **FIGURE 29-1**. It may come as a surprise, then, that Einstein's theory of relativity is based on just two simply stated postulates, and that algebra is all the mathematics required to work out its main results.

The postulates of special relativity put forward by Einstein can be stated as follows:

Equivalence of Physical Laws
The laws of physics are the same in all inertial frames of reference.

Constancy of the Speed of Light
The speed of light in a vacuum, $c = 3.00 \times 10^8$ m/s, is the same in all inertial frames of reference, independent of the motion of the source or the receiver.

All the consequences of relativity explored in this chapter follow as a direct result of these two postulates.

Inertial Frames of Reference   The first postulate is certainly reasonable. Recall from Section 5-2 that an inertial frame of reference is one in which Newton's laws of motion are obeyed. Specifically, an object with no force acting on it has zero acceleration in all inertial frames. Einstein's first postulate simply extends this notion of an inertial frame to cover *all* the known laws of physics, including those dealing with thermodynamics, electricity, magnetism, and electromagnetic waves. For example, a mechanics experiment performed on the surface of the Earth (which is approximately an inertial frame) gives the same results as when the same experiment is carried out in an airplane moving with constant velocity. In addition, the behavior of heat, magnets, and electric circuits is the same in the airplane as on the ground, as indicated in **FIGURE 29-2**.

All inertial frames of reference move with constant velocity (that is, zero acceleration) relative to one another. Hence, the special theory of relativity is "special" in the sense that it restricts our considerations to frames with no acceleration. The more general case, in which accelerated motion is considered, is the subject of the *general* theory of relativity, which we discuss later in this chapter. In the case of Earth, the accelerations associated with its orbital and rotational motions are small enough to be ignored in most experiments. Thus, unless otherwise stated, we shall consider the Earth and objects moving with constant velocity relative to it to be inertial frames of reference.

The Speed of Light   The second postulate of relativity is less intuitive than the first. Specifically, it states that light travels with the same speed, c, regardless of whether the source or the observer is in motion. To understand the implications of this assertion, consider for a moment the case of waves on water. In **FIGURE 29-3 (a)** we see an observer at rest relative to the water, and two moving sources generating waves. The waves produced by both the speedboat and the tugboat travel at the characteristic speed of water waves, v_w, once they are generated. Thus the observer sees a wave speed that is independent of the speed of the source—just as postulated for light and shown in **FIGURE 29-3 (b)**.

On the other hand, suppose the observer is in motion with a speed v with respect to the water. If the observer is moving to the right, and water waves are moving to the left with a speed v_w, as in **FIGURE 29-4 (a)**, the waves move past the observer with a speed $v + v_w$. Similarly, if the water waves are moving to the right, as in **FIGURE 29-4 (b)**, the observer finds them to have a speed $v - v_w$. Clearly, the fact that the observer is in motion with respect to the medium through which the waves are traveling (water

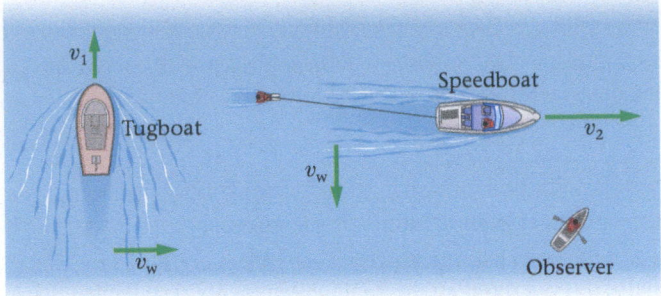

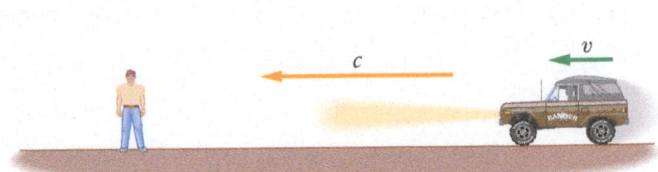

(a) Speed of water waves independent of speed of source

(b) Speed of light waves independent of speed of source

▲ **FIGURE 29-3 Wave speed versus source speed** The speed of a wave is independent of the speed of the source that generates it. **(a)** Water waves produced by a slow-moving tugboat have the same speed as waves produced by a high-powered speedboat. **(b)** The speed of a beam of light, c, is independent of the speed of its source.

in this case) means that the speed of the water waves depends on the speed of the observer.

Before Einstein's theory of relativity, it was generally accepted that a similar situation would apply to light waves. In particular, light was thought to propagate through a hypothetical medium, referred to as the *luminiferous ether*, or the ether for short, that permeates all space. Because the Earth rotates about its axis with a speed of roughly 1000 mi/h at the equator, and orbits the Sun with a speed of about 67,000 mi/h, it follows that it must move relative to the ether. If this is the case, it should be possible to detect this motion by measuring differences in the speed of light propagating in different directions—just as in the case of water waves. Extremely precise experiments were carried out to this end by the American physicists A. A. Michelson (1852–1931) and E. W. Morley (1838–1923) from 1883 to 1887. They were unable to detect *any* difference in the speed of light. More recent and accurate experiments have come to precisely the same conclusion; namely, the second postulate of relativity is an accurate description of the way light behaves.

To see how counterintuitive the second postulate can be, consider the situation illustrated in **FIGURE 29-5**. In this case a ray of light is propagating to the right with a speed c relative to observer 1. A second observer is moving to the right as well, with a speed of $0.9c$. Although it seems natural to think that observer 2 should see the ray of light passing with a speed of only $0.1c$, this is not the case. Observer 2, like observers in all inertial frames of reference, sees the ray go by with the speed of light, c.

For the observations given in Figure 29-5 to be valid—that is, for both observers to measure the same speed of light—the behavior of space and time must differ from our everyday experience when speeds approach the speed of light. This is indeed the case, as we shall see in considerable detail in the next few sections. In everyday circumstances, however, the physics described by Newton's laws are perfectly adequate. In fact, Newton's laws are valid in the limit of very small speeds, whereas Einstein's theory of relativity gives correct results for all speeds from zero up to the speed of light.

All inertial observers measure the same speed for light, and hence they are all equally correct in claiming that they are at rest. For example, observer 1 in Figure 29-5 may say that he is at rest and that observer 2 is moving to the right with a speed of $0.9c$. Observer 2, however, is equally justified in saying that she is at rest and that observer 1 is in motion with a speed of $0.9c$ to the left. From the point of view of relativity, both observers are equally correct. There is no absolute rest or absolute motion, only motion relative to something else.

Finally, note that it would not make sense for observer 2 in Figure 29-5 to have a speed greater than that of light. If this were the case, it would not be possible for the light ray to pass the observer, much less to pass with the speed c. Thus we conclude that *the ultimate speed in the universe is the speed of light in a vacuum*. In the next several sections of this chapter, we shall see several more ways of arriving at precisely the same conclusion.

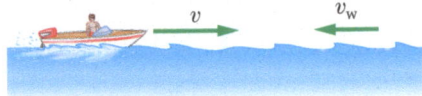

(a) High relative speed

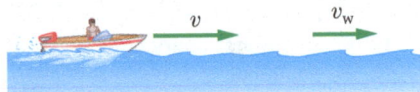

(b) Low relative speed

▲ **FIGURE 29-4 Wave speed versus observer speed** The speed of a wave depends on the speed of the observer relative to the medium through which the wave propagates. **(a)** The water waves move relative to the observer with speed $v + v_w$. **(b)** In this case, the waves move relative to the observer with speed $v - v_w$.

Big Idea **2** Light travels through the vacuum of space with a speed c that is greater than the speed of any object of finite mass.

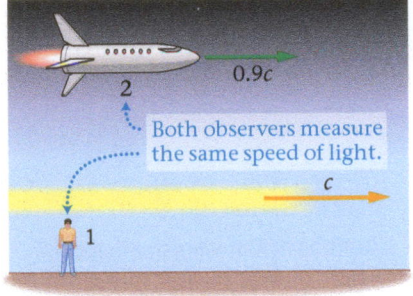

Both observers measure the same speed of light.

▲ **FIGURE 29-5 The speed of light for different observers** A beam of light is moving to the right with a speed c relative to observer 1. Observer 2 is moving to the right with a speed of $0.9c$. Still, from the point of view of observer 2, the beam of light is moving to the right with a speed of c, in agreement with the second postulate of relativity.

Enhance Your Understanding   (Answers given at the end of the chapter)

1. Observer 1 shines a beam of light toward observer 2, who approaches observer 1 with a speed of $c/2$. Observer 1 measures the speed of the light to be c. What speed does observer 2 measure for the light?

Section Review

*   The laws of nature are the same in all inertial frames of reference.

*   The speed of light is the ultimate speed in the universe—an object of finite mass can approach the speed of light, but can never attain it. All observers in inertial reference frames measure the same speed of light, regardless of the motion of the source or the receiver.

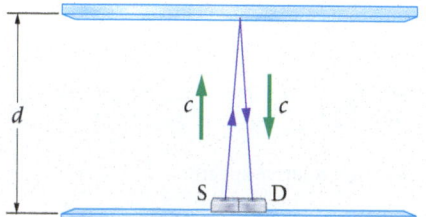

▲ **FIGURE 29-6 A stationary light clock** Light emitted by the source S travels to a mirror a distance d away and is reflected back into the detector D. The time between emission and detection is one cycle, or one "tick," of the clock.

29-2 The Relativity of Time and Time Dilation

We generally think of time as moving forward at a constant rate, as suggested by our everyday experience. This is simply not the case, however, when dealing with speeds approaching the speed of light. If you were to observe a spaceship moving past you with a speed of $0.5c$, for example, you would notice that the clocks on the ship run slow compared with your clocks—even if they were identical in all other respects.

The Light Clock   To calculate the difference between the rates of a moving clock and one at rest, consider the "light clock" shown in **FIGURE 29-6**. In this clock, a cycle begins when a burst of light is emitted from the light source S. The light then travels a distance d to a mirror, where it is reflected. It travels back a distance d to the detector D, and triggers the next burst of light. Each round trip of light can be thought of as one "tick" of the clock. (The points S and D are separated slightly in Figure 29-6 for clarity.)

We begin by calculating the time interval between the ticks of this clock when it is at rest—that is, when its speed relative to the observer making the measurement is zero. Because the light covers a total distance $2d$ with a constant speed c, the time between ticks is simply

$$\Delta t_0 = \frac{2d}{c} \qquad \text{29-1}$$

The subscript 0 indicates the clock is at rest ($v = 0$) when the measurement is made.

In contrast, consider the same light clock moving with a finite speed v, as in **FIGURE 29-7**. Notice that the light must now follow a zigzag path in order to complete a tick of the clock. Because this path is clearly longer than $2d$, and the speed of the light is still the same—according to the second postulate of relativity—the time between ticks, Δt, must be greater than Δt_0. With more time elapsing between ticks, the clock runs slow. We refer to this phenomenon as **time dilation**, because the time interval for one tick has been increased—or dilated—from Δt_0 for a clock at rest relative to an observer to $\Delta t > \Delta t_0$ for a clock in motion relative to an observer.

Time Dilation   To calculate the dilated time, Δt, notice that in the time $\Delta t/2$ the clock moves a horizontal distance $v\Delta t/2$, which is halfway to its position at the end of the tick. The distance traveled by the light in this time is $c\Delta t/2$, which is the hypotenuse

▶ **FIGURE 29-7 A moving light clock** A moving light clock requires a time Δt to complete one cycle. Note that the light follows a zigzag path that is longer than $2d$; hence, the time between ticks is greater for the moving clock than it is for the clock at rest.

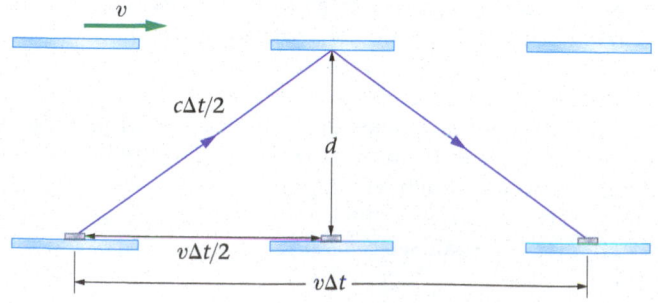

of the right triangle shown in Figure 29-7. Applying the Pythagorean theorem to this triangle, we find the following relationship:

$$\left(\frac{v\Delta t}{2}\right)^2 + d^2 = \left(\frac{c\Delta t}{2}\right)^2$$

Solving for the time Δt, we find

$$\Delta t = \frac{2d}{\sqrt{c^2 - v^2}} = \frac{2d}{c\sqrt{1 - v^2/c^2}}$$

Recalling that $\Delta t_0 = 2d/c$ is the time between ticks for a clock at rest, we can relate the two time intervals as follows:

Time Dilation

$$\Delta t = \frac{\Delta t_0}{\sqrt{1 - v^2/c^2}}$$

29-2

SI unit: s

Notice that $\Delta t = \Delta t_0$ for $v = 0$, as expected. For speeds v that are greater than zero but less than c, the denominator in $\Delta t = \Delta t_0/\sqrt{1 - v^2/c^2}$ is less than 1. As a result, it follows that Δt is greater than Δt_0. Finally, in the limit that the speed v approaches the speed of light, c, the denominator vanishes, and the time interval Δt goes to infinity. This behavior is illustrated in **FIGURE 29-8**, where we show the ratio $\Delta t/\Delta t_0$ as a function of the speed v. The fact that Δt goes to infinity means that it takes an infinite amount of time for one tick—in other words, as v approaches the speed of light, a clock slows to the point of stopping. Clearly, then, the speed of light provides a natural upper limit to the possible speed of an object.

Big Idea 3 Moving clocks run slower than clocks at rest. This applies to any type of "clock," including the aging process of animals and the decay rate of unstable particles. The rate of a clock approaches zero as the clock's speed approaches the speed of light.

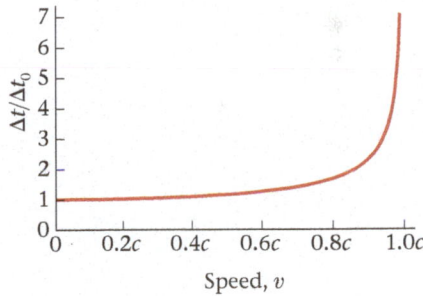

EXERCISE 29-1   ELAPSED TIME

A spaceship carrying a light clock moves with a speed of $0.500c$ relative to an observer on Earth. According to this observer, how much time does it take for the spaceship's clock to advance 1.00 s?

REASONING AND SOLUTION

Substituting $\Delta t_0 = 1.00$ s and $v = 0.500c$ in $\Delta t = \Delta t_0/\sqrt{1 - v^2/c^2}$, we obtain

$$\Delta t = \frac{\Delta t_0}{\sqrt{1 - v^2/c^2}} = \frac{1.00 \text{ s}}{\sqrt{1 - (0.500c)^2/c^2}} = \frac{1.00 \text{ s}}{\sqrt{1 - 0.250}} = 1.15 \text{ s}$$

Even at this high speed, the relativistic effect is relatively small—only about 15%.

▲ **FIGURE 29-8 Time dilation as a function of speed** As the speed of a clock increases, the time required for it to advance by 1 s increases slowly at first and then rapidly near the speed of light.

Time dilation applies to any type of clock, not just the light clock. If this were not the case—if different clocks ran at different rates when in motion with constant velocity—the first postulate of relativity would be violated.

CONCEPTUAL EXAMPLE 29-2   THE RATE OF TIME

A clock moving with a finite speed v is observed to run slow. If the speed of light were twice as great as it actually is, would the factor by which the clock runs slow be increased, decreased, or unchanged?

REASONING AND DISCUSSION

As Figure 29-8 shows, the factor by which time is dilated increases as the speed of a clock approaches the speed of light. If the speed of light were twice as great, the speed of a moving clock would be a smaller fraction of the speed of light; hence, the time dilation factor would be less.

Extending the preceding argument, it follows that in the limit of an infinite speed of light there would be no time dilation at all. This conclusion is verified by noting that in the limit $c \to \infty$ the denominator in $\Delta t = \Delta t_0/\sqrt{1 - v^2/c^2}$ is equal to 1, so that Δt and Δt_0 are equal. Because the speed of light is practically infinite in everyday terms, the relativistic time dilation effect is negligible for everyday objects.

ANSWER

The factor by which the clock runs slow would be decreased.

Events, Proper Time, and Everyday Speeds Before we proceed, it is useful to introduce some of the terms commonly employed in relativity. First, an **event** is a physical occurrence that happens at a *specified location* and at a *specified time*. In three dimensions, for example, we specify an event by giving the values of the coordinates x, y, and z as well as the time t. If two events occur at the same location but at different times, the time between the events is referred to as the **proper time**:

> The *proper time* is the amount of time separating two events that occur at the *same location*.

As an example, the proper time between ticks in a light clock is the time between the emission of light (event 1) and its detection (event 2) when the clock is at rest relative to the observer. Thus, Δt_0 in $\Delta t = \Delta t_0/\sqrt{1 - v^2/c^2}$ is the proper time, and Δt is the corresponding time when the clock moves relative to the observer with a speed v.

PROBLEM-SOLVING NOTE

Identifying the Proper Time

The key to solving problems involving time dilation is to correctly identify the proper time, Δt_0. Simply put, the proper time is the time between events that occur at the same location.

In Exercise 29-1 we calculated the relativistic effect of time dilation for the case of a clock moving with a speed equal to half the speed of light. In everyday circumstances, however, speeds are never so great. In fact, the greatest speed a human might reasonably attain today is the speed of a space station in orbit. As we saw in Chapter 12, this speed is only about 7700 m/s, or 17,000 mi/h. Although this is a rather large speed, it is still only 1/39,000th the speed of light.

To find the time dilation in a case like this we cannot simply substitute $v = 7700$ m/s into $\Delta t = \Delta t_0/\sqrt{1 - v^2/c^2}$, because a typical calculator does not have enough decimal places to give the correct answer. You might want to try it yourself; your calculator will probably give the incorrect result that Δt is equal to Δt_0. To find the correct answer, we must use the binomial expansion (Appendix A) to reexpress $\Delta t = \Delta t_0/\sqrt{1 - v^2/c^2}$ as follows:

$$\Delta t = \frac{\Delta t_0}{\sqrt{1 - v^2/c^2}} \approx \Delta t_0\left(1 + \frac{1}{2}\frac{v^2}{c^2} + \cdots\right)$$

Substituting the speed of the space station and the speed of light, we find

$$\Delta t \approx \Delta t_0\left[1 + \frac{1}{2}\left(\frac{7700 \text{ m/s}}{3.00 \times 10^8 \text{ m/s}}\right)^2\right] = \Delta t_0(1 + 3.3 \times 10^{-10})$$

Thus, a clock on board a space station runs slow by a factor of 1.00000000033. At this rate, it would take almost 100 years for the clock on the station to lose 1 s compared with a clock on Earth.

RWP* Clearly, such small differences in time cannot be measured with an ordinary clock. In recent years, however, atomic clocks have been constructed that have sufficient accuracy to put relativity to a direct test. The physicists J. C. Hafele and R. E. Keating conducted such a test in 1971 by placing one atomic clock on board a jet airplane and leaving an identical clock at rest in the laboratory. After flying the moving clock at high speed for many hours, the experimenters found that it had run slower (by a fraction of a microsecond) than the clock left in the lab. The discrepancy in the times agreed with the predictions of relativity. In subsequent years, and as recently as 2010, the results of the experiment have been confirmed with increased precision. Time slows down when you fly in an airplane!

Another aspect of time dilation is the fact that different observers disagree on *simultaneity*. For example, suppose observer 1 notes that two events at different locations occur at the same time. For observer 2, who is moving with a speed v relative to observer 1, these same two events are not simultaneous. Thus, relativity not only changes the rate at which time progresses for different observers, it also changes the amount of time that separates events.

Space Travel and Biological Aging

To this point, our discussion of time dilation has been applied solely to clocks. Clocks are not the only objects that show time dilation, however. In fact, we can make the following statement:

---

*Real World Physics applications are denoted by the acronym RWP.

Relativistic time dilation applies equally to *all physical processes*, including chemical reactions and biological functions.

Thus an astronaut in a moving spaceship *ages more slowly* than one who remains on Earth, and by precisely the same factor that a clock on the spaceship runs more slowly than one at rest. To the astronaut, however, time progresses as usual. The implications of time dilation for a high-speed trip to a nearby star are considered in the next Example.

EXAMPLE 29-3   BENNY AND JENNY—SEPARATED AT LAUNCH

Astronaut Benny travels to Vega, the fifth brightest star in the night sky, leaving his 35.0-year-old twin sister Jenny behind on Earth. Benny travels with a speed of 0.990c, and Vega is 25.3 light-years from Earth. **(a)** How much time does the trip take from the point of view of Jenny? **(b)** How much has Benny aged when he arrives at Vega?

PICTURE THE PROBLEM

Our sketch shows the spacecraft as it travels in a straight line from Earth to Vega. The speed of the spacecraft relative to Earth is 99% of the speed of light, $v = 0.990c$. In addition, the distance to Vega is 25.3 light-years (ly); that is, $d = 25.3$ ly. [*Note:* One light-year is the distance light travels with a speed c in one year. Specifically, 1 light-year = 1 ly = $c(1\,\text{y})$.]

REASONING AND STRATEGY

The two events of interest in this problem are (1) leaving Earth and (2) arriving at Vega. For Jenny, these two events occur at different locations. It follows that the time interval for her is Δt, and not the proper time, Δt_0.

For Benny, however, the two events occur at the same location—namely, just outside the spacecraft door. (In fact, from Benny's point of view the spacecraft is at rest, and the stars are in motion.) Therefore, the time interval measured by Benny is the proper time, Δt_0.

Finally, we recall that Δt and Δt_0 are related by $\Delta t = \Delta t_0/\sqrt{1 - v^2/c^2}$.

Known   Benny's speed, $v = 0.990c$; distance to Vega, $d = 25.3$ ly.
Unknown   **(a)** Time of trip for Jenny, $\Delta t = ?$ **(b)** Time of trip for Benny, $\Delta t_0 = ?$

SOLUTION

Part (a)

1. The spacecraft covers a distance d in a time Δt with a speed $v = 0.990c$. Use $v = d/\Delta t$ to solve for the time, Δt:

$$v = \frac{d}{\Delta t}$$
$$\Delta t = \frac{d}{v} = \frac{25.3\,\text{ly}}{0.990c} = \frac{25.3c(1\text{y})}{0.990c} = 25.6\,\text{y}$$

Part (b)

2. Rearrange $\Delta t = \Delta t_0/\sqrt{1 - v^2/c^2}$ (Equation 29-2) to solve for the proper time, Δt_0:

$$\Delta t = \frac{\Delta t_0}{\sqrt{1 - v^2/c^2}}$$
$$\Delta t_0 = \Delta t \sqrt{1 - v^2/c^2}$$

3. Substitute $v = 0.990c$ and $\Delta t = 25.6$ y to find Δt_0:

$$\Delta t_0 = \Delta t\sqrt{1 - v^2/c^2} = (25.6\,\text{y})\sqrt{1 - \frac{(0.990c)^2}{c^2}} = 3.61\,\text{y}$$

INSIGHT

When Benny reaches Vega he is only 38.6 y old, whereas Jenny, who stayed behind on Earth, is 60.6 y old.

From the point of view of Benny, the trip took 3.61 y at a speed of 0.990c. As a result, he would say that the distance covered in traveling to Vega was only (3.61 y)(0.990c) = 3.57 ly. To Benny, the distance to Vega has been reduced. We consider this result in greater detail in the next section.

PRACTICE PROBLEM — PREDICT/CALCULATE

(a) Suppose Benny would like to age only 2.00 y during his trip to Vega. Should his speed be greater than, less than, or equal to 0.990c? Explain. **(b)** How fast must Benny travel to age only 2.00 y on his trip to Vega? [**Answer: (a)** Benny must travel faster than 0.990c if he is to age less than 3.61 y. **(b)** $v = 0.997c$]

Some related homework problems: Problem 12, Problem 14

In the following Quick Example, we determine the speed of an astronaut by considering the change in her heart rate.

QUICK EXAMPLE 29-4   HEARTTHROB: FIND THE ASTRONAUT'S SPEED

An astronaut traveling with a speed v relative to Earth takes her pulse and finds that her heart beats once every 0.925 s. Mission control on Earth, which monitors her heart activity, observes one heartbeat every 1.57 s. What is the astronaut's speed relative to Earth?

REASONING AND SOLUTION

The astronaut's heart is at rest relative to her (good!), and hence she measures the proper time, Δt_0. Mission control measures the dilated time, Δt. These times are related by $\Delta t = \Delta t_0/\sqrt{1 - v^2/c^2}$, which can be solved for the speed, v.

1. Identify the proper time, Δt_0, and the dilated time, Δt:

$$\Delta t_0 = 0.925 \text{ s}$$
$$\Delta t = 1.57 \text{ s}$$

2. Solve the time-dilation expression, $\Delta t = \Delta t_0/\sqrt{1 - v^2/c^2}$, for the speed, v:

$$v = c\sqrt{1 - \Delta t_0^2/\Delta t^2}$$

3. Substitute numerical values:

$$v = c\sqrt{1 - (0.925 \text{ s})^2/(1.57 \text{ s})^2} = 0.808c$$

The Decay of the Muon

A particularly interesting example of time dilation involves subatomic particles called *muons* that are created by cosmic radiation high in Earth's atmosphere. A muon is an unstable particle; in fact, a muon at rest exists for only about 2.20×10^{-6} s, on average, before it decays. Suppose, for example, that a muon is created at an altitude of 5000 m above the surface of the Earth. If this muon travels toward the ground with a speed of $0.995c$ for 2.20×10^{-6} s, it will cover a distance of only 657 m before decaying. Thus, without time dilation, one would conclude that the muons produced at high altitude should not reach the surface of the Earth. In fact, large numbers of muons do reach the ground. The reason is that they age more slowly due to their motion—just like the astronaut traveling to Vega considered in Example 29-3. The next Example examines this time dilation effect.

EXAMPLE 29-5   THE LIFE AND TIMES OF A MUON

Consider muons traveling toward Earth with a speed of $0.995c$ from their point of creation at a height of 5000 m.
(a) Find the average lifetime of these muons, assuming that a muon at rest has an average lifetime of 2.20×10^{-6} s.
(b) Calculate the average distance these muons can cover before decaying.

PICTURE THE PROBLEM

Our sketch shows a muon that has just been created at an altitude of 5000 m. The muon is heading straight for the ground with a speed $v = 0.995c$. As a result of its high speed, the muon's "internal clock" runs slow, allowing it to live longer as seen by an observer on Earth.

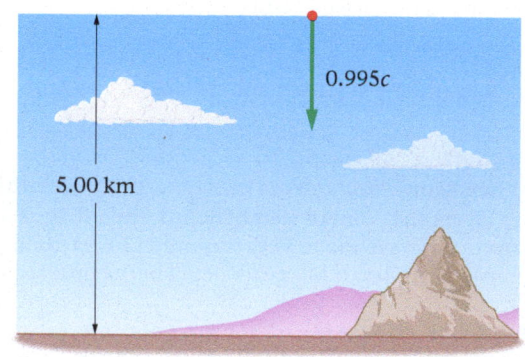

0.995c

5.00 km

REASONING AND STRATEGY

The two events to be considered in this case are (1) the creation of a muon and (2) the decay of a muon. From the point of view of an observer on Earth, these two events occur at different locations. It follows that the corresponding time is Δt.

From the muon's point of view, the two events occur at the same location, and the time between them is $\Delta t_0 = 2.20 \times 10^{-6}$ s. We can find the time Δt by using time dilation: $\Delta t = \Delta t_0/\sqrt{1 - v^2/c^2}$.

Known   Speed of muons, $v = 0.995c$; lifetime of muon at rest, $\Delta t_0 = 2.20 \times 10^{-6}$ s.
Unknown   (a) Lifetime of moving muons, $\Delta t = ?$ (b) Average distance traveled by muons, $d_{av} = ?$

CONTINUED

SOLUTION

Part (a)

1. Substitute $v = 0.995c$ and $\Delta t_0 = 2.20 \times 10^{-6}$ s into $\Delta t = \Delta t_0 / \sqrt{1 - v^2/c^2}$:

$$\Delta t = \frac{\Delta t_0}{\sqrt{1 - v^2/c^2}}$$
$$= \frac{2.20 \times 10^{-6} \text{ s}}{\sqrt{1 - (0.995c)^2/c^2}} = 22.0 \times 10^{-6} \text{ s}$$

Part (b)

2. Multiply $v = 0.995c$ by the time 22.0×10^{-6} s to find the average distance covered:

$$d_{av} = (0.995c)(22.0 \times 10^{-6} \text{ s}) = 6570 \text{ m}$$

INSIGHT

Thus relativistic time dilation allows the muons to travel about 10 times farther (6570 m instead of 657 m) than would have been expected from nonrelativistic physics. As a result, muons are detected on Earth's surface.

PRACTICE PROBLEM

Muons produced at the laboratory of the European Organization for Nuclear Research (CERN) in Geneva are accelerated to high speeds. At these speeds, the muons are observed to have a lifetime 30.00 times greater than the lifetime of muons at rest. What is the speed of the muons at CERN?   [**Answer:** $v = 0.9994c$]

Some related homework problems: Problem 11, Problem 17

Enhance Your Understanding   (Answers given at the end of the chapter)

2. Two identical atomic clocks are manufactured at a factory. One clock remains at the factory, and the other is transported across the country to a physics lab. When the clock arrives at the physics lab, it reads 9:00 A.M. At this same time, does the clock at the factory read a time that is before 9:00 A.M., equal to 9:00 A.M., or after 9:00 A.M.? Explain.

Section Review

- The time elapsed on an object moving with a speed v is $\Delta t = \Delta t_0 / \sqrt{1 - v^2/c^2}$. In this expression, Δt_0 is the proper time—that is, the time between events that occur at the same location. The time Δt is greater than the proper time for all speeds greater than zero and less than c.

29-3 The Relativity of Length and Length Contraction

Just as time is altered for an observer moving with a speed close to the speed of light, so too is distance. For example, a meterstick moving with a speed of $0.5c$ would appear noticeably shorter than a meterstick at rest. As the speed of the meterstick approaches c, its length diminishes toward zero.

To see why lengths contract, and to calculate the amount of contraction, recall the example of the twins Benny and Jenny and the trip to Vega (Example 29-3). From Jenny's point of view on Earth, Benny's trip took 25.6 y and covered a distance of $(0.990c)(25.6 \text{ y}) = 25.3$ ly. From Benny's point of view, however, the trip took only 3.61 y. Both twins agree on their relative velocity, however, and hence it follows that as far as Benny is concerned, his trip covered a distance of only $(0.990c)(3.61 \text{ y}) = 3.57$ ly. This is an example of length contraction, as illustrated in **FIGURE 29-9**.

Proper Length and Contracted Length   In general, we would like to determine the contracted length L of an object moving with a speed v. When the object is at rest ($v = 0$), we say that its length is the **proper length,** L_0:

> The proper length is the distance between two points as measured by an observer who is at rest with respect to them.

In the Benny and Jenny Example, Jenny is at rest with respect to Earth and Vega. As a result, the distance between them as measured by Jenny is the proper length; that is, $L_0 = 25.3$ ly. The contracted length, $L = 3.57$ ly, is measured by Benny.

PROBLEM-SOLVING NOTE

Identifying the Proper Length

The key to solving problems involving length contraction is to correctly identify the proper length, L_0. Specifically, the proper length is the distance between two points that are at rest relative to the observer.

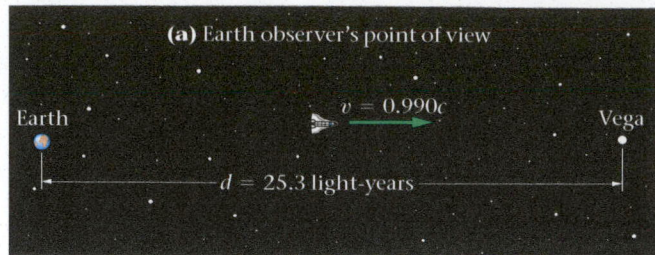
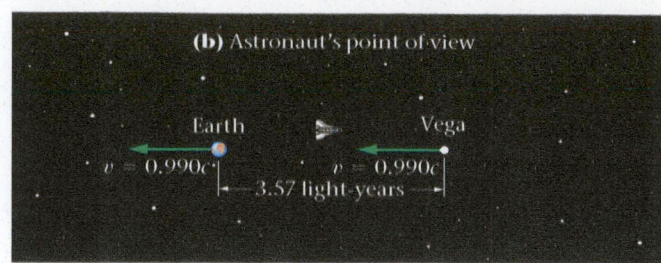

▲ **FIGURE 29-9 A relativistic trip to Vega** **(a)** From the Earth observer's point of view the spaceship is traveling with a speed of 0.990c, covering a distance of 25.3 ly in a time of 25.6 y. **(b)** From the astronaut's point of view the spaceship is at rest, and Earth and Vega are moving with a speed of 0.990c. For the astronaut the trip takes only 3.61 y and covers a contracted distance of only 3.57 ly.

PROBLEM-SOLVING NOTE

Measuring Proper Length and Proper Time

Keep in mind that the observer who measures the proper length is *not* necessarily the same observer who measures the proper time.

Big Idea **4** Moving objects are shorter in the direction of motion than their counterparts at rest. The length of an object approaches zero as its speed approaches the speed of light.

As for the times measured for the trip, from Jenny's point of view the two events (event 1, departing Earth; event 2, arriving at Vega) occur at different locations. As a result, she measures the *dilated time*, $\Delta t = 25.6$ y, even though she also measures the *proper length*, L_0. In contrast, Benny measures the *proper time*, $\Delta t_0 = 3.61$ y, and the *contracted length*, L. Notice that one must be careful to determine from the definitions given previously which observer measures the proper time and which observer measures the proper length—*it should never be assumed*, for example, that just because one observer measures the proper time that observer also measures the proper length.

We now use these observations to obtain a general expression relating L and L_0. To begin, notice that both observers measure the same relative speed, v. For Jenny the speed is $v = L_0/\Delta t$, and for Benny it is $v = L/\Delta t_0$. Setting these speeds equal, we obtain

$$v = \frac{L_0}{\Delta t} = \frac{L}{\Delta t_0}$$

Solving for L in terms of L_0, we find $L = L_0(\Delta t_0/\Delta t)$. Finally, using Equation 29-2 ($\Delta t = \Delta t_0/\sqrt{1 - v^2/c^2}$) to express Δt in terms of Δt_0, we obtain the following:

Length Contraction

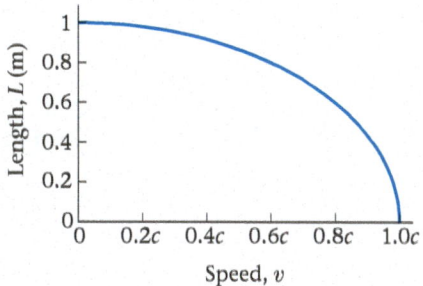

$$L = L_0\sqrt{1 - \frac{v^2}{c^2}}$$ 29-3

SI unit: m

Substituting the numerical values from the Benny and Jenny Example gives a contracted length of $L = (25.3 \text{ ly})\sqrt{1 - (0.990c)^2/c^2} = 3.57$ ly, as expected.

Notice that if $v = 0$ in $L = L_0\sqrt{1 - v^2/c^2}$, we find that $L = L_0$; that is, there is no change in length. On the other hand, as v approaches the speed of light, the contracted length L approaches zero. In general, the length of a moving object is always less than its proper length. **FIGURE 29-10** shows L as a function of speed v for a meterstick, where we see again that the speed of light is the ultimate speed possible.

▲ **FIGURE 29-10 Length contraction** The length of a meterstick as a function of its speed. Note that the length shrinks to zero in the limit that its speed approaches the speed of light.

EXAMPLE 29-6 **HALF A METER**

Find the speed for which the length of a meterstick is 0.500 m.

PICTURE THE PROBLEM
Our sketch shows a moving meterstick with a contracted length $L = 0.500$ m. Also shown is a meterstick that is at rest relative to the observer, which has its proper length, $L_0 = 1.00$ m.

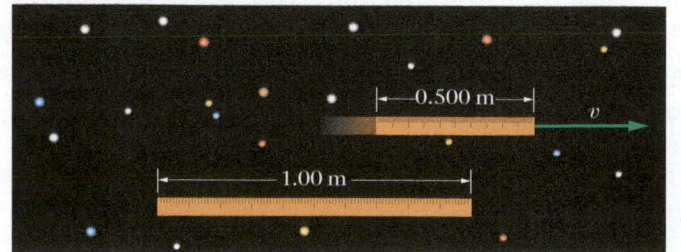

REASONING AND STRATEGY
We can find the desired speed by applying length contraction to the moving meterstick. In particular, given L and L_0 we can solve $L = L_0\sqrt{1 - v^2/c^2}$ for the speed, v.

Known Length of meterstick at rest, $L_0 = 1.00$ m; length of moving meter stick, $L = 0.500$ m.
Unknown Speed of moving meterstick, $v = ?$

CONTINUED

SOLUTION

1. Rearrange $L = L_0\sqrt{1 - v^2/c^2}$ to solve for the speed v:

$$v = c\sqrt{1 - \frac{L^2}{L_0^2}}$$

2. Substitute numerical values:

$$v = c\sqrt{1 - \frac{L^2}{L_0^2}} = c\sqrt{1 - \frac{(0.500 \text{ m})^2}{(1.00 \text{ m})^2}} = 0.866c$$

INSIGHT

An observer traveling along with the moving meterstick would see it as having its proper length of 1.00 m. From this observer's point of view, it is the *other* meterstick that is only 0.500 m long. Thus, lengths—like times—are relative, and depend on the observer making the measurement.

PRACTICE PROBLEM

Find the length of a meterstick that is moving with half the speed of light.   [**Answer:** $L = 0.866$ m]

Some related homework problems: Problem 22, Problem 27

▶ **FIGURE 29-11 A muon reaches Earth's surface (a)** From Earth's point of view a muon travels downward with a speed of 0.995c for a distance of 5.00 km. The muon can reach Earth's surface only if it ages slowly due to its motion. **(b)** From the muon's point of view Earth is moving upward with a speed of 0.995c, and the distance to Earth's surface is only 499 m. From this point of view the muon reaches Earth's surface because the distance is contracted, not because the muon lives longer.

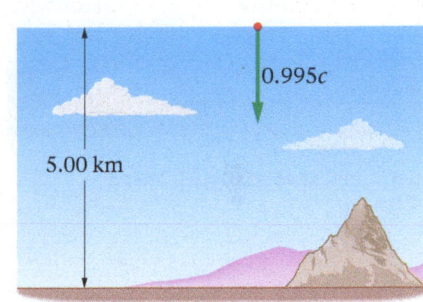

(a) Earth observer's point of view

In **FIGURE 29-11** we illustrate the effects of length contraction for the case of a muon traveling toward Earth's surface, as discussed in Example 29-5. From the point of view of the Earth, as illustrated in Figure 29-11 (a), the muon travels a distance of 5.00 km at a speed of 0.995c. To cover this distance, the muon must live about 10 times longer than it would at rest. From the point of view of the muon, as illustrated in Figure 29-11 (b), the Earth is moving upward at a speed of 0.995c, and the distance the Earth must travel to reach the muon is only $L = (5.00 \text{ km})\sqrt{1 - (0.995c)^2/c^2} = 499$ m. The Earth can easily cover this distance during the muon's resting lifetime of 2.20 μs.

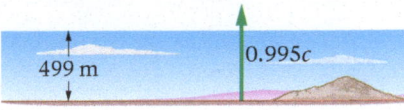

(b) Muon's point of view

One final comment: The length contraction calculated in Equation 29-3 *pertains only to lengths in the direction of the relative motion*. Lengths at right angles to the relative motion are unaffected.

CONCEPTUAL EXAMPLE 29-7   ANGLE OF REPOSE

An astronaut is resting on a bed inclined at an angle θ above the floor of a spaceship, as shown in the first sketch. From the point of view of an observer who sees the spaceship moving to the right with a speed approaching c, is the angle the bed makes with the floor greater than, less than, or equal to the angle θ observed by the astronaut in the spaceship?

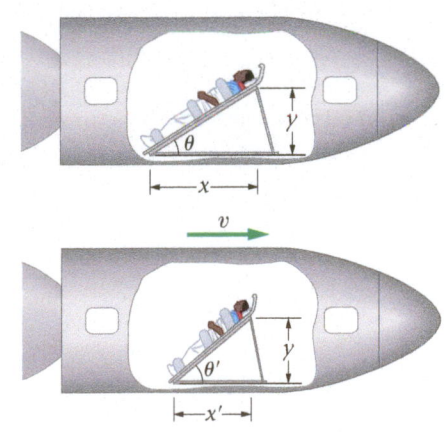

REASONING AND DISCUSSION

A person observing the spaceship moving with a speed v notices a contracted length, x', in the direction of motion, but an unchanged length, y, perpendicular to the direction of motion, as shown in the second sketch. As a result of the contraction in just one direction, the bed is inclined at an angle greater than the angle θ measured by the astronaut.

ANSWER

The angle is greater than θ.

The fact that lengths in different directions contract differently has interesting effects on the way a rapidly moving object appears to the eye. Additional effects are related to the finite time it takes for light to propagate to the eye from different parts of

(a) Streetcar at rest

(b) Streetcar at relativistic speed

◀ **FIGURE 29-12 Relativistic distortions** If the streetcar depicted at rest in part **(a)** moved to the left at nearly the speed of light, it would look like the computer-simulated image in part **(b)**. The streetcar would appear compressed in the direction of its motion. Because the streetcar would move significantly in the time it took for light from it to reach you, however, you would also observe other odd effects—for example, you would be able to see its back surface even from a position directly alongside it.

an object. An example of the way an object would actually look if it moved past us at nearly the speed of light is shown in **FIGURE 29-12**.

RWP The effects of relativity are both an advantage and a disadvantage for the prospect of interstellar space travel. On the one hand, if a new propulsion technology were to allow humans to accelerate a spacecraft to a speed nearly that of light, relative to the stars, the travel distance would be contracted and time would be dilated, making the trip appear shorter to those in the spacecraft. On the other hand, time would advance at its normal rate on Earth, requiring great patience for those awaiting news from the explorers. For example, if the spacecraft were to travel to the nearest star, Alpha Centauri, at a speed of $0.995c$, the 4.37-light-year trip would appear to take only 159 days to those on the spacecraft. The trip would take more than 4 years from the point of view of those on Earth, however, and round-trip messages between the Earth and Alpha Centauri would take longer than 8 years.

Enhance Your Understanding   (Answers given at the end of the chapter)

3. A horizontal meterstick moving to the right is less than 1 m in length. If the meterstick moves to the left instead, is its length greater than, less than, or equal to 1 m? Explain.

Section Review

*   The length of an object moving with a speed v is $L = L_0\sqrt{1 - v^2/c^2}$. In this expression, L_0 is the proper length—that is, the length as measured by an observer at rest relative to the object.

29-4 The Relativistic Addition of Velocities

Suppose you're piloting a spaceship in deep space, moving toward an asteroid with a speed of 25 mi/h. To signal a colleague on the asteroid, you activate a beam of light on the nose of the ship and point it in the direction of motion. Because light in a vacuum travels with the same speed c relative to all inertial observers, the speed of the light beam relative to the asteroid is simply c, *not* c + 25 mi/h. Clearly, then, simple addition of velocities—which seems to work just fine for everyday speeds— no longer applies.

The correct way to add velocities, valid for all speeds from zero to the speed of light, was obtained by Einstein. Consider, for example, a one-dimensional system consisting of a spaceship (1), a probe (2), and an asteroid (3), as indicated in **FIGURE 29-13**. Suppose the spaceship moves toward the asteroid with the velocity v_{13}, where the subscripts mean "the velocity of object 1 relative to object 3." Because the spaceship moves to the right in Figure 29-13 (a), its velocity is positive ($v_{13} > 0$); if it moved to the left, its velocity would be negative ($v_{13} < 0$). Now, imagine that the ship sends out a probe, whose velocity relative to the ship is v_{21}, as shown in Figure 29-13 (b). The question is, how do we add velocities correctly to get the velocity of the probe relative to the asteroid, v_{23}?

The "classical" answer is to simply add the velocities, as we did in Chapter 3; that is, $v_{23} = v_{21} + v_{13}$. Einstein showed that the correct result, taking into account relativity, contains an additional factor. The correct result is as follows:

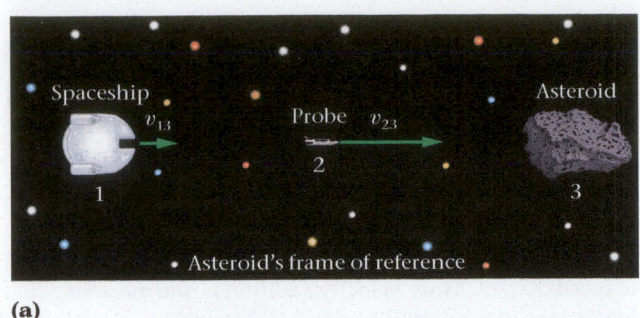

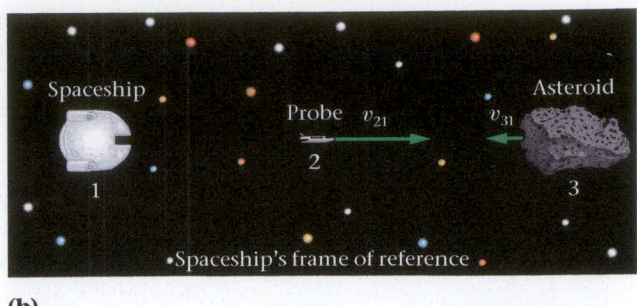

(a)

(b)

▲ **FIGURE 29-13 The velocity of a probe: Two frames of reference** A spaceship approaches an asteroid and sends a probe to investigate it. **(a)** In the asteroid's frame of reference, where the asteroid is at rest, the spaceship approaches with the velocity v_{13} and the probe approaches with the velocity v_{23}. **(b)** In the spaceship's frame of reference, the probe is launched with the velocity v_{21} and the asteroid approaches with the velocity $v_{31} = -v_{13}$. The relativistic sum of the velocities v_{13} and v_{21}, using Equation 29-4, yields the probe's velocity relative to the asteroid, v_{23}.

Relativistic Addition of Velocities

$$v_{23} = \frac{v_{21} + v_{13}}{1 + \dfrac{v_{21}v_{13}}{c^2}}$$

29-4

SI unit: m/s

If the speed of light were infinite, $c \to \infty$, the denominator would be $1 + v_{13}v_{21}/(\infty)^2 = 1$ and we would recover classical velocity addition. Thus, as we've seen before, it is the *finite* speed of light that is responsible for relativistic effects.

   To return to the example at the beginning of this section, what if the probe is actually a beam of light, with the velocity $v_{21} = c$? In this case, relativistic velocity addition says that the speed of the probe (light) relative to the asteroid is

$$v_{23} = \frac{v_{21} + v_{13}}{1 + \dfrac{v_{21}v_{13}}{c^2}} = \frac{c + v_{13}}{1 + \dfrac{c\,v_{13}}{c^2}} = \frac{c\left(1 + \dfrac{v_{13}}{c}\right)}{1 + \dfrac{v_{13}}{c}} = c$$

Thus, as expected, both the observer in the spaceship (1) and the observer on the asteroid (3) measure the same velocity for the light beam, $v_{21} = v_{23} = c$, regardless of the velocity of the ship relative to the asteroid, v_{13}.

   We've seen that $v_{23} = (v_{21} + v_{13})/(1 + v_{21}v_{13}/c^2)$ gives the correct result for a beam of light, but how does it work when applied to speeds much smaller than the speed of light? As an example, suppose a person on a spaceship moving at 25 mi/h throws a ball with a speed of 15 mi/h in the direction of the asteroid. According to the classical addition of velocities, the velocity of the ball relative to the asteroid is 15 mi/h + 25 mi/h = 40 mi/h. Application of relativistic velocity addition gives the following result:

$$v = 39.99999999999997 \text{ mi/h}$$

Thus, in any practical measurement, the velocity of the ball relative to the asteroid will be 40 mi/h. We conclude that the classical result, $v_{23} = v_{21} + v_{13}$, although not strictly correct, is appropriate for small speeds.

EXERCISE 29-8   SPEED OF A PROBE

Suppose the spaceship described previously is approaching an asteroid with a speed of 0.825c. If the spaceship launches a probe toward the asteroid with a speed of 0.759c relative to the ship, what is the speed of the probe relative to the asteroid?

CONTINUED

PROBLEM-SOLVING NOTE

Adding Velocities

To add velocities correctly, it is important to identify each velocity in terms of the object or observer relative to which it is measured, just as for velocity addition in nonrelativistic physics (see Section 3-6).

REASONING AND SOLUTION

Substituting $v_{13} = 0.825c$ and $v_{21} = 0.759c$ in $v_{23} = (v_{21} + v_{13})/(1 + v_{21}v_{13}/c^2)$, we obtain

$$v_{23} = \frac{v_{21} + v_{13}}{1 + \dfrac{v_{21}v_{13}}{c^2}} = \frac{0.759c + 0.825c}{1 + \dfrac{(0.759c)(0.825c)}{c^2}} = 0.974c$$

As expected, the speed relative to the asteroid is less than c. Notice, however, that classical velocity addition gives a strikingly different prediction for the speed relative to the asteroid. In fact, classical velocity addition gives $v_{23} = v_{21} + v_{13} = 0.759c + 0.825c = 1.584c$—a result that is greater than the speed of light. Thus, velocity addition is one more way that experiments can verify the predictions of relativity.

In the following Example, we apply relativistic velocity addition to a physical system.

EXAMPLE 29-9　GENERATION NEXT

At starbase Faraway Point, you observe two spacecraft approaching from the same direction. The *Picard* is approaching with a speed of 0.806c, and the *La Forge* is approaching with a speed of 0.906c. **(a)** Find the velocity of the *La Forge* relative to the *Picard*. **(b)** Find the velocity of the *La Forge* relative to the *Picard* if the *La Forge*'s direction of motion is reversed.

PICTURE THE PROBLEM

Our sketch shows the two spacecraft approaching the starbase along the same line, moving in the positive direction. The speed of the *La Forge* is 0.906c, and the speed of the *Picard* is 0.806c. In addition, note that we have numbered the objects in this system as follows: *Picard* (1), *La Forge* (2), Faraway Point (3). Finally, we show the *La Forge* ahead of the *Picard* in our sketch, but the final result is the same even if their positions are reversed.

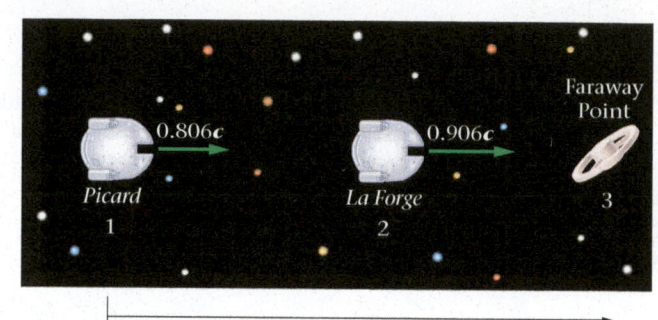

REASONING AND STRATEGY

a. The key to solving a problem like this is to choose the velocities v_{13}, v_{21}, and v_{23} in a way that is consistent with our numbering scheme. In this case, we know that $v_{13} = 0.806c$ and $v_{23} = 0.906c$. The velocity we want to find, *La Forge* relative to *Picard*, is v_{21}. This can be determined by substituting the known quantities into $v_{23} = (v_{21} + v_{13})/(1 + v_{21}v_{13}/c^2)$.

b. In this case the *La Forge* is traveling in the negative direction; therefore, $v_{23} = -0.906c$.

Known　Speed of *Picard*, $v_{13} = 0.806c$; speed of *La Forge*, $v_{23} = 0.906c$; velocity of *La Forge* for part (b), $v_{23} = -0.906c$.
Unknown　**(a)** Velocity of *La Forge* relative to *Picard*, $v_{21} = ?$ **(b)** Velocity of *La Forge* relative to *Picard* when *La Forge* reverses direction, $v_{21} = ?$

SOLUTION

Part (a)

1. Rearrange the expression for relativistic addition of velocities, $v_{23} = (v_{21} + v_{13})/(1 + v_{21}v_{13}/c^2)$, to solve for v_{21}:

$$v_{21} = \frac{v_{23} - v_{13}}{1 - \dfrac{v_{23}v_{13}}{c^2}}$$

2. Substitute $v_{13} = 0.806c$ and $v_{23} = 0.906c$ to find v_{21}, the velocity of the *La Forge* relative to the *Picard*:

$$v_{21} = \frac{v_{23} - v_{13}}{1 - \dfrac{v_{23}v_{13}}{c^2}} = \frac{0.906c - 0.806c}{1 - \dfrac{(0.906c)(0.806c)}{c^2}} = 0.371c$$

Part (b)

3. Use the result from part (a), only this time with $v_{23} = -0.906c$:

$$v_{21} = \frac{v_{23} - v_{13}}{1 - \dfrac{v_{23}v_{13}}{c^2}} = \frac{(-0.906c) - 0.806c}{1 - \dfrac{(-0.906c)(0.806c)}{c^2}} = -0.989c$$

INSIGHT

In part (a), a nonrelativistic calculation of the velocity of the *La Forge* relative to the *Picard* gives 0.100c, considerably less than the correct value of 0.371c obtained from relativistic velocity addition. In part (b), a nonrelativistic calculation gives a speed of 1.712c, as opposed to the correct speed, 0.989c, which is less than c. Also, notice that the negative value of the velocity v_{21} ($-0.989c$) means the *Picard* sees the *La Forge* moving to the left.

CONTINUED

Finally, when we originally set up Equation 29-4, $v_{23} = (v_{21} + v_{13})/(1 + v_{21}v_{13}/c^2)$, we arbitrarily chose the *Picard, La Forge,* and Faraway Point to be 1, 2, and 3, respectively. It's important to note, however, that any assignment of 1, 2, and 3 is equally valid and will yield precisely the same results.

PRACTICE PROBLEM

Find the velocity of the *La Forge* relative to the *Picard* if the *Picard*'s direction of motion is reversed but the *La Forge* is still moving toward Faraway Point. [**Answer:** In this case $v_{13} = -0.806c$; hence, $v_{21} = 0.989c$. The *Picard* sees the *La Forge* moving to the right.]

Some related homework problems: Problem 37, Problem 38, Problem 39

To get a better feeling for relativistic velocity addition, consider a spacecraft, initially at rest, that increases its speed by $0.1c$ when it fires its rockets. At first, the speed of the spacecraft increases linearly with the number of times the rockets are fired, as indicated by the line labeled "Classical velocity addition" in **FIGURE 29-14**. As the spacecraft's speed approaches the speed of light, however, further rocket firings have less and less effect, as seen in the curve labeled "Relativistic velocity addition." In the limit of infinite time—and an infinite number of rocket firings—the speed of the spacecraft approaches c, without ever attaining that value.

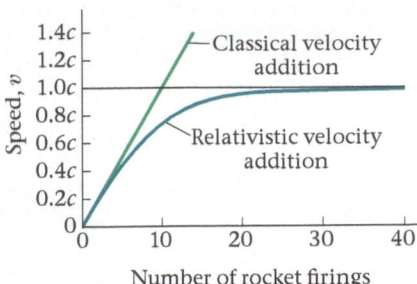

▲ **FIGURE 29-14 Relativistic velocity addition** A comparison of classical velocity addition and relativistic velocity addition. Classically, a spacecraft that continues to fire its rockets attains a greater and greater speed. The correct, relativistic, result is that the spacecraft approaches the speed of light in the limit of an infinite number of rocket firings.

Enhance Your Understanding   (Answers given at the end of the chapter)

4. A passenger jogs toward the front of a train with a speed of 2 m/s. The train itself moves with a speed of 8 m/s relative to the ground. In classical physics, we would say that the speed of the passenger relative to the ground is 10 m/s. In principle, is the correct relativistic speed of the passenger relative to the ground greater than, less than, or equal to 10 m/s? Explain.

Section Review

* When the velocities v_{21} and v_{13} are added, the result is given by relativistic velocity addition, $v_{23} = (v_{21} + v_{13})/(1 + v_{21}v_{13}/c^2)$. In the limit that c is infinite, which is a good approximation for everyday speeds, this reduces to the familiar classical velocity addition, $v_{23} = v_{21} + v_{13}$.

29-5 Relativistic Momentum

The first postulate of relativity states that the laws of physics are the same for all observers in all inertial frames of reference. Among the most fundamental of these laws are the conservation of momentum and the conservation of energy for an isolated system. In this section we consider the relativistic expression for momentum, leaving energy conservation for the next section.

When one considers the unusual way that relativistic velocities add, as given in Equation 29-4, it comes as no surprise that the classical expression for momentum, $p = mv$, is not valid for all speeds. As an example, we saw in Chapter 9 that if a large mass with a speed v collides elastically with a small mass at rest, the small mass is given a speed $2v$. Clearly, this cannot happen if the speed of the large mass is greater than $0.5c$, because the small mass cannot have a speed greater than the speed of light. Thus the nonrelativistic relationship $p = mv$ must be modified for speeds comparable to c.

A detailed analysis shows that the correct relativistic expression for the magnitude of momentum is the following:

Relativistic Momentum

$$p = \frac{mv}{\sqrt{1 - \dfrac{v^2}{c^2}}}$$ 29-5

SI unit: kg · m/s

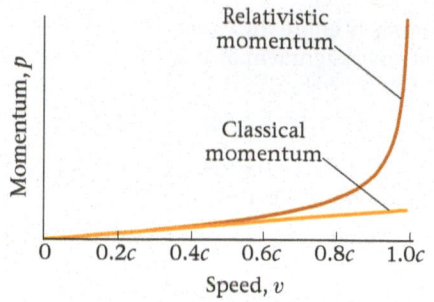

▲ **FIGURE 29-15 Relativistic momentum** The magnitude of momentum as a function of speed. Classically, the momentum increases linearly with speed, as shown by the straight line. The correct relativistic momentum increases to infinity as the speed of light is approached.

As v approaches the speed of light, the relativistic momentum becomes significantly greater than the classical momentum, eventually diverging to infinity as $v \rightarrow c$, as shown in **FIGURE 29-15**. For low speeds the classical and relativistic results agree.

EXERCISE 29-10 MOMENTUM

Find **(a)** the classical and **(b)** the relativistic momentum of a 3.98-kg mass moving with a speed of 0.965c.

REASONING AND SOLUTION

a. The classical momentum is given by the expression $p = mv$, as we saw in Chapter 9:

$$p = mv = (3.98 \text{ kg})(0.965)(3.00 \times 10^8 \text{ m/s}) = 1.15 \times 10^9 \text{ kg} \cdot \text{m/s}$$

b. For relativistic speeds, as in this case, the correct expression for the momentum is $p = mv/\sqrt{1 - v^2/c^2}$:

$$p = \frac{mv}{\sqrt{1 - \dfrac{v^2}{c^2}}} = \frac{(3.98 \text{ kg})(0.965)(3.00 \times 10^8 \text{ m/s})}{\sqrt{1 - \dfrac{(0.965c)^2}{c^2}}} = 4.39 \times 10^9 \text{ kg} \cdot \text{m/s}$$

As expected, the relativistic momentum is larger in magnitude than the classical momentum.

Next we consider a system in which the relativistic momentum is conserved.

EXAMPLE 29-11 THE UNKNOWN MASS

A satellite, initially at rest in deep space, explodes into two pieces. One piece has a mass of 241 kg and moves away from the explosion with a speed of 0.690c. The other piece moves away from the explosion in the opposite direction with a speed of 0.762c. Find the mass of the second piece of the satellite. (The speeds of the pieces of the satellite in this Example are rather large in order to more clearly illustrate the effects of relativity.)

PICTURE THE PROBLEM

The two pieces of the satellite and their speeds (0.690c and 0.762c) are shown in the sketch. Notice that the two pieces move in opposite directions, which means that they move away from one another along the same line.

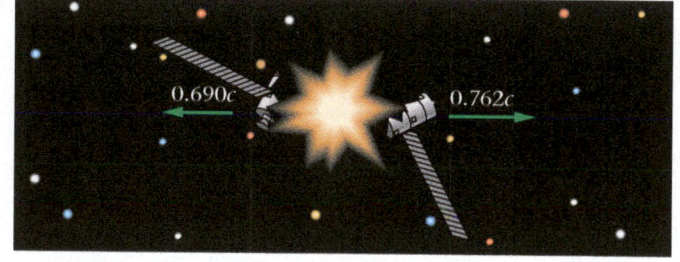

REASONING AND STRATEGY

The basic idea in this system is that because no external forces act on the satellite, its total momentum must be conserved. The initial momentum is zero; hence, the final momentum must be zero as well. This means that the pieces will move in opposite directions, as mentioned, and with momenta of equal magnitude.

   Thus, we begin by calculating the magnitude of the momentum for the first piece of the satellite. Next, we set the momentum of the second piece equal to the same magnitude and solve for the mass.

Known   Mass of piece 1 of satellite, $m_1 = 241$ kg; speed of piece 1, $v_1 = 0.690c$; speed of piece 2, $v_2 = 0.762c$.
Unknown   Mass of piece 2 of satellite, $m_2 = ?$

SOLUTION

1. Calculate the magnitude of the momentum for the piece of the satellite with a mass $m_1 = 241$ kg and a speed $v_1 = 0.690c$:

$$p_1 = \frac{m_1 v_1}{\sqrt{1 - \dfrac{v_1^2}{c^2}}} = \frac{(241 \text{ kg})(0.690)(3.00 \times 10^8 \text{ m/s})}{\sqrt{1 - \dfrac{(0.690c)^2}{c^2}}}$$
$$= 6.89 \times 10^{10} \text{ kg} \cdot \text{m/s}$$

2. Set the magnitude of the momentum of the second piece of the satellite equal to the magnitude of the momentum of the first piece:

$$p_2 = \frac{m_2 v_2}{\sqrt{1 - \dfrac{v_2^2}{c^2}}} = p_1$$

3. Solve the above relationship for m_2, the mass of the second piece of the satellite, and substitute the numerical values $v_2 = 0.762c$ and $p_1 = 6.89 \times 10^{10}$ kg · m/s:

$$m_2 = \left(\frac{p_1}{v_2}\right)\sqrt{1 - \frac{v_2^2}{c^2}} = \left[\frac{6.89 \times 10^{10} \text{ kg} \cdot \text{m/s}}{(0.762)(3.00 \times 10^8 \text{ m/s})}\right]\sqrt{1 - \frac{(0.762c)^2}{c^2}}$$
$$= 195 \text{ kg}$$

CONTINUED

Equation 29-5 is sometimes thought of in terms of a mass that increases with speed. Suppose, for example, that an object has the mass m_0 when it is at rest; that is, its **rest mass** is m_0. When the speed of the object is v, its momentum is

$$p = \frac{m_0 v}{\sqrt{1 - \dfrac{v^2}{c^2}}} = \left(\frac{m_0}{\sqrt{1 - \dfrac{v^2}{c^2}}} \right) v$$

If we now make the identification $m = m_0/\sqrt{1 - v^2/c^2}$ we can write the momentum as follows:

$$p = mv$$

Thus the classical expression for momentum can be used for all speeds if we simply interpret the mass as increasing with speed according to the expression

$$m = \frac{m_0}{\sqrt{1 - \dfrac{v^2}{c^2}}} \qquad \text{29-6}$$

Notice that m approaches infinity as $v \to c$. Hence a constant force acting on an object generates less and less acceleration, $a = F/m$, as the speed of light is approached. This gives one further way of seeing that the speed of light cannot be exceeded.

Though Equation 29-6 helps to show why an object of finite mass can never be accelerated to speeds exceeding the speed of light, it must be noted that the concept of a relativistic mass has its limitations. For example, the relativistic kinetic energy of an object is *not* obtained by simply replacing m in $\frac{1}{2} mv^2$ with the expression given in Equation 29-6, as we shall see in detail in the next section.

Enhance Your Understanding   (Answers given at the end of the chapter)

5. Is the relativistic momentum of an object moving with the speed v greater than, less than, or equal to the classical momentum, mv? Explain.

Section Review

- The relativistic momentum of an object with speed v and mass m is $p = mv/\sqrt{1 - v^2/c^2}$.

29-6 Relativistic Energy and $E = mc^2$

We have just seen that, from the point of view of momentum, an object's mass increases as its speed increases. Therefore, when work is done on an object, part of the work goes into increasing its speed, and part goes into increasing its mass. It follows, then, that *mass is another form of energy*. This result, like time dilation, was completely unanticipated before the introduction of the theory of relativity.

Big Idea 5 Energy and mass are two forms of the same thing, and can be converted from one to the other according to the relationship $E = mc^2$.

PHYSICS IN CONTEXT
Looking Ahead

The energy released in nuclear reactions is due to a conversion of mass to energy, in accordance with $E = mc^2$. This will be discussed in detail in Chapter 32.

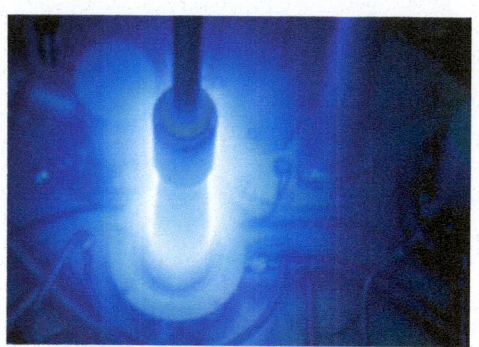

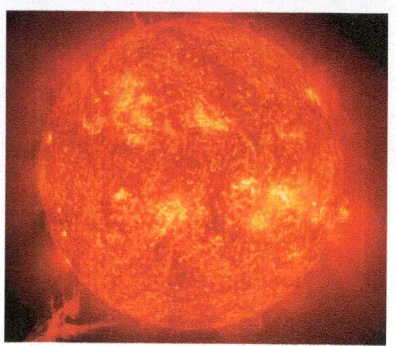

▲ **FIGURE 29-16 Visualizing Concepts Nuclear Reactions** A nuclear reactor (top) converts mass to energy by means of fission reactions, in which large nuclei (such as those of uranium or plutonium) are split into smaller fragments. This photo shows the blue glow referred to as Cherenkov radiation in a nuclear reactor at the Saclay Nuclear Research Centre in Saclay, France. The Sun (bottom) and other stars are powered by fusion reactions, in which light nuclei (such as those of hydrogen) combine to form heavier ones (such as those of helium). Although this process is utilized in the hydrogen bomb, it has not yet been successfully harnessed as a practical source of power on Earth.

Consider, for example, an object whose mass while at rest is m_0. Einstein was able to show that when the object moves with a speed v, its **total energy**, E, is given by the following expression:

Relativistic Energy

$$E = \frac{m_0 c^2}{\sqrt{1 - \dfrac{v^2}{c^2}}} = mc^2$$ 29-7

SI unit: J

This is Einstein's most famous result from relativity; that is, $E = mc^2$, where m is the relativistic mass given in Equation 29-6.

Notice that the total energy, E, does not vanish when the speed goes to zero, as does the classical kinetic energy. Instead, the energy of an object at rest—its **rest energy**, E_0—is

Rest Energy

$$E_0 = m_0 c^2$$ 29-8

SI unit: J

Because the speed of light is so large, it follows that the mass of an object times the *speed of light squared* is an enormous amount of energy, as illustrated in the following Exercise.

EXERCISE 29-12 REST ENERGY

Find the rest energy of a 0.19-kg apple.

REASONING AND SOLUTION

Substituting $m_0 = 0.19$ kg and $c = 3.00 \times 10^8$ m/s in the expression for the rest energy, $E_0 = m_0 c^2$, we find

$$E_0 = m_0 c^2 = (0.19 \text{ kg})(3.00 \times 10^8 \text{ m/s})^2 = 1.7 \times 10^{16} \text{ J}$$

This is an enormous amount of energy, as we detail below.

Converting Mass to Energy To put the result of Exercise 29-12 in context, let's compare it with the total yearly energy usage in the United States, which is about 10^{20} J. This means that if the rest energy of the apple could be converted entirely to usable forms of energy, it could supply the energy needs of the entire United States for about 90 minutes. Put another way, if the rest energy of the apple could be used to light a 100-W lightbulb, the bulb would stay lit for about 5 million years.

RWP This example illustrates the basic principle behind the operation of nuclear power plants, in which a small decrease in mass—due to various nuclear reactions—is used to generate electrical energy. The reactions that power such plants are referred to as **fission reactions,** in which a large nucleus splits into smaller nuclei and neutrons, as described in Section 32-5. For example, the nucleus of a uranium-235 atom may decay into two smaller nuclei and a number of neutrons. Because the mass of the uranium nucleus is greater than the sum of the masses of the fragments of the decay, the reaction releases an enormous amount of energy. In fact, 1 lb of uranium can produce about 3×10^6 kWh of electrical energy, compared with the 1 kWh that can be produced by the combustion of 1 lb of coal.

RWP The Sun is also powered by the conversion of mass to energy. In this case, however, the energy is released by **fusion reactions,** in which two light nuclei combine to form a heavier nucleus. The detailed reactions are presented in Section 32-6. In the following Example, we determine the amount of mass lost by the Sun per second.

Examples of fission and fusion reactions are illustrated in **FIGURE 29-16**.

EXAMPLE 29-13   THE PRODIGAL SUN

Energy is radiated by the Sun at the rate of about 3.92×10^{26} W. Find the corresponding decrease in the Sun's mass for every second that it radiates.

PICTURE THE PROBLEM

Our sketch indicates energy radiated continuously by the Sun at the rate $P = 3.92 \times 10^{26}$ W $= 3.92 \times 10^{26}$ J/s. We label the energy given off in the time interval $\Delta t = 1.00$ s with ΔE. The corresponding decrease in mass is Δm.

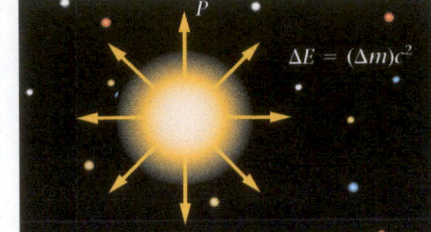

REASONING AND STRATEGY

If the energy radiated by the Sun in 1.00 s is ΔE, the corresponding decrease in mass, according to the relationship $E = mc^2$, is given by $\Delta m = \Delta E/c^2$.

To find ΔE, we simply recall (Chapter 7) that power is energy per time, $P = \Delta E/\Delta t$. Thus the energy radiated by the Sun in 1.00 s is $\Delta E = P\Delta t$, with $\Delta t = 1.00$ s.

Known   Rate of energy given off by the Sun, $P = 3.92 \times 10^{26}$ W; time of radiation, $\Delta t = 1.00$ s.
Unknown Decrease in mass of Sun, $\Delta m = $?

SOLUTION

1. Calculate the energy radiated by the Sun in 1.00 s:

$$\Delta E = P\Delta t = (3.92 \times 10^{26} \text{ J/s})(1.00 \text{ s}) = 3.92 \times 10^{26} \text{ J}$$

2. Divide ΔE by the speed of light squared, c^2, to find the decrease in mass:

$$\Delta m = \frac{\Delta E}{c^2} = \frac{3.92 \times 10^{26} \text{ J}}{(3.00 \times 10^8 \text{ m/s})^2} = 4.36 \times 10^9 \text{ kg}$$

INSIGHT

Thus, the Sun loses a rather large amount of mass each second—roughly the equivalent of 40 cruise ships. Because the Sun has a mass of 1.99×10^{30} kg, however, the mass it loses in 1500 y is only 10^{-10} of its total mass. Even after 1.5 billion years of radiating at its present rate, the Sun will lose a mere 0.01% of its mass. Clearly, the Sun will not evaporate into space anytime soon.

PRACTICE PROBLEM

Find the power radiated by a star whose mass decreases by the mass of the Moon (7.35×10^{22} kg) in half a million years (5.00×10^5 y).   [**Answer:** $P = 4.20 \times 10^{26}$ W, slightly more than the power of the Sun]

Some related homework problems: Problem 54, Problem 61

CONCEPTUAL EXAMPLE 29-14   COMPARE THE MASS   PREDICT/EXPLAIN

(a) When you compress a spring between your fingers, does its mass increase, decrease, or stay the same? (b) Which of the following is the *best explanation* for your prediction?

  I. Compressing the spring increases its potential energy, which in turn increases its mass.

 II. The mass of the spring decreases because some of its mass has been converted to spring potential energy.

III. The spring has the same mass whether it is compressed or not—it's still the same spring.

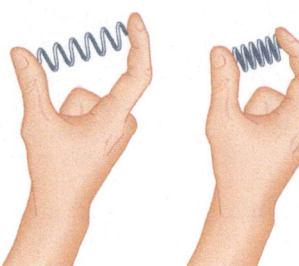

REASONING AND DISCUSSION

When the spring is compressed by an amount x, its energy is increased by the amount $\Delta E = \frac{1}{2}kx^2$, as we saw in Chapter 8. Because the energy of the spring has increased, its mass increases as well, by the amount $\Delta m = \Delta E/c^2$.

ANSWER

(a) The mass of the spring increases. (b) The best explanation is I.

As one might expect, the increase in the mass of a compressed spring is generally too small to be measured. For example, if the energy of a spring increases by 1.00 J, its mass increases by only $\Delta m = (1.00 \text{ J})/c^2 = 1.11 \times 10^{-17}$ kg.

Matter and Antimatter

A particularly interesting aspect of the equivalence of mass and energy is the existence of **antimatter.** For every elementary particle known to exist, there is a corresponding

Before annihilation

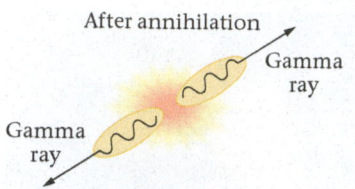

After annihilation

Gamma ray

Gamma ray

▲ **FIGURE 29-17 Electron-positron annihilation** An electron and a positron annihilate each other when they come into contact. The result is the emission of two energetic gamma rays with no mass. The mass of the original particles has been converted into the energy of the gamma rays.

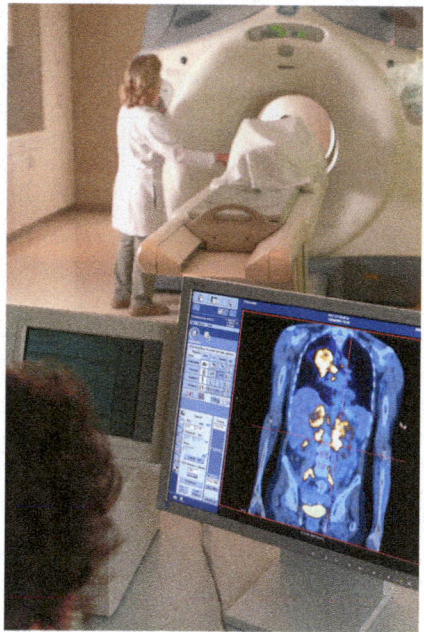

▲ **FIGURE 29-18 A PET scanner** In a PET scan, positrons emitted by a radioactively labeled metabolite are annihilated when they collide with electrons. The mass of the two particles is converted to energy in the form of a pair of gamma rays, which are always emitted in diametrically opposite directions. A ring of detectors surrounding the patient records the radiation and uses it to construct an image. In this way it is possible to map the location of particular metabolic activities in the body.

PROBLEM-SOLVING NOTE

Rest Mass

To correctly evaluate the relativistic kinetic energy, it is necessary to use the rest mass in Equation 29-9.

antimatter particle that has precisely the same mass but exactly the opposite charge. For example, an *electron* has a mass $m_e = 9.11 \times 10^{-31}$ kg and a charge $-e = -1.60 \times 10^{-19} C$; an *antielectron* has a mass 9.11×10^{-31} kg and a charge $+1.60 \times 10^{-19}$ C. Because an antielectron has a positive charge, it is generally referred to as a **positron.**

Antimatter is frequently created in accelerators, where particles collide at speeds approaching the speed of light. In fact, it is possible to create antiatoms in the lab made entirely of antimatter. An intriguing possibility is that the universe may actually contain entire antigalaxies of antimatter.

If this is indeed the case, one would have to be a bit careful about visiting such a galaxy, because particles of matter and antimatter have a rather interesting behavior when they meet—they **annihilate** one another. This situation is illustrated in **FIGURE 29-17**, where we show an electron and a positron coming into contact. The result is that the particles cease to exist, which satisfies charge conservation, since the net charge of the system is zero before and after the annihilation. As for energy conservation, the mass of the two particles is converted into two gamma rays, which are similar to X-rays only more energetic. Each of the gamma rays must have an energy that is at least $E = m_e c^2$. Thus, in matter-antimatter annihilation the particles vanish in a burst of radiation.

BIO Electron-positron annihilation is the basis for the diagnostic imaging technique called positron emission tomography (PET), which is often used to examine biological processes within the brain, heart, and other organs. In a typical PET brain scan, for example, a patient is injected with glucose (the primary energy source for brain activity) that has been "tagged" with radioactive tracers. These tracers emit positrons in a nuclear reaction described in Section 32-7, and the resulting positrons, in turn, encounter electrons in the brain and undergo annihilation. The resulting gamma rays exit through the patient's skull and are monitored by the PET scanner, similar to the one shown in **FIGURE 29-18**, which converts them into false-color images showing the glucose metabolism levels within the brain. Thus, surprising as it may seem, this powerful diagnostic tool actually relies on the annihilation of matter and antimatter inside a person's brain!

The conversion between mass and energy can go the other way as well. That is, an energetic gamma ray, which has zero mass, can be converted into a particle-antiparticle pair. For example, a gamma ray with an energy of at least $(2m_e)c^2$ can be converted into an electron and a positron; that is, the energy of the gamma ray can be converted into the rest energy of the two particles. Thus we can see that the equivalence of mass and energy, as given by the relationship $E = mc^2$, has far-reaching implications.

Relativistic Kinetic Energy

When work is done on an isolated object, accelerating it from rest to a finite speed v, its total energy increases, as given by Equation 29-7 ($E = m_0 c^2 / \sqrt{1 - v^2/c^2}$). We refer to the *increase* in the object's energy as its *kinetic energy*. Thus the total energy, E, of an object is the sum of its rest energy, $m_0 c^2$, and its kinetic energy, K. In particular,

$$E = \frac{m_0 c^2}{\sqrt{1 - \dfrac{v^2}{c^2}}} = m_0 c^2 + K$$

Solving for the kinetic energy, we find:

Relativistic Kinetic Energy

$$K = \frac{m_0 c^2}{\sqrt{1 - \dfrac{v^2}{c^2}}} - m_0 c^2 \qquad\qquad 29\text{-}9$$

SI unit: J

As a check, notice that the kinetic energy is zero when the speed is zero, as expected. A comparison between the relativistic and classical kinetic energies is presented in **FIGURE 29-19**.

Although the kinetic-energy expression in Equation 29-9 looks nothing like the familiar classical kinetic energy, $\frac{1}{2}mv^2$, it does approach this value in the limit of small speeds. To see this, we can expand Equation 29-9 for small v using the binomial expansion given in Appendix A. The result is as follows:

$$K = \frac{m_0 c^2}{\sqrt{1 - \dfrac{v^2}{c^2}}} - m_0 c^2 = m_0 c^2 \left[\frac{1}{\sqrt{1 - \dfrac{v^2}{c^2}}} \right] - m_0 c^2$$

$$= m_0 c^2 \left[1 + \frac{1}{2}\left(\frac{v^2}{c^2}\right) + \frac{3}{8}\left(\frac{v^2}{c^2}\right)^2 + \cdots \right] - m_0 c^2$$

The second term in the square brackets is much smaller than the first term for everyday speeds. For example, if the speed v is equal to $0.00001c$ (a rather large everyday speed of roughly 6000 mi/h), the second term is only about a billionth of a percent of the first term. Later terms in the expansion are smaller still. Hence, for all practical purposes, the kinetic energy for low speeds is

$$K = m_0 c^2 \left[1 + \frac{1}{2}\left(\frac{v^2}{c^2}\right) \right] - m_0 c^2 = m_0 c^2 + \frac{1}{2}m_0 v^2 - m_0 c^2$$

Canceling the rest energy, $m_0 c^2$, we find

$$K = \frac{1}{2}m_0 v^2$$

Notice that the subscript 0 in this expression simply emphasizes the fact that the mass to be used is the rest mass.

We apply the relativistic kinetic energy in the next Example.

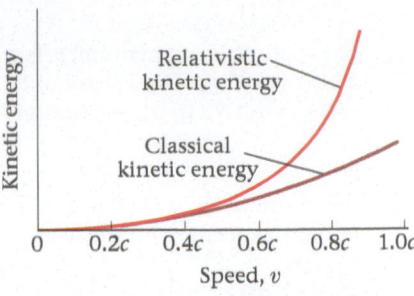

▲ **FIGURE 29-19 Relativistic and classical kinetic energies** The relativistic kinetic energy (upper curve) goes to infinity as the speed of light is approached. The classical kinetic energy (lower curve) agrees with the relativistic result when the speed is small compared with the speed of light.

PHYSICS IN CONTEXT
Looking Back

Kinetic energy $\left(K = \frac{1}{2}mv^2\right)$, introduced originally in Chapter 7, is shown in this chapter to be valid only for speeds that are small compared to the speed of light. The generalized kinetic energy, valid for any speed, is given in Equation 29-9.

EXAMPLE 29-15 RELATIVISTIC MOMENTUM AND KINETIC ENERGY

Scientists on a space station observe a fast-moving asteroid whose relativistic momentum is $1.77m_0 v$. The asteroid's rest mass is $m_0 = 9.12 \times 10^{17}$ kg. **(a)** What is the speed of the asteroid, v, expressed as a multiple of the speed of light? **(b)** What is the kinetic energy of the asteroid?

PICTURE THE PROBLEM
Our sketch shows the high-speed asteroid moving past the space station. The asteroid moves with a speed close to the speed of light, and hence its momentum is significantly greater than the classical value of $m_0 v$.

REASONING AND STRATEGY

a. We are given that the relativistic momentum of the asteroid is $p = 1.77m_0 v$. Since the relativistic momentum is given by $p = m_0 v / \sqrt{1 - v^2/c^2}$ (Equation 29-6), we can solve for the speed, v.

b. From part (a), we know that $p = 1.77m_0 v$, or equivalently that $1/\sqrt{1 - v^2/c^2} = 1.77$. We can find the relativistic kinetic energy by substituting this result directly into the expression $K = m_0 c^2 / \sqrt{1 - v^2/c^2} - m_0 c^2$ (Equation 29-9).

Known Momentum of asteroid, $p = 1.77\,m_0 v$; rest mass of asteroid, $m_0 = 9.12 \times 10^{17}$ kg.
Unknown (a) Speed of asteroid in terms of c, $v = ?$ (b) Kinetic energy of asteroid, $K = ?$

SOLUTION

Part (a)

1. Use relativistic momentum to solve for the speed, v:

$$p = \frac{m_0 v}{\sqrt{1 - \dfrac{v^2}{c^2}}}$$

$$v = c\sqrt{1 - \left(\frac{m_0 v}{p}\right)^2}$$

CONTINUED

2. The relativistic momentum, p, is greater than the classical momentum, $m_0 v$, by a factor of 1.77. Therefore, we substitute $m_0 v/p = 1/1.77$ in the expression for v from Step 1:

$$v = c\sqrt{1 - \left(\frac{1}{1.77}\right)^2} = 0.825c$$

Part (b)

3. Substitute $1/\sqrt{1 - v^2/c^2} = 1.77$ and $m_0 = 9.12 \times 10^{17}$ kg into the expression for the relativistic kinetic energy:

$$K = \frac{m_0 c^2}{\sqrt{1 - \dfrac{v^2}{c^2}}} - m_0 c^2 = m_0 c^2\left(\frac{1}{\sqrt{1 - \dfrac{v^2}{c^2}}} - 1\right) = m_0 c^2(1.77 - 1)$$

$$= (9.12 \times 10^{17}\ \text{kg})(3.00 \times 10^8\ \text{m/s})^2(0.77) = 6.3 \times 10^{34}\ \text{J}$$

INSIGHT

In comparison, the classical kinetic energy at this speed is $\frac{1}{2}m_0 v^2 = 2.79 \times 10^{34}$ J, considerably less than the relativistic value. In fact, the classical kinetic energy is *always* less than the relativistic kinetic energy at any given speed, as we can see in Figure 29-19.

It should be noted that typical asteroids in the solar system have speeds of roughly 30 km/s, or about $0.0001c$. The high speed in this Example illustrates the difference between classical and relativistic systems.

PRACTICE PROBLEM

How fast must the asteroid be moving if its relativistic kinetic energy is to be 7.14×10^{34} J?   [**Answer:** $v = 0.845c$]

Some related homework problems: Problem 52, Problem 63

Once again, we see that the speed of light is the ultimate speed possible for an object of finite rest mass. As is clear from Figure 29-19, the kinetic energy of an object goes to infinity as its speed approaches c. Thus to accelerate an object to the speed of light would require an infinite amount of energy. Any finite amount of work will increase the speed only to a speed less than c.

Enhance Your Understanding   (Answers given at the end of the chapter)

6. An object of mass m moves with speed v. Rank the following energies in order of increasing value: **(A)** relativistic energy; **(B)** relativistic kinetic energy; **(C)** classical kinetic energy.

Section Review

- The relativistic energy of an object with rest mass m_0 and speed v is $E = m_0 c^2/\sqrt{1 - v^2/c^2} = mc^2$.

- The rest energy of an object with rest mass m_0 is $E_0 = m_0 c^2$.

- The relativistic kinetic energy of an object with rest mass m_0 and speed v is $K = m_0 c^2/\sqrt{1 - v^2/c^2} - m_0 c^2$.

29-7 The Relativistic Universe

It is a profound understatement to say that relativity has revolutionized our understanding of the universe. If we think back over the results presented in the last several sections—time dilation, length contraction, mass–energy equivalence—it's clear that relativity reveals to us a universe that is far richer and more varied in its behavior than was ever imagined before. In fact, it is often said that the universe is not only stranger than we imagine, but also stranger than we *can* imagine.

By way of analogy, it is almost as if we had spent our lives on a small island on the equator. We would have no knowledge of snow or deserts or mountain ranges. Although our knowledge of Earth would be valid for our small island, we would have an incomplete picture of the world. Our situation with respect to relativity is similar. Before relativity, our knowledge of the physical universe seemed complete, with only minor details to be worked out. After all, Newton's laws and other fundamental principles of physics gave correct predictions for virtually everything we experienced. What Einstein revealed with his theory of relativity, however, was that we were seeing only a small part of the whole, and that the behavior we experience at low speeds cannot be extended to high speeds. We weren't missing small details—instead, like the mid-ocean islanders, we were missing most of the picture!

Now, it might seem that relativity plays no significant role in our daily lives, since we do not move at speeds approaching the speed of light. In many respects this observation is correct—relativity is not used to design better cars or airplanes, nor is it used to calculate the orbits needed to send astronauts to the Moon or spacecraft to Mars. On the other hand, every major hospital has a particle accelerator in its basement to produce radioactive elements for various types of treatments, and the accelerator brings particles to speeds very close to the speed of light—relativistic effects cannot be ignored in such cases. In addition, GPS technology can provide accurate positional information only by taking into account relativistic effects—both those having to do with speed, as we have seen in the previous sections, as well as those due to gravity, to be discussed in the next section. Thus, as these examples show, we now live in a world where relativity is truly an important part of our everyday lives.

Enhance Your Understanding   (Answers given at the end of the chapter)

7. If the speed of light were infinite, would the effects of relativity be greater than, less than, or the same as the effects we observe with a finite speed of light? Explain.

Section Review

- Relativity plays a significant role in everyday life.

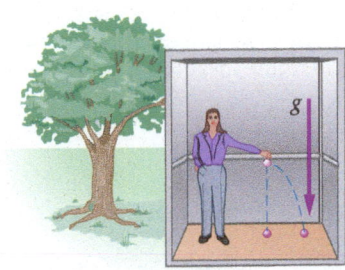

▲ **FIGURE 29-20 A frame of reference in a gravitational field** An observer in an elevator at rest on Earth's surface. If the observer drops or throws a ball, it falls with a downward acceleration of g.

29-8 General Relativity

Einstein's **general theory of relativity** applies to accelerated frames of reference and to gravitation. In fact, the theory provides a link between these two types of physical processes that leads to a new interpretation of gravity.

Consider two observers, both standing within closed elevators. Observer 1 is in an elevator that is at rest on the surface of the Earth, as **FIGURE 29-20** shows. If this observer drops or throws an object, it falls toward the floor of the elevator with an acceleration equal to the acceleration due to gravity.

Observer 2 stands in an identical elevator located in deep space. If this elevator is at rest, or moving with a constant velocity, the observer experiences weightlessness within the elevator, as shown in **FIGURE 29-21**. If an object is released, it remains in place. Now suppose the elevator is given an upward acceleration equal to the acceleration due to gravity, g, as indicated in **FIGURE 29-22**. An object that is released now remains at rest relative to the background stars while the floor of the elevator accelerates upward toward it with the acceleration g. Similarly, if observer 2 throws the ball horizontally it will follow a parabolic path to the floor, just as for observer 1. In addition, the floor of the elevator exerts a force mg on the feet of observer 1 to give that observer (whose mass is m) an upward acceleration g.

We conclude, then, that when observer 2 conducts an experiment in his accelerating elevator, the results are the same as those obtained by observer 1 in her elevator at rest on Earth. Einstein extended these observations to a general principle, the **principle of equivalence:**

Principle of Equivalence

All physical experiments conducted in a uniform gravitational field and in an accelerated frame of reference give identical results.

Thus the two observers cannot tell, without looking outside the elevator, whether they are at rest in a gravitational field or in deep space in an accelerating elevator.

Big Idea 6 An experiment conducted in an accelerating frame of reference gives the same results as it would when conducted in a gravitational field. Thus, there is no physical way to distinguish between a gravitational field and an accelerating frame of reference.

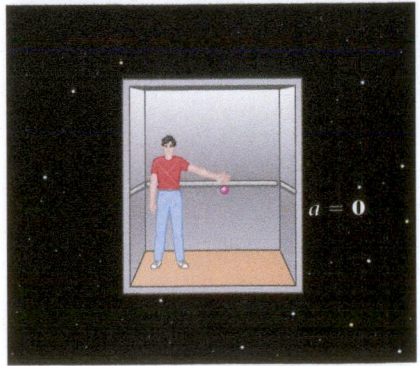

▲ **FIGURE 29-21 An inertial frame of reference with no gravitational field** An observer in an elevator in deep space experiences weightlessness. If the observer releases an object, it remains at rest.

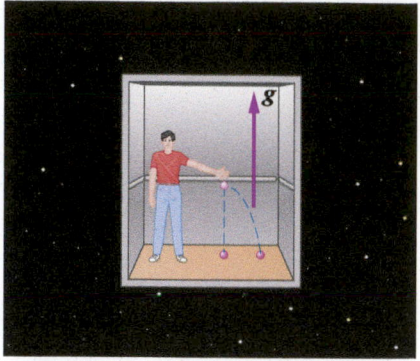

▲ **FIGURE 29-22 An accelerated frame of reference** If the elevator in Figure 29-21 is given an upward acceleration of magnitude *g*, the observer in the elevator will note that objects that are dropped or thrown fall toward the floor of the elevator with an acceleration *g*, just as for the observer in Figure 29-20.

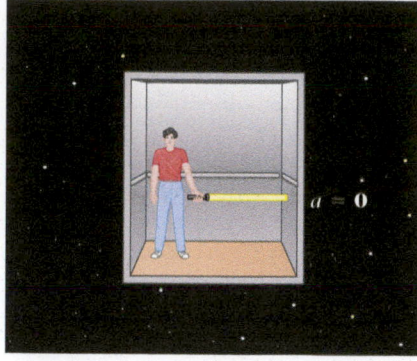

(a) Nonaccelerated elevator

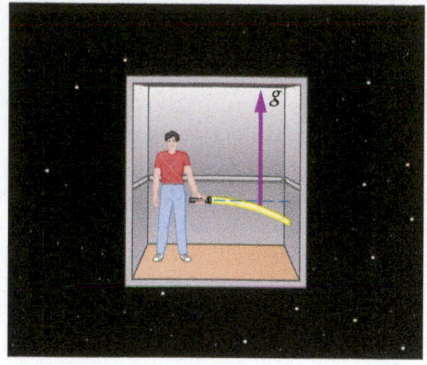

(b) Accelerated elevator

▲ **FIGURE 29-23 A light experiment in two different frames of reference (a)** In a nonaccelerating elevator, a beam of light travels on a straight line as it crosses the elevator. **(b)** In an accelerated elevator, the elevator moves upward as the light crosses the elevator; hence, the light strikes the opposite wall at a lower level. The path of the light in this case appears parabolic to the observer riding in the elevator.

▲ **FIGURE 29-24 The principle of equivalence** By the principle of equivalence, a beam of light in a gravitational field should follow a parabolic path, just as in the accelerated elevator in Figure 29-23 (b). The amount of bending of the light's path has been exaggerated here for clarity.

The Effect of Gravity on Light Now, let's apply the principle of equivalence to a simple experiment involving light. If the observer in **FIGURE 29-23 (a)** shines a flashlight toward the opposite wall of the elevator (which has zero acceleration), the light strikes the wall at its initial height. If the same experiment is conducted in an accelerated elevator, as in **FIGURE 29-23 (b)**, the elevator is accelerating upward during the time the light travels across the elevator. Thus, by the time the light reaches the far wall, it strikes it at a lower level. In fact, the light has followed a parabolic path, just as one would expect for a ball that was projected horizontally.

Applying the principle of equivalence, Einstein concluded that a beam of light in a gravitational field must also bend downward, just as it does in an accelerated elevator; that is, *gravity bends light*. This phenomenon is illustrated in **FIGURE 29-24**, where the amount of bending has been exaggerated for clarity.

In order to put Einstein's prediction to the test, it is necessary to increase the amount of bending as much as possible to make it large enough to be measured. Thus, we need to use the strongest gravitational field available. In our solar system, the strongest gravitational field is provided by the Sun; hence, experiments were planned to look for the bending of light produced by the Sun.

To see what effect the Sun's gravitational field might have on light, consider the Sun and a ray of light from a distant star, as shown in **FIGURE 29-25**. As the light passes the Sun, it is bent, as indicated. It follows that an observer on Earth must look in a direction that is farther from the Sun than the actual direction of the distant star—the Sun's gravitational field displaces the distant stars to apparent positions farther from the Sun. If we imagine the Sun moving in front of a background field of stars, as in **FIGURE 29-26**, the stars near the Sun are displaced outward. It is almost as if the Sun were a lens, slightly distorting the scene behind it.

Because the Sun is so bright, it is possible to carry out an experiment like the one shown in Figures 29-25 and 29-26 only during a total eclipse of the Sun, when the Sun's light is blocked by the Moon. During the eclipse, photographs can be taken to show the positions of the background stars. Later, these photographs can be compared with photographs of the same star field taken 6 months later, when the Sun is on the other side of Earth. Comparing the photographs allows one to measure the displacement of the stars. This experiment was carried out during an expedition to Africa in 1919 by Sir Arthur Eddington. His results confirmed the predictions of the general theory of relativity and made Einstein a household name.

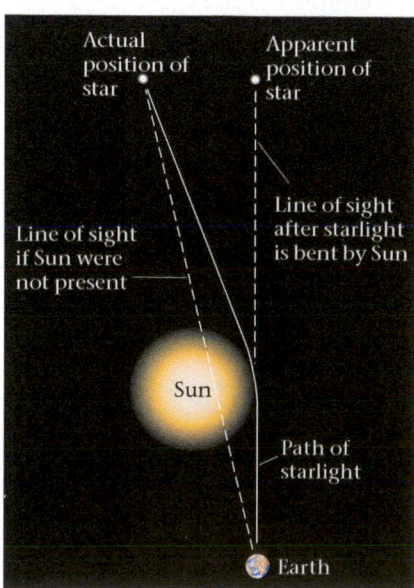

◀ **FIGURE 29-25 Gravitational bending of light** As light from a distant star passes close to the Sun, it is bent. The result is that an observer on Earth sees the star along a line of sight that is displaced away from the center of the Sun.

▲ **FIGURE 29-26 Bending of light near the Sun** As the Sun moves across a starry background, the stars near it appear to be displaced outward, away from the center of the Sun. This is the effect that was used in the first experimental confirmation of general relativity.

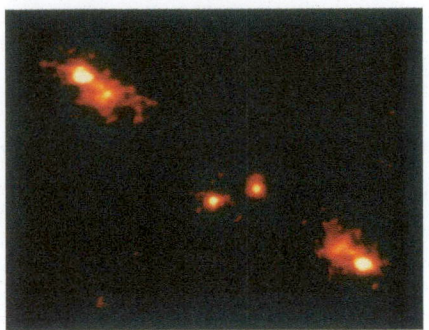

◄ **FIGURE 29-27 Gravitational lensing** The images created by gravitational lenses take a number of forms. Sometimes the light from a distant object is stretched out into an arc or even a complete ring. In other cases, such as that of the distant quasar at left, we may see a pair of images. (The quasar appears at the upper left and lower right. The lensing galaxy is not visible in the photo—the small dots at center are unrelated objects.) In still other instances, four images may be produced, as in the famous "Einstein cross" shown at right. The lensing galaxy at the center is some 400 million light-years from us, while the quasar whose multiple images surround it is about 20 times farther away.

RWP Because gravity can bend light, the more powerful the gravitational force, the more the bending, and the more dramatic the results. In Figure 12-29 we saw what can happen when a large galaxy or a cluster of galaxies, with its immense gravitational field, lies between Earth and a more distant galaxy. The intermediate galaxy can produce significant bending of light, resulting in multiple images of the distant galaxy that can form arcs or crosses. Examples of such *gravitational lensing* are shown in **FIGURE 29-27**.

RWP An intense gravitational field can also be produced when a star burns up its nuclear fuel and collapses to a very small size. In such a case, the gravitational field can become strong enough to actually trap light—that is, to bend it around to the point that it cannot escape the star. Because such a star cannot emit light of its own, it is referred to as a **black hole.** Black holes, by definition, cannot be directly observed; however, their presence can be inferred by their gravitational effects on other bodies. It is also possible to detect the intense radiation emitted by ionized matter as it falls into a black hole. (Recall from Section 25-1 that accelerated charges produce electromagnetic radiation.) By these and other means, the existence of black holes in the centers of many galaxies has been firmly established. In fact, it is now thought that black holes may be relatively common in the universe. Two examples of the effects black holes can have on their surroundings are given in **FIGURE 29-28**.

Recall that in Chapter 12 we calculated the escape speed for Earth. The result was

$$v_e = \sqrt{\frac{2GM_E}{R_E}}$$

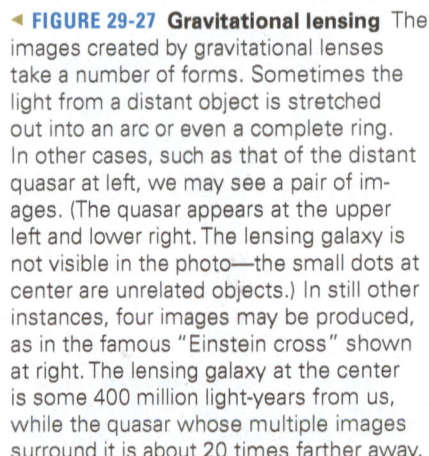

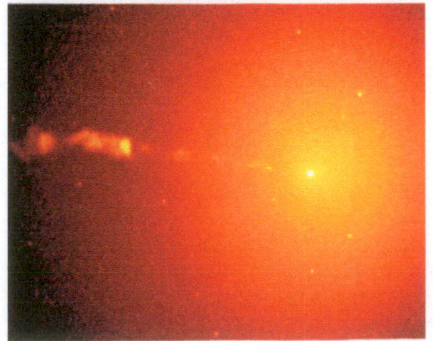

▶ **FIGURE 29-28 Visualizing Concepts Black holes** Although black holes are invisible, they can be detected indirectly. If one member of a binary star collapses to become a black hole, an accretion disk may form around it, as shown in the artist's conception at top. The accretion disk is a ring of matter wrenched from its companion, whirling around at ever greater speeds as it spirals into the black hole. The radiation emitted by this matter is the signature of a black hole. The same mechanism, on a much vaster scale, probably accounts for the enormous radiation from active galaxies, such as M87 (bottom), a giant elliptical galaxy with an enormous jet of matter emanating from its nucleus.

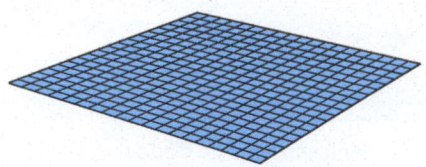

(a) Flat space, away from massive objects

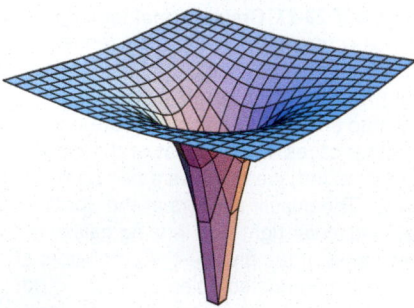

(b) Warped space, near a massive object

▲ **FIGURE 29-29 Warped space and black holes (a)** Regions of space that are far from any large masses can be thought of as flat. In these regions, light propagates in straight lines. **(b)** Near a large concentrated mass, such as a black hole, space can be thought of as "warped." In these regions, the paths of light rays are bent.

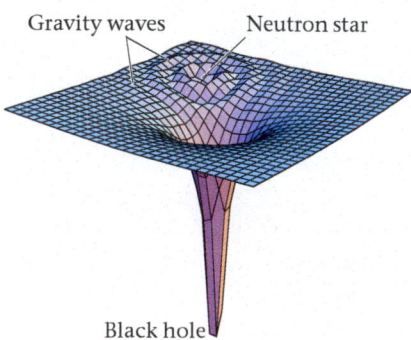

Gravity waves   Neutron star

Black hole

▲ **FIGURE 29-30 Gravity waves** Gravity waves can be thought of as "ripples" in the warped space described in Figure 29-29. In the case illustrated here, a neutron star orbits a black hole. As a result of its acceleration, the neutron star emits gravity waves.

Replacing M_E with M and R_E with R gives the escape speed for any spherical body of mass M and radius R. Now, if we set the escape speed equal to the speed of light, $v_e = c$, we find the following:

$$v_e = \sqrt{\frac{2GM}{R}} = c$$

Solving for R, we find that for an astronomical body to be a black hole, its radius must be no greater than

$$R = \frac{2GM}{c^2} \qquad \text{29-10}$$

Although this calculation is too simplistic and not entirely correct, the final result given in Equation 29-10, known as the **Schwarzschild radius,** does indeed agree with the results of general relativity. In the following Exercise, we calculate the radius of a black hole for a specific case.

EXERCISE 29-16 BLACK HOLE

To what radius must Earth be condensed for it to become a black hole?

REASONING AND SOLUTION

Substituting G, M_E, and c in Equation 29-10 ($R = 2GM/c^2$), we find the following radius:

$$R = \frac{2GM_E}{c^2} = \frac{2(6.67 \times 10^{-11}\,\text{N} \cdot \text{m}^2/\text{kg}^2)(5.98 \times 10^{24}\,\text{kg})}{(3.00 \times 10^8\,\text{m/s})^2} = 8.86\,\text{mm}$$

Thus Earth would have to be compressed to roughly the size of a walnut for it to become a black hole.

Gravity Waves A convenient way to visualize the effects of intense gravitational fields is to think of space as a sheet of rubber with a square array of grid lines, as shown in **FIGURE 29-29**. In the absence of mass, the sheet is flat and the grid lines are straight, as in Figure 29-29 (a). In this case, a beam of light follows a straight-line path, parallel to a grid line. If a large mass is present, however, the sheet of rubber is deformed, as in Figure 29-29 (b), and light rays follow the curved paths of the grid lines. In cases where one large mass orbits another, as in **FIGURE 29-30**, the result is a series of ripples moving outward through the rubber sheet. These ripples represent **gravity waves,** one of the many intriguing predictions of general relativity.

When a gravity wave passes through a given region of space, it causes a local deformation of space, as Figure 29-30 suggests. Early attempts to detect gravity waves were based on measuring the distortion a gravity wave would produce in a large metal bar. Unfortunately, the sensitivity of these devices was too low to detect the weak waves that are thought to pass through the Earth all the time. Ironically, these detectors *were* sensitive enough to detect the relatively strong gravity waves that must have accompanied the 1987 supernova explosion in the nearby Large Magellanic Cloud (a small satellite galaxy of our Milky Way). As luck would have it, however, none of the detectors were operating at the time.

The next generation of gravity wave detectors have now been built. These detectors, which go by the name of Laser Interferometer Gravitational Wave Observatory, or LIGO for short, should be sensitive enough (once upgrades are completed) to detect several gravitational wave events per year. One type of event that LIGO will be looking for is the final death spiral of a neutron star as it plunges into a black hole. The neutron star might orbit the black hole for 150,000 years, but only in the last 15 minutes of its life does it find fame, because during those few minutes its acceleration is great enough to produce gravity waves detectable by LIGO.

The basic operating principle of LIGO is to send a laser beam in two different directions along 4-km-long vacuum tubes. At the far end of each tube the beam is reflected back to its starting point. If the two beams travel equal distances, they will interfere constructively when reunited, just like the two rays that meet at the center of a two-slit interference pattern, as discussed in Section 28-2. If the path lengths along the two tunnels are slightly different, however, the beams will have at least partial destructive interference when they combine. The resulting difference in intensity is what LIGO will measure.

▶ **FIGURE 29-31 How LIGO detects a gravity wave** When a gravity wave passes through the Earth, it "distorts" everything in its path. The idea behind LIGO is to detect these distortions using interference effects similar to those observed in Young's two-slit experiment (Chapter 28). In particular, LIGO sends laser light along two 4-km vacuum tubes oriented at right angles to one another. When a gravity wave passes, one arm will be lengthened and the other arm will be shortened. The resulting difference in the path lengths changes the interference pattern, allowing us to "see" the gravity wave.

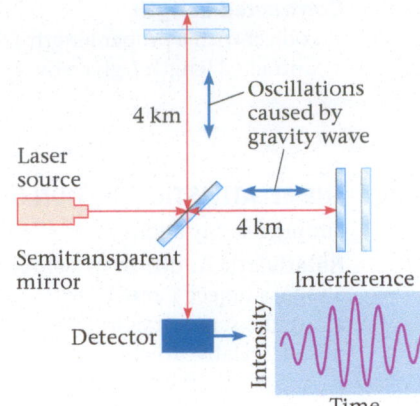

The connection with gravity waves is indicated in **FIGURE 29-31**. Here we see that as a gravity wave passes a LIGO facility, it will cause one tube to increase in length and the other tube to decrease in length. Hence, if the observatory can detect a small enough change in length, it will "see" the gravity wave. Just how much change in length is a gravity wave expected to produce? Incredibly, a typical strong gravity wave will change the length of a tube by less than the diameter of an atomic nucleus. Thus, the task for LIGO is to measure this small change in length in a tube that is 4 km long—no easy task, but the LIGO scientists are confident it can be done.

Enhance Your Understanding   (Answers given at the end of the chapter)

8. When the Sun passes near the line of sight to a distant star, does the position of the star in the sky shift away from the Sun, closer to the Sun, or stay the same? Explain.

Section Review

- The principle of equivalence states that the behavior of a physical system in a gravitational field is identical to its behavior in an accelerated frame of reference.

- The radius of a black hole of mass M is $R = 2GM/c^2$.

CHAPTER 29 REVIEW

CHAPTER SUMMARY

29-1 THE POSTULATES OF SPECIAL RELATIVITY

Einstein's theory of relativity is based on just two postulates.

Equivalence of Physical Laws
The laws of physics are the same in all inertial frames of reference.

Constancy of the Speed of Light
The speed of light in a vacuum, $c = 3.00 \times 10^8$ m/s, is the same in all inertial frames of reference, independent of the motion of the source or the receiver.

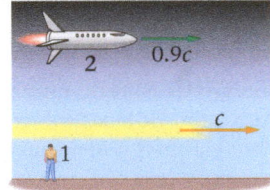

29-2 THE RELATIVITY OF TIME AND TIME DILATION

Clocks that move relative to one another keep time at different rates. In particular, a moving clock runs slower than one that is at rest relative to a given observer.

Proper Time
The proper time, Δt_0, is the amount of time separating two events that occur at the same location.

Time Dilation
If two events separated by a proper time Δt_0 occur in a frame of reference moving with a speed v relative to an observer, the dilated time measured by the observer, Δt, is given by

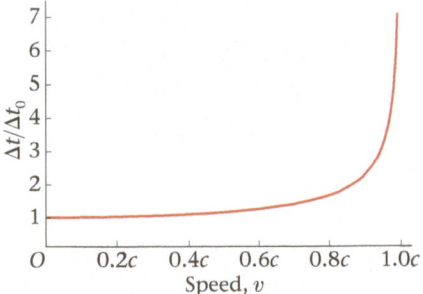

$$\Delta t = \frac{\Delta t_0}{\sqrt{1 - v^2/c^2}} \qquad 29\text{-}2$$

29-3 THE RELATIVITY OF LENGTH AND LENGTH CONTRACTION

The length of an object depends on its speed relative to a given observer.

Proper Length
The proper length, L_0, is the distance between two points as measured by an observer who is at rest with respect to them.

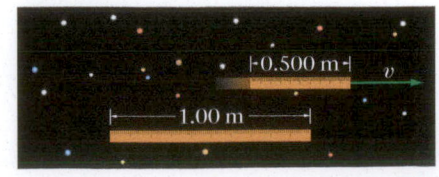

Contracted Length

An object with a proper length L_0 moving with a speed v relative to an observer has a contracted length L given by

$$L = L_0\sqrt{1 - \frac{v^2}{c^2}}$$

29-3

29-4 THE RELATIVISTIC ADDITION OF VELOCITIES

Simple velocity addition, $v = v_1 + v_2$, is valid only in the limit of very small speeds.

Relativistic Addition of Velocities

Suppose object 1 moves with a velocity v_{13} relative to object 3. If object 2 moves along the same straight line with a velocity v_{21} relative to object 1, the velocity of object 2 relative to object 3, v_{23}, is

$$v_{23} = \frac{v_{21} + v_{13}}{1 + \frac{v_{21}v_{13}}{c^2}}$$

29-4

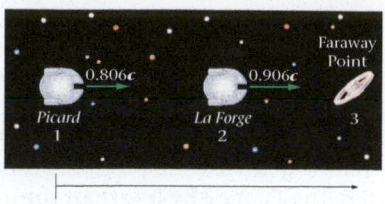

29-5 RELATIVISITIC MOMENTUM

The momentum of an object of mass m and speed v is

$$p = \frac{mv}{\sqrt{1 - \frac{v^2}{c^2}}}$$

29-5

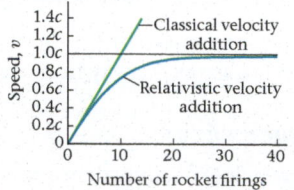

29-6 RELATIVISITIC ENERGY AND $E = mc^2$

One of the most important results of relativity is that mass is another form of energy.

Relativistic Energy

The total energy, E, of an object with rest mass m_0 and speed v is

$$E = \frac{m_0c^2}{\sqrt{1 - \frac{v^2}{c^2}}} = mc^2$$

29-7

Rest Energy

When an object is at rest, its energy E_0 is

$$E_0 = m_0c^2$$

29-8

Relativistic Kinetic Energy

The relativistic kinetic energy, K, of an object of rest mass m_0 moving with a speed v is its total energy, E, minus its rest energy, E_0. In particular,

$$K = \frac{m_0c^2}{\sqrt{1 - \frac{v^2}{c^2}}} - m_0c^2$$

29-9

29-7 THE RELATIVISITIC UNIVERSE

Relativity is important in understanding the universe, and plays a fundamental role in technologies like GPS.

29-8 GENERAL RELATIVITY

General relativity deals with accelerated frames of reference and with gravity.

Principle of Equivalence

One of the basic principles on which general relativity is founded is the following: All physical experiments conducted in a gravitational field and in an accelerated frame of reference give identical results.

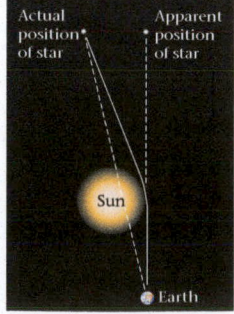

ANSWERS TO ENHANCE YOUR UNDERSTANDING QUESTIONS

1. Observer 2 also measures the speed of light to be c.

2. The clock at the factory is at rest, and hence it runs faster than the clock that has been moving. As a result, the clock in the factory reads a time that is after 9:00 A.M.

3. The moving meterstick is less than 1 m long, regardless of whether it moves to the right or to the left.

4. The correct speed of the passenger relative to the ground is less than 10 m/s, because relativistic velocity addition always gives a reduced result so that the final speed can never exceed the speed of light.

5. The relativistic momentum is greater than mv (for speeds greater than zero), because the denominator in the expres-

sion for the relativistic momentum is less than 1. In fact, the relativistic momentum increases to infinity as the speed approaches c.

6. $C < B < A$.

7. If the speed of light were infinite, the effects of relativity would be less than they are now. In fact, the effects of relativity would be nonexistent.

8. The position of the star in the sky shifts away from the Sun, because the Sun bends the light from the star, as shown in Figure 29-25.

CONCEPTUAL QUESTIONS

For instructor-assigned homework, go to www.masteringphysics.com.

(Answers to odd-numbered Conceptual Questions can be found in the back of the book.)

1. Some distant galaxies are moving away from us at speeds greater than $0.5c$. What is the speed of the light received on Earth from these galaxies? Explain.

2. Describe some of the everyday consequences that would follow if the speed of light were 35 mi/h.

3. When we view a distant galaxy, we notice that the light coming from it has a longer wavelength (it is "red-shifted") than the corresponding light here on Earth. Is this consistent with the

postulate that all observers measure the same speed of light? Explain.

4. According to the theory of relativity, the maximum speed for any particle is the speed of light. Is there a similar restriction on the maximum energy of a particle? Is there a maximum momentum? Explain.

5. Give an argument that shows that an object of finite mass cannot be accelerated from rest to a speed greater than the speed of light in a vacuum.

PROBLEMS AND CONCEPTUAL EXERCISES

Answers to odd-numbered Problems and Conceptual Exercises can be found in the back of the book. **BIO** *identifies problems of biological or medical interest;* **CE** *indicates a conceptual exercise.* **Predict/Explain** *problems ask for two responses: (a) your prediction of a physical outcome, and (b) the best explanation among three provided; and* **Predict/Calculate** *problems ask for a prediction of a physical outcome, based on fundamental physics concepts, and follow that with a numerical calculation to verify the prediction. On all problems, bullets (•, ••, •••) indicate the level of difficulty.*

SECTION 29-1   THE POSTULATES OF SPECIAL RELATIVITY

1. • **CE Predict/Explain** You are in a spaceship, traveling directly away from the Moon with a speed of $0.9c$. A light signal is sent in your direction from the surface of the Moon. (a) As the signal passes your ship, do you measure its speed to be greater than, less than, or equal to $0.1c$? (b) Choose the *best explanation* from among the following:

   I. The speed you measure will be greater than $0.1c$; in fact, it will be c, since all observers in inertial frames measure the same speed of light.

   II. You will measure a speed less than $0.1c$ because of time dilation, which causes clocks to run slow.

   III. When you measure the speed you will find it to be $0.1c$, which is the difference between c and $0.9c$.

2. • Albert is piloting his spaceship, heading east with a speed of $0.95c$ relative to Earth. Albert's ship sends a light beam in the forward (eastward) direction, which travels away from his ship at a speed c. Meanwhile, Isaac is piloting his ship in the westward direction, also at $0.95c$, toward Albert's ship. With what speed does Isaac see Albert's light beam pass his ship?

SECTION 29-2   THE RELATIVITY OF TIME AND TIME DILATION

3. • **CE** A street performer tosses a ball straight up into the air (event 1) and then catches it in his mouth (event 2). For each of the following observers, state whether the time they measure between

these two events is the proper time or the dilated time: **(a)** the street performer; **(b)** a stationary observer on the other side of the street; **(c)** a person sitting at home watching the performance on TV; **(d)** a person observing the performance from a moving car.

4. • **CE Predict/Explain** A clock in a moving rocket is observed to run slow. **(a)** If the rocket reverses direction, does the clock run slow, fast, or at its normal rate? **(b)** Choose the *best explanation* from among the following:

   I. The clock will run slow, just as before. The rate of the clock depends only on relative speed, not on direction of motion.

   II. When the rocket reverses direction the rate of the clock reverses too, and this makes it run fast.

   III. Reversing the direction of the rocket undoes the time dilation effect, and so the clock will now run at its normal rate.

5. • **CE Predict/Explain** Suppose you are a traveling salesman for SSC, the Spacely Sprockets Company. You travel on a spaceship that reaches speeds near the speed of light, and you are paid by the hour. **(a)** When you return to Earth after a sales trip, would you prefer to be paid according to the clock at Spacely Sprockets universal headquarters on Earth, according to the clock on the spaceship in which you travel, or would your pay be the same in either case? **(b)** Choose the *best explanation* from among the following:

   I. You want to be paid according to the clock on Earth, because the clock on the spaceship runs slow when it approaches the speed of light.

II. Collect your pay according to the clock on the spaceship because according to you the clock on Earth has run slow.

III. Your pay would be the same in either case because motion is relative, and all inertial observers will agree on the amount of time that has elapsed.

6. • A neon sign in front of a café flashes on and off once every 3.5 s, as measured by the head cook. How much time elapses between flashes of the sign as measured by an astronaut in a spaceship moving toward Earth with a speed of $0.88c$?

7. • A lighthouse sweeps its beam of light around in a circle once every 8.5 s. To an observer in a spaceship moving away from Earth, the beam of light completes one full circle every 16 s. What is the speed of the spaceship relative to Earth?

8. • As a spaceship flies past with speed v, you observe that 1.0000 s elapses on the ship's clock in the same time that 1.0000 min elapses on Earth. How fast is the ship traveling, relative to the Earth? (Express your answer as a fraction of the speed of light.)

9. • How fast should your spacecraft travel so that clocks on board will advance 10.0 times slower than clocks on Earth?

10. • Usain Bolt set a world record for the 100-m dash on August 16, 2009. If observers on a spaceship moving with a speed of $0.9250c$ relative to Earth saw Usain Bolt's run and measured his time to be 25.21 s, find the time that was recorded on Earth.

11. •• (a) Find the average distance (in the Earth's frame of reference) covered by the muons in Example 29-5 if their speed relative to Earth is $0.850c$. (b) What is the average lifetime of these muons?

12. •• Referring to Example 29-3, (a) how much does Benny age if he travels to Vega with a speed of $0.9995c$? (b) How much time is required for the trip according to Jenny?

13. •• **The Pi Meson** An elementary particle called a pi meson (or pion for short) has an average lifetime of 2.6×10^{-8} s when at rest. If a pion moves with a speed of $0.89c$ relative to Earth, find (a) the average lifetime of the pion as measured by an observer on Earth and (b) the average distance traveled by the pion as measured by the same observer. (c) How far would the pion have traveled relative to Earth if relativistic time dilation did not occur?

14. •• **Predict/Calculate** (a) Is it possible for you to travel far enough and fast enough so that when you return from a trip, you are younger than your stay-at-home sister, who was born 5.0 y after you? (b) Suppose you fly on a rocket with a speed $v = 0.99c$ for 1 y, according to the ship's clocks and calendars. How much time elapses on Earth during your 1-y trip? (c) If you were 22 y old when you left home and your sister was 17, what are your ages when you return?

15. •• In order to cross the galaxy quickly, a spaceship leaves Earth traveling at $0.9999995c$. After 15 minutes a radio message is sent from Earth to the spacecraft. (a) In the Earth–galaxy frame of reference, how far from Earth is the spaceship when the message is sent? (b) How much time elapses on spaceship clocks between when the ship leaves Earth and when the message is sent? (c) How much time elapses on Earth clocks between when the message is sent and when it is received? (d) If the spaceship replies instantly with another radio message, what total time elapses on Earth clocks between when the message is sent and when the reply is received?

16. •• An observer moving toward Earth with a speed of $0.85c$ notices that it takes 5.0 min for a person on Earth to fill her car with gas. Suppose, instead, that the observer had been moving away from Earth with a speed of $0.90c$. How much time would the observer have measured for the car to be filled in this case?

17. •• **Predict/Calculate** An astronaut moving with a speed of $0.75c$ relative to Earth measures her heart rate to be 71 beats per minute.

(a) When an Earth-based observer measures the astronaut's heart rate, is the result greater than, less than, or equal to 71 beats per minute? Explain. (b) Calculate the astronaut's heart rate as measured on Earth.

18. •• **BIO** Newly sprouted sunflowers can grow at the rate of 0.30 in. per day. One such sunflower is left on Earth, and an identical one is placed on a spacecraft that is traveling away from Earth with a speed of $0.97c$. How tall is the sunflower on the spacecraft when a person on Earth says his is 2.0 in. high?

19. •• As measured in Earth's frame of reference, the Earth and the Moon are 384,000 km apart. A spaceship flies past the Earth and the Moon with a constant velocity, and the clocks on the ship show that the trip lasts only 1.00 s. How fast is the ship traveling?

20. •• Captain Jean-Luc is piloting the USS *Enterprise XXIII* at a constant speed $v = 0.825c$. As the *Enterprise* passes the planet Vulcan, he notices that his watch and the Vulcan clocks both read 1:00 P.M. At 3:00 P.M., according to his watch, the *Enterprise* passes the planet Endor. If the Vulcan and Endor clocks are synchronized with each other, what time do the Endor clocks read when the *Enterprise* passes by?

21. •• **Predict/Calculate** A plane flies with a constant velocity of 222 m/s. The clocks on the plane show that it takes exactly 2.00 h to travel a certain distance. (a) According to ground-based clocks, will the flight take slightly more or slightly less than 2.00 h? (b) Calculate how much longer or shorter than 2.00 h this flight will last, according to clocks on the ground.

SECTION 29-3 THE RELATIVITY OF LENGTH AND LENGTH CONTRACTION

22. • How fast does a 275-m spaceship move relative to an observer who measures the ship's length to be 155 m?

23. • Suppose the speed of light in a vacuum were only 25.0 mi/h. Find the length of a bicycle being ridden at a speed of 20.0 mi/h as measured by an observer sitting on a park bench, given that its proper length is 1.89 m.

24. • A rectangular painting is $W = 117$ cm wide and $H = 76.2$ cm high, as indicated in **FIGURE 29-32**. At what speed, v, must the painting move parallel to its width if it is to appear to be square?

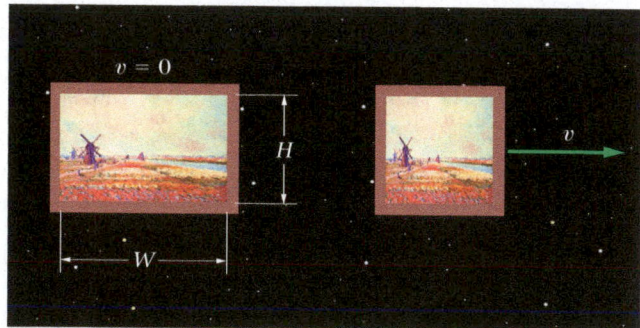

FIGURE 29-32 Problem 24

25. • The particle accelerator at CERN in Geneva, Switzerland, has a circumference of 27 km. Inside it protons are accelerated to a speed of $0.999999972c$. What is the circumference of the accelerator in the frame of reference of the protons?

26. •• A cubical box is 0.75 m on a side. (a) What are the dimensions of the box as measured by an observer moving with a speed of $0.82c$ parallel to one of the edges of the box? (b) What is the volume of the box, as measured by this observer?

27. •• When parked, your car is 5.8 m long. Unfortunately, your garage is only 3.7 m long. (a) How fast would your car have to be moving

for an observer on the ground to find your car shorter than your garage? **(b)** When you are driving at this speed, how long is your garage, as measured in the car's frame of reference?

28. •• An astronaut travels to a distant star with a speed of $0.75c$ relative to Earth. From the astronaut's point of view, the star is 7.5 ly from Earth. On the return trip, the astronaut travels with a speed of $0.89c$ relative to Earth. What is the distance covered on the return trip, as measured by the astronaut? Give your answer in light-years.

29. •• **Predict/Calculate** Laboratory measurements show that an electron traveled 3.50 cm in a time of 0.200 ns. **(a)** In the rest frame of the electron, did the lab travel a distance greater than or less than 3.50 cm? Explain. **(b)** What is the electron's speed? **(c)** In the electron's frame of reference, how far did the laboratory travel?

30. •• You and a friend travel through space in identical spaceships. Your friend informs you that he has made some length measurements and that his ship is 150 m long but that yours is only 110 m long. From your point of view, **(a)** how long is your friend's ship, **(b)** how long is your ship, and **(c)** what is the speed of your friend's ship relative to yours?

31. •• A ladder 5.0 m long leans against a wall inside a spaceship. From the point of view of a person on the ship, the base of the ladder is 3.0 m from the wall, and the top of the ladder is 4.0 m above the floor. The spaceship moves past the Earth with a speed of $0.90c$ in a direction parallel to the floor of the ship. Find the angle the ladder makes with the floor as seen by an observer on Earth.

32. ••• When traveling past an observer with a relative speed v, a rocket is measured to be 9.00 m long. When the rocket moves with a relative speed $2v$, its length is measured to be 5.00 m. **(a)** What is the speed v? **(b)** What is the proper length of the rocket?

33. ••• **Predict/Calculate** The starships *Picard* and *La Forge* are traveling in the same direction toward the Andromeda galaxy. The *Picard* moves with a speed of $0.90c$ relative to the *La Forge*. A person on the *La Forge* measures the length of the two ships and finds the same value. **(a)** If a person on the *Picard* also measures the lengths of the two ships, which of the following is observed: (i) the *Picard* is longer; (ii) the *La Forge* is longer; or (iii) both ships have the same length? Explain. **(b)** Calculate the ratio of the proper length of the *Picard* to the proper length of the *La Forge*.

SECTION 29-4   THE RELATIVISTIC ADDITION OF VELOCITIES

34. • A spaceship moving toward Earth with a speed of $0.85c$ launches a probe in the forward direction with a speed of $0.15c$ relative to the ship. Find the speed of the probe relative to Earth.

35. • Suppose the probe in Problem 34 is launched in the opposite direction to the motion of the spaceship. Find the speed of the probe relative to Earth in this case.

36. • Suppose the speed of light is 35 mi/h. A paper girl riding a bicycle at 22 mi/h throws a rolled-up newspaper in the forward direction, as shown in **FIGURE 29-33**. If the paper is thrown with a speed of 19 mi/h relative to the bike, what is its speed, v, with respect to the ground?

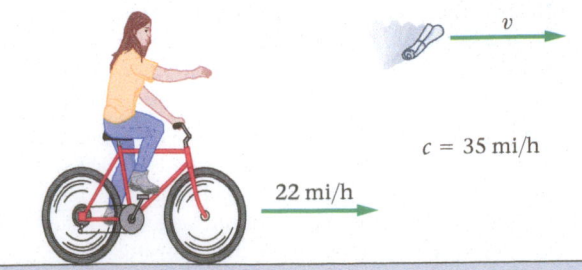

FIGURE 29-33 Problem 36

37. •• Two asteroids head straight for Earth from the same direction. Their speeds relative to Earth are $0.80c$ for asteroid 1 and $0.60c$ for asteroid 2. Find the speed of asteroid 1 relative to asteroid 2.

38. •• Two rocket ships approach Earth from opposite directions, each with a speed of $0.85c$ relative to Earth. What is the speed of one ship relative to the other?

39. •• A spaceship and an asteroid are moving in the same direction away from Earth with speeds of $0.77c$ and $0.41c$, respectively. What is the relative speed between the spaceship and the asteroid?

40. •• An electron moves to the right in a laboratory accelerator with a speed of $0.84c$. A second electron in a different accelerator moves to the left with a speed of $0.43c$ relative to the first electron. Find the speed of the second electron relative to the lab.

41. •• A uranium nucleus is traveling at $0.95c$ in the positive direction relative to the laboratory when it suddenly splits into two pieces. Piece A is propelled in the forward direction with a speed of $0.45c$ relative to the original nucleus. Piece B is sent backward at $0.33c$ relative to the original nucleus. Find the velocities of pieces A and B as measured by an observer in the laboratory.

42. •• **Predict/Calculate** Two rocket ships are racing toward Earth, as shown in **FIGURE 29-34**. Ship A is in the lead, approaching the Earth at $0.80c$ and separating from ship B with a relative speed of $0.50c$. **(a)** As seen from Earth, what is the speed, v, of ship B? **(b)** If ship A increases its speed by $0.10c$ relative to the Earth, does the relative speed between ship A and ship B increase by $0.10c$, by more than $0.10c$, or by less than $0.10c$? Explain. **(c)** Find the relative speed between ships A and B for the situation described in part (b).

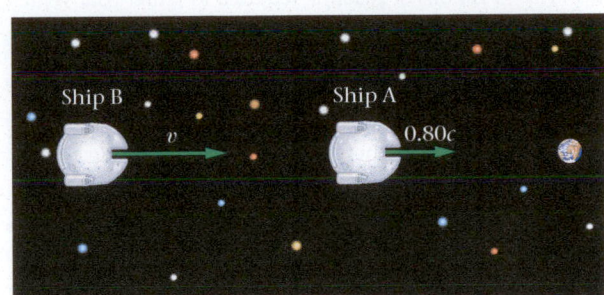

FIGURE 29-34 Problem 42

SECTION 29-5   RELATIVISTIC MOMENTUM

43. • A 2.8×10^6-kg spaceship moves away from Earth with a speed of $0.86c$. What is the magnitude of the ship's **(a)** classical and **(b)** relativistic momentum?

44. • An asteroid with a mass of 8.2×10^{11} kg is observed to have a relativistic momentum of magnitude 7.74×10^{20} kg·m/s. What is the speed of the asteroid relative to the observer?

45. •• An object has a relativistic momentum that is 8.50 times greater than its classical momentum. What is its speed?

46. •• A football player with a mass of 88 kg and a speed of 2.0 m/s collides head-on with a player from the opposing team whose mass is 120 kg. The players stick together and are at rest after the collision. Find the speed of the second player, assuming the speed of light is 3.0 m/s.

47. •• A space probe with a rest mass of 8.2×10^7 kg and a speed of $0.50c$ smashes into an asteroid at rest and becomes embedded within it. If the speed of the probe–asteroid system is $0.26c$ after the collision, what is the rest mass of the asteroid?

48. •• At what speed does the classical momentum, $p = mv$, give an error, when compared with the relativistic momentum, of **(a)** 1.00% and **(b)** 5.00%?

49. ••• A proton has 1836 times the rest mass of an electron. At what speed will an electron have the same momentum as a proton moving at $0.0100c$?

50. ••• Star A has a mass of 3.0×10^{28} kg and is traveling in the positive direction at $0.35c$ relative to the galaxy. Star B has mass 2.0×10^{29} kg and is traveling in the negative direction at $0.15c$. The two stars collide and stick together. What is the velocity of the resulting two-star system after the collision?

SECTION 29-6 RELATIVISTIC ENERGY AND $E = mc^2$

51. • **CE** Particles A through D have the following rest energies and total energies:

Particle	Rest Energy	Total Energy
A	$6E$	$6E$
B	$2E$	$4E$
C	$4E$	$6E$
D	$3E$	$4E$

Rank these particles in order of increasing **(a)** rest mass, **(b)** kinetic energy, and **(c)** speed. Indicate ties where appropriate.

52. • Find the work that must be done on a proton to accelerate it from rest to a speed of $0.95c$.

53. • A spring with a force constant of 595 N/m is compressed a distance of 37 cm. Find the resulting increase in the spring's mass.

54. • The 15 gallons of gasoline required to refuel your vehicle has a mass of about 41 kg and releases 1.8×10^9 J when burned in the engine. If its rest energy could be converted entirely to usable forms of energy, what equivalent mass would release the same amount of energy as burning the 41 kg of gasoline?

55. • What minimum energy must a gamma ray have to create an electron-antielectron pair?

56. • When a proton encounters an antiproton, the two particles annihilate each other, producing two gamma rays. Assuming the particles were at rest when they annihilated, find the energy of each of the two gamma rays produced. (*Note:* The rest energies of an antiproton and a proton are identical.)

57. •• If a neutron moves with a speed of $0.99c$, what are its **(a)** total energy, **(b)** rest energy, and **(c)** kinetic energy?

58. •• A rocket with a mass of 2.7×10^6 kg has a relativistic kinetic energy of 2.7×10^{23} J. **(a)** How fast is the rocket moving? **(b)** What is the rocket's total energy?

59. •• An object has a total energy that is 4.8 times its rest energy. **(a)** What is the speed of the object? **(b)** What is the ratio of the object's relativistic kinetic energy to its rest energy?

60. •• At the CERN particle accelerator in Geneva, Switzerland, protons are accelerated to a speed of $0.999999972c$. What is the total energy of each proton?

61. •• A nuclear power plant converts fuel energy at an average rate of 2.8×10^3 MW during a year of operation. Find the corresponding change in mass of reactor fuel over the entire year.

62. •• A helium atom has a rest mass of $m_{\text{He}} = 4.002603$ u. When disassembled into its constituent particles (2 protons, 2 neutrons, 2 electrons), the well-separated individual particles have the following masses: $m_{\text{p}} = 1.007276$ u, $m_{\text{n}} = 1.008665$ u, $m_{\text{e}} = 0.000549$ u. How much work is required to completely disassemble a helium atom? (*Note:* 1 u of mass has a rest energy of 931.49 MeV.)

63. •• What is the percent difference between the classical kinetic energy, $K_{\text{cl}} = \frac{1}{2}m_0v^2$, and the correct relativistic kinetic energy, $K = m_0c^2/\sqrt{1 - v^2/c^2} - m_0c^2$, at a speed of **(a)** $0.10c$ and **(b)** $0.90c$?

64. •• Predict/Calculate Consider a baseball with a rest mass of 0.145 kg. **(a)** How much work is required to increase the speed of the baseball from 25.0 m/s to 35.0 m/s? **(b)** Is the work required to increase the speed of the baseball from 200,000,025 m/s to 200,000,035 m/s greater than, less than, or the same as the amount found in part (a)? Explain. **(c)** Calculate the work required for the increase in speed indicated in part (b).

65. ••• A lump of putty with a mass of 0.240 kg and a speed of $0.980c$ collides head-on and sticks to an identical lump of putty moving with the same speed. After the collision the system is at rest. What is the mass of the system after the collision?

SECTION 29-8 GENERAL RELATIVITY

66. • Find the radius to which the Sun must be compressed for it to become a black hole.

67. •• **The Black Hole in the Center of the Milky Way** Recent measurements show that the black hole at the center of the Milky Way galaxy, which is believed to coincide with the powerful radio source Sagittarius A*, is 2.6 million times more massive than the Sun; that is, $M = 5.2 \times 10^{36}$ kg. **(a)** What is the maximum radius of this black hole? **(b)** Find the acceleration of gravity at the Schwarzschild radius of this black hole, using the expression for R given in Equation 29-10. **(c)** How does your answer to part (b) change if the mass of the black hole is doubled? Explain.

GENERAL PROBLEMS

68. • **CE** Two observers are moving relative to one another. Which of the following quantities will they always measure to have the same value: **(a)** their relative speed; **(b)** the time between two events; **(c)** the length of an object; **(d)** the speed of light in a vacuum; **(e)** the speed of a third observer?

69. • **CE** You are standing next to a runway as an airplane lands. **(a)** If you and the pilot observe a clock in the cockpit, which of you measures the proper time? **(b)** If you and the pilot observe a large clock on the control tower, which of you measures the proper time? **(c)** Which of you measures the proper length of the airplane? **(d)** Which of you measures the proper length of the runway?

70. • **CE** An apple drops from the bough of a tree to the ground. Is the mass of the apple near the top of its fall greater than, less than, or the same as its mass after it has landed? Explain.

71. • **CE** Predict/Explain Consider two apple pies that are identical in every respect, except that pie 1 is piping hot and pie 2 is at room temperature. **(a)** If identical forces are applied to the two pies, is the acceleration of pie 1 greater than, less than, or equal to the acceleration of pie 2? **(b)** Choose the *best explanation* from among the following:

   I. The acceleration of pie 1 is greater because the fact that it is hot means it has the greater energy.

   II. The fact that pie 1 is hot means it behaves as if it has more mass than pie 2, and therefore it has a smaller acceleration.

   III. The pies have the same acceleration regardless of their temperature because they have identical rest masses.

72. • **CE** Predict/Explain An uncharged capacitor is charged by moving some electrons from one plate of the capacitor to the other plate. **(a)** Is the mass of the charged capacitor greater than, less than, or the same as the mass of the uncharged capacitor? **(b)** Choose the *best explanation* from among the following:

   I. The charged capacitor has more mass because it is storing energy within it, just like a compressed spring.

II. The charged capacitor has less mass because some of its mass now appears as the energy of the electric field between its plates.

III. The capacitor has the same mass whether it is charged or not because charging it only involves moving electrons from one plate to the other without changing the total number of electrons.

73. •• **Cosmic Rays** Protons in cosmic rays have been observed with kinetic energies as large as 1.0×10^{20} eV. (a) How fast are these protons moving? Give your answer as a fraction of the speed of light. (b) Show that the kinetic energy of a single one of these protons is much greater than the kinetic energy of a 15-mg ant walking with a speed of 8.8 mm/s.

74. •• At the CERN particle accelerator in Geneva, Switzerland, protons are accelerated to a speed of $0.999999972c$. In the tunnel next to the beamline a worker drops a wrench from a height of 1.1 m above the floor. (a) How much time elapses before the tool hits the floor from the worker's frame of reference? (b) How much time elapses as measured by a proton in the accelerator?

75. •• A ^{14}C nucleus, initially at rest, emits a beta particle. The beta particle is an electron with 156 keV of kinetic energy. (a) What is the speed of the beta particle? (b) What is the momentum of the beta particle? (c) What is the momentum of the nucleus after it emits the beta particle? (d) What is the speed of the nucleus after it emits the beta particle?

76. •• A clock at rest has a rectangular shape, with a width of 24 cm and a height of 12 cm. When this clock moves parallel to its width with a certain speed v its width and height are the same. Relative to a clock at rest, how much time does it take for the moving clock to advance by 1.0 s?

77. •• A starship moving toward Earth with a speed of $0.850c$ launches a shuttle craft in the forward direction. The shuttle, which has a proper length of 12.5 m, is only 6.55 m long as viewed from Earth. What is the speed of the shuttle relative to the starship?

78. •• When a particle of charge q and momentum p enters a uniform magnetic field at right angles it follows a circular path of radius $R = p/qB$, as shown in **FIGURE 29-35**. What radius does this expression predict for a proton traveling with a speed $v = 0.99c$ through a magnetic field $B = 0.20$ T if you use (a) the nonrelativistic momentum $(p = mv)$ or (b) the relativistic momentum $(p = mv/\sqrt{1 - v^2/c^2})$?

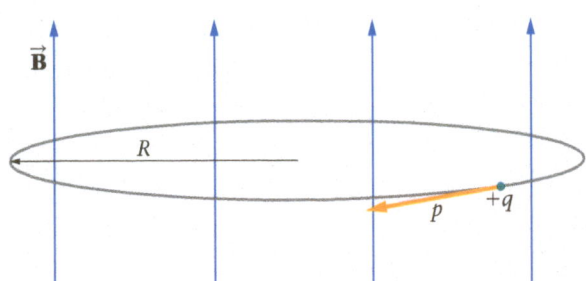

FIGURE 29-35 Problem 78

79. •• A 2.5-m titanium rod in a moving spacecraft is at an angle of 45° with respect to the direction of motion. The craft moves directly toward Earth at $0.98c$. As viewed from Earth, (a) how long is the rod and (b) what angle does the rod make with the direction of motion?

80. •• Electrons are accelerated from rest through a potential difference of 276,000 V. What is the final speed predicted (a) classically and (b) relativistically?

81. •• The rest energy, m_0c^2, of a particle with a kinetic energy K and a momentum p can be determined as follows:

$$m_0c^2 = \frac{(pc)^2 - K^2}{2K}$$

Suppose a pion (a subatomic particle) is observed to have a kinetic energy $K = 35.0$ MeV and a momentum $p = 5.61 \times 10^{-20}$ kg³ m/s = 105 MeV/c. What is the rest energy of the pion? Give your answer in MeV.

82. •• **Predict/Calculate** Consider a "relativistic air track" on which two identical air carts undergo a completely inelastic collision. One cart is initially at rest; the other has an initial speed of $0.650c$. (a) In classical physics, the speed of the carts after the collision would be $0.325c$. Do you expect the final speed in this relativistic collision to be greater than or less than $0.325c$? Explain. (b) Use relativistic momentum conservation to find the speed of the carts after they collide and stick together.

83. ••• **Predict/Calculate** In Conceptual Example 29-7 we considered an astronaut at rest on an inclined bed inside a moving spaceship. From the point of view of observer 1, on board the ship, the astronaut has a length L_0 and is inclined at an angle θ_0 above the floor. Observer 2 sees the spaceship moving to the right with a speed v. (a) Show that the length of the astronaut as measured by observer 2 is

$$L = L_0\sqrt{1 - \left(\frac{v^2}{c^2}\right)\cos^2 \theta_0}$$

(b) Show that the angle θ the astronaut makes with the floor of the ship, as measured by observer 2, is given by

$$\tan \theta = \frac{\tan \theta_0}{\sqrt{1 - v^2/c^2}}$$

84. ••• A pulsar is a collapsed, rotating star that sends out a narrow beam of radiation, like the light from a lighthouse. With each revolution, we see a brief, intense pulse of radiation from the pulsar. Suppose a pulsar is receding directly away from Earth with a speed of $0.800c$, and the starship *Endeavor* is sent out toward the pulsar with a speed of $0.950c$ relative to Earth. If an observer on Earth finds that 153 pulses are emitted by the pulsar every second, at what rate does an observer on the *Endeavor* see pulses emitted?

85. ••• Show that an object with momentum p and rest mass m_0 has a speed given by

$$v = \frac{c}{\sqrt{1 + (m_0c/p)^2}}$$

86. ••• **Decay of the Σ^- Particle** When at rest, the Σ^- particle has a lifetime of 0.15 ns before it decays into a neutron and a pion. One particular Σ^- particle is observed to travel 3.0 cm in the lab before decaying. What was its speed? (*Hint:* Its speed was not $\frac{2}{3}c$.)

PASSAGE PROBLEMS

Relativity in a TV Set

The first televisions used cathode-ray tubes, or CRTs, to form a picture. Today, liquid-crystal displays (LCD), light-emitting diodes (LED), and flexible organic LEDs (OLED) have largely replaced it, but the CRT still serves its purpose in certain niche applications.

The basic idea behind a CRT is fairly simple: use a beam of electrons to "paint" a picture on a fluorescent screen. This is illustrated in **FIGURE 29-36**. First, a heated coil at the negative terminal of the tube (the cathode) produces electrons which are accelerated toward the positive terminal (the anode) to form a beam of electrons—the so-called "cathode ray." A series of horizontal and vertical deflecting plates then direct the beam to any desired spot on a fluorescent screen to produce a glowing dot that can be seen. Moving the glowing dot rapidly around the screen, and varying its intensity with the control grid, allows one to produce a glowing image of any desired object. The first televised image—a dollar sign—was transmitted by Philo T. Farnsworth in 1927, and television inventors have been seeing dollar signs ever since.

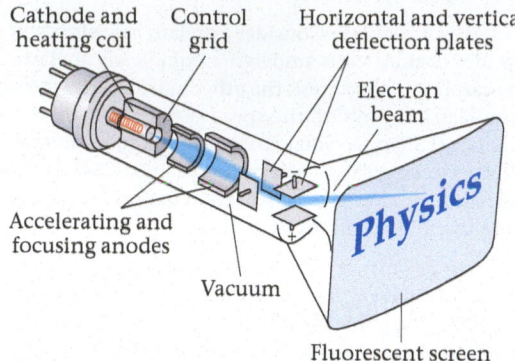

Cathode and heating coil
Control grid
Horizontal and vertical deflection plates
Electron beam
Accelerating and focusing anodes
Vacuum
Physics
Fluorescent screen

FIGURE 29-36 A cathode-ray tube. (Problems 87, 88, 89, and 90)

The interior of a CRT must be a very good vacuum, typically 10^{-7} of an atmosphere or less, to ensure electrons aren't scattered by air molecules on their way to the screen. Electrons in a television set are accelerated through a potential difference of 25.0 kV, which is sufficient to give them speeds comparable to the speed of light. As a result, relativity must be used to accurately determine their behavior. Thus, even in something as commonplace as a CRT, Einstein's theory of relativity proves itself to be of great practical value.

87. • Find the speed of an electron accelerated through a voltage of 25.0 kV—*ignoring* relativity. Express your answer as a fraction times the speed of light. (Speeds over about $0.1c$ are generally regarded as relativistic.)

A. $0.221c$ B. $0.281c$ C. $0.312c$ D. $0.781c$

88. • When relativistic effects are included, do you expect the speed of the electrons to be greater than, less than, or the same as the result found in the previous problem?

89. • Find the speed of the electrons in Problem 87, this time using a correct relativistic calculation. As before, express your answer as a fraction times the speed of light.

A. $0.301c$ B. $0.312c$ C. $0.412c$ D. $0.953c$

90. • Suppose the accelerating voltage in Problem 87 is increased by a factor of 10. What is the correct relativistic speed of an electron in this case?

A. $0.205c$ B. $0.672c$ C. $0.740c$ D. $0.862c$

91. •• **REFERRING TO EXAMPLE 29-9** The *Picard* approaches starbase Faraway Point with a speed of $0.806c$, and the *La Forge* approaches the starbase with a speed of $0.906c$. Suppose the *Picard* now launches a probe toward the starbase. (a) What velocity must the probe have relative to the *Picard* if it is to be at rest relative to the *La Forge*? (b) What velocity must the probe have relative to the *Picard* if its velocity relative to the *La Forge* is to be $0.100c$? (c) For the situation described in part (b), what is the velocity of the probe relative to the Faraway Point starbase?

92. •• **REFERRING TO EXAMPLE 29-9** Faraway Point starbase launches a probe toward the approaching starships. The probe has a velocity relative to the *Picard* of $-0.906c$. The *Picard* approaches starbase Faraway Point with a speed of $0.806c$, and the *La Forge* approaches the starbase with a speed of $0.906c$. (a) What is the velocity of the probe relative to the *La Forge*? (b) What is the velocity of the probe relative to Faraway Point starbase?

30

Quantum Physics

Big Ideas

1 Energy is quantized—it comes in discrete amounts.

2 Light is quantized—it comes in discrete packets called photons.

3 Light has particle-like properties, and particles have wavelike properties.

4 It is impossible to simultaneously know both a particle's position and its momentum.

5 The wavelike nature of particles allows them to "tunnel" through energy barriers.

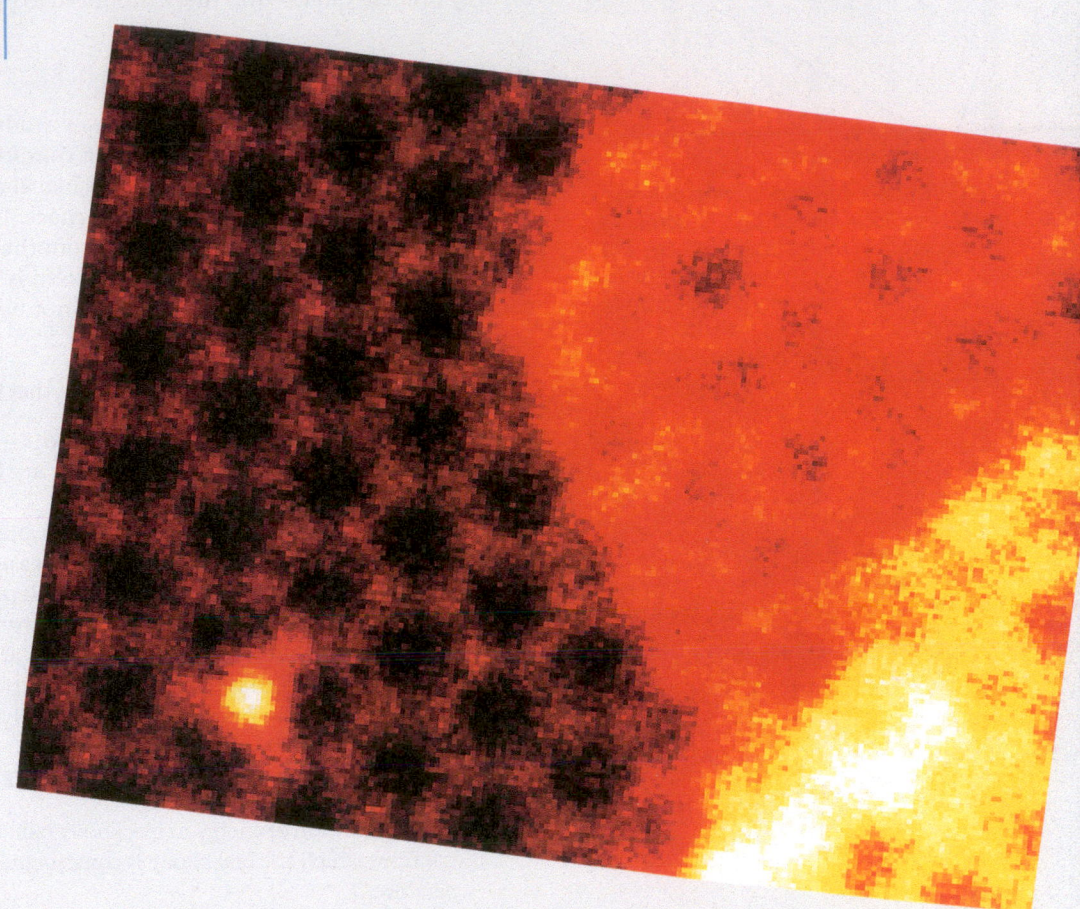

▲ Most of the images we encounter are made with visible light. Even those that are not, such as thermograms and X-rays, employ other kinds of electromagnetic radiation. Until about 100 years ago, the idea of making a picture by means of particles rather than radiation would have seemed absurd—like trying to create a portrait by bouncing paintballs off a subject and onto a canvas. Yet by the 1920s, physicists discovered that particles often act just like waves. Indeed, the wave properties of electrons can be exploited to create remarkably detailed images, such as this scanning transmission electron micrograph (STEM) of graphene—basically a two-dimensional sheet of carbon atoms arranged in a hexagonal lattice. This chapter explores the sometimes odd-seeming laws that describe the behavior of nature in the atomic and subatomic realms, and the series of revolutions in physics between the 1890s and the 1930s that uncovered them.

To understand the behavior of nature at the atomic level, it is necessary to introduce a number of new concepts to physics and to modify many others. In this chapter we consider the basic ideas of quantum physics and show that they lead to a deeper understanding of microscopic systems—in much the same way that relativity extends physics into the realm of high speeds.

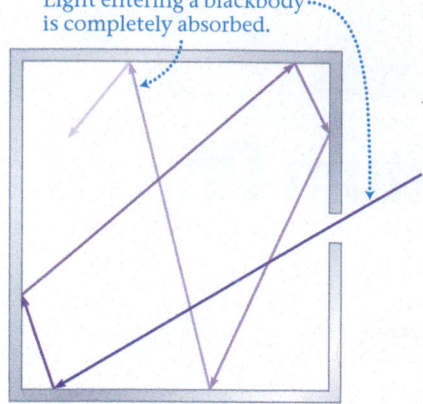

Light entering a blackbody is completely absorbed.

▲ **FIGURE 30-1 An ideal blackbody** In an ideal blackbody, incident light is completely absorbed. In the case shown here, the absorption occurs as the result of multiple reflections within a cavity. The blackbody, and the electromagnetic radiation it contains, are in thermal equilibrium at a temperature T.

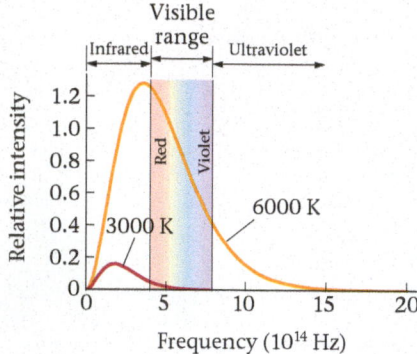

(a)

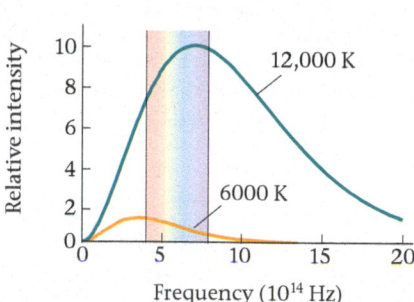

(b)

▲ **FIGURE 30-2 Blackbody radiation** Blackbody radiation as a function of frequency for various temperatures: **(a)** 3000 K and 6000 K; **(b)** 6000 K and 12,000 K. Note that as the temperature is increased, the peak in the radiation shifts toward higher frequency.

30-1 Blackbody Radiation and Planck's Hypothesis of Quantized Energy

If you've ever looked through a small opening into a hot furnace, you've seen the glow of light associated with its high temperature. As unlikely as it may seem, this light played a central role in the revolution of physics that occurred in the early 1900s. It was through the study of such systems that the idea of *energy quantization*—energy taking on only discrete values—was first introduced to physics.

Blackbody Radiation

In the late 1800s, physicists were actively studying the electromagnetic radiation given off by a physical system known as a **blackbody.** An example of a blackbody is illustrated in **FIGURE 30-1**. Notice that this blackbody has a cavity with a small opening to the outside world—much like a furnace. Light that enters the cavity through the opening is reflected multiple times from the interior walls until it is completely absorbed. It is for this reason that the system is referred to as "black," even though the material from which it is made need not be black at all. Thus, we can define a blackbody as follows:

> An ideal blackbody absorbs all the light that is incident on it.

Objects that absorb much of the incident light—though not all of it—are reasonable approximations to a blackbody; objects that are highly reflective and shiny are poor representations of a blackbody.

The basic experiment performed with a blackbody is the following: Heat the blackbody to a fixed temperature, T, and measure the amount of electromagnetic radiation it gives off at a given frequency, f. Repeat this measurement for a number of different frequencies, then plot the intensity of radiation versus the frequency. The results of a typical blackbody experiment are shown in **FIGURE 30-2** for a variety of temperatures. Notice that there is little radiation at low frequencies, a peak in the radiation at intermediate frequencies, and finally a fall-off to little radiation again at high frequencies.

Now, what is truly remarkable about the blackbody experiment is the following:

> The distribution of energy in blackbody radiation is *independent* of the material from which the blackbody is constructed—it depends only on the temperature, T.

Therefore, a blackbody of steel and one of wood give precisely the same results when held at the same temperature. When physicists observe a phenomenon that is independent of the details of the system, it is a clear signal that they are observing something of fundamental significance. This was certainly the case with blackbody radiation.

Areas and Peaks for Blackbody Curves Two aspects of the blackbody curves in Figure 30-2 are of particular importance. First, notice that as the temperature is increased, the area under the curve increases. The total area under the curve is a measure of the total energy emitted by the blackbody, and hence it follows that an object radiates more energy as it becomes hotter, as one would expect.

Second, the peak in the blackbody curve moves to higher frequency as the absolute temperature T is increased. This movement, or "displacement," of the peak with temperature is described by **Wien's displacement law:**

Wien's Displacement Law

$$f_{\text{peak}} = (5.88 \times 10^{10}\ \text{s}^{-1} \cdot \text{K}^{-1})T$$

SI unit: Hz = s^{-1}

30-1

Thus, there is a direct connection between the temperature of an object and the frequency of radiation it emits most strongly.

CONCEPTUAL EXAMPLE 30-1   COMPARE TEMPERATURES

Betelgeuse is a red-giant star in the constellation Orion; Rigel is a bluish-white star in the same constellation. **(a)** Is the surface temperature of Betelgeuse greater than, less than, or the same as the surface temperature of Rigel? **(b)** Which of the following is the *best explanation* for your prediction?

   I. The surface temperature of Betelgeuse is greater than that of Rigel because red is a hotter color than blue.

   II. The two stars have the same surface temperature because the color of light has nothing to do with temperature.

   III. Red light has a lower frequency than blue light, and hence the surface temperature of Betelgeuse is lower than that of Rigel.

REASONING AND DISCUSSION

Recall from Chapter 25 that red light has a lower frequency than blue light. It follows from Wien's displacement law ($f_{peak} = (5.88 \times 10^{10} \, \text{s}^{-1} \cdot \text{K}^{-1})T$) that a red star has a lower temperature than a blue star. Therefore, Betelgeuse has the lower surface temperature.

ANSWER

(a) The surface temperature of Betelgeuse is lower than that of Rigel. **(b)** The best explanation is III.

RWP* To be more specific about the conclusion given in Conceptual Example 30-1, let's consider Figure 30-2 in greater detail. At the lowest temperature shown, 3000 K, the radiation is more intense at the red end of the visible spectrum than at the blue end. An object at this temperature—like the heating coil on a stove or the bolt in **FIGURE 30-3**, for example–appears "red hot" to the eye. Even so, most of the radiation at this temperature is in the infrared, and thus is not visible to the eye at all. A blackbody at 6000 K, like the surface of the Sun, gives out strong radiation throughout the visible spectrum, although there is still more radiation at the red end than at the blue end. As a result, the light of the Sun appears somewhat yellowish. Finally, at 12,000 K a blackbody appears bluish white, and most of its radiation is in the ultraviolet. The temperature of the star Rigel is determined from the location of its radiation peak in the following Exercise.

EXERCISE 30-2   SURFACE TEMPERATURE

Find the surface temperature of Rigel, given that its radiation peak occurs at a frequency of 1.17×10^{15} Hz.

REASONING AND SOLUTION

Solving $f_{peak} = (5.88 \times 10^{10} \, \text{s}^{-1} \cdot \text{K}^{-1})T$ for the temperature, T, we find the following:

$$T = \frac{f_{peak}}{5.88 \times 10^{10} \, \text{s}^{-1} \cdot \text{K}^{-1}} = \frac{1.17 \times 10^{15} \, \text{Hz}}{5.88 \times 10^{10} \, \text{s}^{-1} \cdot \text{K}^{-1}} = 19,900 \, \text{K}$$

This is a little more than three times the surface temperature of the Sun. Thus blackbody radiation allows us to determine the temperature of a distant star that we may never visit.

▲ **FIGURE 30-3** All objects emit electromagnetic radiation over a range of frequencies. The frequency that is radiated most intensely depends on the object's temperature, as specified by Wien's law. The glowing bolt in this picture radiates primarily in the infrared part of the spectrum, but it is hot enough (a few thousand kelvin) that a significant portion of its radiation falls within the red end of the visible region. The other bolts are too cool to radiate any detectable amount of visible light.

Planck's Quantum Hypothesis

Although experimental understanding of blackbody radiation was quite extensive in the late 1800s, there was a problem. Attempts to explain the blackbody curves of Figure 30-2 with classical physics failed—and failed miserably. To see the problem, consider the curves shown in **FIGURE 30-4**. The green curve is the experimental result for a blackbody at a given temperature. In contrast, the blue curve shows the prediction of classical physics. Clearly, the classical result cannot be valid. After all, its curve diverges to infinity at high frequency, which in turn implies that the blackbody radiates an infinite amount of energy. This unphysical divergence at high frequencies is referred to as the *ultraviolet catastrophe*.

---

*Real World Physics applications are denoted by the acronym RWP.

▶ **FIGURE 30-4 The ultraviolet catastrophe** Classical physics predicts a blackbody radiation curve that rises without limit as the frequency increases. This outcome is referred to as the ultraviolet catastrophe. By assuming energy quantization, Planck was able to derive a curve that agreed with experimental results.

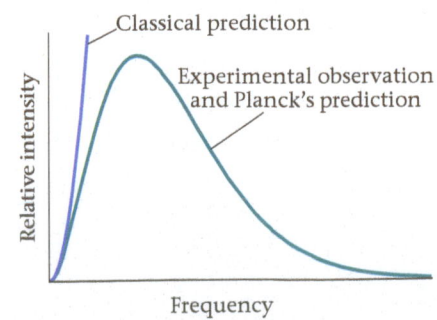

The German physicist Max Planck (1858–1947) worked long and hard on this problem. Eventually, he was able to construct a mathematical formula that agreed with experiment for all frequencies. His next problem was to "derive" the equation. The only way he could do this, it turned out, was to make the following bold and unprecedented assumption:

> The radiation energy in a blackbody at the frequency f must be an integral multiple of a constant (h) times the frequency.

That is, Planck assumed that the energy in a blackbody is *quantized*. We can express this mathematically as follows:

Quantized Energy

$$E_n = nhf \quad n = 0, 1, 2, 3, \ldots \qquad \text{30-2}$$

The constant, h, in this expression is known as **Planck's constant,** and it has the following value:

Planck's Constant, h

$$h = 6.63 \times 10^{-34} \, \text{J} \cdot \text{s} \qquad \text{30-3}$$
SI unit: J · s

Big Idea 1 The results from experiments on blackbody radiation cannot be explained by classical physics. In particular, these experiments show that energy is quantized—that is, it comes in discrete amounts—rather than being continuous, as had been previously thought.

This constant is recognized today as one of the fundamental constants of nature, on an equal footing with other such constants as the speed of light in a vacuum and the rest mass of an electron.

Quantized Energy   The assumption of energy quantization is quite a departure from classical physics, in which energy can take on any value at all and is related to the amplitude of a wave rather than its frequency. In Planck's calculation, the energy can have only the discrete values hf, $2hf$, $3hf$, and so on. Because of this quantization, it follows that the energy can change only in *quantum jumps* of energy no smaller than hf as the system goes from one quantum state to another. The fundamental increment, or *quantum*, of energy, hf, is incredibly small, as can be seen from the small magnitude of Planck's constant. The next Example explores the size of the quantum and the value of the *quantum number, n*, for a typical macroscopic system.

EXAMPLE 30-3   QUANTUM NUMBERS

Suppose the maximum speed of a 1.9-kg mass attached to a spring with a force constant of 47 N/m is 0.84 m/s. (a) Find the frequency of oscillation. (b) What is total energy of this mass–spring system? (c) Determine the size of one quantum of energy in this system. (d) Assuming the energy of this system satisfies $E_n = nhf$, find the quantum number, n.

PICTURE THE PROBLEM
Our sketch shows a 1.9-kg mass oscillating on a spring with a force constant of 47 N/m. The mass has its maximum speed of $v_{\text{max}} = 0.84$ m/s when it passes through the equilibrium position. At this moment, the total energy of the system is simply the kinetic energy of the mass.

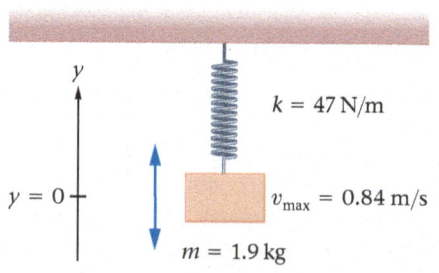

REASONING AND STRATEGY

a. We can find the frequency f of oscillation using $\omega = \sqrt{k/m}$ (Equation 13-10) and $\omega = 2\pi f$ (Equation 13-5).

b. The total energy is simply the kinetic energy as the mass passes through equilibrium: $E = K_{\text{max}} = \frac{1}{2}mv_{\text{max}}^2$.

c. The energy of one quantum is hf, where f is the frequency found in part (a).

d. We determine the quantum number by solving $E_n = nhf$ for n.

Known   Mass attached to spring, $m = 1.9$ kg; force constant of spring, $k = 47$ N/m; maximum speed of mass, $v_{\text{max}} = 0.84$ m/s.

Unknown   (a) Frequency of oscillation, $f = ?$ (b) Total energy, $E = ?$ (c) Quantum of energy, $hf = ?$ (d) Quantum number, $n = ?$

CONTINUED

SOLUTION

Part (a)

1. Calculate the frequency of oscillation using $\omega = \sqrt{k/m} = 2\pi f$:

$$\omega = \sqrt{\frac{k}{m}} = 2\pi f$$

$$f = \frac{1}{2\pi}\sqrt{\frac{k}{m}} = \frac{1}{2\pi}\sqrt{\frac{47 \text{ N/m}}{1.9 \text{ kg}}} = 0.79 \text{ Hz}$$

Part (b)

2. Calculate the maximum kinetic energy of the mass $\left(\frac{1}{2}mv_{max}^2\right)$ to find the total energy, E, of the system:

$$E = \frac{1}{2}mv_{max}^2 = \frac{1}{2}(1.9 \text{ kg})(0.84 \text{ m/s})^2 = 0.67 \text{ J}$$

Part (c)

3. The energy of one quantum is hf, where $f = 0.79$ Hz:

$$hf = (6.63 \times 10^{-34} \text{ J} \cdot \text{s})(0.79 \text{ Hz}) = 5.2 \times 10^{-34} \text{ J}$$

Part (d)

4. Set $E_n = nhf$ equal to the total energy, E, of the system and solve for n:

$$E_n = nhf = E$$

$$n = \frac{E}{hf} = \frac{0.67 \text{ J}}{5.2 \times 10^{-34} \text{ J}} = 1.3 \times 10^{33}$$

INSIGHT

The numbers found in parts (c) and (d) are incredible for their size. For example, the quantum of energy is on the order of 10^{-34} J. For comparison, the energy required to break a bond in a DNA molecule is on the order of 10^{-20} J. Thus, the energy quantum for a macroscopic system is smaller than the energy needed to affect a molecule by a factor of about 10^{14} (a hundred trillion). Similarly, the number of quanta in the system, roughly 10^{33}, is comparable to the number of atoms in four Olympic-size swimming pools.

PRACTICE PROBLEM — PREDICT/CALCULATE

Suppose the force constant of the spring is increased to 52 N/m, while everything else in the system remains the same. (a) Does the quantum of energy increase, decrease, or stay the same? Explain. (b) Calculate the quantum of energy for this case. [**Answer: (a)** Increasing the force constant increases the frequency. This, in turn, increases the quantum of energy. **(b)** The quantum of energy is increased to 5.5×10^{-34} J.]

Some related homework problems: Problem 11, Problem 85, Problem 86

Clearly, then, the quantum numbers in typical macroscopic systems are incredibly large. As a result, a change of one in the quantum number is completely insignificant and undetectable. Similarly, the change in energy from one quantum state to the next is so small that it cannot be measured in a typical experiment; hence, for all practical purposes, the energy of a macroscopic system seems to change continuously, even though it actually changes by small increments. In contrast, in an atomic system, the energy jumps are of great importance, as we shall see in the next section.

Returning to the ultraviolet catastrophe for a moment, we can now see how Planck's hypothesis removes the unphysical divergence at high frequency predicted by classical physics. In Planck's theory, the higher the frequency f, the greater the quantum of energy, hf. Therefore, as the frequency is increased, the amount of energy required for even the smallest quantum jump increases as well. A blackbody has only a finite amount of energy, however, and hence it simply cannot supply the large amount of energy required to produce an extremely high-frequency quantum jump. As a result, the amount of radiation at high frequency drops off toward zero.

Planck's theory of energy quantization leads to an adequate description of the experimental results for blackbody radiation. Still, the theory was troubling and somewhat unsatisfying to Planck and to many other physicists as well. Although the idea of energy quantization worked, at least in this case, it seemed ad hoc and more of a mathematical trick than a true representation of nature. With the work of Einstein, however, which we present in the next section, the well-founded misgivings about quantum theory began to fade away.

Enhance Your Understanding   (Answers given at the end of the chapter)

1. If the temperature of a blackbody is increased, does the (a) peak frequency and (b) total energy radiated increase, decrease, or stay the same?

CONTINUED

Section Review

- The location of the peak in blackbody radiation is given by $f_{peak} = (5.88 \times 10^{10} \text{ s}^{-1} \cdot \text{K}^{-1})T$.

- To correctly describe blackbody radiation, Planck found it necessary to assume that the energy of a blackbody is quantized. Thus, he assumed that $E_n = nhf$, where $n = 1, 2, 3, \ldots$. The constant h in this expression is known as Planck's constant, and its value is $h = 6.63 \times 10^{-34} \text{ J} \cdot \text{s}$.

30-2 Photons and the Photoelectric Effect

From Max Planck's point of view, energy quantization in a blackbody was probably related to quantized vibrations of atoms in the walls of the blackbody. We are familiar, for example, with the fact that a string tied at both ends can produce standing waves at only certain discrete frequencies (Chapter 14), so perhaps atoms vibrating in a blackbody behave in a similar way, vibrating only with certain discrete energies (this is the basic idea behind string theory today). Certainly, Planck didn't think the *light* in a blackbody had a quantized energy, because most physicists at the time thought of light as being a wave, which can have any energy.

Big Idea ❷ Light is quantized— it comes in discrete packets called photons. Photons have an energy that is proportional to their frequency.

Einstein's Photon Model of Light A brash young physicist named Albert Einstein took the idea of quantized energy seriously, however, and applied it to the radiation in a blackbody. Einstein proposed that *light itself* comes in bundles of energy called **photons**, and that photons obey Planck's hypothesis of energy quantization. In Einstein's model, light of frequency f consists of photons with an energy given by the following relationship:

Energy of a Photon of Frequency f

$$E = hf$$

30-4

SI unit: J

Thus the energy in a beam of light of frequency f can have only the values hf, $2hf$, $3hf$, and so on. Planck's initial reaction to Einstein's suggestion was that he had gone too far with the idea of quantization. As it turns out, nothing could have been further from the truth.

In Einstein's photon model, a beam of light can be thought of as a beam of particles, each carrying the energy hf, as indicated in **FIGURE 30-5**. If the beam of light is made more intense while the frequency remains the same, the result is that the photons in the beam are more tightly packed, so that more photons pass a given point in a given time. In this way, more photons shine on a given surface in a given time, increasing the energy delivered to the surface per time. Even so, each photon in the more intense beam has exactly the same amount of energy as a photon in the less intense beam. The energy of a typical photon of visible light is calculated in the next Exercise.

Low-intensity light beam

High-intensity light beam

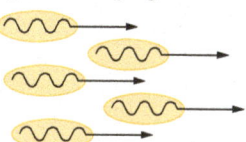

▲ **FIGURE 30-5 The photon model of light** In the photon model of light, a beam of light consists of many photons, each with an energy hf. The more intense the beam, the more tightly packed the photons.

EXERCISE 30-4 PHOTON ENERGY

Calculate the energy of a photon of green light that has a frequency of 5.39×10^{14} Hz (which corresponds to a wavelength of 557 nm). Give the energy in both joules and electron volts.

REASONING AND SOLUTION

Applying $E = hf$ with the given frequency, we find

$$E = hf = (6.63 \times 10^{-34} \text{ J} \cdot \text{s})(5.39 \times 10^{14} \text{ s}^{-1}) = 3.57 \times 10^{-19} \text{ J}$$

$$= (3.57 \times 10^{-19} \text{ J})\left(\frac{1 \text{ eV}}{1.60 \times 10^{-19} \text{ J}}\right) = 2.23 \text{ eV}$$

All photons of this frequency of light have precisely the same energy.

Notice that the energy of a visible photon is on the order of an electron volt (eV). This is significant because the eV is also the typical energy scale for atomic and molecular systems, as we show in the next Example.

EXAMPLE 30-5   WHEN OXYGENS SPLIT

Molecular oxygen (O_2) is a diatomic molecule. The energy required to dissociate 1 mol of O_2 to form 2 mol of atomic oxygen (O) is 118 kcal. **(a)** Find the energy (in joules) required to dissociate one O_2 molecule into two O atoms. **(b)** Assuming the dissociation energy for one molecule is supplied by a single photon, find the frequency of the photon.

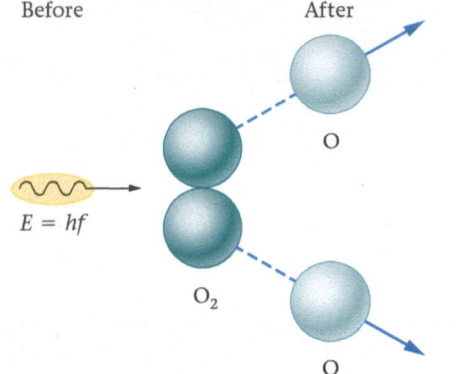
Before   After

PICTURE THE PROBLEM
In our sketch, we show a single photon dissociating an O_2 molecule to produce two O atoms. The energy of the photon is $E = hf$.

REASONING AND STRATEGY

a. This part of the problem is simply a matter of converting from kcal per mole to joules per molecule. This can be done by using the facts that 1 kcal = 4186 J and Avogadro's number (Section 17-1) is 6.02×10^{23} molecules/mol.

b. We can find the frequency of the photon by setting hf equal to the energy E found in part (a).

Known   Energy of dissociation for 1 mol = 118 kcal.

Unknown   **(a)** Energy to dissociate one O_2 molecule = ? **(b)** Frequency of photon, f = ?

SOLUTION

Part (a)

1. Convert 118 kcal/mol to J/molecule:

$$\left(118 \frac{\text{kcal}}{\text{mol}}\right)\left(\frac{4186\,\text{J}}{1\,\text{kcal}}\right)\left(\frac{1}{6.02 \times 10^{23}\,\text{molecules/mol}}\right)$$
$$= 8.21 \times 10^{-19}\,\text{J/molecule}$$

Part (b)

2. Use $E = 8.21 \times 10^{-19}$ J = hf to solve for the frequency, f:

$$f = \frac{E}{h} = \frac{8.21 \times 10^{-19}\,\text{J}}{6.63 \times 10^{-34}\,\text{J} \cdot \text{s}} = 1.24 \times 10^{15}\,\text{Hz}$$

INSIGHT
This frequency is in the ultraviolet. In fact, ultraviolet rays in Earth's upper atmosphere cause O_2 molecules to dissociate, freeing up atomic oxygen, which can then combine with other oxygen atoms or O_2 molecules to form ozone, O_3.

Converting the energy found in part (a) to eV, using 1 eV = 1.60×10^{-19} J, we find 5.13 eV/molecule. Thus, as expected, the eV is a typical measure of energy in atomic and molecular systems.

PRACTICE PROBLEM
An infrared photon has a frequency of 1.00×10^{13} Hz. How much energy is carried by 1 mol of these photons?   [**Answer:** 3990 J = 0.953 kcal]

Some related homework problems: Problem 19, Problem 26

BIO As we've seen, photons typically have small amounts of energy on a macroscopic scale. As a result, enormous numbers of photons are involved in most everyday situations. For example, a typical lightbulb gives off roughly 10^{18} photons per second, even though we need only about 10^3 photons per second to see. Amateur astronomers push the limits of low-light human vision, and routinely view astronomical objects like distant galaxies with only a few thousand photons reaching their eyes per second. At this rate, the photons are separated by about 60 miles—it follows that only one photon at a time from the distant galaxy traverses the astronomer's telescope.

The Photoelectric Effect Einstein applied his photon model of light to the **photoelectric effect**, in which a beam of light (*photo-*) hits the surface of a metal and ejects an electron (*-electric*). The effect can be measured using a device like that pictured in FIGURE 30-6. In this apparatus, incoming light ejects an electron—referred to as a photoelectron—from a metal plate called the emitter (E); the electron is then attracted to a collector plate (C), which is at a positive potential relative to the emitter. The result is an electric current that can be measured with an ammeter.

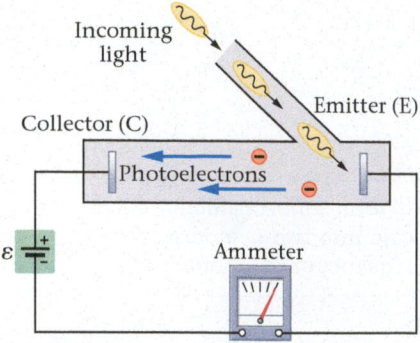

▲ **FIGURE 30-6 The photoelectric effect**
The photoelectric effect can be studied with a device like the one shown. Light shines on a metal plate, ejecting electrons, which are then attracted to a positively charged "collector" plate. The result is an electric current that can be measured with an ammeter.

The minimum amount of energy necessary to eject an electron from a particular metal is referred to as the **work function,** W_0, for that metal. Work functions vary from metal to metal but are typically on the order of a few electron volts. If an electron is given an energy E by the beam of light that is greater than W_0, the excess energy goes into kinetic energy of the ejected electron. The maximum kinetic energy (K) a photo-electron can have, then, is

$$K_{max} = E - W_0 \qquad \text{30-5}$$

Just as with blackbody radiation, the photoelectric effect exhibits behavior that is at odds with classical physics. Two of the main areas of disagreement are the following:

- Classical physics predicts that a beam of light of *any* color (frequency) can eject electrons, as long as the beam has sufficient intensity. That is, if a beam is intense enough, the energy it delivers to an electron will exceed the work function and cause it to be ejected.
- Classical physics also predicts that the maximum kinetic energy of an ejected electron should increase as the intensity of the light beam is increased. In particular, the more energy the beam delivers to the metal, the more energy any given electron can have as it is ejected.

Although both of these predictions are reasonable—necessary, in fact, from the classical physics point of view—they simply do not agree with experiments on the photoelectric effect. In fact, experiments show the following behavior:

- To eject electrons, the incident light beam must have a frequency greater than a certain minimum value, referred to as the **cutoff frequency,** f_0. If the frequency of the light is less than f_0, it will not eject electrons, no matter how intense the beam.
- If the frequency of light is greater than the cutoff frequency, f_0, the effect of increasing the intensity is to increase the *number* of electrons that are emitted per second. The maximum kinetic energy of the electrons does not increase with the intensity of the light; the kinetic energy depends only on the frequency of the light.

As we shall see, these observations are explained quite naturally in terms of photons.

In Einstein's model each photon has an energy determined solely by its frequency. Therefore, making a beam of a given frequency more intense simply means increasing the number of photons hitting the metal in a given time—not increasing the energy carried by a photon. An electron, then, is ejected only if an incoming photon has an energy that is at least equal to the work function: $E = hf_0 = W_0$. The *cutoff frequency* is thus defined as follows:

Cutoff Frequency, f_0

$$f_0 = \frac{W_0}{h} \qquad \text{30-6}$$

SI unit: Hz = s^{-1}

If the frequency of the light is greater than f_0, the electron can leave the metal with a finite kinetic energy; if the frequency is less than f_0, no electrons are ejected, no matter how intense the beam. We determine a typical cutoff frequency in the next Exercise.

EXERCISE 30-6 CUTOFF FREQUENCY

The work function for an aluminum surface is 4.08 eV. Find the cutoff frequency, f_0, for aluminum.

REASONING AND SOLUTION

Substitution in $f_0 = W_0/h$ yields

$$f_0 = \frac{W_0}{h} = \frac{(4.08 \text{ eV})\left(\dfrac{1.60 \times 10^{-19} \text{ J}}{1 \text{ eV}}\right)}{6.63 \times 10^{-34} \text{ J} \cdot \text{s}} = 9.85 \times 10^{14} \text{ Hz}$$

This frequency is in the near-ultraviolet portion of the electromagnetic spectrum.

▶ **FIGURE 30-7 The kinetic energy of photoelectrons** The maximum kinetic energy of photoelectrons as a function of the frequency of light. Note that sodium and gold have different cutoff frequencies, as one might expect for different materials. On the other hand, the slope of the two lines is the same, h, as predicted by Einstein's photon model of light.

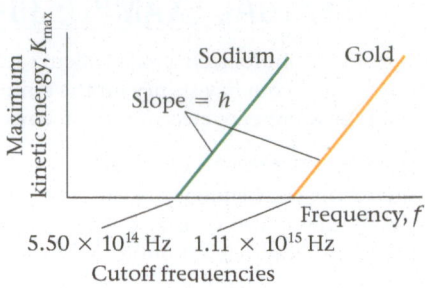

The fact that a more intense beam of monochromatic light delivers more photons per time to the metal just means that more electrons are ejected per time. Because each electron receives precisely the same amount of energy, however, the maximum kinetic energy is the same regardless of the intensity. In fact, if we return to Equation 30-5 ($K_{max} = E - W_0$) and replace the energy, E, with the energy of a photon, hf, we find

$$K_{max} = hf - W_0 \qquad \text{30-7}$$

Notice that K_{max} depends linearly on the frequency but is independent of the intensity. A plot of K_{max} for sodium (Na) and gold (Au) is given in **FIGURE 30-7**. Clearly, both lines have the same slope, h, as expected from Equation 30-7, but they have different cutoff frequencies. Therefore, with the result given in Equation 30-7, Einstein was able to show that Planck's constant, h, appears in a natural way in the photoelectric effect and is not limited in applicability to the blackbody.

EXAMPLE 30-7 | WHITE LIGHT ON POTASSIUM

A beam of white light containing photons with frequencies between 4.25×10^{14} Hz and 7.46×10^{14} Hz is incident on a potassium surface, which has a work function of 2.29 eV. **(a)** What is the cutoff frequency for a potassium surface? (Photons with frequencies less than the cutoff will not eject electrons.) **(b)** What is the maximum kinetic energy of the electrons that are ejected from this surface?

PICTURE THE PROBLEM

Our sketch shows a beam of white light, represented by photons with different frequencies, incident on a potassium surface. Electrons are ejected from this surface with a kinetic energy that depends on the frequency of the photon that was absorbed.

REASONING AND STRATEGY

a. We can find the cutoff frequency, f_0, for potassium using the relationship $f_0 = W_0/h$, with $W_0 = 2.29$ eV.

b. We can obtain the maximum kinetic energy for a given frequency, f, from Equation 30-7: $K_{max} = hf - W_0$. Clearly, the higher the frequency, the greater the maximum kinetic energy. It follows that the greatest possible kinetic energy corresponds to the highest frequency photons in the beam, 7.46×10^{14} Hz.

Known Frequency range, 4.25×10^{14} Hz to 7.46×10^{14} Hz; work function, $W_0 = 2.29$ eV.

Unknown **(a)** Cutoff frequency, $f_0 = ?$ **(b)** Maximum kinetic energy of ejected electrons, $K_{max} = ?$

SOLUTION

Part (a)

1. Use $f_0 = W_0/h$, with $W_0 = 2.29$ eV, to calculate the cutoff frequency for potassium. Be sure to convert W_0 to joules:

$$f_0 = \frac{W_0}{h} = \frac{(2.29 \text{ eV})\left(\dfrac{1.60 \times 10^{-19} \text{ J}}{1 \text{ eV}}\right)}{6.63 \times 10^{-34} \text{ J} \cdot \text{s}} = 5.53 \times 10^{14} \text{ Hz}$$

Part (b)

2. Using $K_{max} = hf - W_0$, calculate K_{max} for the maximum frequency in the beam, $f = 7.46 \times 10^{14}$ Hz:

$$\begin{aligned} K_{max} &= hf - W_0 \\ &= (6.63 \times 10^{-34} \text{ J} \cdot \text{s})(7.46 \times 10^{14} \text{ Hz}) \\ &\quad -(2.29 \text{ eV})\left(\frac{1.60 \times 10^{-19} \text{ J}}{1 \text{ eV}}\right) = 1.28 \times 10^{-19} \text{ J} \end{aligned}$$

INSIGHT

The cutoff frequency for potassium is in the green part of the visible spectrum. The maximum kinetic energy of an ejected electron for this beam of light is about 0.80 eV.

PRACTICE PROBLEM

What frequency of light would be necessary to give a maximum kinetic energy of 1.80 eV to the electrons ejected from this surface? **[Answer: 9.87×10^{14} Hz]**

Some related homework problems: Problem 31, Problem 32, Problem 33

CONCEPTUAL EXAMPLE 30-8 EJECTED ELECTRONS

Consider a photoelectric experiment such as the one illustrated in Figure 30-6. A beam of light with a frequency greater than the cutoff frequency shines on the emitter. If the frequency of this beam is increased while the intensity is held constant, does the number of electrons ejected per second from the metal surface increase, decrease, or stay the same?

REASONING AND DISCUSSION

Increasing the frequency of the beam means that each photon carries more energy; however, we know that the intensity of the beam remains constant. It follows, then, that fewer photons hit the surface per time—otherwise, the intensity would increase. Because fewer photons hit the surface per time, fewer electrons are ejected per time.

ANSWER

The number of electrons ejected per second decreases.

▲ **FIGURE 30-8 Visualizing Concepts Solar Energy Panels** The photoelectric effect is the basic mechanism used by photovoltaic cells, which are now used to power both terrestrial devices such as emergency call stations (top) and the solar panels that supply electricity to the International Space Station (bottom).

RWP Applications of the photoelectric effect are common all around us. For example, if you have ever dashed into an elevator as its doors were closing, you were probably saved from being crushed by the photoelectric effect. Many elevators and garage-door systems use a beam of light and a photoelectric device known as a *photocell* as a safety feature. As long as the beam of light strikes the photocell, the photoelectric effect generates enough ejected electrons to produce a detectable electric current. When the light beam is blocked—by a late arrival at the elevator, for example—the electric current produced by the photocell is interrupted and the doors are signaled to open. Similar photocells automatically turn on streetlights at dusk, measure the amount of light entering a camera, and function as proximity sensors for touchless faucets, soap dispensers, and hand dryers.

RWP Photocells are also the basic unit in the *solar energy panels* that convert some of the energy in sunlight into electrical energy, as shown in **FIGURE 30-8**. A small version of a solar energy panel can be found on many pocket calculators. These panels are efficient enough to operate their calculators with nothing more than dim indoor lighting. Larger outdoor panels can operate billboards and safety lights in remote areas far from commercial power lines. Truly large solar panels, 240 ft in length, power the International Space Station, and make the station visible to the naked eye from Earth's surface. On an even larger scale, power stations consisting of millions of solar panels that generate hundreds of megawatts of electric power have been built at various locations on Earth. These applications of solar panels may only hint at the potential for solar energy in the future, especially when one considers that sunlight delivers about 200,000 times more energy to Earth each day than the entire world's electrical energy production.

Finally, it's interesting to note that though Einstein is best known for his development of the theory of relativity, he was awarded the Nobel Prize in Physics not for relativity, but for the photoelectric effect.

Enhance Your Understanding (Answers given at the end of the chapter)

2. **(a)** If the wavelength of a beam of light is increased, does the energy of photons in the beam increase, decrease, or stay the same? Explain. **(b)** If the intensity of a beam of light is increased, does the energy of photons in the beam increase, decrease, or stay the same? Explain.

Section Review

- Einstein's photon model of light says that light is composed of photons, each with an energy given by $E = hf$.

- The cutoff frequency for the photoelectric effect is $f_0 = W_0/h$, where W_0 is the work function—the energy required to remove an electron from a material.

- The maximum kinetic energy of electrons ejected from a material by light of frequency f is $K_{max} = hf - W_0$.

30-3 The Mass and Momentum of a Photon

When we say that a photon is like a "particle" of light, we put the word *particle* in quotation marks because a photon is quite different from everyday particles in many important respects. For example, a typical particle can be held at rest in your hand. It can also be placed on a scale to have its mass determined. These operations are not possible with a photon.

First, photons travel with the speed of light, which means that all observers see them as having the same speed. It is not possible to stop a photon and hold it in your hand. In contrast, particles with a finite rest mass can never attain the speed of light. It follows, then, that photons must have *zero rest mass*. This condition can be seen mathematically by rewriting Equation 29-7 for the total energy, E, as follows:

$$E\sqrt{1 - \frac{v^2}{c^2}} = m_0 c^2 \qquad\qquad 30\text{-}8$$

Photons travel at the speed of light, $v = c$, and thus the left side of the equation is zero. As a result, the right side of the equation must also be zero, which can happen only if the rest mass is zero:

Rest Mass of a Photon

$$m_0 = 0 \qquad\qquad 30\text{-}9$$

Second, photons differ from everyday particles in that they have a finite momentum even though they have no mass. To see how this can be, notice that the relativistic equation for momentum, Equation 29-5, can be rewritten as follows:

$$p\sqrt{1 - \frac{v^2}{c^2}} = m_0 v \qquad\qquad 30\text{-}10$$

Dividing Equation 30-10 by Equation 30-8 yields

$$\frac{p}{E} = \frac{v}{c^2}$$

Once again setting $v = c$ for a photon, we find that the momentum of a photon is related to its total energy as follows:

$$p = \frac{E}{c}$$

Finally, recalling that $E = hf$, and that $f = c/\lambda$, we obtain the following result:

Momentum of a Photon

$$p = \frac{hf}{c} = \frac{h}{\lambda} \qquad\qquad 30\text{-}11$$

Notice that a photon's momentum increases with its frequency, and thus with its energy.

As one might expect, the momentum of a typical photon of visible light is quite small, as is illustrated in the next Exercise.

EXERCISE 30-9   PHOTON MOMENTUM

Calculate the momentum of a photon of green light that has a frequency of 5.39×10^{14} Hz.

REASONING AND SOLUTION

Substituting $f = 5.39 \times 10^{14}$ Hz in $p = \dfrac{hf}{c} = \dfrac{h}{\lambda}$, we find

$$p = \frac{hf}{c} = \frac{(6.63 \times 10^{-34}\,\text{J} \cdot \text{s})(5.39 \times 10^{14}\,\text{Hz})}{3.00 \times 10^{8}\,\text{m/s}} = 1.19 \times 10^{-27}\,\text{kg} \cdot \text{m/s}$$

PHYSICS IN CONTEXT
Looking Back

The expression for total relativistic energy (Chapter 29) is used here to find the momentum of a photon.

PHYSICS IN CONTEXT
Looking Ahead

Photons are fascinating in that they have zero mass and yet have finite momentum and energy. They also travel at the speed of light, which is not possible for any finite-mass object. We use the concept of photons again in Chapter 31 when we study atomic radiation.

PROBLEM-SOLVING NOTE

The Momentum of a Photon

The relationship $p = mv$ is not valid for photons. In fact, photons have a finite momentum even though their rest mass is zero.

RWP As small as the momentum of a photon is, it can still have a significant impact if the number of photons is large. For example, NASA is studying the feasibility of constructing spaceships that would be powered by huge "light sails" that transfer the momentum of photons from the Sun to the ship. A sail would reflect photons, creating a change in momentum, and hence a reaction force on the sail due to the **radiation pressure** of the photons. With a large enough sail, it may one day be possible to cruise among the stars with space-faring sailboats. In fact, the designs under consideration by NASA would result in the largest and fastest spacecraft ever constructed.

RWP A more "down to Earth," though no less exotic, application of radiation pressure is the *optical tweezer*. Basically, an optical tweezer is a laser beam that is made more intense in the middle than near the edges. When a small, translucent object is placed in this beam, the recoil produced by photons passing through the object exerts a small force on it directed toward the center of the beam. The situation is similar to a ball suspended in the "beam" of air produced by a hair dryer. Although the force exerted by such tweezers is typically on the order of only 10^{-12} N, it is still large enough to manipulate cells, DNA, and other subcellular particles. In fact, optical tweezers can even capture, lift, and separate a single bacterium from a bacterial culture for further study.

Enhance Your Understanding
(Answers given at the end of the chapter)

3. If the wavelength of a beam of light is increased, does the momentum of photons in the beam increase, decrease, or stay the same? Explain.

Section Review

- The momentum of a photon of frequency f is $p = hf/c = h/\lambda$. Photons have momentum even though their mass is zero.

30-4 Photon Scattering and the Compton Effect

Einstein's photon model of the photoelectric effect, published in 1905, focused on the energy of a photon, $E = hf$. The momentum of a photon, $p = hf/c = h/\lambda$, plays a key role in a different type of experiment, in which an X-ray photon undergoes a collision with an electron initially at rest. The result of this collision is that the photon is scattered, which changes its direction and energy. This type of process, referred to as the **Compton effect,** was explained in terms of the photon model of light by the American physicist Arthur Holly Compton (1892–1962) in 1923. The success of the photon model in explaining both the Compton effect and the photoelectric effect gained it widespread acceptance.

Compton Scattering   In **FIGURE 30-9** we see an X-ray photon striking an electron at rest and scattering at an angle θ with respect to the incident direction. To understand the behavior of a system like this, we treat the photon as a particle, with a certain energy and momentum, that collides with another particle (the electron) of mass m_e and initial speed zero. If we assume the electron is free to move, this collision conserves both energy and momentum.

► **FIGURE 30-9 The Compton effect** An X-ray photon scattering from an electron at rest can be thought of as a collision between two particles. The result is a change of wavelength for the scattered photon. This is referred to as the Compton effect.

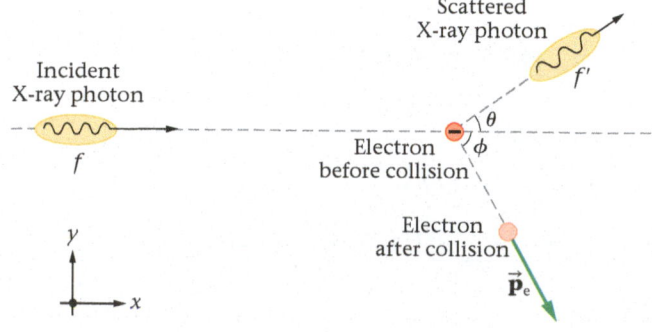

First, to conserve energy the following relationship must be satisfied:

energy of incident photon = energy of scattered photon + final kinetic energy of electron

$$hf = hf' + K \qquad \text{30-12}$$

In general, the frequency of the scattered photon, f', is less than the frequency of the incident photon, f, because part of the energy of the incident photon has gone into kinetic energy of the electron. Because the frequency of the scattered photon is reduced, its wavelength, $\lambda' = c/f'$, is increased.

Next, we conserve both the x and y components of momentum, using $p = h/\lambda$ as the magnitude of the incident photon's momentum, and $p' = h/\lambda'$ as the magnitude of the scattered photon's momentum. For the x component of momentum we have the following relationship:

$$\frac{h}{\lambda} = \frac{h}{\lambda'}\cos\theta + p_e\cos\phi \qquad \text{30-13}$$

Note that we assume the electron has a momentum p_e at an angle ϕ to the incident direction. Conserving the y component of momentum, which initially is zero, we have

$$0 = \frac{h}{\lambda'}\sin\theta - p_e\sin\phi \qquad \text{30-14}$$

Given the initial wavelength and frequency, $\lambda = c/f$, and the scattering angle, θ, we can use the three relationships, Equations 30-12, 30-13, and 30-14, to solve for the three unknowns: λ', ϕ, and K.

Of particular interest is the change in wavelength produced by the scattering. We find the following result:

Compton Shift Formula

$$\Delta\lambda = \lambda' - \lambda = \frac{h}{m_e c}(1 - \cos\theta) \qquad \text{30-15}$$

SI unit: m

In this expression, the quantity $h/m_e c = 2.43 \times 10^{-12}$ m is referred to as the *Compton wavelength of an electron*. The maximum change in the photon's wavelength occurs when it scatters in the reverse direction ($\theta = 180°$), in which case the change is twice the Compton wavelength, $\Delta\lambda = 2(h/m_e c)$. On the other hand, if the scattering angle is zero ($\theta = 0$)—in which case there really is no scattering at all—the change in wavelength is zero; $\Delta\lambda = 0$.

PHYSICS IN CONTEXT
Looking Back

We study collisions involving photons in the same way we studied collisions involving massive objects in Chapter 9.

PROBLEM-SOLVING NOTE

Scattering Angle

The angle θ in the Compton shift formula is the angle between the incident direction of an X-ray and its direction of propagation after scattering.

CONCEPTUAL EXAMPLE 30-10   CHANGE IN WAVELENGTH

If X-rays are scattered from protons instead of from electrons, is the change in their wavelength for a given angle of scattering increased, decreased, or unchanged?

REASONING AND DISCUSSION

As can be seen from Equation 30-15 [$\Delta\lambda = \lambda' - \lambda = (h/m_e c)(1 - \cos\theta)$], the change in wavelength for a given angle is proportional to $h/m_e c$. If a proton is substituted for the electron, the change in wavelength will be proportional to $h/m_p c$ instead. Because protons have about 2000 times the mass of electrons, the change in wavelength will be reduced by a factor of about 2000.

ANSWER

The change in wavelength is decreased.

The next Example gives a detailed analysis of a photon-electron collision.

EXAMPLE 30-11   SCATTERING X-RAYS

An X-ray photon with a wavelength of 0.650 nm scatters from a free electron at rest. After scattering, the photon moves at an angle of 152° relative to the incident direction. Find **(a)** the wavelength and **(b)** the energy of the scattered photon. **(c)** Determine the kinetic energy of the recoiling electron.

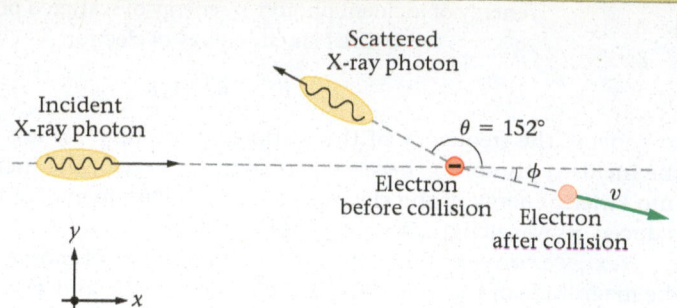

PICTURE THE PROBLEM

As shown in the sketch, the incoming photon is scattered at an angle of 152° relative to its initial direction; thus, it almost heads back the way it came. After the collision, the electron moves with a speed v at an angle ϕ relative to the forward direction.

REASONING AND STRATEGY

a. We can find the wavelength after scattering, λ', by using the Compton shift formula, $\lambda' - \lambda = (h/m_e c)(1 - \cos \theta)$. We are given the initial wavelength, λ, as well as the scattering angle, θ.

b. The energy of the scattered photon is $E' = hf' = hc/\lambda'$.

c. Because energy is conserved, the kinetic energy of the electron is given by $hf = hf' + K$. We can solve this relationship to yield the kinetic energy, K.

Known   Wavelength, $\lambda = 0.650$ nm; scattering angle, $\theta = 152°$.
Unknown   **(a)** Wavelength after scattering, $\lambda' = ?$ **(b)** Energy of scattered photon, $E' = ?$ **(c)** Kinetic energy of recoiling electron, $K = ?$

SOLUTION

Part (a)

1. Use $\lambda' - \lambda = (h/m_e c)(1 - \cos \theta)$ to find the wavelength of the scattered photon:

$$\lambda' = \lambda + \frac{h}{m_e c}(1 - \cos \theta) = 0.650 \times 10^{-9} \text{ m}$$

$$+ \left(\frac{6.63 \times 10^{-34} \text{ J} \cdot \text{s}}{(9.11 \times 10^{-31} \text{ kg})(3.00 \times 10^8 \text{ m/s})} \right)(1 - \cos 152°)$$

$$= 6.55 \times 10^{-10} \text{ m} = 0.655 \text{ nm}$$

Part (b)

2. The energy of the scattered photon is given by $E' = hf' = hc/\lambda'$:

$$E' = hf' = \frac{hc}{\lambda'} = \frac{(6.63 \times 10^{-34} \text{ J} \cdot \text{s})(3.00 \times 10^8 \text{ m/s})}{6.55 \times 10^{-10} \text{ m}}$$

$$= 3.04 \times 10^{-16} \text{ J}$$

Part (c)

3. Find the initial energy of the photon using $E = hf = hc/\lambda$:

$$E = hf = \frac{hc}{\lambda} = \frac{(6.63 \times 10^{-34} \text{ J} \cdot \text{s})(3.00 \times 10^8 \text{ m/s})}{6.50 \times 10^{-10} \text{ m}}$$

$$= 3.06 \times 10^{-16} \text{ J}$$

4. Subtract the final energy of the photon from its initial energy to find the kinetic energy of the electron:

$$K = hf - hf' = 3.06 \times 10^{-16} \text{ J} - 3.04 \times 10^{-16}$$

$$= 2 \times 10^{-18} \text{ J} = 10 \text{ eV}$$

INSIGHT

The energy of the X-ray photon is roughly 1900 eV = 1.9 keV both before and after scattering. Only a very small fraction of its energy is delivered to the electron. Still, the electron acquires about 10 eV of energy—enough to ionize an atom.

PRACTICE PROBLEM

At what scattering angle will the photon have a wavelength of 0.652 nm?   **[Answer: $\theta = 79.9°$]**

Some related homework problems: Problem 56, Problem 57

As we've seen in both the photoelectric effect and the Compton effect, a photon can behave like a particle with a well-defined energy and momentum. We also know, however, that light exhibits wavelike behavior, as when it produces interference fringes in Young's two-slit experiment (Chapter 28). That light can have such seemingly opposite attributes is one of the deepest mysteries of quantum physics and, at the same time, one of its most significant insights. In the next section we expand on this insight and extend it to the behavior of matter.

Enhance Your Understanding (Answers given at the end of the chapter)

4. A photon is scattered by an electron. Is the change in wavelength of the photon greater when the angle of scattering is 0° or 180°? Explain.

Section Review

• The change in wavelength of a photon scattering through an angle from an electron of mass m_e is $\Delta\lambda = \lambda' - \lambda = (h/m_e c)(1 - \cos\theta)$.

30-5 The de Broglie Hypothesis and Wave–Particle Duality

As we've seen, the Compton effect was explained in terms of photons in 1923. Another significant advance in quantum physics occurred that same year, when a French graduate student, Louis de Broglie (1892–1987), put forward a most remarkable hypothesis that would later win him the Nobel Prize in Physics. His suggestion was basically the following:

> Because light, which we usually think of as a wave, can exhibit particle-like behavior, perhaps a particle of matter, like an electron, can exhibit wavelike behavior.

In particular, de Broglie proposed that the *same* relationship between wavelength and momentum that Compton applied to the photon, $p = h/\lambda$, should apply to particles as well. Thus, if the momentum of a particle is p, its de Broglie wavelength is:

de Broglie Wavelength

$$\lambda = \frac{h}{p} \qquad\qquad 30\text{-}16$$

SI unit: m

The greater a particle's momentum, the smaller its de Broglie wavelength.

How can the idea of a wavelength for matter make sense, however, when we know that objects like baseballs and cars behave like particles, not like waves? To see how this is possible, we calculate the de Broglie wavelength of a typical macroscopic object: a 0.13-kg apple moving with a speed of 5.0 m/s. Substituting these values into $p = mv$, and using $\lambda = h/p$, we find the following wavelength: $\lambda = 1.0 \times 10^{-33}$ m. Clearly, this wavelength, which is smaller than the diameter of an atom by a factor of 10^{23}, is much too small to be observed in any macroscopic experiment. Thus, an apple could have a wavelength as given by the de Broglie relationship, and we would never notice it.

In contrast, consider an electron with a kinetic energy of 10.0 eV, a typical atomic energy. Using $K = \frac{1}{2}mv^2 = p^2/2m$ to solve for the momentum p, we find that the de Broglie wavelength in this case is $\lambda = 3.88 \times 10^{-10}$ m = 3.88 Å. Now this wavelength, which is comparable to the size of an atom or molecule, would clearly be significant in such systems. Therefore, the de Broglie wavelength may be unobservable in macroscopic systems but all-important in atomic systems.

PHYSICS IN CONTEXT
Looking Ahead

De Broglie waves appear again in Chapter 31 when we study the Bohr model of the atom. As we shall see, the allowed Bohr orbits correspond to standing waves formed from the de Broglie waves of electrons.

QUICK EXAMPLE 30-12 THE SPEED AND WAVELENGTH OF AN ELECTRON

How fast is an electron moving if its de Broglie wavelength is 5.16×10^{-7} m?

REASONING AND SOLUTION

We can start by replacing the momentum in the de Broglie relationship, $\lambda = h/p$, with the classical expression $p = mv$. We then rearrange to solve for the speed, v.

1. Write Equation 30-16 ($\lambda = h/p$) in terms of the electron's mass m and speed v, using $p = mv$: $\qquad \lambda = \dfrac{h}{mv}$

CONTINUED

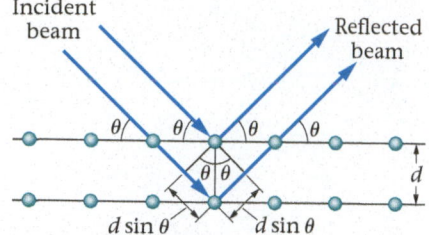

▲ FIGURE 30-10 Scattering from a crystal Scattering of X-rays or particles from a crystal. Note that waves reflecting from the lower plane of atoms have a path length that is longer than the path of the upper waves by the amount $2d \sin\theta$.

**PHYSICS
IN CONTEXT
Looking Back**

Diffraction and two-slit interference (Chapter 28) play a key role in understanding neutron diffraction and electron interference.

Big Idea 3 Light is generally thought of as a wave, but it also has particle-like properties, such as coming in discrete units called photons. Similarly, particles (like electrons and protons) can have wavelike properties, such as showing interference and diffraction patterns.

▶ FIGURE 30-11 Diffraction patterns Diffraction patterns can be observed by passing a beam of X-rays or particles through a crystal. The beams emerging from the crystal are at specific angles, due to constructive interference, and can be recorded on photographic film.

▶ FIGURE 30-12 X-ray diffraction Because the spacing of atoms in a crystal is of the same order of magnitude as the wavelengths of X-rays, the planes of a crystal can serve as a diffraction grating for X-rays. When properly interpreted, the resulting diffraction patterns can provide remarkably detailed information about the structure of the crystal. The historic photograph at left, made by two of the pioneers of X-ray crystallography in 1912, records the diffraction pattern from a simple inorganic salt. The photograph at right, made nearly a century later, shows the much more intricate pattern produced by a large protein molecule.

2. Rearrange to solve for the speed, v: $\qquad v = \dfrac{h}{m\lambda}$

3. Substitute numerical values:

$$v = \frac{h}{m\lambda} = \frac{(6.63 \times 10^{-34}\,\text{J}\cdot\text{s})}{(9.11 \times 10^{-31}\,\text{kg})(5.16 \times 10^{-7}\,\text{m})}$$
$$= 1410\,\text{m/s}$$

The relatively small value obtained for the electron's speed justifies using the nonrelativistic expression for momentum, $p = mv$.

In order for the de Broglie wavelength to be taken seriously, however, it must be observed experimentally. We next consider ways in which this can be done.

Diffraction of X-Rays and Particles by Crystals

An especially powerful way to investigate wave properties is through the study of interference patterns. Consider, for example, directing a coherent beam of X-rays onto a crystalline substance composed of regularly spaced planes of atoms, as indicated in **FIGURE 30-10**. Notice that the reflected beam combines rays that have followed different paths, with different path lengths. If the difference in path lengths is half a wavelength, destructive interference occurs; if the difference in path lengths is one wavelength, constructive interference results, and so on. From Figure 30-10 it is clear that the difference in path lengths for rays reflecting from adjacent planes that have a spacing d is $2d \sin\theta$. Thus, constructive interference occurs when the following conditions are met:

Constructive Interference When Scattering from a Crystal

$$2d \sin\theta = m\lambda \quad m = 1, 2, 3, \ldots \qquad \qquad 30\text{-}17$$

This is very similar to the condition for constructive interference in a diffraction grating, as derived in Equation 28-16. The resulting pattern of interference maxima—referred to as a **diffraction pattern**—can be projected onto photographic film, as indicated in **FIGURE 30-11**, and used to determine the geometrical properties of a crystal. Examples of diffraction patterns are shown in **FIGURE 30-12**.

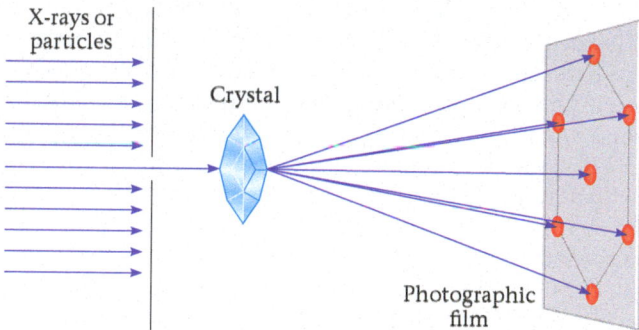

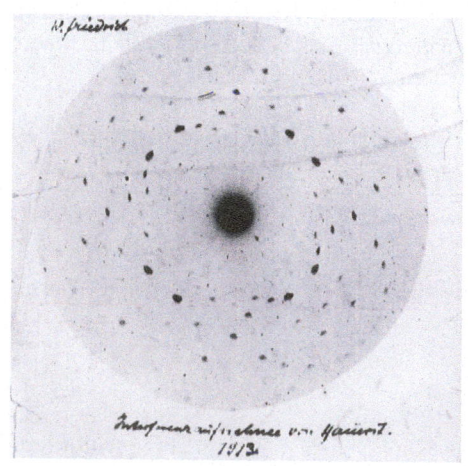

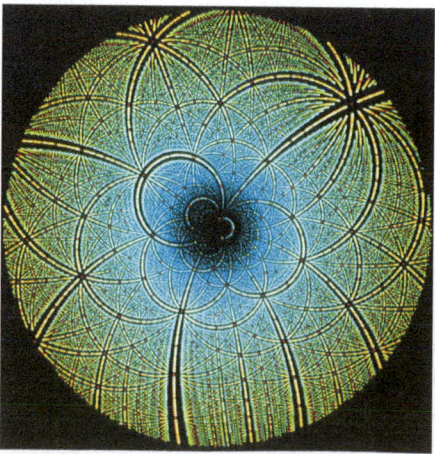

▶ **FIGURE 30-13 Neutron diffraction** The neutron diffraction pattern of the protein lysozyme, a digestive enzyme.

Now, if particles—like electrons, for example—have a wavelength comparable to atomic distances, it should be possible to produce diffraction patterns with them in much the same way as with X-rays. This is indeed the case, as was shown by the American physicists C. J. Davisson and L. H. Germer, who, in 1928, produced diffraction patterns by scattering low-energy electrons (about 54 eV) from crystals of nickel. The spacing between spots in the electron diffraction pattern allowed the researchers to determine the electron's wavelength, which verified in detail the de Broglie relationship, $\lambda = h/p$. Similar diffraction patterns have since been observed with neutrons, hydrogen atoms, and helium atoms. An example of a neutron diffraction pattern is shown in **FIGURE 30-13**.

EXAMPLE 30-13   NEUTRON DIFFRACTION

A beam of neutrons moving with a speed of 1450 m/s is diffracted from a crystal of table salt (sodium chloride), which has an interplanar spacing of $d = 0.282$ nm. **(a)** Find the de Broglie wavelength of the neutrons. **(b)** Find the angle of the first interference maximum.

PICTURE THE PROBLEM
Our sketch shows a beam of neutrons reflecting from two adjacent atomic planes of table salt. The distance d between the planes is 0.282 nm.

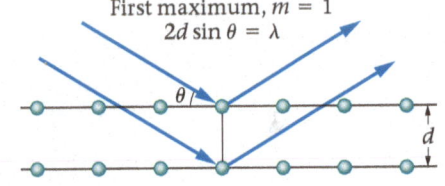

First maximum, $m = 1$
$2d \sin \theta = \lambda$

REASONING AND STRATEGY
a. We can find the de Broglie wavelength using Equation 30-16: $\lambda = h/p$. Because the speed of the neutrons is much less than the speed of light, we can write the momentum as $p = m_n v$, where m_n is the mass of a neutron.

b. Referring to Equation 30-17 ($2d \sin \theta = m\lambda$, where $m = 1, 2, 3, \dots$), we see that the first interference maximum occurs when $m = 1$ in the relationship $2d \sin \theta = m\lambda$. Thus we set $2d \sin \theta$ equal to λ and solve for θ.

Known   Speed of neutrons, $v = 1450$ m/s; interplanar spacing, $d = 0.282$ nm.
Unknown   **(a)** de Broglie wavelength, $\lambda = ?$ **(b)** Angle of first interference maximum, $\theta = ?$

SOLUTION

Part (a)

1. Calculate the de Broglie wavelength using $\lambda = h/p = h/m_n v$:

$$\lambda = \frac{h}{p} = \frac{h}{m_n v} = \frac{6.63 \times 10^{-34} \, \text{J} \cdot \text{s}}{(1.67 \times 10^{-27} \, \text{kg})(1450 \, \text{m/s})} = 0.274 \, \text{nm}$$

Part (b)

2. Set $m = 1$ in $2d \sin \theta = m\lambda$:

$$2d \sin \theta = \lambda$$

3. Solve for the angle θ:

$$\theta = \sin^{-1}\left(\frac{\lambda}{2d}\right) = \sin^{-1}\left(\frac{0.274 \, \text{nm}}{2(0.282 \, \text{nm})}\right) = 29.1°$$

INSIGHT
If we change the speed of the neutrons, we will change their wavelength. This, in turn, changes the angle of the interference maxima. This connection between the neutron speed and the interference maxima provides a detailed and precise way of verifying the de Broglie relationship.

PRACTICE PROBLEM
Find the angle of the first interference maximum if the neutron speed is increased to 2250 m/s.   [**Answer: 18.2°**]

Some related homework problems: Problem 63, Problem 66

RWP In *electron microscopes*, a beam of electrons is used to form an image of an object in much the same way that a beam of light is used in a light microscope. One of the differences, however, is that the wavelength of the electrons can be much shorter than the wavelength of visible light. For example, the shortest visible wavelength is that of blue light, which is about 380 nm. In comparison, we have seen that the de Broglie wavelength of an electron with an energy of only 10.0 eV is 0.388 nm—about 1000 times smaller than the wavelength of blue light. Because the ability to resolve small objects depends on using a wavelength that is smaller than the object to be imaged, an electron microscope can see much finer detail than a light microscope, as **FIGURE 30-14** indicates.

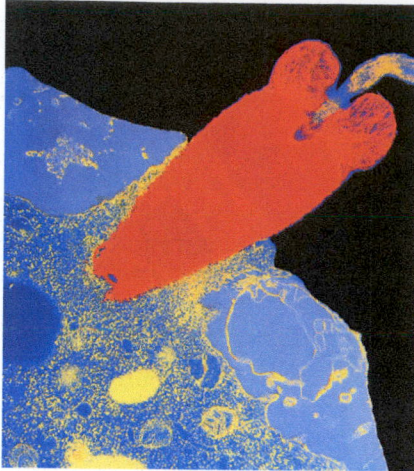
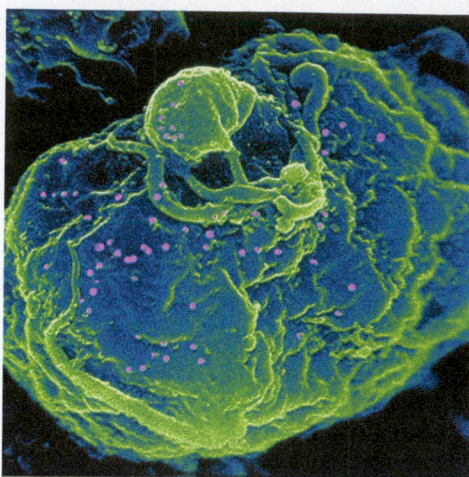

▶ **FIGURE 30-14 Electron micrography** Electron micrographs can be produced in several different ways. In a transmission electron micrograph (TEM), such as the image at left, the beam of electrons passes through the specimen, as in an ordinary light microscope. This colorized photo shows a sea urchin sperm cell penetrating the membrane of an egg. In a scanning electron micrograph (SEM), like the image at right, the electrons are reflected from the surface of the specimen (which is usually coated with a thin layer of metal atoms first) and used to produce a startlingly detailed three-dimensional image. This photo, also colorized for greater clarity, shows particles of the AIDS virus HIV (purple) emerging from an infected human cell.

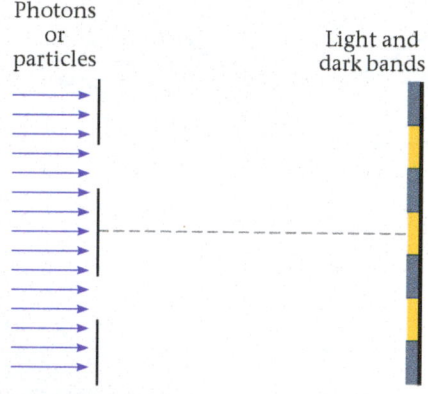

▲ **FIGURE 30-15 The two-slit experiment** When photons or particles are passed through a screen with two slits, the result is an interference pattern of light and dark bands.

Wave–Particle Duality

Notice that we have now come full circle in our study of waves and particles, from considering light as a wave, and then noting that it has particle-like properties, to investigating particles of matter like electrons, and finding that they have wavelike properties. This type of behavior is referred to as the **wave–particle duality.** Thus, as strange as it may seem, light is a wave, but it also comes in discrete units called photons. Electrons come in discrete units of well-defined mass and charge, but they also have wave properties.

To illustrate the wave–particle duality, we consider a two-slit experiment, as shown in **FIGURE 30-15**. If light is passed through these slits, it forms interference patterns of dark and light fringes, as we saw in Chapter 28. If the intensity of the light is reduced to a very low level, it is possible to have only a single photon at a time passing through the apparatus. This photon lands on the distant screen. Eventually, as more and more photons land on the screen, the interference pattern emerges.

Similar behavior can be observed with electrons passing through a pair of slits. The results, as shown in **FIGURE 30-16**, are the direct analog of the results obtained with light. Notice that the dark portions of the interference pattern are those places where electron waves passing through the two slits have combined to produce destructive interference—that is, at these special points an electron is essentially able to "cancel itself out" to produce a dark fringe.

▶ **FIGURE 30-16 Creation of an interference pattern by electrons passing through two slits** At first, the electrons seem to arrive at the screen in random locations. As the number of electrons increases, however, the interference pattern becomes more evident. (Compare Figures 28-4 and 28-5.)

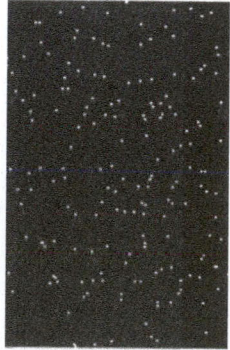

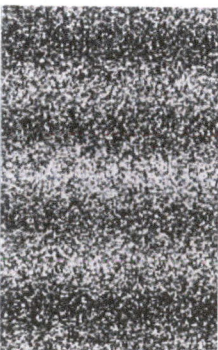

Clearly, then, subatomic particles and light are quite different from objects and waves that we observe on a macroscopic level. In fact, one of the most profound and unexpected insights of quantum physics is simply that even though baseballs, apples, and people are composed of electrons, protons, and neutrons, the behavior of these subatomic particles is nothing like the behavior of baseballs, apples, and people. In short, an electron is not like a metal ball reduced in size. An electron has properties that are different from those of any object we experience on the macroscopic level. To try to force light and electrons into categories like waves and particles is to miss the essence of their existence—they are neither one nor the other, though they have characteristics of both. Again, one is reminded of the saying "The universe is not only stranger than we imagine, it is stranger than we *can* imagine."

Enhance Your Understanding (Answers given at the end of the chapter)

5. The de Broglie waves for two particles of equal mass are shown in **FIGURE 30-17**. Is the kinetic energy of particle 1 greater than, less than, or equal to the kinetic energy of particle 2? Explain.

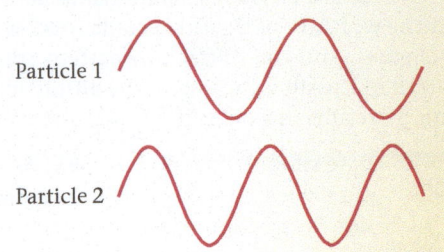

Particle 1

Particle 2

▲ **FIGURE 30-17** Enhance Your Understanding 5.

Section Review

- The de Broglie wavelength of a particle with a momentum p is $\lambda = h/p$. This is the same relationship that is satisfied by photons.

30-6 The Heisenberg Uncertainty Principle

One of the more interesting aspects of the interference pattern in Figure 30-16 is that the dots, corresponding to observations of individual electrons, appear on the screen in random order. To be specific, each time this experiment is run, the dots appear in different locations and in different sequence—all that remains the same is the final pattern that emerges after a large number of electrons are observed. The point is that as any given electron passes through the two-slit apparatus, *it is not possible to predict exactly where that one electron will land on the screen.* We can give the probability that it will land at different locations, but the fate of each individual electron is uncertain.

This kind of "uncertainty" is inherent in quantum physics and is due to the fact that matter has wavelike properties. As a simple example, consider a beam of electrons moving in the x direction and passing through a single slit, as in **FIGURE 30-18**. The result of this experiment is a diffraction pattern similar to that found for light in Chapter 28 (see Figure 28-24). In particular, if the beam passes through a slit of width W, it produces a large central maximum—where the probability of detecting an electron is high—with a dark fringe on either side, at an angle θ given by Equation 28-12 with $m = 1$:

$$\sin \theta = \frac{\lambda}{W} \qquad \text{30-18}$$

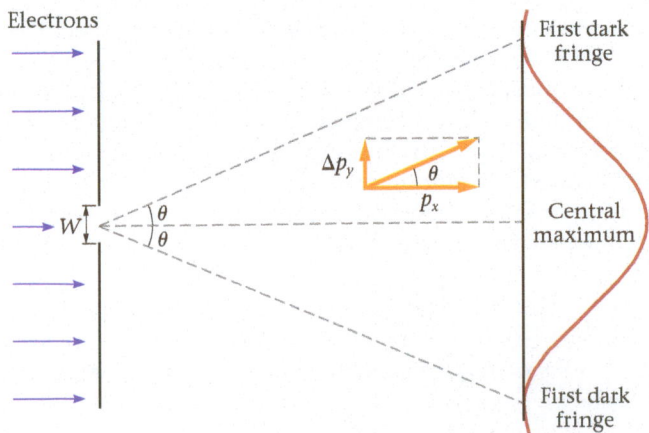

Electrons

W

Δp_y
θ
p_x

First dark fringe

Central maximum

First dark fringe

◀ **FIGURE 30-18 Diffraction pattern of electrons** Central region of the diffraction pattern formed by electrons passing through a single slit in a screen. The curve to the right is a measure of the number of electrons detected at any given location. Note that no electrons are detected at the dark fringes. This diffraction pattern is identical in form with that produced by light passing through a single slit.

**PHYSICS
IN CONTEXT
Looking Back**

Diffraction (Chapter 28) forms the basis for our understanding of the uncertainty principle.

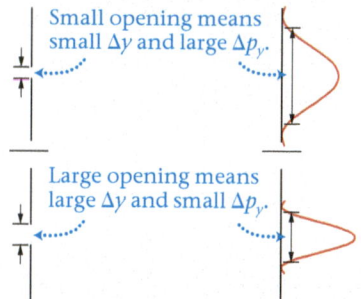

▲ **FIGURE 30-19 Uncertainty in position and momentum** Reciprocal relationship between the uncertainty in position (Δy) and the uncertainty in momentum (Δp_y). As in Figure 30-18, the curves at the right indicate the number of electrons detected at any given location.

Big Idea **4** One consequence of the wavelike properties of matter is that it is impossible to simultaneously know both a particle's position and its momentum. This implies an "uncertainty" in these quantities that is inherent in their quantum nature, and cannot be removed.

If the incoming electrons have a momentum p_x, the wavelength in Equation 30-18 is given by the de Broglie relationship: $\lambda = h/p_x$.

We interpret this experiment as follows: By passing the beam of electrons through the slit, we have determined their location in the y direction to within an uncertainty of roughly $\Delta y \sim W$. After passing through the slit, however, the beam spreads out to form a diffraction pattern—with some electrons acquiring a y component of momentum. Thus there is now an uncertainty in the y component of momentum, Δp_y.

There is a reciprocal relationship between Δy and Δp_y, as we show in **FIGURE 30-19**. As the width W of the slit is made smaller, to decrease Δy, the angle in Equation 30-18 increases, and the diffraction pattern spreads out, increasing Δp_y. Conversely, if the width of the slit is increased, the diffraction pattern becomes narrower; that is, increasing Δy results in a decrease in Δp_y. To summarize:

> If we know the position of a particle with greater precision, its momentum is more uncertain; if we know the momentum of a particle with greater precision, its position is more uncertain.

We can give this conclusion in mathematical form by returning to Equation 30-18 and Figure 30-18. First, assuming small angles, θ, we can use the approximation that $\sin \theta \sim \theta$ to write Equation 30-18 as $\theta \sim \lambda/W$. Similarly, from Figure 30-18 we see that $\tan \theta = \Delta p_y/p_x$. Because $\tan \theta \sim \theta$ for small angles, it follows that $\theta \sim \Delta p_y/p_x$. Equating these two expressions for θ, we get

$$\theta \sim \frac{\lambda}{W} \sim \frac{\Delta p_y}{p_x}$$

Setting $p_x = h/\lambda$ and $W \sim \Delta y$ yields

$$\frac{\lambda}{\Delta y} \sim \frac{\Delta p_y}{h/\lambda}$$

Finally, canceling λ and rearranging, we obtain

$$\Delta p_y\, \Delta y \sim h$$

Thus, the product of uncertainties in position and momentum cannot be less than a certain finite amount that is approximately equal to Planck's constant.

A thorough treatment of this system yields the more precise relationship first given by the German physicist Werner Heisenberg (1901–1976) in 1927, known as the **Heisenberg uncertainty principle:**

The Heisenberg Uncertainty Principle: Momentum and Position

$$\Delta p_y\, \Delta y \geq \frac{h}{2\pi} \qquad\qquad 30\text{-}19$$

In fact, Heisenberg showed that this relationship is a general principle and not restricted in any way to the single-slit system considered here. There is simply an unremovable, intrinsic uncertainty in nature that is the result of the wave behavior of matter. Because the wavelength of matter, $\lambda = h/p$, depends directly on the magnitude of h, so too does the uncertainty.

One way of stating the uncertainty principle is that it is impossible to know both the position and the momentum of a particle with arbitrary precision at any given time. For example, if the position is known precisely, so that Δy approaches zero, it follows that Δp_y must approach infinity, as shown in Figure 30-19. This result implies that the y component of momentum is completely uncertain. Similarly, complete knowledge of $p_y (\Delta p_y \to 0)$ implies that the position y is completely uncertain.

As one might expect, the uncertainty restrictions given in Equation 30-19 have negligible impact on macroscopic systems, but are all-important in atomic and nuclear systems. This is illustrated in the following Example.

EXAMPLE 30-14 WHAT IS YOUR POSITION?

If the speed of an object is 35.0 m/s, with an uncertainty of 5.00%, what is the minimum uncertainty in the object's position if it is **(a)** an electron or **(b)** a volleyball ($m = 0.350$ kg)?

CONTINUED

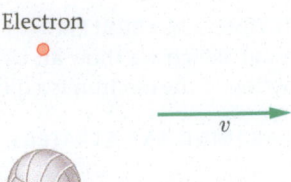

Electron

v

Volleyball

PICTURE THE PROBLEM
As shown in the sketch, the electron and the volleyball have the same speed. The fact that the electron has the smaller mass means that the values of its momentum, and of its uncertainty in momentum, are less than the corresponding values for the volleyball. As a result, the uncertainty in position will be larger for the electron.

REASONING AND STRATEGY
The 5.00% uncertainty in speed means that the magnitude of the momentum (mass times speed) of the electron and the volleyball are also uncertain by 5.00%. The minimum uncertainty in position, then, is given by the Heisenberg uncertainty principle: $\Delta y = h/(2\pi \, \Delta p_y)$.

Known   Speed, $v = 35.0$ m/s; uncertainty $= 5.00\% = 0.0500$.

Unknown (a) Uncertainty in momentum for electron ($m_e = 9.11 \times 10^{-31}$ kg), $\Delta p_y = ?$

   (b) Uncertainty in momentum for volleyball ($m = 0.350$ kg), $\Delta p_y = ?$

SOLUTION

Part (a)

1. Calculate the uncertainty in the electron's momentum:

$$\Delta p_y = 0.0500 m_e v = 0.0500(9.11 \times 10^{-31} \text{ kg})(35.0 \text{ m/s})$$
$$= 1.59 \times 10^{-30} \text{ kg} \cdot \text{m/s}$$

2. Use the uncertainty principle to find the minimum uncertainty in position:

$$\Delta y = \frac{h}{2\pi \, \Delta p_y} = \frac{6.63 \times 10^{-34} \text{ J} \cdot \text{s}}{2\pi(1.59 \times 10^{-30} \text{ kg} \cdot \text{m/s})} = 6.64 \times 10^{-5} \text{ m}$$

Part (b)

3. Calculate the uncertainty in the volleyball's momentum:

$$\Delta p_y = 0.0500 m_{\text{volleyball}} v$$
$$= 0.0500(0.350 \text{ kg})(35.0 \text{ m/s}) = 0.613 \text{ kg} \cdot \text{m/s}$$

4. As before, use the uncertainty principle to find the minimum uncertainty in position:

$$\Delta y = \frac{h}{2\pi \, \Delta p_y} = \frac{6.63 \times 10^{-34} \text{ J} \cdot \text{s}}{2\pi(0.613 \text{ kg} \cdot \text{m/s})} = 1.72 \times 10^{-34} \text{ m}$$

INSIGHT
To put these results in perspective, the minimum uncertainty in the position of the electron is roughly 100,000 times the size of an atom. Clearly, this is a significant uncertainty on the atomic level. On the other hand, the minimum uncertainty of the volleyball is smaller than an atom by a factor of about 10^{24}. It follows, then, that the uncertainty principle does not have measurable consequences on typical macroscopic objects.

PRACTICE PROBLEM — PREDICT/CALCULATE
Suppose the uncertainty in speed (and momentum) is reduced by a factor of two to 2.50%. (a) As a result, does the uncertainty in position increase, decrease, or stay the same? Explain. (b) Find the uncertainties in position for this case.   [**Answer:** (a) The uncertainty in position varies inversely with the uncertainty in momentum, and hence it increases by a factor of two. (b) Electron, $\Delta y = 1.33 \times 10^{-4}$ m; volleyball, $\Delta y = 3.44 \times 10^{-34}$ m]

Some related homework problems: Problem 72, Problem 73

The Heisenberg uncertainty principle also sets the typical energy scales in atomic and nuclear systems. For example, if an electron is known to be confined to an atom, the uncertainty in its position, Δy, will be roughly 1 Å = 0.1 nm. This implies a finite uncertainty for Δp_y, which in turn implies a finite kinetic energy—even though the average value of p_y is zero.

EXAMPLE 30-15   AN ELECTRON IN A BOX

Suppose an electron is confined to a box that is about 0.50 Å $= 0.50 \times 10^{-10}$ m on a side. If this distance is taken as the uncertainty in the position of the electron, (a) calculate the corresponding minimum uncertainty in the momentum. (b) Because the electron is confined to a stationary box, its average momentum is zero. The magnitude of the electron's momentum is nonzero, however. Assuming the magnitude of the electron's momentum is the same as its uncertainty in momentum, calculate the corresponding kinetic energy.

CONTINUED

PICTURE THE PROBLEM

In our sketch we show an electron bouncing around inside a box that is 0.50 Å on a side. The rapid motion of the electron is a quantum effect, due to the uncertainty principle.

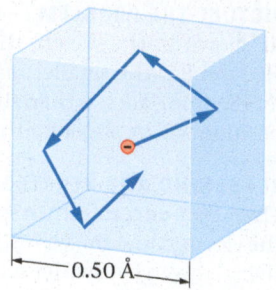

REASONING AND STRATEGY

a. Letting $\Delta y = 0.50$ Å $= 0.50 \times 10^{-10}$ m, we can find the minimum uncertainty in the momentum by using the Heisenberg uncertainty principle: $\Delta p_y = h/(2\pi \Delta y)$.

b. We can calculate the kinetic energy of the electron from its momentum by using $K = p^2/2m$. For the momentum, p, we will use the value of Δp_y obtained in part (a).

Known Uncertainty in position, $\Delta y = 0.50$ Å $= 0.50 \times 10^{-10}$ m; mass of electron, $m_e = 9.11 \times 10^{-31}$ kg.

Unknown (a) Uncertainty in momentum of electron, $\Delta p_y = ?$ (b) Kinetic energy of electron, $K = ?$

0.50 Å

SOLUTION

Part (a)

1. Use $\Delta y = 0.50$ Å $= 0.50 \times 10^{-10}$ m and the uncertainty principle to find Δp_y:

$$\Delta p_y = \frac{h}{2\pi \Delta y} = \frac{6.63 \times 10^{-34}\,\text{J}\cdot\text{s}}{2\pi(0.50 \times 10^{-10}\,\text{m})} = 2.1 \times 10^{-24}\,\text{kg}\cdot\text{m/s}$$

Part (b)

2. Set $p = \Delta p_y$, and use $K = p^2/2m$ to find the kinetic energy:

$$K = \frac{p^2}{2m} = \frac{(2.1 \times 10^{-24}\,\text{kg}\cdot\text{m/s})^2}{2(9.11 \times 10^{-31}\,\text{kg})} = 2.4 \times 10^{-18}\,\text{J} = 15\,\text{eV}$$

INSIGHT

The important point of this rough estimate of an electron's kinetic energy is that it is of the order of typical atomic energies. Thus, the fact that an electron is confined to an object the size of an atom means that its energy must be on the order of 10 eV. This is why the electron in our sketch is bouncing around inside the box rather than resting on its floor—as would be the case for a tennis ball inside a cardboard box. The uncertainty principle implies that the electron must have about 10 eV of kinetic energy when it is confined in this way; hence, it must be moving quite rapidly. We consider the case of a tennis ball in a cardboard box in the following Practice Problem.

PRACTICE PROBLEM

A tennis ball ($m = 0.06$ kg) is confined within a cardboard box 0.5 m on a side. Estimate the kinetic energy of this ball and its corresponding speed.   [**Answer:** $\Delta K \sim 4 \times 10^{-67}$ J, $\Delta v \sim 4 \times 10^{-33}$ m/s. For all practical purposes, the tennis ball appears to be at rest in the box.]

Some related homework problems: Problem 73, Problem 77, Problem 78

Finally, the uncertainty relationship $\Delta y \Delta p_y \geq h/2\pi$ is just one of many forms the uncertainty principle takes. There are a number of quantities like y and p_y that satisfy the same type of relationship between their uncertainties. Perhaps the most important of these other uncertainty principles is the one that relates the uncertainty in energy to the uncertainty in time:

The Heisenberg Uncertainty Principle: Energy and Time

$$\Delta E \Delta t \geq \frac{h}{2\pi}$$

30-20

For example, the shorter the half-life of an unstable particle, the greater the uncertainty in its energy.

CONCEPTUAL EXAMPLE 30-16   UNCERTAINTIES

If Planck's constant were magically reduced to zero, would the position-versus-momentum and energy-versus-time uncertainties increase or decrease?

REASONING AND DISCUSSION

If h were zero, the product of the uncertainties—$\Delta y \Delta p_y$ and $\Delta E \Delta t$—would be reduced to zero as well. The result is that position and momentum, for example, could be determined simultaneously with arbitrary accuracy—that is, with zero uncertainty. In this limit, particles would behave as predicted in classical physics, with well-defined positions and momenta at all times.

ANSWER

The uncertainties would decrease; in fact, they would decrease to zero, as in classical physics.

Thus, if h were zero, the classical description of particles moving along well-defined trajectories, and having no wavelike properties, would be valid. Notice, however, that Planck's constant has an extremely small magnitude, on the order of 10^{-34} in SI units. It is this incredibly small difference from zero that is responsible for all the quantum behavior seen on the atomic scale.

If h were relatively large, however, the wavelike properties of matter would be apparent even on the macroscopic level. For example, consider pitching a baseball toward the catcher's glove in a universe where h is rather large. If the ball is thrown with a well-defined momentum toward the glove, its position will be uncertain, and the ball may end up anywhere. Similarly, if the catcher gives the glove a relatively small uncertainty in position, its uncertainty in momentum is large, meaning that the glove is moving rapidly about its average position. So, even if the pitcher could aim the ball to go where desired—which is not possible—there is no way to know where to aim it! Clearly, if h were significantly larger than it actually is, our experience of the natural world would be significantly different.

Enhance Your Understanding   (Answers given at the end of the chapter)

6. Which of the following has the greater uncertainty in momentum: (a) an electron free to move throughout the length of a wire or (b) an electron confined to a particular atom?

Section Review

- The Heisenberg uncertainty principle as applied to position and momentum is $\Delta p_y \Delta y \geq h/2\pi$. Thus, for example, a reduced uncertainty in position implies an increased uncertainty in momentum.

- The Heisenberg uncertainty principle for time and energy is $\Delta E \Delta t \geq h/2\pi$. Thus, for example, a reduced uncertainty in time implies an increased uncertainty in energy.

Big Idea 5 The wavelike nature of matter allows particles to "tunnel" through regions that would be forbidden to them in classical physics.

30-7 Quantum Tunneling

Because particles have wavelike properties, any behavior seen in waves can be found in particles as well, under the right conditions. An example is the phenomenon known as **tunneling**, in which a wave, or a quantum particle, "tunnels" through a region of space that would be forbidden to it if it were simply a small piece of matter like a classical particle.

FIGURE 30-20 shows a case of tunneling by light. In Figure 30-20 (a) we see a beam of light propagating within glass and undergoing total internal reflection (Chapter 26) when it encounters the glass–air interface. If light were composed of classical particles, this would be the end of the story; the particles of light would simply change direction at the interface and continue propagating within the glass.

What can actually happen in such a system is more interesting, however. In Figure 30-20 (b) we show a second piece of glass brought near the first piece, but with a small vacuum gap between them. If the gap is small, but still finite, a weak beam of light is observed in the second piece of glass, propagating in the same direction as the original beam. We say that the light waves have "tunneled" across the gap. The strength of the beam of light in the second piece of glass depends very sensitively on the size of the gap through which it tunnels, decreasing exponentially as the width of the gap is increased.

Just as with light, electrons and other particles can tunnel across gaps that would be forbidden to them if they were classical particles. An electron, for instance, is not simply a point of mass; instead, it has wave properties that extend outward from it like ripples on a pond. These waves, like those of light, can "feel" the surroundings of an electron, allowing it to "tunnel" through a barrier to a region on the other side where it can propagate.

RWP An example of electron tunneling is shown in **FIGURE 30-21**, where we illustrate the operation of a scanning tunneling microscope, or STM. In the lower part of the figure we see the material, or specimen, to be investigated with the microscope.

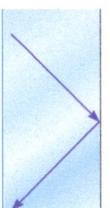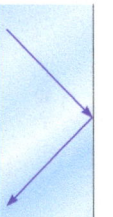

(a) Total internal reflection   **(b)** Light tunnels across the gap

▲ **FIGURE 30-20 Optical tunneling (a)** An incident beam of light undergoes total internal reflection from the glass–air interface. **(b)** If a second piece of glass is brought near the first piece, a weak beam of light is observed continuing in the same direction as the incident beam. We say that the light has "tunneled" across the gap.

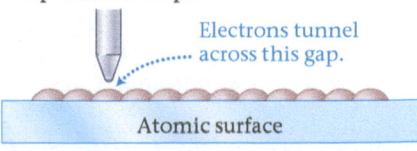

Tip of microscope

Electrons tunnel across this gap.

Atomic surface

▲ **FIGURE 30-21 Operation of a scanning tunneling microscope**

The upper portion of the figure shows the key element of the microscope—a small, pointed tip of metal that can be moved up and down with piezoelectric supports. This tip is brought very close to the specimen being observed, leaving a small vacuum gap between them. Classically, electrons in the specimen are not able to move across the gap to the tip, but in reality they can tunnel to the tip and create a small electric current. The number of electrons that tunnel, and thus the magnitude of the tunneling current, depends on the width of the gap.

In one version of the STM, the tunneling current between the specimen and the tip is held constant. Imagine, for example, that the tip in Figure 30-21 is scanned horizontally from left to right. Because the tunneling current is very sensitive to the size of the gap between the tip and the specimen, the tip must be moved up and down with the contours of the specimen in order to maintain a gap of constant width. The tip is moved by sending electrical voltage to the piezoelectric supports; thus, the voltage going to the supports is a measure of the height of the surface being scanned and can be converted to a visual image of the surface, as in **FIGURE 30-22**. The resolution of these microscopes is on the atomic level, showing the hills and valleys created by atoms on the surface of the material being examined.

▶ **FIGURE 30-22 Scanning tunneling microscopy** STM images are particularly good at recording the "terrain" of surfaces at the atomic level. The image at left shows the hexagonal structure of graphene, a form of carbon. At right, a DNA molecule sits on a substrate of graphite. Three turns of the double helix are visible, magnified about 2,000,000 times.

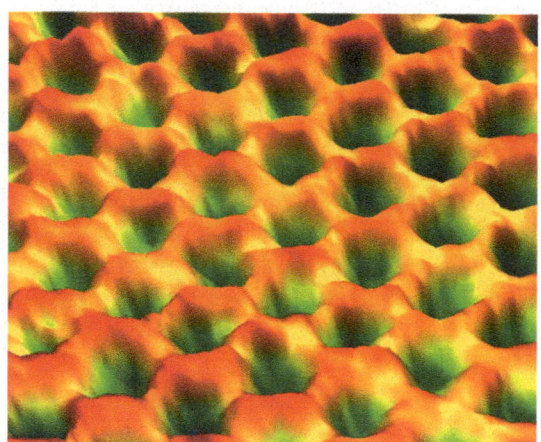
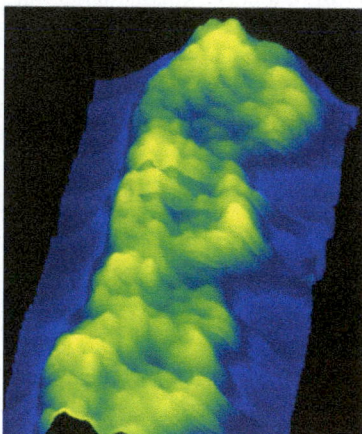

Enhance Your Understanding   (Answers given at the end of the chapter)

7. If the de Broglie wavelength of a particle is increased, does the probability that it will tunnel through a given barrier increase, decrease, or stay the same? Explain.

Section Review

- The wavelike nature of particles allows them to move, or "tunnel," through regions of space that would be forbidden to them if they were classical particles.

CHAPTER 30 REVIEW

CHAPTER SUMMARY

30-1 BLACKBODY RADIATION AND PLANCK'S HYPOTHESIS OF QUANTIZED ENERGY
An ideal blackbody is an object that absorbs all the light incident on it.
Wien's Displacement Law
The frequency at which the radiation from a blackbody is maximum is given by the following relationship:

$$f_{peak} = (5.88 \times 10^{10} \, s^{-1} \cdot K^{-1})T \qquad\qquad 30\text{-}1$$

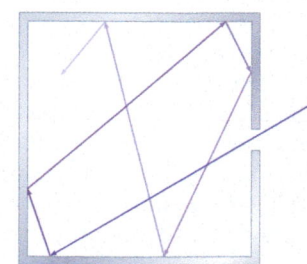

Planck's Quantum Hypothesis

Planck hypothesized that the energy in a blackbody at a frequency f must be an integer multiple of the constant $h = 6.63 \times 10^{-34}$ J·s; that is,

$$E_n = nhf \quad n = 0, 1, 2, 3, \dots \qquad \text{30-2}$$

The constant h is known as Planck's constant.

30-2 PHOTONS AND THE PHOTOELECTRIC EFFECT

Light is composed of particle-like photons, which carry energy in discrete amounts.

Energy of a Photon

The energy of a photon depends on its frequency. A photon with the frequency f has the energy

$$E = hf \qquad \text{30-4}$$

Noting the relationship $\lambda f = c$, we can also express the energy of a photon in terms of its wavelength, $E = hc/\lambda$.

Low-intensity light beam

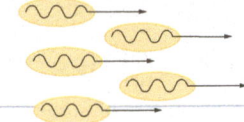

High-intensity light beam

30-3 THE MASS AND MOMENTUM OF A PHOTON

A photon is like a "typical" particle in some respects but different in others. In particular, a photon has zero rest mass, yet it still has a nonzero momentum.

Momentum of a Photon

The momentum, p, of a photon of frequency f and wavelength $\lambda = c/f$ is given by

$$p = \frac{hf}{c} = \frac{h}{\lambda} \qquad \text{30-11}$$

30-4 PHOTON SCATTERING AND THE COMPTON EFFECT

Photons can undergo collisions with particles, in much the same way that particles can collide with other particles.

The Compton Effect

If a photon of wavelength λ undergoes a collision with an electron (mass $= m_e$) and scatters into a new direction at an angle θ from its incident direction, its new wavelength, λ', is given by the Compton shift formula:

$$\Delta\lambda = \lambda' - \lambda = \frac{h}{m_e c}(1 - \cos\theta) \qquad \text{30-15}$$

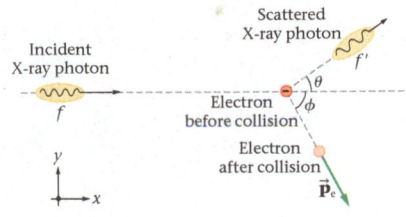

30-5 THE DE BROGLIE HYPOTHESIS AND WAVE–PARTICLE DUALITY

de Broglie Wavelength

According to de Broglie, the relationship between momentum and wavelength should be the same for both light and particles. Thus, the de Broglie wavelength of a particle of momentum p is

$$\lambda = \frac{h}{p} \qquad \text{30-16}$$

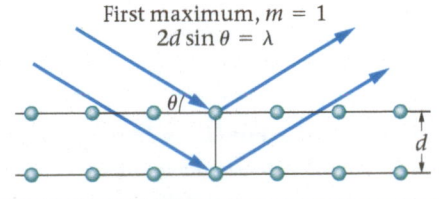

30-6 THE HEISENBERG UNCERTAINTY PRINCIPLE

Because particles have wavelengths and can behave as waves, their position and momentum cannot be determined simultaneously with arbitrary precision.

The Heisenberg Uncertainty Principle: Momentum and Position

In terms of momentum and position, the Heisenberg uncertainty principle states the following:

$$\Delta p_y \Delta y \geq \frac{h}{2\pi} \qquad \text{30-19}$$

Thus, as momentum is determined more precisely, the position becomes more uncertain, and vice versa.

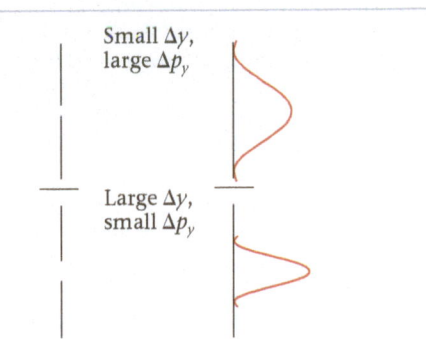

30-7 QUANTUM TUNNELING

Particles, because of their wavelike behavior, can pass through regions of space that would be forbidden to a classical particle. This phenomenon is referred to as tunneling.

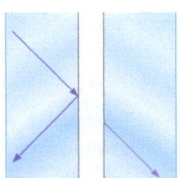

ANSWERS TO ENHANCE YOUR UNDERSTANDING QUESTIONS

1. **(a)** The peak frequency increases with temperature. **(b)** The total energy radiated increases with temperature.

2. **(a)** Increasing the wavelength of light decreases its frequency, which, in turn, decreases the energy of a photon. **(b)** The energy of the photons stays the same. Increasing the intensity means the beam of light delivers more photons per second.

3. The momentum of a photon is inversely proportional to its wavelength. Therefore, increasing the wavelength decreases the momentum.

4. The change in wavelength is greater when the photon is reversed in direction—that is, when the scattering angle is 180°.

5. Particle 2 has the smaller wavelength, and hence it has the greater momentum. It follows that the kinetic energy ($p^2/2m$) of particle 1 is less than the kinetic energy of particle 2.

6. **(b)** An electron confined to an atom has a fairly small uncertainty in position, and hence it has a greater uncertainty in momentum.

7. The longer the de Broglie wavelength, the more likely a particle can tunnel through a barrier because tunneling is a quantum phenomenon that depends on the wavelike nature of particles. As the de Broglie wavelength gets smaller and goes toward zero, tunneling will no longer occur.

CONCEPTUAL QUESTIONS

(Answers to odd-numbered Conceptual Questions can be found in the back of the book.)

For instructor-assigned homework, go to www.masteringphysics.com.

1. Give a brief description of the "ultraviolet catastrophe."

2. How does Planck's hypothesis of energy quantization resolve the "ultraviolet catastrophe"?

3. Is there a lowest temperature below which blackbody radiation is no longer given off by an object? Explain.

4. How can an understanding of blackbody radiation allow us to determine the temperature of distant stars?

5. **Differential Fading** Many vehicles in the United States have a small American flag decal in one of their windows. If the decal has been in place for a long time, the colors will show some fading from exposure to the Sun, as shown in **FIGURE 30-23**. In fact, the red stripes are generally more faded than the blue background for the stars,

FIGURE 30-23 Differential fading. (Conceptual Question 5)

as shown in the photo. Photographs and posters react in the same way, with red colors showing the most fading. Explain this effect in terms of the photon model of light.

6. A source of light is monochromatic. What can you say about the photons emitted by this source?

7. **(a)** Is it possible for a photon from a green source of light to have more energy than a photon from a blue source of light? Explain. **(b)** Is it possible for a photon from a green source of light to have more energy than a photon from a red source of light? Explain.

8. Why does the existence of a cutoff frequency in the photoelectric effect argue in favor of the photon model of light?

9. Why can an electron microscope resolve smaller objects than a light microscope?

10. A proton is about 2000 times more massive than an electron. Is it possible for an electron to have the same de Broglie wavelength as a proton? Explain.

PROBLEMS AND CONCEPTUAL EXERCISES

Answers to odd-numbered Problems and Conceptual Exercises can be found in the back of the book. **BIO** *identifies problems of biological or medical interest;* **CE** *indicates a conceptual exercise.* **Predict/Explain** *problems ask for two responses: (a) your prediction of a physical outcome, and (b) the best explanation among three provided; and* **Predict/Calculate** *problems ask for a prediction of a physical outcome, based on fundamental physics concepts, and follow that with a numerical calculation to verify the prediction. On all problems, bullets (•, ••, •••) indicate the level of difficulty.*

SECTION 30-1 BLACKBODY RADIATION AND PLANCK'S HYPOTHESIS OF QUANTIZED ENERGY

1. • **CE Predict/Explain** The blackbody spectrum of blackbody A peaks at a longer wavelength than that of blackbody B. **(a)** Is the temperature of blackbody A higher than or lower than the temperature of blackbody B? **(b)** Choose the *best explanation* from among the following:
   I. Blackbody A has the higher temperature because the higher the temperature the longer the wavelength.
   II. Blackbody B has the higher temperature because an increase in temperature means an increase in frequency, which corresponds to a decrease in wavelength.

2. • **The Surface Temperature of Betelgeuse** Betelgeuse, a red-giant star in the constellation Orion, has a peak in its radiation at a

frequency of 1.82×10^{14} Hz. What is the surface temperature of Betelgeuse?

3. • **Terrestrial Radiation** The average surface temperature of Earth is 288 K. What is the wavelength of the most intense radiation emitted by Earth into outer space? This radiation plays a major role in the cooling of the surface and atmosphere at night.

4. • The Sun has a surface temperature of about 5800 K. At what frequency does the Sun emit the most radiation?

5. •• **(a)** What is the frequency of the most intense radiation emitted by your body? Assume a skin temperature of 91 °F. **(b)** What is the wavelength of this radiation?

6. •• **The Cosmic Background Radiation** Outer space is filled with a sea of photons, created in the early moments of the universe. The

frequency distribution of this "cosmic background radiation" matches that of a blackbody at a temperature near 2.7 K. **(a)** What is the peak frequency of this radiation? **(b)** What is the wavelength that corresponds to the peak frequency?

7. •• **(a)** By what factor does the peak frequency change if the Kelvin temperature of an object is doubled from 20.0 K to 40.0 K? **(b)** By what factor does the peak frequency change if the Celsius temperature of an object is doubled from 20.0 °C to 40.0 °C?

8. •• **Predict/Calculate A Famous Double Star** Albireo in the constellation Cygnus, which appears as a single star to the naked eye, is actually a beautiful double-star system, as shown in **FIGURE 30-24.** The brighter of the two stars is referred to as A (or Beta-01 Cygni), with a surface temperature of $T_A = 4700$ K; its companion is B (or Beta-02 Cygni), with a surface temperature of $T_B = 13,000$ K. **(a)** When viewed through a telescope, one star is a brilliant blue color, and the other has a warm golden color, as shown in the photo. Is the blue star A or B? Explain. **(b)** What is the ratio of the peak frequencies emitted by the two stars, (f_A/f_B)?

FIGURE 30-24 The double star Albireo in the constellation Cygnus. (Problem 8)

9. •• **Predict/Calculate Halogen Lightbulbs** Halogen lightbulbs allow their filaments to operate at a higher temperature than the filaments in standard incandescent bulbs. For comparison, the filament in a standard lightbulb operates at about 2900 K, whereas the filament in a halogen bulb may operate at 3400 K. **(a)** Which bulb has the higher peak frequency? **(b)** Calculate the ratio of peak frequencies (f_{hal}/f_{std}). **(c)** The human eye is most sensitive to a frequency around 5.5×10^{14} Hz. Which bulb produces a peak frequency closer to this value?

10. •• **Predict/Calculate** An incandescent lightbulb contains a tungsten filament that reaches a temperature of about 3020 K, roughly half the surface temperature of the Sun. **(a)** Treating the filament as a blackbody, determine the frequency for which its radiation is a maximum. **(b)** Do you expect the lightbulb to radiate more energy in the visible or in the infrared part of the spectrum? Explain.

11. •• **Exciting an Oxygen Molecule** An oxygen molecule (O_2) vibrates with an energy identical to that of a single particle of mass $m = 1.340 \times 10^{-26}$ kg attached to a spring with a force constant of $k = 1215$ N/m. The energy levels of the system are uniformly spaced, as indicated in **FIGURE 30-25**, with a separation given by hf. **(a)** What is the vibration frequency of this molecule? **(b)** How much energy must be added to the molecule to excite it from one energy level to the next higher level?

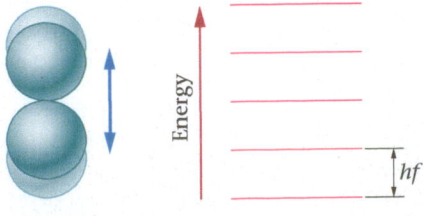

FIGURE 30-25 Problem 11

SECTION 30-2   PHOTONS AND THE PHOTOELECTRIC EFFECT

12. • **CE** A source of red light, a source of green light, and a source of blue light each produce beams of light with the same power. Rank these sources in order of increasing **(a)** wavelength of light, **(b)** frequency of light, and **(c)** number of photons emitted per second. Indicate ties where appropriate.

13. • **CE** Predict/Explain A source of red light has a higher wattage than a source of green light. **(a)** Is the energy of photons emitted by the red source greater than, less than, or equal to the energy of photons emitted by the green source? **(b)** Choose the *best explanation* from among the following:
    I. The photons emitted by the red source have the greater energy because that source has the greater wattage.
    II. The red-source photons have less energy than the green-source photons because they have a lower frequency. The wattage of the source doesn't matter.
    III. Photons from the red source have a lower frequency, but that source also has a greater wattage. The two effects cancel, so the photons have equal energy.

14. • **CE** Predict/Explain A source of yellow light has a higher wattage than a source of blue light. **(a)** Is the number of photons emitted per second by the yellow source greater than, less than, or equal to the number of photons emitted per second by the blue source? **(b)** Choose the *best explanation* from among the following:
    I. The yellow source emits more photons per second because (i) it emits more energy per second than the blue source, and (ii) its photons have less energy than those of the blue source.
    II. The yellow source has the higher wattage, which means its photons have higher energy than the blue-source photons. Therefore, the yellow source emits fewer photons per second.
    III. The two sources emit the same number of photons per second because the higher wattage of the yellow source compensates for the higher energy of the blue photons.

15. • **CE** Predict/Explain Light of a particular wavelength does not eject electrons from the surface of a given metal. **(a)** Should the wavelength of the light be increased or decreased in order to cause electrons to be ejected? **(b)** Choose the *best explanation* from among the following:
    I. The photons have too little energy to eject electrons. To increase their energy, their wavelength should be increased.
    II. The energy of a photon is proportional to its frequency; that is, inversely proportional to its wavelength. To increase the energy of the photons so they can eject electrons, one must decrease their wavelength.

16. • **CE** Light of a particular wavelength and intensity does not eject electrons from the surface of a given metal. Can electrons be ejected from the metal by increasing the intensity of the light? Explain.

17. • When people visit the local tanning salon, they absorb photons of ultraviolet (UV) light to get the desired tan. What are the frequency and wavelength of a UV photon whose energy is 5.9×10^{-19} J?

18. • An AM radio station operating at a frequency of 810 kHz radiates 290 kW of power from its antenna. How many photons are emitted by the antenna every second?

19. • A photon with a wavelength of less than 50.4 nm can ionize a helium atom. What is the ionization potential of helium?

20. • A flashlight emits 2.9 W of light energy. Assuming a frequency of 5.2×10^{14} Hz for the light, determine the number of photons given off by the flashlight per second.

21. • Light of frequency 9.95×10^{14} Hz ejects electrons from the surface of silver. If the maximum kinetic energy of the ejected electrons is 0.180×10^{-19} J, what is the work function of silver?

22. •• The work function of platinum is 6.35 eV. What frequency of light must be used to eject electrons from a platinum surface with a maximum kinetic energy of 2.88×10^{-19} J?

23. •• **(a)** How many 350-nm (UV) photons are needed to provide a total energy of 2.5 J? **(b)** How many 750-nm (red) photons are needed to provide the same energy?

24. •• **(a)** How many photons per second are emitted by a monochromatic lightbulb ($\lambda = 650$ nm) that emits 45 W of power? **(b)** If you stand 15 m from this bulb, how many photons enter each of your eyes per second? Assume your pupil is 5.0 mm in diameter and that the bulb radiates uniformly in all directions.

25. •• **Dissociating the Hydrogen Molecule** The energy required to separate a hydrogen molecule into its individual atoms is 104.2 kcal per mole of H_2. **(a)** If the dissociation energy for a single H_2 molecule is provided by one photon, determine its frequency and wavelength. **(b)** In what region of the electromagnetic spectrum does the photon found in part (a) lie? (Refer to the spectrum shown in Figure 25-12.)

26. •• **(a)** How many photons are emitted per second by a He-Ne laser that emits 2.2 mW of power at a wavelength $\lambda = 632.8$ nm? **(b)** What is the frequency of the electromagnetic waves emitted by a He-Ne laser?

27. •• **Predict/Calculate** You have two lightbulbs of different power and color, as indicated in **FIGURE 30-26**. One is a 150-W red bulb, and the other is a 40-W blue bulb. **(a)** Which bulb emits more photons per second? **(b)** Which bulb emits photons of higher energy? **(c)** Calculate the number of photons emitted per second by each bulb. Take $\lambda_{red} = 640$ nm and $\lambda_{blue} = 430$ nm. (Most of the electromagnetic radiation given off by incandescent lightbulbs is in the infrared portion of the spectrum. For the purposes of this problem, however, assume that all of the radiated power is at the wavelengths indicated.)

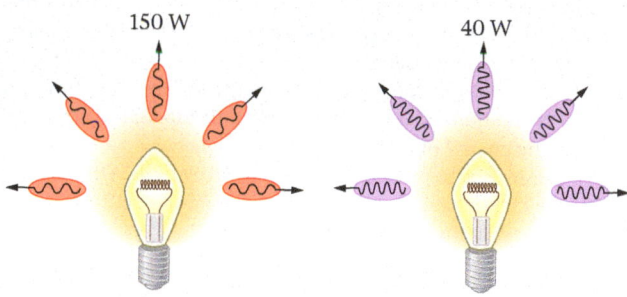

150 W   40 W

FIGURE 30-26 Problem 27

28. •• The maximum wavelength an electromagnetic wave can have and still eject an electron from a copper surface is 264 nm. What is the work function of a copper surface?

29. •• **Predict/Calculate** Aluminum and calcium have photoelectric work functions of $W_{Al} = 4.08$ eV and $W_{Ca} = 2.87$ eV, respectively. **(a)** Which metal requires higher-frequency light to produce photoelectrons? Explain. **(b)** Calculate the minimum frequency that will produce photoelectrons from each surface.

30. •• **Predict/Calculate** Two beams of light with different wavelengths ($\lambda_A > \lambda_B$) are used to produce photoelectrons from a given metal surface. **(a)** Which beam produces photoelectrons with greater kinetic energy? Explain. **(b)** Find K_{max} for cesium ($W_0 = 2.14$ eV) if $\lambda_A = 620$ nm and $\lambda_B = 410$ nm.

31. •• **Predict/Calculate** Zinc and cadmium have photoelectric work functions given by $W_{Zn} = 4.33$ eV and $W_{Cd} = 4.22$ eV, respectively. **(a)** If both metals are illuminated by UV radiation of the same wavelength, which one gives off photoelectrons with the greater maximum kinetic energy? Explain. **(b)** Calculate the maximum kinetic energy of photoelectrons from each surface if $\lambda = 275$ nm.

32. •• White light, with frequencies ranging from 4.00×10^{14} Hz to 7.90×10^{14} Hz, is incident on a barium surface. Given that the work function of barium is 2.52 eV, find **(a)** the maximum kinetic energy of electrons ejected from this surface and **(b)** the range of frequencies for which no electrons are ejected.

33. •• Electromagnetic waves, with frequencies ranging from 4.00×10^{14} Hz to 9.00×10^{16} Hz, are incident on an aluminum surface. Given that the work function of aluminum is 4.08 eV, find **(a)** the maximum kinetic energy of electrons ejected from this surface and **(b)** the range of frequencies for which no electrons are ejected.

34. •• **Predict/Calculate** Platinum has a work function of 6.35 eV, and iron has a work function of 4.50 eV. Light of frequency 1.88×10^{15} Hz ejects electrons from both of these surfaces. **(a)** From which surface will the ejected electrons have a greater maximum kinetic energy? Explain. **(b)** Calculate the maximum kinetic energy of ejected electrons for each surface.

35. •• When light with a frequency $f_1 = 547.5$ THz illuminates a metal surface, the most energetic photoelectrons have 1.260×10^{-19} J of kinetic energy. When light with a frequency $f_2 = 738.8$ THz is used instead, the most energetic photoelectrons have 2.480×10^{-19} J of kinetic energy. Using these experimental results, determine the approximate value of Planck's constant.

36. •• **BIO Owl Vision** Owls have large, sensitive eyes for good night vision. Typically, the pupil of an owl's eye can have a diameter of 8.5 mm (as compared with a maximum diameter of about 7.0 mm for humans). In addition, an owl's eye is about 100 times more sensitive to light of low intensity than a human eye, allowing owls to detect light with an intensity as small as 5.0×10^{-13} W/m². Find the minimum number of photons per second an owl can detect, assuming a frequency of 7.0×10^{14} Hz for the light.

37. ••• **BIO Thermal Photon Emission** A child with a fever has an inner-ear temperature of 37.9 °C. Treat the eardrum as a blackbody with a radius of 2.5 mm. **(a)** At what frequency f_{peak} is the most intense radiation emitted by the ear? **(b)** At what rate is energy emitted by the child's ear into surroundings with the temperature 293 K? Use Equation 16-19 and assume the emissivity of the eardrum is 1.00. **(c)** How many photons per second are emitted by the eardrum if they all have the same frequency as you found in part (a)?

SECTION 30-3 THE MASS AND MOMENTUM OF A PHOTON

38. • The photons used in microwave ovens have a momentum of 5.1×10^{-33} kg·m/s. **(a)** What is their wavelength? **(b)** How does the wavelength of the microwaves compare with the size of the holes in the metal screen on the door of the oven?

39. • What speed must an electron have if its momentum is to be the same as that of an X-ray photon with a wavelength of 0.25 nm?

40. • What is the wavelength of a photon that has the same momentum as an electron moving with a speed of 1500 m/s?

41. • What is the frequency of a photon that has the same momentum as a neutron moving with a speed of 1200 m/s?

42. •• A hydrogen atom, initially at rest, emits an ultraviolet photon with a wavelength of $\lambda = 122$ nm. What is the recoil speed of the atom after emitting the photon?

43. •• A blue-green photon ($\lambda = 486$ nm) is absorbed by a free hydrogen atom, initially at rest. What is the recoil speed of the hydrogen atom after absorbing the photon?

44. •• A hydrogen atom, initially at rest, absorbs an ultraviolet photon with a wavelength of $\lambda = 102.6$ nm. What is the atom's final speed if it now emits an identical photon in a direction that is **(a)** perpendicular or **(b)** opposite to the direction of motion of the original photon?

45. •• A plutonium atom with a mass of 4.05×10^{-25} kg emits a gamma photon with a wavelength of 0.0282 nm from its nucleus. What is the recoil speed of the atom after emitting the photon?

46. •• **Human Recoil** A person with a surface temperature of 35 °C emits about 1.36×10^{22} photons per second into an environment at 20 °C. Suppose the person has a mass of 70.0 kg and stands on a frictionless surface. Find the recoil speed of the person after 1.00 h assuming all the photons are emitted in the same direction with a wavelength of 16 μm.

47. •• **Laser Cooling** Lasers can cool a group of atoms by slowing them down, because the slower the atoms, the lower their temperature(see Equation 17-12). A rubidium atom of mass 1.42×10^{-25} kg and speed 212 m/s undergoes laser cooling when it absorbs a photon of wavelength 781 nm that is traveling in a direction opposite to the motion of the atom. **(a)** This occurs a total of 7500 times in rapid succession. What is the atom's new speed after the 7500 absorption events? **(b)** How many such absorption events are required to bring the rubidium atom to rest from its initial speed of 212 m/s?

48. ••• A laser produces a 5.00-mW beam of light, consisting of photons with a wavelength of 632.8 nm. **(a)** How many photons are emitted by the laser each second? **(b)** The laser beam strikes a black surface and is absorbed. What is the change in the momentum of each photon that is absorbed? **(c)** What force does the laser beam exert on the black surface?

49. ••• A laser produces a 7.50-mW beam of light, consisting of photons with a wavelength of 632.8 nm. **(a)** How many photons are emitted by the laser each second? **(b)** The laser beam strikes a mirror at normal incidence and is reflected. What is the change in momentum of each reflected photon? Give the magnitude only. **(c)** What force does the laser beam exert on the mirror?

SECTION 30-4   PHOTON SCATTERING AND THE COMPTON EFFECT

50. • **CE** Predict/Explain An X-ray photon scatters from a free electron at an angle θ from its initial direction of motion. As the scattering angle θ increases, does the change in wavelength $\Delta\lambda$ of the photon increase, decrease, or remain the same? **(b)** Choose the *best explanation* from among the following:
  I. A larger scattering angle means a larger transfer of momentum and thus a larger change in wavelength of the photon.
  II. A larger scattering angle means a larger change in frequency and thus a smaller change in wavelength.
  III. Photons always scatter elastically, so that the energy of the scattered photon does not change and neither does its wavelength.

51. • An X-ray photon has 38.0 keV of energy before it scatters from a free electron, and 33.5 keV after it scatters. What is the kinetic energy of the recoiling electron?

52. • In the Compton effect, an X-ray photon scatters from a free electron. Find the *change* in the photon's wavelength if it scatters at an angle of **(a)** $\theta = 30.0°$, **(b)** $\theta = 90.0°$, and **(c)** $\theta = 180.0°$ relative to the incident direction.

53. • An X-ray scattering from a free electron is observed to change its wavelength by 3.03 pm. At what angle to the incident direction does the scattered X-ray move?

54. •• The maximum Compton shift in wavelength occurs when a photon is scattered through 180°. What scattering angle will produce a wavelength shift of one-fourth the maximum?

55. •• Predict/Calculate Consider two different photons that scatter through an angle of 180° from a free electron. One is a visible-light photon with $\lambda = 520$ nm, the other is an X-ray photon with $\lambda = 0.030$ nm. **(a)** Which (if either) photon experiences the greater change in wavelength as a result of the scattering? Explain. **(b)** Which photon experiences the greater percentage change in wavelength? Explain. **(c)** Calculate the percentage change in wavelength of each photon.

56. •• An X-ray photon with a wavelength of 0.240 nm scatters from a free electron at rest. The scattered photon moves at an angle of 125° relative to its incident direction. Find **(a)** the initial momentum and **(b)** the final momentum of the photon.

57. •• An X-ray photon scatters from a free electron at rest at an angle of 115° relative to the incident direction. **(a)** If the scattered photon has a wavelength of 0.320 nm, what is the wavelength of the incident photon? **(b)** Determine the energy of the incident and scattered photons. **(c)** Find the kinetic energy of the recoil electron.

58. ••• A photon has an energy E and wavelength λ before scattering from a free electron. After scattering through a 135° angle, the photon's wavelength has increased by 10.0%. Find the initial wavelength and energy of the photon.

59. ••• Find the direction of propagation of the scattered electron in Problem 53, given that the incident X-ray has a wavelength of 0.525 nm and propagates in the positive x direction.

SECTION 30-5   THE DE BROGLIE HYPOTHESIS AND WAVE — PARTICLE DUALITY

60. • **CE** Predict/Explain **(a)** As you accelerate your car away from a stoplight, does the de Broglie wavelength of the car increase, decrease, or stay the same? **(b)** Choose the *best explanation* from among the following:
  I. The de Broglie wavelength will increase because the momentum of the car has increased.
  II. The momentum of the car increases. It follows that the de Broglie wavelength will decrease, because it is inversely proportional to the wavelength.
  III. The de Broglie wavelength of the car depends only on its mass, which doesn't change by pulling away from the stoplight. Therefore, the de Broglie wavelength stays the same.

61. • **CE** By what factor does the de Broglie wavelength of a particle change if **(a)** its momentum is doubled or **(b)** its kinetic energy is doubled? Assume the particle is nonrelativistic.

62. • A particle with a mass of 6.69×10^{-27} kg has a de Broglie wavelength of 7.42 pm. What is the particle's speed?

63. • What speed must a neutron have if its de Broglie wavelength is to be equal to the interionic spacing of table salt (0.282 nm)?

64. • A 71-kg jogger runs with a speed of 4.2 m/s. If the jogger is considered to be a particle, what is her de Broglie wavelength?

65. • Find the kinetic energy of an electron whose de Broglie wavelength is 2.5 Å.

66. •• A beam of neutrons with a de Broglie wavelength of 0.250 nm diffracts from a crystal of table salt, which has an interionic spacing of 0.282 nm. (a) What is the speed of the neutrons? (b) What is the angle of the second interference maximum?

67. •• **Predict/Calculate** An electron and a proton have the same speed. (a) Which has the longer de Broglie wavelength? Explain. (b) Calculate the ratio (λ_e/λ_p).

68. •• **Predict/Calculate** An electron and a proton have the same de Broglie wavelength. (a) Which has the greater kinetic energy? Explain. (b) Calculate the ratio of the electron's kinetic energy to the kinetic energy of the proton.

69. •• **Atom Trap** A researcher uses laser cooling (see Problem 47) to build an atomic trap that contains rubidium atoms of mass 1.42×10^{-25} kg. If the atoms are separated by 75 nm on average, what speed should they have so that their de Broglie wavelengths are equal to their average separation? (In general, overlapping de Broglie waves lead to quantum behavior. This is the basic technique used to produce a *Bose-Einstein condensate*, which is a quantum state of matter.)

70. •• Diffraction effects become significant when the width of an aperture is comparable to the wavelength of the waves being diffracted. (a) At what speed will the de Broglie wavelength of a 65-kg student be equal to the 0.76-m width of a doorway? (b) At this speed, how much time will it take the student to travel a distance of 1.0 mm? (For comparison, the age of the universe is approximately 4×10^{17} s.)

71. ••• A particle has a mass m and an electric charge q. The particle is accelerated from rest through a potential difference V. What is the particle's de Broglie wavelength, expressed in terms of m, q, and V?

SECTION 30-6 THE HEISENBERG UNCERTAINTY PRINCIPLE

72. • A baseball (0.15 kg) and an electron both have a speed of 38 m/s Find the uncertainty in position of each of these objects, given that the uncertainty in their speed is 5.0%.

73. • The uncertainty in position of a proton confined to the nucleus of an atom is roughly the diameter of the nucleus. If this diameter is 5.7×10^{-15} m, what is the uncertainty in the proton's momentum?

74. • The measurement of an electron's energy requires a time interval of 1.0×10^{-8} s. What is the smallest possible uncertainty in the electron's energy?

75. • An excited state of a particular atom has a mean lifetime of 0.60×10^{-9} s, which we may take as the uncertainty Δt. What is the minimum uncertainty in any measurement of the energy of this state?

76. •• **Quantum Well** A quantum well is a region of semiconductor or metal film that confines particles to a small region, resulting in quantized energies that are useful for the operation of a variety of electronic devices. If an electron in a quantum well is confined to an area that is 510 nm wide, what is the uncertainty in the electron's velocity along the same direction?

77. •• The uncertainty in an electron's position is 0.010 nm. (a) What is the minimum uncertainty Δp in its momentum? (b) What is the kinetic energy of an electron whose momentum is equal to this uncertainty $(\Delta p = p)$?

78. •• The uncertainty in a proton's position is 0.010 nm. (a) What is the minimum uncertainty Δp in its momentum? (b) What is the kinetic energy of a proton whose momentum is equal to this uncertainty $(\Delta p = p)$?

79. •• An electron has a momentum $p \approx 1.7 \times 10^{-25}$ kg·m/s. What is the minimum uncertainty in its position that will keep the relative uncertainty in its momentum $(\Delta p/p)$ below 1.0%?

GENERAL PROBLEMS

80. • **CE** Suppose you perform an experiment on the photoelectric effect using light with a frequency high enough to eject electrons. If the intensity of the light is increased while the frequency is held constant, describe whether the following quantities increase, decrease, or stay the same: (a) The maximum kinetic energy of an ejected electron; (b) the minimum de Broglie wavelength of an electron; (c) the number of electrons ejected per second; (d) the electric current in the phototube.

81. • **CE** Suppose you perform an experiment on the photoelectric effect using light with a frequency high enough to eject electrons. If the frequency of the light is increased while the intensity is held constant, describe whether the following quantities increase, decrease, or stay the same: (a) The maximum kinetic energy of an ejected electron; (b) the minimum de Broglie wavelength of an electron; (c) the number of electrons ejected per second; (d) the electric current in the phototube.

82. • **CE** An electron that is accelerated from rest through a potential difference V_0 has a de Broglie wavelength λ_0. What potential difference will double the electron's wavelength? (Express your answer in terms of V_0.)

83. • **BIO Human Vision** Studies have shown that some people can detect 545-nm light with as few as 100 photons entering the eye per second. What is the power delivered by such a beam of light?

84. •• You want to construct a photocell that works with visible light. Three materials are readily available: aluminum ($W_0 = 4.08$ eV), lead ($W_0 = 4.25$ eV), and cesium ($W_0 = 2.14$ eV). Which material(s) would be suitable?

85. •• The latent heat for converting ice at 0 °C to water at 0 °C is 80.0 kcal/kg (Chapter 17). (a) How many photons of frequency 6.0×10^{14} Hz must be absorbed by a 1.0-kg block of ice at 0 °C to melt it to water at 0 °C? (b) How many molecules of H_2O can one photon convert from ice to water?

86. •• A microwave oven can heat 205 mL of water from 20.0 °C to 90.0 °C in 2.00 min. If the wavelength of the microwaves is $\lambda = 12.2$ cm, how many photons were absorbed by the water? (Assume no loss of heat by the water.)

87. •• **Lunar Ranging Experiment** At the Apache Point Observatory in New Mexico, laser pulses of 532-nm light containing 0.115 J of energy are aimed at the Moon. (a) How many photons of light are in each pulse? (b) The laser pulse spreads out to a spot of radius 2.0 km at the Moon's surface. The retroreflector mirror array left by the *Apollo 15* astronauts has an area of about 0.60 m². How many laser photons hit the mirror array? (c) The reflected light spreads out to a spot of radius 2.0 km back at the Apache Point Observatory. How many reflected photons are collected by the 3.5-m-diameter telescope? (In practice only about 0.25 photon per pulse is collected.)

88. •• Light with a frequency of 1.14×10^{15} Hz ejects electrons from the surface of sodium, which has a work function of 2.28 eV. What is the minimum de Broglie wavelength of the ejected electrons?

89. •• An electron moving with a speed of 2.1×10^6 m/s has the same momentum as a photon. Find (a) the de Broglie wavelength of the electron and (b) the wavelength of the photon.

90. ••**BIO The Cold Light of Fireflies** Fireflies are often said to give off "cold light." Given that the peak in a firefly's radiation occurs at

about 5.4×10^{14} Hz, determine the temperature of a blackbody that would have the same peak frequency. From your result, would you say that firefly radiation is well approximated by blackbody radiation? Explain.

91. •• **Predict/Calculate** When light with a wavelength of 545 nm shines on a metal surface, electrons are ejected with speeds of 3.10×10^5 m/s or less. (a) Give a strategy that allows you to use the preceding information to calculate the work function and cutoff frequency for this surface. (b) Carry out your strategy and determine the work function and cutoff frequency.

92. •• When a beam of atoms emerges from an oven at the absolute temperature T, the most probable de Broglie wavelength for a given atom is

$$\lambda_{mp} = \frac{h}{\sqrt{5mkT}}$$

In this expression, m is the mass of an atom, and k is Boltzmann's constant (Chapter 17). What is the most probable speed of a hydrogen atom emerging from an oven at 450 K?

93. •• **Millikan's Photoelectric Experiment** Robert A. Millikan (1868–1953), although best known for his "oil-drop experiment," which measured the charge of an electron, also performed pioneering research on the photoelectric effect. In experiments on lithium, for example, Millikan observed a maximum kinetic energy of 0.550 eV when electrons were ejected with 433.9-nm light. When light of 253.5 nm was used, he observed a maximum kinetic energy of 2.57 eV. (a) What is the work function, W_0, for lithium, as determined from Millikan's results? (b) What maximum kinetic energy do you expect Millikan found when he used light with a wavelength of 365.0 nm?

94. •• A jar is filled with monatomic helium gas at a temperature of 25 °C. The pressure inside the jar is one atmosphere; that is, 101 kPa. (a) Find the average de Broglie wavelength of the helium atoms. (b) Calculate the average separation between helium atoms in the jar. (*Note:* The fact that the spacing between atoms is much greater than the de Broglie wavelength means quantum effects are negligible, and the atoms can be treated as particles.)

95. •• **The Compton Wavelength** The *Compton wavelength*, λ_C, of a particle of mass m is defined as follows: $\lambda_C = h/mc$. (a) Calculate the Compton wavelength of a proton. (b) Calculate the energy of a photon that has the same wavelength as found in part (a). (c) Show, in general, that a photon with a wavelength equal to the Compton wavelength of a particle has an energy that is equal to the rest energy of the particle.

96. ••• **Predict/Calculate** Light of frequency 8.22×10^{14} Hz ejects electrons from surface A with a maximum kinetic energy that is 2.00×10^{-19} J greater than the maximum kinetic energy of electrons ejected from surface B. (a) If the frequency of the light is increased, does the difference in maximum kinetic energy observed from the two surfaces increase, decrease, or stay the same? Explain. (b) Calculate the difference in work function for these two surfaces.

PASSAGE PROBLEMS

Hydrothermal Vents

Hydrothermal vents are the undersea equivalent of hot springs or geysers on land, in which water heated by magma in volcanically active areas emerges through the Earth's crust. The existence of hydrothermal vents was confirmed in 1976, when a research vehicle 2500 m below the surface of the ocean at the Galapagos Rift detected plumes of hot water and photographed the seafloor in the area, revealing a rich community of life dependent on the vents. At those depths, the pressure is so high that even though the water coming from the vents may have a temperature of several hundred degrees Celsius, the water doesn't boil. The hot water from a vent carries hydrogen sulfide and other minerals into the surrounding cold water. These sulfide compounds support populations of bacteria, which in turn support a rich web of life around the vent, including creatures such as crabs, tubeworms, and shrimp.

Some types of vents are called "black smokers" after the dark metal sulfide particles carried by the plume of hot water. The mineral deposits that precipitate from the hot water form a chimney around the plume. Near some of these vents researchers found an "eyeless" shrimp (*Rimicaris exoculata*), so called because it doesn't have the typical eyestalks and image-forming eyes found in other shrimp—not surprisingly, given that no sunlight penetrates to the depths where it lives. But the shrimp do have specialized dorsal organs that contain rhodopsin, a visual pigment. Researchers have found that even though the peak radiation emitted by the hot vents is outside the visual range of the electromagnetic spectrum, enough photons in the visual range come from the vents for the shrimp to be able to "see" the vents with their dorsal organs. In this way, they can locate the energy-producing bacteria but avoid the hot-enough-to-cook-shrimp water coming from the vents.

97. • At a typical vent temperature of 350 °C, what's the peak frequency of the radiation emitted by the vent, assuming the vent is a blackbody?

    A. 1.68×10^8 Hz   **B.** 4.53×10^{12} Hz
    C. 2.06×10^{13} Hz   **D.** 3.66×10^{13} Hz

98. • In which region of the electromagnetic spectrum is the peak frequency of the radiation emitted by the vent in the previous problem?

    A. ultraviolet   **B.** visible   **C.** infrared   **D.** microwave

99. • Rhodopsin is most sensitive to light with a vacuum wavelength of 500 nm. Does this light have a higher, lower, or the same frequency as the peak frequency of the vent radiation?

100. • For light emitted from a different vent, the energy of a photon at the peak frequency is 0.160 eV. What is the temperature of the vent, assuming this vent is also a black body?

    A. 160 °C   **B.** 280 °C   **C.** 380 °C   **D.** 660 °C

101. •• **Predict/Calculate** **Referring to Example 30-11** An X-ray photon with $\lambda = 0.6500$ nm scatters from an electron, giving the electron a kinetic energy of 7.750 eV. (a) Is the scattering angle of the photon greater than, less than, or equal to 152°? (b) Find the scattering angle.

102. •• **Predict/Calculate** **Referring to Example 30-11** An X-ray photon with $\lambda = 0.6500$ nm scatters from an electron. The wavelength of the scattered photon is 0.6510 nm. (a) Is the scattering angle in this case greater than, less than, or equal to 152°? (b) Find the scattering angle.

31

Atomic Physics

Big Ideas

1 Matter is composed of atoms—the smallest units of a given element.

2 Bohr's model of the hydrogen atom accounts for the observed spectrum of hydrogen.

3 The wavelike nature of matter finds support in the Bohr model of an atom.

4 Quantum mechanics provides a complete description of hydrogen as well as all other atoms.

5 The Pauli exclusion principle says that only one electron may occupy a given state.

6 The radiation given off by atoms is due to electrons moving from one orbital shell to another.

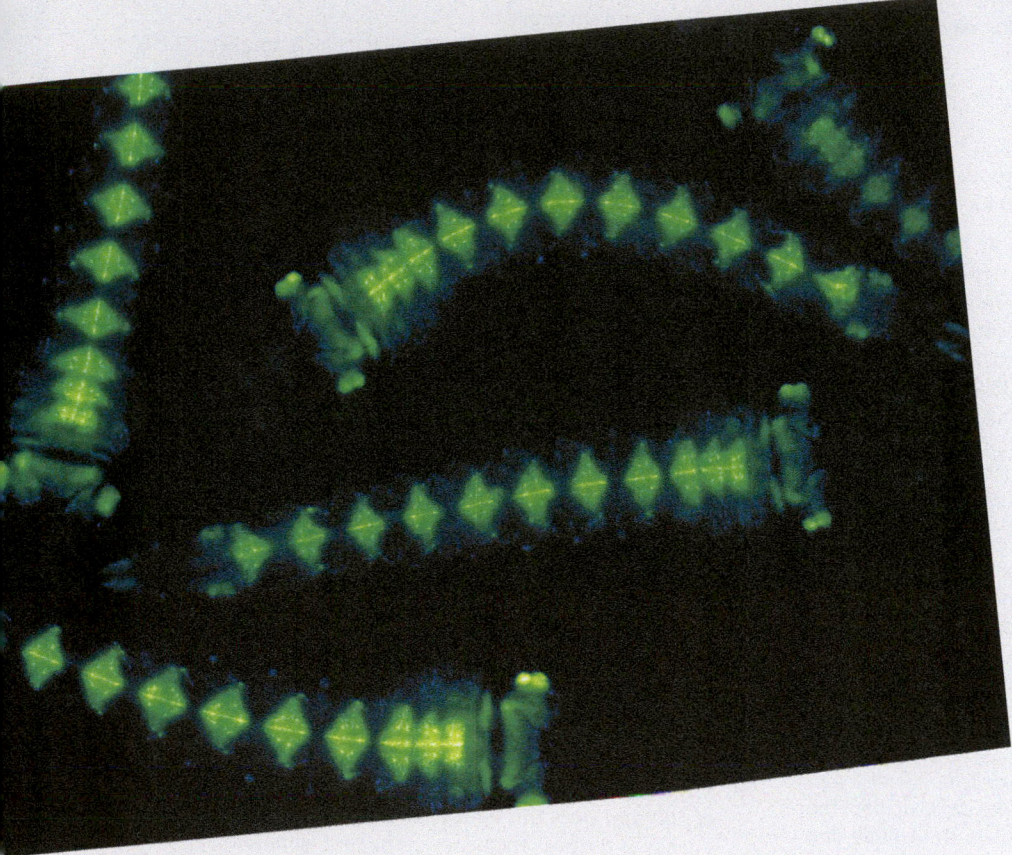

▲ A number of animals, including many species of coral, sponges, and even scorpions, emit beautiful and sometimes eerie colors when illuminated by ultraviolet light. We do not ordinarily expect mosquito larvae, like those shown here, to fall into this category. However, scientists have recently inserted a jellyfish gene that directs the production of green fluorescent protein (GFP) into mosquito larvae, which raises the hope that additional genes can be introduced to make them unable to transmit malaria. This chapter explores our modern understanding of the structure of the atom, which makes it possible to explain not only the phenomenon of fluorescence but many others as well, including the properties of the chemical elements and the generation of laser light.

In this chapter we begin by developing the quantum model of the simplest of all atoms—the hydrogen atom. We then show that the basic features of hydrogen apply to more complex atoms as well. As a result, we are able to understand—in detail—the arrangement of elements in the periodic table.

31-1 Early Models of the Atom

Speculations about the microscopic structure of matter have intrigued humankind for thousands of years. Ancient Greek philosophers, including Leucippus and Democritus, considered the question of what would happen if you took a small object, like a block of copper, and cut it in half, then cut it in half again, and again, for many subsequent divisions. They reasoned that eventually you would reduce the block to a single speck of copper that could not be divided further. This smallest piece of an element was called the **atom** (a + tom), which means, literally, "without division."

It was not until the late nineteenth century, however, that the question of atoms began to yield to direct scientific investigation. We now consider some of the more important early developments in atomic models that helped lead to our current understanding.

The Thomson Model: Plum Pudding

In 1897 the English physicist J. J. Thomson (1856–1940) discovered a "particle" that is smaller in size and thousands of times less massive than even the smallest atom. The **electron,** as this particle was named, was also found to have a negative electric charge—in contrast with atoms, which are electrically neutral. Thomson proposed, therefore, that atoms have an internal structure that includes both electrons and a quantity of positively charged matter. The latter would account for most of the mass of an atom, and would have a charge equal in magnitude to the charge on the electrons.

The picture of an atom that Thomson settled on is one he referred to as the "plum-pudding model." In this model, electrons are embedded in a more or less uniform distribution of positive charge—like raisins spread throughout a pudding. This model is illustrated in **FIGURE 31-1**. Although the plum-pudding model was in agreement with everything Thomson knew about atoms at the time, new experiments were soon to rule out this model and replace it with one that was more like the solar system than a pudding.

The Rutherford Model: A Miniature Solar System

Inspired by the findings and speculations of Thomson, other physicists began to investigate atomic structure. In 1909, Ernest Rutherford (1871–1937) and his coworkers Hans Geiger (1882–1945) and Ernest Marsden (1889–1970) (at that time a 20-year-old undergraduate) decided to test Thomson's model by firing a beam of positively charged particles, known as **alpha particles,** at a thin gold foil. Alpha particles—which were later found to be the nuclei of helium atoms—carry a positive charge, and hence they should be deflected as they pass through the positively charged "pudding" in the gold foil. The deflection should have the following properties: (i) It should be relatively small, because the alpha particles have a substantial mass and the positive charge in the atom is spread out; and (ii) all the alpha particles should be deflected in roughly the same way, because the positive pudding fills virtually all the space.

When Geiger and Marsden performed the experiment, their results did not agree with these predictions. In fact, most of the alpha particles passed right through the foil as if it were not there—as if the atoms in the foil were mostly empty space. Because the results were rather surprising, Rutherford suggested that the experiment be modified to look not only for alpha particles with small angles of deflection—as originally expected—but for ones with large deflections as well.

This suggestion turned out to be an inspired hunch. Not only were large-angle deflections observed, but some of the alpha particles, in fact, were found to have practically reversed their direction of motion. Rutherford was stunned. In his own words, "It was almost as incredible as if you fired a fifteen-inch shell at a piece of tissue paper and it came back and hit you."

To account for the results of these experiments, Rutherford proposed that an atom has a structure similar to that of the solar system, as illustrated in **FIGURE 31-2**. In particular, he imagined that the lightweight, negatively charged electrons orbit a small, positively charged **nucleus** containing almost all the atom's mass. In this nuclear model of the atom, most of the atom is indeed empty space, allowing the majority of the

Big Idea **1** Matter is composed of atoms—the smallest units of a given element. Atoms have internal structure, and contain both positive and negative electric charges. Early models of the atom contained fatal flaws.

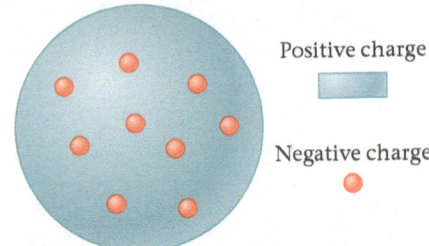

▲ **FIGURE 31-1 The plum-pudding model of an atom** The model of an atom proposed by J. J. Thomson consists of a uniform positive charge, which accounts for most of the mass of an atom, with small negatively charged electrons scattered throughout, like raisins in a pudding.

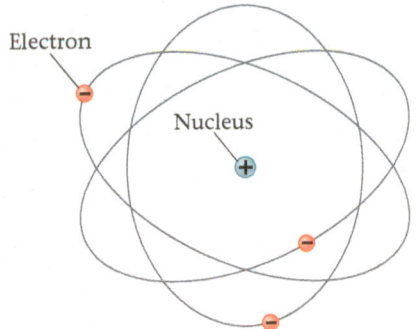

▲ **FIGURE 31-2 The solar-system model of an atom** Ernest Rutherford proposed that an atom is like a miniature solar system, with a massive, positively charged nucleus orbited by lightweight, negatively charged electrons.

alpha particles to pass right through. Furthermore, the positive charge of the atom is now highly concentrated in a small nucleus, rather than spread throughout the atom. This means that an alpha particle that happens to make a head-on collision with the nucleus can actually be turned around, as observed in the experiments.

To see just how small the nucleus must be in his model, Rutherford combined the experimental data with detailed theoretical calculations. His result was that the radius of a nucleus must be smaller than the diameter of the atom by a factor of about 10,000. To put this value in perspective, imagine an atom magnified until its nucleus is as large as the Sun. At what distance would an electron orbit in this "atomic" solar system? Using the factor given by Rutherford, we find that the orbit of the electron would have a radius essentially the same as the orbit of Pluto—inside this radius would be only empty space and the nucleus. Thus an atom must have an even larger fraction of empty space than the solar system!

Flaws in the Model   Although Rutherford's nuclear model of the atom seems reasonable, it contains fatal flaws. First, an orbiting electron undergoes a centripetal acceleration toward the nucleus (Chapter 6). As we know from Section 25-1, however, any electric charge that accelerates gives off energy in the form of electromagnetic radiation. Thus, an orbiting electron would be losing energy constantly; in fact, the situation would be similar to a satellite losing energy to air resistance when it orbits too close to the Earth's atmosphere. Just as in the case of a satellite, an electron would spiral inward and eventually plunge into the nucleus. The entire process of collapse would occur in a fraction of a second (about 10^{-11} s in fact), and therefore the atoms in Rutherford's model would simply not be stable—in contrast with the observed stability of atoms in nature.

Even if we ignore the stability problem for a moment, there is another serious discrepancy between Rutherford's model and experiment. Maxwell's equations state that the frequency of radiation from an orbiting electron should be the same as the frequency of its orbit. In the case of an electron spiraling inward, the frequency would increase continuously. Thus, if we look at light coming from an atom, the Rutherford model indicates that we should see a continuous range of frequencies. This prediction is in striking contrast with experiments, which show that light coming from an atom has only certain discrete frequencies and wavelengths, as we discuss in the next section.

Enhance Your Understanding   (Answers given at the end of the chapter)

1.   The concept of a nucleus was introduced in **(a)** the Thomson model of the atom, **(b)** the Rutherford model of the atom, **(c)** both of these models, or **(d)** neither of these models.

Section Review

*   Atoms are electrically neutral, but contain both positive and negative charges.

*   Early models of the atom contained serious flaws.

31-2 The Spectrum of Atomic Hydrogen

A red-hot piece of metal glows with a ruddy light that represents only a small fraction of its total radiation output. As we saw in Chapter 30, the metal gives off blackbody radiation that extends in a continuous distribution over all possible frequencies. This blackbody distribution, or spectrum, of radiation is characteristic of the entire collection of atoms that make up the metal—it is not characteristic of the spectrum of light that would be given off by a single, isolated metal atom.

To see the light produced by an isolated atom, we turn our attention from a solid—where the atoms are close together and strongly interacting—to a low-pressure gas—where the atoms are far apart and have little interaction with one another. Consider, then, an experiment in which we seal a low-pressure gas in a tube. If we apply a large voltage between the ends of the tube, the gas will emit electromagnetic radiation characteristic of the individual gas atoms. When this radiation is passed through a

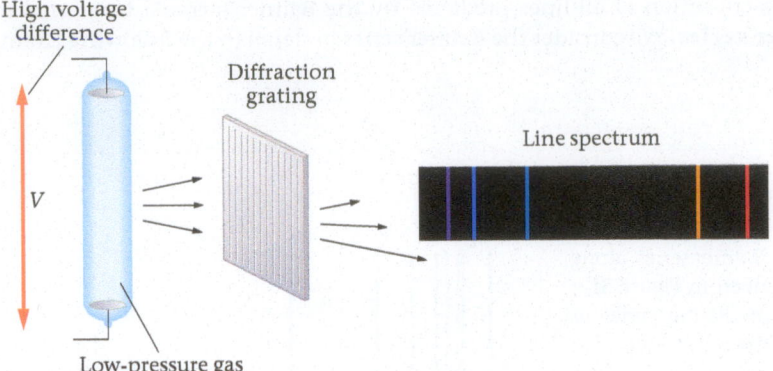

High voltage difference

V

Low-pressure gas

Diffraction grating

Line spectrum

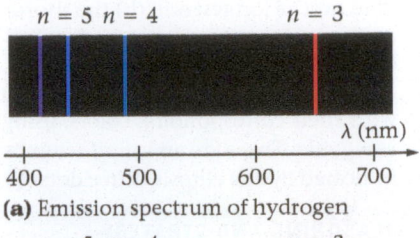

◀ **FIGURE 31-3 The line spectrum of an atom** The light given off by individual atoms, as in a low-pressure gas, consists of a series of discrete wavelengths corresponding to different colors.

$n = 5$ $n = 4$ $\qquad$ $n = 3$

λ (nm)

400 $\qquad$ 500 $\qquad$ 600 $\qquad$ 700

(a) Emission spectrum of hydrogen

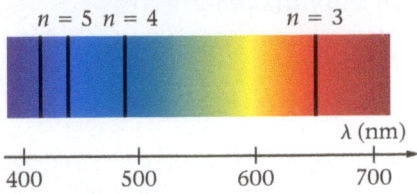

$n = 5$ $n = 4$ $\qquad$ $n = 3$

λ (nm)

400 $\qquad$ 500 $\qquad$ 600 $\qquad$ 700

(b) Absorption spectrum of hydrogen

▲ **FIGURE 31-4 The line spectrum of hydrogen** The emission **(a)** and absorption **(b)** spectra of hydrogen. Note that the wavelengths absorbed by hydrogen (dark lines) are the same as those emitted by hydrogen (colored lines). The location of these lines is predicted by the Balmer series (Equation 31-1) with the appropriate values of n.

Illuminating Sodium Vapor with Sodium and Mercury Lamps

PROBLEM-SOLVING NOTE

Calculating Wavelengths for the Balmer Series

Note that the formula for the Balmer series gives the inverse of the wavelength, rather than the wavelength itself.

diffraction grating (Chapter 28), it is separated into its various wavelengths, as indicated in **FIGURE 31-3**. The result of such an experiment is that a series of bright "lines" is observed, reminiscent of the barcodes used in supermarkets. The precise wavelength associated with each of these lines provides a sort of "fingerprint" identifying a particular type of atom, just as each product in a supermarket has its own unique barcode.

This type of spectrum, with its bright lines in different colors, is referred to as a **line spectrum.** As an example, we show the visible part of the line spectrum of atomic hydrogen in **FIGURE 31-4 (a)**. Hydrogen produces additional lines in the infrared and ultraviolet parts of the electromagnetic spectrum.

The line spectrum shown in Figure 31-4 (a) is an *emission spectrum*; it shows light that is emitted by the hydrogen atoms. Similarly, if light of all colors is passed through a tube of hydrogen gas, some wavelengths will be absorbed by the atoms, giving rise to an *absorption spectrum*, which consists of dark lines (where the atoms absorb the radiation) against an otherwise bright background. The absorption lines occur at precisely the same wavelengths as the emission lines. **FIGURE 31-4 (b)** shows the absorption spectrum of hydrogen.

The Balmer Series The first step in developing a quantitative understanding of the hydrogen spectrum occurred in 1885, when Johann Jakob Balmer (1825–1898), a Swiss schoolteacher, used trial-and-error methods to discover the following simple formula that gives the wavelength of the visible lines in the spectrum:

$$\frac{1}{\lambda} = R\left(\frac{1}{2^2} - \frac{1}{n^2}\right) \qquad n = 3, 4, 5, \ldots \text{(Balmer series)} \qquad 31\text{-}1$$

The constant, R, in this expression is known as the *Rydberg constant*. Its value is

$$R = 1.097 \times 10^7 \, \text{m}^{-1}$$

Each integer value of n $(3, 4, 5, \ldots)$ in Balmer's formula corresponds to the wavelength, λ, of a different spectral line. For example, if we set $n = 5$ in Equation 31-1, we find

$$\frac{1}{\lambda} = (1.097 \times 10^7 \, \text{m}^{-1})\left(\frac{1}{2^2} - \frac{1}{5^2}\right)$$

Solving for the wavelength, we have

$$\lambda = 4.341 \times 10^{-7} \, \text{m} = 434.1 \, \text{nm}$$

This is the bluish line, second from the left in Figure 31-4 (a). The reddish line corresponding to $n = 3$ is responsible for much of the red light emitted by the Lagoon Nebula in **FIGURE 31-5**.

▶ **FIGURE 31-5** Emission nebulas, like the Lagoon Nebula in Sagittarius shown here, are masses of glowing interstellar gas. The gas is excited by high-energy radiation from nearby stars and emits light at wavelengths characteristic of the atoms present, chiefly hydrogen. Much of the visible light from such nebulas is contributed by the red Balmer line of hydrogen with a wavelength of 656.3 nm, known as H-alpha.

The collection of all lines predicted by the Balmer formula is referred to as the **Balmer series.** We consider the Balmer series in detail in the following Example.

EXAMPLE 31-1 THE BALMER SERIES

Find **(a)** the longest and **(b)** the shortest wavelengths in the Balmer series of spectral lines.

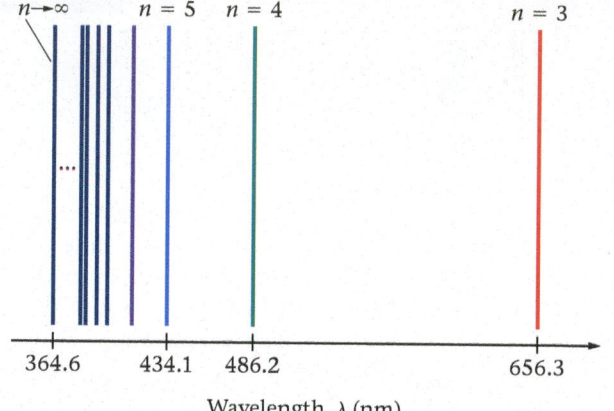

PICTURE THE PROBLEM

In our sketch we indicate the first several lines in the Balmer series, along with their corresponding colors, using the results given in Figure 31-4 as a guide. There are an infinite number of lines in the Balmer series, as indicated by the ellipsis (three dots) to the right of the $n \to \infty$ line.

REASONING AND STRATEGY

By substituting the values $n = 3, n = 4$, and $n = 5$ in the Balmer series (Equation 31-1, $1/\lambda = R(1/2^2 - 1/n^2)$), we find that the wavelength decreases with increasing n. Hence, **(a)** the longest wavelength corresponds to $n = 3$, and **(b)** the shortest wavelength corresponds to $n \to \infty$.

Known Wavelengths in the Balmer series are given by the expression $1/\lambda = R(1/2^2 - 1/n^2)$.

Unknown **(a)** Longest wavelength in Balmer series, $\lambda_{max} = ?$ **(b)** Shortest wavelength in Balmer series, $\lambda_{min} = ?$

SOLUTION

Part (a)

1. To find the longest wavelength in the Balmer series, substitute $n = 3$ in $1/\lambda = R(1/2^2 - 1/n^2)$:

$$\frac{1}{\lambda_{max}} = R\left(\frac{1}{2^2} - \frac{1}{3^2}\right) = (1.097 \times 10^7 \text{ m}^{-1})\left(\frac{5}{36}\right)$$

2. Invert the result in Step 1 to find the wavelength:

$$\lambda_{max} = \frac{36}{5(1.097 \times 10^7 \text{ m}^{-1})} = 656.3 \text{ nm}$$

Part (b)

3. The shortest wavelength is found in the limit $n \to \infty$ or, equivalently, $(1/n^2) \to 0$. Make this substitution in $1/\lambda = R(1/2^2 - 1/n^2)$:

$$\frac{1}{\lambda_{min}} = R\left(\frac{1}{2^2} - 0\right) = (1.097 \times 10^7 \text{ m}^{-1})\left(\frac{1}{4}\right)$$

4. Invert the result in Step 3 to find the wavelength:

$$\lambda_{min} = \frac{4}{(1.097 \times 10^7 \text{ m}^{-1})} = 364.6 \text{ nm}$$

INSIGHT

The longest wavelength corresponds to visible light with a reddish hue, whereas the shortest wavelength is well within the ultraviolet portion of the electromagnetic spectrum—it is invisible to our eyes.

PRACTICE PROBLEM

Which value of n corresponds to a wavelength of 377.1 nm in the Balmer series? **Answer:** $n = 11$]

Some related homework problems: Problem 6, Problem 7

Other Series of Spectral Lines in Hydrogen In **FIGURE 31-6** we show that the Balmer series is not the only series of lines produced by atomic hydrogen. The series with the shortest wavelengths is the **Lyman series**—all of its lines are in the ultraviolet. Similarly, the series with wavelengths just longer than those in the Balmer series is the **Paschen series.** The lines in this series are all in the infrared. The formula that gives the wavelengths in all the series of hydrogen is

$$\frac{1}{\lambda} = R\left(\frac{1}{n'^2} - \frac{1}{n^2}\right) \quad n' = 1, 2, 3, \ldots$$

$$n = n' + 1, n' + 2, n' + 3, \ldots \qquad \text{31-2}$$

Referring to Equation 31-1 ($1/\lambda = R(1/2^2 - 1/n^2)$), we can see that the Balmer series corresponds to the choice $n' = 2$. Similarly, the Lyman series is given with $n' = 1$, and the Paschen series corresponds to $n' = 3$. As one might expect, there are an

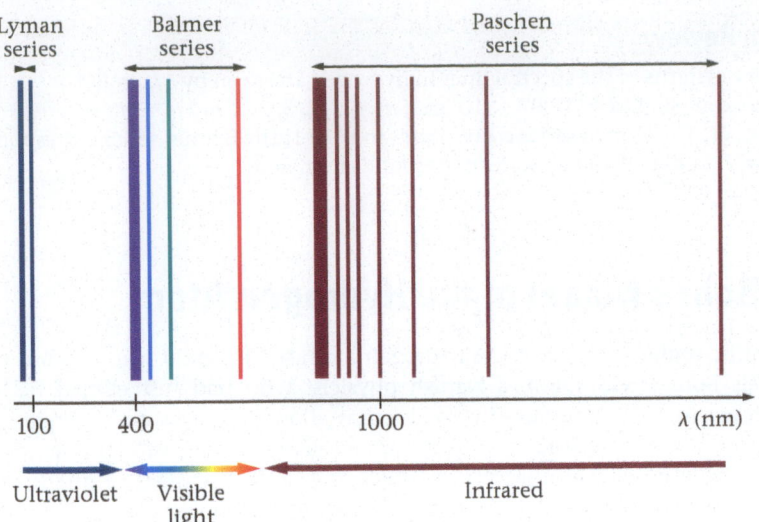

infinite number of series in hydrogen, each corresponding to a different choice for the integer n'. The names of the most common spectral series of hydrogen are listed in Table 31-1.

EXERCISE 31-2 LYMAN AND PASCHEN SERIES

Find **(a)** the shortest wavelength in the Lyman series and **(b)** the longest wavelength in the Paschen series.

REASONING AND SOLUTION

a. Substitute $n' = 1$ (which corresponds to the Lyman series) and $n \to \infty$ (which yields the smallest wavelength in a series) in the expression $1/\lambda = R(1/n'^2 - 1/n^2)$:

$$\frac{1}{\lambda} = R\left(\frac{1}{1^2} - 0\right) = (1.097 \times 10^7 \,\text{m}^{-1})$$

$$\lambda = \frac{1}{(1.097 \times 10^7 \,\text{m}^{-1})} = 91.16 \,\text{nm}$$

Any other choice for n will give a longer wavelength.

b. Substitute $n' = 3$ (for the Paschen series) and $n = 4$ (for the longest wavelength in this series) in the expression $1/\lambda = R(1/n'^2 - 1/n^2)$:

$$\frac{1}{\lambda} = R\left(\frac{1}{3^2} - \frac{1}{4^2}\right) = (1.097 \times 10^7 \,\text{m}^{-1})\left(\frac{7}{144}\right)$$

$$\lambda = \frac{144}{7(1.097 \times 10^7 \,\text{m}^{-1})} = 1875 \,\text{nm}$$

Any other choice for n will give a shorter wavelength.

As successful as $1/\lambda = R(1/n'^2 - 1/n^2)$ is in giving the various wavelengths of radiation produced by hydrogen, it is still just an empirical formula. It gives no insight as to *why* these particular wavelengths, and no others, are produced. The goal of atomic physicists in the early part of the twentieth century was to *derive* $1/\lambda = R(1/n'^2 - 1/n^2)$ from basic physical principles. The first significant step in that direction is the topic of the next section.

TABLE 31-1 Common Spectral Series of Hydrogen

n'	Series name
1	Lyman
2	Balmer
3	Paschen
4	Brackett
5	Pfund

PROBLEM-SOLVING NOTE

Correctly Applying Equation 31-2

Notice that n and n' are integers in Equation 31-2 and that the integer n must always be greater than n'.

Enhance Your Understanding   (Answers given at the end of the chapter)

2. For a given value of n', do the wavelengths of spectral lines increase, decrease, or stay the same with increasing n?

CONTINUED

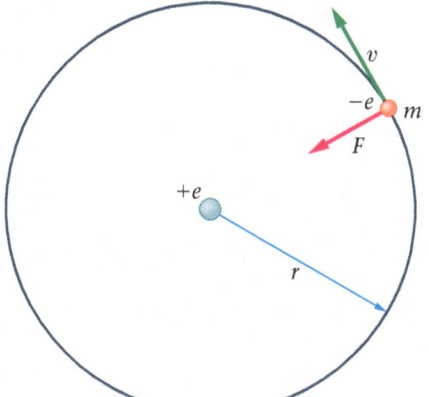

▲ **FIGURE 31-7** Niels Bohr, applying the principles of classical mechanics, with some members of his family.

31-3 Bohr's Model of the Hydrogen Atom

Our scientific understanding of the hydrogen atom took a giant leap forward in 1913, when Niels Bohr (1885–1962), a Danish physicist who had just earned his doctorate in physics in 1911, introduced a model that allowed him to derive Equation 31-2 $(1/\lambda = R(1/n'^2 - 1/n^2))$. Bohr's model combined elements of classical physics, which he is happily applying to a motorcycle in **FIGURE 31-7**, with the ideas of quantum physics introduced by Planck and Einstein about ten years earlier. As such, his model is a hybrid that spanned the gap between the classical physics of Newton and Maxwell and the newly emerging quantum physics.

Assumptions of the Bohr Model

Bohr's model of the hydrogen atom is based on four assumptions. Two are specific to his model and do not apply to the full quantum mechanical picture of hydrogen that will be introduced in Section 31-5. The remaining two assumptions are quite general—they apply not only to hydrogen, but to all atoms.

The two specific assumptions of the Bohr model are as follows:

- The electron in a hydrogen atom moves in a circular orbit about the nucleus.
- Only certain circular orbits are allowed. In these orbits the angular momentum of the electron is equal to an integer times Planck's constant divided by 2π. That is, the angular momentum of an electron in the nth allowed orbit is $L_n = nh/2\pi$, where $n = 1, 2, 3, \ldots$.

The next two assumptions are more general:

- Electrons do not give off electromagnetic radiation when they are in an allowed orbit. Thus, the orbits are stable.
- Electromagnetic radiation is given off or absorbed only when an electron changes from one allowed orbit to another. If the energy difference between two allowed orbits is ΔE, the frequency, f, of the photon that is emitted or absorbed is given by the relationship $|\Delta E| = hf$.

Notice that Bohr's model retains the classical picture of an electron orbiting a nucleus, as in Rutherford's model. It adds the stipulations, however, that only certain orbits are allowed and that no radiation is given off from these orbits. Radiation is given off *only* when an electron shifts from one orbit to another, and then the radiation is in the form of a photon that obeys Einstein's quantum relationship $E = hf$. Thus, as mentioned before, the Bohr model is a hybrid that includes ideas from both classical and quantum physics. We now use this model to determine the behavior of hydrogen.

Bohr Orbits

We begin by determining the radii of the allowed orbits in the Bohr model and the speed of the electrons in these orbits. There are two conditions that we must apply. First, for the electron to move in a circular orbit of radius r with a speed v, as depicted in **FIGURE 31-8**, the electrostatic force of attraction between the electron and the nucleus must be equal in magnitude to the mass of the electron times its centripetal acceleration, mv^2/r. Recalling Coulomb's law (Equation 19-5), we see that the electrostatic force between the electron (with charge $-e$) and the nucleus (with charge $+e$) has a magnitude given by ke^2/r^2. It follows that $mv^2/r = ke^2/r^2$, or, canceling one power of r,

$$mv^2 = k\frac{e^2}{r}$$

 31-3

▲ **FIGURE 31-8 A Bohr orbit** In the Bohr model of hydrogen, electrons orbit the nucleus in circular orbits. The centripetal acceleration of the electron, v^2/r, is produced by the Coulomb force of attraction between the electron and the nucleus.

This relationship is completely analogous to the one that was used to derive Kepler's third law (Chapter 12), except in that case the force of attraction was provided by gravity.

The second condition for an allowed orbit is that the angular momentum of the electron must be a nonzero integer n times $h/2\pi$. Because the electron moves with a speed v in a circular path of radius r, its angular momentum is $L = rmv$ (Equation 11-12). Hence, the condition for the nth allowed orbit is $L_n = r_n m v_n = nh/2\pi$, or, solving for v_n, we have

$$v_n = \frac{nh}{2\pi m r_n} \qquad n = 1, 2, 3, \ldots \qquad \text{31-4}$$

Combining these two conditions allows us to solve for the two unknowns, r_n and v_n.

For example, if we substitute v_n from Equation 31-4 into Equation 31-3, we can solve for r_n. Specifically, we find the following:

$$m\left(\frac{nh}{2\pi m r_n}\right)^2 = k\frac{e^2}{r_n}$$

Rearranging and solving for r_n, we obtain

$$r_n = \left(\frac{h^2}{4\pi^2 mke^2}\right)n^2 \qquad n = 1, 2, 3, \ldots \qquad \text{31-5}$$

The quantity in parentheses is the radius for the smallest ($n = 1$) orbit. Substitution of numerical values for h, m, e, and k gives the following value for r_1:

$$r_1 = \frac{h^2}{4\pi^2 mke^2}$$

$$= \frac{(6.626 \times 10^{-34}\,\text{J}\cdot\text{s})^2}{4\pi^2(9.109 \times 10^{-31}\,\text{kg})(8.988 \times 10^9\,\text{N}\cdot\text{m}^2/\text{C}^2)(1.602 \times 10^{-19}\,\text{C})^2}$$

$$= 5.29 \times 10^{-11}\,\text{m}$$

This radius, which is about half an angstrom and is referred to as the **Bohr radius**, is in agreement with the observed size of hydrogen atoms. Notice the n^2 dependence in the radii of allowed orbits: $r_n = r_1 n^2 = (5.29 \times 10^{-11}\,\text{m})n^2$. This dependence is illustrated in **FIGURE 31-9**.

To complete the solution, we can substitute our result for r_n (Equation 31-5) into the expression for v_n (Equation 31-4). This yields

$$v_n = \frac{nh}{2\pi m}\left(\frac{4\pi^2 mke^2}{n^2 h^2}\right) = \frac{2\pi ke^2}{nh} \qquad n = 1, 2, 3, \ldots \qquad \text{31-6}$$

Thus, the speed of the electron is smaller in orbits farther from the nucleus.

PHYSICS IN CONTEXT
Looking Back

Bohr orbits for the hydrogen atom are calculated in the same way that gravitational orbits of planets and satellites were calculated in Chapter 12.

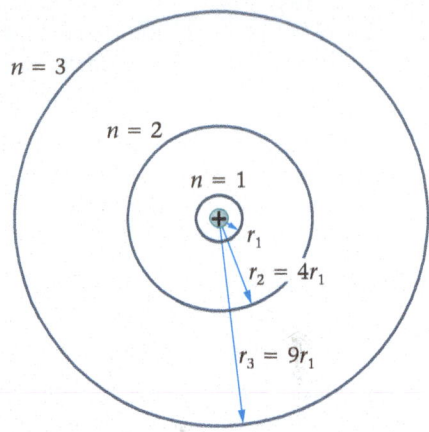

▲ **FIGURE 31-9 The first three Bohr orbits** The first Bohr orbit has a radius $r_1 = 5.29 \times 10^{-11}$m. The second and third Bohr orbits have radii $r_2 = 2^2 r_1 = 4r_1$ and $r_3 = 3^2 r_1 = 9r_1$, respectively. (*Note:* For clarity, the nucleus is drawn larger than its true scale relative to the size of the atom.)

EXAMPLE 31-3   FIRST AND SECOND BOHR ORBITS

Find the kinetic energy of the electron in **(a)** the first Bohr orbit ($n = 1$) and **(b)** the second Bohr orbit ($n = 2$).

PICTURE THE PROBLEM
The first two orbits of the Bohr model are shown in the sketch. Notice that the second orbit has a radius four times greater than the radius of the first orbit. In addition, the speed of the electron in the second orbit is half its value in the first orbit.

REASONING AND STRATEGY
The speed of the electron can be determined by direct substitution in $v_n = 2\pi ke^2/nh$ (Equation 31-6). Once the speed is determined, the kinetic energy is simply $K = \frac{1}{2}mv^2$.

Known   Speed of electron in nth Bohr orbit, $v_n = 2\pi ke^2/nh$; kinetic energy, $K = \frac{1}{2}mv^2$.

Unknown   **(a)** Kinetic energy of electron for $n = 1$, $K_1 = ?$ **(b)** Kinetic energy of electron for $n = 2$, $K_2 = ?$

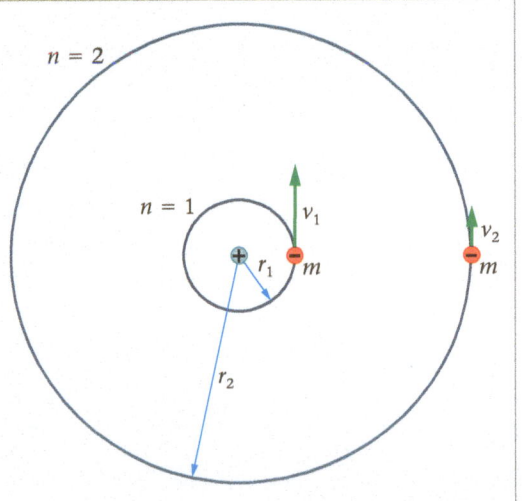

CONTINUED

SOLUTION

Part (a)

1. Substitute $n = 1$ in Equation 31-6 ($v_n = 2\pi ke^2/nh$) to obtain the speed of the electron:

$$v_1 = \frac{2\pi ke^2}{h}$$

$$= \frac{2\pi(8.99 \times 10^9 \, \text{N} \cdot \text{m}^2/\text{C}^2)(1.60 \times 10^{-19} \, \text{C})^2}{(6.63 \times 10^{-34} \, \text{J} \cdot \text{s})}$$

$$= 2.18 \times 10^6 \, \text{m/s}$$

2. The corresponding kinetic energy is $K_1 = \frac{1}{2}mv_1^2$:

$$K_1 = \frac{1}{2}mv_1^2 = \frac{1}{2}(9.11 \times 10^{-31} \, \text{kg})(2.18 \times 10^6 \, \text{m/s})^2$$

$$= 2.16 \times 10^{-18} \, \text{J}$$

Part (b)

3. The speed in the $n = 2$ orbit is half the value of the speed in the $n = 1$ orbit. The corresponding kinetic energy is one-fourth the kinetic energy in part (a):

$$K_2 = \frac{1}{2}mv_2^2 = \frac{1}{2}m(v_1/2)^2$$

$$= \frac{1}{4}K_1 = \frac{1}{4}(2.16 \times 10^{-18} \, \text{J}) = 5.40 \times 10^{-19} \, \text{J}$$

INSIGHT

The speed of the electron in the first Bohr orbit is less than the speed of light by a factor of about 137. In higher orbits, the electron's speed is even less. It follows that relativistic effects are small for the hydrogen atom.

In addition, notice that the kinetic energy of the electron in the first Bohr orbit is approximately 13.6 eV. We shall encounter this particular value for energy again later in the section.

PRACTICE PROBLEM

An electron in a Bohr orbit has a kinetic energy of 8.64×10^{-20} J. Find the radius of this orbit. [**Answer:** $r = 1.32 \times 10^{-9}$ m, which corresponds to $n = 5$]

Some related homework problems: Problem 15, Problem 20, Problem 21

Applying Bohr's Model to Other Atoms Bohr's model applies equally well to singly ionized helium, doubly ionized lithium, and other ions with only a single electron. In the case of singly ionized helium (one electron removed) the charge on the nucleus is $+2e$, for doubly ionized lithium (two electrons removed) it is $+3e$, and so on. In the general case, we may consider a nucleus that contains Z protons and has a charge of $+Ze$, where Z is the atomic number associated with that nucleus. Hydrogen, which has only a single proton in its nucleus, corresponds to $Z = 1$.

To be more explicit about the Z dependence, notice that the electrostatic force between an electron and a nucleus with Z protons has a magnitude of $k(e)(Ze)/r^2 = kZe^2/r^2$. Thus the results derived earlier in this section can be applied to the more general case if we simply replace e^2 with Ze^2. For example, Equation 31-3 becomes $mv^2 = kZe^2/r$. Similarly, the radius of an allowed orbit is

$$r_n = \left(\frac{h^2}{4\pi^2 mkZe^2}\right)n^2 \qquad n = 1, 2, 3, \ldots \qquad \text{31-7}$$

For example, the radius of the $n = 1$ orbit of singly ionized helium is half the radius of the $n = 1$ orbit of hydrogen.

The Energy of a Bohr Orbit

To find the energy of an electron in a Bohr orbit, we simply note that its total mechanical energy, E, is the sum of its kinetic and potential energies:

$$E = K + U = \frac{1}{2}mv^2 + U$$

Using the fact that $mv^2 = kZe^2/r$ for a hydrogen-like atom of atomic number Z (Equation 31-3), and the fact that the electrostatic potential energy of a charge $-e$ and a charge $+Ze$ a distance r apart is $U = -kZe^2/r$ (Equation 20-8), we find that the total mechanical energy is

$$E = \frac{1}{2}\left(\frac{kZe^2}{r}\right) - \frac{kZe^2}{r} = -\frac{kZe^2}{2r}$$

Finally, substituting the radius of a Bohr orbit, as given in Equation 31-7 $(r_n = (h^2/4\pi^2 mkZe^2)n^2)$, we obtain the corresponding energy for the nth orbit:

$$E_n = -\frac{kZe^2}{2r_n} = -\left(\frac{kZe^2}{2}\right)\left(\frac{4\pi^2 mkZe^2}{h^2}\right)\frac{1}{n^2}$$

$$= -\left(\frac{2\pi^2 mk^2e^4}{h^2}\right)\frac{Z^2}{n^2} \qquad n = 1, 2, 3, \ldots \qquad \text{31-8}$$

Using the numerical values for m, k, e, and h, as we did in calculating the Bohr radius, we have

$$E_n = -(13.6\text{ eV})\frac{Z^2}{n^2} \qquad n = 1, 2, 3, \ldots \qquad \text{31-9}$$

Let's first consider the specific case of hydrogen. With $Z = 1$ we find that the energies of the orbits in hydrogen are given by the relationship $E_n = -(13.6\text{ eV})/n^2$. We plot these energies in **FIGURE 31-10** for various values of n. This type of plot is referred to as an **energy-level diagram.** Notice that the **ground state** $(n = 1)$ corresponds to the lowest possible energy of the system. The higher energy levels are referred to as **excited states.** As the integer n tends to infinity, the energy of the excited states approaches zero—the energy the electron and proton would have if they were at rest and separated by an infinite distance. Thus to **ionize** hydrogen—that is, to remove the electron from the atom—requires a minimum energy of 13.6 eV. This value, which is a specific prediction of the Bohr model, is in complete agreement with experiment.

EXERCISE 31-4 IONIZATION ENERGY

In singly ionized helium, a single electron orbits a helium nucleus. Calculate the minimum energy required to remove this electron.

REASONING AND SOLUTION

The nucleus of helium has a charge of $+2e$. Substitution of $Z = 2$ and $n = 1$ in $E_n = -(13.6\text{ eV})\,Z^2/n^2$ yields

$$E_1 = -(13.6\text{ eV})\left(\frac{2^2}{1^2}\right) = -54.4\text{ eV}$$

Therefore, 54.4 eV must be added to remove the electron. We can understand this result as follows: First, the electron in singly ionized helium experiences a stronger attractive force than the single electron in hydrogen. In fact, the force is greater by a factor of two. Second, the $n = 1$ orbit of singly ionized helium has one-half the radius of the $n = 1$ orbit in hydrogen. As a result, four times as much energy is required to remove the electron from the helium ion.

Room-Temperature Hydrogen At room temperature, most hydrogen atoms are in the ground state. This is because the typical thermal energy of such atoms is too small to cause even the lowest-energy excitation from the ground state. Specifically, a typical thermal energy, $k_B T$, corresponding to room temperature $(T \sim 300\text{ K})$ is only about $\frac{1}{40}$ eV. In comparison, the energy required to excite an electron from the ground state of hydrogen to the first excited state is roughly 10 eV; that is, $\Delta E = E_2 - E_1 = (-3.40\text{ eV}) - (-13.6\text{ eV}) = 10.2\text{ eV}$. Excitations to higher excited states require even more energy. As a result, typical intermolecular collisions are simply not energetic enough to produce an excited state in hydrogen.

The Spectrum of Hydrogen

To find a formula describing the spectrum of hydrogen, we use Bohr's assumption that the frequency of emitted radiation for a change in energy equal to ΔE is given by $|\Delta E| = hf$. Because $\lambda f = c$ for electromagnetic radiation (Equation 25-4), we can rewrite this relationship in terms of the wavelength as $|\Delta E| = hc/\lambda$. To find $|\Delta E|$, we recall that the energy for hydrogen is given by $E_n = -(13.6\text{ eV})\,Z^2/n^2$ with $Z = 1$. Therefore, the *change* in energy as the electron moves from an excited outer orbit with $n = n_i$ to a lower orbit with $n = n_f < n_i$ has the following magnitude:

$$|\Delta E| = \left(\frac{2\pi^2 mk^2e^4}{h^2}\right)\left(\frac{1}{n_f^2} - \frac{1}{n_i^2}\right)$$

PROBLEM-SOLVING NOTE

The Z Dependence of Hydrogen-like Ions

Note that the energy of a hydrogen-like ion depends on Z^2, the square of the atomic number.

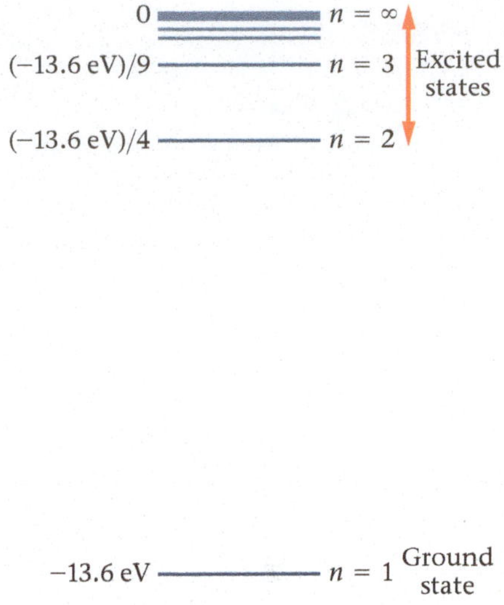

▲ **FIGURE 31-10 Energy-level diagram for the Bohr model of hydrogen** The energy of the ground state of hydrogen is −13.6 eV. Excited states of hydrogen approach zero energy. Note that the difference in energy from the ground state to the first excited state is $\frac{3}{4}$(13.6 eV), and the energy difference from the first excited state to the zero level is only $\frac{1}{4}$(13.6 eV).

PHYSICS IN CONTEXT
Looking Back

The energy of a Bohr orbit is determined using the electric potential energy introduced in Chapter 20.

Big Idea 2 Bohr's model of the hydrogen atom accounts for the observed spectrum of hydrogen. Even so, it has serious flaws and cannot be extended to more complex atoms.

PHYSICS IN CONTEXT
Looking Back

Photons emitted from a hydrogen atom in the Bohr model obey the relationship $E = hf$, which was used to understand blackbody radiation and the photoelectric effect in Chapter 30.

Using $|\Delta E| = hc/\lambda$, we can now solve for $1/\lambda$:

$$\frac{1}{\lambda} = \left(\frac{2\pi^2 m k^2 e^4}{h^3 c}\right)\left(\frac{1}{n_f^2} - \frac{1}{n_i^2}\right) \qquad \text{31-10}$$

Comparing this with Equation 31-2 ($1/\lambda = R(1/n'^2 - 1/n^2)$), we see that the expressions have precisely the same form, provided we identify n_f with n', and n_i with n. In addition, we see that the Rydberg constant in Equation 31-2, $R = 1.097 \times 10^7$ m^{-1}, has been replaced by the rather unusual constant $2\pi^2 m k^2 e^4/h^3 c$ in Equation 31-10. It is remarkable that when the known values of the fundamental constants m, k, e, h, and c are substituted into $2\pi^2 m k^2 e^4/h^3 c$, the resulting value is precisely 1.097×10^7 m^{-1}. This completes the derivation of $1/\lambda = R(1/n'^2 - 1/n^2)$, one of the most significant accomplishments of the Bohr model.

The origin of the line spectrum of hydrogen can be visualized in **FIGURE 31-11**. Notice that transitions that involve an electron jumping from an excited state ($n_i = n > 1$) to the ground state ($n_f = n' = 1$) result in the Lyman series of lines in the ultraviolet. Jumps ending in the $n = 2$ level give rise to the Balmer series, and jumps ending in the $n = 3$ level give the Paschen series. The largest energy jump in each series occurs when an electron falls from $n = \infty$ to the final level. Thus each series of spectral lines has a well-defined shortest wavelength.

RWP* One way to make practical use of the ability of atoms to emit and absorb photons of characteristic energies is to measure time. As mentioned in Chapters 12 and 13, Global Positioning System (GPS) satellites use extremely accurate atomic clocks to keep track of signal time intervals to determine the position of a receiver. So, how does an atomic clock work? In most atomic clocks, either cesium or rubidium gas is exposed to microwave radiation that varies in frequency over a small range. When the energy of the microwave photons matches the energy difference between two closely spaced energy levels in the atoms, the atoms absorb the photons and the transparency of the gas is slightly reduced. This signal is used to calibrate a crystal oscillator that determines the output of the clock. In fact, the transition in cesium atoms, which occurs at a frequency of exactly 9,192,631,770 Hz, is used as the basis for the international definition of one second of time, as mentioned in Chapter 1 . Commercial atomic clocks are accurate to within a few nanoseconds per day, or one second in several million years. Even more accurate atomic clocks are being developed.

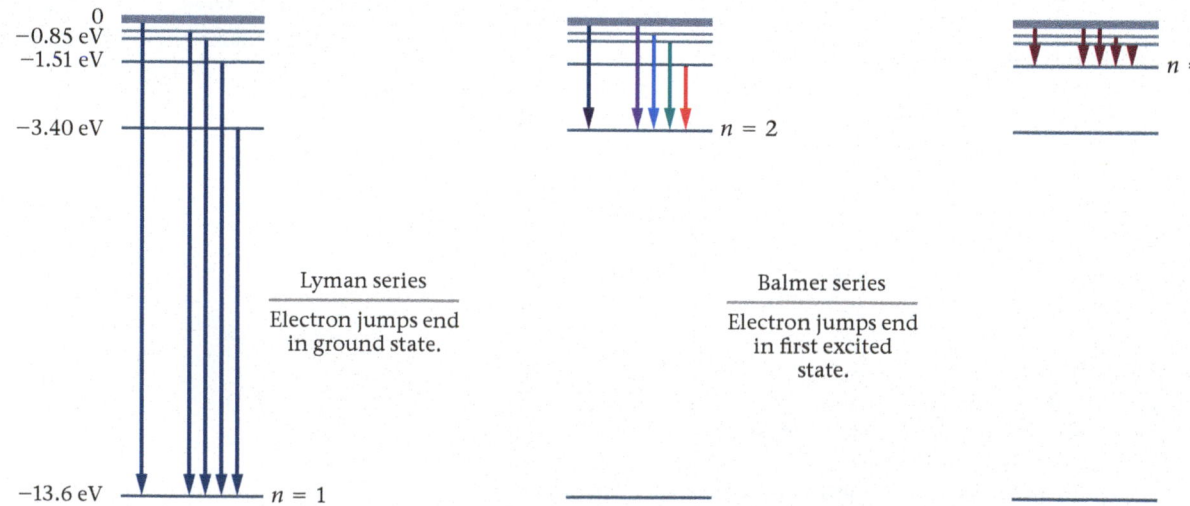

▲ **FIGURE 31-11 The origin of spectral series in hydrogen** Each series of spectral lines in hydrogen is the result of electrons jumping from an excited state to a particular lower level. For the Lyman series the lower level is the ground state. The lower level for the Balmer series is the first excited state ($n = 2$), and the lower level for the Paschen series is the second excited state ($n = 3$).

*Real World Physics applications are denoted by the acronym RWP.

CONCEPTUAL EXAMPLE 31-5   COMPARE WAVELENGTHS

The wavelength of the photon emitted when an electron in hydrogen jumps from the $n_i = 100$ state to the $n_f = 2$ state is (a) greater than, (b) less than, or (c) equal to the wavelength of the photon when the electron jumps from the $n_i = 2$ state to the $n_f = 1$ state.

REASONING AND DISCUSSION

We begin by noting that the wavelength is inversely proportional to the energy difference between levels ($|\Delta E| = hc/\lambda$); thus the smaller the energy difference, the greater the wavelength.

Next, we note that the full range of energies in the hydrogen atom extends from -13.6 eV to 0, and that the difference in energy between $n_i = 2$ and $n_f = 1$ is three-quarters of this range. It follows that the energy difference between $n_i = 2$ and $n_f = 1$ is greater than the energy difference between *any* state with $n_i > 2$ and $n_f = 2$. In fact, the maximum energy difference ending in the state 2 is $\frac{1}{4}(-13.6$ eV$)$, corresponding to $n_i = \infty$ and $n_f = 2$.

Because the energy difference for $n_i = 100$ and $n_f = 2$ is less than the energy difference for $n_i = 2$ and $n_f = 1$, the wavelength for $n_i = 100$ and $n_f = 2$ is the greater of the two.

ANSWER

(a) The wavelength for $n_i = 100$, $n_f = 2$ is greater than the wavelength for $n_i = 2$, $n_f = 1$.

QUICK EXAMPLE 31-6   FIND THE WAVELENGTHS

An electron in a hydrogen atom is in the initial state $n_i = 4$. Calculate the wavelength of the photon emitted by this electron if it jumps to the final state (a) $n_f = 3$, (b) $n_f = 2$, or (c) $n_f = 1$. (*Note:* To simplify the calculations, use $1/\lambda = R(1/n_f^2 - 1/n_i^2)$, with the Rydberg constant $R = 1.097 \times 10^7$ m^{-1}).

PROBLEM-SOLVING NOTE

The Frequency and Wavelength of a Photon

Recall that the frequency of a photon is given by $hf = |\Delta E|$, and its wavelength is given by $hc/\lambda = |\Delta E|$.

REASONING AND SOLUTION

The wavelengths can be found with the expression $1/\lambda = (1.097 \times 10^7$ m$^{-1})(1/n_f^2 - 1/n_i^2)$, using the appropriate values of n_f and n_i for each case.

Part (a)

1. Substitute $n_i = 4$ and $n_f = 3$ in $1/\lambda = R(1/n_f^2 - 1/n_i^2)$:

$$\frac{1}{\lambda} = R\left(\frac{1}{3^2} - \frac{1}{4^2}\right) = (1.097 \times 10^7\, \text{m}^{-1})\left(\frac{7}{144}\right)$$
$$\lambda = 1875 \text{ nm}$$

Part (b)

2. Repeat with $n_f = 2$:

$$\frac{1}{\lambda} = R\left(\frac{1}{2^2} - \frac{1}{4^2}\right) = (1.097 \times 10^7\, \text{m}^{-1})\left(\frac{12}{64}\right)$$
$$\lambda = 486.2 \text{ nm}$$

Part (c)

3. Finally, use $n_f = 1$:

$$\frac{1}{\lambda} = R\left(\frac{1}{1^2} - \frac{1}{4^2}\right) = (1.097 \times 10^7\, \text{m}^{-1})\left(\frac{15}{16}\right)$$
$$\lambda = 97.23 \text{ nm}$$

Other wavelengths are possible when an electron in the $n_i = 4$ state jumps to a lower state. For example, after dropping from the $n = 4$ state to the $n = 3$ state, the electron might then jump from the $n = 3$ state to the $n = 2$ state, and finally from the $n = 2$ state to the $n = 1$ state. Alternatively, the electron might first jump from the $n = 4$ state to the $n = 2$ state, and then from the $n = 2$ state to the $n = 1$ state. Thus an electron in an excited state may result in the emission of a variety of different wavelengths.

Just as an electron can *emit* a photon when it jumps to a lower level, it can also *absorb* a photon and jump to a higher level. This process occurs, however, only if the photon has the proper energy. In particular, the photon must have an energy that precisely matches the energy difference between the lower level of the electron and the higher level to which it is raised. This situation is explored in more detail in the next Quick Example.

QUICK EXAMPLE 31-7   ABSORBING A PHOTON: WHAT IS THE FREQUENCY?

Find the frequency a photon must have if it is to raise an electron in a hydrogen atom from the $n = 2$ state to the $n = 6$ state.

CONTINUED

REASONING AND SOLUTION

We begin by finding the energy of the two states in question. Next, we set the difference in energy between these two states equal to the energy of a photon, hf, and solve for the corresponding frequency.

1. Calculate the energy of the $n = 6$ state in joules:

$$E_6 = -(13.6\,\text{eV})\frac{1}{6^2} = -0.378\,\text{eV} = -6.05 \times 10^{-20}\,\text{J}$$

2. Calculate the energy of the $n = 2$ state:

$$E_2 = -(13.6\,\text{eV})\frac{1}{2^2} = -3.40\,\text{eV} = -5.44 \times 10^{-19}\,\text{J}$$

3. Calculate the difference in energy between these two states:

$$(-6.05 \times 10^{-20}\,\text{J}) - (-5.44 \times 10^{-19}\,\text{J}) = 4.84 \times 10^{-19}\,\text{J}$$

4. Set the energy of a photon equal to this energy difference:

$$hf = 4.84 \times 10^{-19}\,\text{J}$$

5. Solve for the frequency of the photon:

$$f = \frac{4.84 \times 10^{-19}\,\text{J}}{h} = \frac{4.84 \times 10^{-19}\,\text{J}}{6.63 \times 10^{-34}\,\text{J} \cdot \text{s}} = 7.30 \times 10^{14}\,\text{Hz}$$

This frequency corresponds to one of the dark lines in the *absorption* spectrum of hydrogen. In fact, it has the same frequency as the violet emission line in hydrogen's visible spectrum.

Enhance Your Understanding   (Answers given at the end of the chapter)

3. If the quantum number n is doubled in the Bohr model, by what factor do the following quantities change? **(a)** The speed of an electron in an orbit. **(b)** The radius of an orbit. **(c)** The ionization energy of an orbit.

Section Review

- The energy of an allowed orbit in the Bohr model for hydrogen is $E_n = -(13.6\,\text{eV})\,1/n^2$, where $n = 1, 2, 3, \ldots$.

- The wavelength of radiation associated with two different energy levels in the Bohr model for hydrogen is $1/\lambda = R(1/n_f^2 - 1/n_i^2)$, where R is the Rydberg constant, $R = 1.097 \times 10^7\,\text{m}^{-1}$.

31-4 de Broglie Waves and the Bohr Model

The fact that hydrogen emits radiation only at certain well-defined wavelengths is reminiscent of the harmonics of standing waves on a string. Recall that a vibrating string tied down at both ends produces a standing wave only if an integral number of half-wavelengths fit within its length (Section 14-8). Perhaps the behavior of hydrogen can be understood in similar terms.

In 1923 de Broglie used his idea of matter waves (Section 30-5) to show that one of Bohr's assumptions could indeed be thought of as a condition for standing waves. As we saw earlier in this chapter, Bohr assumed that the angular momentum of an electron in an allowed orbit must be a nonzero integer times $h/2\pi$. Specifically,

$$rmv = n\frac{h}{2\pi} \quad n = 1, 2, 3, \ldots$$

In Bohr's model there is no particular reason for this condition other than that it produces results that agree with experiment.

Now, de Broglie imagined his matter waves as analogous to a wave on a string— except that in this case the "string" is not tied down at both ends. Instead, it is formed into a circle of radius r representing an electron's orbit about the nucleus, as illustrated in **FIGURE 31-12**. The condition for a standing wave in this case is that an integral number

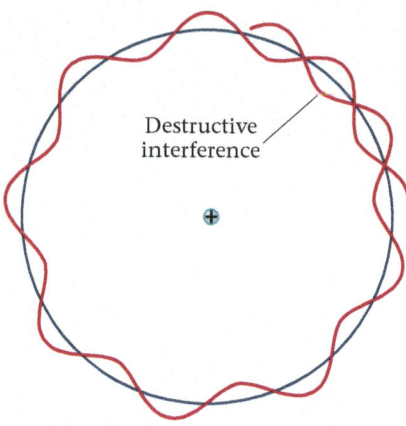

(a)

(b)

▲ **FIGURE 31-12 de Broglie wavelengths and Bohr orbits** Bohr's condition that the angular momentum of an allowed orbit must be an integer n times $h/2\pi$ is equivalent to the condition that n de Broglie wavelengths must fit into the circumference of an orbit. **(a)** The de Broglie waves for the $n = 7$ and $n = 8$ orbits. **(b)** If an integral number of wavelengths do not fit the circumference of an orbit, the result is destructive interference.

▶ **FIGURE 31-13** Although we can't see de Broglie waves, the standing waves they produce are important because they correspond to allowed states in atoms and molecules. One way to visualize de Broglie waves is to make an analogy with mechanical standing waves. In these photos, a loop of wire is oscillated vertically about the support point at the bottom of the loop. The oscillations set up waves that travel on the circumference of the wire—like de Broglie waves on a Bohr orbit in hydrogen. If the wavelength of the mechanical wave is tuned to an appropriate value—by adjusting the frequency of the oscillator—the result is a variety of different standing wave patterns, analogous to different energy levels of de Broglie waves in the Bohr model.

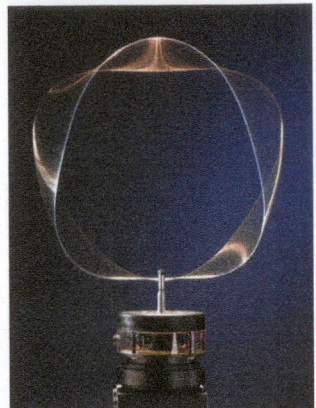

of wavelengths fit into the circumference of the orbit, as in the mechanical analog shown in **FIGURE 31-13**. Stated mathematically, the condition is $n\lambda = 2\pi r$, as illustrated in Figure 31-12 (a) for the cases $n = 7$ and $n = 8$. Other wavelengths would result in destructive interference, as Figure 31-12 (b) shows.

Finally, de Broglie combined the standing wave condition with his matter–wave relationship $p = h/\lambda$ (Equation 30-16). The result is as follows:

$$p = mv = \frac{h}{\lambda} = \frac{h}{(2\pi r/n)} = \frac{nh}{2\pi r} \quad n = 1, 2, 3, \ldots$$

Multiplying both sides of this equation by r to obtain the angular momentum, we find

$$L = rmv = \frac{nh}{2\pi} \quad n = 1, 2, 3, \ldots$$

This is precisely the Bohr orbital condition, now understood as a reflection of the wave nature of matter.

EXAMPLE 31-8   THE WAVELENGTH OF AN ELECTRON

Find the wavelength associated with an electron in the $n = 4$ state of hydrogen.

PICTURE THE PROBLEM
Our sketch shows that four wavelengths fit around the circumference of the $n = 4$ orbit. Recall that the radius of this orbit is $4^2 = 16$ times the radius of the ground-state orbit.

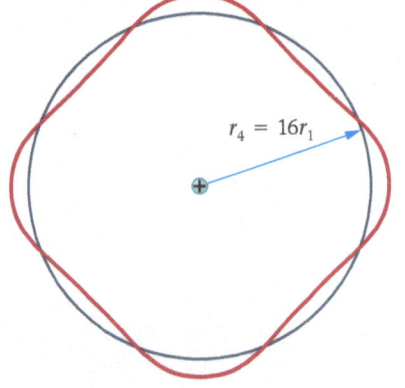

$r_4 = 16r_1$

REASONING AND SOLUTION
To find the wavelength of this matter wave we simply calculate the circumference of the $n = 4$ orbit, and then divide by 4.

Known   State of hydrogen, $n = 4$.
Unknown Wavelength of electron, $\lambda = ?$

SOLUTION

1. Calculate the radius of the $n = 4$ orbit: $r_4 = 4^2 r_1 = 16(5.29 \times 10^{-11}\,\text{m}) = 8.46 \times 10^{-10}\,\text{m}$

2. Use this result to find the circumference of the orbit: $2\pi r_4 = 2\pi(8.46 \times 10^{-10}\,\text{m}) = 5.32 \times 10^{-9}\,\text{m}$

3. Divide the circumference by 4 to find the wavelength: $\lambda_4 = \frac{1}{4}(2\pi r_4) = \frac{1}{4}(5.32 \times 10^{-9}\,\text{m}) = 1.33 \times 10^{-9}\,\text{m}$

INSIGHT
An equivalent way of determining the wavelength is to use the de Broglie relationship, $p = mv = h/\lambda$. Solving for the wavelength yields $\lambda_4 = h/mv_4$, and substituting v_4 from Equation 31-6 ($v_n = 2\pi ke^2/nh$) yields the same result given in Step 3.

PRACTICE PROBLEM — PREDICT/CALCULATE
(a) If the de Broglie wavelength is increased, does the corresponding value of n increase, decrease, or stay the same? Explain. (b) Find the value of n for which the electron in a hydrogen atom has a de Broglie wavelength of

CONTINUED

2.66 × 10⁻⁹ m.   [**Answer: (a)** The de Broglie wavelength is proportional to n, so increasing λ corresponds to an increase in n. **(b)** The wavelength is doubled from its value in this Example; thus, the value of n doubles to $n = 8$.]

Some related homework problems: Problem 35, Problem 38, Problem 39

Big Idea **3** The wavelike nature of matter finds support in the Bohr model of an atom. In particular, the allowed Bohr orbits correspond to standing wave patterns of the de Broglie matter waves.

PHYSICS IN CONTEXT
Looking Back

Bohr orbits can be understood in terms of standing waves, just like the standing waves on a string studied in Chapter 14, only this time using the de Broglie waves introduced in Chapter 30.

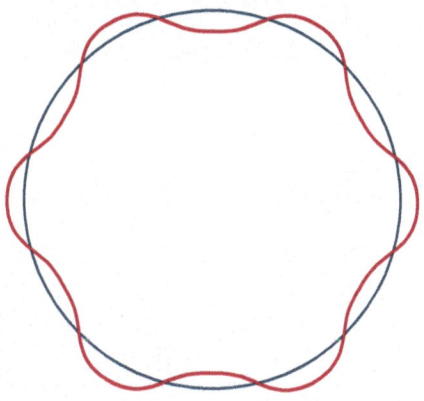

▲ **FIGURE 31-14** Enhance Your Understanding 4.

Taking Matter Waves Seriously   The striking success of de Broglie's matter waves in deriving Bohr's angular momentum condition encouraged physicists to give the idea of matter waves serious consideration. If we accept matter waves as being real, however, a large number of new questions must be addressed. For example, if particles like electrons can be described by matter waves, how do the matter waves behave? What determines the value of a matter wave at a particular location? What is the physical significance of a matter wave that has a large value at one location and a small value at another location?

These questions were answered by Erwin Schrödinger (1887–1961), Max Born (1882–1970), and others. In particular, Schrödinger introduced an equation—similar in many respects to the equation that describes sound waves—to describe the behavior of matter waves. Today, this equation, known as **Schrödinger's equation,** forms the basis for quantum mechanics, which is the quantum physics version of classical mechanics. In fact, Schrödinger's equation plays the same role in quantum mechanics that Newton's laws play in classical mechanics and Maxwell's equations play in electromagnetism.

After the introduction of the Schrödinger equation, Born developed an interpretation of the matter waves that was quite different from that for mechanical waves. For example, in the case of a wave on a string, the amplitude of a wave simply represents the displacement of the string from its equilibrium position. For a matter wave, on the other hand, the amplitude is related to the *probability* of finding a particle in a particular location. Thus, matter waves do not tell us precisely where a particle is located; rather, they give the probability of finding the particle at a given place, as we shall see in detail in the next section.

Enhance Your Understanding   (Answers given at the end of the chapter)

4. What is the value of n for the Bohr orbit shown in **FIGURE 31-14**?

Section Review

• Bohr's condition for the angular momentum of an allowed orbit in hydrogen, $L = rmv = nh/2\pi$, is equivalent to the condition that an integer number of de Broglie wavelengths fit in the circumference of the orbit; that is, $n\lambda = 2\pi r$, where $\lambda = h/p$.

Big Idea **4** Quantum mechanics provides a full and accurate description of hydrogen as well as all the other atoms. In quantum mechanics, the orbits of the Bohr model are replaced with probability "clouds" of various sizes and shapes.

31-5 The Quantum Mechanical Hydrogen Atom

Although Schrödinger's equation and its solution for the hydrogen atom are beyond the scope of this text, we present here some of the main features obtained by this analysis. Other than relativistic effects, the Schrödinger equation presents our most complete understanding of the hydrogen atom and of behavior at the atomic level in general. As we shall see, many aspects of the Bohr model survive in this analysis, though there are also significant differences.

To begin, we note that the Bohr model was characterized by a single quantum number, n. In contrast, the quantum mechanical description of the hydrogen atom requires four quantum numbers. They are as follows:

• **The principal quantum number, n:** The quantum number $n = 1, 2, 3, \ldots$ plays a similar role in the quantum mechanical hydrogen atom and in Bohr's model. In Bohr's model, n is the only quantum number, and it determines the radius of an orbit, its angular momentum, and its energy. In particular, the energy in the Bohr model is $E = (-13.6 \text{ eV})/n^2$. The energy given by Schrödinger's equation is precisely the same, if we neglect small relativistic effects and small magnetic interactions within the atom.

- **The orbital angular momentum quantum number, ℓ:** In the Bohr model an electron's orbital angular momentum is determined by the quantum number n. In particular, $L_n = nh/2\pi$, where $n = 1, 2, 3, \ldots$. In the quantum mechanical solution, there is a separate quantum number, ℓ, for the orbital angular momentum. This quantum number can take on the following values for any given value of the principal quantum number, n:

$$\ell = 0, 1, 2, \ldots, (n - 1) \qquad \text{31-11}$$

The magnitude of the angular momentum for any given value of ℓ is given by the following relationship:

$$L = \sqrt{\ell(\ell + 1)}\,\frac{h}{2\pi} \qquad \text{31-12}$$

Notice that the angular momentum of the electron can have a range of values for a given n, in contrast with the Bohr model, where the angular momentum has just a single value. In particular, the electron in a hydrogen atom can have zero angular momentum; an orbiting electron in the Bohr model always has a nonzero angular momentum.

Finally, although the energy of the hydrogen atom does not depend on ℓ, the energy of more complex atoms does have an ℓ dependence, as we shall see in the next section.

- **The magnetic quantum number, m_ℓ:** If a hydrogen atom is placed in an external magnetic field, its energy is found to depend not only on n but on an additional quantum number as well. This quantum number, m_ℓ, is referred to as the magnetic quantum number. The allowed values of m_ℓ are as follows:

$$m_\ell = -\ell, -\ell + 1, -\ell + 2, \ldots, -1, 0, 1, \ldots, \ell - 2, \ell - 1, \ell \qquad \text{31-13}$$

This quantum number gives the component of the orbital angular momentum vector along a specified direction, usually chosen to be the z axis. With this choice, L_z has the following values:

$$L_z = m_\ell \frac{h}{2\pi}$$

Only a single component of the orbital angular momentum can be known precisely; it is not possible to know all three components of the angular momentum simultaneously, due to the Heisenberg uncertainty principle.

- **The electron spin quantum number, m_s:** The final quantum number needed to describe the hydrogen atom is related to the angular momentum of the electron itself. Just as Earth spins on its axis at the same time that it orbits the Sun, the electron can be thought of as having both an orbital and a "spin" angular momentum. The spin quantum number for an electron takes on just two values:

$$m_s = -\tfrac{1}{2}, \tfrac{1}{2}$$

These two values correspond to the electron's spin being "up" $\left(m_s = \tfrac{1}{2}\right)$ or "down" $\left(m_s = -\tfrac{1}{2}\right)$ with respect to the z axis.

It should be noted that the spin angular momentum is an *intrinsic* property of an electron, like its mass and its charge—*all* electrons have exactly the *same* mass, the *same* charge, and the *same* spin angular momentum. Thus, we do not imagine the electron to be a small spinning sphere, like a microscopic planet. You can't speed up or slow down the spin of an electron. Instead, spin is simply one of the properties that defines an electron.

Energy Levels and States
The energy-level structure of hydrogen in zero magnetic field is shown in **FIGURE 31-15**, along with the corresponding quantum numbers. The energies are the same as in the Bohr model, and hence the spectrum will be the same as in the Bohr model, and experiment.

We define a **state** of hydrogen to be a specific assignment of values for each of the four quantum numbers. For example, there are two states that correspond to the

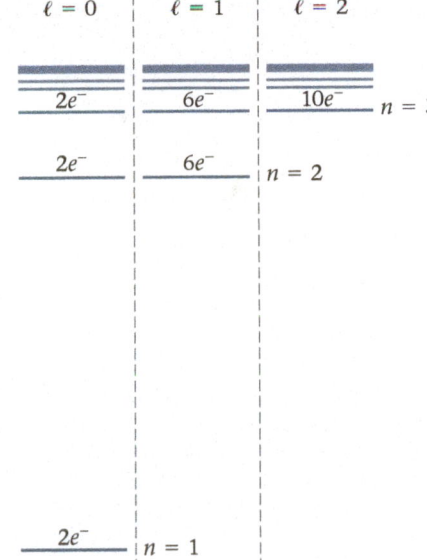

▲ **FIGURE 31-15 Energy-level structure of hydrogen** The values of the quantum mechanical energy levels for hydrogen are in complete agreement with the Bohr model. In the quantum model, however, each energy level has associated with it a specific number of "quantum states" determined by the values of all four quantum numbers, as specified in Table 31-2. In multielectron atoms, these states lead to the formation of the periodic table of elements, as we show in Section 31-6.

lowest possible energy level of hydrogen. These are $n = 1, \ell = 0, m_\ell = 0, m_s = \frac{1}{2}$; and $n = 1, \ell = 0, m_\ell = 0, m_s = -\frac{1}{2}$. Similarly, there are two states corresponding to the $n = 2, \ell = 0$ energy level, and six states corresponding to the $n = 2, \ell = 1$ level. These states are listed in Table 31-2. When we consider multielectron atoms in the next section, we shall see that the number of states associated with a given energy level determines the number of electrons (e^-) it can accommodate. Thus, the notation $2e^-$ in Figure 31-15 indicates an energy level that can hold two electrons, $6e^-$ indicates an energy level that can hold six electrons, and so on. Once an energy level is "filled," additional electrons must occupy other levels. This progressive filling of energy levels ultimately leads to the periodic table of elements.

TABLE 31-2 States of Hydrogen for $n = 1$ and $n = 2$

$n = 1, \ell = 0$			**Two states**
$n = 1$	$\ell = 0$	$m_\ell = 0$	$m_s = \frac{1}{2}$
$n = 1$	$\ell = 0$	$m_\ell = 0$	$m_s = -\frac{1}{2}$
$n = 2, \ell = 0$			**Two states**
$n = 2$	$\ell = 0$	$m_\ell = 0$	$m_s = \frac{1}{2}$
$n = 2$	$\ell = 0$	$m_\ell = 0$	$m_s = -\frac{1}{2}$
$n = 2, \ell = 1$			**Six states**
$n = 2$	$\ell = 1$	$m_\ell = 1$	$m_s = \frac{1}{2}$
$n = 2$	$\ell = 1$	$m_\ell = 1$	$m_s = -\frac{1}{2}$
$n = 2$	$\ell = 1$	$m_\ell = 0$	$m_s = \frac{1}{2}$
$n = 2$	$\ell = 1$	$m_\ell = 0$	$m_s = -\frac{1}{2}$
$n = 2$	$\ell = 1$	$m_\ell = -1$	$m_s = \frac{1}{2}$
$n = 2$	$\ell = 1$	$m_\ell = -1$	$m_s = -\frac{1}{2}$

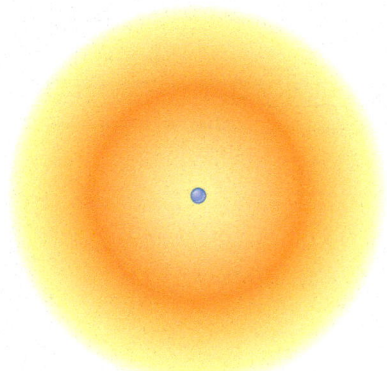

▲ **FIGURE 31-16 Probability cloud for the ground state of hydrogen** In the quantum mechanical model of hydrogen, the electron can be found at any distance from the nucleus. The probability of finding the electron at a given location is proportional to the density of the "probability cloud."

Electron Probability Clouds: Three-Dimensional Standing Waves

As mentioned in the previous section, the solution to Schrödinger's equation gives a matter wave, or **wave function,** as it is known, corresponding to a particular physical system. The wave function for hydrogen gives the probability of finding the electron at a particular location. The best way to visualize this probability distribution is in terms of a "probability cloud," as shown in **FIGURE 31-16**. The probability of finding the electron is greatest where the cloud is densest. In the case shown in Figure 31-16, corresponding to the ground state, $n = 1, \ell = 0, m_\ell = 0$, the probability of finding the electron is distributed with spherical symmetry. Notice that the probability decreases rapidly far from the nucleus, as one would expect.

FIGURE 31-17 gives a different way of looking at the probability distribution for the ground state. Here we plot the probability versus distance from the nucleus. It is interesting to note that the maximum probability occurs at a distance from the nucleus equal to the Bohr radius. Thus certain aspects of the Bohr model find their way into the final solution for hydrogen. The difference, however, is that in the Bohr model the electron is always at a particular distance from the nucleus as it moves in its circular orbit. In the quantum solution to hydrogen, the electron can be found at virtually any distance from the nucleus, not just one distance.

States of higher quantum number have increasingly complex probability distributions, as indicated in **FIGURE 31-18**. As the quantum number n is increased, for example,

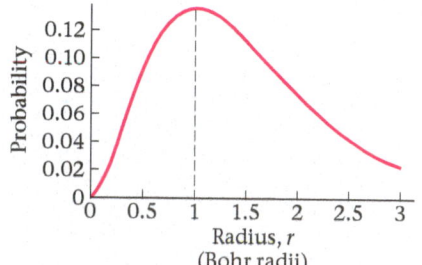

◀ **FIGURE 31-17 Probability as a function of distance** This plot shows the probability of finding an electron in the ground state of hydrogen at a given distance from the nucleus. Note that the probability is greatest at a distance equal to the Bohr radius, r_1.

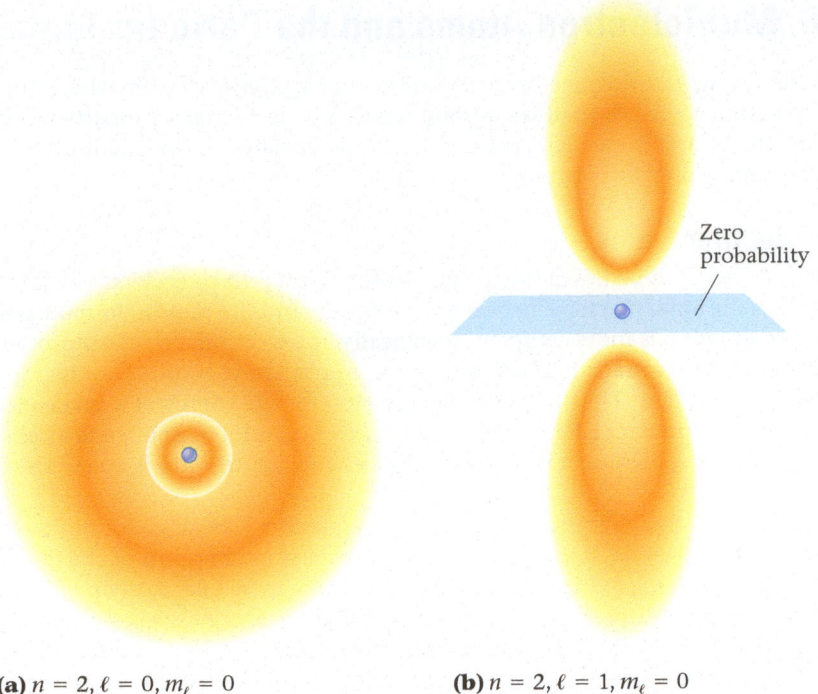

◄ **FIGURE 31-18 Probability clouds for excited states of hydrogen** The "probability cloud" for an electron in hydrogen increases in complexity as the quantum numbers increase. **(a)** $n = 2, \ell = 0$; **(b)** $n = 2, \ell = 1$.

Zero probability

(a) $n = 2, \ell = 0, m_\ell = 0$ **(b)** $n = 2, \ell = 1, m_\ell = 0$

the basic shape of the distribution remains the same, but additional nodes appear. This is illustrated in Figure 31-18 (a) for the case $n = 2, \ell = 0, m_\ell = 0$. Notice that the distribution is spherically symmetric, as in the state $n = 1, \ell = 0, m_\ell = 0$, but now there is a node—where the probability is zero—separating an inner and an outer portion of the distribution. As the quantum number ℓ is increased, the distributions become more complex in their shape. An example is shown in Figure 31-18 (b) for the state $n = 2, \ell = 1, m_\ell = 0$.

CONCEPTUAL EXAMPLE 31-9   FINDING THE ELECTRON

The probability cloud for the $n = 2, \ell = 1, m_\ell = 0$ state of hydrogen is shown in Figure 31-18 (b). This cloud consists of two lobes of high probability separated by a plane of zero probability. Given that an electron in this state can never be halfway between the two lobes, how is it possible that the electron is as likely to be found in the upper lobe as in the lower lobe?

REASONING AND DISCUSSION
The probability lobes of an electron are the result of a standing wave pattern, analogous to the standing waves found on a string tied down at both ends. Both the node on a string and the region of zero probability of an electron are the result of destructive interference. For example, the displacement of a string may be of equal amplitude on either side of a node—where the displacement is always zero—just as the electron probability may be high on either side of a probability node and zero in the middle. In summary, an electron does not simply move from place to place in an atom, like a small ball of charge; instead, it forms a standing wave pattern with nodes in certain locations.

ANSWER
The zero probability plane is a node of the matter wave for the electron—basically the electron "cancels itself out" on this plane as it moves back and forth from the upper lobe to the lower lobe.

Enhance Your Understanding   (Answers given at the end of the chapter)

5. How many states correspond to the $n = 3, \ell = 2$ level of hydrogen?

Section Review

* The quantum mechanical model of hydrogen introduces a number of new "quantum numbers" that characterize the allowed states. The energy levels are the same as in the Bohr model, and hence the predicted spectrum is the same as well. The quantum model is more complex than the Bohr model, and allows for a full description not only of hydrogen, but of all other atoms as well.

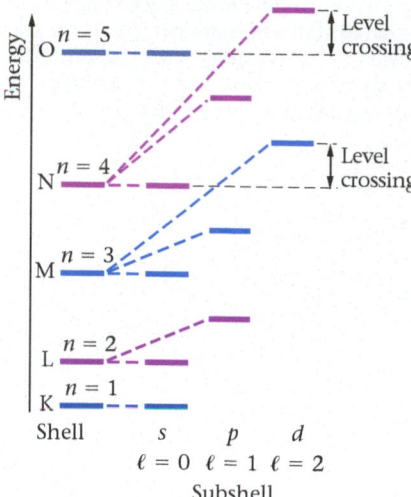

▲ **FIGURE 31-19 Energy levels in multi-electron atoms** In multielectron atoms the energy increases with n for fixed ℓ and increases with ℓ for fixed n. Note the possibility of energy-level crossing. For example, the $n = 3$, $\ell = 2$ energy level is higher than the $n = 4$, $\ell = 0$ energy level.

31-6 Multielectron Atoms and the Periodic Table

In this section we extend our considerations of atomic physics to atoms that have more than one electron. As we shall see, certain regularities arise in the properties of multielectron atoms, and these regularities are intimately related to the quantum numbers described in the previous section.

Multielectron Atoms

One of the great simplicities of the hydrogen atom is that the only electrostatic force in the atom is the attractive force between the electron and the proton. In multielectron atoms the situation is more complex. Specifically, the electrons in such atoms experience repulsive electrostatic interactions with one another, in addition to their interaction with the nucleus. Thus the simple expression for the energy of a single-electron atom given in Equation 31-9 ($E_n = -(13.6 \text{ eV}) Z^2/n^2$) cannot be applied to atoms with multiple electrons, because it does not include energy contributions due to the forces between electrons.

Although no simple formula analogous to Equation 31-9 exists for multielectron atoms, the energy levels of these atoms can be understood in terms of the four quantum numbers (n, ℓ, m_ℓ, m_s) used to describe hydrogen. In fact, by applying Schrödinger's equation to such atoms, we have discovered that the energy levels of multielectron atoms depend on the principal quantum number, n, and on the orbital quantum number, ℓ. For example, increasing n for a fixed value of ℓ results in an increase in energy. This relationship is illustrated in **FIGURE 31-19**, where we see that the energy for $n = 2$ and $\ell = 0$ is greater than the energy for $n = 1$ and $\ell = 0$, and the energy for $n = 3$ and $\ell = 1$ is greater than the energy for $n = 2$ and $\ell = 1$. It is also found that the energy increases with increasing ℓ for fixed n. Thus the energy for $n = 2$ and $\ell = 1$ is greater than the energy for $n = 2$ and $\ell = 0$.

Figure 31-19 also shows that in some cases the energy levels corresponding to different values of n can cross. For example, the energy for $n = 3$ and $\ell = 2$ is greater than the energy for $n = 4$ and $\ell = 0$. A similar crossing occurs with the $n = 4$, $\ell = 2$ state and the $n = 5$, $\ell = 0$ state. In all cases, however, the energy still increases with n for fixed ℓ, and increases with ℓ for fixed n. We shall see the effect of these energy-level crossings in terms of the periodic table later in this section.

Because the energy levels of multielectron atoms depend on n and ℓ, we give specific names and designations to the various values of these quantum numbers. For example, all electrons that have the same value of n are said to be in the same **shell.** Specifically, electrons with $n = 1$ are said to be in the K shell, those with $n = 2$ are in the L shell, those with $n = 3$ are in the M shell, and so on. These designations are summarized in Table 31-3 and displayed in Figure 31-19.

Similarly, electrons in a given shell with the same value of ℓ are said to be in the same **subshell,** and different values of ℓ have different alphabetical designations. For example, electrons with $\ell = 0$ are said to be in the s subshell, those with $\ell = 1$ are in the p subshell, and those with $\ell = 2$ are in the d subshell. These names, though not particularly logical, are used for historical reasons. After the f subshell ($\ell = 3$), the names of subsequent subshells continue in alphabetical order, as listed in Table 31-3.

The Pauli Exclusion Principle

As mentioned in Section 31-3, most hydrogen atoms are in their ground state at room temperature, because typical thermal energies are not great enough to excite the electron to higher energy levels. The same is true of multielectron atoms—they too are generally found in their ground state at room temperature. The question is this: What is the ground state of a multielectron atom?

The answer to this question involves an entirely new fundamental principle of physics put forward by the Austrian physicist Wolfgang Pauli (1900–1958) in 1925. Pauli's "exclusion principle" states that no two electrons in an atom can be in the same state at the same time. That is, once an electron occupies a given state, as defined by the values of its quantum numbers, other electrons are *excluded* from occupying the same state:

TABLE 31-3 Shell and Subshell Designations

n	Shell
1	K
2	L
3	M
4	N
...	...

ℓ	Subshell
0	s
1	p
2	d
3	f
4	g
...	...

The Pauli Exclusion Principle

Only one electron at a time may have a particular set of quantum numbers, n, ℓ, m_ℓ, and m_s. Once a particular state is occupied, other electrons are excluded from that state.

Because of the exclusion principle, the ground state of a multielectron atom is *not* obtained by placing all the electrons in the lowest possible energy state, as one might at first suppose. Once the lowest-energy states are occupied, additional electrons in the atom must occupy levels of higher energy. As more electrons are added to an atom, they fill up one subshell after another until all the electrons are accommodated. The situation is analogous to placing marbles into a jar, as illustrated in **FIGURE 31-20**. The first few marbles occupy the lowest level of the jar, but as more marbles are added, they must occupy levels of higher gravitational potential energy, simply because the lower levels are already occupied.

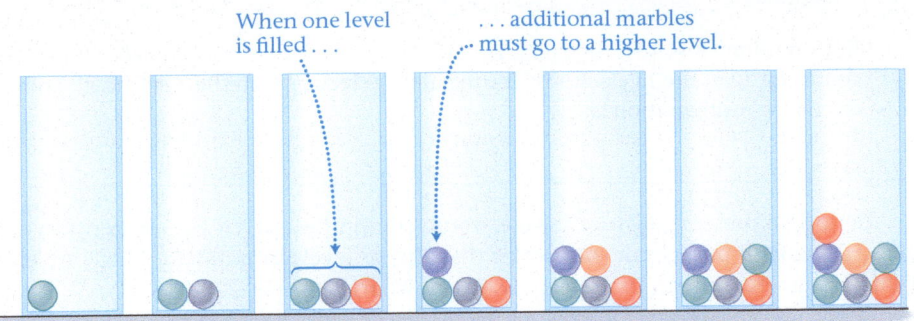

When one level is filled . . .

. . . additional marbles must go to a higher level.

◀ **FIGURE 31-20 Filling a jar with marbles** As marbles are added to a jar, they fill in one level, then another, and another. Once a given level is filled with marbles, additional marbles are excluded from that level, analogous to the filling of energy levels in multielectron atoms.

In the case of atoms, the lowest energy level corresponds to $n = 1$ and $\ell = 0$—that is, to the s subshell of the K shell. To completely define a state, however, we must specify all four quantum numbers: n, ℓ, m_ℓ, and m_s. First, recall that $m_\ell = 0, \pm1, \pm2, \ldots, \pm\ell$ for general ℓ. It follows that in the $\ell = 0$ state the only possible value of m_ℓ is 0. The quantum number m_s, however, can always take on two values: $m_s = +\frac{1}{2}$ and $m_s = -\frac{1}{2}$. Therefore, two electrons can occupy the $n = 1$, $\ell = 0$ energy level, because two different states—$n = 1$, $\ell = 0$, $m_\ell = 0$, $m_s = -\frac{1}{2}$; and $n = 1$, $\ell = 0$, $m_\ell = 0$, $m_s = +\frac{1}{2}$—correspond to that level, as indicated in **FIGURE 31-21**.

The next higher energy level corresponds to $n = 2$ and $\ell = 0$. Again, two states correspond to this level, allowing it to hold two electrons. Lest we conclude that all levels can hold two electrons, consider the next level up: $n = 2$ and $\ell = 1$. In this case, m_ℓ can take on three values: $m_\ell = 0, \pm1$. For each of these three values, m_s can take on two values. Therefore, the $n = 2$, $\ell = 1$ energy level can accommodate 6 electrons, as shown in Figure 31-21. For general ℓ, the number of possible values of m_ℓ $(0, \pm1, \pm2, \ldots, \pm\ell)$ is $2\ell + 1$. When we multiply by 2 (for the number of values of m_s), we find a total number of states equal to $2(2\ell + 1)$. For example, notice in Figure 31-21 that the $n = 3$, $\ell = 2$ energy level can hold $2(2 \cdot 2 + 1) = 10$ electrons.

FIGURE 31-22 presents the ground-state electron arrangements for the following elements: hydrogen (1 electron), helium (2 electrons), lithium (3 electrons), beryllium (4 electrons), boron (5 electrons), carbon (6 electrons), nitrogen (7 electrons), and oxygen (8 electrons). Notice that the energy levels are filled from the bottom upward, like marbles in a jar, with each level having a predetermined maximum number of electrons.

Big Idea 5 The Pauli exclusion principle states that only a certain number of electrons may occupy a given energy level. As a result, electrons in multielectron atoms are forced to occupy higher energy levels, which results in different atomic properties as recognized in the periodic table of elements.

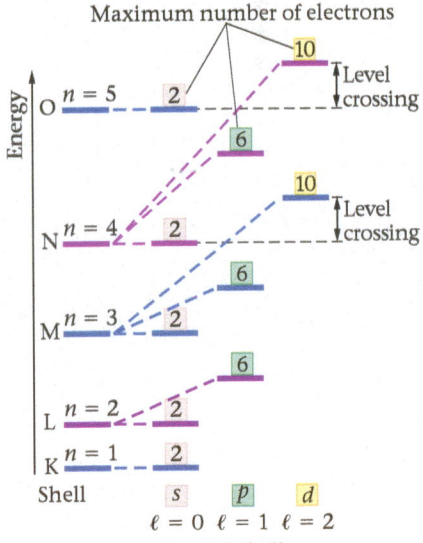

Maximum number of electrons

▲ **FIGURE 31-21 Maximum number of electrons in an energy level** The maximum number of electrons that can occupy a given energy level in a multielectron atom is $2(2\ell + 1)$. Occupancy by more than this number of electrons would violate the Pauli exclusion principle.

Electronic Configurations

Indicating the arrangements of electrons as in Figure 31-22 is instructive, but also somewhat cumbersome. This is especially so when we consider elements with a large number of electrons. To streamline the process, we introduce a shorthand notation that can be applied to all elements.

As an example of this notation, consider the element lithium, which has two electrons in the $n = 1$, $\ell = 0$ state and one electron in the $n = 2$, $\ell = 0$ state. This arrangement of electrons—referred to as an **electronic configuration**—is abbreviated as follows:

$$1s^2 2s^1$$

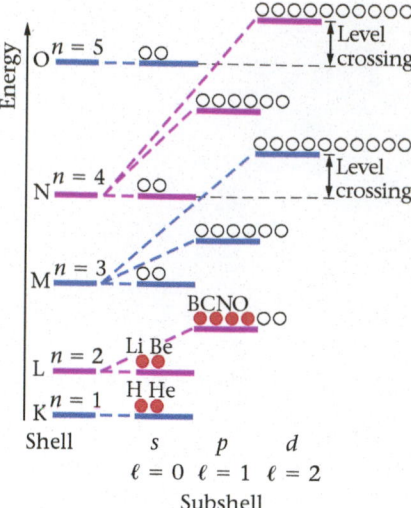

▲ **FIGURE 31-22 First eight elements of the periodic table** The elements hydrogen and helium fill the $n = 1$, $\ell = 0$ level; the elements lithium and beryllium fill the $n = 2$, $\ell = 0$ level; the elements boron, carbon, nitrogen, and oxygen fill four of the six available states in the $n = 2$, $\ell = 1$ level.

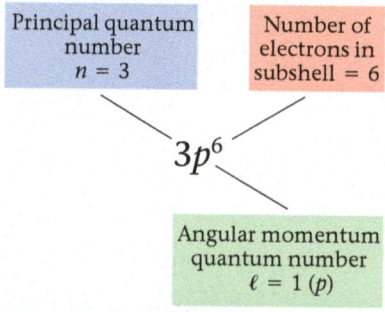

▲ **FIGURE 31-23 Designation of electronic configuration** The example shown here indicates six electrons in the $n = 3$, $\ell = 1$ (p) energy level.

In this expression, the $1s^2$ part indicates $n = 1$ ($1s^2$), $\ell = 0$ ($1s^2$), and an occupancy of two electrons ($1s^2$). Similarly, the $2s^1$ part indicates one electron ($2s^1$) in the $n = 2$ ($2s^1$) and $\ell = 0$ ($2s^1$) state. The general labeling is shown in **FIGURE 31-23**.

EXERCISE 31-10   SHORTHAND CONFIGURATION

Use the shorthand notation just introduced to write the electronic configuration for the ground state of **(a)** oxygen and **(b)** aluminum.

REASONING AND SOLUTION

a. The $1s$ and $2s$ subshells of oxygen ($Z = 8$) are filled, which accounts for 4 of the 8 electrons. The remaining 4 electrons are in the $2p$ subshell. Therefore, the electronic configuration is

$$1s^2\,2s^2\,2p^4$$

b. In aluminum ($Z = 13$), the $1s$, $2s$, $2p$, and $3s$ subshells are fully occupied. This accounts for 12 of the 13 electrons. The remaining electron is in the $3p$ subshell. The electronic configuration is

$$1s^2 2s^2 2p^6 3s^2 3p^1$$

Table 31-4 presents a list of the electronic configurations for the elements hydrogen (H) through potassium (K). Notice that the $3d$ levels in potassium have no electrons, in agreement with the level crossing shown in Figure 31-22.

The Periodic Table

Referring to the elements listed in Table 31-4, we observe a number of interesting patterns. For example, notice that the elements hydrogen, lithium, sodium, and potassium all have the same type of electronic configuration for the final (and outermost) electron in the atom. In particular, hydrogen's outermost (and only) electron is the $1s^1$ electron. In the case of lithium, we see that the outermost electron is $2s^1$, the outermost

TABLE 31-4 Electronic Configurations of the Elements Hydrogen Through Potassium

Atomic number	Element	Electronic configuration
1	Hydrogen (H)	$1s^1$
2	Helium (He)	$1s^2$
3	Lithium (Li)	$1s^2 2s^1$
4	Beryllium (Be)	$1s^2 2s^2$
5	Boron (B)	$1s^2 2s^2 2p^1$
6	Carbon (C)	$1s^2 2s^2 2p^2$
7	Nitrogen (N)	$1s^2 2s^2 2p^3$
8	Oxygen (O)	$1s^2 2s^2 2p^4$
9	Fluorine (F)	$1s^2 2s^2 2p^5$
10	Neon (Ne)	$1s^2 2s^2 2p^6$
11	Sodium (Na)	$1s^2 2s^2 2p^6 3s^1$
12	Magnesium (Mg)	$1s^2 2s^2 2p^6 3s^2$
13	Aluminum (Al)	$1s^2 2s^2 2p^6 3s^2 3p^1$
14	Silicon (Si)	$1s^2 2s^2 2p^6 3s^2 3p^2$
15	Phosphorus (P)	$1s^2 2s^2 2p^6 3s^2 3p^3$
16	Sulfur (S)	$1s^2 2s^2 2p^6 3s^2 3p^4$
17	Chlorine (Cl)	$1s^2 2s^2 2p^6 3s^2 3p^5$
18	Argon (Ar)	$1s^2 2s^2 2p^6 3s^2 3p^6$
19	Potassium (K)	$1s^2 2s^2 2p^6 3s^2 3p^6 4s^1$

electron of sodium is $3s^1$, and for potassium it is $4s^1$. Continuing through the list of elements, we note that rubidium has a $5s^1$ outer electron, cesium has a $6s^1$ outer electron, and francium has a $7s^1$ outer electron. In each case the outermost electron is a single electron in an otherwise empty s subshell.

What makes this similarity in the *electronic configurations* of particular interest is that these elements have similar *chemical properties* as well. In particular, the metallic members of this group of elements (lithium, sodium, potassium, rubidium, cesium, and francium) are referred to as the *alkali metals*. In each of these metals, the outermost electron is easily removed from the atom, leading to a stable, positively charged ion, as in the familiar case of the sodium ion, Na^+. Thus the regular pattern in the filling of shells leads to a regular pattern in the properties of the elements.

Grouping various elements with similar chemical properties was the motivation behind the development of the **periodic table** of elements by the Russian chemist Dmitri Mendeleev (1834–1907). The periodic table is presented in Appendix E. Notice that the elements just mentioned form Group I in the leftmost column of the table. Although Mendeleev grouped these elements strictly on the basis of their chemical properties, we can now see that the grouping also corresponds to the filling of shells, in accordance with the Pauli exclusion principle.

Many groups of elements appear in the periodic table. As another example, Group VII consists of the *halogens*: fluorine, chlorine, bromine, and iodine. Note that these elements also have similar configurations of their outer electrons. In this case, the outer electrons are $2p^5$, $3p^5$, and so on. Thus we see that these elements are just one electron short of filling one of the p subshells. As a result, halogens are highly reactive—they can readily acquire a single electron from another element to form a stable negative ion, as in the case of chlorine, which forms the chloride ion, Cl^-.

Group VIII consists of the *noble gases*. These elements all have completely filled subshells. Thus they do not readily gain or lose an electron. It is for this reason that the noble gases are relatively inert.

Finally, the *transition elements* represent those cases where a crossing of energy levels occurs. For example, the $4s$ subshell is filled at the element calcium ($Z = 20$). The next electron, rather than going into the $4p$ subshell, goes into the $3d$ subshell. In fact, the 10 elements from scandium ($Z = 21$) to zinc ($Z = 30$) correspond to filling of the $3d$ subshell. After this subshell is filled, additional electrons go into the $4p$ subshell in the elements gallium ($Z = 31$) to krypton ($Z = 36$).

FIGURE 31-24 shows the meaning of the various symbols used in the periodic table. Notice that each box in the table gives the symbol for the element, its atomic mass, and its atomic number. Each box also includes the configuration of the outermost electrons in the element. In the case of iron, shown in Figure 31-24, this configuration is $3d^6 4s^2$.

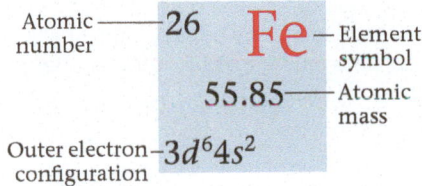

▲ **FIGURE 31-24 Designation of elements in the periodic table** An explanation of the various entries to be found for each element in the periodic table of Appendix E. The example shown here is for the element iron (Fe).

CONCEPTUAL EXAMPLE 31-11   COMPARE THE ENERGY   PREDICT/EXPLAIN

The energy required to remove the outermost electron from sodium is 5.1 eV. **(a)** Is the energy required to remove the outermost electron from potassium greater than, less than, or equal to 5.1 eV? **(b)** Which of the following is the *best explanation* for your prediction?

 I. The energy required to remove an outermost electron is the same for all elements.

 II. The outermost electron in potassium is in the $n = 4$ level. This is a higher level than the $n = 3$ level for the outermost electron in sodium, and hence less energy is required to remove the electron.

 III. The outermost electron in potassium is in the $n = 4$ level. This is a higher level than the $n = 3$ level for the outermost electron in sodium, and hence more energy is required to remove the electron.

REASONING AND DISCUSSION
Referring to the periodic table, we see that potassium has one electron in its outermost shell, just like sodium. In this respect the two elements are alike. The difference between the elements is that the outermost electron in potassium is in a higher energy state; that is, its electron is in an $n = 4$ state, as opposed to the $n = 3$ state in sodium. Less energy is required to remove an electron from a state of higher energy—that is, from a state that is closer to the zero energy of a free electron. Therefore, the outermost electron of potassium can be removed with less energy than 5.1 eV. In fact, the required energy is only 4.3 eV.

ANSWER
(a) Less energy is required to remove the outermost electron from potassium. **(b)** The best explanation is II.

▲ **FIGURE 31-25** Neon (atomic number 10), one of the "noble gases," is most widely known for its role in neon tubes. Electrons in neon atoms are excited by an electrical discharge through the tube. When they return to the ground state, they emit electromagnetic radiation, much of it in the red part of the visible spectrum. The other colors familiarly found in "neon signs" or "neon lights" are produced by adding other elements of the same chemical family: argon, krypton, or xenon.

Big Idea **6** The radiation given off by atoms can be understood as the result of electrons moving from one electron shell to another. The difference in energy between the shells appears as the energy of the emitted photon.

▶ **FIGURE 31-26** **X-ray tube** An X-ray tube accelerates electrons in a vacuum through a potential difference and then directs the electrons onto a metal target. X-rays are produced when the electrons decelerate in the target, and as a result of the excitations they cause when they collide with the metal atoms of the target.

Enhance Your Understanding (Answers given at the end of the chapter)

6. How many electrons are in the element whose electronic configuration is $1s^2 2s^2 2p^6 3s^2 3p^6$? What is this element?

Section Review

- The Pauli exclusion principle results in electrons occupying progressively higher energy levels in multielectron atoms. As more energy levels are filled, the chemical behaviors of the elements show similar characteristics, as indicated in the periodic table of elements.

31-7 Atomic Radiation

We conclude this chapter with a brief investigation of various types of radiation associated with multielectron atoms. Examples range from X-rays that are energetic enough to pass through a human body, to the soft white light of a fluorescent lightbulb, and the brilliant colors of neon lights, like those shown in **FIGURE 31-25**.

X-Rays

X-rays were discovered quite by accident by the German physicist Wilhelm Roentgen (1845–1923) on November 8, 1895. Within months of their discovery, X-rays were being used in medical applications, and they have played an important role in medicine ever since. Today, the X-rays used to give diagnostic images in hospitals and dentist offices are produced by an X-ray tube similar to the one shown in **FIGURE 31-26**. The basic operating principle of this device is that an energetic beam of electrons is generated and directed at a metal target—when the electrons collide with the target, X-rays are emitted. **FIGURE 31-27** shows a typical plot of the X-ray intensity per wavelength versus the wavelength for such a device.

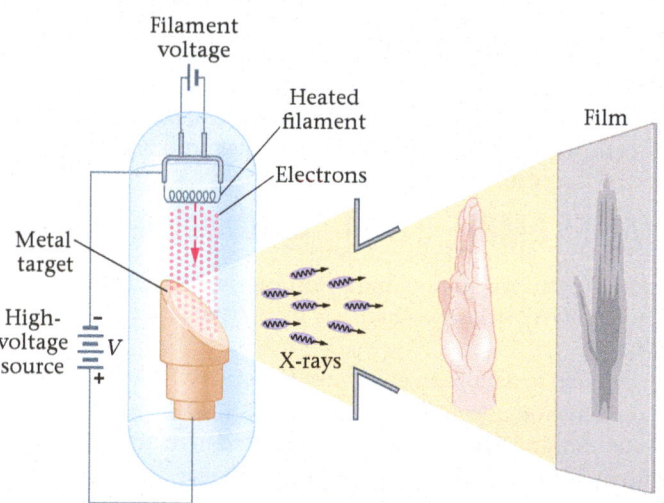

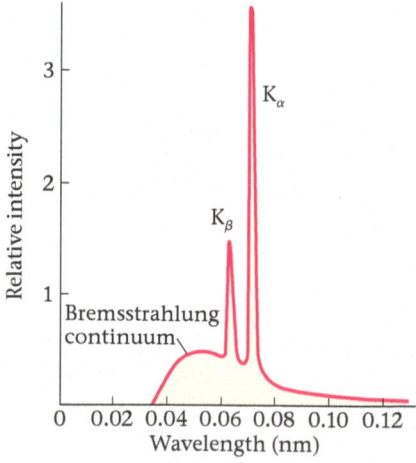

RWP The radiation produced by an X-ray tube is created by two completely different physical mechanisms. The first mechanism is referred to as **bremsstrahlung**, which is German for "braking radiation." What is meant by this expression is that as the energetic electrons impact the target, they undergo a rapid deceleration. This is the "braking." As we know from Chapter 25, an accelerated charge gives off electromagnetic

◀ **FIGURE 31-27** **X-ray spectrum** The spectrum produced by an X-ray tube in which electrons are accelerated from rest through a potential difference of 35,000 V and directed against a molybdenum target.

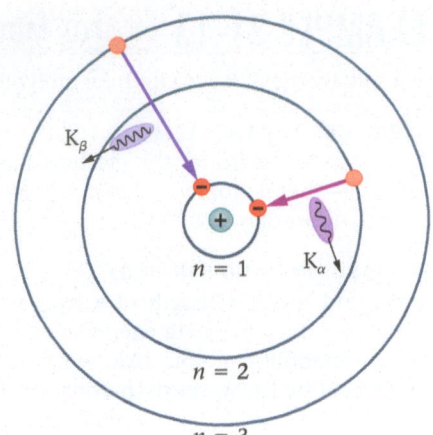

▶ **FIGURE 31-28 Production of characteristic X-rays** When an electron strikes a metal atom in the target of an X-ray tube, it may knock one of the two K-shell ($n = 1$) electrons out of the atom. The resulting vacancy in the K shell will be filled by an electron dropping from a higher shell. If an electron drops from the $n = 2$ shell to the $n = 1$ shell, we say that the resulting photon is a K_α X-ray. Similarly, if an electron in the $n = 3$ shell drops to the $n = 1$ shell, the result is a K_β X-ray. Clearly, the K_β X-ray has the greater energy and the shorter wavelength, as we see in Figure 31-27.

radiation; hence, as the electrons suddenly come to rest in the target they give off high-energy radiation in the form of X-rays. These X-rays cover a wide range of wavelengths, giving rise to the continuous part of the spectrum shown in Figure 31-27.

The sharp peaks in Figure 31-27 are produced by the second physical mechanism. To understand their origin, imagine what happens if one of the electrons in the incident beam is energetic enough to knock an electron out of a target atom. In addition, suppose the ejected electron comes from the lowest energy level of the atom—that is, from the K shell. This "vacancy" is filled almost immediately when an electron from an outer shell drops to the K shell, with the energy difference given off as a photon. In an atom of large atomic number [molybdenum ($Z = 42$) is often used for a target, as is tungsten ($Z = 74$)], the photon that emerges is an X-ray. If an electron drops from the $n = 2$ level to the K shell, the sharp peak of radiation that results is called the K_α line. Similarly, if an electron drops from the $n = 3$ level to the K shell, the resulting peak is called the K_β line. These lines are shown in Figure 31-27, and the corresponding electron jumps are indicated in **FIGURE 31-28**. Because the wavelengths of these lines vary from element to element—that is, they are characteristic of a certain element—they are referred to as **characteristic X-rays.** Similar characteristic X-rays can be emitted when electrons drop in energy to fill a vacancy in the L shell, the M shell, and so on.

The energy an incoming electron must have to dislodge a K-shell electron from an atom can be estimated using results from the Bohr model. The basic idea is that the energy of a K-shell electron in an atom of atomic number Z is given approximately by Equation 31-9 ($E_n = -(13.6 \text{ eV})Z^2/n^2$), with one minor modification: Because there are two electrons in the K shell, each electron shields the other from the nucleus. That is, the negative charge on one electron, $-e$, partially cancels the positive charge of the nucleus, $+Ze$, giving an effective charge experienced by the second electron of $+(Z - 1)e$. Thus, replacing Z with $Z - 1$ in Equation 31-9, and setting $n = 1$, we obtain a reasonable estimate for the energy of a K-shell electron:

$$E_K = -(13.6 \text{ eV})\frac{(Z - 1)^2}{1^2} \qquad 31\text{-}14$$

We apply this result in the following Quick Example.

QUICK EXAMPLE 31-12   FIND THE VOLTAGE OF AN X-RAY TUBE

Estimate the minimum energy an incoming electron must have to knock a K-shell electron out of a tungsten atom ($Z = 74$).

REASONING AND SOLUTION
The energy required to knock a K-shell electron from a tungsten atom is the energy required to increase its energy from $E_K = -(13.6 \text{ eV})(Z - 1)^2/1^2$ to 0, which corresponds to the energy of an electron at rest at infinite separation.

1. Calculate $(Z - 1)^2$ for $Z = 74$: $(74 - 1)^2 = 5329$
2. Multiply -13.6 eV by $(Z - 1)^2$: $(-13.6 \text{ eV})(5329) = -72,500 \text{ eV}$

Thus, an incoming electron must have an energy of at least 72.5 keV to remove a K-shell electron from tungsten and produce characteristic X-rays. This means that the incoming electrons must be accelerated through a potential difference of 72.5 kV. Typical X-ray tubes, like those in dental offices, operate in the 100-kV range.

Equation 31-14 ($E_K = -(13.6 \text{ eV})(Z - 1)^2/1^2$) can also be used to obtain an estimate of the K_α wavelength for a given element. We show how this can be done for the case of molybdenum in the next Example.

EXAMPLE 31-13 K_α FOR MOLYBDENUM

Estimate the K_α wavelength for molybdenum ($Z = 42$).

PICTURE THE PROBLEM

Our sketch indicates the electron jump that is responsible for the K_α X-ray—that is, from $n = 2$ to $n = 1$. Notice that the net charge from the K shell outward is $+(Z - 1)e$.

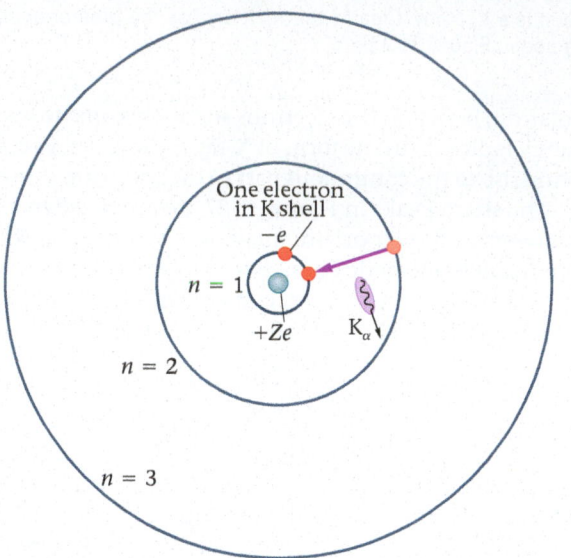

REASONING AND STRATEGY

To find the wavelength of a K_α X-ray, we start with the relationship between the change in energy of an electron and the wavelength of the corresponding photon: $|\Delta E| = hf = hc/\lambda$. Once we have calculated ΔE, we can find the wavelength using $\lambda = hc/|\Delta E|$.

To find ΔE, we first calculate the energy of an electron in the K shell of molybdenum, using $E_K = -(13.6 \text{ eV})(Z - 1)^2/1^2$ with $Z = 42$.

Next, we calculate the energy of an electron in the L shell ($n = 2$) of molybdenum, because it is an L-shell electron that fills the vacancy in the K shell and emits a K_α X-ray. Notice that an L-shell electron sees a nucleus with an effective charge of $+(Z - 1)e$. This follows because there is one electron in the K shell, and this electron partially screens the nucleus. As a result, the energy of an L-shell electron is given by the following expression: $E_L = -(13.6 \text{ eV})(Z - 1)^2/2^2$.

Finally, with these two energies determined, the change in energy is simply $\Delta E = E_K - E_L$.

Known   Atomic number for molybdenum, $Z = 42$.
Unknown K_α wavelength, $\lambda = ?$

SOLUTION

1. Calculate the energy of a K-shell electron, E_K:

$$E_K = -(13.6 \text{ eV})\frac{(Z - 1)^2}{1^2} = -(13.6 \text{ eV})\frac{(42 - 1)^2}{1^2}$$
$$= -22{,}900 \text{ eV}$$

2. Calculate the energy of an L-shell electron, E_L:

$$E_L = -(13.6 \text{ eV})\frac{(Z - 1)^2}{2^2} = -(13.6 \text{ eV})\frac{(42 - 1)^2}{2^2}$$
$$= -5720 \text{ eV}$$

3. Determine the change in energy, ΔE, of an electron that jumps from the L shell to the K shell:

$$\Delta E = E_K - E_L = -22{,}900 \text{ eV} - (-5720 \text{ eV}) = -17{,}200 \text{ eV}$$

4. Calculate the wavelength corresponding to ΔE:

$$\lambda = \frac{hc}{|\Delta E|} = \frac{(6.63 \times 10^{-34} \text{ J} \cdot \text{s})(3.00 \times 10^8 \text{ m/s})}{(17{,}200 \text{ eV})(1.60 \times 10^{-19} \text{ J/eV})}$$
$$= 7.23 \times 10^{-11} \text{ m} = 0.0723 \text{ nm}$$

INSIGHT

Comparing our result with Figure 31-27, which shows the X-ray spectrum for molybdenum, we see that our approximate wavelength of 0.0723 nm is in good agreement with experiment.

PRACTICE PROBLEM

Which element has a K_α peak at a wavelength of approximately 0.155 nm?   [**Answer:** Copper, $Z = 29$]

Some related homework problems: Problem 60, Problem 61, Problem 78

Medical Use of X-Rays As mentioned earlier in this section, X-rays were put to medical use as soon as people found out about their properties and how to produce them. In fact, Roentgen himself, whose first article about X-rays was published in late December 1895, produced an X-ray image of his wife's left hand, clearly showing the bones in her fingers and her wedding ring. Less than two months later, in February 1896, American physicians were starting to test X-rays on patients. One of the earliest patients was a young boy named Eddie McCarthy, who had his broken forearm X-rayed. A New Yorker by the name of Tolson Cunningham had a bullet removed from his leg after its position was determined by a 45-minute X-ray exposure.

BIO Standard X-rays can be difficult to interpret, however, because they cast shadows of all the body materials they pass through onto a single sheet of film. It is somewhat like placing several transparencies on top of one another and trying to decipher their individual contents. With the advent of high-speed computers, a new type of X-ray image is now possible. In a **computerized axial tomography scan** (CAT scan), thin beams of X-rays are directed through the body from a variety of directions. The intensity of the transmitted beam is detected for each direction, and the results are sent to a computer for processing. The result is an image that shows the physician a "cross-sectional slice" through the body. In this way, each part of the body can be viewed individually and with clarity. If a series of such slices are stacked together in the computer, they can give a three-dimensional view of the body's interior. Examples of CAT scans are shown in **FIGURE 31-29**.

Lasers

The production of light by humans advanced significantly when flames were replaced by the lightbulb. An even greater advancement occurred in 1960, however, when the first laser was developed. Lasers produce light that is intense, highly collimated (unidirectional), and pure in its color. Because of these properties, lasers are used in a multitude of technological applications, ranging from supermarket scanners to CDs, from laser pointers to eye surgery. In fact, lasers are now almost as common in everyday life as the lightbulb.

To understand just what a laser is and what makes it so special, we start with its name. The word **laser** is an acronym for light amplification by the stimulated emission of radiation. As we shall see, the properties of stimulated emission lead directly to the amplification of light.

To begin, consider two energy levels in an atom, and suppose an electron occupies the higher of the two levels. If the electron is left alone, it will eventually drop to the lower level in a time that is typically about 10^{-8} s, giving off a photon in the process referred to as **spontaneous emission.** The photon given off in this process can propagate in any direction.

In contrast, suppose the electron in the excited state just described is not left alone. For example, a photon with an energy equal to the energy difference between the two energy levels might pass near the electron, which *enhances* the probability that the electron will drop to the lower level. That is, the incident photon can *stimulate* the emission of a second photon by the electron. The photon given off in this process of **stimulated emission** has the same energy as the incident photon, the same phase, and propagates in the same direction, as indicated in **FIGURE 31-30**. This accounts for the fact that laser light is highly focused and of a single color.

As for the amplification of light, notice that a single photon entering an excited atom can cause two identical photons to exit the atom. If each of these two photons encounters another excited atom and undergoes the same process, the number of photons increases to four. Continuing in this manner, the photons undergo a sort of "chain reaction" that doubles the number of photons with each generation. It is this property of stimulated emission that results in **light amplification.**

In order for the light amplification process to work, it is necessary that photons continue to encounter atoms with electrons in excited states. Under ordinary conditions this will not be the case, because most electrons are in the lowest possible energy levels. For laser action to occur, atoms must first be prepared in an excited state. Then, before the electrons have a chance to drop to a lower level by way of spontaneous emission, the process of stimulated emission can proceed. This requires what is known as a **population inversion,** in which more electrons are found in the excited state than in a lower state. In addition, the excited state must be one that lasts for a relatively long time so that photons will continue to encounter excited atoms. A long-lived excited state is referred to as a **metastable state.** So to produce a laser, one needs a metastable excited state with a population inversion.

RWP A specific example of a laser is the **helium-neon laser,** shown schematically in **FIGURE 31-31 (a),** in which the neon atoms produce the laser light. The appropriate energy-level diagrams for neon are shown in **FIGURE 31-31 (b).** The excited state E_3 is metastable—electrons promoted to that level stay in the level for a relatively long time. Electrons are excited to this level by an electrical power supply connected to the

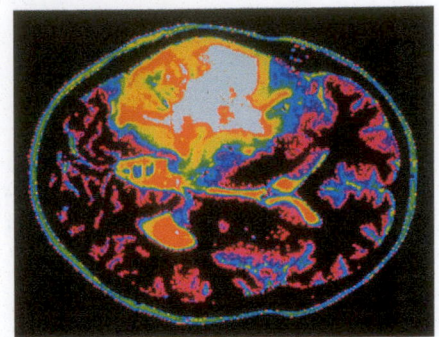

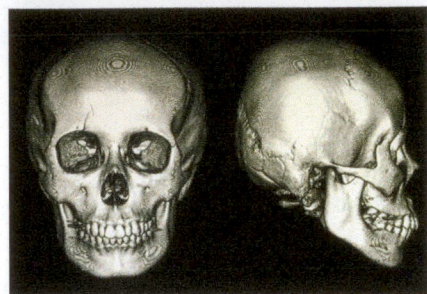

▲ **FIGURE 31-29 Visualizing Concepts CAT Scans** The false-color CAT scan (top) represents a horizontal section through the brain, revealing a large benign tumor (the white and orange area at the top). A series of CAT scans can also be combined to create remarkable three-dimensional images such as the one on the bottom.

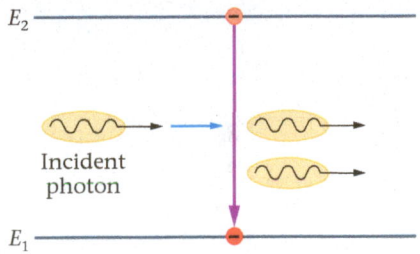

▲ **FIGURE 31-30 Stimulated emission** A photon with an energy equal to the difference $E_2 - E_1$ can enhance the probability that an electron in the state E_2 will drop to the state E_1 and emit a photon. When a photon stimulates the emission of a second photon in this way, the new photon has the same frequency, direction, and phase as the incident photon. If each of the two photons resulting from this process in turn produces two photons, the total number of photons can increase exponentially, resulting in an intense beam of light.

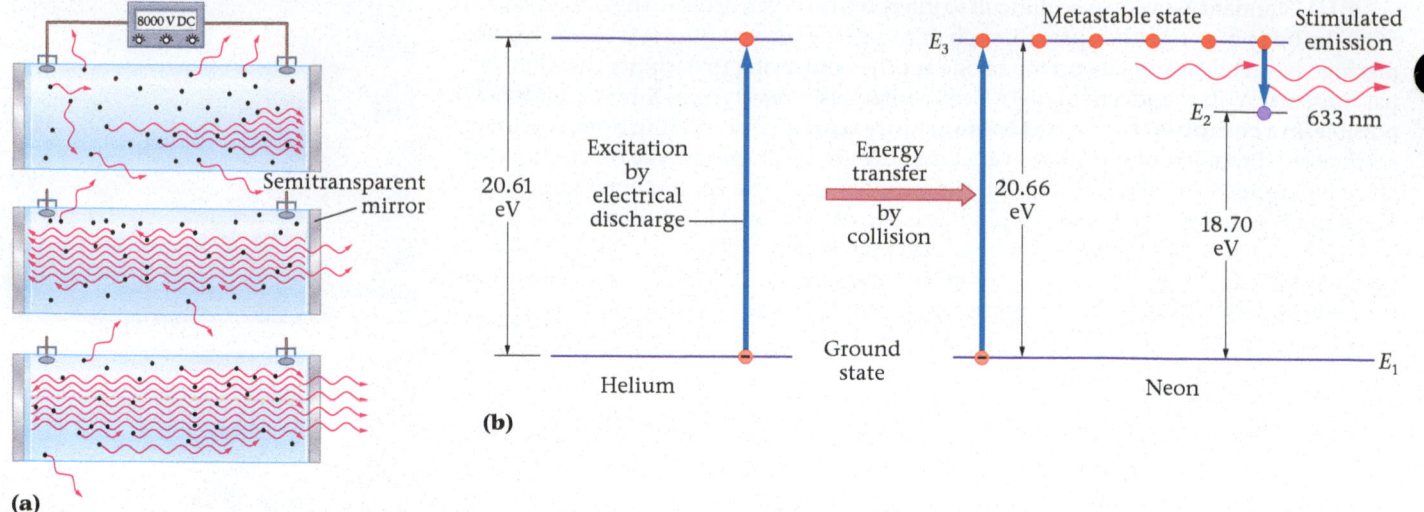

▲ **FIGURE 31-31 The helium-neon laser (a)** A schematic representation of the basic features of a helium-neon laser. **(b)** The relevant energy levels in helium and neon that result in the laser action.

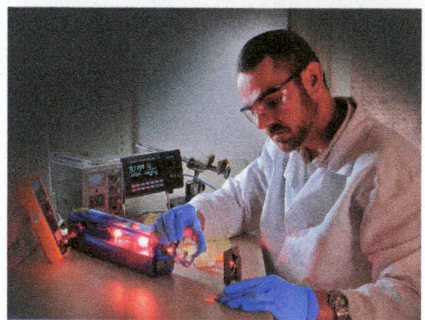

▲ **FIGURE 31-32** An operating helium-neon laser, showing the glowing gas plasma produced when high voltage is placed across the tube containing the helium-neon mixture.

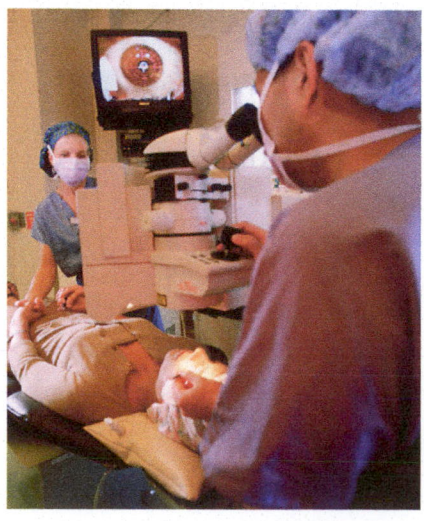

◄ **FIGURE 31-33** A doctor uses a microscope and video monitor to ensure precise cutting of the corneal flap in the LASIK procedure. A UV laser will then remove some of the material under the flap, flattening the cornea and correcting the patient's nearsightedness.

tube containing the helium-neon mixture. The power supply causes electrons to move through the tube, colliding with neon atoms and exciting them. Similarly, excited helium atoms colliding with the neon atoms also cause excitations to the E_3 level. Because this level is metastable, the excitation processes can cause a population inversion, setting the stage for laser action. Stimulated emission then occurs, allowing the electrons to drop to the lower level, E_2. The electrons subsequently proceed through various intermediate steps to the ground state. It is the emission of light between the levels E_3 and E_2 that results in laser light. The difference in energy between these levels is 1.96 eV, and hence the light coming out of the laser is red, with a wavelength of 633 nm.

To enhance the output of a laser, the light is reflected back and forth between mirrors. In fact, this reflection produces a resonant condition in the light, much like standing waves in an organ pipe. An operating laser is shown in **FIGURE 31-32**. Helium-neon lasers revolutionized barcode scanning for retail checkout counters in the early 1970s, although they were later replaced with less expensive solid-state diode lasers. Today helium-neon lasers are still widely used in educational and laboratory science applications.

Laser Therapy Lasers are used today in a variety of medical procedures. A particularly common example is *laser eye surgery*. In these techniques, a laser that emits high-energy photons in the UV range (typically at wavelengths of 193 nm) is used to reshape the cornea and correct nearsightedness. For example, in LASIK (**la**ser **i**n **s**itu **k**eratomileusis) eye surgery, shown in **FIGURE 31-33**, the procedure begins with a small mechanical shaver known as a microkeratome cutting a flap in the cornea, leaving a portion of the cornea uncut to serve as a hinge. After the mechanical cut is made, the corneal flap is folded back, exposing the middle portion of the cornea as shown in **FIGURE 31-34 (a)**. Next, an *excimer laser* sends pulses of UV light onto the cornea, each pulse vaporizing a small layer of corneal material (0.1 to 0.5 μm in thickness) with no heating. This process continues until the cornea is flattened just enough to correct the nearsightedness, after which the corneal flap is put back into place.

BIO *Photorefractive keratectomy* (PRK) is similar to LASIK eye surgery, except that material is removed directly from the surface of the cornea, without the use of a corneal flap, as shown in **FIGURE 31-34 (b)**. To correct nearsightedness, the laser beam is directed onto the central portion of the cornea (left), resulting in a flattening of the cornea. To correct farsightedness, it is necessary to increase the curvature of the cornea. This is

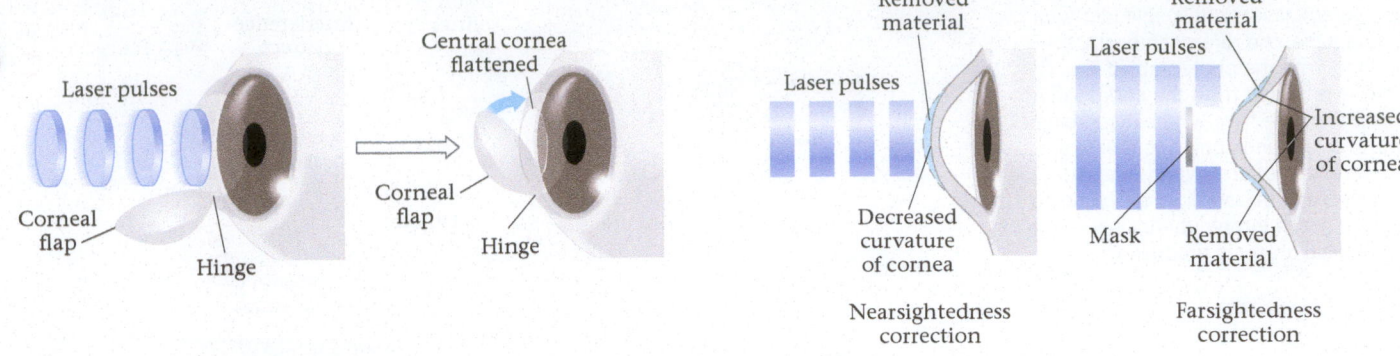

(a) LASIK

(b) PRK

▲ **FIGURE 31-34 Laser vision correction** **(a)** In LASIK eye surgery, a flap of the cornea is cut and folded back. Next, an ultraviolet excimer laser is used to vaporize some of the underlying corneal material. When the flap is replaced, the cornea is flatter than it was, correcting the patient's nearsightedness. **(b)** In the PRK procedure, the laser removes material directly from the corneal surface. If the cornea is too curved, which produces nearsightedness, the laser beam is directed at its center and the cornea is flattened. If the cornea is too flat, which produces farsightedness, the central region is masked and material is removed from the periphery, increasing the curvature.

accomplished by masking the central portion of the cornea so that the laser removes only peripheral portions of the cornea (right). In both cases, it is necessary to keep the beam focused at the desired location on the eye. This is difficult, because the eye routinely moves by small amounts roughly every 15 ms. In the most sophisticated application of PRK, these eye movements can be tracked and the aiming of the laser beam corrected accordingly.

BIO Another medical application of lasers is known as *photodynamic therapy,* or PDT. In this type of therapy, light-sensitive chemicals (such as porphyrins) are injected into the bloodstream and are taken up by cells throughout the body. These chemicals are found to remain in cancerous cells for longer periods of time than in normal cells. Thus, after an appropriate time interval, the light-sensitive chemicals are preferentially concentrated in cancerous cells. If a laser beam with the precise wavelength absorbed by the light-sensitive chemicals illuminates the cancer cells, the resulting chemical reactions kill the cancer cells without damaging the adjacent normal cells. The laser beam can be directed to the desired location using a flexible fiber-optic cable in conjunction with a bronchoscope to treat lung cancer or with an endoscope to treat esophageal cancer.

PDT is also used to treat certain types of age-related macular degeneration (AMD). For most people, macular degeneration results when abnormal blood vessels behind the retina leak fluid and blood into the central region of the retina, or macula, causing it to degenerate. In this case, light-sensitive chemicals are preferentially taken up by the abnormal blood vessels; hence, a laser beam of the proper wavelength can destroy these blood vessels without damaging the normal structures in the retina.

RWP Lasers can also be used to take three-dimensional photographs known as **holograms.** A typical setup for taking a hologram is shown in **FIGURE 31-35**. Notice that the hologram is produced with no focusing lenses, in contrast with normal photography. The basic procedure in holography begins with the splitting of a laser beam into two separate beams. One beam, the reference beam, is directed onto the photographic film. The second beam, called the object beam, is directed onto the object to be recorded in the hologram. The object beam reflects from various parts of the object and then combines with the reference beam on the film. Because the laser light is coherent, and because the object and reference beams travel different distances, the combined light results in an interference pattern. In fact, if you look at a hologram in normal light it simply looks like a confusing mass of swirls and lines. When a laser beam illuminates the hologram, however, the interference pattern causes the laser light to propagate away from the hologram in exactly the same way as the light that originally produced the interference pattern. Thus, a person viewing the hologram

▶ **FIGURE 31-35 Holography** To create a hologram, laser light is split into two beams. One, the reference beam, is directed onto the photographic film. The other is reflected from the surface of an object onto the film, where it combines with the reference beam to create an interference pattern. When this pattern is illuminated with laser light of the same wavelength, a three-dimensional image of the original object is produced.

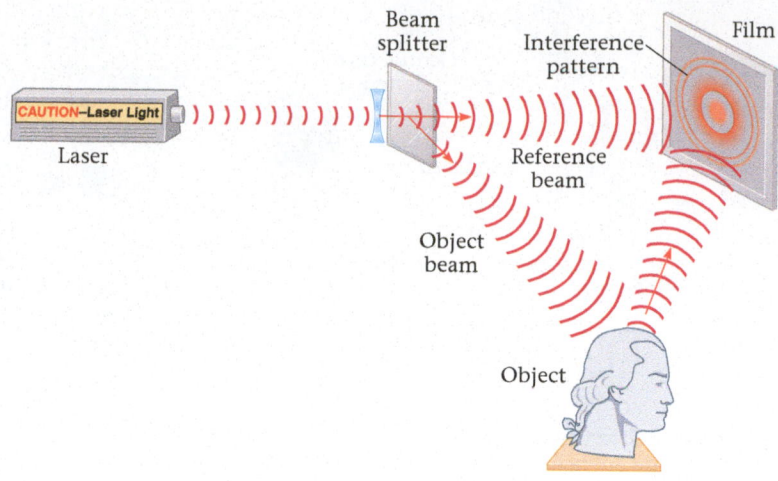

▲ **FIGURE 31-36** A hologram creates a three-dimensional image in empty space. The image can be viewed from different angles to reveal different parts of the original subject.

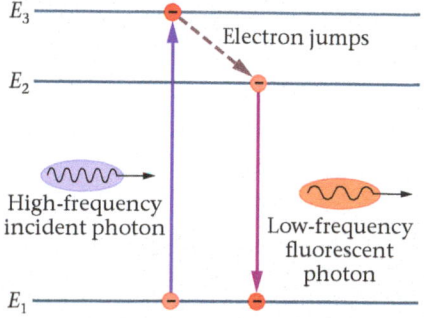

▲ **FIGURE 31-37 The mechanism of fluorescence** In fluorescence, a high-frequency photon raises an electron to an excited state. When the electron drops back to the ground state, it may do so by way of various intermediate states. The jumps between intermediate states produce photons of lower frequency, which are observed as the phenomenon of fluorescence.

sees precisely the same patterns of light that would have been observed when the hologram was recorded.

Holograms give a true three-dimensional image, as illustrated in **FIGURE 31-36**. When you view a hologram, you can move your vantage point to see different parts of the scene that is recorded. In particular, viewing the hologram from one angle may obscure an object in the background, but by moving your head, you can look around the foreground objects to get a clear view of the obscured object. In addition, you have to adjust your focus as you shift your gaze from foreground to background objects, just as in the real world. Finally, if you cut a hologram into pieces, each piece still shows the entire scene! This is analogous to your ability to see everything in your front yard through a small window just as you can through a large window—both show the entire scene.

The holograms on your credit cards are referred to as *rainbow holograms*, because they are designed to be viewed with white light, which contains all the colors of the rainbow. Although these holograms give an impression of three-dimensionality, they do not compare in quality to a hologram viewed with a laser.

Fluorescence and Phosphorescence

In Quick Example 31-6 it was noted that an electron in an excited state can emit photons of various energies as it falls to the ground state. This type of behavior is at the heart of fluorescence and phosphorescence.

Consider the energy levels shown in **FIGURE 31-37**. If an atom with these energy levels absorbs a photon of energy $E_3 - E_1$, it can excite an electron from state E_1 to state E_3. In some atoms, the most likely way for the electron to return to the ground state is by first jumping to level E_2 and then jumping to level E_1. The photons emitted in these jumps have less energy than the photon that caused the excitation in the first place. Hence, in a system like this, an atom is illuminated with a photon of one frequency, and it subsequently emits photons of lower energy and lower frequency. The emission of light of lower frequency after illumination by a higher frequency is referred to as **fluorescence.** In essence, fluorescence can be thought of as a conversion process, in which photons of high frequency are converted to photons of lower frequency.

RWP Perhaps the most common example of fluorescence is the *fluorescent lightbulb*. This device uses fluorescence to convert high-frequency ultraviolet light to lower-frequency visible light. In particular, the tube of a fluorescent light contains mercury vapor. When electricity is applied to one of these tubes, a filament is heated, which produces electrons. These electrons are accelerated by an applied voltage. The electrons strike mercury atoms in the tube, exciting them, and they give off ultraviolet light as they decay to their ground state. This process is not particularly useful in itself, because the ultraviolet light is invisible to us. However, the inside of the tube is coated with a phosphor that absorbs the ultraviolet light and then emits a lower-frequency light that is visible. Thus, several different physical processes must take place before a fluorescent lightbulb produces visible light. A *black light* is a fluorescent lightbulb with a modified coating that allows some of the less harmful ultraviolet light out so that fluorescent pigments and objects can be clearly seen.

RWP Fluorescence is also utilized in white light-emitting diode (LED) lamp technology. The LED in these devices actually emits violet light with a wavelength near 400 nm, and then a phosphor coating converts the violet light to white light. This technology has made highly efficient and long-lasting LED lamps an affordable alternative to compact fluorescent bulbs, and in 2014 the Nobel Prize in physics was awarded to Isamu Akasaki, Hiroshi Amano, and Shuji Nakamura for "the invention of efficient blue light-emitting diodes which has enabled bright and energy-saving white light sources."

RWP Fluorescence finds many other less familiar applications as well. In forensics, the analysis of a crime scene is aided by the fact that human bones and teeth are fluorescent. Thus, illuminating a crime scene with ultraviolet light can make items of interest stand out for easy identification. In addition, the use of a fluorescent dye can make fingerprints visible with great clarity.

BIO Many creatures produce fluorescence in their bodies as well. For example, several types of coral glow brightly when illuminated with ultraviolet light. It is also well known that scorpions are strongly fluorescent, giving off a distinctive green light. In fact, it is often possible to discern a greenish cast when viewing a scorpion in sunlight. At night in the desert, scorpions stand out with a bright green glow when a person illuminates the area with a portable ultraviolet light. This aids researchers who would like to find certain scorpions for study, and campers who are just as interested in avoiding scorpions altogether. Some examples of fluorescence in different situations, including forensics and geology, are shown in **FIGURE 31-38**.

BIO Another example is the green fluorescence produced by the jellyfish *Aequorea victoria*, which finds many uses in biological experiments. The gene that produces the green fluorescent protein (GFP) can serve as a marker to identify whether an organism has incorporated a new segment of DNA into its genome. For example, bacterial colonies that incorporate the GFP gene can be screened by eye simply by viewing the colony under an ultraviolet light. Recently, GFP has been inserted into the genome of a white rabbit, giving rise to the "GFP bunny." The bunny appears normal in white light, but when viewed under light with a wavelength of 392 nm, it glows

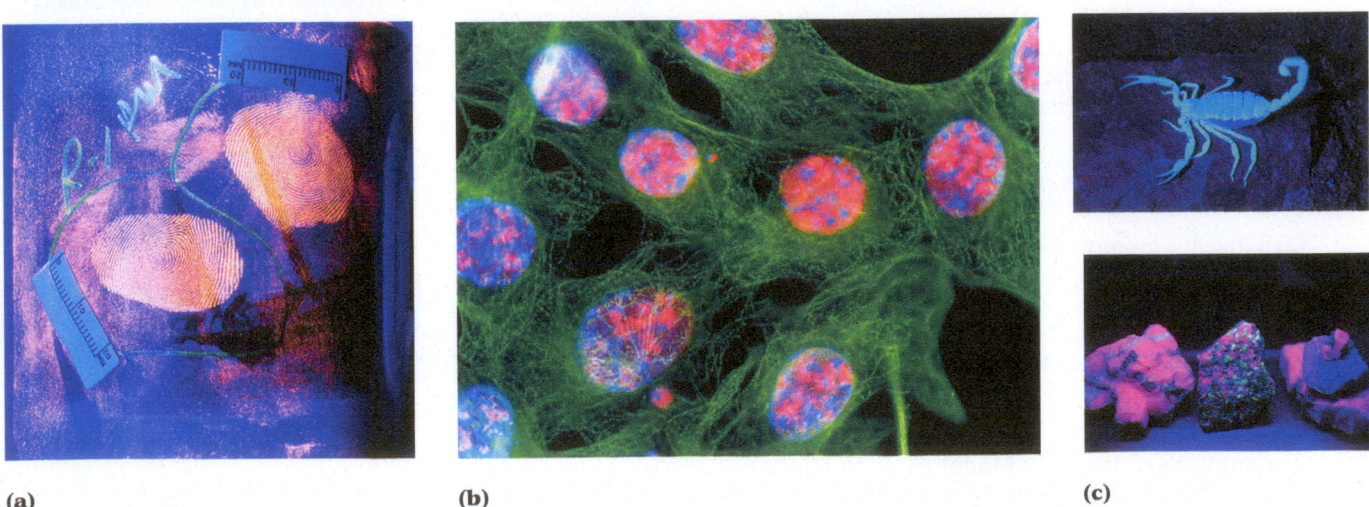

(a)   (b)   (c)

▲ **FIGURE 31-38 Visualizing Concepts Fluorescence (a)** The fingerprints on this mug become clearly visible when treated with fluorescent dye and illuminated with ultraviolet light. **(b)** In the technique known as immunofluorescence, an antibody molecule is linked to a fluorescent dye molecule, making it possible to visualize cellular structures and components that are otherwise largely invisible. **(c)** A variety of creatures, including the scorpion at the top, are naturally fluorescent when illuminated with ultraviolet light. So too are many minerals, such as the ones at the bottom.

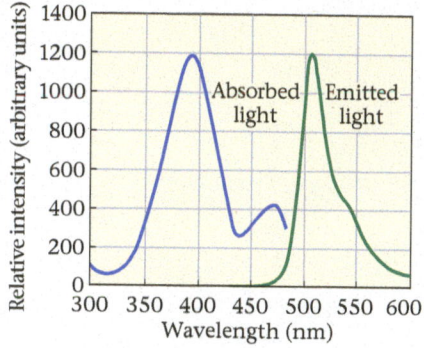

▲ FIGURE 31-39 The fluorescence spectrum of GFP The green fluorescent protein (GFP) strongly absorbs light with a wavelength of about 400 nm (violet). It reemits green light with a wavelength of 509 nm.

with a bright green light at 509 nm. The fluorescence spectrum of GFP is shown in **FIGURE 31-39**.

Phosphorescence is similar to fluorescence, except that phosphorescent materials continue to give off a secondary glow long after the initial illumination that excited the atoms. In fact, phosphorescence may persist for periods of time ranging from a few seconds to several hours, as on the hands of a watch that glow in the dark.

Enhance Your Understanding   (Answers given at the end of the chapter)

7. Is the wavelength of a K_β X-ray for a given element longer than, shorter than, or the same as the wavelength of a K_α X-ray for that element? Explain.

Section Review

- Radiation given off or absorbed by atoms results from electrons moving from one energy level to another.

CHAPTER 31 REVIEW

CHAPTER SUMMARY

31-1   EARLY MODELS OF THE ATOM

An atom is the smallest unit of a given element.

The Thomson Model: Plum Pudding
In Thomson's model, an atom is imagined to be like a positively charged pudding with negatively charged electrons scattered throughout.

The Rutherford Model: A Miniature Solar System
Rutherford discovered that an atom is somewhat like an atomic-scale solar system: mostly empty space, with most of its mass concentrated in the nucleus. The electrons were thought to orbit the nucleus.

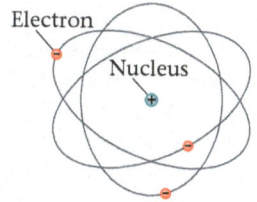

31-2   THE SPECTRUM OF ATOMIC HYDROGEN

Excited atoms of hydrogen in a low-pressure gas give off light of specific wavelengths. This is referred to as the spectrum of hydrogen.

Line Spectra
The spectrum of hydrogen is a series of bright lines of well-defined wavelengths.

Series
Hydrogen's line spectrum is formed by a series of lines that are grouped together. The wavelengths of these series are given by the expression

$$\frac{1}{\lambda} = R\left(\frac{1}{n'^2} - \frac{1}{n^2}\right) \quad n' = 1, 2, 3, \ldots$$
$$n = n' + 1, n' + 2, n' + 3, \ldots \qquad \textbf{31-2}$$

Each line in a given series corresponds to a different value of n. The different series correspond to different values of n'. For example, $n' = 1$ is the Lyman series, $n' = 2$ is the Balmer series, and $n' = 3$ is the Paschen series.

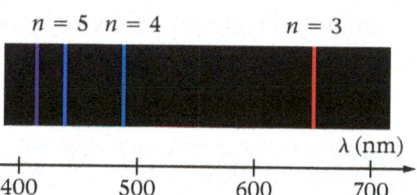

31-3   BOHR'S MODEL OF THE HYDROGEN ATOM

Bohr's model of hydrogen is basically a solar-system model, with only certain orbits allowed.

Assumptions of the Bohr Model
The Bohr model assumes the following: (i) Electrons move in circular orbits about the nucleus; (ii) allowed orbits must have an angular momentum equal to $L_n = nh/2\pi$, where $n = 1, 2, 3, \ldots$; (iii) electrons in allowed orbits do not give off electromagnetic radiation; and (iv) radiation is emitted only when electrons jump from one orbit to another.

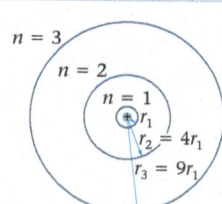

Bohr Orbits
The radii of allowed orbits in the Bohr model are given by

$$r_n = \left(\frac{h^2}{4\pi^2 mkZe^2}\right)n^2 = (5.29 \times 10^{-11}\,\text{m})n^2 \quad n = 1, 2, 3 \ldots \qquad 31\text{-}7$$

The Energy of a Bohr Orbit
The energy of an allowed Bohr orbit is

$$E_n = -(13.6\,\text{eV})\frac{Z^2}{n^2} \qquad n = 1, 2, 3, \ldots \qquad 31\text{-}9$$

These expressions correspond to hydrogen when $Z = 1$.

31-4   DE BROGLIE WAVES AND THE BOHR MODEL

De Broglie showed that the allowed orbits of the Bohr model correspond to standing matter waves of the electrons.

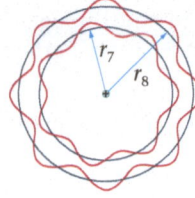

31-5   THE QUANTUM MECHANICAL HYDROGEN ATOM

The correct description of the hydrogen atom is derived from Schrödinger's equation.

Quantum Numbers
The quantum mechanical hydrogen atom is described by four quantum numbers: (i) the principal quantum number, n, which is analogous to n in the Bohr model; (ii) the orbital angular momentum quantum number, ℓ, which takes on the values $\ell = 0, 1, 2, \ldots, (n-1)$; the orbital angular momentum has a magnitude given by $L = \sqrt{\ell(\ell+1)}(h/2\pi)$; (iii) the magnetic quantum number, m_ℓ, for which the allowed values are $m_\ell = -\ell, -\ell+1, -\ell+2, \ldots, -1, 0, 1, \ldots, \ell-2, \ell-1, \ell$; the z component of the orbital angular momentum is $L_z = m_\ell(h/2\pi)$; and (iv) the electron spin quantum number, m_s, which can have the values $m_s = -\frac{1}{2}, \frac{1}{2}$.

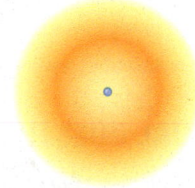

Electron Probability Clouds: Three-Dimensional Standing Waves
In the quantum mechanical hydrogen atom, the electron does not orbit at a precise distance from the nucleus. Instead, the electron distribution is represented by a probability cloud, where the densest regions of the cloud correspond to regions of highest probability.

31-6   MULTIELECTRON ATOMS AND THE PERIODIC TABLE

As electrons are added to atoms, the properties of the atoms change in a regular and predictable way.

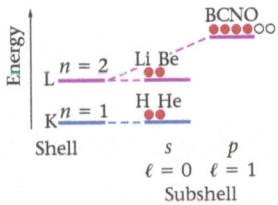

Multielectron Atoms
Energy levels in a multielectron atom depend on n and ℓ. The energy increases with increasing n for fixed ℓ, and with increasing ℓ for fixed n.

Shells and Subshells
Electrons that have the same value of n are said to be in the same shell. Electrons in a given shell that have the same value of ℓ are said to be in the same subshell.

The Pauli Exclusion Principle
The Pauli exclusion principle states that only a single electron may have a particular set of quantum numbers. This means that it is not possible for all the electrons in a multielectron atom to occupy the lowest energy level.

Electronic Configurations
The arrangement of electrons is indicated by the electronic configuration. For example, the configuration $1s^2$ indicates that 2 electrons ($1s^2$) are in the $n = 1$ ($1s^2$), $\ell = 0$ ($1s^2$) state.

The Periodic Table
As electrons fill subshells of progressively higher energy, they produce the elements of the periodic table. Atoms with the same configuration of outermost electrons generally have similar chemical properties.

31-7 ATOMIC RADIATION

Atoms can give off radiation ranging from X-rays to visible light to infrared rays.

X-Rays

X-rays characteristic of a particular element are given off when an electron in an inner shell is knocked out of the atom, and an electron from an outer shell drops down to take its place.

Lasers

A laser is a device that produces light amplification by the stimulated emission of radiation.

Fluorescence and Phosphorescence

When an electron in an atom is excited to a high energy level, it may return to the ground state through a series of lower-energy jumps. These jumps give off radiation of longer wavelength than the radiation that caused the original excitation.

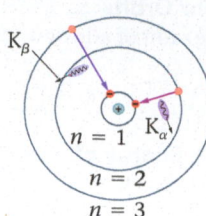

ANSWERS TO ENHANCE YOUR UNDERSTANDING QUESTIONS

1. **(b)** The Rutherford model of the atom introduced the concept of a nucleus.

2. The wavelengths decrease with increasing n for fixed n'.

3. **(a)** The speed is halved. **(b)** The radius quadruples. **(c)** The ionization energy is reduced by a factor of four.

4. $n = 6$, because six wavelengths fit around the orbit.

5. There are 10 states in the $n = 3$, $\ell = 2$ level of hydrogen.

6. There are 18 electrons in this element. The element is argon (Ar).

7. The wavelength of the K_β X-ray is shorter than the wavelength of the K_α X-ray because the energy difference between levels is greater for the K_β X-ray.

CONCEPTUAL QUESTIONS

(Answers to odd-numbered Conceptual Questions can be found in the back of the book.)

For instructor-assigned homework, go to www.masteringphysics.com. (MP)

1. What observation led Rutherford to propose that atoms have a small nucleus containing most of the atom's mass?

2. Do you expect the light given off by **(a)** a neon sign or **(b)** an incandescent lightbulb to be continuous in distribution or in the form of a line spectrum? Explain.

3. In principle, how many spectral lines are there in any given series of hydrogen? Explain.

4. Is there an upper limit to the radius of an allowed Bohr orbit? Explain.

5. **(a)** Is there an upper limit to the wavelength of lines in the spectrum of hydrogen? Explain. **(b)** Is there a lower limit? Explain.

6. The principal quantum number, n, can increase without limit in the hydrogen atom. Does this mean that the energy of the hydrogen atom also can increase without limit? Explain.

7. For each of the following configurations of outermost electrons, state whether the configuration is allowed by the rules of quantum mechanics. If the configuration is not allowed, give the rule or rules that are violated. **(a)** $2d^1$, **(b)** $1p^7$, **(c)** $3p^5$, **(d)** $4g^6$.

8. **(a)** In the quantum mechanical model of the hydrogen atom, there is one value of n for which the angular momentum of the electron must be zero. What is this value of n? **(b)** Can the angular momentum of the electron be zero in states with other values of n? Explain.

9. The elements fluorine, chlorine, and bromine are found to exhibit similar chemical properties. Explain.

PROBLEMS AND CONCEPTUAL EXERCISES

Answers to odd-numbered Problems and Conceptual Exercises can be found in the back of the book. **BIO** *identifies problems of biological or medical interest;* **CE** *indicates a conceptual exercise.* **Predict/Explain** *problems ask for two responses: (a) your prediction of a physical outcome, and (b) the best explanation among three provided; and* **Predict/Calculate** *problems ask for a prediction of a physical outcome, based on fundamental physics concepts, and follow that with a numerical calculation to verify the prediction. On all problems, bullets (•, ••, •••) indicate the level of difficulty.*

SECTION 31-1 EARLY MODELS OF THE ATOM

1. • The electron in a hydrogen atom is typically found at a distance of about 5.3×10^{-11} m from the nucleus, which has a diameter of about 1.0×10^{-15} m. If you assume the hydrogen atom to be a sphere of radius 5.3×10^{-11} m, what fraction of its volume is occupied by the nucleus?

2. • Referring to Problem 1, suppose the nucleus of the hydrogen atom were enlarged to the size of a baseball (diameter = 7.3 cm). At what typical distance from the center of the baseball would you expect to find the electron?

3. •• In the Thomson model of the atom, the mass of an atom is distributed uniformly throughout its volume in a material analogous

to pudding. **(a)** Calculate the density of the "pudding" of a gold atom that has a mass of 3.27×10^{-25} kg, modeling it as a sphere of radius 0.144 nm. **(b)** Compare the density in part (a) with the density of an alpha particle of mass 6.64×10^{-27} kg and radius 1.68×10^{-15} m.

4. •• Copper atoms have 29 protons in their nuclei. If the copper nucleus is a sphere with a diameter of 4.8×10^{-15} m, find the work required to bring an alpha particle (charge = $+2e$) from rest at infinity to the "surface" of the nucleus.

5. •• In Rutherford's scattering experiments, alpha particles (charge = $+2e$) were fired at a gold foil. Consider an alpha particle with an initial kinetic energy K heading directly for the

nucleus of a gold atom (charge = +79e). The alpha particle will come to rest when all its initial kinetic energy has been converted to electric potential energy. Find the distance of closest approach between the alpha particle and the gold nucleus for the case $K = 5.3$ MeV.

SECTION 31-2 THE SPECTRUM OF ATOMIC HYDROGEN

6. • Find the wavelength of the Balmer series spectral line corresponding to $n = 12$.

7. • What is the smallest value of n for which the wavelength of a Balmer series line is less than 400 nm?

8. • Find the wavelength of the three longest-wavelength lines of the Lyman series.

9. • Find the wavelength of the longest-wavelength lines of the Paschen series.

10. •• Find (a) the longest wavelength in the Lyman series and (b) the shortest wavelength in the Paschen series.

SECTION 31-3 BOHR'S MODEL OF THE HYDROGEN ATOM

11. • CE Predict/Explain (a) If the mass of the electron were magically doubled, would the ionization energy of hydrogen increase, decrease, or stay the same? (b) Choose the *best explanation* from among the following:
    I. The ionization energy would increase because the increased mass would mean the electron would orbit closer to the nucleus and would require more energy to move to infinity.
    II. The ionization energy would decrease because a more massive electron is harder to hold in orbit, and therefore it is easier to remove the electron and leave the hydrogen ionized.
    III. The ionization energy would be unchanged because, just like in gravitational orbits, the orbit of the electron is independent of its mass. As a result, there is no change in the energy required to move it to infinity.

12. • CE Consider the Bohr model as applied to the following three atoms: (A) neutral hydrogen in the state $n = 2$; (B) singly ionized helium in the state $n = 1$; (C) doubly ionized lithium in the state $n = 3$. Rank these three atoms in order of increasing Bohr radius. Indicate ties where appropriate.

13. • CE Consider the Bohr model as applied to the following three atoms: (A) neutral hydrogen in the state $n = 3$; (B) singly ionized helium in the state $n = 2$; (C) doubly ionized lithium in the state $n = 1$. Rank these three atoms in order of increasing energy. Indicate ties where appropriate.

14. • CE An electron in the $n = 1$ Bohr orbit has the kinetic energy K_1. In terms of K_1, what is the kinetic energy of an electron in the $n = 2$ Bohr orbit?

15. • Find the ratio v/c for an electron in the first excited state ($n = 2$) of hydrogen.

16. • Find the magnitude of the force exerted on an electron in the ground-state orbit of the Bohr model.

17. • How much energy is required to ionize hydrogen when it is in the $n = 3$ state?

18. • Find the energy of the photon required to excite a hydrogen atom from the $n = 1$ state to the $n = 4$ state.

19. • Atomic Clock Transition The global standard for time is based on a transition in cesium atoms that occurs when a microwave photon of frequency 9,192,631,770 Hz is absorbed. What is the energy difference in eV between the two levels of cesium that correspond with this transition?

20. •• A hydrogen atom is in its second excited state, $n = 3$. Using the Bohr model of hydrogen, find (a) the linear momentum and (b) the angular momentum of the electron in this atom.

21. •• Referring to Problem 20, find (a) the kinetic energy of the electron, (b) the potential energy of the atom, and (c) the total energy of the atom. Give your results in eV.

22. •• Initially, an electron is in the $n = 2$ state of hydrogen. If this electron acquires an additional 2.86 eV of energy, what is the value of n in the final state of the electron?

23. •• Identify the initial and final states if an electron in hydrogen emits a photon with a wavelength of 656 nm.

24. •• Predict/Calculate An electron in hydrogen absorbs a photon and jumps to a higher orbit. (a) Find the energy the photon must have if the initial state is $n = 3$ and the final state is $n = 5$. (b) If the initial state was $n = 5$ and the final state $n = 7$, would the energy of the photon be greater than, less than, or the same as that found in part (a)? Explain. (c) Calculate the photon energy for part (b).

25. •• Predict/Calculate Consider the following four transitions in a hydrogen atom:
    (i) $n_i = 2, n_f = 6$ (ii) $n_i = 2, n_f = 8$
    (iii) $n_i = 7, n_f = 8$ (iv) $n_i = 6, n_f = 2$

    Find (a) the longest- and (b) the shortest-wavelength photon that can be emitted or absorbed by these transitions. Give the value of the wavelength in each case. (c) For which of these transitions does the atom lose energy? Explain.

26. •• Predict/Calculate Muonium Muonium is a hydrogen-like atom in which the electron is replaced with a muon, a fundamental particle with a charge of $-e$ and a mass equal to $207m_e$. (The muon is sometimes referred to loosely as a "heavy electron.") (a) What is the Bohr radius of muonium? (b) Will the wavelengths in the Balmer series of muonium be greater than, less than, or the same as the wavelengths in the Balmer series of hydrogen? Explain. (c) Calculate the longest wavelength of the Balmer series in muonium.

27. •• Predict/Calculate (a) Find the radius of the $n = 4$ Bohr orbit of a doubly ionized lithium atom (Li^{2+}, $Z = 3$). (b) Is the energy required to raise an electron from the $n = 4$ state to the $n = 5$ state in Li^{2+} greater than, less than, or equal to the energy required to raise an electron in hydrogen from the $n = 4$ state to the $n = 5$ state? Explain. (c) Verify your answer to part (b) by calculating the relevant energies.

28. •• Applying the Bohr model to a triply ionized beryllium atom (Be^{3+}, $Z = 4$), find (a) the shortest wavelength of the Lyman series for Be^{3+} and (b) the ionization energy required to remove the final electron in Be^{3+}.

29. •• (a) Calculate the time required for an electron in the $n = 2$ state of hydrogen to complete one orbit about the nucleus. (b) The typical "lifetime" of an electron in the $n = 2$ state is roughly 10^{-8} s— after this time the electron is likely to have dropped back to the $n = 1$ state. Estimate the number of orbits an electron completes in the $n = 2$ state before dropping to the ground state.

30. •• Predict/Calculate The kinetic energy of an electron in a particular Bohr orbit of hydrogen is 1.35×10^{-19} J. (a) Which Bohr orbit does the electron occupy? (b) Suppose the electron moves away from the nucleus to the next higher Bohr orbit. Does the kinetic energy of the electron increase, decrease, or stay the same? Explain. (c) Calculate the kinetic energy of the electron in the orbit referred to in part (b).

31. •• Predict/Calculate The potential energy of a hydrogen atom in a particular Bohr orbit is -1.20×10^{-19} J. (a) Which Bohr orbit does the electron occupy in this atom? (b) Suppose the electron moves away from the nucleus to the next higher Bohr orbit. Does

the potential energy of the atom increase, decrease, or stay the same? Explain. **(c)** Calculate the potential energy of the atom for the orbit referred to in part (b).

32. ••• Consider a head-on collision between two hydrogen atoms, both initially in their ground state and moving with the same aspeed. Find the minimum speed necessary to leave both atoms in their $n = 4$ state after the collision.

33. ••• A hydrogen atom is in the initial state $n_i = n$, where $n > 1$. **(a)** Find the frequency of the photon that is emitted when the electron jumps to state $n_f = n - 1$. **(b)** Find the frequency of the electron's orbital motion in the state n. **(c)** Compare your results for parts (a) and (b) in the limit of large n.

SECTION 31-4 DE BROGLIE WAVES AND THE BOHR MODEL

34. • **CE Predict/Explain** **(a)** Is the de Broglie wavelength of an electron in the $n = 2$ Bohr orbit of hydrogen greater than, less than, or equal to the de Broglie wavelength in the $n = 1$ Bohr orbit? **(b)** Choose the *best explanation* from among the following:
I. The de Broglie wavelength in the nth state is $2\pi r/n$, where r is proportional to n^2. Therefore, the wavelength increases with increasing n, and is greater for $n = 2$ than for $n = 1$.
II. The de Broglie wavelength of an electron in the nth state is such that n wavelengths fit around the circumference of the orbit. Therefore, $\lambda = 2\pi r/n$ and the wavelength for $n = 2$ is less than for $n = 1$.
III. The de Broglie wavelength depends on the mass of the electron, and that is the same regardless of which state of the hydrogen atom the electron occupies.

35. • Find the de Broglie wavelength of an electron in the ground state of the hydrogen atom.

36. •• The de Broglie wavelength of an electron in a hydrogen atom is 1.66 nm. Identify the integer n that corresponds to its orbit.

37. •• The orbital radius of an electron in a hydrogen atom is 0.846 nm. What is its de Broglie wavelength?

38. •• Find an expression, in terms of n and the fundamental constants, for the de Broglie wavelength of an electron in the nth state of the hydrogen atom.

39. •• What is the radius of the hydrogen-atom Bohr orbit shown in **FIGURE 31-40**?

40. •• **Predict/Calculate** **(a)** An electron in a hydrogen atom jumps from the $n = 6$ orbit to the $n = 2$ orbit and emits a photon. Will its de Broglie wavelength increase, decrease, or stay the same after the jump? **(b)** Calculate its de Broglie wavelength before and after the jump.

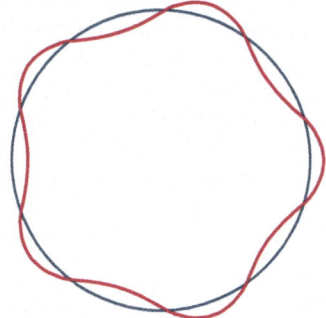

FIGURE 31-40 Problem 39

41. •• **(a)** Find the kinetic energy (in eV) of an electron whose de Broglie wavelength is equal to 0.5 Å = 0.05 nm, a typical atomic size. **(b)** If the wavelength of the electron is reduced to 10^{-15} m, a typical nuclear size, is its kinetic energy greater than, less than, or equal to the value found in part (a)? Explain.

SECTION 31-5 THE QUANTUM MECHANICAL HYDROGEN ATOM

42. • What are the allowed values of ℓ when the principal quantum number is $n = 5$?

43. • How many different values of m_ℓ are possible when the principal quantum number is $n = 5$?

44. •• Give the value of the quantum number ℓ, if one exists, for a hydrogen atom whose orbital angular momentum has a magnitude of **(a)** $\sqrt{6}(h/2\pi)$, **(b)** $\sqrt{15}(h/2\pi)$, **(c)** $\sqrt{30}(h/2\pi)$, or **(d)** $\sqrt{36}(h/2\pi)$.

45. •• **Predict/Calculate** Hydrogen atom number 1 is known to be in the $4f$ state. **(a)** What is the energy of this atom? **(b)** What is the magnitude of this atom's orbital angular momentum? **(c)** Hydrogen atom number 2 is in the $5d$ state. Is this atom's energy greater than, less than, or the same as that of atom 1? Explain. **(d)** Is the magnitude of the orbital angular momentum of atom 1 greater than, less than, or the same as that of atom 2? Explain.

46. •• **Predict/Calculate** An electron in a hydrogen atom makes a transition from a $5p$ state to a $4d$ state. **(a)** Does the energy of the electron increase, decrease, or stay the same? Explain. **(b)** Does the magnitude of the electron's angular momentum increase, decrease, or stay the same? Explain. **(c)** Calculate the change in the electron's energy in eV. **(d)** Calculate the change in the magnitude of the electron's angular momentum.

47. •• **Predict/Calculate** A hydrogen atom has an orbital angular momentum with a magnitude of $10\sqrt{57}(h/2\pi)$. **(a)** Determine the value of the quantum number ℓ for this atom. **(b)** What is the minimum possible value of this atom's principal quantum number, n? Explain. **(c)** If $10\sqrt{57}(h/2\pi)$ is the *maximum* orbital angular momentum this atom can have, what is its energy?

48. •• **Predict/Calculate** The electron in a hydrogen atom with an energy of -0.544 eV is in a subshell with 18 states. **(a)** What is the principal quantum number, n, for this atom? **(b)** What is the maximum possible orbital angular momentum this atom can have? **(c)** Is the number of states in the subshell with the next lowest value of ℓ equal to 16, 14, or 12? Explain.

49. ••• **Predict/Calculate** Consider two different states of a hydrogen atom. In state I the maximum value of the magnetic quantum number is $m_\ell = 3$; in state II the corresponding maximum value is $m_\ell = 2$. Let L_I and L_{II} represent the magnitudes of the orbital angular momentum of an electron in states I and II, respectively. **(a)** Is L_I greater than, less than, or equal to L_{II}? Explain. **(b)** Calculate the ratio L_I/L_{II}.

SECTION 31-6 MULTIELECTRON ATOMS AND THE PERIODIC TABLE

50. • **CE** How many electrons can occupy **(a)** the $2p$ subshell and **(b)** the $3p$ subshell?

51. • **CE** **(a)** How many electrons can occupy the $3d$ subshell? **(b)** How many electrons can occupy the $n = 2$ shell?

52. • **CE** The electronic configuration of a given atom is $1s^2 2s^2 2p^6 3s^2 3p^1$. How many electrons are in this atom?

53. • Give the electronic configuration for the ground state of carbon.

54. • Give a list of all possible sets of the four quantum numbers (n, ℓ, m_ℓ, m_s) for electrons in the $3p$ subshell.

55. •• List the values of the four quantum numbers (n, ℓ, m_ℓ, m_s) for each of the electrons in the ground state of neon.

56. •• List the values of the four quantum numbers (n, ℓ, m_ℓ, m_s) for each of the electrons in the ground state of magnesium.

57. •• The configuration of the outer electrons in Ni is $3d^8 4s^2$. Write out the complete electronic configuration for Ni.

58. •• Determine the number of different sets of quantum numbers possible for each of the following shells: **(a)** $n = 2$, **(b)** $n = 3$, **(c)** $n = 4$.

SECTION 31-7 ATOMIC RADIATION

59. • **CE** Predict/Explain **(a)** In an X-ray tube, do you expect the wavelength of the characteristic X-rays to increase, decrease, or stay the same if the energy of the electrons striking the target is increased? **(b)** Choose the *best explanation* from among the following:
  I. Increasing the energy of the incoming electrons will increase the wavelength of the emitted X-rays.
  II. When the energy of the incoming electrons is increased, the energy of the X-rays is also increased; this, in turn, decreases the wavelength.
  III. The wavelength of characteristic X-rays depends only on the material used in the metal target, and does not change if the energy of incoming electrons is increased.

60. • Using the Bohr model, estimate the wavelength of the K_α X-ray in nickel ($Z = 28$).

61. • Using the Bohr model, estimate the energy of a K_α X-ray emitted by lead ($Z = 82$).

62. •• The K-shell ionization energy of iron is 8500 eV, and its L-shell ionization energy is 2125 eV. What is the wavelength of K_α X-rays emitted by iron?

63. •• An electron drops from the L shell to the K shell and gives off an X-ray with a wavelength of 0.0279 nm. What is the atomic number of this atom?

64. •• Consider an X-ray tube that uses platinum ($Z = 78$) as its target. **(a)** Use the Bohr model to estimate the minimum kinetic energy electrons must have in order for K_α X-rays to just appear in the X-ray spectrum of the tube. **(b)** Assuming the electrons are accelerated from rest through a voltage V, estimate the minimum voltage necessary to produce the K_α X-rays.

65. •• **BIO** Photorefractive Keratectomy A person's vision may be improved significantly by having the cornea reshaped with a laser beam, in a procedure known as photorefractive keratectomy. The excimer laser used in these treatments produces ultraviolet light with a wavelength of 193 nm. **(a)** What is the difference in energy between the two levels that participate in stimulated emission in the excimer laser? **(b)** How many photons from this laser are required to deliver an energy of 1.58×10^{-13} J to the cornea?

66. •• Predict/Calculate An ultraviolet photon is absorbed by a fluorescent molecule, which then emits two photons. **(a)** Are the wavelengths of the emitted photons longer than, shorter than, or the same as the wavelength of the ultraviolet photon? Explain. **(b)** If the ultraviolet photon has a wavelength of 254 nm and one of the photons emitted by the fluorescent material has a wavelength of 415 nm, what is the wavelength of the other emitted photon? Assume that none of the energy of the absorbed ultraviolet photon is converted to other forms.

67. •• Fluorescent Lightbulbs The most common fluorescent lightbulb converts ultraviolet light from mercury atoms to white light. If an ultraviolet photon of wavelength 185 nm is emitted by a mercury atom and absorbed by the phosphor, what is the maximum number of red 655-nm photons that the phosphor can emit?

GENERAL PROBLEMS

68. • **CE** Consider the following three transitions in a hydrogen atom: (A) $n_i = 5, n_f = 2$; (B) $n_i = 7, n_f = 2$; (C) $n_i = 7, n_f = 6$. Rank the transitions in order of increasing **(a)** wavelength and **(b)** frequency of the emitted photon. Indicate ties where appropriate.

69. • **CE** Suppose an electron is in the ground state of hydrogen. **(a)** What is the highest-energy photon this system can absorb without dissociating the electron from the proton? Explain. **(b)** What is the lowest-energy photon this system can absorb? Explain.

70. • **CE** The electronic configuration of a particular carbon atom is $1s^2 2s^2 2p^1 3s^1$. Is this atom in its ground state or in an excited state? Explain.

71. • **CE** Do you expect the ionization energy of sodium (Na) to be greater than, less than, or equal to the ionization energy of lithium (Li)? Explain.

72. • Find the minimum frequency a photon must have if it is to ionize the ground state of the hydrogen atom.

73. •• The electron in a hydrogen atom makes a transition from the $n = 4$ state to the $n = 2$ state, as indicated in **FIGURE 31-41**. **(a)** Determine the linear momentum of the photon emitted as a result of this transition. **(b)** Using your result to part (a), find the recoil speed of the hydrogen atom, assuming it was at rest before the photon was emitted.

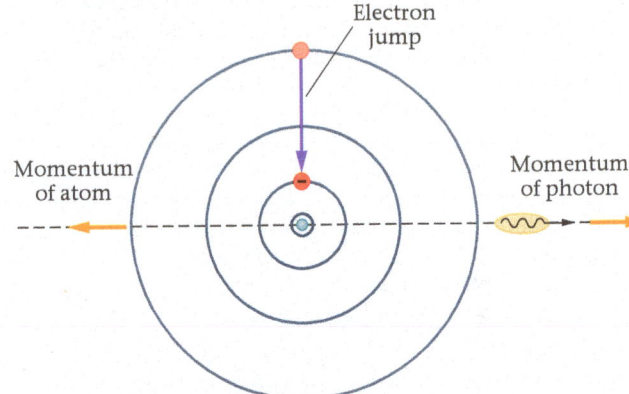

FIGURE 31-41 Problem 73

74. •• **BIO** Laser Eye Surgery In laser eye surgery, the laser emits a 1.45-ns pulse focused on a spot that is 32.5 μm in diameter. **(a)** If the energy contained in the pulse is 2.95 mJ, what is the power per square meter (the irradiance) associated with this beam? **(b)** Suppose a molecule with a diameter of 0.500 nm is irradiated by the laser beam. How much energy does the molecule receive in one pulse from the laser? (The energy obtained in part (b) is more than enough to dissociate a molecule.)

75. •• Consider an electron in the ground-state orbit of the Bohr model of hydrogen. **(a)** Find the time required for the electron to complete one orbit about the nucleus. **(b)** Calculate the current (in amperes) corresponding to the electron's motion.

76. •• A particular Bohr orbit in a hydrogen atom has a total energy of -0.85 eV. What are **(a)** the kinetic energy of the electron in this orbit and **(b)** the electric potential energy of the system?

77. •• An ionized atom has only a single electron. The $n = 6$ Bohr orbit of this electron has a radius of 2.72×10^{-10} m. Find **(a)** the atomic number Z of this atom and **(b)** the total energy E of its $n = 3$ Bohr orbit.

78. •• Find the approximate wavelength of K_β X-rays emitted by molybdenum ($Z = 42$), and compare your result with Figure 31-27. (*Hint*: An electron in the M shell is shielded from the nucleus by the single electron in the K shell, plus all the electrons in the L shell.)

79. •• **CE** Diamagnetism Suppose **FIGURE 31-42** shows an electron in an atom that is orbiting a nucleus counterclockwise (that is, the charge is moving to the right at the moment pictured in the diagram). The indicated magnetic field has just been turned on from an initial value of zero. **(a)** Treating the electron as a circular ring of electric current, use Faraday's law to determine whether the electron will speed up or slow down in response to turning on the magnetic field. **(b)** The circulation of the electron produces a magnetic field at the center of its orbit. Will the magnitude of its magnetic field increase or decrease in response to turning on the external magnetic field? **(c)** Does the magnetic field produced by the orbiting electron point in the same direction as the external magnetic field, or in the opposite direction? (This question offers a crude explanation for the effect of *diamagnetism*, presented in Chapter 22 .)

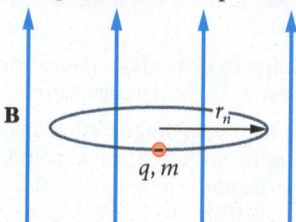

FIGURE 31-42 Problems 79 and 82

80. •• **Predict/Calculate Rydberg Atoms** There is no limit to the size a hydrogen atom can attain, provided it is free from disruptive outside influences. In fact, radio astronomers have detected radiation from large, so-called "Rydberg atoms" in the diffuse hydrogen gas of interstellar space. **(a)** Find the smallest value of n such that the Bohr radius of a single hydrogen atom is greater than 8.0 microns, the size of a typical single-celled organism. **(b)** Find the wavelength of radiation this atom emits when its electron drops from level n to level $n - 1$. **(c)** If the electron drops one more level, from $n - 1$ to $n - 2$, is the emitted wavelength greater than or less than the value found in part (b)? Explain.

81. ••• **Predict/Calculate The Pickering Series** In 1896, the American astronomer Edward C. Pickering (1846–1919) discovered an unusual series of spectral lines in light from the hot star Zeta Puppis. After some time, it was determined that these lines are produced by singly ionized helium. In fact, the "Pickering series" is produced when electrons drop from higher levels to the $n = 4$ level of He$^+$. Spectral lines in the Pickering series have wavelengths given by

$$\frac{1}{\lambda} = C\left(\frac{1}{16} - \frac{1}{n^2}\right).$$

In this expression, $n = 5, 6, 7, \ldots$ **(a)** Do you expect the constant C to be greater than, less than, or equal to the Rydberg constant R? Explain. **(b)** Find the numerical value of C. **(c)** Pickering lines with $n = 6, 8, 10, \ldots$ correspond to Balmer lines in hydrogen with $n = 3, 4, 5, \ldots$. Verify this assertion for the $n = 6$ Pickering line.

82. ••• Consider a particle of mass m, charge q, and constant speed v moving perpendicular to a uniform magnetic field of magnitude B, as shown in Figure 31-42. The particle follows a circular path. Suppose the angular momentum of the particle about the center of its circular motion is quantized in the following way: $mvr = n\hbar$, where $n = 1, 2, 3, \ldots$, and $\hbar = h/2\pi$.

a. Show that the radii of its allowed orbits have the following values:

$$r_n = \sqrt{\frac{n\hbar}{qB}}$$

b. Find the speed of the particle in each allowed orbit.

83. ••• Consider a particle of mass m confined in a one-dimensional box of length L. In addition, suppose the matter wave associated with this particle is analogous to a wave on a string of length L that is fixed at both ends. Using the de Broglie relationship, show that **(a)** the quantized values of the linear momentum of the particle are

$$p_n = \frac{nh}{2L} \qquad n = 1, 2, 3, \ldots$$

and **(b)** the allowed energies of the particle are

$$E_n = n^2\left(\frac{h^2}{8mL^2}\right) \qquad n = 1, 2, 3, \ldots$$

PASSAGE PROBLEMS

BIO Welding a Detached Retina

As a person ages, a normal part of the process is a shrinkage of the vitreous gel—the gelatinous substance that fills the interior of the eye. When this happens, the usual result is that the gel pulls away cleanly from the retina, with little or no adverse effect on the person's vision. This is referred to as posterior vitreous detachment. In some cases, however, the vitreous membrane that surrounds the vitreous gel pulls on the retina as the gel contracts, eventually creating a hole or a tear in the retina itself. At this point, fluid can seep through the hole in the retina and separate it from the underlying supporting cells—the retinal pigment epithelium. This process, known as rhegmatogenous retinal detachment, causes a blind spot in the person's vision. If not treated immediately, a retinal detachment can lead to permanent vision loss.

One way to treat a detached retina is to "weld" it back in place using a laser beam. This type of operation is performed with an argon laser because the blue-green light it produces passes through the vitreous gel with little absorption or damage, but is strongly absorbed by the red pigments in the retina and the retinal epithelium. An argon laser produces light consisting primarily of two wavelengths, 488.0 nm (blue-green) and 514.5 nm (green), and has a power output ranging between 1 W and 20 W. Other types of lasers, including a krypton laser that produces red light at 647.1 nm, or a krypton laser that produces yellow light at 568.2 nm, are sometimes used for this procedure.

84. •• Suppose an argon laser emits 1.49×10^{19} photons per second, half with a wavelength of 488.0 nm and half with a wavelength of 514.5 nm. What is the power output of this laser in watts?

A. 1.49 W	**B.** 5.76 W
C. 5.92 W	**D.** 6.07 W

85. •• A red krypton laser also emits 1.49×10^{19} photons per second. Is its power output greater than, less than, or equal to the power output of the argon laser in Problem 84?

86. •• The energy difference (in eV) between two states of a particular atom is 2.541 eV. What are the type and wavelength of the laser that corresponds to this transition?

A. Kr, 647.1 nm	**B.** Ar, 514.5 nm
C. Kr, 568.2 nm	**D.** Ar, 488.0 nm

87. •• How would the number of photons emitted per second by a yellow Kr laser compare to the number emitted per second by a red Kr laser that has the same power output?

A. The yellow Kr laser emits more photons per second.
B. The yellow Kr laser emits fewer photons per second.
C. The yellow Kr laser emits the same number of photons per second.

88. •• **Predict/Calculate REFERRING TO EXAMPLE 31-8** Suppose the electron is in a state whose standing wave consists of two wavelengths. **(a)** Is the wavelength of this standing wave greater than or less than 1.33×10^{-9} m? **(b)** Find the wavelength of this standing wave.

89. •• **Referring to Example 31-8 (a)** Which state has a de Broglie wavelength of 3.99×10^{-9} m? **(b)** What is the Bohr radius of this state?

Nuclear Physics and Nuclear Radiation

Big Ideas

1 Nuclei consist of protons and neutrons, each with a mass almost 2000 times that of an electron.

2 Radioactivity results when an unstable nucleus decays and emits either a particle or a photon.

3 Radioactive decay proceeds at a rate that is proportional to the amount of radioactive material.

4 Nuclear fission occurs when a large nucleus breaks up into smaller pieces.

5 Nuclear fusion occurs when two light nuclei fuse together to form a larger nucleus.

6 The fundamental forces of nature are gravity, weak nuclear, electromagnetic, and strong nuclear.

▲ The energy radiated by the Sun is produced by nuclear reactions deep in its interior. In this chapter, we explore the atomic nucleus, its basic constituents, and the types of reactions that transform one nucleus to another.

Atoms consist of electrons orbiting a nucleus. In this chapter we turn our attention to the nucleus, and show that its behavior is responsible for reactions that power the Sun, generate electricity in nuclear power plants, and provide the means for treating cancer and other ailments.

TABLE 32-1 Numbers That Characterize a Nucleus

Z	Atomic number = number of protons in nucleus
N	Neutron number = number of neutrons in nucleus
A	Mass number = number of nucleons in nucleus

32-1 The Constituents and Structure of Nuclei

The simplest nucleus is that of the hydrogen atom. This nucleus consists of a single **proton,** whose mass is about 1836 times greater than the mass of an electron and whose electric charge is $+e$. All other nuclei contain neutrons in addition to protons. The **neutron** is an electrically neutral particle (its electric charge is zero) with a mass just slightly greater than that of the proton. No other particles are found in nuclei. Collectively, protons and neutrons are referred to as **nucleons.**

Nuclear Notation Nuclei are characterized by the number and type of nucleons they contain. First, the **atomic number,** Z, is defined as the number of protons in a nucleus. In an electrically neutral atom, the number of electrons is also equal to Z. Next, the number of neutrons in a nucleus is designated by the **neutron number,** N. Finally, the total number of nucleons in a nucleus is the **mass number,** A. These definitions are summarized in Table 32-1. Clearly, the mass number is the sum of the atomic number and the neutron number:

$$A = Z + N \qquad \qquad 32\text{-}1$$

A special notation is used to indicate the composition of a nucleus. Consider, for example, an unstable but very useful form of carbon known as carbon-14. The nucleus of carbon-14 is written as follows:

$$^{14}_{6}\text{C}$$

In this expression, C represents the chemical element carbon. The number 6 is the atomic number of carbon, $Z = 6$, and the number 14 is the mass number of this nucleus, $A = 14$. This means that carbon-14 has 14 nucleons in its nucleus. The neutron number can be found by solving $A = Z + N$ for N: that is, $N = A - Z = 14 - 6 = 8$. Thus the nucleus of carbon-14 consists of 6 protons and 8 neutrons. The most common form of carbon is carbon-12, whose nucleus is designated as $^{12}_{6}\text{C}$. This nucleus has 6 protons and 6 neutrons.

In general, the nucleus of an arbitrary element, X, with atomic number Z and mass number A, is represented as

$$^{A}_{Z}X$$

Once a given element is specified, the value of Z is known. As a result, the subscript Z is sometimes omitted.

EXERCISE 32-1 NUCLEAR SYMBOLS

  a. Give the symbol for a nucleus of phosphorus that contains 16 neutrons.
  b. Deuterium is a type of "heavy hydrogen." The nucleus of deuterium can be written as $^{2}_{1}\text{H}$. What is the number of protons and neutrons in a deuterium nucleus?

REASONING AND SOLUTION

  a. Looking up phosphorus in the periodic table in Appendix E, we find that its atomic number is $Z = 15$. In addition, we are given that $N = 16$. Therefore, $A = Z + N = 31$, and hence the symbol for this nucleus is $^{31}_{15}\text{P}$.
  b. We obtain the number of protons (the atomic number) from the subscript in the symbol $^{2}_{1}\text{H}$; therefore, $Z = 1$. The number of neutrons is given by $N = A - Z$, where A is the superscript. Therefore, the number of neutrons is $N = 2 - 1 = 1$.

Isotopes All nuclei of a given element have the same number of protons, Z. They may have different numbers of neutrons, N, however. Nuclei with the same value of Z but different values of N are referred to as **isotopes**. For example, $^{12}_{6}\text{C}$ and $^{13}_{6}\text{C}$ are two isotopes of carbon, with $^{12}_{6}\text{C}$ being the most common one, constituting about 98.89% of naturally occurring carbon. About 1.11% of natural carbon is $^{13}_{6}\text{C}$. Values for the percentage abundance of various isotopes can be found in Appendix F.

Also given in Appendix F are the atomic masses of many common isotopes. These masses are given in terms of the **atomic mass unit,** u, defined so that the mass of one atom of $^{12}_{6}\text{C}$ is exactly 12 u. The value of u is as follows:

Definition of Atomic Mass Unit, u

$$1\,u = 1.660540 \times 10^{-27}\,kg$$

32-2

SI unit: kg

Protons have a mass just slightly greater than 1 u, and the neutron is slightly more massive than the proton. The precise masses of the proton and neutron are given in Table 32-2, along with the mass of the electron.

TABLE 32-2 **Mass and Charge of Particles in the Atom**

Particle	Mass (kg)	Mass (MeV/c^2)	Mass (u)	Charge (C)
Proton	1.672623×10^{-27}	938.28	1.007276	$+1.6022 \times 10^{-19}$
Neutron	1.674929×10^{-27}	939.57	1.008665	0
Electron	9.109390×10^{-31}	0.511	0.0005485799	-1.6022×10^{-19}

When we consider nuclear reactions later in this chapter, an important consideration will be the energy equivalent of a given mass, as given in Einstein's famous relationship, $E = mc^2$ (Equation 29-8). The energy equivalent of one atomic mass unit is

$$E = mc^2 = (1\,u)c^2$$

$$= (1.660540 \times 10^{-27}\,kg)(2.998 \times 10^8\,m/s)^2\left(\frac{1\,eV}{1.6022 \times 10^{-19}\,J}\right)$$

$$= 931.5\,MeV$$

In this expression, $1\,MeV = 10^6\,eV$. When we recall that the ionization energy of hydrogen is only 13.6 eV, it is clear that the energy equivalent of nucleons is enormous compared with typical atomic energies. In general, energies involving the nucleus are on the order of MeV, and energies associated with the electrons in an atom are on the order of eV. Finally, because mass and energy can be converted from one form to the other, it is common to express the atomic mass unit in terms of energy as follows:

$$1\,u = 931.5\,MeV/c^2$$

32-3

The masses of the proton, neutron, and electron are also given in units of MeV/c^2 in Table 32-2.

Nuclear Size and Density

To obtain an estimate for the size of a nucleus, Rutherford did a simple calculation using energy conservation. He considered the case of a particle of charge $+q$ and mass m approaching a nucleus of charge $+Ze$ with a speed v. He further assumed that the approach was head-on. At some distance, d, from the center of the nucleus, the incoming particle comes to rest instantaneously, before turning around. It follows that the radius of the nucleus is less than d. In the next Example, we obtain a symbolic expression for the distance d.

EXAMPLE 32-2   SETTING A LIMIT ON THE RADIUS OF A NUCLEUS

A particle of mass m, charge $+q$, and speed v heads directly toward a distant, stationary nucleus of charge $+Ze$. Find a symbolic expression for the distance of closest approach between the incoming particle and the center of the nucleus.

PICTURE THE PROBLEM

The incoming particle moves on a line that passes through the center of the nucleus. Far from the nucleus, the particle's speed is v. The particle turns around (comes to rest instantaneously) a distance d from the center of the nucleus.

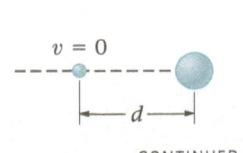

REASONING AND STRATEGY

We can find the distance d by applying energy conservation. In particular, the initial energy of the system is the kinetic

CONTINUED

energy of the particle, $\frac{1}{2}mv^2$, assuming the particle approaches the nucleus from infinity. The final energy is the electric potential energy, $U = kq_1q_2/r = k(Ze)q/d$. Setting these energies equal to each other allows us to solve for d.

Known   Mass of particle $= m$; charge of particle $= q$; speed of particle at infinity $= v$; charge of nucleus $= +Ze$.

Unknown   Distance of closest approach, $d = ?$ (symbolic answer desired)

SOLUTION

1. Write an expression for the initial energy of the system:

    $$E_i = \tfrac{1}{2}mv^2$$

2. Write an expression for the final energy of the system:

    $$E_f = \frac{k(Ze)q}{d}$$

3. Set the final energy equal to the initial energy and solve for d:

    $$\tfrac{1}{2}mv^2 = \frac{k(Ze)q}{d}$$

    $$d = \frac{k(Ze)q}{\left(\tfrac{1}{2}mv^2\right)}$$

INSIGHT

We see that the distance of closest approach is inversely proportional to the initial kinetic energy of the incoming particle, and directly proportional to the charge on the nucleus.

PRACTICE PROBLEM

Find the distance $d_{1/2}$ at which the speed of the incoming particle is equal to $\frac{1}{2}v$. $\left[\text{Answer: } d_{1/2} = kZeq\middle/\tfrac{3}{4}\left(\tfrac{1}{2}mv^2\right)\right]$

Some related homework problems: Problem 6, Problem 7

Using the result obtained in Example 32-2, Rutherford found that for an alpha particle approaching a gold nucleus in one of his experiments, the distance of closest approach was 3.2×10^{-14} m. A similar calculation for alpha particles fired at silver atoms gives a closest approach distance of 2.0×10^{-14} m. This suggests that the size of the nucleus varies from element to element; in particular, the nucleus of silver is smaller than that of gold. In fact, more careful measurements since Rutherford's time have established that the average radius of a nucleus of mass number A is given approximately by the following expression:

$$r = (1.2 \times 10^{-15}\text{ m})A^{1/3} \qquad \text{32-4}$$

Notice that the length scale of the nucleus is on the order of 10^{-15} m, as opposed to the length scale of an atom, which is on the order of 10^{-10} m. Recall that 10^{-15} m is referred to as a *femtometer* (fm). To honor the pioneering work of Enrico Fermi (1901–1954) in the field of nuclear physics, the femtometer is often referred to as the **fermi:**

Definition of the Fermi, fm

1 fermi = 1 fm = 10^{-15} m

SI unit: m

We now use Equation 32-4 to find the radius of a particular nucleus.

EXERCISE 32-3   RADIUS OF A NUCLEUS

Find the radius of a $^{16}_{8}$O nucleus.

REASONING AND SOLUTION

Substitute $A = 16$ in the expression for the radius of a nucleus, $r = (1.2 \times 10^{-15}\text{ m})A^{1/3}$:

$$r = (1.2 \times 10^{-15}\text{ m})(16)^{1/3} = 3.0 \times 10^{-15}\text{ m} = 3.0\text{ fm}$$

The fact that the radius of a nucleus depends on $A^{1/3}$ has interesting consequences for the density of the nucleus, as we see in the next Example.

EXAMPLE 32-4   NUCLEAR DENSITY

Using the expression $r = (1.2 \times 10^{-15} \text{ m})A^{1/3}$, calculate the density of a nucleus with mass number A. (For this problem, let $m = 1.67 \times 10^{-27}$ kg be the mass of a proton or a neutron.)

PICTURE THE PROBLEM
Our sketch shows a collection of neutrons and protons in a densely packed nucleus of radius r.

REASONING AND STRATEGY
To find the density of a nucleus, we must divide its mass, M, by its volume, V. Ignoring the small difference in mass between a neutron and a proton, we can express the mass of a nucleus as $M = Am$, where $m = 1.67 \times 10^{-27}$ kg. The volume of a nucleus is simply the volume of a sphere of radius r; that is, $V = 4\pi r^3/3$.

 Proton
 Neutron

Known   Mass number of nucleus $= A$; mass of proton or neutron, $m = 1.67 \times 10^{-27}$ kg.
Unknown   Density of nucleus, $\rho = ?$

SOLUTION

1. Write an expression for the mass of a nucleus:

$$M = Am = A(1.67 \times 10^{-27} \text{ kg})$$

2. Write an expression for the volume of a nucleus:

$$V = \tfrac{4}{3}\pi r^3 = \tfrac{4}{3}\pi[(1.2 \times 10^{-15} \text{ m})A^{1/3}]^3$$
$$= \tfrac{4}{3}\pi(1.7 \times 10^{-45} \text{ m}^3)A$$

3. Divide the mass by the volume to find the density:

$$\rho = \frac{M}{V} = \frac{Am}{\tfrac{4}{3}\pi(1.7 \times 10^{-45} \text{ m}^3)A}$$
$$= \frac{(1.67 \times 10^{-27} \text{ kg})}{\tfrac{4}{3}\pi(1.7 \times 10^{-45} \text{ m}^3)} = 2.3 \times 10^{17} \text{ kg/m}^3$$

INSIGHT
Notice that the density of a nucleus is found to be *independent* of the mass number, A. This means that a nucleus can be thought of as a collection of closely packed nucleons, much like a group of marbles in a bag. The neutrons in a nucleus serve to separate the protons, thereby reducing their mutual electrostatic repulsion.

The density of a nucleus is incredibly large. For example, a single teaspoon of nuclear matter would weigh about a billion tons.

PRACTICE PROBLEM
Find the surface area of a nucleus in terms of the mass number, A, assuming it to be a sphere.
[**Answer:** area $= (1.8 \times 10^{-29} \text{ m}^2)A^{2/3}$]

Some related homework problems: Problem 5, Problem 9

Nuclear Stability

We know that like charges repel one another, and that the force of repulsion increases rapidly with decreasing distance. It follows that protons in a nucleus, with a separation of only about a fermi, must exert relatively large forces on one another. Applying Coulomb's law (Equation 19-5, $F = k|q_1||q_2|/r^2$), we find the following force for two protons (charge $+e$) separated by a distance of 10^{-15} m:

$$F = \frac{ke^2}{r^2} = 230 \text{ N}$$

The acceleration such a force would give to a proton is $a = F/m = (230 \text{ N})/ (1.67 \times 10^{-27} \text{ kg}) = 1.4 \times 10^{29} \text{ m/s}^2$, which is about 10^{28} times greater than the acceleration of gravity! Thus, if protons in the nucleus experienced only the electrostatic force, the nucleus would fly apart in an instant. It follows that large attractive forces must also act within the nucleus.

Big Idea **1** Nuclei are composed of protons and neutrons, each of which has almost 2000 times the mass of the electron. The nucleus itself has a density that is roughly the same for all elements.

The attractive force that holds a nucleus together is called the **strong nuclear force.** This force has the following properties:

- The strong force is short range, acting only to distances of a couple fermis.
- The strong force is attractive and acts with nearly equal strength between protons and protons, protons and neutrons, and neutrons and neutrons.

In addition, the strong nuclear force does not act on electrons. As a result, it has no effect on the chemical properties of an atom.

It is the competition between the repulsive electrostatic forces and the attractive strong nuclear forces that determines whether a given nucleus is stable. **FIGURE 32-1** shows the neutron number, N, and atomic number (proton number), Z, for nuclei that are stable. The nuclei that have relatively small atomic numbers are most stable when the numbers of protons and neutrons in the nucleus are approximately equal—that is, when $N = Z$. For example, $^{12}_{6}C$ and $^{13}_{6}C$ are both stable. As the atomic number increases, however, we see that the points corresponding to stable nuclei deviate from the line $N = Z$. In fact, we see that large stable nuclei tend to contain significantly more neutrons than protons, as in the case of $^{185}_{75}Re$. Because all nucleons experience the strong nuclear force, but only the protons experience the electrostatic force, the neutrons effectively "dilute" the nuclear charge density, reducing the effect of the repulsive forces that otherwise would make the nucleus disintegrate.

▶ **FIGURE 32-1** *N* and *Z* for stable and unstable nuclei Stable nuclei with proton numbers less than 104 are indicated by small red dots. Notice that large nuclei have significantly more neutrons, *N*, than protons, *Z*. The inset shows unstable nuclei and their decay modes for proton numbers between 65 and 80.

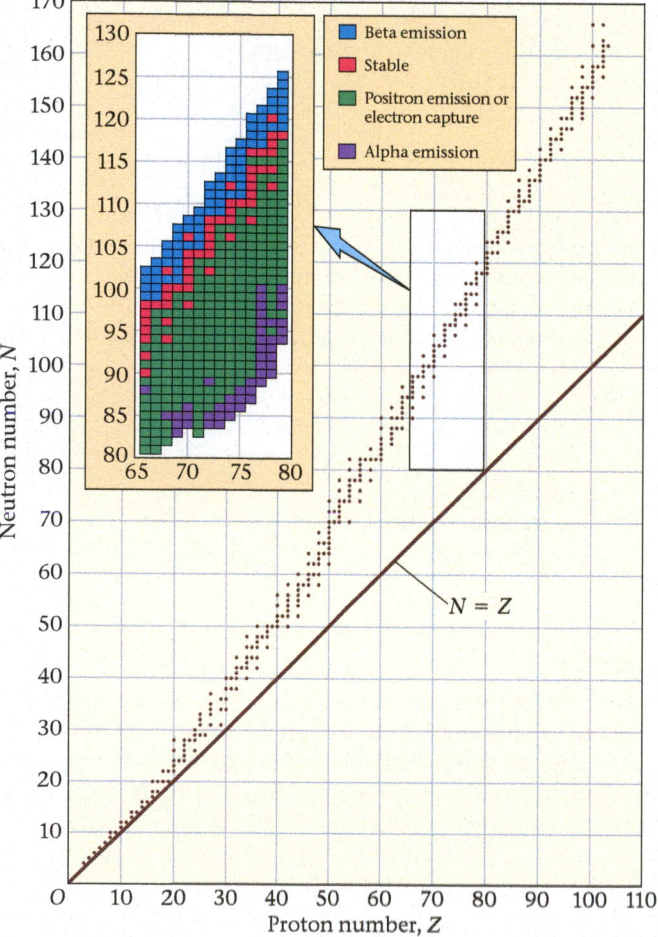

As the number of protons in a nucleus increases, however, a point is reached at which the strong nuclear forces are no longer able to compensate for the repulsive forces between protons. In fact, the largest number of protons in a stable nucleus is $Z = 83$, corresponding to the element bismuth. Nuclei with more than 83 protons are simply not stable, as can be seen by noting that all the elements with $Z > 83$

in Appendix F decay in a finite time—that is, they have a finite half-life. (We shall discuss the half-life of unstable nuclei in detail in Section 32-3.) The nuclei of many well-known elements, such as radon and uranium, disintegrate—decay—in a finite time. We turn now to a discussion of the various ways in which an unstable nucleus decays.

Enhance Your Understanding   (Answers given at the end of the chapter)

1. Consider a stable nucleus with a proton number of roughly 70. Rank the numbers N, Z, and A for this nucleus in order of increasing value. Indicate ties where appropriate.

Section Review

• The atomic number of an element is the number of protons Z in its nucleus. In a neutral atom, this is also the number of electrons. The number of neutrons in a nucleus is N, and the mass number A is defined as the total number of neutrons and protons in the nucleus, $A = Z + N$. Neutrons and protons are referred to as nucleons.

• The atomic mass unit is defined as $1 \text{ u} = 1.660540 \times 10^{-27}$ kg or, equivalently, $1 \text{ u} = 931.5 \text{ MeV}/c^2$.

32-2 **Radioactivity**

Nothing lasts forever, it seems. An unstable nucleus certainly doesn't—sooner or later it changes its composition by emitting a particle of one type or another. Alternatively, a nucleus in an excited state may rearrange its nucleons into a lower-energy state and emit a high-energy photon. We refer to such processes as the **decay** of a nucleus, and the various emissions that result are known collectively as **radioactivity.**

When a nucleus undergoes radioactive decay, the mass of the system decreases. That is, the mass of the initial nucleus before decay is greater than the mass of the resulting nucleus plus the mass of the emitted particle or photon. The difference in mass, $\Delta m < 0$, appears as a release of energy, according to the relationship $E = |\Delta m| c^2$. The mass difference for any given decay can be determined by referring to Appendix F. Notice that the atomic masses listed in Appendix F are the masses of neutral atoms; that is, the values given in the table include the mass of the electrons in an atom. This factor must be considered whenever the mass difference of a reaction is calculated, as we shall see later in this section.

Three types of particles with mass are given off during the various processes of radioactive decay. They are as follows:

• Alpha (α) particles, which are the nuclei of $^4_2$He. An alpha particle consists of two protons and two neutrons. When a nucleus decays by giving off alpha particles, we say that it emits α rays.

• Electrons, which are also referred to as beta (β) particles. The electrons given off by a nucleus are called β rays, or β^- rays to be more precise. (The minus sign is a reminder that the charge of an electron is $-e$.)

• **Positrons,** which have the same mass as an electron but a charge of $+e$. If a nucleus gives off positrons, we say that it emits β^+ rays. (A positron, which is short for "positive electron," is the **antiparticle** of the ordinary electron. Positrons will be considered in greater detail in Section 32-7.)

Finally, radioactivity may take the form of a photon rather than a particle with a finite mass:

• A nucleus in an excited state may emit a high-energy photon, a gamma (γ) ray, and drop to a lower-energy state.

The following Conceptual Example examines the behavior of radioactivity in a magnetic field.

Big Idea **2** Radioactivity results when an unstable nucleus decays and emits either a particle or a photon.

CONCEPTUAL EXAMPLE 32-5 | IDENTIFY THE RADIATION

A sample of radioactive material is placed at the bottom of a small hole drilled into a piece of lead. The sample emits α rays, β^- rays, and γ rays into a region of constant magnetic field. It is observed that the radiation follows three distinct paths, 1, 2, and 3, as shown in the sketch. Identify each path with the corresponding type of radiation.

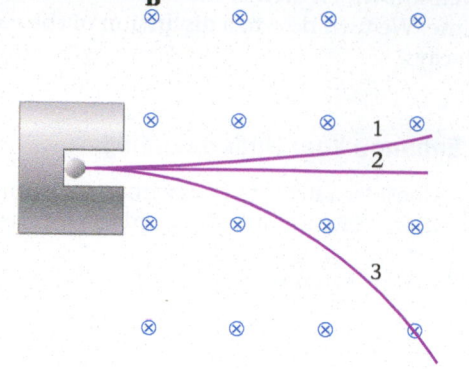

REASONING AND DISCUSSION

First, because γ rays are uncharged, they are not deflected by the magnetic field. It follows that path 2 corresponds to γ rays.

Next, the right-hand rule for the magnetic force (Section 22-2) indicates that positively charged particles will be deflected upward, and negatively charged particles will be deflected downward. As a result, path 1 corresponds to α rays, and path 3 corresponds to β^- rays.

ANSWER

Path 1, α rays; path 2, γ rays; path 3, β^- rays.

Radioactivity was discovered by the French physicist Antoine Henri Becquerel (1852–1908) in 1896, when he observed that uranium was able to expose photographic emulsion, even when the emulsion was covered. Thus radioactivity has the ability to penetrate various materials. In fact, the various types of radioactivity were initially named according to their ability to penetrate, starting with α rays, which are the least penetrating. Typical penetrating abilities for α, β, and γ rays are as follows:

- α rays can barely penetrate a sheet of paper.
- β rays (both β^- and β^+) can penetrate a few millimeters of aluminum.
- γ rays can penetrate several centimeters of lead.

We turn now to a detailed examination of each of these types of decay.

Alpha Decay

When a nucleus decays by giving off an α particle ($^4_2$He), it loses two protons and two neutrons. As a result, its atomic number, Z, decreases by 2, and its mass number decreases by 4. Symbolically, we can write this process as follows:

$$^A_Z X \longrightarrow \, ^{A-4}_{Z-2} Y + \, ^4_2 He.$$

In this expression, X is referred to as the **parent nucleus,** and Y is the **daughter nucleus.** Notice that the sum of the atomic numbers on the right side of this process is equal to the atomic number on the left side; a similar relationship applies to the mass numbers.

The next Example considers the alpha decay of uranium-238. We first use conservation of atomic number and mass number to determine the identity of the daughter nucleus. Next, we use the mass difference to calculate the amount of energy released by the decay.

EXAMPLE 32-6 | URANIUM DECAY

Determine **(a)** the daughter nucleus and **(b)** the energy released when $^{238}_{92}$U undergoes alpha decay.

PICTURE THE PROBLEM

Our sketch shows the specified decay of $^{238}_{92}$U into a daughter nucleus plus an α particle. The number of neutrons and protons is indicated for $^{238}_{92}$U. The α particle consists of two neutrons and two protons.

REASONING AND STRATEGY

a. We can identify the daughter nucleus by requiring that the total number of neutrons and protons be the same before and after the decay.

CONTINUED

b. To find the energy, we first calculate the mass before and after the decay. The magnitude of the difference in mass, $|\Delta m|$, times the speed of light squared, c^2, gives the amount of energy released.

Known Alpha decay of $^{238}_{92}$U, which removes two protons and two neutrons from the nucleus.

Unknown **(a)** Daughter nucleus, $^A_Z Y = ?$ **(b)** Energy released, $E = ?$

SOLUTION

Part (a)

1. Determine the number of neutrons and protons in the daughter nucleus. Add these numbers to obtain the mass number of the daughter nucleus:

$$N = 146 - 2 = 144$$
$$Z = 92 - 2 = 90$$
$$A = N + Z = 144 + 90 = 234$$

2. Referring to Appendix F, we see that the daughter nucleus is thorium-234:

$$^A_Z Y = {}^{234}_{90}\text{Th}$$

Part (b)

3. Use Appendix F to find the initial mass of the system—that is, the mass of a $^{238}_{92}$U atom:

$$m_i = 238.050786 \text{ u}$$

4. Use Appendix F to find the final mass of the system—that is, the mass of a $^{234}_{90}$Th atom plus the mass of a $^4_2$He atom:

$$m_f = 234.043596 \text{ u} + 4.002603 \text{ u} = 238.046199 \text{ u}$$

5. Calculate the mass difference and the corresponding energy release (recall that $1 \text{ u} = 931.5 \text{ MeV}/c^2$):

$$\Delta m = m_f - m_i$$
$$= 238.046199 \text{ u} - 238.050786 \text{ u} = -0.004587 \text{ u}$$
$$E = |\Delta m|c^2 = (0.004587 \text{ u})\left(\frac{931.5 \text{ MeV}/c^2}{1 \text{ u}}\right)c^2$$
$$= 4.273 \text{ MeV}$$

INSIGHT

Each of the masses used in this decay includes the electrons of the corresponding neutral atom. Because the number of electrons initially (92) is the same as the number of electrons in a thorium atom (90) plus the number of electrons in a helium atom (2), the electrons make no contribution to the total mass difference, Δm.

PRACTICE PROBLEM

Find the daughter nucleus and the energy released when $^{226}_{88}$Ra undergoes alpha decay. **[Answer:** $^{222}_{86}$Rn; 4.871 MeV**]**

Some related homework problems: Problem 23, Problem 71

A considerable amount of energy is released in the alpha decay of uranium-238. As indicated in **FIGURE 32-2**, this energy appears as kinetic energy of the daughter nucleus and the α particle as they move off in opposite directions. The following Conceptual Example compares the kinetic energy of the daughter nucleus with that of the α particle.

CONCEPTUAL EXAMPLE 32-7 COMPARE KINETIC ENERGIES

When a stationary $^{238}_{92}$U nucleus decays into a $^{234}_{90}$Th nucleus and an α particle, is the kinetic energy of the α particle greater than, less than, or the same as the kinetic energy of the $^{234}_{90}$Th nucleus?

REASONING AND DISCUSSION

Because no external forces are involved in the decay process, it follows that the momentum of the system is conserved. Letting subscript 1 refer to the $^{234}_{90}$Th nucleus, and subscript 2 to the α particle, we can express the condition for momentum conservation as $m_1 v_1 = m_2 v_2$. Solving for the speed of the α particle, we have $v_2 = (m_1/m_2)v_1$; that is, the α particle has the greater speed because m_1 is greater than m_2.

The fact that the α particle has the greater speed does not, in itself, ensure that its kinetic energy is the greater of the two. After all, the α particle also has the smaller mass. To compare the kinetic energies, note that the kinetic energy of the $^{234}_{90}$Th nucleus is $\frac{1}{2}m_1 v_1^2$. The kinetic energy of the α particle is $\frac{1}{2}m_2 v_2^2 = \frac{1}{2}m_2[(m_1/m_2)v_1]^2 = \frac{1}{2}m_1 v_1^2(m_1/m_2) > \frac{1}{2}m_1 v_1^2$. Therefore, the α particle carries away the majority of the kinetic energy released in the decay. In this particular case, the α particle has a kinetic energy that is $m_1/m_2 = 234/4 = 58.5$ times greater than the kinetic energy of the $^{234}_{90}$Th nucleus.

ANSWER

The α particle has the greater kinetic energy.

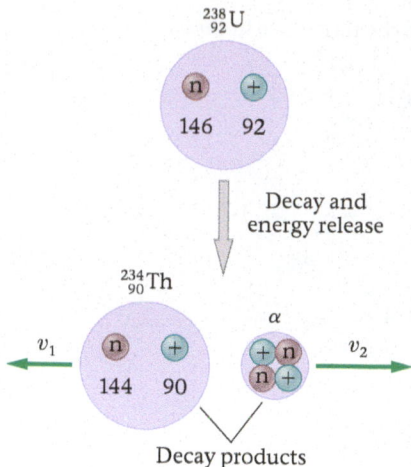

◀ **FIGURE 32-2 Alpha decay of uranium-238** When $^{238}_{92}$U decays to $^{234}_{90}$Th and an alpha particle, the mass of the system decreases. The "lost" mass is actually converted into energy; it appears as the kinetic energy of the $^{234}_{90}$Th nucleus and the alpha particle.

RWP* Although you may not be aware of it, many homes are protected from the hazards of fire by a small radioactive device—a smoke detector, like the one shown in **FIGURE 32-3**—that uses the alpha decay of a man-made radioactive isotope, $^{241}_{95}$Am. In this type of smoke detector, a minute quantity of $^{241}_{95}$Am is placed between two metal plates connected to a battery. The α particles emitted by the radioactive source ionize the air, allowing a measurable electric current to flow between the plates. As long as this current flows, the smoke detector remains silent. When smoke enters the detector, however, the ionized air molecules tend to stick to the smoke particles and become neutralized. This reduces the current and triggers the alarm. These "ionization" smoke detectors are more sensitive than the "photoelectric" detectors that rely on the thickness of smoke to dim a beam of light.

Beta Decay

The basic process that occurs in beta decay is the conversion of a neutron to a proton and an electron:

$$^1_0n \longrightarrow {}^1_1p + e^-$$

When a nucleus decays by giving off an electron, its mass number is unchanged (since protons and neutrons count equally in determining A), but its atomic number increases by 1. This process can be represented as follows:

$$^A_ZX \longrightarrow {}^A_{Z+1}Y + e^-$$

Similarly, if a nucleus undergoes a different type of decay in which it gives off a positron, the process can be written as

$$^A_ZX \longrightarrow {}^A_{Z-1}Y + e^+$$

In the next Example we determine the energy that is released as carbon-14 undergoes beta decay.

◀ **FIGURE 32-3** Smoke detectors like this one make use of a synthetic radioactive isotope, americium-241. The alpha particles emitted when this isotope decays ionize air molecules, making them able to conduct a small current. Smoke particles neutralize the ions, which interrupts the current and sets off an alarm.

*Real World Physics applications are denoted by the acronym RWP.

EXAMPLE 32-8 | BETA DECAY OF CARBON-14

Find **(a)** the daughter nucleus and **(b)** the energy released when $^{14}_6$C undergoes β^- decay.

PICTURE THE PROBLEM
In our sketch we show the nucleus of $^{14}_6$C giving off a β^- particle and converting into a daughter nucleus. The number of neutrons and protons in the $^{14}_6$C nucleus is indicated. The β^- particle is not a nucleon.

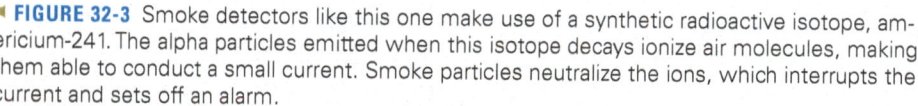

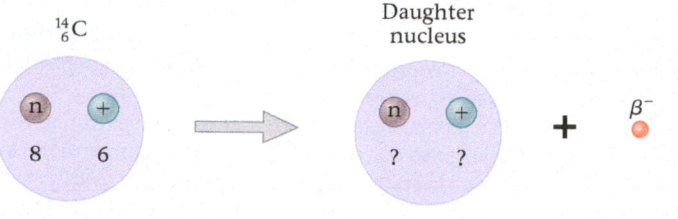

REASONING AND STRATEGY

a. We can identify the daughter nucleus by requiring that the total number of nucleons be the same before and after the decay. The number of neutrons will be decreased by 1, and the number of protons will be increased by 1.

b. To find the energy, we begin by calculating the mass before and after the decay. The magnitude of the difference in mass, $|\Delta m|$, times the speed of light squared, c^2, gives the amount of energy released.

CONTINUED

Known   Beta decay of $^{14}_{6}C$, which converts a neutron to a proton in the nucleus, and emits an electron.

Unknown   (a) Daughter nucleus, $^{A}_{Z}Y$ = ? (b) Energy released, E = ?

SOLUTION

Part (a)

1. Determine the number of neutrons and protons in the daughter nucleus. Add these numbers to obtain the mass number of the daughter nucleus:

$$N = 8 - 1 = 7$$
$$Z = 6 + 1 = 7$$
$$A = N + Z = 7 + 7 = 14$$

2. Referring to Appendix F, we see that the daughter nucleus is nitrogen-14:

$$^{A}_{Z}Y = {}^{14}_{7}N$$

Part (b)

3. Use Appendix F to find the initial mass of the system—that is, the mass of a $^{14}_{6}C$ atom:

$$m_i = 14.003242 \text{ u}$$

4. Use Appendix F to find the final mass of the system, which is simply the mass of a $^{14}_{7}N$ atom (the mass of the β^- particle is included in the mass of $^{14}_{7}N$, as we point out in the Insight):

$$m_f = 14.003074 \text{ u}$$

5. Calculate the mass difference and the corresponding energy release (recall that 1 u = 931.5 MeV/c^2):

$$\Delta m = m_f - m_i$$
$$= 14.003074 \text{ u} - 14.003242 \text{ u} = -0.000168 \text{ u}$$
$$E = |\Delta m| c^2 = (0.000168 \text{ u}) \left(\frac{931.5 \text{ MeV}/c^2}{1 \text{ u}} \right) c^2$$
$$= 0.156 \text{ MeV}$$

INSIGHT

With regard to the masses used in this calculation, notice that the mass of $^{14}_{6}C$ includes the mass of its 6 electrons. Similarly, the mass of $^{14}_{7}N$ includes the mass of 7 electrons in the neutral $^{14}_{7}N$ atom. However, when the $^{14}_{6}C$ nucleus converts to a $^{14}_{7}N$ nucleus, the number of electrons orbiting the nucleus is still 6. In effect, the newly created $^{14}_{7}N$ atom is missing one electron; that is, the mass of the $^{14}_{7}N$ atom includes the mass of one too many electrons. Therefore, it is not necessary to add the mass of an electron (representing the β^- particle) to the final mass of the system, because this extra electron mass is already included in the mass of the $^{14}_{7}N$ atom.

PRACTICE PROBLEM

Find the daughter nucleus and the energy released when $^{234}_{90}Th$ undergoes β^- decay.   [**Answer:** $^{234}_{91}Pa$; 0.274 MeV]

Some related homework problems: Problem 24, Problem 26

Discovery of the Neutrino   Referring to Conceptual Example 32-7, we would expect the kinetic energy of an electron emitted during beta decay to account for most of the energy released by the decay process. In fact, energy conservation allows us to predict the precise amount of kinetic energy the electron should have. It turns out, however, that when the kinetic energy of emitted electrons is measured, a range of values is obtained, as indicated in **FIGURE 32-4**. Specifically, we find that all electrons given off in beta decay have energies that are less than would be predicted by energy conservation. On closer examination it is found that beta decay seems to violate conservation of linear and angular momentum as well! For these reasons, beta decay was an interesting and intriguing puzzle for physicists.

The resolution of this puzzle was given by Pauli in 1930, when he proposed that the "missing" energy and momentum were actually carried off by a particle that was not observed in the experiments. For this particle to have been unobserved, it must have zero charge and little or no mass. Fermi dubbed Pauli's hypothetical particle the **neutrino,** meaning, literally, "little neutral one." We now know that neutrinos do in fact exist and that they account exactly for the missing energy and momentum. They interact so weakly with matter, however, that it wasn't until 1956 that they were observed experimentally. Recent experiments on neutrinos given off by the Sun provide the best evidence yet that the mass of a neutrino is in fact finite—though extremely small. In fact, the best estimate of the neutrino mass at this time is that it is less than about 2 eV/c^2. For comparison, the mass of the electron is 511,000 eV/c^2.

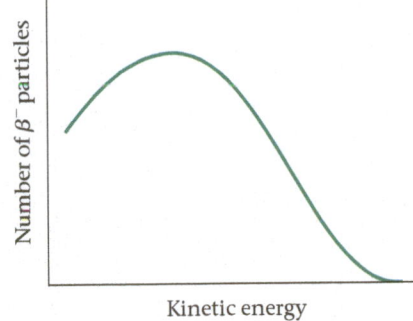

▲ **FIGURE 32-4 Energy of electrons in β^- decay** When electrons are emitted during β^- decay, they come off with a range of energies. This indicates that another particle (the neutrino) must also be taking away some of the energy.

PROBLEM-SOLVING NOTE

The Effects of Beta Decay

In β^- decay, the number of neutrons decreases by 1, and the number of protons increases by 1. The mass number is unchanged.

PHYSICS IN CONTEXT
Looking Back

In considering the decay of a nucleus, we use the concepts of electrostatic repulsion, momentum conservation, and energy conservation. For example, we compared the kinetic energies of nuclear decay products in Conceptual Example 32-7 with the same methods we used to study air carts in Chapter 9.

To give an indication of just how weakly neutrinos interact with matter, only one in every 200 million neutrinos that pass through the Earth interacts with it in any way. As far as the neutrinos are concerned, it is almost as if the Earth did not exist. Right now, in fact, billions of neutrinos are passing through your body every second without the slightest effect.

We can now write the correct expression for the decay of a neutron. Indicating the electron neutrino with the symbol ν_e, we have the following:

$$^1_0n \longrightarrow ^1_1p + e^- + \bar{\nu}_e \qquad \text{32-5}$$

The bar over the neutrino symbol indicates that the neutrino given off in β^- decay is actually an **antineutrino,** the antiparticle counterpart of the neutrino (just as the positron is the antiparticle of the electron). The neutrino itself is given off in β^+ decay.

Gamma Decay

An atom in an excited state can emit a photon when one of its electrons drops to a lower-energy level. Similarly, a nucleus in an excited state can emit a photon as it decays to a state of lower energy. Because nuclear energies are so much greater than typical atomic energies, the photons given off by a nucleus are highly energetic. In fact, these photons have energies that place them well beyond X-rays in the electromagnetic spectrum. We refer to such high-energy photons as **gamma (γ) rays.**

As an example of a situation in which a γ ray can be given off, consider the following beta decay:

$$^{14}_6C \longrightarrow ^{14}_7N^* + e^- + \bar{\nu}_e$$

The asterisk on the nitrogen symbol indicates that the nitrogen nucleus has been left in an excited state as a result of the beta decay. Subsequently, the nitrogen nucleus may decay to its ground state with the emission of a γ ray:

$$^{14}_7N^* \longrightarrow ^{14}_7N + \gamma$$

Notice that neither the atomic number nor the mass number is changed by the emission of a γ ray.

QUICK EXAMPLE 32-9 GAMMA-RAY EMISSION: FIND THE CHANGE IN MASS

A $^{226}_{88}Ra$ nucleus in an excited state emits a γ ray with a wavelength of 5.66×10^{-12} m. Find the decrease in mass of the $^{226}_{88}Ra$ nucleus as a result of this process.

REASONING AND SOLUTION

The γ ray is a photon, and it moves with the speed of light. Therefore, we can convert its wavelength to frequency with the relationship $\lambda f = c$. The energy of the γ ray is $E = hf$, and this energy divided by c^2 gives the change in mass.

1. Find the frequency of the γ ray:

$$f = \frac{c}{\lambda} = \frac{3.00 \times 10^8 \text{ m/s}}{5.66 \times 10^{-12} \text{ m}} = 5.30 \times 10^{19} \text{ Hz}$$

2. Calculate the energy of the γ ray:

$$E = hf = (6.63 \times 10^{-34} \text{ J·s})(5.30 \times 10^{19} \text{ Hz})$$
$$= 3.51 \times 10^{-14} \text{ J} = 0.219 \text{ MeV}$$

3. Determine the mass difference corresponding to the energy of the γ ray:

$$|\Delta m| = \frac{E}{c^2} = \frac{0.219 \text{ MeV}}{c^2}\left(\frac{1 \text{ u}}{931.5 \text{ MeV}/c^2}\right)$$
$$= 0.000235 \text{ u}$$

As a result of emitting this γ ray, the mass of the $^{226}_{88}Ra$ nucleus decreases by an amount that is a bit less than half the mass of an electron.

Radioactive Decay Series

Consider an unstable nucleus that decays and produces a daughter nucleus. If the daughter nucleus is also unstable, it will eventually decay and produce its own daughter nucleus, which may in turn be unstable. In such cases, an original parent nucleus

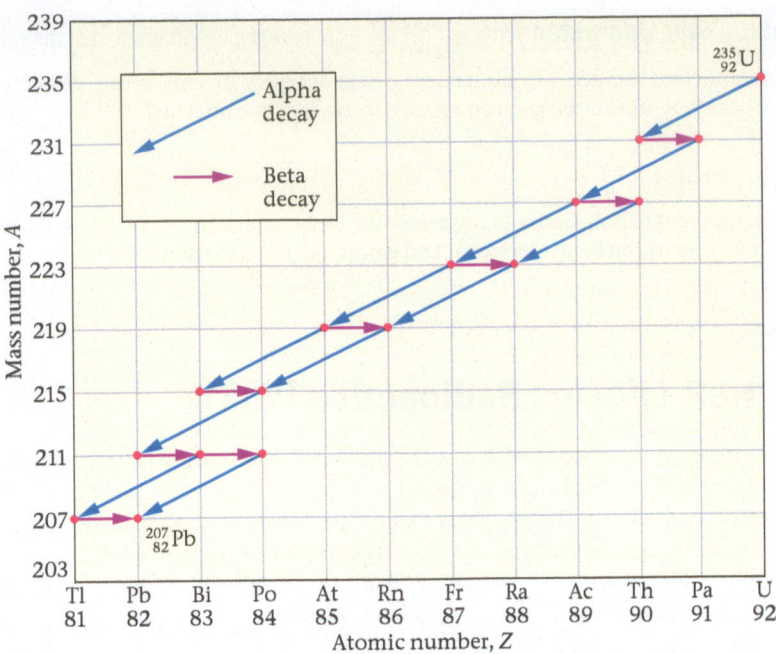

◄ **FIGURE 32-5 Radioactive decay series** **of** $^{235}_{92}U$ When $^{235}_{92}U$ decays, it passes through a number of intermediate nuclei before reaching the stable end of the series, $^{207}_{82}Pb$. Note that some intermediary nuclei can decay in only one way, whereas others have two decay possibilities.

can produce a series of related nuclei referred to as a **radioactive decay series.** An example of a radioactive decay series is shown in **FIGURE 32-5**. In this case, the parent nucleus is $^{235}_{92}U$, and the final nucleus of the series is $^{207}_{82}Pb$, which is stable.

Notice that several of the intermediate nuclei in this series can decay in two different ways—either by alpha decay or by beta decay. Thus there are various "paths" a $^{235}_{92}U$ nucleus can follow as it transforms into a $^{207}_{82}Pb$ nucleus. In addition, the intermediate nuclei in this series decay fairly rapidly, at least on a geological time scale. For example, any actinium-227 $\left(^{227}_{89}Ac\right)$ that was present when the Earth formed would have decayed away long ago. The fact that actinium-227 is still found on the Earth today in natural uranium deposits is due to its continual production in this and other decay series.

Activity

The rate at which nuclear decay occurs—that is, the number of decays per second—is referred to as the **activity.** A highly active material has many nuclear decays occurring every second. For example, a typical sample of radium (usually a fraction of a gram) might have 10^5 to 10^{10} decays per second.

The unit we use to measure activity is the curie, named in honor of Pierre (1859–1906) and Marie (1867–1934) Curie, pioneers in the study of radioactivity. The **curie** (Ci) is defined as follows:

$$1 \text{ curie} = 1 \text{ Ci} = 3.7 \times 10^{10} \text{ decays/s} \qquad \text{32-6}$$

The reason for this choice is that 1 Ci is roughly the activity of 1 g of radium. In SI units, we measure activity in terms of the **becquerel** (Bq):

$$1 \text{ becquerel} = 1 \text{ Bq} = 1 \text{ decay/s} \qquad \text{32-7}$$

The units of activity most often encountered in practical applications are the millicurie $(1 \text{ mCi} = 10^{-3} \text{ Ci})$ and the microcurie $(1 \text{ }\mu\text{Ci} = 10^{-6} \text{ Ci})$.

EXERCISE 32-10 RADIUM ACTIVITY

A sample of radium has an activity of 6.5 μCi. How many decays per second occur in this sample?

REASONING AND SOLUTION

Using the definition 1 curie = 1 Ci = 3.7×10^{10} decays/s, we find

$$6.5 \text{ }\mu\text{Ci} = 6.5 \times 10^{-6} \text{ Ci}$$

$$= (6.5 \times 10^{-6} \text{ Ci})\left(\frac{3.7 \times 10^{10} \text{ decays/s}}{1 \text{ Ci}}\right) = 2.4 \times 10^5 \text{ decays/s}$$

32-3 Half-Life and Radioactive Dating

The phenomenon of radioactive decay, though fundamentally random in its behavior, has certain properties that make it useful as a type of "nuclear clock." In fact, it has been discovered that radioactive decay can be used to date numerous items of interest from the recent—and not so recent—past. In this section we consider the behavior of radioactive decay as a function of time, introduce the concept of a half-life, and show explicitly how these concepts can be applied to dating.

An Analogy for Nuclear Decay To begin, consider an analogy in which coins represent nuclei, and the side that comes up when a given coin is tossed determines whether the corresponding nucleus decays. Suppose, for example, that we toss a group of 64 coins and remove any coin that comes up tails. We expect that—on average—32 coins will be removed after the first round of tosses. Which coins will be removed—and the precise number that will be removed—cannot be known, because the flip of a coin, like the decay of a nucleus, is a random process. When we toss the remaining 32 coins, we expect an average of 16 more to be removed, and so on, with each round of tosses decreasing the number of coins by a factor of 2. The results after the first few rounds are shown by the points in **FIGURE 32-6**.

Also shown in Figure 32-6 is a smooth curve representing the following mathematical function:

$$N = (64)e^{-(\ln 2)t} = (64)e^{-(0.693)t} \qquad \text{32-8}$$

▶ **FIGURE 32-6 Tossing coins as an analogy for nuclear decay** The points on this graph show the number of coins that remain (on average) if one starts with 64 coins and removes half with each round of tosses. The curve is a plot of the mathematical function $64e^{-(0.693)t}$, where $t = 1$ means one round of tosses, $t = 2$ means two rounds, and so on.

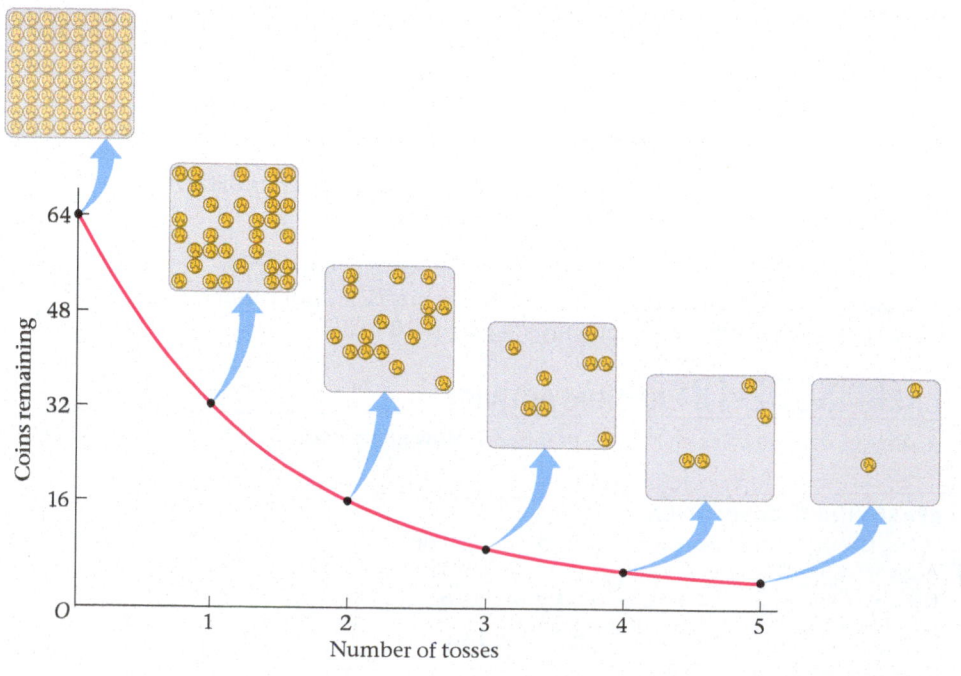

In this expression, N represents the number of coins, and the time variable, t, represents the number of rounds of tosses. For example, if we set $t = 1$ in Equation 32-8, we find $N = 32$, and if we set $t = 2$, we find $N = 16$. This type of "exponential dependence" is a general feature whenever the number of some quantity increases or decreases by a constant factor with each constant interval of time. Examples include the balance in a bank account with compounding interest and the population of the Earth as a function of time.

When nuclei decay, their behavior is much like that of the coins in our analogy. Which nucleus decays in a given interval of time, and the precise number that decay, are controlled by a random process that causes the decay—on average—of a given fraction of the original number of nuclei. Thus the number of nuclei, N, remaining at time t is given by an expression analogous to the one for the coins:

$$N = N_0 e^{-\lambda t} \qquad \text{32-9}$$

Here, N_0 is the number of nuclei present at time $t = 0$, and the constant λ is referred to as the **decay constant**. (*Note: λ does not* represent a wavelength in this context.) In the analogy of the coins, $N_0 = 64$ and $\lambda = 0.693 \text{ s}^{-1}$. Notice that the larger the value of the decay constant, the more rapidly the number of nuclei decreases with time. **FIGURE 32-7** shows the dependence on λ graphically.

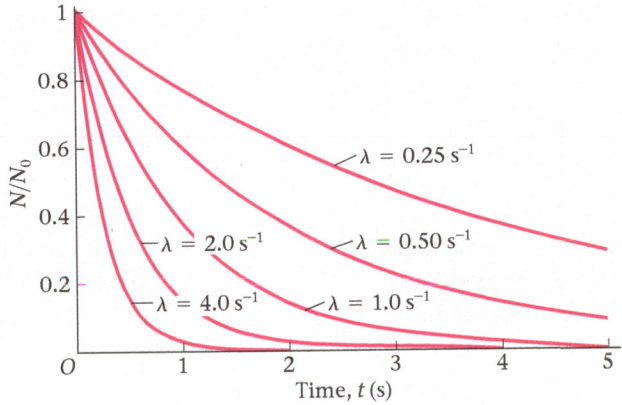

PROBLEM-SOLVING NOTE

Consistent Units and the Decay Constant

When calculating the number of nuclei present at a given time, be sure to use consistent units for the time, t, and the decay constant, λ. For example, if you express λ in units of y^{-1}, measure the time in units of y.

Big Idea 3 Radioactive decay proceeds at a rate determined by the amount of radioactive material present in a sample. Measurement of decay rates can be used to date materials.

◄ **FIGURE 32-7 Dependence on the decay constant** The larger the decay constant, λ, the more rapidly the population of a group of nuclei decreases. In this plot, the value of λ doubles as we move downward from one curve to the next.

QUICK EXAMPLE 32-11 FIND THE RADON LEVEL

Radon can pose a health risk when high levels become trapped in the basement of a house. Test kits, like the one shown in **FIGURE 32-8**, are used by homeowners to monitor the radiation level. Suppose 3.38×10^7 radon atoms are in a basement at a time when it is sealed to prevent any additional radon from entering. Given that the decay constant of radon is 0.181 d^{-1}, how many radon atoms remain in the basement after **(a)** 7 d and **(b)** 14 d?

REASONING AND SOLUTION

Given the decay constant ($\lambda = 0.181 \text{ d}^{-1}$) and the initial number of radioactive atoms ($N_0 = 3.38 \times 10^7$), we can use the expression $N = N_0 e^{-\lambda t}$ to find the number of atoms remaining at any given time.

Part (a)

1. Evaluate $N = N_0 e^{-\lambda t}$ 	 $N = N_0 e^{-\lambda t}$
   with $t = 7$ d: 	 $= (3.38 \times 10^7 \text{ atoms})e^{-(0.181 \text{ d}^{-1})(7 \text{ d})} = 9.52 \times 10^6 \text{ atoms}$

Part (b)

2. Evaluate $N = N_0 e^{-\lambda t}$ 	 $N = N_0 e^{-\lambda t}$
   with $t = 14$ d: 	 $= (3.38 \times 10^7 \text{ atoms})e^{-(0.181 \text{ d}^{-1})(14 \text{ d})} = 2.68 \times 10^6 \text{ atoms}$

After 7 d the number of radon atoms has decreased by a factor of about 3.55. After 14 d, the number has decreased by a factor of about $3.55^2 = 12.6$. It follows that every 7 d the number of radon atoms decreases by another factor of 3.55.

▲ **FIGURE 32-8** Radon-222 ($^{222}_{86}$Rn) is an isotope produced in a radioactive decay series that includes uranium-238 and radium-226. Because uranium is naturally present in certain kinds of rocks and the soils derived from them, radon, which is a gas, can accumulate in basements and similar enclosed underground spaces that lack adequate ventilation. Radon-222 is itself radioactive, undergoing alpha decay with a half-life of about 4 days, and although its concentration is generally small, it may produce radiation levels great enough to be a health hazard if exposure is prolonged. Homeowners in many parts of the country use test kits to monitor the radiation levels produced by radon.

Half-Life A useful way to characterize the rate at which a given type of nucleus decays is in terms of its half-life, which is defined as follows:

The **half-life** of a given type of radioactive nucleus is the time required for the number of such nuclei to decrease by a factor of 2; that is, for the number to decrease from N_0 to $\frac{1}{2}N_0$, from $\frac{1}{2}N_0$ to $\frac{1}{4}N_0$, and so on.

We can solve for this time, call it $T_{1/2}$, by setting $N = \frac{1}{2}N_0$ in Equation 32-9 ($N = N_0 e^{-\lambda t}$):

$$\frac{1}{2}N_0 = N_0 e^{-\lambda T_{1/2}}$$

Canceling N_0 and taking the natural logarithm of both sides of the equation, we find

$$T_{1/2} = \frac{\ln 2}{\lambda} = \frac{0.693}{\lambda} \qquad \text{32-10}$$

A large decay constant λ corresponds to a short half-life, in agreement with the plots shown in Figure 32-7.

CONCEPTUAL EXAMPLE 32-12   HOW MANY NUCLEI?   PREDICT/EXPLAIN

A system consists of N_0 radioactive nuclei at time $t = 0$. **(a)** Is the number of nuclei that remain after *half* of a half-life (that is, at time $t = \frac{1}{2}T_{1/2}$) equal to $\frac{1}{4}N_0$, $\frac{3}{4}N_0$, or $(1/\sqrt{2})N_0$? **(b)** Which of the following is the *best explanation* for your prediction?

   I. In half of a half-life, the number of nuclei remaining will be half of one half N_0—that is, $\frac{1}{4}N_0$.
   II. In half of a half-life, one-quarter of the nuclei will decay, which means that $\frac{3}{4}N_0$ nuclei remain.
   III. In the first half of a half-life, the number of nuclei remaining is reduced to $(1/\sqrt{2})N_0$. The number is reduced by another factor of $1/\sqrt{2}$ in the second half of a half-life, for a total reduction to $(1/2)N_0$.

REASONING AND DISCUSSION

Referring to Quick Example 32-11, we note that if the number of nuclei decreases by a factor f in the time $\frac{1}{2}T_{1/2}$, it will decrease by the factor f^2 in the time $2\left(\frac{1}{2}T_{1/2}\right) = T_{1/2}$. We know, however, that the number of nuclei remaining at the time $T_{1/2}$ is $\frac{1}{2}N_0$. It follows that $f^2 = \frac{1}{2}$, or that $f = 1/\sqrt{2}$. Therefore, the number of nuclei remaining at the time $\frac{1}{2}T_{1/2}$ is $(1/\sqrt{2})N_0$.

ANSWER

(a) At half a half-life, the number of nuclei remaining is $(1/\sqrt{2})N_0$. **(b)** The best explanation is III.

Activity The property that makes radioactivity so useful as a clock is that its **decay rate**, R, or **activity**, depends on time in a straightforward way. To see this, think back to the analogy of the coins. The number of coins that decay (are removed) on the first round of tosses is $32 = \frac{1}{2}(64)$, the number that decay on the second round is $16 = \frac{1}{2}(32)$, and so on. That is, the number that decay in any given interval of time is proportional to the number present at the beginning of the interval.

The same type of analysis applies to nuclei. Therefore, the number of nuclei that decay, ΔN, in a given time interval, Δt, is proportional to the number, N, that are present at time t. In fact, a detailed analysis gives the following result for the activity, R:

$$R = \left|\frac{\Delta N}{\Delta t}\right| = \lambda N \qquad \text{32-11}$$

There are two points to note regarding this equation. First, observe that the number of nuclei is decreasing, $\Delta N < 0$. It is for this reason that we take the absolute value of the quantity $\Delta N/\Delta t$. Second, notice that the proportionality constant is simply λ, the decay constant. That λ is the correct constant of proportionality can be shown using the methods of calculus.

Combining Equation 32-11 ($R = \lambda N$) with Equation 32-9 ($N = N_0 e^{-\lambda t}$), we obtain the time dependence of the activity, R:

$$R = \lambda N_0 e^{-\lambda t} = R_0 e^{-\lambda t} \qquad \text{32-12}$$

In this equation, $R_0 = \lambda N_0$ is the initial value of the activity.

QUICK EXAMPLE 32-13   FIND THE ACTIVITY OF RADON

Referring to Quick Example 32-11, calculate the number of radon atoms that disintegrate per second (a) initially and (b) after 7 d.

REASONING AND SOLUTION

(a) The initial activity (number of decays per second) is determined with $R_0 = \lambda N_0$, where the initial number of atoms is given in Quick Example 32-11 as $N_0 = 3.38 \times 10^7$ atoms, and the decay constant is $\lambda = 0.181 \text{ d}^{-1}$. (b) After 7 d the number of radon atoms has diminished to $N = 9.52 \times 10^6$ atoms. Using this value in $R = \lambda N$ yields the desired activity.

Part (a)

1. Use $R_0 = \lambda N_0$ to calculate the activity (in atoms decaying per second) for $N_0 = 3.38 \times 10^7$:

$$R_0 = \lambda N_0 = (0.181 \text{ d}^{-1})(3.38 \times 10^7 \text{ atoms})$$
$$= 6.12 \times 10^6 \text{ atoms/d} = 70.8 \text{ atoms/s}$$

Part (b)

2. Repeat the calculation for $N = 9.52 \times 10^6$:

$$R = \lambda N = (0.181 \text{ d}^{-1})(9.52 \times 10^6 \text{ atoms})$$
$$= 1.72 \times 10^6 \text{ atoms/d} = 19.9 \text{ atoms/s}$$

The initial activity (number of atoms decaying per time) is 70.8 Bq. In terms of the curie, the initial activity is 0.00191 μCi.

Referring to Equation 32-12 ($R = R_0 e^{-\lambda t}$), we see that the basic idea of radioactive dating is simply this: If we know the initial activity of a sample, R_0, and we also know the sample's activity now, R, we can find the corresponding time, t, as follows:

$$\frac{R}{R_0} = e^{-\lambda t}$$

32-13

$$t = -\frac{1}{\lambda} \ln \frac{R}{R_0} = \frac{1}{\lambda} \ln \frac{R_0}{R}$$

The current activity, R, can be measured in the lab, but how can we know the initial activity, R_0? We address this question next for the specific case of carbon-14.

Carbon-14 Dating

To determine the initial activity of carbon-14 requires a basic knowledge of the role it plays in Earth's biosphere. First, we note that carbon-14 is unstable, with a half-life of 5730 y. It follows that the carbon-14 initially present in any *closed* system will decay away to practically nothing in a time of several half-lives. Even so, the ratio of carbon-14 to carbon-12 in Earth's atmosphere remains approximately constant over time at the value 1.20×10^{-12}. Evidently, Earth's atmosphere is not a closed system, at least as far as carbon-14 is concerned.

This is indeed the case. Cosmic rays, which are high-energy particles from outer space, are continuously entering Earth's upper atmosphere and initiating nuclear reactions in nitrogen-14 (a stable isotope). These reactions result in a steady production of carbon-14. Thus, the steady level of carbon-14 in the atmosphere is a result of the balance between the *production rate* due to cosmic rays and the *decay rate* due to the properties of the carbon-14 nucleus.

Next, we note that living organisms have the same ratio of carbon-14 to carbon-12 as the atmosphere, because they continuously exchange carbon with their surroundings. When an organism dies, however, the exchange of carbon ceases and the carbon-14 in the organism (wood, bone, shell, etc.) begins to decay. This process is illustrated in **FIGURE 32-9**, where we see that the carbon-14 activity of an organism is constant until it dies, at which point it decreases exponentially with a half-life of 5730 y.

All that remains to implement carbon-14 dating is to determine the initial activity, R_0. For convenience, we calculate R_0 for a 1-g sample of carbon. As we know from Chapter 17, 12 g of carbon-12 consists of Avogadro's number of atoms, 6.02×10^{23}. Therefore, 1 g consists of $(6.02 \times 10^{23})/12 = 5.02 \times 10^{22}$ carbon-12 atoms. Multiplying this number of atoms by the ratio of carbon-14 to carbon-12, 1.20×10^{-12}, shows that the number of carbon-14 atoms in the 1-g sample is 6.02×10^{10}. Next, we

▶ **FIGURE 32-9 Activity of carbon-14**
While an organism is living and exchanging carbon with the atmosphere, its carbon-14 activity remains constant. When the organism dies, the carbon-14 activity decays exponentially with a half-life of 5730 years.

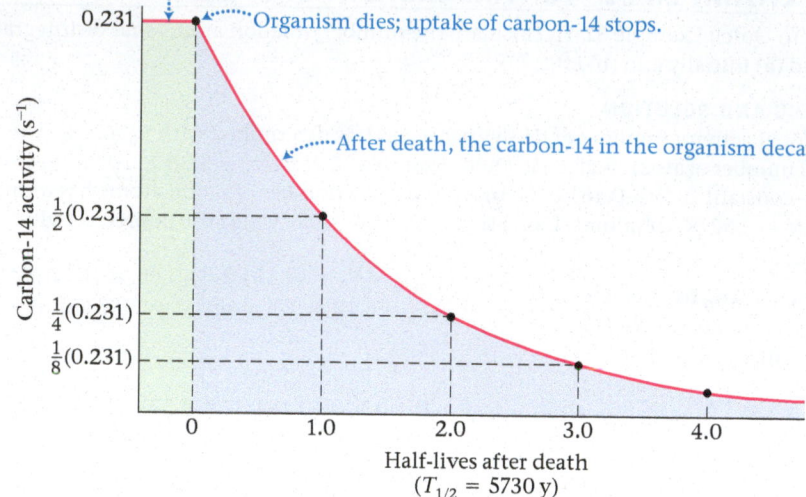

need to know the decay constant, which we obtain by rearranging Equation 32-10 ($T_{1/2} = \ln 2/\lambda = 0.693/\lambda$). This yields

$$\lambda = \frac{\ln 2}{T_{1/2}} = \frac{0.693}{T_{1/2}}$$

Using $T_{1/2} = 5730\,\text{y} = 1.81 \times 10^{11}\,\text{s}$, we find $\lambda = 3.83 \times 10^{-12}\,\text{s}^{-1}$. Combining these results, we find that the initial activity of a 1-g sample of carbon is

$$R_0 = \lambda N_0 = (3.83 \times 10^{-12}\,\text{s}^{-1})(6.02 \times 10^{10}) = 0.231\,\text{Bq} \qquad \text{32-14}$$

This is the initial activity used in Figure 32-9. It follows that 5730 y after an organism dies, its carbon-14 activity per gram of carbon will have decreased to about $\frac{1}{2}(0.231\,\text{Bq}) = 0.116\,\text{Bq}$.

RWP The next Example applies this basic idea to a real-world case of some interest—the Iceman of the Alps.

EXAMPLE 32-14   AGE OF THE ICEMAN: YOU DON'T LOOK A DAY OVER 5000

On September 19, 1991, a German couple hiking in the Italian Alps discovered a body trapped in the ice. Subsequent investigation revealed the remarkably well-preserved body to be that of a Stone Age man. When the carbon-14 dating method was applied to the remains of the Iceman and some of the materials he had carried with him, it was found that the carbon-14 activity was about 0.121 Bq per gram of carbon (**FIGURE 32-10**). Using this information, date the remains of the Iceman.

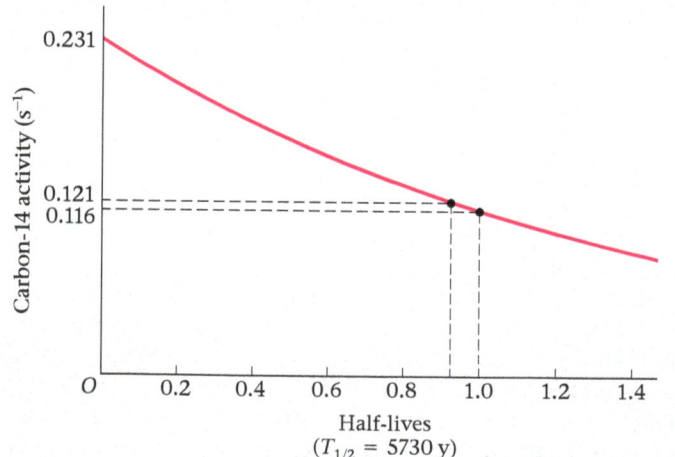

▲ **FIGURE 32-10**

PICTURE THE PROBLEM
Our sketch shows the decay of the carbon-14 activity of a gram of carbon as a function of time. The initial activity of carbon-14 in such a sample is 0.231 Bq.

REASONING AND STRATEGY
We can obtain the age of the remains directly from Equation 32-13, $t = (1/\lambda) \ln(R_0/R)$. In this case, $R_0 = 0.231\,\text{Bq}$ and $R = 0.121\,\text{Bq}$. Because an answer in years would be most useful, we express the decay constant as $\lambda = 0.693/T_{1/2}$, with $T_{1/2} = 5730\,\text{y}$.

The observed activity of 0.121 Bq is slightly greater than $\frac{1}{2}(0.231\,\text{Bq}) = 0.116\,\text{Bq}$; hence, we expect the age of the remains to be slightly less than the half-life of 5730 y.

CONTINUED

Known Activity of 1 g of carbon from the Iceman, $R = 0.121$ Bq.
Unknown Age of the Iceman, $t = ?$

SOLUTION

1. Determine the value of the decay constant, λ, in units of y^{-1}:

$$\lambda = \frac{0.693}{5730 \text{ y}} = 1.21 \times 10^{-4} \text{ y}^{-1}$$

2. Substitute λ, R_0, and R into $t = (1/\lambda) \ln(R_0/R)$:

$$t = \frac{1}{\lambda} \ln\left(\frac{R_0}{R}\right)$$

$$= \frac{1}{(1.21 \times 10^{-4} \text{ y}^{-1})} \ln\left(\frac{0.231 \text{ Bq}}{0.121 \text{ Bq}}\right) = 5340 \text{ y}$$

INSIGHT

We conclude that the Iceman, who has been dubbed Ötzi, died in the mountains during the Stone Age, some 5340 y ago. Detailed examination of Ötzi's body and possessions indicates he was probably an itinerant sheepherder and/or hunter. He met his end in a violent fashion, however. Recent CT scans confirm that Ötzi was killed by an arrow that entered through his shoulder blade and lodged less than an inch from his left lung.

PRACTICE PROBLEM — PREDICT/CALCULATE

(a) If the remains of the Iceman's sister are found in the year 2991, will the carbon-14 activity of 1 g of carbon in her body be greater than, less than, or equal to 0.121 Bq? Explain. **(b)** Determine the carbon-14 activity of 1 g of carbon from the Iceman's sister. **[Answer: (a)** The activity will be less than 0.121 Bq because more time will have elapsed. **(b)** The activity in the year 2991 will be 0.107 Bq.]

Some related homework problems: Problem 38, Problem 39

As useful as carbon-14 dating is, it is limited to time spans of only a few half-lives— say, 10,000 to 15,000 y. Beyond that range, the current activity will be so small that accurate measurements will be difficult. To measure dates on different time scales, different radioactive isotopes must be used. Other frequently used isotopes and their half-lives are $^{210}_{82}\text{Pb}$ (22.3 y), $^{40}_{19}\text{K}$ (1.28×10^9 y), and $^{238}_{92}\text{U}$ (4.468×10^9 y).

Enhance Your Understanding (Answers given at the end of the chapter)

3. Nucleus 1 has a greater decay constant than nucleus 2. Is the half-life of nucleus 1 greater than, less than, or equal to the half-life of nucleus 2?

Section Review

- The number of radioactive atoms N at time t is given by $N = N_0 e^{-\lambda t}$. In this expression, λ is the decay constant, and N_0 is the initial number of atoms.

- The half-life of a radioactive element is $T_{1/2} = \ln 2/\lambda = 0.693/\lambda$.

- The activity, R, or rate of decay, of a radioactive element is $R = R_0 e^{-\lambda t}$. In this expression, R_0 is the initial value of the activity.

32-4 Nuclear Binding Energy

An α particle consists of two protons and two neutrons. Does it follow that the mass of an α particle is twice the mass of a proton plus twice the mass of a neutron? One would certainly think so, but in fact this is not the case. Alpha particles, and all other stable nuclei that contain more than one nucleon, have a mass that is *less* than the mass of the individual nucleons added together.

As strange as this result may seem, it is just one more manifestation of Einstein's theory of relativity. In particular, the reduction in mass of a nucleus, compared with

PHYSICS IN CONTEXT
Looking Back

The most famous equation in physics, $E = mc^2$ (Chapter 29), plays a key role in nuclear reactions. In general, the mass of nuclear decay products does not add up to the mass of the initial nucleus—instead, the mass difference appears as energy given off during the decay. $E = mc^2$ also occurs in the study of matter/antimatter annihilation.

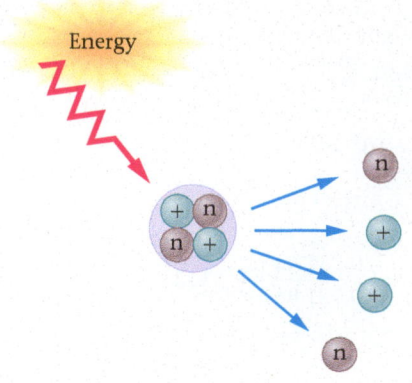

◄ **FIGURE 32-11 The concept of binding energy** The minimum energy that must be supplied to a stable nucleus to break it into its constituent nucleons is referred to as the binding energy. Because the stable nucleus is the lower-energy state, it has less mass than the sum of the masses of its individual constituents.

the mass of its constituents, corresponds to a reduction in its energy, according to the relationship $E = mc^2$. This reduction in energy is referred to as the **binding energy** of the nucleus. To separate a nucleus into its individual nucleons requires an energy at least as great as the binding energy; therefore, the binding energy indicates how firmly a given nucleus is held together. We illustrate this concept in **FIGURE 32-11**.

The next Example uses $E = (\Delta m)c^2$ and the atomic masses given in Appendix F to calculate the binding energy of tritium.

EXAMPLE 32-15 THE BINDING ENERGY OF TRITIUM

Using the information given in Appendix F, calculate the binding energy of tritium, $^3_1$H.

PICTURE THE PROBLEM

In our sketch we show energy being added to the nucleus of tritium, which consists of one proton and two neutrons. The result is one hydrogen atom and two neutrons.

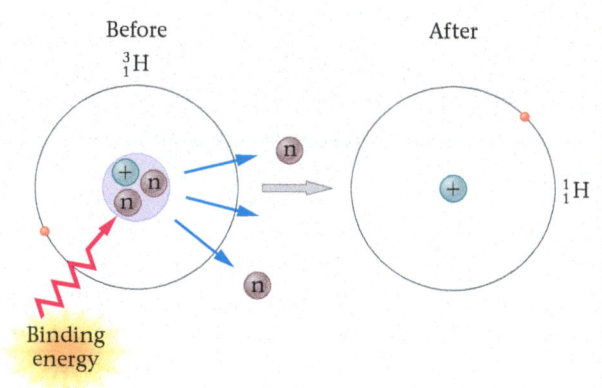

REASONING AND STRATEGY

To find the binding energy, we simply calculate Δm and multiply by c^2. The only point to be careful about is that, with the exception of the neutron, all the masses given in Appendix F are for neutral atoms. This means that the mass of tritium includes the mass of one electron. The same is true of the hydrogen atom, however. It follows that the electron mass will cancel when we calculate Δm.

Known   Mass of tritium = 3.016049 u; mass of hydrogen = 1.007825 u; mass of neutron = 1.008665 u.

Unknown   Binding energy of tritium, E = ?

SOLUTION

1. The initial mass is the mass of tritium:

$m_i = 3.016049$ u

2. The final mass is the mass of hydrogen plus two neutrons:

$m_f = 1.007825$ u $+ 2(1.008665$ u$) = 3.025155$ u

3. Calculate the change in mass, Δm:

$\Delta m = m_f - m_i$
$= 3.025155$ u $- 3.016049$ u $= 0.009106$ u

4. Find the corresponding binding energy (recall that 1 u = 931.5 MeV/c^2):

$E = (\Delta m)c^2 = (0.009106 \text{ u})\left(\dfrac{931.5 \text{ MeV}/c^2}{1 \text{ u}}\right)c^2$

$= 8.482$ MeV

INSIGHT

Thus, it takes about 8.5 MeV to separate the nucleons in a tritium nucleus. Compare this with the fact that only 13.6 eV is required to remove the electron from tritium. As we've noted before, nuclear processes typically involve energies in the MeV range, whereas atomic processes require energies in the eV range.

PRACTICE PROBLEM

Calculate the binding energy of an α particle.   [**Answer: 28.296 MeV**]

Some related homework problems: Problem 44, Problem 45

Binding Energy per Nucleon   In addition to knowing the binding energy of a given nucleus, it is also of interest to know the binding energy per nucleon. In the case of tritium, there are three nucleons; hence, the binding energy per nucleon is $\frac{1}{3}(8.482 \text{ MeV}) = 2.827$ MeV. **FIGURE 32-12** presents the binding energy per nucleon for various stable nuclei as a function of the mass number, A. Notice that the binding-energy

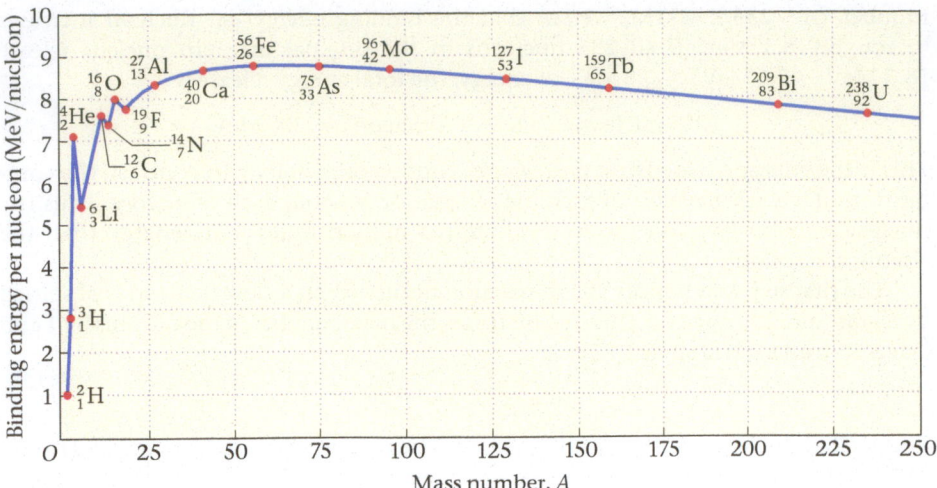

◄ **FIGURE 32-12 The curve of binding energy** This graph shows the binding energy per nucleon for a variety of nuclei. Notice that the binding energy per nucleon for tritium is approximately 2.8 MeV, in agreement with Example 32-15.

curve rises rapidly to a maximum near $A = 60$, and then decreases slowly to a value near 7.4 MeV per nucleon for larger A.

It follows that nuclei with A in the range $A = 50$ to $A = 75$ are the most stable nuclei in nature. In addition, the fact that the binding energy per nucleon changes very little for large A means that the strong nuclear force "saturates," in the sense that adding more nucleons does not add to the binding energy per nucleon. This phenomenon can be understood as a consequence of the short range of the nuclear force. Because of this short range, each nucleon interacts with only a few nucleons that are close neighbors to it in the nucleus. Nucleons on the other side of the nucleus are too far away to interact. Thus, from the point of view of a given nucleon, the attractive energy it feels due to other nucleons in the nucleus is essentially the same whether the nucleus has 150 nucleons or 200 nucleons—only the nearby nucleons interact.

The fact that the binding-energy curve has a maximum has important implications, as will be shown in the next two sections. For example, energy can be released when large nuclei split into smaller nuclei (fission), or when small nuclei combine to form a larger nucleus (fusion).

PROBLEM-SOLVING NOTE

Interpreting the Mass Difference

If the mass difference due to a reaction is negative, $\Delta m = m_f - m_i < 0$, it follows that energy is *released* by the reaction. If Δm is positive, energy must be *supplied* to the system to make the reaction occur.

Enhance Your Understanding   (Answers given at the end of the chapter)

4. Is the binding energy per nucleon of $^{238}_{92}U$ greater than, less than, or equal to the binding energy per nucleon of $^{96}_{42}Mo$?

Section Review

- All stable nuclei that contain more than one nucleon have a mass that is *less* than the mass of the individual nucleons added together. The difference in mass corresponds to the binding energy—that is, the energy required to separate the nucleons.

32-5 Nuclear Fission

A new type of physical phenomenon was discovered in 1939 when Otto Hahn and Fritz Strassman found that, under certain conditions, a uranium nucleus can split apart into two smaller nuclei. This process is called **nuclear fission.** As we shall see, nuclear fission releases an amount of energy that is many orders of magnitude greater than the energy released in chemical reactions. This fact has had a profound impact on the course of human events over the last several decades.

To obtain a rough estimate of just how much energy is released as a result of a fission reaction, consider the binding-energy curve in Figure 32-12. For a large nucleus like uranium-235, $^{235}_{92}U$, we see that the binding energy per nucleon is approximately 7.5 MeV. If this nucleus splits into two nuclei with roughly half the mass

Big Idea **4** Nuclear fission occurs when a large nucleus breaks up into smaller pieces. The reduction in mass that occurs during fission is converted to energy.

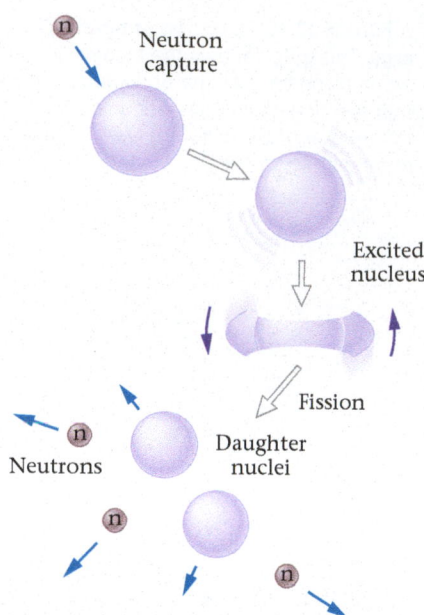

▲ **FIGURE 32-13 Nuclear fission** When a large nucleus captures a neutron, it may become excited and ultimately split into two smaller nuclei. This is the process of fission.

number $(A \sim 235/2 \sim 115)$, we see that the binding energy per nucleon increases to roughly 8.3 MeV. If all 235 nucleons in the original uranium nucleus release $(8.3 \text{ MeV} - 7.5 \text{ MeV}) = 0.8 \text{ MeV}$ of energy, the total energy release is

$$(235 \text{ nucleons})(0.8 \text{ MeV/nucleon}) = 200 \text{ MeV}$$

This is the energy released by a *single nucleus* undergoing fission. In contrast, the combustion of a *single molecule* of gasoline releases only about 2 eV of energy. Thus the nuclear reaction gives off about one hundred million times more energy than the chemical reaction!

The first step in a typical fission reaction occurs when a slow neutron is absorbed by a uranium-235 nucleus. This step increases the mass number of the nucleus by 1 and leaves it in an excited state:

$$_{0}^{1}\text{n} + _{92}^{235}\text{U} \longrightarrow _{92}^{236}\text{U}^{*}$$

The excited nucleus oscillates wildly and becomes highly distorted, as depicted in **FIGURE 32-13**. In many respects, the nucleus behaves like a spinning drop of water. Like a drop of water, the nucleus can distort only so much before it breaks apart into smaller pieces—that is, before it undergoes fission.

There are about 90 different ways in which the uranium-235 nucleus can fission. Typically, 2 or 3 neutrons (2.47 on average) are released during the fission process, in addition to the 2 smaller nuclei that are formed. The reason that neutrons are released can be seen by examining Figure 32-1, where we show N and Z for various nuclei. A large nucleus, like uranium-235, contains a higher percentage of neutrons than a smaller nucleus—that is, the larger nuclei deviate more from the $N = Z$ line. Thus, if $_{92}^{235}\text{U}$ were to simply break in two—keeping the same percentage of neutrons and protons in each piece—the smaller nuclei would have too many neutrons to be stable. As a result, neutrons are typically given off in a fission reaction.

One of the possible fission reactions for $_{92}^{235}\text{U}$ is considered in the following Example.

EXAMPLE 32-16 | A FISSION REACTION OF URANIUM-235

When uranium-235 captures a neutron, it may undergo the following fission reaction: $_{0}^{1}\text{n} + _{92}^{235}\text{U} \longrightarrow _{92}^{236}\text{U}^{*} \longrightarrow _{56}^{141}\text{Ba} + ? + 3_{0}^{1}\text{n}$. **(a)** Complete the reaction (that is, identify the nucleus represented by?) and **(b)** determine the energy it releases.

PICTURE THE PROBLEM
The indicated fission reaction is shown in our sketch, starting with the $_{92}^{236}\text{U}^{*}$ nucleus. The known numbers of protons and neutrons are indicated for both $_{92}^{236}\text{U}^{*}$ and $_{56}^{141}\text{Ba}$. The corresponding numbers for the unidentified nucleus are to be determined.

REASONING AND STRATEGY

a. The total number of protons must be the same before and after the reaction, as must the number of neutrons. By conserving protons and neutrons, we can determine the missing numbers on the unidentified nucleus.

b. As in other reactions, we calculate the difference in mass, $|\Delta m|$, and multiply by c^2. The same number of electrons appears on both sides of the reaction; hence, the electron mass will cancel when we calculate $|\Delta m|$.

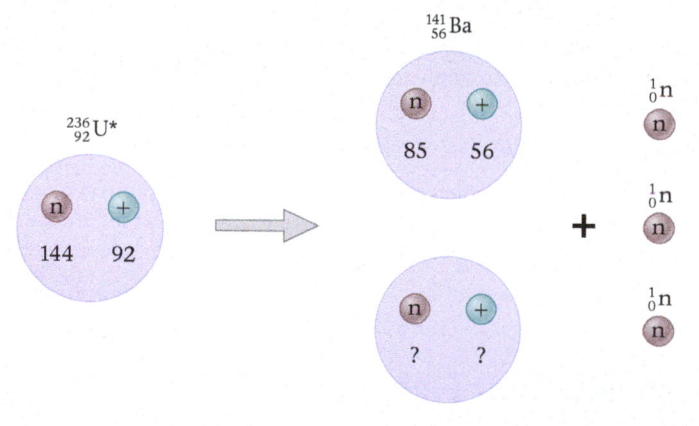

Known Fission reaction of uranium-235, $_{0}^{1}\text{n} + _{92}^{235}\text{U} \longrightarrow _{92}^{236}\text{U}^{*} \longrightarrow _{56}^{141}\text{Ba} + ? + 3_{0}^{1}\text{n}$.
Unknown **(a)** Unidentified nucleus in reaction. **(b)** Energy released, $E = ?$

SOLUTION

Part (a)

1. Determine the number of protons, Z, neutrons, N, and nucleons, A, in the unidentified nucleus. When calculating the number of neutrons, be sure to include the three individual neutrons given off in the reaction:

$Z = 92 - 56 = 36$
$N = 144 - 85 - 3 = 56$
$A = Z + N = 36 + 56 = 92$

CONTINUED

2. Use Z and A to determine the unidentified nucleus:

unidentified nucleus $= {}^{92}_{36}\text{Kr}$

Part (b)

3. Use Appendix F to find the initial mass of the system; that is, the mass of a ${}^{235}_{92}\text{U}$ atom plus the mass of a neutron:

$m_i = 235.043925 \text{ u} + 1.008665 \text{ u} = 236.052590 \text{ u}$

4. Use Appendix F to find the final mass of the system, including both nuclei and the three neutrons:

$m_f = 140.914406 \text{ u} + 91.926111 \text{ u} + 3(1.008665 \text{ u})$
$= 235.866512 \text{ u}$

5. Calculate the mass difference and the corresponding energy release (recall that 1 u = 931.5 MeV/c^2):

$\Delta m = m_f - m_i$
$= 235.866512 \text{ u} - 236.052590 \text{ u} = -0.186078 \text{ u}$

$$E = |\Delta m|c^2 = (0.186078 \text{ u})\left(\frac{931.5 \text{ MeV}/c^2}{1 \text{ u}}\right)c^2$$

$$= 173.3 \text{ MeV}$$

INSIGHT
Thus, as suggested by our crude calculation earlier in this section, the energy given off by a typical fission reaction is on the order of 200 MeV.

PRACTICE PROBLEM
Complete the following reaction, and determine the energy it releases: ${}^{1}_{0}\text{n} + {}^{235}_{92}\text{U} \longrightarrow {}^{236}_{92}\text{U}^* \longrightarrow {}^{140}_{54}\text{Xe} + ? + 2{}^{1}_{0}\text{n}$. [**Answer:** ${}^{94}_{38}\text{Sr}$; 184.7 MeV]

Some related homework problems: Problem 51, Problem 52

Chain Reactions

The fact that fission reactions of ${}^{235}_{92}\text{U}$ give off more than one neutron on average has significant implications. To see why, recall that the fission of ${}^{235}_{92}\text{U}$ is initiated by the absorption of a neutron in the first place. So the neutrons given off by one ${}^{235}_{92}\text{U}$ fission reaction may cause additional fission reactions in other ${}^{235}_{92}\text{U}$ nuclei. A reaction that proceeds from one nucleus to another in this fashion is referred to as a **chain reaction.**

Suppose, for the sake of discussion, that a fission reaction gives off two neutrons and that both neutrons induce additional fissions in other nuclei. These nuclei in turn give off two neutrons. Starting with one nucleus to begin the chain reaction, we have two nuclei in the second generation of the chain, four nuclei in the third generation, and so on, as indicated in **FIGURE 32-14**. After only 100 generations, the number of nuclei undergoing fission is 1.3×10^{30}. If each of these reactions gives off 200 MeV of energy, the total energy release after just 100 generations is 4.1×10^{19} J. To put this into everyday terms, this is enough energy to supply the needs of the entire United States for half a year. Clearly, a rapidly developing **runaway** chain reaction, like the one just described, would result in the explosive release of an enormous amount of energy.

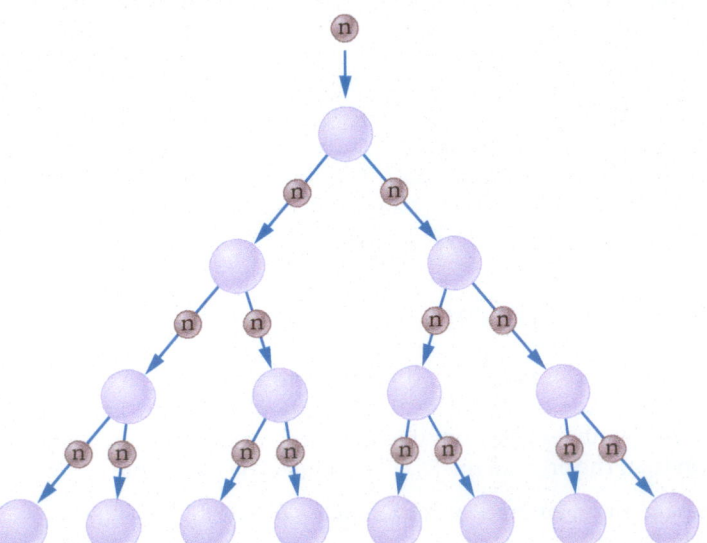

◄ **FIGURE 32-14 A chain reaction** A chain reaction occurs when a neutron emitted by one fission reaction induces a fission reaction in another nucleus. In a runaway chain reaction, the number of neutrons given off by one reaction that cause an additional reaction is greater than one. In the case shown here, two neutrons from one reaction induce additional reactions. In a controlled chain reaction, like those used in nuclear power plants, only one neutron, on average, induces additional reactions.

◀ **FIGURE 32-15 Visualizing Concepts Nuclear Reactors** (Top) The core of a nuclear reactor at a power plant. The core sits in a pool of water, which provides cooling while absorbing stray radiation. The blue glow suffusing the pool is radiation from electrons traveling through the water at relativistic velocities. Above the core is a crane used to replace the fuel rods when their radioactive material becomes depleted. (Center) A closer view of a reactor core. The structures projecting above the core house the mechanism that lowers and raises the control rods, thereby regulating the rate of fission in the reactor. The tubes that are nearly flush with the top of the core house the fuel rods. (Bottom) A technician loads pellets of fissionable material into fuel rods.

A great deal of effort has gone into controlling the chain reaction of $^{235}_{92}$U. If the release of energy can be kept to a manageable and usable level—avoiding explosions or meltdowns—it follows that a powerful source of energy is at our disposal.

RWP The first controlled nuclear chain reaction was achieved by Enrico Fermi in 1942, using a racquetball court at the University of Chicago for his improvised laboratory. His reactor consisted of blocks of uranium (the fuel) stacked together with blocks of graphite (the moderator) to form a large "pile." In such a **nuclear reactor,** the **moderator** slows the neutrons given off during fission, making it more likely that they will be captured by other uranium nuclei and cause additional fissions. The reaction rate is adjusted with **control rods** made of a material (such as cadmium) that is very efficient at absorbing neutrons. With the control rods fully inserted into the pile, any reaction that begins quickly dies out because neutrons are absorbed rather than allowed to cause additional fissions. As the control rods are pulled partway out of the pile, more neutrons become available to induce reactions. When, on average, one neutron given off by any fission reaction produces an additional reaction, the reactor is said to be **critical.** If the control rods are pulled out even farther from the pile, the number of neutrons causing fission reactions in the next generation becomes greater than one, and a runaway reaction begins to occur. Nuclear reactors that produce power for practical applications are operated near their critical condition by continuous adjustment of the placement of the control rods.

Fermi's original reactor, which was designed to test the basic scientific principles underlying fission reactions, produced a power of only about 0.5 W when operating near its critical condition. This is the power necessary to operate a flashlight bulb. In comparison, modern-day nuclear power reactors, like the ones shown in **FIGURE 32-15,** typically produce 1000 MW of power, enough to power an entire city. Because of their ability to produce a stable source of electrical power for a long period of time, nuclear reactors are also used to power space probes. For instance, both the *Voyager 1* spacecraft, the first man-made object to leave the solar system, and the *Cassini* spacecraft, in orbit around Saturn, are powered by the heat produced in the radioactive decay of plutonium-238, which is used to produce electricity by means of the *thermoelectric effect.*

Enhance Your Understanding   (Answers given at the end of the chapter)

5. Identify the missing nucleus in the following reaction: $^{1}_{0}$n + $^{235}_{92}$U ⟶ $^{236}_{92}$U* ⟶ $^{144}_{56}$Ba + ? + 3$^{1}_{0}$n.

Section Review

- Nuclear fission occurs when a large, unstable nucleus splits into two smaller nuclei. Neutrons are usually released during the fission process as well. The neutrons can cause additional fission reactions, resulting in a chain reaction.

32-6 Nuclear Fusion

When two light nuclei combine to form a more massive nucleus, the reaction is referred to as **nuclear fusion.** The binding-energy curve in Figure 32-12 shows that when light nuclei undergo fusion, the resulting nucleus has a greater binding energy per nucleon than the original nuclei. This implies the following:

> The large nucleus formed by fusion has less mass than the sum of the masses of the original small nuclei.

The mass difference appears as the energy ($E = mc^2$) given off by the reaction.

Thermonuclear Fusion   It's not easy to initiate a fusion reaction. The reason is that to combine two small nuclei—say, two protons—into a single nucleus requires that the initial nuclei have a kinetic energy great enough to overcome the Coulomb repulsion that protons exert on each other. The temperature required to give protons the needed kinetic energy is about 10^7 K. When the temperature is high enough to initiate fusion, we say that the resulting process is a **thermonuclear fusion reaction.**

One place where such high temperatures may be encountered is at the core of a star. In fact, all stars generate energy by the process of thermonuclear reactions. Most stars (including the Sun) fuse hydrogen to produce helium. At this very moment, the Sun is converting roughly 600,000,000 tons of hydrogen into helium every second. Some stars also fuse helium or other heavier elements.

In its early stages, a star begins as an enormous cloud of gas and dust. As the cloud begins to fall inward under the influence of its own gravity, it converts gravitational potential energy into kinetic energy. This means that the gas heats up. When the temperature becomes high enough to begin the fusion process, the resulting release of energy tends to stabilize the star, preventing its further collapse. A star like the Sun can burn its hydrogen for about 10 billion years, producing a remarkably stable output of energy so important to life on Earth. When its hydrogen fuel is depleted, the Sun will enter a new phase of its life, becoming a red giant and converting its fusion process to one that fuses helium. When the Sun enters its red-giant phase, it will expand greatly in size, eventually engulfing the Earth and vaporizing it.

RWP The Sun is powered by the **proton-proton cycle** of fusion reactions, as first described by Hans Bethe (1906–2005). This cycle consists of three steps. The first two steps are as follows:

$$^1_1H + {}^1_1H \longrightarrow {}^2_1H + e^+ + \nu_e$$

$$^1_1H + {}^2_1H \longrightarrow {}^3_2He + \gamma$$

A number of reactions are possible for the third step, but the dominant one is

$$^3_2He + {}^3_2He \longrightarrow {}^4_2He + {}^1_1H + {}^1_1H$$

The overall energy production during this cycle is about 27 MeV.

In the next Example we show how the mass difference in a deuterium-deuterium reaction can be used to calculate the amount of energy released.

QUICK EXAMPLE 32-17   THE DEUTERIUM-DEUTERIUM REACTION

Find the energy released in the deuterium-deuterium reaction, $^2_1H + {}^2_1H \longrightarrow {}^3_1H + {}^1_1H$.

REASONING AND SOLUTION

As is usual in studying nuclear reactions, we first calculate the change in mass, Δm, and then use the relativistic result, $E = \Delta mc^2$, to convert the change in mass to the corresponding energy.

1. Calculate the initial mass: $m_i = 2(2.014102\ u) = 4.028204\ u$

2. Calculate the final mass: $m_f = 3.016049\ u + 1.007825\ u = 4.023874\ u$

3. Find the difference in mass: $\Delta m = 4.023874\ u - 4.028204\ u = -0.004330\ u$

4. Convert the mass difference $E = |\Delta m|c^2 = (0.004330\ u)\left(\dfrac{931.5\ \text{MeV}/c^2}{1\ u}\right)c^2$
    to the corresponding
    energy release: $= 4.033\ \text{MeV}$

The tritium produced in this reaction has a half-life of 12.33 y.

RWP   Just as nuclear fission reactions can be controlled, allowing for the generation of usable power, it would be desirable to control fusion as well. In fact, there are many potential advantages of fusion over fission. For example, fusion reactions yield more energy from a given mass of fuel than fission reactions. In addition, one type of fuel for a fission reactor, ^2_1H, is readily obtained from seawater. To date, however, more energy

Big Idea 5 Nuclear fusion occurs when two light nuclei fuse together to form a larger nucleus. The reduction in mass that occurs during fusion is converted to energy.

▲ **FIGURE 32-16** Nuclear fusion, which powers the stars, may eventually turn out to be the clean, inexpensive, and renewable energy source that our society needs. But fusion requires enormous temperatures and pressures, and so far the problems involved in creating a practical fusion technology have not been overcome. Several different approaches have been explored in the effort to produce sustained nuclear fusion. One of them, embodied in the Z machine at Sandia National Laboratories in New Mexico, shown here, involves the compression of a plasma by an intense burst of X-rays. So far, temperatures of nearly 2 billion Kelvin have been attained, but only for very brief periods of time.

is required to produce sustained fusion reactions in the lab than is released by the reactions. Still, researchers are close to the break-even point, and many are confident that fusion reactors are feasible. An example of an experimental fusion reactor is shown in **FIGURE 32-16**.

Most attempts at controlled nuclear fusion employ one of two basic methods. In both methods the basic idea is to overcome the repulsive electrical forces that keep nuclei apart by giving them sufficiently large kinetic energies, which can be accomplished by heating the fuel to temperatures on the order of 10^7 K.

At temperatures like these, all atoms are completely ionized, forming a gas of electrons and nuclei known as a *plasma*. To initiate fusion in the plasma, one must maintain the plasma long enough for collisions to occur between the appropriate nuclei. This can be accomplished using a technique known as **magnetic confinement,** in which powerful magnetic fields confine a plasma within a "magnetic bottle." The trapped plasma is kept away from the walls of the container—preventing their melting—and is heated until fusion begins.

Another approach to reaching the high temperatures and pressures required for fusion is by way of **inertial confinement.** In this technique, one begins by dropping a small solid pellet of fuel into a vacuum chamber. When the pellet reaches the center of the chamber, it is bombarded from all sides by high-power laser beams. These beams heat and vaporize the surface of the pellet in an almost instantaneous event. The heating that causes fusion is still to come, however. As the vaporized exterior of the pellet expands rapidly outward, it exerts an inward "thrust" that causes the remainder of the pellet to implode. This implosion is so violent that the pellet's temperature and pressure rise to levels sufficient to ignite the desired fusion reactions. Global efforts have improved both magnetic confinement and inertial confinement techniques, but many obstacles remain to the production of electrical power by nuclear fusion.

Enhance Your Understanding   (Answers given at the end of the chapter)

6. Identify the missing nucleus in the following reaction: $^2_1H + ^2_1H \longrightarrow ? + ^1_1H + \gamma$.

Section Review

*   Nuclear fusion occurs when small nuclei combine to form a larger nucleus. Energy can be released in fusion reactions. In fact, the power generated by the Sun and other stars is produced by fusion reactions.

32-7 Practical Applications of Nuclear Physics

Nuclear physics, though it involves objects none of us will ever touch or see, nevertheless has significant impact on our everyday life. In this section we consider a number of ways in which nuclear physics affects us all, either directly or indirectly.

Biological Effects of Radiation

Although nuclear radiation can be beneficial when used to image our bodies, or to treat a variety of diseases, it can also be harmful to living tissues. For example, the high-energy photons, electrons, and α particles given off by radioactive decay can ionize a neutral atom by literally "knocking" one or more of its electrons out of the atom. In fact, because typical nuclear decays give off energies in the MeV range, and typical ionization energies are in the 10 eV range, it follows that a single α, β, or γ particle can ionize thousands of atoms or molecules.

RWP Such ionization can be harmful to a living cell by altering the structure of some of its molecules. The result can be a cell that no longer functions or behaves normally, or even a cell that will soon die. This effect is the basis for radiation treatments of cancer, as in **FIGURE 32-17 (a)**, which seek to concentrate high doses of radiation on a cancerous tumor in order to kill the malignant cells. But just what is a "high dose," and what units do we use to measure a person's radiation dose with a radiation badge like the one shown in **FIGURE 32-17 (b)**?

The first radiation unit to be defined, called the **roentgen** (R), is directly related to the amount of ionization caused by X-rays or γ rays. Suppose such radiation is sent through a mass, m, of dry air at standard temperature and pressure (STP). If the radiation creates ions in the air with a total charge of q, we say that the dose of radiation delivered to the air is proportional to q/m. Specifically, we say that the dose of X-rays or γ rays is 1 R if an ionization charge of 2.58×10^{-4} C is produced in 1 kg of dry air.

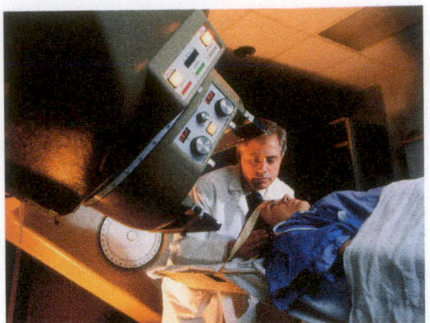

(a)

Definition of the Roentgen, R

$$1 \text{ R} = 2.58 \times 10^{-4} \text{ C/kg} \quad \text{(X-rays or } \gamma \text{ rays in dry air at STP)} \qquad \text{32-15}$$

SI unit: C/kg

The roentgen is one key measure of radiation dose, but there are others. The focus of the **rad,** an acronym for **rad**iation **a**bsorbed **d**ose, is the amount of energy that is absorbed by the irradiated material, regardless of the type of radiation that delivers the energy. For example, if a 1-kg sample of any material absorbs 0.01 J of energy, we say that it has received a dose of 1 rad:

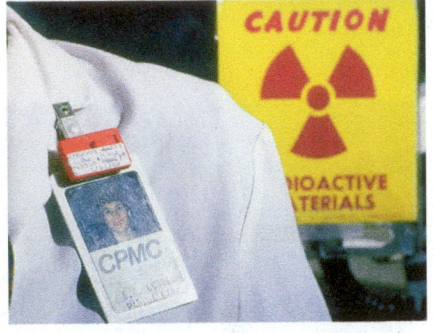

(b)

Definition of the Rad

$$1 \text{ rad} = 0.01 \text{ J/kg} \quad \text{(any type of radiation)} \qquad \text{32-16}$$

SI unit: J/kg

Because the rad depends only on the amount of energy absorbed, and not on the type of radiation, more information is needed if we are to have an indication of the biological effect a certain dose will produce. For example, a 1-rad dose of X-rays is far less likely to cause a cataract in the cornea of the eye than a 1-rad dose of neutrons. To take into account such differences, we introduce a quantity called the **relative biological effectiveness,** or **RBE.** The standard used for comparison is the biological effect produced by a 1-rad dose of 200-keV X-rays. Thus, the RBE is defined as follows:

Definition of Relative Biological Effectiveness, RBE

$$\text{RBE} = \frac{\text{the dose of 200-keV X-rays necessary to produce a given biological effect}}{\text{the dose of a particular type of radiation necessary to produce the same biological effect}} \qquad \text{32-17}$$

SI unit: dimensionless

▲ **FIGURE 32-17** Ironically, the fact that radiation can be lethal to cells also makes it a valuable therapeutic tool in the fight against cancer. **(a)** Radiation from cobalt-60 is being used to attack a malignant tumor. Today, many hospitals have their own particle accelerators to manufacture short-lived radioactive isotopes for the treatment of different kinds of cancer or to use as tracers. **(b)** Because radiation can kill or damage living cells, people who work with radioactive materials or other sources of ionizing radiation (such as X-ray tubes) must be careful to monitor their cumulative exposure. One way to do this is with a simple film badge like the one shown here. When the photographic film is developed, the degree of darkening indicates the amount of radiation it has received.

Representative RBE values are given in Table 32-3. Notice that the larger the RBE for a certain type of radiation, the greater the biological damage caused by a given dose of that radiation.

Combining the dose in rad and the RBE value for a given radiation yields a new unit of radiation referred to as the **biologically equivalent dose.** This unit is measured in **rem**, which stands for **r**oentgen **e**quivalent in **m**an. To be specific, the dose in rem is defined as follows:

Definition of Roentgen Equivalent in Man, rem

$$\text{dose in rem} = \text{dose in rad} \times \text{RBE} \qquad \text{32-18}$$

SI unit: J/kg

Defined in this way, 1 rem of radiation produces the same amount of biological damage no matter which type of radiation is involved. In addition, the larger the dose in rem, the greater the biological damage. Referring to Table 32-3, we see that 1 rad of 200-keV X-rays produces a radiation dose of 1 rem, whereas 1 rad of protons produces a dose of 10 rem. The dose of radiation we receive per year from cosmic rays is about 28 mrem = 0.028 rem. Other typical values of radiation doses are presented in Table 32-4. (For comparison, a dose of 50 to 100 rem damages blood-forming tissues, whereas a dose of 500 rem usually results in death.)

TABLE 32-3 Relative Biological Effectiveness, RBE, for Different Types of Radiation

Type of radiation	RBE
Heavy ions	20
α rays	10–20
Protons	10
Fast neutrons	10
Slow neutrons	4–5
β rays	1.0–1.7
γ rays	1
200-kev X-rays	1

TABLE 32-4 Typical Radiation Doses

Source of radiation	Radiation dose (rem/y)
Inhaled radon	~0.200
Medical/dental examinations	0.040
Cosmic rays	0.028
Natural radioactivity in the Earth and atmosphere	0.028
Nuclear medicine	0.015

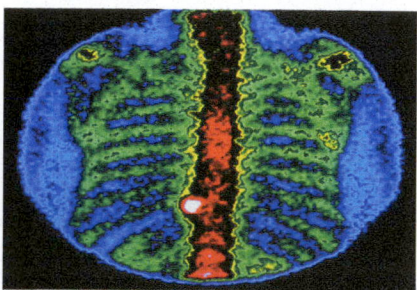

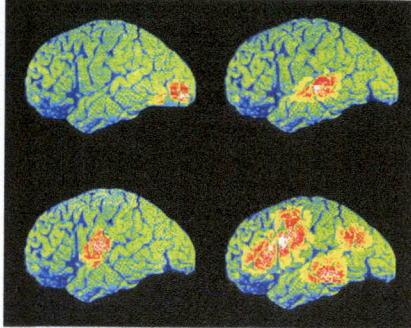

▲ **FIGURE 32-18** Whereas X-ray images are created by passing a beam of radiation through the body, other imaging techniques make use of radiation produced within the body. One method involves the use of radioactive isotopes that tend to become concentrated in particular tissues or structures, such as the thyroid gland or the bones. Radiation from that component can then be measured, or even used to create an image, as in the bone scan at top, where areas of abnormally high radiation (white) indicate the presence of malignancy. A sophisticated variant of this approach is used to produce PET scans, which are most commonly employed to visualize areas of the body where cellular energy production is most intense. The series of images below records brain activity associated with seeing words (upper left), hearing words (upper right), speaking words (lower left), and thinking about words while uttering them (lower right). (The image of the brain, superimposed for reference, is not part of the PET scan.)

EXERCISE 32-18 FIND THE DOSE

A biological sample receives a dose of 436 rad from neutrons with an RBE of 6.17. **(a)** Find the dose in rem. **(b)** If this same dose in rem is to be delivered by α rays with an RBE of 12.5, what dose in rad is required?

REASONING AND SOLUTION

a. The dose in rem is the product of the dose in rad and the RBE:

$$(436\text{ rad})(6.17) = 2690\text{ rem}$$

b. The dose in rad is the dose in rem divided by the RBE:

$$\frac{2690\text{ rem}}{12.5} = 215\text{ rad}$$

Radioactive Tracers

RWP The phenomenon of radioactivity has found applications of many types in medicine and biology. One of these applications employs a radioactive isotope as a sort of "identification tag" or "tracer" to determine the location and quantity of a substance of interest.

For example, an artificially produced radioactive isotope of iodine, $^{131}_{53}\text{I}$, is used to determine the condition of a person's thyroid gland. A healthy thyroid plays an important role in the distribution of iodine—a necessary nutrient—throughout the body. To see if the thyroid is functioning as it should, a patient drinks a small quantity of radioactive sodium iodide, which incorporates the isotope $^{131}_{53}\text{I}$. A couple hours later, the radiation given off by the patient's thyroid gland is measured, giving a direct indication of the amount of sodium iodide the gland has processed.

Because radioactive tracers differ from their nonradioactive counterparts only in the composition of their nuclei, they undergo the same chemical and metabolic reactions. This makes them useful in diagnosing and treating a variety of different conditions. For example, chromium-51 is used to label red blood cells and to quantify gastrointestinal protein loss, copper-64 is used to study genetic diseases affecting copper metabolism, and yttrium-90 is used for liver cancer therapy.

PET Scans

RWP Perhaps the most amazing medical use of radioactive decay is in the diagnostic technique known as **positron emission tomography** (PET). Like the images made using radioactive tracers, PET scans are produced with radiation that *emerges from within* the body, as opposed to radiation that is generated externally and is then passed through the body. Remarkably, the radiation in a PET scan is produced by the *annihilation of matter and antimatter* within the patient's body! It sounds like science fiction, but it is, in fact, a valuable medical tool.

To produce a PET scan, a patient is given a radiopharmaceutical, like fluorodeoxyglucose (FDG), which contains an atom that decays by giving off a positron (the antiparticle to the electron). In the case of FDG, a fluorine-18 atom is attached to a molecule of glucose, the basic energy fuel of cells. The fluorine-18 atom undergoes the following decay with a half-life of 110 minutes:

$$^{18}_{9}\text{F} \longrightarrow {}^{18}_{8}\text{O} + e^+ + \nu_e$$

Almost immediately after the positron is emitted in this decay, it encounters an electron, and the two particles annihilate each other in a burst of energy. In fact, the annihilation process generates two powerful γ rays moving in opposite directions. As these γ rays emerge from the body, specialized detectors observe them and determine their point of origin. The resulting computerized image shows the areas in the body where glucose metabolism is most intense.

PET scans are particularly useful in examining the brain. For example, a PET scan using FDG can show which regions of the brain are most active when a person is performing a specific mental task, like counting, speaking, or translating a foreign language. An example is shown in **FIGURE 32-18**. The scans can also show abnormality in brain function, as when one side of a brain becomes more active than the other during an epileptic seizure. Finally, PET scans can be used to locate tumors in the brain and other parts of the body, and to monitor the progress of their treatment.

Magnetic Resonance Imaging

RWP A diagnostic technique that gives images similar to those obtained with computerized tomography (Section 31-7) is **magnetic resonance imaging,** or MRI. The basic physical mechanism used in MRI is the interaction of a nucleus with an externally applied magnetic field.

Consider the simplest nucleus—a single proton. Protons have a magnetic moment, like a small bar magnet, and when they are placed in a constant magnetic field, they precess about the field with a frequency proportional to the field strength. In addition, protons have a spin of one-half, like electrons; hence, their spin can be either "up" or "down" relative to the direction of the constant magnetic field. When an oscillating magnetic field is applied perpendicular to the constant magnetic field, it can cause protons to "flip" from one spin state to the other if the frequency of oscillation is in resonance with the proton's precessional frequency. These spin flips result in the absorption or release of energy in the radio portion of the electromagnetic spectrum, which can be detected electronically.

By varying the strength of the constant magnetic field as a function of position, spin flips can be detected in various parts of the body being examined. Using a computer to combine signals from various positions allows for the generation of detailed cross-sectional images, as illustrated in **FIGURE 32-19**.

One of the advantages of MRI is that the photons associated with the magnetic fields used in the imaging process are very low in energy. In fact, typical energies are about 10^{-7} eV, much less than typical ionization energies; hence, MRI causes very little cellular damage. In contrast, photons used in CAT scans have energies that range from 10^4 to 10^6 eV, more than enough to produce cellular damage.

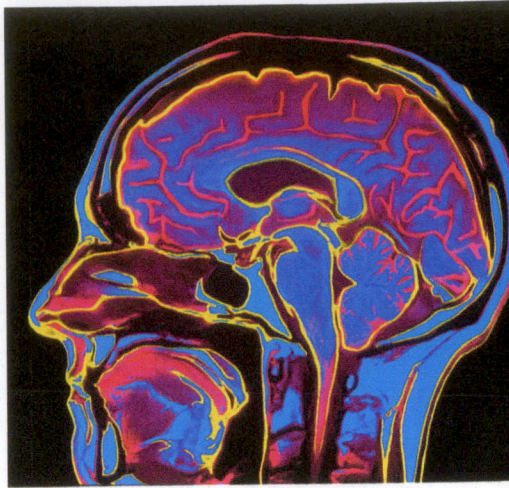

▲ **FIGURE 32-19** Magnetic resonance imaging (MRI) is a safe, noninvasive technique for visualizing internal body structures. No ionizing radiation is involved, so the risk of tissue damage is minimal. MRI images generally show soft tissue with greater clarity than do ordinary X-rays.

Enhance Your Understanding   (Answers given at the end of the chapter)

7. Is the biological damage caused by 1 rem of alpha radiation greater than, less than, or equal to the biological damage caused by 1 rem of gamma radiation?

Section Review

- Radiation from nuclear decay can have beneficial effects, but can also be harmful. The unit used to measure the damage caused by radiation is the rem.

32-8 Elementary Particles

Scientists have long sought to identify the fundamental building blocks of all matter, the **elementary particles.** At one point, it was thought that atoms were elementary particles—one for each element. As we saw in the previous chapter, however, this idea was put to rest when atoms were discovered to be made up of electrons, protons, and neutrons. Of these three particles, only the electron is presently considered to be elementary— protons and neutrons are now known to be composed of still smaller particles. In addition, approximately 300 new particles were discovered in the last half of the twentieth century, most of which are unstable and have lifetimes of only 10^{-6} to 10^{-23} s.

Although a complete accounting of the current theories of elementary particles is beyond the scope of this text, we shall outline the basic insights that have been derived from these theories and from related experiments. We begin by describing the four fundamental forces of nature, since particles can be categorized according to which of these forces they experience.

The Fundamental Forces of Nature

Although nature presents us with a myriad of different physical phenomena—from tornadoes and volcanoes to sunspots and comets to galaxies and black holes—all are

TABLE 32-5 The Fundamental Forces

Force	Relative Strength	Range
Strong nuclear	1	≈ 1 fm
Electromagnetic	10^{-2}	Infinite $(\alpha\ 1/r^2)$
Weak nuclear	10^{-6}	$\approx 10^{-3}$ fm
Gravitational	10^{-43}	Infinite $(\alpha\ 1/r^2)$

the result of just *four* fundamental forces. This is one example of the simplicity that physicists see in nature. These forces, in order of diminishing strength, are the strong nuclear force, the electromagnetic force, the weak nuclear force, and gravity. If we assign a strength of 1 to the strong nuclear force, for purposes of comparison, then the strength of the electromagnetic force is 10^{-2}, the strength of the weak nuclear force is 10^{-6}, and the strength of the gravitational force is an incredibly tiny 10^{-43}. These results are summarized in Table 32-5.

All objects of finite mass experience gravitational forces. This is one reason that gravity is such an important force in the universe, even though it is spectacularly weak. Similarly, objects with a finite charge experience electromagnetic forces. As for the weak and strong nuclear forces, some particles experience only the weak force, whereas others experience both the weak and the strong forces. We turn now to a discussion of particles that fall into these latter two categories.

Big Idea **6** There are four fundamental forces known in nature—gravity, the weak nuclear force, the electromagnetic force, and the strong nuclear force. The weak and strong nuclear forces act within nuclei; the gravitational and electromagnetic forces have infinite range.

Leptons

Particles that are acted on by the weak nuclear force but not by the strong nuclear force are referred to as **leptons.** There are only six leptons known to exist, all of which are listed in Table 32-6. The most familiar of these are the electron and its corresponding neutrino—both of which are stable. No internal structure has ever been detected in any of these particles. As a result, all six leptons have the status of true elementary particles.

TABLE 32-6 Leptons

Particle	Particle Symbol	Antiparticle Symbol	Rest Energy (MeV)	Lifetime (s)
Electron	e^- or β^-	e^+ or β^+	0.511	Stable
Muon	μ^-	μ^+	105.7	2.2×10^{-6}
Tau	τ^-	τ^+	1777	10^{-13}
Electron neutrino	ν_e	$\bar{\nu}_e$	≈ 0	Stable
Muon neutrino	ν_μ	$\bar{\nu}_\mu$	≈ 0	Stable
Tau neutrino	ν_τ	$\bar{\nu}_\tau$	≈ 0	Stable

The weak nuclear force is responsible for most radioactive decay processes, such as beta decay. It is also a force of extremely short range. In fact, the weak force can be felt only by particles that are separated by roughly one-thousandth the diameter of a nucleus. Beyond that range, the weak force has practically no effect at all.

Hadrons

Hadrons are particles that experience both the weak and the strong nuclear forces. They are also acted on by gravity, since all hadrons have finite mass. The two most familiar hadrons are the proton and the neutron. A partial list of the hundreds of hadrons known to exist is given in Table 32-7. Notice that the proton is the only stable hadron (though some theories suggest that even it may decay with the incredibly long half-life of 10^{35} y).

The strong nuclear force is the only force powerful enough to hold a nucleus together. It is a short-range force, extending only to distances comparable to the diameter of a nucleus, but within that range it is strong enough to counteract the intense electromagnetic repulsion between positively charged protons. Outside the nucleus, however, the strong nuclear force is of negligible strength.

In striking contrast with leptons, none of the hadrons are elementary particles. In fact, all hadrons are composed of either two or three smaller particles called **quarks.** Hadrons formed from two quarks are referred to as **mesons;** those formed from three quarks are **baryons.** The properties of quarks will be considered next.

Quarks

To account for the internal structure observed in hadrons, Murray Gell-Mann (1929–) and George Zweig (1937–) independently proposed in 1964 that all hadrons are composed of

TABLE 32-7 Hadrons

Particle		Particle Symbol	Antiparticle Symbol	Rest Energy (MeV)	Lifetime (s)
MESONS					
	Pion	π^+	π^-	139.6	2.6×10^{-8}
		π^0	π^0	135.0	0.8×10^{-16}
	Kaon	K^+	K^-	493.7	1.2×10^{-8}
		K^0_S	$\overline{K}^0_S$	497.7	0.9×10^{-10}
		K^0_L	$\overline{K}^0_L$	497.7	5.1×10^{-8}
	Eta	η^0	η^0	547.9	$<10^{-18}$
BARYONS					
	Proton	p	$\overline{p}$	938.3	Stable
	Neutron	n	$\overline{n}$	939.6	880
	Sigma	Σ^+	$\overline{\Sigma}^-$	1189	0.8×10^{-10}
		Σ^0	$\overline{\Sigma}^0$	1193	7×10^{-20}
		Σ^-	$\overline{\Sigma}^+$	1197	1.5×10^{-10}
	Omega	Ω^-	Ω^+	1672	0.8×10^{-10}

a number of truly elementary particles that Gell-Mann dubbed quarks. Originally, it was proposed that there are three types of quarks, arbitrarily named up (u), down (d), and strange (s). Discoveries of new and more massive hadrons, such as the J/ψ particle discovered in 1974, have necessitated the addition of three more quarks. The equally whimsical names for these new quarks are charmed (c), top or truth (t), and bottom or beauty (b). Table 32-8 lists the six quarks, along with some of their more important properties.

TABLE 32-8 Quarks and Antiquarks

Name	Rest Energy (MeV)	Quarks		Antiquarks	
		Symbol	Charge	Symbol	Charge
Up	360	u	$+\frac{2}{3}e$	$\overline{u}$	$-\frac{2}{3}e$
Down	360	d	$-\frac{1}{3}e$	$\overline{d}$	$+\frac{1}{3}e$
Charmed	1500	c	$+\frac{2}{3}e$	$\overline{c}$	$-\frac{2}{3}e$
Strange	540	s	$-\frac{1}{3}e$	$\overline{s}$	$+\frac{1}{3}e$
Top	173,000	t	$+\frac{2}{3}e$	$\overline{t}$	$-\frac{2}{3}e$
Bottom	5000	b	$-\frac{1}{3}e$	$\overline{b}$	$+\frac{1}{3}e$

The antiparticles to the quarks are also given in Table 32-8. Notice that antiquarks are indicated with a bar over the symbol for the corresponding quark. For example, the symbol for the up quark is u; the symbol for the corresponding antiquark is $\overline{u}$.

Quarks are unique among the elementary particles in a number of ways. For example, they all have charges that are fractions of the charge of the electron. As can be seen in Table 32-8, some quarks have a charge of $+\left(\frac{2}{3}\right)e$ or $-\left(\frac{2}{3}\right)e$; others have a charge of $+\left(\frac{1}{3}\right)e$ or $-\left(\frac{1}{3}\right)e$. No other particles are known to have charges that differ from integer multiples of the electron's charge.

Now it might seem that the fractional charge of a quark would make it easy to identify experimentally. In fact, a number of experiments have searched for quarks in just that way, by looking for particles with fractional charge. No such particle has ever been observed, however. It is now believed that a free, independent quark cannot exist; quarks must always be bound with other quarks. This concept is referred to as **quark confinement.** The physical reason behind confinement is that the force between two quarks

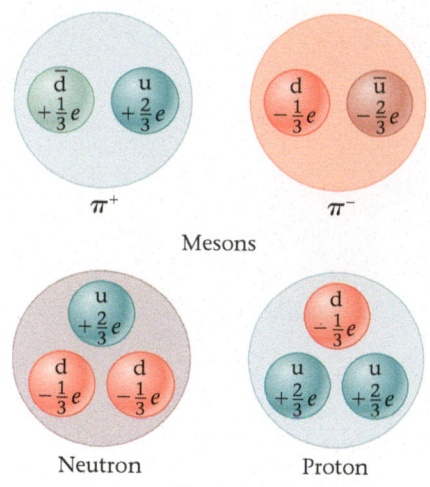

▲ **FIGURE 32-20 The quark composition of mesons and baryons** Mesons and baryons are composed of various quark combinations—mesons always have a quark and an antiquark, baryons always have three quarks. Note that even though quarks have fractional charges (in units of the electron charge, e), the resulting mesons and baryons always have integer charges.

TABLE 32-9 Quark Composition of Some Hadrons

Particle	Quark Composition
MESONS	
π^+	$u\bar{d}$
π^-	$\bar{u}d$
K^+	$u\bar{s}$
K^-	$\bar{u}s$
K^0	$d\bar{s}$
BARYONS	
p	uud
n	udd
Σ^+	uus
Σ^0	uds
Σ^-	dds
Ξ^0	uss
Ξ^-	dss
Ω^-	sss

increases with separation—like two particles connected by a spring. Hence, an infinite amount of energy is required to increase the separation between two quarks to infinity.

The smallest system of bound quarks that can be observed as an independent particle is a pair of quarks. In fact, **mesons** consist of bound pairs of quarks and antiquarks, as illustrated schematically in **FIGURE 32-20**. For example, the π^+ meson is composed of a $u\bar{d}$ pair of quarks. Note that this combination of quarks gives the π^+ meson a net charge of $+e$. The π^- meson, the antiparticle to the π^+ meson, consists of a $\bar{u}d$ pair with a charge of $-e$. Quarks are always bound in configurations that result in charges that are integer multiples of e.

Baryons are bound systems consisting of three quarks, as shown in Figure 32-20. The proton, for example, has the composition uud, with a net charge of $+e$. The neutron, on the other hand, is formed from the combination udd, with a net charge of 0. A variety of hadrons and their corresponding quark compositions are given in Table 32-9.

Finally, not long after the quark model of elementary particles was introduced, it was found that some quark compositions implied a violation of the Pauli exclusion principle. To resolve these discrepancies, it was suggested that quarks must come in three different varieties, which were given the completely arbitrary but colorful names red, green, and blue. Though these quark "colors" have nothing to do with visible colors in the electromagnetic spectrum, they bring quarks into agreement with the exclusion principle and explain other experimental observations that were difficult to understand before the introduction of this new property. The theory of how colored quarks interact with one another is called **quantum chromodynamics**, or QCD, in analogy with the theory describing interactions between charged particles, which is known as **quantum electrodynamics**, or QED.

RWP One prediction of the theory of elementary particles is the existence of a particle called the *Higgs boson*. First predicted in 1964, the existence of the Higgs boson would solve a mystery about why certain other particles have mass when aspects of particle theory suggest they should be massless. It would also help explain why the weak nuclear force has a much shorter range than the electromagnetic force. However, the Higgs boson requires a very large energy to become observable, and it lasts for only 10^{-22} second before decaying into an assortment of other particles. After a search spanning several decades, on March 14, 2013, scientists at the European Organization for Nuclear Research (CERN) announced that they had indeed observed a particle consistent with the predicted properties of the Higgs boson. Nuclear physicists around the world celebrated the discovery as a triumph of the theory of elementary particles.

Enhance Your Understanding   (Answers given at the end of the chapter)

8. How many quarks make up the following particles: **(a)** electron, **(b)** proton, **(c)** neutron, **(d)** pion?

Section Review

- Elementary particles come in two basic varieties. Leptons experience the weak nuclear force, but not the strong nuclear force. Hadrons experience both the strong and weak nuclear forces. Hadrons are composed of smaller particles known as quarks. Mesons are formed from two quarks. Baryons, like protons and neutrons, are formed from three quarks.

32-9 Unified Forces and Cosmology

As discussed earlier, the universe as we see it today has four fundamental forces through which various particles interact with one another. This has not always been the case, however. Shortly after the Big Bang these four forces were combined into a single force sometimes referred to as the **unified force.** This situation lasted for only a brief interval of time. As the early universe expanded and cooled, it eventually underwent a type

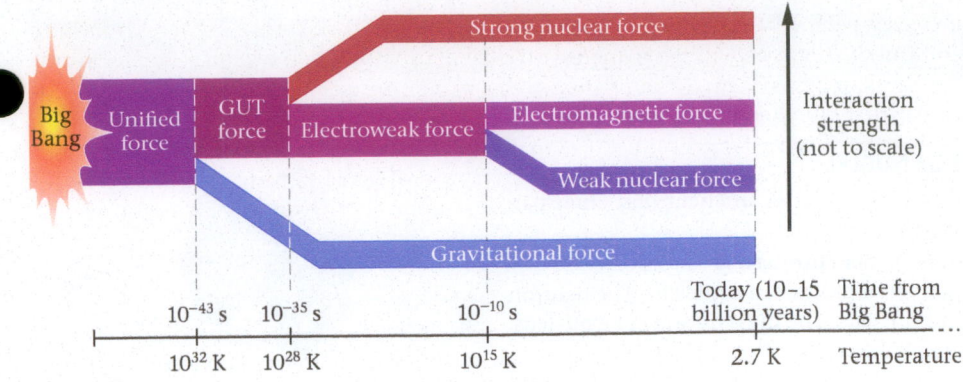

of "phase transition" in which the gravitational force took on a separate identity. This transition occurred at a time of approximately 10^{-43} s after the Big Bang, when the temperature of the universe was about 10^{32} K.

The phase transition just described was the first of three such transitions to occur in the early universe, as we see in **FIGURE 32-21**. At 10^{-35} s, when the temperature was 10^{28} K, the strong nuclear force became a separate force. Similarly, the weak nuclear force became a separate force at 10^{-10} s, when the temperature was 10^{15} K. From 10^{-10} s until the present, the situation has remained the same, even as the temperature of the universe has dipped to a chilly 2.7 K.

Let's look at these forces and the transitions between them more carefully. First, the electromagnetic force combines the forces associated with both electricity and magnetism. Although electricity and magnetism were originally thought to be separate forces, the work of Maxwell and others showed that these forces are simply different aspects of the same underlying force. For example, changing electric fields generate magnetic fields, and changing magnetic fields generate electric fields. In fact, the theory of electromagnetism can be thought of as the first *unified field theory*, in which seemingly different forces are combined into one all-encompassing theory.

At times earlier than 10^{-10} s the weak nuclear force was indistinguishable from the electromagnetic force. Thus, even though these forces seem very different today, we can recognize them as different aspects of the same underlying force—much like the two faces of a coin look very different but are part of the same physical object. The theory that encompasses the weak nuclear force and the electromagnetic force is called the **electroweak theory.** It was developed by Sheldon Glashow, Abdus Salam, and Steven Weinberg.

Going further back in time, the strong nuclear force was indistinguishable from the electroweak force before 10^{-35} s. Although no one has yet succeeded in producing a theory that combines the electroweak force and the strong nuclear force, most physicists feel confident that such a theory exists. This hypothetical theory is referred to as the **grand unified theory,** or GUT.

Finally, a theory that encompasses gravity along with the other forces of nature is one of the ultimate goals of physics. Many physicists, including Einstein in his later years, have worked long and hard toward such an end, but so far with little success.

PHYSICS IN CONTEXT
Looking Back

The evolution of the universe has been marked by a series of phase transitions as it cooled after the Big Bang. These phase transitions are similar to those studied in Chapter 17, and have resulted in an original single force splitting into the four distinct types of forces we see in the universe today.

Enhance Your Understanding (Answers given at the end of the chapter)

9. How many fundamental forces of nature existed immediately after the Big Bang?

Section Review

- Electromagnetism combines the effects of electricity and magnetism in one theory. The electroweak theory combines electromagnetism with the weak nuclear force. The hypothetical grand unified theory would combine the electroweak theory with the strong nuclear force. A variety of theories have been introduced in an attempt to combine gravity with the other forces of nature.

CHAPTER 32 REVIEW

CHAPTER SUMMARY

32-1 THE CONSTITUENTS AND STRUCTURE OF NUCLEI

Nuclei are composed of just two types of particles: protons and neutrons. These particles are referred to collectively as nucleons.

Atomic Number, Z; Neutron Number, N; Mass Number, A

The atomic number, Z, is equal to the number of protons in a nucleus. The neutron number, N, is the number of neutrons in a nucleus. The mass number of a nucleus, A, is the total number of nucleons it contains. Thus, $A = N + Z$.

Designation of Nuclei

A nucleus with atomic number Z and mass number A is designated as

$$^{A}_{Z}X$$

Isotopes

Isotopes are nuclei with the same atomic number but different neutron number.

32-2 RADIOACTIVITY

Radioactivity refers to the emissions observed when an unstable nucleus changes its composition or when an excited nucleus decays to a lower-energy state.

Alpha Decay

In alpha decay, an α particle (which consists of two protons and two neutrons) is emitted.

Beta Decay

Beta decay refers to the emission of an electron, as when a neutron decays into a proton, an electron, and an antineutrino.

Gamma Decay

Gamma decay occurs when an excited nucleus drops to a lower-energy state and emits a photon.

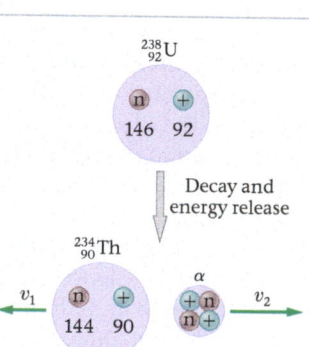

32-3 HALF-LIFE AND RADIOACTIVE DATING

Radioactive nuclei decay with time in a well-defined way. As a result, many radioactive nuclei can be used as a type of "nuclear clock."

Nuclei as a Function of Time and the Decay Constant, λ

If the number of radioactive nuclei in a sample at time $t = 0$ is N_0, then the number, N, at a later time is

$$N = N_0 e^{-\lambda t} \qquad \text{32–9}$$

The constant, λ, in this expression is referred to as the decay constant.

Half-life, $T_{1/2}$

The half-life of a radioactive material is the time required for half of its nuclei to decay. In terms of the decay constant, the half-life is

$$T_{1/2} = \frac{\ln 2}{\lambda} = \frac{0.693}{\lambda} \qquad \text{32–10}$$

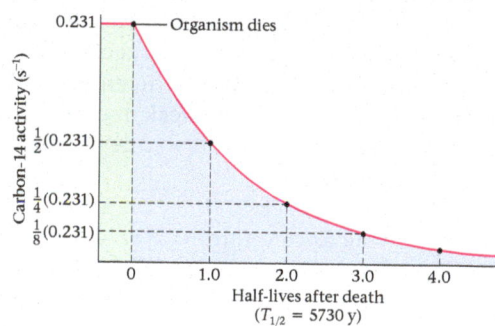

32-4 NUCLEAR BINDING ENERGY

The binding energy of a nucleus is the energy that must be supplied to separate it into its component nucleons.

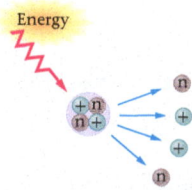

32-5 NUCLEAR FISSION

Nuclear fission is the process in which a large nucleus captures a neutron and then divides into two smaller "daughter" nuclei.

Chain Reactions

When neutrons given off by one fission reaction initiate additional fission reactions, we refer to the process as a chain reaction.

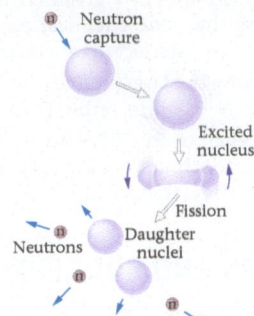

32-6 NUCLEAR FUSION

Nuclear fusion occurs when two small nuclei merge to form a larger nucleus.

32-7 PRACTICAL APPLICATIONS OF NUCLEAR PHYSICS

Nuclear radiation can have both harmful and useful effects.

32-8 ELEMENTARY PARTICLES

Elementary particles are the fundamental building blocks of all matter.

The Fundamental Forces of Nature

There are just four fundamental forces in nature. In order of decreasing strength, they are the strong nuclear force, the electromagnetic force, the weak nuclear force, and the gravitational force.

Leptons

Leptons are elementary particles that experience the weak nuclear force but not the strong nuclear force.

Hadrons

Hadrons are composite particles that experience both the weak and the strong nuclear forces.

Quarks

Quarks are elementary particles that combine to form hadrons. Mesons are formed from quark-antiquark pairs; baryons are formed from combinations of three quarks.

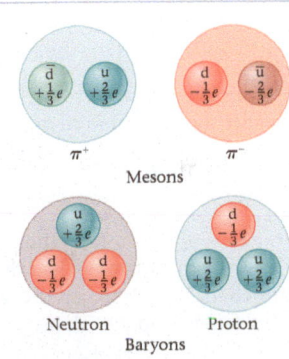

32-9 UNIFIED FORCES AND COSMOLOGY

The four fundamental forces observed in the universe today began as a single force at the time of the Big Bang. As the universe expanded and cooled, the single force split into four different forces with different characteristics.

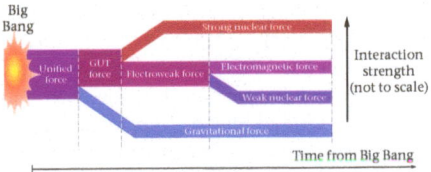

ANSWERS TO ENHANCE YOUR UNDERSTANDING QUESTIONS

1. $Z < N < A$.
2. Alpha decay results in the greatest mass change.
3. The half-life of nucleus 1 is less than the half-life of nucleus 2, because the half-life is inversely proportional to the decay constant.
4. The binding energy per nucleon of $^{238}_{92}\text{U}$ is less than the binding energy per nucleon of $^{96}_{42}\text{Mo}$.
5. $^{89}_{36}\text{Kr}$
6. $^{3}_{1}\text{H}$
7. The biological damage is the same for 1 rem of any kind of radiation.
8. (a) 0 (b) 3 (c) 3 (d) 2
9. There was just one fundamental force originally, before the universe cooled and the forces split to produce the four that are observed today.

CONCEPTUAL QUESTIONS

For instructor-assigned homework, go to www.masteringphysics.com.

(Answers to odd-numbered Conceptual Questions can be found in the back of the book.)

1. Nucleus A and nucleus B have different numbers of protons and different numbers of neutrons. Explain how it is still possible for these nuclei to have equal radii.

2. Is it possible for a form of heavy hydrogen to decay by emitting an α particle? Explain.

3. Which is more likely to expose film kept in a cardboard box, α particles or β particles? Explain.

4. Different isotopes of a given element have different masses, but they have the same chemical properties. Explain why chemical properties are unaffected by a change of isotope.

5. Explain why the large, stable nuclei in Figure 32-1 are found to lie above the $N = Z$ line, rather than below the line.

6. Suppose each of the following items is about 10,000 years old: a feather, a tooth, an obsidian arrowhead, a deer hide moccasin. Which of these items cannot be dated with carbon-14? Explain.

7. Two different samples contain the same radioactive isotope. Is it possible for these samples to have different activities? Explain.

8. Two samples contain different radioactive isotopes. Is it possible for these samples to have the same activity? Explain.

9. Two different types of radiation deliver the same amount of energy to a sample of tissue. Does it follow that each of these types of radiation has the same RBE? Explain.

PROBLEMS AND CONCEPTUAL EXERCISES

Answers to odd-numbered Problems and Conceptual Exercises can be found in the back of the book. **BIO** *identifies problems of biological or medical interest;* **CE** *indicates a conceptual exercise.* **Predict/Explain** *problems ask for two responses: (a) your prediction of a physical outcome, and (b) the best explanation among three provided; and* **Predict/Calculate** *problems ask for a prediction of a physical outcome, based on fundamental physics concepts, and follow that with a numerical calculation to verify the prediction. On all problems, bullets (•, ••, •••) indicate the level of difficulty.*

SECTION 32-1 THE CONSTITUENTS AND STRUCTURE OF NUCLEI

1. • Identify Z, N, and A for the following isotopes: (a) $^{238}_{92}$U, (b) $^{239}_{94}$Pu, (c) $^{144}_{60}$Nd.

2. • Identify Z, N, and A for the following isotopes: (a) $^{32}_{15}$P, (b) $^{99}_{43}$Tc, (c) $^{222}_{86}$Rn.

3. • What are the nuclear radii of (a) $^{197}_{79}$Au and (b) $^{60}_{27}$Co?

4. • A certain chlorine nucleus has a radius of approximately 4.0×10^{-15} m. How many neutrons are in this nucleus?

5. •• **Predict/Calculate** (a) What is the nuclear density of $^{228}_{90}$Th? (b) Do you expect the nuclear density of an alpha particle to be greater than, less than, or the same as that of $^{228}_{90}$Th? Explain. (c) Calculate the nuclear density of an alpha particle.

6. •• **Predict/Calculate** (a) What initial kinetic energy must an alpha particle have if it is to approach a stationary gold nucleus to within a distance of 25.5 fm? (b) If the initial speed of the alpha particle is reduced by a factor of 2, by what factor is the distance of closest approach changed? Explain.

7. •• **Predict/Calculate** An α particle with a kinetic energy of 0.85 MeV approaches a stationary gold nucleus. (a) What is the speed of the α particle? (b) What is the distance of closest approach between the α particle and the gold nucleus? (c) If this same α particle were fired at a copper nucleus instead, would its distance of closest approach be greater than, less than, or the same as that found in part (b)? Explain. (To obtain the mass of an alpha particle, refer to Appendix F and subtract the mass of two electrons from the mass of $^4_2$He.)

8. •• Suppose a marble with a radius of 1.2 cm has the density of a nucleus, as given in Example 32-4. (a) What is the mass of this marble? (b) How many of these marbles would be required to have a mass equal to the mass of Earth?

9. •• **Predict/Calculate** (a) Find the nuclear radius of $^{30}_{15}$P. (b) What mass number would be required for a nucleus to have twice the radius found in part (a)? (c) Verify your answer to part (b) with an explicit calculation.

10. •• **Predict/Calculate** Suppose a uranium-236 nucleus undergoes fission by splitting into two smaller nuclei of equal size. (a) Is the radius of each of the smaller nuclei one-half, more than one-half, or less than one-half the radius of the uranium-236 nucleus? Explain. (b) Calculate the radius of the uranium-236 nucleus. (c) Calculate the radii of the two smaller nuclei.

11. •• A hypothetical nucleus weighs 1 lb. (a) How many nucleons are in this nucleus? (b) What is the radius of this nucleus?

SECTION 32-2 RADIOACTIVITY

12. • **CE** **Predict/Explain** Consider a nucleus that undergoes α decay. (a) Is the radius of the resulting daughter nucleus greater than, less than, or equal to the radius of the original nucleus? (b) Choose the *best explanation* from among the following:
   - I. The decay adds an alpha particle to the nucleus, causing its radius to increase.
   - II. When the nucleus undergoes α decay it ejects two neutrons and two protons. This decreases the number of nucleons in the nucleus, and therefore its radius will decrease.
   - III. An α decay leaves the number of nucleons unchanged. As a result, the radius of the nucleus stays the same.

13. • **CE** **Predict/Explain** Consider a nucleus that undergoes β decay. (a) Is the radius of the resulting daughter nucleus greater than, less than, or the same as that of the original nucleus? (b) Choose the *best explanation* from among the following:
   - I. Capturing a β particle will cause the radius of a nucleus to increase. Therefore, the daughter nucleus has the greater radius.
   - II. The original nucleus emits a β particle, and anytime a particle is emitted from a nucleus the result is a smaller radius. Therefore, the radius of the daughter nucleus is less than the radius of the original nucleus.
   - III. When a nucleus emits a β particle a neutron is converted to a proton, but the number of nucleons is unchanged. As a result, the radius of the daughter nucleus is the same as that of the original nucleus.

14. • **CE** Which of the three decay processes (α, β, or γ) results in a new element? Explain.

15. • Identify the daughter nucleus when the following undergo alpha decay: (a) $^{228}_{90}$Th (b) $^{212}_{83}$Bi, and (c) $^{243}_{96}$Cm.

16. • Identify the daughter nucleus when the following undergo β^- decay: (a) $^{24}_{11}$Na, (b) $^{40}_{19}$K, and (c) $^{89}_{36}$Kr.

17. • Identify the daughter nucleus when the following undergo gamma decay: (a) $^{137}_{56}$Ba, (b) $^{99}_{43}$Tc, and (c) $^{206}_{87}$Fr.

18. • Complete the following nuclear reaction: $^7_3$Li $+ ^1_1$H $\rightarrow ^4_2$He $+$?

19. • Complete the following nuclear reaction: $^{212}_{84}$Po $\rightarrow ^{208}_{82}$Pb $+$?

20. • Complete the following nuclear reaction: ? $\rightarrow ^{14}_7$N $+$ e$^-$ $+ \bar{\nu}_e$

21. •• **CE** One possible decay series for $^{238}_{92}$U is $^{234}_{90}$Th, $^{234}_{91}$Pa, $^{234}_{92}$U, $^{230}_{90}$Th, $^{226}_{88}$Ra, $^{222}_{86}$Rn, $^{218}_{84}$Po, $^{218}_{85}$At, $^{218}_{86}$Rn, $^{214}_{84}$Po, $^{210}_{82}$Pb, $^{210}_{83}$Bi, $^{206}_{81}$Tl, and $^{206}_{82}$Pb. Identify, in the order given, each of the 14 decays that occur in this series.

22. •• Complete the following nuclear reaction and determine the amount of energy it releases: $^3_1$H $\rightarrow ^3_2$He $+$? $+$? Be sure to take into account the mass of the electrons associated with the neutral atoms.

23. •• The following nuclei are observed to decay by emitting an α particle: (a) $^{212}_{84}$Po and (b) $^{239}_{94}$Pu. Write out the decay process for each of these nuclei, and determine the energy released in each reaction. Be sure to take into account the mass of the electrons associated with the neutral atoms.

24. •• The following nuclei are observed to decay by emitting a β^- particle: (a) $^{35}_{16}$S and (b) $^{212}_{82}$Pb. Write out the decay process for each of these nuclei, and determine the energy released in each reaction. Be sure to take into account the mass of the electrons associated with the neutral atoms.

25. •• The following nuclei are observed to decay by emitting a β^+ particle: (a) $^{18}_{9}$F and (b) $^{22}_{11}$Na. Write out the decay process for each of these nuclei, and determine the energy released in each reaction. Be sure to take into account the mass of the electrons associated with the neutral atoms.

26. •• Find the energy released when $^{211}_{82}$Pb undergoes β^- decay to become $^{211}_{83}$Bi. Be sure to take into account the mass of the electrons associated with the neutral atoms.

27. ••• It is observed that $^{66}_{28}$Ni, with an atomic mass of 65.9291 u, decays by β^- emission. (a) Identify the nucleus that results from this decay. (b) If the nucleus found in part (a) has an atomic mass of 65.9289 u, what is the maximum kinetic energy of the emitted electron?

SECTION 32-3 HALF-LIFE AND RADIOACTIVE DATING

28. • **CE** The half-life of carbon-14 is 5730 y. (a) Is it possible for a particular nucleus in a sample of carbon-14 to decay after only 1 s has passed? Explain. (b) Is it possible for a particular nucleus to decay after 10,000 y? Explain.

29. • **CE** Suppose we were to discover that the ratio of carbon-14 to carbon-12 in the atmosphere was significantly smaller 10,000 years ago than it is today. How would this affect the ages we have assigned to objects on the basis of carbon-14 dating? In particular, would the true age of an object be greater than or less than the age we had previously assigned to it? Explain.

30. • **CE** A radioactive sample is placed in a closed container. Two days later only one-quarter of the sample is still radioactive. What is the half-life of this sample?

31. • Radon gas has a half-life of 3.82 d. What is the decay constant for radon?

32. • A radioactive substance has a decay constant equal to $4.4 \times 10^{-8}\,\text{s}^{-1}$. What is the half-life of this substance?

33. • The number of radioactive nuclei in a particular sample decreases over a period of 18 d to one-sixteenth the original number. What is the half-life of these nuclei?

34. •• The half-life of $^{15}_{8}$O is 122 s. How much time does it take for the number of $^{15}_{8}$O nuclei in a given sample to decrease by a factor of 10^{-4}?

35. •• **BIO** **A Radioactive Tag** A drug prepared for a patient is tagged with $^{99}_{43}$Tc, which has a half-life of 6.05 h. (a) What is the decay constant of this isotope? (b) How many $^{99}_{43}$Tc nuclei are required to give an activity of 1.50 μCi?

36. •• **BIO** Referring to Problem 35, suppose the drug containing $^{99}_{43}$Tc with an activity of 1.50 μCi is injected into the patient 2.05 h after it is prepared. What is its activity at the time it is injected?

37. •• **BIO** **Radioactive Tracer** The isotope $^{95}_{43}$Tc has a half-life of 61 d and is sometimes used as a radioactive tracer in plant and animal systems. If a spiny lobster is given an activity of 0.852 μCi with $^{95}_{43}$Tc, how much time will elapse before its activity falls below a 0.100 μCi detection limit?

38. •• An archeologist on a dig finds a fragment of an ancient basket woven from grass. Later, it is determined that the carbon-14 content of the grass in the basket is 9.05% that of an equal carbon sample from present-day grass. What is the age of the basket?

39. •• The bones of a saber-toothed tiger are found to have an activity per gram of carbon that is 12.5% of what would be found in a similar live animal. How old are these bones?

40. •• Charcoal from an ancient fire pit is found to have a carbon-14 content that is only 4.8% that of an equivalent sample of carbon from a living tree. What is the age of the fire pit?

41. •• One of the many isotopes used in cancer treatment is $^{198}_{79}$Au, with a half-life of 2.70 d. Determine the mass of this isotope that is required to give an activity of 225 Ci.

42. •• **Smoke Detectors** The radioactive isotope $^{241}_{95}$Am, with a half-life of 432 y, is the active element in many smoke detectors. Suppose such a detector will no longer function if the activity of the $^{241}_{95}$Am it contains drops below $\frac{1}{512}$ of its initial activity. For what length of time will this smoke detector work?

43. •• **BIO** **Radioactivity in the Bones** Because of its chemical similarity to calcium, $^{90}_{38}$Sr can collect in the bones and present a health risk. What percentage of $^{90}_{38}$Sr present initially still exists after a period of (a) 50.0 y, (b) 60.0 y, and (c) 70.0 y?

SECTION 32-4 NUCLEAR BINDING ENERGY

44. • The atomic mass of gold-197 is 196.96654 u. How much energy is required to completely separate the nucleons in a gold-197 nucleus?

45. • The atomic mass of carbon-12 is 12.000000 u. How much energy is required to completely separate the nucleons in a carbon-12 nucleus?

46. •• Calculate the average binding energy per nucleon of (a) $^{56}_{26}$Fe and (b) $^{238}_{92}$U.

47. •• Calculate the average binding energy per nucleon of (a) $^{4}_{2}$He and (b) $^{64}_{30}$Zn.

48. •• Find the energy required to remove one neutron from $^{23}_{11}$Na.

49. ••• **Predict/Calculate** (a) Consider the following nuclear process, in which a proton is removed from an oxygen nucleus:

$$^{16}_{8}\text{O} + \text{energy} \rightarrow {}^{15}_{7}\text{N} + {}^{1}_{1}\text{H}$$

Find the energy required for this process to occur.
(b) Now consider a process in which a neutron is removed from an oxygen nucleus:

$$^{16}_{8}\text{O} + \text{energy} \rightarrow {}^{15}_{8}\text{O} + {}^{1}_{0}\text{n}$$

Find the energy required for this process to occur.
(c) Which particle, the proton or the neutron, do you expect to be more tightly bound in the oxygen nucleus? Verify your answer.

SECTION 32-5 NUCLEAR FISSION

50. • Find the number of neutrons released by the following fission reaction:

$$^{1}_{0}\text{n} + {}^{235}_{92}\text{U} \rightarrow {}^{132}_{50}\text{Sn} + {}^{101}_{42}\text{Mo} + (?)\,\text{neutrons}$$

51. •• Complete the following fission reaction and determine the amount of energy it releases:

$$^{1}_{0}\text{n} + {}^{235}_{92}\text{U} \rightarrow {}^{133}_{51}\text{Sb} + ? + 5{}^{1}_{0}\text{n}$$

52. •• Complete the following fission reaction and determine the amount of energy it releases:

$$^{1}_{0}\text{n} + {}^{235}_{92}\text{U} \rightarrow {}^{88}_{38}\text{Sr} + {}^{136}_{54}\text{Xe} + (?)\,\text{neutrons}$$

53. •• A gallon of gasoline releases about 2.0×10^8 J of energy when it is burned. How many gallons of gas must be burned to release the same amount of energy as is released when 1.0 lb of $^{235}_{92}$U undergoes fission? (Assume that each fission reaction in $^{235}_{92}$U releases 173 MeV.)

54. •• Assuming a release of 173 MeV per fission reaction, determine the minimum mass of $^{235}_{92}$U that must undergo fission to supply the annual energy needs of the United States. (The amount of energy consumed in the United States each year is 8.4×10^{19} J.)

55. •• Assuming a release of 173 MeV per fission reaction, calculate how many reactions must occur per second to produce a power output of 150 MW.

SECTION 32-6 NUCLEAR FUSION

56. • Consider a fusion reaction in which two deuterium nuclei fuse to form a tritium nucleus and a proton. How much energy is released in this reaction?

57. • Consider a fusion reaction in which a proton fuses with a neutron to form a deuterium nucleus. How much energy is released in this reaction?

58. • Find the energy released in the following fusion reaction:

$$^1_1\text{H} + {}^2_1\text{H} \rightarrow {}^3_2\text{He} + \gamma$$

59. •• (a) Complete the following fusion reaction and determine the energy it releases:

$$^2_1\text{H} + {}^3_1\text{H} \rightarrow ? + {}^1_0\text{n}$$

(b) How many of these reactions must occur per second to produce a power output of 750 MW?

60. ••• **The Evaporating Sun** The Sun radiates energy at the prodigious rate of 3.90×10^{26} W. (a) At what rate, in kilograms per second, does the Sun convert mass into energy? (b) Assuming that the Sun has radiated at this same rate for its entire lifetime of 4.50×10^9 y, and that the current mass of the Sun is 2.00×10^{30} kg, what percentage of its original mass has been converted to energy?

SECTION 32-7 PRACTICAL APPLICATIONS OF NUCLEAR PHYSICS

61. • **BIO Radiation Damage** A sample of tissue absorbs a 55-rad dose of α particles (RBE = 20). How many rad of protons (RBE = 10) cause the same amount of damage to the tissue?

62. • **BIO X-ray Damage** How many rad of 200-keV X-rays cause the same amount of biological damage as 50 rad of heavy ions?

63. •• **BIO** A patient undergoing radiation therapy for cancer receives a 225-rad dose of radiation. (a) Assuming the cancerous growth has a mass of 0.17 kg, calculate how much energy it absorbs. (b) Assuming the growth to have the specific heat of water, determine its increase in temperature.

64. •• **BIO** Alpha particles with an RBE of 13 deliver a 48-mrad radiation dose to a 82-kg patient. (a) What dosage, in rem, does the patient receive? (b) How much energy is absorbed by the patient?

65. ••• **BIO A Radioactive Pharmaceutical** As part of a treatment program, a patient ingests a radioactive pharmaceutical containing $^{32}_{15}$P, which emits β rays with an RBE of 1.50. The half-life of $^{32}_{15}$P is 14.28 d, and the initial activity of the medication is 1.34 MBq. (a) How many electrons are emitted over the period of 7.00 d? (b) If the β rays have an energy of 705 keV, what is the total amount of energy absorbed by the patient's body in 7.00 d? (c) Find the absorbed dosage in rem, assuming the radiation is absorbed by 125 g of tissue.

SECTION 32-8 ELEMENTARY PARTICLES

66. •• The pion π^+ has a rest energy of 139.6 MeV and an average lifetime of 2.6×10^{-8} s. (a) What is its mass in kg? (b) Neglecting the effects of relativity, at what speed will the pion travel 1.0 m during an average lifetime?

67. ••• **The Higgs Boson** The first evidence for the Higgs boson suggests it has a rest energy of 125 GeV and an average lifetime of 1.56×10^{-22} s. (a) What is its mass in kg? (b) Neglecting the effects of relativity, at what speed will the Higgs boson travel 1.0 m during an average lifetime? (c) When relativistic time dilation is taken into account, it can be shown that the speed of a particle that travels a distance d in the laboratory frame of reference while a time t_0 elapses in the particle's frame of reference is given by

$$v = \frac{d/t_0}{\sqrt{1 + d^2/c^2 t_0^2}}$$

Use this expression to find the required speed for the Higgs boson to travel 1.0 m in the laboratory while an average lifetime elapses in the boson's frame of reference. Report your answer as the difference between the boson's speed and the speed of light—that is, find δ, where $v = (1 - \delta)c$.

GENERAL PROBLEMS

68. • **CE** An α particle (charge $+2e$) and a β particle (charge $-e$) deflect in opposite directions when they pass through a magnetic field. Which particle deflects by a greater amount, given that both particles have the same speed? Explain.

69. • **CE** The initial activity of sample A is twice that of sample B. After two half-lives of sample A have elapsed, the two samples have the same activity. What is the ratio of the half-life of B to the half-life of A?

70. • Determine the number of neutrons and protons in (a) $^{232}_{90}$Th, (b) $^{211}_{82}$Pb, and (c) $^{60}_{27}$Co.

71. • Identify the daughter nucleus that results when (a) $^{210}_{82}$Pb undergoes α decay, (b) $^{239}_{92}$U undergoes β^- decay, and (c) $^{11}_6$C undergoes β^+ decay.

72. • A patient is exposed to 260 rad of gamma rays. What is the dose the patient receives in rem?

73. • **CE** The two radioactive decay series that begin with $^{232}_{90}$Th and end with $^{208}_{82}$Pb are shown in **FIGURE 32-22**. Identify the ten intermediary nuclei that appear in these series.

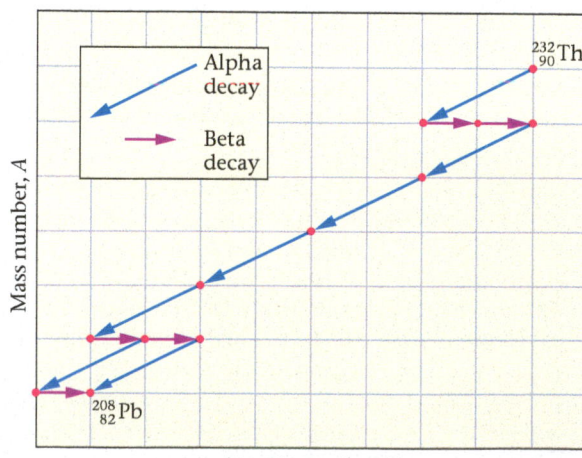

FIGURE 32-22 Problem 73

74. •• **Moon Rocks** In one of the rocks brought back from the Moon, it is found that 80.5% of the initial potassium-40 in the rock has decayed to argon-40. **(a)** If the half-life for this decay is 1.20×10^9 years, how old is the rock? **(b)** How much longer will it take before only 10.0% of the original potassium-40 is still present in the rock?

75. •• **Predict/Calculate Mantles in Gas Lanterns** Gas lanterns used on camping trips have mantles (small lacy bags that give off light) made from a rayon mesh impregnated with thorium and other materials. Thorium is used, even though it is radioactive, because it forms an oxide that can withstand being incandescent for long periods of time. Almost all natural thorium is $^{232}_{90}$Th, which has a half-life of 1.405×10^{10} y. **(a)** A typical mantle contains 325 mg of thorium. What is the activity of the mantle? **(b)** If the half-life of thorium had been double its actual value, by what factor would the activity of a mantle be changed? Explain.

76. •• **Voyager 1 Power** The *Voyager 1* spacecraft, the first man-made object to leave our solar system, is powered by electricity generated by the heat produced when $^{238}_{94}$Pu decays with a half-life of 87.75 y. Engineers predict that the spacecraft will have insufficient power to operate its instruments when the amount of $^{238}_{94}$Pu in its power generators drops below 68% of the amount it had at launch on September 5, 1977. In what year will its instruments cease to function?

77. •• An α particle fired head-on at a stationary nickel nucleus approaches to a radius of 15 fm before being turned around. **(a)** What is the maximum Coulomb force exerted on the α particle? **(b)** What is the electric potential energy of the α particle at its point of closest approach? **(c)** Find the initial kinetic energy of the α particle.

78. •• Calculate the number of disintegrations per second that one would expect from a 1.9-g sample of $^{226}_{88}$Ra. What is the activity of this sample in curies?

79. •• **Predict/Calculate** The activity of sample A is 16 Ci and that of sample B is 4.0 Ci. After 8.0 days the two samples have the same activity of 0.25 Ci. **(a)** Is the half-life of type A nuclei greater than, less than, or the same as the half-life of type B nuclei? **(b)** Calculate the half-lives of type A and type B nuclei.

80. •• **Radius of a Neutron Star** Neutron stars are so named because they are composed of neutrons and have a density the same as that of a nucleus. Referring to Example 32-4 for the nuclear density, find the radius of a neutron star whose mass is 0.50 that of the Sun.

81. •• A specimen taken from the wrappings of a mummy contains 7.82 g of carbon and has an activity of 1.38 Bq. How old is the mummy? (Refer to Section 32-3 for relevant information regarding the isotopes of carbon.)

82. •• **(a)** How many fission reactions are required to light a 120-W lightbulb for 31 d? Assume an energy release of 212 MeV per fission reaction and a 32% conversion efficiency. **(b)** What mass of $^{235}_{92}$U corresponds to the number of fission reactions found in part (a)?

83. •• **Predict/Calculate** Energy is released when three α particles fuse to form carbon-12. **(a)** Is the mass of carbon-12 greater than, less than, or the same as the mass of three α particles? Explain. **(b)** Calculate the energy given off in this fusion reaction.

84. •• **Predict/Calculate (a)** What dosage (in rad) must a 1.5-kg sample of water absorb to increase its temperature by 1.0 C°? **(b)** If the mass of the water sample is increased, does the dosage found in part (a) increase, decrease, or stay the same? Explain.

85. •• **BIO Chest X-rays** A typical chest X-ray uses X-rays with an RBE of 0.85. If the radiation dosage is 2.5 mrem, find the energy absorbed

by a 78-kg patient, assuming one-quarter of the patient's body is exposed to the X-rays.

86. •• **BIO Airplane Flight** During a typical 2.5-h airplane flight, passengers are exposed to 65 μJ/h of radiation above the background radiation they receive on the ground, mostly from cosmic ray protons that are otherwise absorbed by the atmosphere before they reach the surface. **(a)** What dose in rad is received by a 75-kg person during the flight? **(b)** What is the person's equivalent dose in rem? Compare your answer to the 2.0-mrem dose typical of a chest X-ray.

87. ••• A γ ray photon emitted by $^{226}_{88}$Ra has an energy of 0.186 MeV. Use conservation of linear momentum to calculate the recoil speed of a $^{226}_{88}$Ra nucleus after such a γ ray is emitted. Assume that the nucleus is at rest initially, and that relativistic effects can be ignored.

88. ••• The energy released by α decay in a 50.0-g sample of $^{239}_{94}$Pu is to be used to heat 4.75 kg of water. Assuming all the energy released by the radioactive decay goes into heating the water, find how much the temperature of the water increases in 1.00 h.

PASSAGE PROBLEMS

BIO Treating a Hyperactive Thyroid

Of the many endocrine glands in the body, the thyroid is one of the most important. Weighing only an ounce and situated just below the "Adam's apple," the thyroid produces hormones that regulate the metabolic rate of every cell in the body. To produce these hormones the thyroid uses iodine from the food we eat—in fact, the thyroid specializes in absorbing iodine.

The central role played by the thyroid is evidenced by the symptoms produced when it ceases to function properly. For example, a person experiencing hyperthyroidism (an overactive thyroid) presents the internist with a wide range of indicators, including weight loss, ravenous appetite, anxiety, fatigue, hyperactivity, apathy, palpitations, arrhythmias, and nausea, just to mention a few.

The most common treatment for hyperthyroidism is to destroy the overactive thyroid tissues with radioactive iodine-131. This treatment takes advantage of the fact that only thyroid cells absorb and concentrate iodine. To begin the treatment, a patient swallows a single, small capsule containing iodine-131. The radioactive isotope quickly enters the bloodstream and is taken up by the overactive thyroid cells, which are destroyed as the iodine-131 decays with a half-life of 8.04 d. Other cells in the body experience very little radiation damage, which minimizes side effects. In one or two months the thyroid activity is reduced to an acceptable level. Sometimes too much—or even all—of the thyroid is killed, which can result in hypothyroidism, or underactive thyroid. This is easily treated, however, with dietary supplements to replace the missing thyroid hormones.

89. • What is the decay constant, λ, for iodine-131?

   A. $9.98 \times 10^{-7}\,\text{s}^{-1}$ **B.** $1.44 \times 10^{-6}\,\text{s}^{-1}$
   C. $2.39 \times 10^{-5}\,\text{s}^{-1}$ **D.** $5.99 \times 10^{-5}\,\text{s}^{-1}$

90. • If a sample of iodine-131 contains 4.5×10^{16} nuclei, what is the activity of the sample? Express your answer in curies.

   A. 0.27 Ci   **B.** 1.2 Ci
   C. 1.7 Ci   **D.** 4.5 Ci

91. • If the half-life of iodine-131 were only half of its actual value, would the activity of the sample in Problem 90 be increased or decreased?

This text is designed for students with a working knowledge of basic algebra and trigonometry. Even so, it is useful to review some of the mathematical tools that are of particular importance in the study of physics. In this Appendix we cover a number of topics related to mathematical notation, trigonometry, algebra, mathematical expansions, and vector multiplication.

Mathematical Notation

COMMON MATHEMATICAL SYMBOLS

In Table A-1 we present some of the more common mathematical symbols, along with a translation into English. Though these symbols are probably completely familiar, it is worthwhile to be sure we all interpret them in the same way.

TABLE A-1 Mathematical Symbols

$=$	is equal to
$\neq$	is not equal to
$\approx$	is approximately equal to
$\propto$	is proportional to
$>$	is greater than
$\geq$	is greater than or equal to
$\gg$	is much greater than
$<$	is less than
$\leq$	is less than or equal to
$\ll$	is much less than
$\pm$	plus or minus
$\mp$	minus or plus
x_{av} or $\bar{x}$	average value of x
Δx	change in $x (x_f - x_i)$
$|x|$	absolute value of x
Σ	sum of
$\rightarrow 0$	approaches 0
∞	infinity

A couple of the symbols in Table A-1 warrant further discussion. First, Δx, which means "change in x," is used frequently, and in many different contexts. Pronounced "delta x," it is defined as the final value of x, x_f, minus the initial value of x, x_i:

$$\Delta x = x_f - x_i \qquad \text{A-1}$$

Thus, Δx is not Δ times x; it is a shorthand way of writing $x_f - x_i$. The same delta notation can be applied to any quantity—it does not have to be x. In general, we can say that

$$\Delta(anything) = (anything)_f - (anything)_i$$

For example, $\Delta t = t_f - t_i$ is the change in time, $\Delta\vec{v} = \vec{v}_f - \vec{v}_i$ is the change in velocity, and so on. Throughout this text, we use the delta notation whenever we want to indicate the change in a given quantity.

Second, the Greek letter Σ (capital sigma) is also encountered frequently. In general, Σ is shorthand for "sum." For example, suppose we have a system comprised of nine masses, m_1 through m_9. The total mass of the system, M, is simply

$$M = m_1 + m_2 + m_3 + m_4 + m_5 + m_6 + m_7 + m_8 + m_9$$

This is a rather tedious way to write M, however, and would be even more so if the number of masses were larger. To simplify our equation, we use the Σ notation:

$$M = \sum_{i=1}^{9} m_i \qquad \text{A-2}$$

With this notation we could sum over any number of masses, simply by changing the upper limit of the sum.

In addition, Σ is often used to designate a general summation, where the number of terms in the sum may not be known, or may vary from one system to another. In a case like this we would simply write Σ without specific upper and lower limits. Thus, a general way of writing the total mass of a system is as follows:

$$M = \sum m \qquad \text{A-3}$$

VECTOR NOTATION

When we draw a vector to represent a physical quantity, we typically use an arrow whose length is proportional to the magnitude of the quantity, and whose direction is the direction of the quantity. (This and other aspects of vector notation are discussed in Chapter 3.) A slight problem arises, however, when a physical quantity points into or out of the page. In such a case, we use the conventions illustrated in **Figure A-1**.

Figure A-1 (a) shows a vector pointing out of the page. Note that we see only the tip. Below, we show the corresponding convention, which is a dot set off by a circle. The dot represents the point of the vector's arrow coming out of the page toward you.

A similar convention is employed in Figure A-1 (b) for a vector pointing into the page. In this case, the arrow moves directly away from you, giving a view of its "tail feathers." The feathers are placed in an X-shaped pattern, so we represent the vector as an X set off by a circle.

These conventions are used in Chapter 22 to represent the magnetic field vector, $\vec{B}$, and in other locations in the text as well.

(a)　　　　　　**(b)**

▲ **FIGURE A-1 Vectors pointing out of and into the page (a)** A vector pointing out of the page is represented by a dot in a circle. The dot indicates the tip of the vector's arrow. **(b)** A vector pointing into the page is represented by an X in a circle. The X indicates the "tail feathers" of the vector's arrow.

SCIENTIFIC NOTATION

In physics, the numerical value of a physical quantity can cover an enormous range, from the astronomically large to the microscopically small. For example, the mass of the Earth is roughly

$$M_E = 5970000000000000000000000 \text{ kg}$$

In contrast, the mass of a hydrogen atom is approximately

$$M_{\text{hydrogen}} = 0.00000000000000000000000000167 \text{ kg}$$

Clearly, representing such large and small numbers with a long string of zeros is clumsy and prone to error.

The preferred method for handling such numbers is to replace the zeros with the appropriate power of ten. For example, the mass of the Earth can be written as follows:

$$M_E = 5.97 \times 10^{24} \text{ kg}$$

The factor of 10^{24} simply means that the decimal point for the mass of the Earth is 24 places to the right of its location in 5.97. Similarly, the mass of a hydrogen atom is

$$M_{\text{hydrogen}} = 1.67 \times 10^{-27} \text{ kg}$$

In this case, the correct location of the decimal point is 27 places to the left of its location in 1.67. This type of representation, using powers of ten, is referred to as **scientific notation**.

Scientific notation also simplifies various mathematical operations, such as multiplication and division. For example, the product of the mass of the Earth and the mass of a hydrogen atom is

$$
\begin{aligned}
M_E M_{\text{hydrogen}} &= (5.97 \times 10^{24} \text{ kg})(1.67 \times 10^{-27} \text{ kg}) \\
&= (5.97 \times 1.67)(10^{24} \times 10^{-27}) \text{ kg}^2 \\
&= 9.99 \times 10^{24-27} \text{ kg}^2 \\
&= 9.99 \times 10^{-3} \text{ kg}^2
\end{aligned}
$$

Similarly, the mass of a hydrogen atom divided by the mass of the Earth is

$$
\begin{aligned}
\frac{M_{\text{hydrogen}}}{M_E} &= \frac{1.67 \times 10^{-27} \text{ kg}}{5.97 \times 10^{24} \text{ kg}} = \frac{1.67}{5.97} \times \frac{10^{-27}}{10^{24}} \\
&= 0.280 \times 10^{-27-24} \\
&= 0.280 \times 10^{-51} = 2.80 \times 10^{-52}
\end{aligned}
$$

Note the change in location of the decimal point in the last two expressions, and the corresponding change in the power of ten.

Exponents and their manipulation are discussed in greater detail later in this Appendix.

Trigonometry

DEGREES AND RADIANS

We all know the definition of a degree; there are 360 degrees in a circle. The definition of a radian is somewhat less well known; there are 2π radians in a circle. An equivalent definition of the radian is the following:

A radian is the angle for which the corresponding arc length is equal to the radius.

To visualize this definition, consider a pie with a piece cut out, as shown in **Figure A-2 (a)**. Note that a piece of pie has three sides—two radial lines from the center, and an arc of crust. If a piece of pie is cut with an angle of one radian, all three sides are equal in

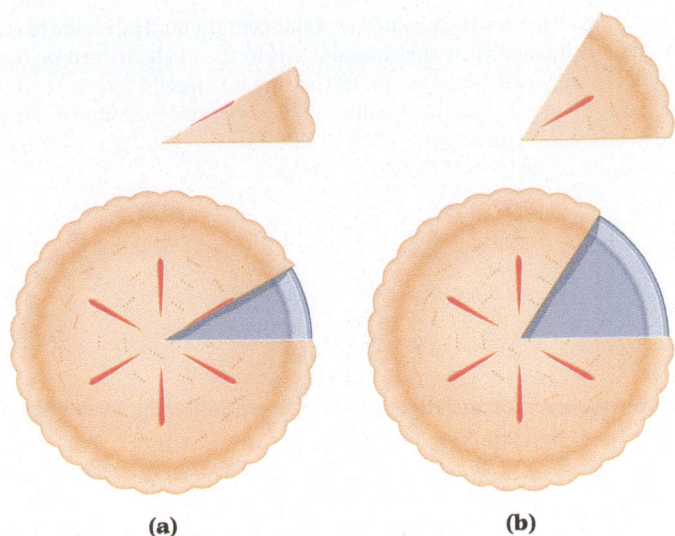

(a)   **(b)**

▲ **FIGURE A-2 The definition of a radian (a)** This piece of pie is cut with an angle less than a radian. Thus, the two radial sides (coming out from the center) are longer than the arc of crust. **(b)** The angle for this piece of pie is equal to one radian (about 57.3°). Thus, all three sides of the piece are of equal length.

length, as shown in **Figure A-2 (b)**. Since a radian is about 57.3°, this amounts to a fairly good-sized piece of pie. Thus, if you want a healthy helping of pie, just tell the server, "One radian, please."

Now, radians are particularly convenient when we are interested in the length of an arc. In **Figure A-3** we show a circular arc corresponding to the radius r and the angle θ. If the angle θ is measured in radians, the length of the arc, s, is given by

$$s = r\theta \qquad \text{A-4}$$

Note that this simple relation is *not valid* when θ is measured in degrees. For a full circle, in which case $\theta = 2\pi$, the length of the arc (which is the circumference of the circle) is $2\pi r$, as expected.

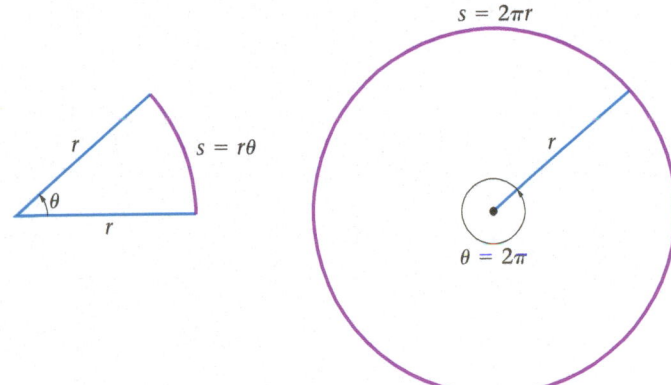

▲ **FIGURE A-3 The Length of an arc** The arc created by a radius *r* rotated through an angle θ has a length $s = r\theta$. If the radius is rotated through a full circle, the angle is $\theta = 2\pi$, and the arc length is the circumference of a circle, $s = 2\pi r$.

TRIGONOMETRIC FUNCTIONS AND THE PYTHAGOREAN THEOREM

Next, we consider some of the more important and frequently used results from trigonometry. We start with the right triangle, shown in **Figure A-4**, and the basic **trigonometric functions**, sin θ (sine theta), cos θ (cosine theta), and tan θ (tangent theta). The cosine of an angle θ is defined to be the side adjacent to the angle divided by the hypotenuse; cos θ = x/r. Similarly, the sine is defined to be the opposite side divided by the hypotenuse, sin θ = y/r, and the tangent is the opposite side divided by the adjacent side, tan θ = y/x. These relations are summarized in the following equations:

$$\cos \theta = \frac{x}{r}$$
$$\sin \theta = \frac{y}{r} \qquad \text{A-5}$$
$$\tan \theta = \frac{y}{x} = \frac{\sin \theta}{\cos \theta}$$

Note that each of the trigonometric functions is the ratio of two lengths, and hence is dimensionless.

According to the **Pythagorean theorem,** the sides of the right triangle in Figure A-4 are related as follows:

$$x^2 + y^2 = r^2 \qquad \text{A-6}$$

Dividing by r^2 yields

$$\frac{x^2}{r^2} + \frac{y^2}{r^2} = 1$$

This can be re-written in terms of sine and cosine to give

$$\sin^2 \theta + \cos^2 \theta = 1$$

Figure A-4 also shows how sine and cosine are used in a typical calculation. In many cases, the hypotenuse of a triangle, r, and one of its angles, θ, are given. To find the short sides of the triangle we rearrange the relations given in Equation A-5. For example, in Figure A-4 we see that x = r cos θ is the length of the short side adjacent to the angle, θ, and y = r sin θ is the length of the short side opposite the angle. The following Example applies this type of calculation to the case of an inclined roadway.

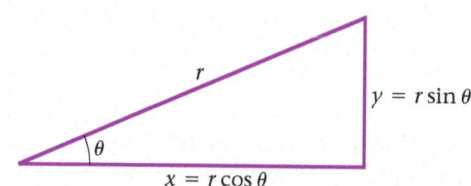

▲ **FIGURE A-4 Relating the short sides of a right triangle to its hypotenuse** The trigonometric functions sin θ and cos θ and the Pythagorean theorem are useful in relating the lengths of the short sides of a right triangle to the length of its hypotenuse.

EXAMPLE A–1 | HIGHWAY TO HEAVEN

You are driving on a long straight road that slopes uphill at an angle of 6.4° above the horizontal. At one point you notice a sign that reads, "Elevation 1500 feet." What is your elevation after you have driven another 1.0 mi?

PICTURE THE PROBLEM
From our sketch, we see that the car is moving along the hypotenuse of a right triangle. The length of the hypotenuse is one mile.

REASONING AND STRATEGY
The elevation gain is the vertical side of the triangle, y. We find y by multiplying the hypotenuse, r, by the sine of theta. That is, since sin θ = y/r it follows that y = r sin θ.

Known Angle of slope, q = 6.4°; driving distance, r = 10 mi; initial elevation = 1500 ft.

Unknown Gain in elevation, y = ?; final elevation = initial elevation + change in elevation = ?

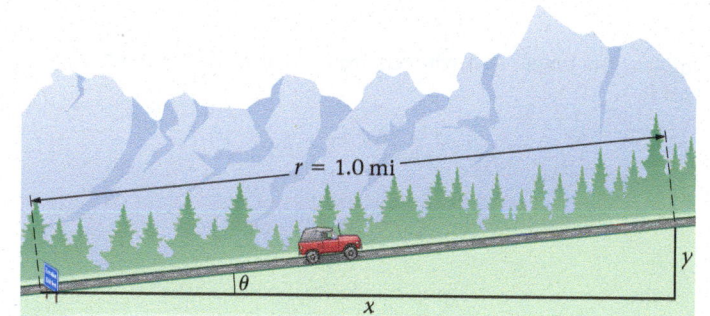

SOLUTION

1. Calculate the elevation gain, y:

$$y = r \sin \theta$$
$$= (1.0 \text{ mi}) \sin 6.4° = (1.0 \text{ mi})(0.11) = 0.11 \text{ mi}$$

2. Convert y from miles to feet:

$$y = (0.11 \text{ mi})\left(\frac{5280 \text{ ft}}{1 \text{ mi}}\right) = 580 \text{ ft}$$

3. Add the elevation gain to the original elevation to obtain the new elevation:

$$\text{elevation} = 1500 \text{ ft} + 580 \text{ ft} = 2100 \text{ ft}$$

INSIGHT
As surprising as it may seem, the horizontal distance covered by the car is r cos θ = (5280 ft) cos 6.4° = 5200 ft, only about 80 ft less than the total distance driven by the car. At the same time, the car rises a distance of 580 ft.

PRACTICE PROBLEM
How far up the road from the first sign should the road crew put another sign reading "Elevation 3500 ft"?
[Answer: 18,000 ft = 3.4 mi]

In some problems, the sides of a triangle (x and y) are given and it is desired to find the corresponding hypotenuse, r, and angle, θ. For example, suppose that $x = 5.0$ m and $y = 2.0$ m. Using the Pythagorean theorem, we find $r = \sqrt{x^2 + y^2} = \sqrt{(5.0\,\text{m})^2 + (2.0\,\text{m})^2} = 5.4$ m. Similarly, to find the angle we use the definition of tangent: $\tan\theta = y/x$. The inverse of this relation is $\theta = \tan^{-1}(y/x) = \tan^{-1}(2.0\,\text{m}/5.0\,\text{m}) = \tan^{-1}(0.40)$. Note that the expression $\tan^{-1}$ is the *inverse tangent function*—it does not mean 1 divided by tangent, but rather "the angle whose tangent is——." Your calculator should have a button on it labeled $\tan^{-1}$. If you enter 0.40 and then press $\tan^{-1}$, you should get 22° (to two significant figures), which means that $\tan 22° = 0.40$. Inverse sine and cosine functions work in the same way.

TRIGONOMETRIC IDENTITIES

In addition to the basic definitions of sine, cosine, and tangent just given, there are a number of useful relationships involving these functions referred to as **trigonometric identities**. First, consider changing the sign of an angle. This corresponds to flipping the triangle in Figure A-4 upside-down, which changes the sign of y but leaves x unaffected. The result is that sine changes its sign, but cosine does not. Specifically, for a general angle A we find the following:

$$\sin(-A) = -\sin A$$
$$\cos(-A) = \cos A \qquad \text{A-7}$$

Next, we consider trigonometric identities relating to the sum or difference of two angles. For example, consider two general angles A and B. The sine and cosine of the sum of these angles, $A + B$, are given below:

$$\sin(A + B) = \sin A \cos B + \sin B \cos A$$
$$\cos(A + B) = \cos A \cos B - \sin A \sin B \qquad \text{A-8}$$

By changing the sign of B, and using the results given in Equation A-7, we obtain the corresponding results for the difference between two angles:

$$\sin(A - B) = \sin A \cos B - \sin B \cos A$$
$$\cos(A - B) = \cos A \cos B + \sin A \sin B \qquad \text{A-9}$$

Applications of these relations can be found in Chapters 4, 14, 23, and 24.

To see how one might use a relation like $\sin(A + B) = \sin A \cos B + \sin B \cos A$, consider the case where $A = B = \theta$. With this substitution we find

$$\sin(\theta + \theta) = \sin\theta \cos\theta + \sin\theta \cos\theta$$

Simplifying somewhat yields the commonly used double-angle formula

$$\sin 2\theta = 2\sin\theta \cos\theta \qquad \text{A-10}$$

This expression is used in deriving Equation 4-12.

As a final example of using trigonometric identities, let $A = 90°$ and $B = \theta$. Making these substitutions in Equations A-9 yields

$$\sin(90° - \theta) = \sin 90° \cos\theta - \sin\theta \cos 90° = \cos\theta$$
$$\cos(90° - \theta) = \cos 90° \cos\theta + \sin 90° \sin\theta = \sin\theta \qquad \text{A-11}$$

Algebra

THE QUADRATIC EQUATION

A well-known result that finds many uses in physics is the solution to the **quadratic equation**

$$ax^2 + bx + c = 0 \qquad \text{A-12}$$

In this equation, a, b, and c are constants and x is a variable. When we refer to the solution of the quadratic equation, we mean the values of x that satisfy Equation A-12. These values are given by the following expression:

> ### Solutions to the Quadratic Equation
>
> $$x = \frac{-b \pm \sqrt{b^2 - 4ac}}{2a} \qquad \text{A-13}$$

Note that there are two solutions to the quadratic equation, in general, corresponding to the plus and minus sign in front of the square root. In the special case that the quantity under the square root vanishes, there will be only a single solution. If the quantity under the square root is negative the result for x is not physical, which means a mistake has probably been made in the calculation.

To illustrate the use of the quadratic equation and its solution, we consider a standard one-dimensional kinematics problem, such as one might encounter in Chapter 2:

A ball is thrown straight upward with an initial speed of 11 m/s. How long does it take for the ball to first reach a height of 4.5 m above its launch point?

The first step in solving this problem is to write the equation giving the height of the ball, y, as a function of time. Referring to Equation 2-11, and replacing x with y, we have

$$y = y_0 + v_0 t - \tfrac{1}{2}gt^2$$

To make this look more like a quadratic equation, we move all the terms onto the left-hand side, which yields

$$\tfrac{1}{2}gt^2 - v_0 t + y - y_0 = 0$$

This is the same as Equation A-12 if we make the following identifications: $x = t$; $a = \tfrac{1}{2}g$; $b = -v_0$; $c = y - y_0$. The desired solution, then, is given by making these substitutions in Equation A-13:

$$t = \frac{v_0 \pm \sqrt{v_0{}^2 - 2g(y - y_0)}}{g}$$

The final step is to use the appropriate numerical values; $g = 9.81$ m/s^2, $v_0 = 11$ m/s, $y - y_0 = 4.5$ m. Straightforward calculation gives $t = 0.54$ s and $t = 1.7$ s. Therefore, the time it takes to first reach a height of 4.5 m is 0.54 s; the second solution is the time when the ball is again at a height of 4.5 m, this time on its way down.

TWO EQUATIONS IN TWO UNKNOWNS

In some problems, two unknown quantities are determined by two interlinked equations. In such cases it often seems at first that you have not been given enough information to obtain a solution. By patiently writing out what is known, however, you can generally use straightforward algebra to solve the problem.

As an example, consider the following problem: A father and daughter share the same birthday. On one birthday the father announces to his daughter, "Today I am four times older than you, but in 5 years I will be only three times older." How old are the father and daughter now?

You might be able to solve this problem by guessing, but here's how to approach it systematically. First, write what is given in the form of equations. Letting F be the father's age in years, and D the daughter's age in years, we know that on this birthday

$$F = 4D \qquad \text{A-14}$$

In 5 years, the father's age will be $F + 5$, the daughter's age will be $D + 5$, and the following will be true:

$$F + 5 = 3(D + 5)$$

Multiplying through the parenthesis gives

$$F + 5 = 3D + 15 \qquad \text{A-15}$$

Now if we subtract Equation A-15 from Equation A-14 we can eliminate one of the unknowns, F:

$$
\begin{aligned}
F &= 4D \\
-(F + 5 &= 3D + 15) \\
\hline
-5 &= D - 15
\end{aligned}
$$

The solution to this new equation shows that the daughter's age now is $D = 10$, and hence the father's age now is $F = 4D = 40$.

The following Example investigates a similar problem. In this case, we use the fact that if you drive with a speed v for a time t the distance covered is $d = vt$.

EXAMPLE A–2   HIT THE ROAD

It takes 1.50 h to drive with a speed v from home to a nearby town, a distance d away. Later, on the way back, the traffic is lighter, and you are able to increase your speed by 15 mi/h. With this higher speed, you get home in just 1.00 h. Find your initial speed v, and the distance to the town, d.

PICTURE THE PROBLEM
Our sketch shows home and the town, separated by a distance d. Going to town the speed is v, returning home the speed is $v + 15$ mi/h.

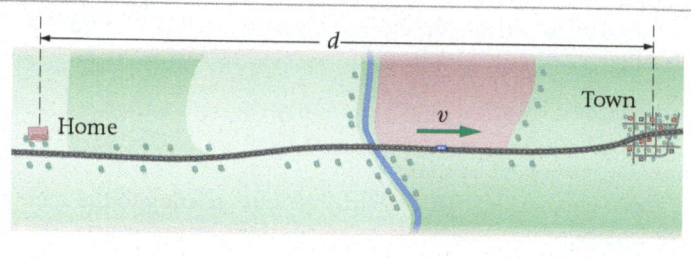

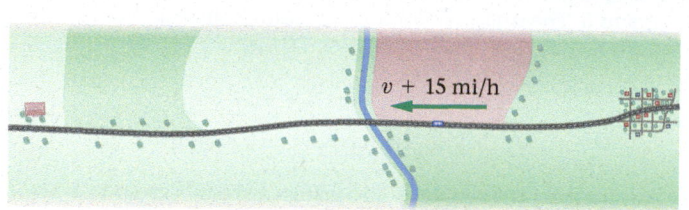

REASONING AND STRATEGY
To determine the two unknowns, v and d, we need two separate equations. One equation corresponds to what we know about the trip to the town, the second equation corresponds to what we know about the return trip.

Known   Drive time outbound, $t = 1.50$ h; increase in speed $= 15$ mi/h; drive time going home $= 1.00$ h.

Unknown Initial speed, $v = ?$; distance to town, $d = ?$

SOLUTION

1. Write an equation for the trip to the town. Recall that this trip takes one and a half hours:

$$d = vt = v(1.50\,\text{h})$$

2. Write an equation for the trip home. This trip takes one hour, and covers the same distance d:

$$d = (v + 15\,\text{mi/h})t = (v + 15\,\text{mi/h})(1.00\,\text{h})$$

3. Subtract these two equations to eliminate d:

$$
\begin{aligned}
d &= v(1.50\,\text{h}) \\
-\,d &= (v + 15\,\text{mi/h})(1.00\,\text{h}) \\
\hline
0 &= v(1.50\,\text{h}) - v(1.00\,\text{h}) - (15\,\text{mi/h})(1.00\,\text{h})
\end{aligned}
$$

4. Solve this new equation for v:

$$0 = v(1.50\,\text{h}) - v(1.00\,\text{h}) - (15\,\text{mi/h})(1.00\,\text{h})$$
$$0 = v(0.50\,\text{h}) - (15\,\text{mi/h})(1.00\,\text{h})$$
$$v = \frac{(15\,\text{mi/h})(1.00\,\text{h})}{(0.50\,\text{h})} = 30\,\text{mi/h}$$

5. Use the first equation to solve for d:

$$d = vt = (30\,\text{mi/h})(1.50\,\text{h}) = 45\,\text{mi}$$

INSIGHT
We could also use the second equation to solve for d. The algebra is a bit messier, but it provides a good double check on our results: $d = (30\,\text{mi/h} + 15\,\text{mi/h})(1.00\,\text{h}) = (45\,\text{mi/h})(1.00\,\text{h}) = 45\,\text{mi}$.

PRACTICE PROBLEM
Suppose the speed going home is 12 mi/h faster than the speed on the way to town, but the times are the same as above. What are v and d in this case?   [**Answer:** $v = 24$ mi/h, $d = 36$ mi]

EXPONENTS AND LOGARITHMS

An **exponent** is the power to which a number is raised. For example, in the expression 10^3, we say that the exponent of 10 is 3. To evaluate 10^3 we simply multiply 10 by itself three times:

$$10^3 = 10 \times 10 \times 10 = 1000$$

Similarly, a negative exponent implies an inverse, as in the relation $10^{-1} = 1/10$. Thus, to evaluate a number like 10^{-4}, for example, we multiply 1/10 by itself four times:

$$10^{-4} = \frac{1}{10} \times \frac{1}{10} \times \frac{1}{10} \times \frac{1}{10} = \frac{1}{10,000} = 0.0001$$

The relations just given apply not just to powers of 10, of course, but to any number at all. Thus, x^4 is

$$x^4 = x \times x \times x \times x$$

and x^{-3} is

$$x^{-3} = \frac{1}{x} \times \frac{1}{x} \times \frac{1}{x} = \frac{1}{x^3}$$

Using these basic rules, it follows that exponents add when two or more numbers are multiplied together:

$$x^2 x^3 = (x \times x)(x \times x \times x)$$
$$= x \times x \times x \times x \times x = x^5 = x^{2+3}$$

On the other hand, exponents multiply when a number is raised to a power:

$$(x^2)^3 = (x \times x) \times (x \times x) \times (x \times x)$$
$$= x \times x \times x \times x \times x \times x = x^6 = x^{2 \times 3}$$

In general, the rules obeyed by exponents can be summarized as follows:

$$x^n x^m = x^{n+m}$$
$$x^{-n} = \frac{1}{x^n}$$
$$\frac{x^n}{x^m} = x^{n-m} \qquad \text{A-16}$$
$$(xy)^n = x^n y^n$$
$$(x^n)^m = x^{nm}$$

Fractional exponents, such as $1/n$, indicate the nth root of a number. Specifically, the square root of x is written as

$$\sqrt{x} = x^{1/2}$$

For n greater than 2 we write the nth root in the following form:

$$\sqrt[n]{x} = x^{1/n} \qquad \text{A-17}$$

Thus, the nth root of a number, x, is the value that gives x when multiplied by itself n times: $(x^{1/n})^n = x^{n/n} = x^1 = x$.

A general method for calculating the exponent of a number is provided by the **logarithm**. For example, suppose x is equal to 10 raised to the power n:

$$x = 10^n$$

In this expression, 10 is referred to as the base. The exponent, n, is equal to the logarithm (log) of x:

$$n = \log x$$

The notation "log" is known as the *common logarithm*, and it refers specifically to base 10.

As an example, suppose that $x = 1000 = 10^n$. Clearly, we can write x as 10^3, which means that the exponent of x is 3:

$$\log x = \log 1000 = \log 10^3 = 3$$

When dealing with a number this simple, the exponent can be determined without a calculator. Suppose, however, that $x = 1205 = 10^n$. To find the exponent for this value of x we use the "log" button on a calculator. The result is

$$n = \log 1205 = 3.081$$

Thus, 10 raised to the 3.081 power gives 1205.

Another base that is frequently used for calculating exponents is $e = 2.718 \ldots$. To represent $x = 1205$ in this base we write

$$x = 1205 = e^m$$

The logarithm to base e is known as the *natural logarithm*, and it is represented by the notation "ln." Using the "ln" button on a calculator, we find

$$m = \ln 1205 = 7.094$$

Thus, e raised to the 7.094 power gives 1205. The connection between the common and natural logarithms is as follows:

$$\ln x = 2.3026 \log x \qquad \text{A-18}$$

In the example just given, we have $\ln 1205 = 7.094 = 2.3026 \log 1205 = 2.3026(3.081)$.

The basic rules obeyed by logarithms follow directly from the rules given for exponents in Equation A-16. In particular,

$$\ln(xy) = \ln x + \ln y$$
$$\ln\left(\frac{x}{y}\right) = \ln x - \ln y \qquad \text{A-19}$$
$$\ln x^n = n \ln x$$

Though these rules are stated in terms of natural logarithms, they are satisfied by logarithms with any base.

Mathematical Expansions

We conclude with a brief consideration of small quantities in mathematics. Consider the following equation:

$$(1 + x)^3 = 1 + 3x + 3x^2 + x^3$$

This expression is valid for all values of x. However, if x is much smaller than one, $x \ll 1$, we can say to a good approximation that

$$(1 + x)^3 \approx 1 + 3x$$

Now, just how good is this approximation? After all, it ignores two terms that would need to be included to produce an equality. In the case $x = 0.001$, for example, the two terms that are neglected, $3x^2$ and x^3, have a combined contribution of only about 3 ten-thousandths of a percent! Clearly, then, little error is made in the approximation $(1 + 0.001)^3 \sim 1 + 3(0.001) = 1.003$. This can be seen visually in **Figure A-5 (a)**, where we plot $(1 + x)^3$ and $1 + 3x$ for x ranging from 0 to 1. Note that there is little difference in the two expressions for x less than about 0.1.

This is just one example of a general result in mathematics that can be derived from the **binomial expansion.** In general, we can say that the following approximation is valid for $x \ll 1$:

$$(1 + x)^n \approx 1 + nx \qquad \text{A-20}$$

This result holds for arbitrary n, not just for the case of $n = 3$. For example, if $n = -1$ we have

$$(1 + x)^{-1} = \frac{1}{1 + x} \approx 1 - x$$

We plot $(1 + x)^{-1}$ and $1 - x$ in **Figure A-5 (b)**, and again we see that the results are in good agreement for x less than about 0.1.

An example of an expansion that arises in the study of relativity concerns the following quotient:

$$\frac{1}{\sqrt{1 - \dfrac{v^2}{c^2}}}$$

In this expression v is the speed of an object and c is the speed of light. Since objects we encounter generally have speeds much less than the speed of light, the ratio v/c is much less than one, and v^2/c^2 is even smaller than v/c. Thus, if we let $x = v^2/c^2$ we have

$$\frac{1}{\sqrt{1 - x}}$$

We can apply the binomial expansion to this result if we replace n with $-1/2$ and x with $-x$ in Equation A-20. This yields

$$\frac{1}{\sqrt{1 - \dfrac{v^2}{c^2}}} \approx 1 + \frac{1}{2}\frac{v^2}{c^2}$$

The two sides of this approximate equality are plotted in **Figure A-5 (c)**, showing the accuracy of the approximation for small v/c.

Another type of mathematical expansion leads to the following useful results:

$$\begin{aligned} \sin\theta &\approx \theta \\ \cos\theta &\approx 1 - \tfrac{1}{2}\theta^2 \end{aligned} \qquad \text{A-21}$$

These expansions are valid for small angles θ measured in radians. Note that the result $\sin\theta \approx \theta$ is used to derive Equations 6-14 and 13-19. (See Table 6-2, p. 170, and Figure 13-21, p. 435, for more details on this expansion.)

Vector Multiplication

There are two distinct ways to multiply vectors, referred to as the **dot product** and the **cross product**. The difference between these two types of multiplication is that the dot product yields a scalar (a number) as its result, whereas the cross product results in a vector. Both types of product have important applications in physics. In what follows, we present the basic techniques associated with dot and cross products, and point out places in the text where they are used.

THE DOT PRODUCT

Consider two vectors, $\vec{A}$ and $\vec{B}$, as shown in **Figure A-6 (a)**. The magnitudes of these vectors are A and B, respectively, and the angle

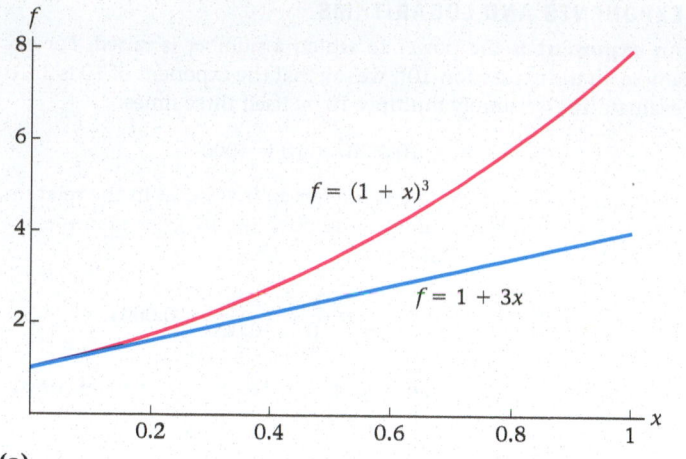

(a)

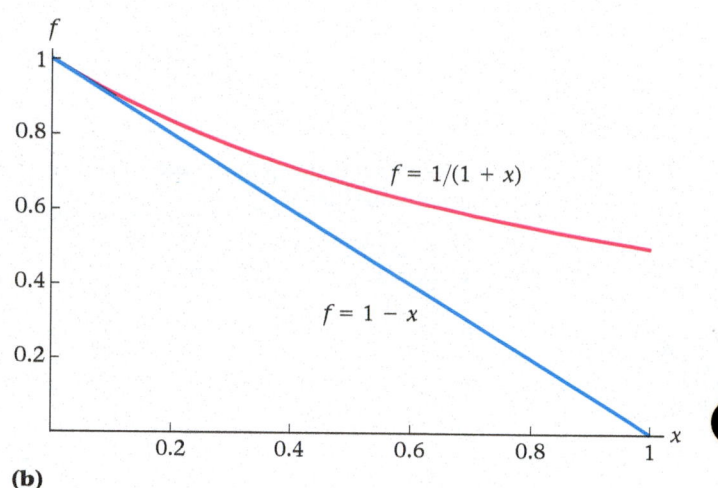

(b)

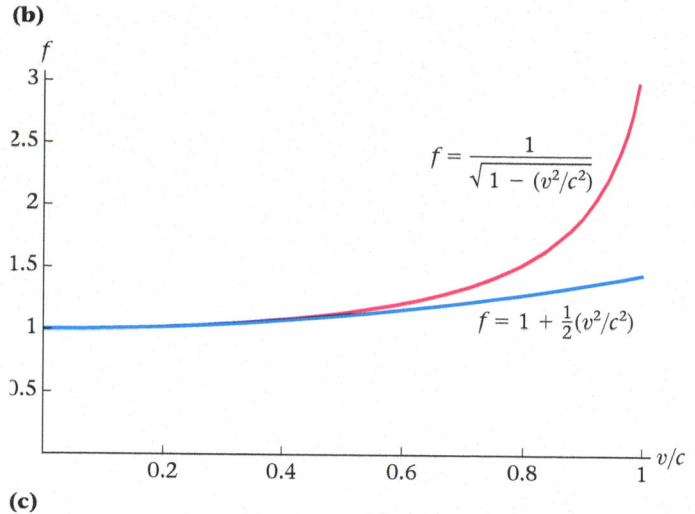

(c)

▲ **FIGURE A-5 Examples of mathematical expansions (a)** A comparison between $(1 + x)^3$ and the result obtained from the binomial expansion, $1 + 3x$. **(b)** A comparison between $1/(1 + x)$ and the result obtained from the binomial expansion, $1 - x$. **(c)** A comparison between $1/\sqrt{1 - (v^2/c^2)}$ and the result obtained from the binomial expansion, $1 + \frac{1}{2}(v^2/c^2)$.

between them is θ. We define the dot product of $\vec{A}$ and $\vec{B}$ as follows:

$$\vec{A} \cdot \vec{B} = AB\cos\theta \qquad \text{A-22}$$

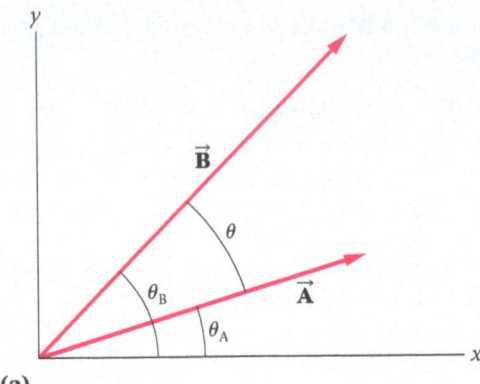

(a)

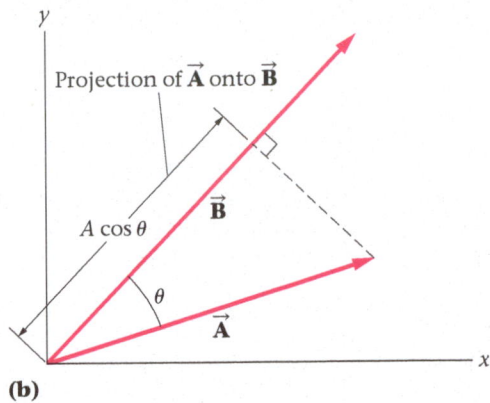

(b)

▲ **FIGURE A-6** The dot product between vectors $\vec{\mathbf{A}}$ and $\vec{\mathbf{B}}$.

In words, the dot product of two vectors is a scalar equal to the magnitude of one vector times the magnitude of the second vector times the cosine of the angle between them.

A geometric interpretation of the dot product is presented in **Figure A-6 (b)**. We begin by projecting the vector $\vec{\mathbf{A}}$ onto the direction of vector $\vec{\mathbf{B}}$. This is done by dropping a perpendicular from the tip of $\vec{\mathbf{A}}$ onto the line that passes through $\vec{\mathbf{B}}$, as shown in **Figure A-6 (b)**. Note that the projection of $\vec{\mathbf{A}}$ on the direction of $\vec{\mathbf{B}}$ has a length given by $A \cos \theta$. It follows that the dot product is simply the projection of $\vec{\mathbf{A}}$ onto $\vec{\mathbf{B}}$ times the magnitude of $\vec{\mathbf{B}}$; that is, $(A \cos \theta)B = AB \cos \theta = \vec{\mathbf{A}} \cdot \vec{\mathbf{B}}$. Equivalently, the dot product can be thought of as the projection of $\vec{\mathbf{B}}$ onto $\vec{\mathbf{A}}$ times the magnitude of $\vec{\mathbf{A}}$.

A few special cases will help to clarify the dot product. Suppose, for example, that $\vec{\mathbf{A}}$ and $\vec{\mathbf{B}}$ are parallel. In this case, $\theta = 0$ and $\vec{\mathbf{A}} \cdot \vec{\mathbf{B}} = AB$. Therefore, when vectors are parallel, the dot product is simply the product of their magnitudes. On the other hand, suppose $\vec{\mathbf{A}}$ and $\vec{\mathbf{B}}$ point in opposite directions. Now we have $\theta = 180$, and therefore $\vec{\mathbf{A}} \cdot \vec{\mathbf{B}} = -AB$. In general, the sign of $\vec{\mathbf{A}} \cdot \vec{\mathbf{B}}$ is positive if the angle between $\vec{\mathbf{A}}$ and $\vec{\mathbf{B}}$ is less than 90°, and is negative if the angle between them is greater than 90°. Finally, if $\vec{\mathbf{A}}$ and $\vec{\mathbf{B}}$ are perpendicular to one another—that is, if $\theta = 90°$—we see that $\vec{\mathbf{A}} \cdot \vec{\mathbf{B}} = AB \cos 90° = 0$. In this case, neither vector has a nonzero projection onto the other vector.

Dot products have a particularly simple form when applied to unit vectors. Recall, for example, that $\hat{\mathbf{x}}$ and $\hat{\mathbf{y}}$ have unit magnitude and are perpendicular to one another. It follows that

$$\hat{\mathbf{x}} \cdot \hat{\mathbf{x}} = 1, \quad \hat{\mathbf{x}} \cdot \hat{\mathbf{y}} = \hat{\mathbf{y}} \cdot \hat{\mathbf{x}} = 0, \quad \hat{\mathbf{y}} \cdot \hat{\mathbf{y}} = 1 \qquad \text{A-23}$$

These results can be applied to the general two-dimensional vectors $\vec{\mathbf{A}} = A_x \hat{\mathbf{x}} + A_y \hat{\mathbf{y}}$ and $\vec{\mathbf{B}} = B_x \hat{\mathbf{x}} + B_y \hat{\mathbf{y}}$ to give

$$
\begin{aligned}
\vec{\mathbf{A}} \cdot \vec{\mathbf{B}} &= (A_x \hat{\mathbf{x}} + A_y \hat{\mathbf{y}}) \cdot (B_x \hat{\mathbf{x}} + B_y \hat{\mathbf{y}}) \\
&= A_x B_x \hat{\mathbf{x}} \cdot \hat{\mathbf{x}} + A_x B_y \hat{\mathbf{x}} \cdot \hat{\mathbf{y}} + A_y B_x \hat{\mathbf{y}} \cdot \hat{\mathbf{x}} + A_y B_y \hat{\mathbf{y}} \cdot \hat{\mathbf{y}} \qquad \text{A-24} \\
&= A_x B_x + A_y B_y
\end{aligned}
$$

Thus, the dot product of tpwo-dimensional vectors is simply the product of their x components plus the product of their y components.

At first glance the result $\vec{\mathbf{A}} \cdot \vec{\mathbf{B}} = AB \cos \theta$ looks quite different from the result $\vec{\mathbf{A}} \cdot \vec{\mathbf{B}} = A_x B_x + A_y B_y$. They are identical, however, as we now show. Suppose that $\vec{\mathbf{A}}$ is at an angle θ_A to the positive x axis, and that $\vec{\mathbf{B}}$ is at an angle $\theta_B > \theta_A$ to the x axis, from which it follows that the angle between $\vec{\mathbf{A}}$ and $\vec{\mathbf{B}}$ is $\theta = \theta_B - \theta_A$. Noting that $A_x = A \cos \theta_A$ and $A_y = A \sin \theta_A$, and similarly for B_x and B_y, we have

$$\vec{\mathbf{A}} \cdot \vec{\mathbf{B}} = AB(\cos \theta_A \cos \theta_B + \sin \theta_A \sin \theta_B)$$

The second trigonometric identity in Equation A-9 can be applied to the quantity in brackets, with the result that $\vec{\mathbf{A}} \cdot \vec{\mathbf{B}} = A_x B_x + A_y B_y = AB \cos(\theta_B - \theta_A) = AB \cos \theta$, as desired.

The most prominent application of dot products in this text is in Chapter 7, where in Equation 7-3 we define the work to be $W = Fd \cos \theta$, with θ the angle between $\vec{\mathbf{F}}$ and $\vec{\mathbf{d}}$. Clearly, this is simply a statement that work is the dot product of force and displacement:

$$W = \vec{\mathbf{F}} \cdot \vec{\mathbf{d}} = Fd \cos \theta$$

Later in the text, in Equation 19-11, we define the electric flux to be $\Phi = EA \cos \theta$. If we let $\vec{\mathbf{A}}$ represent a vector that has a magnitude equal to the area, A, and points in the direction of the normal to the area, we can write the electric flux as a dot product:

$$\Phi = \vec{\mathbf{E}} \cdot \vec{\mathbf{A}} = EA \cos \theta$$

Similar remarks apply to the magnetic flux, defined in Equation 23-1.

THE CROSS PRODUCT

When two vectors are multiplied with the cross product, the result is a third vector that is perpendicular to both original vectors. An example is shown in **Figure A-7**, where we see a vector $\vec{\mathbf{A}}$, a vector $\vec{\mathbf{B}}$, and their cross product, $\vec{\mathbf{C}}$:

$$\vec{\mathbf{C}} = \vec{\mathbf{A}} \times \vec{\mathbf{B}} \qquad \text{A-25}$$

Notice that $\vec{\mathbf{C}}$ is perpendicular to the plane formed by the vectors $\vec{\mathbf{A}}$ and $\vec{\mathbf{B}}$. In addition, the direction of $\vec{\mathbf{C}}$ is given by the following right-hand rule:

To find the direction of $\vec{\mathbf{C}} = \vec{\mathbf{A}} \times \vec{\mathbf{B}}$, point the fingers of your right hand in the direction of $\vec{\mathbf{A}}$ and curl them toward $\vec{\mathbf{B}}$. Your thumb is now pointing in the direction of $\vec{\mathbf{C}}$.

It is clear from this rule that if $\vec{\mathbf{A}} \times \vec{\mathbf{B}} = \vec{\mathbf{C}}$, then $\vec{\mathbf{B}} \times \vec{\mathbf{A}} = -\vec{\mathbf{C}}$.

The magnitude of $\vec{\mathbf{C}} = \vec{\mathbf{A}} \times \vec{\mathbf{B}}$ depends on the magnitudes of the vectors $\vec{\mathbf{A}}$ and $\vec{\mathbf{B}}$, and on the angle θ between them. In particular,

$$C = AB \sin \theta \qquad \text{A-26}$$

Comparing with Equation A-22, we see that the cross product involves a sin θ, whereas the dot product depends on cos θ. As a result, it follows that the cross product has zero magnitude when $\vec{\mathbf{A}}$ and

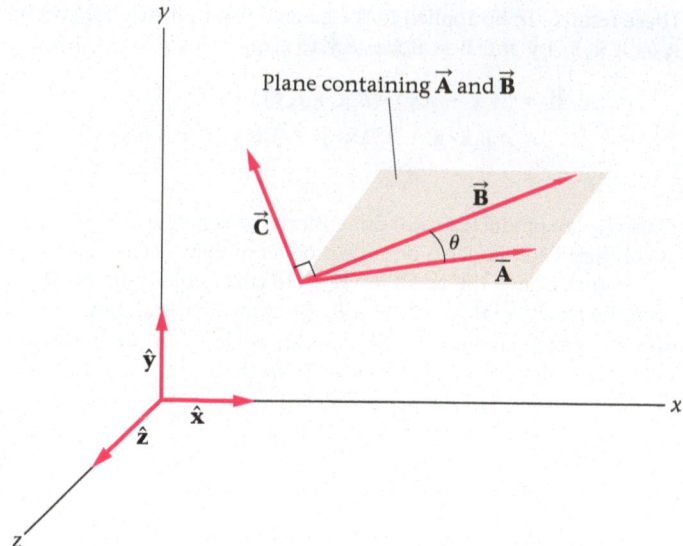

▲ **FIGURE A-7** The vector cross product of $\vec{A}$ and $\vec{B}$.

$\vec{B}$ point in the same direction ($\theta = 0°$) or in opposite directions ($\theta = 180°$). On the other hand, the cross product has its greatest magnitude, $C = AB$, when $\vec{A}$ and $\vec{B}$ are perpendicular to one another ($\theta = 90°$).

When we apply these rules to unit vectors, which are at right angles to one another and of unit magnitude, the results are particularly simple. For example, consider the cross product $\hat{x} \times \hat{y}$. Referring to Figure A-7, we see that this cross product points in the positive z direction. In addition, the magnitude of $\hat{x} \times \hat{y}$ is $(1)(1)\sin 90° = 1$. It follows, therefore, that $\hat{x} \times \hat{y} = \hat{z}$. On the other hand, $\hat{x} \times \hat{x} = \hat{y} \times \hat{y} = \hat{z} \times \hat{z} = 0$ because $\theta = 0°$ in each of these cases. To summarize:

$$\hat{x} \times \hat{y} = \hat{z}, \quad \hat{y} \times \hat{z} = \hat{x}, \quad \hat{z} \times \hat{x} = \hat{y}$$
$$\hat{y} \times \hat{x} = -\hat{z}, \quad \hat{z} \times \hat{y} = -\hat{x}, \quad \hat{x} \times \hat{z} = -\hat{y} \qquad \text{A-27}$$
$$\hat{x} \times \hat{x} = 0, \quad \hat{y} \times \hat{y} = 0, \quad \hat{z} \times \hat{z} = 0$$

As an example of how to use these unit-vector results, consider the cross product of the two-dimensional vectors shown in Figure A-6,

$\vec{A} = A_x\hat{x} + A_y\hat{y}$ and $\vec{B} = B_x\hat{x} + B_y\hat{y}$. Straightforward application of Equation A-27 yields

$$\begin{aligned}\vec{C} = \vec{A} \times \vec{B} &= (A_x\hat{x} + A_y\hat{y}) \times (B_x\hat{x} + B_y\hat{y}) \\ &= A_xB_x(\hat{x} \times \hat{x}) + A_xB_y(\hat{x} \times \hat{y}) \\ &\quad + A_yB_x(\hat{y} \times \hat{x}) + A_yB_y(\hat{y} \times \hat{y}) \\ &= (A_xB_y - A_yB_x)\hat{z} \end{aligned} \qquad \text{A-28}$$

Notice that $\vec{C}$ is perpendicular to both $\vec{A}$ and $\vec{B}$, as required for a cross product. In addition, the magnitude of $\vec{C}$ is

$$A_xB_y - A_yB_x = AB(\cos\theta_A \sin\theta_B - \sin\theta_A \cos\theta_B)$$

If we now apply the first trigonometric identity in Equation A-9, we recover the result given in Equation A-26:

$$\begin{aligned}C &= AB(\cos\theta_A \sin\theta_B - \sin\theta_A \cos\theta_B) \\ &= AB\sin(\theta_B - \theta_A) = AB\sin\theta\end{aligned}$$

The first application of cross products in this text is torque, which is discussed in Chapter 11. In fact, Equation 11-2 defines the magnitude of the torque, τ, as follows: $\tau = r(F\sin\theta)$. As one might expect by referring to Equations A-25 and A-26, the torque vector, $\vec{\tau}$, can be written as the following cross product:

$$\vec{\tau} = \vec{r} \times \vec{F}$$

Similarly, the angular momentum vector, $\vec{L}$, whose magnitude is given in Equation 11-13, is simply

$$\vec{L} = \vec{r} \times \vec{p}$$

Finally, the cross product appears again in magnetism. In fact, the magnitude of the magnetic force on a charge q with a velocity $\vec{v}$ in a magnetic field $\vec{B}$ is $F = qvB\sin\theta$, as given in Equation 22-1. As one might guess, the vector form of this force is

$$\vec{F} = q\vec{v} \times \vec{B}$$

The advantage of a cross product expression like this is that it contains both the direction and magnitude of a vector in one compact equation. In fact, we can now see the origin of the right-hand rule for magnetic forces given in Section 22-2.

APPENDIX B   TYPICAL VALUES

Mass

Sun	2.00×10^{30} kg
Earth	5.97×10^{24} kg
Moon	7.35×10^{22} kg
747 airliner (maximum takeoff weight)	3.5×10^{5} kg
blue whale	178,000 kg = 197 tons
elephant	5400 kg
mountain gorilla	180 kg
human	70 kg
bowling ball	7 kg
half gallon of milk	1.81 kg = 4 lbs
baseball	0.141–0.148 kg
golf ball	0.045 kg
female calliope hummingbird (smallest bird in North America)	3.5×10^{-3} kg $= \frac{1}{8}$ oz
raindrop	3×10^{-5} kg
antibody molecule (IgG)	2.5×10^{-22} kg
hydrogen atom	1.67×10^{-27} kg

Length

orbital radius of Earth (around Sun)	1.5×10^{8} km
orbital radius of Moon (around Earth)	3.8×10^{5} km
altitude of geosynchronous satellite	35,800 km = 22,300 mi
radius of Earth	6370 km
altitude of Earth's ozone layer	50 km
height of Mt. Everest	8848 m
height of Washington Monument	169 m = 555 ft
pitcher's mound to home plate	18.44 m
baseball bat	1.067 m
CD (diameter)	120 mm
aorta (diameter)	18 mm
period in sentence (diameter)	0.5 mm
red blood cell	7.8 μm $= \frac{1}{3300}$ in.
typical bacterium (E. coli)	2 μm
wavelength of green light	550 nm
virus	20–300 nm
large protein molecule	25 nm
diameter of DNA molecule	2.0 nm
radius of hydrogen atom	5.29×10^{-11} m

Time

estimated age of Earth	approx. 4.6 billion y $\approx 10^{17}$ s
estimated age of human species	approx. 150,000 y $\approx 5 \times 10^{12}$ s
half life of carbon-14	5730 y = 1.81×10^{11} s
period of Halley's comet	76 y = 2.40×10^{9} s
half life of technetium-99	6 h = 2.16×10^{4} s
time for driver of car to apply brakes	0.46 s
human reaction time	60–180 ms
air bag deployment time	10 ms

period of middle C sound wave	3.9 ms
collision time for batted ball	2 ms
decay of excited atomic state	10^{-8} s
period of green light wave	1.8×10^{-15} s

Speed

light	3×10^8 m/s
meteor	35–95 km/s
space shuttle (orbital velocity)	8.5 km/s = 19,000 mi/h
rifle bullet	700–750 m/s
sound in air (STP)	340 m/s
fastest human nerve impulses	140 m/s
747 at takeoff	80.5 m/s
kangaroo	18.1 m/s = 40.5 mi/h
200-m dash (Olympic record)	10.1 m/s
butterfly	1 m/s
blood speed in aorta	0.35 m/s
giant tortoise	0.076 m/s = 0.170 mi/h
Mer de Glace glacier (French Alps)	4×10^{-6} m/s

Acceleration

protons in particle accelerator	9×10^{13} m/s^2
ultracentrifuge	3×10^6 m/s^2
meteor impact	10^5 m/s^2
baseball struck by bat	3×10^4 m/s^2
loss of consciousness	$7.14g = 70$ m/s^2
acceleration of gravity on Earth (g)	9.81 m/s^2
braking auto	8 m/s^2
acceleration of gravity on the moon	1.62 m/s^2
rotation of Earth at equator	3.4×10^{-2} m/s^2

APPENDIX C PLANETARY DATA

Name	Equatorial Radius (km)	Mass (Relative to Earth's)*	Mean Density (kg/m^3)	Surface Gravity (Relative to Earth's)	Orbital Semimajor Axis × 10^6 km	Orbital Semimajor Axis A. U.	Escape Speed (km/s)	Orbital Period (Years)	Orbital Eccentricity
Mercury	2440	0.0553	5430	0.38	57.9	0.387	4.2	0.240	0.206
Venus	6052	0.816	5240	0.91	108.2	0.723	10.4	0.615	0.007
Earth	6370	1	5510	1	149.6	1	11.2	1.000	0.017
Mars	3394	0.108	3930	0.38	227.9	1.523	5.0	1.881	0.093
Jupiter	71,492	318	1360	2.53	778.4	5.203	60	11.86	0.048
Saturn	60,268	95.1	690	1.07	1427.0	9.539	36	29.42	0.054
Uranus	25,559	14.5	1270	0.91	2871.0	19.19	21	83.75	0.047
Neptune	24,776	17.1	1640	1.14	4497.1	30.06	24	163.7	0.009
Pluto	1137	0.0021	2060	0.07	5906	39.84	1.2	248.0	0.249

*Mass of Earth = 5.97×10^{24} kg

APPENDIX D   ELEMENTS OF ELECTRICAL CIRCUITS

Circuit Element	Symbol	Physical Characteristics
resistor		Resists the flow of electric current. Converts electrical energy to thermal energy.
capacitor		Stores electrical energy in the form of an electric field.
inductor		Stores electrical energy in the form of a magnetic field.
incandescent lightbulb		A device containing a resistor that gets hot enough to give off visible light.
battery		A device that produces a constant difference in electrical potential between its terminals.
ac generator		A device that produces a potential difference between its terminals that oscillates with time.
switches (open and closed)		Devices to control whether electric current is allowed to flow through a portion of a circuit.
ground		Sets the electric potential at a point in a circuit equal to a constant value usually taken to be $V = 0$.

APPENDIX E   PERIODIC TABLE OF THE ELEMENTS

Key:

Atomic number	26 **Fe**	Symbol
	55.85	Atomic mass
Outer electron configuration	$3d^64s^2$	

Transition elements

GROUP I

1 **H** 1.01 $1s^1$	
3 **Li** 6.94 $2s^1$	
11 **Na** 22.99 $3s^1$	
19 **K** 39.10 $4s^1$	
37 **Rb** 85.47 $5s^1$	
55 **Cs** 132.91 $6s^1$	
87 **Fr** (223) $7s^1$	

GROUP II

4 **Be** 9.01 $2s^2$
12 **Mg** 24.31 $3s^2$
20 **Ca** 40.08 $4s^2$
38 **Sr** 87.62 $5s^2$
56 **Ba** 137.33 $6s^2$
88 **Ra** 226.03 $7s^2$

Transition elements (d):

III	IV	V	VI	VII	VIII					
21 **Sc** 44.96 $3d^14s^2$	22 **Ti** 47.88 $3d^24s^2$	23 **V** 50.94 $3d^34s^2$	24 **Cr** 52.00 $3d^54s^1$	25 **Mn** 54.94 $3d^54s^2$	26 **Fe** 55.85 $3d^64s^2$	27 **Co** 58.93 $3d^74s^2$	28 **Ni** 58.69 $3d^84s^2$	29 **Cu** 63.55 $3d^{10}4s^1$	30 **Zn** 65.39 $3d^{10}4s^2$	
39 **Y** 88.96 $4d^15s^2$	40 **Zr** 91.22 $4d^25s^2$	41 **Nb** 92.91 $4d^45s^1$	42 **Mo** 95.94 $4d^55s^1$	43 **Tc** (98) $4d^55s^2$	44 **Ru** 101.07 $4d^75s^1$	45 **Rh** 102.91 $4d^85s^1$	46 **Pd** 106.42 $4d^{10}5s^0$	47 **Ag** 107.87 $4d^{10}5s^1$	48 **Cd** 112.41 $4d^{10}5s^2$	
57 **La** 138.91 $5d^16s^2$ *	72 **Hf** 178.49 $5d^26s^2$	73 **Ta** 180.95 $5d^36s^2$	74 **W** 183.85 $5d^46s^2$	75 **Re** 186.21 $5d^56s^2$	76 **Os** 190.2 $5d^66s^2$	77 **Ir** 192.22 $5d^76s^2$	78 **Pt** 195.08 $5d^96s^1$	79 **Au** 196.97 $5d^{10}6s^1$	80 **Hg** 200.59 $5d^{10}6s^2$	
89 **Ac** 227.03 $6d^17s^2$ †	104 **Rf** (261) $6d^27s^2$	105 **Db** (262) $6d^37s^2$	106 **Sg** (266) $6d^47s^2$	107 **Bh** (264) $6d^57s^2$	108 **Hs** (269) $6d^67s^2$	109 **Mt** (268) $6d^77s^2$	110 (271)	111 (272)	112 (277)	

Main group (p):

III	IV	V	VI	VII	VIII
					2 **He** 4.00 $1s^2$
5 **B** 10.81 $2p^1$	6 **C** 12.01 $2p^2$	7 **N** 14.01 $2p^3$	8 **O** 16.00 $2p^4$	9 **F** 19.00 $2p^5$	10 **Ne** 20.18 $2p^6$
13 **Al** 26.98 $3p^1$	14 **Si** 28.09 $3p^2$	15 **P** 30.97 $3p^3$	16 **S** 32.07 $3p^4$	17 **Cl** 35.45 $3p^5$	18 **Ar** 39.95 $3p^6$
31 **Ga** 69.72 $4p^1$	32 **Ge** 72.61 $4p^2$	33 **As** 74.92 $4p^3$	34 **Se** 78.96 $4p^4$	35 **Br** 79.90 $4p^5$	36 **Kr** 83.80 $4p^6$
49 **In** 114.82 $5p^1$	50 **Sn** 118.71 $5p^2$	51 **Sb** 121.76 $5p^3$	52 **Te** 127.60 $5p^4$	53 **I** 126.90 $5p^5$	54 **Xe** 131.29 $5p^6$
81 **Tl** 204.36 $6p^1$	82 **Pb** 207.2 $6p^2$	83 **Bi** 208.98 $6p^3$	84 **Po** (209) $6p^4$	85 **At** (210) $6p^5$	86 **Rn** (222) $6p^6$
	114 (289)		116 (289)		118 (293)

Lanthanides (f):

58 **Ce** 140.12 $4f^15d^16s^2$	59 **Pr** 140.91 $4f^36s^2$	60 **Nd** 144.24 $4f^46s^2$	61 **Pm** (145) $4f^56s^2$	62 **Sm** 150.36 $4f^66s^2$	63 **Eu** 151.96 $4f^76s^2$	64 **Gd** 157.25 $5d^14f^76s^2$	65 **Tb** 158.93 $4f^96s^2$	66 **Dy** 162.50 $4f^{10}6s^2$	67 **Ho** 164.93 $4f^{11}6s^2$	68 **Er** 167.26 $4f^{12}6s^2$	69 **Tm** 168.93 $4f^{13}6s^2$	70 **Yb** 173.04 $4f^{14}6s^2$	71 **Lu** 174.97 $5d^14f^{14}6s^2$

Actinides (f):

90 **Th** 232.04 $6d^27s^2$	91 **Pa** 231.04 $5f^26d^17s^2$	92 **U** 238.03 $5f^36d^17s^2$	93 **Np** 237.05 $5f^46d^17s^2$	94 **Pu** (244) $5f^66d^07s^2$	95 **Am** (243) $5f^76d^07s^2$	96 **Cm** (247) $5f^76d^17s^2$	97 **Bk** (247) $5f^86d^17s^2$	98 **Cf** (251) $5f^{10}6d^07s^2$	99 **Es** (252) $5f^{11}6d^07s^2$	100 **Fm** (257) $5f^{12}6d^07s^2$	101 **Md** (258) $5f^{13}6d^07s^2$	102 **No** (259) $5f^{14}6d^07s^2$	103 **Lr** (262) $5f^{14}6d^17s^2$

PERIODS: 1, 2, 3, 4, 5, 6, 7

APPENDIX F   PROPERTIES OF SELECTED ISOTOPES

Atomic Number (Z)	Element	Symbol	Mass Number (A)	Atomic Mass*	Abundance (%) or Decay Mode† (if radioactive)	Half-Life (if radioactive)
0	(Neutron)	n	1	1.008665	β^-	10.6 min
1	Hydrogen	H	1	1.007825	99.985	
	Deuterium	D	2	2.014102	0.015	
	Tritium	T	3	3.016049	β^-	12.33 y
2	Helium	He	3	3.016029	0.00014	
			4	4.002603	≈ 100	
3	Lithium	Li	6	6.015123	7.5	
			7	7.016003	92.5	
4	Beryllium	Be	7	7.016930	EC, γ	53.3 d
			8	8.005305	2α	6.7×10^{-17} s
			9	9.012183	100	
5	Boron	B	10	10.012938	19.9	
			11	11.009305	80.1	
			12	12.014353	β^-	20.2 ms
6	Carbon	C	11	11.011433	β^+, EC	20.3 min
			12	12.000000	98.89	
			13	13.003355	1.11	
			14	14.003242	β^-	5730 y
7	Nitrogen	N	13	13.005739	β^-	9.96 min
			14	14.003074	99.63	
			15	15.000109	0.37	
8	Oxygen	O	15	15.003065	β^+, EC	122 s
			16	15.994915	99.76	
			18	17.999159	0.204	
9	Fluorine	F	19	18.998403	100	
			18	18.000938	EC	109.77 min
10	Neon	Ne	20	19.992439	90.51	
			22	21.991384	9.22	
11	Sodium	Na	22	21.994435	β^+, EC, γ	2.602 y
			23	22.989770	100	
			24	23.990964	β^-, γ	15.0 h
12	Magnesium	Mg	24	23.985045	78.99	
13	Aluminum	Al	27	26.981541	100	
14	Silicon	Si	28	27.976928	92.23	
			31	30.975364	β^-, γ	2.62 h
15	Phosphorus	P	31	30.973763	100	
			32	31.973908	β^-	14.28 d
16	Sulfur	S	32	31.972072	95.0	
			35	34.969033	β^-	87.4 d
17	Chlorine	Cl	35	34.968853	75.77	
			37	36.965903	24.23	
18	Argon	Ar	40	39.962383	99.60	
19	Potassium	K	39	38.963708	93.26	
			40	39.964000	β^-, EC, γ, β^+	1.28×10^9 y

Atomic Number (Z)	Element	Symbol	Mass Number (A)	Atomic Mass*	Abundance (%) or Decay Mode† (if radioactive)	Half-Life (if radioactive)
20	Calcium	Ca	30	39.962591	96.94	
24	Chromium	Cr	52	51.940510	83.79	
25	Manganese	Mn	55	54.938046	100	
26	Iron	Fe	56	55.934939	91.8	
27	Cobalt	Co	59	58.933198	100	
			60	59.933820	β^-, γ	5.271 y
28	Nickel	Ni	58	57.935347	68.3	
			60	59.930789	26.1	
			64	63.927968	0.91	
29	Copper	Cu	63	62.929599	69.2	
			64	63.929766	β^-, β^+	12.7 h
			65	64.927792	30.8	
30	Zinc	Zn	64	63.929145	48.6	
			66	65.926035	27.9	
33	Arsenic	As	75	74.921596	100	
35	Bromine	Br	79	78.918336	50.69	
36	Krypton	Kr	84	83.911506	57.0	
			89	88.917563	β^-	3.2 min
			92	91.926153	β^-	1.84 s
38	Strontium	Sr	86	85.909273	9.8	
			88	87.905625	82.6	
			90	89.907746	β^-	28.8 y
39	Yttrium	Y	89	89.905856	100	
41	Niobium	Nb	98	97.910331	β^-	2.86 s
43	Technetium	Tc	98	97.907210	β^-, γ	4.2×10^6 y
47	Silver	Ag	107	106.905095	51.83	
			109	108.904754	48.17	
48	Cadmium	Cd	114	113.903361	28.7	
49	Indium	In	115	114.90388	95.7; β^-	5.1×10^{14} y
50	Tin	Sn	120	119.902199	32.4	
51	Antimony	Sb	133	132.915237	β^-	2.5 min
53	Iodine	I	127	126.904477	100	
			131	130.906118	β^-, γ	8.04 d
54	Xenon	Xe	132	131.90415	26.9	
			136	135.90722	8.9	
55	Cesium	Cs	133	132.90543	100	
56	Barium	Ba	137	136.90582	11.2	
			138	137.90524	71.7	
			141	140.914406	β^-	18.27 min
			144	143.92273	β^-	11.9 s
61	Promethium	Pm	145	144.91275	EC, α, γ	17.7 y
74	Tungsten (Wolfram)	W	184	183.95095	30.7	
76	Osmium	Os	191	190.96094	β^-, γ	15.4 d
			192	191.96149	41.0	
78	Platinum	Pt	195	194.96479	33.8	
79	Gold	Au	197	196.96656	100	

Atomic Number (Z)	Element	Symbol	Mass Number (A)	Atomic Mass*	Abundance (%) or Decay Mode† (if radioactive)	Half-Life (if radioactive)
81	Thallium	Tl	205	204.97441	70.5	
			210	209.990069	β^-	1.3 min
82	Lead	Pb	204	203.973044	β^-, 1.48	1.4×10^{17} y
			206	205.97446	24.1	
			207	206.97589	22.1	
			208	207.97664	52.3	
			210	209.98418	α, β^-, γ	22.3 y
			211	210.98874	β^-, γ	36.1 min
			212	211.99188	β^-, γ	10.64 h
			214	213.99980	β^-, γ	26.8 min
83	Bismuth	Bi	209	208.98039	100	
			211	210.98726	α, β^-, γ	2.15 min
			212	211.991272	α	60.55 min
84	Polonium	Po	210	209.98286	α, γ	138.38 d
			212	211.988852	α	0.299 μs
			214	213.99519	α, γ	164 μs
86	Radon	Rn	222	222.017574	α, β	3.8235 d
87	Francium	Fr	223	223.019734	α, β^-, γ	21.8 min
88	Radium	Ra	226	226.025406	α, γ	1.60×10^3 y
			228	228.031069	β^-	5.76 y
89	Actinium	Ac	227	227.027751	α, β^-, γ	21.773 y
90	Thorium	Th	228	228.02873	α, γ	1.9131 y
			231	231.036297	α, β^-	25.52 h
			232	232.038054	100; α, γ	1.41×10^{10} y
			234	234.043596	β^-	24.10 d
91	Protactium	Pa	234	234.043302	β^-	6.70 h
92	Uranium	U	232	232.03714	α, γ	72 y
			233	233.039629	α, γ	1.592×10^5 y
			235	235.043925	0.72; α, γ	7.038×10^8 y
			236	236.045563	α, γ	2.342×10^7 y
			238	238.050786	99.275; α, γ	4.468×10^9 y
			239	239.054291	β^-, γ	23.5 min
93	Neptunium	Np	239	239.052932	β^-, γ	2.35 d
94	Plutonium	Pu	239	239.052158	α, γ	2.41×10^4 y
95	Americium	Am	243	243.061374	α, γ	7.37×10^3 y
96	Curium	Cm	245	245.065487	α, γ	8.5×10^3 y
97	Berkelium	Bk	247	247.07003	α, γ	1.4×10^3 y
98	Californium	Cf	249	249.074849	α, γ	351 y
99	Einsteinium	Es	254	254.08802	α, γ, β^-	276 d
100	Fermium	Fm	253	253.08518	EC, α, γ	3.0 d
101	Mendelevium	Md	255	255.0911	EC, α	27 min
102	Nobelium	No	255	255.0933	EC, α	3.1 min
103	Lawrencium	Lr	257	257.0998	α	$\approx$35 s

*The masses given throughout this table are those for the neutral atom, including the Z electrons.
†EC stands for electron capture.

CHAPTER 1

1. (a) 0.152 gigadollars (b) 1.52×10^{-4} teradollars
3. 3×10^8 m/s
5. (a) Yes (b) No (a) Yes
7. (a) Yes (b) Yes (c) No (d) Yes
9. $p = 1$
11. $p = \frac{1}{2}$
13. $[M]/[T]^2$
15. 330.7 m
17. (a) three (b) four
19. (a) 3.14 (b) 3.1416 (c) 3.141593
21. 18 ft^3
23. 0.9788 km
25. 32.9 m
27. 322 ft/s^2
29. (a) 75 ft/s (b) 51 mi/h
31. (a) 67 mutchkin (b) 0.031 gal
33. 32.2 ft/s^2
35. 10^5 seats
37. (a) 10^{10} gal/y (b) 10^9 lb/y
39. (a) Yes (b) No (c) No (d) Yes
41. (a) 6.75×10^{-4} mm (b) 2.66×10^{-5} in
43. 1.2×10^3 lb
45. 55 shakes/s
47. (a) 4.7×10^{-4} mm/s (b) 0.94 ft/week
49. (a) greater than (b) 46 ft/s^2
51. (a) 310 mi/h (b) 0.70 m
53. (a) 51 rev (b) 8.7 ft/rev
55. $p = 1$; $q = -1$
57. 43 mi/h
59. D. 52 s
61. C. 4.78×10^9

CHAPTER 2

1. (a) 1.95 mi (b) 0.75 mi
3. (a) distance = 5 m, Δx = 5 m
   (b) distance = 2 m, Δx = −2 m
5. (a) distance = 130 m, Δx = 100 m
   (b) distance = 260 m, Δx = 0
7. (a) less than (b) I
9. 23.32 mi/h
11. (a) 0.098 m/s (b) 0.22 mi/h
13. 2.2 km
15. 5.5 m/s
17. 8.5 m
19. (a) less than (b) 0.75 m/s
21. (a) positive (b) zero (c) positive
   (d) negative (e) 2.0 m/s (f) 0 m/s
   (g) 1.0 m/s (h) −1.5 m/s

23. (a) x (m)
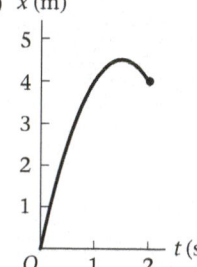

(b) 4.0 m/s (c) 4.0 m/s
25. 35 mi/h

27. (a)
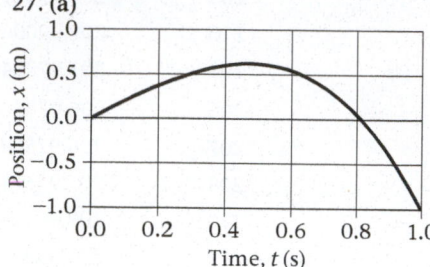

(b) 0.55 m/s (c) 0.56 m/s
   (d) closer to 0.56 m/s
29. (a) greater than (b) I
31. (a) 3.8 m/s (b) 9.9 m/s
33. (a) 32.8 m/s north (b) 15.4 m/s north
35. (a) 10 m (b) 20 m (c) 40 m
37. (a) factor of two (b) 4.3 s (c) 8.6 s
39. 6.50 m/s
41. 9.0 m/s to the west
43. (a) 2.41 m/s (b) 11.5 m
45. (a) 0 m; 5.0 m/s; −20 m/s^2
   (b)
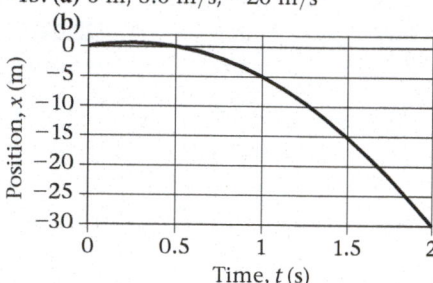

(c) −5.0 m/s (d) 25 m/s

47. (a)
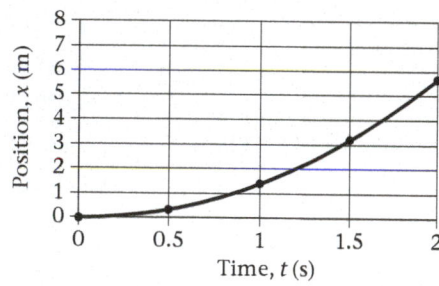

(b) 2.2 m/s (c) 2.8 m/s^2
49. 9.1 m/s^2
51. (a) 0.077 s (b) 0.46 μm
53. 3.8×10^4 m/s^2
55. (a) 21 m (b) greater than 6.0 m/s
   (c) 8.49 m/s
57. (a) 32 m/s^2 (b) less than 8.0 cm
   (c) 4.0 cm
59. (a) 5.7 s (b) −0.175 m/s^2 (c) 2.2 m/s
61. (a) 4.8 m/s^2 toward third base (b) 3.4 m
63. (a) 8.8 m (b) 4.4 m (c) 4.4 m/s
65. (a) a_A = 1.71 m/s^2, a_B = 3.36 m/s^2,
   a_C = 4.91 m/s^2 (b) θ_A = 10.0°,
   θ_B = 20.0°, θ_C = 30.0°
67. 2.7 s
69. 11 m/s
71. (a) 8.4 m/s (b) −1.4 m/s
73. 0.10 s

75. (a) equal to (b) II
77. (a) 11 m (b) 15 m/s (c) 2.1 s
79. 13 m
81. (a) first (dropped) ball (b) ball 1: 24.5 m/s; ball 2: 15.7 m/s
83. (a) 0.49 s (b) 4.8 m/s
85. (a) 65 m (b) 36 m/s (c) 8.5 s
87. (a) 6.7 m/s (b) −5.9 m/s (c) 9.6 m/s
89. 1.01 m/s
91. 5.9 km
93. ball A v negative, a negative; ball B v positive, a negative
95. (a) greater than (b) less than
97. (a) 4.0 m/s^2 (b) 14.5 m (c) 64.5 m
99. (a) 9.81 m/s^2 downward (b) 13.9 m
   (c) 2.21 s (d) 16.5 m/s
101. 3.0 m
103. (a) 12.4 m/s^2 (b) −9.20 m/s^2
105. (a) $\sqrt{2gh}$ (b) $-gh/d$
107. (a) 0.60 s (b) 0.44 m
109. B. 6.78 ft/s
111. plot C
113. 4.3 m/s^2
115. (a) 2.3 s (b) 20 m/s

CHAPTER 3

1. (a) factor of 2 (b) factor of 1
3. $D < C < B < A$
5. 187 ft
7. 3°
9. (a) $(75 \text{ m})\hat{\mathbf{x}} + (47 \text{ m})\hat{\mathbf{y}}$
   (b) $(39 \text{ m})\hat{\mathbf{x}} + (79 \text{ m})\hat{\mathbf{y}}$
11. (a) 56 m (b) 130°
13. 42 ft
15. (a) −34° (b) 17 m (c) direction remains same; magnitude doubles (d) −34°; 34 m
17. (a) $\vec{\mathbf{A}}$ (b) $\vec{\mathbf{A}}$
19. (a) 73.5 m (b) 202 m
21. (a) less than (b) equal to
23. (a)
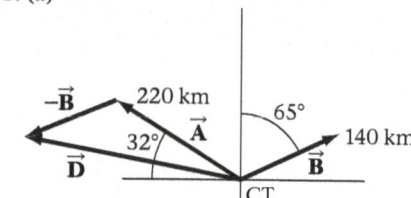

(b) 320 km, 10° north of west
25. (a)
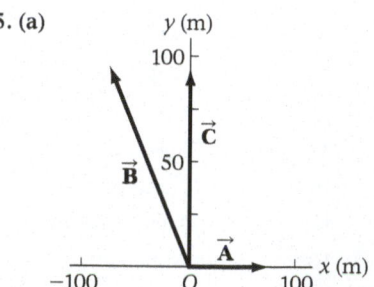

(b) about 120 m; about 130° (c) 121 m; 128°
27. (a) −27°; 11 units (b) 207°; 11 units
   (c) 27°; 11 units

29. $(40\text{ m})\hat{\mathbf{x}} + (-36\text{ m})\hat{\mathbf{y}}$
31. (a) 23 m (b) 23 m
33. (a) $-22°$; 5.4 m (b) 110°; 5.4 m
    (c) 45°; 4.2 m
35. (a) $(20\text{ m})\hat{\mathbf{x}} + (-27\text{ m})\hat{\mathbf{y}}$
    (b) $(-20\text{ m})\hat{\mathbf{x}} + (27\text{ m})\hat{\mathbf{y}}$
37. $(2.1\text{ m})\hat{\mathbf{x}} + (0.74\text{ m})\hat{\mathbf{y}}$
39. 40° south of west; 5.3 km
41. (a) 1.4 km
    (b) $(-490\text{ m})\hat{\mathbf{x}} + (-950\text{ m})\hat{\mathbf{y}}$
    (c) $(490\text{ m})\hat{\mathbf{x}} + (950\text{ m})\hat{\mathbf{y}}$
43. 59° north of east; 4.9 m/s
45. $(-9.8\text{ m/s}^2)\hat{\mathbf{y}}$
47. 22.0 m/s
49. 15.3 m/s
51. (a) $(-4.5\text{ m/s})\hat{\mathbf{x}}$ (b) $(4.5\text{ m/s})\hat{\mathbf{x}}$
53. 18 s
55. (a) 11° west of north
    (b)

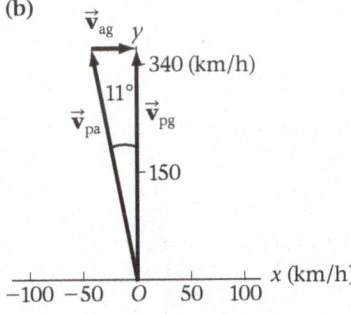

    (c) increased
57. 11 m/s
59. (a) Jet Ski A (b) $\Delta t_A/\Delta t_B = 0.82$
61. (a) equal to (b) II
63. $-53.1°$; 40.3 m
65. between 27° and 360°
67. (a) 326 km (b) 46.4° south of east
    (c) 40.3 min
69. (a) 10.4 m/s (b) 70.4° below horizontal
71. (a) $(16.6\text{ m/s})\hat{\mathbf{x}} - (17.2\text{ m/s})\hat{\mathbf{y}}$
    (b) 23.9 m/s; 46.0° below horizontal
73. (a) 53° north of east; 12 m/s (b) 53°
    south of west; 12 m/s
75. (a) 18 cm; 215° counterclockwise from
    $\vec{\mathbf{A}}$ (b) 0.0034 mm/s at 215°
77. (a) 3.7 (b) 67 m/s
79. (a) 3.5 cm (b) less than (c) 1.7 cm
81. $A = 28\text{ m}; B = 19\text{ m}$
83. C. 1.00 m/s
85. D. 52.1°
87. (a) 12° upstream (b) decrease

CHAPTER 4

1. (a) straight upward (b) III
3. (a) 12.0 s (b) 55.5 s
5. (a) $x = -55\text{ m}; y = 31\text{ m}$
    (b) $v_x = -22\text{ m/s}; v_y = 6.2\text{ m/s}$
    (c) increase with time
7. (a) 48 s (b) 0.021 m/s^2
    (c) $v_x = 2.5\text{ m/s}; v_y = 1.0\text{ m/s}$
9. (a) 27° (b) 0.87 m/s
11. (a) greater than (b) II
13. 46.2 m/s
15. 1.77 m/s^2
17. (a) 9.7 m (b) 1.4 m
19. (a) 3.1 m (b) 3.4 m beyond the far edge
21. 2.6 m/s
23. (a) 1.41 m (b) the same time

25. 1.20 m
27. C. 60°
29. (a) A < B < C (b) A = B = C
31. (a) 14.3 m/s (b) 2.24 s
33. 1.3 m
35. $\theta_A = -90°; \theta_B = -47°$
37. 67.3 m
39. (a) 0.26 s (b) 2.2 m
41. 75.5°
43. 3.03 m
45. (a) 44.5° (b) 30.1 m/s
47. (a) 21.50 m/s (b) 0.18 m/s
49. (a) 849 m/s (b) less than
51. 63.1°
53. equal to
55. equal to
57. (a) less than (b) II
59. 0.786 m
61. 1.2 s
63. (a) $(15.6\text{ m/s})\hat{\mathbf{x}} + (3.7\text{ m/s})\hat{\mathbf{y}}$ (b) 17 m
65. (a) 1.52 m (b) $(4.03\text{ m/s})\hat{\mathbf{y}}$
    (c) x (m)

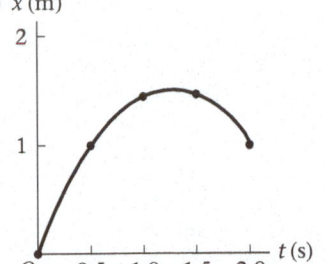

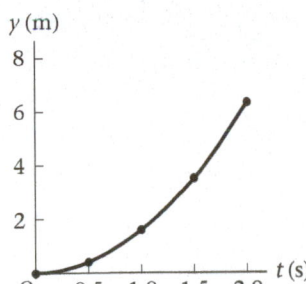

67. 24.0 m
69. (a) $x = 0.950\text{ m}; y = 0.950\text{ m}$
    (b) $x = 0.950\text{ m}; y = 0.699\text{ m}$
71. 17 m
73. (a) 12.1 m/s; at the peak (b) 13.2 m/s
    at 23.5° above horizontal (c) 1.40 m
    (d) 1.40 m
75. 5.8 m
77. 5.8 m/s
79. (a) 6.53 m/s (b) 0.270 s
    (c) $(6.53\text{ m/s})\hat{\mathbf{x}} + (-2.65\text{ m/s})\hat{\mathbf{y}}$
81. (a) $v_1\dfrac{X}{Y}$ (b) $\left(v_1\dfrac{X}{Y}t - X\right)\hat{\mathbf{x}} + (Y - v_1 t)\hat{\mathbf{y}}$
83. 76.0°
87. (a) 5.60 m (b) 4.98 m (c) 5.20 m
89. D. 6.0 mm
91. C. 40 cm
93. (a) 18.2 m (b) 2.5 m/s
95. (a) 1.40 m (b) stay the same (c) 1.40 m

CHAPTER 5

1. $\frac{1}{2}T$
3. 2.72 m
5. 1.6 kN

7.

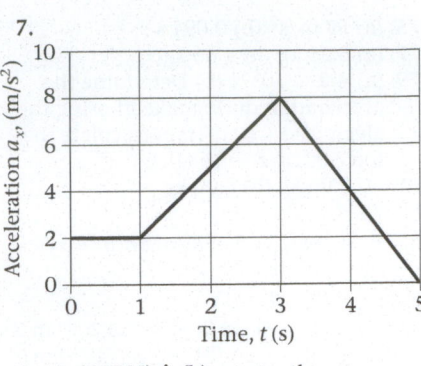

9. (a) $(827\text{ N})\hat{\mathbf{y}}$ (b) greater than
11. 0.14 kg
13. (a) 54.3 m/s (b) 646 m
15. (a) $(-18.7\text{ kN})\hat{\mathbf{x}}$ (b) Determine the
    acceleration from the speeds and
    displacement, and the force from the
    acceleration and mass.
17. (a) greater than (b) II
19. 9.2 km/s^2
21. (a) the same as (b) more than
    (c) 0.93 m/s^2
23. (a) 1.04 N (b) 3.59 N
25. (a) sometimes true (b) never true
    (c) always true (d) sometimes true
27. 7.8 kN
29. 0.10 kN
31. 1.5 m/s^2
33. 146°; 7.8 N
35. (a) 24.5° (b) 4.77×10^{20} N
    (c) 0.00649 m/s^2
37. (a) 0.023 N (b) 6.6 days
39. 2.4 m/s^2 upward
41. (a) upward (b) 1.9 m/s^2 (c) The eleva-
    tor is either speeding up if it is travel-
    ing upward, or slowing down if it is
    traveling downward.
43. (a) $(0.0119\text{ N})\hat{\mathbf{y}}$ (b) the same as
45. 0.11 kN
47. (a) 91 N (b) 0.13 kN
49. (a) $\frac{1}{2}mg$ (b) mg (c) $2mg$
51. (a) 23 N (b) increase
53. 60°
55. (a) higher than (b) I
57. (a) greater than (b) III
59. 69 m/s^2
61.

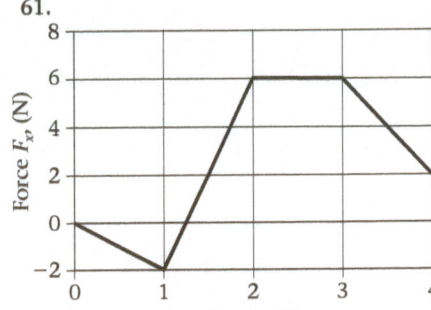

63. 1.57×10^5 N
65. A = B = C
67. $(5.29\text{ N})\hat{\mathbf{x}} + (-6.00\text{ N})\hat{\mathbf{y}}$
69. (a) 3.5 (b) 25 m/s^2 upward
    (c) 0.25 m/s
71. 2.70×10^4 kg
73. (a) $(-0.42\text{ kN})\hat{\mathbf{x}}$ (b) 2

75. (a) 59 m/s² (b) 0.091 s
77. (a) 5.25 kg (b) 4.66 kg
79. (a) 4.2×10^5 N (b) Determine the acceleration during takeoff using the given data (x, t). Then calculate the force using $F = ma$.
81. (a) 2.06 kg (b) 5.00 kg
83. 74 kg
85. $F_1 = \frac{1}{2}m(a_1 + a_2)$; $F_2 = \frac{1}{2}m(a_1 - a_2)$
87. C. 1.05×10^4 N
89. D. 4.21×10^4 N
91. (a) decreased (b) 2.00 kg (c) 2.50 m/s²
93. (a) greater than (b) 36° (c) 0.079 m/s²

CHAPTER 6

1. (a) equal to (b) II
3. (a) 0.29 (b) equal to
5. 0.10 kN
7. box 3 < box 2 < box 1
9. (a) 0.42 (b) 23°
11. (a) 0.64 N (b) 0.25
13. $v^2/2\mu g$ (b) quadruples (c) the same
15. (a) greater than (b) greater than
17. 2.13 kN/m
19. 9.4 cm
21. (a) 4.8 cm (b) Yes; the spring compression would double.
23. (a) 0.33 kN (b) 0.66 kN (c) 0.66 kN
25. (a) 5.1 N (b) 15 N
27. 4.5 kg
29. (a) Graph B (b) II
31. $F_1 = F_2 = 0.85$ N
33. (a) 0.020 kN (b) remains the same
35. (a) stay the same (b) increase
37. (a) $T_1 < T_2 < T_3$ (b) $T_1 < T_2 < T_3$
39. (a) decrease (b) 1.4 m/s²
41. (a) less than (b) 4.36 m/s² (c) 15.3 N
43. (a) 3.2 m/s² (b) greater than (c) 4.7 m/s²
45. Path B
47. 6.1 kN
49. 101 m
51. (a) Your apparent weight is greater at the bottom of the Ferris wheel, where the normal force must both support your weight and provide the upward centripetal acceleration. (b) 0.52 kN at top; 0.56 kN at bottom
53. less than
55. inward
57. (a) 35 N (b) greater than (c) 40 N
59. 0.48 km/s²
61. 0.285
63.

applied force	friction force magnitude	motion
0 N	0 N	at rest
5.0 N	5.0 N	at rest
11 N	11 N	at rest
15 N	8.8 N	accelerating
11 N	8.8 N	accelerating
8.0 N	8.8 N → 8.0 N	decelerating → at rest
5.0 N	5.0 N	at rest

65. (a) 58° (b) 1.4 N

67. (a) 32 N (b) increase
69. (a) static (b) 0.50
71. (a) 0.39 N (b) less than
73. 28°
75. 5.4 cm
77. $v = \sqrt{Mrg/m}$
79. (a) 0.163 s (b) 0.102 m
81. C. 2.7 N/m
83. A. 0.23 N
85. (a) 0.315 (b) 0.127
87. (a) 15 m/s (b) equal to

CHAPTER 7

1. (a) positive (b) zero
3. 0.34 kJ
5. (a) 38 J (b) zero
7. 1.2 J
9. 0.15 kJ
11. (a) 14 J (b) decrease
13. 35 m
15. (a) 0.55 J (b) 0.28 J (c) −1.4 J
17. (a) positive (b) negative
19. 12.7 MJ
21. (a) 1500 kg (b) 16 m/s
23. C < A = D < B
25. (a) −10 J (b) 0.63 N upward
27. (a) −587 J (b) 0.284
29. (a) −7.2 kJ (b) 28 m (c) 0.26 kN
31. (a) 44 J (b) 44 J
33. (a) 0.45 J (b) 0.24 J
35. Spring 1 is stiffer
37. (a) 0.80 m/s (b) 0.21 m
39. $F_4 < F_3 < F_1 < F_2$
41. 3.2 μW
43. 1.1 kN
45. (a) 0.22 W (b) 0.57 N
47. (a) 15.6 N (b) double the speed
49. (a) less than (b) 1.8 kW (c) 5.4 kW
51. (a) 3T (b) $v\sqrt{2}$
53. (a) $3 \Delta x$ (b) III
55. 1.2×10^{-20} J
57. (a) 0.18 kW (b) 6.3 kJ
59. (a) $W_0/4$ (b) $3W_0/4$
61. $K/2$
63. (a) 13.1 J (b) 13.1 J (c) stay the same
65. 0.348
67. (a) 0.80 J (b) 1.4 J (c) −1.3 J
69. (a) 1.04 kN/m (b) 563 N
71. (a) 12 W (b) 1.1 kJ
73. (a) 15 h (b) 0.62 m/s (c) 1.6 s
75. (a) kg/m (b) $\sqrt[3]{p/b}$ (c) 1.26
77. $\frac{1}{2}(k_1 + k_2)x^2$
79. B. 7.7 − 15 m/s
81. C. 490 J
83. (a) 0.93 kN/m (b) 1.7 cm
85. (a) 7.8 s (b) less than (c) 3.9 s

CHAPTER 8

1. (a) −3 J (b) −4 J
3. $W_1 = -0.10$ kJ; $W_2 = -47$ J; $W_3 = -66$ J
5. (a) $W_1 = 51$ J; $W_2 = 51$ J (b) increase
7. (a) equal to (b) II
9. 7.6 MJ
11. 538 kJ
13. (a) factor of 4 (b) 3.85 J
15. (a) 0.59 mm (b) 5.7 W
17. −0.70 J
19. (a) K (b) I

21. 6.73 m/s
23. 5.9 m/s
25.

y (m)	4.0	3.0	2.0	1.0	0
U (J)	8.2	6.2	4.1	2.1	0
K (J)	0	2.1	4.1	6.2	8.2
E (J)	8.2	8.2	8.2	8.2	8.2

27. (a) 15 m/s (b) 43 m
29. (a) 0; 113 J; 113 J (b) 113 J; 226 J; 113 J
31. (a) $v_f = \sqrt{2gh\left(\dfrac{m_2 - m_1}{m_1 + m_2}\right)}$ (b) 1.1 m/s
33. (a) less than (b) III
35. 2.7 m/s
37. −45 MJ
39. (a) decrease (b) stay the same (c) decrease
41.

	W_{nc}(J)	U (J)	K (J)	E (J)
(a)	0	34	0	34
(b)	−2.3	24	7.0	31
(c)	−4.6	15	14	29

43. (a) 0.76 kN (b) −38 kJ (c) 30 m/s
45. As the object travels from point A to point B, its speed increases. From point B to point C, its speed decreases, from point C to point D, the speed increases again, and from point D to point E, the speed decreases. At point E it comes to rest momentarily before retracing its path all the way back to point A, at which time the cycle begins again.
47. 0.6 m and 4.6 m
49. (a)

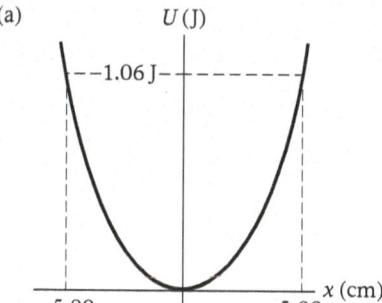

(b) ± 4.5 cm
51. (a) 500 N/m (b) 0.29 m
53. (a) disagree (b) agree (c) agree
55. 7.65 m/s
57. (a) 1.5 mm (b) 0.46 J
59. 1.04 m
61. −58 GJ
63. $\Delta y = h\left(\dfrac{m_2 - m_1}{m_1 + m_2}\right) = 6$ cm
65. 1.1 kN
67. 0.85 m
69. 6.3 cm
71. 48.2°
73. (a) $5r/2$ (b) Although a larger mass would have a larger gravitational force on it, it would also have a larger inertia, hence the effect of the extra mass cancels out.

75. 0.097 m
77. C. 290 N/m
79. A. 0.766 mJ
81. (a) 1.03 m/s (b) 1.19 m/s

CHAPTER 9

1. 1.2 kg
3. 53 kg·m/s
5. (a) 0.42 m/s (b) No (c) 0.19 J
7. (a) 1.3 kg·m/s (b) -0.14 kg·m/s
   (c) The quantity in part (a)
9. inelastic
11. (a) less than (b) I
13. D < C < B < A
15. 9.0 ms
17. (a) 0.33 kN (b) 73.6 kg
19. 2.7 kg·m/s
21. (a) 27° above horizontal; 5.6 kg·m/s
    (b) double in magnitude; no change in
    direction (c) no change
23. 31 kg
25. 2; The piece with the smaller kinetic
    energy has the larger mass.
27. -0.80 m/s
29. 12 m/s; 11 m/s
31. -0.067 J
33. (a) 5.5 m/s (b) 3.8 m/s (c) -0.38 kN
35. 1.32 m/s; 107°
37. 1.1 J
39. (a) 0.9978 (b) 1×10^{-6} (c) 0.9807
41. $v_{1,f} = 0.40$ m/s; $v_{2,f} = 0.16$ m/s;
    $v_{1,f} = 0.64$ m/s
43. less than
45. 0.50 m
47. 1.05 m
49. $X_{cm} = 0$; $Y_{cm} = 0.036$ nm
51.

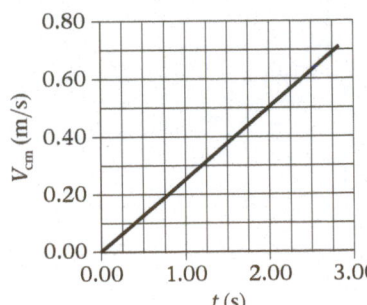

53. (a) zero (b) zero
55. 63 rocks/min
57. 0.23 kN
59. (a) 0.25 N (b) more than 2.5 N (c) 2.8 N
61. C < A < B
63. -0.10 kg·m/s
65. (a) green path (b) II
67. (a) 0.57 (b) -3.1×10^3 kg·m/s

(c) 1.4×10^3 kg·m/s
(d) 4.5×10^3 kg·m/s
69. 0.24 N
71. 0.48 kg
73. $R/2$
75. 3.80 cm/s
77. (a) No (b) 6.71 cm
79. (a) 65 N (b) 3.1 m/s² (c) 9.4 m/s
81. (a) $3v_0/7$ (b) 3/7
87. (b) The net force on the two masses
    must point downward to accelerate the
    center of mass downward, hence the
    upward scale force must be less than
    the total weight.
89. A 0.95 knot
91. B. 270 kN
93. (a) 0.160 km/s (b) 1.58 m/s
95. (a) 0.19 m/s (b) -0.022 J

CHAPTER 10

1. $\frac{\pi}{6}$rad; $\frac{\pi}{4}$rad; $\frac{\pi}{2}$rad; π rad
3. (a) 60 rev/h (b) 360 deg/min
   (c) $\pi/30$ rad/s
5. 9.4 rad
7. 0.018 s
9. (a) 31 rad/s (b) 8.3 rad/s (c) -4.4 rad/s
   (d) -43 rad/s
11. (a) 1.3×10^2 rad/s (b) 2.1×10^2 rad/s
    (c) 3.0×10^2 rad/s (d) positive (e) 85
    rad/s²; 85 rad/s²
13. $\omega/2$
15. (a) -0.226 rad/s² (b) 28.1 s
17. (a) 1.4 rev (b) 2.9 s
19. (a) 6.29 rev/s² (b) 327 rev
21. (a) -334 rad/s² (b) 33.1 m (c) 10.0 in
23. (a) equal to (b) III
25. 26 rad/s
27. (a) 1.3×10^2 rad/s (b) 690 m/s²
29. (a) 0.503 m/s² (b) 0.503 m/s²
31. (a) 0.29 m/s (b) The linear speed would
    also double.
33. 172° below the direction of motion;
    2.1 m/s²
35. (a) 4.12×10^{16} rad/s
    (b) 6.56×10^{15} orbits
    (c) 8.98×10^{22} m/s²
37. 0.5 rev
39. 1.12 m/s
41. (a) 0.069 s (b) 91 rad
43. 69 page/min
45. (a) greater than (b) II
47. 0.36 m
49. 1.4 mJ
51. (a) 12 mJ (b) doubled (c) 68 rad/s
53. (a) 1.2 MJ (b) 1.8 min
55. (a) 1.1 kJ (b) 0.17 km
57. 0.17 kg·m²
59. (a) Equate the initial and final mechan-
    ical energies, then solve the resulting
    expression for the mass of the pulley.
    (b) 2.2 kg
61. (a) 0.53 m/s (b) stay the same
63. 5.3 m/s
65. (a) 18 J (b) 5.3 J (c) 13 J
67. greater than
69. 874 m
71. (a) 2.0 kJ (b) 3.5% (c) 86%

73. $I_3 < I_2 < I_1$
75. (a) 2.07 rev (b) 0.501 rev
77. (a) decrease (b) 50.0 rad/s
    (c) 20.8 rad/s (d) -7.31×10^{-3} rad/s²
79. (a) 8.6 kJ (b) 3.5 kJ
81. 0.20 m/s
83. (a) 6.9 rad/s² (b) It doubles.
85. (a) 15 rad/s² (b) 4.0×10^3 rad
87. 750 rev
89. (a) $\sqrt{3g/L}$ (b) $\sqrt{3gL}$
91. (a) 0.50 m (b) 2.4 rotations
93. (a) 1.5 m (b) 2.7 rev (c) increase
95. C. 30.6 rpm
97. C. 3
99. (a) Object 2 first, object 1 second,
    object 3 third (b) $K_1 = K_2 = K_3$

CHAPTER 11

1. 60 N
3. 25 N·m
5. 68 N·m
7. (a) counterclockwise (b) counterclockwise (c) clockwise
9. C < B < A
11. 1.09 kg·m²
13. 4.05 N·m
15. (a) -0.25 N·m (b) decrease
17. C < B < A
19. (a) Greatest about the x axis; least
    about the z axis. (b) 7.4 rad/s²
    (c) 5.1 rad/s² (d) 3.0 rad/s²
21. 0.290 N·m
23. (a) more than (b) 90.3 N
25. 1.1 m
27. 9.14 cm
29. (a) $(-21.3$ N$) \hat{y}$ (b) $(32.1$ N$) \hat{y}$
31. (a) $F/\sqrt{2}$ (b) $-F/2$ (c) $F/2$
33. $F_H = 330$ N; $F_j = 534$ N
35. 49.7 cm from the milk
37. (a) less than (b) 0.080 kg
39. 1.05 kg
41. 208 kg
43. 18 kN·m
45. 2.5 kg·m²/s
47. (a) point B (b) the same as
    (c) $L_A = 0$ kg·m²/s;
    $L_B = 2.55 \times 10^3$ kg·m²/s;
    $L_O = 2.55 \times 10^3$ kg·m²/s
49. 0.20 kN·m
51. (a) stay the same (b) III
53. (a) stay the same (b) decrease
55. 2.84 rad/s
57. (a) faster (b) 38 rev/min
59. (a) $\dfrac{Iv}{I + mR^2}$ (b) As $I \to 0$, $v_{c,g} \to 0$
    (c) As $I \to \infty$, $v_{c,g} \to v$
61. 0.089 N·m
63. 24.8 J
65. (a) 33 J (b) 1.4 W
67. (a) 0.85 W (b) 1.2 W (c) 2.1 W
69. (a) counterclockwise (b) move to the right
71. (a) move toward the left (b) II
73. 5.5 mJ
75. (a) increase (b) $2v$
77. 4.4 kN/m
79. (a) $f_1 = 0.89$ kN; $f_2 = 0.15$ kN;
    $f_3 = 0.15$ kN (b) $f_1 = 0.89$ kN;
    $f_2 = 0.22$ kN; $f_3 = 0.22$ kN

81. (a) 2.49 m (b) 1.49 kN
83. 0.16 N
85. 962 N
87. $\sqrt{15}\,Mg$
89. 60°
91. 2L/3
93. A. 0.0023 N·m
95. A. $F_{1y} = -1.7$ N, $F_{2y} = 3.5$ N
97. (a) increase (b) 1.0 N
99. 121°

CHAPTER 12

1. B < A < C < D
3. (a) 5.1 kN (b) 0.17 kN
5. (a) 3.4×10^{-12} N (b) 3.4×10^{-12} N
7. (a) 3.56×10^{22} N toward the Sun
   (b) 2.40×10^{20} N toward the Sun
   (c) 3.58×10^{22} N toward the Earth-
   Moon system
9. 3.58×10^{22} N at 0.147° toward Earth
   from the Sun-Moon line
11. (a) 4.7×10^{-8} N at 16° to the left of
    downward (b) All magnitudes reduced
    by a factor of 4, directions unchanged.
13. (a) 3.70 m/s² (b) 8.87 m/s²
15. 5.6×10^{-8} N
17. 13.5 m/s²
19. $M_E/96$
21. (a) Use $\frac{1}{2}mv_i^2 = mgh_f$ to find g,
    and use $g = GM/R^2$ to find M.
    (b) 8.94×10^{22} kg
23. (a) increase (b) stay the same
25. 1.98 h
27. 1.4×10^{11} m
29. (a) less than (b) 1.5×10^{17} kg
31. (a) satellite 2 (b) 5.59 km/s (c) 4.56 km/s
33. (a) farther from (b) 2.36×10^7 m
35. 1.91×10^{30} kg
37. 36.3 GJ
39. (a) -1.5×10^{-8} J (b) quadruple
    (c) double
41. -48.8 nJ
43. (a) 9.0×10^{10} J (b) 2.0×10^{12} J
45. less than
47. 5.03 km/s
49. 1.05 km/s
51. (a) less than (b) 783 m/s
53. 5.73×10^7 m
55. 2.7×10^5 m
57. (a) 2.07×10^{-7} m/s (b) increase
59. (a) 8.2 N (b) 1.7×10^8 m
63. increase
65. B < C < A
67. -6.34×10^{-10} J
69. (a) 1.98×10^{20} N (b) 4.36×10^{20} N
    (c) Orbiting the Sun, with a small effect
    due to the Earth.
71. 7.40×10^{-5} m/s
73. (a) more than (b) 39.2 m/s²
75. (a) 0.18 N (b) 1.7 m/s
77. $T = \sqrt{\dfrac{72\pi^2 r_1^3}{Gm_1}}$
79. 2.3 m/s
81. (a) greater than (b) westward (c) 25.4 h
83. $C = 1.00$
85. (a) No (b) 7.66 km/s (c) 1.55 h
87. (a) $\sqrt{GM_E/r}$ (c) Yes

89. $T = 2\pi\sqrt{\dfrac{\sqrt{3}\,R^3}{GM}}$
91. C. 2.7×10^{12} J
93. more than
95. 8.3×10^{10} m
97. (a) increase (b) 15.8 km/s
    (c) stay the same

CHAPTER 13

1. 1.75 s; 0.57 Hz
3. 6.3 s
5. (a) 0.029 s (b) 34 Hz
7. (a) 0.018 s; 55 Hz (b) 1800 rpm
9. (a) T/2 (b) 3T/4
11. (a) 1.5 Hz (b) 0.34 s
13. (a) v_{max} (b) zero (c) zero (d) $-a_{max}$
15. (a) $x = (3.50$ nm$)\cos[(1.26 \times 10^{15}$ rad/s$)\,t]$ (b) sine function
17. (a) 0.88 s (b) -2.7 cm
19. 10.7 s
21. (a) 20 m (b) 30 s
23. (a) 0.0495g (b) 0.128 m/s (c) minimum
25. 12g
27. (a) 1.1 km/s² (b) 6.2 m/s
29. (a) 0.017 J (b) 0.34 N
31. 2T/3
33. (a) greater than (b) III
35. (a) equal to (b) II
37. 2.28 kg
39. Block 1: 0.708 s; block 2: 0.708 s
41. (a) 13 rad/s (b) 0.45 m/s (c) 0.48 s
43. (a) 1.06×10^3 kg (b) 9.3×10^2 kg
45. 6.31 cm
47. (a) greater than (b) $\sqrt{2}\,T$
49. 1.3 J
51. (a) 0.26 m/s (b) 2.8 cm
53. (a) 13.1 N/m (b) 0.309 s (c) decrease
55. downward
57. 8.95 m
59. 2.4 s
61. $T = 2\sqrt{2}\,\pi\sqrt{\dfrac{R}{g}}$
63. (a) greater than (b) 0.667 m
65. (a) 1.2 s (b) 0.91 m/s
67. 2D
69. (a) remain the same (b) remain the same
71. 0.52 m/s
73. 2.9×10^{-9} Hz
75. (a) zero (b) zero (c) positive
    (d) negative
77. (a) 0.384 Hz (b) 0.125 m/s (c) 5.20 cm
79. (a) less than (b) 0.240 km/s (c) 0.433 s
81. 0.34 s
83. (a) 0.107 m (b) 0.125 s
85. (a) greater than 0.50 m/s (b) 0.79 m/s
    (c) 1.2 J
87. (a) 7.9 N/m (b) 4.6 kg
89. (a) less than (b) $T = \pi\left(\sqrt{\dfrac{\ell}{g}} + \sqrt{\dfrac{L}{g}}\right)$
    (c) 1.5 s
91. $T = \pi\sqrt{\dfrac{2L}{g}\left(\dfrac{m_1 + m_2}{m_1 - m_2}\right)}$
93. B. 10
95. B. 2.2 Hz
97. 11 s
99. (a) stay the same (b) 0.199 s

CHAPTER 14

1. (a) 44 cm (b) 8.0 cm
3. 1.2 m
5. (a) 0.43 Hz (b) 3.7 m/s
7. (a) 2.7 m/s (b) 0.59 Hz
9. (a) 1.00 (b) 0.50
11. (a) less than (b) II
13. (a) B < C < A (b) A < C < B
    (c) A = B = C
15. (a) less time (b) 0.10 s (c) 0.083 s
17. (a) zero (b) zero (c) 18 N (d) 6.6 m/s
    (e) 1.4 s
19. $v \propto \sqrt{\dfrac{F}{R^2\rho}}$
21. $y = (0.11\text{ m})\cos\left(\dfrac{\pi}{1.3\text{ m}}x - \dfrac{\pi}{0.091\text{ s}}t\right)$
23. 1.94×10^8 m/s
25. (a)

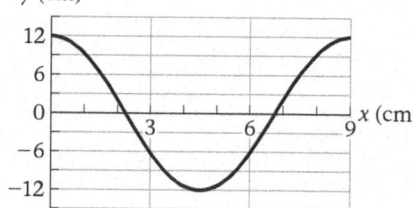

(b)

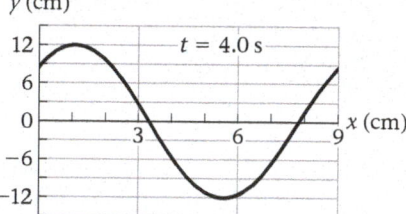

(c)

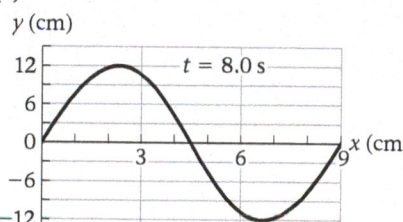

    (d) 8.0 s
27. 266 m
29. lowest note 27.50 Hz, 12.5 m; highest
    note 4.187 kHz, 8.19 cm
31. (a) 0.847 m (b) decrease (c) 0.707 m
33. 1.01 m/s
35. 6.3×10^{-4} W/m²
37. 107 W/m²
39. (a) 104 dB (b) 99.6 dB (c) 2.0×10^6 m
41. (a) 69.5 dB (b) less than
43. (a) greater than (b) III
45. 19.5 m/s
47. (a) 35.3 kHz (b) higher (c) 35.4 kHz
49. 4.2 min
51. (a) 0.33 kHz (b) (i)
53. 15.7 m/s
55. (a) 2.0 cm (b) -2.0 cm
57. 2.1 kHz
59. 0.322 m

61. 147 Hz
63. 49 Hz
65. 2.6 kHz
67. (a) 50 Hz (b) 150 Hz
69. (a) third (b) 44 cm
71. (a) 2.60 Hz (b) 5.21 Hz (c) increase
73. (a) 92.5 Hz (b) 1.85 m
75. string B
77. 3 Hz, 4 Hz, 7 Hz
79. 469 Hz
81. 1.20 m
83. A, III; B, I; C, I; D, III; E, II; F, II
85. 6.72×10^6 m
87. 4 antinodes
89. 1.0×10^{-10} W/m^2
91. 43 m/s
93. (a) No (b) 875 Hz
95. 36.1 m/s
97. 6.73 m/s approaching, 7.00 m/s receding
99. (a) 2.00 (b) 1/3 (c) 2.00
101. (a) 0.0032 J (b) 0.0081 J (c) No
103. 7.06 Hz
105. (a) increase
107. C. 64 Hz
109. B. 2.3 m
111. (a) greater than (b) 30.1 m/s
113. (a) increase (b) 464 Hz

CHAPTER 15

1. 10^3 N
3. No
5. 5.8×10^8 N
7. 1.9×10^6 Pa
9. 601 N
11. (a) 145 cm^2 (b) decrease (c) 43.7 lb/in^2
13. A = B = C = D
15. 2.5 kN
17. 1.1 m
19. (a) 3.2 m (b) 129 kPa
21. (a) 1.45×10^5 Pa (b) 1.54×10^5 Pa
    (c) Because pressure increases with depth.
23. 1.03×10^5 Pa
25. (a) The pressure difference between the
    liquid surface and your mouth creates an
    upward force on the liquid. (b) 10.3 m
27. 0.74 m
29. (a) equal to (b) No
31. 8.4 kN
33. (a) remain the same (b) zero
35. 1.08 kg/m^3
37. (a) decrease (b) III
39. less than
41. at the same level
43. (a) 3.1×10^{-4} m^3 (b) 7.1×10^3 kg/m^3
45. (a) 1.04×10^3 kg/m^3 (b) 0.0744 m^3
    (c) 26 N
47. (a) 0.008 m^3 (b) 31 N
49. 2.1 cm
51. 62 m/s
53. 16 min
55. (a) 1.9 cm/s (b) 0.039 cm/s
57. 7.9 m/s
59. (a) 32 cm/s (b) 43 Pa
61. (a) 6.1 m/s (b) 1.4 cm
63. (a) The tube with the greater diameter
    (b) The tube with the smaller diameter
    (c) 0.90 m/s
65. 7.5 m/s

67. (a) 19 kPa (b) 2.0 kN
69. (a) 980 kN (b) upward
71. (a) less than (b) $P_{narrow} - P_{wide} = -\frac{15}{2}\rho v^2$
73. (a) 20 cm^3 (b) 2.2
75. (a) 2.0 cm^3/s (b) 1.4 L/s
77. move up
79. (a) higher (b) stay the same
81. greater than
83. 95.3 kPa
85. (a) 8.97×10^3 kg/m^3 (b) 820 kg/m^3
87. (a) 0.00141 m^3 (b) decrease
    (c) 0.00140 m^3
89. (a) 10.6 N
91. (a) 4.1 kPa (b) increase (c) 4.3 kPa
93. (a) 3.1×10^{-4} m^3 (b) drop (c) 1.5 cm
95. (a) 5.9 m/s (b) 119 kPa
97. (a) 9.58×10^{-3} m^3 (b) 10.1 kg
99. (a) 94 Pa (b) 0.0025 m^3/s
101. (a) 17 μm (b) yes (c) 21 kg
103. 0.28
107. B. 360 kg/m^3
109. C. 0.28H
111. (a) decrease (b) 0.20 cm
113. (a) 0.715 m (b) 0.104 m

CHAPTER 16

1. (a) less than (b) II
3. 2530°C
5. (a) 5.7×10^3 °C (b) 1.0×10^4 °F
7. (a) 79.3 kPa (b) 151°C
9. 0.23 K/s
11. 320°F = 160°C
13. (a) strip B (b) lead
15. (a) decreased (b) I
17. 1.6 m
19. (a) 96.23 mm (b) −45°C
21. (a) 83°C (b) −82°C
23. 0.0014 in
25. (a) 19×10^{-6} K^{-1} (b) brass
27. (a) water will overflow (b) 24 cm^3
29. 98 W, 0.13 hp
31. 2.6×10^4
33. 4.3 h
35. (a) greater than (b) I
37. 15 kJ
39. 13 K
41. 22.9°C
43. (a) 118 kJ (b) 16.4 kW
45. (a) 837 J/(kg·K); glass
47. 70.5°C
49. $\Delta T_B < \Delta T_A < \Delta T_C$
51. (a) less than (b) III
53. 63 kJ/min
55. 1.6 J
57. (a) 0.800 m (b) decrease by a factor of 2
59. (a) greater than (b) 98°C
61. 49°C
63. (a) 9P/4 (b) 9P
65. (a) No (b) Yes (c) No
67. (a) greater than (b) II
69. 1 K
71. 43 J/s
73. (a) 591 steps (b) 18°C
75. 77.9°F
77. (a) heated (b) 860°C
79. 0.17 C°
81. (a) 637 kW (b) 664 kW
    (c) $T_{f,1} = 33.9$°C, $T_{f,2} = 34.1$°C

83. 2.4 min
85. (a) 3.9×10^{26} W (b) 1.4 kW/m^2
87. 9:59:56.8 A.M.
89. 1.1 m
91. C. Using cooled blankets above and below
93. A. 95 W
95. 0.93 kg
97. (a) increase (b) 35.2 J

CHAPTER 17

1. (a) equal to (b) less than
3. 1/4
5. 0.55 L
7. 1.33×10^3 kg
9. C < A < B < D
11. (a) 3.9×10^{24} atoms (b) 1.26
13. 15.6 g/mol
15. 0.8328
17. 1.60×10^5 Pa
19. (a) less than (b) III
21. 19.1 K
23. (a) 4.0×10^6 Pa (b) 1.18×10^{-20} J
    (c) The pressure decreases by a factor of
    2 and the average kinetic energy of O_2
    remains the same.
25. (a) decrease (b) 920 m/s
27. (a) 2.01% (b) 2.01%
29. 2.6×10^{-19} N
31. 51 kg
33. 1.1×10^6 N/m^2
35. $Y_3 < Y_1 = Y_2 < Y_4$
37. (a) 1.8×10^{-3} (b) 0.60 cm
39. 233 N
41. 4.2 kPa
43. (a) 2.3 kPa (b) 15°C (59°F)
45. (a) 3.5 MPa (b) increase
47. (a) gas (b) solid (c) 5707 kPa
49. 6.6×10^5 J
51. 528 kJ
53. D < C < B < A
55. (a) No (b) 0°C; 0.4 kg
57. (a) 27.3 s (b) 61.5 s (c) 246 s
    (d) The water is boiling.
59. 2.9 MJ
61. 44.2 g
63. 2.3 kg
65. (a) 3.6°C (b) No ice is present.
67. 50.3 h
69. (a) less than (b) III
71. 2.0×10^{22} molecules
73. (a) 0.21 kg (b) 93 kJ
75. 133°C
77. (a) 2.07×10^7 Pa (b) 369°C
79. (a) 0.034 (b) 0.034 (c) 0.043
81. (a) 4.2×10^{-7} m (b) 2.7×10^{-6} m
83. 1.1×10^6 students
85. (a) 25 W (b) 19 h
87. 15 s
89. A. 65.2 mol
91. A. 47 cm
93. (a) 0°C (b) 1.8°C

CHAPTER 18

1. (a) zero (b) zero (c) −100 J
3. $\Delta U = -11.6 \times 10^5$ J, $W = 7.7 \times 10^5$ J,
    $Q = -3.9 \times 10^5$ J
5. (a) 119 J (b) 35 J (c) zero
7. (a) −822 kJ (b) 196 C

9. (a) -4.24 MJ/mi (b) decrease
11. (a) zero (b) -136 J (c) -205 J (d) 341 J (e) 494 J
13. The pressure decreases by a factor of 3.
15. 60 Pa
17. (a) on the system (b) -670 J
19. (a) 1200 kJ (b) No
21. (a) 408 K (b) 555 kJ
23. (a) added to the system (b) 55 kJ
25. (a) 150 kJ (b) zero (c) 150 kJ
27. (a) $T_A = 267$ K, $T_B = 357$ K, $T_C = 89.1$ K (b) A→B heat enters, B→C heat leaves, C→A heat enters (c) A→B 376 kJ, B→C -375 kJ, C→A 150 kJ
29. (a) increase (b) 310 K
31. (a) (i) (b) (iii) (c) (i) (d) (iii)
33. (a) 3.1 kJ (b) 1.9 kJ
35. (a) 23 kJ (b) 14 kJ
37. (a) 8000 K (b) 2.88×10^7 Pa
39. (a) 0.315 (b) 0.630 (c) 1.0×10^5 Pa; 222 K
41. (a) Process 1: 159 kJ; process 2: 1060 kJ (b) 424 kJ (c) 795 kJ
43. 0.30
45. (a) 8.5 kJ (b) 6.0 kJ
47. (a) 1.16 GW (b) 1.71 GW
49. (a) 407 K (b) decreased (c) 339 K
51. (a) 0.660 (b) 872 MW (c) 0.297
53. (a) 1/3 (b) 2/3
55. (a) 0.59 kJ (b) 0.19
57. (a) 1.1×10^8 J (b) \$340
59. 0.41 kW
61. 1.5×10^5 J
63. (a) increase (b) I
65. 1.12×10^4 J/K
67. $D < C < B < A$
69. 0.11 kJ/K
71. 2.6 J/K
73. (a) negative (b) positive (c) negative
75. 0.250
77. (a) 1900 m^3 (b) 2.6×10^7 J
79. (a) Engine A (b) 0.32
81. (a) 751 kJ (b) 194 kJ (c) 945 kJ
83. (a) $nR\Delta T$ (b) zero (c) $\frac{3}{2}nR\Delta T$ (d) zero (e) $\frac{3}{2}nR\Delta T$ (f) $nRT\ln(V_f/V_i)$ (g) zero
85. (a) -503 J/K (b) 168 kJ (c) 569 J/K (d) 66 J/K
87. (a) decreases (b) -0.65 kJ/K (c) 0.66 kJ/K (d) 12 J/K
89.

	Q	W	ΔU
A → B	806 J	806 J	0
B → C	-938 J	-375 J	-563 J
C → A	563 J	0	563 J

93. A. 6.10%
95. C. 500 m
97. (a) greater than (b) 0.488 (c) 512 J

CHAPTER 19

1. (a) decrease (b) I
3. (a) rabbit fur positive, glass negative (b) glass positive, rubber negative (c) glass and rubber
5. (a) -4.4×10^{11} C (b) 2.4×10^8 C
7. 21 cm
9. (a) 933 (b) 9.91×10^{-25} kg
11. $C < A < B$

13. 1.32 m
15. (a) 1.93 N (b) the same as
17. 83 N to the right
19. (a) $F_1 = F_3 < F_2$ (b) 0° (c) 150° (d) 300°
21. 4.7×10^{-10} m
23. (a) 0.14 kN to the right (b) The direction would not change but the magnitude would be cut to a ninth.
25. $(x, y) = (0, -6.9$ km$)$
27. 174.7°; 1.5 N
29. (a) 70 cm (b) No
31. (a) 248°; 58 N (b) The direction would not change but the magnitude would be cut to a fourth.
33. (a) greater than (b) 3.09×10^6 m/s
35. 3.7×10^{-7} C (b) No (c) The tension will be zero.
37. $q_1 = 9\, q_2$
39. (a) 4.05×10^4 N/C (b) 4.50×10^3 N/C
41. (a) $(-3.0 \times 10^7$ N/C$)\,\hat{\mathbf{x}}$ (b) $(5.9 \times 10^7$ N/C$)\,\hat{\mathbf{x}}$
43. Regions 1 and 3
45. (a) 6.89×10^7 N/C (b) greater than (c) 4.19×10^8 N/C
47. (a) 1.3×10^6 N/C (b) less than
49. (a) $(x, y) = (32$ cm, 0$)$ (b) -81 pC
51. $q_1 = 1.8\ \mu$C; $q_2 = -3.6\ \mu$C; $q_3 = 1.8\ \mu$C
53.

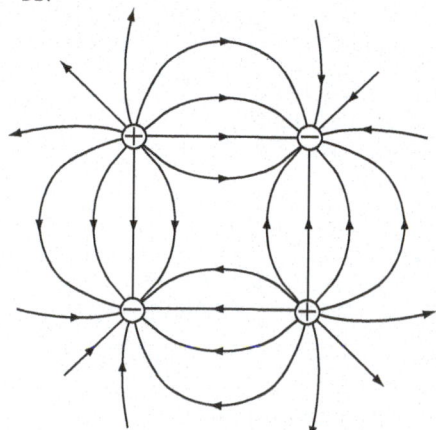

55.

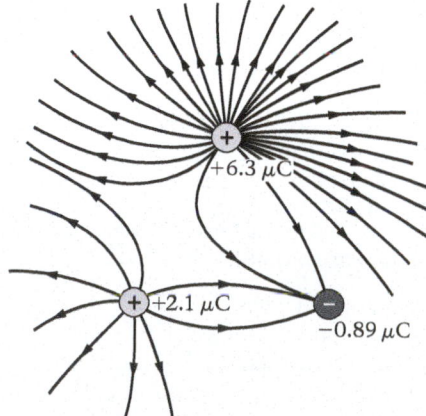

57. (a) the same as (b) II
59. $D < C < B < A$
61. 9.5 cm
63. 6.2×10^{-6} C/m^2
65. (a) -1.41×10^5 N·m^2/C (b) 1.82×10^5 N·m^2/C (c) 4.11×10^5 N·m^2/C (d) $-3.64\ \mu$C

67. 9.7×10^{-6} C/m^2
69. (a) greater than (b) II
71. negative
73. (a) 6.46×10^5 m/s (b) 2.04×10^6 m/s
75. $-2.6\ \mu$C
77. 5.8×10^{-5} N
79. (a) $-Q$ (b) $+Q$ (c) zero
81. (a) 2.6×10^4 N/C (b) 0.066 N
83. (a) 0.55 N (b) greater than (c) $(-4.4$ N$)\,\hat{\mathbf{y}}$
85. $-1.15\ \mu$C
87. 0.254 m
89. 8.85×10^{-6} C/m^2
91. (a) 1.2×10^3 N/C (b) 0.042 N
93. (a) $(6.5 \times 10^4$ N/C$)\hat{\mathbf{x}}$ (b) zero (c) The electric dipole field is very complex, making it difficult to sum the electric flux.
95. (a) 6.5×10^{-7} C (b) 2.0×10^{-6} C and -6.8×10^{-7} C
97. C. 5.40×10^{-7} N
99. B. 1.4 cm
101. (a) increase (b) decrease (c) $3.48\ \mu$C (d) 12.8°
103. (a) 2.00×10^4 N/C (b) 14.2°

CHAPTER 20

1. increasing
3. (a) zero (b) -4.1×10^6 V (c) -4.1×10^6 V
5. 2.36×10^6 V/m
7. $-2.00e$
9. (a) -5.72 mJ (b) 5.72 mJ (c) zero
11. (a) zero (b) -48 V (c) 48 V (d) No
13. (a) 1.1×10^{-20} J (b) stay the same
15. (a) 515 V (b) 183 eV
17. (a) 2.61×10^5 m/s (b) 1.12×10^7 m/s
19. 5 MV
21. 30 m/s
23. (a) 28 V (b) -74 V
25. (a) $-Q/\sqrt{2}$ (b) positive
27. $6.8\ \mu$C
29. 27.2 V
31. $C < A = B < D$
33. (a) 222 V (b) 0.286 m
35. (a) 8.82 m/s (b) 1.83 m
37. (a) 2.51 kV (b) $(x, y) = (-0.141$ m, 1.927 m$)$
39. -0.27 J
41. $\left(4 + \sqrt{2}\right)\dfrac{kQ^2}{a}$
43. (a) greater than (b) III
45. (a)

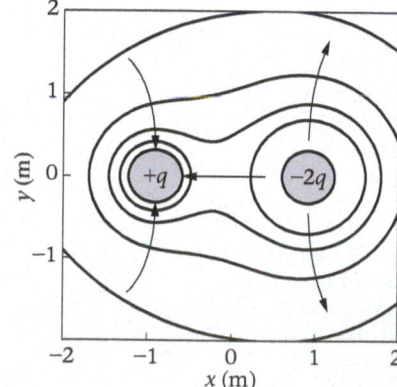

(b)

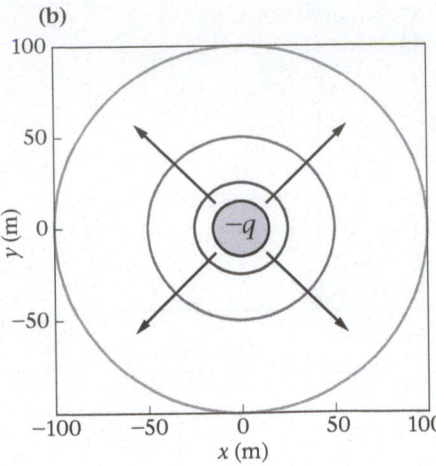

47. (a) parallel to the yz plane (b) increase
    (c) 0.3 mm
49. (a) 15.0 V (b) −1.5 J
51. 1.5 V
53. 709 μF
55. 5.4
57. (a) increased (b) 84 μm
59. (a) 2.10 μm (b) 7.8 μm
61. 15 kV
63. (a) 4.0 nF (b) 6.6 C
65. 0.29 mJ
67. (a) 5.8 × 10^{-14} J (b) decrease
69. 10.6 J/m^3
71. (a) 0.27 C (b) 49 J
73. (a) decrease (b) I
75. (a) decrease (b) decrease (c) decrease
    (d) decrease
77. (a) remain the same (b) increase
    (c) increase (d) increase
79. (a) 4.9 μm (b) 68 V
81. 13.6 eV
83. 2.9 MeV
85. 46 km/s
87. 0.915 J
89. (a) −2.50 × 10^2 V/m
    (b) −5.00 × 10^2 V/m (c) 559 V/m; 243°
91. 4.09 F
93. 1.0 kV
95. (a) 7.0 × 10^{-13} J
    (b) 7.3 × 10^6 molecules
97. 3.58 × 10^{-14} m
99. (a)

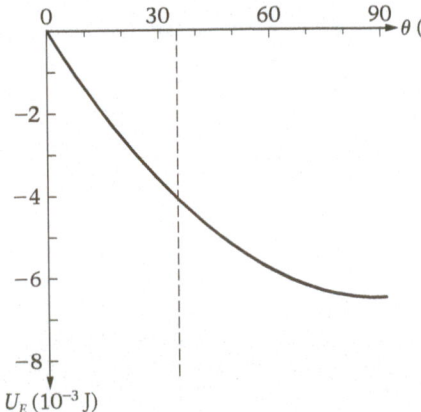

(b)

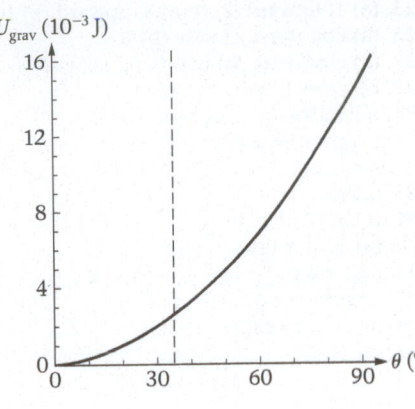

(c)

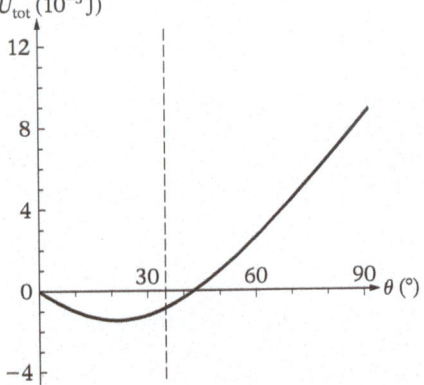

101. 4.15 μF
103. B. toward the tail
105. C. 4.9 × 10^{-9} C
107. (a) negative (b) 0.250 m
     (c) Yes; −0.500 m
109. (a) No (b) 0.714 m (c) 1.67 m

CHAPTER 21

1. 37 C; 2.3 × 10^{20} electrons
3. 6.25 × 10^4 electrons/s
5. (a) 1.5 kC (b) 8.5 y
7. material A
9. 0.22 Ω
11. 0.5 mm
13. 8.7 kΩ
15. (a) 2.5 mV (b) increase
17. (a) 7.2 × 10^{-13} A (b) decrease by a
    factor of 2
19. (a) 0.14 Ω/m (b) decrease (c) 0.10 Ω/m
21. $\left(\dfrac{C}{A}\right)^2 I_{AB}$
23. (a) bulb A (b) 4
25. 74 A
27. 0.65 kW
29. 3.47
31. 155-minute reserve capacity
33. (a) $I_1 = I_2 = I_3$ (b) $V_1 < V_2 < V_3$
35. 6 resistors
37. 2.7 V across 35 Ω; 6.3 V across 82 Ω;
    9.00 V across 45 Ω
39. (a) 3.1 W (b) 1.5 kW
41. (a) 86 Ω (b) $V_{11\,\Omega} = 1.8$ V; $V_{53\,\Omega} = 8.5$ V;
    $V_R = 14$ V (c) larger

43. (a) 77 Ω (b) $I_{22\,\Omega} = 0.55$ A;
    $I_{67\,\Omega} = 0.18$ A; $I_R = 0.16$ A (c) smaller
45. 34 Ω
47. 12 lightbulbs
49. (a) $I_{12\,\Omega} = 0.26$ A; $I_{55\,\Omega} = 0.16$ A; $I_R = $
    0.10 A (b) The currents through the
    12-Ω resistor and R decrease, and
    the current through the 55-Ω resistor
    increases.
51. (a) 1.4 A (b) 11.6 V (c) decrease
53. (a) stay the same (b) I_0
55. (a) decrease (b) 0.12 A clockwise
57. (a) $I_{11\,\Omega} = 0.92$ A; $I_{6.2\,\Omega} = $
    $I_{12\,\Omega} = 0.27$ A; $I_{7.5\,\Omega} = 0.65$ A
    (b) same as (a)
59. (a) $I_{3.9\,\Omega} = I_{9.8\,\Omega} = 0.72$ A; $I_{1.2\,\Omega} = $
    1.8 A; $I_{6.7\,\Omega} = 1.0$ A (b) greater than
    (c) 2.2 V
61. (a) C_1 (b) C_2
63. (a) increase (b) II
65. 54 μF
67. 1.4 V
69. $U_{15\,\mu F} = 2.4$ mJ; $U_{8.2\,\mu F} = 0.71$ mJ;
    $U_{22\,\mu F} = 0.27$ mJ
71. (a) 23 μF
    (b) $Q_{7.5\,\mu F} = 0.11$ mC; $Q_{15\,\mu F} = 0.23$ mC
    (c) $U_{7.5\,\mu F} = 0.84$ mJ; $U_{15\,\mu F} = 1.7$ mJ
73. 6.47 V
75. (a) 2.6 × 10^{-4} C (b) 28 mA
77. (a) 8.91 ms (b) 548 μC (c) 61.5 mA
79. 6.1 kΩ
81. (a) 0.125 m/s (b) 25.0 m/s
83. (a) decrease (b) increase
85. (a) increase (b) stay the same
87. $P_{12\,\Omega} = 0.99$ W; $P_{55\,\Omega} = 1.3$ W;
    $P_{65\,\Omega} = 1.1$ W
89. (a) A = B = C (b) A < C < B
91. (a) 0.91 J (b) 4.0 min
93. (a) greater than (b) 0.82 A (c) 0.54 A
95. 0.26 MΩ
97. (a) greater than (b) 0.27 A
    $I_{35\,\Omega} = I_{82\,\Omega} = 0.103$ A; $I_{45\,\Omega} = 0.27$ A
99. (a) 7.5 A (b) 22 V
    (c) $P_A = P_E = 0$; $P_B = 66$ W; $P_C = 130$ W;
    $P_D = 65$ W
101. $R_1 = 27$ Ω; $R_2 = 13$ Ω
103. A. 3.70 × 10^{-6} A
105. B. 1.22 V
107. (a) $\Delta V_1 = 12.0$ V; $\Delta V_2 = 4.80$ V;
     $\Delta V_3 = 7.20$ V (b) $I_1 = 0.120$ A;
     $I_2 = I_3 = 24.0$ mA
109. (a) stay the same (b) decrease
     (c) 99.0 ms

CHAPTER 22

1. (a) less than (b) II
3. A < B < C
5. 1.0 × 10^9 m/s^2
7. zero
9. (a) 81° (b) 38° (c) 1.2°
11. 4.4 × 10^{-16} N
13. (a) particle 2 (b) 1/4
15. (a) (5.0 × 10^6 N/C)$\hat{\mathbf{x}}$ (b) (−0.2 T)$\hat{\mathbf{z}}$
17. 2.5 km/s
19. (a) 4.3 μm (b) 7.9 mm
21. 0.34 mT
23. (a) negatively charged (b) 1.5 u

25. (a) 4.21×10^{-7} s (b) stay the same
(c) 4.21×10^{-7} s
27. (a) 70 m (b) 1.9 ms (c) 0.68 km
29. 2.3 N
31. $(-0.046$ T$)\hat{z}$
33. 2.3 A
35. (a) north, 18° above horizontal; 1.6 N
(b) east; 1.5 N
37. (a) counterclockwise (b) clockwise
(c) not rotate at all
39. 0.20 N·m
41. (a) away from you (down into the table) (b) 2.4×10^{-6} N·m
43. 1.88×10^{-5} T
47. 3.6 N toward each other
49. 13 A
51. 8.9 μT at 63° below the x axis
53. (a) stay the same (b) III
55. 0.015 T
57. 1.3×10^5 turns
59. 5.2 km
61. out of the page
63. 60° counterclockwise from the $+x$ axis
65. toward wire 3
67. 7.0 μm
69. $F_{top} = F_{bottom} = 0$; $F_{left} = F_{right} = 1.4$ N
71. (a) $B_3 < B_1 < B_2$
(b) B_1, out of the page; B_2, into the page; B_3, out of the page
73. 2.6 kA
75. 2.8 mN; 63° measured from the positive z-axis toward the negative y-axis in the yz-plane
77. $\vec{B}_A = 7.8\ \mu$T out of page; $\vec{B}_B = 0$; $\vec{B}_C = 7.8\ \mu$T into page
79. 0.0019 N·m (b) 6.8 s
81. (a) 1.3 mT (b) 0.29 kN/m
83. (a) 1.6×10^{11} N (b) The force from the magnetar is 280,000 times greater than the electron-proton force within a hydrogen atom
85. (a) 4.68 mT (b) 297 A
87. (a) 2.6 kT/s (b) increase by a factor of $\sqrt{2}$
89. (a) clockwise (b) $\dfrac{2I}{3\pi}$

91. 4.6 V
93. 64 mT
95. A. 1.3×10^{-13} T
97. C. 7.3×10^{10}
99. (a) decreased (b) 104 km/s
101. 0.99 A to the left

CHAPTER 23

1. 4.2×10^{-4} Wb
3. 0.027 T
5. 1.7×10^{-2} Wb
7. (a) 11.7 A (b) The current would be cut to a fourth.
9. (a) zero (b) III
11. (a) $t = 0$ s, 0.2 s, 0.4 s, and 0.6 s
(b) $t = 0.1$ s, 0.3 s, and 0.5 s
13. 0.52 mV
15. 1, counterclockwise; 2, zero; 3, zero; 4, clockwise
17. (a) near $t = 0.5$ s (b) $t = 0.1$ s, 0.3 s, and 0.5 s (c) -0.06 kV, 0, 0.06 kV
19. 36 V

21. 2.3×10^3 turns
23. (a) 1, upward; 2, zero; 3, upward (b) III
25. (a) counterclockwise (b) II
27. (a) clockwise (b) counterclockwise
29. (a) zero (b) clockwise
31. A, clockwise; B, zero; C, counterclockwise
33. $-975\ \mu$C
35. 0.29 T
37. (a) 4.67 m/s (b) No
39. (a) 0.48 A (b) 6.5 m/s
41. 330 rpm
43. 17 mV
45. (a) horizontal (b) 29.3 mV
47. -1.66 V
49. (a) 6.9×10^{-3} m² (b) 0.42 V
51. -50.9 mV
53. 3.6×10^{-4} s
55. A = B = C < D
57. (a) 0.75 s (b) 0.37 Ω
59. (a) 3.8 H (b) 0.25 s (c) 1.6 A
61. 0.012 m²
63. (a) 0.067 H (b) 470 m
65. (a) 0.075 J (b) 0.14 J (c) decrease
67. $2\,V_s$
69. (a) less than (b) 9.4 turns
71. 92
73. 0.38 A; 0.19 kV
75. less than
77. 8.0 μV
79. (a) counterclockwise (b) counterclockwise
81. 0.51 V
83. 0.26 kV
85. (a) 0.10 Wb (b) 0.081 V
87. (a) zero (b) 1 V (c) zero (d) -0.3 V
89. (a) 0.12 ms (b) 39 μJ
91. 1.1 μV
93. (a) zero (b) $-vWB$ (c) zero (d) zero; counterclockwise; zero
95. (a) $1/\sqrt{\varepsilon_0\mu_0}$ (b) 3.00×10^8 m/s
97. C. 2.1×10^{-4} V
99. B. 0.11 s
101. (a) zero (b) zero (c) counterclockwise (d) to the left
103. (a) to the right (b) clockwise (c) 2.2 T

CHAPTER 24

1. 53 V
3. 41 Ω
5. (a) 8.72 W (b) 17.4 W
7. (a) 1.67 kW (b) 0.923
9. 28 μF
11. 2.68 A
13. (a) decrease (b) 0.620 A
15. (a) 50.9 mA (b) multiplicative factor of 2 (c) 102 mA
17. $0 \le f < 60$ Hz
19. (a) 0.979 A (b) in parallel (c) 11.7 μF
21. 3.79 mA
23. 7.6 μF
25. (a) 66.3 Hz (b) 3.48 W
27. (a) 0.80 (b) increase
29. (a) less than (b) III
31. 3.67 A
33. 0.64 mH
35. 37.3 mA
37. (a) 0.575 (b) 71.6 mW

39. 8.7 GHz $< f < \infty$
41. (a)

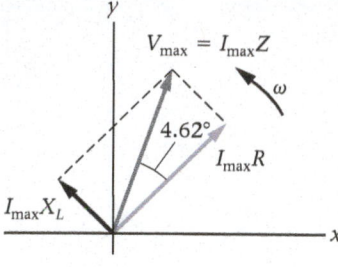

(b) 136 W
43.

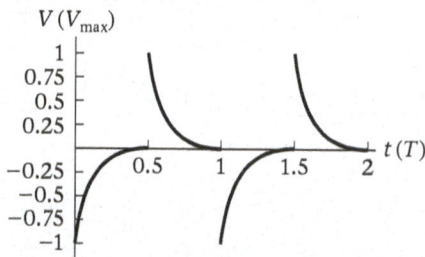

45. in parallel
47. (a) less than (b) II
49. $V_{rms, R} = 43$ V; $V_{rms, L} = 0.051$ V; $V_{rms, C} = 0.11$ kV
51. (a) 4.3 A (b) 8.7 A
53. (a) 88 V (b) 95 V (c) 7.1 V (d) 9.0 V
55. (a)

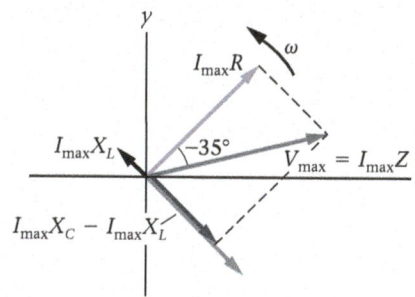

(b) 91.7 W
57. (a) 51.5 Ω (b) 0.486 (c) 9.42 A (d) 58.9 μF (e) 4.58 A
59. 54 μF
61. 21 Ω
63. (a) higher (b) positive (c) 31°
65. (a) 44.5 Hz (b) 44.5 Hz
67. (a) 494 Hz (b) 180 Ω (c) 0.58
69. (a) 9.7 pH (b) 10 Ω
71. equal to
73. (a) increase (b) II
75. 572 mA
77. (a) 44.2 Ω (b) 268 Ω
79. (a) 1.5 μH (b) 0.11 pF (c) 0.21 m
81. 4.51 W
83. (a) 606 Ω (b) increase
85. (a) 29 Ω (b) 71 μF (c) decrease
87. (a) 23.1 W (b) 0.672 W (c) 0.672 W
89. (a) 0.60 (b) Either 85 μF or 300 μF
91. (a) $V_{max}/2$ (b) $V_{max}/\sqrt{2}$
93. (a) 0.24 kHz (b) 93 V (c) 76 V
95. D. 1.7 MΩ

97. D. $+33°$

99. (a) increase (b) 87.2 Hz (c) 0.634 A

CHAPTER 25

1. increasing
3. (a) x direction (b) y direction (c) positive z direction
5. (a) up (b) down (c) east (d) west
7. case 1, $+x$; case 2, $+z$; case 3, $-x$
9. 4.1×10^{16} m
11. -697 Hz
13. (a) away from (b) $0.15c$
15. 2×10^8 m
17. 17 h
19. (a) 2.2×10^7 m/s (b) toward
21. 672.0 nm
23. -3.44 kHz
25. (a) moving away (b) 15.8 m/s
27. 5.41×10^{14} Hz
29. 2.5×10^6 wavelengths
31. 0.122 m
33. (a) 9.38×10^{14} Hz $< f_{\text{UV-B}} < 1.07 \times 10^{15}$ Hz (b) UV-A
35. (a) decrease (b) 3/4
37. 97 s
39. 93 m
41. (a) 0.2 km (b) 2 m
43. (a) multiplicative factor of 2 (b) multiplicative factor of 4
45. 2.25×10^{-10} T
47. (a) 0.87 kV/m (b) 2.0 kW/m² (c) 1.0 kW/m²
49. 75.4 V/m
51. (a) 95 mW/m² (b) 0.95 mW/m²
53. 6.6×10^{-4}
55. 2.6×10^{-11} J
57. 2.3 mN
59. (a) NIF (b) NIF (c) NIF
61. (a) 0.15 mJ (b) 3.8 kW/cm² (c) 3.8×10^5
63. (a) 0.28 kW/m² (b) 4.6 cm
65. (a) 0.030 J (b) 0.92 μPa
67. (a) 1.56×10^{12} W (b) 4.95×10^{17} W/m² (c) 1.37×10^{10} V/m
69. (a) greater than (b) III
71. 0.0042 W/m²
73. 0.375
75. graph B
77. (a) A = C < B (b) Case A, 6.94 W/m²; case B, 40.4 W/m²; case C, 6.94 W/m²
79. (a) $\frac{1}{2}I_0$ (b) $0.375I_0$ (c) $0.0938I_0$ (d) zero
81. (a) multiplicative factor of 2 (b) multiplicative factor of 4
83. 2.00×10^{-11} m
85. (a) zero (b) $\vec{B} = (-9.57 \times 10^{-9}\,\text{T})\,\hat{x} + (2.07 \times 10^{-8}\,\text{T})\,\hat{y}$
87. B < A = C < D
89. 9.0 mm
91. (a) increased (b) 915 rev/s
93. (a) 53.5° (b) 501 W/m²
95. 6.6
97. (a) 1.0 mW/m² (b) 0.71 kN/C
99. 209 W/m²
101. 15.6 W
103. (a) 1.4×10^{-8} N (b) 1.5×10^{-10} m/s² (c) 4.6 d (d) 63 s

105. 3.85×10^{26} W
107. 49°
109. C. blue
111. C. 1.33×10^{-5} N/m²
113. (a) 48.0° (b) decrease

CHAPTER 26

1. 38°
3. 65°
5. (a) afternoon (b) $-8.8°$
7. 31.7°
9. (a) 5 times (b) 4 times
11. 7.25 ft
13. 1.1 m
15. (a) 61 m (b) increase
17. concave
19. -7.98 cm
21. (a) upright (b) reduced (c) virtual
23. toward the right
25. 72.0 cm
27. (a) 55 cm behind the mirror; 9 cm tall (b) upright
29. -0.55 m; 0.27
31. -1.14 m
33. (a) -4.3 cm (b) 9.0 cm
35. (a) 0.11 km in front of the mirror (b) real (c) -5.45
37. (a) 66 cm (b) 17 cm
39. 19 m
41. $d_o = 1.4$ m; $d_i = 0.48$ m
43. (a) point 1 (b) point 4
45. (a) shallower (b) II
47. 1.5
49. 6% error; 4% error
51. 14.4 ns
53. 21.6°
55. 4.1 ft
57. (a) 1.4 (b) decreased
59. 1.4
61. 30.5°
63. (a) $\frac{2}{3}|f|$ (b) upright (c) virtual
65. (a) $2f$ (b) inverted (c) real
67. (a) Just beyond F_2 to the right of lens 2 (b) inverted (c) real
69. $d_i = -17$ cm; $m = 0.55$
71. (a) 0.98 m (b) 68 cm
73. (a) farther away (b) 2.8 cm
75. (a) 34 cm (b) -1.2
77. (a) farther away (b) 34 cm
79. 3.0 cm
81. (a) virtual (b) 69 cm (c) convex
83. (a) greater than (b) I
85. 13.92°
87. convex
89. (a) decrease (b) II
91. 0.25 m
93. (a) 11.5 cm and 34.5 cm (b) 11.5 cm, virtual and upright; 34.5 cm, real and inverted
95. (a) 1.8 (b) No
97. (a) 61.8°
99. (a) -0.667 (b) 0.45
101. $\dfrac{f_1 f_2}{f_1 + f_2}$
103. $d = t \sin\theta \left(1 - \dfrac{\cos\theta}{\sqrt{n^2 - \sin^2\theta}}\right)$
107. C. -12.8 cm

109. C. 91.0 cm
111. (a) 0.83 cm (b) 4.2 cm
113. (a) 2.5 cm (b) -5.0 cm

CHAPTER 27

1. (a) farther from (b) II
3. the flower
5. (a) 2.67 cm (b) 2.44 cm
7. D < B < C < A
9. (a) $f/2.8$ (b) 23 mm, 16 mm, 8.1 mm, 5.9 mm, 4.1 mm
11. (a) $f/11$ (b) 2.1 mm
13. (a) 1/25 s (b) $f/5.6$
15. (a) $f/3.33$ (b) $f/30.6$
17. farsighted
19. (a) nearsighted (b) No
21. 0.95 m
23. 1.33 m
25. (a) 4.4 mm in front of the lens closest to the object (b) 0.18
27. (a) increase (b) 3.5 mm
29. (a) 15.0 cm to the left of the converging lens (b) -0.167
31. -125 cm
33. 44.3 cm
35. -3.8 m
37. (a) -2.75 m (b) -2.77 m
39. (a) nearsighted (b) far point, 15 cm
41. 24.2 cm
43. (a) diverging (b) converging (c) -0.20 diopters; $+2.3$ diopters
45. near point 32.3 cm; far point 16.0 m
47. 12.0 cm
49. (a) 9.042×10^{-3} rad (b) 2.1 m
51. 3.33 cm
53. 12.7 mm
55. 2.96
57. 13 cm
59. (a) objective A (b) 1.1 cm (c) 0.65 cm
61. 3.3×10^{-3} rad
63. (a) 3.83 mm (b) 10.0 cm
65. 9.1 mm
67. (a) 7.85 mm (b) 10 cm (c) 5.1 cm
69. 2.6 m
71. (a) 6.00 (b) 35.0 cm
73. (a) 35.0 cm (b) 36.9 cm
75. (a) right lens (b) 2.5
77. 20°
79. less than
81. (a) decrease (b) II
83. (a) positive (b) I
85. (a)

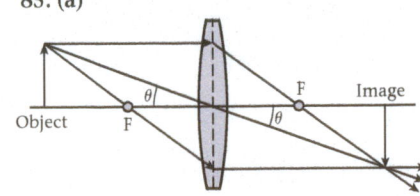

(b)

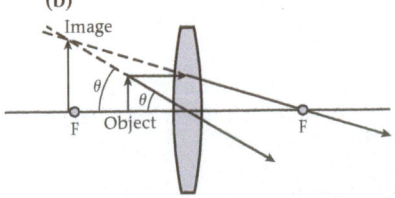

(c)

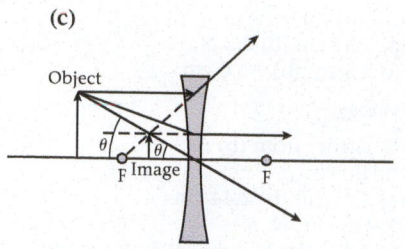

Object

θ θ

F Image F

(d) A simple magnifier helps the eye view an object that is closer than the near point, thus making it appear larger.

87. (a) 2.42 cm (b) 2.42 cm
89. 4.6 cm
91. (a) 73.8 cm (b) 42.0
93. 9.24 cm
95. (a) 16 cm to the left of the lens (b) real (c) -0.80 (d) inverted
97. (a) 0.88 m (b) upright; 0.40
101. C. 4.35 diopters
103. A. 1.9 mm
105. (a) greater than (b) -351 cm
107. (a) greater than (b) 2.81 diopters

CHAPTER 28

1. destructively
3. 16.0 m
5. (a) 300 m (b) 13 s
7. minimum
9. 203 Hz and 609 Hz
11. 0.68 kHz and 2.0 kHz
13. (a) 4λ (b) 2.5λ
15. (a) decrease (b) II
17. 785 nm
19. (a) 415 nm (b) increase (c) 587 nm
21. 128 μm
23. 95.1 μm
25. 69.1 cm
27. (a) 2.24 cm (b) increased
29. greater than
31. 488 nm and 627 nm
33. 493 nm and 658 nm
35. (a) destructive (b) 3.5 m
37. (a) destructive (b) constructive (c) constructive
39. (a) 102 nm (b) thinner
41. (a) 110 nm and 330 nm (b) 220 nm and 440 nm
43. (a) 2.00 (b) 0.500 (c) 2.00
45. 30°
47. 26.9 cm
49. 14 dark fringes
51. 3.20 cm
53. (a) 0.78 m (b) 1.3 m
55. 5.5 cm
57. 7.0 mm
59. (a) 1.1×10^{-5} rad (b) 7.0 μm
61. 2.95°, 5.90°, 8.87°
63. 3.9°
65. 356 nm
67. one
69. 20.3°
71. (a) The $m = 3$ maximum of the 420-nm light overlaps the $m = 2$ maximum of the 630-nm light. (b) 35°
73. 3.29 m
75. (a) bright (b) I

77. 32 μm
79. (a) 98.7 nm (b) 98.7 nm
81. 76 μm
83. 491 nm
85. 4.0×10^5 m
87. 12 μm
89. (c) narrower
91. 6.74 μm
93. C. 0.369 mm
95. C. 11.5 ft
97. (a) longer (b) 0.63 μm
99. (a) increase (b) 15.1°

CHAPTER 29

1. (a) greater than (b) I
3. (a) proper time (b) proper time (c) proper time (d) dilated time
5. (a) according to the clock on Earth (b) I
7. 0.85c
9. 0.995c
11. (a) 1.06 km (b) 4.18×10^{-6} s
13. (a) 5.7×10^{-8} s (b) 15 m (c) 6.9 m
15. (a) 2.70×10^{11} m (b) 0.900 s (c) 57.0 y (d) 114 y
17. (a) less than (b) 47 beats/min
19. 0.788c
21. (a) slightly more (b) 1.97 ns
23. 1.13 m
25. 6.4 m
27. (a) 0.77c (b) 2.4 m
29. (a) less than (b) 0.583c (c) 2.84 cm
31. 72°
33. (a) (i) the *Picard* is longer (b) 2.3
35. 0.80c
37. 0.38c
39. 0.53c
41. piece A 0.98c; piece B 0.90c
43. (a) 7.2×10^{14} kg·m/s (b) 1.4×10^{15} kg·m/s
45. 0.993c
47. 9.4×10^7 kg
49. 0.999c
51. (a) B < D < C < A (b) A < D < B = C (c) A < D < C < B
53. 4.5×10^{-16} kg
55. 1.02 MeV
57. (a) 1.07 nJ (b) 0.151 nJ (c) 0.92 nJ
59. (a) 0.98c (b) 3.8
61. 0.98 kg
63. (a) 0.75% (b) 69%
65. 2.4 kg
67. (a) 7.7×10^6 km (b) 5.8×10^6 m/s² (c) decreases by a factor of 2
69. (a) the pilot (b) you (c) the pilot (d) you
71. (a) less than (b) II
73. (a) 1.00c
75. (a) 0.643c (b) 2.29×10^{-22} kg·m/s (c) -2.29×10^{-22} kg·m/s (d) 9.85 km/s
77. 0.007c
79. (a) 1.8 m (b) 79°
81. 140 MeV
87. C. 0.312c
89. A. 0.301c
91. (a) 0.371c (b) 0.454c (c) 0.922c

CHAPTER 30

1. (a) lower than (b) II
3. 17.7 μm
5. (a) 1.80×10^{13} Hz (b) 16.7 μm
7. (a) 2.00 (b) 1.068
9. (a) halogen (b) 1.2 (c) halogen
11. (a) 4.792×10^{13} Hz (b) 3.18×10^{-20} J
13. (a) less than (b) II
15. (a) decreased (b) II
17. 8.9×10^{14} Hz; 0.34 μm
19. 3.95×10^{-18} J
21. 6.42×10^{-19} J
23. (a) 4.4×10^{18} photons (b) 9.4×10^{18} photons
25. (a) 1.09×10^{15} Hz; 275 nm (b) ultraviolet
27. (a) red (b) blue (c) red, 4.8×10^{20} photons/s; blue, 8.6×10^{19} photons/s
29. (a) aluminum (b) Al, 9.85×10^{14} Hz; Ca, 6.93×10^{14} Hz
31. (a) cadmium (b) Zn, 0.19 eV; Cd, 0.30 eV
33. (a) 369 eV (b) 4.00×10^{14} Hz $\leq f < 9.85 \times 10^{14}$ Hz
35. 6.377×10^{-34} J·s
37. (a) 1.83×10^{13} Hz (b) 2.22 mW (c) 1.83×10^{17} photons/s
39. 2.9×10^6 m/s
41. 9.1×10^{17} Hz
43. 0.815 m/s
45. 58.0 m/s
47. (a) 167 m/s (b) 3.55×10^4
49. (a) 2.39×10^{16} photons/s (b) 2.10×10^{-27} kg·m/s (c) 5.00×10^{-11} N
51. 4.5 keV
53. 104°
55. (a) The change in wavelength is the same for both photons. (b) X-ray (c) visible, 9.3×10^{-4} %; X-ray, 16%
57. (a) 0.317 nm (b) incident, 3.92 keV; scattered, 3.88 keV (c) 40 eV
59. $-37.7°$
61. (a) 1/2 (b) $1/\sqrt{2}$
63. 1.40 km/s
65. 3.9×10^{-18} J
67. (a) electron (b) 1836
69. 0.062 m/s
71. $h/\sqrt{2mqV}$
73. 1.9×10^{-20} kg·m/s
75. 1.8×10^{-25} J
77. (a) 1.1×10^{-23} kg·m/s (b) 0.39 keV
79. 62 nm
81. (a) increase (b) decrease (c) decrease (d) decrease
83. 3.65×10^{-17} W
85. (a) 8.4×10^{23} photons (b) 40 molecules/photon
87. (a) 3.08×10^{17} photons (b) 1.47×10^{10} photons (c) 1.13×10^4 photons
89. (a) 0.35 nm (b) 0.35 nm
91. (a) Substitute $\frac{1}{2}mv_{max}^2$ for K_{max} in the equation $K_{max} = hc/\lambda - W_0$ and solve for W_0. Then calculate $f_0 = W_0/h$. (b) 2.01 eV; 4.84×10^{14} Hz
93. (a) 2.29 eV (b) 1.08 eV

95. (a) 1.32 fm (b) 1.51×10^{-10} J
97. D. 3.66×10^{13} Hz
99. higher
101. (a) less than (b) 94°

CHAPTER 31

1. 8.4×10^{-16}
3. (a) 2.61×10^4 kg/m^3
  (b) 3.34×10^{17} kg/m^3
5. 43 fm
7. $n = 7$
9. 1875 nm, 1282 nm, 1094 nm
11. (a) increase (b) I
13. C < B < A
15. 3.64×10^{-3}
17. 1.51 eV
19. 3.81×10^{-5} eV
21. (a) 1.51 eV (b) −3.02 eV· (c) −1.51 eV
23. $n_i = 3 \rightarrow n_f = 2$
25. (a) (iii), 19.06 μm (b) (ii), 388.9 nm
  (c) (iv)
27. (a) 2.83×10^{-10} m (b) greater than
  (c) H, 0.306 eV; Li^{2+}, 2.75 eV
29. (a) 1.22×10^{-15} s (b) 8×10^6 orbits
31. (a) $n = 6$ (b) increase
  (c) $- 0.888 \times 10^{-19}$ J

33. (a) $f_{\text{photon}} = \dfrac{2\pi^2 mk^2 e^4}{h^3}\left[\dfrac{1}{(n-1)^2} - \dfrac{1}{n^2}\right]$

  (b) $f_{\text{orbit}} = \dfrac{4\pi^2 mk^2 e^4}{n^3 h^3}$

  (c) For very large n, $f_{\text{photon}} = f_{\text{orbit}}$
35. 0.332 nm
37. 1.33 nm
39. 1.32 nm
41. (a) 600 eV (b) greater than
43. 9
45. (a) −0.850 eV (b) 3.66×10^{-34} J·s
  (c) greater than (d) greater than
47. (a) $\ell = 75$ (b) $n = 76$ (c) −2.35 meV
49. (a) greater than (b) $\sqrt{2}$
51. (a) 10 electrons (b) 8 electrons
53. $1s^2\,2s^2\,2p^2$
55.

electron	n	ℓ	m_ℓ	m_s
$1s^1$	1	0	0	$-\frac{1}{2}$
$1s^2$	1	0	0	$\frac{1}{2}$
$2s^1$	2	0	0	$-\frac{1}{2}$
$2s^2$	2	0	0	$\frac{1}{2}$
$2p^1$	2	1	−1	$-\frac{1}{2}$
$2p^2$	2	1	−1	$\frac{1}{2}$
$2p^3$	2	1	0	$-\frac{1}{2}$
$2p^4$	2	1	0	$\frac{1}{2}$
$2p^5$	2	1	1	$-\frac{1}{2}$
$2p^6$	2	1	1	$\frac{1}{2}$

57. $1s^2 2s^2 2p^6 3s^2 3p^6 3d^8 4s^2$
59. (a) stay the same (b) III

61. 66.9 keV
63. $Z = 67$
65. (a) 6.44 eV (b) 1.53×10^5 photons
67. 3 photons
69. (a) 13.6 eV (b) 10.2 eV
71. less than
73. (a) 1.36×10^{-27} kg·m/s (b) 0.815 m/s
75. (a) 1.52×10^{-16} s (b) 0.00105 A
77. (a) $Z = 7$ (b) −74.0 eV
79. (a) speed up (b) increase (c) opposite
81. (a) greater than (b) 4.38×10^7 m^{-1}
85. less than
87. B. The yellow Kr laser emits fewer photons per second.
89. (a) $n = 12$ (b) 7.62 nm

CHAPTER 32

1. (a) $Z = 92; N = 146; A = 238$
  (b) $Z = 94; N = 145; A = 239$
  (c) $Z = 60; N = 84; A = 144$
3. (a) 7.0 fm (b) 4.7 fm
5. (a) 2.3×10^{17} kg/m^3 (b) the same as
  (c) 2.3×10^{17} kg/m^3
7. (a) 6.4×10^6 m/s (b) 0.27 pm
  (c) less than
9. (a) 3.7 fm (b) 240 (c) 240
11. (a) 3×10^{26} nucleons (b) 800 nm
13. (a) the same as (b) III
15. (a) $^{224}_{88}$Ra (b) $^{208}_{81}$Tl (c) $^{239}_{94}$Pu
17. (a) $^{137}_{56}$Ba (b) $^{99}_{43}$Tc (c) $^{206}_{87}$Fr
19. $^4_2$He
21. α decay; β decay; β decay; α decay; α decay; α decay; α decay; β decay; β decay; α decay; α decay; β decay; α decay; β decay
23. (a) $^{212}_{84}$Po $\rightarrow$ $^{208}_{82}$Pb + $^4_2$He; 8.95 MeV
  (b) $^{239}_{94}$Pu $\rightarrow$ $^{235}_{92}$U + $^4_2$He; 5.244 MeV
25. (a) $^{18}_9$F $\rightarrow$ $^{18}_8$O + e$^+$ + v_e; 0.634 MeV
  (b) $^{22}_{11}$Na $\rightarrow$ $^{22}_{10}$Ne + e$^+$ + v_e; 1.819 MeV
27. (a) $^{66}_{29}$Cu (b) 0.2 MeV
29. less than
31. 0.181 d^{-1}
33. 4.5 d
35. (a) 0.115 h^{-1} (b) 1.7×10^9 nuclei
37. 189 d
39. 1.72×10^4 y
41. 0.92 mg
43. (a) 30.0% (b) 23.6% (c) 18.5%
45. 92.16 MeV
47. (a) 7.074 MeV/nucleon
  (b) 8.736 MeV/nucleon
49. (a) 12.13 MeV
  (b) 15.66 MeV
  (c) neutron
51. $^{98}_{41}$Nb; 171.1 MeV
53. 1.6×10^5 gal
55. 5.4×10^{18} s^{-1}
57. 2.224 MeV
59. (a) $^4_2$He; 17.59 MeV (b) 2.7×10^{20} s^{-1}
61. 110 rad
63. (a) 0.38 J (b) 0.54 mK
65. (a) 6.9×10^{11} (b) 0.078 J (c) 93 rem
67. (a) 2.23×10^{-25} kg (b) 6.41×10^{21} m/s
  (c) $\delta = 1.1 \times 10^{-27}$
69. 2.00
71. (a) $^{206}_{80}$Hg (b) $^{239}_{93}$Np (c) $^{11}_5$B
73. $^{228}_{88}$Ra, $^{228}_{89}$Ac, $^{228}_{90}$Th, $^{224}_{88}$Ra, $^{220}_{86}$Rn, $^{216}_{84}$Po, $^{212}_{82}$Pb, $^{212}_{83}$Bi, $^{212}_{84}$Po, and $^{208}_{81}$Tl

75. (a) 1.32×10^3 Bq (b) reduced by a factor of 2
77. (a) 57 N (b) 5.4 MeV (c) 5.4 MeV
79. (a) less than (b) type A, 1.3 d; type B, 2.0 d
81. 2220 y
83. (a) less than (b) 7.274 MeV
85. 0.57 mJ
87. 264 m/s
89. A. 9.98×10^{-7} s^{-1}
91. increased

CHAPTER 1

1. No. The factor of 2 is dimensionless.
3. **(a)** Not possible, since units have dimensions. For example, seconds can only have the dimension of time. **(b)** Possible, since different units can be used to measure the same dimensions. For example, time can be measured in seconds, minutes, or hours.
5. **(a)** Both Albert and Isaac are running with a speed of 3.4 m/s; thus, Albert's speed is equal to Isaac's speed. **(b)** Albert and Isaac have different velocities, because they are running in different directions.
7. To the nearest power of ten: **(a)** 1 m; **(b)** 10^{-2} m; **(c)** 10 m; **(d)** 100 m; **(e)** 10^{7} m.

CHAPTER 2

1. The displacement is the same for you and your dog; the distance covered by the dog is greater.
3. No. After one complete orbit the astronaut's displacement is zero. The distance traveled, however, is roughly 25,000 miles.
5. A speedometer measures speed, not velocity. For example, if you drive with constant speed in a circular path, your speedometer maintains the same reading, even though your velocity is constantly changing.
7. Constant-velocity motion; that is, straight-line motion with constant speed.
9. **(a)** Yes. The object might simply be at rest. **(b)** Yes. An example would be a ball thrown straight upward; at the top of its trajectory its velocity is zero, but it has a nonzero acceleration downward.
11. Yes. A ball thrown straight upward and caught when it returns to its release point has zero average velocity, but it has been accelerating the entire time.
13. When she returns to her original position, her speed is the same as it was initially; that is, 4.5 m/s.

CHAPTER 3

1. **(a)** scalar; **(b)** vector; **(c)** vector; **(d)** scalar.
3. **(a)** $\vec{\mathbf{A}}$ and $\vec{\mathbf{B}}$ have the same magnitude. **(b)** $\vec{\mathbf{A}}$ and $\vec{\mathbf{B}}$ have opposite directions.
5. Yes, if they have the same magnitude and point in opposite directions.
7. Note that the magnitudes A, B, and C satisfy the Pythagorean theorem. It follows that $\vec{\mathbf{A}}$, $\vec{\mathbf{B}}$, and $\vec{\mathbf{C}}$ form a right triangle, with $\vec{\mathbf{C}}$ as the hypotenuse. Thus, $\vec{\mathbf{A}}$ and $\vec{\mathbf{B}}$ are perpendicular to one another.
9. $\vec{\mathbf{A}}$ and $\vec{\mathbf{B}}$ must be collinear and point in opposite directions. In addition,

the magnitude of $\vec{\mathbf{A}}$ must be greater than the magnitude of $\vec{\mathbf{B}}$, that is, $|A| > |B|$.
11. The vector $\vec{\mathbf{A}}$ points in one of two directions: (1) 135° counterclockwise from the x axis; (2) 45° clockwise from the x axis. Case (1) corresponds to $A_x < 0$ and case (2) corresponds to $A_x > 0$.
13. Tilt the umbrella forward so that it points in the opposite direction of the rain's velocity relative to you.

CHAPTER 4

1. Ignoring air resistance, the acceleration of a projectile is vertically downward at all times.
3. The projectile was launched at an angle of 30° above the positive x axis; when it landed its direction of motion was 30° below the positive x axis. Hence, its change in direction was 60° clockwise.
5. At its highest point, the projectile is moving horizontally. This means that gravity has reduced its y component of velocity to zero. The x component of velocity is unchanged by gravity, however. Therefore, the projectile has a velocity equal to $\vec{\mathbf{v}} = (4\ \text{m/s})\hat{\mathbf{x}}$ at the highest point in its trajectory.
7. Less than 45°. See Figure 4-14.
9. **(a)** From the child's point of view the scoop of ice cream falls straight downward. **(b)** From the point of view of the parents the scoop of ice cream follows a parabolic trajectory.
11. **(a)** At its highest point the projectile moves horizontally. Therefore, its launch angle was 50°. **(b)** In this case, the launch angle was 150°, or, equivalently, 30° above the negative x axis.

CHAPTER 5

1. The force exerted on the car by the brakes causes it to slow down, but your body continues to move forward with the same velocity (due to inertia) until the seat belt exerts a force on it to decrease its speed.
3. No. You are at rest relative to your immediate surroundings, but you are in motion relative to other objects in the universe.
5. When the magnitude of the force exerted on the girl by the rope equals the magnitude of her weight, the net force acting on her is zero. As a result, she moves with constant velocity.
7. The astronaut should push the jet pack away from him, in the opposite direction from the spaceship. As a result, the reaction force exerted on him by the pack will accelerate him toward the ship.

9. Mr. Ed's reasoning is incorrect because he is adding two action-reaction forces that act on *different* objects. Wilbur should point out that the net force exerted on the cart is simply the force exerted on it by Mr. Ed. Thus the cart will accelerate. The equal and opposite reaction force acts on Mr. Ed, and does not cancel the force acting on the cart.
11. The whole brick has twice the force acting on it, but it also has twice the mass. Since the acceleration of an object is proportional to the force exerted on it and inversely proportional to its mass, the whole brick has the same acceleration as the half brick.
13. Yes. An object with zero net force acting on it has a constant velocity. This velocity may or may not have zero magnitude.
15. The acceleration of an object is inversely proportional to its mass. The diver and the Earth experience the same force, but the Earth—with its much larger mass—has a much smaller acceleration.
17. An astronaut can tell which of two objects is more massive by pushing on both with the same force. Since acceleration is inversely proportional to mass, the more massive object can be recognized by its smaller acceleration.
19. On solid ground you come to rest in a much smaller distance than when you plunge into water. The smaller the distance over which you come to rest the greater the acceleration, and the greater the acceleration the greater the force exerted on you. Thus, when you land on solid ground the large force it exerts on you may be enough to cause injury.
21. Yes, it would still hurt. The reason is that even though the bucket is "weightless," it still has nonzero mass. Therefore, it has inertia, and when it is kicked it will resist having its state of motion changed. The way it resists a change in its motion is by exerting an equal-but-opposite reaction force on the object trying to make the change; in this case, your foot. This is the force that can make kicking the bucket painful.
23. In everyday units, 1 N is approximately a quarter pound. Objects with a weight of roughly a quarter pound include a quarter-pound hamburger, an apple, a glass of water, and half a dozen Fig Newtons.

CHAPTER 6

1. The clothesline has a finite mass, and so the tension in the line must have an upward component to oppose the

downward force of gravity. Thus, the line sags much the same as if a weight were hanging from it.

3. The braking distance of a skidding car depends on its initial speed and the coefficient of kinetic friction. Thus, if the coefficient of friction is known reasonably well, the initial speed can be determined from the length of the skid marks.

5. The force that ultimately is responsible for stopping a train is the frictional force between its metal wheels and the metal track. These are fairly smooth surfaces. In contrast, the frictional force that stops a car is between the rubber tires and the concrete roadway. These are rougher surfaces with a greater coefficient of friction.

7. As you brake harder your car has a greater acceleration. The greater the acceleration of the car, the greater the force required to give the flat of strawberries the same acceleration. When the required force exceeds the maximum force of static friction the strawberries begin to slide.

9. For a drop of water to stay on a rotating wheel an inward force is required to give the drop the necessary centripetal acceleration. Since the force between a drop of water and the wheel is small, the drop will separate from the wheel rather than follow its circular path.

11. A centripetal force is required to make the motorcycle follow a circular path, and this force increases rapidly with the speed of the cycle. If the necessary centripetal force exceeds the weight of the motorcycle, because its speed is high enough, then the track must exert a downward force on the cycle at the top of the circle. This keeps the cycle in firm contact with the track.

13. Since the passengers are moving in a circular path a centripetal force must be exerted on them. This force, which is radially inward, is supplied by the wall of the cylinder.

15. This helps because the students sitting on the trunk increase the normal force between your tires and the road. Since the force of friction is proportional to the normal force, this increases the frictional force enough (one hopes) to allow your car to move.

17. The normal force exerted on a gecko by a vertical wall is zero. If the gecko is to stay in place, however, the force of static friction must exert an upward force equal to the gecko's weight. For this to happen when the normal force is zero would require an infinite coefficient of static friction. Thus, we conclude that the physics of gecko feet is more complex than our simple models of friction.

19. Yes. The steering wheel can accelerate a car—even if its speed remains the same—by changing its direction of motion.

21. When a bicycle rider leans inward on a turn, the force applied to the wheels of the bicycle by the ground is both upward and inward. It is this inward force that produces the centripetal acceleration of the rider.

CHAPTER 7

1. No. Work requires that a force acts through a distance.

3. True. To do work on an object a force must have a nonzero component along its direction of motion.

5. The frictional force between your shoes and the ground does positive work on you whenever you begin to walk.

7. Gravity exerts an equal and opposite force on the package, and hence the net work done on it is zero. The result is no change in kinetic energy.

9. No, we must also know how much time it takes for engine 1 to do the work. For example, if engine 1 takes twice as much time to do twice the work of engine 2, it has the same power output as engine 2. If it takes more than twice as much time, then engine 1 actually produces less power than engine 2.

CHAPTER 8

1. The kinetic energy cannot be negative, since m and v^2 are always positive or zero. The gravitational potential energy can be negative since any level can be chosen to be zero.

3. The initial mechanical energy of the system is the gravitational potential energy of the mass-Earth system. As the mass moves downward, the gravitational potential energy of the system decreases. At the same time, the potential energy of the spring increases as it is compressed. Initially, the decrease in gravitational potential energy is greater than the increase in spring potential energy, which means that the mass gains kinetic energy. Eventually, the increase in spring energy equals the decrease in gravitational energy and the mass comes to rest.

5. The object's kinetic energy is a maximum when it is released, and a minimum when it reaches its greatest height. The gravitational potential of the system is a minimum when the object is released, and a maximum when the object reaches its greatest height.

7. The jumper's initial kinetic energy is largely converted to a compressional, spring-like potential energy as the pole bends. The pole straightens out, converting its potential energy into

gravitational potential energy. As the jumper falls, the gravitational potential energy is converted into kinetic energy, and finally, the kinetic energy is converted to compressional potential energy as the cushioning pad on the ground is compressed.

9. Zero force implies that the rate of change in the potential energy with distance is zero—that is, the potential energy is constant—but the value of the potential energy can be anything at all. Similarly, if the potential energy is zero, it does not mean that the force is zero. Again, what matters is the rate of change of the potential energy with distance.

11. The dive begins with the diver climbing the ladder to the diving board, which converts chemical energy in the muscles into an increased gravitational potential energy. Next, by jumping on the board the diver causes the board to flex and to store potential energy. As the board rebounds, the diver springs into the air, using the kinetic energy derived from the leg muscles and the potential energy released by the board. The diver's kinetic energy is then converted into an increased gravitational potential energy until the highest point of the dive is reached. After that, gravitational potential energy is converted back to kinetic energy as the diver moves downward. Finally, the kinetic energy of the diver is converted into heat, sound, and flowing water as splashdown occurs.

CHAPTER 9

1. The momentum of the keys increases as they fall because a net force acts on them. The momentum of the universe is unchanged because an equal and opposite force acts on the Earth.

3. If the kinetic energy is zero the speeds must be zero as well. This means that the momentum is zero.

5. Yes, in much the same way that a propeller in water can power a speedboat.

7. (a) The force due to braking—which ultimately comes from friction with the road—reduces the momentum of the car. The momentum lost by the car does not simply disappear, however. Instead, it shows up as an increase in the momentum of the Earth. (b) As with braking, the ultimate source of the force accelerating the car is the force of static friction between the tires and the road.

9. (a) Yes. Suppose two objects have momenta of equal magnitude. If these objects collide in a head-on, completely inelastic collision, they will be at rest after the collision. In this case, all of the initial kinetic energy is converted to other forms of energy. (b) No. In

order for its momentum to change, an external force must act on the system. We are given, however, that the system is isolated. Therefore, the only forces acting on it are internal forces.

11. The kinetic energy of the bullet is much greater than that of the gun. Thus, less energy is dissipated in stopping the gun.

13. The plane weighs the same whether the fly lands on the dashboard or flies about the cockpit. The reason is that the fly must exert a downward force on the air in the cockpit equal in magnitude to its own weight in order to stay aloft. This force ultimately acts downward on the plane, just as if the fly had landed.

15. Your center of mass is somewhere directly above the area of contact between your foot and the ground.

CHAPTER 10

1. All points on the rigid object have the same angular speed. Not all points have the same linear speed, however. The farther a given point is from the axis of rotation the greater its linear speed.

3. No. As long as you are driving in a circular path you will have a nonzero centripetal acceleration.

5. (a) No, because your direction of motion is constantly changing. (b) Yes, your linear speed is simply the radius of the wheel times your angular speed. (c) Yes. It is equal to your linear speed squared divided by the radius of the wheel. (d) No. Your centripetal acceleration is always toward the center of the wheel, but this is in a different direction as you rotate to different locations.

7. (a) A basketball thrown with no spin. (b) A spinning airplane propeller on a plane that is at rest. (c) A bicycle wheel on a moving bicycle.

9. When the chunky stew is rolled down the aisle, all of the contents of the can roll together with approximately the same angular speed. This is because the chunky stew is thick and almost solid. The beef broth, however, is little more than water. Therefore, when the broth is rolled down the aisle, almost all that is actually rolling is the metal can itself. It follows that the stew has the greater initial kinetic energy, and hence it rolls a greater distance.

CHAPTER 11

1. No. Torque depends both on the magnitude of the force and on the distance from the axis of rotation, or moment arm, at which it is applied. A small force can produce the same torque as a large force if it is applied farther from the axis of rotation.

3. The long pole has a large moment of inertia, which means that for a given applied torque the walker and pole have a small angular acceleration. This allows more time for the walker to "correct" his balance.

5. A force applied radially to a wheel produces zero torque, though the net force is nonzero.

7. No. In most cars the massive engine is located in the front, thus the car's center of mass is not in the middle of the car, but is closer to the front end. This means that the force exerted on the front tires is greater than the force exerted on the rear tires. (This situation is analogous to Active Example 11-1.)

9. You are in static equilibrium as you sit in your chair; so is the building where you have your physics class.

11. The angular speed of the dust cloud increases, just like a skater pulling in her arms, due to conservation of angular momentum.

13. Yes. Imagine turning on a ceiling fan. This increases the fan's angular momentum, without changing its linear momentum.

CHAPTER 12

1. No. The force of Earth's gravity is practically as strong in orbit as it is on the surface of the Earth. The astronauts experience weightlessness because they are in constant free fall.

3. As the tips of the fingers approach one another, we can think of them as two small spheres (or we can replace the finger tips with two small marbles if we like). As we know, the net gravitational attraction outside a sphere of mass is the same as that of an equivalent point mass at its center. Therefore, the two fingers simply experience the finite force of two point masses separated by a finite distance.

5. No. A satellite must be moving relative to the center of the Earth to maintain its orbit, but the North Pole is at rest relative to the center of the Earth. Therefore, a satellite cannot remain fixed above the North Pole.

7. Yes. The rotational motion of the Earth is to the east, and therefore if you launch in that direction you are adding the speed of the Earth's rotation to the speed of your rocket.

9. As the astronauts approach a mascon, its increased gravitational attraction would increase the speed of the spacecraft. Similarly, as they pass the mascon, its gravitational attraction would now be in the backward direction, which would decrease their speed.

11. It makes more sense to think of the Moon as orbiting the Sun, with the Earth providing a smaller force that

makes the Moon "wobble" back and forth in its solar orbit.

CHAPTER 13

1. The motion is periodic. It is not simple harmonic, however, because the position and velocity of the ball do not vary sinusoidally with time.

3. The motion of the air cart is periodic; it repeats after a fixed length of time. It is not simple harmonic motion, however, because the position and velocity of the cart do not vary sinusoidally with time.

5. The period remains the same because, even though the distance traveled by the object is doubled, its speed at any given time is also doubled.

7. The period of a pendulum is independent of the mass of its bob. Therefore, the period should be unaffected.

CHAPTER 14

1. To generate a longitudinal wave, hit the nail on the head in a direction parallel to its length. To generate a transverse wave, hit the nail in a direction that is perpendicular to its length.

3. This wave is longitudinal, because each cat moves in the same direction as the wave.

5. The Doppler effect applies to radar waves as well as to sound waves. In particular, the ball sees a Doppler-shifted radar frequency for the waves coming from the radar gun. Then, the ball acts as a moving source for waves of this frequency, giving a second Doppler shift to the echoes that are picked up by the radar gun. This provides a one-to-one correspondence between the final observed frequency and the speed of the ball.

7. The sliding part of a trombone varies the length of the vibrating air column that produces the trombone's sound. By adjusting this length, the player controls the resonant frequencies of the instrument. This, in turn, varies the frequency of sound produced by the trombone.

9. You hear no beats because the difference in frequency between these notes is too great to produce detectable beats.

CHAPTER 15

1. To draw a liquid up a straw you expand your lungs, which reduces the air pressure inside your mouth to less than atmospheric pressure. The resulting difference in pressure produces a net upward force on the liquid in the straw.

3. The pressure in a tank of water increases with depth, hence the pressure is greatest near the bottom. To provide sufficient support there, the metal

bands must be spaced more closely together.

5. This experiment shows that a certain pressure is needed at the bottom of the water column and not just a certain weight of water. To blow the top off the barrel it is necessary to increase the pressure in the barrel enough so that the increase in pressure times the surface area of the top exceeds 400 N. Thus, the required height of water is the height that gives the necessary increase in pressure. But the increase in pressure, $\rho g h$, depends only on the height of the water in the tube, not on its weight.

7. Two quantities are unknown; the object's density and its volume. The two weight measurements provide two independent conditions that can be solved for the two unknowns.

9. The Great Salt Lake has water with a higher salinity, and hence a higher density, than ocean water. In fact, the density of its water is somewhat greater than the density of a typical human body. This means that a person can float in the Great Salt Lake much like a block of wood floats in fresh water.

11. The problem is that as you go deeper into the water, the pressure pushing against your chest and lungs increases rapidly. Even if you had a long tube on your snorkel, you would find it difficult to expand your lungs to take a breath. The air coming through the snorkel is at atmospheric pressure, but the water pushing against your chest might have twice that pressure or more. Thus, scuba gear not only holds air for you to breathe in a tank, it also feeds this air to you under pressure.

13. As the water falls it speeds up. Still, the amount of water that passes a given point in a given time is the same at any height. If the thickness of the water stayed the same, and its speed increased, the amount of water per time would increase. Hence, the thickness of the water must decrease to offset the increase in speed.

15. If you takeoff into the wind the air speed over the wings is greater than if you takeoff with the wind. This means that more lift is produced when taking off into the wind, which is clearly the preferable situation.

17. The ball should spin so that the side facing the batter is moving upward and the side facing the pitcher is moving downward.

CHAPTER 16

1. The coffee is not in equilibrium because its temperature is different from that of its surroundings. Over time the temperature of the coffee will decrease, until finally it is the same as room temperature. At this point it will be in equilibrium—as long as the room stays at the same temperature.

3. No. Heat is not a quantity that one object has more of than another. Heat is the energy that is transferred between objects of different temperatures.

5. The mercury level drops at the beginning because the glass is the first to increase its temperature when it comes into contact with the hot liquid. Therefore, the glass expands before the mercury, leading to a drop in level. As the mercury attains the same temperature few moments later, its level will increase.

7. If the objects have different masses, the less massive object will have a greater temperature change, since it has the smaller heat capacity. On the other hand, the objects may have the same mass but differ in the material from which they are made. In this case, the object with the smaller specific heat will have the greater temperature change.

9. Both the metal and the wood are at a lower temperature than your skin. Therefore, heat will flow from your skin to both the metal and the wood. The metal feels cooler, however, because it has a greater thermal conductivity. This allows the heat from your skin to flow to a larger effective volume than is the case with the wood.

11. Two important factors work in favor of the water-filled balloon. First, the water has a large heat capacity, hence it can take on a large amount of heat with little change in temperature. Second, water is a better conductor of heat than air, hence the heat from the flame is conducted into a large volume of water—which gives it a larger effective heat capacity.

13. The hollow fibers of hair are effective insulators because the gas within the fibers has a low thermal conductivity. This is analogous to double-pane windows, which trap a layer of gas between the panes for a greatly enhanced insulating effect.

CHAPTER 17

1. If the temperature of the air in a house is increased, and the amount of air in the house remains constant, it follows from the ideal gas law that the pressure will increase as well.

3. Yes. If the pressure and volume are changed in such a way that their product remains the same, it follows from the ideal gas law that the temperature of the gas will remain the same. If the temperature of the gas is the same, the average kinetic energy of its molecules will not change.

5. No. The temperature at which water boils on a mountaintop is less than its boiling temperature at sea level due to the low atmospheric pressure on the mountain. Therefore, if the stove is barely able to boil water on the mountain, it will not be able to boil it at sea level, where the required temperature is greater.

7. If we look at the phase diagram in Figure 17-16, we can see that in order to move upward in the graph from the sublimation curve to the fusion curve, the pressure acting on the system must be increased.

9. As high-speed molecules leave the drop, it draws heat from its surroundings to keep its temperature constant. As long as the temperature of the drop is constant, it will continue to have the same fraction of molecules moving quickly enough to escape.

CHAPTER 18

1. **(a)** Yes. Heat can flow into the system if at the same time the system expands, as in an isothermal expansion of a gas. **(b)** Yes. Heat can flow out of the system if at the same time the system is compressed, as in an isothermal compression of a gas.

3. Yes. In an isothermal expansion, all the heat added to the system to keep its temperature constant appears as work done by the system.

5. **(a)** No; **(b)** yes; **(c)** no; **(d)** no; **(e)** yes; **(f)** no.

7. Yes, this is possible. The problem is that you would need low-temperature reservoirs of ever lower temperature to keep the process going.

9. The law of thermodynamics most pertinent to this situation is the second law, which states that physical processes move in the direction of increasing disorder. To decrease the disorder in one region of space requires work to be done, and a larger increase in disorder in another region of space.

CHAPTER 19

1. No, the basic physics of electric charges would not have been affected at all by an opposite assignment of positive and negative labels. The use of + and − signs, as opposed to labels such as A and B, has the distinct advantage that it gives zero net charge to an object that contains equal amounts of positive and negative charge.

3. Initially, the bits of paper are uncharged and are attracted to the comb by polarization effects. (See Figure 19-5 and the accompanying discussion.) When one of the bits of paper comes into contact with the comb, it acquires charge from the comb. Now the piece of paper and the comb have charge of the same sign,

and hence there is a repulsive force between them.

5. No. Even uncharged objects are attracted to a charged rod, due to polarization effects. See Figure 19-5 and the accompanying discussion.

7. The proton can be moving in any direction at all relative to the direction of the electric field. On the other hand, the direction of the proton's acceleration *must* be in the same direction as the electric field.

9. No. The electric field of this system is nonzero, unless the separation d vanishes.

11. Electric fields can exist in a vacuum, just as light can propagate through a vacuum. In fact, we shall see in Chapter 25 that light is simply a wave of oscillating electric and magnetic fields. Therefore, it also follows that magnetic fields can exist in a vacuum as well.

CHAPTER 20

1. The electric field is a measure of how much the electric potential changes from one position to another. Therefore, the electric field in each of these regions is zero.

3. An equipotential surface must always cross an electric field line at right angles. Therefore, the equipotential surface at point 1 is concave to the right, the equipotential surface at point 2 is vertical, and the equipotential surface at point 3 is concave to the left.

5. The electric field is either zero on this surface, or it is nonzero and perpendicular to the surface. If the electric field had a nonzero component parallel to the surface, the electric potential would decrease as one moved in the direction of the parallel component.

7. Not necessarily. The electric potential energy of the two charges would be the same only if the charges are also equal. In general, the electric potential energy is the charge times the electric potential. Therefore, just having equal potentials does not guarantee equal potential energy.

9. The capacitance of a capacitor depends on (b) the separation of the plates and (e) the area of the plates. The capacitance does not depend on (a) the charge on the plates, (c) the voltage difference between the plates, or (d) the electric field between the plates.

CHAPTER 21

1. No. An electric current is produced when a net charge moves. If your body is electrically neutral, no current is produced when you walk.

3. No. An electron may have a fairly large velocity at any given time, but because its direction of motion keeps changing—due to its collisions with

atoms in the wire—its average velocity is almost zero.

5. Yes. Just connect two of these resistors in parallel and you will have an equivalent resistance of $R/2$.

7. Resistors connected in parallel have the same potential difference across their terminals.

9. Resistivity is an intrinsic property of a particular substance. In this sense it is similar to density, which has a particular value for each particular substance. Resistance, however, is a property associated with a given resistor. For example, the resistance of a given wire can be large because its resistivity is large, or because it is long. Similarly, the weight of a ball can be large because its density is large, or because it has a large radius.

11. The light shines brightest immediately after the switch is closed. With time, the intensity of the light diminishes. Eventually, the light stops glowing altogether.

CHAPTER 22

1. No. The particles may have charge of the same sign but move in opposite directions along the same line. In this way, they would both move perpendicular to the field, but would deflect in opposite directions.

3. No. If the electron moves in the same direction as the magnetic field, or opposite to the direction of the field, the magnetic force exerted on it will be zero. As a result, its velocity will remain constant.

5. The radius of curvature is proportional to the speed of the particle. It follows that the particle moving in a circle of large radius (and large circumference) has a proportionally larger speed than the particle moving in a circle of small radius (and small circumference). Therefore, the time required for an orbit ($t = d/v$) is the same for both particles.

CHAPTER 23

1. The magnetic field indicates the strength and direction of the magnetic force that a charged particle moving with a certain velocity would experience at a given point in space. The magnetic flux, on the other hand, can be thought of as a measure of the "amount" of magnetic field that passes through a given area.

3. The metal plate moving between the poles of a magnet experiences eddy currents that retard its motion. This helps to damp out oscillations in the balance, resulting in more accurate readings.

5. Initially, the rod accelerates to the left, due to the downward current it

carries. As it speeds up, however, the motional emf it generates will begin to counteract the emf of the battery. Eventually the two emfs balance one another, and current stops flowing in the rod. From this point on, the rod continues to move with constant speed.

7. When the angular speed of the coil in an electric generator is increased, the rate at which the magnetic flux changes increases as well. As a result, the magnitude of the induced emf produced by the generator increases. Of course, the frequency of the induced emf increases as well.

CHAPTER 24

1. Current and voltage are not always in phase in an ac circuit because capacitors and inductors respond not to the current itself—as a resistor does—but to the charge (capacitor) or to the rate of change of the current (inductor). The charge takes time to build up; therefore, a capacitor's voltage lags behind the current. The rate of change of current is greatest when the current is least; therefore, an inductor's voltage leads the current. Resistors, of course, are always in phase with the current.

3. In the phasor diagram for an LC circuit, the impedance is always perpendicular to the current—the only question, therefore, is whether ϕ is $+90°$ or $-90°$. At a frequency less than the resonance frequency the capacitive reactance ($X_C = 1/\omega C$) is greater than the inductive reactance ($X_L = \omega L$). Therefore, the situation is similar to that of a circuit with only a capacitor, in which case the phase angle is $\phi = -90°$.

5. As frequency is increased there is no change in resistance, R. On the other hand, the capacitive reactance ($X_C = 1/\omega C$) decreases and the inductive reactance ($X_L = \omega L$) increases.

7. Yes. Recall that the impedance of an RLC circuit is $Z = \sqrt{R^2 + (X_L - X_C)^2}$. Because the quantity in parentheses is squared, we get the same impedance when X_L is greater than X_C by a certain amount as when X_C is greater than X_L by the same amount.

CHAPTER 25

1. Presumably, an "invisible man" would be invisible because light passes through his body unimpeded, just as if it were passing through thin air. If some light were deflected or absorbed, we would see this effect and the person would no longer be invisible. For a person to see, however, some light must be absorbed by the retina. This absorption would cause the invisible man to be visible.

3. Your watch must have an LCD display. In such watches, the light coming from

the display is linearly polarized. If the polarization direction of the display and the sunglasses align, you can read the time. If these directions are at 90° to one another no light will pass through the sunglasses, and the display will appear black.

5. View the light reflected from a horizontal surface, such as a tabletop. This light is polarized primarily in the horizontal direction. Therefore, if you rotate the sheet of polarizing material until you receive a maximum amount of reflected light, you will know that its transmission axis is horizontal.

7. As mentioned in the answer to conceptual question 4, the light reflected from a horizontal surface is polarized primarily in the horizontal direction. If the glasses are merely tinted, reflected light will have the same intensity no matter how the glasses are rotated. If they are Polaroid, however, you will notice a striking difference in reflected intensity as you rotate the glasses.

CHAPTER 26

1. Three images are formed of object A. One extends from $(-2\,\text{m}, 2\,\text{m})$ to $(-1\,\text{m}, 2\,\text{m})$ to $(-1\,\text{m}, 3\,\text{m})$. Another image forms an "L" from $(1\,\text{m}, -3\,\text{m})$ to $(1\,\text{m}, -2\,\text{m})$ to $(2\,\text{m}, -2\,\text{m})$. Finally, the third image extends from $(-1\,\text{m}, -3\,\text{m})$ to $(-1\,\text{m}, -2\,\text{m})$ to $(-2\,\text{m}, -2\,\text{m})$.

3. A plane mirror is flat, which means that its radius of curvature is infinite. This means that the focal length of the mirror is also infinite. Whether you consider the focal length to be positive infinity (the limit of a concave mirror) or negative infinity (the limit of a convex mirror) doesn't matter, because in either case the term $1/f$ in the mirror equation will be zero.

5. We can consider the Sun to be infinitely far from the mirror. As a result, its parallel rays will be focused at the focal point of the mirror. Therefore, the distance from the mirror to the paper should be $f = \frac{1}{2}R$.

7. The key to this system is the fact that the lifeguards can run much faster on the sand than they can swim in the water. Therefore, the path ACB—even though it is longer—has a shorter travel time because more time is spent on the sand and less time is spent in the water.

9. The Sun is already well below the horizon when you see it setting. The reason is that as the Sun's light enters the atmosphere from the vacuum of space, it is bent toward the normal; that is, toward the surface of Earth. Therefore, the light from the Sun can still reach us even when a straight line from our eyes to the Sun would go below the horizon.

11. No. In order to see, a person's eyes must first bring light to a focus, and then must absorb the light to convert it to nervous impulses that can travel to the brain. Both the bending of the light and its absorption would give away the presence of the invisible man. Therefore, if the man were truly invisible, he would be unable to see.

CHAPTER 27

1. When you focus on an object at infinity your ciliary muscles are relaxed. It takes muscular effort to focus your eyes on nearby objects.

3. If you are a distance D in front of a mirror your image is a distance D behind the mirror. Therefore, you can see your image clearly if the distance from you to your image, $2D$, is equal to N. In other words, the minimum distance to the mirror is $D = N/2$.

5. Yes, it matters. A simple magnifier is nothing more than a convex lens. As we can see from Figure 26-35, a convex lens forms an enlarged (magnified) image only when the object is closer to the lens than its focal length.

CHAPTER 28

1. At a point where destructive interference is complete, there is no energy. The total energy in the system is unchanged, however, because those regions with constructive interference have increased amounts of energy. One can simply think of the energy as being redistributed.

3. No. The net signal could be near zero if the waves from the two antennas interfere destructively.

5. Submerging the two-slit experiment in water would reduce the wavelength of the light from λ to λ/n, where $n = 1.33$ is the index of refraction of water. Therefore, the angles to all the bright fringes would be reduced, as we can see from Equation 28-1. It follows that the two-slit pattern of bright fringes would be more tightly spaced in this case.

7. A cat's eye would give greater resolution in the vertical direction, because the effective aperture is greater in that direction. As shown in Equation 28-14, the greater the aperture, D, the smaller the angle, θ, and the greater the resolution.

9. In an iridescent object, the color one sees is determined by constructive and destructive interference. The conditions for interference, however, depend on path length, and path length depends on the angle from which one views the system. This is analogous to viewing a two-slit system from different angles and seeing alternating regions of constructive and destructive interference. A painted object, on the

other hand, simply reflects light of a given color in all directions.

CHAPTER 29

1. The light received from these galaxies moves with the speed c, as is true for all light in a vacuum.

3. Yes. All light, regardless of its wavelength, has the same speed in a vacuum. The frequency of this "red shifted" light will be affected, however. Recalling that $v = \lambda f$, we see that a longer wavelength also implies a smaller frequency.

5. Since the total energy of an object of finite mass goes to infinity in the limit that $v \to c$, an infinite amount of energy is required to accelerate an object to the speed c.

CHAPTER 30

1. The "ultraviolet catastrophe" refers to the classical prediction that the intensity of light emitted by a blackbody increases without limit as the frequency is increased.

3. No, all objects with finite temperature give off blackbody radiation. Only an object at absolute zero—which is unattainable—gives off no blackbody radiation.

5. If you look at a painting, photo, decal, or similar object, you are viewing light that it reflects to you. Therefore, when an area appears red, it is because the pigments there absorb blue photons and reflect red photons. Similarly, a blue area absorbs red photons. Blue photons carry considerably more energy than red photons, however, and hence blue photons are more likely to cause damage to the pigment molecules, or to alter their structure. It follows that red pigments (which absorb blue photons) are more likely to become faded when exposed to intense light.

7. **(a)** A photon from a green light source *always* has less energy than a photon from a blue light source. **(b)** A photon from a green light source *always* has more energy than a photon from a red light source. The reason for these results is that the energy of a photon depends linearly on the frequency of light; that is, $E = hf$.

9. The resolution of a microscope is determined by the wavelength of the imaging radiation—the smaller the wavelength the greater the resolution. Since the typical wavelength of an electron in an electron microscope is much smaller than the wavelength of visible light, the electron microscope has the greater resolution.

CHAPTER 31

1. The observation that alpha particles are sometimes reversed in direction

when they strike a thin gold foil led to the idea that there must be a great concentration of positive charge and mass within an atom. This became the nucleus in Rutherford's model.

3. In principle, there are an infinite number of spectral lines in any given series. The lines become more closely spaced as one moves higher in the series, which makes them hard to distinguish in practice.

5. (a) There is no upper limit to the wavelength of lines in the spectrum of hydrogen. The reason is that the wavelength is inversely proportional to the energy difference between successive energy levels. The spacing between these levels goes to zero as one moves to higher levels, and therefore the corresponding wavelengths go to infinity. (b) There is a lower limit to the wavelength, however, because there is an upper limit of 13.6 eV to the energy difference between any two energy levels.

7. All of these questions can be answered by referring to Figure 31-17 and Table

31-3. (a) Not allowed; there is no d subshell in the $n = 2$ shell. (b) Not allowed for two reasons. First, there is no p subshell in the $n = 1$ shell. Second, a p subshell cannot hold seven electrons. (c) Allowed. (d) Not allowed; the $n = 4$ shell does not have a g subshell.

9. These elements all have similar configurations of their outermost electrons. In fact, the outermost electrons in fluorine, chlorine, and bromine are $2p^5, 3p^5$, and $4p^5$, respectively. Therefore, each of these atoms is one electron shy of completing the p subshell. This accounts for their similar chemical behavior.

CHAPTER 32

1. The radius of a nucleus is given by the following expression: $r = (1.2 \times 10^{-15}\,\text{m})A^{1/3}$. Therefore, the radius depends only on the total number of nucleons in the nucleus, A, and not on the number of protons and neutrons separately. If the number of protons plus the number of neutrons is the same for two nuclei, their radii will

be equal as well.

3. Alpha particles, which can barely penetrate a sheet of paper, are very unlikely to expose film in a cardboard box. Beta particles, on the other hand, are able to penetrate a few millimeters of aluminum. Therefore, beta particles are more likely to expose the film than alpha particles.

5. Above the $N = Z$ line, a nucleus contains more neutrons than protons. This helps to make the nucleus stable, by spreading out the positive charge of the protons. If a nucleus were below the $N = Z$ line, it would have more protons than neutrons, and electrostatic repulsion would blow the nucleus apart.

7. Yes. The two samples may contain different quantities of the radioactive isotope, and hence their activities may be different.

9. No. The RBE is related to the amount of biological effect produced by a given type of radiation, not to the amount of energy it delivers.

Photo Credits

Chapter 1—p. 1 Andrew Cowie/AFP/GettyImage p. 3 **(center right)** John F. Kennedy Space Center/NASA p. 3 **(top right)** Oliver Meckes/Science Source p. 4 **(bottom)** U.S. Department of Commerce p. 4 **(top)** Bureau International des Poids et Mesures p. 6 Jeff Greenberg/PhotoEdit, Inc p. 8 Antonin Thuillier/AFP/GettyImages p. 9 **(top)** David R. Frazier/Science Source p. 10 CNRI/Science Photo Library/Science Source p. 11 US Department of Energy p. 15 **(center right)** Jeff Greenberg/PhotoEdit, Inc p. 15 **(top)** Andrew Cowie/AFP/GettyImage p. 17 **(bottom)** USDA Forest Service p. 19 **(left)** Vadim Petrakov/Shutterstock, Visions of America, LLC/Alamy, Eric Grave/Science Source

Chapter 2—p. 19 Dave Thompson-British Athletics/Contributor/Getty Images p. 26 James Bowyer/Getty Images p. 29 NASA p. 32 Chris Urso/AP Images p. 33 David Scharf/Science Source p. 43 **(bottom right)** Rainer Albiez/Shutterstock p. 43 **(center right)** Steve Kazlowski/Danita Delimont Photography/Newscom p. 43 **(top right)** Jim Sugar/Corbis p. 45 Julia Davila-Lampe/Flickr Open/Getty Images p. 54 Jill Morgan/Alamy p. 55 Stephen Dalton/Science Source, Pearson Education/PH College

Chapter 3—p. 60 Maksym Yemelyanov/Fotolia p. 61 Jeff Greenberg/Science Source p. 66 **(bottom left)** Delmas Lehman/Fotolia p. 72 George Whiteley/Science Source p. 76 Laurent Rebours/AP Images p. 84 US Department of Defense p. 86 **(center right)** Ezra Shaw/Getty Images p. 86 **(top left)** PIXOLOGICSTUDIO/Science Photo Library/Alamy

Chapter 4—p. 88 Richard Megna/Fundamental Photographs p. 94 **(top center)** NASA p. 94 **(top left)** Richard Megna/Fundamental Photographs p. 94 **(top right)** Richard Megna/Fundamental Photographs p. 97 **(top left)** Nick Turner/Alamy p. 97 **(top right)** Jon Bilous/Alamy p. 100 **(center)** John Terence Turner/Alamy p. 100 **(left)** NaturePL/Superstock p. 100 **(right)** Bernd Zoller/ImageBroker/Alamy p. 107 G. I. Bernard/Science Source, Gregory G. Dimijian/Science Source

Chapter 5—p. 112 **(top)** Tcly/Shutterstock p. 117 Alessandro Bianchi/Reuters p. 119 James S. Walker p. 122 **(bottom)** NASA p. 122 **(center)** Bruce Bennett/Getty Images p. 128 Federal Aviation Administration p. 136 Pete Saloutos/Corbis p. 140 Johnson Space Center/NASA p. 153 Frank Greenaway/Dorling kindersly, Ltd, United States Marines

Chapter 6—p. 155 Greg Epperson/iStock/Getty Image p. 156 **(left)** Ian Aitken/Dorling Kindersley, Ltd. p. 156 **(right)** Ernest Braun/Stone Allstock/Getty Images p. 160 James Marvin Phelps/Shutterstock p. 162 **(left)** Photodisc/Getty Images p. 162 **(right)** Andreas Bauer/iStock/Getty Images p. 164 **(bottom)** Richard Megna/Fundamental Photographs p. 164 **(top)** Richard Megna/Fundamental Photographs p. 168 **(bottom)** Science Museum/Science & Society/The Image Works p. 168 **(top)** Oote boe 3/Alamy p. 172 Oleg Zabielin/Shutterstock p. 173 Warner Bros pictures/Everett Collection p. 178 Brian McEntire/Shutterstock p. 181 **(bottom)** © Bombardier p. 181 **(center)** Mark Meyer/Navy Visual News Service p. 181 **(top)** Dave Martin/AP Images p. 182 Vasiliy Koval/Fotolia p. 190 Tsuneo Nakamura/Volvox Inc/Alamy p. 191 V. J. Caiozzo, University of California, Irvine, AF archive/Alamy

Chapter 7—p. 195 VisualCommunications/iStock/Getty Images p. 198 **(bottom)** AFP/Getty Images p. 198 **(center)** AFP/Getty Images p. 198 **(top)** holbox/Shutterstock p. 209 Dr T J McMaster, Wellcome Images p. 213 National Geographic/Getty Images

Chapter 8—p. 223 Alex Emanuel Koch/Shutterstock p. 225 **(left)** Herbert Esser/Fotolia p. 225 **(right)** Westend61/Getty Images p. 229 Wales News Service/Splash News/Newscom p. 231 Barry Bland/Alamy p. 233 Todd Winner/Stocktrek Images/Alamy p. 234 **(bottom)** Brisbane/Shutterstock p. 234 **(top)** hatchapong/Fotolia p. 239 I Love Images/Cultura/Getty Images p. 246 Federal Highway Administration, Colorado Division p. 249 **(center)** Stephen Derr/The Image Bank/Getty Images p. 249 **(right)** Dmitry Kalinovsky/Shutterstock

Chapter 9—p. 257 Sasa Huzjak/Alamy, National Oceanic and Atmospheric Administration p. 257 stockshoppe/Shutterstock p. 263 **(left)** Kathy Willens/ AP Images p. 263 **(right)** Wally McNamee/Corbis p. 266 Oxford Scientific/Getty Images p. 268 NASA p. 270 NASA p. 272 **(bottom)** Franck Seguin/Corbis p. 272 **(bottom)** Jacqueline Larma/AP Images p. 272 **(top)** Bruce Bennett/Getty Images p. 280 **(bottom left)** Richard Megna/Fundamental Photographs p. 280 **(top left)** Richard Megna/Fundamental Photographs p. 280 **(top right)** Richard Megna/Fundamental Photographs p. 287 **(bottom)** Richard Megna/Fundamental Photographs p. 287 **(top)** Richard Megna/Fundamental Photographs p. 289 **(bottom)** Vittorio Bruno/Shutterstock p. 289 **(top)** US Air Force p. 290 Richard Megna/Fundamental Photographs p. 290 Richard Megna/Fundamental Photographs

p. 291 **(bottom)** David J. Phillip/AP Images p. 291 **(top)** Vittorio Bruno/Shutterstock

Chapter 10—p. 299 European Southern Observatory p. 300 **(bottom center)** Novastock/PhotoEdit Inc. p. 300 **(bottom left)** NASA p. 300 **(bottom right)** A. Barry Dowsett/Science Source p. 301 Daniel Dempster/Alamy p. 308 **(bottom)** Karl Weatherly/Photodisc/Getty Images p. 308 **(top)** Michael Putland/Getty Images p. 309 **(bottom)** Vasiliy Koval/Fotolia p. 309 **(top)** Pavel Losevsky/Fotolia p. 309 **(top)** Pavel Losevsky/Fotolia p. 312 Richard Megna/Fundamental Photographs p. 317 Mehau Kulyk/Science Photo Library/Science Source p. 318 Mark Scheuern/Alamy p. 326 Jeff Hester/Arizona State University/NASA p. 327 **(right)** Farmer/Fotolia p. 327 **(top left)** Paul Bricknell/Dorling Kindersley, Ltd p. 331 Yoshikazu Tsuno/AFP Photo/Getty Images p. 333 Ames Research Center/Ligo Project/NASA

Chapter 11—p. 334 Ursula Düren/picture-alliance/dpa/AP Images p. 335 Kevin M. Law/Alamy p. 336 Fred Winner/Science Source p. 337 US Coast Guard p. 340 Siegfried Haasch/Getty Images p. 342 Richard Megna/Fundamental Photographs p. 348 EggHeadPhoto/Shutterstock p. 353 **(top center)** Janelle Lugge/Shutterstock p. 353 **(top left)** altrendo images/Getty Images p. 353 **(top right)** Andreas Edelmann/Fotolia p. 360 Rogan Thomson/ActionPlus/Corbis p. 362 **(bottom)** NASA p. 362 **(top left)** NASA Earth Observatory p. 362 **(top right)** Travis Heying/The Wichita Eagle/AP Images p. 366 **(left)** Pearson Education/PH College p. 366 **(right)** Science Source p. 368 NASA p. 371 Polka Dot Images/Getty Images

Chapter 12—p. 381 Denis Sarrazin/NASA Earth Observatory p. 384 MShieldsPhotos/Alamy p. 387 **(bottom right)** European Space Agency p. 387 **(left)** Ralph White/NASA Earth Observing System p. 387 **(top right)** Philip A. Dwyer/MCT/Newscom p. 388 **(bottom)** NASA p. 388 **(top)** NASA Headquarters p. 394 **(left)** NASA Headquarters p. 394 **(right)** NASA Headquarters p. 396 **(left)** NASA Headquarters p. 396 **(right)** NASA/Johnson Space Center p. 402 **(bottom)** AP Images p. 402 **(center)** Mark Pilkington/Geological Survey of Canada/SPL/Science Source p. 402 **(top)** USGS p. 404 David Nunuk/Science Source p. 408 **(bottom)** David R. Frazier Photolibrary, Inc./Alamy p. 408 **(top)** David R. Frazier Photolibrary, Inc./Alamy p. 415 NASA/Jet Propulsion Laboratory, NASA/Johnson Space Center, NASA Headquarters

Chapter 13—USGS p. 418 Sergii Shcherbakov/Fotolia p. 419 Tewan Banditrukkanka/Shutterstock p. 420 FloridaStock/Shutterstock p. 431 NASA Headquarters p. 433 **(left)** Ted Kinsman/Science Source p. 433 **(right)** Terje Rakke/Getty Images p. 443 **(bottom)** Julianwphoto/Fotolia p. 443 **(top)** Marc Tielemans/Alamy p. 446 **(bottom)** AP Images p. 446 **(bottom)** michaklootwijk/Fotolia p. 446 **(top)** Balfour Studios/Alamy p. 450 **(left)** Pearson Education p. 450 **(right)** Pearson Education p. 451 Dmitry Vereshchagin/Fotolia p. 453 **(top left)** Dr. Harold G. Craighead

Chapter 14—p. 456 Jeremy Hoare/Alamy p. 457 **(bottom)** CoverSpot Photography/Alamy p. 457 **(top)** Laurent_Imagery/Flickr State/Getty Images p. 464 Leonard Lessin/Science Source p. 466 **(center)** Tsekhmister/Shutterstock p. 466 **(left)** Oxford Scientific/Getty Images p. 466 **(right)** Phanie/Science Source p. 467 **(bottom)** AP Images p. 467 **(center)** Patricio Robles Gil/AgeFotostock p. 467 **(top)** francescodemarco/Fotolia p. 477 **(bottom)** ESA&Hubble/NASA p. 477 **(center)** Chris Gallagher/Science Source p. 477 **(top)** NOAA p. 483 **(bottom)** Richard Megna/Fundamental Photographs p. 483 **(center)** Richard Megna/Fundamental Photographs p. 483 **(top)** Richard Megna/Fundamental Photographs p. 484 **(top center)** Richard Megna/Fundamental Photographs p. 484 **(top left)** Richard Megna/Fundamental Photographs p. 484 **(top right)** Richard Megna/Fundamental Photographs p. 485 **(left)** Netfalls/Fotolia p. 485 **(right)** kataijudit/Shutterstock p. 489 **(bottom)** epitavi/Fotolia p. 489 **(top)** Dave King/Dorling Kindersley/Getty Images, Daniel Ruta/iStock/Getty Images,

Chapter 15—p. 502 Violetstar/Fotolia p. 505 Michael Fritzen/Fotolia p. 506 **(left)** Richard Megna/Fundamental Photographs p. 506 **(right)** Michal Heron/Pearson Education/PH College p. 510 Richard Megna/Fundamental Photographs p. 516 Joe Traver/The LIFE Images Collection/Getty Images p. 517 Ruslan Dashinsky/iStockphoto/Getty Images p. 518 **(bottom)** UPPA/ZUMA Press/Newscom p. 518 **(top left)** Gary Blakeley/Shutterstock p. 518 **(top right)** Steve Bower/Shutterstock p. 519 Seth Resnick/RGB Ventures/SuperStock/Alamy p. 522 Bluestone Productions/Taxi/Getty Images p. 528 **(bottom)** Kim Schott/iStockphoto/Getty Images p. 528 **(top)** Judy Kennamer/Shutterstock p. 533 AlasdairJames/Vetta/Getty Images p. 540 Fabrication Enterprises Inc.

Chapter 16—p. 544 Fred Olivier/Nature Picture Library p. 552 **(top left)** wajan/Fotolia p. 553 **(center)** Joe Sohm/The Image Works p. 553 **(top)** Richard Choy/Photolibrary/Getty Images p. 556 **(bottom)** NHPA/Science Source p. 556

Text and Illustration Credits

Note: Page numbers followed by f indicate figures; those followed by t indicate tables.

A

A (mass number), 1120, 1120t, 1128
a waves, 711
Aberration, spherical, 921, 921f
Aberrations, 974–975, 974f, 975f
Absolute value. *See* Magnitude
Absolute zero, 548, 549f, 550, 647–648
Absorption spectrum, 1085, 1085f
ac (alternating current). *See* Alternating current; Alternating-current (ac) circuits
Accelerated frames of reference, 1039, 1039f, 1040f
Acceleration, 28–33
  angular, 303–304, 304f, 338–342
  apparent weight and, 139–140, 139f
  automobiles and, 207, 213–215
  average, 28–29, 30f, 73–76, 74f, 75f
  of center of mass, 285–287
  centripetal, 177–182, 177f, 309–311, 309f
    in centrifuges, 182, 182f
    vs. tangential acceleration, 310, 310f
  constant, 33–43, 33f, 91–92, 91t
    angular velocity and, 304–305, 304f
    free fall and, 43–44
  force and, 121–123, 121f, 125–130
  free fall and, 43–44
  graphical interpretation of, 29–30, 29f
  instantaneous, 29–31, 30f, 73–74, 74f
  instantaneous angular, 303–304
  linear, 29f
  mass and, 121, 121f, 122
  maximum, in simple harmonic motion, 427
  notation for, 5, 6t
  of rotating object
    centripetal, 309
    tangential, 310
  signs of, and change in speed, 31–33, 32f
  in simple harmonic motion, 426–428
  tangential, 310–311, 310f
  torque and, 338–342
  typical magnitudes of, 29t
  typical values for, A-10
  units of, 123t
  velocity and, 119–120
  work and, 204–206
Acceleration due to gravity (g)
  altitude and, 386–387
  in connected objects, 175–176
  on Moon, 388
  notation for, 44
  projectile motion and, 91–92
  sample values of, 44f
Acceleration vectors, 73–74, 74f, 75f, 76f
Accelerometer, 33, 33f
Accommodation, in human eye, 955, 955f
Accretion disks, 1041f
Achromatic lenses, 974, 975f
Acousto-optic modulation (AOM), 1007, 1007f
Actinium-227, decay of, 1131
Action-reaction force pairs, 129–131, 130f
Active noise reduction, 481
Activity, in nuclear decay
  definition of, 1131
  time and, 1134–1135, 1136f
  units of, 1131
Adair, Robert, 499
Addition
  of vectors, 65–68
  of velocity, 1028–1031, 1029f, 1031f

Adiabatic processes, 625–627, 626f, 627f, 630–631, 630f
Adiabats, 626, 630f
Age-related macular degeneration (AMD), 1109
Aging, time dilation and, 1023
Air, vibrating columns of, 485–489
Air conditioners, 638, 638f
Air flow, 527–528, 527f, 528f
Air pressure, 581–582
Air resistance
  free fall and, 43
  projectile motion and, 92–93, 104, 104f, 106f
Air splints, 505, 506f
Air tracks, 119, 119f
Air wedges, 991–993, 992f
Airplanes
  heads-up display in, 918
  human-powered, 213, 213f
  lift in wings of, 527
  takeoff distance for, 38–39
  turbulence and, 428
Airport metal detectors, resonance and, 870, 871f
Akasaki, Isamu, 1111
Algebra, A-3–A-6
Algorimeter, 540, 540f
Alkali metals, 1103
Alpha decay, 1126–1128, 1128f
Alpha particles, 1083, 1125, 1137
Alpha rays, 1126
Alternating voltage, 846, 846f
Alternating-current (ac) circuits, 822, 822f, 845–872, 846
  capacitors in, 851–854, 851f, 853f
  definition of, 731
  elements, properties of, 865t
  in generators, 822, 822f
  high and low frequency limits of, 866f
  inductors in, 859–863, 861f
  RC, 754–757, 855–862
  resistors in, 846, 847f
  RL, 861f
  RLC, 869–870
  root mean square (rms) values, 847–848
  voltages and currents in, 846–847, 846f
Altitude
  of geosynchronous satellites, 396, 396f
  gravitational potential energy and, 229–231, 398–400
  gravity and, 386–387
Alveoli, surface tension in, 533
Amano, Hiroshi, 1111
Amber, electrical properties of, 658–659, 658f
Ambulatory oscillation, 442–443
Ammeters, 758
Ampere (A), 731
Ampère, André-Marie, 731
Ampère's law, 789, 789f
Amplitude, of simple harmonic motion, 422, 426–428, 432f
  damped oscillations and, 444
  driven oscillations and, 445–446, 445f
  pendulums and, 439–440
  period and, 431
Amusement park rides, 140
Andromeda galaxy, 886, 887f
Angle(s)
  of incidence, 914–915, 914f, 935, 935f
  launch
    general (arbitrary), 99–103, 99f, 100f
    range and, 104, 106, 106f
    zero, 94–95, 94f, 95f
  of reflection, 914–915, 914f, 915f
  of refraction, 930, 930f, 933, 933f

scattering, 1063
  units for, 301
  of vector direction, 62, 62f, 63f, 65
  work and, 198–199, 198f, 201
Angular acceleration, 303–304, 304f, 338–342
Angular displacement,, 301, 302f
Angular frequency, 425
Angular magnification
  definition of, 967
  of magnifying glass, 967
  of microscope eyepiece, 969
  of telescope, 971
Angular momentum, 356–364
  angular speed and, 356–357
  in Bohr model of the hydrogen atom, 1094–1096
  in circular motion, 356–357, 356f, 358, 358f
  conservation of. *See* Conservation of angular momentum
  definition of, 356
  kinetic energy and, 363
  moment arm and, 357
  of nontangential motion, 356, 356f
  orbital motion and, 392
  of point particle, 356–357, 356f
  rate of change of, 359
  right-hand rule for, 366–367, 366f, 367f
  rotational collisions and, 363
  sign conventions for, 358
  units of, 356
Angular position, 300, 300f
Angular separation
  index of refraction and, 1002
  resolution and, 1002–1005, 1002f
Angular size
  object distance and, 966–967, 966f
  in telescopes, 971
Angular speed, 302–303, 302f, 303f
  angular momentum and, 356–357
  conservation of angular momentum and, 361–363
  kinetic energy and, 313–314, 313f
  linear speed and, 307–310, 307f
Angular velocity, 301–303, 302f, 303f
  constant acceleration and, 304–305, 304f
  instantaneous, 301–303
  period and, 302–303
  right-hand rule for, 366
  as vectors, 366
Animals
  communication by, 467
  fluorescence in, 1111
  iridescence and, 1008, 1008f
  Newton's third law of motion and, 133
Annihilation, matter-antimatter, 1036, 1036f, 1146–1147
Antennas, 881–882, 882f, 890f, 984–985, 985f
Antilock braking system (ABS), 163, 163f
Antimatter, 1035–1036
Antineutrinos, 1130
Antinodes, 483, 483f, 484f, 486, 486f
Antiparticles, 1125
Antiquarks, 1149, 1149t
Antiscalding devices, 553
Antitheft devices, magnetic, 821
AOM (Acousto-optic modulation), 1007, 1007f
Apertures
  in cameras, 958
  circular, 1002, 1002f
  in telescopes, 973
Apochromatic lenses, 974
Apparent depth, 931–932, 932f
Apparent size
  angular magnification and, 966–967
  Galileo thermometer and, 951

Apparent weight, 138–141, 139f
Aquatic creatures, 675
Aqueous humor, 955, 955f
Arc length, 301, 301f
Archimedes' principle, 513–521
  applications of, 514–521
  balloons and, 518, 518f
  in body fat measurement, 515–516
  buoyant force in, 513–514
  in complete submersion, 514–516
  floatation and, 517–521, 517f, 518f
  statement of, 513
  submersion and, 514–516
Area
  notation for, 5, 6t
  shear modulus and, 595
  thermal expansion of, 553
  in Young's modulus, 592
Arm, human
  center of mass in, 284–285, 284f
  tension and torque for, 348–349
Arrow of time, 632
Arteries, blood flow in, 531–533
Asteroids, 401–403
Astigmatism, 965
Astronauts
  aging of, time dilation and, 1023
  centrifuges for, 309, 309f
  golfing on Moon by, 104
  jet packs for, 127, 127f, 134, 134f
  mass of, measurement of, 451
  tethered, 173, 173f
  weightlessness in, 332–333, 431, 431f, 450
Astronomical telescopes, 973, 973f
Atmosphere
  depth of, 540–541
  escape speeds and, 405
  evaporation of, 601–602
  visible light and, 892
Atmospheric pressure
  barometers and, 509, 509f
  and boiling temperature of water, 598
  mercury and, 509
  units of, 505, 509t
  visualization of, 506f
Atom(s), 659, 659f
  early models of, 1083–1084, 1083f
  electric charge of, 659, 659f
  electric force and, 664–665
  energy levels in multielectron, 1100–1102, 1100f
  hydrogen. *See* Hydrogen atom
  line spectrums of, 1084–1088, 1085f
  magnetic fields and, 795
  mass and charge of particles in, 1121t
  neon, 1104f
  nucleus of, 1083–1084, 1083f
  radiation and. *See* Atomic radiation
  states of
    excited, 1091, 1098–1099, 1099f, 1107
    exclusion principle for, 1100–1101
    ground, 1091, 1098, 1098f, 1100
  structure of, 659, 659f
Atomic clocks, 4, 1022, 1092
Atomic mass, 583
Atomic mass unit, 1120–1121
Atomic number (Z), 1120
Atomic radiation, 1104–1112. *See also* Radiation
  fluorescence and, 1110–1111
  phosphorescence and, 1112
  shells and subshells in, 1104–1105, 1105f
  in x-rays, 1104–1105, 1104f
Atomic-force microscopy, 209, 209f
Atomizers, 528, 528f
Atwood's machine, 175–176, 329
Aurora australis, 782
Aurora borealis, 782, 782f
Autoclaves, 598

Automobiles
  acceleration and, 74, 74f, 207, 213–215
  banked curves for, 180, 181f
  braking in, 163, 163f
  circular motion and, 178–181
  coasting, 201–202
  collisions of, 154, 276–277
  crumple zones for, 154
  electric motors in, 824
  headlight differentiation on, 1002
  headlight differentiation with, 1002f
  heads-up display in, 918, 918f
  skidding of, 179–180
  speedometers in, 821
  stopping distance for, 39–41, 41f
  sudden stops in, 347, 347f
Average acceleration, 28–29, 30f
Average acceleration vector, 73–76, 74f, 75f, 76f
Average force, 210–211, 262–265, 262f
Average power, 212–213
  in ac circuits, 847–848, 847f, 865t
  for RC circuits, 858
Average speed, 22–23
Average velocity, 23–24
  equation of motion for, 35, 35f
  graphical interpretation of, 24–25, 25f, 28, 28f
  slope and, 25
Average velocity vectors, 72–73, 73f
Avogadro, Amedeo, 582
Avogadro's number, 582
Axis of rotation, 300, 300f
  direction of, 366–367
  Earth and, 317, 317f
  moment of inertia and, 313–318
  precession and, 368
  static equilibrium and, 343–346
  torque and, 344

B

b waves, 711
Balance, center of mass and, 350f, 353f
Ballistic pendulums, 275
Ballistocardiograph, 269
Balloons, buoyancy of, 518
Balmer, John Jakob, 1085
Balmer series, 1085–1086, 1085f, 1087f, 1087t, 1092, 1092f
Banked turns, 180, 181f
Bar (unit of measure), 505
Bar magnets, 770, 770f
Barometers, 509, 509f
Barringer Meteor Crater, 402
Barton, Otis, 537, 616
Baryons, 1148, 1149t, 1150, 1150f
Baseball
  batting in, 262–263, 262f, 263f, 271, 499
  catching, 237
  pitching in, 128–129, 128f, 305–306, 1073
Bathysphere, 537, 616
Bats, navigation by, 466, 466f, 468, 469f
Batteries, 732–734
  capacitors and, 712–715
  check meters for, 740, 740f
  electromotive force and, 732
  energy and power in, 738–739
  internal resistance in, 733, 743, 743f
  work done by, 733
Batting, in baseball, 262–263, 262f, 263f, 271, 499
Baumgartner, Felix, 518, 518f
Beats/beat frequency, 489–492, 490f
Becquerel (Bq), 1131
Becquerel, Antoine Henri, 1126
Beebe, William, 537, 616
Bees, pollen collection by, 693–694
Bell, Alexander Graham, 470
Bernoulli effect, 527–529, 529f
Bernoulli's equation, 523–530
  applications of, 527–530
  statement of
    with constant height, 524
    general case, 526

visualization of, 528f
work-energy theorem for and, 523–524
Beta (β) particles, 1125
Beta decay, 1128–1130, 1129f
(β) rays, 1125, 1126
Bethe, Hans, 1143
Bicycle reflectors, 918, 918f
Bimetallic strips
  in circuit breakers, 850, 850f
  in thermal expansion, 552–553, 552f
Binding energy of nuclei, 1138–1139, 1138f
Binoculars, 934, 934f
Binoculars, nonreflective coating on, 996, 996f
Binomial expansion, A-6
Biologically equivalent dose, 1145
Bioluminescence, 536
Black holes, 407, 1041–1042, 1042f
Black lights, 1111
Blackbodies, 1052–1055, 1052f
Blackbody radiation, 569, 1052–1055, 1052f
  line spectrums and, 1084–1085, 1085f
  temperature of Sun and, 1053
Blood flow
  in arteries, 531–533
  in countercurrent exchange, 566, 566f
  speed of, measurement of, 477, 778–779, 778f
Blue shift, in galaxies, 472
Blu-ray technology, 996–997, 996f
Boats
  gravity escape system for, 199–200
  river crossing by, 78–79
Bod Pod, 516, 516f
Body fat, measurement of, 515–516
Body Mass Measurement Device, 431, 431f
Bohr, Niels, 366f, 665, 1088, 1088f
Bohr model of hydrogen atom, 1088–1094
  angular momentum in, 1094–1096
  assumptions of, 1088
  de Broglie waves and, 1094
  energy levels in, 1091, 1091f
  line spectrums and, 1092
  magnetic fields in, 806
  orbit in, 665, 665f, 1088–1089, 1088f, 1090–1091
Bohr radius, 1089, 1089f
Boiling point, 598
Bolometers, 737
Boltzmann, Ludwig, 580
Boltzmann constant, 580–581
Born, Max, 1096
Bows and arrows, potential energy in, 232–233
Boyle, Robert, 583
Boyle's law, 584
Brahe, Tycho, 391
Brain
  electrical activity in, 711, 806
  magnetic fields in, 805, 806
  PET scans and, 1146–1147
Brakes
  antilock, 163, 163f
  sudden stops and, 347, 347f
Braking, magnetic, 817
Breathing, surface tension in, 533
Breezes, sea and land, 567, 567f
Bremsstrahlung, 1104
Brewster, David, 935
Brewster's angle, 935, 935f
Bridges
  collapse of, resonance in, 446, 446f
  expansion joints in, 553
  thermal expansion in, 553
Bright fringe, 986–989, 987f
  conditions for, 987–988, 987f
  in diffraction grating, 1005–1006, 1005f
  numbering systems for, 988f
  in single-slit diffraction, 999–1000
Bright spot, in circular shadow, 1001, 1001f

BTUs (British thermal units), 557
Bubbles
  in Cartesian diver, 518
  soap, interference and, 993, 993f
  volume of, 585
Bulk modulus, 595f, 596, 596f
Buoyant force, 513–520, 513f, 518f
  in Archimedes' principle, 513–514
  of floating objects, 517–520
  string tension and, 516–517

C

C ray (center-of-curvature ray), 922–923 922f,
Calculation(s)
  dimensional consistency in, 5
  guidelines for, 13–14
  mathematical expansions for, A-5–A-6
  order-of-magnitude, 11–12
  round-off error in, 8
  significant figures in, 6–9
Calculators, use of, direction angles and, 65
Calorie (C), 557
Calorie (cal), 557
Calorimeters, 560
Calorimetry, 560
Cameras
  apertures in, 958
  electronic flash unit of, 719, 719f
  focusing in, 957
  lenses of, 957–958, 957f
Cancer treatments
  lasers in, use of, 1109
  radiation and, 1144–1145, 1145f
Canoes, separation of, 267, 267f, 268–270
Cantilever, atomic-force, 209, 209f
Capacitance, 712–718
  applications of, 717, 717f
  in computer keyboards, 717, 717f
  definition of, 712
  dielectrics and, 715–718, 715f
  equivalent, for capacitors, 751, 753
  of parallel-plate capacitor, 713–714, 713f, 716
  units of measure for, 712
  voltage and, 718–720
Capacitive reactance, 851–852
Capacitor(s), 712–715
  in alternating-current (ac) circuits, 851–854, 851f, 853f
  batteries and, 712–715
  capacitive reactance in, 851–852
  charging, 754–755
  connection of
    in parallel, 751–752, 751f
    in series, 752–754, 753f
  definition of, 712
  dielectrics and, 715f
  dielectrics in, 715–718
  discharging, 756, 756f
  in electric circuits, 751–754, 751f
  energy storage in, 718–719, 718–720
  equivalent capacitance for, 751, 753
  in flash units, 719, 719f
  at large and small frequencies, 866–867, 866f
  in LC circuits, 867–869, 868f
  parallel-plate. See Parallel-plate capacitor
  phasor diagrams for, 853–854
  power consumption in, 854
  in RC circuits, 754–755, 757f
  rms current in, 851–853
  visualization of, 714f
  voltage phasors and, 855, 855f, 863, 863f
Captioning device, for hearing impaired, 918
Carbon, notation for nucleus of, 1120
Carbon-14 dating, 1135–1137, 1136f
Carnot, Sadi, 634
Carnot's theorem, 634–636
Cartesian diver, 518, 518f
CAT scan (computerized axial tomography scan), 1107, 1107f

Cathode-ray tubes, 776f
Cavendish, Henry, 388–389
Cavitation, 500
CDs
  interference in, 996–997, 996f, 997f
  reflection grating and, 1008, 1008f
Celsius, Anders, 546
Celsius scale, 546
  conversion to
    Fahrenheit scale, 550
    Kelvin scale, 550
Center of curvature, 920, 920f, 923–924, 928
Center of mass, 283–288
  acceleration of, 285–287
  balance and, 350f, 353f
  coordinates for, 284–285
  definition of, 283
  in human arm, 284–285, 284f
  of irregularly shaped object, 352, 352f
  location of, 283–285, 283f, 284f
  motion of, 285–288, 287f
  for multiple or continuous objects, 283–284, 284f
  of object on surface, 353, 353f
  of suspended object, 350f, 352, 352f
  torque and, 346–347, 350–353, 350f
  for two objects, 283–284, 283f
  velocity of, 285–288
Center-of-curvature ray (C ray), 922–923, 922f
Centimeters, 4
Centrifuges, 181–182, 182f, 191, 309–311, 309f, 332
Centripetal acceleration, 177–182, 177f, 309–311, 309f, 310f, 393
Centripetal force, 178, 393
Cesium fountain atomic clock, 4
Cgs system, 4
Chain reactions, 1141–1142, 1141f
Characteristic x-rays, 1105, 1105f
Charge conservation, junction rule and, 747–748, 747f, 748f
Charge distribution, 662, 669, 679, 679f
Charge separation, 660, 703
Charged particles, helical motion of, 782
Charging
  in electric circuits, 754–755
  by induction, 681–682, 681f
Charles, Jacques, 585
Charles's law, 585, 585f
Chicxulub impact crater, 402
Chromatic aberrations, 974, 974f, 975f
Chromatic musical scale, 485, 485t
Ciliary muscles, 955, 955f, 956
Circuit breakers, 849–850, 850f
Circular motion, 177–182
  angular momentum in, 356–357, 356f, 358, 358f
  in magnetic fields, 779–780, 779f
  uniform, simple harmonic motion and, 424–430, 424f, 425f
  with variable speed, 182, 182f
Clausius, Rudolf, 641
Climate, adiabatic processes and, 631
Closed circuits, 732, 747–748, 748f
Coatings, quarter-wave, 996
Coefficient of kinetic friction, 157, 157t
Coefficient of linear expansion, 551, 552t
Coefficient of performance
  for heat pumps, 640
  for refrigerators, 637
Coefficient of static friction, 160f, 161–162
Coefficient of viscosity, 531
Coefficient of volume expansion, 552t, 554
Coherent light sources, 984
Coils
  inductance of, 825, 825f, 827
  rotating, 822, 822f
Collector plates, 1057, 1058f
Collisions
  asteroid-Earth, 401–402
  automobile, 154, 276–277

average force in, 262f, 263–265
center of mass and, 286–288
change in momentum and, 262–265
completely inelastic, 273
crumple zones for, 154
definition of, 271
elastic, 271, 277–282
impulse and, 262
inelastic, 271–277
molecular, 586–591
photon scattering and, 1063
recoil and, 269
rotational, 363–364, 363f
visualization of, 272f
Color of light
dispersion and, 940–941, 943f
thin film interference and, 993–995, 993f
Column of air
ear canal as, 487, 487f, 498, 498f
harmonics of, 486–489
Comets, 404f, 414, 415
Communications. See Telecommunication systems
Component method, for vector addition, 67–68, 67f
Compound microscopes, 969–971, 969f
Compressible fluids, 522
Compression of springs, 208–212, 208f, 231–232
Compton, Arthur Holly, 1062
Compton effect, 1062–1063, 1062f
Compton shift formula, 1063
Computer(s)
keyboards, capacitance in, 717, 717f
processing speed of, 419
Computerized axial tomography scan (CAT scan), 1107, 1107f
Computerized jaw tracking, 842
Concave lenses, 936, 936f, 959, 959f, 961, 961f. See also Diverging lenses
imaging characteristics of, 938t
ray tracing for, 936–938, 937f
Concave mirrors, 918–919
focal length and, 920–921
image formation and, 923, 923f
imaging characteristics of, 924t
Conduction, 562–566, 562f, 563f
Conduction electrons, 662
Conductors, 562, 662–663, 662f. See also Superconductors
arbitrarily shaped, 710, 710f
ideal, 709–710, 709f
shielding by, 679–681, 680f, 681f
Cones, retinal, 955
Connected objects, 172–177
Conservation of angular momentum, 360–364
angular speed and, 361–363
definition of, 360
in nature, 361–362
rotational collisions and, 363, 363f
Conservation of electric charge, 660
Conservation of energy. See Energy conservation
Conservation of momentum, 266–271, 1032–1033
in elastic collisions, 277–282
in inelastic collisions, 272–273, 275–276
internal vs. external forces and, 266–268
net force and, 266
propulsion and, 288–289
for system of objects, 267–269
Conservative forces, 224–233, 227t
definitions of, 224
height and, 225–226, 226f
mechanical energy and, 233–234
vs. nonconservative forces, 224, 225f
path and, 225–227, 225f–227f
springs as, 231–233, 231f
total work and, 225
work done by, 224, 226, 226f
Constancy, of speed of light, 1018

Constant acceleration, 33–43, 33f, 91–92, 91t
angular velocity and, 304–305, 304f
free fall and, 43–44
Constant forces, 121–122, 196–207, 207f
Constant mass, second law of motion and, 261
Constant pressure, molar specific heat for gas at, 629
Constant temperature, 624, 624f
Constant velocity, 26f, 119–120
inertial frames of reference and, 1018
motion and, 89–91, 89f
Constant volume, molar specific heat for gas at, 629
Constant-pressure processes, 621–622, 621f, 622f, 628–629, 628f
Constant-velocity, straight-line drift motion, 779
Constant-volume gas thermometer, 548–549, 548f, 580, 580f
Constant-volume processes, 623–624, 623f, 628f, 629
Constructive interference, 478, 478f, 479f, 984–995, 984f
air wedge and, 991–992, 992f
conditions for, 987–988, 993–995
de Broglie waves and Bohr orbits, 1094f
in diffraction gratings, 1006
opposite phase sources and, 481
in phase sources and, 479
when scattering from crystals, 1066
Contact forces, 131–133
Contact lenses, 965
Continuity, of fluid flow, 521–523
Contour maps, 247, 247f
Contracted length, 1025–1026
Control rods, in nuclear chain reaction, 1142
Convection, 566–570, 566f, 567f
Converging lenses, 936, 936f, 963, 963f. See also Convex lenses
chromatic aberration and, 974, 974f, 975f
in microscopes, 969, 969f
in telescopes, 971, 971f
Conversion
of basic units, 9–11
of temperature
Celsius to Fahrenheit, 547
Celsius to Kelvin, 550
Fahrenheit to Celsius, 547
of vectors
from magnitude and angle, 62, 62f, 63f
to magnitude and direction, 63, 63f, 64
Convex lenses, 936, 936f, 959, 959f, 963f. See also Converging lenses
imaging characteristics of, 938t
ray tracing for, 936–938, 937f, 938f
Convex mirrors, 918–919, 920f
focal length and, 920
image formation and, 922–923, 922f, 924f
imaging characteristics of, 924t
ray tracing for, 922–923, 922f
Cooling
adiabatic, 631
through evaporation, 600, 600f, 601f
Cooling devices, 637–638, 638f
Coordinate systems, 20, 20f
choice of
circular motion, 178
connected objects, 174
incline, 158
for forces in two dimensions, 134–137
for free-body diagrams, 124–125, 124f
for inclined surfaces, 143
for momentum, 259
for normal force, 142, 142f
two-dimensional, 62, 62f
Copernicus, 391
Cornea, shape of, 955, 955f
in astigmatism, 965
in nearsightedness, 963, 963f

surgical correction of, 963, 963f, 1108–1109, 1109f
Corner reflectors, 918–919, 918f
Cosmic rays, 1135
Cosmology, 1150–1151
Coulomb (C), 659
Coulomb, Charles-Augustin de, 659
Coulomb's law, 663–670
Coulombs per volt, 712
Countercurrent exchange, 566, 566f
Crab Nebula, 326, 362, 362f
Crest, of wave, 459, 459f
Crick, F. H. C., 1007
Crickets, chirp rate in, 454–455, 576
Critical angles, 933
Critical nuclear reactors, 1142
Critical point, 599
Critical temperatures, 737
Critically damped oscillations, 444
Cross products, vector, 337, 775, A-7
Crossed polarizers, 900
Crumple zones, 154
Crystals
diffraction gratings and, 1007–1008
diffraction pattern from, 1066, 1066f
Cunningham, Tolson, 1106
Curie (Ci), 1131
Curie, Paul and Marie, 1131
Current. See Electric current
Current phasors
in ac capacitor circuit, 853, 853f
in ac resistor circuit, 846, 847f
in RL circuits, 861, 861f
in RLC circuits, 863, 863f
Curvature. See Center of curvature
Cutoff frequency, 1058, 1059f

D

D waves, 711
DaVinci code, 191
Da Vinci, Leonardo, 916f
Damped oscillations, 444–445, 444f
Damping constant, 444
Dark fringe, 986–989, 986f
conditions for, 987–988, 988t
location of, determination of
in circular opening, 1002
in single-slit diffraction, 998–999, 998f, 999f
numbering systems for, 988f
Dating. See Carbon-14 dating; Radioactive dating
Daughter nucleus, 1126, 1130–1131, 1140f
Davisson, C. J., 1067
de Broglie, Louis, 1065, 1094
de Broglie wavelength, 1065–1066
de Broglie waves, 1094, 1095f
Decay. See Radioactive decay
Decay constant, 1133, 1133f
Decay rate (R), 1134–1135
Deceleration, 31–32
Decibels, 470–472, 472t
Decimal places, in addition and subtraction, 7
Deep sea diving, bouyancy in, 537, 616
Deep View, 537
Defibrillators, 695, 719–720, 720f
Deformation
elastic, 596
shear, 594, 594f
of solids, 592–596
Degrees, 300, 301, 546–547
Delay circuits, 757
Delta notation, 21
Deltoid muscle, 378
Density
of Earth, 317f, 389–390
of fluids
buoyancy and, 518–519
in common substances, 503t
definition of, 503
floatation and, 517–518, 517f, 518f
in ice, 519–520, 556
measurement of, 503
of ice, 519–520, 556
of Moon, 389

of nucleus, 1123
surface charge, 669, 709–710, 709f
of water, 555–556, 555f
Depth
apparent, 931–932, 932f
of atmosphere, 540–541
fluid pressure and, 507–509, 508f
Destructive interference, 478–479, 479f, 984–997, 984f
air wedge and, 992–993, 992f
in CDs and DVDs, 996–997, 996f
conditions for, 988, 993–996
de Broglie waves and Bohr orbits, 1094f
phase sources and, 479, 481
Detectors, gravitational wave, 1042
Dewar, James, 569
Diamagnetism, 796–797, 797f
Diatomic molecules, 591
Dielectric breakdown, 717–718
Dielectric constant, 716, 716t
Dielectric strength, 717, 718t
Dielectrics, 715–718, 715f
Diesel engines, 627
Diffraction, 998–1001
gratings. See Diffraction gratings
patterns of
for circular openings, 1002, 1002f
interference, 986–987, 986f, 987f, 1005–1006, 1005f
x-ray, 1007, 1007f
Poisson's bright spot and, 1001, 1001f
shadows and, 1000
single-slit, 998–999, 998f
smearing and, 1002
telescopes and, 1005
x-ray, 1007, 1007f
Diffraction gratings, 1005–1008
acousto-optic modulation using, 1007, 1007f
constructive interference in, 1006
crystals and, 1007–1008
interference patterns formed by, 1005–1006, 1005f
path-length differences in, 1006, 1006f
on reflecting surfaces, 1008
Diffraction pattern
from crystals, 1066, 1066f
of electrons, 1069–1070
of neutrons, 1067, 1067f
of x-rays, 1066, 1066f
Diffraction-limited resolution, angular separation and, 1002–1005, 1002f, 1005f
Diffuse reflection, 915, 915f
Digital cameras, 957
Digital Micromirror Device (DMD), 915, 915f
Dimensional consistency, 5
Dimmer switches, 862–863
Dinosaurs, 501
Diopters, 962
Dipoles, electric, 677, 677f, 709, 709f
Direct-current (dc) circuits, 731
Direction
average velocity and, 23–24
of axis of rotation, 366–367
change in, of polarization, 898–899, 898f
of electric current, 733, 733f
of electric field lines, 680–681, 680f
of electric fields, 671–672, 672f, 707f, 708, 708f
of excess charge, 679
of fields, in electromagnetic waves, 882, 882f
of force, 118, 119, 121–123, 343–350
of induced current, 816, 816f
of magnetic fields
for current-carrying wire, 788, 788f
for magnets, 771
of magnetic force, 774–776, 774f
magnitude and, 63, 63f, 64, 68–69
momentum and, 258–259
of phasors, 846, 847f
of pressure, 505, 505f

Direction (*continued*)
  of propagation
    in electromagnetic waves, 882,
      882–883, 884–889
    refraction and, 929–930, 942
  of tension, 165–166, 165f
  of vectors, 13, 13f, 61, 62, 64
  of weight, 137–138
Discharge, electric, 680, 680f, 756, 756f
Disorder, entropy and, 645–647
Dispersion, of light, 940–941, 943f
Displacement, 20–21
  angular, 301, 302f
  for blackbody curves, 1052–1053
  definition of, 49
  of fluid, 512, 513–514, 517
  force at angle to, 198–199, 198f
  of interference fringes, 993
  positive/negative, 21, 49
  in projectile motion, 93
  time and, 93
  velocity and, 38, 93
  visualization of, 198f
  work and, 196, 196f, 198–199, 201
Displacement vectors, 61, 61f, 71–72,
  72f
Distance, 20
  average speed and, 22–24
  intensity and, 468–469, 469t
  notation for, 5, 6t
  in orbital motion, 393–394
  orbital motion and, 393f
  stopping, 163, 163f
  stopping, calculation of, 39–41, 41f
  typical length of, 3t
  universal gravitation and, 382
  zero, and work, 198
Distortions, 1027–1028, 1028f
Distribution, of electric charge, 662, 669,
  679, 709
Diverging lenses, 936, 936f, 974. *See also*
  Concave lenses
  chromatic aberration and, 974–975,
    974f, 975f
  nearsightedness correction and, 961,
    961f
  in telescopes, 972
Diving, buoyancy in deep sea, 537, 616
Doloriometer, 540, 540f
Dolphins, navigation by, 466, 496
Domains, magnetic, 796, 796f
Doppler, Christian, 472
Doppler effect, 472–478
  for electromagnetic waves, 887–889,
    887f
  frequency and, 476
  light and, 472, 477
  for moving observer, 472–473, 472f,
    476–477
  for moving source, 473–475, 474f,
    476–477
  radar and, 887
  uses of, 477, 477f
Doppler radar, 477, 887
Dot products, 199, A-6–A-7
Double-slit experiment. *See* Two-slit
  diffraction
Dragonflies, 256, 256f
Drift speed, 733
Driven oscillations, 445–446, 445f
DVDs
  interference in, 996–997, 996f, 997f
  lasers and, 996–997, 997f
  reflection grating and, 1008, 1008f
Dynamic microphones, 813, 813f
Dynamo theory of magnetism, 772

E

Ear
  frequency range of, 485, 487f
  sound perception by, 468–469, 487
Ear canal, as column of air, 487, 487f,
  498, 498f
Earth
  atmosphere of, 601–602
  axis of rotation and, 317, 317f
  density of, 389–390

electric field of, 693
escape speed for, 404–407
gravitational force between Moon
  and, 388–389, 388f
gravitational force of, on Moon, 408
gravitational force on, 381–385
impacts with, by astronomical bodies,
  401–402, 402f
inertial frames of reference and, 1018
magnetic field of, 771–772, 772f, 774,
  795–796, 795f
mass of, 388–389
moment of inertia of, 316–317, 317f
orbital speed of, 392–393
precession of, 368
rotation of, 327, 362–363, 368
tidal deformation on, 408–409, 408f
weight on, 137
ECG. *See* Electrocardiograph (EKG)
Echolocation, 466–467, 469f
Eddington, Arthur, 1040
Eddy currents, 817–818, 817f
EEG. *See* Electronencephalograph
Efficiency, of heat engines, 633–636
Einstein, Albert, 1018, 1018f, 1028,
  1034, 1056–1057, 1060. *See also*
  Relativity
Einthoven's triangle, 711
EKG (Electrocardiograph), 710–711,
  710f, 711f
Elastic collisions, 277–282
  definition of, 271, 277
  in one dimension, 277–280, 279f
  in two dimensions, 280–282, 281f
Elastic deformation, 596
Electric charge(s), 658–662
  acceleration of, and electromagnetic
    waves, 883–884, 884f
  of amber, 658–659, 658f
  of atoms, 659, 659f
  conductors and, 662–663, 662f
  conservation of, 660
  distribution of, 662, 669, 679, 709
  excess, 661, 677, 679
  grounding of, 681, 681f
  on ideal conductors, 709–710, 709f
  induction of, 681–682, 681f
  insulators and, 662–663
  near sharp points, 680, 681f
  negative, 658
    electric potential and, 701, 704
    induced, 680, 680f
    magnetic force for, 775
  net force of, 666–668
  polarization of, 661–662, 661f
  in pollination, 693–694
  positive, 658
    electric potential and, 701, 704
    magnetic force for, 775
  quantized, 661
  rubbing and, 660, 660t
  semiconductors and, 662
  separation of, 660
  in static cling, 693
  strength of, 663–665
  transfer of, 660, 660f
  triboelectric charging and, 660, 660t
  types of, 658–659
  unit of, 659
Electric circuits
  alternating-current. *See* Alternating-
    current (ac) circuits
  antennas and, 881–882, 882f, 890f
  capacitors in. *See* Capacitor(s)
  charging in, 754–755
  closed, 732, 747–748, 748f
  complex, 745, 745f, 747, 747f
  direct-current (dc), 731
  discharge in, 756, 756f
  elements of, A-11
  energy dissipation in, 738–739, 738f,
    740, 745
  in flashlight, 732, 732f, 733
  flow of electrons in, 733–734, 733f
  household, safety devices for,
    849–851, 850f
  inductors and, 826

inverter, 849
Kirchhoff's rules for, 747–750, 747
  f–749f
LC, 867–868, 868f
measurement devices for, 757–758
notation in, 734
open, 732
oscillating, 867–872
power in, 738–741
primary, 808, 808f
RC, 754–757
rectifying, 849
resistors in, 739, 741–747, 741f, 743f,
  744f. *See also* Resistors
resonance in, 867–872
RL, 828–829, 828f, 830f
RLC, 863–867, 863f
secondary, 808, 808f
short, 744, 744f
Electric current
  in ac resistor circuit, 846f
  alternating. *See* Alternating-current
    (ac) circuits
  circulating, 795, 817, 817f
  definition of, 731
  eddy, 817–818, 817f
  flow of, 733, 733f
  induced, 808, 808f, 809f
    direction of, 816, 816f
    Lenz's law for, 814–818, 814f, 815f
    magnets and, 814–818, 815f
    motional electromotive force and,
      816, 816f
  in inductors, 860–861
  loops of, 785–786
    circular, 792–793, 792f
    general, 786
    magnetic flux and, 810
    magnetic force on, 785, 785f
    magnetic moments of, 786
    rectangular, 785–786, 785f
    torque and, 785–787, 786f
  magnetic fields and, 788–792
  measurement of, 758, 758f
  phasors. *See* Current phasors
  resistance in. *See* Electrical resistance
  in *RL* circuits, 828f, 830f
  root mean square values of, 847, 848
    in capacitors, 851–853
    in inductors, 860, 860f
    in *RC* circuits, 862, 862f
    in *RL* circuits, 862, 862f
    in *RLC* circuits, 872, 872f
  units of, 731
  visualization of, 732f
  voltage and, in capacitors, 853–854
Electric dipoles, 677, 677f, 709, 709f
Electric field(s), 670–675
  of aquatic creatures, 675
  of capacitors, dielectrics and,
    715–718, 715f
  within conductors, 679
  definition of, 670
  direction of, 671–672, 672f, 707f,
    708, 708f
  of Earth, 693
  electric force per charge and, 671
  electric potential and, 697–698, 697f,
    698f
  in electromagnetic waves, 881–882,
    882f
  energy storage in, 720
  equipotential surfaces of, 707–711,
    708f, 709f
  Gauss's law and, 684, 684f
  magnitude of, 707–708, 707f
  in nerve cells, 691
  of point charge, 672, 672f
  shielding and, 679–681, 680f
  superposition of, 672–674, 673f
Electric field lines, 675–679, 677f
  at conductor surfaces, 680
  direction of, 680–681, 680f
  for electric dipole, 677, 677f
  for multiple charges, 676–677
  near sharp points, 680, 681f
  properties of, 676

rules for drawing, 676
shielding and, 680f
uniform plate charge, 677, 677f
Electric flux, 682–683, 682f
Electric force(s)
  calculation of, 672–673, 673f
  differences from magnetic force,
    777–778, 777f
  superposition of, 666
Electric force per charge, 671
Electric generators, 822–824, 822f
  alternating-current, 822, 822f
  magnetic flux and, 821
  shake motion, 823, 824f
Electric guitars, 813, 814f
Electric motors, 824–825, 824f
  in automobiles, 824
  magnetic flux and, 821
  magnetic torque and, 824
  torque and, 787
Electric potential
  in batteries, 732
  change in, 697, 700, 703
  dependence of, on algebraic sign, 704
  electric fields and, 697–698, 697f, 698f
  for point charge, 703, 704f
  properties of, 696t
  superposition of, 704
  units of measure for, 697
Electric potential energy, 696–700
  change in, 696–697, 696f
  in closed loop, 748, 748f
  definition of, 696
  electric circuits and, 738–739
  kinetic energy and, 700
  properties of, 696t
  speed of, 700–701
  superposition of, 706
Electrical energy
  in defibrillators, 719–720
  production of, 822f, 823
  storage of, 718–720
Electrical plugs, 850–851, 850f
Electrical power, mechanical power and,
  819–821
Electrical resistance, 734–750
  in batteries, 733, 743, 743f
  electrons and, 734–735
  equivalent, 741–745, 741f, 746f,
    749–750, 749f
  footwear safety and, 767
  heating, 740, 740f
  internal, 733, 743, 743f
  Ohm's law of, 734–735
  Resistivity, resistance and, 735
  resistivity and, 735
  superconductivity and, 737
  units of measure for, 734
Electricity
  generation of, 822, 822f
  household, 849–850
  kilowatt-hours of, 740
  transmission of, 834, 834f
Electrocardiograph (EKG), 710–711,
  710f, 711f
Electrodialysis, 672
Electrognathography, 842
Electromagnet. *See* Solenoids
Electromagnetic flowmeter, 778, 778f
Electromagnetic force, 1148
Electromagnetic spectrum, 889–892,
  890f
Electromagnetic waves
  electric charge acceleration and,
    883–884, 884f
  electric fields and, 882
  energy in, 892–893
  intensity of, 893–894, 893f
  linearly polarized, 897, 897f
  magnetic fields and, 882
  momentum in, 895
  oscillation and, 882–884
  phase changes and, 990–991, 991f
  polarization of, 896–903
  production of, 881–884, 882f
  propagation of, 884–889
    direction of, 882, 883f

Doppler effect and, 887–889
  speed of light and, 884–885
radiation from, 567
radiation pressure and, 895–896, 896f
speed of, 883, 884–886
Electromotive force (emf), 732, 733
  in electric generators, 822
  induced
    magnetic fields and, 808, 809f
    magnitude of, 812
    values of, maximum and mini-
      mum, 821
  motional, 816, 816f, 818–819, 821
  of rotating coils, 822, 822f
Electron(s), 659–661
  charge on, 659f
  conduction, 662
  diffraction pattern of, 1069–1070
  display of, uncertainty principle of,
    1072
  drift speed of, 733–734
  in early atomic models, 1083–1084,
    1083f
  ejection of, 1057–1059, 1058f
  electric charge of, 659
  electronic configurations for,
    1101–1103, 1102f, 1102t
  flow of, in electric circuit, 733–734,
    733f
  intrinsic properties of, 659
  as leptons, 1148t
  magnetic field in, 795
  magnitude of charge of, 659
  mass of, 659
  matter waves of, 1073–1074,
    1094–1096, 1095f, 1098
  moving, with magnetic forces, 776,
    776f
  notation for, 1101–1102
  population inversion and, 1107
  probability clouds of, 1098–1099,
    1098f, 1099f
  properties of, 1097
  in radioactive delay, 1125
  resistance and, 734–735
  shells and subshells for, 1100, 1103
  speed of, 1065–1066
  tunneling by, 1073–1074, 1073f,
    1074f
  wave-particle duality of, 1068–1069
Electron microscopes, 1067–1068, 1068f
Electron neutrino, 1148t
Electron spin quantum number, 1097
Electron volt (eV), 697, 1057
Electronencephalograph (EEG), 711,
  711f
Electronic configurations, 1101–1103,
  1102f, 1102t
Electronic flash unit, 719, 719f
Electrostatic force, 658, 658f
  Coulomb's law for, 663
  magnitude of, between point charges,
    663, 664f
  in nucleus, 1124
  strength of, 665f
Electrostatic precipitators, 681
Electrostatic shielding, 679, 680f, 681f
Electroweak theory, 1151
Elementary particles, 1147–1150
Elements
  electronic configurations of,
    1101–1103, 1102f, 1102t
  moles of, 582, 582f
  periodic table of, 1103, 1103f, A-12
Elephants, infrasonic communication
  by, 467
Ellipse, 391, 391f, 392f
Emission nebulas, 1085f
Emission spectrum, 1085, 1085f
Emissivity, 568–569
Emitters, 1057, 1058f
Endoscopes, 934
Endoscopic surgery, 662
Energy
  of Bohr orbits, 1090–1091
  conservation of. *See* Energy
    conservation

conversion of mass to, 1034
degradation of, entropy and, 645
dissipation of, in resistors, 738–739,
  738f, 740, 745
electric potential, 696–700
of electromagnetic waves, 892–893
excited state of, 1091
ground state of, 1091
internal, of gas, 591
kinetic. *See* Kinetic energy
mass and, 1034
mechanical, 233–246
notation for, 5, 6t
from oceans, 656
of photons, 1056
potential, 228–233
quantization, 1052–1055, 1054
release in nuclear fission, 1139–1140
rest, 1034
thermal, 545
time and, Heisenberg uncertainty
  principle on, 1072
total, 1034
units of, 196–197
Energy conservation
  in astronomical systems, 401–407
  calorimetry and, 560
  electric potential energy and,
    700–702
  in electrical systems, 703, 703f
  in fluids, 523–526
  heat and, 556–557
  impact speed of astronomical bodies
    and, 401–403
  kinematics and, 234–235, 235f
  loop rule and, 748–749, 748f, 749f
  for mechanical energy, 233–241
  in oscillatory motion, 435–438, 435f
  phase changes and, 606–608
  rolling motion and, 318–323
  speed and, 234–241
  springs and, 240–241
  surface tension and, 533
  in thermodynamic systems, 618–619
  in transformers, 833
  visualization of, 234f
Energy density, 720
Energy levels, in multielectron atoms,
  1100–1102, 1101f
Energy recovery technology, 824
Energy-level diagrams, 1091, 1091f
Engines
  diesel, 627
  heat, 632–636, 632f
  steam, 632–634, 633f
Entropy, 640–647
  change in, 641–645
  disorder and, 645–647
  heat engines and, 642
  living systems and, 646–647
  in Universe, 642–643, 646
Equation of state, for ideal gas, 580,
  580f, 582
Equations of continuity, 522
Equations of motion
  for angular motion, 305
  constant acceleration, 91t
  for one-dimensional kinematics,
    33–43
  for parabolas, 96–97
  for projectiles, 93, 94–95, 99
Equilibrium
  phase, 597–600, 597f
  spherical charge distributions and,
    679, 679f
  static, in fluids, 507–512
  thermal, 545
  torque and, 343–350
  translational, 169–172
Equilibrium vapor pressure, 597, 597f
Equipotential plots, 707–708, 707f
Equipotential surfaces, 707–711
  definition of, 707
  for dipoles, 709, 709f
  of electric fields, 707–711, 708f, 709f
  ideal conductors as, 709–710, 709f
  for point charge, 707, 707f

Equipotentials, 247
Equivalence
  mass-energy, 1033–1038, 1121
  of physical laws, 1018
  principle of, in general relativity,
    1039, 1040f
Equivalent capacitance, for capacitors
  in parallel, 751
  in series, 753
Equivalent resistance, for resistors,
  741–745, 741f
  in combination circuits, 745, 746f
  in parallel, 743
  in series, 742
Escape speed, 404–407, 407f
Eta, 1149t
European Organization for Nuclear
  Research (CERN), 1150
eV (Electron volt), 697, 1057
Evaporation, 600–602
  of atmosphere, 601–602
  cooling through, 600, 600f, 601f
Events, in relativity, 1022
Excimer laser, 1108, 1109f
Excited state of atoms, 1091
  metastable, 1107
  probability clouds for, 1098–1099,
    1099f
Exclusion principle
  quarks and, 1150
  for states of atoms, 1100–1101
Expansion joints, 553
Explosions
  center of mass and, 287, 288f
  momentum in, 270
  stellar, 361–362, 362f
Exponents, A-5
External forces
  center of mass and, 286–288
  driven oscillations and, 445
  momentum and, 266–268
Eye. *See* Human eye
Eyepiece, of microscope, 969, 969f

F
f -number, 958
F ray (focal-point ray), 922–923, 922f,
  923f, 937, 937f, 938f
Fahrenheit, Daniel Gabriel, 547
Fahrenheit scale, 547
Far point, visual, 956
Farad (F), 712
Faraday, Michael, 712, 808
Faraday's law of induction, 812–814
Farsightedness, 963, 963f
Fat, body, measurement of, 515–516
Femtometer (fm), 1122
Fermi (fm), 1122
Fermi, Enrico, 10f, 1122, 1129, 1142
Ferromagnetism, 795–796, 796f
Fifth harmonic, 486, 486f
Filters, polarizing, 898, 898f, 903f
Fireworks, center of mass of, 287–288,
  288f
First harmonic
  in columns of air, 486–488, 486f,
    487f, 488f
  on a string, 482–483, 482f, 483f
First law of thermodynamics, 618–620
Fission reactions, 1034, 1034f
Fizeau, Arman, 885
Flash units, electronic, 719, 719f
Flashlights
  circuit in, 732, 732f, 733
  shake, 823, 824f
Flea jump, potential energy in,
  232, 232f
Fleas, propulsion by, 133
Flight
  of dragonflies, 256, 256f
  evolution of, 221–222
  human-powered, 213, 213f
  time of, 103–104
  turbulence during, 428
Floatation, 517–521, 517f, 518f
Floaters, 957
Flowmeter, electromagnetic, 778, 778f

Fluid(s). *See also* Water
  Archimedes' principle of, 513–521
  Bernoulli's equation for, 523–530
  boiling point of, 598
  buoyancy and, 513–519, 513f, 517f
  buoyant force due to, 513–514, 513f
  complete submersion in, 514–516
  compressible vs. incompressible, 522
  definition of, 502
  density of. *See* Fluid density
  displacement of, 512, 513–514, 517
  flow of. *See* Fluid flow
  ideal, 530–531, 531f
  incompressible, 522, 522f
  Pascal's principle and, 511–512, 512f
  seeking own level, 509–510, 510f
  static equilibrium in, 507–512
  surface tension of, 533, 533f
  viscosity of, 531–533, 531f, 531t
  weight of, pressure and, 507, 507f
Fluid density, 503
  buoyancy and, 518–519
  of common substances, 503t
  definition of, 503
  floatation and, 517–518, 517f, 518f
  in ice, 519–520, 556
  measurement of, 503
  temperature and, 555–556, 556f
Fluid displacement, 512, 513–514, 517
Fluid flow, 523–524, 525
  in arterioles, 532
  Bernoulli's equation for, 518–527
  equation of continuity for, 522
  fluid pressure and, 523–526
  height of, change in, 525, 526
  speed of, 521–523, 521f
  speed of, change in, 523–525, 526
  Torricelli's law and, 529
  viscosity and, 531–533, 531f
  volume flow rate in, 532–533
  work-energy theorem for and,
    523–524
Fluid pressure, 500–512, 523–526
  with depth, 507–509, 508f
  weight and, 507, 507f
Fluorescence, 1110–1111, 1110f, 1111f
Flux. *See* Electric flux; Magnetic flux
Foamcrete, 127–128
Focal length
  of contact lens, 965
  in farsightedness, 963
  in human eye, 955
  in lenses, 939, 939f
  of magnifying glass, 967
  in microscope eyepieces, 969
  in mirrors, 920–921, 920f
  in nearsightedness, 961
  refractive power and, 962, 962f
  sign conventions for, 926, 939
  of telescopes, 971
Focal point
  aberrations and, 921, 921f, 974, 974f
  of lenses, 936–938, 936f–938f
  in microscopes, 969, 969f
  of mirrors, 920–923, 920f–923f
  in telescopes, 971, 971f
Focal-point ray (F ray), 922–923, 922f,
  923f, 937, 937f, 938f
Focusing
  in cameras, 957
  in human eye, 955–956, 956
Food, potential energy of, 230
Food preservation, gamma rays for, 892f
Foot, 4
Footwear safety, electrical resistance
  and, 767
Force(s)
  acceleration and, 121–123, 121f,
    125–130
  action-reaction, 129–131, 130f
  at angle to displacement, 198–199,
    198f
  atmospheric pressure and, 505–506
  buoyant, 513–520, 513f, 518f
  centripetal, 178, 393
  circular motion and, 177–182
  conservative. *See* Conservative forces

Force(s) (continued)
   constant, 121–122
   contact, 131–133
   deforming, 592–596
   direction of, 118, 119, 121–123
   electromotive, 732
   electrostatic. See Electrostatic force
   exerted by spheres, 669
   free-body diagrams of external,
      124–125, 124f, 125f, 130
   frictional, 119, 156–164
   fundamental, of nature, 1147–1148,
      1148t, 1150–1151, 1151f
   gravitational. See Gravity
   with horizontal components,
      343–350, 348f
   impulse and, 262–263, 262f
   internal vs. external, effect on
      momentum of, 266–268
   magnetic. See Magnetic force
   magnitude of, 118, 142, 142f
   measurement of, 122
   net, 118, 119, 120, 122, 132f
   nontangential, and torque, 336–337
   normal, 124, 141–145, 141f, 142f
   pairs, 129–130, 130f
   path and, 224–225
   per area, pressure and, 504
   between point charges, 663, 664f
   reduction of, 119
   restoring, 421
   spring, 166–169, 208–212
   in strings, 164–166
   strong nuclear, 1124
   superposition of, 666
   tangential, and torque, 335–336,
      335f, 336f
   units of, 123t
   variable, work done by, 207–212
   as vectors, 118, 122, 134–137, 134f
   with vertical components, 343–350,
      348f
   visualization of, 122f, 198f
   work done by. See Work
   zero net, 343–350, 343f
Force constant, 167, 168
Force constant, for mass on a spring,
   430–435
Force fields, 670–671. See also Electric
   field(s)
Force pairs
   action-reaction, 129–131, 130f
   internal vs. external forces and,
      266–268
Force per charge, electric, 671
Forensics, fluorescence use in, 1111
Fosbury flop, 285, 291f
Fountain jets, 97f
Frames of reference
   accelerated, 1039, 1039f, 1040f
   definition of, 119–120
   inertial, 1018, 1018f
   noninertial, 120
Franklin, Benjamin, 669, 693
Free fall
   acceleration in, 43, 43f, 93, 93f
   air resistance and, 43
   characteristics of, 43–44
   definition of, 43
   examples of, 43f
   gravity and, 44
   of objects, 43–49
   orbital motion and, 382, 382f
   position and, 44, 44f
   from rest, 45, 45f
   separation, 45
   speed and, 46–48
   symmetry of, 43f, 47, 47–49
   velocity and, 43f, 44
   weight and, 137–138, 137f
   weightlessness and, 140
Free space
   permeability of, 789
   permittivity of, 683
Free-body diagrams, 124–125, 124f,
   125f, 130
Frequency

angular, 425
beat, 489–492, 490f
cutoff, 1058, 1059f
definition of, 419
harmonic, 482–488
natural, 445
of periodic motion, 419–420
range, of human ear, 485
RLC circuits and, 866
of sound waves, 466–467
typical values for, 420t
units of, 419
wavelength and, 459
   for electromagnetic waves,
      889–890, 890f
   for sound waves, 459, 459f
Frets, guitar, 485, 485f
Friction, 156–164
   calculation of, 156–157, 160–161
   definition of, 155
   functions of, 156
   kinetic, 156–160
   as nonconservative force, 224, 224f
   origin of, 156, 156f
   reduction of, 119
   static, 160–164
   surface and, 156, 156f
   visualization of, 156f
   work done by, 224, 225–227, 225f
Fringes. See Bright fringe; Dark fringe;
   Interference fringes
From the Earth to the Moon (Verne), 54,
   404
Frost heaving, 556, 556f
Frost wedging, 599
Fundamental mode, 482–484, 482f,
   483f, 486–489
Fuses, 849
Fusion, latent heat of, 602, 603t
Fusion curve, 598–599, 599f
Fusion reactions, 1034, 1034f
Fusion reactors, 1143–1144, 1144f

G
Galaxies
   red shifted, 887, 887f
   speed of, Doppler effect and, 472, 477
Galileo Galilei, 43, 407, 438–439, 885,
   972
Galileo thermometer, 951
Gallinule, 505, 505f
Galvanometers, 787, 787f
Gamma decay, 1130
Gamma rays, 890f, 891, 1126, 1130
Gas(es)
   equations of state for, 582, 582f
   ideal, 580–586
      definition of, 580
      equation of state for, 580, 580f, 582
      internal energy of, 591
      isotherms and, 583–584, 624, 624f
      molar specific heat for, 628–629,
         628f
      moles in, 582–583
      pressure of, 580–581, 580f, 583,
         583f, 585–586
      volume of, 585–586
   isotherms for, 583–584, 583f
   kinetic theory of, 586–592
   molecular speed of, 587–590, 588f
   moles in, 582
   noble, 1103
   pressure of, calculation of, 580–581
   thermal behavior of, 548–549
   volume of, temperature and, 585,
      585f
   work done by expanding, 621–622,
      621f, 623, 623f, 624, 624f
Gauge pressure, 506
Gauss (G), 774
Gaussian surface, 684–685
Gauss's law, 683–684
Gay-Lussac, Joseph, 585
Geiger, Hans, 1083
Gell-Mann, Murray, 1148, 1149
General launch angle, 99–103, 99f, 100f
General theory of relativity, 1039–1043

gravity waves and, 1042–1043, 1042f
principle of equivalence in, 1039,
   1040f
Schwarzschild radius, 1042
Generators. See Electric generators
Geographic poles, 771–772, 772f
Geomagnetism, 771–772, 772f
Geometrical optics, 914, 954–975
Geosynchronous satellites, 396, 396f
Germer, L. H., 1067
GFCI (Ground fault circuit interrupter),
   850f, 851, 851f
GFP (Green fluorescent protein),
   1111–1112, 1112f
Glare, 902, 902f, 903f, 915
Glashow, Sheldon, 1151
Glass
   electrical properties of, 658–659
   light tunneling and, 1073, 1073f
   slabs and refraction, 932f, 933
Global Positioning System (GPS), 395,
   395f
Golf
   on Moon, 104
   putt judgement in, 244
Gossamer Albatross, 213, 213f
GPS (Global Positioning System), 395,
   395f
GPS technology, 1039
Grams, 4
Grand unified theory (GUT), 1151
Graphical method, for vector addition,
   66–67, 66f, 67f
Graphs
   acceleration, 29, 29f
   average acceleration, 30f
   average velocity, 24–25, 25f, 28, 28f, 35f
   instantaneous acceleration, 30f
   instantaneous velocity, 27f, 28f
   one-dimensional motion, 25f
   v-vs.-t, 29–30
   x-vs.-t, 24
Grating spectroscope, 1007–1008, 1008f
Gratings. See Diffraction gratings;
   Reflection grating
Gravimeters, 387, 387f, 441
Gravitational field, of Sun, 1040, 1041f
Gravitational lensing, 407, 407f, 1041,
   1041f
Gravitational potential energy, 247f,
   398–401, 398f
   altitude and, 229–231, 398–400
   of fluids, 510, 510f
   kinetic energy and, 402, 403f
   of mass, 400–401, 400f
   potential well of, 402, 403f
   as scalar, 400
   zero level of, 410
Gravitropism, 190, 384, 384f
Gravity, 381–409
   altitude and, 386–387
   atoms and, 664–665
   as attractive force, 664
   compared to electric force, 664
   distance and, 393–394, 393f
   on Earth, 381–385, 385f
   effect of, on light, 1040
   force of, 120–121
   free fall and, 44–45
   as fundamental force, 1148
   inverse square dependence and, 383
   Isaac Newton and, 382
   law of universal gravitation and,
      381–385
   magnitude of, 382
   on Moon, 137, 387–388
   orbital motion and, 391–394
   radiation pressure and, 896, 896f
   range and, 104
   spherical bodies and, 385–391
   superposition and, 383
   tides and, 407–409, 408f
   universal gravitational constant and,
      388–389, 389f
   values of, based on latitude, 44f
   vector addition and, 383

waves, 1042–1043, 1042f, 1043f
weight and, 137–139
work done by, 199–200, 203, 203f,
   205, 224, 224f–226f, 225–227
Gravity escape system, 199–200
Gravity Field and Steady-State Ocean
   Circulation Explorer, 387, 387f
Gravity maps, 441
Green fluorescent protein (GFP),
   1111–1112, 1112f
Ground fault circuit interrupter (GFCI),
   850f, 851, 851f
Ground state, of atoms, 1091
   of multielectron atoms, 1100–1101
   probability clouds for, 1098, 1098f
Grounded plugs, 850–851, 850f
Grounding, 681, 681f
Guitar frets, 485, 485f
Guitar pickups, 813, 814f
Gyroscopes, 367–368, 367f

H
Hadrons, 1148, 1149t, 1150t
Hafele, J. C., 1022
Hahn, Otto, 1139
Hale-Bopp comet, 404f
Half-life, of radioactive material, 1133f,
   1134
Halley's comet, 414
Halogens, 1103
Hang time, concept of, 100, 100f
Harmonic motion. See Simple harmonic
   motion
Harmonic waves, 457, 457f, 458f,
   463–464, 463f
Harmonics
   of columns of air, 485–489, 486f,
      487f, 488f
   on a string, 483–485, 483f
Headlights, resolution and, 1002
Headphones, active noise reduction, 481
Heads-up display, 918, 918f
Hearing
   frequency range in, 466, 485, 487f
   sound perception in, 468–469, 470,
      487
Heart
   electrical activity in, 710–711, 710f,
      711f
   pacemakers for, 757, 757f
Heartbeat recoil detectors, 269
Heat
   calories in, 557
   definition of, 545
   energy and, 545–546
   energy conservation and, 556–557
   flow of, 562–566, 562f
   latent, 602–606
   measurement of, 558–560
   mechanical equivalent of, 557, 557f
   sign conventions for, 618
   specific, 559–560
   in thermal equilibrium, 559–560
   thermal expansion and, 550–556
   in thermodynamic systems, 618–620
   types of, 558–562
   units of, 557
   work and, 556–557
Heat capacity, 558–559
Heat death, of Universe, 646
Heat engines, 632–638, 632f, 633f, 638f
   Carnot's theorem and, 634–636
   efficiency of, 633–636
   entropy change and, 642
   living systems as, 646–647
   maximum efficiency of, 635
Heat exchange
   by conduction, 562–566, 562f, 563f,
      566
   by convection, 566–570, 566f, 567f
   phase changes and, 606–608
   by radiation, 567–568, 567f, 568f
Heat flow
   in adiabatic processes, 624–626, 626f
   entropy and, 645–646, 645f
   spontaneous, 632, 637
   in thermodynamic systems, 618–620

Heat pumps, 639–640, 639f
Heating devices, 637, 638f
Heating elements, 738, 738f
Height
  conservative forces and, 225–226, 226f
  difference, and speed, 236–239
  of fluid flow, change in, 525, 526
  loss of, from walking, 123
  nonconservative forces and, 243–244
Heisenberg, Werner, 1070
Heisenberg uncertainty principle
  for energy and time, 1072
  for momentum and position, 1070–1071, 1070f
Helical motion, 782, 782f
Helioseismology, 453
Helium-neon laser, 1107–1108, 1108f
Henry (H), 826
Henry, Joseph, 826
Hero of Alexandria, 626f
Hero's engine, 632f
Hertz (Hz), 419
Hertz, Heinrich, 419, 881
Higgs boson particle, 1150
Hohmann transfer, 397
Holiday lights, construction of, 745
Holograms, 1109–1110, 1110f
Hooke's law, 167–169
Horizontal motion
  independence of, 93–94, 93f, 94f
  launch angle and, 94–95, 99–103, 99f, 100f
Horsepower (hp), 213
Horseshoe magnets, 770, 770f
Hubble Space Telescope, 921, 922f
Hubble's law, 477
Human body, electric potential and, 710–711
Human eye, 955–957
  accommodation in, 955, 955f
  accommodation of, and work, 210
  focal length in, 955
  focusing in, 955–956, 956f
  refraction in, 955
  saccadic motion in, 821
  shape of
    in farsightedness, 963, 963f
    in nearsightedness, 961, 961f
  structure of, 955, 955f
Human-powered centrifuge, 191, 332
Human-powered flight, 213, 213f
Hurricanes, conservation of angular momentum in, 361, 362f
Huygen's principle, 987, 987f
Hydraulic lift, 511–512, 512f
Hydroelectric power, 822f
Hydrogen atom. *See also* Atom(s); Bohr model of hydrogen atom
  absorption spectrum of, 1085, 1085f
  Bohr orbits for, 1088–1089, 1088f
  energy-level structure of, 1097, 1097–1098, 1097f
  ionization of, 1091
  line spectral series of, 1085–1086, 1085f, 1087t, 1091–1092, 1092f
  nucleus of, 1120
  probability clouds for, 1098–1100, 1098f, 1099f
  quantum numbers for, 1096–1097
  at room-temperature, 1091
  wave function for, 1098–1100
Hydrometers, 538, 538f
Hydrostatic paradox, 541
Hyperopia. *See* Farsightedness

I

Ice
  density of, 519–520, 556
  formation of, 606–607, 606f
  melting of, 599–606, 599f
Ice cream, homemade, 605
Icebergs, buoyancy of, 519–520, 519f
Ideal conductors, 709–710, 709f
Ideal fluids, 530–531, 531f
Ideal gases, 580–586
  definition of, 580

equation of state for, 580, 580f, 582
internal energy of, 591
isotherms and, 583–584, 624, 624f
molar specific heat for, 628–629, 628f
moles in, 582–583
pressure of, 580–581, 580f, 583, 583f, 585–586
volume of, 585–586
Ideal heat pump, 639–640
Ideal reflector, 569
Ideal springs, 168
Image(s)
  characteristics of, 924t, 938t
  location of, 922–928, 924t, 936–941, 938t
  orientation of, 922–928, 924t, 936–938, 938t
  real, 923, 923f, 927, 940
  size of. *See* Image size
  virtual, 922f, 923, 927, 940
Image distance
  in mirror equation, 924–925
  negative, 926
  in plane mirrors, 916, 916f
  sign conventions for
    in lenses, 940
    in mirrors, 927
  in thin-lens equation, 939
Image formation
  in human eye, 955, 955f
  in magnifying glass, 967, 967f
  in plane mirrors, 916–919, 916f
  in spherical mirrors, 922–923, 922f, 923f, 924f
  in telescopes, 971
  in two-lens system, 959–961, 960f
Image size
  in lenses, 936–941, 937f, 938t
  magnifying glass and, 966–967, 966f
  in mirrors, 917, 922–928, 923, 923f, 924t
  telescopes and, 971
  in two-lens system, 959, 959f–960
Immunofluorescence, 1111f
Impact craters, 402, 402f
Impedance, 855–856, 865t
  in *RC* circuits, 855
  in *RL* circuits, 861–862
  in *RLC* circuits, 863–864
  unit of measure for, 855
Impulse, 262–266
  average force and, 262–263, 262f
  definition of, 262
  magnitude of, 262
  net, calculation of, 265
  visualization of, 263f
In phase wave sources, 479
Incidence, angle of, 914–915, 914f, 935, 935f
Incidence wave, 914, 914f
Inclined surface(s), 143–145
  coordinate system for, 158
  energy conservation and, 246f
  kinetic friction and, 159–162
  mechanical energy and, 238–239
  normal force and, 143, 143f
  power and speed on, 214–215, 214f
  rolling object on, 320, 320f
  weight on, 143–144, 144f
Incoherent light sources, 984
Incompressible fluids, 522
Index of refraction
  angular separation and, 1002
  for common substances, 930t
  definition of, 929
  dispersion and, 941–942
  in human eye, 955
  phase changes and, 991, 991f
  thin film phase change and, 995–996, 996f
Induced current. *See* under Electric current
Induced electromotive force (emf). *See* under Electromotive force (emf)
Induced negative charge, 680, 680f
Inductance, 825–827
  definition of, 825

finite, 827
mutual, 825
of solenoids, 826
units of, 826
Induction
  charging by, 681–682, 681f
  Faraday's law of, 812–814
  magnetic, 808, 808f, 809f
Induction stoves, 817
Inductive reactance, 860, 860f
Inductors, 825–832, 825f
  circuits in, 860–861
  electric circuits and, 826
  energy stored in, 830–832
  magnetic field and, 830–831
  notation for, 828
  phasor diagrams for, 861f
  in *RL* circuits, 828, 828f
  voltage phasors and, 863, 863f
Inelastic collisions, 271–277
  definition of, 271–272
  in one dimension, 272–275, 272f
  in two dimensions, 275–277
Inertia
  law of, 119
  moments of, 313–318, 314f, 315f, 316t
Inertial confinement, 1144
Inertial frames of reference, 120, 1018, 1018f
Infrared radiation, 567, 567f
Infrared waves, 890, 890f, 891f, 1086
Infrasound, 466–467, 467f
Instantaneous acceleration, 29–31, 30f
Instantaneous acceleration vectors, 73–74, 74f
Instantaneous angular acceleration, 303–304
Instantaneous angular velocity, 301–303
Instantaneous speed, 26
Instantaneous velocity, 26–28, 26f, 27f, 27t, 28f
Instantaneous velocity vectors, 73, 73f
Insulated windows, 566
Insulators, 562, 662–663, 662f
Intact (eye implantation), 963
Intensity
  definition of, 468
  distance and, 468–469, 469t
  of electromagnetic waves, 893–894, 893f
  vs. intensity level, 470
  of light, 897–899, 898f
  of sound, 468–472
  units of, 468
  of waves, 468f
Intensity levels, 470–471, 472t
Interference, 478–482, 486f
  in air wedges, 991–993, 992f
  beats/beat frequency, 489–492
  in CDs and DVDs, 996–997, 996f, 997f
  color of light and, 993–995, 993f
  conditions for, 984
  constructive, 478, 478f, 479, 479f, 481, 984–995, 984f, 992f, 1006
  destructive, 478–479, 479, 479f, 481, 984–997, 984f
  Huygen's principle and, 987, 987f
  Newton's rings and, 993, 993f
  opposite phase sources and, 481
  patterns of, 986–987, 986f, 987f
    diffraction grating and, 1005–1006, 1005f
    in water waves, 984, 984f
  in phase sources and, 479
  reflection and, 990–997, 991f
  in single-slit diffraction, 998f, 999–1000
  soap bubbles and, 993, 993f
  superposition and, 984–986
  in thin films, 993–996, 996f
  thin films and, 993–996, 994f
  in two-slit diffraction, 987–988, 987f, 988f
  in waves, 984–985, 984f
Interference fringes, 987–990, 988f

angular vs. linear position of, 988
circular, 993, 993f
conditions for, 987–988, 987f
diffraction grating and, 1005–1006
displacement of, 993
linear distance from central, 988–989
location of, determination of, 988
numbering systems for, 987f
in single-slit diffraction, 998–999, 998f
spacing of
  in air wedge, 992f
  change of medium and, 989–990
  in Newton's rings, 993, 993f
Interference patterns, 479–480, 479f, 1066, 1068, 1068f
Interferomety, 1005
Internal energy. *See* Thermal energy
  of gas, 591
  of systems, 618, 618f, 619f, 620
Internal forces, momentum and, 266–268
Internal resistance, 733, 743, 743f
International Space Station, 1060
Intracorneal ring, for vision correction, 963
Inverse square dependence, 383, 383f
Inverter circuits, 849
Io (planet)
  orbit of, 394
  tidal locking and, 408–409
Ions
  negative, 660
  positive, 660
  Z dependence of hydrogen-like, 1090–1091, 1091f
Iridescence, 1008, 1008f
Iris, 956, 956f
Irreversible processes, entropy and, 642
Irreversible thermal processes, 621
Isothermal processes, 624, 624f
Isotherms, 583–584, 624, 624f, 630, 630f
Isotopes
  definition of, 1120
  properties of selected, A-13–A-14
  radioactive, 1146, 1146f

J

Jet packs, 127, 127f, 134, 134f
Joule (J), 196–197, 203, 557
Joule, James Prescott, 196, 557
Joules per coulomb, 697
Joules per kelvin (J/K), 558
Junction rule, Kirchhoff's, 747–748, 747f, 748f
Junctions, 747
Jupiter, moons of, tidal locking and, 408–409

K

Kaons, 1149t
Keating, R. E., 1022
Keck reflecting telescopes, 973, 973f
Kelvin scale, 550, 589
Kepler, Johannes, 391
Kepler Space Telescope, 895
Kepler's laws of orbital motion, 391–394
  first law, 391–392, 391f, 392f
  second law, 392–393, 392f
  third law, 393–394, 394f
Keratometer, 965
Keyboards, computer, capacitance and, 717, 717f
Kilocalorie, 557
Kilogram, 2, 3
Kilowatt-hour (kWh), 740
Kinematics, 19, 41f
  energy conservation and, 234–235, 235f
  linear, 305
  Newton's second law of motion and, 127–129
  rotational, 304–307
  two-dimensional. *See* Motion; Projectile motion
Kinetic energy
  angular momentum and, 363

Kinetic energy (continued)
conservative vs. nonconservative
forces and, 224, 225f
in elastic collisions, 277–282
electric potential energy and, 700
gravitational potential energy and,
402, 403f
of inelastic collisions, 272–274
mechanical energy and, 236–241
molecular collisions and, 586–591
motion and, 246
oscillation and, 435–436
of photoelectrons, 1058–1059,
1059f
vs. potential energy, 228
potential energy curve and, 247f
of pulleys, 321
relativistic, 1036–1038, 1037f
of rolling motion, 318–322, 318–323
rotational, 313–314, 317–318
in simple harmonic motion, 418–419,
421, 421t, 435–436, 436f
of springs, 211, 211f
temperature and, 589
translational motion and, 313–318
work and, 203–207
Kinetic friction, 156–160
calculation of, 156–157
coefficients of, 157, 157t
on inclined surface, 159–162
magnitude of, 158
as nonconservative force, 224,
244–245
normal force and, 156–158, 157f
rules of thumb for, 157
vs. static friction, 160, 164f
Kinetic theory of gases, 586–592
Kirchhoff, Gustave, 747
Kirchhoff's rules, 748–751
compared to equivalent resistance,
749–750
junction rule of, 747–748, 747f
loop rule of, 748–749, 748f, 749f
Knuckle cracking, 500

L
Lagoon Nebula, 1085, 1085f
Land breezes, 567, 567f
Landing, of projectiles, 97–98, 103, 105
Laser(s), 1107–1110
in DVD technology, 996–997, 997f
excimer, 1108, 1109f
helium-neon, 1107–1108, 1108f
holograms and, 1109–1110, 1110f
medical use of, 1108–1109
as optical tweezers, 1062
use of lasers for, 1109f
vision correction using, 1108–1109
Laser Interferometer Gravitational Wave
Observatory (LIGO), 1042,
1043f
Laser therapy, 1108
LASIK eye surgery, 1108, 1108f, 1109f
Latent heat, 602–606
applications of, 604
definition of, 602
types of, 602–604
for various materials, 603t
Launch angle
general, 99–103, 99f, 100f
range and, 104, 106, 106f
zero, 94–95, 94f, 95f
Lava bombs, 46–47, 46f, 47f, 97f
Law of induction, 812–814
Law of Malus, 898
Law of universal gravitation, 381–385
Laws of nature, 2
LC circuits, 867–868, 868f
mass-string analogies with, 869t
natural frequency of, 868–869
in radio/television wave reception,
883, 883f
LCD (liquid-crystal display), 901, 901f,
903, 903f
LEDs (light-emitting diodes), 734f, 1111
Leeuwenhoek, Antonie van, 968
Leg, as physical pendulum, 442–443

Length, 2–3
change in, of solids, 592–594, 594f
contracted, 1025–1028
proper, 1025
of typical distances, 3t
typical values for, A-9
Length contraction
calculation of, 1026
muons and, 1027, 1027f
Lens(es)
aberrations in, 974–975, 974f, 975f
achromatic, 974, 975f
apochromatic, 974
camera, 957, 957f
concave, 959, 959f, 961, 961f
converging, 963, 963f
chromatic aberration and, 974,
974f, 975f
in microscopes, 969, 969f
in telescopes, 971, 971f
convex, 959, 959f, 963f
corrective, 959–962, 959f
diverging
chromatic aberration and,
974–975, 974f, 975f
nearsightedness correction and,
961, 961f
in telescopes, 972
focal length in, 939, 939f
in human eye, 955, 955f
in microscopes, 969–970, 969f
Newton's rings and, 993, 993f
principal axis of, 937, 937f
ray tracing for, 936–939, 937f, 938f
refraction in, 913, 937–938, 937f
refractive power of, 962, 962f
thin-lens equation for, 939–941,
939f
Lensing, gravitational, 407, 407f, 1041,
1041f
Lenz, Heinrich, 814
Lenz's law, 814–818, 814f, 815f
Leptons, 1148, 1148t
Levitation, magnetic, 796–797, 797f
Lewis, Gilbert Newton, 17
Lift, air flow and, 527, 527f
Lifting, 200, 205
Light. See also Light waves
amplification of, 1107
color of, 896, 940–941, 943f, 993–995,
993f
conductivity and, 662–663
dispersion of, 940–941, 943f
effect of gravity on, 1040
intensity of, 897–899, 898f
interference and sources of, 984
linearly polarized, 903
monochromatic, 984, 987–988, 987f
photon model of, 1056–1057,
1056f
polarized vs. unpolarized, 897, 897f
refraction of, 928–936
sources of, coherent and incoherent, 984
ultraviolet, 890f, 891, 892f, 907
visible, 890–891, 892
wave-particle duality of, 896, 1064,
1068–1069, 1068f
Light clock, 1020, 1020f
Light dimmer switches, 862–863
Light waves. See also Light
diffraction of, 998–1001
Doppler effect and, 472, 477,
887–889, 887f
gravitational bending of, 1040, 1040f,
1041f
Huygen's principle and, 987, 987f
Poisson's bright spot and, 1001
reflection of, 914–915, 914f, 990–997,
991f, 993f
speed of, 884–886, 1018–1019, 1019f
tunneling by, 1073, 1073f
two-slit diffraction with, 986–990,
987f
Lightbulbs, fluorescent, 1110
Light-emitting diodes (LEDs), 734f, 1111
Lightning rods, 680, 680f, 693

LIGO (Laser Interferometer Gravitational
Wave Observatory), 1042, 1043f
Line spectrum, 1084–1088, 1085f,
1091–1092, 1092f
Linear expansion, 551
Linear kinematics, 305
Linear momentum. See Momentum
Linear motion, 304–311
Linear reasoning, 206
Linnaeus, Carolus, 546
Liquid crystal, 901
Liquid-crystal display (LCD), 901, 901f,
903, 903f
Liquid(s)
boiling point of, 590–591
phase equilibrium and, 597–600, 597f
Lithotripsy, shock wave, 467
Living organisms
entropy and, 646–647
ferromagnetism in, 796, 796f
Logarithms, A-5
Longitudinal waves, 458, 458f, 464, 464f
Loop detection, on roadways, 844
Loop rule, Kirchhoff's, 748–749, 748f,
749f
Loudness. See Intensity; Intensity levels
Loudspeakers, 784, 784f. See also Speak-
ers
Lungs
gas pressure in, 581
spring-like property of, 168
surface tension in, 533
Lyman series, 1086–1087, 1087f, 1087t,
1092, 1092f

M
M ray (midpoint ray), 937, 937f–939f
Macular degeneration, 1109
Magnet(s)
induced currents and, 814–818, 815f
magnetic field lines for, 770f, 771
poles on, 770–771, 770f
in shake motion power generators,
823, 824f
types of, 770
Magnetars, 805
Magnetic antitheft devices, 821
Magnetic braking, 817
Magnetic confinement, 1144
Magnetic domains, 796, 796f
Magnetic energy per volume, 831
Magnetic field(s), 770–772, 788–791, 806
atoms and, 795
in Bohr model of hydrogen atom, 806
in the brain, 805, 806
circular motion in, 779–780
of current-carrying wire, 788–789,
788f, 789f
diamagnetic effect in, 796–797, 797f
direction of
for current-carrying wire, 788, 788f
for magnets, 771
of Earth, 771–772, 772f, 774,
795–796, 795f
effects of, 770–782
electromagnetic waves and, 881–882,
882f
in electrons, 795
energy storage in, 830–832
force on moving electric charge by,
772–773, 773f
induced electromotive force and,
808, 809f
inductors and, 830–831
of magnetars, 805
magnetic domains and, 796, 796f
magnetic flux and, 809–812
magnitude of, 773
magnitudes of typical, 774t
microscopic origin of, 795
motion in, of charged particles,
777–782
notation in, 770
paramagnetism and, 796
production of, 788–792
right-hand rule, 788
of solenoids, 793–795, 793f

temperature and, 795
units for, 774
work and, 819–821
Magnetic field lines, 770–771, 771f
inside solenoids, 793, 793f
magnetic flux and, 809, 810f
for magnets, 770f, 771, 771f
Magnetic flux, 809–812
in conversion of work to energy,
819–821
definition of, 810
electric generators and motors and,
821
magnetic field lines and, 809, 810f
transformers and, 832–833
unit of, 810
Magnetic force
on current-carrying wires,
783–785
between current-carrying wires, 791,
791f
differences from electric force,
777–778, 777f
direction of, 774–776, 774f
on electric current loops, 785, 785f
in loudspeakers, 784, 784f
magnitude of, 772–773
for positive/negative charges, 775
right-hand rule for, 774–776, 775f
vector cross products and, 775
Magnetic induction, 808, 808f, 809f
Magnetic moment, of electric current
loops, 786
Magnetic poles, 770–771, 770f
Magnetic quantum number, 1097
Magnetic resonance imaging (MRI), 794,
1147, 1147f
Magnetic tape recording, 814
Magnetic torque, 785–787, 786f, 824
Magnetism
dynamo theory of, 772
levitation and, 796–797, 797f
in matter, 795–797
recycling and, 796, 796f
units of measure for, 773, 774
Magnetite, 796
Magnetoencephalography, 806
Magnification, 966–969
angular, 967, 969
apparent size and, 966–967
in lenses, 939
in microscopes, 970–971
in mirrors, 926
sign conventions for, 926, 939
in telescopes, 971
in two-lens system, 960
units of, 967
Magnifying glass, 966–969, 966f, 967f
angular magnification in, 967
focal length of, 967
image formation in, 967, 967f
image size and, 966–967, 966f
Magnitude, 13
conversion of, 62–63, 63f
direction and, 63, 63f, 64, 68–69
of electric fields, 707–708, 707f
of electron's charge, 659
of electrostatic forces between point
charges, 663
of force, 118, 142, 142f
of gravitational force, 382
of impulse, 262
of induced electromotive force, 812
of magnetic fields, 773
of magnetic force, 772–773
of momentum, 258–259, 1031
of torque, 336–337
of vectors, 61–63
of weight, 137–138
Malus, Etienne-Louis, 898
Malus's law, 898
Maps
contour, 247, 247f
gravity, 441
Marconi, Guglielmo, 881
Marine tides, 407–409, 408f
Mars

mass of, 390
meteorites from, 413–414
Marsden, Ernest, 1083
Mass, 2, 3
 acceleration and, 121, 121f, 122
 of alpha particles, 1137
 atomic, 583
 center of, 283–288
 conservion of, to energy, 1034
 constant, second law of motion and, 261
 definition of, 118
 of Earth, 388–389
 of electons, 659
 energy and, 1034
 gravitational force and, 382–383
 gravitational potential energy of, 400–401, 400f
 of Jupiter, 394
 molecular, 583
 of moles, 583
 moments of inertia for, 313–318, 315f, 316f, 316t
 momentum and, 258, 261
 orbital motion and, 394, 394f
 of Pluto, 411
 of protons, 659
 of rotating object, 313–318
 speed of wave on a spring and, 460–462
 of string, 460
 on strings, periods of, 430–435
 typical, in kilograms, 118t
 typical values for, A-9
 units of, 123t
 on vertical spring, 433–434, 433f
 visualization of, 122f
 weight vs., 137–138, 137f
Mass number (A), 1120
Mass per length, of string, 460
Mass spectrometer, 780, 780f
Masseter muscle, 377
Mathematics, expansions in, A-5–A-6
Matter and antimatter, 1035–1036
Matter waves, 1065, 1073–1074, 1094–1096, 1095f, 1098
Maximum acceleration, in simple harmonic motion, 427
Maximum efficiency, of heat engines, 635–636
Maximum range, 104
Maximum speed, in simple harmonic motion, 426
Maximum value, vs. root mean square (rms) values, 848
Maximum voltage, in RC circuit, 855
Maxwell, James Clerk, 588, 881, 885–886
Maxwell speed distribution, 588, 588f, 600, 601f
McCarthy, Eddie, 1106
Measurement
 accuracy of, 6–7
 of body fat, 515–516
 units of, 2–4
Mechanical energy, 233–246
 conservation of, 233–241. See also Energy conservation
 definition of, 233
 food energy conversion to, 230
 kinetic energy and, 236–241
 nonconservative forces and, 241–246
 total, of gravitational system, 401
Mechanical equivalent of heat, 557, 557f
Mechanical work, conversion of, to electrical energy, 818–821
Mechanics, 19
Mendeleev, Dmitri, 1103
Mercury
 atmospheric pressure and, 509
 orbital motion of, 394–395
Mesons, 1148, 1149t, 1150, 1150f
Metal(s)
 alkali, 1103
 thermal expansion of, 552–553
 work function and, 1058
Metal detectors, resonance and, 870, 870f

Metastable state, 1107
Meteorites, from Mars, 413–414
Meteors, infrasound waves and, 467
Meter, 2, 3
Michelson, Albert A., 907, 1019
Microcoulomb, 659
Microcurie, 1131
Microfarad, 712
Microhematocrit centrifuge, 309
Microphones, dynamic, 813, 813f
Microscopes
 electron, 1067–1068, 1068f
 optics in, 969–971, 969f
 scanning tunneling, 1073–1074, 1073f, 1074f
Microwaves, 890, 890f
Midpoint ray (M ray), 937, 937f
Millicurie, 1131
Millikan, Robert A., 661
Mirages, 932, 932f
Mirror(s)
 corner reflector, 918, 918f
 parabolic, 921, 921f
 plane, 916–919, 916f, 919f
 principal axis of, 920, 920f
 ray tracing for, 922–923, 922f, 923f
 sign conventions for, 920, 926–927
 spherical. See Spherical mirrors
Mirror equation, 922, 924–926, 925f
Mirror images, 916
Mirror writing, 916f
Mks system, 2
Moderator, in nuclear chain reaction, 1142
Molar specific heat, 628–629
Molecular collisions, 586–591
Molecular mass, 583
Molecular oscillations, 449
Molecules
 collisions of, kinetic energy and, 586–591
 diatomic, 591
 moles and, 582, 582f
 speed distribution of, 587–590, 588f
 units of measure for, 582
Moles, 582, 582f, 583
Moment arm, 337–338, 337f, 357
Moment of inertia, 313–318, 314f, 315f, 316t
 definition of, 314
 of Earth, 316–317, 317f
Momentum
 angular. See Angular momentum
 average force and, 263–265
 calculation of
  changes in, 258–259
  for system of objects, 259
 changes in
  calculation of, 258–259, 259f
  changing mass and, 288–289
  impulse and, 262–265, 262f, 263f
  internal vs. external forces and, 266–268
  second law of motion and, 261
 conservation of. See Conservaton of momentum
 coordinate system for, 259
 definition of, 258
 direction and, 258–259
 in electromagnetic waves, 895
 external forces and, 266–268
 internal forces and, 266–268
 magnitude of, 258–259
 mass and, 258, 261
 of photons, 1061–1062
 position and, Heisenberg uncertainty principle on, 1070–1071, 1070f
 relativistic, 1031–1033, 1032f
 rocket propulsion and, 288–289
 second law of motion and, 261
 total, 258–259
 units of, 258
 as vector, 258–259
 velocity and, 258
Momentum-impulse theory, 262
Monochromatic light, 984, 987–988, 987f

Moon
 acceleration due to gravity on, 388
 corner reflectors on, 918
 density of, 389
 escape speed for, 405
 golfing on, 104
 gravitational force between Earth and, 387–388, 388f
 gravity on, 137, 387–388
 tides and, 408, 408f
 weight vs. mass on, 137
Morley, E. W., 1019
Motion. See also Projectile motion
 of center of mass, 285–288, 287f
 circular, 177–182
  angular momentum of, 356–357, 356f, 358, 358f
  simple harmonic motion and, 424–430
 with constant acceleration, 33–34
 constant velocity and, 89–91, 89f
 constant-velocity, straight-line drift, 779
 equations of, 33–43, 39t
 factors affecting, of mass on a string, 432f
 force reduction and, 119
 helical, 782, 782f
 horizontal
  independence of, 93–94, 93f, 94f
  launch angle and, 94–95, 99–103, 99f, 100f
 independence of, 89–92, 93–94, 93f, 94f
 in magnetic fields, 779
 nontangential, 356, 356f
 oscillatory, 247
 periodic, 419–420
 projectile. See Projectile motion
 properties of, in one-dimension, 20–21, 25f
 relative, 76–79
 retrograde, 391
 rolling, 311–312, 311f, 312f
 rotational. See also Rotational motion
 saccadic, 821
 simple harmonic, 417?430
 speed and, 89–91, 93–103
 translational, 311–312, 311f, 312f
 turning points of, 246–247, 425
 in two-dimensions, 89–103
 velocity and, 89–94, 89f
 vertical, independence of, 93–94, 93f, 94f
 of waves, 453f, 457, 457f
Motorcycles, loop detection and, 844
Motors, electric. See Electric motors
Movie captioning device, for hearing impaired, 918
Moving observer/source
 Doppler effect for, 472–477, 472f, 474f
 relativistic length and, 1025–1026, 1026f
 simultaneity and, 1022
 speed of, vs. wave speed, 1018–1019, 1019f
MRI (Magnetic resonance imaging), 794, 1147, 1147f
Multielectron atoms, 1100–1102
 energy levels in, 1100–1102, 1101f
 ground state of, 1100–1101
 radiation and, 1104–1112
Multimeters, 758, 758f
Multiplication of vectors, by scalars, 70–71, 70f
Muons, 1148t
 length contraction and, 1027, 1027f
 time dilation and, 1024
Muscae volitantes, 957
Muscle(s)
 biceps, 377
 ciliary, 955, 955f, 956
 deltoid, 378
 masseter, 377
 triceps, 378
Music, chromatic scale in, 485, 485t

Musical instruments, 485, 488–489
Mutchkin, 17
Mutual inductance, 825
Myopia. See Nearsightedness

N

N (neutron number), 1120
Nakamura, Shuji, 1111
NASA, 1062
Nasal strips, 168–169
National Electric Code, 767
Natural frequency, 445, 868–869
Nature, laws of, 2, 1018–1019
Navigation, by echolocation, 466, 466f, 496
Near point, visual
 definition of, 956
 in farsightedness, 963, 963f
 magnification and, 968
Nearsightedness, 960–961, 961f, 963, 1108–1109
Nebula
 Crab, 326, 362, 362f
 emission, 1085f
 Lagoon, 1085, 1085f
Negative charges, 658
 electric potential and, 701, 704
 induced, 680, 680f
 magnetic force for, 775
Negative work, 200, 200f
Neon, 1104f
Nerve cells, electric fields in, 691
Net force, 132f
 center of mass and, 286–288, 287f
 change of momentum and, 261
 conservation of momentum and, 266
 definition of, 118
 of electric charges, 666–668
 first law of motion and, 119
 internal vs. external forces and, 266–268
 second law of motion and, 122
 transitional equilibrium and, 169–172, 169f
 velocity and, 120
Neutral electric charge, 659
Neutrinos, 1129–1130, 1148t
Neutron(s)
 in atoms, 659
 definition of, 1120
 as hadrons, 1148, 1149t
 mass of, 1121, 1121t
Neutron diffraction, 1067, 1067f
Neutron number (N), 1120
Neutron stars, 362
Newton, Isaac, 973
Newtonian reflectors, 973, 973f
Newtons (N), 122–123, 123t
Newton's cradle, 280f
Newton's law of universal gravitation, 381–385, 664
Newton's laws of motion
 first law, 118–120
 second law, 120–129, 339–342
  acceleration in, 120–122, 121f
  application of, to force components, 134–137
  momentum and, 261
  net force in, 122
  for rotational motion, 354–355, 359
  for a system of particles, 286–288
  in vector components, 122, 126–127
 third law, 129–134
Newton's rings, 993, 993f
Nexrad, 888
Nijmegen High Field Magnet Laboratory, 796
Noble gases, 1103
Nodes, 483, 483f, 484f, 486, 486f
Noggin, 17
Nonconservative forces, 224, 227t, 241–246
 vs. conservative forces, 224, 225f
 distance and, 244–246

Nonconservative forces (*continued*)
  energy conservation and, 246f
  height and, 243–244
  path and, 224, 225f
  work done by, 241–246
Noninertial frames of reference, 120
Nonomic materials, 735
Nonreflective coatings, 995–996, 996f
Nonzero net force, 287–288, 287f
Normal force, 124, 141–145
  on inclined surfaces, 143, 143f
  kinetic friction and, 156–158, 157f
  magnitude of, 141f, 142, 142f
  weight and, 141–142
North pole
  geographic, 771–772, 772f
  magnetic, 770–771, 770f
Notation
  for acceleration, 5, 6t
  acceleration due to gravity, 44
  for antiquarks, 1149
  for area, 5, 6t
  for batteries, 732
  for Celsius scale, 546
  delta, 21
  for distance, 5, 6t
  in electric circuits, 734
  for electrons, 1101–1102
  for energy, 5, 6t
  for grounding, 681
  for inductors, 828
  in magnetic field, 770
  in periodic table of elements, 1103, 1103f
  scientific, A-1
  subscripts in, 71–72
  for vectors, 61, 61f, 62, 69, 122
  for velocity, 5, 6t
  for volume, 5, 6t
Nuclear fission, 1139–1142, 1140f
  chain reactions in, 1141–1142, 1141f
  nuclear reactors and, 1142
Nuclear force, 1124, 1139, 1148, 1151
Nuclear fusion, 1142–1144
Nuclear physics, practical applications of, 1144–1147
Nuclear reactors, 1034, 1034f, 1142, 1142f
Nucleons, 1120, 1138–1139, 1139f
Nucleus (Nuclei)
  of atoms, 1083–1084, 1083f
  binding energy of, 1138–1139, 1138f
  combining of, 1142–1144
  decay of
    randomness of, 1132, 1132f
    types of, 1125–1132
  density of, 1123
  notation for, 1120, 1120t
  parent and daughter, 1126, 1130–1131, 1140f
  radioactive decay series of, 1131, 1131f
  size of, 1121–1122
  splitting of, 1139–1142
  stability of, 1123–1124
  structure of, 1120
Numerical values, 6–9

O

Object distance
  angular size and, 966–967, 966f
  center of curvature and, 923–924, 928
  close in, 928
  in mirror equation, 924
  negative, 927
  in plane mirrors, 916, 916f
  sign conventions for
    in lenses, 940
    in mirrors, 927
  in thin-lens equation, 939
Objective, of microscope, 969, 969f
Ocean Thermal Energy Conversion (OTEC), 656
Ocean tides, 407–409, 408f
Oceans, bathysphere exploration of, 616
Oersted, Hans Christian, 788
Ohm (Ω), 734
Ohm, Georg Simon, 734

Ohmmeters, 734, 758
Ohm's law, 734, 735
Omega particle, 1149t
One-dimensional kinematic systems, 125, 125f
Onnes, Heike Kamerlingh, 737
Open circuits, 732
Opposite phase wave sources, 481
Optic nerve, 955, 955f
Optical fibers, 934, 934f, 935f
Optical illusions, 931–932, 932f
Optical pyrometer, 567
Optical tweezers, 1062
Optically active compounds, 901, 912
Optics
  corrective, 958–966
  geometrical, 914, 954–975
  human eye of, 955–957
  in microscopes, 969–971, 969f
  physical, 983–1008
  in telescopes, 971–974
Optics (Ptolemy), 948–949
Orbital angular momentum quantum number, 1097
Orbital maneuvers, 397–398, 397f
Orbital motion
  angular momentum and, 392
  in Bohr model of hydrogen atom, 665, 665f
  centripetal acceleration and, 393
  centripetal force and, 393, 393f
  circular, 393, 393f
  distance and, 393–394
  elliptical, 391–392, 391f, 392f
  free fall, 382, 382f
  gravity and, 393
  Kepler's laws of, 391–394
  mass and, 394, 394f
  of Mercury, 394–395
  period of, 393–394, 393f
  of spacecraft, 395–397, 396f, 397f
  Sun and, 391–394
  visualization of, 394f
Orbital speed, of Earth, 392–393
Order-of-magnitude calculations, 11–12
Organ pipes, 488–489
Origin, launch from. See General launch angle
Oscillations, 415–451
  ambulatory, 442–443
  damped, 444–445, 444f
  driven, 445–446, 445f
  in electric circuits, 867–872
  energy conservation and, 435–438, 435f
  kinetic energy and, 435–436
  in *LC* circuits, 868f
  of mass on a spring, 418, 418f, 420–423, 421f, 430–435
  of mass on a string, 868–869, 869t
  molecular, 449
  natural frequency of, 445
  overdamped, 444
  pendulums and, 438–444
  period of, 419–420
  potential energy and, 435–436
  resonance and, 445–446, 445f
  in simple harmonic motion, 417–420, 418f, 421f
  in walking, 442–443
Oscillatory motion, 247
Overdamped oscillations, 444

P

P ray (parallel ray), 920, 920f, 922–923, 922f, 923f, 937, 937f, 938f
P wave, 711
Pacemakers, for heart, 757, 757f
Paleomagnetism, 772
Parabolic path, of projectiles, 96–97, 97f, 105, 105f
Parallel ray (P ray), 920, 920f, 922–923, 922f, 923f, 937, 937f, 938f
Parallel-plate capacitor, 677, 677f, 713f
  capacitance of, 713–714, 713f, 716
  dielectrics in, 715–718, 715f
  touchscreens and, 717

Paramagnetism, 796
Paraxial rays, 921
Parent nucleus, 1126, 1130–1131, 1140f
Particle accelerators, 1039
Particles
  alpha, 1125, 1137
  charged, helical motion of, 785
  elementary, 1147–1150
  sub-atomic
    mass and charge of particles in, 1121f
    radioactive decay and, 1125–1132
    wavelength and momentum in, 1065
    wavelike properties of, 1068f, 1073
Pascal (Pa), 505
Pascal, Blaise, 505
Pascal's principle, 511–512
Paschen series, 1086–1087, 1087f, 1087t, 1092, 1092f
Path
  conservative forces and, 225f–227f
  forces on closed, 224–227
  nonconservative forces and, 224, 225f
  speed and, 236–237, 236f
  work and, 226–227, 226f
Patterned ground, 599, 600f
Paul, Les, 813
Pauli, Wolfgang, 366f, 1100, 1129
Pauli exclusion principle, 1150
Pauli exclusion principle says that only one electron may occupy a given state, 1100–1101
PDT (Photodynamic therapy), 1109
Pendulums, 438–444
  ballistic, 275
  Galileo's studies of, 438–439
  leg as, 442–443
  motion of, 439, 439f
  period of, 416
  periods of, 440, 442
  physical, 442–443
  potential energy of, 439–440, 439f
  restoring force for, 439
  simple, 439–442
Period, 416–430
  amplitude and, 431
  angular velocity and, 302–303
  definition of, 419
  of mass on spring, 430–435, 432f
  of orbital motion, 393–394, 393f
  of pendulum, 440, 442
  time, expression of, 428
  typical values for, 420t
  units of, 419
  of wave and wavelength, 459, 459f
Periodic motion, 419–420
Periodic table of elements, 1103, 1103f, A-12
Permeability of free space, 789
Permittivity of free space, 683
Perspiration, evaporation of, 600, 600f, 601f
PET (positron emission tomography) scans, 1036, 1036f, 1146–1147, 1146f
Phase angle, 858, 858f, 865t
Phase changes, 598–599, 606–608
  boiling point and, 598
  electromagnetic waves and, 990–991, 991f
  energy conservation in, 606–608
  equilibrium in, 597–600
  equilibrium vapor pressure in, 597–598, 597f
  fusion curve and, 598–599, 599f
  index of refraction and, 991, 991f
  latent heat and, 602–606
  phase diagrams and, 598–599, 599f
  reflection and, 990–991, 991f
  sublimation curve and, 599, 599f
  temperature and, 603, 603f, 604f
  in thin films, 995–996, 996f
Phase diagrams, 598–599, 599f, 613
Phase equilibrium, 597–600, 597f
Phase relationship, of waves, 984–985, 984f

Phasor diagrams
  ac resistor circuit, 846, 847f
  for capacitor circuits, 853, 853f
  circuit elements and, 865t
  for inductor circuits, 860–861, 861f
  power factor in, 858
  for *RC* circuits, 855f, 858, 858f
  for *RL* circuits, 861f
  for *RLC* circuits, 863–864, 863f
Phasors. See also Current phasors; Phasor diagrams; Voltage phasors
  direction of, 846, 847f
  drawing, 852, 861
  function of, 846
Phosphorescence, 1112
Photocells, 1060
Photoconductors, 662
Photocopiers, 663
Photodynamic therapy (PDT), 1109
Photoelastic stress analysis, crossed polarizers and, 900–901, 901f
Photoelectric effect, 1057–1058, 1058f, 1060, 1060f
Photoelectrons, 1057–1059, 1058f, 1059f
Photons
  absorption of, 1093
  energy of frequency *f*, 1056
  light amplification and, 1107
  light and, 1056–1057, 1056f
  momentum of, 1061–1062
  radiation pressure of, 1062
  radioactivity and, 1125
  rest mass of, 1061
  scattering of. See Compton effect
  spontaneous emissions of, 1107
  stimulated emission of, 1107, 1107f
  wave-particle duality of, 1064, 1068–1069, 1068f
Photorefractive keratectomy (PRK), 1108, 1109f, 1117
Physical laws, equivalence of, 1018
Physical optics, 983–1008
Physical quantities
  dimensions of, 5–6, 6t
  magnitude of, 13
  measurement units of, 2–4
Physics
  definition of, 2
  goal of, 1
  problem solving in, 13–14
Piano, shape of, 485
Pickering, Edward C., 1118
Pickering series, 1118
Pickups, guitar, 813, 814f
Picofarad (pF), 712
Pions, 1149t
Pipes
  columns of air in, 486, 488–489
  fluid flow through, 521–523, 531–532
  thermal expansion in, 556
Pitch, 466–467
  Doppler effect and, 472–478
  human perception of, 466
Pitching, in baseball, 128–129, 128f, 305–306, 1073
Pixels, 1005, 1005f
Planck, Max, 1054
Planck's constant, 1054
Planck's hypothesis of quantized energy, 1054, 1055
Plane mirrors, 916–919, 916f, 919f
Plane waves, 914, 914f
Planets. See also specific planets
  data table for, A-10
  orbits of, 391–394
Plasma, in nuclear fusion, 1144
Plates. See Conductors
Plimsoll mark, 518f, 519
Plugs, electrical, 850–851, 850f
Plum pudding model of atoms, 1083, 1083f
Pluto, mass of, 411
Point charges
  in Coulomb's law, 663–664, 664f
  electric field lines for, 676, 676f
  electric field of, 672, 672f
  electric flux for, 683, 683f

electric potential of, 703–707, 704f
equipotential surfaces for, 707, 707f
magnitude of electrostatic forces between, 669–670
over uniform plate, 677, 677f
spheres and, 669
Pointillism, 1005, 1005f
Poise, 531
Poiseuille, Jean Louis Marie, 531
Poiseuille's equation, 532
Poisson, Siméon D., 1001
Poisson's bright spot, 1001, 1001f
Polaris, 368
Polarization
definition of, 897, 897f
direction of, change in, 898–899, 898f
of electric charges, 661–662, 661f
electromagnetic waves and, 896–903, 897f
filters and, 898, 898f, 903f
light intensity and, 897–899, 898f
in optically active compounds, 901, 912
by reflection, 902, 902f
by scattering, 901–902, 902f, 903f
total, 934–935, 935f
Polarized plugs, 850, 850f
Polarizers, 897–898, 897f, 898f
crossed, 900
direction of light wave in, 897, 897f, 898f
transmission axis of, 898, 898f
Poles
geographic, 771–772, 772f
magnetic, 770–771, 770f
Pollination, 693–694
Population inversion, 1107
Porro prisms, 934, 934f
Position
vs. acceleration, in simple harmonic motion, 426–427, 426f
acceleration and, 36–37, 36f
angular, 300, 300f
energy and, in simple harmonic motion, 435–436, 435f
harmonic waves and, 463–464
of launched projectile, 99
momentum and, Heisenberg uncertainty principle on, 1070–1071, 1070f
in one-dimensional coordinate system, 20
potential energy and, 246–247
time and, 34–35, 35f, 36–37, 36f, 38
for mass on string, 431, 432f
in simple harmonic motion, 421–422, 422f, 424–425, 425f
velocity and, 34–35, 35f, 38
Position vectors, 71, 71f
Positive charges, 658
electric potential and, 701, 704
magnetic force for, 775
Positive work, 200, 200f, 211, 211f
Positron, 1036, 1125
Positron emission tomography (PET) scans, 1036, 1036f, 1146–1147, 1146f
Potential. *See* Electric potential; Potential energy
Potential energy, 228–233
in bows and arrows, 232–233
definition of, 228
of flea jump, 232, 232f
of food, 230
gravitational, 228–231, 228f, 229f, 398–401, 398f
vs. kinetic energy, 228
motion and, 246–247, 246f, 247f
in simple harmonic motion, 418–419, 421, 421t, 435–436, 435f
of springs, 231–233, 231f
zero level of, 228
Potential energy curves, 246–247, 246f, 247f
Potential well, of gravitational potential energy, 402, 403f
Power, 212–215

definition of, 212
dissipation of, in resistors, 745
electrical, 738–741
in batteries, 738
conversion of ac and dc, 849
dissipation of, in resistors, 738–739, 738f, 740, 745
units of measure for, 738
hydroelectric, 822f
radiated, 568
refractive, 962, 962f
speed and, 214–215, 214f
from torque, 365
units of, 213
Power factor, 858
Prairie dog burrows, air flow in, 528, 528f
Precession, 368
Precipitators, electrostatic, 681
Prefixes, 4, 5t
Premature infants, respiratory distress in, 533
Presbyopia, 956
Pressure
air, 581–582
airflow speed and, 527–528, 527f
atmospheric, 505, 509, 509f, 509t
constant, in thermodynamic processes, 621–622, 621f, 622f
dependence on speed of, 527
equilibrium vapor, 597, 597f
fluid, 500–512, 523–526
definition of, 504
with depth, 507–509, 508f
direction of, 505, 505f
proportionality of, 505
units of, 504
weight and, 507, 507f
gas, 580, 583, 583f
calculation of, 580–581
constant, 585, 585f
isotherms and, 583–584, 583f
in kinetic theory, 588
origin of, 587, 587f
gauge, 506
phase equilibrium and, 597–600, 597f
Primary electric circuit, 808, 808f
Principal axis, 920, 920f, 937, 937f
Principal maxima, 1005–1006, 1005f
Principal quantum number, 1096–1097
Prisms
in binoculars, 934, 934f
concave lenses and, 937, 937f
convex lenses and, 936, 936f
PRK (Photorefractive keratectomy), 1108, 1109f, 1117
Probability clouds, electron, 1098–1099, 1098f, 1099f
Problem solving, guidelines for, 13–14
Projectile(s), 92–94
collision of, 101–102
definition of, 92
equations of motion for, 93, 94–95, 99
independence of motion and, 93–94
landing of, 97–98, 103, 105
launch angles of, 94–95, 94f, 99–103, 99f, 100f
motion of. *See* Projectile motion
parabolic path of, 96–97, 97f, 105, 105f
speed and, 104–107, 104f
Projectile motion
acceleration and, 91–92, 91t, 93, 93f
air resistance and, 92–93, 104, 104f, 106f
characteristics of, 103–107
displacement and, 93
equations for, 93, 94–95, 99
launch angle in. *See* Launch angle
parabolic path in, 96–97, 97f, 105, 105f
symmetry in, 104–106
time and, 93
in two-dimensions, 103–107
velocity and, 94–99, 104–105
Proper length, definition of, 1025

Proper time, 1022
Propulsion
by fleas, 133
rocket, 288–289, 289f
visualization of, 289f
Proton(s), 1120
in atoms, 659
as hadrons, 1148, 1149t
mass of, 659
nuclear stability and, 1123–1124, 1124f
Proton-proton cycle of fusion reactions, 1143
Ptolemy, 391, 948–949
Pulleys
angular acceleration and velocity of, 304f
angular acceleration of, 304
in Atwood's machine, 175–176, 176f
connected objects and, 173–174
energy conservation and, 321
string tension and, 165–166
torque and, 354–355, 354f
transitional equilibrium and, 169–170, 169f
Pulmonary surfactant, 533
Pulsars, 362, 362f
Pupil, of human eye, 956, 956f, 1002
Pythagorean theorem, A-1–A-3

Q
q waves, 711
QRS complex, 711
Quadratic equation, A-3
Quantities
small, mathematical expansion for, A-5–A-6
unknown, solving for, A-3–A-4
Quantity. *See* Physical quantities
Quantized electric charge, 661
Quantized energy, 1054
Quantum, definition of, 1054
Quantum chromodynamics (QCD), 1150
Quantum electrodynamics (QED), 1150
Quantum mechanics, 1096–1099
Quantum numbers, 1054–1055, 1096–1097
Quantum particles, tunneling by, 1073–1074
Quantum theory, 1054–1055
Quantum tunneling, 1073–1074
Quark(s), 1148–1150, 1149t, 1150t
Quark confinement, 1149–1150
Quarter-wave coatings, 996
Quasi-static thermal processes, 621

R
Racetracks, banked curves for, 180, 181f
Rad (radiation absorbed dose), 1145
Radar, Doppler effect and, 477, 477f, 887
Radial keratotomy, 963, 963f
Radians, 301, A-1
Radiated power, 568
Radiation, 567–568. *See also* Atomic radiation
biological effects of, 1144–1147
blackbody, 569, 1052–1055, 1052f
dosages of, 1145, 1146t
emissivity and, 568–569
temperature and, 1052–1053
ultraviolet light, 907
visualization of
infrared, 567f
visible, 567f
Radiation absorbed dose (Rad), 1145
Radiation pressure, 895–896, 896f, 1062
Radio tuners, 445, 870, 883, 883f
Radio waves, 890, 890f
Radioactive dating, 1135–1137, 1136f
Radioactive decay, 1125–1132
Radioactive decay series, 1131, 1131f
Radioactive tracers, 1146
Radioactivity, 1125–1132
as a clock, 1134–1135
discovery of, 1126
half-life of, 1133, 1133f
Radius, 301, 301f

Bohr, 1089, 1089f
moment of inertia and, 315–316, 315f, 316t
Schwarzschild, 1042
turning, 179–180
Radon, 1133, 1133f
Rails (bird) and pressure, 505, 505f
Rainbow holograms, 1110
Rainbows, 942, 943f
Raindrops, 943f
Ramp. *See* Inclined surfaces
Range, in projectile motion, 103–104, 103f
gravity and, 104
launch angles and, 104, 106, 106f
maximum, 104
Ray tracing
for lenses, 936–939, 937f, 938f
for mirrors, 922–923, 922f, 923f
properties of, 937
Rayleigh's criterion, 1002
RBE (relative biological effectiveness), 1145, 1145t
RC circuits, 754–757, 754f, 855–862
applications of, 757
capacitors in, 855, 855f, 857
electric current in, 855, 858f
impedance in, 855
maximum current in, 855
phase angle for, 858, 858f
phasor diagrams for, 855f
power factor in, 857–859
resistors in, 855, 855f
rms current in, 862, 862f
voltage in, 855, 858f
Reactance
capacitive, 851–852
inductive, 860, 860f
Real images, 923, 923f, 927, 940
Reasoning, linear, 206
Recoil, 269
Rectifying circuits, 849
Recycling and magnetism, 796f
Red shift, 472, 477, 887, 887f
Reflected wave, 914, 914f
Reflection, 462–463, 914–915, 914f
angles and, 914–915, 914f, 915f, 933
diffuse, 915, 915f
dispersion and, 942–943, 943f
interference and, 990–997, 991f
law of, 914–915, 914f
mirror images and, 918
phase changes and, 990–991, 991f
by plane mirrors, 916–917, 916f
polarization and, 902, 902f
radiation and, 569
specular, 915, 915f
in spherical mirrors, 920–921, 921f
total internal, 933–935, 933f, 934f, 935f
of waves
of light, 914–915, 914f, 990–997, 991f, 993f
of sound, 462–463, 462f, 463f
Reflection grating, 1008, 1008f
Reflections on the Motive Power of Fire (Carnot), 634
Reflector, ideal, 569
Reflectors, 973
Reflectors, bicycle, 918
Refraction, 928–936, 955
angle of, 929–930
and apparent depth, 931–932, 932f
definition of, 928
direction of propagation and, 929–930, 929f, 942
index of, 929, 930t, 941–942, 955, 991, 991f, 995–996, 996f, 1002
in lenses, 913, 937–938, 937f
optical illusions due to, 931–932, 932f
properties of, 930
Snell's law and, 929–930
total internal reflection and, 933–934, 933f, 935f
total polarization and, 935, 935f
Refractive power, 962, 962f
Refractors, 972, 973f

Refrigerator magnets, 771, 771f
Refrigerators, 637, 638f
Relative biological effectiveness (RBE), 1145, 1145t
Relative motion, 76–79
Relativistic addition of velocities, 1028–1031, 1029f, 1031f
Relativistic energy, 1033–1038
Relativistic kinetic energy, 1036–1038, 1037f
Relativistic momentum, 1031–1033, 1032f
Relativity
    distortions and, 1027–1028, 1028f
    effects of, on space travel, 1028
    events in, 1022
    general theory of, 1039–1043
    of length, 1025–1028
    proper time in, 1022
    special theory of, 1018–1020
    universe and, 1033–1038
REM (roentgen equivalent in man), 1145
Resistance. See Air resistance; Electrical resistance
Resistance heating, 740
Resistivity, 735–737
    of common substances, 735t
    electrical resistance and, 735
    in semiconductors, 737
    temperature and, 737
    units of measure for, 735
Resistors
    in combination circuits, 745, 746f, 747, 747f
    dissipation of power in, 738–739, 738f, 740, 745
    in electric circuits, 739
    in parallel, 743–745, 743f
    phasors and, 855, 855f, 863, 863f
    power dissipation in, 738–739, 738f, 740, 745, 847–848
    in RC circuits, 754–755, 756f
    in series, 741–743, 741f
Resolution, 1002–1005
    angular separation and, 1002–1005, 1002f
    circular apertures and, 1002, 1002f
    diffraction-limited, 1002–1005, 1002f, 1005f
    pointillism and, 1005
Resonance
    in electric circuits, 867–872
    metal detectors and, 870, 870f
    oscillations and, 445–446, 445f
    RLC circuits and, 870, 870f
Resonance curves, 445–446, 445f
Respiration, surface tension in, 533
Rest energy, 1034
Rest mass, 1033, 1061
Restoring force, 421, 439–440
Retina, 916, 955, 955f
Retrograde motion, 391
Retroreflectors, 918, 918f
Reversible thermal processes, 621, 621f
Reversible thermal processes, entropy and, 640–645
Revolutions, 300, 301
Right-hand rule, 366–367
    for angular velocity, 366, 366f
    for direction of propagation for elec-
        tromagnetic waves, 882, 883f
    magnetic fields and, 788–789
    for magnetic force, 774–776, 775f
    for torque, 367, 367f
Rings of Saturn, 409
River, crossing of, 78–79
RL circuits, 828–829
    electric current in, 828f, 830f
    impedance in, 861–862
    inductors in, 828, 828f
    phasor diagrams for, 861f
    rms current in, 862f
RLC circuits, 863–867
    frequency and, 866
    impedance in, 863–864
    phase angle for, 865

phasor diagrams for, 863–864, 863f
    resonance and, 870, 870f
    rms current in, 872, 872f
rms. See Root mean square
Roche limit, 409
Rockets
    escape speed of, 404–407, 407f
    orbital motion of, 394–397, 397f
    propulsion and, 288–289, 289f
Rods and cones, retinal, 955
Roentgen (R), 1145
Roentgen, Wilhelm, 1104–1105, 1106
Roentgen equivalent in man (REM), 1145
Rolling motion, 311–312, 311f, 312, 318–323
Romer, Ole, 885
Root mean square (rms) speed, 589
Root mean square (rms) values
    in alternating-current (ac) circuits, 847–848
    vs. maximum value, 848
Rotation
    axis of, 300, 300f, 362–363, 368
    period of, 302–303
Rotational collisions, angular momen-
    tum and, 363–364, 363f
Rotational kinematics, 304–307
Rotational motion, 300–315
    angular acceleration and, 303–304, 304f
    angular displacement and, 301, 302f
    angular position and, 300, 300f
    angular speed and, 302–303, 302f, 303f, 313–314, 313f
    angular velocity and, 301–303, 302f, 303f
    of Earth, 362–363, 368
    equations of, 305
    of gyroscopes, 367–368, 367f
    linear motion and, 304–311
    Newton's second law for, 339–342, 359
    Newton's second law of motion and, 354–355
    rolling motion and, 311–312, 311f, 312f
    translational motion and, 311–312, 311f, 312f
    vectors and, 366–368
    visualization of, 300f
Rotational work, 365, 365f
Round-off error, 8
Rubbing, electric charges and, 660, 660t
Rumford, Count, 556
Runaway chain reaction, 1141
Rutherford, Ernest, 1083, 1121–1122
Rutherford solar system model of atom, 1083–1084, 1083f
Rydberg constant, 1085

S
Saccadic motion, 821
Safety devices
    for footwear, electrical resistance and, 767
    for household electric circuits, 849–851, 850f
    photocells and, 1060, 1060f
Salam, Abdus, 1151
Salt
    boiling point and, 591
    in ice-cream making, 605–606
Satellite dishes, 921
Satellites
    geosynchronous, 396, 396f
    GPS (global positioning system), 1092
    orbital motion of, 394–397, 396f, 397f
Saturn, rings of, 409
Scalar components, of vector, 62, 62f
Scalars, 12, 61
    definition of, 79
    gravitational potential energy and, 400
Scale, chromatic musical, 485, 485t
Scanning electron micrograph (SEM),

1068f
Scanning tunneling microscope (STM), 1073–1074, 1073f, 1074f
Scattering, of photons. See Compton effect
Schrödinger, Erwin, 1096
Schrödinger's equation, 1096
Schwarzschild radius, 1042
Scientific notation, 8, A-1
Scorpions, fluorescence in, 1111
Scott, David, 43, 59
Sea breezes, 567, 567f
Second(s), 2, 4
Second harmonic, 479f, 483, 483f, 486, 488, 488f
Second law of thermodynamics, 631–632, 646
Secondary electric circuit, 808, 808f, 812
Seismographs, 813, 814f
Selenium, 662
Self-inductance, 825, 825f
SEM (Scanning electron micrograph), 1068f
Semiconductors, 662, 737
Shadows, diffraction and, 1000
Shape
    center of mass and, 352, 352f
    change in, of solids, 594–595, 594f
    moments of inertia for, 313–316, 314f, 315f, 316f
Shear deformation, 594, 594f
Shear modulus, 594, 594f
Shells, for electrons, 1100, 1100t, 1103, 1104–1105, 1105f
Shepard, Alan, 104
Shielding, electrostatic, 679–681, 680f, 681f
Ships, Plimsoll mark on, 518f, 519
Shock wave lithotripsy, 467
Short circuits, 744, 744f
Shrimp, peacock mantis, spring poten-
    tial energy in, 233, 233f
Shutter speed, 958
SI units, 2
Sigma particle, 1149t
Sign conventions
    for angular acceleration, 303
    for angular momentum, 358
    for angular position, 300
    for angular velocity, 302
    of current flow, 733
    for Doppler effect, 473, 475
    for focal length, 926, 939
    for heat, 619t
    for heat capacity, 559
    for image distance
        in lenses, 940
        in mirrors, 927
    for magnification, 926, 939
    for mirrors, 920, 926–927
    for object distance
        in lenses, 940
        in mirrors, 927
    for rotational motion, 300
    for torque, 337
    for work, 618, 619t
Significant figures, 6–9
Simple harmonic motion, 420–430
    acceleration in, 426–428, 426f
    amplitude in, 422, 426–428, 432f
    definition of, 420
    energy conservation in, 435f
    maximum speed in, 426
    period of mass on spring in, 430–435
    position vs. time in, 421–423, 422f, 431, 432f
    sine and cosine functions in, 422f
    uniform circular motion and, 424–430, 424f, 425f
    velocity in, 425–426, 426f
Simple harmonic motions
    energy conservation in, 435–438
    springs and, 418f, 421f
Simultaneity, time dilation and, 1022
Single-slit diffraction, 998–999, 998f

Siphons, 541
Siri's formula, 516
Size. See Apparent size; Image size
Skylab, 414–415
Slinky, as sound wave model, 464, 464f
Slope, 25
Slug, 4
Smearing, diffraction-induced, 1002
Smoke detectors, alpha decay use in, 1128, 1128f
Snell's law, 929–930
Soap bubbles, interference and, 993, 993f
Solar energy, 678
Solar energy panels
    photocells in, 1060, 1060f
    photovoltaic, 678, 678f
Solar system
    elliptical orbital motion in, 391–393
    gravitational force in, 385–390
Solar system model of atoms, 1083–1084, 1083f
Solenoids, 793–795, 793f, 826
Solid(s)
    deformation of, 592–596
        length change in, 592–594, 594f
        shape change in, 594–595, 594f
        stress and strain on, 596, 596f
        volume change in, 595–596, 595f, 596f
    to liquid phase changes, 598–599, 599f
Sound. See also Sound wave(s)
    of dinosaurs, 501
    Doppler effect and, 472–478, 472f. 474f, 474f
    human perception of, 470
    infrasonic, 466–467, 467f
    intensity of, 468–472, 468f, 468t
    loudness of. See Intensity
    pitch of, 466–467
    speed of, 464–465, 465t
    ultrasonic, 466, 466f
Sound wave(s), 464–467. See also Sound
    Doppler effect and, 472–478, 472f, 474f
    frequency of, 466–467
    longitudinal waves as, 464, 464f
    mechanical model of, 464, 464f
    properties of, 464–465, 464f
    reflection of, 462f, 463, 464f
    shape of, 464, 464t
    speed of, 460–461
South pole, magnetic, 770–771, 770f
Space travel, effects of relativity on, 1028
Spacecraft
    comet sampling by, 415
    escape speed of, 404–407, 407f
    maneuvers, 397–398, 397f
    orbital motion of, 394–397, 397f
Special theory of relativity, 1018–1120
    equivalence of physical laws in, 1018
    inertial frames of reference in, 1018
    speed of light in, 1018–1019
Specific heat, 559–560, 604
    for common substances, 559t
    for constant-pressure processes, 628–629, 628f
    for constant-volume processes, 628–629, 628f
    definition of, 559
    molar, 628–629
Spectroscope, grating, 1007–1008, 1008f
Specular beams, 1008
Specular reflection, 915, 915f
Speed
    acceleration and, 29, 30
    angular, 302–303, 302f, 303f, 307–310
        angular momentum and, 356–357
        conservation of angular momen-
            tum and, 361–363
    average, 22–23
    of blood flow, measurement of, 778–779, 778f
    change in, 31–32, 32f, 33f
    change in, work and, 203–207
    definition of, 12

dependence of pressure on, 527
distribution of, in gases, 587–590
in elastic collisions, 280–282
of electric potential energy, 700–701
of electrons, 1065–1066
energy conservation and, 234–241
escape, 404–407, 407f
of fluid, 523–524
of fluid flow, 521–523, 521f
of fluid flow, change in, 523–525, 526
free fall and, 46–48
of galaxies, Doppler effect and, 472, 477
height difference and, 236–239
in inelastic collisions, 272–273
instantaneous, 26
of light, 884–886
    astronomical perspective on, 886
    constancy of, 1018–1019, 1019f
    measurement of, 884, 885f
    in a vacuum, 884
linear, 307–310, 307f
maximum, in simple harmonic motion, 426
motion and, 89–91, 93–103
orbital, of Earth, 392–393
particle, and magnetic force, 773–774
path independence of, 236–237, 236f
projectiles and, 104–107, 104f
of rotating object, 308
rotational motion and, 302–303, 302f, 303f, 307–311, 307f
of sound, 464–465, 465t
of sound waves, 460–461
tangential, 307–310, 308f
translational, 311–312, 312f
typical values for, A-10
velocity and, 41–43
walking, 442–443
wave vs. source, 1018–1019, 1019f
of waves, 459, 459f, 460–462
Speed distribution, of molecules, 587–590, 588f, 601, 601f
Speedometers, 817, 821
Spheres
    charge distribution on, 669, 679, 679f
    as Gaussian surface, 684, 684f
    gravitational attraction of, 385–391
    point charge and, 669
Spherical aberration, 974, 974f
Spherical charge distributions, 669–670
Spherical mirror images, focal length in, 920f
Spherical mirrors, 918–919, 920f
    aberrations in, 921, 921f
    focal length in, 920
    imaging characteristics of, 924t
    mirror equation for, 922, 924–926, 925f
    ray tracing for, 922–923, 922f, 923f
Spherical wave fronts, 914, 914f
Sphygmomanometer, 540
Spider webs, resonance and, 446
Spontaneous emission, 1107
Spontaneous heat flow, 632, 637
Spring(s)
    energy conservation and, 240–241, 435–436
    force, 166–169, 167f, 208–212
    ideal, 168
    kinetic energy of, 435–436
    period of mass on, 430–435
    potential energy of, 231–233, 231f
    rules of thumb for, 168
    turning points and, 247, 247f
    visualization of, 168f
    work done by, 208–212, 208f, 210f, 211f
Springs
    potential energy of, 435–436
    simple harmonic motion and, 417–420, 418f, 421f, 430–435
    thermal expansion in, 553
    vertical, oscillation and, 433–434, 433f
Standing waves, 482–489, 484f
    in columns of air, 485–489

on a string, 482–486
in Sun, 489
three-dimensional, 1094–1095, 1095f
Star(s)
    explosion of, 361–362, 362f
    neutron, 362
    pole, 368
    pulsar, 362, 362f
    radiation and, 1053
    thermonuclear fusion in, 1143
Stardust spacecraft, 415
State function, definition of, 620
State of a system
    definition of, 620
    entropy and, 641
Static cling, 693
Static equilibrium, 343–346, 343f
    axis of rotation and, 344
    conditions for, 343
    in fluids, 507–512
Static friction, 160–164, 160f
    calculation of, 160–161
    coefficients of, 160f, 161–162
    vs. kinetic friction, 160, 164f
    magnitude of, 162–163
    rules of thumb for, 161
    visualization of, 162f
Steam engines, 632–634, 633f
Stefan-Boltzmann constant, 568
Stefan-Boltzmann law, 568
Stenosis, arterial, 533
Stimulated emission, 1107, 1107f
STM (Scanning tunneling microscope), 1073–1074, 1073f, 1074f
Strain, 596, 596f
Strassman, Fritz, 1139
Stretching
    potential energy of spring, 231–232
    of solids, 592–594, 592f
    of springs, work done in, 208–210, 208f
Strings
    connected objects and, 172–177, 172f–174f
    mass per length of, 460
    in musical instruments, 485
    pulleys and, 165–166, 175–176
    tension and, 164–166, 164f, 165f, 460
    transitional equilibrium and, 169–172, 169f
    waves on, 457, 457f, 458f, 460–463
        harmonics of, 482–486, 483f
        motion of, 458f
        speed of, 460–462
Strong nuclear force, 1124, 1139, 1148, 1151
Sublimation, latent heat of, 603
Sublimation curve, 599, 599f
Submarines, tourist, 537
Submerged volume, 519–520, 519f
Submersion, 514–516
Subscripts
    in notation, 71–72
    relative motion and, 78, 78f
    velocity and, 78
Subshells, for electrons, 1100, 1100t, 1103, 1104–1105, 1105f
Subtraction, of vectors, 69–70, 69f
Sun
    mass to energy conversion and, 1034
    orbital motion and, 391–394
    proton-proton cycle of fusion reactions in, 1143
    standing waves in, 489
    surface vibrations on, 453
    temperature and blackbody radiation of, 1053
Sunglasses, polarized light and, 897, 902–903, 902f, 903f
Sunsets, 902, 902f, 903f
Superconductors, 737–738, 737f
Superposition, 383
    of electric fields, 672–674, 673f
    of electric forces, 666
    of electric potential, 704
    of electric potential energy, 706

interference and, 984–986
of waves, 478, 478f
Surface
    friction and, 156, 156f
    inclined. See Inclined surface(s)
Surface charge density, 669, 709–710, 709f
Surface tension, 533, 533f
Surfactant, 533
Sweating, evaporation and, 600, 600f, 601f
Swim bladders, buoyancy of, 518
Switches, dimmer, 862–863
Symmetry, in projectile motion, 104–106
Synchrotron undulator, 781, 781f
Synchrotrons, 884
Système International d'Unités (SI), 2
Systems
    internal energy of, 618, 618f, 619f, 620
    state of, 620, 641

T

T wave, 711, 712
Tacoma Narrows bridge, 446, 446f
Talus slope, 162f
Tangent lines, 27
Tangential acceleration, 310–311, 310f
Tangential speed, 307–310, 308f
Tau particle, 1148t
TCAS (Traffic Collision Avoidance System), 91–92
Telecommunication systems
    electromagnetic waves and, 881
    optical fibers and, 934
Telescopes, 971–974
    angular size in, 971
    astronomical, 973, 973f
    converging lenses in, 971, 971f
    diffraction and, 1005
    diverging lenses in, 972
    focal length in, 971
    image formation in, 971
    image size and, 971
    magnification in, 971
    parabolic mirrors in, 921
    reflecting, 973, 973f
    refractor, 973, 973f
Television tuners, 870, 883, 884f
Television waves, 890
Televisions, screen pixels in, 1005, 1005f
TEM (Transmission electron micrograph), 1068f
Temperature, 545
    absolute zero and, 548
    boiling point and, 598
    constant
        in isothermal processes, 624, 624f
        isotherms and, 583–584, 583f
    cricket chirp rate and, 454–455, 576
    critical, 737
    equilibrium vapor pressure and, 597–598, 597f
    fluid density and, 555–556, 555f
    kinetic energy and, 589
    in kinetic theory of gases, 589
    magnetic field and, 795
    phase changes and, 603, 603f, 604f
    in phase equilibrium, 597–600, 597f
    radiation and, 1052–1053
    resistivity and, 737
    scales for, 546–550, 550f
    speed of sound, 465
    thermal equilibrium and, 618
    units for, 546–547
Tension
    in connected objects, 173–175
    direction of, 165–166, 165f
    as nonconservative force, 227t
    speed of wave on a spring and, 460–462
    strings and, 164–166, 164f
    surface, 533, 533f
    transitional equilibrium and, 169–172, 169f, 170f, 171f, 172f
Terminals, battery, 732
Termometro Lentos, 951

Tesla (T), 773, 774
Tesla, Nikola, 773
Theremin, capacitance and, 717, 718f
Thermal conductivity, 563, 563t
    applications of, 566
    heat flow and, 564–565
Thermal contact, 545
Thermal energy, 545
Thermal equilibrium, 545
    heat flow in, 545–546, 618
    temperature and, 618
Thermal expansion, 550–556, 599, 600f
    area, 553
    coefficients of, 551, 554, 554t
    examples of, 552–553
    linear, 551, 552t
    visualization of, 553f
    volume, 554, 554f
    water and, 555–556, 555f
Thermal processes, 620–627
    adiabatic, 625–627, 626f, 627f, 630–631, 630f
    characteristics of, 627t
    at constant pressure, 621f, 622f
    at constant volume, 623–624
    constant-pressure, 621–622, 621f, 622f
    constant-volume, 623–624, 623f, 628–629
    irreversible, 621
    isothermal, 624, 624f
    quasi-static, 621
    reversible, 621, 621f
Thermistors, 737
Thermodynamics, 546f
    definition of, 545
    first law of, 618–620
    second law of, 631–632, 646
    third law of, 647–648
    zeroth law of, 545–546, 618
Thermograms, 891f
Thermometer
    constant-volume gas, 548–549, 548f, 580, 580f
    cricket, 454–455, 576
    Galileo, 951
Thermonuclear fusion reaction, 1143
Thermos bottles, 569, 570f
Thin films, interference and
    conditions for interference in, 993–996
    phase changes in, 996f
Thin-lens equation, 939–941, 939f
Third harmonic, 483, 483f, 486, 486f, 488f
Third law of thermodynamics, 647–648
Thompson, Benjamin, 556
Thomson, J. J., 1083
Thomson, William (Lord Kelvin), 550
Thomson model of atoms, 1083, 1083f
Three-dimensional standing waves, 1094–1095, 1095f
Thrust, 289
Tidal bulges, 408
Tidal deformation, 408–409, 408f
Tidal locking, 408
Tides, 407–409, 408f
Time, 2, 4
    vs. acceleration, in simple harmonic motion, 426–427, 426f
    displacement and, 93
    energy and, Heisenberg uncertainty principle on, 1072
    energy and, in simple harmonic motion, 435–436, 436f
    expression of, period and, 428
    of flight, 103–104
    harmonic waves and, 463–464
    position and, 34–35, 35f
        for mass on string, 431, 432f
        in simple harmonic motion, 421–422, 422f, 425, 425f
    typical intervals of, 5t
    typical values for, A-9–A-10
    velocity and, 33–34, 35f, 44, 89–91, 93
Time constant, in *RC* circuit, 755, 755f

Time dilation, 1020–1021, 1021f
  airplane travel and, 1022
  muons and, 1024
  physical processes and, 1022–1023
  simultaneity and, 1022
Tomography, computerized axial, 1107, 1107f
Torque
  acceleration and, 338–342
  applications of, 335–336, 335f, 336f, 354–356
  axis of rotation and, 344
  center of mass and, 346–347, 350–353, 350f
  conservation of angular momentum and, 360
  definition of, 335, 337
  electric motors and, 787
  equilibrium and, 343–350
  force direction and, 343–350
  for general force, 337
  magnetic, 785–787, 786f, 824
  magnitude of, 336–337
  moment arm and, 337–338, 337f
  in nature, examples of, 336
  nontangential forces and, 336–337
  power produced by, 365
  pulleys and, 354–355
  right-hand rule for, 367, 367f
  sign convention for, 337
  for tangential force, 335–336, 335f, 336f
  units of, 335
  as vector, 367
  work done by, 365
  zero, 347–350
  zero, static equilibrium and, 343–346
Torricelli, Evangelista, 509, 540–541
Torricelli's law, 529
Torsiversion, 379
Total acceleration, of rotating object, 311
Total energy, 1034
Total internal reflection, 933–935
  binoculars and, 934, 934f, 935f
  critical angles for, 933
  optical fibers and, 934, 934f, 935f
Total work
  calculation of, 201–203
  conservative forces and, 225–226
  kinetic energy and, 203–207
Touchscreens, capacitance and, 717, 717f
Tourist submarines, 537
Traction, 165–166, 187, 187f
Traffic Collision Avoidance System (TCAS), 91–92
Traffic lights, 844
Transcranial magnetic stimulation, 805
Transformers, 832–834, 832f, 834f
  energy conservation in, 833
  equations for, 832, 833
  magnetic flux and, 832–833
Transition elements, 1103
Translational equilibrium, 169–172
Translational motion, 311–318, 312f
Transmission axis, in polarizers, 898
Transmission axis, of polarized light, 898f
Transmission electron micrograph (TEM), 1068f
Transverse waves, 457, 457f, 882. See also Light; Radio waves
Triboelectric charging, 660, 660t
Triceps muscle, 378
Trigonometric functions, A-1–A-3
Trigonometric identities, A-3
Trigonometry, A-1
Triple point, 599
Trough, of wave, 459, 459f
Tuners, radio/television, 445, 883, 883f
Tuning forks, in neurologic testing, 450
Tunneling, 1073–1074
Turn signals, 757
Turning points, 246–247, 247f, 425
Turning radius, 179–180
Twenty Thousand Leagues Under the Sea (Verne), 656

Two-lens system
  image formation in, 959–961, 960f
  image size in, 959–960
  magnification in, 959–960, 960
  vision correction using, 959, 959f, 960–961, 961f, 963f
Two-slit diffraction, 986–990, 986f, 987f
  change of medium and, 989–990
  interference in, 987–988, 987f, 988f
Two-slit experiment, 1068, 1068f
Tyrannosaurus rex, 377

U
Ultracentrifuges, 182
Ultrasound, 466, 466f, 477
Ultraviolet catastrophe, 1053, 1053f, 1055
Ultraviolet light, 890f, 891, 892f, 907, 1086
Ultraviolet radiation, 567
Ultraviolet rays, 890f, 891
Uncertainty principle, 1068–1072, 1070f
Underdamped oscillations, 444
Unified force, of Universe, 1150–1151, 1151f
Uniform circular motion, simple harmonic motion and, 424–430, 424f, 425f
Unit vectors, 70–71, 70f, 71f
Units of measure
  for activity rate of nuclear decay, 1131
  for angles, 301
  for angular momentum, 356
  for angular position, 300
  for atmospheric pressure, 505,509t
  for atoms, 582
  basic, 2–4
  for capacitance, 712
  in cgs system, 2–4
  for electric charge, 659
  of electric current, 731
  for electric potential, 697
  for electrical power, 738
  for electrical resistance, 734
  for electricity usage, 740
  for electromotive force, 732
  for entropy change, 641
  for frequency, 419
  for heat, 557
  for heat capacity, 558
  for impulse, 262
  intensity level, 470
  intensity of, 468
  for magnetic fields, 774
  for magnification, 967
  in mks system, 2
  for molecules, 582
  for momentum, 258
  for period, 419
  for pressure, 504
  for resistivity, 735
  for temperature, 546–547
  for torque, 335
  for viscosity, 531
Universal gas constant, 582
Universal gravitation, 381–385
Universal gravitational constant, 379–380, 382, 388–390, 389f
Universe
  entropy in, 642–643, 646
  heat death of, 646
  momentum in, 270
  relativity and, 1033–1038
  unified force of, 1150–1151, 1151f
Unpolarized light
  definition of, 897, 897f
  intensity of, 898–899, 898f
  scattering of, 901–902, 902f
  transmission of, 900–903, 900f, 902f, 903f
Uranium
  enrichment of, 611
  splitting of nucleus of, 1139–1140

Uranium-238, alpha decay of, 1126–1128, 1128f
UV. See Ultraviolet light

V
Vacuum (environment)
  speed of light in, 884–886, 885f
  in Thermos bottle, 569, 570f
Vampire bats, flight of, 265, 266f
Vapor, phase equilibrium and, 597–600, 597f
Vapor pressure curve, 597–598, 597f, 613
Vaporization, latent heat of, 603, 603f, 603f
Vapor-pressure curve, 597f, 598
Variable air capacitors, 714f
Vector(s), 12–13
  acceleration, 73–74, 74f, 75f, 76f
  addition of, 65–68
    component method, 67–68, 67f
    graphical method, 66–67, 66f, 67f
  angular momentum, 366–367
  angular velocity, 366
  average velocity, 72–73, 73f
  components of. See Vector components
  conversion of, 61–63
  coordinate system for, 62, 62f
  cross products, 337, 775
  definition of, 61, 79
  direction of, 13, 13f, 61, 62, 64
  displacement, 61, 61f, 71–72, 72f
  dot products of, 199
  force as, 118, 122, 134–137, 134f
  impulse as, 262
  magnitude of, 68–69
  momentum, 366–367
  momentum as, 258–259
  notation for, 61, 61f, 62, 69, 122
  one-dimensional, 12–13, 13f
  position, 71, 71f
  relative motion and, 76–79
  rotational motion and, 366–368
  vs. scalars, 61
  subtraction of, 69–70, 69f
  3-D, 775–776, 775f
  unit, 70–71, 70f, 71f
  velocity, 12, 72–73, 75–76, 75f, 76f
  velocity and, 12–13, 13f
  weight as, 137–138
Vector components, 61–66
  multiplication of, 70–71, 70f
  Newton's laws of motion in, 126–127
  Newton's second law of motion in, 122
  signs of, determination of, 64–65, 64f, 65f
Vector multiplication, A-6–A-7
Vector sum, 66
Velocity
  acceleration and, 119–120
  angular, 301–303, 302f, 303f
    right-hand rule for, 366, 366f
    as vector, 366
  average, 23–24, 24–25, 25, 25f, 28, 28f
  of center of mass, 285–288
  constant, 26f, 89–91, 89f, 119–120, 1018
  constant acceleration and, 34
  displacement and, 38, 93
  in elastic collisions, 277–279
  free fall and, 43f, 44
  in inelastic collisions, 272–274
  instantaneous, 26–28, 26f, 27f, 27t, 28f
  of launched projectile, 99–100
  motion and, 89–94, 89f
  net force and, 120
  notation for, 5, 6t
  position and, 34–35, 35f, 38
  projectile motion and, 83-85, 83f, 85t, 104–105
  relative, 76–79
  relativistic addition of, 1028–1031, 1029f, 1031f
  in rolling motion, 312, 312f
  signs of, and change in speed, 31–33, 32f

  in simple harmonic motion, 425–426, 426f
  speed and, 41–43
  subscripts and, 78
  time and, 12, 33–34, 35f, 44, 89–91, 93
  vectors and, 12–13, 13f
Velocity selectors, 778
Velocity vectors, 72–73, 73f, 75–76, 75f, 76f
Verne, Jules, 54, 404, 656
Vertical springs, oscillation and, 433–434, 433f
Vibrating columns of air, 485–489
Virtual images, 922f, 923, 927, 940
Viscosity, 531–533, 531f, 531t
Visible light, 890–891, 892
Vision correction
  for astigmatism, 965
  for farsightedness, 963, 963f
  for nearsightedness, 961, 961f, 963
  use of lasers for, 1108–1109, 1109f
Vitreous humor, 955, 955f
Volt (V), 697
Volta, Alessandro, 697
Voltage
  in ac resistor circuit, 846f
  calculation of, 734–738
  capacitance and, 718–720
  conversion of, 832–834
  electric current and, in capacitors, 853–854
  in inductors, 860–861
  measurement of, 758, 758f
  rms value of, in ac circuit, 847, 848
  transformers and, 832–833
Voltage phasors
  in ac capacitor circuit, 853, 853f
  in ac resistor circuit, 846, 847f
  in RL circuits, 861, 861f
  in RLC circuits, 863, 863f
Voltmeters, 758
Volume
  change in, of solids, 595–596, 595f, 596f
  expansion of, 554, 554f
  flow rate, 532–533
  gas, isotherms and, 583–584, 583f
  isotherms and, 624, 624f
  notation for, 5, 6t
  submerged, 519–520, 519f
Vomit comet, 140, 140f, 450
v-vs.-t graphs, 29–30, 29f, 30, 30f, 33f

W
Walking
  loss of height from, 123
  oscillatory motion in, 442–443
Walking speed, 442–443
Water
  boiling point of, 598
  density of, 555–556, 555f
  equilibrium vapor pressure for, 597, 597f
  fusion curve for, 599
  specific heat of, 559–560
  speed distribution for, 600, 601f
  thermal behavior of, 555–556, 555f, 600–601
  triple point of, 599
  vapor-pressure curve for, 597f, 598
Water waves, 458, 458f
  diffraction of, 998f
  interference pattern of, 984, 984f
  speed of, 474
Watson, J. D., 1007
Watt (W), 213
Watt, James, 213
Wave(s), 456–492
  crest of, 459, 459f
  definition of, 457
  gravity and, 1042–1043, 1042f, 1043f
  harmonic, 457, 457f, 458f, 463–464, 463f
  infrared, 890, 890f, 891f
  interference in, 478–482, 984–985, 984f

light. *See* Light waves
linearly polarized, 897, 897f
longitudinal, 458, 458f, 464, 464f
matter. *See* Matter waves
motion of, 453f, 457, 457f
period of, 459, 459f
phase relationship of, 984–985,
    984f
radio, 890, 890f
shape of, 457–459, 457f–459f,
    463–464
sound. *See* Sound wave(s)
spherical, 914, 914f
standing, 482–489, 1094–1095
on a string, 457, 457f, 460–463
    harmonics and, 482–486
    motion of, 458f
    speed of, 460–462
    superposition of, 478, 478f
television, 890
transverse, 457, 457f, 882
trough of, 459, 459f
types of, 457–459
water. *See* Water waves
Wave fronts, 914, 914f, 929
Wave function
    harmonic, 463–464, 463f, 1093
    matter wave as, 1096, 1098
Wavelength, 459
    change in, Compton shift formula
        for, 1063
    de Broglie, 1065–1066
    frequency and
        for electromagnetic waves,
            889–890, 890f
        for sound waves, 459, 459f
    in hydrogen spectral series, 1085–
        1087, 1085f, 1091–1092, 1092f
    line spectrum for, 1085, 1085f
Wave-particle duality, 896, 1064,
    1068–1069, 1068f
Weak nuclear force, 1148, 1151
Weather glass, 540
Weber (Wb), 810
Weber, Wilhelm, 810
Wedgwood, Joseph, 567
Weighing the Earth experiemnt,
    388–389
Weight, 3, 137–145

apparent, 138–141, 139f
buoyant force and, 513–514, 513f
direction of, 137–138
fluid pressure and, 507, 507f
gravity and, 137–139
on inclined surface, 143–144, 144f
magnitude of, 138
vs. mass, 3
mass vs., 137–138, 137f
normal forces and, 141–142
as vector, 137–138
Weight vectors, 137–138
Weightlessness, 140, 332–333
Weinberg, Steven, 1151
Whales, communication by, 467
Wheels, rolling, friction of, 163, 163f,
    164f
Wien's displacement law, 1052–1053,
    1053f
Windows, insulated, 566
Windshield wipers, delay circuits in, 757
Wings
    airplane, lift in, 527
    of dragonflies, 256, 256f
Wires, current-carrying, 783–793
    circular, 792f
    helical, 793–795, 793f
    magnetic field of, 788–789, 788f,
        789f
    magnetic force between, 791, 791f
    magnetic force on, 783–785
Woodpeckers, drumming of, 420
Work
    acceleration and, 204–206
    angles and, 198–199, 198f, 201
    calculation of, 196
        at angle to displacement, 198
        in direction of displacement, 198
        total work, 201
        using average force, 210–211
    change in speed and, 203–207
    definition of, 196, 198
    displacement and, 196–203, 196f,
        198–199, 201
    done by batteries, 733
    done by conservative force, 224, 226,
        226f
    done by constant force, 196–207,
        207, 207f

at angle to displacement, 198–202,
    198f
in direction of displacement,
    196–198, 196f
done by constant-pressure processes,
    621–622, 621f, 622f
done by constant-volume processes,
    623–624, 623f
done by friction, 224, 225–227, 225f
done by gravity, 224, 224f–226f,
    225–227
done by heat engine, 632–636
done by nonconservative forces, 224,
    241–246
done by pressure, 525
done by springs, 208–212, 208f, 210f,
    211f
done by torque, 365
done by variable force, 207–212,
    208f
as dot products of vectors, 199
entropy and, 644–645
heat as, 556–557
mechanical, conversion of, to electri-
    cal energy, 818–821
negative, 200, 200f, 211, 211f
path dependence of, 200, 226–227,
    226f
positive, 200, 200f, 211, 211f
sign conventions for, 618, 619t
signs of, determination of, 200
in thermodynamic systems, 618–619,
    619f, 620
total, 201–207, 225–226
typical values of, 197t
units of, 196–197
visualization of, 198f
zero, 200
zero distance and, 198
Work function, 1058
Work-energy theorem, 203, 204–207
    Bernoulli's equation and, 523–524
    torque and, 365
Wristwatches, shake motion generators
    in, 823, 824f

X
x component, of vectors, 62, 62f, 63f,
    64, 64f, 65f, 67f, 80

x coordinate of the center of mass,
    284–285
x unit vector, 70–71, 71f
X-ray diffraction, 1007, 1007f, 1066, 1066f
X-ray photons, 1062–1067
X-rays
    atomic radiation in, 1104–1105, 1104f
    characteristic, 1105, 1105f
    discovery of, 1104
    medical use of, 1106–1107
x-vs.-*t* graphs, 24, 26f, 26t, 27f

Y
y component, of vectors, 62, 63f, 64, 64f,
    65f, 67f, 80
y coordinate of the center of mass,
    284–285
y unit vector, 70–71, 71f
Yerkes refractor, 973, 973f
Young, Thomas, 592, 986
Young's modulus, 592, 592t
Young's two-slit experiment, 986–988,
    986f
Yo-yos, 322–323, 331

Z
Z (atomic number), 1120
Zero, absolute, 548, 549f, 550
Zero launch angle, 94–95, 94f, 95f
Zero level, of potential energy, 228
Zero net force
    center of mass and, 286–287
    of electric charges, 667–668
    static equilibrium and, 343–346, 343f
    with vertical and horizontal force
        components, 347–350, 348f
Zero net torque
    static equilibrium and, 343–346
    with vertical and horizontal force
        components, 347–350, 348f
Zero torque
    center of mass and, 350–351
    static equilibrium and, 343–346, 343f
    with vertical and horizontal force
        components, 347–350, 348f
Zero work, 200, 200f
Zeroth law of thermodynamics,
    545–546, 546f, 618
Zweig, George, 1148